Lines

Slope of line through (x_1, y_1) and (x_2, y_2):

$$m = \frac{y_2 - y_1}{x_2 - x_1}$$

Point-slope equation of line through (x_1, y_1) with slope m:

$$y - y_1 = m(x - x_1)$$

Slope-intercept equation of line with slope m and y-intercept b:

$$y = b + mx$$

Rules of Exponents

$$a^x a^t = a^{x+t}$$
$$\frac{a^x}{a^t} = a^{x-t}$$
$$(a^x)^t = a^{xt}$$

Definition of Natural Log

$y = \ln x$ means $e^y = x$

ex: $\ln 1 = 0$ since $e^0 = 1$

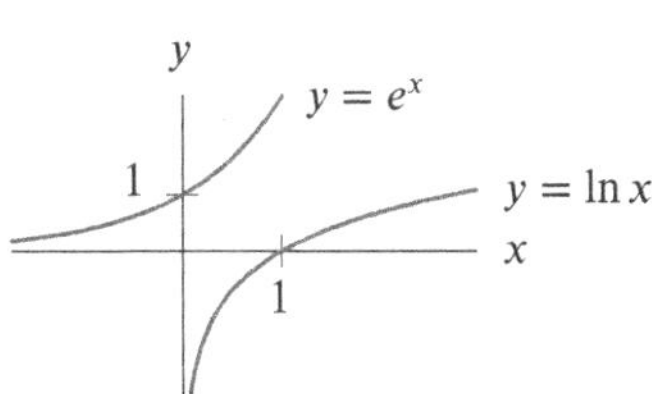

Identities

$$\ln e^x = x$$
$$e^{\ln x} = x$$

Rules of Natural Logarithms

$$\ln(AB) = \ln A + \ln B$$
$$\ln\left(\frac{A}{B}\right) = \ln A - \ln B$$
$$\ln A^p = p \ln A$$

Distance and Midpoint Formulas

Distance D between (x_1, y_1) and (x_2, y_2):

$$D = \sqrt{(x_2 - x_1)^2 + (y_2 - y_1)^2}$$

Midpoint of (x_1, y_1) and (x_2, y_2):

$$\left(\frac{x_1 + x_2}{2}, \frac{y_1 + y_2}{2}\right)$$

Quadratic Formula

If $ax^2 + bx + c = 0$, then

$$x = \frac{-b \pm \sqrt{b^2 - 4ac}}{2a}$$

Factoring Special Polynomials

$$x^2 - y^2 = (x + y)(x - y)$$
$$x^3 + y^3 = (x + y)(x^2 - xy + y^2)$$
$$x^3 - y^3 = (x - y)(x^2 + xy + y^2)$$

Circles

Center (h, k) and radius r:

$$(x - h)^2 + (y - k)^2 = r^2$$

Ellipse

$$\frac{x^2}{a^2} + \frac{y^2}{b^2} = 1$$

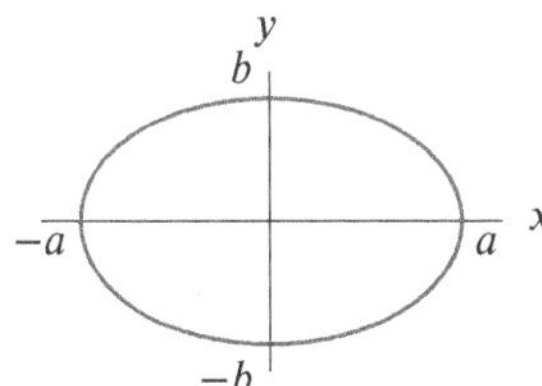

Hyperbola

$$\frac{x^2}{a^2} - \frac{y^2}{b^2} = 1$$

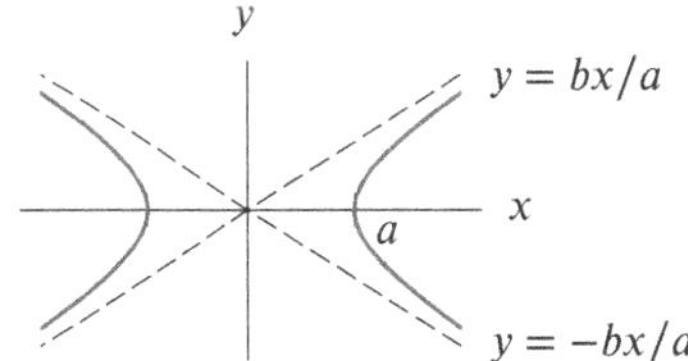

Geometric Formulas

Conversion Between Radians and Degrees: π radians $= 180°$

Triangle

$A = \frac{1}{2}bh$

$= \frac{1}{2}ab\sin\theta$

Circle

$A = \pi r^2$

$C = 2\pi r$

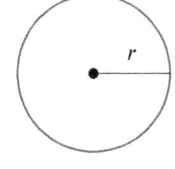

Sector of Circle

$A = \frac{1}{2}r^2\theta$ (θ in radians)

$s = r\theta$ (θ in radians)

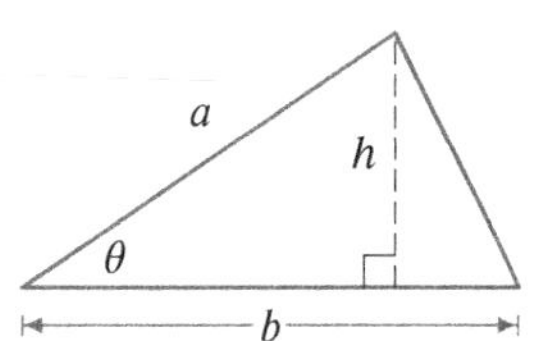

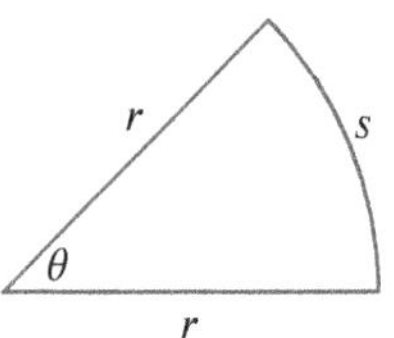

Sphere

$V = \frac{4}{3}\pi r^3 \quad A = 4\pi r^2$

Cylinder

$V = \pi r^2 h$

Cone

$V = \frac{1}{3}\pi r^2 h$

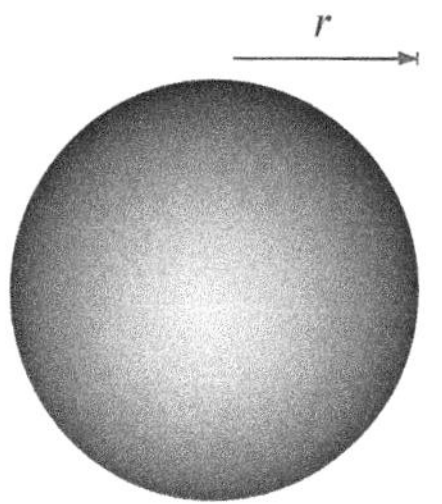

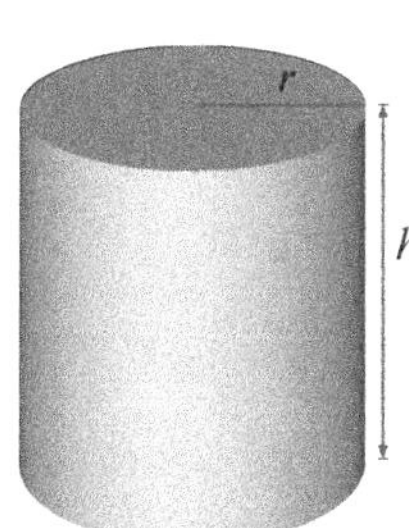

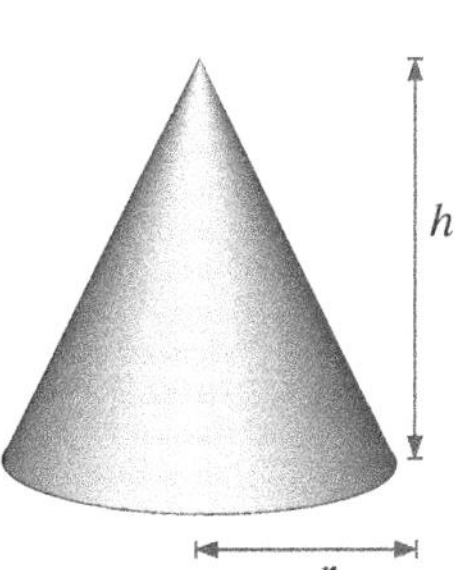

Trigonometric Functions

$\sin\theta = \dfrac{y}{r}$

$\cos\theta = \dfrac{x}{r}$

$\tan\theta = \dfrac{y}{x}$

$\tan\theta = \dfrac{\sin\theta}{\cos\theta}$

$\cos^2\theta + \sin^2\theta = 1$

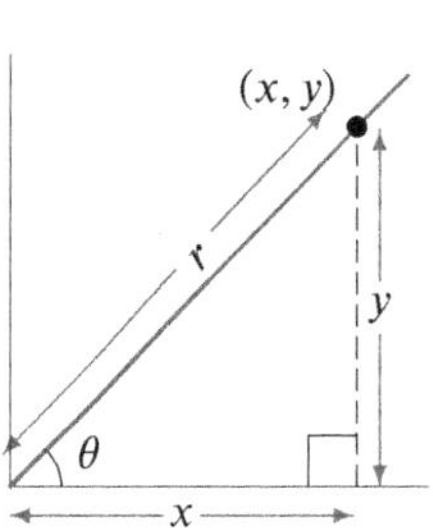

$\sin(A \pm B) = \sin A\cos B \pm \cos A\sin B$

$\cos(A \pm B) = \cos A\cos B \mp \sin A\sin B$

$\sin(2A) = 2\sin A\cos A$

$\cos(2A) = 2\cos^2 A - 1 = 1 - 2\sin^2 A$

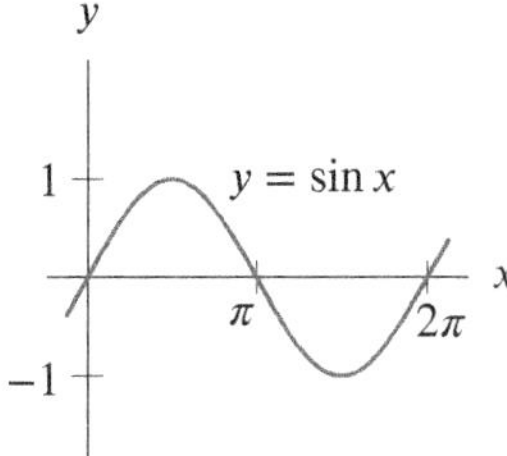

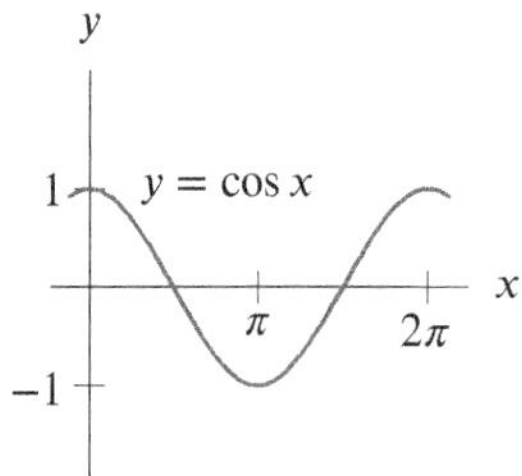

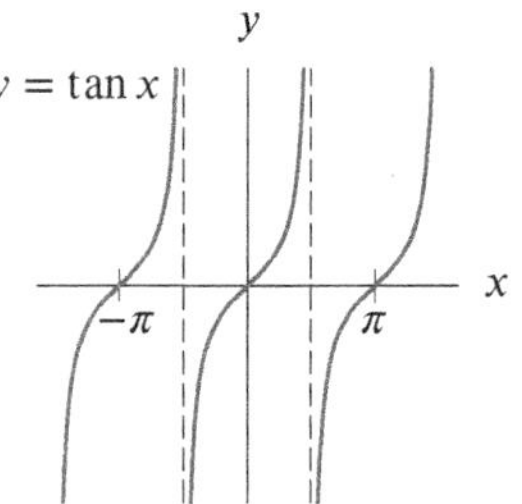

The Binomial Theorem

$$(x+y)^n = x^n + nx^{n-1}y + \frac{n(n-1)}{1\cdot 2}x^{n-2}y^2 + \frac{n(n-1)(n-2)}{1\cdot 2\cdot 3}x^{n-3}y^3 + \cdots + nxy^{n-1} + y^n$$

$$(x-y)^n = x^n - nx^{n-1}y + \frac{n(n-1)}{1\cdot 2}x^{n-2}y^2 - \frac{n(n-1)(n-2)}{1\cdot 2\cdot 3}x^{n-3}y^3 + \cdots \pm nxy^{n-1} \mp y^n$$

CALCULUS

Seventh Edition

EMEA EDITION

We dedicate this book to Andrew M. Gleason.

His brilliance and the extraordinary kindness and dignity with which he treated others made an enormous difference to us, and to many, many people. Andy brought out the best in everyone.

Deb Hughes Hallett
for the Calculus Consortium

CALCULUS

Seventh Edition

EMEA EDITION

Produced by the Calculus Consortium and initially funded by a National Science Foundation Grant.

Deborah Hughes-Hallett
University of Arizona

William G. McCallum
University of Arizona

Andrew M. Gleason
Harvard University

Eric Connally
Harvard University Extension

David Lovelock
University of Arizona

Douglas Quinney
University of Keele

Daniel E. Flath
Macalester College

Guadalupe I. Lozano
University of Arizona

Karen Rhea
University of Michigan

Selin Kalaycıoğlu
New York University

Jerry Morris
Sonoma State University

Ayşe Şahin
Wright State University

Brigitte Lahme
Sonoma State University

David Mumford
Brown University

Adam H. Spiegler
Loyola University Chicago

Patti Frazer Lock
St. Lawrence University

Brad G. Osgood
Stanford University

Jeff Tecosky-Feldman
Haverford College

David O. Lomen
University of Arizona

Cody L. Patterson
University of Texas at San Antonio

Thomas W. Tucker
Colgate University

Aaron D. Wootton
University of Portland

with the assistance of

Otto K. Bretscher
Colby College

Adrian Iovita
University of Washington

David E. Sloane, MD
Harvard Medical School

Coordinated by
Elliot J. Marks

ACQUISITIONS EDITOR	Shannon Corliss
VICE PRESIDENT AND DIRECTOR	Laurie Rosatone
DEVELOPMENT EDITOR	Adria Giattino
FREELANCE DEVELOPMENTAL EDITOR	Anne Scanlan-Rohrer/Two Ravens Editorial
MARKETING MANAGER	John LaVacca
SENIOR PRODUCT DESIGNER	David Dietz
SENIOR PRODUCTION EDITOR	Laura Abrams
COVER DESIGNER	Maureen Eide
CHAPTER OPENING PHOTO	©Patrick Zephyr/Patrick Zephyr Nature Photography

Problems from Calculus: The Analysis of Functions, by Peter D. Taylor (Toronto: Wall & Emerson, Inc., 1992). Reprinted with permission of the publisher.

This book was set in Times Roman by the Consortium using TEX, Mathematica, and the package ASTEX, which was written by Alex Kasman. It was printed and bound by TJ International. The cover was printed by TJ International.

Founded in 1807, John Wiley & Sons, Inc. has been a valued source of knowledge and understanding for more than 200 years, helping people around the world meet their needs and fulfill their aspirations. Our company is built on a foundation of principles that include responsibility to the communities we serve and where we live and work. In 2008, we launched a Corporate Citizenship Initiative, a global effort to address the environmental, social, economic, and ethical challenges we face in our business. Among the issues we are addressing are carbon impact, paper specifications and procurement, ethical conduct within our business and among our vendors, and community and charitable support. For more information, please visit our website: www.wiley.com/go/citizenship.

This material is based upon work supported by the National Science Foundation under Grant No. DUE-9352905. Opinions expressed are those of the authors and not necessarily those of the Foundation.

ISBN: 9781119585817

The inside back cover will contain printing identification and country of origin if omitted from this page. In addition, if the ISBN on the back cover dffers from the ISBN on this page, the one on the back cover is correct.

Printed and bound by CPI Group (UK) Ltd, Croydon, CR0 4YY

C9781119585817_090724

PREFACE

Calculus is one of the greatest achievements of the human intellect. Inspired by problems in astronomy, Newton and Leibniz developed the ideas of calculus 300 years ago. Since then, each century has demonstrated the power of calculus to illuminate questions in mathematics, the physical sciences, engineering, and the social and biological sciences.

Calculus has been so successful both because its central theme—change—is pivotal to an analysis of the natural world and because of its extraordinary power to reduce complicated problems to simple procedures. Therein lies the danger in teaching calculus: it is possible to teach the subject as nothing but procedures—thereby losing sight of both the mathematics and of its practical value. This edition of *Calculus* continues our effort to promote courses in which understanding and computation reinforce each other. It reflects the input of users at research universities, four-year colleges, community colleges, and secondary schools, as well as of professionals in partner disciplines such as engineering and the natural and social sciences.

Mathematical Thinking Supported by Theory and Modeling

The first stage in the development of mathematical thinking is the acquisition of a clear intuitive picture of the central ideas. In the next stage, the student learns to reason with the intuitive ideas in plain English. After this foundation has been laid, there is a choice of direction. All students benefit from both theory and modeling, but the balance may differ for different groups. Some students, such as mathematics majors, may prefer more theory, while others may prefer more modeling. For instructors wishing to emphasize the connection between calculus and other fields, the text includes:

- A variety of problems from the **physical sciences** and **engineering**.
- Examples from the **biological sciences** and **economics**.
- Models from the **health sciences** and of **population growth**.
- Problems on **sustainability**.
- Case studies on **medicine** by David E. Sloane, MD.

Active Learning: Good Problems

As instructors ourselves, we know that interactive classrooms and well-crafted problems promote student learning. Since its inception, the hallmark of our text has been its innovative and engaging problems. These problems probe student understanding in ways often taken for granted. Praised for their creativity and variety, these problems have had influence far beyond the users of our textbook.

The Seventh Edition continues this tradition. Under our approach, which we call the "Rule of Four," ideas are presented graphically, numerically, symbolically, and verbally, thereby encouraging students to deepen their understanding. Graphs and tables in this text are assumed to show all necessary information about the functions they represent, including direction of change, local extrema, and discontinuities.

Problems in this text include:

- **Strengthen Your Understanding** problems at the end of every section. These problems ask students to reflect on what they have learned by deciding "What is wrong?" with a statement and to "Give an example" of an idea.
- **ConcepTests** promote active learning in the classroom. These can be used with or without personal response systems (*e.g.*, clickers), and have been shown to dramatically improve student learning. Available in a book or on the web at www.wiley.com/college/hughes-hallett.
- **Class Worksheets** allow instructors to engage students in individual or group class-work. Samples are available in the Instructor's Manual, and all are on the web at www.wiley.com/college/hughes-hallett.
- **Data and Models** Many examples and problems throughout the text involve data-driven models. For example, Section 11.7 has a series of problems studying the spread of the chikungunya virus that arrived

in the US in 2013. Projects at the end of each chapter of the E-Text (at www.wiley.com/college/hughes-hallett) provide opportunities for sustained investigation of real-world situations that can be modeled using calculus.

- **Drill Exercises** build student skill and confidence.

Enhancing Learning Online

This Seventh Edition provides opportunities for students to experience the concepts of calculus in ways that would not be possible in a traditional textbook. The E-Text of *Calculus*, powered by VitalSource, provides interactive demonstrations of concepts, embedded videos that illustrate problem-solving techniques, and built-in assessments that allow students to check their understanding as they read. The E-Text also contains additional content not found in the print edition:

- Worked example **videos** by Donna Krawczyk at the University of Arizona, which provide students the opportunity to see and hear hundreds of the book's examples being explained and worked out in detail
- Embedded **Interactive Explorations**, applets that present and explore key ideas graphically and dynamically—especially useful for display of three-dimensional graphs
- Material that reviews and extends the major ideas of each chapter: Chapter Summary, Review Exercises and Problems, CAS Challenge Problems, and Projects
- Challenging problems that involve further exploration and application of the mathematics in many sections
- Section on the ϵ, δ definition of limit (1.10)
- Appendices that include preliminary ideas useful in this course

Problems Available in WileyPLUS

Students and instructors can access a wide variety of problems through WileyPLUS with ORION, Wiley's digital learning environment. ORION Learning provides an adaptive, personalized learning experience that delivers easy-to-use analytics so instructors and students can see exactly where they're excelling and where they need help. WileyPLUS with ORION features the following resources:

- Online version of the text, featuring hyperlinks to referenced content, applets, videos, and supplements.
- Homework management tools, which enable the instructor to assign questions easily and grade them automatically, using a rich set of options and controls.
- QuickStart pre-designed reading and homework assignments. Use them as-is or customize them to fit the needs of your classroom.
- Intelligent Tutoring questions, in which students are prompted for responses as they step through a problem solution and receive targeted feedback based on those responses.
- Algebra & Trigonometry Refresher material, delivered through ORION, Wiley's personalized, adaptive learning environment that assesses students' readiness and provides students with an opportunity to brush up on material necessary to master Calculus, as well as to determine areas that require further review.

Flexibility and Adaptability: Varied Approaches

The Seventh Edition of *Calculus* is designed to provide flexibility for instructors who have a range of preferences regarding inclusion of topics and applications and the use of computational technology. For those who prefer the lean topic list of earlier editions, we have kept clear the main conceptual paths. For example,

- The Key Concept chapters on the derivative and the definite integral (Chapters 2 and 5) can be covered at the outset of the course, right after Chapter 1.

- Limits and continuity (Sections 1.7, 1.8, and 1.9) can be covered in depth before the introduction of the derivative (Sections 2.1 and 2.2), or after.
- Approximating Functions Using Series (Chapter 10) can be covered before, or without, Chapter 9.
- In Chapter 4 (Using the Derivative), instructors can select freely from Sections 4.3–4.8.
- Chapter 8 (Using the Definite Integral) contains a wide range of applications. Instructors can select one or two to do in detail.

To use calculus effectively, students need skill in both symbolic manipulation and the use of technology. The balance between the two may vary, depending on the needs of the students and the wishes of the instructor. The book is adaptable to many different combinations.

The book does not require any specific software or technology. It has been used with graphing calculators, graphing software, and computer algebra systems. Any technology with the ability to graph functions and perform numerical integration will suffice. Students are expected to use their own judgment to determine where technology is useful.

Content

This content represents our vision of how calculus can be taught. It is flexible enough to accommodate individual course needs and requirements. Topics can easily be added or deleted, or the order changed.

Changes to the text in the Seventh Edition are in italics. In all chapters, problems were added and others were updated. In total, there are more than 1300 new problems.

Chapter 1: A Library of Functions

This chapter introduces all the elementary functions to be used in the book. Although the functions are probably familiar, the graphical, numerical, verbal, and modeling approach to them may be new. We introduce exponential functions at the earliest possible stage, since they are fundamental to the understanding of real-world processes.

The content on limits and continuity in this chapter has been revised and expanded to emphasize the limit as a central idea of calculus. Section 1.7 gives an intuitive introduction to the ideas of limit and continuity. Section 1.8 introduces one-sided limits and limits at infinity and presents properties of limits of combinations of functions, such as sums and products. The new Section 1.9 gives a variety of algebraic techniques for computing limits, together with many new exercises and problems applying those techniques, and introduces the Squeeze Theorem. The new online Section 1.10 contains the ϵ, δ *definition of limit, previously in Section 1.8.*

Chapter 2: Key Concept: The Derivative

The purpose of this chapter is to give the student a practical understanding of the definition of the derivative and its interpretation as an instantaneous rate of change. The power rule is introduced; other rules are introduced in Chapter 3.

Chapter 3: Short-Cuts to Differentiation

The derivatives of all the functions in Chapter 1 are introduced, as well as the rules for differentiating products; quotients; and composite, inverse, hyperbolic, and implicitly defined functions.

Chapter 4: Using the Derivative

The aim of this chapter is to enable the student to use the derivative in solving problems, including optimization, graphing, rates, parametric equations, and indeterminate forms. It is not necessary to cover all the sections in this chapter.

Chapter 5: Key Concept: The Definite Integral

The purpose of this chapter is to give the student a practical understanding of the definite integral as a limit of Riemann sums and to bring out the connection between the derivative and the definite integral in the Fundamental Theorem of Calculus.

The difference between total distance traveled during a time interval is contrasted with the change in position.

Chapter 6: Constructing Antiderivatives

This chapter focuses on going backward from a derivative to the original function, first graphically and numerically, then analytically. It introduces the Second Fundamental Theorem of Calculus and the concept of a differential equation.

Chapter 7: Integration

This chapter includes several techniques of integration, including substitution, parts, partial fractions, and trigonometric substitutions; others are included in the table of integrals. There are discussions of numerical methods and of improper integrals.

Chapter 8: Using the Definite Integral

This chapter emphasizes the idea of subdividing a quantity to produce Riemann sums which, in the limit, yield a definite integral. It shows how the integral is used in geometry, physics, economics, and probability; polar coordinates are introduced. It is not necessary to cover all the sections in this chapter.

Distance traveled along a parametrically defined curve during a time interval is contrasted with arc length.

Chapter 9: Sequences and Series

This chapter focuses on sequences, series of constants, and convergence. It includes the integral, ratio, comparison, limit comparison, and alternating series tests. It also introduces geometric series and general power series, including their intervals of convergence.

Rearrangement of the terms of a conditionally convergent series is discussed.

Chapter 10: Approximating Functions

This chapter introduces Taylor Series and Fourier Series using the idea of approximating functions by simpler functions.

The term Maclaurin series is introduced for a Taylor series centered at 0. *Term-by-term differentiation of a Taylor series within its interval of convergence is introduced without proof. This term-by-term differentiation allows us to show that a power series is its own Taylor series.*

Chapter 11: Differential Equations

This chapter introduces differential equations. The emphasis is on qualitative solutions, modeling, and interpretation.

Chapter 12: Functions of Several Variables

This chapter introduces functions of many variables from several points of view, using surface graphs, contour diagrams, and tables. We assume throughout that functions of two or more variables are defined on regions with piecewise smooth boundaries. We conclude with a section on continuity.

Chapter 13: A Fundamental Tool: Vectors

This chapter introduces vectors geometrically and algebraically and discusses the dot and cross product.

An application of the cross product to angular velocity is given.

Chapter 14: Differentiating Functions of Several Variables

Partial derivatives, directional derivatives, gradients, and local linearity are introduced. The chapter also discusses higher order partial derivatives, quadratic Taylor approximations, and differentiability.

Chapter 15: Optimization

The ideas of the previous chapter are applied to optimization problems, both constrained and unconstrained.

Chapter 16: Integrating Functions of Several Variables

This chapter discusses double and triple integrals in Cartesian, polar, cylindrical, and spherical coordinates.

Chapter 17: Parameterization and Vector Fields

This chapter discusses parameterized curves and motion, vector fields and flowlines.

Additional problems are provided on parameterizing curves in 3-space that are not contained in a coordinate plane.

Chapter 18: Line Integrals

This chapter introduces line integrals and shows how to calculate them using parameterizations. Conservative fields, gradient fields, the Fundamental Theorem of Calculus for Line Integrals, and Green's Theorem are discussed.

Chapter 19: Flux Integrals and Divergence

This chapter introduces flux integrals and shows how to calculate them over surface graphs, portions of cylinders, and portions of spheres. The divergence is introduced and its relationship to flux integrals discussed in the Divergence Theorem.

We calculate the surface area of the graph of a function using flux.

Chapter 20: The Curl and Stokes' Theorem

The purpose of this chapter is to give students a practical understanding of the curl and of Stokes' Theorem and to lay out the relationship between the theorems of vector calculus.

Chapter 21: Parameters, Coordinates, and Integrals

This chapter covers parameterized surfaces, the change of variable formula in a double or triple integral, and flux though a parameterized surface.

Appendices

There are online appendices on roots, accuracy, and bounds; complex numbers; Newton's method; and vectors in the plane. The appendix on vectors can be covered at any time, but may be particularly useful in the conjunction with Section 4.8 on parametric equations.

Supplementary Materials and Additional Resources

Supplements for the instructor can be obtained online at the book companion site or by contacting your Wiley representative. The following supplementary materials are available for this edition:

- **Instructor's Manual** containing teaching tips, calculator programs, overhead transparency masters, sample worksheets, and sample syllabi.
- **Computerized Test Bank**, comprised of nearly 7,000 questions, mostly algorithmically-generated, which allows for multiple versions of a single test or quiz.

- **Instructor's Solution Manual** with complete solutions to all problems.
- **Student Solution Manual** with complete solutions to half the odd-numbered problems.
- **Graphing Calculator Manual**, to help students get the most out of their graphing calculators, and to show how they can apply the numerical and graphing functions of their calculators to their study of calculus.
- **Additional Material**, elaborating specially marked points in the text and password-protected electronic versions of the instructor ancillaries, can be found on the web at www.wiley.com/college/hughes-hallett.

ConcepTests

ConcepTests, modeled on the pioneering work of Harvard physicist Eric Mazur, are questions designed to promote active learning during class, particularly (but not exclusively) in large lectures. Our evaluation data show students taught with ConcepTests outperformed students taught by traditional lecture methods 73% versus 17% on conceptual questions, and 63% versus 54% on computational problems.

Advanced Placement (AP) Teacher's Guide

The AP Guide, written by a team of experienced AP teachers, provides tips, multiple-choice questions, and free-response questions that align to each chapter of the text. It also features a collection of labs designed to complement the teaching of key AP Calculus concepts.

New material has been added to reflect recent changes in the learning objectives for AB and BC Calculus, including extended coverage of limits, continuity, sequences, and series. Also new to this edition are grids that align multiple choice and free-response questions to the College Board's Enduring Understandings, Learning Objectives, and Essential Knowledge.

Acknowledgements

First and foremost, we want to express our appreciation to the National Science Foundation for their faith in our ability to produce a revitalized calculus curriculum and, in particular, to our program officers, Louise Raphael, John Kenelly, John Bradley, and James Lightbourne. We also want to thank the members of our Advisory Board, Benita Albert, Lida Barrett, Simon Bernau, Robert Davis, M. Lavinia DeConge-Watson, John Dossey, Ron Douglas, Eli Fromm, William Haver, Seymour Parter, John Prados, and Stephen Rodi.

In addition, a host of other people around the country and abroad deserve our thanks for their contributions to shaping this edition. They include: Huriye Arikan, Pau Atela, Ruth Baruth, Paul Blanchard, Lewis Blake, David Bressoud, Stephen Boyd, Lucille Buonocore, Matthew Michael Campbell, Jo Cannon, Ray Cannon, Phil Cheifetz, Scott Clark, Jailing Dai, Ann Davidian, Tom Dick, Srdjan Divac, Tevian Dray, Steven Dunbar, Penny Dunham, David Durlach, John Eggers, Wade Ellis, Johann Engelbrecht, Brad Ernst, Sunny Fawcett, Paul Feehan, Sol Friedberg, Melanie Fulton, Tom Gearhart, David Glickenstein, Chris Goff, Sheldon P. Gordon, Salim Haïdar, Elizabeth Hentges, Rob Indik, Adrian Iovita, David Jackson, Sue Jensen, Alex Kasman, Matthias Kawski, Christopher Kennedy, Mike Klucznik, Donna Krawczyk, Stephane Lafortune, Andrew Lawrence, Carl Leinert, Daniel Look, Andrew Looms, Bin Lu, Alex Mallozzi, Corinne Manogue, Jay Martin, Eric Mazur, Abby McCallum, Dan McGee, Ansie Meiring, Lang Moore, Jerry Morris, Hideo Nagahashi, Kartikeya Nagendra, Alan Newell, Steve Olson, John Orr, Arnie Ostebee, Andrew Pasquale, Scott Pilzer, Wayne Raskind, Maria Robinson, Laurie Rosatone, Ayse Sahin, Nataliya Sandler, Ken Santor, Anne Scanlan-Rohrer, Ellen Schmierer, Michael Sherman, Pat Shure, David Smith, Ernie Solheid, Misha Stepanov, Steve Strogatz, Carl Swenson, Peter Taylor, Dinesh Thakur, Sally Thomas, Joe Thrash, Alan Tucker, Doug Ulmer, Ignatios Vakalis, Bill Vélez, Joe Vignolini, Stan Wagon, Hannah Winkler, Debra Wood, Deane Yang, Bruce Yoshiwara, Kathy Yoshiwara, and Paul Zorn.

Reports from the following reviewers were most helpful for the sixth edition:

Barbara Armenta, James Baglama, Jon Clauss, Ann Darke, Marcel Finan, Dana Fine, Michael Huber, Greg Marks, Wes Ostertag, Ben Smith, Mark Turner, Aaron Weinberg, and Jianying Zhang.

Reports from the following reviewers were most helpful for the seventh edition:

Scott Adamson, Janet Beery, Tim Biehler, Lewis Blake, Mark Booth, Tambi Boyle, David Brown, Jeremy Case, Phil Clark, Patrice Conrath, Pam Crawford, Roman J. Dial, Rebecca Dibbs, Marcel B. Finan, Vauhn

Foster-Grahler, Jill Guerra, Salim M. Haidar, Ryan A. Hass, Firas Hindeleh, Todd King, Mary Koshar, Dick Lane, Glenn Ledder, Oscar Levin, Tom Linton, Erich McAlister, Osvaldo Mendez, Cindy Moss, Victor Padron, Michael Prophet, Ahmad Rajabzadeh, Catherine A. Roberts, Kari Rothi, Edward J. Soares, Diana Staats, Robert Talbert, James Vicich, Wendy Weber, Mina Yavari, and Xinyun Zhu.

Finally, we extend our particular thanks to Jon Christensen for his creativity with our three-dimensional figures.

Deborah Hughes-Hallett	David O. Lomen	Douglas Quinney
Andrew M. Gleason	David Lovelock	Karen Rhea
William G. McCallum	Guadalupe I. Lozano	Ayşe Şahin
Eric Connally	Jerry Morris	Adam Spiegler
Daniel E. Flath	David O. Mumford	Jeff Tecosky-Feldman
Selin Kalaycıoğlu	Brad G. Osgood	Thomas W. Tucker
Brigitte Lahme	Cody L. Patterson	Aaron D. Wootton
Patti Frazer Lock		

To Students: How to Learn from this Book

- This book may be different from other math textbooks that you have used, so it may be helpful to know about some of the differences in advance. This book emphasizes at every stage the *meaning* (in practical, graphical or numerical terms) of the symbols you are using. There is much less emphasis on "plug-and-chug" and using formulas, and much more emphasis on the interpretation of these formulas than you may expect. You will often be asked to explain your ideas in words or to explain an answer using graphs.
- The book contains the main ideas of calculus in plain English. Your success in using this book will depend on your reading, questioning, and thinking hard about the ideas presented. Although you may not have done this with other books, you should plan on reading the text in detail, not just the worked examples.
- There are very few examples in the text that are exactly like the homework problems. This means that you can't just look at a homework problem and search for a similar–looking "worked out" example. Success with the homework will come by grappling with the ideas of calculus.
- Many of the problems that we have included in the book are open-ended. This means that there may be more than one approach and more than one solution, depending on your analysis. Many times, solving a problem relies on common sense ideas that are not stated in the problem but which you will know from everyday life.
- Some problems in this book assume that you have access to a graphing calculator or computer. There are many situations where you may not be able to find an exact solution to a problem, but you can use a calculator or computer to get a reasonable approximation.
- This book attempts to give equal weight to four methods for describing functions: graphical (a picture), numerical (a table of values) algebraic (a formula), and verbal. Sometimes you may find it easier to translate a problem given in one form into another. The best idea is to be flexible about your approach: if one way of looking at a problem doesn't work, try another.
- Students using this book have found discussing these problems in small groups very helpful. There are a great many problems which are not cut-and-dried; it can help to attack them with the other perspectives your colleagues can provide. If group work is not feasible, see if your instructor can organize a discussion session in which additional problems can be worked on.
- You are probably wondering what you'll get from the book. The answer is, if you put in a solid effort, you will get a real understanding of one of the most important accomplishments of the millennium—calculus—as well as a real sense of the power of mathematics in the age of technology.

CONTENTS

For online material, see www.wiley.com/college/hughes-hallett.

10 APPROXIMATING FUNCTIONS USING SERIES 513

11 DIFFERENTIAL EQUATIONS 561

12 FUNCTIONS OF SEVERAL VARIABLES 651

13 A FUNDAMENTAL TOOL: VECTORS 701

14 DIFFERENTIATING FUNCTIONS OF SEVERAL VARIABLES 739

15 OPTIMIZATION: LOCAL AND GLOBAL EXTREMA 805

16 INTEGRATING FUNCTIONS OF SEVERAL VARIABLES 839

Chapter One

FOUNDATION FOR CALCULUS: FUNCTIONS AND LIMITS

Contents

1.1 FUNCTIONS AND CHANGE

In mathematics, a *function* is used to represent the dependence of one quantity upon another.

Let's look at an example. In 2015, Boston, Massachusetts, had the highest annual snowfall, 110.6 inches, since recording started in 1872. Table 1.1 shows one 14-day period in which the city broke another record with a total of 64.4 inches.[1]

Table 1.1 *Daily snowfall in inches for Boston, January 27 to February 9, 2015*

Day	1	2	3	4	5	6	7	8	9	10	11	12	13	14
Snowfall	22.1	0.2	0	0.7	1.3	0	16.2	0	0	0.8	0	0.9	7.4	14.8

You may not have thought of something so unpredictable as daily snowfall as being a function, but it *is* a function of day, because each day gives rise to one snowfall total. There is no formula for the daily snowfall (otherwise we would not need a weather bureau), but nevertheless the daily snowfall in Boston does satisfy the definition of a function: Each day, t, has a unique snowfall, S, associated with it.

We define a function as follows:

> A **function** is a rule that takes certain numbers as inputs and assigns to each a definite output number. The set of all input numbers is called the **domain** of the function and the set of resulting output numbers is called the **range** of the function.

The input is called the *independent variable* and the output is called the *dependent variable*. In the snowfall example, the domain is the set of days $\{1, 2, 3, 4, 5, 6, 7, 8, 9, 10, 11, 12, 13, 14\}$ and the range is the set of daily snowfalls $\{0, 0.2, 0.7, 0.8, 0.9, 1.3, 7.4, 14.8, 16.2, 22.1\}$. We call the function f and write $S = f(t)$. Notice that a function may have identical outputs for different inputs (Days 8 and 9, for example).

Some quantities, such as a day or date, are *discrete*, meaning they take only certain isolated values (days must be integers). Other quantities, such as time, are *continuous* as they can be any number. For a continuous variable, domains and ranges are often written using interval notation:

The set of numbers t such that $a \le t \le b$ is called a *closed interval* and written $[a, b]$.

The set of numbers t such that $a < t < b$ is called an *open interval* and written (a, b).

The Rule of Four: Tables, Graphs, Formulas, and Words

Functions can be represented by tables, graphs, formulas, and descriptions in words. For example, the function giving the daily snowfall in Boston can be represented by the graph in Figure 1.1, as well as by Table 1.1.

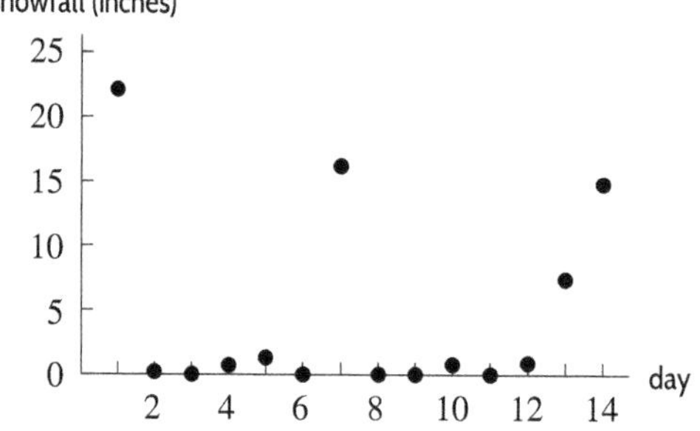

Figure 1.1: Boston snowfall, starting January 27, 2015

As another example of a function, consider the snowy tree cricket. Surprisingly enough, all such crickets chirp at essentially the same rate if they are at the same temperature. That means that the chirp rate is a function of temperature. In other words, if we know the temperature, we can determine

[1] http://w2.weather.gov/climate/xmacis.php?wfo=box. Accessed June 2015.

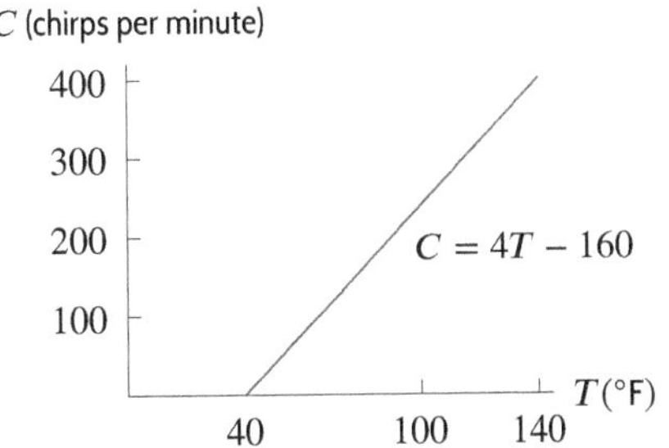

Figure 1.2: Cricket chirp rate versus temperature

the chirp rate. Even more surprisingly, the chirp rate, C, in chirps per minute, increases steadily with the temperature, T, in degrees Fahrenheit, and can be computed by the formula

$$C = 4T - 160$$

to a fair level of accuracy. We write $C = f(T)$ to express the fact that we think of C as a function of T and that we have named this function f. The graph of this function is in Figure 1.2.

Notice that the graph of $C = f(T)$ in Figure 1.2 is a solid line. This is because $C = f(T)$ is a *continuous function*. Roughly speaking, a continuous function is one whose graph has no breaks, jumps, or holes. This means that the independent variable must be continuous. (We give a more precise definition of continuity of a function in Section 1.7.)

Examples of Domain and Range

If the domain of a function is not specified, we usually take it to be the largest possible set of real numbers. For example, we usually think of the domain of the function $f(x) = x^2$ as all real numbers. However, the domain of the function $g(x) = 1/x$ is all real numbers except zero, since we cannot divide by zero.

Sometimes we restrict the domain to be smaller than the largest possible set of real numbers. For example, if the function $f(x) = x^2$ is used to represent the area of a square of side x, we restrict the domain to nonnegative values of x.

Example 1 The function $C = f(T)$ gives chirp rate as a function of temperature. We restrict this function to temperatures for which the predicted chirp rate is positive, and up to the highest temperature ever recorded at a weather station, 134°F. What is the domain of this function f?

Solution If we consider the equation

$$C = 4T - 160$$

simply as a mathematical relationship between two variables C and T, any T value is possible. However, if we think of it as a relationship between cricket chirps and temperature, then C cannot be less than 0. Since $C = 0$ leads to $0 = 4T - 160$, and so $T = 40$°F, we see that T cannot be less than 40°F. (See Figure 1.2.) In addition, we are told that the function is not defined for temperatures above 134°. Thus, for the function $C = f(T)$ we have

$$\begin{aligned}\text{Domain} &= \text{All } T \text{ values between 40°F and 134°F}\\ &= \text{All } T \text{ values with } 40 \le T \le 134\\ &= [40, 134].\end{aligned}$$

Example 2 Find the range of the function f, given the domain from Example 1. In other words, find all possible values of the chirp rate, C, in the equation $C = f(T)$.

Solution Again, if we consider $C = 4T - 160$ simply as a mathematical relationship, its range is all real C values. However, when thinking of the meaning of $C = f(T)$ for crickets, we see that the function predicts cricket chirps per minute between 0 (at $T = 40$°F) and 376 (at $T = 134$°F). Hence,

$$\begin{aligned}\text{Range} &= \text{All } C \text{ values from 0 to 376}\\ &= \text{All } C \text{ values with } 0 \le C \le 376\\ &= [0, 376].\end{aligned}$$

In using the temperature to predict the chirp rate, we thought of the temperature as the *independent variable* and the chirp rate as the *dependent variable*. However, we could do this backward, and calculate the temperature from the chirp rate. From this point of view, the temperature is dependent on the chirp rate. Thus, which variable is dependent and which is independent may depend on your viewpoint.

Linear Functions

The chirp-rate function, $C = f(T)$, is an example of a *linear function*. A function is linear if its slope, or rate of change, is the same at every point. The rate of change of a function that is not linear may vary from point to point.

Olympic and World Records

During the early years of the Olympics, the height of the men's winning pole vault increased approximately 8 inches every four years. Table 1.2 shows that the height started at 130 inches in 1900, and increased by the equivalent of 2 inches a year. So the height was a linear function of time from 1900 to 1912. If y is the winning height in inches and t is the number of years since 1900, we can write

$$y = f(t) = 130 + 2t.$$

Since $y = f(t)$ increases with t, we say that f is an *increasing function.* The coefficient 2 tells us the rate, in inches per year, at which the height increases.

Table 1.2 *Men's Olympic pole vault winning height (approximate)*

Year	1900	1904	1908	1912
Height (inches)	130	138	146	154

This rate of increase is the *slope* of the line in Figure 1.3. The slope is given by the ratio

$$\text{Slope} = \frac{\text{Rise}}{\text{Run}} = \frac{146-138}{8-4} = \frac{8}{4} = 2 \text{ inches/year.}$$

Calculating the slope (rise/run) using any other two points on the line gives the same value.

What about the constant 130? This represents the initial height in 1900, when $t = 0$. Geometrically, 130 is the *intercept* on the vertical axis.

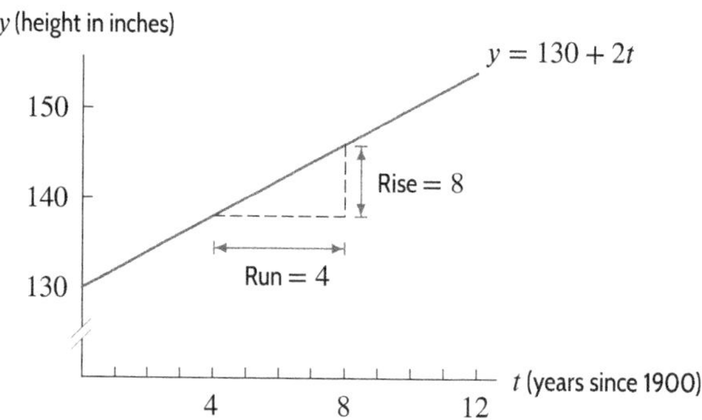

Figure 1.3: Olympic pole vault records

You may wonder whether the linear trend continues beyond 1912. Not surprisingly, it does not exactly. The formula $y = 130 + 2t$ predicts that the height in the 2012 Olympics would be 354 inches or 29 feet 6 inches, which is considerably higher than the actual value of 19 feet 7.05 inches. There is clearly a danger in *extrapolating* too far from the given data. You should also observe that the data in Table 1.2 is discrete, because it is given only at specific points (every four years). However, we have treated the variable t as though it were continuous, because the function $y = 130 + 2t$ makes sense for all values of t. The graph in Figure 1.3 is of the continuous function because it is a solid line, rather than four separate points representing the years in which the Olympics were held.

As the pole vault heights have increased over the years, the time to run the mile has decreased. If y is the world record time to run the mile, in seconds, and t is the number of years since 1900, then records show that, approximately,

$$y = g(t) = 260 - 0.39t.$$

The 260 tells us that the world record was 260 seconds in 1900 (at $t = 0$). The slope, -0.39, tells us that the world record decreased by about 0.39 seconds per year. We say that g is a *decreasing function*.

Difference Quotients and Delta Notation

We use the symbol Δ (the Greek letter capital delta) to mean "change in," so Δx means change in x and Δy means change in y.

The slope of a linear function $y = f(x)$ can be calculated from values of the function at two points, given by x_1 and x_2, using the formula

$$m = \frac{\text{Rise}}{\text{Run}} = \frac{\Delta y}{\Delta x} = \frac{f(x_2) - f(x_1)}{x_2 - x_1}.$$

The quantity $(f(x_2) - f(x_1))/(x_2 - x_1)$ is called a *difference quotient* because it is the quotient of two differences. (See Figure 1.4.) Since $m = \Delta y/\Delta x$, the units of m are y-units over x-units.

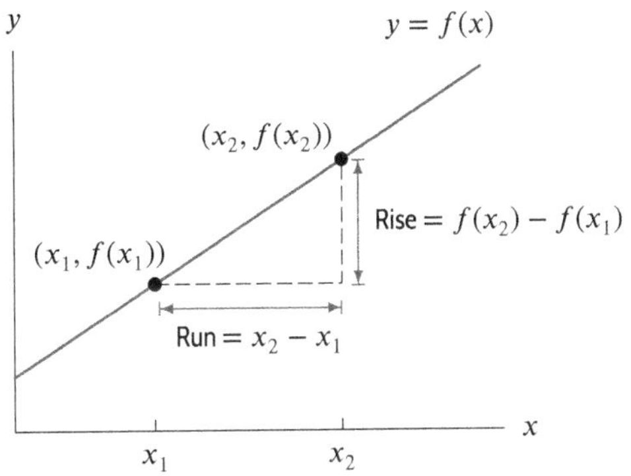

Figure 1.4: Difference quotient $= \dfrac{f(x_2) - f(x_1)}{x_2 - x_1}$

Families of Linear Functions

> A **linear function** has the form
>
> $$y = f(x) = b + mx.$$
>
> Its graph is a line such that
> - m is the **slope**, or rate of change of y with respect to x.
> - b is the **vertical intercept**, or value of y when x is zero.

Notice that if the slope, m, is zero, we have $y = b$, a horizontal line.

> To recognize that a table of x and y values comes from a linear function, $y = b + mx$, look for differences in y-values that are constant for equally spaced x-values.

Formulas such as $f(x) = b + mx$, in which the constants m and b can take on various values, give a *family of functions*. All the functions in a family share certain properties—in this case, all the graphs are straight lines. The constants m and b are called *parameters*; their meaning is shown in Figures 1.5 and 1.6. Notice that the greater the magnitude of m, the steeper the line.

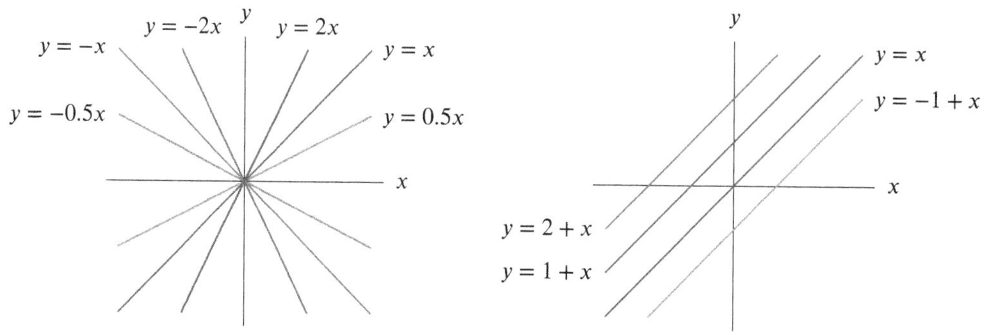

Figure 1.5: The family $y = mx$ (with $b = 0$)

Figure 1.6: The family $y = b + x$ (with $m = 1$)

Increasing versus Decreasing Functions

The terms increasing and decreasing can be applied to other functions, not just linear ones. See Figure 1.7. In general,

> A function f is **increasing** if the values of $f(x)$ increase as x increases.
> A function f is **decreasing** if the values of $f(x)$ decrease as x increases.
>
> The graph of an *increasing* function *climbs* as we move from left to right.
> The graph of a *decreasing* function *falls* as we move from left to right.
>
> A function $f(x)$ is **monotonic** if it increases for all x or decreases for all x.

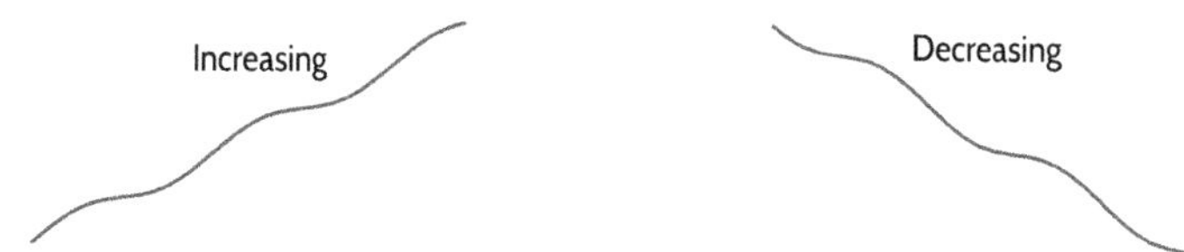

Figure 1.7: Increasing and decreasing functions

Proportionality

A common functional relationship occurs when one quantity is *proportional* to another. For example, the area, A, of a circle is proportional to the square of the radius, r, because

$$A = f(r) = \pi r^2.$$

> We say y is (directly) **proportional** to x if there is a nonzero constant k such that
>
> $$y = kx.$$
>
> This k is called the constant of proportionality.

We also say that one quantity is *inversely proportional* to another if one is proportional to the reciprocal of the other. For example, the speed, v, at which you make a 50-mile trip is inversely proportional to the time, t, taken, because v is proportional to $1/t$:

$$v = 50\left(\frac{1}{t}\right) = \frac{50}{t}.$$

Exercises and Problems for Section 1.1

EXERCISES

1. The population of a city, P, in millions, is a function of t, the number of years since 2010, so $P = f(t)$. Explain the meaning of the statement $f(5) = 7$ in terms of the population of this city.

2. The pollutant PCB (polychlorinated biphenyl) can affect the thickness of pelican eggshells. Thinking of the thickness, T, of the eggshells, in mm, as a function of the concentration, P, of PCBs in ppm (parts per million), we have $T = f(P)$. Explain the meaning of $f(200)$ in terms of thickness of pelican eggs and concentration of PCBs.

3. Describe what Figure 1.8 tells you about an assembly line whose productivity is represented as a function of the number of workers on the line.

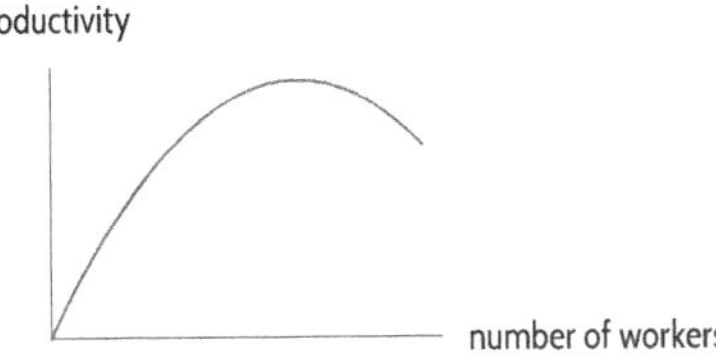

Figure 1.8

■ For Exercises 4–7, find an equation for the line that passes through the given points.

4. $(0, 0)$ and $(1, 1)$
5. $(0, 2)$ and $(2, 3)$
6. $(-2, 1)$ and $(2, 3)$
7. $(-1, 0)$ and $(2, 6)$

■ For Exercises 8–11, determine the slope and the y-intercept of the line whose equation is given.

8. $2y + 5x - 8 = 0$
9. $7y + 12x - 2 = 0$
10. $-4y + 2x + 8 = 0$
11. $12x = 6y + 4$

12. Match the graphs in Figure 1.9 with the following equations. (Note that the x and y scales may be unequal.)

   (a) $y = x - 5$
   (b) $-3x + 4 = y$
   (c) $5 = y$
   (d) $y = -4x - 5$
   (e) $y = x + 6$
   (f) $y = x/2$

(I)
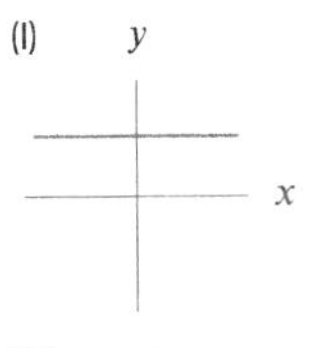

(II)
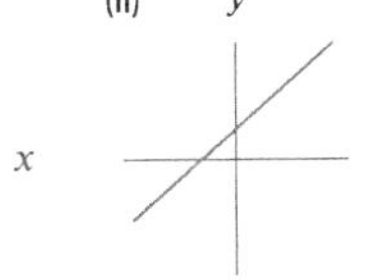

(III)
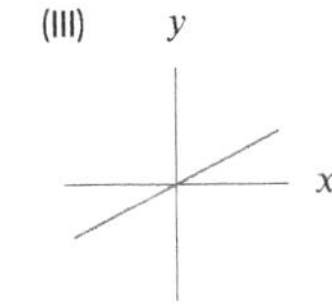

(IV)
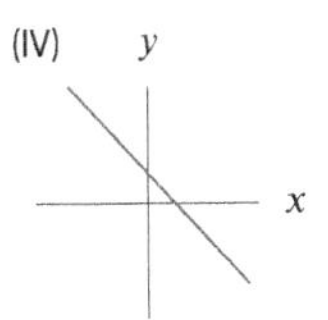

(V)
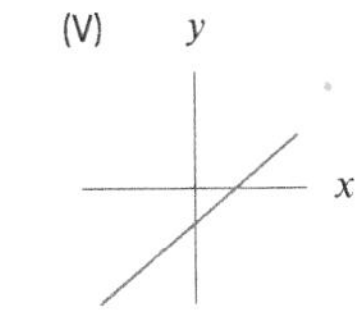

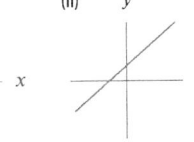

(VI)
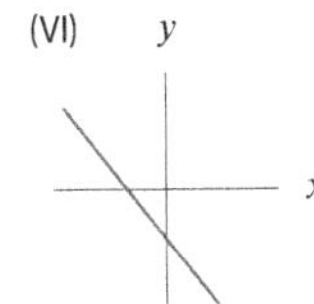

Figure 1.9

13. Match the graphs in Figure 1.10 with the following equations. (Note that the x and y scales may be unequal.)

   (a) $y = -2.72x$
   (b) $y = 0.01 + 0.001x$
   (c) $y = 27.9 - 0.1x$
   (d) $y = 0.1x - 27.9$
   (e) $y = -5.7 - 200x$
   (f) $y = x/3.14$

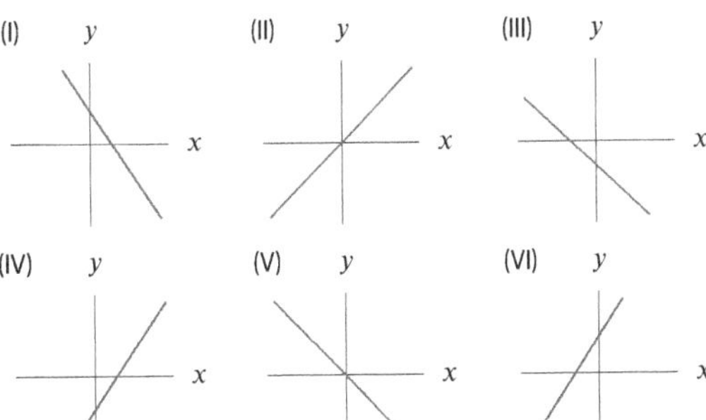

Figure 1.10

14. Estimate the slope and the equation of the line in Figure 1.11.

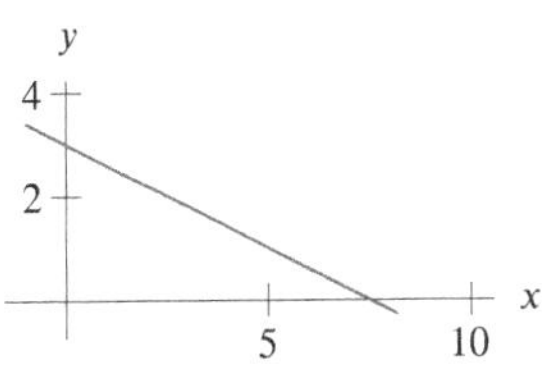

Figure 1.11

15. Find an equation for the line with slope m through the point (a, c).

16. Find a linear function that generates the values in Table 1.3.

Table 1.3

x	5.2	5.3	5.4	5.5	5.6
y	27.8	29.2	30.6	32.0	33.4

■ For Exercises 17–19, use the facts that parallel lines have equal slopes and that the slopes of perpendicular lines are negative reciprocals of one another.

17. Find an equation for the line through the point $(2, 1)$ which is perpendicular to the line $y = 5x - 3$.

18. Find equations for the lines through the point $(1, 5)$ that are parallel to and perpendicular to the line with equation $y + 4x = 7$.

19. Find equations for the lines through the point (a, b) that are parallel and perpendicular to the line $y = mx + c$, assuming $m \neq 0$.

For Exercises 20–23, give the approximate domain and range of each function. Assume the entire graph is shown.

20.

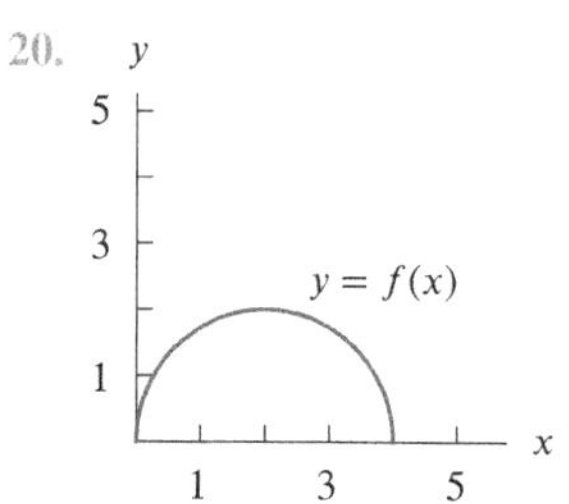

21.

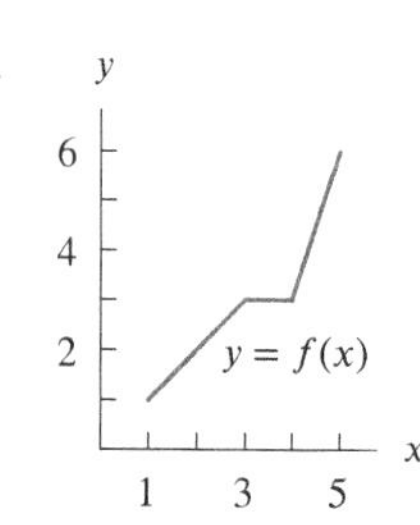

22.

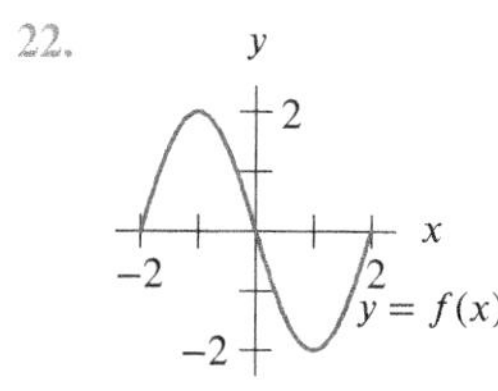

23.

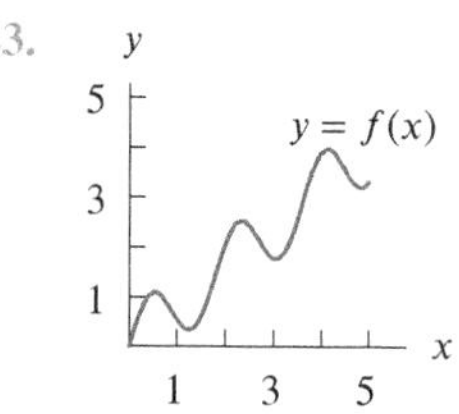

Find the domain and range in Exercises 24–25.

24. $y = x^2 + 2$

25. $y = \dfrac{1}{x^2 + 2}$

26. If $f(t) = \sqrt{t^2 - 16}$, find all values of t for which $f(t)$ is a real number. Solve $f(t) = 3$.

In Exercises 27–31, write a formula representing the function.

27. The volume of a sphere is proportional to the cube of its radius, r.

28. The average velocity, v, for a trip over a fixed distance, d, is inversely proportional to the time of travel, t.

29. The strength, S, of a beam is proportional to the square of its thickness, h.

30. The energy, E, expended by a swimming dolphin is proportional to the cube of the speed, v, of the dolphin.

31. The number of animal species, N, of a certain body length, l, is inversely proportional to the square of l.

PROBLEMS

32. In December 2010, the snowfall in Minneapolis was unusually high,[2] leading to the collapse of the roof of the Metrodome. Figure 1.12 gives the snowfall, S, in Minneapolis for December 6–15, 2010.

(a) How do you know that the snowfall data represents a function of date?
(b) Estimate the snowfall on December 12.
(c) On which day was the snowfall more than 10 inches?
(d) During which consecutive two-day interval was the increase in snowfall largest?

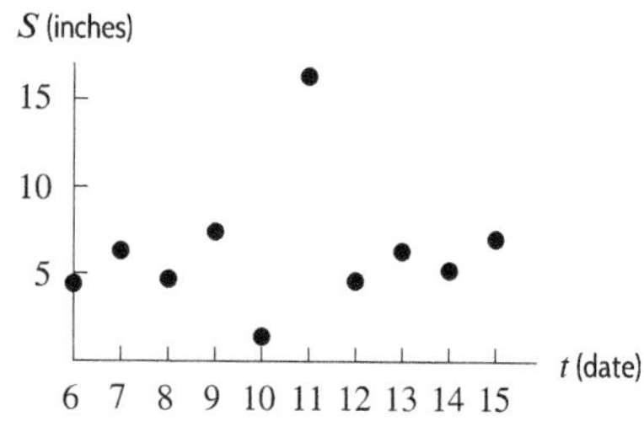

Figure 1.12

33. The value of a car, $V = f(a)$, in thousands of dollars, is a function of the age of the car, a, in years.

(a) Interpret the statement $f(5) = 6$.
(b) Sketch a possible graph of V against a. Is f an increasing or decreasing function? Explain.
(c) Explain the significance of the horizontal and vertical intercepts in terms of the value of the car.

34. Which graph in Figure 1.13 best matches each of the following stories?[3] Write a story for the remaining graph.

(a) I had just left home when I realized I had forgotten my books, so I went back to pick them up.
(b) Things went fine until I had a flat tire.
(c) I started out calmly but sped up when I realized I was going to be late.

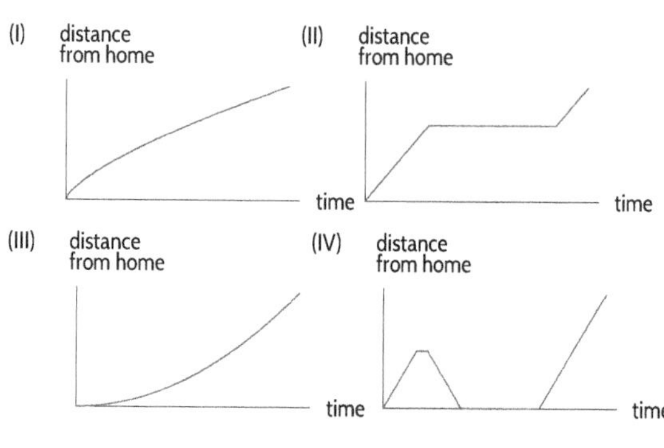

Figure 1.13

In Problems 35–38 the function $S = f(t)$ gives the average annual sea level, S, in meters, in Aberdeen, Scotland,[4]

[2]http://www.crh.noaa.gov/mpx/Climate/DisplayRecords.php

[3]Adapted from Jan Terwel, "Real Math in Cooperative Groups in Secondary Education." *Cooperative Learning in Mathematics*, ed. Neal Davidson, p. 234 (Reading: Addison Wesley, 1990).

[4]www.gov.uk, accessed January 7, 2015.

as a function of t, the number of years before 2012. Write a mathematical expression that represents the given statement.

35. In 2000 the average annual sea level in Aberdeen was 7.049 meters.

36. The average annual sea level in Aberdeen in 2012.

37. The average annual sea level in Aberdeen was the same in 1949 and 2000.

38. The average annual sea level in Aberdeen decreased by 8 millimeters from 2011 to 2012.

■ Problems 39–42 ask you to plot graphs based on the following story: "As I drove down the highway this morning, at first traffic was fast and uncongested, then it crept nearly bumper-to-bumper until we passed an accident, after which traffic flow went back to normal until I exited."

39. Driving speed against time on the highway

40. Distance driven against time on the highway

41. Distance from my exit vs time on the highway

42. Distance between cars vs distance driven on the highway

43. An object is put outside on a cold day at time $t = 0$. Its temperature, $H = f(t)$, in °C, is graphed in Figure 1.14.

(a) What does the statement $f(30) = 10$ mean in terms of temperature? Include units for 30 and for 10 in your answer.
(b) Explain what the vertical intercept, a, and the horizontal intercept, b, represent in terms of temperature of the object and time outside.

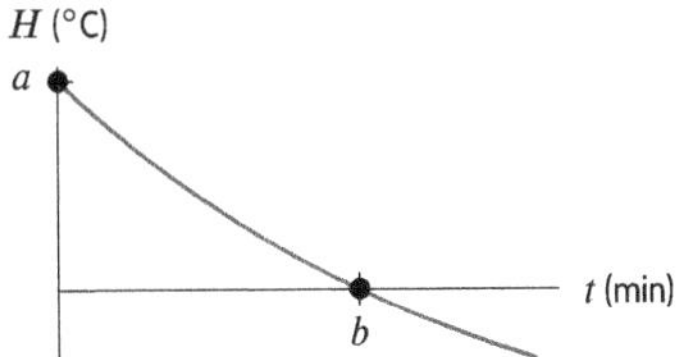

Figure 1.14

44. A rock is dropped from a window and falls to the ground below. The height, s (in meters), of the rock above ground is a function of the time, t (in seconds), since the rock was dropped, so $s = f(t)$.

(a) Sketch a possible graph of s as a function of t.
(b) Explain what the statement $f(7) = 12$ tells us about the rock's fall.
(c) The graph drawn as the answer for part (a) should have a horizontal and vertical intercept. Interpret each intercept in terms of the rock's fall.

45. You drive at a constant speed from Chicago to Detroit, a distance of 275 miles. About 120 miles from Chicago you pass through Kalamazoo, Michigan. Sketch a graph of your distance from Kalamazoo as a function of time.

46. US imports of crude oil and petroleum have been increasing.[5] There have been many ups and downs, but the general trend is shown by the line in Figure 1.15.

(a) Find the slope of the line. Include its units of measurement.
(b) Write an equation for the line. Define your variables, including their units.
(c) Assuming the trend continues, when does the linear model predict imports will reach 18 million barrels per day? Do you think this is a reliable prediction? Give reasons.

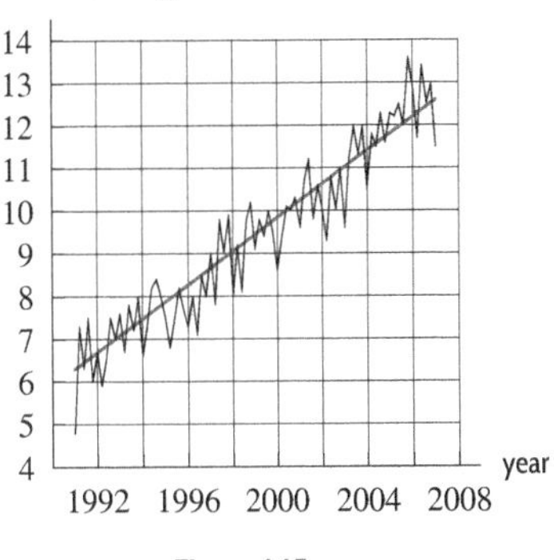

Figure 1.15

■ Problems 47–49 use Figure 1.16 showing how the quantity, Q, of grass (kg/hectare) in different parts of Namibia depended on the average annual rainfall, r, (mm), in two different years.[6]

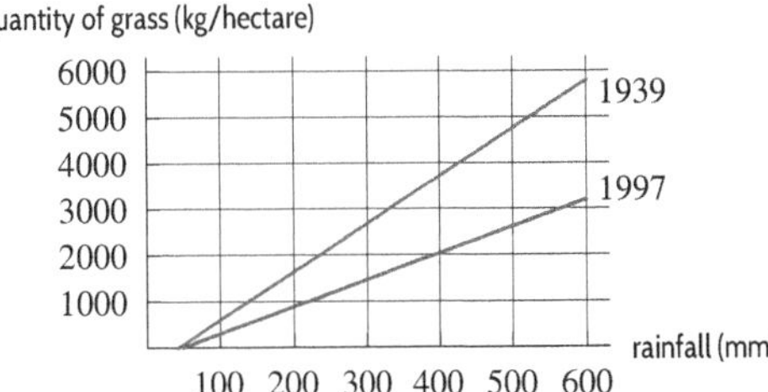

Figure 1.16

47. (a) For 1939, find the slope of the line, including units.
(b) Interpret the slope in this context.
(c) Find the equation of the line.

48. (a) For 1997, find the slope of the line, including units.
(b) Interpret the slope in this context.
(c) Find the equation of the line.

49. Which of the two functions in Figure 1.16 has the larger difference quotient $\Delta Q/\Delta r$? What does this tell us about grass in Namibia?

[5] http://www.theoildrum.com/node/2767. Accessed May 2015.

[6] David Ward and Ben T. Ngairorue, "Are Namibia's Grasslands Desertifying?", *Journal of Range Management* 53, 2000, 138–144.

50. Marmots are large squirrels that hibernate in the winter and come out in the spring. Figure 1.17 shows the date (days after Jan 1) that they are first sighted each year in Colorado as a function of the average minimum daily temperature for that year.[7]

(a) Find the slope of the line, including units.
(b) What does the sign of the slope tell you about marmots?
(c) Use the slope to determine how much difference 6°C warming makes to the date of first appearance of a marmot.
(d) Find the equation of the line.

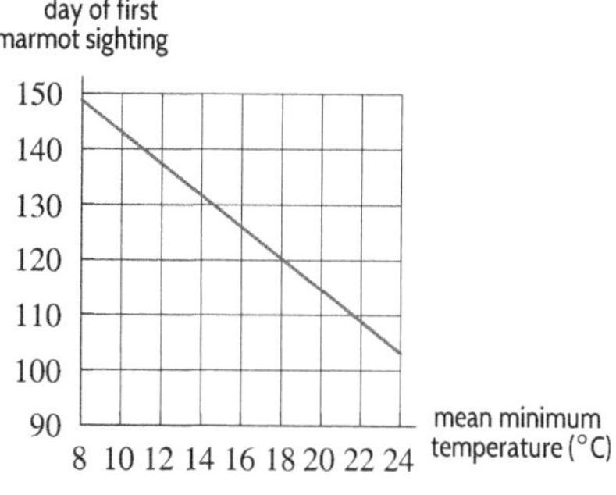

Figure 1.17

51. In Colorado spring has arrived when the bluebell first flowers. Figure 1.18 shows the date (days after Jan 1) that the first flower is sighted in one location as a function of the first date (days after Jan 1) of bare (snow-free) ground.[8]

(a) If the first date of bare ground is 140, how many days later is the first bluebell flower sighted?
(b) Find the slope of the line, including units.
(c) What does the sign of the slope tell you about bluebells?
(d) Find the equation of the line.

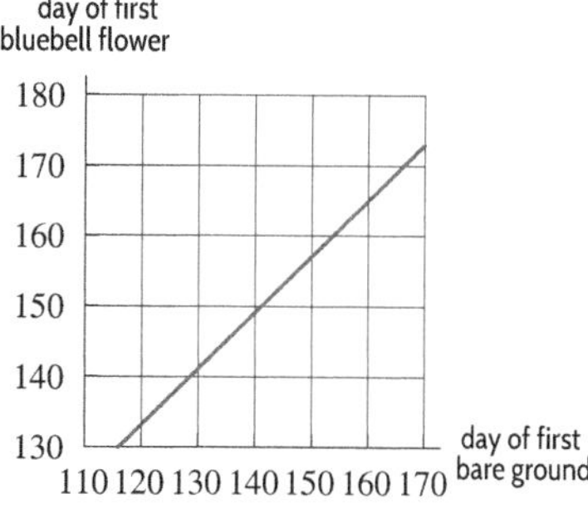

Figure 1.18

52. On March 5, 2015, Capracotta, Italy, received 256 cm (100.787 inches) of snow in 18 hours.[9]

(a) Assuming the snow fell at a constant rate and there were already 100 cm of snow on the ground, find a formula for $f(t)$, in cm, for the depth of snow as a function of t hours since the snowfall began on March 5.
(b) What are the domain and range of f?

53. In a California town, the monthly charge for waste collection is \$8 for 32 gallons of waste and \$12.32 for 68 gallons of waste.

(a) Find a linear formula for the cost, C, of waste collection as a function of the number of gallons of waste, w.
(b) What is the slope of the line found in part (a)? Give units and interpret your answer in terms of the cost of waste collection.
(c) What is the vertical intercept of the line found in part (a)? Give units and interpret your answer in terms of the cost of waste collection.

54. For tax purposes, you may have to report the value of your assets, such as cars or refrigerators. The value you report drops with time. "Straight-line depreciation" assumes that the value is a linear function of time. If a \$950 refrigerator depreciates completely in seven years, find a formula for its value as a function of time.

55. Residents of the town of Maple Grove who are connected to the municipal water supply are billed a fixed amount monthly plus a charge for each cubic foot of water used. A household using 1000 cubic feet was billed \$40, while one using 1600 cubic feet was billed \$55.

(a) What is the charge per cubic foot?
(b) Write an equation for the total cost of a resident's water as a function of cubic feet of water used.
(c) How many cubic feet of water used would lead to a bill of \$100?

56. A controversial 1992 Danish study[10] reported that men's average sperm count decreased from 113 million per milliliter in 1940 to 66 million per milliliter in 1990.

(a) Express the average sperm count, S, as a linear function of the number of years, t, since 1940.
(b) A man's fertility is affected if his sperm count drops below about 20 million per milliliter. If the linear model found in part (a) is accurate, in what year will the average male sperm count fall below this level?

[7] David W. Inouye, Billy Barr, Kenneth B. Armitage, and Brian D. Inouye, "Climate change is affecting altitudinal migrants and hibernating species", *PNAS* 97, 2000, 1630–1633.
[8] David W. Inouye, Billy Barr, Kenneth B. Armitage, and Brian D. Inouye, "Climate change is affecting altitudinal migrants and hibernating species", *PNAS* 97, 2000, 1630–1633.
[9] http://iceagenow.info/2015/03/official-italy-captures-world-one-day-snowfall-record/
[10] "Investigating the Next Silent Spring," *US News and World Report*, pp. 50–52 (March 11, 1996).

57. Let $f(t)$ be the number of US billionaires in year t.

(a) Express the following statements[11] in terms of f.
  (i) In 2001 there were 272 US billionaires.
  (ii) In 2014 there were 525 US billionaires.

(b) Find the average yearly increase in the number of US billionaires from 2001 to 2014. Express this using f.

(c) Assuming the yearly increase remains constant, find a formula predicting the number of US billionaires in year t.

58. The cost of planting seed is usually a function of the number of acres sown. The cost of the equipment is a *fixed cost* because it must be paid regardless of the number of acres planted. The costs of supplies and labor vary with the number of acres planted and are called *variable costs*. Suppose the fixed costs are \$10,000 and the variable costs are \$200 per acre. Let C be the total cost, measured in thousands of dollars, and let x be the number of acres planted.

(a) Find a formula for C as a function of x.
(b) Graph C against x.
(c) Which feature of the graph represents the fixed costs? Which represents the variable costs?

59. An airplane uses a fixed amount of fuel for takeoff, a (different) fixed amount for landing, and a third fixed amount per mile when it is in the air. How does the total quantity of fuel required depend on the length of the trip? Write a formula for the function involved. Explain the meaning of the constants in your formula.

60. For the line $y = f(x)$ in Figure 1.19, evaluate

(a) $f(423) - f(422)$ **(b)** $f(517) - f(513)$

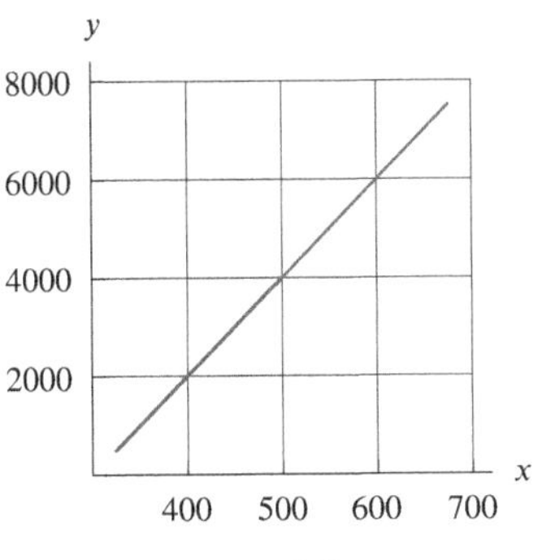

Figure 1.19

61. For the line $y = g(x)$ in Figure 1.20, evaluate

(a) $g(4210) - g(4209)$ **(b)** $g(3760) - g(3740)$

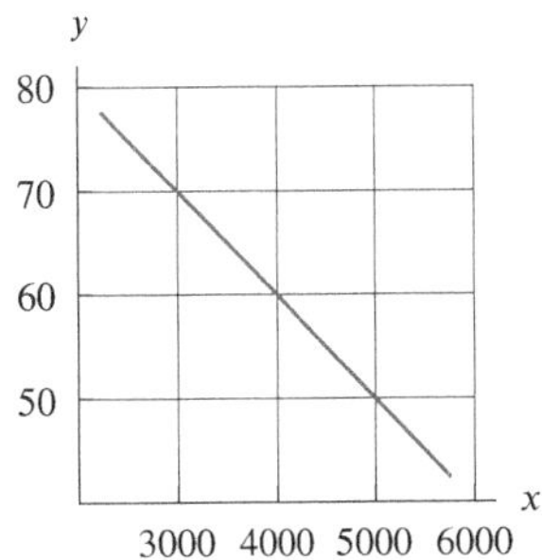

Figure 1.20

62. An alternative to petroleum-based diesel fuel, biodiesel, is derived from renewable resources such as food crops, algae, and animal oils. The table shows the recent annual percent growth in US biodiesel exports.[12]

(a) Find the largest time interval over which the percentage growth in the US exports of biodiesel was an increasing function of time. Interpret what increasing means, practically speaking, in this case.

(b) Find the largest time interval over which the actual US exports of biodiesel was an increasing function of time. Interpret what increasing means, practically speaking, in this case.

Year	2010	2011	2012	2013	2014
% growth over previous yr	−60.5	−30.5	69.9	53.0	−57.8

63. Hydroelectric power is electric power generated by the force of moving water. Figure 1.21 shows[13] the annual percent growth in hydroelectric power consumption by the US industrial sector between 2006 and 2014.

(a) Find the largest time interval over which the percentage growth in the US consumption of hydroelectric power was an increasing function of time. Interpret what increasing means, practically speaking, in this case.

(b) Find the largest time interval over which the actual US consumption of hydroelectric power was a decreasing function of time. Interpret what decreasing means, practically speaking, in this case.

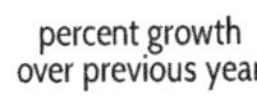

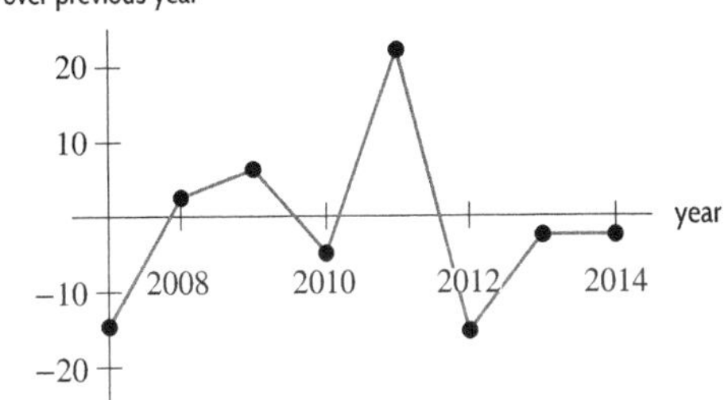

Figure 1.21

[11] www.statista.com, accessed March 18, 2015.
[12] www.eia.doe.gov, accessed March 29, 2015.
[13] Yearly values have been joined with line segments to highlight trends in the data; however, values in between years should not be inferred from the segments. From www.eia.doe.gov, accessed March 29, 2015.

64. Solar panels are arrays of photovoltaic cells that convert solar radiation into electricity. The table shows the annual percent change in the US price per watt of a solar panel.[14]

Year	2005	2006	2007	2008	2009	2010
% growth over previous yr	6.7	9.7	−3.7	3.6	−20.1	−29.7

(a) Find the largest time interval over which the percentage growth in the US price per watt of a solar panel was a decreasing function of time. Interpret what decreasing means, practically speaking, in this case.

(b) Find the largest time interval over which the actual price per watt of a solar panel was a decreasing function of time. Interpret what decreasing means, practically speaking, in this case.

65. Table 1.4 shows the average annual sea level, S, in meters, in Aberdeen, Scotland,[15] as a function of time, t, measured in years before 2008.

Table 1.4

t	0	25	50	75	100	125
S	7.094	7.019	6.992	6.965	6.938	6.957

(a) What was the average sea level in Aberdeen in 2008?

(b) In what year was the average sea level 7.019 meters? 6.957 meters?

(c) Table 1.5 gives the average sea level, S, in Aberdeen as a function of the year, x. Complete the missing values.

Table 1.5

x	1883	?	1933	1958	1983	2008
S	?	6.938	?	6.992	?	?

66. The table gives the required standard weight, w, in kilograms, of American soldiers, aged between 21 and 27, for height, h, in centimeters.[16]

(a) How do you know that the data in this table could represent a linear function?

(b) Find weight, w, as a linear function of height, h. What is the slope of the line? What are the units for the slope?

(c) Find height, h, as a linear function of weight, w. What is the slope of the line? What are the units for the slope?

h (cm)	172	176	180	184	188	192	196
w (kg)	79.7	82.4	85.1	87.8	90.5	93.2	95.9

67. A company rents cars at \$40 a day and 15 cents a mile. Its competitor's cars are \$50 a day and 10 cents a mile.

(a) For each company, give a formula for the cost of renting a car for a day as a function of the distance traveled.

(b) On the same axes, graph both functions.

(c) How should you decide which company is cheaper?

68. A \$25,000 vehicle depreciates \$2000 a year as it ages. Repair costs are \$1500 per year.

(a) Write formulas for each of the two linear functions at time t, value, $V(t)$, and repair costs to date, $C(t)$. Graph them.

(b) One strategy is to replace a vehicle when the total cost of repairs is equal to the current value. Find this time.

(c) Another strategy is to replace the vehicle when the value of the vehicle is some percent of the original value. Find the time when the value is 6%.

69. A bakery owner knows that customers buy a total of q cakes when the price, p, is no more than $p = d(q) = 20 - q/20$ dollars. She is willing to make and supply as many as q cakes at a price of $p = s(q) = 11 + q/40$ dollars each. (The graphs of the functions $d(q)$ and $s(q)$ are called a *demand curve* and a *supply curve*, respectively.) The graphs of $d(q)$ and $s(q)$ are in Figure 1.22.

(a) Why, in terms of the context, is the slope of $d(q)$ negative and the slope of $s(q)$ positive?

(b) Is each of the ordered pairs (q, p) a solution to the inequality $p \leq 20 - q/20$? Interpret your answers in terms of the context.

$$(60, 18) \qquad (120, 12)$$

(c) Graph in the qp-plane the solution set of the system of inequalities $p \leq 20 - q/20$, $p \geq 11 + q/40$. What does this solution set represent in terms of the context?

(d) What is the rightmost point of the solution set you graphed in part (c)? Interpret your answer in terms of the context.

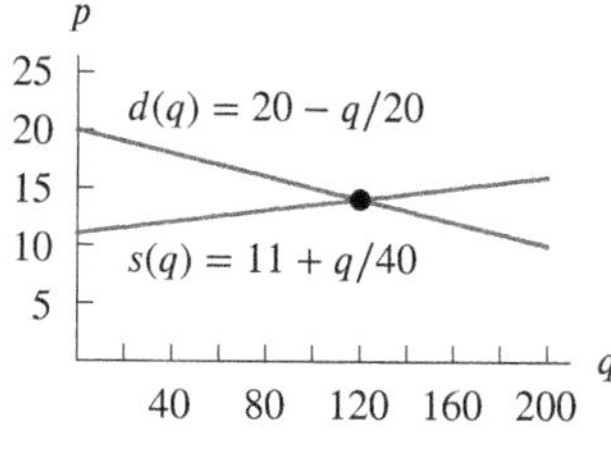

Figure 1.22

[14] We use the official price per peak watt, which uses the maximum number of watts a solar panel can produce under ideal conditions. From www.eia.doe.gov, accessed March 29, 2015.

[15] www.decc.gov.uk, accessed June 2011.

[16] Adapted from usmilitary.about.com, accessed March 29, 2015.

70. **(a)** Consider the functions graphed in Figure 1.23(a). Find the coordinates of C.
(b) Consider the functions in Figure 1.23(b). Find the coordinates of C in terms of b.

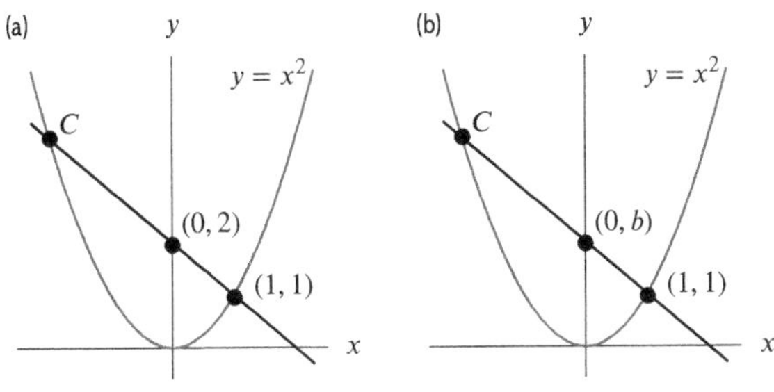

Figure 1.23

71. When Galileo was formulating the laws of motion, he considered the motion of a body starting from rest and falling under gravity. He originally thought that the velocity of such a falling body was proportional to the distance it had fallen. What do the experimental data in Table 1.6 tell you about Galileo's hypothesis? What alternative hypothesis is suggested by the two sets of data in Table 1.6 and Table 1.7?

Table 1.6

Distance (ft)	0	1	2	3	4
Velocity (ft/sec)	0	8	11.3	13.9	16

Table 1.7

Time (sec)	0	1	2	3	4
Velocity (ft/sec)	0	32	64	96	128

Strengthen Your Understanding

■ In Problems 72–76, explain what is wrong with the statement.

72. For constants m and b, the slope of the linear function $y = b + mx$ is $m = \Delta x/\Delta y$.
73. The lines $x = 3$ and $y = 3$ are both linear functions of x.
74. The line $y - 3 = 0$ has slope 1 in the xy-plane.
75. Values of y on the graph of $y = 0.5x - 3$ increase more slowly than values of y on the graph of $y = 0.5 - 3x$.
76. The equation $y = 2x + 1$ indicates that y is directly proportional to x with a constant of proportionality 2.

■ In Problems 77–78, give an example of:

77. A linear function with a positive slope and a negative x-intercept.
78. A formula representing the statement "q is inversely proportional to the cube root of p and has a positive constant of proportionality."

■ Are the statements in Problems 79–84 true or false? Give an explanation for your answer.

79. For any two points in the plane, there is a linear function whose graph passes through them.
80. If $y = f(x)$ is a linear function, then increasing x by 1 unit changes the corresponding y by m units, where m is the slope.
81. The linear functions $y = -x + 1$ and $x = -y + 1$ have the same graph.
82. The linear functions $y = 2 - 2x$ and $x = 2 - 2y$ have the same graph.
83. If y is a linear function of x, then the ratio y/x is constant for all points on the graph at which $x \neq 0$.
84. If $y = f(x)$ is a linear function, then increasing x by 2 units adds $m + 2$ units to the corresponding y, where m is the slope.
85. Which of the following functions has its domain identical with its range?

(a) $f(x) = x^2$ **(b)** $g(x) = \sqrt{x}$
(c) $h(x) = x^3$ **(d)** $i(x) = |x|$

1.2 EXPONENTIAL FUNCTIONS

Population Growth

The population of Burkina Faso, a sub-Saharan African country,[17] from 2007 to 2013 is given in Table 1.8. To see how the population is growing, we look at the increase in population in the third column. If the population had been growing linearly, all the numbers in the third column would be the same.

[17]dataworldbank.org, accessed March 29, 2015.

Table 1.8 *Population of Burkina Faso (estimated), 2007–2013*

Year	Population (millions)	Change in population (millions)
2007	14.235	
		0.425
2008	14.660	
		0.435
2009	15.095	
		0.445
2010	15.540	
		0.455
2011	15.995	
		0.465
2012	16.460	
		0.474
2013	16.934	

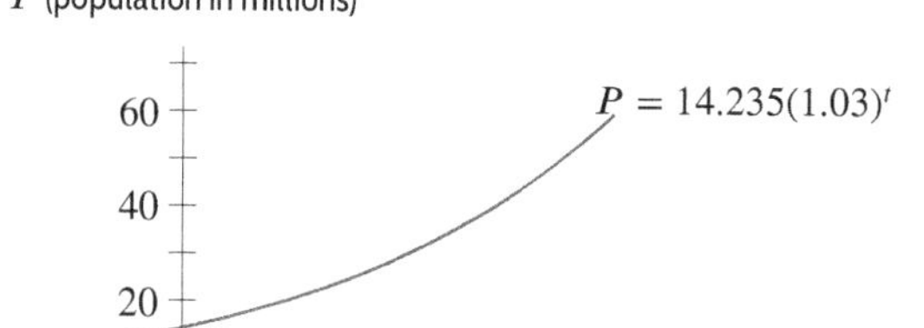

Figure 1.24: Population of Burkina Faso (estimated): Exponential growth

Suppose we divide each year's population by the previous year's population. For example,

$$\frac{\text{Population in 2008}}{\text{Population in 2007}} = \frac{14.660 \text{ million}}{14.235 \text{ million}} = 1.03$$

$$\frac{\text{Population in 2009}}{\text{Population in 2008}} = \frac{15.095 \text{ million}}{14.660 \text{ million}} = 1.03.$$

The fact that both calculations give 1.03 shows the population grew by about 3% between 2007 and 2008 *and* between 2008 and 2009. Similar calculations for other years show that the population grew by a factor of about 1.03, or 3%, every year. Whenever we have a constant growth factor (here 1.03), we have exponential growth. The population t years after 2007 is given by the *exponential* function

$$P = 14.235(1.03)^t.$$

If we assume that the formula holds for 50 years, the population graph has the shape shown in Figure 1.24. Since the population is growing faster and faster as time goes on, the graph is bending upward; we say it is *concave up*. Even exponential functions which climb slowly at first, such as this one, eventually climb extremely quickly.

> To recognize that a table of t and P values comes from an exponential function, look for ratios of P values that are constant for equally spaced t values.

Concavity

We have used the term concave up[19] to describe the graph in Figure 1.24. In words:

> The graph of a function is **concave up** if it bends upward as we move left to right; it is **concave down** if it bends downward. (See Figure 1.25 for four possible shapes.) A line is neither concave up nor concave down.

Figure 1.25: Concavity of a graph

[19] In Chapter 2 we consider concavity in more depth.

Elimination of a Drug from the Body

Now we look at a quantity which is decreasing exponentially instead of increasing. When a patient is given medication, the drug enters the bloodstream. As the drug passes through the liver and kidneys, it is metabolized and eliminated at a rate that depends on the particular drug. For the antibiotic ampicillin, approximately 40% of the drug is eliminated every hour. A typical dose of ampicillin is 250 mg. Suppose $Q = f(t)$, where Q is the quantity of ampicillin, in mg, in the bloodstream at time t hours since the drug was given. At $t = 0$, we have $Q = 250$. Since every hour the amount remaining is 60% of the previous amount, we have

$$f(0) = 250$$
$$f(1) = 250(0.6)$$
$$f(2) = (250(0.6))(0.6) = 250(0.6)^2,$$

and after t hours,

$$Q = f(t) = 250(0.6)^t.$$

This is an *exponential decay function*. Some values of the function are in Table 1.9; its graph is in Figure 1.26.

Notice the way in which the function in Figure 1.26 is decreasing. Each hour a smaller quantity of the drug is removed than in the previous hour. This is because as time passes, there is less of the drug in the body to be removed. Compare this to the exponential growth in Figure 1.24, where each step upward is larger than the previous one. Notice, however, that both graphs are concave up.

Table 1.9 *Drug elimination*

t (hours)	Q (mg)
0	250
1	150
2	90
3	54
4	32.4
5	19.4

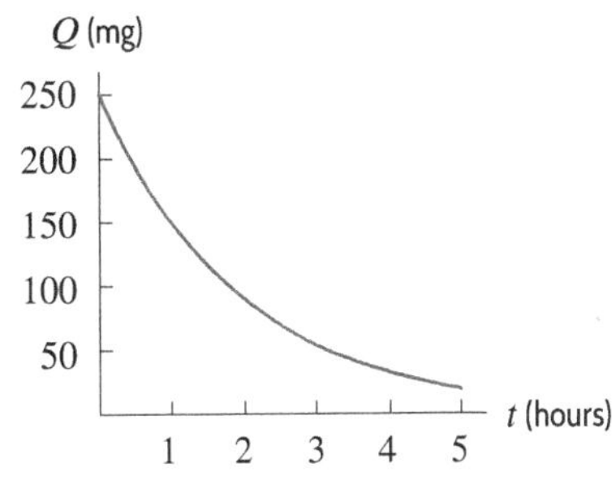

Figure 1.26: Drug elimination: Exponential decay

The General Exponential Function

> We say P is an **exponential function** of t with base a if
>
> $$P = P_0 a^t,$$
>
> where P_0 is the initial quantity (when $t = 0$) and a is the factor by which P changes when t increases by 1.
> If $a > 1$, we have exponential growth; if $0 < a < 1$, we have exponential decay.

Provided $a > 0$, the largest possible domain for the exponential function is all real numbers. The reason we do not want $a \leq 0$ is that, for example, we cannot define $a^{1/2}$ if $a < 0$. Also, we do not usually have $a = 1$, since $P = P_0 1^t = P_0$ is then a constant function.

The value of a is closely related to the percent growth (or decay) rate. For example, if $a = 1.03$, then P is growing at 3%; if $a = 0.94$, then P is decaying at 6%, so the growth rate is $r = a - 1$.

Example 1 Suppose that $Q = f(t)$ is an exponential function of t. If $f(20) = 88.2$ and $f(23) = 91.4$:
(a) Find the base. (b) Find the growth rate. (c) Evaluate $f(25)$.

Solution (a) Let

$$Q = Q_0 a^t.$$

Substituting $t = 20, Q = 88.2$ and $t = 23, Q = 91.4$ gives two equations for Q_0 and a:

$$88.2 = Q_0 a^{20} \quad \text{and} \quad 91.4 = Q_0 a^{23}.$$

Dividing the two equations enables us to eliminate Q_0:

$$\frac{91.4}{88.2} = \frac{Q_0 a^{23}}{Q_0 a^{20}} = a^3.$$

Solving for the base, a, gives

$$a = \left(\frac{91.4}{88.2}\right)^{1/3} = 1.012.$$

(b) Since $a = 1.012$, the growth rate is $1.012 - 1 = 0.012 = 1.2\%$.
(c) We want to evaluate $f(25) = Q_0 a^{25} = Q_0(1.012)^{25}$. First we find Q_0 from the equation

$$88.2 = Q_0(1.012)^{20}.$$

Solving gives $Q_0 = 69.5$. Thus,

$$f(25) = 69.5(1.012)^{25} = 93.6.$$

Half-Life and Doubling Time

Radioactive substances, such as uranium, decay exponentially. A certain percentage of the mass disintegrates in a given unit of time; the time it takes for half the mass to decay is called the *half-life* of the substance.

A well-known radioactive substance is carbon-14, which is used to date organic objects. When a piece of wood or bone was part of a living organism, it accumulated small amounts of radioactive carbon-14. Once the organism dies, it no longer picks up carbon-14. Using the half-life of carbon-14 (about 5730 years), we can estimate the age of the object. We use the following definitions:

> The **half-life** of an exponentially decaying quantity is the time required for the quantity to be reduced by a factor of one half.
> The **doubling time** of an exponentially increasing quantity is the time required for the quantity to double.

The Family of Exponential Functions

The formula $P = P_0 a^t$ gives a family of exponential functions with positive parameters P_0 (the initial quantity) and a (the base, or growth/decay factor). The base tells us whether the function is increasing ($a > 1$) or decreasing ($0 < a < 1$). Since a is the factor by which P changes when t is increased by 1, large values of a mean fast growth; values of a near 0 mean fast decay. (See Figures 1.27 and 1.28.) All members of the family $P = P_0 a^t$ are concave up.

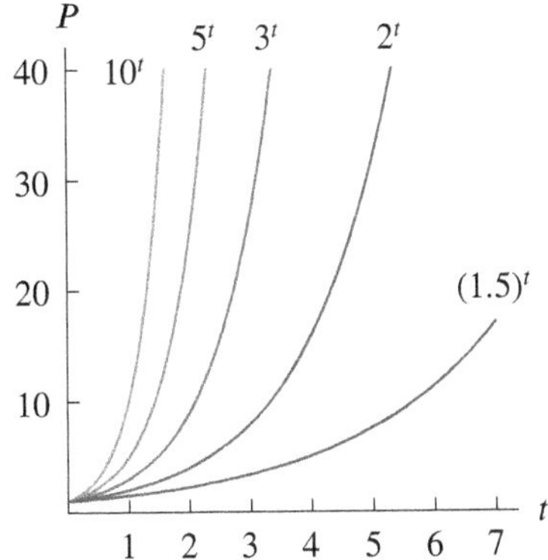

Figure 1.27: Exponential growth: $P = a^t$, for $a > 1$

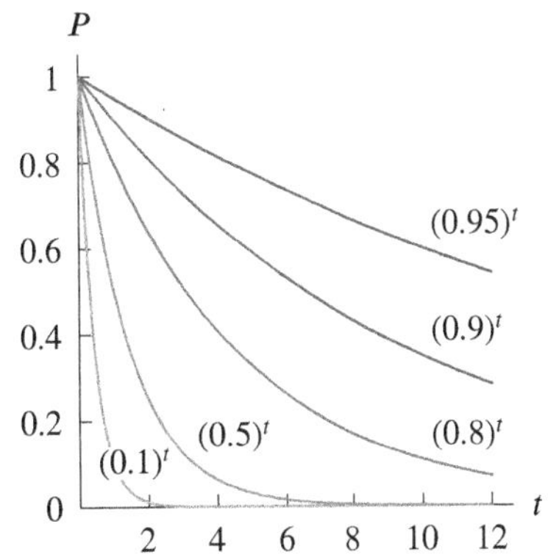

Figure 1.28: Exponential decay: $P = a^t$, for $0 < a < 1$

Example 2 Figure 1.29 is the graph of three exponential functions. What can you say about the values of the six constants a, b, c, d, p, q?

Solution All the constants are positive. Since a, c, p represent y-intercepts, we see that $a = c$ because these graphs intersect on the y-axis. In addition, $a = c < p$, since $y = p \cdot q^x$ crosses the y-axis above the other two.

Since $y = a \cdot b^x$ is decreasing, we have $0 < b < 1$. The other functions are increasing, so $1 < d$ and $1 < q$.

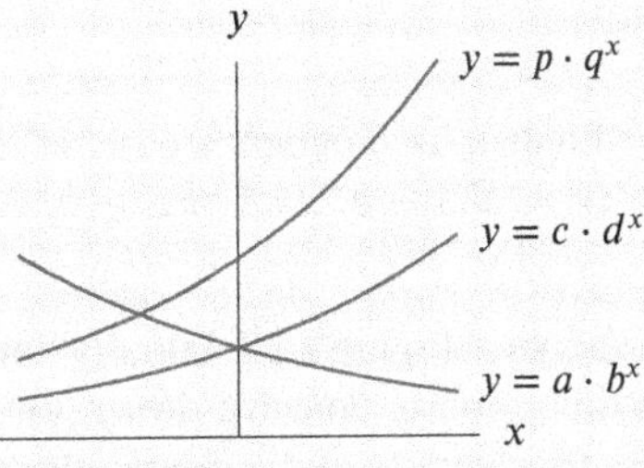

Figure 1.29: Three exponential functions

Exponential Functions with Base *e*

The most frequently used base for an exponential function is the famous number $e = 2.71828\ldots$. This base is used so often that you will find an e^x button on most scientific calculators. At first glance, this is all somewhat mysterious. Why is it convenient to use the base $2.71828\ldots$? The full answer to that question must wait until Chapter 3, where we show that many calculus formulas come out neatly when e is used as the base. We often use the following result:

Any **exponential growth** function can be written, for some $a > 1$ and $k > 0$, in the form

$$P = P_0 a^t \quad \text{or} \quad P = P_0 e^{kt}$$

and any **exponential decay** function can be written, for some $0 < a < 1$ and $-k < 0$, as

$$Q = Q_0 a^t \quad \text{or} \quad Q = Q_0 e^{-kt},$$

where P_0 and Q_0 are the initial quantities.

We say that P and Q are growing or decaying at a *continuous*[19] *rate* of k. (For example, $k = 0.02$ corresponds to a continuous rate of 2%.)

[19]The reason that k is called the continuous rate is explored in detail in Chapter 11.

Example 3 Convert the functions $P = e^{0.5t}$ and $Q = 5e^{-0.2t}$ into the form $y = y_0 a^t$. Use the results to explain the shape of the graphs in Figures 1.30 and 1.31.

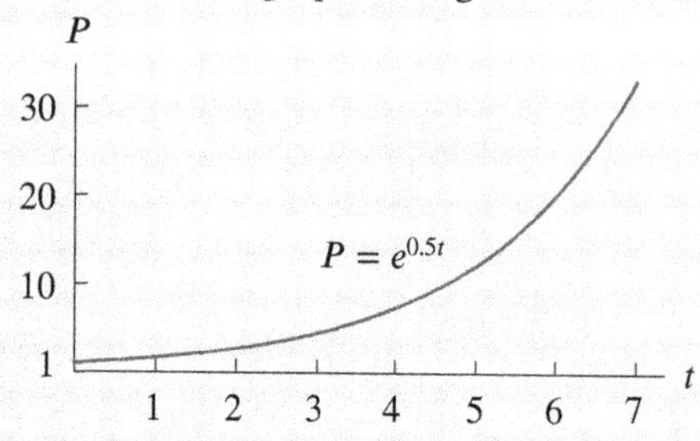

Figure 1.30: An exponential growth function

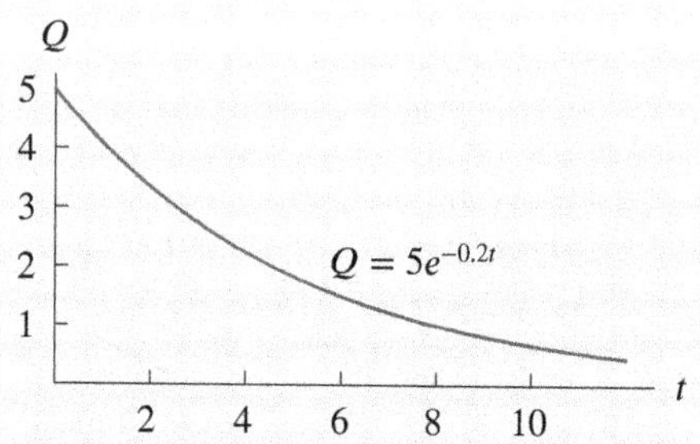

Figure 1.31: An exponential decay function

Solution We have

$$P = e^{0.5t} = (e^{0.5})^t = (1.65)^t.$$

Thus, P is an exponential growth function with $P_0 = 1$ and $a = 1.65$. The function is increasing and its graph is concave up, similar to those in Figure 1.27. Also,

$$Q = 5e^{-0.2t} = 5(e^{-0.2})^t = 5(0.819)^t,$$

so Q is an exponential decay function with $Q_0 = 5$ and $a = 0.819$. The function is decreasing and its graph is concave up, similar to those in Figure 1.28.

Example 4 The quantity, Q, of a drug in a patient's body at time t is represented for positive constants S and k by the function $Q = S(1 - e^{-kt})$. For $t \geq 0$, describe how Q changes with time. What does S represent?

Solution The graph of Q is shown in Figure 1.32. Initially none of the drug is present, but the quantity increases with time. Since the graph is concave down, the quantity increases at a decreasing rate. This is realistic because as the quantity of the drug in the body increases, so does the rate at which the body excretes the drug. Thus, we expect the quantity to level off. Figure 1.32 shows that S is the saturation level. The line $Q = S$ is called a *horizontal asymptote*.

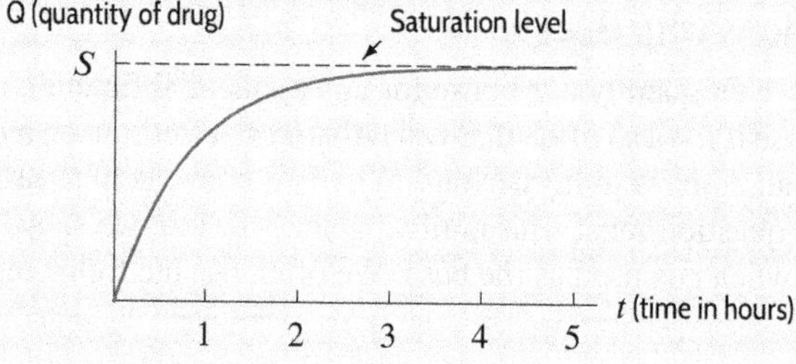

Figure 1.32: Buildup of the quantity of a drug in body

Exercises and Problems for Section 1.2

EXERCISES

In Exercises 1–4, decide whether the graph is concave up, concave down, or neither.

1. 

2. 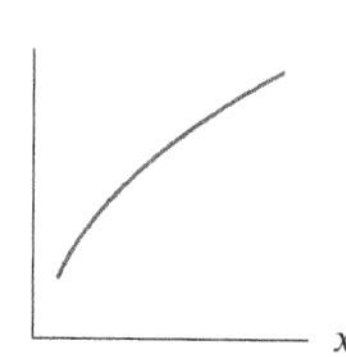

3. 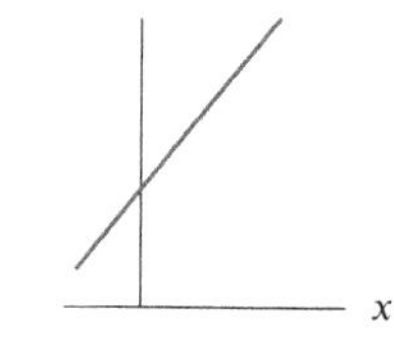

4. 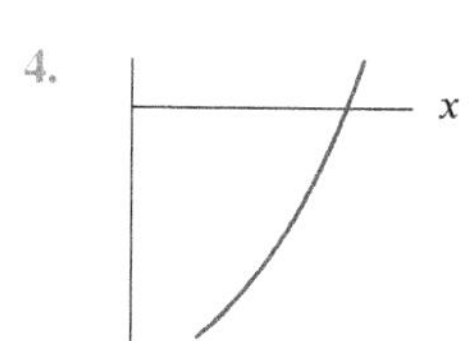

■ The functions in Exercises 5–8 represent exponential growth or decay. What is the initial quantity? What is the growth rate? State if the growth rate is continuous.

5. $P = 5(1.07)^t$

6. $P = 7.7(0.92)^t$

7. $P = 3.2e^{0.03t}$

8. $P = 15e^{-0.06t}$

■ Write the functions in Exercises 9–12 in the form $P = P_0 a^t$. Which represent exponential growth and which represent exponential decay?

9. $P = 15e^{0.25t}$

10. $P = 2e^{-0.5t}$

11. $P = P_0 e^{0.2t}$

12. $P = 7e^{-\pi t}$

■ In Exercises 13–14, let $f(t) = Q_0 a^t = Q_0(1 + r)^t$.

(a) Find the base, a.

(b) Find the percentage growth rate, r.

13. $f(5) = 75.94$ and $f(7) = 170.86$

14. $f(0.02) = 25.02$ and $f(0.05) = 25.06$

15. A town has a population of 1000 people at time $t = 0$. In each of the following cases, write a formula for the population, P, of the town as a function of year t.

(a) The population increases by 50 people a year.
(b) The population increases by 5% a year.

16. An air-freshener starts with 30 grams and evaporates over time. In each of the following cases, write a formula for the quantity, Q grams, of air-freshener remaining t days after the start and sketch a graph of the function. The decrease is:

(a) 2 grams a day **(b)** 12% a day

17. For which pairs of consecutive points in Figure 1.33 is the function graphed:

(a) Increasing and concave up?
(b) Increasing and concave down?
(c) Decreasing and concave up?
(d) Decreasing and concave down?

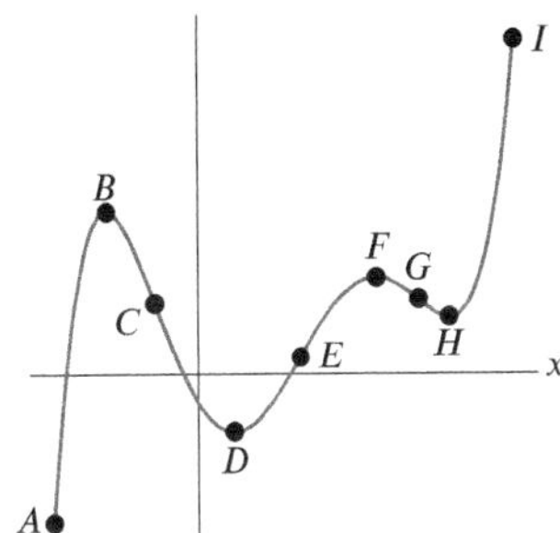

Figure 1.33

18. The table gives the average temperature in Wallingford, Connecticut, for the first 10 days in March.

(a) Over which intervals was the average temperature increasing? Decreasing?
(b) Find a pair of consecutive intervals over which the average temperature was increasing at a decreasing rate. Find another pair of consecutive intervals over which the average temperature was increasing at an increasing rate.

Day	1	2	3	4	5	6	7	8	9	10
°F	42°	42°	34°	25°	22°	34°	38°	40°	49°	49°

PROBLEMS

19. **(a)** Which (if any) of the functions in the following table could be linear? Find formulas for those functions.
(b) Which (if any) of these functions could be exponential? Find formulas for those functions.

x	$f(x)$	$g(x)$	$h(x)$
−2	12	16	37
−1	17	24	34
0	20	36	31
1	21	54	28
2	18	81	25

■ In Problems 20–21, find all the tables that have the given characteristic.

(A)

x	0	40	80	160
y	2.2	2.2	2.2	2.2

(B)

x	−8	−4	0	8
y	51	62	73	95

(C)

x	−4	−3	4	6
y	18	0	4.5	−2.25

(D)

x	3	4	5	6
y	18	9	4.5	2.25

20. y could be a linear function of x.

21. y could be an exponential function of x.

22. Table 1.10 shows some values of a linear function f and an exponential function g. Find exact values (not decimal approximations) for each of the missing entries.

Table 1.10

x	0	1	2	3	4
$f(x)$	10	?	20	?	?
$g(x)$	10	?	20	?	?

23. Match the functions $h(s)$, $f(s)$, and $g(s)$, whose values are in Table 1.11, with the formulas

$$y = a(1.1)^s\,, \quad y = b(1.05)^s\,, \quad y = c(1.03)^s,$$

assuming a, b, and c are constants. Note that the function values have been rounded to two decimal places.

Table 1.11

s	$h(s)$	s	$f(s)$	s	$g(s)$
2	1.06	1	2.20	3	3.47
3	1.09	2	2.42	4	3.65
4	1.13	3	2.66	5	3.83
5	1.16	4	2.93	6	4.02
6	1.19	5	3.22	7	4.22

24. Each of the functions g, h, k in Table 1.12 is increasing, but each increases in a different way. Which of the graphs in Figure 1.34 best fits each function?

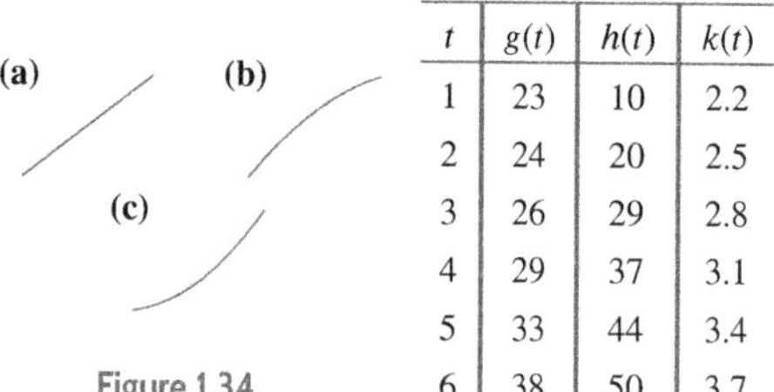

Figure 1.34

Table 1.12

t	$g(t)$	$h(t)$	$k(t)$
1	23	10	2.2
2	24	20	2.5
3	26	29	2.8
4	29	37	3.1
5	33	44	3.4
6	38	50	3.7

25. Each of the functions in Table 1.13 decreases, but each decreases in a different way. Which of the graphs in Figure 1.35 best fits each function?

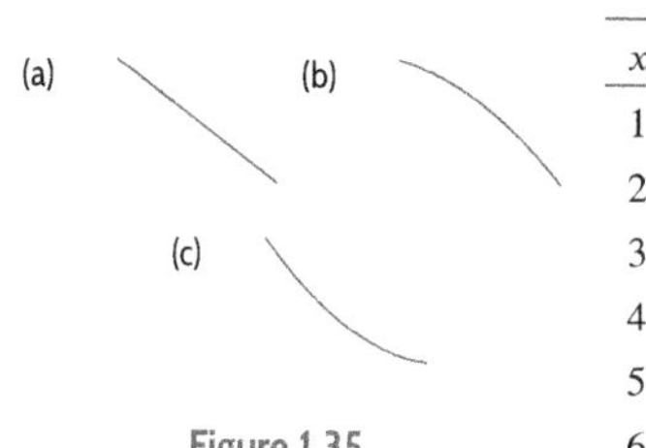

Figure 1.35

Table 1.13

x	$f(x)$	$g(x)$	$h(x)$
1	100	22.0	9.3
2	90	21.4	9.1
3	81	20.8	8.8
4	73	20.2	8.4
5	66	19.6	7.9
6	60	19.0	7.3

26. Figure 1.36 shows $Q = 50(1.2)^t$, $Q = 50(0.6)^t$, $Q = 50(0.8)^t$, and $Q = 50(1.4)^t$. Match each formula to a graph.

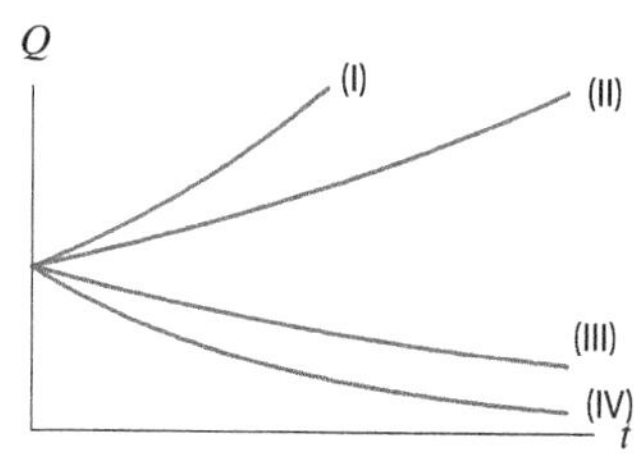

Figure 1.36

■ In Problems 27–32, give a possible formula for the function.

27.

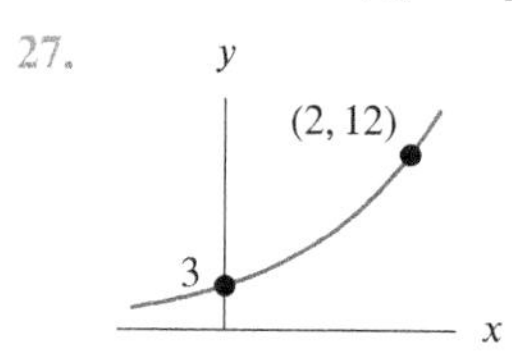

28.

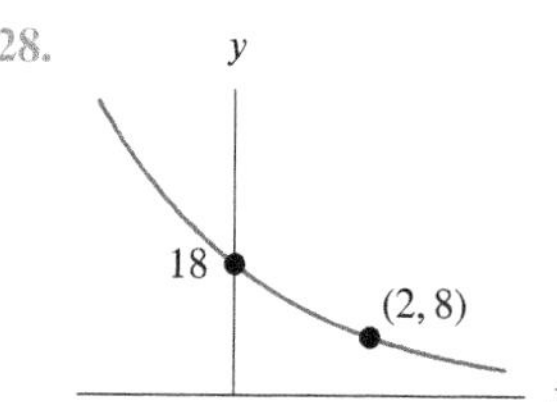

29.

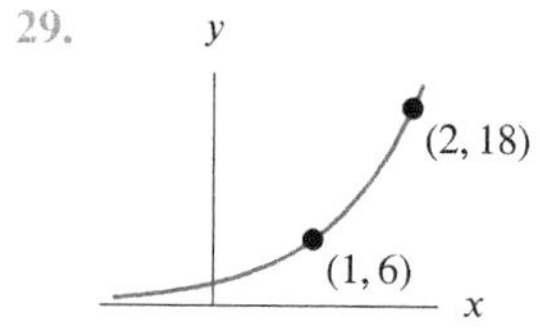

30.

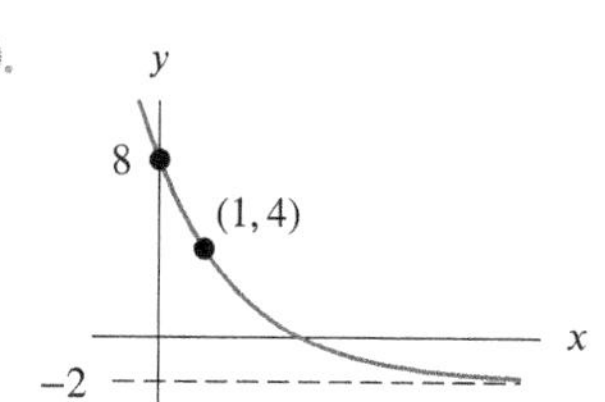

31.

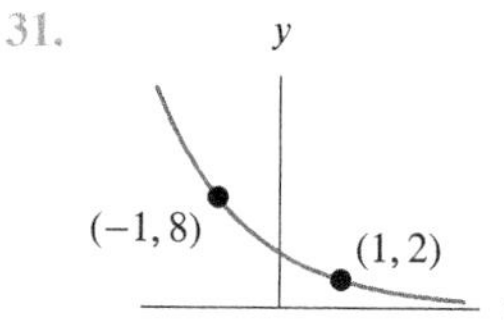

32.

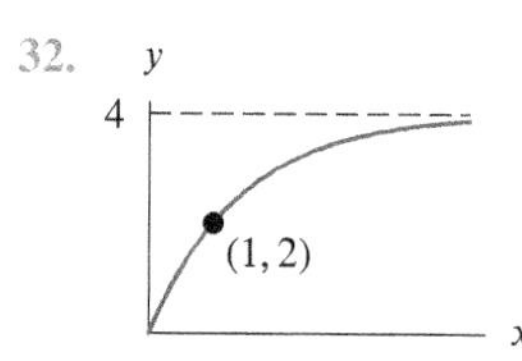

33. The table gives the number of North American houses (millions) with analog cable TV.[20]

(a) Plot the number of houses, H, in millions, with cable TV versus year, Y.
(b) Could H be a linear function of Y? Why or why not?
(c) Could H be an exponential function of Y? Why or why not?

Year	2010	2011	2012	2013	2014	2015
Houses	18.3	13	7.8	3.9	1	0.5

34. When a new product is advertised, more and more people try it. However, the rate at which new people try it slows as time goes on.

(a) Graph the total number of people who have tried such a product against time.
(b) What do you know about the concavity of the graph?

[20]http://www.statista.com. Accessed May 2015.

35. Sketch reasonable graphs for the following. Pay particular attention to the concavity of the graphs.

(a) The total revenue generated by a car rental business, plotted against the amount spent on advertising.
(b) The temperature of a cup of hot coffee standing in a room, plotted as a function of time.

36. (a) A population, P, grows at a continuous rate of 2% a year and starts at 1 million. Write P in the form $P = P_0e^{kt}$, with P_0, k constants.
(b) Plot the population in part (a) against time.

37. A 2008 study of 300 oil fields producing a total of 84 million barrels per day reported that daily production was decaying at a continuous rate of 9.1% per year.[21] Find the estimated production in these fields in 2025 if the decay continues at the same rate.

38. In 2014, the world's population reached 7.17 billion[22] and was increasing at a rate of 1.1% per year. Assume that this growth rate remains constant. (In fact, the growth rate has decreased since 2008.)

(a) Write a formula for the world population (in billions) as a function of the number of years since 2014.
(b) Estimate the population of the world in the year 2020.
(c) Sketch world population as a function of years since 2014. Use the graph to estimate the doubling time of the population of the world.

39. Aircraft require longer takeoff distances, called takeoff rolls, at high-altitude airports because of diminished air density. The table shows how the takeoff roll for a certain light airplane depends on the airport elevation. (Takeoff rolls are also strongly influenced by air temperature; the data shown assume a temperature of 0° C.) Determine a formula for this particular aircraft that gives the takeoff roll as an exponential function of airport elevation.

Elevation (ft)	Sea level	1000	2000	3000	4000
Takeoff roll (ft)	670	734	805	882	967

40. One of the main contaminants of a nuclear accident, such as that at Chernobyl, is strontium-90, which decays exponentially at a continuous rate of approximately 2.47% per year. After the Chernobyl disaster, it was suggested that it would be about 100 years before the region would again be safe for human habitation. What percent of the original strontium-90 would still remain then?

41. The decrease in the number of colonies of honey bees, essential to pollinate crops providing one quarter of US food consumption, worries policy makers. US beekeepers say that over the three winter months a 5.9% decline in the number of colonies per month is economically sustainable, but higher rates are not.[23]

(a) Assuming a constant percent colony loss, which function I–III could describe a winter monthly colony loss trend that is economically sustainable? Assume y is the number of US bee colonies, t is time in months, and a is a positive constant.

I. $y = a(1.059)^t$ II. $y = a(0.962)^t$
III. $y = a(0.935)^t$

(b) What is the annual bee colony trend described by each of the functions in part (a)?

42. A certain region has a population of 10,000,000 and an annual growth rate of 2%. Estimate the doubling time by guessing and checking.

43. According to the EPA, sales of electronic devices in the US doubled between 1997 and 2009, when 438 million electronic devices sold.[24]

(a) Find an exponential function, $S(t)$, to model sales in millions since 1997.
(b) What was the annual percentage growth rate between 1997 and 2009?

44. (a) Estimate graphically the doubling time of the exponentially growing population shown in Figure 1.37. Check that the doubling time is independent of where you start on the graph.
(b) Show algebraically that if $P = P_0a^t$ doubles between time t and time $t + d$, then d is the same number for any t.

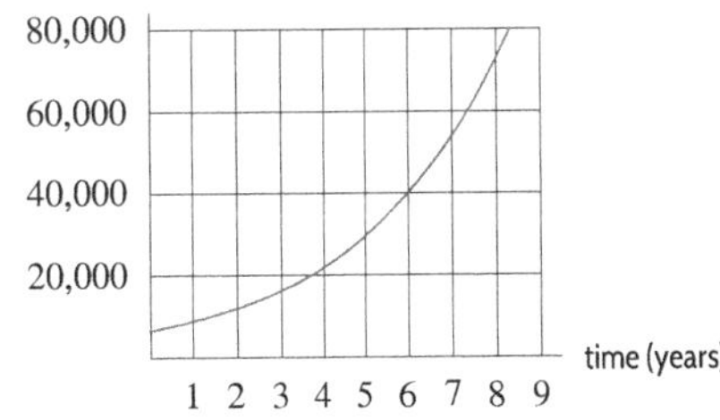

Figure 1.37

45. A deposit of P_0 into a bank account has a doubling time of 50 years. No other deposits or withdrawals are made.

(a) How much money is in the bank account after 50 years? 100 years? 150 years? (Your answer will involve P_0.)
(b) How many times does the amount of money double in t years? Use this to write a formula for P, the amount of money in the account after t years.

[21] International Energy Agency, *World Energy Outlook*, 2008.
[22] www.indexmundi.com, accessed June 14, 2015.
[23] "A Key to America's Crops Is Disappearing at a Staggering Rate," www.businessinsider.com, accessed January 2016.
[24] http://www.epa.gov/osw/conserve/materials/ecycling/docs /summarybaselinereport2011.pdf. Accessed DATE

46. A 325 mg aspirin has a half-life of H hours in a patient's body.

(a) How long does it take for the quantity of aspirin in the patient's body to be reduced to 162.5 mg? To 81.25 mg? To 40.625 mg? (Note that $162.5 = 325/2$, etc. Your answers will involve H.)

(b) How many times does the quantity of aspirin, A mg, in the body halve in t hours? Use this to give a formula for A after t hours.

47. (a) The half-life of radium-226 is 1620 years. If the initial quantity of radium is Q_0, explain why the quantity, Q, of radium left after t years, is given by

$$Q = Q_0\,(0.99572)^t.$$

(b) What percentage of the original amount of radium is left after 500 years?

48. In the early 1960s, radioactive strontium-90 was released during atmospheric testing of nuclear weapons and got into the bones of people alive at the time. If the half-life of strontium-90 is 29 years, what fraction of the strontium-90 absorbed in 1960 remained in people's bones in 2010? [Hint: Write the function in the form $Q = Q_0(1/2)^{t/29}$.]

49. Food bank usage in Britain has grown dramatically over the past decade. The number of users, in thousands, of the largest food bank in year t is estimated to be $N(t) = 1.3e^{0.81t}$, where t is the number of years since 2006.[25]

(a) What does the 1.3 represent in this context? Give units.

(b) What is the continuous growth rate of users per year?

(c) What is the annual percent growth rate of users per year?

(d) Using only your answer for part (c), decide if the doubling time is more or less than 1 year.

■ Problems 50–51 concern biodiesel, a fuel derived from renewable resources such as food crops, algae, and animal oils. The table shows the percent growth over the previous year in US biodiesel consumption.[26]

Year	2008	2009	2010	2011	2012	2013	2014
% growth	−14.1	5.9	−19.3	240.8	1.0	56.9	1.1

50. (a) According to the US Department of Energy, the US consumed 322 million gallons of biodiesel in 2009. Approximately how much biodiesel (in millions of gallons) did the US consume in 2010? In 2011?

(b) Graph the points showing the annual US consumption of biodiesel, in millions of gallons of biodiesel, for the years 2009 to 2014. Label the scales on the horizontal and vertical axes.

51. (a) True or false: The annual US consumption of biodiesel grew exponentially from 2008 to 2010. Justify your answer without doing any calculations.

(b) According to this data, during what single year(s), if any, did the US consumption of biodiesel at least triple?

52. Hydroelectric power is electric power generated by the force of moving water. The table shows the annual percent change in hydroelectric power consumption by the US industrial sector.[27]

Year	2009	2010	2011	2012	2013	2014
% growth over previous yr	6.3	−4.9	22.4	−15.5	−3.0	−3.4

(a) According to the US Department of Energy, the US industrial sector consumed about 2.65 quadrillion[28] BTUs of hydroelectric power in 2009. Approximately how much hydroelectric power (in quadrillion BTUs) did the US consume in 2010? In 2011?

(b) Graph the points showing the annual US consumption of hydroelectric power, in quadrillion BTUs, for the years 2009 to 2014. Label the scales on the horizontal and vertical axes.

(c) According to this data, when did the largest yearly decrease, in quadrillion BTUs, in the US consumption of hydroelectric power occur? What was this decrease?

■ Problems 53–54 concern wind power, which has been used for centuries to propel ships and mill grain. Modern wind power is obtained from windmills that convert wind energy into electricity. Figure 1.38 shows the annual percent growth in US wind power consumption[29] between 2009 and 2014.

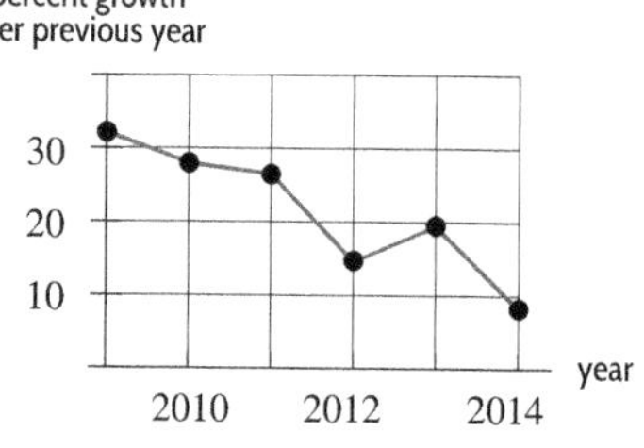

Figure 1.38

[25] Estimates for the Trussell Trust: http://www.bbc.com/news/education-30346060 and http://www.trusselltrust.org/stats. Accessed November 2015.

[26] www.eia.doe.gov, accessed March 29, 2015.

[27] From www.eia.doe.gov, accessed March 31, 2015.

[28] 1 quadrillion BTU=10^{15} BTU.

[29] Yearly values have been joined with line segments to highlight trends in the data. Actual values in between years should not be inferred from the segments. From www.eia.doe.gov, accessed April 1, 2015.

53. **(a)** According to the US Department of Energy, the US consumption of wind power was 721 trillion BTUs in 2009. How much wind power did the US consume in 2010? In 2011?
(b) Graph the points showing the annual US consumption of wind power, in trillion BTUs, for the years 2009 to 2014. Label the scales on the horizontal and vertical axes.
(c) Based on this data, in what year did the largest yearly increase, in trillion BTUs, in the US consumption of wind power occur? What was this increase?

54. **(a)** According to Figure 1.38, during what single year(s), if any, did the US consumption of wind power energy increase by at least 25%? Decrease by at least 25%?
(b) True or false: The US consumption of wind power energy doubled from 2008 to 2011?

55. **(a)** The exponential functions in Figure 1.39 have b, d, q positive. Which of the constants a, c, and p must be positive?
(b) Which of the constants a, b, c, d, p, and q must be between 0 and 1?
(c) Which two of the constants a, b, c, d, p, and q must be equal?
(d) What information about the constants a and b does the point $(1, 1)$ provide?

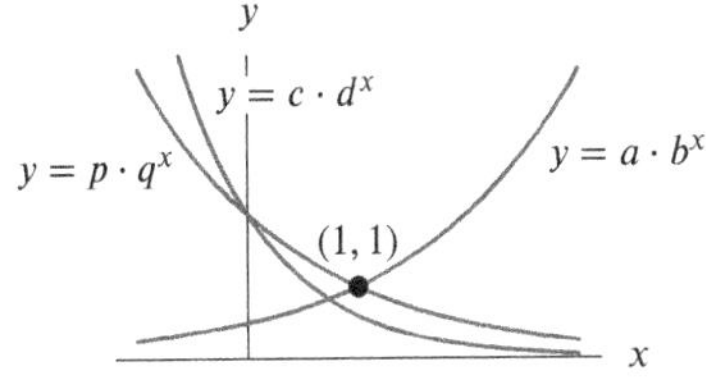

Figure 1.39

Strengthen Your Understanding

In Problems 56–59, explain what is wrong with the statement.

56. A quantity that doubles daily has an exponential growth rate of 200% per day.

57. The function $y = e^{-0.25x}$ is decreasing and its graph is concave down.

58. The function $y = 2x$ is increasing, and its graph is concave up.

59. The points $(0, 1)$, $(1, e)$, $(2, 2e)$, and $(3, 3e)$ are all on the graph of $y = e^x$

In Problems 60–64, give an example of:

60. A decreasing exponential function with a vertical intercept of π.

61. A formula representing the statement "q decreases at a constant percent rate, and $q = 2.2$ when $t = 0$."

62. A function that is increasing at a constant percent rate and that has the same vertical intercept as $f(x) = 0.3x + 2$.

63. A function with a horizontal asymptote at $y = -5$ and range $y > -5$.

64. An exponential function that grows slower than $y = e^x$ for $x > 0$.

Are the statements in Problems 65–72 true or false? Give an explanation for your answer.

65. The function $f(x) = e^{2x}/(2e^{5x})$ is exponential.

66. The function $y = 2 + 3e^{-t}$ has a y-intercept of $y = 3$.

67. The function $y = 5 - 3e^{-4t}$ has a horizontal asymptote of $y = 5$.

68. If $y = f(x)$ is an exponential function and if increasing x by 1 increases y by a factor of 5, then increasing x by 2 increases y by a factor of 10.

69. If $y = Ab^x$ and increasing x by 1 increases y by a factor of 3, then increasing x by 2 increases y by a factor of 9.

70. An exponential function can be decreasing.

71. If a and b are positive constants, $b \neq 1$, then $y = a + ab^x$ has a horizontal asymptote.

72. The function $y = 20/(1 + 2e^{-kt})$ with $k > 0$, has a horizontal asymptote at $y = 20$.

1.3 NEW FUNCTIONS FROM OLD

Shifts and Stretches

The graph of a constant multiple of a given function is easy to visualize: each y-value is stretched or shrunk by that multiple. For example, consider the function $f(x)$ and its multiples $y = 3f(x)$ and $y = -2f(x)$. Their graphs are shown in Figure 1.40. The factor 3 in the function $y = 3f(x)$ stretches each $f(x)$ value by multiplying it by 3; the factor -2 in the function $y = -2f(x)$ stretches $f(x)$ by multiplying by 2 and reflects it across the x-axis. You can think of the multiples of a given function as a family of functions.

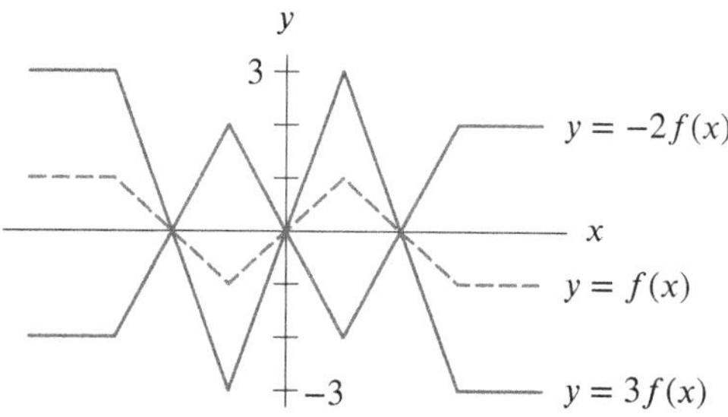

Figure 1.40: Multiples of the function $f(x)$

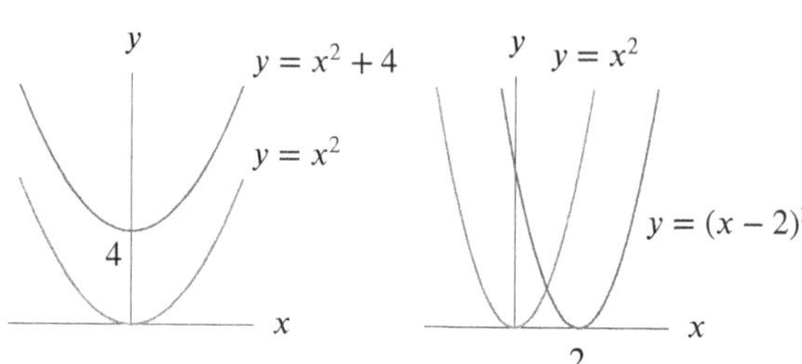

Figure 1.41: Graphs of $y = x^2$ with $y = x^2 + 4$ and $y = (x-2)^2$

It is also easy to create families of functions by shifting graphs. For example, $y - 4 = x^2$ is the same as $y = x^2 + 4$, which is the graph of $y = x^2$ shifted up by 4. Similarly, $y = (x-2)^2$ is the graph of $y = x^2$ shifted right by 2. (See Figure 1.41.)

- Multiplying a function by a constant, c, stretches the graph vertically (if $c > 1$) or shrinks the graph vertically (if $0 < c < 1$). A negative sign (if $c < 0$) reflects the graph across the x-axis, in addition to shrinking or stretching.
- Replacing y by $(y - k)$ moves a graph up by k (down if k is negative).
- Replacing x by $(x - h)$ moves a graph to the right by h (to the left if h is negative).

Composite Functions

If oil is spilled from a tanker, the area of the oil slick grows with time. Suppose that the oil slick is always a perfect circle. Then the area, A, of the oil slick is a function of its radius, r:

$$A = f(r) = \pi r^2.$$

The radius is also a function of time, because the radius increases as more oil spills. Thus, the area, being a function of the radius, is also a function of time. If, for example, the radius is given by

$$r = g(t) = 1 + t,$$

then the area is given as a function of time by substitution:

$$A = \pi r^2 = \pi(1+t)^2.$$

We are thinking of A as a *composite function* or a "function of a function," which is written

$$A = \underbrace{f(g(t))}_{\substack{\text{Composite function;}\\ f \text{ is outside function,}\\ g \text{ is inside function}}} = \pi(g(t))^2 = \pi(1+t)^2.$$

To calculate A using the formula $\pi(1+t)^2$, the first step is to find $1 + t$, and the second step is to square and multiply by π. The first step corresponds to the inside function $g(t) = 1 + t$, and the second step corresponds to the outside function $f(r) = \pi r^2$.

Example 1 If $f(x) = x^2$ and $g(x) = x - 2$, find each of the following:

(a) $f(g(3))$ (b) $g(f(3))$ (c) $f(g(x))$ (d) $g(f(x))$

Solution (a) Since $g(3) = 1$, we have $f(g(3)) = f(1) = 1$.

(b) Since $f(3) = 9$, we have $g(f(3)) = g(9) = 7$. Notice that $f(g(3)) \neq g(f(3))$.

(c) $f(g(x)) = f(x-2) = (x-2)^2$.

(d) $g(f(x)) = g(x^2) = x^2 - 2$. Again, notice that $f(g(x)) \neq g(f(x))$.

Notice that the horizontal shift in Figure 1.41 can be thought of as a composition $f(g(x)) = (x-2)^2$.

Example 2 Express each of the following functions as a composition:

(a) $h(t) = (1+t^3)^{27}$ (b) $k(y) = e^{-y^2}$ (c) $l(y) = -(e^y)^2$

Solution In each case think about how you would calculate a value of the function. The first stage of the calculation gives you the inside function, and the second stage gives you the outside function.

(a) For $(1+t^3)^{27}$, the first stage is cubing and adding 1, so an inside function is $g(t) = 1+t^3$. The second stage is taking the 27th power, so an outside function is $f(y) = y^{27}$. Then

$$f(g(t)) = f(1+t^3) = (1+t^3)^{27}.$$

In fact, there are lots of different answers: $g(t) = t^3$ and $f(y) = (1+y)^{27}$ is another possibility.

(b) To calculate e^{-y^2} we square y, take its negative, and then take e to that power. So if $g(y) = -y^2$ and $f(z) = e^z$, then we have

$$f(g(y)) = e^{-y^2}.$$

(c) To calculate $-(e^y)^2$, we find e^y, square it, and take the negative. Using the same definitions of f and g as in part (b), the composition is

$$g(f(y)) = -(e^y)^2.$$

Since parts (b) and (c) give different answers, we see the order in which functions are composed is important.

Odd and Even Functions: Symmetry

There is a certain symmetry apparent in the graphs of $f(x) = x^2$ and $g(x) = x^3$ in Figure 1.42. For each point (x, x^2) on the graph of f, the point $(-x, x^2)$ is also on the graph; for each point (x, x^3) on the graph of g, the point $(-x, -x^3)$ is also on the graph. The graph of $f(x) = x^2$ is symmetric about the y-axis, whereas the graph of $g(x) = x^3$ is symmetric about the origin. The graph of any polynomial involving only even powers of x has symmetry about the y-axis, while polynomials with only odd powers of x are symmetric about the origin. Consequently, any functions with these symmetry properties are called *even* and *odd*, respectively.

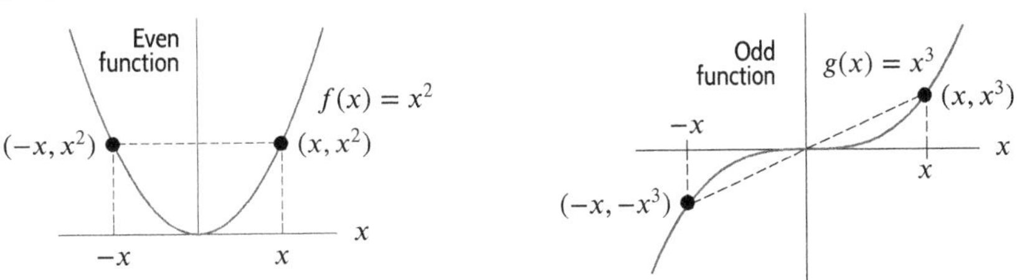

Figure 1.42: Symmetry of even and odd functions

> For any function f,
> f is an **even** function if $f(-x) = f(x)$ for all x.
> f is an **odd** function if $f(-x) = -f(x)$ for all x.

For example, $g(x) = e^{x^2}$ is even and $h(x) = x^{1/3}$ is odd. However, many functions do not have any symmetry and are neither even nor odd.

Inverse Functions

On August 26, 2005, the runner Kenenisa Bekele[30] of Ethiopia set a world record for the 10,000-meter race. His times, in seconds, at 2000-meter intervals are recorded in Table 1.14, where $t = f(d)$ is the number of seconds Bekele took to complete the first d meters of the race. For example, Bekele ran the first 4000 meters in 629.98 seconds, so $f(4000) = 629.98$. The function f was useful to athletes planning to compete with Bekele.

Let us now change our point of view and ask for distances rather than times. If we ask how far Bekele ran during the first 629.98 seconds of his race, the answer is clearly 4000 meters. Going backward in this way from numbers of seconds to numbers of meters gives f^{-1}, the *inverse function*[31] of f. We write $f^{-1}(629.98) = 4000$. Thus, $f^{-1}(t)$ is the number of meters that Bekele ran during the first t seconds of his race. See Table 1.15, which contains values of f^{-1}.

The independent variable for f is the dependent variable for f^{-1}, and vice versa. The domains and ranges of f and f^{-1} are also interchanged. The domain of f is all distances d such that $0 \le d \le 10000$, which is the range of f^{-1}. The range of f is all times t, such that $0 \le t \le 1577.53$, which is the domain of f^{-1}.

Table 1.14 *Bekele's running time*

d (meters)	$t = f(d)$ (seconds)
0	0.00
2000	315.63
4000	629.98
6000	944.66
8000	1264.63
10000	1577.53

Table 1.15 *Distance run by Bekele*

t (seconds)	$d = f^{-1}(t)$ (meters)
0.00	0
315.63	2000
629.98	4000
944.66	6000
1264.63	8000
1577.53	10000

Which Functions Have Inverses?

If a function has an inverse, we say it is *invertible*. Let's look at a function which is not invertible. Consider the flight of the Mercury spacecraft *Freedom 7*, which carried Alan Shepard, Jr. into space in May 1961. Shepard was the first American to journey into space. After launch, his spacecraft rose to an altitude of 116 miles, and then came down into the sea. The function $f(t)$ giving the altitude in miles t minutes after lift-off does not have an inverse. To see why it does not, try to decide on a value for $f^{-1}(100)$, which should be the time when the altitude of the spacecraft was 100 miles. However, there are two such times, one when the spacecraft was ascending and one when it was descending. (See Figure 1.43.)

The reason the altitude function does not have an inverse is that the altitude has the same value for two different times. The reason the Bekele time function did have an inverse is that each running time, t, corresponds to a unique distance, d.

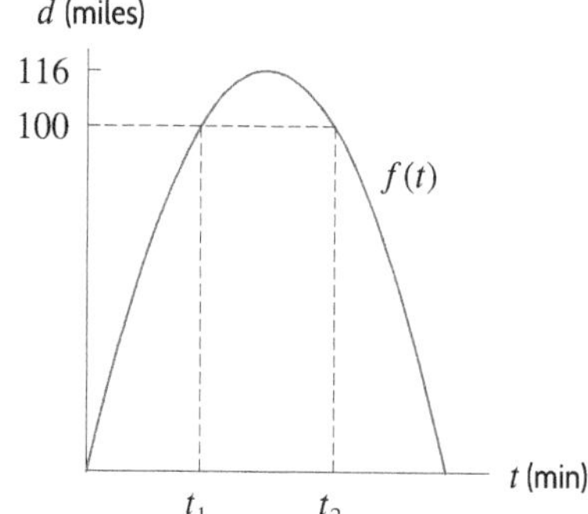

Figure 1.43: Two times, t_1 and t_2, at which altitude of spacecraft is 100 miles

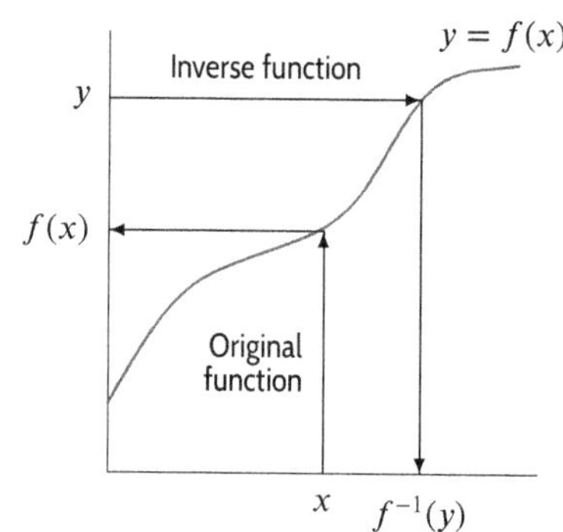

Figure 1.44: A function which has an inverse

[30] kenenisabekelle.com/, accessed January 11, 2011.

[31] The notation f^{-1} represents the inverse function, which is not the same as the reciprocal, $1/f$.

Figure 1.44 suggests when an inverse exists. The original function, f, takes us from an x-value to a y-value, as shown in Figure 1.44. Since having an inverse means there is a function going from a y-value to an x-value, the crucial question is whether we can get back. In other words, does each y-value correspond to a unique x-value? If so, there's an inverse; if not, there is not. This principle may be stated geometrically, as follows:

> A function has an inverse if (and only if) its graph intersects any horizontal line at most once.

For example, the function $f(x) = x^2$ does not have an inverse because many horizontal lines intersect the parabola twice.

Definition of an Inverse Function

If the function f is invertible, its inverse is defined as follows:

$$f^{-1}(y) = x \quad \text{means} \quad y = f(x).$$

Formulas for Inverse Functions

If a function is defined by a formula, it is sometimes possible to find a formula for the inverse function. In Section 1.1, we looked at the snowy tree cricket, whose chirp rate, C, in chirps per minute, is approximated at the temperature, T, in degrees Fahrenheit, by the formula

$$C = f(T) = 4T - 160.$$

So far we have used this formula to predict the chirp rate from the temperature. But it is also possible to use this formula backward to calculate the temperature from the chirp rate.

Example 3 Find the formula for the function giving temperature in terms of the number of cricket chirps per minute; that is, find the inverse function f^{-1} such that

$$T = f^{-1}(C).$$

Solution Since C is an increasing function, f is invertible. We know $C = 4T - 160$. We solve for T, giving

$$T = \frac{C}{4} + 40,$$

so

$$f^{-1}(C) = \frac{C}{4} + 40.$$

Graphs of Inverse Functions

The function $f(x) = x^3$ is increasing everywhere and so has an inverse. To find the inverse, we solve

$$y = x^3$$

for x, giving

$$x = y^{1/3}.$$

The inverse function is

$$f^{-1}(y) = y^{1/3}$$

or, if we want to call the independent variable x,

$$f^{-1}(x) = x^{1/3}.$$

The graphs of $y = x^3$ and $y = x^{1/3}$ are shown in Figure 1.45. Notice that these graphs are the reflections of one another across the line $y = x$. For example, $(8, 2)$ is on the graph of $y = x^{1/3}$ because $2 = 8^{1/3}$, and $(2, 8)$ is on the graph of $y = x^3$ because $8 = 2^3$. The points $(8, 2)$ and $(2, 8)$ are reflections of one another across the line $y = x$.

In general, we have the following result.

> If the x- and y-axes have the same scales, the graph of f^{-1} is the reflection of the graph of f across the line $y = x$.

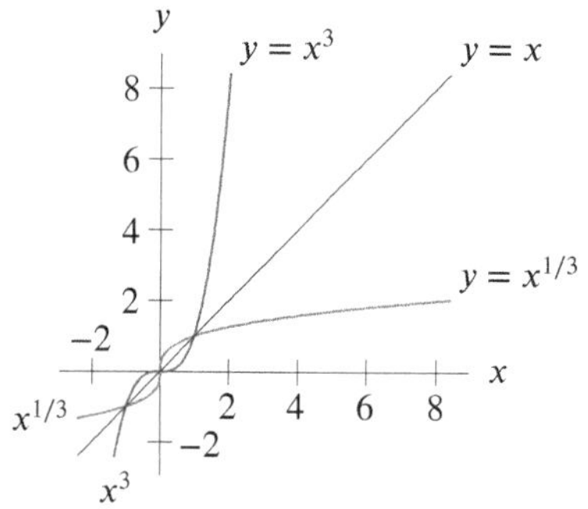

Figure 1.45: Graphs of inverse functions, $y = x^3$ and $y = x^{1/3}$, are reflections across the line $y = x$

Exercises and Problems for Section 1.3

EXERCISES

For the functions f in Exercises 1–3, graph:

(a) $f(x+2)$ (b) $f(x-1)$ (c) $f(x)-4$
(d) $f(x+1)+3$ (e) $3f(x)$ (f) $-f(x)+1$

1.

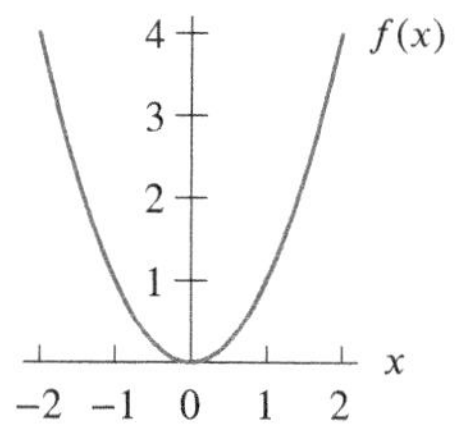

2.

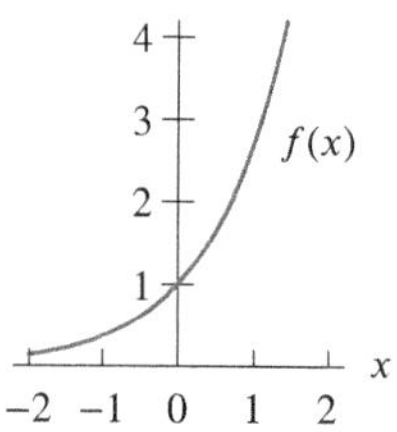

3.

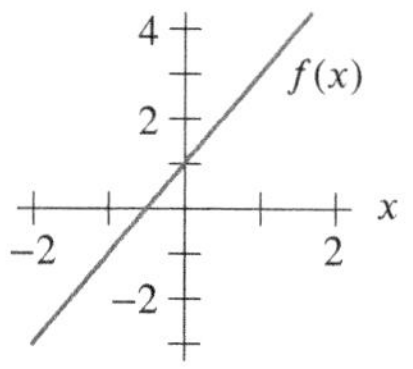

In Exercises 4–7, use Figure 1.46 to graph the functions.

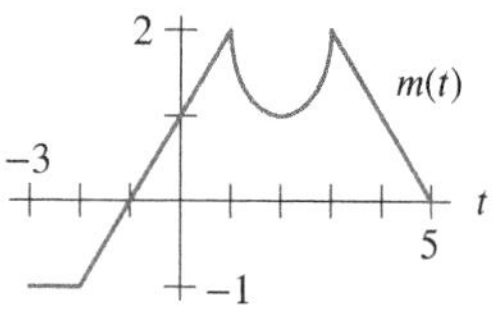

Figure 1.46

4. $n(t) = m(t) + 2$

5. $p(t) = m(t-1)$

6. $k(t) = m(t+1.5)$

7. $w(t) = m(t-0.5) - 2.5$

8. Use Figure 1.47 to graph each of the following. Label any intercepts or asymptotes that can be determined.

(a) $y = f(x) + 3$ (b) $y = 2f(x)$
(c) $y = f(x+4)$ (d) $y = 4 - f(x)$

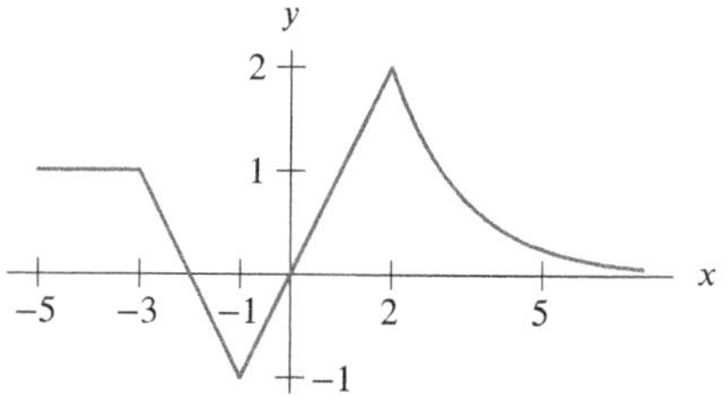

Figure 1.47

For the functions f and g in Exercises 9–12, find

(a) $f(g(1))$ (b) $g(f(1))$ (c) $f(g(x))$
(d) $g(f(x))$ (e) $f(t)g(t)$

9. $f(x) = x^2, g(x) = x + 1$

10. $f(x) = \sqrt{x+4}, g(x) = x^2$

11. $f(x) = e^x, g(x) = x^2$

12. $f(x) = 1/x, g(x) = 3x + 4$

13. If $f(x) = x^2 + 1$, find and simplify:

(a) $f(t+1)$ (b) $f(t^2+1)$ (c) $f(2)$
(d) $2f(t)$ (e) $(f(t))^2 + 1$

14. For $g(x) = x^2 + 2x + 3$, find and simplify:

(a) $g(2+h)$ (b) $g(2)$
(c) $g(2+h) - g(2)$

■ Simplify the quantities in Exercises 15–18 using $m(z) = z^2$.

15. $m(z+1) - m(z)$
16. $m(z+h) - m(z)$
17. $m(z) - m(z-h)$
18. $m(z+h) - m(z-h)$

■ Are the functions in Exercises 19–26 even, odd, or neither?

19. $f(x) = x^6 + x^3 + 1$
20. $f(x) = x^3 + x^2 + x$
21. $f(x) = x^4 - x^2 + 3$
22. $f(x) = x^3 + 1$
23. $f(x) = 2x$
24. $f(x) = e^{x^2-1}$
25. $f(x) = x(x^2 - 1)$
26. $f(x) = e^x - x$

■ For Exercises 27–28, decide if the function $y = f(x)$ is invertible.

27.

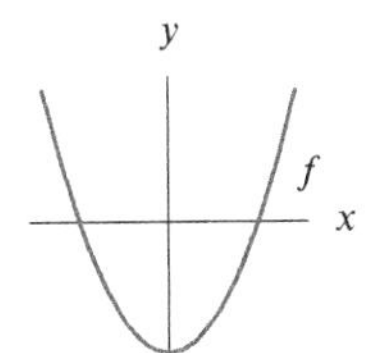

28.

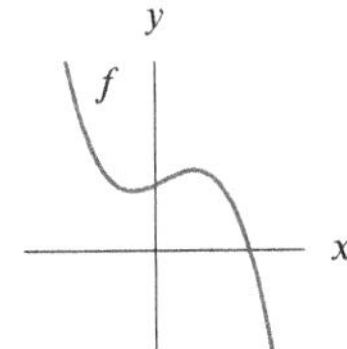

■ For Exercises 29–31, use a graph of the function to decide whether or not it is invertible.

29. $f(x) = x^2 + 3x + 2$
30. $f(x) = x^3 - 5x + 10$
31. $f(x) = x^3 + 5x + 10$

32. Let p be the price of an item and q be the number of items sold at that price, where $q = f(p)$. What do the following quantities mean in terms of prices and quantities sold?

(a) $f(25)$ (b) $f^{-1}(30)$

33. Let $C = f(A)$ be the cost, in dollars, of building a store of area A square feet. In terms of cost and square feet, what do the following quantities represent?

(a) $f(10{,}000)$ (b) $f^{-1}(20{,}000)$

34. Let $f(x)$ be the temperature (°F) when the column of mercury in a particular thermometer is x inches long. What is the meaning of $f^{-1}(75)$ in practical terms?

35. (a) Write an equation for a graph obtained by vertically stretching the graph of $y = x^2$ by a factor of 2, followed by a vertical upward shift of 1 unit. Sketch it.
(b) What is the equation if the order of the transformations (stretching and shifting) in part (a) is interchanged?
(c) Are the two graphs the same? Explain the effect of reversing the order of transformations.

PROBLEMS

36. How does the graph of $Q = S(1 - e^{-kt})$ in Example 4 on page 18 relate to the graph of the exponential decay function, $y = Se^{-kt}$?

■ In Problems 37–38 find possible formulas for the graphs using shifts of x^2 or x^3.

37.

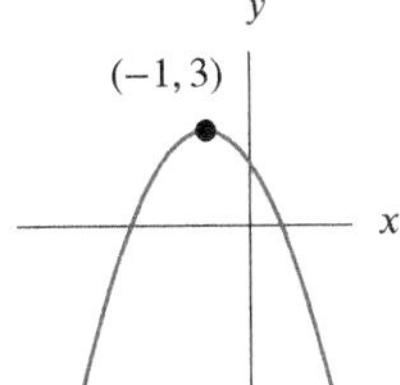

38.

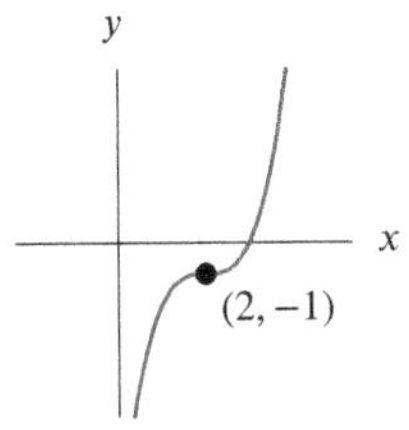

■ In Problems 39–42, use Figure 1.48 to estimate the function value or explain why it cannot be done.

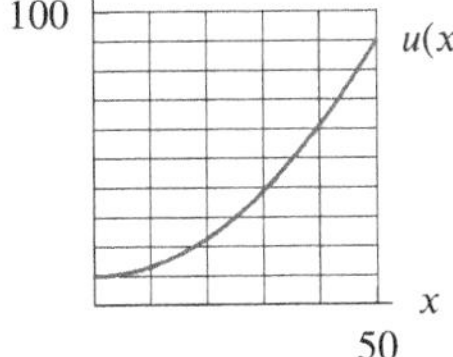

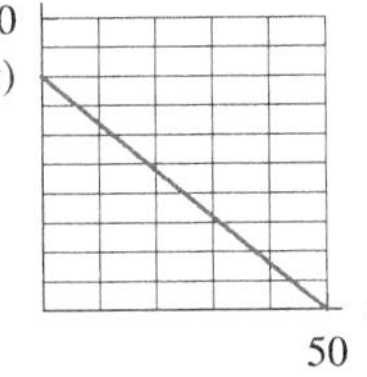

Figure 1.48

39. $u(v(10))$
40. $u(v(40))$
41. $v(u(10))$
42. $v(u(40))$

■ For Problems 43–48, use the graphs in Figure 1.49.

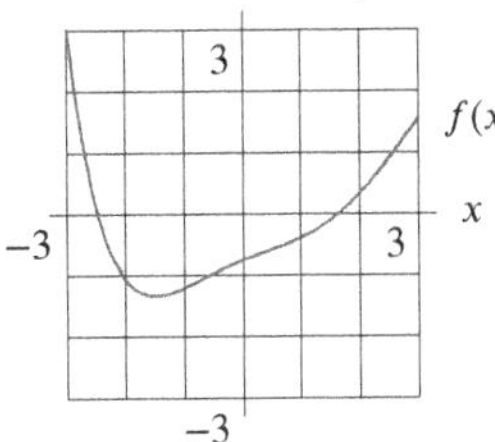

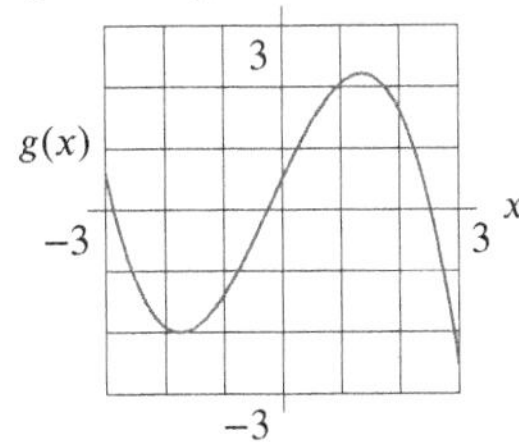

Figure 1.49

43. Estimate $f(g(1))$.
44. Estimate $g(f(2))$.
45. Estimate $f(f(1))$.
46. Graph $f(g(x))$.
47. Graph $g(f(x))$.
48. Graph $f(f(x))$.

■ For Problems 49–52, determine functions f and g such that $h(x) = f(g(x))$. (Note: There is more than one correct answer. Do not choose $f(x) = x$ or $g(x) = x$.)

49. $h(x) = (x+1)^3$
50. $h(x) = x^3 + 1$
51. $h(x) = \sqrt{x^2 + 4}$
52. $h(x) = e^{2x}$

53. A tree of height y meters has, on average, B branches, where $B = y - 1$. Each branch has, on average, n leaves, where $n = 2B^2 - B$. Find the average number of leaves on a tree as a function of height.

54. A spherical balloon is growing with radius $r = 3t + 1$, in centimeters, for time t in seconds. Find the volume of the balloon at 3 seconds.

55. Complete the following table with values for the functions f, g, and h, given that:

(a) f is an even function.
(b) g is an odd function.
(c) h is the composition $h(x) = g(f(x))$.

x	$f(x)$	$g(x)$	$h(x)$
−3	0	0	
−2	2	2	
−1	2	2	
0	0	0	
1			
2			
3			

56. Write a table of values for f^{-1}, where f is as given below. The domain of f is the integers from 1 to 7. State the domain of f^{-1}.

x	1	2	3	4	5	6	7
$f(x)$	3	−7	19	4	178	2	1

57. **(a)** Use Figure 1.50 to estimate $f^{-1}(2)$.
(b) Sketch a graph of f^{-1} on the same axes.

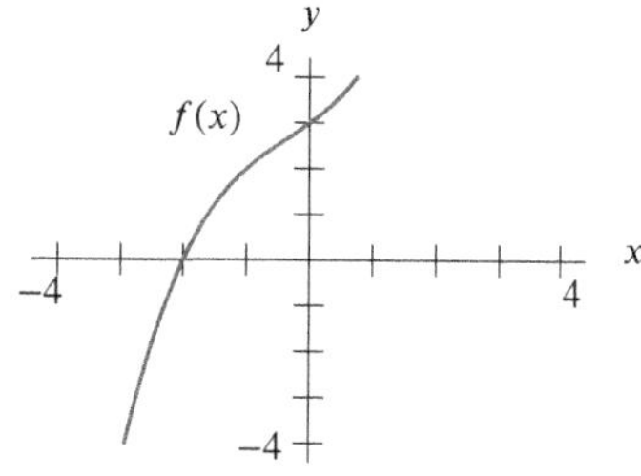

Figure 1.50

■ For Problems 58–61, decide if the function f is invertible.

58. $f(d)$ is the total number of gallons of fuel an airplane has used by the end of d minutes of a particular flight.

59. $f(t)$ is the number of customers in Saks Fifth Avenue at t minutes past noon on December 18, 2014.

60. $f(n)$ is the number of students in your calculus class whose birthday is on the nth day of the year.

61. $f(w)$ is the cost of mailing a letter weighing w grams.

■ In Problems 62–66, interpret the expression in terms of carbon footprint, a measure of the environmental impact in kilograms of green house gas (GHG) emissions. Assume that a bottle of drinking water that travels d km from its production source has a carbon footprint $f(d)$ kg of GHGs.[32]

62. $f(150)$

63. $f(8700) = 0.25$

64. $f^{-1}(1.1)$

65. $f(150) + f(0)$

66. $\dfrac{f(1500) - f(150)}{1500 - 150}$

■ In Problems 67–70 the functions $r = f(t)$ and $V = g(r)$ give the radius and the volume of a commercial hot air balloon being inflated for testing. The variable t is in minutes, r is in feet, and V is in cubic feet. The inflation begins at $t = 0$. In each case, give a mathematical expression that represents the given statement.

67. The volume of the balloon t minutes after inflation began.

68. The volume of the balloon if its radius were twice as big.

69. The time that has elapsed when the radius of the balloon is 30 feet.

70. The time that has elapsed when the volume of the balloon is 10,000 cubic feet.

71. The cost of producing q articles is given by the function $C = f(q) = 100 + 2q$.

(a) Find a formula for the inverse function.
(b) Explain in practical terms what the inverse function tells you.

72. Figure 1.51 shows $f(t)$, the number (in millions) of motor vehicles registered[33] in the world in the year t.

(a) Is f invertible? Explain.
(b) What is the meaning of $f^{-1}(400)$ in practical terms? Evaluate $f^{-1}(400)$.
(c) Sketch the graph of f^{-1}.

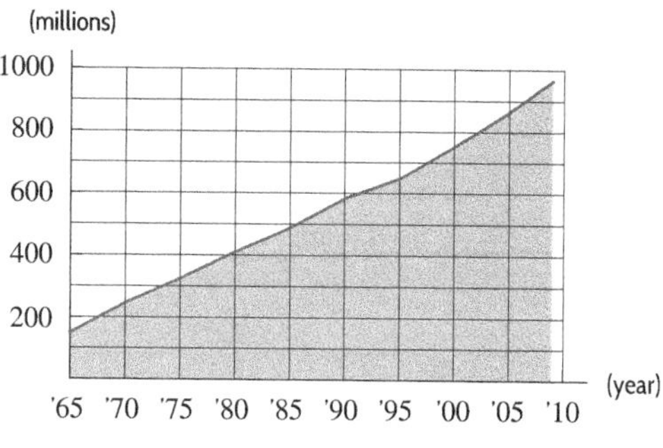

Figure 1.51

[32]For the data in Problem 63, see www.triplepundit.com. Accessed March 1, 2015.
[33]www.earth-policy.org, accessed June 5, 2011. In 2000, about 30% of the registered vehicles were in the US.

73. Figure 1.52 is a graph of the function $f(t)$. Here $f(t)$ is the depth in meters below the Atlantic Ocean floor where t million-year-old rock can be found.[34]

(a) Evaluate $f(15)$, and say what it means in practical terms.
(b) Is f invertible? Explain.
(c) Evaluate $f^{-1}(120)$, and say what it means in practical terms.
(d) Sketch a graph of f^{-1}.

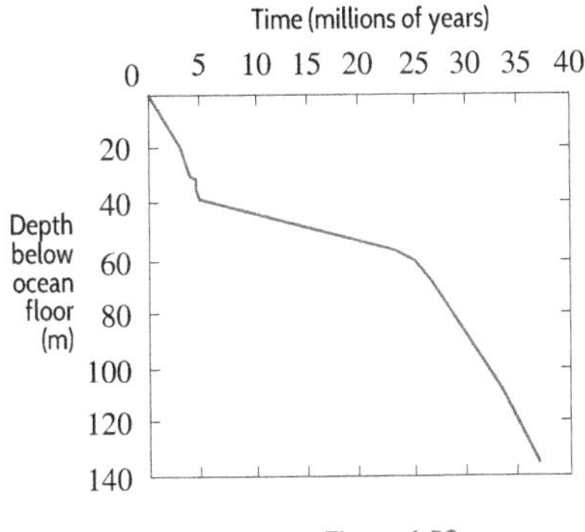

Figure 1.52

74. Figure 1.53 shows graphs of 4 useful functions: the step, the sign, the ramp, and the absolute value. We have

$$\text{step}(x) = \begin{cases} 0 & \text{if } x < 0 \\ 1 & \text{if } x \geq 0 \end{cases}.$$

Match the remaining 3 graphs with the following formulas in terms of the step function.

(a) x step(x)
(b) step(x) − step$(-x)$
(c) x step(x) − x step$(-x)$

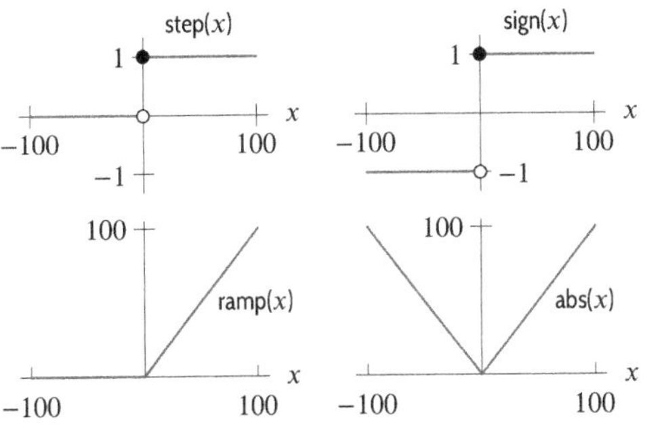

Figure 1.53

Strengthen Your Understanding

In Problems 75–79, explain what is wrong with the statement.

75. The graph of $f(x) = -(x+1)^3$ is the graph of $g(x) = -x^3$ shifted right by 1 unit.

76. The functions $f(x) = 3^x$ and $g(x) = x^3$ are inverses of each other.

77. $f(x) = 3x+5$ and $g(x) = -3x-5$ are inverse functions of each other.

78. The function $f(x) = e^x$ is its own inverse.

79. The inverse of $f(x) = x$ is $f^{-1}(x) = 1/x$.

In Problems 80–83, give an example of:

80. An invertible function whose graph contains the point $(0, 3)$.

81. An even function whose graph does not contain the point $(0, 0)$.

82. An increasing function $f(x)$ whose values are greater than those of its inverse function $f^{-1}(x)$ for $x > 0$.

83. Two functions $f(x)$ and $g(x)$ such that moving the graph of f to the left 2 units gives the graph of g and moving the graph of f up 3 also gives the graph of g.

Are the statements in Problems 84–93 true or false? Give an explanation for your answer.

84. The composition of the exponential functions $f(x) = 2^x$ and $g(x) = 3^x$ is exponential.

85. The graph of $f(x) = 100(10^x)$ is a horizontal shift of the graph of $g(x) = 10^x$.

86. If f is an increasing function, then f^{-1} is an increasing function.

87. If a function is even, then it does not have an inverse.

88. If a function is odd, then it does not have an inverse.

89. The function $f(x) = e^{-x^2}$ is decreasing for all x.

90. If $g(x)$ is an even function then $f(g(x))$ is even for every function $f(x)$.

91. If $f(x)$ is an even function then $f(g(x))$ is even for every function $g(x)$.

92. There is a function which is both even and odd.

93. The composition of two linear functions is linear.

In Problems 94–97, suppose f is an increasing function and g is a decreasing function. Give an example of functions f and g for which the statement is true, or say why such an example is impossible.

94. $f(x) + g(x)$ is decreasing for all x.

95. $f(x) - g(x)$ is decreasing for all x.

96. $f(x)g(x)$ is decreasing for all x.

97. $f(g(x))$ is increasing for all x.

[34]Data of Dr. Murlene Clark based on core samples drilled by the research ship *Glomar Challenger*, taken from *Initial Reports of the Deep Sea Drilling Project*.

1.4 LOGARITHMIC FUNCTIONS

In Section 1.2, we approximated the population of Burkina Faso (in millions) by the function

$$P = f(t) = 14.235(1.03)^t,$$

where t is the number of years since 2007. Now suppose that instead of calculating the population at time t, we ask when the population will reach 20 million. We want to find the value of t for which

$$20 = f(t) = 14.235(1.03)^t.$$

We use logarithms to solve for a variable in an exponent.

Logarithms to Base 10 and to Base e

We define the *logarithm* function, $\log_{10} x$, to be the inverse of the exponential function, 10^x, as follows:

> The **logarithm** to base 10 of x, written $\mathbf{log_{10}\,x}$, is the power of 10 we need to get x. In other words,
>
> $$\log_{10} x = c \quad \text{means} \quad 10^c = x.$$
>
> We often write $\log x$ in place of $\log_{10} x$.

The other frequently used base is e. The logarithm to base e is called the *natural logarithm* of x, written $\ln x$ and defined to be the inverse function of e^x, as follows:

> The **natural logarithm** of x, written $\ln x$, is the power of e needed to get x. In other words,
>
> $$\ln x = c \quad \text{means} \quad e^c = x.$$

Values of $\log x$ are in Table 1.16. Because no power of 10 gives 0, $\log 0$ is undefined. The graph of $y = \log x$ is shown in Figure 1.54. The domain of $y = \log x$ is positive real numbers; the range is all real numbers. In contrast, the inverse function $y = 10^x$ has domain all real numbers and range all positive real numbers. The graph of $y = \log x$ has a vertical asymptote at $x = 0$, whereas $y = 10^x$ has a horizontal asymptote at $y = 0$.

One big difference between $y = 10^x$ and $y = \log x$ is that the exponential function grows extremely quickly whereas the log function grows extremely slowly. However, $\log x$ does go to infinity, albeit slowly, as x increases. Since $y = \log x$ and $y = 10^x$ are inverse functions, the graphs of the two functions are reflections of one another across the line $y = x$, provided the scales along the x- and y-axes are equal.

Table 1.16 *Values for* $\log x$ *and* 10^x

x	$\log x$
0	undefined
1	0
2	0.3
3	0.5
4	0.6
⋮	⋮
10	1

x	10^x
0	1
1	10
2	100
3	10^3
4	10^4
⋮	⋮
10	10^{10}

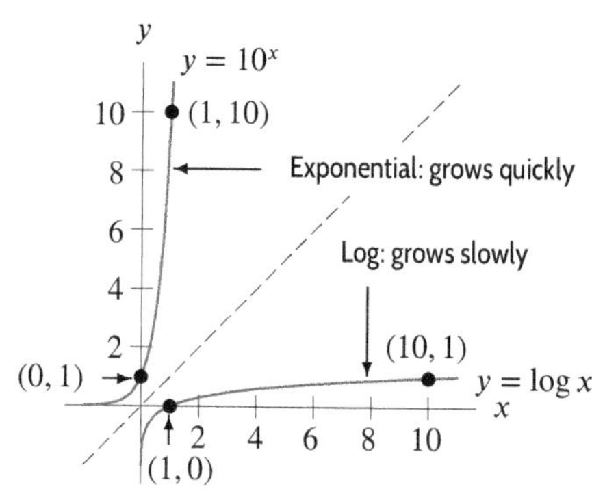

Figure 1.54: Graphs of $\log x$ and 10^x

The graph of $y = \ln x$ in Figure 1.55 has roughly the same shape as the graph of $y = \log x$. The x-intercept is $x = 1$, since $\ln 1 = 0$.The graph of $y = \ln x$ also climbs very slowly as x increases. Both graphs, $y = \log x$ and $y = \ln x$, have *vertical asymptotes* at $x = 0$.

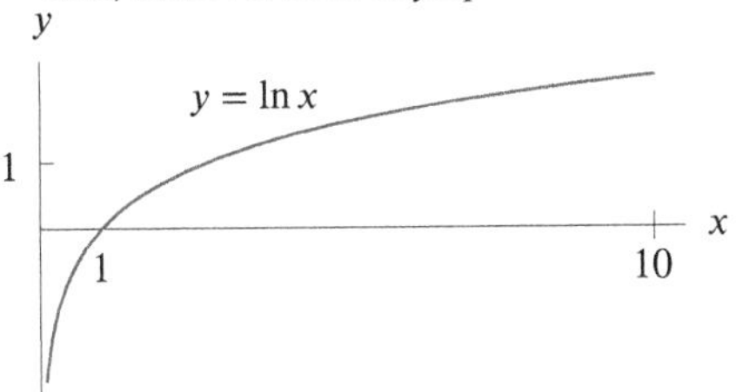

Figure 1.55: Graph of the natural logarithm

The following properties of logarithms may be deduced from the properties of exponents:

Properties of Logarithms

Note that $\log x$ and $\ln x$ are not defined when x is negative or 0.

1. $\log(AB) = \log A + \log B$	1. $\ln(AB) = \ln A + \ln B$
2. $\log\left(\frac{A}{B}\right) = \log A - \log B$	2. $\ln\left(\frac{A}{B}\right) = \ln A - \ln B$
3. $\log(A^p) = p \log A$	3. $\ln(A^p) = p \ln A$
4. $\log(10^x) = x$	4. $\ln e^x = x$
5. $10^{\log x} = x$	5. $e^{\ln x} = x$

In addition, $\log 1 = 0$ because $10^0 = 1$, and $\ln 1 = 0$ because $e^0 = 1$.

Solving Equations Using Logarithms

Logs are frequently useful when we have to solve for unknown exponents, as in the next examples.

Example 1 Find t such that $2^t = 7$.

Solution First, notice that we expect t to be between 2 and 3 (because $2^2 = 4$ and $2^3 = 8$). To calculate t, we take logs to base 10 of both sides. (Natural logs could also be used.)

$$\log(2^t) = \log 7.$$

Then use the third property of logs, which says $\log(2^t) = t \log 2$, and get:

$$t \log 2 = \log 7.$$

Using a calculator to find the logs gives

$$t = \frac{\log 7}{\log 2} \approx 2.81.$$

Example 2 Find when the population of Burkina Faso reaches 20 million by solving $20 = 14.235(1.03)^t$.

Solution Dividing both sides of the equation by 14.235, we get

$$\frac{20}{14.235} = (1.03)^t.$$

Now take logs of both sides:

$$\log\left(\frac{20}{14.235}\right) = \log(1.03^t).$$

Using the fact that $\log(A^t) = t\log A$, we get

$$\log\left(\frac{20}{14.235}\right) = t\log(1.03).$$

Solving this equation using a calculator to find the logs, we get

$$t = \frac{\log(20/14.235)}{\log(1.03)} = 11.5 \text{ years},$$

which is between $t = 11$ and $t = 12$. This value of t corresponds to the year 2018.

Example 3 Traffic pollution is harmful to school-age children. The concentration of carbon monoxide, CO, in the air near a busy road is a function of distance from the road. The concentration decays exponentially at a continuous rate of 3.3% per meter.[36] At what distance from the road is the concentration of CO half what it is on the road?

Solution If C_0 is the concentration of CO on the road, then the concentration x meters from the road is

$$C = C_0 e^{-0.033x}.$$

We want to find the value of x making $C = C_0/2$, that is,

$$C_0 e^{-0.033x} = \frac{C_0}{2}.$$

Dividing by C_0 and then taking natural logs yields

$$\ln\left(e^{-0.033x}\right) = -0.033x = \ln\left(\frac{1}{2}\right) = -0.6931,$$

so

$$x = 21 \text{ meters}.$$

At 21 meters from the road the concentration of CO in the air is half the concentration on the road.

In Example 3 the decay rate was given. However, in many situations where we expect to find exponential growth or decay, the rate is not given. To find it, we must know the quantity at two different times and then solve for the growth or decay rate, as in the next example.

Example 4 The population of Mexico was 100.3 million in 2000 and 120.3 million in 2014.[37] Assuming it increases exponentially, find a formula for the population of Mexico as a function of time.

Solution If we measure the population, P, in millions and time, t, in years since 2000, we can say

$$P = P_0 e^{kt} = 100.3e^{kt},$$

where $P_0 = 100.3$ is the initial value of P. We find k by using the fact that $P = 120.3$ when $t = 14$,

[36] Rickwood, P. and Knight, D. (2009). "The health impacts of local traffic pollution on primary school age children." *State of Australian Cities 2009 Conference Proceedings*, www.be.unsw.edu.au, accessed March 24, 2016.

[37] www.indexmundi.com, accessed January 28, 2016.

so

$$120.3 = 100.3e^{k\cdot 14}.$$

To find k, we divide both sides by 100.3, giving

$$\frac{120.3}{100.3} = 1.1994 = e^{14k}.$$

Now take natural logs of both sides:

$$\ln(1.1994) = \ln(e^{14k}).$$

Using a calculator and the fact that $\ln(e^{14k}) = 14k$, this becomes

$$0.182 = 14k.$$

So

$$k = 0.013,$$

and therefore

$$P = 100.3e^{0.013t}.$$

Since $k = 0.013 = 1.3\%$, the population of Mexico was growing at a continuous rate of 1.3% per year.

In Example 4 we chose to use e for the base of the exponential function representing Mexico's population, making clear that the continuous growth rate was 1.3%. If we had wanted to emphasize the annual growth rate, we could have expressed the exponential function in the form $P = P_0a^t$.

Example 5 Give a formula for the inverse of the following function (that is, solve for t in terms of P):

$$P = f(t) = 14.235(1.029)^t.$$

Solution We want a formula expressing t as a function of P. Take logs:

$$\log P = \log(14.235(1.029)^t).$$

Since $\log(AB) = \log A + \log B$, we have

$$\log P = \log 14.235 + \log((1.029)^t).$$

Now use $\log(A^t) = t\log A$:

$$\log P = \log 14.235 + t\log 1.029.$$

Solve for t in two steps, using a calculator at the final stage:

$$t\log 1.029 = \log P - \log 14.235$$

$$t = \frac{\log P}{\log 1.029} - \frac{\log 14.235}{\log 1.029} = 80.545\log P - 92.898.$$

Thus,

$$f^{-1}(P) = 80.545\log P - 92.898.$$

Note that

$$f^{-1}(20) = 80.545(\log 20) - 92.898 = 11.89,$$

which agrees with the result of Example 2.

Exercises and Problems for Section 1.4

EXERCISES

In Exercises 1–6, simplify the expression completely.

1. $e^{\ln(1/2)}$
2. $10^{\log(AB)}$
3. $5e^{\ln(A^2)}$
4. $\ln(e^{2AB})$
5. $\ln(1/e) + \ln(AB)$
6. $2\ln\left(e^A\right) + 3\ln B^e$

For Exercises 7–18, solve for x using logs.

7. $3^x = 11$
8. $17^x = 2$
9. $20 = 50(1.04)^x$
10. $4 \cdot 3^x = 7 \cdot 5^x$
11. $7 = 5e^{0.2x}$
12. $2^x = e^{x+1}$
13. $50 = 600e^{-0.4x}$
14. $2e^{3x} = 4e^{5x}$
15. $7^{x+2} = e^{17x}$
16. $10^{x+3} = 5e^{7-x}$
17. $2x - 1 = e^{\ln x^2}$
18. $4e^{2x-3} - 5 = e$

For Exercises 19–24, solve for t. Assume a and b are positive, a and $b \neq 1$, and k is nonzero.

19. $a = b^t$
20. $P = P_0\, a^t$
21. $Q = Q_0\, a^{nt}$
22. $P_0\, a^t = Q_0\, b^t$
23. $a = be^t$
24. $P = P_0\, e^{kt}$

In Exercises 25–28, put the functions in the form $P = P_0 e^{kt}$.

25. $P = 15(1.5)^t$
26. $P = 10(1.7)^t$
27. $P = 174(0.9)^t$
28. $P = 4(0.55)^t$

Find the inverse function in Exercises 29–31.

29. $p(t) = (1.04)^t$
30. $f(t) = 50e^{0.1t}$
31. $f(t) = 1 + \ln t$

PROBLEMS

32. The exponential function $y(x) = Ce^{\alpha x}$ satisfies the conditions $y(0) = 2$ and $y(1) = 1$. Find the constants C and α. What is $y(2)$?

For Problems 33–34, find k such that $p = p_0 e^{kt}$ has the given doubling time.

33. 10
34. 0.4

35. A culture of bacteria originally numbers 500. After 2 hours there are 1500 bacteria in the culture. Assuming exponential growth, how many are there after 6 hours?

36. One hundred kilograms of a radioactive substance decay to 40 kg in 10 years. How much remains after 20 years?

37. The population of the US was 281.4 million in 2000 and 316.1 million in 2013.[37] Assuming exponential growth,
   (a) In what year is the population expected to go over 350 million?
   (b) What population is predicted for the 2020 census?

38. The population of a region is growing exponentially. There were 40,000,000 people in 2005 ($t = 0$) and 48,000,000 in 2015. Find an expression for the population at any time t, in years. What population would you predict for the year 2020? What is the doubling time?

39. Oil consumption in China grew exponentially[38] from 8.938 million barrels per day in 2010 to 10.480 million barrels per day in 2013. Assuming exponential growth continues at the same rate, what will oil consumption be in 2025?

40. The concentration of the car exhaust fume nitrous oxide, NO_2, in the air near a busy road is a function of distance from the road. The concentration decays exponentially at a continuous rate of 2.54% per meter.[39] At what distance from the road is the concentration of NO_2 half what it is on the road?

41. For children and adults with diseases such as asthma, the number of respiratory deaths per year increases by 0.33% when pollution particles increase by a microgram per cubic meter of air.[40]
   (a) Write a formula for the number of respiratory deaths per year as a function of quantity of pollution in the air. (Let Q_0 be the number of deaths per year with no pollution.)
   (b) What quantity of air pollution results in twice as many respiratory deaths per year as there would be without pollution?

[37] data.worldbank.org, accessed April 1, 2015.
[38] Based on www.eia.gov/cfapps/ipdbproject, accessed May 2015.
[39] Rickwood, P. and Knight, D. (2009). "The health impacts of local traffic pollution on primary school age children", *State of Australian Cities 2009 Conference Proceedings*, www.be.unsw.edu.au, accessed March 24, 2016.
[40] Brook, R. D., Franklin, B., Cascio, W., Hong, Y., Howard, G., Lipsett, M., Luepker, R., Mittleman, M., Samet, J., and Smith, S. C., "Air pollution and cardiovascular disease." *Circulation,* 2004;109:2655–2671.

42. The number of alternative fuel vehicles[41] running on E85, a fuel that is up to 85% plant-derived ethanol, increased exponentially in the US between 2005 and 2010.

(a) Use this information to complete the missing table values.
(b) How many E85-powered vehicles were there in the US in 2004?
(c) By what percent did the number of E85-powered vehicles grow from 2005 to 2009?

Year	2005	2006	2007	2008	2009	2010
No. E85 vehicles	246,363	?	?	?	?	618,505

43. A cup of coffee contains 100 mg of caffeine, which leaves the body at a continuous rate of 17% per hour.

(a) Write a formula for the amount, A mg, of caffeine in the body t hours after drinking a cup of coffee.
(b) Graph the function from part (a). Use the graph to estimate the half-life of caffeine.
(c) Use logarithms to find the half-life of caffeine.

44. Persistent organic pollutants (POPS) are a serious environmental hazard. Figure 1.56 shows their natural decay over time in human fat.[42]

(a) How long does it take for the concentration to decrease from 100 units to 50 units?
(b) How long does it take for the concentration to decrease from 50 units to 25 units?
(c) Explain why your answers to parts (a) and (b) suggest that the decay may be exponential.
(d) Find an exponential function that models concentration, C, as a function of t, the number of years since 1970.

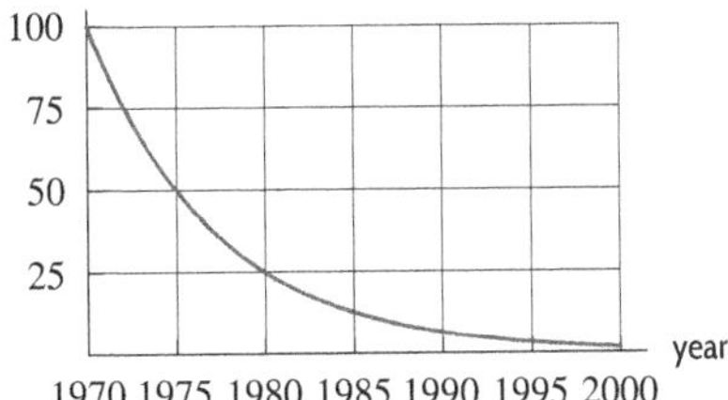

Figure 1.56

45. At time t hours after taking the cough suppressant hydrocodone bitartrate, the amount, A, in mg, remaining in the body is given by $A = 10(0.82)^t$.

(a) What was the initial amount taken?
(b) What percent of the drug leaves the body each hour?
(c) How much of the drug is left in the body 6 hours after the dose is administered?
(d) How long is it until only 1 mg of the drug remains in the body?

46. Different isotopes (versions) of the same element can have very different half-lives. With t in years, the decay of plutonium-240 is described by the formula

$$Q = Q_0 e^{-0.00011t},$$

whereas the decay of plutonium-242 is described by

$$Q = Q_0 e^{-0.0000018t}.$$

Find the half-lives of plutonium-240 and plutonium-242.

47. The size of an exponentially growing bacteria colony doubles in 5 hours. How long will it take for the number of bacteria to triple?

48. Air pressure, P, decreases exponentially with height, h, above sea level. If P_0 is the air pressure at sea level and h is in meters, then

$$P = P_0 e^{-0.00012h}.$$

(a) At the top of Denali, height 6194 meters (about 20,320 feet), what is the air pressure, as a percent of the pressure at sea level?
(b) The maximum cruising altitude of an ordinary commercial jet is around 12,000 meters (about 39,000 feet). At that height, what is the air pressure, as a percent of the sea level value?

49. With time, t, in years since the start of 1980, textbook prices have increased at 6.7% per year while inflation has been 3.3% per year.[43] Assume both rates are continuous growth rates.

(a) Find a formula for $B(t)$, the price of a textbook in year t if it cost $\$B_0$ in 1980.
(b) Find a formula for $P(t)$, the price of an item in year t if it cost $\$P_0$ in 1980 and its price rose according to inflation.
(c) A textbook cost \$50 in 1980. When is its price predicted to be double the price that would have resulted from inflation alone?

50. In November 2010, a "tiger summit" was held in St. Petersburg, Russia.[44] In 1900, there were 100,000 wild tigers worldwide; in 2010 the number was 3200.

(a) Assuming the tiger population has decreased exponentially, find a formula for $f(t)$, the number of wild tigers t years since 1900.
(b) Between 2000 and 2010, the number of wild tigers decreased by 40%. Is this percentage larger or smaller than the decrease in the tiger population predicted by your answer to part (a)?

[41] www.eia.gov, accessed April 1, 2015.
[42] K.C. Jones, P. de Voogt, "Persistent organic pollutants (POPs): state of the science," *Environmental Pollution* 100, 1999, pp. 209–221.
[43] Data from "Textbooks headed for ash heap of history", http://educationtechnews.com, Vol 5, 2010.
[44] "Tigers would be extinct in Russia if unprotected," Yahoo! News, Nov. 21, 2010.

51. In 2014, the populations of China and India were approximately 1.355 and 1.255 billion people,[45] respectively. However, due to central control the annual population growth rate of China was 0.44% while the population of India was growing by 1.25% each year. If these growth rates remain constant, when will the population of India exceed that of China?

52. The revenue of Apple® went from \$54.5 billion in 2013 to \$74.6 billion[46] in 2015. Find an exponential function to model the revenue as a function of years since 2013. What is the continuous percent growth rate, per year, of revenue?

53. The world population was 6.9 billion at the end of 2010 and is predicted to reach 9 billion by the end of 2050.[47]

(a) Assuming the population is growing exponentially, what is the continuous growth rate per year?
(b) The United Nations celebrated the "Day of 6 Billion" on October 12, 1999, and "Day of 7 Billion" on October 31, 2011. Using the growth rate in part (a), when is the "Day of 8 Billion" predicted to be?

54. In the early 1920s, Germany had tremendously high inflation, called hyperinflation. Photographs of the time show people going to the store with wheelbarrows full of money. If a loaf of bread cost $1/4$ marks in 1919 and 2,400,000 marks in 1922, what was the average yearly inflation rate between 1919 and 1922?

55. In 2010, there were about 246 million vehicles (cars and trucks) and about 308.7 million people in the US.[48] The number of vehicles grew 15.5% over the previous decade, while the population has been growing at 9.7% per decade. If the growth rates remain constant, when will there be, on average, one vehicle per person?

56. Tiny marine organisms reproduce at different rates. Phytoplankton doubles in population twice a day, but foraminifera doubles every five days. If the two populations are initially the same size and grow exponentially, how long does it take for

(a) The phytoplankton population to be double the foraminifera population.
(b) The phytoplankton population to be 1000 times the foraminifera population.

57. A picture supposedly painted by Vermeer (1632–1675) contains 99.5% of its carbon-14 (half-life 5730 years). From this information decide whether the picture is a fake. Explain your reasoning.

58. Cyanide is used in solution to isolate gold in a mine.[49] This may result in contaminated groundwater near the mine, requiring the poison be removed, as in the following table, where t is in years since 2012.

(a) Find an exponential model for $c(t)$, the concentration, in parts per million, of cyanide in the groundwater.
(b) Use the model in part (a) to find the number of years it takes for the cyanide concentration to fall to 10 ppm.
(c) The filtering process removing the cyanide is sped up so that the new model is $D(t) = c(2t)$. Find $D(t)$.
(d) If the cyanide removal was started three years earlier, but run at the speed of part (a), find a new model, $E(t)$.

t (years)	0	1	2
$c(t)$ (ppm)	25.0	21.8	19.01

59. In 2015, Nepal had two massive earthquakes, the first measuring 7.8 on the Richter scale and the second measuring 7.3. The Richter scale compares the strength, W, of the seismic waves of an earthquake with the strength, W_0, of the waves of a standard earthquake giving a Richter rating of $R = \log\left(W/W_0\right)$. By what factor were the seismic waves in Nepal's first earthquake stronger than the seismic waves in

(a) A standard earthquake?
(b) The second Nepal earthquake?

60. Find the equation of the line l in Figure 1.57.

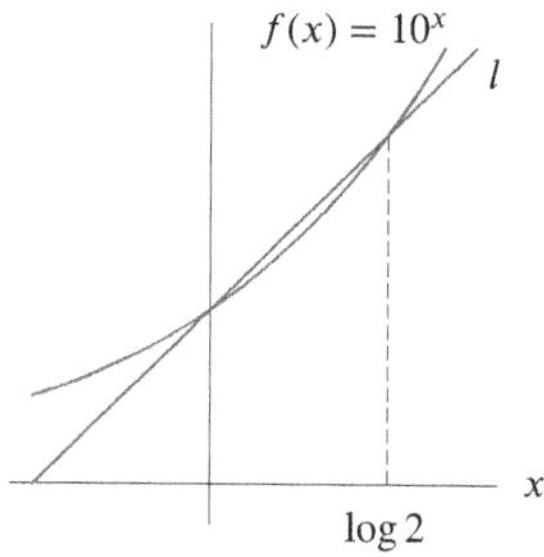

Figure 1.57

61. Without a calculator or computer, match the functions e^x, $\ln x$, x^2, and $x^{1/2}$ to their graphs in Figure 1.58.

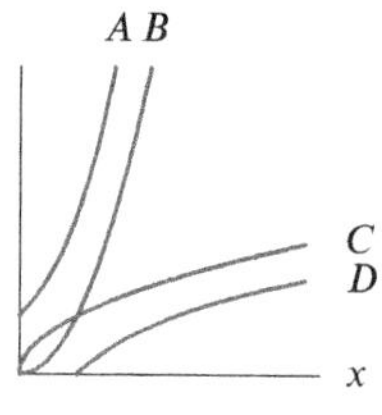

Figure 1.58

[45] www.indexmundi.com, accessed April 11, 2015.
[46] techcrunch.com, accessed April 2, 2015.
[47] "Reviewing the Bidding on the Climate Files", in About Dot Earth, *New York Times*, Nov. 19, 2010.
[48] http://www.autoblog.com/2010/01/04/report-number-of-cars-in-the-u-s-dropped-by-four-million-in-20/ and http://2010.census.gov/news/releases/operations/cb10-cn93.html. Accessed February 2012.
[49] www.www.miningfacts.org/environment/what-is-the-role-of-cyanide-in-mining/r. Accessed June 9, 2015.

62. Is there a difference between $\ln(\ln(x))$ and $\ln^2(x)$? (Note: $\ln^2(x)$ is another way of writing $(\ln x)^2$.)

63. If $h(x) = \ln(x + a)$, where $a > 0$, what is the effect of increasing a on

(a) The y-intercept? **(b)** The x-intercept?

64. If $h(x) = \ln(x + a)$, where $a > 0$, what is the effect of increasing a on the vertical asymptote?

65. If $g(x) = \ln(ax + 2)$, where $a \neq 0$, what is the effect of increasing a on

(a) The y-intercept? **(b)** The x-intercept?

66. If $f(x) = a\ln(x + 2)$, what is the effect of increasing a on the vertical asymptote?

67. If $g(x) = \ln(ax + 2)$, where $a \neq 0$, what is the effect of increasing a on the vertical asymptote?

68. Show that the growth rate k of the exponential function $f(t) = P_0e^{kt}$, with $P_0 > 0$, can be computed from two values of f by a difference quotient of the form:

$$k = \frac{\ln f(t_2) - \ln f(t_1)}{t_2 - t_1}.$$

Strengthen Your Understanding

In Problems 69–74, explain what is wrong with the statement.

69. The function $-\log|x|$ is odd.

70. For all $x > 0$, the value of $\ln(100x)$ is 100 times larger than $\ln x$.

71. $\ln x > \log x$.

72. $\ln(A + B) = (\ln A)(\ln B)$.

73. $\ln(A + B) = \ln A + \ln B$.

74. $\dfrac{1}{\ln x} = e^x$

In Problems 75–77, give an example of:

75. A function that grows slower than $y = \log x$ for $x > 1$.

76. A function $f(x)$ such that $\ln(f(x))$ is only defined for $x < 0$.

77. A function with a vertical asymptote at $x = 3$ and defined only for $x > 3$.

Are the statements in Problems 78–81 true or false? Give an explanation for your answer.

78. The graph of $f(x) = \ln x$ is concave down.

79. The graph of $g(x) = \log(x - 1)$ crosses the x-axis at $x = 1$.

80. The inverse function of $y = \log x$ is $y = 1/\log x$.

81. If a and b are positive constants, then $y = \ln(ax + b)$ has no vertical asymptote.

1.5 TRIGONOMETRIC FUNCTIONS

Trigonometry originated as part of the study of triangles. The name *tri-gon-o-metry* means the measurement of three-cornered figures, and the first definitions of the trigonometric functions were in terms of triangles. However, the trigonometric functions can also be defined using the unit circle, a definition that makes them periodic, or repeating. Many naturally occurring processes are also periodic. The water level in a tidal basin, the blood pressure in a heart, an alternating current, and the position of the air molecules transmitting a musical note all fluctuate regularly. Such phenomena can be represented by trigonometric functions.

Radians

There are two commonly used ways to represent the input of the trigonometric functions: radians and degrees. The formulas of calculus, as you will see, are neater in radians than in degrees.

> An angle of 1 **radian** is defined to be the angle at the center of a unit circle which cuts off an arc of length 1, measured counterclockwise. (See Figure 1.59(a).) A unit circle has radius 1.

An angle of 2 radians cuts off an arc of length 2 on a unit circle. A negative angle, such as $-1/2$ radians, cuts off an arc of length $1/2$, but measured clockwise. (See Figure 1.59(b).)

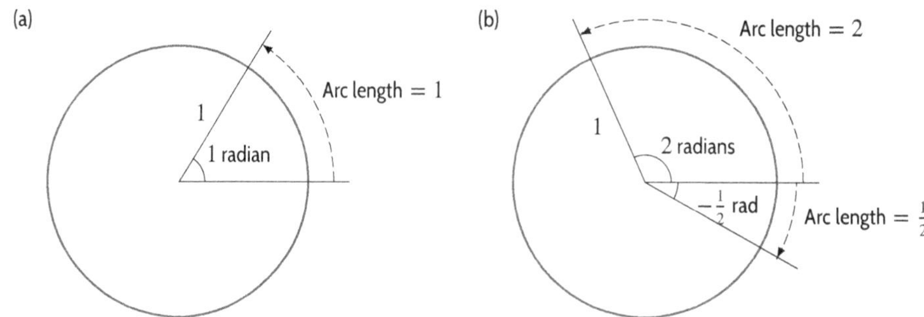

Figure 1.59: Radians defined using unit circle

It is useful to think of angles as rotations, since then we can make sense of angles larger than 360°; for example, an angle of 720° represents two complete rotations counterclockwise. Since one full rotation of 360° cuts off an arc of length 2π, the circumference of the unit circle, it follows that

$$360° = 2\pi \text{ radians}, \quad \text{so} \quad 180° = \pi \text{ radians}.$$

In other words, 1 radian $= 180°/\pi$, so one radian is about 60°. The word radians is often dropped, so if an angle or rotation is referred to without units, it is understood to be in radians.

Radians are useful for computing the length of an arc in any circle. If the circle has radius r and the arc cuts off an angle θ, as in Figure 1.60, then we have the following relation:

$$\text{Arc length} = s = r\theta.$$

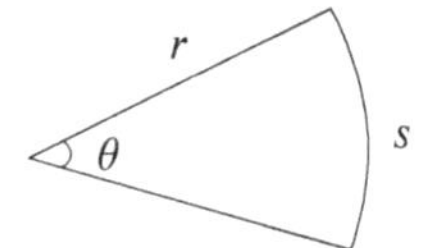

Figure 1.60: Arc length of a sector of a circle

The Sine and Cosine Functions

The two basic trigonometric functions—the sine and cosine—are defined using a unit circle. In Figure 1.61, an angle of t radians is measured counterclockwise around the circle from the point $(1, 0)$. If P has coordinates (x, y), we define

$$\cos t = x \quad \text{and} \quad \sin t = y.$$

We assume that the angles are *always* in radians unless specified otherwise.

Since the equation of the unit circle is $x^2+y^2 = 1$, writing $\cos^2 t$ for $(\cos t)^2$, we have the identity

$$\cos^2 t + \sin^2 t = 1.$$

As t increases and P moves around the circle, the values of $\sin t$ and $\cos t$ oscillate between 1 and -1, and eventually repeat as P moves through points where it has been before. If t is negative, the angle is measured clockwise around the circle.

Amplitude, Period, and Phase

The graphs of sine and cosine are shown in Figure 1.62. Notice that sine is an odd function, and cosine is even. The maximum and minimum values of sine and cosine are $+1$ and -1, because those are the maximum and minimum values of y and x on the unit circle. After the point P has moved around the complete circle once, the values of $\cos t$ and $\sin t$ start to repeat; we say the functions are *periodic*.

For any periodic function of time, the
- **Amplitude** is half the distance between the maximum and minimum values (if it exists).
- **Period** is the smallest time needed for the function to execute one complete cycle.

The amplitude of $\cos t$ and $\sin t$ is 1, and the period is 2π. Why 2π? Because that's the value of t when the point P has gone exactly once around the circle. (Remember that $360° = 2\pi$ radians.)

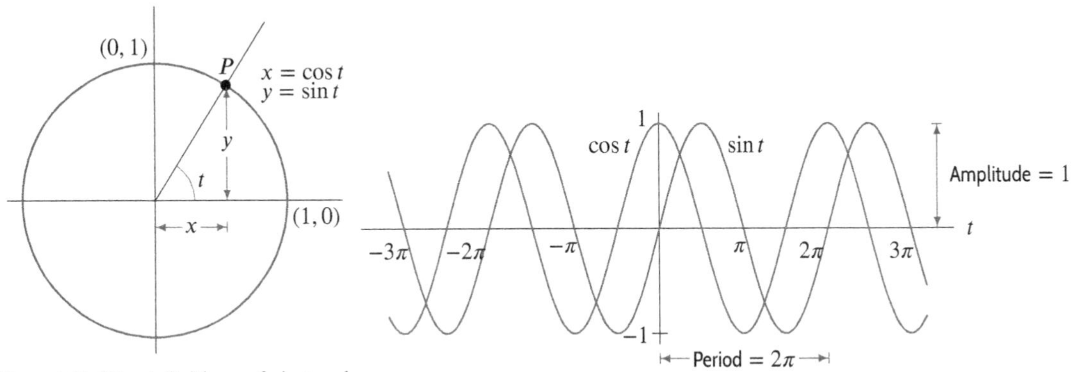

Figure 1.61: The definitions of $\sin t$ and $\cos t$

Figure 1.62: Graphs of $\cos t$ and $\sin t$

In Figure 1.62, we see that the sine and cosine graphs are exactly the same shape, only shifted horizontally. Since the cosine graph is the sine graph shifted $\pi/2$ to the left,

$$\cos t = \sin(t + \pi/2).$$

Equivalently, the sine graph is the cosine graph shifted $\pi/2$ to the right, so

$$\sin t = \cos(t - \pi/2).$$

We say that the *phase difference* or *phase shift* between $\sin t$ and $\cos t$ is $\pi/2$.

Functions whose graphs are the shape of a sine or cosine curve are called *sinusoidal* functions.

To describe arbitrary amplitudes and periods of sinusoidal functions, we use functions of the form

$$f(t) = A\sin(Bt) \quad \text{and} \quad g(t) = A\cos(Bt),$$

where $|A|$ is the amplitude and $2\pi/|B|$ is the period.
The graph of a sinusoidal function is shifted horizontally by a distance $|h|$ when t is replaced by $t - h$ or $t + h$.
Functions of the form $f(t) = A\sin(Bt) + C$ and $g(t) = A\cos(Bt) + C$ have graphs which are shifted vertically by C and oscillate about this value.

Example 1 Find and show on a graph the amplitude and period of the functions

(a) $y = 5\sin(2t)$ (b) $y = -5\sin\left(\frac{t}{2}\right)$ (c) $y = 1 + 2\sin t$

Solution

(a) From Figure 1.63, you can see that the amplitude of $y = 5\sin(2t)$ is 5 because the factor of 5 stretches the oscillations up to 5 and down to −5. The period of $y = \sin(2t)$ is π, because when t changes from 0 to π, the quantity $2t$ changes from 0 to 2π, so the sine function goes through one complete oscillation.

(b) Figure 1.64 shows that the amplitude of $y = -5\sin(t/2)$ is again 5, because the negative sign reflects the oscillations in the t-axis, but does not change how far up or down they go. The period of $y = -5\sin(t/2)$ is 4π because when t changes from 0 to 4π, the quantity $t/2$ changes from 0 to 2π, so the sine function goes through one complete oscillation.

(c) The 1 shifts the graph $y = 2\sin t$ up by 1. Since $y = 2\sin t$ has an amplitude of 2 and a period of 2π, the graph of $y = 1 + 2\sin t$ goes up to 3 and down to -1, and has a period of 2π. (See Figure 1.65.) Thus, $y = 1 + 2\sin t$ also has amplitude 2.

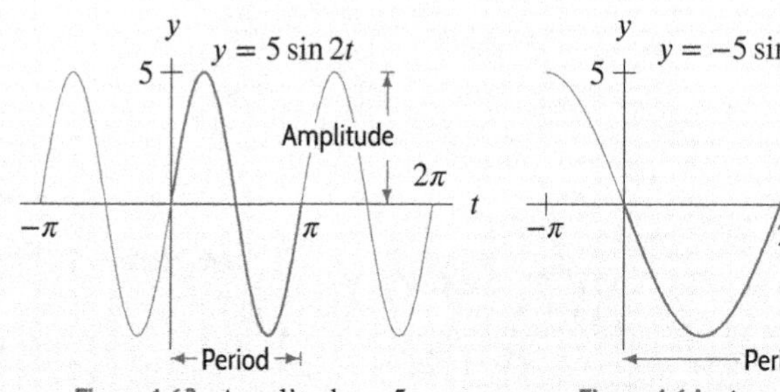

Figure 1.63: Amplitude = 5, period = π

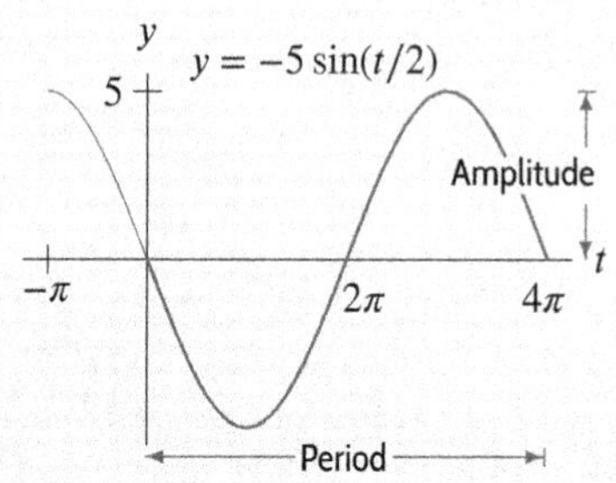

Figure 1.64: Amplitude = 5, period = 4π

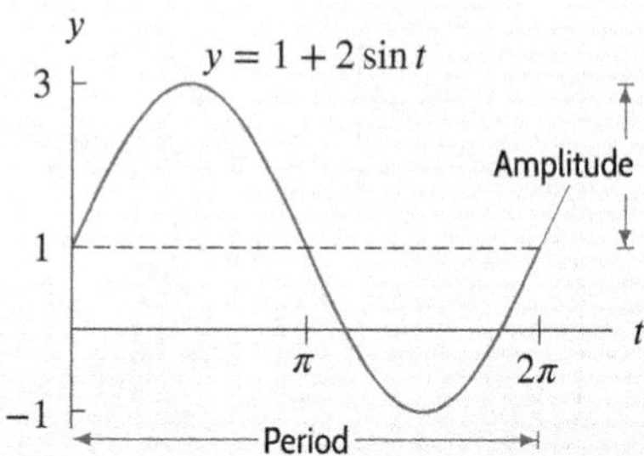

Figure 1.65: Amplitude = 2, period = 2π

Example 2 Find possible formulas for the following sinusoidal functions.

(a)

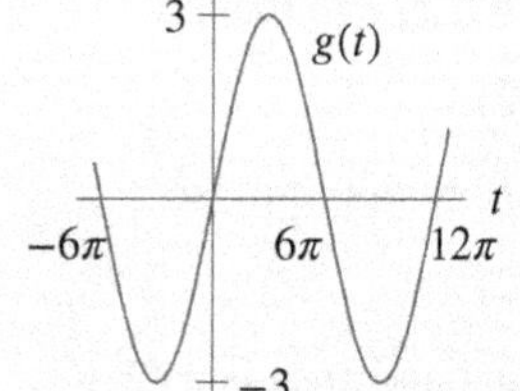

(b)

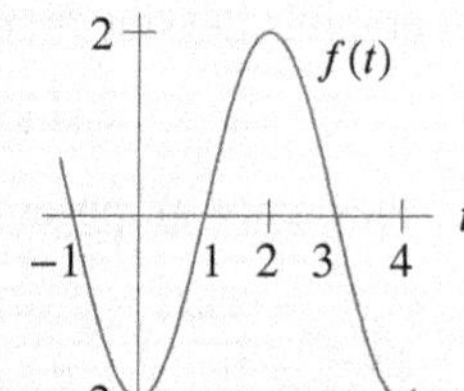

(c)

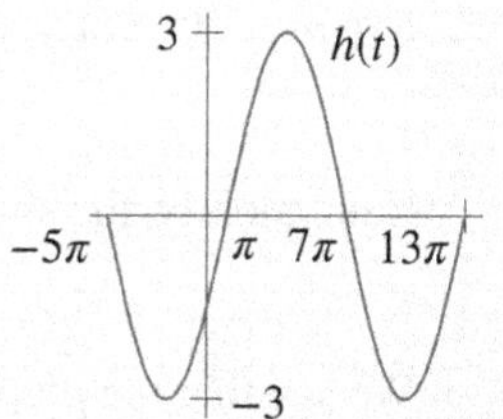

Solution (a) This function looks like a sine function with amplitude 3, so $g(t) = 3\sin(Bt)$. Since the function executes one full oscillation between $t = 0$ and $t = 12\pi$, when t changes by 12π, the quantity Bt changes by 2π. This means $B \cdot 12\pi = 2\pi$, so $B = 1/6$. Therefore, $g(t) = 3\sin(t/6)$ has the graph shown.

(b) This function looks like an upside-down cosine function with amplitude 2, so $f(t) = -2\cos(Bt)$. The function completes one oscillation between $t = 0$ and $t = 4$. Thus, when t changes by 4, the quantity Bt changes by 2π, so $B \cdot 4 = 2\pi$, or $B = \pi/2$. Therefore, $f(t) = -2\cos(\pi t/2)$ has the graph shown.

(c) This function looks like the function $g(t)$ in part (a), but shifted a distance of π to the right. Since $g(t) = 3\sin(t/6)$, we replace t by $(t - \pi)$ to obtain $h(t) = 3\sin[(t - \pi)/6]$.

Example 3 On July 1, 2007, high tide in Boston was at midnight. The water level at high tide was 9.9 feet; later, at low tide, it was 0.1 feet. Assuming the next high tide is at exactly 12 noon and that the height of the water is given by a sine or cosine curve, find a formula for the water level in Boston as a function of time.

Solution Let y be the water level in feet, and let t be the time measured in hours from midnight. The oscillations have amplitude 4.9 feet $(= (9.9 - 0.1)/2)$ and period 12, so $12B = 2\pi$ and $B = \pi/6$. Since the water is highest at midnight, when $t = 0$, the oscillations are best represented by a cosine function. (See Figure 1.66.) We can say

$$\text{Height above average} = 4.9\cos\left(\frac{\pi}{6}t\right).$$

Since the average water level was 5 feet $(= (9.9 + 0.1)/2)$, we shift the cosine up by adding 5:

$$y = 5 + 4.9\cos\left(\frac{\pi}{6}t\right).$$

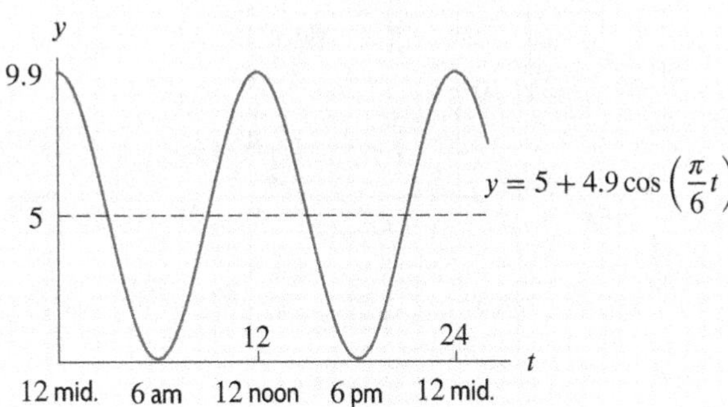

Figure 1.66: Function approximating the tide in Boston on July 1, 2007

Example 4 Of course, there's something wrong with the assumption in Example 3 that the next high tide is at noon. If so, the high tide would always be at noon or midnight, instead of progressing slowly through the day, as in fact it does. The interval between successive high tides actually averages about 12 hours 24 minutes. Using this, give a more accurate formula for the height of the water as a function of time.

Solution The period is 12 hours 24 minutes = 12.4 hours, so $B = 2\pi/12.4$, giving

$$y = 5 + 4.9\cos\left(\frac{2\pi}{12.4}t\right) = 5 + 4.9\cos(0.507t).$$

Example 5 Use the information from Example 4 to write a formula for the water level in Boston on a day when the high tide is at 2 pm.

Solution When the high tide is at midnight,

$$y = 5 + 4.9\cos(0.507t).$$

Since 2 pm is 14 hours after midnight, we replace t by $(t - 14)$. Therefore, on a day when the high tide is at 2 pm,

$$y = 5 + 4.9\cos(0.507(t - 14)).$$

The Tangent Function

If t is any number with $\cos t \neq 0$, we define the tangent function as follows

$$\tan t = \frac{\sin t}{\cos t}.$$

Figure 1.61 on page 41 shows the geometrical meaning of the tangent function: $\tan t$ is the slope of the line through the origin $(0, 0)$ and the point $P = (\cos t, \sin t)$ on the unit circle.

The tangent function is undefined wherever $\cos t = 0$, namely, at $t = \pm\pi/2, \pm 3\pi/2, \ldots$, and it has a vertical asymptote at each of these points. The function $\tan t$ is positive where $\sin t$ and $\cos t$ have the same sign. The graph of the tangent is shown in Figure 1.67.

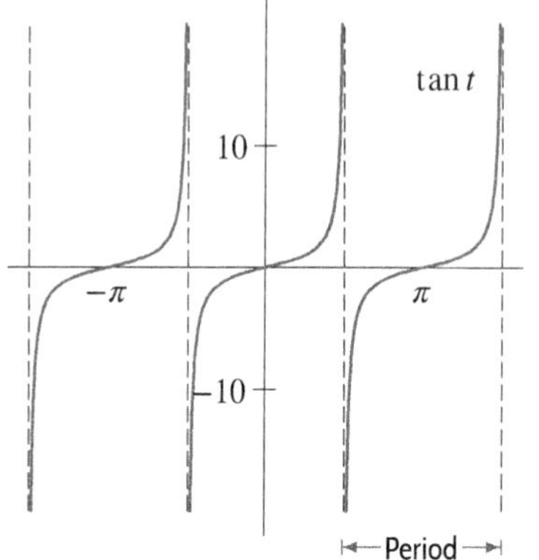

Figure 1.67: The tangent function

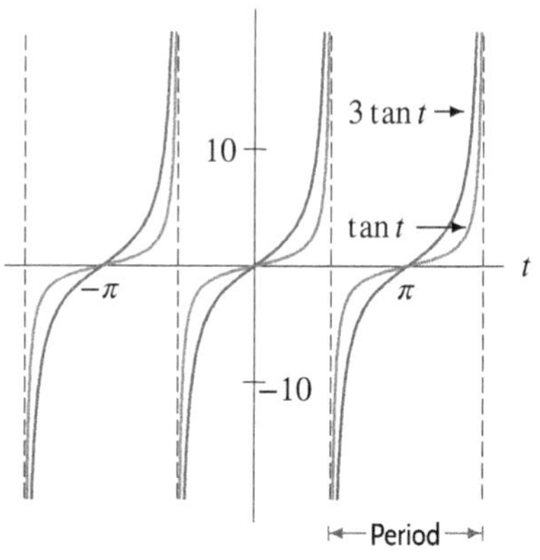

Figure 1.68: Multiple of tangent

The tangent function has period π, because it repeats every π units. Does it make sense to talk about the amplitude of the tangent function? Not if we're thinking of the amplitude as a measure of the size of the oscillation, because the tangent becomes infinitely large near each vertical asymptote. We can still multiply the tangent by a constant, but that constant no longer represents an amplitude. (See Figure 1.68.)

The Inverse Trigonometric Functions

On occasion, you may need to find a number with a given sine. For example, you might want to find x such that

$$\sin x = 0$$

or such that

$$\sin x = 0.3.$$

The first of these equations has solutions $x = 0, \pm\pi, \pm 2\pi, \ldots$. The second equation also has infinitely many solutions. Using a calculator and a graph, we get

$$x \approx 0.305, 2.84, 0.305 \pm 2\pi, 2.84 \pm 2\pi, \ldots.$$

For each equation, we pick out the solution between $-\pi/2$ and $\pi/2$ as the preferred solution. For example, the preferred solution to $\sin x = 0$ is $x = 0$, and the preferred solution to $\sin x = 0.3$ is $x = 0.305$. We define the inverse sine, written "arcsin" or "$\sin^{-1}$," as the function which gives the preferred solution.

For $-1 \le y \le 1$,

$$\arcsin y = x$$

means $\quad \sin x = y \quad \text{with} \quad -\frac{\pi}{2} \le x \le \frac{\pi}{2}.$

Thus the arcsine is the inverse function to the piece of the sine function having domain $[-\pi/2, \pi/2]$. (See Table 1.17 and Figure 1.69.) On a calculator, the arcsine function[50] is usually denoted by $\boxed{\sin^{-1}}$.

[50]Note that $\sin^{-1} x = \arcsin x$ is not the same as $(\sin x)^{-1} = 1/\sin x$.

Table 1.17 *Values of* $\sin x$ *and* $\sin^{-1} x$

x	$\sin x$
$-\frac{\pi}{2}$	-1.000
-1.0	-0.841
-0.5	-0.479
0.0	0.000
0.5	0.479
1.0	0.841
$\frac{\pi}{2}$	1.000

x	$\sin^{-1} x$
-1.000	$-\frac{\pi}{2}$
-0.841	-1.0
-0.479	-0.5
0.000	0.0
0.479	0.5
0.841	1.0
1.000	$\frac{\pi}{2}$

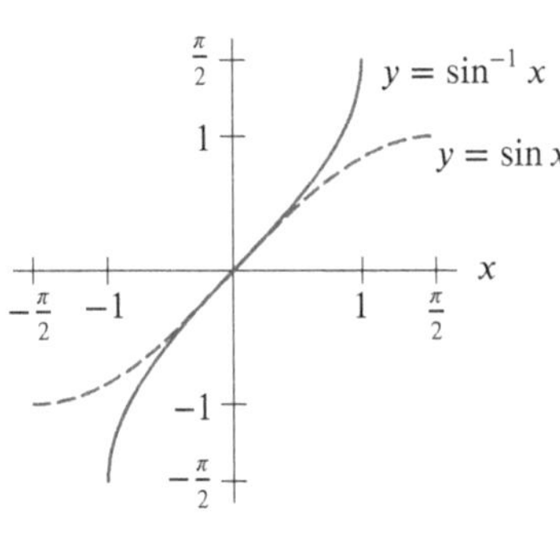

Figure 1.69: The arcsine function

The inverse tangent, written "arctan" or "$\tan^{-1}$," is the inverse function for the piece of the tangent function having the domain $-\pi/2 < x < \pi/2$. On a calculator, the inverse tangent is usually denoted by $\boxed{\tan^{-1}}$. The graph of the arctangent is shown in Figure 1.71.

> For any y,
> $$\arctan y = x$$
> means $$\tan x = y \quad \text{with} \quad -\frac{\pi}{2} < x < \frac{\pi}{2}.$$

The inverse cosine function, written "arccos" or "$\cos^{-1}$," is discussed in Problem 74. The range of the arccosine function is $0 \le x \le \pi$.

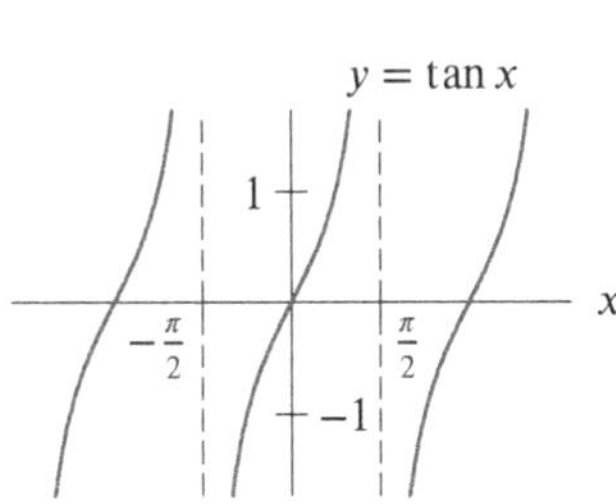

Figure 1.70: The tangent function

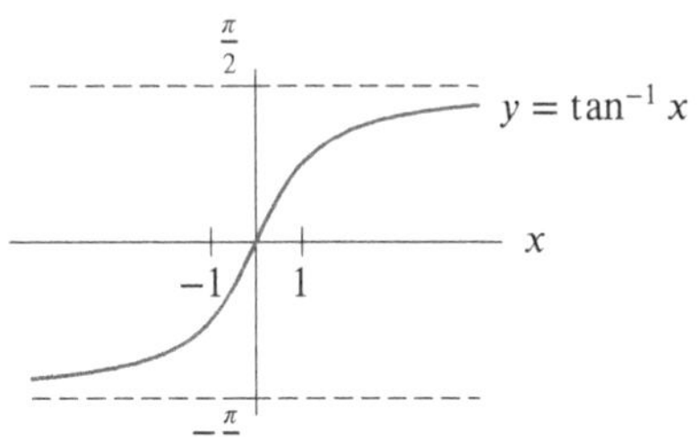

Figure 1.71: The arctangent function

Exercises and Problems for Section 1.5

EXERCISES

■ For Exercises 1–9, draw the angle using a ray through the origin, and determine whether the sine, cosine, and tangent of that angle are positive, negative, zero, or undefined.

1. $\frac{3\pi}{2}$
2. 2π
3. $\frac{\pi}{4}$
4. 3π
5. $\frac{\pi}{6}$
6. $\frac{4\pi}{3}$
7. $\frac{-4\pi}{3}$
8. 4
9. -1

■ Find the period and amplitude in Exercises 10–13.

10. $y = 7\sin(3t)$
11. $z = 3\cos(u/4) + 5$
12. $w = 8 - 4\sin(2x + \pi)$
13. $r = 0.1\sin(\pi t) + 2$

■ For Exercises 14–23, find a possible formula for each graph.

14.
15.

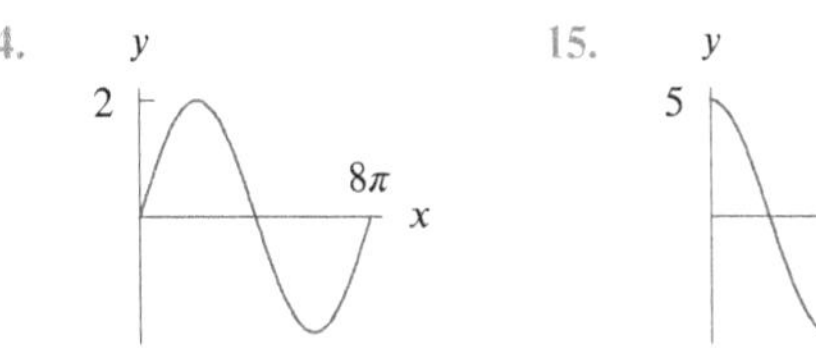

16. y 4 x π

17.

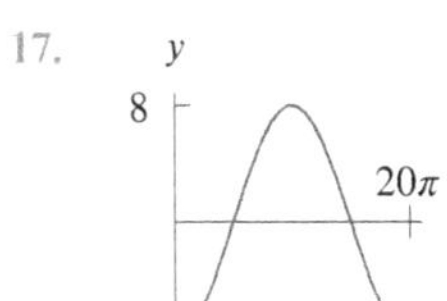

18.

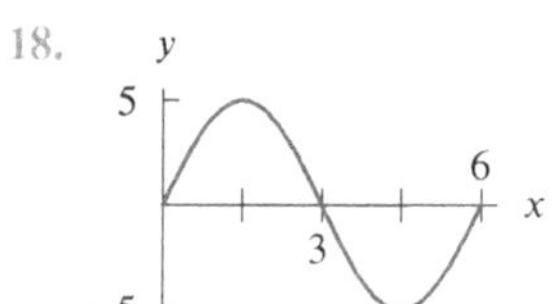

19.

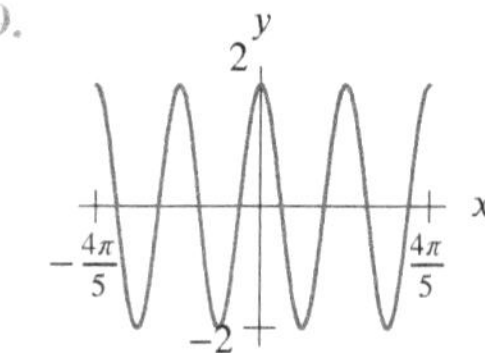

20.

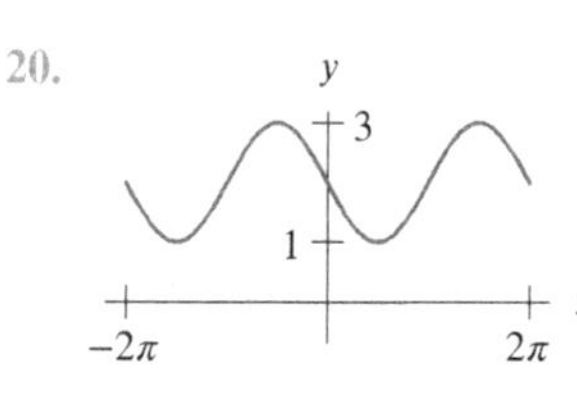

21.

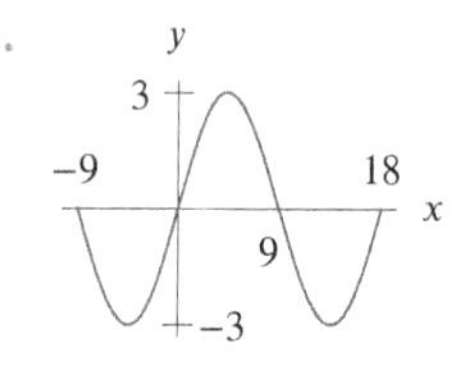

22.

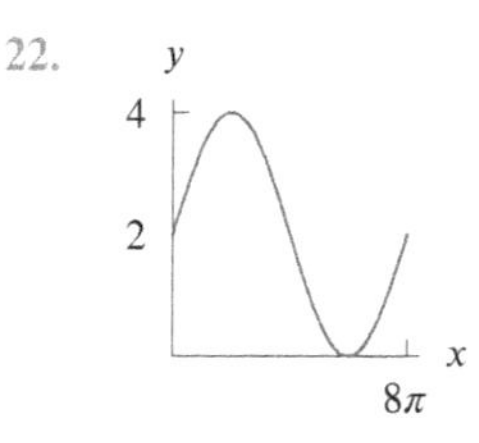

23.

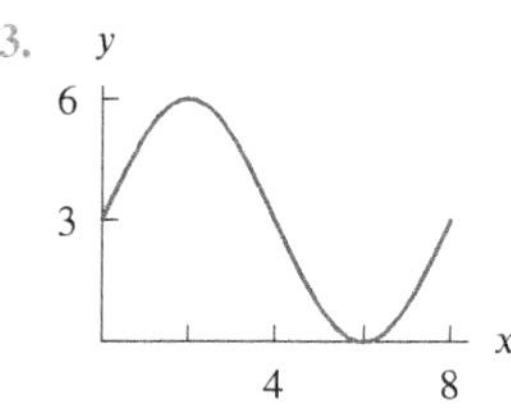

■ In Exercises 24–29, calculate the quantity without using the the trigonometric functions on a calculator. You are given that $\sin(\pi/12) = 0.259$ and $\cos(\pi/5) = 0.809$. You may want to draw a picture showing the angles involved.

24. $\cos\left(-\frac{\pi}{5}\right)$ 25. $\sin\frac{\pi}{5}$ 26. $\cos\frac{\pi}{12}$

27. $\sin(-\pi/12)$ 28. $\tan\pi/12$ 29. $\tan\pi/5$

■ In Exercises 30–34, find a solution to the equation if possible. Give the answer in exact form and in decimal form.

30. $2 = 5\sin(3x)$ 31. $1 = 8\cos(2x + 1) - 3$

32. $8 = 4\tan(5x)$ 33. $1 = 8\tan(2x + 1) - 3$

34. $8 = 4\sin(5x)$

35. What is the period of the earth's revolution around the sun?

36. What is the approximate period of the moon's revolution around the earth?

37. When a car's engine makes less than about 200 revolutions per minute, it stalls. What is the period of the rotation of the engine when it is about to stall?

38. A compact disc spins at a rate of 200 to 500 revolutions per minute. What are the equivalent rates measured in radians per second?

39. Find the angle, in degrees, that a wheelchair ramp makes with the ground if the ramp rises 1 foot over a horizontal distance of

(a) 12 ft, the normal requirement[51]
(b) 8 ft, the steepest ramp legally permitted
(c) 20 ft, the recommendation if snow can be expected on the ramp

PROBLEMS

40. **(a)** Use a graphing calculator or computer to estimate the period of $2\sin\theta + 3\cos(2\theta)$.
(b) Explain your answer, given that the period of $\sin\theta$ is 2π and the period of $\cos(2\theta)$ is π.

41. Without a calculator or computer, match the formulas with the graphs in Figure 1.72.

(a) $y = 2\cos(t - \pi/2)$ **(b)** $y = 2\cos t$
(c) $y = 2\cos(t + \pi/2)$

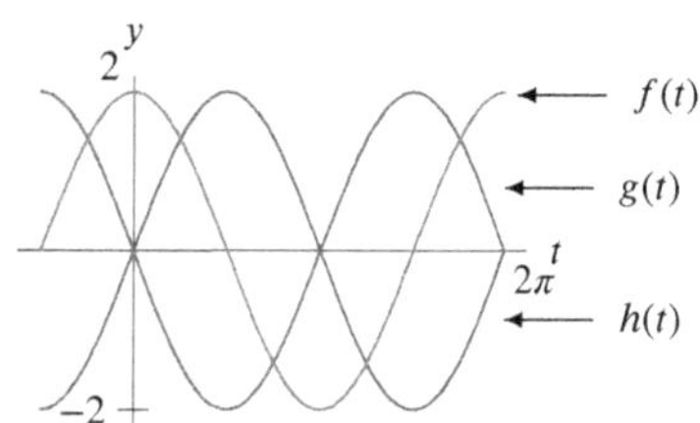

Figure 1.72

42. Figure 1.73 shows four periodic functions of the family $f(t) = A\cos(B(t-h))$, all with the same amplitude and period but with different horizontal shifts.

(a) Find the value of A.
(b) Find the value of B.
(c) Which graph corresponds to each of the following values of h: 0, 20, 50, 60?

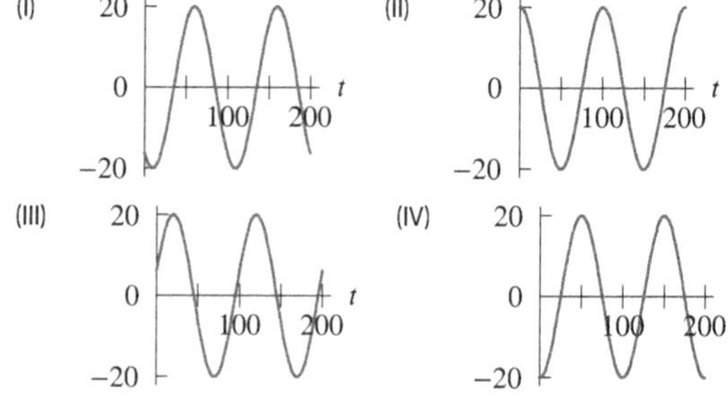

Figure 1.73

[51] http://www.access-board.gov/adaag/html/adaag.htm#4.1.6(3)a, accessed June 6, 2011.

■ In Problems 43–49, graph the given function on the axes in Figure 1.74.

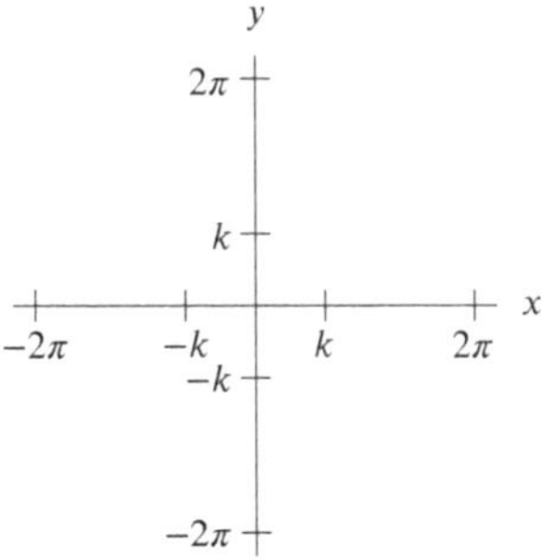

Figure 1.74

43. $y = k \sin x$

44. $y = -k \cos x$

45. $y = k(\cos x) + k$

46. $y = k(\sin x) - k$

47. $y = k \sin(x - k)$

48. $y = k \cos(x + k)$

49. $y = k \sin(2\pi x/k)$

■ For Problems 50–53, use Figure 1.75 to estimate the given value for $f(t) = A\cos(B(t - h)) + C$, which approximates the plotted average monthly water temperature of the Mississippi River in Louisiana.[52]

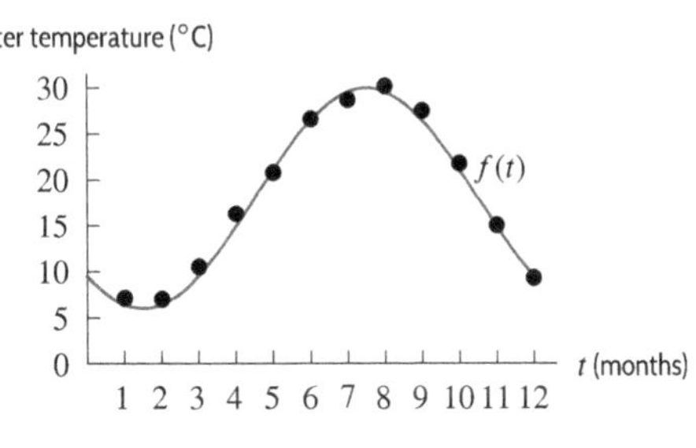

Figure 1.75

50. C

51. A

52. B

53. h

■ For Problems 54–57, use Figure 1.75 to estimate the given value for $f(t) = A\sin(B(t - h)) + C$, which approximates the plotted average monthly water flow of the Mississippi River.[53]

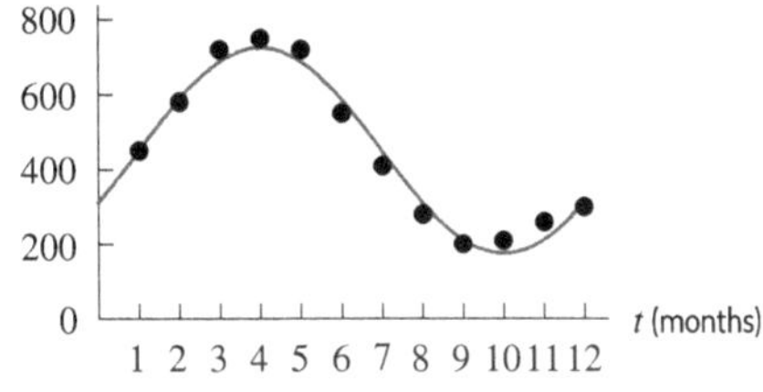

Figure 1.76

54. C

55. A

56. B

57. h

58. The visitors' guide to St. Petersburg, Florida, contains the chart shown in Figure 1.77 to advertise their good weather. Fit a trigonometric function approximately to the data, where H is temperature in degrees Fahrenheit, and the independent variable is time in months. In order to do this, you will need to estimate the amplitude and period of the data, and when the maximum occurs. (There are many possible answers to this problem, depending on how you read the graph.)

H (°F)	Jan	Feb	Mar	Apr	May	June	July	Aug	Sept	Oct	Nov	Dec
100°												
90°												
80°												
70°												
60°												
50°												

Figure 1.77: "St. Petersburg...where we're famous for our wonderful weather and year-round sunshine." (Reprinted with permission)

59. What is the difference between $\sin x^2$, $\sin^2 x$, and $\sin(\sin x)$? Express each of the three as a composition. (Note: $\sin^2 x$ is another way of writing $(\sin x)^2$.)

60. On the graph of $y = \sin x$, points P and Q are at consecutive lowest and highest points. Find the slope of the line through P and Q.

61. A population of animals oscillates sinusoidally between a low of 700 on January 1 and a high of 900 on July 1.

    (a) Graph the population against time.
    (b) Find a formula for the population as a function of time, t, in months since the start of the year.

62. The desert temperature, H, oscillates daily between 40°F at 5 am and 80°F at 5 pm. Write a possible formula for H in terms of t, measured in hours from 5 am.

63. The depth of water in a tank oscillates sinusoidally once every 6 hours. If the smallest depth is 5.5 feet and the largest depth is 8.5 feet, find a possible formula for the depth in terms of time in hours.

64. The voltage, V, of an electrical outlet in a home as a function of time, t (in seconds), is $V = V_0 \cos(120\pi t)$.

    (a) What is the period of the oscillation?
    (b) What does V_0 represent?
    (c) Sketch the graph of V against t. Label the axes.

65. The power output, P, of a solar panel varies with the position of the sun. Let $P = 10 \sin\theta$ watts, where θ is the angle between the sun's rays and the panel, $0 \le \theta \le \pi$. On a typical summer day in Ann Arbor, Michigan, the sun rises at 6 am and sets at 8 pm and the angle is

[52] waterdata.usgs.gov/nwis. Accessed May 2015.
[53] www.americaswetlandresources.com. Accessed May 2015.

$\theta = \pi t/14$, where t is time in hours since 6 am and $0 \le t \le 14$.

(a) Write a formula for a function, $f(t)$, giving the power output of the solar panel (in watts) t hours after 6 am on a typical summer day in Ann Arbor.
(b) Graph the function $f(t)$ in part (a) for $0 \le t \le 14$.
(c) At what time is the power output greatest? What is the power output at this time?
(d) On a typical winter day in Ann Arbor, the sun rises at 8 am and sets at 5 pm. Write a formula for a function, $g(t)$, giving the power output of the solar panel (in watts) t hours after 8 am on a typical winter day.

66. A baseball hit at an angle of θ to the horizontal with initial velocity v_0 has horizontal range, R, given by

$$R = \frac{v_0^2}{g}\sin(2\theta).$$

Here g is the acceleration due to gravity. Sketch R as a function of θ for $0 \le \theta \le \pi/2$. What angle gives the maximum range? What is the maximum range?

67. **(a)** Match the functions $\omega = f(t)$, $\omega = g(t)$, $\omega = h(t)$, $\omega = k(t)$, whose values are in the table, with the functions with formulas:

(i) $\omega = 1.5 + \sin t$ (ii) $\omega = 0.5 + \sin t$
(iii) $\omega = -0.5 + \sin t$ (iv) $\omega = -1.5 + \sin t$

(b) Based on the table, what is the relationship between the values of $g(t)$ and $k(t)$? Explain this relationship using the formulas you chose for g and k.
(c) Using the formulas you chose for g and h, explain why all the values of g are positive, whereas all the values of h are negative.

t	$f(t)$	t	$g(t)$	t	$h(t)$	t	$k(t)$
6.0	−0.78	3.0	1.64	5.0	−2.46	3.0	0.64
6.5	−0.28	3.5	1.15	5.1	−2.43	3.5	0.15
7.0	0.16	4.0	0.74	5.2	−2.38	4.0	−0.26
7.5	0.44	4.5	0.52	5.3	−2.33	4.5	−0.48
8.0	0.49	5.0	0.54	5.4	−2.27	5.0	−0.46

68. For a boat to float in a tidal bay, the water must be at least 2.5 meters deep. The depth of water around the boat, $d(t)$, in meters, where t is measured in hours since midnight, is

$$d(t) = 5 + 4.6\sin(0.5t).$$

(a) What is the period of the tides in hours?
(b) If the boat leaves the bay at midday, what is the latest time it can return before the water becomes too shallow?

69. The Bay of Fundy in Canada has the largest tides in the world. The difference between low and high water levels is 15 meters (nearly 50 feet). At a particular point the depth of the water, y meters, is given as a function of time, t, in hours since midnight by

$$y = D + A\cos\left(B(t - C)\right).$$

(a) What is the physical meaning of D?
(b) What is the value of A?
(c) What is the value of B? Assume the time between successive high tides is 12.4 hours.
(d) What is the physical meaning of C?

70. Match graphs A-D in Figure 1.78 with the functions below. Assume a, b, c and d are positive constants.

$f(t) = \sin t + b$ $\quad$ $h(t) = \sin t + e^{ct} + d$
$g(t) = \sin t + at + b$ $\quad$ $r(t) = \sin t - e^{ct} + b$

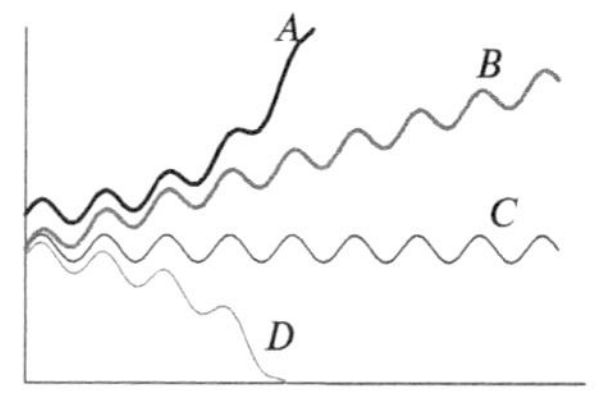

Figure 1.78

71. In Figure 1.79, the blue curve shows monthly mean carbon dioxide (CO_2) concentration, in parts per million (ppm) at Mauna Loa Observatory, Hawaii, as a function of t, in months, since December 2005. The black curve shows the monthly mean concentration adjusted for seasonal CO_2 variation.[54]

(a) Approximately how much did the monthly mean CO_2 increase between December 2005 and December 2010?
(b) Find the average monthly rate of increase of the monthly mean CO_2 between December 2005 and December 2010. Use this information to find a linear function that approximates the black curve.
(c) The seasonal CO_2 variation between December 2005 and December 2010 can be approximated by a sinusoidal function. What is the approximate period of the function? What is its amplitude? Give a formula for the function.
(d) The blue curve may be approximated by a function of the form $h(t) = f(t) + g(t)$, where $f(t)$ is sinusoidal and $g(t)$ is linear. Using your work in parts (b) and (c), find a possible formula for $h(t)$. Graph $h(t)$ using the scale in Figure 1.79.

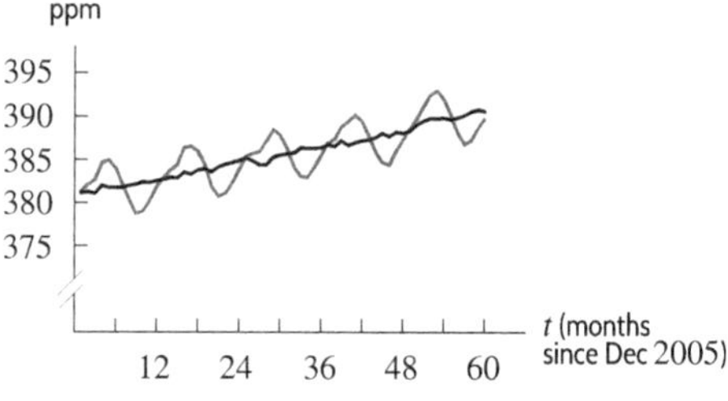

Figure 1.79

[54] www.esrl.noaa.gov/gmd/ccgg/trends/. Accessed March 2011. Monthly means joined by segments to highlight trends.

72. Find the area of the trapezoidal cross-section of the irrigation canal shown in Figure 1.80.

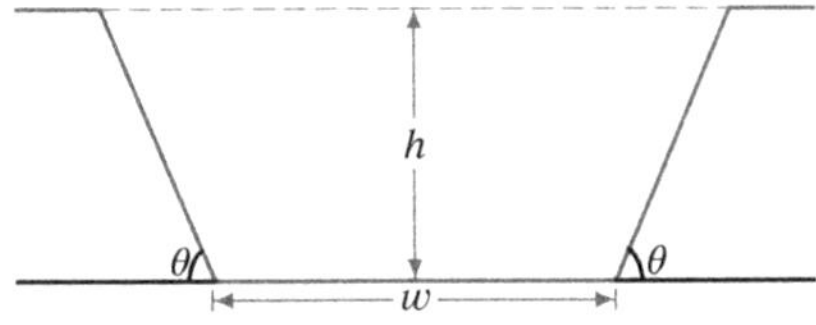

Figure 1.80

73. Graph $y = \sin x$, $y = 0.4$, and $y = -0.4$.

(a) From the graph, estimate to one decimal place all the solutions of $\sin x = 0.4$ with $-\pi \leq x \leq \pi$.

(b) Use a calculator to find arcsin(0.4). What is the relation between arcsin(0.4) and each of the solutions you found in part (a)?

(c) Estimate all the solutions to $\sin x = -0.4$ with $-\pi \leq x \leq \pi$ (again, to one decimal place).

(d) What is the relation between arcsin(0.4) and each of the solutions you found in part (c)?

74. This problem introduces the arccosine function, or inverse cosine, denoted by $\boxed{\cos^{-1}}$ on most calculators.

(a) Using a calculator set in radians, make a table of values, to two decimal places, of $g(x) = \arccos x$, for $x = -1, -0.8, -0.6, \ldots, 0, \ldots, 0.6, 0.8, 1$.

(b) Sketch the graph of $g(x) = \arccos x$.

(c) Why is the domain of the arccosine the same as the domain of the arcsine?

(d) What is the range of the arccosine?

(e) Why is the range of the arccosine *not* the same as the range of the arcsine?

Strengthen Your Understanding

In Problems 75–78, explain what is wrong with the statement.

75. The functions $f(x) = 3\cos x$ and $g(x) = \cos 3x$ have the same period.

76. For the function $f(x) = \sin(Bx)$ with $B > 0$, increasing the value of B increases the period.

77. The function $y = \sin x \cos x$ is periodic with period 2π.

78. For positive A, B, C, the maximum value of the function $y = A\sin(Bx) + C$ is $y = A$.

In Problems 79–80, give an example of:

79. A sine function with period 23.

80. A cosine function which oscillates between values of 1200 and 2000.

Are the statements in Problems 81–97 true or false? Give an explanation for your answer.

81. The family of functions $y = a\sin x$, with a a positive constant, all have the same period.

82. The family of functions $y = \sin ax$, a a positive constant, all have the same period.

83. The function $f(\theta) = \cos\theta - \sin\theta$ is increasing on $0 \leq \theta \leq \pi/2$.

84. The function $f(t) = \sin(0.05\pi t)$ has period 0.05.

85. If t is in seconds, $g(t) = \cos(200\pi t)$ executes 100 cycles in one second.

86. The function $f(\theta) = \tan(\theta - \pi/2)$ is not defined at $\theta = \pi/2, 3\pi/2, 5\pi/2 \ldots$.

87. $\sin|x| = \sin x$ for $-2\pi < x < 2\pi$

88. $\sin|x| = |\sin x|$ for $-2\pi < x < 2\pi$

89. $\cos|x| = |\cos x|$ for $-2\pi < x < 2\pi$

90. $\cos|x| = \cos x$ for $-2\pi < x < 2\pi$

91. The function $f(x) = \sin(x^2)$ is periodic, with period 2π.

92. The function $g(\theta) = e^{\sin\theta}$ is periodic.

93. If $f(x)$ is a periodic function with period k, then $f(g(x))$ is periodic with period k for every function $g(x)$.

94. If $g(x)$ is a periodic function, then $f(g(x))$ is periodic for every function $f(x)$.

95. The function $f(x) = |\sin x|$ is even.

96. $\sin^{-1}(\sin t) = t$ for all t.

97. The function $f(t) = \sin^{-1}(\sin t)$ is periodic with period 2π.

1.6 POWERS, POLYNOMIALS, AND RATIONAL FUNCTIONS

Power Functions

A *power function* is a function in which the dependent variable is proportional to a power of the independent variable:

> A **power function** has the form
>
> $f(x) = kx^p$, where k and p are constant.

For example, the volume, V, of a sphere of radius r is given by

$$V = g(r) = \frac{4}{3}\pi r^3.$$

As another example, the gravitational force, F, on a unit mass at a distance r from the center of the earth is given by Newton's Law of Gravitation, which says that, for some positive constant k,

$$F = \frac{k}{r^2} \quad \text{or} \quad F = kr^{-2}.$$

We consider the graphs of the power functions x^n, with n a positive integer. Figures 1.81 and 1.82 show that the graphs fall into two groups: odd and even powers. For n greater than 1, the odd powers have a "seat" at the origin and are increasing everywhere else. The even powers are first decreasing and then increasing. For large x, the higher the power of x, the faster the function climbs.

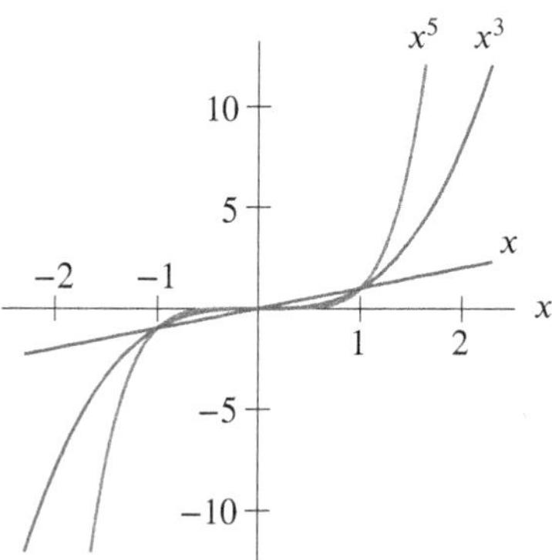

Figure 1.81: Odd powers of x: "Seat" shaped for $n > 1$

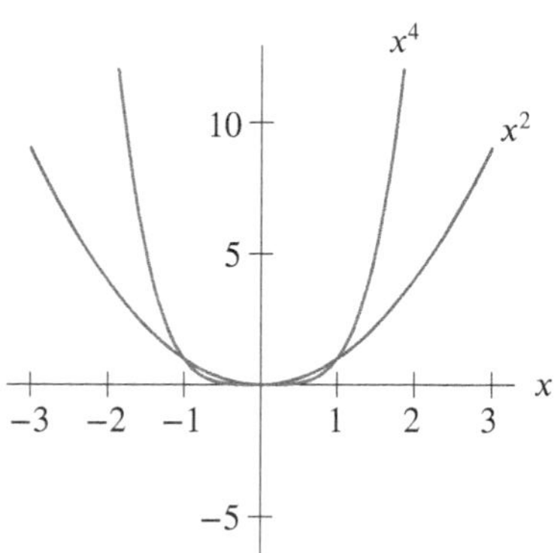

Figure 1.82: Even powers of x: ⋃-shaped

Exponentials and Power Functions: Which Dominate?

In everyday language, the word exponential is often used to imply very fast growth. But do exponential functions always grow faster than power functions? To determine what happens "in the long run," we often want to know which functions *dominate* as x gets arbitrarily large.

Let's consider $y = 2^x$ and $y = x^3$. The close-up view in Figure 1.83(a) shows that between $x = 2$ and $x = 4$, the graph of $y = 2^x$ lies below the graph of $y = x^3$. The far-away view in Figure 1.83(b) shows that the exponential function $y = 2^x$ eventually overtakes $y = x^3$. Figure 1.83(c), which gives a very far-away view, shows that, for large x, the value of x^3 is insignificant compared to 2^x. Indeed, 2^x is growing so much faster than x^3 that the graph of 2^x appears almost vertical in comparison to the more leisurely climb of x^3.

We say that Figure 1.83(a) gives a *local* view of the functions' behavior, whereas Figure 1.83(c) gives a *global* view.

In fact, *every* exponential growth function eventually dominates *every* power function. Although an exponential function may be below a power function for some values of x, if we look at large enough x-values, a^x (with $a > 1$) will eventually dominate x^n, no matter what n is.

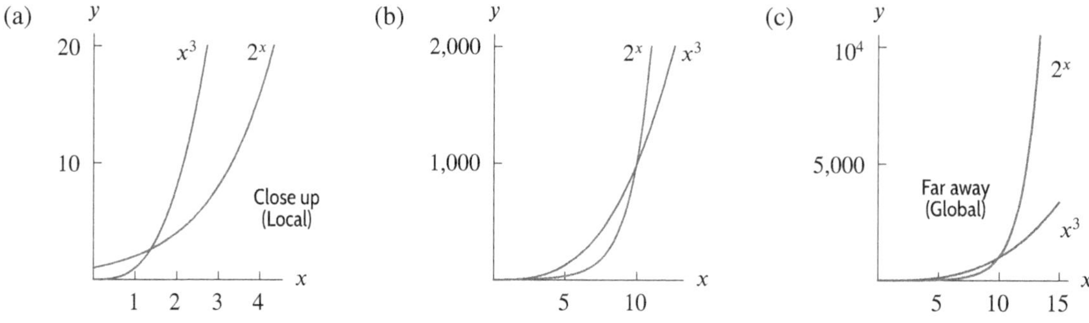

Figure 1.83: Comparison of $y = 2^x$ and $y = x^3$: Notice that $y = 2^x$ eventually dominates $y = x^3$

Polynomials

Polynomials are the sums of power functions with nonnegative integer exponents:

$$y = p(x) = a_n x^n + a_{n-1}x^{n-1} + \cdots + a_1 x + a_0.$$

Here n is a nonnegative integer called the *degree* of the polynomial, and $a_n, a_{n-1}, \ldots, a_1, a_0$ are constants, with leading coefficient $a_n \neq 0$. An example of a polynomial of degree $n = 3$ is

$$y = p(x) = 2x^3 - x^2 - 5x - 7.$$

In this case $a_3 = 2, a_2 = -1, a_1 = -5$, and $a_0 = -7$. The shape of the graph of a polynomial depends on its degree; typical graphs are shown in Figure 1.84. These graphs correspond to a positive coefficient for x^n; a negative leading coefficient turns the graph upside down. Notice that the quadratic "turns around" once, the cubic "turns around" twice, and the quartic (fourth degree) "turns around" three times. An nth-degree polynomial "turns around" at most $n - 1$ times (where n is a positive integer), but there may be fewer turns.

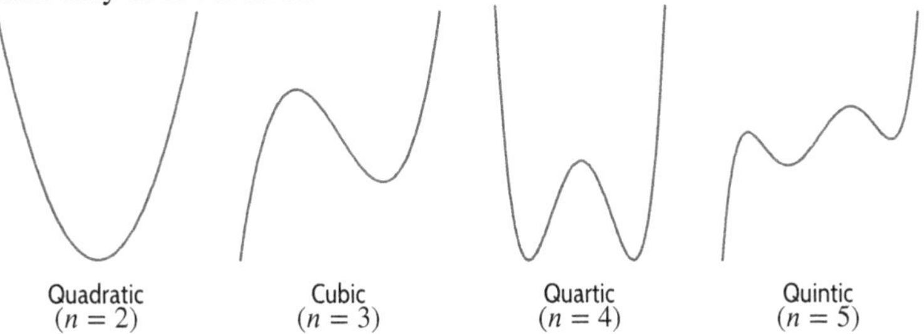

Figure 1.84: Graphs of typical polynomials of degree n

Example 1 Find possible formulas for the polynomials whose graphs are in Figure 1.85.

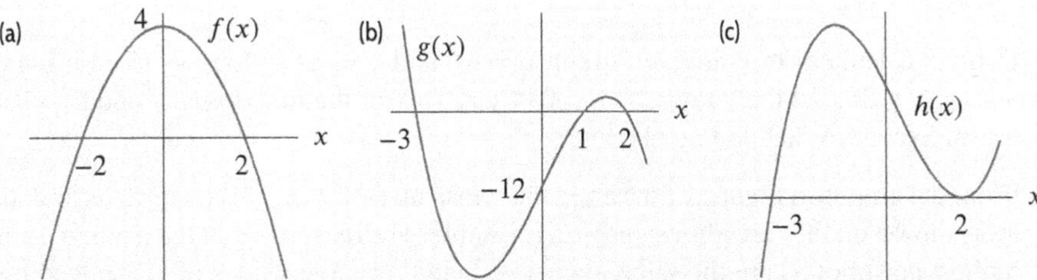

Figure 1.85: Graphs of polynomials

Solution

(a) This graph appears to be a parabola, turned upside down, and moved up by 4, so

$$f(x) = -x^2 + 4.$$

The negative sign turns the parabola upside down and the +4 moves it up by 4. Notice that this formula does give the correct x-intercepts since $0 = -x^2 + 4$ has solutions $x = \pm 2$. These values of x are called *zeros* of f.

We can also solve this problem by looking at the x-intercepts first, which tell us that $f(x)$ has factors of $(x + 2)$ and $(x - 2)$. So

$$f(x) = k(x + 2)(x - 2).$$

To find k, use the fact that the graph has a y-intercept of 4, so $f(0) = 4$, giving

$$4 = k(0 + 2)(0 - 2),$$

or $k = -1$. Therefore, $f(x) = -(x + 2)(x - 2)$, which multiplies out to $-x^2 + 4$.

Note that $f(x) = 4 - x^4/4$ also has the same basic shape, but is flatter near $x = 0$. There are many possible answers to these questions.

(b) This looks like a cubic with factors $(x + 3)$, $(x - 1)$, and $(x - 2)$, one for each intercept:

$$g(x) = k(x + 3)(x - 1)(x - 2).$$

Since the y-intercept is -12, we have

$$-12 = k(0+3)(0-1)(0-2).$$

So $k = -2$, and we get the cubic polynomial

$$g(x) = -2(x+3)(x-1)(x-2).$$

(c) This also looks like a cubic with zeros at $x = 2$ and $x = -3$. Notice that at $x = 2$ the graph of $h(x)$ touches the x-axis but does not cross it, whereas at $x = -3$ the graph crosses the x-axis. We say that $x = 2$ is a *double zero*, but that $x = -3$ is a single zero.

To find a formula for $h(x)$, imagine the graph of $h(x)$ to be slightly lower down, so that the graph has one x-intercept near $x = -3$ and two near $x = 2$, say at $x = 1.9$ and $x = 2.1$. Then a formula would be

$$h(x) \approx k(x+3)(x-1.9)(x-2.1).$$

Now move the graph back to its original position. The zeros at $x = 1.9$ and $x = 2.1$ move toward $x = 2$, giving

$$h(x) = k(x+3)(x-2)(x-2) = k(x+3)(x-2)^2.$$

The double zero leads to a repeated factor, $(x-2)^2$. Notice that when $x > 2$, the factor $(x-2)^2$ is positive, and when $x < 2$, the factor $(x-2)^2$ is still positive. This reflects the fact that $h(x)$ does not change sign near $x = 2$. Compare this with the behavior near the single zero at $x = -3$, where h does change sign.

We cannot find k, as no coordinates are given for points off of the x-axis. Any positive value of k stretches the graph vertically but does not change the zeros, so any positive k works.

Example 2 Using a calculator or computer, graph $y = x^4$ and $y = x^4 - 15x^2 - 15x$ for $-4 \le x \le 4$ and for $-20 \le x \le 20$. Set the y range to $-100 \le y \le 100$ for the first domain, and to $-100 \le y \le 200{,}000$ for the second. What do you observe?

Solution From the graphs in Figure 1.86 we see that close up ($-4 \le x \le 4$) the graphs look different; from far away, however, they are almost indistinguishable. The reason is that the leading terms (those with the highest power of x) are the same, namely x^4, and for large values of x, the leading term dominates the other terms.

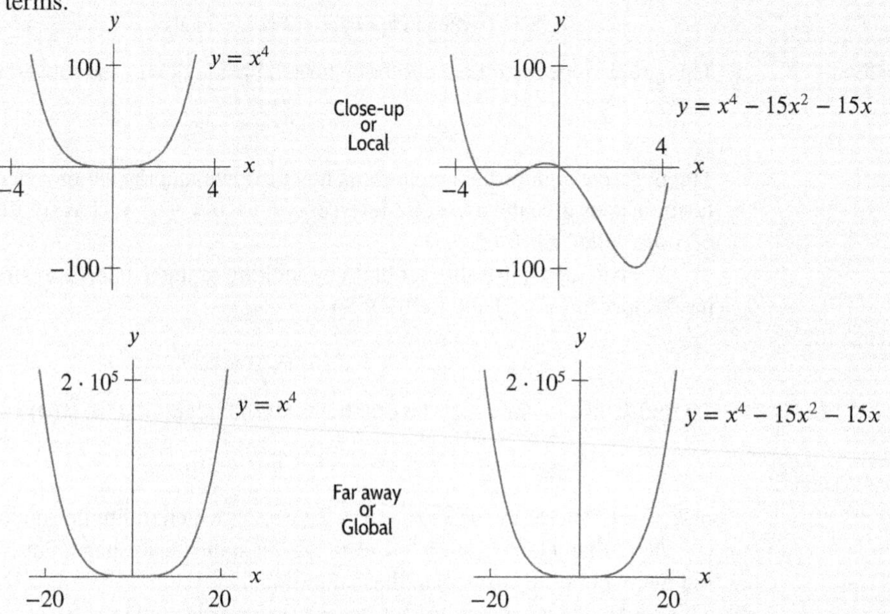

Figure 1.86: Local and global views of $y = x^4$ and $y = x^4 - 15x^2 - 15x$

Rational Functions

Rational functions are ratios of polynomials, p and q:

$$f(x) = \frac{p(x)}{q(x)}.$$

Example 3 Look at a graph and explain the behavior of $y = \dfrac{1}{x^2+4}$.

Solution The function is even, so the graph is symmetric about the y-axis. As x gets larger, the denominator gets larger, making the value of the function closer to 0. Thus the graph gets arbitrarily close to the x-axis as x increases without bound. See Figure 1.87.

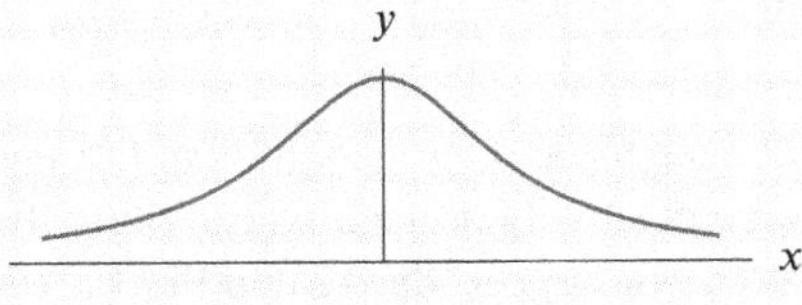

Figure 1.87: Graph of $y = \frac{1}{x^2+4}$

In the previous example, we say that $y = 0$ (i.e. the x-axis) is a *horizontal asymptote*. Writing "→" to mean "tends to," we have $y \to 0$ as $x \to \infty$ and $y \to 0$ as $x \to -\infty$.

> If the graph of $y = f(x)$ approaches a horizontal line $y = L$ as $x \to \infty$ or $x \to -\infty$, then the line $y = L$ is called a **horizontal asymptote.**[55] This occurs when
>
> $$f(x) \to L \quad \text{as} \quad x \to \infty \qquad \text{or} \qquad f(x) \to L \quad \text{as} \quad x \to -\infty.$$
>
> If the graph of $y = f(x)$ approaches the vertical line $x = K$ as $x \to K$ from one side or the other, that is, if
>
> $$y \to \infty \quad \text{or} \quad y \to -\infty \quad \text{when} \quad x \to K,$$
>
> then the line $x = K$ is called a **vertical asymptote.**

The graphs of rational functions may have vertical asymptotes where the denominator is zero. For example, the function in Example 3 has no vertical asymptotes as the denominator is never zero. The function in Example 4 has two vertical asymptotes corresponding to the two zeros in the denominator.

Rational functions have horizontal asymptotes if $f(x)$ approaches a finite number as $x \to \infty$ or $x \to -\infty$. We call the behavior of a function as $x \to \pm\infty$ its *end behavior*.

Example 4 Look at a graph and explain the behavior of $y = \dfrac{3x^2-12}{x^2-1}$, including end behavior.

Solution Factoring gives

$$y = \frac{3x^2-12}{x^2-1} = \frac{3(x+2)(x-2)}{(x+1)(x-1)}$$

so $x = \pm 1$ are vertical asymptotes. If $y = 0$, then $3(x+2)(x-2) = 0$ or $x = \pm 2$; these are the

[55] We are assuming that $f(x)$ gets arbitrarily close to L as $x \to \infty$.

x-intercepts. Note that zeros of the denominator give rise to the vertical asymptotes, whereas zeros of the numerator give rise to x-intercepts. Substituting $x = 0$ gives $y = 12$; this is the y-intercept. The function is even, so the graph is symmetric about the y-axis.

Table 1.18 *Values of* $y = \frac{3x^2-12}{x^2-1}$

x	$y = \frac{3x^2-12}{x^2-1}$
± 10	2.909091
± 100	2.999100
± 1000	2.999991

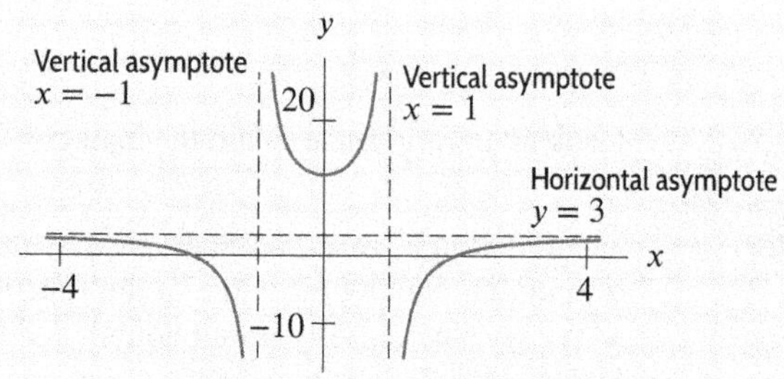

Figure 1.88: Graph of the function $y = \frac{3x^2-12}{x^2-1}$

To see what happens as $x \to \pm\infty$, look at the y-values in Table 1.18. Clearly y is getting closer to 3 as x gets large positively or negatively. Alternatively, realize that as $x \to \pm\infty$, only the highest powers of x matter. For large x, the 12 and the 1 are insignificant compared to x^2, so

$$y = \frac{3x^2 - 12}{x^2 - 1} \approx \frac{3x^2}{x^2} = 3 \quad \text{for large } x.$$

So $y \to 3$ as $x \to \pm\infty$, and therefore the horizontal asymptote is $y = 3$. See Figure 1.88. Since, for $x > 1$, the value of $(3x^2 - 12)/(x^2 - 1)$ is less than 3, the graph lies *below* its asymptote. (Why doesn't the graph lie below $y = 3$ when $-1 < x < 1$?)

Exercises and Problems for Section 1.6 Online Resource: Additional Problems for Section 1.6

EXERCISES

For Exercises 1–2, what happens to the value of the function as $x \to \infty$ and as $x \to -\infty$?

1. $y = 0.25x^3 + 3$
2. $y = 2 \cdot 10^{4x}$

In Exercises 3–10, determine the end behavior of each function as $x \to +\infty$ and as $x \to -\infty$.

3. $f(x) = -10x^4$
4. $f(x) = 3x^5$
5. $f(x) = 5x^4 - 25x^3 - 62x^2 + 5x + 300$
6. $f(x) = 1000 - 38x + 50x^2 - 5x^3$
7. $f(x) = \dfrac{3x^2 + 5x + 6}{x^2 - 4}$
8. $f(x) = \dfrac{10 + 5x^2 - 3x^3}{2x^3 - 4x + 12}$
9. $f(x) = 3x^{-4}$
10. $f(x) = e^x$

In Exercises 11–16, which function dominates as $x \to \infty$?

11. $1000x^4$ or $0.2x^5$
12. $10e^{0.1x}$ or $5000x^2$
13. $100x^5$ or 1.05^x
14. $2x^4$ or $10x^3 + 25x^2 + 50x + 100$
15. $20x^4 + 100x^2 + 5x$ or $25 - 40x^2 + x^3 + 3x^5$
16. $\sqrt{x}$ or $\ln x$
17. Each of the graphs in Figure 1.89 is of a polynomial. The windows are large enough to show end behavior.
    - **(a)** What is the minimum possible degree of the polynomial?
    - **(b)** Is the leading coefficient of the polynomial positive or negative?

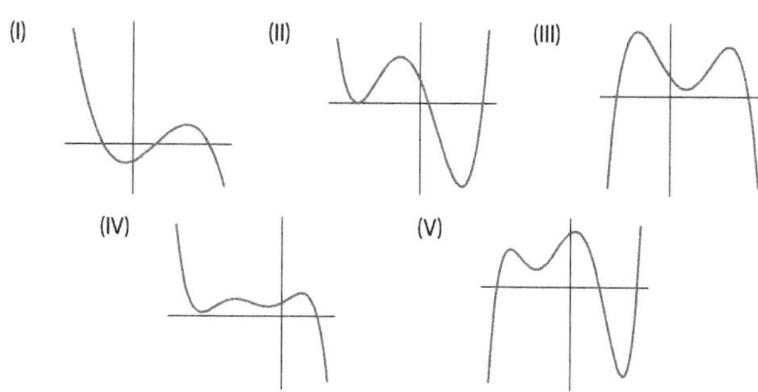

Figure 1.89

■ Find cubic polynomials for the graphs in Exercises 18–19.

18.

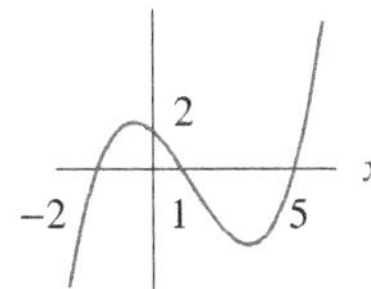

19.

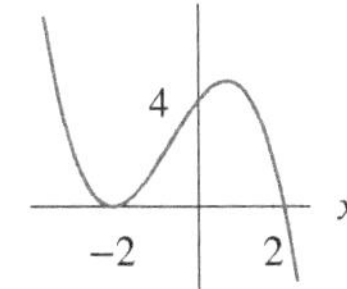

■ Find possible formulas for the graphs in Exercises 20–23.

20.

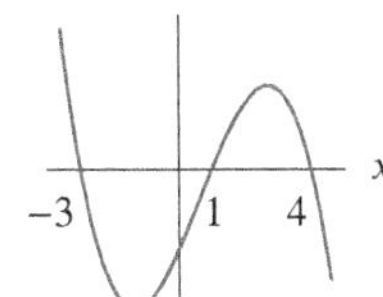

21.

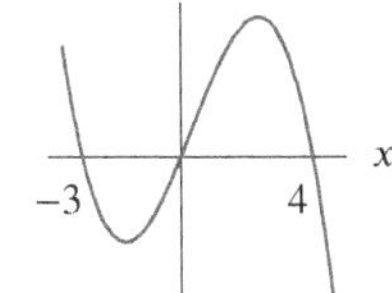

22.

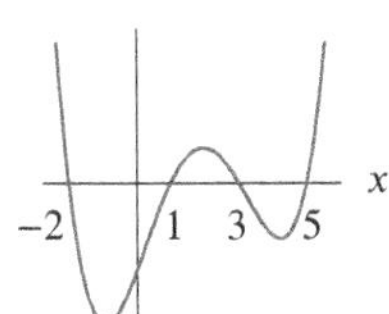

23.

■ In Exercises 24–26, choose the functions that are in the given family, assuming a, b, and c are positive constants.

$$f(x) = \sqrt{x^4 + 16} \qquad g(x) = ax^{23}$$
$$h(x) = -\frac{1}{5^{x-2}} \qquad p(x) = \frac{a^3 b^x}{c}$$
$$q(x) = \frac{ab^2}{c} \qquad r(x) = -x + b - \sqrt{cx^4}$$

24. Exponential 25. Quadratic 26. Linear

■ In Exercises 27–32, choose each of the families the given function is in, assuming a is a positive integer and b and c are positive constants.

I. Exponential II. Power
III. Polynomial IV. Rational

27. $f(x) = \dfrac{ax}{b} + c$

28. $g(x) = ax^2 + \dfrac{b}{x^2}$

29. $h(x) = b\left(\dfrac{x}{c}\right)^a$

30. $k(x) = bx^a$

31. $j(x) = ax^{-1} + \dfrac{b}{x}$

32. $l(x) = \left(\dfrac{a+b}{c}\right)^{2x}$

■ In Exercises 33–40, which of the following functions have the given property?

I. $y = \dfrac{x^2 - 2}{x^2 + 2}$ II. $y = \dfrac{x^2 + 2}{x^2 - 2}$
III. $y = (x-1)(1-x)(x+1)^2$ IV. $y = x^3 - x$
V. $y = x - \dfrac{1}{x}$ VI. $y = (x^2 - 2)(x^2 + 2)$

33. A polynomial of degree 3.

34. A polynomial of degree 4.

35. A rational function that is not a polynomial.

36. Exactly two distinct zeros.

37. Exactly one vertical asymptote.

38. More than two distinct zeros.

39. Exactly two vertical asymptotes.

40. A horizontal asymptote.

■ For Exercises 41–44, assuming the window is large enough to show end behavior, identify the graph as that of a rational function, exponential function or logarithmic function.

41.

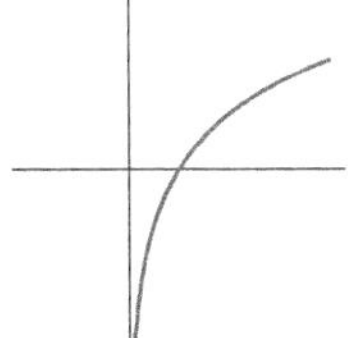

Figure 1.90

42.

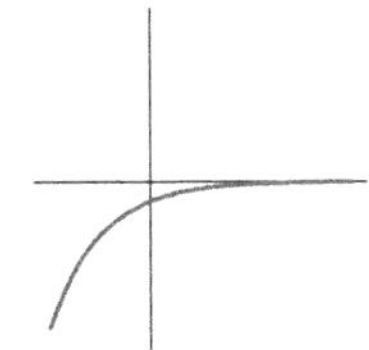

Figure 1.91

43.

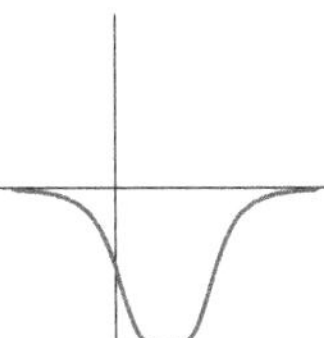

Figure 1.92

44.

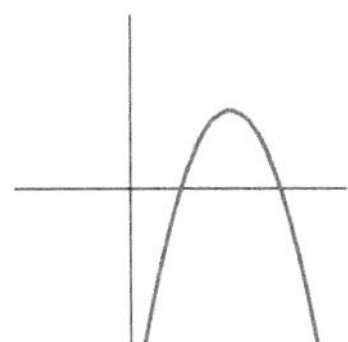

Figure 1.93

PROBLEMS

45. How many distinct roots can a polynomial of degree 5 have? (List all possibilities.) Sketch a possible graph for each case.

46. Find a calculator window in which the graphs of $f(x) = x^3 + 1000x^2 + 1000$ and $g(x) = x^3 - 1000x^2 - 1000$ appear indistinguishable.

47. A cubic polynomial with positive leading coefficient is shown in Figure 1.94 for $-10 \le x \le 10$ and $-10 \le y \le 10$. What can be concluded about the total number of zeros of this function? What can you say about the location of each of the zeros? Explain.

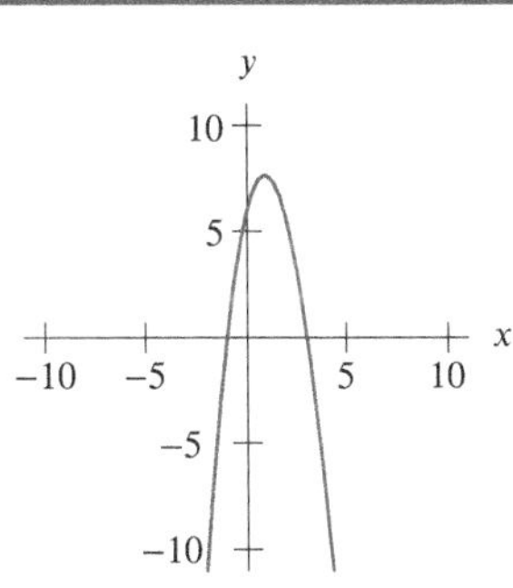

Figure 1.94

48. **(a)** If $f(x) = ax^2 + bx + c$, what can you say about the values of a, b, and c if:

(i) $(1, 1)$ is on the graph of $f(x)$?

(ii) $(1, 1)$ is the vertex of the graph of $f(x)$? (Hint: The axis of symmetry is $x = -b/(2a)$.)

(iii) The y-intercept of the graph is $(0, 6)$?

(b) Find a quadratic function satisfying all three conditions.

49. A box of fixed volume V has a square base with side length x. Write a formula for the height, h, of the box in terms of x and V. Sketch a graph of h versus x.

50. A closed cylindrical can of fixed volume V has radius r.

(a) Find the surface area, S, as a function of r.
(b) What happens to the value of S as $r \to \infty$?
(c) Sketch a graph of S against r, if $V = 10$ cm^3.

51. The DuBois formula relates a person's surface area s, in m^2, to weight w, in kg, and height h, in cm, by

$$s = 0.01w^{0.25}h^{0.75}.$$

(a) What is the surface area of a person who weighs 65 kg and is 160 cm tall?
(b) What is the weight of a person whose height is 180 cm and who has a surface area of 1.5 m^2?
(c) For people of fixed weight 70 kg, solve for h as a function of s. Simplify your answer.

52. According to *Car and Driver*, an Alfa Romeo going at 70 mph requires 150 feet to stop.[56] Assuming that the stopping distance is proportional to the square of velocity, find the stopping distances required by an Alfa Romeo going at 35 mph and at 140 mph.

53. Poiseuille's Law gives the rate of flow, R, of a gas through a cylindrical pipe in terms of the radius of the pipe, r, for a fixed drop in pressure between the two ends of the pipe.

(a) Find a formula for Poiseuille's Law, given that the rate of flow is proportional to the fourth power of the radius.
(b) If $R = 400$ cm^3/sec in a pipe of radius 3 cm for a certain gas, find a formula for the rate of flow of that gas through a pipe of radius r cm.
(c) What is the rate of flow of the same gas through a pipe with a 5 cm radius?

54. A pomegranate is thrown from ground level straight up into the air at time $t = 0$ with velocity 64 feet per second. Its height at time t seconds is $f(t) = -16t^2 + 64t$. Find the time it hits the ground and the time it reaches its highest point. What is the maximum height?

55. The height of an object above the ground at time t is given by

$$s = v_0 t - \frac{g}{2}t^2,$$

where v_0 is the initial velocity and g is the acceleration due to gravity.

(a) At what height is the object initially?
(b) How long is the object in the air before it hits the ground?
(c) When will the object reach its maximum height?
(d) What is that maximum height?

56. The rate, R, at which a population in a confined space increases is proportional to the product of the current population, P, and the difference between the carrying capacity, L, and the current population. (The carrying capacity is the maximum population the environment can sustain.)

(a) Write R as a function of P.
(b) Sketch R as a function of P.

■ In Problems 57–61, the length of a plant, L, is a function of its mass, M. A unit increase in a plant's mass stretches the plant's length more when the plant is small, and less when the plant is large.[57] Assuming $M > 0$, decide if the function agrees with the description.

57. $L = 2M^{1/4}$

58. $L = 0.2M^3 + M^4$

59. $L = 2M^{-1/4}$

60. $L = \dfrac{4(M+1)^2 - 1}{(M+1)^2}$

61. $L = \dfrac{10(M+1)^2 - 1}{(M+1)^3}$

■ In Problems 62–64, find all horizontal and vertical asymptotes for each rational function.

62. $f(x) = \dfrac{5x - 2}{2x + 3}$

63. $f(x) = \dfrac{x^2 + 5x + 4}{x^2 - 4}$

64. $f(x) = \dfrac{5x^3 + 7x - 1}{x^3 - 27}$

65. For each function, fill in the blanks in the statements:
$f(x) \to$ __________ as $x \to -\infty$,
$f(x) \to$ __________ as $x \to +\infty$.

(a) $f(x) = 17 + 5x^2 - 12x^3 - 5x^4$
(b) $f(x) = \dfrac{3x^2 - 5x + 2}{2x^2 - 8}$
(c) $f(x) = e^x$

66. A rational function $y = f(x)$ is graphed in Figure 1.95. If $f(x) = g(x)/h(x)$ with $g(x)$ and $h(x)$ both quadratic functions, give possible formulas for $g(x)$ and $h(x)$.

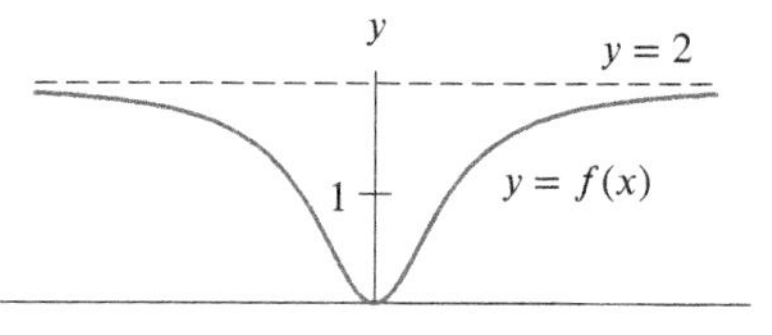

Figure 1.95

[56] http://www.caranddriver.com/alfa-romeo/4c. Accessed February 2016.

[57] Niklas, K. and Enquist, B., "Invariant scaling relationships for interspecific plant biomass production rates and body size", *PNAS*, Feb 27, 2001.

67. After running 3 miles at a speed of x mph, a man walked the next 6 miles at a speed that was 2 mph slower. Express the total time spent on the trip as a function of x. What horizontal and vertical asymptotes does the graph of this function have?

68. Which of the functions I–III meet each of the following descriptions? There may be more than one function for each description, or none at all.

 (a) Horizontal asymptote of $y = 1$.
 (b) The x-axis is a horizontal asymptote.
 (c) Symmetric about the y-axis.
 (d) An odd function.
 (e) Vertical asymptotes at $x = \pm 1$.

I. $y = \dfrac{x-1}{x^2+1}$ II. $y = \dfrac{x^2-1}{x^2+1}$ III. $y = \dfrac{x^2+1}{x^2-1}$

69. Values of three functions are given in Table 1.19, rounded to two decimal places. One function is of the form $y = ab^t$, one is of the form $y = ct^2$, and one is of the form $y = kt^3$. Which function is which?

Table 1.19

t	$f(t)$	t	$g(t)$	t	$h(t)$
2.0	4.40	1.0	3.00	0.0	2.04
2.2	5.32	1.2	5.18	1.0	3.06
2.4	6.34	1.4	8.23	2.0	4.59
2.6	7.44	1.6	12.29	3.0	6.89
2.8	8.62	1.8	17.50	4.0	10.33
3.0	9.90	2.0	24.00	5.0	15.49

70. Use a graphing calculator or a computer to graph $y = x^4$ and $y = 3^x$. Determine approximate domains and ranges that give each of the graphs in Figure 1.96.

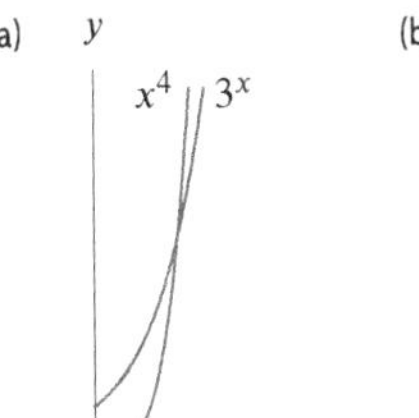

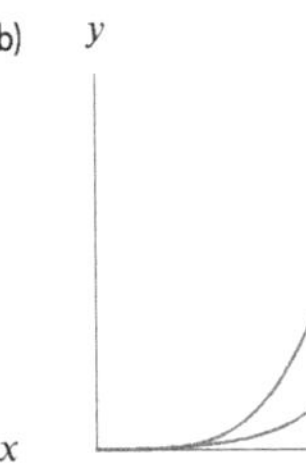

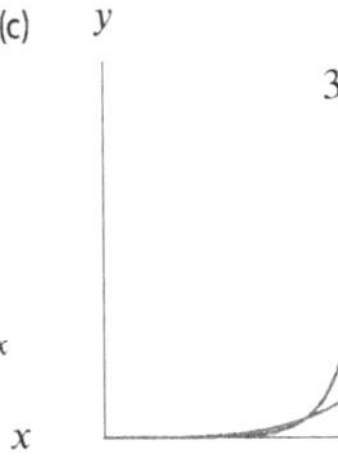

Figure 1.96

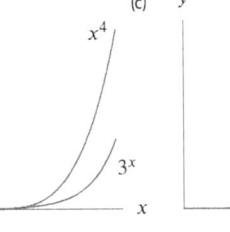

71. Consider the point P at the intersection of the circle $x^2+y^2 = 2a^2$ and the parabola $y = x^2/a$ in Figure 1.97. If a is increased, the point P traces out a curve. For $a > 0$, find the equation of this curve.

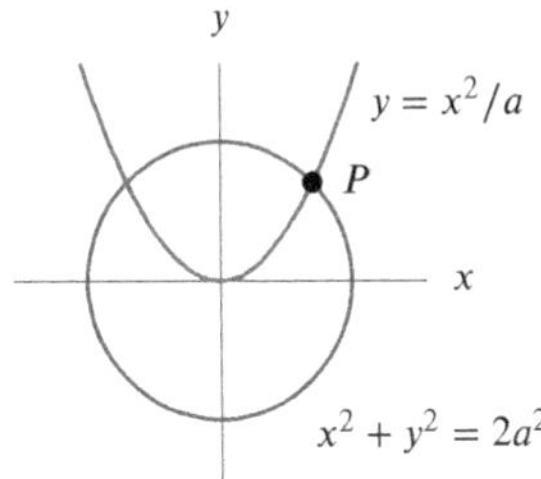

Figure 1.97

72. When an object of mass m moves with a velocity v that is small compared to the velocity of light, c, its energy is given approximately by

$$E \approx \frac{1}{2}mv^2.$$

If v is comparable in size to c, then the energy must be computed by the exact formula

$$E = mc^2\left(\frac{1}{\sqrt{1-v^2/c^2}} - 1\right).$$

 (a) Plot a graph of both functions for E against v for $0 \le v \le 5 \cdot 10^8$ and $0 \le E \le 5 \cdot 10^{17}$. Take $m = 1$ kg and $c = 3 \cdot 10^8$ m/sec. Explain how you can predict from the exact formula the position of the vertical asymptote.
 (b) What do the graphs tell you about the approximation? For what values of v does the first formula give a good approximation to E?

73. If $y = 100x^{-0.2}$ and $z = \ln x$, explain why y is an exponential function of z.

Strengthen Your Understanding

In Problems 74–79, explain what is wrong with the statement.

74. The graph of a polynomial of degree 5 cuts the horizontal axis five times.

75. A fourth-degree polynomial tends to infinity as $x \to \infty$.

76. A rational function has a vertical asymptote.

77. $x^5 > x^3$ for $x > 0$

78. Every rational function has a horizontal asymptote.

79. A function cannot cross its horizontal asymptote.

In Problems 80–85, give an example of:

80. A polynomial of degree 3 whose graph cuts the horizontal axis three times to the right of the origin.

81. A rational function with horizontal asymptote $y = 3$.

82. A rational function that is not a polynomial and that has no vertical asymptote.

83. A function that has a vertical asymptote at $x = -7\pi$.

84. A function that has exactly 17 vertical asymptotes.

85. A function that has a vertical asymptote which is crossed by a horizontal asymptote.

Are the statements in Problems 86–89 true or false? Give an explanation for your answer.

86. Every polynomial of even degree has a least one real zero.

87. Every polynomial of odd degree has a least one real zero.

88. The composition of two quadratic functions is quadratic.

89. For $x > 0$ the graph of the rational function $f(x) = \dfrac{5(x^3 - x)}{x^2 + x}$ is a line.

90. List the following functions in order from smallest to largest as $x \to \infty$ (that is, as x increases without bound).

(a) $f(x) = -5x$ **(b)** $g(x) = 10^x$
(c) $h(x) = 0.9^x$ **(d)** $k(x) = x^5$
(e) $l(x) = \pi^x$

1.7 INTRODUCTION TO LIMITS AND CONTINUITY

In this section we switch focus from families of functions to an intuitive introduction to *continuity* and *limits*. Limits are central to a full understanding of calculus.

The Idea of Continuity

Roughly speaking, a function is *continuous on an interval* if its graph has no breaks, jumps, or holes in that interval. A function is *continuous at a point* if nearby values of the independent variable give nearby values of the function. Most real world phenomena are modeled using continuous functions.

The Graphical Viewpoint: Continuity on an Interval

A continuous function has a graph which can be drawn without lifting the pencil from the paper.

Example: The function $f(x) = 3x^3 - x^2 + 2x - 1$ is continuous on any interval. (See Figure 1.98.)

Example: The function $f(x) = 1/x$ is not defined at $x = 0$. It is continuous on any interval not containing 0. (See Figure 1.99.)

Example: A company rents cars for \$7 per hour or fraction thereof, so it costs \$7 for a trip of one hour or less, \$14 for a trip between one and two hours, and so on. If $p(x)$ is the price of trip lasting x hours, then its graph (in Figure 1.100) is a series of steps. This function is not continuous on any open interval containing a positive integer because the graph jumps at these points.

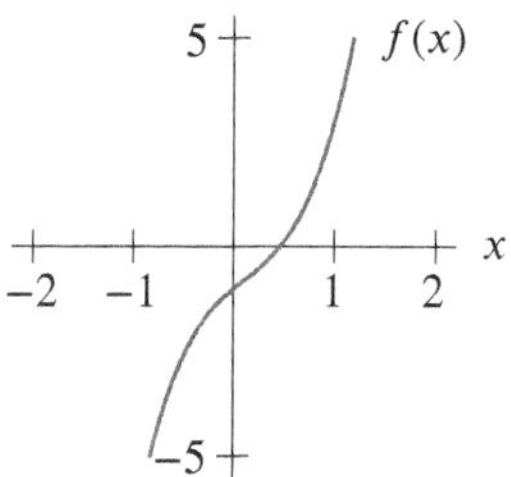

Figure 1.98: The graph of $f(x) = 3x^3 - x^2 + 2x - 1$

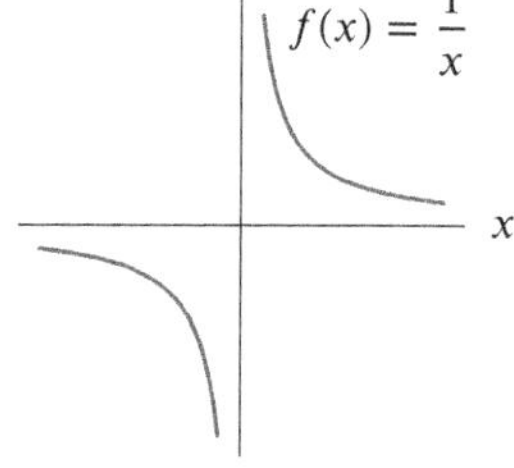

Figure 1.99: Not continuous on any interval containing 0

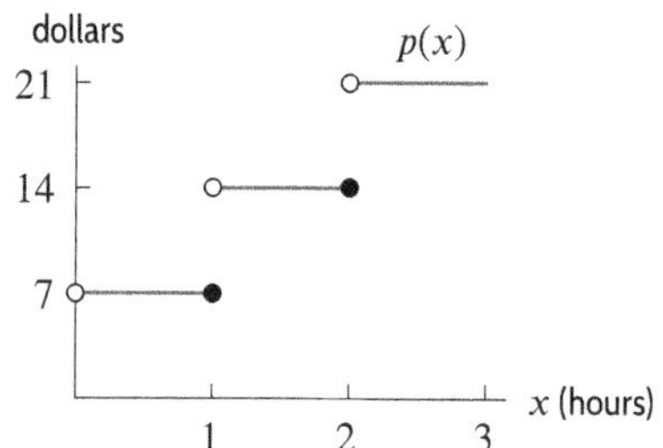

Figure 1.100: Cost of renting a car

The Numerical Viewpoint: Continuity at a Point

In practical work, continuity of a function at a point is important because it means that small errors in the independent variable lead to small errors in the value of the function. Conversely, if a function is not continuous at a point, a small error in input measurement can lead to an enormous error in output.

Example: Suppose that $f(x) = x^2$ and that we want to compute $f(\pi)$. Knowing f is continuous tells us that taking $x = 3.14$ should give a good approximation to $f(\pi)$, and that we can get as accurate an approximation to $f(\pi)$ as we want by using enough decimals of π.

Example: If $p(x)$ is the price of renting a car graphed in Figure 1.100, then $p(0.99) = p(1) =\$7$, whereas $p(1.01) =\$14$, because as soon as we pass one hour, the price jumps to \$14. So a small difference in time can lead to a significant difference in the cost. Hence p is not continuous at $x = 1$. As we see from its graph, this also means it is not continuous on any open interval including $x = 1$.

We express continuity at a point by saying that if $f(x)$ is continuous at $x = c$, the values of $f(x)$ approach $f(c)$ as x approaches c.

Example 1 Investigate the continuity of $f(x) = x^2$ at $x = 2$.

Solution From Table 1.20, it appears that the values of $f(x) = x^2$ approach $f(2) = 4$ as x approaches 2. Thus f appears to be continuous at $x = 2$.

Table 1.20 *Values of x^2 near $x = 2$*

x	1.9	1.99	1.999	2.001	2.01	2.1
x^2	3.61	3.96	3.996	4.004	4.04	4.41

Which Functions Are Continuous?

Most of the functions we have seen are continuous everywhere they are defined. For example, exponential functions, polynomials, and the sine and cosine are continuous everywhere. Rational functions are continuous on any interval in which their denominators are not zero. Functions created by adding, multiplying, or composing continuous functions are also continuous.[58]

The Idea of a Limit

Continuity at a point describes behavior of a function *near* a point, as well as *at* the point. To find the value of a function at a point we can just evaluate the function there. To focus on what happens to the values of a function near a point, we introduce the concept of limit. First some notation:

> We write $\lim_{x\to c} f(x) = L$ if the values of $f(x)$ approach L as x approaches c.

How should we find the limit L, or even know whether such a number exists? We look for trends in the values of $f(x)$ as x gets closer to c, but $x \neq c$. A graph from a calculator or computer often helps.

Example 2 Use a graph to estimate $\lim_{\theta\to 0}\left(\frac{\sin\theta}{\theta}\right)$. (Use radians.)

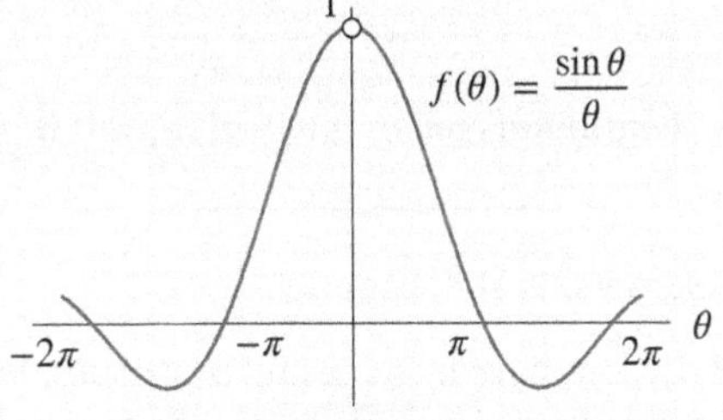

Figure 1.101: Find the limit as $\theta \to 0$

[58]For more details, see Theorem 1.8 in Section 1.8.

Solution Figure 1.101 shows that as θ approaches 0 from either side, the value of $\sin\theta/\theta$ appears to approach 1, suggesting that $\lim_{\theta\to 0}(\sin\theta/\theta) = 1$. Zooming in on the graph near $\theta = 0$ provides further support for this conclusion. Notice that $\sin\theta/\theta$ is undefined at $\theta = 0$.

Figure 1.101 suggests that $\lim_{\theta\to 0}(\sin\theta/\theta) = 1$, but to be sure we need to be more precise about words like "approach" and "close."

Definition of Limit

By the beginning of the 19th century, calculus had proved its worth, and there was no doubt about the correctness of its answers. However, it was not until the work of the French mathematician Augustin Cauchy (1789–1857) that calculus was put on a rigorous footing. Cauchy gave a formal definition of the limit, similar to the following:

> A function f is defined on an interval around c, except perhaps at the point $x = c$. We define the **limit** of the function $f(x)$ as x approaches c, written $\lim_{x\to c} f(x)$, to be a number L (if one exists) such that $f(x)$ is as close to L as we want whenever x is sufficiently close to c (but $x \neq c$). If L exists, we write
> $$\lim_{x\to c} f(x) = L.$$

Note that this definition ensures a limit, if it exists, cannot have more than one value.

Finding Limits

As we saw in Example 2, we can often estimate the value of a limit from its graph. We can also estimate a limit by using a table of values. However, no matter how closely we zoom in on a graph at a point or how close to a point we evaluate a function, there are always points closer, so in using these techniques for estimating limits we are never completely sure we have the exact answer. In Example 4 and Section 1.9 we show how a limit can be found exactly using algebra.

Definition of Continuity

In Example 2 we saw that the values of $\sin\theta/\theta$ approach 1 as θ approaches 0. However, $\sin\theta/\theta$ is not continuous at $\theta = 0$ since its graph has a hole there; see Figure 1.101. This illustrates an important difference between limits and continuity: a limit is only concerned with what happens *near* a point but continuity depends on what happens *near* a point and *at* that point. We now give a more precise definition of continuity using limits.

> The function f is **continuous** at $x = c$ if f is defined at $x = c$ and if
> $$\lim_{x\to c} f(x) = f(c).$$
> In other words, $f(x)$ is as close as we want to $f(c)$ provided x is close enough to c. The function is **continuous on an interval** $[a, b]$ if it is continuous at every point in the interval.[59]

The Intermediate Value Theorem

Continuous functions have many useful properties. For example, to locate the zeros of a continuous function, we can look for intervals where the function changes sign.

[59] If c is an endpoint of the interval, we can define continuity at $x = c$ using one-sided limits at c; see Section 1.8.

Example 3 What do the values in Table 1.21 tell you about the zeros of $f(x) = \cos x - 2x^2$?

Table 1.21

x	$f(x)$
0	1.00
0.2	0.90
0.4	0.60
0.6	0.11
0.8	−0.58
1.0	−1.46

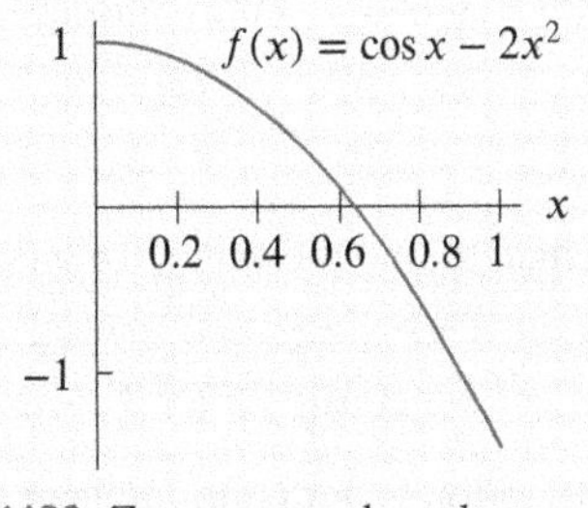

Figure 1.102: Zeros occur where the graph of a continuous function crosses the horizontal axis

Solution Since $f(x)$ is the difference of two continuous functions, it is continuous. Since f is positive at $x = 0.6$ and negative at $x = 0.8$, and its graph has no breaks, the graph must cross the axis between these points. We conclude that $f(x) = \cos x - 2x^2$ has at least one zero in the interval $0.6 < x < 0.8$. Figure 1.102 suggests that there is only one zero in the interval $0 \le x \le 1$, but we cannot be sure of this from a graph or a table of values.

In the previous example, we concluded that $f(x) = \cos x - 2x^2$ has a zero between $x = 0$ and $x = 1$ because it changed from positive to negative without skipping values in between—in other words, because it is continuous. If it were not continuous, the graph could jump across the x-axis, changing sign but not creating a zero. For example, $f(x) = 1/x$ has opposite signs at $x = -1$ and $x = 1$, but no zeros for $-1 \le x \le 1$ because of the break at $x = 0$. (See Figure 1.99.)

More generally, the intuitive notion of continuity tells us that, as we follow the graph of a continuous function f from some point $(a, f(a))$ to another point $(b, f(b))$, then f takes on all intermediate values between $f(a)$ and $f(b)$. (See Figure 1.103.) The formal statement of this property is known as the Intermediate Value Theorem, and is a powerful tool in theoretical proofs in calculus.

Theorem 1.1: Intermediate Value Theorem

Suppose f is continuous on a closed interval $[a, b]$. If k is any number between $f(a)$ and $f(b)$, then there is at least one number c in $[a, b]$ such that $f(c) = k$.

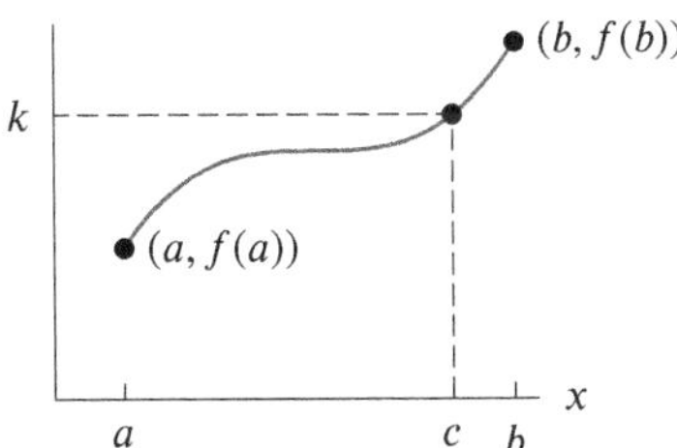

Figure 1.103: The Intermediate Value Theorem guarantees that a continuous f takes on the value k

Finding Limits Exactly Using Continuity and Algebra

The concept of limit is critical to a formal justification of calculus, so an understanding of limit outside the context of continuity is important. We have already seen how to use a graph or a table of values to estimate a limit. We now see how to find the exact value of a limit.

Limits of Continuous Functions

In Example 1, the limit of $f(x) = x^2$ at $x = 2$ is the same as $f(2)$. This is because $f(x)$ is a continuous function at $x = 2$, and this is precisely the definition of continuity:

Limits of Continuous Functions

If a function $f(x)$ is continuous at $x = c$, the limit is the value of $f(x)$ there:

$$\lim_{x \to c} f(x) = f(c).$$

Thus, to find the limits for a continuous function: Substitute c.

Limits of Functions Which Are not Continuous

If a function is not continuous at a point $x = c$ it is still possible for the limit to exist, but it can be hard to find. Sometimes such limits can be computed using algebra.

Example 4 Use a graph to estimate $\lim_{x \to 3} \frac{x^2 - 9}{x - 3}$ and then use algebra to find the limit exactly.

Solution Evaluating $(x^2 - 9)/(x - 3)$ at $x = 3$ gives us $0/0$, so the function is undefined at $x = 3$, shown as a hole in Figure 1.104. However, we see that as x approaches 3 from either side, the value of $(x^2 - 9)/(x - 3)$ appears to approach 6, suggesting the limit is 6.

To find this limit exactly, we first use algebra to rewrite the function. We have

$$\frac{x^2 - 9}{x - 3} = \frac{(x + 3)(x - 3)}{x - 3}.$$

Since $x \neq 3$ in the limit, we can cancel the common factor $x - 3$ to see

$$\lim_{x \to 3} \frac{x^2 - 9}{x - 3} = \lim_{x \to 3} \frac{(x + 3)(x - 3)}{x - 3} = \lim_{x \to 3}(x + 3).$$

Since $x + 3$ is continuous, we have

$$\lim_{x \to 3} \frac{x^2 - 9}{x - 3} = \lim_{x \to 3}(x + 3) = 3 + 3 = 6.$$

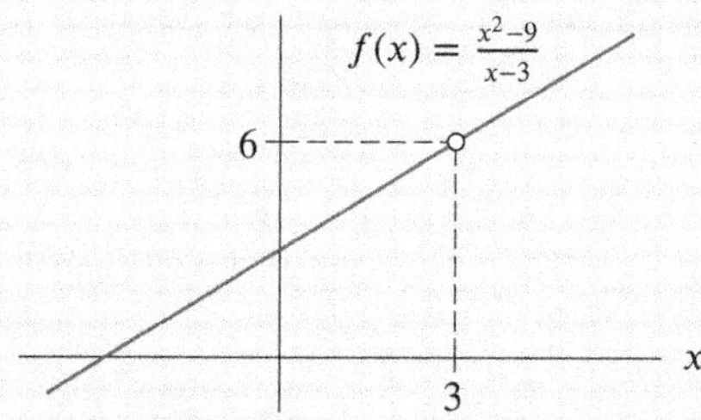

Figure 1.104: Find the limit as $x \to 3$

If a function is continuous at a point, then its limit must exist at that point. Example 4 illustrates that the existence of a limit is not enough to guarantee continuity.

More precisely, as we can see from Figure 1.104, the function $f(x) = (x^2 - 9)/(x - 3)$ is not continuous at $x = 3$ as there is a hole in its graph at that point. However, Example 4 shows that the limit $\lim_{x \to 3} f(x)$ does exist and is equal to 6. This is because for a function $f(x)$ to be continuous at a point $x = c$, the limit has to exist at $x = c$, the function has to be defined at $x = c$ and the limit has to equal the function. In this case, even though the limit does exist, the function does not have a value at $x = 3$, so it cannot be continuous there.

When Limits Do Not Exist

Whenever there is no number L such that $\lim_{x\to c} f(x) = L$, we say $\lim_{x\to c} f(x)$ does not exist. The following three examples show some of the ways in which a limit may fail to exist.

Example 5 Use a graph to explain why $\lim_{x\to 2} \dfrac{|x-2|}{x-2}$ does not exist.

Solution As we see in Figure 1.105, the value of $f(x) = \dfrac{|x-2|}{x-2}$ approaches -1 as x approaches 2 from the left and the value of $f(x)$ approaches 1 as x approaches 2 from the right. This means if $\lim_{x\to 2} \dfrac{|x-2|}{x-2} = L$ then L would have to be both 1 and -1. Since L cannot have two different values, the limit does not exist.

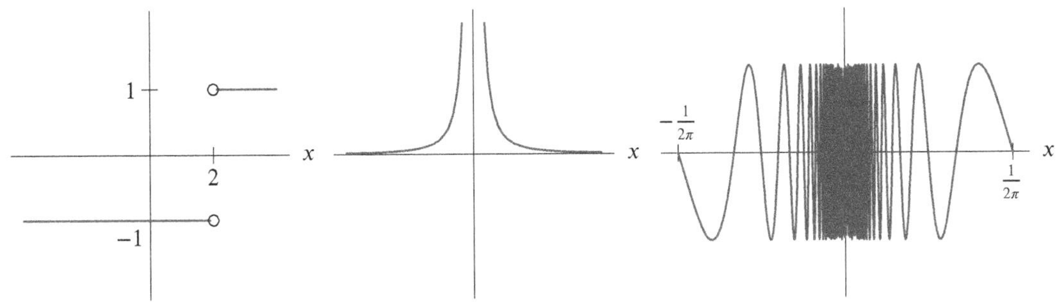

Figure 1.105: Graph of $f(x) = |x-2|/(x-2)$

Figure 1.106: Graph of $g(x) = 1/x^2$

Figure 1.107: Graph of $h(x) = \sin(1/x)$

Example 6 Use a graph to explain why $\lim_{x\to 0} \dfrac{1}{x^2}$ does not exist.

Solution As x approaches zero, $g(x) = 1/x^2$ becomes arbitrarily large, so it cannot approach any finite number L. See Figure 1.106. Therefore we say $1/x^2$ has no limit as $x \to 0$.

Since $1/x^2$ gets arbitrarily large on both sides of $x = 0$, we can write $\lim_{x\to 0} 1/x^2 = \infty$. The limit still does not exist since it does not approach a real number L, but we can use limit notation to describe its behavior. This behavior may also be described as "diverging to infinity."

Example 7 Explain why $\lim_{x\to 0} \sin\left(\dfrac{1}{x}\right)$ does not exist.

Solution The sine function has values between -1 and 1. The graph of $h(x) = \sin(1/x)$ in Figure 1.107 oscillates more and more rapidly as $x \to 0$. There are x-values approaching 0 where $\sin(1/x) = -1$. There are also x-values approaching 0 where $\sin(1/x) = 1$. So if the limit existed, it would have to be both -1 and 1. Thus, the limit does not exist.

Notice that in all three examples, the function is not continuous at the given point. This is because continuity requires the existence of a limit, so failure of a limit to exist at a point automatically means a function is not continuous there.

Exercises and Problems for Section 1.7

EXERCISES

1. **(a)** Using Figure 1.108, find all values of x for which f is not continuous.
   (b) List the largest open intervals on which f is continuous.

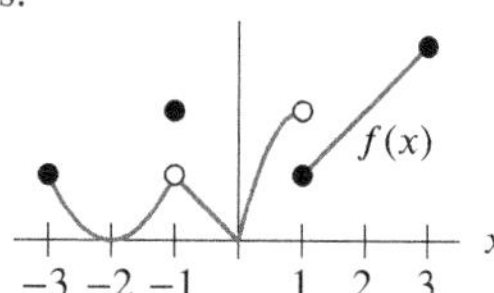

Figure 1.108

2. **(a)** Using Figure 1.109, find all values of x for which f is not continuous.
   (b) List the largest open intervals on which f is continuous.

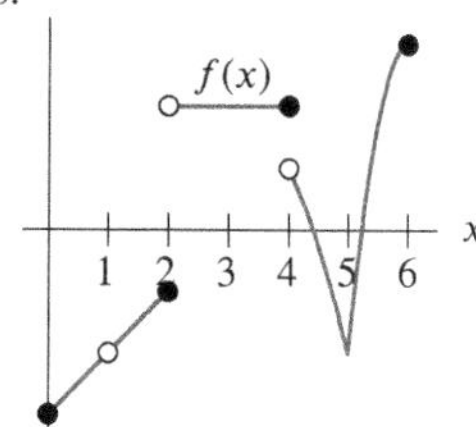

Figure 1.109

3. Use the graph of $f(x)$ in Figure 1.110 to give approximate values for the following limits (if they exist). If the limit does not exist, say so.

   (a) $\lim_{x\to-3} f(x)$ **(b)** $\lim_{x\to-2} f(x)$ **(c)** $\lim_{x\to-1} f(x)$
   (d) $\lim_{x\to0} f(x)$ **(e)** $\lim_{x\to1} f(x)$ **(f)** $\lim_{x\to3} f(x)$

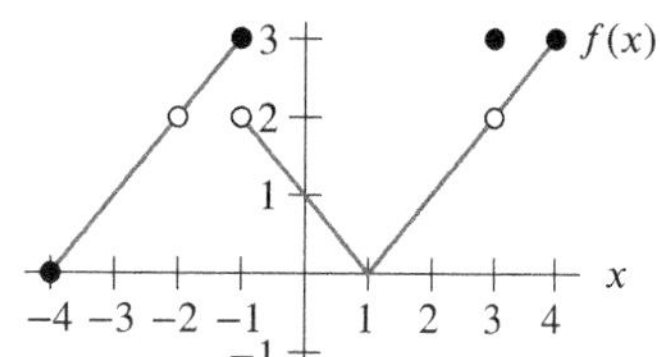

Figure 1.110

In Exercises 4–5, the graph of $y = f(x)$ is given.
(a) Give the x-values where $f(x)$ is not continuous.
(b) Does the limit of $f(x)$ exist at each x-value where $f(x)$ is not continuous? If so, give the value of the limit.

4.

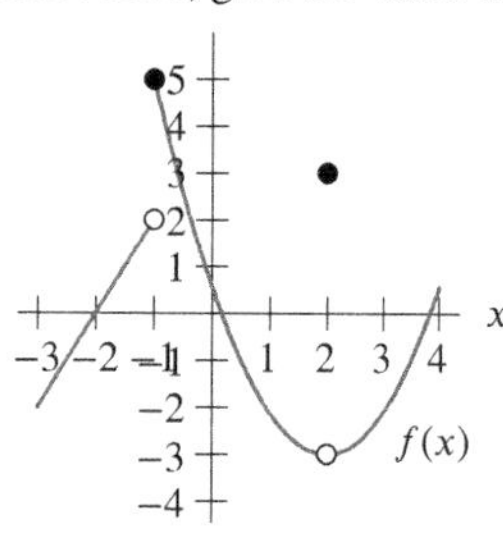

Figure 1.111

5.

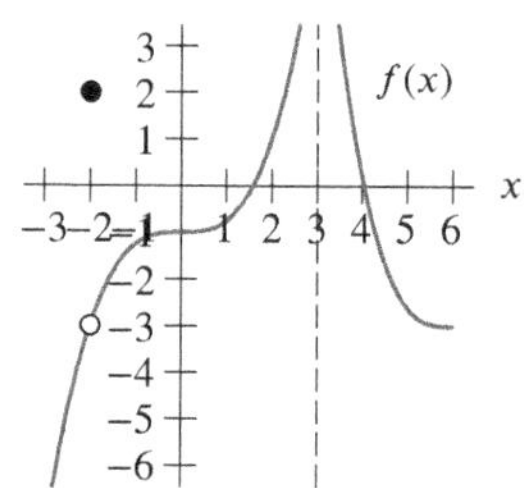

Figure 1.112

6. Assume $f(x)$ is continuous on an interval around $x = 0$, except possibly at $x = 0$. What does the table of values suggest as the value of $\lim_{x\to0} f(x)$? Does the limit definitely have this value?

x	-0.1	-0.01	0.01	0.1
$f(x)$	1.987	1.999	1.999	1.987

7. Assume $g(t)$ is continuous on an interval around $t = 3$, except possibly at $t = 3$. What does the table of values suggest as the value of $\lim_{t\to3} g(t)$? Does the limit definitely have this value?

t	2.9	2.99	3.01	3.1
$g(t)$	0.948	0.995	1.005	1.049

In Exercises 8–9,
(a) Make a table of values of $f(x)$ for $x = -0.1, -0.01, 0.01$, and 0.1.
(b) Use the table to estimate $\lim_{x\to0} f(x)$.

8. $f(x) = \dfrac{\sin(5x)}{x}$ 9. $f(x) = \dfrac{e^{3x} - 1}{x}$

10. Use a table of values to estimate $\lim_{x\to1} (5 + \ln x)$.

In Exercises 11–16, is the function continuous on the interval?

11. $\dfrac{1}{x-2}$ on $[-1, 1]$
12. $\dfrac{1}{x-2}$ on $[0, 3]$
13. $\dfrac{1}{\sqrt{2x-5}}$ on $[3, 4]$
14. $2x + x^{-1}$ on $[-1, 1]$
15. $\dfrac{1}{\cos x}$ on $[0, \pi]$
16. $\dfrac{e^{\sin\theta}}{\cos\theta}$ on $[-\frac{\pi}{4}, \frac{\pi}{4}]$

17. Are the following functions continuous? Explain.

   (a) $f(x) = \begin{cases} x & x \le 1 \\ x^2 & 1 < x \end{cases}$
   (b) $g(x) = \begin{cases} x & x \le 3 \\ x^2 & 3 < x \end{cases}$

18. Let f be the function given by

$$f(x) = \begin{cases} 4 - x & 0 \le x \le 3 \\ x^2 - 8x + 17 & 3 < x < 5 \\ 12 - 2x & 5 \le x \le 6 \end{cases}.$$

(a) Find all values of x for which f is not continuous.
(b) List the largest open intervals on which f is continuous.

■ In Exercises 19–22, show there is a number c, with $0 \le c \le 1$, such that $f(c) = 0$.

19. $f(x) = x^3 + x^2 - 1$
20. $f(x) = e^x - 3x$
21. $f(x) = x - \cos x$
22. $f(x) = 2^x - 1/x$

■ In Exercises 23–28, use algebra to find the limit exactly.

23. $\lim_{x\to 2} \dfrac{x^2 - 4}{x - 2}$
24. $\lim_{x\to -3} \dfrac{x^2 - 9}{x + 3}$
25. $\lim_{x\to 1} \dfrac{x^2 + 4x - 5}{x - 1}$
26. $\lim_{x\to 1} \dfrac{x^2 - 4x + 3}{x^2 + 3x - 4}$
27. $\lim_{x\to 1} \dfrac{x^2 + 4}{x + 8}$
28. $\lim_{h\to 0} \dfrac{(5 + h)^2 - 5^2}{h}$

■ For Exercises 29–30, find the value of the constant k such that

29. $\lim_{x\to 5}(kx + 10) = 20$
30. $\lim_{x\to 2} \dfrac{(x + 6)(x - k)}{x^2 + x} = 4$

■ In Exercises 31–34 find k so that the function is continuous on any interval.

31. $f(x) = \begin{cases} kx & x \le 3 \\ 5 & 3 < x \end{cases}$

32. $f(x) = \begin{cases} kx & 0 \le x < 2 \\ 3x^2 & 2 \le x \end{cases}$

33. $h(x) = \begin{cases} kx & 0 \le x < 1 \\ x + 3 & 1 \le x \le 5. \end{cases}$

34. $g(t) = \begin{cases} t + k & t \le 5 \\ kt & 5 < t \end{cases}$

PROBLEMS

35. Which of the following are continuous functions of time?

(a) The quantity of gas in the tank of a car on a journey between New York and Boston.
(b) The number of students enrolled in a class during a semester.
(c) The age of the oldest person alive.

36. An electrical circuit switches instantaneously from a 6-volt battery to a 12-volt battery 7 seconds after being turned on. Graph the battery voltage against time. Give formulas for the function represented by your graph. What can you say about the continuity of this function?

37. A stone dropped from the top of a cliff falls freely for 5 seconds before it hits the ground. Figure 1.113 shows the speed $v = f(t)$ (in meters/sec) of the stone as a function of time t in seconds for $0 \le t \le 7$.

(a) Is f continuous? Explain your answer in the context of the problem.
(b) Sketch a graph of the height, $h = g(t)$, of the stone for $0 \le t \le 7$. Is g continuous? Explain.

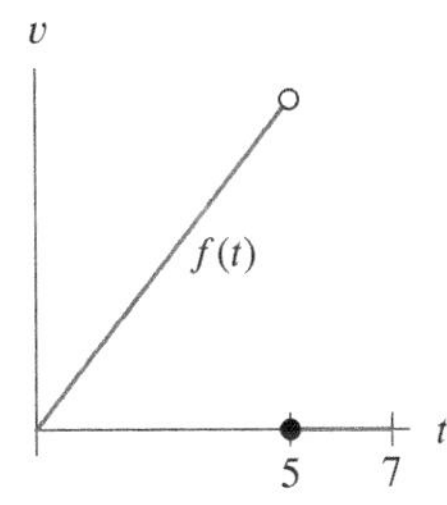

Figure 1.113

38. Beginning at time $t = 0$, a car undergoing a crash test accelerates for two seconds, maintains a constant speed for one second, and then crashes into a test barrier at $t = 3$ seconds.

(a) Sketch a possible graph of $v = f(t)$, the speed of the car (in meters/sec) after t seconds, on the interval $0 \le t \le 4$.
(b) Is the function f in part (a) continuous? Explain your answer in the context of this problem.

39. Discuss the continuity of the function g graphed in Figure 1.114 and defined as follows:

$$g(\theta) = \begin{cases} \dfrac{\sin\theta}{\theta} & \text{for } \theta \ne 0 \\ 1/2 & \text{for } \theta = 0. \end{cases}$$

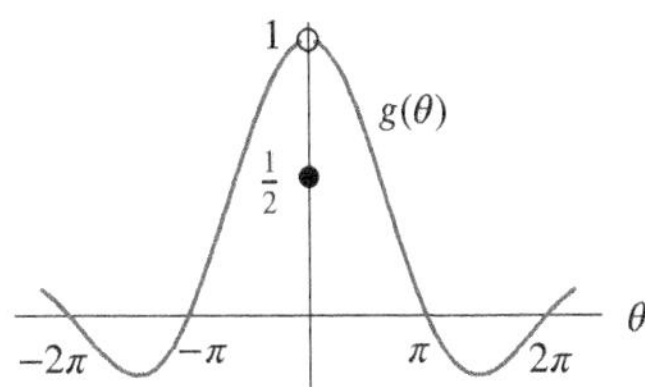

Figure 1.114

40. Is the following function continuous on $[-1, 1]$?

$$f(x) = \begin{cases} \dfrac{x}{|x|} & x \ne 0 \\ 0 & x = 0 \end{cases}$$

Estimate the limits in Problems 41–42 graphically.

41. $\lim_{x\to 0} \dfrac{|x|}{x}$

42. $\lim_{x\to 0} x \ln|x|$

In Problems 43–48, use a graph to estimate the limit. Use radians unless degrees are indicated by $\theta°$.

43. $\lim_{\theta\to 0} \dfrac{\sin(2\theta)}{\theta}$

44. $\lim_{\theta\to 0} \dfrac{\sin\theta°}{\theta°}$

45. $\lim_{h\to 0} \dfrac{e^h - 1}{h}$

46. $\lim_{h\to 0} \dfrac{e^{5h} - 1}{h}$

47. $\lim_{h\to 0} \dfrac{2^h - 1}{h}$

48. $\lim_{h\to 0} \dfrac{\cos(3h) - 1}{h}$

In Problems 49–54, find a value of k, if any, making $h(x)$ continuous on $[0, 5]$.

49. $h(x) = \begin{cases} k\cos x & 0 \le x \le \pi \\ 12 - x & \pi < x \end{cases}$

50. $h(x) = \begin{cases} kx & 0 \le x \le 1 \\ 2kx + 3 & 1 < x \le 5. \end{cases}$

51. $h(x) = \begin{cases} k\sin x & 0 \le x \le \pi \\ x + 4 & \pi < x \le 5. \end{cases}$

52. $h(x) = \begin{cases} e^{kx} & 0 \le x < 2 \\ x + 1 & 2 \le x \le 5. \end{cases}$

53. $h(x) = \begin{cases} 0.5x & 0 \le x < 1 \\ \sin(kx) & 1 \le x \le 5. \end{cases}$

54. $h(x) = \begin{cases} \ln(kx + 1) & 0 \le x \le 2 \\ x + 4 & 2 < x \le 5. \end{cases}$

55. **(a)** Use Figure 1.115 to decide at what points $f(x)$ is not continuous.
(b) At what points is the function $|f(x)|$ not continuous?

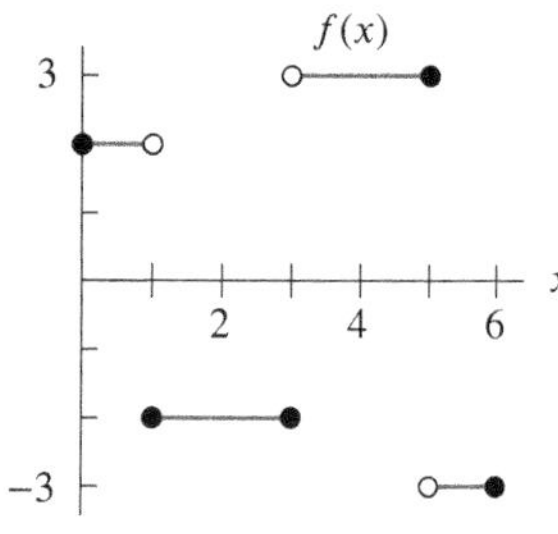

Figure 1.115

56. For t in months, a population, in thousands, is approximated by a continuous function

$$P(t) = \begin{cases} e^{kt} & 0 \le t \le 12 \\ 100 & t > 12. \end{cases}$$

(a) What is the initial value of the population?
(b) What must be the value of k?
(c) Describe in words how the population is changing.

57. A 0.6 ml dose of a drug is injected into a patient steadily for half a second. At the end of this time, the quantity, Q, of the drug in the body starts to decay exponentially at a continuous rate of 0.2% per second. Using formulas, express Q as a continuous function of time, t in seconds.

In Problems 58–61, at what values of x is the function not continuous? If possible, give a value for the function at each point of discontinuity so the function is continuous everywhere.

58. $f(x) = \dfrac{x^2 - 1}{x + 1}$

59. $g(x) = \dfrac{x^2 - 4x - 5}{x - 5}$

60. $f(z) = \dfrac{z^2 - 11z + 18}{2z - 18}$

61. $q(t) = \dfrac{-t^3 + 9t}{t^2 - 9}$

In Problems 62–65, is the function continuous for all x? If not, say where it is not continuous and explain in what way the definition of continuity is not satisfied.

62. $f(x) = 1/x$

63. $f(x) = \begin{cases} |x|/x & x \ne 0 \\ 0 & x = 0 \end{cases}$

64. $f(x) = \begin{cases} x/x & x \ne 0 \\ 1 & x = 0 \end{cases}$

65. $f(x) = \begin{cases} 2x/x & x \ne 0 \\ 3 & x = 0 \end{cases}$

66. Graph three different functions, continuous on $0 \le x \le 1$, and with the values in the table. The first function is to have exactly one zero in $[0, 1]$, the second is to have at least two zeros in the interval $[0.6, 0.8]$, and the third is to have at least two zeros in the interval $[0, 0.6]$.

x	0	0.2	0.4	0.6	0.8	1.0
$f(x)$	1.00	0.90	0.60	0.11	−0.58	−1.46

67. Let $p(x)$ be a cubic polynomial with $p(5) < 0$, $p(10) > 0$, and $p(12) < 0$. What can you say about the number and location of zeros of $p(x)$?

68. **(a)** What does a graph of $y = e^x$ and $y = 4 - x^2$ tell you about the solutions to the equation $e^x = 4 - x^2$?
(b) Evaluate $f(x) = e^x + x^2 - 4$ at $x = -4, -3, -2, -1, 0, 1, 2, 3, 4$. In which intervals do the solutions to $e^x = 4 - x^2$ lie?

69. **(a)** Does $f(x)$ satisfy the conditions for the Intermediate Value Theorem on $0 \le x \le 2$ if

$$f(x) = \begin{cases} e^x & 0 \le x \le 1 \\ 4 + (x - 1)^2 & 1 < x \le 2? \end{cases}$$

(b) What are $f(0)$ and $f(2)$? Can you find a value of k between $f(0)$ and $f(2)$ such that the equation $f(x) = k$ has no solution? If so, what is it?

70. Let $g(x)$ be continuous with $g(0) = 3$, $g(1) = 8$, $g(2) = 4$. Use the Intermediate Value Theorem to explain why $g(x)$ is not invertible.

71. By graphing $y = (1 + x)^{1/x}$, estimate $\lim_{x\to 0}(1 + x)^{1/x}$. You should recognize the answer you get. What does the limit appear to be?

72. Investigate $\lim_{h\to 0}(1 + h)^{1/h}$ numerically.

73. Let $f(x) = \sin(1/x)$.

(a) Find a sequence of x-values that approach 0 such that $\sin(1/x) = 0$.
[Hint: Use the fact that $\sin(\pi) = \sin(2\pi) = \sin(3\pi) = \ldots = \sin(n\pi) = 0$.]

(b) Find a sequence of x-values that approach 0 such that $\sin(1/x) = 1$.
[Hint: Use the fact that $\sin(n\pi/2) = 1$ if $n = 1, 5, 9, \ldots$.]

(c) Find a sequence of x-values that approach 0 such that $\sin(1/x) = -1$.

(d) Explain why your answers to any two of parts (a)–(c) show that $\lim_{x\to 0} \sin(1/x)$ does not exist.

Strengthen Your Understanding

In Problems 74–76, explain what is wrong with the statement.

74. For any function $f(x)$, if $f(a) = 2$ and $f(b) = 4$, the Intermediate Value Theorem says that f takes on the value 3 for some x between a and b.
75. If $f(x)$ is continuous on $0 \le x \le 2$ and if $f(0) = 0$ and $f(2) = 10$, the Intermediate Value Theorem says that $f(1) = 5$.
76. If $\lim_{x\to c} f(x)$ exists, then $f(x)$ is continuous at $x = c$.

In Problems 77–80, give an example of:

77. A function which is defined for all x and continuous everywhere except at $x = 15$.
78. A function to which the Intermediate Value Theorem does not apply on the interval $-1 \le x \le 1$.
79. A function that is continuous on $[0, 1]$ but not continuous on $[1, 3]$.
80. A function that is increasing but not continuous on $[0, 10]$.

Are the statements in Problems 81–83 true or false? Give an explanation for your answer.

81. If a function is not continuous at a point, then it is not defined at that point.
82. If f is continuous on the interval $[0, 10]$ and $f(0) = 0$ and $f(10) = 100$, then $f(c)$ cannot be negative for c in $[0, 10]$.
83. If $f(x)$ is not continuous on the interval $[a, b]$, then $f(x)$ must omit at least one value between $f(a)$ and $f(b)$.

1.8 EXTENDING THE IDEA OF A LIMIT

In Section 1.7, we introduced the idea of limit to describe the behavior of a function close to a point. We now extend limit notation to describe a function's behavior to values on only one side of a point and around an asymptote, and we extend limits to combinations of functions.

One-Sided Limits

When we write

$$\lim_{x\to 2} f(x),$$

we mean the number that $f(x)$ approaches as x approaches 2 *from both sides*. We examine values of $f(x)$ as x approaches 2 through values greater than 2 (such as 2.1, 2.01, 2.003) and values less than 2 (such as 1.9, 1.99, 1.994). If we want x to approach 2 only through values greater than 2, we write

$$\lim_{x\to 2^+} f(x)$$

for the number that $f(x)$ approaches (assuming such a number exists). Similarly,

$$\lim_{x\to 2^-} f(x)$$

denotes the number (if it exists) obtained by letting x approach 2 through values less than 2. We call $\lim_{x\to 2^+} f(x)$ a *right-hand limit* and $\lim_{x\to 2^-} f(x)$ a *left-hand limit*.

For the function graphed in Figure 1.116, we have

$$\lim_{x\to 2^-} f(x) = L_1 \qquad \lim_{x\to 2^+} f(x) = L_2.$$

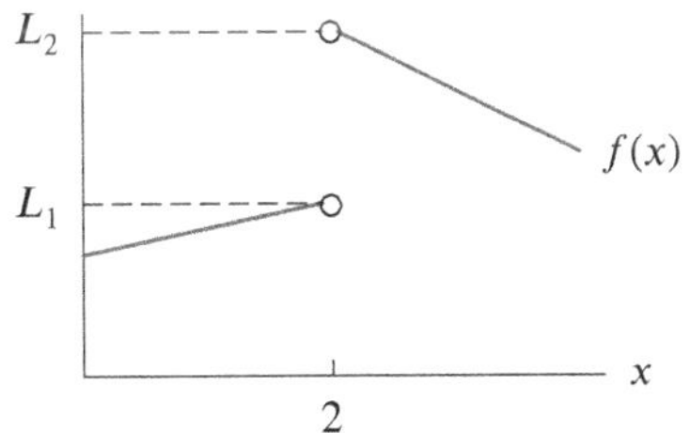

Figure 1.116: Left- and right-hand limits at $x = 2$

Observe that in this example $L_1 \neq L_2$; that is, $f(x)$ approaches different values as x approaches 2 from the left and from the right. Because of this, $\lim_{x\to 2} f(x)$ does not exist, since there is no single value that $f(x)$ approaches for all values of x close to 2.

Example 1 Find each of the following limits or explain why it does not exist:

(a) $\lim_{x\to 2} \dfrac{|x-2|}{x-2}$ (b) $\lim_{x\to 2^+} \dfrac{|x-2|}{x-2}$ (c) $\lim_{x\to 2^-} \dfrac{|x-2|}{x-2}$

Solution (a) In Example 5 of Section 1.7 we saw that $\lim_{x\to 2} \dfrac{|x-2|}{x-2}$ does not exist as it would have to take two different values. (See Figure 1.117.) However, it is still possible that the one-sided limits exist.

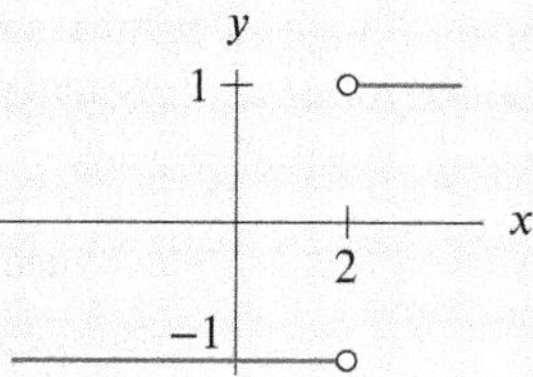

Figure 1.117: Limit of $y = |x-2|/(x-2)$ does not exist at $x = 2$

(b) To determine $\lim_{x\to 2^+} \dfrac{|x-2|}{x-2}$, we look at the values of $|x-2|/(x-2)$ for values of x greater than 2. When $x > 2$,

$$\frac{|x-2|}{x-2} = \frac{x-2}{x-2} = 1.$$

So as x approaches 2 from the right, the value of $|x-2|/(x-2)$ is always 1. Therefore,

$$\lim_{x\to 2^+} \frac{|x-2|}{x-2} = 1.$$

(c) If $x < 2$, then

$$\frac{|x-2|}{x-2} = \frac{-(x-2)}{x-2} = -1.$$

So as x approaches 2 from the left, the value of $|x-2|/(x-2)$ is always -1. Therefore,

$$\lim_{x\to 2^-} \frac{|x-2|}{x-2} = -1.$$

Limits and Asymptotes

We can use limit notation to describe the asymptotes of a function.

Horizontal Asymptotes and Limits

Sometimes we want to know what happens to $f(x)$ as x gets large, that is, the end behavior of f.

If $f(x)$ stays as close to a number L as we please when x is sufficiently large, then we write

$$\lim_{x\to\infty} f(x) = L.$$

Similarly, if $f(x)$ stays as close to L as we please when x is negative and has a sufficiently large absolute value, then we write

$$\lim_{x\to-\infty} f(x) = L.$$

The symbol ∞ does not represent a number. Writing $x \to \infty$ means that we consider arbitrarily large values of x. If the limit of $f(x)$ as $x \to \infty$ or $x \to -\infty$ is L, we say that the graph of f has $y = L$ as a *horizontal asymptote*.

Example 2 Investigate $\lim_{x\to\infty} \frac{1}{x}$ and $\lim_{x\to-\infty} \frac{1}{x}$.

Solution A graph of $f(x) = 1/x$ in a large window shows $1/x$ approaching zero as x grows large in magnitude in either the positive or the negative direction (see Figure 1.118). This is as we would expect, since dividing 1 by larger and larger numbers yields answers which are closer and closer to zero. This suggests that

$$\lim_{x\to\infty} \frac{1}{x} = \lim_{x\to-\infty} \frac{1}{x} = 0,$$

and that $f(x) = 1/x$ has $y = 0$ as a horizontal asymptote as $x \to \pm\infty$.

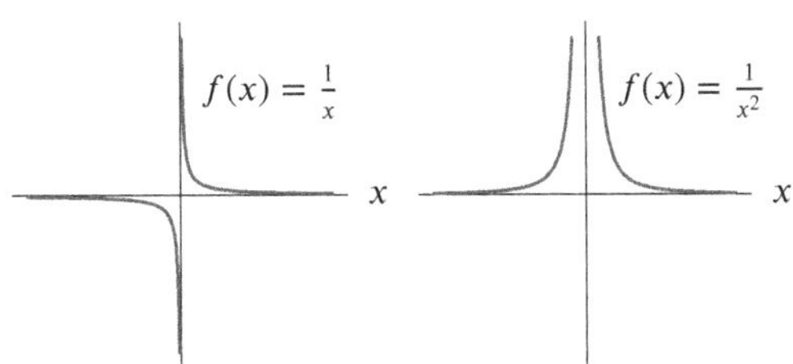

Figure 1.118: Limits describe the asymptotes of $f(x) = 1/x$

Figure 1.119: Vertical asymptote of $f(x) = 1/x^2$ at $x = 0$

There are many quantities in fields such as finance and medicine which change over time, t, and knowing the end behavior, or the limit of the quantity as $t \to \infty$, can be extremely important.

Example 3 Total sales $P(t)$ of an app (in thousands) t months after the app was introduced is given in Table 1.22 and can be modeled by $P(t) = \dfrac{532}{1 + 869e^{-0.8t}}$.

(a) Use Table 1.22 to estimate and interpret $\lim_{t\to\infty} P(t)$.

(b) Find $\lim_{t\to\infty} P(t)$ using the model.

Table 1.22 *Total sales of an app for increasing t*

t (months)	8	12	16	20	24	28
$P(t)$ (sales in 1000s)	217.685	502.429	530.727	531.948	531.998	531.999

Solution (a) From Table 1.22, it appears that as t gets larger, $P(t)$ approaches 532. This makes sense as over time there is a limit on the number of people who are interested in the app. So the maximum potential sales for this app are estimated to be 532 thousand.

(b) As t gets large, $e^{-0.8t}$ gets close to zero, so $P(t)$ gets close to $532/(1 + 0)$. This suggests

$$\lim_{t\to\infty} \frac{532}{1 + 869e^{-0.8t}} = 532.$$

Vertical Asymptotes and Limits

In Section 1.7, we wrote $\lim_{x\to 0} 1/x^2 = \infty$ because $1/x^2$ becomes arbitrarily large on both sides of zero; see Figure 1.119. Similarly, if a function becomes arbitrarily large in magnitude but negative on both sides of a point we can say its limit at that point is $-\infty$. In each case, the limit does not exist since values of $1/x^2$ do not approach a real number; we are just able to use limit notation to describe its behavior.

Though $1/x^2$ also has a vertical asymptote at $x = 0$, we cannot say the same for $1/x$ since the behavior of $1/x$ as x approaches 0 from the right does not match the behavior as x approaches 0 from the left. We can however use one-sided limit notation to describe this type of behavior; that is, we write $\lim_{x\to 0^+} \frac{1}{x} = \infty$ and $\lim_{x\to 0^-} \frac{1}{x} = -\infty$.

Example 4 Describe the behavior of $f(t) = \dfrac{3+4t}{t+2}$ near $t = -2$.

Solution From Figure 1.120, we can see that as t approaches -2 from the right, $f(t)$ approaches $-\infty$. Therefore, we can say that

$$\lim_{t\to -2^+} f(t) = -\infty.$$

On the other hand, as t approaches -2 from the left, $f(t)$ gets arbitrarily large. Therefore,

$$\lim_{t\to -2^-} f(t) = \infty.$$

Since neither of the one-sided limits exists, the limit $\lim_{t\to -2} f(t)$ does not exist either.

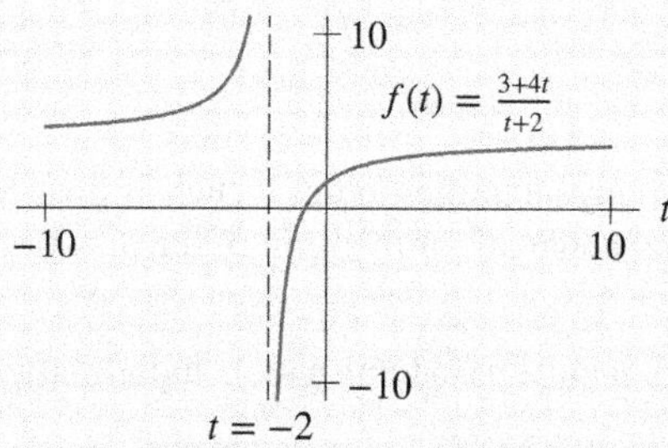

Figure 1.120: Behavior of $f(t)$ near $t = -2$

Limits and Continuity for Combinations of Functions

The following properties of limits allow us to extend our knowledge of the limiting behavior of two functions to their sums and products, and sometimes to their quotients. These properties hold for both one- and two-sided limits, as well as limits at infinity (when $x \to \infty$ or $x \to -\infty$). They underlie many limit calculations, though we may not acknowledge them explicitly.

Theorem 1.2: Properties of Limits

Assuming all the limits on the right-hand side exist:

1. If b is a constant, then $\lim_{x\to c} (bf(x)) = b\left(\lim_{x\to c} f(x)\right)$.
2. $\lim_{x\to c} (f(x) + g(x)) = \lim_{x\to c} f(x) + \lim_{x\to c} g(x)$.
3. $\lim_{x\to c} (f(x)g(x)) = \left(\lim_{x\to c} f(x)\right)\left(\lim_{x\to c} g(x)\right)$.
4. $\lim_{x\to c} \dfrac{f(x)}{g(x)} = \dfrac{\lim_{x\to c} f(x)}{\lim_{x\to c} g(x)}$, provided $\lim_{x\to c} g(x) \neq 0$.
5. For any constant k, $\lim_{x\to c} k = k$.
6. $\lim_{x\to c} x = c$.

Since a function is continuous at a point only when it has a limit there, the properties of limits lead to similar properties of continuity for combinations of functions.

Theorem 1.3: Continuity of Sums, Products, and Quotients of Functions

If f and g are continuous on an interval and b is a constant, then, on that same interval,

1. $bf(x)$ is continuous.
2. $f(x) + g(x)$ is continuous.
3. $f(x)g(x)$ is continuous.
4. $f(x)/g(x)$ is continuous, provided $g(x) \neq 0$ on the interval.

We prove the third of these properties.

Proof Let c be any point in the interval. We must show that $\lim_{x\to c}(f(x)g(x)) = f(c)g(c)$. Since $f(x)$ and $g(x)$ are continuous, we know that $\lim_{x\to c} f(x) = f(c)$ and $\lim_{x\to c} g(x) = g(c)$. So, by the third property of limits in Theorem 1.2,

$$\lim_{x\to c}(f(x)g(x)) = \left(\lim_{x\to c} f(x)\right)\left(\lim_{x\to c} g(x)\right) = f(c)g(c).$$

Since c was chosen arbitrarily, we have shown that $f(x)g(x)$ is continuous at every point in the interval.

Continuity also extends to compositions and inverses of functions.

Theorem 1.4: Continuity of Composite and Inverse Functions

If f and g are continuous, then

1. if the composite function $f(g(x))$ is defined on an interval, then $f(g(x))$ is continuous on that interval.
2. if f has an inverse function f^{-1}, then f^{-1} is continuous.

Example 5 Use Theorems 1.3 and 1.4 to explain why the function is continuous.
(a) $f(x) = x^2 \cos x$ (b) $g(x) = \ln x$ (c) $h(x) = \sin(e^x)$

Solution (a) Since $y = x^2$ and $y = \cos x$ are continuous everywhere, by Theorem 1.3 their product $f(x) = x^2 \cos x$ is continuous everywhere.
(b) Since $y = e^x$ is continuous everywhere it is defined and $\ln x$ is the inverse function of e^x, by Theorem 1.4, $g(x) = \ln x$ is continuous everywhere it is defined.
(c) Since $y = \sin x$ and $y = e^x$ are continuous everywhere, by Theorem 1.4 their composition, $h(x) = \sin(e^x)$, is continuous.

Exercises and Problems for Section 1.8 Online Resource: Additional Problems for Section 1.8

EXERCISES

1. Use Figure 1.121 to find the limits or explain why they don't exist.

(a) $\lim_{x\to -1^+} f(x)$ (b) $\lim_{x\to 0^-} f(x)$ (c) $\lim_{x\to 0} f(x)$
(d) $\lim_{x\to 1^-} f(x)$ (e) $\lim_{x\to 1} f(x)$ (f) $\lim_{x\to 2^-} f(x)$

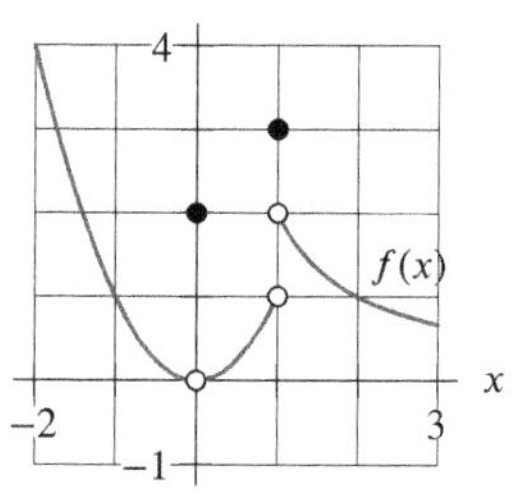

Figure 1.121

2. Use Figure 1.122 to estimate the following limits, if they exist.

(a) $\lim_{x\to 1^-} f(x)$ (b) $\lim_{x\to 1^+} f(x)$ (c) $\lim_{x\to 1} f(x)$
(d) $\lim_{x\to 2^-} f(x)$ (e) $\lim_{x\to 2^+} f(x)$ (f) $\lim_{x\to 2} f(x)$

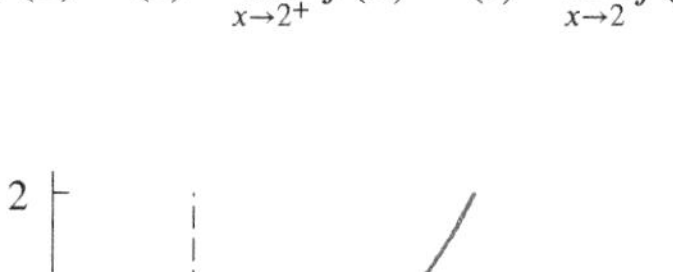

Figure 1.122

3. Use Figure 1.123 to find each of the following or explain why they don't exist.

(a) $f(-2)$ (b) $f(0)$ (c) $\lim_{x\to -4^+} f(x)$
(d) $\lim_{x\to -2^-} f(x)$ (e) $\lim_{x\to -2^+} f(x)$ (f) $\lim_{x\to 0} f(x)$
(g) $\lim_{x\to 2} f(x)$ (h) $\lim_{x\to 4^-} f(x)$

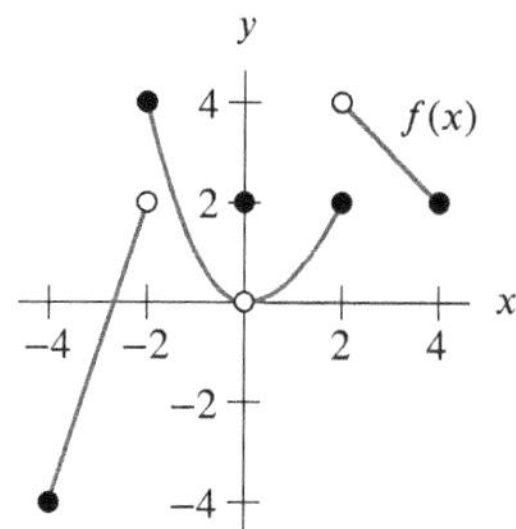

Figure 1.123

4. Use Figure 1.124 to find each of the following or explain why they don't exist.

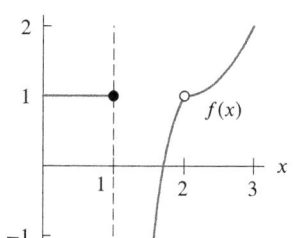

(a) $f(0)$ (b) $f(4)$ (c) $\lim_{x\to -2^-} f(x)$
(d) $\lim_{x\to -2^+} f(x)$ (e) $\lim_{x\to -2} f(x)$ (f) $\lim_{x\to 0} f(x)$
(g) $\lim_{x\to 2^-} f(x)$ (h) $\lim_{x\to 2^+} f(x)$ (i) $\lim_{x\to 2} f(x)$
(j) $\lim_{x\to 4} f(x)$

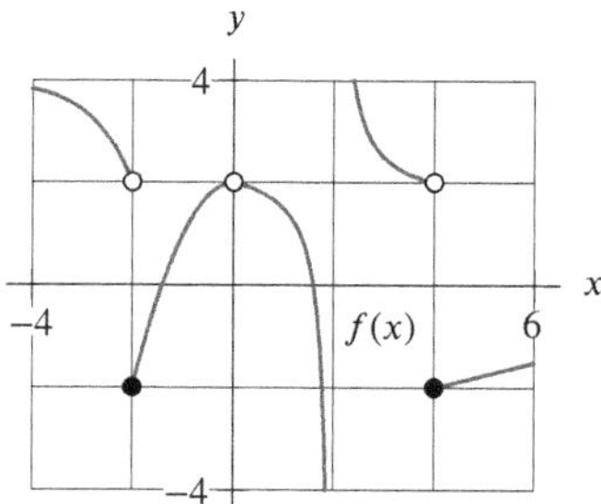

Figure 1.124

5. Use Figure 1.125 to estimate the following limits.

(a) $\lim_{x\to\infty} f(x)$ **(b)** $\lim_{x\to-\infty} f(x)$

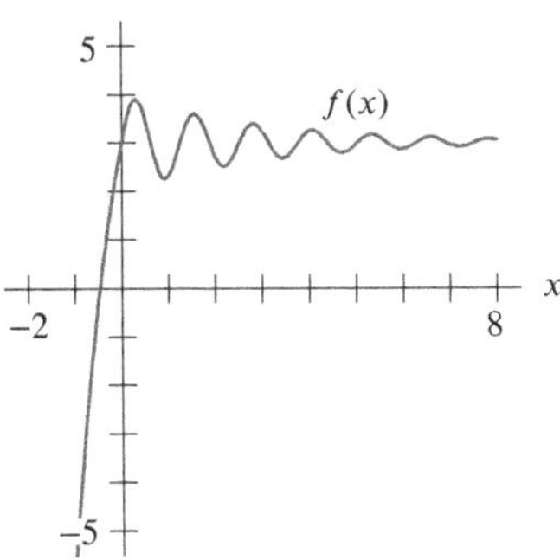

Figure 1.125

■ In Exercises 6–8, calculate the limit using the limit properties and $\lim_{x\to 2} f(x) = 7$, $\lim_{x\to 2} g(x) = -4$, $\lim_{x\to 2} h(x) = \frac{1}{2}$.

6. $\lim_{x\to 2}(f(x) - 2h(x))$

7. $\lim_{x\to 2}(g(x))^2$

8. $\lim_{x\to 2} \dfrac{f(x)}{g(x)\cdot h(x)}$

9. Using Figures 1.126 and 1.127, estimate

(a) $\lim_{x\to 1^-}(f(x) + g(x))$ **(b)** $\lim_{x\to 1^+}(f(x) + 2g(x))$

(c) $\lim_{x\to 1^-} f(x)g(x)$ **(d)** $\lim_{x\to 1^+} \dfrac{f(x)}{g(x)}$

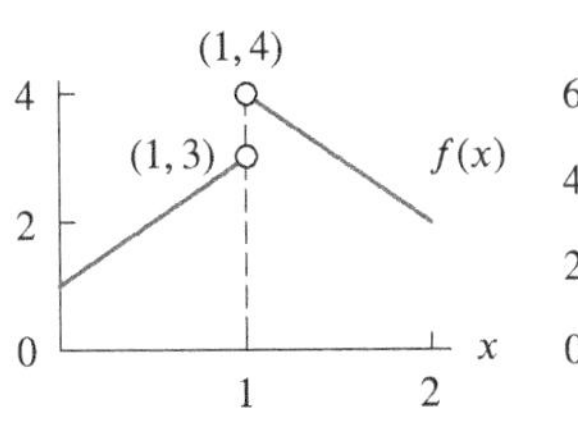

Figure 1.126

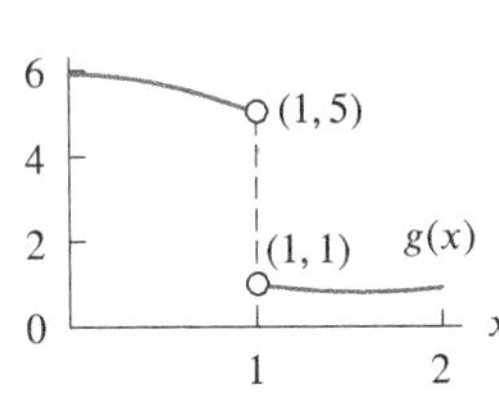

Figure 1.127

■ In Exercises 10–15, draw a possible graph of $f(x)$. Assume $f(x)$ is defined and continuous for all real x.

10. $\lim_{x\to\infty} f(x) = -\infty$ and $\lim_{x\to-\infty} f(x) = -\infty$

11. $\lim_{x\to\infty} f(x) = -\infty$ and $\lim_{x\to-\infty} f(x) = \infty$

12. $\lim_{x\to\infty} f(x) = 1$ and $\lim_{x\to-\infty} f(x) = \infty$

13. $\lim_{x\to\infty} f(x) = -\infty$ and $\lim_{x\to-\infty} f(x) = 3$

14. $\lim_{x\to\infty} f(x) = \infty$ and $\lim_{x\to-1} f(x) = 2$

15. $\lim_{x\to 3} f(x) = 5$ and $\lim_{x\to-\infty} f(x) = \infty$

■ In Exercises 16–28, find the limits using your understanding of the end behavior of each function.

16. $\lim_{x\to\infty} x^2$

17. $\lim_{x\to-\infty} x^2$

18. $\lim_{x\to-\infty} x^3$

19. $\lim_{x\to\infty} x^3$

20. $\lim_{x\to\infty} e^x$

21. $\lim_{x\to\infty} e^{-x}$

22. $\lim_{x\to\infty} 5^{-x}$

23. $\lim_{x\to\infty} \sqrt{x}$

24. $\lim_{x\to\infty} \ln x$

25. $\lim_{x\to\infty} x^{-2}$

26. $\lim_{x\to-\infty} x^{-2}$

27. $\lim_{x\to-\infty} x^{-3}$

28. $\lim_{x\to\infty} \left(\frac{1}{2}\right)^x$

■ In Exercises 29–34, give $\lim_{x\to-\infty} f(x)$ and $\lim_{x\to+\infty} f(x)$.

29. $f(x) = -x^4$

30. $f(x) = 5 + 21x - 2x^3$

31. $f(x) = x^5 + 25x^4 - 37x^3 - 200x^2 + 48x + 10$

32. $f(x) = \dfrac{3x^3 + 6x^2 + 45}{5x^3 + 25x + 12}$

33. $f(x) = 8x^{-3}$

34. $f(x) = 25e^{0.08x}$

35. Does $f(x) = \dfrac{|x|}{x}$ have right or left limits at 0? Is $f(x)$ continuous?

■ In Exercises 36–38, use algebra to evaluate $\lim_{x\to a^+} f(x)$, $\lim_{x\to a^-} f(x)$, and $\lim_{x\to a} f(x)$ if they exist. Sketch a graph to confirm your answers.

36. $a = 4$, $f(x) = \dfrac{|x-4|}{x-4}$

37. $a = 2$, $f(x) = \dfrac{|x-2|}{x}$

38. $a = 3$, $f(x) = \begin{cases} x^2 - 2, & 0 < x < 3 \\ 2, & x = 3 \\ 2x + 1, & 3 < x \end{cases}$

PROBLEMS

39. By graphing $y = (1 + 1/x)^x$, estimate $\lim_{x\to\infty}(1 + 1/x)^x$. You should recognize the answer you get.

40. Investigate $\lim_{x\to\infty}(1 + 1/x)^x$ numerically.

41. **(a)** Sketch $f(x) = e^{1/(x^2+0.0001)}$ around $x = 0$.

(b) Is $f(x)$ continuous at $x = 0$? Use properties of continuous functions to confirm your answer.

42. What does a calculator suggest about $\lim_{x\to 0^+} xe^{1/x}$? Does the limit appear to exist? Explain.

In Problems 43–52, evaluate $\lim_{x\to\infty}$ for the function.

43. $f(x) = \dfrac{x+3}{2-x}$

44. $f(x) = \dfrac{\pi + 3x}{\pi x - 3}$

45. $f(x) = \dfrac{x-5}{5+2x^2}$

46. $f(x) = \dfrac{x^2+2x-1}{3+3x^2}$

47. $f(x) = \dfrac{x^2+4}{x+3}$

48. $f(x) = \dfrac{2x^3-16x^2}{4x^2+3x^3}$

49. $f(x) = \dfrac{x^4+3x}{x^4+2x^5}$

50. $f(x) = \dfrac{3e^x+2}{2e^x+3}$

51. $f(x) = \dfrac{2^{-x}+5}{3^{-x}+7}$

52. $f(x) = \dfrac{2e^{-x}+3}{3e^{-x}+2}$

53. **(a)** Sketch the graph of a continuous function f with *all* of the following properties:
  (i) $f(0) = 2$
  (ii) $f(x)$ is decreasing for $0 \le x \le 3$
  (iii) $f(x)$ is increasing for $3 < x \le 5$
  (iv) $f(x)$ is decreasing for $x > 5$
  (v) $f(x) \to 9$ as $x \to \infty$

(b) Is it possible that the graph of f is concave down for all $x > 6$? Explain.

54. Sketch the graph of a function f with *all* of the following properties:

(i) $f(0) = 2$ (ii) $f(4) = 2$
(iii) $\lim_{x\to-\infty} f(x) = 2$ (iv) $\lim_{x\to 0} f(x) = 0$
(v) $\lim_{x\to 2} f(x) = \infty$ (vi) $\lim_{x\to 4^-} f(x) = 2$
(vii) $\lim_{x\to 4^+} f(x) = -2$

55. Sketch the graph of a function f with *all* of the following properties:

(i) $f(-2) = 1$ (ii) $f(2) = -2$
(iii) $f(3) = 3$ (iv) $\lim_{x\to-\infty} f(x) = -2$
(v) $\lim_{x\to-1^-} f(x) = -\infty$ (vi) $\lim_{x\to-1^+} f(x) = \infty$
(vii) $\lim_{x\to 2} f(x) = 1$ (viii) $\lim_{x\to 3^-} f(x) = 3$
(ix) $\lim_{x\to 3^+} f(x) = 2$ (x) $\lim_{x\to\infty} f(x) = 1$

56. The graph of $f(x)$ has a horizontal asymptote at $y = -4$, a vertical asymptote at $x = 3$, and no other asymptotes.

(a) Find a value of a such that $\lim_{x\to a} f(x)$ does not exist.
(b) If $\lim_{x\to\infty} f(x)$ exists, what is its value?

57. A patient takes a 100 mg dose of a drug once daily for four days starting at time $t = 0$ (t in days). Figure 1.128 shows a graph of $Q = f(t)$, the amount of the drug in the patient's body, in mg, after t days.

(a) Estimate and interpret each of the following:
  (i) $\lim_{t\to 1^-} f(t)$ (ii) $\lim_{t\to 1^+} f(t)$

(b) For what values of t is f not continuous? Explain the meaning of the points of discontinuity.

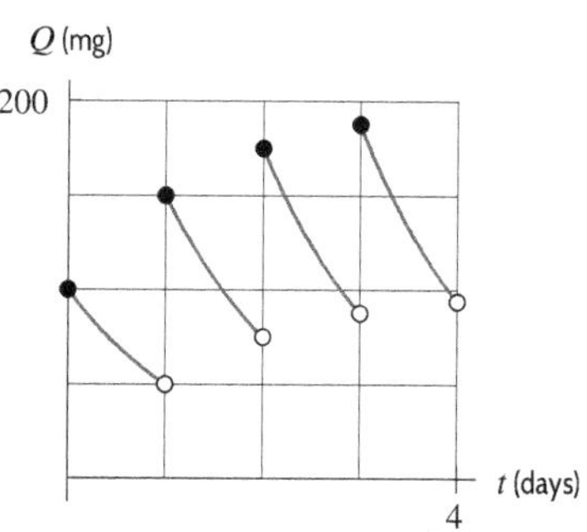

Figure 1.128

58. If $p(x)$ is the function on page 58 giving the price of renting a car, explain why $\lim_{x\to 1} p(x)$ does not exist.

59. Evaluate $\lim_{x\to 3} \dfrac{x^2+5x}{x+9}$ using the limit properties. State the property you use at each step.

60. Let $\lim_{x\to\infty} f(x) = \infty$ and $\lim_{x\to\infty} g(x) = \infty$. Give possible formulas for $f(x)$ and $g(x)$ if

(a) $\lim_{x\to\infty} \dfrac{f(x)}{g(x)} = \infty$ **(b)** $\lim_{x\to\infty} \dfrac{f(x)}{g(x)} = 3$

(c) $\lim_{x\to\infty} \dfrac{f(x)}{g(x)} = 0$

61. **(a)** Rewrite $\dfrac{1}{x-5} - \dfrac{10}{x^2-25}$ in the form $f(x)/g(x)$ for polynomials $f(x)$ and $g(x)$.
(b) Evaluate the limit $\lim_{x\to 5}\left(\dfrac{1}{x-5} - \dfrac{10}{x^2-25}\right)$.
(c) Explain why you cannot use Property 4 of the limit properties to evaluate $\lim_{x\to 5}\left(\dfrac{1}{x-5} - \dfrac{10}{x^2-25}\right)$.

In Problems 62–63, modify the definition of limit on page 60 to give a definition of each of the following.

62. A right-hand limit

63. A left-hand limit

64. Use Theorem 1.2 on page 71 to explain why if f and g are continuous on an interval, then so are $f+g$, fg, and f/g (assuming $g(x) \neq 0$ on the interval).

Strengthen Your Understanding

■ In Problems 65–66, explain what is wrong with the statement.

65. If $P(x)$ and $Q(x)$ are polynomials, $P(x)/Q(x)$ must be continuous for all x.

66. $\lim\limits_{x\to 1} \dfrac{x-1}{|x-1|} = 1$

■ In Problems 67–68, give an example of:

67. A rational function that has a limit at $x = 1$ but is not continuous at $x = 1$.

68. A function $f(x)$ where $\lim\limits_{x\to\infty} f(x) = 2$ and $\lim\limits_{x\to-\infty} f(x) = -2$.

■ Suppose that $\lim_{x\to 3} f(x) = 7$. Are the statements in Problems 69–73 true or false? If a statement is true, explain how you know. If a statement is false, give a counterexample.

69. $\lim\limits_{x\to 3}(xf(x)) = 21$.

70. If $g(3) = 4$, then $\lim\limits_{x\to 3}(f(x)g(x)) = 28$.

71. If $\lim\limits_{x\to 3} g(x) = 5$, then $\lim\limits_{x\to 3}(f(x) + g(x)) = 12$.

72. If $\lim\limits_{x\to 3}(f(x) + g(x)) = 12$, then $\lim\limits_{x\to 3} g(x) = 5$.

73. If $\lim\limits_{x\to 3} g(x)$ does not exist, then $\lim\limits_{x\to 3}(f(x)g(x))$ does not exist.

■ In Problems 74–79, is the statement true or false? Assume that $\lim_{x\to 7^+} f(x) = 2$ and $\lim_{x\to 7^-} f(x) = -2$. Explain.

74. $\lim\limits_{x\to 7} f(x)$ exists.

75. $\lim\limits_{x\to 7}(f(x))^2$ exists.

76. $\dfrac{\lim\limits_{x\to 7^+} f(x)}{\lim\limits_{x\to 7^-} f(x)}$ exists.

77. The function $(f(x))^2$ is continuous at $x = 7$.

78. If $f(7) = 2$, the function $f(x)$ is continuous at $x = 7$.

79. If $f(7) = 2$, the function $(f(x))^2$ is continuous at $x = 7$.

■ In Problems 80–81, let $f(x) = (1/x)\sin(1/x)$. Is the statement true or false? Explain.

80. The function $f(x)$ has horizontal asymptote $y = 0$.

81. The function $f(x)$ has a vertical asymptote at $x = 0$.

1.9 FURTHER LIMIT CALCULATIONS USING ALGEBRA

Sections 1.7 and 1.8 are sufficient for the later chapters of this book. In this optional section we explore further algebraic calculations of limits.

Limits of Quotients

In calculus we often encounter limits of the form $\lim\limits_{x\to c} f(x)/g(x)$ where $f(x)$ and $g(x)$ are continuous. There are three types of behavior for this type of limit:

- When $g(c) \neq 0$, the limit can be evaluated by substitution.
- When $g(c) = 0$ but $f(c) \neq 0$, the limit is undefined.
- When $g(c) = 0$ and $f(c) = 0$, the limit may or may not exist and can take any value.

We explore each these behaviors in more detail. When $g(c) \neq 0$, by Theorem 1.3, the limit can be found by substituting:

$$\lim_{x\to c} \frac{f(x)}{g(x)} = \frac{f(c)}{g(c)}.$$

When $g(c) = 0$, the situation is more complicated as substitution cannot be used.

Example 1 Evaluate the following limit or explain why it does not exist:

$$\lim_{x\to 3} \frac{x+1}{x-3}.$$

Solution If we try to evaluate at $x = 3$, we get $4/0$ which is undefined. Figure 1.129 shows that as x approaches 3 from the right, the function becomes arbitrarily large, and as x approaches 3 from the left, the function becomes arbitrarily large but negative, so this limit does not exist.

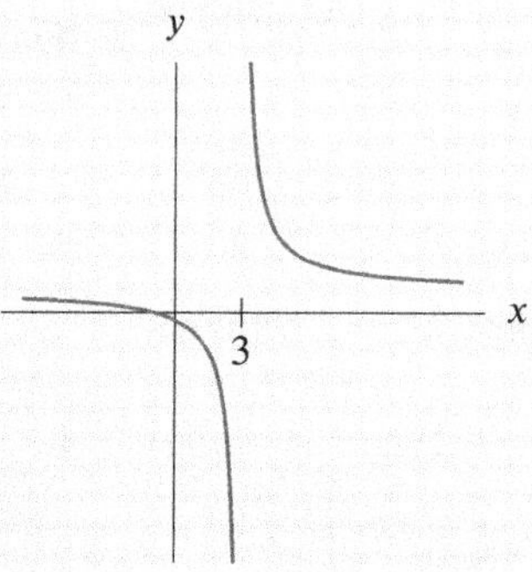

Figure 1.129: Limit of $y = (x + 1)/(x - 3)$ does not exist at $x = 3$

The limit in Example 1 does not exist because as x approaches 3, the denominator gets close to zero and the numerator gets close to 4. This means we are dividing a number close to 4 by a smaller and smaller number, resulting in a larger and larger number. This observation holds in general: for continuous functions, if $g(c) = 0$ but $f(c) \neq 0$, then $\lim_{x \to c} f(x)/g(x)$ does not exist.

Limits of the Form $0/0$ and Holes in Graphs

In Example 4 of Section 1.7 we saw that when both $f(c) = 0$ and $g(c) = 0$, so we have a limit of the form $0/0$, the limit can exist. We now explore limits of this form in more detail.

Example 2 Evaluate the following limit or explain why it does not exist:

$$\lim_{x \to 3} \frac{x^2 - x - 6}{x - 3}.$$

Solution If we try to evaluate at $x = 3$, we get $0/0$ which is undefined. Figure 1.130 suggests that as x approaches 3, the function gets close to 5, which suggests the limit is 5.

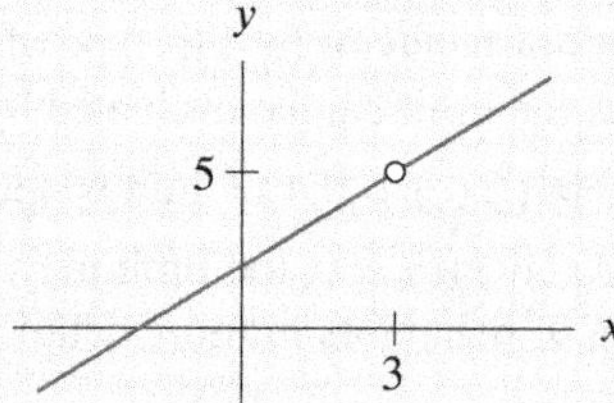

Figure 1.130: Graph of $y = (x^2 - x - 6)/(x - 3)$ is the same as the graph of $y = x + 2$ except at $x = 3$

This limit is similar to the one we saw in Example 4 of Section 1.7, so we check it algebraically using a similar method. Since the numerator factors as $x^2 - x - 6 = (x - 3)(x + 2)$ and $x \neq 3$ in the limit, we can cancel the common factor $x - 3$. We have:

$$\begin{aligned}
\lim_{x \to 3} \frac{x^2 - x - 6}{x - 3} &= \lim_{x \to 3} \frac{(x - 3)(x + 2)}{x - 3} && \text{Factoring the numerator} \\
&= \lim_{x \to 3} (x + 2) && \text{Canceling } (x - 3) \text{ since } x \neq 3 \\
&= 3 + 2 = 5 && \text{Substituting } x = 3 \text{ since } x + 2 \text{ is continuous}
\end{aligned}$$

Even though $f(x) = x^2 - x - 6$ and $g(x) = x - 3$ approach 0 as x approaches 3, the limit in Example 2 exists and is equal to 5 because the values of $f(x)$ are approximately 5 times the value of $g(x)$ near $x = 3$.

Using a Function with the Same Values to Evaluate a Limit

The limit in Example 2 exists because provided $x \neq 3$ we have

$$\frac{x^2 - x - 6}{x - 3} = x + 2,$$

so their limits are the same. This means their graphs are identical except at $x = 3$ where the first has a hole, and the second passes through $(3, 5)$. The key point is that when two functions take the same values close to but not necessarily at $x = c$, then their limits are the same at $x = c$. We use this observation in the following example.

Example 3 Evaluate the following limits or explain why they don't exist:

(a) $\displaystyle\lim_{h\to 0} \frac{(3+h)^2 - 3^2}{h}$ (b) $\displaystyle\lim_{x\to 4} \frac{x-4}{\sqrt{x}-2}$ (c) $\displaystyle\lim_{x\to 1} \frac{x-1}{x^2 - 2x + 1}$

Solution (a) At $h = 0$, we get $0/0$, so the function is undefined. We calculate algebraically:

$$\begin{aligned}
\lim_{h\to 0} \frac{(3+h)^2 - 3^2}{h} &= \lim_{h\to 0} \frac{3^2 + 6h + h^2 - 3^2}{h} && \text{Expanding the numerator} \\
&= \lim_{h\to 0} \frac{6h + h^2}{h} \\
&= \lim_{h\to 0} (6 + h) && \text{Canceling } h \text{ since } h \neq 0 \\
&= 6 + 0 = 6. && \text{Substituting } h = 0 \text{ since } 6 + h \text{ is continuous}
\end{aligned}$$

The limit exists because $y = \dfrac{(3+h)^2 - 3^2}{h}$ has the same values as the continuous function $y = 6 + h$ except at $h = 0$. If we were to sketch their graphs, they would be identical except at $h = 0$ where the rational function has a hole.

(b) At $x = 4$, we have $0/0$, so the function is undefined. We decide to multiply in the numerator and denominator by $\sqrt{x} + 2$ because that creates a factor of $x - 4$ in the denominator:

$$\begin{aligned}
\lim_{x\to 4} \frac{x-4}{\sqrt{x}-2} &= \lim_{x\to 4} \left(\frac{x-4}{\sqrt{x}-2}\right)\left(\frac{\sqrt{x}+2}{\sqrt{x}+2}\right) && \text{Multiplying by 1 does not change the limit} \\
&= \lim_{x\to 4} \frac{(x-4)(\sqrt{x}+2)}{(\sqrt{x}-2)(\sqrt{x}+2)} \\
&= \lim_{x\to 4} \frac{(x-4)(\sqrt{x}+2)}{x-4} && \text{Expanding the denominator} \\
&= \lim_{x\to 4} (\sqrt{x}+2) && \text{Canceling } (x-4) \text{ since } x \neq 4 \\
&= \sqrt{4} + 2 = 4. && \text{Substituting } x = 4 \text{ since } \sqrt{x}+2 \text{ is continuous}
\end{aligned}$$

Once again, we see that the limit exists because the function $y = \dfrac{x-4}{\sqrt{x}-2}$ has the same values as the continuous function $y = \sqrt{x} + 2$ except at $x = 4$. If we were to sketch their graphs, they would be identical except at $x = 4$ where the former has a hole.

(c) At $x = 1$, we get $0/0$, so the function is undefined. We try to calculate algebraically:

$$\begin{aligned}\lim_{x\to 1}\frac{x-1}{x^2-2x+1} &= \lim_{x\to 1}\frac{x-1}{(x-1)^2} && \text{Factoring the denominator}\\ &= \lim_{x\to 1}\frac{1}{(x-1)} && \text{Canceling } (x-1) \text{ since } x\neq 1\end{aligned}$$

So the function $y = \dfrac{x-1}{x^2-2x+1}$ takes the same values as $y = \dfrac{1}{x-1}$ except at $x = 1$, so it must have the same limit. Since the limit of the denominator of $y = 1/(x-1)$ is 0 and the numerator is 1 at $x = 1$, this limit does not exist. The limit is of the same form as in Example 1.

Although the limit in part (c) of Example 3 did not exist, we used the same idea as in the other examples to show this. The values of $y = \dfrac{x-1}{x^2-2x+1}$ are equal to the values of $y = \dfrac{1}{x-1}$ except at $x = 1$, so instead we considered the limit of $1/(x-1)$ at $x = 1$.

Finding Limits Using a New Variable

In order to evaluate the limit $\lim\limits_{x\to 4}\dfrac{x-4}{\sqrt{x}-2}$ algebraically in Example 3, we needed an additional step to eliminate the square root in the denominator. After this step, the process was identical to the other examples. Another way to find this limit is to use a new variable to eliminate the square root.

Example 4 By letting $t = \sqrt{x}$, evaluate the limit

$$\lim_{x\to 4}\frac{x-4}{\sqrt{x}-2}.$$

Solution If we let $t = \sqrt{x}$, we look for the limit as t approaches 2, since as x gets closer to 4, the value of t gets closer to 2. We calculate step by step algebraically:

$$\begin{aligned}\lim_{x\to 4}\frac{x-4}{\sqrt{x}-2} &= \lim_{t\to 2}\frac{t^2-4}{t-2} && \text{Letting } t=\sqrt{x}\\ &= \lim_{t\to 2}\frac{(t-2)(t+2)}{t-2} && \text{Factoring the numerator}\\ &= \lim_{t\to 2}(t+2) && \text{Canceling } (t-2) \text{ since } t\neq 2\\ &= 2+2 = 4. && \text{Evaluating the limit by substituting } t=2 \text{ since } t+2 \text{ is continuous}\end{aligned}$$

Calculating Limits at Infinity

Algebraic techniques can also be used to evaluate limits at infinity and one-sided limits.

Example 5 Evaluate the limit

$$\lim_{y\to\infty}\frac{3+4y}{y+2}.$$

Solution The limits of the numerator and the denominator do not exist as y grows arbitrarily large since both also grow arbitrarily large. However, dividing numerator and denominator by the highest power of y occurring enables us to see that the limit does exist:

$$\lim_{y\to\infty} \frac{3+4y}{y+2} = \lim_{y\to\infty} \frac{3/y+4}{1+2/y} \quad \text{Dividing numerator and denominator by } y$$

$$= \frac{0+4}{1+0} = 4 \quad \text{Since } 1/y \to 0 \text{ as } y \to \infty$$

In Example 5, as y goes to infinity, the limit of neither the numerator nor of the denominator exists; however, the limit of the quotient does exist. The reason for this is that though both numerator and denominator grow without bound, the constants 3 and 2 are insignificant for large y-values. Thus, $3 + 4y$ behaves like $4y$ and $2 + y$ behaves like y, so $(3 + 4y)/(y + 2)$ behaves like $4y/y$.

The Squeeze Theorem

There are some limits for which the techniques we already have cannot be used. For example, if $f(x) = x^2 \cos(1/x)$, then $f(x)$ is undefined at $x = 0$. This means f is not continuous at $x = 0$, so we cannot evaluate the limit by substitution. Also, there is no obvious function $g(x)$ with the same values as $f(x)$ close to $x = 0$ that we could use to calculate the limit. From Figure 1.131, it looks as if the limit does exist and equals 0, so we need a new way to check this is indeed the limit.

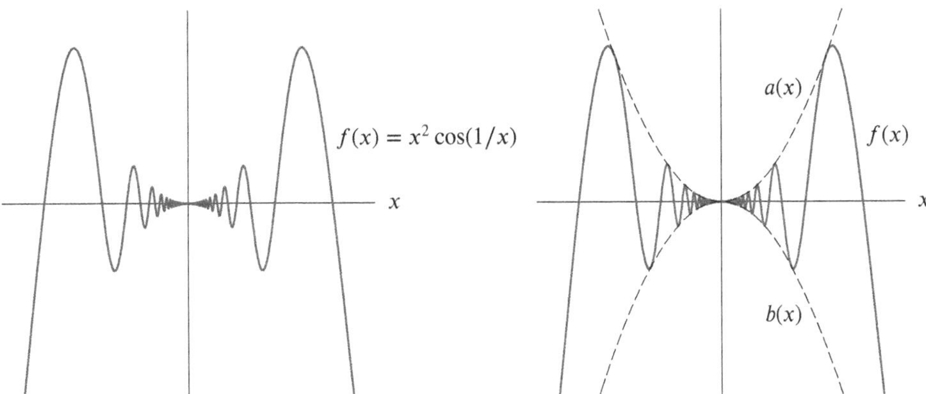

Figure 1.131: Find $\lim_{x\to 0} f(x)$

Figure 1.132: Graph of $f(x)$ is squeezed between $b(x) = -x^2$ and $a(x) = x^2$

From Figure 1.132, we can see that the graph of $a(x) = x^2$ is always above $f(x)$ and $b(x) = -x^2$ is always below $f(x)$, so $f(x)$ is always between $a(x)$ and $b(x)$. Since $a(x)$ and $b(x)$ both get closer and closer to 0 as x gets closer to 0, the values of $f(x)$ are "squeezed" between them and must have the same limit. Thus, we conclude $\lim_{x\to 0} f(x) = 0$.

Our calculation of this limit is an illustration of the Squeeze Theorem:

Theorem 1.5: The Squeeze Theorem

If $b(x) \le f(x) \le a(x)$ for all x close to $x = c$ except possibly at $x = c$, and $\lim_{x\to c} b(x) = L = \lim_{x\to c} a(x)$, then

$$\lim_{x\to c} f(x) = L.$$

The Squeeze Theorem can also be used to find limits at infinity.

Example 6 After driving over a speed bump at $t = 0$ seconds a car bounces up and down so the height of its body from the ground in inches is given by $h(t) = 7 + e^{-0.5t}\sin(2\pi t)$. Find and interpret $\lim_{t\to\infty} h(t)$.

Solution Since $-1 \le \sin(2\pi t) \le 1$ for all t, we have

$$7 - e^{-0.5t} \le h(t) \le 7 + e^{-0.5t},$$

so we try the Squeeze Theorem with $a(t) = 7 + e^{-0.5t}$ and $b(t) = 7 - e^{-0.5t}$. We have:

$$\lim_{t\to\infty} a(t) = \lim_{t\to\infty}(7 + e^{-0.5t}) = \lim_{t\to\infty}\left(7 + \frac{1}{e^{0.5t}}\right) \quad \text{Rewriting } e^{-0.5t} \text{ as } 1/e^{0.5t}$$

$$= \lim_{t\to\infty} 7 = 7. \quad \text{Since } 1/e^{0.5t} \to 0 \text{ as } t \to \infty$$

Similarly $\lim_{t\to\infty} b(t) = 7$, so by the Squeeze Theorem, $\lim_{t\to\infty} h(t) = 7$. This value makes sense as over time the shock absobers of the car will lessen the bouncing until the height stabilizes to 7 inches.

Exercises and Problems for Section 1.9

EXERCISES

In Exercises 1–3, find the limit.

1. $\lim_{x\to 0} \frac{3x^2}{x^2}$
2. $\lim_{x\to 0} \frac{3x^2}{x}$
3. $\lim_{x\to 0} \frac{3x^2}{x^4}$

For Exercises 4–23, use algebra to simplify the expression and find the limit.

4. $\lim_{x\to 3} \frac{x^2 - 3x}{x - 3}$
5. $\lim_{t\to 0} \frac{t^4 + t^2}{2t^3 - 9t^2}$
6. $\lim_{x\to 0} \frac{x^3 - 3x}{x\sqrt{2x+3}}$
7. $\lim_{x\to -4} \frac{x + 4}{2x^2 + 5x - 12}$
8. $\lim_{y\to 1} \frac{y^2 - 5y + 4}{y - 1}$
9. $\lim_{x\to 1} \frac{x^2 + 2x - 3}{x^2 - 3x + 2}$
10. $\lim_{t\to -2} \frac{2t^2 + 3t - 2}{t^2 + 5t + 6}$
11. $\lim_{x\to 3} \frac{x^2 - 9}{x^2 + x - 12}$
12. $\lim_{y\to -1} \frac{2y^2 + y - 1}{3y^2 + 2y - 1}$
13. $\lim_{h\to 0} \frac{(3+h)^2 - 9}{h}$
14. $\lim_{x\to -3} \frac{(x+5)^2 - 4}{x^2 - 9}$
15. $\lim_{x\to\sqrt{3}} \frac{5x^2 - 15}{x^4 - 9}$
16. $\lim_{x\to 2} \frac{2/x - 1}{x - 2}$
17. $\lim_{t\to 3} \frac{1/t - 1/3}{t - 3}$
18. $\lim_{t\to 0} \frac{1/(t+1) - 1}{t}$
19. $\lim_{h\to 0} \frac{1/(4+h) - 1/4}{h}$
20. $\lim_{z\to 1} \frac{\sqrt{z} - 1}{z - 1}$
21. $\lim_{h\to 0} \frac{\sqrt{9+h} - 3}{h}$
22. $\lim_{x\to 0} \frac{4^x - 1}{2^x - 1}$
23. $\lim_{h\to 0} \frac{(1+h)^4 - 1}{h}$

In Exercises 24–26, for the given constant c and functions $f(x)$ and $g(x)$, answer the following:

(a) Is the limit $\lim_{x\to c} \frac{f(x)}{g(x)}$ of the form $0/0$?

(b) Find $\lim_{x\to c} \frac{f(x)}{g(x)}$.

24. $c = 4$, $f(x) = 2x - 1$, $g(x) = x^{-2}$
25. $c = 1$, $f(x) = 2x^3 + x^2 - 3x$, $g(x) = x^2 + 3x - 4$
26. $c = 3$, $f(x) = x^2 - 3x$, $g(x) = \sqrt{x+1} - 2$

In Exercises 27–34, use algebra to evaluate the limit.

27. $\lim_{z\to\infty} \frac{5z^2 + 2z + 1}{2z^3 - z^2 + 9}$
28. $\lim_{x\to\infty} \frac{x + 7x^2 - 11}{3x^2 - 2x}$
29. $\lim_{t\to\infty} \frac{4e^t + 3e^{-t}}{5e^t + 2e}$
30. $\lim_{x\to\infty} \frac{2^{x+1}}{3^{x-1}}$
31. $\lim_{x\to\infty} \frac{2^{3x+2}}{3^{x+3}}$
32. $\lim_{x\to\infty} xe^{-x}$
33. $\lim_{t\to\infty}(4e^t)(7e^{-t})$
34. $\lim_{t\to\infty} t^{-2} \cdot \sin t$

35. Find $\lim_{x\to -1} f(x)$ if, for all x, $-4x+6 \le f(x) \le x^2-2x+7$.

36. Find $\lim_{x\to 0} f(x)$ if, for all x, $4\cos(2x) \le f(x) \le 3x^2 + 4$.

37. Find $\lim_{x\to\infty} f(x)$ if, for $x > 0$, $\frac{4x^2 - 5}{x^2} \le f(x) \le \frac{4x^6 + 3}{x^6}$.

PROBLEMS

In Problems 38–49, find all values for the constant k such that the limit exists.

38. $\lim_{x\to 4} \dfrac{x^2-k^2}{x-4}$

39. $\lim_{x\to 1} \dfrac{x^2-kx+4}{x-1}$

40. $\lim_{x\to -2} \dfrac{x^2+4x+k}{x+2}$

41. $\lim_{x\to 5} \dfrac{x^2-kx+5}{x^2-2x-15}$

42. $\lim_{x\to 0} \dfrac{e^k+2x-8}{e^x-1}$

43. $\lim_{x\to 1} \dfrac{k^2-40x-9}{\ln x}$

44. $\lim_{x\to\infty} \dfrac{x^2+3x+5}{4x+1+x^k}$

45. $\lim_{x\to -\infty} \dfrac{e^{2x}-5}{e^{kx}+3}$

46. $\lim_{x\to\infty} \dfrac{x^3-6}{x^k+3}$

47. $\lim_{x\to\infty} \dfrac{e^{kx}+11}{e^{5x}-3}$

48. $\lim_{x\to\infty} \dfrac{3^{kx}+6}{3^{2x}+4}$

49. $\lim_{x\to -\infty} \dfrac{3^{kx}+6}{3^{2x}+4}$

In Problems 50–55, use the indicated new variable to evaluate the limit.

50. $\lim_{y\to 4} \dfrac{\sqrt{y}-2}{y-4}$, let $t=\sqrt{y}$

51. $\lim_{x\to 9} \dfrac{x-\sqrt{x}-6}{\sqrt{x}-3}$, let $t=\sqrt{x}$

52. $\lim_{h\to 0} \dfrac{\sqrt{1+h}-1}{h}$, let $t=\sqrt{1+h}$

53. $\lim_{x\to 1} \dfrac{\sqrt[3]{x}-1}{x-1}$, let $t=\sqrt[3]{x}$

54. $\lim_{x\to 0} \dfrac{e^{3x}-e^{2x}}{e^x-1}$, let $t=e^x$

55. $\lim_{x\to\infty} \dfrac{2e^{3x}-1}{5e^{3x}+e^x+1}$, let $t=e^x$

56. Use the Squeeze Theorem to prove $\lim_{x\to\infty} \dfrac{\sin x}{x}=0$.

57. Use the Squeeze Theorem to prove $\lim_{x\to\infty} \dfrac{1}{x+e^{-x}}=0$.

In Problems 58–61, use the Squeeze Theorem to calculate the limit.

58. $\lim_{x\to\infty} \dfrac{\cos^2 x}{2x+1}$

59. $\lim_{x\to 0} x^4\sin(1/x)$

60. $\lim_{x\to\infty} \dfrac{x}{\sqrt{x^3+1}}$

61. $\lim_{x\to\infty} \dfrac{1}{x+2\cos^2 x}$

62. Let $\lim_{x\to\infty} f(x)=0$ and $\lim_{x\to\infty} g(x)=0$. Give possible formulas for $f(x)$ and $g(x)$ if

(a) $\lim_{x\to\infty} \dfrac{f(x)}{g(x)}=0$

(b) $\lim_{x\to\infty} \dfrac{f(x)}{g(x)}=1$

(c) $\lim_{x\to\infty} \dfrac{f(x)}{g(x)}=\infty$

In Problems 63–66, for the given constant c and function $f(x)$, find a function $g(x)$ that has a hole in its graph at $x=c$ but $f(x)=g(x)$ everywhere else that $f(x)$ is defined. Give the coordinates of the hole.

63. $f(x)=x^2+1,\ c=3$

64. $f(x)=x^2+1,\ c=0$

65. $f(x)=\ln x,\ c=1$

66. $f(x)=\sin x,\ c=\pi$

In Problems 67–72, for the given m and n, evaluate $\lim_{x\to 1} f(x)$ or explain why it does not exist, where

$$f(x)=\frac{(x-1)^n}{(x-1)^m}.$$

67. $n=3, m=2$

68. $n=2, m=3$

69. $n=2, m=2$

70. n and m are positive integers with $n>m$.

71. n and m are positive integers with $m>n$.

72. n and m are positive integers with $m=n$.

73. For any $f(x)$, where $-\dfrac{1}{x}\le f(x)\le \dfrac{1}{x}$, find values of c and L for which the Squeeze Theorem can be applied.

Strengthen Your Understanding

In Problems 74–75, explain what is wrong with the statement.

74. If $f(x)=\dfrac{x^2-1}{x+1}$ and $g(x)=x-1$, then $f=g$.

75. If $f(1)=0$ and $g(1)=1$, then

$$\lim_{x\to 1}\frac{f(x)}{g(x)}=\frac{0}{1}=0.$$

Are the statements in Problems 76–83 true or false? Explain.

76. If $b(x)\le f(x)\le a(x)$ and $\lim_{x\to 0} b(x)=-1$, $\lim_{x\to 0} a(x)=1$, then $-1\le \lim_{x\to 0} f(x)\le 1$.

77. If $0\le f(x)\le a(x)$ and $\lim_{x\to 0} a(x)=0$, then $\lim_{x\to 0} f(x)=0$.

78. If $b(x)\le f(x)\le a(x)$ and $\lim_{x\to 0} b(x)=\lim_{x\to 0} f(x)$ then $\lim_{x\to 0} f(x)=\lim_{x\to 0} a(x)$.

79. If $\lim_{x\to 0} g(x) = 0$ then $\lim_{x\to 0} \dfrac{f(x)}{g(x)} = \infty$.

80. If $\lim_{x\to 0} g(x) = 0$ and $\lim_{x\to 0} f(x) \neq 0$ then $\lim_{x\to 0} \dfrac{f(x)}{g(x)} = \infty$.

81. If $\lim_{x\to 0} \dfrac{f(x)}{g(x)}$ exists, then $\lim_{x\to 0} f(x)$ exists and $\lim_{x\to 0} g(x)$ exists.

82. If $\lim_{x\to 0} \dfrac{f(x)}{g(x)}$ exists, and $\lim_{x\to 0} g(x)$ exists then $\lim_{x\to 0} f(x)$ exists.

83. If $\lim_{x\to c^+} g(x) = 1$ and $\lim_{x\to c^-} g(x) = -1$ and $\lim_{x\to c} \dfrac{f(x)}{g(x)}$ exists, then $\lim_{x\to c} f(x) = 0$.

Online Resource: Section 1.10, review problems, and projects

Chapter Two

KEY CONCEPT: THE DERIVATIVE

Contents

2.1 HOW DO WE MEASURE SPEED?

The speed of an object at an instant in time is surprisingly difficult to define precisely. Consider the statement: "At the instant it crossed the finish line, the horse was traveling at 42 mph." How can such a claim be substantiated? A photograph taken at that instant will show the horse motionless—it is no help at all. There is some paradox in trying to study the horse's motion at a particular instant in time, since by focusing on a single instant we stop the motion!

Problems of motion were of central concern to Zeno and other philosophers as early as the fifth century B.C. The modern approach, made famous by Newton's calculus, is to stop looking for a simple notion of speed at an instant, and instead to look at speed over small time intervals containing the instant. This method sidesteps the philosophical problems mentioned earlier but introduces new ones of its own.

We illustrate the ideas discussed above with an idealized example, called a thought experiment. It is idealized in the sense that we assume that we can make measurements of distance and time as accurately as we wish.

A Thought Experiment: Average and Instantaneous Velocity

We look at the speed of a small object (say, a grapefruit) that is thrown straight upward into the air at $t = 0$ seconds. The grapefruit leaves the thrower's hand at high speed, slows down until it reaches its maximum height, and then speeds up in the downward direction and finally, "Splat!" (See Figure 2.1.)

Suppose that we want to determine the speed, say, at $t = 1$ second. Table 2.1 gives the height, y, of the grapefruit above the ground as a function of time. During the first second the grapefruit travels $90-6 = 84$ feet, and during the second second it travels only $142-90 = 52$ feet. Hence the grapefruit traveled faster over the first interval, $0 \leq t \leq 1$, than the second interval, $1 \leq t \leq 2$.

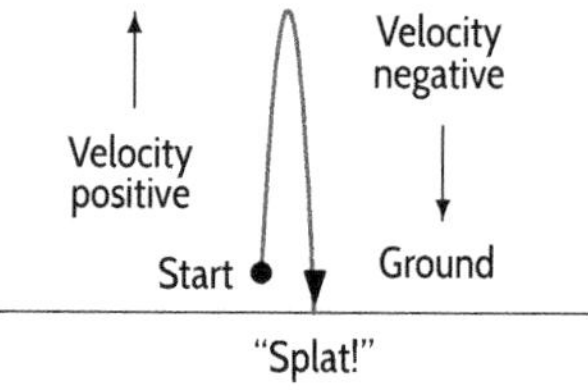

Figure 2.1: The grapefruit's path is straight up and down

Table 2.1 *Height of the grapefruit above the ground*

t (sec)	0	1	2	3	4	5	6
y (feet)	6	90	142	162	150	106	30

Velocity Versus Speed

From now on, we will distinguish between velocity and speed. Suppose an object moves along a line. We pick one direction to be positive and say that the *velocity* is positive if it is in that direction, and negative if it is in the opposite direction. For the grapefruit, upward is positive and downward is negative. (See Figure 2.1.) *Speed* is the magnitude of the velocity and so is always positive or zero.

If $s(t)$ is the position of an object at time t, then the **average velocity** of the object over the interval $a \leq t \leq b$ is

$$\text{Average velocity} = \frac{\text{Change in position}}{\text{Change in time}} = \frac{s(b) - s(a)}{b - a}.$$

In words, the **average velocity** of an object over an interval is the net change in position during the interval divided by the change in time.

Example 1 Compute the average velocity of the grapefruit over the interval $4 \le t \le 5$. What is the significance of the sign of your answer?

Solution During this one-second interval, the grapefruit moves $(106 - 150) = -44$ feet. Therefore the average velocity is $-44/(5 - 4) = -44$ ft/sec. The negative sign means the height is decreasing and the grapefruit is moving downward.

Example 2 Compute the average velocity of the grapefruit over the interval $1 \le t \le 3$.

Solution Average velocity $= (162 - 90)/(3 - 1) = 72/2 = 36$ ft/sec.

The average velocity is a useful concept since it gives a rough idea of the behavior of the grapefruit: If two grapefruits are hurled into the air, and one has an average velocity of 10 ft/sec over the interval $0 \le t \le 1$ while the second has an average velocity of 100 ft/sec over the same interval, the second one is moving faster.

But average velocity over an interval does not solve the problem of measuring the velocity of the grapefruit at *exactly* $t = 1$ second. To get closer to an answer to that question, we have to look at what happens near $t = 1$ in more detail. The data[1] in Figure 2.2 shows the average velocity over small intervals on either side of $t = 1$.

Notice that the average velocity before $t = 1$ is slightly more than the average velocity after $t = 1$. We expect to define the velocity *at* $t = 1$ to be between these two average velocities. As the size of the interval shrinks, the values of the velocity before $t = 1$ and the velocity after $t = 1$ get closer together. In the smallest interval in Figure 2.2, both velocities are 68.0 ft/sec (to one decimal place), so we define the velocity at $t = 1$ to be 68.0 ft/sec (to one decimal place).

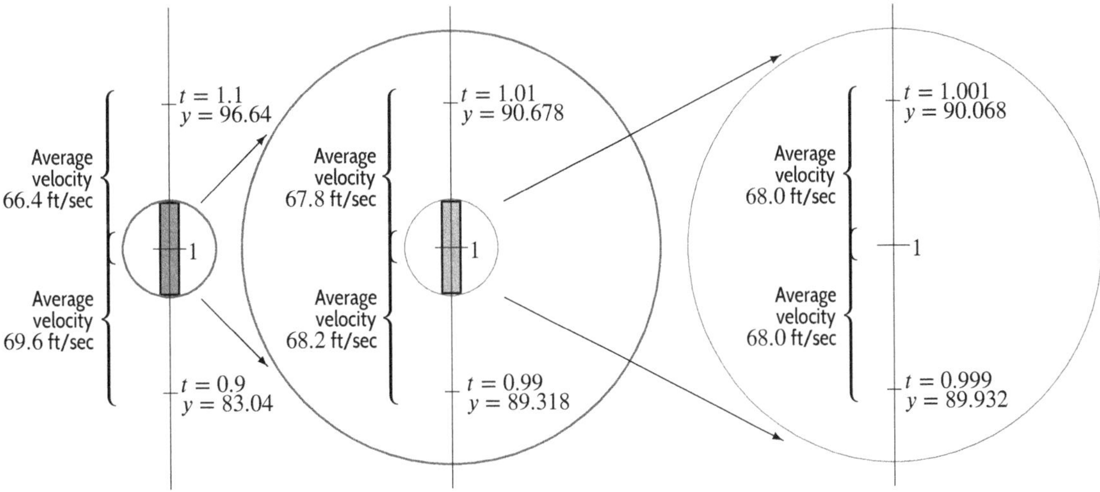

Figure 2.2: Average velocities over intervals on either side of $t = 1$: showing successively smaller intervals

Of course, if we calculate to more decimal places, the average velocities before and after $t = 1$ would no longer agree. To calculate the velocity at $t = 1$ to more decimal places of accuracy, we take smaller and smaller intervals on either side of $t = 1$ until the average velocities agree to the number of decimal places we want. In this way, we can estimate the velocity at $t = 1$ to any accuracy.

[1]The data is in fact calculated from the formula $y = 6 + 100t - 16t^2$.

Defining Instantaneous Velocity Using Limit Notation

When we take smaller and smaller intervals, it turns out that the average velocities get closer and closer to 68 ft/sec. It seems natural, then, to define *instantaneous velocity* at the instant $t = 1$ to be 68 ft/sec. Its definition depends on our being convinced that smaller and smaller intervals give averages that come arbitrarily close to 68; that is, the average speed approaches 68 ft/sec as a limit.

Notice how we have replaced the original difficulty of computing velocity at a point by a search for an argument to convince ourselves that the average velocities approach a limit as the time intervals shrink in size. Showing that the limit is exactly 68 requires the definition of the limit given on page 60 in Section 1.7

To define instantaneous velocity at an arbitrary point $t = a$, we use the same method as for $t = 1$. On small intervals of size h around $t = a$, we calculate

$$\text{Average velocity} = \frac{s(a+h) - s(a)}{h}.$$

The instantaneous velocity is the number that the average velocities approach as the intervals decrease in size, that is, as h becomes smaller. So we make the following definition:

Let $s(t)$ be the position at time t. Then the **instantaneous velocity** at $t = a$ is defined as

$$\begin{array}{c}\text{Instantaneous velocity}\\ \text{at } t = a\end{array} = \lim_{h\to 0} \frac{s(a+h) - s(a)}{h}.$$

In words, the **instantaneous velocity** of an object at time $t = a$ is given by the limit of the average velocity over an interval, as the interval shrinks around a.

This limit refers to the number that the average velocities approach as the intervals shrink. To estimate the limit, we look at intervals of smaller and smaller, but never zero, length.

Visualizing Velocity: Slope of Curve

Now we visualize velocity using a graph of height as a function of time. The cornerstone of the idea is the fact that, on a very small scale, most functions look almost like straight lines. Imagine taking the graph of a function near a point and "zooming in" to get a close-up view. (See Figure 2.3.) The more we zoom in, the more the curve appears to be a straight line. We call the slope of this line the *slope of the curve* at the point.

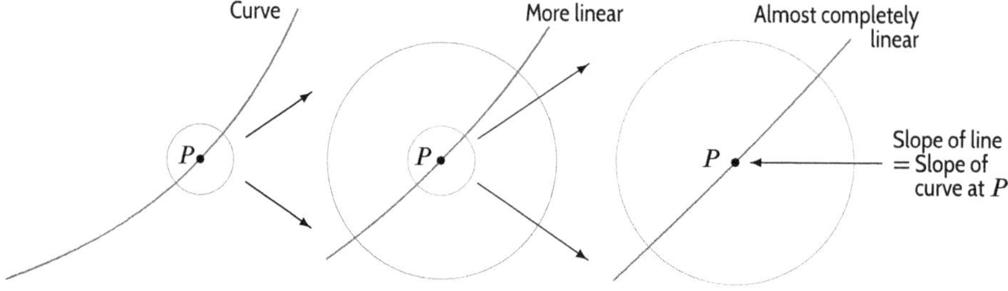

Figure 2.3: Estimating the slope of the curve at the point by "zooming in"

To visualize the instantaneous velocity, we think about how we calculated it. We took average velocities over small intervals containing $t = 1$. Two such velocities are represented by the slopes of the lines in Figure 2.4. As the length of the interval shrinks, the slope of the line gets closer to the slope of the curve at $t = 1$.

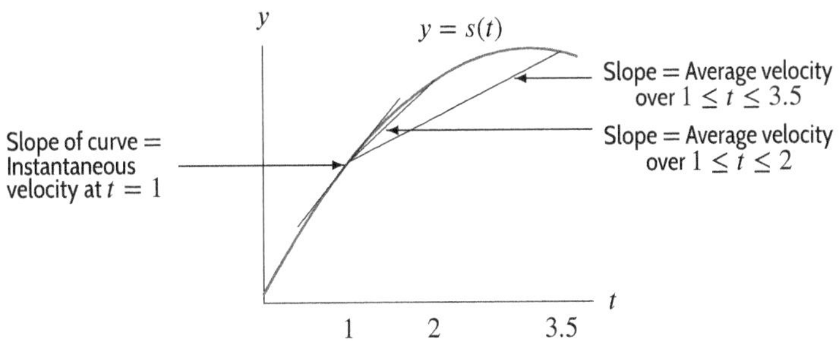

Figure 2.4: Average velocities over small intervals

> The **instantaneous velocity** is the slope of the curve at a point.

Let's go back to the grapefruit. Figure 2.5 shows the height of the grapefruit plotted against time. (Note that this is not a picture of the grapefruit's path, which is straight up and down.)

How can we visualize the average velocity on this graph? Suppose $y = s(t)$. We consider the interval $1 \le t \le 2$ and the expression

$$\text{Average velocity} = \frac{\text{Change in position}}{\text{Change in time}} = \frac{s(2) - s(1)}{2 - 1} = \frac{142 - 90}{1} = 52 \text{ ft/sec.}$$

Now $s(2) - s(1)$ is the change in position over the interval, and it is marked vertically in Figure 2.5. The 1 in the denominator is the time elapsed and is marked horizontally in Figure 2.5. Therefore,

$$\text{Average velocity} = \frac{\text{Change in position}}{\text{Change in time}} = \text{Slope of line joining } B \text{ and } C.$$

(See Figure 2.5.) A similar argument shows the following:

> The **average velocity** over any time interval $a \le t \le b$ is the slope of the line joining the points on the graph of $s(t)$ corresponding to $t = a$ and $t = b$.

Figure 2.5 shows how the grapefruit's velocity varies during its journey. At points A and B the curve has a large positive slope, indicating that the grapefruit is traveling up rapidly. Point D is almost at the top: the grapefruit is slowing down. At the peak, the slope of the curve is zero: the fruit has slowed to zero velocity for an instant in preparation for its return to earth. At point E the curve has a small negative slope, indicating a slow velocity of descent. Finally, the slope of the curve at point G is large and negative, indicating a large downward velocity that is responsible for the "Splat."

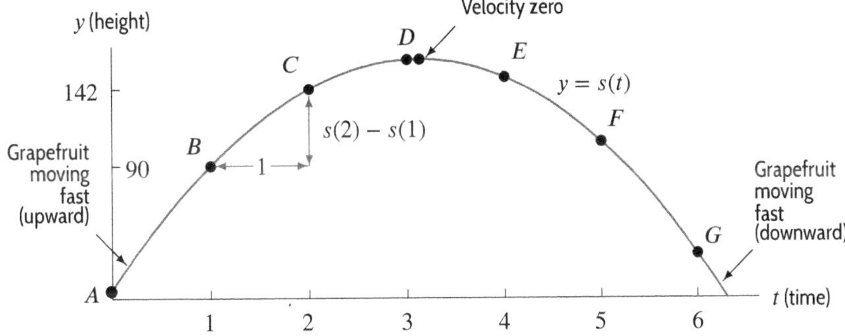

Figure 2.5: The height, y, of the grapefruit at time t

Using Limits to Compute the Instantaneous Velocity

Suppose we want to calculate the instantaneous velocity for $s(t) = t^2$ at $t = 3$. We must find:

$$\lim_{h\to 0} \frac{s(3+h) - s(3)}{h} = \lim_{h\to 0} \frac{(3+h)^2 - 9}{h}.$$

We show two possible approaches.

Example 3 Estimate $\lim_{h\to 0} \frac{(3+h)^2 - 9}{h}$ numerically.

Solution The limit is the value approached by this expression as h approaches 0. The values in Table 2.2 seem to be converging to 6 as $h \to 0$. So it is a reasonable guess that

$$\lim_{h\to 0} \frac{(3+h)^2 - 9}{h} = 6.$$

However, we cannot be sure that the limit is *exactly* 6 by looking at the table. To calculate the limit exactly requires algebra.

Table 2.2 *Values of $\left((3+h)^2 - 9\right)/h$ near $h = 0$*

h	−0.1	−0.01	−0.001	0.001	0.01	0.1
$\left((3+h)^2 - 9\right)/h$	5.9	5.99	5.999	6.001	6.01	6.1

Example 4 Use algebra to find $\lim_{h\to 0} \frac{(3+h)^2 - 9}{h}$.

Solution Expanding the numerator gives

$$\frac{(3+h)^2 - 9}{h} = \frac{9 + 6h + h^2 - 9}{h} = \frac{6h + h^2}{h}.$$

Since taking the limit as $h \to 0$ means looking at values of h near, but not equal, to 0, we can cancel h, giving

$$\lim_{h\to 0} \frac{(3+h)^2 - 9}{h} = \lim_{h\to 0}(6+h).$$

As h approaches 0, the values of $(6+h)$ approach 6, so

$$\lim_{h\to 0} \frac{(3+h)^2 - 9}{h} = \lim_{h\to 0}(6+h) = 6.$$

Exercises and Problems for Section 2.1

EXERCISES

1. The distance, s, a car has traveled on a trip is shown in the table as a function of the time, t, since the trip started. Find the average velocity between $t = 2$ and $t = 5$.

t (hours)	0	1	2	3	4	5
s (km)	0	45	135	220	300	400

2. The table gives the position of a particle moving along the x-axis as a function of time in seconds, where x is in meters. What is the average velocity of the particle from $t = 0$ to $t = 4$?

t	0	2	4	6	8
$x(t)$	−2	4	−6	−18	−14

3. The table gives the position of a particle moving along the x-axis as a function of time in seconds, where x is in angstroms. What is the average velocity of the particle from $t = 2$ to $t = 8$?

t	0	2	4	6	8
$x(t)$	0	14	−6	−18	−4

4. Figure 2.6 shows a particle's distance from a point as a function of time, t. What is the particle's average velocity from $t = 0$ to $t = 3$?

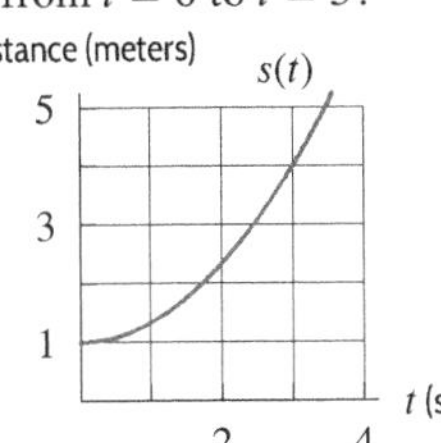

Figure 2.6

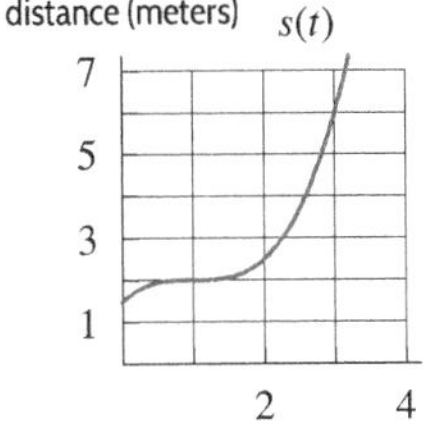

Figure 2.7

5. Figure 2.7 shows a particle's distance from a point as a function of time, t. What is the particle's average velocity from $t = 1$ to $t = 3$?

6. An observer tracks the distance a plane has rolled along the runway after touching down. Figure 2.8 shows this distance, x, in thousands of feet, as a function of time, t, in seconds since touchdown. Find the average velocity of the plane over the following time intervals:

   (a) Between 0 and 20 seconds.
   (b) Between 20 and 40 seconds.

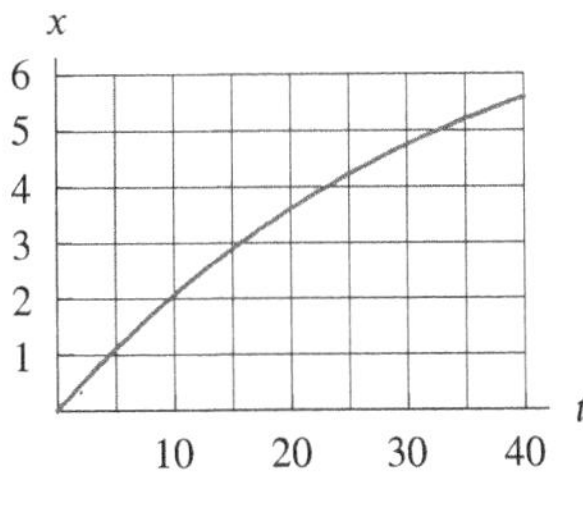

Figure 2.8

7. At time t in seconds, a particle's distance $s(t)$, in micrometers (μm), from a point is given by $s(t) = e^t - 1$. What is the average velocity of the particle from $t = 2$ to $t = 4$?

8. At time t in seconds, a particle's distance $s(t)$, in centimeters, from a point is given by $s(t) = 4+3\sin t$. What is the average velocity of the particle from $t = \pi/3$ to $t = 7\pi/3$?

9. In a time of t seconds, a particle moves a distance of s meters from its starting point, where $s = 3t^2$.

   (a) Find the average velocity between $t = 1$ and $t = 1 + h$ if:

   (i) $h = 0.1$, (ii) $h = 0.01$, (iii) $h = 0.001$.

   (b) Use your answers to part (a) to estimate the instantaneous velocity of the particle at time $t = 1$.

10. In a time of t seconds, a particle moves a distance of s meters from its starting point, where $s = 4t^3$.

   (a) Find the average velocity between $t = 0$ and $t = h$ if:

   (i) $h = 0.1$, (ii) $h = 0.01$, (iii) $h = 0.001$.

   (b) Use your answers to part (a) to estimate the instantaneous velocity of the particle at time $t = 0$.

11. In a time of t seconds, a particle moves a distance of s meters from its starting point, where $s = \sin(2t)$.

   (a) Find the average velocity between $t = 1$ and $t = 1 + h$ if:

   (i) $h = 0.1$, (ii) $h = 0.01$, (iii) $h = 0.001$.

   (b) Use your answers to part (a) to estimate the instantaneous velocity of the particle at time $t = 1$.

12. A car is driven at a constant speed. Sketch a graph of the distance the car has traveled as a function of time.

13. A car is driven at an increasing speed. Sketch a graph of the distance the car has traveled as a function of time.

14. A car starts at a high speed, and its speed then decreases slowly. Sketch a graph of the distance the car has traveled as a function of time.

PROBLEMS

In Problems 15–20, estimate the limit by substituting smaller and smaller values of h. For trigonometric functions, use radians. Give answers to one decimal place.

15. $\lim\limits_{h\to 0} \dfrac{(3+h)^3 - 27}{h}$

16. $\lim\limits_{h\to 0} \dfrac{\cos h - 1}{h}$

17. $\lim\limits_{h\to 0} \dfrac{7^h - 1}{h}$

18. $\lim\limits_{h\to 0} \dfrac{e^{1+h} - e}{h}$

19. $\lim\limits_{h\to 0} \dfrac{\sin h}{h}$

20. $\lim\limits_{h\to 0} \dfrac{\sqrt{1+h} - 1}{h}$

21. Match the points labeled on the curve in Figure 2.9 with the given slopes.

Slope	Point
−3	
−1	
0	
1/2	
1	
2	

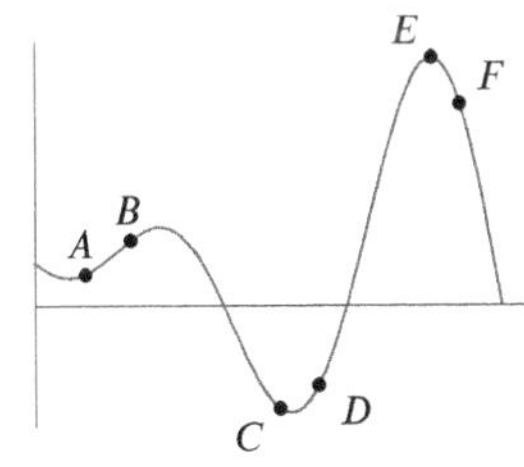

Figure 2.9

22. For the function shown in Figure 2.10, at what labeled points is the slope of the graph positive? Negative? At which labeled point does the graph have the greatest (i.e., most positive) slope? The least slope (i.e., negative and with the largest magnitude)?

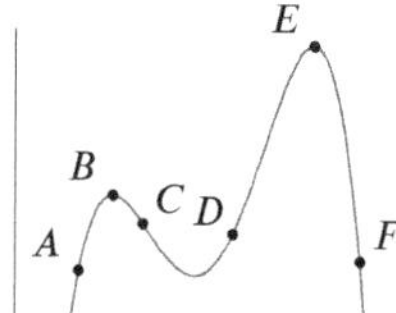

Figure 2.10

23. For the graph $y = f(x)$ in Figure 2.11, arrange the following numbers from smallest to largest:

- The slope of the graph at A.
- The slope of the graph at B.
- The slope of the graph at C.
- The slope of the line AB.
- The number 0.
- The number 1.

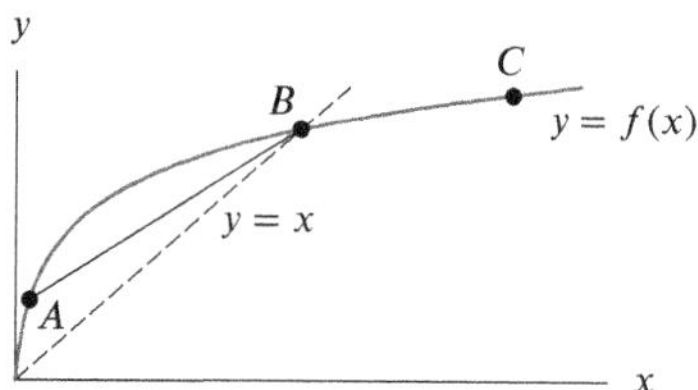

Figure 2.11

24. The graph of $f(t)$ in Figure 2.12 gives the position of a particle at time t. List the following quantities in order, smallest to largest.

- A, average velocity between $t = 1$ and $t = 3$,
- B, average velocity between $t = 5$ and $t = 6$,
- C, instantaneous velocity at $t = 1$,
- D, instantaneous velocity at $t = 3$,
- E, instantaneous velocity at $t = 5$,
- F, instantaneous velocity at $t = 6$.

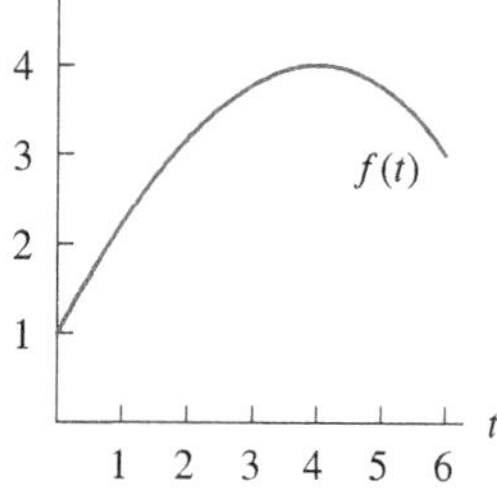

Figure 2.12

25. Find the average velocity over the interval $0 \le t \le 0.2$, and estimate the velocity at $t = 0.2$ of a car whose position, s, is given by the following table.

t (sec)	0	0.2	0.4	0.6	0.8	1.0
s (ft)	0	0.5	1.8	3.8	6.5	9.6

26. Figure 2.13 shows $f(t)$ and $g(t)$, the positions of two cars with respect to time, t, in minutes.

(a) Describe how the velocity of each car is changing during the time shown.
(b) Find an interval over which the cars have the same average velocity.
(c) Which of the following statements are true?

(i) Sometime in the first half minute, the two cars are traveling at the same instantaneous velocity.
(ii) During the second half minute (from $t = 1/2$ to $t = 1$), there is a time that the cars are traveling at the same instantaneous velocity.
(iii) The cars are traveling at the same velocity at $t = 1$ minute.
(iv) There is no time during the period shown that the cars are traveling at the same velocity.

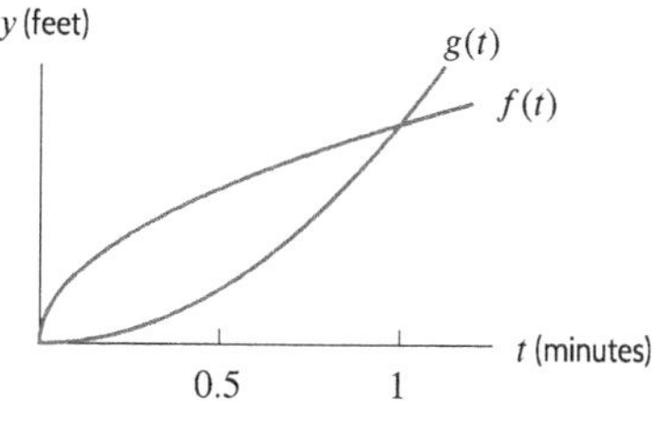

Figure 2.13

27. A particle moves at varying velocity along a line and $s = f(t)$ represents the particle's distance from a point as a function of time, t. Sketch a possible graph for f if the average velocity of the particle between $t = 2$ and $t = 6$ is the same as the instantaneous velocity at $t = 5$.

28. A particle moves along a line with varying velocity. At time t the particle is at a distance $s = f(t)$ from a fixed point on the line. Sketch a possible graph for f if the average velocity of the particle between $t = 0$ and $t = 5$ is the same as its instantaneous velocity at exactly two times between $t = 0$ and $t = 5$.

29. A ball is tossed into the air from a bridge, and its height, y (in feet), above the ground t seconds after it is thrown is given by

$$y = f(t) = -16t^2 + 50t + 36.$$

(a) How high above the ground is the bridge?
(b) What is the average velocity of the ball for the first second?
(c) Approximate the velocity of the ball at $t = 1$ second.
(d) Graph f, and determine the maximum height the ball reaches. What is the velocity at the time the ball is at the peak?
(e) Use the graph to decide at what time, t, the ball reaches its maximum height.

■ In Problems 30–33, use algebra to evaluate the limit.

30. $\lim_{h\to 0} \frac{(2+h)^2-4}{h}$

31. $\lim_{h\to 0} \frac{(1+h)^3-1}{h}$

32. $\lim_{h\to 0} \frac{3(2+h)^2-12}{h}$

33. $\lim_{h\to 0} \frac{(3+h)^2-(3-h)^2}{2h}$

Strengthen Your Understanding

■ In Problems 34–36, explain what is wrong with the statement.

34. Velocity and speed are the same.

35. Since $\lim_{h\to 0}(2+h)^2 = 4$, we have

$$\lim_{h\to 0} \frac{(2+h)^2-2^2}{h} = 0.$$

36. The particle whose position is shown in Figure 2.14 has velocity at time $t = 4$ greater than the velocity at $t = 2$.

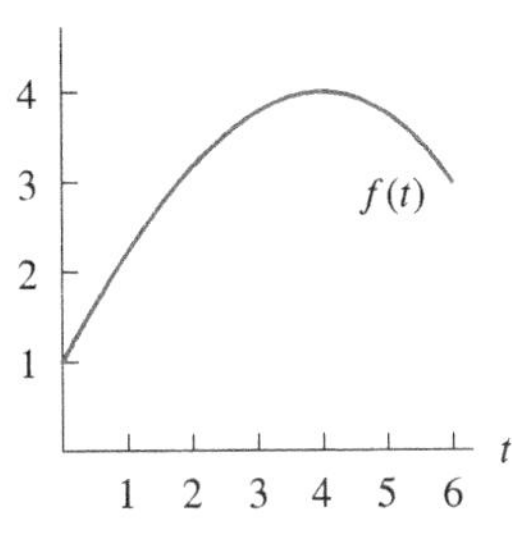

Figure 2.14

■ In Problems 37–38, give an example of:

37. A function which has a negative instantaneous velocity for $t < 0$ and a positive instantaneous velocity for $t > 0$.

38. A function giving the position of a particle that has the same speed at $t = -1$ and $t = 1$ but different velocities.

■ Are the statements in Problems 39–43 true or false? Give an explanation for your answer.

39. If a car is going 50 miles per hour at 2 pm and 60 miles per hour at 3 pm, then it travels between 50 and 60 miles during the hour between 2 pm and 3 pm.

40. If a car travels 80 miles between 2 and 4 pm, then its velocity is close to 40 mph at 2 pm.

41. If the time interval is short enough, then the average velocity of a car over the time interval and the instantaneous velocity at a time in the interval can be expected to be close.

42. If an object moves with the same average velocity over every time interval, then its average velocity equals its instantaneous velocity at any time.

43. The formula Distance traveled = Average velocity × Time is valid for every moving object for every time interval.

2.2 THE DERIVATIVE AT A POINT

Average Rate of Change

In Section 2.1, we looked at the change in height divided by the change in time; this ratio is called the *difference quotient*. Now we define the rate of change of a function f that depends on a variable other than time. We say:

$$\text{Average rate of change of } f \text{ over the interval from } a \text{ to } a+h = \frac{f(a+h)-f(a)}{h}.$$

The numerator, $f(a+h) - f(a)$, measures the change in the value of f over the interval from a to $a+h$. The difference quotient is the change in f divided by the change in the independent variable, which we call x. Although the interval is no longer necessarily a time interval, we still talk about the *average rate of change* of f over the interval. If we want to emphasize the independent variable, we talk about the average rate of change of f *with respect to x*.

Instantaneous Rate of Change: The Derivative

We define the *instantaneous rate of change* of a function at a point in the same way that we defined instantaneous velocity: we look at the average rate of change over smaller and smaller intervals. This instantaneous rate of change is called the *derivative of f at a*, denoted by $f'(a)$.

The **derivative of f at a**, written $f'(a)$, is defined as

$$\text{Rate of change of } f \text{ at } a = f'(a) = \lim_{h\to 0} \frac{f(a+h)-f(a)}{h}.$$

If the limit exists, then f is said to be **differentiable at a**.

To emphasize that $f'(a)$ is the rate of change of $f(x)$ as the variable x changes, we call $f'(a)$ the derivative of f *with respect to* x at $x = a$. When the function $y = s(t)$ represents the position of an object, the derivative $s'(t)$ is the velocity.

Example 1 Eucalyptus trees, common in California and the Pacific Northwest, grow better with more water. Scientists in North Africa, analyzing where to plant trees, found that the volume of wood that grows on a square kilometer, in meters^3, is approximated, for r cm rainfall per year and $60 \le r \le 120$, by[2]

$$V(r) = 0.2r^2 - 20r + 600.$$

(a) Calculate the average rate of change of V with respect to r over the intervals $90 \le r \le 100$ and $100 \le r \le 110$.

(b) By choosing small values for h, estimate the instantaneous rate of change of V with respect to r at $r = 100$ cm.

Solution (a) Using the formula for the average rate of change gives

$$\begin{array}{c}\text{Average rate of change of volume}\\ \text{for } 90 \le r \le 100\end{array} = \frac{V(100)-V(90)}{10} = \frac{600-420}{10} = 18 \text{ meter}^3/\text{cm}.$$

$$\begin{array}{c}\text{Average rate of change of volume}\\ \text{for } 100 \le r \le 110\end{array} = \frac{V(110)-V(100)}{10} = \frac{820-600}{10} = 22 \text{ meter}^3/\text{cm}.$$

So we see that the average rate of change of the volume of wood grown on a square kilometer increases as the rainfall increases.

(b) With $h = 0.1$ and $h = -0.1$, we have the difference quotients

$$\frac{V(100.1)-V(100)}{0.1} = 20.02 \text{ m}^3/\text{cm} \quad \text{and} \quad \frac{V(99.9)-V(100)}{-0.1} = 19.98 \text{ m}^3/\text{cm}.$$

With $h = 0.01$ and $h = -0.01$,

$$\frac{V(100.01)-V(100)}{0.01} = 20.002 \text{ m}^3/\text{cm} \quad \text{and} \quad \frac{V(99.99)-V(100)}{-0.01} = 19.998 \text{ m}^3/\text{cm}.$$

These difference quotients suggest that when the yearly rainfall is 100 cm, the instantaneous rate of change of the volume of wood grown on a square kilometer is about 20 meter^3 per cm of rainfall. To confirm that the instantaneous rate of change of the function is exactly 20, that is, $V'(100) = 20$, we would need to take the limit as $h \to 0$.

Visualizing the Derivative: Slope of Curve and Slope of Tangent

As with velocity, we can visualize the derivative $f'(a)$ as the slope of the graph of f at $x = a$. In addition, there is another way to think of $f'(a)$. Consider the difference quotient $(f(a+h)-f(a))/h$. The numerator, $f(a+h)-f(a)$, is the vertical distance marked in Figure 2.15 and h is the horizontal distance, so

$$\text{Average rate of change of } f = \frac{f(a+h)-f(a)}{h} = \text{Slope of line } AB.$$

[2]Sepp, C., "Is Urban Forestry a Solution to the Energy Crises of Sahelian Cities?", www.nzdl.org, accessed Feb 11, 2012.

As h becomes smaller, the line AB approaches the tangent line to the curve at A. (See Figure 2.16.) We say

$$\text{Instantaneous rate of change of } f \text{ at } a = \lim_{h\to 0} \frac{f(a+h)-f(a)}{h} = \text{Slope of tangent at } A.$$

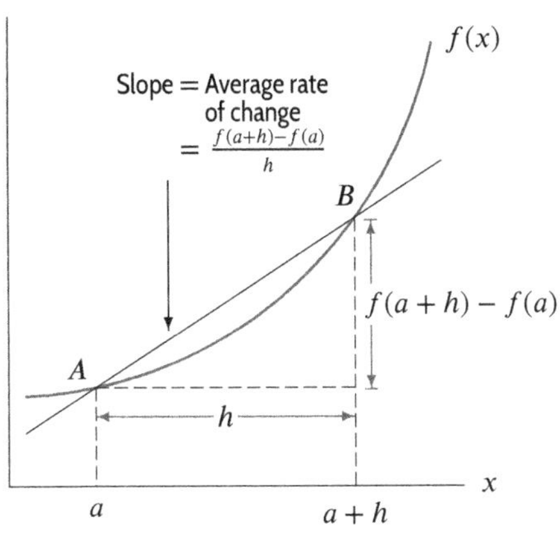

Figure 2.15: Visualizing the average rate of change of f

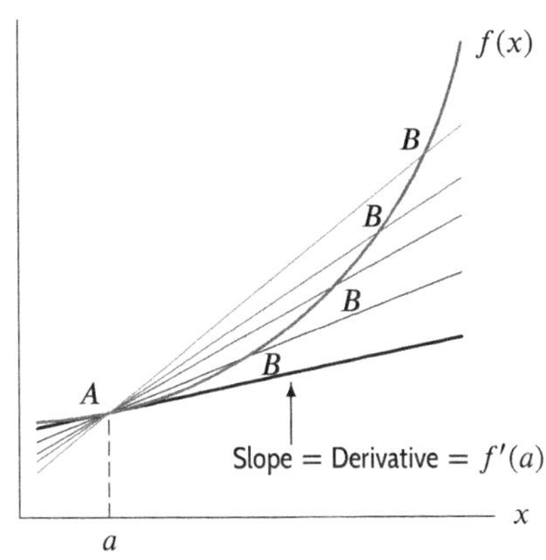

Figure 2.16: Visualizing the instantaneous rate of change of f

> The derivative at point A can be interpreted as:
> - The slope of the curve at A.
> - The slope of the tangent line to the curve at A.

The slope interpretation is often useful in gaining rough information about the derivative, as the following examples show.

Example 2 Is the derivative of $\sin x$ at $x = \pi$ positive or negative?

Solution Looking at a graph of $\sin x$ in Figure 2.17 (remember, x is in radians), we see that a tangent line drawn at $x = \pi$ has negative slope. So the derivative at this point is negative.

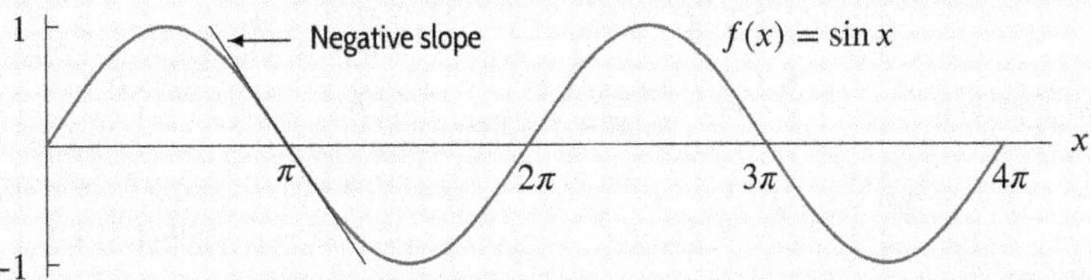

Figure 2.17: Tangent line to $\sin x$ at $x = \pi$

Recall that if we zoom in on the graph of a function $y = f(x)$ at the point $x = a$, we usually find that the graph looks like a straight line with slope $f'(a)$.

Example 3 By zooming in on the point $(0, 0)$ on the graph of the sine function, estimate the value of the derivative of $\sin x$ at $x = 0$, with x in radians.

Solution Figure 2.18 shows graphs of $\sin x$ with smaller and smaller scales. On the interval $-0.1 \le x \le 0.1$, the graph looks like a straight line of slope 1. Thus, the derivative of $\sin x$ at $x = 0$ is about 1.

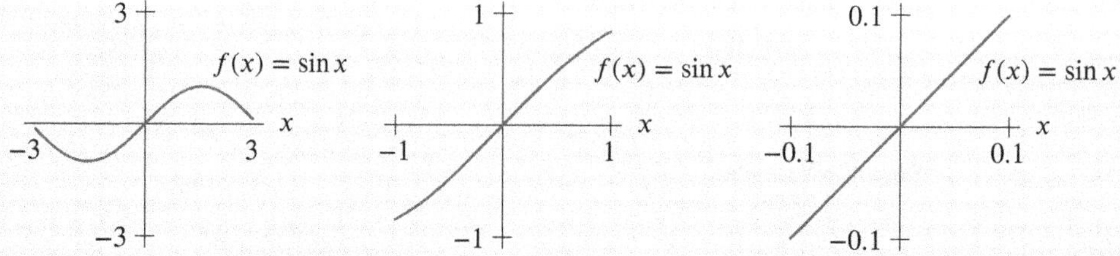

Figure 2.18: Zooming in on the graph of $\sin x$ near $x = 0$ shows the derivative is about 1 at $x = 0$

Later we will show that the derivative of $\sin x$ at $x = 0$ is exactly 1. (See page 161 in Section 3.5.) From now on we will assume that this is so. This simple result is one of the reasons we choose to use radians when doing calculus with trigonometric functions. If we had done Example 3 in degrees, the derivative of $\sin x$ would have turned out to be a much messier number. (See Problem 30, page 98.)

Estimating the Derivative

Example 4 Estimate the value of the derivative of $f(x) = 2^x$ at $x = 0$ graphically and numerically.

Solution Graphically: Figure 2.19 indicates that the graph is concave up. Assuming this, the slope at A is between the slope of BA and the slope of AC. Since

$$\text{Slope of line } BA = \frac{(2^0 - 2^{-1})}{(0-(-1))} = \frac{1}{2} \quad \text{and} \quad \text{Slope of line } AC = \frac{(2^1 - 2^0)}{(1-0)} = 1,$$

we know that at $x = 0$ the derivative of 2^x is between $1/2$ and 1.

Numerically: To estimate the derivative at $x = 0$, we look at values of the difference quotient

$$\frac{f(0+h) - f(0)}{h} = \frac{2^h - 2^0}{h} = \frac{2^h - 1}{h}$$

for small h. Table 2.3 shows some values of 2^h together with values of the difference quotients. (See Problem 43 on page 98 for what happens for very small values of h.)

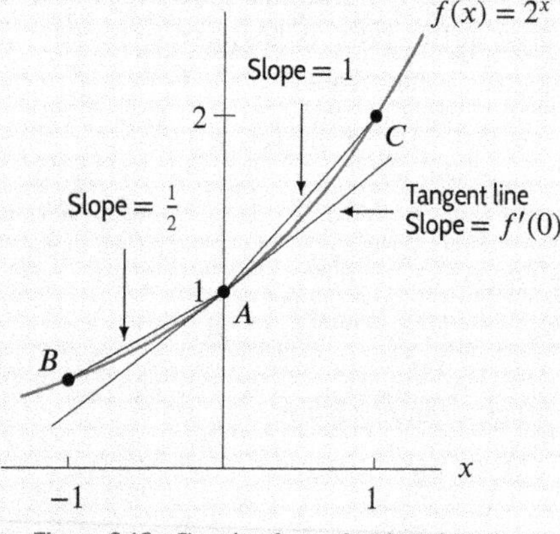

Figure 2.19: Graph of $y = 2^x$ showing the derivative at $x = 0$

Table 2.3 *Numerical values for difference quotient of 2^x at $x = 0$*

h	2^h	Difference quotient: $\frac{2^h-1}{h}$
−0.0003	0.999792078	0.693075
−0.0002	0.999861380	0.693099
−0.0001	0.999930688	0.693123
0	1	
0.0001	1.00006932	0.693171
0.0002	1.00013864	0.693195
0.0003	1.00020797	0.693219

The concavity of the curve tells us that difference quotients calculated with negative h's are smaller than the derivative, and those calculated with positive h's are larger. From Table 2.3 we see that the derivative is between 0.693123 and 0.693171. To three decimal places, $f'(0) = 0.693$.

Example 5 Find an approximate equation for the tangent line to $f(x) = 2^x$ at $x = 0$.

Solution From the previous example, we know the slope of the tangent line is about 0.693 at $x = 0$. Since the tangent line has y-intercept 1, its equation is

$$y = 0.693x + 1.$$

Computing the Derivative Algebraically

The graph of $f(x) = 1/x$ in Figure 2.20 leads us to expect that $f'(2)$ is negative. To compute $f'(2)$ exactly, we use algebra.

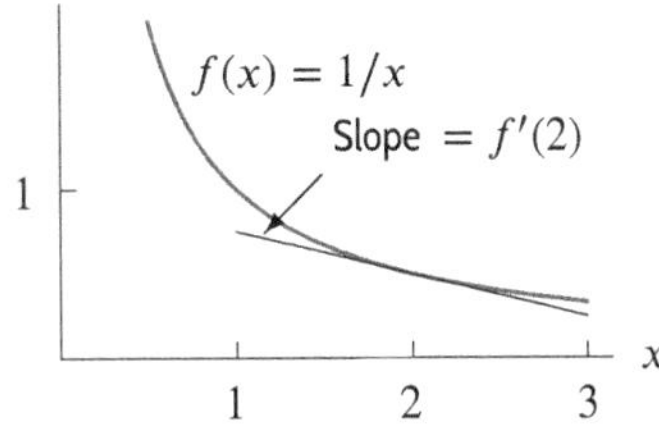

Figure 2.20: Tangent line to $f(x) = 1/x$ at $x = 2$

Example 6 Find the derivative of $f(x) = 1/x$ at the point $x = 2$.

Solution The derivative is the limit of the difference quotient, so we look at

$$f'(2) = \lim_{h \to 0} \frac{f(2+h) - f(2)}{h}.$$

Using the formula for f and simplifying gives

$$f'(2) = \lim_{h \to 0} \frac{1}{h}\left(\frac{1}{2+h} - \frac{1}{2}\right) = \lim_{h \to 0}\left(\frac{2-(2+h)}{2h(2+h)}\right) = \lim_{h \to 0} \frac{-h}{2h(2+h)}.$$

Since the limit only examines values of h close to, but not equal to, zero, we can cancel h. We get

$$f'(2) = \lim_{h \to 0} \frac{-1}{2(2+h)} = -\frac{1}{4}.$$

Thus, $f'(2) = -1/4$. The slope of the tangent line in Figure 2.20 is $-1/4$.

Exercises and Problems for Section 2.2

EXERCISES

1. The table shows values of $f(x) = x^3$ near $x = 2$ (to three decimal places). Use it to estimate $f'(2)$.

x	1.998	1.999	2.000	2.001	2.002
x^3	7.976	7.988	8.000	8.012	8.024

2. By choosing small values for h, estimate the instantaneous rate of change of the function $f(x) = x^3$ with respect to x at $x = 1$.

3. The income that a company receives from selling an item is called the revenue. Production decisions are based, in part, on how revenue changes if the quantity sold changes; that is, on the rate of change of revenue with respect to quantity sold. Suppose a company's revenue, in dollars, is given by $R(q) = 100q - 10q^2$, where q is the quantity sold in kilograms.

   (a) Calculate the average rate of change of R with respect to q over the intervals $1 \le q \le 2$ and $2 \le q \le 3$.

(b) By choosing small values for h, estimate the instantaneous rate of change of revenue with respect to change in quantity at $q = 2$ kilograms.

4. **(a)** Make a table of values rounded to two decimal places for the function $f(x) = e^x$ for $x = 1, 1.5, 2, 2.5$, and 3. Then use the table to answer parts (b) and (c).
(b) Find the average rate of change of $f(x)$ between $x = 1$ and $x = 3$.
(c) Use average rates of change to approximate the instantaneous rate of change of $f(x)$ at $x = 2$.

5. **(a)** Make a table of values, rounded to two decimal places, for $f(x) = \log x$ (that is, log base 10) with $x = 1, 1.5, 2, 2.5, 3$. Then use this table to answer parts (b) and (c).
(b) Find the average rate of change of $f(x)$ between $x = 1$ and $x = 3$.
(c) Use average rates of change to approximate the instantaneous rate of change of $f(x)$ at $x = 2$.

6. If $f(x) = x^3 + 4x$, estimate $f'(3)$ using a table with values of x near 3, spaced by 0.001.

7. Graph $f(x) = \sin x$, and use the graph to decide whether the derivative of $f(x)$ at $x = 3\pi$ is positive or negative.

8. For the function $f(x) = \log x$, estimate $f'(1)$. From the graph of $f(x)$, would you expect your estimate to be greater than or less than $f'(1)$?

9. Estimate $f'(2)$ for $f(x) = 3^x$. Explain your reasoning.

10. The graph of $y = f(x)$ is shown in Figure 2.21. Which is larger in each of the following pairs?

(a) Average rate of change: Between $x = 1$ and $x = 3$, or between $x = 3$ and $x = 5$?
(b) $f(2)$ or $f(5)$?
(c) $f'(1)$ or $f'(4)$?

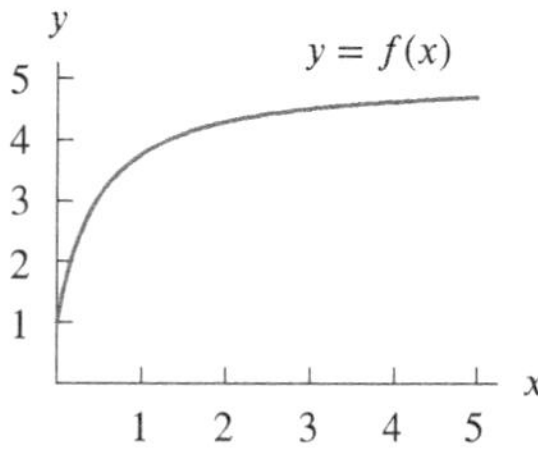

Figure 2.21

11. Use Figure 2.22 to decide which is larger in each of the following pairs

(a) Average rate of change between $x = 0$ and $x = 2$ or between $x = 2$ and $x = 4$
(b) $g(1)$ or $g(4)$
(c) $g'(2)$ or $g'(4)$

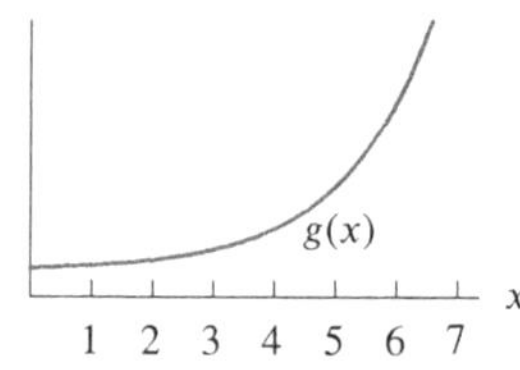

Figure 2.22

12. Figure 2.23 shows the graph of f. Match the derivatives in the table with the points a, b, c, d, e.

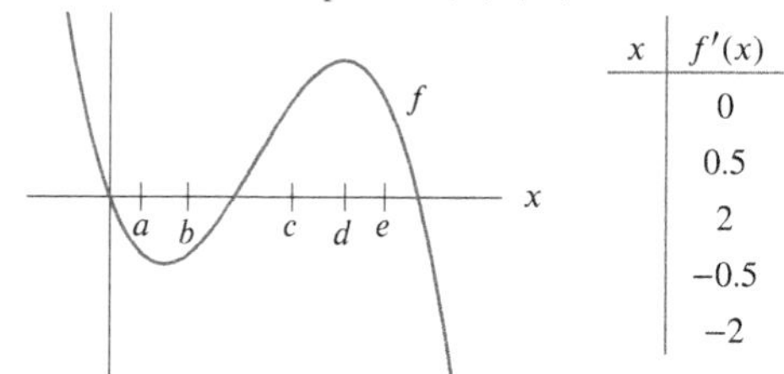

x	$f'(x)$
	0
	0.5
	2
	−0.5
	−2

Figure 2.23

13. Label points A, B, C, D, E, and F on the graph of $y = f(x)$ in Figure 2.24.

(a) Point A is a point on the curve where the derivative is negative.
(b) Point B is a point on the curve where the value of the function is negative.
(c) Point C is a point on the curve where the derivative is largest.
(d) Point D is a point on the curve where the derivative is zero.
(e) Points E and F are different points on the curve where the derivative is about the same.

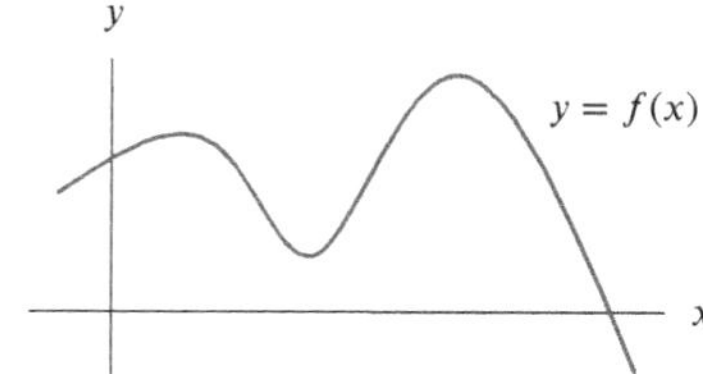

Figure 2.24

In Exercises 14–18, interpret the expression in terms of Arctic Sea ice extent, the area of sea covered by ice.[3] Let $E(x)$ and $F(t)$ be the Arctic Sea ice extent, both in millions of square kilometers, as a function of time, x, in years, since February 1979, and time, t, in days, since December 31, 2014.

14. $E(29) = 15$

15. $E(4) = 16$

16. $\dfrac{E(29) - E(4)}{29 - 4} = -0.04$

17. $F(32) = 13.97$

18. $\dfrac{F(59) - F(32)}{59 - 32} = 0.0159$

[3] Sea ice extent definition and data values from nsidc.org/cryosphere/seaice/data/terminology.html. Accessed March 2015.

PROBLEMS

19. Suppose that $f(x)$ is a function with $f(100) = 35$ and $f'(100) = 3$. Estimate $f(102)$.

20. Show how to represent the following on Figure 2.25.

    (a) $f(4)$ (b) $f(4) - f(2)$

    (c) $\dfrac{f(5) - f(2)}{5 - 2}$ (d) $f'(3)$

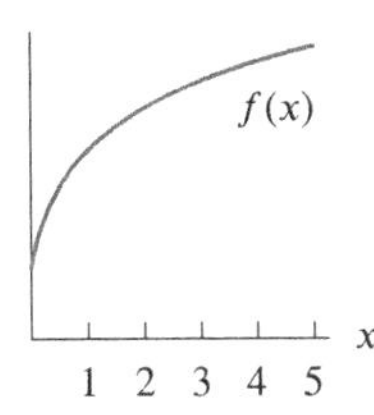

Figure 2.25

21. For each of the following pairs of numbers, use Figure 2.25 to decide which is larger. Explain your answer.

    (a) $f(3)$ or $f(4)$?
    (b) $f(3) - f(2)$ or $f(2) - f(1)$?
    (c) $\dfrac{f(2) - f(1)}{2 - 1}$ or $\dfrac{f(3) - f(1)}{3 - 1}$?
    (d) $f'(1)$ or $f'(4)$?

22. With the function f given by Figure 2.25, arrange the following quantities in ascending order:

$$0, \quad f'(2), \quad f'(3), \quad f(3) - f(2)$$

23. The function in Figure 2.26 has $f(4) = 25$ and $f'(4) = 1.5$. Find the coordinates of the points A, B, C.

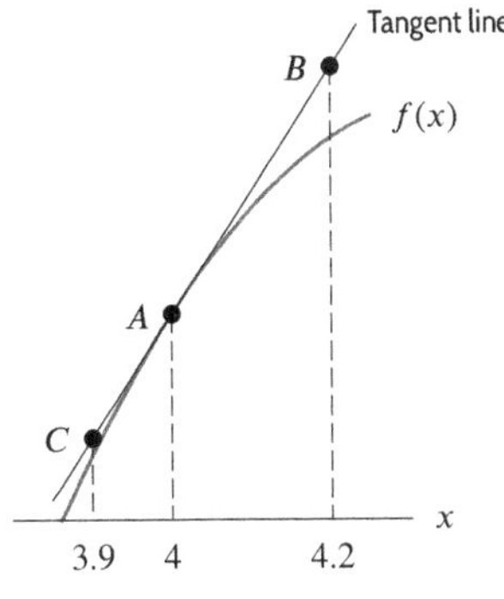

Figure 2.26

24. Use Figure 2.27 to fill in the blanks in the following statements about the function g at point B.

    (a) $g(\_\_) = \_\_$ (b) $g'(\_\_) = \_\_$

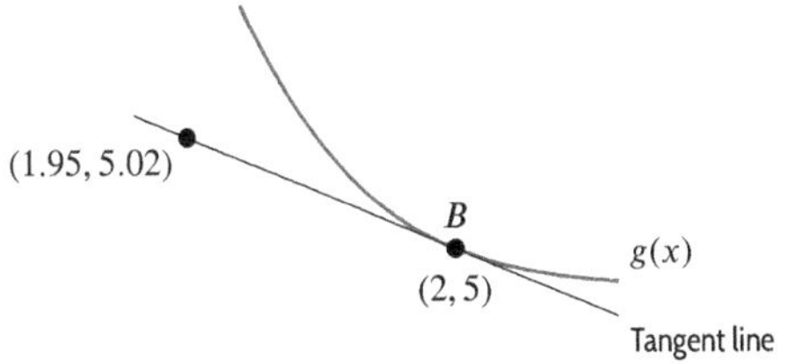

Figure 2.27

25. On a copy of Figure 2.28, mark lengths that represent the quantities in parts (a)–(d). (Pick any positive x and h.)

    (a) $f(x)$ (b) $f(x + h)$

    (c) $f(x + h) - f(x)$ (d) h

    (e) Using your answers to parts (a)–(d), show how the quantity $\dfrac{f(x + h) - f(x)}{h}$ can be represented as the slope of a line in Figure 2.28.

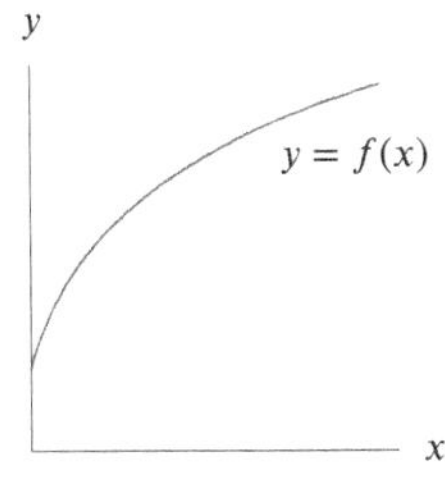

Figure 2.28

26. On a copy of Figure 2.29, mark lengths that represent the quantities in parts (a)–(d). (Pick any convenient x, and assume $h > 0$.)

    (a) $f(x)$ (b) $f(x+h)$ (c) $f(x+h) - f(x)$ (d) h

    (e) Using your answers to parts (a)–(d), show how the quantity $\dfrac{f(x + h) - f(x)}{h}$ can be represented as the slope of a line on the graph.

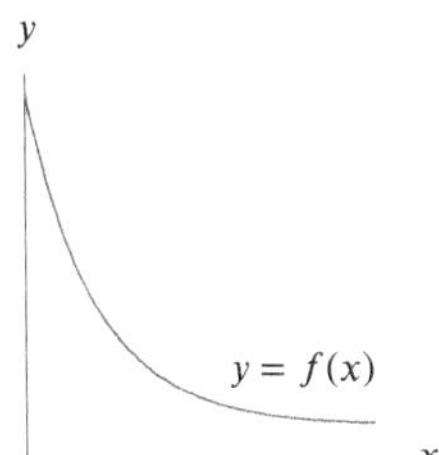

Figure 2.29

27. Consider the function shown in Figure 2.30.

(a) Write an expression involving f for the slope of the line joining A and B.
(b) Draw the tangent line at C. Compare its slope to the slope of the line in part (a).
(c) Are there any other points on the curve at which the slope of the tangent line is the same as the slope of the tangent line at C? If so, mark them on the graph. If not, why not?

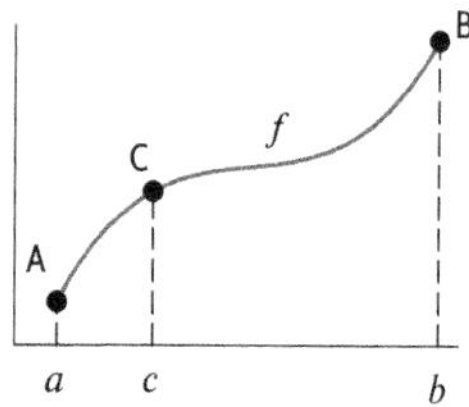

Figure 2.30

28. (a) If f is even and $f'(10) = 6$, what is $f'(-10)$?
(b) If f is any even function and $f'(0)$ exists, what is $f'(0)$?

29. If g is an odd function and $g'(4) = 5$, what is $g'(-4)$?

30. (a) Estimate $f'(0)$ if $f(x) = \sin x$, with x in degrees.
(b) In Example 3 on page 93, we found that the derivative of $\sin x$ at $x = 0$ was 1. Why do we get a different result here? (This problem shows why radians are almost always used in calculus.)

31. Find the equation of the tangent line to $f(x) = x^2 + x$ at $x = 3$. Sketch the function and this tangent line.

32. Estimate the instantaneous rate of change of the function $f(x) = x \ln x$ at $x = 1$ and at $x = 2$. What do these values suggest about the concavity of the graph between 1 and 2?

33. Estimate the derivative of $f(x) = x^x$ at $x = 2$.

34. For $y = f(x) = 3x^{3/2} - x$, use your calculator to construct a graph of $y = f(x)$, for $0 \le x \le 2$. From your graph, estimate $f'(0)$ and $f'(1)$.

35. Let $f(x) = \ln(\cos x)$. Use your calculator to approximate the instantaneous rate of change of f at the point $x = 1$. Do the same thing for $x = \pi/4$. (Note: Be sure that your calculator is set in radians.)

36. The population, $P(t)$, of China, in billions, can be approximated by[4]

$$P(t) = 1.394(1.006)^t,$$

where t is the number of years since the start of 2014. According to this model, how fast was the population growing at the start of 2014 and at the start of 2015? Give your answers in millions of people per year.

37. The US population[5] officially reached 300 million on October 17, 2006 and was gaining 1 person each 11 seconds. If $f(t)$ is the US population in millions t years after October 17, 2006, find $f(0)$ and $f'(0)$.

For Problems 38–41, estimate the change in y for the given change in x.

38. $y = f(x)$, $f'(100) = 0.4$, x increases from 100 to 101

39. $y = f(x)$, $f'(12) = 30$, x increases from 12 to 12.2

40. $y = g(x)$, $g'(250) = -0.5$, x increases from 250 to 251.5

41. $y = p(x)$, $p'(400) = 2$, x decreases from 400 to 398

42. (a) Graph $f(x) = \frac{1}{2}x^2$ and $g(x) = f(x) + 3$ on the same set of axes. What can you say about the slopes of the tangent lines to the two graphs at the point $x = 0$? $x = 2$? Any point $x = x_0$?
(b) Explain why adding a constant value, C, to any function does not change the value of the slope of its graph at any point. [Hint: Let $g(x) = f(x) + C$, and calculate the difference quotients for f and g.]

43. Suppose Table 2.3 on page 94 is continued with smaller values of h. A particular calculator gives the results in Table 2.4. (Your calculator may give slightly different results.) Comment on the values of the difference quotient in Table 2.4. In particular, why is the last value of $(2^h - 1)/h$ zero? What do you expect the calculated value of $(2^h - 1)/h$ to be when $h = 10^{-20}$?

Table 2.4 *Questionable values of difference quotients of 2^x near $x = 0$*

h	Difference quotient: $(2^h - 1)/h$
10^{-4}	0.6931712
10^{-6}	0.693147
10^{-8}	0.6931
10^{-10}	0.69
10^{-12}	0

44. (a) Let $f(x) = x^2$. Explain what Table 2.5 tells us about $f'(1)$.
(b) Find $f'(1)$ exactly.
(c) If x changes by 0.1 near $x = 1$, what does $f'(1)$ tell us about how $f(x)$ changes? Illustrate your answer with a sketch.

Table 2.5

x	x^2	Difference in successive x^2 values
0.998	0.996004	
		0.001997
0.999	0.998001	
		0.001999
1.000	1.000000	
		0.002001
1.001	1.002001	
		0.002003
1.002	1.004004	

[4] www.worldometers.info, accessed April 1, 2015.
[5] www.today.com/id/15298443/ns/today/t/us-population-hits-million-mark/#.VsuG1hgVmV0, accessed February 2015.

■ Use algebra to evaluate the limits in Problems 45–50.

45. $\lim_{h\to 0} \frac{(-3+h)^2-9}{h}$

46. $\lim_{h\to 0} \frac{(2-h)^3-8}{h}$

47. $\lim_{h\to 0} \frac{1/(1+h)-1}{h}$

48. $\lim_{h\to 0} \frac{1/(1+h)^2-1}{h}$

49. $\lim_{h\to 0} \frac{\sqrt{4+h}-2}{h}$ [Hint: Multiply by $\sqrt{4+h}+2$ in numerator and denominator.]

50. $\lim_{h\to 0} \frac{1/\sqrt{4+h}-1/2}{h}$

■ For Problems 51–55, estimate the value of $f'(1)$ by substituting small values for h. Then use algebra to find $f'(1)$ exactly.

51. $f(x)=3x+1$

52. $f(x)=x^2+x+1$

53. $f(x)=\sqrt{x}$

54. $f(x)=\frac{1}{x+1}$

55. $f(x)=2x+x^{-1}$

■ Find the derivatives in Problems 56–61 algebraically.

56. $f(x) = 5x^2$ at $x = 10$

57. $f(x)=x^3$ at $x=-2$

58. $g(t)=t^2+t$ at $t=-1$

59. $f(x)=x^3+5$ at $x=1$

60. $g(x)=1/x$ at $x=2$

61. $g(z)=z^{-2}$, find $g'(2)$

■ For Problems 62–65, find the equation of the line tangent to the function at the given point.

62. $f(x) = 5x^2$ at $x = 10$

63. $f(x)=x^3$ at $x=-2$

64. $f(x)=x$ at $x=20$

65. $f(x)=1/x^2$ at $(1,1)$

Strengthen Your Understanding

■ In Problems 66–67, explain what is wrong with the statement.

66. For the function $f(x)=\log x$ we have $f'(0.5)<0$.

67. The derivative of a function $f(x)$ at $x=a$ is the tangent line to the graph of $f(x)$ at $x=a$.

■ In Problems 68–69, give an example of:

68. A continuous function which is always increasing and positive.

69. A linear function with derivative 2 at $x=0$.

■ Are the statements in Problems 70–72 true or false? Give an explanation for your answer.

70. You cannot be sure of the exact value of a derivative of a function at a point using only the information in a table of values of the function. The best you can do is find an approximation.

71. If you zoom in (with your calculator) on the graph of $y=f(x)$ in a small interval around $x=10$ and see a straight line, then the slope of that line equals the derivative $f'(10)$.

72. If $f(x)$ is concave up, then $f'(a)<(f(b)-f(a))/(b-a)$ for $a<b$.

73. Assume that f is an odd function and that $f'(2)=3$, then $f'(-2)=$

(a) 3 **(b)** −3
(c) 1/3 **(d)** −1/3

2.3 THE DERIVATIVE FUNCTION

In the previous section we looked at the derivative of a function at a fixed point. Now we consider what happens at a variety of points. The derivative generally takes on different values at different points and is itself a function.

First, remember that the derivative of a function at a point tells us the rate at which the value of the function is changing at that point. Geometrically, we can think of the derivative as the slope of the curve or of the tangent line at the point.

Example 1 Estimate the derivative of the function $f(x)$ graphed in Figure 2.31 at $x=-2,-1,0,1,2,3,4,5$.

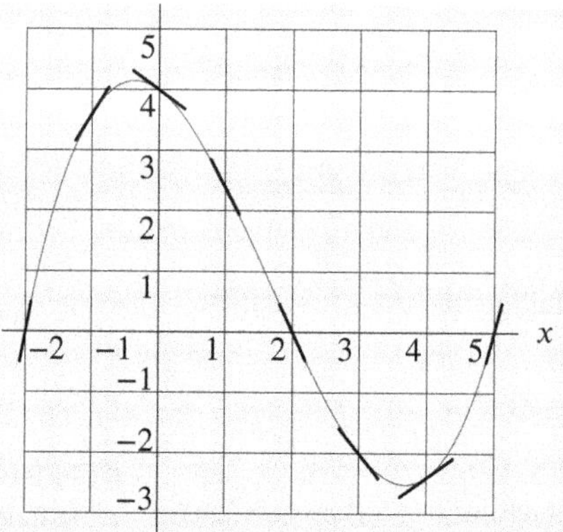

Figure 2.31: Estimating the derivative graphically as the slope of the tangent line

Solution From the graph we estimate the derivative at any point by placing a straightedge so that it forms the tangent line at that point, and then using the grid squares to estimate the slope of the straightedge. For example, the tangent at $x = -1$ is drawn in Figure 2.31, and has a slope of about 2, so $f'(-1) \approx 2$. Notice that the slope at $x = -2$ is positive and fairly large; the slope at $x = -1$ is positive but smaller. At $x = 0$, the slope is negative, by $x = 1$ it has become more negative, and so on. Some estimates of the derivative are listed in Table 2.6. You should check these values. Are they reasonable? Is the derivative positive where you expect? Negative?

Table 2.6 *Estimated values of derivative of function in Figure 2.31*

x	−2	−1	0	1	2	3	4	5
$f'(x)$	6	2	−1	−2	−2	−1	1	4

Notice that for every x-value, there's a corresponding value of the derivative. Therefore, the derivative is itself a function of x.

For any function f, we define the **derivative function**, f', by

$$f'(x) = \text{Rate of change of } f \text{ at } x = \lim_{h\to 0} \frac{f(x+h) - f(x)}{h}.$$

For every x-value for which this limit exists, we say f is *differentiable at* that x-value. If the limit exists for all x in the domain of f, we say f is *differentiable everywhere*. Most functions we meet are differentiable at every point in their domain, except perhaps for a few isolated points.

The Derivative Function: Graphically

Example 2 Sketch the graph of the derivative of the function shown in Figure 2.31.

Solution We plot the values of the derivative in Table 2.6 and connect them with a smooth curve to obtain the estimate of the derivative function in Figure 2.32. Values of the derivative function give slopes of the original graph.

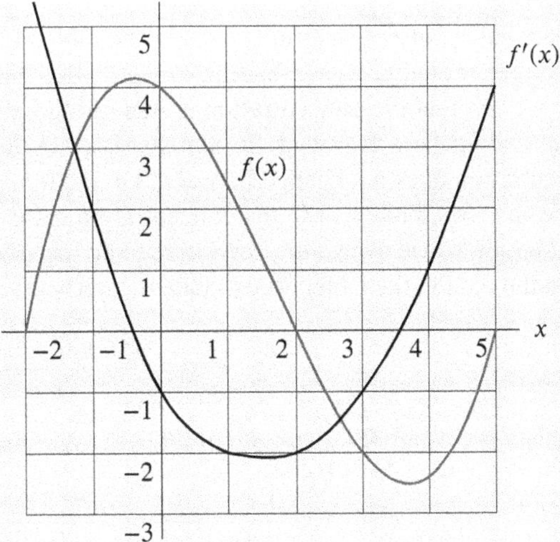

Figure 2.32: Function (colored) and derivative (black) from Example 1

Check that this graph of f' makes sense: Where the values of f' are positive, f is increasing ($x < -0.3$ or $x > 3.8$) and where f' is negative, f is decreasing. Notice that at the points where f has large positive slope, such as $x = -2$, the graph of the derivative is far above the x-axis, as it should be, since the value of the derivative is large there. At points where the slope is gentler, such as $x = -1$, the graph of f' is closer to the x-axis, since the derivative is smaller.

What Does the Derivative Tell Us Graphically?

Where f' is positive, the tangent line to f is sloping up; where f' is negative, the tangent line to f is sloping down. If $f' = 0$ everywhere, then the tangent line to f is horizontal everywhere, and f is constant. We see that the sign of f' tells us whether f is increasing or decreasing.

> If $f' > 0$ on an interval, then f is *increasing* over that interval.
> If $f' < 0$ on an interval, then f is *decreasing* over that interval.

Moreover, the magnitude of the derivative gives us the magnitude of the rate of change; so if f' is large (positive or negative), then the graph of f is steep (up or down), whereas if f' is small the graph of f slopes gently. With this in mind, we can learn about the behavior of a function from the behavior of its derivative.

The Derivative Function: Numerically

If we are given values of a function instead of its graph, we can estimate values of the derivative.

Example 3 Table 2.7 gives values of $c(t)$, the concentration (μg/cm^3) of a drug in the bloodstream at time t (min). Construct a table of estimated values for $c'(t)$, the rate of change of $c(t)$ with respect to time.

Table 2.7 *Concentration as a function of time*

t (min)	0	0.1	0.2	0.3	0.4	0.5	0.6	0.7	0.8	0.9	1.0
$c(t)$ (μg/cm^3)	0.84	0.89	0.94	0.98	1.00	1.00	0.97	0.90	0.79	0.63	0.41

Solution We estimate values of c' using the values in the table. To do this, we have to assume that the data points are close enough together that the concentration does not change wildly between them. From the table, we see that the concentration is increasing between $t = 0$ and $t = 0.4$, so we expect a positive derivative there. However, the increase is quite slow, so we expect the derivative to be small. The concentration does not change between 0.4 and 0.5, so we expect the derivative to be roughly 0 there. From $t = 0.5$ to $t = 1.0$, the concentration starts to decrease, and the rate of decrease gets larger and larger, so we expect the derivative to be negative and of greater and greater magnitude.

Using the data in the table, we estimate the derivative using the difference quotient:

$$c'(t) \approx \frac{c(t+h) - c(t)}{h}.$$

Since the data points are 0.1 apart, we use $h = 0.1$, giving, for example,

$$c'(0) \approx \frac{c(0.1) - c(0)}{0.1} = \frac{0.89 - 0.84}{0.1} = 0.5\ \mu\text{g/cm}^3\text{/min}.$$

$$c'(0.1) \approx \frac{c(0.2) - c(0.1)}{0.1} = \frac{0.94 - 0.89}{0.1} = 0.5\ \mu\text{g/cm}^3\text{/min}.$$

See Table 2.8. Notice that the derivative has small positive values until $t = 0.4$, where it is roughly 0, and then it gets more and more negative, as we expected. The slopes are graphed in Figure 2.33.

Table 2.8 *Estimated derivative of concentration*

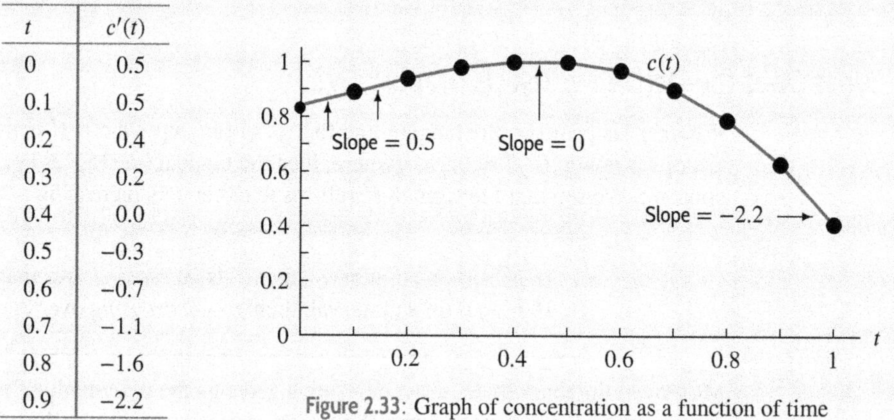

t	$c'(t)$
0	0.5
0.1	0.5
0.2	0.4
0.3	0.2
0.4	0.0
0.5	−0.3
0.6	−0.7
0.7	−1.1
0.8	−1.6
0.9	−2.2

Figure 2.33: Graph of concentration as a function of time

Improving Numerical Estimates for the Derivative

In the previous example, the estimate for the derivative at 0.2 used the interval to the right; we found the average rate of change between $t = 0.2$ and $t = 0.3$. However, we could equally well have gone to the left and used the rate of change between $t = 0.1$ and $t = 0.2$ to approximate the derivative at 0.2. For a more accurate result, we could average these slopes and say

$$c'(0.2) \approx \frac{1}{2}\left(\begin{array}{c}\text{Slope to left}\\ \text{of } 0.2\end{array} + \begin{array}{c}\text{Slope to right}\\ \text{of } 0.2\end{array}\right) = \frac{0.5 + 0.4}{2} = 0.45.$$

In general, averaging the slopes leads to a more accurate answer.

Derivative Function: From a Formula

If we are given a formula for f, can we come up with a formula for f'? We often can, as shown in the next example. Indeed, much of the power of calculus depends on our ability to find formulas for the derivatives of all the functions we described earlier. This is done systematically in Chapter 3.

Derivative of a Constant Function

The graph of a constant function $f(x) = k$ is a horizontal line, with a slope of 0 everywhere. Therefore, its derivative is 0 everywhere. (See Figure 2.34.)

> If $f(x) = k$, then $f'(x) = 0$.

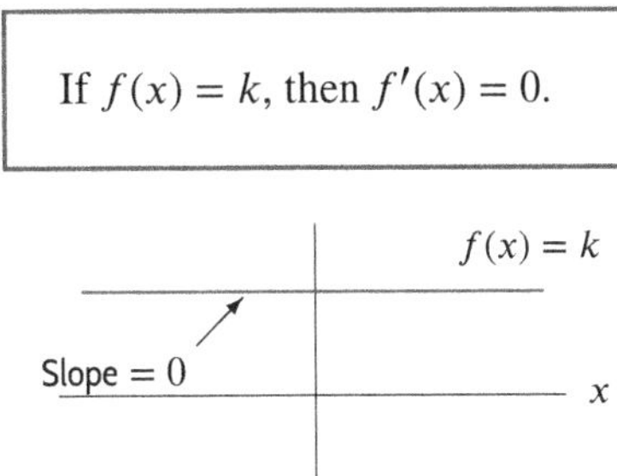

Figure 2.34: A constant function

Derivative of a Linear Function

We already know that the slope of a straight line is constant. This tells us that the derivative of a linear function is constant.

> If $f(x) = b + mx$, then $f'(x) = \text{Slope} = m$.

Derivative of a Power Function

Example 4 Find a formula for the derivative of $f(x) = x^2$.

Solution Before computing the formula for $f'(x)$ algebraically, let's try to guess the formula by looking for a pattern in the values of $f'(x)$. Table 2.9 contains values of $f(x) = x^2$ (rounded to three decimals), which we can use to estimate the values of $f'(1)$, $f'(2)$, and $f'(3)$.

Table 2.9 *Values of $f(x) = x^2$ near $x = 1$, $x = 2$, $x = 3$ (rounded to three decimals)*

x	x^2	x	x^2	x	x^2
0.999	0.998	1.999	3.996	2.999	8.994
1.000	1.000	2.000	4.000	3.000	9.000
1.001	1.002	2.001	4.004	3.001	9.006
1.002	1.004	2.002	4.008	3.002	9.012

Near $x = 1$, the value of x^2 increases by about 0.002 each time x increases by 0.001, so

$$f'(1) \approx \frac{0.002}{0.001} = 2.$$

Similarly, near $x = 2$ and $x = 3$, the value of x^2 increases by about 0.004 and 0.006, respectively, when x increases by 0.001. So

$$f'(2) \approx \frac{0.004}{0.001} = 4 \quad \text{and} \quad f'(3) \approx \frac{0.006}{0.001} = 6.$$

Knowing the value of f' at specific points can never tell us the formula for f', but it certainly can be suggestive: Knowing $f'(1) \approx 2$, $f'(2) \approx 4$, $f'(3) \approx 6$ suggests that $f'(x) = 2x$.

The derivative is calculated by forming the difference quotient and taking the limit as h goes to zero. The difference quotient is

$$\frac{f(x+h) - f(x)}{h} = \frac{(x+h)^2 - x^2}{h} = \frac{x^2 + 2xh + h^2 - x^2}{h} = \frac{2xh + h^2}{h}.$$

Since h never actually reaches zero, we can cancel it in the last expression to get $2x + h$. The limit of this as h goes to zero is $2x$, so

$$f'(x) = \lim_{h \to 0}(2x + h) = 2x.$$

Example 5 Calculate $f'(x)$ if $f(x) = x^3$.

Solution We look at the difference quotient

$$\frac{f(x+h) - f(x)}{h} = \frac{(x+h)^3 - x^3}{h}.$$

Multiplying out gives $(x+h)^3 = x^3 + 3x^2h + 3xh^2 + h^3$, so

$$f'(x) = \lim_{h \to 0} \frac{x^3 + 3x^2h + 3xh^2 + h^3 - x^3}{h} = \lim_{h \to 0} \frac{3x^2h + 3xh^2 + h^3}{h}.$$

Since in taking the limit as $h \to 0$, we consider values of h near, but not equal to, zero, we can cancel h, giving

$$f'(x) = \lim_{h \to 0} \frac{3x^2h + 3xh^2 + h^3}{h} = \lim_{h \to 0}(3x^2 + 3xh + h^2).$$

As $h \to 0$, the value of $(3xh + h^2) \to 0$, so

$$f'(x) = \lim_{h \to 0}(3x^2 + 3xh + h^2) = 3x^2.$$

The previous two examples show how to compute the derivatives of power functions of the form $f(x) = x^n$, when n is 2 or 3. We can use the Binomial Theorem to show the *power rule* for a positive integer n:

$$\text{If } f(x) = x^n \text{ then } f'(x) = nx^{n-1}.$$

This result is in fact valid for any real value of n.

Exercises and Problems for Section 2.3 Online Resource: Additional Problems for Section 2.3

EXERCISES

1. **(a)** Estimate $f'(2)$ using the values of f in the table.
   (b) For what values of x does $f'(x)$ appear to be positive? Negative?

x	0	2	4	6	8	10	12
$f(x)$	10	18	24	21	20	18	15

2. Find approximate values for $f'(x)$ at each of the x-values given in the following table.

x	0	5	10	15	20
$f(x)$	100	70	55	46	40

3. Values of $f(x)$ are in the table. Where in the interval $-12 \le x \le 9$ does $f'(x)$ appear to be the greatest? Least?

x	−12	−9	−6	−3	0	3	6	9
$f(x)$	1.02	1.05	1.12	1.14	1.15	1.14	1.12	1.06

■ For Exercises 4–13, graph the derivative of the given functions.

4.
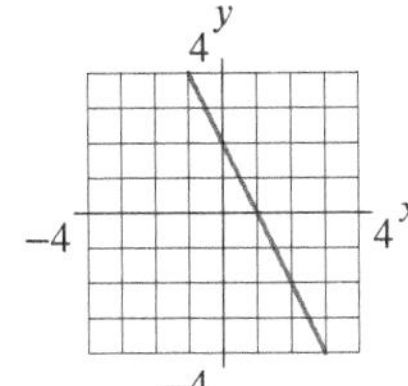

5.
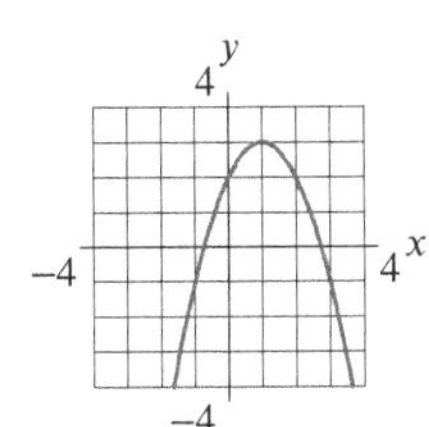

6.
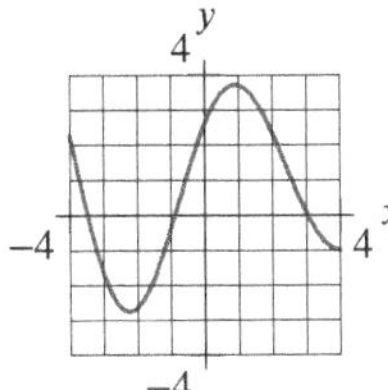

7.
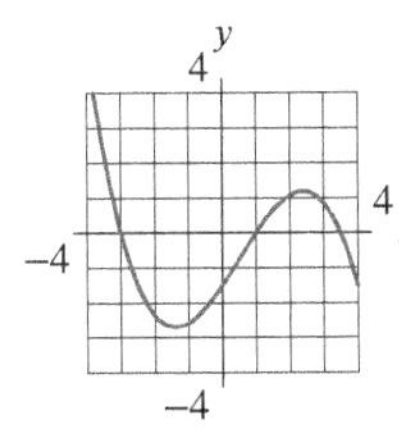

8.
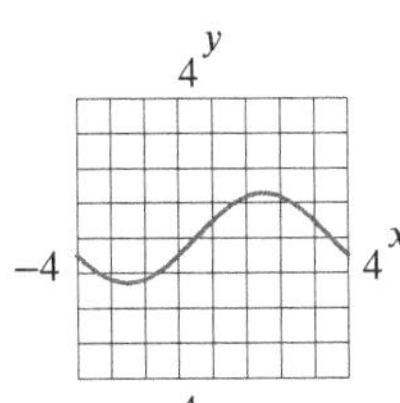

9.
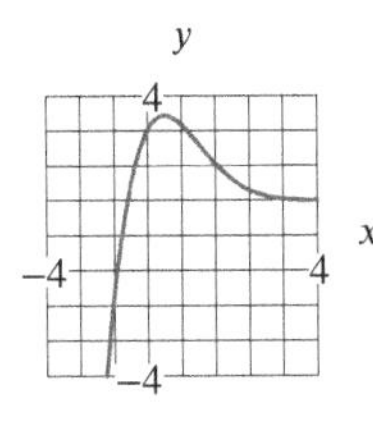

10.
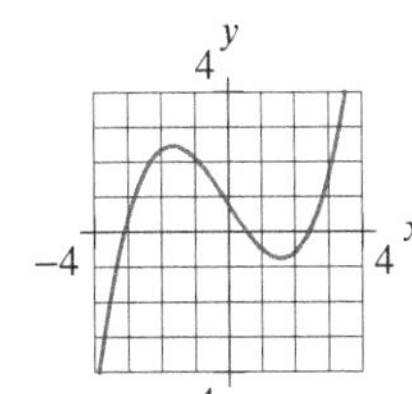

11.
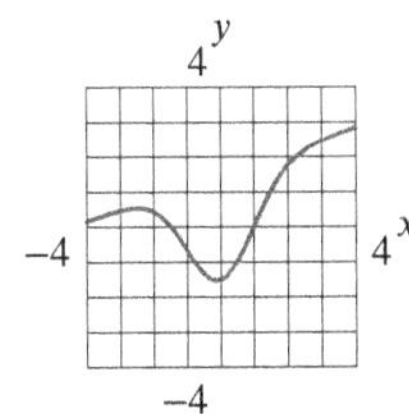

12.
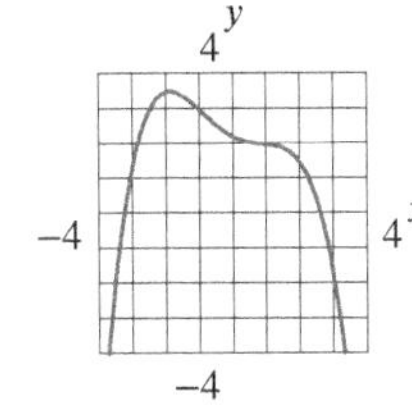

13.
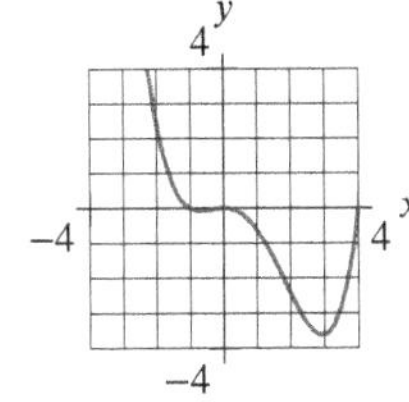

■ For Exercises 14–19, sketch the graph of $f(x)$, and use this graph to sketch the graph of $f'(x)$.

14. $f(x) = 5x$
15. $f(x) = x^2$
16. $f(x) = x(x - 1)$
17. $f(x) = e^x$
18. $f(x) = \cos x$
19. $f(x) = \ln x$

■ In Exercises 20–21, find a formula for the derivative using the power rule. Confirm it using difference quotients.

20. $k(x) = 1/x$
21. $l(x) = 1/x^2$

■ In Exercises 22–27, find a formula for the derivative of the function using the difference quotient.

22. $g(x) = 2x^2 - 3$
23. $m(x) = 1/(x + 1)$
24. $g(x) = 4x - 5$
25. $g(x) = x^2 + 2x + 1$
26. $g(x) = x^3 + 1$
27. $g(x) = 1/\sqrt{x}$

PROBLEMS

28. In each case, graph a smooth curve whose slope meets the condition.
   (a) Everywhere positive and increasing gradually.
   (b) Everywhere positive and decreasing gradually.
   (c) Everywhere negative and increasing gradually (becoming less negative).
   (d) Everywhere negative and decreasing gradually (becoming more negative).

29. Draw a possible graph of $y = f(x)$ given the following information about its derivative.
   - $f'(x) > 0$ for $x < -1$
   - $f'(x) < 0$ for $x > -1$
   - $f'(x) = 0$ at $x = -1$

30. For $f(x) = \ln x$, construct tables, rounded to four decimals, near $x = 1$, $x = 2$, $x = 5$, and $x = 10$. Use the tables to estimate $f'(1)$, $f'(2)$, $f'(5)$, and $f'(10)$. Then guess a general formula for $f'(x)$.

31. Given the numerical values shown, find approximate values for the derivative of $f(x)$ at each of the x-values given. Where is the rate of change of $f(x)$ positive? Where is it negative? Where does the rate of change of $f(x)$ seem to be greatest?

x	0	1	2	3	4	5	6	7	8
$f(x)$	18	13	10	9	9	11	15	21	30

32. Values of x and $g(x)$ are given in the table. For what value of x does $g'(x)$ appear to be closest to 3?

x	2.7	3.2	3.7	4.2	4.7	5.2	5.7	6.2
$g(x)$	3.4	4.4	5.0	5.4	6.0	7.4	9.0	11.0

33. In the graph of f in Figure 2.35, at which of the labeled x-values is

(a) $f(x)$ greatest? **(b)** $f(x)$ least?
(c) $f'(x)$ greatest? **(d)** $f'(x)$ least?

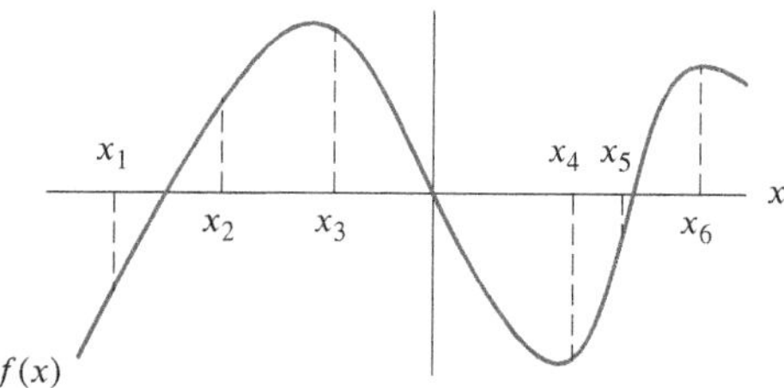

Figure 2.35

■ For Problems 34–43, sketch the graph of $f'(x)$.

34.

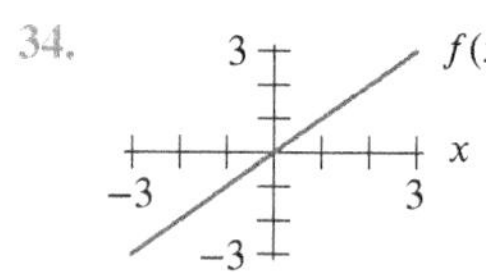

35.

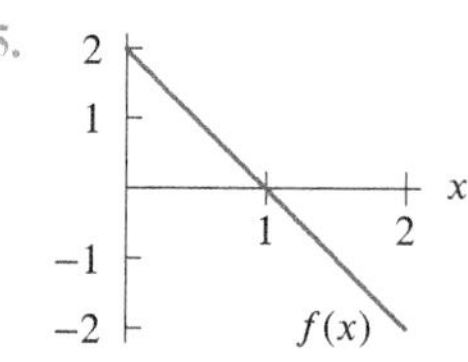

36.

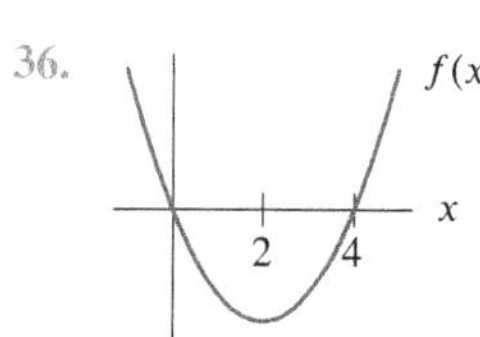

37.

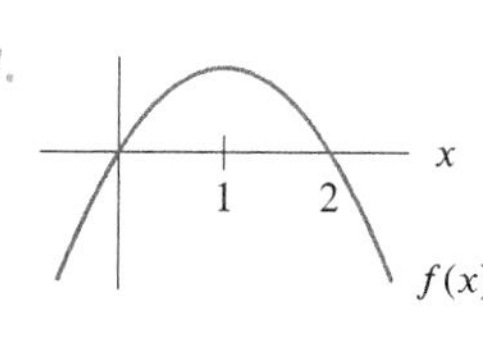

38.

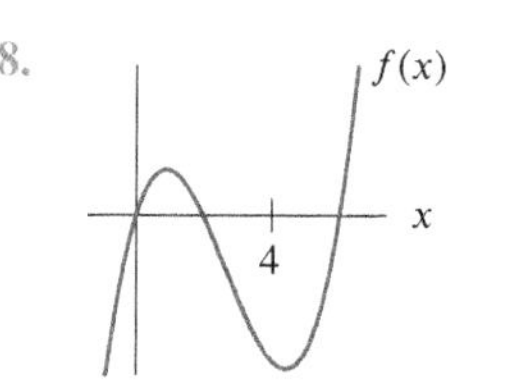

39.

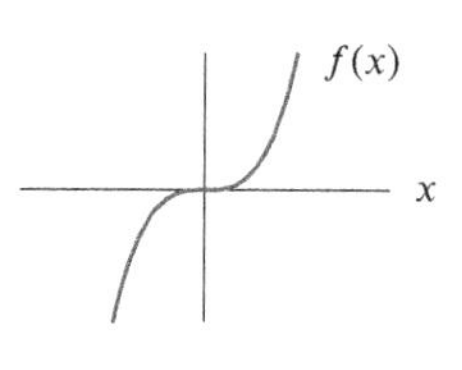

40.

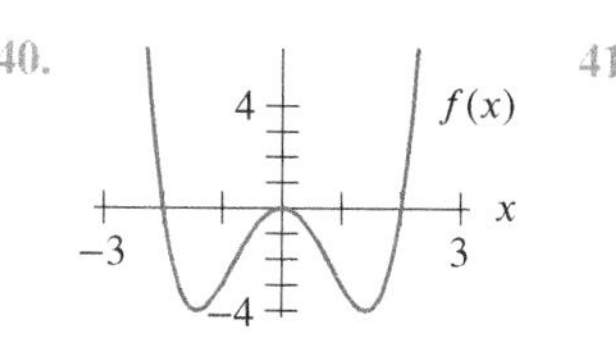

41.

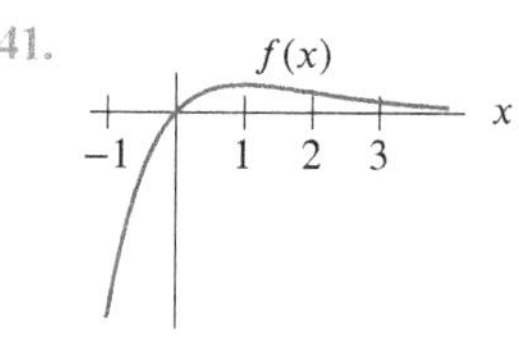

42.

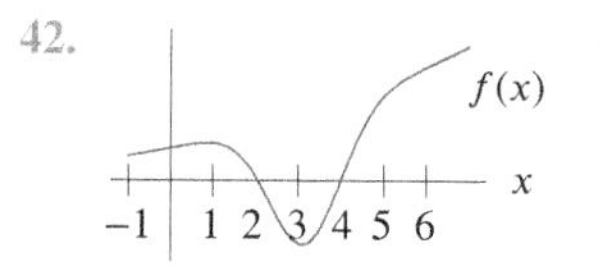

43.

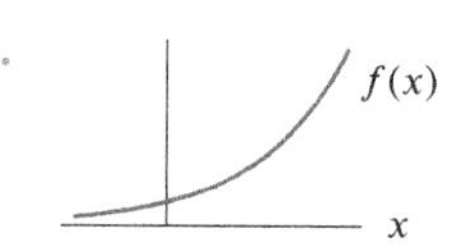

■ In Problems 44–47, match f' with the corresponding f in Figure 2.36.

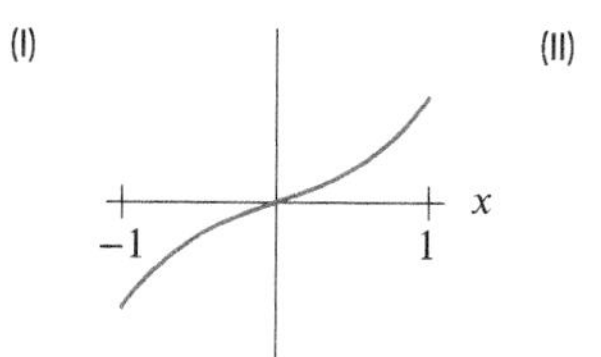

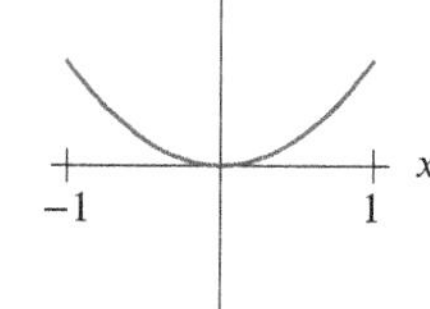

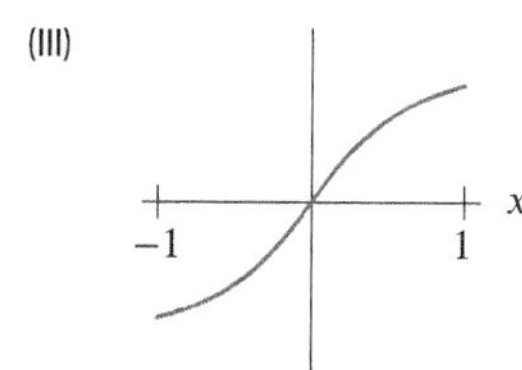

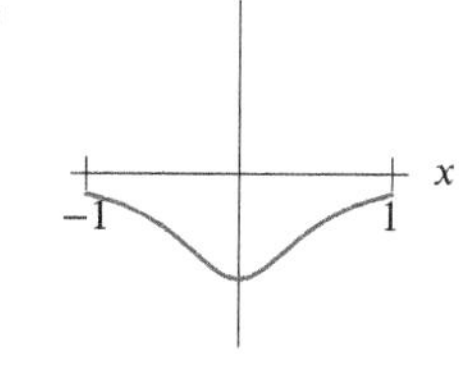

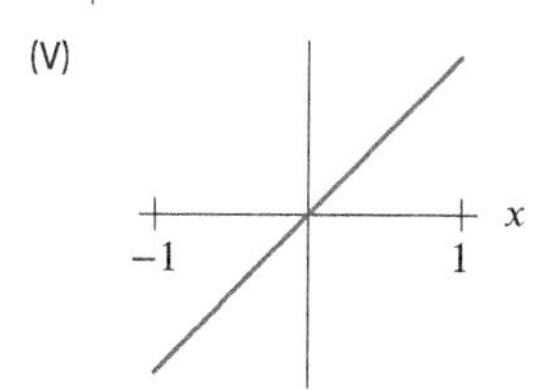

Figure 2.36

44.

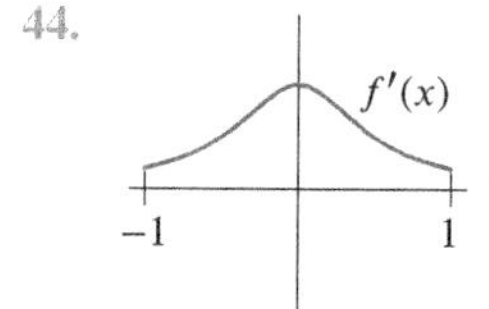

45.

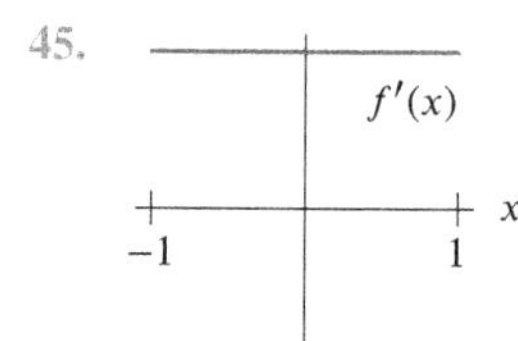

46.

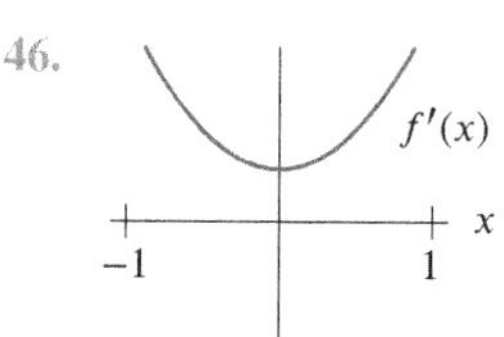

47.

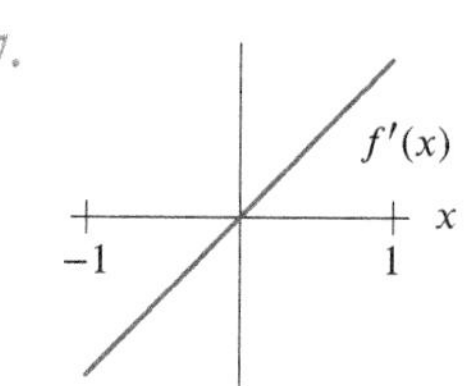

48. Roughly sketch the shape of the graph of a quadratic polynomial, f, if it is known that:

- $(1, 3)$ is on the graph of f.
- $f'(0) = 3$, $f'(2) = 1$, $f'(3) = 0$.

49. A vehicle moving along a straight road has distance $f(t)$ from its starting point at time t. Which of the graphs in Figure 2.37 could be $f'(t)$ for the following scenarios? (Assume the scales on the vertical axes are all the same.)

(a) A bus on a popular route, with no traffic
(b) A car with no traffic and all green lights
(c) A car in heavy traffic conditions

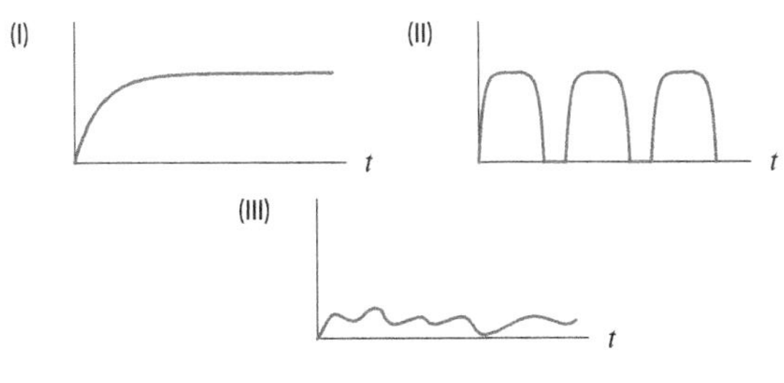

Figure 2.37

50. A child inflates a balloon, admires it for a while and then lets the air out at a constant rate. If $V(t)$ gives the volume of the balloon at time t, then Figure 2.38 shows $V'(t)$ as a function of t. At what time does the child:

(a) Begin to inflate the balloon?
(b) Finish inflating the balloon?
(c) Begin to let the air out?
(d) What would the graph of $V'(t)$ look like if the child had alternated between pinching and releasing the open end of the balloon, instead of letting the air out at a constant rate?

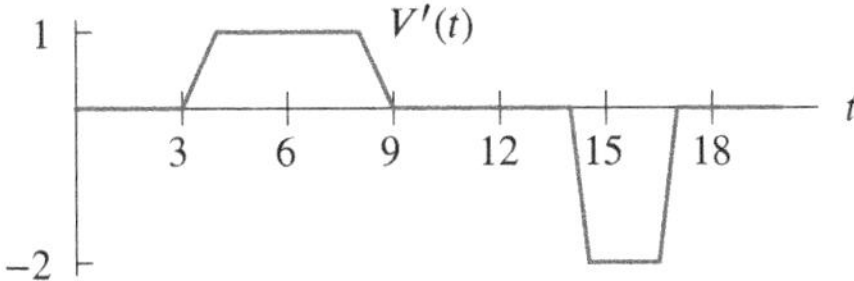

Figure 2.38

51. Figure 2.39 shows a graph of voltage across an electrical capacitor as a function of time. The current is proportional to the derivative of the voltage; the constant of proportionality is positive. Sketch a graph of the current as a function of time.

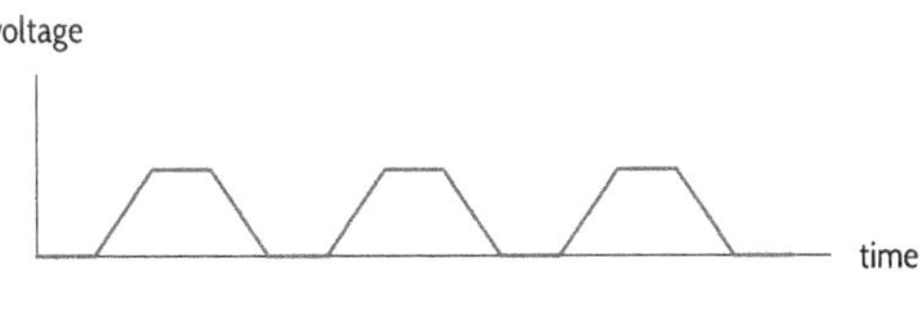

Figure 2.39

52. Figure 2.40 is the graph of f', the derivative of a function f. On what interval(s) is the function f

(a) Increasing? **(b)** Decreasing?

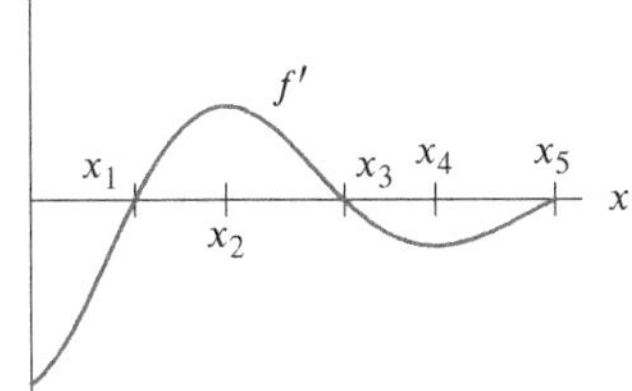

Figure 2.40: Graph of f', not f

53. The derivative of f is the spike function in Figure 2.41. What can you say about the graph of f?

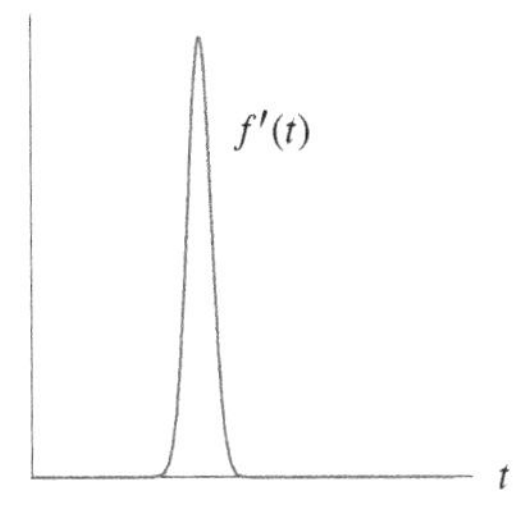

Figure 2.41

54. The population of a herd of deer is modeled by

$$P(t) = 4000 + 500 \sin\left(2\pi t - \frac{\pi}{2}\right)$$

where t is measured in years from January 1.

(a) How does this population vary with time? Sketch a graph of $P(t)$ for one year.
(b) Use the graph to decide when in the year the population is a maximum. What is that maximum? Is there a minimum? If so, when?
(c) Use the graph to decide when the population is growing fastest. When is it decreasing fastest?
(d) Estimate roughly how fast the population is changing on the first of July.

55. The graph in Figure 2.42 shows the accumulated federal debt since 1970. Sketch the derivative of this function. What does it represent?

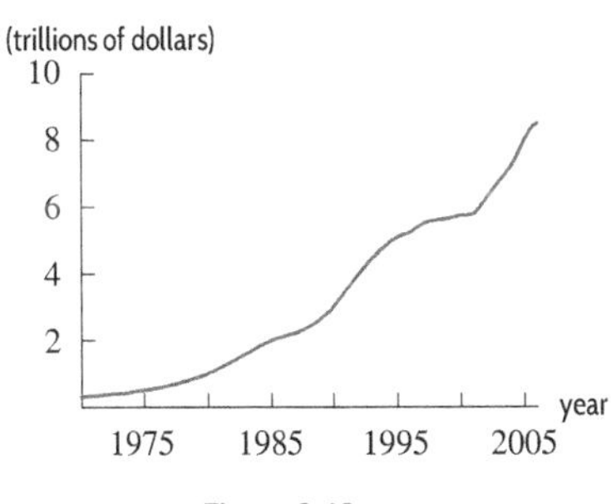

Figure 2.42

56. Draw the graph of a continuous function $y = f(x)$ that satisfies the following three conditions:
    - $f'(x) > 0$ for $x < -2$,
    - $f'(x) < 0$ for $-2 < x < 2$,
    - $f'(x) = 0$ for $x > 2$.
57. Draw a possible graph of a continuous function $y = f(x)$ that satisfies the following three conditions:
    - $f'(x) > 0$ for $1 < x < 3$
    - $f'(x) < 0$ for $x < 1$ and $x > 3$
    - $f'(x) = 0$ at $x = 1$ and $x = 3$
58. If $\lim_{x\to\infty} f(x) = 50$ and $f'(x)$ is positive for all x, what is $\lim_{x\to\infty} f'(x)$? (Assume this limit exists.) Explain your answer with a picture.
59. Using a graph, explain why if $f(x)$ is an even function, then $f'(x)$ is odd.
60. Using a graph, explain why if $g(x)$ is an odd function, then $g'(x)$ is even.

Strengthen Your Understanding

In Problems 61–63, explain what is wrong with the statement.

61. The graph of the derivative of the function $f(x) = \cos x$ is always above the x-axis.
62. A function, f, whose graph is above the x-axis for all x has a positive derivative for all x.
63. If $f'(x) = g'(x)$ then $f(x) = g(x)$.

In Problems 64–65, give an example of:

64. A function representing the position of a particle which has positive velocity for $0 < t < 0.5$ and negative velocity for $0.5 < t < 1$.
65. A family of linear functions all with the same derivative.

Are the statements in Problems 66–69 true or false? Give an explanation for your answer.

66. The derivative of a linear function is constant.
67. If $g(x)$ is a vertical shift of $f(x)$, then $f'(x) = g'(x)$.
68. If $f'(x)$ is increasing, then $f(x)$ is also increasing.
69. If $f(a) \neq g(a)$, then $f'(a) \neq g'(a)$.

2.4 INTERPRETATIONS OF THE DERIVATIVE

We have seen the derivative interpreted as a slope and as a rate of change. In this section, we illustrate the process of obtaining other interpretations.

An Alternative Notation for the Derivative

So far we have used the notation f' to stand for the derivative of the function f. An alternative notation for derivatives was introduced by the German mathematician Gottfried Wilhelm Leibniz (1646–1716). If the variable y depends on the variable x, that is, if

$$y = f(x),$$

then he wrote dy/dx for the derivative, so

$$\frac{dy}{dx} = f'(x).$$

Leibniz's notation is quite suggestive if we think of the letter d in dy/dx as standing for "small difference in … ." For small changes Δx and Δy in x and y, the notation dy/dx reminds us that the derivative is a limit of ratios of the form

$$\frac{\text{Difference in } y\text{-values}}{\text{Difference in } x\text{-values}} = \frac{\Delta y}{\Delta x}.$$

Although not formally correct, it can be helpful to think of dy/dx as a small change in y divided by a small change in x. That is, when Δx is small,

$$\frac{dy}{dx} \approx \frac{\Delta y}{\Delta x}.$$

Formally, the separate entities dy and dx have no independent meaning: they are all part of one notation. On the other hand, many scientists and mathematicians think of dy and dx as separate entities representing "infinitesimally" small differences in y and x, even though it is difficult to say exactly how small "infinitesimal" is.

Another way to view the notation dy/dx is to think of d/dx as a single symbol meaning "the derivative with respect to x of ...." So dy/dx can be viewed as

$$\frac{d}{dx}(y), \quad \text{meaning "the derivative with respect to } x \text{ of } y."$$

Using Units to Interpret the Derivative

The notation dy/dx makes it easy to see the units for the derivative: the units for y divided by the units for x. The units often suggest the interpretation of the derivative.

For example, suppose $s = f(t)$ gives the distance, in meters, of a body from a fixed point as a function of time, t, in seconds. Suppose we know that $f'(2) = 10$. We can write this same fact in Leibniz notation as

$$\left.\frac{ds}{dt}\right|_{t=2} = 10.$$

The Leibniz notation reminds us that the units of this derivative are meters per second. Then

$$\left.\frac{ds}{dt}\right|_{t=2} = f'(2) = 10 \text{ meters/sec}$$

tells us that when $t = 2$ seconds, the body is moving at an instantaneous velocity of 10 meters/sec. This means that if the body continued to move at this speed for one more second, it would move 10 more meters. Thus, for a time increment Δt, the distance moved would be given by

$$\Delta s \approx f'(2) \cdot \Delta t \qquad \text{that is,} \qquad \Delta s \approx 10 \cdot \Delta t.$$

In practice, however, the velocity of the body may not remain 10 meters/sec for long, so the approximation is only reliable for small Δt.

Example 1 The cost of extracting T tons of ore from a copper mine is $C = f(T)$ dollars. What does it mean to say that $f'(2000) = 100$?

Solution In the Leibniz notation,

$$f'(2000) = \left.\frac{dC}{dT}\right|_{T=2000} = 100.$$

Since C is measured in dollars and T is measured in tons, dC/dT is measured in dollars per ton. For small ΔT, we have

$$\Delta C \approx 100\Delta T.$$

So the statement $f'(2000) = 100$ says that when 2000 tons of ore have already been extracted from the mine, the additional cost ΔC of extracting the next ton ($\Delta T = 1$) is approximately \$100.

As another example, if 5 tons is a small increase, we use this relationship to approximate the additional cost of increasing production by 5 tons above 2000:

$$\text{Additional cost } = \Delta C \approx 100\Delta T = 100 \cdot 5 = 500 \text{ dollars}.$$

We see that the derivative, 100, acts as a *multiplier*, converting Δt tons into ΔC in dollars.

Example 2 The cost C (in dollars) of building a house A square feet in area is given by the function $C = f(A)$. What is the practical interpretation of the function $f'(A)$?

Solution In the Leibniz notation,

$$f'(A) = \frac{dC}{dA}.$$

This is a cost divided by an area, so it is measured in dollars per square foot. You can think of dC as the extra cost of building an extra dA square feet of house. Then you can think of dC/dA as the additional cost per square foot. So if you are planning to build a house A square feet in area, $f'(A)$ is the cost per square foot of the *extra* area involved in building a slightly larger house, and is called the *marginal cost*. The marginal cost of adding one more square foot is probably smaller than the average cost per square foot for the entire house, since once you are already set up to build a large house, the cost of adding one square foot is likely to be small.

Example 3 If $q = f(p)$ gives the number of pounds of sugar produced by a manufacturer when the price per pound is p cents, then what are the units and the meaning of the statement $f'(30) = 50$?

Solution We have

$$f'(30) = \left.\frac{dq}{dp}\right|_{p=30} = 50,$$

so the units of $f'(30)$ are pounds per cent. This says that when the price is 30¢, the quantity produced is increasing at 50 pounds/cent. Thus, if the price increased by one cent from 30¢ to 31¢, the quantity produced would increase by approximately 50 pounds.

Example 4 Water is flowing through a pipe at a constant rate of 10 cubic feet per second. Interpret this rate as the derivative of some function.

Solution You might think at first that the statement has something to do with the velocity of the water, but in fact a flow rate of 10 cubic feet per second could be achieved either with very slowly moving water through a large pipe, or with very rapidly moving water through a narrow pipe. If we look at the units—cubic feet per second—we realize that we are being given the rate of change of a quantity measured in cubic feet. But a cubic foot is a measure of volume, so we are being told the rate of change of a volume. One way to visualize this is to imagine all the water that is flowing through the pipe ending up in a tank somewhere. Let $V(t)$ be the volume of water in the tank at time t. Then we are being told that the rate of change of $V(t)$ is 10, or

$$V'(t) = \frac{dV}{dt} = 10.$$

Example 5 Let $N = g(t)$ be the estimated number of alternative-fueled vehicles[6] in use in the US, in thousands, where t is the number of years since 2008. Explain the meaning of the statements:

(a) $g'(3) = 253$ (b) $g^{-1}(1191) = 3$ (c) $(g^{-1})'(1191) = 0.004$

Solution

(a) The units of N are thousands of vehicles, the units of t are years, so the units of $g'(t)$ are thousand vehicles per year. Thus, the statement $g'(3) = 253$ tells us that in the year 2011, the use of alternative-fueled vehicles was increasing at 253,000 per year. Thus, in one more year after 2011 we would have expected the number of alternative-fueled vehicles in use in the US to increase by about 253,000 vehicles.

(b) The statement $g^{-1}(1191) = 3$, which is equivalent to $g(3) = 1191$, tells us that the year in which the number of alternative-fueled vehicles was 1,191,000 was 2011.

(c) The units of $(g^{-1})'(V)$ are years per thousand vehicles. The statement $(g^{-1})'(1191) = 0.004$ tells us that when the number of alternative-fueled vehicles was 1,191,000, it took about 0.004 years, or between 1 and 2 days, for the number of alternative-fueled vehicles to grow by a thousand vehicles.

[6]www.eia.gov, accessed April 2015.

Exercises and Problems for Section 2.4 Online Resource: Additional Problems for Section 2.4

EXERCISES

1. The cost, C (in dollars), to produce g gallons of a chemical can be expressed as $C = f(g)$. Using units, explain the meaning of the following statements in terms of the chemical:

   (a) $f(200) = 1300$ (b) $f'(200) = 6$

2. The time for a chemical reaction, T (in minutes), is a function of the amount of catalyst present, a (in milliliters), so $T = f(a)$.

   (a) If $f(5) = 18$, what are the units of 5? What are the units of 18? What does this statement tell us about the reaction?
   (b) If $f'(5) = -3$, what are the units of 5? What are the units of -3? What does this statement tell us?

3. The temperature, T, in degrees Fahrenheit, of a cold yam placed in a hot oven is given by $T = f(t)$, where t is the time in minutes since the yam was put in the oven.

   (a) What is the sign of $f'(t)$? Why?
   (b) What are the units of $f'(20)$? What is the practical meaning of the statement $f'(20) = 2$?

4. The temperature, H, in degrees Celsius, of a cup of coffee placed on the kitchen counter is given by $H = f(t)$, where t is in minutes since the coffee was put on the counter.

   (a) Is $f'(t)$ positive or negative? Give a reason for your answer.
   (b) What are the units of $f'(20)$? What is its practical meaning in terms of the temperature of the coffee?

5. The cost, C (in dollars), to produce q quarts of ice cream is $C = f(q)$. In each of the following statements, what are the units of the two numbers? In words, what does each statement tell us?

   (a) $f(200) = 600$ (b) $f'(200) = 2$

6. An economist is interested in how the price of a certain item affects its sales. At a price of $\$p$, a quantity, q, of the item is sold. If $q = f(p)$, explain the meaning of each of the following statements:

   (a) $f(150) = 2000$ (b) $f'(150) = -25$

7. Let $S(t)$ be the amount of water, measured in acre-feet,[7] that is stored in a reservoir in week t.

   (a) What are the units of $S'(t)$?
   (b) What is the practical meaning of $S'(t) > 0$? What circumstances might cause this situation?

8. The wind speed, W, in meters per second, at a distance x km from the center of a hurricane is given by $W = h(x)$.

   (a) Give the the units of dW/dx.
   (b) For a certain hurricane, $h'(15) > 0$. What does this tell you about the hurricane?

9. Suppose $C(r)$ is the total cost of paying off a car loan borrowed at an annual interest rate of $r\%$. What are the units of $C'(r)$? What is the practical meaning of $C'(r)$? What is its sign?

10. Let $f(x)$ be the elevation in feet of the Mississippi River x miles from its source. What are the units of $f'(x)$? What can you say about the sign of $f'(x)$?

■ In Exercises 11–15, give the units and sign of the derivative.

11. $h'(t)$, where $h(t)$ is the altitude of a parachutist, in feet, t seconds after he jumps out of a plane.

12. $f'(t)$, where $f(t)$ is the temperature of a room, in degrees Celsius, t minutes after a heater is turned on.

13. $P'(r)$, where $P(r)$ is the monthly payment on a car loan, in dollars, if the annual interest rate is $r\%$.

14. $T'(v)$, where $T(v)$ is the time, in minutes, that it takes to drive from Tucson to Phoenix at a constant speed of v miles per hour.

15. $W'(x)$, where $W(x)$ is the work required, in joules, to stretch a spring x cm beyond its natural length.

16. Suppose $P(t)$ is the monthly payment, in dollars, on a mortgage which will take t years to pay off. What are the units of $P'(t)$? What is the practical meaning of $P'(t)$? What is its sign?

17. After investing \$1000 at an annual interest rate of 2% compounded continuously for t years, your balance is $\$B$, where $B = f(t)$. What are the units of dB/dt? What is the financial interpretation of dB/dt?

18. Investing \$1000 at an annual interest rate of $r\%$, compounded continuously, for 10 years gives you a balance of $\$B$, where $B = g(r)$. Give a financial interpretation of the statements:

    (a) $g(2) \approx 1221$.
    (b) $g'(2) \approx 122$. What are the units of $g'(2)$?

19. Meteorologists define the temperature lapse rate to be $-dT/dz$ where T is the air temperature in Celsius at altitude z kilometers above the ground.

    (a) What are the units of the lapse rate?
    (b) What is the practical meaning of a lapse rate of 6.5?

20. Let $\left.\frac{dV}{dr}\right|_{r=2} = 16$.

    (a) For small Δr, write an approximate equation relating ΔV and Δr near $r = 2$.
    (b) Estimate ΔV if $\Delta r = 0.1$.
    (c) Let $V = 32$ when $r = 2$. Estimate V when $r = 2.1$.

[7] An acre-foot is the amount of water it takes to cover one acre of area with 1 foot of water.

21. Let $R = f(S)$ and $f'(10) = 3$.
   (a) For small ΔS, write an approximate equation relating ΔR and ΔS near $S = 10$.
   (b) Estimate the change in R if S changes from $S = 10$ to $S = 10.2$.
   (c) Let $f(10) = 13$. Estimate $f(10.2)$.

22. Let $y = f(x)$ with $f(3) = 5$ and $f'(3) = 9$.
   (a) For small Δx, write an approximate equation relating Δy and Δx near $x = 3$.
   (b) Estimate $f(2.9)$.

PROBLEMS

23. If t is the number of years since 2014, the population, P, of China,[8] in billions, can be approximated by the function

$$P = f(t) = 1.394(1.006)^t.$$

Estimate $f(6)$ and $f'(6)$, giving units. What do these two numbers tell you about the population of China?

24. A city grew in population throughout the 1980s and into the early 1990s. The population was at its largest in 1995, and then shrank until 2010. Let $P = f(t)$ represent the population of the city t years since 1980. Sketch graphs of $f(t)$ and $f'(t)$, labeling the units on the axes.

25. Let Δx be small, and $y = f(x)$, $y = g(x)$, $y = h(x)$, $y = k(x)$ be functions with

$$f'(1) = 3,\ g'(1) = 0.3,\ h'(1) = -0.3,\ k'(1) = -3.$$

   (a) For Δx positive, for which functions is $\Delta y > \Delta x$?
   (b) For which functions is $|\Delta y| < |\Delta x|$?

26. Analysis of satellite data indicates that the Greenland ice sheet lost approximately 2900 gigatons (gt) of mass between March 2002 and September 2014. The mean mass loss rate for 2013–14 was 6 gt/year; the rate for 2012–13 was 474 gt/year.[9]
   (a) What derivative does this tell us about? Define the function and give units for each variable.
   (b) What numerical statement can you make about the derivative? Give units.

27. A laboratory study investigating the relationship between diet and weight in adult humans found that the weight of a subject, W, in pounds, was a function, $W = f(c)$, of the average number of Calories per day, c, consumed by the subject.
   (a) In terms of diet and weight, interpret the statements $f(1800) = 155$, $f'(2000) = 0.0003$, and $f^{-1}(162) = 2200$.
   (b) What are the units of $f'(c) = dW/dc$?

28. An economist is interested in how the price of a certain commodity affects its sales. Suppose that at a price of \$$p$, a quantity q of the commodity is sold. If $q = f(p)$, explain in economic terms the meaning of the statements $f(10) = 240{,}000$ and $f'(10) = -29{,}000$.

29. On May 9, 2007, CBS Evening News had a 4.3 point rating. (Ratings measure the number of viewers.) News executives estimated that a 0.1 drop in the ratings for the CBS Evening News corresponds to a \$5.5 million drop in revenue.[10] Express this information as a derivative. Specify the function, the variables, the units, and the point at which the derivative is evaluated.

30. Let $W(h)$ be an invertible function which tells how many gallons of water an oak tree of height h feet uses on a hot summer day. Give practical interpretations for each of the following quantities or statements.

   (a) $W(50)$ (b) $W^{-1}(40)$ (c) $W'(5) = 3$

31. The population of Mexico in millions is $P = f(t)$, where t is the number of years since 2014. Explain the meaning of the statements:

   (a) $f'(-6) = 1.35$ (b) $f^{-1}(113.2) = -5$
   (c) $(f^{-1})'(113.2) = 0.74$

32. Let $f(t)$ be the number of centimeters of rainfall that has fallen since midnight, where t is the time in hours. Interpret the following in practical terms, giving units.

   (a) $f(10) = 3.1$ (b) $f^{-1}(5) = 16$
   (c) $f'(10) = 0.4$ (d) $(f^{-1})'(5) = 2$

33. Water is flowing into a tank; the depth, in feet, of the water at time t in hours is $h(t)$. Interpret, with units, the following statements.

   (a) $h(5) = 3$ (b) $h'(5) = 0.7$
   (c) $h^{-1}(5) = 7$ (d) $(h^{-1})'(5) = 1.2$

34. As the altitude of an object increases, the pressure of the earth's atmosphere on the object decreases. Let $P = f(h)$ denote the atmospheric pressure, in kilopascals (kPa), at an altitude of h thousand meters above sea level. Interpret each of the following statements in terms of altitude and atmospheric pressure.

   (a) $f(1) = 88$ (b) $f'(1) = -11.5$
   (c) $f^{-1}(10) = 17.6$ (d) $(f^{-1})'(10) = -0.76$

[8] www.worldometers.info, accessed April 1, 2015.
[9] www.arctic.noaa.gov. Accessed April 3, 2015.
[10] *OC Register*, May 9, 2007; *The New York Times*, May 14, 2007.

35. As the distance of an object from the surface of the earth increases, the weight of the object decreases. Let $w = f(d)$ be the weight of an object, in Newtons, at a distance of d kilometers above the earth's surface. Interpret each of the following quantities in terms of weight and distance from the surface of the earth. Then, decide whether the quantity is positive or negative, and explain your answer.

(a) $f(80)$ **(b)** $f'(80)$
(c) $f^{-1}(200)$ **(d)** $(f^{-1})'(200)$

36. The acceleration a, in meters/second^2, of a child on a merry-go-round depends on her distance, r meters from the center, and her speed v, in meters per second. When v is constant, a is a function of r, so $a = f(r)$. When r is constant, a is a function of v, so $a = g(v)$. Explain what each of the following statements means in terms of the child.

(a) $f'(2) = -1$ **(b)** $g'(2) = 2$

37. Let $p(h)$ be the pressure in dynes per cm^2 on a diver at a depth of h meters below the surface of the ocean. What do each of the following quantities mean to the diver? Give units for the quantities.
(a) $p(100)$ **(b)** h such that $p(h) = 1.2 \cdot 10^6$
(c) $p(h) + 20$ **(d)** $p(h + 20)$
(e) $p'(100)$ **(f)** h such that $p'(h) = 100{,}000$

38. Let $g(t)$ be the height, in inches, of Amelia Earhart (one of the first woman airplane pilots) t years after her birth. What are the units of $g'(t)$? What can you say about the signs of $g'(10)$ and $g'(30)$? (Assume that $0 \le t < 39$, Amelia Earhart's age when her plane disappeared.)

39. If $g(v)$ is the fuel efficiency, in miles per gallon, of a car going at v miles per hour, what are the units of $g'(90)$? What is the practical meaning of the statement $g'(55) = -0.54$?

40. The function $P(d)$ gives the total electricity, in kWh, that a solar array has generated between the start of the year and the end of the dth day of the year. For each statement below, give a mathematical equation in terms of P, its inverse, or derivatives.

(a) The array had generated 3500 kWh of electricity by the end of January 4.
(b) At the end of January 4, the array was generating electricity at a rate of 1000 kWh per day.
(c) When the array had generated 5000 kWh of electricity, it took approximately half a day to generate an additional 1000 kWh of electricity.
(d) At the end of January 30, it took approximately one day to generate an additional 2500 kWh of electricity.

41. Let P be the total petroleum reservoir on Earth in the year t. (In other words, P represents the total quantity of petroleum, including what's not yet discovered, on Earth at time t.) Assume that no new petroleum is being made and that P is measured in barrels. What are the units of dP/dt? What is the meaning of dP/dt? What is its sign? How would you set about estimating this derivative in practice? What would you need to know to make such an estimate?

42. **(a)** If you jump out of an airplane without a parachute, you fall faster and faster until air resistance causes you to approach a steady velocity, called the *terminal* velocity. Sketch a graph of your velocity against time.
(b) Explain the concavity of your graph.
(c) Assuming air resistance to be negligible at $t = 0$, what natural phenomenon is represented by the slope of the graph at $t = 0$?

43. Let W be the amount of water, in gallons, in a bathtub at time t, in minutes.
(a) What are the meaning and units of dW/dt?
(b) Suppose the bathtub is full of water at time t_0, so that $W(t_0) > 0$. Subsequently, at time $t_p > t_0$, the plug is pulled. Is dW/dt positive, negative, or zero:
(i) For $t_0 < t < t_p$?
(ii) After the plug is pulled, but before the tub is empty?
(iii) When all the water has drained from the tub?

44. The temperature T of the atmosphere depends on the altitude a above the surface of Earth. Two measurements in the stratosphere find temperatures to be −36.1°C at 35 km and −34.7°C at 35.5 km.
(a) Estimate the derivative dT/da at 35 km. Include units.
(b) Estimate the temperature at 34 km.
(c) Estimate the temperature at 35.2 km.

45. The Arctic Sea ice extent, the area of sea covered by ice, grows over the winter months, typically from November to March. Let $F(t)$ be the Arctic Sea ice extent, in million of square kilometers, as a function of time, t, in days since November 1, 2014. Then $F'(t) = 0.073$ on January 1, 2015.[11]
(a) Give the units of the 0.073, and interpret the number in practical terms.
(b) Estimate ΔF between January 1 and January 6, 2015. Explain what this tells us about Arctic Sea ice.

46. The depth, h (in mm), of the water runoff down a slope during a steady rain is a function of the distance, x (in meters), from the top of the slope,[12] so $h = f(x)$. We have $f'(15) = 0.02$.
(a) What are the units of the 15?
(b) What are the units of the 0.02?

[11] Sea Ice Index data from nsidc.org/data/seaice_index/archives.html. Accessed March 2015.
[12] R. S. Anderson and S. P. Anderson, *Geomorphology* (Cambridge: Cambridge University Press, 2010), p. 369.

(c) About how much difference in runoff depth is there between two points around 15 meters down the slope if one of them is 4 meters farther from the top of the slope than the other?

47. Average leaf width, w (in mm), in tropical Australia[13] is a function of the average annual rainfall, r (in mm), so $w = f(r)$. We have $f'(1500) = 0.0218$.

(a) What are the units of the 1500?
(b) What are the units of the 0.0218?
(c) About how much difference in average leaf width would you find in two forests whose average annual rainfalls are near 1500 mm but differ by 200 mm?

48. When an ice dam for a glacial lake breaks, the maximal outflow rate, Q in meters^3/sec is a function of V, the volume of the lake (in millions of meters^3).

(a) What are the units of dQ/dV?
(b) Observation shows that $dQ/dV|_{V=12} = 22$. About how much is the difference in maximal outflow when dams break in two lakes with volumes near $12 \cdot 10^6$ meters^3 if one of them has volume 500,000 meters^3 greater than the other?

49. The growth rate, R (in cell divisions per hour), of the *E. coli* bacterium is a function of C (in units of 10^{-4} M), the concentration of glucose in the solution containing the bacteria. (1 M is 1 mole of glucose per liter of solution.)[14]

(a) What are the units of dR/dC?
(b) Observation shows the $dR/dC|_{C=1.5} = 0.1$. What are the units of the 1.5?
(c) Two populations are growing in glucose solutions near $1.5 \cdot 10^{-4}$ M, but one is $0.2 \cdot 10^{-4}$ M more concentrated than the other. How much difference is there between the bacteria cell division rates?

50. A company's revenue from car sales, C (in thousands of dollars), is a function of advertising expenditure, a, in thousands of dollars, so $C = f(a)$.

(a) What does the company hope is true about the sign of f'?
(b) What does the statement $f'(100) = 2$ mean in practical terms? How about $f'(100) = 0.5$?
(c) Suppose the company plans to spend about \$100,000 on advertising. If $f'(100) = 2$, should the company spend more or less than \$100,000 on advertising? What if $f'(100) = 0.5$?

51. In April 2015 in the US, there was one birth every 8 seconds, one death every 12 seconds, and one new international migrant every 32 seconds.[15]

(a) Let $f(t)$ be the population of the US, where t is time in seconds measured from the start of April 2015. Find $f'(0)$. Give units.
(b) To the nearest second, how long did it take for the US population to add one person in April 2015?

52. During the 1970s and 1980s, the buildup of chlorofluorocarbons (CFCs) created a hole in the ozone layer over Antarctica. After the 1987 Montreal Protocol, an agreement to phase out CFC production, the ozone hole has shrunk. The ODGI (ozone depleting gas index) shows the level of CFCs present.[16] Let $O(t)$ be the ODGI for Antarctica in year t; then $O(2016) = 81.4$ and $O'(2016) = -1.5$. Assuming that the ODGI decreases at a constant rate, estimate when the ozone hole will have recovered, which occurs when ODGI = 0.

53. Let $P(x)$ be the number of people of height $\leq x$ inches in the US. What is the meaning of $P'(66)$? What are its units? Estimate $P'(66)$ (using common sense). Is $P'(x)$ ever negative? [Hint: You may want to approximate $P'(66)$ by a difference quotient, using $h = 1$. Also, you may assume the US population is about 300 million, and note that 66 inches = 5 feet 6 inches.]

54. When you breathe, a muscle (called the diaphragm) reduces the pressure around your lungs and they expand to fill with air. The table shows the volume of a lung as a function of the reduction in pressure from the diaphragm. Pulmonologists (lung doctors) define the *compliance* of the lung as the derivative of this function.[17]

(a) What are the units of compliance?
(b) Estimate the maximum compliance of the lung.
(c) Explain why the compliance gets small when the lung is nearly full (around 1 liter).

Pressure reduction (cm of water)	Volume (liters)
0	0.20
5	0.29
10	0.49
15	0.70
20	0.86
25	0.95
30	1.00

[13]H. Shugart, *Terrestrial Ecosystems in Changing Environments* (Cambridge: CUP, 1998), p. 145.
[14]Jacques Monod, "The Growth of Bacterial Cultures", *Annu. Rev. Microbiol.* 1949.3: 371–394.
[15]www.census.gov, accessed April 8, 2015.
[16]www.esrl.noaa.gov/gmd/odgi, accessed April, 2017.
[17]en.wikipedia.org, accessed April 3, 2015.

Strengthen Your Understanding

In Problems 55–57, explain what is wrong with the statement.

55. If the position of a car at time t is given by $s(t)$ then the velocity of the car is $s'(t)$ and the units of s' are meters per second.
56. A spherical balloon originally contains 3 liters of air and it is leaking 1% of its volume per hour. If $r(t)$ is the radius of the balloon at time t then $r'(t) > 0$.
57. A laser printer takes $T(P)$ minutes to produce P pages, so the derivative $\dfrac{dT}{dP}$ is measured in pages per minute.

In Problems 58–59, give an example of:

58. A function whose derivative is measured in years/dollar.
59. A function whose derivative is measured in miles/day.

Are the statements in Problems 60–63 true or false? Give an explanation for your answer.

60. By definition, the instantaneous velocity of an object equals a difference quotient.
61. If $y = f(x)$, then $\left.\dfrac{dy}{dx}\right|_{x=a} = f'(a)$.
62. If $f(t)$ is the quantity in grams of a chemical produced after t minutes and $g(t)$ is the same quantity in kilograms, then $f'(t) = 1000g'(t)$.
63. If $f(t)$ is the quantity in kilograms of a chemical produced after t minutes and $g(t)$ is the quantity in kilograms produced after t seconds, then $f'(t) = 60g'(t)$.

For Problems 64–65, assume $g(v)$ is the fuel efficiency, in miles per gallon, of a car going at a speed of v miles per hour.

64. What are the units of $g'(v) = \dfrac{dg}{dv}$? There may be more than one option.

    (a) $(\text{miles})^2/(\text{gal})(\text{hour})$
    (b) hour/gal
    (c) gal/hour
    (d) $(\text{gal})(\text{hour})/(\text{miles})^2$
    (e) (miles/gallon)/(miles/hour)

65. What is the practical meaning of $g'(55) = -0.54$? There may be more than one option.

    (a) When the car is going 55 mph, the rate of change of the fuel efficiency decreases *to* approximately 0.54 miles/gal.
    (b) When the car is going 55 mph, the rate of change of the fuel efficiency decreases *by* approximately 0.54 miles/gal.
    (c) If the car speeds up from 55 mph to 56 mph, then the fuel efficiency is approximately −0.54 miles per gallon.
    (d) If the car speeds up from 55 mph to 56 mph, then the car becomes less fuel efficient by approximately 0.54 miles per gallon.

2.5 THE SECOND DERIVATIVE

Second and Higher Derivatives

Since the derivative is itself a function, we can consider its derivative. For a function f, the derivative of its derivative is called the *second derivative*, and written f'' (read "f double-prime") or y''. If $y = f(x)$, the second derivative can also be written as $\dfrac{d^2y}{dx^2}$, which means $\dfrac{d}{dx}\left(\dfrac{dy}{dx}\right)$, the derivative of $\dfrac{dy}{dx}$. We can also differentiate the second derivative to get the third derivative written as $f'''(x)$ or $\dfrac{d^3y}{dx^3}$. In fact, we can take the nth derivative written as $f^{(n)}(x)$ or $\dfrac{d^ny}{dx^n}$ for any integer $n \geq 1$.

What Do Derivatives Tell Us?

Recall that the derivative of a function tells you whether a function is increasing or decreasing:

- If $f' > 0$ on an interval, then f is *increasing* over that interval.
- If $f' < 0$ on an interval, then f is *decreasing* over that interval.

If f' is always positive on an interval or always negative on an interval, then f is monotonic over that interval.

Since f'' is the derivative of f',

- If $f'' > 0$ on an interval, then f' is *increasing* over that interval.
- If $f'' < 0$ on an interval, then f' is *decreasing* over that interval.

What does it mean for f' to be increasing or decreasing? An example in which f' is increasing is shown in Figure 2.43, where the curve is bending upward, or is *concave up*. In the example shown in Figure 2.44, in which f' is decreasing, the graph is bending downward, or is *concave down*. These figures suggest the following result:

> If $f'' > 0$ on an interval, then f' is increasing, so the graph of f is concave up there.
> If $f'' < 0$ on an interval, then f' is decreasing, so the graph of f is concave down there.

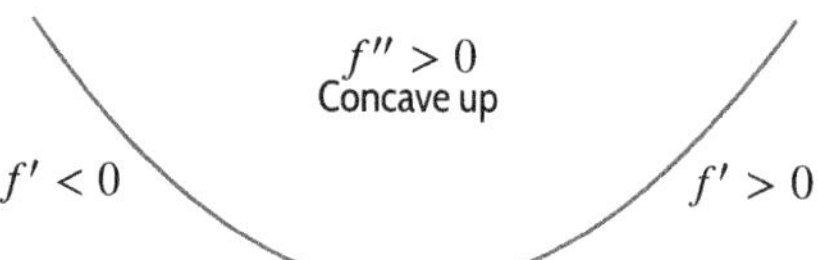

Figure 2.43: Meaning of f'': The slope of f increases from left to right, f'' is positive, and f is concave up

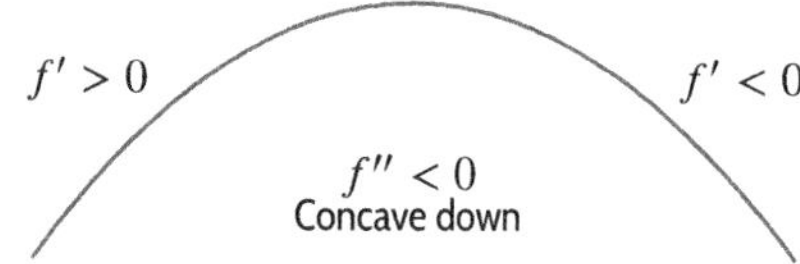

Figure 2.44: Meaning of f'': The slope of f decreases from left to right, f'' is negative, and f is concave down

Warning! The graph of a function f can be concave up everywhere and yet have $f'' = 0$ at some point. For instance, the graph of $f(x) = x^4$ in Figure 2.45 is concave up, but it can be shown that $f''(0) = 0$. If we are told that the graph of a function f is concave up, we can be sure that f'' is not negative, that is $f'' \geq 0$, but not that f'' is positive, $f'' > 0$.

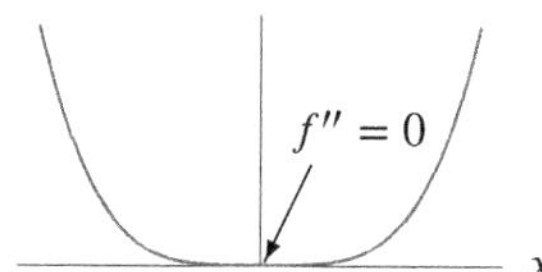

Figure 2.45: Graph of $f(x) = x^4$

> If the graph of f is concave up and f'' exists on an interval, then $f'' \geq 0$ there.
> If the graph of f is concave down and f'' exists on an interval, then $f'' \leq 0$ there.

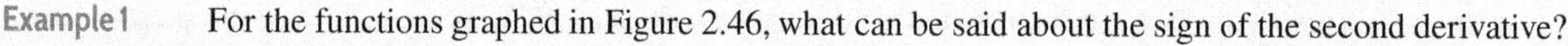

Example 1 For the functions graphed in Figure 2.46, what can be said about the sign of the second derivative?

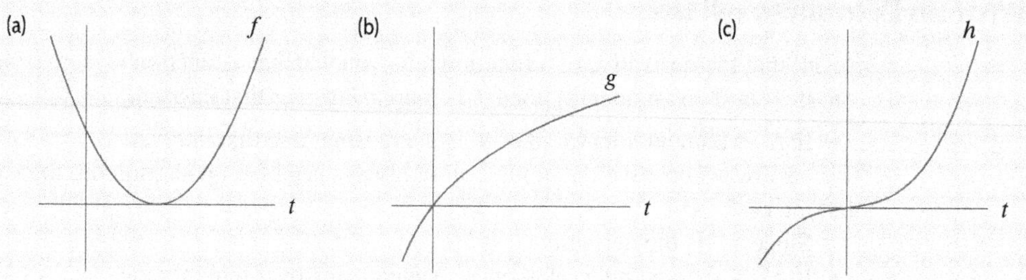

Figure 2.46: What signs do the second derivatives have?

Solution (a) The graph of f is concave up everywhere, so $f'' \geq 0$ everywhere.
(b) The graph of g is concave down everywhere, so $g'' \leq 0$ everywhere.
(c) For $t < 0$, the graph of h is concave down, so $h'' \leq 0$ there. For $t > 0$, the graph of h is concave up, so $h'' \geq 0$ there.

Example 2 Sketch the second derivative f'' for the function f of Example 1 on page 99, graphed with its derivative, f', in Figure 2.47. Relate the resulting graph of f'' to the graphs of f and f'.

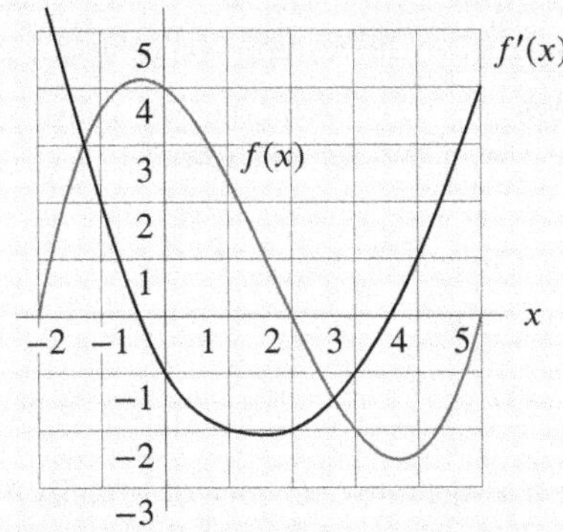

Figure 2.47: Function, f in color; derivative, f', in black

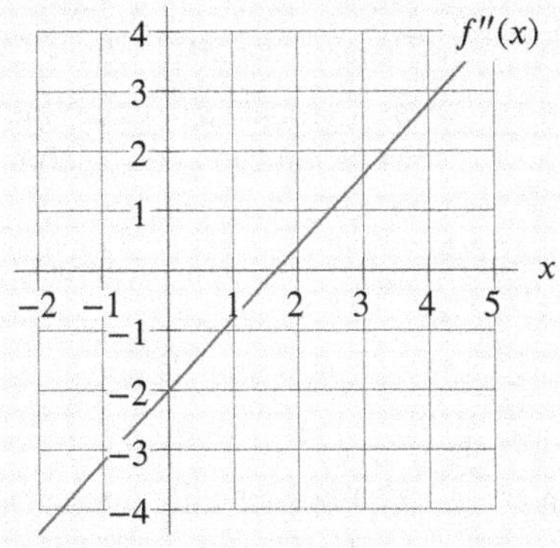

Figure 2.48: Graph of f''

Solution We want to sketch the derivative of f'. We do this by estimating the slopes of f' and plotting them, obtaining Figure 2.48.

We observe that where $f'' > 0$, the graph of f is concave up and f' is increasing, and that where $f'' < 0$, the graph of f is concave down and f' is decreasing. Where $f''(x) = 0$, the graph of f changes from concave down to concave up, and f' changes from decreasing to increasing.

Interpretation of the Second Derivative as a Rate of Change

If we think of the derivative as a rate of change, then the second derivative is a rate of change of a rate of change. If the second derivative is positive, the rate of change of f is increasing; if the second derivative is negative, the rate of change of f is decreasing.

The second derivative can be a matter of practical concern. A 2009 article[18] reported that although the US economy was shrinking, the rate of decrease had slowed. (The derivative of the size of the economy was negative and the second derivative was positive). The article continued "although the economy is spiralling down, it is doing so more slowly."

Example 3 A population, P, growing in a confined environment often follows a *logistic* growth curve, like that shown in Figure 2.49. Relate the sign of d^2P/dt^2 to how the rate of growth, dP/dt, changes over time. What are practical interpretations of t_0 and L?

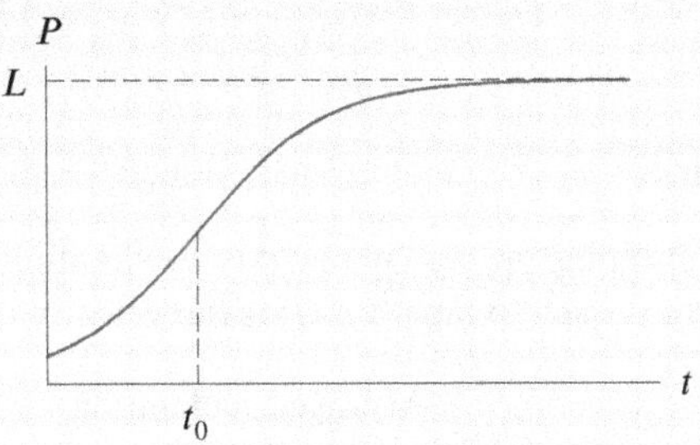

Figure 2.49: Logistic growth curve

[18] *The Economist,* February 19, 2009, Washington, DC, "The second derivative may be turning positive," www.economist.com/node/13145616

Solution For $t < t_0$, the rate of growth, dP/dt, is increasing and $d^2P/dt^2 \geq 0$. At t_0, the rate dP/dt is a maximum. In other words, at time t_0 the population is growing fastest. For $t > t_0$, the rate of growth, dP/dt, is decreasing and $dP^2/dt^2 \leq 0$. At t_0, the curve changes from concave up to concave down, and $d^2P/dt^2 = 0$ there.

The quantity L represents the limiting value of the population as $t \to \infty$. Biologists call L the *carrying capacity* of the environment.

Example 4 Tests on the 2011 Chevy Corvette ZR1 sports car gave the results[19] in Table 2.10.

(a) Estimate dv/dt for the time intervals shown.

(b) What can you say about the sign of d^2v/dt^2 over the period shown?

Table 2.10 *Velocity of 2011 Chevy Corvette ZR1*

Time, t (sec)	0	3	6	9	12
Velocity, v (meters/sec)	0	23	42	59	72

Solution (a) For each time interval we can calculate the average rate of change of velocity. For example, from $t = 0$ to $t = 3$ we have

$$\frac{dv}{dt} \approx \text{Average rate of change of velocity} = \frac{23-0}{3-0} = 7.67\ \frac{\text{m/sec}}{\text{sec}}.$$

Estimated values of dv/dt are in Table 2.11.

(b) Since the values of dv/dt are decreasing between the points shown, we expect $d^2v/dt^2 \leq 0$. The graph of v against t in Figure 2.50 supports this; it is concave down. The fact that $dv/dt > 0$ tells us that the car is speeding up; the fact that $d^2v/dt^2 \leq 0$ tells us that the rate of increase decreased (actually, did not increase) over this time period.

Table 2.11 *Estimates for dv/dt (meters/sec/sec)*

Time interval (sec)	0 − 3	3 − 6	6 − 9	9 − 12
Average rate of change (dv/dt)	7.67	6.33	5.67	4.33

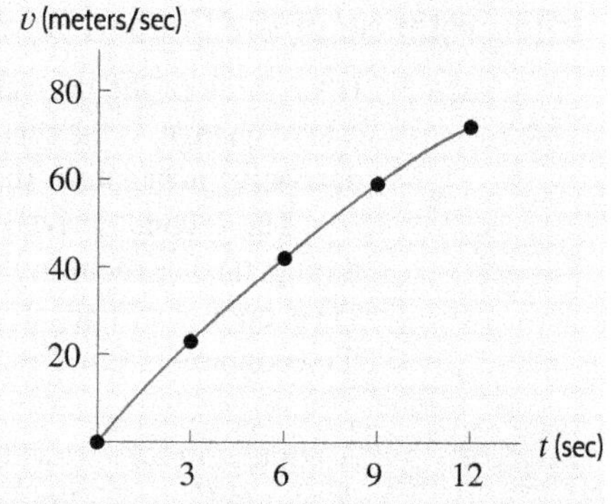

Figure 2.50: Velocity of 2011 Chevy Corvette ZR1

Velocity and Acceleration

When a car is speeding up, we say that it is accelerating. We define *acceleration* as the rate of change of velocity with respect to time. If $v(t)$ is the velocity of an object at time t, we have

$$\begin{matrix}\text{Average acceleration} \\ \text{from } t \text{ to } t+h\end{matrix} = \frac{v(t+h)-v(t)}{h},$$

[19] Adapted from data in http://www.corvette-web-central.com/Corvettetopspeed.html

$$\text{Instantaneous acceleration} = v'(t) = \lim_{h\to 0} \frac{v(t+h) - v(t)}{h}.$$

If the term velocity or acceleration is used alone, it is assumed to be instantaneous. Since velocity is the derivative of position, acceleration is the second derivative of position. Summarizing:

If $y = s(t)$ is the position of an object at time t, then

- Velocity: $v(t) = \dfrac{dy}{dt} = s'(t)$.
- Acceleration: $a(t) = \dfrac{d^2y}{dt^2} = s''(t) = v'(t)$.

Example 5 A particle is moving along a straight line; its acceleration is zero only once. Its distance, s, to the right of a fixed point is given by Figure 2.51. Estimate:

(a) When the particle is moving to the right and when it is moving to the left.

(b) When the acceleration of the particle is zero, when it is negative, and when it is positive.

Solution (a) The particle is moving to the right whenever s is increasing. From the graph, this appears to be for $0 < t < \frac{2}{3}$ and for $t > 2$. For $\frac{2}{3} < t < 2$, the value of s is decreasing, so the particle is moving to the left.

(b) Since the acceleration is zero only once, this must be when the curve changes concavity, at about $t = \frac{4}{3}$. Then the acceleration is negative for $t < \frac{4}{3}$, since the graph is concave down there, and the acceleration is positive for $t > \frac{4}{3}$, since the graph is concave up there.

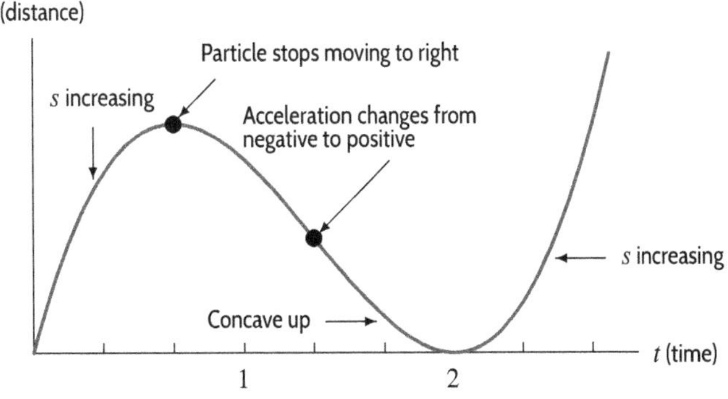

Figure 2.51: Distance of particle to right of a fixed point

Exercises and Problems for Section 2.5 Online Resource: Additional Problems for Section 2.5

EXERCISES

1. Fill in the blanks:

(a) If f'' is positive on an interval, then f' is __________ on that interval, and f is __________ on that interval.

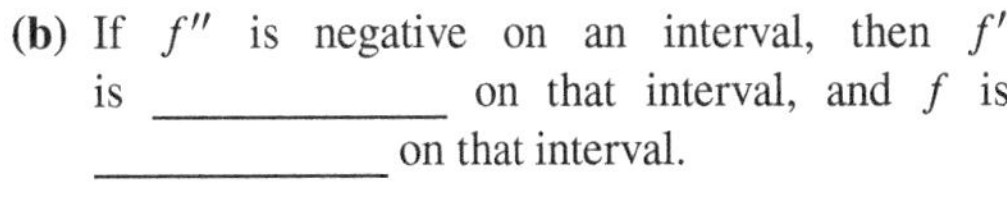

(b) If f'' is negative on an interval, then f' is __________ on that interval, and f is __________ on that interval.

2. For the function graphed in Figure 2.52, are the following nonzero quantities positive or negative?

(a) $f(2)$ **(b)** $f'(2)$ **(c)** $f''(2)$

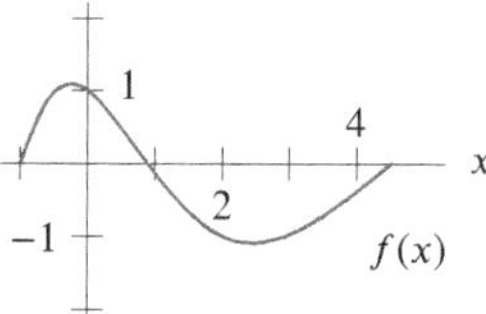

Figure 2.52

3. At one of the labeled points on the graph in Figure 2.53 both dy/dx and d^2y/dx^2 are positive. Which is it?

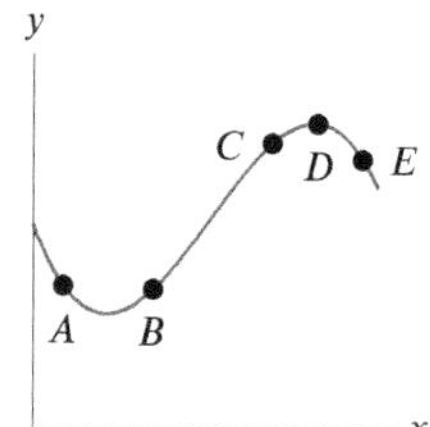

Figure 2.53

4. At exactly two of the labeled points in Figure 2.54, the derivative f' is 0; the second derivative f'' is not zero at any of the labeled points. On a copy of the table, give the signs of f, f', f'' at each marked point.

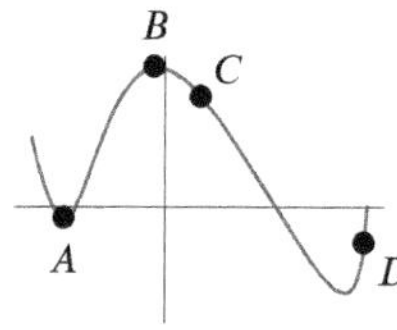

Point	f	f'	f''
A			
B			
C			
D			

Figure 2.54

5. Graph the functions described in parts (a)–(d).

(a) First and second derivatives everywhere positive.
(b) Second derivative everywhere negative; first derivative everywhere positive.
(c) Second derivative everywhere positive; first derivative everywhere negative.
(d) First and second derivatives everywhere negative.

6. Sketch the graph of a function whose first derivative is everywhere negative and whose second derivative is positive for some x-values and negative for other x-values.

7. Sketch the graph of the height of a particle against time if velocity is positive and acceleration is negative.

■ For Exercises 8–13, give the signs of the first and second derivatives for the function. Each derivative is either positive everywhere, zero everywhere, or negative everywhere.

8.

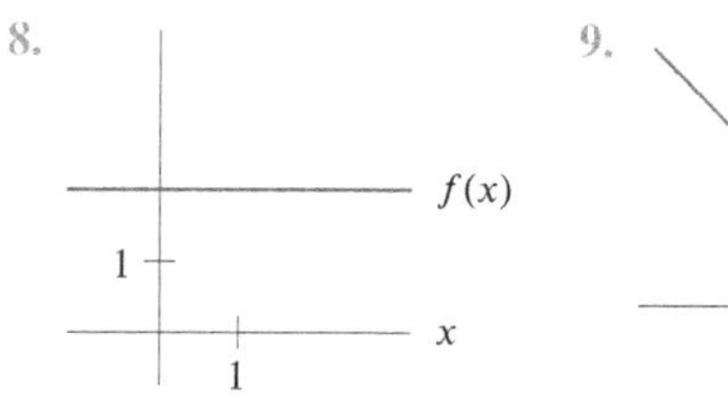

9.

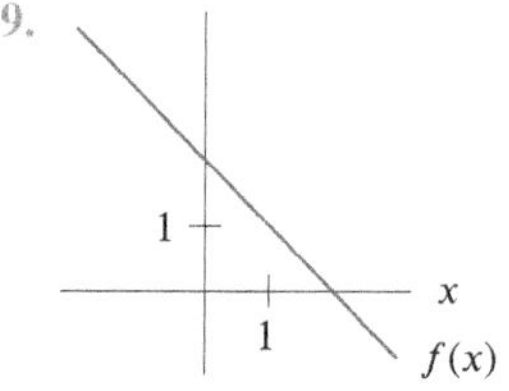

10.

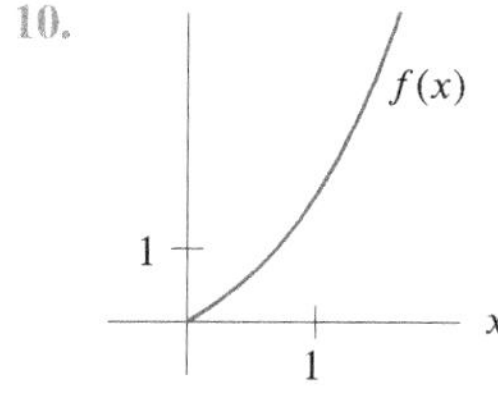

11.

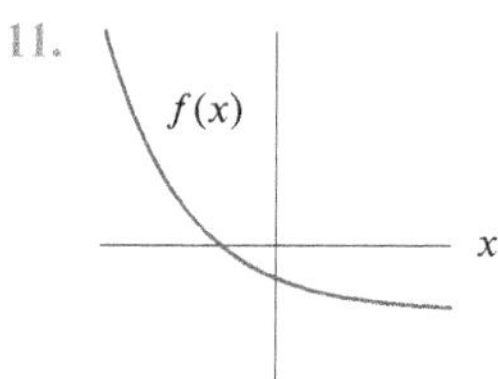

12.

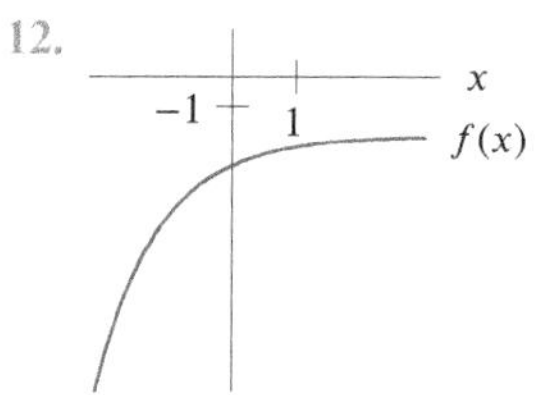

13.

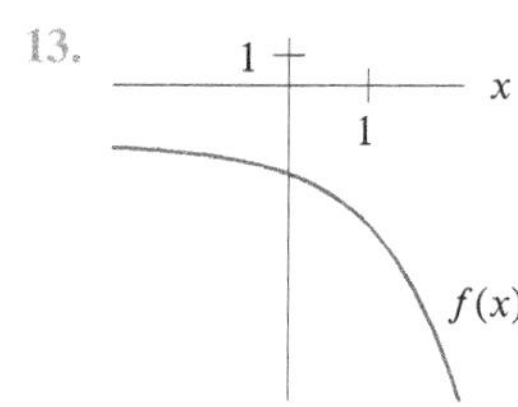

14. The position of a particle moving along the x-axis is given by $s(t) = 5t^2 + 3$. Use difference quotients to find the velocity $v(t)$ and acceleration $a(t)$.

PROBLEMS

15. The table gives the number of passenger cars, $C = f(t)$, in millions,[20] in the US in the year t.

(a) Do $f'(t)$ and $f''(t)$ appear to be positive or negative during the period 1975–1990?
(b) Do $f'(t)$ and $f''(t)$ appear to be positive or negative during the period 1990–2000?
(c) Estimate $f'(2005)$. Using units, interpret your answer in terms of passenger cars.

t	1975	1980	1985	1990	1995	2000	2005
C	106.7	121.6	127.9	133.7	128.4	133.6	136.6

[20] www.bts.gov/publications/national_transportation_statistics/html/table_01_11.html. Accessed April 27, 2011.

16. An accelerating sports car goes from 0 mph to 60 mph in five seconds. Its velocity is given in the following table, converted from miles per hour to feet per second, so that all time measurements are in seconds. (Note: 1 mph is 22/15 ft/sec.) Find the average acceleration of the car over each of the first two seconds.

Time, t (sec)	0	1	2	3	4	5
Velocity, $v(t)$ (ft/sec)	0	30	52	68	80	88

17. A small plane is taking off from a runway at a regional airport. The plane's ground velocity increases throughout the first 20 seconds. Table 2.12 gives the ground velocity, in meters per second, at two-second intervals.

(a) Graph the data.
(b) From the graph, estimate when the plane is accelerating the fastest and estimate that acceleration.

Table 2.12

Time (sec)	0	2	4	6	8	10
Velocity (m/s)	2.7	2.7	4	6.3	8.5	11.6
Time (sec)	12	14	16	18	20	
Velocity (m/s)	13.4	17.4	21.9	29.1	32.6	

18. Sketch the curves described in (a)–(c):

(a) Slope is positive and increasing at first but then is positive and decreasing.
(b) The first derivative of the function whose graph is in part (a).
(c) The second derivative of the function whose graph is in part (a).

In Problems 19–24, graph the second derivative of the function.

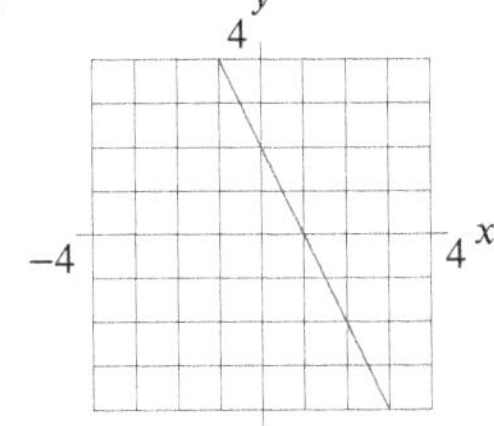

20.

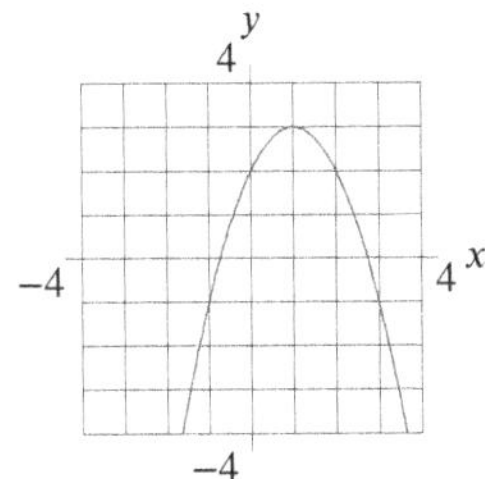

21.

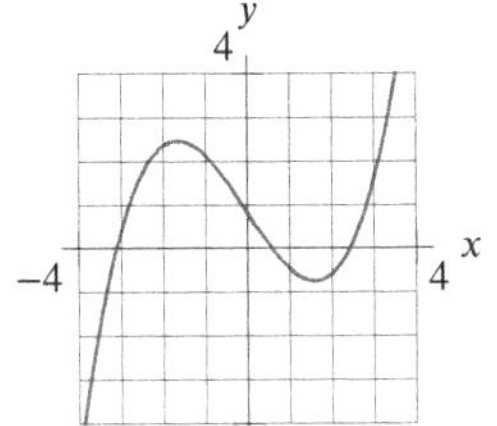

22.

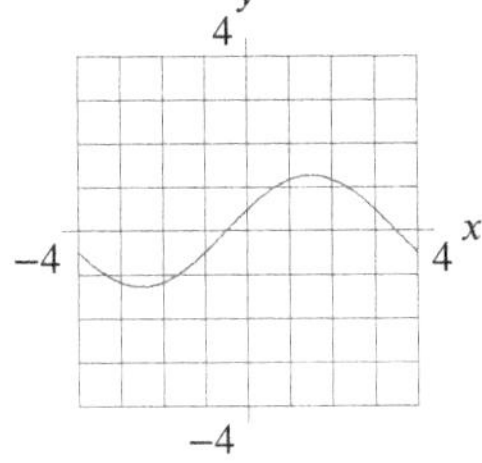

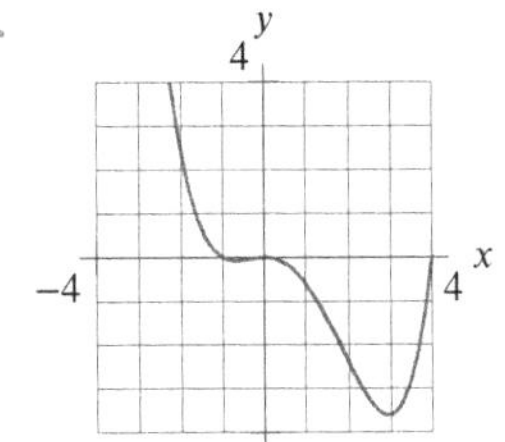

24.

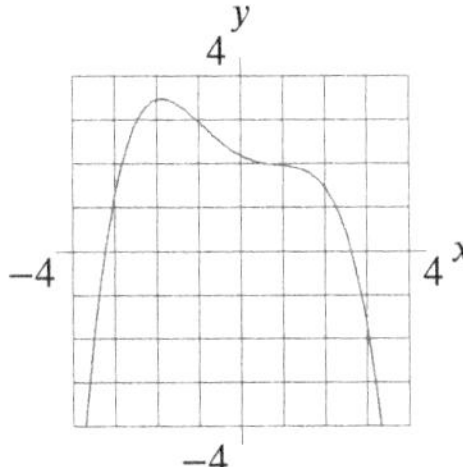

25. Let $P(t)$ represent the price of a share of stock of a corporation at time t. What does each of the following statements tell us about the signs of the first and second derivatives of $P(t)$?

(a) "The price of the stock is rising faster and faster."
(b) "The price of the stock is close to bottoming out."

26. A headline in the New York Times on December 14, 2014, read:[21]

> "A Steep Slide in Law School Enrollment Accelerates"

(a) What function is the author talking about?
(b) Draw a possible graph for the function.
(c) In terms of derivatives, what is the headline saying?

27. In economics, *total utility* refers to the total satisfaction from consuming some commodity. According to the economist Samuelson:[22]

> As you consume more of the same good, the total (psychological) utility increases. However, ... with successive new units of the good, your total utility will grow at a slower and slower rate because of a fundamental tendency for your psychological ability to appreciate more of the good to become less keen.

(a) Sketch the total utility as a function of the number of units consumed.
(b) In terms of derivatives, what is Samuelson saying?

28. "Winning the war on poverty" has been described cynically as slowing the rate at which people are slipping below the poverty line. Assuming that this is happening:

(a) Graph the total number of people in poverty against time.
(b) If N is the number of people below the poverty line at time t, what are the signs of dN/dt and d^2N/dt^2? Explain.

[21] http://www.nytimes.com, accessed January 8, 2015.
[22] From Paul A. Samuelson, *Economics*, 11th edition (New York: McGraw-Hill, 1981).

29. An industry is being charged by the Environmental Protection Agency (EPA) with dumping unacceptable levels of toxic pollutants in a lake. Over a period of several months, an engineering firm makes daily measurements of the rate at which pollutants are being discharged into the lake. The engineers produce a graph similar to either Figure 2.55(a) or Figure 2.55(b). For each case, give an idea of what argument the EPA might make in court against the industry and in the industry's defense.

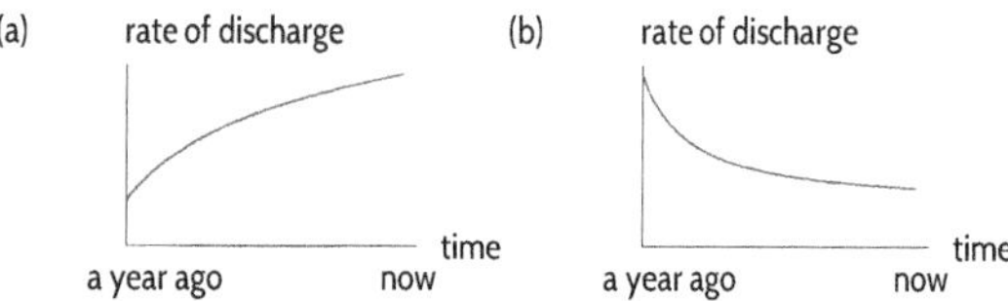

Figure 2.55

In Problems 30–35, the length of a plant, L, is a function of its mass, M, so $L = f(M)$. A unit increase in a plant's mass stretches the plant's length more when the plant is small, and less when the plant is large. Assuming $M > 0$, decide if f' agrees with this description.

30. f' is constant

31. f' is negative

32. f' is decreasing

33. f' is increasing

34. $f'(M) = 0.4M + 0.3M^2$

35. $f'(M) = 0.25/M^{3/4}$

36. At which of the marked x-values in Figure 2.56 can the following statements be true?

(a) $f(x) < 0$
(b) $f'(x) < 0$
(c) $f(x)$ is decreasing
(d) $f'(x)$ is decreasing
(e) Slope of $f(x)$ is positive
(f) Slope of $f(x)$ is increasing

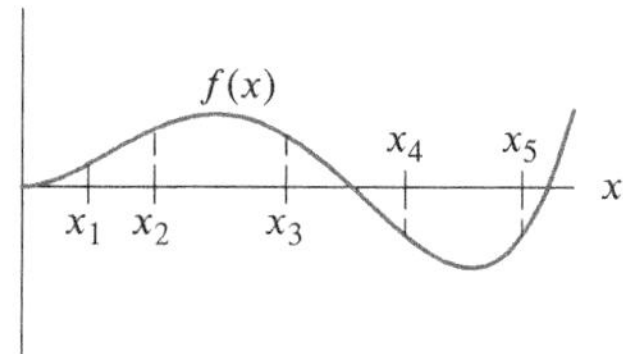

Figure 2.56

37. Figure 2.57 gives the position, $f(t)$, of a particle at time t. At which of the marked values of t can the following statements be true?

(a) The position is positive
(b) The velocity is positive
(c) The acceleration is positive
(d) The position is decreasing
(e) The velocity is decreasing

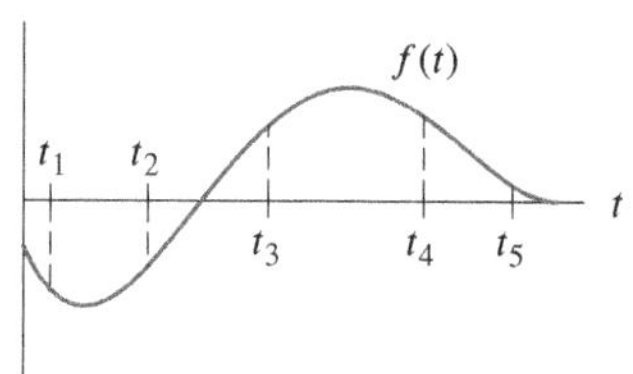

Figure 2.57

38. The graph of f' (not f) is given in Figure 2.58. At which of the marked values of x is

(a) $f(x)$ greatest?
(b) $f(x)$ least?
(c) $f'(x)$ greatest?
(d) $f'(x)$ least?
(e) $f''(x)$ greatest?
(f) $f''(x)$ least?

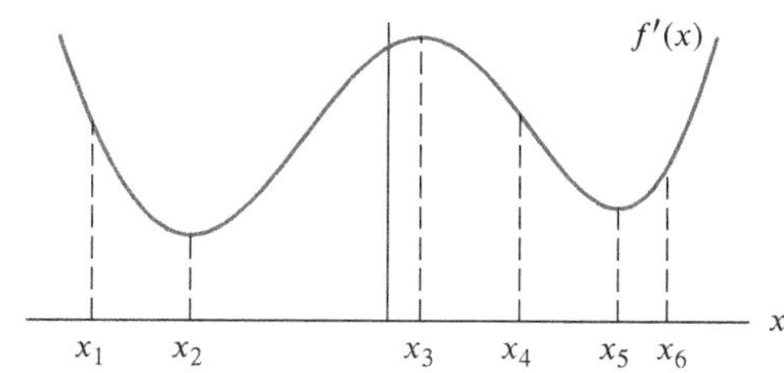

Figure 2.58: Graph of f', not f

For Problems 39–40, sketch the graph of one continuous function f that has all of the indicated properties.

39.
- $f'(x) > 0$ for $0 < x < 3$ and $8 < x < 12$
  $f'(x) < 0$ for $3 < x < 8$
- $f''(x) > 0$ for $5 < x < 9$
  $f''(x) < 0$ for $0 < x < 5$ and $9 < x < 12$
- $\lim_{x\to\infty} f(x) = 2$

40.
- $f'(x) > 0$ for $-2 < x < 2$
  $f'(x) < 0$ for $-6 < x < -2$ and $2 < x < 6$
- $f''(x) > 0$ for $-3 < x < 0$ and $3 < x < 6$
  $f''(x) < 0$ for $-6 < x < -3$ and $0 < x < 3$
- $\lim_{x\to-\infty} f(x) = 2$ and $\lim_{x\to\infty} f(x) = 3$

41. A function f has $f(5) = 20$, $f'(5) = 2$, and $f''(x) < 0$, for $x \geq 5$. Which of the following are possible values for $f(7)$ and which are impossible?

(a) 26 **(b)** 24 **(c)** 22

42. Chlorofluorocarbons (CFCs) were used as propellants in spray cans until their buildup in the atmosphere started destroying the ozone, which protects us from ultraviolet rays. Since the 1987 Montreal Protocol (an agreement to curb CFCs), the CFCs in the atmosphere above the US have been reduced from a high of 1915 parts per trillion (ppt) in 2000 to 1640 ppt in 2014.[23] The reduction has been approximately linear. Let $C(t)$ be the concentration of CFCs in ppt in year t.
   - **(a)** Find $C(2000)$ and $C(2014)$.
   - **(b)** Estimate $C'(2000)$ and $C'(2014)$.
   - **(c)** Assuming $C(t)$ is linear, find a formula for $C(t)$.
   - **(d)** When is $C(t)$ expected to reach 1500 ppt, the level before CFCs were introduced?
   - **(e)** If you were told that in the future, $C(t)$ would not be exactly linear, and that $C''(t) > 0$, would your answer to part (d) be too early or too late?

43. **(a)** Using the graph of f' in Figure 2.59, for what values of x is f increasing?
   - **(b)** For what values of x is f concave up?
   - **(c)** For what values of x is $f'''(x) > 0$?

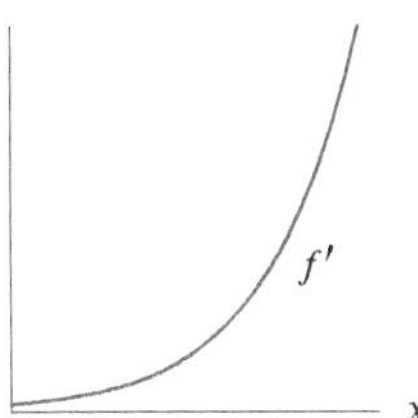

Figure 2.59: Graph of f'

Strengthen Your Understanding

In Problems 44–45, explain what is wrong with the statement.

44. A function that is not concave up is concave down.

45. When the acceleration of a car is zero, the car is not moving.

In Problems 46–47, give an example of:

46. A function that has a nonzero first derivative but zero second derivative.

47. A function for which $f'(0) = 0$ but $f''(0) \neq 0$.

Are the statements in Problems 48–52 true or false? Give an explanation for your answer.

48. If $f''(x) > 0$ then $f'(x)$ is increasing.

49. The instantaneous acceleration of a moving particle at time t is the limit of difference quotients.

50. A function which is monotonic on an interval is either increasing or decreasing on the interval.

51. The function $f(x) = x^3$ is monotonic on any interval.

52. The function $f(x) = x^2$ is monotonic on any interval.

2.6 DIFFERENTIABILITY

What Does It Mean for a Function to Be Differentiable?

A function is differentiable at a point if it has a derivative there. In other words:

> The function f is **differentiable** at x if
>
> $$\lim_{h\to 0} \frac{f(x+h) - f(x)}{h} \quad \text{exists.}$$
>
> Thus, the graph of f has a nonvertical tangent line at x. The value of the limit and the slope of the tangent line are the derivative of f at x.

Occasionally we meet a function which fails to have a derivative at a few points. A function fails to be differentiable at a point if:

- The function is not continuous at the point.
- The graph has a sharp corner at that point.
- The graph has a vertical tangent line.

[23] www.esrl.noaa.gov, accessed April 27, 2015.

Figure 2.60 shows a function which appears to be differentiable at all points except $x = a$ and $x = b$. There is no tangent at A because the graph has a corner there. As x approaches a from the left, the slope of the line joining P to A converges to some positive number. As x approaches a from the right, the slope of the line joining P to A converges to some negative number. Thus the slopes approach different numbers as we approach $x = a$ from different sides. Therefore the function is not differentiable at $x = a$. At B, the graph has a vertical tangent. As x approaches b, the slope of the line joining B to Q does not approach a limit; it just keeps growing larger and larger. Again, the limit defining the derivative does not exist and the function is not differentiable at $x = b$.

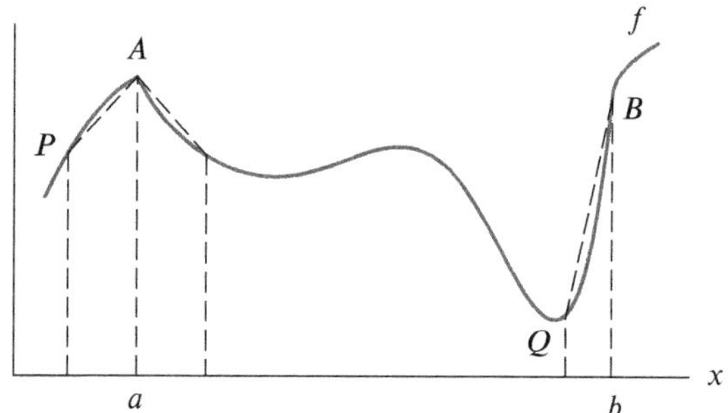

Figure 2.60: A function which is not differentiable at A or B

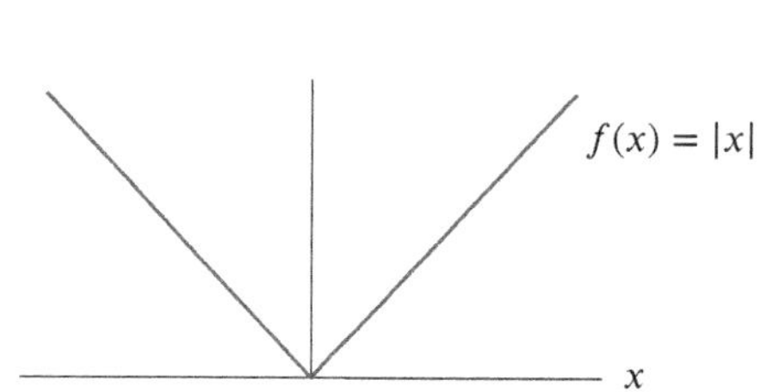

Figure 2.61: Graph of absolute value function, showing point of non-differentiability at $x = 0$

Examples of Nondifferentiable Functions

An example of a function whose graph has a corner is the *absolute value* function defined as follows:

$$f(x) = |x| = \begin{cases} x & \text{if } x \geq 0, \\ -x & \text{if } x < 0. \end{cases}$$

This function is called *piecewise linear* because each part of it is linear. Its graph is in Figure 2.61. Near $x = 0$, even close-up views of the graph of $f(x)$ look the same, so this is a corner which can't be straightened out by zooming in.

Example 1 Try to compute the derivative of the function $f(x) = |x|$ at $x = 0$. Is f differentiable there?

Solution To find the derivative at $x = 0$, we want to look at

$$\lim_{h\to 0} \frac{f(h) - f(0)}{h} = \lim_{h\to 0} \frac{|h| - 0}{h} = \lim_{h\to 0} \frac{|h|}{h}.$$

As h approaches 0 from the right, h is positive, so $|h| = h$, and the ratio is always 1. As h approaches 0 from the left, h is negative, so $|h| = -h$, and the ratio is -1. Since the limits are different from each side, the limit of the difference quotient does not exist. Thus, the absolute value function is not differentiable at $x = 0$. The limits of 1 and -1 correspond to the fact that the slope of the right-hand part of the graph is 1, and the slope of the left-hand part is -1.

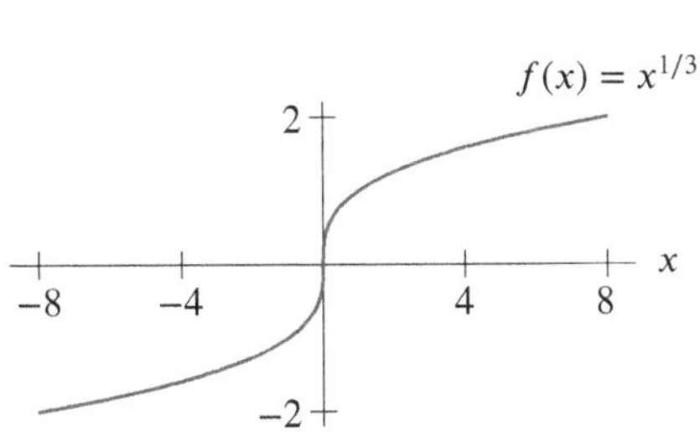

Figure 2.62: Continuous function not differentiable at $x = 0$: Vertical tangent

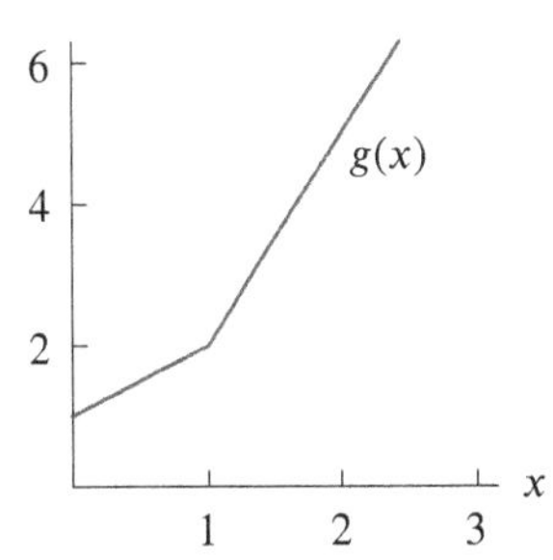

Figure 2.63: Continuous function not differentiable at $x = 1$

Example 2 Investigate the differentiability of $f(x) = x^{1/3}$ at $x = 0$.

Solution This function is smooth at $x = 0$ (no sharp corners) but appears to have a vertical tangent there. (See Figure 2.62.) Looking at the difference quotient at $x = 0$, we see

$$\lim_{h\to 0} \frac{(0+h)^{1/3} - 0^{1/3}}{h} = \lim_{h\to 0} \frac{h^{1/3}}{h} = \lim_{h\to 0} \frac{1}{h^{2/3}}.$$

As $h \to 0$ the denominator becomes small, so the fraction grows without bound. Hence, the function fails to have a derivative at $x = 0$.

Example 3 Consider the function given by the formulas

$$g(x) = \begin{cases} x+1 & \text{if } x \le 1 \\ 3x-1 & \text{if } x > 1. \end{cases}$$

Draw the graph of g. Is g continuous? Is g differentiable at $x = 1$?

Solution The graph in Figure 2.63 has no breaks in it, so the function is continuous. However, the graph has a corner at $x = 1$ which no amount of magnification will remove. To the left of $x = 1$, the slope is 1; to the right of $x = 1$, the slope is 3. Thus, the difference quotient at $x = 1$ has no limit, so the function g is not differentiable at $x = 1$.

The Relationship Between Differentiability and Continuity

The fact that a function which is differentiable at a point has a tangent line suggests that the function is continuous there, as the next theorem shows.

> **Theorem 2.1: A Differentiable Function Is Continuous**
>
> If $f(x)$ is differentiable at a point $x = a$, then $f(x)$ is continuous at $x = a$.

Proof We assume f is differentiable at $x = a$. Then we know that $f'(a)$ exists where

$$f'(a) = \lim_{x\to a} \frac{f(x) - f(a)}{x - a}.$$

To show that f is continuous at $x = a$, we want to show that $\lim_{x\to a} f(x) = f(a)$. We calculate $\lim_{x\to a}(f(x) - f(a))$, hoping to get 0. By algebra, we know that for $x \neq a$,

$$f(x) - f(a) = (x - a) \cdot \frac{f(x) - f(a)}{x - a}.$$

Taking the limits, we have

$$\begin{aligned}
\lim_{x\to a}(f(x) - f(a)) &= \lim_{x\to a}\left((x - a)\frac{f(x) - f(a)}{x - a}\right) \\
&= \left(\lim_{x\to a}(x - a)\right) \cdot \left(\lim_{x\to a}\frac{f(x) - f(a)}{x - a}\right) && \text{(By Theorem 1.2, Property 3)} \\
&= 0 \cdot f'(a) = 0. && \text{(Since } f'(a) \text{ exists)}
\end{aligned}$$

Thus we know that $\lim_{x\to a} f(x) = f(a)$, which means $f(x)$ is continuous at $x = a$.

Exercises and Problems for Section 2.6

EXERCISES

For the graphs in Exercises 1–2, list the x-values for which the function appears to be

(a) Not continuous.  **(b)** Not differentiable. 

1\. 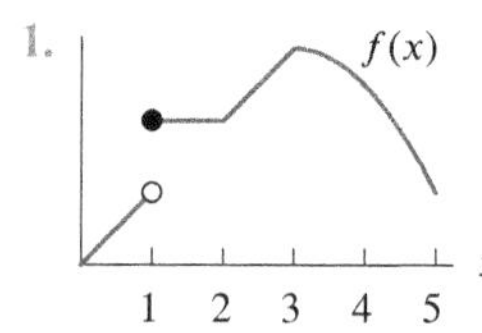

2\. 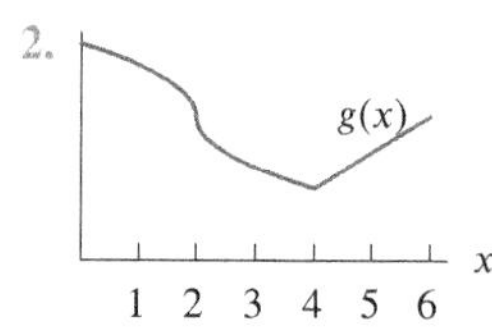

In Exercises 3–4, does the function appear to be differentiable on the interval of x-values shown?

3\. 4. 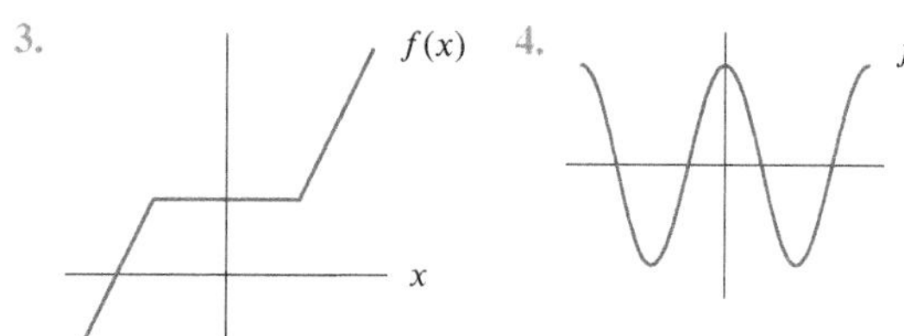

In Exercises 5–7, decide if the function is differentiable at $x = 0$. Try zooming in on a graphing calculator, or calculating the derivative $f'(0)$ from the definition.

5. $f(x) = x \cdot |x|$

6. $f(x) = \begin{cases} -2x & \text{for } x < 0 \\ x^2 & \text{for } x \geq 0 \end{cases}$

7. $f(x) = \begin{cases} (x+1)^2 & \text{for } x < 0 \\ 2x + 1 & \text{for } x \geq 0 \end{cases}$

PROBLEMS

Decide if the functions in Problems 8–10 are differentiable at $x = 0$. Try zooming in on a graphing calculator, or calculating the derivative $f'(0)$ from the definition.

8. $f(x) = (x + |x|)^2 + 1$

9. $f(x) = \begin{cases} x\sin(1/x) + x & \text{for } x \neq 0 \\ 0 & \text{for } x = 0 \end{cases}$

10. $f(x) = \begin{cases} x^2\sin(1/x) & \text{for } x \neq 0 \\ 0 & \text{for } x = 0 \end{cases}$

11. In each of the following cases, sketch the graph of a continuous function $f(x)$ with the given properties.

(a) $f''(x) > 0$ for $x < 2$ and for $x > 2$ and $f'(2)$ is undefined.

(b) $f''(x) > 0$ for $x < 2$ and $f''(x) < 0$ for $x > 2$ and $f'(2)$ is undefined.

12. Look at the graph of $f(x) = (x^2 + 0.0001)^{1/2}$ shown in Figure 2.64. The graph of f appears to have a sharp corner at $x = 0$. Do you think f has a derivative at $x = 0$?

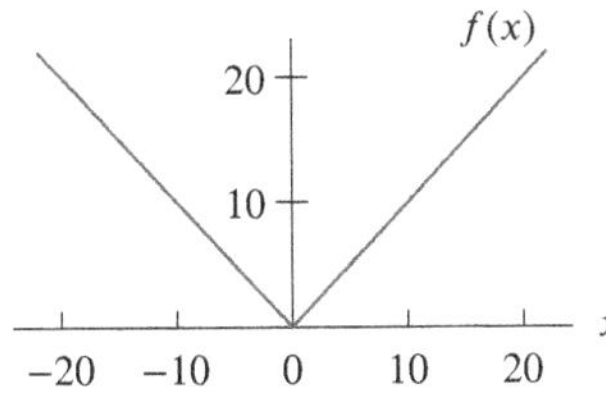

Figure 2.64

13. The acceleration due to gravity, g, varies with height above the surface of the earth, in a certain way. If you go down below the surface of the earth, g varies in a different way. It can be shown that g is given by

$$g = \begin{cases} \dfrac{GMr}{R^3} & \text{for } r < R \\[2ex] \dfrac{GM}{r^2} & \text{for } r \geq R, \end{cases}$$

where R is the radius of the earth, M is the mass of the earth, G is the gravitational constant, and r is the distance to the center of the earth.

(a) Sketch a graph of g against r.
(b) Is g a continuous function of r? Explain your answer.
(c) Is g a differentiable function of r? Explain your answer.

14. An electric charge, Q, in a circuit is given as a function of time, t, by

$$Q = \begin{cases} C & \text{for } t \leq 0 \\ Ce^{-t/RC} & \text{for } t > 0, \end{cases}$$

where C and R are positive constants. The electric current, I, is the rate of change of charge, so

$$I = \frac{dQ}{dt}.$$

(a) Is the charge, Q, a continuous function of time?
(b) Do you think the current, I, is defined for all times, t? [Hint: To graph this function, take, for example, $C = 1$ and $R = 1$.]

15. A magnetic field, B, is given as a function of the distance, r, from the center of a wire as follows:

$$B = \begin{cases} \dfrac{r}{r_0} B_0 & \text{for } r \leq r_0 \\[2ex] \dfrac{r_0}{r} B_0 & \text{for } r > r_0. \end{cases}$$

(a) Sketch a graph of B against r. What is the meaning of the constant B_0?
(b) Is B continuous at $r = r_0$? Give reasons.
(c) Is B differentiable at $r = r_0$? Give reasons.

16. A cable is made of an insulating material in the shape of a long, thin cylinder of radius r_0. It has electric charge distributed evenly throughout it. The electric field, E, at a distance r from the center of the cable is given by

$$E = \begin{cases} kr & \text{for } r \leq r_0 \\ k\dfrac{r_0^2}{r} & \text{for } r > r_0. \end{cases}$$

(a) Is E continuous at $r = r_0$?
(b) Is E differentiable at $r = r_0$?
(c) Sketch a graph of E as a function of r.

17. Graph the function defined by

$$g(r) = \begin{cases} 1 + \cos(\pi r/2) & \text{for } -2 \leq r \leq 2 \\ 0 & \text{for } r < -2 \text{ or } r > 2. \end{cases}$$

(a) Is g continuous at $r = 2$? Explain your answer.
(b) Do you think g is differentiable at $r = 2$? Explain your answer.

18. The potential, ϕ, of a charge distribution at a point on the y-axis is given by

$$\phi = \begin{cases} 2\pi\sigma\left(\sqrt{y^2 + a^2} - y\right) & \text{for } y \geq 0 \\ 2\pi\sigma\left(\sqrt{y^2 + a^2} + y\right) & \text{for } y < 0, \end{cases}$$

where σ and a are positive constants. [Hint: To graph this function, take, for example, $2\pi\sigma = 1$ and $a = 1$.]

(a) Is ϕ continuous at $y = 0$?
(b) Do you think ϕ is differentiable at $y = 0$?

19. Sometimes, odd behavior can be hidden beneath the surface of a rather normal-looking function. Consider the following function:

$$f(x) = \begin{cases} 0 & \text{if } x < 0 \\ x^2 & \text{if } x \geq 0. \end{cases}$$

(a) Sketch a graph of this function. Does it have any vertical segments or corners? Is it differentiable everywhere? If so, sketch the derivative f' of this function. [Hint: You may want to use the result of Example 4 on page 103.]
(b) Is the derivative function, $f'(x)$, differentiable everywhere? If not, at what point(s) is it not differentiable? Draw the second derivative of $f(x)$ wherever it exists. Is the second derivative function, $f''(x)$, differentiable? Continuous?

20. Figure 2.65 shows graphs of four useful but nondifferentiable functions: the step, the sign, the ramp, and the absolute value. Match the graphs with differentiable versions of these functions, called soft versions.

(a) $e^x/(e^x + e^{-x})$
(b) $xe^x/(e^x + e^{-x})$
(c) $(e^x - e^{-x})/(e^x + e^{-x})$
(d) $(xe^x - xe^{-x})/(e^x + e^{-x})$

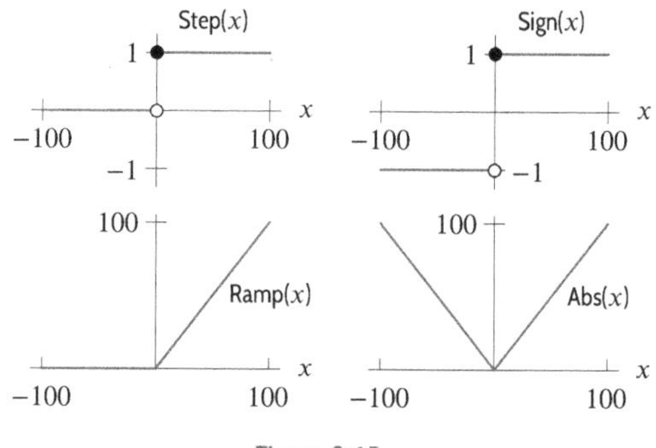

Figure 2.65

Strengthen Your Understanding

In Problems 21–22, explain what is wrong with the statement.

21. A function f that is not differentiable at $x = 0$ has a graph with a sharp corner at $x = 0$.

22. If f is not differentiable at a point then it is not continuous at that point.

In Problems 23–25, give an example of:

23. A continuous function that is not differentiable at $x = 2$.

24. An invertible function that is not differentiable at $x = 0$.

25. A rational function that has zeros at $x = \pm 1$ and is not differentiable at $x = \pm 2$.

Are the statements in Problems 26–30 true or false? If a statement is true, give an example illustrating it. If a statement is false, give a counterexample.

26. There is a function which is continuous on $[1, 5]$ but not differentiable at $x = 3$.

27. If a function is differentiable, then it is continuous.

28. If a function is continuous, then it is differentiable.

29. If a function is not continuous, then it is not differentiable.

30. If a function is not differentiable, then it is not continuous.

31. Which of the following would be a counterexample to the statement: "If f is differentiable at $x = a$ then f is continuous at $x = a$"?

(a) A function which is not differentiable at $x = a$ but is continuous at $x = a$.
(b) A function which is not continuous at $x = a$ but is differentiable at $x = a$.
(c) A function which is both continuous and differentiable at $x = a$.
(d) A function which is neither continuous nor differentiable at $x = a$.

Online Resource: Review problems and Projects

Chapter Three

SHORT-CUTS TO DIFFERENTIATION

Contents

3.1 POWERS AND POLYNOMIALS

Derivative of a Constant Times a Function

Figure 3.1 shows the graph of $y = f(x)$ and of three multiples: $y = 3f(x)$, $y = \frac{1}{2}f(x)$, and $y = -2f(x)$. What is the relationship between the derivatives of these functions? In other words, for a particular x-value, how are the slopes of these graphs related?

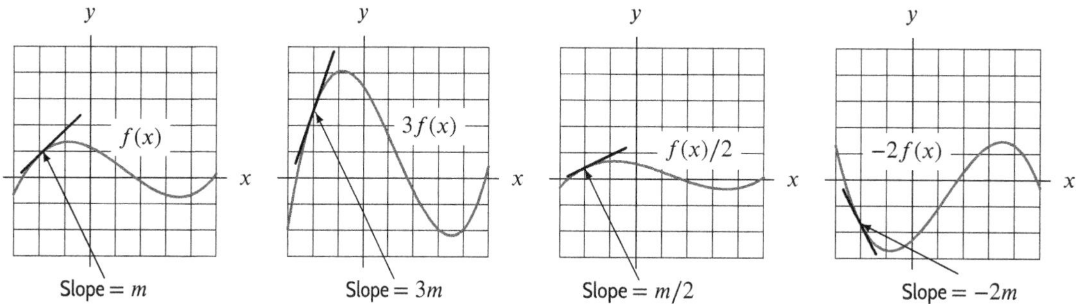

Figure 3.1: A function and its multiples: Derivative of multiple is multiple of derivative

Multiplying the value of a function by a constant stretches or shrinks the original graph (and reflects it across the x-axis if the constant is negative). This changes the slope of the curve at each point. If the graph has been stretched, the "rises" have all been increased by the same factor, whereas the "runs" remain the same. Thus, the slopes are all steeper by the same factor. If the graph has been shrunk, the slopes are all smaller by the same factor. If the graph has been reflected across the x-axis, the slopes will all have their signs reversed. In other words, if a function is multiplied by a constant, c, so is its derivative:

Theorem 3.1: Derivative of a Constant Multiple

If f is differentiable and c is a constant, then

$$\frac{d}{dx}[cf(x)] = cf'(x).$$

Proof Although the graphical argument makes the theorem plausible, to prove it we must use the definition of the derivative:

$$\frac{d}{dx}[cf(x)] = \lim_{h\to 0}\frac{cf(x+h)-cf(x)}{h} = \lim_{h\to 0} c\,\frac{f(x+h)-f(x)}{h}$$
$$= c\lim_{h\to 0}\frac{f(x+h)-f(x)}{h} = cf'(x).$$

We can take c across the limit sign by the properties of limits (part 1 of Theorem 1.2 on page 71).

Derivatives of Sums and Differences

Suppose we have two functions, $f(x)$ and $g(x)$, with the values listed in Table 3.1. Values of the sum $f(x)+g(x)$ are given in the same table.

Table 3.1 *Sum of functions*

x	$f(x)$	$g(x)$	$f(x)+g(x)$
0	100	0	100
1	110	0.2	110.2
2	130	0.4	130.4
3	160	0.6	160.6
4	200	0.8	200.8

We see that adding the increments of $f(x)$ and the increments of $g(x)$ gives the increments of $f(x) + g(x)$. For example, as x increases from 0 to 1, $f(x)$ increases by 10 and $g(x)$ increases by 0.2, while $f(x) + g(x)$ increases by $110.2 - 100 = 10.2$. Similarly, as x increases from 3 to 4, $f(x)$ increases by 40 and $g(x)$ by 0.2, while $f(x) + g(x)$ increases by $200.8 - 160.6 = 40.2$.

From this example, we see that the rate at which $f(x)+g(x)$ is increasing is the sum of the rates at which $f(x)$ and $g(x)$ are increasing. Similar reasoning applies to the difference, $f(x) - g(x)$. In terms of derivatives:

Theorem 3.2: Derivative of Sum and Difference

If f and g are differentiable, then

$$\frac{d}{dx}[f(x)+g(x)] = f'(x)+g'(x) \quad \text{and} \quad \frac{d}{dx}[f(x)-g(x)] = f'(x)-g'(x).$$

Proof Using the definition of the derivative:

$$\begin{aligned}\frac{d}{dx}[f(x)+g(x)] &= \lim_{h\to 0}\frac{[f(x+h)+g(x+h)]-[f(x)+g(x)]}{h}\\ &= \lim_{h\to 0}\left[\underbrace{\frac{f(x+h)-f(x)}{h}}_{\text{Limit of this is } f'(x)}+\underbrace{\frac{g(x+h)-g(x)}{h}}_{\text{Limit of this is } g'(x)}\right]\\ &= f'(x)+g'(x).\end{aligned}$$

We have used the fact that the limit of a sum is the sum of the limits, part 2 of Theorem 1.2 on page 71. The proof for $f(x) - g(x)$ is similar.

Powers of x

In Chapter 2 we showed that

$$f'(x) = \frac{d}{dx}(x^2) = 2x \quad \text{and} \quad g'(x) = \frac{d}{dx}(x^3) = 3x^2.$$

The graphs of $f(x) = x^2$ and $g(x) = x^3$ and their derivatives are shown in Figures 3.2 and 3.3. Notice $f'(x) = 2x$ has the behavior we expect. It is negative for $x < 0$ (when f is decreasing), zero for $x = 0$, and positive for $x > 0$ (when f is increasing). Similarly, $g'(x) = 3x^2$ is zero when $x = 0$, but positive everywhere else, as g is increasing everywhere else.

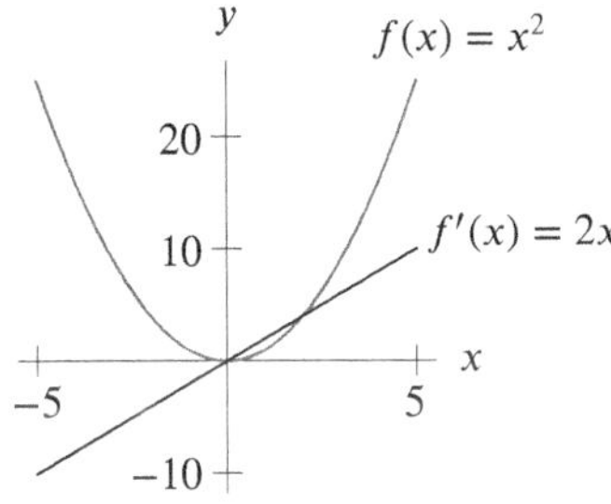

Figure 3.2: Graphs of $f(x) = x^2$ and its derivative $f'(x) = 2x$

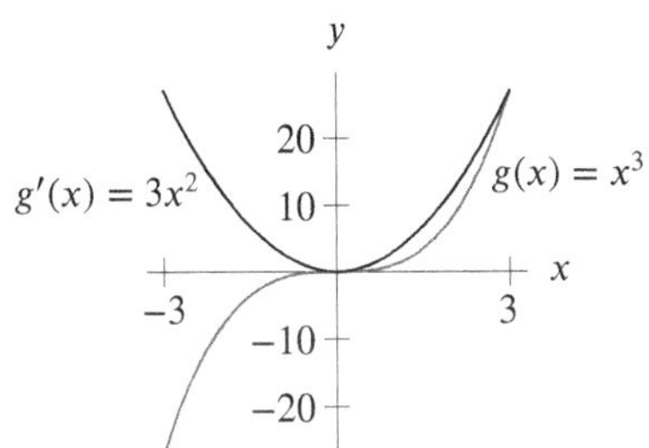

Figure 3.3: Graphs of $g(x) = x^3$ and its derivative $g'(x) = 3x^2$

These examples are special cases of the power rule, which we justify for any positive integer n on page 133:

The Power Rule

For any constant real number n,

$$\frac{d}{dx}(x^n) = nx^{n-1}.$$

Problem 118 (available online) asks you to show that this rule holds for negative integral powers; such powers can also be differentiated using the quotient rule (Section 3.3). In Section 3.6 we indicate how to justify the power rule for powers of the form $1/n$.

Example 1 Use the power rule to differentiate (a) $\frac{1}{x^3}$, (b) $x^{1/2}$, (c) $\frac{1}{\sqrt[3]{x}}$.

Solution

(a) For $n = -3$: $\frac{d}{dx}\left(\frac{1}{x^3}\right) = \frac{d}{dx}(x^{-3}) = -3x^{-3-1} = -3x^{-4} = -\frac{3}{x^4}$.

(b) For $n = 1/2$: $\frac{d}{dx}\left(x^{1/2}\right) = \frac{1}{2}x^{(1/2)-1} = \frac{1}{2}x^{-1/2} = \frac{1}{2\sqrt{x}}$.

(c) For $n = -1/3$: $\frac{d}{dx}\left(\frac{1}{\sqrt[3]{x}}\right) = \frac{d}{dx}\left(x^{-1/3}\right) = -\frac{1}{3}x^{(-1/3)-1} = -\frac{1}{3}x^{-4/3} = -\frac{1}{3x^{4/3}}$.

Example 2 Use the definition of the derivative to justify the power rule for $n = -2$: Show $\frac{d}{dx}(x^{-2}) = -2x^{-3}$.

Solution Provided $x \neq 0$, we have

$$\begin{aligned}
\frac{d}{dx}\left(x^{-2}\right) = \frac{d}{dx}\left(\frac{1}{x^2}\right) = \lim_{h\to 0}\left(\frac{\frac{1}{(x+h)^2} - \frac{1}{x^2}}{h}\right) &= \lim_{h\to 0}\frac{1}{h}\left[\frac{x^2 - (x+h)^2}{(x+h)^2x^2}\right] && \text{(Combining fractions over a common denominator)}\\
&= \lim_{h\to 0}\frac{1}{h}\left[\frac{x^2 - (x^2 + 2xh + h^2)}{(x+h)^2x^2}\right] && \text{(Multiplying out)}\\
&= \lim_{h\to 0}\frac{-2xh - h^2}{h(x+h)^2x^2} && \text{(Simplifying numerator)}\\
&= \lim_{h\to 0}\frac{-2x - h}{(x+h)^2x^2} && \text{(Dividing numerator and denominator by } h)\\
&= \frac{-2x}{x^4} && \text{(Letting } h \to 0)\\
&= -2x^{-3}.
\end{aligned}$$

The graphs of x^{-2} and its derivative, $-2x^{-3}$, are shown in Figure 3.4. Does the graph of the derivative have the features you expect?

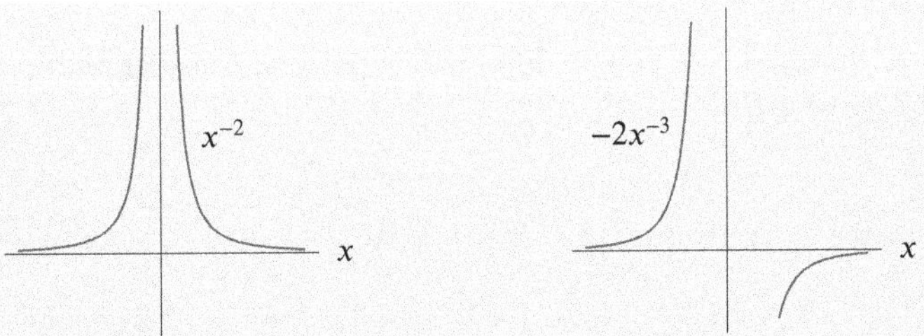

Figure 3.4: Graphs of x^{-2} and its derivative, $-2x^{-3}$

Justification of $\frac{d}{dx}(x^n) = nx^{n-1}$, for n a Positive Integer

To calculate the derivatives of x^2 and x^3, we had to expand $(x+h)^2$ and $(x+h)^3$. To calculate the derivative of x^n, we must expand $(x+h)^n$. Let's look back at the previous expansions:

$$(x+h)^2 = x^2 + 2xh + h^2, \qquad (x+h)^3 = x^3 + 3x^2h + 3xh^2 + h^3,$$

and multiply out a few more examples:

$$(x+h)^4 = x^4 + 4x^3h + 6x^2h^2 + 4xh^3 + h^4,$$
$$(x+h)^5 = x^5 + 5x^4h + \underbrace{10x^3h^2 + 10x^2h^3 + 5xh^4 + h^5}_{\text{Terms involving } h^2 \text{ and higher powers of } h}.$$

In general, we can say

$$(x+h)^n = x^n + nx^{n-1}h + \underbrace{\cdots + h^n}_{\text{Terms involving } h^2 \text{ and higher powers of } h}.$$

We have just seen this is true for $n = 2, 3, 4, 5$. It can be proved in general using the Binomial Theorem (see www.wiley.com/college/hughes-hallett). Now to find the derivative,

$$\begin{aligned}\frac{d}{dx}(x^n) &= \lim_{h\to 0}\frac{(x+h)^n - x^n}{h}\\ &= \lim_{h\to 0}\frac{(x^n + nx^{n-1}h + \cdots + h^n) - x^n}{h}\\ &= \lim_{h\to 0}\frac{nx^{n-1}h + \overbrace{\cdots + h^n}^{\text{Terms involving } h^2 \text{ and higher powers of } h}}{h}.\end{aligned}$$

When we factor out h from terms involving h^2 and higher powers of h, each term will still have an h in it. Factoring and dividing, we get:

$$\frac{d}{dx}(x^n) = \lim_{h\to 0}\frac{h(nx^{n-1} + \cdots + h^{n-1})}{h} = \lim_{h\to 0}(nx^{n-1} + \overbrace{\cdots + h^{n-1}}^{\text{Terms involving } h \text{ and higher powers of } h}).$$

But as $h \to 0$, all terms involving an h will go to 0, so

$$\frac{d}{dx}(x^n) = \lim_{h\to 0}(nx^{n-1} + \underbrace{\cdots + h^{n-1}}_{\text{These terms go to } 0}) = nx^{n-1}.$$

Derivatives of Polynomials

Now that we know how to differentiate powers, constant multiples, and sums, we can differentiate any polynomial.

Example 3 Find the derivatives of (a) $5x^2 + 3x + 2$, (b) $\sqrt{3}x^7 - \dfrac{x^5}{5} + \pi$.

Solution (a)

$$\begin{aligned}\frac{d}{dx}(5x^2+3x+2) &= 5\frac{d}{dx}(x^2) + 3\frac{d}{dx}(x) + \frac{d}{dx}(2)\\ &= 5\cdot 2x + 3\cdot 1 + 0 \qquad \text{(Since the derivative of a constant, } \tfrac{d}{dx}(2)\text{, is zero.)}\\ &= 10x+3.\end{aligned}$$

(b)

$$\begin{aligned}\frac{d}{dx}\left(\sqrt{3}x^7 - \frac{x^5}{5} + \pi\right) &= \sqrt{3}\frac{d}{dx}(x^7) - \frac{1}{5}\frac{d}{dx}(x^5) + \frac{d}{dx}(\pi)\\ &= \sqrt{3}\cdot 7x^6 - \frac{1}{5}\cdot 5x^4 + 0 \qquad \text{(Since } \pi \text{ is a constant, } d\pi/dx = 0.)\\ &= 7\sqrt{3}x^6 - x^4.\end{aligned}$$

We can also use the rules we have seen so far to differentiate expressions that are not polynomials.

Example 4 Differentiate (a) $5\sqrt{x} - \dfrac{10}{x^2} + \dfrac{1}{2\sqrt{x}}$. (b) $0.1x^3 + 2x^{\sqrt{2}}$.

Solution (a)

$$\begin{aligned}\frac{d}{dx}\left(5\sqrt{x} - \frac{10}{x^2} + \frac{1}{2\sqrt{x}}\right) &= \frac{d}{dx}\left(5x^{1/2} - 10x^{-2} + \frac{1}{2}x^{-1/2}\right)\\ &= 5\cdot\frac{1}{2}x^{-1/2} - 10(-2)x^{-3} + \frac{1}{2}\left(-\frac{1}{2}\right)x^{-3/2}\\ &= \frac{5}{2\sqrt{x}} + \frac{20}{x^3} - \frac{1}{4x^{3/2}}.\end{aligned}$$

(b) $$\frac{d}{dx}(0.1x^3 + 2x^{\sqrt{2}}) = 0.1\frac{d}{dx}(x^3) + 2\frac{d}{dx}(x^{\sqrt{2}}) = 0.3x^2 + 2\sqrt{2}x^{\sqrt{2}-1}.$$

Example 5 Find the second derivative and interpret its sign for
(a) $f(x) = x^2$, (b) $g(x) = x^3$, (c) $k(x) = x^{1/2}$.

Solution

(a) If $f(x) = x^2$, then $f'(x) = 2x$, so $f''(x) = \dfrac{d}{dx}(2x) = 2$. Since f'' is always positive, f is concave up, as expected for a parabola opening upward. (See Figure 3.5.)

(b) If $g(x) = x^3$, then $g'(x) = 3x^2$, so $g''(x) = \dfrac{d}{dx}(3x^2) = 3\dfrac{d}{dx}(x^2) = 3\cdot 2x = 6x$. This is positive for $x > 0$ and negative for $x < 0$, which means x^3 is concave up for $x > 0$ and concave down for $x < 0$. (See Figure 3.6.)

(c) If $k(x) = x^{1/2}$, then $k'(x) = \frac{1}{2}x^{(1/2)-1} = \frac{1}{2}x^{-1/2}$, so

$$k''(x) = \frac{d}{dx}\left(\frac{1}{2}x^{-1/2}\right) = \frac{1}{2}\cdot\left(-\frac{1}{2}\right)x^{-(1/2)-1} = -\frac{1}{4}x^{-3/2}.$$

Now k' and k'' are only defined on the domain of k, that is, $x \geq 0$. When $x > 0$, we see that $k''(x)$ is negative, so k is concave down. (See Figure 3.7.)

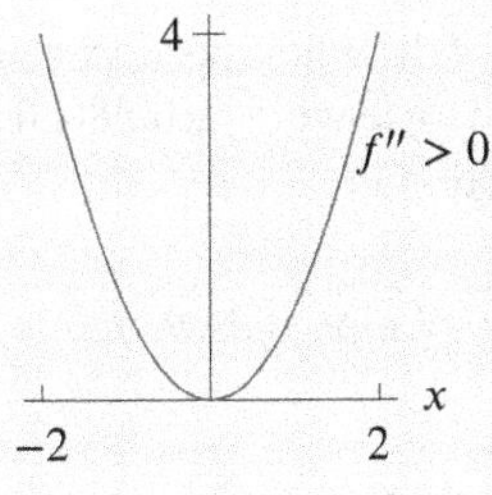

Figure 3.5: $f(x) = x^2$ has $f''(x) = 2$

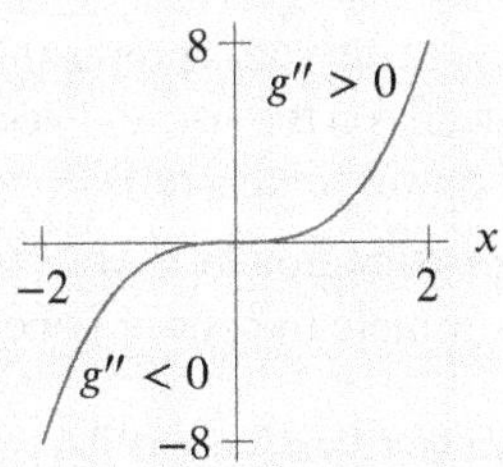

Figure 3.6: $g(x) = x^3$ has $g''(x) = 6x$

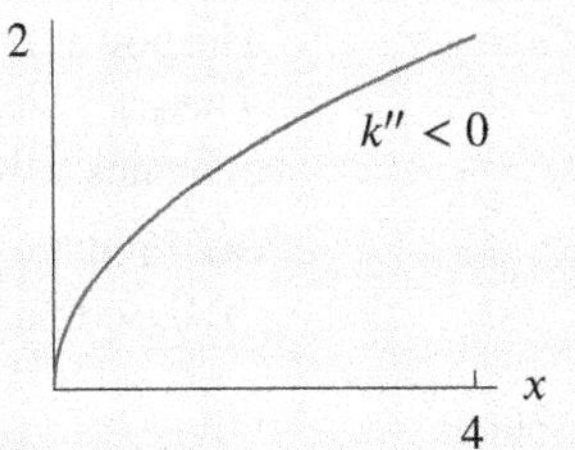

Figure 3.7: $k(x) = x^{1/2}$ has $k''(x) = -\frac{1}{4}x^{-3/2}$

Example 6 If the position of a body, in meters, is given as a function of time t, in seconds, by

$$s = -4.9t^2 + 5t + 6,$$

find the velocity and acceleration of the body at time t.

Solution The velocity, v, is the derivative of the position:

$$v = \frac{ds}{dt} = \frac{d}{dt}(-4.9t^2 + 5t + 6) = -9.8t + 5,$$

and the acceleration, a, is the derivative of the velocity:

$$a = \frac{dv}{dt} = \frac{d}{dt}(-9.8t + 5) = -9.8.$$

Note that v is in meters/second and a is in meters/second^2.

The Tangent Line Approximation for Power Functions

The derivative $f'(a)$ is the slope of the function $y = f(x)$ at $x = a$—that is, $f'(a)$ is the slope of the tangent line to the graph of $y = f(x)$ at $x = a$. Thus the equation of the tangent line is, for some constant b,

$$y = b + f'(a)x.$$

This equation is usually written in a form obtained from the difference quotient

$$f'(a) = \frac{y - f(a)}{x - a}.$$

Multiplying by $(x - a)$ gives $f'(a)(x - a) = y - f(a)$, so we can write

$$y = \underbrace{f(a)}_{\text{Value at } a} + \underbrace{f'(a)(x - a)}_{\text{Change from } a \text{ to } x}.$$

Since the tangent line lies close to the graph of the function for values of x close to a, we approximate the function value, $f(x)$, by the y-value on the tangent line, giving the following:

> **Tangent Line Approximation**
>
> For x near a:
>
> $$f(x) \approx f(a) + f'(a)(x - a).$$

In the following example, we see how the tangent line can be used to approximate the wind power generated at different wind speeds.

Example 7 The power P generated by a wind turbine is proportional to the cube of the wind speed, v, so $P = kv^3$, where k is a constant that depends on the size and design. The North Hoyle wind turbine[1] off the coast of Wales, UK, has $k = 2$ for power in kilowatts (kw) and wind speed in meters/sec (mps).

(a) Find the tangent line approximation to $P = 2v^3$ at $v = 10$ mps.

(b) Use this tangent line to estimate the power generated when $v = 12$ mps and when $v = 9.5$ mps.

Solution Since $P = 2v^3$, the derivative is $dP/dv = 6v^2$.

(a) At $v = 10$ we have $P = 2 \cdot 10^3 = 2000$ kw and $dP/dv = 6 \cdot 10^2 = 600$ kw/mps. Thus, the tangent line approximation is

$$P \approx 2000 + 600(v - 10).$$

(b) When $v = 12$ and $v = 9.5$, substituting gives

$$P \approx 2000 + \underbrace{600(12 - 10)}_{600(2)=1200} = 3200 \text{ kw.}$$

$$P \approx 2000 + \underbrace{600(9.5 - 10)}_{600(-0.5)=-300} = 1700 \text{ kw.}$$

Notice that the derivative, 600 kw/mps, acts as a multiplier, converting the increase in wind speed, $\Delta v = 2$ mps, into an increase in power, $\Delta P = 600 \cdot 2 = 1200$ kw. Similarly, the decrease in wind speed $\Delta v = -0.5$ mps is converted to a decrease in power $\Delta P = -300$ kw.

For comparison, the exact values are $P = 2 \cdot 12^3 = 3456$ kw and $P = 2 \cdot 9.5^3 = 1714.75$ kw.

Exercises and Problems for Section 3.1 Online Resource: Additional Problems for Section 3.1

EXERCISES

1. Let $f(x) = 7$. Using the definition of the derivative, show that $f'(x) = 0$ for all values of x.

2. Let $f(x) = 17x+11$. Use the definition of the derivative to calculate $f'(x)$.

■ For Exercises 3–5, determine if the derivative rules from this section apply. If they do, find the derivative. If they don't apply, indicate why.

3. $y = 3^x$

4. $y = x^3$

5. $y = x^\pi$

■ For Exercises 6–49, find the derivatives of the given functions. Assume that a, b, c, and k are constants.

6. $y = x^{12}$

7. $y = x^{11}$

8. $y = -x^{-11}$

9. $y = x^{-12}$

10. $y = x^{3.2}$

11. $y = x^{-3/4}$

12. $y = x^{4/3}$

13. $y = x^{3/4}$

14. $y = x^2 + 5x + 7$

15. $f(t) = t^3 - 3t^2 + 8t - 4$

16. $f(x) = \dfrac{1}{x^4}$

17. $g(t) = \dfrac{1}{t^5}$

18. $f(z) = -\dfrac{1}{z^{6.1}}$

19. $y = \dfrac{1}{r^{7/2}}$

20. $y = \sqrt{x}$

21. $f(x) = \sqrt[4]{x}$

22. $h(\theta) = \dfrac{1}{\sqrt[3]{\theta}}$

23. $f(x) = \sqrt{\dfrac{1}{x^3}}$

24. $h(x) = \ln e^{ax}$

25. $y = 4x^{3/2} - 5x^{1/2}$

26. $f(t) = 3t^2 - 4t + 1$

27. $y = 17x + 24x^{1/2}$

28. $y = z^2 + \dfrac{1}{2z}$

29. $f(x) = 5x^4 + \dfrac{1}{x^2}$

[1] http://www.raeng.org.uk/publications/other/23-wind-turbine, Royal Academy of Engineering. Accessed January 7, 2016.

30. $h(w) = -2w^{-3} + 3\sqrt{w}$

31. $y = -3x^4 - 4x^3 - 6x$

32. $y = 3t^5 - 5\sqrt{t} + \dfrac{7}{t}$

33. $y = 3t^2 + \dfrac{12}{\sqrt{t}} - \dfrac{1}{t^2}$

34. $y = \sqrt{x}(x+1)$

35. $y = t^{3/2}(2 + \sqrt{t})$

36. $h(t) = \dfrac{3}{t} + \dfrac{4}{t^2}$

37. $h(\theta) = \theta(\theta^{-1/2} - \theta^{-2})$

38. $y = \dfrac{x^2+1}{x}$

39. $f(z) = \dfrac{z^2+1}{3z}$

40. $g(x) = \dfrac{x^2 + \sqrt{x} + 1}{x^{3/2}}$

41. $y = \dfrac{\theta - 1}{\sqrt{\theta}}$

42. $g(t) = \dfrac{\sqrt{t}(1+t)}{t^2}$

43. $j(x) = \dfrac{x^3}{a} + \dfrac{a}{b}x^2 - cx$

44. $f(x) = \dfrac{ax+b}{x}$

45. $h(x) = \dfrac{ax+b}{c}$

46. $V = \frac{4}{3}\pi r^2 b$

47. $w = 3ab^2 q$

48. $y = ax^2 + bx + c$

49. $P = a + b\sqrt{t}$

■ In Exercises 50–57, use the tangent line approximation.

50. Given $f(4) = 5, f'(4) = 7$, approximate $f(4.02)$.

51. Given $f(4) = 5, f'(4) = 7$, approximate $f(3.92)$.

52. Given $f(5) = 3, f'(5) = -2$, approximate $f(5.03)$.

53. Given $f(2) = -4, f'(2) = -3$, approximate $f(1.95)$.

54. Given $f(-3) = -4, f'(-3) = 2$, approximate $f(-2.99)$.

55. Given $f(3) = -4, f'(3) = -2$ approximate $f(2.99)$.

56. Given $f(x) = x^4 - x^2 + 3$ approximate $f(1.04)$.

57. Given $f(x) = x^3 + x^2 - 6$, approximate $f(0.97)$.

PROBLEMS

58. If $f(x) = 6x^4 - 2x + 7$, find $f'(x)$, $f''(x)$, and $f'''(x)$.

59. **(a)** Let $f(x) = x^7$. Find $f'(x)$, $f''(x)$, and $f'''(x)$.
(b) Find the smallest n such that $f^{(n)}(x) = 0$.

60. If $p(t) = t^3 + 2t^2 - t + 4$, find d^2p/dt^2 and d^3p/dt^3.

61. If $w(x) = \sqrt{x} + (1/\sqrt{x})$, find d^2w/dx^2 and d^3w/dx^3.

■ For Problems 62–67, determine if the derivative rules from this section apply. If they do, find the derivative. If they don't apply, indicate why.

62. $y = (x+3)^{1/2}$

63. $y = \pi^x$

64. $g(x) = x^\pi - x^{-\pi}$

65. $y = 3x^2 + 4$

66. $y = \dfrac{1}{3x^2+4}$

67. $y = \dfrac{1}{3z^2} + \dfrac{1}{4}$

68. If $f(x) = (3x+8)(2x-5)$, find $f'(x)$ and $f''(x)$.

69. Find the equation of the line tangent to the graph of f at $(1,1)$, where f is given by $f(x) = 2x^3 - 2x^2 + 1$.

70. Find the equation of the line tangent to $y = x^2 + 3x - 5$ at $x = 2$.

71. Find the equation of the line tangent to $f(x)$ at $x = 2$, if

$$f(x) = \frac{x^3}{2} - \frac{4}{3x}.$$

72. **(a)** Find the equation of the tangent line to $f(x) = x^3$ at the point where $x = 2$.
(b) Graph the tangent line and the function on the same axes. If the tangent line is used to estimate values of the function near $x = 2$, will the estimates be overestimates or underestimates?

73. **(a)** Use Figure 3.8 to rank the quantities $f'(-1), f'(0), f'(1), f'(4)$ from smallest to largest.
(b) Confirm your answer by calculating the quantities using the formula, $f(x) = x^3 - 3x^2 + 2x + 10$.

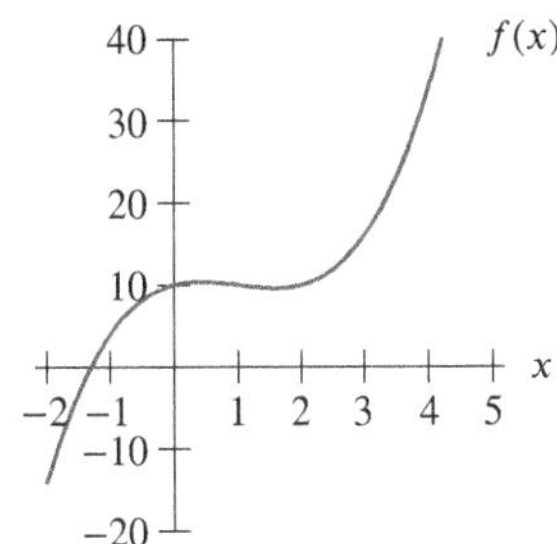

Figure 3.8

74. The graph of $y = x^3 - 9x^2 - 16x + 1$ has a slope of 5 at two points. Find the coordinates of the points.

75. On what intervals is the graph of $f(x) = x^4 - 4x^3$ both decreasing and concave up?

76. For what values of x is the function $y = x^5 - 5x$ both increasing and concave up?

77. **(a)** Find the *eighth* derivative of $f(x) = x^7 + 5x^5 - 4x^3 + 6x - 7$. Think ahead!
(The nth derivative, $f^{(n)}(x)$, is the result of differentiating $f(x)$ n times.)
(b) Find the seventh derivative of $f(x)$.

■ For the functions in Problems 78–81:
(a) Find the derivative at $x = -1$.
(b) Find the second derivative at $x = -1$.

(c) Use your answers to parts (a) and (b) to match the function to one of the graphs in Figure 3.9, each of which is shown centered on the point $(-1, -1)$.

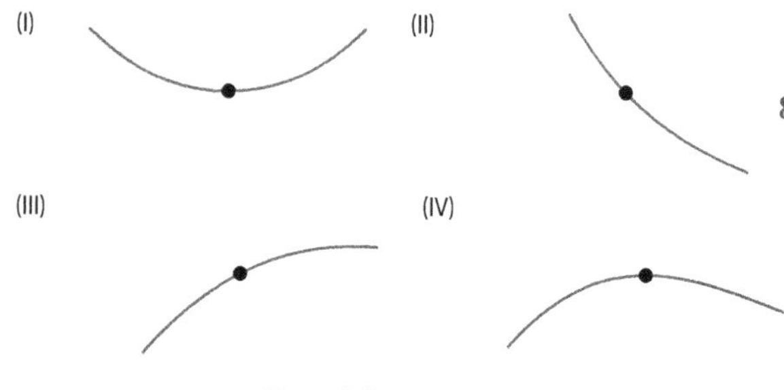

Figure 3.9

78. $k(x) = x^3 - x - 1$

79. $f(x) = 2x^3 + 3x^2 - 2$

80. $g(x) = x^4 - x^2 - 2x - 3$

81. $h(x) = 2x^4 + 8x^3 + 15x^2 + 14x + 4$

82. With t in years since 2016, the height of a sand dune (in centimeters) is $f(t) = 700 - 3t^2$. Find $f(5)$ and $f'(5)$. Using units, explain what each means in terms of the sand dune.

83. A rubber balloon contains neon. As the air pressure, P (in atmospheres), outside the balloon increases, the volume of gas, V (in liters), in the balloon decreases according to $V = f(P) = 25/P$.

   (a) Evaluate and interpret $f(2)$, including units.
   (b) Evaluate and interpret $f'(2)$, including units.
   (c) Assuming that the pressure increases at a constant rate, does the volume of the balloon decrease faster when the pressure is 1 atmosphere or when the pressure is 2 atmospheres? Justify your answer.

84. A ball is dropped from the top of the Empire State building to the ground below. The height, y, of the ball above the ground (in feet) is given as a function of time, t (in seconds), by

$$y = 1250 - 16t^2.$$

   (a) Find the velocity of the ball at time t. What is the sign of the velocity? Why is this to be expected?
   (b) Show that the acceleration of the ball is a constant. What are the value and sign of this constant?
   (c) When does the ball hit the ground, and how fast is it going at that time? Give your answer in feet per second and in miles per hour (1 ft/sec = 15/22 mph).

85. At a time t seconds after it is thrown up in the air, a tomato is at a height of $f(t) = -4.9t^2 + 25t + 3$ meters.

   (a) What is the average velocity of the tomato during the first 2 seconds? Give units.
   (b) Find (exactly) the instantaneous velocity of the tomato at $t = 2$. Give units.
   (c) What is the acceleration at $t = 2$?
   (d) How high does the tomato go?
   (e) How long is the tomato in the air?

86. Let $f(t)$ and $g(t)$ give, respectively, the amount of water (in acre-feet) in two different reservoirs on day t. Suppose that $f(0) = 2000, g(0) = 1500$ and that $f'(0) = 11, g'(0) = 13.5$. Let $h(t) = f(t) - g(t)$.

   (a) Evaluate $h(0)$ and $h'(0)$. What do these quantities tell you about the reservoir?
   (b) Assume h' is constant for $0 \le t \le 250$. Does h have any zeros? What does this tell you about the reservoirs?

87. A jökulhlaup is the rapid draining of a glacial lake when an ice dam bursts. The maximum outflow rate, Q (in m^3/sec), during a jökulhlaup is given[2] in terms of its volume, v (in km^3), before the dam-break by $Q = 7700v^{0.67}$.

   (a) Find $\dfrac{dQ}{dv}$.
   (b) Evaluate $\left.\dfrac{dQ}{dv}\right|_{v=0.1}$. Include units. What does this derivative mean for glacial lakes?

88. (a) For $y = kx^n$, show that near any point $x = a$, we have $\Delta y/y \approx n\Delta x/a$.
   (b) Interpret this relationship in terms of percent change in y and x.

In Problems 89–92, use the fact that for a power function $y = kx^n$, for small changes, the percent change in output y is approximately n times the percent change in input x. (See Problem 88.)

89. An error of 5% in the measurement of the radius r of a circle leads to what percent error in the area A?

90. If we want to measure the volume V of a sphere accurate to 3%, how accurately must we measure the radius r?

91. The stopping distance s, in feet, of a car traveling v mph is $s = v^2/20$. An increase of 10% in the speed of the car leads to what percent increase in the stopping distance?

92. The average wind speed in Hyannis, MA, in August is 9 mph. In nearby Nantucket, it is 10 mph. What percent increase in power, P, is there for a wind turbine in Nantucket compared to Hyannis in August? Assume $P = kv^3$, where v is wind speed.

[2] J. J. Clague and W. H. Mathews, "The magnitude of joulhlaups." *Journal of Glaciology*, vol.12, issue 66, pp. 501–504.

93. The depth, h (in mm), of the water runoff down a slope during steady rain[3] is a function of the distance, x (in meters), from the top of the slope, $h = f(x) = 0.07x^{2/3}$.

(a) Find $f'(x)$.
(b) Find $f'(30)$. Include units.
(c) Explain how you can use your answer to part (b) to estimate the difference in runoff depths between a point 30 meters down the slope and a point 6 meters farther down.

94. If M is the mass of the earth and G is a constant, the acceleration due to gravity, g, at a distance r from the center of the earth is given by

$$g = \frac{GM}{r^2}.$$

(a) Find dg/dr.
(b) What is the practical interpretation (in terms of acceleration) of dg/dr? Why would you expect it to be negative?
(c) You are told that $M = 6 \cdot 10^{24}$ and $G = 6.67 \cdot 10^{-20}$ where M is in kilograms, r in kilometers, and g in km/sec^2. What is the value of dg/dr at the surface of the earth ($r = 6400$ km)? Include units.
(d) What does this tell you about whether or not it is reasonable to assume g is constant near the surface of the earth?

95. The period, T, of a pendulum is given in terms of its length, l, by

$$T = 2\pi\sqrt{\frac{l}{g}},$$

where g is the acceleration due to gravity (a constant).

(a) Find dT/dl.
(b) What is the sign of dT/dl? What does this tell you about the period of pendulums?

96. (a) Use the formula for the area of a circle of radius r, $A = \pi r^2$, to find dA/dr.
(b) The result from part (a) should look familiar. What does dA/dr represent geometrically?
(c) Use the difference quotient to explain the observation you made in part (b).

97. Show that for any power function $f(x) = x^n$, we have $f'(1) = n$.

98. Suppose W is proportional to r^3. The derivative dW/dr is proportional to what power of r?

99. Given a power function of the form $f(x) = ax^n$, with $f'(2) = 3$ and $f'(4) = 24$, find n and a.

100. (a) Find the value of a making $f(x)$ continuous at $x = 1$:

$$f(x) = \begin{cases} ax & 0 \le x \le 1 \\ x^2 + 3 & 1 < x \le 2. \end{cases}$$

(b) With the value of a you found in part (a), does $f(x)$ have a derivative at every point in $0 < x < 2$? Explain.

101. Let $f(x) = x^3+3x^2-2x+1$ and $g(x) = x^3+3x^2-2x-4$.

(a) Show that the derivatives of $f(x)$ and $g(x)$ are the same.
(b) Use the graphs of $f(x)$ and $g(x)$ to explain why their derivatives are the same.
(c) Are there other functions which share the same derivative as $f(x)$ and $g(x)$?

102. Let $f(x) = x^4 - 3x^2 + 1$.

(a) Show that $f(x)$ is an even function.
(b) Show that $f'(x)$ is an odd function.
(c) If $g(x)$ is a polynomial that is an even function, will its derivative always be an odd function?

Strengthen Your Understanding

■ In Problems 103–104, explain what is wrong with the statement.

103. The only function that has derivative $2x$ is x^2.

104. The derivative of $f(x) = 1/x^2$ is $f'(x) = 1/(2x)$.

■ In Problems 105–107, give an example of:

105. Two functions $f(x)$ and $g(x)$ such that

$$\frac{d}{dx}(f(x) + g(x)) = 2x + 3.$$

106. A function whose derivative is $g'(x) = 2x$ and whose graph has no x-intercepts.

107. A function which has second derivative equal to 6 everywhere.

■ In Problems 108–110, is the statement true or false? Give an explanation for your answer.

108. The derivative of a polynomial is always a polynomial.

109. The derivative of π/x^2 is $-\pi/x$.

110. If $f'(2) = 3.1$ and $g'(2) = 7.3$, then the graph of $f(x) + g(x)$ has slope 10.4 at $x = 2$.

■ In Problems 111–112, is the statement true or false? You are told that f'' and g'' exist and that f and g are concave up for all x. If a statement is true, explain how you know. If a statement is false, give a counterexample.

111. $f(x) + g(x)$ is concave up for all x.

112. $f(x) - g(x)$ cannot be concave up for all x.

[3] R. S. Anderson and S. P. Anderson, *Geomorphology*, Cambridge University Press, p. 369.

3.2 THE EXPONENTIAL FUNCTION

What do we expect the graph of the derivative of the exponential function $f(x) = a^x$ to look like? The exponential function in Figure 3.10 increases slowly for $x < 0$ and more rapidly for $x > 0$, so the values of f' are small for $x < 0$ and larger for $x > 0$. Since the function is increasing for all values of x, the graph of the derivative must lie above the x-axis. It appears that the graph of f' may resemble the graph of f itself.

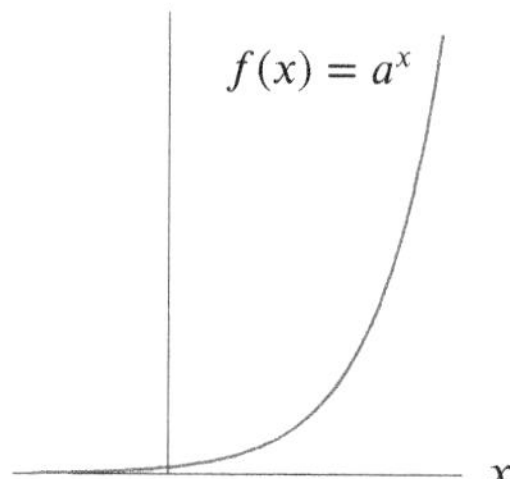

Figure 3.10: $f(x) = a^x$, with $a > 1$

In this section we see that $f'(x) = k \cdot a^x$, where k is a constant, so in fact $f'(x)$ is proportional to $f(x)$. This property of exponential functions makes them particularly useful in modeling because many quantities have rates of change which are proportional to themselves. For example, the simplest model of population growth has this property.

Derivatives of Exponential Functions and the Number e

We start by calculating the derivative of $g(x) = 2^x$, which is given by

$$g'(x) = \lim_{h\to 0}\left(\frac{2^{x+h} - 2^x}{h}\right) = \lim_{h\to 0}\left(\frac{2^x 2^h - 2^x}{h}\right) = \lim_{h\to 0} 2^x\left(\frac{2^h - 1}{h}\right)$$

$$= \lim_{h\to 0}\left(\frac{2^h - 1}{h}\right)\cdot 2^x. \qquad \text{(Since } x \text{ and } 2^x \text{ are fixed during this calculation)}$$

To find $\lim_{h\to 0}(2^h - 1)/h$, see Table 3.2, where we have substituted values of h near 0. The table suggests that the limit exists and has value 0.693. Let us call the limit k, so $k = 0.693$. Then

$$\frac{d}{dx}(2^x) = k \cdot 2^x = 0.693 \cdot 2^x.$$

So the derivative of 2^x is proportional to 2^x with constant of proportionality 0.693. A similar calculation shows that the derivative of $f(x) = a^x$ is

$$f'(x) = \lim_{h\to 0}\left(\frac{a^{x+h} - a^x}{h}\right) = \lim_{h\to 0}\left(\frac{a^h - 1}{h}\right)\cdot a^x.$$

Table 3.2

h	$(2^h - 1)/h$
−0.1	0.6697
−0.01	0.6908
−0.001	0.6929
0.001	0.6934
0.01	0.6956
0.1	0.7177

Table 3.3

a	$k = \lim_{h\to 0}\frac{a^h-1}{h}$
2	0.693
3	1.099
4	1.386
5	1.609
6	1.792
7	1.946

Table 3.4

h	$(1+h)^{1/h}$
−0.001	2.7196422
−0.0001	2.7184178
−0.00001	2.7182954
0.00001	2.7182682
0.0001	2.7181459
0.001	2.7169239

The quantity $\lim_{h\to 0}(a^h - 1)/h$ is also a constant, although the value of the constant depends on a. Writing $k = \lim_{h\to 0}(a^h - 1)/h$, we see that the derivative of $f(x) = a^x$ is proportional to a^x:

$$\frac{d}{dx}(a^x) = k \cdot a^x.$$

For particular values of a, we can estimate k by substituting values of h near 0 into the expression $(a^h - 1)/h$. Table 3.3 shows the results. Notice that for $a = 2$, the value of k is less than 1, while for $a = 3, 4, 5, \ldots$, the values of k are greater than 1. The values of k appear to be increasing, so we guess that there is a value of a between 2 and 3 for which $k = 1$. If so, we have found a value of a with the remarkable property that the function a^x is equal to its own derivative.

So let us look for such an a. This means we want to find a such that

$$\lim_{h\to 0} \frac{a^h - 1}{h} = 1, \qquad \text{or, for small } h, \qquad \frac{a^h - 1}{h} \approx 1.$$

Solving for a, we can estimate a as follows:

$$a^h - 1 \approx h, \qquad \text{or} \qquad a^h \approx 1 + h, \qquad \text{so} \qquad a \approx (1 + h)^{1/h}.$$

Taking small values of h, as in Table 3.4, we see $a \approx 2.718\ldots$. This is the number e introduced in Chapter 1. In fact, it can be shown that if

$$e = \lim_{h\to 0}(1 + h)^{1/h} = 2.718\ldots \qquad \text{then} \qquad \lim_{h\to 0} \frac{e^h - 1}{h} = 1.$$

This means that e^x is its own derivative:

$$\boxed{\frac{d}{dx}(e^x) = e^x.}$$

Figure 3.11 shows the graphs 2^x, 3^x, and e^x together with their derivatives. Notice that the derivative of 2^x is below the graph of 2^x, since $k < 1$ there, and the graph of the derivative of 3^x is above the graph of 3^x, since $k > 1$ there. With $e \approx 2.718$, the function e^x and its derivative are identical.

Note on Round-Off Error and Limits

If we try to evaluate $(1 + h)^{1/h}$ on a calculator by taking smaller and smaller values of h, the values of $(1 + h)^{1/h}$ at first get closer to $2.718\ldots$. However, they will eventually move away again because of the *round-off error* (that is, errors introduced by the fact that the calculator can only hold a certain number of digits).

As we try smaller and smaller values of h, how do we know when to stop? Unfortunately, there is no fixed rule. A calculator can only suggest the value of a limit, but can never confirm that this value is correct. In this case, it looks like the limit is $2.718\ldots$ because the values of $(1 + h)^{1/h}$ approach this number for a while. To be sure this is correct, we have to find the limit analytically.

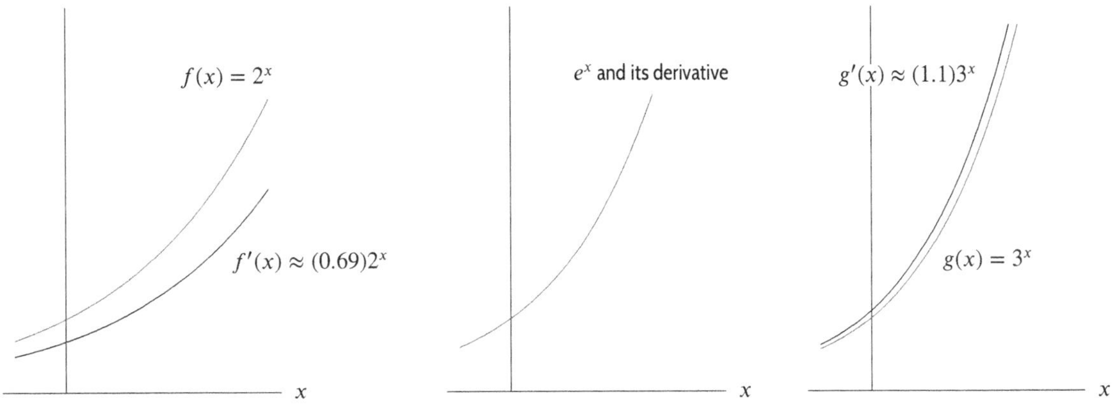

Figure 3.11: Graphs of the functions 2^x, e^x, and 3^x and their derivatives

A Formula for the Derivative of a^x

To get a formula for the derivative of a^x, we must calculate

$$f'(x) = \lim_{h\to 0} \frac{a^{x+h} - a^x}{h} = \underbrace{\left(\lim_{h\to 0} \frac{a^h - 1}{h}\right)}_{k} a^x.$$

However, without knowing the value of a, we can't use a calculator to estimate k. We take a different approach, rewriting $a = e^{\ln a}$, so

$$\lim_{h\to 0} \frac{a^h - 1}{h} = \lim_{h\to 0} \frac{(e^{\ln a})^h - 1}{h} = \lim_{h\to 0} \frac{e^{(\ln a)h} - 1}{h}.$$

To evaluate this limit, we use a limit that we already know:

$$\lim_{h\to 0} \frac{e^h - 1}{h} = 1.$$

In order to use this limit, we substitute $t = (\ln a)h$. Since t approaches 0 as h approaches 0, we have

$$\lim_{h\to 0} \frac{e^{(\ln a)h} - 1}{h} = \lim_{t\to 0} \frac{e^t - 1}{(t/\ln a)} = \lim_{t\to 0}\left(\ln a \cdot \frac{e^t - 1}{t}\right) = \ln a \left(\lim_{t\to 0} \frac{e^t - 1}{t}\right) = (\ln a) \cdot 1 = \ln a.$$

Thus, we have

$$f'(x) = \lim_{h\to 0} \frac{a^{x+h} - a^x}{h} = \left(\lim_{h\to 0} \frac{a^h - 1}{h}\right) a^x = (\ln a)a^x.$$

In Section 3.6 we obtain the same result by another method. We conclude that:

$$\boxed{\frac{d}{dx}(a^x) = (\ln a)a^x.}$$

Thus, for any a, the derivative of a^x is proportional to a^x. The constant of proportionality is $\ln a$. The derivative of a^x is equal to a^x if the constant of proportionality is 1, that is, if $\ln a = 1$, then $a = e$. The fact that the constant of proportionality is 1 when $a = e$ makes e a particularly convenient base for exponential functions.

Example 1 Differentiate $2 \cdot 3^x + 5e^x$.

Solution

$$\frac{d}{dx}(2 \cdot 3^x + 5e^x) = 2\frac{d}{dx}(3^x) + 5\frac{d}{dx}(e^x) = 2\ln 3 \cdot 3^x + 5e^x \approx (2.1972)3^x + 5e^x.$$

We can now use the new differentiation formula to compute rates.

Example 2 The population of the world in billions can be modeled by the function $f(t) = 6.91(1.011)^t$, where t is years since 2010. Find and interpret $f(0)$ and $f'(0)$.

Solution We have $f(t) = 6.91(1.011)^t$ so $f'(t) = 6.91(\ln 1.011)(1.011)^t = 0.0756(1.011)^t$. Therefore,

$$f(0) = 6.91 \text{ billion people}$$

and

$$f'(0) = 0.0756 \text{ billion people per year.}$$

In 2010, the population of the world was 6.91 billion people and was increasing at a rate of 0.0756 billion, or 75.6 million, people per year.

Exercises and Problems for Section 3.2 Online Resource: Additional Problems for Section 3.2

EXERCISES

In Exercises 1–25, find the derivatives of the functions . Assume that a and k are constants.

1. $f(x) = 2e^x + x^2$
2. $y = 5t^2 + 4e^t$
3. $f(x) = a^{5x}$
4. $f(x) = 12e^x + 11^x$
5. $y = 5x^2 + 2^x + 3$
6. $f(x) = 2^x + 2 \cdot 3^x$
7. $y = 4 \cdot 10^x - x^3$
8. $z = (\ln 4)e^x$
9. $y = \dfrac{3^x}{3} + \dfrac{33}{\sqrt{x}}$
10. $y = 2^x + \dfrac{2}{x^3}$
11. $z = (\ln 4)4^x$
12. $f(t) = (\ln 3)^t$
13. $y = 5 \cdot 5^t + 6 \cdot 6^t$
14. $h(z) = (\ln 2)^z$
15. $f(x) = e^2 + x^e$
16. $y = \pi^2 + \pi^x$
17. $f(x) = e^\pi + \pi^x$
18. $f(x) = \pi^x + x^\pi$
19. $f(x) = e^k + k^x$
20. $f(x) = e^{1+x}$
21. $f(t) = e^{t+2}$
22. $f(\theta) = e^{k\theta} - 1$
23. $y(x) = a^x + x^a$
24. $f(x) = x^{\pi^2} + (\pi^2)^x$
25. $g(x) = 2x - \dfrac{1}{\sqrt[3]{x}} + 3^x - e$

In Exercises 26–28, find formulas for f'' and f'''.

26. $f(x) = 2^x$
27. $f(t) = 5^{t+1}$
28. $f(v) = \sqrt{v} + 3^v$

PROBLEMS

In Problems 29–39, can the functions be differentiated using the rules developed so far? Differentiate if you can; otherwise, indicate why the rules discussed so far do not apply.

29. $y = x^2 + 2^x$
30. $y = \sqrt{x} - (\frac{1}{2})^x$
31. $y = x^2 \cdot 2^x$
32. $f(s) = 5^s e^s$
33. $y = e^{x+5}$
34. $y = e^{5x}$
35. $y = 4^{(x^2)}$
36. $f(z) = (\sqrt{4})^z$
37. $f(\theta) = 4^{\sqrt{\theta}}$
38. $f(x) = 4^{(3^x)}$
39. $y = \dfrac{2^x}{x}$

40. **(a)** Use Figure 3.12 to rank the quantities $f'(1), f'(2), f'(3)$ from smallest to largest.
(b) Confirm your answer by calculating the quantities using the formula, $f(x) = 2e^x - 3x^2\sqrt{x}$.

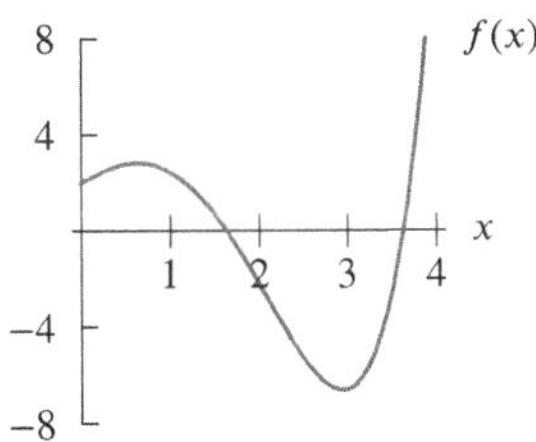

Figure 3.12

41. An animal population is given by $P(t) = 300(1.044)^t$ where t is the number of years since the study of the population began. Find $P'(5)$ and interpret your result.

42. With a yearly inflation rate of 2%, prices are given by

$$P = P_0(1.02)^t,$$

where P_0 is the price in dollars when $t = 0$ and t is time in years. Suppose $P_0 = 1$. How fast (in cents/year) are prices rising when $t = 10$?

43. After a storm, a high-pressure system moves into Duluth, Minnesota. The air pressure t hours after noon is given by $P(t) = 1050 - 44(0.94)^t$ mb (millibars). Find the air pressure and the rate at which it is increasing at 3 pm. Include units.

44. The value of an automobile purchased in 2014 can be approximated by the function $V(t) = 30(0.85)^t$, where t is the time, in years, from the date of purchase, and $V(t)$ is the value, in thousands of dollars.

(a) Evaluate and interpret $V(4)$, including units.
(b) Find an expression for $V'(t)$, including units.
(c) Evaluate and interpret $V'(4)$, including units.
(d) Use $V(t)$, $V'(t)$, and any other considerations you think are relevant to write a paragraph in support of or in opposition to the following statement: "From a monetary point of view, it is best to keep this vehicle as long as possible."

45. With t in years since the start of 2014, worldwide annual extraction of copper is $17.9(1.025)^t$ million tonnes.[4]

(a) How fast is the annual extraction changing at time t? Give units.
(b) How fast is the annual extraction changing at the start of 2025?
(c) Suppose annual extraction changes at the rate found in part (b) for the five years 2025–2030. By how much does the annual extraction change over this period?
(d) Is your answer to part (c) larger or smaller than the change in annual extraction predicted by the model $17.9(1.025)^t$?

46. Food bank usage in Britain has grown dramatically over the past decade. The number of users, in thousands, of the largest bank is estimated[5] to be $N(t) = 1.3(2.25)^t$, where t is the number of years since 2006.

(a) At what rate is the number of food bank users changing at time t? Give units.
(b) Does this rate of change increase or decrease with time?

47. In 2012, the population of Mexico was 115 million and growing 1.09% annually, while the population of the US was 314 million and growing 0.9% annually.[6]

(a) Find the Mexican growth rate in people/year in 2012.
(b) Find the US growth rate, measured the same way, and use it determine which population was growing faster in 2012.

48. Some antique furniture increased very rapidly in price over the past decade. For example, the price of a particular rocking chair is well approximated by

$$V = 75(1.35)^t,$$

where V is in dollars and t is in years since 2000. Find the rate, in dollars per year, at which the price is increasing at time t.

49. Find the quadratic polynomial $g(x) = ax^2 + bx + c$ which best fits the function $f(x) = e^x$ at $x = 0$, in the sense that

$$g(0) = f(0), \text{ and } g'(0) = f'(0), \text{ and } g''(0) = f''(0).$$

Using a computer or calculator, sketch graphs of f and g on the same axes. What do you notice?

Strengthen Your Understanding

In Problems 50–51, explain what is wrong with the statement.

50. The derivative of $f(x) = 2^x$ is $f'(x) = x2^{x-1}$.
51. The derivative of $f(x) = \pi^e$ is $f'(x) = e\pi^{e-1}$.

In Problems 52–53, give an example of:

52. An exponential function for which the derivative is always negative.
53. A function f such that $f'''(x) = f(x)$.

Are the statements in Problems 54–56 true or false? Give an explanation for your answer.

54. If $f(x)$ is increasing, then $f'(x)$ is increasing.
55. There is no function such that $f'(x) = f(x)$ for all x.
56. If $f(x)$ is defined for all x, then $f'(x)$ is defined for all x.

3.3 THE PRODUCT AND QUOTIENT RULES

We now know how to find derivatives of powers and exponentials, and of sums and constant multiples of functions. This section shows how to find the derivatives of products and quotients.

Using Δ Notation

To express the difference quotients of general functions, some additional notation is helpful. We write Δf, read "delta f," for a small change in the value of f at the point x,

$$\Delta f = f(x+h) - f(x).$$

In this notation, the derivative is the limit of the ratio $\Delta f/h$:

$$f'(x) = \lim_{h\to 0} \frac{\Delta f}{h}.$$

[4]Data from http://minerals.usgs.gov/minerals/pubs/commodity/ Accessed February 8, 2015.
[5]Estimates for the Trussell Trust http://www.bbc.com/news/education-30346060 and http://www.trusselltrust.org/stats. Accessed November 2015.
[6]www.indexmundi.com, accessed April 9, 2015.

The Product Rule

Suppose we know the derivatives of $f(x)$ and $g(x)$ and want to calculate the derivative of the product, $f(x)g(x)$. The derivative of the product is calculated by taking the limit, namely,

$$\frac{d[f(x)g(x)]}{dx} = \lim_{h\to 0} \frac{f(x+h)g(x+h) - f(x)g(x)}{h}.$$

To picture the quantity $f(x+h)g(x+h) - f(x)g(x)$, imagine the rectangle with sides $f(x+h)$ and $g(x+h)$ in Figure 3.13, where $\Delta f = f(x+h) - f(x)$ and $\Delta g = g(x+h) - g(x)$. Then

$$\begin{aligned} f(x+h)g(x+h) - f(x)g(x) &= (\text{Area of whole rectangle}) - (\text{Unshaded area}) \\ &= \text{Area of the three shaded rectangles} \\ &= \Delta f \cdot g(x) + f(x) \cdot \Delta g + \Delta f \cdot \Delta g. \end{aligned}$$

Now divide by h:

$$\frac{f(x+h)g(x+h) - f(x)g(x)}{h} = \frac{\Delta f}{h} \cdot g(x) + f(x) \cdot \frac{\Delta g}{h} + \frac{\Delta f \cdot \Delta g}{h}.$$

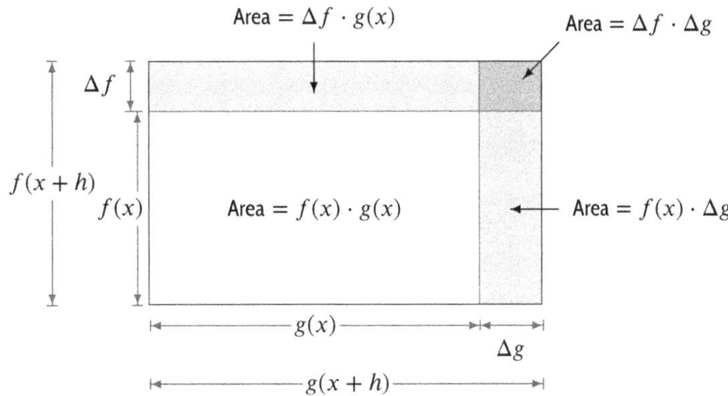

Figure 3.13: Illustration for the product rule (with Δf, Δg positive)

To evaluate the limit as $h \to 0$, we examine the three terms on the right separately. Notice that

$$\lim_{h\to 0} \frac{\Delta f}{h} \cdot g(x) = f'(x)g(x) \quad \text{and} \quad \lim_{h\to 0} f(x) \cdot \frac{\Delta g}{h} = f(x)g'(x).$$

In the third term we multiply the top and bottom by h to get $\frac{\Delta f}{h} \cdot \frac{\Delta g}{h} \cdot h$. Then,

$$\lim_{h\to 0} \frac{\Delta f \cdot \Delta g}{h} = \lim_{h\to 0} \frac{\Delta f}{h} \cdot \frac{\Delta g}{h} \cdot h = \lim_{h\to 0} \frac{\Delta f}{h} \cdot \lim_{h\to 0} \frac{\Delta g}{h} \cdot \lim_{h\to 0} h = f'(x) \cdot g'(x) \cdot 0 = 0.$$

Therefore, we conclude that

$$\begin{aligned} \lim_{h\to 0} \frac{f(x+h)g(x+h) - f(x)g(x)}{h} &= \lim_{h\to 0} \left(\frac{\Delta f}{h} \cdot g(x) + f(x) \cdot \frac{\Delta g}{h} + \frac{\Delta f \cdot \Delta g}{h} \right) \\ &= \lim_{h\to 0} \frac{\Delta f}{h} \cdot g(x) + \lim_{h\to 0} f(x) \cdot \frac{\Delta g}{h} + \lim_{h\to 0} \frac{\Delta f \cdot \Delta g}{h} \\ &= f'(x)g(x) + f(x)g'(x). \end{aligned}$$

Thus we have proved the following rule:

Theorem 3.3: The Product Rule

If $u = f(x)$ and $v = g(x)$ are differentiable, then

$$(fg)' = f'g + fg'.$$

The product rule can also be written

$$\frac{d(uv)}{dx} = \frac{du}{dx} \cdot v + u \cdot \frac{dv}{dx}.$$

In words:

The derivative of a product is the derivative of the first times the second plus the first times the derivative of the second.

Another justification of the product rule is given in Problem 49 on page 185.

Example 1 Differentiate (a) x^2e^x, (b) $(3x^2+5x)e^x$, (c) $\frac{e^x}{x^2}$.

Solution (a)

$$\frac{d(x^2e^x)}{dx} = \left(\frac{d(x^2)}{dx}\right)e^x + x^2\frac{d(e^x)}{dx} = 2xe^x + x^2e^x = (2x+x^2)e^x.$$

(b)

$$\begin{aligned}\frac{d((3x^2+5x)e^x)}{dx} &= \left(\frac{d(3x^2+5x)}{dx}\right)e^x + (3x^2+5x)\frac{d(e^x)}{dx}\\ &= (6x+5)e^x + (3x^2+5x)e^x = (3x^2+11x+5)e^x.\end{aligned}$$

(c) First we must write $\frac{e^x}{x^2}$ as the product $x^{-2}e^x$:

$$\begin{aligned}\frac{d}{dx}\left(\frac{e^x}{x^2}\right) = \frac{d(x^{-2}e^x)}{dx} &= \left(\frac{d(x^{-2})}{dx}\right)e^x + x^{-2}\frac{d(e^x)}{dx}\\ &= -2x^{-3}e^x + x^{-2}e^x = (-2x^{-3}+x^{-2})e^x.\end{aligned}$$

The Quotient Rule

Suppose we want to differentiate a function of the form $Q(x) = f(x)/g(x)$. (Of course, we have to avoid points where $g(x) = 0$.) We want a formula for Q' in terms of f' and g'.

Assuming that $Q(x)$ is differentiable, we can use the product rule on $f(x) = Q(x)g(x)$:

$$\begin{aligned}f'(x) &= Q'(x)g(x) + Q(x)g'(x)\\ &= Q'(x)g(x) + \frac{f(x)}{g(x)}g'(x).\end{aligned}$$

Solving for $Q'(x)$ gives

$$Q'(x) = \frac{f'(x) - \frac{f(x)}{g(x)}g'(x)}{g(x)}.$$

Multiplying the top and bottom by $g(x)$ to simplify gives

$$\frac{d}{dx}\left(\frac{f(x)}{g(x)}\right) = \frac{f'(x)g(x) - f(x)g'(x)}{(g(x))^2}.$$

So we have the following rule:

Theorem 3.4: The Quotient Rule

If $u = f(x)$ and $v = g(x)$ are differentiable, then

$$\left(\frac{f}{g}\right)' = \frac{f'g - fg'}{g^2},$$

or equivalently,

$$\frac{d}{dx}\left(\frac{u}{v}\right) = \frac{\frac{du}{dx}\cdot v - u\cdot\frac{dv}{dx}}{v^2}.$$

In words:

The derivative of a quotient is the derivative of the numerator times the denominator minus the numerator times the derivative of the denominator, all over the denominator squared.

Example 2 Differentiate (a) $\dfrac{5x^2}{x^3+1}$, (b) $\dfrac{1}{1+e^x}$, (c) $\dfrac{e^x}{x^2}$.

Solution (a)

$$\frac{d}{dx}\left(\frac{5x^2}{x^3+1}\right) = \frac{\left(\frac{d}{dx}(5x^2)\right)(x^3+1) - 5x^2\frac{d}{dx}(x^3+1)}{(x^3+1)^2} = \frac{10x(x^3+1) - 5x^2(3x^2)}{(x^3+1)^2}$$

$$= \frac{-5x^4+10x}{(x^3+1)^2}.$$

(b)

$$\frac{d}{dx}\left(\frac{1}{1+e^x}\right) = \frac{\left(\frac{d}{dx}(1)\right)(1+e^x) - 1\frac{d}{dx}(1+e^x)}{(1+e^x)^2} = \frac{0(1+e^x) - 1(0+e^x)}{(1+e^x)^2}$$

$$= \frac{-e^x}{(1+e^x)^2}.$$

(c) This is the same as part (c) of Example 1, but this time we do it by the quotient rule.

$$\frac{d}{dx}\left(\frac{e^x}{x^2}\right) = \frac{\left(\frac{d(e^x)}{dx}\right)x^2 - e^x\left(\frac{d(x^2)}{dx}\right)}{(x^2)^2} = \frac{e^x x^2 - e^x(2x)}{x^4}$$

$$= e^x\left(\frac{x^2-2x}{x^4}\right) = e^x\left(\frac{x-2}{x^3}\right).$$

This is, in fact, the same answer as before, although it looks different. Can you show that it is the same?

Exercises and Problems for Section 3.3 Online Resource: Additional Problems for Section 3.3

EXERCISES

1. If $f(x) = x^2(x^3+5)$, find $f'(x)$ two ways: by using the product rule and by multiplying out before taking the derivative. Do you get the same result? Should you?

2. If $f(x) = 2^x \cdot 3^x$, find $f'(x)$ two ways: by using the product rule and by using the fact that $2^x \cdot 3^x = 6^x$. Do you get the same result?

For Exercises 3–30, find the derivative. It may be to your advantage to simplify first. Assume that a, b, c, and k are constants.

3. $f(x) = xe^x$
4. $y = x \cdot 2^x$
5. $y = \sqrt{x} \cdot 2^x$
6. $y = (t^2 + 3)e^t$
7. $f(x) = (x^2 - \sqrt{x})3^x$
8. $y = (t^3 - 7t^2 + 1)e^t$
9. $f(x) = \dfrac{x}{e^x}$
10. $g(x) = \dfrac{25x^2}{e^x}$
11. $y = \dfrac{t+1}{2^t}$
12. $g(w) = \dfrac{w^{3.2}}{5^w}$
13. $q(r) = \dfrac{3r}{5r+2}$
14. $g(t) = \dfrac{t-4}{t+4}$
15. $z = \dfrac{3t+1}{5t+2}$
16. $z = \dfrac{t^2+5t+2}{t+3}$
17. $f(t) = 2te^t - \dfrac{1}{\sqrt{t}}$
18. $f(x) = \dfrac{x^2+3}{x}$
19. $w = \dfrac{y^3 - 6y^2 + 7y}{y}$
20. $g(t) = \dfrac{4}{3+\sqrt{t}}$
21. $f(z) = \dfrac{z^2+1}{\sqrt{z}}$
22. $w = \dfrac{5-3z}{5+3z}$
23. $h(r) = \dfrac{r^2}{2r+1}$
24. $f(z) = \dfrac{3z^2}{5z^2+7z}$
25. $w(x) = \dfrac{17e^x}{2^x}$
26. $h(p) = \dfrac{1+p^2}{3+2p^2}$
27. $f(x) = \dfrac{x^2+3x+2}{x+1}$
28. $f(x) = \dfrac{ax+b}{cx+k}$
29. $f(x) = (2 - 4x - 3x^2)(6x^e - 3\pi)$
30. $f(x) = (3x^2 + \pi)(e^x - 4)$

PROBLEMS

In Problems 31–33, use Figure 3.14 and the product or quotient rule to estimate the derivative, or state why the rules of this section do not apply. The graph of $f(x)$ has a sharp corner at $x = 2$.

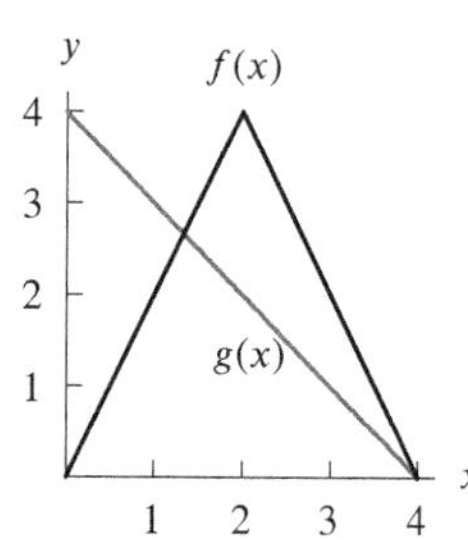

Figure 3.14

31. Let $h(x) = f(x) \cdot g(x)$. Find:
(a) $h'(1)$ (b) $h'(2)$ (c) $h'(3)$
32. Let $k(x) = f(x)/g(x)$. Find:
(a) $k'(1)$ (b) $k'(2)$ (c) $k'(3)$
33. Let $j(x) = g(x)/f(x)$. Find:
(a) $j'(1)$ (b) $j'(2)$ (c) $j'(3)$

For Problems 34–39, let $h(x) = f(x) \cdot g(x)$, and $k(x) = f(x)/g(x)$, and $l(x) = g(x)/f(x)$. Use Figure 3.15 to estimate the derivatives.

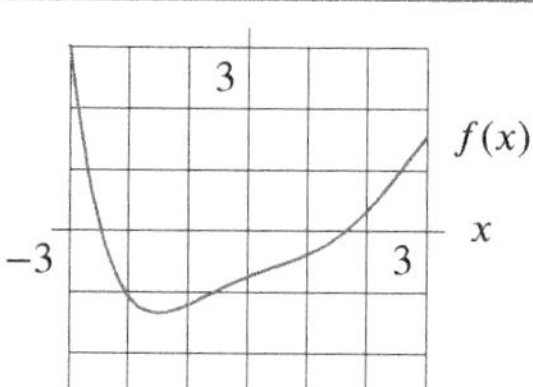

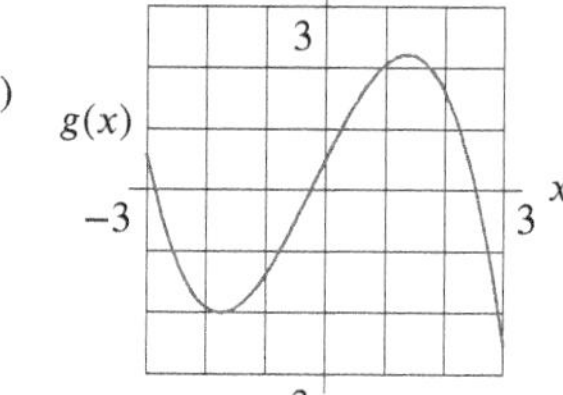

Figure 3.15

34. $h'(1)$
35. $k'(1)$
36. $h'(2)$
37. $k'(2)$
38. $l'(1)$
39. $l'(2)$
40. Differentiate $f(t) = e^{-t}$ by writing it as $f(t) = \dfrac{1}{e^t}$.
41. Differentiate $f(x) = e^{2x}$ by writing it as $f(x) = e^x \cdot e^x$.
42. Differentiate $f(x) = e^{3x}$ by writing it as $f(x) = e^x \cdot e^{2x}$ and using the result of Problem 41.
43. For what intervals is $f(x) = xe^x$ concave up?
44. For what intervals is $g(x) = \dfrac{1}{x^2+1}$ concave down?
45. Find the equation of the tangent line to the graph of $f(x) = 5xe^x$ at the point at which $x = 0$.
46. Find the equation of the tangent line to the graph of $f(x) = x^3e^x$ at the point at which $x = 2$.
47. Find the equation of the tangent line to the graph of $f(x) = \dfrac{2x-5}{x+1}$ at the point at which $x = 0$.

In Problems 48–51, the functions $f(x)$, $g(x)$, and $h(x)$ are differentiable for all values of x. Find the derivative of each of the following functions, using symbols such as $f(x)$ and $f'(x)$ in your answers as necessary.

48. $x^2 f(x)$

49. $4^x(f(x)+g(x))$

50. $\dfrac{f(x)}{g(x)+1}$

51. $\dfrac{f(x)g(x)}{h(x)}$

52. The differentiable functions f and g have the values in the table. For each of the following functions h, find $h'(2)$.

(a) $h(x) = f(x)+g(x)$ **(b)** $h(x) = f(x)g(x)$
(c) $h(x) = \dfrac{f(x)}{g(x)}$

x	$f(x)$	$g(x)$	$f'(x)$	$g'(x)$
2	3	4	5	−2

53. If $H(3) = 1, H'(3) = 3, F(3) = 5, F'(3) = 4$, find:

(a) $G'(3)$ if $G(z) = F(z) \cdot H(z)$
(b) $G'(3)$ if $G(w) = F(w)/H(w)$

54. Find the slope of the line tangent to $h(x) = f(x)g(x)$ at $x = 3$, given that the line tangent to the graph of $f(x)$ at $x = 3$ is $y = 2x - 1$, and the line tangent to the graph of $g(x)$ at $x = 3$ is $y = 13 - 3x$.

55. Find a possible formula for a function $y = f(x)$ such that $f'(x) = 10x^9e^x + x^{10}e^x$.

56. The density of veins on leaves tells us about a region's past climate. Scientists measure vein density, V, in mm per mm^2, by estimating the average distance, x, in mm, between veins on a leaf, and using the formula:[7]

$$V = f(x) = \frac{0.629}{x} + 1.073.$$

(a) Calculate $f'(x)$ using the power rule.
(b) Calculate $f'(x)$ using the quotient rule.
(c) What are the units of $f'(x)$?
(d) Calculate $f'(1)$ and interpret the meaning of your answer in practical terms.

57. The quantity, q, of a skateboard sold depends on the selling price, p, in dollars, so we write $q = f(p)$. You are given that $f(140) = 15{,}000$ and $f'(140) = -100$.

(a) What do $f(140) = 15{,}000$ and $f'(140) = -100$ tell you about the sales of skateboards?
(b) The total revenue, R, earned by the sale of skateboards is given by $R = pq$. Find $\left.\dfrac{dR}{dp}\right|_{p=140}$.
(c) What is the sign of $\left.\dfrac{dR}{dp}\right|_{p=140}$? If the skateboards are currently selling for \$140, what happens to revenue if the price is increased to \$141?

58. A museum has decided to sell one of its paintings and to invest the proceeds. If the picture is sold between the years 2015 and 2025 and the money from the sale is invested in a bank account earning 2% interest per year compounded annually, then $B(t)$, the balance in the year 2025, depends on the year, t, in which the painting is sold and the sale price $P(t)$. If t is measured from the year 2015 so that $0 \le t \le 10$ then

$$B(t) = P(t)(1.02)^{10-t}.$$

(a) Explain why $B(t)$ is given by this formula.
(b) Show that the formula for $B(t)$ is equivalent to

$$B(t) = (1.02)^{10}\frac{P(t)}{(1.02)^t}.$$

(c) Find $B'(5)$, given that $P(5) = 150{,}000$ dollars and $P'(5) = 2000$ dollars/year.

59. Let $f(v)$ be the gas consumption (in liters/km) of a car going at velocity v (in km/hr). In other words, $f(v)$ tells you how many liters of gas the car uses to go one kilometer, if it is going at velocity v. You are told that

$$f(80) = 0.05 \text{ and } f'(80) = 0.0005.$$

(a) Let $g(v)$ be the distance the same car goes on one liter of gas at velocity v. What is the relationship between $f(v)$ and $g(v)$? Find $g(80)$ and $g'(80)$.
(b) Let $h(v)$ be the gas consumption in liters per hour. In other words, $h(v)$ tells you how many liters of gas the car uses in one hour if it is going at velocity v. What is the relationship between $h(v)$ and $f(v)$? Find $h(80)$ and $h'(80)$.
(c) How would you explain the practical meaning of the values of these functions and their derivatives to a driver who knows no calculus?

60. The manager of a political campaign uses two functions to predict fund-raising: $N(t)$ gives the number of donors on day t of the campaign, and $A(t)$ gives the average donation, in dollars, per donor on day t.

(a) Let $P(t)$ be the total money raised on day t of the campaign. How are $P(t)$, $N(t)$, and $A(t)$ related?
(b) Explain how it is possible for the number of donors to decrease, yet the total money raised to increase. Interpret your answer using the product rule.
(c) On the third day of the campaign there are 300 donors, each contributing an average of 100 dollars. If each day there are 3 fewer donors, use the product rule to explain what the campaign manager must do to ensure the total money raised does not change from the third day to the fourth day.

[7]http://www.ncbi.nlm.nih.gov/pubmed/24725225. Blonder and Enquist, "Inferring Climate from Angiosperm Leaf Venation Networks." Accessed March, 2015.

61. A mutual fund holds $N(t)$ million shares of a company with total value $A(t)$ million dollars, where t is in days after January 1, 2016.

 (a) Express $P(t)$, the price per share, in terms of $N(t)$ and $A(t)$.
 (b) What should the mutual fund do to increase $A(t)$ over time even if the value of each share, $P(t)$, decreases?
 (c) On January 1, 2016, the fund holds 2 million shares with a total value of 32 million dollars. If the price per share is dropping at a rate of \$0.23 per day, use the quotient rule to explain how the fund should set the rate of sale or purchase of shares so that the total value does not change.

62. A patient's total cholesterol level, $T(t)$, and good cholesterol level, $G(t)$, at t weeks after January 1, 2016, are measured in milligrams per deciliter of blood (mg/dl). The cholesterol ratio, $R(t) = G(t)/T(t)$ is used to gauge the safety of a patient's cholesterol, with risk of cholesterol-related illnesses being minimized when $R(t) > 1/5$ (that is, good cholesterol is at least $1/5$ of total cholesterol).

 (a) Explain how it is possible for total cholesterol of the patient to increase but the cholesterol ratio to remain constant.
 (b) On January 1, the patient's total cholesterol level is 120 mg/dl and good cholesterol level is 30 mg/dl. Though $R > 1/5$, the doctor prefers that the patient's good cholesterol increase to 40 mg/dl, so prescribes a diet starting January 1 which increases good cholesterol by 1 mg/dl per week without changing the cholesterol ratio. What is the rate of change of total cholesterol the first week of the diet?

For Problems 63–66, use the graphs of $h(x)$ and $g(x)$ to find $f'(0)$ where $f(x) = g(x)h(x)$. If this is not possible, identify the missing values needed to find $f'(0)$.

63.
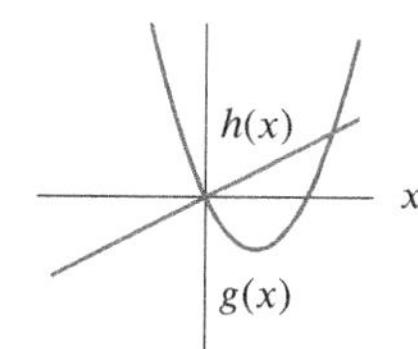

64.
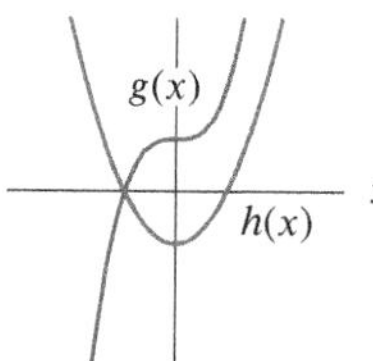

65.
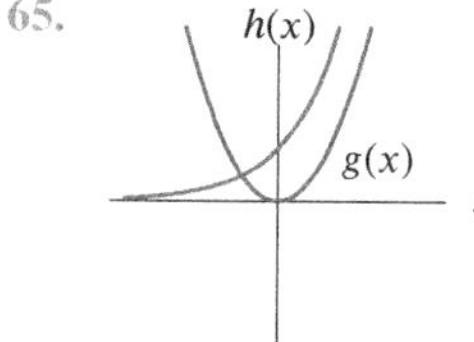

66.
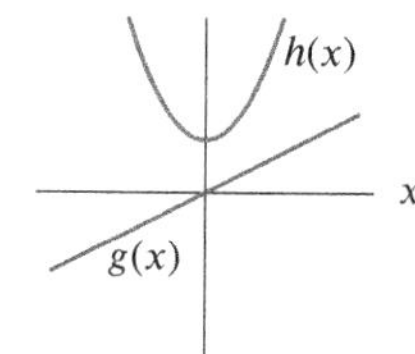

Strengthen Your Understanding

In Problems 67–69, explain what is wrong with the statement.

67. The derivative of $f(x) = x^2e^x$ is $f'(x) = 2xe^x$.

68. Differentiating $f(x) = x/(x+1)$ by the quotient rule gives

$$f'(x) = \frac{x\frac{d}{dx}(x+1) - (x+1)\frac{d}{dx}(x)}{(x+1)^2}.$$

69. The quotient $f(x) = (x+1)/e^{-x}$ cannot be differentiated using the product rule.

In Problems 70–71, give an example of:

70. A function involving a sine and an exponential that can be differentiated using the product rule or the quotient rule.

71. A function $f(x)$ that can be differentiated both using the product rule and in some other way.

Are the statements in Problems 72–74 true or false? Give an explanation for your answer.

72. Let f and g be two functions whose second derivatives are defined. Then

$$(fg)'' = fg'' + f''g.$$

73. If the function $f(x)/g(x)$ is defined but not differentiable at $x = 1$, then either $f(x)$ or $g(x)$ is not differentiable at $x = 1$.

74. Suppose that f'' and g'' exist and f and g are concave up for all x, then $f(x)g(x)$ is concave up for all x.

75. Which of the following would be a counterexample to the product rule?

 (a) Two differentiable functions f and g satisfying $(fg)' = f'g'$.
 (b) A differentiable function f such that $(xf(x))' = xf'(x) + f(x)$.
 (c) A differentiable function f such that $(f(x)^2)' = 2f(x)$.
 (d) Two differentiable functions f and g such that $f'(a) = 0$ and $g'(a) = 0$ and fg has positive slope at $x = a$.

In Problems 76–79, let $f(x) = g(x)h(x)$, with $g'(2) = 0.5$ and $h'(2) = 0.7$. Either explain why the given statement is true, or provide values for the unknown terms in the product rule to show that it can be false.

76. Since $g'(2) < 1$ and $h'(2) < 1$, we must have $f'(2) < 1$.

77. Since $g'(2) \neq 0$ and $h'(2) \neq 0$, the only way $f'(2) = 0$ is if both $g(2) = 0$ and $h(2) = 0$.

78. Since $g'(2)$ and $h'(2)$ are both positive, $f'(2)$ must also be positive.

79. If $g(2) = 0$, then $f'(2)$ is half the value of $h(2)$.

3.4 THE CHAIN RULE

The chain rule enables us to differentiate composite functions such as $\sqrt{x^2+1}$ or e^{-x^2}. Before seeing a formula, let's think about the derivative of a composite function in a practical situation.

Intuition Behind the Chain Rule

Imagine we are moving straight upward in a hot air balloon. Let y be our distance from the ground. The temperature, H, is changing as a function of altitude, so $H = f(y)$. How does our temperature change with time?

The rate of change of our temperature is affected both by how fast the temperature is changing with altitude (about 16°F per mile), and by how fast we are climbing (say 2 mph). Then our temperature changes by 16° for every mile we climb, and since we move 2 miles in an hour, our temperature changes by $16 \cdot 2 = 32$ degrees in an hour.

Since temperature is a function of height, $H = f(y)$, and height is a function of time, $y = g(t)$, we can think of temperature as a composite function of time, $H = f(g(t))$, with f as the outside function and g as the inside function. The example suggests the following result, which turns out to be true:

Rate of change of composite function	=	Rate of change of outside function	×	Rate of change of inside function

The Derivative of a Composition of Functions

We now obtain a formula for the chain rule. Suppose $f(g(x))$ is a composite function, with f being the outside function and g being the inside. Let us write

$$z = g(x) \quad \text{and} \quad y = f(z), \quad \text{so} \quad y = f(g(x)).$$

Then a small change in x, called Δx, generates a small change in z, called Δz. In turn, Δz generates a small change in y called Δy. Provided Δx and Δz are not zero, we can say:

$$\frac{\Delta y}{\Delta x} = \frac{\Delta y}{\Delta z} \cdot \frac{\Delta z}{\Delta x}.$$

Since $\frac{dy}{dx} = \lim_{\Delta x \to 0} \frac{\Delta y}{\Delta x}$, this suggests that in the limit as Δx, Δy, and Δz get smaller and smaller, we have:

The Chain Rule

$$\frac{dy}{dx} = \frac{dy}{dz} \cdot \frac{dz}{dx}.$$

In other words:

The rate of change of a composite function is the product of the rates of change of the outside and inside functions.

Since $\dfrac{dy}{dz} = f'(z)$ and $\dfrac{dz}{dx} = g'(x)$, we can also write

$$\frac{d}{dx} f(g(x)) = f'(z) \cdot g'(x).$$

Substituting $z = g(x)$, we can rewrite this as follows:

Theorem 3.5: The Chain Rule

If f and g are differentiable functions, then

$$\frac{d}{dx} f(g(x)) = f'(g(x)) \cdot g'(x).$$

In words:

The derivative of a composite function is the product of the derivatives of the outside and inside functions. The derivative of the outside function must be evaluated at the inside function.

A justification of the chain rule is given in Problem 50 on page 185. The following example shows how units confirm that the rate of change of a composite function is the product of the rates of change of the outside and inside functions.

Example 1 The length, L, in micrometers (μm), of steel depends on the air temperature, H°C, and the temperature H depends on time, t, measured in hours. If the length of a steel bridge increases by 0.2 μm for every degree increase in temperature, and the temperature is increasing at 3°C per hour, how fast is the length of the bridge increasing? What are the units for your answer?

Solution We want to know how much the length of the bridge changes in one hour; this rate is in μm/hr. We are told that the length of the bridge changes by 0.2 μm for each degree that the temperature changes, and that the temperature changes by 3°C each hour. Thus, in one hour, the length of the bridge changes by $0.2 \cdot 3 = 0.6$ μm.

Now we do the same calculation using derivative notation and the chain rule. We know that

$$\text{Rate length increasing with respect to temperature} = \frac{dL}{dH} = 0.2\ \mu\text{m}/°\text{C}$$

$$\text{Rate temperature increasing with respect to time} = \frac{dH}{dt} = 3°\text{C/hr}.$$

We want to calculate the rate at which the length is increasing with respect to time, or dL/dt. We think of L as a function of H, and H as a function of t. The chain rule tells us that

$$\frac{dL}{dt} = \frac{dL}{dH} \cdot \frac{dH}{dt} = \left(0.2 \frac{\mu\text{m}}{°\text{C}}\right) \cdot \left(3 \frac{°\text{C}}{\text{hr}}\right) = 0.6\ \mu\text{m/hr}.$$

Thus, the length is increasing at 0.6 μm/hr. Notice that the units work out as we expect.

Example 1 shows us how to interpret the chain rule in practical terms. The next examples show how the chain rule is used to compute derivatives of functions given by formulas.

Example 2 Find the derivatives of the following functions:

(a) $(x^2+1)^{100}$ (b) $\sqrt{3x^2+5x-2}$ (c) $\dfrac{1}{x^2+x^4}$ (d) e^{3x} (e) e^{x^2}

Solution (a) Here $z = g(x) = x^2 + 1$ is the inside function; $f(z) = z^{100}$ is the outside function. Now $g'(x) = 2x$ and $f'(z) = 100z^{99}$, so

$$\frac{d}{dx}((x^2+1)^{100}) = 100z^{99} \cdot 2x = 100(x^2+1)^{99} \cdot 2x = 200x(x^2+1)^{99}.$$

(b) Here $z = g(x) = 3x^2 + 5x - 2$ and $f(z) = \sqrt{z}$, so $g'(x) = 6x + 5$ and $f'(z) = \dfrac{1}{2\sqrt{z}}$. Hence

$$\frac{d}{dx}(\sqrt{3x^2+5x-2}) = \frac{1}{2\sqrt{z}} \cdot (6x+5) = \frac{1}{2\sqrt{3x^2+5x-2}} \cdot (6x+5).$$

(c) Let $z = g(x) = x^2 + x^4$ and $f(z) = 1/z$, so $g'(x) = 2x + 4x^3$ and $f'(z) = -z^{-2} = -\dfrac{1}{z^2}$. Then

$$\frac{d}{dx}\left(\frac{1}{x^2+x^4}\right) = -\frac{1}{z^2}(2x+4x^3) = -\frac{2x+4x^3}{(x^2+x^4)^2}.$$

We could have done this problem using the quotient rule. Try it and see that you get the same answer!

(d) Let $z = g(x) = 3x$ and $f(z) = e^z$. Then $g'(x) = 3$ and $f'(z) = e^z$, so

$$\frac{d}{dx}\left(e^{3x}\right) = e^z \cdot 3 = 3e^{3x}.$$

(e) To figure out which is the inside function and which is the outside, notice that to evaluate e^{x^2} we first evaluate x^2 and then take e to that power. This tells us that the inside function is $z = g(x) = x^2$ and the outside function is $f(z) = e^z$. Therefore, $g'(x) = 2x$, and $f'(z) = e^z$, giving

$$\frac{d}{dx}(e^{x^2}) = e^z \cdot 2x = e^{x^2} \cdot 2x = 2xe^{x^2}.$$

To differentiate a complicated function, we may have to use the chain rule more than once, as in the following example.

Example 3 Differentiate: (a) $\sqrt{e^{-x/7}+5}$ (b) $\left(1 - e^{2\sqrt{t}}\right)^{19}$

Solution (a) Let $z = g(x) = e^{-x/7} + 5$ be the inside function; let $f(z) = \sqrt{z}$ be the outside function. Now $f'(z) = \dfrac{1}{2\sqrt{z}}$, but we need the chain rule to find $g'(x)$.

We choose inside and outside functions whose composition is $g(x)$. Let $u = h(x) = -x/7$ and $k(u) = e^u + 5$ so $g(x) = k(h(x)) = e^{-x/7} + 5$. Then $h'(x) = -1/7$ and $k'(u) = e^u$, so

$$g'(x) = e^u \cdot \left(-\frac{1}{7}\right) = -\frac{1}{7}e^{-x/7}.$$

Using the chain rule to combine the derivatives of $f(z)$ and $g(x)$, we have

$$\frac{d}{dx}(\sqrt{e^{-x/7}+5}) = \frac{1}{2\sqrt{z}}\left(-\frac{1}{7}e^{-x/7}\right) = -\frac{e^{-x/7}}{14\sqrt{e^{-x/7}+5}}.$$

(b) Let $z = g(t) = 1 - e^{2\sqrt{t}}$ be the inside function and $f(z) = z^{19}$ be the outside function. Then $f'(z) = 19z^{18}$ but we need the chain rule to differentiate $g(t)$.

Now we choose $u = h(t) = 2\sqrt{t}$ and $k(u) = 1 - e^u$, so $g(t) = k(h(t))$. Then $h'(t) = 2 \cdot \frac{1}{2}t^{-1/2} = \frac{1}{\sqrt{t}}$ and $k'(u) = -e^u$, so

$$g'(t) = -e^u \cdot \frac{1}{\sqrt{t}} = -\frac{e^{2\sqrt{t}}}{\sqrt{t}}.$$

Using the chain rule to combine the derivatives of $f(z)$ and $g(t)$, we have

$$\frac{d}{dx}(1 - e^{2\sqrt{t}})^{19} = 19z^{18}\left(-\frac{e^{2\sqrt{t}}}{\sqrt{t}}\right) = -19\frac{e^{2\sqrt{t}}}{\sqrt{t}}\left(1 - e^{2\sqrt{t}}\right)^{18}.$$

It is often faster to use the chain rule without introducing new variables, as in the following examples.

Example 4 Differentiate $\sqrt{1 + e^{\sqrt{3+x^2}}}$.

Solution The chain rule is needed three times:

$$\begin{aligned}
\frac{d}{dx}\left(\sqrt{1 + e^{\sqrt{3+x^2}}}\right) &= \frac{1}{2}\left(1 + e^{\sqrt{3+x^2}}\right)^{-1/2} \cdot \frac{d}{dx}\left(1 + e^{\sqrt{3+x^2}}\right) \\
&= \frac{1}{2}\left(1 + e^{\sqrt{3+x^2}}\right)^{-1/2} \cdot e^{\sqrt{3+x^2}} \cdot \frac{d}{dx}\left(\sqrt{3+x^2}\right) \\
&= \frac{1}{2}\left(1 + e^{\sqrt{3+x^2}}\right)^{-1/2} \cdot e^{\sqrt{3+x^2}} \cdot \frac{1}{2}\left(3+x^2\right)^{-1/2} \cdot \frac{d}{dx}\left(3+x^2\right) \\
&= \frac{1}{2}\left(1 + e^{\sqrt{3+x^2}}\right)^{-1/2} \cdot e^{\sqrt{3+x^2}} \cdot \frac{1}{2}\left(3+x^2\right)^{-1/2} \cdot 2x.
\end{aligned}$$

Example 5 Find the derivative of e^{2x} by the chain rule and by the product rule.

Solution Using the chain rule, we have

$$\frac{d}{dx}(e^{2x}) = e^{2x} \cdot \frac{d}{dx}(2x) = e^{2x} \cdot 2 = 2e^{2x}.$$

Using the product rule, we write $e^{2x} = e^x \cdot e^x$. Then

$$\frac{d}{dx}(e^{2x}) = \frac{d}{dx}(e^x e^x) = \left(\frac{d}{dx}(e^x)\right)e^x + e^x\left(\frac{d}{dx}(e^x)\right) = e^x \cdot e^x + e^x \cdot e^x = 2e^{2x}.$$

Using the Product and Chain Rules to Differentiate a Quotient

If you prefer, you can differentiate a quotient by the product and chain rules, instead of by the quotient rule. The resulting formulas may look different, but they will be equivalent.

Example 6 Find $k'(x)$ if $k(x) = \dfrac{x}{x^2 + 1}$.

Solution One way is to use the quotient rule:

$$k'(x) = \frac{1 \cdot (x^2+1) - x \cdot (2x)}{(x^2+1)^2}$$
$$= \frac{1-x^2}{(x^2+1)^2}.$$

Alternatively, we can write the original function as a product,

$$k(x) = x\frac{1}{x^2+1} = x \cdot (x^2+1)^{-1},$$

and use the product rule:

$$k'(x) = 1 \cdot (x^2+1)^{-1} + x \cdot \frac{d}{dx}\left[(x^2+1)^{-1}\right].$$

Now use the chain rule to differentiate $(x^2+1)^{-1}$, giving

$$\frac{d}{dx}\left[(x^2+1)^{-1}\right] = -(x^2+1)^{-2} \cdot 2x = \frac{-2x}{(x^2+1)^2}.$$

Therefore,

$$k'(x) = \frac{1}{x^2+1} + x \cdot \frac{-2x}{(x^2+1)^2} = \frac{1}{x^2+1} - \frac{2x^2}{(x^2+1)^2}.$$

Putting these two fractions over a common denominator gives the same answer as the quotient rule.

Exercises and Problems for Section 3.4 Online Resource: Additional Problems for Section 3.4

EXERCISES

In Exercises 1–57, find the derivatives. Assume that a, b, and c are constants.

1. $f(x) = (x+1)^{99}$
2. $w = (t^3+1)^{100}$
3. $g(x) = (4x^2+1)^7$
4. $f(x) = \sqrt{1-x^2}$
5. $y = \sqrt{e^x+1}$
6. $w = (\sqrt{t}+1)^{100}$
7. $h(w) = (w^4-2w)^5$
8. $s(t) = (3t^2+4t+1)^3$
9. $w(r) = \sqrt{r^4+1}$
10. $k(x) = (x^3+e^x)^4$
11. $f(x) = e^{2x}\left(x^2+5^x\right)$
12. $y = e^{3w/2}$
13. $g(x) = e^{\pi x}$
14. $B = 15e^{0.20t}$
15. $w = 100e^{-x^2}$
16. $f(\theta) = 2^{-\theta}$
17. $y = \pi^{(x+2)}$
18. $g(x) = 3^{(2x+7)}$
19. $f(t) = te^{5-2t}$
20. $p(t) = e^{4t+2}$
21. $v(t) = t^2e^{-ct}$
22. $g(t) = e^{(1+3t)^2}$
23. $w = e^{\sqrt{s}}$
24. $y = e^{-4t}$
25. $y = \sqrt{s^3+1}$
26. $y = te^{-t^2}$
27. $f(z) = \sqrt{z}e^{-z}$
28. $z(x) = \sqrt[3]{2^x+5}$
29. $z = 2^{5t-3}$
30. $w = \sqrt{(x^2 \cdot 5^x)^3}$
31. $f(y) = \sqrt{10^{(5-y)}}$
32. $f(z) = \dfrac{\sqrt{z}}{e^z}$
33. $y = \dfrac{\sqrt{z}}{2^z}$
34. $y = \left(\dfrac{x^2+2}{3}\right)^2$
35. $h(x) = \sqrt{\dfrac{x^2+9}{x+3}}$
36. $y = \dfrac{e^x-e^{-x}}{e^x+e^{-x}}$
37. $y = \dfrac{1}{e^{3x}+x^2}$
38. $h(z) = \left(\dfrac{b}{a+z^2}\right)^4$
39. $f(x) = \dfrac{1}{\sqrt{x^3+1}}$
40. $f(z) = \dfrac{1}{(e^z+1)^2}$
41. $w = (t^2+3t)(1-e^{-2t})$
42. $h(x) = 2^{e^{3x}}$
43. $f(x) = 6e^{5x}+e^{-x^2}$
44. $f(x) = e^{-(x-1)^2}$
45. $f(w) = (5w^2+3)e^{w^2}$
46. $f(\theta) = (e^{\theta}+e^{-\theta})^{-1}$

47. $y = \sqrt{e^{-3t^2} + 5}$

48. $z = (te^{3t} + e^{5t})^9$

49. $f(y) = e^{e^{(y^2)}}$

50. $f(t) = 2e^{-2e^{2t}}$

51. $f(x) = (ax^2 + b)^3$

52. $f(t) = ae^{bt}$

53. $f(x) = axe^{-bx}$

54. $g(\alpha) = e^{\alpha e^{-2\alpha}}$

55. $y = ae^{-be^{-cx}}$

56. $y = (e^x - e^{-x})^2$

57. $y = \left(x^2+5\right)^3 \left(3x^3-2\right)^2$

PROBLEMS

■ In Problems 58–61, use Figure 3.16 and the chain rule to estimate the derivative, or state why the chain rule does not apply. The graph of $f(x)$ has a sharp corner at $x = 2$.

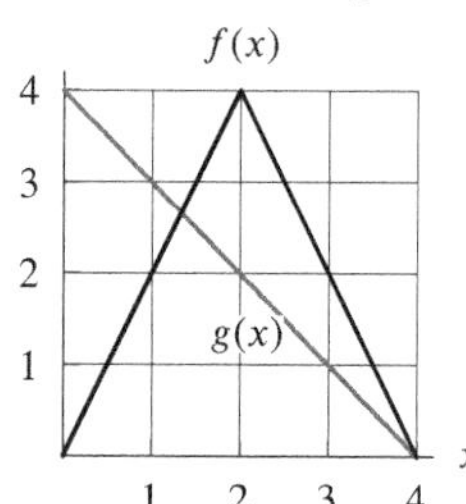

Figure 3.16

58. Let $h(x) = f(g(x))$. Find:

(a) $h'(1)$ (b) $h'(2)$ (c) $h'(3)$

59. Let $u(x) = g(f(x))$. Find:

(a) $u'(1)$ (b) $u'(2)$ (c) $u'(3)$

60. Let $v(x) = f(f(x))$. Find:

(a) $v'(1)$ (b) $v'(2)$ (c) $v'(3)$

61. Let $w(x) = g(g(x))$. Find:

(a) $w'(1)$ (b) $w'(2)$ (c) $w'(3)$

■ In Problems 62–65, use Figure 3.17 to evaluate the derivative.

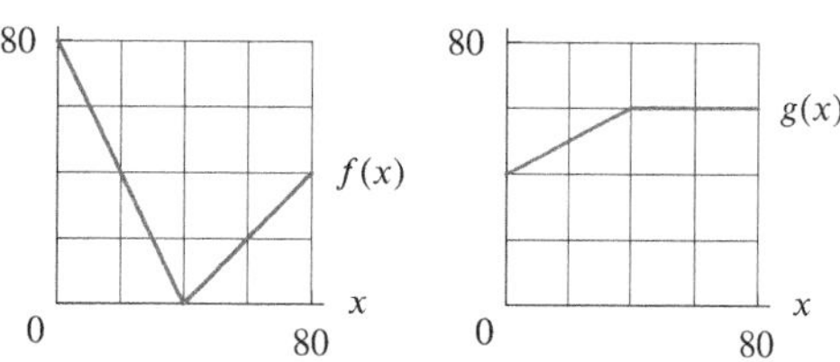

Figure 3.17

62. $\left.\frac{d}{dx} f(g(x))\right|_{x=30}$

63. $\left.\frac{d}{dx} f(g(x))\right|_{x=70}$

64. $\left.\frac{d}{dx} g(f(x))\right|_{x=30}$

65. $\left.\frac{d}{dx} g(f(x))\right|_{x=70}$

66. Find the equation of the tangent line to $f(x) = (x-1)^3$ at the point where $x = 2$.

67. Find the equation of the line tangent to $y = f(x)$ at $x = 1$, where $f(x) = 6e^{5x} + e^{-x^2}$.

68. Find the equation of the line tangent to $f(t) = 100e^{-0.3t}$ at $t = 2$.

69. For what values of x is the graph of $y = e^{-x^2}$ concave down?

70. For what intervals is $f(x) = xe^{-x}$ concave down?

71. Suppose $f(x) = (2x+1)^{10}(3x-1)^7$. Find a formula for $f'(x)$. Decide on a reasonable way to simplify your result, and find a formula for $f''(x)$.

72. Find a possible formula for a function $m(x)$ such that $m'(x) = x^5 \cdot e^{(x^6)}$.

73. Given $F(2) = 1, F'(2) = 5, F(4) = 3, F'(4) = 7$ and $G(4) = 2, G'(4) = 6, G(3) = 4, G'(3) = 8$, find:

(a) $H(4)$ if $H(x) = F(G(x))$
(b) $H'(4)$ if $H(x) = F(G(x))$
(c) $H(4)$ if $H(x) = G(F(x))$
(d) $H'(4)$ if $H(x) = G(F(x))$
(e) $H'(4)$ if $H(x) = F(x)/G(x)$

74. Given $f(x)$ with $f(1) = 2$ and $f'(1) = 5$, find

(a) $g'(1)$ if $g(x) = (f(x))^3$
(b) $h'(1)$ if $h(x) = f(x^3)$

75. Given $f(x)$ with $f(2) = 7$ and $f'(2) = 3$ and $f'(4) = -2$, find

(a) $g'(2)$ if $g(x) = (f(x))^2$
(b) $h'(2)$ if $h(x) = f(x^2)$

76. A particle is moving on the x-axis, where x is in centimeters. Its velocity, v, in cm/sec, when it is at the point with coordinate x is given by

$$v = x^2 + 3x - 2.$$

Find the acceleration of the particle when it is at the point $x = 2$. Give units in your answer.

77. A fish population is approximated by $P(t) = 10e^{0.6t}$, where t is in months. Calculate and use units to explain what each of the following tells us about the population:

(a) $P(12)$ **(b)** $P'(12)$

78. The world's population[8] is about $f(t) = 7.17e^{0.011t}$ billion, where t is time in years since July 2014. Find $f(0)$, $f'(0)$, $f(10)$, and $f'(10)$. Using units, interpret your answers in terms of population.

[8]www.indexmundi.com, accessed April 9, 2015.

79. Fourth-quarter net sales for the Hershey Company[9], in billion dollars, in t years from 2012 can be approximated by $f(t) = 1.75e^{0.12t}$. Find $f(3)$ and $f'(3)$. Give units and interpret in terms of Hershey sales.

80. For t in years since 2010, daily oil consumption in China, in thousands of barrels, was approximated by[10]

$$B = 8938e^{0.05t}.$$

(a) Is daily oil consumption increasing or decreasing with time?
(b) How fast is oil consumption changing at time t?

81. The balance in a bank account t years after money is deposited is given by $f(t) = 5000e^{0.02t}$ dollars.

(a) How much money was deposited? What is the interest rate of the account?
(b) Find $f(10)$ and $f'(10)$. Give units and interpret in terms of balance in the account.

82. For $t \geq 0$ in minutes, the temperature, H, of a pot of soup in degrees Celsius is[11]

$$H = 5 + 95e^{-0.054t}.$$

(a) Is the temperature increasing or decreasing with time?
(b) How fast is the temperature changing at time t? Give units.

83. A yam is put in a hot oven, maintained at a constant temperature 200°C. At time $t = 30$ minutes, the temperature T of the yam is 120° and is increasing at an (instantaneous) rate of 2°/min. Newton's law of cooling (or, in our case, warming) tells us that the temperature at time t is

$$T(t) = 200 - ae^{-bt}.$$

Find a and b.

84. The 2010 census[12] determined the population of the US was 308.75 million on April 1, 2010. If the population was increasing exponentially at a rate of 2.85 million per year on that date, find a formula for the population as a function of time, t, in years since that date.

85. If you invest P dollars in a bank account at an annual interest rate of $r\%$, then after t years you will have B dollars, where

$$B = P\left(1 + \frac{r}{100}\right)^t.$$

(a) Find dB/dt, assuming P and r are constant. In terms of money, what does dB/dt represent?
(b) Find dB/dr, assuming P and t are constant. In terms of money, what does dB/dr represent?

86. The theory of relativity predicts that an object whose mass is m_0 when it is at rest will appear heavier when moving at speeds near the speed of light. When the object is moving at speed v, its mass m is given by

$$m = \frac{m_0}{\sqrt{1 - (v^2/c^2)}}, \quad \text{where } c \text{ is the speed of light.}$$

(a) Find dm/dv.
(b) In terms of physics, what does dm/dv tell you?

87. Since the 1950s, the carbon dioxide concentration in the air has been recorded at the Mauna Loa Observatory in Hawaii.[13] A graph of this data is called the Keeling Curve, after Charles Keeling, who started recording the data. With t in years since 1950, fitting functions to the data gives three models for the carbon dioxide concentration in parts per million (ppm):

$$f(t) = 303 + 1.3t$$
$$g(t) = 304e^{0.0038t}$$
$$h(t) = 0.0135t^2 + 0.5133t + 310.5.$$

(a) What family of function is used in each model?
(b) Find the rate of change of carbon dioxide in 2010 in each of the three models. Give units.
(c) Arrange the three models in increasing order of the rates of change they give for 2010. (Which model predicts the largest rate of change in 2010? Which predicts the smallest?)
(d) Consider the same three models for all positive time t. Will the ordering in part (c) remain the same for all t? If not, how will it change?

88. Instruments on a plane measure the distance traveled, x (in km), and the quantity of fuel in the tank, q (in liters), after t minutes of flight.

(a) Give the units of dx/dt. Explain its meaning for the flight. Is it positive or negative?
(b) Give the units of dq/dt. Explain its meaning for the flight. Is it positive or negative?
(c) The quantity of fuel, q, is also a function of the distance traveled, x. Give the units and meaning of dq/dx. Is it positive or negative?
(d) Use the chain rule to express dq/dx in terms of dx/dt and dq/dt.

For Problems 89–92, if $f(x)$ has a positive slope everywhere, and $g(x)$ has a negative slope everywhere, decide if $h(x)$ is increasing, or decreasing, or neither.

89. $h(x) = f(g(x))$

90. $h(x) = g(f(x))$

91. $h(x) = g(g(x))$

92. $h(x) = f(f(x))$

[9] www.thehersheycompany.com, accessed April 10, 2015.
[10] Based on /www.eia.gov/cfapps/ipdbproject/ Accessed May 2015.
[11] Based on http://www.ugrad.math.ubc.ca/coursedoc/math100/notes/diffeqs/cool.html. Accessed May 2015.
[12] http://2010.census.gov/2010census/, accessed March 10, 2013.
[13] www.esrl.noaa.gov/gmd/ccgg/, accessed March 10, 2013.

Strengthen Your Understanding

■ In Problems 93–95, explain what is wrong with the statement.

93. The derivative of $g(x) = (e^x + 2)^5$ is $g'(x) = 5(e^x + 2)^4$.
94. The derivative of $w(x) = e^{x^2}$ is $w'(x) = e^{x^2}$.
95. If $f(x) = h(g(x))$ and $h'(3) = 0$, then $f'(3) = 0$.

■ In Problems 96–97, give an example of:

96. A function involving a sine and an exponential that requires the chain rule to differentiate.
97. A function that can be differentiated both using the chain rule and by another method.

■ Are the statements in Problems 98–101 true or false? If a statement is true, explain how you know. If a statement is false, give a counterexample.

98. If $f(x) = h(g(x))$ and $g'(2) = 0$, then $f'(2) = 0$.
99. $(fg)'(x)$ is never equal to $f'(x)g'(x)$.
100. If the derivative of $f(g(x))$ is equal to the derivative of $f(x)$ for all x, then $g(x) = x$ for all x.
101. Suppose that f'' and g'' exist and that f and g are concave up for all x, then $f(g(x))$ is concave up for all x.

3.5 THE TRIGONOMETRIC FUNCTIONS

Derivatives of the Sine and Cosine

Since the sine and cosine functions are periodic, their derivatives must be periodic also. (Why?) Let's look at the graph of $f(x) = \sin x$ in Figure 3.18 and estimate the derivative function graphically.

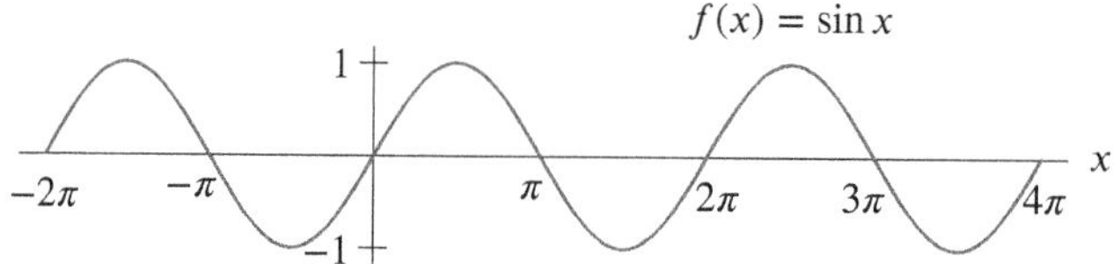

Figure 3.18: The sine function

First we might ask where the derivative is zero. (At $x = \pm\pi/2, \pm 3\pi/2, \pm 5\pi/2$, etc.) Then ask where the derivative is positive and where it is negative. (Positive for $-\pi/2 < x < \pi/2$; negative for $\pi/2 < x < 3\pi/2$, etc.) Since the largest positive slopes are at $x = 0, 2\pi$, and so on, and the largest negative slopes are at $x = \pi, 3\pi$, and so on, we get something like the graph in Figure 3.19.

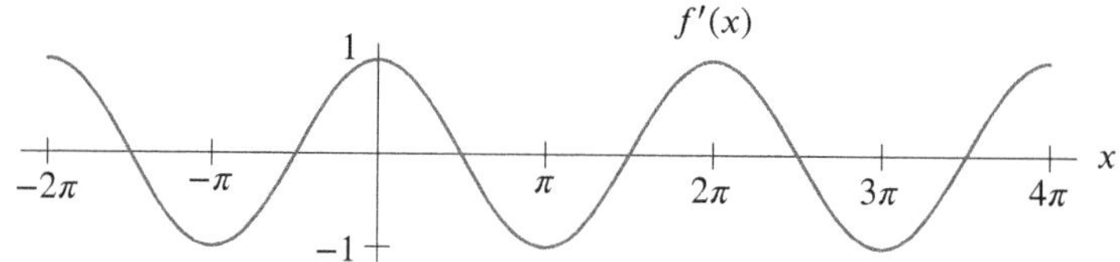

Figure 3.19: Derivative of $f(x) = \sin x$

The graph of the derivative in Figure 3.19 looks suspiciously like the graph of the cosine function. This might lead us to conjecture, quite correctly, that the derivative of the sine is the cosine. Of course, we cannot be sure, just from the graphs, that the derivative of the sine really is the cosine. However, for now we'll assume that the derivative of the sine *is* the cosine and confirm the result at the end of the section.

One thing we can do now is to check that the derivative function in Figure 3.19 has amplitude 1 (as it ought to if it is the cosine). That means we have to convince ourselves that the derivative of $f(x) = \sin x$ is 1 when $x = 0$. The next example suggests that this is true when x is in radians.

Example 1 Using a calculator set in radians, estimate the derivative of $f(x) = \sin x$ at $x = 0$.

Solution Since $f(x) = \sin x$,

$$f'(0) = \lim_{h\to 0} \frac{\sin(0+h) - \sin 0}{h} = \lim_{h\to 0} \frac{\sin h}{h}.$$

Table 3.5 contains values of $(\sin h)/h$ which suggest that this limit is 1, so we estimate

$$f'(0) = \lim_{h\to 0} \frac{\sin h}{h} = 1.$$

Table 3.5

h (radians)	-0.1	-0.01	-0.001	-0.0001	0.0001	0.001	0.01	0.1
$(\sin h)/h$	0.99833	0.99998	1.0000	1.0000	1.0000	1.0000	0.99998	0.99833

Warning: It is important to notice that in the previous example h was in *radians*; any conclusions we have drawn about the derivative of $\sin x$ are valid *only* when x is in radians. If you find the derivative with h in degrees, you get a different result.

Example 2 Starting with the graph of the cosine function, sketch a graph of its derivative.

Solution The graph of $g(x) = \cos x$ is in Figure 3.20(a). Its derivative is 0 at $x = 0, \pm\pi, \pm 2\pi$, and so on; it is positive for $-\pi < x < 0$, $\pi < x < 2\pi$, and so on; and it is negative for $0 < x < \pi$, $2\pi < x < 3\pi$, and so on. The derivative is in Figure 3.20(b).

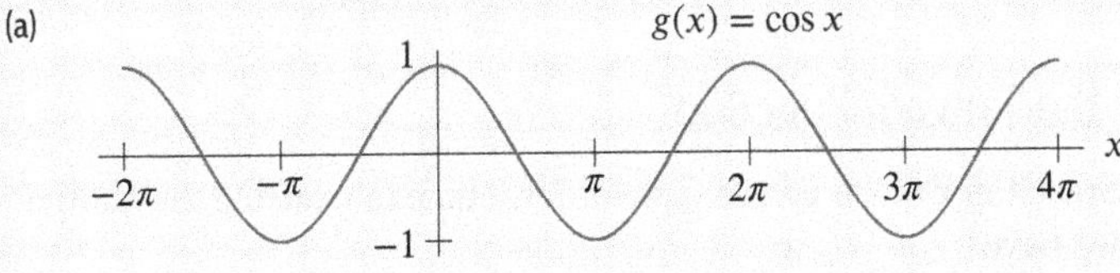

Figure 3.20: $g(x) = \cos x$ and its derivative, $g'(x)$

As we did with the sine, we use the graphs to make a conjecture. The derivative of the cosine in Figure 3.20(b) looks exactly like the graph of sine, except it is reflected across the x-axis. But how can we be sure that the derivative is $-\sin x$?

Example 3 Use the relation $\frac{d}{dx}(\sin x) = \cos x$ to show that $\frac{d}{dx}(\cos x) = -\sin x$.

Solution Since the cosine function is the sine function shifted to the left by $\pi/2$ (that is, $\cos x = \sin(x + \pi/2)$), we expect the derivative of the cosine to be the derivative of the sine, shifted to the left by $\pi/2$. Differentiating using the chain rule:

$$\frac{d}{dx}(\cos x) = \frac{d}{dx}\left(\sin\left(x + \frac{\pi}{2}\right)\right) = \cos\left(x + \frac{\pi}{2}\right).$$

But $\cos(x + \pi/2)$ is the cosine shifted to the left by $\pi/2$, which gives a sine curve reflected across the x-axis. So we have

$$\frac{d}{dx}(\cos x) = \cos\left(x + \frac{\pi}{2}\right) = -\sin x.$$

At the end of this section and in Problems 88 and 89 (available online), we show that our conjectures for the derivatives of sin x and cos x are correct. Thus, we have:

> For x in radians, $\dfrac{d}{dx}(\sin x) = \cos x$ and $\dfrac{d}{dx}(\cos x) = -\sin x$.

Example 4 Differentiate (a) $2\sin(3\theta)$, (b) $\cos^2 x$, (c) $\cos(x^2)$, (d) $e^{-\sin t}$.

Solution Use the chain rule:

(a) $\dfrac{d}{d\theta}(2\sin(3\theta)) = 2\dfrac{d}{d\theta}(\sin(3\theta)) = 2(\cos(3\theta))\dfrac{d}{d\theta}(3\theta) = 2(\cos(3\theta))3 = 6\cos(3\theta).$

(b) $\dfrac{d}{dx}(\cos^2 x) = \dfrac{d}{dx}\left((\cos x)^2\right) = 2(\cos x)\cdot\dfrac{d}{dx}(\cos x) = 2(\cos x)(-\sin x) = -2\cos x \sin x.$

(c) $\dfrac{d}{dx}\left(\cos(x^2)\right) = -\sin(x^2)\cdot\dfrac{d}{dx}(x^2) = -2x\sin(x^2).$

(d) $\dfrac{d}{dt}(e^{-\sin t}) = e^{-\sin t}\dfrac{d}{dt}(-\sin t) = -(\cos t)e^{-\sin t}.$

Derivative of the Tangent Function

Since $\tan x = \sin x/\cos x$, we differentiate $\tan x$ using the quotient rule. Writing $(\sin x)'$ for $d(\sin x)/dx$, we have:

$$\frac{d}{dx}(\tan x) = \frac{d}{dx}\left(\frac{\sin x}{\cos x}\right) = \frac{(\sin x)'(\cos x) - (\sin x)(\cos x)'}{\cos^2 x} = \frac{\cos^2 x + \sin^2 x}{\cos^2 x} = \frac{1}{\cos^2 x}.$$

> For x in radians, $\dfrac{d}{dx}(\tan x) = \dfrac{1}{\cos^2 x}$.

The graphs of $f(x) = \tan x$ and $f'(x) = 1/\cos^2 x$ are in Figure 3.21. Is it reasonable that f' is always positive? Are the asymptotes of f' where we expect?

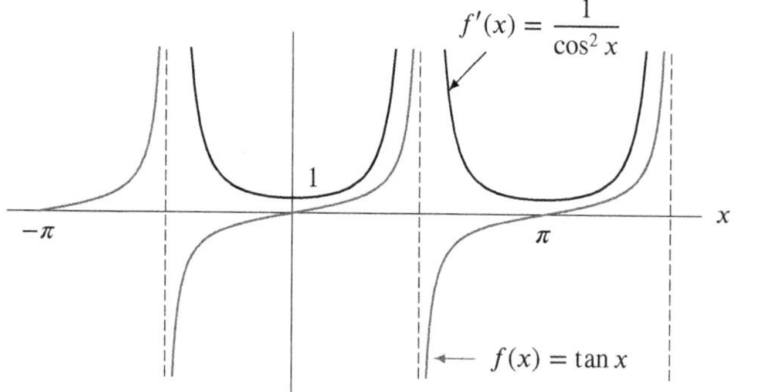

Figure 3.21: The function $\tan x$ and its derivative

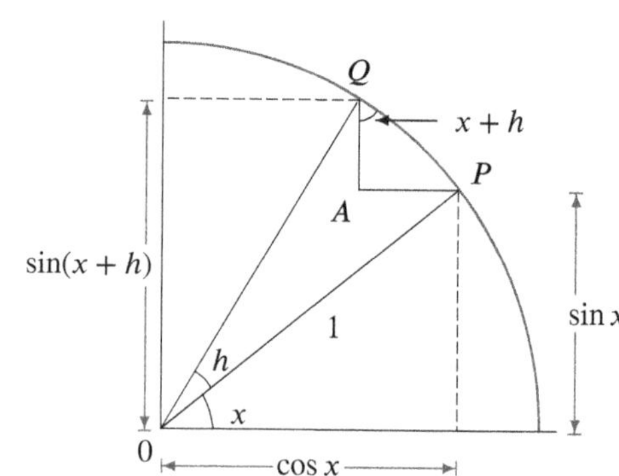

Figure 3.22: Unit circle showing $\sin(x+h)$ and $\sin x$

Example 5 Differentiate (a) $2\tan(3t)$, (b) $\tan(1-\theta)$, (c) $\dfrac{1+\tan t}{1-\tan t}$.

Solution (a) Use the chain rule:

$$\frac{d}{dt}(2\tan(3t)) = 2\frac{1}{\cos^2(3t)}\frac{d}{dt}(3t) = \frac{6}{\cos^2(3t)}.$$

(b) Use the chain rule:

$$\frac{d}{d\theta}(\tan(1-\theta)) = \frac{1}{\cos^2(1-\theta)}\cdot\frac{d}{d\theta}(1-\theta) = -\frac{1}{\cos^2(1-\theta)}.$$

(c) Use the quotient rule:

$$\begin{aligned}\frac{d}{dt}\left(\frac{1+\tan t}{1-\tan t}\right) &= \frac{\frac{d}{dt}(1+\tan t)(1-\tan t)-(1+\tan t)\frac{d}{dt}(1-\tan t)}{(1-\tan t)^2}\\ &= \frac{\frac{1}{\cos^2 t}(1-\tan t)-(1+\tan t)\left(-\frac{1}{\cos^2 t}\right)}{(1-\tan t)^2}\\ &= \frac{2}{\cos^2 t\cdot(1-\tan t)^2}.\end{aligned}$$

Example 6 The Bay of Fundy in Canada is known for extreme tides. The depth of the water, y, in meters can be modeled as a function of time, t, in hours after midnight, by

$$y = 10 + 7.5\cos(0.507t).$$

How quickly is the depth of the water rising or falling at 6:00 am and at 9:00 am?

Solution To find how fast the water depth is changing, we compute the derivative of y, using the chain rule:

$$\frac{dy}{dt} = -7.5(0.507)\sin(0.507t) = -3.8025\sin(0.507t).$$

When $t = 6$, we have $\frac{dy}{dt} = -3.8025\sin(0.507\cdot 6) = -0.378$ meters/hour. So the tide is falling at 0.378 meters/hour.

When $t = 9$, we have $\frac{dy}{dt} = -3.8025\sin(0.507\cdot 9) = 3.760$ meters/hour. So the tide is rising at 3.760 meters/hour.

Informal Justification of $\frac{d}{dx}(\sin x) = \cos x$

Consider the unit circle in Figure 3.22. To find the derivative of $\sin x$, we need to estimate

$$\frac{\sin(x+h)-\sin x}{h}.$$

In Figure 3.22, the quantity $\sin(x+h)-\sin x$ is represented by the length QA. The arc QP is of length h, so

$$\frac{\sin(x+h)-\sin x}{h} = \frac{QA}{\text{Arc } QP}.$$

Now, if h is small, QAP is approximately a right triangle because the arc QP is almost a straight line. Furthermore, using geometry, we can show that angle $AQP = x + h$. For small h, we have

$$\frac{\sin(x+h)-\sin x}{h} = \frac{QA}{\text{Arc } QP} \approx \cos(x+h).$$

As $h \to 0$, the approximation gets better, so

$$\frac{d}{dx}(\sin x) = \lim_{h\to 0} \frac{\sin(x+h) - \sin x}{h} = \cos x.$$

Other derivations of this result are given in Problems 88 and 89 (available online).

Exercises and Problems for Section 3.5 Online Resource: Additional Problems for Section 3.5

EXERCISES

1. Construct a table of values for $\cos x$, $x = 0, 0.1, 0.2, \ldots, 0.6$. Using the difference quotient, estimate the derivative at these points (use $h = 0.001$), and compare it with $-\sin x$.

In Exercises 2–49, find the derivatives of the functions. Assume a, b, and c are constants.

2. $r(\theta) = \sin\theta + \cos\theta$
3. $s(\theta) = \cos\theta \sin\theta$
4. $z = \cos(4\theta)$
5. $f(x) = \sin(3x)$
6. $y = 5\sin(3t)$
7. $P = 4\cos(2t)$
8. $g(x) = \sin(2 - 3x)$
9. $R(x) = 10 - 3\cos(\pi x)$
10. $g(\theta) = \sin^2(2\theta) - \pi\theta$
11. $g(t) = (2 + \sin(\pi t))^3$
12. $f(x) = x^2 \cos x$
13. $w = \sin(e^t)$
14. $f(x) = e^{\cos x}$
15. $f(y) = e^{\sin y}$
16. $z = \theta e^{\cos\theta}$
17. $R(\theta) = e^{\sin(3\theta)}$
18. $g(\theta) = \sin(\tan\theta)$
19. $w(x) = \tan(x^2)$
20. $f(x) = \sqrt{1 - \cos x}$
21. $f(x) = \sqrt{3 + \sin(8x)}$
22. $f(x) = \cos(\sin x)$
23. $f(x) = \tan(\sin x)$
24. $k(x) = \sqrt{(\sin(2x))^3}$
25. $f(x) = 2x\sin(3x)$
26. $y = e^{\theta}\sin(2\theta)$
27. $f(x) = e^{-2x} \cdot \sin x$
28. $z = \sqrt{\sin t}$
29. $y = \sin^5\theta$
30. $g(z) = \tan(e^z)$
31. $z = \tan(e^{-3\theta})$
32. $w = e^{-\sin\theta}$
33. $Q = \cos(e^{2x})$
34. $h(t) = t\cos t + \tan t$
35. $f(\alpha) = \cos\alpha + 3\sin\alpha$
36. $k(\alpha) = \sin^5\alpha\cos^3\alpha$
37. $f(\theta) = \theta^3\cos\theta$
38. $y = \cos^2 w + \cos(w^2)$
39. $y = \sin(\sin x + \cos x)$
40. $y = \sin(2x) \cdot \sin(3x)$
41. $P = \dfrac{\cos t}{t^3}$
42. $t(\theta) = \dfrac{\cos\theta}{\sin\theta}$
43. $f(x) = \sqrt{\dfrac{1 - \sin x}{1 - \cos x}}$
44. $r(y) = \dfrac{y}{\cos y + a}$
45. $G(x) = \dfrac{\sin^2 x + 1}{\cos^2 x + 1}$
46. $y = a\sin(bt) + c$
47. $P = a\cos(bt + c)$
48. $y = x^2 e^x \sin x$
49. $y = x^3 e^{5x}\sin(2x)$

PROBLEMS

In Problems 50–52, find formulas for f'' and f'''.

50. $f(x) = \sin(2x)$
51. $f(\theta) = \theta\cos\theta$
52. $f(u) = e^{2u}\sin u$

53. Is the graph of $y = \sin(x^4)$ increasing or decreasing when $x = 10$? Is it concave up or concave down?

54. Find the line tangent to $f(t) = 3\sin(2t) + 5$ at the point where $t = \pi$.

55. Find the line tangent to $f(x) = 3x + \cos(5x)$ at the point where $x = 0$.

56. Find the line tangent to $f(t) = 8 + \sin(3t)$ at the point where $t = 0$.

57. Find the 50th derivative of $y = \cos x$.

58. Find a function $F(x)$ satisfying $F'(x) = \sin(4x)$.

59. Let $f(x) = \sin^2 x + \cos^2 x$.
   (a) Find $f'(x)$ using the formula for $f(x)$ and derivative formulas from this section. Simplify your answer.
   (b) Use a trigonometric identity to check your answer to part (a). Explain.

60. On page 42 the depth, y, in feet, of water in Boston Harbor is given in terms of t, the number of hours since midnight, by

$$y = 5 + 4.9\cos\left(\frac{\pi}{6}t\right).$$

   (a) Find dy/dt. What does dy/dt represent, in terms of water level?

(b) For $0 \leq t \leq 24$, when is dy/dt zero? (Figure 1.66 on page 43 may be helpful.) Explain what it means (in terms of water level) for dy/dt to be zero.

61. A boat at anchor is bobbing up and down in the sea. The vertical distance, y, in feet, between the sea floor and the boat is given as a function of time, t, in minutes, by

$$y = 15 + \sin(2\pi t).$$

(a) Find the vertical velocity, v, of the boat at time t.
(b) Make rough sketches of y and v against t.

62. The voltage, V, in volts, in an electrical outlet is given as a function of time, t, in seconds, by the function $V = 156\cos(120\pi t)$.

(a) Give an expression for the rate of change of voltage with respect to time.
(b) Is the rate of change ever zero? Explain.
(c) What is the maximum value of the rate of change?

63. An oscillating mass of m gm at the end of a spring is at a distance y from its equilibrium position given by

$$y = A\sin\left(\left(\sqrt{\frac{k}{m}}\right)t\right).$$

The constant k measures the stiffness of the spring.

(a) Find a time at which the mass is farthest from its equilibrium position. Find a time at which the mass is moving fastest. Find a time at which the mass is accelerating fastest.
(b) What is the period, T, of the oscillation?
(c) Find dT/dm. What does the sign of dT/dm tell you?

64. With t in years, the population of a herd of deer is represented by

$$P(t) = 4000 + 500\sin\left(2\pi t - \frac{\pi}{2}\right).$$

(a) How does this population vary with time? Graph $P(t)$ for one year.
(b) When in the year the population is a maximum? What is that maximum? Is there a minimum? If so, when?
(c) When is the population growing fastest? When is it decreasing fastest?
(d) How fast is the population changing on July 1?

65. An environmentalist reports that the depth of the water in a new reservoir is approximated by

$$h = d(t) = \begin{cases} kt & 0 \leq t \leq 2 \\ 50 + \sin(0.1t) & t > 2, \end{cases}$$

where t is in weeks since the date the reservoir was completed and h is in meters.

(a) During what period was the reservoir filling at a constant rate? What was that rate?
(b) In this model, is the rate at which the water level is changing defined for all times $t > 0$? Explain.

66. Normal human body temperature fluctuates with a rhythm tied to our sleep cycle.[14] If $H(t)$ is body temperature in degrees Celsius at time t in hours since 9 am, then $H(t)$ may be modeled by

$$H(t) = 36.8 + 0.6\sin\left(\frac{\pi}{12}t\right).$$

(a) Calculate $H'(t)$ and give units.
(b) Calculate $H'(4)$ and $H'(12)$, then interpret the meaning of your answers in everyday terms.

In Problems 67–70, find and interpret the value of the expression in practical terms. Let $C(t)$ be the concentration of carbon dioxide in parts per million (ppm) in the air as a function of time, t, in months since December 1, 2005:[15]

$$C(t) = 3.5\sin\left(\frac{\pi t}{6}\right) + 381 + \frac{t}{6}.$$

67. $C'(36)$

68. $C'(60)$

69. $C'(30)$

70. $\dfrac{C(60) - C(0)}{60}$

71. A rubber duck bounces up and down in a pool after a stone is dropped into the water. The height of the duck, in inches, above the equilibrium position of the water is given as a function of time t, in seconds, by

$$d(t) = e^{-t}(\cos t + \sin t).$$

(a) Find and interpret the practical meaning of the derivative $d'(t)$.
(b) Determine when $d'(t) = 0$ for $t \geq 0$. What can you say about the duck when $d'(t) = 0$?
(c) Determine $\lim_{t\to\infty} d(t)$ and explain why this limit makes sense in practical terms.

72. The metal bar of length l in Figure 3.23 has one end attached at the point P to a circle of radius a. Point Q at the other end can slide back and forth along the x-axis.

(a) Find x as a function of θ.
(b) Assume lengths are in centimeters and the angular velocity ($d\theta/dt$) is 2 radians/second counterclockwise. Find the velocity at which the point Q is moving when

(i) $\theta = \pi/2$, (ii) $\theta = \pi/4$.

[14] Model based on data from circadian.org. Accessed January 2014.
[15] Based on data from Mauna Loa, Hawaii, at esrl.noaa.gov/gmd/ccgg/trends/. Accessed March 2011.

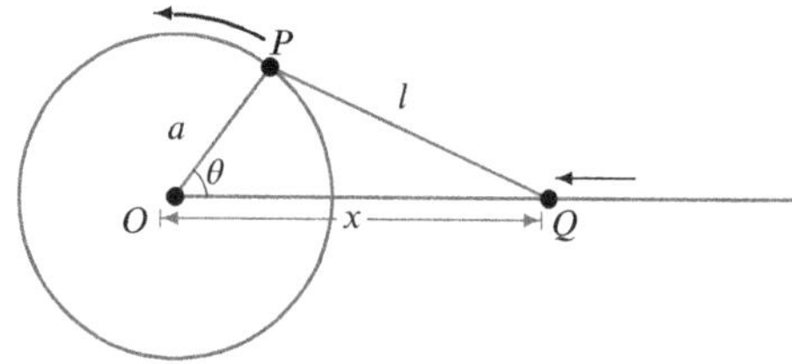

Figure 3.23

73. Let $f(x) = e^{-x}\sin x$.
   (a) Find the derivative $f'(x)$.
   (b) Explain why $f'(x) = 0$ at precisely the same points where $\tan x = 1$.

Strengthen Your Understanding

In Problems 74–75, explain what is wrong with the statement.

74. The derivative of $n(x) = \sin(\cos x)$ is $n'(x) = \cos(-\sin x)$.
75. The derivative of $f(x) = \sin(\sin x)$ is $f'(x) = (\cos x)(\sin x) + (\sin x)(\cos x)$.

In Problems 76–77, give an example of:

76. A trigonometric function whose derivative must be calculated using the chain rule.
77. A function $f(x)$ such that $f''(x) = -f(x)$.

Are the statements in Problems 78–80 true or false? Give an explanation for your answer.

78. The derivative of $\tan\theta$ is periodic.
79. If a function is periodic, with period c, then so is its derivative.
80. The only functions whose fourth derivatives are equal to $\cos t$ are of the form $\cos t + C$, where C is any constant.

3.6 THE CHAIN RULE AND INVERSE FUNCTIONS

In this section we will use the chain rule to calculate the derivatives of fractional powers, logarithms, exponentials, and the inverse trigonometric functions.[16] The same method is used to obtain a formula for the derivative of a general inverse function.

Finding the Derivative of an Inverse Function: Derivative of $x^{1/2}$

Earlier we calculated the derivative of x^n with n an integer, but we have been using the result for non-integer values of n as well. We now confirm that the power rule holds for $n = 1/2$ by calculating the derivative of $f(x) = x^{1/2}$ using the chain rule. Since

$$(f(x))^2 = x,$$

the derivative of $(f(x))^2$ and the derivative of x must be equal, so

$$\frac{d}{dx}(f(x))^2 = \frac{d}{dx}(x).$$

We can use the chain rule with $f(x)$ as the inside function to obtain:

$$\frac{d}{dx}(f(x))^2 = 2f(x)\cdot f'(x) = 1.$$

Solving for $f'(x)$ gives

$$f'(x) = \frac{1}{2f(x)} = \frac{1}{2x^{1/2}},$$

or

$$\frac{d}{dx}(x^{1/2}) = \frac{1}{2x^{1/2}} = \frac{1}{2}x^{-1/2}.$$

A similar calculation can be used to obtain the derivative of $x^{1/n}$ where n is a positive integer.

[16]It requires a separate justification, not given here, that these functions are differentiable.

Derivative of $\ln x$

We use the chain rule to differentiate an identity involving $\ln x$. Since $e^{\ln x} = x$, we have

$$\frac{d}{dx}(e^{\ln x}) = \frac{d}{dx}(x),$$

$$e^{\ln x} \cdot \frac{d}{dx}(\ln x) = 1. \qquad \text{(Since } e^x \text{ is outside function and } \ln x \text{ is inside function)}$$

Solving for $d(\ln x)/dx$ gives

$$\frac{d}{dx}(\ln x) = \frac{1}{e^{\ln x}} = \frac{1}{x},$$

so

$$\boxed{\frac{d}{dx}(\ln x) = \frac{1}{x}.}$$

Example 1 Differentiate (a) $\ln(x^2+1)$ (b) $t^2 \ln t$ (c) $\sqrt{1+\ln(1-y)}$.

Solution (a) Using the chain rule:

$$\frac{d}{dx}\left(\ln(x^2+1)\right) = \frac{1}{x^2+1}\frac{d}{dx}(x^2+1) = \frac{2x}{x^2+1}.$$

(b) Using the product rule:

$$\frac{d}{dt}(t^2 \ln t) = \frac{d}{dt}(t^2)\cdot \ln t + t^2\frac{d}{dt}(\ln t) = 2t\ln t + t^2 \cdot \frac{1}{t} = 2t\ln t + t.$$

(c) Using the chain rule:

$$\begin{aligned}
\frac{d}{dy}\left(\sqrt{1+\ln(1-y)}\right) &= \frac{d}{dy}(1+\ln(1-y))^{1/2} \\
&= \frac{1}{2}(1+\ln(1-y))^{-1/2}\cdot\frac{d}{dy}(1+\ln(1-y)) && \text{(Using the chain rule)} \\
&= \frac{1}{2\sqrt{1+\ln(1-y)}}\cdot\frac{1}{1-y}\cdot\frac{d}{dy}(1-y) && \text{(Using the chain rule again)} \\
&= \frac{-1}{2(1-y)\sqrt{1+\ln(1-y)}}.
\end{aligned}$$

Derivative of a^x

In Section 3.2, we saw that the derivative of a^x is proportional to a^x. Now we see another way of calculating the constant of proportionality. We use the identity

$$\ln(a^x) = x\ln a.$$

Differentiating both sides, using $\frac{d}{dx}(\ln x) = \frac{1}{x}$ and the chain rule, and remembering that $\ln a$ is a constant, we obtain:

$$\frac{d}{dx}(\ln a^x) = \frac{1}{a^x}\cdot\frac{d}{dx}(a^x) = \ln a.$$

Solving gives the result we obtained earlier:

$$\boxed{\frac{d}{dx}(a^x) = (\ln a)a^x.}$$

Derivatives of Inverse Trigonometric Functions

In Section 1.5 we defined $\arcsin x$ as the angle between $-\pi/2$ and $\pi/2$ (inclusive) whose sine is x. Similarly, $\arctan x$ as the angle strictly between $-\pi/2$ and $\pi/2$ whose tangent is x. To find $\frac{d}{dx}(\arctan x)$ we use the identity $\tan(\arctan x) = x$. Differentiating using the chain rule gives

$$\frac{1}{\cos^2(\arctan x)} \cdot \frac{d}{dx}(\arctan x) = 1,$$

so

$$\frac{d}{dx}(\arctan x) = \cos^2(\arctan x).$$

Using the identity $1 + \tan^2\theta = \frac{1}{\cos^2\theta}$, and replacing θ by $\arctan x$, we have

$$\cos^2(\arctan x) = \frac{1}{1 + \tan^2(\arctan x)} = \frac{1}{1 + x^2}.$$

Thus we have

$$\boxed{\frac{d}{dx}(\arctan x) = \frac{1}{1 + x^2}.}$$

By a similar argument, we obtain the result:

$$\boxed{\frac{d}{dx}(\arcsin x) = \frac{1}{\sqrt{1 - x^2}}.}$$

Example 2 Differentiate (a) $\arctan(t^2)$ (b) $\arcsin(\tan\theta)$.

Solution Use the chain rule:

(a) $\frac{d}{dt}\left(\arctan(t^2)\right) = \frac{1}{1 + (t^2)^2} \cdot \frac{d}{dt}(t^2) = \frac{2t}{1 + t^4}.$

(b) $\frac{d}{dt}(\arcsin(\tan\theta)) = \frac{1}{\sqrt{1 - (\tan\theta)^2}} \cdot \frac{d}{d\theta}(\tan\theta) = \frac{1}{\sqrt{1 - \tan^2\theta}} \cdot \frac{1}{\cos^2\theta}.$

Derivative of a General Inverse Function

Each of the previous results gives the derivative of an inverse function. In general, if a function f has a differentiable inverse, f^{-1}, we find its derivative by differentiating $f(f^{-1}(x)) = x$ by the chain rule:

$$\frac{d}{dx}\left(f\left(f^{-1}(x)\right)\right) = 1$$
$$f'\left(f^{-1}(x)\right) \cdot \frac{d}{dx}\left(f^{-1}(x)\right) = 1$$

so

$$\boxed{\frac{d}{dx}\left(f^{-1}(x)\right) = \frac{1}{f'(f^{-1}(x))}.}$$

Thus, the derivative of the inverse is the reciprocal of the derivative of the original function, but evaluated at the point $f^{-1}(x)$ instead of the point x.

Example 3 Figure 3.24 shows $f(x)$ and $f^{-1}(x)$. Using Table 3.6, find

(a) (i) $f(2)$ (ii) $f^{-1}(2)$ (iii) $f'(2)$ (iv) $(f^{-1})'(2)$

(b) The equation of the tangent lines at the points P and Q.

(c) What is the relationship between the two tangent lines?

Table 3.6

x	$f(x)$	$f'(x)$
0	1	0.7
1	2	1.4
2	4	2.8
3	8	5.5

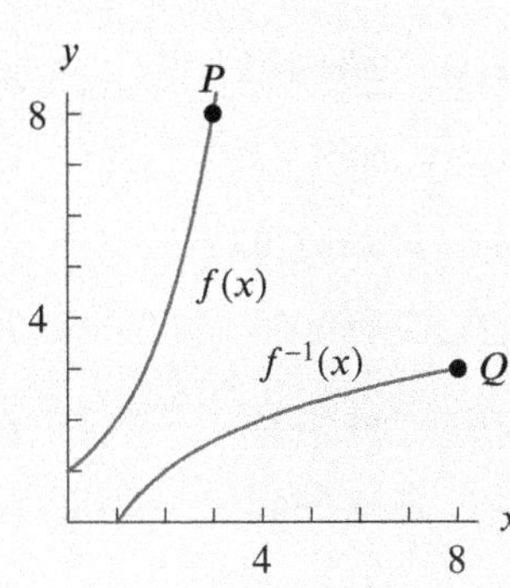

Solution (a) Reading from the table, we have

(i) $f(2) = 4$.

(ii) $f^{-1}(2) = 1$.

(iii) $f'(2) = 2.8$.

(iv) To find the derivative of the inverse function, we use

$$(f^{-1})'(2) = \frac{1}{f'(f^{-1}(2))} = \frac{1}{f'(1)} = \frac{1}{1.4} = 0.714.$$

Notice that the derivative of f^{-1} is the reciprocal of the derivative of f. However, the derivative of f^{-1} is evaluated at 2, while the derivative of f is evaluated at 1, where $f^{-1}(2) = 1$ and $f(1) = 2$.

(b) At the point P, we have $f(3) = 8$ and $f'(3) = 5.5$, so the equation of the tangent line at P is

$$y - 8 = 5.5(x - 3).$$

At the point Q, we have $f^{-1}(8) = 3$, so the slope at Q is

$$(f^{-1})'(8) = \frac{1}{f'(f^{-1}(8))} = \frac{1}{f'(3)} = \frac{1}{5.5}.$$

Thus, the equation of the tangent line at Q is

$$y - 3 = \frac{1}{5.5}(x - 8).$$

(c) The two tangent lines have reciprocal slopes, and the points (3, 8) and (8, 3) are reflections of one another across the line $y = x$. Thus, the two tangent lines are reflections of one another across the line $y = x$.

Exercises and Problems for Section 3.6 Online Resource: Additional Problems for Section 3.6

EXERCISES

■ For Exercises 1–41, find the derivative. It may be to your advantage to simplify before differentiating. Assume a, b, c, and k are constants.

1. $f(t) = \ln(t^2 + 1)$

2. $f(x) = \ln(1 - x)$

3. $f(x) = \ln(5x^2 + 3)$

4. $y = 2x^2 + 3\ln x$

5. $y = \arcsin(x+1)$

6. $f(x) = \arctan(3x)$

7. $P = 3\ln(x^2+5x+3)$

8. $Q = a\ln(bx+c)$

9. $f(x) = \ln(e^{2x})$

10. $f(x) = e^{\ln(e^{2x^2+3})}$

11. $f(x) = \ln(1-e^{-x})$

12. $f(\alpha) = \ln(\sin\alpha)$

13. $f(x) = \ln(e^x+1)$

14. $y = x\ln x - x + 2$

15. $j(x) = \ln(e^{ax}+b)$

16. $y = x^3\ln x$

17. $h(w) = w^3\ln(10w)$

18. $f(x) = \ln(e^{7x})$

19. $f(x) = e^{(\ln x)+1}$

20. $f(\theta) = \ln(\cos\theta)$

21. $f(t) = \ln(e^{\ln t})$

22. $f(y) = \arcsin(y^2)$

23. $s(x) = \arctan(2-x)$

24. $g(\alpha) = \sin(\arcsin\alpha)$

25. $g(t) = e^{\arctan(3t^2)}$

26. $g(t) = \cos(\ln t)$

27. $h(z) = z^{\ln 2}$

28. $h(w) = w\arcsin w$

29. $f(x) = e^{\ln(kx)}$

30. $r(t) = \arcsin(2t)$

31. $j(x) = \cos\left(\sin^{-1}x\right)$

32. $f(x) = \cos(\arctan 3x)$

33. $f(z) = \dfrac{1}{\ln z}$

34. $g(t) = \dfrac{\ln(kt)+t}{\ln(kt)-t}$

35. $f(x) = \ln(\sin x+\cos x)$

36. $f(t) = \ln(\ln t)+\ln(\ln 2)$

37. $f(w) = 6\sqrt{w} + \dfrac{1}{w^2} + 5\ln w$

38. $y = 2x(\ln x + \ln 2) - 2x + e$

39. $f(x) = \cos(\arcsin(x+1))$

40. $a(t) = \ln\left(\dfrac{1-\cos t}{1+\cos t}\right)^4$

41. $T(u) = \arctan\left(\dfrac{u}{1+u}\right)$

PROBLEMS

42. Let $f(x) = \ln(3x)$.

(a) Find $f'(x)$ and simplify your answer.
(b) Use properties of logs to rewrite $f(x)$ as a sum of logs.
(c) Differentiate the result of part (b). Compare with the result in part (a).

43. On what intervals is $\ln(x^2+1)$ concave up?

44. Use the chain rule to obtain the formula for $\dfrac{d}{dx}(\arcsin x)$.

45. Using the chain rule, find $\dfrac{d}{dx}(\log x)$. (Recall $\log x = \log_{10} x$.)

46. To compare the acidity of different solutions, chemists use the pH (which is a single number, not the product of p and H). The pH is defined in terms of the concentration, x, of hydrogen ions in the solution as

$$\text{pH} = -\log x.$$

Find the rate of change of pH with respect to hydrogen ion concentration when the pH is 2. [Hint: Use the result of Problem 45.]

47. The number of years, T, it takes an investment of \$1000 to grow to \$$F$ in an account which pays 5% interest compounded continuously is given by

$$T = g(F) = 20\ln(0.001F).$$

Find $g(5000)$ and $g'(5000)$. Give units with your answers and interpret them in terms of money in the account.

48. A firm estimates that the total revenue, R, in dollars, received from the sale of q goods is given by

$$R = \ln(1+1000q^2).$$

The marginal revenue, MR, is the rate of change of the total revenue as a function of quantity. Calculate the marginal revenue when $q = 10$.

49. Average leaf width, w (in mm), in tropical Australia[17] is a function of the average annual rainfall, x (in mm). We have $w = f(x) = 32.7\ln(x/244.5)$.

(a) Find $f'(x)$.
(b) Find $f'(2000)$. Include units.
(c) Explain how you can use your answer to part (b) to estimate the difference in average leaf widths in a forest whose average annual rainfall is 2000 mm and one whose annual rainfall is 150 mm more.

50. **(a)** Find the equation of the tangent line to $y = \ln x$ at $x = 1$.
(b) Use it to calculate approximate values for $\ln(1.1)$ and $\ln(2)$.
(c) Using a graph, explain whether the approximate values are smaller or larger than the true values. Would the same result have held if you had used the tangent line to estimate $\ln(0.9)$ and $\ln(0.5)$? Why?

51. **(a)** For $x > 0$, find and simplify the derivative of $f(x) = \arctan x + \arctan(1/x)$.
(b) What does your result tell you about f?

52. **(a)** Given that $f(x) = x^3$, find $f'(2)$.
(b) Find $f^{-1}(x)$.
(c) Use your answer from part (b) to find $(f^{-1})'(8)$.
(d) How could you have used your answer from part (a) to find $(f^{-1})'(8)$?

[17] H. H. Shugart, *Terrestrial Ecosystems in Changing Environments* (Cambridge: Cambridge University Press, 1998), p. 145.

53. **(a)** For $f(x) = 2x^5 + 3x^3 + x$, find $f'(x)$.
(b) How can you use your answer to part (a) to determine if $f(x)$ is invertible?
(c) Find $f(1)$.
(d) Find $f'(1)$.
(e) Find $(f^{-1})'(6)$.

54. Imagine you are zooming in on the graph of each of the following functions near the origin:

$$y = x \qquad y = \sqrt{x}$$
$$y = x^2 \qquad y = \sin x$$
$$y = x\sin x \qquad y = \tan x$$
$$y = \sqrt{x/(x+1)} \qquad y = x^3$$
$$y = \ln(x+1) \qquad y = \frac{1}{2}\ln(x^2+1)$$
$$y = 1 - \cos x \qquad y = \sqrt{2x - x^2}$$

Which of them look the same? Group together those functions which become indistinguishable, and give the equations of the lines they look like.

In Problems 55–58, use Figure 3.25 to find a point x where $h(x) = n(m(x))$ has the given derivative.

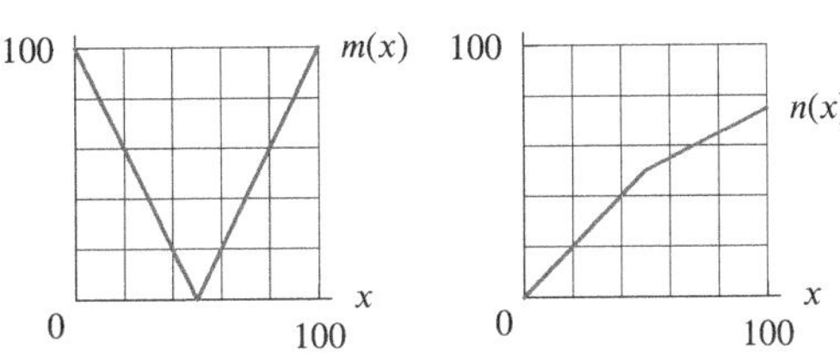

Figure 3.25

55. $h'(x) = -2$

56. $h'(x) = 2$

57. $h'(x) = 1$

58. $h'(x) = -1$

In Problems 59–61, use Figure 3.26 to estimate the derivatives.

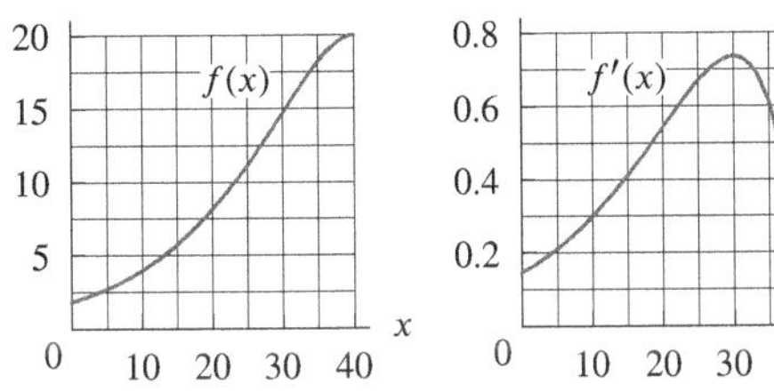

Figure 3.26

59. $(f^{-1})'(5)$

60. $(f^{-1})'(10)$

61. $(f^{-1})'(15)$

[18]www.indexmundi.com, accessed April 10, 2015.

In Problems 62–64, use Figure 3.27 to calculate the derivative.

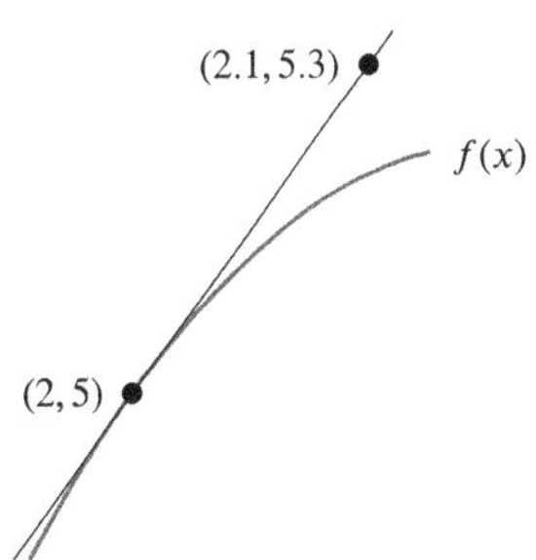

Figure 3.27

62. $h'(2)$ if $h(x) = (f(x))^3$

63. $k'(2)$ if $k(x) = (f(x))^{-1}$

64. $g'(5)$ if $g(x) = f^{-1}(x)$

65. Use the table and the fact that $f(x)$ is invertible and differentiable everywhere to find $(f^{-1})'(3)$.

x	$f(x)$	$f'(x)$
3	1	7
6	2	10
9	3	5

66. At a particular location, $f(p)$ is the number of gallons of gas sold when the price is p dollars per gallon.

(a) What does the statement $f(2) = 4023$ tell you about gas sales?
(b) Find and interpret $f^{-1}(4023)$.
(c) What does the statement $f'(2) = -1250$ tell you about gas sales?
(d) Find and interpret $(f^{-1})'(4023)$

67. Let $P = f(t)$ give the US population[18] in millions in year t.

(a) What does the statement $f(2014) = 319$ tell you about the US population?
(b) Find and interpret $f^{-1}(319)$. Give units.
(c) What does the statement $f'(2014) = 2.44$ tell you about the population? Give units.
(d) Evaluate and interpret $(f^{-1})'(319)$. Give units.

68. Figure 3.28 shows the number of motor vehicles,[19] $f(t)$, in millions, registered in the world t years after 1965. With units, estimate and interpret

(a) $f(20)$ (b) $f'(20)$
(c) $f^{-1}(500)$ (d) $(f^{-1})'(500)$

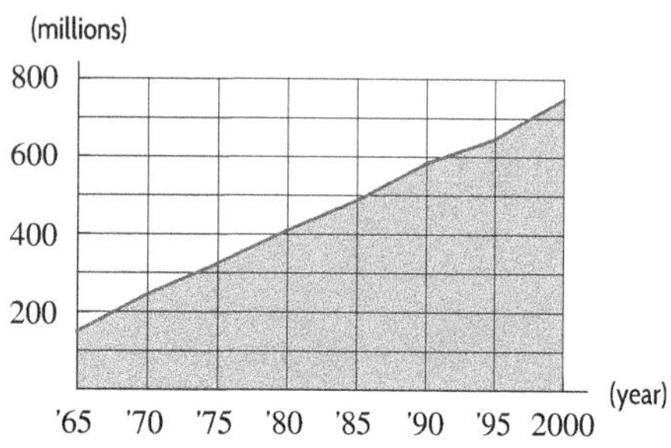

Figure 3.28

69. Using Figure 3.29, where $f'(2) = 2.1$, $f'(4) = 3.0$, $f'(6) = 3.7$, $f'(8) = 4.2$, find $(f^{-1})'(8)$.

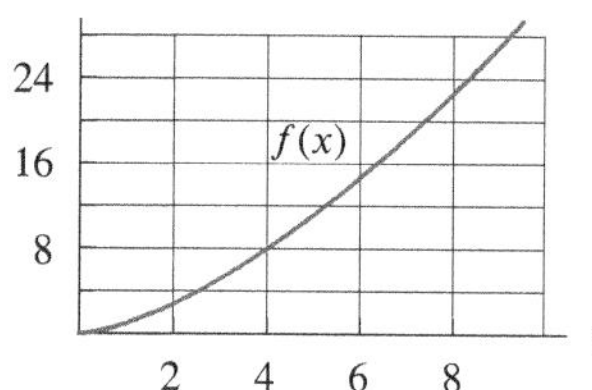

Figure 3.29

70. Figure 3.30 shows an invertible function $f(x)$.

(a) Sketch a graph of $f^{-1}(x)$.
(b) Find each x-value where $f^{-1}(x)$ has a horizontal tangent line.
(c) Find each x-value where $f^{-1}(x)$ has a vertical tangent line.
(d) Explain how you could have answered (b) and (c) just using the graph of $f(x)$.

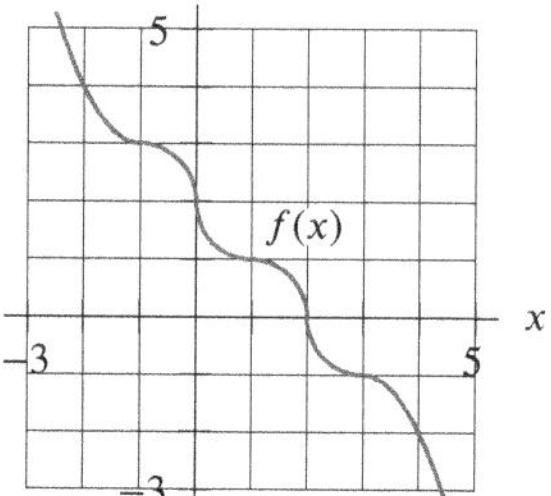

Figure 3.30

71. An invertible function $f(x)$ has values in the table. Evaluate

(a) $f'(a) \cdot (f^{-1})'(A)$ (b) $f'(b) \cdot (f^{-1})'(B)$
(c) $f'(c) \cdot (f^{-1})'(C)$

x	a	b	c	d
$f(x)$	A	B	C	D

72. Assuming $g(x) > 0$, use the chain rule to explain why the derivative of $f(x) = \ln(g(x))$ is zero exactly where the derivative of $g(x)$ is zero.

73. If f is continuous, invertible, and defined for all x, why must at least one of the statements $(f^{-1})'(10) = 8$, $(f^{-1})'(20) = -6$ be wrong?

74. Suppose $f(x)$ is an invertible function that is differentiable at every point of its domain and let $g(x) = f^{-1}(x)$.

(a) Is $g(x)$ also differentiable at every point of its domain? Explain your answer.
(b) Explain why there is no value $x = a$ with $g'(a) = 0$.

Strengthen Your Understanding

In Problems 75–77, explain what is wrong with the statement.

75. If $w(x) = \ln(1 + x^4)$ then $w'(x) = 1/(1 + x^4)$.

76. The derivative of $f(x) = \ln(\ln x)$ is

$$f'(x) = \frac{1}{x}\ln x + \ln x\frac{1}{x} = \frac{2\ln x}{x}.$$

77. Given $f(2) = 6$, $f'(2) = 3$, and $f^{-1}(3) = 4$, we have $(f^{-1})'(2) = \frac{1}{f^{-1}(f'(2))} = \frac{1}{f^{-1}(3)} = \frac{1}{4}$.

In Problems 78–81, give an example of:

78. A function that is equal to a constant multiple of its derivative but that is not equal to its derivative.

79. A function whose derivative is c/x, where c is a constant.

80. A function $f(x)$ for which $f'(x) = f'(cx)$, where c is a constant.

81. A function f such that $\frac{d}{dx}\left(f^{-1}(x)\right) = \frac{1}{f'(x)} = 1$.

Are the statements in Problems 82–83 true or false? Give an explanation for your answer.

82. The graph of $\ln(x^2)$ is concave up for $x > 0$.

83. If $f(x)$ has an inverse function, $g(x)$, then the derivative of $g(x)$ is $1/f'(x)$.

[19] www.earth-policy.org, accessed May 18, 2007.

3.7 IMPLICIT FUNCTIONS

In earlier chapters, most functions were written in the form $y = f(x)$; here y is said to be an *explicit* function of x. An equation such as

$$x^2 + y^2 = 4$$

is said to give y as an *implicit* function of x. Its graph is the circle in Figure 3.31. Since there are x-values which correspond to two y-values, y is not a function of x on the whole circle. Solving for y gives

$$y = \pm\sqrt{4 - x^2},$$

where $y = \sqrt{4 - x^2}$ represents the top half of the circle and $y = -\sqrt{4 - x^2}$ represents the bottom half. So y is a function of x on the top half, and y is a different function of x on the bottom half.

But let's consider the circle as a whole. The equation does represent a curve which has a tangent line at each point. The slope of this tangent can be found by differentiating the equation of the circle with respect to x:

$$\frac{d}{dx}(x^2) + \frac{d}{dx}(y^2) = \frac{d}{dx}(4).$$

If we think of y as a function of x and use the chain rule, we get

$$2x + 2y\frac{dy}{dx} = 0.$$

Solving gives

$$\frac{dy}{dx} = -\frac{x}{y}.$$

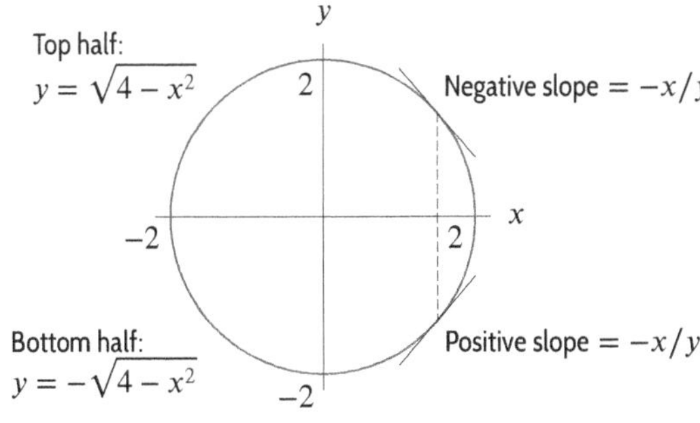

Figure 3.31: Graph of $x^2 + y^2 = 4$

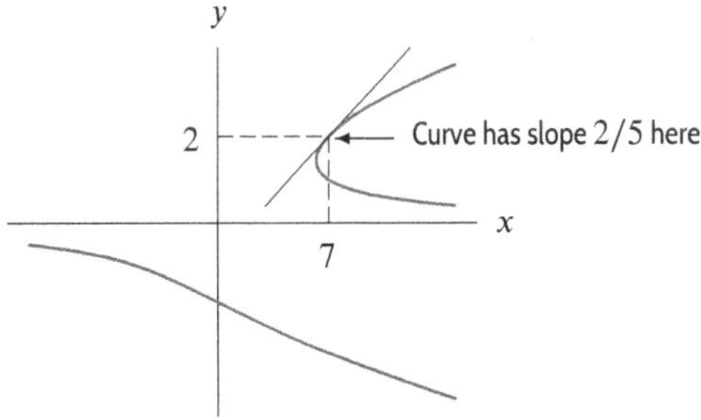

Figure 3.32: Graph of $y^3 - xy = -6$ and its tangent line at $(7, 2)$

The derivative here depends on both x and y (instead of just on x). This is because for many x-values there are two y-values, and the curve has a different slope at each one. Figure 3.31 shows that for x and y both positive, we are on the top right quarter of the curve and the slope is negative (as the formula predicts). For x positive and y negative, we are on the bottom right quarter of the curve and the slope is positive (as the formula predicts).

Differentiating the equation of the circle has given us the slope of the curve at all points except $(2, 0)$ and $(-2, 0)$, where the tangent is vertical. In general, this process of *implicit differentiation* leads to a derivative whenever the expression for the derivative does not have a zero in the denominator.

Example 1 Make a table of x and approximate y-values for the equation $y^3 - xy = -6$ near $x = 7$, $y = 2$. Your table should include the x-values 6.8, 6.9, 7.0, 7.1, and 7.2.

Solution We would like to solve for y in terms of x, but we cannot isolate y by factoring. There is a formula for solving cubics, somewhat like the quadratic formula, but it is too complicated to be useful here. Instead, first observe that $x = 7$, $y = 2$ does satisfy the equation. (Check this!) Now find dy/dx by implicit differentiation:

$$\begin{aligned}
\frac{d}{dx}(y^3) - \frac{d}{dx}(xy) &= \frac{d}{dx}(-6) \\
3y^2\frac{dy}{dx} - 1\cdot y - x\frac{dy}{dx} &= 0 \quad \text{(Differentiating with respect to } x\text{)} \\
3y^2\frac{dy}{dx} - x\frac{dy}{dx} &= y \\
(3y^2 - x)\frac{dy}{dx} &= y \quad \text{(Factoring out } \tfrac{dy}{dx}\text{)} \\
\frac{dy}{dx} &= \frac{y}{3y^2 - x}.
\end{aligned}$$

When $x = 7$ and $y = 2$, we have

$$\frac{dy}{dx} = \frac{2}{12 - 7} = \frac{2}{5}.$$

(See Figure 3.32.) The equation of the tangent line at $(7, 2)$ is

$$y - 2 = \frac{2}{5}(x - 7)$$

or

$$y = 0.4x - 0.8.$$

Since the tangent lies very close to the curve near the point $(7, 2)$, we use the equation of the tangent line to calculate the following approximate y-values:

x	6.8	6.9	7.0	7.1	7.2
Approximate y	1.92	1.96	2.00	2.04	2.08

Notice that although the equation $y^3 - xy = -6$ leads to a curve which is difficult to deal with algebraically, it still looks like a straight line locally.

Example 2 Find all points where the tangent line to $y^3 - xy = -6$ is either horizontal or vertical.

Solution From the previous example, $\dfrac{dy}{dx} = \dfrac{y}{3y^2 - x}$. The tangent is horizontal when the numerator of dy/dx equals 0, so $y = 0$. Since we also must satisfy $y^3 - xy = -6$, we get $0^3 - x\cdot 0 = -6$, which is impossible. We conclude that there are no points on the curve where the tangent line is horizontal.

The tangent is vertical when the denominator of dy/dx is 0, giving $3y^2 - x = 0$. Thus, $x = 3y^2$ at any point with a vertical tangent line. Again, we must also satisfy $y^3 - xy = -6$, so

$$\begin{aligned}
y^3 - (3y^2)y &= -6, \\
-2y^3 &= -6, \\
y &= \sqrt[3]{3} \approx 1.442.
\end{aligned}$$

We can then find x by substituting $y = \sqrt[3]{3}$ in $y^3 - xy = -6$. We get $3 - x(\sqrt[3]{3}) = -6$, so $x = 9/(\sqrt[3]{3}) \approx 6.240$. So the tangent line is vertical at $(6.240, 1.442)$.

Using implicit differentiation and the expression for dy/dx to locate the points where the tangent is vertical or horizontal, as in the previous example, is a first step in obtaining an overall picture of the curve $y^3 - xy = -6$. However, filling in the rest of the graph, even roughly, by using the sign of dy/dx to tell us where the curve is increasing or decreasing can be difficult.

Exercises and Problems for Section 3.7 Online Resource: Additional Problems for Section 3.7

EXERCISES

For Exercises 1–21, find dy/dx. Assume a, b, c are constants.

1. $x^2 + y^2 = \sqrt{7}$
2. $x^2 + y^3 = 8$
3. $x^2 + xy - y^3 = xy^2$
4. $x^2 + y^2 + 3x - 5y = 25$
5. $xy + x + y = 5$
6. $x^2y - 2y + 5 = 0$
7. $x^2y^3 - xy = 6$
8. $\sqrt{x} = 5\sqrt{y}$
9. $\sqrt{x} + \sqrt{y} = 25$
10. $xy - x - 3y - 4 = 0$
11. $6x^2 + 4y^2 = 36$
12. $ax^2 - by^2 = c^2$
13. $\ln x + \ln(y^2) = 3$
14. $x \ln y + y^3 = \ln x$
15. $\sin(xy) = 2x + 5$
16. $e^{\cos y} = x^3 \arctan y$
17. $\arctan(x^2y) = xy^2$
18. $e^{x^2} + \ln y = 0$
19. $(x - a)^2 + y^2 = a^2$
20. $x^{2/3} + y^{2/3} = a^{2/3}$
21. $\sin(ay) + \cos(bx) = xy$

In Exercises 22–25, find the slope of the tangent to the curve at the point specified.

22. $x^2 + y^2 = 1$ at $(0, 1)$
23. $\sin(xy) = x$ at $(1, \pi/2)$
24. $x^3 + 2xy + y^2 = 4$ at $(1, 1)$
25. $x^3 + 5x^2y + 2y^2 = 4y + 11$ at $(1, 2)$

For Exercises 26–30, find the equations of the tangent lines to the following curves at the indicated points.

26. $xy^2 = 1$ at $(1, -1)$
27. $\ln(xy) = 2x$ at $(1, e^2)$
28. $y^2 = \dfrac{x^2}{xy - 4}$ at $(4, 2)$
29. $y = \dfrac{x}{y + a}$ at $(0, 0)$
30. $x^{2/3} + y^{2/3} = a^{2/3}$ at $(a, 0)$

PROBLEMS

31. **(a)** Find dy/dx given that $x^2 + y^2 - 4x + 7y = 15$.
    (b) Under what conditions on x and/or y is the tangent line to this curve horizontal? Vertical?

32. **(a)** Find the slope of the tangent line to the ellipse $\dfrac{x^2}{25} + \dfrac{y^2}{9} = 1$ at the point (x, y).
    (b) Are there any points where the slope is not defined?

33. **(a)** Find all points on $y^2 + xy + x^2 = 1$ with $x = 1$.
    (b) Find dy/dx for $y^2 + xy + x^2 = 1$.
    (c) Find the slope of the tangent line to $y^2+xy+x^2 = 1$ at each point with $x = 1$.

34. Find the equations of the tangent lines at $x = 2$ to the ellipse

$$\frac{(x-2)^2}{16} + \frac{y^2}{4} = 1.$$

35. **(a)** Find the equations of the tangent lines to the circle $x^2 + y^2 = 25$ at the points where $x = 4$.
    (b) Find the equations of the normal lines to this circle at the same points. (The normal line is perpendicular to the tangent line at that point.)
    (c) At what point do the two normal lines intersect?

36. Find the equation of the tangent line to the curve $y = x^2$ at $x = 1$. Show that this line is also a tangent to a circle centered at $(8, 0)$ and find the equation of this circle.

37. If $y = \arcsin x$ then $x = \sin y$. Use implicit differentiation on $x = \sin y$ to show that

$$\frac{d}{dx} \arcsin x = \frac{1}{\sqrt{1 - x^2}}.$$

38. Show that the power rule for derivatives applies to rational powers of the form $y = x^{m/n}$ by raising both sides to the nth power and using implicit differentiation.

39. At pressure P atmospheres, a certain fraction f of a gas decomposes. The quantities P and f are related, for some positive constant K, by the equation

$$\frac{4f^2P}{1 - f^2} = K.$$

    (a) Find df/dP.
    (b) Show that $df/dP < 0$ always. What does this mean in practical terms?

40. For constants a, b, n, R, Van der Waal's equation relates the pressure, P, to the volume, V, of a fixed quantity of a gas at constant temperature T:

$$\left(P + \frac{n^2 a}{V^2}\right)(V - nb) = nRT.$$

Find the rate of change of volume with pressure, dV/dP.

Strengthen Your Understanding

In Problems 41–42, explain what is wrong with the statement.

41. If $y = \sin(xy)$ then $dy/dx = y\cos(xy)$.
42. The formula $dy/dx = -x/y$ gives the slope of the circle $x^2+y^2 = 10$ at every point in the plane except where $y = 0$.

In Problems 43–44, give an example of:

43. A formula for dy/dx leading to a vertical tangent at $y = 2$ and a horizontal tangent at $x = \pm 2$.
44. A curve that has two horizontal tangents at the same x-value, but no vertical tangents.
45. True or false? Explain your answer: If y satisfies the equation $y^2 + xy - 1 = 0$, then dy/dx exists everywhere.

3.8 HYPERBOLIC FUNCTIONS

There are two combinations of e^x and e^{-x} which are used so often in engineering that they are given their own name. They are the *hyperbolic sine*, abbreviated sinh, and the *hyperbolic cosine*, abbreviated cosh. They are defined as follows:

Hyperbolic Sine and Cosine

$$\cosh x = \frac{e^x + e^{-x}}{2} \qquad \sinh x = \frac{e^x - e^{-x}}{2}$$

Properties of Hyperbolic Functions

The graphs of $\cosh x$ and $\sinh x$ are given in Figures 3.33 and 3.34 together with the graphs of multiples of e^x and e^{-x}. The graph of $\cosh x$ is called a *catenary*; it is the shape of a hanging cable.

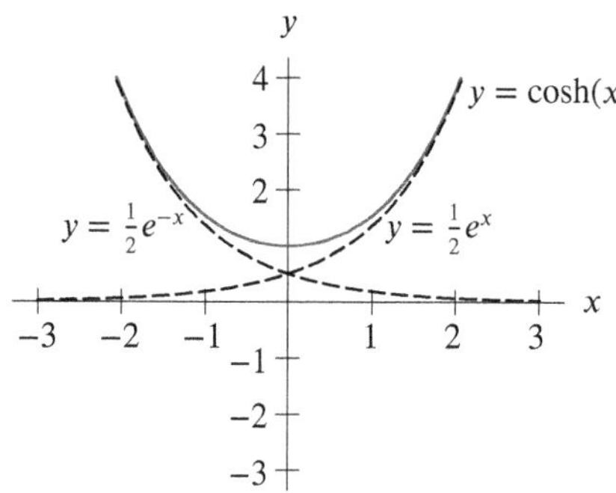

Figure 3.33: Graph of $y = \cosh x$

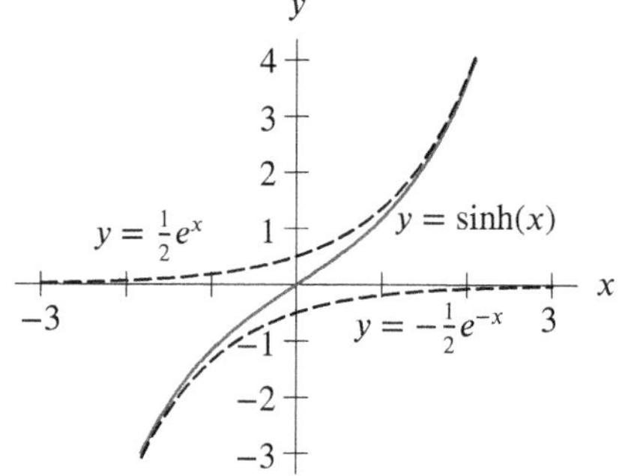

Figure 3.34: Graph of $y = \sinh x$

The graphs suggest that the following results hold:

$$\cosh 0 = 1 \qquad \sinh 0 = 0$$

$$\cosh(-x) = \cosh x \qquad \sinh(-x) = -\sinh x$$

To show that the hyperbolic functions really do have these properties, we use their formulas.

Example 1 Show that (a) $\cosh(0) = 1$ (b) $\cosh(-x) = \cosh x$

Solution (a) Substituting $x = 0$ into the formula for $\cosh x$ gives

$$\cosh 0 = \frac{e^0 + e^{-0}}{2} = \frac{1+1}{2} = 1.$$

(b) Substituting $-x$ for x gives

$$\cosh(-x) = \frac{e^{-x} + e^{-(-x)}}{2} = \frac{e^{-x} + e^{x}}{2} = \cosh x.$$

Thus, we know that $\cosh x$ is an even function.

Example 2 Describe and explain the behavior of $\cosh x$ as $x \to \infty$ and $x \to -\infty$.

Solution From Figure 3.33, it appears that as $x \to \infty$, the graph of $\cosh x$ resembles the graph of $\frac{1}{2}e^x$. Similarly, as $x \to -\infty$, the graph of $\cosh x$ resembles the graph of $\frac{1}{2}e^{-x}$. This behavior is explained by using the formula for $\cosh x$ and the facts that $e^{-x} \to 0$ as $x \to \infty$ and $e^x \to 0$ as $x \to -\infty$:

$$\text{As } x \to \infty, \qquad \cosh x = \frac{e^x + e^{-x}}{2} \to \frac{1}{2}e^x.$$

$$\text{As } x \to -\infty, \qquad \cosh x = \frac{e^x + e^{-x}}{2} \to \frac{1}{2}e^{-x}.$$

Identities Involving cosh x and sinh x

The reason the hyperbolic functions have names that remind us of the trigonometric functions is that they share similar properties. A familiar identity for trigonometric functions is

$$(\cos x)^2 + (\sin x)^2 = 1.$$

To discover an analogous identity relating $(\cosh x)^2$ and $(\sinh x)^2$, we first calculate

$$(\cosh x)^2 = \left(\frac{e^x + e^{-x}}{2}\right)^2 = \frac{e^{2x} + 2e^x e^{-x} + e^{-2x}}{4} = \frac{e^{2x} + 2 + e^{-2x}}{4}$$

$$(\sinh x)^2 = \left(\frac{e^x - e^{-x}}{2}\right)^2 = \frac{e^{2x} - 2e^x e^{-x} + e^{-2x}}{4} = \frac{e^{2x} - 2 + e^{-2x}}{4}.$$

If we add these expressions, the resulting right-hand side contains terms involving both e^{2x} and e^{-2x}. If, however, we subtract the expressions for $(\cosh x)^2$ and $(\sinh x)^2$, we obtain a simple result:

$$(\cosh x)^2 - (\sinh x)^2 = \frac{e^{2x} + 2 + e^{-2x}}{4} - \frac{e^{2x} - 2 + e^{-2x}}{4} = \frac{4}{4} = 1.$$

Thus, writing $\cosh^2 x$ for $(\cosh x)^2$ and $\sinh^2 x$ for $(\sinh x)^2$, we have the identity

$$\boxed{\cosh^2 x - \sinh^2 x = 1}$$

This identity shows us how the hyperbolic functions got their name. Suppose (x, y) is a point in the plane and $x = \cosh t$ and $y = \sinh t$ for some t. Then the point (x, y) lies on the hyperbola $x^2 - y^2 = 1$.

Extending the analogy to the trigonometric functions, we define

Hyperbolic Tangent

$$\tanh x = \frac{\sinh x}{\cosh x} = \frac{e^x - e^{-x}}{e^x + e^{-x}}$$

Derivatives of Hyperbolic Functions

We calculate the derivatives using the fact that $\frac{d}{dx}(e^x) = e^x$. The results are again reminiscent of the trigonometric functions. For example,

$$\frac{d}{dx}(\cosh x) = \frac{d}{dx}\left(\frac{e^x + e^{-x}}{2}\right) = \frac{e^x - e^{-x}}{2} = \sinh x.$$

We find $\frac{d}{dx}(\sinh x)$ similarly, giving the following results:

$$\frac{d}{dx}(\cosh x) = \sinh x \qquad \frac{d}{dx}(\sinh x) = \cosh x$$

Example 3 Compute the derivative of $\tanh x$.

Solution Using the quotient rule gives

$$\frac{d}{dx}(\tanh x) = \frac{d}{dx}\left(\frac{\sinh x}{\cosh x}\right) = \frac{(\cosh x)^2 - (\sinh x)^2}{(\cosh x)^2} = \frac{1}{\cosh^2 x}.$$

Exercises and Problems for Section 3.8 Online Resource: Additional Problems for Section 3.8

EXERCISES

In Exercises 1–11, find the derivative of the function.

1. $y = \sinh(3z + 5)$
2. $y = \cosh(2x)$
3. $g(t) = \cosh^2 t$
4. $f(t) = \cosh(\sinh t)$
5. $f(t) = t^3 \sinh t$
6. $y = \cosh(3t)\sinh(4t)$
7. $y = \tanh(12 + 18x)$
8. $f(t) = \cosh(e^{t^2})$
9. $g(\theta) = \ln(\cosh(1 + \theta))$
10. $f(y) = \sinh(\sinh(3y))$
11. $f(t) = \cosh^2 t - \sinh^2 t$
12. Show that $d(\sinh x)/dx = \cosh x$.
13. Show that $\sinh 0 = 0$.
14. Show that $\sinh(-x) = -\sinh(x)$.

In Exercises 15–16, simplify the expressions.

15. $\cosh(\ln t)$
16. $\sinh(\ln t)$

PROBLEMS

17. Describe and explain the behavior of $\sinh x$ as $x \to \infty$ and as $x \to -\infty$.

18. If $x = \cosh t$ and $y = \sinh t$, explain why the point (x, y) always lies on the curve $x^2 - y^2 = 1$. (This curve is called a hyperbola and gave this family of functions its name.)

19. Is there an identity analogous to $\sin(2x) = 2\sin x \cos x$ for the hyperbolic functions? Explain.

20. Is there an identity analogous to $\cos(2x) = \cos^2 x - \sin^2 x$ for the hyperbolic functions? Explain.

■ Prove the identities in Problems 21–22.

21. $\cosh(A + B) = \cosh A \cosh B + \sinh A \sinh B$

22. $\sinh(A + B) = \sinh A \cosh B + \cosh A \sinh B$

■ In Problems 23–26, find the limit of the function as $x \to \infty$.

23. $\dfrac{\sinh(2x)}{\cosh(3x)}$

24. $\dfrac{e^{2x}}{\sinh(2x)}$

25. $\dfrac{\sinh(x^2)}{\cosh(x^2)}$

26. $\dfrac{\cosh(2x)}{\sinh(3x)}$

27. For what values of k is $\lim_{x\to\infty} e^{-3x}\cosh kx$ finite?

28. For what values of k is $\lim_{x\to\infty} \dfrac{\sinh kx}{\cosh 2x}$ finite?

29. The cable between the two towers of a power line hangs in the shape of the curve

$$y = \frac{T}{w}\cosh\left(\frac{wx}{T}\right),$$

where T is the tension in the cable at its lowest point and w is the weight of the cable per unit length. This curve is called a *catenary*.

(a) Suppose the cable stretches between the points $x = -T/w$ and $x = T/w$. Find an expression for the "sag" in the cable. (That is, find the difference between the height of the cable at the highest and lowest points.)

(b) Show that the shape of the cable satisfies the equation

$$\frac{d^2y}{dx^2} = \frac{w}{T}\sqrt{1 + \left(\frac{dy}{dx}\right)^2}.$$

30. The Saint Louis Arch can be approximated by using a function of the form $y = b - a\cosh(x/a)$. Putting the origin on the ground in the center of the arch and the y-axis upward, find an approximate equation for the arch given the dimensions shown in Figure 3.35. (In other words, find a and b.)

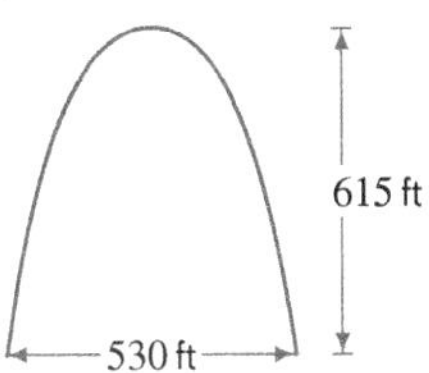

Figure 3.35

31. (a) Using a calculator or computer, sketch the graph of $y = 2e^x + 5e^{-x}$ for $-3 \le x \le 3$, $0 \le y \le 20$. Observe that it looks like the graph of $y = \cosh x$. Approximately where is its minimum?

(b) Show algebraically that $y = 2e^x + 5e^{-x}$ can be written in the form $y = A\cosh(x - c)$. Calculate the values of A and c. Explain what this tells you about the graph in part (a).

32. The following problem is a generalization of Problem 31. Show that any function of the form

$$y = Ae^x + Be^{-x}, \quad A > 0,\ B > 0,$$

can be written, for some K and c, in the form

$$y = K\cosh(x - c).$$

What does this tell you about the graph of $y = Ae^x + Be^{-x}$?

33. (a) Find $\tanh 0$.

(b) For what values of x is $\tanh x$ positive? Negative? Explain your answer algebraically.

(c) On what intervals is $\tanh x$ increasing? Decreasing? Use derivatives to explain your answer.

(d) Find $\lim_{x\to\infty} \tanh x$ and $\lim_{x\to-\infty} \tanh x$. Show this information on a graph.

(e) Does $\tanh x$ have an inverse? Justify your answer using derivatives.

Strengthen Your Understanding

■ In Problems 34–37, explain what is wrong with the statement.

34. The function $f(x) = \cosh x$ is periodic.

35. The derivative of the function $f(x) = \cosh x$ is $f'(x) = -\sinh x$.

36. $\cosh^2 x + \sinh^2 x = 1$.

37. $\tanh x \to \infty$ as $x \to \infty$.

■ In Problems 38–40, give an example of:

38. A hyperbolic function which is concave up.

39. A value of k such that $\lim_{x\to\infty} e^{kx}\cosh x$ does not exist.

40. A function involving the hyperbolic cosine that passes through the point $(1, 3)$.

Are the statements in Problems 41–45 true or false? Give an explanation for your answer.

41. The function $\tanh x$ is odd, that is, $\tanh(-x) = -\tanh x$.

42. The 100th derivative of $\sinh x$ is $\cosh x$.

43. $\sinh x + \cosh x = e^x$.

44. The function $\sinh x$ is periodic.

45. The function $\sinh^2 x$ is concave down everywhere.

3.9 LINEAR APPROXIMATION AND THE DERIVATIVE

The Tangent Line Approximation

When we zoom in on the graph of a differentiable function, it looks like a straight line. In fact, the graph is not exactly a straight line when we zoom in; however, its deviation from straightness is so small that it can't be detected by the naked eye. Let's examine what this means. The straight line that we think we see when we zoom in on the graph of $f(x)$ at $x = a$ has slope equal to the derivative, $f'(a)$, so the equation is

$$y = f(a) + f'(a)(x - a).$$

The fact that the graph looks like a line means that y is a good approximation to $f(x)$. (See Figure 3.36.) This suggests the following definition:

The Tangent Line Approximation

Suppose f is differentiable at a. Then, for values of x near a, the tangent line approximation to $f(x)$ is

$$f(x) \approx f(a) + f'(a)(x - a).$$

The expression $f(a) + f'(a)(x - a)$ is called the *local linearization* of f near $x = a$. We are thinking of a as fixed, so that $f(a)$ and $f'(a)$ are constant.

The **error**, $E(x)$, in the approximation is defined by

$$E(x) = f(x) - f(a) - f'(a)(x - a).$$

It can be shown that the tangent line approximation is the best linear approximation to f near a. See Problem 51.

Example 1 What is the tangent line approximation for $f(x) = \sin x$ near $x = 0$?

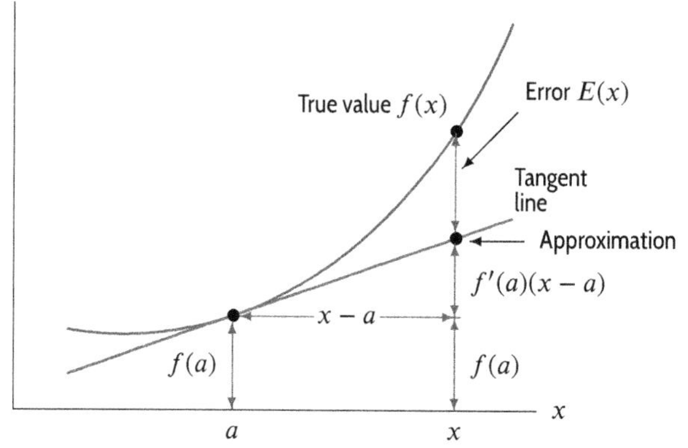

Figure 3.36: The tangent line approximation and its error

Solution The tangent line approximation of f near $x = 0$ is

$$f(x) \approx f(0) + f'(0)(x - 0).$$

If $f(x) = \sin x$, then $f'(x) = \cos x$, so $f(0) = \sin 0 = 0$ and $f'(0) = \cos 0 = 1$, and the approximation is

$$\sin x \approx x.$$

This means that, near $x = 0$, the function $f(x) = \sin x$ is well approximated by the function $y = x$. If we zoom in on the graphs of the functions $\sin x$ and x near the origin, we won't be able to tell them apart. (See Figure 3.37.)

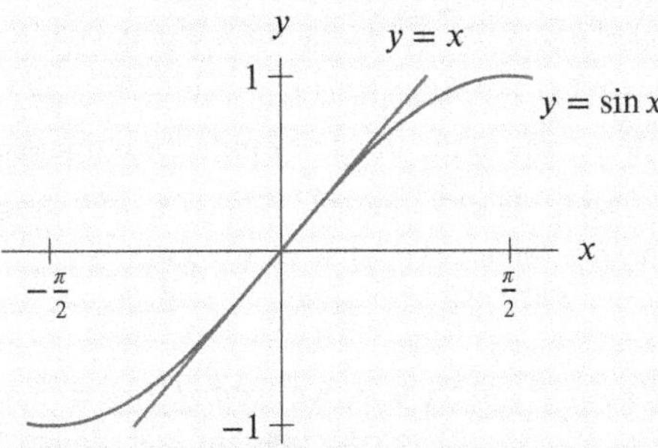

Figure 3.37: Tangent line approximation to $y = \sin x$

Example 2 What is the local linearization of e^{kx} near $x = 0$?

Solution If $f(x) = e^{kx}$, then $f(0) = 1$ and, by the chain rule, $f'(x) = ke^{kx}$, so $f'(0) = ke^{k\cdot 0} = k$. Thus

$$f(x) \approx f(0) + f'(0)(x - 0)$$

becomes

$$e^{kx} \approx 1 + kx.$$

This is the tangent line approximation to e^{kx} near $x = 0$. In other words, if we zoom in on the functions $f(x) = e^{kx}$ and $y = 1 + kx$ near the origin, we won't be able to tell them apart.

Estimating the Error in the Approximation

Let us look at the error, $E(x)$, which is the difference between $f(x)$ and the local linearization. (Look back at Figure 3.36.) The fact that the graph of f looks like a line as we zoom in means that not only is $E(x)$ small for x near a, but also that $E(x)$ is small relative to $(x - a)$. To demonstrate this, we prove the following theorem about the ratio $E(x)/(x - a)$.

Theorem 3.6: Differentiability and Local Linearity

Suppose f is differentiable at $x = a$ and $E(x)$ is the error in the tangent line approximation, that is:

$$E(x) = f(x) - f(a) - f'(a)(x - a).$$

Then

$$\lim_{x\to a} \frac{E(x)}{x - a} = 0.$$

Proof Using the definition of $E(x)$, we have

$$\frac{E(x)}{x-a} = \frac{f(x) - f(a) - f'(a)(x-a)}{x-a} = \frac{f(x)-f(a)}{x-a} - f'(a).$$

Taking the limit as $x \to a$ and using the definition of the derivative, we see that

$$\lim_{x\to a} \frac{E(x)}{x-a} = \lim_{x\to a}\left(\frac{f(x)-f(a)}{x-a} - f'(a)\right) = f'(a) - f'(a) = 0.$$

Theorem 3.6 says that $E(x)$ approaches 0 faster than $(x-a)$. For the function in Example 3, we see that $E(x) \approx k(x-a)^2$ for constant k if x is near a.

Example 3 Let $E(x)$ be the error in the tangent line approximation to $f(x) = x^3 - 5x + 3$ for x near 2.

(a) What does a table of values for $E(x)/(x-2)$ suggest about $\lim_{x\to 2} E(x)/(x-2)$?

(b) Make another table to see that $E(x) \approx k(x-2)^2$. Estimate the value of k. Check that a possible value is $k = f''(2)/2$.

Solution (a) Since $f(x) = x^3 - 5x + 3$, we have $f'(x) = 3x^2 - 5$, and $f''(x) = 6x$. Thus, $f(2) = 1$ and $f'(2) = 3 \cdot 2^2 - 5 = 7$, so the tangent line approximation for x near 2 is

$$f(x) \approx f(2) + f'(2)(x-2)$$
$$f(x) \approx 1 + 7(x-2).$$

Thus,

$$E(x) = \text{True value} - \text{Approximation} = (x^3 - 5x + 3) - (1 + 7(x-2)).$$

The values of $E(x)/(x-2)$ in Table 3.7 suggest that $E(x)/(x-2)$ approaches 0 as $x \to 2$.

(b) Notice that if $E(x) \approx k(x-2)^2$, then $E(x)/(x-2)^2 \approx k$. Thus we make Table 3.8 showing values of $E(x)/(x-2)^2$. Since the values are all approximately 6, we guess that $k = 6$ and $E(x) \approx 6(x-2)^2$.

Since $f''(2) = 12$, our value of k satisfies $k = f''(2)/2$.

Table 3.7

x	$E(x)/(x-2)$
2.1	0.61
2.01	0.0601
2.001	0.006001
2.0001	0.00060001

Table 3.8

x	$E(x)/(x-2)^2$
2.1	6.1
2.01	6.01
2.001	6.001
2.0001	6.0001

The relationship between $E(x)$ and $f''(x)$ that appears in Example 3 holds more generally. If $f(x)$ satisfies certain conditions, it can be shown that the error in the tangent line approximation behaves near $x = a$ as

$$E(x) \approx \frac{f''(a)}{2}(x-a)^2.$$

This is part of a general pattern for obtaining higher-order approximations called Taylor polynomials, which are studied in Chapter 10.

Why Differentiability Makes a Graph Look Straight

We use the properties of the error $E(x)$ to understand why differentiability makes a graph look straight when we zoom in.

Example 4 Consider the graph of $f(x) = \sin x$ near $x = 0$, and its linear approximation computed in Example 1. Show that there is an interval around 0 with the property that the distance from $f(x) = \sin x$ to the linear approximation is less than $0.1|x|$ for all x in the interval.

Solution The linear approximation of $f(x) = \sin x$ near 0 is $y = x$, so we write

$$\sin x = x + E(x).$$

Since $\sin x$ is differentiable at $x = 0$, Theorem 3.6 tells us that

$$\lim_{x\to 0} \frac{E(x)}{x} = 0.$$

If we take $\epsilon = 1/10$, then the definition of limit guarantees that there is a $\delta > 0$ such that

$$\left|\frac{E(x)}{x}\right| < 0.1 \quad \text{for all} \quad |x| < \delta.$$

In other words, for x in the interval $(-\delta, \delta)$, we have $|x| < \delta$, so

$$|E(x)| < 0.1|x|.$$

(See Figure 3.38.)

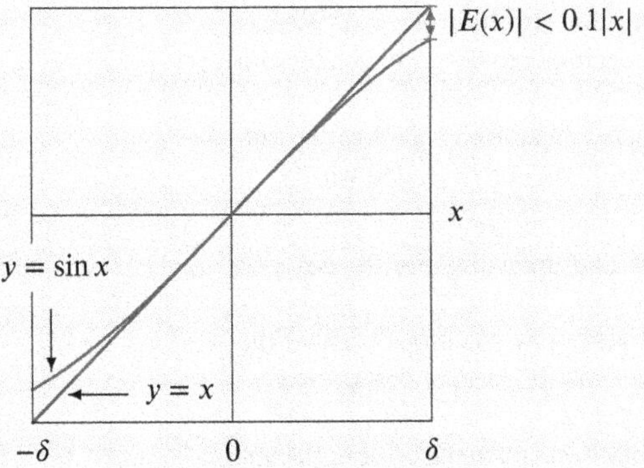

Figure 3.38: Graph of $y = \sin x$ and its linear approximation $y = x$, showing a window in which the magnitude of the error, $|E(x)|$, is less than $0.1|x|$ for all x in the window

We can generalize from this example to explain why differentiability makes the graph of f look straight when viewed over a small graphing window. Suppose f is differentiable at $x = a$. Then we know $\lim_{x\to a}\left|\frac{E(x)}{x-a}\right| = 0$. So, for any $\epsilon > 0$, we can find a δ small enough so that

$$\left|\frac{E(x)}{x-a}\right| < \epsilon, \qquad \text{for} \quad a-\delta < x < a+\delta.$$

So, for any x in the interval $(a-\delta, a+\delta)$, we have

$$|E(x)| < \epsilon|x-a|.$$

Thus, the error, $E(x)$, is less than ϵ times $|x-a|$, the distance between x and a. So, as we zoom in on the graph by choosing smaller ϵ, the deviation, $|E(x)|$, of f from its tangent line shrinks, even relative to the scale on the x-axis. So, zooming makes a differentiable function look straight.

Exercises and Problems for Section 3.9

EXERCISES

1. Find the tangent line approximation for $\sqrt{1+x}$ near $x=0$.
2. What is the tangent line approximation to e^x near $x=0$?
3. Find the tangent line approximation to $1/x$ near $x=1$.
4. Find the local linearization of $f(x)=x^2$ near $x=1$.
5. What is the local linearization of e^{x^2} near $x=1$?
6. Show that $1-x/2$ is the tangent line approximation to $1/\sqrt{1+x}$ near $x=0$.
7. Show that $e^{-x}\approx 1-x$ near $x=0$.
8. Local linearization gives values too small for the function x^2 and too large for the function $\sqrt{x}$. Draw pictures to explain why.
9. Using a graph like Figure 3.37, estimate to one decimal place the magnitude of the error in approximating $\sin x$ by x for $-1\le x\le 1$. Is the approximation an over- or an underestimate?
10. For x near 0, local linearization gives

$$e^x \approx 1+x.$$

Using a graph, decide if the approximation is an over- or underestimate, and estimate to one decimal place the magnitude of the error for $-1\le x\le 1$.

PROBLEMS

11. **(a)** Find the best linear approximation, $L(x)$, to $f(x)=e^x$ near $x=0$.
    (b) What is the sign of the error, $E(x)=f(x)-L(x)$ for x near 0?
    (c) Find the true value of the function at $x=1$. What is the error? (Give decimal answers.) Illustrate with a graph.
    (d) Before doing any calculations, explain which you expect to be larger, $E(0.1)$ or $E(1)$, and why.
    (e) Find $E(0.1)$.
12. **(a)** Find the tangent line approximation to $\cos x$ at $x=\pi/4$.
    (b) Use a graph to explain how you know whether the tangent line approximation is an under- or overestimate for $0\le x\le \pi/2$.
    (c) To one decimal place, estimate the error in the approximation for $0\le x\le \pi/2$.
13. Suppose $f(x)$ has $f(50)=99.5$ and $f'(50)=0.2$.
    (a) Find a linear function that approximates $f(x)$ for x near 50.
    (b) Give an estimate for x such that $f(x)=100$.
    (c) What should be true of the graph of f for your estimate in part (b) to be fairly accurate?
14. The graphs in Figure 3.39 have the same window and the same scale. Use local linearization at $x=0$ to match each graph with a formula (a)–(d).
    (a) $h(x)=x^3-3x^2+3x$
    (b) $g(x)=x^{2/3}$
    (c) $k(x)=2x^2-4x$
    (d) $f(x)=e^x-1$

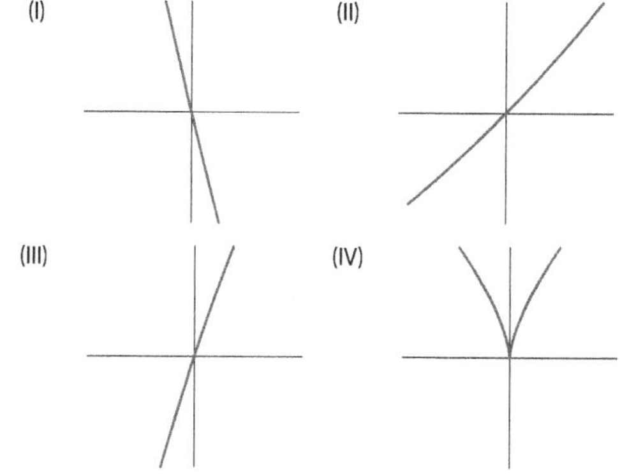

Figure 3.39

15. **(a)** Graph $f(x)=x^3-3x^2+3x+1$.
    (b) Find and add to your sketch the local linearization to $f(x)$ at $x=2$.
    (c) Mark on your sketch the true value of $f(1.5)$, the tangent line approximation to $f(1.5)$ and the error in the approximation.
16. **(a)** Show that $1+kx$ is the local linearization of $(1+x)^k$ near $x=0$.
    (b) Someone claims that the square root of 1.1 is about 1.05. Without using a calculator, do you think that this estimate is about right?
    (c) Is the actual number above or below 1.05?

17. Figure 3.40 shows $f(x)$ and its local linearization at $x = a$. What is the value of a? Of $f(a)$? Is the approximation an under- or overestimate? Use the linearization to approximate the value of $f(1.2)$.

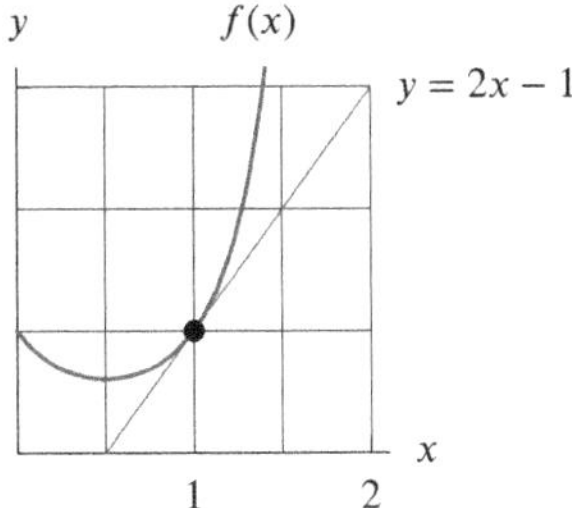

Figure 3.40

■ In Problems 18–19, the equation has a solution near $x = 0$. By replacing the left side of the equation by its linearization, find an approximate value for the solution.

18. $e^x + x = 2$

19. $x + \ln(1 + x) = 0.2$

20. (a) Given that $f(7) = 13$ and $f'(7) = -0.38$, estimate $f(7.1)$.
    (b) Suppose also $f''(x) < 0$ for all x. Does this make your answer to part (a) an under- or overestimate?

21. A function g has $g(3) = 7$ and $g(3.001) - g(3) = 0.0025$.
    (a) Estimate $g'(3)$.
    (b) Using the estimate from part (a), find the the tangent line approximation to $g(x)$ at $x = 3$ and use it to estimate $g(3.1)$.
    (c) If $g''(x) > 0$ for $3 \le x \le 3.1$, is the value for $g(3.1)$ found in part (b) an underestimate, an overestimate, exact or is there not enough information to decide?

22. (a) Explain why the following equation has a solution near 0:
$$e^t = 0.02t + 1.098.$$
    (b) Replace e^t by its linearization near 0. Solve the new equation to get an approximate solution to the original equation.

23. The speed of sound in dry air is
$$f(T) = 331.3\sqrt{1 + \frac{T}{273.15}} \text{ meters/second}$$
where T is the temperature in degrees Celsius. Find a linear function that approximates the speed of sound for temperatures near 0°C.

24. Live phytoplankton of diameter x micrometers sink in the ocean at a rate of $u = 0.021x^{1.177}$ meters per day.[20] The rate is important, because when phytoplankton are too deep they do not have enough light to carry on photosynthesis.
    (a) Find du/dx.
    (b) Evaluate $du/dx|_{x=300}$. Include units. Explain the meaning of this derivative for phytoplankton.
    (c) Use the derivative to estimate the difference in the sinking rates of phytoplankton of diameters 280 and 310 micrometers.
    (d) Give the tangent line approximation for the phytoplankton sinking rate for diameters near 300 micrometers.

25. The generation time for an organism is the time from its birth until it begins to reproduce. For marine organisms from bacteria to whales,[21] the generation time, G, in days, is a function of their length, L, in cm: $G = 40.9L^{0.579}$.
    (a) Find dG/dL.
    (b) Evaluate $dG/dL|_{L=1}$. Include units. Explain the meaning of this derivative for generation times.
    (c) Use the derivative to estimate the difference in generation times of organisms of lengths 0.9 and 1.1 cm.
    (d) Give the tangent line approximation for the generation time for lengths near 1 cm.

26. Air pressure at sea level is 30 inches of mercury. At an altitude of h feet above sea level, the air pressure, P, in inches of mercury, is given by
$$P = 30e^{-3.23\times10^{-5}h}$$
    (a) Sketch a graph of P against h.
    (b) Find the equation of the tangent line at $h = 0$.
    (c) A rule of thumb used by travelers is that air pressure drops about 1 inch for every 1000-foot increase in height above sea level. Write a formula for the air pressure given by this rule of thumb.
    (d) What is the relation between your answers to parts (b) and (c)? Explain why the rule of thumb works.
    (e) Are the predictions made by the rule of thumb too large or too small? Why?

27. On October 7, 2010, the *Wall Street Journal*[22] reported that Android cell phone users had increased to 10.9 million by the end of August 2010 from 866,000 a year earlier. During the same period, iPhone users increased to 13.5 million, up from 7.8 million a year earlier. Let $A(t)$ be the number of Android users, in millions, at time t in years since the end of August 2009. Let $P(t)$ be the number of iPhone users in millions.

[20] Adapted from M. Denny, *How the Ocean Works* (Princeton: Princeton University Press, 2008), pp. 76 and 104.
[21] Adapted from M. Denny, *How the Ocean Works* (Princeton: Princeton University Press, 2008), p. 124.
[22] "Apple Readies Verizon iPhone", WSJ, Oct 7, 2010.

(a) Estimate $A'(0)$. Give units.
(b) Estimate $P'(0)$. Give units.
(c) Using the tangent line approximation, when are the numbers of Android and iPhone users predicted to be the same?
(d) What assumptions did you make in part (c)?

28. Table 3.9 shows the water stored $S(t)$, in acre-feet,[23] of Lake Sonoma, a reservoir in Northern California, from March 2014 until April 2015[24]. With t in months, let $t = 0$ be March 2014.

(a) Find a linear approximation for the water stored near $t = 2$ and use it to approximate the water stored in October 2014.
(b) Find and use an appropriate linear approximation to estimate the water stored in October 2015.

Table 3.9

t	0	1	2	3	4	5	6
$S(t)$	182,566	185,569	179,938	171,770	163,150	154,880	147,391
t	7	8	9	10	11	12	13
$S(t)$	141,146	136,553	191,296	189,093	218,354	216,019	212,740

29. Small water bugs swim in groups as protection against attacks from fish. In one observational study, the number, a, of attacks per individual bug per hour depended on the group size, s, according to the model[25] $a = 120s^{-1.118}$. The model is appropriate for group sizes from 1 to 100.

(a) How many attacks are expected on an individual bug per hour if swimming in a group of size 10? 20? 50?
(b) Find da/ds.
(c) Evaluate $da/ds|_{s=50}$. Include units.
(d) Find a linear approximation for a as a function of s for group sizes near 50 bugs.
(e) Use your linear approximation to estimate the difference in number of attacks between groups of 48 and 53 bugs. Compare your answer with the difference obtained using the original power function model.

30. If C (in units of 10^{-4} molar) is the concentration of glucose in a solution, then *E. coli* bacterium in the solution grows at a rate, R (in cell divisions per hour), given by[26]

$$R = \frac{1.35C}{0.22 + C} \text{ cell divisions per hour.}$$

(a) Find the growth rate of bacteria growing in $2 \cdot 10^{-4}$ molar glucose solution. Include units.
(b) Find dR/dC.
(c) Find $dR/dC|_{C=2}$. Include units.
(d) Find the tangent line approximation of the growth rate for bacteria growing in glucose concentrations near $2 \cdot 10^{-4}$ molar.
(e) Use the tangent line approximation to estimate the growth rate in a $2.2 \cdot 10^{-4}$ molar glucose solution. Compare it with the growth rate from the original model.

31. Writing g for the acceleration due to gravity, the period, T, of a pendulum of length l is given by

$$T = 2\pi\sqrt{\frac{l}{g}}.$$

(a) Show that if the length of the pendulum changes by Δl, the change in the period, ΔT, is given by

$$\Delta T \approx \frac{T}{2l}\Delta l.$$

(b) If the length of the pendulum increases by 2%, by what percent does the period change?

32. Suppose now the length of the pendulum in Problem 31 remains constant, but that the acceleration due to gravity changes.

(a) Use the method of the preceding problem to relate ΔT approximately to Δg, the change in g.
(b) If g increases by 1%, find the percent change in T.

33. Suppose f has a continuous positive second derivative for all x. Which is larger, $f(1+\Delta x)$ or $f(1)+f'(1)\Delta x$? Explain.

34. Suppose $f'(x)$ is a differentiable decreasing function for all x. In each of the following pairs, which number is the larger? Give a reason for your answer.

(a) $f'(5)$ and $f'(6)$
(b) $f''(5)$ and 0
(c) $f(5 + \Delta x)$ and $f(5) + f'(5)\Delta x$

[23] An acre-foot is the amount of water it takes to cover one acre of area with 1 foot of water.
[24] http://cdec.water.ca.gov, accessed May 28, 2015.
[25] W. A. Foster and J. E. Treherne, "Evidence for the Dilution Effect in the Selfish Herd from Fish Predation on a Marine Insect," *Nature* 293 (October, 1981), pp. 466–467.
[26] Jacques Monod, "The Growth of Bacterial Cultures," *Annual Review of Microbiology* 1949.3, pp. 371–394.

■ Problems 35–37 investigate the motion of a projectile shot from a cannon. The fixed parameters are the acceleration of gravity, $g = 9.8\ \text{m/sec}^2$, and the muzzle velocity, $v_0 = 500\ \text{m/sec}$, at which the projectile leaves the cannon. The angle θ, in degrees, between the muzzle of the cannon and the ground can vary.

35. The range of the projectile is

$$f(\theta) = \frac{v_0^2}{g}\sin\frac{\pi\theta}{90}wq = 25{,}510\sin\frac{\pi\theta}{90}\ \text{meters.}$$

(a) Find the range with $\theta = 20°$.
(b) Find a linear function of θ that approximates the range for angles near 20°.
(c) Find the range and its approximation from part (b) for 21°.

36. The time that the projectile stays in the air is

$$t(\theta) = \frac{2v_0}{g}\sin\frac{\pi\theta}{180} = 102\sin\frac{\pi\theta}{180}\ \text{seconds.}$$

(a) Find the time in the air for $\theta = 20°$.
(b) Find a linear function of θ that approximates the time in the air for angles near 20°.
(c) Find the time in air and its approximation from part (b) for 21°.

37. At its highest point the projectile reaches a peak altitude given by

$$h(\theta) = \frac{v_0^2}{2g}\sin^2\frac{\pi\theta}{180} = 12{,}755\sin^2\frac{\pi\theta}{180}\ \text{meters.}$$

(a) Find the peak altitude for $\theta = 20°$.
(b) Find a linear function of θ that approximates the peak altitude for angles near 20°.
(c) Find the peak altitude and its approximation from part (b) for 21°.

■ In Problems 38–40, find the local linearization of $f(x)$ near 0 and use this to approximate the value of a.

38. $f(x) = (1+x)^r, \quad a = (1.2)^{3/5}$

39. $f(x) = e^{kx}, \quad a = e^{0.3}$

40. $f(x) = \sqrt{b^2 + x}, \quad a = \sqrt{26}$

■ In Problems 41–45, find a formula for the error $E(x)$ in the tangent line approximation to the function near $x = a$. Using a table of values for $E(x)/(x-a)$ near $x = a$, find a value of k such that $E(x)/(x-a) \approx k(x-a)$. Check that, approximately, $k = f''(a)/2$ and that $E(x) \approx (f''(a)/2)(x-a)^2$.

41. $f(x) = x^4, \ a = 1$

42. $f(x) = \cos x, \ a = 0$

43. $f(x) = e^x, \ a = 0$

44. $f(x) = \sqrt{x}, \ a = 1$

45. $f(x) = \ln x, \ a = 1$

46. Multiply the local linearization of e^x near $x = 0$ by itself to obtain an approximation for e^{2x}. Compare this with the actual local linearization of e^{2x}. Explain why these two approximations are consistent, and discuss which one is more accurate.

47. **(a)** Show that $1 - x$ is the local linearization of $\dfrac{1}{1+x}$ near $x = 0$.
(b) From your answer to part (a), show that near $x = 0$,

$$\frac{1}{1+x^2} \approx 1 - x^2.$$

(c) Without differentiating, what do you think the derivative of $\dfrac{1}{1+x^2}$ is at $x = 0$?

48. From the local linearizations of e^x and $\sin x$ near $x = 0$, write down the local linearization of the function $e^x \sin x$. From this result, write down the derivative of $e^x \sin x$ at $x = 0$. Using this technique, write down the derivative of $e^x \sin x/(1+x)$ at $x = 0$.

49. Use local linearization to derive the product rule,

$$[f(x)g(x)]' = f'(x)g(x) + f(x)g'(x).$$

[Hint: Use the definition of the derivative and the local linearizations $f(x+h) \approx f(x) + f'(x)h$ and $g(x+h) \approx g(x) + g'(x)h$.]

50. Derive the chain rule using local linearization. [Hint: In other words, differentiate $f(g(x))$, using $g(x+h) \approx g(x) + g'(x)h$ and $f(z+k) \approx f(z) + f'(z)k$.]

51. Consider a function f and a point a. Suppose there is a number L such that the linear function g

$$g(x) = f(a) + L(x-a)$$

is a good approximation to f. By good approximation, we mean that

$$\lim_{x\to a}\frac{E_L(x)}{x-a} = 0,$$

where $E_L(x)$ is the approximation error defined by

$$f(x) = g(x) + E_L(x) = f(a) + L(x-a) + E_L(x).$$

Show that f is differentiable at $x = a$ and that $f'(a) = L$. Thus the tangent line approximation is the only good linear approximation.

52. Consider the graph of $f(x) = x^2$ near $x = 1$. Find an interval around $x = 1$ with the property that throughout any smaller interval, the graph of $f(x) = x^2$ never differs from its local linearization at $x = 1$ by more than $0.1|x-1|$.

Strengthen Your Understanding

In Problems 53–54, explain what is wrong with the statement.

53. To approximate $f(x) = e^x$, we can always use the linear approximation $f(x) = e^x \approx x + 1$.

54. The linear approximation for $F(x) = x^3 + 1$ near $x = 0$ is an underestimate for the function F for all x, $x \neq 0$.

In Problems 55–57, give an example of:

55. Two different functions that have the same linear approximation near $x = 0$.

56. A non-polynomial function that has the tangent line approximation $f(x) \approx 1$ near $x = 0$.

57. A function that does not have a linear approximation at $x = -1$.

58. Let f be a differentiable function and let L be the linear function $L(x) = f(a) + k(x - a)$ for some constant a. Decide whether the following statements are true or false for all constants k. Explain your answer.

(a) L is the local linearization for f near $x = a$,
(b) If $\lim_{x \to a}(f(x) - L(x)) = 0$, then L is the local linearization for f near $x = a$.

3.10 THEOREMS ABOUT DIFFERENTIABLE FUNCTIONS

A Relationship Between Local and Global: The Mean Value Theorem

We often want to infer a global conclusion (for example, f is increasing on an interval) from local information (for example, f' is positive at each point on an interval). The following theorem relates the average rate of change of a function on an interval (global information) to the instantaneous rate of change at a point in the interval (local information).

Theorem 3.7: The Mean Value Theorem

If f is continuous on $a \leq x \leq b$ and differentiable on $a < x < b$, then there exists a number c, with $a < c < b$, such that
$$f'(c) = \frac{f(b) - f(a)}{b - a}.$$
In other words, $f(b) - f(a) = f'(c)(b - a)$.

To understand this theorem geometrically, look at Figure 3.41. Join the points on the curve where $x = a$ and $x = b$ with a secant line and observe that
$$\text{Slope of secant line } = \frac{f(b) - f(a)}{b - a}.$$
Now consider the tangent lines drawn to the curve at each point between $x = a$ and $x = b$. In general, these lines have different slopes. For the curve shown in Figure 3.41, the tangent line at $x = a$ is flatter than the secant line. Similarly, the tangent line at $x = b$ is steeper than the secant line. However, there appears to be at least one point between a and b where the slope of the tangent line to the curve is precisely the same as the slope of the secant line. Suppose this occurs at $x = c$. Then
$$\text{Slope of tangent line } = f'(c) = \frac{f(b) - f(a)}{b - a}.$$
The Mean Value Theorem tells us that the point $x = c$ exists, but it does not tell us how to find c.

Problems 73 and 74 (available online) in Section 4.2 show how the Mean Value Theorem can be derived.

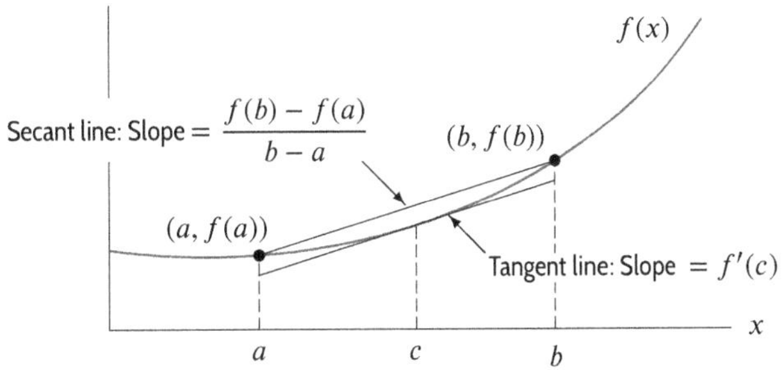

Figure 3.41: The point c with $f'(c) = \frac{f(b)-f(a)}{b-a}$

If f satisfies the conditions of the Mean Value Theorem on $a < x < b$ and $f(a) = f(b) = 0$, the Mean Value Theorem tells us that there is a point c, with $a < c < b$, such that $f'(c) = 0$. This result is called **Rolle's Theorem**.

The Increasing Function Theorem

We say that a function f is *increasing* on an interval if, for any two numbers x_1 and x_2 in the interval such that $x_1 < x_2$, we have $f(x_1) < f(x_2)$. If instead we have $f(x_1) \leq f(x_2)$, we say f is *nondecreasing*.

Theorem 3.8: The Increasing Function Theorem

Suppose that f is continuous on $a \leq x \leq b$ and differentiable on $a < x < b$.

- If $f'(x) > 0$ on $a < x < b$, then f is increasing on $a \leq x \leq b$.
- If $f'(x) \geq 0$ on $a < x < b$, then f is nondecreasing on $a \leq x \leq b$.

Proof Suppose $a \leq x_1 < x_2 \leq b$. By the Mean Value Theorem, there is a number c, with $x_1 < c < x_2$, such that

$$f(x_2) - f(x_1) = f'(c)(x_2 - x_1).$$

If $f'(c) > 0$, this says $f(x_2) - f(x_1) > 0$, which means f is increasing. If $f'(c) \geq 0$, this says $f(x_2) - f(x_1) \geq 0$, which means f is nondecreasing.

It may seem that something as simple as the Increasing Function Theorem should follow immediately from the definition of the derivative, and you may be surprised that the Mean Value Theorem is needed.

The Constant Function Theorem

If f is constant on an interval, then we know that $f'(x) = 0$ on the interval. The following theorem is the converse.

Theorem 3.9: The Constant Function Theorem

Suppose that f is continuous on $a \leq x \leq b$ and differentiable on $a < x < b$. If $f'(x) = 0$ on $a < x < b$, then f is constant on $a \leq x \leq b$.

Proof The proof is the same as for the Increasing Function Theorem, only in this case $f'(c) = 0$ so $f(x_2) - f(x_1) = 0$. Thus $f(x_2) = f(x_1)$ for $a \leq x_1 < x_2 \leq b$, so f is constant.

A proof of the Constant Function Theorem using the Increasing Function Theorem is given in Problems 17 and 25.

The Racetrack Principle

Theorem 3.10: The Racetrack Principle[27]

Suppose that g and h are continuous on $a \leq x \leq b$ and differentiable on $a < x < b$, and that $g'(x) \leq h'(x)$ for $a < x < b$.

- If $g(a) = h(a)$, then $g(x) \leq h(x)$ for $a \leq x \leq b$.
- If $g(b) = h(b)$, then $g(x) \geq h(x)$ for $a \leq x \leq b$.

The Racetrack Principle has the following interpretation. We can think of $g(x)$ and $h(x)$ as the positions of two racehorses at time x, with horse h always moving faster than horse g. If they start together, horse h is ahead during the whole race. If they finish together, horse g was ahead during the whole race.

Proof Consider the function $f(x) = h(x) - g(x)$. Since $f'(x) = h'(x) - g'(x) \geq 0$, we know that f is nondecreasing by the Increasing Function Theorem. So $f(x) \geq f(a) = h(a) - g(a) = 0$. Thus $g(x) \leq h(x)$ for $a \leq x \leq b$. This proves the first part of the Racetrack Principle. Problem 24 asks for a proof of the second part.

Example 1 Explain graphically why $e^x \geq 1 + x$ for all values of x. Then use the Racetrack Principle to prove the inequality.

Solution The graph of the function $y = e^x$ is concave up everywhere and the equation of its tangent line at the point $(0, 1)$ is $y = x + 1$. (See Figure 3.42.) Since the graph always lies above its tangent, we have the inequality

$$e^x \geq 1 + x.$$

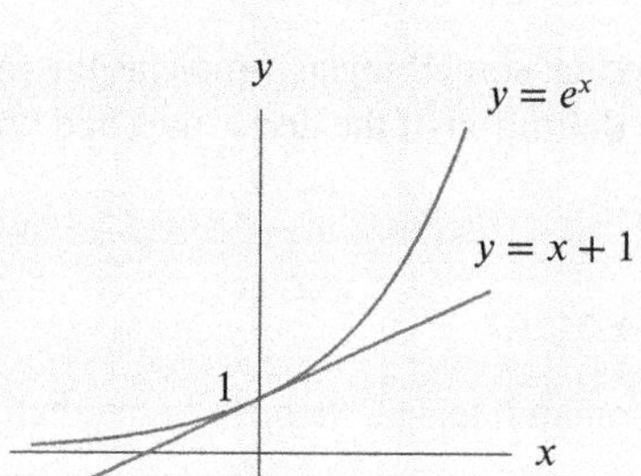

Figure 3.42: Graph showing that $e^x \geq 1 + x$

Now we prove the inequality using the Racetrack Principle. Let $g(x) = 1 + x$ and $h(x) = e^x$.

[27]Based on the Racetrack Principle in *Calculus & Mathematica*, by William Davis, Horacio Porta, Jerry Uhl (Reading: Addison Wesley, 1994).

Then $g(0) = h(0) = 1$. Furthermore, $g'(x) = 1$ and $h'(x) = e^x$. Hence $g'(x) \leq h'(x)$ for $x \geq 0$. So by the Racetrack Principle, with $a = 0$, we have $g(x) \leq h(x)$, that is, $1 + x \leq e^x$.

For $x \leq 0$ we have $h'(x) \leq g'(x)$. So by the Racetrack Principle, with $b = 0$, we have $g(x) \leq h(x)$, that is, $1 + x \leq e^x$.

Exercises and Problems for Section 3.10

EXERCISES

In Exercises 1–5, decide if the statements are true or false. Give an explanation for your answer.

1. Let $f(x) = [x]$, the greatest integer less than or equal to x. Then $f'(x) = 0$, so $f(x)$ is constant by the Constant Function Theorem.
2. If $a < b$ and $f'(x)$ is positive on $[a, b]$ then $f(a) < f(b)$.
3. If $f(x)$ is increasing and differentiable on the interval $[a, b]$, then $f'(x) > 0$ on $[a, b]$.
4. The Racetrack Principle can be used to justify the statement that if two horses start a race at the same time, the horse that wins must have been moving faster than the other throughout the race.
5. Two horses start a race at the same time and one runs slower than the other throughout the race. The Racetrack Principle can be used to justify the fact that the slower horse loses the race.

Do the functions graphed in Exercises 6–9 appear to satisfy the hypotheses of the Mean Value Theorem on the interval $[a, b]$? Do they satisfy the conclusion?

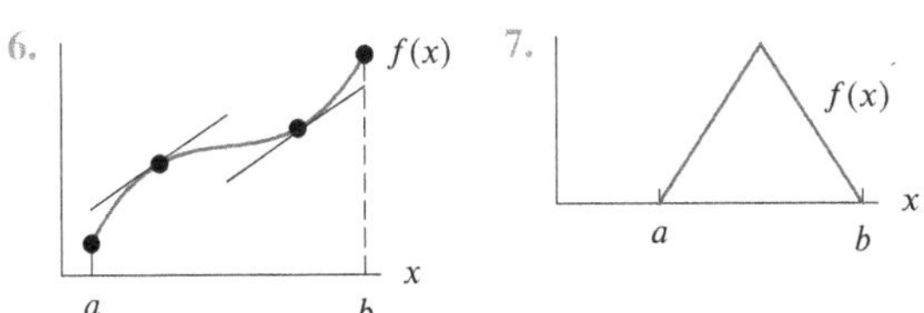

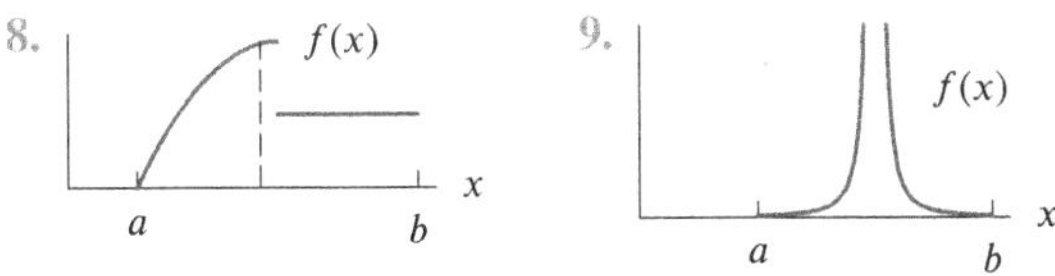

PROBLEMS

10. Applying the Mean Value Theorem with $a = 2$, $b = 7$ to the function in Figure 3.43 leads to $c = 4$. What is the equation of the tangent line at 4?

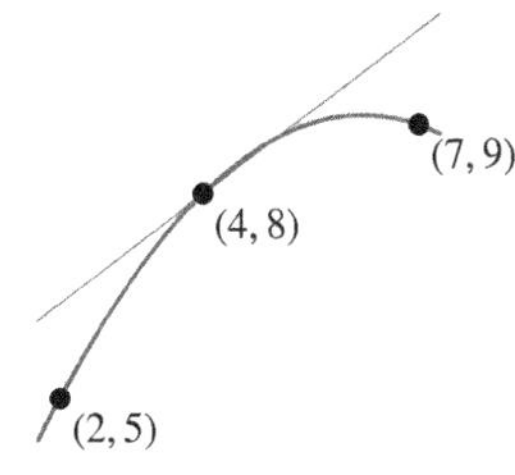

Figure 3.43

11. Applying the Mean Value Theorem with $a = 3$, $b = 13$ to the function in Figure 3.44 leads to the point c shown. What is the value of $f'(c)$? What can you say about the values of $f'(x_1)$ and $f'(x_2)$?

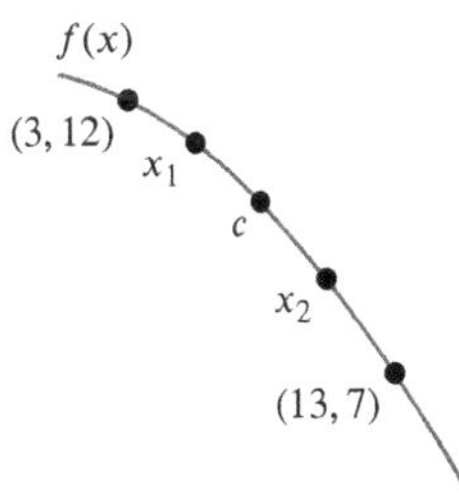

Figure 3.44

12. Let $p(x) = x^5 + 8x^4 - 30x^3 + 30x^2 - 31x + 22$. What is the relationship between $p(x)$ and $f(x) = 5x^4 + 32x^3 - 90x^2 + 60x - 31$? What do the values of $p(1)$ and $p(2)$ tell you about the values of $f(x)$?
13. Let $p(x)$ be a seventh-degree polynomial with 7 distinct zeros. How many zeros does $p'(x)$ have?
14. Use the Racetrack Principle and the fact that $\sin 0 = 0$ to show that $\sin x \leq x$ for all $x \geq 0$.
15. Use the Racetrack Principle to show that $\ln x \leq x - 1$.

16. Use the fact that $\ln x$ and e^x are inverse functions to show that the inequalities $e^x \geq 1+x$ and $\ln x \leq x-1$ are equivalent for $x > 0$.

17. State a Decreasing Function Theorem, analogous to the Increasing Function Theorem. Deduce your theorem from the Increasing Function Theorem. [Hint: Apply the Increasing Function Theorem to $-f$.]

18. Dominic drove from Phoenix to Tucson on Interstate 10, a distance of 116 miles. The speed limit on this highway varies between 55 and 75 miles per hour. He started his trip at 11:44 pm and arrived in Tucson at 1:12 am. Prove that Dominic was speeding at some point during his trip.

In Problems 19–22, use one of the theorems in this section to prove the statements.

19. If $f'(x) \leq 1$ for all x and $f(0) = 0$, then $f(x) \leq x$ for all $x \geq 0$.

20. If $f''(t) \leq 3$ for all t and $f(0) = f'(0) = 0$, then $f(t) \leq \frac{3}{2}t^2$ for all $t \geq 0$.

21. If $f'(x) = g'(x)$ for all x and $f(5) = g(5)$, then $f(x) = g(x)$ for all x.

22. If f is differentiable and $f(0) < f(1)$, then there is a number c, with $0 < c < 1$, such that $f'(c) > 0$.

23. The position of a particle on the x-axis is given by $s = f(t)$; its initial position and velocity are $f(0) = 3$ and $f'(0) = 4$. The acceleration is bounded by $5 \leq f''(t) \leq 7$ for $0 \leq t \leq 2$. What can we say about the position $f(2)$ of the particle at $t = 2$?

24. Suppose that g and h are continuous on $[a, b]$ and differentiable on (a, b). Prove that if $g'(x) \leq h'(x)$ for $a < x < b$ and $g(b) = h(b)$, then $h(x) \leq g(x)$ for $a \leq x \leq b$.

25. Deduce the Constant Function Theorem from the Increasing Function Theorem and the Decreasing Function Theorem. (See Problem 17.)

26. Prove that if $f'(x) = g'(x)$ for all x in (a, b), then there is a constant C such that $f(x) = g(x) + C$ on (a, b). [Hint: Apply the Constant Function Theorem to $h(x) = f(x) - g(x)$.]

27. Suppose that $f'(x) = f(x)$ for all x. Prove that $f(x) = Ce^x$ for some constant C. [Hint: Consider $f(x)/e^x$.]

28. Suppose that f is continuous on $[a, b]$ and differentiable on (a, b) and that $m \leq f'(x) \leq M$ on (a, b). Use the Racetrack Principle to prove that $f(x) - f(a) \leq M(x-a)$ for all x in $[a, b]$, and that $m(x-a) \leq f(x) - f(a)$ for all x in $[a, b]$. Conclude that $m \leq (f(b)-f(a))/(b-a) \leq M$. This is called the Mean Value Inequality. In words: If the instantaneous rate of change of f is between m and M on an interval, so is the average rate of change of f over the interval.

29. Suppose that $f''(x) \geq 0$ for all x in (a, b). We will show the graph of f lies above the tangent line at $(c, f(c))$ for any c with $a < c < b$.
   (a) Use the Increasing Function Theorem to prove that $f'(c) \leq f'(x)$ for $c \leq x < b$ and that $f'(x) \leq f'(c)$ for $a < x \leq c$.
   (b) Use (a) and the Racetrack Principle to conclude that $f(c) + f'(c)(x - c) \leq f(x)$, for $a < x < b$.

Strengthen Your Understanding

In Problems 30–32, explain what is wrong with the statement.

30. The Mean Value Theorem applies to $f(x) = |x|$, for $-2 < x < 2$.

31. The following function satisfies the conditions of the Mean Value Theorem on the interval $[0, 1]$:

$$f(x) = \begin{cases} x & \text{if } 0 < x \leq 1 \\ 1 & \text{if } x = 0. \end{cases}$$

32. If $f'(x) = 0$ on $a < x < b$, then by the Constant Function Theorem f is constant on $a \leq x \leq b$.

In Problems 33–37, give an example of:

33. An interval where the Mean Value Theorem applies when $f(x) = \ln x$.

34. An interval where the Mean Value Theorem does not apply when $f(x) = 1/x$.

35. A continuous function f on the interval $[-1, 1]$ that does not satisfy the conclusion of the Mean Value Theorem.

36. A function f that is differentiable on the interval $(0, 2)$, but does not satisfy the conclusion of the Mean Value Theorem on the interval $[0, 2]$.

37. A function that is differentiable on $(0, 1)$ and not continuous on $[0, 1]$, but which satisfies the conclusion of the Mean Value Theorem.

Are the statements in Problems 38–41 true or false for a function f whose domain is all real numbers? If a statement is true, explain how you know. If a statement is false, give a counterexample.

38. If $f'(x) \geq 0$ for all x, then $f(a) \leq f(b)$ whenever $a \leq b$.

39. If $f'(x) \leq g'(x)$ for all x, then $f(x) \leq g(x)$ for all x.

40. If $f'(x) = g'(x)$ for all x, then $f(x) = g(x)$ for all x.

41. If $f'(x) \leq 1$ for all x and $f(0) = 0$, then $f(x) \leq x$ for all x.

Online Resource: Review problems and Projects

Chapter Four

USING THE DERIVATIVE

Contents

4.1 USING FIRST AND SECOND DERIVATIVES

What Derivatives Tell Us About a Function and Its Graph

In Chapter 2, we saw the following connection between the derivatives of a function and the function itself:

- If $f' > 0$ on an interval, then f is increasing on that interval.
- If $f' < 0$ on an interval, then f is decreasing on that interval.

If f' is always positive on an interval or always negative on an interval, then f is monotonic over that interval.

- If $f'' > 0$ on an interval, then the graph of f is concave up on that interval.
- If $f'' < 0$ on an interval, then the graph of f is concave down on that interval.

Now that we have formulas for the derivatives of the elementary functions, we can do more with these principles than we could in Chapter 2.

When we graph a function on a computer or calculator, we see only part of the picture, and we may miss some significant features. Information given by the first and second derivatives can help identify regions with interesting behavior.

Example 1 Use the first and second derivatives to analyze the function $f(x) = x^3 - 9x^2 - 48x + 52$.

Solution Since f is a cubic polynomial, we expect a graph that is roughly S-shaped. We use the derivative to determine where the function is increasing and where it is decreasing. The derivative of f is

$$f'(x) = 3x^2 - 18x - 48.$$

To find where $f' > 0$ or $f' < 0$, we first find where $f' = 0$, that is, where

$$3x^2 - 18x - 48 = 0.$$

Factoring, we get

$$3(x + 2)(x - 8) = 0, \qquad \text{so} \qquad x = -2 \text{ or } x = 8.$$

Since $f' = 0$ *only* at $x = -2$ and $x = 8$, and since f' is continuous, f' cannot change sign on any of the three intervals $x < -2$, or $-2 < x < 8$, or $8 < x$.

How can we tell the sign of f' on each of these intervals? One way is to pick a point and substitute into f'. For example, since $f'(-3) = 33 > 0$, we know f' is positive for $x < -2$, so f is increasing for $x < -2$. Similarly, since $f'(0) = -48$ and $f'(10) = 72$, we know that f decreases between $x = -2$ and $x = 8$ and increases for $x > 8$. Summarizing:

f increasing ↗	$x = -2$	f decreasing ↘	$x = 8$	f increasing ↗	x
$f' > 0$	$f' = 0$	$f' < 0$	$f' = 0$	$f' > 0$	

We find that $f(-2) = 104$ and $f(8) = -396$. Hence on the interval $-2 < x < 8$ the function decreases from a high of 104 to a low of -396. One more point on the graph is easy to get: the y intercept, $f(0) = 52$. This gives us three points on the graph and tells us where it slopes up and where it slopes down. See Figure 4.1.

To determine concavity, we use the second derivative. We have

$$f''(x) = 6x - 18.$$

Thus, $f''(x) < 0$ when $x < 3$ and $f''(x) > 0$ when $x > 3$, so the graph of f is concave down for $x < 3$ and concave up for $x > 3$. At $x = 3$, we have $f''(x) = 0$. See Figure 4.1. Summarizing:

$x = 3$

f concave down $\cap$ | f concave up $\cup$

$f'' < 0$ | $f'' = 0$ | $f'' > 0$

As expected, the graph in Figure 4.1 is concave down for $x < 3$ and concave up for $x > 3$.

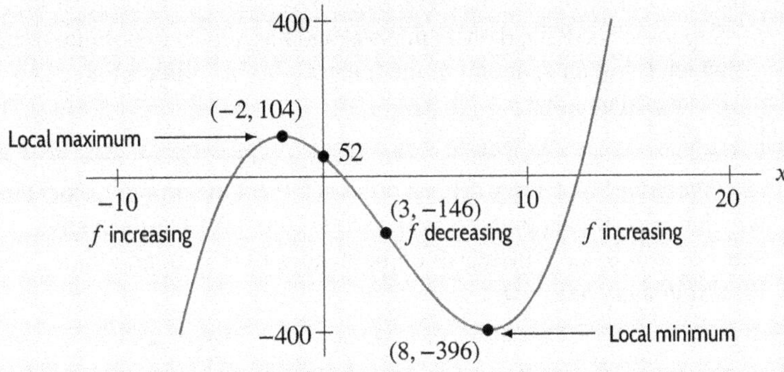

Figure 4.1: Useful graph of $f(x) = x^3 - 9x^2 - 48x + 52$

Local Maxima and Minima

We are often interested in points such as those marked in Figure 4.1 as local maximum and local minimum. We have the following definition:

> Suppose p is a point in the domain of f:
> - f has a **local minimum** at p if $f(p)$ is less than or equal to the values of f for points near p.
> - f has a **local maximum** at p if $f(p)$ is greater than or equal to the values of f for points near p.

We use the adjective "local" because we are describing only what happens near p. Local maxima and minima are sometimes called *local extrema*.

How Do We Detect a Local Maximum or Minimum?

In the preceding example, the points $x = -2$ and $x = 8$, where $f'(x) = 0$, played a key role in leading us to local maxima and minima. We give a name to such points:

> For any function f, a point p in the domain of f where $f'(p) = 0$ or $f'(p)$ is undefined is called a **critical point** of the function. In addition, the point $(p, f(p))$ on the graph of f is also called a critical point. A **critical value** of f is the value, $f(p)$, at a critical point, p.

Notice that "critical point of f" can refer either to points in the domain of f or to points on the graph of f. You will know which meaning is intended from the context.

Geometrically, at a critical point where $f'(p) = 0$, the line tangent to the graph of f at p is horizontal. At a critical point where $f'(p)$ is undefined, there is no horizontal tangent to the graph—there's either a vertical tangent or no tangent at all. (For example, $x = 0$ is a critical point for the absolute value function $f(x) = |x|$.) However, most of the functions we work with are differentiable everywhere, and therefore most of our critical points are of the $f'(p) = 0$ variety.

The critical points divide the domain of f into intervals within which the sign of the derivative remains the same, either positive or negative. Therefore, if f is defined on the interval between two successive critical points, its graph cannot change direction on that interval; it is either increasing or decreasing. The following result, which is proved on page 198, tells us that all local maxima and minima which are not at endpoints occur at critical points.

Theorem 4.1: Local Extrema and Critical Points

Suppose f is defined on an interval and has a local maximum or minimum at the point $x = a$, which is not an endpoint of the interval. If f is differentiable at $x = a$, then $f'(a) = 0$. Thus, a is a critical point.

Warning! Not every critical point is a local maximum or local minimum. Consider $f(x) = x^3$, which has a critical point at $x = 0$. (See Figure 4.2.) The derivative, $f'(x) = 3x^2$, is positive on both sides of $x = 0$, so f increases on both sides of $x = 0$, and there is neither a local maximum nor a local minimum at $x = 0$.

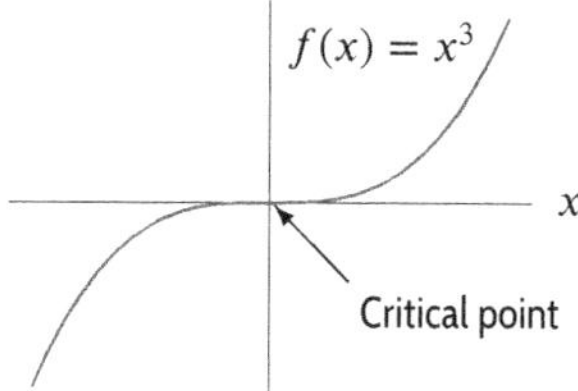

Figure 4.2: Critical point which is not a local maximum or minimum

Testing for Local Maxima and Minima at a Critical Point

If f' has different signs on either side of a critical point p, with $f'(p) = 0$, then the graph changes direction at p and looks like one of those in Figure 4.3. So we have the following criterion:

The First-Derivative Test for Local Maxima and Minima

Suppose p is a critical point of a continuous function f. Moving from left to right:

- If f' changes from negative to positive at p, then f has a local minimum at p.
- If f' changes from positive to negative at p, then f has a local maximum at p.

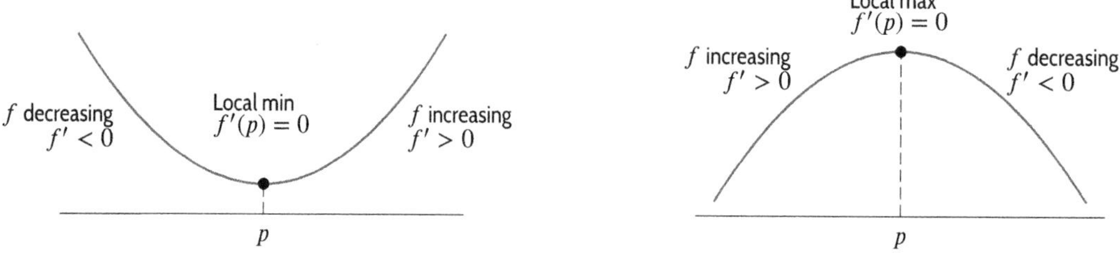

Figure 4.3: Changes in direction at a critical point, p: Local maxima and minima

Example 2 Use a graph of the function $f(x) = \dfrac{1}{x(x-1)}$ to observe its local maxima and minima. Confirm your observation analytically.

Solution The graph in Figure 4.4 suggests that this function has no local minima but that there is a local maximum at about $x = 1/2$. Confirming this analytically means using the formula for the derivative to show that what we expect is true. Since $f(x) = (x^2 - x)^{-1}$, we have

$$f'(x) = -1(x^2 - x)^{-2}(2x - 1) = -\frac{2x - 1}{(x^2 - x)^2}.$$

So $f'(x) = 0$ where $2x - 1 = 0$. Thus, the only critical point in the domain of f is $x = 1/2$.

Furthermore, $f'(x) > 0$ where $0 < x < 1/2$, and $f'(x) < 0$ where $1/2 < x < 1$. Thus, f increases for $0 < x < 1/2$ and decreases for $1/2 < x < 1$. According to the first-derivative test, the critical point $x = 1/2$ is a local maximum.

For $-\infty < x < 0$ or $1 < x < \infty$, there are no critical points and no local maxima or minima. Although $1/(x(x - 1)) \to 0$ both as $x \to \infty$ and as $x \to -\infty$, we don't say 0 is a local minimum because $1/(x(x - 1))$ never actually *equals* 0.

Notice that although $f' > 0$ everywhere that it is defined for $x < 1/2$, the function f is not increasing throughout this interval. The problem is that f and f' are not defined at $x = 0$, so we cannot conclude that f is increasing when $x < 1/2$.

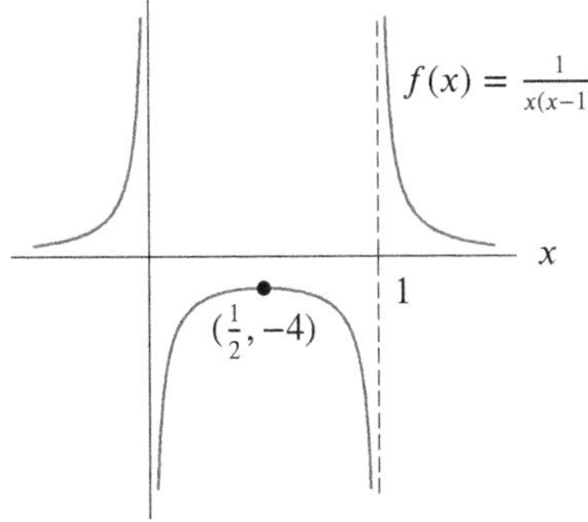

Figure 4.4: Find local maxima and minima

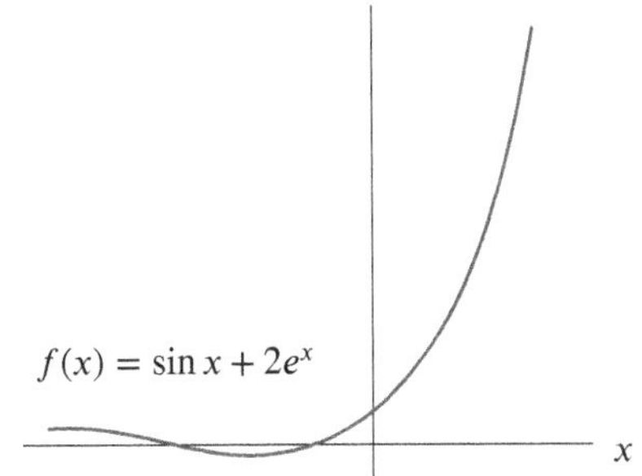

Figure 4.5: Explain the absence of local maxima and minima for $x \geq 0$

Example 3 The graph of $f(x) = \sin x + 2e^x$ is in Figure 4.5. Using the derivative, explain why there are no local maxima or minima for $x > 0$.

Solution Local maxima and minima can occur only at critical points. Now $f'(x) = \cos x + 2e^x$, which is defined for all x. We know $\cos x$ is always between -1 and 1, and $2e^x > 2$ for $x > 0$, so $f'(x)$ cannot be 0 for any $x > 0$. Therefore there are no local maxima or minima there.

The Second-Derivative Test for Local Maxima and Minima

Knowing the concavity of a function can also be useful in testing if a critical point is a local maximum or a local minimum. Suppose p is a critical point of f, with $f'(p) = 0$, so that the graph of f has a horizontal tangent line at p. If the graph is concave up at p, then f has a local minimum at p. Likewise, if the graph is concave down, f has a local maximum. (See Figure 4.6.) This suggests:

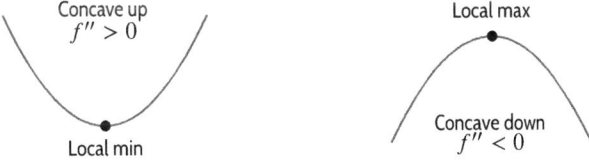

Figure 4.6: Local maxima and minima and concavity

The Second-Derivative Test for Local Maxima and Minima

- If $f'(p) = 0$ and $f''(p) > 0$ then f has a local minimum at p.
- If $f'(p) = 0$ and $f''(p) < 0$ then f has a local maximum at p.
- If $f'(p) = 0$ and $f''(p) = 0$ then the test tells us nothing.

Example 4 Classify as local maxima or local minima the critical points of $f(x) = x^3 - 9x^2 - 48x + 52$.

Solution As we saw in Example 1 on page 192,

$$f'(x) = 3x^2 - 18x - 48$$

and the critical points of f are $x = -2$ and $x = 8$. We have

$$f''(x) = 6x - 18.$$

Thus $f''(8) = 30 > 0$, so f has a local minimum at $x = 8$. Since $f''(-2) = -30 < 0$, f has a local maximum at $x = -2$.

Warning! The second-derivative test does not tell us anything if both $f'(p) = 0$ and $f''(p) = 0$. For example, if $f(x) = x^3$ and $g(x) = x^4$, both $f'(0) = f''(0) = 0$ and $g'(0) = g''(0) = 0$. The point $x = 0$ is a minimum for g but is neither a maximum nor a minimum for f. However, the first-derivative test is still useful. For example, g' changes sign from negative to positive at $x = 0$, so we know g has a local minimum there.

Concavity and Inflection Points

Investigating points where the slope changes sign led us to critical points. Now we look at points where the concavity changes.

A point, p, at which the graph of a continuous function, f, changes concavity is called an **inflection point** of f.

The words "inflection point of f" can refer either to a point in the domain of f or to a point on the graph of f. The context of the problem will tell you which is meant.

How Do We Detect an Inflection Point?

To identify candidates for an inflection point of a continuous function, we often use the second derivative.

Suppose f'' is defined on both sides of a point p:
- If f'' is zero or undefined at p, then p is a possible inflection point.
- To test whether p is an inflection point, check whether f'' changes sign at p.

The following example illustrates how local maxima and minima and inflection points are found.

Example 5 For $x \geq 0$, find the local maxima and minima and inflection points for $g(x) = xe^{-x}$ and graph g.

Solution Taking derivatives and simplifying, we have

$$g'(x) = (1 - x)e^{-x} \quad \text{and} \quad g''(x) = (x - 2)e^{-x}.$$

So $x = 1$ is a critical point, and $g' > 0$ for $x < 1$ and $g' < 0$ for $x > 1$. Hence g increases to a local maximum at $x = 1$ and then decreases. Since $g(0) = 0$ and $g(x) > 0$ for $x > 0$, there is a local minimum at $x = 0$. Also, $g(x) \to 0$ as $x \to \infty$. There is an inflection point at $x = 2$ since $g'' < 0$ for $x < 2$ and $g'' > 0$ for $x > 2$. The graph is in Figure 4.7.

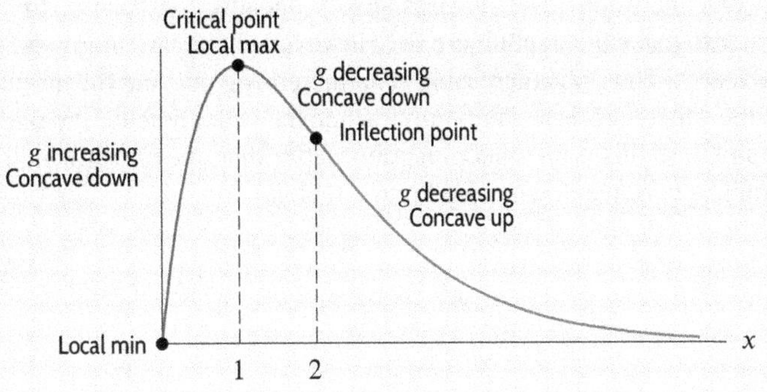

Figure 4.7: Graph of $g(x) = xe^{-x}$

Warning! Not every point x where $f''(x) = 0$ (or f'' is undefined) is an inflection point (just as not every point where $f' = 0$ is a local maximum or minimum). For instance $f(x) = x^4$ has $f''(x) = 12x^2$ so $f''(0) = 0$, but $f'' > 0$ when $x < 0$ and when $x > 0$, so there is *no* change in concavity at $x = 0$. See Figure 4.8.

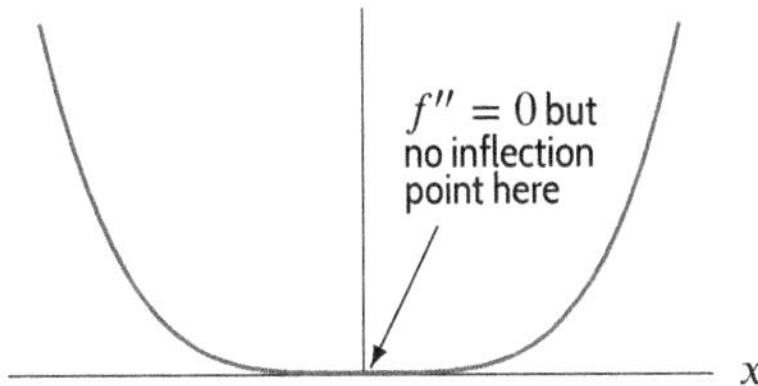

Figure 4.8: Graph of $f(x) = x^4$

Inflection Points and Local Maxima and Minima of the Derivative

Inflection points can also be interpreted in terms of first derivatives. Applying the first-derivative test for local maxima and minima to f', we obtain the following result:

> Suppose a function f has a continuous derivative. If f'' changes sign at p, then f has an inflection point at p, and f' has a local minimum or a local maximum at p.

Figure 4.9 shows two inflection points. Notice that the curve crosses the tangent line at these points and that the slope of the curve is a maximum or a minimum there.

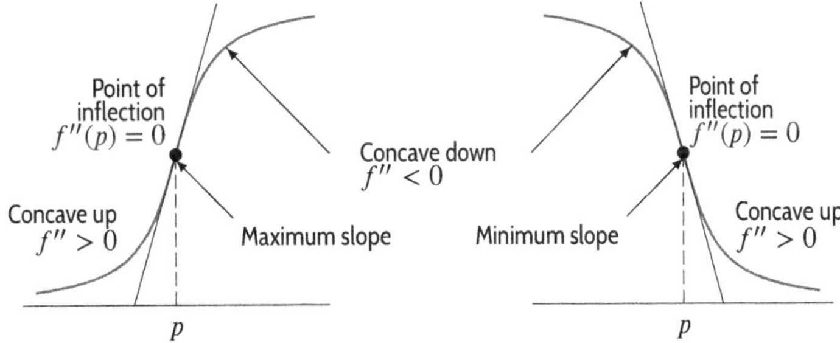

Figure 4.9: Change in concavity at p: Points of inflection

Example 6 Water is being poured into the vase in Figure 4.10 at a constant rate, measured in volume per unit time. Graph $y = f(t)$, the depth of the water against time, t. Explain the concavity and indicate the inflection points.

Solution At first the water level, y, rises slowly because the base of the vase is wide, and it takes a lot of water to make the depth increase. However, as the vase narrows, the rate at which the water is rising increases. Thus, y is increasing at an increasing rate and the graph is concave up. The rate of increase in the water level is at a maximum when the water reaches the middle of the vase, where the diameter is smallest; this is an inflection point. After that, the rate at which y increases decreases again, so the graph is concave down. (See Figure 4.11.)

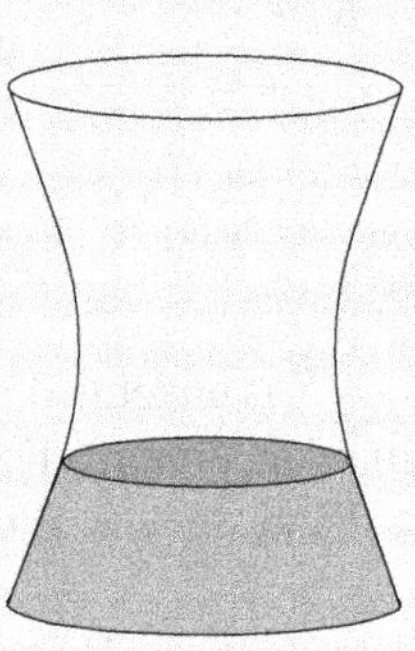

Figure 4.10: A vase

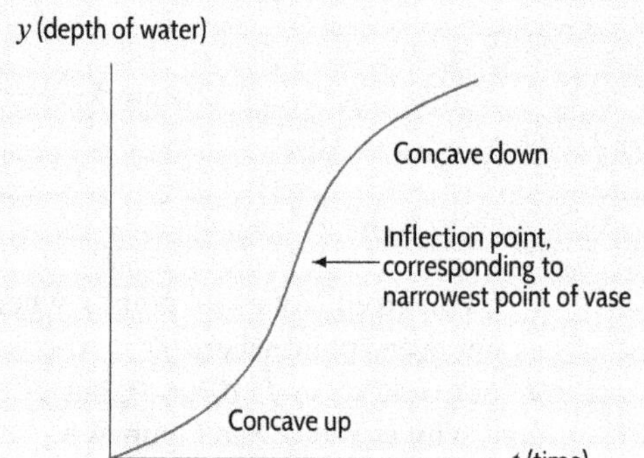

Figure 4.11: Graph of depth of water in the vase, y, against time, t

Showing Local Extrema Are at Critical Points

We now prove Theorem 4.1, which says that inside an interval, local maxima and minima can only occur at critical points. Suppose that f has a local maximum at $x = a$. Assuming that $f'(a)$ is defined, the definition of the derivative gives

$$f'(a) = \lim_{h\to 0} \frac{f(a+h) - f(a)}{h}.$$

Since this is a two-sided limit, we have

$$f'(a) = \lim_{h\to 0^-} \frac{f(a+h) - f(a)}{h} = \lim_{h\to 0^+} \frac{f(a+h) - f(a)}{h}.$$

By the definition of local maximum, $f(a+h) \le f(a)$ for all sufficiently small h. Thus $f(a+h) - f(a) \le 0$ for sufficiently small h. The denominator, h, is negative when we take the limit from the left and positive when we take the limit from the right. Thus

$$\lim_{h\to 0^-} \frac{f(a+h) - f(a)}{h} \ge 0 \quad \text{and} \quad \lim_{h\to 0^+} \frac{f(a+h) - f(a)}{h} \le 0.$$

Since both these limits are equal to $f'(a)$, we have $f'(a) \ge 0$ and $f'(a) \le 0$, so we must have $f'(a) = 0$. The proof for a local minimum at $x = a$ is similar.

Exercises and Problems for Section 4.1 Online Resource: Additional Problems for Section 4.1

EXERCISES

1. Indicate all critical points on the graph of f in Figure 4.12 and determine which correspond to local maxima of f, which to local minima, and which to neither.

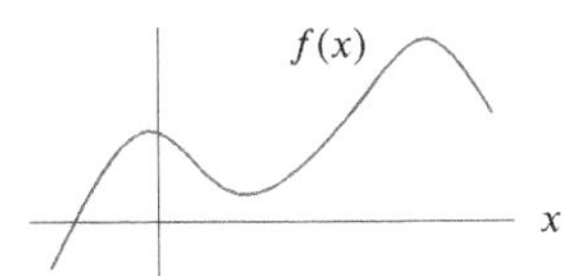

Figure 4.12

2. Graph a function which has exactly one critical point, at $x = 2$, and exactly one inflection point, at $x = 4$.

3. Graph a function with exactly two critical points, one of which is a local minimum and the other is neither a local maximum nor a local minimum.

■ In Exercises 4–8, use derivatives to find the critical points and inflection points.

4. $f(x) = x^3 - 9x^2 + 24x + 5$

5. $f(x) = x^5 - 10x^3 - 8$

6. $f(x) = x^5 + 15x^4 + 25$

7. $f(x) = 5x - 3\ln x$

8. $f(x) = 4xe^{3x}$

■ In Exercises 9–12, find all critical points and then use the first-derivative test to determine local maxima and minima. Check your answer by graphing.

9. $f(x) = 3x^4 - 4x^3 + 6$

10. $f(x) = (x^2 - 4)^7$

11. $f(x) = (x^3 - 8)^4$

12. $f(x) = \dfrac{x}{x^2 + 1}$

■ In Exercises 13–16, find all critical points and then use the second-derivative test to determine local maxima and minima.

13. $f(x) = 9 + 6x^2 - x^3$

14. $f(x) = x^4 - 18x^2 + 10$

15. $f(x) = e^{-2x^2}$

16. $f(x) = 2x - 5\ln x$

■ In Exercises 17–20, find the critical points of the function and classify them as local maxima or minima or neither.

17. $g(x) = xe^{-3x}$

18. $h(x) = x + 1/x$

19. $f(x) = \sin x - 0.5x$

20. $f(x) = \sin x + \cos x$

21. **(a)** Use a graph to estimate the x-values of any critical points and inflection points of $f(x) = e^{-x^2}$.
(b) Use derivatives to find the x-values of any critical points and inflection points exactly.

■ In Exercises 22–24, use Figure 4.13 to determine which of the two values is greater.

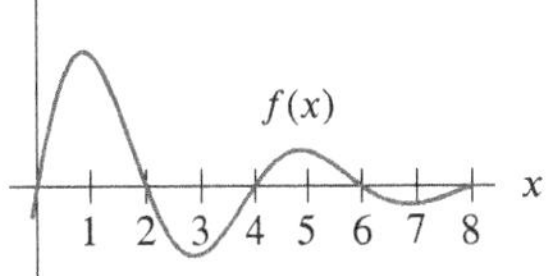

Figure 4.13

22. $f'(0)$ or $f'(4)$?

23. $f'(2)$ or $f'(6)$?

24. $f''(1)$ or $f''(3)$?

■ In Exercises 25–28, the function f is defined for all x. Use the graph of f' to decide:
(a) Over what intervals is f increasing? Decreasing?
(b) Does f have local maxima or minima? If so, which, and where?

25.

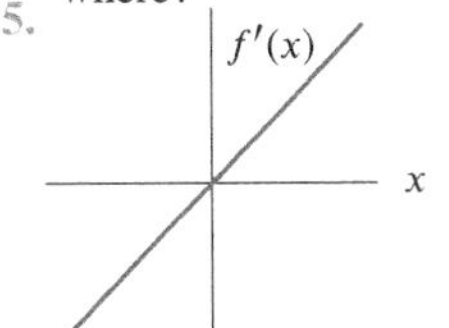

26.

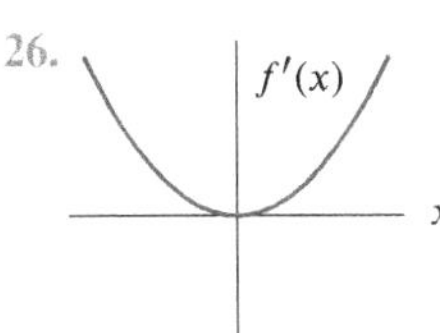

27.

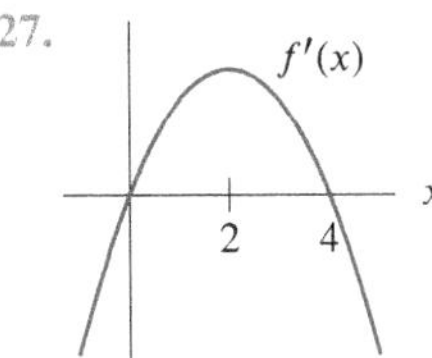

28.

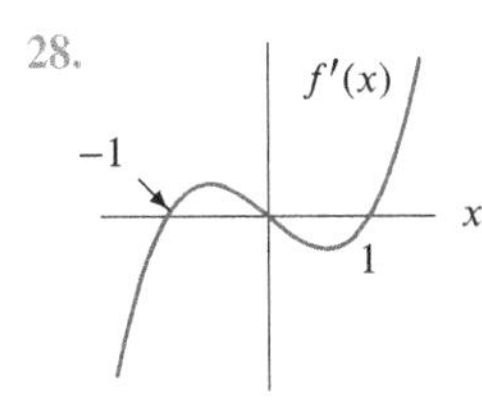

PROBLEMS

29. **(a)** If a is a positive constant, find all critical points of $f(x) = x^3 - ax$.
(b) Find the value of a so that f has local extrema at $x = \pm 2$.

30. **(a)** If a is a constant, find all critical points of $f(x) = 5ax - 2x^2$.
(b) Find the value of a so that f has a local maximum at $x = 6$.

31. **(a)** If b is a positive constant and $x > 0$, find all critical points of $f(x) = x - b\ln x$.
(b) Use the second-derivative test to determine whether the function has a local maximum or local minimum at each critical point.

32. **(a)** If a is a nonzero constant, find all critical points of

$$f(x) = \frac{a}{x^2} + x.$$

(b) Use the second-derivative test to show that if a is positive then the graph has a local minimum, and if a is negative then the graph has a local maximum.

33. **(a)** Show that if a is a positive constant, then $x = 0$ is the only critical point of $f(x) = x + a\sqrt{x}$.

(b) Use derivatives to show that f is increasing and its graph is concave down for all $x > 0$.

34. If U and V are positive constants, find all critical points of

$$F(t) = Ue^t + Ve^{-t}.$$

35. Figure 4.14 is the graph of a derivative f'. On the graph, mark the x-values that are critical points of f. At which critical points does f have local maxima, local minima, or neither?

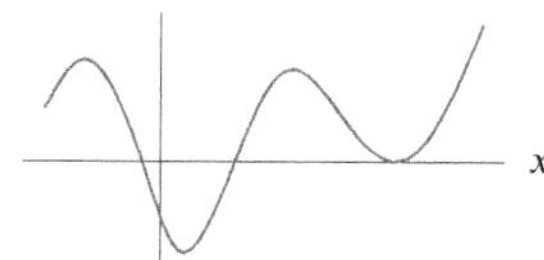

Figure 4.14

36. Figure 4.14 is the graph of a derivative f'. On the graph, mark the x-values that are inflection points of f.

37. Figure 4.14 is the graph of a second derivative f''. On the graph, mark the x-values that are inflection points of f.

38. **(a)** Figure 4.15 shows the graph of f. Which of the x-values A, B, C, D, E, F, and G appear to be critical points of f?

(b) Which appear to be inflection points of f?

(c) How many local maxima does f appear to have? How many local minima?

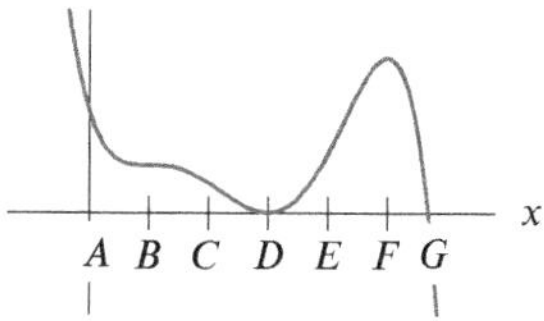

Figure 4.15

39. Figure 4.15 shows the graph of the **derivative**, f'.

(a) Which of the x-values A, B, C, D, E, F, and G appear to be critical points of f?

(b) Which appear to be inflection points of f?

(c) How many local maxima does f appear to have? How many local minima?

■ For Problems 40–43, sketch a possible graph of $y = f(x)$, using the given information about the derivatives $y' = f'(x)$ and $y'' = f''(x)$. Assume that the function is defined and continuous for all real x.

40.

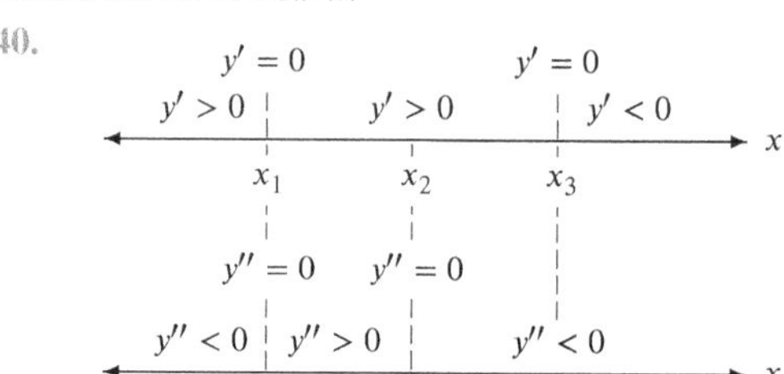

41.

$y' < 0$

x_1 x_2 x_3 x

$y'' = 0$ $y'' = 0$ $y'' = 0$

$y'' > 0$ $y'' < 0$ $y'' > 0$ $y'' < 0$ x

42.

y' undefined $y' = 0$

$y' > 0$ $y' < 0$ $y' > 0$ x

x_1 x_2

y'' undefined

$y'' > 0$ $y'' > 0$ x

43.

$y' = 2$

$y' = 2$ $y' > 0$ x

x_1

$y'' = 0$

$y'' = 0$ $y'' > 0$ x

44. Suppose f has a continuous derivative whose values are given in the following table.

(a) Estimate the x-coordinates of critical points of f for $0 \le x \le 10$.

(b) For each critical point, indicate if it is a local maximum of f, local minimum, or neither.

x	0	1	2	3	4	5	6	7	8	9	10
$f'(x)$	5	2	1	−2	−5	−3	−1	2	3	1	−1

45. **(a)** The following table gives values of the differentiable function $y = f(x)$. Estimate the x-values of critical points of $f(x)$ on the interval $0 < x < 10$. Classify each critical point as a local maximum, local minimum, or neither.

(b) Now assume that the table gives values of the continuous function $y = f'(x)$ (instead of $f(x)$). Estimate and classify critical points of the function $f(x)$.

x	0	1	2	3	4	5	6	7	8	9	10
y	1	2	1	−2	−5	−3	−1	2	3	1	−1

46. If water is flowing at a constant rate (i.e., constant volume per unit time) into the vase in Figure 4.16, sketch a graph of the depth of the water against time. Mark on the graph the time at which the water reaches the corner of the vase.

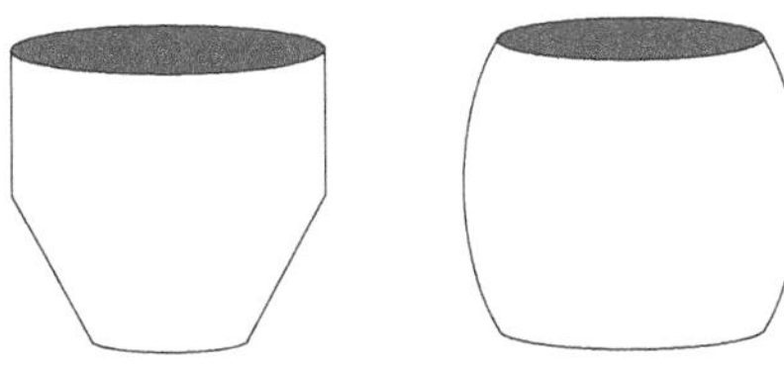

Figure 4.16 Figure 4.17

47. If water is flowing at a constant rate (i.e., constant volume per unit time) into the Grecian urn in Figure 4.17, sketch a graph of the depth of the water against time. Mark on the graph the time at which the water reaches the widest point of the urn.

48. Find and classify the critical points of $f(x) = x^3(1-x)^4$ as local maxima and minima.

49. If $m, n \geq 2$ are integers, find and classify the critical points of $f(x) = x^m(1-x)^n$.

50. The rabbit population on a small Pacific island is approximated by

$$P = \frac{2000}{1 + e^{5.3-0.4t}}$$

with t measured in years since 1774, when Captain James Cook left 10 rabbits on the island.

(a) Graph P. Does the population level off?
(b) Estimate when the rabbit population grew most rapidly. How large was the population at that time?
(c) What natural causes could lead to the shape of the graph of P?

51. The Arctic Sea ice extent, the area of the sea covered by ice, grows seasonally over the winter months each year, typically from November to March, and is modeled by $G(t)$, in millions of square kilometers, t months after November 1, 2014.[1]

(a) What is the sign of $G'(t)$ for $0 < t < 4$?
(b) Suppose $G''(t) < 0$ for $0 < t < 4$. What does this tell us about how the Arctic Sea ice extent grows?
(c) Sketch a graph of $G(t)$ for $0 \leq t \leq 4$, given that $G(0) = 10.3$, $G(4) = 14.4$, and G'' is as in part (b). Label your axes, including units.

52. Find values of a and b so that $f(x) = x^2 + ax + b$ has a local minimum at the point $(6, -5)$.

53. Find the value of a so that $f(x) = xe^{ax}$ has a critical point at $x = 3$.

54. Find values of a and b so that $f(x) = axe^{bx}$ has $f(1/3) = 1$ and f has a local maximum at $x = 1/3$.

55. For a function f and constant $k \neq 0$, we have

$$f'(x) = -k^2 x e^{-0.5x^2/k^2}$$
$$f''(x) = (x^2 - k^2)e^{-0.5x^2/k^2}.$$

(a) What is the only critical point of f?
(b) How many inflection points does the graph of f have?
(c) Is the critical point of f a local maximum, local minimum, or neither?

56. Graph $f(x) = x + \sin x$, and determine where f is increasing most rapidly and least rapidly.

57. You might think the graph of $f(x) = x^2 + \cos x$ should look like a parabola with some waves on it. Sketch the actual graph of $f(x)$ using a calculator or computer. Explain what you see using $f''(x)$.

■ Problems 58–59 show graphs of the three functions f, f', f''. Identify which is which.

58.

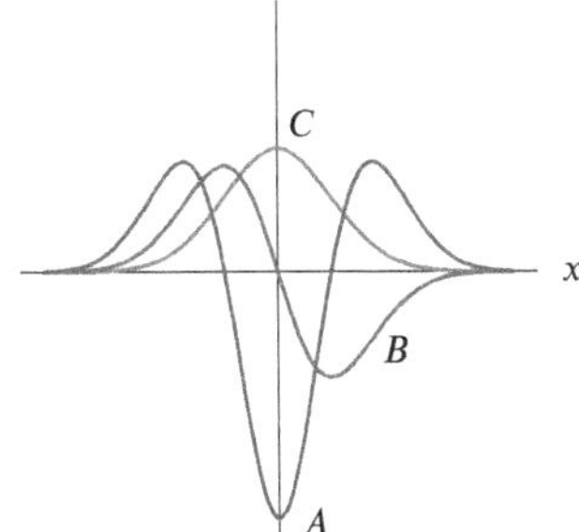

59.

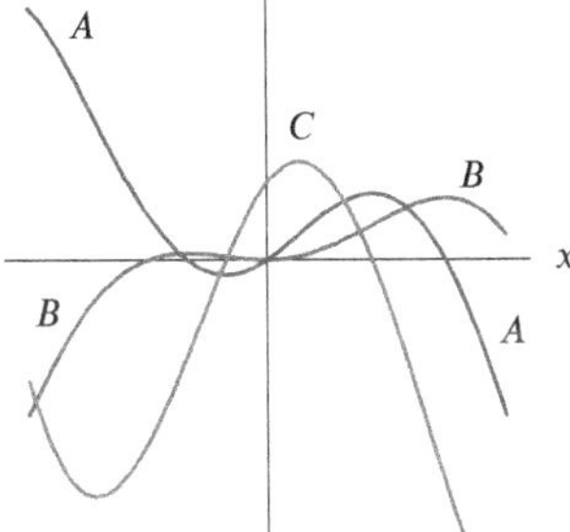

[1] Data on the Arctic Sea ice extent was recorded daily in 2014 and is archived at nsidc.org/arcticseaicenews/2015/03/. Accessed September 2015.

■ Problems 60–61 show graphs of f, f', f''. Each of these three functions is either odd or even. Decide which functions are odd and which are even. Use this information to identify which graph corresponds to f, which to f', and which to f''.

60.

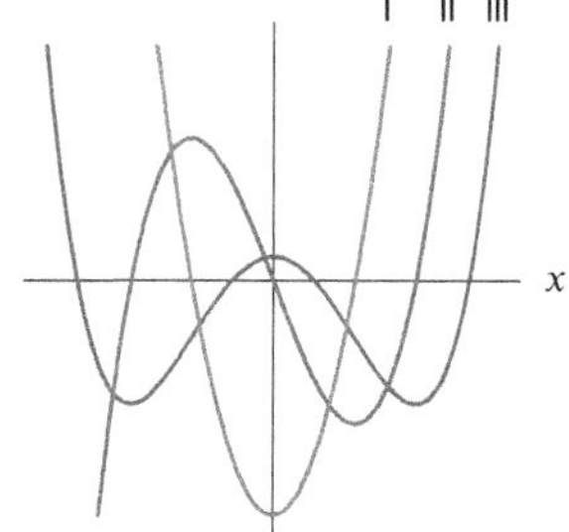

61.

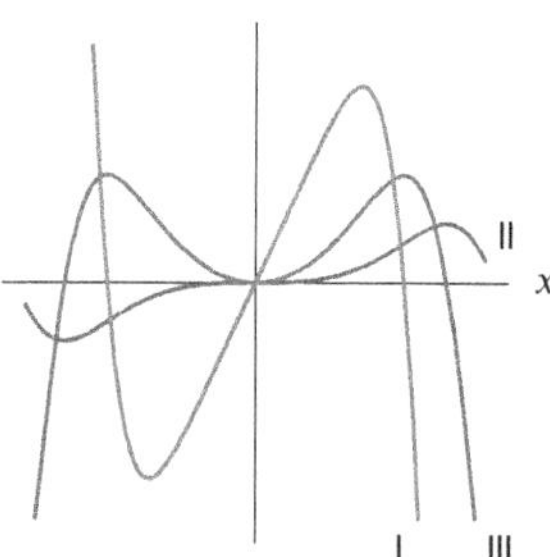

62. Use the derivative formulas and algebra to find the intervals where $f(x) = (x+50)/(x^2+525)$ is increasing and the intervals where it is decreasing. It is possible, but difficult, to solve this problem by graphing f; describe the difficulty.

Strengthen Your Understanding

■ In Problems 63–66, explain what is wrong with the statement.

63. An increasing function has no inflection points.

64. For any function f, if $f''(0) = 0$, there is an inflection point at $x = 0$.

65. If p is a critical point, and f' is negative to the left of p and positive to the right of p, and if $f''(p)$ exists, then $f''(p) > 0$.

66. If f has exactly two critical points, then one is a local maximum and the other is a local minimum.

■ In Problems 67–70, give an example of:

67. A function which has no critical points on the interval between 0 and 1.

68. A function, f, which has a critical point at $x = 1$ but for which $f'(1) \neq 0$.

69. A function with local maxima and minima at an infinite number of points.

70. A function f that has a local maximum at $x = a$ and for which $f''(a)$ is not negative.

■ Are the statements in Problems 71–79 true or false for a function f whose domain is all real numbers? If a statement is true, explain how you know. If a statement is false, give a counterexample.

71. A local minimum of f occurs at a critical point of f.

72. If $x = p$ is not a critical point of f, then $x = p$ is not a local maximum of f.

73. A local maximum of f occurs at a point where

$$f'(x) = 0.$$

74. If $x = p$ is not a local maximum of f, then $x = p$ is not a critical point of f.

75. If $f'(p) = 0$, then f has a local minimum or local maximum at $x = p$.

76. If f' is continuous and f has no critical points, then f is everywhere increasing or everywhere decreasing.

77. If f'' is continuous and the graph of f has an inflection point at $x = p$, then $f''(p) = 0$.

78. A critical point of f must be a local maximum or minimum of f.

79. Every cubic polynomial has an inflection point.

■ In Problems 80–83, give an example of a function f that makes the statement true, or say why such an example is impossible. Assume that f'' exists everywhere.

80. f is concave up and $f(x)$ is positive for all x.

81. f is concave down and $f(x)$ is positive for all x.

82. f is concave down and $f(x)$ is negative for all x.

83. f is concave up and $f(x)$ is negative for all x.

84. Given that $f'(x)$ is continuous everywhere and changes from negative to positive at $x = a$, which of the following statements must be true?

(a) a is a critical point of $f(x)$
(b) $f(a)$ is a local maximum
(c) $f(a)$ is a local minimum
(d) $f'(a)$ is a local maximum
(e) $f'(a)$ is a local minimum

4.2 OPTIMIZATION

The largest and smallest values of a quantity often have practical importance. For example, automobile engineers want to construct a car that uses the least amount of fuel, scientists want to calculate which wavelength carries the maximum radiation at a given temperature, and urban planners want to design traffic patterns to minimize delays. Such problems belong to the field of mathematics called *optimization*. The next three sections show how the derivative provides an efficient way of solving many optimization problems.

Global Maxima and Minima

The single greatest (or least) value of a function f over a specified domain is called the *global maximum* (or *minimum*) of f. Recall that the local maxima and minima tell us where a function is locally largest or smallest. Now we are interested in where the function is absolutely largest or smallest in a given domain. We make the following definition:

Suppose p is a point in the domain of f:

- f has a **global minimum** at p if $f(p)$ is less than or equal to all values of f.
- f has a **global maximum** at p if $f(p)$ is greater than or equal to all values of f.

Global maxima and minima are sometimes called *extrema* or *optimal values*.

Existence of Global Extrema

The following theorem describes when global extrema are guaranteed to exist:

Theorem 4.2: The Extreme Value Theorem

If f is continuous on the closed interval $a \leq x \leq b$, then f has a global maximum and a global minimum on that interval.

For a proof of Theorem 4.2, see www.wiley.com/college/hughes-hallett.

How Do We Find Global Maxima and Minima?

If f is a continuous function defined on a closed interval $a \leq x \leq b$ (that is, an interval containing its endpoints), then Theorem 4.2 guarantees that global maxima and minima exist. Figure 4.18 illustrates that the global maximum or minimum of f occurs either at a critical point or at an endpoint of the interval, $x = a$ or $x = b$. These points are the candidates for global extrema.

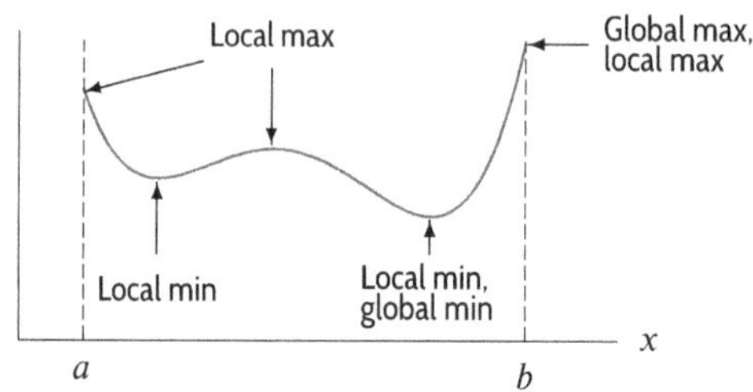

Figure 4.18: Global maximum and minimum on a closed interval $a \le x \le b$

Global Maxima and Minima on a Closed Interval: Test the Candidates

For a continuous function f on a closed interval $a \le x \le b$:

- Find the critical points of f in the interval.
- Evaluate the function at the critical points and at the endpoints, a and b. The largest value of the function is the global maximum; the smallest value is the global minimum.

If the function is defined on an open interval $a < x < b$ (that is, an interval not including its endpoints) or on all real numbers, there may or may not be a global maximum or minimum. For example, there is no global maximum in Figure 4.19 because the function has no actual largest value. The global minimum in Figure 4.19 coincides with the local minimum. There is a global minimum but no global maximum in Figure 4.20.

Global Maxima and Minima on an Open Interval or on All Real Numbers

For a continuous function, f, find the value of f at all the critical points and sketch a graph. Look at values of f when x approaches the endpoints of the interval, or approaches $\pm\infty$, as appropriate. If there is only one critical point, look at the sign of f' on either side of the critical point.

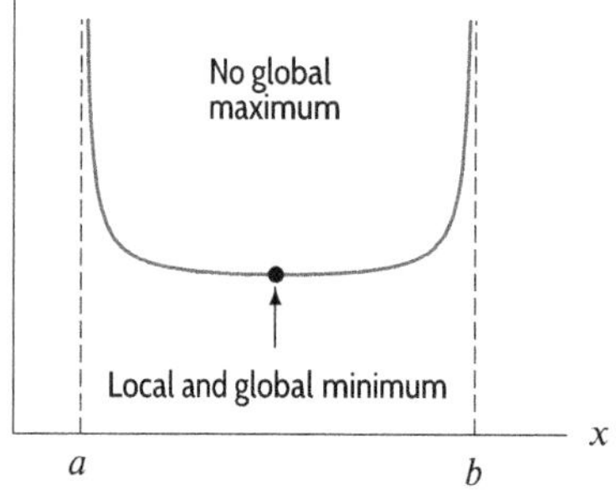

Figure 4.19: Global minimum on $a < x < b$

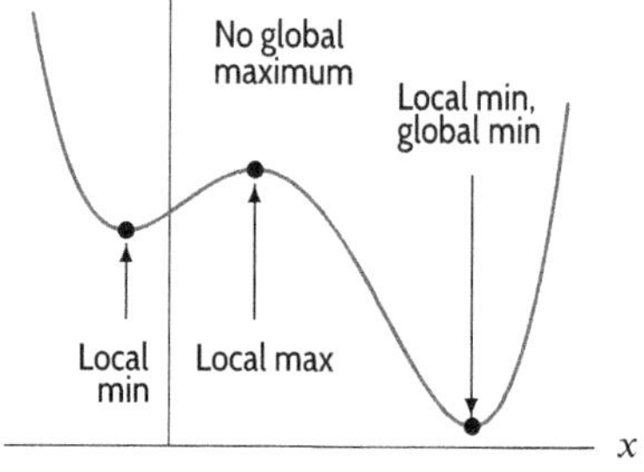

Figure 4.20: Global minimum when the domain is all real numbers

Example 1 Find the global maxima and minima of $f(x) = x^3 - 9x^2 - 48x + 52$ on the following intervals:

(a) $-5 \le x \le 12$ (b) $-5 \le x \le 14$ (c) $-5 \le x < \infty$.

Solution (a) We have previously obtained the critical points $x = -2$ and $x = 8$ using

$$f'(x) = 3x^2 - 18x - 48 = 3(x+2)(x-8).$$

We evaluate f at the critical points and the endpoints of the interval:

$$\begin{aligned} f(-5) &= (-5)^3 - 9(-5)^2 - 48(-5) + 52 = -58 \\ f(-2) &= 104 \\ f(8) &= -396 \\ f(12) &= -92. \end{aligned}$$

Comparing these function values, we see that the global maximum on $[-5, 12]$ is 104 and occurs at $x = -2$, and the global minimum on $[-5, 12]$ is -396 and occurs at $x = 8$.

(b) For the interval $[-5, 14]$, we compare

$$f(-5) = -58, \quad f(-2) = 104, \quad f(8) = -396, \quad f(14) = 360.$$

The global maximum is now 360 and occurs at $x = 14$, and the global minimum is still -396 and occurs at $x = 8$. Since the function is increasing for $x > 8$, changing the right-hand end of the interval from $x = 12$ to $x = 14$ alters the global maximum but not the global minimum. See Figure 4.21.

(c) Figure 4.21 shows that for $-5 \le x < \infty$ there is no global maximum, because we can make $f(x)$ as large as we please by choosing x large enough. The global minimum remains -396 at $x = 8$.

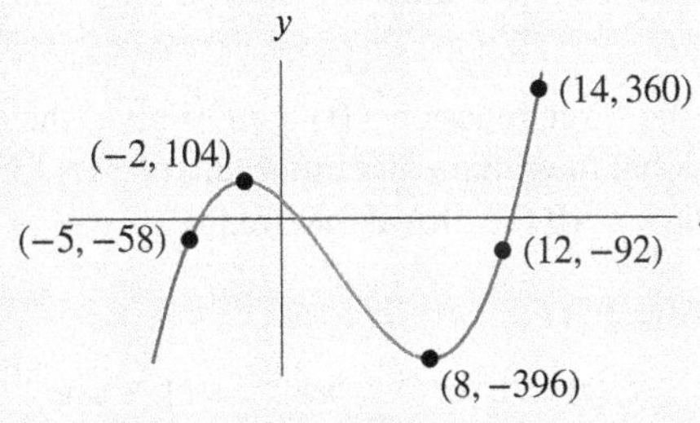

Figure 4.21: Graph of $f(x) = x^3 - 9x^2 - 48x + 52$

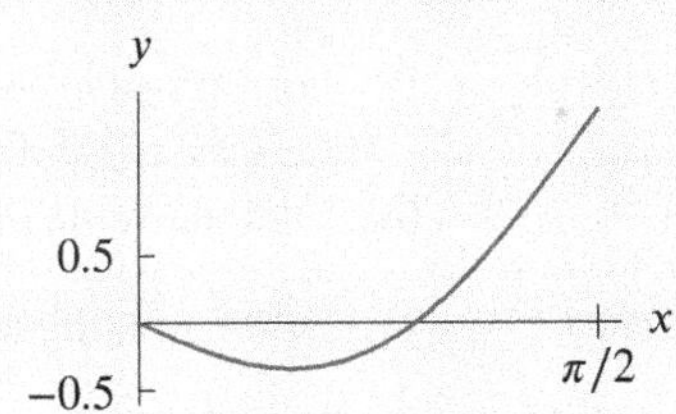

Figure 4.22: Graph of $g(x) = x - \sin(2x)$

Example 2 Find the global maximum and minimum of $g(x) = x - \sin(2x)$ on the interval $0 \le x \le \pi/2$.

Solution Since $g(x) = x - \sin(2x)$ is continuous and the interval $0 \le x \le \pi/2$ is closed, there must be a global maximum and minimum. The possible candidates are critical points in the interval and endpoints. Since there are no points where $g'(x)$ is undefined, we solve $g'(x) = 0$ to find all the critical points:

$$g'(x) = 1 - 2\cos(2x) = 0,$$

so $\cos(2x) = 1/2$. Therefore $2x = \pi/3, 5\pi/3, \ldots$. Thus the only critical point in the interval is $x = \pi/6$. We compare values of g at the critical points and the endpoints:

$$g(0) = 0, \quad g(\pi/6) = \pi/6 - \sqrt{3}/2 = -0.342, \quad g(\pi/2) = \pi/2 = 1.571.$$

Thus the global maximum is 1.571 at $x = \pi/2$ and the global minimum is -0.342 at $x = \pi/6$. See Figure 4.22.

Example 3 Jared is coughing. The speed, $v(r)$, in meters/sec, with which he expels air depends on the radius, r, of his windpipe, given for $0 \le r \le 9$ in millimeters (mm) by

$$v(r) = 0.1(9 - r)r^2.$$

What value of r maximizes the speed? For what value is the speed minimized?

Solution Notice that $v(0) = v(9) = 0$, and that $v(r)$ is positive for $0 < r < 9$. Therefore the maximum occurs somewhere between $r = 0$ and $r = 9$. Since

$$v(r) = 0.1(9 - r)r^2 = 0.9r^2 - 0.1r^3,$$

the derivative is

$$v'(r) = 1.8r - 0.3r^2 = 0.3r(6 - r).$$

The derivative is zero if $r = 0$ or $r = 6$. These are the critical points of v. We already know $v(0) = v(9) = 0$, and

$$v(6) = 0.1(9 - 6)6^2 = 10.8 \text{ meters/sec.}$$

Thus, v has a global maximum at $r = 6$ mm. The global minimum of $v = 0$ meters/sec occurs at both endpoints $r = 0$ mm and $r = 9$ mm.

In applications, the function being optimized often contains a parameter whose value depends on the situation, and the maximum or minimum depends on the parameter.

Example 4

(a) For a positive constant b, the surge function $f(t) = te^{-bt}$ gives the quantity of a drug in the body for time $t \ge 0$. Find the global maximum and minimum of $f(t)$ for $t \ge 0$.
(b) Find the value of b making $t = 10$ the global maximum.

Solution

(a) Differentiating and factoring gives

$$f'(t) = 1 \cdot e^{-bt} - bte^{-bt} = (1 - bt)e^{-bt},$$

so there is a critical point at $t = 1/b$.

The sign of f' is determined by the sign of $(1 - bt)$, so f' is positive to the left of $t = 1/b$ and negative to the right of $t = 1/b$. Since f increases to the left of $t = 1/b$ and decreases to the right of $t = 1/b$, the global maximum occurs at $t = 1/b$. In addition, $f(0) = 0$ and $f(t) \ge 0$ for all $t \ge 0$, so the global minimum occurs at $t = 0$. Thus

The global maximum value is $f\left(\frac{1}{b}\right) = \frac{1}{b}e^{-b(1/b)} = \frac{e^{-1}}{b}$.

The global minimum value is $f(0) = 0$.

(b) Since $t = 10$ gives the global maximum, we have $1/b = 10$, so $b = 0.1$. See Figure 4.23.

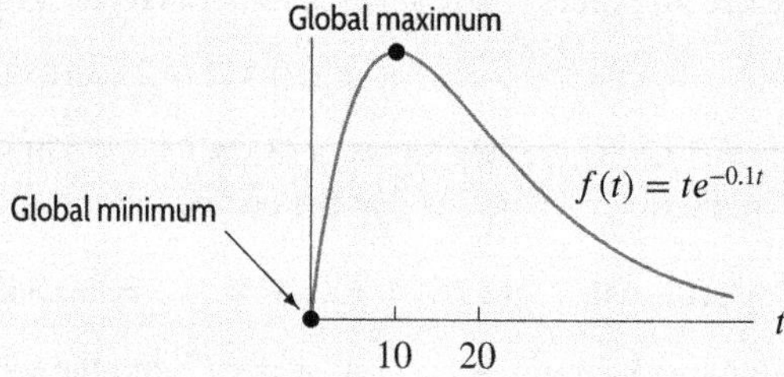

Figure 4.23: Graph of $f(t) = te^{-bt}$ for $b = 0.1$

Example 5 When an arrow is shot into the air, its range, R, is defined as the horizontal distance from the archer to the point where the arrow hits the ground. If the ground is horizontal and we neglect air resistance, it can be shown that

$$R = \frac{{v_0}^2 \sin(2\theta)}{g},$$

where v_0 is the initial velocity of the arrow, g is the (constant) acceleration due to gravity, and θ is the angle above horizontal, so $0 \le \theta \le \pi/2$. (See Figure 4.24.) What angle θ maximizes R?

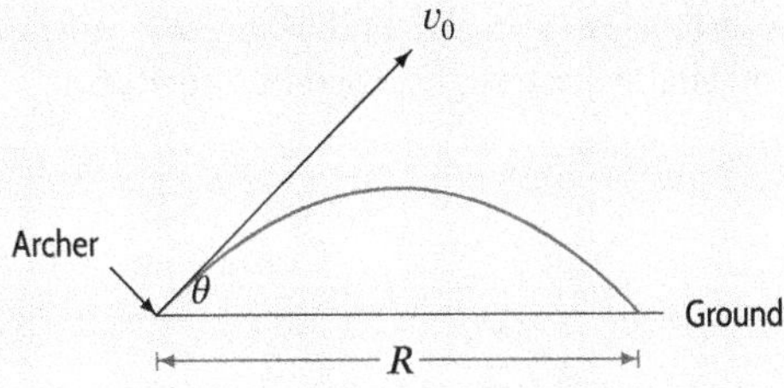

Figure 4.24: Arrow's path

Solution We can find the maximum of this function without using calculus. The maximum value of R occurs when $\sin(2\theta) = 1$, so $\theta = \arcsin(1)/2 = \pi/4$, giving $R = v_0^2/g$.

Let's see how we can do the same problem with calculus. We want to find the global maximum of R for $0 \le \theta \le \pi/2$. First we look for critical points:

$$\frac{dR}{d\theta} = 2\frac{v_0^2 \cos(2\theta)}{g}.$$

Setting $dR/d\theta$ equal to 0, we get

$$0 = \cos(2\theta), \quad \text{or} \quad 2\theta = \pm\frac{\pi}{2}, \pm\frac{3\pi}{2}, \pm\frac{5\pi}{2}, \ldots$$

so $\pi/4$ is the only critical point in the interval $0 \le \theta \le \pi/2$. The range at $\theta = \pi/4$ is $R = {v_0}^2/g$.

Now we check the value of R at the endpoints $\theta = 0$ and $\theta = \pi/2$. Since $R = 0$ at each endpoint (the arrow is shot horizontally or vertically), the critical point $\theta = \pi/4$ is both a local and a global maximum on $0 \le \theta \le \pi/2$. Therefore, the arrow goes farthest if shot at an angle of $\pi/4$, or 45°.

Finding Upper and Lower Bounds

The problem of finding the *bounds* of a function is closely related to finding maxima and minima. In Example 1 on page 204, the value of $f(x)$ on the interval $[-5, 12]$ ranges from -396 to 104. Thus

$$-396 \le f(x) \le 104,$$

and we say that -396 is a *lower bound* for f and 104 is an *upper bound* for f on $[-5, 12]$. (See Appendix A for more on bounds.) Of course, we could also say that

$$-400 \le f(x) \le 150,$$

so that f is also bounded below by -400 and above by 150 on $[-5, 12]$. However, we consider the -396 and 104 to be the *best possible bounds* because they describe most accurately how the function $f(x)$ behaves on $[-5, 12]$.

Example 6 An object on a spring oscillates about its equilibrium position at $y = 0$. See Figure 4.25. Its displacement, y, from equilibrium is given as a function of time, t, by

$$y = e^{-t}\cos t.$$

Find the greatest distance the object goes above and below the equilibrium for $t \geq 0$.

Solution We are looking for bounds for y as a function of t. What does the graph look like? We think of it as a cosine curve with a decreasing amplitude of e^{-t}; in other words, it is a cosine curve squashed between the graphs of $y = e^{-t}$ and $y = -e^{-t}$, forming a wave with lower and lower crests and shallower and shallower troughs. (See Figure 4.26.)

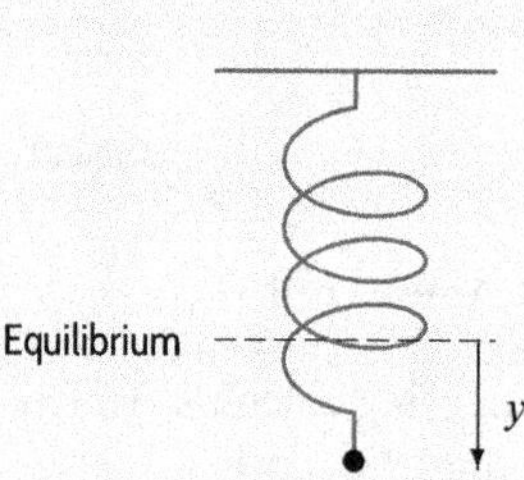

Figure 4.25: Object on spring ($y < 0$ below equilibrium)

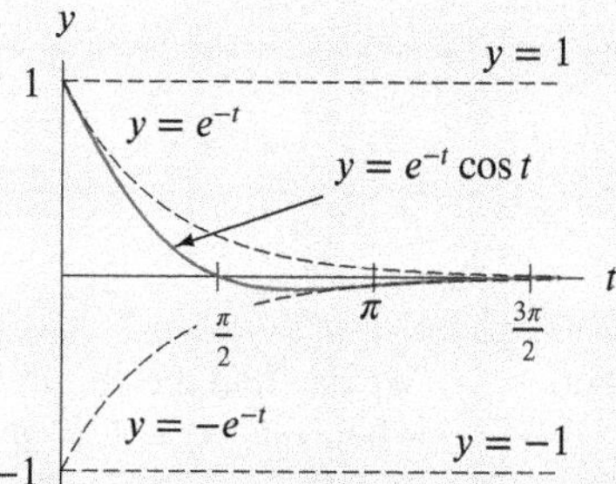

Figure 4.26: $f(t) = e^{-t}\cos t$ for $t \geq 0$

From the graph, we see that for $t \geq 0$, the curve lies between the horizontal lines $y = -1$ and $y = 1$. This means that -1 and 1 are bounds:

$$-1 \leq e^{-t}\cos t \leq 1.$$

The line $y = 1$ is the best possible upper bound because the graph comes up that high at $t = 0$. However, we can find a better lower bound if we find the global minimum value of y for $t \geq 0$; this minimum occurs in the first trough between $t = \pi/2$ and $t = 3\pi/2$ because later troughs are squashed closer to the t-axis. At the minimum, $dy/dt = 0$. The product rule gives

$$\frac{dy}{dt} = (-e^{-t})\cos t + e^{-t}(-\sin t) = -e^{-t}(\cos t + \sin t) = 0.$$

Since e^{-t} is never 0, we have

$$\cos t + \sin t = 0, \quad \text{so} \quad \frac{\sin t}{\cos t} = -1.$$

The smallest positive solution of

$$\tan t = -1 \quad \text{is} \quad t = \frac{3\pi}{4}.$$

Thus, the global minimum we see on the graph occurs at $t = 3\pi/4$. The value of y there is

$$y = e^{-3\pi/4}\cos\left(\frac{3\pi}{4}\right) \approx -0.07.$$

The greatest distance the object goes below equilibrium is 0.07. Thus, for all $t \geq 0$,

$$-0.07 \leq e^{-t}\cos t \leq 1.$$

Notice how much smaller in magnitude the lower bound is than the upper. This is a reflection of how quickly the factor e^{-t} causes the oscillation to die out.

Exercises and Problems for Section 4.2 Online Resource: Additional Problems for Section 4.2

EXERCISES

For Exercises 1–2, indicate all critical points on the given graphs. Determine which correspond to local minima, local maxima, global minima, global maxima, or none of these. (Note that the graphs are on closed intervals.)

1.
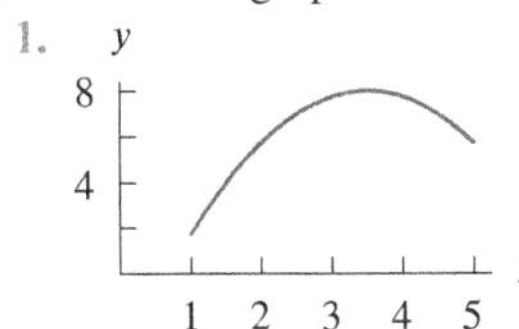

2.
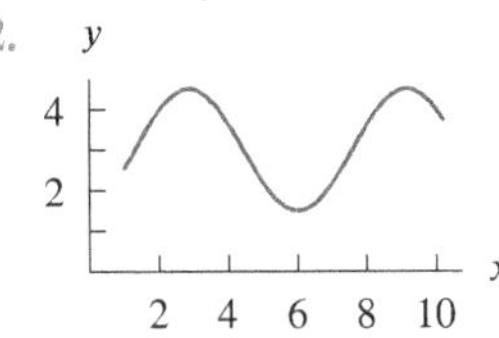

3. For $x > 0$, find the x-value and the corresponding y-value that maximizes $y = 25 + 6x^2 - x^3$, by
   (a) Estimating the values from a graph of y.
   (b) Finding the values using calculus.

In Exercises 4–10, find the global maximum and minimum for the function on the closed interval.

4. $f(x) = x^3 - 3x^2 + 20, \quad -1 \le x \le 3$
5. $f(x) = x^4 - 8x^2, \quad -3 \le x \le 1$
6. $f(x) = xe^{-x^2/2}, \quad -2 \le x \le 2$
7. $f(x) = 3x^{1/3} - x, \quad -1 \le x \le 8$
8. $f(x) = x - 2\ln(x+1), \quad 0 \le x \le 2$
9. $f(x) = x^2 - 2|x|, \quad -3 \le x \le 4$
10. $f(x) = \dfrac{x+1}{x^2+3}, \quad -1 \le x \le 2$

In Exercises 11–13, find the value(s) of x for which:
(a) $f(x)$ has a local maximum or local minimum. Indicate which are maxima and which are minima.
(b) $f(x)$ has a global maximum or global minimum.

11. $f(x) = x^{10} - 10x$, and $0 \le x \le 2$
12. $f(x) = x - \ln x$, and $0.1 \le x \le 2$
13. $f(x) = \sin^2 x - \cos x$, and $0 \le x \le \pi$

In Exercises 14–21, find the exact global maximum and minimum values of the function. The domain is all real numbers unless otherwise specified.

14. $g(x) = 4x - x^2 - 5$
15. $f(x) = x + 1/x$ for $x > 0$
16. $g(t) = te^{-t}$ for $t > 0$
17. $f(x) = x - \ln x$ for $x > 0$
18. $f(t) = \dfrac{t}{1+t^2}$
19. $f(t) = (\sin^2 t + 2)\cos t$
20. $f(x) = 2e^x + 3e^{-x}$
21. $f(x) = e^{3x} - e^{2x}$

In Exercises 22–27, find the best possible bounds for the function.

22. $x^3 - 4x^2 + 4x, \quad$ for $\quad 0 \le x \le 4$
23. $e^{-x^2}, \quad$ for $|x| \le 0.3$
24. $x^3e^{-x}, \quad$ for $x \ge 0$
25. $x + \sin x, \quad$ for $0 \le x \le 2\pi$
26. $\ln(1+x), \quad$ for $x \ge 0$
27. $\ln(1+x^2), \quad$ for $-1 \le x \le 2$

PROBLEMS

28. Figure 4.27 shows a function f. Does f have a global maximum? A global minimum? If so, where? Assume that $f(x)$ is defined for all x and that the graph does not change concavity outside the window shown.

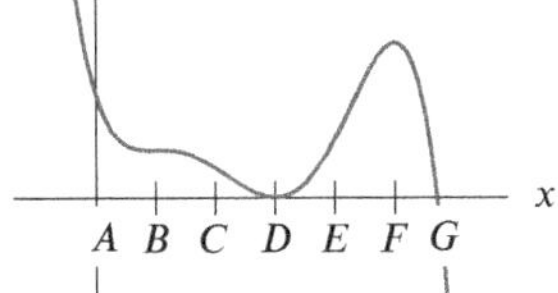

Figure 4.27

29. Figure 4.27 shows a derivative, f'. Does the function f have a global maximum? A global minimum? If so, where? Assume that $f(x)$ and $f'(x)$ are defined for all x and that the graph of $f'(x)$ does not change concavity outside the window shown.

30. A grapefruit is tossed straight up with an initial velocity of 50 ft/sec. The grapefruit is 5 feet above the ground when it is released. Its height, in feet, at time t seconds is given by

$$y = -16t^2 + 50t + 5.$$

How high does it go before returning to the ground?

31. Find the value(s) of x that give critical points of $y = ax^2 + bx + c$, where a, b, c are constants. Under what conditions on a, b, c is the critical value a maximum? A minimum?

32. What value of w minimizes S if $S - 5pw = 3qw^2 - 6pq$ and p and q are positive constants?

33. At what value(s) of T does $Q = AT(S - T)$ have a critical point? Assume A and S are nonzero constants.

34. For a positive constant a, find the values of x that give critical points of y if

$$\frac{1}{y} = \frac{x}{a + 4x^2}.$$

35. Let $y = at^2e^{-bt}$ with a and b positive constants. For $t \geq 0$, what value of t maximizes y? Sketch the curve if $a = 1$ and $b = 1$.

36. For some positive constant C, a patient's temperature change, T, due to a dose, D, of a drug is given by

$$T = \left(\frac{C}{2} - \frac{D}{3}\right) D^2.$$

(a) What dosage maximizes the temperature change?
(b) The sensitivity of the body to the drug is defined as dT/dD. What dosage maximizes sensitivity?

37. A warehouse selling cement has to decide how often and in what quantities to reorder. It is cheaper, on average, to place large orders, because this reduces the ordering cost per unit. On the other hand, larger orders mean higher storage costs. The warehouse always reorders cement in the same quantity, q. The total weekly cost, C, of ordering and storage is given by

$$C = \frac{a}{q} + bq, \quad \text{where } a, b \text{ are positive constants.}$$

(a) Which of the terms, a/q and bq, represents the ordering cost and which represents the storage cost?
(b) What value of q gives the minimum total cost?

38. A company has 100 units to spend for equipment and labor combined. The company spends x on equipment and $100 - x$ on labor, enabling it to produce Q items where

$$Q = 5x^{0.3}(100 - x)^{0.8}.$$

How much should the company spend on equipment to maximize production Q? On labor? What is the maximum production Q?

39. A chemical reaction converts substance A to substance Y. At the start of the reaction, the quantity of A present is a grams. At time t seconds later, the quantity of Y present is y grams. The rate of the reaction, in grams/sec, is given by

$$\text{Rate} = ky(a - y), \quad k \text{ is a positive constant.}$$

(a) For what values of y is the rate nonnegative? Graph the rate against y.
(b) For what values of y is the rate a maximum?

40. The potential energy, U, of a particle moving along the x-axis is given by

$$U = b\left(\frac{a^2}{x^2} - \frac{a}{x}\right),$$

where a and b are positive constants and $x > 0$. What value of x minimizes the potential energy?

41. For positive constants A and B, the force between two atoms in a molecule is given by

$$f(r) = -\frac{A}{r^2} + \frac{B}{r^3},$$

where $r > 0$ is the distance between the atoms. What value of r minimizes the force between the atoms?

42. When an electric current passes through two resistors with resistance r_1 and r_2, connected in parallel, the combined resistance, R, can be calculated from the equation

$$\frac{1}{R} = \frac{1}{r_1} + \frac{1}{r_2},$$

where R, r_1, and r_2 are positive. Assume that r_2 is constant.

(a) Show that R is an increasing function of r_1.
(b) Where on the interval $a \leq r_1 \leq b$ does R take its maximum value?

43. The bending moment M of a beam, supported at one end, at a distance x from the support is given by

$$M = \tfrac{1}{2}wLx - \tfrac{1}{2}wx^2,$$

where L is the length of the beam, and w is the uniform load per unit length. Find the point on the beam where the moment is greatest.

44. As an epidemic spreads through a population, the number of infected people, I, is expressed as a function of the number of susceptible people, S, by

$$I = k\ln\left(\frac{S}{S_0}\right) - S + S_0 + I_0, \quad \text{for } k, S_0, I_0 > 0.$$

(a) Find the maximum number of infected people.
(b) The constant k is a characteristic of the particular disease; the constants S_0 and I_0 are the values of S and I when the disease starts. Which of the following affects the maximum possible value of I? Explain.

- The particular disease, but not how it starts.
- How the disease starts, but not the particular disease.
- Both the particular disease and how it starts.

45. Two points on the curve $y = \dfrac{x^3}{1 + x^4}$ have opposite x-values, x and $-x$. Find the points making the slope of the line joining them greatest.

46. The function $y = t(x)$ is positive and continuous with a global maximum at the point $(3, 3)$. Graph $t(x)$ if $t'(x)$ and $t''(x)$ have the same sign for $x < 3$, but opposite signs for $x > 3$.

47. Figure 4.28 gives the derivative of $g(x)$ on $-2 \le x \le 2$.

(a) Write a few sentences describing the behavior of $g(x)$ on this interval.
(b) Does the graph of $g(x)$ have any inflection points? If so, give the approximate x-coordinates of their locations. Explain your reasoning.
(c) What are the global maxima and minima of g on $[-2, 2]$?
(d) If $g(-2) = 5$, what do you know about $g(0)$ and $g(2)$? Explain.

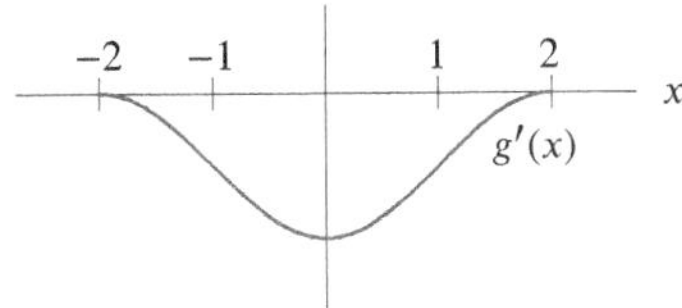

Figure 4.28

48. Figure 4.29 shows the second derivative of $h(x)$ for $-2 \le x \le 1$. If $h'(-1) = 0$ and $h(-1) = 2$,

(a) Explain why $h'(x)$ is never negative on this interval.
(b) Explain why $h(x)$ has a global maximum at $x = 1$.
(c) Sketch a possible graph of $h(x)$ for $-2 \le x \le 1$.

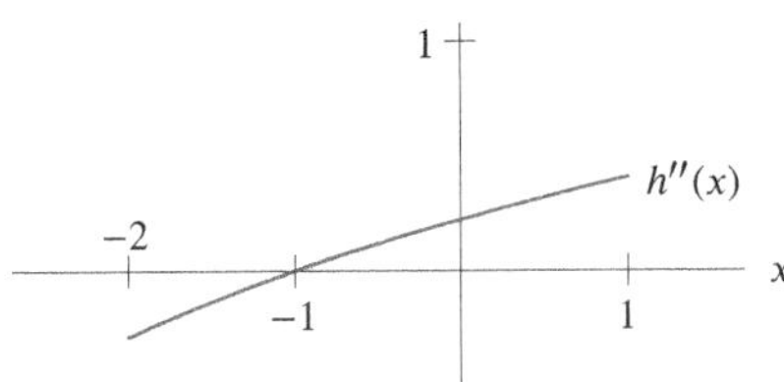

Figure 4.29

49. Figure 4.30 shows $f(x) = e^{3x}$ and $g(x) = e^{-2x}$.

(a) Explain why the graphs intersect only at $x = 0$.
(b) Use your answer to part (a) to find the minimum value of $h(x) = 2e^{3x} + 3e^{-2x}$.

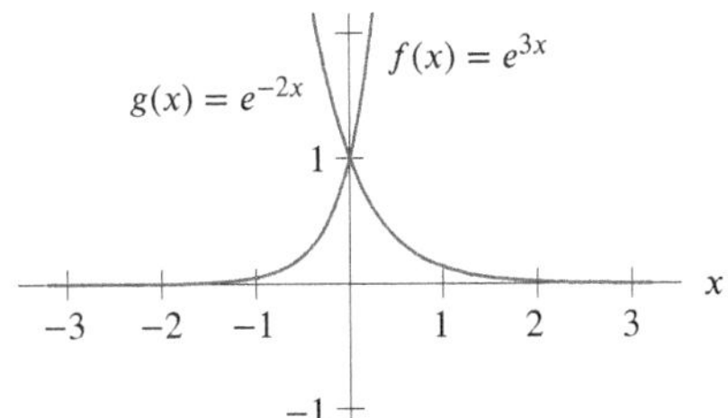

Figure 4.30

50. Figure 4.31 shows $f(x) = \cos(3x)$ and $g(x) = -\cos(2x)$ on the interval $0 \le x \le \pi$.

(a) Show that the graphs intersect at $x = \pi/5$, $x = 3\pi/5$, and $x = \pi$.
(b) Use the graphs and part (a) to find the maximum and minimum value of $h(x) = 2\sin(3x)+3\sin(2x)$ on $0 \le x \le \pi$.

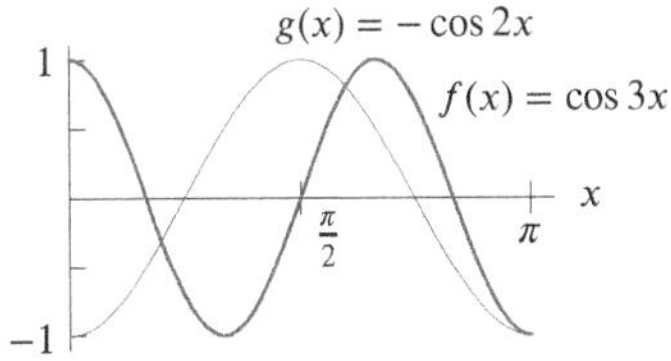

Figure 4.31

51. You are given the n numbers $a_1, a_2, a_3, \cdots, a_n$. Show that the average of these numbers gives the minimum of

$$D(x) = (x-a_1)^2 + (x-a_2)^2 + (x-a_3)^2 + \cdots + (x-a_n)^2.$$

Strengthen Your Understanding

In Problems 52–54, explain what is wrong with the statement.

52. The function $f(x) = (x-1)^2(x-2)$, $0 \le x \le 3$ has a global maximum at $x = 1$.

53. The global minimum of $f(x) = x^4$ on any closed interval $a \le x \le b$ occurs at $x = 0$.

54. The best possible bounds for $f(x) = 1/(1-x)$ on the interval $0 \le x \le 2$ are $f(0) \le f(x) \le f(2)$.

In Problems 55–58, give an example of:

55. A function which has a global maximum at $x = 0$ and a global minimum at $x = 1$ on the interval $0 \le x \le 1$ but no critical points in between $x = 0$ and $x = 1$.

56. A function for which the global maximum is equal to the global minimum.

57. An interval where the best possible bounds for $f(x) = x^2$ are $2 \le f(x) \le 5$.

58. A differentiable function f with best possible bounds $-1 \le f(x) \le 1$ on the interval $-4 \le x \le 4$.

In Problems 59–63, let $f(x) = x^2$. Decide if the following statements are true or false. Explain your answer.

59. f has an upper bound on the interval $(0, 2)$.

60. f has a global maximum on the interval $(0, 2)$.
61. f does not have a global minimum on the interval $(0, 2)$.
62. f does not have a global minimum on any interval (a, b).
63. f has a global minimum on any interval $[a, b]$.
64. Which of the following statements is implied by the statement "If f is continuous on $[a, b]$ then f has a global maximum on $[a, b]$?"
    (a) If f has a global maximum on $[a, b]$ then f must be continuous on $[a, b]$.
    (b) If f is not continuous on $[a, b]$ then f does not have a global maximum on $[a, b]$.
    (c) If f does not have a global maximum on $[a, b]$ then f is not continuous on $[a, b]$.

■ Are the statements in Problems 65–69 true of false? Give an explanation for your answer.

65. Since the function $f(x) = 1/x$ is continuous for $x > 0$ and the interval $(0, 1)$ is bounded, f has a maximum on the interval $(0, 1)$.
66. The Extreme Value Theorem says that only continuous functions have global maxima and minima on every closed, bounded interval.
67. The global maximum of $f(x) = x^2$ on every closed interval is at one of the endpoints of the interval.
68. A function can have two different upper bounds.
69. If a differentiable function $f(x)$ has a global maximum on the interval $0 \le x \le 10$ at $x = 0$, then $f'(0) \le 0$.

4.3 OPTIMIZATION AND MODELING

Finding global maxima and minima is often made possible by having a formula for the function to be maximized or minimized. The process of translating a problem into a function with a known formula is called *mathematical modeling*. The examples that follow give the flavor of modeling.

Example 1 What are the dimensions of an aluminum can that holds 40 in^3 of juice and that uses the least material? Assume that the can is cylindrical, and is capped on both ends.

Solution It is often a good idea to think about a problem in general terms before trying to solve it. Since we're trying to use as little material as possible, why not make the can very small, say, the size of a peanut? We can't, since the can must hold 40 in^3. If we make the can short, to try to use less material in the sides, we'll have to make it fat as well, so that it can hold 40 in^3. See Figure 4.32(a).

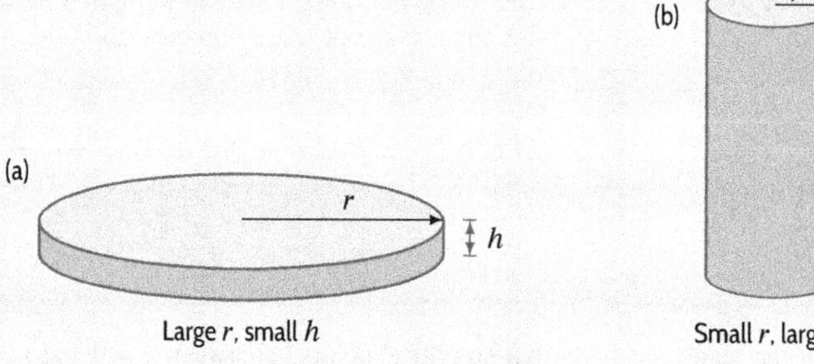

Figure 4.32: Various cylindrical-shaped cans

Table 4.1 *Height, h, and material, M, used in can for various choices of radius, r*

r (in)	h (in)	M (in^2)
0.2	318.31	400.25
1	12.73	86.28
2	3.18	65.13
3	1.41	83.22
4	0.80	120.53
10	0.13	636.32

If we try to save material by making the top and bottom small, the can has to be tall to accommodate the 40 in^3 of juice. So any savings we get by using a small top and bottom may be outweighed by the height of the sides. See Figure 4.32(b).

Table 4.1 gives the height h and amount of material M used in the can for some choices of the radius, r. You can see that r and h change in opposite directions, and that more material is used at the extremes (very large or very small r) than in the middle. It appears that the radius needing the smallest amount of material, M, is somewhere between 1 and 3 inches. Thinking of M as a function of the radius, r, we get the graph in Figure 4.33. The graph shows that the global minimum we want is at a critical point.

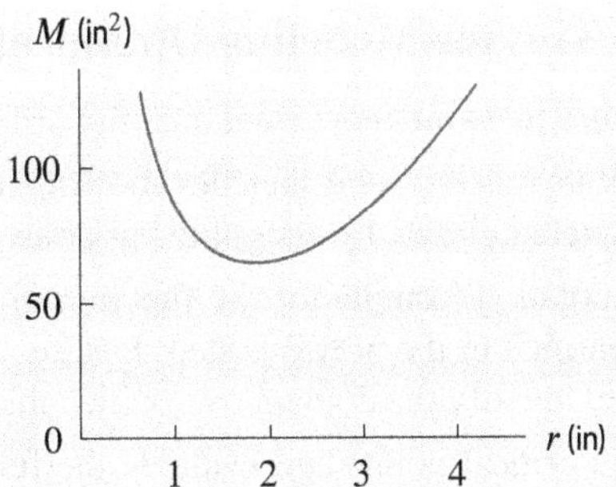

Figure 4.33: Total material used in can, M, as a function of radius, r

Both the table and the graph were obtained from a mathematical model, which in this case is a formula for the material used in making the can. Finding such a formula depends on knowing the geometry of a cylinder, in particular its area and volume. We have

$$M = \text{Material used in the can} = \text{Material in ends} + \text{Material in the side}$$

where

$$\text{Material in ends} = 2 \cdot \text{Area of a circle with radius } r = 2 \cdot \pi r^2,$$
$$\text{Material in the side} = \text{Area of curved side of cylinder with height } h \text{ and radius } r = 2\pi rh.$$

We have

$$M = 2\pi r^2 + 2\pi rh.$$

However, h is not independent of r: if r grows, h shrinks, and vice versa. To find the relationship, we use the fact that the volume of the cylinder, $\pi r^2 h$, is equal to the constant 40 in^3:

$$\text{Volume of can } = \pi r^2 h = 40, \quad \text{giving} \quad h = \frac{40}{\pi r^2}.$$

This means

$$\text{Material in the side} = 2\pi rh = 2\pi r \frac{40}{\pi r^2} = \frac{80}{r}.$$

Thus we obtain the formula for the total material, M, used in a can of radius r if the volume is 40 in^3:

$$M = 2\pi r^2 + \frac{80}{r}.$$

The domain of this function is all $r > 0$ because the radius of the can cannot be negative or zero.

To find the minimum of M, we look for critical points:

$$\frac{dM}{dr} = 4\pi r - \frac{80}{r^2} = 0 \quad \text{at a critical point,} \quad \text{so} \quad 4\pi r = \frac{80}{r^2}.$$

Therefore,

$$\pi r^3 = 20, \quad \text{so} \quad r = \left(\frac{20}{\pi}\right)^{1/3} = 1.85 \text{ inches,}$$

which agrees with the graph in Figure 4.33. We also have

$$h = \frac{40}{\pi r^2} = \frac{40}{\pi(1.85)^2} = 3.7 \text{ inches.}$$

The material used is $M = 2\pi(1.85)^2 + 80/1.85 = 64.7$ in^2.

To confirm that we have found the global minimum, we look at the formula for dM/dr. For small r, the $-80/r^2$ term dominates and for large r, the $4\pi r$ term dominates, so dM/dr is negative for $r < 1.85$ and positive for $r > 1.85$. Thus, M is decreasing for $r < 1.85$ and increasing for $r > 1.85$, so the global minimum occurs at $r = 1.85$.

Practical Tips for Modeling Optimization Problems

1. Make sure that you know what quantity or function is to be optimized.
2. If possible, make several sketches showing how the elements that vary are related. Label your sketches clearly by assigning variables to quantities which change.
3. Try to obtain a formula for the function to be optimized in terms of the variables that you identified in the previous step. If necessary, eliminate from this formula all but one variable. Identify the domain over which this variable varies.
4. Find the critical points and evaluate the function at these points and the endpoints (if relevant) to find the global maxima and/or minima.

The next example, another problem in geometry, illustrates this approach.

Example 2 Alaina wants to get to the bus stop as quickly as possible. The bus stop is across a grassy park, 2000 feet west and 600 feet north of her starting position. Alaina can walk west along the edge of the park on the sidewalk at a speed of 6 ft/sec. She can also travel through the grass in the park, but only at a rate of 4 ft/sec. What path gets her to the bus stop the fastest?

Solution

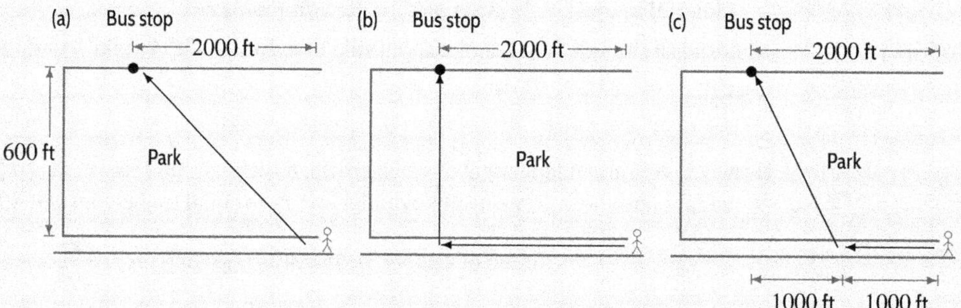

Figure 4.34: Three possible paths to the bus stop

We might first think that she should take a path that is the shortest distance. Unfortunately, the path that follows the shortest distance to the bus stop is entirely in the park, where her speed is slow. (See Figure 4.34(a).) That distance is $\sqrt{2000^2 + 600^2} = 2088$ feet, which takes her about 522 seconds to traverse. She could instead walk quickly the entire 2000 feet along the sidewalk, which leaves her just the 600-foot northward journey through the park. (See Figure 4.34(b).) This method would take $2000/6 + 600/4 \approx 483$ seconds total walking time.

But can she do even better? Perhaps another combination of sidewalk and park gives a shorter travel time. For example, what is the travel time if she walks 1000 feet west along the sidewalk and the rest of the way through the park? (See Figure 4.34(c).) The answer is about 458 seconds.

We make a model for this problem. We label the distance that Alaina walks west along the sidewalk x and the distance she walks through the park y, as in Figure 4.35. Then the total time, t, is

$$t = t_{\text{sidewalk}} + t_{\text{park}}.$$

Since

$$\text{Time} = \text{Distance}/\text{Speed},$$

and she can walk 6 ft/sec on the sidewalk and 4 ft/sec in the park, we have

$$t = \frac{x}{6} + \frac{y}{4}.$$

Now, by the Pythagorean Theorem, $y = \sqrt{(2000 - x)^2 + 600^2}$. Therefore

$$t = \frac{x}{6} + \frac{\sqrt{(2000 - x)^2 + 600^2}}{4} \qquad \text{for } 0 \le x \le 2000.$$

We can find the critical points of this function analytically. (See Problem 16 on page 219.) Alternatively, we can graph the function on a calculator and estimate the critical point, which is $x \approx 1463$ feet. This gives a minimum total time of about 445 seconds.

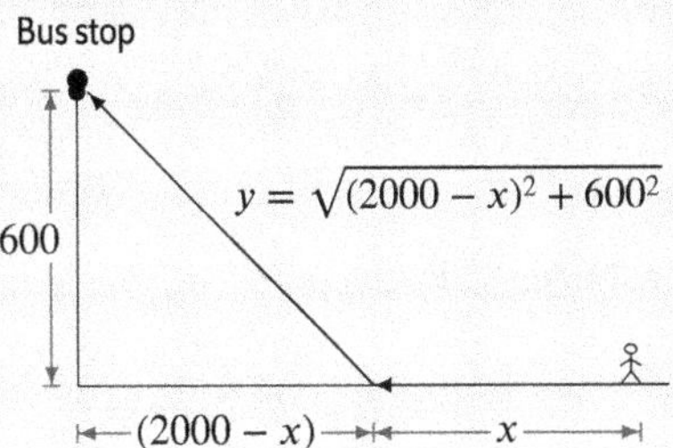

Figure 4.35: Modeling time to bus stop

Example 3 Figure 4.36 shows the curves $y = \sqrt{x}$, $x = 9$, $y = 0$, and a rectangle with vertical sides at $x = a$ and $x = 9$. Find the dimensions of the rectangle having the maximum possible area.

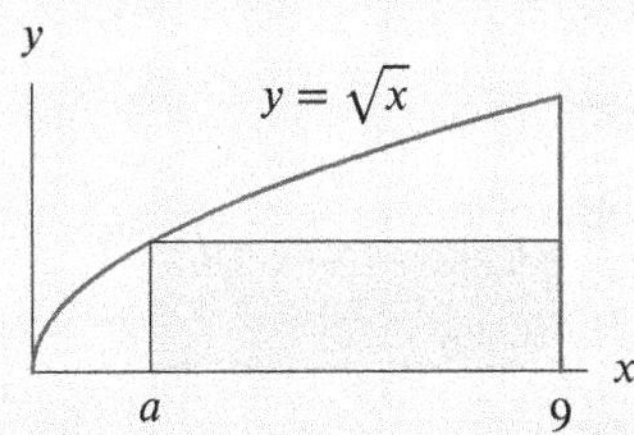

Figure 4.36: Find the rectangle of maximum area with one corner on $y = \sqrt{x}$

Solution We want to choose a to maximize the area of the rectangle with corners at $(a, \sqrt{a})$ and $(9, \sqrt{a})$. The area of this rectangle is given by

$$R = \text{Height} \cdot \text{Length} = \sqrt{a}(9 - a) = 9a^{1/2} - a^{3/2}.$$

We are restricted to $0 \le a \le 9$. To maximize this area, we first set $dR/da = 0$ to find critical points:

$$\begin{aligned} \frac{dR}{da} = \frac{9}{2}a^{-1/2} - \frac{3}{2}a^{1/2} &= 0 \\ \frac{9}{2\sqrt{a}} &= \frac{3\sqrt{a}}{2} \\ 18 &= 6a \\ a &= 3. \end{aligned}$$

Notice that $R = 0$ at the endpoints $a = 0$ and $a = 9$, and R is positive between these values. Since $a = 3$ is the only critical point, the rectangle with the maximum area has length $9 - 3 = 6$ and height $\sqrt{3}$.

Example 4 A closed box has a fixed surface area A and a square base with side x.

(a) Find a formula for the volume, V, of the box as a function of x. What is the domain of V?
(b) Graph V as a function of x.
(c) Find the maximum value of V.

Solution (a) The height of the box is h, as shown in Figure 4.37. The box has six sides, four with area xh and two, the top and bottom, with area x^2. Thus,

$$4xh + 2x^2 = A.$$

So

$$h = \frac{A - 2x^2}{4x}.$$

Then, the volume, V, is given by

$$V = x^2h = x^2\left(\frac{A - 2x^2}{4x}\right) = \frac{x}{4}\left(A - 2x^2\right) = \frac{A}{4}x - \frac{1}{2}x^3.$$

Since the area of the top and bottom combined must be less than A, we have $2x^2 \le A$. Thus, the domain of V is $0 \le x \le \sqrt{A/2}$.

(b) Figure 4.38 shows the graph for $x \ge 0$. (Note that A is a positive constant.)

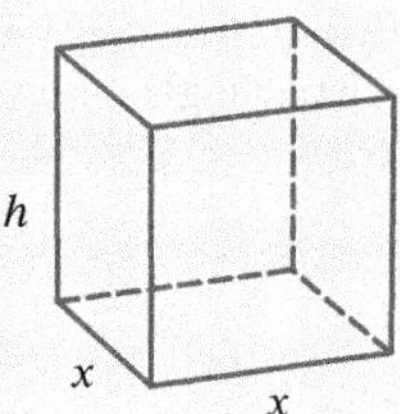

Figure 4.37: Box with base of side x, height h, surface area A, and volume V

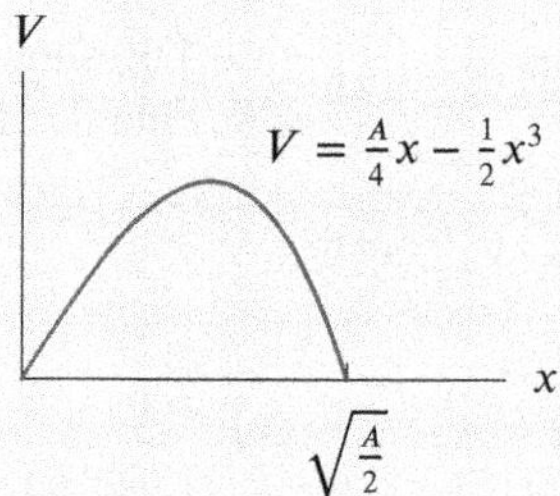

Figure 4.38: Volume, V, against length of side of base, x

(c) To find the maximum, we differentiate, regarding A as a constant:

$$\frac{dV}{dx} = \frac{A}{4} - \frac{3}{2}x^2 = 0$$

so

$$x = \pm\sqrt{\frac{A}{6}}.$$

Since $x \ge 0$ in the domain of V, we use $x = \sqrt{A/6}$. Figure 4.38 indicates that at this value of x, the volume is a maximum.

From the formula, we see that $dV/dx > 0$ for $x < \sqrt{A/6}$, so V is increasing, and that $dV/dx < 0$ for $x > \sqrt{A/6}$, so V is decreasing. Thus, $x = \sqrt{A/6}$ gives the global maximum.

Evaluating V at $x = \sqrt{A/6}$ and simplifying, we get

$$V = \frac{A}{4}\sqrt{\frac{A}{6}} - \frac{1}{2}\left(\sqrt{\frac{A}{6}}\right)^3 = \left(\frac{A}{6}\right)^{3/2}.$$

Example 5 A light is suspended at a height h above the floor. (See Figure 4.39.) The illumination at the point P is inversely proportional to the square of the distance from the point P to the light and directly proportional to the cosine of the angle θ. How far from the floor should the light be to maximize the illumination at the point P?

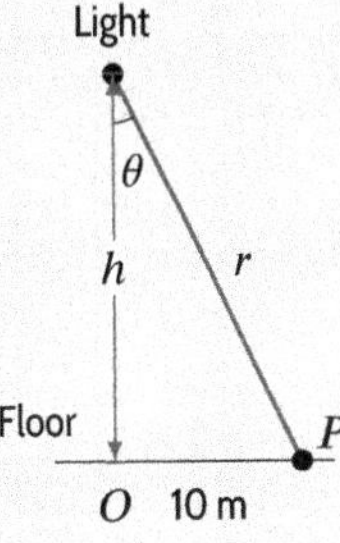

Figure 4.39: How high should the light be?

Solution If the illumination is represented by I and r is the distance from the light to the point P, then we know that for some $k \geq 0$,

$$I = \frac{k\cos\theta}{r^2}.$$

Since $r^2 = h^2 + 10^2$ and $\cos\theta = h/r = h/\sqrt{h^2+10^2}$, we have, for $h \geq 0$,

$$I = \frac{kh}{(h^2+10^2)^{3/2}}.$$

To find the height at which I is maximized, we differentiate using the quotient rule:

$$\begin{aligned}\frac{dI}{dh} &= \frac{k(h^2+10^2)^{3/2} - kh(\frac{3}{2}(h^2+10^2)^{1/2}(2h))}{[(h^2+10^2)^{3/2}]^2}\\ &= \frac{(h^2+10^2)^{1/2}[k(h^2+10^2) - 3kh^2]}{(h^2+10^2)^3}\\ &= \frac{k(h^2+10^2) - 3kh^2}{(h^2+10^2)^{5/2}}\\ &= \frac{k(10^2 - 2h^2)}{(h^2+10^2)^{5/2}}.\end{aligned}$$

Setting $dI/dh = 0$ for $h \geq 0$ gives

$$\begin{aligned}10^2 - 2h^2 &= 0\\ h &= \sqrt{50} \text{ meters.}\end{aligned}$$

Since $dI/dh > 0$ for $h < \sqrt{50}$ and $dI/dh < 0$ for $h > \sqrt{50}$, there is a local maximum at $h = \sqrt{50}$ meters. There is only one critical point, so the global maximum of I occurs at that point. Thus, the illumination is greatest if the light is suspended at a height of $\sqrt{50} \approx 7$ meters above the floor.

A Graphical Example: Minimizing Gas Consumption

Next we look at an example in which a function is given graphically and the optimum values are read from a graph. We already know how to estimate the optimum values of $f(x)$ from a graph of $f(x)$—read off the highest and lowest values. In this example, we see how to estimate the optimum value of the quantity $f(x)/x$ from a graph of $f(x)$ against x. The question we investigate is how to set driving speeds to maximize fuel efficiency.[2]

Example 6 Gas consumption, g (in gallons/hour), as a function of velocity, v (in mph), is given in Figure 4.40. What velocity minimizes the gas consumption per mile, represented by g/v?

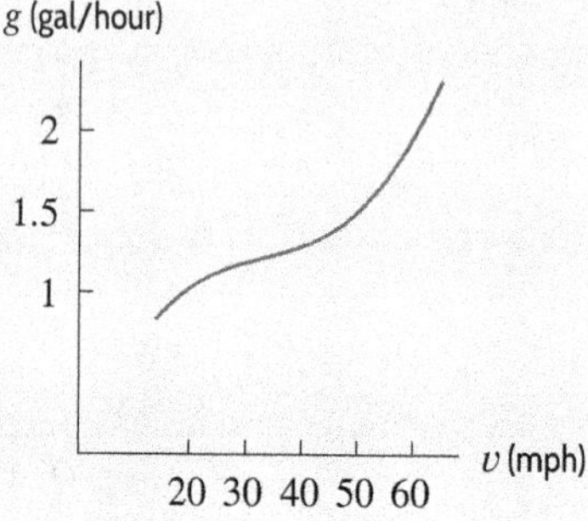

Figure 4.40: Gas consumption versus velocity

Solution Figure 4.41 shows that g/v is the slope of the line from the origin to the point P. Where on the curve should P be to make the slope a minimum? From the possible positions of the line shown in Figure 4.41, we see that the slope of the line is both a local and global minimum when the line is tangent to the curve. From Figure 4.42, we can see that the velocity at this point is about 50 mph. Thus to minimize gas consumption per mile, we should drive about 50 mph.

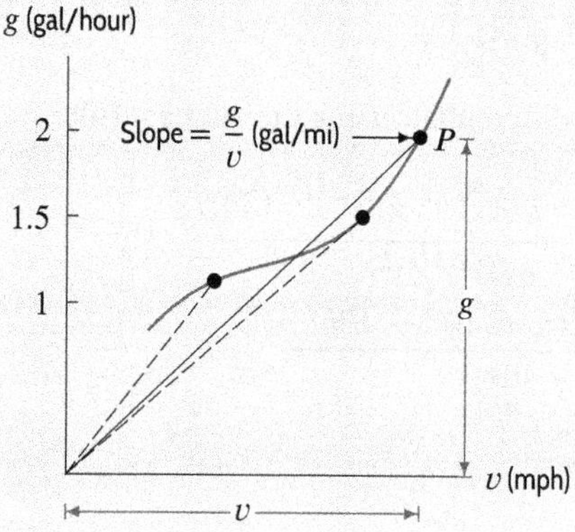

Figure 4.41: Graphical representation of gas consumption per mile, g/v

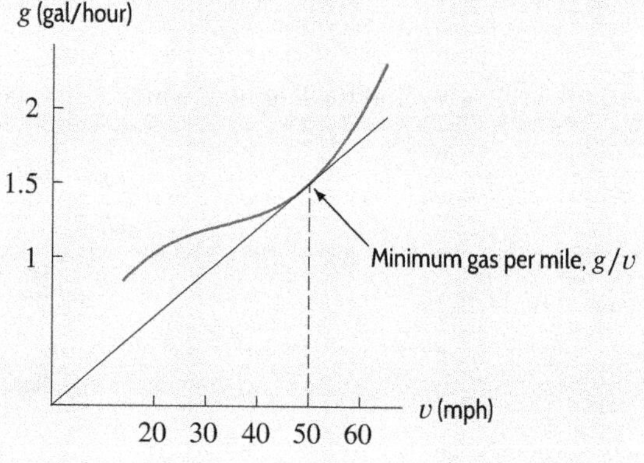

Figure 4.42: Velocity for maximum fuel efficiency

Exercises and Problems for Section 4.3

EXERCISES

1. The sum of two nonnegative numbers is 100. What is the maximum value of the product of these two numbers?

2. The product of two positive numbers is 784. What is the minimum value of their sum?

3. The sum of two times one nonnegative number and five times another is 600. What is the maximum value of the product of these two numbers?

4. The sum of three nonnegative numbers is 36, and one of the numbers is twice one of the other numbers. What is the maximum value of the product of these three numbers?

5. The perimeter of a rectangle is 64 cm. Find the lengths of the sides of the rectangle giving the maximum area.

6. If you have 100 feet of fencing and want to enclose a rectangular area up against a long, straight wall, what is the largest area you can enclose?

[2]Adapted from Peter D. Taylor, *Calculus: The Analysis of Functions* (Toronto: Wall & Emerson, 1992).

■ For the solids in Exercises 7–10, find the dimensions giving the minimum surface area, given that the volume is 8 cm^3.

7. A closed rectangular box, with a square base x by x cm and height h cm.

8. An open-topped rectangular box, with a square base x by x cm and height h cm.

9. A closed cylinder with radius r cm and height h cm.

10. A cylinder open at one end with radius r cm and height h cm.

■ In Exercises 11–12, find the x-value maximizing the shaded area. One vertex is on the graph of $f(x) = x^2/3 - 50x + 1000$, where $0 \le x \le 20$.

11.

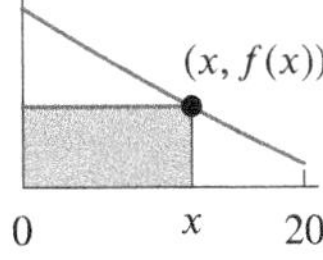

12.

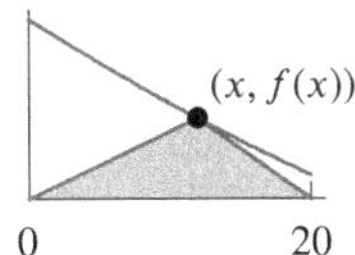

13. A rectangle has one side on the x-axis and two vertices on the curve
$$y = \frac{1}{1+x^2}.$$
Find the vertices of the rectangle with maximum area.

14. A right triangle has one vertex at the origin and one vertex on the curve $y = e^{-x/3}$ for $1 \le x \le 5$. One of the two perpendicular sides is along the x-axis; the other is parallel to the y-axis. Find the maximum and minimum areas for such a triangle.

15. A rectangle has one side on the x-axis, one side on the y-axis, one vertex at the origin and one on the curve $y = e^{-2x}$ for $x \ge 0$. Find the

(a) Maximum area **(b)** Minimum perimeter

PROBLEMS

16. Find analytically the exact critical point of the function which represents the time, t, to walk to the bus stop in Example 2. Recall that t is given by
$$t = \frac{x}{6} + \frac{\sqrt{(2000-x)^2 + 600^2}}{4}.$$

17. Of all rectangles with given area, A, which has the shortest diagonals?

18. A rectangular beam is cut from a cylindrical log of radius 30 cm. The strength of a beam of width w and height h is proportional to wh^2. (See Figure 4.43.) Find the width and height of the beam of maximum strength.

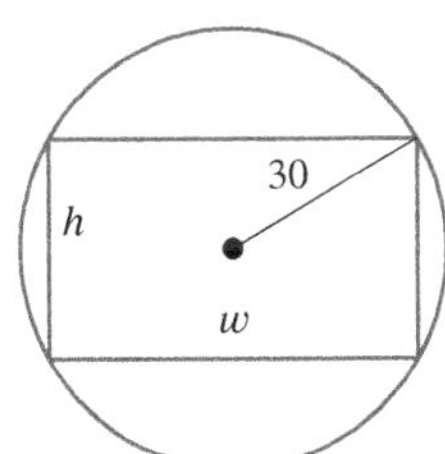

Figure 4.43

■ In Problems 19–20 a vertical line divides a region into two pieces. Find the value of the coordinate x that maximizes the product of the two areas.

19. 20.

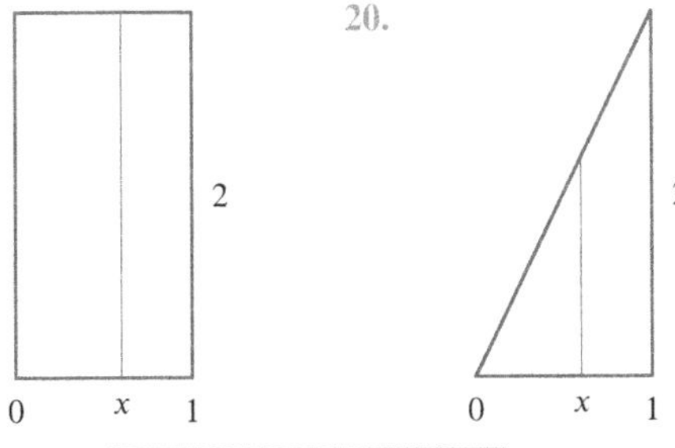

■ In Problems 21–23 the figures are made of rectangles and semicircles.

(a) Find a formula for the area.

(b) Find a formula for the perimeter.

(c) Find the dimensions x and y that maximize the area given that the perimeter is 100.

21.

22.

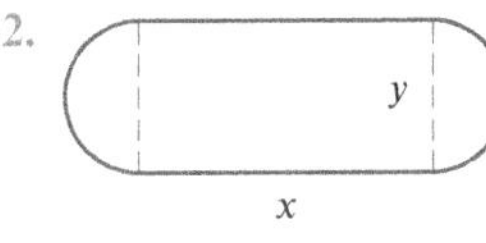

23.

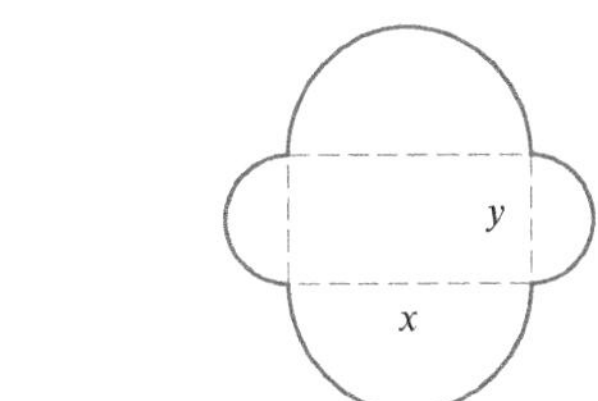

24. A piece of wire of length L cm is cut into two pieces. One piece, of length x cm, is made into a circle; the rest is made into a square.

(a) Find the value of x that makes the sum of the areas of the circle and square a minimum. Find the value of x giving a maximum.

(b) For the values of x found in part (a), show that the ratio of the length of wire in the square to the length of wire in the circle is equal to the ratio of the area of the square to the area of the circle.[3]

(c) Are the values of x found in part (a) the only values of x for which the ratios in part (b) are equal?

[3]From Sally Thomas.

■ In Problems 25–28, find the minimum and maximum values of the expression where x and y are lengths in Figure 4.44 and $0 \le x \le 10$.

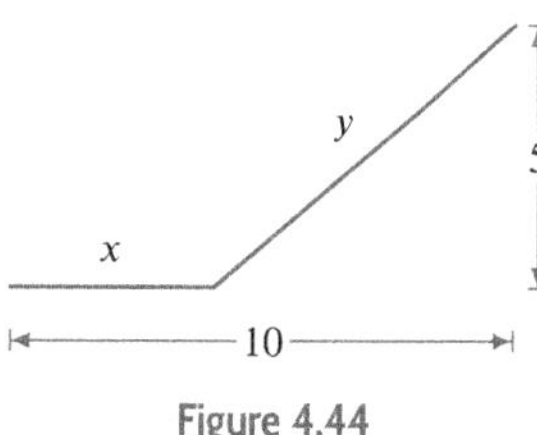

Figure 4.44

25. x 26. y 27. $x + 2y$ 28. $2x + y$

29. Which point on the curve $y = \sqrt{1 - x}$ is closest to the origin?

30. Which point on the curve $y = \sqrt{x + 2}$ is closest to the origin? What is the minimum distance from the origin?

31. Which point on the curve $y = \sqrt{5x + 15}$ is closest to the origin? What is the minimum distance from the origin?

32. Find the point(s) on the ellipse

$$\frac{x^2}{9} + y^2 = 1$$

(a) Closest to the point $(2, 0)$.
(b) Closest to the focus $(\sqrt{8}, 0)$.

[Hint: Minimize the square of the distance—this avoids square roots.]

33. What are the dimensions of the closed cylindrical can that has surface area 280 square centimeters and contains the maximum volume?

34. A hemisphere of radius 1 sits on a horizontal plane. A cylinder stands with its axis vertical, the center of its base at the center of the sphere, and its top circular rim touching the hemisphere. Find the radius and height of the cylinder of maximum volume.

35. A smokestack deposits soot on the ground with a concentration inversely proportional to the square of the distance from the stack. With two smokestacks 20 miles apart, the concentration of the combined deposits on the line joining them, at a distance x from one stack, is given by

$$S = \frac{k_1}{x^2} + \frac{k_2}{(20 - x)^2}$$

where k_1 and k_2 are positive constants which depend on the quantity of smoke each stack is emitting. If $k_1 = 7k_2$, find the point on the line joining the stacks where the concentration of the deposit is a minimum.

36. In a chemical reaction, substance A combines with substance B to form substance Y. At the start of the reaction, the quantity of A present is a grams, and the quantity of B present is b grams. At time t seconds after the start of the reaction, the quantity of Y present is y grams. Assume $a < b$ and $y \le a$. For certain types of reactions, the rate of the reaction, in grams/sec, is given by

$$\text{Rate} = k(a - y)(b - y), \quad k \text{ is a positive constant.}$$

(a) For what values of y is the rate nonnegative? Graph the rate against y.
(b) Use your graph to find the value of y at which the rate of the reaction is fastest.

37. A wave of wavelength λ traveling in deep water has speed, v, given for positive constants c and k, by

$$v = k\sqrt{\frac{\lambda}{c} + \frac{c}{\lambda}}$$

As λ varies, does such a wave have a maximum or minimum velocity? If so, what is it? Explain.

38. A circular ring of wire of radius r_0 lies in a plane perpendicular to the x-axis and is centered at the origin. The ring has a positive electric charge spread uniformly over it. The electric field in the x-direction, E, at the point x on the axis is given by

$$E = \frac{kx}{\left(x^2 + r_0^2\right)^{3/2}} \quad \text{for} \quad k > 0.$$

At what point on the x-axis is the field greatest? Least?

39. A woman pulls a sled which, together with its load, has a mass of m kg. If her arm makes an angle of θ with her body (assumed vertical) and the coefficient of friction (a positive constant) is μ, the least force, F, she must exert to move the sled is given by

$$F = \frac{mg\mu}{\sin\theta + \mu\cos\theta}.$$

If $\mu = 0.15$, find the maximum and minimum values of F for $0 \le \theta \le \pi/2$. Give answers as multiples of mg.

40. Four equally massive particles can be made to rotate, equally spaced, around a circle of radius r. This is physically possible provided the radius and period T of the rotation are chosen so that the following *action* function is at its global minimum:

$$A(r) = \frac{r^2}{T} + \frac{T}{r}, \quad r > 0.$$

(a) Find the radius r at which $A(r)$ has a global minimum.
(b) If the period of the rotation is doubled, determine whether the radius of the rotation increases or decreases, and by approximately what percentage.

41. You run a small furniture business. You sign a deal with a customer to deliver up to 400 chairs, the exact number to be determined by the customer later. The price will be \$90 per chair up to 300 chairs, and above 300, the price will be reduced by \$0.25 per chair (on the whole order) for every additional chair over 300 ordered. What are the largest and smallest revenues your company can make under this deal?

42. The cost of fuel to propel a boat through the water (in dollars per hour) is proportional to the cube of the speed. A certain ferry boat uses \$100 worth of fuel per hour when cruising at 10 miles per hour. Apart from fuel, the cost of running this ferry (labor, maintenance, and so on) is \$675 per hour. At what speed should it travel so as to minimize the cost *per mile* traveled?

43. A business sells an item at a constant rate of r units per month. It reorders in batches of q units, at a cost of $a + bq$ dollars per order. Storage costs are k dollars per item per month, and, on average, $q/2$ items are in storage, waiting to be sold. [Assume r, a, b, k are positive constants.]

(a) How often does the business reorder?
(b) What is the average monthly cost of reordering?
(c) What is the total monthly cost, C of ordering and storage?
(d) Obtain Wilson's lot size formula, the optimal batch size which minimizes cost.

44. A bird such as a starling feeds worms to its young. To collect worms, the bird flies to a site where worms are to be found, picks up several in its beak, and flies back to its nest. The *loading curve* in Figure 4.45 shows how the number of worms (the load) a starling collects depends on the time it has been searching for them.[4] The curve is concave down because the bird can pick up worms more efficiently when its beak is empty; when its beak is partly full, the bird becomes much less efficient. The traveling time (from nest to site and back) is represented by the distance PO in Figure 4.45. The bird wants to maximize the rate at which it brings worms to the nest, where

$$\text{Rate worms arrive} = \frac{\text{Load}}{\text{Traveling time} + \text{Searching time}}$$

(a) Draw a line in Figure 4.45 whose slope is this rate.
(b) Using the graph, estimate the load which maximizes this rate.
(c) If the traveling time is increased, does the optimal load increase or decrease? Why?

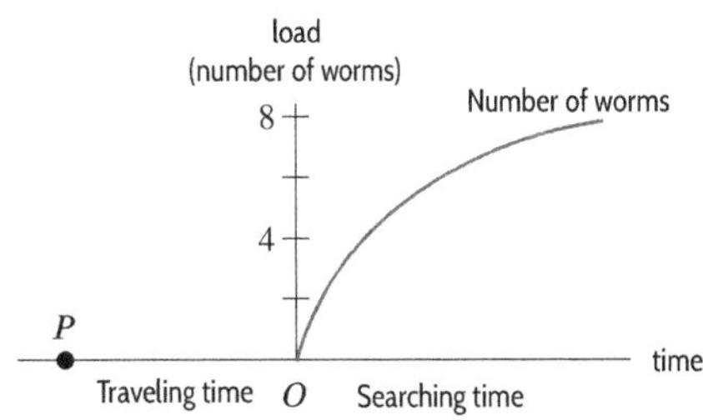

Figure 4.45

45. On the same side of a straight river are two towns, and the townspeople want to build a pumping station, S. See Figure 4.46. The pumping station is to be at the river's edge with pipes extending straight to the two towns. Where should the pumping station be located to minimize the total length of pipe?

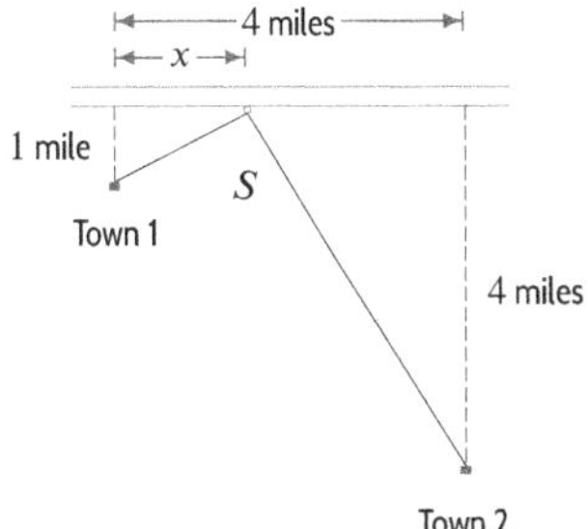

Figure 4.46

46. A pigeon is released from a boat (point B in Figure 4.47) floating on a lake. Because of falling air over the cool water, the energy required to fly one meter over the lake is twice the corresponding energy e required for flying over the bank ($e = 3$ joule/meter). To minimize the energy required to fly from B to the loft, L, the pigeon heads to a point P on the bank and then flies along the bank to L. The distance $\overline{AL}$ is 2000 m, and $\overline{AB}$ is 500 m.

(a) Express the energy required to fly from B to L via P as a function of the angle θ (the angle BPA).
(b) What is the optimal angle θ?
(c) Does your answer change if $\overline{AL}$, $\overline{AB}$, and e have different numerical values?

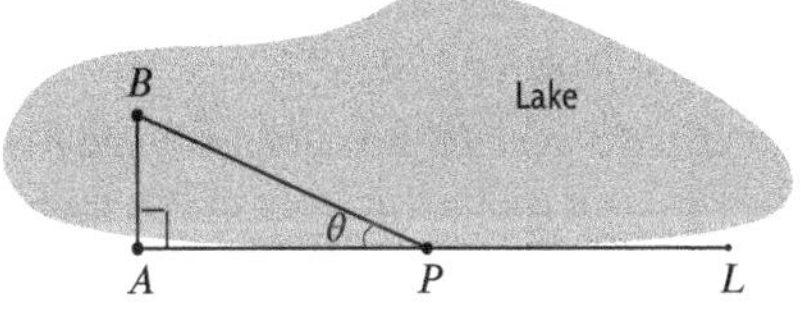

Figure 4.47

[4]Alex Kacelnick (1984). Reported by J. R. Krebs and N. B. Davies, *An Introduction to Behavioural Ecology* (Oxford: Blackwell, 1987).

47. To get the best view of the Statue of Liberty in Figure 4.48, you should be at the position where θ is a maximum. If the statue stands 92 meters high, including the pedestal, which is 46 meters high, how far from the base should you be? [Hint: Find a formula for θ in terms of your distance from the base. Use this function to maximize θ, noting that $0 \le \theta \le \pi/2$.]

Figure 4.48

48. A light ray starts at the origin and is reflected off a mirror along the line $y = 1$ to the point $(2, 0)$. See Figure 4.49. Fermat's Principle says that light's path minimizes the time of travel.[5] The speed of light is a constant.

 (a) Using Fermat's Principle, find the optimal position of P.
 (b) Using your answer to part (a), derive the Law of Reflection, that $\theta_1 = \theta_2$.

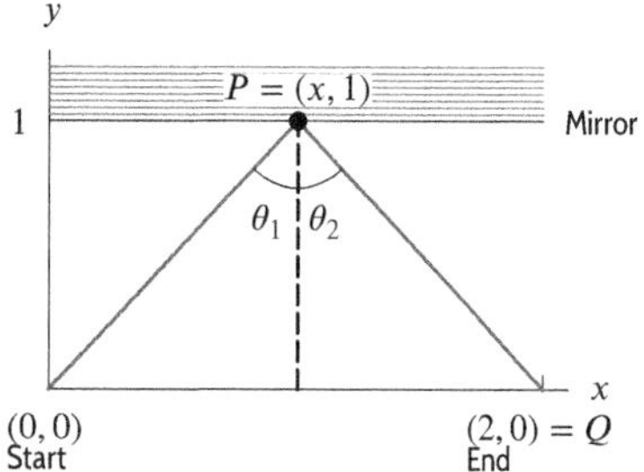

Figure 4.49

49. (a) For which positive number x is $x^{1/x}$ largest? Justify your answer.
 [Hint: You may want to write $x^{1/x} = e^{\ln(x^{1/x})}$.]
 (b) For which positive integer n is $n^{1/n}$ largest? Justify your answer.
 (c) Use your answer to parts (a) and (b) to decide which is larger: $3^{1/3}$ or $\pi^{1/\pi}$.

50. The *arithmetic mean* of two numbers a and b is defined as $(a + b)/2$; the *geometric mean* of two positive numbers a and b is defined as $\sqrt{ab}$.

 (a) For two positive numbers, which of the two means is larger? Justify your answer.
 [Hint: Define $f(x) = (a+x)/2 - \sqrt{ax}$ for fixed a.]
 (b) For three positive numbers a, b, c, the arithmetic and geometric mean are $(a + b + c)/3$ and $\sqrt[3]{abc}$, respectively. Which of the two means of three numbers is larger? [Hint: Redefine $f(x)$ for fixed *a and b*.]

51. A line goes through the origin and a point on the curve $y = x^2e^{-3x}$, for $x \ge 0$. Find the maximum slope of such a line. At what x-value does it occur?

52. The distance, s, traveled by a cyclist who starts at 1 pm is given in Figure 4.50. Time, t, is in hours since noon.

 (a) Explain why the quantity s/t is represented by the slope of a line from the origin to the point (t, s) on the graph.
 (b) Estimate the time at which the quantity s/t is a maximum.
 (c) What is the relationship between the quantity s/t and the instantaneous speed of the cyclist at the time you found in part (b)?

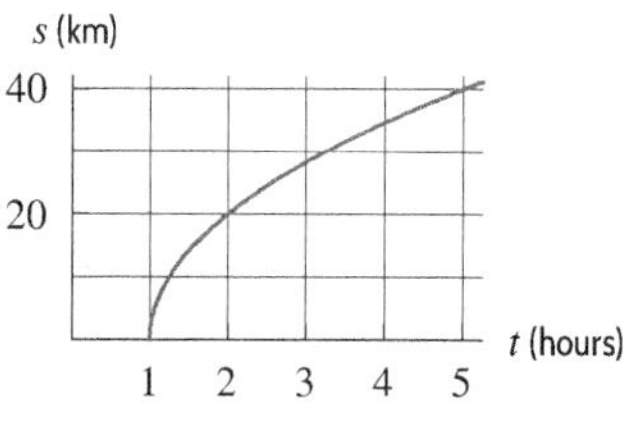

Figure 4.50

53. When birds lay eggs, they do so in clutches of several at a time. When the eggs hatch, each clutch gives rise to a brood of baby birds. We want to determine the clutch size which maximizes the number of birds surviving to adulthood per brood. If the clutch is small, there are few baby birds in the brood; if the clutch is large, there are so many baby birds to feed that most die of starvation. The number of surviving birds per brood as a function of clutch size is shown by the benefit curve in Figure 4.51.[6]

 (a) Estimate the clutch size which maximizes the number of survivors per brood.
 (b) Suppose also that there is a biological cost to having a larger clutch: the female survival rate is reduced by large clutches. This cost is represented by the dotted line in Figure 4.51. If we take cost into account by assuming that the optimal clutch size in fact maximizes the vertical distance between the curves, what is the new optimal clutch size?

[5] http://en.wikipedia.org, accessed April 10, 2015.
[6] Data from C. M. Perrins and D. Lack, reported by J. R. Krebs and N. B. Davies in *An Introduction to Behavioural Ecology* (Oxford: Blackwell, 1987).

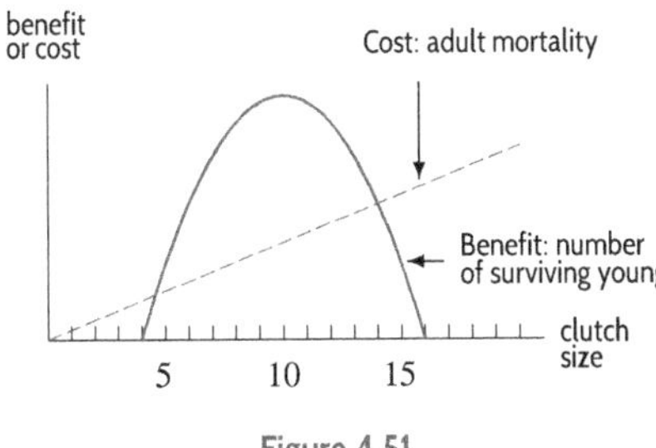

Figure 4.51

54. Let $f(v)$ be the amount of energy consumed by a flying bird, measured in joules per second (a joule is a unit of energy), as a function of its speed v (in meters/sec). Let $a(v)$ be the amount of energy consumed by the same bird, measured in joules per meter.

 (a) Suggest a reason in terms of the way birds fly for the shape of the graph of $f(v)$ in Figure 4.52.
 (b) What is the relationship between $f(v)$ and $a(v)$?
 (c) Where on the graph is $a(v)$ a minimum?
 (d) Should the bird try to minimize $f(v)$ or $a(v)$ when it is flying? Why?

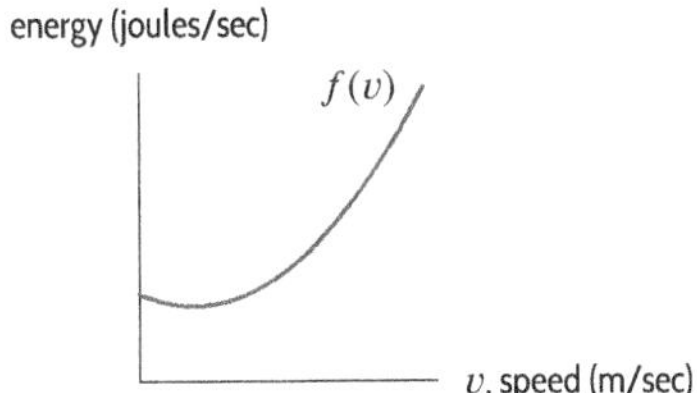

Figure 4.52

55. The forward motion of an aircraft in level flight is reduced by two kinds of forces, known as *induced drag* and *parasite drag*. Induced drag is a consequence of the downward deflection of air as the wings produce lift. Parasite drag results from friction between the air and the entire surface of the aircraft. Induced drag is inversely proportional to the square of speed and parasite drag is directly proportional to the square of speed. The sum of induced drag and parasite drag is called total drag. The graph in Figure 4.53 shows a certain aircraft's induced drag and parasite drag functions.

 (a) Sketch the total drag as a function of air speed.
 (b) Estimate two different air speeds which each result in a total drag of 1000 pounds. Does the total drag function have an inverse? What about the induced and parasite drag functions?
 (c) Fuel consumption (in gallons per hour) is roughly proportional to total drag. Suppose you are low on fuel and the control tower has instructed you to enter a circular holding pattern of indefinite duration to await the passage of a storm at your landing field. At what air speed should you fly the holding pattern? Why?

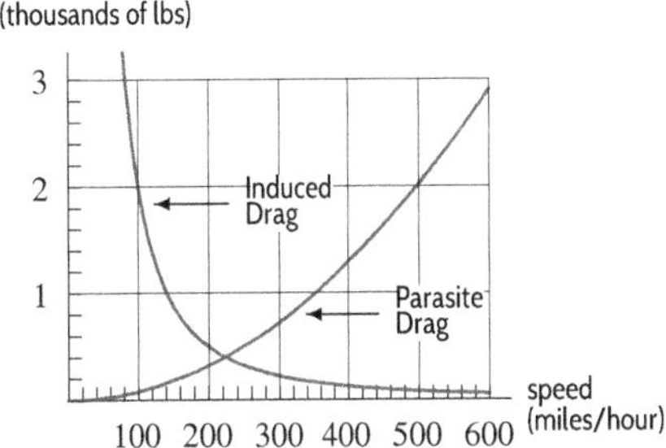

Figure 4.53

56. Let $f(v)$ be the fuel consumption, in gallons per hour, of a certain aircraft as a function of its airspeed, v, in miles per hour. A graph of $f(v)$ is given in Figure 4.54.

 (a) Let $g(v)$ be the fuel consumption of the same aircraft, but measured in gallons per mile instead of gallons per hour. What is the relationship between $f(v)$ and $g(v)$?
 (b) For what value of v is $f(v)$ minimized?
 (c) For what value of v is $g(v)$ minimized?
 (d) Should a pilot try to minimize $f(v)$ or $g(v)$?

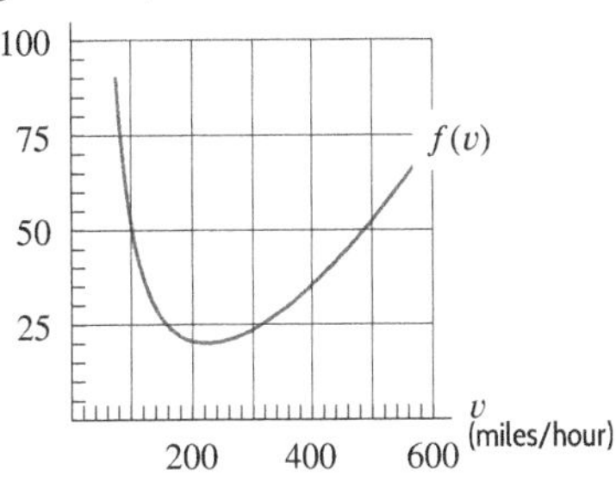

Figure 4.54

■ Problems 57–58 use the fact that a physical system is in stable equilibrium if the total energy, E, is a local minimum.[7]

57. A mass m hanging on the end of a spring extends its length by y. See Figure 4.55. For g, the acceleration due to gravity, and positive constant k, the total energy is

$$E = \frac{1}{2}ky^2 - mgy.$$

Is there a length that gives a stable equilibrium position for a constant mass m? If so, what is it?

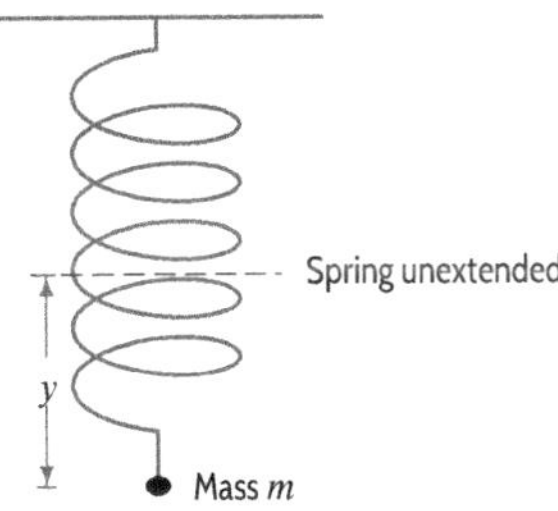

Figure 4.55

[7]Based on J. Meriam and L. Kraige, *Engineering Mechanics: Statics* (New York: Wiley, 1992)

58. The top of a rod of mass m and length l slides vertically, while the bottom end is attached to a spring and only slides horizontally. See Figure 4.56, where $0 \leq \theta \leq \pi/2$. For g, the acceleration due to gravity, and positive constant k, the total energy in terms of the angle θ is given by

$$E = \frac{1}{2}kl^2 \sin^2\theta + \frac{1}{2}mgl\cos\theta.$$

(a) Find the critical points for E.

(b) Show that if $k > mg/(2l)$ there is a stable equilibrium at $\theta = 0$.

(c) Show that if $\cos\theta = mg/(2kl)$, then

$$\frac{d^2E}{d\theta^2} = kl^2\left(\left(\frac{mg}{2kl}\right)^2 - 1\right).$$

What does this tell you about a stable equilibrium at this point?

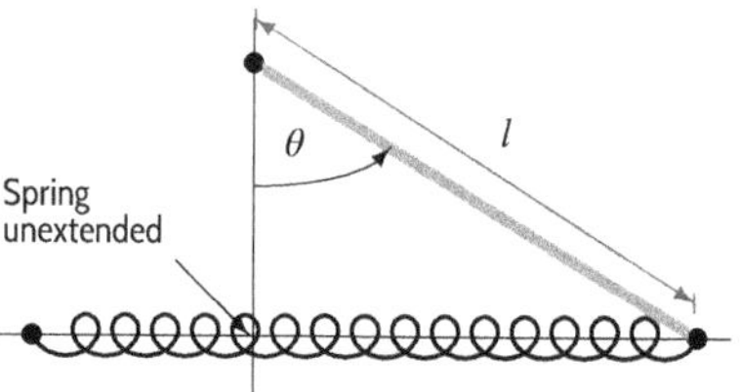

Figure 4.56

Strengthen Your Understanding

In Problems 59–61, explain what is wrong with the statement.

59. If A is the area of a rectangle of sides x and $2x$, for $0 \leq x \leq 10$, the maximum value of A occurs where $dA/dx = 0$.

60. An open box is made from a 20-inch-square piece of cardboard by cutting squares of side h from the corners and folding up the edges, giving the box in Figure 4.57. To find the maximum volume of such a box, we work on the domain $h \geq 0$.

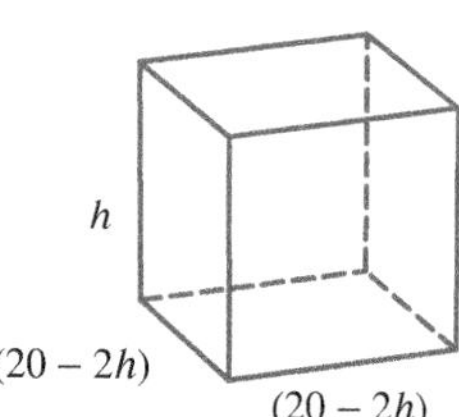

Figure 4.57: Box of volume $V = h(20 - 2h)^2$

61. The solution of an optimization problem modeled by a quadratic function occurs at the vertex of the quadratic.

In Problems 62–64, give an example of:

62. The sides of a rectangle with perimeter 20 cm and area smaller than 10 cm^2.

63. A context for a modeling problem where you are given that $xy = 120$ and you are minimizing the quantity $2x + 6y$.

64. A modeling problem where you are minimizing the cost of the material in a cylindrical can of volume 250 cubic centimeters.

4.4 FAMILIES OF FUNCTIONS AND MODELING

In Chapter 1, we saw that the graph of one function can tell us about the graphs of many others. The graph of $y = x^2$ tells us, indirectly, about the graphs of $y = x^2 + 2$, $y = (x + 2)^2$, $y = 2x^2$, and countless other functions. We say that all functions of the form $y = a(x + b)^2 + c$ form a *family of functions*; their graphs are similar to that of $y = x^2$, except for shifts and stretches determined by the values of a, b, and c. The constants a, b, c are called *parameters*. Different values of the parameters give different members of the family.

The Bell-Shaped Curve: $y = e^{-(x-a)^2/b}$

The family of bell-shaped curves includes the family of *normal density* functions, used in probability and statistics.[8] We assume that $b > 0$. See Section 8.8 for applications of the normal distribution. First we let $b = 1$ and examine the role of a.

[8]Probabilists divide our function by a constant, $\sqrt{\pi b}$, to get the normal density.

Example 1 Graph $y = e^{-(x-a)^2}$ for $a = -2, 0, 2$ and explain the role of a in the shape of the graph.

Solution See Figure 4.58. The role of the parameter a is to shift the graph of $y = e^{-x^2}$ to the right or left. Notice that the value of y is always positive. Since $y \to 0$ as $x \to \pm\infty$, the x-axis is a horizontal asymptote. Thus $y = e^{-(x-a)^2}$ is the family of horizontal shifts of the bell-shaped curve $y = e^{-x^2}$.

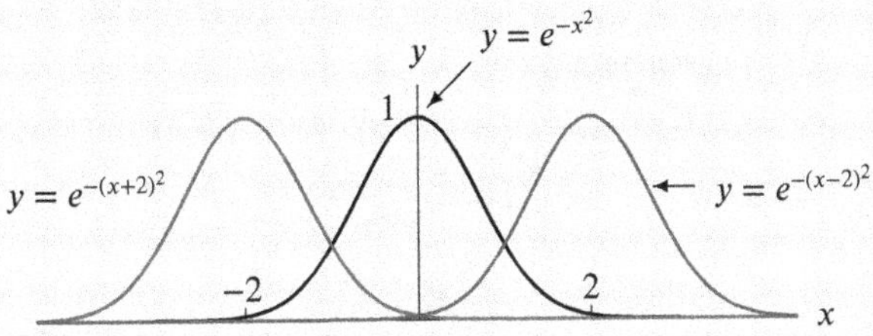

Figure 4.58: Bell-shaped curves with center a: Family $y = e^{-(x-a)^2}$

We now consider the role of the parameter b by studying the family with $a = 0$.

Example 2 Find the critical points and points of inflection of $y = e^{-x^2/b}$.

Solution To investigate the critical points and points of inflection, we calculate

$$\frac{dy}{dx} = -\frac{2x}{b}e^{-x^2/b}$$

and, using the product rule, we get

$$\frac{d^2y}{dx^2} = -\frac{2}{b}e^{-x^2/b} - \frac{2x}{b}\left(-\frac{2x}{b}e^{-x^2/b}\right) = \frac{2}{b}\left(\frac{2x^2}{b} - 1\right)e^{-x^2/b}.$$

Critical points occur where $dy/dx = 0$, that is, where

$$\frac{dy}{dx} = -\frac{2x}{b}\, e^{-x^2/b} = 0.$$

Since $e^{-x^2/b}$ is never zero, the only critical point is $x = 0$. At that point, $y = 1$ and $d^2y/dx^2 < 0$. Hence, by the second-derivative test, there is a local maximum at $x = 0$, and this is also a global maximum.

Inflection points occur where the second derivative changes sign; thus, we start by finding values of x for which $d^2y/dx^2 = 0$. Since $e^{-x^2/b}$ is never zero, $d^2y/dx^2 = 0$ when

$$\frac{2x^2}{b} - 1 = 0.$$

Solving for x gives

$$x = \pm\sqrt{\frac{b}{2}}.$$

Looking at the expression for d^2y/dx^2, we see that d^2y/dx^2 is negative for $x = 0$, and positive as $x \to \pm\infty$. Therefore the concavity changes at $x = -\sqrt{b/2}$ and at $x = \sqrt{b/2}$, so we have inflection points at $x = \pm\sqrt{b/2}$.

Returning to the two-parameter family $y = e^{-(x-a)^2/b}$, we conclude that there is a maximum at $x = a$, obtained by horizontally shifting the maximum at $x = 0$ of $y = e^{-x^2/b}$ by a units. There are inflection points at $x = a \pm \sqrt{b/2}$ obtained by shifting the inflection points $x = \pm\sqrt{b/2}$ of $y = e^{-x^2/b}$ by a units. (See Figure 4.59.) At the inflection points $y = e^{-1/2} \approx 0.6$.

With this information we can see the effect of the parameters. The parameter a determines the location of the center of the bell and the parameter b determines how narrow or wide the bell is. (See Figure 4.60.) If b is small, then the inflection points are close to a and the bell is sharply peaked near a; if b is large, the inflection points are farther away from a and the bell is spread out.

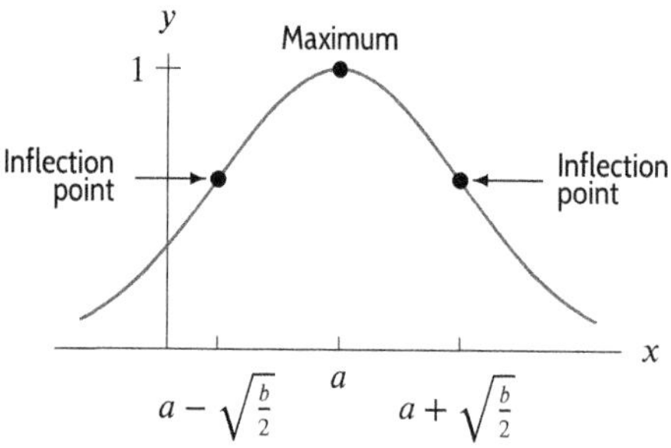

Figure 4.59: Graph of $y = e^{-(x-a)^2/b}$: bell-shaped curve with peak at $x = a$

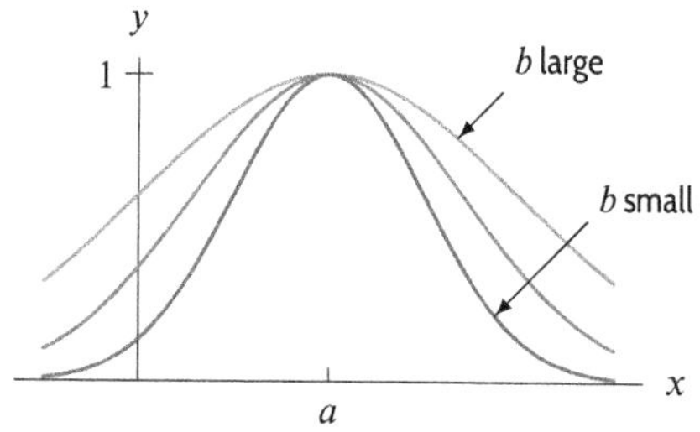

Figure 4.60: Graph of $y = e^{-(x-a)^2/b}$ for fixed a and various b

Modeling with Families of Functions

One reason for studying families of functions is their use in mathematical modeling. Confronted with the problem of modeling some phenomenon, a crucial first step involves recognizing families of functions which might fit the available data.

Motion Under Gravity: $y = -4.9t^2 + v_0 t + y_0$

The position of an object moving vertically under the influence of gravity can be described by a function in the two-parameter family

$$y = -4.9t^2 + v_0 t + y_0$$

where t is time in seconds and y is the distance in meters above the ground. Why do we need the parameters v_0 and y_0? Notice that at time $t = 0$ we have $y = y_0$. Thus the parameter y_0 gives the height above ground of the object at time $t = 0$. Since $dy/dt = -9.8t + v_0$, the parameter v_0 gives the velocity of the object at time $t = 0$. From this equation we see that $dy/dt = 0$ when $t = v_0/9.8$. This is the time when the object reaches its maximum height.

Example 3 Give a function describing the position of a projectile launched upward from ground level with an initial velocity of 50 m/sec. How high does the projectile rise?

Solution We have $y_0 = 0$ and $v_0 = 50$, so the height of the projectile after t seconds is $y = -4.9t^2 + 50t$. It reaches its maximum height when $t = 50/9.8 = 5.1$ seconds, and its height at that time is $-4.9(5.1)^2 + 50(5.1) = 127.5$, or about 128 meters.

Exponential Model with a Limit: $y = a(1 - e^{-bx})$

We consider $a, b > 0$. The graph of one member, with $a = 2$ and $b = 1$, is in Figure 4.61. Such a graph represents a quantity which is increasing but leveling off. For example, a body dropped in a thick fluid speeds up initially, but its velocity levels off as it approaches a terminal velocity. Similarly, if a pollutant pouring into a lake builds up toward a saturation level, its concentration may be described in this way. The graph also represents the temperature of an object in an oven.

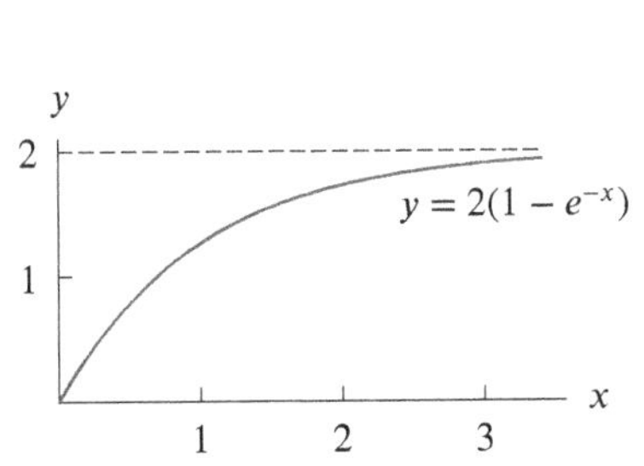

Figure 4.61: One member of the family $y = a(1 - e^{-bx})$, with $a = 2, b = 1$

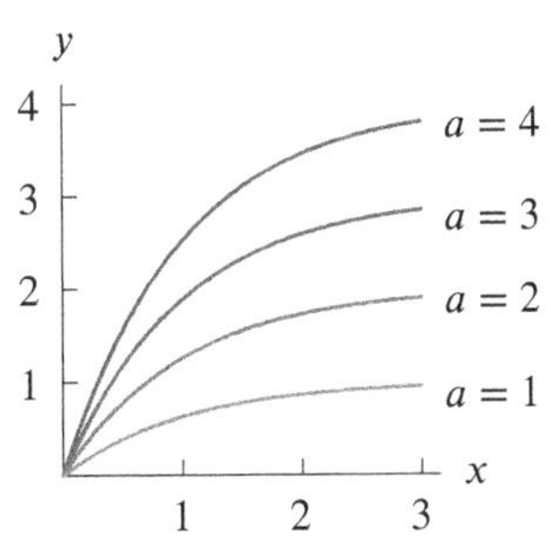

Figure 4.62: Fixing $b = 1$ gives $y = a(1 - e^{-x})$, graphed for various a

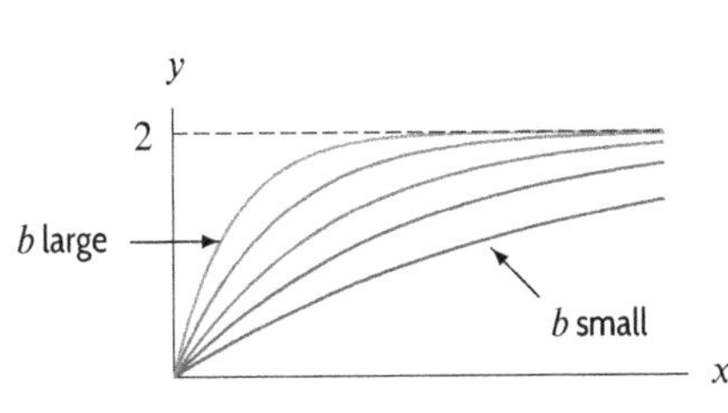

Figure 4.63: Fixing $a = 2$ gives $y = 2(1 - e^{-bx})$, graphed for various b

Example 4 Describe the effect of varying the parameters a and b on the graph of $y = a(1 - e^{-bx})$.

Solution First examine the effect on the graph of varying a. Fix b at some positive number, say $b = 1$. Substitute different values for a and look at the graphs in Figure 4.62. We see that as x gets larger, y approaches a from below, so a is an upper bound for this function. Analytically, this follows from the fact that $e^{-bx} \to 0$ as $x \to \infty$. Physically, the value of a represents the terminal velocity of a falling body or the saturation level of the pollutant in the lake.

Now examine the effect of varying b on the graph. Fix a at some positive number, say $a = 2$. Substitute different values for b and look at the graphs in Figure 4.63. The parameter b determines how sharply the curve rises and how quickly it gets close to the line $y = a$.

Let's confirm the last observation in Example 4 analytically. For $y = a(1 - e^{-bx})$, we have $dy/dx = abe^{-bx}$, so the slope of the tangent to the curve at $x = 0$ is ab. For larger b, the curve rises more rapidly at $x = 0$. How long does it take the curve to climb halfway up from $y = 0$ to $y = a$? When $y = a/2$, we have

$$a(1 - e^{-bx}) = \frac{a}{2}, \qquad \text{which leads to} \qquad x = \frac{\ln 2}{b}.$$

If b is large then $(\ln 2)/b$ is small, so in a short distance the curve is already half way up to a. If b is small, then $(\ln 2)/b$ is large and we have to go a long way out to get up to $a/2$. See Figure 4.64.

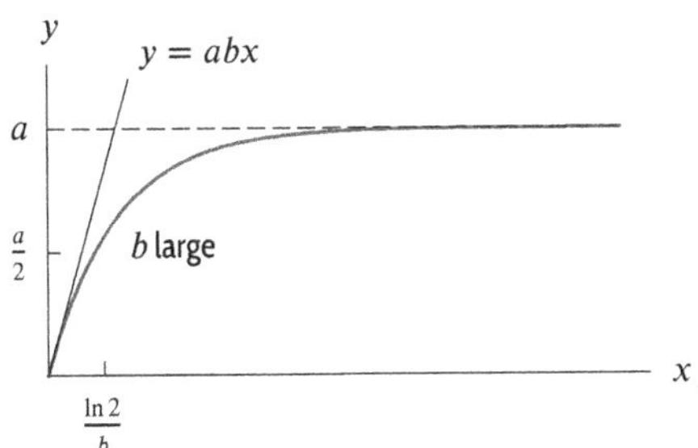

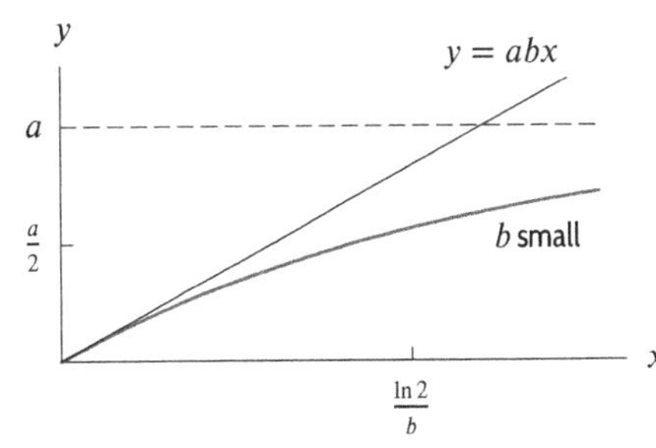

Figure 4.64: Tangent at $x = 0$ to $y = a(1 - e^{-bx})$, with fixed a, and large and small b

The following example illustrates an application of this family.

Example 5 The number, N, of people who have heard a rumor spread by mass media by time t is modeled by

$$N(t) = a(1 - e^{-bt}).$$

There are 200,000 people in the population who hear the rumor eventually. If 10% of them heard it the first day, find a and b, assuming t is measured in days.

Solution Since $\lim_{t\to\infty} N(t) = a$, we have $a = 200{,}000$. When $t = 1$, we have $N = 0.1(200{,}000) = 20{,}000$ people, so substituting into the formula gives

$$N(1) = 20{,}000 = 200{,}000\left(1 - e^{-b(1)}\right).$$

Solving for b gives

$$\begin{aligned} 0.1 &= 1 - e^{-b} \\ e^{-b} &= 0.9 \\ b &= -\ln 0.9 = 0.105. \end{aligned}$$

The Logistic Model: $y = L/(1 + Ae^{-kt})$

The *logistic* family models the growth of a population limited by the environment. (See Section 11.7.) We assume that $L, A, k > 0$ and we look at the roles of each of the three parameters in turn.

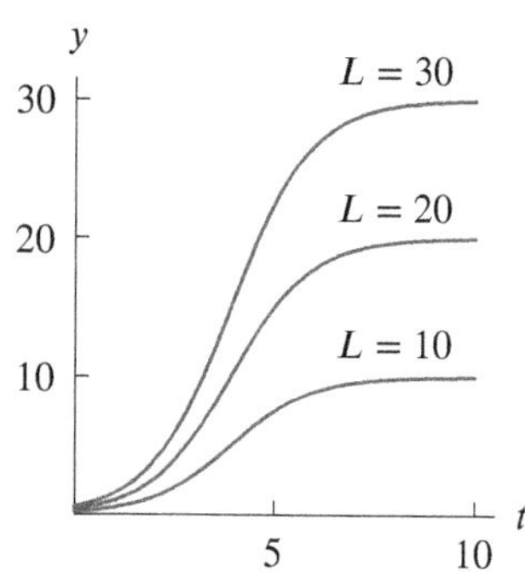

Figure 4.65: Varying L: Graph of $y = L/(1 + Ae^{-kt})$ with $A = 50$, $k = 1$

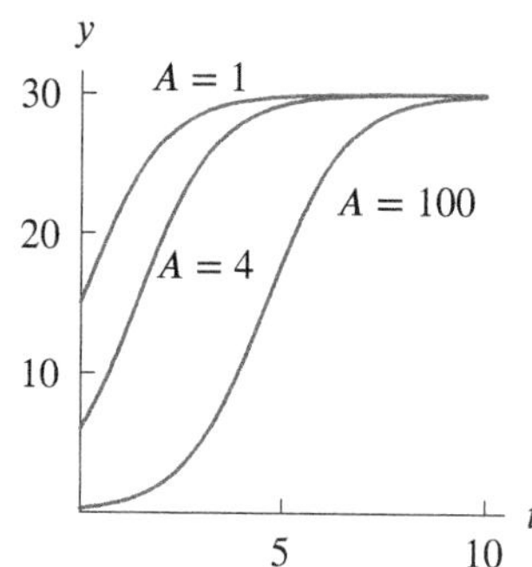

Figure 4.66: Varying A: Graph of $y = L/(1 + Ae^{-kt})$ with $L = 30$, $k = 1$

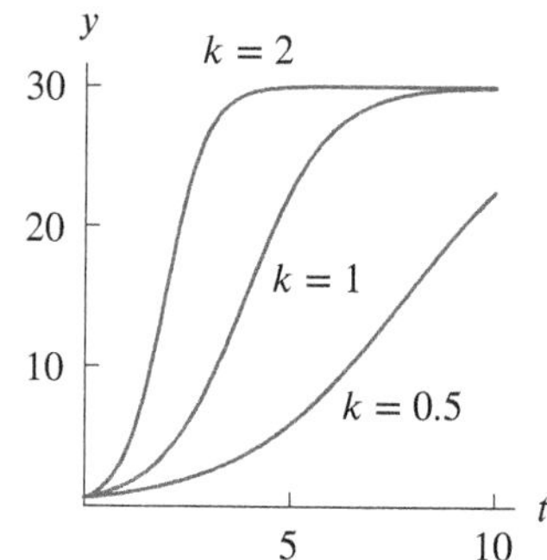

Figure 4.67: Varying k: Graph of $y = L/(1 + Ae^{-kt})$ with $L = 30$, $A = 50$

Logistic curves with varying values of L are shown in Figure 4.65. The values of y level off as $t \to \infty$ because $Ae^{-kt} \to 0$ as $t \to \infty$. Thus, as t increases, the values of y approach L. The line $y = L$ is a horizontal asymptote, called the *limiting value* or *carrying capacity*, and representing the maximum sustainable population. The parameter L stretches or shrinks the graph vertically.

In Figure 4.66 we investigate the effect of the parameter A, with k and L fixed. The parameter A alters the point at which the curve cuts the y-axis—larger values of A move the y-intercept closer to the origin. At $t = 0$ we have $y = L/(1 + A)$, confirming that increasing A decreases the value of y at $t = 0$.

Figure 4.67 shows the effect of the parameter k. With L and A fixed, we see that varying k affects the rate at which the function approaches the limiting value L. If k is small, the graph rises slowly; if k is large, the graph rises steeply. At $t = 0$, we have $dy/dt = LAk/(1 + A)^2$, so the initial slope of a logistic curve is proportional to k.

The graphs suggest that none of the curves has a critical point for $t > 0$. Some curves appear to have a point of inflection; others have none. To investigate, we take derivatives:

$$\frac{dy}{dt} = \frac{LAke^{-kt}}{(1 + Ae^{-kt})^2}.$$

Since every factor of dy/dt is positive, the first derivative is always positive. Thus, there are no critical points and the function is always increasing.

Using a computer algebra system or the quotient rule, we find

$$\frac{d^2y}{dt^2} = \frac{LAk^2e^{-kt}(-1 + Ae^{-kt})}{(1 + Ae^{-kt})^3}.$$

Since L, A, k, e^{-kt}, and the denominator are always positive, the sign of d^2y/dt^2 is determined by the sign of $(-1+Ae^{-kt})$. Points of inflection may occur where $d^2y/dt^2 = 0$. This is where $-1+Ae^{-kt} = 0$, or

$$Ae^{-kt} = 1.$$

At this value of t,

$$y = \frac{L}{1 + Ae^{-kt}} = \frac{L}{1+1} = \frac{L}{2}.$$

In Problem 46 on page 232, we see that d^2y/dt^2 changes sign at $y = L/2$. Since the concavity changes at $y = L/2$, there is a point of inflection when the population is half the carrying capacity. If the initial population is $L/2$ or above, there is no inflection point. (See the top graph in Figure 4.66.)

To find the value of t at the inflection point, we solve for t in the equation

$$Ae^{-kt} = 1$$
$$t = \frac{\ln(1/A)}{-k} = \frac{\ln A}{k}.$$

Thus, increasing the value of A moves the inflection point to the right. (See the bottom two graphs in Figure 4.66.)

Exercises and Problems for Section 4.4

EXERCISES

In Exercises 1–6, investigate the one-parameter family of functions. Assume that a is positive.

(a) Graph $f(x)$ using three different values for a.

(b) Using your graph in part (a), describe the critical points of f and how they appear to move as a increases.

(c) Find a formula for the x-coordinates of the critical point(s) of f in terms of a.

1. $f(x) = (x - a)^2$
2. $f(x) = x^3 - ax$
3. $f(x) = ax^3 - x$
4. $f(x) = x - a\sqrt{x}$
5. $f(x) = x^2e^{-ax}$
6. $f(x) = \dfrac{a}{x^2} + x$ for $x > 0$
7. Figure 4.68 shows $f(x) = 1+e^{-ax}$ for $a = 1, 2, 5$. Without a calculator, identify the graphs by looking at $f'(0)$.

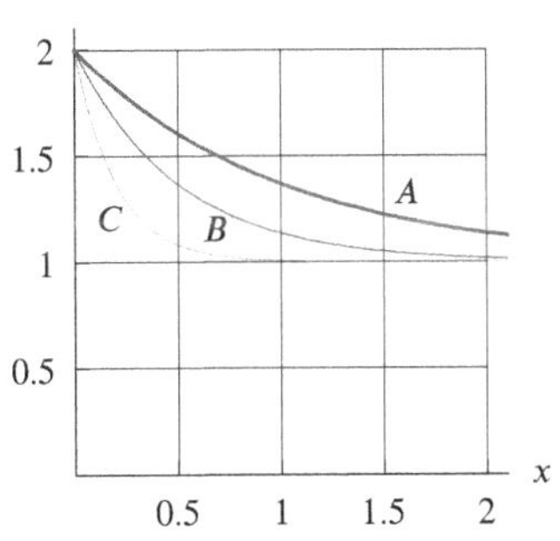

Figure 4.68

8. Figure 4.69 shows $f(x) = xe^{-ax}$ for $a = 1, 2, 3$. Without a calculator, identify the graphs by locating the critical points of $f(x)$.

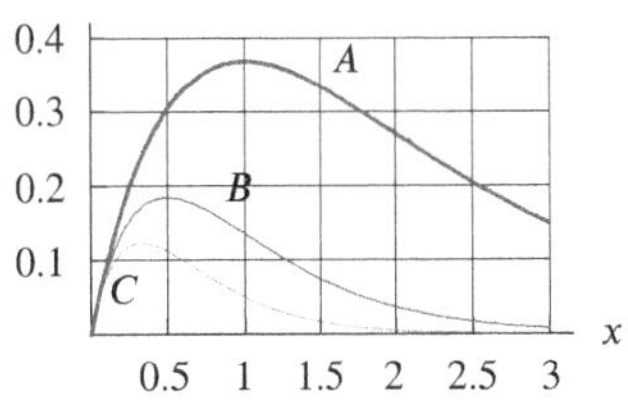

Figure 4.69

9. Consider the family

$$y = \frac{A}{x + B}.$$

(a) If $B = 0$, what is the effect of varying A on the graph?

(b) If $A = 1$, what is the effect of varying B?

(c) On one set of axes, graph the function for several values of A and B.

10. If A and B are positive constants, find all critical points of

$$f(w) = \frac{A}{w^2} - \frac{B}{w}.$$

■ In Exercises 11–16, investigate the given two parameter family of functions. Assume that a and b are positive.

(a) Graph $f(x)$ using $b = 1$ and three different values for a.
(b) Graph $f(x)$ using $a = 1$ and three different values for b.
(c) In the graphs in parts (a) and (b), how do the critical points of f appear to move as a increases? As b increases?
(d) Find a formula for the x-coordinates of the critical point(s) of f in terms of a and b.

11. $f(x) = (x - a)^2 + b$

12. $f(x) = x^3 - ax^2 + b$

13. $f(x) = ax(x - b)^2$

14. $f(x) = \dfrac{ax}{x^2 + b}$

15. $f(x) = \sqrt{b - (x - a)^2}$

16. $f(x) = \dfrac{a}{x} + bx$ for $x > 0$

PROBLEMS

17. Figure 4.70 shows $f(x) = x + a^2/x$ for $a = 1, 2$, and a third positive integer value of a. Without a calculator, identify the graphs and the third value of a.

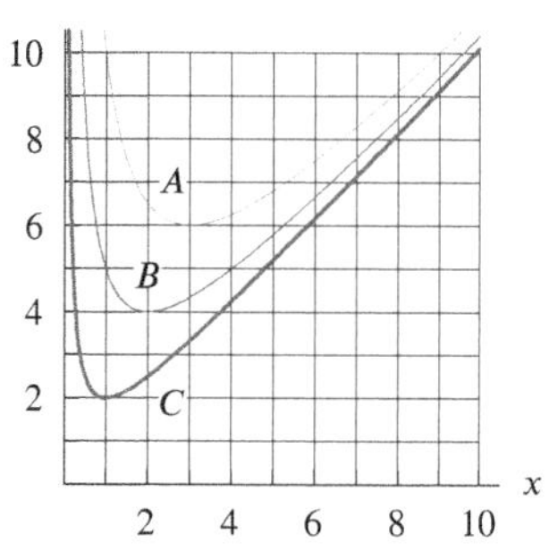

Figure 4.70

18. Figure 4.71 shows $f(x) = x + a \sin x$ for positive values of a. Explain why any two of the curves intersect at the same points on $0 \le x \le 7$.

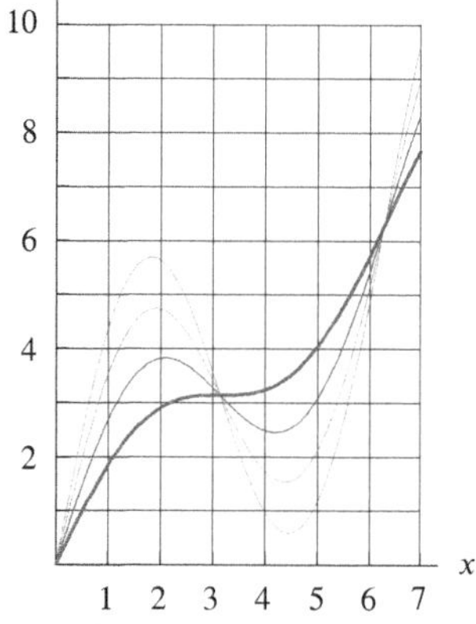

Figure 4.71

19. **(a)** Sketch graphs of $y = xe^{-bx}$ for $b = 1, 2, 3, 4$. Describe the graphical significance of b.
(b) Find the coordinates of the critical point of $y = xe^{-bx}$ and use it to confirm your answer to part (a).

20. **(a)** Graph $y = f(x) = \sqrt{x + A}$ using three different values for the constant $A \ge 1$.
(b) What is the domain of $f(x)$?
(c) Find the value of x that gives the point on the curve closest to the origin. Does the answer depend on A?

21. For b a positive constant and n an even integer, $n \ge 2$, find the critical points of $f(x) = x(x - b)^n$. Illustrate the effects of the parameters b and n with graphs.

22. **(a)** Graph $f(x) = x + a \sin x$ for $a = 0.5$ and $a = 3$.
(b) For what values of a is $f(x)$ increasing for all x?

23. **(a)** Graph $f(x) = x^2 + a \sin x$ for $a = 1$ and $a = 20$.
(b) For what values of a is $f(x)$ concave up for all x?

24. Let $y = a \cosh(x/a)$ for $a > 0$. Sketch graphs for $a = 1, 2, 3$. Describe in words the effect of increasing a.

25. Let $f(x) = bxe^{1+bx}$, where b is constant and $b > 0$.

(a) What is the x-coordinate of the critical point of f?
(b) Is the critical point a local maximum or a local minimum?
(c) Show that the y-coordinate of the critical point does not depend on the value of b.

26. Let $h(x) = e^{-x} + kx$, where k is any constant. For what value(s) of k does h have

(a) No critical points?
(b) One critical point?
(c) A horizontal asymptote?

27. Let $g(x) = x - ke^x$, where k is any constant. For what value(s) of k does the function g have a critical point?

28. Show that $f(x) = x - k\sqrt{x}$, with k a positive constant and $x \ge 0$, has a local minimum at a point whose x-coordinate is $1/4$ of the way between its x-intercepts.

29. For any constant a, let $f(x) = ax - x \ln x$ for $x > 0$.

(a) What is the x-intercept of the graph of $f(x)$?
(b) Graph $f(x)$ for $a = -1$ and $a = 1$.
(c) For what values of a does $f(x)$ have a critical point for $x > 0$? Find the coordinates of the critical point and decide if it is a local maximum, a local minimum, or neither.

30. Let $f(x) = x^2 + \cos(kx)$, for $k > 0$.
 (a) Graph f for $k = 0.5, 1, 3, 5$. Find the smallest number k at which you see points of inflection in the graph of f.
 (b) Explain why the graph of f has no points of inflection if $k \le \sqrt{2}$, and infinitely many points of inflection if $k > \sqrt{2}$.
 (c) Explain why f has only a finite number of critical points, no matter what the value of k.

31. Let $f(x) = e^x - kx$, for $k > 0$.
 (a) Graph f for $k = 1/4, 1/2, 1, 2, 4$. Describe what happens as k changes.
 (b) Show that f has a local minimum at $x = \ln k$.
 (c) Find the value of k for which the local minimum is the largest.

32. Let $f(x) = e^{ax} - e^{-bx}$ for positive constants a and b. Explain why f is always increasing.

33. Let $f(x) = e^{-ax} + e^{bx}$ for non-zero constants a and b. Explain why the graph of f is always concave up.

34. Let $f(x) = ax^4 - bx$ for positive constants a and b. Explain why the graph of f is always concave up.

35. Let $f(x) = a\ln x - bx$, for positive constants a and b. Explain why there is an interval on which f is increasing and an interval on which f is decreasing.

36. Let $f(x) = ax^3 - bx$ for positive constants a and b. Explain why there is an interval on which f is increasing and another interval on which it is decreasing.

37. Let $f(x) = ax^4 - bx^2$ for positive constants a and b. Explain why there is an interval on which the graph of f is concave up and an interval on which the graph of f is concave down.

38. (a) Find all critical points of $f(x) = x^3 - ax + b$.
 (b) Under what conditions on a and b does this function have no critical points?
 (c) Under what conditions on a and b does this function have exactly one critical point? What is the one critical point, and is it a local maximum, a local minimum, or neither?
 (d) Under what conditions on a and b does this function have exactly two critical points? What are they? Which are local maxima, which are local minima, and which are neither?
 (e) Is it ever possible for this function to have more than two critical points? Explain.

39. Figure 4.72 shows graphs of four members of the family

$$f(x) = \frac{Ae^{-x} + Be^x}{e^{-x} + e^x}.$$

 (a) What value does $f(x)$ approach as $x \to \infty$? As $x \to -\infty$?
 (b) Explain the graphical significance of A and B.
 (c) Which graph corresponds to each pair of values for (A, B): $(-2, 5), (7, 0), (8, -4), (2, 6)$?

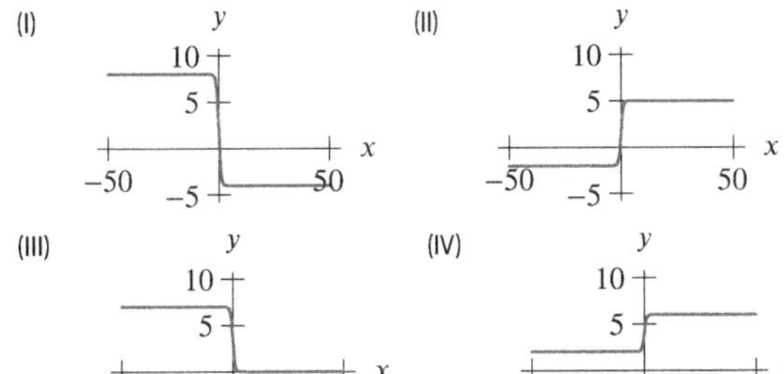

Figure 4.72

40. Figure 4.73 shows graphs of four members of the family

$$f(x) = \frac{Axe^{-x} + Bxe^x}{e^{-x} + e^x}.$$

 (a) What value does $f(x)$ approach as $x \to \infty$? As $x \to -\infty$?
 (b) Explain the graphical significance of A and B.
 (c) Which graph corresponds to each pair of values for (A, B): $(2, 1), (-1, 2), (-2, 0), (-1, -2)$?

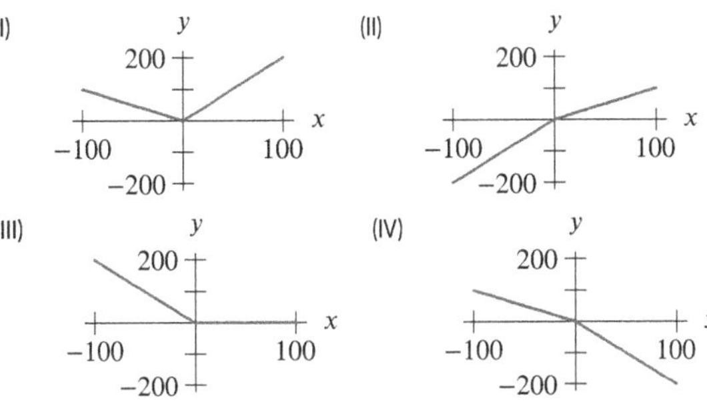

Figure 4.73

41. Consider the surge function $y = axe^{-bx}$ for $a, b > 0$.
 (a) Find the local maxima, local minima, and points of inflection.
 (b) How does varying a and b affect the shape of the graph?
 (c) On one set of axes, graph this function for several values of a and b.

42. Sketch several members of the family $y = e^{-ax}\sin bx$ for $b = 1$, and describe the graphical significance of the parameter a.

43. Sketch several members of the family $y = e^{-ax}\sin bx$ for $a = 1$, and describe the graphical significance of the parameter b.

44. If $a > 0$, $b > 0$, show that $f(x) = a(1 - e^{-bx})$ is everywhere increasing and everywhere concave down.

45. Find a formula for the family of cubic polynomials with an inflection point at the origin. How many parameters are there?

46. **(a)** Derive formulas for the first and second derivatives of the logistic function:

$$y = \frac{L}{1 + Ae^{-kt}} \quad \text{for } L, A, \text{ and } k \text{ positive constants.}$$

(b) Derive a formula for the t value of any inflection point(s).
(c) Use the second derivative to determine the concavity on either side of any inflection points.

Find formulas for the functions described in Problems 47–58.

47. A function of the form $y = a(1 - e^{-bx})$ with $a, b > 0$ and a horizontal asymptote of $y = 5$.

48. A function of the form $y = be^{-(x-a)^2/2}$ with its maximum at the point $(0, 3)$.

49. A curve of the form $y = e^{-(x-a)^2/b}$ for $b > 0$ with a local maximum at $x = 2$ and points of inflection at $x = 1$ and $x = 3$.

50. A logistic curve with carrying capacity of 12, y-intercept of 4, and point of inflection at $(0.5, 6)$.

51. A function of the form $y = \dfrac{a}{1 + be^{-t}}$ with y-intercept 2 and an inflection point at $t = 1$.

52. A cubic polynomial with a critical point at $x = 2$, an inflection point at $(1, 4)$, and a leading coefficient of 1.

53. A fourth-degree polynomial whose graph is symmetric about the y-axis, has a y-intercept of 0, and global maxima at $(1, 2)$ and $(-1, 2)$.

54. A function of the form $y = a\sin(bt^2)$ whose first critical point for positive t occurs at $t = 1$ and whose derivative is 3 when $t = 2$.

55. A function of the form $y = a\cos(bt^2)$ whose first critical point for positive t occurs at $t = 1$ and whose derivative is -2 when $t = 1/\sqrt{2}$.

56. A function of the form $y = ae^{-x} + bx$ with the global minimum at $(1, 2)$.

57. A function of the form $y = bxe^{-ax}$ with a local maximum at $(3, 6)$.

58. A function of the form $y = at + b/t$, with a local minimum $(3, 12)$ and a local maximum at $(-3, -12)$.

59. A family of functions is given by

$$r(x) = \frac{1}{a + (x - b)^2}.$$

(a) For what values of a and b does the graph of r have a vertical asymptote? Where are the vertical asymptotes in this case?
(b) Find values of a and b so that the function r has a local maximum at the point $(3, 5)$.

60. **(a)** Find all critical points of $f(x) = x^4 + ax^2 + b$.
(b) Under what conditions on a and b does this function have exactly one critical point? What is the one critical point, and is it a local maximum, a local minimum, or neither?
(c) Under what conditions on a and b does this function have exactly three critical points? What are they? Which are local maxima and which are local minima?
(d) Is it ever possible for this function to have two critical points? No critical points? More than three critical points? Give an explanation in each case.

61. Let $y = Ae^x + Be^{-x}$ for any constants A, B.

(a) Sketch the graph of the function for
(i) $A = 1,\ B = 1$ (ii) $A = 1,\ B = -1$
(iii) $A = 2,\ B = 1$ (iv) $A = 2,\ B = -1$
(v) $A = -2,\ B = -1$ (vi) $A = -2,\ B = 1$
(b) Describe in words the general shape of the graph if A and B have the same sign. What effect does the sign of A have on the graph?
(c) Describe in words the general shape of the graph if A and B have different signs. What effect does the sign of A have on the graph?
(d) For what values of A and B does the function have a local maximum? A local minimum? Justify your answer using derivatives.

62. The temperature, T, in °C, of a yam put into a 200°C oven is given as a function of time, t, in minutes, by

$$T = a(1 - e^{-kt}) + b.$$

(a) If the yam starts at 20°C, find a and b.
(b) If the temperature of the yam is initially increasing at 2°C per minute, find k.

63. For positive a, b, the potential energy, U, of a particle is

$$U = b\left(\frac{a^2}{x^2} - \frac{a}{x}\right) \quad \text{for } x > 0.$$

(a) Find the intercepts and asymptotes.
(b) Compute the local maxima and minima.
(c) Sketch the graph.

64. The force, F, on a particle with potential energy U is given by

$$F = -\frac{dU}{dx}.$$

Using the expression for U in Problem 63, graph F and U on the same axes, labeling intercepts and local maxima and minima.

65. The Lennard-Jones model predicts the potential energy $V(r)$ of a two-atom molecule as a function of the distance r between the atoms to be

$$V(r) = \frac{A}{r^{12}} - \frac{B}{r^6}, \quad r > 0,$$

where A and B are positive constants.

(a) Evaluate $\lim_{r\to 0^+} V(r)$, and interpret your answer.

(b) Find the critical point of $V(r)$. Is it a local maximum or local minimum?
(c) The inter-atomic force is given by $F(r) = -V'(r)$. At what distance r is the inter-atomic force zero? (This is called the equilibrium size of the molecule.)
(d) Describe how the parameters A and B affect the equilibrium size of the molecule.

66. For positive A, B, the force between two atoms is a function of the distance, r, between them:

$$f(r) = -\frac{A}{r^2} + \frac{B}{r^3} \qquad r > 0.$$

(a) What are the zeros and asymptotes of f?
(b) Find the coordinates of the critical points and inflection points of f.
(c) Graph f.

(d) Illustrating your answers with a sketch, describe the effect on the graph of f of:
(i) Increasing B, holding A fixed
(ii) Increasing A, holding B fixed

67. An organism has size W at time t. For positive constants A, b, and c, the Gompertz growth function gives

$$W = Ae^{-e^{b-ct}}, \quad t \geq 0.$$

(a) Find the intercepts and asymptotes.
(b) Find the critical points and inflection points.
(c) Graph W for various values of A, b, and c.
(d) A certain organism grows fastest when it is about 1/3 of its final size. Would the Gompertz growth function be useful in modeling its growth? Explain.

Strengthen Your Understanding

In Problems 68–69, explain what is wrong with the statement.

68. Every function of the form $f(x) = x^2 + bx + c$, where b and c are constants, has two zeros.

69. Every function of the form $f(x) = a/x + bx$, where a and b are nonzero constants, has two critical points.

In Problems 70–73, give an example of:

70. A family of quadratic functions which has zeros at $x = 0$ and $x = b$.

71. A member of the family $f(x) = ax^3 - bx$ that has no critical points.

72. A family of functions, $f(x)$, depending on a parameter a, such that each member of the family has exactly one critical point.

73. A family of functions, $g(x)$, depending on two parameters, a and b, such that each member of the family has exactly two critical points and one inflection point. You may want to restrict a and b.

74. Let $f(x) = ax + b/x$. Suppose a and b are positive. What happens to $f(x)$ as b increases?
(a) The critical points move further apart.
(b) The critical points move closer together.
(c) The critical values move further apart.
(d) The critical values move closer together.

75. Let $f(x) = ax + b/x$. Suppose a and b are positive. What happens to $f(x)$ as a increases?
(a) The critical points move further apart.
(b) The critical points move closer together.
(c) The critical values move further apart.
(d) The critical values move closer together.

4.5 APPLICATIONS TO MARGINALITY

Management decisions within a particular business usually aim at maximizing profit for the company. In this section we see how the derivative can be used to maximize profit. Profit depends on both production cost and revenue (or income) from sales. We begin by looking at the cost and revenue functions.

> The **cost function**, $C(q)$, gives the total cost of producing a quantity q of some good.

What sort of function do we expect C to be? The more goods that are made, the higher the total cost, so C is an increasing function. In fact, cost functions usually have the general shape shown in Figure 4.74. The intercept on the C-axis represents the *fixed costs*, which are incurred even if nothing is produced. (This includes, for instance, the machinery needed to begin production.) The cost function increases quickly at first and then more slowly because producing larger quantities of a good is usually more efficient than producing smaller quantities—this is called *economy of scale*. At still higher production levels, the cost function starts to increase faster again as resources become scarce, and sharp increases may occur when new factories have to be built. Thus, the graph of $C(q)$ may start out concave down and become concave up later on.

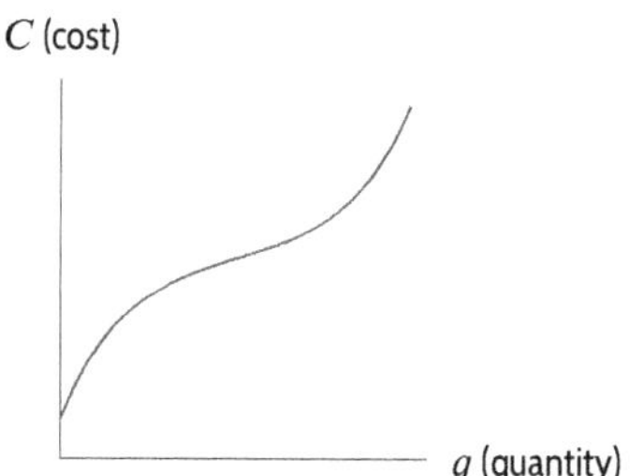

Figure 4.74: Cost as a function of quantity

> The **revenue function**, $R(q)$, gives the total revenue received by a firm from selling a quantity q of some good.

Revenue is income obtained from sales. If the price per item is p, and the quantity sold is q, then

$$\text{Revenue} = \text{Price} \times \text{Quantity}, \quad \text{so} \quad R = pq.$$

If the price per item does not depend on the quantity sold, then the graph of $R(q)$ is a straight line through the origin with slope equal to the price p. See Figure 4.75. In practice, for large values of q, the market may become glutted, causing the price to drop, giving $R(q)$ the shape in Figure 4.76.

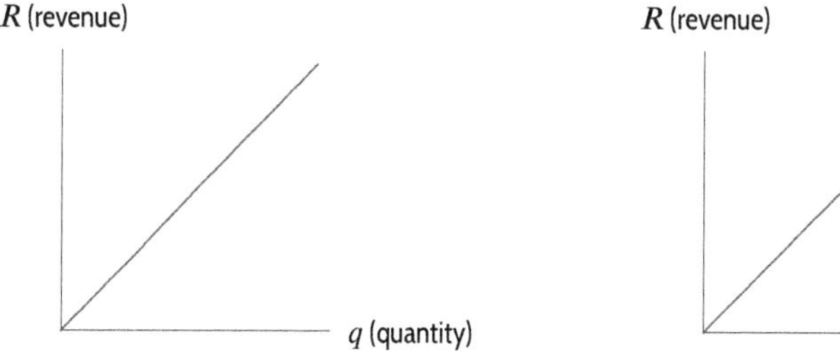

Figure 4.75: Revenue: Constant price

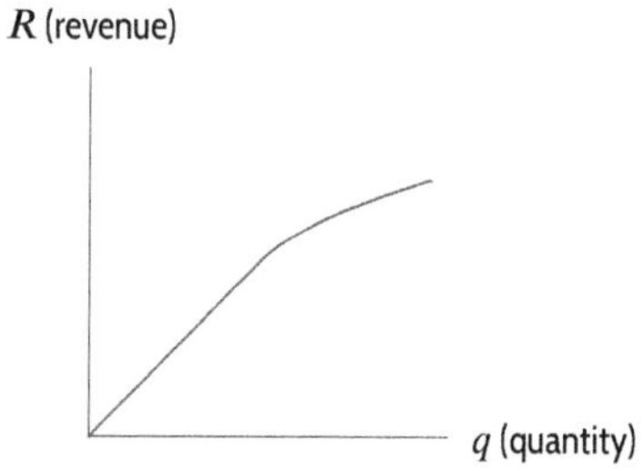

Figure 4.76: Revenue: Decreasing price

The profit is usually written as π. (Economists use π to distinguish it from the price, p; this π has nothing to do with the area of a circle, and merely stands for the Greek equivalent of the letter "p.") The profit resulting from producing and selling q items is defined by

$$\text{Profit} = \text{Revenue} - \text{Cost}, \quad \text{so} \quad \pi(q) = R(q) - C(q).$$

Example 1 If cost, C, and revenue, R, are given by the graph in Figure 4.77, for what production quantities, q, does the firm make a profit? Approximately what production level maximizes profit?

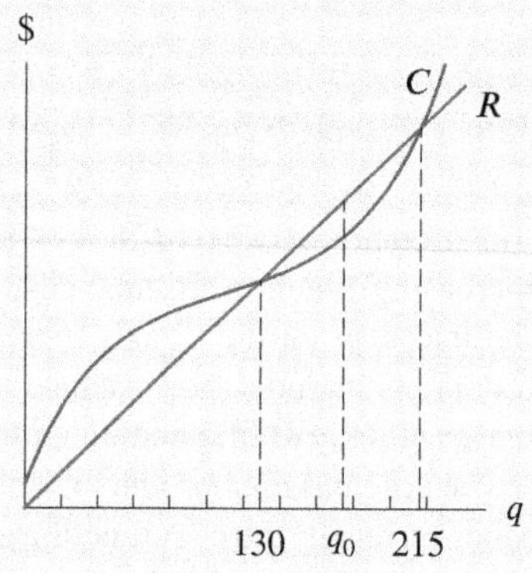

Figure 4.77: Costs and revenues for Example 1

Solution The firm makes a profit whenever revenues are greater than costs, that is, when $R > C$. The graph of R is above the graph of C approximately when $130 < q < 215$. Production between $q = 130$ units and $q = 215$ units generates a profit. The vertical distance between the cost and revenue curves is largest at q_0, so q_0 units gives maximum profit.

Marginal Analysis

Many economic decisions are based on an analysis of the costs and revenues "at the margin." Let's look at this idea through an example.

Suppose we are running an airline and we are trying to decide whether to offer an additional flight. How should we decide? We'll assume that the decision is to be made purely on financial grounds: if the flight will make money for the company, it should be added. Obviously we need to consider the costs and revenues involved. Since the choice is between adding this flight and leaving things the way they are, the crucial question is whether the *additional costs* incurred are greater or smaller than the *additional revenues* generated by the flight. These additional costs and revenues are called the *marginal costs* and *marginal revenues*.

Suppose $C(q)$ is the function giving the total cost of running q flights. If the airline had originally planned to run 100 flights, its costs would be $C(100)$. With the additional flight, its costs would be $C(101)$. Therefore,

$$\text{Additional cost "at the margin"} = C(101) - C(100).$$

Now

$$C(101) - C(100) = \frac{C(101) - C(100)}{101 - 100},$$

and this quantity is the average rate of change of cost between 100 and 101 flights. In Figure 4.78 the average rate of change is the slope of the line joining the $C(100)$ and $C(101)$ points on the graph. If the graph of the cost function is not curving fast near the point, the slope of this line is close to the slope of the tangent line there. Therefore, the average rate of change is close to the instantaneous rate of change. Since these rates of change are not very different, many economists choose to define marginal cost, MC, as the instantaneous rate of change of cost with respect to quantity:

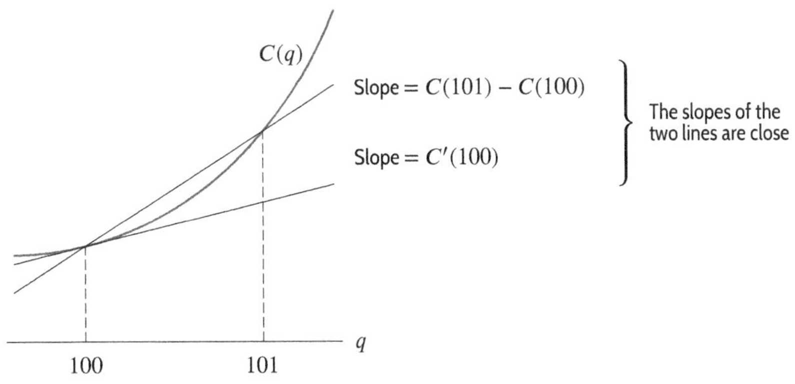

Figure 4.78: Marginal cost: Slope of one of these lines

$$\text{Marginal cost} = MC = C'(q) \qquad \text{so} \qquad \text{Marginal cost} \approx C(q+1) - C(q).$$

Similarly, if the revenue generated by q flights is $R(q)$ and the number of flights increases from 100 to 101, then

$$\text{Additional revenue "at the margin"} = R(101) - R(100).$$

Now $R(101) - R(100)$ is the average rate of change of revenue between 100 and 101 flights. As before, the average rate of change is usually almost equal to the instantaneous rate of change, so economists often define:

$$\text{Marginal revenue} = MR = R'(q) \quad \text{so} \quad \text{Marginal revenue} \approx R(q+1) - R(q).$$

We often refer to total cost, C, and total revenue, R, to distinguish them from marginal cost, MC, and marginal revenue, MR. If the words cost and revenue are used alone, they are understood to mean total cost and total revenue.

Example 2 If $C(q)$ and $R(q)$ for the airline are given in Figure 4.79, should the company add the 101st flight?

Solution The marginal revenue is the slope of the revenue curve, and the marginal cost is the slope of the cost curve at the point 100. From Figure 4.79, you can see that the slope at the point A is smaller than the slope at B, so $MC < MR$. This means that the airline will make more in extra revenue than it will spend in extra costs if it runs another flight, so it should go ahead and run the 101st flight.

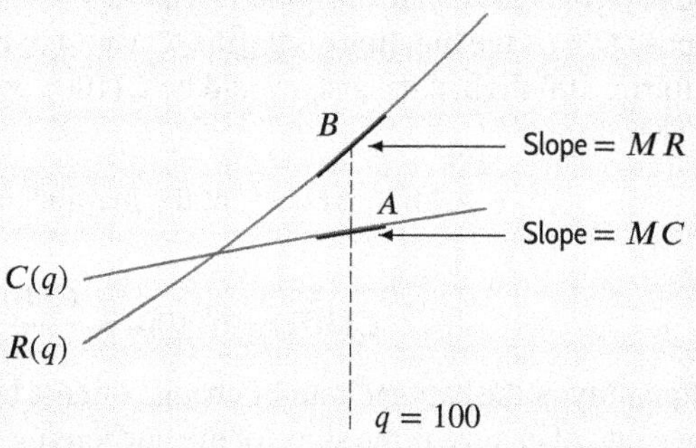

Figure 4.79: Should the company add the 101st flight?

Since MC and MR are derivative functions, they can be estimated from the graphs of total cost and total revenue.

Example 3 If R and C are given by the graphs in Figure 4.80, sketch graphs of $MR = R'(q)$ and $MC = C'(q)$.

Figure 4.80: Total revenue and total cost for Example 3

Solution The revenue graph is a line through the origin, with equation

$$R = pq$$

where p is the price, which is a constant. The slope is p and

$$MR = R'(q) = p.$$

The total cost is increasing, so the marginal cost is always positive (above the q-axis). For small q values, the total cost curve is concave down, so the marginal cost is decreasing. For larger q, say $q > 100$, the total cost curve is concave up and the marginal cost is increasing. Thus the marginal cost has a minimum at about $q = 100$. (See Figure 4.81.)

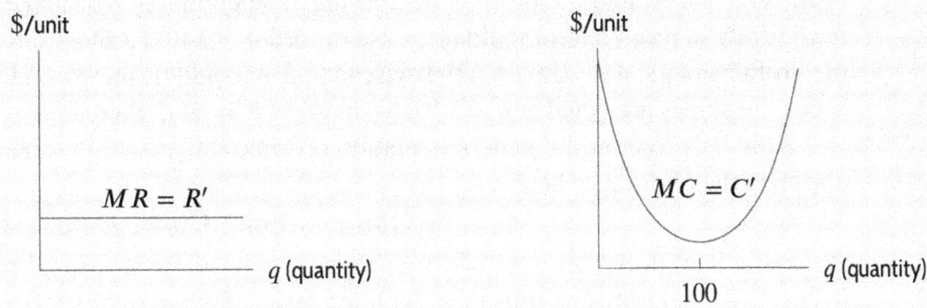

Figure 4.81: Marginal revenue and costs for Example 3

Maximizing Profit

Now let's look at how to maximize profit, given functions for total revenue and total cost.

Example 4 Find the maximum profit if the total revenue and total cost are given, for $0 \leq q \leq 200$, by the curves R and C in Figure 4.82.

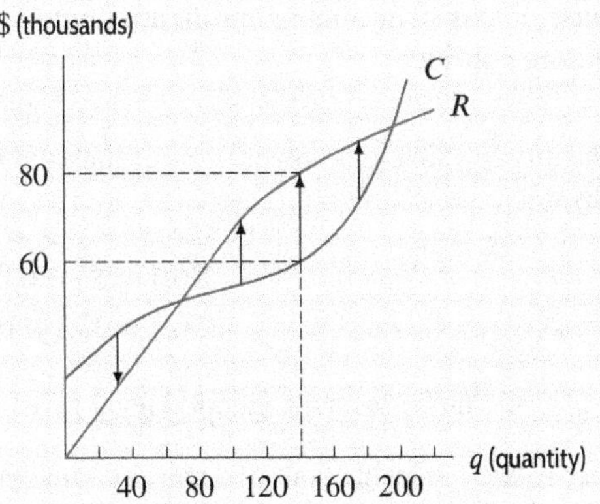

Figure 4.82: Maximum profit at $q = 140$

Solution The profit is represented by the vertical difference between the curves and is marked by the vertical arrows on the graph. When revenue is below cost, the company is taking a loss; when revenue is above cost, the company is making a profit. We can see that the profit is largest at about $q = 140$, so this is the production level we're looking for. To be sure that the local maximum is a global maximum, we need to check the endpoints. At $q = 0$ and $q = 200$, the profit is negative, so the global maximum is indeed at $q = 140$.

To find the actual maximum profit, we estimate the vertical distance between the curves at $q = 140$. This gives a maximum profit of \$80,000 − \$60,000 = \$20,000.

Suppose we wanted to find the minimum profit. In this example, we must look at the endpoints, when $q = 0$ or $q = 200$. We see the minimum profit is negative (a loss), and it occurs at $q = 0$.

Maximum Profit Occurs Where $MR = MC$

In Example 4, observe that at $q = 140$ the slopes of the two curves in Figure 4.82 are equal. To the left of $q = 140$, the revenue curve has a larger slope than the cost curve, and the profit increases as q increases. The company will make more money by producing more units, so production should increase toward $q = 140$. To the right of $q = 140$, the slope of the revenue curve is less than the slope of the cost curve, and the profit is decreasing. The company will make more money by producing fewer units so production should decrease toward $q = 140$. At the point where the slopes are equal, the profit has a local maximum; otherwise the profit could be increased by moving toward that point. Since the slopes are equal at $q = 140$, we have $MR = MC$ there.

Now let's look at the general situation. To maximize or minimize profit over an interval, we optimize the profit, π, where

$$\pi(q) = R(q) - C(q).$$

We know that global maxima and minima can only occur at critical points or at endpoints of an interval. To find critical points of π, look for zeros of the derivative:

$$\pi'(q) = R'(q) - C'(q) = 0.$$

So

$$R'(q) = C'(q),$$

that is, the slopes of the revenue and cost curves are equal. This is the same observation that we made in the previous example. In economic language,

> The maximum (or minimum) profit can occur where
>
> Marginal cost = Marginal revenue.

Of course, maximal or minimal profit does not *have* to occur where $MR = MC$; there are also the endpoints to consider.

Example 5 Find the quantity q which maximizes profit if the total revenue, $R(q)$, and total cost, $C(q)$, are given in dollars by

$$R(q) = 5q - 0.003q^2$$
$$C(q) = 300 + 1.1q,$$

where $0 \leq q \leq 800$ units. What production level gives the minimum profit?

Solution We look for production levels that give marginal revenue = marginal cost:

$$MR = R'(q) = 5 - 0.006q$$
$$MC = C'(q) = 1.1.$$

So $5 - 0.006q = 1.1$, giving

$$q = 3.9/0.006 = 650 \text{ units}.$$

Does this value of q represent a local maximum or minimum of π? We can tell by looking at production levels of 649 units and 651 units. When $q = 649$ we have $MR = \$1.106$, which is greater than the (constant) marginal cost of \$1.10. This means that producing one more unit will bring in more revenue than its cost, so profit will increase. When $q = 651$, $MR = \$1.094$, which is *less* than MC, so it is not profitable to produce the 651st unit. We conclude that $q = 650$ is a local maximum for the profit function π. The profit earned by producing and selling this quantity is $\pi(650) = R(650) - C(650) = \$967.50$.

To check for global maxima we need to look at the endpoints. If $q = 0$, the only cost is \$300 (the fixed costs) and there is no revenue, so $\pi(0) = -\$300$. At the upper limit of $q = 800$, we have $\pi(800) = \$900$. Therefore, the maximum profit is at the production level of 650 units, where $MR = MC$. The minimum profit (a loss) occurs when $q = 0$ and there is no production at all.

Exercises and Problems for Section 4.5 Online Resource: Additional Problems for Section 4.5

EXERCISES

1. Total cost and revenue are approximated by the functions $C = 5000 + 2.4q$ and $R = 4q$, both in dollars. Identify the fixed cost, marginal cost per item, and the price at which this item is sold.

2. Total cost is $C = 8500 + 4.65q$ and total revenue is $R = 5.15q$, both in dollars, where q represents the quantity produced.

   (a) What is the fixed cost?
   (b) What is the marginal cost per item?
   (c) What is the price at which this item is sold?
   (d) For what production levels does this company make a profit?
   (e) How much does the company make for each additional unit sold?

3. When production is 4500, marginal revenue is \$8 per unit and marginal cost is \$9.25 per unit. Do you expect maximum profit to occur at a production level above or below 4500? Explain.

4. Figure 4.83 shows cost and revenue. For what production levels is the profit function positive? Negative? Estimate the production at which profit is maximized.

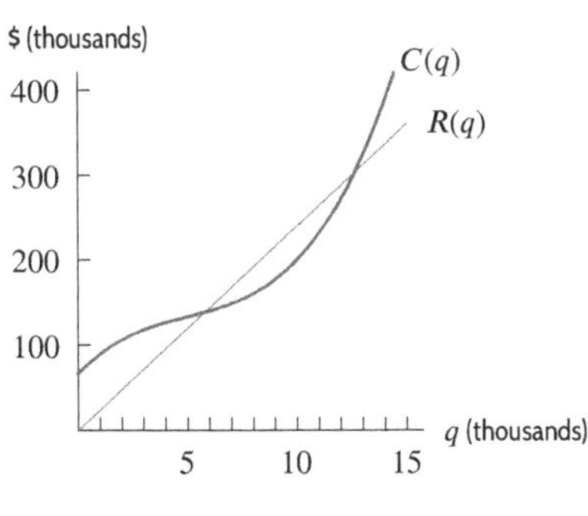

Figure 4.83

5. Figure 4.84 gives cost and revenue. What are fixed costs? What quantity maximizes profit, and what is the maximum profit earned?

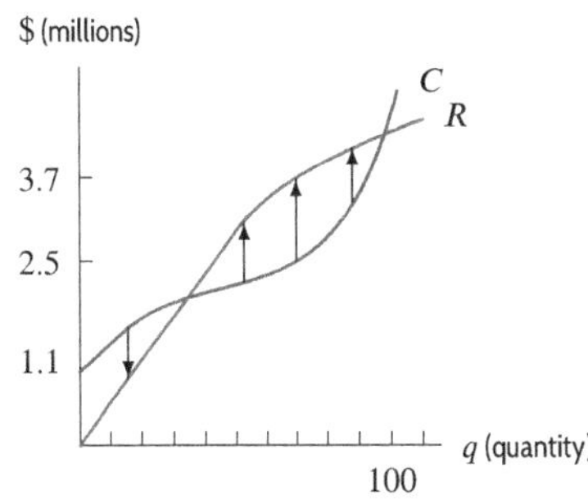

Figure 4.84

6. Figure 4.85 shows cost and revenue for producing q units. For the production levels in (a)–(d), is the company making or losing money? Should the company be increasing or decreasing production to increase profits?

   (a) $q = 75$
   (b) $q = 150$
   (c) $q = 225$
   (d) $q = 300$

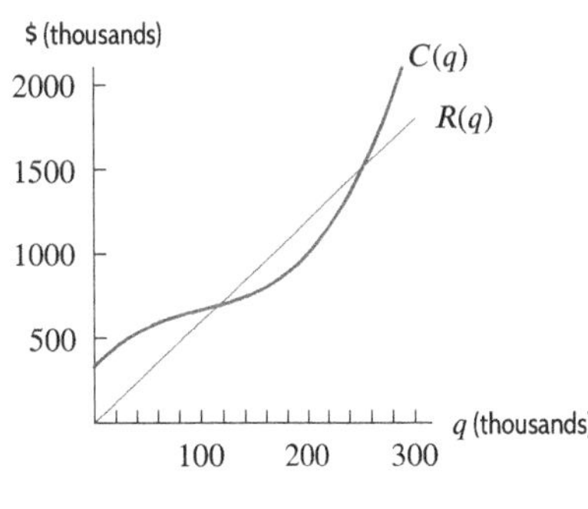

Figure 4.85

In Exercises 7–10, give the cost, revenue, and profit functions.

7. An online seller of T-shirts pays \$500 to start up the website and \$6 per T-shirt, then sells the T-shirts for \$12 each.

8. A car wash operator pays \$35,000 for a franchise, then spends \$10 per car wash, which costs the consumer \$15.

9. A couple running a house-cleaning business invests \$5000 in equipment, and they spend \$15 in supplies to clean a house, for which they charge \$60.

10. A lemonade stand operator sets up the stand for free in front of the neighbor's house, makes 5 quarts of lemonade for $4, then sells each 8-oz cup for 25 cents.

11. The revenue from selling q items is $R(q) = 500q - q^2$, and the total cost is $C(q) = 150 + 10q$. Write a function that gives the total profit earned, and find the quantity which maximizes the profit.

12. Revenue is given by $R(q) = 450q$ and cost is given by $C(q) = 10{,}000 + 3q^2$. At what quantity is profit maximized? What is the total profit at this production level?

13. A company estimates that the total revenue, R, in dollars, received from the sale of q items is $R = \ln(1 + 1000q^2)$. Calculate and interpret the marginal revenue if $q = 10$.

14. Table 4.2 shows cost, $C(q)$, and revenue, $R(q)$.

(a) At approximately what production level, q, is profit maximized? Explain your reasoning.
(b) What is the price of the product?
(c) What are the fixed costs?

Table 4.2

q	0	500	1000	1500	2000	2500	3000
$R(q)$	0	1500	3000	4500	6000	7500	9000
$C(q)$	3000	3800	4200	4500	4800	5500	7400

15. Table 4.3 shows marginal cost, MC, and marginal revenue, MR.

(a) Use the marginal cost and marginal revenue at a production of $q = 5000$ to determine whether production should be increased or decreased from 5000.
(b) Estimate the production level that maximizes profit.

Table 4.3

q	5000	6000	7000	8000	9000	10000
MR	60	58	56	55	54	53
MC	48	52	54	55	58	63

PROBLEMS

16. Let $C(q)$ be the total cost of producing a quantity q of a certain product. See Figure 4.86.

(a) What is the meaning of $C(0)$?
(b) Describe in words how the marginal cost changes as the quantity produced increases.
(c) Explain the concavity of the graph (in terms of economics).
(d) Explain the economic significance (in terms of marginal cost) of the point at which the concavity changes.
(e) Do you expect the graph of $C(q)$ to look like this for all types of products?

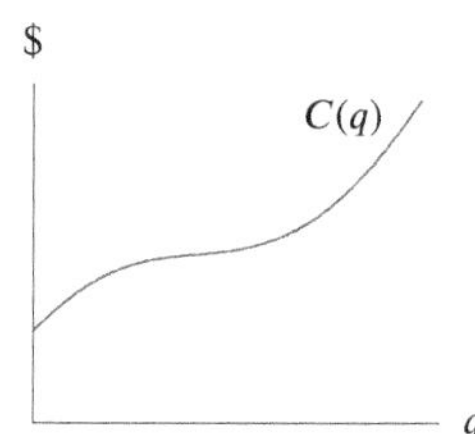

Figure 4.86

17. When production is 2000, marginal revenue is $4 per unit and marginal cost is $3.25 per unit. Do you expect maximum profit to occur at a production level above or below 2000? Explain.

18. If $C'(500) = 75$ and $R'(500) = 100$, should the quantity produced be increased or decreased from $q = 500$ in order to increase profits?

19. An online seller of knitted sweaters finds that it costs $35 to make her first sweater. Her cost for each additional sweater goes down until it reaches $25 for her 100th sweater, and after that it starts to rise again. If she can sell each sweater for $35, is the quantity sold that maximizes her profit less than 100? Greater than 100?

20. The marginal revenue and marginal cost for a certain item are graphed in Figure 4.87. Do the following quantities maximize profit for the company? Explain your answer.

(a) $q = a$ (b) $q = b$

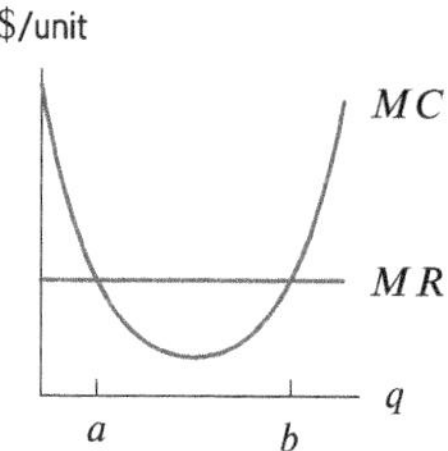

Figure 4.87

21. The total cost $C(q)$ of producing q goods is given by:

$$C(q) = 0.01q^3 - 0.6q^2 + 13q.$$

(a) What is the fixed cost?
(b) What is the maximum profit if each item is sold for \$7? (Assume you sell everything you produce.)
(c) Suppose exactly 34 goods are produced. They all sell when the price is \$7 each, but for each \$1 increase in price, 2 fewer goods are sold. Should the price be raised, and if so by how much?

22. A company manufactures only one product. The quantity, q, of this product produced per month depends on the amount of capital, K, invested (i.e., the number of machines the company owns, the size of its building, and so on) and the amount of labor, L, available each month. We assume that q can be expressed as a *Cobb-Douglas production function*:

$$q = cK^{\alpha}L^{\beta},$$

where c, α, β are positive constants, with $0 < \alpha < 1$ and $0 < \beta < 1$. In this problem we will see how the Russian government could use a Cobb-Douglas function to estimate how many people a newly privatized industry might employ. A company in such an industry has only a small amount of capital available to it and needs to use all of it, so K is fixed. Suppose L is measured in man-hours per month, and that each man-hour costs the company w rubles (a ruble is the unit of Russian currency). Suppose the company has no other costs besides labor, and that each unit of the good can be sold for a fixed price of p rubles. How many man-hours of labor per month should the company use in order to maximize its profit?

23. An agricultural worker in Uganda is planting clover to increase the number of bees making their home in the region. There are 100 bees in the region naturally, and for every acre put under clover, 20 more bees are found in the region.

(a) Draw a graph of the total number, $N(x)$, of bees as a function of x, the number of acres devoted to clover.
(b) Explain, both geometrically and algebraically, the shape of the graph of:
    (i) The marginal rate of increase of the number of bees with acres of clover, $N'(x)$.
    (ii) The average number of bees per acre of clover, $N(x)/x$.

24. If you invest x dollars in a certain project, your return is $R(x)$. You want to choose x to maximize your return per dollar invested,[9] which is

$$r(x) = \frac{R(x)}{x}.$$

(a) The graph of $R(x)$ is in Figure 4.88, with $R(0) = 0$. Illustrate on the graph that the maximum value of $r(x)$ is reached at a point at which the line from the origin to the point is tangent to the graph of $R(x)$.
(b) Also, the maximum of $r(x)$ occurs at a point at which the slope of the graph of $r(x)$ is zero. On the same axes as part (a), sketch $r(x)$. Illustrate that the maximum of $r(x)$ occurs where its slope is 0.
(c) Show, by taking the derivative of the formula for $r(x)$, that the conditions in part (a) and (b) are equivalent: the x-value at which the line from the origin is tangent to the graph of R is the same as the x-value at which the graph of r has zero slope.

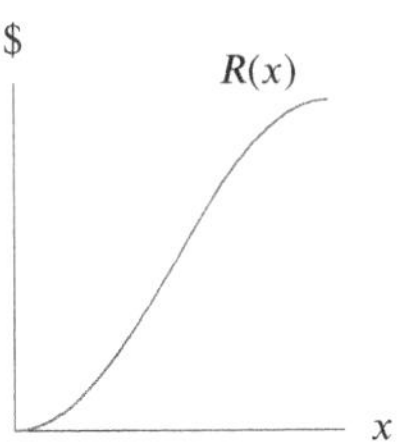

Figure 4.88

Problems 25–26 involve the *average cost* of manufacturing a quantity q of a good, which is defined to be

$$a(q) = \frac{C(q)}{q}.$$

25. Figure 4.89 shows the cost of production, $C(q)$, as a function of quantity produced, q.

(a) For some q_0, sketch a line whose slope is the marginal cost, MC, at that point.
(b) For the same q_0, explain why the average cost $a(q_0)$ can be represented by the slope of the line from that point on the curve to the origin.
(c) Use the method of Example 6 on page 218 to explain why at the value of q which minimizes $a(q)$, the average and marginal costs are equal.

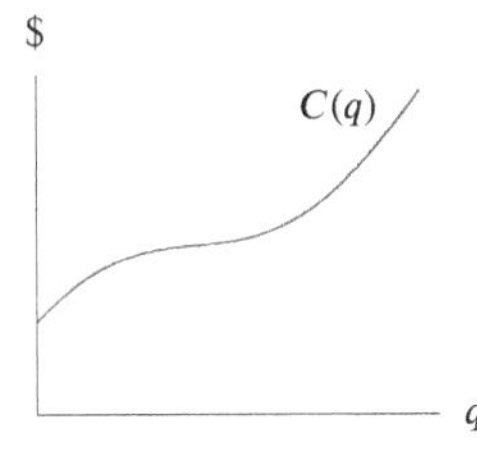

Figure 4.89

[9]From Peter D. Taylor, *Calculus: The Analysis of Functions* (Toronto: Wall & Emerson, 1992).

26. The average cost per item to produce q items is given by

$$a(q) = 0.01q^2 - 0.6q + 13, \quad \text{for} \quad q > 0.$$

(a) What is the total cost, $C(q)$, of producing q goods?
(b) What is the minimum marginal cost? What is the practical interpretation of this result?
(c) At what production level is the average cost a minimum? What is the lowest average cost?
(d) Compute the marginal cost at $q \overset{!}{=} 30$. How does this relate to your answer to part (c)? Explain this relationship both analytically and in words.

In many applications, we want to maximize or minimize some quantity subject to a condition. Such constrained optimization problems are solved using Lagrange multipliers in multivariable calculus; Problems 27–29 show an alternate method.[10]

27. Minimize $x^2 + y^2$ while satisfying $x + y = 4$ using the following steps.

(a) Graph $x + y = 4$. On the same axes, graph $x^2 + y^2 = 1$, $x^2 + y^2 = 4$, $x^2 + y^2 = 9$.
(b) Explain why the minimum value of $x^2 + y^2$ on $x + y = 4$ occurs at the point at which a graph of $x^2 + y^2 =$ Constant is tangent to the line $x + y = 4$.
(c) Using your answer to part (b) and implicit differentiation to find the slope of the circle, find the minimum value of $x^2 + y^2$ such that $x + y = 4$.

28. The quantity Q of an item which can be produced from quantities x and y of two raw materials is given by $Q = 10xy$ at a cost of $C = x + 2y$ thousand dollars. If the budget for raw materials is 10 thousand dollars, find the maximum production using the following steps.

(a) Graph $x + 2y = 10$ in the first quadrant. On the same axes, graph $Q = 10xy = 100$, $Q = 10xy = 200$, and $Q = 10xy = 300$.
(b) Explain why the maximum production occurs at a point at which a production curve is tangent to the cost line $C = 10$.
(c) Using your answer to part (b) and implicit differentiation to find the slope of the curve, find the maximum production under this budget.

29. With quantities x and y of two raw materials available, $Q = x^{1/2}y^{1/2}$ thousand items can be produced at a cost of $C = 2x + y$ thousand dollars. Using the following steps, find the minimum cost to produce 1 thousand items.

(a) Graph $x^{1/2}y^{1/2} = 1$. On the same axes, graph $2x + y = 2$, $2x + y = 3$, and $2x + y = 4$.
(b) Explain why the minimum cost occurs at a point at which a cost line is tangent to the production curve $Q = 1$.
(c) Using your answer to part (b) and implicit differentiation to find the slope of the curve, find the minimum cost to meet this production level.

Strengthen Your Understanding

In Problems 30–32, explain what is wrong with the statement.

30. If $C(100) = 90$ and $R(100) = 150$, increasing the quantity produced from 100 increases profit.

31. If $MC(200) = 10$ and $MR(200) = 10$, the company is maximizing profit.

32. For the cost, C, and revenue, R, in Figure 4.90, profit is maximized when the quantity produced is about 3,500 units.

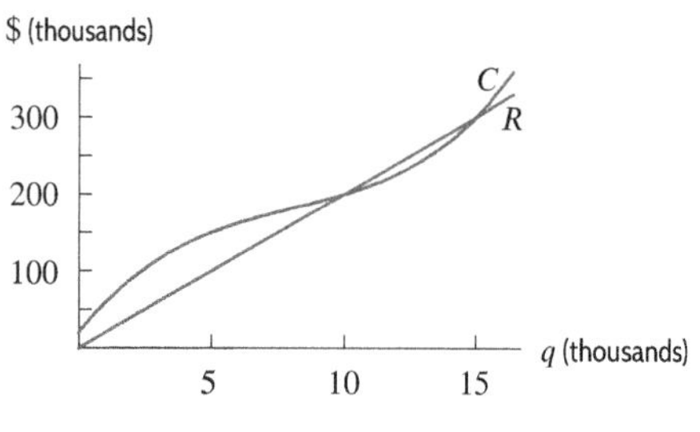

Figure 4.90

In Problems 33–34, give an example of:

33. A quantity, q, in Figure 4.90 where $MC > MR$.

34. Cost and revenue curves such that the item can never be sold for a profit.

35. Which is correct? A company generally wants to

(a) Maximize revenue
(b) Maximize marginal revenue
(c) Minimize cost
(d) Minimize marginal cost
(e) None of the above

36. Which is correct? A company can increase its profit by increasing production if, at its current level of production,

(a) Marginal revenue − Marginal cost > 0
(b) Marginal revenue − Marginal cost = 0
(c) Marginal revenue − Marginal cost < 0
(d) Marginal revenue − Marginal cost is increasing

[10]Kelly Black, "Putting Constraints in Optimization for First-Year Calculus Students," pp. 310–312, *SIAM Review*, Vol. 39, No. 2, June 1997.

4.6 RATES AND RELATED RATES

Derivatives represent rates of change. In this section, we see how to calculate rates in a variety of situations.

Example 1 A spherical snowball is melting. Its radius decreases at a constant rate of 2 cm per minute from an initial value of 70 cm. How fast is the volume decreasing half an hour later?

Solution The radius, r, starts at 70 cm and decreases at 2 cm/min. At time t minutes since the start,

$$r = 70 - 2t \text{ cm}.$$

The volume of the snowball is given by

$$V = \frac{4}{3}\pi r^3 = \frac{4}{3}\pi(70 - 2t)^3 \text{ cm}^3.$$

The rate at which the volume is changing at time t is

$$\frac{dV}{dt} = \frac{4}{3}\pi \cdot 3(70 - 2t)^2(-2) = -8\pi(70 - 2t)^2 \text{ cm}^3/\text{min}.$$

The volume is measured in cm^3, and time is in minutes, so after half an hour $t = 30$, and

$$\left.\frac{dV}{dt}\right|_{t=30} = -8\pi(70 - 2 \cdot 30)^2 = -800\pi \text{ cm}^3/\text{min}.$$

Thus, the rate at which the volume is changing is $-800\pi \approx -2500$ cm^3/min, so it is decreasing at a rate of about 2500 cm^3/min.

Example 2 A skydiver of mass m jumps from a plane at time $t = 0$. Under certain assumptions, the distance, $s(t)$, he has fallen in time t is given by

$$s(t) = \frac{m^2 g}{k^2}\left(\frac{kt}{m} + e^{-kt/m} - 1\right) \quad \text{for some positive constant } k.$$

(a) Find $s'(0)$ and $s''(0)$ and interpret in terms of the skydiver.
(b) Relate the units of $s'(t)$ and $s''(t)$ to the units of t and $s(t)$.

Solution (a) Differentiating using the chain rule gives

$$s'(t) = \frac{m^2 g}{k^2}\left(\frac{k}{m} + e^{-kt/m}\left(-\frac{k}{m}\right)\right) = \frac{mg}{k}\left(1 - e^{-kt/m}\right)$$

$$s''(t) = \frac{mg}{k}(-e^{kt/m})\left(-\frac{k}{m}\right) = ge^{-kt/m}.$$

Since $e^{-k \cdot 0/m} = 1$, evaluating at $t = 0$ gives

$$s'(0) = \frac{mg}{k}(1 - 1) = 0 \quad \text{and} \quad s''(0) = g.$$

The first derivative of distance is velocity, so the fact that $s'(0) = 0$ tells us that the skydiver starts with zero velocity. The second derivative of distance is acceleration, so the fact that $s''(0) = g$ tells us that the skydiver's initial acceleration is g, the acceleration due to gravity.

(b) The units of velocity, $s'(t)$, and acceleration, $s''(t)$, are given by

$$\text{Units of } s'(t) \text{ are } \frac{\text{Units of } s(t)}{\text{Units of } t} = \frac{\text{Units of distance}}{\text{Units of time}}; \quad \text{for example, meters/sec.}$$

$$\text{Units of } s''(t) \text{ are } \frac{\text{Units of } s'(t)}{\text{Units of } t} = \frac{\text{Units of distance}}{(\text{Units of time})^2}; \quad \text{for example, meters/sec}^2.$$

Related Rates

In Example 1, the radius of the snowball decreased at a constant rate. A more realistic scenario is that the radius decreases at a varying rate. In this case, we may not be able to write a formula for V as a function of t. However, we may still be able to calculate dV/dt, as in the following example.

Example 3 A spherical snowball melts in such a way that the instant at which its radius is 20 cm, its radius is decreasing at 3 cm/min. At what rate is the volume of the ball of snow changing at that instant?

Solution Since the snowball is spherical, we again have that

$$V = \frac{4}{3}\pi r^3.$$

We can no longer write a formula for r in terms of t, but we know that

$$\frac{dr}{dt} = -3 \quad \text{when} \quad r = 20.$$

We want to know dV/dt when $r = 20$. Think of r as an (unknown) function of t and differentiate the expression for V with respect to t using the chain rule:

$$\frac{dV}{dt} = \frac{4}{3}\pi \cdot 3r^2\frac{dr}{dt} = 4\pi r^2\frac{dr}{dt}.$$

At the instant at which $r = 20$ and $dr/dt = -3$, we have

$$\frac{dV}{dt} = 4\pi \cdot 20^2 \cdot (-3) = -4800\pi \approx -15{,}080 \text{ cm}^3/\text{min}.$$

So the volume of the ball is decreasing at a rate of 15,080 cm^3 per minute at the moment when $r = 20$ cm. Notice that we have sidestepped the problem of not knowing r as a function of t by calculating the derivatives only at the moment we are interested in.

Example 4 Figure 4.91 shows the fuel consumption, g, in miles per gallon (mpg), of a car traveling at v mph. At one moment, the car was going 70 mph and its deceleration was 8000 miles/hour^2. How fast was the fuel consumption changing at that moment? Include units.

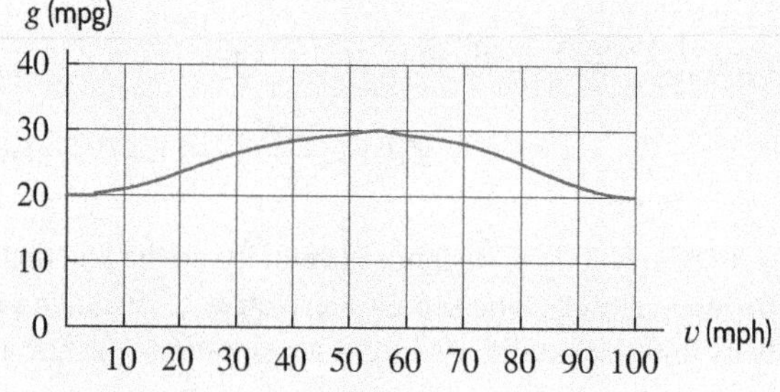

Figure 4.91: Fuel consumption versus velocity

Solution Acceleration is rate of change of velocity, dv/dt, and we are told that the deceleration is 8000 miles/hour^2, so we know $dv/dt = -8000$ when $v = 70$. We want dg/dt. The chain rule gives

$$\frac{dg}{dt} = \frac{dg}{dv} \cdot \frac{dv}{dt}.$$

The value of dg/dv is the slope of the curve in Figure 4.91 at $v = 70$. Since the points (30, 40) and (100, 20) lie approximately on the tangent to the curve at $v = 70$, we can estimate the derivative

$$\frac{dg}{dv} \approx \frac{20 - 40}{100 - 30} = -\frac{2}{7} \text{ mpg/mph.}$$

Thus,

$$\frac{dg}{dt} = \frac{dg}{dv} \cdot \frac{dv}{dt} \approx -\frac{2}{7}(-8000) \approx 2300 \text{ mpg/hr.}$$

In other words, fuel consumption is increasing at about $2300/60 \approx 38$ mpg per minute. Since we approximated dg/dv, we can only get a rough estimate for dg/dt.

A famous problem involves the rate at which the top of a ladder slips down a wall as the foot of the ladder moves.

Example 5

(a) A 3-meter ladder stands against a high wall. The foot of the ladder moves outward at a constant speed of 0.1 meter/sec. When the foot is 1 meter from the wall, how fast is the top of the ladder falling? What about when the foot is 2 meters from the wall?

(b) If the foot of the ladder moves out at a constant speed, how does the speed at which the top falls change as the foot gets farther out?

Solution

(a) Let the foot be x meters from the base of the wall and let the top be y meters from the base. See Figure 4.92. Then, since the ladder is 3 meters long, by Pythagoras' Theorem,

$$x^2 + y^2 = 3^2 = 9.$$

Thinking of x and y as functions of t, we differentiate with respect to t using the chain rule:

$$2x\frac{dx}{dt} + 2y\frac{dy}{dt} = 0.$$

We are interested in the moment at which $dx/dt = 0.1$ and $x = 1$. We want to know dy/dt, so we solve, giving

$$\frac{dy}{dt} = -\frac{x}{y}\frac{dx}{dt}.$$

When the foot of the ladder is 1 meter from the wall, $x = 1$ and $y = \sqrt{9 - 1^2} = \sqrt{8}$, so

$$\frac{dy}{dt} = -\frac{1}{\sqrt{8}}0.1 = -0.035 \text{ meter/sec.}$$

Thus, the top falls at 0.035 meter/sec.

When the foot is 2 meters from the wall, $x = 2$ and $y = \sqrt{9 - 2^2} = \sqrt{5}$, so

$$\frac{dy}{dt} = -\frac{2}{\sqrt{5}}0.1 = -0.089 \text{ meter/sec.}$$

Thus, the top falls at 0.089 meter/sec. Notice that the top falls faster when the base of the ladder is farther from the wall.

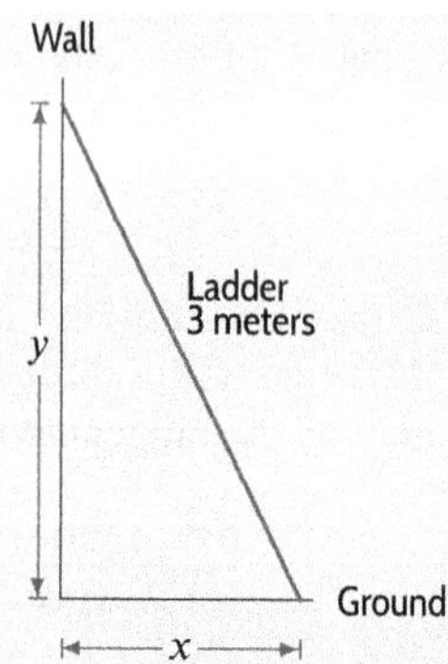

Figure 4.92: Side view of ladder standing against wall (x, y in meters)

(b) As the foot of the ladder moves out, x increases and y decreases. Looking at the expression

$$\frac{dy}{dt} = -\frac{x}{y}\frac{dx}{dt},$$

we see that if dx/dt is constant, the magnitude of dy/dt increases as the foot gets farther out. Thus, the top falls faster and faster.

Example 6 An airplane, flying at 450 km/hr at a constant altitude of 5 km, is approaching a camera mounted on the ground. Let θ be the angle of elevation above the ground at which the camera is pointed. See Figure 4.93. When $\theta = \pi/3$, how fast does the camera have to rotate in order to keep the plane in view?

Solution Suppose the camera is at point C and the plane is vertically above point B. Let x km be the distance between B and C. The fact that the plane is moving horizontally toward C at 450 km/hr means that x is decreasing and $dx/dt = -450$ km/hr. From Figure 4.93, we see that $\tan\theta = 5/x$.

Differentiating $\tan\theta = 5/x$ with respect to t and using the chain rule gives

$$\frac{1}{\cos^2\theta}\frac{d\theta}{dt} = -5x^{-2}\frac{dx}{dt}.$$

We want to calculate $d\theta/dt$ when $\theta = \pi/3$. At that moment, $\cos\theta = 1/2$ and $\tan\theta = \sqrt{3}$, so $x = 5/\sqrt{3}$. Substituting gives

$$\frac{1}{(1/2)^2}\frac{d\theta}{dt} = -5\left(\frac{5}{\sqrt{3}}\right)^{-2} \cdot (-450)$$

$$\frac{d\theta}{dt} = 67.5 \text{ radians/hour.}$$

Since there are 60 minutes in an hour, the camera must turn at roughly 1 radian per minute if it is to remain pointed at the plane. With 1 radian $\approx 60°$, this is a rotation of about one degree per second.

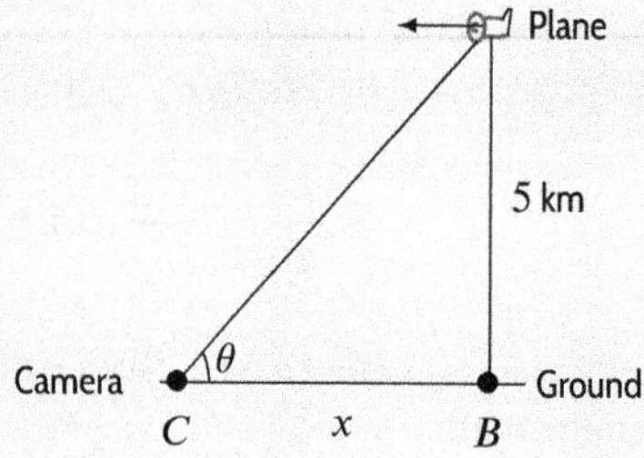

Figure 4.93: Plane approaching a camera at C (side view; x in km)

Exercises and Problems for Section 4.6 Online Resource: Additional Problems for Section 4.6

EXERCISES

1. With time, t, in minutes, the temperature, H, in degrees Celsius, of a bottle of water put in the refrigerator at $t = 0$ is given by

$$H = 4 + 16e^{-0.02t}.$$

How fast is the water cooling initially? After 10 minutes? Give units.

2. The world population[11] P, in billions, is approximately

$$P = 7.17e^{0.01064t},$$

where t is in years since April 20, 2014. At what rate was the world's population increasing on that date? Give your answer in millions of people per year.

3. If $x^2 + y^2 = 25$ and $dx/dt = 6$, find dy/dt when y is positive and

(a) $x = 0$ **(b)** $x = 3$ **(c)** $x = 4$

4. If $xy = 100$ and $dx/dt = 5$, find dy/dt when

(a) $x = 10$ **(b)** $x = 25$ **(c)** $x = 50$

5. With length, l, in meters, the period T, in seconds, of a pendulum is given by

$$T = 2\pi\sqrt{\frac{l}{9.8}}.$$

(a) How fast does the period increase as l increases?
(b) Does this rate of change increase or decrease as l increases?

6. The Dubois formula relates a person's surface area, s, in meters^2, to weight, w, in kg, and height, h, in cm, by

$$s = 0.01w^{0.25}h^{0.75}.$$

(a) What is the surface area of a person who weighs 60 kg and is 150 cm tall?
(b) The person in part (a) stays constant height but increases in weight by 0.5 kg/year. At what rate is his surface area increasing when his weight is 62 kg?

7. A plane is climbing at 500 feet per minute, and the air temperature outside the plane is falling at 2°C per 1000 feet. What is the rate of change (as a function of time) of the air temperature just outside the plane?

8. If θ is the angle between a line through the origin and the positive x-axis, the area, in cm^2, of part of a rose petal is

$$A = \frac{9}{16}(4\theta - \sin(4\theta)).$$

If the angle θ is increasing at a rate of 0.2 radians per minute, at what rate is the area changing when $\theta = \pi/4$?

[11] www.indexmundi.com, accessed April 20, 2015.

9. Atmospheric pressure decays exponentially as altitude increases. With pressure, P, in inches of mercury and altitude, h, in feet above sea level, we have

$$P = 30e^{-3.23\times10^{-5}h}.$$

(a) At what altitude is the atmospheric pressure 25 inches of mercury?
(b) A glider measures the pressure to be 25 inches of mercury and experiences a pressure increase of 0.1 inches of mercury per minute. At what rate is it changing altitude?

10. The gravitational force, F, on a rocket at a distance, r, from the center of the earth is given by

$$F = \frac{k}{r^2},$$

where $k = 10^{13}$ newton · km^2. When the rocket is 10^4 km from the center of the earth, it is moving away at 0.2 km/sec. How fast is the gravitational force changing at that moment? Give units. (A newton is a unit of force.)

11. The power, P, dissipated when a 9-volt battery is put across a resistance of R ohms is given by

$$P = \frac{81}{R}.$$

What is the rate of change of power with respect to resistance?

12. A voltage V across a resistance R generates a current

$$I = \frac{V}{R}$$

in amps. A constant voltage of 9 volts is put across a resistance that is increasing at a rate of 0.2 ohms per second when the resistance is 5 ohms. At what rate is the current changing?

13. The potential, ϕ, of a charge distribution at a point on the positive x-axis is given, for x in centimeters, by

$$\phi = 2\pi\left(\sqrt{x^2+4} - x\right).$$

A particle at $x = 3$ is moving to the left at a rate of 0.2 cm/sec. At what rate is its potential changing?

14. The average cost per item, C, in dollars, of manufacturing a quantity q of cell phones is given by

$$C = \frac{a}{q} + b \qquad \text{where } a, b \text{ are positive constants.}$$

(a) Find the rate of change of C as q increases. What are its units?
(b) If production increases at a rate of 100 cell phones per week, how fast is the average cost changing? Is the average cost increasing or decreasing?

15. A pyramid has height h and a square base with side x. The volume of a pyramid is $V = \frac{1}{3}x^2h$. If the height remains fixed and the side of the base is decreasing by 0.002 meter/yr, at what rate is the volume decreasing when the height is 120 meters and the width is 150 meters?

16. A thin uniform rod of length l cm and a small particle lie on a line separated by a distance of a cm. If K is a positive constant and F is measured in newtons, the gravitational force between them is

$$F = \frac{K}{a(a+l)}.$$

(a) If a is increasing at the rate 2 cm/min when $a = 15$ and $l = 5$, how fast is F decreasing?

(b) If l is decreasing at the rate 2 cm/min when $a = 15$ and $l = 5$, how fast is F increasing?

PROBLEMS

In Problems 17–20, use Figure 4.94 showing the altitude of a plane as a function of the time since take-off and air pressure as a function of altitude.[12] How fast is air pressure changing outside the plane at the given moment in its flight?

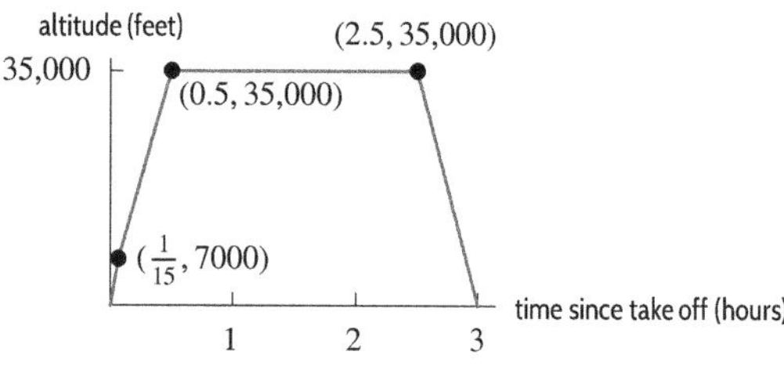

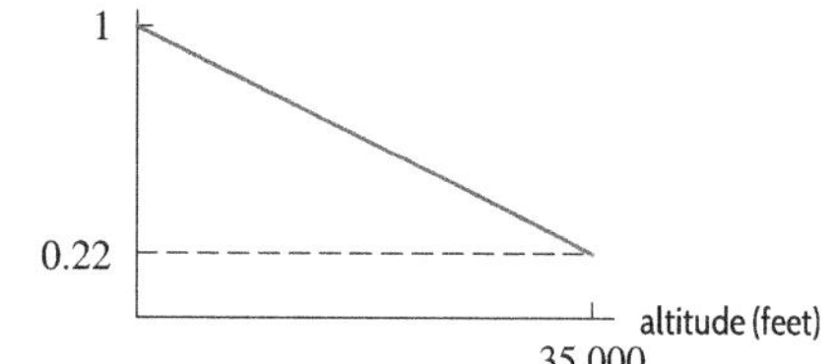

Figure 4.94

17. 1 hour after takeoff

18. 2 minutes after takeoff

19. 10 minutes after takeoff

20. 2.75 hours after takeoff

21. A rectangle has one side of 10 cm. How fast is the area of the rectangle changing at the instant when the other side is 12 cm and increasing at 3 cm per minute?

22. A rectangle has one side of 8 cm. How fast is the diagonal of the rectangle changing at the instant when the other side is 6 cm and increasing at 3 cm per minute?

23. A right triangle has one leg of 7 cm. How fast is its area changing at the instant that the other leg has length 10 cm and is decreasing at 2 cm per second?

24. The area, A, of a square is increasing at 3 cm^2 per minute. How fast is the side length of the square changing when $A = 576$ cm^2?

25. Car A is driving east toward an intersection. Car B has already gone through the same intersection and is heading north. At what rate is the distance between the cars changing at the instant when car A is 40 miles from the intersection and traveling at 50 mph and car B is 30 miles from the intersection and traveling at 60 mph? Are the cars getting closer together or farther apart at this time?

26. Car A is driving south, away from an intersection. Car B is approaching the intersection and is moving west. At what rate is the distance between the cars changing at the instant when car A is 40 miles from the intersection and traveling at 55 mph and car B is 30 miles from the intersection and traveling at 45 mph? Are the cars getting closer together or farther apart at this time?

27. A dose, D, of a drug causes a temperature change, T, in a patient. For C a positive constant, T is given by

$$T = \left(\frac{C}{2} - \frac{D}{3}\right)D^2.$$

(a) What is the rate of change of temperature change with respect to dose?

(b) For what doses does the temperature change increase as the dose increases?

28. When a company spends L thousand dollars on labor and K thousand dollars on equipment, it produces P units:

$$P = 500L^{0.3}K^{0.7}.$$

(a) How many units are produced when labor spending is $L = 85$ and equipment spending is $K = 50$?

(b) The company decides to keep equipment expenditures constant at $K = 50$, but to increase labor expenditures by 2 thousand dollars per year. At what rate is the quantity of items produced changing when labor spending is $L = 90$?

29. An item costs \$500 at time $t = 0$ and costs $\$P$ in year t. When inflation is $r\%$ per year, the price is given by

$$P = 500e^{rt/100}.$$

(a) If r is a constant, at what rate is the price rising (in dollars per year)

(i) Initially? (ii) After 2 years?

(b) Now suppose that r is increasing by 0.3 per year when $r = 4$ and $t = 2$. At what rate (dollars per year) is the price increasing at that time?

[12] Change in air pressure with respect to altitude is not quite linear, but this approximation is fairly close to real air pressure.

30. A 10 m ladder leans against a vertical wall and the bottom of the ladder slides away from the wall at a rate of 0.5 m/sec. How fast is the top of the ladder sliding down the wall when the bottom of the ladder is

(a) 4 m from the wall? (b) 8 m from the wall?

31. Gasoline is pouring into a vertical cylindrical tank of radius 3 feet. When the depth of the gasoline is 4 feet, the depth is increasing at 0.2 ft/sec. How fast is the volume of gasoline changing at that instant?

32. Water is being pumped into a vertical cylinder of radius 5 meters and height 20 meters at a rate of 3 meters^3/min. How fast is the water level rising when the cylinder is half full?

33. A spherical snowball is melting. Its radius is decreasing at 0.2 cm per hour when the radius is 15 cm. How fast is its volume decreasing at that time?

34. The radius of a spherical balloon is increasing by 2 cm/sec. At what rate is air being blown into the balloon at the moment when the radius is 10 cm? Give units in your answer.

35. If two electrical resistances, R_1 and R_2, are connected in parallel, their combined resistance, R, is given by

$$\frac{1}{R} = \frac{1}{R_1} + \frac{1}{R_2}.$$

Suppose R_1 is held constant at 10 ohms, and that R_2 is increasing at 2 ohms per minute when R_2 is 20 ohms. How fast is R changing at that moment?

36. For positive constants A and B, the force, F, between two atoms in a molecule at a distance r apart is given by

$$F = -\frac{A}{r^2} + \frac{B}{r^3}.$$

(a) How fast does force change as r increases? What type of units does it have?
(b) If at some time t the distance is changing at a rate k, at what rate is the force changing with time? What type of units does this rate of change have?

37. For positive constants k and g, the velocity, v, of a particle of mass m at time t is given by

$$v = \frac{mg}{k}\left(1 - e^{-kt/m}\right).$$

At what rate is the velocity changing at time 0? At $t = 1$? What do your answers tell you about the motion?

38. The circulation time of a mammal (that is, the average time it takes for all the blood in the body to circulate once and return to the heart) is proportional to the fourth root of the body mass of the mammal. The constant of proportionality is 17.40 if circulation time is in seconds and body mass is in kilograms. The body mass of a growing child is 45 kg and is increasing at a rate of 0.1 kg/month. What is the rate of change of the circulation time of the child?

39. A certain quantity of gas occupies a volume of 20 cm^3 at a pressure of 1 atmosphere. The gas expands without the addition of heat, so, for some constant k, its pressure, P, and volume, V, satisfy the relation

$$PV^{1.4} = k.$$

(a) Find the rate of change of pressure with volume. Give units.
(b) The volume is increasing at 2 cm^3/min when the volume is 30 cm^3. At that moment, is the pressure increasing or decreasing? How fast? Give units.

40. The metal frame of a rectangular box has a square base. The horizontal rods in the base are made out of one metal and the vertical rods out of a different metal. If the horizontal rods expand at a rate of 0.001 cm/hr and the vertical rods expand at a rate of 0.002 cm/hr, at what rate is the volume of the box expanding when the base has an area of 9 cm^2 and the volume is 180 cm^3?

41. A ruptured oil tanker causes a circular oil slick on the surface of the ocean. When its radius is 150 meters, the radius of the slick is expanding by 0.1 meter/minute and its thickness is 0.02 meter. At that moment:

(a) How fast is the area of the slick expanding?
(b) The circular slick has the same thickness everywhere, and the volume of oil spilled remains fixed. How fast is the thickness of the slick decreasing?

42. A potter forms a piece of clay into a cylinder. As he rolls it, the length, L, of the cylinder increases and the radius, r, decreases. If the length of the cylinder is increasing at 0.1 cm per second, find the rate at which the radius is changing when the radius is 1 cm and the length is 5 cm.

43. A cone-shaped coffee filter of radius 6 cm and depth 10 cm contains water, which drips out through a hole at the bottom at a constant rate of 1.5 cm^3 per second.

(a) If the filter starts out full, how long does it take to empty?
(b) Find the volume of water in the filter when the depth of the water is h cm.
(c) How fast is the water level falling when the depth is 8 cm?

44. Water is being poured into the cone-shaped container in Figure 4.95. When the depth of the water is 2.5 in, it is increasing at 3 in/min. At that time, how fast is the surface area, A, that is covered by water increasing? [Hint: $A = \pi rs$, where r, s are as shown.]

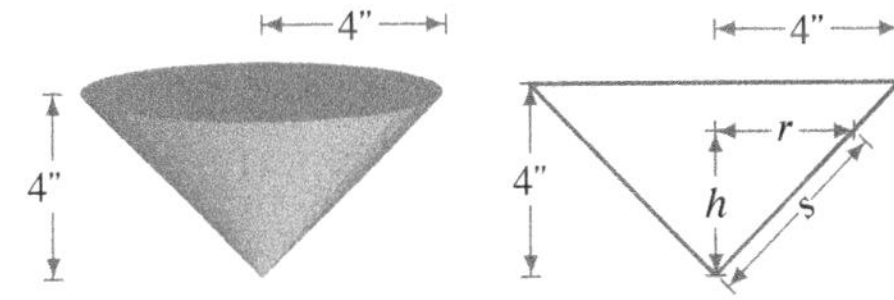

Figure 4.95: Cone and cross section

45. Grit, which is spread on roads in winter, is stored in mounds which are the shape of a cone. As grit is added to the top of a mound at 2 cubic meters per minute, the angle between the slant side of the cone and the vertical remains 45°. How fast is the height of the mound increasing when it is half a meter high? [Hint: Volume $V = \pi r^2 h/3$, where r is radius and h is height.]

46. A gas station stands at the intersection of a north-south road and an east-west road. A police car is traveling toward the gas station from the east, chasing a stolen truck which is traveling north away from the gas station. The speed of the police car is 100 mph at the moment it is 3 miles from the gas station. At the same time, the truck is 4 miles from the gas station going 80 mph. At this moment:

(a) Is the distance between the car and truck increasing or decreasing? How fast? (Distance is measured along a straight line joining the car and the truck.)
(b) How does your answer change if the truck is going 70 mph instead of 80 mph?

47. The London Eye is a large Ferris wheel that has diameter 135 meters and revolves continuously. Passengers enter the cabins at the bottom of the wheel and complete one revolution in about 27 minutes. One minute into the ride a passenger is rising at 0.06 meters per second. How fast is the horizontal motion of the passenger at that moment?

48. Point P moves around the unit circle.[13] (See Figure 4.96.) The angle θ, in radians, changes with time as shown in Figure 4.97.

(a) Estimate the coordinates of P when $t = 2$.
(b) When $t = 2$, approximately how fast is the point P moving in the x-direction? In the y-direction?

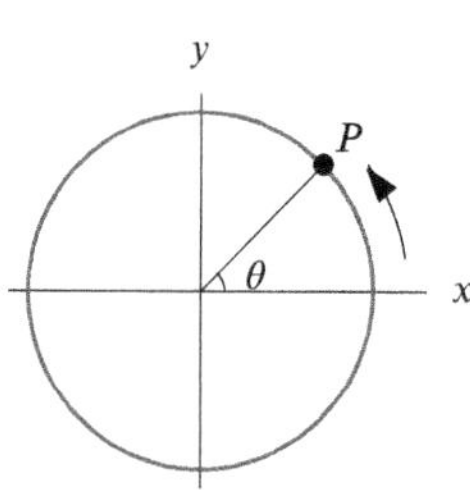

Figure 4.96

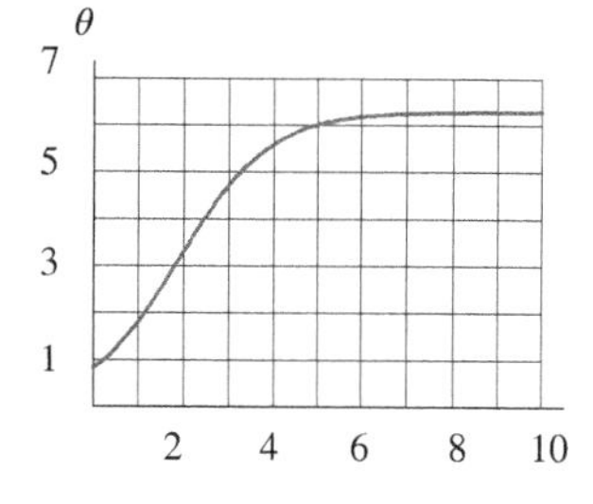

Figure 4.97

49. On February 16, 2007, paraglider Eva Wisnierska[14] was caught in a freak thunderstorm over Australia and carried upward at a speed of about 3000 ft/min. Table 4.4 gives the temperature at various heights. Approximately how fast (in °F/ per minute) was her temperature decreasing when she was at 4000 feet?

Table 4.4

y (thousand ft)	2	4	6	8	10	12	14	16
H (°F)	60	52	38	31	23	16	9	2

[13]Based on an idea from Caspar Curjel.
[14]www.sciencedaily.com, accessed May 15, 2007.

50. Coroners estimate time of death using the rule of thumb that a body cools about 2°F during the first hour after death and about 1°F for each additional hour. Assuming an air temperature of 68°F and a living body temperature of 98.6°F, the temperature $T(t)$ in °F of a body at a time t hours since death is given by

$$T(t) = 68 + 30.6e^{-kt}.$$

(a) For what value of k will the body cool by 2°F in the first hour?
(b) Using the value of k found in part (a), after how many hours will the temperature of the body be decreasing at a rate of 1°F per hour?
(c) Using the value of k found in part (a), show that, 24 hours after death, the coroner's rule of thumb gives approximately the same temperature as the formula.

51. A train is traveling at 0.8 km/min along a long straight track, moving in the direction shown in Figure 4.98. A movie camera, 0.5 km away from the track, is focused on the train.

(a) Express z, the distance between the camera and the train, as a function of x.
(b) How fast is the distance from the camera to the train changing when the train is 1 km from the camera? Give units.
(c) How fast is the camera rotating (in radians/min) at the moment when the train is 1 km from the camera?

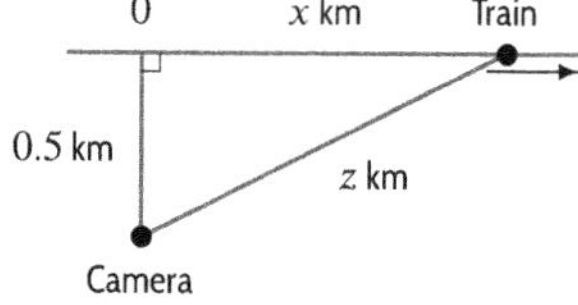

Figure 4.98

52. A lighthouse is 2 km from the long, straight coastline shown in Figure 4.99. Find the rate of change of the distance of the spot of light from the point O with respect to the angle θ.

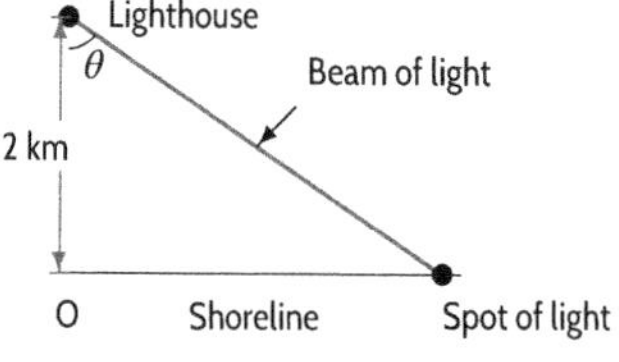

Figure 4.99

53. When the growth of a spherical cell depends on the flow of nutrients through the surface, it is reasonable to assume that the growth rate, dV/dt, is proportional to the surface area, S. Assume that for a particular cell $dV/dt = \frac{1}{3}S$. At what rate is its radius r increasing?

54. A circular region is irrigated by a 20 meter long pipe, fixed at one end and rotating horizontally, spraying water. One rotation takes 5 minutes. A road passes 30 meters from the edge of the circular area. See Figure 4.100.

(a) How fast is the end of the pipe, P, moving?
(b) How fast is the distance PQ changing when θ is $\pi/2$? When θ is 0?

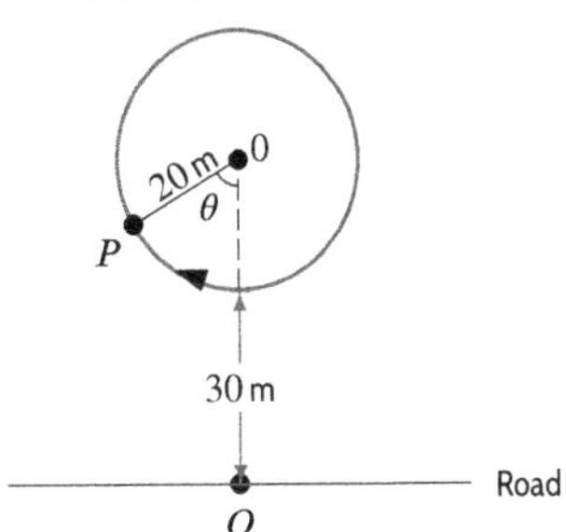

Figure 4.100

55. For the amusement of the guests, some hotels have elevators on the outside of the building. One such hotel is 300 feet high. You are standing by a window 100 feet above the ground and 150 feet away from the hotel, and the elevator descends at a constant speed of 30 ft/sec, starting at time $t = 0$, where t is time in seconds. Let θ be the angle between the line of your horizon and your line of sight to the elevator. (See Figure 4.101.)

(a) Find a formula for $h(t)$, the elevator's height above the ground as it descends from the top of the hotel.
(b) Using your answer to part (a), express θ as a function of time t and find the rate of change of θ with respect to t.
(c) The rate of change of θ is a measure of how fast the elevator appears to you to be moving. At what height is the elevator when it appears to be moving fastest?

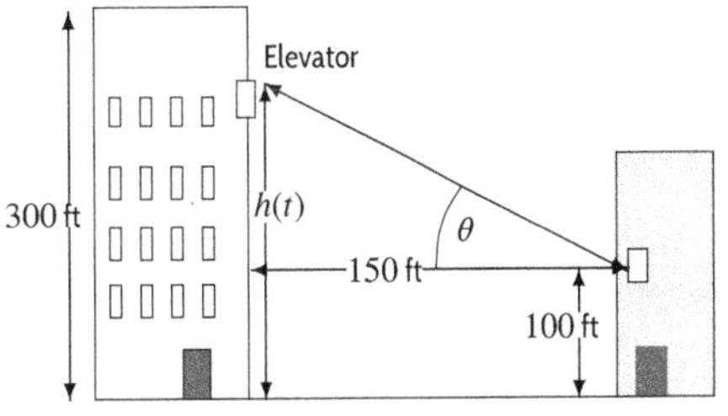

Figure 4.101

56. In a romantic relationship between Angela and Brian, who are unsuited for each other, $a(t)$ represents the affection Angela has for Brian at time t days after they meet, while $b(t)$ represents the affection Brian has for Angela at time t. If $a(t) > 0$ then Angela likes Brian; if $a(t) < 0$ then Angela dislikes Brian; if $a(t) = 0$ then Angela neither likes nor dislikes Brian. Their affection for each other is given by the relation $(a(t))^2 + (b(t))^2 = c$, where c is a constant.

(a) Show that $a(t) \cdot a'(t) = -b(t) \cdot b'(t)$.
(b) At any time during their relationship, the rate per day at which Brian's affection for Angela changes is $b'(t) = -a(t)$. Explain what this means if Angela

(i) Likes Brian, (ii) Dislikes Brian.

(c) Use parts (a) and (b) to show that $a'(t) = b(t)$. Explain what this means if Brian

(i) Likes Angela, (ii) Dislikes Angela.

(d) If $a(0) = 1$ and $b(0) = 1$ who first dislikes the other?

Strengthen Your Understanding

■ In Problems 57–58, explain what is wrong with the statement.

57. If the radius, R, of a circle increases at a constant rate, its diameter, D, increases at the same constant rate.

58. If two variables x and y are functions of t and are related by the equation $y = 1 - x^2$ then $dy/dt = -2x$.

■ In Problems 59–60, give an example of:

59. Two functions f and g where $y = f(x)$ and $x = g(t)$ such that dy/dt and dx/dt are both constant.

60. Two functions g and f where $x = g(t)$ and $y = f(x)$ such that dx/dt is constant and dy/dt is not constant.

■ Are the statements in Problems 61–62 true of false? Give an explanation for your answer.

61. If the radius of a circle is increasing at a constant rate, then so is the circumference.

62. If the radius of a circle is increasing at a constant rate, then so is the area.

63. The light in the lighthouse in Figure 4.102 rotates at 2 revolutions per minute. To calculate the speed at which the spot of light moves along the shore, it is best to differentiate:

(a) $r^2 = 5^2 + x^2$ **(b)** $x = r\sin\theta$
(c) $x = 5\tan\theta$ **(d)** $r^2 = 2^2 + x^2$

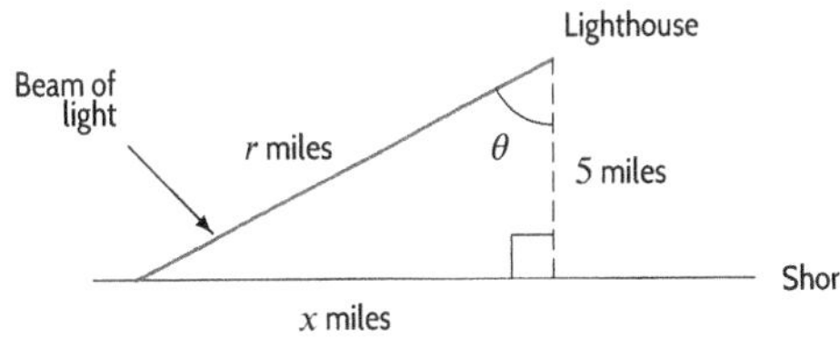

Figure 4.102

4.7 L'HOPITAL'S RULE, GROWTH, AND DOMINANCE

Suppose we want to calculate the exact value of the limit

$$\lim_{x\to 0}\frac{e^{2x}-1}{x}.$$

Substituting $x = 0$ gives us $0/0$, which is undefined:

$$\frac{e^{2(0)}-1}{0}=\frac{1-1}{0}=\frac{0}{0}.$$

Substituting values of x near 0 gives us an approximate value for the limit.

However, the limit can be calculated exactly using local linearity. Suppose we let $f(x)$ be the numerator, so $f(x) = e^{2x} - 1$, and $g(x)$ be the denominator, so $g(x) = x$. Then $f(0) = 0$ and $f'(x) = 2e^{2x}$, so $f'(0) = 2$. When we zoom in on the graph of $f(x) = e^{2x} - 1$ near the origin, we see its tangent line $y = 2x$ shown in Figure 4.103. We are interested in the ratio $f(x)/g(x)$, which is approximately the ratio of the y-values in Figure 4.103. So, for x near 0,

$$\frac{f(x)}{g(x)}=\frac{e^{2x}-1}{x}\approx\frac{2x}{x}=\frac{2}{1}=\frac{f'(0)}{g'(0)}.$$

As $x \to 0$, this approximation gets better, and we have

$$\lim_{x\to 0}\frac{e^{2x}-1}{x}=2.$$

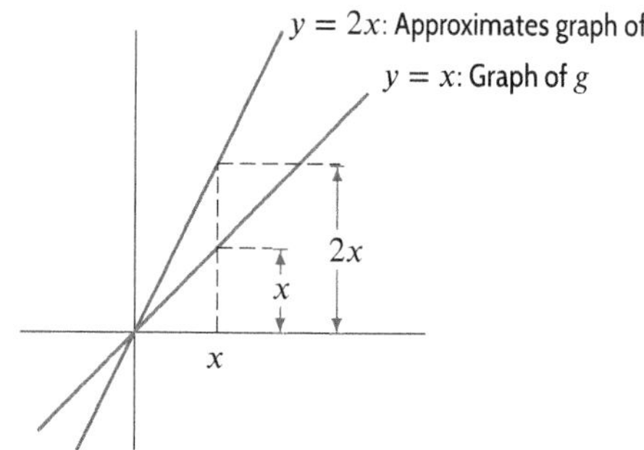

Figure 4.103: Ratio $(e^{2x}-1)/x$ is approximated by ratio of slopes as we zoom in near the origin

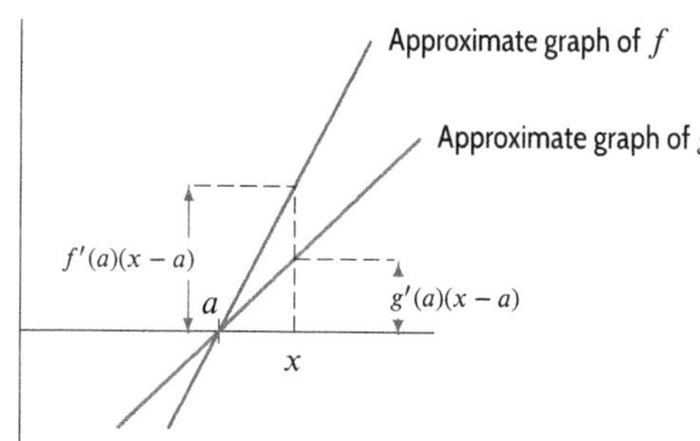

Figure 4.104: Ratio $f(x)/g(x)$ is approximated by ratio of slopes, $f'(a)/g'(a)$, as we zoom in at a

L'Hopital's Rule

If $f(a) = g(a) = 0$, we can use the same method to investigate limits of the form

$$\lim_{x\to a}\frac{f(x)}{g(x)}.$$

As in the previous case, we zoom in on the graphs of $f(x)$ and $g(x)$. Figure 4.104 shows that both graphs cross the x-axis at $x = a$. This suggests that the limit of $f(x)/g(x)$ as $x \to a$ is the ratio of slopes, giving the following result:

> **L'Hopital's rule:**[15] If f and g are differentiable, $f(a) = g(a) = 0$, and $g'(a) \neq 0$, then
>
> $$\lim_{x\to a}\frac{f(x)}{g(x)}=\frac{f'(a)}{g'(a)}.$$

Example 1 Use l'Hopital's rule to confirm that $\lim_{x\to 0}\frac{\sin x}{x}=1$.

[15] The marquis Guillaume de l'Hospital (1661–1704) was a French nobleman who wrote the first calculus text.

Solution Let $f(x) = \sin x$ and $g(x) = x$. Then $f(0) = g(0) = 0$ and $f'(x) = \cos x$ and $g'(x) = 1$. Thus,

$$\lim_{x\to 0} \frac{\sin x}{x} = \frac{\cos 0}{1} = 1.$$

If we also have $f'(a) = g'(a) = 0$, then we can use the following result:

> **More general form of l'Hopital's rule:** If f and g are differentiable and $f(a) = g(a) = 0$, then
>
> $$\lim_{x\to a} \frac{f(x)}{g(x)} = \lim_{x\to a} \frac{f'(x)}{g'(x)},$$
>
> provided the limit on the right exists.

Example 2 Calculate $\lim_{t\to 0} \dfrac{e^t - 1 - t}{t^2}$.

Solution Let $f(t) = e^t - 1 - t$ and $g(t) = t^2$. Then $f(0) = e^0 - 1 - 0 = 0$ and $g(0) = 0$, and $f'(t) = e^t - 1$ and $g'(t) = 2t$. So

$$\lim_{t\to 0} \frac{e^t - 1 - t}{t^2} = \lim_{t\to 0} \frac{e^t - 1}{2t}.$$

Since $f'(0) = g'(0) = 0$, the ratio $f'(0)/g'(0)$ is not defined. So we use l'Hopital's rule again:

$$\lim_{t\to 0} \frac{e^t - 1 - t}{t^2} = \lim_{t\to 0} \frac{e^t - 1}{2t} = \lim_{t\to 0} \frac{e^t}{2} = \frac{1}{2}.$$

We can also use l'Hopital's rule in cases involving infinity.

> **L'Hopital's rule applies to limits involving infinity**, provided f and g are differentiable. For a any real number or $\pm\infty$:
>
> - When $\lim_{x\to a} f(x) = \pm\infty$ and $\lim_{x\to a} g(x) = \pm\infty$,
>
> or
>
> - When $\lim_{x\to\infty} f(x) = \lim_{x\to\infty} g(x) = 0$.
>
> it can be shown that:
>
> $$\lim_{x\to a} \frac{f(x)}{g(x)} = \lim_{x\to a} \frac{f'(x)}{g'(x)},$$
>
> provided the limit on the right-hand side exists.

The next example shows how this version of l'Hopital's rule is used.

Example 3 Calculate $\lim_{x\to\infty} \dfrac{5x + e^{-x}}{7x}$.

Solution Let $f(x) = 5x + e^{-x}$ and $g(x) = 7x$. Then $\lim_{x\to\infty} f(x) = \lim_{x\to\infty} g(x) = \infty$, and $f'(x) = 5 - e^{-x}$ and $g'(x) = 7$, so

$$\lim_{x\to\infty} \frac{5x + e^{-x}}{7x} = \lim_{x\to\infty} \frac{(5 - e^{-x})}{7} = \frac{5}{7}.$$

L'Hopital's rule can also be applied to one-sided limits.

Dominance: Powers, Polynomials, Exponentials, and Logarithms

In Chapter 1, we see that some functions are much larger than others as $x \to \infty$. For positive functions f and g, we say that g *dominates* f as $x \to \infty$ if $\lim_{x\to\infty} \frac{f(x)}{g(x)} = 0$. L'Hopital's rule gives us an easy way of checking this.

Example 4 Check that $x^{1/2}$ dominates $\ln x$ as $x \to \infty$.

Solution We apply l'Hopital's rule to $(\ln x)/x^{1/2}$:

$$\lim_{x\to\infty} \frac{\ln x}{x^{1/2}} = \lim_{x\to\infty} \frac{1/x}{\frac{1}{2}x^{-1/2}}.$$

To evaluate this limit, we simplify and get

$$\lim_{x\to\infty} \frac{1/x}{\frac{1}{2}x^{-1/2}} = \lim_{x\to\infty} \frac{2x^{1/2}}{x} = \lim_{x\to\infty} \frac{2}{x^{1/2}} = 0.$$

Therefore we have

$$\lim_{x\to\infty} \frac{\ln x}{x^{1/2}} = 0,$$

which tells us that $x^{1/2}$ dominates $\ln x$ as $x \to \infty$.

Example 5 Check that any exponential function of the form e^{kx} (with $k > 0$) dominates any power function of the form Ax^p (with A and p positive) as $x \to \infty$.

Solution We apply l'Hopital's rule repeatedly to Ax^p/e^{kx}:

$$\lim_{x\to\infty} \frac{Ax^p}{e^{kx}} = \lim_{x\to\infty} \frac{Apx^{p-1}}{ke^{kx}} = \lim_{x\to\infty} \frac{Ap(p-1)x^{p-2}}{k^2e^{kx}} = \cdots$$

Keep applying l'Hopital's rule until the power of x is no longer positive. Then the limit of the numerator must be a finite number, while the limit of the denominator must be ∞. Therefore we have

$$\lim_{x\to\infty} \frac{Ax^p}{e^{kx}} = 0,$$

so e^{kx} dominates Ax^p.

Recognizing the Form of a Limit

Although expressions like $0/0$ and ∞/∞ have no numerical value, they are useful in describing the form of a limit. We can also use l'Hopital's rule to calculate some limits of the form $0 \cdot \infty$, $\infty - \infty$, 1^∞, 0^0, and ∞^0.

Example 6 Calculate $\lim_{x\to\infty} xe^{-x}$.

Solution Since $\lim_{x\to\infty} x = \infty$ and $\lim_{x\to\infty} e^{-x} = 0$, we see that

$$xe^{-x} \to \infty \cdot 0 \quad \text{as } x \to \infty.$$

Rewriting

$$\infty \cdot 0 \quad \text{as} \quad \frac{\infty}{\frac{1}{0}} = \frac{\infty}{\infty}$$

gives a form whose value can be determined using l'Hopital's rule, so we rewrite the function xe^{-x} as

$$xe^{-x} = \frac{x}{e^x} \to \frac{\infty}{\infty} \qquad \text{as } x \to \infty.$$

Taking $f(x) = x$ and $g(x) = e^x$ gives $f'(x) = 1$ and $g'(x) = e^x$, so

$$\lim_{x\to\infty} xe^{-x} = \lim_{x\to\infty} \frac{x}{e^x} = \lim_{x\to\infty} \frac{1}{e^x} = 0.$$

A Famous Limit

In the following example, l'Hopital's rule is applied to calculate a limit that can be used to define e.

Example 7 Evaluate $\lim\limits_{x\to\infty} \left(1 + \dfrac{1}{x}\right)^x$.

Solution As $x \to \infty$, we see that $\left(1 + \dfrac{1}{x}\right)^x \to 1^\infty$, a form whose value in this context is to be determined. Since $\ln 1^\infty = \infty \cdot \ln 1 = \infty \cdot 0$, we write

$$y = \left(1 + \frac{1}{x}\right)^x$$

and find the limit of $\ln y$:

$$\lim_{x\to\infty} \ln y = \lim_{x\to\infty} \ln\left(1 + \frac{1}{x}\right)^x = \lim_{x\to\infty} x \ln\left(1 + \frac{1}{x}\right) = \infty \cdot 0.$$

As in the previous example, we rewrite

$$\infty \cdot 0 \quad \text{as} \quad \frac{0}{\frac{1}{\infty}} = \frac{0}{0},$$

which suggests rewriting

$$\lim_{x\to\infty} x \ln\left(1 + \frac{1}{x}\right) \quad \text{as} \quad \lim_{x\to\infty} \frac{\ln(1 + 1/x)}{1/x}.$$

Since $\lim_{x\to\infty} \ln(1 + 1/x) = 0$ and $\lim_{x\to\infty}(1/x) = 0$, we can use l'Hopital's rule with $f(x) = \ln(1 + 1/x)$ and $g(x) = 1/x$. We have

$$f'(x) = \frac{1}{1 + 1/x}\left(-\frac{1}{x^2}\right) \qquad \text{and} \qquad g'(x) = -\frac{1}{x^2},$$

so

$$\lim_{x\to\infty} \ln y = \lim_{x\to\infty} \frac{\ln(1 + 1/x)}{1/x} = \lim_{x\to\infty} \frac{1}{1 + 1/x}\left(-\frac{1}{x^2}\right) \Big/ \left(-\frac{1}{x^2}\right)$$
$$= \lim_{x\to\infty} \frac{1}{1 + 1/x} = 1.$$

Since $\lim_{x\to\infty} \ln y = 1$, we have

$$\lim_{x\to\infty} y = e^1 = e.$$

Example 8 Put the following limits in a form that can be evaluated using l'Hopital's rule:

(a) $\lim_{x\to 0^+} x\ln x$ (b) $\lim_{x\to\infty} x^{1/x}$ (c) $\lim_{x\to 0^+} x^x$ (d) $\lim_{x\to 0} \frac{1}{x} - \frac{1}{\sin x}$

Solution (a) We have

$$\lim_{x\to 0^+} x\ln x = 0\cdot\infty.$$

We can rewrite

$$0\cdot\infty \quad \text{as} \quad \frac{\infty}{1/0} = \frac{\infty}{\infty}.$$

This corresponds to rewriting

$$\lim_{x\to 0^+} x\ln x \quad \text{as} \quad \lim_{x\to 0^+} \frac{\ln x}{1/x}.$$

This is an ∞/∞ form that can be evaluated using l'Hopital's rule. (Note that we could also have written

$$0\cdot\infty \quad \text{as} \quad \frac{0}{1/\infty} = \frac{0}{0},$$

but this leads to rewriting

$$\lim_{x\to 0^+} x\ln x \quad \text{as} \quad \lim_{x\to 0^+} \frac{x}{1/\ln x}.$$

It turns out that l'Hopital's rule fails to simplify this limit.)

(b) In this case we have a ∞^0 form, so we take the logarithm and get a $0\cdot\infty$ form:

$$\lim_{x\to\infty} \ln(x^{1/x}) = \lim_{x\to\infty} \frac{1}{x}\ln x = \lim_{x\to\infty} \frac{\ln x}{x}.$$

This is an ∞/∞ form that can be evaluated using l'Hopital's rule. Once we get the answer, we exponentiate it to get the original limit.

(c) Since $\lim_{x\to 0^+} x = 0$, this is a 0^0 form. If we take the logarithm, we get

$$\ln 0^0 = 0\cdot\ln 0 = 0\cdot\infty.$$

This corresponds to the limit

$$\lim_{x\to 0^+} \ln x^x = \lim_{x\to 0^+} x\ln x,$$

which is considered in part (a).

(d) We have

$$\lim_{x\to 0} \frac{1}{x} - \frac{1}{\sin x} = \frac{1}{0} - \frac{1}{0} = \infty - \infty.$$

Limits like this can often be calculated by adding the fractions:

$$\lim_{x\to 0} \frac{1}{x} - \frac{1}{\sin x} = \lim_{x\to 0} \frac{\sin x - x}{x\sin x},$$

giving a $0/0$ form that can be evaluated using l'Hopital's rule twice.

Exercises and Problems for Section 4.7

EXERCISES

■ In Exercises 1–12, for each limit, indicate whether l'Hopital's rule applies. You do not have to evaluate the limits.

1. $\lim_{x\to 0} \frac{\sin x}{5x}$
2. $\lim_{x\to 2} \frac{2e^x - x}{x}$
3. $\lim_{x\to 1} \frac{x^2 - 3x + 2}{x^2 - 1}$
4. $\lim_{x\to 0} \frac{e^{2x} - 1}{\cos x}$
5. $\lim_{x\to 0} \frac{\cos x}{x}$
6. $\lim_{x\to 0} \frac{e^{3x} - 1}{\sin x}$

7. $\lim_{x\to 1} \dfrac{x^2 - 2x + 5}{x^2 - 1}$

8. $\lim_{x\to 2} \dfrac{x^2 - 4}{3x - 6}$

9. $\lim_{x\to 0} \dfrac{2e^x - 2}{\cos x - 1}$

10. $\lim_{x\to 0} \dfrac{\cos x - 1}{e^x}$

11. $\lim_{x\to 0} \dfrac{e^x}{\sin x}$

12. $\lim_{x\to 0} \dfrac{\sin x - x}{x^2}$

■ In Exercises 13–38, find the limit. Use l'Hopital's rule if it applies.

13. $\lim_{x\to 2} \dfrac{x-2}{x^2 - 4}$

14. $\lim_{x\to 1} \dfrac{x^2 + 3x - 4}{x - 1}$

15. $\lim_{x\to 1} \dfrac{x^6 - 1}{x^4 - 1}$

16. $\lim_{x\to 0} \dfrac{e^x - 1}{\sin x}$

17. $\lim_{x\to 0} \dfrac{\sin x}{e^x}$

18. $\lim_{x\to 1} \dfrac{\ln x}{x - 1}$

19. $\lim_{x\to 0} \dfrac{e^{4x} - 1}{\cos x}$

20. $\lim_{x\to 1} \dfrac{x^a - 1}{x^b - 1}, b \neq 0$

21. $\lim_{x\to a} \dfrac{\sqrt[3]{x} - \sqrt[3]{a}}{x - a}, a \neq 0$

22. $\lim_{x\to 3} \dfrac{x^2 - 9}{x - 3}$

23. $\lim_{x\to 1} \dfrac{x^3 - 5x^2 + 4}{3x^2 + 4x - 7}$

24. $\lim_{x\to 2} \dfrac{x^3 - x^2 - 4}{5x^2 - 10x}$

25. $\lim_{x\to 1} \dfrac{2x^5 - 2}{3x^4 - 3x}$

26. $\lim_{x\to 0} \dfrac{\cos x - 1}{3x}$

27. $\lim_{x\to 0} \dfrac{\sin x + 5x}{8x}$

28. $\lim_{x\to 0} \dfrac{e^{3x} - 1}{5x}$

29. $\lim_{x\to 0} \dfrac{\sin(5x)}{3x^2}$

30. $\lim_{x\to 1} \dfrac{x^2 - 1}{4x^3 - 2}$

31. $\lim_{x\to 0} \dfrac{e^{5x}}{\cos x - 1}$

32. $\lim_{x\to 0} \dfrac{3x}{\cos x}$

33. $\lim_{x\to 1} \dfrac{3x^2 + 4}{x^2 + 3x + 5}$

34. $\lim_{x\to \infty} \dfrac{\sin x}{x}$

35. $\lim_{x\to \infty} \dfrac{\ln x}{x}$

36. $\lim_{x\to \infty} \dfrac{(\ln x)^3}{x^2}$

37. $\lim_{x\to \infty} \dfrac{5x + 1}{e^x}$

38. $\lim_{x\to \infty} \dfrac{x^2 - 4}{3x^2 + 5x + 1}$

■ In Exercises 39–42, which function dominates as $x \to \infty$?

39. x^5 and $0.1x^7$

40. $0.01x^3$ and $50x^2$

41. $\ln(x + 3)$ and $x^{0.2}$

42. x^{10} and $e^{0.1x}$

PROBLEMS

43. The functions f and g and their tangent lines at $(4, 0)$ are shown in Figure 4.105. Find $\lim_{x\to 4} \dfrac{f(x)}{g(x)}$.

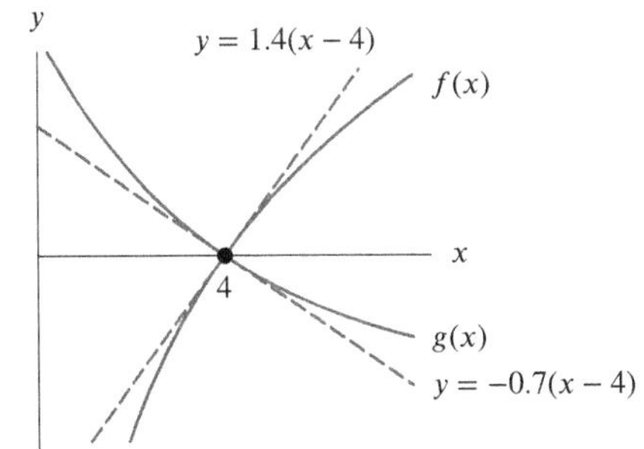

Figure 4.105

■ For Problems 44–47, find the sign of $\lim_{x\to a} \dfrac{f(x)}{g(x)}$ from the figure.

44. 45.

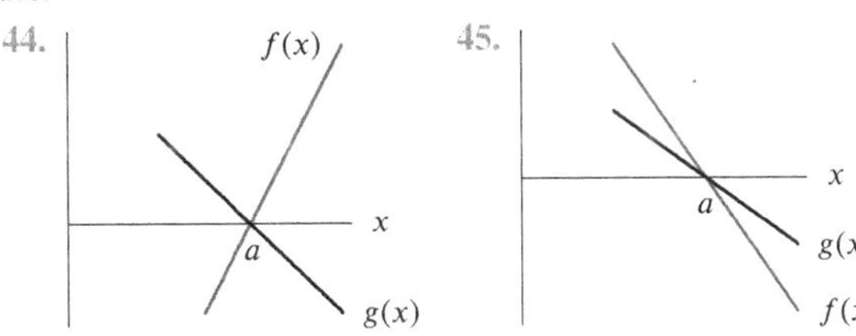

46. 47.

f(x) x a g(x) Assume $f''(a) \neq 0, g''(a) \neq 0$

f(x) x a g(x) Assume $f'''(a) \neq 0, g'''(a) \neq 0$

■ In Problems 48–51, based on your knowledge of the behavior of the numerator and denominator, predict the value of the limit. Then find each limit using l'Hopital's rule.

48. $\lim_{x\to 0} \dfrac{x^2}{\sin x}$

49. $\lim_{x\to 0} \dfrac{\sin^2 x}{x}$

50. $\lim_{x\to 0} \dfrac{\sin x}{x^{1/3}}$

51. $\lim_{x\to 0} \dfrac{x}{(\sin x)^{1/3}}$

■ In Problems 52–57, describe the form of the limit ($0/0$, ∞/∞, $\infty \cdot 0$, $\infty - \infty$, 1^∞, 0^0, ∞^0, or none of these). Does l'Hopital's rule apply? If so, explain how.

52. $\lim_{x\to \infty} \dfrac{x}{e^x}$

53. $\lim_{x\to 1} \dfrac{x}{x - 1}$

54. $\lim_{t\to \infty} \left(\dfrac{1}{t} - \dfrac{2}{t^2}\right)$

55. $\lim_{t\to 0^+} \dfrac{1}{t} - \dfrac{1}{e^t - 1}$

56. $\lim_{x\to 0} (1 + x)^x$

57. $\lim_{x\to \infty} (1 + x)^{1/x}$

In Problems 58–71 determine whether the limit exists, and where possible evaluate it.

58. $\lim_{x\to 1} \frac{\ln x}{x^2 - 1}$

59. $\lim_{t\to \pi} \frac{\sin^2 t}{t - \pi}$

60. $\lim_{n\to\infty} \sqrt[n]{n}$

61. $\lim_{x\to 0^+} x \ln x$

62. $\lim_{x\to 0} \frac{\sinh(2x)}{x}$

63. $\lim_{x\to 0} \frac{1 - \cosh(3x)}{x}$

64. $\lim_{x\to 1^-} \frac{\cos^{-1} x}{x - 1}$

65. $\lim_{x\to 0} \left(\frac{1}{x} - \frac{1}{\sin x}\right)$

66. $\lim_{t\to 0^+} \left(\frac{2}{t} - \frac{1}{e^t - 1}\right)$

67. $\lim_{t\to 0} \left(\frac{1}{t} - \frac{1}{e^t - 1}\right)$

68. $\lim_{x\to\infty} \left(1 + \sin\left(\frac{3}{x}\right)\right)^x$

69. $\lim_{t\to 0} \frac{\sin^2 At}{\cos At - 1}, A \neq 0$

70. $\lim_{t\to\infty} e^t - t^n$, where n is a positive integer

71. $\lim_{x\to 0^+} x^a \ln x$, where a is a positive constant.

In Problems 72–74, explain why l'Hopital's rule cannot be used to calculate the limit. Then evaluate the limit if it exists.

72. $\lim_{x\to 1} \frac{\sin(2x)}{x}$

73. $\lim_{x\to 0} \frac{\cos x}{x}$

74. $\lim_{x\to\infty} \frac{e^{-x}}{\sin x}$

In Problems 75–77, evaluate the limit using the fact that

$$\lim_{n\to\infty} \left(1 + \frac{1}{n}\right)^n = e.$$

75. $\lim_{x\to 0^+} (1 + x)^{1/x}$

76. $\lim_{n\to\infty} \left(1 + \frac{2}{n}\right)^n$

77. $\lim_{x\to 0^+} (1 + kx)^{t/x}; k > 0$

78. Show that $\lim_{n\to\infty} \left(1 - \frac{1}{n}\right)^n = e^{-1}$.

79. Use the result of Problem 78 to evaluate

$$\lim_{n\to\infty} \left(1 - \frac{\lambda}{n}\right)^n.$$

In Problems 80–82, evaluate the limits where

$$f(t) = \left(\frac{3^t + 5^t}{2}\right)^{1/t} \quad \text{for } t \neq 0.$$

80. $\lim_{t\to -\infty} f(t)$

81. $\lim_{t\to +\infty} f(t)$

82. $\lim_{t\to 0} f(t)$

In Problems 83–86, evaluate the limits as x approaches 0.

83. $\frac{\sinh(2x)}{x}$

84. $\frac{1 - \cosh(3x)}{x}$

85. $\frac{1 - \cosh(5x)}{x^2}$

86. $\frac{x - \sinh(x)}{x^3}$

Problems 87–89 are examples Euler used to illustrate l'Hopital's rule. Find the limit.

87. $\lim_{x\to 0} \frac{e^x - 1 - \ln(1 + x)}{x^2}$

88. $\lim_{x\to \pi/2} \frac{1 - \sin x + \cos x}{\sin x + \cos x - 1}$

89. $\lim_{x\to 1} \frac{x^x - x}{1 - x + \ln x}$

Strengthen Your Understanding

In Problems 90–91, explain what is wrong with the statement.

90. There is a positive integer n such that function x^n dominates e^x as $x \to \infty$.

91. L'Hopital's rule shows that

$$\lim_{x\to\infty} \frac{5x + \cos x}{x} = 5.$$

In Problems 92–93, give an example of:

92. A limit of a rational function for which l'Hopital's rule cannot be applied.

93. A function f such that L'Hopital's rule can be applied to find

$$\lim_{x\to\infty} \frac{f(x)}{\ln x}.$$

94. Is the following statement true of false? If $g'(a) \neq 0$, then $\lim_{x\to a} \frac{f(x)}{g(x)} = \frac{f'(a)}{g'(a)}$. Give an explanation for your answer.

95. Which of the limits cannot be computed with l'Hopital's rule?

(a) $\lim_{x\to 0} \frac{\sin x}{x}$

(b) $\lim_{x\to 0} \frac{\cos x}{x}$

(c) $\lim_{x\to 0} \frac{x}{\sin x}$

(d) $\lim_{x\to\infty} \frac{x}{e^x}$

4.8 PARAMETRIC EQUATIONS

Representing Motion in the Plane

To represent the motion of a particle in the xy-plane we use two equations, one for the x-coordinate of the particle, $x = f(t)$, and another for the y-coordinate, $y = g(t)$. Thus at time t the particle is at the point $(f(t), g(t))$. The equation for x describes the right-left motion; the equation for y describes the up-down motion. The equations for x and y are called *parametric equations* with *parameter* t.

Example 1 Describe the motion of the particle whose coordinates at time t are $x = \cos t$ and $y = \sin t$.

Solution Since $(\cos t)^2 + (\sin t)^2 = 1$, we have $x^2 + y^2 = 1$. That is, at any time t the particle is at a point (x, y) on the unit circle $x^2 + y^2 = 1$. We plot points at different times to see how the particle moves on the circle. (See Figure 4.106 and Table 4.5.) The particle completes one full trip counterclockwise around the circle every 2π units of time. Notice how the x-coordinate goes repeatedly back and forth from -1 to 1 while the y-coordinate goes repeatedly up and down from -1 to 1. The two motions combine to trace out a circle.

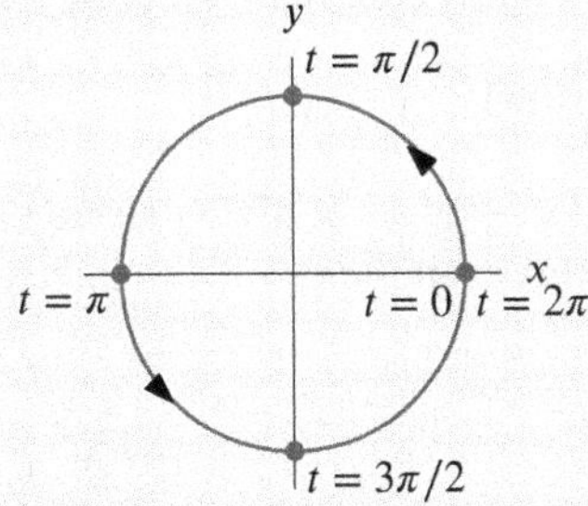

Figure 4.106: The circle parameterized by $x = \cos t$, $y = \sin t$

Table 4.5 *Points on the circle with* $x = \cos t$, $y = \sin t$

t	x	y
0	1	0
$\pi/2$	0	1
π	−1	0
$3\pi/2$	0	−1
2π	1	0

Example 2 Figure 4.107 shows the graphs of two functions, $f(t)$ and $g(t)$. Describe the motion of the particle whose coordinates at time t are $x = f(t)$ and $y = g(t)$.

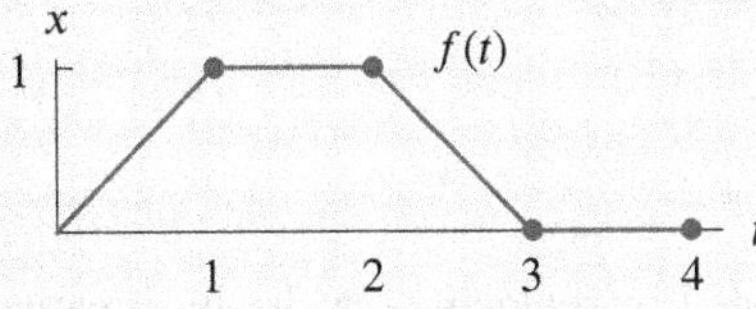

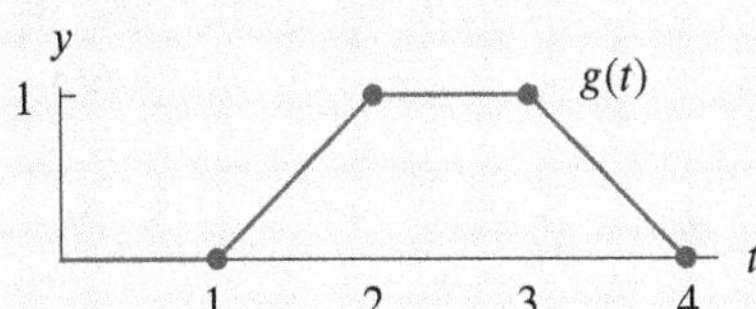

Figure 4.107: Graphs of $x = f(t)$ and $y = g(t)$ used to trace out the path $(f(t), g(t))$ in Figure 4.108

Solution Between times $t = 0$ and $t = 1$, the x-coordinate goes from 0 to 1, while the y-coordinate stays fixed at 0. So the particle moves along the x-axis from $(0, 0)$ to $(1, 0)$. Then, between times $t = 1$ and $t = 2$, the x-coordinate stays fixed at $x = 1$, while the y-coordinate goes from 0 to 1. Thus, the particle moves along the vertical line from $(1, 0)$ to $(1, 1)$. Similarly, between times $t = 2$ and $t = 3$, it moves horizontally back to $(0, 1)$, and between times $t = 3$ and $t = 4$ it moves down the y-axis to $(0, 0)$. Thus, it traces out the square in Figure 4.108.

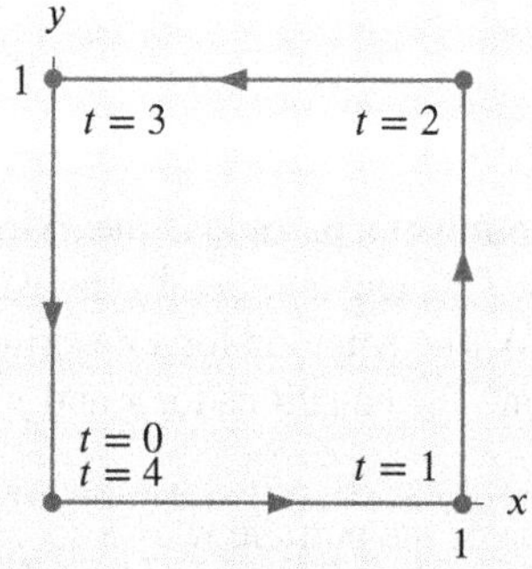

Figure 4.108: The square parameterized by $(f(t), g(t))$

Different Motions Along the Same Path

Example 3 Describe the motion of the particle whose x and y coordinates at time t are given by the equations

$$x = \cos(3t), \quad y = \sin(3t).$$

Solution Since $(\cos(3t))^2 + (\sin(3t))^2 = 1$, we have $x^2 + y^2 = 1$, giving motion around the unit circle. But from Table 4.6, we see that the particle in this example is moving three times as fast as the particle in Example 1. (See Figure 4.109.)

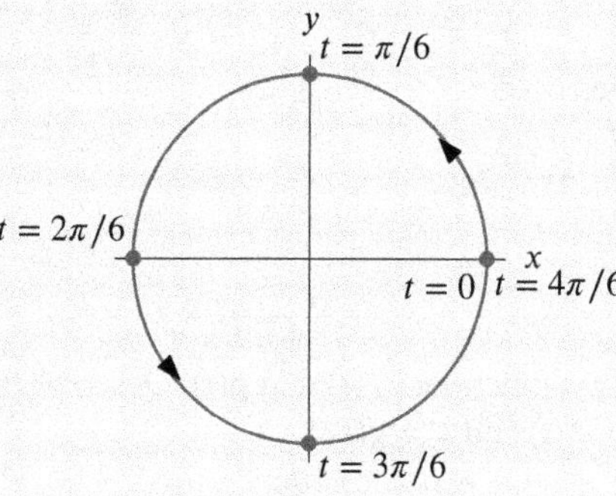

Figure 4.109: The circle parameterized by $x = \cos(3t)$, $y = \sin(3t)$

Table 4.6 *Points on circle with* $x = \cos(3t)$, $y = \sin(3t)$

t	x	y
0	1	0
$\pi/6$	0	1
$2\pi/6$	−1	0
$3\pi/6$	0	−1
$4\pi/6$	1	0

Example 3 is obtained from Example 1 by replacing t by $3t$; this is called a *change in parameter.* If we make a change in parameter, the particle traces out the same curve (or a part of it) but at a different speed or in a different direction.

Example 4 Describe the motion of the particle whose x and y coordinates at time t are given by

$$x = \cos(e^{-t^2}), \quad y = \sin(e^{-t^2}).$$

Solution As in Examples 1 and 3, we have $x^2 + y^2 = 1$, so the motion lies on the unit circle. As time t goes from $-\infty$ (way back in the past) to 0 (the present) to ∞ (way off in the future), e^{-t^2} goes from near 0 to 1 back to near 0. So $(x, y) = (\cos(e^{-t^2}), \sin(e^{-t^2}))$ goes from near $(1, 0)$ to $(\cos 1, \sin 1)$ and back to near $(1, 0)$. The particle does not actually reach the point $(1, 0)$. (See Figure 4.110 and Table 4.7.)

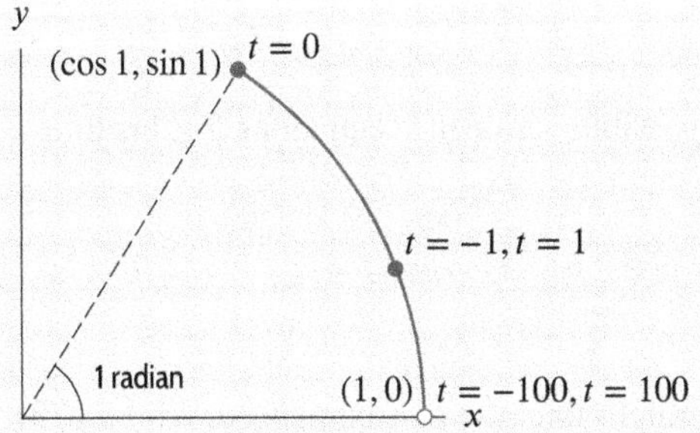

Figure 4.110: The circle parameterized by $x = \cos(e^{-t^2})$, $y = \sin(e^{-t^2})$

Table 4.7 *Points on circle with* $x = \cos(e^{-t^2})$, $y = \sin(e^{-t^2})$

t	x	y
−100	~ 1	~ 0
−1	0.93	0.36
0	0.54	0.84
1	0.93	0.36
100	~ 1	~ 0

Motion in a Straight Line

An object moves with constant speed along a straight line through the point (x_0, y_0). Both the x- and y-coordinates have a constant rate of change. Let $a = dx/dt$ and $b = dy/dt$. Then at time t the object has coordinates $x = x_0 + at$, $y = y_0 + bt$. (See Figure 4.111.) Notice that a represents the change in x in one unit of time, and b represents the change in y. Thus the line has slope $m = b/a$.

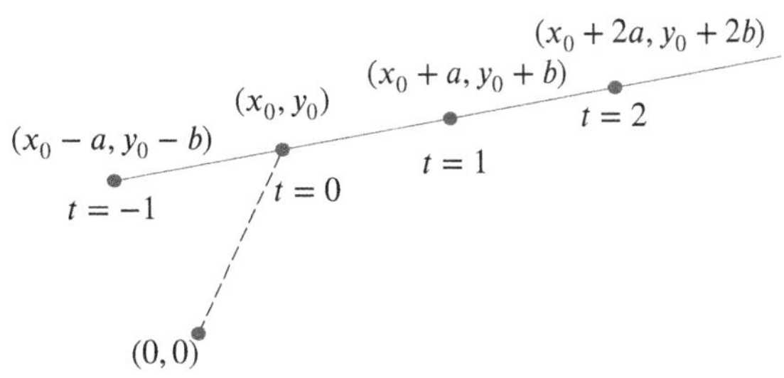

Figure 4.111: The line $x = x_0 + at$, $y = y_0 + bt$

This yields the following:

Parametric Equations for a Straight Line

An object moving along a line through the point (x_0, y_0), with $dx/dt = a$ and $dy/dt = b$, has parametric equations

$$x = x_0 + at, \quad y = y_0 + bt.$$

The slope of the line is $m = b/a$.

Example 5 Find parametric equations for:

(a) The line passing through the points $(2, -1)$ and $(-1, 5)$.
(b) The line segment from $(2, -1)$ to $(-1, 5)$.

Solution (a) Imagine an object moving with constant speed along a straight line from $(2, -1)$ to $(-1, 5)$, making the journey from the first point to the second in one unit of time. Then $dx/dt = ((-1) - 2)/1 = -3$ and $dy/dt = (5 - (-1))/1 = 6$. Thus the parametric equations are

$$x = 2 - 3t, \quad y = -1 + 6t.$$

(b) In the parameterization in part (a), $t = 0$ corresponds to the point $(2, -1)$ and $t = 1$ corresponds to the point $(-1, 5)$. So the parameterization of the segment is

$$x = 2 - 3t, \quad y = -1 + 6t, \qquad 0 \le t \le 1.$$

There are many other possible parametric equations for this line.

Speed and Velocity

An object moves along a straight line at a constant speed, with $dx/dt = a$ and $dy/dt = b$. In one unit of time, the object moves a units horizontally and b units vertically. Thus, by the Pythagorean Theorem, it travels a distance $\sqrt{a^2 + b^2}$. So its speed is

$$\text{Speed} = \frac{\text{Distance traveled}}{\text{Time taken}} = \frac{\sqrt{a^2 + b^2}}{1} = \sqrt{a^2 + b^2}.$$

For general motion along a curve with varying speed, we make the following definition:

The **instantaneous speed** of a moving object is defined to be

$$\text{Speed} = \sqrt{\left(\frac{dx}{dt}\right)^2 + \left(\frac{dy}{dt}\right)^2}.$$

The quantity $v_x = dx/dt$ is the **instantaneous velocity** in the x-direction; $v_y = dy/dt$ is the **instantaneous velocity** in the y-direction. The **velocity vector** $\vec{v}$ is written $\vec{v} = v_x \vec{i} + v_y \vec{j}$.

The quantities v_x and v_y are called the *components* of the velocity in the x- and y-directions. The velocity vector $\vec{v}$ is a useful way to keep track of the velocities in both directions using one mathematical object. The symbols $\vec{i}$ and $\vec{j}$ represent vectors of length one in the positive x and y-directions, respectively. For more about vectors, see Appendix D.

Example 6 A particle moves along a curve in the xy-plane with $x(t) = 2t + e^t$ and $y(t) = 3t - 4$, where t is time. Find the velocity vector and speed of the particle when $t = 1$.

Solution Differentiating gives

$$\frac{dx}{dt} = 2 + e^t, \quad \frac{dy}{dt} = 3.$$

When $t = 1$ we have $v_x = 2 + e$, $v_y = 3$. So the velocity vector is $\vec{v} = (2 + e)\vec{i} + 3\vec{j}$ and the speed is $\sqrt{(2 + e)^2 + 3^2} = \sqrt{13 + 4e + e^2} = 5.591$.

Example 7 A particle moves in the xy-plane with $x = 2t^3 - 9t^2 + 12t$ and $y = 3t^4 - 16t^3 + 18t^2$, where t is time.

(a) At what times is the particle

(i) Stopped? (ii) Moving parallel to the x- or y- axis?

(b) Find the speed of the particle at time t.

Solution (a) Differentiating gives

$$\frac{dx}{dt} = 6t^2 - 18t + 12, \quad \frac{dy}{dt} = 12t^3 - 48t^2 + 36t.$$

We are interested in the points at which $dx/dt = 0$ or $dy/dt = 0$. Solving gives

$$\frac{dx}{dt} = 6(t^2 - 3t + 2) = 6(t-1)(t-2) \quad \text{so} \quad \frac{dx}{dt} = 0 \quad \text{if } t = 1 \text{ or } t = 2.$$

$$\frac{dy}{dt} = 12t(t^2 - 4t + 3) = 12t(t-1)(t-3) \quad \text{so} \quad \frac{dy}{dt} = 0 \quad \text{if } t = 0,\ t = 1, \text{ or } t = 3.$$

(i) The particle is stopped if both dx/dt and dy/dt are 0, which occurs at $t = 1$.

(ii) The particle is moving parallel to the x-axis if $dy/dt = 0$ but $dx/dt \neq 0$. This occurs at $t = 0$ and $t = 3$. The particle is moving parallel to the y-axis if $dx/dt = 0$ but $dy/dt \neq 0$. This occurs at $t = 2$.

(b) We have

$$\text{Speed} = \sqrt{\left(\frac{dx}{dt}\right)^2 + \left(\frac{dy}{dt}\right)^2} = \sqrt{(6t^2 - 18t + 12)^2 + (12t^3 - 48t^2 + 36t)^2}.$$

Example 8 A child is sitting on a Ferris wheel of diameter 10 meters, making one revolution every 2 minutes. Find the speed of the child

(a) Using geometry. (b) Using a parameterization of the motion.

Solution (a) The child moves at a constant speed around a circle of radius 5 meters, completing one revolution every 2 minutes. One revolution around a circle of radius 5 is a distance of 10π, so the child's speed is $10\pi/2 = 5\pi \approx 15.7$ m/min. See Figure 4.112.

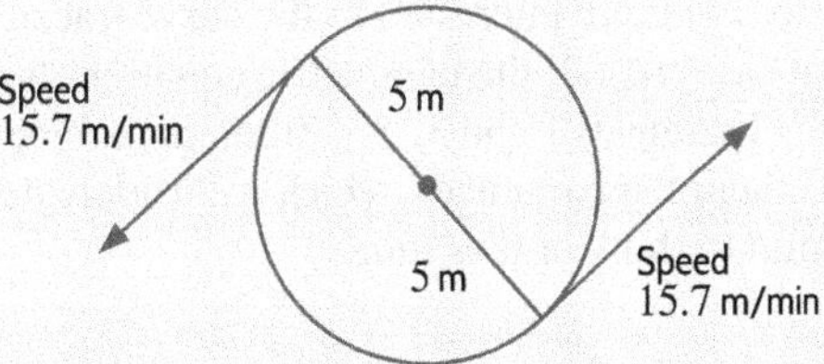

Figure 4.112: Motion of a child on a Ferris wheel at two different times is represented by the arrows. The direction of each arrow is the direction of motion at that time.

(b) The Ferris wheel has radius 5 meters and completes 1 revolution counterclockwise every 2 minutes. If the origin is at the center of the circle and we measure x and y in meters, the motion is parameterized by equations of the form

$$x = 5\cos(\omega t), \quad y = 5\sin(\omega t),$$

where ω is chosen to make the period 2 minutes. Since the period of $\cos(\omega t)$ and $\sin(\omega t)$ is $2\pi/\omega$, we must have

$$\frac{2\pi}{\omega} = 2, \quad \text{so} \quad \omega = \pi.$$

Thus, for t in minutes, the motion is described by the equations

$$x = 5\cos(\pi t), \quad y = 5\sin(\pi t).$$

So the speed is given by

$$\begin{aligned}\text{Speed} &= \sqrt{\left(\frac{dx}{dt}\right)^2 + \left(\frac{dy}{dt}\right)^2} \\ &= \sqrt{(-5\pi)^2 \sin^2(\pi t) + (5\pi)^2 \cos^2(\pi t)} = 5\pi\sqrt{\sin^2(\pi t) + \cos^2(\pi t)} = 5\pi \approx 15.7 \text{ m/min},\end{aligned}$$

which agrees with the speed we calculated in part (a).

Tangent Lines

To find the tangent line at a point (x_0, y_0) to a curve given parametrically, we find the straight line motion through (x_0, y_0) with the same velocity in the x and y directions as a particle moving along the curve.

Example 9 Find the tangent line at the point $(1, 2)$ to the curve defined by the parametric equations

$$x = t^3, \quad y = 2t.$$

Solution At time $t = 1$ the particle is at the point $(1, 2)$. The velocity in the x-direction at time t is $v_x = dx/dt = 3t^2$, and the velocity in the y-direction is $v_y = dy/dt = 2$. So at $t = 1$ the velocity in the x-direction is 3 and the velocity in the y-direction is 2. Thus the tangent line has parametric equations

$$x = 1 + 3t, \quad y = 2 + 2t.$$

Parametric Representations of Curves in the Plane

Sometimes we are more interested in the curve traced out by the particle than we are in the motion itself. In that case we call the parametric equations a *parameterization* of the curve. As we can see by comparing Examples 1 and 3, two different parameterizations can describe the same curve in the xy-plane. Though the parameter, which we usually denote by t, may not have physical meaning, it is often helpful to think of it as time.

Example 10 Give a parameterization of the semicircle of radius 1 shown in Figure 4.113.

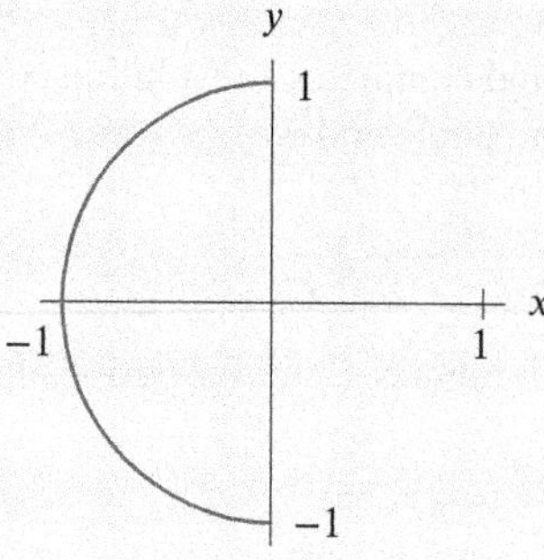

Figure 4.113: Parameterization of semicircle for Example 10

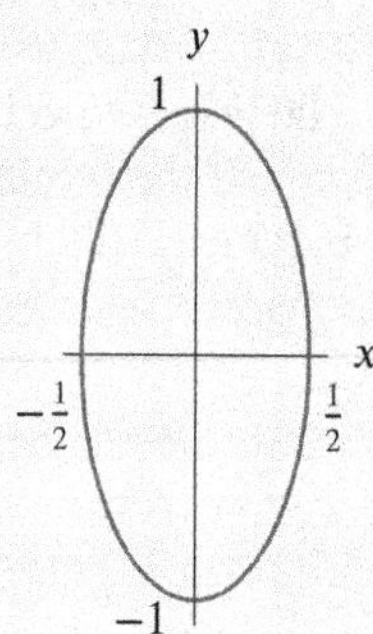

Figure 4.114: Parameterization of the ellipse $4x^2 + y^2 = 1$ for Example 11

Solution We can use the equations $x = \cos t$ and $y = \sin t$ for counterclockwise motion in a circle, from Example 1 on page 259. The particle passes $(0, 1)$ at $t = \pi/2$, moves counterclockwise around the circle, and reaches $(0, -1)$ at $t = 3\pi/2$. So a parameterization is

$$x = \cos t, \ y = \sin t, \quad \frac{\pi}{2} \le t \le \frac{3\pi}{2}.$$

To find the xy-equation of a curve given parametrically, we eliminate the parameter t in the parametric equations. In the previous example, we use the Pythagorean identity, so

$$\cos^2 t + \sin^2 t = 1 \quad \text{gives} \quad x^2 + y^2 = 1.$$

Example 11 Give a parameterization of the ellipse $4x^2 + y^2 = 1$ shown in Figure 4.114.

Solution Since $(2x)^2 + y^2 = 1$, we adapt the parameterization of the circle in Example 1. Replacing x by $2x$ gives the equations $2x = \cos t, y = \sin t$. A parameterization of the ellipse is thus

$$x = \tfrac{1}{2}\cos t, \qquad y = \sin t, \qquad 0 \le t \le 2\pi.$$

We usually require that the parameterization of a curve go from one end of the curve to the other without retracing any portion of the curve. This is different from parameterizing the motion of a particle, where, for example, a particle may move around the same circle many times.

Parameterizing the Graph of a Function

The graph of any function $y = f(x)$ can be parameterized by letting the parameter t be x:

$$x = t, \quad y = f(t).$$

Example 12 Give parametric equations for the curve $y = x^3 - x$. In which direction does this parameterization trace out the curve?

Solution Let $x = t, \ y = t^3 - t$. Thus, $y = t^3 - t = x^3 - x$. Since $x = t$, as time increases the x-coordinate moves from left to right, so the particle traces out the curve $y = x^3 - x$ from left to right.

Curves Given Parametrically

Some complicated curves can be graphed more easily using parametric equations; the next example shows such a curve.

Example 13 Assume t is time in seconds. Sketch the curve traced out by the particle whose motion is given by

$$x = \cos(3t), \quad y = \sin(5t).$$

Solution The x-coordinate oscillates back and forth between 1 and -1, completing 3 oscillations every 2π seconds. The y-coordinate oscillates up and down between 1 and -1, completing 5 oscillations every 2π seconds. Since both the x- and y-coordinates return to their original values every 2π seconds, the curve is retraced every 2π seconds. The result is a pattern called a Lissajous figure. (See Figure 4.115.)

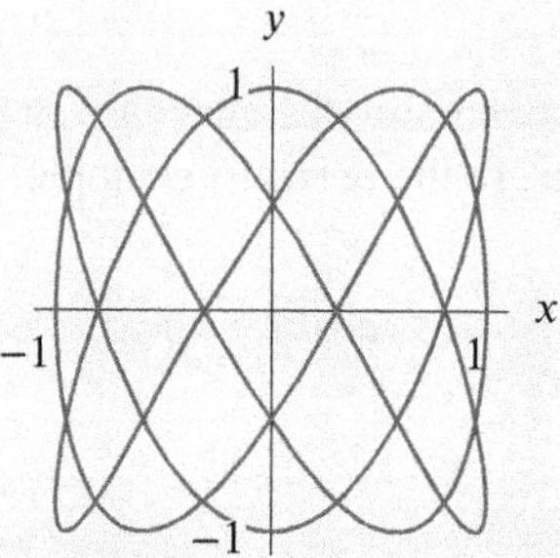

Figure 4.115: A Lissajous figure: $x = \cos(3t)$, $y = \sin(5t)$

Slope and Concavity of Parametric Curves

Suppose we have a curve traced out by the parametric equations $x = f(t)$, $y = g(t)$. To find the slope at a point on the curve, we could, in theory, eliminate the parameter t and then differentiate the function we obtain. However, the chain rule gives us an easier way.

Suppose the curve traced out by the parametric equations is represented by $y = h(x)$. (It may be represented by an implicit function.) Thinking of x and y as functions of t, the chain rule gives

$$\frac{dy}{dt} = \frac{dy}{dx} \cdot \frac{dx}{dt},$$

so we obtain the slope of the curve as a function of t:

$$\text{Slope of curve} = \frac{dy}{dx} = \frac{dy/dt}{dx/dt}.$$

We can find the second derivative, d^2y/dx^2, by a similar method and use it to investigate the concavity of the curve. The chain rule tells us that if w is any differentiable function of x, then

$$\frac{dw}{dx} = \frac{dw/dt}{dx/dt}.$$

For $w = dy/dx$, we have

$$\frac{dw}{dx} = \frac{d}{dx}\left(\frac{dy}{dx}\right) = \frac{d^2y}{dx^2},$$

so the chain rule gives the second derivative at any point on a parametric curve:

$$\frac{d^2y}{dx^2} = \frac{d}{dt}\left(\frac{dy}{dx}\right) \Big/ \frac{dx}{dt}.$$

Example 14 If $x = \cos t$, $y = \sin t$, find the point corresponding to $t = \pi/4$, the slope of the curve at the point, and d^2y/dx^2 at the point.

Solution The point corresponding to $t = \pi/4$ is $(\cos(\pi/4), \sin(\pi/4)) = \left(1/\sqrt{2}, 1/\sqrt{2}\right)$.

To find the slope, we use

$$\frac{dy}{dx} = \frac{dy/dt}{dx/dt} = \frac{\cos t}{-\sin t},$$

so when $t = \pi/4$,

$$\text{Slope} = \frac{\cos(\pi/4)}{-\sin(\pi/4)} = -1.$$

Thus, the curve has slope -1 at the point $(1/\sqrt{2}, 1/\sqrt{2})$. This is as we would expect, since the curve traced out is the circle of Example 1.

To find d^2y/dx^2, we use $w = dy/dx = -(\cos t)/(\sin t)$, so

$$\frac{d^2y}{dx^2} = \frac{d}{dt}\left(-\frac{\cos t}{\sin t}\right) \Big/ (-\sin t) = -\frac{(-\sin t)(\sin t) - (\cos t)(\cos t)}{\sin^2 t} \cdot \left(-\frac{1}{\sin t}\right) = -\frac{1}{\sin^3 t}.$$

Thus, at $t = \pi/4$

$$\left.\frac{d^2y}{dx^2}\right|_{t=\pi/4} = -\frac{1}{(\sin(\pi/4))^3} = -2\sqrt{2}.$$

Since the second derivative is negative, the concavity is negative. This is as expected, since the point is on the top half of the circle where the graph is bending downward.

Exercises and Problems for Section 4.8 Online Resource: Additional Problems for Section 4.8

EXERCISES

For Exercises 1–4, use the graphs of f and g to describe the motion of a particle whose position at time t is given by $x = f(t)$, $y = g(t)$.

1.

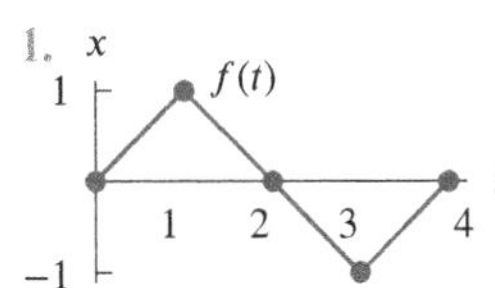

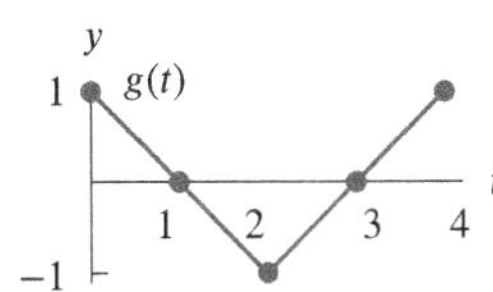

2.

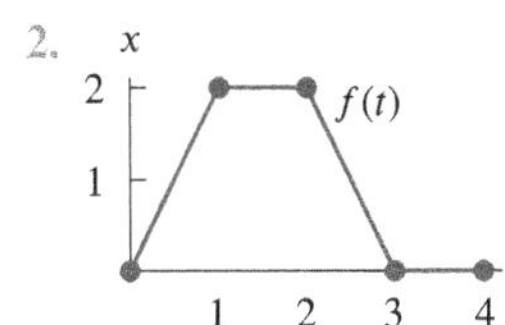

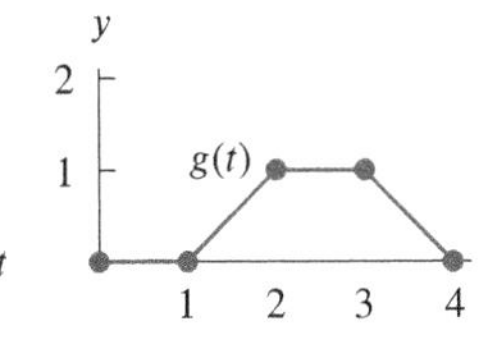

3.

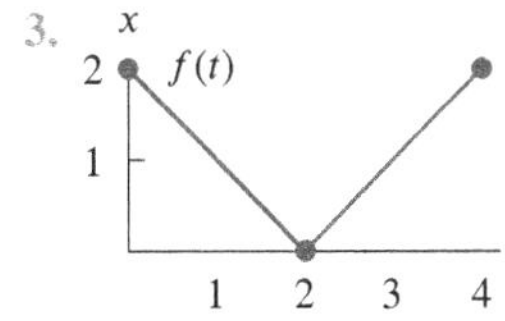

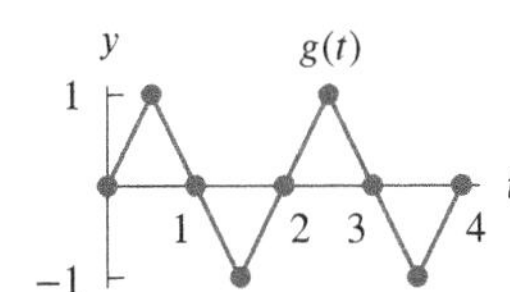

4.

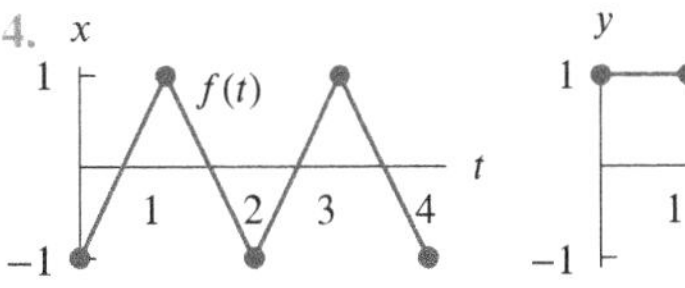

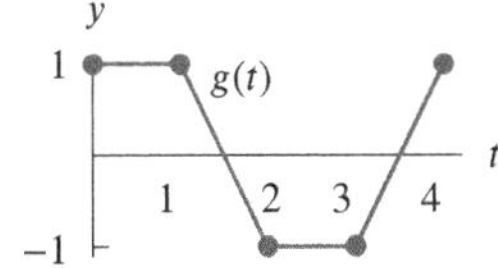

In Exercises 5–11, write a parameterization for the curves in the xy-plane.

5. A circle of radius 3 centered at the origin and traced out clockwise.

6. A vertical line through the point $(-2, -3)$.

7. A circle of radius 5 centered at the point $(2, 1)$ and traced out counterclockwise.

8. A circle of radius 2 centered at the origin traced clockwise starting from $(-2, 0)$ when $t = 0$.

9. The line through the points $(2, -1)$ and $(1, 3)$.

10. An ellipse centered at the origin and crossing the x-axis at ± 5 and the y-axis at ± 7.

11. An ellipse centered at the origin, crossing the x-axis at ± 3 and the y-axis at ± 7. Start at the point $(-3, 0)$ and trace out the ellipse counterclockwise.

■ Exercises 12–17 give parameterizations of the unit circle or a part of it. Describe in words how the circle is traced out, including when and where the particle is moving clockwise and when and where the particle is moving counterclockwise.

12. $x = \sin t, \quad y = \cos t$

13. $x = \cos t, \quad y = -\sin t$

14. $x = \cos(t^2), \quad y = \sin(t^2)$

15. $x = \cos(t^3 - t), \quad y = \sin(t^3 - t)$

16. $x = \cos(\ln t), \quad y = \sin(\ln t)$

17. $x = \cos(\cos t), \quad y = \sin(\cos t)$

■ In Exercises 18–20, what curves do the parametric equations trace out? Find the equation for each curve.

18. $x = 2 + \cos t,\ y = 2 - \sin t$

19. $x = 2 + \cos t,\ y = 2 - \cos t$

20. $x = 2 + \cos t,\ y = \cos^2 t$

■ In Exercises 21–26, the parametric equations describe the motion of a particle. Find an equation of the curve along which the particle moves.

21. $x = 3t + 1$, $y = t - 4$

22. $x = t^2 + 3$, $y = t^2 - 2$

23. $x = t + 4$, $y = t^2 - 3$

24. $x = \cos 3t$, $y = \sin 3t$

25. $x = 3\cos t$, $y = 3\sin t$

26. $x = 2 + 5\cos t$, $y = 7 + 5\sin t$

■ In Exercises 27–29, find an equation of the tangent line to the curve for the given value of t.

27. $x = t^3 - t, \quad y = t^2 \quad$ when $t = 2$

28. $x = t^2 - 2t, \quad y = t^2 + 2t \quad$ when $t = 1$

29. $x = \sin(3t), \quad y = \sin(4t) \quad$ when $t = \pi$

■ For Exercises 30–33, find the speed for the given motion of a particle. Find any times when the particle comes to a stop.

30. $x = t^2, \quad y = t^3$

31. $x = \cos(t^2), \quad y = \sin(t^2)$

32. $x = \cos 2t, \quad y = \sin t$

33. $x = t^2 - 4t,\ y = t^3 - 12t$

34. Find parametric equations for the tangent line at $t = 2$ for Problem 30.

PROBLEMS

■ Problems 35–36 show motion twice around a square, beginning at the origin at time $t = 0$ and parameterized by $x = f(t)$, $y = g(t)$. Sketch possible graphs of f and g consisting of line segments.

35.

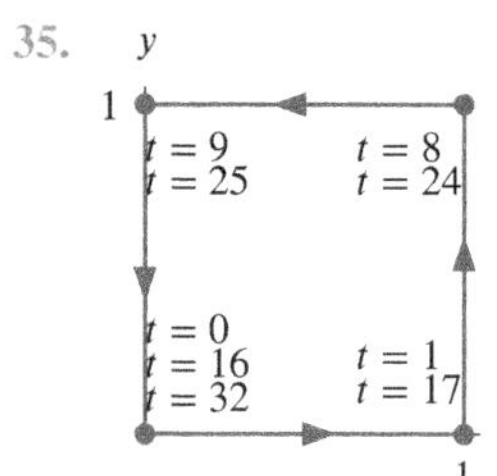

36.

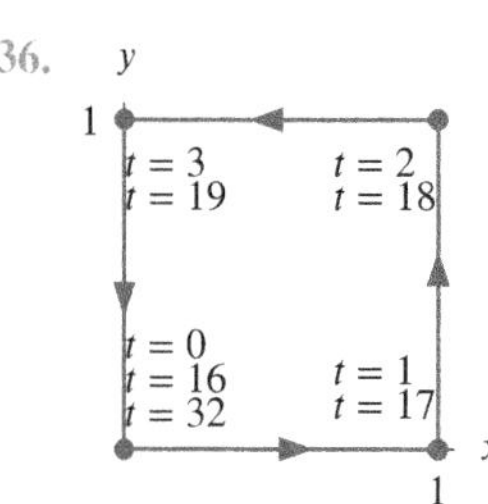

37. A line is parameterized by $x = 10 + t$ and $y = 2t$.

(a) What part of the line do we get by restricting t to $t < 0$?

(b) What part of the line do we get by restricting t to $0 \le t \le 1$?

38. A line is parameterized by $x = 2 + 3t$ and $y = 4 + 7t$.

(a) What part of the line is obtained by restricting t to nonnegative numbers?

(b) What part of the line is obtained if t is restricted to $-1 \le t \le 0$?

(c) How should t be restricted to give the part of the line to the left of the y-axis?

39. (a) Explain how you know that the following two pairs of equations parameterize the same line:

$$x = 2 + t, y = 4 + 3t \text{ and } x = 1 - 2t, y = 1 - 6t.$$

(b) What are the slope and y intercept of this line?

40. Describe the similarities and differences among the motions in the plane given by the following three pairs of parametric equations:
(a) $x = t, \quad y = t^2$ (b) $x = t^2, \quad y = t^4$
(c) $x = t^3, \quad y = t^6$.

41. What can you say about the values of a, b and k if the equations

$$x = a + k\cos t, \quad y = b + k\sin t, \quad 0 \le t \le 2\pi,$$

trace out the following circles in Figure 4.116?
(a) C_1 (b) C_2 (c) C_3

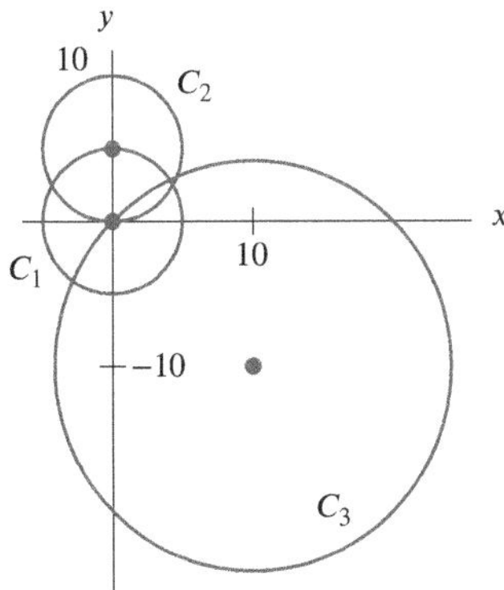

Figure 4.116

42. Suppose $a, b, c, d, m, n, p, q > 0$. Match each pair of parametric equations with one of the lines l_1, l_2, l_3, l_4 in Figure 4.117.

I. $\begin{cases} x = a + ct, \\ y = -b + dt. \end{cases}$ II. $\begin{cases} x = m + pt, \\ y = n - qt. \end{cases}$

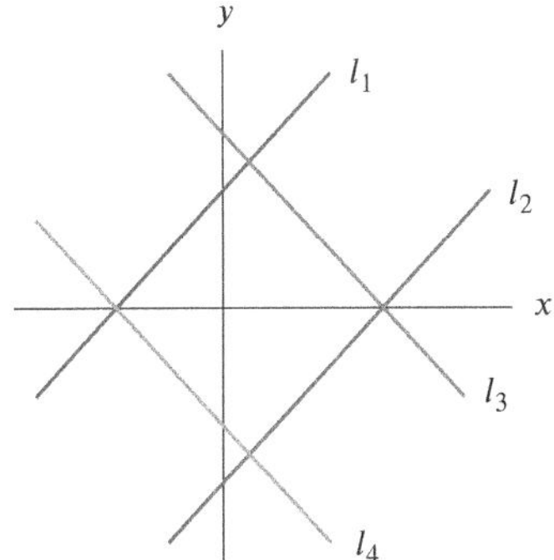

Figure 4.117

43. Describe in words the curve represented by the parametric equations

$$x = 3 + t^3, \quad y = 5 - t^3.$$

44. **(a)** Sketch the parameterized curve $x = t\cos t$, $y = t\sin t$ for $0 \le t \le 4\pi$.
(b) By calculating the position at $t = 2$ and $t = 2.01$, estimate the speed at $t = 2$.
(c) Use derivatives to calculate the speed at $t = 2$ and compare your answer to part (b).

45. The position of a particle at time t is given by $x = e^t$ and $y = 2e^{2t}$.

(a) Find dy/dx in terms of t.
(b) Eliminate the parameter and write y in terms of x.
(c) Using your answer to part (b), find dy/dx in terms of x.

46. At time t, the position of a particle moving on a curve is given by $x(t) = t^2 + 4$ and $y(t) = 3t + 5$. At $t = 1$:

(a) What is the position of the particle?
(b) What is the slope of the curve?
(c) What is the speed of the particle?

47. At time t, the position of a particle moving on a curve is given by $x(t) = 3t^2 - 1$ and $y(t) = t^2 - 3t$. At $t = 2$:

(a) What is the position of the particle?
(b) What is the slope of the curve?
(c) What is the speed of the particle?

48. There are many ways to parameterize the same line.

(a) Give the constant speed parameterization of a line if $t = 0$ and $t = 1$ correspond to the first and second points, respectively, on the line:

(i) $(2, 1); (5, 10)$ (ii) $(3, 4); (-1, -8)$

(b) Show that parts (i) and (ii) parameterize the same line by finding the equation of each line in slope-intercept form $y = b + mx$.

49. For x and y in meters, the motion of a particle is given by

$$x = t^3 - 3t, \quad y = t^2 - 2t,$$

where the y-axis is vertical and the x-axis is horizontal.

(a) Does the particle ever come to a stop? If so, when and where?
(b) Is the particle ever moving straight up or down? If so, when and where?
(c) Is the particle ever moving straight horizontally right or left? If so, when and where?

50. At time t, the position of a particle moving on a curve is given by $x = e^{2t} - e^{-2t}$ and $y = 3e^{2t} + e^{-2t}$.

(a) Find all values of t at which the curve has

(i) A horizontal tangent.

(ii) A vertical tangent.

(b) Find dy/dx in terms of t.
(c) Find $\lim_{t\to\infty} dy/dx$.

51. Figure 4.118 shows the graph of a parameterized curve $x = f(t)$, $y = f'(t)$ for a function $f(t)$.

(a) Is $f(t)$ an increasing or decreasing function?
(b) As t increases, is the curve traced from P to Q or from Q to P?
(c) Is $f(t)$ concave up or concave down?

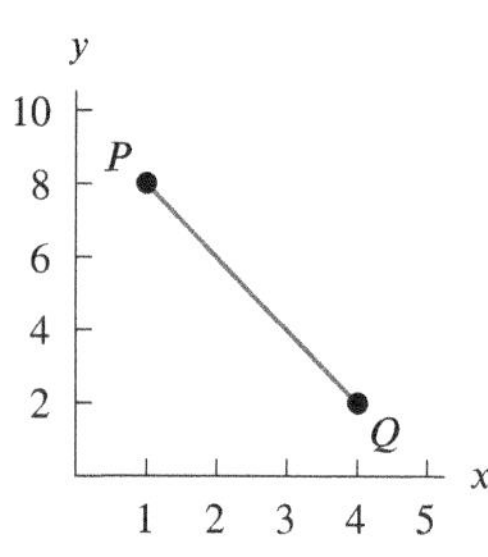

Figure 4.118

52. At time t, the position of a particle is $x(t) = 5\sin(2t)$ and $y(t) = 4\cos(2t)$, with $0 \le t < 2\pi$.

    (a) Graph the path of the particle for $0 \le t < 2\pi$, indicating the direction of motion.
    (b) Find the position and velocity of the particle when $t = \pi/4$.
    (c) How many times does the particle pass through the point found in part (b)?
    (d) What does your answer to part (b) tell you about the direction of the motion relative to the coordinate axes when $t = \pi/4$?
    (e) What is the speed of the particle at time $t = \pi$?

53. Two particles move in the xy-plane. At time t, the position of particle A is given by $x(t) = 4t - 4$ and $y(t) = 2t - k$, and the position of particle B is given by $x(t) = 3t$ and $y(t) = t^2 - 2t - 1$.

    (a) If $k = 5$, do the particles ever collide? Explain.
    (b) Find k so that the two particles do collide.
    (c) At the time that the particles collide in part (b), which particle is moving faster?

54. **(a)** Find d^2y/dx^2 for $x = t^3 + t, y = t^2$.
    (b) Is the curve concave up or down at $t = 1$?

55. **(a)** An object moves along the path $x = 3t$ and $y = \cos(2t)$, where t is time. Write the equation for the line tangent to this path at $t = \pi/3$.
    (b) Find the smallest positive value of t for which the y-coordinate is a local maximum.
    (c) Find d^2y/dx^2 when $t = 2$. What does this tell you about the concavity of the graph at $t = 2$?

56. The position of a particle at time t is given by $x = e^t + 3$ and $y = e^{2t} + 6e^t + 9$.

    (a) Find dy/dx in terms of t.
    (b) Find d^2y/dx^2. What does this tell you about the concavity of the graph?
    (c) Eliminate the parameter and write y in terms of x.
    (d) Using your answer from part (c), find dy/dx and d^2y/dx^2 in terms of x. Show that these answers are the same as the answers to parts (a) and (b).

57. A particle moves in the xy-plane with position at time t given by $x = \sin t$ and $y = \cos(2t)$ for $0 \le t < 2\pi$.

    (a) At what time does the particle first touch the x-axis? What is the speed of the particle at that time?
    (b) Is the particle ever at rest?
    (c) Discuss the concavity of the graph.

58. At time t, a projectile launched with angle of elevation α and initial velocity v_0 has position $x(t) = (v_0 \cos \alpha)t$ and $y(t) = (v_0 \sin \alpha)t - \frac{1}{2}gt^2$, where g is the acceleration due to gravity.

    (a) A football player kicks a ball at an angle of 36° above the ground with an initial velocity of 60 feet per second. Write the parametric equations for the position of the football at time t seconds. Use $g = 32$ ft/sec^2.
    (b) Graph the path that the football follows.
    (c) How long does it take for the football to hit the ground? How far is it from the spot where the football player kicked it?
    (d) What is the maximum height the football reaches during its flight?
    (e) At what speed is the football traveling 1 second after it was kicked?

Strengthen Your Understanding

■ In Problems 59–60, explain what is wrong with the statement.

59. The line segment from $(2, 2)$ to $(0, 0)$ is parameterized by $x = 2t$, $y = 2t$, $0 \le t \le 1$.

60. A circle of radius 2 centered at $(0, 1)$ is parameterized by $x = 2\cos \pi t$, $y = 2\sin \pi t$, $0 \le t \le 2$.

■ In Problems 61–62, give an example of:

61. A parameterization of a quarter circle centered at the origin of radius 2 in the first quadrant.

62. A parameterization of the line segment between $(0, 0)$ and $(1, 2)$.

■ Are the statements in Problems 63–64 true of false? Give an explanation for your answer.

63. The curve given parametrically by $x = f(t)$ and $y = g(t)$ has no sharp corners if f and g are differentiable.

64. If a curve is given parametrically by $x = \cos(t^2)$, $y = \sin(t^2)$, then its slope is $\tan(t^2)$.

Online Resource: Review problems and Projects

Chapter Five

KEY CONCEPT: THE DEFINITE INTEGRAL

Contents

5.1 HOW DO WE MEASURE DISTANCE TRAVELED?

For positive constant velocities, we can find the distance a moving object travels using the formula

$$\text{Distance} = \text{Velocity} \times \text{Time}.$$

In this section we see how to estimate the distance when the velocity is not a constant.

A Thought Experiment: How Far Did the Car Go?

Velocity Data Every Two Seconds

A car is moving with increasing velocity along a straight road. Table 5.1 shows the velocity every two seconds:

Table 5.1 *Velocity of car every two seconds*

Time (sec)	0	2	4	6	8	10
Velocity (ft/sec)	20	30	38	44	48	50

How far has the car traveled? Since we don't know how fast the car is moving at every moment, we can't calculate the distance exactly, but we can make an estimate. The velocity is increasing, so the car is going at least 20 ft/sec for the first two seconds. Since Distance = Velocity × Time, the car goes at least $20 \cdot 2 = 40$ feet during the first two seconds. Likewise, it goes at least $30 \cdot 2 = 60$ feet during the next two seconds, and so on. During the ten-second period it goes at least

$$20 \cdot 2 + 30 \cdot 2 + 38 \cdot 2 + 44 \cdot 2 + 48 \cdot 2 = 360 \text{ feet}.$$

Thus, 360 feet is an underestimate of the total distance traveled during the ten seconds.

To get an overestimate, we can reason this way: Since the velocity is increasing, during the first two seconds, the car's velocity is at most 30 ft/sec, so it moved at most $30 \cdot 2 = 60$ feet. In the next two seconds it moved at most $38 \cdot 2 = 76$ feet, and so on. Thus, over the ten-second period it moved at most

$$30 \cdot 2 + 38 \cdot 2 + 44 \cdot 2 + 48 \cdot 2 + 50 \cdot 2 = 420 \text{ feet}.$$

Therefore,

$$360 \text{ feet} \leq \text{Total distance traveled} \leq 420 \text{ feet}.$$

There is a difference of 60 feet between the upper and lower estimates.

Velocity Data Every One Second

What if we want a more accurate estimate? Then we make more frequent velocity measurements, say every second, as in Table 5.2.

As before, we get a lower estimate for each second by using the velocity at the beginning of that second. During the first second the velocity is at least 20 ft/sec, so the car travels at least $20 \cdot 1 = 20$ feet. During the next second the car moves at least 26 feet, and so on. We have

$$\begin{aligned} \text{New lower estimate} &= 20 \cdot 1 + 26 \cdot 1 + 30 \cdot 1 + 34 \cdot 1 + 38 \cdot 1 \\ &\quad + 41 \cdot 1 + 44 \cdot 1 + 46 \cdot 1 + 48 \cdot 1 + 49 \cdot 1 \\ &= 376 \text{ feet}. \end{aligned}$$

Table 5.2 *Velocity of car every second*

Time (sec)	0	1	2	3	4	5	6	7	8	9	10
Velocity (ft/sec)	20	26	30	34	38	41	44	46	48	49	50

Notice that this lower estimate is greater than the old lower estimate of 360 feet.

We get a new upper estimate by considering the velocity at the end of each second. During the first second the velocity is at most 26 ft/sec, so the car moves at most $26 \cdot 1 = 26$ feet; in the next second it moves at most 30 feet, and so on.

$$\begin{aligned}\text{New upper estimate} &= 26 \cdot 1 + 30 \cdot 1 + 34 \cdot 1 + 38 \cdot 1 + 41 \cdot 1 \\ &\quad + 44 \cdot 1 + 46 \cdot 1 + 48 \cdot 1 + 49 \cdot 1 + 50 \cdot 1 \\ &= 406 \text{ feet.}\end{aligned}$$

This is less than the old upper estimate of 420 feet. Now we know that

$$376 \text{ feet} \leq \text{Total distance traveled} \leq 406 \text{ feet.}$$

The difference between upper and lower estimates is now 30 feet, half of what it was before. By halving the interval of measurement, we have halved the difference between the upper and lower estimates.

Visualizing Distance on the Velocity Graph: Two-Second Data

We can represent both upper and lower estimates on a graph of the velocity. The graph also shows how changing the time interval between velocity measurements changes the accuracy of our estimates.

The velocity can be graphed by plotting the two-second data in Table 5.1 and drawing a curve through the data points. (See Figure 5.1.) The area of the first dark rectangle is $20 \cdot 2 = 40$, the lower estimate of the distance moved during the first two seconds. The area of the second dark rectangle is $30 \cdot 2 = 60$, the lower estimate for the distance moved in the next two seconds. The total area of the dark rectangles represents the lower estimate for the total distance moved during the ten seconds.

If the dark and light rectangles are considered together, the first area is $30 \cdot 2 = 60$, the upper estimate for the distance moved in the first two seconds. The second area is $38 \cdot 2 = 76$, the upper estimate for the next two seconds. The upper estimate for the total distance is represented by the sum of the areas of the dark and light rectangles. Therefore, the area of the light rectangles alone represents the difference between the two estimates.

To visualize the difference between the two estimates, look at Figure 5.1 and imagine the light rectangles all pushed to the right and stacked on top of each other. This gives a rectangle of width 2 and height 30. The height, 30, is the difference between the initial and final values of the velocity: $30 = 50 - 20$. The width, 2, is the time interval between velocity measurements.

Visualizing Distance on the Velocity Graph: One-Second Data

Figure 5.2 shows the velocities measured every second. The area of the dark rectangles again represents the lower estimate, and the area of the dark and light rectangles together represent the upper

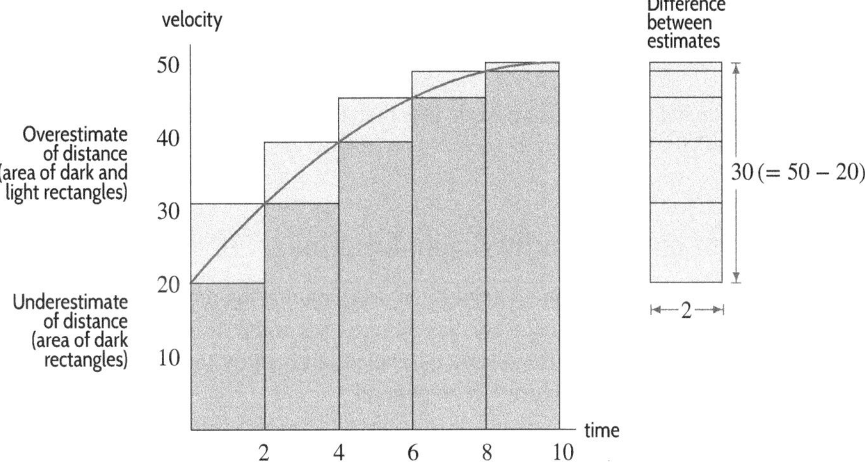

Figure 5.1: Velocity measured every 2 seconds

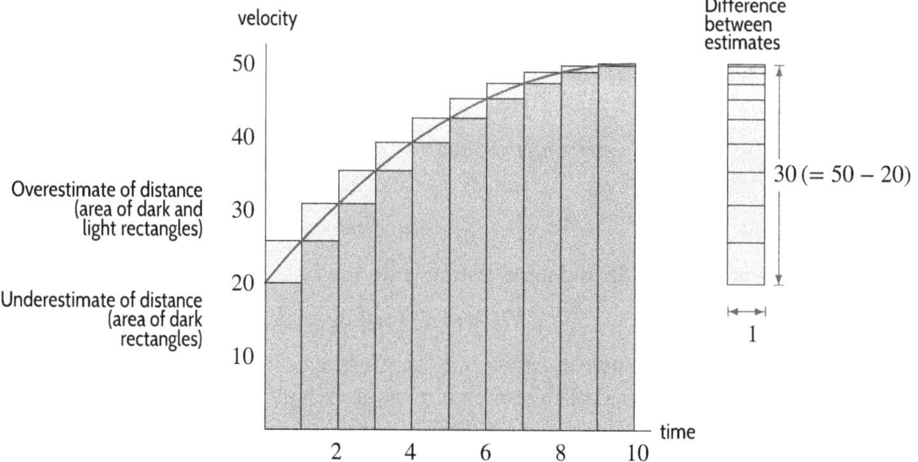

Figure 5.2: Velocity measured every second

estimate. As before, the difference between the two estimates is represented by the area of the light rectangles. This difference can be calculated by stacking the light rectangles vertically, giving a rectangle of the same height as before but of half the width. Its area is therefore half what it was before. Again, the height of this stack is $50 - 20 = 30$, but its width is now 1.

Example 1 What would be the difference between the upper and lower estimates if the velocity were given every tenth of a second? Every hundredth of a second? Every thousandth of a second?

Solution Every tenth of a second: Difference between estimates $= (50 - 20)(1/10) = 3$ feet.
Every hundredth of a second: Difference between estimates $= (50 - 20)(1/100) = 0.3$ feet.
Every thousandth of a second: Difference between estimates $= (50 - 20)(1/1000) = 0.03$ feet.

Example 2 How frequently must the velocity be recorded in order to estimate the total distance traveled to within 0.1 feet?

Solution The difference between the velocity at the beginning and end of the observation period is $50 - 20 = 30$. If the time between successive measurements is Δt, then the difference between the upper and lower estimates is $(30)\Delta t$. We want

$$(30)\Delta t < 0.1,$$

or

$$\Delta t < \frac{0.1}{30} = 0.0033.$$

So if the measurements are made less than 0.0033 seconds apart, the distance estimate is accurate to within 0.1 feet.

Visualizing Distance on the Velocity Graph: Area Under Curve

As we make more frequent velocity measurements, the rectangles used to estimate the distance traveled fit the curve more closely. See Figures 5.3 and 5.4. In the limit, as the number of subdivisions increases, we see that the distance traveled is given by the area between the velocity curve and the horizontal axis. See Figure 5.5. In general:

> If the velocity is positive, the total distance traveled is the area under the velocity curve.

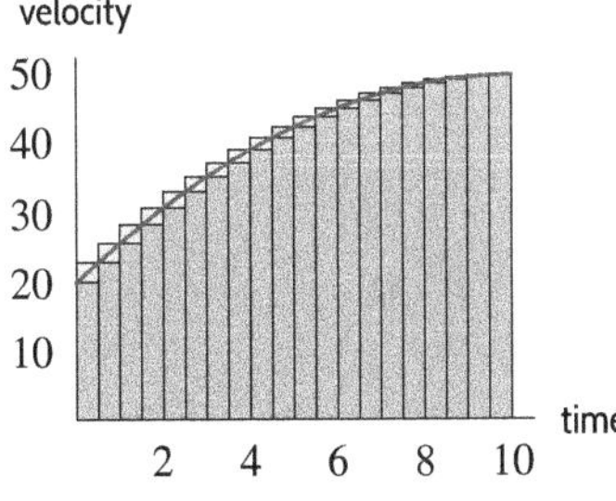

Figure 5.3: Velocity measured every 1/2 second

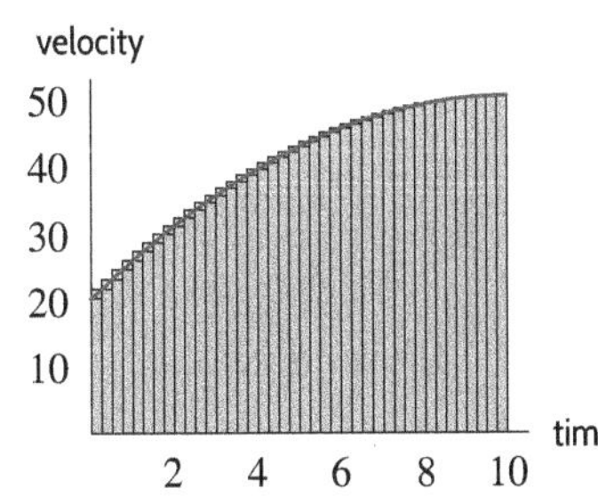

Figure 5.4: Velocity measured every 1/4 second

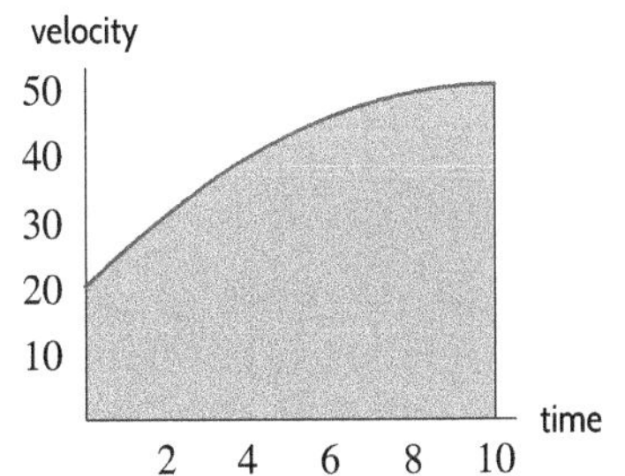

Figure 5.5: Distance traveled is area under curve

Example 3 With time t in seconds, the velocity of a bicycle, in feet per second, is given by $v(t) = 5t$. How far does the bicycle travel in 3 seconds?

Solution The velocity is linear. See Figure 5.6. The distance traveled is the area between the line $v(t) = 5t$ and the t-axis. Since this region is a triangle of height 15 and base 3,

$$\text{Distance traveled} = \text{Area of triangle} = \frac{1}{2} \cdot 15 \cdot 3 = 22.5 \text{ feet.}$$

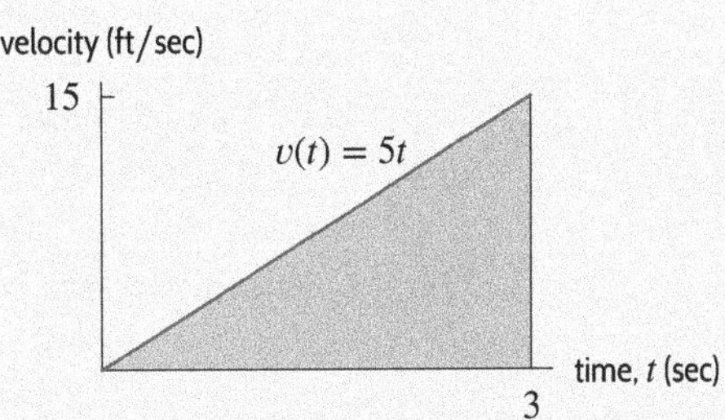

Figure 5.6: Shaded area represents distance traveled

Positive and Negative Velocity: Change in Position Versus Total Distance Traveled

In our thought experiment, we considered only motion in one direction; velocity was always positive. However if the car turns around and heads in the opposite direction, we say the velocity is negative. Then some terms in the sum are negative: Positive terms represent motion in the original direction, while negative terms represent motion in the opposite direction.

The sum now gives the *change in position of the car*, rather than the distance traveled. The change in position tells us how far the car is from its starting point and in which direction. To find the *distance traveled*, we take the absolute value of all terms in the sum.

Example 4 A particle moves along the y-axis with velocity 30 cm/sec for 5 seconds and velocity -10 cm/sec for the next 5 seconds. Positive velocity indicates upward motion; negative velocity represents downward motion. Explain what is represented by the sums

(a) $30 \cdot 5 + (-10) \cdot 5$ (b) $30 \cdot 5 + 10 \cdot 5$

Solution The particle moves upward at 30 cm/sec for 5 seconds, then turns around and moves downward at 10 cm/sec for 5 seconds. Thus the particle

$$\text{Moves upward:} \quad 30 \cdot 5 = 150 \text{ cm}$$
$$\text{Moves downward:} \quad 10 \cdot 5 = 50 \text{ cm.}$$

(a) The sum $30 \cdot 5 + (-10) \cdot 5$ is the distance moved upward *minus* the distance moved downward, representing

$$\text{Change in position} = 30 \cdot 5 + (-10) \cdot 5 = 150 - 50 = 100 \text{ cm upward.}$$

(b) The sum $30 \cdot 5 + 10 \cdot 5$ is the distance moved upward *plus* the distance moved downward, representing

$$\text{Distance traveled} = 30 \cdot 5 + 10 \cdot 5 = 150 + 50 = 200 \text{ cm.}$$

Figure 5.7 shows the velocity versus time. The area of the rectangle above the t-axis represents the distance moved upward, while the area of the rectangle below the t-axis represents the distance moved downward. The difference in their areas gives the change in position; the sum of their areas gives the total distance traveled.

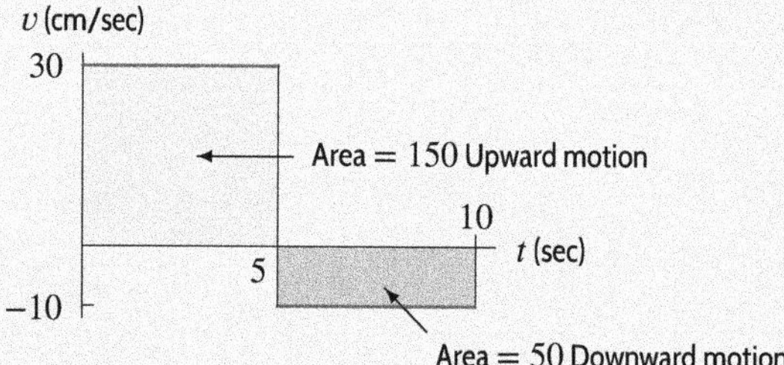

Figure 5.7: Areas give change in position and distance traveled

As we saw in Example 4, when velocity is negative as well as positive, there are two different sums. The sum using the velocity (positive and negative terms) gives change in position; the sum using the absolute value of velocity, that is, speed, (all positive terms) gives total distance traveled.

In terms of area:

> If the velocity can be negative as well as positive, then
> - **Change in position** is given by the area above axis *minus* area below the axis
> - **Distance traveled** is given by the area above axis *plus* area below the axis

Left and Right Sums

We now write the estimates for the distance traveled by the car in new notation. Let $v = f(t)$ denote any nonnegative velocity function. We want to find the distance traveled between times $t = a$ and $t = b$. We take measurements of $f(t)$ at equally spaced times $t_0, t_1, t_2, \ldots, t_n$, with time $t_0 = a$ and time $t_n = b$. The time interval between any two consecutive measurements is

$$\Delta t = \frac{b - a}{n},$$

where Δt means the change, or increment, in t.

During the first time interval, from t_0 and t_1, the velocity can be approximated by $f(t_0)$, so the distance traveled is approximately

$$f(t_0)\Delta t.$$

During the second time interval, the velocity is about $f(t_1)$, so the distance traveled is about

$$f(t_1)\Delta t.$$

Continuing in this way and adding all the estimates, we get an estimate for the total distance traveled. In the last interval, the velocity is approximately $f(t_{n-1})$, so the last term is $f(t_{n-1})\Delta t$:

$$\begin{array}{c}\text{Total distance traveled}\\ \text{between } t = a \text{ and } t = b\end{array} \approx f(t_0)\Delta t + f(t_1)\Delta t + f(t_2)\Delta t + \cdots + f(t_{n-1})\Delta t.$$

This is called a *left-hand sum* because we used the value of velocity from the left end of each time interval. It is represented by the sum of the areas of the rectangles in Figure 5.8.

We can also calculate a *right-hand sum* by using the value of the velocity at the right end of each time interval. In that case the estimate for the first interval is $f(t_1)\Delta t$, for the second interval it is $f(t_2)\Delta t$, and so on. The estimate for the last interval is now $f(t_n)\Delta t$, so

$$\begin{array}{c}\text{Total distance traveled}\\ \text{between } t = a \text{ and } t = b\end{array} \approx f(t_1)\Delta t + f(t_2)\Delta t + f(t_3)\Delta t + \cdots + f(t_n)\Delta t.$$

The right-hand sum is represented by the area of the rectangles in Figure 5.9.

If f is an increasing function, as in Figures 5.8 and 5.9, the left-hand sum is an underestimate and the right-hand sum is an overestimate of the total distance traveled. If f is decreasing, as in Figure 5.10, then the roles of the two sums are reversed.

Accuracy of Estimates

For either increasing or decreasing velocity functions, the exact value of the distance traveled lies somewhere between the two estimates. Thus, the accuracy of our estimate depends on how close these two sums are. For a function which is increasing throughout or decreasing throughout the interval $[a, b]$:

$$\left| \begin{array}{c}\text{Difference between}\\ \text{upper and lower estimates}\end{array} \right| = \left| \begin{array}{c}\text{Difference between}\\ f(a) \text{ and } f(b)\end{array} \right| \cdot \Delta t = |f(b) - f(a)| \cdot \Delta t.$$

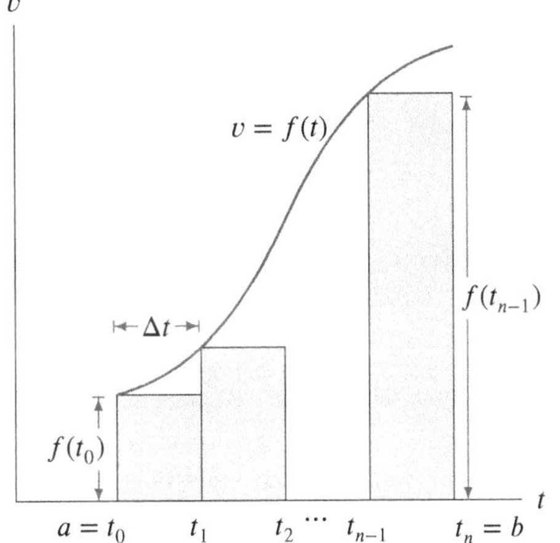

Figure 5.8: Left-hand sums

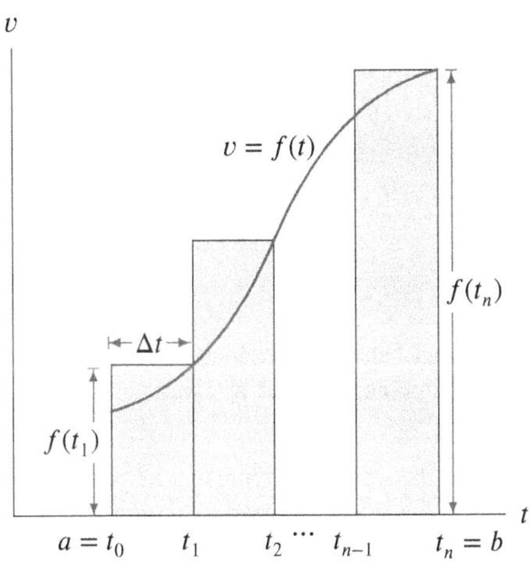

Figure 5.9: Right-hand sums

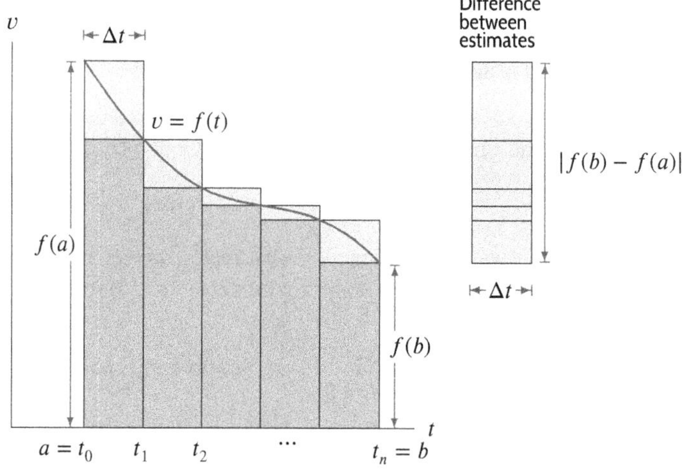

Figure 5.10: Left and right sums if f is decreasing

(Absolute values make the differences nonnegative.) In Figure 5.10, the area of the light rectangles is the difference between estimates. By making the time interval, Δt, between measurements small enough, we can make this difference between lower and upper estimates as small as we like.

Exercises and Problems for Section 5.1 Online Resource: Additional Problems for Section 5.1

EXERCISES

1. Figure 5.11 shows the velocity of a car for $0 \leq t \leq 12$ and the rectangles used to estimate the distance traveled.

   (a) Do the rectangles represent a left or a right sum?
   (b) Do the rectangles lead to an upper or a lower estimate?
   (c) What is the value of n?
   (d) What is the value of Δt?
   (e) Give an approximate value for the estimate.

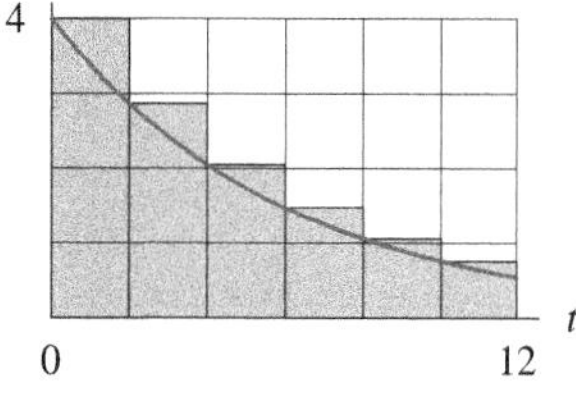

Figure 5.11

2. Figure 5.12 shows the velocity of a car for $0 \leq t \leq 24$ and the rectangles used to estimate the distance traveled.

   (a) Do the rectangles represent a left or a right sum?
   (b) Do the rectangles lead to an upper or a lower estimate?
   (c) What is the value of n?
   (d) What is the value of Δt?
   (e) Estimate the distance traveled.

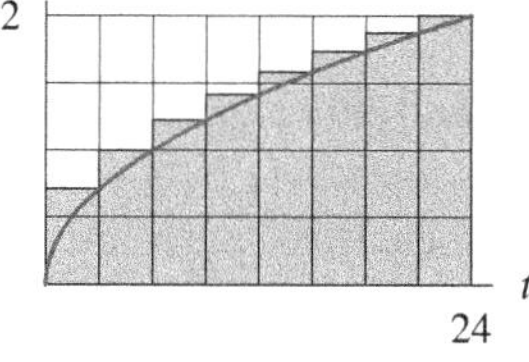

Figure 5.12

3. Figure 5.13 shows the velocity of an object for $0 \leq t \leq 6$. Calculate the following estimates of the distance the object travels between $t = 0$ and $t = 6$, and indicate whether each result is an upper or lower estimate of the distance traveled.

   (a) A left sum with $n = 2$ subdivisions
   (b) A right sum with $n = 2$ subdivisions

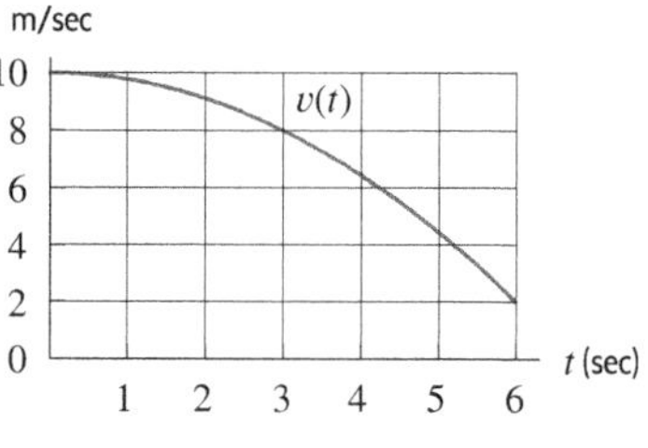

Figure 5.13

4. Figure 5.14 shows the velocity of an object for $0 \le t \le 8$. Calculate the following estimates of the distance the object travels between $t = 0$ and $t = 8$, and indicate whether each is an upper or lower estimate of the distance traveled.

   (a) A left sum with $n = 2$ subdivisions
   (b) A right sum with $n = 2$ subdivisions

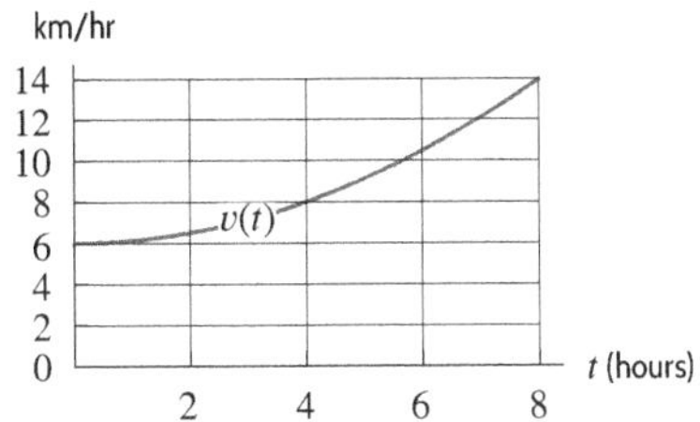

Figure 5.14

5. Figure 5.15 shows the velocity, v, of an object (in meters/sec). Estimate the total distance the object traveled between $t = 0$ and $t = 6$.

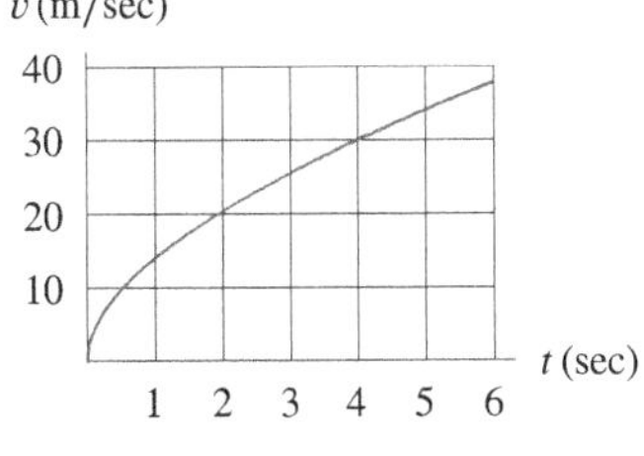

Figure 5.15

6. A bicyclist traveling at 20 ft/sec puts on the brakes to slow down at a constant rate, coming to a stop in 3 seconds.

   (a) Figure 5.16 shows the velocity of the bike during braking. What are the values of a and b in the figure?
   (b) How far does the bike travel while braking?

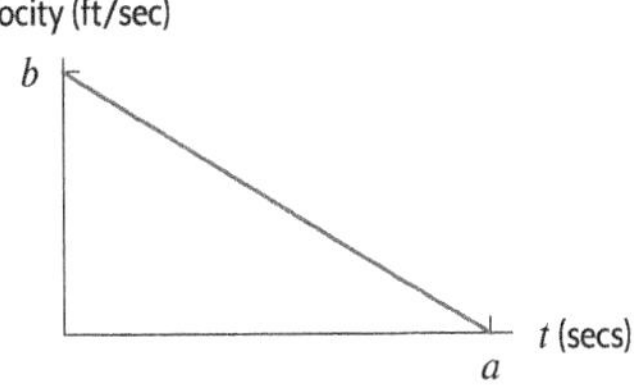

Figure 5.16

7. A bicyclist accelerates at a constant rate, from 0 ft/sec to 15 ft/sec in 10 seconds.

   (a) Figure 5.17 shows the velocity of the bike while it is accelerating. What is the value of b in the figure?
   (b) How far does the bike travel while it is accelerating?

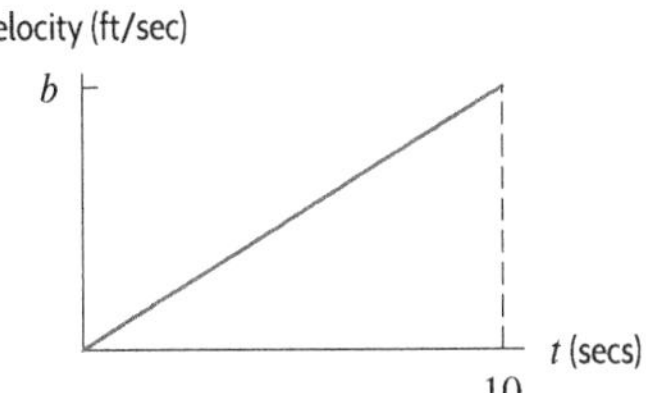

Figure 5.17

8. A car accelerates at a constant rate from 44 ft/sec to 88 ft/sec in 5 seconds.

   (a) Figure 5.18 shows the velocity of the car while it is accelerating. What are the values of a, b and c in the figure?
   (b) How far does the car travel while it is accelerating?

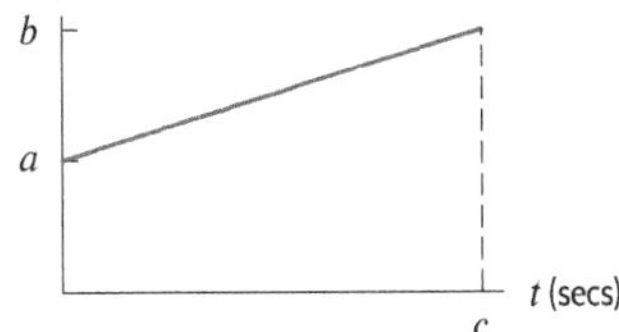

Figure 5.18

9. A car slows down at a constant rate from 90 ft/sec to 20 ft/sec in 12 seconds.

   (a) Figure 5.19 shows the velocity of the car while it is slowing down. What are the values of a, b and c in the figure?
   (b) How far does the car travel while it is slowing down?

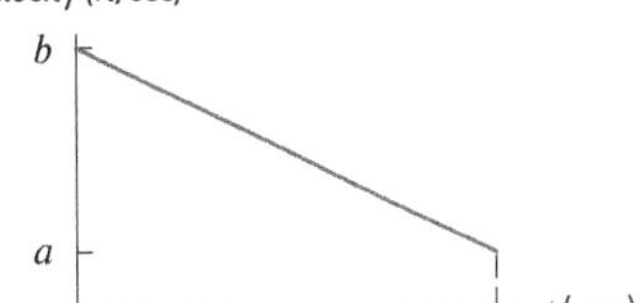

Figure 5.19

10. The velocity $v(t)$ in Table 5.3 is increasing, $0 \le t \le 12$.

   (a) Find an upper estimate for the total distance traveled using

   (i) $n = 4$ (ii) $n = 2$

   (b) Which of the two answers in part (a) is more accurate? Why?
   (c) Find a lower estimate of the total distance traveled using $n = 4$.

Table 5.3

t	0	3	6	9	12
$v(t)$	34	37	38	40	45

11. The velocity $v(t)$ in Table 5.4 is decreasing, $2 \le t \le 12$. Using $n = 5$ subdivisions to approximate the total distance traveled, find

(a) An upper estimate **(b)** A lower estimate

Table 5.4

t	2	4	6	8	10	12
$v(t)$	44	42	41	40	37	35

12. A car comes to a stop five seconds after the driver applies the brakes. While the brakes are on, the velocities in the table are recorded.

(a) Give lower and upper estimates of the distance the car traveled after the brakes were applied.
(b) On a sketch of velocity against time, show the lower and upper estimates of part (a).
(c) Find the difference between the estimates. Explain how this difference can be visualized on the graph in part (b).

Time since brakes applied (sec)	0	1	2	3	4	5
Velocity (ft/sec)	88	60	40	25	10	0

13. Table 5.5 gives the ground speed of a small plane accelerating for takeoff. Find upper and lower estimates for the distance traveled by the plane during takeoff.

Table 5.5

Time (sec)	0	2	4	6	8	10
Speed (m/s)	2.7	2.7	4	6.3	8.5	11.6
Time (sec)	12	14	16	18	20	
Speed (m/s)	13.4	17.4	21.9	29.1	32.6	

14. Figure 5.20 shows the velocity of a particle, in cm/sec, along a number line for time $-3 \le t \le 3$.

(a) Describe the motion in words: Is the particle changing direction or always moving in the same direction? Is the particle speeding up or slowing down?
(b) Make over and underestimates of the distance traveled for $-3 \le t \le 3$.

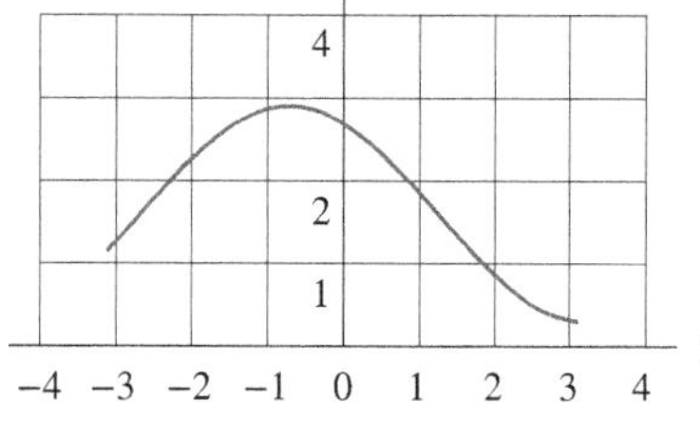

Figure 5.20

15. At time, t, in seconds, your velocity, v, in meters/second, is given by

$$v(t) = 1 + t^2 \quad \text{for} \quad 0 \le t \le 6.$$

Use $\Delta t = 2$ to estimate the distance traveled during this time. Find the upper and lower estimates, and then average the two.

16. For time, t, in hours, $0 \le t \le 1$, a bug is crawling at a velocity, v, in meters/hour given by

$$v = \frac{1}{1+t}.$$

Use $\Delta t = 0.2$ to estimate the distance that the bug crawls during this hour. Find an overestimate and an underestimate. Then average the two to get a new estimate.

■ Exercises 17–20 show the velocity, in cm/sec, of a particle moving along a number line. (Positive velocities represent movement to the right; negative velocities to the left.) Find the change in position and total distance traveled between times $t = 0$ and $t = 5$ seconds.

17. **18.**

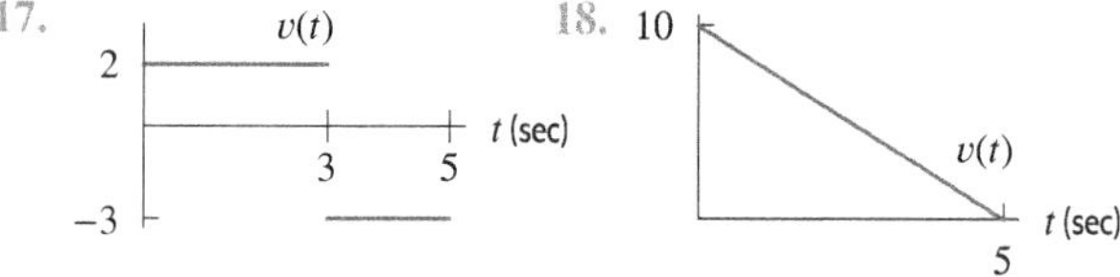

19. **20.**

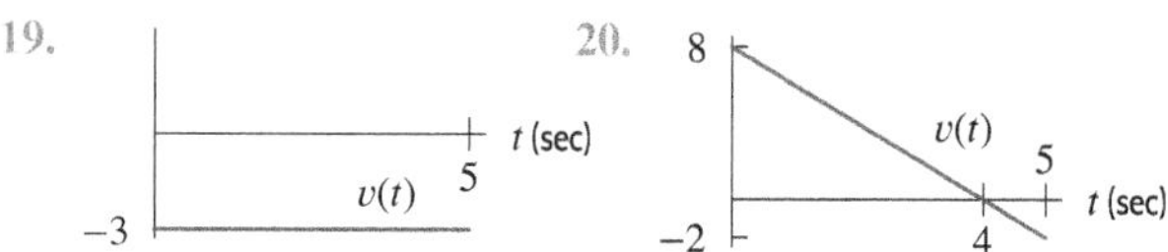

21. The velocity of a car, in ft/sec, is $v(t) = 10t$ for t in seconds, $0 \le t \le 6$.

(a) Use $\Delta t = 2$ to give upper and lower estimates for the distance traveled. What is their average?
(b) Find the distance traveled using the area under the graph of $v(t)$. Compare it to your answer for part (a).

22. A particle moves with velocity $v(t) = 3 - t$ along the x-axis, with time t in seconds, $0 \le t \le 4$.

(a) Use $\Delta t = 1$ to give upper and lower estimates for the total displacement. What is their average?
(b) Graph $v(t)$. Give the total displacement.

23. Use the expressions for left and right sums on page 277 and Table 5.6.

 (a) If $n = 4$, what is Δt? What are t_0, t_1, t_2, t_3, t_4? What are $f(t_0), f(t_1), f(t_2), f(t_3), f(t_4)$?
 (b) Find the left and right sums using $n = 4$.
 (c) If $n = 2$, what is Δt? What are t_0, t_1, t_2? What are $f(t_0), f(t_1), f(t_2)$?
 (d) Find the left and right sums using $n = 2$.

Table 5.6

t	15	17	19	21	23
$f(t)$	10	13	18	20	30

24. Use the expressions for left and right sums on page 277 and Table 5.7.

 (a) If $n = 4$, what is Δt? What are t_0, t_1, t_2, t_3, t_4? What are $f(t_0), f(t_1), f(t_2), f(t_3), f(t_4)$?
 (b) Find the left and right sums using $n = 4$.
 (c) If $n = 2$, what is Δt? What are t_0, t_1, t_2? What are $f(t_0), f(t_1), f(t_2)$?
 (d) Find the left and right sums using $n = 2$.

Table 5.7

t	0	4	8	12	16
$f(t)$	25	23	22	20	17

PROBLEMS

25. Roger runs a marathon. His friend Jeff rides behind him on a bicycle and clocks his speed every 15 minutes. Roger starts out strong, but after an hour and a half he is so exhausted that he has to stop. Jeff's data follow:

Time since start (min)	0	15	30	45	60	75	90
Speed (mph)	12	11	10	10	8	7	0

 (a) Assuming that Roger's speed is never increasing, give upper and lower estimates for the distance Roger ran during the first half hour.
 (b) Give upper and lower estimates for the distance Roger ran in total during the entire hour and a half.
 (c) How often would Jeff have needed to measure Roger's speed in order to find lower and upper estimates within 0.1 mile of the actual distance he ran?

26. The velocity of a particle moving along the x-axis is given by $f(t) = 6 - 2t$ cm/sec. Use a graph of $f(t)$ to find the exact change in position of the particle from time $t = 0$ to $t = 4$ seconds.

In Problems 27–30, find the difference between the upper and lower estimates of the distance traveled at velocity $f(t)$ on the interval $a \le t \le b$ for n subdivisions.

27. $f(t) = 5t + 8, a = 1, b = 3, n = 100$

28. $f(t) = 25 - t^2, a = 1, b = 4, n = 500$

29. $f(t) = \sin t, a = 0, b = \pi/2, n = 100$

30. $f(t) = e^{-t^2/2}, a = 0, b = 2, n = 20$

31. A 2015 Porsche 918 Spyder accelerates from 0 to 88 ft/sec (60 mph) in 2.2 seconds, the fastest acceleration of any car available for retail sale in 2015.[1]

 (a) Assuming that the acceleration is constant, graph the velocity from $t = 0$ to $t = 2.2$.
 (b) How far does the car travel during this time?

32. A baseball thrown directly upward at 96 ft/sec has velocity $v(t) = 96 - 32t$ ft/sec at time t seconds.

 (a) Graph the velocity from $t = 0$ to $t = 6$.
 (b) When does the baseball reach the peak of its flight? How high does it go?
 (c) How high is the baseball at time $t = 5$?

33. Figure 5.21 gives your velocity during a trip starting from home. Positive velocities take you away from home and negative velocities take you toward home. Where are you at the end of the 5 hours? When are you farthest from home? How far away are you at that time?

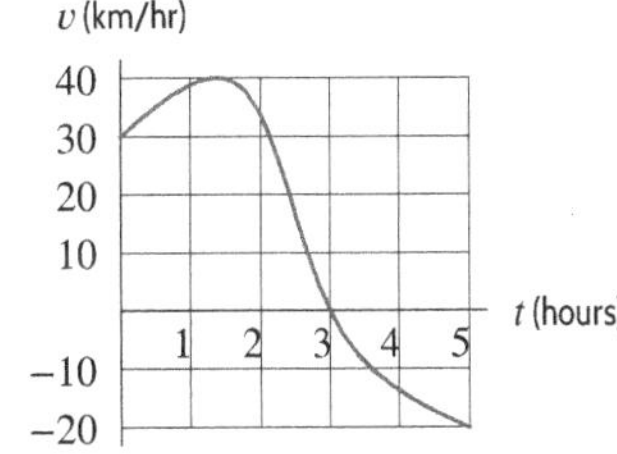

Figure 5.21

34. When an aircraft attempts to climb as rapidly as possible, its climb rate decreases with altitude. (This occurs because the air is less dense at higher altitudes.) The table shows performance data for a single-engine aircraft.

Altitude (1000 ft)	0	1	2	3	4	5
Climb rate (ft/min)	925	875	830	780	730	685
Altitude (1000 ft)	6	7	8	9	10	
Climb rate (ft/min)	635	585	535	490	440	

 (a) Calculate upper and lower estimates for the time required for this aircraft to climb from sea level to 10,000 ft.
 (b) If climb rate data were available in increments of 500 ft, what would be the difference between a lower and upper estimate of climb time based on 20 subdivisions?

[1]K. C. Colwell, "First Test: 2015 Porsche 918 Spyder," *Car and Driver*, August 1, 2014.

35. Table 5.8 shows the upward vertical velocity $v(t)$, in ft/min, of a small plane at time t seconds during a short flight.

(a) When is the plane going up? Going down?
(b) If the airport is located 110 ft above sea level, estimate the maximum altitude the plane reaches during the flight.

Table 5.8

t	$v(t)$	t	$v(t)$
0	10	100	−140
10	20	110	−180
20	60	120	0
30	490	130	−820
40	890	140	−1270
50	980	150	−780
60	830	160	−940
70	970	170	−540
80	300	180	−230
90	10	190	0

36. A bicyclist is pedaling along a straight road for one hour with a velocity v shown in Figure 5.22. She starts out five kilometers from the lake and positive velocities take her toward the lake. [Note: The vertical lines on the graph are at 10-minute (1/6-hour) intervals.]

(a) Does the cyclist ever turn around? If so, at what time(s)?
(b) When is she going the fastest? How fast is she going then? Toward the lake or away?
(c) When is she closest to the lake? Approximately how close to the lake does she get?
(d) When is she farthest from the lake? Approximately how far from the lake is she then?
(e) What is the total distance she traveled?

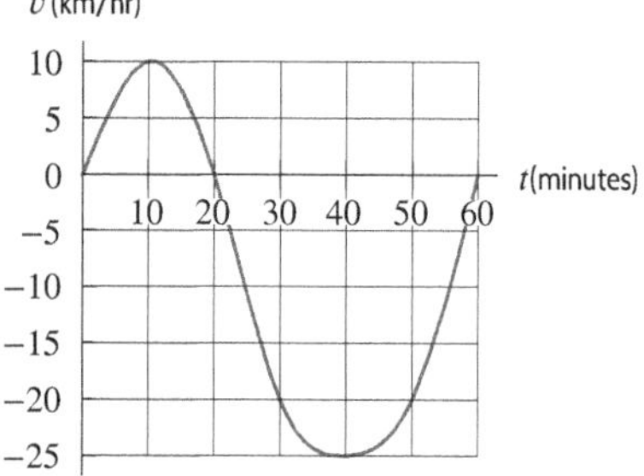

Figure 5.22

37. Two cars travel in the same direction along a straight road. Figure 5.23 shows the velocity, v, of each car at time t. Car B starts 2 hours after car A and car B reaches a maximum velocity of 50 km/hr.

(a) For approximately how long does each car travel?
(b) Estimate car A's maximum velocity.
(c) Approximately how far does each car travel?

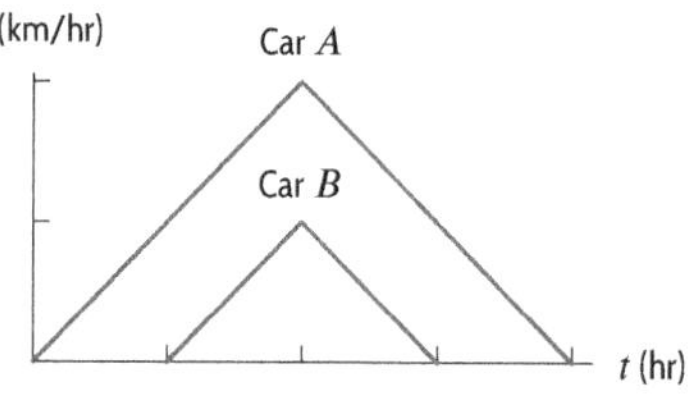

Figure 5.23

38. Two cars start at the same time and travel in the same direction along a straight road. Figure 5.24 gives the velocity, v, of each car as a function of time, t. Which car:

(a) Attains the larger maximum velocity?
(b) Stops first?
(c) Travels farther?

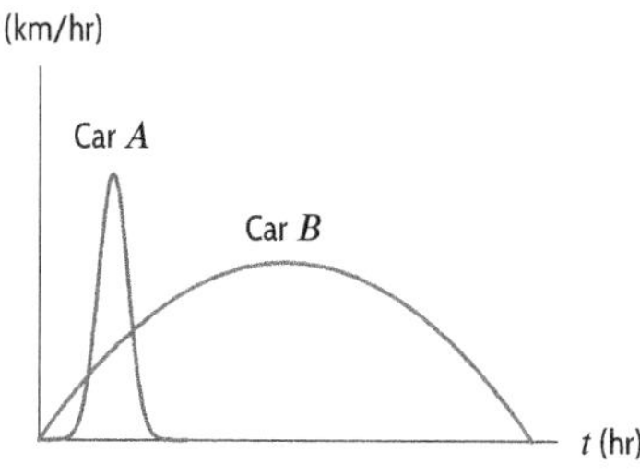

Figure 5.24

39. A car initially going 50 ft/sec brakes at a constant rate (constant negative acceleration), coming to a stop in 5 seconds.

(a) Graph the velocity from $t = 0$ to $t = 5$.
(b) How far does the car travel?
(c) How far does the car travel if its initial velocity is doubled, but it brakes at the same constant rate?

40. A car moving with velocity v has a stopping distance proportional to v^2.

(a) If a car going 20 mi/hr has a stopping distance of 50 feet, what is its stopping distance going 40 mi/hr? What about 60 mi/hr?
(b) After applying the brakes, a car going 30 ft/sec stops in 5 seconds and has $v = 30 - 6t$. Explain why the stopping distance is given by the area under the graph of v against t.
(c) By looking at areas under graphs of v, explain why a car with the same deceleration as the car in part (b) but an initial speed of 60 ft/sec has a stopping distance 4 times as far.

41. A woman drives 10 miles, accelerating uniformly from rest to 60 mph. Graph her velocity versus time. How long does it take for her to reach 60 mph?

42. An object has zero initial velocity and a constant acceleration of 32 ft/sec^2. Find a formula for its velocity as a function of time. Use left and right sums with $\Delta t = 1$ to find upper and lower bounds on the distance that the object travels in four seconds. Find the precise distance using the area under the curve.

Strengthen Your Understanding

■ In Problems 43–44, explain what is wrong with the statement.

43. If a car accelerates from 0 to 50 ft/sec in 10 seconds, then it travels 250 ft.

44. For any acceleration, you can estimate the total distance traveled by a car in 1 second to within 0.1 feet by recording its velocity every 0.1 second.

■ In Problems 45–46, give an example of:

45. A velocity function f and an interval $[a, b]$ such that the distance denoted by the right-hand sum for f on $[a, b]$ is less than the distance denoted by the left-hand sum, no matter what the number of subdivisions.

46. A velocity $f(t)$ and an interval $[a, b]$ such that at least 100 subdivisions are needed in order for the difference between the upper and lower estimates to be less than or equal to 0.1.

■ Are the statements in Problems 47–49 true or false? Give an explanation for your answer.

47. For an increasing velocity function on a fixed time interval, the left-hand sum with a given number of subdivisions is always less than the corresponding right-hand sum.

48. For a decreasing velocity function on a fixed time interval, the difference between the left-hand sum and right-hand sum is halved when the number of subdivisions is doubled.

49. For a given velocity function on a given interval, the difference between the left-hand sum and right-hand sum gets smaller as the number of subdivisions gets larger.

5.2 THE DEFINITE INTEGRAL

In Section 5.1, we saw how distance traveled can be approximated by a sum of areas of rectangles. We also saw how the approximation improves as the width of the rectangles gets smaller. In this section, we construct these sums for any function f, whether or not it represents a velocity.

Sigma Notation

Suppose $f(t)$ is a continuous function for $a \le t \le b$. We divide the interval from a to b into n equal subdivisions, and we call the width of an individual subdivision Δt, so

$$\Delta t = \frac{b-a}{n}.$$

Let $t_0, t_1, t_2, \ldots, t_n$ be endpoints of the subdivisions. Both the left-hand and right-hand sums can be written more compactly using *sigma*, or summation, notation. The symbol $\sum$ is a capital sigma, or Greek letter "S." We write

$$\text{Right-hand sum} = f(t_1)\Delta t + f(t_2)\Delta t + \cdots + f(t_n)\Delta t = \sum_{i=1}^{n} f(t_i)\Delta t.$$

The $\sum$ tells us to add terms of the form $f(t_i)\Delta t$. The "$i = 1$" at the base of the sigma sign tells us to start at $i = 1$, and the "n" at the top tells us to stop at $i = n$.

In the left-hand sum we start at $i = 0$ and stop at $i = n - 1$, so we write

$$\text{Left-hand sum} = f(t_0)\Delta t + f(t_1)\Delta t + \cdots + f(t_{n-1})\Delta t = \sum_{i=0}^{n-1} f(t_i)\Delta t.$$

Taking the Limit to Obtain the Definite Integral

Now we take the limit of these sums as n goes to infinity. If f is continuous for $a \le t \le b$, the limits of the left- and right-hand sums exist and are equal. The *definite integral* is the limit of these sums. A formal definition of the definite integral is given in the online supplement to the text at www.wiley.com/college/hughes-hallett.

Suppose f is continuous for $a \leq t \leq b$. The **definite integral** of f from a to b, written

$$\int_a^b f(t)\,dt,$$

is the limit of the left-hand or right-hand sums with n subdivisions of $a \leq t \leq b$ as n gets arbitrarily large, that is, as $\Delta t \to 0$. In other words,

$$\int_a^b f(t)\,dt = \lim_{n\to\infty} (\text{Left-hand sum}) = \lim_{n\to\infty} \left(\sum_{i=0}^{n-1} f(t_i)\Delta t \right)$$

and

$$\int_a^b f(t)\,dt = \lim_{n\to\infty} (\text{Right-hand sum}) = \lim_{n\to\infty} \left(\sum_{i=1}^{n} f(t_i)\Delta t \right).$$

Each of these sums is called a *Riemann sum*, f is called the *integrand*, and a and b are called the *limits of integration*.

The "$\int$" notation comes from an old-fashioned "S," which stands for "sum" in the same way that $\sum$ does. The "dt" in the integral comes from the factor Δt. Notice that the limits on the $\sum$ symbol are 0 and $n - 1$ for the left-hand sum, and 1 and n for the right-hand sum, whereas the limits on the $\int$ sign are a and b.

Computing a Definite Integral

In practice, we often approximate definite integrals numerically using a calculator or computer. They use programs which compute sums for larger and larger values of n, and eventually give a value for the integral. Some (but not all) definite integrals can be computed exactly. However, any definite integral can be approximated numerically.

In the next example, we see how numerical approximation works. For each value of n, we show an over and an underestimate for the integral $\int_1^2 (1/t)\,dt$. As we increase the value of n, the over and underestimates get closer together, trapping the value of the integral between them. By increasing the value of n sufficiently, we can calculate the integral to any desired accuracy.

Example 1 Calculate the left-hand and right-hand sums with $n = 2$ for $\int_1^2 \frac{1}{t}\,dt$. What is the relation between the left- and right-hand sums for $n = 10$ and $n = 250$ and the integral?

Solution Here $a = 1$ and $b = 2$, so for $n = 2$, $\Delta t = (2 - 1)/2 = 0.5$. Therefore, $t_0 = 1$, $t_1 = 1.5$ and $t_2 = 2$. (See Figure 5.25.) We have

$$\text{Left-hand sum} = f(1)\Delta t + f(1.5)\Delta t = 1(0.5) + \frac{1}{1.5}(0.5) = 0.8333,$$

$$\text{Right-hand sum} = f(1.5)\Delta t + f(2)\Delta t = \frac{1}{1.5}(0.5) + \frac{1}{2}(0.5) = 0.5833.$$

In Figure 5.25 we see that the left-hand sum is bigger than the area under the curve and the right-hand sum is smaller. So the area under the curve $f(t) = 1/t$ from $t = 1$ to $t = 2$ is between them:

$$0.5833 < \int_1^2 \frac{1}{t}\,dt < 0.8333.$$

Since $1/t$ is decreasing, when $n = 10$ in Figure 5.26 we again see that the left-hand sum is larger than the area under the curve, and the right-hand sum smaller. A calculator or computer gives

$$0.6688 < \int_1^2 \frac{1}{t}\,dt < 0.7188.$$

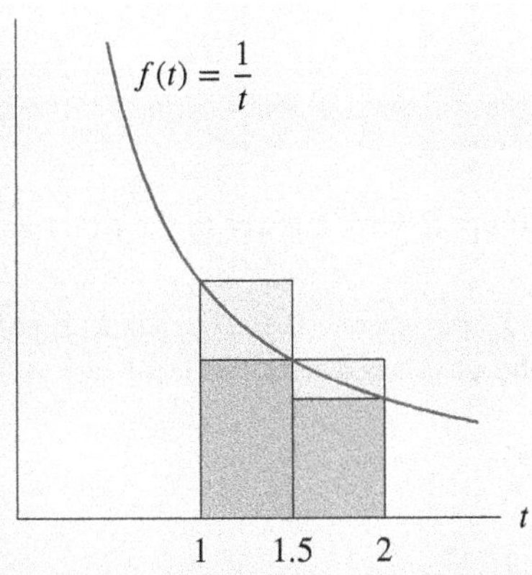

Figure 5.25: Approximating $\int_1^2 \frac{1}{t}\,dt$ with $n = 2$

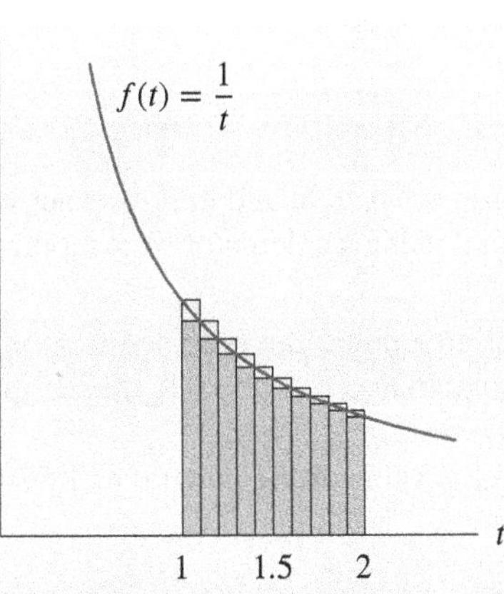

Figure 5.26: Approximating $\int_1^2 \frac{1}{t}\,dt$ with $n = 10$

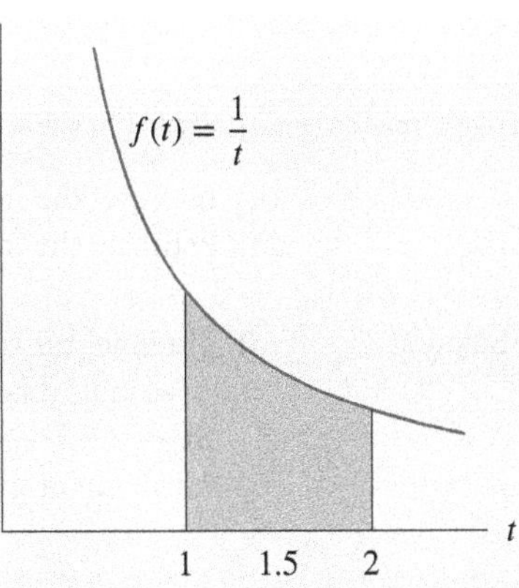

Figure 5.27: Shaded area is exact value of $\int_1^2 \frac{1}{t}\,dt$

The left- and right-hand sums trap the exact value of the integral between them. As the subdivisions become finer, the left- and right-hand sums get closer together.

When $n = 250$, a calculator or computer gives

$$0.6921 < \int_1^2 \frac{1}{t}\,dt < 0.6941.$$

So, to two decimal places, we can say that

$$\int_1^2 \frac{1}{t}\,dt \approx 0.69.$$

The exact value is known to be $\displaystyle\int_1^2 \frac{1}{t}\,dt = \ln 2 = 0.693147\ldots$. See Figure 5.27.

The Definite Integral as an Area

If $f(x)$ is positive we can interpret each term $f(x_0)\Delta x, f(x_1)\Delta x, \ldots$ in a left- or right-hand Riemann sum as the area of a rectangle. See Figure 5.28. As the width Δx of the rectangles approaches zero, the rectangles fit the curve of the graph more exactly, and the sum of their areas gets closer and closer to the area under the curve shaded in Figure 5.29. This suggests that:

> When $f(x) \geq 0$ and $a < b$:
>
> $$\text{Area under graph of } f \text{ and above } x\text{-axis between } a \text{ and } b = \int_a^b f(x)\,dx.$$

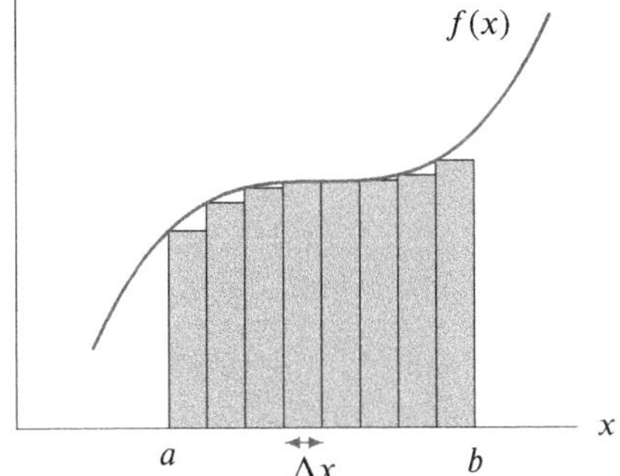

Figure 5.28: Area of rectangles approximating the area under the curve

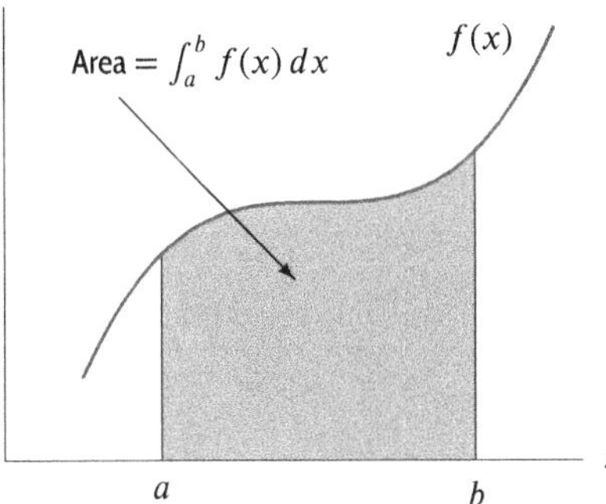

Figure 5.29: The definite integral $\int_a^b f(x)\,dx$

Example 2 Consider the integral $\int_{-1}^{1} \sqrt{1-x^2}\,dx$.

(a) Interpret the integral as an area, and find its exact value.
(b) Estimate the integral using a calculator or computer. Compare your answer to the exact value.

Solution (a) The integral is the area under the graph of $y = \sqrt{1-x^2}$ between -1 and 1. See Figure 5.30. Rewriting this equation as $x^2 + y^2 = 1$, we see that the graph is a semicircle of radius 1 and area $\pi/2 = 1.5707963\ldots$.
(b) A calculator gives the value of the integral as $1.5707963\ldots$.

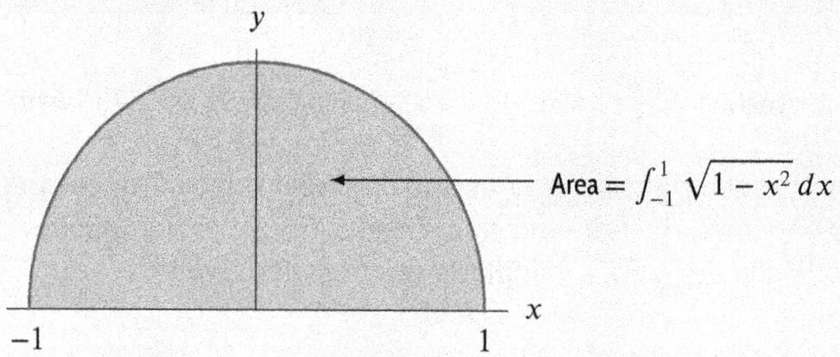

Figure 5.30: Area interpretation of $\int_{-1}^{1} \sqrt{1-x^2}\,dx$

When $f(x)$ Is Not Positive

We have assumed in drawing Figure 5.29 that the graph of $f(x)$ lies above the x-axis. If the graph lies below the x-axis, then each value of $f(x)$ is negative, so each $f(x)\Delta x$ is negative, and the area gets counted negatively. In that case, the definite integral is the negative of the area.

> When $f(x)$ is positive for some x values and negative for others, and $a < b$:
> $\int_a^b f(x)\,dx$ is the sum of areas above the x-axis, counted positively, and areas below the x-axis, counted negatively.

Example 3 How does the definite integral $\int_{-1}^{1} (x^2-1)\,dx$ relate to the area between the parabola $y = x^2 - 1$ and the x-axis?

Solution A calculator gives $\int_{-1}^{1}(x^2-1)\,dx = -1.33$. Since the parabola lies below the axis between $x = -1$ and $x = 1$ (see Figure 5.31), the area between the parabola and the x-axis is approximately 1.33.

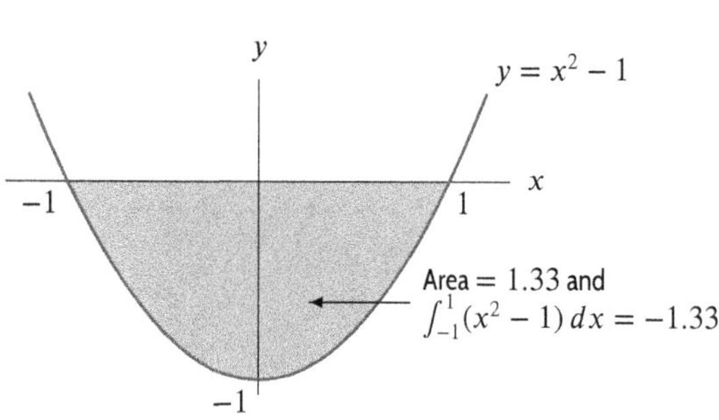

Figure 5.31: Integral $\int_{-1}^{1}(x^2-1)\,dx$ is negative of shaded area

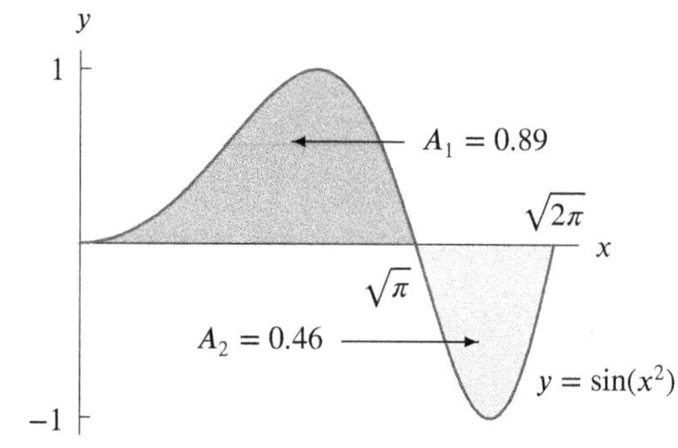

Figure 5.32: Integral $\int_0^{\sqrt{2\pi}} \sin(x^2)\,dx = A_1 - A_2$

Example 4 Consider the definite integral $\int_0^{\sqrt{2\pi}} \sin(x^2)\,dx$.

(a) Interpret the integral in terms of areas.
(b) Find the total area between $\sin(x^2)$ and the x-axis for $0 \le x \le \sqrt{2\pi}$.

Solution (a) The integral is the area above the x-axis, A_1, minus the area below the x-axis, A_2. See Figure 5.32. Estimating the integral with a calculator gives

$$\int_0^{\sqrt{2\pi}} \sin(x^2)\,dx = 0.43.$$

The graph of $y = \sin(x^2)$ crosses the x-axis where $x^2 = \pi$, that is, at $x = \sqrt{\pi}$. The next crossing is at $x = \sqrt{2\pi}$. Breaking the integral into two parts and calculating each one separately gives

$$\int_0^{\sqrt{\pi}} \sin(x^2)\,dx = 0.89 \quad \text{and} \quad \int_{\sqrt{\pi}}^{\sqrt{2\pi}} \sin(x^2)\,dx = -0.46.$$

So $A_1 = 0.89$ and $A_2 = 0.46$. Then, as we would expect,

$$\int_0^{\sqrt{2\pi}} \sin(x^2)\,dx = A_1 - A_2 = 0.89 - 0.46 = 0.43.$$

(b) The total area between $\sin(x^2)$ and the x-axis is the sum of the two areas

$$A_1 + A_2 = 0.89 + 0.46 = 1.35.$$

More General Riemann Sums

Left- and right-hand sums are special cases of Riemann sums. For a general Riemann sum we allow subdivisions to have different lengths. Also, instead of evaluating f only at the left or right endpoint of each subdivision, we allow it to be evaluated anywhere in the subdivision. Thus, a general Riemann sum has the form

$$\sum_{i=1}^{n} \text{Value of } f(t) \text{ at some point in } i^{\text{th}} \text{ subdivision} \times \text{Length of } i^{\text{th}} \text{ subdivision}.$$

(See Figure 5.33.) As before, we let $t_0, t_1, \ldots, t_n$ be the endpoints of the subdivisions, so the length of the ith subdivision is $\Delta t_i = t_i - t_{i-1}$. For each i we choose a point c_i in the ith subinterval at which to evaluate f, leading to the following definition:

> A general Riemann sum for f on the interval $[a, b]$ is a sum of the form
>
> $$\sum_{i=1}^{n} f(c_i)\Delta t_i,$$
>
> where $a = t_0 < t_1 < \cdots < t_n = b$, and, for $i = 1, \ldots, n$, $\Delta t_i = t_i - t_{i-1}$, and $t_{i-1} \le c_i \le t_i$.

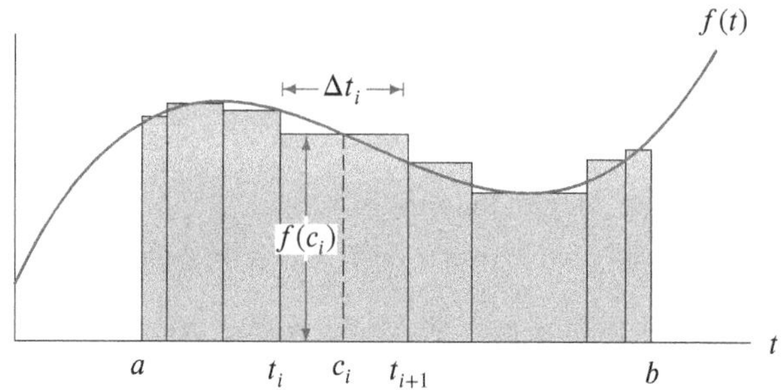

Figure 5.33: A general Riemann sum approximating $\int_a^b f(t)\,dt$

The definite integral is defined as the limit of the Riemann sum as the longest subinterval tends to 0. If f is continuous, we can make a general Riemann sum as close as we like to the value of the definite integral by making the interval lengths small enough. Thus, in approximating definite integrals or in proving theorems about them, we can use general Riemann sums rather than left- or right-hand sums. Generalized Riemann sums are especially useful in establishing properties of the definite integral; see www.wiley.com/college/hughes-hallett.

Exercises and Problems for Section 5.2 Online Resource: Additional Problems for Section 5.2

EXERCISES

In Exercises 1–2, rectangles have been drawn to approximate $\int_0^6 g(x)\,dx$.

(a) Do the rectangles represent a left or a right sum?
(b) Do the rectangles lead to an upper or a lower estimate?
(c) What is the value of n?
(d) What is the value of Δx?

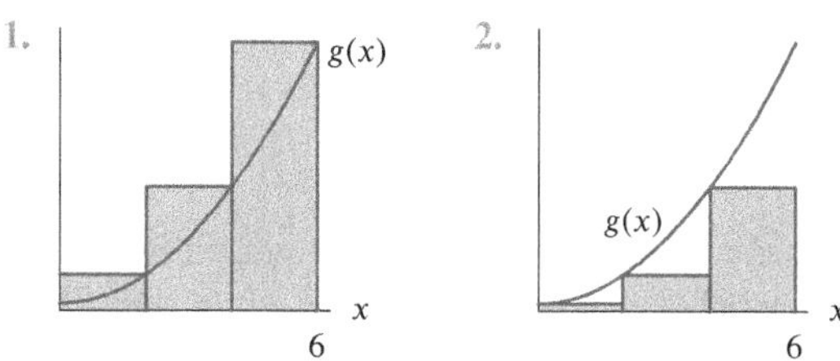

3. Figure 5.34 shows a Riemann sum approximation with n subdivisions to $\int_a^b f(x)\,dx$.

(a) Is it a left- or right-hand approximation? Would the other one be larger or smaller?
(b) What are a, b, n and Δx?

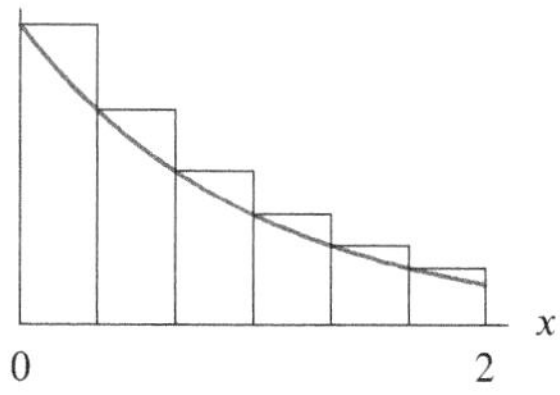

Figure 5.34

4. Figure 5.35 shows a Riemann sum approximation with n subdivisions to $\int_a^b f(x)\,dx$.

(a) Is it a left- or right-hand approximation? Would the other one be larger or smaller?
(b) What are a, b, n, and Δx?

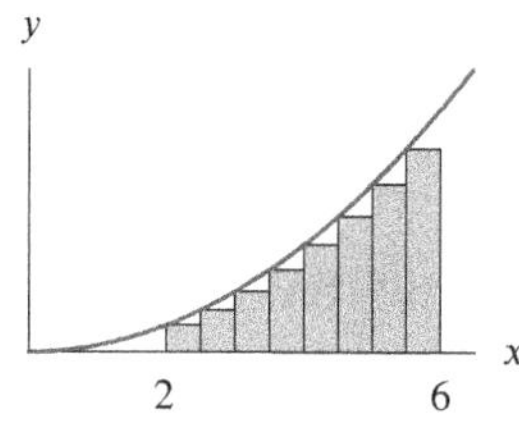

Figure 5.35

5. Using Figure 5.36, draw rectangles representing each of the following Riemann sums for the function f on the interval $0 \le t \le 8$. Calculate the value of each sum.

(a) Left-hand sum with $\Delta t = 4$
(b) Right-hand sum with $\Delta t = 4$
(c) Left-hand sum with $\Delta t = 2$
(d) Right-hand sum with $\Delta t = 2$

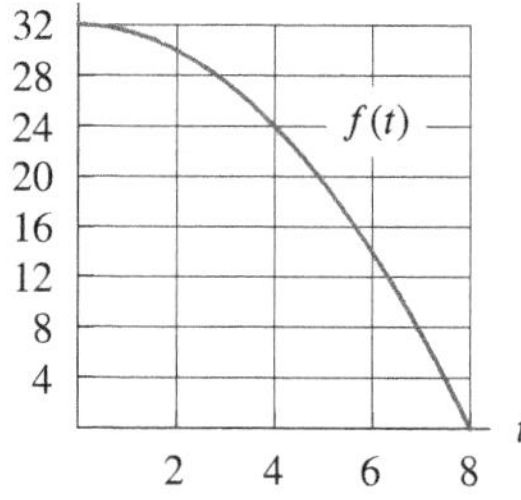

Figure 5.36

6. Use the table to estimate $\int_0^{40} f(x)dx$. What values of n and Δx did you use?

x	0	10	20	30	40
$f(x)$	350	410	435	450	460

7. Use the table to estimate $\int_0^{12} f(x)\,dx$.

x	0	3	6	9	12
$f(x)$	32	22	15	11	9

8. Use the table to estimate $\int_0^{15} f(x)\,dx$.

x	0	3	6	9	12	15
$f(x)$	50	48	44	36	24	8

9. Write out the terms of the right-hand sum with $n = 5$ that could be used to approximate $\int_3^7 \frac{1}{1+x}\,dx$. Do not evaluate the terms or the sum.

10. Use Figure 5.37 to estimate $\int_0^{20} f(x)\,dx$.

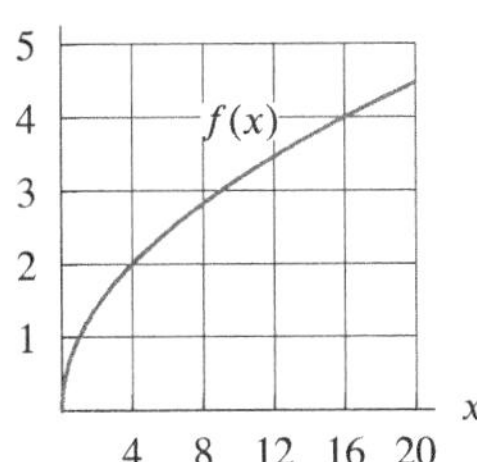

Figure 5.37

11. Use Figure 5.38 to estimate $\int_{-10}^{15} f(x)dx$.

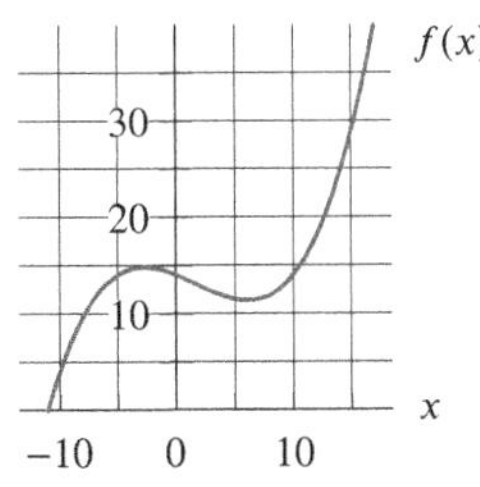

Figure 5.38

12. Using Figure 5.39, estimate $\int_{-3}^{5} f(x)dx$.

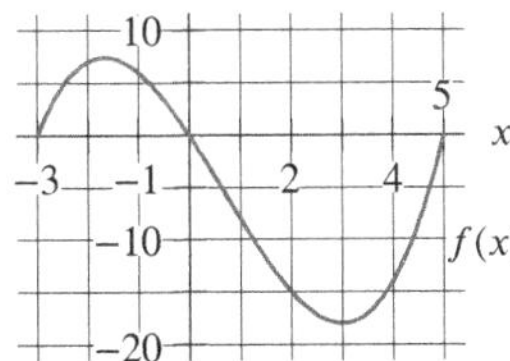

Figure 5.39

■ In Exercises 13–18, use a calculator or a computer to find the value of the definite integral.

13. $\int_1^4 (x^2 + x)\,dx$

14. $\int_0^3 2^x\,dx$

15. $\int_{-1}^1 e^{-x^2}\,dx$

16. $\int_0^3 \ln(y^2 + 1)\,dy$

17. $\int_0^1 \sin(t^2)dt$

18. $\int_3^4 \sqrt{e^z + z}\,dz$

19. For $f(x) = 3x + 2$, evaluate the Riemann sums:

(a) $\sum_{i=0}^{3} f(x_i)\,\Delta x$ where $\Delta x = 2, x_0 = 3$

(b) $\sum_{i=1}^{2} f(x_i)\,\Delta x$ where $\Delta x = 3, x_0 = 7$

(c) $\sum_{i=1}^{4} f(x_i)\,\Delta x$ where $\Delta x = 0.5, x_0 = -5$

20. For $g(x) = 4x - 1$, evaluate the Riemann sums:

(a) $\sum_{i=0}^{3} g(x_i)\,\Delta x$ where $\Delta x = 3, x_0 = 2$

(b) $\sum_{i=1}^{4} g(x_i)\,\Delta x$ where $\Delta x = 2, x_0 = 4$

(c) $\sum_{i=2}^{5} g(x_i)\,\Delta x$ where $\Delta x = 3, x_0 = 1$

21. For $h(x) = \frac{1}{2}x + 5$ evaluate the Riemann sums:

(a) $\sum_{i=0}^{4} h(x_i)\,\Delta x$ where $\Delta x = 2, x_0 = 2$

(b) $\sum_{i=2}^{5} h(x_i)\,\Delta x$ where $\Delta x = 3, x_0 = 0$

(c) $\sum_{i=4}^{7} h(x_i)\,\Delta x$ where $\Delta x = 2, x_0 = 1$

22. Use Table 5.9 to evaluate the Riemann sums:

Table 5.9

t	3	4	5	6	7	8	9
$f(t)$	−40	−17	4	23	40	55	68
t	10	11	12	13	14	15	
$f(t)$	79	88	95	100	103	104	

(a) $\sum_{i=0}^{n-1} f(t_i)\,\Delta t$ where $t_0 = 3, t_n = 15, n = 4$

(b) $\sum_{i=1}^{n} f(t_i)\,\Delta t$ where $t_0 = 3, t_n = 15, n = 3$

(c) $\sum_{i=1}^{n} f(t_i)\,\Delta t$ where $t_0 = 5, t_n = 13, n = 4$

PROBLEMS

23. The graph of $f(t)$ is in Figure 5.40. Which of the following four numbers could be an estimate of $\int_0^1 f(t)dt$ accurate to two decimal places? Explain your choice.

I. −98.35 II. 71.84 III. 100.12 IV. 93.47

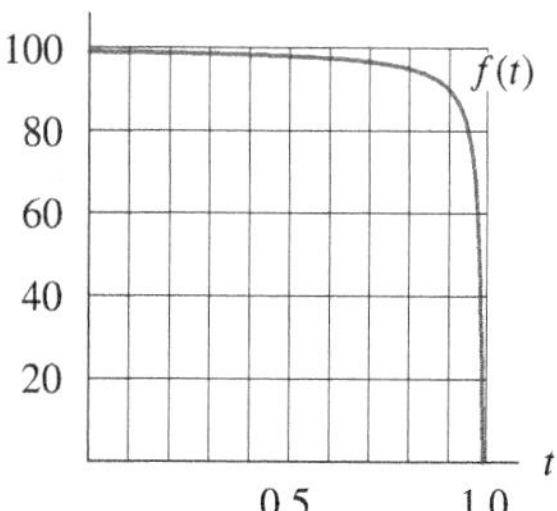

Figure 5.40

24. (a) What is the total area between the graph of $f(x)$ in Figure 5.41 and the x-axis, between $x = 0$ and $x = 5$?
(b) What is $\int_0^5 f(x)\,dx$?

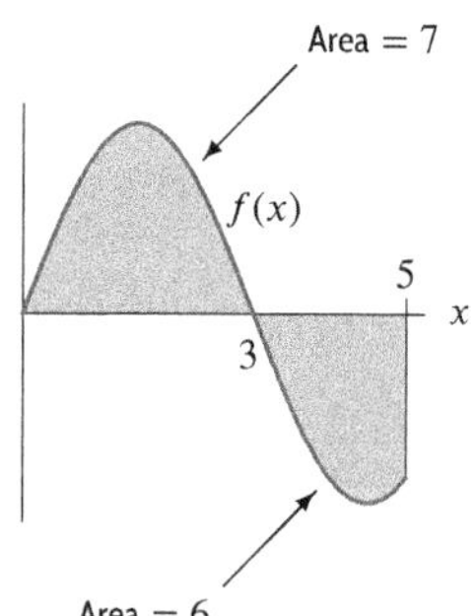

Figure 5.41

25. Find the total area between $y = 4 - x^2$ and the x-axis for $0 \le x \le 3$.

26. (a) Find the total area between $f(x) = x^3 - x$ and the x-axis for $0 \le x \le 3$.
(b) Find $\int_0^3 f(x)dx$.
(c) Are the answers to parts (a) and (b) the same? Explain.

■ In Problems 27–33, find the area of the region between the curve and the horizontal axis

27. Under $y = 6x^3 - 2$ for $5 \le x \le 10$.

28. Under $y = \cos t$ for $0 \le t \le \pi/2$.

29. Under $y = \ln x$ for $1 \le x \le 4$.

30. Under $y = 2\cos(t/10)$ for $1 \le t \le 2$.

31. Under $y = \cos\sqrt{x}$ for $0 \le x \le 2$.

32. Under the curve $y = 7 - x^2$ and above the x-axis.

33. Above the curve $y = x^4 - 8$ and below the x-axis.

34. Use Figure 5.42 to find the values of

(a) $\int_a^b f(x)\,dx$ (b) $\int_b^c f(x)\,dx$
(c) $\int_a^c f(x)\,dx$ (d) $\int_a^c |f(x)|\,dx$

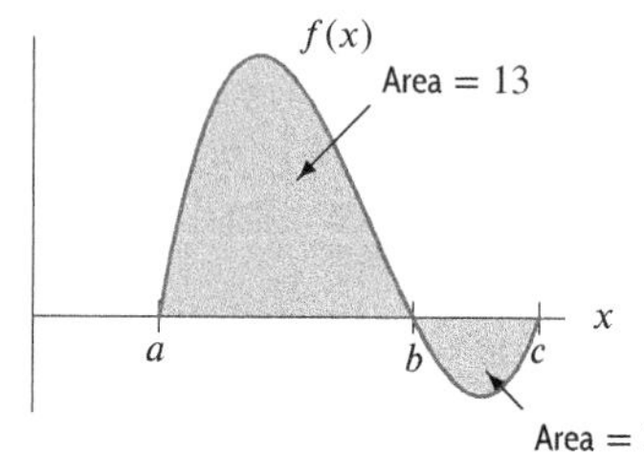

Figure 5.42

35. Given $\int_{-2}^0 f(x)\,dx = 4$ and Figure 5.43, estimate:

(a) $\int_0^2 f(x)\,dx$ (b) $\int_{-2}^2 f(x)\,dx$
(c) The total shaded area.

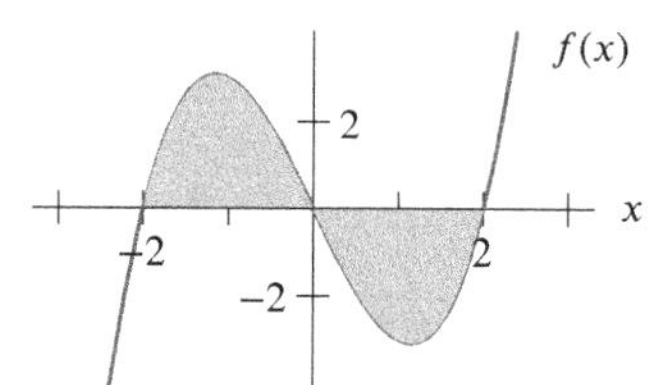

Figure 5.43

36. (a) Using Figure 5.44, find $\int_{-3}^0 f(x)\,dx$.
(b) If the area of the shaded region is A, estimate $\int_{-3}^4 f(x)\,dx$.

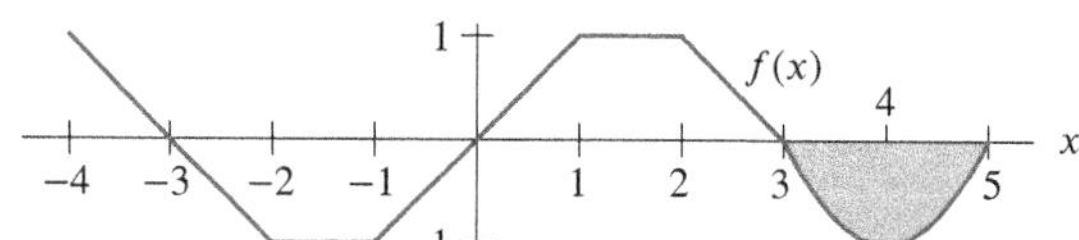

Figure 5.44

37. Use Figure 5.45 to find the values of

(a) $\int_0^2 f(x)\,dx$ (b) $\int_3^7 f(x)\,dx$
(c) $\int_2^7 f(x)\,dx$ (d) $\int_5^8 f(x)\,dx$

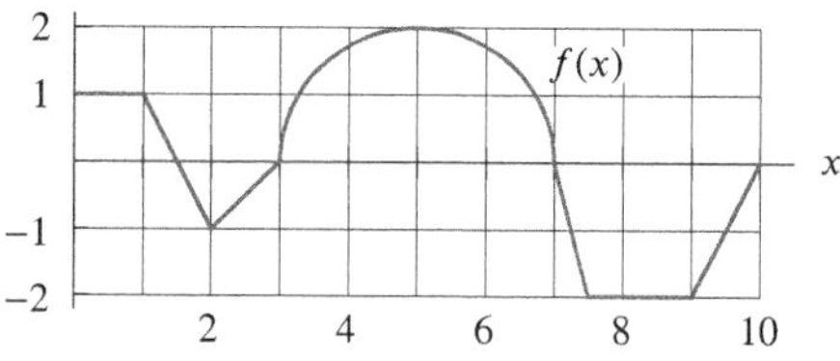

Figure 5.45: Graph consists of a semicircle and line segments

In Problems 38–41, find the integral by finding the area of the region between the curve and the horizontal axis.

38. $\int_0^{10} (x-5)\,dx$

39. $\int_0^{8} (6-2x)\,dx$

40. $\int_{-8}^{6} \left(\frac{1}{2}x+3\right) dx$

41. $\int_{-10}^{1} \frac{-4x-16}{3}\,dx$

42. (a) Graph $f(x)=x(x+2)(x-1)$.
(b) Find the total area between the graph and the x-axis between $x=-2$ and $x=1$.
(c) Find $\int_{-2}^{1} f(x)\,dx$ and interpret it in terms of areas.

43. Compute the definite integral $\int_0^4 \cos\sqrt{x}\,dx$ and interpret the result in terms of areas.

44. Without computation, decide if $\int_0^{2\pi} e^{-x}\sin x\,dx$ is positive or negative. [Hint: Sketch $e^{-x}\sin x$.]

45. Estimate $\int_0^1 e^{-x^2}\,dx$ using $n=5$ rectangles to form a

(a) Left-hand sum (b) Right-hand sum

In Problems 46–53, estimate the integral using a left-hand sum and a right-hand sum with the given value of n.

46. $\int_0^{12} x^2\,dx,\ n=4$

47. $\int_{-2}^{8} \frac{1}{4}x^4\,dx,\ n=5$

48. $\int_{-1}^{8} 2^x\,dx,\ n=3$

49. $\int_{-4}^{4} \left(\frac{1}{2^x}+1\right) dx,$ $n=4$

50. $\int_1^3 \frac{3}{x}\,dx,\ n=4$

51. $\int_0^{\pi} \sin(x)\,dx,\ n=4$

52. $\int_{5.5}^{7} \left(x^2+3x\right)\,dx,\ n=3$

53. $\int_1^4 \sqrt{x}\,dx,\ n=3$

54. (a) On a sketch of $y=\ln x$, represent the left Riemann sum with $n=2$ approximating $\int_1^2 \ln x\,dx$. Write out the terms in the sum, but do not evaluate it.
(b) On another sketch, represent the right Riemann sum with $n=2$ approximating $\int_1^2 \ln x\,dx$. Write out the terms in the sum, but do not evaluate it.
(c) Which sum is an overestimate? Which sum is an underestimate?

55. (a) Draw the rectangles that give the left-hand sum approximation to $\int_0^{\pi}\sin x\,dx$ with $n=2$.
(b) Repeat part (a) for $\int_{-\pi}^{0}\sin x\,dx$.
(c) From your answers to parts (a) and (b), what is the value of the left-hand sum approximation to $\int_{-\pi}^{\pi}\sin x\,dx$ with $n=4$?

56. (a) Use a calculator or computer to find $\int_0^6 (x^2+1)\,dx$. Represent this value as the area under a curve.
(b) Estimate $\int_0^6 (x^2+1)\,dx$ using a left-hand sum with $n=3$. Represent this sum graphically on a sketch of $f(x)=x^2+1$. Is this sum an overestimate or underestimate of the true value found in part (a)?
(c) Estimate $\int_0^6 (x^2+1)\,dx$ using a right-hand sum with $n=3$. Represent this sum on your sketch. Is this sum an overestimate or underestimate?

57. (a) Graph $f(x)=\begin{cases}1-x & 0\le x\le 1\\ x-1 & 1<x\le 2.\end{cases}$
(b) Find $\int_0^2 f(x)\,dx$.
(c) Calculate the 4-term left Riemann sum approximation to the definite integral. How does the approximation compare to the exact value?

58. Estimate $\int_1^2 x^2\,dx$ using left- and right-hand sums with four subdivisions. How far from the true value of the integral could your estimate be?

59. Without computing the sums, find the difference between the right- and left-hand Riemann sums if we use $n=500$ subintervals to approximate $\int_{-1}^{1}(2x^3+4)\,dx$.

In Problems 60–65, the limit is either a right-hand or left-hand Riemann sum $\sum f(t_i)\Delta t$. For the given choice of t_i, write the limit as a definite integral.

60. $\lim_{n\to\infty}\sum_{i=1}^{n}\left(\frac{i}{n}\right)^2\left(\frac{1}{n}\right);\quad t_i=\frac{i}{n}$

61. $\lim_{n\to\infty}\sum_{i=0}^{n-1}2\left(\frac{3i}{n}\right)\left(\frac{3}{n}\right);\quad t_i=\frac{3i}{n}$

62. $\lim_{n\to\infty}\sum_{i=0}^{n-1}\frac{1}{n}e^{1+i/n};\quad t_i=1+\frac{i}{n}$

63. $\lim_{n\to\infty}\sum_{i=1}^{n}\left(7\left(\frac{i}{2n}\right)^2+3\right)\frac{1}{2n};\quad t_i=\frac{i}{2n}$

64. $\lim_{n\to\infty}\sum_{i=0}^{n-1}5\cos\left(\frac{\pi i}{n}\right)\frac{\pi}{n};\quad t_i=\frac{\pi i}{n}$

65. $\lim_{n\to\infty}\sum_{i=1}^{n}\frac{3}{n}\sqrt{\left(1+\frac{3i}{n}\right)^2+\left(1+\frac{3i}{n}\right)};\quad t_i=1+\frac{3i}{n}$

In Problems 66–69, express the given limit of a Riemann sum as a definite integral and then evaluate the integral.

66. $\lim_{n\to\infty}\sum_{i=1}^{n}2\left(\frac{3i}{n}\right)\cdot\frac{3}{n}$

67. $\lim_{n\to\infty}\sum_{i=0}^{n-1}\sqrt{\frac{4i}{n}}\cdot\frac{4}{n}$

68. $\lim_{n\to\infty}\sum_{i=0}^{n-1}\sqrt{4-\left(\frac{2i}{n}\right)^2}\cdot\frac{2}{n}$

69. $\lim_{n\to\infty}\sum_{i=1}^{n}\left(8\left(1+\frac{i}{n}\right)-8\right)\cdot\frac{1}{n}$

In Problems 70–73, use Figure 5.46 to find limits a and b in the interval $[0, 5]$ with $a < b$ satisfying the given condition.

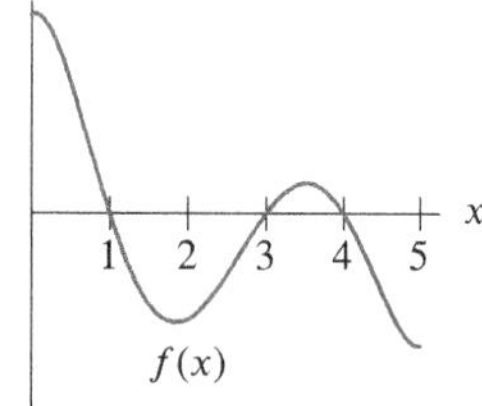

Figure 5.46

70. $\int_0^b f(x)\,dx$ is largest

71. $\int_a^4 f(x)\,dx$ is smallest

72. $\int_a^b f(x)\,dx$ is largest

73. $\int_a^b f(x)\,dx$ is smallest

Strengthen Your Understanding

In Problems 74–76, explain what is wrong with the statement.

74. For any function, $\int_1^3 f(x)\,dx$ is the area between the the graph of f and the x-axis on $1 \le x \le 3$.

75. The left-hand sum with 10 subdivisions for the integral $\int_1^2 \sin(x)\,dx$ is

$$0.1\,(\sin(1) + \sin(1.1) + \cdots + \sin(2))\,.$$

76. $\int_1^1 e^x \ln(x^2 + \cos(x^2))\,dx = e \ln 2.$

In Problems 77–78, give an example of:

77. A function f and an interval $[a, b]$ such that $\int_a^b f(x)\,dx$ is negative.

78. A function f such that $\int_1^3 f(x)\,dx < \int_1^2 f(x)\,dx$.

In Problems 79–81 decide whether the statement is true or false. Justify your answer.

79. On the interval $a \le t \le b$, the integral of the velocity is the total distance traveled from $t = a$ to $t = b$.

80. A 4-term left-hand Riemann sum approximation cannot give the exact value of a definite integral.

81. If $f(x)$ is decreasing and $g(x)$ is increasing, then $\int_a^b f(x)\,dx \neq \int_a^b g(x)\,dx$.

In Problems 82–84, is the statement true for all continuous functions $f(x)$ and $g(x)$? Explain your answer.

82. $\int_0^2 f(x)\,dx \le \int_0^3 f(x)\,dx$.

83. $\int_0^2 f(x)\,dx = \int_0^2 f(t)\,dt$.

84. If $\int_2^6 f(x)\,dx \le \int_2^6 g(x)\,dx$, then $f(x) \le g(x)$ for $2 \le x \le 6$.

In Problems 85–86, graph a continuous function $f(x) \ge 0$ on $[0, 10]$ with the given properties.

85. The maximum value taken on by $f(x)$ for $0 \le x \le 10$ is 1. In addition $\int_0^{10} f(x)\,dx = 5$.

86. The maximum value taken on by $f(x)$ for $0 \le x \le 10$ is 5. In addition $\int_0^{10} f(x)\,dx = 1$.

5.3 THE FUNDAMENTAL THEOREM AND INTERPRETATIONS

The Notation and Units for the Definite Integral

Just as the Leibniz notation dy/dx for the derivative reminds us that the derivative is the limit of a ratio of differences, the notation for the definite integral recalls the meaning of the integral. The symbol

$$\int_a^b f(x)\,dx$$

reminds us that an integral is a limit of sums of terms of the form "$f(x)$ times a small difference in x." Officially, dx is not a separate entity, but a part of the whole integral symbol. Just as one thinks of d/dx as a single symbol meaning "the derivative with respect to x of…," one can think of $\int_a^b \ldots\, dx$ as a single symbol meaning "the integral of … with respect to x."

However, many scientists and mathematicians informally think of dx as an "infinitesimally" small bit of x multiplied by $f(x)$. This viewpoint is often the key to interpreting the meaning of a definite integral.

For example, if $f(t)$ is the velocity of a moving particle at time t, then $f(t)\,dt$ may be thought of informally as velocity × time, giving the distance traveled by the particle during a small bit of time dt. The integral $\int_a^b f(t)\,dt$ may then be thought of as the sum of all these small distances, giving the net change in position of the particle between $t = a$ and $t = b$. The notation for the integral suggests units for the value of the integral. Since the terms being added up are products of the form "$f(x)$ times a difference in x," the unit of measurement for $\int_a^b f(x)\,dx$ is the product of the units for $f(x)$ and the units for x. For example, if $f(t)$ is velocity in meters/second and t is time in seconds, then

$$\int_a^b f(t)\,dt$$

has units of (meters/sec) ×(sec) = meters. This is what we expect, since the value of this integral represents change in position.

As another example, graph $y = f(x)$ with the same units of measurement of length along the x- and y-axes, say cm. Then $f(x)$ and x are measured in the same units, so

$$\int_a^b f(x)\,dx$$

is measured in square units of cm × cm = cm^2. Again, this is what we would expect since in this context the integral represents an area.

The Fundamental Theorem of Calculus

We have seen that change in position can be calculated as the limit of Riemann sums of the velocity function $v = f(t)$. Thus, change in position is given by the definite integral $\int_a^b f(t)\,dt$. If we let $F(t)$ denote the position function, then the change in position can also be written as $F(b) - F(a)$. Thus we have:

$$\int_a^b f(t)\,dt = \begin{array}{c}\text{Change in position from}\\ t = a \text{ to } t = b\end{array} = F(b) - F(a)$$

We also know that the position F and velocity f are related using derivatives: $F'(t) = f(t)$. Thus, we have uncovered a connection between the integral and derivative, which is so important that it is called the Fundamental Theorem of Calculus. It applies to any function F with a continuous derivative $f = F'$.

Theorem 5.1: The Fundamental Theorem of Calculus[2]

If f is continuous on the interval $[a, b]$ and $f(t) = F'(t)$, then

$$\int_a^b f(t)\,dt = F(b) - F(a).$$

To understand the Fundamental Theorem of Calculus, think of $f(t) = F'(t)$ as the rate of change of the quantity $F(t)$. To calculate the total change in $F(t)$ between times $t = a$ and $t = b$, we divide the interval $a \le t \le b$ into n subintervals, each of length Δt. For each small interval, we estimate the change in $F(t)$, written ΔF, and add these. In each subinterval we assume the rate of change of $F(t)$ is approximately constant, so that we can say

$$\Delta F \approx \text{Rate of change of } F \times \text{Time elapsed}.$$

For the first subinterval, from t_0 to t_1, the rate of change of $F(t)$ is approximately $F'(t_0)$, so

$$\Delta F \approx F'\left(t_0\right)\Delta t.$$

Similarly, for the second interval

$$\Delta F \approx F'\left(t_1\right)\Delta t.$$

[2]This result is sometimes called the First Fundamental Theorem of Calculus, to distinguish it from the Second Fundamental Theorem of Calculus discussed in Section 6.4.

Summing over all the subintervals, we get

$$\begin{array}{c}\text{Total change in } F(t) \\ \text{between } t=a \text{ and } t=b\end{array} = \sum_{i=0}^{n-1} \Delta F \approx \sum_{i=0}^{n-1} F'\left(t_i\right) \Delta t.$$

We have approximated the change in $F(t)$ as a left-hand sum.

However, the total change in $F(t)$ between the times $t = a$ and $t = b$ is simply $F(b) - F(a)$. Taking the limit as n goes to infinity converts the Riemann sum to a definite integral and suggests the following interpretation of the Fundamental Theorem of Calculus:[3]

$$F(b) - F(a) = \begin{array}{c}\text{Total change in } F(t) \\ \text{between } t=a \text{ and } t=b\end{array} = \int_a^b F'(t)\,dt.$$

In words, the definite integral of a rate of change gives the total change.

This argument does not, however, constitute a proof of the Fundamental Theorem. The errors in the various approximations must be investigated using the definition of limit. A proof is given in Section 6.4 where we learn how to construct antiderivatives using the Second Fundamental Theorem of Calculus.

Example 1 If $F'(t) = f(t)$ and $f(t)$ is velocity in miles/hour, with t in hours, what are the units of $\int_a^b f(t)\,dt$ and $F(b) - F(a)$?

Solution Since the units of $f(t)$ are miles/hour and the units of t are hours, the units of $\int_a^b f(t)\,dt$ are (miles/hour) $\times$ hours $=$ miles. Since F measures change in position, the units of $F(b) - F(a)$ are miles. As expected, the units of $\int_a^b f(t)\,dt$ and $F(b) - F(a)$ are the same.

The Definite Integral of a Rate of Change: Applications of the Fundamental Theorem

Many applications are based on the Fundamental Theorem, which tells us that the definite integral of a rate of change gives the total change.

Example 2 Let $F(t)$ represent a bacteria population which is 5 million at time $t = 0$. After t hours, the population is growing at an instantaneous rate of 2^t million bacteria per hour. Estimate the total increase in the bacteria population during the first hour, and the population at $t = 1$.

Solution Since the rate at which the population is growing is $F'(t) = 2^t$, we have

$$\text{Change in population} = F(1) - F(0) = \int_0^1 2^t\,dt.$$

Using a calculator to evaluate the integral, we find

$$\text{Change in population} = \int_0^1 2^t\,dt = 1.44 \text{ million bacteria.}$$

Since $F(0) = 5$, the population at $t = 1$ is given by

$$\text{Population} = F(1) = F(0) + \int_0^1 2^t\,dt = 5 + 1.44 = 6.44 \text{ million.}$$

[3]We could equally well have used a right-hand sum, since the definite integral is their common limit.

The following example uses the fact that the definite integral of the velocity gives the change in position, or, if the velocity is positive, the total distance traveled.

Example 3 Two cars start from rest at a traffic light and accelerate for several minutes. Figure 5.47 shows their velocities as a function of time.

(a) Which car is ahead after one minute? (b) Which car is ahead after two minutes?

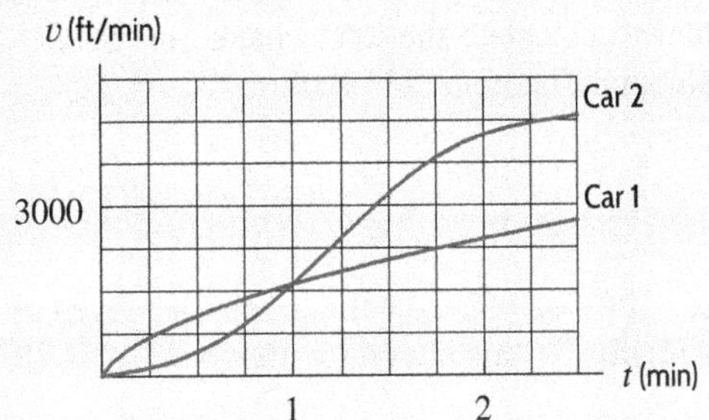

Figure 5.47: Velocities of two cars in Example 3. Which is ahead when?

Solution

(a) For the first minute car 1 goes faster than car 2, and therefore car 1 must be ahead at the end of one minute.

(b) At the end of two minutes the situation is less clear, since car 1 was going faster for the first minute and car 2 for the second. However, if $v = f(t)$ is the velocity of a car after t minutes, then, since the integral of velocity is distance traveled, we know that

$$\text{Distance traveled in two minutes} = \int_0^2 f(t)\,dt,$$

This definite integral may also be interpreted as the area under the graph of f between 0 and 2. Since the area in Figure 5.47 representing the distance traveled by car 2 is clearly larger than the area for car 1, we know that car 2 has traveled farther than car 1.

Example 4 Biological activity in water is reflected in the rate at which carbon dioxide, CO_2, is added or removed. Plants take CO_2 out of the water during the day for photosynthesis and put CO_2 into the water at night. Animals put CO_2 into the water all the time as they breathe. Figure 5.48 shows the rate of change of the CO_2 level in a pond.[4] At dawn, there were 2.600 mmol of CO_2 per liter of water.

(a) At what time was the CO_2 level lowest? Highest?

(b) Estimate how much CO_2 enters the pond during the night ($t = 12$ to $t = 24$).

(c) Estimate the CO_2 level at dusk (twelve hours after dawn).

(d) Does the CO_2 level appear to be approximately in equilibrium?

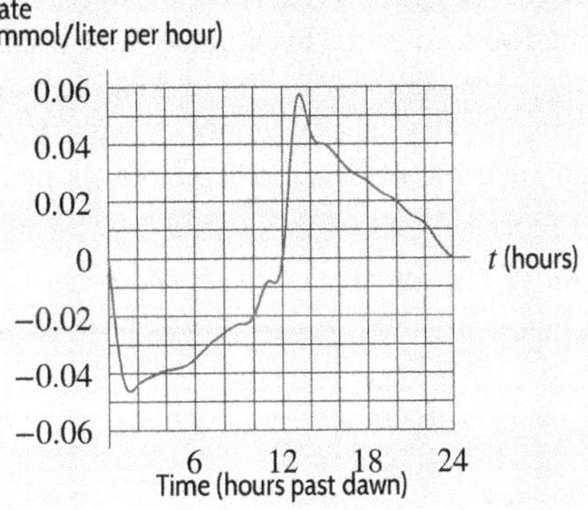

Figure 5.48: Rate at which CO_2 enters a pond over a 24-hour period

[4]Data from R. J. Beyers, *The Pattern of Photosynthesis and Respiration in Laboratory Microsystems* (Mem. 1st. Ital. Idrobiol., 1965).

Solution Let $f(t)$ be the rate at which CO_2 is entering the water at time t and let $F(t)$ be the concentration of CO_2 in the water at time t, so $F'(t) = f(t)$.

(a) From Figure 5.48, we see $f(t)$ is negative for $0 \le t \le 12$, so the CO_2 level is decreasing during this interval (daytime). Since $f(t)$ is positive for $12 < t < 24$, the CO_2 level is increasing during this interval (night). The CO_2 is lowest at $t = 12$ (dusk) and highest at $t = 0$ and $t = 24$ (dawn).

(b) We want to calculate the total change in the CO_2 level in the pond, $F(24) - F(12)$. By the Fundamental Theorem of Calculus,

$$F(24) - F(12) = \int_{12}^{24} f(t)\,dt.$$

We use values of $f(t)$ from the graph (displayed in Table 5.10) to construct a left Riemann sum approximation to this integral with $n = 6$, $\Delta t = 2$:

$$\begin{aligned}\int_{12}^{24} f(t)\,dt &\approx f(12)\cdot 2 + f(14)\cdot 2 + f(16)\cdot 2 + \cdots + f(22)\cdot 2\\ &\approx (0.000)2 + (0.045)2 + (0.035)2 + \cdots + (0.012)2 = 0.278.\end{aligned}$$

Thus, between $t = 12$ and $t = 24$,

$$\text{Change in } CO_2 \text{ level } = F(24) - F(12) = \int_{12}^{24} f(t)\,dt \approx 0.278 \text{ mmol/liter}.$$

(c) To find the CO_2 level at $t = 12$, we use the Fundamental Theorem to estimate the change in CO_2 level during the day:

$$F(12) - F(0) = \int_{0}^{12} f(t)\,dt$$

Using a left Riemann sum as in part (b), we have

$$F(12) - F(0) = \int_{0}^{12} f(t)\,dt \approx -0.328.$$

Since initially there were $F(0) = 2.600$ mmol/liter, we have

$$F(12) = F(0) - 0.328 = 2.272 \text{ mmol/liter}.$$

(d) The amount of CO_2 removed during the day is represented by the area of the region below the t-axis; the amount of CO_2 added during the night is represented by the area above the t-axis. These areas look approximately equal, so the CO_2 level is approximately in equilibrium.

Using Riemann sums to estimate these areas, we find that about 0.278 mmol/l of CO_2 was released into the pond during the night and about 0.328 mmol/l of CO_2 was absorbed from the pond during the day. These quantities are sufficiently close that the difference could be due to measurement error, or to errors from the Riemann sum approximation.

Table 5.10 *Rate, $f(t)$, at which CO_2 is entering or leaving water (read from Figure 5.48)*

t	$f(t)$	t	$f(t)$	t	$f(t)$	t	$f(t)$	t	$f(t)$	t	$f(t)$
0	0.000	4	−0.039	8	−0.026	12	0.000	16	0.035	20	0.020
2	−0.044	6	−0.035	10	−0.020	14	0.045	18	0.027	22	0.012

Calculating Definite Integrals: Computational Use of the Fundamental Theorem

The Fundamental Theorem of Calculus owes its name to its central role in linking rates of change (derivatives) to total change. However, the Fundamental Theorem also provides an exact way of computing certain definite integrals.

Example 5 Compute $\int_1^3 2x\,dx$ by two different methods.

Solution Using left- and right-hand sums, we can approximate this integral as accurately as we want. With $n = 100$, for example, the left-sum is 7.96 and the right sum is 8.04. Using $n = 500$ we learn

$$7.992 < \int_1^3 2x\,dx < 8.008.$$

The Fundamental Theorem, on the other hand, allows us to compute the integral exactly. We take $f(x) = 2x$. We know that if $F(x) = x^2$, then $F'(x) = 2x$. So we use $f(x) = 2x$ and $F(x) = x^2$ and obtain

$$\int_1^3 2x\,dx = F(3) - F(1) = 3^2 - 1^2 = 8.$$

Notice that to use the Fundamental Theorem to calculate a definite integral, we need to know the antiderivative, F. Chapter 6 discusses how antiderivatives are computed.

Exercises and Problems for Section 5.3 Online Resource: Additional Problems for Section 5.3

EXERCISES

1. If $f(t)$ is measured in dollars per year and t is measured in years, what are the units of $\int_a^b f(t)\,dt$?
2. If $f(t)$ is measured in meters/second^2 and t is measured in seconds, what are the units of $\int_a^b f(t)\,dt$?
3. If $f(x)$ is measured in pounds and x is measured in feet, what are the units of $\int_a^b f(x)\,dx$?

■ In Exercises 4–7, explain in words what the integral represents and give units.

4. $\int_1^3 v(t)\,dt$, where $v(t)$ is velocity in meters/sec and t is time in seconds.
5. $\int_0^6 a(t)\,dt$, where $a(t)$ is acceleration in km/hr^2 and t is time in hours.
6. $\int_{2005}^{2011} f(t)\,dt$, where $f(t)$ is the rate at which world population is growing in year t, in billion people per year.
7. $\int_0^5 s(x)\,dx$, where $s(x)$ is rate of change of salinity (salt concentration) in gm/liter per cm in sea water, and where x is depth below the surface of the water in cm.

8. For the two cars in Example 3, page 295, estimate:
   (a) The distances moved by car 1 and car 2 during the first minute.
   (b) The time at which the two cars have gone the same distance.

■ In Exercises 9–14, let $f(t) = F'(t)$. Write the integral $\int_a^b f(t)\,dt$ and evaluate it using the Fundamental Theorem of Calculus.

9. $F(t) = t^2$; $a = 1$, $b = 3$
10. $F(t) = 3t^2 + 4t$; $a = 2$, $b = 5$
11. $F(t) = \ln t$; $a = 1$, $b = 5$
12. $F(t) = \sin t$; $a = 0$, $b = \pi/2$
13. $F(t) = 7 \cdot 4^t$; $a = 2$, $b = 3$
14. $F(t) = \tan t$; $a = -\pi/4$, $b = \pi/4$

PROBLEMS

15. (a) Differentiate $x^3 + x$.
    (b) Use the Fundamental Theorem of Calculus to find $$\int_0^2 (3x^2 + 1)\,dx.$$

16. (a) What is the derivative of $\sin t$?
    (b) The velocity of a particle at time t is $v(t) = \cos t$. Use the Fundamental Theorem of Calculus to find the total distance traveled by the particle between $t = 0$ and $t = \pi/2$.

17. (a) If $F(t) = \frac{1}{2}\sin^2 t$, find $F'(t)$.
(b) Find $\int_{0.2}^{0.4} \sin t \cos t\, dt$ two ways:
(i) Numerically.
(ii) Using the Fundamental Theorem of Calculus.

18. (a) If $F(x) = e^{x^2}$, find $F'(x)$.
(b) Find $\int_0^1 2xe^{x^2}\, dx$ two ways:
(i) Numerically.
(ii) Using the Fundamental Theorem of Calculus.

In Problems 19–20, find the area under the graph of $f(t)$ for $0 \le t \le 5$ using the Fundamental Theorem of Calculus. Compare your answer with what you get using areas of triangles.

19. $f(t) = 6t$

20. $f(t) = 10 - 2t$

21. Pollution is removed from a lake at a rate of $f(t)$ kg/day on day t.
(a) Explain the meaning of the statement $f(12) = 500$.
(b) If $\int_5^{15} f(t)\, dt = 4000$, give the units of the 5, the 15, and the 4000.
(c) Give the meaning of $\int_5^{15} f(t)\, dt = 4000$.

22. Oil leaks out of a tanker at a rate of $r = f(t)$ gallons per minute, where t is in minutes. Write a definite integral expressing the total quantity of oil which leaks out of the tanker in the first hour.

23. Water is leaking out of a tank at a rate of $R(t)$ gallons/hour, where t is measured in hours.
(a) Write a definite integral that expresses the total amount of water that leaks out in the first two hours.
(b) In Figure 5.49, shade the region whose area represents the total amount of water that leaks out in the first two hours.
(c) Give an upper and lower estimate of the total amount of water that leaks out in the first two hours.

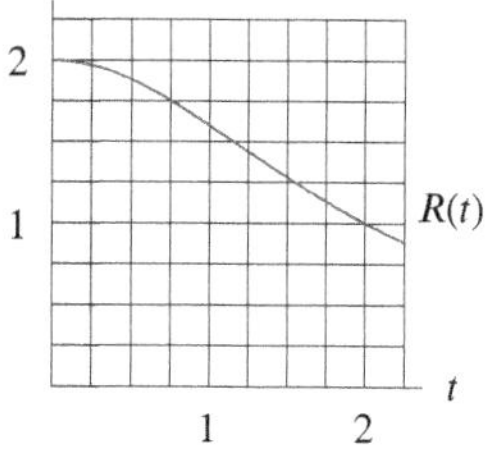

Figure 5.49

24. As coal deposits are depleted, it becomes necessary to strip-mine larger areas for each ton of coal. Figure 5.50 shows the number of acres of land per million tons of coal that will be defaced during strip-mining as a function of the number of million tons removed, starting from the present day.
(a) Estimate the total number of acres defaced in extracting the next 4 million tons of coal (measured from the present day). Draw four rectangles under the curve, and compute their area.
(b) Re-estimate the number of acres defaced using rectangles above the curve.
(c) Use your answers to parts (a) and (b) to get a better estimate of the actual number of acres defaced.

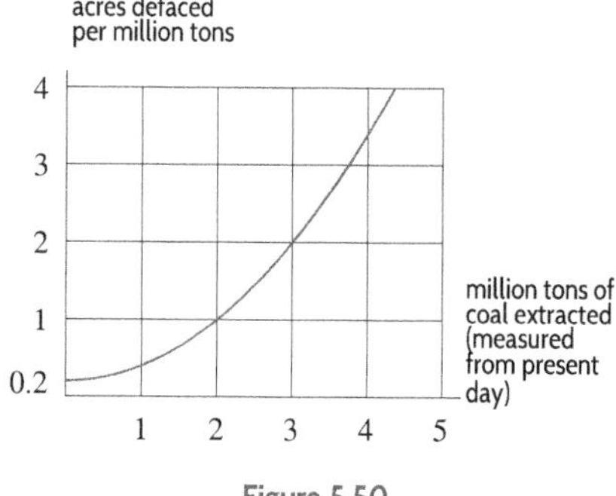

Figure 5.50

25. The rate at which the world's oil is consumed (in billions of barrels per year) is given by $r = f(t)$, where t is in years and $t = 0$ is the start of 2004.
(a) Write a definite integral representing the total quantity of oil consumed between the start of 2004 and the start of 2009.
(b) Between 2004 and 2009, the rate was modeled by $r = 32e^{0.05t}$. Using a left-hand sum with five subdivisions, find an approximate value for the total quantity of oil consumed between the start of 2004 and the start of 2009.
(c) Interpret each of the five terms in the sum from part (b) in terms of oil consumption.

26. A bungee jumper leaps off the starting platform at time $t = 0$ and rebounds once during the first 5 seconds. With velocity measured downward, for t in seconds and $0 \le t \le 5$, the jumper's velocity is approximated[5] by $v(t) = -4t^2 + 16t$ meters/sec.
(a) How many meters does the jumper travel during the first five seconds?
(b) Where is the jumper relative to the starting position at the end of the five seconds?
(c) What does $\int_0^5 v(t)\, dt$ represent in terms of the jump?

[5]Based on www.itforus.oeiizk.waw.pl/tresc/activ//modules/bj.pdf. Accessed February 12, 2012.

27. The table gives annual US emissions, $H(t)$, of hydrofluorocarbons, or "super greenhouse gases," in millions of metric tons of carbon-dioxide equivalent. Let t be in years since 2000.[6]

    (a) What are the units and meaning of $\int_0^{12} H(t)\,dt$?

    (b) Estimate $\int_0^{12} H(t)\,dt$.

Year	2000	2002	2004	2006	2008	2010	2012
$H(t)$	158.9	154.4	152.1	153.9	162.7	167.0	173.6

28. An old rowboat has sprung a leak. Water is flowing into the boat at a rate, $r(t)$, given in the table.

    (a) Compute upper and lower estimates for the volume of water that has flowed into the boat during the 15 minutes.

    (b) Draw a graph to illustrate the lower estimate.

t minutes	0	5	10	15
$r(t)$ liters/min	12	20	24	16

29. Annual coal production in the US (in billion tons per year) is given in the table.[7] Estimate the total amount of coal produced in the US between 1997 and 2009. If $r = f(t)$ is the rate of coal production t years since 1997, write an integral to represent the 1997–2009 coal production.

Year	1997	1999	2001	2003	2005	2007	2009
Rate	1.090	1.094	1.121	1.072	1.132	1.147	1.073

30. The amount of waste a company produces, W, in tons per week, is approximated by $W = 3.75e^{-0.008t}$, where t is in weeks since January 1, 2016. Waste removal for the company costs \$150/ton. How much did the company pay for waste removal during the year 2016?

31. A two-day environmental cleanup started at 9 am on the first day. The number of workers fluctuated as shown in Figure 5.51. If the workers were paid \$10 per hour, how much was the total personnel cost of the cleanup?

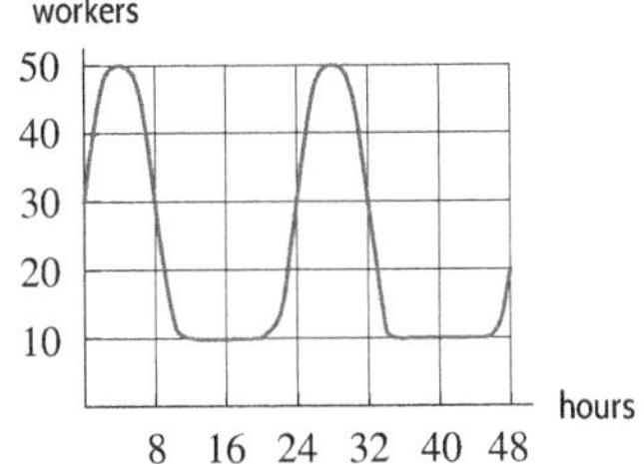

Figure 5.51

32. In Problem 31, suppose workers were paid \$10 per hour for work between 9 am and 5 pm and \$15 per hour for work during the rest of the day. What would the total personnel costs have been under these conditions?

33. Figure 5.52 shows solar radiation, in watts per square meter (w/m^2), in Santa Rosa, California, throughout a typical January day.[8] Estimate the daily energy produced, in kwh, by a 20-square-meter solar array located in Santa Rosa if it converts 18% of solar radiation into energy.

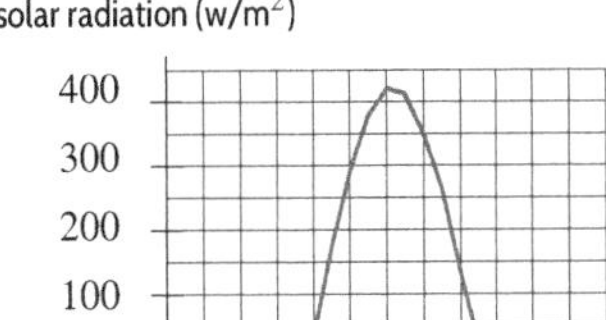

Figure 5.52

34. A warehouse charges its customers \$5 per day for every 10 cubic feet of space used for storage. Figure 5.53 records the storage used by one company over a month. How much will the company have to pay?

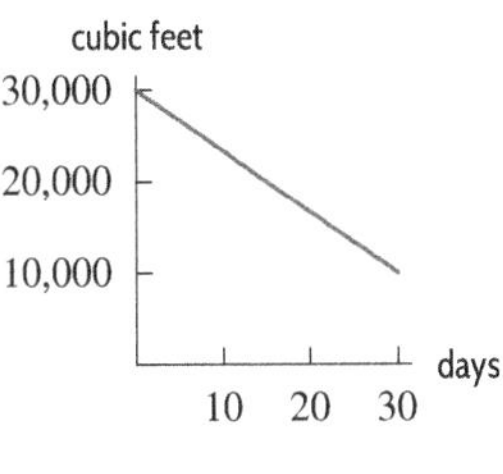

Figure 5.53

35. A cup of coffee at 90°C is put into a 20°C room when $t = 0$. The coffee's temperature is changing at a rate of $r(t) = -7e^{-0.1t}$ °C per minute, with t in minutes. Estimate the coffee's temperature when $t = 10$.

36. Water is pumped out of a holding tank at a rate of $5 - 5e^{-0.12t}$ liters/minute, where t is in minutes since the pump is started. If the holding tank contains 1000 liters of water when the pump is started, how much water does it hold one hour later?

37. The concentration of a medication in the plasma changes at a rate of $h(t)$ mg/ml per hour, t hours after the delivery of the drug.

    (a) Explain the meaning of the statement $h(1) = 50$.

    (b) There is 250 mg/ml of the medication present at time $t = 0$ and $\int_0^3 h(t)\,dt = 480$. What is the plasma concentration of the medication present three hours after the drug is administered?

[6]www.epa.gov/climatechange/ghgemissions/inventoryexplorer/#allsectors/allgas/gas/all. Accessed April 2015.
[7]http://www.eia.doe.gov/cneaf/coal/page/special/tbl1.html. Accessed May 2011.
[8]http://cdec.water.ca.gov/, accessed January 8, 2015.

Problems 38–39 concern the graph of f' in Figure 5.54.

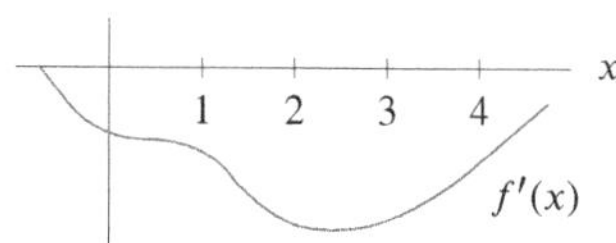

Figure 5.54: Graph of f', not f

38. Which is greater, $f(0)$ or $f(1)$?

39. List the following in increasing order:
$$\frac{f(4)-f(2)}{2}, \quad f(3)-f(2), \quad f(4)-f(3).$$

40. A force F parallel to the x-axis is given by the graph in Figure 5.55. Estimate the work, W, done by the force, where $W = \int_0^{16} F(x)\,dx$.

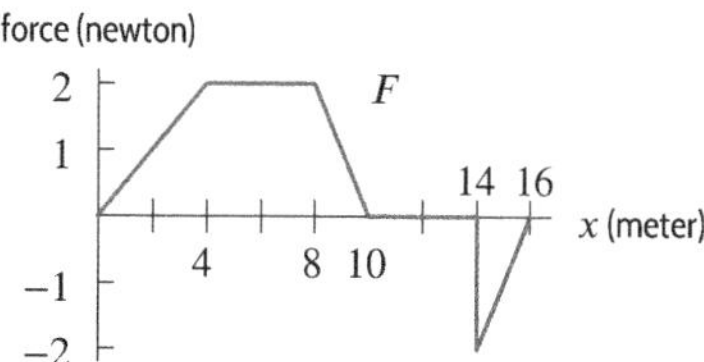

Figure 5.55

41. Let $f(1) = 7, f'(t) = e^{-t^2}$. Use left- and right-hand sums of 5 rectangles each to estimate $f(2)$.

42. Figure 5.56 shows a continuous function f. Rank the following integrals in ascending numerical order.

(i) $\int_0^2 f(x)\,dx$ (ii) $\int_0^1 f(x)\,dx$

(iii) $\int_0^2 (f(x))^{1/2}\,dx$ (iv) $\int_0^2 (f(x))^2\,dx$.

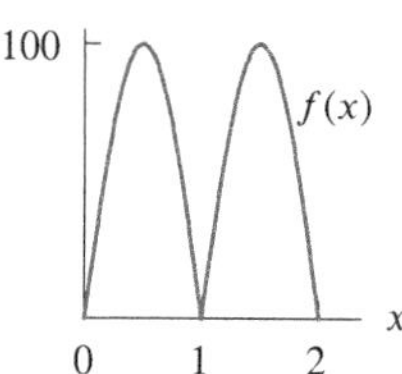

Figure 5.56

43. The graphs in Figure 5.57 represent the velocity, v, of a particle moving along the x-axis for time $0 \le t \le 5$. The vertical scales of all graphs are the same. Identify the graph showing which particle:

(a) Has a constant acceleration.
(b) Ends up farthest to the left of where it started.
(c) Ends up the farthest from its starting point.
(d) Experiences the greatest initial acceleration.
(e) Has the greatest average velocity.
(f) Has the greatest average acceleration.

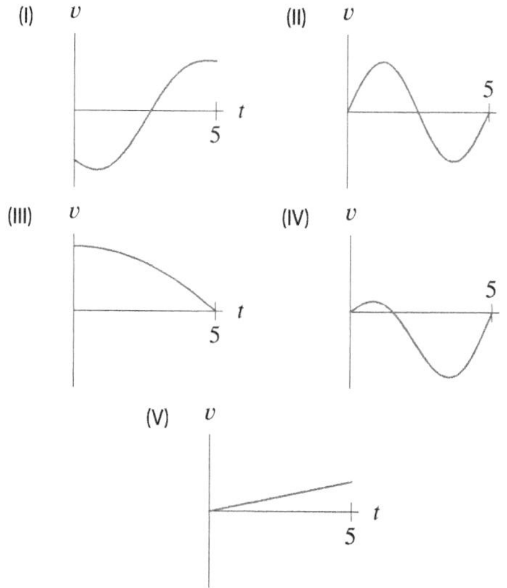

Figure 5.57

44. A mouse moves back and forth in a straight tunnel, attracted to bits of cheddar cheese alternately introduced to and removed from the ends (right and left) of the tunnel. The graph of the mouse's velocity, v, is given in Figure 5.58, with positive velocity corresponding to motion toward the right end. Assume that the mouse starts ($t = 0$) at the center of the tunnel.

(a) Use the graph to estimate the time(s) at which:
 (i) The mouse changes direction.
 (ii) The mouse is moving most rapidly to the right; to the left.
 (iii) The mouse is farthest to the right of center; farthest to the left.
 (iv) The mouse's speed (i.e., the magnitude of its velocity) is decreasing.
 (v) The mouse is at the center of the tunnel.

(b) What is the total distance the mouse traveled?

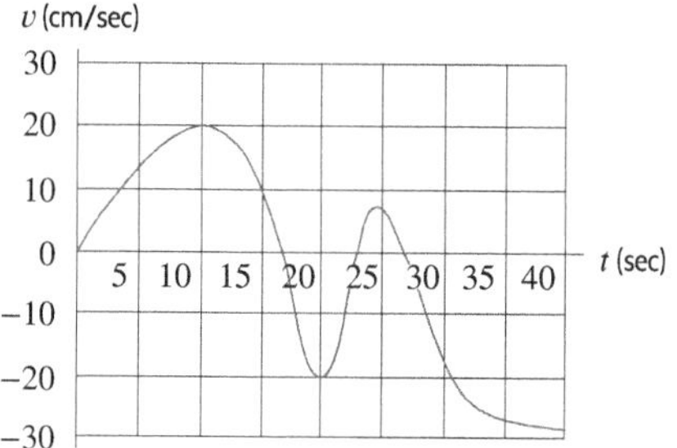

Figure 5.58

In Problems 45–46, oil is pumped from a well at a rate of $r(t)$ barrels per day, with t in days. Assume $r'(t) < 0$ and $t_0 > 0$.

45. What does the value of $\int_0^{t_0} r(t)\,dt$ tell us about the oil well?

46. Rank in order from least to greatest:
$$\int_0^{2t_0} r(t)\,dt, \quad \int_{t_0}^{2t_0} r(t)\,dt, \quad \int_{2t_0}^{3t_0} r(t)\,dt.$$

47. Height velocity graphs are used by endocrinologists to follow the progress of children with growth deficiencies. Figure 5.59 shows the height velocity curves of an average boy and an average girl between ages 3 and 18.

    (a) Which curve is for girls and which is for boys? Explain how you can tell.
    (b) About how much does the average boy grow between ages 3 and 10?
    (c) The growth spurt associated with adolescence and the onset of puberty occurs between ages 12 and 15 for the average boy and between ages 10 and 12.5 for the average girl. Estimate the height gained by each average child during this growth spurt.
    (d) When fully grown, about how much taller is the average man than the average woman? (The average boy and girl are about the same height at age 3.)

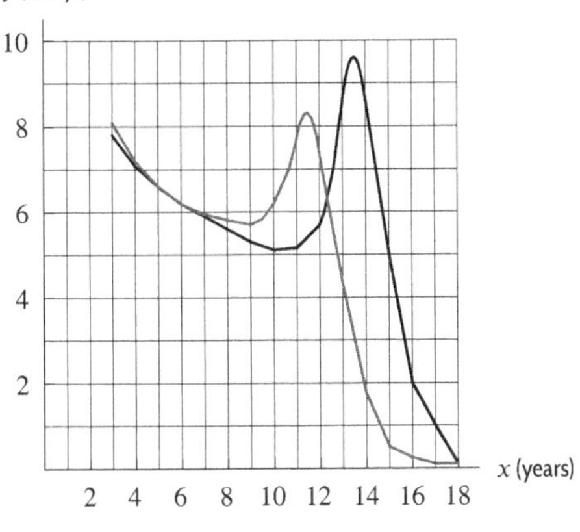

Figure 5.59

48. Figure 5.60 shows the rate of change in the average plasma concentration of the drug Omeprazole (in ng/ml per hour) for six hours after the first dose is administered using two different capsules: immediate-release and delayed-release.[9]

    (a) Which graph corresponds to which capsule?
    (b) Do the two capsules provide the same maximum concentration? If not, which provides the larger maximum concentration?

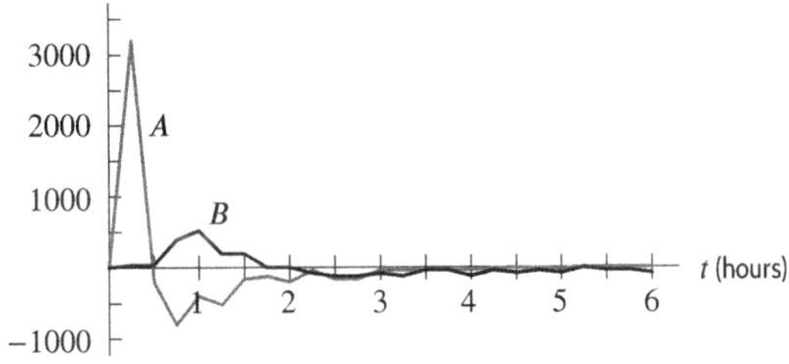

Figure 5.60

49. Table 5.11 shows the monthly change in water stored in Lake Sonoma, California, from March through November 2014. The change is measured in acre-feet per month.[10] On March 1, the water stored was 182,566 acre-feet. Let $S(t)$ be the total water, in acre-feet, stored in month t, where $t = 0$ is March.

    (a) Find and interpret $S(0)$ and $S(3)$.
    (b) Approximately when do maximum and minimum values of $S(t)$ occur?
    (c) Does $S(t)$ appear to have inflection points? If so, approximately when?

Table 5.11

Month	Mar	Apr	May	June	July
Change in water	3003	−5631	−8168	−8620	−8270
Month	Aug	Sept	Oct	Nov	
Change in water	−7489	−6245	−4593	54,743	

Change in water in acre-feet per month

■ In Problems 50–52, evaluate the expressions using Table 5.12. Give exact values if possible; otherwise, make the best possible estimates using left-hand Riemann sums.

Table 5.12

t	0.0	0.1	0.2	0.3	0.4	0.5
$f(t)$	0.3	0.2	0.2	0.3	0.4	0.5
$g(t)$	2.0	2.9	5.1	5.1	3.9	0.8

50. $\int_0^{0.5} f(t)\,dt$

51. $\int_{0.2}^{0.5} g'(t)\,dt$

52. $\int_0^{0.3} g(f(t))\,dt$

■ In Problems 53–55, let $C(n)$ be a city's cost, in millions of dollars, for plowing the roads when n inches of snow have fallen. Let $c(n) = C'(n)$. Evaluate the expressions and interpret your answers in terms of the cost of plowing snow, given

$$c'(n) < 0, \quad \int_0^{15} c(n)\,dn = 7.5, \quad c(15) = 0.7,$$
$$c(24) = 0.4, \quad C(15) = 8, \quad C(24) = 13.$$

53. $\int_{15}^{24} c(n)\,dn$

54. $C(0)$

55. $c(15) + \int_{15}^{24} c'(n)\,dn$

[9] Data adapted from C. W. Howden, "Review article: immediate-release proton-pump inhibitor therapy—potential advantages", *Aliment Pharmacol Ther.* (2005).

[10] Date from http://cdec.water.ca.gov/cgi-progs/stationInfo?station_id=WRS, accessed June, 2015. An acre-foot is the amount of water it takes to cover one acre of area with 1 foot of water.

■ Problems 56–58 refer to a May 2, 2010, article:[11]

"The crisis began around 10 am yesterday when a 10-foot wide pipe in Weston sprang a leak, which worsened throughout the afternoon and eventually cut off Greater Boston from the Quabbin Reservoir, where most of its water supply is stored... Before water was shut off to the ruptured pipe [at 6:40 pm], brown water had been roaring from a massive crater [at a rate of] 8 million gallons an hour rushing into the nearby Charles River."

Let $r(t)$ be the rate in gallons/hr that water flowed from the pipe t hours after it sprang its leak.

56. Which is larger: $\int_0^2 r(t)\,dt$ or $\int_2^4 r(t)\,dt$?

57. Which is larger: $\int_0^4 r(t)\,dt$ or $4r(4)$?

58. Give a reasonable overestimate of $\int_0^8 r(t)\,dt$.

■ In Problems 59–61, list the expressions (I)–(III) in order from smallest to largest, where $r(t)$ is the hourly rate that an animal burns calories and $R(t)$ is the total number of calories burned since time $t = 0$. Assume $r(t) > 0$ and $r'(t) < 0$ for $0 \le t \le 12$.

59. Letting $t_0 = 0, t_{100} = 12$:

I. $\sum_{i=0}^{99} r(t_i)\,\Delta t$ II. $\sum_{i=1}^{100} r(t_i)\,\Delta t$ III. $\int_0^{12} r(t)\,dt$

60. I. $R(10)$ II. $R(12)$ III. $R(10)+r(10)\cdot 2$

61. I. $\int_5^8 r(t)\,dt$ II. $\int_8^{11} r(t)\,dt$ III. $R(12)-R(9)$

Strengthen Your Understanding

■ In Problems 62–63, explain what is wrong with the statement.

62. If $f(t)$ represents the rate, in lbs per year, at which a dog gains weight t years after it is born, then $\int_0^4 f(t)dt$ represents the weight of the dog when the dog is four years old.

63. If $f(x) = \sqrt{x}$ the Fundamental Theorem of Calculus states that $\int_4^9 \sqrt{x}\,dx = \sqrt{9} - \sqrt{4}$.

■ In Problems 64–65, give an example of:

64. A function $f(x)$ and limits of integration a and b such that $\int_a^b f(x)\,dx = e^4 - e^2$.

65. The graph of a velocity function of a car that travels 200 miles in 4 hours.

66. True or False? The units for an integral of a function $f(x)$ are the same as the units for $f(x)$.

5.4 THEOREMS ABOUT DEFINITE INTEGRALS

Properties of the Definite Integral

For the definite integral $\int_a^b f(x)\,dx$, we have so far only considered the case $a < b$. We now allow $a \ge b$. We still set $x_0 = a$, $x_n = b$, and $\Delta x = (b-a)/n$. As before, we have $\int_a^b f(x)dx = \lim_{n\to\infty} \sum_{i=1}^{n} f(x_i)\Delta x$.

Theorem 5.2: Properties of Limits of Integration

If a, b, and c are any numbers and f is a continuous function, then

1. $\int_b^a f(x)\,dx = -\int_a^b f(x)\,dx.$
2. $\int_a^c f(x)\,dx + \int_c^b f(x)\,dx = \int_a^b f(x)\,dx.$

In words:

1. The integral from b to a is the negative of the integral from a to b.
2. The integral from a to c plus the integral from c to b is the integral from a to b.

By interpreting the integrals as areas, we can justify these results for $f \ge 0$. In fact, they are true for all functions for which the integrals make sense.

[11] "A catastrophic rupture hits region's water system," *The Boston Globe*, May 2, 2010.

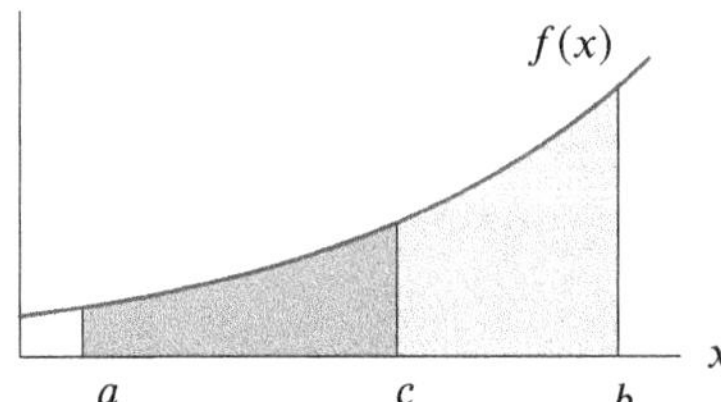

Figure 5.61: Additivity of the definite integral ($a < c < b$)

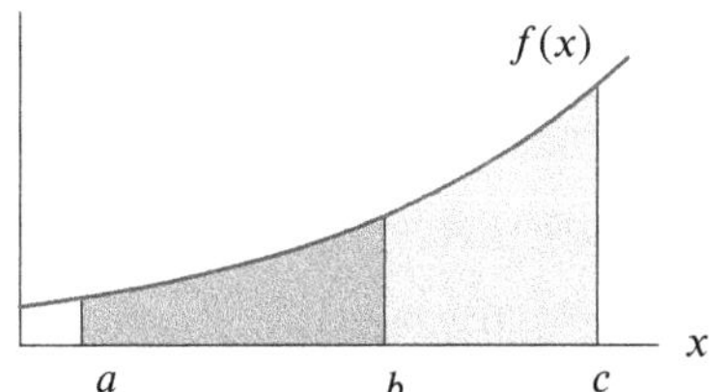

Figure 5.62: Additivity of the definite integral ($a < b < c$)

Why is $\int_b^a f(x)\,dx = -\int_a^b f(x)\,dx$?

By definition, both integrals are approximated by sums of the form $\sum f(x_i)\Delta x$. The only difference in the sums for $\int_b^a f(x)\,dx$ and $\int_a^b f(x)\,dx$ is that in the first $\Delta x = (a-b)/n = -(b-a)/n$ and in the second $\Delta x = (b-a)/n$. Since everything else about the sums is the same, we must have $\int_b^a f(x)\,dx = -\int_a^b f(x)\,dx$.

Why is $\int_a^c f(x)\,dx + \int_c^b f(x)\,dx = \int_a^b f(x)\,dx$?

Suppose $a < c < b$. Figure 5.61 suggests that $\int_a^c f(x)\,dx + \int_c^b f(x)\,dx = \int_a^b f(x)\,dx$ since the area under f from a to c plus the area under f from c to b together make up the whole area under f from a to b.

This property holds for all numbers a, b, and c, not just those satisfying $a < c < b$. (See Figure 5.62.) For example, the area under f from 3 to 6 is equal to the area from 3 to 8 *minus* the area from 6 to 8, so

$$\int_3^6 f(x)\,dx = \int_3^8 f(x)\,dx - \int_6^8 f(x)\,dx = \int_3^8 f(x)\,dx + \int_8^6 f(x)\,dx.$$

Example 1 Given that $\int_0^{1.25} \cos(x^2)\,dx = 0.98$ and $\int_0^1 \cos(x^2)\,dx = 0.90$, what are the values of the following integrals? (See Figure 5.63.)

(a) $\displaystyle\int_1^{1.25} \cos(x^2)\,dx$ (b) $\displaystyle\int_{-1}^{1} \cos(x^2)\,dx$ (c) $\displaystyle\int_{1.25}^{-1} \cos(x^2)\,dx$

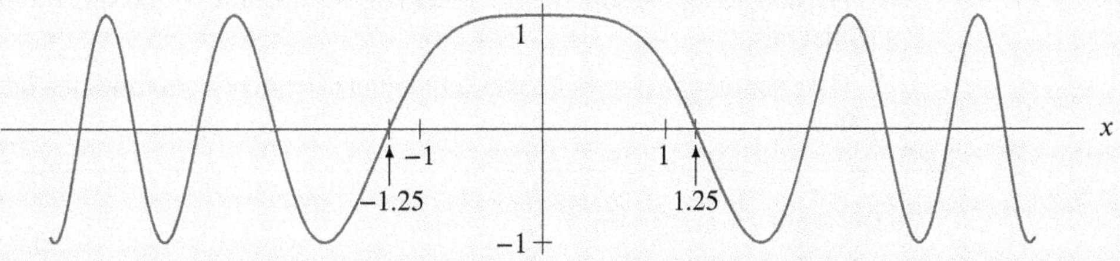

Figure 5.63: Graph of $f(x) = \cos(x^2)$

Solution (a) Since, by the additivity property,

$$\int_0^{1.25} \cos(x^2)\,dx = \int_0^1 \cos(x^2)\,dx + \int_1^{1.25} \cos(x^2)\,dx,$$

we get

$$0.98 = 0.90 + \int_1^{1.25} \cos(x^2)\,dx,$$

so

$$\int_{1}^{1.25} \cos(x^2)\,dx = 0.08.$$

(b) By the additivity property, we have

$$\int_{-1}^{1} \cos(x^2)\,dx = \int_{-1}^{0} \cos(x^2)\,dx + \int_{0}^{1} \cos(x^2)\,dx.$$

By the symmetry of $\cos(x^2)$ about the y-axis,

$$\int_{-1}^{0} \cos(x^2)\,dx = \int_{0}^{1} \cos(x^2)\,dx = 0.90.$$

So

$$\int_{-1}^{1} \cos(x^2)\,dx = 0.90 + 0.90 = 1.80.$$

(c) Using both properties in Theorem 5.2, we have

$$\int_{1.25}^{-1} \cos(x^2)\,dx = -\int_{-1}^{1.25} \cos(x^2)\,dx = -\left(\int_{-1}^{0} \cos(x^2)\,dx + \int_{0}^{1.25} \cos(x^2)\,dx\right)$$
$$= -(0.90 + 0.98) = -1.88.$$

Theorem 5.3: Properties of Sums and Constant Multiples of the Integrand

Let f and g be continuous functions and let c be a constant.

1. $\displaystyle\int_{a}^{b} (f(x) \pm g(x))\,dx = \int_{a}^{b} f(x)\,dx \pm \int_{a}^{b} g(x)\,dx.$
2. $\displaystyle\int_{a}^{b} cf(x)\,dx = c\int_{a}^{b} f(x)\,dx.$

In words:

1. The integral of the sum (or difference) of two functions is the sum (or difference) of their integrals.
2. The integral of a constant times a function is that constant times the integral of the function.

Why Do These Properties Hold?

Both can be visualized by thinking of the definite integral as the limit of a sum of areas of rectangles.

For property 1, suppose that f and g are positive on the interval $[a, b]$ so that the area under $f(x)+g(x)$ is approximated by the sum of the areas of rectangles like the one shaded in Figure 5.64. The area of this rectangle is

$$(f(x_i) + g(x_i))\Delta x = f(x_i)\Delta x + g(x_i)\Delta x.$$

Since $f(x_i)\Delta x$ is the area of a rectangle under the graph of f, and $g(x_i)\Delta x$ is the area of a rectangle under the graph of g, the area under $f(x) + g(x)$ is the sum of the areas under $f(x)$ and $g(x)$.

For property 2, notice that multiplying a function by c stretches or shrinks the graph in the vertical direction by a factor of c. Thus, it stretches or shrinks the height of each approximating rectangle by c, and hence multiplies the area by c.

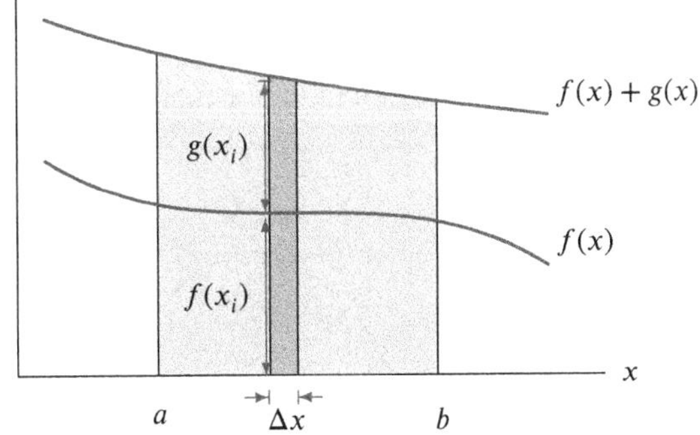

Figure 5.64: Area $= \int_a^b [f(x)+g(x)]\,dx = \int_a^b f(x)\,dx + \int_a^b g(x)\,dx$

Example 2 Evaluate the definite integral $\int_0^2 (1+3x)\,dx$ exactly.

Solution We can break this integral up as follows:

$$\int_0^2 (1+3x)\,dx = \int_0^2 1\,dx + \int_0^2 3x\,dx = \int_0^2 1\,dx + 3\int_0^2 x\,dx.$$

From Figures 5.65 and 5.66 and the area interpretation of the integral, we see that

$$\int_0^2 1\,dx = \begin{array}{c}\text{Area of}\\ \text{rectangle}\end{array} = 2 \quad \text{and} \quad \int_0^2 x\,dx = \begin{array}{c}\text{Area of}\\ \text{triangle}\end{array} = \frac{1}{2}\cdot 2\cdot 2 = 2.$$

Therefore,

$$\int_0^2 (1+3x)\,dx = \int_0^2 1\,dx + 3\int_0^2 x\,dx = 2 + 3\cdot 2 = 8.$$

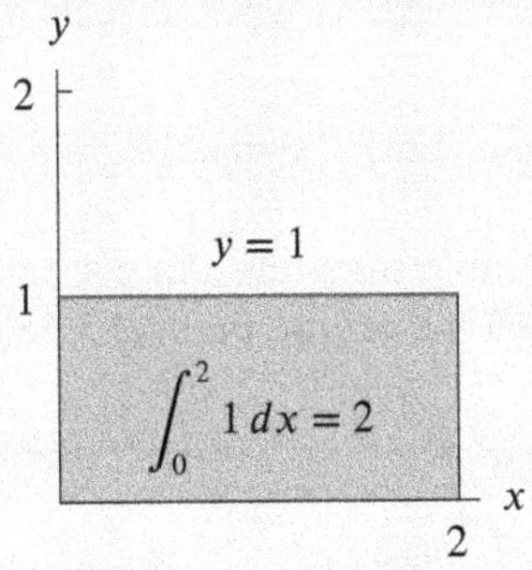

Figure 5.65: Area representing $\int_0^2 1\,dx$

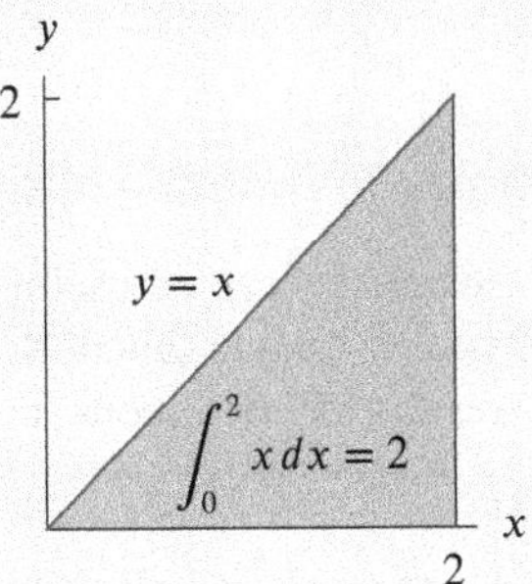

Figure 5.66: Area representing $\int_0^2 x\,dx$

Area Between Curves

Theorem 5.3 enables us to find the area of a region between curves. We have the following result:

> If the graph of $f(x)$ lies above the graph of $g(x)$ for $a \le x \le b$, then
>
> $$\begin{array}{c}\text{Area between } f \text{ and } g\\ \text{for } a \le x \le b\end{array} = \int_a^b (f(x) - g(x))\,dx.$$

Example 3 Find the area of the shaded region in Figure 5.67.

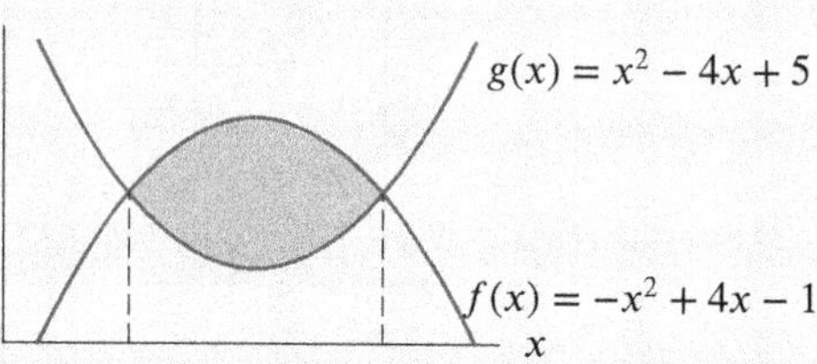

Figure 5.67: Area between two parabolas

Solution The curves cross where

$$\begin{aligned} x^2 - 4x + 5 &= -x^2 + 4x - 1 \\ 2x^2 - 8x + 6 &= 0 \\ 2(x-1)(x-3) &= 0 \\ x &= 1, 3. \end{aligned}$$

Since $f(x) = -x^2 + 4x - 1$ is above $g(x) = x^2 - 4x + 5$ for x between 1 and 3, we find the shaded area by subtraction:

$$\begin{aligned} \text{Area} = \int_1^3 f(x)\,dx - \int_1^3 g(x)\,dx &= \int_1^3 (f(x) - g(x))\,dx \\ &= \int_1^3 ((-x^2 + 4x - 1) - (x^2 - 4x + 5))\,dx \\ &= \int_1^3 (-2x^2 + 8x - 6)\,dx = 2.667. \end{aligned}$$

Using Symmetry to Evaluate Integrals

Symmetry can be useful in evaluating definite integrals. An even function is symmetric about the y-axis. An odd function is symmetric about the origin. Figures 5.68 and 5.69 suggest the following results for continuous functions:

> If f is even, then $\displaystyle\int_{-a}^{a} f(x)\,dx = 2\int_0^a f(x)\,dx.$ If g is odd, then $\displaystyle\int_{-a}^{a} g(x)\,dx = 0.$

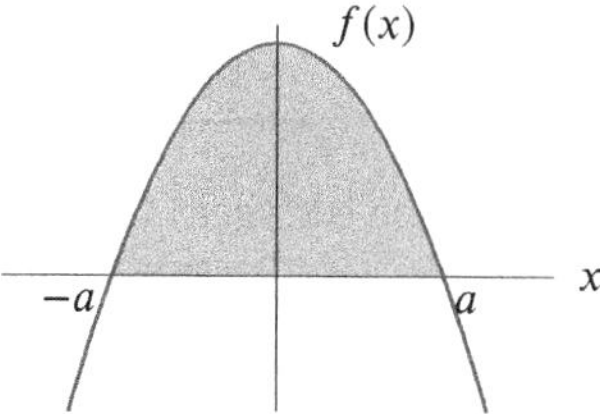

Figure 5.68: For an even function, $\int_{-a}^{a} f(x)\,dx = 2\int_0^a f(x)\,dx$

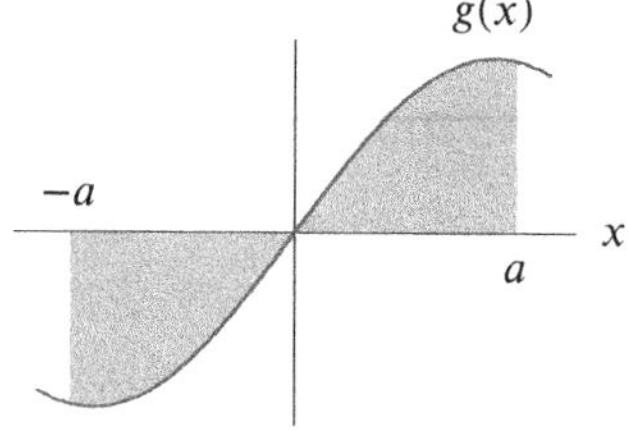

Figure 5.69: For an odd function, $\int_{-a}^{a} g(x)\,dx = 0$

Example 4 Given that $\int_0^\pi \sin t\,dt = 2$, find (a) $\displaystyle\int_{-\pi}^{\pi} \sin t\,dt$ (b) $\displaystyle\int_{-\pi}^{\pi} |\sin t|\,dt$

Solution Graphs of $\sin t$ and $|\sin t|$ are in Figures 5.70 and 5.71.

(a) Since $\sin t$ is an odd function

$$\int_{-\pi}^{\pi} \sin t\,dt = 0.$$

(b) Since $|\sin t|$ is an even function

$$\int_{-\pi}^{\pi} |\sin t|\,dt = 2\int_0^{\pi} |\sin t|\,dt = 4.$$

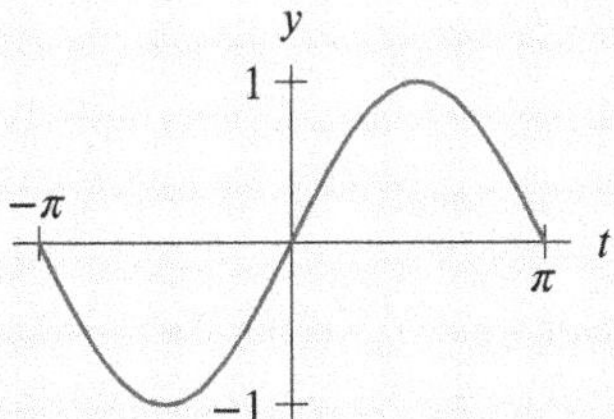

Figure 5.70

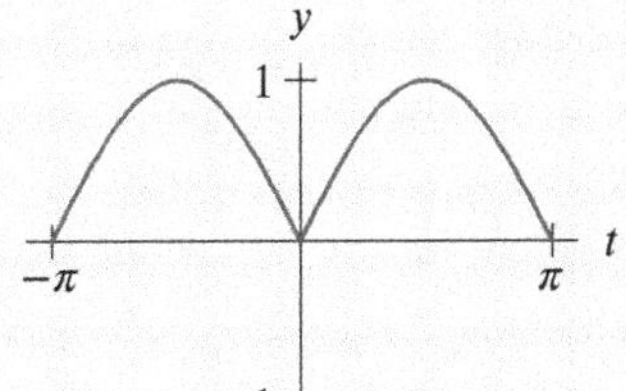

Figure 5.71

Comparing Integrals

Suppose we have constants m and M such that $m \le f(x) \le M$ for $a \le x \le b$. We say f is *bounded above* by M and *bounded below* by m. Then the graph of f lies between the horizontal lines $y = m$ and $y = M$. So the definite integral lies between $m(b-a)$ and $M(b-a)$. See Figure 5.72.

Suppose $f(x) \le g(x)$ for $a \le x \le b$, as in Figure 5.73. Then the definite integral of f is less than or equal to the definite integral of g. This leads us to the following results:

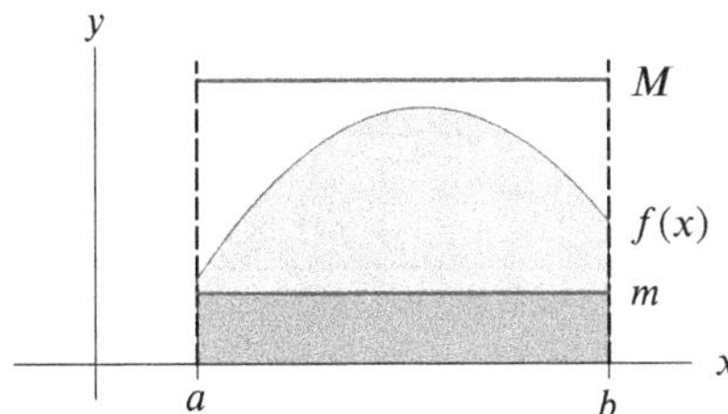

Figure 5.72: The area under the graph of f lies between the areas of the rectangles

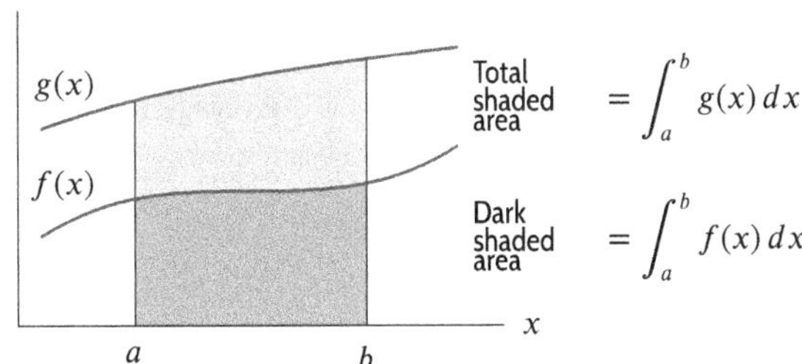

Figure 5.73: If $f(x) \le g(x)$ then $\int_a^b f(x)\,dx \le \int_a^b g(x)\,dx$

Theorem 5.4: Comparison of Definite Integrals

Let f and g be continuous functions.

1. If $m \le f(x) \le M$ for $a \le x \le b$, then $\displaystyle m(b-a) \le \int_a^b f(x)\,dx \le M(b-a).$

2. If $f(x) \le g(x)$ for $a \le x \le b$, then $\displaystyle \int_a^b f(x)\,dx \le \int_a^b g(x)\,dx.$

Example 5 Explain why $\int_0^{\sqrt{\pi}} \sin(x^2)\,dx \le \sqrt{\pi}$.

Solution Since $\sin(x^2) \le 1$ for all x (see Figure 5.74), part 2 of Theorem 5.4 gives

$$\int_0^{\sqrt{\pi}} \sin(x^2)\,dx \le \int_0^{\sqrt{\pi}} 1\,dx = \sqrt{\pi}.$$

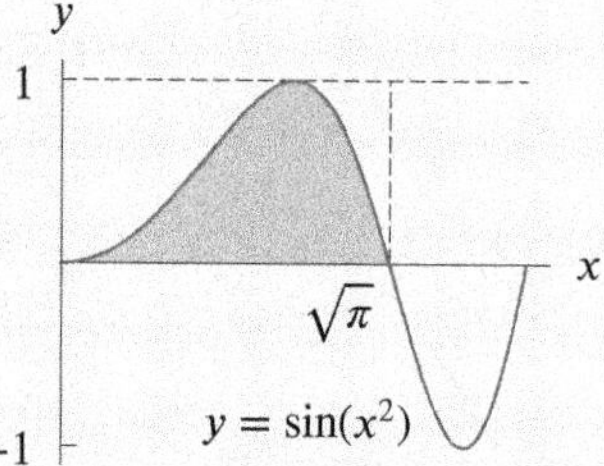

Figure 5.74: Graph showing that $\int_0^{\sqrt{\pi}} \sin(x^2)\,dx < \sqrt{\pi}$

The Definite Integral as an Average

We know how to find the average of n numbers: Add them and divide by n. But how do we find the average value of a continuously varying function? Let us consider an example. Suppose $f(t)$ is the temperature at time t, measured in hours since midnight, and that we want to calculate the average temperature over a 24-hour period. One way to start is to average the temperatures at n equally spaced times, $t_1, t_2, \dots, t_n$, during the day.

$$\text{Average temperature} \approx \frac{f(t_1) + f(t_2) + \cdots + f(t_n)}{n}.$$

The larger we make n, the better the approximation. We can rewrite this expression as a Riemann sum over the interval $0 \le t \le 24$ if we use the fact that $\Delta t = 24/n$, so $n = 24/\Delta t$:

$$\begin{aligned}\text{Average temperature} &\approx \frac{f(t_1) + f(t_2) + \cdots + f(t_n)}{24/\Delta t}\\ &= \frac{f(t_1)\Delta t + f(t_2)\Delta t + \cdots + f(t_n)\Delta t}{24}\\ &= \frac{1}{24}\sum_{i=1}^{n} f(t_i)\Delta t.\end{aligned}$$

As $n \to \infty$, the Riemann sum tends toward an integral, and $1/24$ of the sum also approximates the average temperature better. It makes sense, then, to write

$$\text{Average temperature} = \lim_{n\to\infty} \frac{1}{24}\sum_{i=1}^{n} f(t_i)\Delta t = \frac{1}{24}\int_0^{24} f(t)\,dt.$$

We have found a way of expressing the average temperature over an interval in terms of an integral. Generalizing for any function f, if $a < b$, we define

$$\text{Average value of } f \text{ from } a \text{ to } b = \frac{1}{b-a}\int_a^b f(x)\,dx.$$

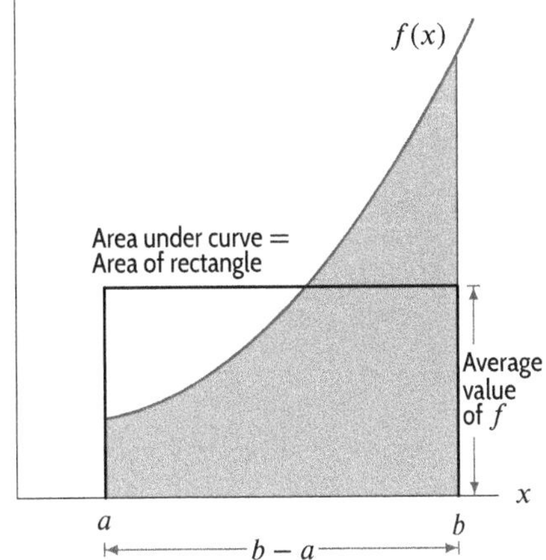

Figure 5.75: Area and average value

How to Visualize the Average on a Graph

The definition of average value tells us that

$$(\text{Average value of } f) \cdot (b-a) = \int_a^b f(x)\,dx.$$

Let's interpret the integral as the area under the graph of f. Then the average value of f is the height of a rectangle whose base is $(b-a)$ and whose area is the same as the area under the graph of f. (See Figure 5.75.)

Example 6 Suppose that $C(t)$ represents the daily cost of heating your house, measured in dollars per day, where t is time measured in days and $t = 0$ corresponds to January 1, 2015. Interpret $\int_0^{90} C(t)\,dt$ and $\frac{1}{90-0}\int_0^{90} C(t)\,dt$.

Solution The units for the integral $\int_0^{90} C(t)\,dt$ are (dollars/day)×(days) = dollars. The integral represents the total cost in dollars to heat your house for the first 90 days of 2015, namely the months of January, February, and March. The second expression is measured in (1/days)(dollars) or dollars per day, the same units as $C(t)$. It represents the average cost per day to heat your house during the first 90 days of 2015.

Example 7 In the year 2014, the population of Nevada[12] was modeled by the function

$$P = f(t) = 2.839(1.017)^t,$$

where P is in millions of people and t is in years since 2014. Use this function to predict the average population of Nevada between the years 2010 and 2020.

Solution We want the average value of $f(t)$ between $t = -6$ and $t = 4$. This is given by

$$\text{Average population} = \frac{1}{4-(-6)}\int_{-6}^{4} f(t)\,dt = \frac{1}{10}(28.91) = 2.891.$$

We used a calculator to evaluate the integral. The average population of Nevada between 2010 and 2020 is predicted to be about 2.891 million people.

[12] quickfacts.census.gov, accessed April 20, 2015.

Exercises and Problems for Section 5.4 Online Resource: Additional Problems for Section 5.4

EXERCISES

In Exercises 1–6, find the integral, given that $\int_a^b f(x)\,dx = 8$, $\int_a^b (f(x))^2\,dx = 12$, $\int_a^b g(t)\,dt = 2$, and $\int_a^b (g(t))^2\,dt = 3$.

1. $\int_a^b (f(x) + g(x))\,dx$
2. $\int_a^b cf(z)\,dz$
3. $\int_a^b \left((f(x))^2 - (g(x))^2\right)dx$
4. $\int_a^b (f(x))^2\,dx - (\int_a^b f(x)\,dx)^2$
5. $\int_a^b \left(c_1 g(x) + (c_2 f(x))^2\right)\,dx$
6. $\int_{a+5}^{b+5} f(x-5)\,dx$

In Exercises 7–10, find the average value of the function over the given interval.

7. $g(t) = 1+t$ over $[0, 2]$
8. $g(t) = e^t$ over $[0, 10]$
9. $f(x) = 2$ over $[a, b]$
10. $f(x)=4x+7$ over $[1,3]$

11. **(a)** Using Figure 5.76, find $\int_1^6 f(x)\,dx$.
    (b) What is the average value of f on $[1, 6]$?

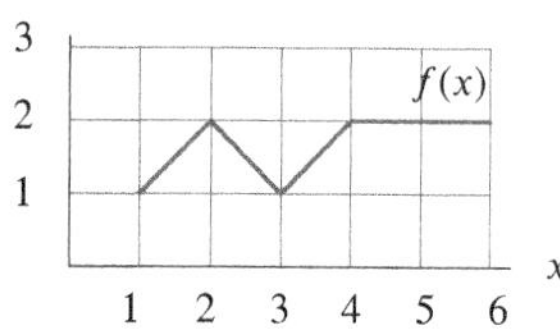

Figure 5.76

12. How do the units for the average value of f relate to the units for $f(x)$ and the units for x?

Find the area of the regions in Exercises 13–20.

13. Under $y = e^x$ and above $y = 1$ for $0 \le x \le 2$.
14. Under $y = 5\ln(2x)$ and above $y = 3$ for $3 \le x \le 5$.
15. Between $y = x^2$ and $y = x^3$ for $0 \le x \le 1$.
16. Between $y = x^{1/2}$ and $y = x^{1/3}$ for $0 \le x \le 1$.
17. Between $y = \sin x + 2$ and $y = 0.5$ for $6 \le x \le 10$.
18. Between $y = \cos t$ and $y = \sin t$ for $0 \le t \le \pi$.
19. Between $y = e^{-x}$ and $y = 4(x - x^2)$.
20. Between $y = e^{-x}$ and $y = \ln x$ for $1 \le x \le 2$.

In Exercises 21–24, without evaluating them, decide which of the two definite integrals is smaller.

21. $\int_0^1 x\,dx$ and $\int_0^1 x^2\,dx$
22. $\int_1^2 x\,dx$ and $\int_1^2 x^2\,dx$
23. $\int_2^3 \cos(x)\,dx$ and $\int_2^3 x\,dx$
24. $\int_{-2}^{-1} x^3\,dx$ and $\int_{-2}^{-1} e^x\,dx$

PROBLEMS

25. Figure 5.77 shows an even function $f(x)$. Find
    (a) $\int_0^4 f(x)\,dx$ **(b)** $\int_{-2}^2 f(x)\,dx$ **(c)** $\int_{-4}^2 f(x)\,dx$

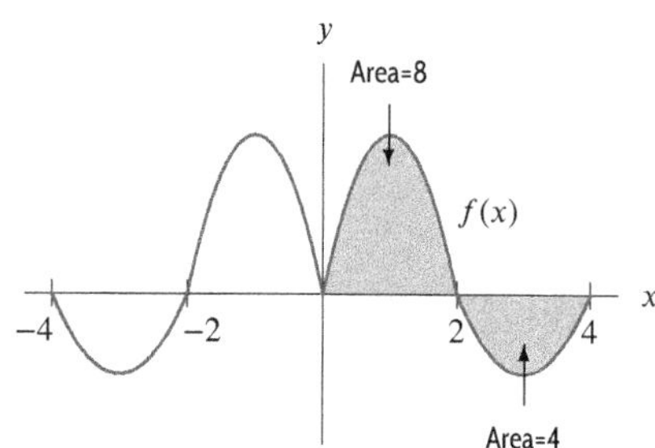

Figure 5.77

26. **(a)** Let $\int_0^3 f(x)dx = 6$. What is the average value of $f(x)$ on the interval $x = 0$ to $x = 3$?
    (b) If $f(x)$ is even, what is $\int_{-3}^3 f(x)dx$? What is the average value of $f(x)$ on the interval $x = -3$ to $x = 3$?
    (c) If $f(x)$ is odd, what is $\int_{-3}^3 f(x)dx$? What is the average value of $f(x)$ on the interval $x = -3$ to $x = 3$?

27. Using Figure 5.78, write $\int_0^3 f(x)\,dx$ in terms of $\int_{-1}^1 f(x)\,dx$ and $\int_1^3 f(x)\,dx$.

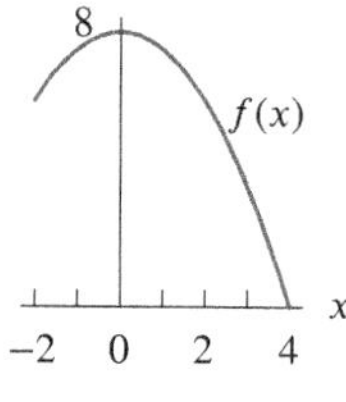

Figure 5.78

28. **(a)** Assume $a \le b$. Use geometry to construct a formula in terms of a and b for

$$\int_a^b 1\,dx.$$

    (b) Use the result of part (a) to find:
    (i) $\int_2^5 1\,dx$ (ii) $\int_{-3}^8 1\,dx$ (iii) $\int_1^3 23\,dx$

29. Given $\int_1^5 f(x)\,dx = 5$, $\int_1^9 f(x)\,dx = 3$, $\int_3^9 f(x)\,dx = 4$, evaluate:

(a) $\int_5^9 f(x)\,dx$ (b) $\int_3^5 f(x)\,dx$ (c) $\int_3^1 f(x)\,dx$

30. If $\int_2^5 (2f(x)+3)\,dx = 17$, find $\int_2^5 f(x)\,dx$.

31. (a) Given that $\int_3^5 g(x)\,dx = 7$, find

(i) $\int_3^5 2g(x)\,dx$ (ii) $\int_5^3 g(x)\,dx$

(b) What values of a and b allow you to calculate $\int_a^b g(x-6)\,dx$ from the information in part (a)?

32. The value, V, of a Tiffany lamp, worth \$225 in 1975, increases at 15% per year. Its value in dollars t years after 1975 is given by

$$V = 225(1.15)^t.$$

Find the average value of the lamp over the period 1975–2010.

33. (a) Assume that $0 \le a \le b$. Use geometry to construct a formula in terms of a and b for

$$\int_a^b x\,dx.$$

(b) Use the result of part (a) to find:

(i) $\int_2^5 x\,dx$ (ii) $\int_{-3}^8 x\,dx$ (iii) $\int_1^3 5x\,dx$

34. If $f(x)$ is odd and $\int_{-2}^3 f(x)\,dx = 30$, find $\int_2^3 f(x)\,dx$.

35. If $f(x)$ is even and $\int_{-2}^2 (f(x)-3)\,dx = 8$, find $\int_0^2 f(x)\,dx$.

36. Without any computation, find $\displaystyle\int_{-\pi/4}^{\pi/4} x^3 \cos x^2\,dx$.

37. If the average value of f on the interval $2 \le x \le 5$ is 4, find $\int_2^5 (3f(x)+2)\,dx$.

38. Suppose $\int_1^3 3x^2\,dx = 26$ and $\int_1^3 2x\,dx = 8$. What is $\int_1^3 (x^2 - x)\,dx$?

39. Figure 5.79 shows the rate, $f(x)$, in thousands of algae per hour, at which a population of algae is growing, where x is in hours.

(a) Estimate the average value of the rate over the interval $x = -1$ to $x = 3$.

(b) Estimate the total change in the population over the interval $x = -3$ to $x = 3$.

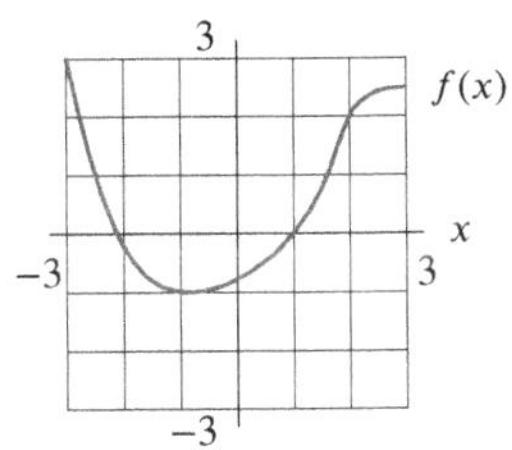

Figure 5.79

[13]www.indexmundi.com, accessed April 20, 2015.

40. (a) Using Figure 5.80, estimate $\int_{-3}^3 f(x)\,dx$.

(b) Which of the following average values of $f(x)$ is larger?

(i) Between $x = -3$ and $x = 3$

(ii) Between $x = 0$ and $x = 3$

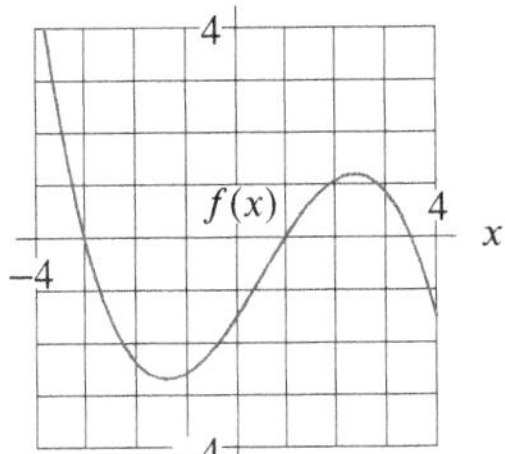

Figure 5.80

41. A bar of metal is cooling from 1000°C to room temperature, 20°C. The temperature, H, of the bar t minutes after it starts cooling is given, in °C, by

$$H = 20 + 980e^{-0.1t}.$$

(a) Find the temperature of the bar at the end of one hour.

(b) Find the average value of the temperature over the first hour.

(c) Is your answer to part (b) greater or smaller than the average of the temperatures at the beginning and the end of the hour? Explain this in terms of the concavity of the graph of H.

42. In 2014, the population of Mexico[13] was growing at 1.21% a year. Assuming that this growth rate continues into the future and that t is in years since 2014, the Mexican population, P, in millions, is given by

$$P = 120(1.0121)^t.$$

(a) Predict the average population of Mexico between 2014 and 2050.

(b) Find the average of the population in 2014 and the predicted population in 2050.

(c) Explain, in terms of the concavity of the graph of P, why your answer to part (b) is larger or smaller than your answer to part (a).

43. (a) Using a graph, decide if the area under $y = e^{-x^2/2}$ between 0 and 1 is more or less than 1.

(b) Find the area.

44. Without computation, show that $2 \leq \int_0^2 \sqrt{1+x^3}\, dx \leq 6$.

In Problems 45–47, let $f(x) = \sqrt{25 - x^2}$. Without evaluating the definite integral, give upper and lower bounds.

45. $\int_0^3 f(x)\, dx$ 46. $\int_3^4 f(x)\, dx$ 47. $\int_0^4 f(x)\, dx$

48. Without calculating the integral, explain why the following statements are false.

(a) $\int_{-2}^{-1} e^{x^2}\, dx = -3$ **(b)** $\int_{-1}^{1} \left| \frac{\cos(x+2)}{1+\tan^2 x} \right| dx = 0$

For Problems 49–52, mark the quantity on a copy of the graph of f in Figure 5.81.

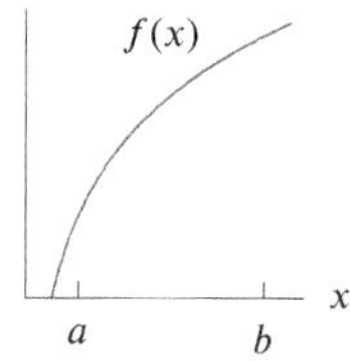

Figure 5.81

49. A length representing $f(b) - f(a)$.

50. A slope representing $\dfrac{f(b) - f(a)}{b - a}$.

51. An area representing $F(b) - F(a)$, where $F' = f$.

52. A length roughly approximating

$$\frac{F(b) - F(a)}{b - a}, \text{ where } F' = f.$$

53. Using Figure 5.82, is each expression positive, negative, zero, or is there not enough information to decide?

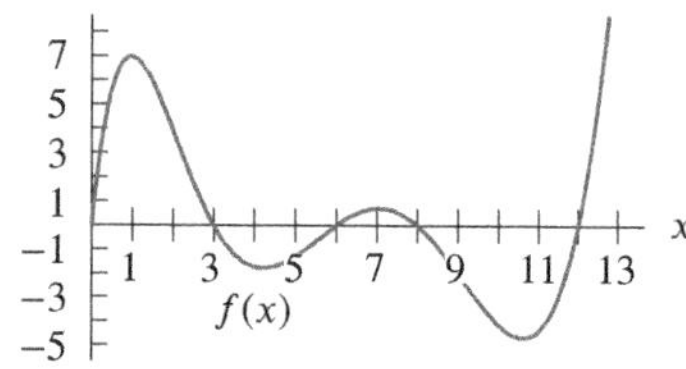

Figure 5.82

(a) $\int_3^8 f(x)\, dx$ **(b)** $\int_6^0 f(x)\, dx$

(c) $\int_8^{12} (f(x) + 6)\, dx$ **(d)** $\int_0^3 f(x+3)\, dx$

54. Using the graph of f in Figure 5.83, arrange the following quantities in increasing order, from least to greatest.

(i) $\int_0^1 f(x)\, dx$ (ii) $\int_1^2 f(x)\, dx$

(iii) $\int_0^2 f(x)\, dx$ (iv) $\int_2^3 f(x)\, dx$

(v) $-\int_1^2 f(x)\, dx$ (vi) The number 0

(vii) The number 20 (viii) The number -10

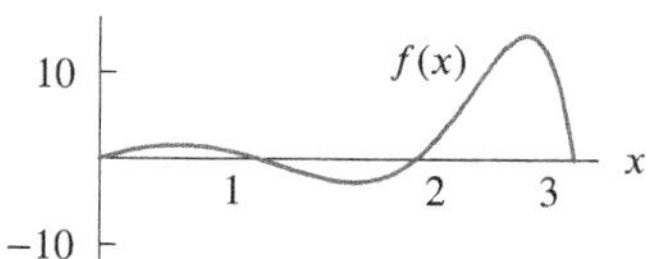

Figure 5.83

55. **(a)** Using Figures 5.84 and 5.85, find the average value on $0 \leq x \leq 2$ of

(i) $f(x)$ (ii) $g(x)$ (iii) $f(x) \cdot g(x)$

(b) Is the following statement true? Explain your answer.

$$\text{Average}(f) \cdot \text{Average}(g) = \text{Average}(f \cdot g)$$

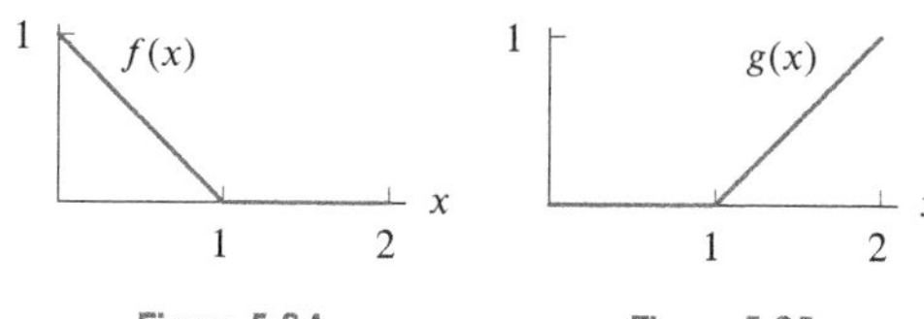

Figure 5.84 Figure 5.85

56. **(a)** Without computing any integrals, explain why the average value of $f(x) = \sin x$ on $[0, \pi]$ must be between 0.5 and 1.

(b) Compute this average.

57. Figure 5.86 shows the *standard normal distribution* from statistics, which is given by

$$\frac{1}{\sqrt{2\pi}} e^{-x^2/2}.$$

Statistics books often contain tables such as the following, which show the area under the curve from 0 to b for various values of b.

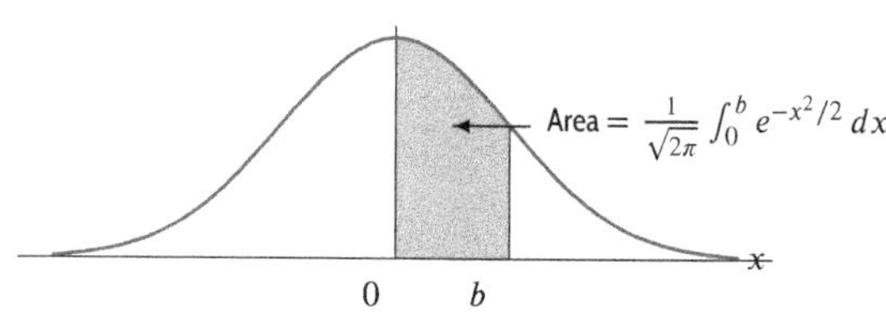

Figure 5.86

b	$\frac{1}{\sqrt{2\pi}}\int_0^b e^{-x^2/2}\,dx$
1	0.3413
2	0.4772
3	0.4987
4	0.5000

Use the information given in the table and the symmetry of the curve about the y-axis to find:

(a) $\displaystyle\frac{1}{\sqrt{2\pi}}\int_1^3 e^{-x^2/2}\,dx$ **(b)** $\displaystyle\frac{1}{\sqrt{2\pi}}\int_{-2}^3 e^{-x^2/2}\,dx$

In Problems 58–59, evaluate the expression, if possible, or say what additional information is needed, given that $\int_{-4}^4 g(x)\,dx = 12$.

58. $\displaystyle\int_0^4 g(x)\,dx$

59. $\displaystyle\int_{-4}^4 g(-x)\,dx$

In Problems 60–63, evaluate the expression if possible, or say what extra information is needed, given $\int_0^7 f(x)\,dx = 25$.

60. $\displaystyle\sqrt{\int_0^7 f(x)\,dx}$

61. $\displaystyle\int_0^{3.5} f(x)\,dx$

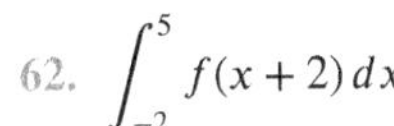

62. $\displaystyle\int_{-2}^5 f(x+2)\,dx$

63. $\displaystyle\int_0^7 (f(x)+2)\,dx$

For Problems 64–66, assuming $F' = f$, mark the quantity on a copy of Figure 5.87.

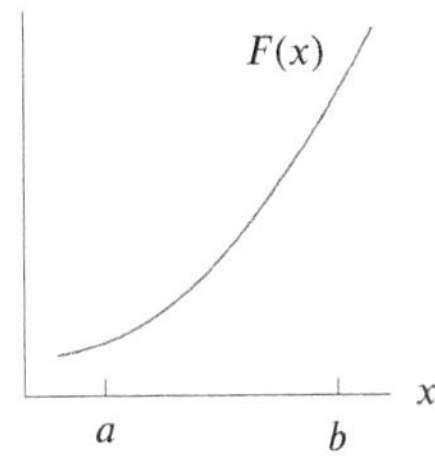

Figure 5.87

64. A slope representing $f(a)$.

65. A length representing $\displaystyle\int_a^b f(x)\,dx$.

66. A slope representing $\displaystyle\frac{1}{b-a}\int_a^b f(x)\,dx$.

Strengthen Your Understanding

In Problems 67–69, explain what is wrong with the statement.

67. If $f(x)$ is a continuous function on $[a, b]$ such that $\int_a^b f(x)dx \geq 0$, then $f(x) \geq 0$ for all x in $[a, b]$.

68. If $f(x)$ is a continuous function on the interval $[a, b]$, then $\int_a^b (5 + 3f(x))\,dx = 5 + 3\int_a^b f(x)\,dx$.

69. If $f(t)$ is the population of fish in a lake on day t, then the average population over a 6-month period is given by

$$\frac{1}{6}\int_0^6 f(t)\,dt.$$

In Problems 70–72, give an example of:

70. A continuous function $f(x)$ on the interval $[0, 1]$ such that $\int_0^1 2f(x)\,dx < \int_0^1 f(x)\,dx$.

71. A continuous function $f(x)$ on the interval $[0, 4]$ such that $\int_0^4 f(x)\,dx = 0$, but $f(x)$ is not equal to 0 everywhere on $[0, 4]$.

72. An expression involving a definite integral that can be interpreted as the average speed for a car over a 5-hour journey.

In Problems 73–88, are the statements true for all continuous functions $f(x)$ and $g(x)$? Give an explanation for your answer.

73. If $\int_0^2 (f(x) + g(x))\,dx = 10$ and $\int_0^2 f(x)\,dx = 3$, then $\int_0^2 g(x)\,dx = 7$.

74. If $\int_0^2 (f(x) + g(x))\,dx = 10$, then $\int_0^2 f(x)\,dx = 3$ and $\int_0^2 g(x)\,dx = 7$.

75. If $\int_0^2 f(x)\,dx = 6$, then $\int_0^4 f(x)\,dx = 12$.

76. If $\int_0^2 f(x)\,dx = 6$ and $g(x) = 2f(x)$, then $\int_0^2 g(x)\,dx = 12$.

77. If $\int_0^2 f(x)\,dx = 6$ and $h(x) = f(5x)$, then $\int_0^2 h(x)\,dx = 30$.

78. If $a = b$, then $\int_a^b f(x)\,dx = 0$.

79. If $a \neq b$, then $\int_a^b f(x)\,dx \neq 0$.

80. $\int_1^2 f(x)\,dx + \int_2^3 g(x)\,dx = \int_1^3 (f(x) + g(x))\,dx$.

81. $\int_{-1}^1 f(x)\,dx = 2\int_0^1 f(x)\,dx$.

82. If $f(x) \leq g(x)$ on the interval $[a, b]$, then the average value of f is less than or equal to the average value of g on the interval $[a, b]$.

83. The average value of f on the interval $[0, 10]$ is the average of the average value of f on $[0, 5]$ and the average value of f on $[5, 10]$.

84. If $a < c < d < b$, then the average value of f on the interval $[c, d]$ is less than the average value of f on the interval $[a, b]$.

85. Suppose that A is the average value of f on the interval $[1, 4]$ and B is the average value of f on the interval $[4, 9]$. Then the average value of f on $[1, 9]$ is the weighted average $(3/8)A + (5/8)B$.

86. On the interval $[a, b]$, the average value of $f(x) + g(x)$ is the average value of $f(x)$ plus the average value of $g(x)$.

87. The average value of the product, $f(x)g(x)$, of two functions on an interval equals the product of the average values of $f(x)$ and $g(x)$ on the interval.

88. The units of the average value of a function f on an interval are the same as the units of f.

89. Which of the following statements follow directly from the rule

$$\int_a^b (f(x) + g(x))\, dx = \int_a^b f(x)\, dx + \int_a^b g(x)\, dx?$$

(a) If $\int_a^b (f(x)+g(x))\, dx = 5+7$, then $\int_a^b f(x)\, dx = 5$ and $\int_a^b g(x)\, dx = 7$.

(b) If $\int_a^b f(x)\, dx = \int_a^b g(x)\, dx = 7$, then $\int_a^b (f(x) + g(x))\, dx = 14$.

(c) If $h(x) = f(x) + g(x)$, then $\int_a^b (h(x) - g(x))\, dx = \int_a^b h(x)\, dx - \int_a^b g(x)\, dx$.

Online Resource: Review problems and Projects

Chapter Six

CONSTRUCTING ANTIDERIVATIVES

Contents

6.1 ANTIDERIVATIVES GRAPHICALLY AND NUMERICALLY

The Family of Antiderivatives

If the derivative of F is f, we call F an *antiderivative* of f. For example, since the derivative of x^2 is $2x$, we say that

$$x^2 \text{ is an antiderivative of } 2x.$$

Notice that $2x$ has many antiderivatives, since $x^2 + 1$, $x^2 + 2$, and $x^2 + 3$, all have derivative $2x$. In fact, if C is any constant, we have

$$\frac{d}{dx}(x^2 + C) = 2x + 0 = 2x,$$

so any function of the form $x^2 + C$ is an antiderivative of $2x$. The function $f(x) = 2x$ has a *family of antiderivatives*.

Let us look at another example. If v is the velocity of a car and s is its position, then $v = ds/dt$ and s is an antiderivative of v. As before, $s + C$ is an antiderivative of v for any constant C. In terms of the car, adding C to s is equivalent to adding C to the odometer reading. Adding a constant to the odometer reading simply means measuring distance from a different starting point, which does not alter the car's velocity.

Visualizing Antiderivatives Using Slopes

Suppose we have the graph of f', and we want to sketch an approximate graph of f. We are looking for the graph of f whose slope at any point is equal to the value of f' there. Where f' is above the x-axis, f is increasing; where f' is below the x-axis, f is decreasing. If f' is increasing, f is concave up; if f' is decreasing, f is concave down.

Example 1 The graph of f' is given in Figure 6.1. Sketch a graph of f in the cases when $f(0) = 0$ and $f(0) = 1$.

Solution For $0 \le x \le 2$, the function f has a constant slope of 1, so the graph of f is a straight line. For $2 \le x \le 4$, the function f is increasing but more and more slowly; it has a maximum at $x = 4$ and decreases thereafter. (See Figure 6.2.) The solutions with $f(0) = 0$ and $f(0) = 1$ start at different points on the vertical axis but have the same shape.

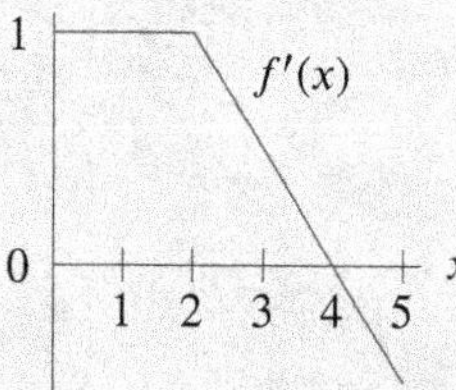

Figure 6.1: Graph of f'

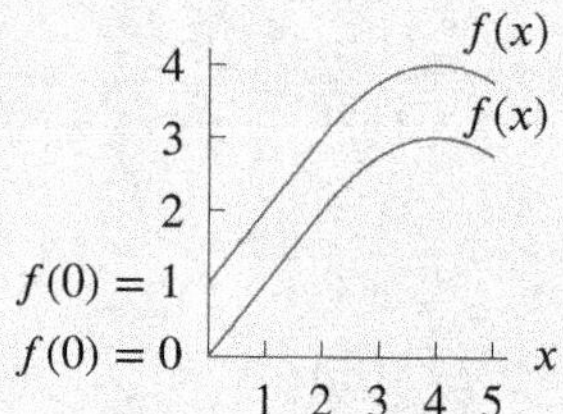

Figure 6.2: Two different f's which have the same derivative f'

Example 2 Sketch a graph of the antiderivative F of $f(x) = e^{-x^2}$ satisfying $F(0) = 0$.

Solution The graph of $f(x) = e^{-x^2}$ is shown in Figure 6.3. The slope of the antiderivative $F(x)$ is given by $f(x)$. Since $f(x)$ is always positive, the antiderivative $F(x)$ is always increasing. Since $f(x)$ is increasing for negative x, we know that $F(x)$ is concave up for negative x. Since $f(x)$ is decreasing for positive x, we know that $F(x)$ is concave down for positive x. Since $f(x) \to 0$ as $x \to \pm\infty$, the graph of $F(x)$ levels off at both ends. See Figure 6.4.

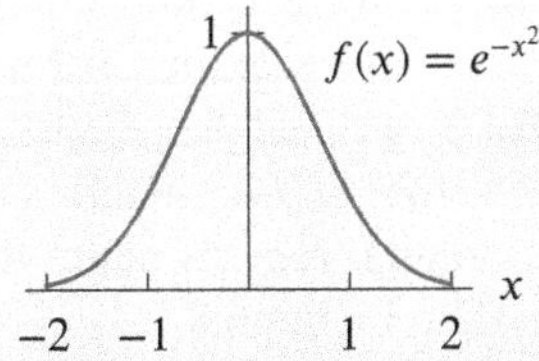

Figure 6.3: Graph of $f(x) = e^{-x^2}$

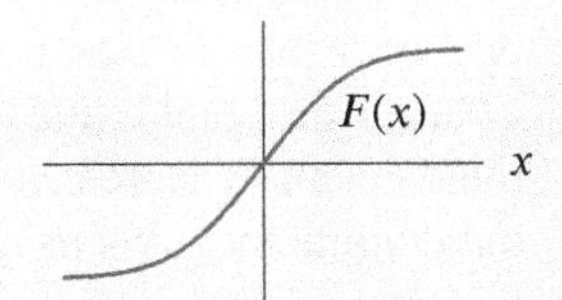

Figure 6.4: An antiderivative $F(x)$ of $f(x) = e^{-x^2}$

Example 3 For the function f' given in Figure 6.5, sketch a graph of three antiderivative functions f, one with $f(0) = 0$, one with $f(0) = 1$, and one with $f(0) = 2$.

Solution To graph f, start at the point on the vertical axis specified by the initial condition and move with slope given by the value of f' in Figure 6.5. Different initial conditions lead to different graphs for f, but for a given x-value they all have the same slope (because the value of f' is the same for each). Thus, the different f curves are obtained from one another by a vertical shift. See Figure 6.6.

- Where f' is positive $(1 < x < 3)$, we see f is increasing; where f' is negative $(0 < x < 1$ or $3 < x < 4)$, we see f is decreasing.
- Where f' is increasing $(0 < x < 2)$, we see f is concave up; where f' is decreasing $(2 < x < 4)$, we see f is concave down.
- Where $f' = 0$, we see f has a local maximum at $x = 3$ and a local minimum at $x = 1$.
- Where f' has a maximum $(x = 2)$, we see f has a point of inflection.

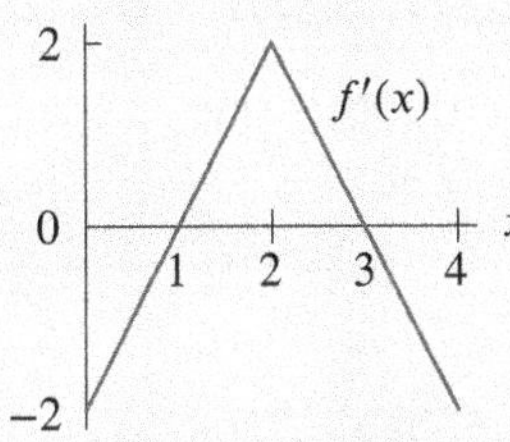

Figure 6.5: Slope function, f'

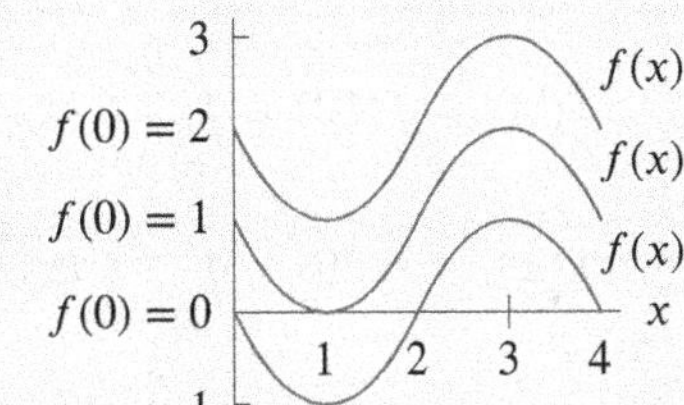

Figure 6.6: Antiderivatives f

Computing Values of an Antiderivative Using Definite Integrals

A graph of f' shows where f is increasing and where f is decreasing. We can calculate the actual value of the function f using the Fundamental Theorem of Calculus (Theorem 5.1 on page 293): If f' is continuous, then

$$\int_a^b f'(x)\,dx = f(b) - f(a).$$

If we know $f(a)$, we can estimate $f(b)$ by computing the definite integral using area or Riemann sums.

Example 4 Figure 6.7 is the graph of the derivative $f'(x)$ of a function $f(x)$. It is given that $f(0) = 100$. Sketch the graph of $f(x)$, showing all critical points and inflection points of f and giving their coordinates.

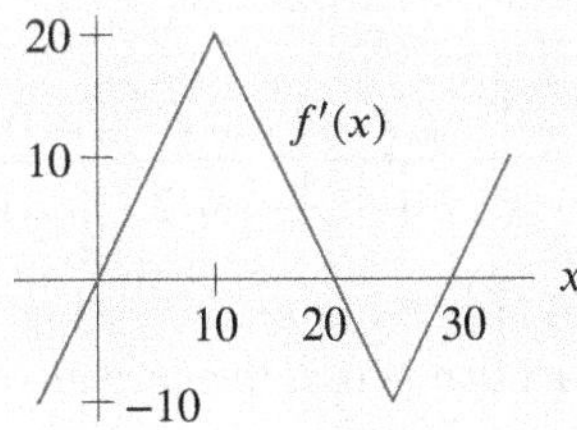

Figure 6.7: Graph of derivative

Solution The critical points of f occur at $x = 0$, $x = 20$, and $x = 30$, where $f'(x) = 0$. The inflection points of f occur at $x = 10$ and $x = 25$, where $f'(x)$ has a maximum or minimum. To find the coordinates of the critical points and inflection points of f, we evaluate $f(x)$ for $x = 0, 10, 20, 25, 30$. Using the Fundamental Theorem, we can express the values of $f(x)$ in terms of definite integrals. We evaluate the definite integrals using the areas of triangular regions under the graph of $f'(x)$, remembering that areas below the x-axis are subtracted. (See Figure 6.8.)

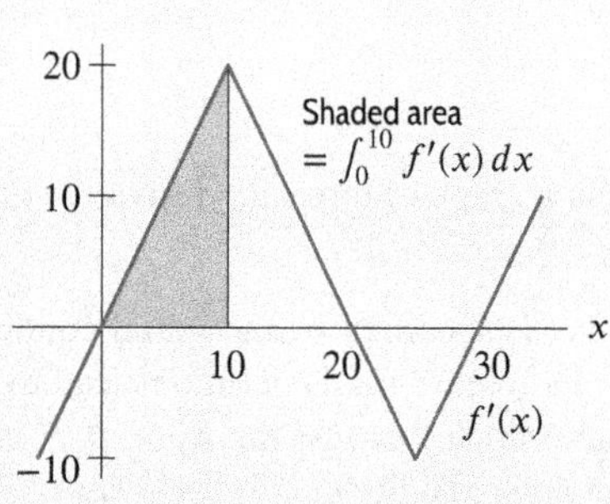

Figure 6.8: Finding $f(10) = f(0) + \int_0^{10} f'(x)\,dx$

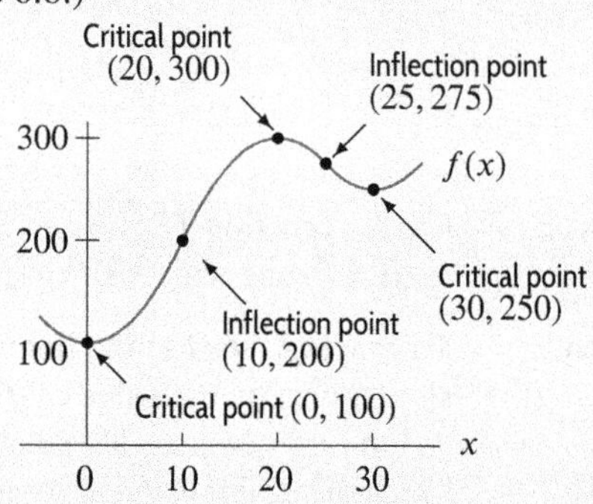

Figure 6.9: Graph of $f(x)$

Since $f(0) = 100$, the Fundamental Theorem gives us the following values of f, which are marked in Figure 6.9.

$$f(10) = f(0) + \int_0^{10} f'(x)\,dx = 100 + (\text{Shaded area in Figure 6.8}) = 100 + \frac{1}{2}(10)(20) = 200,$$

$$f(20) = f(10) + \int_{10}^{20} f'(x)\,dx = 200 + \frac{1}{2}(10)(20) = 300,$$

$$f(25) = f(20) + \int_{20}^{25} f'(x)\,dx = 300 - \frac{1}{2}(5)(10) = 275,$$

$$f(30) = f(25) + \int_{25}^{30} f'(x)\,dx = 275 - \frac{1}{2}(5)(10) = 250.$$

Example 5 Suppose $F'(t) = t\cos t$ and $F(0) = 2$. Find $F(b)$ at the points $b = 0, 0.1, 0.2, \ldots, 1.0$.

Solution We apply the Fundamental Theorem with $f(t) = t\cos t$ and $a = 0$ to get values for $F(b)$:

$$F(b) - F(0) = \int_0^b F'(t)\,dt = \int_0^b t\cos t\,dt.$$

Since $F(0) = 2$, we have

$$F(b) = 2 + \int_0^b t\cos t\,dt.$$

Calculating the definite integral $\int_0^b t\cos t\,dt$ numerically for $b = 0, 0.1, 0.2, \ldots, 1.0$ gives the values for F in Table 6.1.

Table 6.1 *Approximate values for F*

b	0	0.1	0.2	0.3	0.4	0.5	0.6	0.7	0.8	0.9	1.0
$F(b)$	2.000	2.005	2.020	2.044	2.077	2.117	2.164	2.216	2.271	2.327	2.382

Notice that $F(b)$ appears to be increasing between $b = 0$ and $b = 1$. This could have been predicted from the fact that $t\cos t$, the derivative of $F(t)$, is positive for t between 0 and 1.

Exercises and Problems for Section 6.1 Online Resource: Additional Problems for Section 6.1

EXERCISES

1. Fill in the blanks in the following statements, assuming that $F(x)$ is an antiderivative of $f(x)$:
   (a) If $f(x)$ is positive over an interval, then $F(x)$ is _______ over the interval.
   (b) If $f(x)$ is increasing over an interval, then $F(x)$ is _______ over the interval.

2. Use Figure 6.10 and the fact that $P = 0$ when $t = 0$ to find values of P when $t = 1, 2, 3, 4$ and 5.

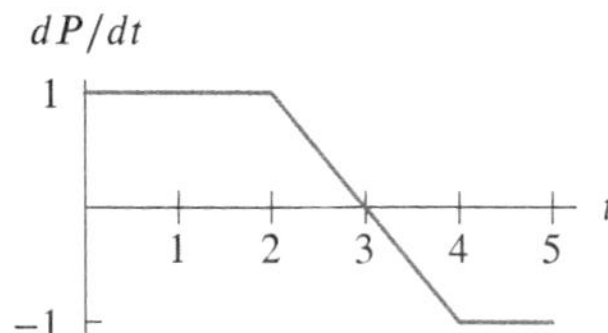

Figure 6.10

3. Use Figure 6.11 and the fact that $P = 2$ when $t = 0$ to find values of P when $t = 1, 2, 3, 4$ and 5.

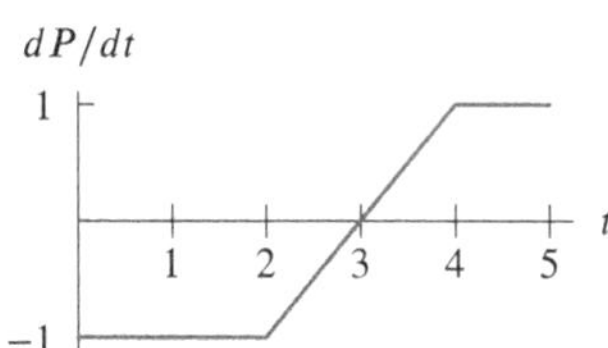

Figure 6.11

4. Let $G'(t) = g(t)$ and $G(0) = 4$. Use Figure 6.12 to find the values of $G(t)$ at $t = 5, 10, 20, 25$.

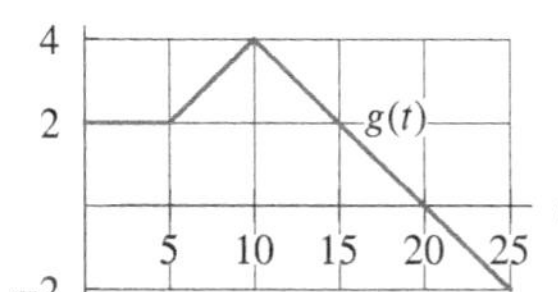

Figure 6.12

■ In Exercises 5–12, sketch two functions F such that $F' = f$. In one case let $F(0) = 0$ and in the other, let $F(0) = 1$.

5. 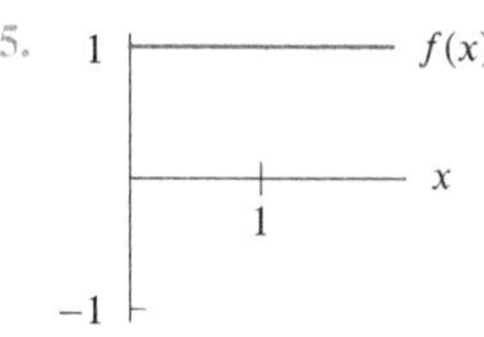

6. 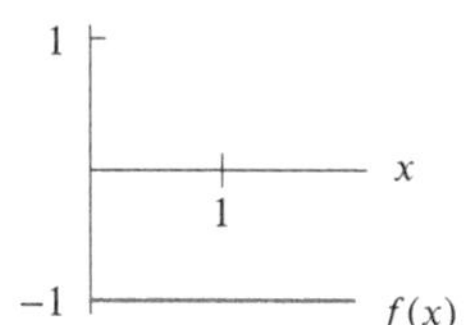

7. 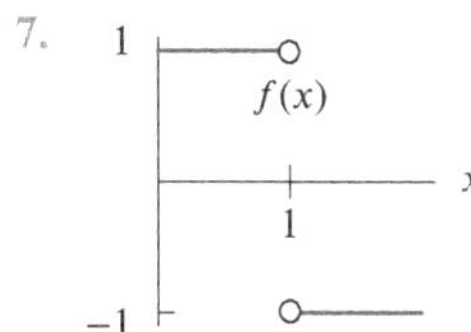

8. 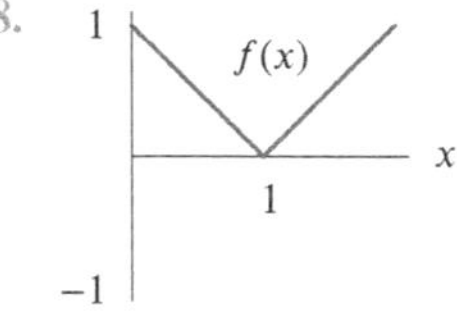

9. 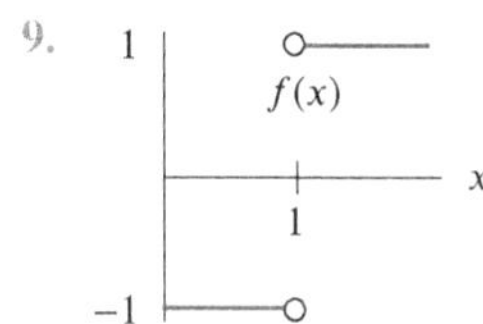

10. 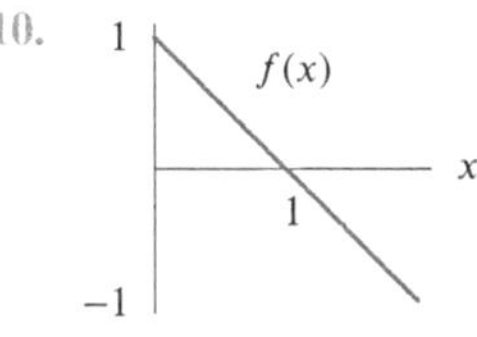

11. 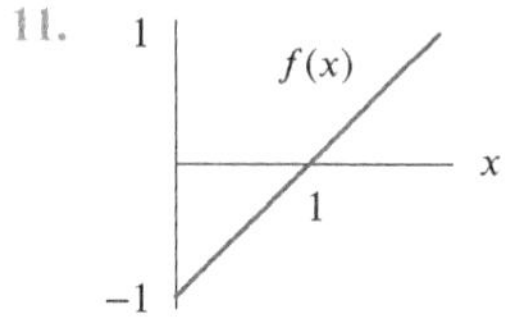

12. 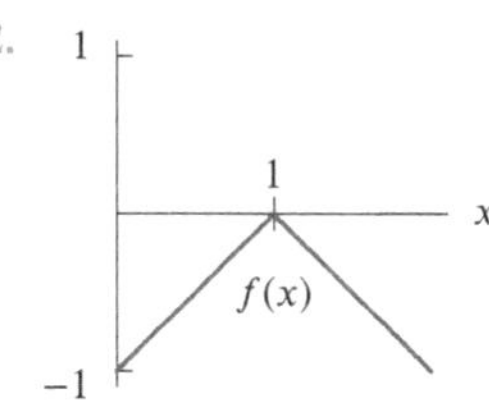

13. Let $F(x)$ be an antiderivative of $f(x)$, with $F(0) = 50$ and $\int_0^5 f(x)\,dx = 12$. What is $F(5)$?

14. Let $F(x)$ be an antiderivative of $f(x)$, with $F(1) = 20$ and $\int_1^4 f(x)\,dx = -7$. What is $F(4)$?

PROBLEMS

15. Let $F(x)$ be an antiderivative of $f(x)$.
    (a) If $\int_2^5 f(x)\,dx = 4$ and $F(5) = 10$, find $F(2)$.
    (b) If $\int_0^{100} f(x)\,dx = 0$, what is the relationship between $F(100)$ and $F(0)$?

16. **(a)** Estimate $\int_0^4 f(x)\,dx$ for $f(x)$ in Figure 6.13.
    (b) Let $F(x)$ be an antiderivative of $f(x)$. Is $F(x)$ increasing or decreasing on the interval $0 \le x \le 4$?
    (c) If $F(0) = 20$, what is $F(4)$?

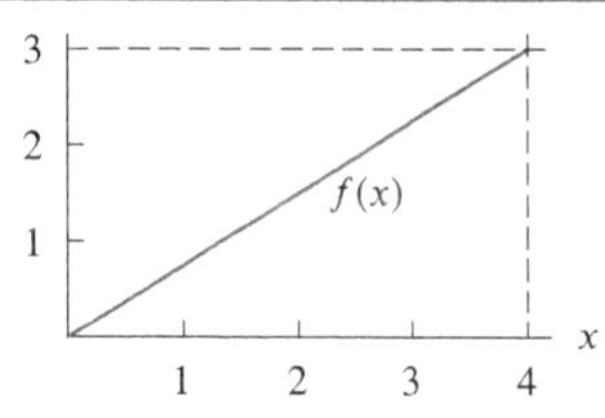

Figure 6.13

17. **(a)** Estimate $\int_0^4 f(x)\,dx$ for $f(x)$ in Figure 6.14.
(b) Let $F(x)$ be an antiderivative of $f(x)$. If $F(0) = 100$, what is $F(4)$?

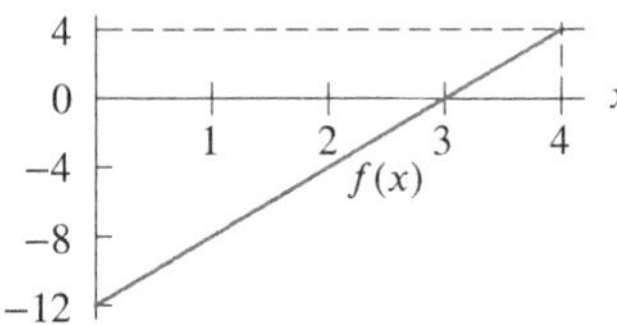

Figure 6.14

18. If $F(0) = 5$ and $F(x)$ is an antiderivative of $f(x) = 3e^{-x^2}$, use a calculator to find $F(2)$.

19. If $G(1) = 50$ and $G(x)$ is an antiderivative of $g(x) = \ln x$, use a calculator to find $G(4)$.

20. Estimate $f(x)$ for $x = 2, 4, 6$, using the given values of $f'(x)$ and the fact that $f(0) = 100$.

x	0	2	4	6
$f'(x)$	10	18	23	25

21. Estimate $f(x)$ for $x = 2, 4, 6$, using the given values of $f'(x)$ and the fact that $f(0) = 50$.

x	0	2	4	6
$f'(x)$	17	15	10	2

22. Using Figure 6.15, sketch a graph of an antiderivative $G(t)$ of $g(t)$ satisfying $G(0) = 5$. Label each critical point of $G(t)$ with its coordinates.

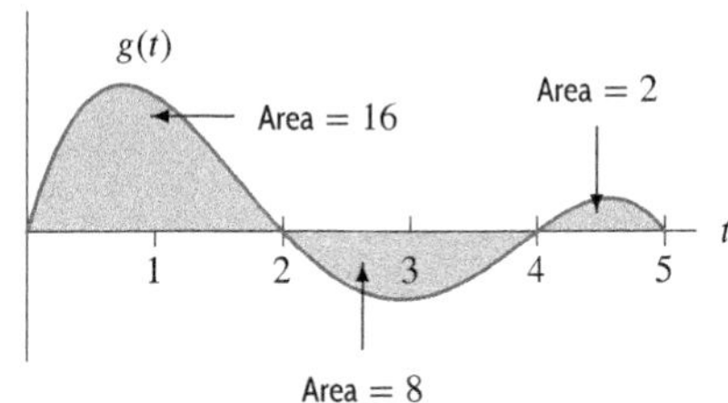

Figure 6.15

23. Use Figure 6.16 and the fact that $F(2) = 3$ to sketch the graph of $F(x)$. Label the values of at least four points.

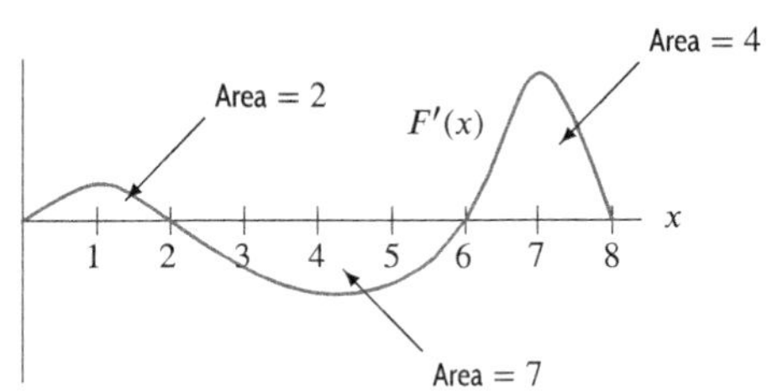

Figure 6.16

24. Figure 6.17 shows the rate of change of the concentration of adrenaline, in micrograms per milliliter per minute, in a person's body. Sketch a graph of the concentration of adrenaline, in micrograms per milliliter, in the body as a function of time, in minutes.

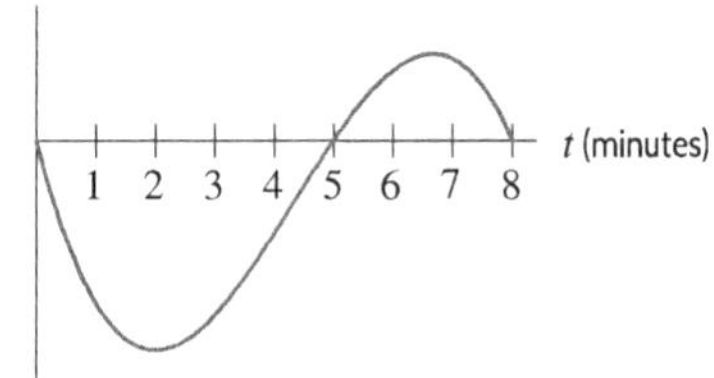

Figure 6.17

25. The graph in Figure 6.18 records the spillage rate at a toxic waste treatment plant over the 50 minutes it took to plug the leak.

(a) Complete the table for the total quantity spilled in liters in time t minutes since the spill started.

Time t (min)	0	10	20	30	40	50
Quantity (liters)	0					

(b) Graph the total quantity leaked against time for the entire fifty minutes. Label axes and include units.

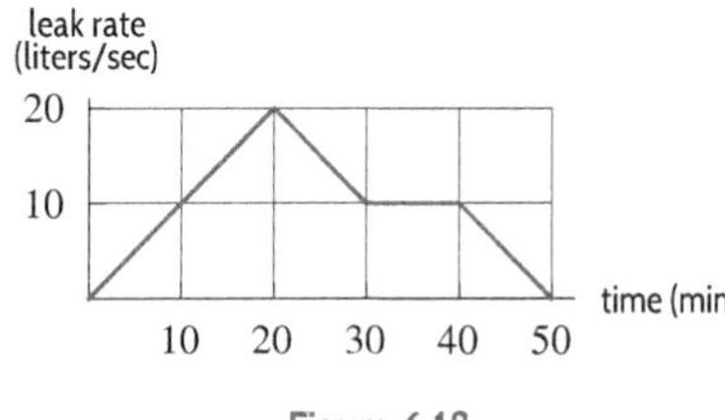

Figure 6.18

26. Two functions, $f(x)$ and $g(x)$, are shown in Figure 6.19. Let F and G be antiderivatives of f and g, respectively. On the same axes, sketch graphs of the antiderivatives $F(x)$ and $G(x)$ satisfying $F(0) = 0$ and $G(0) = 0$. Compare F and G, including a discussion of zeros and x- and y-coordinates of critical points.

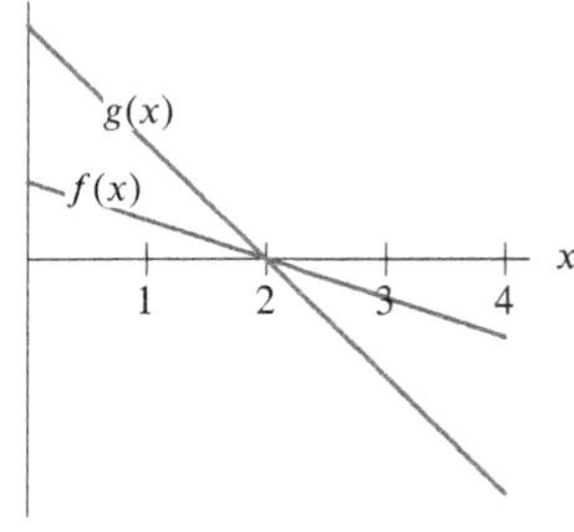

Figure 6.19

27. Let $F(x)$ be an antiderivative of $f(x) = 1 - x^2$.

(a) On what intervals is $F(x)$ increasing?
(b) On what intervals is the graph of $F(x)$ concave up?

In Problems 28–31, sketch two functions F with $F'(x) = f(x)$. In one, let $F(0) = 0$; in the other, let $F(0) = 1$. Mark x_1, x_2, and x_3 on the x-axis of your graph. Identify local maxima, minima, and inflection points of $F(x)$.

28.

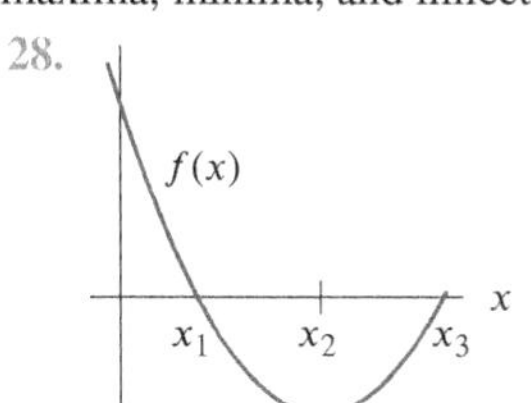

29.

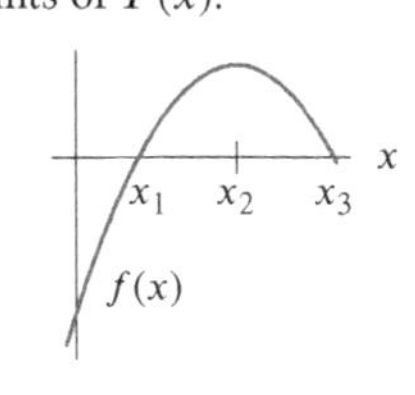

30.

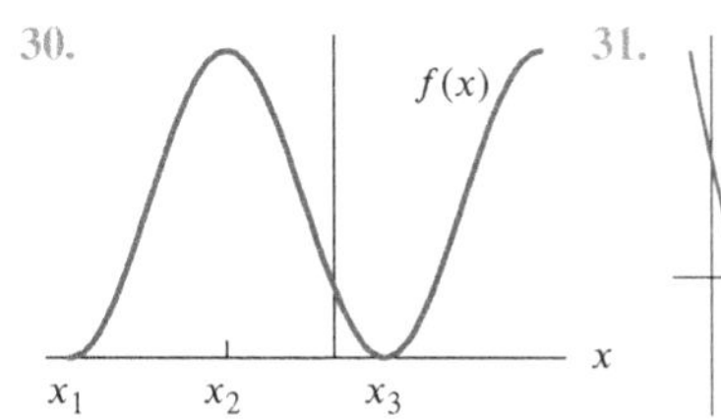

31.

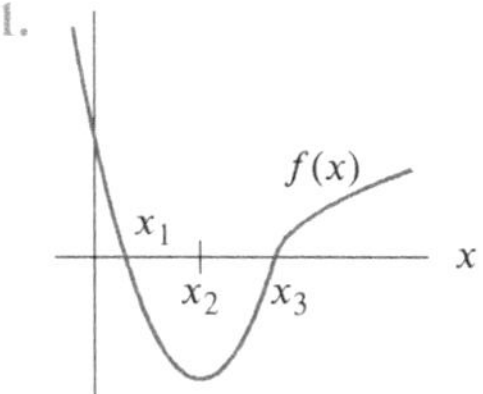

32. A particle moves back and forth along the x-axis. Figure 6.20 approximates the velocity of the particle as a function of time. Positive velocities represent movement to the right and negative velocities represent movement to the left. The particle starts at the point $x = 5$. Graph the distance of the particle from the origin, with distance measured in kilometers and time in hours.

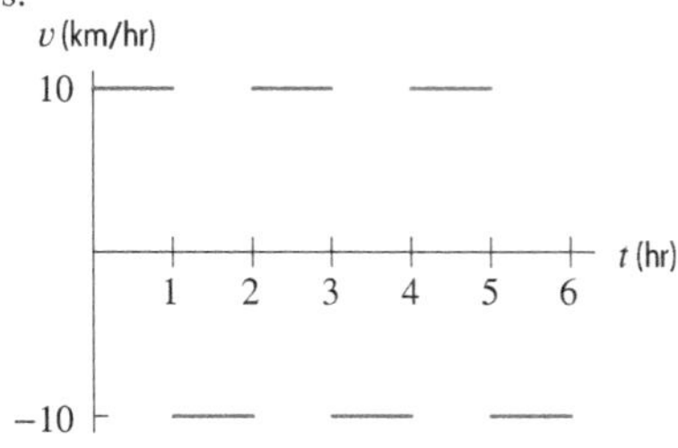

Figure 6.20

33. Assume f' is given by the graph in Figure 6.21. Suppose f is continuous and that $f(0) = 0$.

(a) Find $f(3)$ and $f(7)$.
(b) Find all x with $f(x) = 0$.
(c) Sketch a graph of f over the interval $0 \leq x \leq 7$.

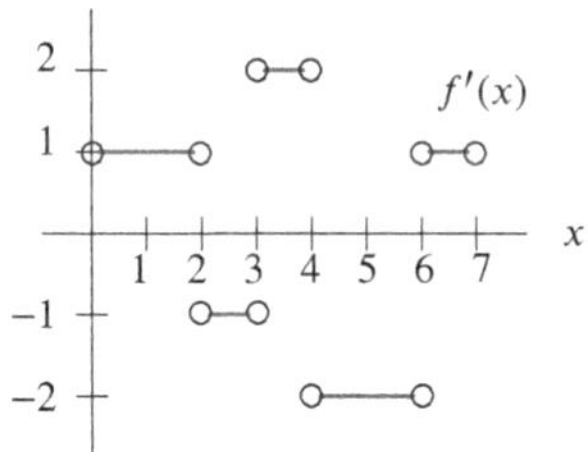

Figure 6.21

34. Urologists are physicians who specialize in the health of the bladder. In a common diagnostic test, urologists monitor the emptying of the bladder using a device that produces two graphs. In one of the graphs the flow rate (in milliliters per second) is measured as a function of time (in seconds). In the other graph, the volume emptied from the bladder is measured (in milliliters) as a function of time (in seconds). See Figure 6.22.

(a) Which graph is the flow rate and which is the volume?
(b) Which one of these graphs is an antiderivative of the other?

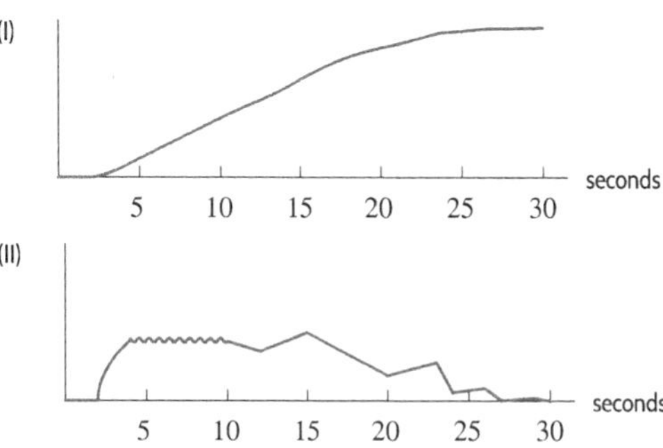

Figure 6.22

35. The Quabbin Reservoir in the western part of Massachusetts provides most of Boston's water. The graph in Figure 6.23 represents the flow of water in and out of the Quabbin Reservoir throughout 2016.

(a) Sketch a graph of the quantity of water in the reservoir, as a function of time.
(b) When, in the course of 2016, was the quantity of water in the reservoir largest? Smallest? Mark and label these points on the graph you drew in part (a).
(c) When was the quantity of water increasing most rapidly? Decreasing most rapidly? Mark and label these times on both graphs.
(d) By July 2017 the quantity of water in the reservoir was about the same as in January 2016. Draw plausible graphs for the flow into and the flow out of the reservoir for the first half of 2017.

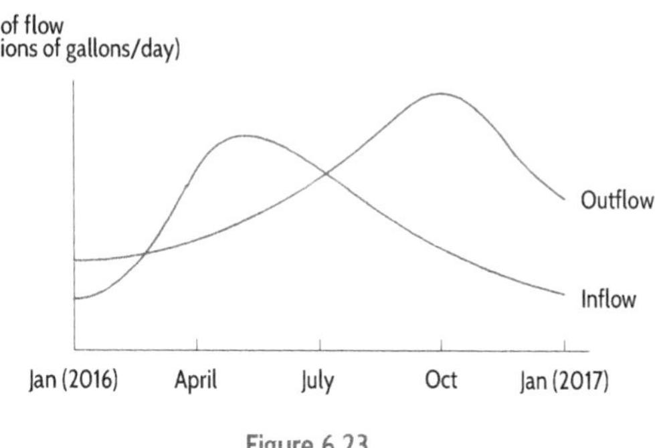

Figure 6.23

36. The birth rate, B, in births per hour, of a bacteria population is given in Figure 6.24. The curve marked D gives the death rate, in deaths per hour, of the same population.

(a) Explain what the shape of each of these graphs tells you about the population.
(b) Use the graphs to find the time at which the net rate of increase of the population is at a maximum.
(c) At time $t = 0$ the population has size N. Sketch the graph of the total number born by time t. Also sketch the graph of the number alive at time t. Estimate the time at which the population is a maximum.

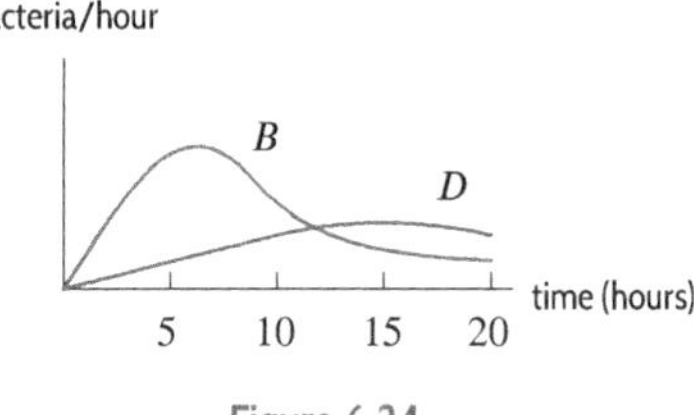

Figure 6.24

Strengthen Your Understanding

In Problems 37–38, explain what is wrong with the statement.

37. Let $F(x)$ be an antiderivative of $f(x)$. If $f(x)$ is everywhere increasing, then $F(x) \geq 0$.
38. If $F(x)$ and $G(x)$ are both antiderivatives of $f(x)$, then $H(x) = F(x) + G(x)$ must also be an antiderivative of $f(x)$.

In Problems 39–40, give an example of:

39. A graph of a function $f(x)$ such that $\int_0^2 f(x)\,dx = 0$.
40. A graph of a function $f(x)$ whose antiderivative is increasing everywhere.

Are the statements in Problems 41–42 true or false? Give an explanation for your answer.

41. A function $f(x)$ has at most one derivative.
42. If $f(t)$ is a linear function with positive slope, then an antiderivative, F, is a linear function.

6.2 CONSTRUCTING ANTIDERIVATIVES ANALYTICALLY

What Is an Antiderivative of $f(x) = 0$?

A function whose derivative is zero everywhere on an interval must have a horizontal tangent line at every point of its graph, and the only way this can happen is if the function is constant. Alternatively, if we think of the derivative as a velocity, and if the velocity is always zero, then the object is standing still; the position function is constant. A rigorous proof of this result using the definition of the derivative is surprisingly subtle. (See the Constant Function Theorem on page 187.)

> If $F'(x) = 0$ on an interval, then $F(x) = C$ on this interval, for some constant C.

What Is the Most General Antiderivative of f?

We know that if a function f has an antiderivative F, then it has a family of antiderivatives of the form $F(x) + C$, where C is any constant. You might wonder if there are any others. To decide, suppose that we have two functions F and G with $F' = f$ and $G' = f$: that is, F and G are both antiderivatives of the same function f. Since $F' = G'$ we have $(G - F)' = 0$. But this means that we must have $G - F = C$, so $G(x) = F(x) + C$, where C is a constant. Thus, any two antiderivatives of the same function differ only by a constant.

> If F and G are both antiderivatives of f on an interval, then $G(x) = F(x) + C$.

The Indefinite Integral

All antiderivatives of $f(x)$ are of the form $F(x) + C$. We introduce a notation for the general antiderivative that looks like the definite integral without the limits and is called the *indefinite integral*:

$$\int f(x)\,dx = F(x) + C.$$

It is important to understand the difference between

$$\int_a^b f(x)\,dx \qquad \text{and} \qquad \int f(x)\,dx.$$

The first is a number and the second is a family of *functions*. The word "integration" is frequently used for the process of finding the antiderivative as well as of finding the definite integral. The context usually makes clear which is intended.

What Is an Antiderivative of $f(x) = k$?

If k is a constant, the derivative of kx is k, so we have

An antiderivative of k is kx.

Using the indefinite integral notation, we have

> If k is constant,
> $$\int k\,dx = kx + C.$$

Finding Antiderivatives

Finding antiderivatives of functions is like taking square roots of numbers: if we pick a number at random, such as 7 or 493, we may have trouble finding its square root without a calculator. But if we happen to pick a number such as 25 or 64, which we know is a perfect square, then we can find its square root exactly. Similarly, if we pick a function which we recognize as a derivative, then we can find its antiderivative easily.

For example, to find an antiderivative of $f(x) = x$, notice that $2x$ is the derivative of x^2; this tells us that x^2 is an antiderivative of $2x$. If we divide by 2, then we guess that

$$\text{An antiderivative of } x \text{ is } \frac{x^2}{2}.$$

To check this statement, take the derivative of $x^2/2$:

$$\frac{d}{dx}\left(\frac{x^2}{2}\right) = \frac{1}{2}\cdot\frac{d}{dx}x^2 = \frac{1}{2}\cdot 2x = x.$$

What about an antiderivative of x^2? The derivative of x^3 is $3x^2$, so the derivative of $x^3/3$ is $3x^2/3 = x^2$. Thus,

$$\text{An antiderivative of } x^2 \text{ is } \frac{x^3}{3}.$$

The pattern looks like

$$\text{An antiderivative of } x^n \text{ is } \frac{x^{n+1}}{n+1}.$$

(We assume $n \neq -1$, or we would have $x^0/0$, which does not make sense.) It is easy to check this formula by differentiation:

$$\frac{d}{dx}\left(\frac{x^{n+1}}{n+1}\right) = \frac{(n+1)x^n}{n+1} = x^n.$$

In indefinite integral notation, we have shown that

> $$\int x^n\,dx = \frac{x^{n+1}}{n+1} + C, \quad n \neq -1.$$

What about when $n = -1$? In other words, what is an antiderivative of $1/x$? Fortunately, we know a function whose derivative is $1/x$, namely, the natural logarithm. Thus, since

$$\frac{d}{dx}(\ln x) = \frac{1}{x},$$

we know that

$$\int \frac{1}{x}\,dx = \ln x + C, \quad \text{for } x > 0.$$

If $x < 0$, then $\ln x$ is not defined, so it can't be an antiderivative of $1/x$. In this case, we can try $\ln(-x)$:

$$\frac{d}{dx}\ln(-x) = (-1)\frac{1}{-x} = \frac{1}{x}$$

so

$$\int \frac{1}{x}\,dx = \ln(-x) + C, \quad \text{for } x < 0.$$

This means $\ln x$ is an antiderivative of $1/x$ if $x > 0$, and $\ln(-x)$ is an antiderivative of $1/x$ if $x < 0$. Since $|x| = x$ when $x > 0$ and $|x| = -x$ when $x < 0$, we can collapse these two formulas into:

$$\text{An antiderivative of } \frac{1}{x} \text{ is } \ln|x|.$$

Therefore

$$\boxed{\int \frac{1}{x}\,dx = \ln|x| + C.}$$

Since the exponential function is its own derivative, it is also its own antiderivative; thus

$$\boxed{\int e^x\,dx = e^x + C.}$$

Also, antiderivatives of the sine and cosine are easy to guess. Since

$$\frac{d}{dx}\sin x = \cos x \quad \text{and} \quad \frac{d}{dx}\cos x = -\sin x,$$

we get

$$\boxed{\int \cos x\,dx = \sin x + C} \quad \text{and} \quad \boxed{\int \sin x\,dx = -\cos x + C.}$$

Example 1 Find $\int (3x + x^2)\,dx$.

Solution We know that $x^2/2$ is an antiderivative of x and that $x^3/3$ is an antiderivative of x^2, so we expect

$$\int (3x + x^2)\,dx = 3\left(\frac{x^2}{2}\right) + \frac{x^3}{3} + C.$$

You should always check your antiderivatives by differentiation—it's easy to do. Here

$$\frac{d}{dx}\left(\frac{3}{2}x^2 + \frac{x^3}{3} + C\right) = \frac{3}{2}\cdot 2x + \frac{3x^2}{3} = 3x + x^2.$$

The preceding example illustrates that the sum and constant multiplication rules of differentiation work in reverse:

Theorem 6.1: Properties of Antiderivatives: Sums and Constant Multiples

In indefinite integral notation,

1. $\int (f(x) \pm g(x))\,dx = \int f(x)\,dx \pm \int g(x)\,dx$
2. $\int cf(x)\,dx = c\int f(x)\,dx.$

In words,

1. An antiderivative of the sum (or difference) of two functions is the sum (or difference) of their antiderivatives.
2. An antiderivative of a constant times a function is the constant times an antiderivative of the function.

These properties are analogous to the properties for definite integrals given on page 304 in Section 5.4, even though definite integrals are numbers and antiderivatives are functions.

Example 2 Find $\int (\sin x + 3\cos x)\,dx$.

Solution We break the antiderivative into two terms:

$$\int (\sin x + 3\cos x)\,dx = \int \sin x\,dx + 3\int \cos x\,dx = -\cos x + 3\sin x + C.$$

Check by differentiating:

$$\frac{d}{dx}(-\cos x + 3\sin x + C) = \sin x + 3\cos x.$$

Using Antiderivatives to Compute Definite Integrals

As we saw in Section 5.3, the Fundamental Theorem of Calculus gives us a way of calculating definite integrals. Denoting $F(b) - F(a)$ by $F(x)\big|_a^b$, the theorem says that if $F' = f$ and f is continuous, then

$$\int_a^b f(x)\,dx = F(x)\Big|_a^b = F(b) - F(a).$$

To find $\int_a^b f(x)\,dx$, we first find F, and then calculate $F(b) - F(a)$. This method of computing definite integrals gives an exact answer. However, the method only works in situations where we can find the antiderivative $F(x)$. This is not always easy; for example, none of the functions we have encountered so far is an antiderivative of $\sin(x^2)$.

Example 3 Compute $\int_1^2 3x^2\,dx$ using the Fundamental Theorem.

Solution Since $F(x) = x^3$ is an antiderivative of $f(x) = 3x^2$,

$$\int_1^2 3x^2\,dx = F(x)\Big|_1^2 = F(2) - F(1),$$

gives

$$\int_1^2 3x^2\,dx = x^3\Big|_1^2 = 2^3 - 1^3 = 7.$$

Notice in this example we used the antiderivative x^3, but $x^3 + C$ works just as well because the constant C cancels out:

$$\int_1^2 3x^2\,dx = (x^3 + C)\Big|_1^2 = (2^3 + C) - (1^3 + C) = 7.$$

Example 4 Compute $\displaystyle\int_0^{\pi/4} \frac{1}{\cos^2\theta}\,d\theta$ exactly.

Solution We use the Fundamental Theorem. Since $F(\theta) = \tan\theta$ is an antiderivative of $f(\theta) = 1/\cos^2\theta$, we get

$$\int_0^{\pi/4} \frac{1}{\cos^2\theta}\,d\theta = \tan\theta\Big|_0^{\pi/4} = \tan\left(\frac{\pi}{4}\right) - \tan(0) = 1.$$

Exercises and Problems for Section 6.2 Online Resource: Additional Problems for Section 6.2

EXERCISES

1. If $p'(x) = q(x)$, write a statement involving an integral sign giving the relationship between $p(x)$ and $q(x)$.
2. If $u'(x) = v(x)$, write a statement involving an integral sign giving the relationship between $u(x)$ and $v(x)$.
3. Which of (I)–(V) are antiderivatives of $f(x) = e^{x/2}$?

   I. $e^{x/2}$ II. $2e^{x/2}$ III. $2e^{(1+x)/2}$
   IV. $2e^{x/2} + 1$ V. $e^{x^2/4}$

4. Which of (I)–(V) are antiderivatives of $f(x) = 1/x$?

   I. $\ln x$ II. $-1/x^2$ III. $\ln x + \ln 3$
   IV. $\ln(2x)$ V. $\ln(x + 1)$

5. Which of (I)–(V) are antiderivatives of

$$f(x) = 2\sin x\cos x?$$

   I. $-2\sin x\cos x$ II. $2\cos^2 x - 2\sin^2 x$
   III. $\sin^2 x$ IV. $-\cos^2 x$
   V. $2\sin^2 x + \cos^2 x$

■ In Exercises 6–21, find an antiderivative.

6. $f(x) = 5$
7. $f(t) = 5t$
8. $f(x) = x^2$
9. $g(t) = t^2 + t$
10. $g(z) = \sqrt{z}$
11. $h(z) = \frac{1}{z}$
12. $r(t) = \frac{1}{t^2}$
13. $h(t) = \cos t$
14. $g(z) = \frac{1}{z^3}$
15. $q(y) = y^4 + \frac{1}{y}$
16. $f(z) = e^z$
17. $g(t) = \sin t$
18. $f(t) = 2t^2 + 3t^3 + 4t^4$
19. $f(x) = 5x - \sqrt{x}$
20. $f(t) = \frac{t^2 + 1}{t}$
21. $p(t) = t^3 - \frac{t^2}{2} - t$

■ In Exercises 22–33, find the general antiderivative.

22. $f(t) = 6t$
23. $h(x) = x^3 - x$
24. $f(x) = x^2 - 4x + 7$
25. $g(t) = \sqrt{t}$
26. $r(t) = t^3 + 5t - 1$
27. $f(z) = z + e^z$
28. $g(x) = \sin x + \cos x$
29. $h(x) = 4x^3 - 7$
30. $g(x) = \frac{5}{x^3}$
31. $p(t) = 2 + \sin t$
32. $p(t) = \frac{1}{\sqrt{t}}$
33. $h(t) = \frac{7}{\cos^2 t}$

■ In Exercises 34–41, find an antiderivative $F(x)$ with $F'(x) = f(x)$ and $F(0) = 0$. Is there only one possible solution?

34. $f(x) = 3$
35. $f(x) = 2x$
36. $f(x) = -7x$
37. $f(x) = 2 + 4x + 5x^2$
38. $f(x) = \frac{1}{4}x$
39. $f(x) = x^2$
40. $f(x) = \sqrt{x}$
41. $f(x) = \sin x$

■ In Exercises 42–55, find the indefinite integrals.

42. $\int (5x+7)\,dx$

43. $\int \left(4t+\frac{1}{t}\right)\,dt$

44. $\int (2+\cos t)\,dt$

45. $\int 7e^x\,dx$

46. $\int (3e^x+2\sin x)\,dx$

47. $\int (4e^x-3\sin x)\,dx$

48. $\int \left(5x^2+2\sqrt{x}\right)\,dx$

49. $\int (x+3)^2\,dx$

50. $\int \frac{8}{\sqrt{x}}\,dx$

51. $\int \left(\frac{3}{t}-\frac{2}{t^2}\right)\,dt$

52. $\int (e^x+5)\,dx$

53. $\int t^3(t^2+1)\,dt$

54. $\int \left(\sqrt{x^3}-\frac{2}{x}\right)\,dx$

55. $\int \left(\frac{x+1}{x}\right)\,dx$

■ In Exercises 56–65, evaluate the definite integrals exactly [as in $\ln(3\pi)$], using the Fundamental Theorem, and numerically [$\ln(3\pi)\approx 2.243$]:

56. $\int_0^3 (x^2+4x+3)\,dx$

57. $\int_1^3 \frac{1}{t}\,dt$

58. $\int_0^{\pi/4} \sin x\,dx$

59. $\int_0^1 2e^x\,dx$

60. $\int_0^2 3e^x\,dx$

61. $\int_2^5 (x^3-\pi x^2)\,dx$

62. $\int_0^1 \sin\theta\,d\theta$

63. $\int_1^2 \frac{1+y^2}{y}\,dy$

64. $\int_0^2 \left(\frac{x^3}{3}+2x\right)\,dx$

65. $\int_0^{\pi/4} (\sin t+\cos t)\,dt$

■ In Exercises 66–75, decide if the statement is True or False by differentiating the right-hand side.

66. $\int (8x^2-4x+3)\,dx=\frac{8}{3}x^3-2x^2+3x+C$

67. $\int \left(3\sqrt{x}+\frac{5}{x^2}\right)\,dx=2x^{3/2}+\frac{5}{x}+C$

68. $\int \sqrt{x^2+1}\,dx=\left(x^2+1\right)^{3/2}+C$

69. $\int x^{-2}\,dx=-\frac{1}{x}+C$

70. $\int x^{-1}\,dx=-1x^0+C=-1+C$

71. $\int e^{2x}\,dx=2e^{2x}+C$

72. $\int e^{x^2}\,dx=2x\cdot e^{x^2}+C$

73. $\int 3\cos x\,dx=3\sin x+C$

74. $\int 5\sin\left(\frac{x}{5}\right)\,dx=5\cos\left(\frac{x}{5}\right)+C$

75. $\int x\cos x\,dx=\frac{x^2}{2}\sin x+C$

PROBLEMS

76. Use the Fundamental Theorem to find the area under $f(x)=x^2$ between $x=0$ and $x=3$.

77. Find the exact area of the region bounded by the x-axis and the graph of $y=x^3-x$.

78. Calculate the exact area above the graph of $y=\frac{1}{2}\left(\frac{3}{\pi}x\right)^2$ and below the graph of $y=\cos x$. The curves intersect at $x=\pm\pi/3$.

79. Find the exact area of the shaded region in Figure 6.25 between $y=3x^2-3$ and the x-axis.

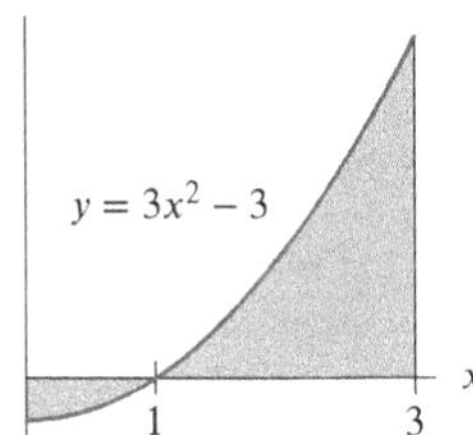

Figure 6.25

80. **(a)** Find the exact area between $f(x)=x^3-7x^2+10x$, the x-axis, $x=0$, and $x=5$.
(b) Find $\int_0^5 (x^3-7x^2+10x)\,dx$ exactly and interpret this integral in terms of areas.

81. Find the exact area between the curve $y=e^x-2$ and the x-axis for $0\le x\le 2$.

82. Find the exact area between the curves $y=x^2$ and $y=2-x^2$.

83. Find the exact area between the x-axis and the graph of $f(x)=(x-1)(x-2)(x-3)$.

84. The area under $1/\sqrt{x}$ on the interval $1\le x\le b$ is equal to 6. Find the value of b using the Fundamental Theorem.

85. Use the Fundamental Theorem to find the value of b if the area under the graph of $f(x)=8x$ between $x=1$ and $x=b$ is equal to 192. Assume $b>1$.

86. Find the exact positive value of c which makes the area under the graph of $y = c(1 - x^2)$ and above the x-axis equal to 1.

87. Sketch the parabola $y = x(x - \pi)$ and the curve $y = \sin x$, showing their points of intersection. Find the exact area between the two graphs.

88. Find the exact average value of $f(x) = \sqrt{x}$ on the interval $0 \le x \le 9$. Illustrate your answer on a graph of $f(x) = \sqrt{x}$.

89. **(a)** What is the average value of $f(t) = \sin t$ over $0 \le t \le 2\pi$? Why is this a reasonable answer?
(b) Find the average of $f(t) = \sin t$ over $0 \le t \le \pi$.

90. Let $\int q(x)\,dx = Q(x)+C$ where $Q(3) = 12$. Given that $\int_3^8 q(x)\,dx = 5$, find $Q(8)$.

91. Water is pumped into a cylindrical tank, standing vertically, at a decreasing rate given at time t minutes by

$$r(t) = 120 - 6t \text{ ft}^3/\text{min} \quad \text{for } 0 \le t \le 10.$$

The tank has radius 5 ft and is empty when $t = 0$. Find the depth of water in the tank at $t = 4$.

92. A car moves along a straight line with velocity, in feet/second, given by

$$v(t) = 6 - 2t \quad \text{for } t \ge 0.$$

(a) Describe the car's motion in words. (When is it moving forward, backward, and so on?)
(b) The car's position is measured from its starting point. When is it farthest forward? Backward?
(c) Find s, the car's position measured from its starting point, as a function of time.

93. In drilling an oil well, the total cost, C, consists of fixed costs (independent of the depth of the well) and marginal costs, which depend on depth; drilling becomes more expensive, per meter, deeper into the earth. Suppose the fixed costs are 1,000,000 riyals (the riyal is the unit of currency of Saudi Arabia), and the marginal costs are

$$C'(x) = 4000 + 10x \text{ riyals/meter},$$

where x is the depth in meters. Find the total cost of drilling a well x meters deep.

94. A helicopter rotor slows down at a constant rate from 350 revs/min to 260 revs/min in 1.5 minutes.

(a) Find the angular acceleration (i.e. change in rev/min) during this time interval. What are the units of this acceleration?
(b) Assuming the angular acceleration remains constant, how long does it take for the rotor to stop? (Measure time from the moment when speed was 350 revs/min.)
(c) How many revolutions does the rotor make between the time the angular speed was 350 revs/min and stopping?

95. Use the fact that $(x^x)' = x^x(1+\ln x)$ to evaluate exactly:

$$\int_1^3 x^x(1 + \ln x)\,dx.$$

96. Assuming that $\int g(x)\,dx = G(x)+C$, where $G(4) = 9$, $G(6) = 4$, and $G(9) = 6$, evaluate the definite integral:

(a) $\int_6^4 g(x)\,dx$ **(b)** $\int_6^9 7g(x)\,dx$

(c) $\int_4^9 (g(x) + 3)\,dx$

■ For Problems 97–99, let $\int g(x)\,dx = G(x) + C$. Which of (I)–(III), if any, is equal to the given integral?

97. $\int g(2x)\,dx$

I. $0.5G(0.5x) + C$ II. $0.5G(2x) + C$
III. $2G(0.5x) + C$

98. $\int \cos(G(x))\,g(x)\,dx$

I. $\sin(G(x))\,g(x) + C$ II. $\sin(G(x))\,G(x) + C$
III. $\sin(G(x)) + C$

99. $\int xg(x)\,dx$

I. $G\left(x^2\right) + C$ II. $xG(x) + C$ III. $\frac{1}{2}x^2G(x)+C$

Strengthen Your Understanding

■ In Problems 100–101, explain what is wrong with the statement.

100. $\int \frac{3x^2+1}{2x}\,dx = \frac{x^3+x}{x^2} + C$

101. For all n, $\int x^n\,dx = \frac{x^{n+1}}{n+1} + C.$

■ In Problems 102–103, give an example of:

102. Two different functions $F(x)$ and $G(x)$ that have the same derivative.

103. A function $f(x)$ whose antiderivative $F(x)$ has a graph which is a line with negative slope.

■ Are the statements in Problems 104–112 true or false? Give an explanation for your answer.

104. An antiderivative of $3\sqrt{x+1}$ is $2(x+1)^{3/2}$.

105. An antiderivative of $3x^2$ is $x^3 + \pi$.

106. An antiderivative of $1/x$ is $\ln|x| + \ln 2$.

107. An antiderivative of e^{-x^2} is $-e^{-x^2}/2x$.

108. $\int f(x)\,dx = (1/x)\int xf(x)\,dx$.

109. If $F(x)$ is an antiderivative of $f(x)$ and $G(x) = F(x)+2$, then $G(x)$ is an antiderivative of $f(x)$.

110. If $F(x)$ and $G(x)$ are two antiderivatives of $f(x)$ for $-\infty < x < \infty$ and $F(5) > G(5)$, then $F(10) > G(10)$.

111. If $F(x)$ is an antiderivative of $f(x)$ and $G(x)$ is an antiderivative of $g(x)$, then $F(x)\cdot G(x)$ is an antiderivative of $f(x)\cdot g(x)$.

112. If $F(x)$ and $G(x)$ are both antiderivatives of $f(x)$ on an interval, then $F(x) - G(x)$ is a constant function.

6.3 DIFFERENTIAL EQUATIONS AND MOTION

An equation of the form

$$\frac{dy}{dx} = f(x)$$

is an example of a *differential equation*. Finding the *general solution* to the differential equation means finding the general antiderivative $y = F(x) + C$ with $F'(x) = f(x)$. Chapter 11 gives more details.

Example 1 Find and graph the general solution of the differential equation

$$\frac{dy}{dx} = 2x.$$

Solution We are asking for a function whose derivative is $2x$. One antiderivative of $2x$ is

$$y = x^2.$$

The general solution is therefore

$$y = x^2 + C,$$

where C is any constant. Figure 6.26 shows several curves in this family.

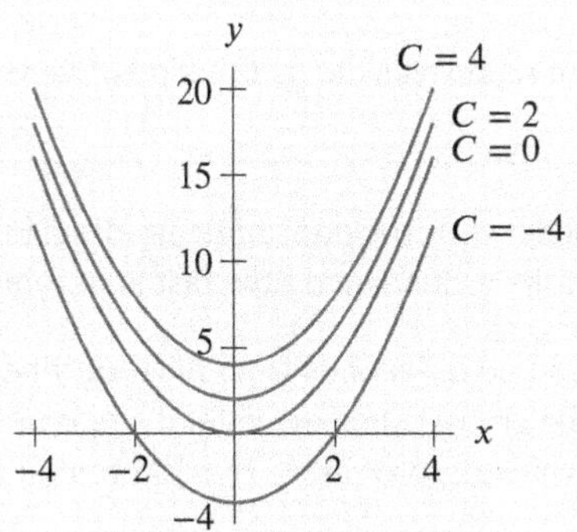

Figure 6.26: Solution curves of $dy/dx = 2x$

How Can We Pick One Solution to the Differential Equation $\frac{dy}{dx} = f(x)$?

Picking one antiderivative is equivalent to selecting a value of C. To do this, we need an extra piece of information, usually that the solution curve passes through a particular point (x_0, y_0). The differential equation plus the extra condition

$$\frac{dy}{dx} = f(x), \qquad y(x_0) = y_0$$

is called an *initial value problem*. (The *initial condition* $y(x_0) = y_0$ is shorthand for $y = y_0$ when $x = x_0$.) An initial value problem usually has a unique solution, called the *particular solution*.

Example 2 Find the solution of the initial value problem

$$\frac{dy}{dx} = 2x, \quad y(3) = 5.$$

Solution We have already seen that the general solution to the differential equation is $y = x^2 + C$. The initial condition allows us to determine the constant C. Substituting $y(3) = 5$ gives

$$5 = y(3) = 3^2 + C,$$

so C is given by

$$C = -4.$$

Thus, the (unique) solution is

$$y = x^2 - 4$$

Figure 6.26 shows this particular solution, marked $C = -4$.

Equations of Motion

We now use differential equations to analyze the motion of an object falling freely under the influence of gravity. It has been known since Galileo's time that an object moving under the influence of gravity (ignoring air resistance) has constant acceleration, g. In the most frequently used units, its value is approximately

$$g = 9.8 \text{ m/sec}^2, \quad \text{or} \quad g = 32 \text{ ft/sec}^2.$$

Thus, if v is the upward velocity and t is the time, we have the differential equation

$$\frac{dv}{dt} = -g.$$

The negative sign represents the fact that positive velocity is measured upward, whereas gravity acts downward.

Example 3 A stone is dropped from a 100-foot-high building. Find, as functions of time, its position and velocity. When does it hit the ground, and how fast is it going at that time?

Solution Suppose t is measured in seconds from the time when the stone was dropped. If we measure distance, s, in feet above the ground, then the velocity, v, is in ft/sec upward, and the acceleration due to gravity is 32 ft/sec^2 downward, so we have the differential equation

$$\frac{dv}{dt} = -32.$$

From what we know about antiderivatives, the general solution is

$$v = -32t + C,$$

where C is some constant. Since $v = C$ when $t = 0$, the constant C represents the initial velocity, v_0. The fact that the stone is dropped rather than thrown tells us that the initial velocity is zero, so the initial condition is $v_0 = 0$. Substituting gives

$$0 = -32(0) + C \quad \text{so} \quad C = 0.$$

Thus,

$$v = -32t.$$

But now we can write a second differential equation:

$$v = \frac{ds}{dt} = -32t.$$

The general solution is

$$s = -16t^2 + K,$$

where K is another constant.

Since the stone starts at the top of the building, we have the initial condition $s = 100$ when $t = 0$. Substituting gives

$$100 = -16(0^2) + K, \quad \text{so} \quad K = 100,$$

and therefore

$$s = -16t^2 + 100.$$

Thus, we have found both v and s as functions of time.

The stone hits the ground when $s = 0$, so we must solve

$$0 = -16t^2 + 100$$

giving $t^2 = 100/16$ or $t = \pm 10/4 = \pm 2.5$ sec. Since t must be positive, $t = 2.5$ sec. At that time, $v = -32(2.5) = -80$ ft/sec. (The velocity is negative because we are considering moving up as positive and down as negative.) After the stone hits the ground, the differential equation no longer applies.

Example 4 An object is thrown vertically upward with a speed of 10 m/sec from a height of 2 meters above the ground. Find the highest point it reaches and the time when it hits the ground.

Solution We must find the position as a function of time. In this example, the velocity is in m/sec, so we use $g = 9.8$ m/sec^2. Measuring distance in meters upward from the ground, we have the differential equation

$$\frac{dv}{dt} = -9.8.$$

As before, v is a function whose derivative is constant, so

$$v = -9.8t + C.$$

Since the initial velocity is 10 m/sec upward, we know that $v = 10$ when $t = 0$. Substituting gives

$$10 = -9.8(0) + C \quad \text{so} \quad C = 10.$$

Thus,

$$v = -9.8t + 10.$$

To find s, we use

$$v = \frac{ds}{dt} = -9.8t + 10$$

and look for a function that has $-9.8t + 10$ as its derivative. The general solution is

$$s = -4.9t^2 + 10t + K,$$

where K is any constant. To find K, we use the fact that the object starts at a height of 2 meters above the ground, so $s = 2$ when $t = 0$. Substituting gives

$$2 = -4.9(0)^2 + 10(0) + K, \quad \text{so} \quad K = 2,$$

and therefore

$$s = -4.9t^2 + 10t + 2.$$

The object reaches its highest point when the velocity is 0, so at that time

$$v = -9.8t + 10 = 0.$$

This occurs when

$$t = \frac{10}{9.8} \approx 1.02 \text{ sec.}$$

When $t = 1.02$ seconds,

$$s = -4.9(1.02)^2 + 10(1.02) + 2 \approx 7.10 \text{ meters.}$$

So the maximum height reached is 7.10 meters. The object reaches the ground when $s = 0$:

$$0 = -4.9t^2 + 10t + 2.$$

Solving this using the quadratic formula gives

$$t \approx -0.18 \text{ and } t \approx 2.22 \text{ sec.}$$

Since the time at which the object hits the ground must be positive, $t \approx 2.22$ seconds.

History of the Equations of Motion

The problem of a body moving freely under the influence of gravity near the surface of the earth intrigued mathematicians and philosophers from Greek times onward and was finally solved by Galileo and Newton. The question to be answered was: How do the velocity and the position of the body vary with time? We define s to be the position, or height, of the body above a fixed point (often ground level), v is the velocity of the body measured upward, and a is the acceleration. The velocity and position at time $t = 0$ are represented by v_0 and s_0 respectively. We assume that the acceleration of the body is a constant, $-g$ (the negative sign means that the acceleration is downward), so

$$\frac{dv}{dt} = a = -g.$$

Problem 34 asks you to show that the motion satisfies

$$v = -gt + v_0,$$
$$s = -\frac{g}{2}t^2 + v_0 t + s_0.$$

Our derivation of the formulas for the velocity and the position of the body hides an almost 2000-year struggle to understand the mechanics of falling bodies, from Aristotle's *Physics* to Galileo's *Dialogues Concerning Two New Sciences*.

Though it is an oversimplification of his ideas, we can say that Aristotle's conception of motion was primarily in terms of *change of position*. This seems entirely reasonable; it is what we commonly observe, and this view dominated discussions of motion for centuries. But it misses a subtlety that was brought to light by Descartes, Galileo, and, with a different emphasis, by Newton. That subtlety is now usually referred to as the *principle of inertia*.

This principle holds that a body traveling undisturbed at constant velocity in a straight line will continue in this motion indefinitely. Stated another way, it says that one cannot distinguish in any absolute sense (that is, by performing an experiment) between being at rest and moving with constant velocity in a straight line. If you are reading this book in a closed room and have no external reference points, there is no experiment that will tell you, one way or the other, whether you are at rest or whether you, the room, and everything in it are moving with constant velocity in a straight line. Therefore, as Newton saw, an understanding of motion should be based on *change of velocity* rather than change of position. Since acceleration is the rate of change of velocity, it is acceleration that must play a central role in the description of motion.

Newton placed a new emphasis on the importance of *forces*. Newton's laws of motion do not say what a force *is*, they say how it *acts*. His first law is the principle of inertia, which says what happens in the *absence* of a force—there is no change in velocity. His second law says that a force acts to produce a change in velocity, that is, an acceleration. It states that $F = ma$, where m is the mass of the object, F is the net force, and a is the acceleration produced by this force.

Galileo demonstrated that a body falling under the influence of gravity does so with constant acceleration. Assuming we can neglect air resistance, this constant acceleration is independent of the mass of the body. This last fact was the outcome of Galileo's famous observation around 1600 that a heavy ball and a light ball dropped off the Leaning Tower of Pisa hit the ground at the same time. Whether or not he actually performed this experiment, Galileo presented a very clear thought experiment in the *Dialogues* to prove the same point. (This point was counter to Aristotle's more common-sense notion that the heavier ball would reach the ground first.) Galileo showed that the mass of the object did not appear as a variable in the equation of motion. Thus, the same constant-acceleration equation applies to all bodies falling under the influence of gravity.

Nearly a hundred years after Galileo's experiment, Newton formulated his laws of motion and gravity, which gave a theoretical explanation of Galileo's experimental observation that the acceleration due to gravity is independent of the mass of the body. According to Newton, acceleration is caused by force, and in the case of falling bodies, that force is the force of gravity.

Exercises and Problems for Section 6.3

Online Resource: Additional Problems for Section 6.3

EXERCISES

1. Show that $y = xe^{-x} + 2$ is a solution of the initial value problem
$$\frac{dy}{dx} = (1-x)e^{-x}, \quad y(0) = 2.$$

2. Show that $y = \sin(2t)$, for $0 \le t < \pi/4$, is a solution to the initial value problem
$$\frac{dy}{dt} = 2\sqrt{1-y^2}, \quad y(0) = 0.$$

In Exercises 3–8, find the general solution to the differential equation.

3. $\dfrac{dy}{dx} = 2x$

4. $\dfrac{dy}{dt} = t^2$

5. $\dfrac{dy}{dx} = x^3 + 5x^4$

6. $\dfrac{dy}{dt} = e^t$

7. $\dfrac{dy}{dx} = \cos x$

8. $\dfrac{dy}{dx} = \dfrac{1}{x}, \quad x > 0$

In Exercises 9–12, find the solution of the initial value problem.

9. $\dfrac{dy}{dx} = 3x^2, \quad y(0) = 5$

10. $\dfrac{dy}{dx} = x^5 + x^6, \quad y(1) = 2$

11. $\dfrac{dy}{dx} = e^x, \quad y(0) = 7$

12. $\dfrac{dy}{dx} = \sin x, \quad y(0) = 3$

PROBLEMS

13. A rock is thrown downward with velocity 10 ft/sec from a bridge 100 ft above the water. How fast is the rock going when it hits the water?

14. A water balloon launched from the roof of a building at time $t = 0$ has vertical velocity $v(t) = -32t + 40$ feet/sec at time t seconds, with $v > 0$ corresponding to upward motion.
    (a) If the roof of the building is 30 feet above the ground, find an expression for the height of the water balloon above the ground at time t.
    (b) What is the average velocity of the balloon between $t = 1.5$ and $t = 3$ seconds?
    (c) A 6-foot person is standing on the ground. How fast is the water balloon falling when it strikes the person on the top of the head?

15. A car starts from rest at time $t = 0$ and accelerates at $-0.6t + 4$ meters/sec^2 for $0 \le t \le 12$. How long does it take for the car to go 100 meters?

16. Ice is forming on a pond at a rate given by
$$\frac{dy}{dt} = k\sqrt{t},$$
where y is the thickness of the ice in inches at time t measured in hours since the ice started forming, and k is a positive constant. Find y as a function of t.

17. A revenue $R(p)$ is obtained by a farmer from selling

grain at price p dollars/unit. The marginal revenue is given by $R'(p) = 25 - 2p$.

(a) Find $R(p)$. Assume the revenue is zero when the price is zero.
(b) For what prices does the revenue increase as the price increases? For what prices does the revenue decrease as price increases?

18. A firm's marginal cost function is $MC = 3q^2 + 6q + 9$.

(a) Write a differential equation for the total cost, $C(q)$.
(b) Find the total cost function if the fixed costs are 400.

19. A tomato is thrown upward from a bridge 25 m above the ground at 40 m/sec.

(a) Give formulas for the acceleration, velocity, and height of the tomato at time t.
(b) How high does the tomato go, and when does it reach its highest point?
(c) How long is it in the air?

20. A car going 80 ft/sec (about 55 mph) brakes to a stop in five seconds. Assume the deceleration is constant.

(a) Graph the velocity against time, t, for $0 \le t \le 5$ seconds.
(b) Represent, as an area on the graph, the total distance traveled from the time the brakes are applied until the car comes to a stop.
(c) Find this area and hence the distance traveled.
(d) Now find the total distance traveled using antidifferentiation.

21. An object is shot vertically upward from the ground with an initial velocity of 160 ft/sec.

(a) At what rate is the velocity decreasing? Give units.
(b) Explain why the graph of velocity of the object against time (with upward positive) is a line.
(c) Using the starting velocity and your answer to part (b), find the time at which the object reaches the highest point.
(d) Use your answer to part (c) to decide when the object hits the ground.
(e) Graph the velocity against time. Mark on the graph when the object reaches its highest point and when it lands.
(f) Find the maximum height reached by the object by considering an area on the graph.
(g) Now express velocity as a function of time, and find the greatest height by antidifferentiation.

22. A stone thrown upward from the top of a 320-foot cliff at 128 ft/sec eventually falls to the beach below.

(a) How long does the stone take to reach its highest point?
(b) What is its maximum height?
(c) How long before the stone hits the beach?
(d) What is the velocity of the stone on impact?

23. A 727 jet needs to attain a speed of 200 mph to take off. If it can accelerate from 0 to 200 mph in 30 seconds, how long must the runway be? (Assume constant acceleration.)

24. A cat, walking along the window ledge of a New York apartment, knocks off a flower pot, which falls to the street 200 feet below. How fast is the flower pot traveling when it hits the street? (Give your answer in ft/sec and in mph, given that 1 ft/sec = 15/22 mph.)

25. An Acura NSX going at 70 mph stops in 157 feet. Find the acceleration, assuming it is constant.

26. (a) Find the general solution of the differential equation $dy/dx = 2x + 1$.
(b) Sketch a graph of at least three solutions.
(c) Find the solution satisfying $y(1) = 5$. Graph this solution with the others from part (b).

27. (a) Find and graph the general solution of the differential equation

$$\frac{dy}{dx} = \sin x + 2.$$

(b) Find the solution of the initial value problem

$$\frac{dy}{dx} = \sin x + 2, \quad y(3) = 5.$$

28. On the moon, the acceleration due to gravity is about 1.6 m/sec^2 (compared to $g \approx 9.8$ m/sec^2 on earth). If you drop a rock on the moon (with initial velocity 0), find formulas for:

(a) Its velocity, $v(t)$, at time t.
(b) The distance, $s(t)$, it falls in time t.

29. (a) Imagine throwing a rock straight up in the air. What is its initial velocity if the rock reaches a maximum height of 100 feet above its starting point?
(b) Now imagine being transplanted to the moon and throwing a moon rock vertically upward with the same velocity as in part (a). How high will it go? (On the moon, $g = 5$ ft/sec^2.)

30. An object is dropped from a 400-foot tower. When does it hit the ground and how fast is it going at the time of the impact?

31. The object in Problem 30 falls off the same 400-foot tower. What would the acceleration due to gravity have to be to make it reach the ground in half the time?

32. A ball that is dropped from a window hits the ground in five seconds. How high is the window? (Give your answer in feet.)

33. On the moon the acceleration due to gravity is 5 ft/sec^2. An astronaut jumps into the air with an initial upward velocity of 10 ft/sec. How high does he go? How long is the astronaut off the ground?

34. Assume the acceleration of a moving body is $-g$ and its initial velocity and position are v_0 and s_0 respectively. Find velocity, v, and position, s, as a function of t.

35. A particle of mass, m, acted on by a force, F, moves in a straight line. Its acceleration, a, is given by Newton's Law:

$$F = ma.$$

The work, W, done by a constant force when the particle moves through a displacement, d, is

$$W = Fd.$$

The velocity, v, of the particle as a function of time, t, is given in Figure 6.27. What is the sign of the work done during each of the time intervals: $[0, t_1]$, $[t_1, t_2]$, $[t_2, t_3]$, $[t_3, t_4]$, $[t_2, t_4]$?

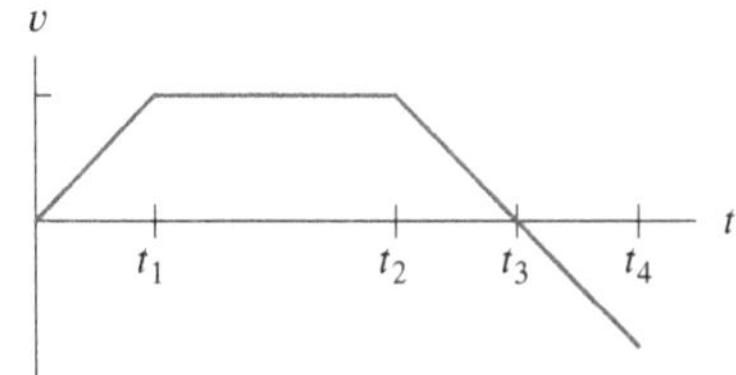

Figure 6.27

Strengthen Your Understanding

In Problems 36–39, explain what is wrong with the statement.

36. A rock dropped from a 400-foot cliff takes twice as long to hit the ground as it would if it were dropped from a 200-foot cliff.

37. The function $y = \cos(t^2)$ is a solution to the initial value problem

$$\frac{dy}{dt} = -\sin(t^2), \quad y(0) = 1.$$

38. Two solutions to a differential equation $dy/dx = f(x)$ have graphs which cross at the initial value.

39. A differential equation cannot have a constant solution.

In Problems 40–42, give an example of:

40. Two different solutions to the differential equation

$$\frac{dy}{dt} = t + 3.$$

41. A differential equation that has solution $y = \cos(5x)$.

42. A differential equation that has a solution that is decreasing for all t.

Are the statements in Problems 43–51 true or false? Give an explanation for your answer.

43. If $F(x)$ is an antiderivative of $f(x)$, then $y = F(x)$ is a solution to the differential equation $dy/dx = f(x)$.

44. If $y = F(x)$ is a solution to the differential equation $dy/dx = f(x)$, then $F(x)$ is an antiderivative of $f(x)$.

45. If an object has constant nonzero acceleration, then the position of the object as a function of time is a quadratic polynomial.

46. In an initial value problem for the differential equation $dy/dx = f(x)$, the value of y at $x = 0$ is always specified.

47. If $f(x)$ is positive for all x, then there is a solution of the differential equation $dy/dx = f(x)$ where $y(x)$ is positive for all x.

48. If $f(x) > 0$ for all x then every solution of the differential equation $dy/dx = f(x)$ is an increasing function.

49. If two solutions of a differential equation $dy/dx = f(x)$ have different values at $x = 3$ then they have different values at every x.

50. If the function $y = f(x)$ is a solution of the differential equation $dy/dx = \sin x/x$, then the function $y = f(x) + 5$ is also a solution.

51. There is only one solution $y(t)$ to the initial value problem $dy/dt = 3t^2$, $y(1) = \pi$.

6.4 SECOND FUNDAMENTAL THEOREM OF CALCULUS

Suppose f is an elementary function, that is, a combination of constants, powers of x, $\sin x$, $\cos x$, e^x, and $\ln x$. Then we have to be lucky to find an antiderivative F which is also an elementary function. But if we can't find F as an elementary function, how can we be sure that F exists at all? In this section we use the definite integral to construct antiderivatives.

Construction of Antiderivatives Using the Definite Integral

Consider the function $f(x) = e^{-x^2}$. Its antiderivative, F, is not an elementary function, but we would still like to find a way of calculating its values. Assuming F exists, we know from the Fundamental

Theorem of Calculus that

$$F(b) - F(a) = \int_a^b e^{-t^2}\, dt.$$

Setting $a = 0$ and replacing b by x, we have

$$F(x) - F(0) = \int_0^x e^{-t^2}\, dt.$$

Suppose we want the antiderivative that satisfies $F(0) = 0$. Then we get

$$F(x) = \int_0^x e^{-t^2}\, dt.$$

This is a formula for F. For any value of x, there is a unique value for $F(x)$, so F is a function. For any fixed x, we can calculate $F(x)$ numerically. For example,

$$F(2) = \int_0^2 e^{-t^2}\, dt = 0.88208\ldots.$$

Notice that our expression for F is not an elementary function; we have *created* a new function using the definite integral. The next theorem says that this method of constructing antiderivatives works in general. This means that if we define F by

$$F(x) = \int_a^x f(t)\, dt,$$

then F must be an antiderivative of f.

Theorem 6.2: Construction Theorem for Antiderivatives

(Second Fundamental Theorem of Calculus) If f is a continuous function on an interval, and if a is any number in that interval, then the function F defined on the interval as follows is an antiderivative of f:

$$F(x) = \int_a^x f(t)\, dt.$$

Proof Our task is to show that F, defined by this integral, is an antiderivative of f. We want to show that $F'(x) = f(x)$. By the definition of the derivative,

$$F'(x) = \lim_{h\to 0} \frac{F(x+h) - F(x)}{h}.$$

To gain some geometric insight, let's suppose f is positive and h is positive. Then we can visualize

$$F(x) = \int_a^x f(t)\, dt \quad \text{and} \quad F(x+h) = \int_a^{x+h} f(t)\, dt$$

as areas, which leads to representing

$$F(x+h) - F(x) = \int_x^{x+h} f(t)\, dt$$

as a difference of two areas. From Figure 6.28, we see that $F(x+h)-F(x)$ is roughly the area of a rectangle of height $f(x)$ and width h (shaded darker in Figure 6.28), so we have

$$F(x+h)-F(x)\approx f(x)h,$$

hence

$$\frac{F(x+h)-F(x)}{h}\approx f(x).$$

More precisely, we can use Theorem 5.4 on comparing integrals on page 307 to conclude that

$$mh\le\int_x^{x+h} f(t)\,dt\le Mh,$$

where m is the greatest lower bound for f on the interval from x to $x+h$ and M is the least upper bound on that interval. (See Figure 6.29.) Hence

$$mh\le F(x+h)-F(x)\le Mh,$$

so

$$m\le\frac{F(x+h)-F(x)}{h}\le M.$$

Since f is continuous, both m and M approach $f(x)$ as h approaches zero. Thus

$$f(x)\le\lim_{h\to 0}\frac{F(x+h)-F(x)}{h}\le f(x).$$

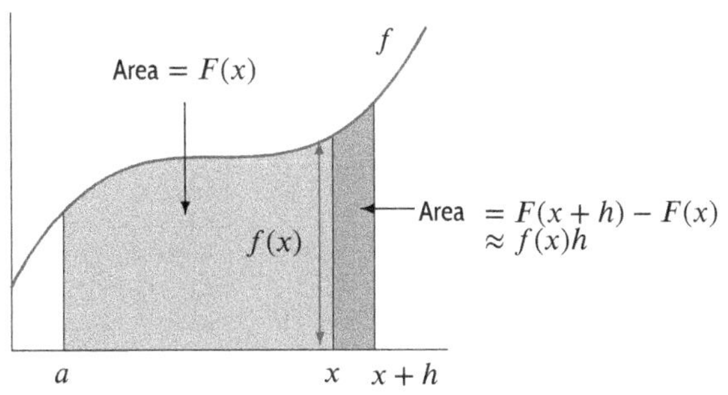

Figure 6.28: $F(x+h)-F(x)$ is area of roughly rectangular region

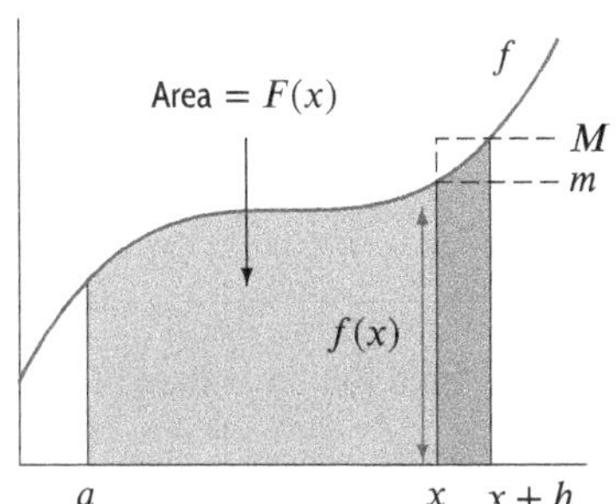

Figure 6.29: Upper and lower bounds for $F(x+h)-F(x)$

Thus both inequalities must actually be equalities, so we have the result we want:

$$f(x)=\lim_{h\to 0}\frac{F(x+h)-F(x)}{h}=F'(x).$$

Relationship Between the Construction Theorem and the Fundamental Theorem of Calculus

If F is constructed as in Theorem 6.2, then we have just shown that $F'=f$. Suppose G is any other antiderivative of f, so $G'=f$, and therefore $F'=G'$. Since the derivative of $F-G$ is zero, the Constant Function Theorem on page 187 tells us that $F-G$ is a constant, so $F(x)=G(x)+C$.

Since we know $F(a)=0$ (by the definition of F), we can write

$$\int_a^b f(t)\,dt=F(b)=F(b)-F(a)=(G(b)+C)-(G(a)+C)=G(b)-G(a).$$

This result, that the definite integral $\int_a^b f(t)\,dt$ can be evaluated using any antiderivative of f, is the (First) Fundamental Theorem of Calculus. Thus, we have shown that the First Fundamental Theorem of Calculus can be obtained from the Construction Theorem (the Second Fundamental Theorem of Calculus).

Using the Construction Theorem for Antiderivatives

The construction theorem enables us to write down antiderivatives of functions that do not have elementary antiderivatives. For example, an antiderivative of $(\sin x)/x$ is

$$F(x) = \int_0^x \frac{\sin t}{t}\, dt.$$

Notice that F is a function; we can calculate its values to any degree of accuracy.[1] This function already has a name: it is called the *sine-integral*, and it is denoted $\mathrm{Si}(x)$.

Example 1 Construct a table of values of $\mathrm{Si}(x)$ for $x = 0, 1, 2, 3$.

Solution Using numerical methods, we calculate the values of $\mathrm{Si}(x) = \int_0^x (\sin t)/t\, dt$ given in Table 6.2. Since the integrand is undefined at $t = 0$, we took the lower limit as 0.00001 instead of 0.

Table 6.2 *A table of values of Si(x)*

x	0	1	2	3
$\mathrm{Si}(x)$	0	0.95	1.61	1.85

The reason the sine-integral has a name is that some scientists and engineers use it all the time (for example, in optics). For them, it is just another common function like sine or cosine. Its derivative is given by

$$\frac{d}{dx}\,\mathrm{Si}(x) = \frac{\sin x}{x}.$$

Example 2 Find the derivative of $x\,\mathrm{Si}(x)$.

Solution Using the product rule,

$$\begin{aligned}\frac{d}{dx}(x\,\mathrm{Si}(x)) &= \left(\frac{d}{dx}x\right)\mathrm{Si}(x) + x\left(\frac{d}{dx}\,\mathrm{Si}(x)\right)\\ &= 1\cdot \mathrm{Si}(x) + x\frac{\sin x}{x}\\ &= \mathrm{Si}(x) + \sin x.\end{aligned}$$

Exercises and Problems for Section 6.4 Online Resource: Additional Problems for Section 6.4

EXERCISES

1. For $x = 0, 0.5, 1.0, 1.5$, and 2.0, make a table of values for $I(x) = \int_0^x \sqrt{t^4+1}\, dt$.

2. Assume that $F'(t) = \sin t \cos t$ and $F(0) = 1$. Find $F(b)$ for $b = 0,\ 0.5,\ 1,\ 1.5,\ 2,\ 2.5$, and 3.

3. **(a)** Continue the table of values for $\mathrm{Si}(x) = \int_0^x (\sin t/t)\, dt$ on page 338 for $x = 4$ and $x = 5$.
   (b) Why is $\mathrm{Si}(x)$ decreasing between $x = 4$ and $x = 5$?

■ In Exercises 4–6, write an expression for the function, $f(x)$, with the given properties.

4. $f'(x) = \sin(x^2)$ and $f(0) = 7$

5. $f'(x) = (\sin x)/x$ and $f(1) = 5$

6. $f'(x) = \mathrm{Si}(x)$ and $f(0) = 2$

[1] You may notice that the integrand, $(\sin t)/t$, is undefined at $t = 0$; such improper integrals are treated in more detail in Chapter 7.

■ In Exercises 7–10, let $F(x) = \int_0^x f(t)\,dt$. Graph $F(x)$ as a function of x.

7. 8.

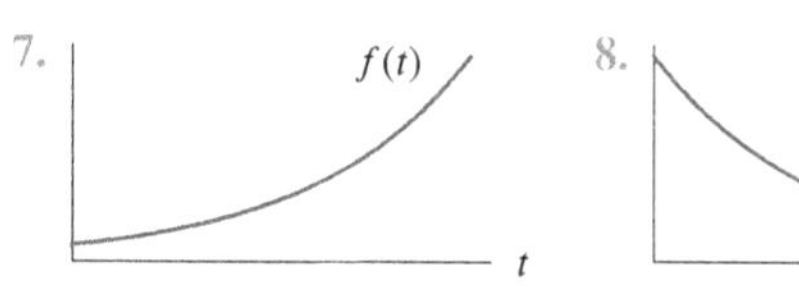

9.

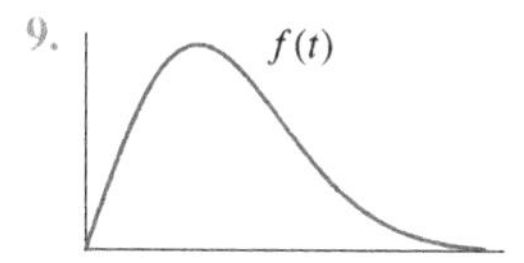

10.

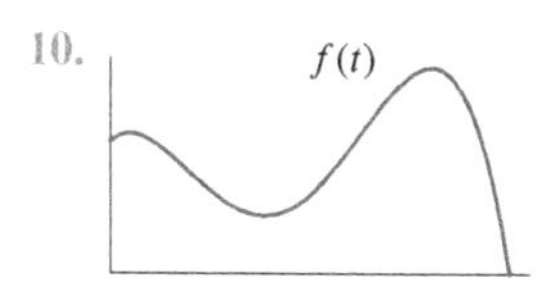

■ Find the derivatives in Exercises 11–16.

11. $\dfrac{d}{dx}\displaystyle\int_0^x \cos(t^2)\,dt$

12. $\dfrac{d}{dt}\displaystyle\int_4^t \sin(\sqrt{x})\,dx$

13. $\dfrac{d}{dx}\displaystyle\int_1^x (1+t)^{200}\,dt$

14. $\dfrac{d}{dx}\displaystyle\int_2^x \ln(t^2+1)\,dt$

15. $\dfrac{d}{dx}\displaystyle\int_{0.5}^x \arctan(t^2)\,dt$

16. $\dfrac{d}{dx}\left[\mathrm{Si}(x^2)\right]$

PROBLEMS

17. Find intervals where the graph of $F(x) = \int_0^x e^{-t^2}\,dt$ is concave up and concave down.

18. Use properties of the function $f(x) = xe^{-x}$ to determine the number of values x that make $F(x) = 0$, given $F(x) = \int_1^x f(t)\,dt$.

■ For Problems 19–21, let $F(x) = \int_0^x \sin(t^2)\,dt$.

19. Approximate $F(x)$ for $x = 0, 0.5, 1, 1.5, 2, 2.5$.

20. Using a graph of $F'(x)$, decide where $F(x)$ is increasing and where $F(x)$ is decreasing for $0 \le x \le 2.5$.

21. Does $F(x)$ have a maximum value for $0 \le x \le 2.5$? If so, at what value of x does it occur, and approximately what is that maximum value?

22. Use Figure 6.30 to sketch a graph of $F(x) = \int_0^x f(t)\,dt$. Label the points x_1, x_2, x_3.

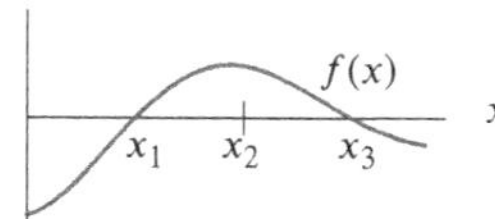

Figure 6.30

23. The graph of the derivative F' of some function F is given in Figure 6.31. If $F(20) = 150$, estimate the maximum value attained by F.

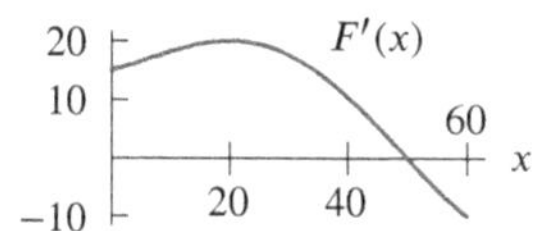

Figure 6.31

24. **(a)** Let $F(x) = \int_5^x f(t)\,dt$, with f in Figure 6.32. Find the x-value at which the maximum value of F occurs.
(b) What is the maximum value of F?

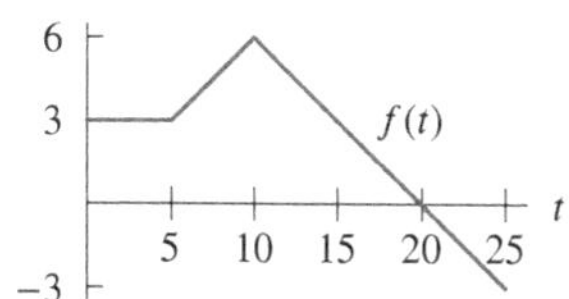

Figure 6.32

25. Let $g(x) = \int_0^x f(t)\,dt$. Using Figure 6.33, find
(a) $g(0)$ **(b)** $g'(1)$
(c) The interval where g is concave up.
(d) The value of x where g takes its maximum on the interval $0 \le x \le 8$.

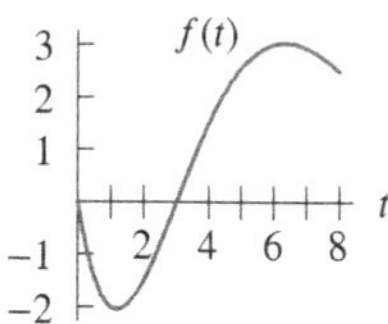

Figure 6.33

26. Let $F(x) = \int_0^x \sin(2t)\,dt$.
(a) Evaluate $F(\pi)$.
(b) Draw a sketch to explain geometrically why the answer to part (a) is correct.
(c) For what values of x is $F(x)$ positive? negative?

27. Let $F(x) = \int_2^x (1/\ln t)\,dt$ for $x \ge 2$.
(a) Find $F'(x)$.
(b) Is F increasing or decreasing? What can you say about the concavity of its graph?
(c) Sketch a graph of $F(x)$.

■ In Problems 28–29, find the value of the function with the given properties.

28. $F(1)$, where $F'(x) = e^{-x^2}$ and $F(0) = 2$

29. $G(-1)$, where $G'(x) = \cos(x^2)$ and $G(0) = -3$

In Problems 30–33, estimate the value of each expression, given $w(t) = \int_0^t q(x)\,dx$ and $v(t) = \int_0^t q'(x)\,dx$. Table 6.3 gives values for $q(x)$, a function with negative first and second derivatives. Are your answers under- or overestimates?

Table 6.3

x	0.0	0.1	0.2	0.3	0.4	0.5
$q(x)$	5.3	5.2	4.9	4.5	3.9	3.1

30. $w(0.4)$
31. $v(0.4)$
32. $w'(0.4)$
33. $v'(0.4)$

In Problems 34–37, use the chain rule to calculate the derivative.

34. $\frac{d}{dx}\int_0^{x^2} \ln(1+t^2)\,dt$
35. $\frac{d}{dt}\int_1^{\sin t} \cos(x^2)\,dx$
36. $\frac{d}{dt}\int_{2t}^{4} \sin(\sqrt{x})\,dx$
37. $\frac{d}{dx}\int_{-x^2}^{x^2} e^{t^2}\,dt$

In Problems 38–41, find the given quantities. The *error function*, erf(x), is defined by

$$\operatorname{erf}(x) = \frac{2}{\sqrt{\pi}}\int_0^x e^{-t^2}\,dt.$$

38. $\frac{d}{dx}(x\operatorname{erf}(x))$
39. $\frac{d}{dx}(\operatorname{erf}(\sqrt{x}))$
40. $\frac{d}{dx}\left(\int_0^{x^3} e^{-t^2}\,dt\right)$
41. $\frac{d}{dx}\left(\int_x^{x^3} e^{-t^2}\,dt\right)$

42. In year x, the forested area in hectares of an island is

$$f(x) = 820 + \int_0^x r(t)\,dt,$$

where $r(t)$ has only one zero, at $t = 40$. Let $r'(40) < 0$, $f(40) = 1170$ and $\int_{30}^{40} r(x)\,dx = 82$.

(a) What is the maximum forested area?
(b) How large is the forested area in year $t = 30$?

43. Let $F(x) = \int_0^x p(t)\,dt$ and $G(x) = \int_3^x p(t)\,dt$, where $F(3) = 5$ and $G(5) = 7$. Find each of the following.

(a) $G(0)$
(b) $F(0)$
(c) $F(5)$
(d) k where $G(x) = F(x)+k$

Strengthen Your Understanding

In Problems 44–46, explain what is wrong with the statement.

44. $\frac{d}{dx}\int_0^5 t^2\,dt = x^2.$

45. $F(x) = \int_{-2}^{x} t^2\,dt$ has a local minimum at $x = 0$.

46. For the function $f(x)$ shown in Figure 6.34, $F(x) = \int_0^x f(t)\,dt$ has a local minimum at $x = 2$.

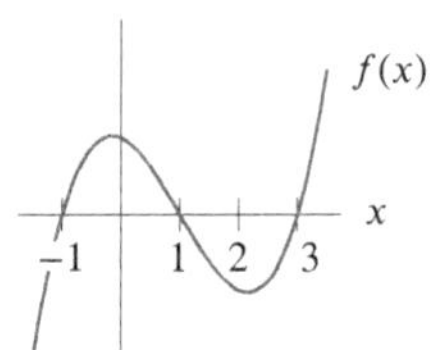

Figure 6.34:

In Problems 47–48, give an example of:

47. A function, $F(x)$, constructed using the Second Fundamental Theorem of Calculus such that F is a nondecreasing function and $F(0) = 0$.

48. A function $G(x)$, constructed using the Second Fundamental Theorem of Calculus, such that G is concave up and $G(7) = 0$.

Are the statements in Problems 49–54 true or false? Give an explanation for your answer.

49. Every continuous function has an antiderivative.

50. $\int_0^x \sin(t^2)\,dt$ is an antiderivative of $\sin(x^2)$.

51. If $F(x) = \int_0^x f(t)\,dt$, then $F(5) - F(3) = \int_3^5 f(t)dt$.

52. If $F(x) = \int_0^x f(t)\,dt$, then $F(x)$ must be increasing.

53. If $F(x) = \int_0^x f(t)\,dt$ and $G(x) = \int_2^x f(t)\,dt$, then $F(x) = G(x) + C$.

54. If $F(x) = \int_0^x f(t)\,dt$ and $G(x) = \int_0^x g(t)\,dt$, then $F(x) + G(x) = \int_0^x (f(t) + g(t))\,dt$.

Online Resource: Review problems and Projects

Chapter Seven

INTEGRATION

Contents

7.1 INTEGRATION BY SUBSTITUTION

In Chapter 3, we learned rules to differentiate any function obtained by combining constants, powers of x, $\sin x$, $\cos x$, e^x, and $\ln x$, using addition, multiplication, division, or composition of functions. Such functions are called *elementary*.

In this chapter, we introduce several methods of antidifferentiation. However, there is a great difference between looking for derivatives and looking for antiderivatives. Every elementary function has elementary derivatives, but most elementary functions—such as $\sqrt{x^3+1}$, $(\sin x)/x$, and e^{-x^2}—do not have elementary antiderivatives.

All commonly occurring antiderivatives can be found with a computer algebra system (CAS). However, just as it is useful to be able to calculate $3+4$ without a calculator, we usually calculate some antiderivatives by hand.

The Guess-and-Check Method

A good strategy for finding simple antiderivatives is to *guess* an answer (using knowledge of differentiation rules) and then *check* the answer by differentiating it. If we get the expected result, then we're done; otherwise, we revise the guess and check again.

The method of guess-and-check is useful in reversing the chain rule. According to the chain rule,

$$\frac{d}{dx}(f(g(x))) = \underbrace{f'}_{\text{Derivative of outside}}(\overbrace{g(x)}^{\text{Inside}}) \cdot \underbrace{g'(x)}_{\text{Derivative of inside}}.$$

Thus, any function which is the result of applying the chain rule is the product of two factors: the "derivative of the outside" and the "derivative of the inside." If a function has this form, its antiderivative is $f(g(x))$.

Example 1 Find $\int 3x^2\cos(x^3)\,dx$.

Solution The function $3x^2\cos(x^3)$ looks like the result of applying the chain rule: there is an "inside" function x^3 and its derivative $3x^2$ appears as a factor. Since the outside function is a cosine which has a sine as an antiderivative, we guess $\sin(x^3)$ for the antiderivative. Differentiating to check gives

$$\frac{d}{dx}(\sin(x^3)) = \cos(x^3)\cdot(3x^2).$$

Since this is what we began with, we know that

$$\int 3x^2\cos(x^3)\,dx = \sin(x^3) + C.$$

The basic idea of this method is to try to find an inside function whose derivative appears as a factor. This works even when the derivative is missing a constant factor, as in the next example.

Example 2 Find $\int x^3\sqrt{x^4+5}\,dx$.

Solution Here the inside function is x^4+5, and its derivative appears as a factor, with the exception of a missing 4. Thus, the integrand we have is more or less of the form

$$g'(x)\sqrt{g(x)},$$

with $g(x) = x^4 + 5$. Since $x^{3/2}/(3/2)$ is an antiderivative of the outside function $\sqrt{x}$, we might guess that an antiderivative is

$$\frac{(g(x))^{3/2}}{3/2} = \frac{(x^4+5)^{3/2}}{3/2}.$$

Let's check and see:

$$\frac{d}{dx}\left(\frac{(x^4+5)^{3/2}}{3/2}\right) = \frac{3}{2}\frac{(x^4+5)^{1/2}}{3/2} \cdot 4x^3 = 4x^3(x^4+5)^{1/2},$$

so $\dfrac{(x^4+5)^{3/2}}{3/2}$ is too big by a factor of 4. The correct antiderivative is

$$\frac{1}{4}\frac{(x^4+5)^{3/2}}{3/2} = \frac{1}{6}(x^4+5)^{3/2}.$$

Thus

$$\int x^3\sqrt{x^4+5}\,dx = \frac{1}{6}(x^4+5)^{3/2} + C.$$

As a final check:

$$\frac{d}{dx}\left(\frac{1}{6}(x^4+5)^{3/2}\right) = \frac{1}{6}\cdot\frac{3}{2}(x^4+5)^{1/2}\cdot 4x^3 = x^3(x^4+5)^{1/2}.$$

As we see in the preceding example, antidifferentiating a function often involves "correcting for" constant factors: if differentiation produces an extra factor of 2, antidifferentiation will require a factor of $\frac{1}{2}$.

The Method of Substitution

When the integrand is complicated, it helps to formalize this guess-and-check method as follows:

To Make a Substitution

Let w be the "inside function" and $dw = w'(x)\,dx = \dfrac{dw}{dx}dx$.

Let's redo the first example using a substitution.

Example 3 Find $\displaystyle\int 3x^2\cos(x^3)\,dx$.

Solution As before, we look for an inside function whose derivative appears—in this case x^3. We let $w = x^3$. Then $dw = w'(x)\,dx = 3x^2\,dx$. The original integrand can now be completely rewritten in terms of the new variable w:

$$\int 3x^2\cos(x^3)\,dx = \int \cos\underbrace{(x^3)}_{w}\cdot\underbrace{3x^2\,dx}_{dw} = \int \cos w\,dw = \sin w + C = \sin(x^3) + C.$$

By changing the variable to w, we can simplify the integrand. We now have $\cos w$, which can be antidifferentiated more easily. The final step, after antidifferentiating, is to convert back to the original variable, x.

Why Does Substitution Work?

The substitution method makes it look as if we can treat dw and dx as separate entities, even canceling them in the equation $dw = (dw/dx)dx$. Let's see why this works. Suppose we have an integral of the form $\int f(g(x))g'(x)\,dx$, where $g(x)$ is the inside function and $f(x)$ is the outside function. If F is an antiderivative of f, then $F' = f$, and by the chain rule $\frac{d}{dx}(F(g(x))) = f(g(x))g'(x)$. Therefore,

$$\int f(g(x))g'(x)\,dx = F(g(x)) + C.$$

Now write $w = g(x)$ and $dw/dx = g'(x)$ on both sides of this equation:

$$\int f(w)\frac{dw}{dx}\,dx = F(w) + C.$$

On the other hand, knowing that $F' = f$ tells us that

$$\int f(w)\,dw = F(w) + C.$$

Thus, the following two integrals are equal:

$$\int f(w)\frac{dw}{dx}\,dx = \int f(w)\,dw.$$

Substituting w for the inside function and writing $dw = w'(x)dx$ leaves the indefinite integral unchanged.

Let's revisit the second example that we did by guess-and-check.

Example 4 Find $\int x^3\sqrt{x^4+5}\,dx$.

Solution The inside function is $x^4 + 5$, with derivative $4x^3$. The integrand has a factor of x^3, and since the only thing missing is a constant factor, we try

$$w = x^4 + 5.$$

Then

$$dw = w'(x)\,dx = 4x^3\,dx,$$

giving

$$\frac{1}{4}dw = x^3\,dx.$$

Thus,

$$\int x^3\sqrt{x^4+5}\,dx = \int \sqrt{w}\,\frac{1}{4}dw = \frac{1}{4}\int w^{1/2}\,dw = \frac{1}{4}\cdot\frac{w^{3/2}}{3/2} + C = \frac{1}{6}(x^4+5)^{3/2} + C.$$

Once again, we get the same result as with guess-and-check.

Warning

We saw in the preceding example that we can apply the substitution method when a *constant* factor is missing from the derivative of the inside function. However, we may not be able to use substitution if anything other than a constant factor is missing. For example, setting $w = x^4 + 5$ to find

$$\int x^2\sqrt{x^4+5}\,dx$$

does us no good because $x^2\,dx$ is not a constant multiple of $dw = 4x^3\,dx$. Substitution works if the integrand contains the derivative of the inside function, *to within a constant factor*.

Some people prefer the substitution method over guess-and-check since it is more systematic, but both methods achieve the same result. For simple problems, guess-and-check can be faster.

Example 5 Find $\int e^{\cos\theta} \sin\theta \, d\theta$.

Solution We let $w = \cos\theta$ since its derivative is $-\sin\theta$ and there is a factor of $\sin\theta$ in the integrand. This gives

$$dw = w'(\theta)\, d\theta = -\sin\theta \, d\theta,$$

so

$$-dw = \sin\theta \, d\theta.$$

Thus

$$\int e^{\cos\theta} \sin\theta \, d\theta = \int e^{w}\,(-dw) = (-1)\int e^{w}\, dw = -e^{w} + C = -e^{\cos\theta} + C.$$

Example 6 Find $\int \frac{e^t}{1+e^t}\, dt$.

Solution Observing that the derivative of $1 + e^t$ is e^t, we see $w = 1 + e^t$ is a good choice. Then $dw = e^t\, dt$, so that

$$\begin{aligned}\int \frac{e^t}{1+e^t}\, dt = \int \frac{1}{1+e^t}\, e^t\, dt = \int \frac{1}{w}\, dw &= \ln|w| + C \\ &= \ln|1+e^t| + C \\ &= \ln(1+e^t) + C. \qquad \text{(Since } (1+e^t) \text{ is always positive.)}\end{aligned}$$

Since the numerator is $e^t\, dt$, we might also have tried $w = e^t$. This substitution leads to the integral $\int (1/(1+w))dw$, which is better than the original integral but requires another substitution, $u = 1+w$, to finish. There are often several different ways of doing an integral by substitution.

Notice the pattern in the previous example: having a function in the denominator and its derivative in the numerator leads to a natural logarithm. The next example follows the same pattern.

Example 7 Find $\int \tan\theta \, d\theta$.

Solution Recall that $\tan\theta = (\sin\theta)/(\cos\theta)$. If $w = \cos\theta$, then $dw = -\sin\theta \, d\theta$, so

$$\int \tan\theta \, d\theta = \int \frac{\sin\theta}{\cos\theta}\, d\theta = \int \frac{-dw}{w} = -\ln|w| + C = -\ln|\cos\theta| + C.$$

One way to think of integration is in terms of standard forms, whose antiderivatives are known. Substitution is useful for putting a complicated integral in a standard form.

Example 8 Give a substitution w and constants k, n so that the following integral has the form $\int kw^n\, dw$:

$$\int e^x \cos^3(e^x)\sin(e^x)\, dx.$$

Solution We notice that one of the factors in the integrand is $(\cos(e^x))^3$, so if we let $w = \cos(e^x)$, this factor is w^3. Then $dw = (-\sin(e^x))e^x\,dx$, so

$$\int e^x \cos^3(e^x)\sin(e^x)\,dx = \int \cos^3(e^x)(\sin(e^x))e^x\,dx = \int w^3\,(-dw).$$

Therefore, we choose $w = \cos(e^x)$ and then $k = -1, n = 3$.

Definite Integrals by Substitution

Example 9 Compute $\displaystyle\int_0^2 xe^{x^2}\,dx.$

Solution To evaluate this definite integral using the Fundamental Theorem of Calculus, we first need to find an antiderivative of $f(x) = xe^{x^2}$. The inside function is x^2, so we let $w = x^2$. Then $dw = 2x\,dx$, so $\frac{1}{2}\,dw = x\,dx$. Thus,

$$\int xe^{x^2}\,dx = \int e^w \frac{1}{2}\,dw = \frac{1}{2}e^w + C = \frac{1}{2}e^{x^2} + C.$$

Now we find the definite integral

$$\int_0^2 xe^{x^2}\,dx = \frac{1}{2}e^{x^2}\bigg|_0^2 = \frac{1}{2}(e^4 - e^0) = \frac{1}{2}(e^4 - 1).$$

There is another way to look at the same problem. After we established that

$$\int xe^{x^2}\,dx = \frac{1}{2}e^w + C,$$

our next two steps were to replace w by x^2, and then x by 2 and 0. We could have directly replaced the original limits of integration, $x = 0$ and $x = 2$, by the corresponding w limits. Since $w = x^2$, the w limits are $w = 0^2 = 0$ (when $x = 0$) and $w = 2^2 = 4$ (when $x = 2$), so we get

$$\int_{x=0}^{x=2} xe^{x^2}\,dx = \frac{1}{2}\int_{w=0}^{w=4} e^w\,dw = \frac{1}{2}e^w\bigg|_0^4 = \frac{1}{2}\left(e^4 - e^0\right) = \frac{1}{2}(e^4 - 1).$$

As we would expect, both methods give the same answer.

To Use Substitution to Find Definite Integrals

Either

- Compute the indefinite integral, expressing an antiderivative in terms of the original variable, and then evaluate the result at the original limits,

or

- Convert the original limits to new limits in terms of the new variable and do not convert the antiderivative back to the original variable.

Example 10 Evaluate $\displaystyle\int_0^{\pi/4} \frac{\tan^3\theta}{\cos^2\theta}\,d\theta.$

Solution To use substitution, we must decide what w should be. There are two possible inside functions, $\tan\theta$ and $\cos\theta$. Now

$$\frac{d}{d\theta}(\tan\theta) = \frac{1}{\cos^2\theta} \quad \text{and} \quad \frac{d}{d\theta}(\cos\theta) = -\sin\theta,$$

and since the integral contains a factor of $1/\cos^2\theta$ but not of $\sin\theta$, we try $w = \tan\theta$. Then $dw = (1/\cos^2\theta)d\theta$. When $\theta = 0$, $w = \tan 0 = 0$, and when $\theta = \pi/4$, $w = \tan(\pi/4) = 1$, so

$$\int_0^{\pi/4} \frac{\tan^3\theta}{\cos^2\theta}\,d\theta = \int_0^{\pi/4} (\tan\theta)^3 \cdot \frac{1}{\cos^2\theta}\,d\theta = \int_0^1 w^3\,dw = \frac{1}{4}w^4\Big|_0^1 = \frac{1}{4}.$$

Example 11 Evaluate $\displaystyle\int_1^3 \frac{dx}{5-x}$.

Solution Let $w = 5 - x$, so $dw = -dx$. When $x = 1$, $w = 4$, and when $x = 3$, $w = 2$, so

$$\int_1^3 \frac{dx}{5-x} = \int_4^2 \frac{-dw}{w} = -\ln|w|\Big|_4^2 = -(\ln 2 - \ln 4) = \ln\left(\frac{4}{2}\right) = \ln 2 = 0.693.$$

Notice that we write the limit $w = 4$ at the bottom of the second integral, even though it is larger than $w = 2$, because $w = 4$ corresponds to the lower limit $x = 1$.

More Complex Substitutions

In the examples of substitution presented so far, we guessed an expression for w and hoped to find dw (or some constant multiple of it) in the integrand. What if we are not so lucky? It turns out that it often works to let w be some messy expression contained inside, say, a cosine or under a root, even if we cannot see immediately how such a substitution helps.

Example 12 Find $\displaystyle\int \sqrt{1+\sqrt{x}}\,dx$.

Solution This time, the derivative of the inside function is nowhere to be seen. Nevertheless, we try $w = 1 + \sqrt{x}$. Then $w - 1 = \sqrt{x}$, so $(w-1)^2 = x$. Therefore $2(w-1)\,dw = dx$. We have

$$\begin{aligned}\int \sqrt{1+\sqrt{x}}\,dx &= \int \sqrt{w}\,2(w-1)\,dw = 2\int w^{1/2}(w-1)\,dw \\ &= 2\int (w^{3/2} - w^{1/2})\,dw = 2\left(\frac{2}{5}w^{5/2} - \frac{2}{3}w^{3/2}\right) + C \\ &= 2\left(\frac{2}{5}(1+\sqrt{x})^{5/2} - \frac{2}{3}(1+\sqrt{x})^{3/2}\right) + C.\end{aligned}$$

Notice that the substitution in the preceding example again converts the inside of the messiest function into something simple. In addition, since the derivative of the inside function is not waiting for us, we have to solve for x so that we can get dx entirely in terms of w and dw.

Example 13 Find $\displaystyle\int (x+7)\sqrt[3]{3-2x}\,dx$.

Solution The inside function is $3-2x$, with derivative -2. However, instead of a factor of -2, the integrand contains the factor $(x+7)$, but substituting $w=3-2x$ turns out to help anyway. Then $dw=-2\,dx$, so $(-1/2)\,dw=dx$. Now we must convert everything to w, including $x+7$. If $w=3-2x$, then $2x=3-w$, so $x=3/2-w/2$, and therefore we can write $x+7$ in terms of w. Thus

$$\begin{aligned}\int (x+7)\sqrt[3]{3-2x}\,dx &= \int \left(\frac{3}{2}-\frac{w}{2}+7\right)\sqrt[3]{w}\left(-\frac{1}{2}\right)dw\\ &= -\frac{1}{2}\int\left(\frac{17}{2}-\frac{w}{2}\right)w^{1/3}\,dw\\ &= -\frac{1}{4}\int (17-w)w^{1/3}\,dw\\ &= -\frac{1}{4}\int (17w^{1/3}-w^{4/3})\,dw\\ &= -\frac{1}{4}\left(17\frac{w^{4/3}}{4/3}-\frac{w^{7/3}}{7/3}\right)+C\\ &= -\frac{1}{4}\left(\frac{51}{4}(3-2x)^{4/3}-\frac{3}{7}(3-2x)^{7/3}\right)+C.\end{aligned}$$

Looking back over the solution, the reason this substitution works is that it converts $\sqrt[3]{3-2x}$, the messiest part of the integrand, to $\sqrt[3]{w}$, which can be combined with the other factor and then integrated.

Exercises and Problems for Section 7.1 Online Resource: Additional Problems for Section 7.1

EXERCISES

1. Use substitution to express each of the following integrals as a multiple of $\int_a^b (1/w)\,dw$ for some a and b. Then evaluate the integrals.

   (a) $\displaystyle\int_0^1 \frac{x}{1+x^2}\,dx$ **(b)** $\displaystyle\int_0^{\pi/4} \frac{\sin x}{\cos x}\,dx$

2. **(a)** Find the derivatives of $\sin(x^2+1)$ and $\sin(x^3+1)$.
   (b) Use your answer to part (a) to find antiderivatives of:
   (i) $x\cos(x^2+1)$ (ii) $x^2\cos(x^3+1)$
   (c) Find the general antiderivatives of:
   (i) $x\sin(x^2+1)$ (ii) $x^2\sin(x^3+1)$

■ Find the integrals in Exercises 3–50. Check your answers by differentiation.

3. $\displaystyle\int e^{3x}\,dx$
4. $\displaystyle\int te^{t^2}\,dt$
5. $\displaystyle\int e^{-x}\,dx$
6. $\displaystyle\int 25e^{-0.2t}\,dt$
7. $\displaystyle\int \sin(2x)\,dx$
8. $\displaystyle\int t\cos(t^2)\,dt$
9. $\displaystyle\int \sin(3-t)\,dt$
10. $\displaystyle\int xe^{-x^2}\,dx$
11. $\displaystyle\int (r+1)^3\,dr$
12. $\displaystyle\int y(y^2+5)^8\,dy$
13. $\displaystyle\int x^2(1+2x^3)^2\,dx$
14. $\displaystyle\int t^2(t^3-3)^{10}\,dt$
15. $\displaystyle\int x(x^2+3)^2\,dx$
16. $\displaystyle\int x(x^2-4)^{7/2}\,dx$
17. $\displaystyle\int y^2(1+y)^2\,dy$
18. $\displaystyle\int (2t-7)^{73}\,dt$
19. $\displaystyle\int x^2e^{x^3+1}\,dx$
20. $\displaystyle\int (x^2+3)^2\,dx$
21. $\displaystyle\int \frac{1}{\sqrt{4-x}}\,dx$
22. $\displaystyle\int \frac{dy}{y+5}$
23. $\displaystyle\int e^{-0.1t+4}\,dt$
24. $\displaystyle\int \tan(\theta+1)\,d\theta$
25. $\displaystyle\int \sin\theta(\cos\theta+5)^7\,d\theta$
26. $\displaystyle\int \sqrt{\cos 3t}\sin 3t\,dt$
27. $\displaystyle\int \sin^6\theta\cos\theta\,d\theta$
28. $\displaystyle\int \sin^3\alpha\cos\alpha\,d\alpha$
29. $\displaystyle\int \sin^6(5\theta)\cos(5\theta)\,d\theta$
30. $\displaystyle\int \tan(2x)\,dx$
31. $\displaystyle\int \frac{(\ln z)^2}{z}\,dz$
32. $\displaystyle\int \frac{e^t+1}{e^t+t}\,dt$
33. $\displaystyle\int \frac{(t+1)^2}{t^2}\,dt$
34. $\displaystyle\int \frac{y}{y^2+4}\,dy$
35. $\displaystyle\int \frac{dx}{1+2x^2}$
36. $\displaystyle\int \frac{dx}{\sqrt{1-4x^2}}$
37. $\displaystyle\int \frac{\cos\sqrt{x}}{\sqrt{x}}\,dx$
38. $\displaystyle\int \frac{e^{\sqrt{y}}}{\sqrt{y}}\,dy$

39. $\int \frac{1+e^x}{\sqrt{x+e^x}}\, dx$

40. $\int \frac{e^x}{2+e^x}\, dx$

41. $\int \frac{x+1}{x^2+2x+19}\, dx$

42. $\int \frac{t}{1+3t^2}\, dt$

43. $\int \frac{e^x - e^{-x}}{e^x + e^{-x}}\, dx$

44. $\int \frac{x\cos(x^2)}{\sqrt{\sin(x^2)}}\, dx$

45. $\int \sinh 3t\, dt$

46. $\int \cosh x\, dx$

47. $\int \cosh(2w+1)\, dw$

48. $\int (\sinh z)e^{\cosh z}\, dz$

49. $\int \cosh^2 x \sinh x\, dx$

50. $\int x\cosh x^2\, dx$

■ For the functions in Exercises 51–58, find the general antiderivative. Check your answers by differentiation.

51. $p(t) = \pi t^3 + 4t$

52. $f(x) = \sin 3x$

53. $f(x) = 2x\cos(x^2)$

54. $r(t) = 12t^2\cos(t^3)$

55. $f(x) = \sin(2-5x)$

56. $f(x) = e^{\sin x}\cos x$

57. $f(x) = \frac{x}{x^2+1}$

58. $f(x) = \frac{1}{3\cos^2(2x)}$

■ For Exercises 59–66, use the Fundamental Theorem to calculate the definite integrals.

59. $\int_0^{\pi} \cos(x+\pi)\, dx$

60. $\int_0^{1/2} \cos(\pi x)\, dx$

61. $\int_0^{\pi/2} e^{-\cos\theta}\sin\theta\, d\theta$

62. $\int_1^2 2xe^{x^2}\, dx$

63. $\int_1^4 \frac{e^{\sqrt{x}}}{\sqrt{x}}\, dx$

64. $\int_{-1}^{e-2} \frac{1}{t+2}\, dt$

65. $\int_1^4 \frac{\cos\sqrt{x}}{\sqrt{x}}\, dx$

66. $\int_0^2 \frac{x}{(1+x^2)^2}\, dx$

■ For Exercises 67–72, evaluate the definite integrals. Whenever possible, use the Fundamental Theorem of Calculus, perhaps after a substitution. Otherwise, use numerical methods.

67. $\int_{-1}^3 (x^3+5x)\, dx$

68. $\int_{-1}^1 \frac{1}{1+y^2}\, dy$

69. $\int_1^3 \frac{1}{x}\, dx$

70. $\int_1^3 \frac{dt}{(t+7)^2}$

71. $\int_{-1}^2 \sqrt{x+2}\, dx$

72. $\int_1^2 \frac{\sin t}{t}\, dt$

■ Find the integrals in Exercises 73–80.

73. $\int y\sqrt{y+1}\, dy$

74. $\int z(z+1)^{1/3}\, dz$

75. $\int \frac{t^2+t}{\sqrt{t+1}}\, dt$

76. $\int \frac{dx}{2+2\sqrt{x}}$

77. $\int x^2\sqrt{x-2}\, dx$

78. $\int (z+2)\sqrt{1-z}\, dz$

79. $\int \frac{t}{\sqrt{t+1}}\, dt$

80. $\int \frac{3x-2}{\sqrt{2x+1}}\, dx$

PROBLEMS

81. Show that the following integral can be calculated by multiplying the numerator and denominator by e^t and using a substitution:

$$\int \frac{1}{1+e^{-t}}\, dt.$$

■ In Problems 82–85, show the two integrals are equal using a substitution.

82. $\int_0^{\pi/3} 3\sin^2(3x)\, dx = \int_0^{\pi} \sin^2(y)\, dy$

83. $\int_1^2 2\ln(s^2+1)\, ds = \int_1^4 \frac{\ln(t+1)}{\sqrt{t}}\, dt$

84. $\int_1^e (\ln w)^3\, dw = \int_0^1 z^3 e^z\, dz$

85. $\int_0^{\pi} x\cos(\pi - x)\, dx = \int_0^{\pi} (\pi - t)\cos t\, dt$

86. Using the substitution $w = x^2$, find a function $g(w)$ such that $\int_{\sqrt{a}}^{\sqrt{b}} dx = \int_a^b g(w)\, dw$ for all $0 < a < b$.

87. Using the substitution $w = e^x$, find a function $g(w)$ such that $\int_a^b e^{-x} dx = \int_{e^a}^{e^b} g(w)\, dw$ for all $a < b$.

■ In Problems 88–94, evaluate the integral. Your answer should not contain f, which is a differentiable function with the following values:

x	0	1	$\pi/2$	e	3
$f(x)$	5	7	8	10	11
$f'(x)$	2	4	6	9	12

88. $\int_0^1 f'(x)\sin f(x)\, dx$

89. $\int_1^3 f'(x)e^{f(x)}\, dx$

90. $\int_1^3 \frac{f'(x)}{f(x)}\, dx$

91. $\int_0^1 e^x f'(e^x)\, dx$

92. $\int_1^e \frac{f'(\ln x)}{x}\, dx$

93. $\int_0^1 f'(x)(f(x))^2\, dx$

94. $\int_0^{\pi/2} \sin x \cdot f'(\cos x)\, dx$

■ In Problems 95–99, explain why the two antiderivatives are really, despite their apparent dissimilarity, different expressions of the same problem. You do not need to evaluate the integrals.

95. $\int \frac{e^x\,dx}{1+e^{2x}}$ and $\int \frac{\cos x\,dx}{1+\sin^2 x}$

96. $\int \frac{\ln x}{x}\,dx$ and $\int x\,dx$

97. $\int e^{\sin x}\cos x\,dx$ and $\int \frac{e^{\arcsin x}}{\sqrt{1-x^2}}\,dx$

98. $\int (\sin x)^3 \cos x\,dx$ and $\int (x^3+1)^3 x^2\,dx$

99. $\int \sqrt{x+1}\,dx$ and $\int \frac{\sqrt{1+\sqrt{x}}}{\sqrt{x}}\,dx$

■ In Problems 100–103, find an expression for the integral which contains g but no integral sign.

100. $\int g'(x)(g(x))^4\,dx$

101. $\int g'(x)e^{g(x)}\,dx$

102. $\int g'(x)\sin g(x)\,dx$

103. $\int g'(x)\sqrt{1+g(x)}\,dx$

■ In Problems 104–106, find a substitution w and constants k, n so that the integral has the form $\int kw^n\,dw$.

104. $\int x^2\sqrt{1-4x^3}\,dx$

105. $\int \frac{\cos t}{\sin t}\,dt$

106. $\int \frac{2x\,dx}{(x^2-3)^2}$

■ In Problems 107–109, find constants k, n, w_0, w_1 so the the integral has the form $\int_{w_0}^{w_1} kw^n\,dw$.

107. $\int_1^5 \frac{3x\,dx}{\sqrt{5x^2+7}}, \quad w = 5x^2+7$

108. $\int_0^5 \frac{2^x\,dx}{2^x+3}, \quad w = 2^x+3$

109. $\int_{\pi/12}^{\pi/4} \sin^7(2x)\cos(2x)\,dx, \quad w = \sin 2x$

■ In Problems 110–114, find a substitution w and a constant k so that the integral has the form $\int ke^w\,dw$.

110. $\int xe^{-x^2}\,dx$

111. $\int e^{\sin\phi}\cos\phi\,d\phi$

112. $\int \sqrt{e^r}\,dr$

113. $\int \frac{z^2\,dz}{e^{-z^3}}$

114. $\int e^{2t}e^{3t-4}\,dt$

■ In Problems 115–116, find a substitution w and constants a, b, A so that the integral has the form $\int_a^b Ae^w\,dw$.

115. $\int_3^7 e^{2t-3}\,dt$

116. $\int_0^1 e^{\cos(\pi t)}\sin(\pi t)\,dt$

117. Integrate:

(a) $\int \frac{1}{\sqrt{x}}\,dx$ (b) $\int \frac{1}{\sqrt{x+1}}\,dx$

(c) $\int \frac{1}{\sqrt{x}+1}\,dx$

118. If appropriate, evaluate the following integrals by substitution. If substitution is not appropriate, say so, and do not evaluate.

(a) $\int x\sin(x^2)\,dx$ (b) $\int x^2\sin x\,dx$

(c) $\int \frac{x^2}{1+x^2}\,dx$ (d) $\int \frac{x}{(1+x^2)^2}\,dx$

(e) $\int x^3e^{x^2}\,dx$ (f) $\int \frac{\sin x}{2+\cos x}\,dx$

■ In Problems 119–125, find the exact area.

119. Under $f(x) = xe^{x^2}$ between $x = 0$ and $x = 2$.

120. Under $f(x) = 1/(x+1)$ between $x = 0$ and $x = 2$.

121. Under $f(x) = \sinh(x/2)$ between $x = 0$ and $x = 2$.

122. Under $f(\theta) = (e^{\theta+1})^3$ for $0 \le \theta \le 2$.

123. Between e^t and e^{t+1} for $0 \le t \le 2$.

124. Between $y = e^x$, $y = 3$, and the y-axis.

125. Under one arch of the curve $V(t) = V_0\sin(\omega t)$, where $V_0 > 0$ and $\omega > 0$.

126. Find the area inside the asteroid in Figure 7.1 given that

$$\int_0^1 (1-x^{2/3})^{3/2}\,dx = \frac{3\pi}{32}.$$

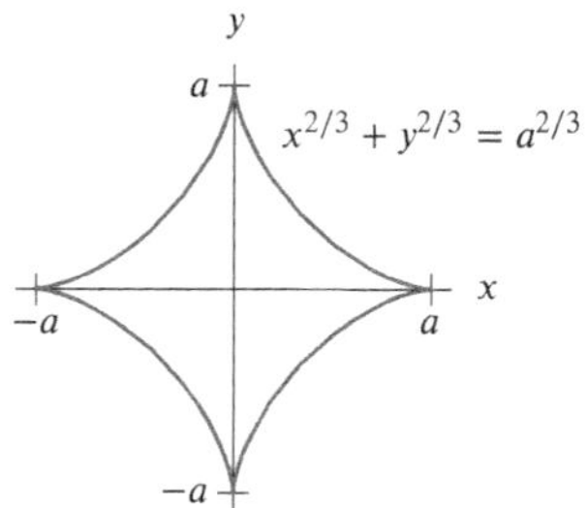

Figure 7.1

127. Find the exact average value of $f(x) = 1/(x+1)$ on the interval $x = 0$ to $x = 2$. Sketch a graph showing the function and the average value.

128. Let $g(x) = f(2x)$. Show that the average value of f on the interval $[0, 2b]$ is the same as the average value of g on the interval $[0, b]$.

129. Suppose $\int_0^2 g(t)\,dt = 5$. Calculate the following:

(a) $\int_0^4 g(t/2)\,dt$ (b) $\int_0^2 g(2-t)\,dt$

130. Suppose $\int_0^1 f(t)\,dt = 3$. Calculate the following:

(a) $\displaystyle\int_0^{0.5} f(2t)\,dt$ (b) $\displaystyle\int_0^1 f(1-t)\,dt$

(c) $\displaystyle\int_1^{1.5} f(3-2t)\,dt$

131. (a) Calculate exactly: $\int_{-\pi}^{\pi} \cos^2\theta \sin\theta\,d\theta$.
(b) Calculate the exact area under the curve $y = \cos^2\theta \sin\theta$ between $\theta = 0$ and $\theta = \pi$.

132. Find $\int 4x(x^2+1)\,dx$ using two methods:

(a) Do the multiplication first, and then antidifferentiate.
(b) Use the substitution $w = x^2 + 1$.
(c) Explain how the expressions from parts (a) and (b) are different. Are they both correct?

133. (a) Find $\int \sin\theta\cos\theta\,d\theta$.
(b) You probably solved part (a) by making the substitution $w = \sin\theta$ or $w = \cos\theta$. (If not, go back and do it that way.) Now find $\int \sin\theta\cos\theta\,d\theta$ by making the *other* substitution.
(c) There is yet another way of finding this integral which involves the trigonometric identities

$$\sin(2\theta) = 2\sin\theta\cos\theta$$
$$\cos(2\theta) = \cos^2\theta - \sin^2\theta.$$

Find $\int \sin\theta\cos\theta\,d\theta$ using one of these identities and then the substitution $w = 2\theta$.
(d) You should now have three different expressions for the indefinite integral $\int \sin\theta\cos\theta\,d\theta$. Are they really different? Are they all correct? Explain.

■ For Problems 134–137, find a substitution w and constants a, b, k so that the integral has the form $\int_a^b k f(w)\,dw$.

134. $\displaystyle\int_{-2}^{5} f\left(x^2\right) x\,dx$

135. $\displaystyle\int_0^{\pi} f(\cos x)\sin x\,dx$

136. $\displaystyle\int_1^{9} f\left(6x\sqrt{x}\right)\sqrt{x}\,dx$

137. $\displaystyle\int_2^{5} \frac{f\left(\ln\left(x^2+1\right)\right)x}{x^2+1}\,dx$

■ In Problems 138–141, given that $f(x) = F'(x)$, use Table 7.1 to evaluate the expression correct to three decimals places.

Table 7.1

x	0	0.5	1	1.5	2	2.5	3
$F(x)$	2	3	16	21	24	26	31

138. $\displaystyle\int_1^{\sqrt{3}} f\left(x^2\right) x\,dx$

139. $\displaystyle\int_{\pi/6}^{\pi/2} f(\sin x)\cos x\,dx$

140. $\displaystyle\int_0^{0.5} e^{F(x)} f(x)\,dx$

141. $\displaystyle\int_1^{3} \frac{f(x)}{F(x)}\,dx$

142. Find possible formulas for $f(x)$ and $g(x)$ given that

$$\int g(f(x))\,g(x)\,dx = \sin(\sin x) + C.$$

143. Find possible formulas for $f(x)$ and $g(x)$ given that

$$\int f(g(x))\,g(x)\cdot\frac{1}{\sqrt{x}}\,dx = \sin\left(e^{\sqrt{x}}\right) + C.$$

144. Find the solution of the initial value problem

$$y' = \tan x + 1, \quad y(0) = 1.$$

145. Let $I_{m,n} = \int_0^1 x^m(1-x)^n dx$ for constant m, n. Show that $I_{m,n} = I_{n,m}$.

146. Let $f(t)$ be the velocity in meters/second of a car at time t in seconds. Give an integral for the change in the position of the car

(a) Over the time interval $0 \le t \le 60$.
(b) In terms of time T in minutes, over the same time interval.

147. With t in years since 1790, the US population in millions can be approximated by

$$P = \frac{192}{1 + 48e^{-0.0317t}}.$$

Based on this model, what was the average US population between 1900 and 1950?

148. Over the past fifty years the carbon dioxide level in the atmosphere has increased. Carbon dioxide is believed to drive temperature, so predictions of future carbon dioxide levels are important. If $C(t)$ is carbon dioxide level in parts per million (ppm) and t is time in years since 1950, three possible models are:[1]

I $C'(t) = 1.3$
II $C'(t) = 0.5 + 0.03t$
III $C'(t) = 0.5e^{0.02t}$

(a) Given that the carbon dioxide level was 311 ppm in 1950, find $C(t)$ for each model.
(b) Find the carbon dioxide level in 2020 predicted by each model.

149. Let $f(t)$ be the rate of flow, in cubic meters per hour, of a flooding river at time t in hours. Give an integral for the total flow of the river

(a) Over the 3-day period $0 \le t \le 72$.
(b) In terms of time T in days over the same 3-day period.

[1]Based on data from www.esrl.noaa.gov/gmd/ccgg. Accessed March 13, 2013.

150. With t in years since April 20th, 2015, the population, P, of the world in billions[2] can be modeled by $P = 7.17e^{0.01064t}$.

(a) What does this model predict for the world population on April 20th, 2020? In 2025?
(b) Use the Fundamental Theorem to predict the average population of the world between April 2015 and April 2025.

151. Oil is leaking out of a ruptured tanker at the rate of $r(t) = 50e^{-0.02t}$ thousand liters per minute.

(a) At what rate, in liters per minute, is oil leaking out at $t = 0$? At $t = 60$?
(b) How many liters leak out during the first hour?

152. At the start of 2014, the world's known copper reserves were 690 million tons. With t in years since the start of 2014, copper has been mined at a rate given by $17.9e^{0.025t}$ million tons per year.[3]

(a) Assuming that copper continues to be extracted at the same rate, write an expression for the total quantity mined in the first T years after the start of 2014.
(b) Under these assumptions, when is the world predicted to run out of copper?

153. Throughout much of the 20th century, the yearly consumption of electricity in the US increased exponentially at a continuous rate of 7% per year. Assume this trend continues and that the electrical energy consumed in 1900 was 1.4 million megawatt-hours.

(a) Write an expression for yearly electricity consumption as a function of time, t, in years since 1900.
(b) Find the average yearly electrical consumption throughout the 20th century.
(c) During what year was electrical consumption closest to the average for the century?
(d) Without doing the calculation for part (c), how could you have predicted which half of the century the answer would be in?

154. An electric current, $I(t)$, flowing out of a capacitor, decays according to $I(t) = I_0 e^{-t}$, where t is time. Find the charge, $Q(t)$, remaining in the capacitor at time t. The initial charge is Q_0 and $Q(t)$ is related to $I(t)$ by

$$Q'(t) = -I(t).$$

155. **(a)** Between 2005 and 2015, ACME Widgets sold widgets at a continuous rate of $R = R_0 e^{0.125t}$ widgets per year, where t is time in years since January 1, 2005. Suppose they were selling widgets at a rate of 1000 per year on January 1, 2005. How many widgets did they sell between 2005 and 2015? How many did they sell if the rate on January 1, 2005 was 1,000,000 widgets per year?
(b) In the first case (1000 widgets per year on January 1, 2005), how long did it take for half the widgets in the ten-year period to be sold? In the second case (1,000,000 widgets per year on January 1, 2005), when had half the widgets in the ten-year period been sold?
(c) In 2015, ACME advertised that half the widgets it had sold in the previous ten years were still in use. Based on your answer to part (b), how long must a widget last in order to justify this claim?

156. The rate at which water is flowing into a tank is $r(t)$ gallons/minute, with t in minutes.

(a) Write an expression approximating the amount of water entering the tank during the interval from time t to time $t + \Delta t$, where Δt is small.
(b) Write a Riemann sum approximating the total amount of water entering the tank between $t = 0$ and $t = 5$. Write an exact expression for this amount.
(c) By how much has the amount of water in the tank changed between $t = 0$ and $t = 5$ if $r(t) = 20e^{0.02t}$?
(d) If $r(t)$ is as in part (c), and if the tank contains 3000 gallons initially, find a formula for $Q(t)$, the amount of water in the tank at time t.

Strengthen Your Understanding

■ In Problems 157–159, explain what is wrong with the statement.

157. $\int (f(x))^2\,dx = (f(x))^3/3 + C$.
158. $\int \cos(x^2)\,dx = \sin(x^2)/(2x) + C$.
159. $\int_0^{\pi/2} \cos(3x)\,dx = (1/3)\int_0^{\pi/2} \cos w\,dw$.

■ In Problems 160–161, give an example of:

160. A possible $f(\theta)$ so that the following integral can be integrated by substitution:

$$\int f(\theta)e^{\cos\theta}\,d\theta.$$

161. An indefinite integral involving $\sin(x^3 - 3x)$ that can be evaluated by substitution.

■ In Problems 162–164, decide whether the statements are true or false. Give an explanation for your answer.

162. $\int f'(x)\cos(f(x))\,dx = \sin(f(x)) + C$.
163. $\int (1/f(x))\,dx = \ln|f(x)| + C$.
164. $\int t\sin(5 - t^2)\,dt$ can be evaluated using substitution.

[2]www.indexmundi.com, accessed April 20, 2015.
[3]Data from http://minerals.usgs.gov/minerals/pubs/commodity/ Accessed February 8, 2015.

7.2 INTEGRATION BY PARTS

The method of substitution reverses the chain rule. Now we introduce *integration by parts*, which is based on the product rule.

Example 1 Find $\int xe^x\,dx$.

Solution We are looking for a function whose derivative is xe^x. The product rule might lead us to guess xe^x, because we know that the derivative has two terms, one of which is xe^x:

$$\frac{d}{dx}(xe^x) = \frac{d}{dx}(x)e^x + x\frac{d}{dx}(e^x) = e^x + xe^x.$$

Of course, our guess is wrong because of the extra e^x. But we can adjust our guess by subtracting e^x; this leads us to try $xe^x - e^x$. Let's check it:

$$\frac{d}{dx}(xe^x - e^x) = \frac{d}{dx}(xe^x) - \frac{d}{dx}(e^x) = e^x + xe^x - e^x = xe^x.$$

It works, so $\int xe^x\,dx = xe^x - e^x + C$.

Example 2 Find $\int \theta\cos\theta\,d\theta$.

Solution We guess the antiderivative is $\theta\sin\theta$ and use the product rule to check:

$$\frac{d}{d\theta}(\theta\sin\theta) = \frac{d(\theta)}{d\theta}\sin\theta + \theta\frac{d}{d\theta}(\sin\theta) = \sin\theta + \theta\cos\theta.$$

To correct for the extra $\sin\theta$ term, we must subtract from our original guess something whose derivative is $\sin\theta$. Since $\frac{d}{d\theta}(\cos\theta) = -\sin\theta$, we try:

$$\frac{d}{d\theta}(\theta\sin\theta + \cos\theta) = \frac{d}{d\theta}(\theta\sin\theta) + \frac{d}{d\theta}(\cos\theta) = \sin\theta + \theta\cos\theta - \sin\theta = \theta\cos\theta.$$

Thus, $\int \theta\cos\theta\,d\theta = \theta\sin\theta + \cos\theta + C$.

The General Formula for Integration by Parts

We can formalize the process illustrated in the last two examples in the following way. We begin with the product rule:

$$\frac{d}{dx}(uv) = u'v + uv'$$

where u and v are functions of x with derivatives u' and v', respectively. We rewrite this as:

$$uv' = \frac{d}{dx}(uv) - u'v$$

and then integrate both sides:

$$\int uv'\,dx = \int \frac{d}{dx}(uv)\,dx - \int u'v\,dx.$$

Since an antiderivative of $\frac{d}{dx}(uv)$ is just uv, we get the following formula:

Integration by Parts

$$\int uv' \, dx = uv - \int u'v \, dx.$$

This formula is useful when the integrand can be viewed as a product and when the integral on the right-hand side is simpler than that on the left. In effect, we were using integration by parts in the previous two examples. In Example 1, we let $xe^x = (x) \cdot (e^x) = uv'$, and choose $u = x$ and $v' = e^x$. Thus, $u' = 1$ and $v = e^x$, so

$$\int \underbrace{(x)}_{u} \underbrace{(e^x)}_{v'} \, dx = \underbrace{(x)}_{u} \underbrace{(e^x)}_{v} - \int \underbrace{(1)}_{u'} \underbrace{(e^x)}_{v} \, dx = xe^x - e^x + C.$$

So uv represents our first guess, and $\int u'v \, dx$ represents the correction to our guess.

Notice what would have happened if, instead of $v = e^x$, we took $v = e^x + C_1$. Then

$$\begin{aligned}\int xe^x \, dx &= x(e^x + C_1) - \int (e^x + C_1) \, dx \\ &= xe^x + C_1x - e^x - C_1x + C \\ &= xe^x - e^x + C,\end{aligned}$$

as before. Thus, it is not necessary to include an arbitrary constant in the antiderivative for v; any antiderivative will do.

What would have happened if we had picked u and v' the other way around? If $u = e^x$ and $v' = x$, then $u' = e^x$ and $v = x^2/2$. The formula for integration by parts then gives

$$\int xe^x \, dx = \frac{x^2}{2}e^x - \int \frac{x^2}{2} \cdot e^x \, dx,$$

which is true but not helpful since the integral on the right is worse than the one on the left. To use this method, we must choose u and v' to make the integral on the right easier to find than the integral on the left.

How to Choose u and v'

- Whatever you let v' be, you need to be able to find v.
- It helps if u' is simpler than u (or at least no more complicated than u).
- It helps if v is simpler than v' (or at least no more complicated than v').

If we pick $v' = x$ in Example 1, then $v = x^2/2$, which is certainly "worse" than v'.

There are some examples which don't look like good candidates for integration by parts because they don't appear to involve products, but for which the method works well. Such examples often involve $\ln x$ or the inverse trigonometric functions. Here is one:

Example 3 Find $\int_2^3 \ln x \, dx$.

Solution This does not look like a product unless we write $\ln x = (1)(\ln x)$. Then we might say $u = 1$ so $u' = 0$, which certainly makes things simpler. But if $v' = \ln x$, what is v? If we knew, we would not need integration by parts. Let's try the other way: if $u = \ln x$, $u' = 1/x$ and if $v' = 1$, $v = x$, so

$$\int_2^3 \underbrace{(\ln x)}_{u} \underbrace{(1)}_{v'} \, dx = \underbrace{(\ln x)}_{u} \underbrace{(x)}_{v} \Big|_2^3 - \int_2^3 \underbrace{\left(\frac{1}{x}\right)}_{u'} \cdot \underbrace{(x)}_{v} \, dx$$

$$= x \ln x \Big|_2^3 - \int_2^3 1 \, dx = (x \ln x - x) \Big|_2^3$$

$$= 3 \ln 3 - 3 - 2 \ln 2 + 2 = 3 \ln 3 - 2 \ln 2 - 1.$$

Notice that when doing a definite integral by parts, we must remember to put the limits of integration (here 2 and 3) on the uv term (in this case $x \ln x$) as well as on the integral $\int u'v \, dx$.

Example 4 Find $\displaystyle\int x^6 \ln x \, dx$.

Solution View $x^6 \ln x$ as uv' where $u = \ln x$ and $v' = x^6$. Then $v = \frac{1}{7}x^7$ and $u' = 1/x$, so integration by parts gives us:

$$\int x^6 \ln x \, dx = \int (\ln x) x^6 \, dx = (\ln x)\left(\frac{1}{7}x^7\right) - \int \frac{1}{7}x^7 \cdot \frac{1}{x} \, dx$$

$$= \frac{1}{7}x^7 \ln x - \frac{1}{7}\int x^6 \, dx$$

$$= \frac{1}{7}x^7 \ln x - \frac{1}{49}x^7 + C.$$

In Example 4 we did not choose $v' = \ln x$, because it is not immediately clear what v would be. In fact, we used integration by parts in Example 3 to find the antiderivative of $\ln x$. Also, using $u = \ln x$, as we have done, gives $u' = 1/x$, which can be considered simpler than $u = \ln x$. This shows that u does not have to be the first factor in the integrand (here x^6).

Example 5 Find $\displaystyle\int x^2 \sin 4x \, dx$.

Solution If we let $v' = \sin 4x$, then $v = -\frac{1}{4}\cos 4x$, which is no worse than v'. Also letting $u = x^2$, we get $u' = 2x$, which is simpler than $u = x^2$. Using integration by parts:

$$\int x^2 \sin 4x \, dx = x^2\left(-\frac{1}{4}\cos 4x\right) - \int 2x\left(-\frac{1}{4}\cos 4x\right) dx$$

$$= -\frac{1}{4}x^2 \cos 4x + \frac{1}{2}\int x \cos 4x \, dx.$$

The trouble is we still have to grapple with $\int x \cos 4x \, dx$. This can be done by using integration by parts again with a new u and v, namely $u = x$ and $v' = \cos 4x$:

$$\int x \cos 4x \, dx = x\left(\frac{1}{4}\sin 4x\right) - \int 1 \cdot \frac{1}{4}\sin 4x \, dx$$

$$= \frac{1}{4}x \sin 4x - \frac{1}{4} \cdot \left(-\frac{1}{4}\cos 4x\right) + C$$

$$= \frac{1}{4}x \sin 4x + \frac{1}{16}\cos 4x + C.$$

Thus,

$$\begin{aligned}\int x^2 \sin 4x\,dx &= -\frac{1}{4}x^2\cos 4x + \frac{1}{2}\int x\cos 4x\,dx\\ &= -\frac{1}{4}x^2\cos 4x + \frac{1}{2}\left(\frac{1}{4}x\sin 4x + \frac{1}{16}\cos 4x + C\right)\\ &= -\frac{1}{4}x^2\cos 4x + \frac{1}{8}x\sin 4x + \frac{1}{32}\cos 4x + C.\end{aligned}$$

Notice that, in this example, each time we used integration by parts, the exponent of x went down by 1. In addition, when the arbitrary constant C is multiplied by $\frac{1}{2}$, it is still represented by C.

Example 6 Find $\int \cos^2\theta\,d\theta$.

Solution Using integration by parts with $u = \cos\theta$, $v' = \cos\theta$ gives $u' = -\sin\theta$, $v = \sin\theta$, so we get

$$\int \cos^2\theta\,d\theta = \cos\theta\sin\theta + \int \sin^2\theta\,d\theta.$$

Substituting $\sin^2\theta = 1 - \cos^2\theta$ leads to

$$\begin{aligned}\int \cos^2\theta\,d\theta &= \cos\theta\sin\theta + \int (1-\cos^2\theta)\,d\theta\\ &= \cos\theta\sin\theta + \int 1\,d\theta - \int \cos^2\theta\,d\theta.\end{aligned}$$

Looking at the right side, we see that the original integral has reappeared. If we move it to the left, we get

$$2\int \cos^2\theta\,d\theta = \cos\theta\sin\theta + \int 1\,d\theta = \cos\theta\sin\theta + \theta + C.$$

Dividing by 2 gives

$$\int \cos^2\theta\,d\theta = \frac{1}{2}\cos\theta\sin\theta + \frac{1}{2}\theta + C.$$

Problem 59 asks you to do this integral by another method.

The previous example illustrates a useful technique: Use integration by parts to transform the integral into an expression containing another copy of the same integral, possibly multiplied by a coefficient, then solve for the original integral.

Example 7 Use integration by parts twice to find $\int e^{2x}\sin(3x)\,dx$.

Solution Using integration by parts with $u = e^{2x}$ and $v' = \sin(3x)$ gives $u' = 2e^{2x}$, $v = -\frac{1}{3}\cos(3x)$, so we get

$$\int e^{2x}\sin(3x)\,dx = -\frac{1}{3}e^{2x}\cos(3x) + \frac{2}{3}\int e^{2x}\cos(3x)\,dx.$$

On the right side we have an integral similar to the original one, with the sine replaced by a cosine. Using integration by parts on that integral in the same way gives

$$\int e^{2x}\cos(3x)\,dx = \frac{1}{3}e^{2x}\sin(3x) - \frac{2}{3}\int e^{2x}\sin(3x)\,dx.$$

Substituting this into the expression we obtained for the original integral gives

$$\begin{aligned}\int e^{2x}\sin(3x)\,dx &= -\frac{1}{3}e^{2x}\cos(3x)+\frac{2}{3}\left(\frac{1}{3}e^{2x}\sin(3x)-\frac{2}{3}\int e^{2x}\sin(3x)\,dx\right)\\ &= -\frac{1}{3}e^{2x}\cos(3x)+\frac{2}{9}e^{2x}\sin(3x)-\frac{4}{9}\int e^{2x}\sin(3x)\,dx.\end{aligned}$$

The right side now has a copy of the original integral, multiplied by $-4/9$. Moving it to the left, we get

$$\left(1+\frac{4}{9}\right)\int e^{2x}\sin(3x)\,dx = -\frac{1}{3}e^{2x}\cos(3x)+\frac{2}{9}e^{2x}\sin(3x).$$

Dividing through by the coefficient on the left, $(1+4/9)=13/9$ and adding a constant of integration C, we get

$$\begin{aligned}\int e^{2x}\sin(3x)\,dx &= \frac{9}{13}\left(-\frac{1}{3}e^{2x}\cos(3x)+\frac{2}{9}e^{2x}\sin(3x)\right)+C\\ &= \frac{1}{13}e^{2x}\left(2\sin(3x)-3\cos(3x)\right)+C.\end{aligned}$$

Example 8 Use a computer algebra system to investigate $\int \sin(x^2)\,dx$.

Solution It can be shown that $\sin(x^2)$ has no elementary antiderivative. A computer algebra system gives an antiderivative involving a non-elementary function, the Fresnel Integral, which you may not recognize.

Exercises and Problems for Section 7.2

Online Resource: Additional Problems for Section 7.2

EXERCISES

1. Use integration by parts to express $\int x^2e^x\,dx$ in terms of

   (a) $\int x^3e^x\,dx$ **(b)** $\int xe^x\,dx$

2. Write $\arctan x = 1\cdot\arctan x$ to find $\int \arctan x\,dx$.

■ Find the integrals in Exercises 3–38.

3. $\int t\sin t\,dt$
4. $\int t^2\sin t\,dt$
5. $\int te^{5t}\,dt$
6. $\int t^2e^{5t}\,dt$
7. $\int pe^{-0.1p}\,dp$
8. $\int (z+1)e^{2z}\,dz$
9. $\int x\ln x\,dx$
10. $\int x^3\ln x\,dx$
11. $\int q^5\ln 5q\,dq$
12. $\int \theta^2\cos 3\theta\,d\theta$
13. $\int \sin^2\theta\,d\theta$
14. $\int \cos^2(3\alpha+1)\,d\alpha$
15. $\int (\ln t)^2\,dt$
16. $\int \ln(x^2)\,dx$
17. $\int y\sqrt{y+3}\,dy$
18. $\int (t+2)\sqrt{2+3t}\,dt$
19. $\int (\theta+1)\sin(\theta+1)\,d\theta$
20. $\int \frac{z}{e^z}\,dz$
21. $\int \frac{\ln x}{x^2}\,dx$
22. $\int \frac{y}{\sqrt{5-y}}\,dy$
23. $\int \frac{t+7}{\sqrt{5-t}}\,dt$
24. $\int x(\ln x)^4\,dx$
25. $\int \sqrt{x}\ln x\,dx$
26. $\int y\sqrt{1-y}\,dy$
27. $\int r(\ln r)^2\,dr$
28. $\int \arcsin w\,dw$
29. $\int \arctan 7z\,dz$
30. $\int x\arctan x^2\,dx$

31. $\int x^3 e^{x^2}\,dx$

32. $\int x^5 \cos x^3\,dx$

33. $\int x \sinh x\,dx$

34. $\int (x-1)\cosh x\,dx$

35. $\int e^{\sqrt{x}}\,dx$

36. $\int \theta^5 \cos\theta^3\,d\theta$

37. $\int (2x+1)^2 \ln(2x+1)\,dx$

38. $\int \frac{e^{1/x}}{x^3}\,dx$

■ Evaluate the integrals in Exercises 39–46 both exactly [e.g. $\ln(3\pi)$] and numerically [e.g. $\ln(3\pi) \approx 2.243$].

39. $\int_1^5 \ln t\,dt$

40. $\int_3^5 x\cos x\,dx$

41. $\int_0^{10} ze^{-z}\,dz$

42. $\int_1^3 t\ln t\,dt$

43. $\int_0^1 \arctan y\,dy$

44. $\int_0^5 \ln(1+t)\,dt$

45. $\int_0^1 \arcsin z\,dz$

46. $\int_0^1 u \arcsin u^2\,du$

47. For each of the following integrals, indicate whether integration by substitution or integration by parts is more appropriate. Do not evaluate the integrals.

(a) $\int x\sin x\,dx$ **(b)** $\int \frac{x^2}{1+x^3}\,dx$

(c) $\int xe^{x^2}\,dx$ **(d)** $\int x^2\cos(x^3)\,dx$

(e) $\int \frac{1}{\sqrt{3x+1}}\,dx$ **(f)** $\int x^2\sin x\,dx$

(g) $\int \ln x\,dx$

48. Find $\int_1^2 \ln x\,dx$ numerically. Find $\int_1^2 \ln x\,dx$ using antiderivatives. Check that your answers agree.

PROBLEMS

■ In Problems 49–51, using properties of ln, find a substitution w and constant k so that the integral has the form $\int k\ln w\,dw$.

49. $\int \ln\left((5-3x)^2\right)\,dx$

50. $\int \ln\left(\frac{1}{\sqrt{4-5x}}\right)\,dx$

51. $\int \frac{\ln\left((\ln x)^3\right)}{x}\,dx$

■ In Problems 52–57, find the exact area.

52. Under $y = te^{-t}$ for $0 \le t \le 2$.

53. Under $f(z) = \arctan z$ for $0 \le z \le 2$.

54. Under $f(y) = \arcsin y$ for $0 \le y \le 1$.

55. Between $y = \ln x$ and $y = \ln(x^2)$ for $1 \le x \le 2$.

56. Between $f(t) = \ln(t^2-1)$ and $g(t) = \ln(t-1)$ for $2 \le t \le 3$.

57. Under the first arch of $f(x) = x\sin x$.

58. In Exercise 13, you evaluated $\int \sin^2\theta\,d\theta$ using integration by parts. (If you did not do it by parts, do so now!) Redo this integral using the identity $\sin^2\theta = (1-\cos 2\theta)/2$. Explain any differences in the form of the answer obtained by the two methods.

59. Compute $\int \cos^2\theta\,d\theta$ in two different ways and explain any differences in the form of your answers. (The identity $\cos^2\theta = (1+\cos 2\theta)/2$ may be useful.)

60. Use integration by parts twice to find $\int e^x \sin x\,dx$.

61. Use integration by parts twice to find $\int e^\theta \cos\theta\,d\theta$.

62. Use the results from Problems 60 and 61 and integration by parts to find $\int xe^x \sin x\,dx$.

63. Use the results from Problems 60 and 61 and integration by parts to find $\int \theta e^\theta \cos\theta\,d\theta$.

64. If f is a twice differentiable function, find

$$\int f''(x)\ln x\,dx + \int \frac{f(x)}{x^2}\,dx$$

(Your answer should contain f, but no integrals.)

65. If f is a twice differentiable function, find $\int xf''(x)\,dx$. (Your answer should contain f, but no integrals.)

■ In Problems 66–69, derive the given formulas.

66. $\int x^n e^x\,dx = x^n e^x - n\int x^{n-1}e^x\,dx$

67. $\int x^n \cos ax\,dx = \frac{1}{a}x^n \sin ax - \frac{n}{a}\int x^{n-1}\sin ax\,dx$

68. $\int x^n \sin ax\,dx = -\frac{1}{a}x^n \cos ax + \frac{n}{a}\int x^{n-1}\cos ax\,dx$

69. $\int \cos^n x\,dx = \frac{1}{n}\cos^{n-1}x\sin x + \frac{n-1}{n}\int \cos^{n-2}x\,dx$

70. Integrating $e^{ax}\sin bx$ by parts twice gives

$$\int e^{ax}\sin bx\,dx = e^{ax}(A\sin bx + B\cos bx) + C.$$

(a) Find the constants A and B in terms of a and b. [Hint: Don't actually perform the integration.]

(b) Evaluate $\int e^{ax}\cos bx\,dx$ by modifying the method in part (a). [Again, do not perform the integration.]

71. Use the table with $f(x) = F'(x)$ to find $\int_0^5 xf'(x)\,dx$.

x	0	1	2	3	4	5
$f(x)$	2	−5	−6	−1	10	27
$F(x)$	10	8	2	−2	2	20

72. Find $\int_2^3 f(x)g'(x)\,dx$, given that $\int_2^3 f'(x)g(x)\,dx = 1.3$ and the values in the table:

x	$f(x)$	$f'(x)$	$g(x)$	$g'(x)$
2	5	−1	0.2	3
3	7	−2	0.1	2

73. Find possible formulas for $f(x)$ and $g(x)$ given that

$$\int x^3 g'(x)\,dx = f(x)g(x) - \int x^2 \cos x\,dx.$$

74. Let f be a function with $f(0) = 6$, $f(1) = 5$, and $f'(1) = 2$. Evaluate the integral $\int_0^1 xf''(x)\,dx$.

75. Given $h(x) = f(x)\sqrt{x}$ and $g'(x) = f(x)/\sqrt{x}$, rewrite in terms of $h(x)$ and $g(x)$:

$$\int f'(x)\sqrt{x}\,dx.$$

Your answer should not include integrals, $f(x), h'(x)$, or $g'(x)$.

76. Given that $f(7) = 0$ and $\int_0^7 f(x)\,dx = 5$, evaluate

$$\int_0^7 xf'(x)\,dx.$$

77. Let $F(a)$ be the area under the graph of $y = x^2e^{-x}$ between $x = 0$ and $x = a$, for $a > 0$.

(a) Find a formula for $F(a)$.
(b) Is F an increasing or decreasing function?
(c) Is F concave up or concave down for $0 < a < 2$?

78. The concentration, C, in ng/ml, of a drug in the blood as a function of the time, t, in hours since the drug was administered is given by $C = 15te^{-0.2t}$. The area under the concentration curve is a measure of the overall effect of the drug on the body, called the bioavailability. Find the bioavailability of the drug between $t = 0$ and $t = 3$.

79. Given $h(x) = f(x)\ln|x|$ and $g'(x) = \dfrac{f(x)}{x}$, rewrite $\int f'(x)\ln|x|\,dx$. in terms of $h(x)$ and $g(x)$.

80. The *error function*, $\text{erf}(x)$, is defined by

$$\text{erf}(x) = \frac{2}{\sqrt{\pi}}\int_0^x e^{-t^2}\,dt.$$

(a) Let $u = \text{erf}(x)$. Use integration by parts to write $\int \text{erf}(x)\,dx = uv - \int v\,u'\,dx$. Give u' and v'.
(b) Evaluate the integral $\int v\,u'\,dx$ from part (a) by making a substitution w. Give the values of w and dw.
(c) Use your answers to parts (a) and (b) to find $\int \text{erf}(x)\,dx$. Your answer may involve $\text{erf}(x)$.

81. The *Eulerian logarithmic integral* $\text{Li}(x)$ is defined[4] as $\text{Li}(x) = \int_2^x \frac{1}{\ln t}\,dt$. Letting $u = \text{Li}(x)$ and $v = \ln x$, use integration by parts to evaluate $\int \text{Li}(x)x^{-1}\,dx$. Your answer will involve $\text{Li}(x)$.

Strengthen Your Understanding

In Problems 82–84, explain what is wrong with the statement.

82. To integrate $\int t\ln t\,dt$ by parts, use $u = t$, $v' = \ln t$.

83. The integral $\int \arctan x\,dx$ cannot be evaluated using integration by parts since the integrand is not a product of two functions.

84. Using integration by parts, we can show that

$$\int f(x)\,dx = xf'(x) - \int xf'(x)\,dx.$$

In Problems 85–87, give an example of:

85. An integral using only powers of θ and $\sin\theta$ which can be evaluated using integration by parts twice.

86. An integral that requires three applications of integration by parts.

87. An integral of the form $\int f(x)g(x)\,dx$ that can be evaluated using integration by parts either with $u = f(x)$ or with $u = g(x)$.

In Problems 88–90, decide whether the statements are true or false. Give an explanation for your answer.

88. $\int t\sin(5-t)\,dt$ can be evaluated by parts.

89. The integral $\int t^2e^{3-t}\,dt$ can be done by parts.

90. When integrating by parts, it does not matter which factor we choose for u.

[4]http://en.wikipedia.org/wiki/Logarithmic_integral_function#Offset_logarithmic_integral, accessed February 17, 2011.

7.3 TABLES OF INTEGRALS

Today, many integrals are done using a CAS. Traditionally, the antiderivatives of commonly used functions were compiled in a table, such as the one in the back of this book. Other tables include *CRC Standard Mathematical Tables* (Boca Raton, Fl: CRC Press). The key to using these tables is being able to recognize the general class of function that you are trying to integrate, so you can know in what section of the table to look.

Warning: This section involves long division of polynomials and completing the square. You may want to review these topics!

Using the Table of Integrals

Part I of the table inside the back cover gives the antiderivatives of the basic functions x^n, a^x, $\ln x$, $\sin x$, $\cos x$, and $\tan x$. (The antiderivative for $\ln x$ is found using integration by parts and is a special case of the more general formula III-13.) Most of these are already familiar.

Part II of the table contains antiderivatives of functions involving products of e^x, $\sin x$, and $\cos x$. All of these antiderivatives were obtained using integration by parts.

Example 1 Find $\int \sin 7z \sin 3z\, dz$.

Solution Since the integrand is the product of two sines, we should use II-10 in the table,

$$\int \sin 7z \sin 3z\, dz = -\frac{1}{40}(7\cos 7z \sin 3z - 3\cos 3z \sin 7z) + C.$$

Part III of the table contains antiderivatives for products of a polynomial and e^x, $\sin x$, or $\cos x$. It also has an antiderivative for $x^n \ln x$, which can easily be used to find the antiderivatives of the product of a general polynomial and $\ln x$. Each *reduction formula* is used repeatedly to reduce the degree of the polynomial until a zero-degree polynomial is obtained.

Example 2 Find $\int (x^5 + 2x^3 - 8)e^{3x}\, dx$.

Solution Since $p(x) = x^5 + 2x^3 - 8$ is a polynomial multiplied by e^{3x}, this is of the form in III-14. Now $p'(x) = 5x^4 + 6x^2$ and $p''(x) = 20x^3 + 12x$, and so on, giving

$$\int (x^5 + 2x^3 - 8)e^{3x}\, dx = e^{3x}\left(\frac{1}{3}(x^5 + 2x^3 - 8) - \frac{1}{9}(5x^4 + 6x^2) + \frac{1}{27}(20x^3 + 12x)\right.$$
$$\left. - \frac{1}{81}(60x^2 + 12) + \frac{1}{243}(120x) - \frac{1}{729}\cdot 120\right) + C.$$

Here we have the successive derivatives of the original polynomial $x^5 + 2x^3 - 8$, occurring with alternating signs and multiplied by successive powers of 1/3.

Part IV of the table contains reduction formulas for the antiderivatives of $\cos^n x$ and $\sin^n x$, which can be obtained by integration by parts. When n is a positive integer, formulas IV-17 and IV-18 can be used repeatedly to reduce the power n until it is 0 or 1.

Example 3 Find $\int \sin^6 \theta\, d\theta$.

Solution Use IV-17 repeatedly:

$$\int \sin^6\theta\,d\theta = -\frac{1}{6}\sin^5\theta\cos\theta + \frac{5}{6}\int \sin^4\theta\,d\theta$$

$$\int \sin^4\theta\,d\theta = -\frac{1}{4}\sin^3\theta\cos\theta + \frac{3}{4}\int \sin^2\theta\,d\theta$$

$$\int \sin^2\theta\,d\theta = -\frac{1}{2}\sin\theta\cos\theta + \frac{1}{2}\int 1\,d\theta.$$

Calculate $\int \sin^2\theta\,d\theta$ first, and use this to find $\int \sin^4\theta\,d\theta$; then calculate $\int \sin^6\theta\,d\theta$. Putting this all together, we get

$$\int \sin^6\theta\,d\theta = -\frac{1}{6}\sin^5\theta\cos\theta - \frac{5}{24}\sin^3\theta\cos\theta - \frac{15}{48}\sin\theta\cos\theta + \frac{15}{48}\theta + C.$$

The last item in **Part IV** of the table is not a formula: it is advice on how to antidifferentiate products of integer powers of $\sin x$ and $\cos x$. There are various techniques to choose from, depending on the nature (odd or even, positive or negative) of the exponents.

Example 4 Find $\displaystyle\int \cos^3 t\sin^4 t\,dt$.

Solution Here the exponent of $\cos t$ is odd, so IV-23 recommends making the substitution $w = \sin t$. Then $dw = \cos t\,dt$. To make this work, we'll have to separate off one of the cosines to be part of dw. Also, the remaining even power of $\cos t$ can be rewritten in terms of $\sin t$ by using $\cos^2 t = 1-\sin^2 t = 1 - w^2$, so that

$$\begin{aligned}\int \cos^3 t\sin^4 t\,dt &= \int \cos^2 t\sin^4 t\cos t\,dt \\ &= \int (1-w^2)w^4\,dw = \int (w^4 - w^6)\,dw \\ &= \frac{1}{5}w^5 - \frac{1}{7}w^7 + C = \frac{1}{5}\sin^5 t - \frac{1}{7}\sin^7 t + C.\end{aligned}$$

Example 5 Find $\displaystyle\int \cos^2 x\sin^4 x\,dx$.

Solution In this example, both exponents are even. The advice given in IV-23 is to convert to all sines or all cosines. We'll convert to all sines by substituting $\cos^2 x = 1 - \sin^2 x$, and then we'll multiply out the integrand:

$$\int \cos^2 x\sin^4 x\,dx = \int (1-\sin^2 x)\sin^4 x\,dx = \int \sin^4 x\,dx - \int \sin^6 x\,dx.$$

In Example 3 we found $\int \sin^4 x\,dx$ and $\int \sin^6 x\,dx$. Put them together to get

$$\begin{aligned}\int \cos^2 x\sin^4 x\,dx &= -\frac{1}{4}\sin^3 x\cos x - \frac{3}{8}\sin x\cos x + \frac{3}{8}x \\ &\quad -\left(-\frac{1}{6}\sin^5 x\cos x - \frac{5}{24}\sin^3 x\cos x - \frac{15}{48}\sin x\cos x + \frac{15}{48}x\right) + C \\ &= \frac{1}{6}\sin^5 x\cos x - \frac{1}{24}\sin^3 x\cos x - \frac{3}{48}\sin x\cos x + \frac{3}{48}x + C.\end{aligned}$$

The last two parts of the table are concerned with quadratic functions: **Part V** has expressions with quadratic denominators; **Part VI** contains square roots of quadratics. The quadratics that appear in these formulas are of the form $x^2 \pm a^2$ or $a^2 - x^2$, or in factored form $(x-a)(x-b)$, where a and b are different constants. Quadratics can be converted to these forms by factoring or completing the square.

Preparing to Use the Table: Transforming the Integrand

To use the integral table, we often need to manipulate or reshape integrands to fit entries in the table. The manipulations that tend to be useful are factoring, long division, completing the square, and substitution.

Factoring

Example 6 Find $\int \frac{3x+7}{x^2+6x+8}\,dx$.

Solution In this case we factor the denominator to get it into a form in the table:

$$x^2+6x+8=(x+2)(x+4).$$

Now in V-27 we let $a=-2$, $b=-4$, $c=3$, and $d=7$, to obtain

$$\int \frac{3x+7}{x^2+6x+8}\,dx=\frac{1}{2}(\ln|x+2|-(-5)\ln|x+4|)+C.$$

Long Division

Example 7 Find $\int \frac{x^2}{x^2+4}\,dx$.

Solution A good rule of thumb when integrating a rational function whose numerator has a degree greater than or equal to that of the denominator is to start by doing *long division*. This results in a polynomial plus a simpler rational function as a remainder. Performing long division here, we obtain:

$$\frac{x^2}{x^2+4}=1-\frac{4}{x^2+4}.$$

Then, by V-24 with $a=2$, we obtain:

$$\int \frac{x^2}{x^2+4}\,dx=\int 1\,dx-4\int \frac{1}{x^2+4}\,dx=x-4\cdot\frac{1}{2}\arctan\frac{x}{2}+C.$$

Completing the Square to Rewrite the Quadratic in the Form w^2+a^2

Example 8 Find $\int \frac{1}{x^2+6x+14}\,dx$.

Solution By completing the square, we can get this integrand into a form in the table:

$$\begin{aligned}x^2+6x+14&=(x^2+6x+9)-9+14\\&=(x+3)^2+5.\end{aligned}$$

Let $w = x + 3$. Then $dw = dx$ and so the substitution gives

$$\int \frac{1}{x^2 + 6x + 14}\,dx = \int \frac{1}{w^2 + 5}\,dw = \frac{1}{\sqrt{5}}\arctan\frac{w}{\sqrt{5}} + C = \frac{1}{\sqrt{5}}\arctan\frac{x+3}{\sqrt{5}} + C,$$

where the antidifferentiation uses V-24 with $a^2 = 5$.

Substitution

Getting an integrand into the right form to use a table of integrals involves substitution and a variety of algebraic techniques.

Example 9 Find $\int e^t \sin(5t + 7)\,dt$.

Solution This looks similar to II-8. To make the correspondence more complete, let's try the substitution $w = 5t + 7$. Then $dw = 5\,dt$, so $dt = \frac{1}{5}\,dw$. Also, $t = (w - 7)/5$. Then the integral becomes

$$\begin{aligned}\int e^t \sin(5t+7)\,dt &= \int e^{(w-7)/5}\sin w\,\frac{dw}{5} \\ &= \frac{e^{-7/5}}{5}\int e^{w/5}\sin w\,dw. \qquad \text{(Since } e^{(w-7)/5} = e^{w/5}e^{-7/5} \text{ and } e^{-7/5} \text{ is a constant)}\end{aligned}$$

Now we can use II-8 with $a = \frac{1}{5}$ and $b = 1$ to write

$$\int e^{w/5}\sin w\,dw = \frac{1}{(1/5)^2 + 1^2}e^{w/5}\left(\frac{\sin w}{5} - \cos w\right) + C,$$

so

$$\begin{aligned}\int e^t \sin(5t+7)\,dt &= \frac{e^{-7/5}}{5}\left(\frac{25}{26}e^{(5t+7)/5}\left(\frac{\sin(5t+7)}{5} - \cos(5t+7)\right)\right) + C \\ &= \frac{5e^t}{26}\left(\frac{\sin(5t+7)}{5} - \cos(5t+7)\right) + C.\end{aligned}$$

Example 10 Find a substitution w and constants k, n so that the following integral has the form $\int kw^n \ln w\,dw$, found in III-13:

$$\int \frac{\ln(x+1) + \ln(x-1)}{\sqrt{x^2 - 1}}x\,dx$$

Solution First we use properties of ln to simplify the integral:

$$\int \frac{\ln(x+1) + \ln(x-1)}{\sqrt{x^2-1}}x\,dx = \int \frac{\ln((x+1)(x-1))}{\sqrt{x^2-1}}x\,dx = \int \frac{\ln(x^2-1)}{\sqrt{x^2-1}}x\,dx.$$

Let $w = x^2 - 1$, $dw = 2x\,dx$, so that $x\,dx = (1/2)\,dw$. Then

$$\int \frac{\ln(x^2-1)}{\sqrt{x^2-1}}x\,dx = \int w^{-1/2}\ln w\,\frac{1}{2}dw = \int \frac{1}{2}w^{-1/2}\ln w\,dw,$$

so $k = 1/2, n = -1/2$.

Exercises and Problems for Section 7.3 Online Resource: Additional Problems for Section 7.3

EXERCISES

For Exercises 1–14, say which formula, if any, to apply from the table of integrals. Give the values of any constants.

1. $\int x^{1/2}e^{3x}\,dx$
2. $\int \cos 3x \sin 4x\,dx$
3. $\int \sin^4 x\,dx$
4. $\int e^{5x}\sin 2x\,dx$
5. $\int e^x \ln x\,dx$
6. $\int \ln x\,dx$
7. $\int \frac{1}{x^2-3x-4}\,dx$
8. $\int \frac{\cos x}{x}\,dx$
9. $\int \sqrt{x^2-9}\,dx$
10. $\int \sqrt{x^3+8}\,dx$
11. $\int \sin e^x\,dx$
12. $\int x^2 \sin 2x\,dx$
13. $\int \frac{4x-2}{x^2+9}\,dx$
14. $\int \frac{4x-2}{x^2-9}\,dx$

For Exercises 15–54, antidifferentiate using the table of integrals. You may need to transform the integrand first.

15. $\int x^5 \ln x\,dx$
16. $\int e^{-3\theta}\cos\theta\,d\theta$
17. $\int x^3 \sin 5x\,dx$
18. $\int (x^2+3)\ln x\,dx.$
19. $\int (x^3+5)^2\,dx.$
20. $\int \sin w \cos^4 w\,dw$
21. $\int \sin^4 x\,dx$
22. $\int x^3 e^{2x}\,dx$
23. $\int x^2 e^{3x}\,dx$
24. $\int x^2 e^{x^3}\,dx$
25. $\int x^4 e^{3x}\,dx$
26. $\int u^5 \ln(5u)\,du$
27. $\int \frac{1}{3+y^2}\,dy$
28. $\int \frac{dx}{9x^2+16}$
29. $\int \frac{dx}{\sqrt{25-16x^2}}$
30. $\int \frac{dx}{\sqrt{9x^2+25}}$
31. $\int \sin 3\theta \cos 5\theta\,d\theta$
32. $\int \sin 3\theta \sin 5\theta\,d\theta$
33. $\int \frac{1}{\cos^3 x}\,dx$
34. $\int \frac{t^2+1}{t^2-1}\,dt$
35. $\int e^{5x}\sin 3x\,dx$
36. $\int \cos 2y \cos 7y\,dy$
37. $\int y^2 \sin 2y\,dy$
38. $\int x^3 \sin x^2\,dx$
39. $\int \frac{1}{\cos^4 7x}\,dx$
40. $\int \frac{1}{\sin^3 3\theta}\,d\theta$
41. $\int \frac{1}{\sin^2 2\theta}\,d\theta$
42. $\int \frac{1}{\cos^5 x}\,dx.$
43. $\int \frac{1}{x^2+4x+3}\,dx$
44. $\int \frac{1}{x^2+4x+4}\,dx$
45. $\int \frac{dz}{z(z-3)}$
46. $\int \frac{dy}{4-y^2}$
47. $\int \frac{1}{1+(z+2)^2}\,dz$
48. $\int \frac{1}{y^2+4y+5}\,dy$
49. $\int \sin^3 x\,dx$
50. $\int \tan^4 x\,dx$
51. $\int \sinh^3 x \cosh^2 x\,dx$
52. $\int \sinh^2 x \cosh^3 x\,dx$
53. $\int \sin^3 3\theta \cos^2 3\theta\,d\theta$
54. $\int z e^{2z^2}\cos(2z^2)\,dz$

For Exercises 55–64, evaluate the definite integrals. Whenever possible, use the Fundamental Theorem of Calculus, perhaps after a substitution. Otherwise, use numerical methods.

55. $\int_0^{\pi/12} \sin(3\alpha)\,d\alpha$
56. $\int_{-\pi}^{\pi} \sin 5x \cos 6x\,dx$
57. $\int_1^2 (x-2x^3)\ln x\,dx$
58. $\int_0^1 \sqrt{3-x^2}\,dx$
59. $\int_0^1 \frac{1}{x^2+2x+1}\,dx$
60. $\int_0^1 \frac{dx}{x^2+2x+5}$
61. $\int_0^{1/\sqrt{2}} \frac{x}{\sqrt{1-x^4}}\,dx$
62. $\int_0^1 \frac{(x+2)}{(x+2)^2+1}\,dx$
63. $\int_{\pi/4}^{\pi/3} \frac{dx}{\sin^3 x}$
64. $\int_{-3}^{-1} \frac{dx}{\sqrt{x^2+6x+10}}$

PROBLEMS

In Problems 65–66, using properties of ln, find a substitution w and constants k, n so that the integral has the form

$$\int kw^n \ln w\,dw.$$

65. $\int (2x+1)^3 \ln(2x+1)\,dx$
66. $\int (2x+1)^3 \ln \frac{1}{\sqrt{2x+1}}\,dx$

In Problems 67–69, find constants a, b, c, m, n so that the integral is in one of the following forms from a table of integrals.[5] Give the form (i)–(iii) you use.

(i) $\displaystyle\int \frac{dx}{ax^2+bx+c}$ (ii) $\displaystyle\int \frac{mx+n}{ax^2+bx+c}\,dx$

(iii) $\displaystyle\int \frac{dx}{\left(ax^2+bx+c\right)^n},\ n>0$

67. $\displaystyle\int \frac{dx}{5-\frac{x}{4}-\frac{x^2}{6}}$

68. $\displaystyle\int \frac{dx}{2x+\frac{5}{7+3x}}$

69. $\displaystyle\int \frac{dx}{(x^2-5x+6)^3(x^2-4x+4)^2(x^2-6x+9)^2}$

In Problems 70–71, find constants a, b, λ so that the integral has the form found in some tables of integrals:[6]

$$\int \frac{e^{2\lambda x}}{ae^{\lambda x}+b}\,dx.$$

70. $\displaystyle\int \frac{e^{6x}}{4+e^{3x+1}}\,dx$

71. $\displaystyle\int \frac{e^{8x}}{4e^{4x}+5e^{6x}}\,dx$

72. According to a table of integrals,[7]

$$\int x^2e^{bx}\,dx = e^{bx}\left(\frac{x^2}{b}-\frac{2x}{b^2}+\frac{2}{b^3}\right)+C.$$

(a) Find a substitution w and constant k so that the integral $\int x^5 e^{bx^2}\,dx$ can be rewritten in the form

$$\int kw^2e^{bw}\,dw.$$

(b) Evaluate the integral in terms of x. Your answer may involve the constant b.

73. Show that for all integers m and n, with $m \neq \pm n$, $\int_{-\pi}^{\pi} \sin m\theta \sin n\theta\, d\theta = 0$.

74. Show that for all integers m and n, with $m \neq \pm n$, $\int_{-\pi}^{\pi} \cos m\theta \cos n\theta\, d\theta = 0$.

75. Water pipelines sometimes spring leaks, and water escapes until the leak is repaired.

(a) At t days after a leak is detected, water leaks from a pipeline at an estimated rate of

$$r(t) = 1 - \frac{1}{\sqrt{t^2+1}} \text{ thousands of gallons per day.}$$

By finding a derivative or looking at a graph, explain why this rate could represent a new leak. In the long run, if the leak is not fixed, what happens to $r(t)$?

(b) What is the shape of the graph of $V(t)$, the total volume of water that has escaped by time t?

(c) Find a formula for $V(t)$.

Strengthen Your Understanding

In Problems 76–80, explain what is wrong with the statement.

76. The table of integrals cannot be used to find $\displaystyle\int \frac{dt}{7-t^2}$.

77. If $a > 0$, then $\int 1/(x^2+4x+a)\,dx$ always involves arctan.

78. By Formula II-8 of the table with $a = 1, b = 1$,

$$\int e^x \sin x\,dx = \frac{1}{2}e^x(\sin x - \cos x) + C.$$

Therefore

$$\int e^{2x+1}\sin(2x+1)\,dx = \frac{1}{2}e^{2x+1}(\sin(2x+1) - \cos(2x+1)) + C.$$

79. The integral $\int \sin x \cos x\,dx$ with $a = 1, b = 1$ is undefined according to Table Formula II-12 since, for $a \neq b$,

$$\int \sin(ax)\cos(bx)\,dx = \frac{1}{b^2-a^2}(b\sin(ax)\sin(bx) + a\cos(ax)\cos(bx)) + C.$$

80. The table can be used to evaluate $\int \sin x/x\,dx$.

In Problems 81–82, give an example of:

81. An indefinite integral involving a square root that can be evaluated by first completing a square.

82. An indefinite integral involving $\sin x$ that can be evaluated with a reduction formula

In Problems 83–86, decide whether the statements are true or false. Give an explanation for your answer.

83. $\int \sin^7\theta \cos^6\theta\, d\theta$ can be written as a polynomial with $\cos\theta$ as the variable.

84. $\int 1/(x^2+4x+5)\,dx$ involves a natural logarithm.

85. $\int 1/(x^2+4x-5)\,dx$ involves an arctangent.

86. $\int x^{-1}((\ln x)^2 + (\ln x)^3)\,dx$ is a polynomial with $\ln x$ as the variable.

[5] http://en.wikipedia.org/wiki/List_of_integrals_of_rational_functions, page accessed February 24, 2010.
[6] http://en.wikipedia.org/wiki/List_of_integrals_of_exponential_functions, page accessed May 5, 2010.
[7] http://en.wikipedia.org/wiki/List_of_integrals_of_exponential_functions, page accessed February 17, 2011.

7.4 ALGEBRAIC IDENTITIES AND TRIGONOMETRIC SUBSTITUTIONS

Although not all functions have elementary antiderivatives, many do. In this section we introduce two powerful methods of integration which show that large classes of functions have elementary antiderivatives. The first is the method of partial fractions, which depends on an algebraic identity and allows us to integrate rational functions. The second is the method of trigonometric substitutions, which allows us to handle expressions involving the square root of a quadratic polynomial. Some of the formulas in the table of integrals can be derived using the techniques of this section.

Method of Partial Fractions

The integral of some rational functions can be obtained by splitting the integrand into *partial fractions*. For example, to find

$$\int \frac{1}{(x-2)(x-5)}\,dx,$$

the integrand is split into partial fractions with denominators $(x-2)$ and $(x-5)$. We write

$$\frac{1}{(x-2)(x-5)} = \frac{A}{x-2} + \frac{B}{x-5},$$

where A and B are constants that need to be found. Multiplying by $(x-2)(x-5)$ gives the identity

$$1 = A(x-5) + B(x-2)$$

so

$$1 = (A+B)x - 5A - 2B.$$

Since this equation holds for all x, the constant terms on both sides must be equal.[8] Similarly, the coefficients of x on both sides must be equal. So

$$\begin{aligned} -5A - 2B &= 1 \\ A + B &= 0. \end{aligned}$$

Solving these equations gives $A = -1/3$, $B = 1/3$. Thus,

$$\frac{1}{(x-2)(x-5)} = \frac{-1/3}{x-2} + \frac{1/3}{x-5}.$$

(Check the answer by writing the right-hand side over the common denominator $(x-2)(x-5)$.)

Example 1 Use partial fractions to integrate $\displaystyle\int \frac{1}{(x-2)(x-5)}\,dx$.

Solution We split the integrand into partial fractions, each of which can be integrated:

$$\int \frac{1}{(x-2)(x-5)}\,dx = \int \left(\frac{-1/3}{x-2} + \frac{1/3}{x-5}\right) dx = -\frac{1}{3}\ln|x-2| + \frac{1}{3}\ln|x-5| + C.$$

You can check that using formula V-26 in the integral table gives the same result.

This method can be used to derive formulas V-29 and V-30 in the integral table. A similar method works on rational functions whenever the denominator of the integrand factors into distinct linear factors and the numerator has degree less than the denominator.

Example 2 Find $\displaystyle\int \frac{x+2}{x^2+x}\,dx$.

[8] We have not shown that the equation holds for $x = 2$ and $x = 5$, but these values do not affect the argument.

Solution We factor the denominator and split the integrand into partial fractions:

$$\frac{x+2}{x^2+x} = \frac{x+2}{x(x+1)} = \frac{A}{x} + \frac{B}{x+1}.$$

Multiplying by $x(x+1)$ gives the identity

$$\begin{aligned} x+2 &= A(x+1) + Bx \\ &= (A+B)x + A. \end{aligned}$$

Equating constant terms and coefficients of x gives $A = 2$ and $A + B = 1$, so $B = -1$. Then we split the integrand into two parts and integrate:

$$\int \frac{x+2}{x^2+x}\,dx = \int \left(\frac{2}{x} - \frac{1}{x+1}\right) dx = 2\ln|x| - \ln|x+1| + C.$$

The next example illustrates what to do if there is a repeated factor in the denominator.

Example 3 Calculate $\int \frac{10x - 2x^2}{(x-1)^2(x+3)}\,dx$ using partial fractions of the form $\frac{A}{x-1}, \frac{B}{(x-1)^2}, \frac{C}{x+3}$.

Solution We are given that the squared factor, $(x-1)^2$, leads to partial fractions of the form:

$$\frac{10x - 2x^2}{(x-1)^2(x+3)} = \frac{A}{x-1} + \frac{B}{(x-1)^2} + \frac{C}{x+3}.$$

Multiplying through by $(x-1)^2(x+3)$ gives

$$\begin{aligned} 10x - 2x^2 &= A(x-1)(x+3) + B(x+3) + C(x-1)^2 \\ &= (A+C)x^2 + (2A + B - 2C)x - 3A + 3B + C. \end{aligned}$$

Equating the coefficients of x^2 and x and the constant terms, we get the simultaneous equations:

$$\begin{aligned} A + C &= -2 \\ 2A + B - 2C &= 10 \\ -3A + 3B + C &= 0. \end{aligned}$$

Solving gives $A = 1, B = 2, C = -3$. Thus, we obtain three integrals which can be evaluated:

$$\begin{aligned} \int \frac{10x - 2x^2}{(x-1)^2(x+3)}\,dx &= \int \left(\frac{1}{x-1} + \frac{2}{(x-1)^2} - \frac{3}{x+3}\right) dx \\ &= \ln|x-1| - \frac{2}{(x-1)} - 3\ln|x+3| + K. \end{aligned}$$

For the second integral, we use the fact that $\int 2/(x-1)^2 dx = 2\int (x-1)^{-2} dx = -2(x-1)^{-1} + K$.

If there is a quadratic in the denominator which cannot be factored, we need an expression of the form $Ax + B$ in the numerator, as the next example shows.

Example 4 Find $\int \frac{2x^2 - x - 1}{(x^2+1)(x-2)}\,dx$ using partial fractions of the form $\frac{Ax+B}{x^2+1}$ and $\frac{C}{x-2}$.

Solution We are given that the quadratic denominator, $(x^2 + 1)$, which cannot be factored further, has a numerator of the form $Ax + B$, so we have

$$\frac{2x^2 - x - 1}{(x^2+1)(x-2)} = \frac{Ax+B}{x^2+1} + \frac{C}{x-2}.$$

Multiplying by $(x^2 + 1)(x - 2)$ gives

$$\begin{aligned} 2x^2 - x - 1 &= (Ax + B)(x - 2) + C(x^2 + 1) \\ &= (A + C)x^2 + (B - 2A)x + C - 2B. \end{aligned}$$

Equating the coefficients of x^2 and x and the constant terms gives the simultaneous equations

$$\begin{aligned} A + C &= 2 \\ B - 2A &= -1 \\ C - 2B &= -1. \end{aligned}$$

Solving gives $A = B = C = 1$, so we rewrite the integral as follows:

$$\int \frac{2x^2 - x - 1}{(x^2+1)(x-2)}\,dx = \int \left(\frac{x+1}{x^2+1} + \frac{1}{x-2}\right)dx.$$

This identity is useful provided we can perform the integration on the right-hand side. The first integral can be done if it is split into two; the second integral is similar to those in the previous examples. We have

$$\int \frac{2x^2 - x - 1}{(x^2+1)(x-2)}\,dx = \int \frac{x}{x^2+1}\,dx + \int \frac{1}{x^2+1}\,dx + \int \frac{1}{x-2}\,dx.$$

To calculate $\int (x/(x^2 + 1))\,dx$, substitute $w = x^2 + 1$, or guess and check. The final result is

$$\int \frac{2x^2 - x - 1}{(x^2+1)(x-2)}\,dx = \frac{1}{2}\ln|x^2+1| + \arctan x + \ln|x-2| + K.$$

The next example shows what to do if the numerator has degree larger than the denominator.

Example 5 Calculate $\displaystyle\int \frac{x^3 - 7x^2 + 10x + 1}{x^2 - 7x + 10}\,dx$ using long division before integrating.

Solution The degree of the numerator is greater than the degree of the denominator, so we divide first:

$$\frac{x^3 - 7x^2 + 10x + 1}{x^2 - 7x + 10} = \frac{x(x^2 - 7x + 10) + 1}{x^2 - 7x + 10} = x + \frac{1}{x^2 - 7x + 10}.$$

The remainder, in this case $1/(x^2 - 7x + 10)$, is a rational function on which we try to use partial fractions. We have

$$\frac{1}{x^2 - 7x + 10} = \frac{1}{(x-2)(x-5)}$$

so in this case we use the result of Example 1 to obtain

$$\int \frac{x^3 - 7x^2 + 10x + 1}{x^2 - 7x + 10}\,dx = \int \left(x + \frac{1}{(x-2)(x-5)}\right)dx = \frac{x^2}{2} - \frac{1}{3}\ln|x-2| + \frac{1}{3}\ln|x-5| + C.$$

Many, though not all, rational functions can be integrated by the strategy suggested by the previous examples.

Strategy for Integrating a Rational Function, $\frac{P(x)}{Q(x)}$

- If degree of $P(x) \geq$ degree of $Q(x)$, try long division and the method of partial fractions on the remainder.
- If $Q(x)$ is the product of distinct linear factors, use partial fractions of the form
$$\frac{A}{(x-c)}.$$
- If $Q(x)$ contains a repeated linear factor, $(x-c)^n$, use partial fractions of the form
$$\frac{A_1}{(x-c)} + \frac{A_2}{(x-c)^2} + \cdots + \frac{A_n}{(x-c)^n}.$$
- If $Q(x)$ contains an unfactorable quadratic $q(x)$, try a partial fraction of the form
$$\frac{Ax+B}{q(x)}.$$

To use this method, we must be able to integrate each partial fraction. We can integrate terms of the form $A/(x-c)^n$ using the power rule (if $n > 1$) and logarithms (if $n = 1$). Next we see how to integrate terms of the form $(Ax+B)/q(x)$, where $q(x)$ is an unfactorable quadratic.

Trigonometric Substitutions

Section 7.1 showed how substitutions could be used to transform complex integrands. Now we see how substitution of $\sin\theta$ or $\tan\theta$ can be used for integrands involving square roots of quadratics or unfactorable quadratics.

Sine Substitutions

Substitutions involving $\sin\theta$ make use of the Pythagorean identity, $\cos^2\theta + \sin^2\theta = 1$, to simplify an integrand involving $\sqrt{a^2 - x^2}$.

Example 6 Find $\displaystyle\int \frac{1}{\sqrt{1-x^2}}\,dx$ using the substitution $x = \sin\theta$.

Solution If $x = \sin\theta$, then $dx = \cos\theta\,d\theta$, and substitution converts $1 - x^2$ to a perfect square:

$$\int \frac{1}{\sqrt{1-x^2}}\,dx = \int \frac{1}{\sqrt{1-\sin^2\theta}}\cos\theta\,d\theta = \int \frac{\cos\theta}{\sqrt{\cos^2\theta}}\,d\theta.$$

Now either $\sqrt{\cos^2\theta} = \cos\theta$ or $\sqrt{\cos^2\theta} = -\cos\theta$ depending on the values taken by θ. If we choose $-\pi/2 \leq \theta \leq \pi/2$, then $\cos\theta \geq 0$, so $\sqrt{\cos^2\theta} = \cos\theta$. Then

$$\int \frac{\cos\theta}{\sqrt{\cos^2\theta}}\,d\theta = \int \frac{\cos\theta}{\cos\theta}\,d\theta = \int 1\,d\theta = \theta + C = \arcsin x + C.$$

The last step uses the fact that $\theta = \arcsin x$ if $x = \sin\theta$ and $-\pi/2 \leq \theta \leq \pi/2$.

From now on, when we substitute $\sin\theta$, we assume that the interval $-\pi/2 \le \theta \le \pi/2$ has been chosen. Notice that the previous example is the case $a = 1$ of formula VI-28. The next example illustrates how to choose the substitution when $a \ne 1$.

Example 7 Use a trigonometric substitution to find $\displaystyle\int \frac{1}{\sqrt{4-x^2}}\,dx$.

Solution This time we choose $x = 2\sin\theta$, with $-\pi/2 \le \theta \le \pi/2$, so that $4 - x^2$ becomes a perfect square:

$$\sqrt{4-x^2} = \sqrt{4-4\sin^2\theta} = 2\sqrt{1-\sin^2\theta} = 2\sqrt{\cos^2\theta} = 2\cos\theta.$$

Then $dx = 2\cos\theta\,d\theta$, so substitution gives

$$\int \frac{1}{\sqrt{4-x^2}}\,dx = \int \frac{1}{2\cos\theta}2\cos\theta\,d\theta = \int 1\,d\theta = \theta + C = \arcsin\left(\frac{x}{2}\right) + C.$$

The general rule for choosing a sine substitution is:

> To simplify $\sqrt{a^2 - x^2}$, for constant a, try $x = a\sin\theta$, with $-\pi/2 \le \theta \le \pi/2$.

Notice $\sqrt{a^2-x^2}$ is only defined on the interval $[-a, a]$. Assuming that the domain of the integrand is $[-a, a]$, the substitution $x = a\sin\theta$, with $-\pi/2 \le \theta \le \pi/2$, is valid for all x in the domain, because its range is $[-a, a]$ and it has an inverse $\theta = \arcsin(x/a)$ on $[-a, a]$.

Example 8 Find the area of the ellipse $4x^2 + y^2 = 9$.

Solution Solving for y shows that $y = \sqrt{9-4x^2}$ gives the upper half of the ellipse. From Figure 7.2, we see by symmetry that

$$\text{Area } = 4\int_0^{3/2} \sqrt{9-4x^2}\,dx.$$

To decide which trigonometric substitution to use, we write the integrand as

$$\sqrt{9-4x^2} = 2\sqrt{\frac{9}{4} - x^2} = 2\sqrt{\left(\frac{3}{2}\right)^2 - x^2}.$$

This suggests that we should choose $x = (3/2)\sin\theta$, so that $dx = (3/2)\cos\theta\,d\theta$ and

$$\sqrt{9-4x^2} = 2\sqrt{\left(\frac{3}{2}\right)^2 - \left(\frac{3}{2}\right)^2\sin^2\theta} = 2\left(\frac{3}{2}\right)\sqrt{1-\sin^2\theta} = 3\cos\theta.$$

When $x = 0$, $\theta = 0$, and when $x = 3/2$, $\theta = \pi/2$, so

$$4\int_0^{3/2} \sqrt{9-4x^2}\,dx = 4\int_0^{\pi/2} 3\cos\theta\left(\frac{3}{2}\right)\cos\theta\,d\theta = 18\int_0^{\pi/2} \cos^2\theta\,d\theta.$$

Using Example 6 on page 356 or formula IV-18 we find

$$\int \cos^2\theta\,d\theta = \frac{1}{2}\cos\theta\sin\theta + \frac{1}{2}\theta + C.$$

So we have

$$\text{Area} = 4\int_0^{3/2} \sqrt{9-4x^2}\,dx = \frac{18}{2}(\cos\theta\sin\theta+\theta)\bigg|_0^{\pi/2} = 9\left(0+\frac{\pi}{2}\right) = \frac{9\pi}{2}.$$

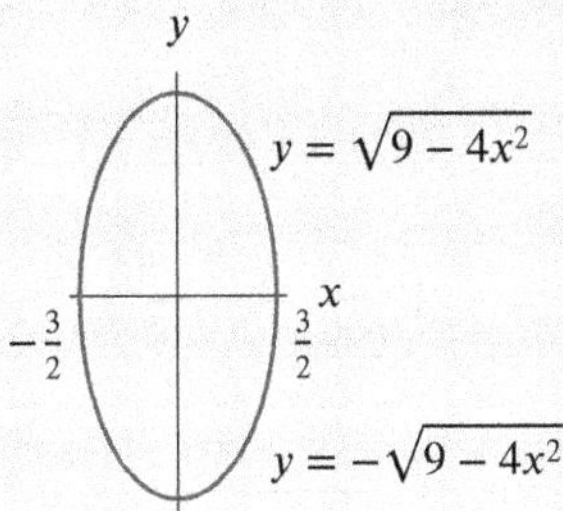

Figure 7.2: The ellipse $4x^2+y^2=9$

In Example 8, we did not return to the original variable x after making the substitution because we had also converted the limits of the definite integral. However, if we are calculating an indefinite integral, we have to return to the original variable. In the next example, we see how a triangle representing the substitution can be useful.

Example 9 Find the indefinite integral $\int \sqrt{9-4x^2}\,dx$ corresponding to Example 8.

Solution From Example 8, we know if $x=(3/2)\sin\theta$, then $dx=(3/2)\cos\theta\,d\theta$, so

$$\int \sqrt{9-4x^2}\,dx = \int 3\cos\theta\cdot\frac{3}{2}\cos\theta\,d\theta = \frac{9}{2}\left(\frac{1}{2}\cos\theta\sin\theta+\frac{1}{2}\theta\right)+C.$$

To rewrite the antiderivative in terms of the original variable x, we use the fact that $\sin\theta=2x/3$ to write $\theta=\arcsin(2x/3)$. To express $\cos\theta$ in terms of x, we draw the right triangle in Figure 7.3 with opposite side $2x$ and hypotenuse 3, so $\sin\theta=2x/3$. Then we use the Pythagorean Theorem to see that $\cos\theta=\sqrt{9-4x^2}/3$, so

$$\begin{aligned}\int \sqrt{9-4x^2}\,dx &= \frac{9}{4}\cos\theta\sin\theta+\frac{9}{4}\theta+C\\ &= \frac{9}{4}\cdot\frac{2x}{3}\cdot\frac{\sqrt{9-4x^2}}{3}+\frac{9}{4}\arcsin\frac{2x}{3}+C = \frac{x\sqrt{9-4x^2}}{2}+\frac{9}{4}\arcsin\frac{2x}{3}+C.\end{aligned}$$

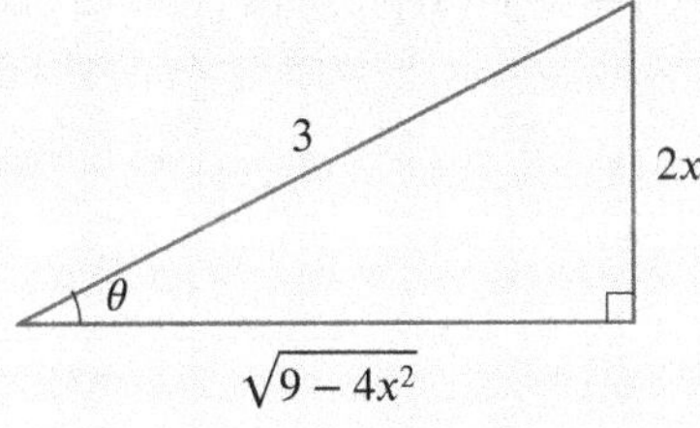

Figure 7.3: Triangle with $\sin\theta=2x/3$

Tangent Substitutions

Integrals involving a^2+x^2 may be simplified by a substitution involving $\tan\theta$ and the trigonometric identities $\tan\theta = \sin\theta/\cos\theta$ and $\cos^2\theta + \sin^2\theta = 1$.

Example 10 Find $\int \frac{1}{x^2+9}\,dx$ using the substitution $x = 3\tan\theta$.

Solution If $x = 3\tan\theta$, then $dx = (3/\cos^2\theta)\,d\theta$, so

$$\int \frac{1}{x^2+9}\,dx = \int \left(\frac{1}{9\tan^2\theta+9}\right)\left(\frac{3}{\cos^2\theta}\right)\,d\theta = \frac{1}{3}\int \frac{1}{\left(\frac{\sin^2\theta}{\cos^2\theta}+1\right)\cos^2\theta}\,d\theta$$

$$= \frac{1}{3}\int \frac{1}{\sin^2\theta+\cos^2\theta}\,d\theta = \frac{1}{3}\int 1\,d\theta = \frac{1}{3}\theta + C = \frac{1}{3}\arctan\left(\frac{x}{3}\right) + C.$$

> To simplify a^2+x^2 or $\sqrt{a^2+x^2}$, for constant a, try $x = a\tan\theta$, with $-\pi/2 < \theta < \pi/2$.

Note that a^2+x^2 and $\sqrt{a^2+x^2}$ are defined on $(-\infty,\infty)$. Assuming that the domain of the integrand is $(-\infty,\infty)$, the substitution $x = a\tan\theta$, with $-\pi/2 < \theta < \pi/2$, is valid for all x in the domain, because its range is $(-\infty,\infty)$ and it has an inverse $\theta = \arctan(x/a)$ on $(-\infty,\infty)$.

Example 11 Use a tangent substitution to show that the following two integrals are equal:

$$\int_0^1 \sqrt{1+x^2}\,dx = \int_0^{\pi/4} \frac{1}{\cos^3\theta}\,d\theta.$$

What area do these integrals represent?

Solution We put $x = \tan\theta$, with $-\pi/2 < \theta < \pi/2$, so that $dx = (1/\cos^2\theta)\,d\theta$, and

$$\sqrt{1+x^2} = \sqrt{1+\frac{\sin^2\theta}{\cos^2\theta}} = \sqrt{\frac{\cos^2\theta+\sin^2\theta}{\cos^2\theta}} = \frac{1}{\cos\theta}.$$

When $x = 0$, $\theta = 0$, and when $x = 1$, $\theta = \pi/4$, so

$$\int_0^1 \sqrt{1+x^2}\,dx = \int_0^{\pi/4} \left(\frac{1}{\cos\theta}\right)\left(\frac{1}{\cos^2\theta}\right)\,d\theta = \int_0^{\pi/4} \frac{1}{\cos^3\theta}\,d\theta.$$

The integral $\int_0^1 \sqrt{1+x^2}\,dx$ represents the area under the hyperbola $y^2 - x^2 = 1$ in Figure 7.4.

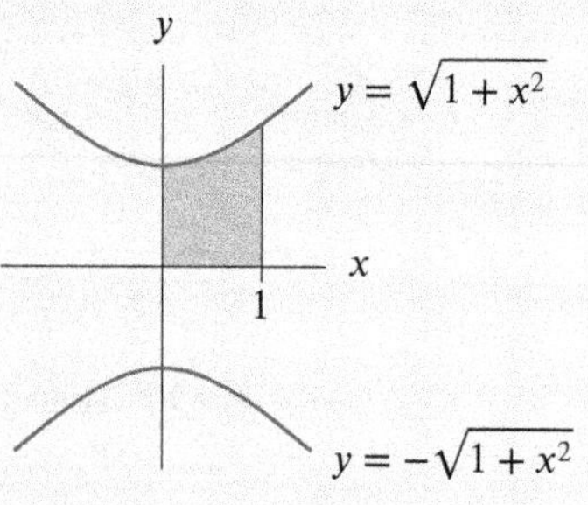

Figure 7.4: The hyperbola $y^2 - x^2 = 1$

Completing the Square to Use a Trigonometric Substitution

To make a trigonometric substitution, we may first need to complete the square.

Example 12 Find $\displaystyle\int \frac{3}{\sqrt{2x-x^2}}\,dx$.

Solution To use a sine or tangent substitution, the expression under the square root sign should be in the form a^2+x^2 or a^2-x^2. Completing the square, we get

$$2x - x^2 = 1 - (x-1)^2.$$

This suggests we substitute $x - 1 = \sin\theta$, or $x = \sin\theta + 1$. Then $dx = \cos\theta\, d\theta$, and

$$\begin{aligned}\int \frac{3}{\sqrt{2x-x^2}}\,dx &= \int \frac{3}{\sqrt{1-(x-1)^2}}\,dx = \int \frac{3}{\sqrt{1-\sin^2\theta}}\cos\theta\,d\theta\\ &= \int \frac{3}{\cos\theta}\cos\theta\,d\theta = \int 3\,d\theta = 3\theta + C.\end{aligned}$$

Since $x - 1 = \sin\theta$, we have $\theta = \arcsin(x-1)$, so

$$\int \frac{3}{\sqrt{2x-x^2}}\,dx = 3\arcsin(x-1) + C.$$

Example 13 Find $\displaystyle\int \frac{1}{x^2+x+1}\,dx$.

Solution Completing the square, we get

$$x^2 + x + 1 = \left(x+\frac{1}{2}\right)^2 + \frac{3}{4} = \left(x+\frac{1}{2}\right)^2 + \left(\frac{\sqrt{3}}{2}\right)^2.$$

This suggests we substitute $x + 1/2 = (\sqrt{3}/2)\tan\theta$, or $x = -1/2 + (\sqrt{3}/2)\tan\theta$. Then $dx = (\sqrt{3}/2)(1/\cos^2\theta)\,d\theta$, so

$$\begin{aligned}\int \frac{1}{x^2+x+1}\,dx &= \int \left(\frac{1}{(x+\frac{1}{2})^2+\frac{3}{4}}\right)\left(\frac{\sqrt{3}}{2}\frac{1}{\cos^2\theta}\right)d\theta\\ &= \frac{\sqrt{3}}{2}\int \left(\frac{1}{\frac{3}{4}\tan^2\theta+\frac{3}{4}}\right)\left(\frac{1}{\cos^2\theta}\right)d\theta = \frac{2}{\sqrt{3}}\int \frac{1}{(\tan^2\theta+1)\cos^2\theta}\,d\theta\\ &= \frac{2}{\sqrt{3}}\int \frac{1}{\sin^2\theta+\cos^2\theta}\,d\theta = \frac{2}{\sqrt{3}}\int 1\,d\theta = \frac{2}{\sqrt{3}}\theta + C.\end{aligned}$$

Since $x + 1/2 = (\sqrt{3}/2)\tan\theta$, we have $\theta = \arctan((2/\sqrt{3})x + 1/\sqrt{3})$, so

$$\int \frac{1}{x^2+x+1}\,dx = \frac{2}{\sqrt{3}}\arctan\left(\frac{2}{\sqrt{3}}x + \frac{1}{\sqrt{3}}\right) + C.$$

Alternatively, using a computer algebra system gives[9]

$$\frac{2\tan^{-1}\left(\frac{2x+1}{\sqrt{3}}\right)}{\sqrt{3}},$$

essentially the same as we obtained by hand. You can check either answer by differentiation.

Exercises and Problems for Section 7.4 Online Resource: Additional Problems for Section 7.4

EXERCISES

Split the functions in Exercises 1–7 into partial fractions.

1. $\dfrac{x+1}{6x+x^2}$

2. $\dfrac{20}{25-x^2}$

3. $\dfrac{1}{w^4-w^3}$

4. $\dfrac{2y}{y^3-y^2+y-1}$

5. $\dfrac{8}{y^3-4y}$

6. $\dfrac{2(1+s)}{s(s^2+3s+2)}$

7. $\dfrac{2}{s^4-1}$

In Exercises 8–14, find the antiderivative of the function in the given exercise.

8. Exercise 1

9. Exercise 2

10. Exercise 3

11. Exercise 4

12. Exercise 5

13. Exercise 6

14. Exercise 7

In Exercises 15–19, evaluate the integral.

15. $\displaystyle\int \frac{3x^2-8x+1}{x^3-4x^2+x+6}\,dx$; use $\dfrac{A}{x-2}+\dfrac{B}{x+1}+\dfrac{C}{x-3}$.

16. $\displaystyle\int \frac{dx}{x^3-x^2}$; use $\dfrac{A}{x}+\dfrac{B}{x^2}+\dfrac{C}{x-1}$.

17. $\displaystyle\int \frac{10x+2}{x^3-5x^2+x-5}\,dx$; use $\dfrac{A}{x-5}+\dfrac{Bx+C}{x^2+1}$.

18. $\displaystyle\int \frac{x^4+12x^3+15x^2+25x+11}{x^3+12x^2+11x}\,dx$; use division and $\dfrac{A}{x}+\dfrac{B}{x+1}+\dfrac{C}{x+11}$.

19. $\displaystyle\int \frac{x^4+3x^3+2x^2+1}{x^2+3x+2}\,dx$; use division.

In Exercises 20–22, use the substitution to find the integral.

20. $\displaystyle\int \frac{1}{\sqrt{9-4x^2}}\,dx, \quad x=\frac{3}{2}\sin t$

21. $\displaystyle\int \frac{1}{\sqrt{4x-3-x^2}}\,dx, \quad x=\sin t+2$

22. $\displaystyle\int \frac{1}{x^2+4x+5}\,dx, \quad x=\tan t-2$

23. Which of the following integrals are best done by a trigonometric substitution, and what substitution?

(a) $\displaystyle\int \sqrt{9-x^2}\,dx$ **(b)** $\displaystyle\int x\sqrt{9-x^2}\,dx$

24. Give a substitution (not necessarily trigonometric) which could be used to compute the following integrals:

(a) $\displaystyle\int \frac{x}{\sqrt{x^2+10}}\,dx$ **(b)** $\displaystyle\int \frac{1}{\sqrt{x^2+10}}\,dx$

PROBLEMS

25. Find a value of k and a substitution w such that

$$\int \frac{12x-2}{(3x+2)(x-1)}\,dx = k\int \frac{dw}{w}.$$

26. Find values of A and B such that

$$\int \frac{12x-2}{(3x+2)(x-1)}\,dx = \int \frac{A\,dx}{3x+2}+\int \frac{B\,dx}{x-1}.$$

27. Write the integral $\displaystyle\int \frac{2x+9}{(3x+5)(4-5x)}\,dx$ in the form $\displaystyle\int \frac{cx+d}{(x-a)(x-b)}\,dx$. Give the values of the constants a, b, c, d. You need not evaluate the integral.

28. Write the integral $\displaystyle\int \frac{dx}{\sqrt{12-4x^2}}$ in the form $\displaystyle\int \frac{k\,dx}{\sqrt{a^2-x^2}}$. Give the values of the positive constants a and k. You need not evaluate the integral.

[9]wolframalpha.com, accessed January 11, 2011.

29. **(a)** Evaluate $\int \frac{3x+6}{x^2+3x}\,dx$ by partial fractions.
(b) Show that your answer to part (a) agrees with the answer you get by using the integral tables.

■ Complete the square and give a substitution (not necessarily trigonometric) which could be used to compute the integrals in Problems 30–37.

30. $\int \frac{1}{x^2+2x+2}\,dx$

31. $\int \frac{1}{x^2+6x+25}\,dx$

32. $\int \frac{dy}{y^2+3y+3}$

33. $\int \frac{x+1}{x^2+2x+2}\,dx$

34. $\int \frac{4}{\sqrt{2z-z^2}}\,dz$

35. $\int \frac{z-1}{\sqrt{2z-z^2}}\,dz$

36. $\int (t+2)\sin(t^2+4t+7)\,dt$

37. $\int (2-\theta)\cos(\theta^2-4\theta)d\theta$

■ Calculate the integrals in Problems 38–53.

38. $\int \frac{1}{(x-5)(x-3)}\,dx$

39. $\int \frac{1}{(x+2)(x+3)}\,dx$

40. $\int \frac{1}{(x+7)(x-2)}\,dx$

41. $\int \frac{x}{x^2-3x+2}\,dx$

42. $\int \frac{dz}{z^2+z}$

43. $\int \frac{dx}{x^2+5x+4}$

44. $\int \frac{dP}{3P-3P^2}$

45. $\int \frac{3x+1}{x^2-3x+2}\,dx$

46. $\int \frac{y+2}{2y^2+3y+1}\,dy$

47. $\int \frac{x+1}{x^3+x}\,dx$

48. $\int \frac{x-2}{x^2+x^4}\,dx$

49. $\int \frac{y^2}{25+y^2}\,dy$

50. $\int \frac{dz}{(4-z^2)^{3/2}}$

51. $\int \frac{10}{(s+2)(s^2+1)}\,ds$

52. $\int \frac{1}{x^2+4x+13}\,dx$

53. $\int \frac{e^x\,dx}{(e^x-1)(e^x+2)}$

■ In Problems 54–65, evaluate the indefinite integral, using a trigonometric substitution and a triangle to express the answer in terms of x.

54. $\int \frac{1}{x^2\sqrt{1+x^2}}\,dx$

55. $\int \frac{x^2}{\sqrt{9-x^2}}\,dx$

56. $\int \frac{\sqrt{1-4x^2}}{x^2}\,dx$

57. $\int \frac{\sqrt{25-9x^2}}{x}\,dx$

58. $\int \frac{1}{x\sqrt{9-4x^2}}\,dx$

59. $\int \frac{1}{x\sqrt{1+16x^2}}\,dx$

60. $\int \frac{1}{x^2\sqrt{4-x^2}}\,dx$

61. $\int \frac{1}{(25+4x^2)^{3/2}}\,dx$

62. $\int \frac{1}{(16-x^2)^{3/2}}\,dx$

63. $\int \frac{x^2}{(1+9x^2)^{3/2}}\,dx$

64. $\int \frac{\sqrt{x^2+4}}{x^4}\,dx$

65. $\int \frac{x^3}{\sqrt{4-x^2}}\,dx$

■ Find the exact area of the regions in Problems 66–71.

66. Bounded by $y = 3x/((x-1)(x-4))$, $y = 0$, $x = 2$, $x = 3$.

67. Bounded by $y = (3x^2+x)/((x^2+1)(x+1))$, $y = 0$, $x = 0$, $x = 1$.

68. Bounded by $y = x^2/\sqrt{1-x^2}$, $y = 0$, $x = 0$, $x = 1/2$.

69. Bounded by $y = x^3/\sqrt{4-x^2}$, $y = 0$, $x = 0$, $x = \sqrt{2}$.

70. Bounded by $y = 1/\sqrt{x^2+9}$, $y = 0$, $x = 0$, $x = 3$.

71. Bounded by $y = 1/(x\sqrt{x^2+9})$, $y = 0$, $x = \sqrt{3}$, $x = 3$.

■ Calculate the integrals in Problems 72–74 by partial fractions and then by using the indicated substitution. Show that the results you get are the same.

72. $\int \frac{dx}{1-x^2}$; substitution $x = \sin\theta$.

73. $\int \frac{2x}{x^2-1}\,dx$; substitution $w = x^2-1$.

74. $\int \frac{3x^2+1}{x^3+x}\,dx$; substitution $w = x^3+x$.

75. **(a)** Show $\int \frac{1}{\sin^2\theta}\,d\theta = -\frac{1}{\tan\theta} + C$.
(b) Calculate $\int \frac{dy}{y^2\sqrt{5-y^2}}$.

■ Solve Problems 76–78 without using integral tables.

76. Calculate the integral $\int \frac{1}{(x-a)(x-b)}\,dx$ for
(a) $a \neq b$ **(b)** $a = b$

77. Calculate the integral $\int \frac{x}{(x-a)(x-b)}\,dx$ for
(a) $a \neq b$ **(b)** $a = b$

78. Calculate the integral $\int \frac{1}{x^2-a}\,dx$ for
(a) $a > 0$ **(b)** $a = 0$ **(c)** $a < 0$

79. A rumor is spread in a school. For $0 < a < 1$ and $b > 0$, the time t at which a fraction p of the school population has heard the rumor is given by

$$t(p) = \int_a^p \frac{b}{x(1-x)}\,dx.$$

(a) Evaluate the integral to find an explicit formula for $t(p)$. Write your answer so it has only one ln term.

(b) At time $t = 0$ one percent of the school population ($p = 0.01$) has heard the rumor. What is a?

(c) At time $t = 1$ half the school population ($p = 0.5$) has heard the rumor. What is b?

(d) At what time has 90% of the school population ($p = 0.9$) heard the rumor?

Strengthen Your Understanding

In Problems 80–81, explain what is wrong with the statement.

80. To integrate

$$\int \frac{1}{(x-1)^2(x-2)}\,dx$$

using a partial fraction decomposition, let

$$\frac{1}{(x-1)^2(x-2)} = \frac{A}{(x-1)^2} + \frac{B}{x-2}.$$

81. Use the substitution $x = 2\sin\theta$ to integrate the following integral:

$$\int \frac{1}{(x^2+4)^{3/2}}\,dx.$$

In Problems 82–85, give an example of:

82. A rational function whose antiderivative is not a rational function.

83. An integral whose evaluation requires factoring a cubic.

84. A linear polynomial $P(x)$ and a quadratic polynomial $Q(x)$ such that the rational function $P(x)/Q(x)$ does not have a partial fraction decomposition of the form

$$\frac{P(x)}{Q(x)} = \frac{A}{x-r} + \frac{B}{x-s}$$

for some constants A, B, r, and s.

85. An integral that can be made easier to evaluate by using the trigonometric substitution $x = \frac{3}{2}\sin\theta$.

In Problems 86–87, decide whether the statements are true or false. Give an explanation for your answer.

86. The integral $\int \frac{1}{\sqrt{9-t^2}}\,dt$ can be made easier to evaluate by using the substitution $t = 3\tan\theta$.

87. To calculate $\int \frac{1}{x^3+x^2}\,dx$, we can split the integrand into

$$\int \left(\frac{A}{x} + \frac{B}{x^2} + \frac{C}{x+1}\right)dx$$

For Problems 88–89, which technique is useful in evaluating the integral?

(a) Integration by parts **(b)** Partial fractions
(c) Long division **(d)** Completing the square
(e) A trig substitution **(f)** Other substitutions

88. $\displaystyle\int \frac{x^2}{\sqrt{1-x^2}}\,dx$

89. $\displaystyle\int \frac{x^2}{1-x^2}\,dx$

7.5 NUMERICAL METHODS FOR DEFINITE INTEGRALS

Many functions do not have elementary antiderivatives. To evaluate the definite integrals of such functions, we cannot use the Fundamental Theorem; we must use numerical methods. We know how to approximate a definite integral numerically using left- and right-hand Riemann sums; in this section, we introduce more accurate methods.

The Midpoint Rule

In the left- and right-hand Riemann sums, the heights of the rectangles are found using the left-hand or right-hand endpoints, respectively, of the subintervals. For the *midpoint rule*, we use the midpoint of each of the subintervals.

For example, in approximating $\int_1^2 f(x)\,dx$ by a Riemann sum with two subdivisions, we first divide the interval $1 \le x \le 2$ into two pieces. The midpoint of the first subinterval is 1.25 and the midpoint of the second is 1.75. The heights of the two rectangles are $f(1.25)$ and $f(1.75)$, respectively. (See Figure 7.5.) The Riemann sum is

$$f(1.25)0.5 + f(1.75)0.5.$$

Figure 7.5 shows that evaluating f at the midpoint of each subdivision often gives a better approximation to the area under the curve than evaluating f at either end.

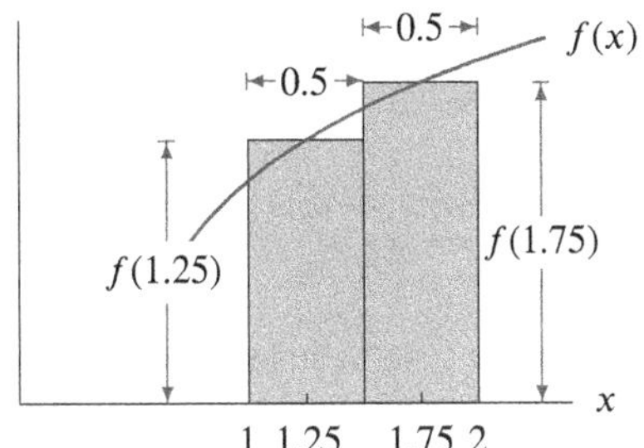

Figure 7.5: Midpoint rule with two subdivisions

Thus, we have three ways of estimating an integral using a Riemann sum:

1. The **left rule** uses the left endpoint of each subinterval.
2. The **right rule** uses the right endpoint of each subinterval.
3. The **midpoint rule** uses the midpoint of each subinterval.

We write LEFT(n), RIGHT(n), and MID(n) to denote the results obtained by using these rules with n subdivisions.

Example 1 For $\int_1^2 \frac{1}{x}\,dx$, compute LEFT(2), RIGHT(2) and MID(2), and compare your answers with the exact value of the integral.

Solution For $n = 2$ subdivisions of the interval $[1, 2]$, we use $\Delta x = 0.5$. Then, to four decimal places,

$$\text{LEFT}(2) = f(1)(0.5) + f(1.5)(0.5) = \frac{1}{1}(0.5) + \frac{1}{1.5}(0.5) = 0.8333$$
$$\text{RIGHT}(2) = f(1.5)(0.5) + f(2)(0.5) = \frac{1}{1.5}(0.5) + \frac{1}{2}(0.5) = 0.5833$$
$$\text{MID}(2) = f(1.25)(0.5) + f(1.75)(0.5) = \frac{1}{1.25}(0.5) + \frac{1}{1.75}(0.5) = 0.6857.$$

All three Riemann sums in this example are approximating

$$\int_1^2 \frac{1}{x}\,dx = \ln x\Big|_1^2 = \ln 2 - \ln 1 = \ln 2 = 0.6931.$$

With only two subdivisions, the left and right rules give quite poor approximations but the midpoint rule is already fairly close to the exact answer. Figures 7.6(a) and (b) show that the midpoint rule is more accurate than the left and right rules because the error to the left of the midpoint tends to cancel the error to the right of the midpoint.

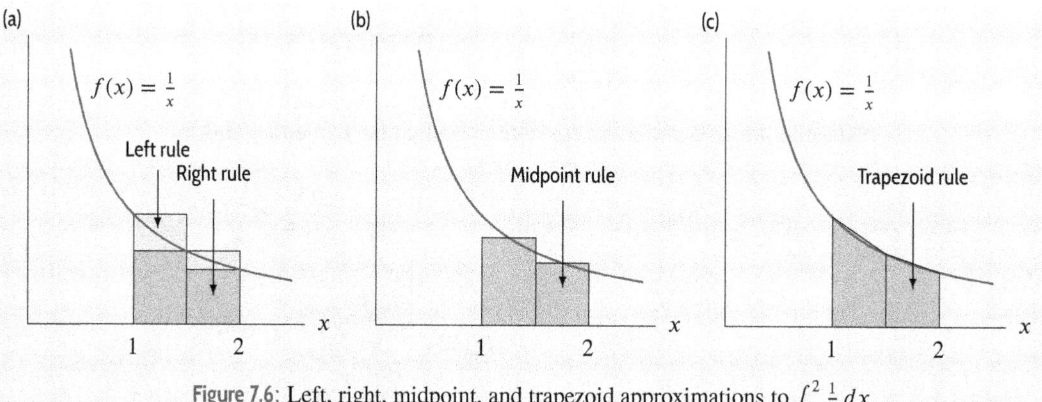

Figure 7.6: Left, right, midpoint, and trapezoid approximations to $\int_1^2 \frac{1}{x}\,dx$

The Trapezoid Rule

We have just seen how the midpoint rule can have the effect of balancing out the errors of the left and right rules. Another way of balancing these errors is to average the results from the left and right rules. This approximation is called the *trapezoid rule*:

$$\text{TRAP}(n) = \frac{\text{LEFT}(n) + \text{RIGHT}(n)}{2}.$$

The trapezoid rule averages the values of f at the left and right endpoints of each subinterval and multiplies by Δx. This is the same as approximating the area under the graph of f in each subinterval by a trapezoid (see Figure 7.7).

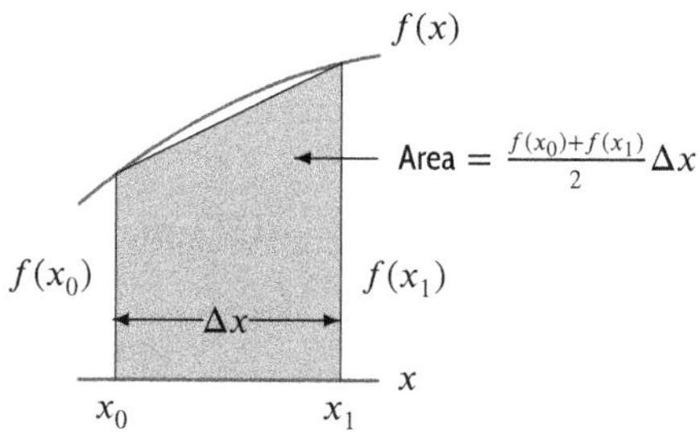

Figure 7.7: Area used in the trapezoid rule

Example 2 For $\displaystyle\int_1^2 \frac{1}{x}\,dx$, compare the trapezoid rule with two subdivisions with the left, right, and midpoint rules.

Solution In the previous example we got LEFT(2) = 0.8333 and RIGHT(2) = 0.5833. The trapezoid rule is the average of these, so TRAP(2) = 0.7083. (See Figure 7.6(c).) The exact value of the integral is 0.6931, so the trapezoid rule is better than the left or right rules. The midpoint rule is still the best, in this example, since MID(2) = 0.6857.

Is the Approximation an Over- or Underestimate?

It is useful to know when a rule is producing an overestimate and when it is producing an underestimate. In Chapter 5 we saw that if the integrand is increasing, the left rule underestimates and the right rule overestimates the integral. If the integrand is decreasing, the roles reverse. Now we see how concavity relates to the errors in the trapezoid and midpoint rules.

The Trapezoid Rule

If the graph of the function is concave down on $[a, b]$, then each trapezoid lies below the graph and the trapezoid rule underestimates. If the graph is concave up on $[a, b]$, the trapezoid rule overestimates. (See Figure 7.8.)

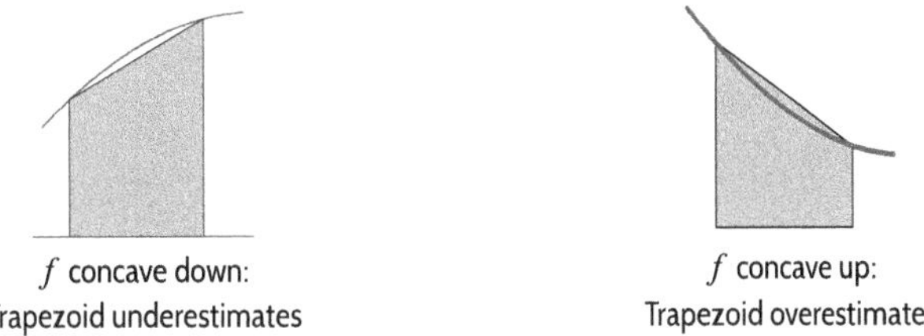

Figure 7.8: Error in the trapezoid rule

The Midpoint Rule

To understand the relationship between the midpoint rule and concavity, take a rectangle whose top intersects the curve at the midpoint of a subinterval. Draw a tangent to the curve at the midpoint; this gives a trapezoid. See Figure 7.9. (This is *not* the same trapezoid as in the trapezoid rule.) The midpoint rectangle and the new trapezoid have the same area, because the shaded triangles in Figure 7.9 are congruent. Hence, if the graph of the function is concave down, the midpoint rule overestimates; if the graph is concave up, the midpoint rule underestimates. (See Figure 7.10.)

If the graph of f is concave down on $[a, b]$, then

$$\text{TRAP}(n) \le \int_a^b f(x)\,dx \le \text{MID}(n).$$

If the graph of f is concave up on $[a, b]$, then

$$\text{MID}(n) \le \int_a^b f(x)\,dx \le \text{TRAP}(n).$$

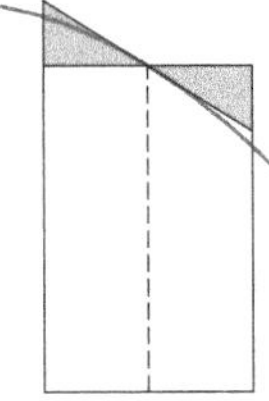

Figure 7.9: Midpoint rectangle and trapezoid with same area

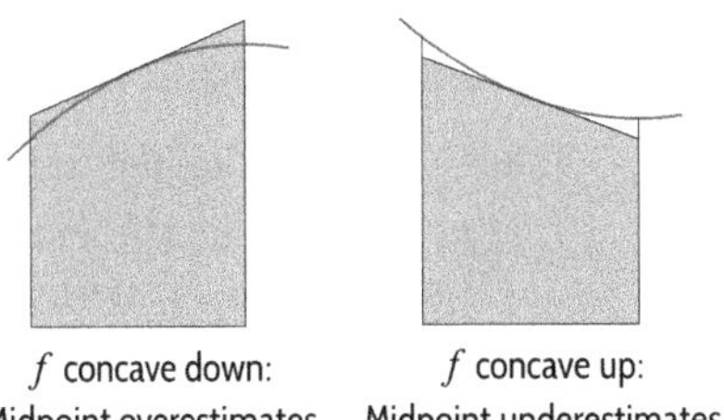

Figure 7.10: Error in the midpoint rule

When we compute an approximation, we are always concerned about the error, namely the difference between the exact answer and the approximation. We usually do not know the exact error; if we did, we would also know the exact answer. We take

$$\text{Error} = \text{Actual value} - \text{Approximate value}.$$

The errors for some methods are much smaller than those for others. In general, the midpoint and trapezoid rules are more accurate than the left or right rules. Comparing the errors in the midpoint and trapezoid rules suggests an even better method, called Simpson's rule.

Error in Left and Right Rules

We work with the example $\int_1^2 (1/x)\,dx$ because we know the exact value of this integral ($\ln 2$) and we can investigate the behavior of the errors.

Let us see what happens to the error in the left and right rules as we increase n. The results are in Table 7.2. A positive error indicates that the Riemann sum is less than the exact value, $\ln 2$. Notice that the errors for the left and right rules have opposite signs but are approximately equal in magnitude. (See Figure 7.11.) This leads us to want to average the left and right rules; this average is the trapezoid rule.

Table 7.2 *Errors for the left and right rule approximation to* $\int_1^2 \frac{1}{x}\,dx = \ln 2 \approx 0.6931471806$

n	Error in left rule	Error in right rule
2	−0.1402	0.1098
10	−0.0256	0.0244
50	−0.0050	0.0050
250	−0.0010	0.0010

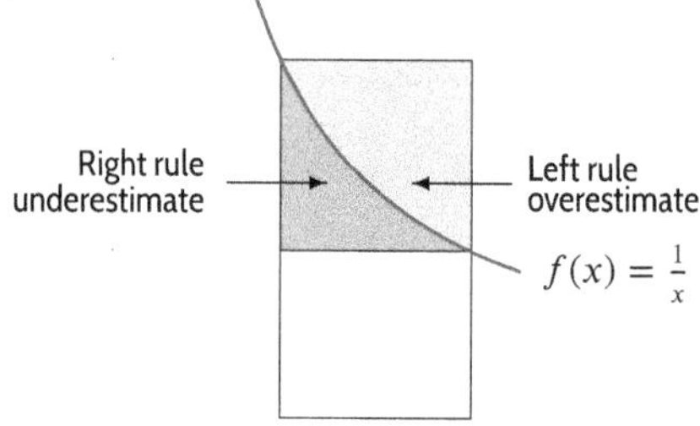

Figure 7.11: Errors in left and right sums

There is another pattern to the errors in Table 7.2. If we compute the *ratio* of the errors in Table 7.3, we see that the error[10] in both the left and right rules decreases by a factor of about 5 as n increases by a factor of 5.

There is nothing special about the number 5; the same holds for any factor. To get one extra digit of accuracy in any calculation, we must make the error $1/10$ as big, so we must increase n by a factor of 10. In fact, *for the left or right rules, each extra digit of accuracy requires about 10 times the work.*

Table 7.3 *Ratio of the errors as n increases for* $\int_1^2 \frac{1}{x}\,dx$

	Ratio of errors in left rule	Ratio of errors in right rule
Error(2) / Error(10)	5.47	4.51
Error(10) / Error(50)	5.10	4.90
Error(50) / Error(250)	5.02	4.98

Error in Trapezoid and Midpoint Rules

Table 7.4 shows that the trapezoid and midpoint rules generally produce better approximations to $\int_1^2 (1/x)\,dx$ than the left and right rules.

Again there is a pattern to the errors. For each n, the midpoint rule is noticeably better than the trapezoid rule; the error for the midpoint rule, in absolute value, seems to be about half the error of the trapezoid rule. To see why, compare the shaded areas in Figure 7.12. Also, notice in Table 7.4 that the errors for the two rules have opposite signs; this is due to concavity.

Table 7.4 *The errors for the trapezoid and midpoint rules for* $\int_1^2 \frac{1}{x}\,dx$

n	Error in trapezoid rule	Error in midpoint rule
2	−0.0152	0.0074
10	−0.00062	0.00031
50	−0.0000250	0.0000125
250	−0.0000010	0.0000005

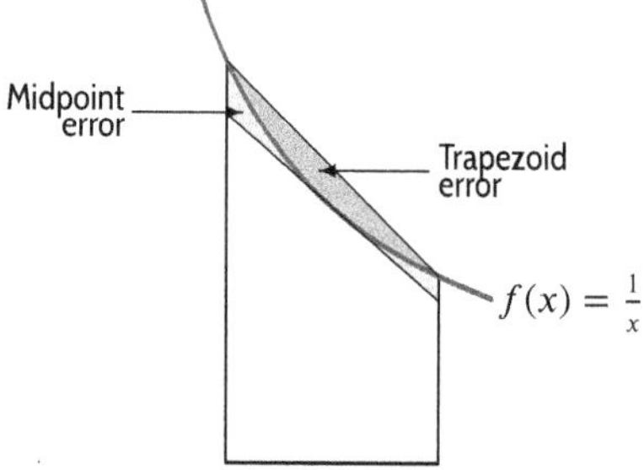

Figure 7.12: Errors in the midpoint and trapezoid rules

[10]The values in Table 7.2 are rounded to 4 decimal places; those in Table 7.3 were computed using more decimal places and then rounded.

We are interested in how the errors behave as n increases. Table 7.5 gives the ratios of the errors for each rule. For each rule, we see that as n increases by a factor of 5, the error decreases by a factor of about $25 = 5^2$. In fact, it can be shown that this squaring relationship holds for any factor, so increasing n by a factor of 10 will decrease the error by a factor of about $100 = 10^2$. Reducing the error by a factor of 100 is equivalent to adding two more decimal places of accuracy to the result. In other words: *In the trapezoid or midpoint rules, each extra 2 digits of accuracy requires about 10 times the work.*

Table 7.5 *Ratios of the errors as n increases for* $\int_1^2 \frac{1}{x}\,dx$

	Ratio of errors in trapezoid rule	Ratio of errors in midpoint rule
Error(2) / Error(10)	24.33	23.84
Error(10) / Error(50)	24.97	24.95
Error(50) / Error(250)	25.00	25.00

Simpson's Rule

Observing that the trapezoid error has the opposite sign and about twice the magnitude of the midpoint error, we may guess that a weighted average of the two rules, with the midpoint rule weighted twice the trapezoid rule, has a smaller error. This approximation is called *Simpson's rule*[11]:

$$\mathrm{SIMP}(n) = \frac{2 \cdot \mathrm{MID}(n) + \mathrm{TRAP}(n)}{3}.$$

Table 7.6 gives the errors for Simpson's rule. Notice how much smaller the errors are than the previous errors. Of course, it is a little unfair to compare Simpson's rule at $n = 50$, say, with the previous rules, because Simpson's rule must compute the value of f at both the midpoint and the endpoints of each subinterval and hence involves evaluating the function at twice as many points.

We see in Table 7.6 that as n increases by a factor of 5, the errors decrease by a factor of about 600, or about 5^4. Again this behavior holds for any factor, so increasing n by a factor of 10 decreases the error by a factor of about 10^4. In other words: *In Simpson's rule, each extra 4 digits of accuracy requires about 10 times the work.*

Table 7.6 *The errors for Simpson's rule and the ratios of the errors*

n	Error	Ratio
2	−0.0001067877	
		550.15
10	−0.0000001940	
		632.27
50	−0.0000000003	

Alternate Approach to Numerical Integration: Approximating by Lines and Parabolas

These rules for numerical integration can be obtained by approximating $f(x)$ on subintervals by a function:

- The left and right rules use constant functions.
- The trapezoid and midpoint rules use linear functions.
- Simpson's rule uses quadratic functions.

Problems 70 and 71 (available online) show how a quadratic approximation leads to Simpson's rule.

[11]Some books and computer programs use slightly different terminology for Simpson's rule; what we call $n = 50$, they call $n = 100$.

Exercises and Problems for Section 7.5

Online Resource: Additional Problems for Section 7.5

EXERCISES

In Exercises 1–6, sketch the area given by the following approximations to $\int_a^b f(x)dx$. Identify each approximation as an overestimate or an underestimate.

(a) LEFT(2) (b) RIGHT(2)
(c) TRAP(2) (d) MID(2)

1.

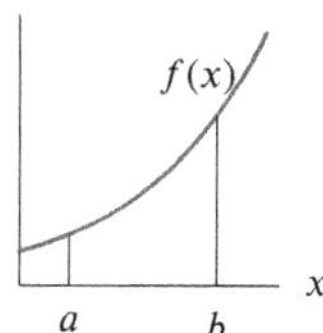

2.

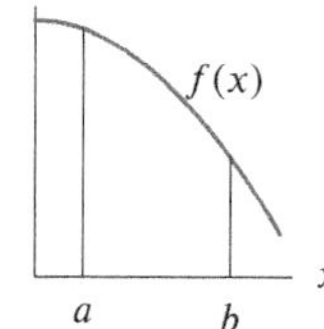

3.

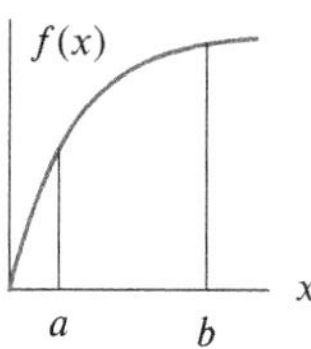

4.

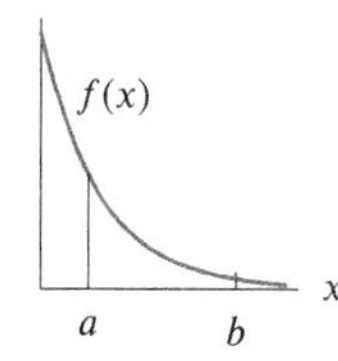

5.

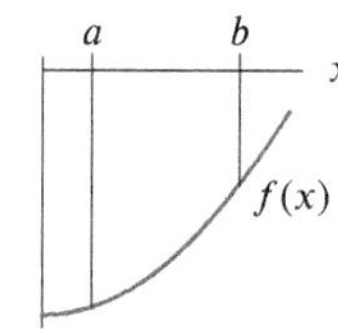

6.

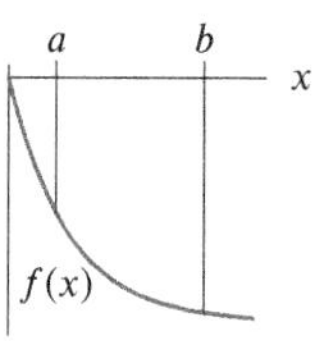

7. Calculate the following approximations to $\int_0^6 x^2 dx$.

(a) LEFT(2) (b) RIGHT(2)
(c) TRAP(2) (d) MID(2)

8. Estimate $\int_0^6 x^2 dx$ using SIMP(2).

9. (a) Find LEFT(2) and RIGHT(2) for $\int_0^4 (x^2+1)\,dx$.
(b) Illustrate your answers to part (a) graphically. Is each approximation an underestimate or overestimate?

10. (a) Find MID(2) and TRAP(2) for $\int_0^4 (x^2+1)\,dx$.
(b) Illustrate your answers to part (a) graphically. Is each approximation an underestimate or overestimate?

11. Calculate the following approximations to $\int_0^\pi \sin\theta\, d\theta$.

(a) LEFT(2) (b) RIGHT(2)
(c) TRAP(2) (d) MID(2)

PROBLEMS

12. Use Table 7.7 to estimate $\int_1^2 g(t)\,dt$ by MID(5).

Table 7.7

t	1.0	1.1	1.2	1.3	1.4	1.5
$g(t)$	−2.1	−2.9	−3.4	−3.7	−3.6	−3.2
t	1.6	1.7	1.8	1.9	2.0	2.1
$g(t)$	−2.5	−1.7	−0.7	0.5	2.1	4.1

13. Compute MID(4) for the integral $\int_0^2 f(x)\,dx$ using the values in Table 7.8.

Table 7.8

x	0	0.25	0.50	0.75	1.00	1.25
$f(x)$	2.3	5.8	7.8	9.3	10.3	10.8
x	1.50	1.75	2.00	2.25	2.50	2.75
$f(x)$	10.8	10.3	9.3	7.8	5.8	3.3

In Problems 14–17, use Table 7.9.

Table 7.9

t	0.0	0.1	0.2	0.3	0.4
$g(t)$	1.87	2.64	3.34	3.98	4.55
t	0.5	0.6	0.7	0.8	0.9
$g(t)$	5.07	5.54	5.96	6.35	6.69

14. Estimate $\int_{0.2}^{0.6} g(t)\,dt$ using a left-hand sum with $n=4$.

15. Estimate $\int_0^{0.9} g(t)\,dt$ using a right-hand sum with $n=3$.

16. Estimate $\int_0^{0.6} g(t)\,dt$ using the midpoint rule with $n=3$.

17. Assuming $g(t)$ is increasing and does not change concavity, which methods underestimate $\int_0^{0.9} g(t)\,dt$?

I. Left-hand sum II. Right-hand sum
III. Midpoint rule IV. Trapezoid rule

In Problems 18–19, compute approximations to $\int_2^3 (1/x^2)\,dx$.

18. TRAP(2) 19. MID(2)

20. (a) Estimate $\int_0^1 1/(1+x^2)\,dx$ by subdividing the interval into eight parts using:
(i) the left Riemann sum
(ii) the right Riemann sum
(iii) the trapezoidal rule
(b) Since the exact value of the integral is $\pi/4$, you can estimate the value of π using part (a). Explain why your first estimate is too large and your second estimate too small.

21. Using the table, estimate the total distance traveled from time $t = 0$ to time $t = 6$ using LEFT, RIGHT, and TRAP.

Time, t	0	1	2	3	4	5	6
Velocity, v	3	4	5	4	7	8	11

22. Using Figure 7.13, order the following approximations to the integral $\int_0^3 f(x)dx$ and its exact value from smallest to largest:
LEFT(n), RIGHT(n), MID(n), TRAP(n), Exact value.

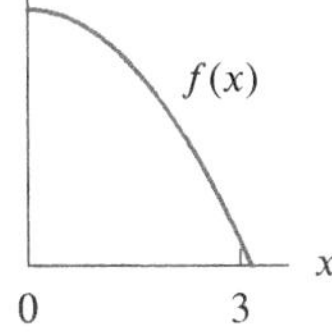

Figure 7.13

23. (a) What is the exact value of $\int_0^2 (x^3 + 3x^2)\,dx$?
(b) Find SIMP(n) for $n = 2, 4, 100$. What do you notice?

24. The results from the left, right, trapezoid, and midpoint rules used to approximate $\int_0^1 g(t)\,dt$, with the same number of subdivisions for each rule, are as follows: 0.601, 0.632, 0.633, 0.664.

(a) Using Figure 7.14, match each rule with its approximation.
(b) Between which two consecutive approximations does the true value of the integral lie?

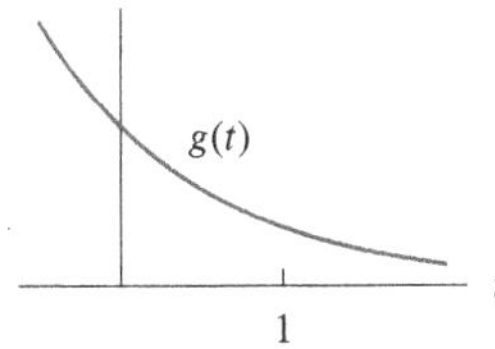

Figure 7.14

25. The graph of $y = e^{-x^2/2}$ is concave down for $-1 < x < 1$.

(a) Without calculation, identify the following as the values of $\int_0^1 e^{-x^2/2}dx$, and the approximations LEFT(1), RIGHT(1), MID(1), TRAP(1):

0.368, 0.684, 0.856, 0.882, 1.

(b) Without calculation, find the values of LEFT(1), RIGHT(1), MID(1), TRAP(1) for $\int_{-1}^0 e^{-x^2/2}dx$.

26. (a) Without calculation, explain which of the following values must be equal to each other: $\int_0^7 x\,dx$, the approximations LEFT(2), RIGHT(2), MID(2), TRAP(2).
(b) Find the values of LEFT(2), RIGHT(2), MID(2), TRAP(2), $\int_0^7 x\,dx$.

27. The table shows approximations to $\int_1^2 f(x)\,dx$ for an increasing f.

(a) The value of TRAP(5) or MID(5) in the table is incorrect. Which one? Find the correct value.
(b) The concavity of the graph of f does not change on $[1, 2]$. Is it concave up or down?
(c) Find the value SIMP(5).

LEFT(5)	RIGHT(5)	TRAP(5)	MID(5)	SIMP(5)
0.3153	0.4539	0.3890	0.3871	

■ In Problems 28–31, decide which approximation—left, right, trapezoid, or midpoint—is guaranteed to give an overestimate for $\int_0^5 f(x)\,dx$, and which is guaranteed to give an underestimate. (There may be more than one.)

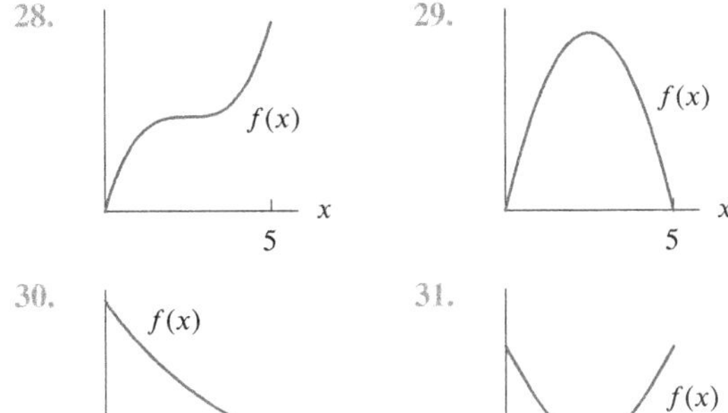

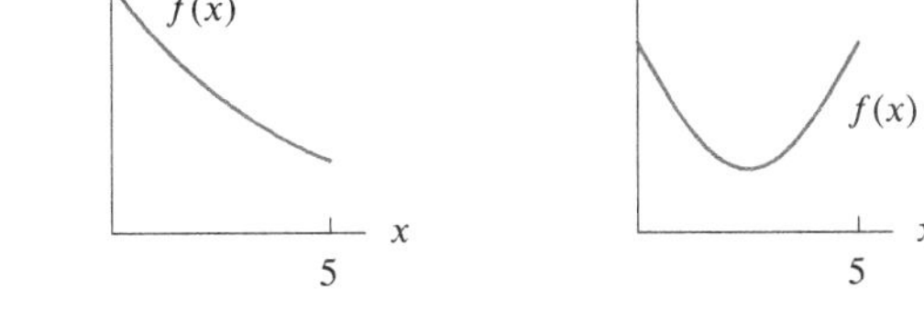

32. Consider the integral $\int_0^4 3\sqrt{x}dx$.

(a) Estimate the value of the integral using MID(2).
(b) Use the Fundamental Theorem of Calculus to find the exact value of the definite integral.
(c) What is the error for MID(2)?
(d) Use your knowledge of how errors change and your answer to part (c) to estimate the error for MID(20).
(e) Use your answer to part (d) to estimate the approximation MID(20).

33. Using a fixed number of subdivisions, we approximate the integrals of f and g on the interval in Figure 7.15.

(a) For which function, f or g, is LEFT more accurate? RIGHT? Explain.
(b) For which function, f or g, is TRAP more accurate? MID? Explain.

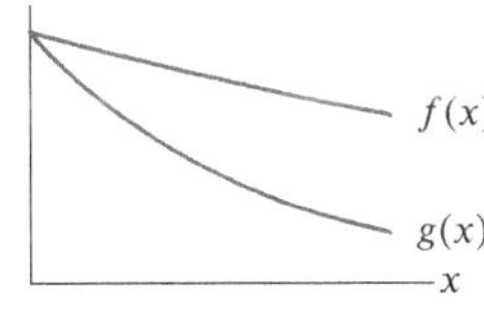

Figure 7.15

34. **(a)** Values for $f(x)$ are in the table. Of LEFT, RIGHT, TRAP, which is most likely to give the best estimate of $\int_0^{12} f(x)\,dx$? Estimate the integral using this method.
(b) Assume $f(x)$ is continuous with no critical points or points of inflection on the interval $0 \le x \le 12$. Is the estimate found in part (a) an over- or under-estimate? Explain.

x	0	3	6	9	12
$f(x)$	100	97	90	78	55

35. Table 7.10 gives approximations to an integral whose true value is 7.621372.
(a) Does the integrand appear to be increasing or decreasing? Concave up or concave down?
(b) Fill in the errors for $n = 3$ in the middle column in Table 7.10.
(c) Estimate the errors for $n = 30$ and fill in the right-hand column in Table 7.10.

Table 7.10

	Approximation $n = 3$	Error $n = 3$	Error $n = 30$
LEFT	5.416101		
RIGHT	9.307921		
TRAP	7.362011		
MID	7.742402		
SIMP	7.615605		

36. Approximations to a definite integral are given in Table 7.11; the exact value of the integral is 0.69315. Fill in the errors for $n = 2$ and estimate the errors when $n = 20$.

Table 7.11

	Approximation $n = 2$	Error $n = 2$	Error $n = 20$
LEFT	0.83333		
RIGHT	0.58333		
TRAP	0.70833		
MID	0.68571		
SIMP	0.69325		

37. **(a)** What is the exact value of $\int_0^1 7x^6\,dx$?
(b) Find LEFT(5), RIGHT(5), TRAP(5), MID(5), and SIMP(5), and compute the error for each.
(c) Repeat part (b) with $n = 10$ (instead of $n = 5$).
(d) For each rule in part (b), compute the ratio of the error for $n = 5$ divided by the error for $n = 10$. Are these values expected?

38. **(a)** What is the exact value of $\int_0^4 e^x\,dx$?
(b) Find LEFT(2), RIGHT(2), TRAP(2), MID(2), and SIMP(2). Compute the error for each.
(c) Repeat part (b) with $n = 4$ (instead of $n = 2$).
(d) For each rule in part (b), as n goes from $n = 2$ to $n = 4$, does the error go down approximately as you would expect?

39. **(a)** Find the exact value of $\int_0^{2\pi} \sin\theta\,d\theta$.
(b) Explain, using pictures, why the MID(1) and MID(2) approximations to this integral give the exact value.
(c) Does MID(3) give the exact value of this integral? How about MID(n)? Explain.

40. To investigate the relationship between the integrand and the errors in the left and right rules, imagine integrating a linear function. For one subinterval of integration, sketch lines with small f' and large f'. How do the errors compare?

41. To investigate the relationship between the integrand and the errors in the midpoint and trapezoid rules, imagine an integrand whose graph is concave down over one subinterval of integration. Sketch graphs where f'' has small magnitude and where f'' has large magnitude. How do the errors compare?

42. A computer takes 3 seconds to compute a particular definite integral accurate to 2 decimal places. Approximately how long does it take the computer to get 10 decimal places of accuracy using each of the following rules? Give your answer in seconds and in appropriate time units (minutes, hours, days, or years).

(a) LEFT **(b)** MID **(c)** SIMP

43. The width, in feet, at various points along the fairway of a hole on a golf course is given in Figure 7.16. If one pound of fertilizer covers 200 square feet, estimate the amount of fertilizer needed to fertilize the fairway.

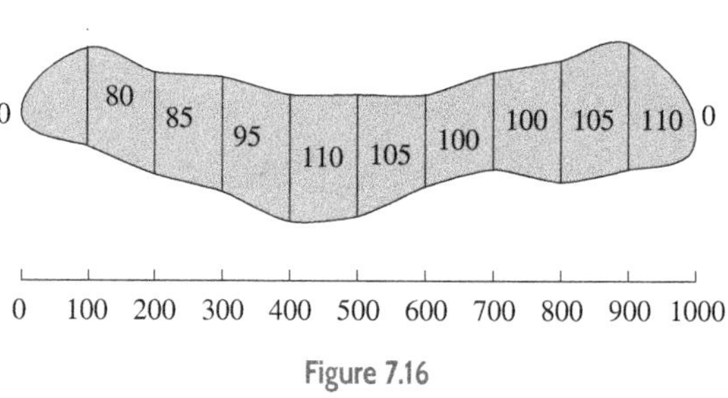

Figure 7.16

■ Problems 44–48 involve approximating $\int_a^b f(x)\,dx$.

44. Show RIGHT(n) = LEFT(n) + $f(b)\Delta x - f(a)\Delta x$.

45. Show TRAP(n) = LEFT(n) + $\frac{1}{2}(f(b) - f(a))\,\Delta x$.

46. Show LEFT($2n$) = $\frac{1}{2}$(LEFT(n) + MID(n)).

47. Check that the equations in Problems 44 and 45 hold for $\int_1^2 (1/x)\,dx$ when $n = 10$.

48. Suppose that $a = 2$, $b = 5$, $f(2) = 13$, $f(5) = 21$ and that LEFT(10) = 3.156 and MID(10) = 3.242. Use Problems 44–46 to compute RIGHT(10), TRAP(10), LEFT(20), RIGHT(20), and TRAP(20).

Strengthen Your Understanding

In Problems 49–52, explain what is wrong with the statement.

49. The midpoint rule never gives the exact value of a definite integral.
50. TRAP(n) $\to 0$ as $n \to \infty$.
51. For any integral, TRAP(n) $\geq$ MID(n).
52. If, for a certain integral, it takes 3 nanoseconds to improve the accuracy of TRAP from one digit to three digits, then it also takes 3 nanoseconds to improve the accuracy from 8 digits to 10 digits.

In Problems 53–54, give an example of:

53. A continuous function $f(x)$ on the interval $[0, 1]$ such that RIGHT(10) $< \int_0^1 f(x)dx <$ MID(10).
54. A continuous function $f(x)$ on the interval $[0, 10]$ such that TRAP(40) $>$ TRAP(80).

In Problems 55–56, decide whether the statements are true or false. Give an explanation for your answer.

55. The midpoint rule approximation to $\int_0^1 (y^2 - 1)\, dy$ is always smaller than the exact value of the integral.
56. The trapezoid rule approximation is never exact.

The left and right Riemann sums of a function f on the interval $[2, 6]$ are denoted by LEFT(n) and RIGHT(n), respectively, when the interval is divided into n equal parts.

In Problems 57–67, decide whether the statements are true for all continuous functions, f. Give an explanation for your answer.

57. If $n = 10$, then the subdivision size is $\Delta x = 1/10$.
58. If we double the value of n, we make Δx half as large.
59. LEFT(10) $\leq$RIGHT(10)
60. As n approaches infinity, LEFT(n) approaches 0.
61. LEFT(n) − RIGHT(n) $= (f(2) - f(6))\Delta x$.
62. Doubling n decreases the difference LEFT(n) − RIGHT(n) by exactly the factor $1/2$.
63. If LEFT(n) = RIGHT(n) for all n, then f is a constant function.
64. The trapezoid estimate TRAP(n) = (LEFT(n) + RIGHT(n))/2 is always closer to $\int_2^6 f(x)dx$ than LEFT(n) or RIGHT(n).
65. $\int_2^6 f(x)\, dx$ lies between LEFT(n) and RIGHT(n).
66. If LEFT(2) $< \int_a^b f(x)\, dx$, then LEFT(4) $< \int_a^b f(x)\, dx$.
67. If $0 < f' < g'$ everywhere, then the error in approximating $\int_a^b f(x)\, dx$ by LEFT(n) is less than the error in approximating $\int_a^b g(x)\, dx$ by LEFT(n).

7.6 IMPROPER INTEGRALS

Our original discussion of the definite integral $\int_a^b f(x)\, dx$ assumed that the interval $a \leq x \leq b$ was of finite length and that f was continuous. Integrals that arise in applications do not necessarily have these nice properties. In this section we investigate a class of integrals, called *improper* integrals, in which one limit of integration is infinite or the integrand is unbounded. As an example, to estimate the mass of the earth's atmosphere, we might calculate an integral which sums the mass of the air up to different heights. In order to represent the fact that the atmosphere does not end at a specific height, we let the upper limit of integration get larger and larger, or tend to infinity.

We consider improper integrals with positive integrands since they are the most common.

One Type of Improper Integral: When the Limit of Integration Is Infinite

Here is an example of an improper integral:

$$\int_1^\infty \frac{1}{x^2}\, dx.$$

To evaluate this integral, we first compute the definite integral $\int_1^b (1/x^2)\, dx$:

$$\int_1^b \frac{1}{x^2}\, dx = -x^{-1}\Big|_1^b = -\frac{1}{b} + \frac{1}{1}.$$

Now take the limit as $b \to \infty$. Since

$$\lim_{b\to\infty} \int_1^b \frac{1}{x^2}\, dx = \lim_{b\to\infty} \left(-\frac{1}{b} + 1\right) = 1,$$

we say that the improper integral $\int_1^\infty (1/x^2)\, dx$ *converges* to 1.

If we think in terms of areas, the integral $\int_1^\infty (1/x^2)\,dx$ represents the area under $f(x) = 1/x^2$ from $x = 1$ extending infinitely far to the right. (See Figure 7.17(a).) It may seem strange that this region has finite area. What our limit computations are saying is that

$$\text{When } b = 10: \quad \int_1^{10} \frac{1}{x^2}\,dx = -\frac{1}{x}\bigg|_1^{10} = -\frac{1}{10} + 1 = 0.9$$

$$\text{When } b = 100: \quad \int_1^{100} \frac{1}{x^2}\,dx = -\frac{1}{100} + 1 = 0.99$$

$$\text{When } b = 1000: \quad \int_1^{1000} \frac{1}{x^2}\,dx = -\frac{1}{1000} + 1 = 0.999$$

and so on. In other words, as b gets larger and larger, the area between $x = 1$ and $x = b$ tends to 1. See Figure 7.17(b). Thus, it does make sense to declare that $\int_1^\infty (1/x^2)\,dx = 1$.

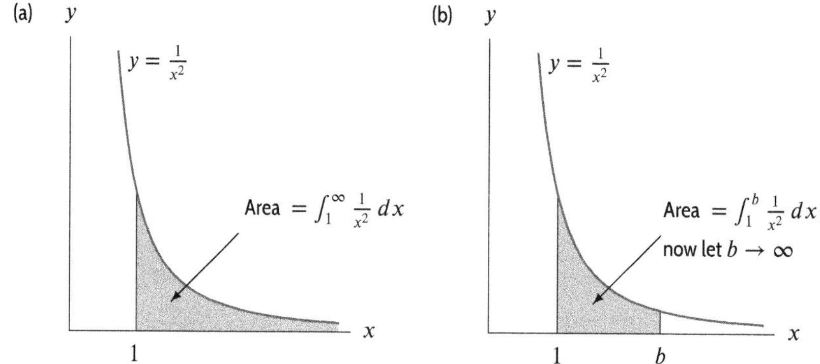

Figure 7.17: Area representation of improper integral

Of course, in another example, we might not get a finite limit as b gets larger and larger. In that case we say the improper integral *diverges*.

> Suppose $f(x)$ is positive for $x \geq a$.
>
> If $\displaystyle\lim_{b\to\infty} \int_a^b f(x)\,dx$ is a finite number, we say that $\displaystyle\int_a^\infty f(x)\,dx$ **converges** and define
>
> $$\int_a^\infty f(x)\,dx = \lim_{b\to\infty} \int_a^b f(x)\,dx.$$
>
> Otherwise, we say that $\displaystyle\int_a^\infty f(x)\,dx$ **diverges**. We define $\displaystyle\int_{-\infty}^b f(x)\,dx$ similarly.

Similar definitions apply if $f(x)$ is negative.

Example 1 Does the improper integral $\displaystyle\int_1^\infty \frac{1}{\sqrt{x}}\,dx$ converge or diverge?

Solution We consider

$$\int_1^b \frac{1}{\sqrt{x}}\,dx = \int_1^b x^{-1/2}\,dx = 2x^{1/2}\bigg|_1^b = 2b^{1/2} - 2.$$

We see that $\int_1^b (1/\sqrt{x})\,dx$ grows without bound as $b \to \infty$. We have shown that the area under the curve in Figure 7.18 is not finite. Thus we say the integral $\int_1^\infty (1/\sqrt{x})\,dx$ *diverges.* We could also say $\int_1^\infty (1/\sqrt{x})\,dx = \infty$.

Notice that $f(x) \to 0$ as $x \to \infty$ does not guarantee convergence of $\int_a^\infty f(x)\,dx$.

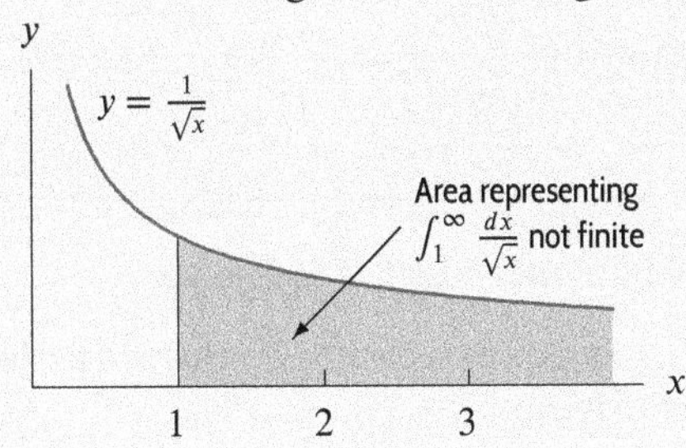

Figure 7.18: $\int_1^\infty \frac{1}{\sqrt{x}}\,dx$ diverges

What is the difference between the functions $1/x^2$ and $1/\sqrt{x}$ that makes the area under the graph of $1/x^2$ approach 1 as $x \to \infty$, whereas the area under $1/\sqrt{x}$ grows very large? Both functions approach 0 as x grows, so as b grows larger, smaller bits of area are being added to the definite integral. The difference between the functions is subtle: the values of the function $1/\sqrt{x}$ *don't shrink fast enough* for the integral to have a finite value. Of the two functions, $1/x^2$ drops to 0 much faster than $1/\sqrt{x}$, and this feature keeps the area under $1/x^2$ from growing beyond 1.

Example 2 Find $\displaystyle\int_0^\infty e^{-5x}\,dx$.

Solution First we consider $\int_0^b e^{-5x}\,dx$:

$$\int_0^b e^{-5x}\,dx = -\frac{1}{5}e^{-5x}\bigg|_0^b = -\frac{1}{5}e^{-5b} + \frac{1}{5}.$$

Since $e^{-5b} = \dfrac{1}{e^{5b}}$, this term tends to 0 as b approaches infinity, so $\int_0^\infty e^{-5x}\,dx$ converges. Its value is

$$\int_0^\infty e^{-5x}\,dx = \lim_{b\to\infty}\int_0^b e^{-5x}\,dx = \lim_{b\to\infty}\left(-\frac{1}{5}e^{-5b} + \frac{1}{5}\right) = 0 + \frac{1}{5} = \frac{1}{5}.$$

Since e^{5x} grows very rapidly, we expect that e^{-5x} will approach 0 rapidly. The fact that the area approaches $1/5$ instead of growing without bound is a consequence of the speed with which the integrand e^{-5x} approaches 0.

Example 3 Determine for which values of the exponent, p, the improper integral $\displaystyle\int_1^\infty \frac{1}{x^p}\,dx$ diverges.

Solution For $p \neq 1$,

$$\int_1^b x^{-p}\,dx = \frac{1}{-p+1}x^{-p+1}\bigg|_1^b = \left(\frac{1}{-p+1}b^{-p+1} - \frac{1}{-p+1}\right).$$

The important question is whether the exponent of b is positive or negative. If it is negative, then as b approaches infinity, b^{-p+1} approaches 0. If the exponent is positive, then b^{-p+1} grows without bound as b approaches infinity. What happens if $p = 1$? In this case we get

$$\int_1^\infty \frac{1}{x}\,dx = \lim_{b\to\infty} \ln x\bigg|_1^b = \lim_{b\to\infty} \ln b - \ln 1.$$

Since $\ln b$ becomes arbitrarily large as b approaches infinity, the integral grows without bound. We conclude that $\int_1^\infty (1/x^p)\,dx$ diverges precisely when $p \le 1$. For $p > 1$ the integral has the value

$$\int_1^\infty \frac{1}{x^p}dx = \lim_{b\to\infty}\int_1^b \frac{1}{x^p}dx = \lim_{b\to\infty}\left(\frac{1}{-p+1}b^{-p+1} - \frac{1}{-p+1}\right) = -\left(\frac{1}{-p+1}\right) = \frac{1}{p-1}.$$

Application of Improper Integrals to Energy

The energy, E, required to separate two charged particles, originally a distance a apart, to a distance b, is given by the integral

$$E = \int_a^b \frac{kq_1q_2}{r^2}\,dr$$

where q_1 and q_2 are the magnitudes of the charges and k is a constant. If q_1 and q_2 are in coulombs, a and b are in meters, and E is in joules, the value of the constant k is $9\cdot 10^9$.

Example 4 A hydrogen atom consists of a proton and an electron, with opposite charges of magnitude $1.6\cdot 10^{-19}$ coulombs. Find the energy required to take a hydrogen atom apart (that is, to move the electron from its orbit to an infinite distance from the proton). Assume that the initial distance between the electron and the proton is the Bohr radius, $R_B = 5.3\cdot 10^{-11}$ meter.

Solution Since we are moving from an initial distance of R_B to a final distance of ∞, the energy is represented by the improper integral

$$E = \int_{R_B}^\infty k\frac{q_1q_2}{r^2}\,dr = kq_1q_2\lim_{b\to\infty}\int_{R_B}^b \frac{1}{r^2}\,dr$$

$$= kq_1q_2\lim_{b\to\infty} -\frac{1}{r}\bigg|_{R_B}^b = kq_1q_2\lim_{b\to\infty}\left(-\frac{1}{b}+\frac{1}{R_B}\right) = \frac{kq_1q_2}{R_B}.$$

Substituting numerical values, we get

$$E = \frac{(9\cdot 10^9)(1.6\cdot 10^{-19})^2}{5.3\cdot 10^{-11}} \approx 4.35\cdot 10^{-18}\text{ joules.}$$

This is about the amount of energy needed to lift a speck of dust 0.000000025 inch off the ground. (In other words, not much!)

What happens if the limits of integration are $-\infty$ and ∞? In this case, we break the integral at any point and write the original integral as a sum of two new improper integrals.

For a positive function $f(x)$, we can use any (finite) number c to define

$$\int_{-\infty}^\infty f(x)\,dx = \int_{-\infty}^c f(x)\,dx + \int_c^\infty f(x)\,dx.$$

If *either* of the two new improper integrals diverges, we say the original integral diverges. Only if both of the new integrals have a finite value do we add the values to get a finite value for the original integral.

It is not hard to show that the preceding definition does not depend on the choice for c.

Another Type of Improper Integral: When the Integrand Becomes Infinite

There is another way for an integral to be improper. The interval may be finite but the function may be unbounded near some points in the interval. For example, consider $\int_0^1 (1/\sqrt{x})\,dx$. Since the graph of $y = 1/\sqrt{x}$ has a vertical asymptote at $x = 0$, the region between the graph, the x-axis, and the lines $x = 0$ and $x = 1$ is unbounded. Instead of extending to infinity in the horizontal direction as in the previous improper integrals, this region extends to infinity in the vertical direction. See Figure 7.19(a). We handle this improper integral in a similar way as before: we compute $\int_a^1 (1/\sqrt{x})\,dx$ for values of a slightly larger than 0 and look at what happens as a approaches 0 from the positive side. (This is written as $a \to 0^+$.)

First we compute the integral:

$$\int_a^1 \frac{1}{\sqrt{x}}\,dx = 2x^{1/2}\Big|_a^1 = 2 - 2a^{1/2}.$$

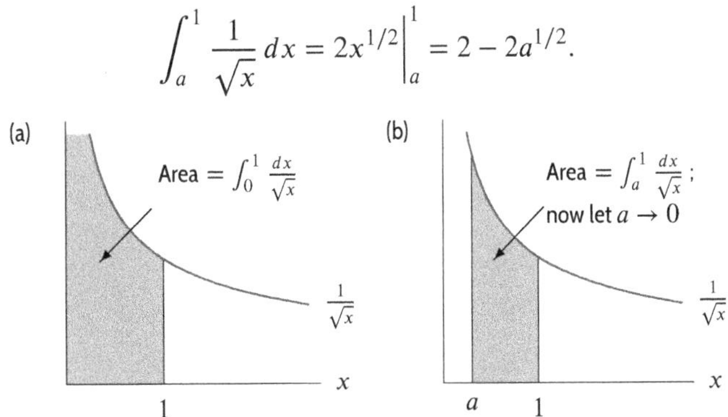

Figure 7.19: Area representation of improper integral

Now we take the limit:

$$\lim_{a\to 0^+} \int_a^1 \frac{1}{\sqrt{x}}\,dx = \lim_{a\to 0^+} (2 - 2a^{1/2}) = 2.$$

Since the limit is finite, we say the improper integral converges, and that

$$\int_0^1 \frac{1}{\sqrt{x}}\,dx = 2.$$

Geometrically, what we have done is to calculate the finite area between $x = a$ and $x = 1$ and take the limit as a tends to 0 from the right. See Figure 7.19(b). Since the limit exists, the integral converges to 2. If the limit did not exist, we would say the improper integral diverges.

Example 5 Investigate the convergence of $\displaystyle\int_0^2 \frac{1}{(x-2)^2}\,dx$.

Solution This is an improper integral since the integrand tends to infinity as x approaches 2 and is undefined at $x = 2$. Since the trouble is at the right endpoint, we replace the upper limit by b, and let b tend to 2 from the left. This is written $b \to 2^-$, with the "–" signifying that 2 is approached from below. See Figure 7.20.

$$\int_0^2 \frac{1}{(x-2)^2}\,dx = \lim_{b\to 2^-} \int_0^b \frac{1}{(x-2)^2}\,dx = \lim_{b\to 2^-} (-1)(x-2)^{-1}\Big|_0^b = \lim_{b\to 2^-} \left(-\frac{1}{(b-2)} - \frac{1}{2}\right).$$

Therefore, since $\displaystyle\lim_{b\to 2^-} \left(-\frac{1}{b-2}\right)$ does not exist, the integral diverges.

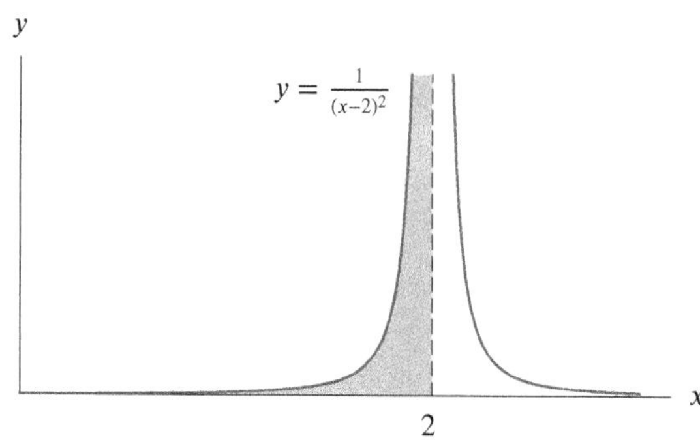

Figure 7.20: Shaded area represents $\int_0^2 \frac{1}{(x-2)^2}\,dx$

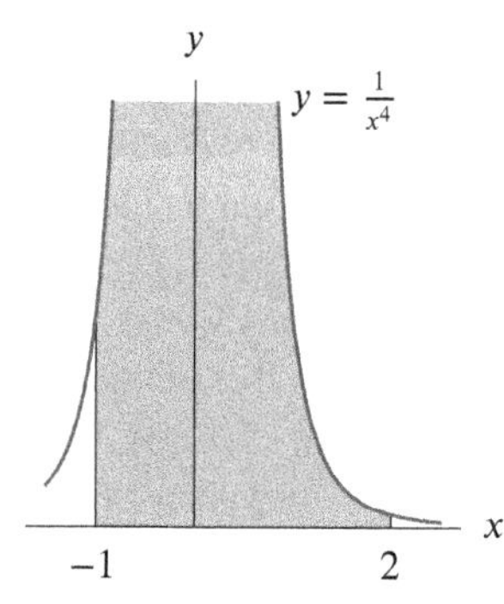

Figure 7.21: Shaded area represents $\int_{-1}^2 \frac{1}{x^4}\,dx$

Suppose $f(x)$ is positive and continuous on $a \le x < b$ and tends to infinity as $x \to b$. If $\lim_{c\to b^-} \int_a^c f(x)\,dx$ is a finite number, we say that $\int_a^b f(x)\,dx$ **converges** and define

$$\int_a^b f(x)\,dx = \lim_{c\to b^-} \int_a^c f(x)\,dx.$$

Otherwise, we say that $\int_a^b f(x)\,dx$ **diverges**.

When $f(x)$ tends to infinity as x approaches a, we define convergence in a similar way. In addition, an integral can be improper because the integrand tends to infinity *inside* the interval of integration rather than at an endpoint. In this case, we break the given integral into two (or more) improper integrals so that the integrand tends to infinity only at endpoints.

Suppose that $f(x)$ is positive and continuous on $[a, b]$ except at the point c. If $f(x)$ tends to infinity as $x \to c$, then we define

$$\int_a^b f(x)\,dx = \int_a^c f(x)\,dx + \int_c^b f(x)\,dx.$$

If *either* of the two new improper integrals diverges, we say the original integral diverges. Only if *both* of the new integrals have a finite value do we add the values to get a finite value for the original integral.

Example 6 Investigate the convergence of $\int_{-1}^2 \frac{1}{x^4}\,dx$.

Solution See Figure 7.21. The trouble spot is $x = 0$, rather than $x = -1$ or $x = 2$. We break the given improper integral into two improper integrals each of which has $x = 0$ as an endpoint:

$$\int_{-1}^2 \frac{1}{x^4}\,dx = \int_{-1}^0 \frac{1}{x^4}\,dx + \int_0^2 \frac{1}{x^4}\,dx.$$

We can now use the previous technique to evaluate the new integrals, if they converge. Since

$$\int_0^2 \frac{1}{x^4}\,dx = \lim_{a\to 0^+} -\frac{1}{3}x^{-3}\Big|_a^2 = \lim_{a\to 0^+}\left(-\frac{1}{3}\right)\left(\frac{1}{8}-\frac{1}{a^3}\right)$$

the integral $\int_0^2 (1/x^4)\,dx$ diverges. Thus, the original integral diverges. A similar computation shows that $\int_{-1}^0 (1/x^4)\,dx$ also diverges.

It is easy to miss an improper integral when the integrand tends to infinity inside the interval. For example, it is fundamentally incorrect to say that $\int_{-1}^2 (1/x^4)\,dx = -\frac{1}{3}x^{-3}\Big|_{-1}^2 = -\frac{1}{24} - \frac{1}{3} = -\frac{3}{8}$.

Example 7 Find $\displaystyle\int_0^6 \frac{1}{(x-4)^{2/3}}\,dx$.

Solution Figure 7.22 shows that the trouble spot is at $x = 4$, so we break the integral at $x = 4$ and consider the separate parts.

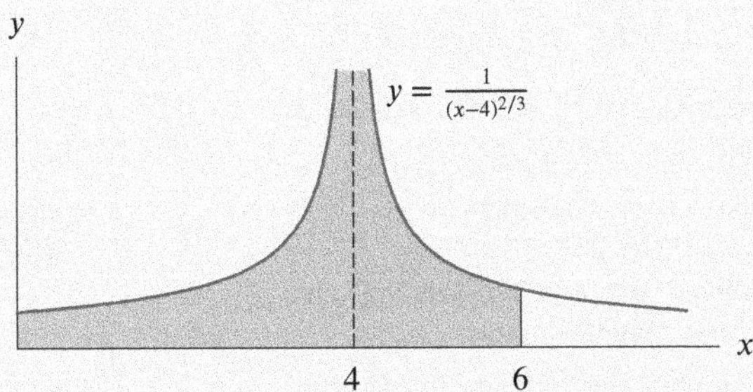

Figure 7.22: Shaded area represents $\int_0^6 \frac{1}{(x-4)^{2/3}}\,dx$

We have

$$\int_0^4 \frac{1}{(x-4)^{2/3}}\,dx = \lim_{b\to 4^-} 3(x-4)^{1/3}\Big|_0^b = \lim_{b\to 4^-}\left(3(b-4)^{1/3} - 3(-4)^{1/3}\right) = 3(4)^{1/3}.$$

Similarly,

$$\int_4^6 \frac{1}{(x-4)^{2/3}}\,dx = \lim_{a\to 4^+} 3(x-4)^{1/3}\Big|_a^6 = \lim_{a\to 4^+}\left(3\cdot 2^{1/3} - 3(a-4)^{1/3}\right) = 3(2)^{1/3}.$$

Since both of these integrals converge, the original integral converges:

$$\int_0^6 \frac{1}{(x-4)^{2/3}}\,dx = 3(4)^{1/3} + 3(2)^{1/3} = 8.542.$$

Finally, there is a question of what to do when an integral is improper at both endpoints. In this case, we just break the integral at any interior point of the interval. The original integral diverges if either or both of the new integrals diverge.

Example 8 Investigate the convergence of $\displaystyle\int_0^\infty \frac{1}{x^2}\,dx$.

Solution This integral is improper both because the upper limit is ∞ and because the function is undefined at $x = 0$. We break the integral into two parts at, say, $x = 1$. We know by Example 3 that $\int_1^\infty (1/x^2)\,dx$ has a finite value. However, the other part, $\int_0^1 (1/x^2)\,dx$, diverges since:

$$\int_0^1 \frac{1}{x^2}\,dx = \lim_{a\to 0^+} -x^{-1}\Big|_a^1 = \lim_{a\to 0^+}\left(\frac{1}{a} - 1\right).$$

Therefore $\displaystyle\int_0^\infty \frac{1}{x^2}\,dx$ diverges as well.

Exercises and Problems for Section 7.6 Online Resource: Additional Problems for Section 7.6

EXERCISES

1. Shade the area represented by:
   (a) $\int_1^\infty (1/x^2)\,dx$ (b) $\int_0^1 (1/\sqrt{x})\,dx$
2. Evaluate the improper integral $\int_0^\infty e^{-0.4x}\,dx$ and sketch the area it represents.
3. (a) Use a calculator or computer to estimate $\int_0^b xe^{-x}\,dx$ for $b = 5, 10, 20$.
   (b) Use your answers to part (a) to estimate the value of $\int_0^\infty xe^{-x}\,dx$, assuming it is finite.
4. (a) Sketch the the area represented by the improper integral $\int_{-\infty}^\infty e^{-x^2}\,dx$.
   (b) Use a calculator or computer to estimate $\int_{-a}^a e^{-x^2}\,dx$ for $a = 1,2,3,4,5$.
   (c) Use the answers to part (b) to estimate the value of $\int_{-\infty}^\infty e^{-x^2}\,dx$, assuming it is finite.

In Exercises 5–39, calculate the integral if it converges. You may calculate the limit by appealing to the dominance of one function over another, or by l'Hopital's rule.

5. $\displaystyle\int_1^\infty \frac{1}{5x+2}\,dx$
6. $\displaystyle\int_1^\infty \frac{1}{(x+2)^2}\,dx$
7. $\displaystyle\int_0^1 \ln x\,dx$
8. $\displaystyle\int_0^\infty e^{-\sqrt{x}}\,dx$
9. $\displaystyle\int_0^\infty xe^{-x^2}\,dx$
10. $\displaystyle\int_1^\infty e^{-2x}\,dx$
11. $\displaystyle\int_0^\infty \frac{x}{e^x}\,dx$
12. $\displaystyle\int_1^\infty \frac{x}{4+x^2}\,dx$
13. $\displaystyle\int_{-\infty}^0 \frac{e^x}{1+e^x}\,dx$
14. $\displaystyle\int_{-\infty}^\infty \frac{dz}{z^2+25}$
15. $\displaystyle\int_1^\infty \frac{z}{(1+z^2)^3}\,dz$
16. $\displaystyle\int_0^\infty \frac{z}{3+z^2}\,dz$
17. $\displaystyle\int_0^4 \frac{1}{\sqrt{x}}\,dx$
18. $\displaystyle\int_{\pi/4}^{\pi/2} \frac{\sin x}{\sqrt{\cos x}}\,dx$
19. $\displaystyle\int_0^1 \frac{1}{v}\,dv$
20. $\displaystyle\int_0^1 \frac{x^4+1}{x}\,dx$
21. $\displaystyle\int_1^\infty \frac{1}{x^2+1}\,dx$
22. $\displaystyle\int_1^\infty \frac{1}{\sqrt{x^2+1}}\,dx$
23. $\displaystyle\int_0^\infty \frac{dt}{\sqrt{t+1}}$
24. $\displaystyle\int_{-1}^1 \frac{dt}{\sqrt{t+1}}$
25. $\displaystyle\int_0^4 \frac{-1}{u^2-16}\,du$
26. $\displaystyle\int_1^\infty \frac{y}{y^4+1}\,dy$
27. $\displaystyle\int_2^\infty \frac{dx}{x\ln x}$
28. $\displaystyle\int_0^1 \frac{\ln x}{x}\,dx$
29. $\displaystyle\int_{16}^{20} \frac{1}{y^2-16}\,dy$
30. $\displaystyle\int_0^\pi \frac{1}{\sqrt{x}}e^{-\sqrt{x}}\,dx$
31. $\displaystyle\int_3^\infty \frac{dx}{x(\ln x)^2}$
32. $\displaystyle\int_0^2 \frac{1}{\sqrt{4-x^2}}\,dx$
33. $\displaystyle\int_4^\infty \frac{dx}{(x-1)^2}$
34. $\displaystyle\int_4^\infty \frac{dx}{x^2-1}$
35. $\displaystyle\int_7^\infty \frac{dy}{\sqrt{y-5}}$
36. $\displaystyle\int_0^3 \frac{y\,dy}{\sqrt{9-y^2}}$
37. $\displaystyle\int_0^\infty te^{-2t}\,dt$
38. $\displaystyle\int_0^{\pi/2} \tan\theta\,d\theta$
39. $\displaystyle\int_3^6 \frac{d\theta}{(4-\theta)^2}$

PROBLEMS

40. Find a formula (not involving integrals) for

$$f(x) = \int_{-\infty}^{x} e^t \, dt.$$

41. In statistics we encounter $P(x)$, a function defined by

$$P(x) = \frac{1}{\sqrt{\pi}} \int_0^x e^{-t^2} \, dt.$$

Use a calculator or computer to evaluate

(a) $P(1)$ **(b)** $P(\infty)$

42. Find the area under the curve $y = xe^{-x}$ for $x \geq 0$.

43. Find the area under the curve $y = 1/\cos^2 t$ between $t = 0$ and $t = \pi/2$.

In Problems 44–47, evaluate $f(3)$.

44. $f(x) = \int_0^{\infty} x^{-t} dt$

45. $f(x) = \int_1^{\infty} t^{-x} dt$

46. $f(x) = \int_0^{\infty} xe^{-xt} \, dt$

47. $f(x) = \int_0^{\infty} 2txe^{-tx^2} dt$

48. The rate, r, at which people get sick during an epidemic of the flu can be approximated by $r = 1000te^{-0.5t}$, where r is measured in people/day and t is measured in days since the start of the epidemic.

(a) Sketch a graph of r as a function of t.
(b) When are people getting sick fastest?
(c) How many people get sick altogether?

49. Find the energy required to separate opposite electric charges of magnitude 1 coulomb. The charges are initially 1 meter apart and one is moved infinitely far from the other. (The definition of energy is on page 388.)

50. The probability that a light bulb manufactured by a company lasts at least a hundred hours is

$$\int_a^{\infty} 0.012e^{-0.012t} \, dt.$$

The CEO claims that 90% of the company's light bulbs last at least 1000 hours. Is this statement accurate?

51. Given that $\int_{-\infty}^{\infty} e^{-x^2} \, dx = \sqrt{\pi}$, calculate the exact value of

$$\int_{-\infty}^{\infty} e^{-(x-a)^2/b} \, dx.$$

52. Assuming $g(x)$ is a differentiable function whose values are bounded for all x, derive Stein's identity, which is used in statistics:

$$\int_{-\infty}^{\infty} g'(x)e^{-x^2/2} \, dx = \int_{-\infty}^{\infty} xg(x)e^{-x^2/2} \, dx.$$

53. Given that

$$\int_0^{\infty} \frac{x^4 e^x}{(e^x - 1)^2} \, dx = \frac{4\pi^4}{15}$$

evaluate

$$\int_0^{\infty} \frac{x^4 e^{2x}}{(e^{2x} - 1)^2} \, dx.$$

54. Let $x_0 = a = \ln 3, x_n = b, \Delta x = (b - a)/n, x_i = x_0 + i\Delta x$. Evaluate

$$\lim_{b\to\infty} \left(\lim_{n\to\infty} \sum_{i=1}^{n} e^{-x_i} \Delta x \right)$$

55. Let $x_0 = a = 4, x_n = b, \Delta x = (b - a)/n, x_i = x_0 + i\Delta x$. Evaluate

$$\lim_{b\to\infty} \left(\lim_{n\to\infty} \sum_{i=1}^{n} \frac{1}{x_i^2} \Delta x \right)$$

Strengthen Your Understanding

In Problems 56–57, explain what is wrong with the statement.

56. If both $\int_1^{\infty} f(x) \, dx$ and $\int_1^{\infty} g(x) \, dx$ diverge, then so does $\int_1^{\infty} f(x)g(x) \, dx$.

57. If $\int_1^{\infty} f(x) \, dx$ diverges, then $\lim_{x\to\infty} f(x) \neq 0$.

In Problems 58–59, give an example of:

58. A function $f(x)$, continuous for $x \geq 1$, such that $\lim_{x\to\infty} f(x) = 0$, but $\int_1^{\infty} f(x) dx$ diverges.

59. A function $f(x)$, continuous at $x = 2$ and $x = 5$, such that the integral $\int_2^5 f(x) \, dx$ is improper and divergent.

In Problems 60–65, decide whether the statements are true or false. Give an explanation for your answer.

60. If f is continuous for all x and $\int_0^{\infty} f(x) \, dx$ converges, then so does $\int_a^{\infty} f(x) \, dx$ for all positive a.

61. If $f(x)$ is a positive periodic function, then $\int_0^{\infty} f(x) \, dx$ diverges.

62. If $f(x)$ is continuous and positive for $x > 0$ and if $\lim_{x\to\infty} f(x) = 0$, then $\int_0^{\infty} f(x) \, dx$ converges.

63. If $f(x)$ is continuous and positive for $x > 0$ and if $\lim_{x\to\infty} f(x) = \infty$, then $\int_0^{\infty} (1/f(x)) \, dx$ converges.

64. If $\int_0^\infty f(x)\,dx$ and $\int_0^\infty g(x)\,dx$ both converge, then $\int_0^\infty (f(x)+g(x))\,dx$ converges.

65. If $\int_0^\infty f(x)\,dx$ and $\int_0^\infty g(x)\,dx$ both diverge, then $\int_0^\infty (f(x)+g(x))\,dx$ diverges.

■ Suppose that f is continuous for all real numbers and that $\int_0^\infty f(x)\,dx$ converges. Let a be any positive number. Decide which of the statements in Problems 66–69 are true and which are false. Give an explanation for your answer.

66. $\int_0^\infty af(x)\,dx$ converges.

67. $\int_0^\infty f(ax)\,dx$ converges.

68. $\int_0^\infty f(a+x)\,dx$ converges.

69. $\int_0^\infty (a+f(x))\,dx$ converges.

7.7 COMPARISON OF IMPROPER INTEGRALS

Making Comparisons

Sometimes it is difficult to find the exact value of an improper integral by antidifferentiation, but it may be possible to determine whether an integral converges or diverges. The key is to *compare* the given integral to one whose behavior we already know. Let's look at an example.

Example 1 Determine whether $\displaystyle\int_1^\infty \frac{1}{\sqrt{x^3+5}}\,dx$ converges.

Solution First, let's see what this integrand does as $x\to\infty$. For large x, the 5 becomes insignificant compared with the x^3, so

$$\frac{1}{\sqrt{x^3+5}} \approx \frac{1}{\sqrt{x^3}} = \frac{1}{x^{3/2}}.$$

Since

$$\int_1^\infty \frac{1}{\sqrt{x^3}}\,dx = \int_1^\infty \frac{1}{x^{3/2}}\,dx = \lim_{b\to\infty}\int_1^b \frac{1}{x^{3/2}}\,dx = \lim_{b\to\infty} -2x^{-1/2}\Big|_1^b = \lim_{b\to\infty}\left(2-2b^{-1/2}\right) = 2,$$

the integral $\int_1^\infty (1/x^{3/2})\,dx$ converges. So we expect our integral to converge as well.

In order to confirm this, we observe that for $0\le x^3\le x^3+5$, we have

$$\frac{1}{\sqrt{x^3+5}} \le \frac{1}{\sqrt{x^3}}.$$

and so for $b\ge 1$,

$$\int_1^b \frac{1}{\sqrt{x^3+5}}\,dx \le \int_1^b \frac{1}{\sqrt{x^3}}\,dx.$$

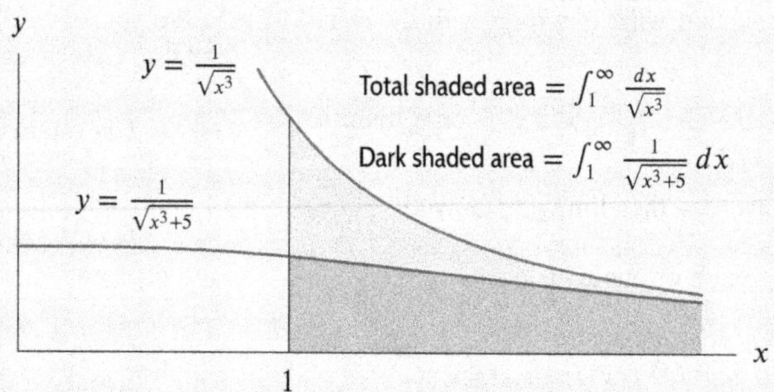

Figure 7.23: Graph showing $\int_1^\infty \frac{1}{\sqrt{x^3+5}}\,dx \le \int_1^\infty \frac{dx}{\sqrt{x^3}}$

(See Figure 7.23.) Since $\int_1^b (1/\sqrt{x^3+5})\,dx$ increases as b approaches infinity but is always smaller than $\int_1^b (1/x^{3/2})\,dx < \int_1^\infty (1/x^{3/2})\,dx = 2$, we know $\int_1^\infty (1/\sqrt{x^3+5})\,dx$ must have a finite value less than 2. Thus,

$$\int_1^\infty \frac{dx}{\sqrt{x^3+5}} \text{ converges to a value less than 2.}$$

Notice that we first looked at the behavior of the integrand as $x \to \infty$. This is useful because the convergence or divergence of the integral is determined by what happens as $x \to \infty$.

The Comparison Test for $\int_a^\infty f(x)\,dx$

Assume $f(x)$ is positive. Making a comparison involves two stages:

1. Guess, by looking at the behavior of the integrand for large x, whether the integral converges or not. (This is the "behaves like" principle.)
2. Confirm the guess by comparison with a positive function $g(x)$:
   - If $f(x) \le g(x)$ and $\int_a^\infty g(x)\,dx$ converges, then $\int_a^\infty f(x)\,dx$ converges.
   - If $g(x) \le f(x)$ and $\int_a^\infty g(x)\,dx$ diverges, then $\int_a^\infty f(x)\,dx$ diverges.

Example 2 Decide whether $\displaystyle\int_4^\infty \frac{dt}{(\ln t) - 1}$ converges or diverges.

Solution Since $\ln t$ grows without bound as $t \to \infty$, the -1 is eventually going to be insignificant in comparison to $\ln t$. Thus, as far as convergence is concerned,

$$\int_4^\infty \frac{1}{(\ln t) - 1}\,dt \quad \text{behaves like} \quad \int_4^\infty \frac{1}{\ln t}\,dt.$$

Does $\int_4^\infty (1/\ln t)\,dt$ converge or diverge? Since $\ln t$ grows very slowly, $1/\ln t$ goes to zero very slowly, and so the integral probably does not converge. We know that $(\ln t) - 1 < \ln t < t$ for all positive t. So, provided $t > e$, we take reciprocals:

$$\frac{1}{(\ln t) - 1} > \frac{1}{\ln t} > \frac{1}{t}.$$

Since $\int_4^\infty (1/t)\,dt$ diverges, we conclude that

$$\int_4^\infty \frac{1}{(\ln t) - 1}\,dt \text{ diverges.}$$

How Do We Know What to Compare With?

In Examples 1 and 2, we investigated the convergence of an integral by comparing it with an easier integral. How did we pick the easier integral? This is a matter of trial and error, guided by any information we get by looking at the original integrand as $x \to \infty$. We want the comparison integrand to be easy and, in particular, to have a simple antiderivative.

Useful Integrals for Comparison

- $\int_1^{\infty} \frac{1}{x^p} dx$ converges for $p > 1$ and diverges for $p \leq 1$.
- $\int_0^{1} \frac{1}{x^p} dx$ converges for $p < 1$ and diverges for $p \geq 1$.
- $\int_0^{\infty} e^{-ax} dx$ converges for $a > 0$.

Of course, we can use any function for comparison, provided we can determine its behavior.

Example 3 Investigate the convergence of $\displaystyle\int_1^{\infty} \frac{(\sin x)+3}{\sqrt{x}} dx$.

Solution Since it looks difficult to find an antiderivative of this function, we try comparison. What happens to this integrand as $x \to \infty$? Since $\sin x$ oscillates between -1 and 1,

$$\frac{2}{\sqrt{x}} = \frac{-1+3}{\sqrt{x}} \leq \frac{(\sin x)+3}{\sqrt{x}} \leq \frac{1+3}{\sqrt{x}} = \frac{4}{\sqrt{x}},$$

the integrand oscillates between $2/\sqrt{x}$ and $4/\sqrt{x}$. (See Figure 7.24.)

What do $\int_1^{\infty}(2/\sqrt{x})\,dx$ and $\int_1^{\infty}(4/\sqrt{x})\,dx$ do? As far as convergence is concerned, they certainly do the same thing, and whatever that is, the original integral does it too. It is important to notice that $\sqrt{x}$ grows very slowly. This means that $1/\sqrt{x}$ gets small slowly, which means that convergence is unlikely. Since $\sqrt{x} = x^{1/2}$, the result in the preceding box (with $p = \frac{1}{2}$) tells us that $\int_1^{\infty}(1/\sqrt{x})\,dx$ diverges. So the comparison test tells us that the original integral diverges.

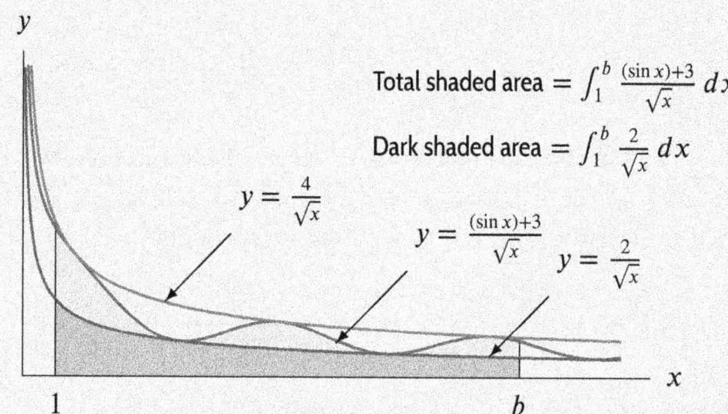

Figure 7.24: Graph showing $\int_1^b \frac{2}{\sqrt{x}} dx \leq \int_1^b \frac{(\sin x)+3}{\sqrt{x}} dx$, for $b \geq 1$

Notice that there are two possible comparisons we could have made in Example 3:

$$\frac{2}{\sqrt{x}} \leq \frac{(\sin x)+3}{\sqrt{x}} \quad \text{or} \quad \frac{(\sin x)+3}{\sqrt{x}} \leq \frac{4}{\sqrt{x}}.$$

Since both $\int_1^{\infty}(2/\sqrt{x})\,dx$ and $\int_1^{\infty}(4/\sqrt{x})\,dx$ diverge, only the first comparison is useful. Knowing that an integral is *smaller* than a divergent integral is of no help whatsoever!

The next example shows what to do if the comparison does not hold throughout the interval of integration.

Example 4 Show $\int_1^\infty e^{-x^2/2}\,dx$ converges.

Solution We know that $e^{-x^2/2}$ goes very rapidly to zero as $x \to \infty$, so we expect this integral to converge. Hence we look for some larger integrand which has a convergent integral. One possibility is $\int_1^\infty e^{-x}\,dx$, because e^{-x} has an elementary antiderivative and $\int_1^\infty e^{-x}\,dx$ converges. What is the relationship between $e^{-x^2/2}$ and e^{-x}? We know that for $x \geq 2$,

$$x \leq \frac{x^2}{2} \quad \text{so} \quad -\frac{x^2}{2} \leq -x,$$

and so, for $x \geq 2$

$$e^{-x^2/2} \leq e^{-x}.$$

Since this inequality holds only for $x \geq 2$, we split the interval of integration into two pieces:

$$\int_1^\infty e^{-x^2/2}\,dx = \int_1^2 e^{-x^2/2}\,dx + \int_2^\infty e^{-x^2/2}\,dx.$$

Now $\int_1^2 e^{-x^2/2}\,dx$ is finite (it is not improper) and $\int_2^\infty e^{-x^2/2}\,dx$ is finite by comparison with $\int_2^\infty e^{-x}\,dx$. Therefore, $\int_1^\infty e^{-x^2/2}\,dx$ is the sum of two finite pieces and therefore must be finite.

The previous example illustrates the following general principle:

> If f is positive and continuous on $[a, b]$,
>
> $$\int_a^\infty f(x)\,dx \text{ and } \int_b^\infty f(x)\,dx$$
>
> either both converge or both diverge.

In particular, when the comparison test is applied to $\int_a^\infty f(x)\,dx$, the inequalities for $f(x)$ and $g(x)$ do not need to hold for all $x \geq a$ but only for x greater than some value, say b.

Exercises and Problems for Section 7.7

EXERCISES

In Exercises 1–9, use the box on page 396 and the behavior of rational and exponential functions as $x \to \infty$ to predict whether the integrals converge or diverge.

1. $\int_1^\infty \frac{x^2}{x^4+1}\,dx$

2. $\int_2^\infty \frac{x^3}{x^4-1}\,dx$

3. $\int_1^\infty \frac{x^2+1}{x^3+3x+2}\,dx$

4. $\int_1^\infty \frac{1}{x^2+5x+1}\,dx$

5. $\int_1^\infty \frac{x}{x^2+2x+4}\,dx$

6. $\int_1^\infty \frac{x^2-6x+1}{x^2+4}\,dx$

7. $\int_1^\infty \frac{5x+2}{x^4+8x^2+4}\,dx$

8. $\int_1^\infty \frac{1}{e^{5t}+2}\,dt$

9. $\int_1^\infty \frac{x^2+4}{x^4+3x^2+11}\,dx$

■ In Exercises 10–13, what, if anything, does the comparison tell us about the convergence of the integral?

10. $\displaystyle\int_1^{\infty} \frac{\sin^2 x}{x^2}\,dx$, compare with $\dfrac{1}{x^2}$

11. $\displaystyle\int_1^{\infty} \frac{\sin^2 x}{x}\,dx$, compare with $\dfrac{1}{x}$

12. $\displaystyle\int_0^{1} \frac{\sin^2 x}{x^2}\,dx$, compare with $\dfrac{1}{x^2}$

13. $\displaystyle\int_0^{1} \frac{\sin^2 x}{\sqrt{x}}\,dx$, compare with $\dfrac{1}{\sqrt{x}}$

■ In Exercises 14–29, decide if the improper integral converges or diverges.

14. $\displaystyle\int_{50}^{\infty} \frac{dz}{z^3}$

15. $\displaystyle\int_1^{\infty} \frac{dx}{1+x}$

16. $\displaystyle\int_1^{\infty} \frac{dx}{x^3+1}$

17. $\displaystyle\int_5^{8} \frac{6}{\sqrt{t-5}}\,dt$

18. $\displaystyle\int_0^{1} \frac{1}{x^{19/20}}\,dx$

19. $\displaystyle\int_{-1}^{5} \frac{dt}{(t+1)^2}$

20. $\displaystyle\int_{-\infty}^{\infty} \frac{du}{1+u^2}$

21. $\displaystyle\int_1^{\infty} \frac{du}{u+u^2}$

22. $\displaystyle\int_1^{\infty} \frac{d\theta}{\sqrt{\theta^2+1}}$

23. $\displaystyle\int_2^{\infty} \frac{d\theta}{\sqrt{\theta^3+1}}$

24. $\displaystyle\int_0^{1} \frac{d\theta}{\sqrt{\theta^3+\theta}}$

25. $\displaystyle\int_0^{\infty} \frac{dy}{1+e^y}$

26. $\displaystyle\int_1^{\infty} \frac{2+\cos\phi}{\phi^2}\,d\phi$

27. $\displaystyle\int_0^{\infty} \frac{dz}{e^z+2^z}$

28. $\displaystyle\int_0^{\pi} \frac{2-\sin\phi}{\phi^2}\,d\phi$

29. $\displaystyle\int_4^{\infty} \frac{3+\sin\alpha}{\alpha}\,d\alpha$

PROBLEMS

30. The graphs of $y = 1/x$, $y = 1/x^2$ and the functions $f(x)$, $g(x)$, $h(x)$, and $k(x)$ are shown in Figure 7.25.

(a) Is the area between $y = 1/x$ and $y = 1/x^2$ on the interval from $x = 1$ to ∞ finite or infinite? Explain.

(b) Using the graph, decide whether the integral of each of the functions $f(x)$, $g(x)$, $h(x)$ and $k(x)$ on the interval from $x = 1$ to ∞ converges, diverges, or whether it is impossible to tell.

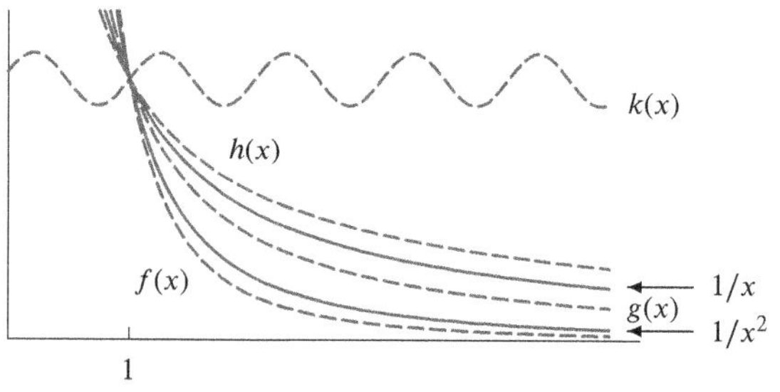

Figure 7.25

31. For $f(x)$ in Figure 7.26, both $\int_a^{\infty} f(x)\,dx$ and $\int_0^{a} f(x)\,dx$ converge. Decide if the following integrals converge, diverge, or could do either. Assume that $0 < g(x) < f(x)$ for $x > a$ and $0 < f(x) < g(x)$ for $x < a$.

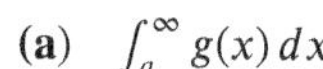

(a) $\int_a^{\infty} g(x)\,dx$ **(b)** $\int_0^{a} g(x)\,dx$

(c) $\int_0^{\infty} f(x)\,dx$ **(d)** $\int_0^{\infty} g(x)\,dx$

(e) $\int_a^{\infty} (f(x)-g(x))\,dx$ **(f)** $\displaystyle\int_a^{\infty} \frac{1}{f(x)}\,dx$

(g) $\int_a^{\infty} (f(x))^2\,dx$

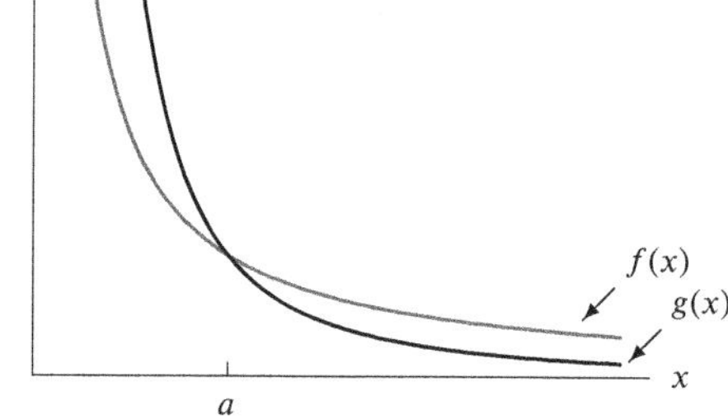

Figure 7.26

32. Suppose $\int_a^{\infty} f(x)\,dx$ converges. What does Figure 7.27 suggest about the convergence of $\int_a^{\infty} g(x)\,dx$?

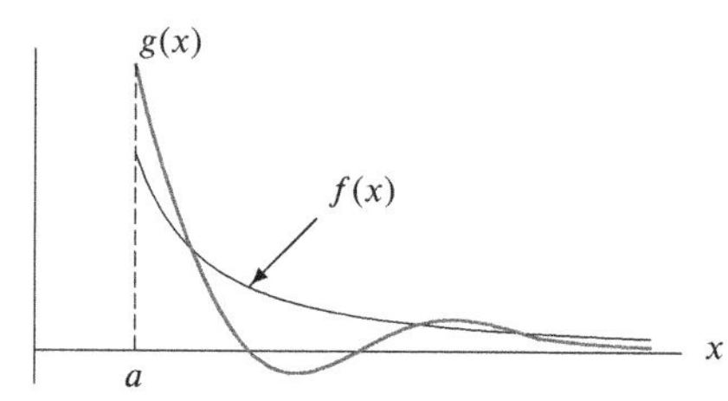

Figure 7.27

■ For what values of p do the integrals in Problems 33–34 converge or diverge?

33. $\displaystyle\int_2^{\infty} \frac{dx}{x(\ln x)^p}$

34. $\displaystyle\int_1^{2} \frac{dx}{x(\ln x)^p}$

35. (a) Find an upper bound for

$$\int_3^\infty e^{-x^2}\,dx.$$

[Hint: $e^{-x^2} \le e^{-3x}$ for $x \ge 3$.]

(b) For any positive n, generalize the result of part (a) to find an upper bound for

$$\int_n^\infty e^{-x^2}\,dx$$

by noting that $nx \le x^2$ for $x \ge n$.

36. In Planck's Radiation Law, we encounter the integral

$$\int_1^\infty \frac{dx}{x^5(e^{1/x}-1)}.$$

(a) Explain why a graph of the tangent line to e^t at $t = 0$ tells us that for all t

$$1 + t \le e^t.$$

(b) Substituting $t = 1/x$, show that for all $x \ne 0$

$$e^{1/x} - 1 > \frac{1}{x}.$$

(c) Use the comparison test to show that the original integral converges.

Strengthen Your Understanding

■ In Problems 37–40, explain what is wrong with the statement.

37. $\int_1^\infty 1/(x^3 + \sin x)\,dx$ converges by comparison with $\int_1^\infty 1/x^3\,dx$.

38. $\int_1^\infty 1/(x^{\sqrt{2}} + 1)\,dx$ is divergent.

39. If $0 \le f(x) \le g(x)$ and $\int_0^\infty g(x)\,dx$ diverges then by the comparison test $\int_0^\infty f(x)\,dx$ diverges.

40. Let $f(x) > 0$. If $\int_1^\infty f(x)\,dx$ is convergent then so is $\int_1^\infty 1/f(x)\,dx$.

■ In Problems 41–42, give an example of:

41. A continuous function $f(x)$ for $x \ge 1$ such that the improper integral $\int_1^\infty f(x)dx$ can be shown to converge by comparison with the integral $\int_1^\infty 3/(2x^2)\,dx$.

42. A positive, continuous function $f(x)$ such that $\int_1^\infty f(x)dx$ diverges and

$$f(x) \le \frac{3}{7x - 2\sin x}, \quad \text{for } x \ge 1.$$

■ In Problems 43–48, decide whether the statements are true or false. Give an explanation for your answer.

43. The integral $\displaystyle\int_0^\infty \frac{1}{e^x + x}\,dx$ converges.

44. The integral $\displaystyle\int_0^1 \frac{1}{x^2 - 3}\,dx$ diverges.

45. If $f(x) < g(x)$ for all x and $a < b$, then

$$\int_a^b f(x)\,dx < \int_a^b g(x)\,dx.$$

46. If $f(x) < g(x)$ for all x and $\int_a^\infty g(x)\,dx$ converges, then $\int_a^\infty f(x)\,dx$ converges.

47. If $|f(x)| < |g(x)|$ for all x and $a < b$, then

$$\int_a^b f(x)\,dx < \int_a^b g(x)\,dx.$$

48. If $|f(x)| < |g(x)|$ for all x and if $\int_a^\infty |g(x)|\,dx$ converges, then $\int_a^\infty |f(x)|\,dx$ converges.

Online Resource: Review problems and Projects

Chapter Eight

USING THE DEFINITE INTEGRAL

Contents

8.1 AREAS AND VOLUMES

In Chapter 5, we calculated areas under graphs using definite integrals. We obtained the integral by slicing up the region, constructing a Riemann sum, and then taking a limit. In this section, we calculate areas of other regions, as well as volumes, using definite integrals. To obtain the integral, we again slice up the region and construct a Riemann sum.

Finding Areas by Slicing

Example 1 Use horizontal slices to set up a definite integral to calculate the area of the isosceles triangle in Figure 8.1.

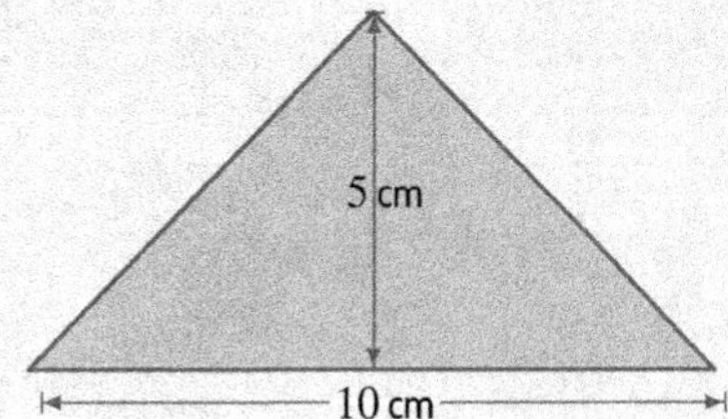

Figure 8.1: Isosceles triangle

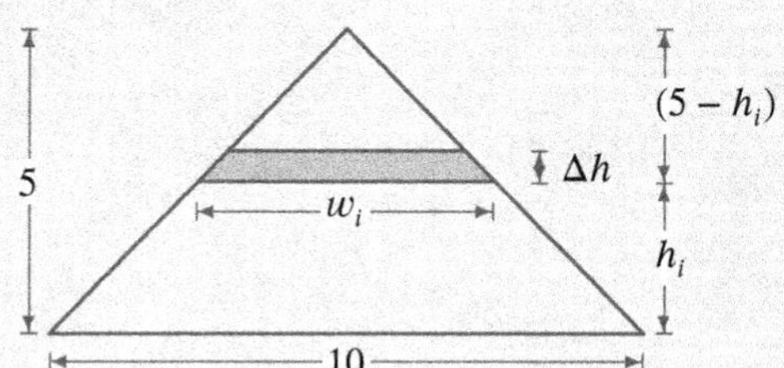

Figure 8.2: Horizontal slices of isosceles triangle

Solution Notice that we can find the area of a triangle without using an integral; we will use this to check the result from integration:

$$\text{Area} = \frac{1}{2}\text{Base} \cdot \text{Height} = 25 \text{ cm}^2.$$

To calculate the area using horizontal slices we divide the region into strips; see Figure 8.2. A typical strip is approximately a rectangle of length w_i and width Δh, so

$$\text{Area of strip} \approx w_i \Delta h \text{ cm}^2.$$

To get w_i in terms of h_i, the height above the base, use the similar triangles in Figure 8.2:

$$\frac{w_i}{10} = \frac{5-h_i}{5}$$
$$w_i = 2(5-h_i) = 10 - 2h_i.$$

Summing the areas of the strips gives the Riemann sum approximation:

$$\text{Area of triangle} \approx \sum_{i=1}^{n} w_i \Delta h = \sum_{i=1}^{n} (10-2h_i)\Delta h \text{ cm}^2.$$

Taking the limit as $n \to \infty$, the width of a strip shrinks, and we get the integral:

$$\text{Area of triangle} = \lim_{n\to\infty} \sum_{i=1}^{n} (10-2h_i)\Delta h = \int_0^5 (10-2h)\, dh \text{ cm}^2.$$

Evaluating the integral gives

$$\text{Area of triangle} = \int_0^5 (10-2h)\, dh = (10h - h^2)\Big|_0^5 = 25 \text{ cm}^2.$$

This agrees with the result we get using Area $= \frac{1}{2}$ Base $\cdot$ Height.

Notice that the limits in the definite integral are the limits for the variable h. Once we decide to slice the triangle horizontally, we know that a typical slice has thickness Δh, so h is the variable in our definite integral, and the limits must be values of h.

Example 2 Use horizontal slices to set up a definite integral representing the area of the semicircle of radius 7 cm in Figure 8.3.

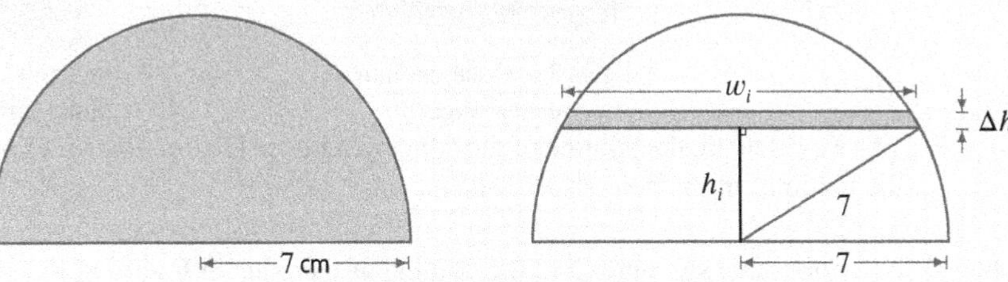

Figure 8.3: Semicircle

Figure 8.4: Horizontal slices of semicircle

Solution As in Example 1, to calculate the area using horizontal slices, we divide the region into strips; see Figure 8.4. A typical strip at height h_i above the base has width w_i and thickness Δh, so

$$\text{Area of strip } \approx w_i \Delta h \text{ cm}^2.$$

To get w_i in terms of h_i, we use the Pythagorean Theorem in Figure 8.4:

$$h_i^2 + \left(\frac{w_i}{2}\right)^2 = 7^2,$$

so

$$w_i = \sqrt{4(7^2 - h_i^2)} = 2\sqrt{49 - h_i^2}.$$

Summing the areas of the strips gives the Riemann sum approximation:

$$\text{Area of semicircle } \approx \sum_{i=1}^{n} w_i \Delta h = \sum_{i=1}^{n} 2\sqrt{49 - h_i^2}\Delta h \text{ cm}^2.$$

Taking the limit as $n \to \infty$, the width of strip shrinks, that is $\Delta h \to 0$, and we get the integral:

$$\text{Area of semicircle } = \lim_{n\to\infty} \sum_{i=1}^{n} 2\sqrt{49 - h_i^2}\Delta h = 2\int_0^7 \sqrt{49 - h^2}\, dh \text{ cm}^2.$$

Using the table of integrals VI-30 and VI-28, or a calculator or computer, gives

$$\text{Area of semicircle } = 2 \cdot \frac{1}{2}\left(h\sqrt{49 - h^2} + 49 \arcsin\left(\frac{h}{7}\right)\right)\Bigg|_0^7 = 49 \arcsin 1 = \frac{49}{2}\pi = 76.97 \text{ cm}^2.$$

As a check, notice that the area of the whole circle of radius 7 is $\pi \cdot 7^2 = 49\pi$ cm^2.

Finding Volumes by Slicing

To calculate the volume of a solid using Riemann sums, we chop the solid into slices whose volumes we can estimate.

Let's see how we might slice a cone standing with the vertex uppermost. We could divide the cone vertically into arch-shaped slices; see Figure 8.5. We could also divide the cone horizontally, giving coin-shaped slices; see Figure 8.6.

Figure 8.5: Cone cut into vertical slices

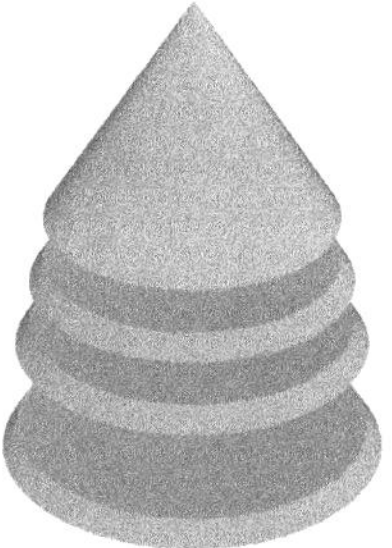

Figure 8.6: Cone cut into horizontal slices

To calculate the volume of the cone, we choose the circular slices because it is easier to estimate the volumes of the coin-shaped slices.

Example 3 Use horizontal slicing to find the volume of the cone in Figure 8.7.

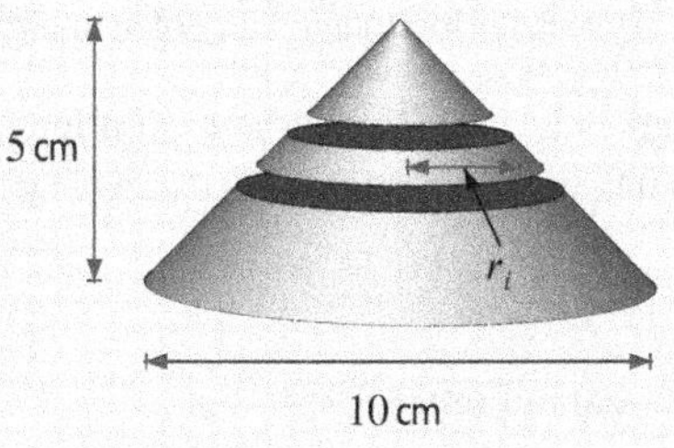

Figure 8.7: Cone

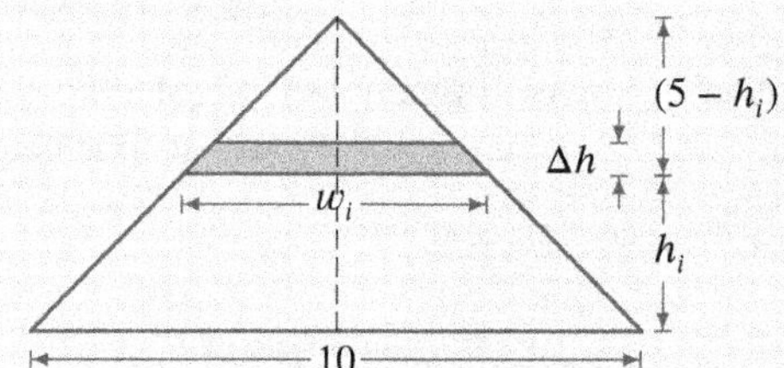

Figure 8.8: Vertical cross-section of cone in Figure 8.7

Solution Each slice is a circular disk of thickness Δh. See Figure 8.7. The disk at height h_i above the base has radius $r_i = \frac{1}{2}w_i$. From Figure 8.8 and the previous example, we have

$$w_i = 10 - 2h_i \quad \text{so} \quad r_i = 5 - h_i.$$

Each slice is approximately a cylinder of radius r_i and thickness Δh, so

$$\text{Volume of slice } \approx \pi r_i^2 \Delta h = \pi(5-h_i)^2 \Delta h \text{ cm}^3.$$

Summing over all slices, we have

$$\text{Volume of cone } \approx \sum_{i=1}^{n} \pi(5-h_i)^2 \Delta h \text{ cm}^3.$$

Taking the limit as $n \to \infty$, so $\Delta h \to 0$, gives

$$\text{Volume of cone } = \lim_{n\to\infty} \sum_{i=1}^{n} \pi(5-h_i)^2 \Delta h = \int_0^5 \pi(5-h)^2\, dh \text{ cm}^3.$$

The integral can be evaluated using the substitution $u = 5 - h$ or by multiplying out $(5-h)^2$. Using the substitution, we have

$$\text{Volume of cone } = \int_0^5 \pi(5-h)^2 dh = -\frac{\pi}{3}(5-h)^3\bigg|_0^5 = \frac{125}{3}\pi \text{ cm}^3.$$

Note that the sum represented by the $\sum$ sign is over all the strips. To simplify the notation, in the future, we will not write limits for $\sum$ or subscripts on the variable, since all we want is the final expression for the definite integral. We now calculate the volume of a hemisphere by slicing.

Example 4 Set up and evaluate an integral giving the volume of the hemisphere of radius 7 cm in Figure 8.9.

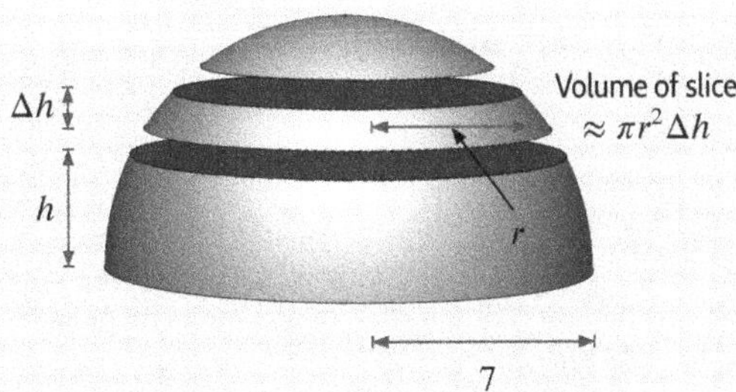

Figure 8.9: Slicing to find the volume of a hemisphere

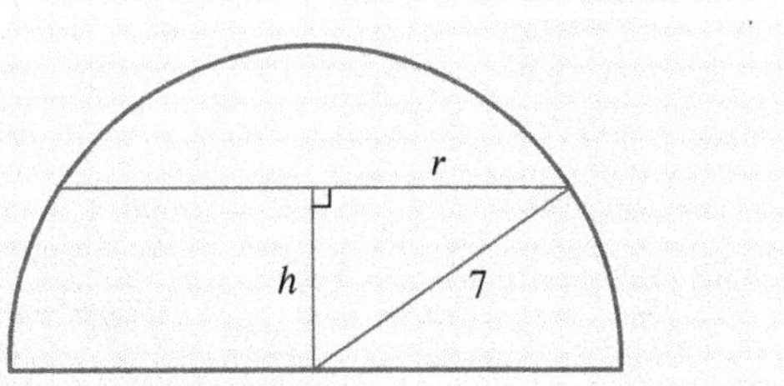

Figure 8.10: Vertical cut through center of hemisphere showing relation between r and h

Solution We will not use the formula $\frac{4}{3}\pi r^3$ for the volume of a sphere. However, our approach can be used to derive that formula.

Divide the hemisphere into horizontal slices of thickness Δh cm. (See Figure 8.9.) Each slice is circular. Let r be the radius of the slice at height h, so

$$\text{Volume of slice} \approx \pi r^2 \Delta h \text{ cm}^3.$$

We express r in terms of h using the Pythagorean Theorem as in Example 2. From Figure 8.10, we have

$$h^2 + r^2 = 7^2,$$

so

$$r = \sqrt{7^2 - h^2} = \sqrt{49 - h^2}.$$

Thus,

$$\text{Volume of slice} \approx \pi r^2 \, \Delta h = \pi(7^2 - h^2)\, \Delta h \text{ cm}^3.$$

Summing the volumes of all slices gives:

$$\text{Volume} \approx \sum \pi r^2 \, \Delta h = \sum \pi(7^2 - h^2)\, \Delta h \text{ cm}^3.$$

As the thickness of each slice tends to zero, the sum becomes a definite integral. Since the radius of the hemisphere is 7, we know that h varies from 0 to 7, so these are the limits of integration:

$$\text{Volume} = \int_0^7 \pi(7^2 - h^2)\, dh = \pi\left(7^2 h - \frac{1}{3}h^3\right)\Bigg|_0^7 = \frac{2}{3}\pi 7^3 = 718.4 \text{ cm}^3.$$

Notice that the volume of the hemisphere is half of $\frac{4}{3}\pi 7^3$ cm^3, as we expected.

We now use slicing to find the volume of a pyramid. We do not use the formula, $V = \frac{1}{3}b^2 \cdot h$, for the volume of a pyramid of height h and square base of side length b, but our approach can be used to derive that formula.

Example 5 When it was built more than 4500 years ago, the Great Pyramid of Egypt had a height of 481 feet and a square base 756 feet by 756 feet.[1] Compute its original volume, in cubic feet.

Solution We slice the pyramid horizontally, creating square slices with thickness Δh. The bottom layer is a square slice 756 feet by 756 feet and volume about $(756)^2\Delta h$ ft^3. As we move up the pyramid, the layers have shorter side lengths. We divide the height into n subintervals of length Δh. See

[1] Its current height is 455 feet. See https://en.wikipedia.org/wiki/Great_Pyramid_of_Giza. Accessed March 2016.

Figure 8.11. Let s be the side length of the slice at height h; then

$$\text{Volume of slice} \approx s^2\,\Delta h\ \text{ft}^3.$$

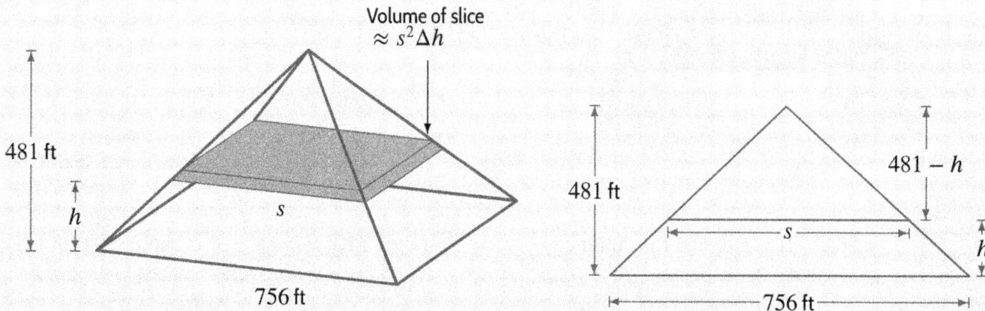

Figure 8.11: The Great Pyramid

Figure 8.12: Cross-section relating s and h

We express s as a function of h using the vertical cross-section in Figure 8.12. By similar triangles, we get

$$\frac{s}{756} = \frac{(481-h)}{481}.$$

Thus,

$$s = \left(\frac{756}{481}\right)(481-h),$$

and the total volume, V, is approximated by adding the volumes of the n layers:

$$V \approx \sum s^2\,\Delta h = \sum \left(\left(\frac{756}{481}\right)(481-h)\right)^2 \Delta h\ \text{ft}^3.$$

As the thickness of each slice tends to zero, the sum becomes a definite integral. Finally, since h varies from 0 to 481, the height of the pyramid, we have

$$V = \int_0^{481} \left(\left(\frac{756}{481}\right)(481-h)\right)^2 dh = \left(\frac{756}{481}\right)^2 \int_0^{481} (481-h)^2\,dh$$

$$= \left(\frac{756}{481}\right)^2 \left(-\frac{(481-h)^3}{3}\right)\Bigg|_0^{481} = \left(\frac{756}{481}\right)^2 \frac{(481)^3}{3} = \frac{1}{3}(756)^2(481) \approx 92 \text{ million ft}^3.$$

Note that $V = \frac{1}{3}(756)^2(481) = \frac{1}{3}b^2 \cdot h$, as expected.

Exercises and Problems for Section 8.1

EXERCISES

1. **(a)** Write a Riemann sum approximating the area of the region in Figure 8.13, using vertical strips as shown.
   (b) Evaluate the corresponding integral.

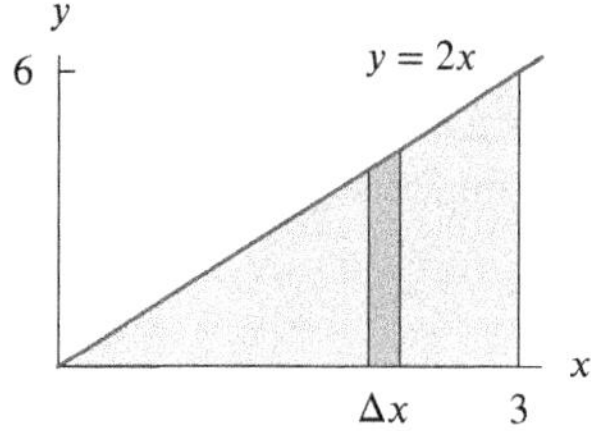

Figure 8.13

2. **(a)** Write a Riemann sum approximating the area of the region in Figure 8.14, using vertical strips as shown.
   (b) Evaluate the corresponding integral.

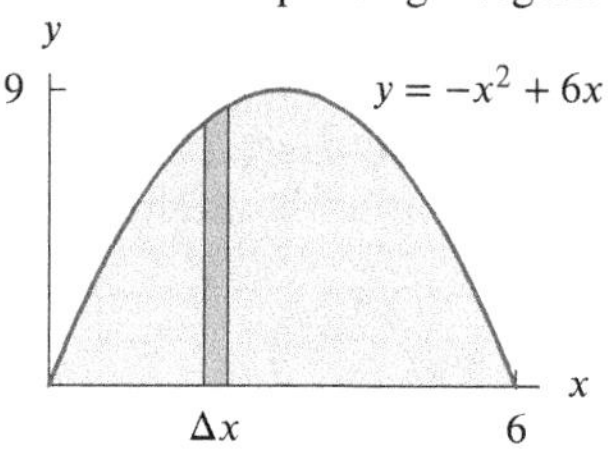

Figure 8.14

3. **(a)** Write a Riemann sum approximating the area of the region in Figure 8.15, using horizontal strips as shown.
(b) Evaluate the corresponding integral.

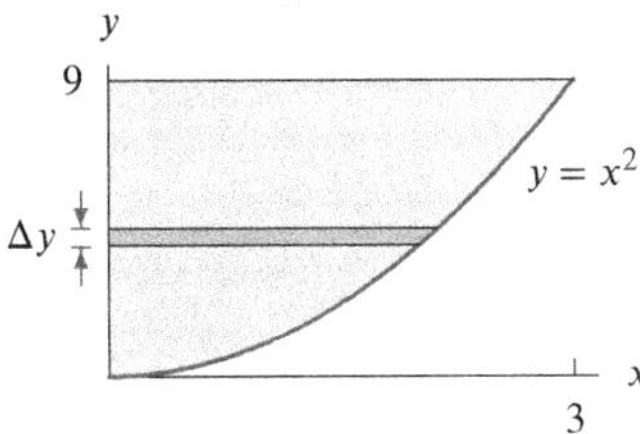

Figure 8.15

4. **(a)** Write a Riemann sum approximating the area of the region in Figure 8.16, using horizontal strips as shown.
(b) Evaluate the corresponding integral.

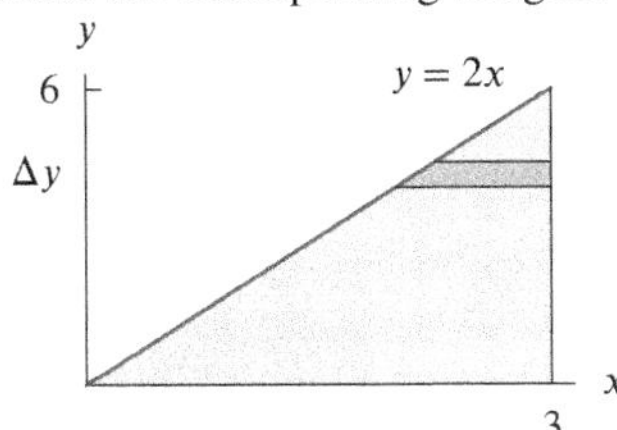

Figure 8.16

In Exercises 5–12, write a Riemann sum and then a definite integral representing the area of the region, using the strip shown. Evaluate the integral exactly.

5.

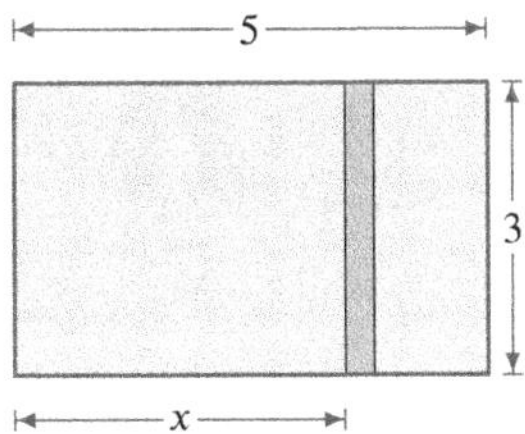

6.

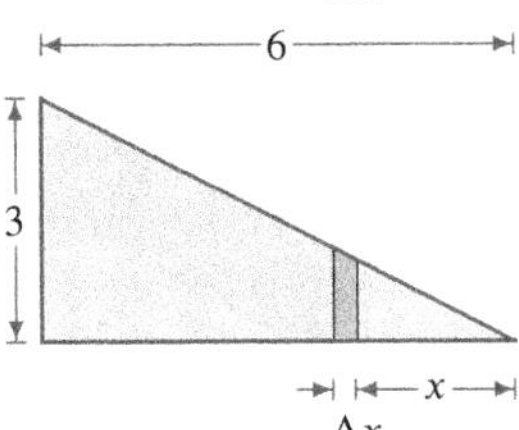

7.

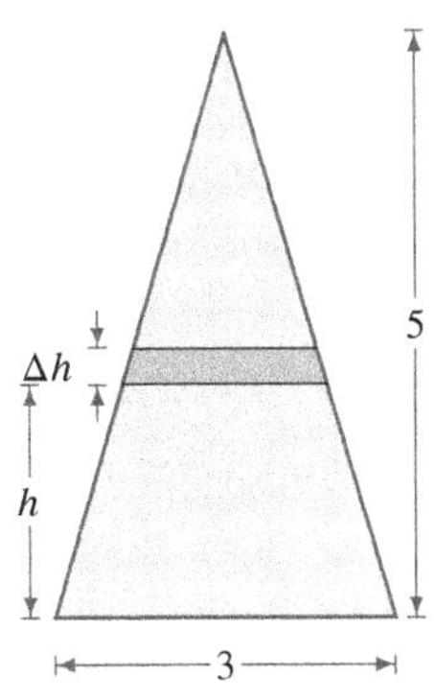

8.

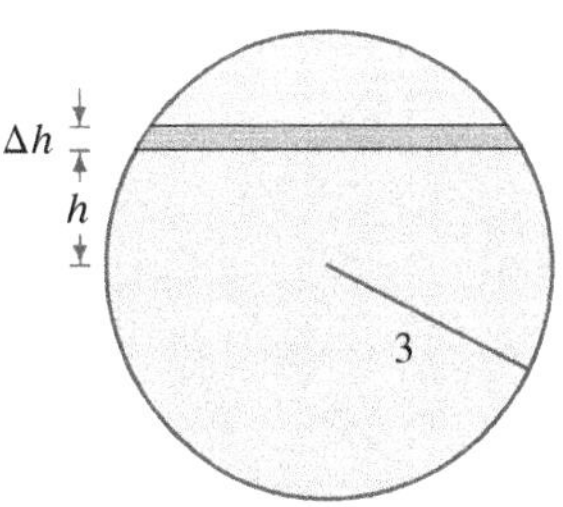

9.

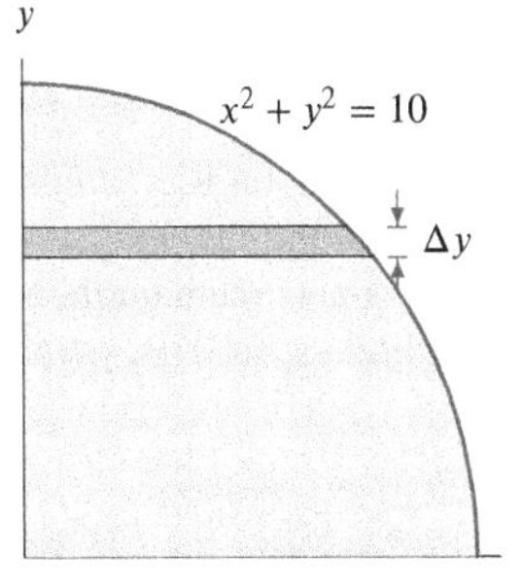

10.

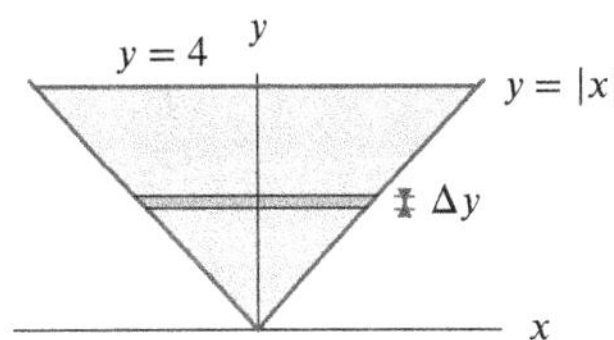

11.

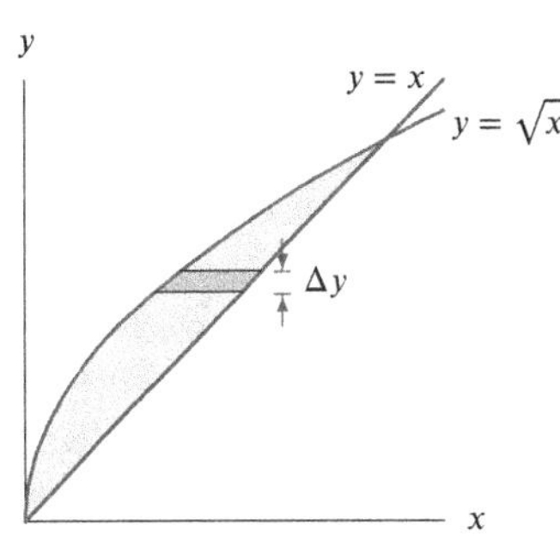

12.

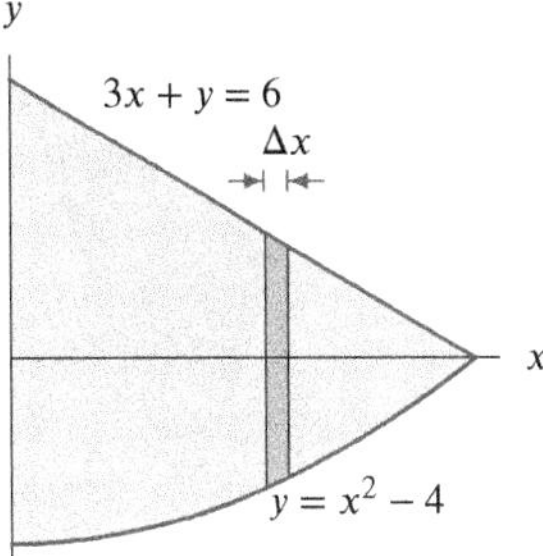

13. **(a)** Match the regions I–IV in Figure 8.17 with the regions $A - D$:
- A: Bounded by $y + x = 2$, $y = x$, $x = 2$
- B: Bounded by $y + x = 2$, $y = x$, $y = 2$
- C: Bounded by $y + x = 2$, $y = x$, $x = 0$
- D: Bounded by $y + x = 2$, $y = x$, $y = 0$

(b) Write integrals representing the areas of the regions II and III using vertical strips. Do not evaluate.

Figure 8.17

In Exercises 14–21, write a Riemann sum and then a definite integral representing the volume of the region, using the slice shown. Evaluate the integral exactly. (Regions are parts of cones, cylinders, spheres, pyramids, and triangular prisms.)

14.

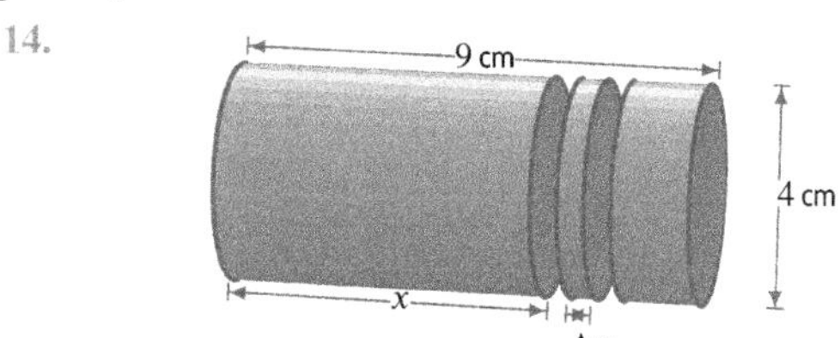

15.

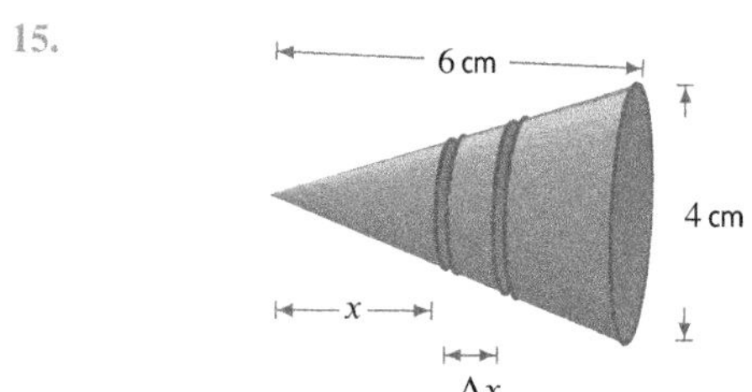

16.

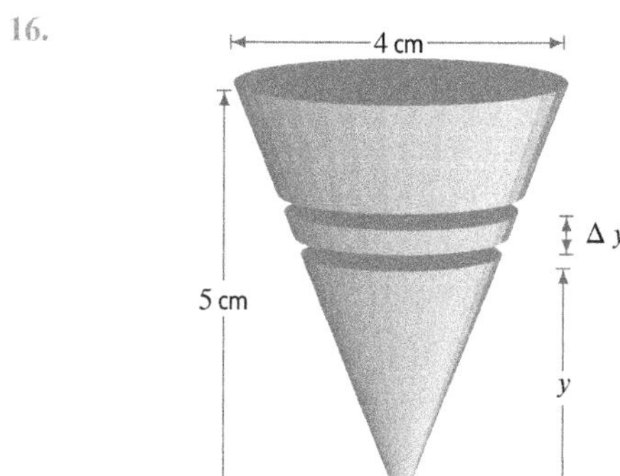

17.

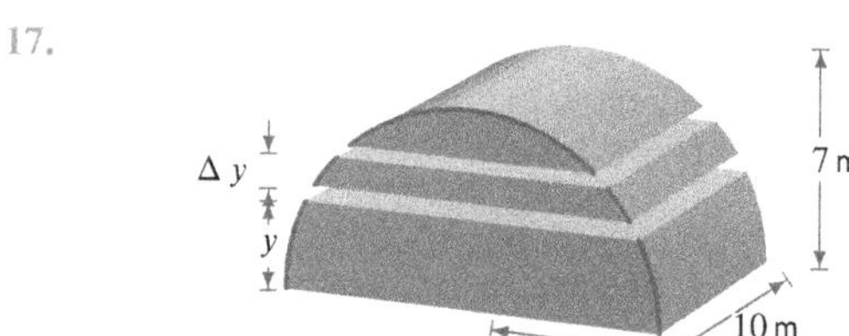

18.

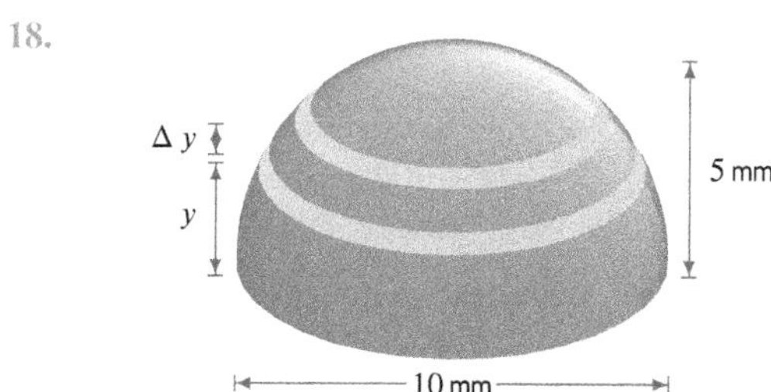

19.

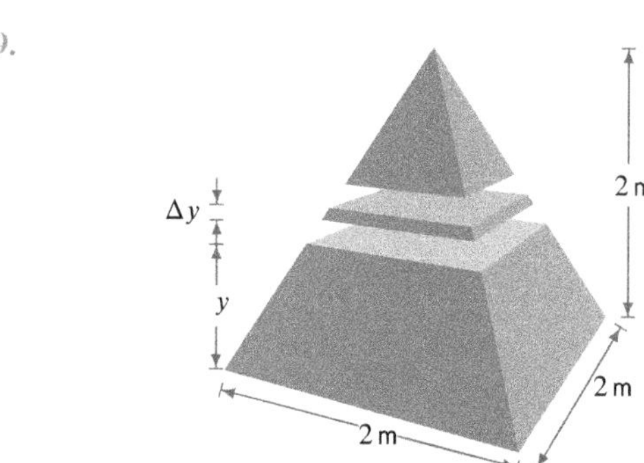

20.

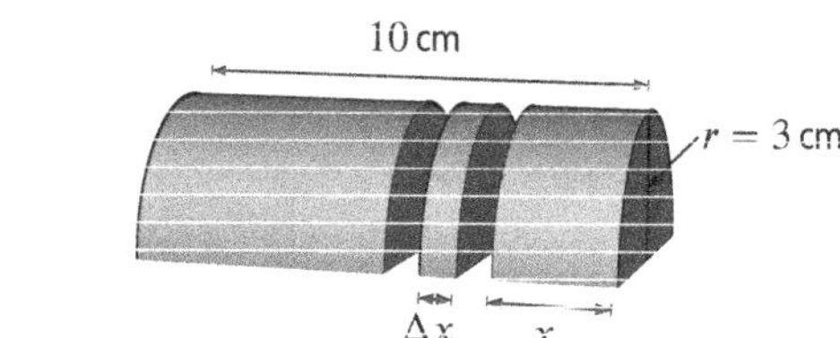

21.

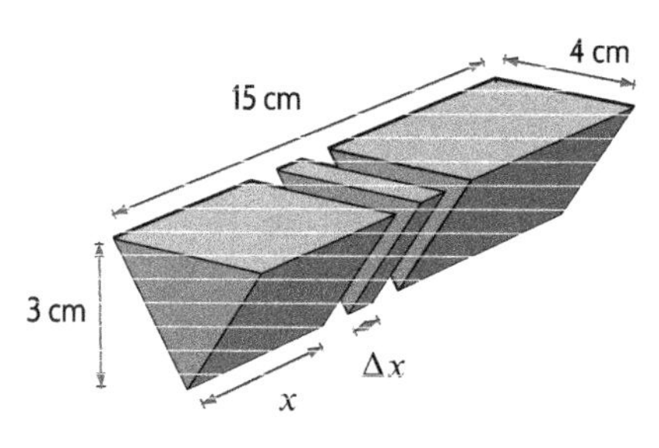

PROBLEMS

The integrals in Problems 22–25 represent the area of either a triangle or part of a circle, and the variable of integration measures a distance. In each case, say which shape is represented, and give the base and height of the triangle or the radius of the circle. Make a sketch to support your answer showing the variable and all other relevant quantities.

22. $\int_0^1 3x\,dx$

23. $\int_{-9}^{9} \sqrt{81 - x^2}\,dx$

24. $\int_0^{\sqrt{15}} \sqrt{15 - h^2}\,dh$

25. $\int_0^7 5\left(1 - \frac{h}{7}\right)\,dh$

26. The integral $\int_0^1 (x - x^2)\,dx$ represents the area of a region between two curves in the plane. Make a sketch of this region.

■ In Problems 27–30, construct and evaluate definite integral(s) representing the area of the region described, using:
(a) Vertical slices **(b)** Horizontal slices

27. Enclosed by $y = x^2$ and $y = 3x$.

28. Enclosed by $y = 2x$ and $y = 12 - x$ and the y-axis.

29. Enclosed by $y = x^2$ and $y = 6 - x$ and the x-axis.

30. Enclosed by $y = 2x$ and $x = 5$ and $y = 6$ and the x-axis.

■ In Problems 31–35, the integral represents the volume of a hemisphere, sphere, or cone, and the variable of integration is a length. Say which shape is represented; give the radius of the hemisphere or sphere or the radius and height of the cone. Make a sketch showing the variable and all relevant quantities.

31. $\int_0^{12} \pi(144 - h^2)\,dh$

32. $\int_0^{12} \pi(x/3)^2 dx$

33. $2\int_0^8 \pi\left(64 - h^2\right)\,dh$

34. $\int_0^6 \pi(3 - y/2)^2 dy$

35. $\int_0^2 \pi(2^2 - (2 - y)^2)\,dy$

36. A cone with base radius 4 and height 16 standing with its vertex upward is cut into 32 horizontal slices of equal thickness Δh.

(a) Find Δh.
(b) Find a formula relating the radius of the cone r at height h above the base.
(c) What is the approximate volume of the bottom slice?
(d) What is the height of the ninth slice from the bottom?
(e) What is the approximate volume of the ninth slice?

37. A hemisphere with a horizontal base of radius 4 is cut parallel to the base into 20 slices of equal thickness.

(a) What is the thickness Δh of each slice?
(b) What is the approximate volume of the slice at height h above the base?

38. Find the volume of a sphere of radius r by slicing.

39. Set up and evaluate an integral to find the volume of a cone of height 12 m and base radius 3 m.

40. Find, by slicing, a formula for the volume of a cone of height h and base radius r.

41. Figure 8.18 shows a solid with both rectangular and triangular cross sections.

(a) Slice the solid parallel to the triangular faces. Sketch one slice and calculate its volume in terms of x, the distance of the slice from one end. Then write and evaluate an integral giving the volume of the solid.
(b) Repeat part (a) for horizontal slices. Instead of x, use h, the distance of a slice from the top.

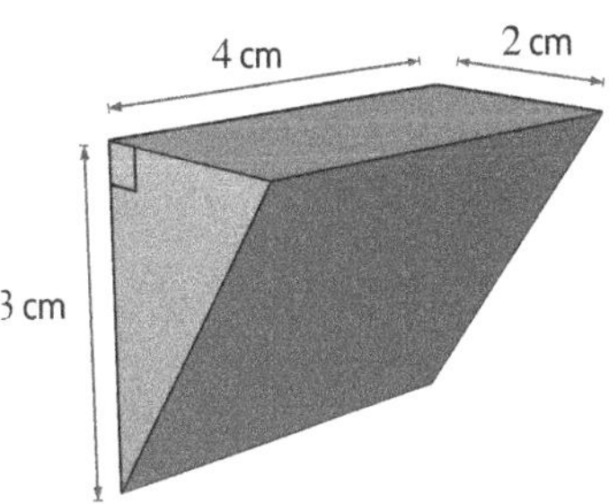

Figure 8.18

42. A rectangular lake is 150 km long and 3 km wide. The vertical cross-section through the lake in Figure 8.19 shows that the lake is 0.2 km deep at the center. (These are the approximate dimensions of Lake Mead, a large reservoir providing water to California, Nevada, and Arizona.) Set up and evaluate a definite integral giving the total volume of water in the lake.

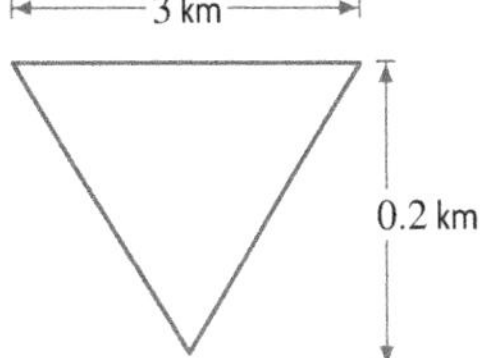

Figure 8.19: Not to scale

43. A dam has a rectangular base 1400 meters long and 160 meters wide. Its cross-section is shown in Figure 8.20. (The Grand Coulee Dam in Washington state is roughly this size.) By slicing horizontally, set up and evaluate a definite integral giving the volume of material used to build this dam.

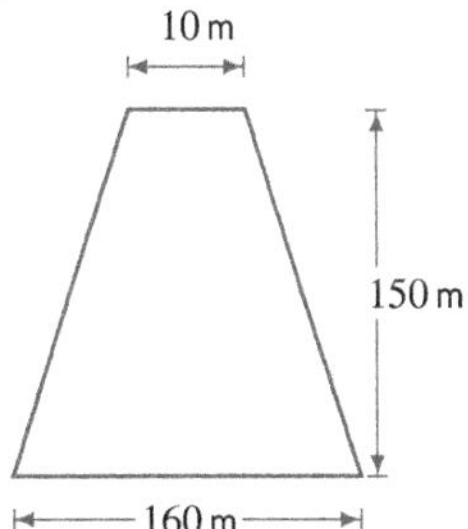

Figure 8.20: Not to scale

44. **(a)** Set up and evaluate an integral giving the volume of a pyramid of height 10 m and square base 8 m by 8 m.
(b) The pyramid in part (a) is cut off at a height of 6 m. See Figure 8.21. Find the volume.

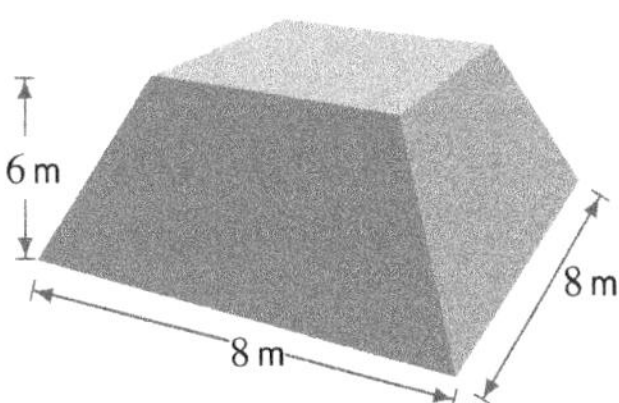

Figure 8.21

45. The exterior of a holding tank is a cylinder with radius 3 m and height 6 m; the interior is cone-shaped; Figure 8.22 shows its cross-section. Using an integral, find the volume of material needed to make the tank.

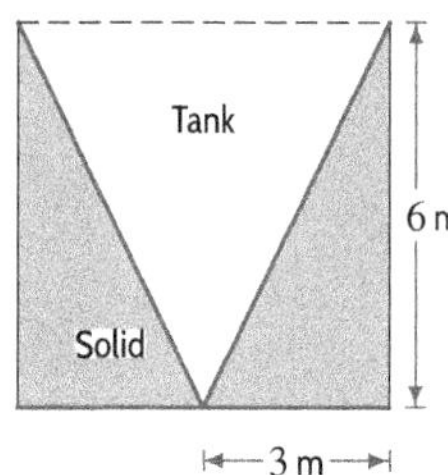

Figure 8.22

In Problems 46–49, the given volume has a horizontal base. Let h be the height above the base of a slice with thickness Δh. Which of (I)–(IV) approximates the volume of this slice?

I. $\pi\sqrt{16-h^2}\Delta h$ II. $\frac{\pi}{25}(20-h)^2\,\Delta h$
III. $25\pi\,(20-h)^2\,\Delta h$ IV. $\pi\left(16-h^2\right)\Delta h$
V. $\frac{1}{25}(20-h)^2\,\Delta h$ VI. $25\,(20-h)^2\,\Delta h$

46. A cone of height $h = 20$ and base radius $r = 4$.

47. A cone of height $h = 20$ and base radius $r = 100$.

48. A pyramid whose base is a square of side $s = 4$ and whose height is $h = 20$.

49. A hemisphere of radius $r = 4$.

Strengthen Your Understanding

In Problems 50–51, explain what is wrong with the statement.

50. To find the area between the line $y = 2x$, the y-axis, and the line $y = 8$ using horizontal slices, evaluate the integral $\int_0^8 2y\,dy$.

51. The volume of the sphere of radius 10 centered at the origin is given by the integral $\int_{-10}^{10}\pi\sqrt{10^2-x^2}\,dx$.

In Problems 52–53, give an example of:

52. A region in the plane where it is easier to compute the area using horizontal slices than it is with vertical slices. Sketch the region.

53. A triangular region in the plane for which both horizontal and vertical slices work just as easily.

In Problems 54–59, are the statements true or false? Give an explanation for your answer.

54. The integral $\int_{-3}^{3}\pi(9-x^2)\,dx$ represents the volume of a sphere of radius 3.

55. The integral $\int_0^h \pi(r-y)\,dy$ gives the volume of a cone of radius r and height h.

56. The integral $\int_0^r \pi\sqrt{r^2-y^2}\,dy$ gives the volume of a hemisphere of radius r.

57. A cylinder of radius r and length l is lying on its side. Horizontal slicing tells us that the volume is given by $\int_{-r}^{r} 2l\sqrt{r^2-y^2}\,dy$.

58. A cone of height 10 has a horizontal base with radius 50. If h is the height of a horizontal slice of thickness Δh, the slice's volume is approximated by $\pi\,(50-5h)^2\,\Delta h$.

59. A semicircle of radius 10 has a horizontal base. If h is the height of a horizontal strip of thickness Δh, the strip's area is approximated by $2\sqrt{100-h^2}\Delta h$

8.2 APPLICATIONS TO GEOMETRY

In Section 8.1, we calculated volumes using slicing and definite integrals. In this section, we use the same method to calculate the volumes of more complicated regions as well as the length of a curve. The method is summarized in the following steps:

To Compute a Volume or Length Using an Integral

- Divide the solid (or curve) into small pieces whose volume (or length) we can easily approximate;
- Add the contributions of all the pieces, obtaining a Riemann sum that approximates the total volume (or length);
- Take the limit as the number of terms in the sum tends to infinity, giving a definite integral for the total volume (or total length);
- Evaluate the definite integral, either exactly or approximately.

In the previous section, all the slices we created were disks or rectangles. We now look at different ways of generating volumes whose cross-sections include circles, rectangles, and also rings.

Volumes of Revolution

One way to create a solid having circular cross-sections is to revolve a region in the plane around a line, giving a *solid of revolution,* as in the following examples.

Example 1 The region bounded by the curve $y = e^{-x}$ and the x-axis between $x = 0$ and $x = 1$ is revolved around the x-axis. Find the volume of this solid of revolution.

Solution We slice the region perpendicular to the x-axis, giving circular disks of thickness Δx. See Figure 8.23. The radius of the disk is $y = e^{-x}$, so:

$$\text{Volume of the slice} \approx \pi y^2\,\Delta x = \pi(e^{-x})^2\,\Delta x,$$

$$\text{Total volume} \approx \sum \pi y^2\,\Delta x = \sum \pi\,(e^{-x})^2\,\Delta x.$$

As the thickness of each slice tends to zero, we get:

$$\begin{aligned}\text{Total volume} &= \int_0^1 \pi(e^{-x})^2\,dx = \pi\int_0^1 e^{-2x}\,dx = \pi\left(-\frac{1}{2}\right)e^{-2x}\Big|_0^1\\ &= \pi\left(-\frac{1}{2}\right)(e^{-2} - e^0) = \frac{\pi}{2}(1 - e^{-2}) \approx 1.36.\end{aligned}$$

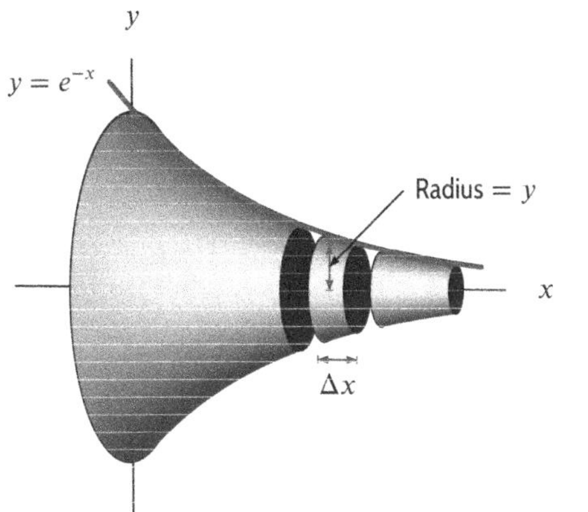

Figure 8.23: A thin strip rotated around the x-axis to form a circular slice

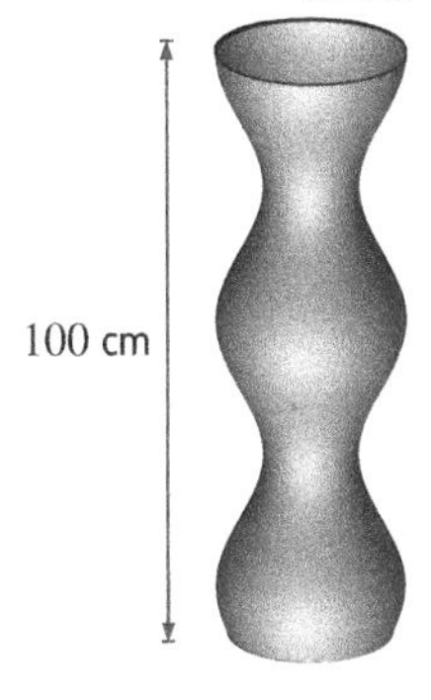

Figure 8.24: A table leg

Example 2 A table leg in Figure 8.24 has a circular cross section with radius r cm at a height of y cm above the ground given by $r = 3 + \cos(\pi y/25)$. Find the volume of the table leg.

Solution The table leg is formed by rotating the curve $r = 3 + \cos(\pi y/25)$ around the y-axis. Slicing the table leg horizontally gives circular disks of thickness Δy and radius $r = 3 + \cos(\pi y/25)$.

To set up a definite integral for the volume, we find the volume of a typical slice:

$$\text{Volume of slice} \approx \pi r^2 \Delta y = \pi\left(3 + \cos\left(\frac{\pi}{25}y\right)\right)^2\,\Delta y.$$

Summing over all slices gives the Riemann sum approximation:

$$\text{Total volume} = \sum \pi\left(3 + \cos\left(\frac{\pi}{25}y\right)\right)^2\,\Delta y.$$

Taking the limit as $\Delta y \to 0$ gives the definite integral:

$$\text{Total volume} = \lim_{\Delta y\to 0}\sum \pi\left(3 + \cos\left(\frac{\pi}{25}y\right)\right)^2\,\Delta y = \int_0^{100} \pi\left(3 + \cos\left(\frac{\pi}{25}y\right)\right)^2\,dy.$$

Evaluating the integral numerically gives:

$$\text{Total volume} = \int_0^{100} \pi\left(3 + \cos\left(\frac{\pi}{25}y\right)\right)^2\,dy = 2984.5\text{ cm}^3.$$

Example 3 The region bounded by the curves $y = x$ and $y = x^2$ is rotated about the line $y = 3$. Compute the volume of the resulting solid.

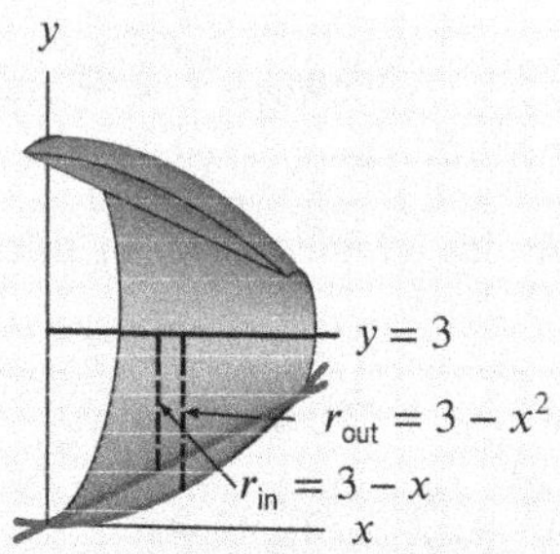

Figure 8.25: Cutaway view of volume showing inner and outer radii

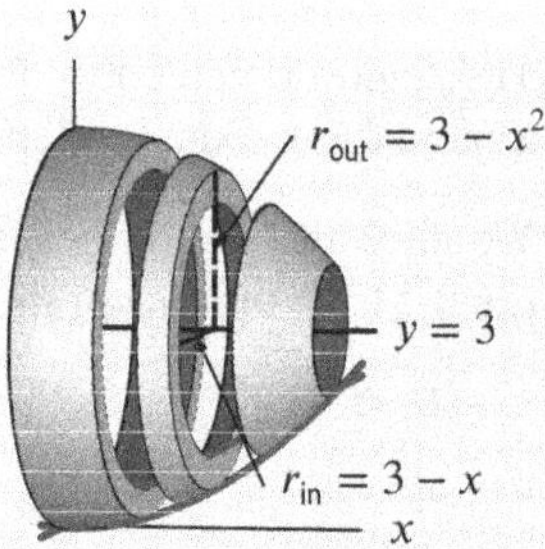

Figure 8.26: One slice (a disk-with-a-hole)

Solution The solid is shaped like a bowl with the base removed. See Figure 8.25. To compute the volume, we divide the area in the xy-plane into thin vertical strips of thickness Δx, as in Figure 8.27.

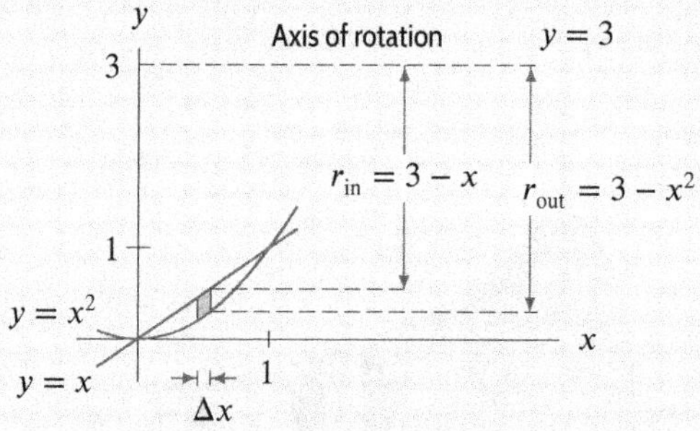

Figure 8.27: The region for Example 3

As each strip is rotated around the line $y = 3$, it sweeps out a slice shaped like a circular disk with a hole in it. See Figure 8.26. This disk-with-a-hole has an inner radius of $r_{in} = 3 - x$ and an outer radius of $r_{out} = 3 - x^2$. Think of the slice as a circular disk of radius r_{out} from which has been removed a smaller disk of radius r_{in}. Then:

$$\text{Volume of slice} \approx \pi r_{out}^2\,\Delta x - \pi r_{in}^2\,\Delta x = \pi(3 - x^2)^2\,\Delta x - \pi(3 - x)^2\,\Delta x.$$

Adding the volumes of all the slices, we have:

$$\text{Total volume} = V \approx \sum \left(\pi r_{out}^2 - \pi r_{in}^2\right)\Delta x = \sum \left(\pi(3 - x^2)^2 - \pi(3 - x)^2\right)\Delta x.$$

We let Δx, the thickness of each slice, tend to zero to obtain a definite integral. Since the curves $y = x$ and $y = x^2$ intersect at $x = 0$ and $x = 1$, these are the limits of integration:

$$\begin{aligned} V &= \int_0^1 \left(\pi(3 - x^2)^2 - \pi(3 - x)^2\right) dx = \pi \int_0^1 \left((9 - 6x^2 + x^4) - (9 - 6x + x^2)\right) dx \\ &= \pi \int_0^1 (6x - 7x^2 + x^4)\, dx = \pi \left(3x^2 - \frac{7x^3}{3} + \frac{x^5}{5}\right)\bigg|_0^1 \approx 2.72. \end{aligned}$$

Volumes of Regions of Known Cross-Section

We now calculate the volume of a solid constructed by a different method. Starting with a region in the xy-plane as a base, the solid is built by standing squares, semicircles, or triangles vertically on edge in this region.

Example 4 Find the volume of the solid whose base is the region in the xy-plane bounded by the curves $y = x^2$ and $y = 8 - x^2$ and whose cross-sections perpendicular to the x-axis are squares with one side in the xy-plane. (See Figure 8.28.)

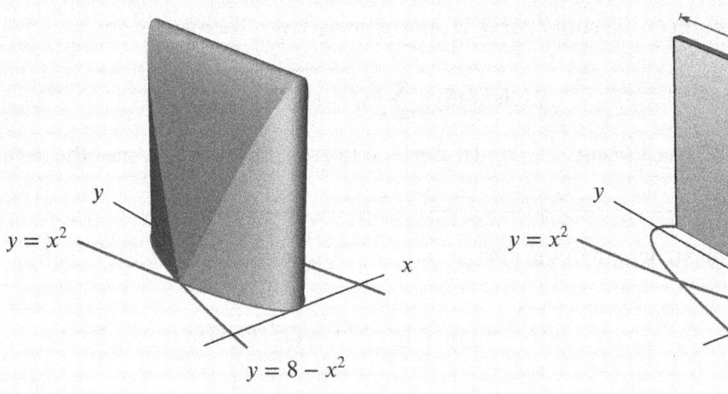

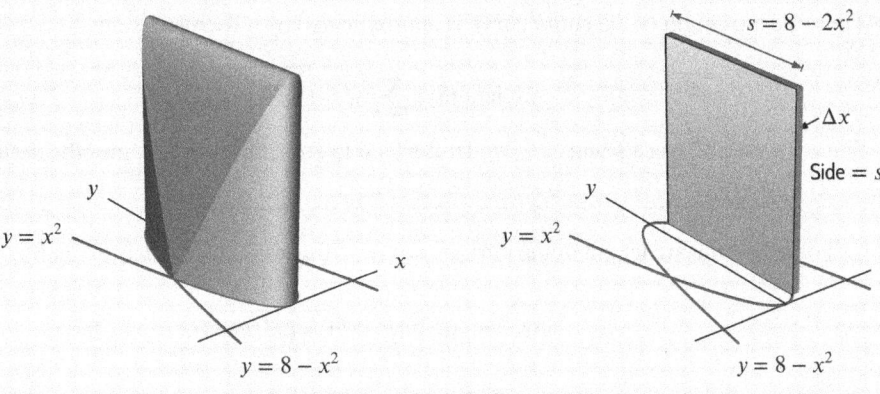

Figure 8.28: The solid for Example 4

Figure 8.29: A slice of the solid for Example 4

Solution We view the solid as a loaf of bread sitting on the xy-plane and made up of square slices. A typical slice of thickness Δx is shown in Figure 8.29. The side length, s, of the square is the distance (in the y direction) between the two curves, so $s = (8 - x^2) - x^2 = 8 - 2x^2$, giving

$$\text{Volume of slice} \approx s^2\,\Delta x = (8 - 2x^2)^2\,\Delta x.$$

Thus

$$\text{Total volume} = V \approx \sum s^2\,\Delta x = \sum (8 - 2x^2)^2\,\Delta x.$$

As the thickness Δx of each slice tends to zero, the sum becomes a definite integral. Since the curves $y = x^2$ and $y = 8 - x^2$ intersect at $x = -2$ and $x = 2$, these are the limits of integration. We have

$$\begin{aligned} V &= \int_{-2}^{2} (8 - 2x^2)^2\,dx = \int_{-2}^{2} (64 - 32x^2 + 4x^4)\,dx \\ &= \left(64x - \frac{32}{3}x^3 + \frac{4}{5}x^5\right)\Bigg|_{-2}^{2} = \frac{2048}{15} \approx 136.5. \end{aligned}$$

Arc Length

A definite integral can be used to compute the *arc length*, or length, of a curve. To compute the length of the curve $y = f(x)$ from $x = a$ to $x = b$, where $a < b$, we divide the curve into small pieces, each one approximately straight.

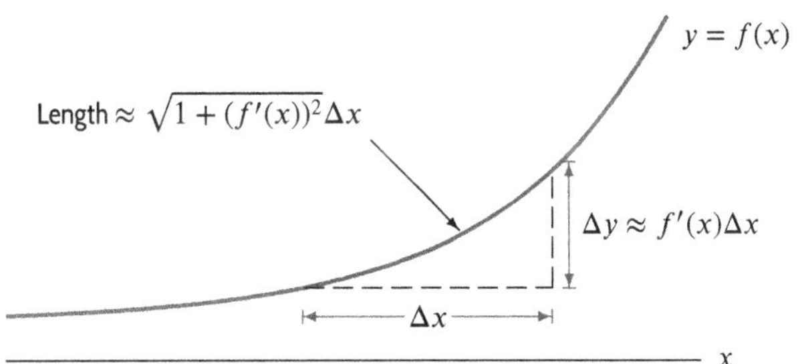

Figure 8.30: Length of a small piece of curve approximated using Pythagoras' theorem

Figure 8.30 shows that a small change Δx corresponds to a small change $\Delta y \approx f'(x)\,\Delta x$. The length of the piece of the curve is approximated by

$$\text{Length} \approx \sqrt{(\Delta x)^2 + (\Delta y)^2} \approx \sqrt{(\Delta x)^2 + (f'(x)\,\Delta x)^2} = \sqrt{1 + (f'(x))^2}\,\Delta x.$$

Thus, the arc length of the entire curve is approximated by a Riemann sum:

$$\text{Arc length} \approx \sum \sqrt{1 + (f'(x))^2}\,\Delta x.$$

Since x varies between a and b, as we let Δx tend to zero, the sum becomes the definite integral:

> For $a < b$, the arc length of the curve $y = f(x)$ from $x = a$ to $x = b$ is given by
>
> $$\text{Arc length} = \int_a^b \sqrt{1 + (f'(x))^2}\,dx.$$

Example 5 Set up and evaluate an integral to compute the length of the curve $y = x^3$ from $x = 0$ to $x = 5$.

Solution If $f(x) = x^3$, then $f'(x) = 3x^2$, so

$$\text{Arc length} = \int_0^5 \sqrt{1 + (3x^2)^2}\,dx.$$

Although the formula for the arc length of a curve is easy to apply, the integrands it generates often do not have elementary antiderivatives. Evaluating the integral numerically, we find the arc length to be 125.68. To check that the answer is reasonable, notice that the curve starts at $(0, 0)$ and goes to $(5, 125)$, so its length must be at least the length of a straight line between these points, or $\sqrt{5^2 + 125^2} = 125.10$. (See Figure 8.31.)

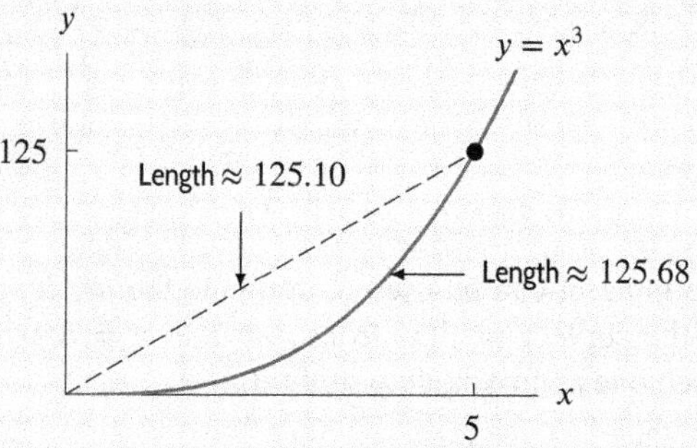

Figure 8.31: Arc length of $y = x^3$ (Note: The picture is distorted because the scales on the two axes are quite different.)

Distance and Arc Length on a Parametric Curve

A particle moving along a curve in the plane given by the parametric equations $x = f(t)$, $y = g(t)$, where t is time, has velocity dx/dt in the x-direction and dy/dt in the y-direction. Thus its speed is given by:

$$\text{Speed} = \sqrt{\left(\frac{dx}{dt}\right)^2 + \left(\frac{dy}{dt}\right)^2}.$$

We find the distance the particle moves along curve by integrating the speed, giving the following result:

> If a particle's position (x, y) is given by differentiable functions of time, t, for $a \le t \le b$, then
>
> $$\text{Distance traveled along curve} = \int_a^b \sqrt{\left(\frac{dx}{dt}\right)^2 + \left(\frac{dy}{dt}\right)^2}\, dt.$$
>
> If the particle does not retrace any part of its path, then distance traveled is also the *arc length of the parametric curve.*

Example 6 Find the circumference of the ellipse given by the parametric equations

$$x = 2\cos t, \quad y = \sin t, \quad 0 \le t \le 2\pi.$$

Solution The parameterization goes once around the ellipse for $0 \le t \le 2\pi$, so using numerical integration we calculate:

$$\text{Circumference} = \int_0^{2\pi} \sqrt{\left(\frac{dx}{dt}\right)^2 + \left(\frac{dy}{dt}\right)^2}\, dt = \int_0^{2\pi} \sqrt{(-2\sin t)^2 + (\cos t)^2}\, dt$$

$$= \int_0^{2\pi} \sqrt{4\sin^2 t + \cos^2 t}\, dt = 9.69.$$

Since the ellipse is inscribed in a circle of radius 2 and circumscribes a circle of radius 1, we expect the length of the ellipse to be between $2\pi(2) \approx 12.57$ and $2\pi(1) \approx 6.28$, so 9.69 is reasonable.

Example 7 At time t, with $0 \le t \le 1$, a particle has position $x = 4(t - t^2)$ and $y = 4(t - t^2)$.

(a) Find the distance traveled by the particle during this time interval.
(b) How does your answer to part (a) relate to the length of the curve?
(c) What is the displacement of the particle between $t = 0$ and $t = 1$?

Solution (a) Since $dx/dt = 4(1 - 2t)$ and $dy/dt = 4(1 - 2t)$, we have

$$\text{Distance traveled} = \int_0^1 \sqrt{(4(1-2t))^2 + (4(1-2t))^2}\, dt$$

$$= 4\sqrt{2}\int_0^1 \sqrt{(1-2t)^2}\, dt = 4\sqrt{2}\int_0^1 |1-2t|\, dt = 2\sqrt{2}.$$

(b) The particle travels along the line $y = x$ between $(0, 0)$ and $(1, 1)$. See Figure 8.32. The length of this curve is $\sqrt{2}$, only half the answer to part (a). This is because the particle travels over the line segment twice, once between $t = 0$ and $t = 1/2$ and again in the other direction between $t = 1/2$ and $t = 1$.

(c) The displacement in both x-direction and y-direction is 0, since the particle is at the same point at $t = 1$ as at $t = 0$.

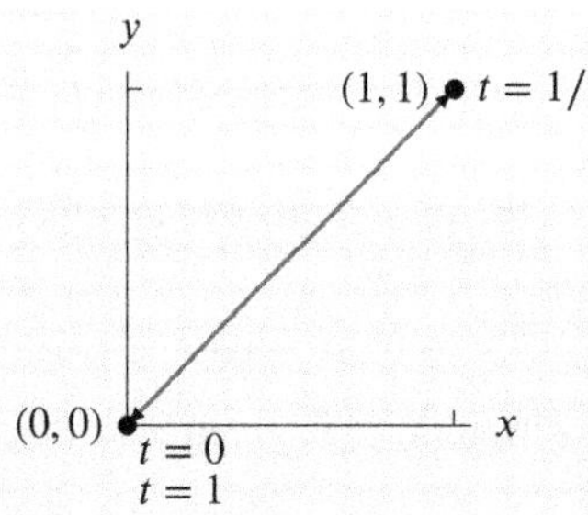

Figure 8.32: Path of particle: $x = y = 4(t - t^2)$ for $0 \leq t \leq 1$

Exercises and Problems for Section 8.2 Online Resource: Additional Problems for Section 8.2

EXERCISES

1. **(a)** The region in Figure 8.33 is rotated around the x-axis. Using the strip shown, write an integral giving the volume.
   (b) Evaluate the integral.

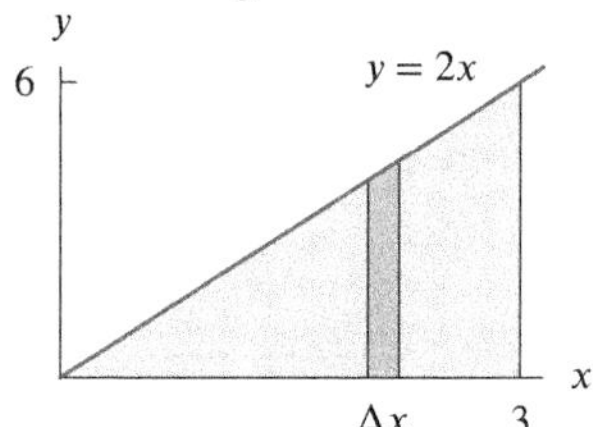

Figure 8.33

2. **(a)** The region in Figure 8.34 is rotated around the x-axis. Using the strip shown, write an integral giving the volume.
   (b) Evaluate the integral.

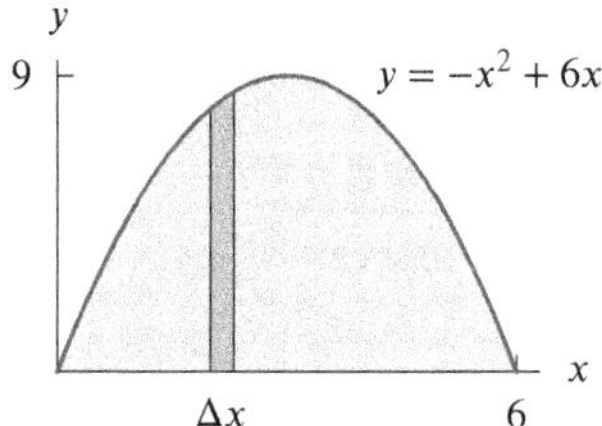

Figure 8.34

3. **(a)** The region in Figure 8.35 is rotated around the y-axis. Using the strip shown, write an integral giving the volume.
   (b) Evaluate the integral.

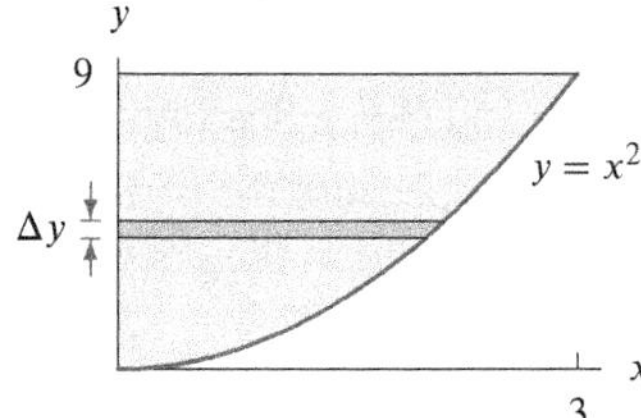

Figure 8.35

4. **(a)** The region in Figure 8.36 is rotated around the y-axis. Write an integral giving the volume.
   (b) Evaluate the integral.

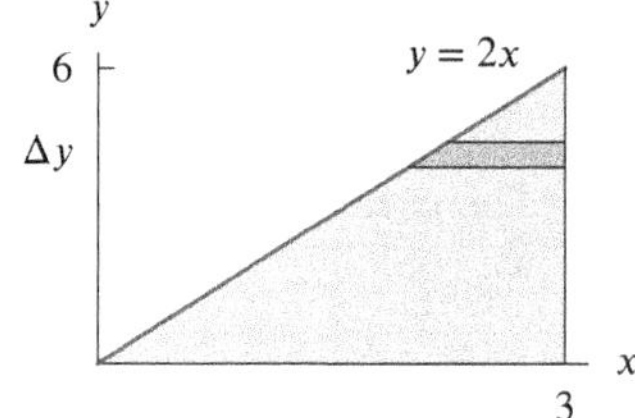

Figure 8.36

In Exercises 5–14, the region is rotated around the x-axis. Find the volume.

5. Bounded by $y = x^2, y = 0, x = 0, x = 1$.
6. Bounded by $y = (x + 1)^2, y = 0, x = 1, x = 2$.
7. Bounded by $y = 4 - x^2, y = 0, x = -2, x = 0$.
8. Bounded by $y = \sqrt{x + 1}, y = 0, x = -1, x = 1$.
9. Bounded by $y = e^x, y = 0, x = -1, x = 1$.
10. Bounded by $y = \cos x, y = 0, x = 0, x = \pi/2$.
11. Bounded by $y = 1/(x + 1), y = 0, x = 0, x = 1$.
12. Bounded by $y = \sqrt{\cosh 2x}, y = 0, x = 0, x = 1$.
13. Bounded by $y = x^2, y = x, x = 0, x = 1$.
14. Bounded by $y = e^{3x}, y = e^x, x = 0, x = 1$.

In Exercises 15–18, the region is rotated around the y-axis. Write, then evaluate, an integral giving the volume.

15. Bounded by $y = 3x$, $x = 0$, $y = 6$.
16. Bounded by $y = 3x$, $y = 0$, $x = 2$.
17. Bounded by $y = \ln x$, $y = 0$, $x = 2$.
18. Bounded by $y = \ln x$, $x = 0$, $y = \ln 2$ and $y = 0$.

For Exercises 19–24, find the arc length of the graph of the function from $x = 0$ to $x = 2$.

19. $f(x) = x^2/2$
20. $f(x) = \cos x$
21. $f(x) = \ln(x + 1)$
22. $f(x) = \sqrt{x^3}$
23. $f(x) = \sqrt{4 - x^2}$
24. $f(x) = \cosh x$

In Exercises 25–28, find the length of the parametric curve.

25. $x = 3 + 5t$, $y = 1 + 4t$ for $1 \le t \le 2$. Explain why your answer is reasonable.
26. $x = \cos(e^t)$, $y = \sin(e^t)$ for $0 \le t \le 1$. Explain why your answer is reasonable.
27. $x = \cos(3t)$, $y = \sin(5t)$ for $0 \le t \le 2\pi$.
28. $x = \cos^3 t$, $y = \sin^3 t$, for $0 \le t \le 2\pi$.

PROBLEMS

In Problems 29–32, set up definite integral(s) to find the volume obtained when the region between $y = x^2$ and $y = 5x$ is rotated about the given axis. Do not evaluate the integral(s).

29. The x-axis
30. The y-axis
31. The line $y = -4$
32. The line $x = -3$

In Problems 33–36 set up, but do not evaluate, an integral that represents the volume obtained when the region in the first quadrant is rotated about the given axis.

33. Bounded by $y = \sqrt[3]{x}$, $x = 4y$. Axis $x = 9$.
34. Bounded by $y = \sqrt[3]{x}$, $x = 4y$. Axis $y = 3$.
35. Bounded by $y = 0$, $x = 9$, $y = \frac{1}{3}x$. Axis $y = -2$.
36. Bounded by $y = 0$, $x = 9$, $y = \frac{1}{3}x$. Axis $x = -1$.
37. Find the length of one arch of $y = \sin x$.
38. Find the perimeter of the region bounded by $y = x$ and $y = x^2$.
39. **(a)** Find (in terms of a) the area of the region bounded by $y = ax^2$, the x-axis, and $x = 2$. Assume $a > 0$.
    (b) If this region is rotated about the x-axis, find the volume of the solid of revolution in terms of a.
40. **(a)** Find (in terms of b) the area of the region between $y = e^{-bx}$ and the x-axis, between $x = 0$ and $x = 1$. Assume $b > 0$.
    (b) If this region is rotated about the x-axis, find the volume of the solid of revolution in terms of b.

For Problems 41–43, sketch the solid obtained by rotating each region around the indicated axis. Using the sketch, show how to approximate the volume of the solid by a Riemann sum, and hence find the volume.

41. Bounded by $y = x^3$, $x = 1$, $y = -1$. Axis: $y = -1$.
42. Bounded by $y = \sqrt{x}$, $x = 1$, $y = 0$. Axis: $x = 1$.
43. Bounded by the first arch of $y = \sin x$, $y = 0$. Axis: x axis.

Problems 44–49 concern the region bounded by $y = x^2$, $y = 1$, and the y-axis, for $x \ge 0$. Find the volume of the solid.

44. The solid obtained by rotating the region around the y-axis.
45. The solid obtained by rotating the region about the x-axis.
46. The solid obtained by rotating the region about the line $y = -2$.
47. The solid whose base is the region and whose cross-sections perpendicular to the x-axis are squares.
48. The solid whose base is the region and whose cross-sections perpendicular to the x-axis are semicircles.
49. The solid whose base is the region and whose cross-sections perpendicular to the y-axis are equilateral triangles.

For Problems 50–54 consider the region bounded by $y = e^x$, the x-axis, and the lines $x = 0$ and $x = 1$. Find the volume of the solid.

50. The solid obtained by rotating the region about the x-axis.
51. The solid obtained by rotating the region about the horizontal line $y = -3$.
52. The solid obtained by rotating the region about the horizontal line $y = 7$.
53. The solid whose base is the given region and whose cross-sections perpendicular to the x-axis are squares.
54. The solid whose base is the given region and whose cross-sections perpendicular to the x-axis are semicircles.

In Problems 55–58, the region is rotated around the y-axis. Write, then evaluate, an integral or sum of integrals giving the volume.

55. Bounded by $y = 2x$, $y = 6 - x$, $y = 0$.
56. Bounded by $y = 2x$, $y = 6 - x$, $x = 0$.
57. Bounded by $y = \sqrt{x}$, $y = 2 - \sqrt{x}$, $y = 0$.
58. Bounded by $y = \sqrt{x}$, $y = 2 - \sqrt{x}$, $x = 0$.
59. Consider the hyperbola $x^2 - y^2 = 1$ in Figure 8.37.
    (a) The shaded region $2 \le x \le 3$ is rotated around the x-axis. What is the volume generated?
    (b) What is the arc length with $y \ge 0$ from $x = 2$ to $x = 3$?

Figure 8.37

60. Rotating the ellipse $x^2/a^2 + y^2/b^2 = 1$ about the x-axis generates an ellipsoid. Compute its volume.

61. Rotating the asteroid $x^{2/3} + y^{2/3} = a^{2/3}$ in Figure 8.38 about the x-axis generates a star-shaped solid. Compute its volume.

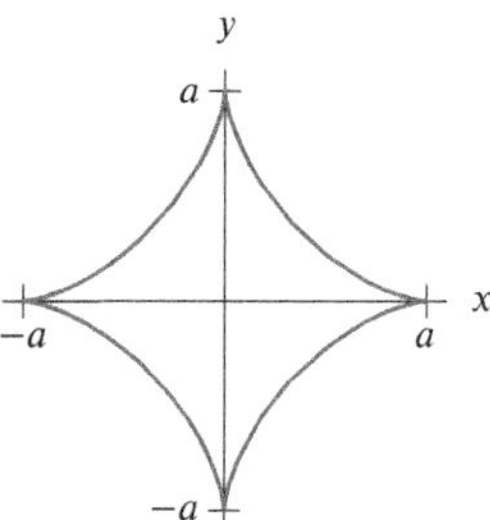

Figure 8.38

62. **(a)** A pie dish with straight sides is 9 inches across the top, 7 inches across the bottom, and 3 inches deep. See Figure 8.39. Compute the volume of this dish.
(b) Make a rough estimate of the volume in cubic inches of a single cut-up apple, and estimate the number of apples needed to make an apple pie that fills this dish.

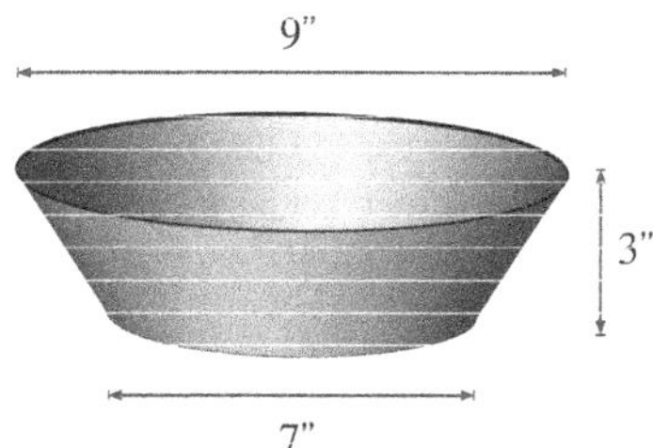

Figure 8.39

63. The design of boats is based on Archimedes' Principle, which states that the buoyant force on an object in water is equal to the weight of the water displaced. Suppose you want to build a sailboat whose hull is parabolic with cross-section $y = ax^2$, where a is a constant. Your boat will have length L and its maximum draft (the maximum vertical depth of any point of the boat beneath the water line) will be H. See Figure 8.40. Every cubic meter of water weighs 10,000 newtons. What is the maximum possible weight for your boat and cargo?

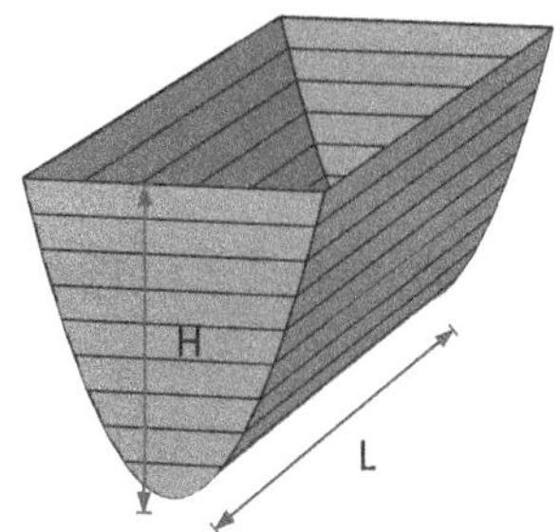

Figure 8.40

64. A tree trunk has a circular cross-section at every height; its circumference is given in the following table. Estimate the volume of the tree trunk using the trapezoid rule.

Height (feet)	0	20	40	60	80	100	120
Circumference (feet)	26	22	19	14	6	3	1

65. The circumference of a tree at different heights above the ground is given in the table below. Assume that all horizontal cross-sections of the tree are circles. Estimate the volume of the tree.

Height (inches)	0	20	40	60	80	100	120
Circumference (inches)	31	28	21	17	12	8	2

66. Compute the perimeter of the region used for the base of the solids in Problems 50–54.

67. Write an integral that represents the arc length of the portion of the graph of $f(x) = -x(x-4)$ that lies above the x-axis. Do not evaluate the integral.

68. Find a curve $y = g(x)$, such that when the region between the curve and the x-axis for $0 \le x \le \pi$ is revolved around the x-axis, it forms a solid with volume given by

$$\int_0^{\pi} \pi(4 - 4\cos^2 x)\,dx.$$

[Hint: Use the identity $\sin^2 x = 1 - \cos^2 x$.]

In Problems 69–70, a hemisphere of radius a has its base on the xy-plane, centered at the origin; the z-axis is vertically upward. Using the given slices,
(a) Write an expression for the volume of a slice.
(b) Write an integral giving the volume of the hemisphere.
(c) Calculate the integral.

69. Vertical slices perpendicular to the x-axis.

70. Horizontal slices perpendicular to the z-axis.

In Problems 71–72, find the volume of the solid whose base is the region in the first quadrant bounded by $y = 4 - x^2$, the x-axis, and the y-axis, and whose cross-section in the given direction is an equilateral triangle. Include a sketch of the region and show how to find the area of a triangular cross-section.

71. Perpendicular to the x-axis.

72. Perpendicular to the y-axis.

73. For $k > 0$, the volume of the solid created by rotating the region bounded by $y = kx(x - 2)$ and the x-axis between $x = 0$ and $x = 2$ around the x-axis is $192\pi/5$. Find k.

74. A particle moves in the xy-plane along the elliptical path $x = 2\cos t$, $y = \sin t$, where x and y are in cm and $0 \le t \le \pi$. Find the distance traveled by the particle.

75. A projectile moves along a path $x = 20t$, $y = 20t - 4.9t^2$, where t is time in seconds, $0 \le t \le 5$, and x and y are in meters. Find the total distance traveled by the projectile.

76. A small plane flies in a sinusoidal pattern at a constant altitude over a river. The path of the plane is $y = 0.1\sin(30x)$, where $0 \le x \le 10$ for x and y in km. Find the total distance traveled by the plane during the flight.

77. The arc length of $y = f(x)$ from $x = 2$ to $x = 12$ is 20. Find the arc length of $g(x) = 4f(0.25x + 1)$ from $x = 4$ to $x = 44$.

78. A particle starts at the origin and moves along the curve $y = 2x^{3/2}/3$ in the positive x-direction at a speed of 3 cm/sec, where x, y are in cm. Find the position of the particle at $t = 6$.

79. A particle's position along a circular path at time t with $0 \le t \le 3$ is given by $x = \cos(\pi t)$ and $y = \sin(\pi t)$.

(a) Find the distance traveled by the particle over this time interval.
(b) How does your answer in part (a) relate to the circumference of the circle?
(c) What is the particle's displacement between $t = 0$ and $t = 3$?

80. A particle moves with velocity dx/dt in the x-direction and dy/dt in the y-direction at time t in seconds, where

$$\frac{dx}{dt} = -\sin t \quad \text{and} \quad \frac{dy}{dt} = \cos t.$$

(a) Find the change in position in the x and y-coordinates between $t = 0$ and $t = \pi$.
(b) If the particle passes through $(2, 3)$ at $t = 0$, find its position at $t = \pi$.
(c) Find the distance traveled by the particle from time $t = 0$ to $t = \pi$.

81. A particle moves with velocity dx/dt in the x-direction and dy/dt in the y-direction at time t in seconds, where

$$\frac{dx}{dt} = 3t^2 \quad \text{and} \quad \frac{dy}{dt} = 12t.$$

(a) Find the change in position in the x and y-coordinates between $t = 0$ and $t = 3$.
(b) If the particle passes through $(-7, 11)$ at $t = 0$, find its position at $t = 3$.
(c) Find the distance traveled by the particle from time $t = 0$ to $t = 3$.

82. An airplane takes off at $t = 0$ hours flying due north. It takes 24 minutes for the plane to reach cruising altitude, and during this time its ground speed (or horizontal velocity), in mph, is

$$\frac{dx}{dt} = 1250 - \frac{1050}{t+1}$$

and its vertical velocity, in mph, is

$$\frac{dy}{dt} = -625t^2 + 250t.$$

(a) What is the cruising altitude of the plane?
(b) What is the ground distance (or horizontal distance) covered from take off until cruising altitude?
(c) Find the total distance traveled by the plane from take off until cruising altitude.

83. Rotate the bell-shaped curve $y = e^{-x^2/2}$ shown in Figure 8.41 around the y-axis, forming a hill-shaped solid of revolution. By slicing horizontally, find the volume of this hill.

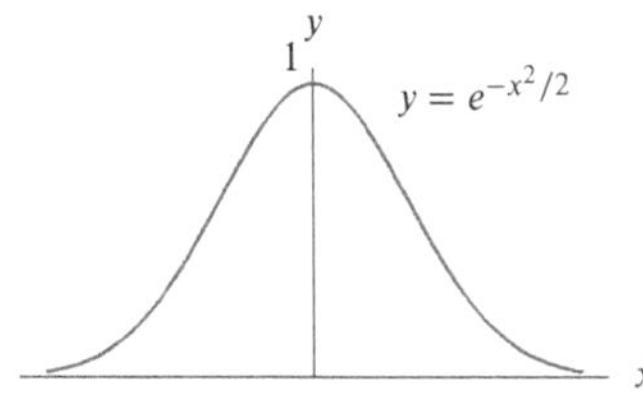

Figure 8.41

84. A bowl has the shape of the graph of $y = x^4$ between the points $(1, 1)$ and $(-1, 1)$ rotated about the y-axis. When the bowl contains water to a depth of h units, it flows out through a hole in the bottom at a rate (volume/time) proportional to $\sqrt{h}$, with constant of proportionality 6.

(a) Show that the water level falls at a constant rate.
(b) Find how long it takes to empty the bowl if it is originally full to the brim.

85. The hull of a boat has widths given by the following table. Reading across a row of the table gives widths at points 0, 10, ..., 60 feet from the front to the back at a certain level below waterline. Reading down a column of the table gives widths at levels 0, 2, 4, 6, 8 feet below waterline at a certain distance from the front. Use the trapezoidal rule to estimate the volume of the hull below waterline.

		Front of boat ⟶ Back of boat						
		0	10	20	30	40	50	60
	0	2	8	13	16	17	16	10
Depth	2	1	4	8	10	11	10	8
below	4	0	3	4	6	7	6	4
waterline	6	0	1	2	3	4	3	2
(in feet)	8	0	0	1	1	1	1	1

86. **(a)** Write an integral which represents the circumference of a circle of radius r.
(b) Evaluate the integral, and show that you get the answer you expect.

87. Find a curve $y = f(x)$ whose arc length from $x = 1$ to $x = 4$ is given by
$$\int_1^4 \sqrt{1+\sqrt{x}}\,dx.$$

88. Write a simplified expression that represents the arc length of the concave-down portion of the graph of $f(x) = e^{-x^2}$. Do not evaluate your answer.

89. Write an expression for the arc length of the portion of the graph of $f(x) = x^4 - 8x^3 + 18x^2 + 3x + 7$ that is concave down. Do not simplify or evaluate the answer.

90. With x and b in meters, a chain hangs in the shape of the catenary $\cosh x = \frac{1}{2}(e^x + e^{-x})$ for $-b \le x \le b$. If the chain is 10 meters long, how far apart are its ends?

91. The path of a robotic camera inspecting a suspension bridge cable is $y = a(e^{x/a} + e^{-x/a})/2$ for $-5 \le x \le 5$ with a constant. Find, but do not evaluate, an integral that gives the distance traveled by the camera.

Strengthen Your Understanding

In Problems 92–95, explain what is wrong with the statement.

92. The solid obtained by rotating the region bounded by the curves $y = 2x$ and $y = 3x$ between $x = 0$ and $x = 5$ around the x-axis has volume $\int_0^5 \pi(3x - 2x)^2\,dx$.

93. The arc length of the curve $y = \sin x$ from $x = 0$ to $x = \pi/4$ is $\int_0^{\pi/4} \sqrt{1+\sin^2 x}\,dx$.

94. The arc length of the curve $y = x^5$ between $x = 0$ and $x = 2$ is less than 32.

95. The circumference of a circle with parametric equations $x = \cos(2\pi t)$, $y = \sin(2\pi t)$ is
$$\int_0^2 \sqrt{(-2\pi\sin(2\pi t))^2 + (2\pi\cos(2\pi t))^2}\,dt.$$

In Problems 96–99, give an example of:

96. A region in the plane which gives the same volume whether rotated about the x-axis or the y-axis.

97. A region where the solid obtained by rotating the region around the x-axis has greater volume than the solid obtained by revolving the region around the y-axis.

98. Two different curves from $(0,0)$ to $(10,0)$ that have the same arc length.

99. A function $f(x)$ whose graph passes through the points $(0,0)$ and $(1,1)$ and whose arc length between $x = 0$ and $x = 1$ is greater than $\sqrt{2}$.

Are the statements in Problems 100–103 true or false? If a statement is true, explain how you know. If a statement is false, give a counterexample.

100. Of two solids of revolution, the one with the greater volume is obtained by revolving the region in the plane with the greater area.

101. If f is differentiable on the interval $[0, 10]$, then the arc length of the graph of f on the interval $[0, 1]$ is less than the arc length of the graph of f on the interval $[1, 10]$.

102. If f is concave up for all x and $f'(0) = 3/4$, then the arc length of the graph of f on the interval $[0, 4]$ is at least 5.

103. If f is concave down for all x and $f'(0) = 3/4$, then the arc length of the graph of f on the interval $[0, 4]$ is at most 5.

8.3 AREA AND ARC LENGTH IN POLAR COORDINATES

Many curves and regions in the plane are easier to describe in polar coordinates than in Cartesian coordinates. Thus their areas and arc lengths are best found using integrals in polar coordinates.

A point, P, in the plane is often identified by its *Cartesian coordinates* (x, y), where x is the horizontal distance to the point from the origin and y is the vertical distance.[2] Alternatively, we can identify the point, P, by specifying its distance, r, from the origin and the angle, θ, shown in Figure 8.42. The angle θ is measured counterclockwise from the positive x-axis to the line joining P to the origin. The labels r and θ are called the *polar coordinates* of point P.

[2]Cartesian coordinates can also be called rectangular coordinates.

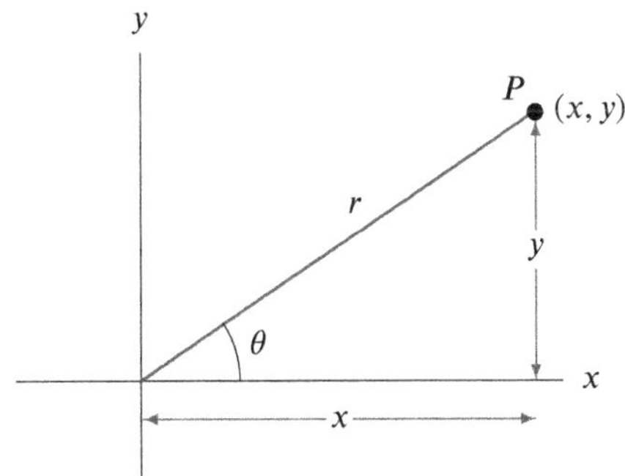

Figure 8.42: Cartesian and polar coordinates for the point P

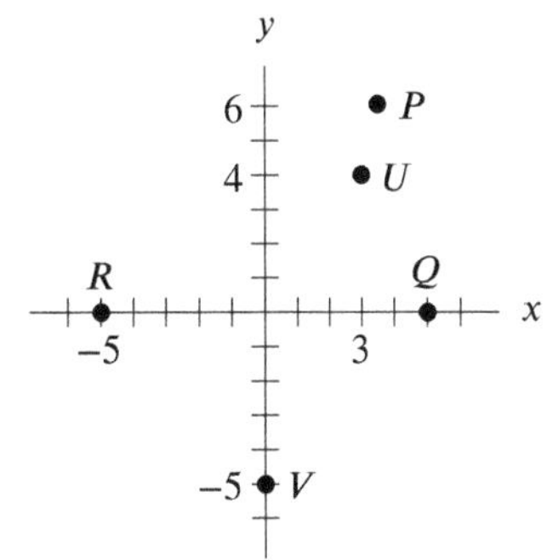

Figure 8.43: Points on the plane for Example 1

Relation Between Cartesian and Polar Coordinates

From the right triangle in Figure 8.42, we see that

- $x = r\cos\theta$ and $y = r\sin\theta$
- $r = \sqrt{x^2 + y^2}$ and $\tan\theta = \dfrac{y}{x}$, $x \neq 0$

The angle θ is determined by the equations $\cos\theta = x/\sqrt{x^2+y^2}$ and $\sin\theta = y/\sqrt{x^2+y^2}$.

Warning: In general $\theta \neq \tan^{-1}(y/x)$. It is not possible to determine which quadrant θ is in from the value of $\tan\theta$ alone.

Example 1

(a) Give Cartesian coordinates for the points with polar coordinates (r, θ) given by $P = (7, \pi/3)$, $Q = (5, 0)$, $R = (5, \pi)$.

(b) Give polar coordinates for the points with Cartesian coordinates (x, y) given by $U = (3, 4)$ and $V = (0, -5)$.

Solution

(a) See Figure 8.43. Point P is a distance of 7 from the origin. The angle θ is $\pi/3$ radians (60°). The Cartesian coordinates of P are

$$x = r\cos\theta = 7\cos\frac{\pi}{3} = \frac{7}{2} \quad \text{and} \quad y = r\sin\theta = 7\sin\frac{\pi}{3} = \frac{7\sqrt{3}}{2}.$$

Point Q is located a distance of 5 units along the positive x-axis with Cartesian coordinates

$$x = r\cos\theta = 5\cos 0 = 5 \quad \text{and} \quad y = r\sin\theta = 5\sin 0 = 0.$$

For point R, which is on the negative x-axis,

$$x = r\cos\theta = 5\cos\pi = -5 \quad \text{and} \quad y = r\sin\theta = 5\sin\pi = 0.$$

(b) For $U = (3, 4)$, we have $r = \sqrt{3^2 + 4^2} = 5$ and $\tan\theta = 4/3$. A possible value for θ is $\theta = \arctan 4/3 = 0.927$ radians, or about 53°. The polar coordinates of U are $(5, 0.927)$. The point V falls on the negative y-axis, so we can choose $r = 5$, $\theta = 3\pi/2$ for its polar coordinates. In this case, we cannot use $\tan\theta = y/x$ to find θ, because $\tan\theta = y/x = -5/0$ is undefined.

Because the angle θ can be allowed to wrap around the origin more than once, there are many possibilities for the polar coordinates of a point. For the point V in Example 1, we can also choose $\theta = -\pi/2$ or $\theta = 7\pi/2$, so that $(5, -\pi/2)$, $(5, 7\pi/2)$, and $(5, 3\pi/2)$ are all polar coordinates for V. However, we often choose θ between 0 and 2π.

Example 2 Give three possible sets of polar coordinates for the point P in Figure 8.44.

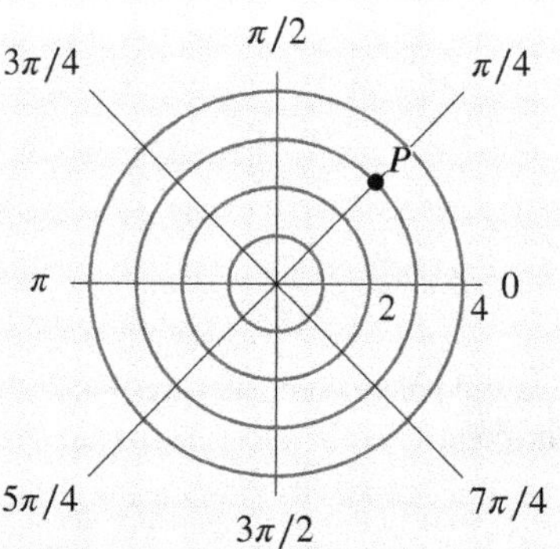

Figure 8.44

Solution Because $r = 3$ and $\theta = \pi/4$, one set of polar coordinates for P is $(3, \pi/4)$. We can also use $\theta = \pi/4 + 2\pi = 9\pi/4$ and $\theta = \pi/4 - 2\pi = -7\pi/4$, to get $(3, 9\pi/4)$ and $(3, -7\pi/4)$.

Graphing Equations in Polar Coordinates

The equations for certain graphs are much simpler when expressed in polar coordinates than in Cartesian coordinates. On the other hand, some graphs that have simple equations in Cartesian coordinates have complicated equations in polar coordinates.

Example 3 (a) Describe in words the graphs of the equation $y = 1$ (in Cartesian coordinates) and the equation $r = 1$ (in polar coordinates).

(b) Write the equation $r = 1$ using Cartesian coordinates. Write the equation $y = 1$ using polar coordinates.

Solution (a) The equation $y = 1$ describes a horizontal line. Since the equation $y = 1$ places no restrictions on the value of x, it describes every point having a y-value of 1, no matter what the value of its x-coordinate. Similarly, the equation $r = 1$ places no restrictions on the value of θ. Thus, it describes every point having an r-value of 1, that is, having a distance of 1 from the origin. This set of points is the unit circle. See Figure 8.45.

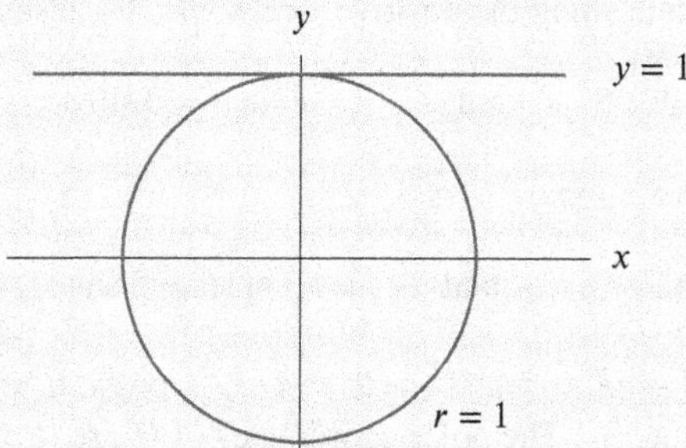

Figure 8.45: The graph of the equation $r = 1$ is the unit circle because $r = 1$ for every point regardless of the value of θ. The graph of $y = 1$ is a horizontal line since $y = 1$ for any x

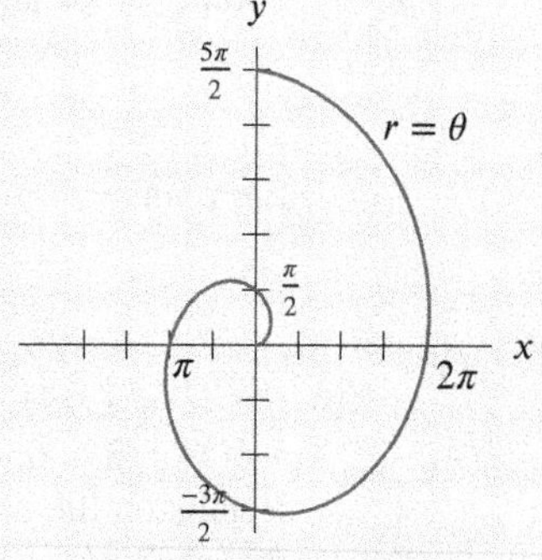

Figure 8.46: A graph of the Archimedean spiral $r = \theta$

(b) Since $r = \sqrt{x^2 + y^2}$, we rewrite the equation $r = 1$ using Cartesian coordinates as $\sqrt{x^2 + y^2} = 1$, or, squaring both sides, as $x^2 + y^2 = 1$. We see that the equation for the unit circle is simpler in polar coordinates than it is in Cartesian coordinates.

On the other hand, since $y = r \sin\theta$, we can rewrite the equation $y = 1$ in polar coordinates as $r \sin\theta = 1$, or, dividing both sides by $\sin\theta$, as $r = 1/\sin\theta$. We see that the equation for this horizontal line is simpler in Cartesian coordinates than in polar coordinates.

Example 4 Graph the equation $r = \theta$. The graph is called an *Archimedean spiral* after the Greek mathematician Archimedes who described its properties (although not by using polar coordinates).

Solution To construct this graph, use the values in Table 8.1. To help us visualize the shape of the spiral, we convert the angles in Table 8.1 to degrees and the r-values to decimals. See Table 8.2.

Table 8.1 *Points on the Archimedean spiral $r = \theta$, with θ in radians*

θ	0	$\frac{\pi}{6}$	$\frac{\pi}{3}$	$\frac{\pi}{2}$	$\frac{2\pi}{3}$	$\frac{5\pi}{6}$	π	$\frac{7\pi}{6}$	$\frac{4\pi}{3}$	$\frac{3\pi}{2}$
r	0	$\frac{\pi}{6}$	$\frac{\pi}{3}$	$\frac{\pi}{2}$	$\frac{2\pi}{3}$	$\frac{5\pi}{6}$	π	$\frac{7\pi}{6}$	$\frac{4\pi}{3}$	$\frac{3\pi}{2}$

Table 8.2 *Points on the Archimedean spiral $r = \theta$, with θ in degrees*

θ	0	30°	60°	90°	120°	150°	180°	210°	240°	270°
r	0.00	0.52	1.05	1.57	2.09	2.62	3.14	3.67	4.19	4.71

Notice that as the angle θ increases, points on the curve move farther from the origin. At 0°, the point is at the origin. At 30°, it is 0.52 units away from the origin, at 60° it is 1.05 units away, and at 90° it is 1.57 units away. As the angle winds around, the point traces out a curve that moves away from the origin, giving a spiral. (See Figure 8.46.)

In our definition, r is positive. However, graphs of curves in polar coordinates are traditionally drawn using negative values of r as well, because this makes the graphs symmetric. If an equation $r = f(\theta)$ gives a negative r-value, it is plotted in the opposite direction to θ. See Examples 5 and 6 and Figures 8.47 and 8.49.

Example 5 For $a > 0$ and n a positive integer, curves of the form $r = a \sin n\theta$ or $r = a \cos n\theta$ are called *roses*. Graph the roses

(a) $r = 3 \sin 2\theta$ (b) $r = 4 \cos 3\theta$

Solution

(a) Using a calculator or making a table of values, we see that the graph is a rose with four petals, each extending a distance of 3 from base to tip. See Figure 8.47. Negative values of r for $\pi/2 < \theta < \pi$ and $3\pi/2 < \theta < 2\pi$ give the petals in Quadrants II and IV. For example, $\theta = 3\pi/4$ gives $r = -3$, which is plotted 3 units from the origin in the direction opposite to $\theta = 3\pi/4$, namely in Quadrant IV.

(b) The graph is a rose with three petals, each extending 4 from base to tip. See Figure 8.48.

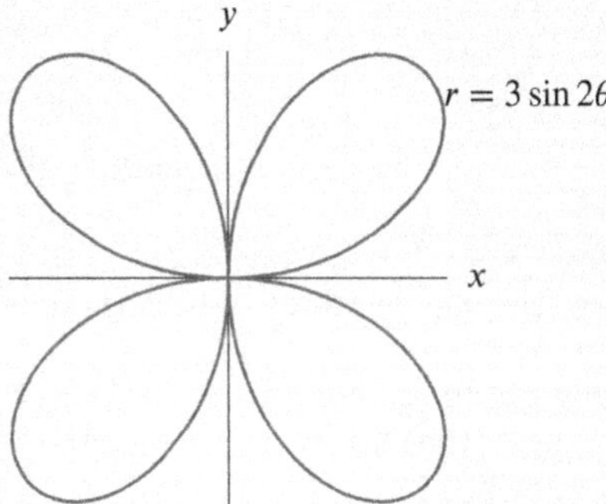

Figure 8.47: Graph of $r = 3 \sin 2\theta$ (petals in Quadrants II and IV have $r < 0$)

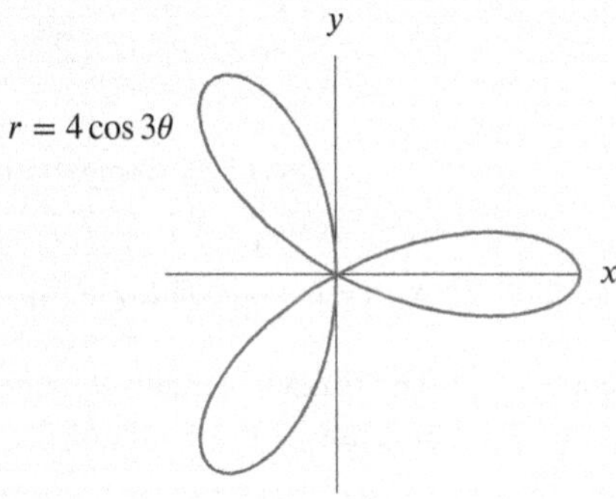

Figure 8.48: Graph of $r = 4 \cos 3\theta$

Example 6 Curves of the form $r = a + b\sin\theta$ or $r = a + b\cos\theta$ are called limaçons. Graph $r = 1 + 2\cos\theta$ and $r = 3 + 2\cos\theta$.

Solution See Figures 8.49 and 8.50. The equation $r = 1 + 2\cos\theta$ leads to negative r values for some θ values between $\pi/2$ and $3\pi/2$; these values give the inner loop in Figure 8.49. For example, $\theta = \pi$ gives $r = -1$, which is plotted 1 unit from the origin in the direction opposite to $\theta = \pi$, namely on the positive x-axis. The equation $r = 3 + 2\cos\theta$ does not lead to negative r-values.

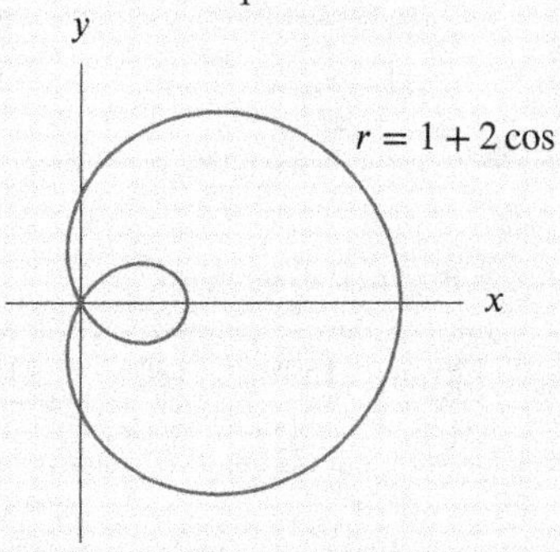

Figure 8.49: Graph of $r = 1 + 2\cos\theta$ (inner loop has $r < 0$)

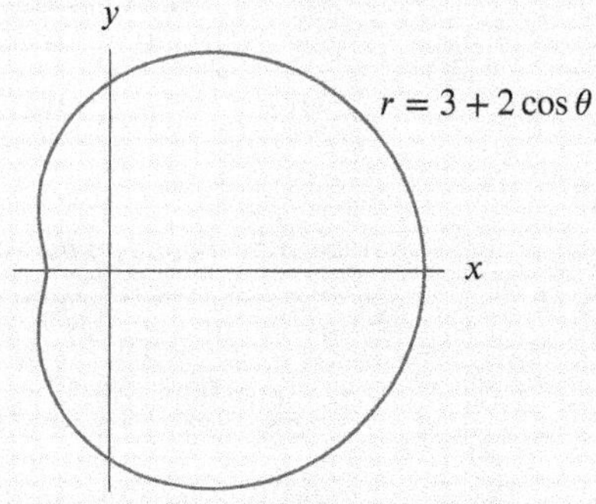

Figure 8.50: Graph of $r = 3 + 2\cos\theta$

Polar coordinates can be used with inequalities to describe regions that are obtained from circles. Such regions are often much harder to represent in Cartesian coordinates.

Example 7 Using inequalities, describe a compact disc with outer diameter 120 mm and inner diameter 15 mm.

Solution The compact disc lies between two circles of radius 7.5 mm and 60 mm. See Figure 8.51. Thus, if the origin is at the center, the disc is represented by

$$7.5 \le r \le 60 \quad \text{and} \quad 0 \le \theta \le 2\pi.$$

Figure 8.51: Compact disc

Figure 8.52: Pizza slice

Example 8 An 18-inch pizza is cut into 12 slices. Use inequalities to describe one of the slices.

Solution The pizza has radius 9 inches; the angle at the center is $2\pi/12 = \pi/6$. See Figure 8.52. Thus, if the origin is at the center of the original pizza, the slice is represented by

$$0 \le r \le 9 \qquad \text{and} \qquad 0 \le \theta \le \frac{\pi}{6}.$$

Area in Polar Coordinates

We can use a definite integral to find the area of a region described in polar coordinates. As previously, we slice the region into small pieces, construct a Riemann sum, and take a limit to obtain the definite integral. In this case, the slices are approximately circular sectors.

To calculate the area of the sector in Figure 8.53, we think of the area of the sector as a fraction $\theta/2\pi$ of the area of the entire circle (for θ in radians). Then

$$\text{Area of sector} = \frac{\theta}{2\pi} \cdot \pi r^2 = \frac{1}{2} r^2 \theta.$$

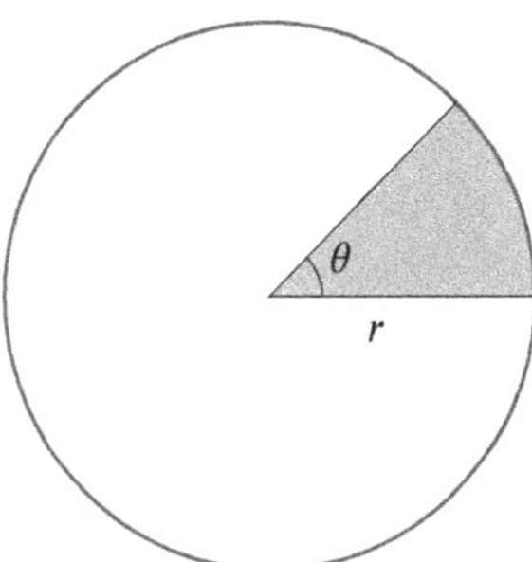

Figure 8.53: Area of shaded sector $= \frac{1}{2} r^2 \theta$ (for θ in radians)

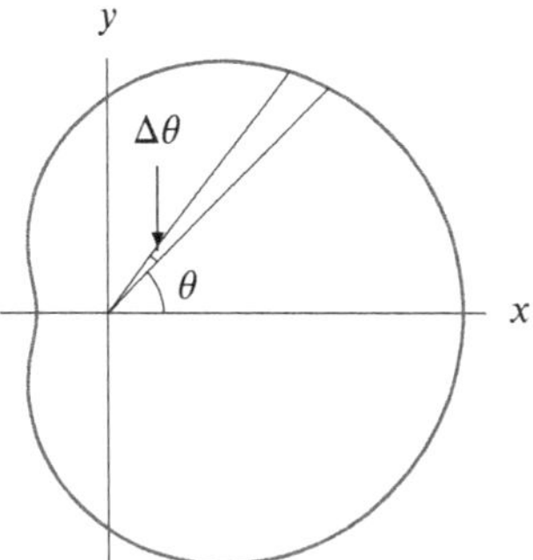

Figure 8.54: Finding the area of the limaçon $r = 3 + 2\cos\theta$

Example 9 Use circular sectors to set up a definite integral to calculate the area of the region bounded by the limaçon $r = 3 + 2\cos\theta$, for $0 \le \theta \le 2\pi$. See Figure 8.54.

Solution The slices are not exactly circular sectors because the radius r depends on θ. However,

$$\text{Area of sector} \approx \frac{1}{2} r^2 \Delta\theta = \frac{1}{2}(3 + 2\cos\theta)^2 \Delta\theta.$$

Thus, for the whole area,

$$\text{Area of region} \approx \sum \frac{1}{2}(3 + 2\cos\theta)^2 \Delta\theta.$$

Taking the limit as $n \to \infty$ and $\Delta\theta \to 0$ gives the integral

$$\text{Area} = \int_0^{2\pi} \frac{1}{2}(3 + 2\cos\theta)^2 \, d\theta.$$

To compute this integral, we expand the integrand and use integration by parts or formula IV-18 from the table of integrals:

$$\begin{aligned}
\text{Area} &= \frac{1}{2}\int_0^{2\pi} (9 + 12\cos\theta + 4\cos^2\theta)\, d\theta \\
&= \frac{1}{2}\left(9\theta + 12\sin\theta + \frac{4}{2}(\cos\theta\sin\theta + \theta)\right)\Bigg|_0^{2\pi} \\
&= \frac{1}{2}(18\pi + 0 + 4\pi) = 11\pi.
\end{aligned}$$

The reasoning in Example 9 suggests a general area formula.

For a curve $r = f(\theta)$, with $f(\theta) \geq 0$, the area in Figure 8.55 is given by

$$\text{Area of region enclosed } = \frac{1}{2}\int_{\alpha}^{\beta} f(\theta)^2 \, d\theta.$$

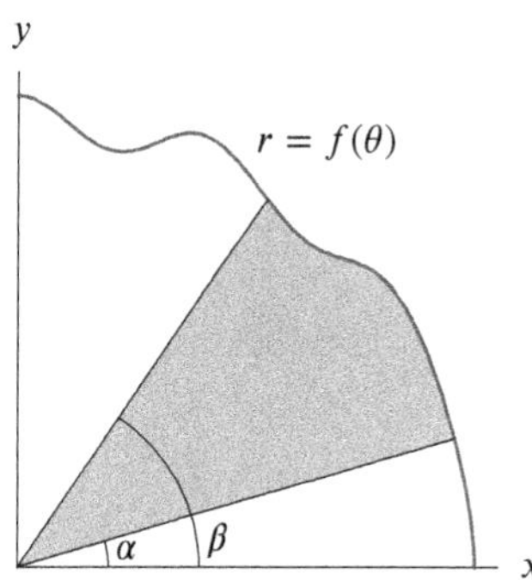

Figure 8.55: Area in polar coordinates

Example 10 Find the area of one petal of the four-petal rose $r = 3\sin 2\theta$ in Figure 8.56.

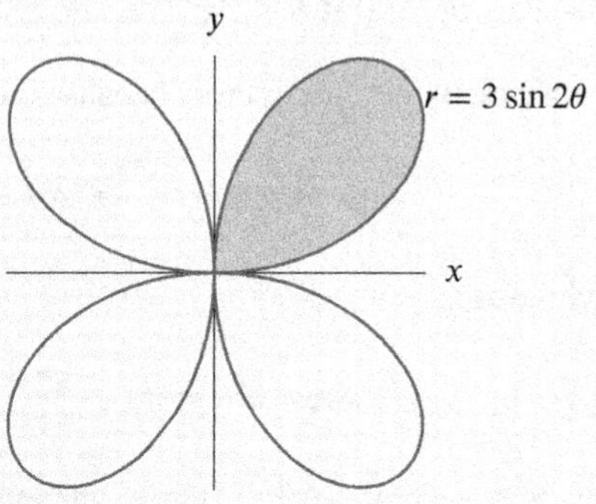

Figure 8.56: One petal of the rose $r = 3\sin 2\theta$ with $0 \leq \theta \leq \pi/2$

Solution The petal in the first quadrant is described by $r = 3\sin 2\theta$ for $0 \leq \theta \leq \pi/2$, so

$$\text{Area of shaded region } = \frac{1}{2}\int_0^{\pi/2} (3\sin 2\theta)^2 \, d\theta = \frac{9}{2}\int_0^{\pi/2} \sin^2 2\theta \, d\theta.$$

Using the substitution $w = 2\theta$, we rewrite the integral and use integration by parts or formula IV-17 from the table of integrals:

$$\begin{aligned}\text{Area } &= \frac{9}{2}\int_0^{\pi/2} \sin^2 2\theta \, d\theta = \frac{9}{4}\int_0^{\pi} \sin^2 w \, dw \\ &= \frac{9}{4}\left(-\frac{1}{2}\cos w \sin w + \frac{1}{2}w\right)\Big|_0^{\pi} = \frac{9}{4}\cdot\frac{\pi}{2} = \frac{9\pi}{8}.\end{aligned}$$

Slope in Polar Coordinates

For a curve $r = f(\theta)$, we can express x and y in terms of θ as a parameter, giving

$$x = r\cos\theta = f(\theta)\cos\theta \quad \text{and} \quad y = r\sin\theta = f(\theta)\sin\theta.$$

To find the slope of the curve, we use the formula for the slope of a parametric curve

$$\frac{dy}{dx} = \frac{dy/d\theta}{dx/d\theta}.$$

Example 11 Find the slope of the curve $r = 3\sin 2\theta$ at $\theta = \pi/3$.

Solution Expressing x and y in terms of θ, we have

$$x = 3\sin(2\theta)\cos\theta \quad \text{and} \quad y = 3\sin(2\theta)\sin\theta.$$

The slope is given by

$$\frac{dy}{dx} = \frac{6\cos(2\theta)\sin\theta + 3\sin(2\theta)\cos\theta}{6\cos(2\theta)\cos\theta - 3\sin(2\theta)\sin\theta}.$$

At $\theta = \pi/3$, we have

$$\left.\frac{dy}{dx}\right|_{\theta=\pi/3} = \frac{6(-1/2)(\sqrt{3}/2) + 3(\sqrt{3}/2)(1/2)}{6(-1/2)(1/2) - 3(\sqrt{3}/2)(\sqrt{3}/2)} = \frac{\sqrt{3}}{5}.$$

Arc Length in Polar Coordinates

We can calculate the arc length of the curve $r = f(\theta)$ by expressing x and y in terms of θ as a parameter

$$x = f(\theta)\cos\theta \qquad y = f(\theta)\sin\theta$$

and using the formula for the arc length of a parametric curve:

$$\text{Arc length} = \int_\alpha^\beta \sqrt{\left(\frac{dx}{d\theta}\right)^2 + \left(\frac{dy}{d\theta}\right)^2}\, d\theta.$$

The calculations may be simplified by using the alternate form of the arc length integral in Problem 45.

Example 12 Find the arc length of one petal of the rose $r = 3\sin 2\theta$ for $0 \le \theta \le \pi/2$. See Figure 8.56.

Solution The curve is given parametrically by

$$x = 3\sin(2\theta)\cos\theta \quad \text{and} \quad y = 3\sin(2\theta)\sin\theta.$$

Thus, calculating $dx/d\theta$ and $dy/d\theta$ and evaluating the integral on a calculator, we have:

$$\begin{aligned}\text{Arc length} &= \int_0^{\pi/2} \sqrt{(6\cos(2\theta)\cos\theta - 3\sin(2\theta)\sin\theta)^2 + (6\cos(2\theta)\sin\theta + 3\sin(2\theta)\cos\theta)^2}\, d\theta \\ &= 7.266.\end{aligned}$$

Exercises and Problems for Section 8.3

EXERCISES

Convert the polar coordinates in Exercises 1–4 to Cartesian coordinates. Give exact answers.

1. $(1, 2\pi/3)$
2. $(\sqrt{3}, -3\pi/4)$
3. $(2\sqrt{3}, -\pi/6)$
4. $(2, 5\pi/6)$

Convert the Cartesian coordinates in Exercises 5–8 to polar coordinates.

5. $(1, 1)$
6. $(-1, 0)$
7. $(\sqrt{6}, -\sqrt{2})$
8. $(-\sqrt{3}, 1)$

9. **(a)** Make a table of values for the equation $r = 1 - \sin\theta$. Include $\theta = 0, \pi/3, \pi/2, 2\pi/3, \pi, \ldots$.
   (b) Use the table to graph the equation $r = 1 - \sin\theta$ in the xy-plane. This curve is called a *cardioid.*
   (c) At what point(s) does the cardioid $r = 1 - \sin\theta$ intersect a circle of radius 1/2 centered at the origin?
   (d) Graph the curve $r = 1 - \sin 2\theta$ in the xy-plane. Compare this graph to the cardioid $r = 1 - \sin\theta$.
10. Graph the equation $r = 1 - \sin(n\theta)$, for $n = 1, 2, 3, 4$. What is the relationship between the value of n and the shape of the graph?
11. Graph the equation $r = 1 - \sin\theta$, with $0 \le \theta \le n\pi$, for $n = 2, 3, 4$. What is the relationship between the value of n and the shape of the graph?
12. Graph the equation $r = 1 - n\sin\theta$, for $n = 2, 3, 4$. What is the relationship between the value of n and the shape of the graph?
13. Graph the equation $r = 1 - \cos\theta$. Describe its relationship to $r = 1 - \sin\theta$.
14. Give inequalities that describe the flat surface of a washer that is one inch in diameter and has an inner hole with a diameter of 3/8 inch.
15. Graph the equation $r = 1 - \sin(2\theta)$ for $0 \le \theta \le 2\pi$. There are two loops. For each loop, give a restriction on θ that shows all of that loop and none of the other loop.
16. A slice of pizza is one eighth of a circle of radius 1 foot. The slice is in the first quadrant, with one edge along the x-axis, and the center of the pizza at the origin. Give inequalities describing this region using:
    (a) Polar coordinates **(b)** Rectangular coordinates

In Exercises 17–19, give inequalities for r and θ which describe the following regions in polar coordinates.

17.

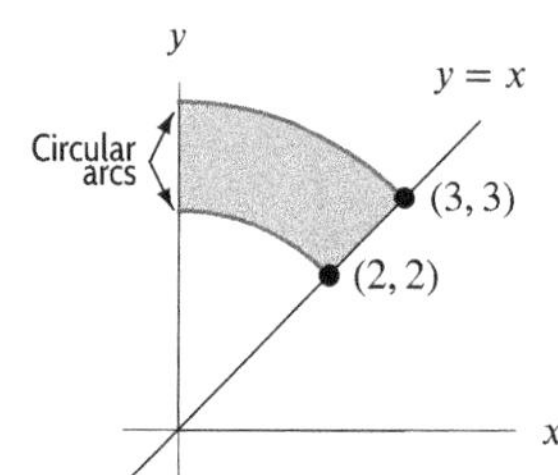

18.

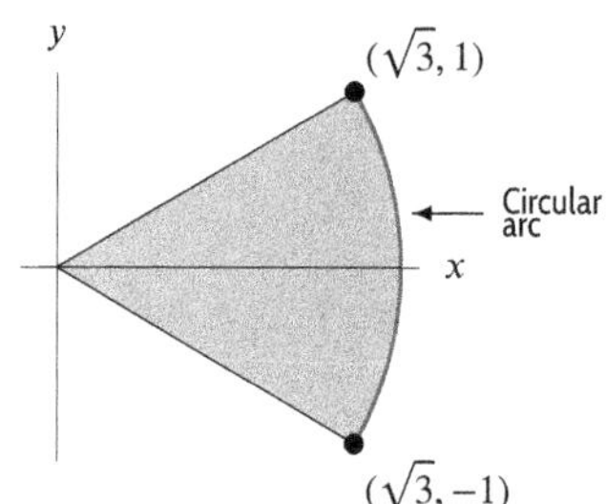

19.

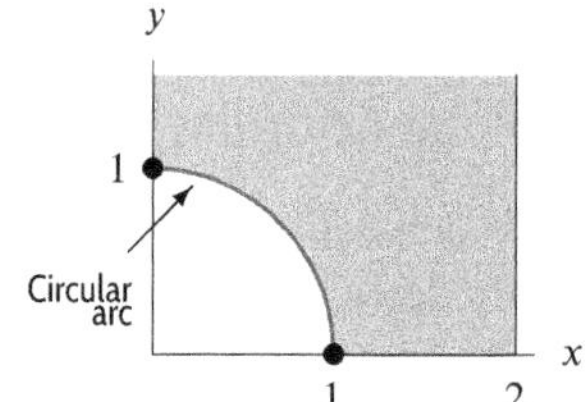

Note: Region extends indefinitely in the y-direction.

20. Find the slope of the curve $r = 2$ at $\theta = \pi/4$.
21. Find the slope of the curve $r = e^{\theta}$ at $\theta = \pi/2$.
22. Find the slope of the curve $r = 1 - \cos\theta$ at $\theta = \pi/2$.
23. Find the arc length of the curve $r = e^{\theta}$ from $\theta = \pi/2$ to $\theta = \pi$.
24. Find the arc length of the curve $r = \theta^2$ from $\theta = 0$ to $\theta = 2\pi$.

PROBLEMS

25. Sketch the polar region described by the following integral expression for area:

$$\frac{1}{2}\int_0^{\pi/3} \sin^2(3\theta)\, d\theta.$$

26. Find the area inside the spiral $r = \theta$ for $0 \le \theta \le 2\pi$.
27. Find the area between the two spirals $r = \theta$ and $r = 2\theta$ for $0 \le \theta \le 2\pi$.

28. Find the area inside the cardioid $r = 1 + \cos\theta$ for $0 \le \theta \le 2\pi$.

29. (a) In polar coordinates, write equations for the line $x = 1$ and the circle of radius 2 centered at the origin.
(b) Write an integral in polar coordinates representing the area of the region to the right of $x = 1$ and inside the circle.
(c) Evaluate the integral.

30. Show that the area formula for polar coordinates gives the expected answer for the area of the circle $r = a$ for $0 \le \theta \le 2\pi$.

31. Show that the arc length formula for polar coordinates gives the expected answer for the circumference of the circle $r = a$ for $0 \le \theta \le 2\pi$.

32. Find the area inside the circle $r = 1$ and outside the cardioid $r = 1 + \sin\theta$.

33. Find the area inside the cardioid $r = 1 - \sin\theta$ and outside the circle $r = 1/2$.

34. Find the area lying outside $r = 2\cos\theta$ and inside $r = 1 + \cos\theta$.

35. (a) Graph $r = 2\cos\theta$ and $r = 2\sin\theta$ on the same axes.
(b) Using polar coordinates, find the area of the region shared by both curves.

36. For what value of a is the area enclosed by $r = \theta, \theta = 0$, and $\theta = a$ equal to 1?

37. (a) Sketch the bounded region inside the lemniscate $r^2 = 4\cos 2\theta$ and outside the circle $r = \sqrt{2}$.
(b) Compute the area of the region described in part (a).

38. Using Example 11 on page 427, find the equation of the tangent line to the curve $r = 3\sin 2\theta$ at $\theta = \pi/3$.

39. Using Example 11 on page 427 and Figure 8.47, find the points where the curve $r = 3\sin 2\theta$ has horizontal and vertical tangents.

40. For what values of θ on the polar curve $r = \theta$, with $0 \le \theta \le 2\pi$, are the tangent lines horizontal? Vertical?

41. (a) In Cartesian coordinates, write an equation for the tangent line to $r = 1/\theta$ at $\theta = \pi/2$.
(b) The graph of $r = 1/\theta$ has a horizontal asymptote as θ approaches 0. Find the equation of this asymptote.

42. Find the maximum value of the y-coordinate of points on the limaçon $r = 1 + 2\cos\theta$.

Find the arc length of the curves in Problems 43–44.

43. $r = \theta, 0 \le \theta \le 2\pi$

44. $r = 1/\theta, \pi \le \theta \le 2\pi$

45. For the curve $r = f(\theta)$ from $\theta = \alpha$ to $\theta = \beta$, show that

$$\text{Arc length} = \int_{\alpha}^{\beta} \sqrt{(f'(\theta))^2 + (f(\theta))^2}\, d\theta.$$

46. Find the arc length of the spiral $r = \theta$ where $0 \le \theta \le \pi$.

47. Find the arc length of part of the cardioid $r = 1 + \cos\theta$ where $0 \le \theta \le \pi/2$.

Strengthen Your Understanding

In Problems 48–51, explain what is wrong with the statement.

48. The point with Cartesian coordinates (x, y) has polar coordinates $r = \sqrt{x^2 + y^2}, \theta = \tan^{-1}(y/x)$.

49. All points of the curve $r = \sin(2\theta)$ for $\pi/2 < \theta < \pi$ are in quadrant II.

50. If the slope of the curve $r = f(\theta)$ is positive, then $dr/d\theta$ is positive.

51. Any polar curve that is symmetric about both the x and y axes must be a circle, centered at the origin.

In Problems 52–55, give an example of:

52. Two different pairs of polar coordinates (r, θ) that correspond to the same point in the plane.

53. The equation of a circle in polar coordinates.

54. A polar curve $r = f(\theta)$ that is symmetric about neither the x-axis nor the y-axis.

55. A polar curve $r = f(\theta)$ other than a circle that is symmetric about the x-axis.

8.4 DENSITY AND CENTER OF MASS

Density and How to Slice a Region

The examples in this section involve the idea of *density*. For example,

- A population density is measured in, say, people per mile (along the edge of a road), or people per unit area (in a city), or bacteria per cubic centimeter (in a test tube).
- The density of a substance (e.g. air, wood, or metal) is the mass of a unit volume of the substance and is measured in, say, grams per cubic centimeter.

Suppose we want to calculate the total mass or total population, but the density is not constant over a region.

> **To find total quantity from density:** Divide the region into small pieces in such a way that the density is approximately constant on each piece, and add the contributions of the pieces.

Example 1 The Massachusetts Turnpike ("the Pike") starts in the middle of Boston and heads west. The number of people living next to it varies as it gets farther from the city. Suppose that, x miles out of town, the population density adjacent to the Pike is $P = f(x)$ people/mile. Express the total population living next to the Pike within 5 miles of Boston as a definite integral.

Solution Divide the Pike into segments of length Δx. The population density at the center of Boston is $f(0)$; let's use that density for the first segment. This gives an estimate of

$$\text{People living in first segment} \approx f(0) \text{ people/ mile} \cdot \Delta x \text{ mile} = f(0)\Delta x \text{ people}.$$

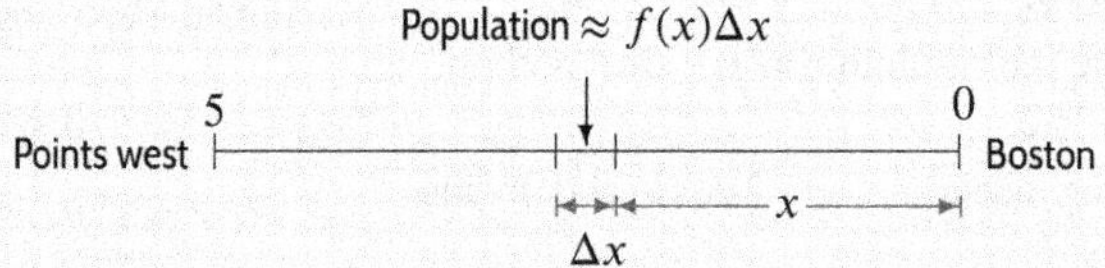

Figure 8.57: Population along the Massachusetts Turnpike

Similarly, the population in a typical segment x miles from the center of Boston is the population density times the length of the interval, or roughly $f(x)\,\Delta x$. (See Figure 8.57.) The sum of all these estimates gives the estimate

$$\text{Total population} \approx \sum f(x)\,\Delta x.$$

Letting $\Delta x \to 0$ gives

$$\text{Total population} = \lim_{\Delta x \to 0} \sum f(x)\,\Delta x = \int_0^5 f(x)\,dx.$$

The 5 and 0 in the limits of the integral are the upper and lower limits of the interval over which we are integrating.

Example 2 The air density h meters above the earth's surface is $f(h)$ kg/m^3. Find the mass of a cylindrical column of air 2 meters in diameter and 25 kilometers high, with base on the surface of the earth.

Solution The column of air is a circular cylinder 2 meters in diameter and 25 kilometers, or 25,000 meters, high. First we must decide how we are going to slice this column. Since the air density varies with altitude but remains constant horizontally, we take horizontal slices of air. That way, the density will be more or less constant over the whole slice, being close to its value at the bottom of the slice. (See Figure 8.58.)

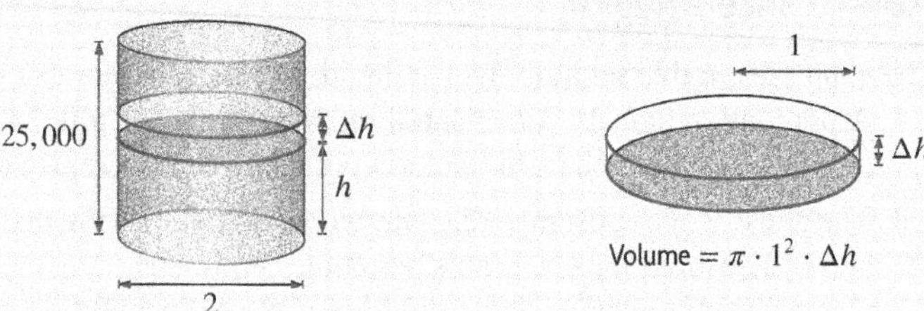

Figure 8.58: Slicing a column of air horizontally

A slice is a cylinder of height Δh and diameter 2 m, so its radius is 1 m. We find the approximate mass of the slice by multiplying its volume by its density. If the thickness of the slice is Δh, then its volume is $\pi r^2 \cdot \Delta h = \pi 1^2 \cdot \Delta h = \pi\, \Delta h\ \text{m}^3$. The density of the slice is given by $f(h)$. Thus,

$$\text{Mass of slice} \approx \text{Volume} \cdot \text{ Density} = (\pi \Delta h\ \text{m}^3)(f(h)\ \text{kg/m}^3) = \pi\, \Delta h \cdot f(h)\ \text{kg}.$$

Adding these slices up yields a Riemann sum:

$$\text{Total mass} \approx \sum \pi f(h)\, \Delta h\ \text{kg}.$$

As $\Delta h \to 0$, this sum approximates the definite integral:

$$\text{Total mass} = \int_0^{25{,}000} \pi f(h)\, dh\ \text{kg}.$$

In order to get a numerical value for the mass of air, we need an explicit formula for the density as a function of height, as in the next example.

Example 3 Find the mass of the column of air in Example 2 if the density of air at height h is given by

$$P = f(h) = 1.28e^{-0.000124h}\ \text{kg/m}^3.$$

Solution Using the result of the previous example, we have

$$\begin{aligned}\text{Mass} &= \int_0^{25{,}000} \pi 1.28 e^{-0.000124h}\, dh = \frac{-1.28\pi}{0.000124}\left(e^{-0.000124h}\Big|_0^{25{,}000}\right)\\ &= \frac{1.28\pi}{0.000124}\left(e^0 - e^{-0.000124(25{,}000)}\right) \approx 31{,}000\ \text{kg}.\end{aligned}$$

It requires some thought to figure out how to slice a region. The key point is that you want the density to be nearly constant within each piece.

Example 4 The population density in Ringsburg is a function of the distance from the city center. At r miles from the center, the density is $P = f(r)$ people per square mile. Ringsburg is circular with radius 5 miles. Write a definite integral that expresses the total population of Ringsburg.

Solution We want to slice Ringsburg up and estimate the population of each slice. If we were to take straight-line slices, the population density would vary on each slice, since it depends on the distance from the city center. We want the population density to be pretty close to constant on each slice. We therefore take slices that are thin rings around the center. (See Figure 8.59.) Since the ring is very thin, we can approximate its area by straightening it into a thin rectangle. (See Figure 8.60.) The width of the rectangle is Δr miles, and its length is approximately equal to the circumference of the ring, $2\pi r$ miles, so its area is about $2\pi r\, \Delta r\ \text{mi}^2$. Since

$$\text{Population on ring} \approx \text{Density} \cdot \text{Area},$$

we get

$$\text{Population on ring} \approx (f(r)\ \text{people/mi}^2)(2\pi r \Delta r\ \text{mi}^2) = f(r) \cdot 2\pi r\, \Delta r\ \text{people}.$$

Adding the contributions from each ring, we get

$$\text{Total population} \approx \sum 2\pi r f(r)\, \Delta r\ \text{people}.$$

So

$$\text{Total population} = \int_0^5 2\pi r f(r)\,dr \text{ people.}$$

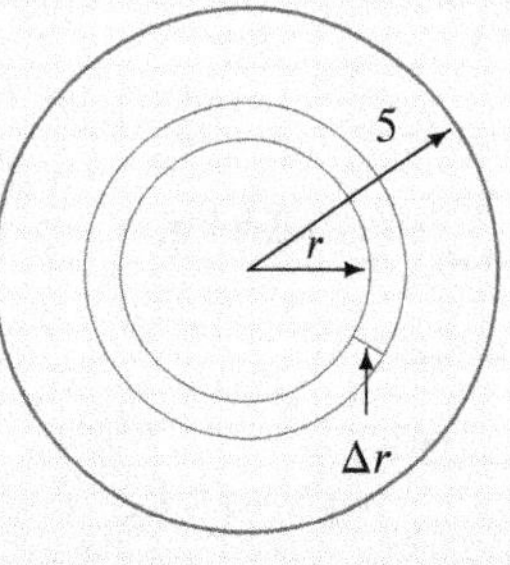

Figure 8.59: Ringsburg

Figure 8.60: Ring from Ringsburg (straightened out)

Note: You may wonder what happens if we calculate the area of the ring by subtracting the area of the inner circle (πr^2) from the area of the outer circle ($\pi(r + \Delta r)^2$), giving

$$\text{Area} = \pi(r + \Delta r)^2 - \pi r^2.$$

Multiplying out and subtracting, we get

$$\begin{aligned}\text{Area} &= \pi(r^2 + 2r\,\Delta r + (\Delta r)^2) - \pi r^2 \\ &= 2\pi r\,\Delta r + \pi(\Delta r)^2.\end{aligned}$$

This expression differs from the one we used before by the $\pi(\Delta r)^2$ term. However, as Δr becomes very small, $\pi(\Delta r)^2$ becomes much, much smaller. We say its smallness is of *second order*, since the power of the small factor, Δr, is 2. In the limit as $\Delta r \to 0$, we can ignore $\pi(\Delta r)^2$.

Center of Mass

The center of mass of a mechanical system is important for studying its behavior when in motion. For example, some sport utility vehicles and light trucks tend to tip over in accidents, because of their high centers of mass.

In this section, we first define the center of mass for a system of point masses on a line. Then we use the definite integral to extend this definition.

Point Masses

Two children on a seesaw, one twice the weight of the other, will balance if the lighter child is twice as far from the pivot as the heavier child. Thus, the balance point is 2/3 of the way from the lighter child and 1/3 of the way from the heavier child. This balance point is the *center of mass* of the mechanical system consisting of the masses of the two children (we ignore the mass of the seesaw itself). See Figure 8.61.

To find the balance point, we use the *displacement* (signed distance) of each child from the pivot to calculate the *moment*, where

$$\text{Moment of mass about pivot} = \text{Mass} \times \text{Displacement from pivot}.$$

A moment represents the tendency of a child to turn the system about the pivot point; the seesaw balances if the total moment is zero. Thus, the center of mass is the point about which the total moment is zero.

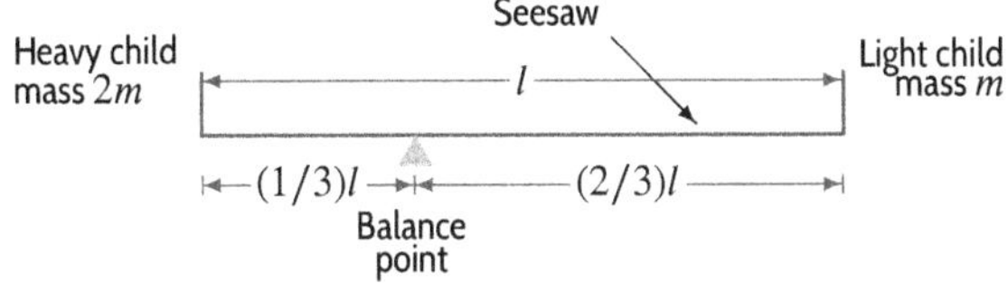

Figure 8.61: Children on seesaw

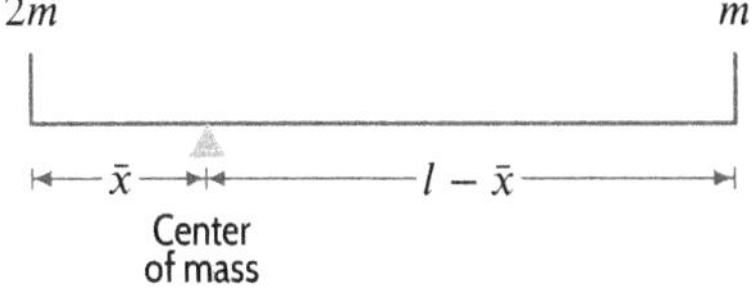

Figure 8.62: Center of mass of point masses

Example 5 Calculate the position of the center of mass of the children in Figure 8.61 using moments.

Solution Suppose the center of mass in Figure 8.62 is at a distance of $\bar{x}$ from the left end. The moment of the left mass about the center of mass is $-2m\bar{x}$ (it is negative since it is to the left of the center of mass); the moment of the right mass about the center of mass is $m(l - \bar{x})$. The system balances if

$$-2m\bar{x} + m(l - \bar{x}) = 0 \quad \text{or} \quad ml - 3m\bar{x} = 0 \quad \text{so} \quad \bar{x} = \frac{1}{3}l.$$

Thus, the center of mass is $l/3$ from the left end.

We use the same method to calculate the center of mass, $\bar{x}$, of the system in Figure 8.63. The sum of the moments of the three masses about $\bar{x}$ is 0, so

$$m_1(x_1 - \bar{x}) + m_2(x_2 - \bar{x}) + m_3(x_3 - \bar{x}) = 0.$$

Solving for $\bar{x}$, we get

$$m_1\bar{x} + m_2\bar{x} + m_3\bar{x} = m_1x_1 + m_2x_2 + m_3x_3$$

$$\bar{x} = \frac{m_1x_1 + m_2x_2 + m_3x_3}{m_1 + m_2 + m_3} = \frac{\sum_{i=1}^{3} m_i x_i}{\sum_{i=1}^{3} m_i}.$$

Generalizing leads to the following formula:

> The **center of mass** of a system of n point masses $m_1, m_2, \ldots, m_n$ located at positions $x_1, x_2, \ldots, x_n$ along the x-axis is given by
>
> $$\bar{x} = \frac{\sum x_i m_i}{\sum m_i}.$$

The numerator is the sum of the moments of the masses about the origin; the denominator is the total mass of the system.

Example 6 Show that the definition of $\bar{x}$ gives the same answer as we found in Example 5.

Solution Suppose the origin is at the left end of the seesaw in Figure 8.61. The total mass of the system is $2m + m = 3m$. We compute

$$\bar{x} = \frac{\sum x_i m_i}{\sum m_i} = \frac{1}{3m}(2m \cdot 0 + m \cdot l) = \frac{ml}{3m} = \frac{l}{3}.$$

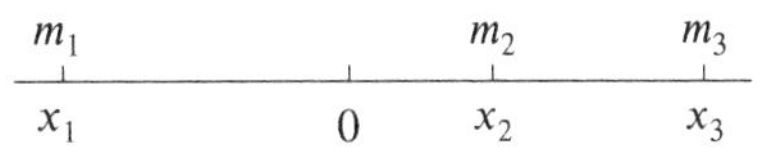

Figure 8.63: Discrete masses m_1, m_2, m_3

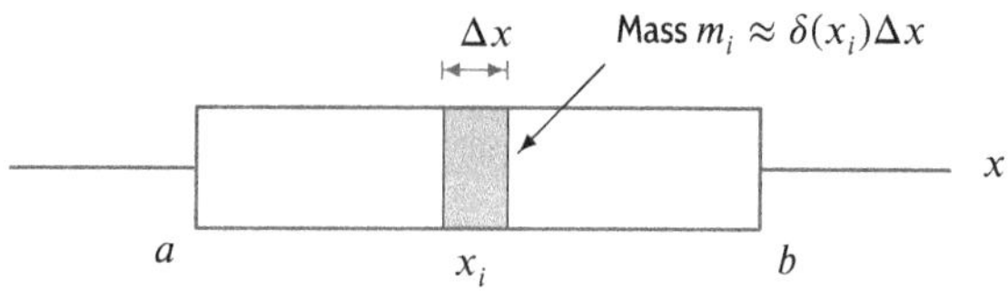

Figure 8.64: Calculating the center of mass of an object of variable density, $\delta(x)$

Continuous Mass Density

Instead of discrete masses arranged along the x-axis, suppose we have an object lying on the x-axis between $x = a$ and $x = b$. At point x, suppose the object has mass density (mass per unit length) of $\delta(x)$. To calculate the center of mass of such an object, divide it into n pieces, each of length Δx. On each piece, the density is nearly constant, so the mass of the piece is given by density times length. See Figure 8.64. Thus, if x_i is a point in the ith piece,

$$\text{Mass of the } i^{\text{th}} \text{ piece, } m_i \approx \delta(x_i)\Delta x.$$

Then the formula for the center of mass, $\bar{x} = \sum x_i m_i / \sum m_i$, applied to the n pieces of the object gives

$$\bar{x} = \frac{\sum x_i \delta(x_i)\Delta x}{\sum \delta(x_i)\Delta x}.$$

In the limit as $n \to \infty$ we have the following formula:

> The **center of mass** $\bar{x}$ of an object lying along the x-axis between $x = a$ and $x = b$ is
>
> $$\bar{x} = \frac{\int_a^b x\delta(x)\,dx}{\int_a^b \delta(x)\,dx},$$
>
> where $\delta(x)$ is the density (mass per unit length) of the object.

As in the discrete case, the denominator is the total mass of the object.

Example 7 Find the center of mass of a 2-meter rod lying on the x-axis with its left end at the origin if:
(a) The density is constant and the total mass is 5 kg. (b) The density is $\delta(x) = 15x^2$ kg/m.

Solution (a) Since the density is constant along the rod, we expect the balance point to be in the middle, that is, $\bar{x} = 1$. To check this, we compute $\bar{x}$. The density is the total mass divided by the length, so $\delta(x) = 5/2$ kg/m. Then

$$\bar{x} = \frac{\text{Moment}}{\text{Mass}} = \frac{\int_0^2 x \cdot \frac{5}{2}\,dx}{5} = \frac{1}{5} \cdot \frac{5}{2} \cdot \frac{x^2}{2}\bigg|_0^2 = 1 \text{ meter.}$$

(b) Since more of the mass of the rod is closer to its right end (the density is greatest there), we expect the center of mass to be in the right half of the rod, that is, between $x = 1$ and $x = 2$. We have

$$\text{Total mass} = \int_0^2 15x^2\,dx = 5x^3\Big|_0^2 = 40 \text{ kg.}$$

Thus,

$$\bar{x} = \frac{\text{Moment}}{\text{Mass}} = \frac{\int_0^2 x \cdot 15x^2 dx}{40} = \frac{15}{40} \cdot \frac{x^4}{4}\bigg|_0^2 = 1.5 \text{ meter.}$$

Two- and Three-Dimensional Regions

For a system of masses that lies in the plane, the center of mass is a point with coordinates $(\bar{x}, \bar{y})$. In three dimensions, the center of mass is a point with coordinates $(\bar{x}, \bar{y}, \bar{z})$. To compute the center of mass in three dimensions, we use the following formulas in which $A_x(x)$ is the area of a slice perpendicular to the x-axis at x, and $A_y(y)$ and $A_z(z)$ are defined similarly. In two dimensions, we use the same formulas for $\bar{x}$ and $\bar{y}$, but we interpret $A_x(x)$ and $A_y(y)$ as the lengths of strips perpendicular to the x- and y-axes, respectively.

> For a region of constant density δ, the center of mass is given by
>
> $$\bar{x} = \frac{\int x\delta A_x(x)\,dx}{\text{Mass}} \qquad \bar{y} = \frac{\int y\delta A_y(y)\,dy}{\text{Mass}} \qquad \bar{z} = \frac{\int z\delta A_z(z)\,dz}{\text{Mass}}.$$

The expression $\delta A_x(x)\Delta x$ is the moment of a slice perpendicular to the x-axis. Thus, these formulas are extensions of the formula for the one-dimensional case. In the two- and three-dimensional case, we are assuming that the density δ is constant. If the density is not constant, finding the center of mass may require a double or triple integral from multivariable calculus.

Example 8 Find the coordinates of the center of mass of the isosceles triangle in Figure 8.65. The triangle has constant density and mass m.

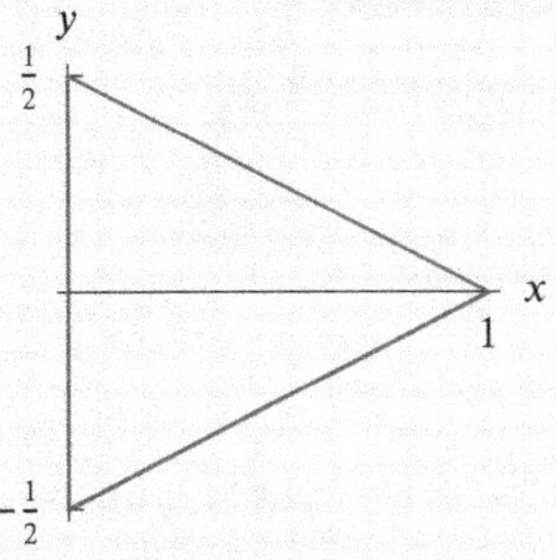

Figure 8.65: Find center of mass of this triangle

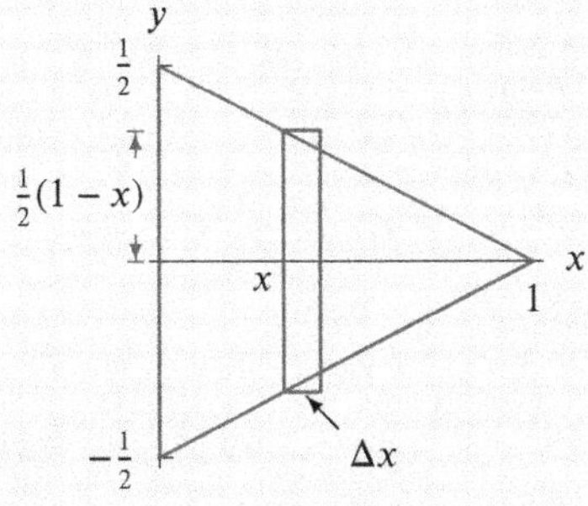

Figure 8.66: Sliced triangle

Solution Because the mass of the triangle is symmetrically distributed with respect to the x-axis, $\bar{y} = 0$. We expect $\bar{x}$ to be closer to $x = 0$ than to $x = 1$, since the triangle is wider near the origin.

The area of the triangle is $\frac{1}{2} \cdot 1 \cdot 1 = \frac{1}{2}$. Thus, Density = Mass/Area = $2m$. If we slice the triangle into strips of width Δx, then the strip at position x has length $A_x(x) = 2 \cdot \frac{1}{2}(1-x) = (1-x)$. See Figure 8.66. So

$$\text{Area of strip} = A_x(x)\Delta x \approx (1-x)\Delta x.$$

Since the density is $2m$, the center of mass is given by

$$\bar{x} = \frac{\int x\delta A_x(x)\,dx}{\text{Mass}} = \frac{\int_0^1 2mx(1-x)\,dx}{m} = 2\left(\frac{x^2}{2} - \frac{x^3}{3}\right)\Bigg|_0^1 = \frac{1}{3}.$$

So the center of mass of this triangle is at the point $(\bar{x}, \bar{y}) = (1/3, 0)$.

Example 9 Find the center of mass of a hemisphere of radius 7 cm and constant density δ.

Solution Stand the hemisphere with its base horizontal in the xy-plane, with the center at the origin. Symmetry tells us that its center of mass lies directly above the center of the base, so $\bar{x} = \bar{y} = 0$. Since the hemisphere is wider near its base, we expect the center of mass to be nearer to the base than the top.

To calculate the center of mass, slice the hemisphere into horizontal disks, as in Figure 8.67. A disk of thickness Δz at height z above the base has

$$\text{Volume of disk} = A_z(z)\Delta z \approx \pi(7^2 - z^2)\Delta z\,\text{cm}^3.$$

So, since the density is δ,

$$\bar{z} = \frac{\int z\delta A_z(z)\,dz}{\text{Mass}} = \frac{\int_0^7 z\delta\pi(7^2 - z^2)\,dz}{\text{Mass}}.$$

Since the total mass of the hemisphere is $(\frac{2}{3}\pi 7^3)\,\delta$, we get

$$\bar{z} = \frac{\delta\pi\int_0^7 (7^2 z - z^3)\,dz}{\text{Mass}} = \frac{\delta\pi\left(7^2 z^2/2 - z^4/4\right)\Big|_0^7}{\text{Mass}} = \frac{\frac{7^4}{4}\delta\pi}{\frac{2}{3}\pi 7^3\delta} = \frac{21}{8} = 2.625\text{ cm}.$$

The center of mass of the hemisphere is 2.625 cm above the center of its base. As expected, it is closer to the base of the hemisphere than its top.

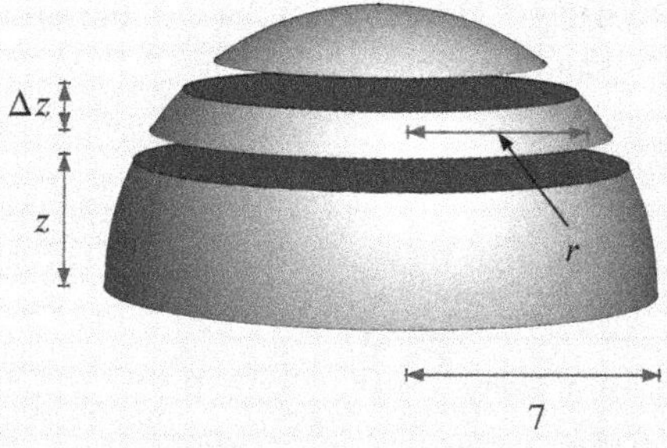

Figure 8.67: Slicing to find the center of mass of a hemisphere

Exercises and Problems for Section 8.4

EXERCISES

1. Find the mass of a rod of length 10 cm with density $\delta(x) = e^{-x}$ gm/cm at a distance of x cm from the left end.

2. A plate occupying the region $0 \le x \le 2, 0 \le y \le 3$ has density $\delta = 5$ gm/cm^2. Set up two integrals giving the mass of the plate, one corresponding to strips in the x-direction and the other corresponding to strips in the y-direction.

3. A rod has length 2 meters. At a distance x meters from its left end, the density of the rod is given by

$$\delta(x) = 2 + 6x\text{ gm/m}.$$

   (a) Write a Riemann sum approximating the total mass of the rod.
   (b) Find the exact mass by converting the sum into an integral.

4. If a rod lies along the x-axis between a and b, the moment of the rod is $\int_a^b x\delta(x)\,dx$, where $\delta(x)$ is its density in grams/meter at a position x meters. Find the moment and center of mass of the rod in Exercise 3.

5. The density of cars (in cars per mile) down a 20-mile stretch of the Pennsylvania Turnpike is approximated by

$$\delta(x) = 300\left(2 + \sin\left(4\sqrt{x + 0.15}\right)\right),$$

   at a distance x miles from the Breezewood toll plaza.

   (a) Sketch a graph of this function for $0 \le x \le 20$.
   (b) Write a Riemann sum that approximates the total number of cars on this 20-mile stretch.
   (c) Find the total number of cars on the 20-mile stretch.

6. (a) Find a Riemann sum which approximates the total mass of a 3×5 rectangular sheet, whose density per unit area at a distance x from one of the sides of length 5 is $1/(1+x^4)$.
   (b) Calculate the mass.

7. A point mass of 2 grams located 3 centimeters to the left of the origin and a point mass of 5 grams located 4 centimeters to the right of the origin are connected by a thin, light rod. Find the center of mass of the system.

8. Find the center of mass of a system containing three point masses of 5 gm, 3 gm, and 1 gm located respectively at $x = -10, x = 1$, and $x = 2$.

9. Find the mass of the block $0 \le x \le 10$, $0 \le y \le 3$, $0 \le z \le 1$, whose density δ, is given by

$$\delta = 2 - z \quad \text{for } 0 \le z \le 1.$$

PROBLEMS

Problems 10–12 refer to a colony of bats which flies out of a cave each night to eat insects. To estimate the colony's size, a naturalist counts samples of bats at different distances from the cave. Table 8.3 gives n, her count per hectare, at a distance r km from the cave. For instance, she counts 300 bats in one hectare at the cave's mouth, and 219 bats in one hectare one kilometer from the cave. The bat count r km from the cave is the same in all directions. [Note that 1 km^2 = 100 hectares, written 100 ha.]

Table 8.3

r	0	1	2	3	4	5
n	300	219	160	117	85	62

10. Give an overestimate of the number of bats between 3 and 4 km from the cave.

11. Give an underestimate of the number of bats between 3 and 4 km from the cave.

12. Letting $n = f(r)$, write an integral in terms of f representing the number of bats in the cave. Assume that bats fly no farther away than 5 km from the cave. Do not evaluate the integral.

13. Find the total mass of the triangular region in Figure 8.68, which has density $\delta(x) = 1 + x$ grams/cm^2.

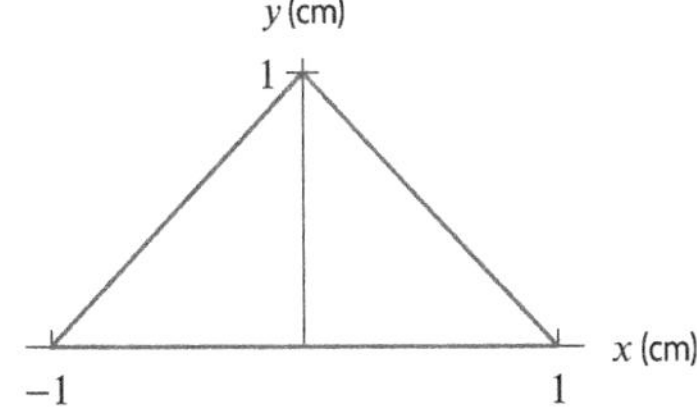

Figure 8.68

14. A rectangular plate is located with vertices at points $(0,0), (2,0), (2,3)$ and $(0,3)$ in the xy-plane. The density of the plate at point (x, y) is $\delta(y) = 2 + y^2$ gm/cm^2 and x and y are in cm. Find the total mass of the plate.

15. A cardboard figure has the shape shown in Figure 8.69. The region is bounded on the left by the line $x = a$, on the right by the line $x = b$, above by $f(x)$, and below by $g(x)$. If the density $\delta(x)$ gm/cm^2 varies only with x, find an expression for the total mass of the figure, in terms of $f(x)$, $g(x)$, and $\delta(x)$.

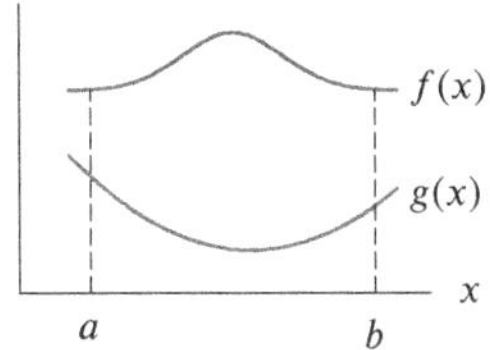

Figure 8.69

16. Circle City, a typical metropolis, is densely populated near its center, and its population gradually thins out toward the city limits. In fact, its population density is 10,000$(3 - r)$ people/square mile at distance r miles from the center.
   (a) Assuming that the population density at the city limits is zero, find the radius of the city.
   (b) What is the total population of the city?

17. The density of oil in a circular oil slick on the surface of the ocean at a distance r meters from the center of the slick is given by $\delta(r) = 50/(1+r)$ kg/m^2.
   (a) If the slick extends from $r = 0$ to $r = 10{,}000$ m, find a Riemann sum approximating the total mass of oil in the slick.
   (b) Find the exact value of the mass of oil in the slick by turning your sum into an integral and evaluating it.
   (c) Within what distance r is half the oil of the slick contained?

18. The soot produced by a garbage incinerator spreads out in a circular pattern. The depth, $H(r)$, in millimeters, of the soot deposited each month at a distance r kilometers from the incinerator is given by $H(r) = 0.115e^{-2r}$.
   (a) Write a definite integral giving the total volume of soot deposited within 5 kilometers of the incinerator each month.
   (b) Evaluate the integral you found in part (a), giving your answer in cubic meters.

19. The concentration of silt in a cylindrical well 2 m across and 20 m deep is 10 g/m^3 at the surface (depth $s = 0$) and increases linearly to 50 g/m^3 at the bottom (depth $s = 20$).
   (a) Find a formula for the silt concentration $C(s)$ as a function of depth s. What is the silt concentration in g/m^3 at a depth of $s = 10$?

(b) How much silt would there be in the bottom meter of the well assuming concentration does not change from 50 g/m^3. Is this an over or underestimate?
(c) Approximate the silt in a horizontal slice of water of thickness Δs at depth s.
(d) Set up and evaluate an integral to calculate the total amount of silt in the well.

20. Three point masses of 4 gm each are placed at $x = -6$, 1 and 3. Where should a fourth point mass of 4 gm be placed to make the center of mass at the origin?

21. A rod of length 3 meters with density $\delta(x) = 1 + x^2$ grams/meter is positioned along the positive x-axis, with its left end at the origin. Find the total mass and the center of mass of the rod.

22. A rod with density $\delta(x) = 2 + \sin x$ lies on the x-axis between $x = 0$ and $x = \pi$. Find the center of mass of the rod.

23. A rod of length 1 meter has density $\delta(x) = 1 + kx^2$ grams/meter, where k is a positive constant. The rod is lying on the positive x-axis with one end at the origin.

(a) Find the center of mass as a function of k.
(b) Show that the center of mass of the rod satisfies $0.5 < \bar{x} < 0.75$.

24. A rod of length 2 meters and density $\delta(x) = 3-e^{-x}$ kilograms per meter is placed on the x-axis with its ends at $x = \pm 1$.

(a) Will the center of mass of the rod be on the left or right of the origin? Explain.
(b) Find the coordinate of the center of mass.

25. One half of a uniform circular disk of radius 1 meter lies in the xy-plane with its diameter along the y-axis, its center at the origin, and $x > 0$. The mass of the half-disk is 3 kg. Find $(\bar{x}, \bar{y})$.

26. A metal plate, with constant density 2 gm/cm^2, has a shape bounded by the curve $y = x^2$ and the x-axis, with $0 \le x \le 1$ and x, y in cm.

(a) Find the total mass of the plate.
(b) Sketch the plate, and decide, on the basis of the shape, whether $\bar{x}$ is less than or greater than $1/2$.
(c) Find $\bar{x}$.

27. A metal plate, with constant density 5 gm/cm^2, has a shape bounded by the curve $y = \sqrt{x}$ and the x-axis, with $0 \le x \le 1$ and x, y in cm.

(a) Find the total mass of the plate.
(b) Find $\bar{x}$ and $\bar{y}$.

28. An isosceles triangle with uniform density, altitude a, and base b is placed in the xy-plane as in Figure 8.70. Show that the center of mass is at $\bar{x} = a/3$, $\bar{y} = 0$. Hence show that the center of mass is independent of the triangle's base.

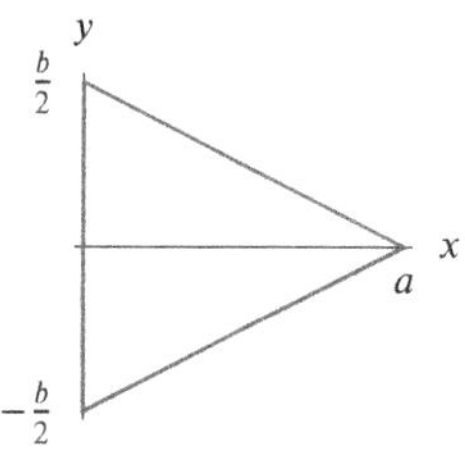

Figure 8.70

29. Find the center of mass of a cone of height 5 cm and base diameter 10 cm with constant density δ gm/cm^3.

30. A solid is formed by rotating the region bounded by the curve $y = e^{-x}$ and the x-axis between $x = 0$ and $x = 1$, around the x-axis. It was shown in Example 1 on page 411 that the volume of this solid is $\pi(1-e^{-2})/2$. Assuming the solid has constant density δ, find $\bar{x}$ and $\bar{y}$.

31. (a) Find the mass of a pyramid of constant density δ gm/cm^3 with a square base of side 40 cm and height 10 cm. [That is, the vertex is 10 cm above the center of the base.]
(b) Find the center of mass of the pyramid.

32. The storage shed in Figure 8.71 is the shape of a half-cylinder of radius r and length l.

(a) What is the volume of the shed?
(b) The shed is filled with sawdust whose density (mass/unit volume) at any point is proportional to the distance of that point from the floor. The constant of proportionality is k. Calculate the total mass of sawdust in the shed.

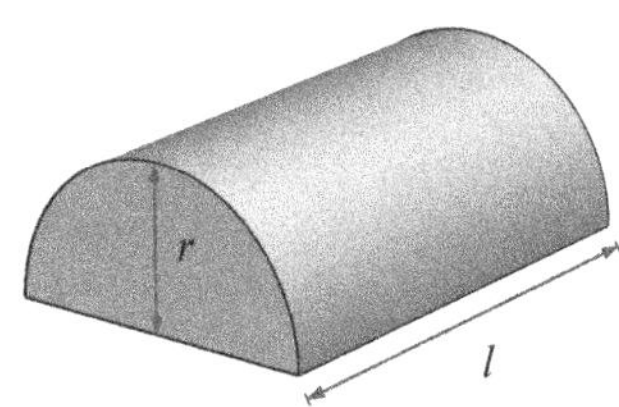

Figure 8.71

33. Water leaks out of a tank through a square hole with 1-inch sides. At time t (in seconds) the velocity of water flowing through the hole is $v = g(t)$ ft/sec. Write a definite integral that represents the total amount of water lost in the first minute.

34. An exponential model for the density of the earth's atmosphere says that if the temperature of the atmosphere were constant, then the density of the atmosphere as a function of height, h (in meters), above the surface of the earth would be given by

$$\delta(h) = 1.28e^{-0.000124h} \text{ kg/m}^3.$$

(a) Write (but do not evaluate) a sum that approximates the mass of the portion of the atmosphere from $h = 0$ to $h = 100$ m (i.e., the first 100 meters above sea level). Assume the radius of the earth is 6400 km.

(b) Find the exact answer by turning your sum in part (a) into an integral. Evaluate the integral.

35. The following table gives the density D (in gm/cm^3) of the earth at a depth x km below the earth's surface. The radius of the earth is about 6370 km. Find an upper and a lower bound for the earth's mass such that the upper bound is less than twice the lower bound. Explain your reasoning; in particular, what assumptions have you made about the density?

x	0	1000	2000	2900	3000	4000	5000	6000	6370
D	3.3	4.5	5.1	5.6	10.1	11.4	12.6	13.0	13.0

Strengthen Your Understanding

In Problems 36–39, explain what is wrong with the statement.

36. A 10 cm rod can have mass density given by $f(x) = x^2 - 5x$ gm/cm, at a point x cm from one end.

37. The center of mass of a rod with density x^2 gm/cm for $0 \le x \le 10$ is given by $\int_0^{10} x^3\, dx$.

38. If the center of mass of a rod is in the center of the rod, then the density of the rod is constant.

39. A disk with radius 3 cm and density $\delta(r) = 3 - r$ gm/cm^2, where r is in centimeters from the center of the disk, has total mass 27π gm.

In Problems 40–42, give an example of:

40. A mass density on a rod such that the rod is most dense at one end but the center of mass is nearer the other end.

41. A rod of length 2 cm, whose density $\delta(x)$ makes the center of mass not at the center of the rod.

42. A rod of length 2 cm, whose density $\delta(x)$ makes the center of mass at the center of the rod.

In Problems 43–49, are the statements true or false? Give an explanation for your answer.

43. To find the total population in a circular city, we always slice it into concentric rings, no matter what the population density function.

44. A city occupies a region in the xy-plane, with population density $\delta(y) = 1 + y$. To set up an integral representing the total population in the city, we should slice the region parallel to the y-axis.

45. The population density in a circular city of radius 2 depends on the distance r from the center by $f(r) = 10-3r$, so that the density is greatest at the center. Then the population of the inner city, $0 \le r \le 1$, is greater than the population of the suburbs, $1 \le r \le 2$.

46. The location of the center of mass of a system of three masses on the x-axis does not change if all the three masses are doubled.

47. The center of mass of a region in the plane cannot be outside the region.

48. Particles are shot at a circular target. The density of particles hitting the target decreases with the distance from the center. To set up a definite integral to calculate the total number of particles hitting the target, we should slice the region into concentric rings.

49. A metal rod of density $f(x)$ lying along the x-axis from $x = 0$ to $x = 4$ has its center of mass at $x = 2$. Then the two halves of the rod on either side of $x = 2$ have equal mass.

8.5 APPLICATIONS TO PHYSICS

Although geometric problems were a driving force for the development of the calculus in the seventeenth century, it was Newton's spectacularly successful applications of the calculus to physics that most clearly demonstrated the power of this new mathematics.

Work

In physics the word "work" has a technical meaning which is different from its everyday meaning. Physicists say that if a constant force, F, is applied to some object to move it a distance, d, then the force has done work on the object. The force must be parallel to the motion (in the same or the opposite direction). We make the following definition:

Work done = Force · Distance or $W = F \cdot d.$

Notice that if we walk across a room holding a book, we do no work on the book, since the force we exert on the book is vertical, but the motion of the book is horizontal. On the other hand, if we lift the book from the floor to a table, we accomplish work.

There are several sets of units in common use. To measure work, we will generally use the two sets of units, International (SI) and British, in the following table.

	Force	Distance	Work
International (SI) units	newton (nt)	meter (m)	joule (j)
British units	pound (lb)	foot (ft)	foot-pound (ft-lb)

Conversions
1 lb = 4.45 nt
1 ft = 0.305 m
1 ft-lb = 1.36 joules

One joule of work is done when a force of 1 newton moves an object through 1 meter, so 1 joule = 1 newton-meter.

Example 1 Calculate the work done on an object when

(a) A force of 2 newtons moves it 12 meters. (b) A 3-lb force moves it 4 feet.

Solution (a) Work done = 2 nt · 12 m = 24 joules. (b) Work done = 3 lb · 4 ft = 12 ft-lb.

In the previous example, the force was constant and we calculated the work by multiplication. In the next example, the force varies, so we need an integral. We divide up the distance moved and sum to get a definite integral representing the work.

Example 2 Hooke's Law says that the force, F, required to compress the spring in Figure 8.72 by a distance x, in meters from its equilibrium position, is given by $F = kx$, for some constant k. Find the work done in compressing the spring by 0.1 m if $k = 8$ nt/m.

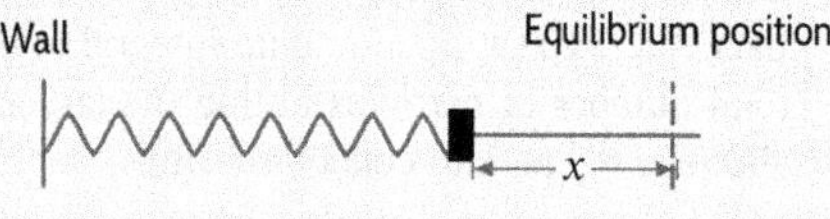

Figure 8.72: Compression of spring: Force is kx

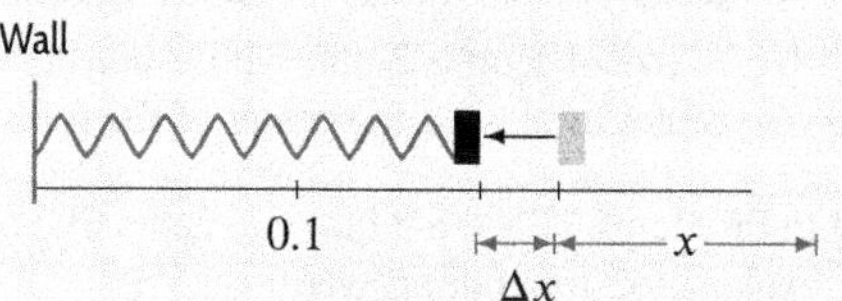

Figure 8.73: Work done in compressing spring a small distance Δx is $kx\Delta x$

Solution Since k is in newtons/meter and x is in meters, we have $F = 8x$ newtons. Since the force varies with x, we divide the distance moved into small increments, Δx, as in Figure 8.73. Then

$$\text{Work done in moving through an increment} \approx F\Delta x = 8x\Delta x \text{ joules.}$$

So, summing over all increments gives the Riemann sum approximation

$$\text{Total work done} \approx \sum 8x\Delta x.$$

Taking the limit as $\Delta x \to 0$ gives

$$\text{Total work done} = \int_0^{0.1} 8x\,dx = 4x^2\Big|_0^{0.1} = 0.04 \text{ joules.}$$

In general, if force is a function $F(x)$ of position x, then in moving from $x = a$ to $x = b$,

$$\text{Work done} = \int_a^b F(x)\,dx.$$

The Force Due to Gravity: Mass Versus Weight

When an object is lifted, work is done against the force exerted by gravity on the object. By Newton's Second Law, the downward gravitational force acting on a mass m is mg, where g is the acceleration due to gravity. To lift the object, we need to exert a force equal to the gravitational force but in the opposite direction.

In International units, $g = 9.8$ m/sec^2, and we usually measure mass, m, in *kilograms*. In British units, mass is seldom used. Instead, we usually talk about the *weight* of an object, which is the force exerted by gravity on the object. Roughly speaking, the mass represents the quantity of matter in an object, whereas the weight represents the force of gravity on it. The mass of an object is the same everywhere, whereas the weight can vary if, for example, the object is moved to outer space where gravitational forces are smaller.

When we are given the weight of an object, we do not multiply by g to find the gravitational force as it has already been done. In British units, a *pound* is a unit of weight. In International units, a kilogram is a unit of mass, and the unit of weight is a newton, where 1 newton $=$ 1 kg $\cdot$ m/sec^2.

Example 3 How much work is done in lifting

(a) A 5-pound book 3 feet off the floor? (b) A 1.5-kilogram book 2 meters off the floor?

Solution (a) The force due to gravity is 5 lb, so $W = F \cdot d = (5 \text{ lb})(3 \text{ ft}) = 15$ foot-pounds.

(b) The force due to gravity is $mg = (1.5 \text{ kg})(g \text{ m/sec}^2)$, so

$$W = F \cdot d = [(1.5 \text{ kg})(9.8 \text{ m/sec}^2)] \cdot (2 \text{ m}) = 29.4 \text{ joules.}$$

In the previous example, work is found by multiplication. In the next example, different parts of the object move different distances, so an integral is needed.

Example 4 A 28-meter uniform chain with a mass density of 2 kilograms per meter is dangling from the roof of a building. How much work is needed to pull the chain up onto the top of the building?

Solution Since 1 meter of the chain has mass density 2 kg, the gravitational force per meter of chain is $(2 \text{ kg})(9.8 \text{ m/sec}^2) = 19.6$ newtons. Let's divide the chain into small sections of length Δy, each requiring a force of $19.6\,\Delta y$ newtons to move it against gravity. See Figure 8.74. If Δy is small, all of this piece is hauled up approximately the same distance, namely y, so

$$\text{Work done on the small piece} \approx (19.6\,\Delta y \text{ newtons})(y \text{ meters}) = 19.6y\,\Delta y \text{ joules.}$$

The work done on the entire chain is given by the total of the work done on each piece:

$$\text{Work done} \approx \sum 19.6y\,\Delta y \text{ joules.}$$

As $\Delta y \to 0$, we obtain a definite integral. Since y varies from 0 to 28 meters, the total work is

$$\text{Work done} = \int_0^{28} (19.6y)\,dy = 9.8y^2\Big|_0^{28} = 7683.2 \text{ joules.}$$

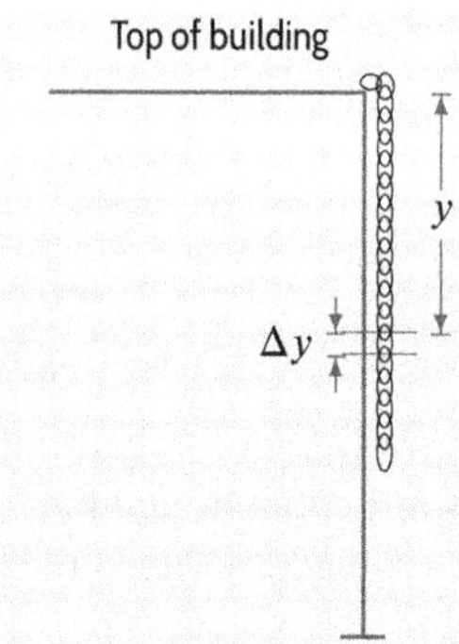

Figure 8.74: Chain for Example 4

Example 5 Calculate the work done in pumping oil from the cone-shaped tank in Figure 8.75 to the rim. The oil has density 800 kg/m^3 and its vertical depth is 10 m.

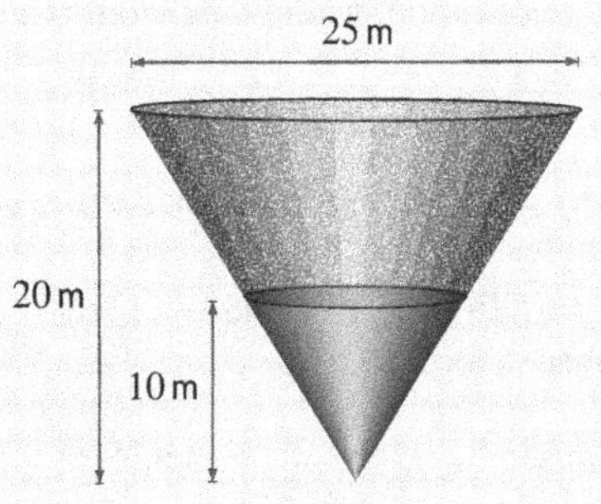

Figure 8.75: Cone-shaped tank containing oil

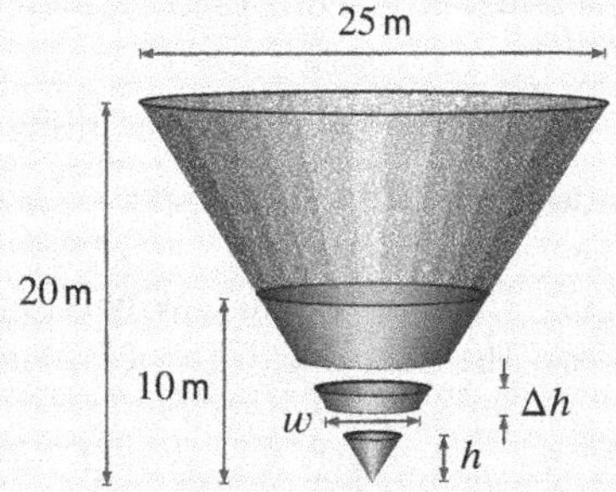

Figure 8.76: Slicing the oil horizontally to compute work

Solution We slice the oil horizontally because each part of such a slice moves the same vertical distance. Each slice is approximately a circular disk with radius $w/2$ m, so, with h in meters,

$$\text{Volume of slice } \approx \pi\left(\frac{w}{2}\right)^2 \Delta h = \frac{\pi}{4} w^2 \Delta h \text{ m}^3.$$

$$\text{Force of gravity on slice } = \text{ Density } \cdot g \cdot \text{ Volume } = 800g\frac{\pi}{4}w^2\Delta h = 200\pi g w^2 \Delta h \text{ nt.}$$

Since each part of the slice has to move a vertical distance of $(20-h)$ m, we have

$$\begin{aligned}\text{Work done on slice } \approx \text{ Force } \cdot \text{ Distance} &= 200\pi g w^2 \Delta h \text{ nt } \cdot (20-h) \text{ m}\\ &= 200\pi g w^2 (20-h)\Delta h \text{ joules.}\end{aligned}$$

To find w in terms of h, we use the similar triangles in Figure 8.76:

$$\frac{w}{h} = \frac{25}{20} \quad \text{so} \quad w = \frac{5}{4}h = 1.25h.$$

Thus,

$$\text{Work done on strip } \approx 200\pi g(1.25h)^2(20-h)\Delta h = 312.5\pi g h^2 (20-h)\Delta h \text{ joules.}$$

Summing and taking the limit as $\Delta h \to 0$ gives an integral with upper limit $h = 10$, the depth of the oil.

$$\text{Total work } = \lim_{\Delta h \to 0} \sum 312.5\pi g h^2 (20-h)\Delta h = \int_0^{10} 312.5\pi g h^2 (20-h)\, dh \text{ joules.}$$

Evaluating the integral using $g = 9.8$ m/sec^2 gives

$$\text{Total work } = 312.5\pi g \left(20\frac{h^3}{3} - \frac{h^4}{4}\right)\Bigg|_0^{10} = 1{,}302{,}083\pi g \approx 4.0 \cdot 10^7 \text{ joules.}$$

In the following example, information is given in British units about the weight of the pyramid, so we do not need to multiply by g to find the gravitational force.

Example 6 It is reported that the Great Pyramid of Egypt was built in 20 years. If the stone making up the pyramid has a density of 200 pounds per cubic foot, find the total amount of work done in building the pyramid. The pyramid is 481 feet high and has a square base 756 feet by 756 feet. Estimate how many workers were needed to build the pyramid.

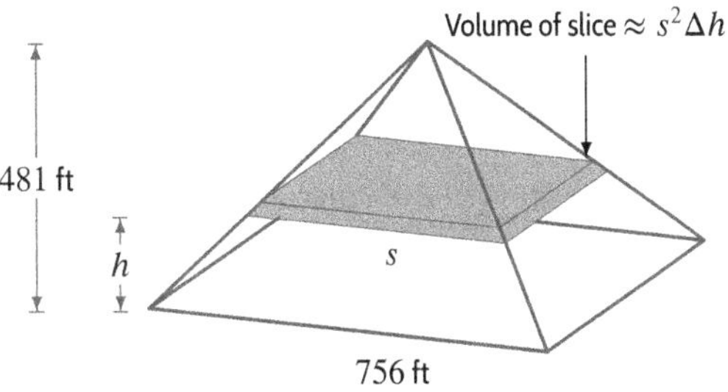

Figure 8.77: Pyramid for Example 6

Solution We assume that the stones were originally located at the approximate height of the construction site. Imagine the pyramid built in layers as we did in Example 5 on page 405.

By similar triangles, the layer at height h has a side length $s = 756(481 - h)/481$ ft. (See Figure 8.77.) The layer at height h has a volume of $s^2\,\Delta h$ ft^3, so its weight is $200s^2\,\Delta h$ lb. This layer is lifted through a height of h, so

$$\text{Work to lift layer } = (200s^2\,\Delta h \text{ lb})(h \text{ ft}) = 200s^2 h\,\Delta h \text{ ft-lb.}$$

Substituting for s in terms of h and summing over all layers gives

$$\text{Total work } \approx \sum 200s^2\, h\Delta h = \sum 200\left(\frac{756}{481}\right)^2 (481 - h)^2 h\,\Delta h \text{ ft-lb.}$$

Since h varies from 0 to 481, as $\Delta h \to 0$, we obtain

$$\text{Total work } = \int_0^{481} 200\left(\frac{756}{481}\right)^2 (481 - h)^2 h\,dh \approx 2.2 \cdot 10^{12} \text{ foot-pounds.}$$

We have calculated the total work done in building the pyramid; now we want to estimate the total number of workers needed. Let's assume every laborer worked 10 hours a day, 300 days a year, for 20 years. Assume that a typical worker lifted ten 50-pound blocks a distance of 4 feet every hour, thus performing 2000 foot-pounds of work per hour (this is a very rough estimate). Then each laborer performed $(10)(300)(20)(2000) = 1.2 \cdot 10^8$ foot-pounds of work over a twenty-year period. Thus, the number of workers needed was about $(2.2 \cdot 10^{12})/(1.2 \cdot 10^8)$, or about 18,000.

Force and Pressure

We can use the definite integral to compute the force exerted by a liquid on a surface, for example, the force of water on a dam. The idea is to get the force from the *pressure*. The pressure in a liquid is the force per unit area exerted by the liquid. Two things you need to know about pressure are:

- At any point, pressure is exerted equally in all directions—up, down, sideways.
- Pressure increases with depth. (That is one of the reasons why deep sea divers have to take much greater precautions than scuba divers.)

At a depth of h meters, the pressure, p, exerted by the liquid, measured in newtons per square meter, is given by computing the total weight of a column of liquid h meters high with a base of 1 square meter. The volume of such a column of liquid is just h cubic meters. If the liquid has density δ (mass per unit volume), then its weight per unit volume is δg, where g is the acceleration due to gravity. The weight of the column of liquid is $\delta g h$, so

$$\boxed{\text{Pressure} = \text{Mass density} \cdot g \cdot \text{Depth} \quad \text{or} \quad p = \delta g h.}$$

Provided the pressure is constant over a given area, we also have the following relation:

$$\boxed{\text{Force} = \text{Pressure} \cdot \text{Area}.}$$

The units and data we will generally use are given in the following table:

	Density of water	Force	Area	Pressure
SI units	1000 kg/m^3 (mass)	newton (nt)	meter^2	pascal (nt/m^2)
British units	62.4 lb/ft^3 (weight)	pound (lb)	foot^2	lb/ft^2

Conversions
1 lb = 4.45 nt
1ft^2 = 0.093 m^2
1 lb/ft^2 = 47.9 pa

In International units, the mass density of water is 1000 kg/m^3, so the pressure at a depth of h meters is $\delta g h = 1000 \cdot 9.8h = 9800h$ nt/m^2. See Figure 8.78.

In British units, the density of the liquid is usually given as a weight per unit volume, rather than a mass per unit volume. In that case, we do not need to multiply by g because it has already been done. For example, water weighs 62.4 lb/ft^3, so the pressure at depth h feet is $62.4h$ lb/ft^2. See Figure 8.79.

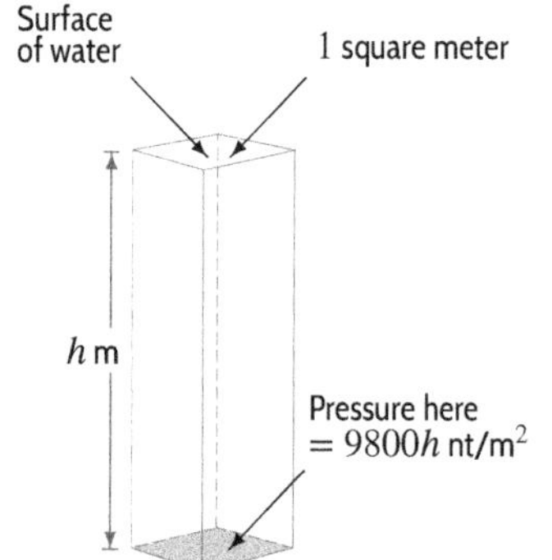

Figure 8.78: Pressure exerted by column of water (International units)

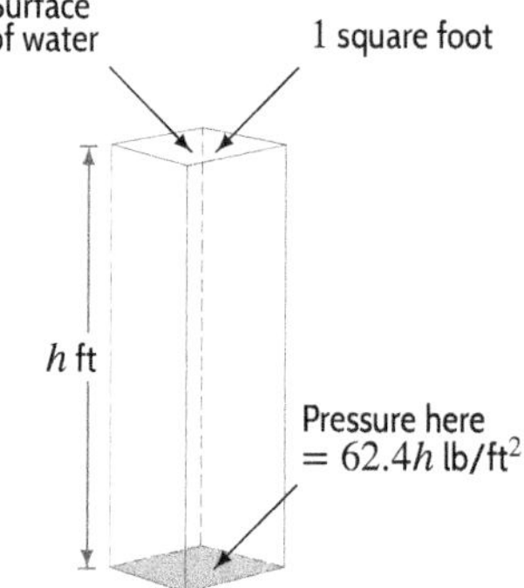

Figure 8.79: Pressure exerted by a column of water (British units)

If the pressure is constant over a surface, we calculate the force on the surface by multiplying the pressure by the area of the surface. If the pressure is not constant, we divide the surface into small pieces *in such a way that the pressure is nearly constant on each one* to obtain a definite integral for the force on the surface. Since the pressure varies with depth, we divide the surface into horizontal strips, each of which is at an approximately constant depth.

Example 7 In 1912, the ocean liner Titanic sank to the bottom of the Atlantic, 12,500 feet (nearly 2.5 miles) below the surface. Find the force on one side of a 100-foot square plate at the depth of the Titanic if the plate is: (a) Lying horizontally (b) Standing vertically.

Solution (a) When the plate is horizontal, the pressure is the same at every point on the plate, so

$$\text{Pressure} = 62.4\ \text{lb/ft}^3 \cdot 12{,}500\ \text{ft} = 780{,}000\ \text{lb/ft}^2.$$

To imagine this pressure, convert to pounds per square inch, giving $780{,}000/144 \approx 5400\ \text{lb/in}^2$. For the horizontal plate

$$\text{Force} = 780{,}000\ \text{lb/ft}^2 \cdot 100^2\ \text{ft}^2 = 7.8 \cdot 10^9\ \text{pounds}.$$

(b) When the plate is vertical, only the bottom is at 12,500 feet; the top is at 12,400 feet. Dividing into horizontal strips of width Δh, as in Figure 8.80, we have

$$\text{Area of strip} = 100\Delta h\ \text{ft}^2.$$

Since the pressure on a strip at a depth of h feet is $62.4h\ \text{lb/ft}^2$,

$$\text{Force on strip} \approx 62.4h \cdot 100\Delta h = 6240h\Delta h\ \text{pounds}.$$

Summing over all strips and taking the limit as $\Delta h \to 0$ gives a definite integral. The strips vary in depth from 12,400 to 12,500 feet, so

$$\text{Total force} = \lim_{\Delta h \to 0} \sum 6240h\Delta h = \int_{12{,}400}^{12{,}500} 6240h\,dh\ \text{pounds}.$$

Evaluating the integral gives

$$\text{Total force} = 6240\frac{h^2}{2}\bigg|_{12{,}400}^{12{,}500} = 3120(12{,}500^2 - 12{,}400^2) = 7.77 \cdot 10^9\ \text{pounds}.$$

Notice that the answer to part (b) is smaller than the answer to part (a). This is because part of the plate is at a smaller depth in part (b) than in part (a).

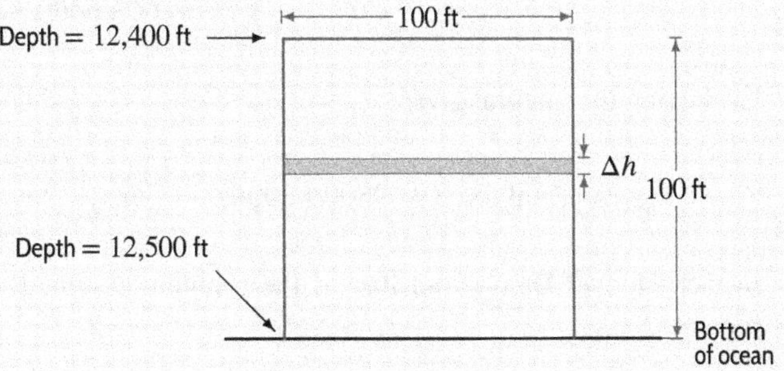

Figure 8.80: Square plate at bottom of ocean; h measured from the surface of water

Example 8 Figure 8.81 shows a dam approximately the size of Hoover Dam, which stores water for California, Nevada, and Arizona. Calculate:

(a) The water pressure at the base of the dam. (b) The total force of the water on the dam.

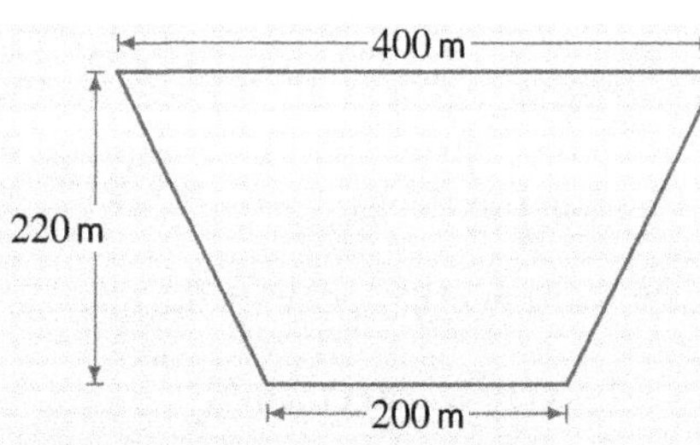

Figure 8.81: Trapezoid-shaped dam

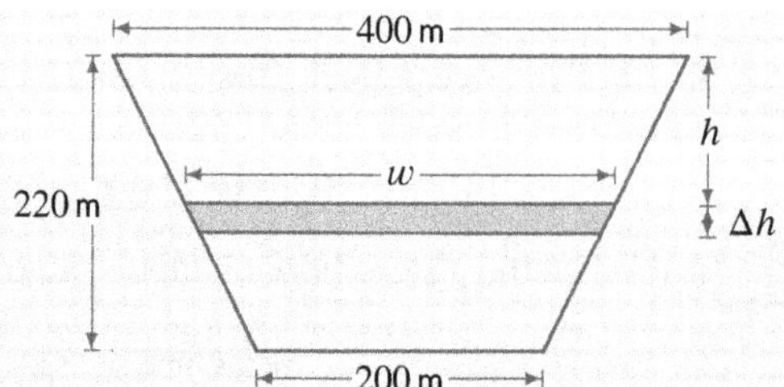

Figure 8.82: Dividing dam into horizontal strips

Solution (a) Since the density of water is $\delta = 1000$ kg/m^3, at the base of the dam,

$$\text{Water pressure } = \delta g h = 1000 \cdot 9.8 \cdot 220 = 2.156 \cdot 10^6 \text{ nt/m}^2.$$

(b) To calculate the force on the dam, we divide the dam into horizontal strips because the pressure along each strip is approximately constant. See Figure 8.82. Since each strip is approximately rectangular,

$$\text{Area of strip } \approx w\Delta h \text{ m}^2.$$

The pressure at a depth of h meters is $\delta g h = 9800h$ nt/m^2. Thus,

$$\text{Force on strip } \approx \text{ Pressure } \cdot \text{ Area } = 9800hw\Delta h \text{ nt}.$$

To find w in terms of h, we use the fact that w decreases linearly from $w = 400$ when $h = 0$ to $w = 200$ when $h = 220$. Thus w is a linear function of h, with slope $(200-400)/220 = -10/11$, so

$$w = 400 - \frac{10}{11}h.$$

Thus

$$\text{Force on strip } \approx 9800h\left(400 - \frac{10}{11}h\right)\Delta h \text{ nt}.$$

Summing over all strips and taking the limit as $\Delta h \to 0$ gives

$$\begin{aligned}\text{Total force on dam } &= \lim_{\Delta h \to 0} \sum 9800h\left(400 - \frac{10}{11}h\right)\Delta h \\ &= \int_0^{220} 9800h\left(400 - \frac{10}{11}h\right) dh \text{ newtons.}\end{aligned}$$

Evaluating the integral gives

$$\text{Total force } = 9800\left(200h^2 - \frac{10}{33}h^3\right)\Bigg|_0^{220} = 6.32 \cdot 10^{10} \text{ newtons.}$$

In fact, Hoover Dam is not flat, as the problem assumed, but arched, to better withstand the pressure.

Exercises and Problems for Section 8.5 Online Resource: Additional Problems for Section 8.5

EXERCISES

1. Find the work done on a 40 lb suitcase when it is raised 9 inches.

2. Find the work done on a 20 kg suitcase when it is raised 30 centimeters.

3. A particle x feet from the origin has a force of $x^2 + 2x$ pounds acting on it. What is the work done in moving the object from the origin a distance of 1 foot?

In Exercises 4–6, the force, F, required to compress a spring by a distance x meters is given by $F = 3x$ newtons.

4. Find the work done in compressing the spring from $x = 1$ to $x = 2$.

5. Find the work done to compress the spring to $x = 3$, starting at the equilibrium position, $x = 0$.

6. **(a)** Find the work done in compressing the spring from $x = 0$ to $x = 1$ and in compressing the spring from $x = 4$ to $x = 5$.
   (b) Which of the two answers is larger? Why?

7. A circular steel plate of radius 20 ft lies flat on the bottom of a lake, at a depth of 150 ft. Find the force on the plate due to the water pressure.

8. A fish tank is 2 feet long and 1 foot wide, and the depth of the water is 1 foot. What is the force on the bottom of the fish tank?

9. A child fills a bucket with sand so that the bucket and sand together weigh 10 lbs, lifts it 2 feet up and then walks along the beach, holding the bucket at a constant height of 2 ft above the ground. How much work is done on the bucket after the child has walked 100 ft?

10. The gravitational force on a 1 kg object at a distance r meters from the center of the earth is $F = 4 \cdot 10^{14}/r^2$ newtons. Find the work done in moving the object from the surface of the earth to a height of 10^6 meters above the surface. The radius of the earth is $6.4 \cdot 10^6$ meters.

PROBLEMS

11. How much work is required to lift a 1000-kg satellite from the surface of the earth to an altitude of $2 \cdot 10^6$ m? The gravitational force is $F = GMm/r^2$, where M is the mass of the earth, m is the mass of the satellite, and r is the distance between them. The radius of the earth is $6.4 \cdot 10^6$ m, its mass is $6 \cdot 10^{24}$ kg, and in these units the gravitational constant, G, is $6.67 \cdot 10^{-11}$.

12. A worker on a scaffolding 75 ft above the ground needs to lift a 500 lb bucket of cement from the ground to a point 30 ft above the ground by pulling on a rope weighing 0.5 lb/ft. How much work is required?

13. An anchor weighing 100 lb in water is attached to a chain weighing 3 lb/ft in water. Find the work done to haul the anchor and chain to the surface of the water from a depth of 25 ft.

14. A 1000-lb weight is being lifted to a height 10 feet off the ground. It is lifted using a rope which weighs 4 lb per foot and which is being pulled up by construction workers standing on a roof 30 feet off the ground. Find the work done to lift the weight.

15. A 2000-lb cube of ice must be lifted 100 ft, and it is melting at a rate of 4 lb per minute. If it can be lifted at a rate of one foot every minute, find the work needed to get the block of ice to the desired height.

16. A cylindrical garbage can of depth 3 ft and radius 1 ft fills with rainwater up to a depth of 2 ft. How much work would be done in pumping the water up to the top edge of the can? (Water weighs 62.4 lb/ft^3.)

17. A rectangular swimming pool 50 ft long, 20 ft wide, and 10 ft deep is filled with water to a depth of 9 ft. Use an integral to find the work required to pump all the water out over the top.

18. A water tank is in the form of a right circular cylinder with height 20 ft and radius 6 ft. If the tank is half full of water, find the work required to pump all of it over the top rim.

19. The tank in Problem 18 is full of water. Find the work required to pump all of it to a point 10 ft above the top of the tank.

20. Water in a cylinder of height 10 ft and radius 4 ft is to be pumped out. Find the work required if
    (a) The tank is full of water and the water is to pumped over the top of the tank.
    (b) The tank is full of water and the water must be pumped to a height 5 ft above the top of the tank.
    (c) The depth of water in the tank is 8 ft and the water must be pumped over the top of the tank.

21. A water tank is in the shape of a right circular cone with height 18 ft and radius 12 ft at the top. If it is filled with water to a depth of 15 ft, find the work done in pumping all of the water over the top of the tank. (The density of water is $\delta = 62.4$ lb/ft^3.)

22. A cone with height 12 ft and radius 4 ft, pointing downward, is filled with water to a depth of 9 ft. Find the work required to pump all the water out over the top.

23. A hemispherical bowl of radius 2 ft contains water to a depth of 1 ft at the center. Let y be measured vertically upward from the bottom of the bowl. Water has density 62.4 lb/ft^3.
    (a) Approximately how much work does it take to move a horizontal slice of water at a distance y from the bottom to the rim of the bowl?
    (b) Write and evaluate an integral giving the work done to move all the water to the rim of the bowl.

24. A bucket of water of mass 20 kg is pulled at constant velocity up to a platform 40 meters above the ground. This takes 10 minutes, during which time 5 kg of water drips out at a steady rate through a hole in the bottom. Find the work needed to raise the bucket to the platform.

25. A gas station stores its gasoline in a tank under the ground. The tank is a cylinder lying horizontally on its side. (In other words, the tank is not standing vertically on one of its flat ends.) If the radius of the cylinder is 4 feet, its length is 12 feet, and its top is 10 feet under the ground, find the total amount of work needed to pump the gasoline out of the tank. (Gasoline weighs 42 lb/ft^3.)

26. **(a)** The trough in Figure 8.83 is full of water. Find the force of the water on a triangular end.
(b) Find the work to pump all the water over the top.

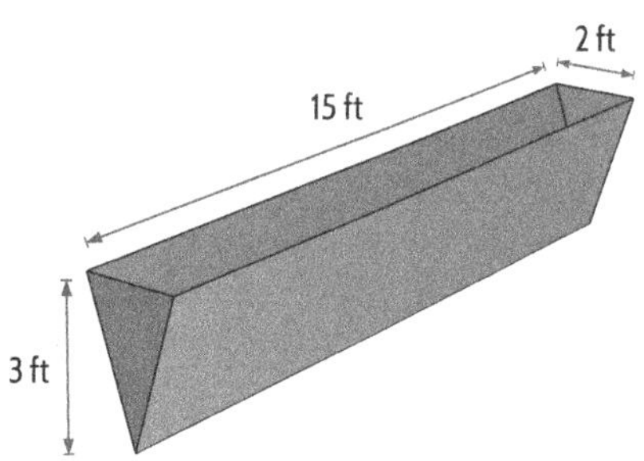

Figure 8.83

27. The dam in Hannawa Falls, NY, on the Raquette River is 60 feet across, 25 feet high, and approximately rectangular. Find the water force on the dam.

28. What is the total force on the bottom and each side of a full rectangular water tank that has length 20 ft, width 10 ft, and depth 15 ft?

29. A rectangular dam is 100 ft long and 50 ft high. If the water is 40 ft deep, find the force of the water on the dam.

30. A lobster tank in a restaurant is 4 ft long by 3 ft wide by 2 ft deep. Find the water force on the bottom and on each of the four sides.

31. The Three Gorges Dam started operation in China in 2008. With the largest electrical generating capacity in the world, the dam is about 2000 m long and 180 m high, and has created a lake longer than Lake Superior.[3] Assume the dam is rectangular in shape.
(a) Estimate the water pressure at the base of the dam.
(b) Set up and evaluate a definite integral giving the total force of the water on the dam.

32. On August 12, 2000, the Russian submarine Kursk sank to the bottom of the sea, 350 feet below the surface. Find the following at the depth of the Kursk.
(a) The water pressure in pounds per square foot and pounds per square inch.
(b) The force on a 5-foot square metal sheet held
(i) Horizontally. (ii) Vertically.

33. The ocean liner Titanic lies under 12,500 feet of water at the bottom of the Atlantic Ocean.
(a) What is the water pressure at the Titanic? Give your answer in pounds per square foot and pounds per square inch.
(b) Set up and calculate an integral giving the total force on a circular porthole (window) of diameter 6 feet standing vertically with its center at the depth of the Titanic.

34. Set up and calculate a definite integral giving the total force on the dam shown in Figure 8.84, which is about the size of the Aswan Dam in Egypt.

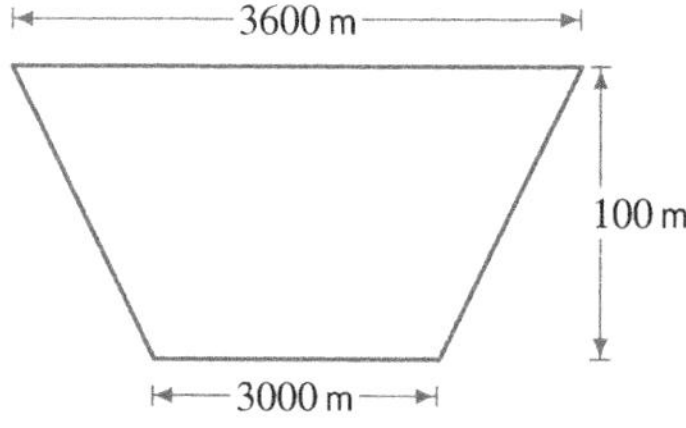

Figure 8.84

35. A climbing plane has vertical velocity $v(h)$ meters/second, where h is its altitude in meters. Write a definite integral for the time it takes the plane to climb 7000 meters from the ground.

36. Climbing a ladder at a constant rate, a painter sprays $1/4$ kg of paint per meter onto a pole from the ground up to a height of 8 meters. Her sprayer starts on the ground with 3 kg of paint and weighs 2 kg when empty.
(a) Find a formula for the mass of the sprayer with paint as a function of h, the height of the sprayer above the ground.
(b) Approximate the work done by the painter in lifting the sprayer from height h to $h + \Delta h$.
(c) Find total work done lifting the sprayer for one coat of paint.

37. Old houses may contain asbestos, now known to be dangerous; removal requires using a special vacuum. A contractor climbs a ladder and sucks up asbestos at a constant rate from a 10 m tall pipe covered by 0.2 kg/m using a vacuum weighing 14 kg with a 1.2 kg capacity.
(a) Let h be the height of the vacuum from the ground. If the vacuum is empty at $h = 0$, find a formula for the mass of the vacuum and the asbestos inside as a function of h.
(b) Approximate the work done by the contractor in lifting the vacuum from height h to $h + \Delta h$.
(c) At what height does the vacuum fill up?
(d) Find total work done lifting the vacuum from height $h = 0$ until the vacuum fills.
(e) Assuming again an empty tank at $h = 0$, find the work done lifting the vacuum when removing the remaining asbestos.

38. A skyrocket firework burns fuel as it climbs. Its mass, $m(h)$ kg, at height h m is given by

$$m(h) = 0.96 + \frac{0.065}{1+h}.$$

(a) What is the mass of the rocket at $h = 0$ just before it is launched?

[3] en.wikipedia.org/wiki/Three_Gorges_Dam. Accessed April 2012.

(b) Show that the mass decreases as the rocket climbs.
(c) What work would be required to lift the rocket 10 m if its mass did not decrease?
(d) Approximate the work done as the rocket goes from height h to $h + \Delta h$.
(e) The rocket explodes at $h = 250$ m. Find total work done by the rocket from launch to explosion.

39. We define the electric potential at a distance r from an electric charge q by q/r. The electric potential of a charge distribution is obtained by adding up the potential from each point. Electric charge is sprayed (with constant density σ in units of charge/unit area) on to a circular disk of radius a. Consider the axis perpendicular to the disk and through its center. Find the electric potential at the point P on this axis at a distance R from the center. (See Figure 8.85.)

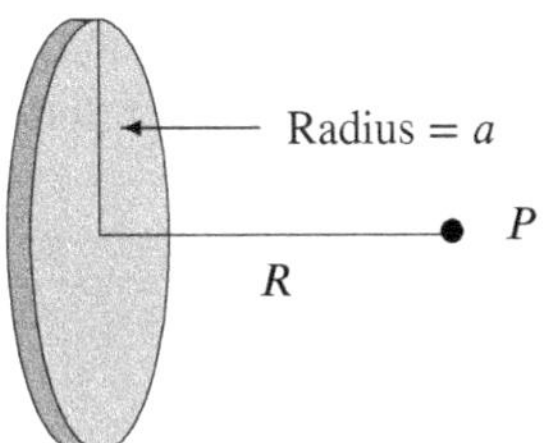

Figure 8.85

40. A uniform, thin, circular disk of radius a and mass M lies on a horizontal plane. The point P lies a distance y directly above O, the center of the disk. Calculate the gravitational force on a mass m at the point P. (See Figure 8.86.) Use the fact that the gravitational force exerted on the mass m by a thin horizontal ring of radius r, mass μ, and center O is toward O and given by

$$F = \frac{G\mu my}{(r^2 + y^2)^{3/2}}, \quad \text{where } G \text{ is constant.}$$

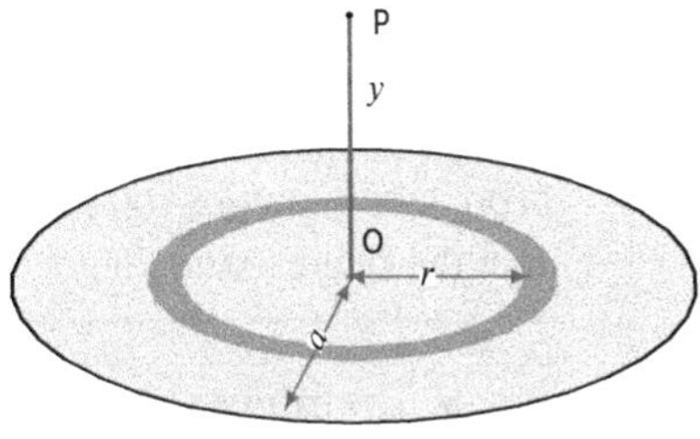

Figure 8.86

Strengthen Your Understanding

■ In Problems 41–43, explain what is wrong with the statement.

41. A 20 meter rope with a mass of 30 kg dangles over the edge of a cliff. Ignoring friction, the work required to pull the rope to the top of the cliff is

$$\text{Work} = (30 \text{ kg})\left(9.8 \frac{\text{m}}{\text{sec}^2}\right)(20 \text{ m}).$$

42. A cylindrical tank is 10 meters deep. It takes twice as much work to pump all the oil out through the top of the tank when the tank is full as when the tank is half full.

43. Lifting a 10 kg rock 2 meters off the ground requires 20 joules of work.

■ In Problems 44–45, give an example of:

44. A situation where work can be computed as Force × Distance without doing an integral.

45. Two cylindrical tanks A and B such that it takes less work to pump the water from tank A to a height of 10 meters than from tank B. Both tanks contain the same volume of water and are less than 10 meters high.

■ In Problems 46–51, are the statements true or false? Give an explanation for your answer.

46. It takes more work to lift a 20 lb weight 10 ft slowly than to lift it the same distance quickly.

47. Work can be negative or positive.

48. The force on a rectangular dam is doubled if its length stays the same and its depth is doubled.

49. To find the force of water on a vertical wall, we always slice the wall horizontally, no matter what the shape of the wall.

50. The force of a liquid on a wall can be negative or positive.

51. If the average value of the force $F(x)$ is 7 on the interval $1 \le x \le 4$, then the work done by the force in moving from $x = 1$ to $x = 4$ is 21.

8.6 APPLICATIONS TO ECONOMICS

Present and Future Value

Many business deals involve payments in the future. If you buy a car or furniture, for example, you may buy it on credit and pay over a period of time. If you are going to accept payment in the future under such a deal, you obviously need to know how much you should be paid. Being paid \$100 in the future is clearly worse than being paid \$100 today for many reasons. If you are given the money today, you can do something else with it—for example, put it in the bank, invest it somewhere, or spend it. Thus, even without considering inflation, if you are to accept payment in the future, you would expect to be paid more to compensate for this loss of potential earnings. The question we will consider now is, how much more?

To simplify matters, we consider only what we would lose by not earning interest; we will not consider the effect of inflation. Let's look at some specific numbers. Suppose you deposit \$100 in an account which earns 3% interest compounded annually, so that in a year's time you will have \$103. Thus, \$100 today will be worth \$103 a year from now. We say that the \$103 is the *future value* of the \$100, and that the \$100 is the *present value* of the \$103. Observe that the present value is smaller than the future value. In general, we say the following:

- The **future value**, $\$B$, of a payment, $\$P$, is the amount to which the $\$P$ would have grown if deposited in an interest-bearing bank account.
- The **present value**, $\$P$, of a future payment, $\$B$, is the amount which would have to be deposited in a bank account today to produce exactly $\$B$ in the account at the relevant time in the future.

With an interest rate of r per year, compounded annually, and a time period of t years, a deposit of $\$P$ grows to a future balance of $\$B$, where

$$B = P(1+r)^t, \quad \text{or equivalently,} \quad P = \frac{B}{(1+r)^t}.$$

Note that for a 3% interest rate, $r = 0.03$. If instead of annual compounding, we have continuous compounding, we get the following result:

$$B = Pe^{rt}, \quad \text{or equivalently,} \quad P = \frac{B}{e^{rt}} = Be^{-rt}.$$

Example 1 You win the lottery and are offered the choice between \$1 million in four yearly installments of \$250,000 each, starting now, and a lump-sum payment of \$920,000 now. Assuming a 6% interest rate per year, compounded continuously, and ignoring taxes, which should you choose?

Solution We do the problem in two ways. First, we assume that you pick the option with the largest present value. The first of the four \$250,000 payments is made now, so

$$\text{Present value of first payment} = \$250{,}000.$$

The second payment is made one year from now, so

$$\text{Present value of second payment} = \$250{,}000e^{-0.06(1)}.$$

Calculating the present value of the third and fourth payments similarly, we find:

$$\begin{aligned}\text{Total present value} &= \$250{,}000 + \$250{,}000e^{-0.06(1)} + \$250{,}000e^{-0.06(2)} + \$250{,}000e^{-0.06(3)} \\ &\approx \$250{,}000 + \$235{,}441 + \$221{,}730 + \$208{,}818 \\ &= \$915{,}989.\end{aligned}$$

Since the present value of the four payments is less than \$920,000, you are better off taking the \$920,000 right now.

Alternatively, we can compare the future values of the two pay schemes. The scheme with the highest future value is the best from a purely financial point of view. We calculate the future value of both schemes three years from now, on the date of the last \$250,000 payment. At that time,

$$\text{Future value of the lump sum payment} = \$920{,}000e^{0.06(3)} \approx \$1{,}101{,}440.$$

Now we calculate the future value of the first \$250,000 payment:

$$\text{Future value of the first payment} = \$250{,}000e^{0.06(3)}.$$

Calculating the future value of the other payments similarly, we find:

$$\begin{aligned}\text{Total future value} &= \$250{,}000e^{0.06(3)} + \$250{,}000e^{0.06(2)} + \$250{,}000e^{0.06(1)} + \$250{,}000 \\ &\approx \$299{,}304 + \$281{,}874 + \$265{,}459 + \$250{,}000 \\ &= \$1{,}096{,}637.\end{aligned}$$

The future value of the \$920,000 payment is greater, so you are better off taking the \$920,000 right now. Of course, since the present value of the \$920,000 payment is greater than the present value of the four separate payments, you would expect the future value of the \$920,000 payment to be greater than the future value of the four separate payments.

(Note: If you read the fine print, you will find that many lotteries do not make their payments right away, but often spread them out, sometimes far into the future. This is to reduce the present value of the payments made, so that the value of the prizes is much less than it might first appear!)

Income Stream

When we consider payments made to or by an individual, we usually think of *discrete* payments, that is, payments made at specific moments in time. However, we may think of payments made by a company as being *continuous*. The revenues earned by a huge corporation, for example, come in essentially all the time and can be represented by a continuous *income stream*, written

$$P(t) \text{ dollars/year.}$$

Notice that $P(t)$ is the *rate* at which deposits are made (its units are dollars per year, for example) and that this rate may vary with time, t.

Present and Future Values of an Income Stream

Just as we can find the present and future values of a single payment, so we can find the present and future values of a stream of payments. We assume that interest is compounded continuously.

Suppose that we want to calculate the present value of the income stream described by a rate of $P(t)$ dollars per year, and that we are interested in the period from now until M years in the future. We divide the stream into many small deposits, each of which is made at approximately one instant. We divide the interval $0 \leq t \leq M$ into subintervals, each of length Δt:

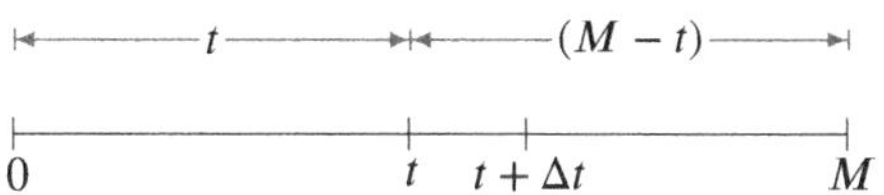

Assuming Δt is small, the rate, $P(t)$, at which deposits are being made will not vary much within one subinterval. Thus, between t and $t + \Delta t$:

$$\begin{aligned}\text{Amount deposited} &\approx \text{Rate of deposits} \times \text{Time} \\ &\approx (P(t) \text{ dollars/year})(\Delta t \text{ years}) \\ &= P(t)\Delta t \text{ dollars.}\end{aligned}$$

Measured from the present, the deposit of $P(t)\Delta t$ is made t years in the future. Thus, with an interest rate r per year, we have

$$\begin{matrix}\text{Present value of money deposited} \\ \text{in interval } t \text{ to } t + \Delta t\end{matrix} \approx P(t)\Delta t e^{-rt}.$$

Summing over all subintervals gives

$$\text{Total present value} \approx \sum P(t)e^{-rt}\Delta t \text{ dollars.}$$

In the limit as $\Delta t \to 0$, we get the following integral:

$$\boxed{\text{Present value } = \int_0^M P(t)e^{-rt}\,dt \text{ dollars.}}$$

In computing future value, the deposit of $P(t)\Delta t$ has a period of $(M - t)$ years to earn interest, and therefore

$$\begin{matrix}\text{Future value of money deposited} \\ \text{in interval } t \text{ to } t + \Delta t\end{matrix} \approx (P(t)\Delta t)\, e^{r(M-t)}.$$

Summing over all subintervals, we get:

$$\text{Total future value } \approx \sum P(t)\Delta t e^{r(M-t)} \text{ dollars.}$$

As the length of the subdivisions tends toward zero, the sum becomes an integral:

$$\boxed{\text{Future value } = \int_0^M P(t)e^{r(M-t)}\,dt \text{ dollars.}}$$

In addition, by writing $e^{r(M-t)} = e^{rM} \cdot e^{-rt}$ and factoring out e^{rM}, we see that

$$\text{Future value } = e^{rM} \cdot \text{ Present value.}$$

Example 2 Find the present and future values of a constant income stream of \$1000 per year over a period of 20 years, assuming an interest rate of 10% per year, compounded continuously.

Solution Using $P(t) = 1000$ and $r = 0.1$, we have

$$\text{Present value } = \int_0^{20} 1000e^{-0.1t}\,dt = 1000\left(-\frac{e^{-0.1t}}{0.1}\right)\Bigg|_0^{20} = 10{,}000(1 - e^{-2}) \approx 8646.65 \text{ dollars.}$$

There are two ways to compute the future value. Using the present value of \$8646.65, we have

$$\text{Future value } = 8646.65e^{0.1(20)} = 63{,}890.58 \text{ dollars.}$$

Alternatively, we can use the integral formula:

$$\text{Future value} = \int_0^{20} 1000e^{0.1(20-t)}dt = \int_0^{20} 1000e^2e^{-0.1t}dt$$
$$= 1000e^2\left(-\frac{e^{-0.1t}}{0.1}\right)\Bigg|_0^{20} = 10{,}000e^2(1-e^{-2}) \approx 63{,}890.58 \text{ dollars.}$$

Notice that the total amount deposited is \$1000 per year for 20 years, or \$20,000. The additional \$43,895.58 of the future value comes from interest earned.

Supply and Demand Curves

In a free market, the quantity of a certain item produced and sold can be described by the supply and demand curves of the item. The *supply curve* shows the quantity of the item the producers will supply at different price levels. It is usually assumed that as the price increases, the quantity supplied will increase. The consumers' behavior is reflected in the *demand curve*, which shows what quantity of goods are bought at various prices. An increase in price is usually assumed to cause a decrease in the quantity purchased. See Figure 8.87.

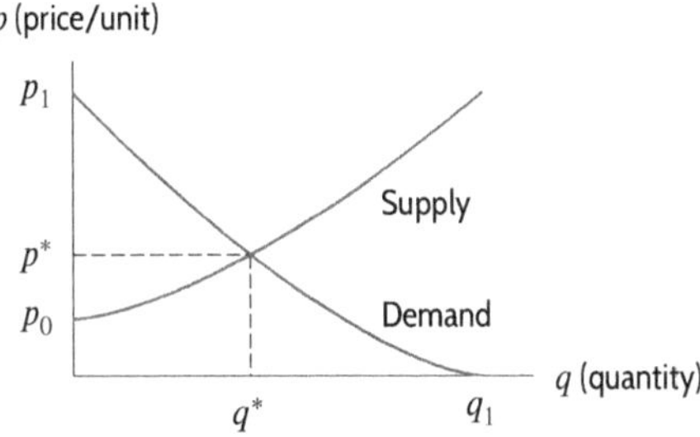

Figure 8.87: Supply and demand curves

It is assumed that the market settles to the *equilibrium price* and *quantity*, p^* and q^*, where the graphs cross. At equilibrium, a quantity q^* of an item is produced and sold for a price of p^* each.

Consumer and Producer Surplus

Notice that at equilibrium, a number of consumers have bought the item at a lower price than they would have been willing to pay. (For example, there are some consumers who would have been willing to pay prices up to p_1.) Similarly, there are some suppliers who would have been willing to produce the item at a lower price (down to p_0, in fact). We define the following terms:

- The **consumer surplus** measures the consumers' gain from trade. It is the total amount gained by consumers by buying the item at the current price rather than at the price they would have been willing to pay.
- The **producer surplus** measures the suppliers' gain from trade. It is the total amount gained by producers by selling at the current price, rather than at the price they would have been willing to accept.

In the absence of price controls, the current price is assumed to be the equilibrium price.

Both consumers and producers are richer for having traded. The consumer and producer surplus measure how much richer they are.

Suppose that all consumers buy the good at the maximum price they are willing to pay. Divide the interval from 0 to q^* into subintervals of length Δq. Figure 8.88 shows that a quantity Δq of items are sold at a price of about p_1, another Δq are sold for a slightly lower price of about p_2, the next Δq for a price of about p_3, and so on. Thus,

$$\text{Consumers' total expenditure} \approx p_1\Delta q + p_2\Delta q + p_3\Delta q + \cdots = \sum p_i\Delta q.$$

If D is the demand function given by $p = D(q)$, and if all consumers who were willing to pay more than p^* paid as much as they were willing, then as $\Delta q \to 0$, we would have

$$\text{Consumer expenditure} = \int_0^{q^*} D(q)dq = \begin{array}{l}\text{Area under demand}\\ \text{curve from 0 to } q^*.\end{array}$$

Now if all goods are sold at the equilibrium price, the consumers' actual expenditure is p^*q^*, the area of the rectangle between the axes and the lines $q = q^*$ and $p = p^*$. Thus, if p^* and q^* are equilibrium price and quantity, the consumer surplus is calculated as follows:

$$\text{Consumer surplus} = \left(\int_0^{q^*} D(q)dq\right) - p^*q^* = \begin{array}{l}\text{Area under demand}\\ \text{curve above } p = p^*.\end{array}$$

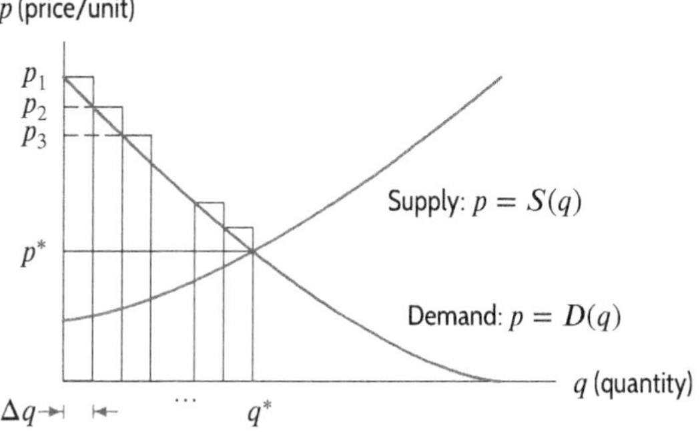

Figure 8.88: Calculation of consumer surplus

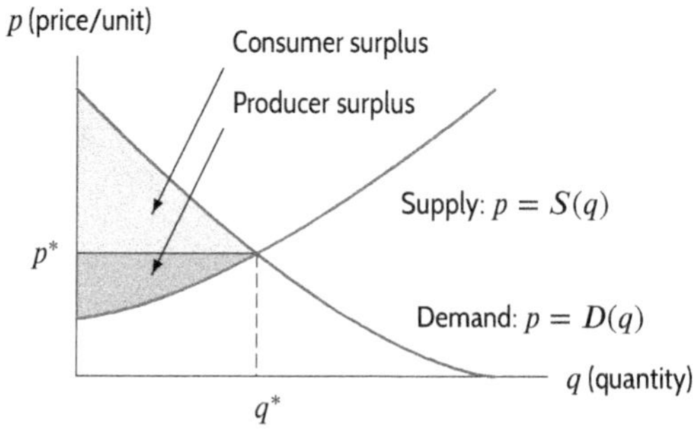

Figure 8.89: Consumer and producer surplus

See Figure 8.89. Similarly, if the supply curve is given by the function $p = S(q)$ and p^* and q^* are equilibrium price and quantity, the producer surplus is calculated as follows:

$$\text{Producer surplus} = p^*q^* - \left(\int_0^{q^*} S(q)dq\right) = \begin{array}{l}\text{Area between supply}\\ \text{curve and line } p = p^*.\end{array}$$

Exercises and Problems for Section 8.6

EXERCISES

In Exercises 1–7 give an expression that represents the statement. Do not simplify your expression.

1. The future value of single \$C deposit, after 20 years, at a 2% interest rate per year, compounded annually.

2. The present value of \$C deposited 20 years from now, at a 2% interest rate per year, compounded annually.

3. The present value of a deposit of \$C, made 10 years from now, with a 2% interest rate per year, compounded continuously.

4. The present value of an income stream paying C dollars/year for a period of 15 years, at a 2% interest rate per year, compounded continuously.

5. The future value at the end of 15 years of an income stream paying C dollars/year throughout the 15 years, at a 2% interest rate per year, compounded continuously.

6. The future value, at the end of C years, of a series of three \$500 deposits, where the first deposit is made now, the second a year from now, and the third two years from now. Assume a 2% interest rate per year, compounded continuously.

7. The continuous interest rate per year for a deposit \$C that will grow to \$25,000 in 30 years.

8. Find the future value of an income stream of \$2000 per year, deposited into an account paying 2% interest per year, compounded continuously, over a 15-year period.

9. Find the present and future values of an income stream of \$5000 a year, for a period of 5 years, if the continuous interest rate is 4% per year.

10. A person deposits money into a retirement account, which pays 5% interest per year compounded continuously, at a rate of \$5000 per year for 15 years. Calculate:
   - **(a)** The balance in the account at the end of the 15 years.
   - **(b)** The amount of money actually deposited into the account.
   - **(c)** The interest earned during the 15 years.

11. **(a)** Find the present and future values of a constant income stream of \$100 per year over a period of 20 years, assuming a 10% annual interest rate per year, compounded continuously.
   (b) How many years will it take for the balance to reach \$5000?

■ Exercises 12–14 concern a single investment of \$10,000. Find the continuous interest rate per year, yielding a future value of \$20,000 in the given time period.

12. 60 years   13. 20 years   14. 5 years

■ Exercises 15–17 concern a constant income stream that pays a total of \$20,000 over a certain time period with an interest rate of 2% per year, compounded continuously. Find the rate at which money is paid, in dollars/year, and the future value of the stream at the end of the given time period.

15. 5 years   16. 10 years   17. 20 years

PROBLEMS

18. A company expects a factory to make \$200 million/year for ten years, beginning in year $t = 5$. With a continuous interest rate of 1% per year, what is the present value of this income stream from $t = 5$ to $t = 15$?

19. With a continuous interest rate of 1.5% per year, a company expects the present value (in \$ millions) of the income stream from a new factory to be given by the integral
$$\int_3^{18} 300e^{-0.015t}\,dt.$$
   - **(a)** What income stream (in \$ millions/year) does the company expect the factory to make, once it comes online?
   - **(b)** For how many years does the company expect the factory to generate an income stream?

20. Draw a graph, with time in years on the horizontal axis, of what an income stream might look like for a company that sells sunscreen in the northeast United States.

21. On March 6, 2007, the Associated Press reported that Ed Nabors had won half of a \$390 million jackpot, the largest lottery prize in US history at the time. Suppose he was given the choice of receiving his \$195 million share paid out continuously over 20 years or one lump sum of \$120 million paid immediately. Based on present value:
   - **(a)** Which option is better if the interest rate is 6% per year, compounded continuously? An interest rate of 3% per year?
   - **(b)** If Mr. Nabors chose the lump-sum option, what assumption was he making about interest rates?

22. Find a constant income stream (in dollars per year) which after 10 years has a future value of \$20,000, assuming a continuous interest rate of 3% per year.

23. **(a)** A bank account earns 2% interest per year compounded continuously. At what (constant, continuous) rate must a parent deposit money into such an account in order to save \$100,000 in 15 years for a child's college expenses?
   (b) If the parent decides instead to deposit a lump sum now in order to attain the goal of \$100,000 in 15 years, how much must be deposited now?

24. **(a)** If you deposit money continuously at a constant rate of \$4000 per year into a bank account that earns 2% interest per year, how many years will it take for the balance to reach \$10,000?
   (b) How many years would it take if the account had \$1000 in it initially?

25. A business associate who owes you \$3000 offers to pay you \$2800 now, or else pay you three yearly installments of \$1000 each, with the first installment paid now. If you use only financial reasons to make your decision, which option should you choose? Justify your answer, assuming a 3% interest rate per year, compounded continuously.

■ In Problems 26–29 find the continuous interest rate per year that yields a future value of \$18,000 in 20 years for each \$9000 investment.

26. A single \$9000 deposit.

27. An initial \$6000 deposit plus a second \$3000 deposit made three years after the first.

28. An initial \$3000 deposit plus a second \$6000 deposit made three years after the first.

29. An income stream of \$300 per year.

30. A family wants to save for college tuition for their daughter. What continuous yearly interest rate $r\%$ is needed in their savings account if their deposits of \$4800 per year are to grow to \$100,000 in 15 years? Assume that they make deposits continuously throughout the year.

31. Big Tree McGee is negotiating his rookie contract with a professional basketball team. They have agreed to a three-year deal which will pay Big Tree a fixed amount at the end of each of the three years, plus a signing bonus at the beginning of his first year. They are still haggling about the amounts and Big Tree must decide between a big signing bonus and fixed payments per year, or a smaller bonus with payments increasing each year. The two options are summarized in the table. All values are payments in millions of dollars.

	Signing bonus	Year 1	Year 2	Year 3
Option #1	6.0	2.0	2.0	2.0
Option #2	1.0	2.0	4.0	6.0

(a) Big Tree decides to invest all income in stock funds which he expects to grow at a rate of 10% per year, compounded continuously. He would like to choose the contract option which gives him the greater future value at the end of the three years when the last payment is made. Which option should he choose?

(b) Calculate the present value of each contract offer.

32. Sales of Version 6.0 of a computer software package start out high and decrease exponentially. At time t, in years, the sales are $s(t) = 100e^{-t}$ thousands of dollars per year. After two years, Version 7.0 of the software is released and replaces Version 6.0. Assuming that all income from software sales is immediately invested in government bonds which pay interest at a 4% rate per year, compounded continuously, calculate the total value of sales of Version 6.0 over the two-year period.

33. The value of some good wine increases with age. Thus, if you are a wine dealer, you have the problem of deciding whether to sell your wine now, at a price of $\$P$ a bottle, or to sell it later at a higher price. Suppose you know that the amount a wine-drinker is willing to pay for a bottle of this wine t years from now is $\$P(1 + 20\sqrt{t})$. Assuming continuous compounding and a prevailing interest rate of 5% per year, when is the best time to sell your wine?

34. An oil company discovered an oil reserve of 100 million barrels. For time $t > 0$, in years, the company's extraction plan is a linear declining function of time as follows:

$$q(t) = a - bt,$$

where $q(t)$ is the rate of extraction of oil in millions of barrels per year at time t and $b = 0.1$ and $a = 10$.

(a) How long does it take to exhaust the entire reserve?

(b) The oil price is a constant \$20 per barrel, the extraction cost per barrel is a constant \$10, and the market interest rate is 10% per year, compounded continuously. What is the present value of the company's profit?

35. You are manufacturing a particular item. After t years, the rate at which you earn a profit on the item is $(2-0.1t)$ thousand dollars per year. (A negative profit represents a loss.) Interest is 3% per year, compounded continuously,

(a) Write a Riemann sum approximating the present value of the total profit earned up to a time M years in the future.

(b) Write an integral representing the present value in part (a). (You need not evaluate this integral.)

(c) For what M is the present value of the stream of profits on this item maximized? What is the present value of the total profit earned up to that time?

36. In May 1991, *Car and Driver* described a Jaguar that sold for \$980,000. At that price only 50 have been sold. It is estimated that 350 could have been sold if the price had been \$560,000. Assuming that the demand curve is a straight line, and that \$560,000 and 350 are the equilibrium price and quantity, find the consumer surplus at the equilibrium price.

37. Using Riemann sums, explain the economic significance of $\int_0^{q^*} S(q)\,dq$ to the producers.

38. Using Riemann sums, give an interpretation of producer surplus, $\int_0^{q^*} (p^* - S(q))\,dq$, analogous to the interpretation of consumer surplus.

39. In Figure 8.89, page 454, mark the regions representing the following quantities and explain their economic meaning:

(a) p^*q^* (b) $\int_0^{q^*} D(q)\,dq$

(c) $\int_0^{q^*} S(q)\,dq$ (d) $\int_0^{q^*} D(q)\,dq - p^*q^*$

(e) $p^*q^* - \int_0^{q^*} S(q)\,dq$ (f) $\int_0^{q^*} (D(q) - S(q))\,dq$

40. The dairy industry is an example of cartel pricing: the government has set milk prices artificially high. On a supply and demand graph, label p^+, a price above the equilibrium price. Using the graph, describe the effect of forcing the price up to p^+ on:

(a) The consumer surplus.

(b) The producer surplus.

(c) The total gains from trade (Consumer surplus + Producer surplus).

41. Rent controls on apartments are an example of price controls on a commodity. They keep the price artificially low (below the equilibrium price). Sketch a graph of supply and demand curves, and label on it a price p^- below the equilibrium price. What effect does forcing the price down to p^- have on:

(a) The producer surplus?

(b) The consumer surplus?

(c) The total gains from trade (Consumer surplus + Producer surplus)?

Strengthen Your Understanding

■ In Problems 42–45, explain what is wrong with the statement.

42. The future value of an income stream of \$2000 per year after 10 years is \$15,000, assuming a 3% continuous interest rate per year.
43. The present value of a lump-sum payment S dollars one year from now is greater with an interest rate of 4% per year, compounded annually, than with an interest rate of 3% per year compounded annually.
44. Payments are made at a constant rate of P dollars per year over a two-year period. The present value of these payments is $2Pe^{-2r}$, where r is the continuous interest rate per year.
45. Producer surplus is measured in the same units as the quantity, q.

■ In Problems 46–49, give an example of:

46. Supply and demand curves where producer surplus is smaller than consumer surplus.
47. A continuous interest rate such that a \$10,000 payment in 10 years' time has a present value of less than \$5000.
48. An interest rate, compounded annually, and a present value that correspond to a future value of \$5000 one year from now.
49. An interest rate, compounded annually, and a table of values that shows how much money you would have to put down in a single deposit t years from now, at $t = 0, 1, 2, 3,$ or 4, if you want to have \$10,000 ten years from now (ignoring inflation).

8.7 DISTRIBUTION FUNCTIONS

Understanding the distribution of various quantities through the population is important to decision makers. For example, the income distribution gives useful information about the economic structure of a society. In this section we look at the distribution of ages in the US. To allocate funding for education, health care, and social security, the government needs to know how many people are in each age group. We will see how to represent such information by a density function.

US Age Distribution

The data in Table 8.4 shows how the ages of the US population were distributed in 2012.[4] To represent this information graphically we use a type of *histogram*, with a vertical bar above each age group constructed so that the *area* of each bar represents the fraction of the population in that age group.[5] The total area of all the rectangles is $100\% = 1$. We only consider people who are less than 100 years old.[6] For the 0–20 age group, the base of the rectangle is 20, and we want the area to be 0.27, so the height must be $0.27/20 = 0.0135$. We treat ages as though they were continuously distributed. The category 0–20, for example, contains people who are just one day short of their twentieth birthday. (See Figure 8.90.)

Table 8.4 *Distribution of ages in the US in 2012*

Age group	Fraction of total population
0–20	27% = 0.27
20–40	27% = 0.27
40–60	28% = 0.28
60–80	15% = 0.15
80–100	3% = 0.03

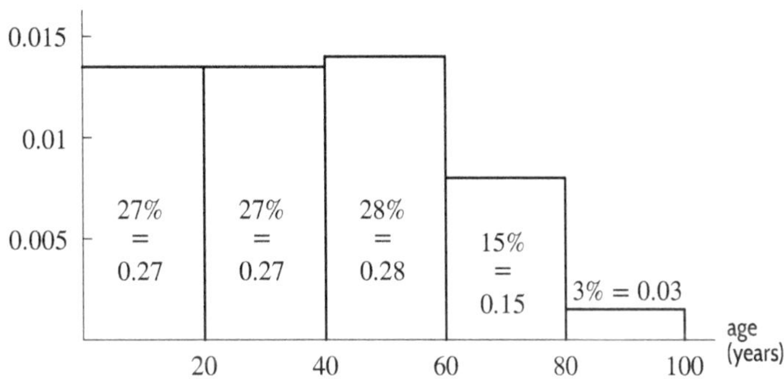

Figure 8.90: How ages were distributed in the US in 2012

[4] www.census.gov, accessed April 22, 2015.
[5] There are other types of histogram which have frequency on the vertical axis.
[6] In fact, 0.02% of the population is over 100, but this is too small to be visible on the histogram.

Example 1 In 2012, estimate what fraction of the US population was:

(a) Between 20 and 60 years old.
(b) Less than 10 years old.
(c) Between 75 and 80 years old.
(d) Between 80 and 85 years old.

Solution

(a) We add the fractions, so $0.27 + 0.28 = 0.55$; that is, 55% of the US population was in this age group.

(b) To find the fraction less than 10 years old, we could assume, for example, that the population was distributed evenly over the 0–20 group. (This means we are assuming that babies were born at a fairly constant rate over the last 20 years, which is probably reasonable.) If we make this assumption, then we can say that the population less than 10 years old was about half that in the 0–20 group, that is, 0.135 of the total population. Notice that we get the same result by computing the area of the rectangle from 0 to 10. (See Figure 8.91.)

(c) To find the population between 75 and 80 years old, since 0.15 of Americans in 2012 were in the 60-80 group, we might apply the same reasoning and say that $\frac{1}{4}(0.15) = 0.0375$ of the population was in this age group. This result is represented as an area in Figure 8.91. The assumption that the population was evenly distributed is not a good one here; certainly there were more people between the ages of 60 and 65 than between 75 and 80. Thus, the estimate of 0.0375 is certainly too high.

(d) Again using the (faulty) assumption that ages in each group were distributed uniformly, we would find that the fraction between 80 and 85 was $\frac{1}{4}(0.03) = 0.0075$. (See Figure 8.91.) This estimate is also poor—there were certainly more people in the 80–85 group than, say, the 95–100 group, and so the 0.0075 estimate is too low.

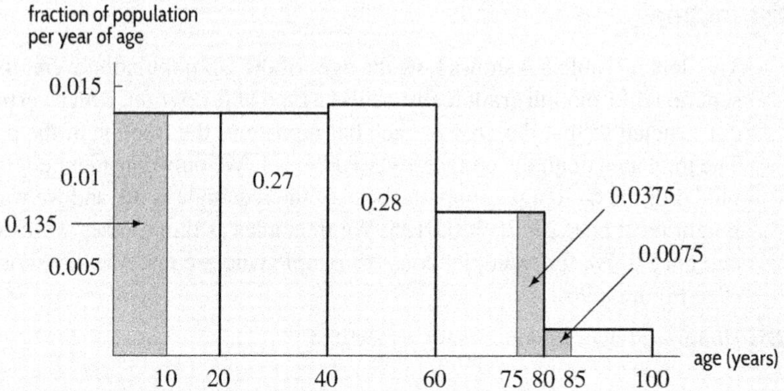

Figure 8.91: Ages in the US in 2012—various subgroups (for Example 1)

Smoothing Out the Histogram

We could get better estimates if we had smaller age groups (each age group in Figure 8.90 is 20 years, which is quite large). The more detailed data in Table 8.5 leads to the new histogram in Figure 8.92. As we get more detailed information, the upper silhouette of the histogram becomes smoother, but the area of any of the bars still represents the percentage of the population in that age group. Imagine, in the limit, replacing the upper silhouette of the histogram by a smooth curve in such a way that area under the curve above one age group is the same as the area in the corresponding rectangle. The total area under the whole curve is again $100\% = 1$. (See Figure 8.92.)

Table 8.5 *Ages in the US in 2012 (more detailed)*

Age group	Fraction of total population
0–10	13% = 0.13
10–20	14% = 0.14
20–30	14% = 0.14
30–40	13% = 0.13
40–50	14% = 0.14
50–60	14% = 0.14
60–70	10% = 0.10
70–80	5% = 0.05
80–90	2% = 0.02
90–100	1% = 0.01

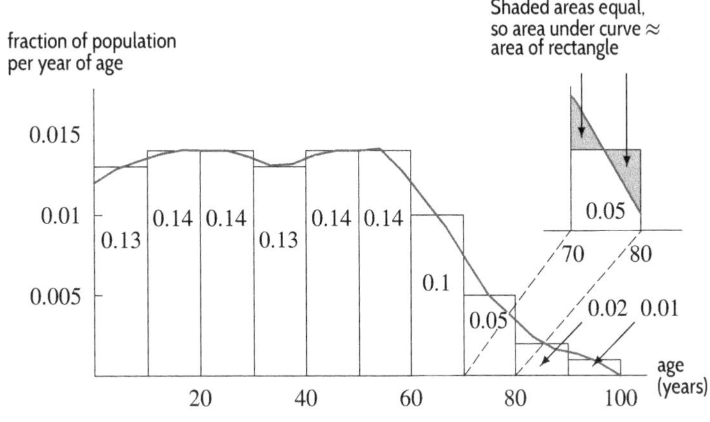

Figure 8.92: Smoothing out the age histogram

The Age Density Function

If t is age in years, we define $p(t)$, the age *density function*, to be a function which "smooths out" the age histogram. This function has the property that

$$\begin{array}{c}\text{Fraction of population}\\ \text{between ages } a \text{ and } b\end{array} = \begin{array}{c}\text{Area under graph of } p\\ \text{between } a \text{ and } b\end{array} = \int_a^b p(t)dt.$$

If a and b are the smallest and largest possible ages (say, $a = 0$ and $b = 100$), so that the ages of all of the population are between a and b, then

$$\int_a^b p(t)dt = \int_0^{100} p(t)dt = 1.$$

What does the age density function p tell us? Notice that we have not talked about the meaning of $p(t)$ itself, but *only* of the integral $\int_a^b p(t)\,dt$. Let's look at this in a bit more detail. Suppose, for example, that $p(10) = 0.015$ per year. This is *not* telling us that 0.015 of the population is precisely 10 years old (where 10 years old means exactly 10, not $10\frac{1}{2}$, not $10\frac{1}{4}$, not 10.1). However, $p(10) = 0.015$ does tell us that for some small interval Δt around 10, the fraction of the population with ages in this interval is approximately $p(10)\,\Delta t = 0.015\,\Delta t$.

The Probability Density Function

Suppose we are interested in how a certain characteristic, x, is distributed through a population. For example, x might be height or age if the population is people, or might be wattage for a population of light bulbs. Then we define a general density function with the following properties:

> The function, $p(x)$, is a **probability density function**, or pdf, if
>
> $$\begin{array}{c}\text{Fraction of population for which}\\ x \text{ is between } a \text{ and } b\end{array} = \begin{array}{c}\text{Area under graph of } p\\ \text{between } a \text{ and } b\end{array} = \int_a^b p(x)dx.$$
>
> $$\int_{-\infty}^{\infty} p(x)\,dx = 1 \quad \text{and} \quad p(x) \geq 0 \quad \text{for all } x.$$

The density function must be nonnegative because its integral always gives a fraction of the population. Also, the fraction of the population with x between $-\infty$ and ∞ is 1 because the entire population has the characteristic x between $-\infty$ and ∞. The function p that was used to smooth out the age histogram satisfies this definition of a density function. We do not assign a meaning to the value $p(x)$ directly, but rather interpret $p(x)\,\Delta x$ as the fraction of the population with the characteristic in a short interval of length Δx around x.

The density function is often approximated by formulas, as in the next example.

Example 2 Find formulas to approximate the density function, p, for the US age distribution. To reflect Figure 8.92, use a continuous function, constant at 0.013 up to age 60 and then dropping linearly.

Solution We have $p(t) = 0.013$ for $0 \le t < 60$. For $t \ge 60$, we need a linear function sloping downward. Because p is continuous, we have $p(60) = 0.013$. Because p is a density function we have $\int_0^{100} p(t)\,dt = 1$. Suppose b is as in Figure 8.93; then

$$\int_0^{100} p(t)\,dt = \int_0^{60} p(t)\,dt + \int_{60}^{100} p(t)\,dt = 60(0.013) + \frac{1}{2}(0.013)b = 1,$$

where $\int_{60}^{100} p(t)\,dt$ is given by the area of the triangle. Solving for b gives $b = 33.85$. Thus the slope of the line is $-0.013/33.85 = -0.00038$, so for $60 \le t \le 60 + 33.85 = 93.85$, we have

$$p(t) - 0.013 = -0.00038(t - 60),$$
$$p(t) = 0.0358 - 0.00038t.$$

According to this way of smoothing the data, there is no one over 93.85 years old, so $p(t) = 0$ for $t > 93.85$.

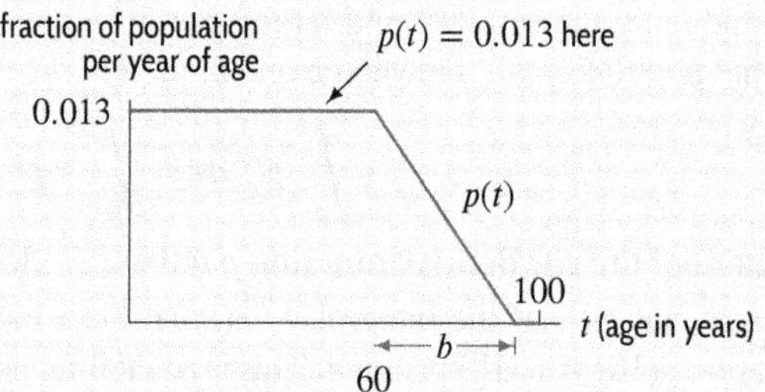

Figure 8.93: Age density function

Cumulative Distribution Function for Ages

Another way of showing how ages are distributed in the US is by using the *cumulative distribution function* $P(t)$, defined by

$$P(t) = \frac{\text{Fraction of population}}{\text{of age less than } t} = \int_0^t p(x)\,dx.$$

Thus, P is the antiderivative of p with $P(0) = 0$, and $P(t)$ gives the area under the density curve between 0 and t.

Notice that the cumulative distribution function is nonnegative and increasing (or at least nondecreasing), since the number of people younger than age t increases as t increases. Another way of seeing this is to notice that $P' = p$, and p is positive (or nonnegative). Thus the cumulative age distribution is a function which starts with $P(0) = 0$ and increases as t increases. $P(t) = 0$ for $t < 0$ because, when $t < 0$, there is no one whose age is less than t. The limiting value of P, as $t \to \infty$, is 1 since as t becomes very large (100 say), everyone is younger than age t, so the fraction of people with age less than t tends toward 1. (See Figure 8.94.) For t less than 60, the graph of P is a line, because p is constant there. For $t > 60$, the graph of P levels off as p tends to 0.

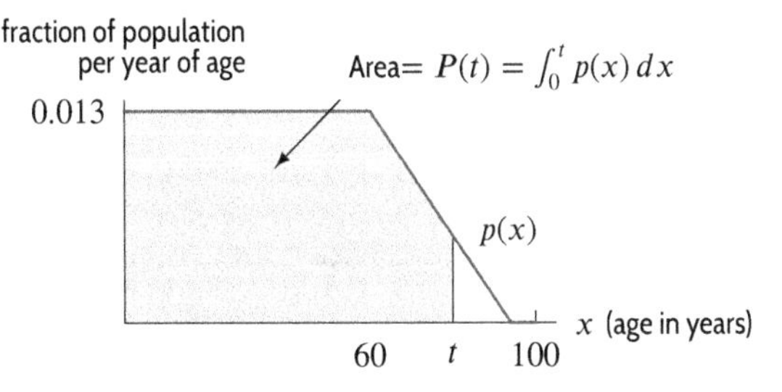

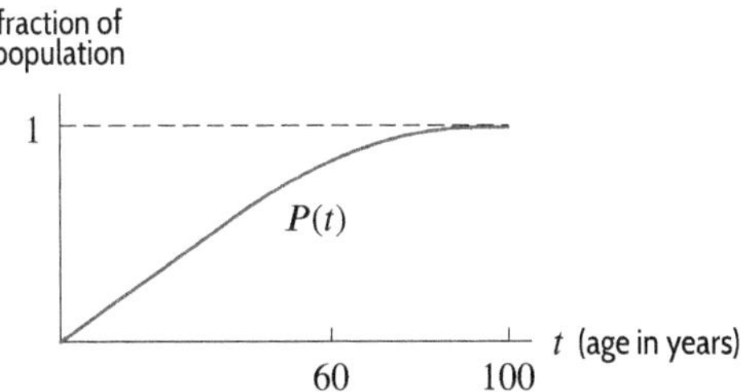

Figure 8.94: $P(t)$, the cumulative age distribution function, and its relation to $p(x)$, the age density function

Cumulative Distribution Function

A **cumulative distribution function**, or cdf, $P(t)$, of a density function p, is defined by

$$P(t) = \int_{-\infty}^{t} p(x)\,dx = \text{Fraction of population having values of } x \text{ below } t.$$

Thus, P is an antiderivative of p, that is, $P' = p$.
Any cumulative distribution has the following properties:

- P is increasing (or nondecreasing).
- $\lim_{t\to\infty} P(t) = 1$ and $\lim_{t\to-\infty} P(t) = 0$.
- $$\text{Fraction of population having values of } x \text{ between } a \text{ and } b = \int_a^b p(x)\,dx = P(b) - P(a).$$

Exercises and Problems for Section 8.7

EXERCISES

1. Match the graphs of the density functions (a), (b), and (c) with the graphs of the cumulative distribution functions I, II, and III.

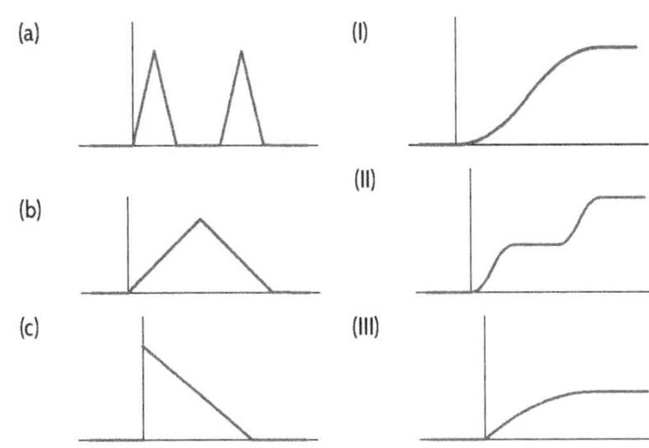

In Exercises 2–4, graph a density function and a cumulative distribution function which could represent the distribution of income through a population with the given characteristics.

2. A large middle class.
3. Small middle and upper classes and many poor people.
4. Small middle class, many poor and many rich people.

Decide if the function graphed in Exercises 5–10 is a probability density function (pdf) or a cumulative distribution function (cdf). Give reasons. Find the value of c. Sketch and label the other function. (That is, sketch and label the cdf if the problem shows a pdf, and the pdf if the problem shows a cdf.)

5. 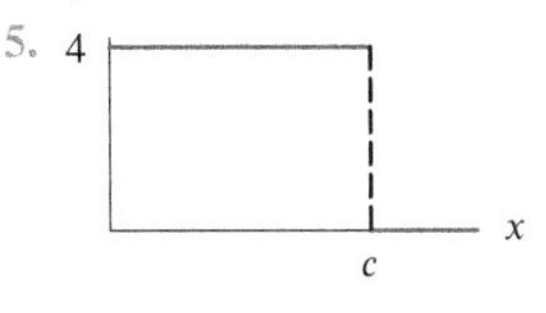

6. 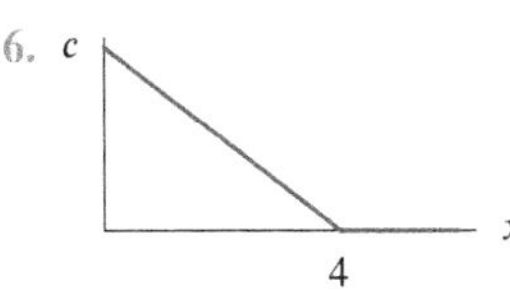

7. 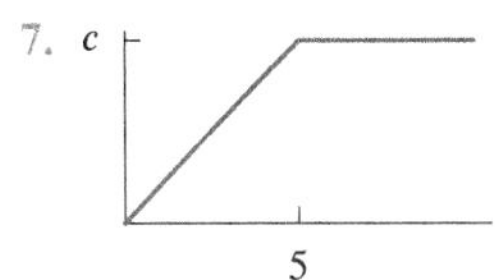

8. 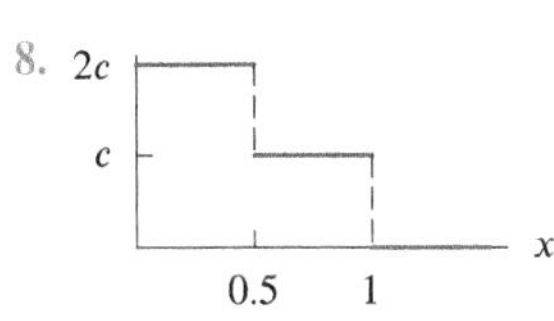

9. 10. 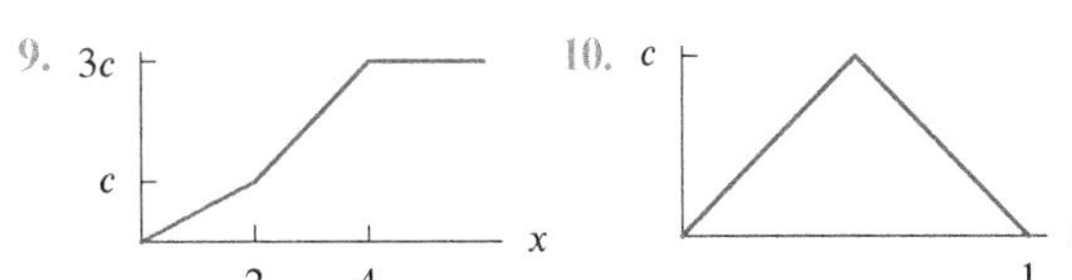

11. Let $p(x)$ be the density function for annual family income, where x is in thousands of dollars. What is the meaning of the statement $p(70) = 0.05$?

12. Find a density function $p(x)$ such that $p(x) = 0$ when $x \geq 5$ and when $x < 0$, and is decreasing when $0 \leq x \leq 5$.

PROBLEMS

13. Figure 8.95 shows the distribution of kinetic energy of molecules in a gas at temperatures 300 kelvins and 500 kelvins. At higher temperatures, more of the molecules in a gas have higher kinetic energies. Which graph corresponds to which temperature?

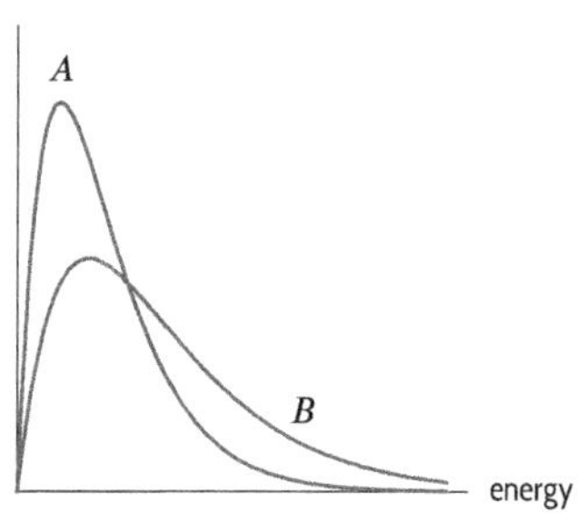

Figure 8.95

14. A large number of people take a standardized test, receiving scores described by the density function p graphed in Figure 8.96. Does the density function imply that most people receive a score near 50? Explain why or why not.

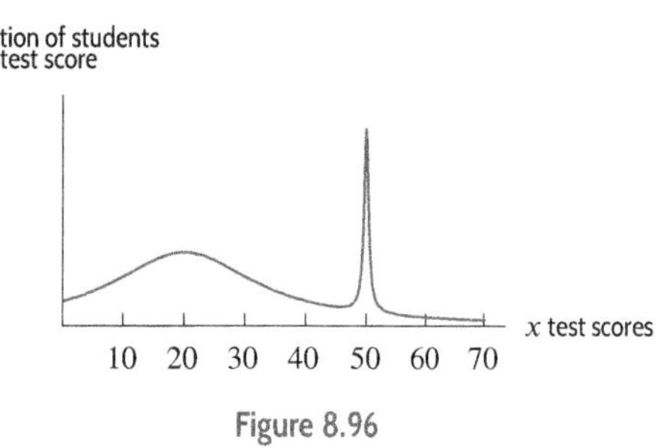

Figure 8.96

15. An experiment is done to determine the effect of two new fertilizers A and B on the growth of a species of peas. The cumulative distribution functions of the heights of the mature peas without treatment and treated with each of A and B are graphed in Figure 8.97.

(a) About what height are most of the unfertilized plants?
(b) Explain in words the effect of the fertilizers A and B on the mature height of the plants.

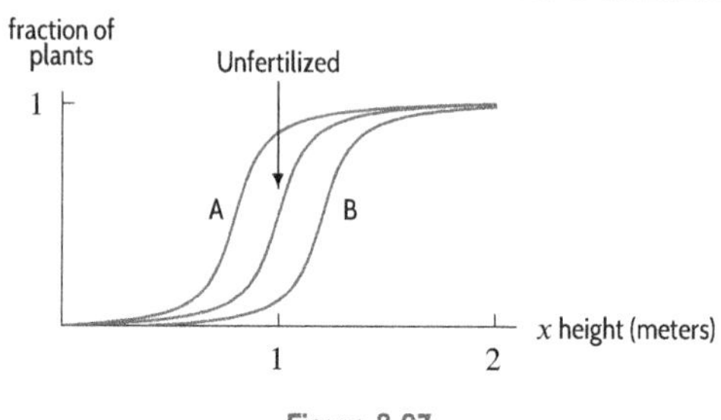

Figure 8.97

16. The cumulative distribution function for heights (in meters) of trees in a forest is $F(x)$.

(a) Explain in terms of trees the meaning of the statement $F(7) = 0.6$.
(b) Which is greater, $F(6)$ or $F(7)$? Justify your answer in terms of trees.

17. The density function for heights of American men, in inches is $p(x)$. What is the meaning of the statement $p(68) = 0.2$?

18. The fraction of the US population of age less than t is $P(t)$. Using Table 8.5 on page 459, make a table of values for $P(t)$.

19. Figure 8.98 shows a density function and the corresponding cumulative distribution function.[7]

(a) Which curve represents the density function and which represents the cumulative distribution function? Give a reason for your choice.
(b) Put reasonable values on the tick marks on each of the axes.

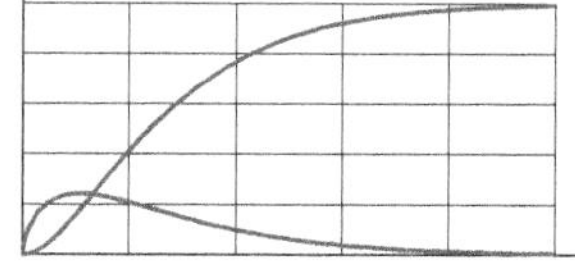

Figure 8.98

20. The density function and cumulative distribution function of heights of grass plants in a meadow are in Figures 8.99 and 8.100, respectively.

(a) There are two species of grass in the meadow, a short grass and a tall grass. Explain how the graph of the density function reflects this fact.

[7]Adapted from *Calculus*, by David A. Smith and Lawrence C. Moore (Lexington, D.C. Heath, 1994).

(b) Explain how the graph of the cumulative distribution function reflects the fact that there are two species of grass in the meadow.
(c) About what percentage of the grasses in the meadow belong to the short grass species?

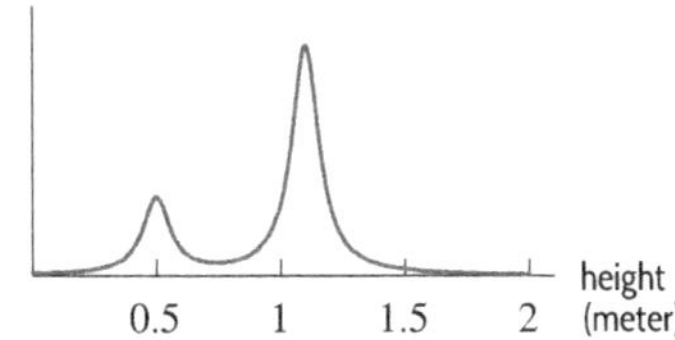

Figure 8.99

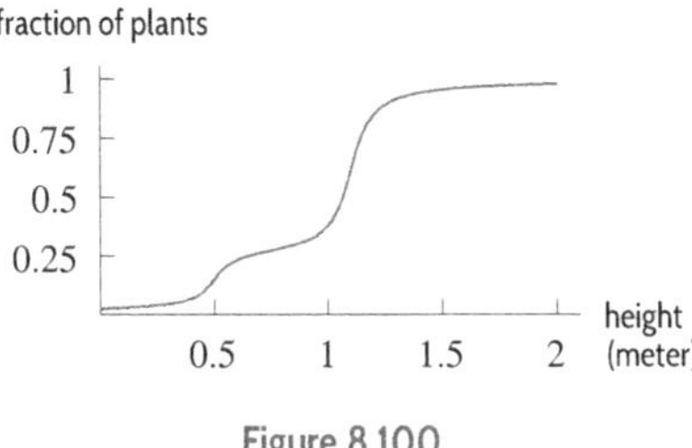

Figure 8.100

21. After measuring the duration of many telephone calls, the telephone company found their data was well approximated by the density function $p(x) = 0.4e^{-0.4x}$, where x is the duration of a call, in minutes.

(a) What percentage of calls last between 1 and 2 minutes?
(b) What percentage of calls last 1 minute or less?
(c) What percentage of calls last 3 minutes or more?
(d) Find the cumulative distribution function.

22. Students at the University of California were surveyed and asked their grade point average. (The GPA ranges from 0 to 4, where 2 is just passing.) The distribution of GPAs is shown in Figure 8.101.[8]

(a) Roughly what fraction of students are passing?
(b) Roughly what fraction of the students have honor grades (GPAs above 3)?
(c) Why do you think there is a peak around 2?
(d) Sketch the cumulative distribution function.

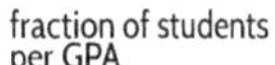

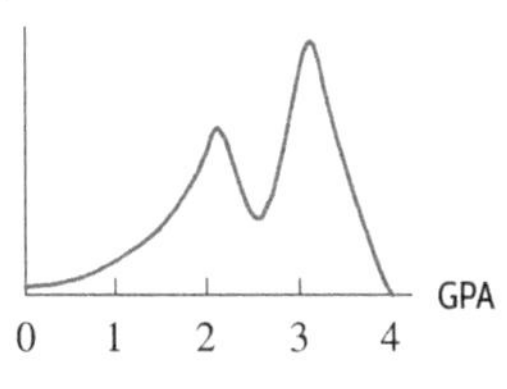

Figure 8.101

23. Figure 8.102[8] shows the distribution of elevation, in miles, across the earth's surface. Positive elevation denotes land above sea level; negative elevation shows land below sea level (i.e., the ocean floor).

(a) Describe in words the elevation of most of the earth's surface.
(b) Approximately what fraction of the earth's surface is below sea level?

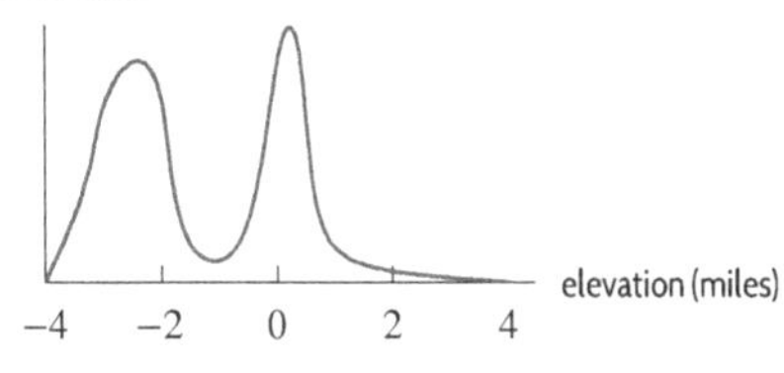

Figure 8.102

24. Consider a population of individuals with a disease. Suppose that t is the number of years since the onset of the disease. The death density function, $f(t) = cte^{-kt}$, approximates the fraction of the sick individuals who die in the time interval $[t, t + \Delta t]$ as follows:

$$\text{Fraction who die} \approx f(t)\Delta t = cte^{-kt}\Delta t$$

where c and k are positive constants whose values depend on the particular disease.

(a) Find the value of c in terms of k.
(b) If 40% of the population dies within 5 years, find c and k.
(c) Find the cumulative death distribution function, $C(t)$. Give your answer in terms of k.

Strengthen Your Understanding

In Problems 25–31, explain what is wrong with the statement.

25. If $p(x)$ is a probability density function with $p(1) = 0.02$, then the probability that x takes the value 1 is 0.02.

26. If $P(x)$ is a cumulative distribution function with $P(5) = 0.4$, then the probability that $x = 5$ is 0.4.

27. The function $p(t) = t^2$ is a density function.

28. The function $p(x) = x^2e^x$ is a density function.

29. The function $P(x) = x^2e^x$ is a cumulative distribution function.

30. The function $P(t) = e^{-t^2}$ is a cumulative distribution function.

31. A probability density function is always increasing.

[8] Adapted from *Statistics*, by Freedman, Pisani, Purves, and Adikhari (New York: Norton, 1991).

In Problems 32–35, give an example of:

32. A density function that is greater than zero on $0 \leq x \leq 20$ and zero everywhere else.

33. A cumulative distribution function that is piecewise linear.

34. A probability density function which is nonzero only between $x = 2$ and $x = 7$.

35. A cumulative distribution function with $P(3) = 0$ and $P(7) = 1$.

In Problems 36–37, are the statements true or false? Give an explanation for your answer.

36. If $p(x) = xe^{-x^2}$ for all x, then $p(x)$ is a probability density function.

37. If $p(x) = xe^{-x^2}$ for all $x > 0$ and $p(x) = 0$ for $x \leq 0$, then $p(x)$ is a probability density function.

8.8 PROBABILITY, MEAN, AND MEDIAN

Probability

Suppose we pick a member of the US population at random and ask what is the probability that the person is between, say, the ages of 70 and 80. We saw in Table 8.5 on page 459 that 5% = 0.05 of the population is in this age group. We say that the probability, or chance, that the person is between 70 and 80 is 0.05. Using any age density function $p(t)$, we can define probabilities as follows:

$$\begin{pmatrix}\text{Probability that a person is}\\ \text{between ages } a \text{ and } b\end{pmatrix} = \begin{pmatrix}\text{Fraction of population}\\ \text{between ages } a \text{ and } b\end{pmatrix} = \int_a^b p(t)\,dt.$$

Since the cumulative distribution function gives the fraction of the population younger than age t, the cumulative distribution can also be used to calculate the probability that a randomly selected person is in a given age group.

$$\begin{pmatrix}\text{Probability that a person is}\\ \text{younger than age } t\end{pmatrix} = \begin{pmatrix}\text{Fraction of population}\\ \text{younger than age } t\end{pmatrix} = P(t) = \int_0^t p(x)\,dx.$$

In the next example, both a density function and a cumulative distribution function are used to describe the same situation.

Example 1 Suppose you want to analyze the fishing industry in a small town. Each day, the boats bring back at least 2 tons of fish, and never more than 8 tons.

(a) Using the density function describing the daily catch in Figure 8.103, find and graph the corresponding cumulative distribution function and explain its meaning.
(b) What is the probability that the catch will be between 5 and 7 tons?

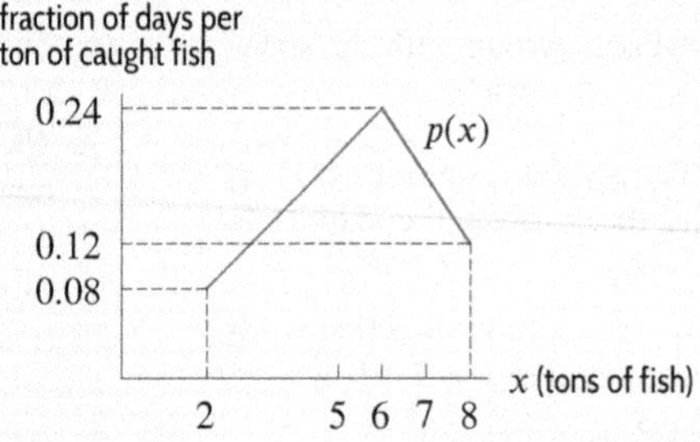

Figure 8.103: Density function of daily catch

Solution (a) The cumulative distribution function $P(t)$ is equal to the fraction of days on which the catch is less than t tons of fish. Since the catch is never less than 2 tons, we have $P(t) = 0$ for $t \leq 2$. Since the catch is always less than 8 tons, we have $P(t) = 1$ for $t \geq 8$. For t in the range $2 < t < 8$, we must evaluate the integral

$$P(t) = \int_{-\infty}^{t} p(x)dx = \int_{2}^{t} p(x)dx.$$

This integral equals the area under the graph of $p(x)$ between $x = 2$ and $x = t$. It can be calculated by noting that $p(x)$ is given by the formula

$$p(x) = \begin{cases} 0.04x & \text{for } 2 \leq x \leq 6 \\ -0.06x + 0.6 & \text{for } 6 < x \leq 8 \end{cases}$$

and $p(x) = 0$ for $x < 2$ or $x > 8$. Thus, for $2 \leq t \leq 6$,

$$P(t) = \int_{2}^{t} 0.04x\,dx = 0.04\frac{x^2}{2}\bigg|_2^t = 0.02t^2 - 0.08.$$

And for $6 \leq t \leq 8$,

$$\begin{aligned} P(t) &= \int_2^t p(x)\,dx = \int_2^6 p(x)\,dx + \int_6^t p(x)\,dx \\ &= 0.64 + \int_6^t (-0.06x + 0.6)\,dx = 0.64 + \left(-0.06\frac{x^2}{2} + 0.6x\right)\bigg|_6^t \\ &= -0.03t^2 + 0.6t - 1.88. \end{aligned}$$

Thus

$$P(t) = \begin{cases} 0.02t^2 - 0.08 & \text{for } 2 \leq t \leq 6 \\ -0.03t^2 + 0.6t - 1.88 & \text{for } 6 < t \leq 8. \end{cases}$$

In addition $P(t) = 0$ for $t < 2$ and $P(t) = 1$ for $8 < t$. (See Figure 8.104.)

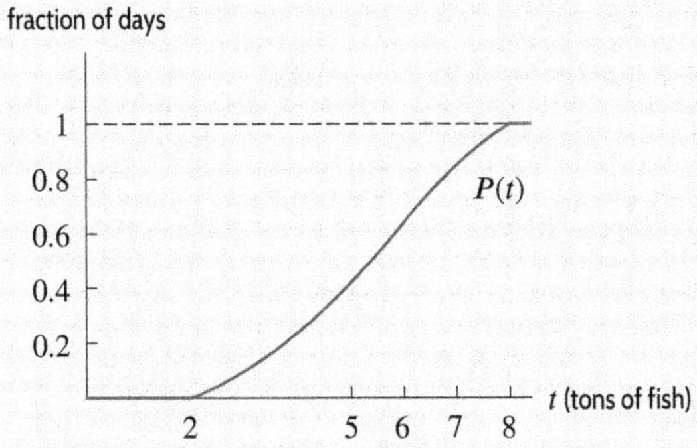

Figure 8.104: Cumulative distribution of daily catch

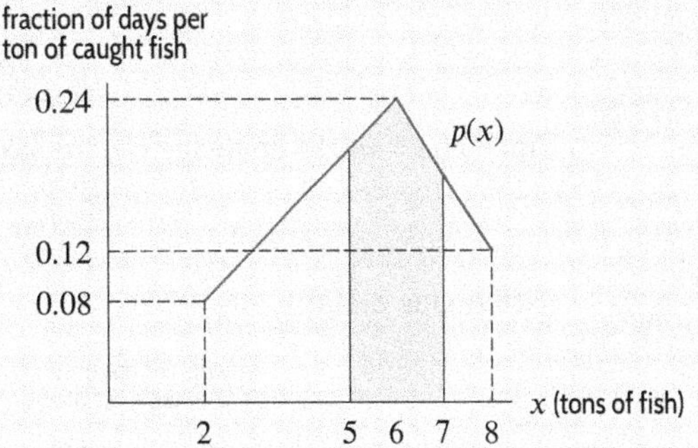

Figure 8.105: Shaded area represents the probability that the catch is between 5 and 7 tons

(b) The probability that the catch is between 5 and 7 tons can be found using either the density function, p, or the cumulative distribution function, P. If we use the density function, this probability can be represented by the shaded area in Figure 8.105, which is about 0.43:

$$\text{Probability catch is between 5 and 7 tons} = \int_5^7 p(x)\,dx = 0.43.$$

The probability can be found from the cumulative distribution as follows:

$$\begin{array}{c}\text{Probability catch is} \\ \text{between 5 and 7 tons}\end{array} = P(7) - P(5) = 0.85 - 0.42 = 0.43.$$

The Median and Mean

It is often useful to be able to give an "average" value for a distribution. Two measures that are in common use are the *median* and the *mean*.

The Median

A **median** of a quantity x distributed through a population is a value T such that half the population has values of x less than (or equal to) T, and half the population has values of x greater than (or equal to) T. Thus, a median T satisfies

$$\int_{-\infty}^{T} p(x)\,dx = 0.5,$$

where p is the density function. In other words, half the area under the graph of p lies to the left of T.

Example 2 Find the median age in the US in 2012, using the age density function given by

$$p(t) = \begin{cases} 0.013 & \text{for } 0 \le t \le 60 \\ 0.0358 - 0.00038t & \text{for } 60 < t \le 93.85. \end{cases}$$

Solution We want to find the value of T such that

$$\int_{-\infty}^{T} p(t)\,dt = \int_{0}^{T} p(t)\,dt = 0.5.$$

Since $p(t) = 0.013$ up to age 60, we have

$$\text{Median} = T = \frac{0.5}{0.013} \approx 38.5 \text{ years.}$$

(See Figure 8.106.)

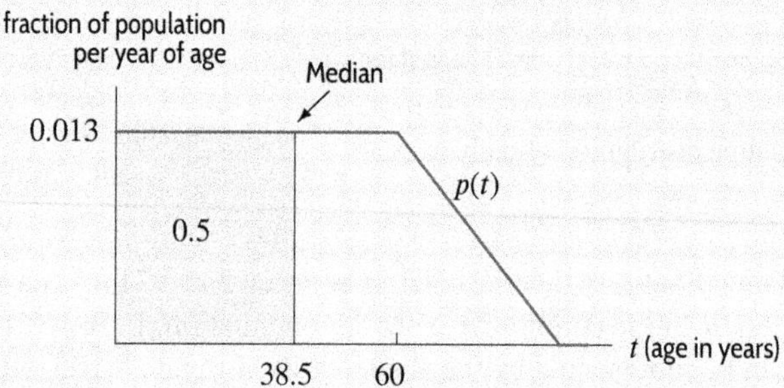

Figure 8.106: Median of age distribution

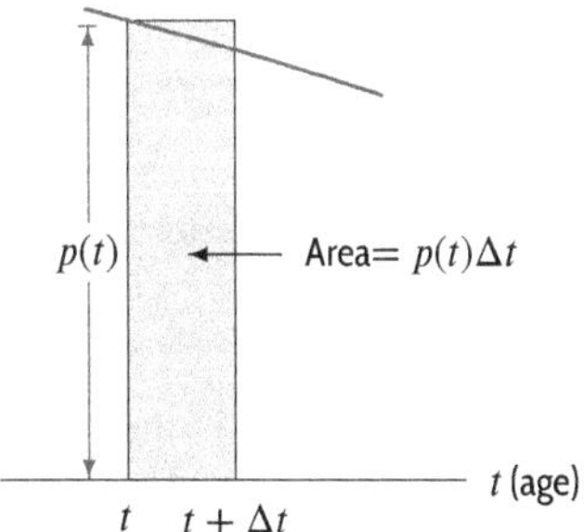

Figure 8.107: Shaded area is percentage of population with age between t and $t + \Delta t$

The Mean

Another commonly used average value is the *mean*. To find the mean of N numbers, you add the numbers and divide the sum by N. For example, the mean of the numbers 1, 2, 7, and 10 is $(1 + 2 + 7 + 10)/4 = 5$. The mean age of the entire US population is therefore defined as

$$\text{Mean age} = \frac{\sum \text{Ages of all people in the US}}{\text{Total number of people in the US}}.$$

Calculating the sum of all the ages directly would be an enormous task; we will approximate the sum by an integral. The idea is to "slice up" the age axis and consider the people whose age is between t and $t + \Delta t$. How many are there?

The fraction of the population between t and $t + \Delta t$ is the area under the graph of p between these points, which is well approximated by the area of the rectangle, $p(t)\Delta t$. (See Figure 8.107.) If the total number of people in the population is N, then

$$\begin{array}{c}\text{Number of people with age} \\ \text{between } t \text{ and } t+\Delta t\end{array} \approx p(t)\Delta t N.$$

The age of all of these people is approximately t:

$$\begin{array}{c}\text{Sum of ages of people} \\ \text{between age } t \text{ and } t+\Delta t\end{array} \approx t p(t)\Delta t N.$$

Therefore, adding and factoring out an N gives us

$$\text{Sum of ages of all people} \approx \left(\sum t p(t)\Delta t\right) N.$$

In the limit, as we allow Δt to shrink to 0, the sum becomes an integral, so

$$\text{Sum of ages of all people} = \left(\int_0^{100} t p(t)\,dt\right) N.$$

Therefore, with N equal to the total number of people in the US, and assuming no person is over 100 years old,

$$\text{Mean age} = \frac{\text{Sum of ages of all people in US}}{N} = \int_0^{100} t p(t)\,dt.$$

We can give the same argument for any[9] density function $p(x)$.

[9]Provided all the relevant improper integrals converge.

If a quantity has density function $p(x)$,

$$\textbf{Mean value} \text{ of the quantity} = \int_{-\infty}^{\infty} xp(x)\,dx.$$

It can be shown that the mean is the point on the horizontal axis where the region under the graph of the density function, if it were made out of cardboard, would balance.

Example 3 Find the mean age of the US population, using the density function of Example 2.

Solution The formula for p is

$$p(t) = \begin{cases} 0 & \text{for } t < 0 \\ 0.013 & \text{for } 0 \leq t \leq 60 \\ 0.0358 - 0.00038t & \text{for } 60 < t \leq 93.85 \\ 0 & \text{for } t > 93.85. \end{cases}$$

Using these formulas, we compute

$$\text{Mean age} = \int_0^{100} tp(t)dt = \int_0^{60} t(0.013)dt + \int_{60}^{93.85} t(0.0358 - 0.00038t)dt$$

$$= 0.013\frac{t^2}{2}\bigg|_0^{60} + 0.0358\frac{t^2}{2}\bigg|_{60}^{93.85} - 0.00038\frac{t^3}{3}\bigg|_{60}^{93.85} \approx 39.3 \text{ years.}$$

The mean is shown is Figure 8.108.

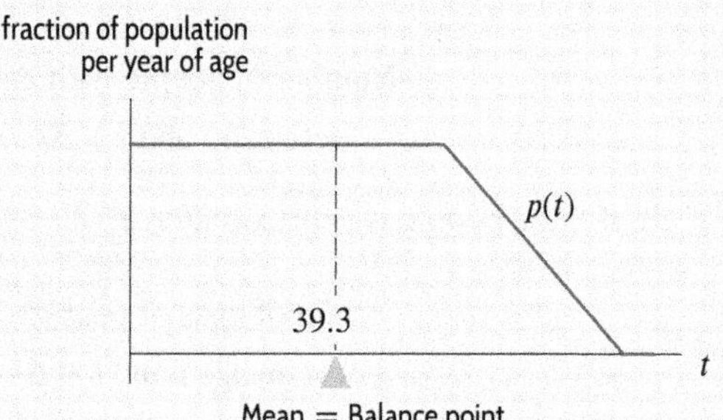

Figure 8.108: Mean of age distribution

Normal Distributions

How much rain do you expect to fall in your home town this year? If you live in Anchorage, Alaska, the answer is something close to 15 inches (including the snow). Of course, you don't expect exactly 15 inches. Some years have more than 15 inches, and some years have less. Most years, however, the amount of rainfall is close to 15 inches; only rarely is it well above or well below 15 inches. What does the density function for the rainfall look like? To answer this question, we look at rainfall data over many years. Records show that the distribution of rainfall is well approximated by a *normal distribution*. The graph of its density function is a bell-shaped curve which peaks at 15 inches and slopes downward approximately symmetrically on either side.

Normal distributions are frequently used to model real phenomena, from grades on an exam to the number of airline passengers on a particular flight. A normal distribution is characterized by its *mean*, μ, and its *standard deviation*, σ. The mean tells us the location of the central peak. The standard deviation tells us how closely the data is clustered around the mean. A small value of σ tells us that the data is close to the mean; a large σ tells us the data is spread out. In the following formula for a normal distribution, the factor of $1/(\sigma\sqrt{2\pi})$ makes the area under the graph equal to 1.

> A **normal distribution** has a density function of the form
> $$p(x) = \frac{1}{\sigma\sqrt{2\pi}} e^{-(x-\mu)^2/(2\sigma^2)},$$
> where μ is the mean of the distribution and σ is the standard deviation, with $\sigma > 0$.

To model the rainfall in Anchorage, we use a normal distribution with $\mu = 15$ and $\sigma = 1$. (See Figure 8.109.)

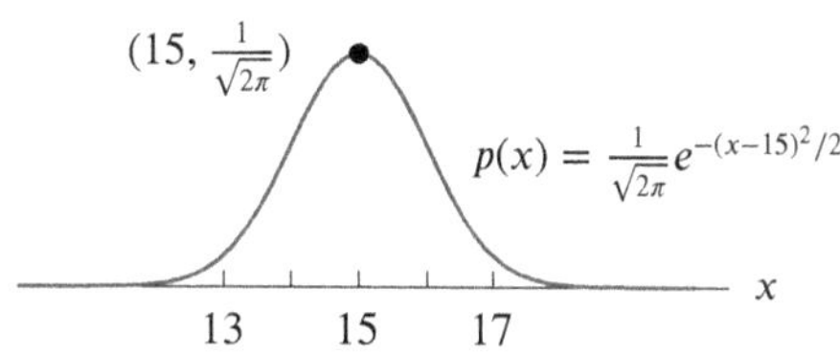

Figure 8.109: Normal distribution with $\mu = 15$ and $\sigma = 1$

Example 4 For Anchorage's rainfall, use the normal distribution with the density function with $\mu = 15$ and $\sigma = 1$ to compute the fraction of the years with rainfall between
(a) 14 and 16 inches, (b) 13 and 17 inches, (c) 12 and 18 inches.

Solution (a) The fraction of the years with annual rainfall between 14 and 16 inches is $\int_{14}^{16} \frac{1}{\sqrt{2\pi}} e^{-(x-15)^2/2}\, dx$. Since there is no elementary antiderivative for $e^{-(x-15)^2/2}$, we find the integral numerically. Its value is about 0.68.

$$\text{Fraction of years with rainfall between 14 and 16 inches} = \int_{14}^{16} \frac{1}{\sqrt{2\pi}} e^{-(x-15)^2/2}\, dx \approx 0.68.$$

(b) Finding the integral numerically again:

$$\text{Fraction of years with rainfall between 13 and 17 inches} = \int_{13}^{17} \frac{1}{\sqrt{2\pi}} e^{-(x-15)^2/2}\, dx \approx 0.95.$$

(c)

$$\text{Fraction of years with rainfall between 12 and 18 inches} = \int_{12}^{18} \frac{1}{\sqrt{2\pi}} e^{-(x-15)^2/2}\, dx \approx 0.997.$$

Since 0.95 is so close to 1, we expect that most of the time the rainfall will be between 13 and 17 inches a year.

Among the normal distributions, the one having $\mu = 0$, $\sigma = 1$ is called the *standard normal distribution*. Values of the corresponding cumulative distribution function are published in tables.

Exercises and Problems for Section 8.8

EXERCISES

1. Show that the area under the fishing density function in Figure 8.103 on page 464 is 1. Why is this to be expected?

2. Find the mean daily catch for the fishing data in Figure 8.103, page 464.

3. **(a)** Using a calculator or computer, sketch graphs of the density function of the normal distribution

$$p(x) = \frac{1}{\sigma\sqrt{2\pi}} e^{-(x-\mu)^2/(2\sigma^2)}.$$

   (i) For fixed μ (say, $\mu = 5$) and varying σ (say, $\sigma = 1, 2, 3$).

   (ii) For varying μ (say, $\mu = 4, 5, 6$) and fixed σ (say, $\sigma = 1$).

   (b) Explain how the graphs confirm that μ is the mean of the distribution and that σ is a measure of how closely the data is clustered around the mean.

In Exercises 4–6, a density function $p(x)$ satisfying (I)–(IV) gives the fraction of years with a given total annual snowfall (in m) for a city.

I. $\int_0^{0.5} p(x)\,dx = 0.1$ II. $\int_0^{2} p(x)\,dx = 0.3$

III. $\int_0^{2.72} p(x)\,dx = 0.5$ IV. $\int_0^{\infty} xp(x)\,dx = 2.65$

4. What is the median annual snowfall (in m)?

5. What is the mean snowfall (in m)?

6. What is the probability of an annual snowfall between 0.5 and 2 m?

PROBLEMS

7. A screening test for susceptibility to diabetes reports a numerical score between 0 to 100. A score greater than 50 indicates a potential risk, with some lifestyle training recommended. Results from 200,000 people who were tested show that :

   - 75% received scores evenly distributed between 0 and 50.
   - 25% received scores evenly distributed between 50 and 100.

   The probability density function (pdf) is in Figure 8.110.

   (a) Find the values of A and B that make this a probability density function.
   (b) Find the median test score.
   (c) Find the mean test score.
   (d) Give a graph of the cumulative distribution function (cdf) for these test scores.

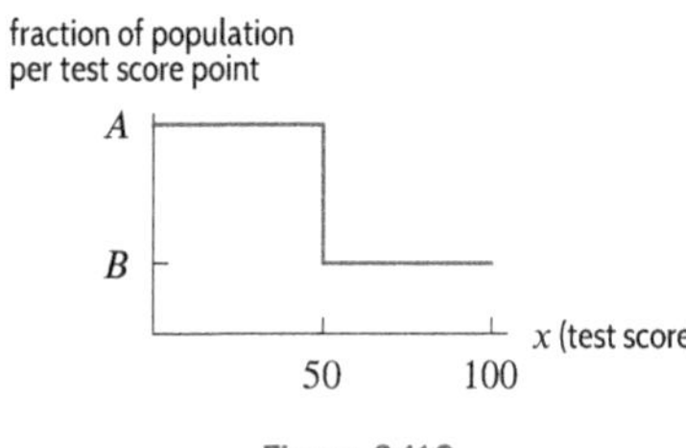

Figure 8.110

In Problems 8–11, use Figure 8.111, a graph of $p(x)$, a density function for the fraction of a region's winters with a given total snowfall (in m).

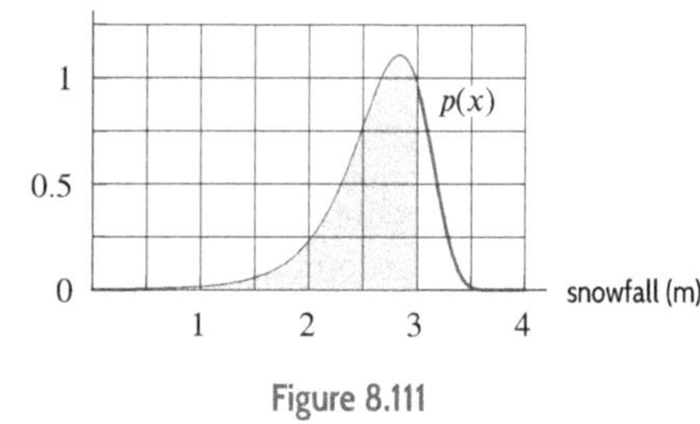

Figure 8.111

8. Which of the following events is most likely?

   I. A winter has 3 m or more of snow?

   II. A winter has 2 m or less of snow?

   III. A winter with between 2 and 3 m of snow?

9. The shaded area is 0.8. What percentage of winters see more than 3 meters of total snowfall?

10. What appears to be the smallest and largest total annual snowfall?

11. If $p(2.8) = 1.1$, approximately what percentage of winters see snowfall totals between 2.8 m and 3.0 m?

12. A quantity x has density function $p(x) = 0.5(2 - x)$ for $0 \le x \le 2$ and $p(x) = 0$ otherwise. Find the mean and median of x.

13. A quantity x has cumulative distribution function $P(x) = x - x^2/4$ for $0 \le x \le 2$ and $P(x) = 0$ for $x < 0$ and $P(x) = 1$ for $x > 2$. Find the mean and median of x.

14. The probability of a transistor failing between $t = a$ months and $t = b$ months is given by $c\int_a^b e^{-ct}dt$, for some constant c.

(a) If the probability of failure within the first six months is 10%, what is c?
(b) Given the value of c in part (a), what is the probability the transistor fails within the second six months?

15. Suppose that x measures the time (in hours) it takes for a student to complete an exam. All students are done within two hours and the density function for x is

$$p(x) = \begin{cases} x^3/4 & \text{if } 0 < x < 2 \\ 0 & \text{otherwise.} \end{cases}$$

(a) What proportion of students take between 1.5 and 2.0 hours to finish the exam?
(b) What is the mean time for students to complete the exam?
(c) Compute the median of this distribution.

16. In 1950 an experiment was done observing the time gaps between successive cars on the Arroyo Seco Freeway.[10] The data show that the density function of these time gaps was given approximately by

$$p(x) = ae^{-0.122x}$$

where x is the time in seconds and a is a constant.

(a) Find a.
(b) Find P, the cumulative distribution function.
(c) Find the median and mean time gap.
(d) Sketch rough graphs of p and P.

17. Consider a group of people who have received treatment for a disease such as cancer. Let t be the *survival time*, the number of years a person lives after receiving treatment. The density function giving the distribution of t is $p(t) = Ce^{-Ct}$ for some positive constant C.

(a) What is the practical meaning for the cumulative distribution function $P(t) = \int_0^t p(x)\,dx$?
(b) The survival function, $S(t)$, is the probability that a randomly selected person survives for at least t years. Find $S(t)$.
(c) Suppose a patient has a 70% probability of surviving at least two years. Find C.

18. While taking a walk along the road where you live, you accidentally drop your glove, but you don't know where. The probability density $p(x)$ for having dropped the glove x kilometers from home (along the road) is

$$p(x) = 2e^{-2x} \quad \text{for } x \geq 0.$$

(a) What is the probability that you dropped it within 1 kilometer of home?
(b) At what distance y from home is the probability that you dropped it within y km of home equal to 0.95?

19. The distribution of IQ scores can be modeled by a normal distribution with mean 100 and standard deviation 15.

(a) Write the formula for the density function of IQ scores.
(b) Estimate the fraction of the population with IQ between 115 and 120.

20. The speeds of cars on a road are approximately normally distributed with a mean $\mu = 58$ km/hr and standard deviation $\sigma = 4$ km/hr.

(a) What is the probability that a randomly selected car is going between 60 and 65 km/hr?
(b) What fraction of all cars are going slower than 52 km/hr?

21. Consider the normal distribution, $p(x)$.

(a) Show that $p(x)$ is a maximum when $x = \mu$. What is that maximum value?
(b) Show that $p(x)$ has points of inflection where $x = \mu + \sigma$ and $x = \mu - \sigma$.
(c) Describe in your own words what μ and σ tell you about the distribution.

22. For a normal population of mean 0, show that the fraction of the population within one standard deviation of the mean does not depend on the standard deviation.
[Hint: Use the substitution $w = x/\sigma$.]

23. Which of the following functions makes the most sense as a model for the probability density representing the time (in minutes, starting from $t = 0$) that the next customer walks into a store?

(a) $p(t) = \begin{cases} \cos t & 0 \leq t \leq 2\pi \\ e^{t-2\pi} & t \geq 2\pi \end{cases}$
(b) $p(t) = 3e^{-3t}$ for $t \geq 0$
(c) $p(t) = e^{-3t}$ for $t \geq 0$
(d) $p(t) = 1/4$ for $0 \leq t \leq 4$

24. Let $P(x)$ be the cumulative distribution function for the household income distribution in the US in 2009.[11] Values of $P(x)$ are in the following table:

Income x (thousand \$)	20	40	60	75	100
$P(x)$ (%)	29.5	50.1	66.8	76.2	87.1

(a) What percent of the households made between \$40,000 and \$60,000? More than \$100,000?
(b) Approximately what was the median income?
(c) Is the statement "More than one-third of households made between \$40,000 and \$75,000" true or false?

[10] Reported by Daniel Furlough and Frank Barnes.
[11] http://www.census.gov/hhes/www/income/income.html, accessed January 7, 2012.

25. If we think of an electron as a particle, the function

$$P(r) = 1 - (2r^2 + 2r + 1)e^{-2r}$$

is the cumulative distribution function of the distance, r, of the electron in a hydrogen atom from the center of the atom. The distance is measured in Bohr radii. (1 Bohr radius $= 5.29 \times 10^{-11}$ m. Niels Bohr (1885–1962) was a Danish physicist.)

For example, $P(1) = 1 - 5e^{-2} \approx 0.32$ means that the electron is within 1 Bohr radius from the center of the atom 32% of the time.

(a) Find a formula for the density function of this distribution. Sketch the density function and the cumulative distribution function.

(b) Find the median distance and the mean distance. Near what value of r is an electron most likely to be found?

(c) The Bohr radius is sometimes called the "radius of the hydrogen atom." Why?

Strengthen Your Understanding

In Problems 26–27, explain what is wrong with the statement.

26. A median T of a quantity distributed through a population satisfies $p(T) = 0.5$ where p is the density function.

27. The following density function has median 1:

$$p(x) = \begin{cases} 0 & \text{for } x < 0 \\ 2(1-x) & \text{for } 0 \le x \le 1 \\ 0 & \text{for } x > 1. \end{cases}$$

In Problems 28–29, give an example of:

28. A distribution with a mean of $1/2$ and standard deviation $1/2$.

29. A distribution with a mean of $1/2$ and median $1/2$.

In Problems 30–34, a quantity x is distributed through a population with probability density function $p(x)$ and cumulative distribution function $P(x)$. Decide if each statement is true or false. Give an explanation for your answer.

30. If $p(10) = 1/2$, then half the population has $x < 10$.

31. If $P(10) = 1/2$, then half the population has $x < 10$.

32. If $p(10) = 1/2$, then the fraction of the population lying between $x = 9.98$ and $x = 10.04$ is about 0.03.

33. If $p(10) = p(20)$, then none of the population has x values lying between 10 and 20.

34. If $P(10) = P(20)$, then none of the population has x values lying between 10 and 20.

Online Resource: Review problems and Projects

Chapter Nine

SEQUENCES AND SERIES

Contents

9.1 SEQUENCES

A sequence[1] is an infinite list of numbers $s_1, s_2, s_3, \ldots, s_n, \ldots$. We call s_1 the first term, s_2 the second term; s_n is the general term. For example, the sequence of squares, $1, 4, 9, \ldots, n^2, \ldots$, can be denoted by the general term $s_n = n^2$. Thus, a sequence is a function whose domain is the positive integers, but it is traditional to denote the terms of a sequence using subscripts, s_n, rather than function notation, $s(n)$. In addition, we may talk about sequences whose general term has no simple formula, such as the sequence 3, 3.1, 3.14, 3.141, 3.1415, ..., in which s_n gives the first n digits of π.

The Numerical, Algebraic, and Graphical Viewpoint

Just as we can view a function algebraically, numerically, graphically, or verbally, we can view sequences in different ways. We may give an algebraic formula for the general term. We may give the numerical values of the first few terms of the sequence, suggesting a pattern for the later terms.

Example 1 Give the first six terms of the following sequences:

(a) $s_n = \dfrac{n(n+1)}{2}$ (b) $s_n = \dfrac{n+(-1)^n}{n}$

Solution (a) Substituting $n = 1, 2, 3, 4, 5, 6$ into the formula for the general term, we get

$$\frac{1\cdot 2}{2}, \frac{2\cdot 3}{2}, \frac{3\cdot 4}{2}, \frac{4\cdot 5}{2}, \frac{5\cdot 6}{2}, \frac{6\cdot 7}{2} = 1,\ 3,\ 6,\ 10,\ 15,\ 21.$$

(b) Substituting $n = 1, 2, 3, 4, 5, 6$ into the formula for the general term, we get

$$\frac{1-1}{1}, \frac{2+1}{2}, \frac{3-1}{3}, \frac{4+1}{4}, \frac{5-1}{5}, \frac{6+1}{6} = 0, \frac{3}{2}, \frac{2}{3}, \frac{5}{4}, \frac{4}{5}, \frac{7}{6}.$$

Example 2 Give a general term for the following sequences:

(a) 1, 2, 4, 8, 16, 32, ... (b) $\dfrac{7}{2}, \dfrac{7}{5}, \dfrac{7}{8}, \dfrac{7}{11}, \dfrac{1}{2}, \dfrac{7}{17}, \ldots$

Solution Although the first six terms do not determine the sequence, we can sometimes use them to guess a possible formula for the general term.

(a) We have powers of 2, so we guess $s_n = 2^n$. When we check by substituting in $n = 1, 2, 3, 4, 5, 6$, we get $2, 4, 8, 16, 32, 64$, instead of $1, 2, 4, 8, 16, 32$. We fix our guess by subtracting 1 from the exponent, so the general term is
$$s_n = 2^{n-1}.$$
Substituting the first six values of n shows that the formula checks.

(b) In this sequence, the fifth term looks different from the others, whose numerators are all 7. We can fix this by rewriting $1/2 = 7/14$. The sequence of denominators is then $2, 5, 8, 11, 14, 17$. This looks like a linear function with slope 3, so we expect the denominator has formula $3n + k$ for some k. When $n = 1$, the denominator is 2, so
$$2 = 3\cdot 1 + k \quad \text{giving} \quad k = -1$$
and the denominator of s_n is $3n - 1$. Our general term is then
$$s_n = \frac{7}{3n-1}.$$
To check this, evaluate s_n for $n = 1, \ldots, 6$.

[1]In everyday English, the words "sequence" and "series" are used interchangeably. In mathematics, they have different meanings and cannot be interchanged.

There are two ways to visualize a sequence. One is to plot points with n on the horizontal axis and s_n on the vertical axis. The other is to label points on a number line $s_1, s_2, s_3, \ldots$. See Figure 9.1 for the sequence $s_n = 1 + (-1)^n/n$.

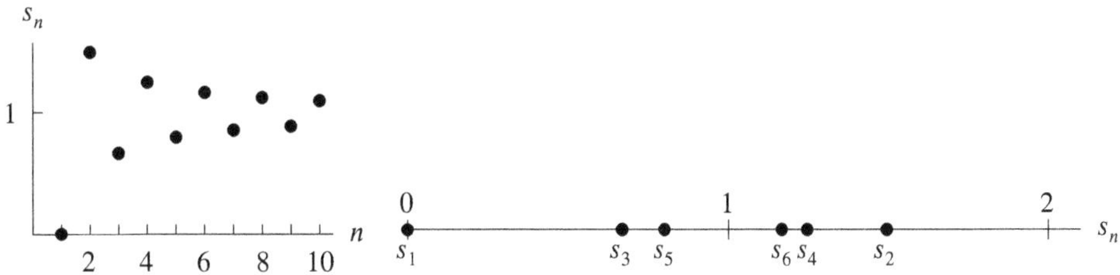

Figure 9.1: The sequence $s_n = 1 + (-1)^n/n$

Defining Sequences Recursively

Sequences can also be defined *recursively*, by giving an equation relating the nth term to the previous terms and as many of the first few terms as are needed to get started.

Example 3 Give the first six terms of the recursively defined sequences.

(a) $s_n = s_{n-1} + 3$ for $n > 1$ and $s_1 = 4$
(b) $s_n = -3s_{n-1}$ for $n > 1$ and $s_1 = 2$
(c) $s_n = \frac{1}{2}(s_{n-1} + s_{n-2})$ for $n > 2$ and $s_1 = 0$, $s_2 = 1$
(d) $s_n = ns_{n-1}$ for $n > 1$ and $s_1 = 1$

Solution (a) When $n = 2$, we obtain $s_2 = s_1 + 3 = 4 + 3 = 7$. When $n = 3$, we obtain $s_3 = s_2 + 3 = 7 + 3 = 10$. In words, we obtain each term by adding 3 to the previous term. The first six terms are

$$4, 7, 10, 13, 16, 19.$$

(b) Each term is -3 times the previous term, starting with $s_1 = 2$. We have $s_2 = -3s_1 = -3 \cdot 2 = -6$ and $s_3 = -3s_2 = -3(-6) = 18$. Continuing, we get

$$2, -6, 18, -54, 162, -486.$$

(c) Each term is the average of the previous two terms, starting with $s_1 = 0$ and $s_2 = 1$. We get $s_3 = (s_2 + s_1)/2 = (1 + 0)/2 = 1/2$. Then $s_4 = (s_3 + s_2)/2 = ((1/2) + 1)/2 = 3/4$. Continuing, we get

$$0, 1, \frac{1}{2}, \frac{3}{4}, \frac{5}{8}, \frac{11}{16}.$$

(d) Here $s_2 = 2s_1 = 2 \cdot 1 = 2$ so $s_3 = 3s_2 = 3 \cdot 2 = 6$ and $s_4 = 4s_3 = 4 \cdot 6 = 24$. Continuing gives

$$1, 2, 6, 24, 120, 720.$$

The general term of part (d) of the previous example is given by $s_n = n(n-1)(n-2) \cdots 3 \cdot 2 \cdot 1$, which is denoted $s_n = n!$ and is called n factorial. We define $0! = 1$.

We can also look at the first few terms of a sequence and try to guess a recursive definition by looking for a pattern.

Example 4 Give a recursive definition of the following sequences.

(a) $1, 3, 7, 15, 31, 63, \ldots$ (b) $1, 4, 9, 16, 25, 36, \ldots$

Solution (a) Each term is twice the previous term plus one; for example, $7 = 2 \cdot 3 + 1$ and $63 = 2 \cdot 31 + 1$. Thus, a recursive definition is

$$s_n = 2s_{n-1} + 1 \text{ for } n > 1 \text{ and } s_1 = 1.$$

There are other ways to define the sequence recursively. We might notice, for example, that the differences of consecutive terms are powers of 2. Thus, we could also use

$$s_n = s_{n-1} + 2^{n-1} \quad \text{for } n > 1 \text{ and } s_1 = 1.$$

(b) We recognize the terms as the squares of the positive integers, but we are looking for a recursive definition which relates consecutive terms. We see that

$$\begin{aligned} s_2 &= s_1 + 3 \\ s_3 &= s_2 + 5 \\ s_4 &= s_3 + 7 \\ s_5 &= s_4 + 9, \end{aligned}$$

so the differences between consecutive terms are consecutive odd integers. The difference between s_n and s_{n-1} is $2n - 1$, so a recursive definition is

$$s_n = s_{n-1} + 2n - 1, \text{ for } n > 1 \text{ and } s_1 = 1.$$

Recursively defined sequences, sometimes called recurrence relations, are powerful tools used frequently in computer science, as well as in differential equations. Finding a formula for the general term can be surprisingly difficult.

Convergence of Sequences

The limit of a sequence s_n as $n \to \infty$ is defined the same way as the limit of a function $f(x)$ as $x \to \infty$; see also Problem 80 (available online).

> The sequence $s_1, s_2, s_3, \ldots, s_n, \ldots$ has a **limit** L, written $\lim_{n\to\infty} s_n = L$, if s_n is as close to L as we please whenever n is sufficiently large. If a limit, L, exists, we say the sequence **converges** to its limit L. If no limit exists, we say the sequence **diverges**.

To calculate the limit of a sequence, we use what we know about the limits of functions, including the properties in Theorem 1.2 and the following facts:

- The sequence $s_n = x^n$ converges to 0 if $|x| < 1$ and diverges if $|x| > 1$
- The sequence $s_n = 1/n^p$ converges to 0 if $p > 0$

Example 5 Do the following sequences converge or diverge? If a sequence converges, find its limit.

(a) $s_n = (0.8)^n$ (b) $s_n = \dfrac{1 - e^{-n}}{1 + e^{-n}}$ (c) $s_n = \dfrac{n^2 + 1}{n}$ (d) $s_n = 1 + (-1)^n$

Solution (a) Since $0.8 < 1$, the sequence converges by the first fact and the limit is 0.

(b) Since $e^{-1} < 1$, we have $\lim_{n\to\infty} e^{-n} = \lim_{n\to\infty} (e^{-1})^n = 0$ by the first fact. Thus, using the properties of limits from Section 1.8, we have

$$\lim_{n\to\infty} \frac{1 - e^{-n}}{1 + e^{-n}} = \frac{1 - 0}{1 + 0} = 1.$$

(c) Since $(n^2 + 1)/n$ grows without bound as $n \to \infty$, the sequence s_n diverges.

(d) Since $(-1)^n$ alternates in sign, the sequence alternates between 0 and 2. Thus the sequence s_n diverges, since it does not get close to any fixed value.

Convergence and Bounded Sequences

A sequence s_n is *bounded* if there are numbers K and M such that $K \leq s_n \leq M$ for all terms. If $\lim_{n\to\infty} s_n = L$, then from some point on, the terms are bounded between $L - 1$ and $L + 1$. Thus we have the following fact:

> A convergent sequence is bounded.

On the other hand, a bounded sequence need not be convergent. In Example 5, we saw that $1+(-1)^n$ diverges, but it is bounded between 0 and 2. To ensure that a bounded sequence converges we need to rule out this sort of oscillation. The following theorem gives a condition that ensures convergence for a bounded sequence. A sequence s_n is called *monotone* if it is either increasing, that is $s_n < s_{n+1}$ for all n, or decreasing, that is $s_n > s_{n+1}$ for all n.

> **Theorem 9.1: Convergence of a Monotone, Bounded Sequence**
>
> If a sequence s_n is bounded and monotone, it converges.

To understand this theorem graphically, see Figure 9.2. The sequence s_n is increasing and bounded above by M, so the values of s_n must "pile up" at some number less than or equal to M. This number is the limit.[2]

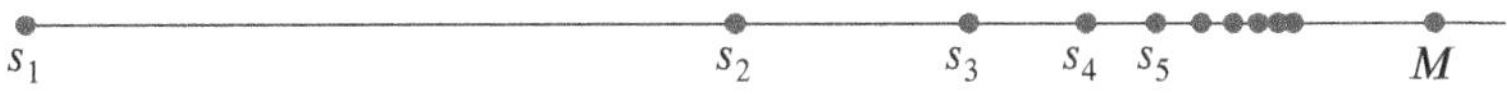

Figure 9.2: Values of s_n for $n = 1, 2, \cdots, 10$

Example 6 The sequence $s_n = (1/2)^n$ is decreasing and bounded below by 0, so it converges. We have already seen that it converges to 0.

Example 7 The sequence $s_n = (1 + 1/n)^n$ can be shown to be increasing and bounded (see Project 2 (available online)). Theorem 9.1 then guarantees that this sequence has a limit, which turns out to be e. (In fact, the sequence can be used to define e.)

Example 8 If $s_n = (1 + 1/n)^n$, find s_{100} and s_{1000}. How many decimal places agree with e?

Solution We have $s_{100} = (1.01)^{100} = 2.7048$ and $s_{1000} = (1.001)^{1000} = 2.7169$. Since $e = 2.7183\ldots$,we see that s_{100} agrees with e to one decimal place and s_{1000} agrees with e to two decimal places.

Exercises and Problems for Section 9.1 Online Resource: Additional Problems for Section 9.1

EXERCISES

■ For Exercises 1–6, find the first five terms of the sequence from the formula for s_n, $n \geq 1$.

1. $2^n + 1$
2. $n + (-1)^n$
3. $\dfrac{2n}{2n+1}$
4. $(-1)^n \left(\dfrac{1}{2}\right)^n$
5. $(-1)^{n+1} \left(\dfrac{1}{2}\right)^{n-1}$
6. $\left(1 - \dfrac{1}{n+1}\right)^{n+1}$

■ In Exercises 7–12, find a formula for s_n, $n \geq 1$.

7. 4, 8, 16, 32, 64, ...
8. 1, 3, 7, 15, 31, ...
9. 2, 5, 10, 17, 26, ...
10. 1, −3, 5, −7, 9, ...
11. 1/3, 2/5, 3/7, 4/9, 5/11, ...
12. 1/2, −1/4, 1/6, −1/8, 1/10, ...

[2]See the online supplement for a proof.

PROBLEMS

13. Match formulas (a)–(d) with graphs (I)–(IV).

(a) $s_n = 1 - 1/n$ **(b)** $s_n = 1 + (-1)^n/n$
(c) $s_n = 1/n$ **(d)** $s_n = 1 + 1/n$

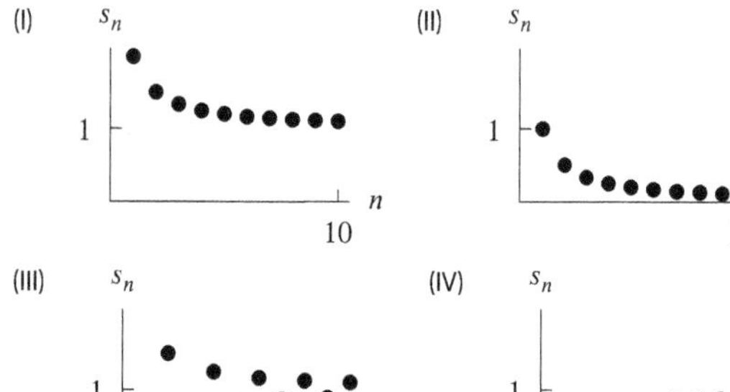

14. Match formulas (a)–(e) with graphs (I)–(V).

(a) $s_n = 2 - 1/n$
(b) $s_n = (-1)^n 2 + 1/n$
(c) $s_n = 2 + (-1)^n/n$
(d) $s_n = 2 + 1/n$
(e) $s_n = (-1)^n 2 + (-1)^n/n$

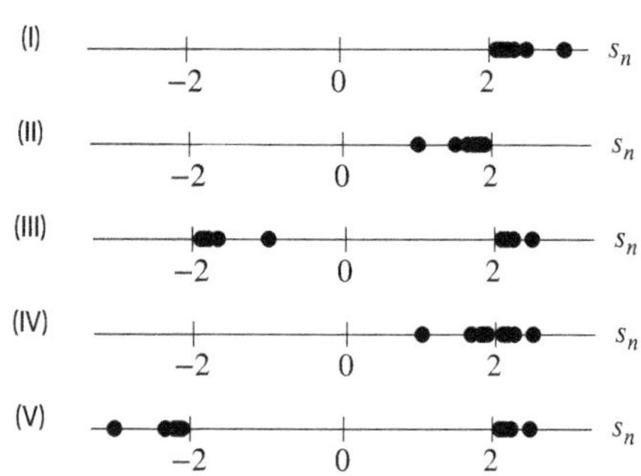

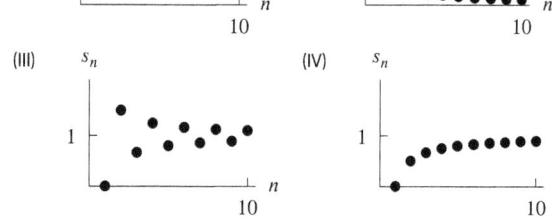

15. Match formulas (a)–(e) with descriptions (I)–(V) of the behavior of the sequence as $n \to \infty$.

(a) $s_n = n(n+1) - 1$
(b) $s_n = 1/(n+1)$
(c) $s_n = 1 - n^2$
(d) $s_n = \cos(1/n)$
(e) $s_n = (\sin n)/n$

(I) Diverges to $-\infty$
(II) Diverges to $+\infty$
(III) Converges to 0 through positive numbers
(IV) Converges to 1
(V) Converges to 0 through positive and negative numbers

■ Do the sequences in Problems 16–27 converge or diverge? If a sequence converges, find its limit.

16. 2^n

17. $(0.2)^n$

18. $3 + e^{-2n}$

19. $(-0.3)^n$

20. $\dfrac{n}{10} + \dfrac{10}{n}$

21. $\dfrac{2^n}{3^n}$

22. $\dfrac{2n+1}{n}$

23. $\dfrac{(-1)^n}{n}$

24. $\dfrac{1}{n} + \ln n$

25. $\dfrac{2n + (-1)^n 5}{4n - (-1)^n 3}$

26. $\dfrac{\sin n}{n}$

27. $\cos(\pi n)$

■ In Problems 28–31, find the first six terms of the recursively defined sequence.

28. $s_n = 2s_{n-1} + 3$ for $n > 1$ and $s_1 = 1$

29. $s_n = s_{n-1} + n$ for $n > 1$ and $s_1 = 1$

30. $s_n = s_{n-1} + \left(\frac{1}{2}\right)^{n-1}$ for $n > 1$ and $s_1 = 0$

31. $s_n = s_{n-1} + 2s_{n-2}$ for $n > 2$ and $s_1 = 1,\ s_2 = 5$

■ In Problems 32–33, let $a_1 = 8, b_1 = 5$, and, for $n > 1$,

$$a_n = a_{n-1} + 3n$$
$$b_n = b_{n-1} + a_{n-1}.$$

32. Give the values of a_2, a_3, a_4.

33. Give the values of b_2, b_3, b_4, b_5.

34. Suppose $s_1 = 0, s_2 = 0, s_3 = 1$, and that $s_n = s_{n-1} + s_{n-2} + s_{n-3}$ for $n \geq 4$. The members of the resulting sequence are called *tribonacci numbers*.[3] Find $s_4, s_5, \ldots, s_{10}$.

■ In Problems 35–40, find a recursive definition for the sequence.

35. $1, 3, 5, 7, 9, \ldots$

36. $2, 4, 6, 8, 10, \ldots$

37. $3, 5, 9, 17, 33, \ldots$

38. $1, 5, 14, 30, 55, \ldots$

39. $1, 3, 6, 10, 15, \ldots$

40. $1, 2, \frac{3}{2}, \frac{5}{3}, \frac{8}{5}, \frac{13}{8}, \ldots$

■ In Problems 41–43, show that the sequence s_n satisfies the recurrence relation.

41. $s_n = 3n - 2$
$s_n = s_{n-1} + 3$ for $n > 1$ and $s_1 = 1$

42. $s_n = n(n+1)/2$
$s_n = s_{n-1} + n$ for $n > 1$ and $s_1 = 1$

43. $s_n = 2n^2 - n$
$s_n = s_{n-1} + 4n - 3$ for $n > 1$ and $s_1 = 1$

[3] http://en.wikipedia.org/wiki/Tribonacci_numbers, accessed March 16, 2011.

In Problems 44–47, use the formula for s_n to give the third term of the sequence, s_3.

44. $s_n = (-1)^n 2^{n-1} \cdot n^2$

45. $s_n = \sum_{k=0}^{n} 3 \cdot 2^k$

46. $s_n = 2 + 3s_{n-1}$, where $s_0 = 5$.

47. $s_n = \int_{1/n}^{1} \frac{1}{x^2}\,dx$

Problems 48–50 concern analog signals in electrical engineering, which are continuous functions $f(t)$, where t is time. To digitize the signal, we sample $f(t)$ every Δt to form the sequence $s_n = f(n\Delta t)$. For example, if $f(t) = \sin t$ with t in seconds, sampling f every $1/10$ second produces the sequence $\sin(1/10), \sin(2/10), \sin(3/10), \ldots$. Give the first 6 terms of a sampling of the signal every Δt seconds.

48. $f(t) = (t-1)^2$, $\Delta t = 0.5$

49. $f(t) = \cos 5t$, $\Delta t = 0.1$

50. $f(t) = \frac{\sin t}{t}$, $\Delta t = 1$

In Problems 51–53, we smooth a sequence, $s_1, s_2, s_3, \ldots$, by replacing each term s_n by t_n, the average of s_n with its neighboring terms

$$t_n = \frac{(s_{n-1} + s_n + s_{n+1})}{3} \quad \text{for } n > 1.$$

Start with $t_1 = (s_1 + s_2)/2$, since s_1 has only one neighbor. Smooth the given sequence once and then smooth the resulting sequence. What do you notice?

51. 18, −18, 18, −18, 18, −18, 18 ...

52. 0, 0, 0, 18, 0, 0, 0, 0 ...

53. 1, 2, 3, 4, 5, 6, 7, 8 ...

In Problems 54–57, for the function f define a sequence recursively by $x_n = f(x_{n-1})$ for $n > 1$ and $x_1 = a$. Depending on f and the starting value a, this sequence may converge to a limit L. If L exists, it has the property that $f(L) = L$. For the functions and starting values given, use a calculator to see if the sequence converges. [To obtain the terms of the sequence, push the function button repeatedly.]

54. $f(x) = \cos x$, $a = 0$

55. $f(x) = e^{-x}$, $a = 0$

56. $f(x) = \sin x$, $a = 1$

57. $f(x) = \sqrt{x}$, $a = 0.5$

58. Let V_n be the number of new SUVs sold in the US in month n, where $n = 1$ is January 2016. In terms of SUVs, what do the following represent?

(a) V_{10}
(b) $V_n - V_{n-1}$
(c) $\sum_{i=1}^{12} V_i$ and $\sum_{i=1}^{n} V_i$

59. **(a)** Let s_n be the number of ancestors a person has n generations ago. (Your ancestors are your parents, grandparents, great-grandparents, etc.) What is s_1? s_2? Find a formula for s_n.
(b) For which n is s_n greater than 6 billion, the current world population? What does this tell you about your ancestors?

60. For $1 \le n \le 10$, find a formula for p_n, the payment in year n on a loan of \$100,000. Interest is 5% per year, compounded annually, and payments are made at the end of each year for ten years. Each payment is \$10,000 plus the interest on the amount of money outstanding.

61. **(a)** Cans are stacked in a triangle on a shelf. The bottom row contains k cans, the row above contains one can fewer, and so on, until the top row, which has one can. How many rows are there? Find a_n, the number of cans in the nth row, $1 \le n \le k$ (where the top row is $n = 1$).
(b) Let T_n be the total number of cans in the top n rows. Find a recurrence relation for T_n in terms of T_{n-1}.
(c) Show that $T_n = \frac{1}{2}n(n+1)$ satisfies the recurrence relation.

62. You are deciding whether to buy a new or a two-year-old car (of the same make) based on which will have cost you less when you resell it at the end of three years. Your cost consists of two parts: the loss in value of the car and the repairs. A new car costs \$20,000 and loses 12% of its value each year. Repairs are \$400 the first year and increase by 18% each subsequent year.

(a) For a new car, find the first three terms of the sequence d_n giving the depreciation (loss of value) in dollars in year n. Give a formula for d_n.
(b) Find the first three terms of the sequence r_n, the repair cost in dollars for a new car in year n. Give a formula for r_n.
(c) Find the total cost of owning a new car for three years.
(d) Find the total cost of owning the two-year-old car for three years. Which should you buy?

63. The Fibonacci sequence, first studied by the thirteenth-century Italian mathematician Leonardo di Pisa, also known as Fibonacci, is defined recursively by

$$F_n = F_{n-1} + F_{n-2} \text{ for } n > 2 \text{ and } F_1 = 1, F_2 = 1.$$

The Fibonacci sequence occurs in many branches of mathematics and can be found in patterns of plant growth (phyllotaxis).

(a) Find the first 12 terms.
(b) Show that the sequence of successive ratios F_{n+1}/F_n appears to converge to a number r satisfying the equation $r^2 = r + 1$. (The number r was known as the golden ratio to the ancient Greeks.)
(c) Let r satisfy $r^2 = r + 1$. Show that the sequence $s_n = Ar^n$, where A is constant, satisfies the Fibonacci equation $s_n = s_{n-1} + s_{n-2}$ for $n > 2$.

64. This problem defines the Calkin-Wilf-Newman sequence of positive rational numbers. The sequence is remarkable because every positive rational number appears as one of its terms and none appears more than once. Every real number x can be written as an integer A plus a number B where $0 \le B < 1$. For example, for $x = 12/5 = 2 + 2/5$ we have $A = 2$ and $B = 2/5$. For $x = 3 = 3 + 0$ we have $A = 3$ and $B = 0$. Define the function $f(x)$ by

$$f(x) = A + (1 - B).$$

For example, $f(12/5) = 2 + (1 - 2/5) = 13/5$ and $f(3) = 3 + (1 - 0) = 4$.

(a) Evaluate $f(x)$ for $x = 25/8$, $13/9$, and π.
(b) Find the first six terms of the recursively defined Calkin-Wilf-Newman sequence: $s_n = 1/f(s_{n-1})$ for $n > 1$ and $s_1 = 1$.

Strengthen Your Understanding

■ In Problems 65–67, explain what is wrong with the statement.

65. The sequence $s_n = \dfrac{3n+10}{7n+3}$, which begins with the terms $\dfrac{13}{10}, \dfrac{16}{17}, \dfrac{19}{24}, \dfrac{22}{31}, \ldots$ converges to 0 because the terms of the sequence get smaller and smaller.

66. A convergent sequence consists entirely of terms greater than 2, then the limit of the sequence is greater than 2.

67. If the convergent sequence has limit L and $s_n < 2$ for all n, then $L < 2$.

■ In Problems 68–69, give an example of:

68. An increasing sequence that converges to 0.

69. A monotone sequence that does not converge.

■ Decide if the statements in Problems 70–78 are true or false. Give an explanation for your answer.

70. You can tell if a sequence converges by looking at the first 1000 terms.

71. If the terms s_n of a convergent sequence are all positive then $\lim_{n\to\infty} s_n$ is positive.

72. If the sequence s_n of positive terms is unbounded, then the sequence has a term greater than a million.

73. If the sequence s_n of positive terms is unbounded, then the sequence has an infinite number of terms greater than a million.

74. If a sequence s_n is convergent, then the terms s_n tend to zero as n increases.

75. A monotone sequence cannot have both positive and negative terms.

76. If a monotone sequence of positive terms does not converge, then it has a term greater than a million.

77. If all terms s_n of a sequence are less than a million, then the sequence is bounded.

78. If a convergent sequence has $s_n \le 5$ for all n, then $\lim_{n\to\infty} s_n \le 5$.

79. Which of the sequences I–IV is monotone and bounded for $n \ge 1$?

I. $s_n = 10 - \dfrac{1}{n}$
II. $s_n = \dfrac{10n+1}{n}$
III. $s_n = \cos n$
IV. $s_n = \ln n$

(a) I
(b) I and II
(c) II and IV
(d) I, II, and III

9.2 GEOMETRIC SERIES

Adding the terms of a sequence produces a *series*. For example, we have the sequence $1, 2, 3, 4, 5, 6, \ldots$ and the series $1 + 2 + 3 + 4 + 5 + 6 + \cdots$. This section introduces infinite series of constants, which are sums of the form

$$1 + \frac{1}{2} + \frac{1}{3} + \frac{1}{4} + \cdots$$
$$0.4 + 0.04 + 0.004 + 0.0004 + \cdots.$$

The individual numbers, $1, \frac{1}{2}, \frac{1}{3}, \ldots$, or $0.4, 0.04, \ldots$, etc., are called *terms* in the series. To talk about the *sum* of the series, we must first explain how to add infinitely many numbers.

Let us look at the repeated administration of a drug. In this example, the terms in the series represent each dose; the sum of the series represents the drug level in the body in the long run.

Repeated Drug Dosage

A person with an ear infection is told to take antibiotic tablets regularly for several days. Since the drug is being excreted by the body between doses, how can we calculate the quantity of the drug remaining in the body at any particular time?

To be specific, let's suppose the drug is ampicillin (a common antibiotic) taken in 250 mg doses four times a day (that is, every six hours). It is known that at the end of six hours, about 4% of the drug is still in the body. What quantity of the drug is in the body right after the tenth tablet? The fortieth?

Let Q_n represent the quantity, in milligrams, of ampicillin in the blood right after the nth tablet. Then

$$Q_1 = 250 \qquad = 250 \text{ mg}$$

$$Q_2 = \underbrace{250(0.04)}_{\text{Remnants of first tablet}} + \underbrace{250}_{\text{New tablet}} \qquad = 260 \text{ mg}$$

$$\begin{aligned} Q_3 &= Q_2(0.04) + 250 = (250(0.04) + 250)(0.04) + 250 \\ &= \underbrace{250(0.04)^2 + 250(0.04)}_{\text{Remnants of first and second tablets}} + \underbrace{250}_{\text{New tablet}} \qquad = 260.4 \text{ mg} \end{aligned}$$

$$\begin{aligned} Q_4 &= Q_3(0.04) + 250 = \left(250(0.04)^2 + 250(0.04) + 250\right)(0.04) + 250 \\ &= \underbrace{250(0.04)^3 + 250(0.04)^2 + 250(0.04) +}_{\text{Remnants of first, second, and third tablets}} \underbrace{250}_{\text{New tablet}} \qquad = 260.416 \text{ mg}. \end{aligned}$$

Looking at the pattern that is emerging, we guess that

$$Q_{10} = 250(0.04)^9 + 250(0.04)^8 + 250(0.04)^7 + \cdots + 250(0.04)^2 + 250(0.04) + 250.$$

Notice that there are 10 terms in this sum—one for every tablet—but that the highest power of 0.04 is the ninth, because no tablet has been in the body for more than 9 six-hour time periods. Now suppose we actually want to find the numerical value of Q_{10}. It seems that we have to add 10 terms—fortunately, there's a better way. Notice the remarkable fact that if you subtract $(0.04)Q_{10}$ from Q_{10}, all but two terms drop out. First, multiplying by 0.04, we get

$$(0.04)Q_{10} = 250(0.04)^{10} + 250(0.04)^9 + 250(0.04)^8 + \cdots + 250(0.04)^3 + 250(0.04)^2 + 250(0.04).$$

Subtracting gives

$$Q_{10} - (0.04)Q_{10} = 250 - 250(0.04)^{10}.$$

Factoring Q_{10} on the left and solving for Q_{10} gives

$$\begin{aligned} Q_{10}(1 - 0.04) &= 250\left(1 - (0.04)^{10}\right) \\ Q_{10} &= \frac{250\left(1 - (0.04)^{10}\right)}{1 - 0.04}. \end{aligned}$$

This is called the *closed-form* expression for Q_{10}. It is easy to evaluate on a calculator, giving $Q_{10} =$ 260.42 (to two decimal places). Similarly, Q_{40} is given in closed form by

$$Q_{40} = \frac{250\left(1 - (0.04)^{40}\right)}{1 - 0.04}.$$

Evaluating this on a calculator shows $Q_{40} = 260.42$, which is the same (to two decimal places) as Q_{10}. Thus after ten tablets, the value of Q_n appears to have stabilized at just over 260 mg.

Looking at the closed forms for Q_{10} and Q_{40}, we can see that, in general, Q_n must be given by

$$Q_n = \frac{250(1 - (0.04)^n)}{1 - 0.04}.$$

What Happens as $n \to \infty$?

What does this closed form for Q_n predict about the long-run level of ampicillin in the body? As $n \to \infty$, the quantity $(0.04)^n \to 0$. In the long run, assuming that 250 mg continue to be taken every six hours, the level right after a tablet is taken is given by

$$Q_n = \frac{250\,(1-(0.04)^n)}{1-0.04} \to \frac{250(1-0)}{1-0.04} = 260.42.$$

The Geometric Series in General

In the previous example we encountered sums of the form $a + ax + ax^2 + \cdots + ax^8 + ax^9$ (with $a = 250$ and $x = 0.04$). Such a sum is called a finite *geometric series*. A geometric series is one in which each term is a constant multiple of the one before. The first term is a, and the constant multiplier, or *common ratio* of successive terms, is x.

A **finite geometric series** has the form

$$a + ax + ax^2 + \cdots + ax^{n-2} + ax^{n-1}.$$

An **infinite geometric series** has the form

$$a + ax + ax^2 + \cdots + ax^{n-2} + ax^{n-1} + ax^n + \cdots.$$

The "$\cdots$" at the end of the second series tells us that the series is going on forever—in other words, that it is infinite.

Sum of a Finite Geometric Series

The same procedure that enabled us to find the closed form for Q_{10} can be used to find the sum of any finite geometric series. Suppose we write S_n for the sum of the first n terms, which means up to the term containing x^{n-1}:

$$S_n = a + ax + ax^2 + \cdots + ax^{n-2} + ax^{n-1}.$$

Multiply S_n by x:

$$xS_n = ax + ax^2 + ax^3 + \cdots + ax^{n-1} + ax^n.$$

Now subtract xS_n from S_n, which cancels out all terms except for two, giving

$$S_n - xS_n = a - ax^n$$
$$(1-x)S_n = a(1-x^n).$$

Provided $x \neq 1$, we can solve to find a closed form for S_n as follows:

The **sum of a finite geometric series** is given by

$$S_n = a + ax + ax^2 + \cdots + ax^{n-1} = \frac{a(1-x^n)}{1-x}, \qquad \text{provided } x \neq 1.$$

Note that the value of n in the formula for S_n is the number of terms in the sum S_n.

Sum of an Infinite Geometric Series

In the ampicillin example, we found the sum Q_n and then let $n \to \infty$. We do the same here. The sum Q_n, which shows the effect of the first n doses, is an example of a *partial sum*. The first three partial sums of the series $a + ax + ax^2 + \cdots + ax^{n-1} + ax^n + \cdots$ are

$$S_1 = a$$
$$S_2 = a + ax$$
$$S_3 = a + ax + ax^2.$$

To find the sum of this infinite series, we consider the partial sum, S_n, of the first n terms. The formula for the sum of a finite geometric series gives

$$S_n = a + ax + ax^2 + \cdots + ax^{n-1} = \frac{a(1-x^n)}{1-x}.$$

What happens to S_n as $n \to \infty$? It depends on the value of x. If $|x| < 1$, then $x^n \to 0$ as $n \to \infty$, so

$$\lim_{n\to\infty} S_n = \lim_{n\to\infty} \frac{a(1-x^n)}{1-x} = \frac{a(1-0)}{1-x} = \frac{a}{1-x}.$$

Thus, provided $|x| < 1$, as $n \to \infty$ the partial sums S_n approach a limit of $a/(1-x)$. When this happens, we define the sum S of the infinite geometric series to be that limit and say the series *converges* to $a/(1-x)$.

> For $|x| < 1$, the **sum of the infinite geometric series** is given by
>
> $$S = a + ax + ax^2 + \cdots + ax^{n-1} + ax^n + \cdots = \frac{a}{1-x}.$$

If, on the other hand, $|x| > 1$, then x^n and the partial sums have no limit as $n \to \infty$ (if $a \neq 0$). In this case, we say the series *diverges*. If $x > 1$, the terms in the series become larger and larger in magnitude, and the partial sums diverge to $+\infty$ (if $a > 0$) or $-\infty$ (if $a < 0$). When $x < -1$, the terms become larger in magnitude, the partial sums oscillate as $n \to \infty$, and the series diverges.

What happens when $x = 1$? The series is

$$a + a + a + a + \cdots,$$

and if $a \neq 0$, the partial sums grow without bound, and the series does not converge. When $x = -1$, the series is

$$a - a + a - a + a - \cdots,$$

and, if $a \neq 0$, the partial sums oscillate between a and 0, and the series does not converge.

Example 1 For each of the following infinite geometric series, find several partial sums and the sum (if it exists).

(a) $1 + \frac{1}{2} + \frac{1}{4} + \frac{1}{8} + \cdots$ (b) $1 + 2 + 4 + 8 + \cdots$ (c) $6 - 2 + \frac{2}{3} - \frac{2}{9} + \frac{2}{27} - \cdots$

Solution (a) This series may be written

$$1 + \frac{1}{2} + \left(\frac{1}{2}\right)^2 + \left(\frac{1}{2}\right)^3 + \cdots$$

which we can identify as a geometric series with $a = 1$ and $x = \frac{1}{2}$, so $S = \dfrac{1}{1-(1/2)} = 2$. Let's check this by finding the partial sums:

$$\begin{aligned}
S_1 &= 1\\
S_2 &= 1 + \frac{1}{2} = \frac{3}{2} = 2 - \frac{1}{2}\\
S_3 &= 1 + \frac{1}{2} + \frac{1}{4} = \frac{7}{4} = 2 - \frac{1}{4}\\
S_4 &= 1 + \frac{1}{2} + \frac{1}{4} + \frac{1}{8} = \frac{15}{8} = 2 - \frac{1}{8}.
\end{aligned}$$

The sequence of partial sums begins

$$1, 2-\frac{1}{2}, 2-\frac{1}{4}, 2-\frac{1}{8}, \ldots.$$

The formula for S_n gives

$$S_n = \frac{1-(\frac{1}{2})^n}{1-\frac{1}{2}} = 2-\left(\frac{1}{2}\right)^{n-1}.$$

Thus, the partial sums are creeping up to the value $S = 2$, so $S_n \to 2$ as $n \to \infty$.

(b) The partial sums of this geometric series (with $a = 1$ and $x = 2$) grow without bound, so the series has no sum:

$$\begin{aligned} S_1 &= 1 \\ S_2 &= 1+2 = 3 \\ S_3 &= 1+2+4 = 7 \\ S_4 &= 1+2+4+8 = 15. \end{aligned}$$

The sequence of partial sums begins

$$1, 3, 7, 15, \ldots.$$

The formula for S_n gives

$$S_n = \frac{1-2^n}{1-2} = 2^n - 1.$$

(c) This is an infinite geometric series with $a = 6$ and $x = -\frac{1}{3}$. The partial sums,

$$S_1 = 6.00, \quad S_2 = 4.00, \quad S_3 \approx 4.67, \quad S_4 \approx 4.44, \quad S_5 \approx 4.52, \quad S_6 \approx 4.49,$$

appear to be converging to 4.5. This turns out to be correct because the sum is

$$S = \frac{6}{1-(-1/3)} = 4.5.$$

Regular Deposits into a Savings Account

People who save money often do so by putting some fixed amount aside regularly. To be specific, suppose \$1000 is deposited every year in a savings account earning 5% a year, compounded annually. What is the balance, B_n, in dollars, in the savings account right after the nth deposit?

As before, let's start by looking at the first few years:

$$\begin{aligned} B_1 &= 1000 \\ B_2 &= B_1(1.05) + 1000 = \underbrace{1000(1.05)}_{\text{Original deposit}} + \underbrace{1000}_{\text{New deposit}} \\ B_3 &= B_2(1.05) + 1000 = \underbrace{1000(1.05)^2 + 1000(1.05)}_{\text{First two deposits}} + \underbrace{1000}_{\text{New deposit}} \end{aligned}$$

Observing the pattern, we see

$$B_n = 1000(1.05)^{n-1} + 1000(1.05)^{n-2} + \cdots + 1000(1.05) + 1000.$$

So B_n is a finite geometric series with $a = 1000$ and $x = 1.05$. Thus we have

$$B_n = \frac{1000\,(1-(1.05)^n)}{1-1.05}.$$

We can rewrite this so that both the numerator and denominator of the fraction are positive:

$$B_n = \frac{1000\,((1.05)^n-1)}{1.05-1}.$$

What Happens as $n \to \infty$?

Common sense tells you that if you keep depositing \$1000 in an account and it keeps earning interest, your balance grows without bound. This is what the formula for B_n shows also: $(1.05)^n \to \infty$ as $n \to \infty$, so B_n has no limit. (Alternatively, observe that the infinite geometric series of which B_n is a partial sum has $x = 1.05$, which is greater than 1, so the series does not converge.)

Exercises and Problems for Section 9.2

EXERCISES

In Exercises 1–7, is a sequence or a series given?

1. $2^2+4^2+6^2+8^2+\cdots$
2. $2^2, 4^2, 6^2, 8^2, \ldots$
3. $1, -2, 3, -4, 5, \ldots$
4. $1+2, 3+4, 5+6, 7+8, \ldots$
5. $1-2+3-4+5-\cdots$
6. $1+2+3+4+5+6+7+8+\cdots$
7. $-S_1+S_2-S_3+S_4-S_5+\cdots$

In Exercises 8–18, decide which of the following are geometric series. For those which are, give the first term and the ratio between successive terms. For those which are not, explain why not.

8. $5-10+20-40+80-\cdots$
9. $2+1+\frac{1}{2}+\frac{1}{4}+\frac{1}{8}+\cdots$
10. $1+\frac{1}{2}+\frac{1}{3}+\frac{1}{4}+\frac{1}{5}+\cdots$
11. $1+x+2x^2+3x^3+4x^4+\cdots$
12. $1-\frac{1}{2}+\frac{1}{4}-\frac{1}{8}+\frac{1}{16}+\cdots$
13. $3+3z+6z^2+9z^3+12z^4+\cdots$
14. $1+2z+(2z)^2+(2z)^3+\cdots$
15. $1-y^2+y^4-y^6+\cdots$
16. $1-x+x^2-x^3+x^4-\cdots$
17. $z^2-z^4+z^8-z^{16}+\cdots$
18. $y^2+y^3+y^4+y^5+\cdots$

In Exercises 19–23, state whether or not the series is geometric. If it is geometric and converges, find the sum of the series.

19. $\frac{1}{3}+\frac{4}{9}+\frac{16}{27}+\frac{64}{81}+\cdots$
20. $\frac{3}{2}-\frac{1}{2}+\frac{1}{6}-\frac{1}{18}+\frac{1}{54}+\cdots$
21. $\frac{1}{2}+\frac{2}{3}+\frac{4}{5}+\frac{5}{6}+\cdots$
22. $\sum_{n=2}^{\infty}\frac{(-1)^n}{2^{2n}}$
23. $\sum_{n=1}^{\infty}\frac{1}{\sqrt{n}}$

In Exercises 24–27, say how many terms are in the finite geometric series and find its sum.

24. $2+2(0.1)+2(0.1)^2+\cdots+2(0.1)^{25}$
25. $2(0.1)+2(0.1)^2+\cdots+2(0.1)^{10}$
26. $2(0.1)^5+2(0.1)^6+\cdots+2(0.1)^{13}$
27. $8+4+2+1+\frac{1}{2}+\cdots+\frac{1}{2^{10}}$

In Exercises 28–30, find the sum of the infinite geometric series.

28. $36+12+4+\frac{4}{3}+\frac{4}{9}+\cdots$
29. $-810+540-360+240-160+\cdots$
30. $80+\frac{80}{\sqrt{2}}+40+\frac{40}{\sqrt{2}}+20+\frac{20}{\sqrt{2}}+\cdots$

In Exercises 31–34, find a closed-form for the geometric series and determine for which values of x it converges.

31. $\sum_{n=0}^{\infty}(5x)^n$
32. $\sum_{n=0}^{\infty}(1-x)^n$
33. $\sum_{n=0}^{\infty}\frac{(1-x/2)^n}{2}$
34. $\sum_{n=0}^{\infty}\frac{(1-x/3)^n}{3}$

PROBLEMS

■ In Problems 35–42, find the sum of the series. For what values of the variable does the series converge to this sum?

35. $1 + z/2 + z^2/4 + z^3/8 + \cdots$

36. $1 + 3x + 9x^2 + 27x^3 + \cdots$

37. $y - y^2 + y^3 - y^4 + \cdots$

38. $2 - 4z + 8z^2 - 16z^3 + \cdots$

39. $3 + x + x^2 + x^3 + \cdots$

40. $4 + y + y^2/3 + y^3/9 + \cdots$

41. $8 + 8\left(x^2 - 5\right) + 8\left(x^2 - 5\right)^2 + 8\left(x^2 - 5\right)^3 + \cdots$

42. $5 + 5\left(z^3 - 8\right) + 5\left(z^3 - 8\right)^2 + 5\left(z^3 - 8\right)^3 + \cdots$

■ In Problems 43–45, for the given value of x determine whether the infinite geometric series converges. If so, find its sum:

$$3 + 3\cos x + 3(\cos x)^2 + 3(\cos x)^3 + \cdots.$$

43. $x = 0$ 44. $x = 2\pi/3$ 45. $x = \pi$

46. Bill invests \$200 at the start of each month for 24 months, starting now. If the investment yields 0.5% per month, compounded monthly, what is its value at the end of 24 months?

47. Once a day, eight tons of pollutants are dumped into a bay. Of this, 25% is removed by natural processes each day. What happens to the quantity of pollutants in the bay over time? Give the long-run quantity right after a dump.

48. **(a)** The total reserves of a non-renewable resource are 400 million tons. Annual consumption, currently 25 million tons per year, is expected to rise by 1% each year. After how many years will the reserves be exhausted?
(b) Instead of increasing by 1% per year, suppose consumption was decreasing by a constant percentage per year. If existing reserves are never to be exhausted, what annual percentage reduction in consumption is required?

49. A repeating decimal can always be expressed as a fraction. This problem shows how writing a repeating decimal as a geometric series enables you to find the fraction.
(a) Write the repeating decimal 0.232323... as a geometric series using the fact that $0.232323\ldots = 0.23 + 0.0023 + 0.000023 + \cdots$.
(b) Use the formula for the sum of a geometric series to show that $0.232323\ldots = 23/99$.

50. One way of valuing a company is to calculate the present value of all its future earnings. A farm expects to sell \$1000 worth of Christmas trees once a year forever, with the first sale in the immediate future. What is the present value of this Christmas tree business? The interest rate is 1% per year, compounded continuously.

51. Around January 1, 1993, Barbra Streisand signed a contract with Sony Corporation for \$2 million a year for 10 years. Suppose the first payment was made on the day of signing and that all other payments were made on the first day of the year. Suppose also that all payments were made into a bank account earning 4% a year, compounded annually.
(a) How much money was in the account
(i) On the night of December 31, 1999?
(ii) On the day the last payment was made?
(b) What was the present value of the contract on the day it was signed?

52. Peter wishes to create a retirement fund from which he can draw \$20,000 when he retires and the same amount at each anniversary of his retirement for 10 years. He plans to retire 20 years from now. What investment need he make today if he can get a return of 5% per year, compounded annually?

53. In 2013, the quantity of copper mined worldwide was 17.9 million tonnes and had been increasing by 0.025% annually. By the end of that year, the world's known copper reserves were 690 million tonnes.[4]
(a) Write and sum a series giving the total quantity of copper mined in the n years since 2013. Assume the quantity mined each year continues to increase at 0.025% per year, with the first increase in 2014.
(b) In what year are the reserves exhausted?
(c) How does the sum constructed in part (a) relate to the integral $\int_0^n 17.9(1.025)^t\,dt$?

54. This problem shows another way of deriving the long-run ampicillin level. (See page 481.) In the long run the ampicillin levels off to Q mg right after each tablet is taken. Six hours later, right before the next dose, there will be less ampicillin in the body. However, if stability has been reached, the amount of ampicillin that has been excreted is exactly 250 mg because taking one more tablet raises the level back to Q mg. Use this to solve for Q.

55. On page 481, you saw how to compute the quantity Q_n mg of ampicillin in the body right after the nth tablet of 250 mg, taken once every six hours.
(a) Do a similar calculation for P_n, the quantity of ampicillin (in mg) in the body right *before* the nth tablet is taken.

[4]Data from http://minerals.usgs.gov/minerals/pubs/commodity, accessed February 8, 2015.

(b) Express P_n in closed form.
(c) What is $\lim_{n\to\infty} P_n$? Is this limit the same as $\lim_{n\to\infty} Q_n$? Explain in practical terms why your answer makes sense.

56. Figure 9.3 shows the quantity of the drug atenolol in the blood as a function of time, with the first dose at time $t = 0$. Atenolol is taken in 50 mg doses once a day to lower blood pressure.

(a) If the half-life of atenolol in the blood is 6.3 hours, what percentage of the atenolol present at the start of a 24-hour period is still there at the end?
(b) Find expressions for the quantities $Q_0, Q_1, Q_2, Q_3, \ldots$, and Q_n shown in Figure 9.3. Write the expression for Q_n in closed form.
(c) Find expressions for the quantities $P_1, P_2, P_3, \ldots$, and P_n shown in Figure 9.3. Write the expression for P_n in closed form.

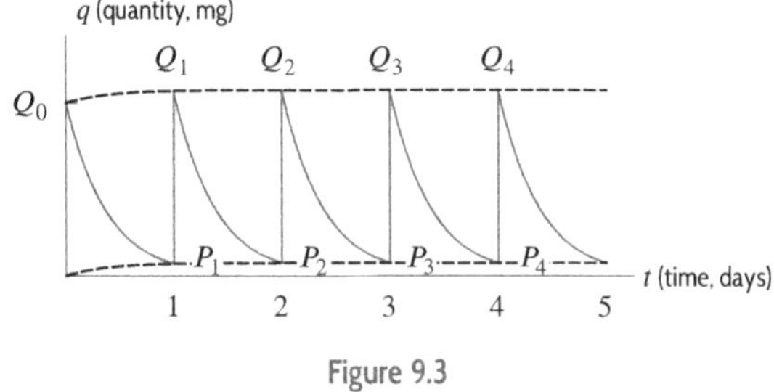

Figure 9.3

57. Draw a graph like that in Figure 9.3 for 250 mg of ampicillin taken every 6 hours, starting at time $t = 0$. Put on the graph the values of $Q_1, Q_2, Q_3, \ldots$ introduced in the text on page 481 and the values of $P_1, P_2, P_3, \ldots$ calculated in Problem 55.

58. In theory, drugs that decay exponentially always leave a residue in the body. However, in practice, once the drug has been in the body for 5 half-lives, it is regarded as being eliminated.[5] If a patient takes a tablet of the same drug every 5 half-lives forever, what is the upper limit to the amount of drug that can be in the body?

59. This problem shows how to estimate the cumulative effect of a tax cut on a country's economy. Suppose the government proposes a tax cut totaling \$100 million. We assume that all the people who have extra money spend 80% of it and save 20%. Thus, of the extra income generated by the tax cut, \$100(0.8) million = \$80 million is spent and becomes extra income to someone else. These people also spend 80% of their additional income, or \$80(0.8) million, and so on. Calculate the total additional spending created by such a tax cut.

60. The government proposes a tax cut of \$100 million as in Problem 59, but that economists now predict that people will spend 90% of their extra income and save only 10%. How much additional spending would be generated by the tax cut under these assumptions?

61. (a) What is the present value of a \$1000 bond which pays \$50 a year for 10 years, starting one year from now? Assume the interest rate is 5% per year, compounded annually.
(b) Since \$50 is 5% of \$1000, this bond is called a 5% bond. What does your answer to part (a) tell you about the relationship between the principal and the present value of this bond if the interest rate is 5%?
(c) If the interest rate is more than 5% per year, compounded annually, which is larger: the principal or the present value of the bond? Why is the bond then described as *trading at a discount*?
(d) If the interest rate is less than 5% per year, compounded annually, why is the bond described as *trading at a premium*?

62. A ball is dropped from a height of 10 feet and bounces. Each bounce is $\frac{3}{4}$ of the height of the bounce before. Thus, after the ball hits the floor for the first time, the ball rises to a height of $10(\frac{3}{4}) = 7.5$ feet, and after it hits the floor for the second time, it rises to a height of $7.5(\frac{3}{4}) = 10(\frac{3}{4})^2 = 5.625$ feet. (Assume that there is no air resistance.)

(a) Find an expression for the height to which the ball rises after it hits the floor for the nth time.
(b) Find an expression for the total vertical distance the ball has traveled when it hits the floor for the first, second, third, and fourth times.
(c) Find an expression for the total vertical distance the ball has traveled when it hits the floor for the nth time. Express your answer in closed form.

63. You might think that the ball in Problem 62 keeps bouncing forever since it takes infinitely many bounces. This is not true!

(a) Show that a ball dropped from a height of h feet reaches the ground in $\frac{1}{4}\sqrt{h}$ seconds. (Assume $g = 32$ ft/sec^2.)
(b) Show that the ball in Problem 62 stops bouncing after

$$\frac{1}{4}\sqrt{10}+\frac{1}{2}\sqrt{10}\sqrt{\frac{3}{4}}\left(\frac{1}{1-\sqrt{3/4}}\right) \approx 11 \text{ seconds.}$$

[5] http://dr.pierce1.net/PDF/half_life.pdf, accessed on May 10, 2003.

Strengthen Your Understanding

In Problems 64–66, explain what is wrong with the statement.

64. The sequence $4, 1, \frac{1}{4}, \frac{1}{16}, \dots$ converges to $\frac{4}{1-1/4} = \frac{16}{3}$.

65. The sum of the infinite geometric series $1 - \frac{3}{2} + \frac{9}{4} - \frac{27}{8} + \cdots$ is $\frac{1}{1+3/2} = \frac{2}{5}$.

66. The following series is convergent:

$$0.000001 + 0.00001 + 0.0001 + 0.001 + \cdots.$$

In Problems 67–72, give an example of:

67. A geometric series that does not converge.

68. A geometric series in which a term appears more than once.

69. A finite geometric series with four distinct terms whose sum is 10.

70. An infinite geometric series that converges to 10.

71. Two geometric series whose sum is geometric.

72. Two geometric series whose sum is not geometric.

73. Which of the following geometric series converge?

(I) $20 - 10 + 5 - 2.5 + \cdots$
(II) $1 - 1.1 + 1.21 - 1.331 + \cdots$
(III) $1 + 1.1 + 1.21 + 1.331 + \cdots$
(IV) $1 + y^2 + y^4 + y^6 + \cdots$, for $-1 < y < 1$

(a) (I) only
(b) (IV) only
(c) (I) and (IV)
(d) (II) and (IV)
(e) None of the other choices is correct.

9.3 CONVERGENCE OF SERIES

We now consider general series in which each term a_n is a number. The series can be written compactly using a $\sum$ sign as follows:

$$\sum_{n=1}^{\infty} a_n = a_1 + a_2 + a_3 + \cdots + a_n + \cdots.$$

For any values of a and x, the geometric series is such a series, with general term $a_n = ax^{n-1}$.

Partial Sums and Convergence of Series

As in Section 9.2, we define the *partial sum*, S_n, of the first n terms of a series as

$$S_n = \sum_{i=1}^{n} a_i = a_1 + a_2 + \cdots + a_n.$$

To investigate the convergence of the series, we consider the sequence of partial sums

$$S_1, S_2, S_3, \dots, S_n, \dots.$$

If S_n has a limit as $n \to \infty$, then we define the sum of the series to be that limit.

> If the sequence S_n of partial sums converges to S, so $\lim_{n\to\infty} S_n = S$, then we say the series $\sum_{n=1}^{\infty} a_n$ **converges** and that its sum is S. We write $\sum_{n=1}^{\infty} a_n = S$. If $\lim_{n\to\infty} S_n$ does not exist, we say that the series **diverges**.

The following example shows how a series leads to a sequence of partial sums and how we use them to determine convergence.

Example 1 Investigate the convergence of the series with $a_n = 1/(n(n+1))$:

$$\sum_{n=1}^{\infty} a_n = \frac{1}{2} + \frac{1}{6} + \frac{1}{12} + \frac{1}{20} + \cdots.$$

Solution In order to determine whether the series converges, we first find the partial sums:

$$\begin{aligned}
S_1 &= \frac{1}{2} \\
S_2 &= \frac{1}{2} + \frac{1}{6} = \frac{2}{3} \\
S_3 &= \frac{1}{2} + \frac{1}{6} + \frac{1}{12} = \frac{3}{4} \\
S_4 &= \frac{1}{2} + \frac{1}{6} + \frac{1}{12} + \frac{1}{20} = \frac{4}{5} \\
&\vdots
\end{aligned}$$

It appears that $S_n = n/(n+1)$ for each positive integer n. We check that this pattern continues by assuming that $S_n = n/(n+1)$ for a given integer n, adding a_{n+1}, and simplifying

$$S_{n+1} = S_n + a_{n+1} = \frac{n}{n+1} + \frac{1}{(n+1)(n+2)} = \frac{n^2+2n+1}{(n+1)(n+2)} = \frac{n+1}{n+2}.$$

Thus the sequence of partial sums has formula $S_n = n/(n+1)$, which converges to 1, so the series $\sum_{n=1}^{\infty} a_n$ converges to 1. That is, we can say that

$$\frac{1}{2} + \frac{1}{6} + \frac{1}{12} + \frac{1}{20} + \cdots = 1.$$

Visualizing Series

We can visualize the terms of the series in Example 1 as the heights of the bars in Figure 9.4. The partial sums of the series are illustrated by stacking the bars on top of each other in Figure 9.5.

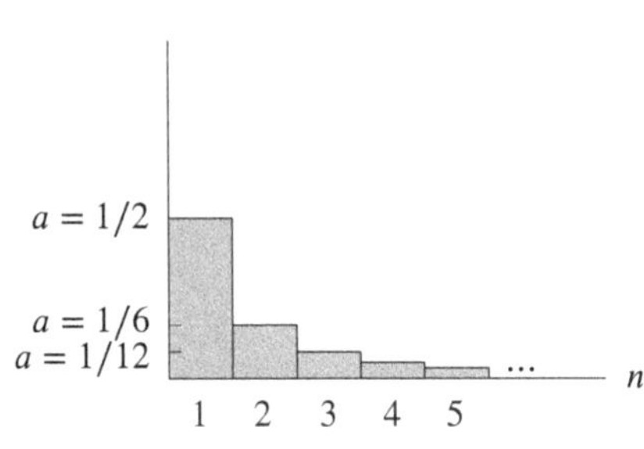

Figure 9.4: Terms of the series with $a_n = 1/(n(n+1))$

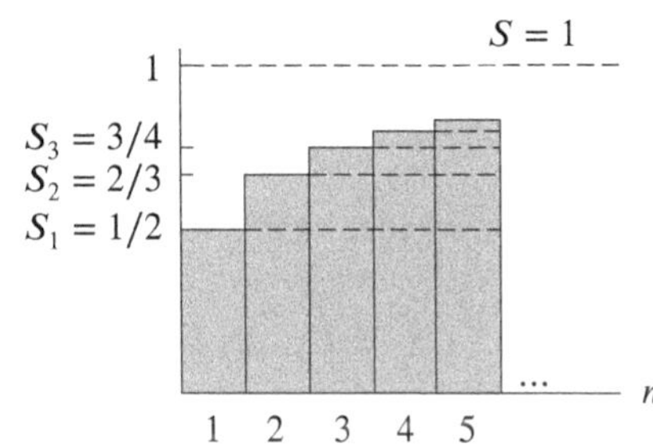

Figure 9.5: Partial sums of the series with $a_n = 1/(n(n+1))$

Here are some properties that are useful in determining whether or not a series converges.

Theorem 9.2: Convergence Properties of Series

1. If $\sum_{n=1}^{\infty} a_n$ and $\sum_{n=1}^{\infty} b_n$ converge and if k is a constant, then
   - $\sum_{n=1}^{\infty}(a_n + b_n)$ converges to $\sum_{n=1}^{\infty} a_n + \sum_{n=1}^{\infty} b_n$.
   - $\sum_{n=1}^{\infty} ka_n$ converges to $k\sum_{n=1}^{\infty} a_n$.
2. Changing a finite number of terms in a series does not change whether or not it converges, although it may change the value of its sum if it does converge.
3. If $\lim_{n\to\infty} a_n \neq 0$ or $\lim_{n\to\infty} a_n$ does not exist, then $\sum_{n=1}^{\infty} a_n$ diverges.
4. If $\sum_{n=1}^{\infty} a_n$ diverges, then $\sum_{n=1}^{\infty} ka_n$ diverges if $k \neq 0$.

For proofs of these properties, see Problems 39–42. As with improper integrals, the convergence of a series is determined by its behavior for large n. (See the "behaves like" principle on page 401.) From Property 2 we see that, if N is a positive integer, then $\sum_{n=1}^{\infty} a_n$ and $\sum_{n=N}^{\infty} a_n$ either both converge or both diverge. Thus, if all we care about is the convergence of a series, we can omit the limits and write $\sum a_n$.

Example 2 Does the series $\sum(1 - e^{-n})$ converge?

Solution Since the terms in the series $a_n = 1 - e^{-n}$ tend to 1, not 0, as $n \to \infty$, the series diverges by Property 3 of Theorem 9.2.

Comparison of Series and Integrals

We investigate the convergence of some series by comparison with an improper integral. The *harmonic series* is the infinite series

$$\sum_{n=1}^{\infty} \frac{1}{n} = 1 + \frac{1}{2} + \frac{1}{3} + \frac{1}{4} + \cdots + \frac{1}{n} + \cdots.$$

Convergence of this sum would mean that the sequence of partial sums

$$S_1 = 1, \quad S_2 = 1 + \frac{1}{2}, \quad S_3 = 1 + \frac{1}{2} + \frac{1}{3}, \quad \ldots, \quad S_n = 1 + \frac{1}{2} + \frac{1}{3} + \cdots + \frac{1}{n}, \quad \ldots$$

tends to a limit as $n \to \infty$. Let's look at some values:

$$S_1 = 1, \quad S_{10} \approx 2.93, \quad S_{100} \approx 5.19, \quad S_{1000} \approx 7.49, \quad S_{10000} \approx 9.79.$$

The growth of these partial sums is slow, but they do in fact grow without bound, so the harmonic series diverges. This is justified in the following example and in Problem 60 (available online).

Example 3 Show that the harmonic series $1 + 1/2 + 1/3 + 1/4 + \cdots$ diverges.

Solution The idea is to approximate $\int_1^\infty (1/x)\,dx$ by a left-hand sum, where the terms $1, 1/2, 1/3, \ldots$ are heights of rectangles of base 1. In Figure 9.6, the sum of the areas of the 3 rectangles is larger than the area under the curve between $x = 1$ and $x = 4$, and the same kind of relationship holds for the first n rectangles. Thus, we have

$$S_n = 1 + \frac{1}{2} + \frac{1}{3} + \cdots + \frac{1}{n} > \int_1^{n+1} \frac{1}{x}\,dx = \ln(n+1).$$

Since $\ln(n+1)$ gets arbitrarily large as $n \to \infty$, so do the partial sums, S_n. Thus, the partial sums have no limit, so the series diverges.

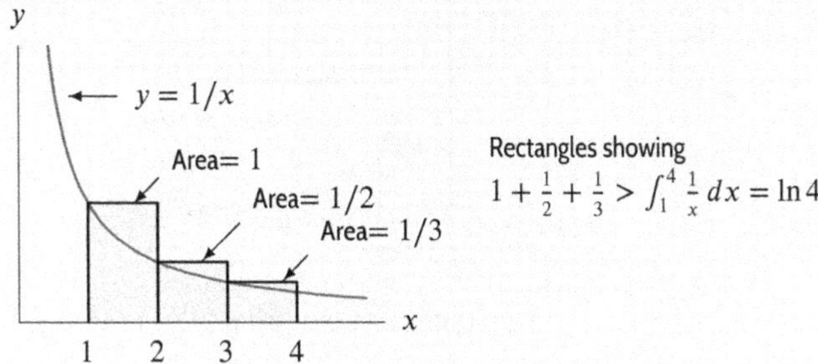

Figure 9.6: Comparing the harmonic series to $\int_1^\infty (1/x)\,dx$

Notice that the harmonic series diverges, even though $\lim_{n\to\infty} a_n = \lim_{n\to\infty} (1/n) = 0$. Although Property 3 of Theorem 9.2 guarantees $\sum a_n$ diverges if $\lim_{n\to\infty} a_n \neq 0$, it is possible for $\sum a_n$ to either converge or diverge if $\lim_{n\to\infty} a_n = 0$. When we have $\lim_{n\to\infty} a_n = 0$, we must investigate the series further to determine whether it converges or diverges.

Example 4 By comparison with the improper integral $\int_1^\infty (1/x^2)\,dx$, show that the following series converges:

$$\sum_{n=1}^{\infty} \frac{1}{n^2} = 1 + \frac{1}{4} + \frac{1}{9} + \cdots.$$

Solution Since we want to show that $\sum_{n=1}^{\infty} 1/n^2$ converges, we want to show that the partial sums of this series tend to a limit. We do this by showing that the sequence of partial sums increases and is bounded above, so Theorem 9.1 applies.

Each successive partial sum is obtained from the previous one by adding one more term in the series. Since all the terms are positive, the sequence of partial sums is increasing.

To show that the partial sums of $\sum_{n=1}^{\infty} 1/n^2$ are bounded, we consider the right-hand sum represented by the area of the rectangles in Figure 9.7. We start at $x = 1$, since the area under the curve is infinite for $0 \le x \le 1$. The shaded rectangles in Figure 9.7 suggest that:

$$\frac{1}{4} + \frac{1}{9} + \frac{1}{16} + \cdots + \frac{1}{n^2} \le \int_1^\infty \frac{1}{x^2}\,dx.$$

The area under the graph is finite, since

$$\int_1^\infty \frac{1}{x^2}\,dx = \lim_{b\to\infty} \int_1^b \frac{1}{x^2}\,dx = \lim_{b\to\infty} \left(-\frac{1}{b} + 1\right) = 1.$$

To get S_n, we add 1 to both sides, giving

$$S_n = 1 + \frac{1}{4} + \frac{1}{9} + \frac{1}{16} + \cdots + \frac{1}{n^2} \le 1 + \int_1^{\infty} \frac{1}{x^2}\,dx = 2.$$

Thus, the increasing sequence of partial sums is bounded above by 2. Hence, by Theorem 9.1 the sequence of partial sums converges, so the series converges.

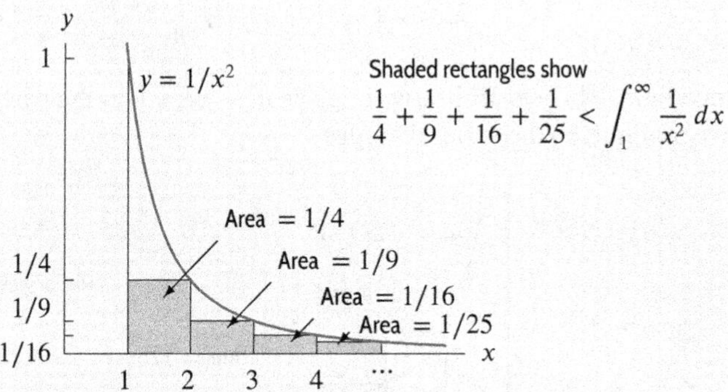

Figure 9.7: Comparing $\sum_{n=1}^{\infty} 1/n^2$ to $\int_1^{\infty}(1/x^2)\,dx$

Notice that we have shown that the series in the Example 4 converges, but we have not found its sum. The integral gives us a bound on the partial sums, but it does not give us the limit of the partial sums. Euler proved the remarkable fact that the sum is $\pi^2/6$.

The method of Examples 3 and 4 can be used to prove the following theorem. See Problem 59 (available online).

Theorem 9.3: The Integral Test

Suppose $a_n = f(n)$, where $f(x)$ is decreasing and positive.

- If $\int_1^{\infty} f(x)\,dx$ converges, then $\sum a_n$ converges.
- If $\int_1^{\infty} f(x)\,dx$ diverges, then $\sum a_n$ diverges.

Suppose $f(x)$ is continuous. Then if $f(x)$ is positive and decreasing for all x beyond some point, say c, the integral test can be used.

The integral test allows us to analyze a family of series, the p-series, and see how convergence depends on the parameter p.

Example 5 For what values of p does the series $\sum_{n=1}^{\infty} 1/n^p$ converge?

Solution If $p \le 0$, the terms in the series $a_n = 1/n^p$ do not tend to 0 as $n \to \infty$. Thus the series diverges for $p \le 0$.

If $p > 0$, we compare $\sum_{n=1}^{\infty} 1/n^p$ to the integral $\int_1^{\infty} 1/x^p\,dx$. In Example 3 of Section 7.6 we saw that the integral converges if $p > 1$ and diverges if $p \le 1$. By the integral test, we conclude that $\sum 1/n^p$ converges if $p > 1$ and diverges if $p \le 1$.

We can summarize Example 5 as follows:

> The **p-series** $\sum_{n=1}^{\infty} 1/n^p$ converges if $p > 1$ and diverges if $p \leq 1$.

Exercises and Problems for Section 9.3 Online Resource: Additional Problems for Section 9.3

EXERCISES

In Exercises 1–3, find the first five terms of the sequence of partial sums.

1. $\sum_{n=1}^{\infty} n$
2. $\sum_{n=1}^{\infty} \frac{(-1)^n}{n}$
3. $\sum_{n=1}^{\infty} \frac{1}{n(n+1)}$

In Exercises 4–7, use the integral test to decide whether the series converges or diverges.

4. $\sum_{n=1}^{\infty} \frac{1}{(n+2)^2}$
5. $\sum_{n=1}^{\infty} \frac{n}{n^2+1}$
6. $\sum_{n=1}^{\infty} \frac{1}{e^n}$
7. $\sum_{n=2}^{\infty} \frac{1}{n(\ln n)^2}$

8. Use comparison with $\int_1^{\infty} x^{-3}\,dx$ to show that $\sum_{n=2}^{\infty} 1/n^3$ converges to a number less than or equal to $1/2$.

9. Use comparison with $\int_0^{\infty} 1/(x^2+1)\,dx$ to show that $\sum_{n=1}^{\infty} 1/(n^2+1)$ converges to a number less than or equal to $\pi/2$.

In Exercises 10–12, explain why the integral test cannot be used to decide if the series converges or diverges.

10. $\sum_{n=1}^{\infty} n^2$
11. $\sum_{n=1}^{\infty} \frac{(-1)^n}{n}$
12. $\sum_{n=1}^{\infty} e^{-n}\sin n$

PROBLEMS

In Problems 13–32, does the series converge or diverge?

13. $\sum_{n=0}^{\infty} \frac{3}{n+2}$
14. $\sum_{n=0}^{\infty} \frac{4}{2n+1}$
15. $\sum_{n=0}^{\infty} \frac{2}{\sqrt{2+n}}$
16. $\sum_{n=0}^{\infty} \frac{2n}{1+n^4}$
17. $\sum_{n=0}^{\infty} \frac{2n}{(1+n^2)^2}$
18. $\sum_{n=0}^{\infty} \frac{2n}{\sqrt{4+n^2}}$
19. $\sum_{n=1}^{\infty} \frac{3}{(2n-1)^2}$
20. $\sum_{n=1}^{\infty} \frac{4}{(2n+1)^3}$
21. $\sum_{n=0}^{\infty} \frac{3}{n^2+4}$
22. $\sum_{n=0}^{\infty} \frac{2}{1+4n^2}$
23. $\sum_{n=1}^{\infty} \frac{n}{n+1}$
24. $\sum_{n=0}^{\infty} \frac{n+1}{2n+3}$
25. $\sum_{n=1}^{\infty} \left(\left(\frac{1}{2}\right)^n + \left(\frac{2}{3}\right)^n\right)$
26. $\sum_{n=1}^{\infty} \left(\left(\frac{3}{4}\right)^n + \frac{1}{n}\right)$
27. $\sum_{n=1}^{\infty} \frac{n+2^n}{n2^n}$
28. $\sum_{n=1}^{\infty} \frac{\ln n}{n}$
29. $\sum_{n=1}^{\infty} \frac{1}{n(1+\ln n)}$
30. $\sum_{n=3}^{\infty} \frac{n+1}{n^2+2n+2}$
31. $\sum_{n=0}^{\infty} \frac{1}{n^2+2n+2}$
32. $\sum_{n=2}^{\infty} \frac{n\ln n+4}{n^2}$

33. Show that $\sum_{n=1}^{\infty} \frac{1}{\ln(2^n)}$ diverges.

34. Show that $\sum_{n=1}^{\infty} \frac{1}{(\ln(2^n))^2}$ converges.

35. **(a)** Find the partial sum, S_n, of $\sum_{n=1}^{\infty} \ln\left(\frac{n+1}{n}\right)$.
    (b) Does the series in part (a) converge or diverge?

36. **(a)** Show $r^{\ln n} = n^{\ln r}$ for positive numbers n and r.
    (b) For what values $r > 0$ does $\sum_{n=1}^{\infty} r^{\ln n}$ converge?

37. Consider the series $\sum_{k=1}^{\infty} \frac{1}{k(k+1)} = \frac{1}{1\cdot 2} + \frac{1}{2\cdot 3} + \cdots$.
    (a) Show that $\frac{1}{k} - \frac{1}{k+1} = \frac{1}{k(k+1)}$.
    (b) Use part (a) to find the partial sums S_3, S_{10}, and S_n.
    (c) Use part (b) to show that the sequence of partial sums S_n, and therefore the series, converges to 1.

38. Consider the series

$$\sum_{k=1}^{\infty} \ln\left(\frac{k(k+2)}{(k+1)^2}\right) = \ln\left(\frac{1\cdot 3}{2\cdot 2}\right) + \ln\left(\frac{2\cdot 4}{3\cdot 3}\right) + \cdots.$$

(a) Show that the partial sum of the first three nonzero terms $S_3 = \ln(5/8)$.

(b) Show that the partial sum $S_n = \ln\left(\frac{n+2}{2(n+1)}\right)$.

(c) Use part (b) to show that the partial sums S_n, and therefore the series, converge to $\ln(1/2)$.

39. Show that if $\sum a_n$ and $\sum b_n$ converge and if k is a constant, then $\sum(a_n+b_n)$, $\sum(a_n-b_n)$, and $\sum ka_n$ converge.

40. Let N be a positive integer. Show that if $a_n = b_n$ for $n \geq N$, then $\sum a_n$ and $\sum b_n$ either both converge, or both diverge.

41. Show that if $\sum a_n$ converges, then $\lim_{n\to\infty} a_n = 0$. [Hint: Consider $\lim_{n\to\infty}(S_n - S_{n-1})$, where S_n is the nth partial sum.]

42. Show that if $\sum a_n$ diverges and $k \neq 0$, then $\sum ka_n$ diverges.

43. The series $\sum a_n$ converges. Explain, by looking at partial sums, why the series $\sum(a_{n+1} - a_n)$ also converges.

44. The series $\sum a_n$ diverges. Give examples that show the series $\sum(a_{n+1} - a_n)$ could converge or diverge.

Strengthen Your Understanding

In Problems 45–48, explain what is wrong with the statement.

45. The series $\sum(1/n)^2$ converges because the terms approach zero as $n \to \infty$.

46. The integral $\int_1^\infty (1/x^3)\,dx$ and the series $\sum_{n=1}^{\infty} 1/n^3$ both converge to the same value, $1/2$.

47. The series $\sum_{n=1}^{\infty} n^k$ converges for $k > 1$ and diverges for $k \leq 1$.

48. Since e^{-x^2} has no elementary antiderivative, the series $\sum_{n=1}^{\infty} e^{-n^2}$ does not converge.

In Problems 49–50, give an example of:

49. A series $\sum_{n=1}^{\infty} a_n$ with $\lim_{n\to\infty} a_n = 0$, but such that $\sum_{n=1}^{\infty} a_n$ diverges.

50. A convergent series $\sum_{n=1}^{\infty} a_n$, whose terms are all positive, such that the series $\sum_{n=1}^{\infty} \sqrt{a_n}$ is not convergent.

Decide if the statements in Problems 51–58 are true or false. Give an explanation for your answer.

51. $\sum_{n=1}^{\infty}(1+(-1)^n)$ is a series of nonnegative terms.

52. If a series converges, then the sequence of partial sums of the series also converges.

53. If $\sum |a_n + b_n|$ converges, then $\sum |a_n|$ and $\sum |b_n|$ converge.

54. The series $\sum_{n=1}^{\infty} 2^{(-1)^n}$ converges.

55. If a series $\sum a_n$ converges, then the terms, a_n, tend to zero as n increases.

56. If the terms, a_n, of a series tend to zero as n increases, then the series $\sum a_n$ converges.

57. If $\sum a_n$ does not converge and $\sum b_n$ does not converge, then $\sum a_n b_n$ does not converge.

58. If $\sum a_n b_n$ converges, then $\sum a_n$ and $\sum b_n$ converge.

9.4 TESTS FOR CONVERGENCE

Comparison of Series

In Section 7.7, we compared two integrals to decide whether an improper integral converged. In Theorem 9.3 we compared an integral and a series. Now we compare two series.

> **Theorem 9.4: Comparison Test**
>
> Suppose $0 \leq a_n \leq b_n$ for all n beyond a certain value.
> - If $\sum b_n$ converges, then $\sum a_n$ converges.
> - If $\sum a_n$ diverges, then $\sum b_n$ diverges.

Since $a_n \leq b_n$, the plot of the a_n terms lies under the plot of the b_n terms. See Figure 9.8. The comparison test says that if the total area for $\sum b_n$ is finite, then the total area for $\sum a_n$ is finite also. If the total area for $\sum a_n$ is not finite, then neither is the total area for $\sum b_n$.

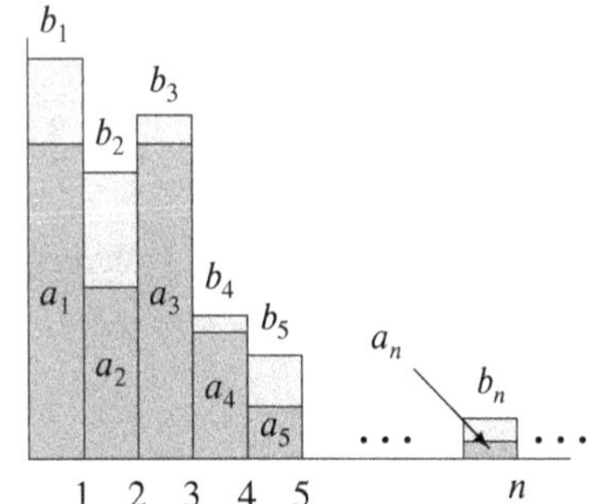

Figure 9.8: Each a_n is represented by the area of a dark rectangle, and each b_n by a dark plus a light rectangle

Example 1 Use the comparison test to determine whether the series $\sum_{n=1}^{\infty} \frac{1}{n^3+1}$ converges.

Solution For $n \geq 1$, we know that $n^3 \leq n^3 + 1$, so

$$0 \leq \frac{1}{n^3+1} \leq \frac{1}{n^3}.$$

Thus, every term in the series $\sum_{n=1}^{\infty} 1/(n^3+1)$ is less than or equal to the corresponding term in $\sum_{n=1}^{\infty} 1/n^3$. Since we saw that $\sum_{n=1}^{\infty} 1/n^3$ converges as a p-series with $p > 1$, we know that $\sum_{n=1}^{\infty} 1/(n^3+1)$ converges.

Example 2 Decide whether the following series converge: (a) $\sum_{n=1}^{\infty} \frac{n-1}{n^3+3}$ (b) $\sum_{n=1}^{\infty} \frac{6n^2+1}{2n^3-1}$.

Solution (a) Since the convergence is determined by the behavior of the terms for large n, we observe that

$$\frac{n-1}{n^3+3} \quad \text{behaves like} \quad \frac{n}{n^3} = \frac{1}{n^2} \quad \text{as} \quad n \to \infty.$$

Since $\sum 1/n^2$ converges, we guess that $\sum (n-1)/(n^3+3)$ converges. To confirm this, we use the comparison test. Since a fraction increases if its numerator is made larger or its denominator is made smaller, we have

$$0 \leq \frac{n-1}{n^3+3} \leq \frac{n}{n^3} = \frac{1}{n^2} \quad \text{for all} \quad n \geq 1.$$

Thus, the series $\sum (n-1)/(n^3+3)$ converges by comparison with $\sum 1/n^2$.

(b) First, we observe that

$$\frac{6n^2+1}{2n^3-1} \quad \text{behaves like} \quad \frac{6n^2}{2n^3} = \frac{3}{n} \quad \text{as} \quad n \to \infty.$$

Since $\sum 1/n$ diverges, so does $\sum 3/n$, and we guess that $\sum (6n^2+1)/(2n^3-1)$ diverges. To confirm this, we use the comparison test. Since a fraction decreases if its numerator is made smaller or its denominator is made larger, we have

$$0 \leq \frac{6n^2}{2n^3} \leq \frac{6n^2+1}{2n^3-1},$$

so

$$0 \leq \frac{3}{n} \leq \frac{6n^2+1}{2n^3-1}.$$

Thus, the series $\sum(6n^2+1)/(2n^3-1)$ diverges by comparison with $\sum 3/n$.

Limit Comparison Test

The comparison test uses the relationship between the terms of two series, $a_n \leq b_n$, which can be difficult to establish. However, the convergence or divergence of a series depends only on the long-run behavior of the terms as $n \to \infty$; this idea leads to the *limit comparison test.*

Example 3 Predict the convergence or divergence of

$$\sum \frac{n^2-5}{n^3+n+2}.$$

Solution As $n \to \infty$, the highest power terms in the numerator and denominator, n^2 and n^3, dominate. Thus the term

$$a_n = \frac{n^2-5}{n^3+n+2}$$

behaves, as $n \to \infty$, like

$$\frac{n^2}{n^3} = \frac{1}{n}.$$

Since the harmonic series $\sum 1/n$ diverges, we guess that $\sum a_n$ also diverges. However, the inequality

$$\frac{n^2-5}{n^3+n+2} \leq \frac{1}{n}$$

is in the wrong direction to use with the comparison test to confirm divergence, since we need the given series to be greater than a known divergent series.

The following test can be used to confirm a prediction of convergence or divergence, as in Example 3, without inequalities.

Theorem 9.5: Limit Comparison Test

Suppose $a_n > 0$ and $b_n > 0$ for all n. If

$$\lim_{n\to\infty} \frac{a_n}{b_n} = c \qquad \text{where } c > 0,$$

then the two series $\sum a_n$ and $\sum b_n$ either both converge or both diverge.

The limit $\lim_{n\to\infty} a_n/b_n = c$ captures the idea that a_n "behaves like" cb_n as $n \to \infty$.

Example 4 Use the limit comparison test to determine if the following series converge or diverge.

(a) $\sum \frac{n^2-5}{n^3+n+2}$ (b) $\sum \sin\left(\frac{1}{n}\right)$

Solution (a) We take $a_n = \dfrac{n^2-5}{n^3+n+2}$. Because a_n behaves like $\dfrac{n^2}{n^3} = \dfrac{1}{n}$ as $n \to \infty$ we take $b_n = 1/n$. We have

$$\lim_{n\to\infty} \frac{a_n}{b_n} = \lim_{n\to\infty} \frac{1}{1/n} \cdot \frac{n^2-5}{n^3+n+2} = \lim_{n\to\infty} \frac{n^3-5n}{n^3+n+2} = 1.$$

The limit comparison test applies with $c = 1$. Since $\sum 1/n$ diverges, the limit comparison test shows that $\sum \dfrac{n^2-5}{n^3+n+2}$ also diverges.

(b) Since $\sin x \approx x$ for x near 0, we know that $\sin(1/n)$ behaves like $1/n$ as $n \to \infty$. Since $\sin(1/n)$ is positive for $n \geq 1$, we can apply the limit comparison test with $a_n = \sin(1/n)$ and $b_n = 1/n$. We have

$$\lim_{n\to\infty} \frac{a_n}{b_n} = \lim_{n\to\infty} \frac{\sin(1/n)}{1/n} = 1.$$

Thus $c = 1$ and since $\sum 1/n$ diverges, the series $\sum \sin(1/n)$ also diverges.

Series of Both Positive and Negative Terms

If $\sum a_n$ has both positive and negative terms, then its plot has rectangles lying both above and below the horizontal axis. See Figure 9.9. The total area of the rectangles is no longer equal to $\sum a_n$. However, it is still true that if the total area of the rectangles above and below the axis is finite, then the series converges. The area of the nth rectangle is $|a_n|$, so we have:

Theorem 9.6: Convergence of Absolute Values Implies Convergence

If $\sum |a_n|$ converges, then so does $\sum a_n$.

Problem 144 (available online) shows how to prove this result.

Example 5 Explain how we know that the following series converges:

$$\sum_{n=1}^{\infty} \frac{(-1)^{n-1}}{n^2} = 1 - \frac{1}{4} + \frac{1}{9} - \cdots.$$

Solution Writing $a_n = (-1)^{n-1}/n^2$, we have

$$|a_n| = \left|\frac{(-1)^{n-1}}{n^2}\right| = \frac{1}{n^2}.$$

The p-series $\sum 1/n^2$ converges, since $p > 1$, so $\sum (-1)^{n-1}/n^2$ converges.

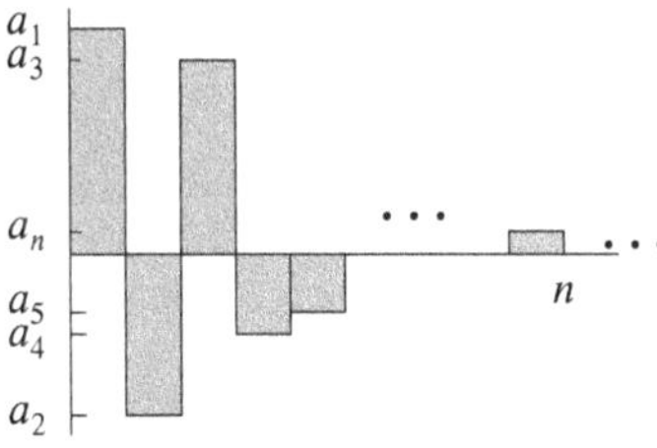

Figure 9.9: Representing a series with positive and negative terms

Comparison with a Geometric Series: The Ratio Test

A geometric series $\sum a_n$ has the property that the ratio a_{n+1}/a_n is constant for all n. For many other series, this ratio, although not constant, tends to a constant as n increases. In some ways, such series behave like geometric series. In particular, a geometric series converges if the ratio $|a_{n+1}/a_n| < 1$. A non-geometric series also converges if the ratio $|a_{n+1}/a_n|$ tends to a limit which is less than 1. This idea leads to the following test.

Theorem 9.7: The Ratio Test

For a series $\sum a_n$, suppose the sequence of ratios $|a_{n+1}|/|a_n|$ has a limit:

$$\lim_{n\to\infty} \frac{|a_{n+1}|}{|a_n|} = L.$$

- If $L < 1$, then $\sum a_n$ converges.
- If $L > 1$, or if L is infinite,[6] then $\sum a_n$ diverges.
- If $L = 1$, the test does not tell us anything about the convergence of $\sum a_n$.

Proof Here are the main steps in the proof. Suppose $\lim\limits_{n\to\infty} \frac{|a_{n+1}|}{|a_n|} = L < 1$. Let x be a number between L and 1. Then for all sufficiently large n, say for all $n \geq k$, we have

$$\frac{|a_{n+1}|}{|a_n|} < x.$$

Then,

$$\begin{aligned}
|a_{k+1}| &< |a_k|x, \\
|a_{k+2}| &< |a_{k+1}|x < |a_k|x^2, \\
|a_{k+3}| &< |a_{k+2}|x < |a_k|x^3,
\end{aligned}$$

and so on. Thus, writing $a = |a_k|$, we have for $i = 1, 2, 3, \ldots$,

$$|a_{k+i}| < ax^i.$$

Now we can use the comparison test: $\sum |a_{k+i}|$ converges by comparison with the geometric series $\sum ax^i$. Since $\sum |a_{k+i}|$ converges, Theorem 9.6 tells us that $\sum a_{k+i}$ converges. So, by property 2 of Theorem 9.2, we see that $\sum a_n$ converges too.

If $L > 1$, then for sufficiently large n, say $n \geq m$,

$$|a_{n+1}| > |a_n|,$$

so the sequence $|a_m|, |a_{m+1}|, |a_{m+2}|, \ldots$, is increasing. Thus, $\lim\limits_{n\to\infty} a_n \neq 0$, so $\sum a_n$ diverges (by Theorem 9.2, property 3). The argument in the case that $|a_{n+1}|/|a_n|$ is unbounded is similar.

Example 6 Show that the following series converges:

$$\sum_{n=1}^{\infty} \frac{1}{n!} = 1 + \frac{1}{2!} + \frac{1}{3!} + \cdots.$$

Solution Since $a_n = 1/n!$ and $a_{n+1} = 1/(n+1)!$, we have

$$\frac{|a_{n+1}|}{|a_n|} = \frac{1/(n+1)!}{1/n!} = \frac{n!}{(n+1)!} = \frac{n(n-1)(n-2)\cdots 2\cdot 1}{(n+1)n(n-1)\cdots 2\cdot 1}.$$

[6] That is, the sequence $|a_{n+1}|/|a_n|$ grows without bound.

We cancel $n(n-1)(n-2)\cdot \cdots \cdot 2\cdot 1$, giving

$$\lim_{n\to\infty}\frac{|a_{n+1}|}{|a_n|}=\lim_{n\to\infty}\frac{n!}{(n+1)!}=\lim_{n\to\infty}\frac{1}{n+1}=0.$$

Because the limit is 0, which is less than 1, the ratio test tells us that $\sum_{n=1}^{\infty} 1/n!$ converges. In Chapter 10, we see that the sum is $e-1$.

Example 7 What does the ratio test tell us about the convergence of the following two series?

$$\sum_{n=1}^{\infty}\frac{1}{n} \quad\text{and}\quad \sum_{n=1}^{\infty}\frac{(-1)^{n-1}}{n}.$$

Solution Because $|(-1)^n|=1$, in both cases we have $\lim_{n\to\infty}|a_{n+1}/a_n|=\lim_{n\to\infty}n/(n+1)=1$. The first series is the harmonic series, which diverges. However, Example 8 will show that the second series converges. Thus, if the ratio test gives a limit of 1, the ratio test does not tell us anything about the convergence of a series.

Alternating Series

A series is called an *alternating series* if the terms alternate in sign. For example,

$$\sum_{n=1}^{\infty}\frac{(-1)^{n-1}}{n}=1-\frac{1}{2}+\frac{1}{3}-\frac{1}{4}+\cdots+\frac{(-1)^{n-1}}{n}+\cdots.$$

The convergence of an alternating series can often be determined using the following test:

Theorem 9.8: Alternating Series Test

An alternating series of the form

$$\sum_{n=1}^{\infty}(-1)^{n-1}a_n=a_1-a_2+a_3-a_4+\cdots+(-1)^{n-1}a_n+\cdots$$

converges if

$$0<a_{n+1}<a_n \quad\text{for all } n \quad\text{and}\quad \lim_{n\to\infty}a_n=0.$$

Although we do not prove this result, we can see why it is reasonable. The first partial sum, $S_1=a_1$, is positive. The second, $S_2=a_1-a_2$, is still positive, since $a_2<a_1$, but S_2 is smaller than S_1. (See Figure 9.10.) The next sum, $S_3=a_1-a_2+a_3$, is greater than S_2 but smaller than S_1. The partial sums oscillate back and forth, and since the distance between them tends to 0, they eventually converge.

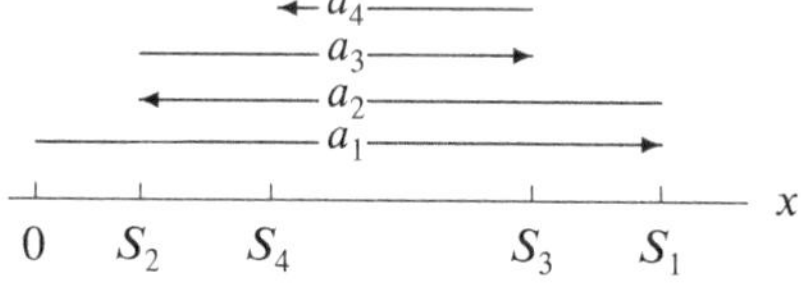

Figure 9.10: Partial sums, S_1, S_2, S_3, S_4 of an alternating series

Example 8 Show that the following alternating harmonic series converges:

$$\sum_{n=1}^{\infty} \frac{(-1)^{n-1}}{n}.$$

Solution We have $a_n = 1/n$ and $a_{n+1} = 1/(n+1)$. Thus,

$$a_{n+1} = \frac{1}{n+1} < \frac{1}{n} = a_n \quad \text{for all } n, \quad \text{and} \quad \lim_{n\to\infty} 1/n = 0.$$

Thus, the hypothesis of Theorem 9.8 is satisfied, so the alternating harmonic series converges.

Suppose S is the sum of an alternating series, so $S = \lim_{n\to\infty} S_n$. Then S is trapped between any two consecutive partial sums, say S_3 and S_4 or S_4 and S_5, so

$$S_2 < S_4 < \cdots < S < \cdots < S_3 < S_1.$$

Thus, the error in using S_n to approximate the true sum S is less than the distance from S_n to S_{n+1}, which is a_{n+1}. Stated symbolically, we have the following result:

Theorem 9.9: Error Bounds for Alternating Series

Let $S_n = \sum_{i=1}^{n} (-1)^{i-1} a_i$ be the nth partial sum of an alternating series and let $S = \lim_{n\to\infty} S_n$. Suppose that $0 < a_{n+1} < a_n$ for all n and $\lim_{n\to\infty} a_n = 0$. Then

$$|S - S_n| < a_{n+1}.$$

Thus, provided S_n converges to S by the alternating series test, the error in using S_n to approximate S is less than the magnitude of the first term of the series which is omitted in the approximation.

Example 9 Estimate the error in approximating the sum of the alternating harmonic series $\sum_{n=1}^{\infty} (-1)^{n-1}/n$ by the sum of the first nine terms.

Solution The ninth partial sum is given by

$$S_9 = 1 - \frac{1}{2} + \frac{1}{3} - \cdots + \frac{1}{9} = 0.7456\ldots.$$

The first term omitted is $-1/10$, with magnitude 0.1. By Theorem 9.9, we know that the true value of the sum differs from $0.7456\ldots$ by less than 0.1.

Absolute and Conditional Convergence

We say that the series $\sum a_n$ is

- **Absolutely convergent** if $\sum a_n$ and $\sum |a_n|$ both converge.
- **Conditionally convergent** if $\sum a_n$ converges but $\sum |a_n|$ diverges.

Conditionally convergent series rely on cancellation between positive and negative terms for their convergence.

Example: The series $\sum_{n=1}^{\infty} \frac{(-1)^{n-1}}{n^2}$ is absolutely convergent because the series converges and the p-series $\sum 1/n^2$ also converges.

Example: The series $\sum_{n=1}^{\infty} \frac{(-1)^{n-1}}{n}$ is conditionally convergent because the series converges but the harmonic series $\sum 1/n$ diverges.

Absolutely and conditionally convergent series behave differently when their terms are reordered:

- **Absolutely** convergent series: Rearranging terms does not change the sum.
- **Conditionally** convergent series: Rearranging terms can change the sum to any number.[7]

See Problem 112 on page 503 and Problems 136 and 137 (available online).

Exercises and Problems for Section 9.4

Online Resource: Additional Problems for Section 9.4

EXERCISES

In Exercises 1–4, use the comparison test to confirm the statement.

1. $\sum_{n=4}^{\infty} \frac{1}{n}$ diverges, so $\sum_{n=4}^{\infty} \frac{1}{n-3}$ diverges.

2. $\sum_{n=1}^{\infty} \frac{1}{n^2}$ converges, so $\sum_{n=1}^{\infty} \frac{1}{n^2+2}$ converges.

3. $\sum_{n=1}^{\infty} \frac{1}{n^2}$ converges, so $\sum_{n=1}^{\infty} \frac{e^{-n}}{n^2}$ converges.

4. $\sum_{n=1}^{\infty} \left(\frac{3}{7}\right)^n$ converges, so $\sum_{n=1}^{\infty} \left(\frac{3n}{7n+1}\right)^n$ converges.

In Exercises 5–8, use end behavior to compare the series to a p-series and predict whether the series converges or diverges.

5. $\sum_{n=1}^{\infty} \frac{n^3+1}{n^4+2n^3+2n}$

6. $\sum_{n=1}^{\infty} \frac{n+4}{n^3+5n-3}$

7. $\sum_{n=1}^{\infty} \frac{1}{n^4+3n^3+7}$

8. $\sum_{n=1}^{\infty} \frac{n-4}{\sqrt{n^3+n^2+8}}$

In Exercises 9–14, use the comparison test to determine whether the series converges.

9. $\sum_{n=1}^{\infty} \frac{n^2}{n^4+1}$

10. $\sum_{n=1}^{\infty} \frac{1}{3^n+1}$

11. $\sum_{n=1}^{\infty} \frac{1}{n^4+e^n}$

12. $\sum_{n=2}^{\infty} \frac{1}{\ln n}$

13. $\sum_{n=1}^{\infty} \frac{n\sin^2 n}{n^3+1}$

14. $\sum_{n=1}^{\infty} \frac{2^n+1}{n2^n-1}$

In Exercises 15–21, use the ratio test to decide whether the series converges or diverges.

15. $\sum_{n=1}^{\infty} \frac{n}{2^n}$

16. $\sum_{n=1}^{\infty} \frac{1}{(2n)!}$

17. $\sum_{n=1}^{\infty} \frac{1}{ne^n}$

18. $\sum_{n=1}^{\infty} \frac{(n!)^2}{(2n)!}$

19. $\sum_{n=1}^{\infty} \frac{n!(n+1)!}{(2n)!}$

20. $\sum_{n=1}^{\infty} \frac{1}{r^n n!}, r>0$

21. $\sum_{n=0}^{\infty} \frac{2^n}{n^3+1}$

In Exercises 22–32, use the limit comparison test to determine whether the series converges or diverges.

22. $\sum_{n=1}^{\infty} \frac{5n+1}{3n^2}$, by comparing to $\sum_{n=1}^{\infty} \frac{1}{n}$

23. $\sum_{n=1}^{\infty} \left(\frac{1+n}{3n}\right)^n$, by comparing to $\sum_{n=1}^{\infty} \left(\frac{1}{3}\right)^n$
[Hint: $\lim_{n\to\infty}(1+1/n)^n = e$.]

24. $\sum_{n=1}^{\infty} \left(1-\cos\frac{1}{n}\right)$, by comparing to $\sum_{n=1}^{\infty} \frac{1}{n^2}$

25. $\sum_{n=1}^{\infty} \frac{1}{n^4-7}$

26. $\sum_{n=1}^{\infty} \frac{n+1}{n^2+2}$

27. $\sum_{n=1}^{\infty} \frac{n^3-2n^2+n+1}{n^4-2}$

28. $\sum_{n=1}^{\infty} \frac{2^n}{3^n-1}$

29. $\sum_{n=1}^{\infty} \frac{1}{2\sqrt{n}+\sqrt{n+2}}$

30. $\sum_{n=1}^{\infty} \left(\frac{1}{2n-1}-\frac{1}{2n}\right)$

31. $\sum_{n=1}^{\infty} \frac{4\sin n+n}{n^2}$

32. $\sum_{n=1}^{\infty} \frac{n}{\cos n+e^n}$

[7] See Walter Rudin, *Principles of Mathematical Analysis*, 3rd ed. (New York: McGraw-Hill, 1976), pp. 76–77.

Which of the series in Exercises 33–36 are alternating?

33. $\sum_{n=1}^{\infty}(-1)^n\left(2-\frac{1}{n}\right)$

34. $\sum_{n=1}^{\infty}\cos(n\pi)$

35. $\sum_{n=1}^{\infty}(-1)^n\cos(n\pi)$

36. $\sum_{n=1}^{\infty}(-1)^n\cos n$

In Exercises 37–42, use the alternating series test to decide whether the series converges.

37. $\sum_{n=1}^{\infty}\frac{(-1)^{n-1}}{\sqrt{n}}$

38. $\sum_{n=1}^{\infty}\frac{(-1)^{n-1}}{2n+1}$

39. $\sum_{n=1}^{\infty}\frac{(-1)^n}{n^2+1}$

40. $\sum_{n=1}^{\infty}\frac{(-1)^n}{n!}$

41. $\sum_{n=1}^{\infty}\frac{(-1)^{n-1}}{n^2+2n+1}$

42. $\sum_{n=1}^{\infty}\frac{(-1)^{n-1}}{e^n}$

43. **(a)** Decide whether the following series is alternating:

$$\sum_{n=1}^{\infty}\frac{\sin n}{n^3}.$$

(b) Use the comparison test to determine whether the following series converges or diverges:

$$\sum_{n=1}^{\infty}\left|\frac{\sin n}{n^3}\right|.$$

(c) Determine whether the following series converges or diverges:

$$\sum_{n=1}^{\infty}\frac{\sin n}{n^3}.$$

In Exercises 44–52, determine whether the series is absolutely convergent, conditionally convergent, or divergent.

44. $\sum_{n=1}^{\infty}\frac{(-1)^n}{2^n}$

45. $\sum_{n=1}^{\infty}\frac{(-1)^n}{2n}$

46. $\sum_{n=1}^{\infty}\frac{(-1)^{n-1}}{\sqrt{n}}$

47. $\sum_{n=1}^{\infty}(-1)^n\frac{n}{n+1}$

48. $\sum_{n=1}^{\infty}\frac{(-1)^n}{n^4+7}$

49. $\sum_{n=2}^{\infty}\frac{(-1)^{n-1}}{n\ln n}$

50. $\sum_{n=1}^{\infty}\frac{\cos n}{n^2}$

51. $\sum_{n=1}^{\infty}(-1)^{n-1}\arcsin\left(\frac{1}{n}\right)$

52. $\sum_{n=1}^{\infty}\frac{(-1)^{n-1}\arctan(1/n)}{n^2}$

PROBLEMS

In Problems 53–54, explain why the comparison test cannot be used to decide if the series converges or diverges.

53. $\sum_{n=1}^{\infty}\frac{(-1)^n}{n^2}$

54. $\sum_{n=1}^{\infty}\sin n$

In Problems 55–56, explain why the ratio test cannot be used to decide if the series converges or diverges.

55. $\sum_{n=1}^{\infty}(-1)^n$

56. $\sum_{n=1}^{\infty}\sin n$

In Problems 57–60, explain why the alternating series test cannot be used to decide if the series converges or diverges.

57. $\sum_{n=1}^{\infty}(-1)^{n-1}n$

58. $\sum_{n=1}^{\infty}(-1)^{n-1}\sin n$

59. $\sum_{n=1}^{\infty}(-1)^{n-1}\left(2-\frac{1}{n}\right)$

60. $\frac{2}{1}-\frac{1}{1}+\frac{2}{2}-\frac{1}{2}+\frac{2}{3}-\frac{1}{3}+\cdots$

In Problems 61–63, use a computer or calculator to investigate the behavior of the partial sums of the alternating series. Which appear to converge? Confirm convergence using the alternating series test. If a series converges, estimate its sum.

61. $1-2+3-4+5+\cdots+(-1)^n(n+1)+\cdots$

62. $1-0.1+0.01-0.001+\cdots+(-1)^n10^{-n}+\cdots$

63. $1-\frac{1}{1!}+\frac{1}{2!}-\frac{1}{3!}+\cdots+(-1)^n\frac{1}{n!}+\cdots$

In Problems 64–92, determine whether the series converges.

64. $\sum_{n=1}^{\infty}\frac{8^n}{n!}$

65. $\sum_{n=1}^{\infty}\frac{1}{4n+3}$

66. $\sum_{n=1}^{\infty}\frac{(-2)^{n-1}}{n^2}$

67. $\sum_{n=1}^{\infty}\frac{(-1)^{n-1}}{2n+1}$

68. $\sum_{n=1}^{\infty}\frac{5+e^n}{3^n}$

69. $\sum_{n=2}^{\infty}\frac{n+2}{n^2-1}$

70. $\sum_{n=1}^{\infty}\frac{n2^n}{3^n}$

71. $\sum_{n=0}^{\infty}\frac{(0.1)^n}{n!}$

72. $\sum_{n=1}^{\infty}\frac{n^2}{n^2+1}$

73. $\sum_{n=1}^{\infty}\frac{1+3^n}{4^n}$

74. $\sum_{n=1}^{\infty}\frac{(n-1)!}{n^2}$

75. $\sum_{n=1}^{\infty}\frac{(2n)!}{(n!)^2}$

76. $\sum_{n=1}^{\infty}e^n$

77. $\sum_{n=0}^{\infty}e^{-n}$

78. $\sum_{n=1}^{\infty}\frac{n+1}{n^3+6}$

79. $\sum_{n=1}^{\infty}\frac{5n+2}{2n^2+3n+7}$

80. $\sum_{n=1}^{\infty}\frac{(-1)^{n-1}}{\sqrt{3n-1}}$

81. $\sum_{n=1}^{\infty}\frac{(-1)^{n-1}2^n}{n^2}$

82. $\sum_{n=1}^{\infty} \frac{1}{\sqrt{n^2(n+2)}}$

83. $\sum_{n=1}^{\infty} \frac{n(n+1)}{\sqrt{n^3+2n^2}}$

84. $\sum_{n=1}^{\infty} \frac{2n^3-1}{n^3+1}$

85. $\sum_{n=1}^{\infty} \frac{n+1}{3n^2-2}$

86. $\sum_{n=1}^{\infty} \frac{6}{n+2^n}$

87. $\sum_{n=1}^{\infty} \frac{2n+1}{\sqrt{3n^3-2}}$

88. $\sum_{n=2}^{\infty} \frac{3}{\ln n^2}$

89. $\sum_{n=1}^{\infty} \frac{\sin n}{n^2}$

90. $\sum_{n=1}^{\infty} \frac{\sin n^2}{n^2}$

91. $\sum_{n=1}^{\infty} \frac{\cos(n\pi)}{n}$

92. $\sum_{n=1}^{\infty} \frac{1}{n^2} \tan\left(\frac{1}{n}\right)$

■ In Problems 93–97, use the expression for a_n to decide:

(a) If the sequence $\{a_n\}_{n=1}^{\infty}$ converges or diverges.

(b) If the series $\sum_{n=1}^{\infty} a_n$ converges or diverges.

93. $a_n = \frac{e^n}{1+e^{2n}}$

94. $a_n = \frac{\sqrt{n}}{1+\sqrt{n}}$

95. $a_n = \frac{n}{1+n^2}$

96. $a_n = \frac{4+2^n}{3^n}$

97. $a_n = 2 + (-1)^n$

■ In Problems 98–101, the series converges. Is the sum affected by rearranging the terms of the series?

98. $\sum_{n=1}^{\infty} \frac{(-1)^n}{n}$

99. $\sum_{n=1}^{\infty} \frac{(-1)^n}{n^2}$

100. $\sum_{n=1}^{\infty} \frac{1}{2^n}$

101. $\sum_{n=2}^{\infty} \frac{(-1)^{n+1}}{\ln n}$

■ In Problems 102–106, for what values of a does the series converge?

102. $\sum_{n=1}^{\infty} \left(\frac{2}{n}\right)^a$

103. $\sum_{n=1}^{\infty} \left(\frac{2}{a}\right)^n, a > 0$

104. $\sum_{n=1}^{\infty} (\ln a)^n, a > 0$

105. $\sum_{n=1}^{\infty} \frac{\ln n}{n^a}$

106. $\sum_{n=1}^{\infty} (-1)^n \arctan\left(\frac{a}{n}\right), a > 0$

■ The series in Problems 107–109 converge by the alternating series test. Use Theorem 9.9 to find how many terms give a partial sum, S_n, within 0.01 of the sum, S, of the series.

107. $\sum_{n=1}^{\infty} \frac{(-1)^{n-1}}{n}$

108. $\sum_{n=1}^{\infty} \left(-\frac{2}{3}\right)^{n-1}$

109. $\sum_{n=1}^{\infty} \frac{(-1)^{n-1}}{(2n)!}$

110. Suppose $0 \le b_n \le 2^n \le a_n$ and $0 \le c_n \le 2^{-n} \le d_n$ for all n. Which of the series $\sum a_n$, $\sum b_n$, $\sum c_n$, and $\sum d_n$ definitely converge and which definitely diverge?

111. Given two convergent series $\sum a_n$ and $\sum b_n$, we know that the term-by-term sum $\sum(a_n + b_n)$ converges. What about the series formed by taking the products of the terms $\sum a_n \cdot b_n$? This problem shows that it may or may not converge.

(a) Show that if $\sum a_n = \sum 1/n^2$ and $\sum b_n = \sum 1/n^3$, then $\sum a_n \cdot b_n$ converges.

(b) Explain why $\sum(-1)^n/\sqrt{n}$ converges.

(c) Show that if $a_n = b_n = (-1)^n/\sqrt{n}$, then $\sum a_n \cdot b_n$ does not converge.

112. Let $\sum a_n$ be a series that converges to a number L and whose terms are all positive. We show that any rearrangement $\sum c_n$ of this series also converges to L.

(a) Show that $S_n < L$ for any partial sum S_n of $\sum c_n$.

(b) Show that for any $K < L$, there is a partial sum S_n of $\sum c_n$ with $S_n > K$.

(c) Use parts (a) and (b) to show $\sum c_n$ converges to L.

Strengthen Your Understanding

■ In Problems 113–115, explain what is wrong with the statement.

113. The series $\sum_{n=1}^{\infty}(-1)^{2n}/n^2$ converges by the alternating series test.

114. The series $\sum_{n=1}^{\infty} 1/(n^2+1)$ converges by the ratio test.

115. The series $\sum_{n=1}^{\infty} 1/n^{3/2}$ converges by comparison with $\sum_{n=1}^{\infty} 1/n^2$.

■ In Problems 116–118, give an example of:

116. A series $\sum_{n=1}^{\infty} a_n$ that converges but $\sum_{n=1}^{\infty} |a_n|$ diverges.

117. An alternating series that does not converge.

118. A series $\sum a_n$ such that

$$\lim_{n\to\infty} \frac{|a_{n+1}|}{|a_n|} = 3.$$

■ Decide if the statements in Problems 119–134 are true or false. Give an explanation for your answer.

119. If the terms s_n of a sequence alternate in sign, then the sequence converges.

120. If $0 \le a_n \le b_n$ for all n and $\sum a_n$ converges, then $\sum b_n$ converges.

121. If $0 \le a_n \le b_n$ for all n and $\sum a_n$ diverges, then $\sum b_n$ diverges.

122. If $b_n \leq a_n \leq 0$ for all n and $\sum b_n$ converges, then $\sum a_n$ converges.

123. If $\sum a_n$ converges, then $\sum |a_n|$ converges.

124. If $\sum a_n$ converges, then $\lim_{n\to\infty} |a_{n+1}|/|a_n| \neq 1$.

125. $\sum_{n=0}^{\infty}(-1)^n \cos(2\pi n)$ is an alternating series.

126. The series $\sum_{n=0}^{\infty}(-1)^n 2^n$ converges.

127. If $\sum a_n$ converges, then $\sum(-1)^n a_n$ converges.

128. If an alternating series converges by the alternating series test, then the error in using the first n terms of the series to approximate the entire series is less in magnitude than the first term omitted.

129. If an alternating series converges, then the error in using the first n terms of the series to approximate the entire series is less in magnitude than the first term omitted.

130. If $\sum |a_n|$ converges, then $\sum(-1)^n |a_n|$ converges.

131. To find the sum of the alternating harmonic series $\sum(-1)^{n-1}/n$ to within 0.01 of the true value, we can sum the first 100 terms.

132. If $\sum a_n$ is absolutely convergent, then it is convergent.

133. If $\sum a_n$ is conditionally convergent, then it is absolutely convergent.

134. If $a_n > 0.5b_n > 0$ for all n and $\sum b_n$ diverges, then $\sum a_n$ diverges.

135. Which test will help you determine if the series converges or diverges?

$$\sum_{k=1}^{\infty} \frac{1}{k^3+1}$$

(a) Integral test
(b) Comparison test
(c) Ratio test

9.5 POWER SERIES AND INTERVAL OF CONVERGENCE

In Section 9.2 we saw that the geometric series $\sum ax^n$ converges for $-1 < x < 1$ and diverges otherwise. This section studies the convergence of more general series constructed from powers. Chapter 10 shows how such power series are used to approximate functions such as e^x, $\sin x$, $\cos x$, and $\ln x$.

A **power series** about $x = a$ is a sum of constants times powers[8] of $(x - a)$:

$$C_0 + C_1(x-a) + C_2(x-a)^2 + \cdots + C_n(x-a)^n + \cdots = \sum_{n=0}^{\infty} C_n(x-a)^n.$$

We think of a as a constant. For any fixed x, the power series $\sum C_n(x-a)^n$ is a series of numbers like those considered in Section 9.3. To investigate the convergence of a power series, we consider the partial sums, which in this case are the polynomials $S_n(x) = C_0 + C_1(x-a) + C_2(x-a)^2 + \cdots + C_n(x-a)^n$. As before, we consider the sequence[9]

$$S_0(x),\ S_1(x),\ S_2(x),\ \ldots\ , S_n(x), \ldots.$$

For a fixed value of x, if this sequence of partial sums converges to a limit S, that is, if $\lim_{n\to\infty} S_n(x) = S$, then we say that the power series **converges** to S for this value of x.

A power series may converge for some values of x and not for others.

Example 1 Find an expression for the general term of the series and use it to write the series using $\sum$ notation:

$$\frac{(x-2)^4}{4} - \frac{(x-2)^6}{9} + \frac{(x-2)^8}{16} - \frac{(x-2)^{10}}{25} + \cdots.$$

[8]For $n = 0$, when $x = a$ we define $(x-a)^0$ to be 1.
[9]Here we call the first term in the sequence $S_0(x)$ rather than $S_1(x)$ so that the last term of $S_n(x)$ is $C_n(x-a)^n$.

Solution The series is about $x = 2$ and the odd terms are zero. We use $(x - 2)^{2n}$ and begin with $n = 2$. Since the series alternates and is positive for $n = 2$, we multiply by $(-1)^n$. For $n = 2$, we divide by 4, for $n = 3$ we divide by 9, and in general, we divide by n^2. One way to write this series is

$$\sum_{n=2}^{\infty} \frac{(-1)^n (x-2)^{2n}}{n^2}.$$

Example 2 Determine whether the power series $\sum_{n=0}^{\infty} \frac{x^n}{2^n}$ converges or diverges for

(a) $x = -1$ (b) $x = 3$

Solution (a) Substituting $x = -1$, we have

$$\sum_{n=0}^{\infty} \frac{x^n}{2^n} = \sum_{n=0}^{\infty} \frac{(-1)^n}{2^n} = \sum_{n=0}^{\infty} \left(-\frac{1}{2}\right)^n.$$

This is a geometric series with ratio $-1/2$, so the series converges to $1/(1 - (-\frac{1}{2})) = 2/3$.

(b) Substituting $x = 3$, we have

$$\sum_{n=0}^{\infty} \frac{x^n}{2^n} = \sum_{n=0}^{\infty} \frac{3^n}{2^n} = \sum_{n=0}^{\infty} \left(\frac{3}{2}\right)^n.$$

This is a geometric series with ratio greater than 1, so it diverges.

Numerical and Graphical View of Convergence

Consider the series

$$(x-1) - \frac{(x-1)^2}{2} + \frac{(x-1)^3}{3} - \frac{(x-1)^4}{4} + \cdots + (-1)^{n-1}\frac{(x-1)^n}{n} + \cdots.$$

To investigate the convergence of this series, we look at the sequence of partial sums graphed in Figure 9.11. The graph suggests that the partial sums converge for x in the interval $(0, 2)$. In Examples 4 and 5, we show that the series converges for $0 < x \leq 2$. This is called the *interval of convergence* of this series.

At $x = 1.4$, which is inside the interval, the series appears to converge quite rapidly:

$$S_5(1.4) = 0.33698\ldots \qquad S_7(1.4) = 0.33653\ldots$$
$$S_6(1.4) = 0.33630\ldots \qquad S_8(1.4) = 0.33645\ldots$$

Table 9.1 shows the results of using $x = 1.9$ and $x = 2.3$ in the power series. For $x = 1.9$, which is inside the interval of convergence but close to an endpoint, the series converges, though rather slowly. For $x = 2.3$, which is outside the interval of convergence, the series diverges: the larger the value of n, the more wildly the series oscillates. In fact, the contribution of the twenty-fifth term is about 28; the contribution of the hundredth term is about $-2{,}500{,}000{,}000$. Figure 9.11 shows the interval of convergence and the partial sums.

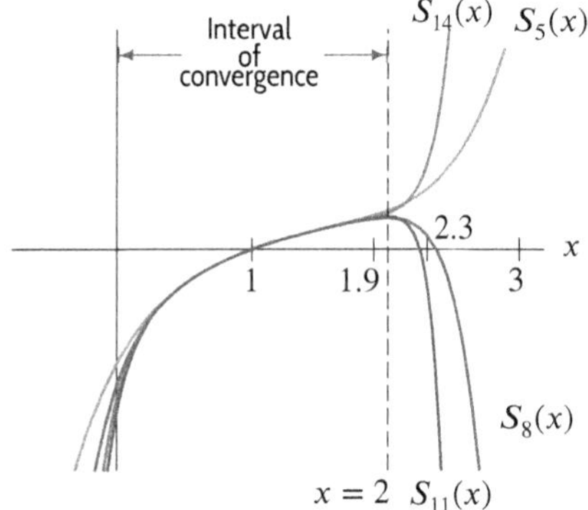

Figure 9.11: Partial sums for series in Example 4 converge for $0 < x \leq 2$

Table 9.1 *Partial sums for series in Example 4 with $x = 1.9$ inside interval of convergence and $x = 2.3$ outside*

n	$S_n(1.9)$	n	$S_n(2.3)$
2	0.495	2	0.455
5	0.69207	5	1.21589
8	0.61802	8	0.28817
11	0.65473	11	1.71710
14	0.63440	14	−0.70701

Notice that the interval of convergence, $0 < x \leq 2$, is centered on $x = 1$. Since the interval extends one unit on either side, we say the *radius of convergence* of this series is 1.

Intervals of Convergence

Each power series falls into one of three cases, characterized by its *radius of convergence*, R. This radius gives an *interval of convergence*.

- The series converges only for $x = a$; the **radius of convergence** is defined to be $R = 0$.
- The series converges for all values of x; the **radius of convergence** is defined to be $R = \infty$.
- There is a positive number R, called the **radius of convergence**, such that the series converges for $|x - a| < R$ and diverges for $|x - a| > R$. See Figure 9.12.

Using the radius of convergence, we make the following definition:

- The **interval of convergence** is the interval between $a - R$ and $a + R$, including any endpoint where the series converges.

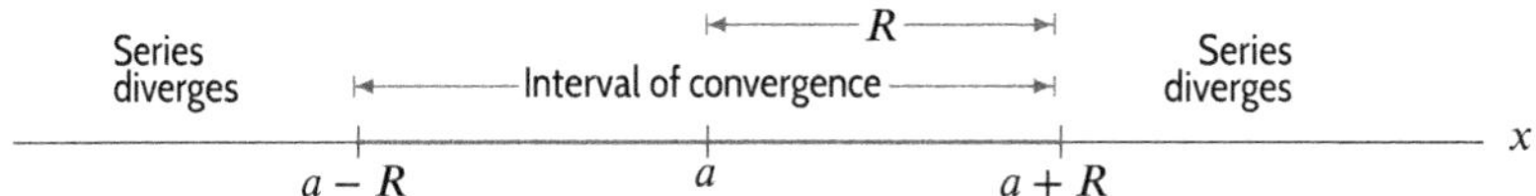

Figure 9.12: Radius of convergence, R, determines an interval, centered at $x = a$, in which the series converges

There are some series whose radius of convergence we already know. For example, the geometric series

$$1 + x + x^2 + \cdots + x^n + \cdots$$

converges for $|x| < 1$ and diverges for $|x| \geq 1$, so its radius of convergence is 1. Similarly, the series

$$1 + \frac{x}{3} + \left(\frac{x}{3}\right)^2 + \cdots + \left(\frac{x}{3}\right)^n + \cdots$$

converges for $|x/3| < 1$ and diverges for $|x/3| \geq 1$, so its radius of convergence is 3.

The next theorem gives a method of computing the radius of convergence for many series. To find the values of x for which the power series $\sum_{n=0}^{\infty} C_n(x-a)^n$ converges, we use the ratio test. Writing $a_n = C_n(x-a)^n$ and assuming $C_n \neq 0$ and $x \neq a$, we have

$$\lim_{n\to\infty} \frac{|a_{n+1}|}{|a_n|} = \lim_{n\to\infty} \frac{|C_{n+1}(x-a)^{n+1}|}{|C_n(x-a)^n|} = \lim_{n\to\infty} \frac{|C_{n+1}||x-a|}{|C_n|} = |x-a| \lim_{n\to\infty} \frac{|C_{n+1}|}{|C_n|}.$$

Case 1. Suppose $\lim_{n\to\infty} |a_{n+1}|/|a_n|$ is infinite. Then the ratio test shows that the power series converges only for $x = a$. The radius of convergence is $R = 0$.
Case 2. Suppose $\lim_{n\to\infty} |a_{n+1}|/|a_n| = 0$. Then the ratio test shows that the power series converges for all x. The radius of convergence is $R = \infty$.
Case 3. Suppose $\lim_{n\to\infty} |a_{n+1}|/|a_n| = K|x-a|$, where $\lim_{n\to\infty} |C_{n+1}|/|C_n| = K$. In Case 1, K does not exist; in Case 2, $K = 0$. Thus, we can assume K exists and $K \neq 0$, and we can define $R = 1/K$. Then we have

$$\lim_{n\to\infty} \frac{|a_{n+1}|}{|a_n|} = K|x-a| = \frac{|x-a|}{R},$$

so the ratio test tells us that the power series:

- Converges for $\dfrac{|x-a|}{R} < 1$; that is, for $|x-a| < R$
- Diverges for $\dfrac{|x-a|}{R} > 1$; that is, for $|x-a| > R$.

The results are summarized in the following theorem.

Theorem 9.10: Method for Computing Radius of Convergence

To calculate the radius of convergence, R, for the power series $\sum_{n=0}^{\infty} C_n(x-a)^n$, use the ratio test with $a_n = C_n(x-a)^n$.

- If $\lim_{n\to\infty} |a_{n+1}|/|a_n|$ is infinite, then $R = 0$.
- If $\lim_{n\to\infty} |a_{n+1}|/|a_n| = 0$, then $R = \infty$.
- If $\lim_{n\to\infty} |a_{n+1}|/|a_n| = K|x-a|$, where K is finite and nonzero, then $R = 1/K$.

Note that the ratio test does not tell us anything if $\lim_{n\to\infty} |a_{n+1}|/|a_n|$ fails to exist, which can occur, for example, if some of the C_ns are zero.

A proof that a power series has a radius of convergence and of Theorem 9.10 is given in the online theory supplement. To understand these facts informally, we can think of a power series as being like a geometric series whose coefficients vary from term to term. The radius of convergence depends on the behavior of the coefficients: if there are constants C and K such that for larger and larger n,

$$|C_n| \approx CK^n,$$

then it is plausible that $\sum C_n x^n$ and $\sum CK^n x^n = \sum C(Kx)^n$ converge or diverge together. The geometric series $\sum C(Kx)^n$ converges for $|Kx| < 1$, that is, for $|x| < 1/K$. We can find K using the ratio test, because

$$\frac{|a_{n+1}|}{|a_n|} = \frac{|C_{n+1}||(x-a)^{n+1}|}{|C_n||(x-a)^n|} \approx \frac{CK^{n+1}|(x-a)^{n+1}|}{CK^n|(x-a)^n|} = K|x-a|.$$

Example 3 Show that the following power series converges for all x:

$$1 + x + \frac{x^2}{2!} + \frac{x^3}{3!} + \cdots + \frac{x^n}{n!} + \cdots.$$

Solution Because $C_n = 1/n!$, none of the C_ns are zero and we can use the ratio test:

$$\lim_{n\to\infty} \frac{|a_{n+1}|}{|a_n|} = |x| \lim_{n\to\infty} \frac{|C_{n+1}|}{|C_n|} = |x| \lim_{n\to\infty} \frac{1/(n+1)!}{1/n!} = |x| \lim_{n\to\infty} \frac{n!}{(n+1)!} = |x| \lim_{n\to\infty} \frac{1}{n+1} = 0.$$

This gives $R = \infty$, so the series converges for all x. We see in Chapter 10 that it converges to e^x.

Example 4 Determine the radius of convergence of the series

$$\frac{(x-1)}{3} - \frac{(x-1)^2}{2\cdot 3^2} + \frac{(x-1)^3}{3\cdot 3^3} - \frac{(x-1)^4}{4\cdot 3^4} + \cdots + (-1)^{n-1}\frac{(x-1)^n}{n\cdot 3^n} + \cdots.$$

What does this tell us about the interval of convergence of this series?

Solution Because $C_n = (-1)^{n-1}/(n \cdot 3^n)$ is never zero we can use the ratio test. We have

$$\lim_{n\to\infty} \frac{|a_{n+1}|}{|a_n|} = |x-1| \lim_{n\to\infty} \frac{|C_{n+1}|}{|C_n|} = |x-1| \lim_{n\to\infty} \frac{\left|\frac{(-1)^n}{(n+1)\cdot 3^{n+1}}\right|}{\left|\frac{(-1)^{n-1}}{n\cdot 3^n}\right|} = |x-1| \lim_{n\to\infty} \frac{n}{3(n+1)} = \frac{|x-1|}{3}.$$

Thus, $K = 1/3$ in Theorem 9.10, so the radius of convergence is $R = 1/K = 3$. The power series converges for $|x-1| < 3$ and diverges for $|x-1| > 3$, so the series converges for $-2 < x < 4$. Notice that the radius of convergence does not tell us what happens at the endpoints, $x = -2$ and $x = 4$. The endpoints are investigated in Example 5.

What Happens at the Endpoints of the Interval of Convergence?

The ratio test does not tell us whether a power series converges at the endpoints of its interval of convergence, $x = a \pm R$. There is no simple theorem that answers this question. Since substituting $x = a \pm R$ converts the power series to a series of numbers, the tests in Sections 9.3 and 9.4 are often useful.

Example 5 Determine the interval of convergence of the series

$$\frac{(x-1)}{3} - \frac{(x-1)^2}{2\cdot 3^2} + \frac{(x-1)^3}{3\cdot 3^3} - \frac{(x-1)^4}{4\cdot 3^4} + \cdots + (-1)^{n-1}\frac{(x-1)^n}{n\cdot 3^n} + \cdots.$$

Solution In Example 4 we showed that this series has $R = 3$; it converges for $-2 < x < 4$ and diverges for $x < -2$ or $x > 4$. We need to determine whether it converges at the endpoints of the interval of convergence, $x = -2$ and $x = 4$. At $x = 4$, we have the series

$$1 - \frac{1}{2} + \frac{1}{3} - \frac{1}{4} + \cdots + \frac{(-1)^{n-1}}{n} + \cdots.$$

This is an alternating series with $a_n = 1/n$, so by the alternating series test (Theorem 9.8), it converges. At $x = -2$, we have the series

$$-1 - \frac{1}{2} - \frac{1}{3} - \frac{1}{4} - \cdots - \frac{1}{n} - \cdots.$$

This is the negative of the harmonic series, so it diverges. Therefore, the interval of convergence is $-2 < x \le 4$. The right endpoint is included and the left endpoint is not.

Series with All Odd, or All Even, Terms

The ratio test requires $\lim_{n\to\infty} |a_{n+1}|/|a_n|$ to exist for $a_n = C_n(x-a)^n$. What happens if the power series has only even or odd powers, so some of the coefficients C_n are zero? Then we use the fact that an infinite series can be written in several ways and pick one in which the terms are nonzero.

Example 6 Find the radius and interval of convergence of the series

$$1 + 2^2x^2 + 2^4x^4 + 2^6x^6 + \cdots.$$

Solution If we take $a_n = 2^n x^n$ for n even and $a_n = 0$ for n odd, $\lim_{n\to\infty} |a_{n+1}|/|a_n|$ does not exist. Therefore, for this series we take

$$a_n = 2^{2n}x^{2n},$$

so that, replacing n by $n+1$, we have

$$a_{n+1} = 2^{2(n+1)}x^{2(n+1)} = 2^{2n+2}x^{2n+2}.$$

Thus,

$$\frac{|a_{n+1}|}{|a_n|} = \frac{\left|2^{2n+2}x^{2n+2}\right|}{\left|2^{2n}x^{2n}\right|} = \left|2^2x^2\right| = 4x^2.$$

We have

$$\lim_{n\to\infty} \frac{|a_{n+1}|}{|a_n|} = 4x^2.$$

The ratio test guarantees that the power series converges if $4x^2 < 1$, that is, if $|x| < \frac{1}{2}$. The radius of convergence is $\frac{1}{2}$. The series converges for $-\frac{1}{2} < x < \frac{1}{2}$ and diverges for $x > \frac{1}{2}$ or $x < -\frac{1}{2}$. At $x = \pm\frac{1}{2}$, all the terms in the series are 1, so the series diverges (by Theorem 9.2, Property 3). Thus, the interval of convergence is $-\frac{1}{2} < x < \frac{1}{2}$.

Example 7 Write the general term a_n of the following series so that none of the terms are zero:

$$x - \frac{x^3}{3!} + \frac{x^5}{5!} - \frac{x^7}{7!} + \frac{x^9}{9!} - \cdots.$$

Solution This series has only odd powers. We can get odd integers using $2n-1$ for $n \ge 1$, since

$$2\cdot 1 - 1 = 1, \quad 2\cdot 2 - 1 = 3, \quad 2\cdot 3 - 1 = 5, \text{ etc.}$$

Also, the signs of the terms alternate, with the first (that is, $n = 1$) term positive, so we include a factor of $(-1)^{n-1}$. Thus we get

$$a_n = (-1)^{n-1}\frac{x^{2n-1}}{(2n-1)!}.$$

We see in Chapter 10 that the series converges to $\sin x$. Exercise 24 shows that the radius of convergence of this series is infinite, so that it converges for all values of x.

Exercises and Problems for Section 9.5 Online Resource: Additional Problems for Section 9.5

EXERCISES

Which of the series in Exercises 1–4 are power series?

1. $x - x^3 + x^6 - x^{10} + x^{15} - \cdots$
2. $\frac{1}{x} + \frac{1}{x^2} + \frac{1}{x^3} + \frac{1}{x^4} + \cdots$
3. $1 + x + (x-1)^2 + (x-2)^3 + (x-3)^4 + \cdots$
4. $x^7 + x + 2$

In Exercises 5–10, find an expression for the general term of the series. Give the starting value of the index (n or k, for example).

5. $\frac{1}{2}x + \frac{1\cdot 3}{2^2\cdot 2!}x^2 + \frac{1\cdot 3\cdot 5}{2^3\cdot 3!}x^3 + \cdots$
6. $px + \frac{p(p-1)}{2!}x^2 + \frac{p(p-1)(p-2)}{3!}x^3 + \cdots$
7. $1 - \frac{(x-1)^2}{2!} + \frac{(x-1)^4}{4!} - \frac{(x-1)^6}{6!} + \cdots$
8. $(x-1)^3 - \frac{(x-1)^5}{2!} + \frac{(x-1)^7}{4!} - \frac{(x-1)^9}{6!} + \cdots$
9. $\frac{x-a}{1} + \frac{(x-a)^2}{2\cdot 2!} + \frac{(x-a)^3}{4\cdot 3!} + \frac{(x-a)^4}{8\cdot 4!} + \cdots$
10. $2(x+5)^3 + 3(x+5)^5 + \frac{4(x+5)^7}{2!} + \frac{5(x+5)^9}{3!} + \cdots$

In Exercises 11–23, find the radius of convergence.

11. $\sum_{n=0}^{\infty} nx^n$
12. $\sum_{n=0}^{\infty} (5x)^n$
13. $\sum_{n=0}^{\infty} n^3x^n$
14. $\sum_{n=0}^{\infty} (2^n + n^2)x^n$
15. $\sum_{n=0}^{\infty} \frac{(n+1)x^n}{2^n + n}$
16. $\sum_{n=1}^{\infty} \frac{2^n(x-1)^n}{n}$
17. $\sum_{n=1}^{\infty} \frac{(x-3)^n}{n2^n}$
18. $\sum_{n=0}^{\infty} (-1)^n \frac{x^{2n}}{(2n)!}$
19. $x - \frac{x^2}{4} + \frac{x^3}{9} - \frac{x^4}{16} + \frac{x^5}{25} - \cdots$
20. $1 + 2x + \frac{4x^2}{2!} + \frac{8x^3}{3!} + \frac{16x^4}{4!} + \frac{32x^5}{5!} + \cdots$
21. $1 + 2x + \frac{4!x^2}{(2!)^2} + \frac{6!x^3}{(3!)^2} + \frac{8!x^4}{(4!)^2} + \frac{10!x^5}{(5!)^2} + \cdots$
22. $3x + \frac{5}{2}x^2 + \frac{7}{3}x^3 + \frac{9}{4}x^4 + \frac{11}{5}x^5 + \cdots$
23. $x - \frac{x^3}{3} + \frac{x^5}{5} - \frac{x^7}{7} + \cdots$
24. Show that the radius of convergence of the power series $x - \frac{x^3}{3!} + \frac{x^5}{5!} - \frac{x^7}{7!} + \cdots$ in Example 7 is infinite.

PROBLEMS

25. (a) Determine the radius of convergence of the series
$$x - \frac{x^2}{2} + \frac{x^3}{3} - \frac{x^4}{4} + \cdots + (-1)^{n-1}\frac{x^n}{n} + \cdots.$$
What does this tell us about the interval of convergence of this series?
(b) Investigate convergence at the end points of the interval of convergence of this series.
26. Show that the series $\sum_{n=1}^{\infty} \frac{(2x)^n}{n}$ converges for $|x| < 1/2$. Investigate whether the series converges for $x = 1/2$ and $x = -1/2$.

In Problems 27–34, find the interval of convergence.

27. $\sum_{n=0}^{\infty} \frac{x^n}{3^n}$
28. $\sum_{n=2}^{\infty} \frac{(x-3)^n}{n}$
29. $\sum_{n=1}^{\infty} \frac{n^2x^{2n}}{2^{2n}}$
30. $\sum_{n=1}^{\infty} \frac{(-1)^n(x-5)^n}{2^n n^2}$
31. $\sum_{n=1}^{\infty} \frac{x^{2n+1}}{n!}$
32. $\sum_{n=0}^{\infty} n!x^n$
33. $\sum_{n=1}^{\infty} \frac{(5x)^n}{\sqrt{n}}$
34. $\sum_{n=1}^{\infty} \frac{(5x)^{2n}}{\sqrt{n}}$

■ In Problems 35–38, use the formula for the sum of a geometric series to find a power series centered at the origin that converges to the expression. For what values does the series converge?

35. $\dfrac{1}{1+2z}$

36. $\dfrac{2}{1+y^2}$

37. $\dfrac{3}{1-z/2}$

38. $\dfrac{8}{4+y}$

39. For all z-values for which it converges, the function f is defined by the series

$$f(z) = 5+5\left(\frac{z-3}{7}\right)+5\cdot\left(\frac{z-3}{7}\right)^2+5\cdot\left(\frac{z-3}{7}\right)^3+\cdots.$$

(a) Find $f(4)$.
(b) Find the interval of convergence of $f(z)$.

40. For all t-values for which it converges, the function f is defined by the series

$$f(t) = \sum_{n=0}^{\infty} \frac{(t-7)^n}{5^n}.$$

(a) Find $f(4)$.
(b) Find the interval of convergence of $f(t)$.

41. The series $\sum C_n x^n$ converges when $x = -4$ and diverges when $x = 7$. Decide whether each of the following statements is true or false, or whether this cannot be determined.

(a) The power series converges when $x = 10$.
(b) The power series converges when $x = 3$.
(c) The power series diverges when $x = 1$.
(d) The power series diverges when $x = 6$.

42. If $\sum C_n(x-3)^n$ converges at $x = 7$ and diverges at $x = 10$, what can you say about the convergence at $x = 11$? At $x = 5$? At $x = 0$?

43. The series $\sum C_n x^n$ converges at $x = -5$ and diverges at $x = 7$. What can you say about its radius of convergence?

44. The series $\sum C_n(x+7)^n$ converges at $x = 0$ and diverges at $x = -17$. What can you say about its radius of convergence?

45. The power series $C_0 + C_1x + C_2x^2 + C_3x^3 + \cdots$ converges to π for $x = k$. Find possible values for k, C_0, C_1, C_2, C_3, where C_i is an integer. [Hint: $\pi = 3 + 0.1 + 0.04 + 0.001 + \cdots$.]

46. For constant p, find the radius of convergence of the binomial power series:[10]

$$1 + px + \frac{p(p-1)x^2}{2!} + \frac{p(p-1)(p-2)x^3}{3!} + \cdots.$$

47. Show that if $C_0 + C_1x + C_2x^2 + C_3x^3 + \cdots$ converges for $|x| < R$ with R given by the ratio test, then so does $C_1 + 2C_2x + 3C_3x^2 + \cdots$. Assume $C_i \neq 0$ for all i.

48. For all x-values for which it converges, the function f is defined by the series

$$f(x) = \sum_{n=0}^{\infty} \frac{x^n}{n!}.$$

(a) What is $f(0)$?
(b) What is the domain of f?
(c) Assuming that f' can be calculated by differentiating the series term-by-term, find the series for $f'(x)$. What do you notice?
(d) Guess what well-known function f is.

49. From Exercise 24 we know the following series converges for all x. We define $g(x)$ to be its sum:

$$g(x) = \sum_{n=1}^{\infty} (-1)^{n-1} \frac{x^{2n-1}}{(2n-1)!}.$$

(a) Is $g(x)$ odd, even, or neither? What is $g(0)$?
(b) Assuming that derivatives can be computed term by term, show that $g''(x) = -g(x)$.
(c) Guess what well-known function g might be. Check your guess using $g(0)$ and $g'(0)$.

Strengthen Your Understanding

■ In Problems 50–52, explain what is wrong with the statement.

50. If $\lim_{n\to\infty} |C_{n+1}/C_n| = 0$, then the radius of convergence for $\sum C_n x^n$ is 0.

51. If $\sum C_n(x-1)^n$ converges at $x = 3$, then

$$\lim_{n\to\infty} \left|\frac{C_n}{C_{n+1}}\right| = 2.$$

52. The series $\sum C_n x^n$ diverges at $x = 2$ and converges at $x = 3$.

■ In Problems 53–56, give an example of:

53. A power series that is divergent at $x = 0$.

54. A power series that converges at $x = 5$ but nowhere else.

55. A power series that converges at $x = 10$ and diverges at $x = 6$.

56. A series $\sum C_n x^n$ with radius of convergence 1 and that converges at $x = 1$ and $x = -1$.

[10]For an explanation of the name, see Section 10.2.

Decide if the statements in Problems 57–69 are true or false. Give an explanation for your answer.

57. $\sum_{n=1}^{\infty}(x-n)^n$ is a power series.

58. If the power series $\sum C_n x^n$ converges for $x = 2$, then it converges for $x = 1$.

59. If the power series $\sum C_n x^n$ converges for $x = 1$, then the power series converges for $x = 2$.

60. If the power series $\sum C_n x^n$ does not converge for $x = 1$, then the power series does not converge for $x = 2$.

61. $\sum C_n(x-1)^n$ and $\sum C_n x^n$ have the same radius of convergence.

62. If $\sum C_n x^n$ and $\sum B_n x^n$ have the same radius of convergence, then the coefficients, C_n and B_n, must be equal.

63. If a power series converges at one endpoint of its interval of convergence, then it converges at the other.

64. A power series always converges at at least one point.

65. If the power series $\sum C_n x^n$ converges at $x = 10$, then it converges at $x = -9$.

66. If the power series $\sum C_n x^n$ converges at $x = 10$, then it converges at $x = -10$.

67. $-5 < x \leq 7$ is a possible interval of convergence of a power series.

68. $-3 < x < 2$ could be the interval of convergence of $\sum C_n x^n$.

69. If $-11 < x < 1$ is the interval of convergence of $\sum C_n(x-a)^n$, then $a = -5$.

70. The power series $\sum C_n x^n$ diverges at $x = 7$ and converges at $x = -3$. At $x = -9$, the series is

(a) Conditionally convergent
(b) Absolutely convergent
(c) Alternating
(d) Divergent
(e) Cannot be determined.

Online Resource: Review problems and Projects

Chapter Ten

APPROXIMATING FUNCTIONS USING SERIES

Contents

10.1 TAYLOR POLYNOMIALS

In this section, we see how to approximate a function by polynomials.

Linear Approximations

We already know how to approximate a function using a degree-1 polynomial, namely the tangent line approximation given in Section 3.9:

$$f(x) \approx f(a) + f'(a)(x - a).$$

The tangent line and the curve have the same slope at $x = a$. As Figure 10.1 suggests, the tangent line approximation to the function is generally more accurate for values of x close to a.

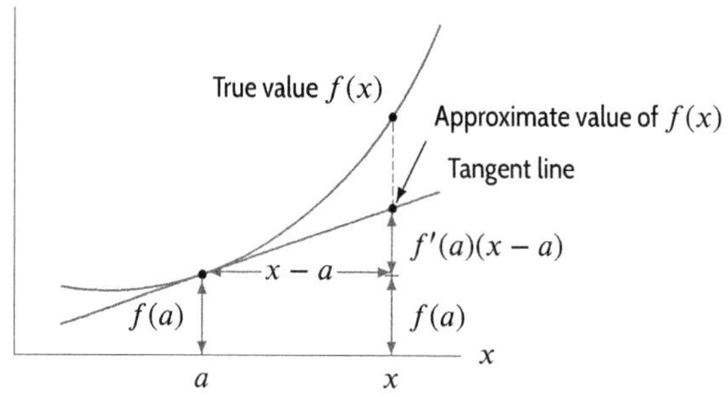

Figure 10.1: Tangent line approximation of $f(x)$ for x near a

We first focus on $a = 0$. The tangent line approximation at $x = 0$ is referred to as the *first Taylor approximation* at $x = 0$, or *first Maclaurin polynomial*, and written as follows:

> **Taylor Polynomial of Degree 1 Approximating $f(x)$ for x near 0**
>
> $$f(x) \approx P_1(x) = f(0) + f'(0)x$$

Example 1 Find the Taylor polynomial of degree 1 for $g(x) = \cos x$, with x in radians, for x near 0.

Solution The tangent line at $x = 0$ is just the horizontal line $y = 1$, as shown in Figure 10.2, so $P_1(x) = 1$. We have

$$g(x) = \cos x \approx 1, \quad \text{for } x \text{ near } 0.$$

If we take $x = 0.05$, then

$$g(0.05) = \cos(0.05) = 0.998\ldots,$$

which is quite close to the approximation $\cos x \approx 1$. Similarly, if $x = -0.1$, then

$$g(-0.1) = \cos(-0.1) = 0.995\ldots$$

is close to the approximation $\cos x \approx 1$. However, if $x = 0.4$, then

$$g(0.4) = \cos(0.4) = 0.921\ldots,$$

so the approximation $\cos x \approx 1$ is less accurate. For x near 0, the graph suggests that the farther x is from 0, the worse the approximation, $\cos x \approx 1$, is likely to be.

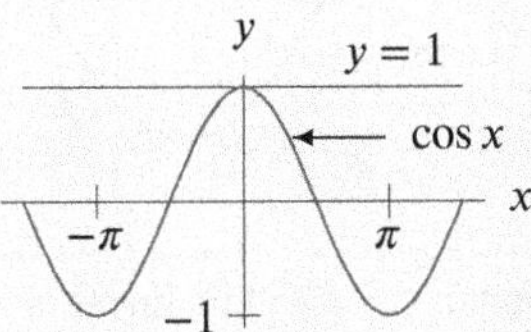

Figure 10.2: Graph of $\cos x$ and its tangent line at $x = 0$

The previous example shows that the Taylor polynomial of degree 1 might actually have degree less than 1.

Quadratic Approximations

To get a more accurate approximation, we use a quadratic function instead of a linear function.

Example 2 Find the quadratic approximation to $g(x) = \cos x$ for x near 0.

Solution To ensure that the quadratic, $P_2(x)$, is a good approximation to $g(x) = \cos x$ at $x = 0$, we require that $\cos x$ and the quadratic have the same value, the same slope, and the same second derivative at $x = 0$. That is, we require $P_2(0) = g(0)$, $P_2'(0) = g'(0)$, and $P_2''(0) = g''(0)$. We take the quadratic polynomial

$$P_2(x) = C_0 + C_1 x + C_2 x^2$$

and determine C_0, C_1, and C_2. Since

$$\begin{aligned} P_2(x) &= C_0 + C_1 x + C_2 x^2 \\ P_2'(x) &= C_1 + 2C_2 x \\ P_2''(x) &= 2C_2 \end{aligned} \qquad \text{and} \qquad \begin{aligned} g(x) &= \cos x \\ g'(x) &= -\sin x \\ g''(x) &= -\cos x, \end{aligned}$$

we have

$$\begin{aligned} C_0 &= P_2(0) = g(0) = \cos 0 = 1 \\ C_1 &= P_2'(0) = g'(0) = -\sin 0 = 0 \\ 2C_2 &= P_2''(0) = g''(0) = -\cos 0 = -1, \end{aligned} \qquad \text{so} \qquad \begin{aligned} C_0 &= 1 \\ C_1 &= 0 \\ C_2 &= -\tfrac{1}{2}. \end{aligned}$$

Consequently, the quadratic approximation is

$$\cos x \approx P_2(x) = 1 + 0 \cdot x - \frac{1}{2}x^2 = 1 - \frac{x^2}{2}, \quad \text{for } x \text{ near } 0.$$

Figure 10.3 suggests that the quadratic approximation $\cos x \approx P_2(x)$ is better than the linear approximation $\cos x \approx P_1(x)$ for x near 0. Let's compare the accuracy of the two approximations. Recall that $P_1(x) = 1$ for all x. At $x = 0.4$, we have $\cos(0.4) = 0.921\ldots$ and $P_2(0.4) = 0.920$, so the quadratic approximation is a significant improvement over the linear approximation. The magnitude of the error is about 0.001 instead of 0.08.

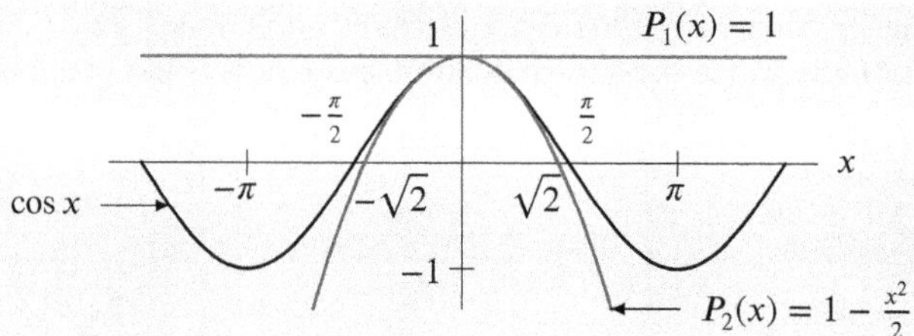

Figure 10.3: Graph of $\cos x$ and its linear, $P_1(x)$, and quadratic, $P_2(x)$, approximations for x near 0

Generalizing the computations in Example 2, we define the *second Taylor approximation* at $x = 0$, or *second Maclaurin polynomial*, as follows:

Taylor Polynomial of Degree 2 Approximating $f(x)$ for x near 0

$$f(x) \approx P_2(x) = f(0) + f'(0)x + \frac{f''(0)}{2}x^2$$

Higher-Degree Polynomials

Figure 10.3 shows that the quadratic approximation can still bend away from the original function for large x. We attempt to fix this by approximating $f(x)$ near 0 using a polynomial of higher degree:

$$f(x) \approx P_n(x) = C_0 + C_1x + C_2x^2 + \cdots + C_{n-1}x^{n-1} + C_nx^n.$$

To find the values of the constants: $C_0, C_1, C_2, \ldots, C_n$, we require that the function $f(x)$ and each of its first n derivatives agree with those of the polynomial $P_n(x)$ at the point $x = 0$. In general, the higher the degree of a Taylor polynomial, the larger the interval on which the function and the polynomial remain close to each other.

To see how to find the constants, let's take $n = 3$ as an example:

$$f(x) \approx P_3(x) = C_0 + C_1x + C_2x^2 + C_3x^3.$$

Substituting $x = 0$ gives

$$f(0) = P_3(0) = C_0.$$

Differentiating $P_3(x)$ yields

$$P_3'(x) = C_1 + 2C_2x + 3C_3x^2,$$

so substituting $x = 0$ shows that

$$f'(0) = P_3'(0) = C_1.$$

Differentiating and substituting again, we get

$$P_3''(x) = 2 \cdot 1C_2 + 3 \cdot 2 \cdot 1C_3x,$$

which gives

$$f''(0) = P_3''(0) = 2 \cdot 1C_2,$$

so that

$$C_2 = \frac{f''(0)}{2 \cdot 1}.$$

The third derivative, denoted by P_3''', is

$$P_3'''(x) = 3 \cdot 2 \cdot 1C_3,$$

so

$$f'''(0) = P_3'''(0) = 3 \cdot 2 \cdot 1C_3,$$

and then

$$C_3 = \frac{f'''(0)}{3 \cdot 2 \cdot 1}.$$

You can imagine a similar calculation starting with $P_4(x)$ and using the fourth derivative $f^{(4)}$, which would give

$$C_4 = \frac{f^{(4)}(0)}{4 \cdot 3 \cdot 2 \cdot 1},$$

and so on. Using factorial notation,[1] we write these expressions as

$$C_3 = \frac{f'''(0)}{3!}, \quad C_4 = \frac{f^{(4)}(0)}{4!}.$$

Writing $f^{(n)}$ for the nth derivative of f, we have, for any positive integer n,

$$C_n = \frac{f^{(n)}(0)}{n!}.$$

So we define the nth *Taylor approximation* at $x = 0$, or n^{th} *Maclaurin polynomial*, as follows:

Taylor Polynomial of Degree n Approximating $f(x)$ for x near 0

$$\begin{aligned} f(x) &\approx P_n(x) \\ &= f(0) + f'(0)x + \frac{f''(0)}{2!}x^2 + \frac{f'''(0)}{3!}x^3 + \frac{f^{(4)}(0)}{4!}x^4 + \cdots + \frac{f^{(n)}(0)}{n!}x^n \end{aligned}$$

We call $P_n(x)$ the Taylor polynomial of degree n centered at $x = 0$ or the Taylor polynomial about (or around) $x = 0$.

Example 3 Construct the Taylor polynomial of degree 7 approximating the function $f(x) = \sin x$ for x near 0. Compare the value of the Taylor approximation with the true value of f at $x = \pi/3$.

Solution We have

$$\begin{aligned} f(x) &= \sin x & \text{giving} \quad f(0) &= 0 \\ f'(x) &= \cos x & f'(0) &= 1 \\ f''(x) &= -\sin x & f''(0) &= 0 \\ f'''(x) &= -\cos x & f'''(0) &= -1 \\ f^{(4)}(x) &= \sin x & f^{(4)}(0) &= 0 \\ f^{(5)}(x) &= \cos x & f^{(5)}(0) &= 1 \\ f^{(6)}(x) &= -\sin x & f^{(6)}(0) &= 0 \\ f^{(7)}(x) &= -\cos x & f^{(7)}(0) &= -1. \end{aligned}$$

Using these values, we see that the Taylor polynomial approximation of degree 7 is

$$\begin{aligned} \sin x \approx P_7(x) &= 0 + 1 \cdot x + \frac{0}{2!} \cdot x^2 - \frac{1}{3!} \cdot x^3 + \frac{0}{4!} \cdot x^4 + \frac{1}{5!} \cdot x^5 + \frac{0}{6!} \cdot x^6 - \frac{1}{7!} \cdot x^7 \\ &= x - \frac{x^3}{3!} + \frac{x^5}{5!} - \frac{x^7}{7!}, \quad \text{for } x \text{ near } 0. \end{aligned}$$

[1] Recall that $k! = k(k-1)\cdots 2 \cdot 1$. In addition, $1! = 1$, and $0! = 1$.

Notice that since $f^{(8)}(0) = 0$, the seventh and eighth Taylor approximations to $\sin x$ are the same.

In Figure 10.4 we show the graphs of the sine function and the approximating polynomial of degree 7 for x near 0. They are indistinguishable where x is close to 0. However, as we look at values of x farther away from 0 in either direction, the two graphs move apart. To check the accuracy of this approximation numerically, we see how well it approximates $\sin(\pi/3) = \sqrt{3}/2 = 0.8660254\ldots$.

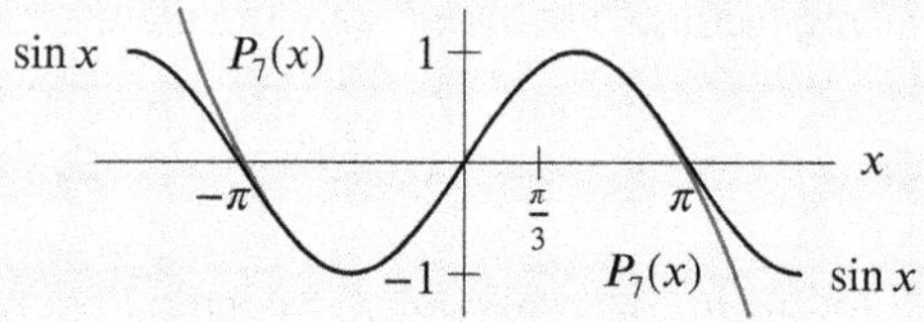

Figure 10.4: Graph of $\sin x$ and its seventh-degree Taylor polynomial, $P_7(x)$, for x near 0

When we substitute $\pi/3 = 1.0471976\ldots$ into the polynomial approximation, we obtain $P_7(\pi/3) = 0.8660213\ldots$, which is extremely accurate—to about four parts in a million.

Example 4 Graph the Taylor polynomial of degree 8 approximating $g(x) = \cos x$ for x near 0.

Solution We find the coefficients of the Taylor polynomial by the method of the preceding example, giving

$$\cos x \approx P_8(x) = 1 - \frac{x^2}{2!} + \frac{x^4}{4!} - \frac{x^6}{6!} + \frac{x^8}{8!}.$$

Figure 10.5 shows that $P_8(x)$ is close to the cosine function for a larger interval of x-values than the quadratic approximation $P_2(x) = 1 - x^2/2$ in Example 2 on page 515.

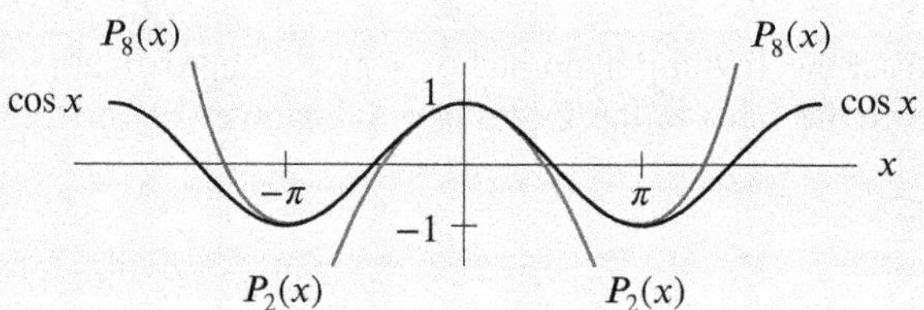

Figure 10.5: $P_8(x)$ approximates $\cos x$ better than $P_2(x)$ for x near 0

Example 5 Construct the Taylor polynomial of degree 10 about $x = 0$ for the function $f(x) = e^x$. Check the accuracy of this approximation at $x = 1$.

Solution We have $f(0) = 1$. Since the derivative of e^x is equal to e^x, all the higher-order derivatives are equal to e^x. Consequently, for any $k = 1, 2, \ldots, 10$, $f^{(k)}(x) = e^x$ and $f^{(k)}(0) = e^0 = 1$. Therefore, the Taylor polynomial approximation of degree 10 is given by

$$e^x \approx P_{10}(x) = 1 + x + \frac{x^2}{2!} + \frac{x^3}{3!} + \frac{x^4}{4!} + \cdots + \frac{x^{10}}{10!}, \quad \text{for } x \text{ near } 0.$$

To check the accuracy of the approximation, we substitute $x = 1$ to get $P_{10}(1) = 2.718281801$ and compare with $e = e^1 = 2.718281828\ldots$. We see P_{10} yields the first seven decimal places for e.

Figure 10.6 shows graphs of $f(x) = e^x$ and the Taylor polynomials of degree $n = 0, 1, 2, 3, 4$. Notice that each successive approximation remains close to the exponential curve for a larger interval of x-values. For any n, the accuracy decreases for large x because e^x grows faster than any polynomial as $x \to \infty$.

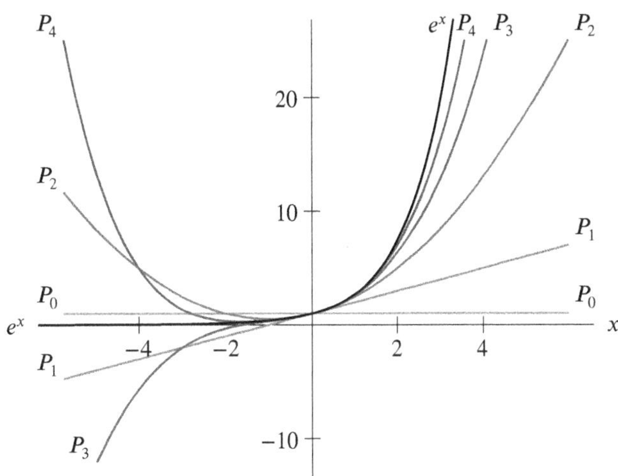

Figure 10.6: For x near 0, the value of e^x is more closely approximated by higher-degree Taylor polynomials

Example 6 Construct the Taylor polynomial of degree n approximating $f(x) = \dfrac{1}{1-x}$ for x near 0.

Solution Differentiating gives $f(0) = 1$, $f'(0) = 1$, $f''(0) = 2$, $f'''(0) = 3!$, $f^{(4)}(0) = 4!$, and so on. This means

$$\frac{1}{1-x} \approx P_n(x) = 1 + x + x^2 + x^3 + x^4 + \cdots + x^n, \quad \text{for } x \text{ near } 0.$$

Let us compare the Taylor polynomial with the formula obtained from the sum of a finite geometric series on page 482:

$$\frac{1-x^{n+1}}{1-x} = 1 + x + x^2 + x^3 + x^4 + \cdots + x^n.$$

If x is close to 0 and x^{n+1} is small enough to neglect, the formula for the sum of a finite geometric series gives us the Taylor approximation of degree n:

$$\frac{1}{1-x} \approx 1 + x + x^2 + x^3 + x^4 + \cdots + x^n.$$

Taylor Polynomials Around $x = a$

Suppose we want to approximate $f(x) = \ln x$ by a Taylor polynomial. This function has no Taylor polynomial about $x = 0$ because the function is not defined for $x \leq 0$. However, it turns out that we can construct a polynomial centered about some other point, $x = a$.

First, let's look at the equation of the tangent line at $x = a$:

$$y = f(a) + f'(a)(x-a).$$

This gives the first Taylor approximation

$$f(x) \approx f(a) + f'(a)(x-a) \quad \text{for } x \text{ near } a.$$

The $f'(a)(x-a)$ term is a correction that approximates the change in f as x moves away from a.

Similarly, the Taylor polynomial $P_n(x)$ centered at $x = a$ is set up as $f(a)$ plus correction terms that are zero for $x = a$. This is achieved by writing the polynomial in powers of $(x - a)$ instead of powers of x:

$$f(x) \approx P_n(x) = C_0 + C_1(x-a) + C_2(x-a)^2 + \cdots + C_n(x-a)^n.$$

If we require n derivatives of the approximating polynomial $P_n(x)$ and the original function $f(x)$ to agree at $x = a$, we get the following result for the nth Taylor approximation at $x = a$:

Taylor Polynomial of Degree n Approximating $f(x)$ for x near a

$$f(x) \approx P_n(x)$$
$$= f(a) + f'(a)(x-a) + \frac{f''(a)}{2!}(x-a)^2 + \cdots + \frac{f^{(n)}(a)}{n!}(x-a)^n$$

We call $P_n(x)$ the Taylor polynomial of degree n centered at $x = a$ or the Taylor polynomial about $x = a$.

The formulas for these coefficients are derived in the same way as we did for $a = 0$. (See Problem 42, page 522.)[2]

Example 7 Construct the Taylor polynomial of degree 4 approximating the function $f(x) = \ln x$ for x near 1.

Solution We have

$$\begin{aligned}
f(x) &= \ln x & \text{so} \quad f(1) &= \ln(1) = 0\\
f'(x) &= 1/x & f'(1) &= 1\\
f''(x) &= -1/x^2 & f''(1) &= -1\\
f'''(x) &= 2/x^3 & f'''(1) &= 2\\
f^{(4)}(x) &= -6/x^4, & f^{(4)}(1) &= -6.
\end{aligned}$$

The Taylor polynomial is therefore

$$\begin{aligned}
\ln x \approx P_4(x) &= 0 + (x-1) - \frac{(x-1)^2}{2!} + 2\frac{(x-1)^3}{3!} - 6\frac{(x-1)^4}{4!}\\
&= (x-1) - \frac{(x-1)^2}{2} + \frac{(x-1)^3}{3} - \frac{(x-1)^4}{4}, \qquad \text{for } x \text{ near } 1.
\end{aligned}$$

Graphs of $\ln x$ and several of its Taylor polynomials are shown in Figure 10.7. Notice that $P_4(x)$ stays reasonably close to $\ln x$ for x near 1, but bends away as x gets farther from 1. Also, note that the Taylor polynomials are defined for $x \leq 0$, but $\ln x$ is not.

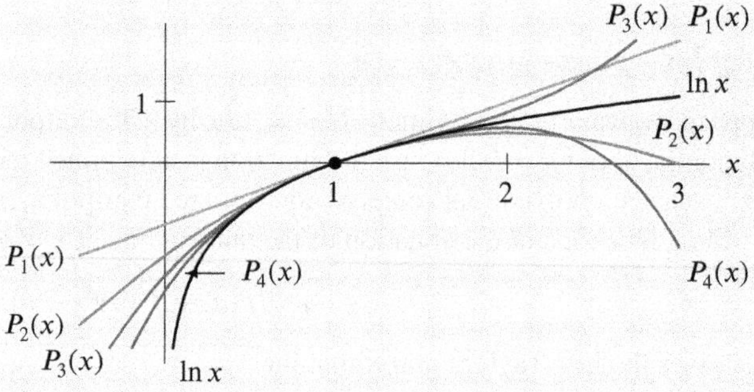

Figure 10.7: Taylor polynomials approximate $\ln x$ closely for x near 1, but not necessarily farther away

[2] A Taylor polynomial is also called a Maclaurin polynomial if $a = 0$.

The examples in this section suggest that the following results are true for common functions:

- Taylor polynomials centered at $x = a$ give good approximations to $f(x)$ for x near a. Farther away, they may or may not be good.
- Typically, a Taylor polynomial of higher degree fits the function closely over a larger interval.

Exercises and Problems for Section 10.1

EXERCISES

For Exercises 1–10, find the Taylor polynomials of degree n approximating the functions for x near 0. (Assume p is a constant.)

1. $\dfrac{1}{1-x}$, $n = 3, 5, 7$
2. $\dfrac{1}{1+x}$, $n = 4, 6, 8$
3. $\sqrt{1+x}$, $n = 2, 3, 4$
4. $\sqrt[3]{1-x}$, $n = 2, 3, 4$
5. $\cos x$, $n = 2, 4, 6$
6. $\ln(1+x)$, $n = 5, 7, 9$
7. $\arctan x$, $n = 3, 4$
8. $\tan x$, $n = 3, 4$
9. $\dfrac{1}{\sqrt{1+x}}$, $n = 2, 3, 4$
10. $(1+x)^p$, $n = 2, 3, 4$

For Exercises 11–16, find the Taylor polynomial of degree n for x near the given point a.

11. $\sqrt{1-x}$, $a = 0$, $n = 3$
12. e^x, $a = 1$, $n = 4$
13. $\dfrac{1}{1+x}$, $a = 2$, $n = 4$
14. $\cos x$, $a = \pi/2$, $n = 4$
15. $\sin x$, $a = -\pi/4$, $n = 3$
16. $\ln(x^2)$, $a = 1$, $n = 4$

PROBLEMS

17. The function $f(x)$ is approximated near $x = 0$ by the third-degree Taylor polynomial
$$P_3(x) = 2 - x - x^2/3 + 2x^3.$$
Give the value of
(a) $f(0)$ **(b)** $f'(0)$
(c) $f''(0)$ **(d)** $f'''(0)$

18. Find the second-degree Taylor polynomial for $f(x) = 4x^2 - 7x + 2$ about $x = 0$. What do you notice?

19. Find the third-degree Taylor polynomial for $f(x) = x^3 + 7x^2 - 5x + 1$ about $x = 0$. What do you notice?

20. **(a)** Based on your observations in Problems 18–19, make a conjecture about Taylor approximations in the case when f is itself a polynomial.
(b) Show that your conjecture is true.

21. The Taylor polynomial of degree 7 of $f(x)$ is given by
$$P_7(x) = 1 - \frac{x}{3} + \frac{5x^2}{7} + 8x^3 - \frac{x^5}{11} + 8x^7.$$
Find the Taylor polynomial of degree 3 of $f(x)$.

22. Find the value of $f^{(5)}(1)$ if $f(x)$ is approximated near $x = 1$ by the Taylor polynomial
$$p(x) = \sum_{n=0}^{10} \frac{(x-1)^n}{n!}.$$

In Problems 23–24, find a simplified formula for $P_5(x)$, the fifth-degree Taylor polynomial approximating f near $x = 0$.

23. Use the values in the table.

$f(0)$	$f'(0)$	$f''(0)$	$f'''(0)$	$f^{(4)}(0)$	$f^{(5)}(0)$
−3	5	−2	0	−1	4

24. Let $f(0) = -1$ and, for $n > 0$, $f^{(n)}(0) = -(-2)^n$.

25. **(a)** From a graph of $f(x) = \ln(2x+4)$ decide if the tangent line approximation to f at $x = 0$ has a positive or negative slope.
(b) Find the tangent line approximation.
(c) Find the second and third Taylor approximations to f about $x = 0$.

For Problems 26–29, suppose $P_2(x) = a + bx + cx^2$ is the second-degree Taylor polynomial for the function f about $x = 0$. What can you say about the signs of a, b, and c if f has the graph given below?

26. 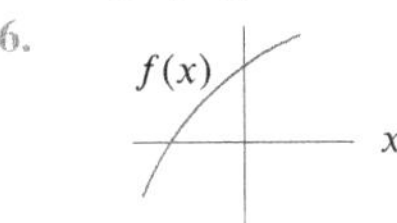

27. 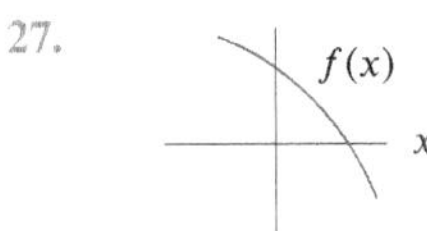

28. 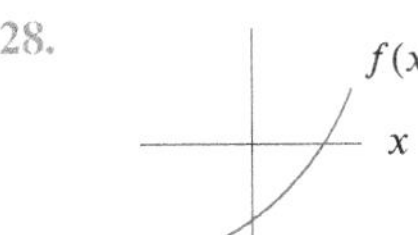

29. 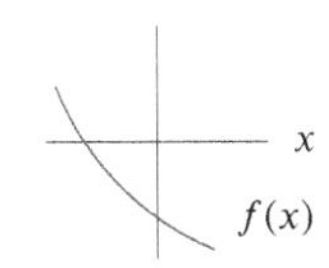

■ In Problems 30–33, use the third degree Taylor polynomial $P_3(x) = 4 + 2(x-1) - 2(x-1)^2 + (x-1)^3/2$ of $f(x)$ about $x = 1$ to find the given value, or explain why you can't.

30. $f(1)$ 31. $f'''(1)$ 32. $f''(0)$ 33. $f^{(4)}(1)$

■ In Problems 34–36, use a Taylor polynomial with the derivatives given to make the best possible estimate of the value.

34. $f(1.1)$, given that $f(1) = 3, f'(1) = 2, f''(1) = -4$.

35. $g(1.8)$, given that $g(2) = -3, g'(2) = -2, g''(2) = 0, g'''(2) = 12$.

36. $h'(0.1)$, given that $h(0) = 6, h'(0) = 2, h''(0) = -4$.

37. Use the Taylor approximation for x near 0,

$$\sin x \approx x - \frac{x^3}{3!},$$

to explain why $\lim_{x\to 0} \frac{\sin x}{x} = 1$.

38. Use the fourth-degree Taylor approximation for x near 0,

$$\cos x \approx 1 - \frac{x^2}{2!} + \frac{x^4}{4!},$$

to explain why $\lim_{x\to 0} \frac{1-\cos x}{x^2} = \frac{1}{2}$.

39. Use a fourth-degree Taylor approximation for e^h, for h near 0, to evaluate the following limits. Would your answer be different if you used a Taylor polynomial of higher degree?

(a) $\lim_{h\to 0} \frac{e^h - 1 - h}{h^2}$

(b) $\lim_{h\to 0} \frac{e^h - 1 - h - \frac{h^2}{2}}{h^3}$

40. If $f(2) = g(2) = h(2) = 0$, and $f'(2) = h'(2) = 0$, $g'(2) = 22$, and $f''(2) = 3$, $g''(2) = 5$, $h''(2) = 7$, calculate the following limits. Explain your reasoning.

(a) $\lim_{x\to 2} \frac{f(x)}{h(x)}$ **(b)** $\lim_{x\to 2} \frac{f(x)}{g(x)}$

41. One of the two sets of functions, f_1, f_2, f_3, or g_1, g_2, g_3, is graphed in Figure 10.8; the other set is graphed in Figure 10.9. Points A and B each have $x = 0$. Taylor polynomials of degree 2 approximating these functions near $x = 0$ are as follows:

$$f_1(x) \approx 2 + x + 2x^2 \qquad g_1(x) \approx 1 + x + 2x^2$$
$$f_2(x) \approx 2 + x - x^2 \qquad g_2(x) \approx 1 + x + x^2$$
$$f_3(x) \approx 2 + x + x^2 \qquad g_3(x) \approx 1 - x + x^2.$$

(a) Which group of functions, the fs or the gs, is represented by each figure?
(b) What are the coordinates of the points A and B?
(c) Match each function with the graphs (I)–(III) in the appropriate figure.

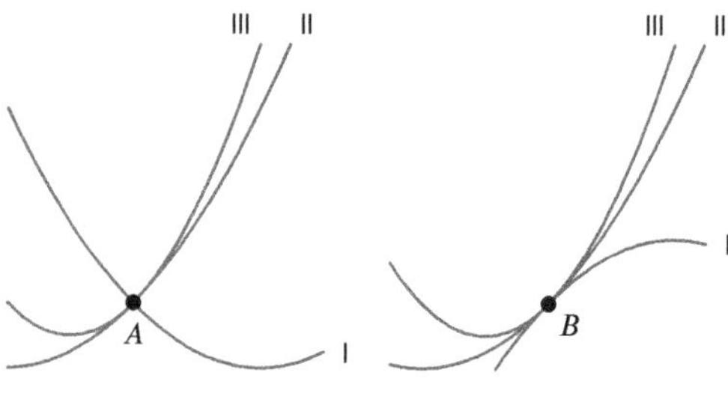

Figure 10.8 Figure 10.9

42. Derive the formulas given in the box on page 520 for the coefficients of the Taylor polynomial approximating a function f for x near a.

43. **(a)** Find and multiply the Taylor polynomials of degree 1 near $x = 0$ for the two functions $f(x) = 1/(1-x)$ and $g(x) = 1/(1-2x)$.
(b) Find the Taylor polynomial of degree 2 near $x = 0$ for the function $h(x) = f(x)g(x)$.
(c) Is the product of the Taylor polynomials for $f(x)$ and $g(x)$ equal to the Taylor polynomial for the function $h(x)$?

44. **(a)** Find and multiply the Taylor polynomials of degree 1 near $x = 0$ for the two functions $f(x)$ and $g(x)$.
(b) Find the Taylor polynomial of degree 2 near $x = 0$ for the function $h(x) = f(x)g(x)$.
(c) Show that the product of the Taylor polynomials for $f(x)$ and $g(x)$ and the Taylor polynomial for the function $h(x)$ are the same if $f''(0)g(0) + f(0)g''(0) = 0$.

45. **(a)** Find the Taylor polynomial approximation of degree 4 about $x = 0$ for the function $f(x) = e^{x^2}$.
(b) Compare this result to the Taylor polynomial approximation of degree 2 for the function $f(x) = e^x$ about $x = 0$. What do you notice?
(c) Use your observation in part (b) to write out the Taylor polynomial approximation of degree 20 for the function in part (a).
(d) What is the Taylor polynomial approximation of degree 5 for the function $f(x) = e^{-2x}$?

46. The integral $\int_0^1 (\sin t/t)\,dt$ is difficult to approximate using, for example, left Riemann sums or the trapezoid rule because the integrand $(\sin t)/t$ is not defined at $t = 0$. However, this integral converges; its value is 0.94608 .... Estimate the integral using Taylor polynomials for $\sin t$ about $t = 0$ of

(a) Degree 3 **(b)** Degree 5

47. Consider the equations $\sin x = 0.2$ and $x - \frac{x^3}{3!} = 0.2$.

(a) How many solutions does each equation have?
(b) Which of the solutions of the two equations are approximately equal? Explain.

48. When we model the motion of a pendulum, we sometimes replace the differential equation

$$\frac{d^2\theta}{dt^2} = -\frac{g}{l}\sin\theta \quad \text{by} \quad \frac{d^2\theta}{dt^2} = -\frac{g}{l}\theta,$$

where θ is the angle between the pendulum and the vertical. Explain why, and under what circumstances, it is reasonable to make this replacement.

49. **(a)** Using a graph, explain why the following equation has a solution at $x = 0$ and another just to the right of $x = 0$:

$$\cos x = 1 - 0.1x.$$

(b) Replace $\cos x$ by its second-degree Taylor polynomial near 0 and solve the equation. Your answers are approximations to the solutions to the original equation at or near 0.

50. Let $P_n(x)$ be the Taylor polynomial of degree n approximating $f(x)$ near $x = 0$ and $Q_n(x)$ the Taylor polynomial of degree n approximating $f'(x)$ near $x = 0$. Show that $Q_{n-1}(x) = P_n'(x)$.

51. Let $P_n(x)$ be the Taylor polynomial of degree n approximating $f(x)$ near $x = 0$ and $Q_n(x)$ the Taylor polynomial of degree n approximating its antiderivative $F(x)$ with $F(0) = 1$. Show that $Q_{n+1}(x) = 1 + \int_0^x P_n(t)\,dt$.

Strengthen Your Understanding

In Problems 52–54, explain what is wrong with the statement.

52. If $f(x) = \ln(2+x)$, then the second-degree Taylor polynomial approximating $f(x)$ near $x = 0$ has a negative constant term.

53. Let $f(x) = \dfrac{1}{1-x}$. The coefficient of the x term of the Taylor polynomial of degree 3 approximating $f(x)$ near $x = 0$ is -1.

54. If the Taylor polynomial of degree 3 approximating f about $x = 0$ is

$$P_3(x) = \frac{x}{2} + \frac{x^2}{3} + \frac{x^3}{4},$$

then the Taylor polynomial of degree 3 approximating f about $x = 1$ is

$$P_3(x) = \frac{(x-1)}{2} + \frac{(x-1)^2}{3} + \frac{(x-1)^3}{4}.$$

In Problems 55–58, give an example of:

55. A function $f(x)$ for which every Taylor polynomial approximation near $x = 0$ involves only odd powers of x.

56. A third-degree Taylor polynomial near $x = 1$ approximating a function $f(x)$ with $f'(1) = 3$.

57. Two different functions having the same Taylor polynomial of degree 1 about $x = 0$.

58. A function whose Taylor polynomial of degree 1 about $x = 0$ is closer to the values of the function for some values of x than its Taylor polynomial of degree 2 about that point.

Decide if the statements in Problems 59–66 are true or false. Give an explanation for your answer.

59. If $f(x)$ and $g(x)$ have the same Taylor polynomial of degree 2 near $x = 0$, then $f(x) = g(x)$.

60. Using $\sin\theta \approx \theta - \theta^3/3!$ with $\theta = 1°$, we have $\sin(1°) \approx 1 - 1^3/6 = 5/6$.

61. The Taylor polynomial of degree 2 for e^x near $x = 5$ is $1 + (x-5) + (x-5)^2/2$.

62. If the Taylor polynomial of degree 2 for $f(x)$ near $x = 0$ is $P_2(x) = 1+x-x^2$, then $f(x)$ is concave up near $x = 0$.

63. The quadratic approximation to $f(x)$ for x near 0 is better than the linear approximation for all values of x.

64. A Taylor polynomial for f near $x = a$ touches the graph of f only at $x = a$.

65. The linear approximation to $f(x)$ near $x = -1$ shows that if $f(-1) = g(-1)$ and $f'(-1) < g'(-1)$, then $f(x) < g(x)$ for all x sufficiently close to -1 (but not equal to -1).

66. The quadratic approximation to $f(x)$ near $x = -1$ shows that if $f(-1) = g(-1)$, $f'(-1) = g'(-1)$, and $f''(-1) < g''(-1)$, then $f(x) < g(x)$ for all x sufficiently close to -1 (but not equal to -1).

10.2 TAYLOR SERIES

In the previous section we saw how to approximate a function near a point by Taylor polynomials. Now we define a Taylor series, which is a power series that can be thought of as a Taylor polynomial that goes on forever.

Taylor Series for $\cos x$, $\sin x$, e^x

We have the following Taylor polynomials centered at $x = 0$ for $\cos x$:

$$\cos x \approx P_0(x) = 1$$
$$\cos x \approx P_2(x) = 1 - \frac{x^2}{2!}$$
$$\cos x \approx P_4(x) = 1 - \frac{x^2}{2!} + \frac{x^4}{4!}$$
$$\cos x \approx P_6(x) = 1 - \frac{x^2}{2!} + \frac{x^4}{4!} - \frac{x^6}{6!}$$
$$\cos x \approx P_8(x) = 1 - \frac{x^2}{2!} + \frac{x^4}{4!} - \frac{x^6}{6!} + \frac{x^8}{8!}.$$

Here we have a sequence of polynomials, $P_0(x)$, $P_2(x)$, $P_4(x)$, $P_6(x)$, $P_8(x)$, ..., each of which is a better approximation to $\cos x$ than the last, for x near 0. When we go to a higher-degree polynomial (say from P_6 to P_8), we add more terms ($x^8/8!$, for example), but the terms of lower degree don't change. Thus, each polynomial includes the information from all the previous ones. We represent the whole sequence of Taylor polynomials by writing the *Taylor series* for $\cos x$:

$$1 - \frac{x^2}{2!} + \frac{x^4}{4!} - \frac{x^6}{6!} + \frac{x^8}{8!} - \cdots.$$

Notice that the partial sums of this series are the Taylor polynomials, $P_n(x)$.

We define the Taylor series for $\sin x$ and e^x similarly. It turns out that, for these functions, the Taylor series converges to the function for all x, so we can write the following:

$$e^x = 1 + x + \frac{x^2}{2!} + \frac{x^3}{3!} + \frac{x^4}{4!} + \cdots$$
$$\sin x = x - \frac{x^3}{3!} + \frac{x^5}{5!} - \frac{x^7}{7!} + \frac{x^9}{9!} - \cdots$$
$$\cos x = 1 - \frac{x^2}{2!} + \frac{x^4}{4!} - \frac{x^6}{6!} + \frac{x^8}{8!} - \cdots$$

These series are also called *Maclaurin series* or *Taylor expansions* of the functions $\sin x$, $\cos x$, and e^x about $x = 0$. Notice that the series for e^x contains both odd and even powers, while $\sin x$ has only odd powers and $\cos x$ has only even powers.[3]

The *general term* of a Taylor series is a formula that gives any term in the series. For example, $x^n/n!$ is the general term in the Taylor expansion for e^x, since substituting $n = 0, 1, 2, \ldots$ gives each of the terms in the series for e^x. Similarly, $(-1)^k x^{2k}/(2k)!$ is the general term in the expansion for $\cos x$ because substituting $k = 0, 1, 2, \ldots$ gives each of its terms. We call n or k the *index*.

Taylor Series in General

Any function f, all of whose derivatives exist at 0, has a Taylor series. However, the Taylor series for f does not necessarily converge to $f(x)$ for all values of x. For the values of x for which the series does converge to $f(x)$, we have the following formula:

Taylor Series for $f(x)$ About $x = 0$

$$f(x) = f(0) + f'(0)x + \frac{f''(0)}{2!}x^2 + \frac{f'''(0)}{3!}x^3 + \cdots + \frac{f^{(n)}(0)}{n!}x^n + \cdots$$

[3] A constant term is an even power.

In addition, just as we have Taylor polynomials centered at points other than 0, we can also have a Taylor series centered at $x = a$ (provided all the derivatives of f exist at $x = a$). For the values of x for which the series converges to $f(x)$, we have the following formula:

Taylor Series for $f(x)$ About $x = a$

$$f(x) = f(a) + f'(a)(x-a) + \frac{f''(a)}{2!}(x-a)^2 + \frac{f'''(a)}{3!}(x-a)^3 + \cdots + \frac{f^{(n)}(a)}{n!}(x-a)^n + \cdots$$

The Taylor series is a power series whose partial sums are the Taylor polynomials. As we saw in Section 9.5, power series converge on an interval centered at $x = a$.

For a given function f and a given x, even if the Taylor series converges, it might not converge to $f(x)$. However, the Taylor series for most common functions, including e^x, $\cos x$, and $\sin x$, do converge to the original function for all x. See Section 10.4.

The Binomial Series Expansion

We now find the Taylor series about $x = 0$ for the function $f(x) = (1+x)^p$, with p a constant, but not necessarily a positive integer. Taking derivatives:

$$\begin{aligned}
f(x) &= (1+x)^p & \text{so}\quad f(0) &= 1\\
f'(x) &= p(1+x)^{p-1} & f'(0) &= p\\
f''(x) &= p(p-1)(1+x)^{p-2} & f''(0) &= p(p-1)\\
f'''(x) &= p(p-1)(p-2)(1+x)^{p-3}, & f'''(0) &= p(p-1)(p-2).
\end{aligned}$$

Thus, the third-degree Taylor polynomial for x near 0 is

$$(1+x)^p \approx P_3(x) = 1 + px + \frac{p(p-1)}{2!}x^2 + \frac{p(p-1)(p-2)}{3!}x^3.$$

Graphing $P_3(x), P_4(x), \ldots$ for various specific values of p suggests that the Taylor polynomials converge to $f(x)$ for $-1 < x < 1$ for all p, and in fact this is correct. (See Problems 38–39, page 529.) The Taylor series for $f(x) = (1+x)^p$ about $x = 0$ is as follows:

The Binomial Series

$$(1+x)^p = 1 + px + \frac{p(p-1)}{2!}x^2 + \frac{p(p-1)(p-2)}{3!}x^3 + \cdots \qquad \text{for } -1 < x < 1.$$

In fact, when p is a positive integer, the binomial series gives the same result as multiplying $(1+x)^p$ out. (Newton discovered that the binomial series can be used for non-integer exponents.)

Example 1 Find the Taylor series about $x = 0$ for $\frac{1}{1+x}$. Explain why the result is a geometric series.

Solution Since $\frac{1}{1+x} = (1+x)^{-1}$, use the binomial series with $p = -1$. Then we have

$$\begin{aligned}
\frac{1}{1+x} = (1+x)^{-1} &= 1 + (-1)x + \frac{(-1)(-2)}{2!}x^2 + \frac{(-1)(-2)(-3)}{3!}x^3 + \cdots\\
&= 1 - x + x^2 - x^3 + \cdots \qquad \text{for } -1 < x < 1.
\end{aligned}$$

This binomial series is also geometric since each term is $-x$ times the previous term. The common ratio is $-x$.

Example 2 Find the Taylor series about $x = 0$ for $\sqrt{1+x}$.

Solution Since $\sqrt{1+x} = (1+x)^{1/2}$, we use the binomial series with $p = 1/2$. Then

$$\begin{aligned} f(x) = \sqrt{1+x} &= 1 + \frac{1}{2}x + \frac{(\frac{1}{2})(-\frac{1}{2})x^2}{2!} + \frac{(\frac{1}{2})(-\frac{1}{2})(-\frac{3}{2})x^3}{3!} + \cdots \\ &= 1 + \frac{1}{2}x + \frac{(-\frac{1}{4})x^2}{2!} + \frac{(\frac{3}{8})x^3}{3!} + \cdots \\ &= 1 + \frac{x}{2} - \frac{x^2}{8} + \frac{x^3}{16} + \cdots \qquad \text{for } -1 < x < 1. \end{aligned}$$

Convergence of Taylor Series

Let us look again at the Taylor polynomial for $\ln x$ about $x = 1$ that we derived in Example 7 on page 520. A similar calculation gives the Taylor series

$$\ln x = (x-1) - \frac{(x-1)^2}{2} + \frac{(x-1)^3}{3} - \frac{(x-1)^4}{4} + \cdots + (-1)^{n-1}\frac{(x-1)^n}{n} + \cdots .$$

Using the ratio test, we get convergence on $0 < x < 2$. Substituting $x = 0$ gives a divergent series (the negative harmonic series); substituting $x = 2$ gives a convergent series by the alternating series test. Thus, this power series has interval of convergence $0 < x \leq 2$.

However, although we know that the series converges in this interval, we do not yet know that its sum is $\ln x$. Figure 10.10 shows the polynomials fit the curve well for $0 < x < 2$, suggesting that the Taylor series does converge to $\ln x$ for $0 < x \leq 2$, as turns out to be the case.

However, when $x > 2$, the polynomials move away from the curve and the approximations get worse as the degree of the polynomial increases. For $x \leq 0$, the function $\ln x$ is not defined. Thus, the Taylor polynomials are effective only as approximations to $\ln x$ for values of x between 0 and 2; outside that interval, they should not be used. Inside the interval, but near the ends, 0 or 2, the polynomials converge very slowly. This means we might have to take a polynomial of very high degree to get an accurate value for $\ln x$.

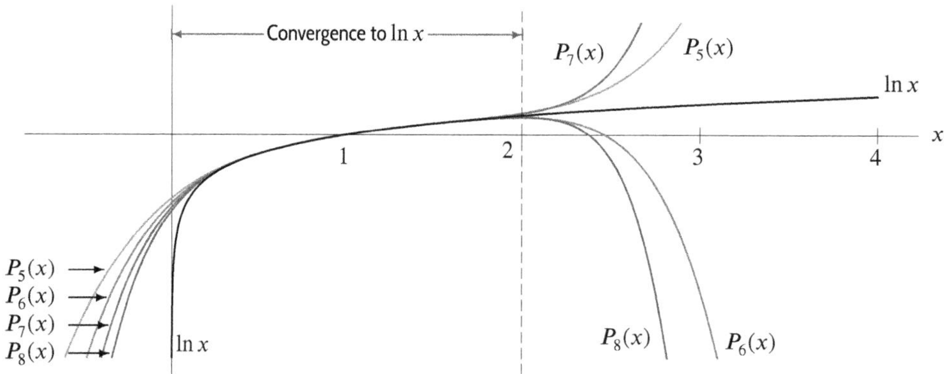

Figure 10.10: Taylor polynomials $P_5(x), P_6(x), P_7(x), P_8(x), \ldots$ converge to $\ln x$ for $0 < x \leq 2$ and diverge outside that interval

Proving that the Taylor series converges to $\ln x$ between 0 and 2, as Figure 10.10 suggests, requires the error term introduced in Section 10.4.

Example 3 Find the Taylor series for $\ln(1+x)$ about $x=0$, and investigate its convergence to $\ln(1+x)$.

Solution Taking derivatives of $\ln(1+x)$ and substituting $x=0$ leads, assuming convergence, to the Taylor series

$$\ln(1+x) = x - \frac{x^2}{2} + \frac{x^3}{3} - \frac{x^4}{4} + \cdots.$$

Notice that this is the same series that we get by substituting $(1+x)$ for x in the series for $\ln x$:

$$\ln x = (x-1) - \frac{(x-1)^2}{2} + \frac{(x-1)^3}{3} - \frac{(x-1)^4}{4} + \cdots \quad \text{for } 0 < x \le 2.$$

Since the series for $\ln x$ about $x=1$ converges to $\ln x$ for x between 0 and 2, the Taylor series for $\ln(1+x)$ about $x=0$ converges to $\ln(1+x)$ for x between -1 and 1. See Figure 10.11. Notice that the series could not possibly converge to $\ln(1+x)$ for $x \le -1$ since $\ln(1+x)$ is not defined there.

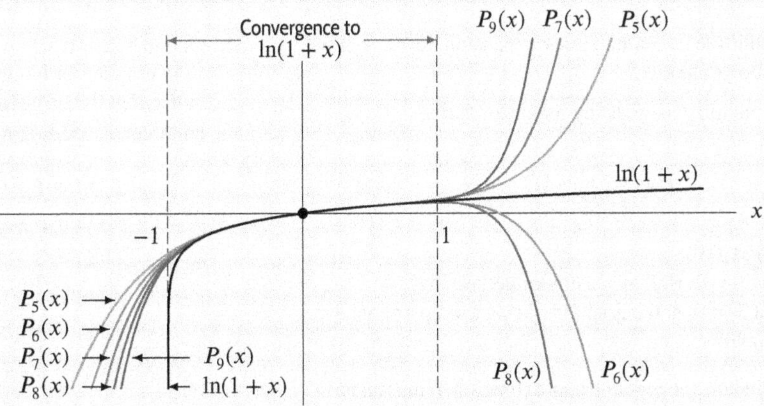

Figure 10.11: Convergence of the Taylor series for $\ln(1+x)$

The following theorem tells us about the relationship between power series and Taylor series; the justification involves calculations similar to those in Problem 60.

Theorem 10.1: A Power Series Is its Own Taylor Series

If a power series about $x=a$ converges to $f(x)$ for $|x-a| < R$, then the power series is the Taylor series for $f(x)$ about $x=a$.

Exercises and Problems for Section 10.2

EXERCISES

For Exercises 1–9, find the first four nonzero terms of the Taylor series for the function about 0.

1. $(1+x)^{3/2}$
2. $\sqrt[4]{x+1}$
3. $\sin(-x)$
4. $\ln(1-x)$
5. $\dfrac{1}{1-x}$
6. $\dfrac{1}{\sqrt{1+x}}$
7. $\sqrt[3]{1-y}$
8. $\tan(t+\pi/4)$
9. $\ln(5+2x)$

For Exercises 10–17, find the first four terms of the Taylor series for the function about the point a.

10. $\sin x$, $a = \pi/4$
11. $\cos\theta$, $a = \pi/4$
12. $\cos t$, $a = \pi/6$
13. $\sin\theta$, $a = -\pi/4$
14. $\tan x$, $a = \pi/4$
15. $1/x$, $a = 1$
16. $1/x$, $a = 2$
17. $1/x$, $a = -1$

In Exercises 18–25, find an expression for the general term of the series and give the range of values for the index (n or k, for example).

18. $\dfrac{1}{1-x} = 1 + x + x^2 + x^3 + x^4 + \cdots$
19. $\dfrac{1}{1+x} = 1 - x + x^2 - x^3 + x^4 - \cdots$
20. $\ln(1-x) = -x - \dfrac{x^2}{2} - \dfrac{x^3}{3} - \dfrac{x^4}{4} - \cdots$
21. $\ln(1+x) = x - \dfrac{x^2}{2} + \dfrac{x^3}{3} - \dfrac{x^4}{4} + \dfrac{x^5}{5} - \cdots$
22. $\sin x = x - \dfrac{x^3}{3!} + \dfrac{x^5}{5!} - \dfrac{x^7}{7!} + \cdots$
23. $\arctan x = x - \dfrac{x^3}{3} + \dfrac{x^5}{5} - \dfrac{x^7}{7} + \cdots$
24. $e^{x^2} = 1 + x^2 + \dfrac{x^4}{2!} + \dfrac{x^6}{3!} + \dfrac{x^8}{4!} + \cdots$
25. $x^2\cos x^2 = x^2 - \dfrac{x^6}{2!} + \dfrac{x^{10}}{4!} - \dfrac{x^{14}}{6!} + \cdots$

PROBLEMS

26. Compute the binomial series expansion for $(1 + x)^3$. What do you notice?

27. The function f is given by its Taylor series

$$f(x) = \sum_{n=1}^{\infty}(-1)^{n-1}\frac{2^n x^n}{n!}.$$

(a) Find $f(0)$.
(b) At $x = 0$, is f increasing or decreasing?
(c) At $x = 0$, is the graph of f concave up or concave down?

28. **(a)** Find the tangent line approximation to the function g whose Taylor series is

$$g(x) = \ln 5 - \sum_{n=1}^{\infty}\frac{2^n\cdot(n-1)!}{5^n}x^n.$$

(b) What are the slope and y-intercept of this line?
(c) Is the tangent line above or below the graph of g?

29. Using the Taylor series for $f(x) = e^x$ around 0, compute the following limit:

$$\lim_{x\to 0}\frac{e^x - 1}{x}.$$

30. Use the fact that the Taylor series of $g(x) = \sin(x^2)$ is

$$x^2 - \frac{x^6}{3!} + \frac{x^{10}}{5!} - \frac{x^{14}}{7!} + \cdots$$

to find $g''(0)$, $g'''(0)$, and $g^{(10)}(0)$. (There is an easy way and a hard way to do this!)

31. The Taylor series of $f(x) = x^2 e^{x^2}$ about $x = 0$ is

$$x^2 + x^4 + \frac{x^6}{2!} + \frac{x^8}{3!} + \frac{x^{10}}{4!} + \cdots.$$

Find $\left.\dfrac{d}{dx}\left(x^2e^{x^2}\right)\right|_{x=0}$ and $\left.\dfrac{d^6}{dx^6}\left(x^2e^{x^2}\right)\right|_{x=0}$.

32. One of the two sets of functions, f_1, f_2, f_3, or g_1, g_2, g_3 is graphed in Figure 10.12; the other set is graphed in Figure 10.13. Taylor series for the functions about a point corresponding to either A or B are as follows:

$$f_1(x) = 3 + (x-1) - (x-1)^2 + \cdots$$
$$f_2(x) = 3 - (x-1) + (x-1)^2 + \cdots$$
$$f_3(x) = 3 - 2(x-1) + (x-1)^2 + \cdots$$
$$g_1(x) = 5 - (x-4) - (x-4)^2 + \cdots$$
$$g_2(x) = 5 - (x-4) + (x-4)^2 + \cdots$$
$$g_3(x) = 5 + (x-4) + (x-4)^2 + \cdots.$$

(a) Which group of functions is represented in each figure?
(b) What are the coordinates of the points A and B?
(c) Match each function with the graphs (I)–(III) in the appropriate figure.

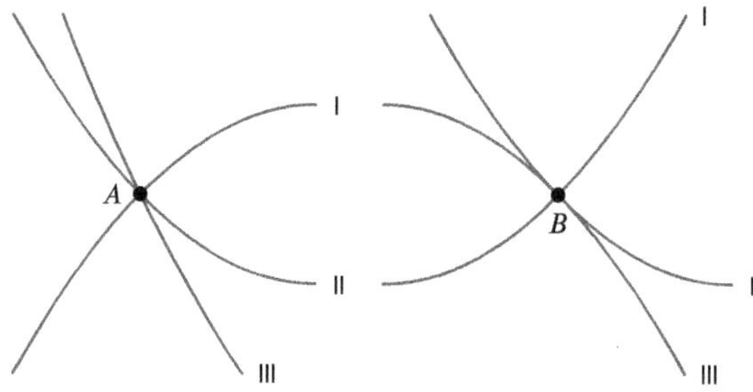

Figure 10.12   Figure 10.13

For Problems 33–36, what must be true of of $f^{(n)}(0)$ to ensure that the Taylor series of $f(x)$ about $x = 0$ has the given property?

33. All coefficients are positive.
34. Only has terms with even exponents.
35. Only has terms with odd exponents.
36. Coefficients for even exponent terms are negative, and coefficients for odd exponent terms are positive.

37. By graphing the function $f(x) = \dfrac{1}{\sqrt{1+x}}$ and several of its Taylor polynomials, estimate the interval of convergence of the series you found in Exercise 6.

38. By graphing the function $f(x) = \sqrt{1+x}$ and several of its Taylor polynomials, estimate where the series we found in Example 2 converges to $\sqrt{1+x}$.

39. (a) By graphing the function $f(x) = \dfrac{1}{1-x}$ and several of its Taylor polynomials, estimate where the series you found in Exercise 5 converges to $1/(1-x)$.
    (b) Compute the radius of convergence analytically.

40. Find the radius of convergence of the Taylor series around $x = 0$ for e^x.

41. Find the radius of convergence of the Taylor series around $x = 0$ for $\ln(1-x)$.

42. (a) Write the general term of the binomial series for $(1+x)^p$ about $x = 0$.
    (b) Find the radius of convergence of this series.

■ By recognizing each series in Problems 43–51 as a Taylor series evaluated at a particular value of x, find the sum of each of the following convergent series.

43. $1 + \frac{2}{1!} + \frac{4}{2!} + \frac{8}{3!} + \cdots + \frac{2^n}{n!} + \cdots$

44. $1 - \frac{1}{3!} + \frac{1}{5!} - \frac{1}{7!} + \cdots + \frac{(-1)^n}{(2n+1)!} + \cdots$

45. $1 + \frac{1}{4} + \left(\frac{1}{4}\right)^2 + \left(\frac{1}{4}\right)^3 + \cdots + \left(\frac{1}{4}\right)^n + \cdots$

46. $1 - \frac{100}{2!} + \frac{10000}{4!} + \cdots + \frac{(-1)^n \cdot 10^{2n}}{(2n)!} + \cdots$

47. $\frac{1}{2} - \frac{(\frac{1}{2})^2}{2} + \frac{(\frac{1}{2})^3}{3} - \frac{(\frac{1}{2})^4}{4} + \cdots + \frac{(-1)^n \cdot (\frac{1}{2})^{n+1}}{(n+1)} + \cdots$

48. $1 - 0.1 + 0.1^2 - 0.1^3 + \cdots$

49. $1 + 3 + \frac{9}{2!} + \frac{27}{3!} + \frac{81}{4!} + \cdots$

50. $1 - \frac{1}{2!} + \frac{1}{4!} - \frac{1}{6!} + \cdots$

51. $1 - 0.1 + \frac{0.01}{2!} - \frac{0.001}{3!} + \cdots$

52. Let

$$f(x) = 1 - 2\cos x + \frac{(-2)(-3)}{2!}(\cos x)^2 + \frac{(-2)(-3)(-4)}{3!}(\cos x)^3 + \cdots.$$

    (a) By letting $z = \cos x$, rewrite $f(x)$ as a rational function in z by recognizing it as a binomial series in z.
    (b) Use part (a) to find the exact value of $f(\pi/3)$.

■ In Problems 53–54 solve exactly for the variable.

53. $1 + x + x^2 + x^3 + \cdots = 5$

54. $x - \frac{1}{2}x^2 + \frac{1}{3}x^3 + \cdots = 0.2$

■ In Problems 55–57, find the described value from the series

$$C(x) = \sum_{n=0}^{\infty} \frac{(-1)^n x^{4n+1}}{(2n)!(4n+1)}.$$

55. The integer a where x^k/a is the term of the series corresponding to $n = 2$.

56. The integer k where x^k/a is the term of the series corresponding to $n = 2$.

57. $C'(0)$

58. By recognizing the limit as a Taylor series, find the exact value of $\lim_{n\to\infty} s_n$, given that $s_0 = 1$ and $s_n = s_{n-1} + \frac{1}{n!} \cdot 2^n$ for $n \geq 0$.

59. Let $i = \sqrt{-1}$. We define $e^{i\theta}$ by substituting $i\theta$ in the Taylor series for e^x. Use this definition[4] to explain Euler's formula

$$e^{i\theta} = \cos\theta + i\sin\theta.$$

60. The power series $\sum C_n x^n$ converges to the function $f(x)$ for $|x| < R$. Assuming that you can differentiate term by term, show that $f^{(n)}(0) = n!C_n$ for all n. Thus the coefficients C_n of the power series are the coefficients $f^{(n)}(0)/n!$ of the Taylor series for $f(x)$.

Strengthen Your Understanding

■ In Problems 61–62, explain what is wrong with the statement.

61. Since

$$\frac{1}{1-x} = 1 + x + x^2 + x^3 + \cdots,$$

we conclude that

$$\frac{1}{1-2} = 1 + 2 + 2^2 + 2^3 + \cdots.$$

62. The radius of convergence is 2 for the following Taylor series: $1 + (x-3) + (x-3)^2 + (x-3)^3 + \cdots$.

■ In Problems 63–64, give an example of:

63. A function with a Taylor series whose third-degree term is zero.

64. A Taylor series that is convergent at $x = -1$.

[4]Complex numbers are discussed in Appendix B.

Decide if the statements in Problems 65–71 are true or false. Assume that the Taylor series for a function converges to that function. Give an explanation for your answer.

65. The Taylor series for $\sin x$ about $x = \pi$ is

$$(x-\pi) - \frac{(x-\pi)^3}{3!} + \frac{(x-\pi)^5}{5!} - \cdots.$$

66. If f is an even function, then the Taylor series for f near $x = 0$ has only terms with even exponents.

67. If f has the following Taylor series about $x = 0$, then $f^{(7)}(0) = -8$:

$$f(x) = 1 - 2x + \frac{3}{2!}x^2 - \frac{4}{3!}x^3 + \cdots.$$

(Assume the pattern of the coefficients continues.)

68. The Taylor series for f converges everywhere f is defined.

69. The graphs of e^x and its Taylor polynomial $P_{10}(x)$ get further and further apart as $x \to \infty$.

70. If the Taylor series for $f(x)$ around $x = 0$ has a finite number of terms and an infinite radius of convergence, then $f(x)$ is a polynomial.

71. If $f(x)$ is a polynomial, then its Taylor series has an infinite radius of convergence.

10.3 FINDING AND USING TAYLOR SERIES

Finding a Taylor series for a function means finding the coefficients. Assuming the function has all its derivatives defined, finding the coefficients can always be done, in theory at least, by differentiation. That is how we derived the four most important Taylor series, those for the functions e^x, $\sin x$, $\cos x$, and $(1+x)^p$.

For many functions, however, computing Taylor series coefficients by differentiation can be a very laborious business. We now introduce easier ways of finding Taylor series, if the series we want is closely related to a series that we already know.

New Series by Substitution

Suppose we want to find the Taylor series for e^{-x^2} about $x = 0$. We could find the coefficients by differentiation. Differentiating e^{-x^2} by the chain rule gives $-2xe^{-x^2}$, and differentiating again gives $-2e^{-x^2} + 4x^2e^{-x^2}$. Each time we differentiate we use the product rule, and the number of terms grows. Finding the tenth or twentieth derivative of e^{-x^2}, and thus the series for e^{-x^2} up to the x^{10} or x^{20} terms, by this method is tiresome (at least without a computer or calculator that can differentiate).

Fortunately, there's a quicker way. Recall that

$$e^y = 1 + y + \frac{y^2}{2!} + \frac{y^3}{3!} + \frac{y^4}{4!} + \cdots \quad \text{for all } y.$$

Substituting $y = -x^2$ tells us that

$$\begin{aligned} e^{-x^2} &= 1 + (-x^2) + \frac{(-x^2)^2}{2!} + \frac{(-x^2)^3}{3!} + \frac{(-x^2)^4}{4!} + \cdots \\ &= 1 - x^2 + \frac{x^4}{2!} - \frac{x^6}{3!} + \frac{x^8}{4!} - \cdots \quad \text{for all } x. \end{aligned}$$

Using this method, it is easy to find the series up to the x^{10} or x^{20} terms. It can be shown that this is the Taylor series for e^{-x^2}.

Example 1 Find the Taylor series about $x = 0$ for $f(x) = \dfrac{1}{1+x^2}$.

Solution The binomial series, or geometric series, tells us that

$$\frac{1}{1+y} = (1+y)^{-1} = 1 - y + y^2 - y^3 + y^4 - \cdots \quad \text{for } -1 < y < 1.$$

Substituting $y = x^2$ gives

$$\frac{1}{1+x^2} = 1 - x^2 + x^4 - x^6 + x^8 - \cdots \quad \text{for } -1 < x < 1,$$

which is the Taylor series for $\dfrac{1}{1+x^2}$.

Notice that substitution can affect the radius of convergence. For example, in Example 1 if we find the Taylor series for $g(x) = 1/(1+4x^2)$ by substituting $y = 4x^2$, the interval of convergence changes from $-1 < y < 1$ to $-1/2 < x < 1/2$.

New Series by Differentiation and Integration

Just as we can get new series by substitution, we can also get new series by differentiation and integration. Term-by-term differentiation of a Taylor series for $f(x)$ gives a Taylor series for $f'(x)$; antidifferentiation works similarly.

Example 2 Find the Taylor series about $x = 0$ for $\dfrac{1}{(1-x)^2}$ from the series for $\dfrac{1}{1-x}$.

Solution We know that $\dfrac{d}{dx}\left(\dfrac{1}{1-x}\right) = \dfrac{1}{(1-x)^2}$, so we start with the geometric series

$$\frac{1}{1-x} = 1 + x + x^2 + x^3 + x^4 + \cdots \quad \text{for } -1 < x < 1.$$

Differentiation term by term gives the binomial series

$$\frac{1}{(1-x)^2} = \frac{d}{dx}\left(\frac{1}{1-x}\right) = 1 + 2x + 3x^2 + 4x^3 + \cdots \quad \text{for } -1 < x < 1.$$

Example 3 Find the Taylor series[5] about $x = 0$ for $\arctan x$ from the series for $\dfrac{1}{1+x^2}$.

Solution We know that $\dfrac{d}{dx}(\arctan x) = \dfrac{1}{1+x^2}$, so we use the series from Example 1:

$$\frac{d}{dx}(\arctan x) = \frac{1}{1+x^2} = 1 - x^2 + x^4 - x^6 + x^8 - \cdots \qquad \text{for } -1 < x < 1.$$

Antidifferentiating term by term and assuming convergence gives

$$\arctan x = \int \frac{1}{1+x^2}\,dx = C + x - \frac{x^3}{3} + \frac{x^5}{5} - \frac{x^7}{7} + \frac{x^9}{9} - \cdots \qquad \text{for } -1 < x < 1,$$

where C is the constant of integration. Since $\arctan 0 = 0$, we have $C = 0$, so

$$\arctan x = x - \frac{x^3}{3} + \frac{x^5}{5} - \frac{x^7}{7} + \frac{x^9}{9} - \cdots \qquad \text{for } -1 < x < 1.$$

Convergence of a Series and Differentiation

Example 2 assumed that the series created by term-by-term differentiation converged, and to the function we wanted. The following theorem, which is proved in more advanced courses, justifies the assumption. Problem 59 confirms that the radius of convergence remains the same under term-by-term differentiation.

[5]The series for $\arctan x$ was discovered by James Gregory (1638–1675).

Theorem 10.2: Convergence of the Term-by-Term Derivative

If a Taylor series for $f(x)$ at $x = a$ converges to $f(x)$ for $|x-a| < R$, then the series found by term-by-term differentiation is the Taylor series for $f'(x)$ and converges to $f'(x)$ on the same interval $|x-a| < R$.

Multiplying and Substituting Taylor Series

We can also form a Taylor series for a product of two functions. In some cases, this is easy; for example, if we want to find the Taylor series about $x = 0$ for the function $f(x) = x^2 \sin x$, we can start with the Taylor series for $\sin x$,

$$\sin x = x - \frac{x^3}{3!} + \frac{x^5}{5!} - \frac{x^7}{7!} + \cdots,$$

and multiply the series by x^2:

$$\begin{aligned} x^2 \sin x &= x^2 \left(x - \frac{x^3}{3!} + \frac{x^5}{5!} - \frac{x^7}{7!} + \cdots \right) \\ &= x^3 - \frac{x^5}{3!} + \frac{x^7}{5!} - \frac{x^9}{7!} + \cdots. \end{aligned}$$

However, in some cases, finding a Taylor series for a product of two functions requires more work.

Example 4 Find the Taylor series about $x = 0$ for $g(x) = \sin x \cos x$.

Solution The Taylor series about $x = 0$ for $\sin x$ and $\cos x$ are

$$\begin{aligned} \sin x &= x - \frac{x^3}{3!} + \frac{x^5}{5!} - \cdots \\ \cos x &= 1 - \frac{x^2}{2!} + \frac{x^4}{4!} - \cdots. \end{aligned}$$

So we have

$$g(x) = \sin x \cos x = \left(x - \frac{x^3}{3!} + \frac{x^5}{5!} - \cdots \right) \left(1 - \frac{x^2}{2!} + \frac{x^4}{4!} - \cdots \right).$$

To multiply these two series, we must multiply each term of the series for $\sin x$ by each term of the series for $\cos x$. Because each series has infinitely many terms, we organize the process by first determining the constant term of the product, then the linear term, and so on.

The constant term of this product is zero because there is no combination of a term from the first series and a term from the second that yields a constant. The linear term of the product is x; we obtain this term by multiplying the x from the first series by the 1 from the second. The degree-2 term of the product is also zero; more generally, we notice that every even-degree term of the product is zero because every combination of a term from the first series and a term from the second yields an odd-degree term. To find the degree-3 term, observe that the combinations of terms that yield degree-3 terms are $x \cdot -\frac{x^2}{2!} = -\frac{1}{2}x^3$ and $-\frac{x^3}{3!} \cdot 1 = -\frac{1}{6}x^3$, and thus the degree-3 term is $-\frac{1}{2}x^3 - \frac{1}{6}x^3 = -\frac{2}{3}x^3$. Continuing in this manner, we find that

$$g(x) = x - \frac{2}{3}x^3 + \frac{2}{15}x^5 - \cdots.$$

There is another way to find this series. Notice that $\sin x \cos x = \frac{1}{2}\sin(2x)$. Substituting $2x$ into the Taylor series about $x = 0$ for the sine function, we get

$$\sin(2x) = 2x - \frac{(2x)^3}{3!} + \frac{(2x)^5}{5!} - \cdots$$

$$= 2x - \frac{4}{3}x^3 + \frac{4}{15}x^5 - \cdots.$$

Therefore, we have

$$g(x) = \frac{1}{2}\sin(2x) = x - \frac{2}{3}x^3 + \frac{2}{15}x^5 - \cdots.$$

We can also obtain a Taylor series for a composite function by substituting a Taylor series into another one, as in the next example.

Example 5 Find the Taylor series about $\theta = 0$ for $g(\theta) = e^{\sin\theta}$.

Solution For all y and θ, we know that

$$e^y = 1 + y + \frac{y^2}{2!} + \frac{y^3}{3!} + \frac{y^4}{4!} + \cdots$$

and

$$\sin\theta = \theta - \frac{\theta^3}{3!} + \frac{\theta^5}{5!} - \cdots.$$

Let's substitute the series for $\sin\theta$ for y:

$$e^{\sin\theta} = 1 + \left(\theta - \frac{\theta^3}{3!} + \frac{\theta^5}{5!} - \cdots\right) + \frac{1}{2!}\left(\theta - \frac{\theta^3}{3!} + \frac{\theta^5}{5!} - \cdots\right)^2 + \frac{1}{3!}\left(\theta - \frac{\theta^3}{3!} + \frac{\theta^5}{5!} - \cdots\right)^3 + \cdots.$$

To simplify, we multiply out and collect terms. The only constant term is the 1, and there's only one θ term. The only θ^2 term is the first term we get by multiplying out the square, namely $\theta^2/2!$. There are two contributors to the θ^3 term: the $-\theta^3/3!$ from within the first parentheses and the first term we get from multiplying out the cube, which is $\theta^3/3!$. Thus the series starts

$$\begin{aligned} e^{\sin\theta} &= 1 + \theta + \frac{\theta^2}{2!} + \left(-\frac{\theta^3}{3!} + \frac{\theta^3}{3!}\right) + \cdots \\ &= 1 + \theta + \frac{\theta^2}{2!} + 0\cdot\theta^3 + \cdots \quad \text{for all } \theta. \end{aligned}$$

Applications of Taylor Series

Example 6 Use the series for $\arctan x$ to estimate the numerical value of π.

Solution Since $\arctan 1 = \pi/4$, we use the series for $\arctan x$ from Example 3. We assume—as is the case—that the series does converge to $\pi/4$ at $x = 1$. Substituting $x = 1$ into the series for $\arctan x$ gives

$$\pi = 4\arctan 1 = 4\left(1 - \frac{1}{3} + \frac{1}{5} - \frac{1}{7} + \frac{1}{9} - \cdots\right).$$

Table 10.1 *Approximating π using the series for* $\arctan x$

n	4	5	25	100	500	1000	10,000
S_n	2.895	3.340	3.182	3.132	3.140	3.141	3.141

Table 10.1 shows the value of the nth partial sum, S_n, obtained by summing the first n nonzero terms. The values of S_n do seem to converge to $\pi = 3.141\ldots$. However, this series converges very slowly, meaning that we have to take a large number of terms to get an accurate estimate for π. So this way of calculating π is not particularly practical. (A better one is given in Project 2 (available online).) However, the expression for π given by this series is surprising and elegant.

A basic question we can ask about two functions is which one gives larger values. Taylor series can often be used to answer this question over a small interval. If the constant terms of the series for two functions are the same, compare the linear terms; if the linear terms are the same, compare the quadratic terms, and so on.

Example 7 By looking at their Taylor series, decide which of the following functions is largest for t near 0.

(a) e^t (b) $\dfrac{1}{1-t}$

Solution The Taylor expansion about $t = 0$ for e^t is

$$e^t = 1 + t + \frac{t^2}{2!} + \frac{t^3}{3!} + \cdots .$$

Viewing $1/(1-t)$ as the sum of a geometric series with initial term 1 and common ratio t, we have

$$\frac{1}{1-t} = 1 + t + t^2 + t^3 + \cdots \text{ for } -1 < t < 1.$$

Notice that these two series have the same constant term and the same linear term. However, their remaining terms are different. For values of t near zero, the quadratic terms dominate all of the subsequent terms,[6] so we can use the approximations

$$e^t \approx 1 + t + \frac{t^2}{2}$$
$$\frac{1}{1-t} \approx 1 + t + t^2.$$

Since

$$1 + t + \frac{1}{2}t^2 < 1 + t + t^2,$$

and since the approximations are valid for t near 0, we conclude that, for t near 0,

$$e^t < \frac{1}{1-t}.$$

See Figure 10.14.

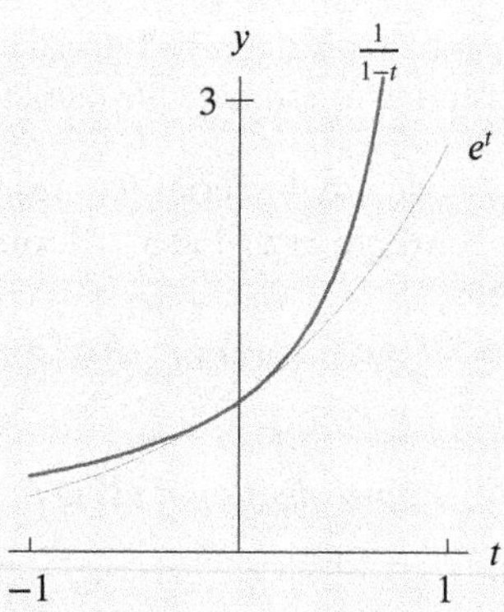

Figure 10.14: Comparing two functions near $t = 0$

[6]To make this argument rigorous, we need the Lagrange error bound given in the next section.

Example 8 Two electrical charges of equal magnitude and opposite signs located near one another are called an electrical dipole. The charges Q and $-Q$ are a distance r apart. (See Figure 10.15.) The electric field, E, at the point P, at a distance R from the dipole is given by

$$E = \frac{Q}{R^2} - \frac{Q}{(R+r)^2}.$$

Use series to investigate the behavior of the electric field far to the left along the line through the dipole. Show that when R is large in comparison to r, the electric field is approximately proportional to $1/R^3$.

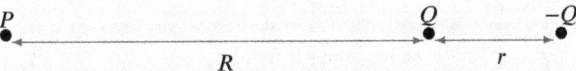

Figure 10.15: Approximating the electric field at P due to a dipole consisting of charges Q and $-Q$ a distance r apart

Solution In order to use a series approximation, we need a variable whose value is small. Although we know that r is much smaller than R, we do not know that r itself is small. The quantity r/R is, however, very small. Hence we expand $1/(R+r)^2$ in powers of r/R so that we can safely use only the first few terms of the Taylor series. First we rewrite using algebra:

$$\frac{1}{(R+r)^2} = \frac{1}{R^2(1+r/R)^2} = \frac{1}{R^2}\left(1+\frac{r}{R}\right)^{-2}.$$

Now we use the binomial expansion for $(1+x)^p$ with $x = r/R$ and $p = -2$:

$$\begin{aligned}\frac{1}{R^2}\left(1+\frac{r}{R}\right)^{-2} &= \frac{1}{R^2}\left(1+(-2)\left(\frac{r}{R}\right)+\frac{(-2)(-3)}{2!}\left(\frac{r}{R}\right)^2+\frac{(-2)(-3)(-4)}{3!}\left(\frac{r}{R}\right)^3+\cdots\right)\\ &= \frac{1}{R^2}\left(1-2\frac{r}{R}+3\frac{r^2}{R^2}-4\frac{r^3}{R^3}+\cdots\right).\end{aligned}$$

So, substituting the series into the expression for E, we have

$$\begin{aligned}E = \frac{Q}{R^2}-\frac{Q}{(R+r)^2} &= Q\left(\frac{1}{R^2}-\frac{1}{R^2}\left(1-2\frac{r}{R}+3\frac{r^2}{R^2}-4\frac{r^3}{R^3}+\cdots\right)\right)\\ &= \frac{Q}{R^2}\left(2\frac{r}{R}-3\frac{r^2}{R^2}+4\frac{r^3}{R^3}-\cdots\right).\end{aligned}$$

Since r/R is smaller than 1, the binomial expansion for $(1+r/R)^{-2}$ converges. We are interested in the electric field far away from the dipole. The quantity r/R is small there, and $(r/R)^2$ and higher powers are smaller still. Thus, we approximate by disregarding all terms except the first, giving

$$E \approx \frac{Q}{R^2}\left(\frac{2r}{R}\right), \quad \text{so} \quad E \approx \frac{2Qr}{R^3}.$$

Since Q and r are constants, this means that E is approximately proportional to $1/R^3$.

In the previous example, we say that E is *expanded in terms of* r/R, meaning that the variable in the expansion is r/R.

Exercises and Problems for Section 10.3

EXERCISES

In Exercises 1–12, using known Taylor series, find the first four nonzero terms of the Taylor series about 0 for the function.

1. e^{-x}
2. $\sqrt{1-2x}$
3. $\cos(\theta^2)$
4. $\ln(1-2y)$
5. $\arcsin x$
6. $t\sin(3t)$
7. $\dfrac{1}{\sqrt{1-z^2}}$
8. $\dfrac{z}{e^{z^2}}$
9. $\phi^3\cos(\phi^2)$
10. $\arctan(r^2)$
11. $\cosh t$
12. $\sinh t$

In Exercises 13–15, find the Taylor series about 0 for the function. Include the general term.

13. $(1+x)^3$
14. $t\sin(t^2)-t^3$
15. $\dfrac{1}{\sqrt{1-y^2}}$

In Exercises 16–17, find the exact value of the expression by identifying it as a known series.

16. $1+0.2+\dfrac{0.2^2}{2!}+\dfrac{0.2^3}{3!}+\dfrac{0.2^4}{4!}+\cdots$
17. $1+0.5+\dfrac{0.5^2}{2}+\dfrac{0.5^3}{3}+\dfrac{0.5^4}{4}+\cdots$

For Exercises 18–23, expand the quantity about 0 in terms of the variable given. Give four nonzero terms.

18. $\dfrac{1}{2+x}$ in terms of $\dfrac{x}{2}$
19. $\sqrt{T+h}$ in terms of $\dfrac{h}{T}$
20. $\dfrac{1}{a-r}$ in terms of $\dfrac{r}{a}$
21. $\dfrac{1}{(a+r)^2}$ in terms of $\dfrac{r}{a}$
22. $\sqrt[3]{P+t}$ in terms of $\dfrac{t}{P}$
23. $\dfrac{a}{\sqrt{a^2+x^2}}$ in terms of $\dfrac{x}{a}$, where $a>0$

PROBLEMS

In Problems 24–27, using known Taylor series, find the first four nonzero terms of the Taylor series about 0 for the function.

24. $\sqrt{(1+t)}\sin t$
25. $e^t\cos t$
26. $\sqrt{1+\sin\theta}$
27. $\dfrac{1}{1-\ln(1+t)}$

28. Let $g(z)$ be the function obtained by substituting $z=x-1$ into $f(x)=1/x$. Use a series for $g(z)$ to get a Taylor series for $f(x)$ around $x=1$.

29. **(a)** Let $g(z)$ be the function obtained by substituting $z=(x-1)/2$ into $f(x)=1/(x+1)$. Use a series for $h(z)$ to find a Taylor series for $f(x)$ around $x=1$.
(b) Find the Taylor series of $g(x)=(x-1)^2/(x+1)$ around $x=1$.

30. The Taylor series for $\ln(1-x)$ about $x=0$ converges for $-1<x<1$. For each of the following functions, use this interval to find the largest interval on which the function's Taylor series about $x=0$ converges.

(a) $\ln(4-x)$ **(b)** $\ln(4+x)$ **(c)** $\ln(1+4x^2)$

31. Use the series for e^x to find the Taylor series for $\sinh 2x$ and $\cosh 2x$.

32. **(a)** Find the first three nonzero terms of the Taylor series for e^x+e^{-x}.
(b) Explain why the graph of e^x+e^{-x} looks like a parabola near $x=0$. What is the equation of this parabola?

33. **(a)** Find the first three nonzero terms of the Taylor series for e^x-e^{-x}.
(b) Explain why the graph of e^x-e^{-x} near $x=0$ looks like the graph of a cubic polynomial symmetric about the origin. What is the equation for this cubic?

34. Find the first three terms of the Taylor series for $f(x)=e^{x^2}$ around 0. Use this information to approximate the integral

$$\int_0^1 e^{x^2}\,dx.$$

35. Find the sum of $\sum_{p=1}^{\infty} px^{p-1}$ for $|x|<1$.

36. For values of y near 0, put the following functions in increasing order, using their Taylor expansions.

(a) $\ln(1+y^2)$ **(b)** $\sin(y^2)$ **(c)** $1-\cos y$

37. For values of θ near 0, put the following functions in increasing order, using their Taylor expansions.

(a) $1+\sin\theta$ **(b)** e^θ **(c)** $\dfrac{1}{\sqrt{1-2\theta}}$

38. A function has the following Taylor series about $x=0$:

$$f(x)=\sum_{n=0}^{\infty}\frac{x^{2n+1}}{2n+1}.$$

Find the ninth-degree Taylor polynomial for $f(2x)$.

39. Figure 10.16 shows the graphs of the four functions below for values of x near 0. Use Taylor series to match graphs and formulas.

(a) $\dfrac{1}{1-x^2}$ **(b)** $(1+x)^{1/4}$

(c) $\sqrt{1+\frac{x}{2}}$ (d) $\frac{1}{\sqrt{1-x}}$

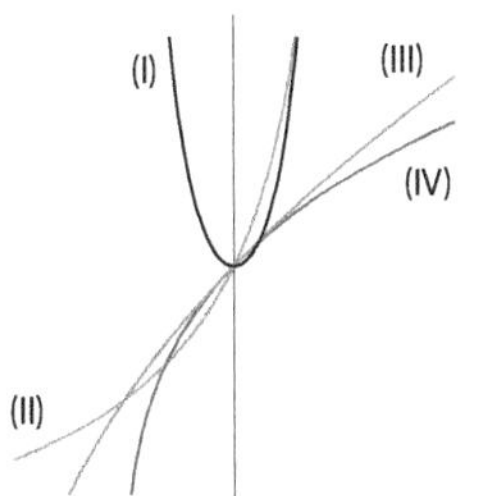

Figure 10.16

40. The *sine integral* function is defined by the improper integral $\text{Si}(x) = \int_0^x \frac{\sin t}{t}\,dt$. Use the Taylor polynomial, $P_7(x)$, of degree 7 about $x = 0$ for the sine function to estimate Si(2).

41. Write out the first four nonzero terms of the Taylor series about $x = 0$ for $f(x) = \int_0^x \sin\left(t^2\right)\,dt$.

42. **(a)** Find the Taylor series for $f(t) = te^t$ about $t = 0$.
(b) Using your answer to part (a), find a Taylor series expansion about $x = 0$ for

$$\int_0^x te^t\,dt.$$

(c) Using your answer to part (b), show that

$$\frac{1}{2}+\frac{1}{3}+\frac{1}{4(2!)}+\frac{1}{5(3!)}+\frac{1}{6(4!)}+\cdots = 1.$$

43. Find the sum of $\sum_{n=1}^{\infty} \frac{k^{n-1}}{(n-1)!}e^{-k}$.

44. Use Taylor series to explain the patterns in the digits in the following expansions:

(a) $\frac{1}{0.98} = 1.02040816\ldots$
(b) $\left(\frac{1}{0.99}\right)^2 = 1.020304050607\ldots$

45. Padé approximants are rational functions used to approximate more complicated functions. In this problem, you will derive the Padé approximant to the exponential function.

(a) Let $f(x) = (1+ax)/(1+bx)$, where a and b are constants. Write down the first three terms of the Taylor series for $f(x)$ about $x = 0$.
(b) By equating the first three terms of the Taylor series about $x = 0$ for $f(x)$ and for e^x, find a and b so that $f(x)$ approximates e^x as closely as possible near $x = 0$.

46. One of Einstein's most amazing predictions was that light traveling from distant stars would bend around the sun on the way to earth. His calculations involved solving for ϕ in the equation

$$\sin\phi + b(1+\cos^2\phi+\cos\phi) = 0,$$

where b is a very small positive constant.

(a) Explain why the equation could have a solution for ϕ which is near 0.
(b) Expand the left-hand side of the equation in Taylor series about $\phi = 0$, disregarding terms of order ϕ^2 and higher. Solve for ϕ. (Your answer will involve b.)

47. A hydrogen atom consists of an electron, of mass m, orbiting a proton, of mass M, where m is much smaller than M. The *reduced mass*, μ, of the hydrogen atom is defined by

$$\mu = \frac{mM}{m+M}.$$

(a) Show that $\mu \approx m$.
(b) To get a more accurate approximation for μ, express μ as m times a series in m/M.
(c) The approximation $\mu \approx m$ is obtained by disregarding all but the constant term in the series. The first-order correction is obtained by including the linear term but no higher terms. If $m \approx M/1836$, by what percentage does including the linear term change the estimate $\mu \approx m$?

48. Resonance in electric circuits leads to the expression

$$\left(\omega L - \frac{1}{\omega C}\right)^2,$$

where ω is the variable and L and C are constants.

(a) Find ω_0, the value of ω making the expression zero.
(b) In practice, ω fluctuates about ω_0, so we are interested in the behavior of this expression for values of ω near ω_0. Let $\omega = \omega_0 + \Delta\omega$ and expand the expression in terms of $\Delta\omega$ up to the first nonzero term. Give your answer in terms of $\Delta\omega$ and L but not C.

49. A pendulum consists of a mass, m, swinging on the end of a string of length l. With the angle between the string and the vertical represented by θ, the motion satisfies the differential equation

$$\theta'' + \frac{g}{l}g\sin\theta = 0.$$

(a) For small swings, we can replace $\sin\theta$ by its lowest nonzero Taylor approximation. What does the differential equation become?
(b) If the amplitude of the oscillation is θ_0, the solutions to the original differential equation are oscillations with[7]

$$\text{Period } = 2\pi\sqrt{\frac{l}{g}}\left(1+\frac{1}{16}\theta_0^2+\cdots\right).$$

[7]http://en.wikipedia.org/wiki/Pendulum. Accessed April, 2015.

The solutions to the approximate differential equation are oscillations with

$$\text{Period} = 2\pi\sqrt{\frac{l}{g}}.$$

If $\theta_0 = 20°$, by what percentage is the more accurate estimate of the period obtained using the solution to the original equation up to the θ_0^2-term different to the approximate estimate using the solution of the approximate equation?

50. Stefan's Law says the rate at which a body's temperature, H, changes with time is proportional to $A^4 - H^4$, where A is the surrounding temperature.[8] Use a Taylor series to expand $A^4 - H^4$ up to the linear term.

51. The Michelson-Morley experiment, which contributed to the formulation of the theory of relativity, involved the difference between the two times t_1 and t_2 that light took to travel between two points. If v is velocity; l_1, l_2, and c are constants; and $v < c$, then t_1 and t_2 are given by

$$t_1 = \frac{2l_2}{c(1 - v^2/c^2)} - \frac{2l_1}{c\sqrt{1 - v^2/c^2}}$$
$$t_2 = \frac{2l_2}{c\sqrt{1 - v^2/c^2}} - \frac{2l_1}{c(1 - v^2/c^2)}.$$

(a) Find an expression for $\Delta t = t_1 - t_2$, and give its Taylor expansion in terms of v^2/c^2 up to the second nonzero term.
(b) For small v, to what power of v is Δt proportional? What is the constant of proportionality?

52. The theory of relativity predicts that when an object moves at speeds close to the speed of light, the object appears heavier. The apparent, or relativistic, mass, m, of the object when it is moving at speed v is given by the formula

$$m = \frac{m_0}{\sqrt{1 - v^2/c^2}},$$

where c is the speed of light and m_0 is the mass of the object when it is at rest.

(a) Use the formula for m to decide what values of v are possible.
(b) Sketch a rough graph of m against v, labeling intercepts and asymptotes.
(c) Write the first three nonzero terms of the Taylor series for m in terms of v.
(d) For what values of v do you expect the series to converge?

53. The potential energy, V, of two gas molecules separated by a distance r is given by

$$V = -V_0\left(2\left(\frac{r_0}{r}\right)^6 - \left(\frac{r_0}{r}\right)^{12}\right),$$

where V_0 and r_0 are positive constants.

(a) Show that if $r = r_0$, then V takes on its minimum value, $-V_0$.
(b) Write V as a series in $(r - r_0)$ up through the quadratic term.
(c) For r near r_0, show that the difference between V and its minimum value is approximately proportional to $(r - r_0)^2$. In other words, show that $V - (-V_0) = V + V_0$ is approximately proportional to $(r - r_0)^2$.
(d) The force, F, between the molecules is given by $F = -dV/dr$. What is F when $r = r_0$? For r near r_0, show that F is approximately proportional to $(r - r_0)$.

54. Van der Waal's equation relates the pressure, P, and the volume, V, of a fixed quantity of a gas at constant temperature T:

$$\left(P + \frac{n^2a}{V^2}\right)(V - nb) = nRT,$$

where a, b, n, R are constants. Find the first two nonzero terms of the Taylor series of P in terms for $1/V$.

55. The hyperbolic sine and cosine are differentiable and satisfy the conditions $\cosh 0 = 1$ and $\sinh 0 = 0$, and

$$\frac{d}{dx}(\cosh x) = \sinh x \quad \frac{d}{dx}(\sinh x) = \cosh x.$$

(a) Using only this information, find the Taylor approximation of degree $n = 8$ about $x = 0$ for $f(x) = \cosh x$.
(b) Estimate the value of cosh 1.
(c) Use the result from part (a) to find a Taylor polynomial approximation of degree $n = 7$ about $x = 0$ for $g(x) = \sinh x$.

In Problems 56–57, find the value of the integer a_i given that the Taylor series of $f(x) = \cos\sqrt{x}$ is:

$$a_0 + \frac{1}{a_1}x + \frac{1}{a_2}x^2 + \frac{1}{a_3}x^3 + \cdots.$$

56. a_2

57. a_3

58. Find $f(42)$ for $f(x) = \int_1^{g(x)} (1/t)\,dt$ where

$$g(x) = \lim_{b\to-\infty}\int_b^x h(t)\,dt, \quad \text{and} \quad h(x) = \sum_{k=0}^{\infty}\frac{x^k}{k!}.$$

59. The ratio test applied to the power series $\sum C_n x^n$ gives R as the radius of convergence. Show that the term-by-term derivative of the power series has the same radius of convergence.

[8]http://en.wikipedia.org/wiki/StefanBoltzmann_law. Accessed May 2015.

Strengthen Your Understanding

In Problems 60–61, explain what is wrong with the statement.

60. Within its radius of convergence,

$$\frac{1}{2+x} = 1 - \frac{x}{2} + \left(\frac{x}{2}\right)^2 - \left(\frac{x}{2}\right)^3 + \cdots.$$

61. Using the Taylor series for $e^x = 1 + x + \frac{x^2}{2!} + \frac{x^3}{3!} + \cdots$, we find that $e^{-x} = 1 - x - \frac{x^2}{2!} - \frac{x^3}{3!} - \cdots$.

In Problems 62–63, give an example of:

62. A function with no Taylor series around 0.

63. A function $f(x)$ that does not have a Taylor series around 0 even though $f(0)$ is defined.

Decide if the statements in Problems 64–68 are true or false. Assume that the Taylor series for a function converges to that function. Give an explanation for your answer.

64. To find the Taylor series for $\sin x + \cos x$ about any point, add the Taylor series for $\sin x$ and $\cos x$ about that point.

65. The Taylor series for $x^3 \cos x$ about $x = 0$ has only odd powers.

66. The Taylor series for $f(x)g(x)$ about $x = 0$ is

$$f(0)g(0) + f'(0)g'(0)x + \frac{f''(0)g''(0)}{2!}x^2 + \cdots.$$

67. If $L_1(x)$ is the linear approximation to $f_1(x)$ near $x = 0$ and $L_2(x)$ is the linear approximation to $f_2(x)$ near $x = 0$, then $L_1(x) + L_2(x)$ is the linear approximation to $f_1(x) + f_2(x)$ near $x = 0$.

68. If $L_1(x)$ is the linear approximation to $f_1(x)$ near $x = 0$ and $L_2(x)$ is the linear approximation to $f_2(x)$ near $x = 0$, then $L_1(x)L_2(x)$ is the quadratic approximation to $f_1(x)f_2(x)$ near $x = 0$.

69. Given that the Taylor series for $\tan x = x + x^3/3 + 21x^5/120 + \cdots$, then that of $3\tan(x/3)$ is

(a) $3x + x^3 + 21x^5/120 + \cdots$
(b) $3x + x^3 + 21x^5/40 + \cdots$
(c) $x + x^3/27 + 7x^5/3240 + \cdots$
(d) $x + x^3/3 + 21x^5/120 + \cdots$

10.4 THE ERROR IN TAYLOR POLYNOMIAL APPROXIMATIONS

In order to use an approximation with confidence, we need to know how big the error could be. The error is the difference between the exact answer and the approximate value. If $P_n(x)$ is the nth-degree Taylor polynomial, the error in approximating $f(x)$ is

$$E_n(x) = f(x) - P_n(x).$$

We have seen that the higher the degree of the Taylor polynomial, in general the larger the interval on which the function and the polynomial are close to each other. For example, Figure 10.17 shows the first three Taylor polynomial approximations about 0 for the function $f(x) = e^x$. If we pick a particular Taylor polynomial, say P_2, the closer x is to 0, the better $P_2(x)$ approximates e^x.

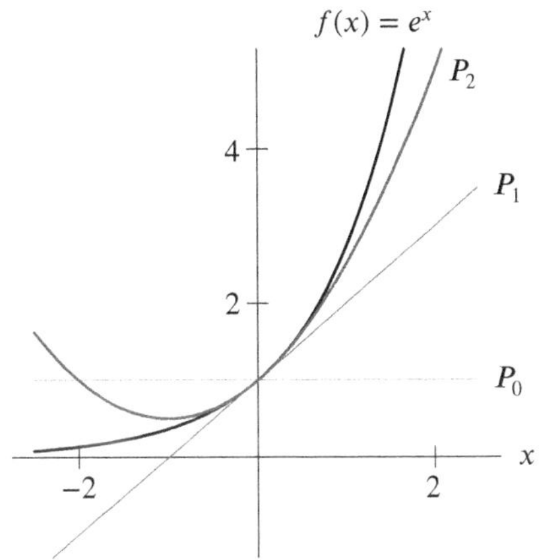

Figure 10.17: Graph of $f(x) = e^x$ and first three Taylor polynomials around $x = 0$

Example 1 Compare the errors in the approximations

$$e^{0.1} \approx 1 + 0.1 + \frac{1}{2!}(0.1)^2 \quad \text{and} \quad e^{0.05} \approx 1 + (0.05) + \frac{1}{2!}(0.05)^2.$$

Solution We are approximating e^x by its second-degree Taylor polynomial about 0. At $x = 0.1$ and $x = 0.05$ the errors are:

$$E_2(0.1) = e^{0.1} - \left(1 + 0.1 + \frac{1}{2!}(0.1)^2\right) = 1.105171\ldots - 1.105000 = 0.000171$$

$$E_2(0.05) = e^{0.05} - \left(1 + 0.05 + \frac{1}{2!}(0.05)^2\right) = 1.051271\ldots - 1.051250 = 0.000021.$$

As expected, the error in the approximation is smaller for $x = 0.05$ than for $x = 0.1$.

In Example 1 we obtained a better approximation by taking x closer 0. Example 2 shows if we are interested in a particular value of x, we can often improve the approximation by choosing a higher-degree polynomial.

Example 2 Compare the errors in the approximations for $e^{0.1}$ using P_2 and P_3.

Solution In the previous example we saw $E_2(0.1) = 0.000171$. We now compute

$$E_3(0.1) = e^{0.1} - P_3(0.1) = e^{0.1} - \left(1 + 0.1 + \frac{1}{2!}(0.1)^2 + \frac{1}{3!}(0.1)^3\right) = 1.105171\ldots - 1.1051666\ldots$$
$$= 0.0000043\ldots$$

Bounds on the Error

In practice, we usually want to find a *bound* on the magnitude of the error—we want to figure out how big it could be—without calculating the error exactly. Lagrange found an expression for an error bound E_n based on the degree-$(n+1)$ term in the Taylor series that is used to make such estimates:

Theorem 10.3: The Lagrange Error Bound for $P_n(x)$

Suppose f and all its derivatives are continuous. If $P_n(x)$ is the degree-n Taylor polynomial for $f(x)$ about a, and $E_n(x) = f(x) - P_n(x)$ is the error function, then

$$|E_n(x)| \le \frac{M}{(n+1)!}|x-a|^{n+1},$$

where M is an upper bound on $\left|f^{(n+1)}\right|$ on the interval between a and x.

To find M in practice, we often find the maximum of $|f^{(n+1)}|$ on the interval and pick any larger value for M. See page 543 for a justification of Theorem 10.3.

Using the Lagrange Error Bound for Taylor Polynomials

Example 3 Give a bound on the error, E_4, when e^x is approximated by its fourth-degree Taylor polynomial about 0 for $-0.5 \le x \le 0.5$.

Solution Let $f(x) = e^x$. We need to find a bound for the fifth derivative, $f^{(5)}(x) = e^x$. Since e^x is positive and increasing, its largest value is at the right endpoint of the interval and:

$$|f^{(5)}(x)| \leq e^{0.5} = \sqrt{e} \quad \text{for } -0.5 \leq x \leq 0.5.$$

Since $\sqrt{e} < 2$, we can take $M = 2$ (or any larger value). Then

$$|E_4| = |f(x) - P_4(x)| \leq \frac{2}{5!}|x|^5.$$

This means, for example, that on $-0.5 \leq x \leq 0.5$, the approximation

$$e^x \approx 1 + x + \frac{x^2}{2!} + \frac{x^3}{3!} + \frac{x^4}{4!}$$

has an error of at most $\frac{2}{120}(0.5)^5 < 0.0006$.

The Lagrange error bound for Taylor polynomials can be used to see how the accuracy of the approximation depends on the value of x and the value of n. Observe that the error bound for a Taylor polynomial of degree n is proportional to $|x - a|^{n+1}$. That means, for example, with a Taylor polynomial of degree n centered at 0, if we decrease x by a factor of 2, the error bound decreases by a factor of 2^{n+1}.

We saw this happen in Example 1 where $E_2(0.1) = 0.000171$ and $E_2(0.05) = 0.000021$. Since

$$\frac{E_2(0.1)}{E_2(0.05)} = \frac{0.000171}{0.000021} = 8.1,$$

we see that as the value of x decreases by a factor of 2, the error decreases by a factor of about $8 = 2^3$.

Convergence of the Taylor Series for cos x

We have already seen that the Taylor polynomials centered at $x = 0$ for $\cos x$ are good approximations for x near 0. (See Figure 10.18.) In fact, for any value of x, if we take a Taylor polynomial centered at $x = 0$ of high enough degree, its graph is nearly indistinguishable from the graph of the cosine function near that point.

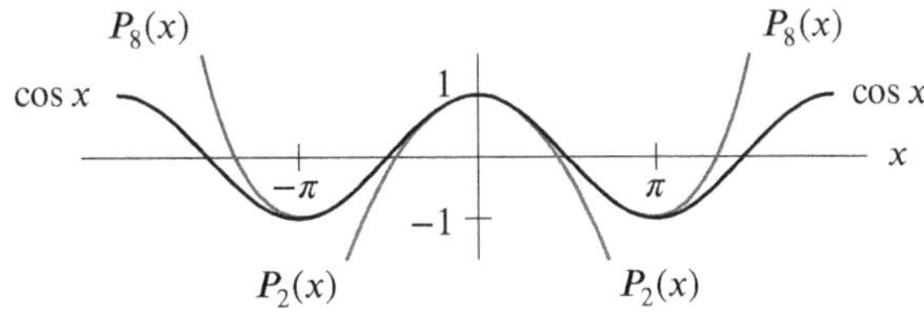

Figure 10.18: Graph of $\cos x$ and two Taylor polynomials for x near 0

Let's see what happens numerically. Let $x = \pi/2$. The successive Taylor polynomial approximations to $\cos(\pi/2) = 0$ about $x = 0$ are

$$\begin{aligned}
P_2(\pi/2) &= \quad 1 - (\pi/2)^2/2! &&= -0.23370\ldots \\
P_4(\pi/2) &= 1 - (\pi/2)^2/2! + (\pi/2)^4/4! &&= \quad 0.01997\ldots \\
P_6(\pi/2) &= \quad \cdots &&= -0.00089\ldots \\
P_8(\pi/2) &= \quad \cdots &&= \quad 0.00002\ldots.
\end{aligned}$$

It appears that the approximations converge to the true value, $\cos(\pi/2) = 0$, very rapidly. Now take a value of x somewhat farther away from 0, say $x = \pi$; then $\cos\pi = -1$ and

$$\begin{aligned}
P_2(\pi) &= 1 - (\pi)^2/2! = -3.93480\ldots \\
P_4(\pi) &= \quad \cdots \quad = \ \ 0.12391\ldots \\
P_6(\pi) &= \quad \cdots \quad = -1.21135\ldots \\
P_8(\pi) &= \quad \cdots \quad = -0.97602\ldots \\
P_{10}(\pi) &= \quad \cdots \quad = -1.00183\ldots \\
P_{12}(\pi) &= \quad \cdots \quad = -0.99990\ldots \\
P_{14}(\pi) &= \quad \cdots \quad = -1.000004\ldots.
\end{aligned}$$

We see that the rate of convergence is somewhat slower; it takes a 14th-degree polynomial to approximate $\cos\pi$ as accurately as an 8th-degree polynomial approximates $\cos(\pi/2)$. If x were taken still farther away from 0, then we would need still more terms to obtain as accurate an approximation of $\cos x$.

Exercise 18 on page 510 uses the ratio test to show that the Taylor series for $\cos x$ converges for all values of x. To show that its sum is indeed $\cos x$, we use Theorem 10.3. This justifies our writing the equality:

$$\cos x = 1 - \frac{x^2}{2!} + \frac{x^4}{4!} - \frac{x^6}{6!} + \frac{x^8}{8!} - \cdots \qquad \text{for all } x.$$

Showing that the Taylor Series for $\cos x$ Converges to $\cos x$

The Lagrange error bound in Theorem 10.3 allows us to see if the Taylor series for a function converges to that function. In the series for $\cos x$, the odd powers are missing, so we assume n is even and write

$$E_n(x) = \cos x - P_n(x) = \cos x - \left(1 - \frac{x^2}{2!} + \cdots + (-1)^{n/2}\frac{x^n}{n!}\right),$$

giving

$$\cos x = 1 - \frac{x^2}{2!} + \cdots + (-1)^{n/2}\frac{x^n}{n!} + E_n(x).$$

Thus, for the Taylor series to converge to $\cos x$, we must have $E_n(x) \to 0$ as $n \to \infty$.

Showing $E_n(x) \to 0$ as $n \to \infty$

Proof Since $f(x) = \cos x$, the $(n+1)^{\text{st}}$ derivative, $f^{(n+1)}(x)$, is $\pm\cos x$ or $\pm\sin x$, no matter what n is. So for all n, we have $|f^{(n+1)}(x)| \le 1$ on the interval between 0 and x.

By the Lagrange error bound with $M = 1$, we have

$$|E_n(x)| = |\cos x - P_n(x)| \le \frac{|x|^{n+1}}{(n+1)!} \quad \text{for every } n.$$

To show that the errors go to zero, we must show that for a fixed x,

$$\frac{|x|^{n+1}}{(n+1)!} \to 0 \quad \text{as} \quad n \to \infty.$$

To see why this is true, consider the ratio of successive terms of this sequence. We have

$$\frac{|x|^{n+2}/(n+2)!}{|x|^{n+1}/(n+1)!} = \frac{|x|}{n+2}.$$

Therefore, we obtain the $(n+1)^{\text{st}}$ term of this sequence, $|x|^{n+2}/(n+2)!$, by multiplying the previous term, $|x|^{n+1}/(n+1)!$, by $|x|/(n+2)$. Since $|x|$ is fixed and $n+2$ is increasing, for sufficiently large n, this ratio is less than $1/2$ (or any other constant between 0 and 1). Thus eventually each term in the sequence is less than $1/2$ the previous term, so the sequence of errors approaches zero. Therefore, the Taylor series $1 - x^2/2! + x^4/4! - \cdots$ does converge to $\cos x$.

Problems 27 and 28 ask you to show that the Taylor series for $\sin x$ and e^x converge to the original function for all x. In each case, you again need the following limit:

$$\lim_{n\to\infty} \frac{x^n}{n!} = 0.$$

Error Bounds Using the Alternating Series Estimate

We saw a method of bounding the error in an alternating series in Theorem 9.9 on page 500. This method can be used for a Taylor series if we know the series converges to the function $f(x)$ and that the series is alternating and has terms whose magnitudes decrease monotonically to zero.

Example 4 As we have shown, the Taylor series for $\cos x$ converges to the function for all x. Use an alternating series to bound the errors in approximating $\cos 1$ by $P_6(1)$ and $P_9(1)$.

Solution The Taylor series is

$$\cos x = 1 - \frac{x^2}{2!} + \frac{x^4}{4!} - \frac{x^6}{6!} + \frac{x^8}{8!} - \frac{x^{10}}{10!} + \cdots,$$

so, substituting $x = 1$, we have

$$\cos 1 = \underbrace{\overbrace{1 - \frac{1}{2!} + \frac{1}{4!} - \frac{1}{6!}}^{P_6(1)} + \frac{1}{8!}}_{P_9(1)} - \frac{1}{10!} + \cdots.$$

This series is alternating and the magnitude of the terms is monotonically decreasing to zero. Thus we can use the alternating series error bound, which tells us that the error between $\cos 1$ and $P_n(1)$ is bounded by the magnitude of the next term in the series. So we have

$$\left|\cos 1 - P_6(1)\right| \le \frac{1}{8!} = 0.0000248,$$

$$\left|\cos 1 - P_9(1)\right| \le \frac{1}{10!} = 0.0000003.$$

Deriving the Lagrange Error Bound

Recall that we constructed $P_n(x)$, the Taylor polynomial of f about 0, so that its first n derivatives equal the corresponding derivatives of $f(x)$. Therefore, $E_n(0) = 0$, $E_n'(0) = 0$, $E_n''(0) = 0, \cdots$, $E_n^{(n)}(0) = 0$. Since $P_n(x)$ is an nth-degree polynomial, its $(n+1)^{\text{st}}$ derivative is 0, so $E_n^{(n+1)}(x) = f^{(n+1)}(x)$. In addition, suppose that $\left|f^{(n+1)}(x)\right|$ is bounded by a positive constant M, for all positive values of x near 0, say for $0 \le x \le d$, so that

$$-M \le f^{(n+1)}(x) \le M \quad \text{for } 0 \le x \le d.$$

This means that

$$-M \le E_n^{(n+1)}(x) \le M \quad \text{for } 0 \le x \le d.$$

Writing t for the variable, we integrate this inequality from 0 to x, giving

$$-\int_0^x M\,dt \le \int_0^x E_n^{(n+1)}(t)\,dt \le \int_0^x M\,dt \quad \text{for } 0 \le x \le d,$$

so

$$-Mx \le E_n^{(n)}(x) \le Mx \quad \text{for } 0 \le x \le d.$$

We integrate this inequality again from 0 to x, giving

$$-\int_0^x Mt\,dt \le \int_0^x E_n^{(n)}(t)\,dt \le \int_0^x Mt\,dt \quad \text{for } 0 \le x \le d,$$

so

$$-\frac{1}{2}Mx^2 \le E_n^{(n-1)}(x) \le \frac{1}{2}Mx^2 \quad \text{for } 0 \le x \le d.$$

By repeated integration, we obtain the following bound:

$$-\frac{1}{(n+1)!}Mx^{n+1} \le E_n(x) \le \frac{1}{(n+1)!}Mx^{n+1} \quad \text{for } 0 \le x \le d,$$

which means that

$$|E_n(x)| = |f(x) - P_n(x)| \le \frac{1}{(n+1)!}Mx^{n+1} \quad \text{for } 0 \le x \le d.$$

When x is to the left of 0, so $-d \le x \le 0$, and when the Taylor series is centered at $a \ne 0$, similar calculations lead to Theorem 10.3.

Exercises and Problems for Section 10.4

EXERCISES

In Exercises 1–8, use Theorem 10.3 to find a bound for the error in approximating the quantity with a third-degree Taylor polynomial for the given function $f(x)$ about $x = 0$. Compare the bound with the actual error.

1. $e^{0.1}$, $f(x) = e^x$
2. $\sin(0.2)$, $f(x) = \sin x$
3. $\cos(-0.3)$, $f(x) = \cos x$
4. $\sqrt{0.9}$, $f(x) = \sqrt{1+x}$
5. $\ln(1.5)$, $f(x) = \ln(1+x)$
6. $1/\sqrt{3}$, $f(x) = (1+x)^{-1/2}$
7. $\tan 1$, $f(x) = \tan x$
8. $0.5^{1/3}$, $f(x) = (1-x)^{1/3}$

In Exercises 9–11, the Taylor polynomial $P_n(x)$ about 0 approximates $f(x)$ with error $E_n(x)$ and the Taylor series converges to $f(x)$. Find the smallest constant K given by the alternating series error bound such that $|E_4(1)| \le K$.

9. $f(x) = e^{-x}$
10. $f(x) = \cos x$
11. $f(x) = \sin x$

In Exercises 12–14, the Taylor polynomial $P_n(x)$ about 0 approximates $f(x)$ with error $E_n(x)$ and the Taylor series converges to $f(x)$ on $|x| \le 1$. Find the smallest constant K given by the alternating series error bound such that $|E_6(1)| \le K$.

12. $\displaystyle\sum_{n=0}^{\infty}(-1)^n\frac{x^n}{2^n}$
13. $\displaystyle\sum_{n=0}^{\infty}\frac{(-x)^n}{n^2+1}$
14. $\displaystyle\sum_{n=0}^{\infty}\frac{(-1)^{n+1}x^n}{(n+1)^n}$

PROBLEMS

15. **(a)** Using a calculator, make a table of values to four decimal places of $\sin x$ for

$$x = -0.5, -0.4, \dots, -0.1, 0, 0.1, \dots, 0.4, 0.5.$$

(b) Add to your table the values of the error $E_1 = \sin x - x$ for these x-values.

(c) Using a calculator or computer, draw a graph of the quantity $E_1 = \sin x - x$ showing that

$$|E_1| < 0.03 \quad \text{for} \quad -0.5 \le x \le 0.5.$$

16. Find a bound on the magnitude of the error if we approximate $\sqrt{2}$ using the Taylor approximation of degree three for $\sqrt{1+x}$ about $x = 0$.

17. **(a)** Let $f(x) = e^x$. Find a bound on the magnitude of the error when $f(x)$ is approximated using $P_3(x)$, its Taylor approximation of degree 3 around 0 over the interval $[-2, 2]$.

(b) What is the actual maximum error in approximating $f(x)$ by $P_3(x)$ over the interval $[-2, 2]$?

18. Let $f(x) = \cos x$ and let $P_n(x)$ be the Taylor approximation of degree n for $f(x)$ around 0. Explain why, for any x, we can choose an n such that

$$|f(x) - P_n(x)| < 10^{-9}.$$

19. Consider the error in using the approximation $\sin\theta \approx \theta$ on the interval $[-1, 1]$.

(a) Reasoning informally, say where the approximation is an overestimate and where it is an underestimate.

(b) Use Theorem 10.3 to bound the error. Check your answer graphically on a computer or calculator.

20. Repeat Problem 19 for the approximation $\sin\theta \approx \theta - \theta^3/3!$.

21. You approximate $f(t) = e^t$ by a Taylor polynomial of degree 0 about $t = 0$ on the interval $[0, 0.5]$.

(a) Reasoning informally, say whether the approximation is an overestimate or an underestimate.

(b) Use Theorem 10.3 to bound the error. Check your answer graphically on a computer or calculator.

22. Repeat Problem 21 using the second-degree Taylor approximation to e^t.

23. (a) Use the graphs of $y = \cos x$ and its Taylor polynomials, $P_{10}(x)$ and $P_{20}(x)$, in Figure 10.19 to bound:

(i) The error in approximating $\cos 6$ by $P_{10}(6)$ and by $P_{20}(6)$.

(ii) The error in approximating $\cos x$ by $P_{20}(x)$ for $|x| \le 9$.

(b) If we want to approximate $\cos x$ by $P_{10}(x)$ to an accuracy of within 0.1, what is the largest interval of x-values on which we can work? Give your answer to the nearest integer.

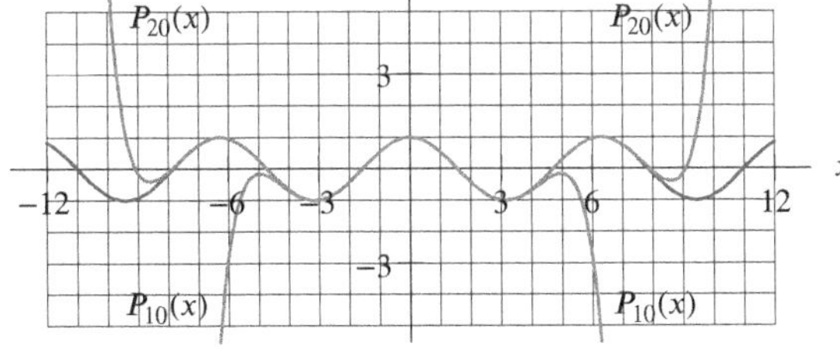

Figure 10.19

24. Give a bound for the error for the nth-degree Taylor polynomial about $x = 0$ approximating $\cos x$ on the interval $[0, 1]$. What is the bound for $\sin x$?

25. What degree Taylor polynomial about $x = 0$ do you need to calculate $\cos 1$ to four decimal places? To six decimal places? Justify your answer using the results of Problem 24.

26. For $|x| \le 0.1$, graph the error

$$E_0 = \cos x - P_0(x) = \cos x - 1.$$

Explain the shape of the graph, using the Taylor expansion of $\cos x$. Find a bound for $|E_0|$ for $|x| \le 0.1$.

27. Show that the Taylor series about 0 for e^x converges to e^x for every x. Do this by showing that the error $E_n(x) \to 0$ as $n \to \infty$.

28. Show that the Taylor series about 0 for $\sin x$ converges to $\sin x$ for every x.

29. To approximate π using a Taylor polynomial, we could use the series for the arctangent or the series for the arcsine. In this problem, we compare the two methods.

(a) Using the fact that $d(\arctan x)/dx = 1/(1+x^2)$ and $\arctan 1 = \pi/4$, approximate the value of π using the third-degree Taylor polynomial of $4\arctan x$ about $x = 0$.

(b) Using the fact that $d(\arcsin x)/dx = 1/\sqrt{1-x^2}$ and $\arcsin 1 = \pi/2$, approximate the value of π using the third-degree Taylor polynomial of $2\arcsin x$ about $x = 0$.

(c) Estimate the maximum error of the approximation you found in part (a).

(d) Explain the problem in estimating the error in the arcsine approximation.

In Problems 30–31, you will compare the Lagrange and alternating series error bounds at $x = 1/2$ in approximating the function $f(x)$ by $P_n(x)$, its Taylor polynomial about $x = 0$. The Taylor series converges to $f(x)$ at $x = 1/2$.

(a) Find the smallest bound for the error using Lagrange's method.

(b) Find a bound for the same error using the alternating series bound.

(c) Compare your answers to parts (a) and (b). What do you notice?

30. $f(x) = 1/(1+x)$ **31.** $f(x) = e^{-x}$,

32. To use the alternating series error bound, the magnitude of the terms must decrease monotonically to 0. Here we use the alternating series error bound on the Taylor series for $\cos x$, whose convergence to $\cos x$ is known.

(a) Find the ratio of the magnitudes of successive terms in the Taylor series for $\cos x$.

(b) Using your answer to part (a), if $x = 1.1$, show that the magnitudes of the terms decrease monotonically to zero.

(c) Bound the error in approximating $\cos x$ by its degree-4 Taylor polynomial when $x = 1.1$.

(d) If $x = 10$, after which term in the Taylor series do the terms of the series decrease monotonically to zero?

(e) Assuming the alternating series error bound applies, bound the error $E_{12}(10)$ using an alternating series.

Strengthen Your Understanding

In Problems 33–34, explain what is wrong with the statement.

33. Let $P_n(x)$ be a Taylor approximation of degree n for a function $f(x)$ about a, where a is a constant. Then $|f(a) - P_n(a)| > 0$ for any n.

34. Let $f(x)$ be a function whose Taylor series about $x = 0$ converges to $f(x)$ for all x. Then there exists a positive integer n such that the nth-degree Taylor polynomial $P_n(x)$ for $f(x)$ about $x = 0$ satisfies the inequality
$$|f(x) - P_n(x)| < 1 \quad \text{for all values of } x.$$

In Problems 35–37, give an example of:

35. A function $f(x)$ whose Taylor series converges to $f(x)$ for all values of x.

36. A polynomial $P(x)$ such that $|1/x - P(x)| < 0.1$ for all x in the interval $[1, 1.5]$.

37. A function $f(x)$ and an interval $[-c, c]$ such that the value of M in the error of the second-degree Taylor polynomial of $f(x)$ centered at 0 on the interval could be 4.

Decide if the statements in Problems 38–42 are true or false. Assume that the Taylor series for a function converges to that function. Give an explanation for your answer.

38. Let $P_n(x)$ be the nth Taylor polynomial for a function f near $x = a$. Although $P_n(x)$ is a good approximation to f near $x = a$, it is not possible to have $P_n(x) = f(x)$ for all x.

39. If $|f^{(n)}(x)| < 10$ for all $n > 0$ and all x, then the Taylor series for f about $x = 0$ converges to $f(x)$ for all x.

40. If $f^{(n)}(0) \geq n!$ for all n, then the Taylor series for f near $x = 0$ diverges at $x = 0$.

41. If $f^{(n)}(0) \geq n!$ for all n, then the Taylor series for f near $x = 0$ diverges at $x = 1$.

42. If $f^{(n)}(0) \geq n!$ for all n, then the Taylor series for f near $x = 0$ diverges at $x = 1/2$.

10.5 FOURIER SERIES

We have seen how to approximate a function by a Taylor polynomial of fixed degree. Such a polynomial is usually very close to the true value of the function near one point (the point at which the Taylor polynomial is centered), but not necessarily at all close anywhere else. In other words, Taylor polynomials are good approximations of a function *locally*, but not necessarily *globally*. In this section, we take another approach: we approximate the function by trigonometric functions, called *Fourier approximations*. The resulting approximation may not be as close to the original function at some points as the Taylor polynomial. However, the Fourier approximation is, in general, close over a larger interval. In other words, a Fourier approximation can be a better approximation globally. In addition, Fourier approximations are useful even for functions that are not continuous. Unlike Taylor approximations, Fourier approximations are periodic, so they are particularly useful for approximating periodic functions.

Many processes in nature are periodic or repeating, so it makes sense to approximate them by periodic functions. For example, sound waves are made up of periodic oscillations of air molecules. Heartbeats, the movement of the lungs, and the electrical current that powers our homes are all periodic phenomena. Two of the simplest periodic functions are the square wave in Figure 10.20 and the triangular wave in Figure 10.21. Electrical engineers use the square wave as the model for the flow of electricity when a switch is repeatedly flicked on and off.

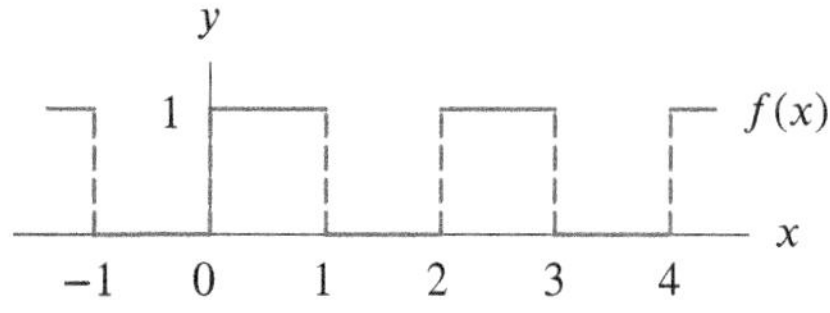

Figure 10.20: Square wave

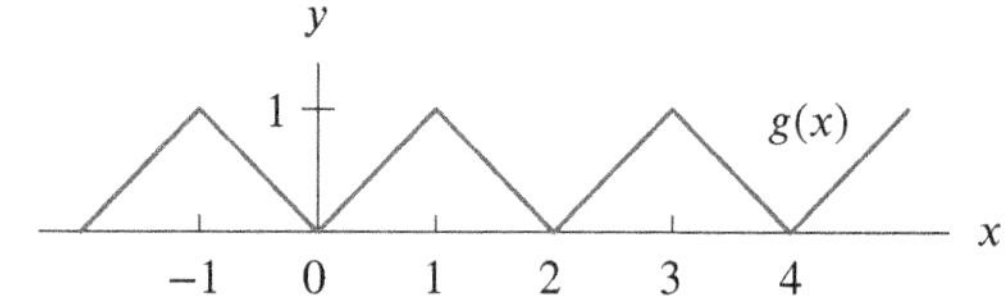

Figure 10.21: Triangular wave

Fourier Polynomials

We can express the square wave and the triangular wave by the formulas

$$f(x) = \begin{cases} \vdots & \vdots \\ 0 & -1 \le x < 0 \\ 1 & 0 \le x < 1 \\ 0 & 1 \le x < 2 \\ 1 & 2 \le x < 3 \\ 0 & 3 \le x < 4 \\ \vdots & \vdots \end{cases} \qquad g(x) = \begin{cases} \vdots & \vdots \\ -x & -1 \le x < 0 \\ x & 0 \le x < 1 \\ 2-x & 1 \le x < 2 \\ x-2 & 2 \le x < 3 \\ 4-x & 3 \le x < 4 \\ \vdots & \vdots \end{cases}$$

However, these formulas are not particularly easy to work with. Worse, the functions are not differentiable at various points. Here we show how to approximate such functions by differentiable, periodic functions.

Since sine and cosine are the simplest periodic functions, they are the building blocks we use. Because they repeat every 2π, we assume that the function f we want to approximate repeats every 2π. (Later, we deal with the case where f has some other period.) We start by considering the square wave in Figure 10.22. Because of the periodicity of all the functions concerned, we only have to consider what happens in the course of a single period; the same behavior repeats in any other period.

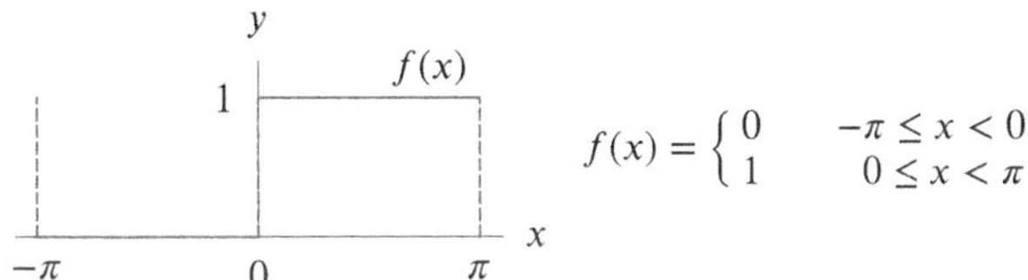

Figure 10.22: Square wave on $[-\pi, \pi]$

We will attempt to approximate f with a sum of trigonometric functions of the form

$$\begin{aligned} f(x) &\approx F_n(x) \\ &= a_0 + a_1 \cos x + a_2 \cos(2x) + a_3 \cos(3x) + \cdots + a_n \cos(nx) \\ &\quad + b_1 \sin x + b_2 \sin(2x) + b_3 \sin(3x) + \cdots + b_n \sin(nx) \\ &= a_0 + \sum_{k=1}^{n} a_k \cos(kx) + \sum_{k=1}^{n} b_k \sin(kx). \end{aligned}$$

$F_n(x)$ is known as a *Fourier polynomial of degree n*, named after the French mathematician Joseph Fourier (1768–1830), who was one of the first to investigate it.[9] The coefficients a_k and b_k are called *Fourier coefficients*. Since each of the component functions $\cos(kx)$ and $\sin(kx)$, $k = 1, 2, \ldots, n$, repeats every 2π, $F_n(x)$ must repeat every 2π and so is a potentially good match for $f(x)$, which also repeats every 2π. The problem is to determine values for the Fourier coefficients that achieve a close match between $f(x)$ and $F_n(x)$. We choose the following values:

The Fourier Coefficients for a Periodic Function f of Period 2π

$$a_0 = \frac{1}{2\pi} \int_{-\pi}^{\pi} f(x)\,dx,$$

$$a_k = \frac{1}{\pi} \int_{-\pi}^{\pi} f(x) \cos(kx)\,dx \quad \text{for } k > 0,$$

$$b_k = \frac{1}{\pi} \int_{-\pi}^{\pi} f(x) \sin(kx)\,dx \quad \text{for } k > 0.$$

Notice that a_0 is just the average value of f over the interval $[-\pi, \pi]$.

[9] The Fourier polynomials are not polynomials in the usual sense of the word.

For an informal justification for the use of these values, see page 555. In addition, the integrals over $[-\pi, \pi]$ for a_k and b_k can be replaced by integrals over any interval of length 2π.

Example 1 Construct successive Fourier polynomials for the square wave function f, with period 2π, given by

$$f(x) = \begin{cases} 0 & -\pi \le x < 0 \\ 1 & 0 \le x < \pi. \end{cases}$$

Solution Since a_0 is the average value of f on $[-\pi, \pi]$, we suspect from the graph of f that $a_0 = \frac{1}{2}$. We can verify this analytically:

$$a_0 = \frac{1}{2\pi}\int_{-\pi}^{\pi} f(x)\,dx = \frac{1}{2\pi}\int_{-\pi}^{0} 0\,dx + \frac{1}{2\pi}\int_{0}^{\pi} 1\,dx = 0 + \frac{1}{2\pi}(\pi) = \frac{1}{2}.$$

Furthermore,

$$a_1 = \frac{1}{\pi}\int_{-\pi}^{\pi} f(x)\cos x\,dx = \frac{1}{\pi}\int_{0}^{\pi} 1\cos x\,dx = 0$$

and

$$b_1 = \frac{1}{\pi}\int_{-\pi}^{\pi} f(x)\sin x\,dx = \frac{1}{\pi}\int_{0}^{\pi} 1\sin x\,dx = \frac{2}{\pi}.$$

Therefore, the Fourier polynomial of degree 1 is given by

$$f(x) \approx F_1(x) = \frac{1}{2} + \frac{2}{\pi}\sin x,$$

and the graphs of the function and the first Fourier approximation are shown in Figure 10.23.

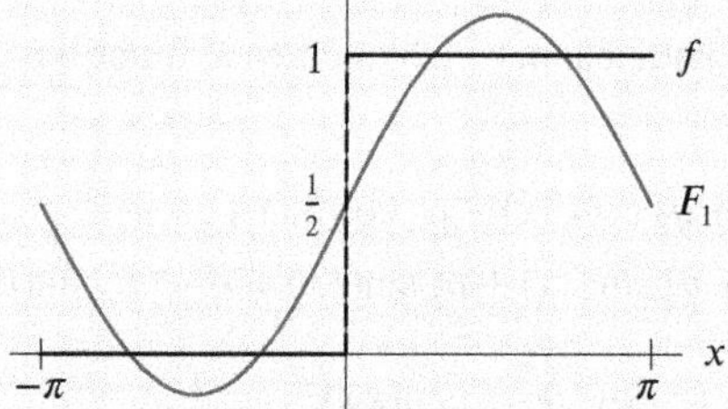

Figure 10.23: First Fourier approximation to the square wave

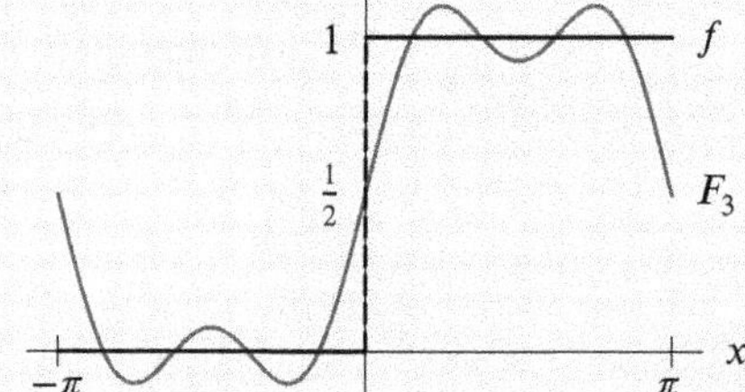

Figure 10.24: Third Fourier approximation to the square wave

We next construct the Fourier polynomial of degree 2. The coefficients a_0, a_1, b_1 are the same as before. In addition,

$$a_2 = \frac{1}{\pi}\int_{-\pi}^{\pi} f(x)\cos(2x)\,dx = \frac{1}{\pi}\int_{0}^{\pi} 1\cos(2x)\,dx = 0$$

and

$$b_2 = \frac{1}{\pi}\int_{-\pi}^{\pi} f(x)\sin(2x)\,dx = \frac{1}{\pi}\int_{0}^{\pi} 1\sin(2x)\,dx = 0.$$

Since $a_2 = b_2 = 0$, the Fourier polynomial of degree 2 is identical to the Fourier polynomial of degree 1. Let's look at the Fourier polynomial of degree 3:

$$a_3 = \frac{1}{\pi}\int_{-\pi}^{\pi} f(x)\cos(3x)\,dx = \frac{1}{\pi}\int_{0}^{\pi} 1\cos(3x)\,dx = 0$$

and

$$b_3 = \frac{1}{\pi}\int_{-\pi}^{\pi} f(x)\sin(3x)\,dx = \frac{1}{\pi}\int_{0}^{\pi} 1\sin(3x)\,dx = \frac{2}{3\pi}.$$

So the approximation is given by

$$f(x) \approx F_3(x) = \frac{1}{2} + \frac{2}{\pi}\sin x + \frac{2}{3\pi}\sin(3x).$$

The graph of F_3 is shown in Figure 10.24. This approximation is better than $F_1(x) = \frac{1}{2} + \frac{2}{\pi}\sin x$, as comparing Figure 10.24 to Figure 10.23 shows.

Without going through the details, we calculate the coefficients for higher-degree Fourier approximations:

$$F_5(x) = \frac{1}{2} + \frac{2}{\pi}\sin x + \frac{2}{3\pi}\sin(3x) + \frac{2}{5\pi}\sin(5x)$$
$$F_7(x) = \frac{1}{2} + \frac{2}{\pi}\sin x + \frac{2}{3\pi}\sin(3x) + \frac{2}{5\pi}\sin(5x) + \frac{2}{7\pi}\sin(7x).$$

Figure 10.25 shows that higher-degree approximations match the step-like nature of the square wave function more and more closely.

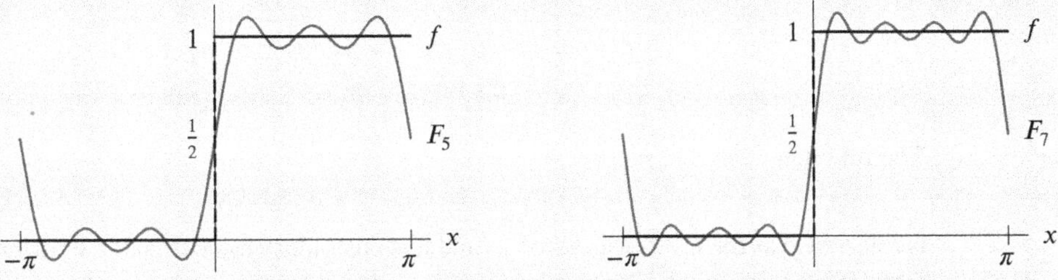

Figure 10.25: Fifth and seventh Fourier approximations to the square wave

We could have used a Taylor series to approximate the square wave, provided we did not center the series at a point of discontinuity. Since the square wave is a constant function on each interval, all its derivatives are zero, and so its Taylor series approximations are the constant functions: 0 or 1, depending on where the Taylor series is centered. They approximate the square wave perfectly on each piece, but they do not do a good job over the whole interval of length 2π. That is what Fourier polynomials succeed in doing: they approximate a curve fairly well everywhere, rather than just near a particular point. The Fourier approximations above look a lot like square waves, so they approximate well *globally*. However, they may not give good values near points of discontinuity. (For example, near $x = 0$, they all give values near $1/2$, which are incorrect.) Thus Fourier polynomials may not be good *local* approximations.

> Taylor polynomials give good *local* approximations to a function;
> Fourier polynomials give good *global* approximations to a function.

Fourier Series

As with Taylor polynomials, the higher the degree of the Fourier approximation, generally the more accurate it is. Therefore, we carry this procedure on indefinitely by letting $n \to \infty$, and we call the resulting infinite series a *Fourier series*.

The Fourier Series for f on $[-\pi, \pi]$

$$f(x) = a_0 + a_1 \cos x + a_2 \cos 2x + a_3 \cos 3x + \cdots$$
$$+ b_1 \sin x + b_2 \sin 2x + b_3 \sin 3x + \cdots$$

where a_k and b_k are the Fourier coefficients.

Thus, the Fourier series for the square wave is

$$f(x) = \frac{1}{2} + \frac{2}{\pi}\sin x + \frac{2}{3\pi}\sin 3x + \frac{2}{5\pi}\sin 5x + \frac{2}{7\pi}\sin 7x + \cdots.$$

Harmonics

Let us start with a function $f(x)$ that is periodic with period 2π, expanded in a Fourier series:

$$f(x) = a_0 + a_1 \cos x + a_2 \cos 2x + a_3 \cos 3x + \cdots$$
$$+ b_1 \sin x + b_2 \sin 2x + b_3 \sin 3x + \cdots$$

The function

$$a_k \cos kx + b_k \sin kx$$

is referred to as the kth *harmonic* of f, and it is customary to say that the Fourier series expresses f in terms of its harmonics. The first harmonic, $a_1 \cos x + b_1 \sin x$, is sometimes called the *fundamental harmonic* of f.

Example 2 Find a_0 and the first four harmonics of a *pulse train* function f of period 2π shown in Figure 10.26:

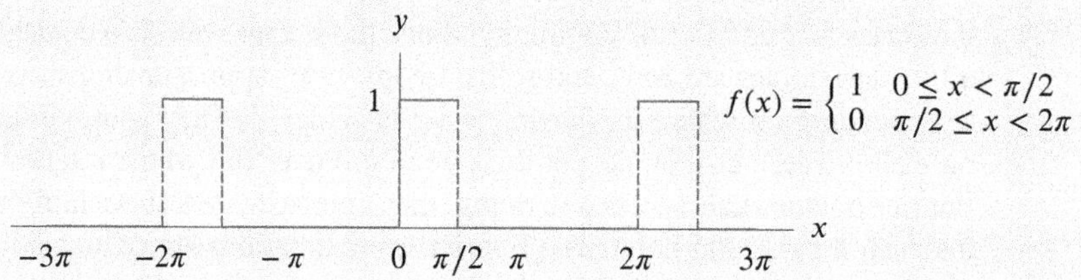

Figure 10.26: A train of pulses with period 2π

Solution First, a_0 is the average value of the function, so

$$a_0 = \frac{1}{2\pi}\int_{-\pi}^{\pi} f(x)\,dx = \frac{1}{2\pi}\int_0^{\pi/2} 1\,dx = \frac{1}{4}.$$

Next, we compute a_k and b_k, $k = 1, 2, 3$, and 4. The formulas

$$a_k = \frac{1}{\pi}\int_{-\pi}^{\pi} f(x)\cos(kx)\,dx = \frac{1}{\pi}\int_0^{\pi/2} \cos(kx)\,dx$$

$$b_k = \frac{1}{\pi}\int_{-\pi}^{\pi} f(x)\sin(kx)\,dx = \frac{1}{\pi}\int_0^{\pi/2} \sin(kx)\,dx$$

lead to the harmonics

$$\begin{aligned}
a_1 \cos x + b_1 \sin x &= \frac{1}{\pi}\cos x + \frac{1}{\pi}\sin x \\
a_2 \cos(2x) + b_2 \sin(2x) &= \frac{1}{\pi}\sin(2x) \\
a_3 \cos(3x) + b_3 \sin(3x) &= -\frac{1}{3\pi}\cos(3x) + \frac{1}{3\pi}\sin(3x) \\
a_4 \cos(4x) + b_4 \sin(4x) &= 0.
\end{aligned}$$

Figure 10.27 shows the graph of the sum of a_0 and these harmonics, which is the fourth Fourier approximation of f.

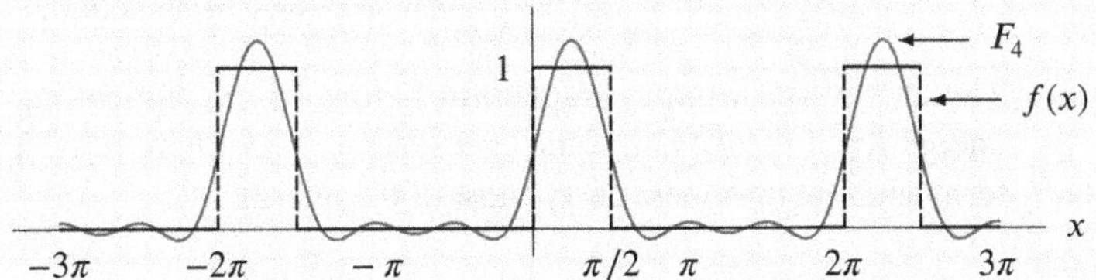

Figure 10.27: Fourth Fourier approximation to pulse train f equals the sum of a_0 and the first four harmonics

Energy and the Energy Theorem

The quantity $A_k = \sqrt{a_k^2 + b_k^2}$ is called the amplitude of the kth harmonic. The square of the amplitude has a useful interpretation. Adopting terminology from the study of periodic waves, we define the *energy* E of a periodic function f of period 2π to be the number

$$E = \frac{1}{\pi}\int_{-\pi}^{\pi} (f(x))^2\, dx.$$

Problem 19 on page 558 asks you to check that for all positive integers k,

$$\frac{1}{\pi}\int_{-\pi}^{\pi} (a_k \cos(kx) + b_k \sin(kx))^2\, dx = a_k^2 + b_k^2 = A_k^2.$$

This shows that the kth harmonic of f has energy A_k^2. The energy of the constant term a_0 of the Fourier series is $\frac{1}{\pi}\int_{-\pi}^{\pi} a_0^2\, dx = 2a_0^2$, so we make the definition

$$A_0 = \sqrt{2}a_0.$$

It turns out that for all reasonable periodic functions f, the energy of f equals the sum of the energies of its harmonics:

The Energy Theorem for a Periodic Function f of Period 2π

$$E = \frac{1}{\pi}\int_{-\pi}^{\pi} (f(x))^2\, dx = A_0^2 + A_1^2 + A_2^2 + \cdots$$

where $A_0 = \sqrt{2}a_0$ and $A_k = \sqrt{a_k^2 + b_k^2}$ (for all integers $k \geq 1$).

The graph of A_k^2 against k is called the *energy spectrum* of f. It shows how the energy of f is distributed among its harmonics.

Example 3 (a) Graph the energy spectrum of the square wave of Example 1.

(b) What fraction of the energy of the square wave is contained in the constant term and first three harmonics of its Fourier series?

Solution (a) We know from Example 1 that $a_0 = 1/2$, $a_k = 0$ for $k \geq 1$, $b_k = 0$ for k even, and $b_k = 2/(k\pi)$ for k odd. Thus

$$A_0^2 = 2a_0^2 = \frac{1}{2}$$

$$A_k^2 = 0 \quad \text{if } k \text{ is even}, \quad k \geq 1,$$

$$A_k^2 = \left(\frac{2}{k\pi}\right)^2 = \frac{4}{k^2\pi^2} \quad \text{if } k \text{ is odd}, \quad k \geq 1.$$

The energy spectrum is graphed in Figure 10.28. Notice that it is customary to represent the energy A_k^2 of the kth harmonic by a vertical line of length A_k^2. The graph shows that the constant term and first harmonic carry most of the energy of f.

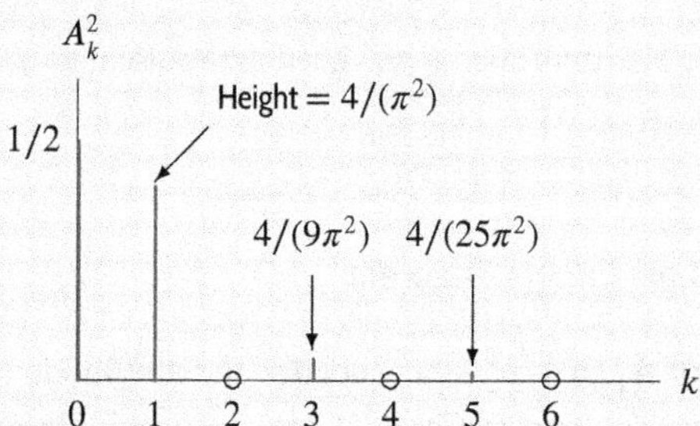

Figure 10.28: The energy spectrum of a square wave

(b) The energy of the square wave $f(x)$ is

$$E = \frac{1}{\pi}\int_{-\pi}^{\pi} (f(x))^2\,dx = \frac{1}{\pi}\int_0^{\pi} 1\,dx = 1.$$

The energy in the constant term and the first three harmonics of the Fourier series is

$$A_0^2 + A_1^2 + A_2^2 + A_3^2 = \frac{1}{2} + \frac{4}{\pi^2} + 0 + \frac{4}{9\pi^2} = 0.950.$$

The fraction of energy carried by the constant term and the first three harmonics is

$$0.95/1 = 0.95, \text{ or } 95\%.$$

Musical Instruments

You may have wondered why different musical instruments sound different, even when playing the same note. A first step might be to graph the periodic deviations from the average air pressure that form the sound waves they produce. This has been done for clarinet and trumpet in Figure 10.29.[10] However, it is more revealing to graph the energy spectra of these functions, as in Figure 10.30. The

[10]Adapted from C.A. Culver, *Musical Acoustics* (New York: McGraw-Hill, 1956), pp. 204, 220.

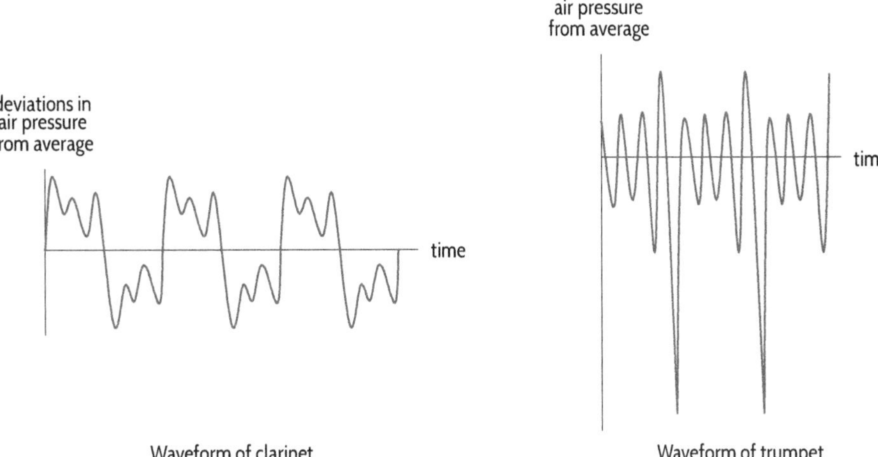

Figure 10.29: Sound waves of a clarinet and trumpet

most striking difference is the relative weakness of the second and fourth harmonics for the clarinet, with the second harmonic completely absent. The trumpet sounds the second harmonic with as much energy as it does the fundamental.

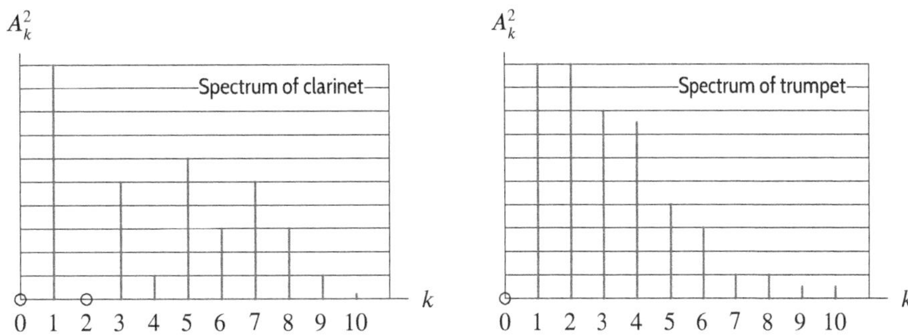

Figure 10.30: Energy spectra of a clarinet and trumpet

What Do We Do If Our Function Does Not Have Period 2π?

We can adapt our previous work to find the Fourier series for a function of period b. Suppose $f(x)$ is given on the interval $[-b/2, b/2]$. In Problem 31, we see how to use a change of variable to get the following result:

The Fourier Series for f on $[-b/2, b/2]$

$$f(x) = a_0 + \sum_{k=1}^{\infty}\left(a_k \cos\left(\frac{2\pi kx}{b}\right) + b_k \sin\left(\frac{2\pi kx}{b}\right)\right)$$

where $a_0 = \dfrac{1}{b}\displaystyle\int_{-b/2}^{b/2} f(x)\,dx$ and, for $k \geq 1$,

$$a_k = \frac{2}{b}\int_{-b/2}^{b/2} f(x)\cos\left(\frac{2\pi kx}{b}\right)dx \quad\text{and}\quad b_k = \frac{2}{b}\int_{-b/2}^{b/2} f(x)\sin\left(\frac{2\pi kx}{b}\right)dx.$$

The constant $2\pi k/b$ is called the angular frequency of the kth harmonic; b is the period of f.

Note that the integrals over $[-b/2, b/2]$ can be replaced by integrals over any interval of length b.

Example 4 Find the fifth-degree Fourier polynomial of the square wave $f(x)$ graphed in Figure 10.31.

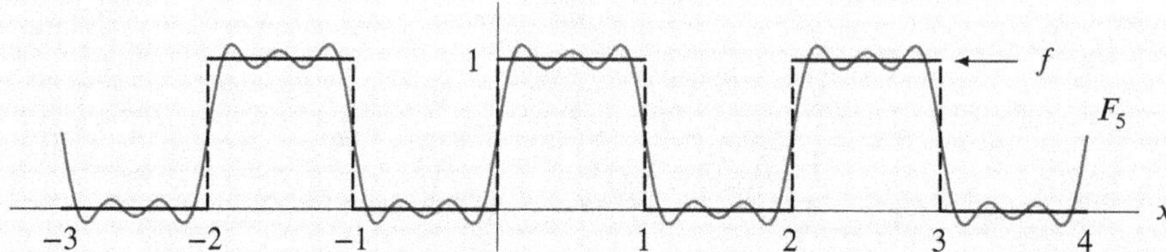

Figure 10.31: Square wave f and its fifth Fourier approximation F_5

Solution Since $f(x)$ repeats outside the interval $[-1, 1]$, we have $b = 2$. As an example of how the coefficients are computed, we find b_1. Since $f(x) = 0$ for $-1 < x < 0$,

$$b_1 = \frac{2}{2}\int_{-1}^{1} f(x)\sin\left(\frac{2\pi x}{2}\right)dx = \int_0^1 \sin(\pi x)dx = -\frac{1}{\pi}\cos(\pi x)\Big|_0^1 = \frac{2}{\pi}.$$

Finding the other coefficients by a similar method, we have

$$f(x) \approx \frac{1}{2} + \frac{2}{\pi}\sin(\pi x) + \frac{2}{3\pi}\sin(3\pi x) + \frac{2}{5\pi}\sin(5\pi x).$$

Notice that the coefficients in this series are the same as those in Example 1. This is because the graphs in Figures 10.25 and 10.31 are the same except with different scales on the x-axes.

Seasonal Variation in the Incidence of Measles

Example 5 Fourier approximations have been used to analyze the seasonal variation in the incidence of diseases. One study[11] done in Baltimore, Maryland, for the years 1901–1931, studied $I(t)$, the average number of cases of measles per 10,000 susceptible children in the tth month of the year. The data points in Figure 10.32 show $f(t) = \log I(t)$. The curve in Figure 10.32 shows the second Fourier approximation of $f(t)$. Figure 10.33 contains the graphs of the first and second harmonics of $f(t)$, plotted separately as deviations about a_0, the average logarithmic incidence rate. Describe what these two harmonics tell you about incidence of measles.

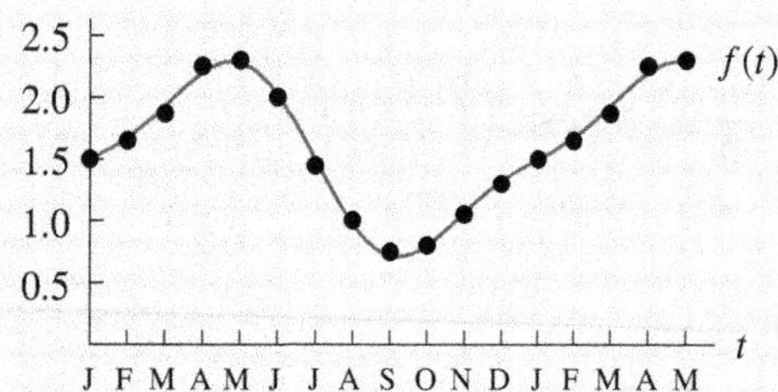

Figure 10.32: Logarithm of incidence of measles per month (dots) and second Fourier approximation (curve)

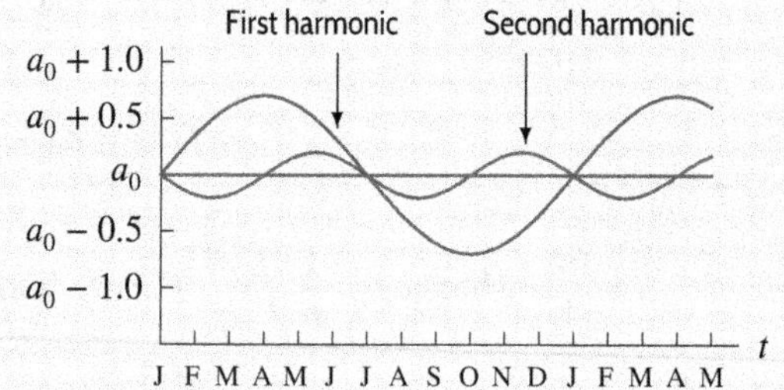

Figure 10.33: First and second harmonics of $f(t)$ plotted as deviations from average log incidence rate, a_0

[11]From C. I. Bliss and D. L. Blevins, *The Analysis of Seasonal Variation in Measles* (Am. J. Hyg. 70, 1959), reported by Edward Batschelet, *Introduction to Mathematics for the Life Sciences* (Springer-Verlag, Berlin, 1979).

Solution Taking the log of $I(t)$ has the effect of reducing the amplitude of the oscillations. However, since the log of a function increases when the function increases and decreases when it decreases, oscillations in $f(t)$ correspond to oscillations in $I(t)$.

Figure 10.33 shows that the first harmonic in the Fourier series has a period of one year (the same period as the original function); the second harmonic has a period of six months. The graph in Figure 10.33 shows that the first harmonic is approximately a sine function with amplitude about 0.7; the second harmonic is approximately the negative of a sine function with amplitude about 0.2. Thus, for t in months ($t = 0$ in January),

$$\log I(t) = f(t) \approx a_0 + 0.7\sin\left(\frac{\pi}{6}t\right) - 0.2\sin\left(\frac{\pi}{3}t\right),$$

where $\pi/6$ and $\pi/3$ are introduced to make the periods 12 and 6 months, respectively. We can estimate a_0 from the original graph of f: it is the average value of f, approximately 1.5. Thus

$$f(t) \approx 1.5 + 0.7\sin\left(\frac{\pi}{6}t\right) - 0.2\sin\left(\frac{\pi}{3}t\right).$$

Figure 10.32 shows that the second Fourier approximation of $f(t)$ is quite good. The harmonics of $f(t)$ beyond the second must be rather insignificant. This suggests that the variation in incidence in measles comes from two sources, one with a yearly cycle that is reflected in the first harmonic and one with a half-yearly cycle reflected in the second harmonic. At this point the mathematics can tell us no more; we must turn to the epidemiologists for further explanation.

Informal Justification of the Formulas for the Fourier Coefficients

Recall that the coefficients in a Taylor series (which is a good approximation locally) are found by differentiation. In contrast, the coefficients in a Fourier series (which is a good approximation globally) are found by integration.

We want to find the constants $a_0, a_1, a_2, \ldots$ and $b_1, b_2, \ldots$ in the expression

$$f(x) = a_0 + \sum_{k=1}^{\infty} a_k\cos(kx) + \sum_{k=1}^{\infty} b_k\sin(kx).$$

Consider the integral

$$\int_{-\pi}^{\pi} f(x)\,dx = \int_{-\pi}^{\pi}\left(a_0 + \sum_{k=1}^{\infty} a_k\cos(kx) + \sum_{k=1}^{\infty} b_k\sin(kx)\right)dx.$$

Splitting the integral into separate terms, and assuming we can interchange integration and summation, we get

$$\begin{aligned}\int_{-\pi}^{\pi} f(x)\,dx &= \int_{-\pi}^{\pi} a_0\,dx + \int_{-\pi}^{\pi}\sum_{k=1}^{\infty} a_k\cos(kx)\,dx + \int_{-\pi}^{\pi}\sum_{k=1}^{\infty} b_k\sin(kx)\,dx \\ &= \int_{-\pi}^{\pi} a_0\,dx + \sum_{k=1}^{\infty}\int_{-\pi}^{\pi} a_k\cos(kx)\,dx + \sum_{k=1}^{\infty}\int_{-\pi}^{\pi} b_k\sin(kx)\,dx.\end{aligned}$$

But for $k \ge 1$, thinking of the integral as an area shows that

$$\int_{-\pi}^{\pi}\sin(kx)\,dx = 0 \quad\text{and}\quad \int_{-\pi}^{\pi}\cos(kx)\,dx = 0,$$

so all terms drop out except the first, giving

$$\int_{-\pi}^{\pi} f(x)\,dx = \int_{-\pi}^{\pi} a_0\,dx = a_0x\Big|_{-\pi}^{\pi} = 2\pi a_0.$$

Thus, we get the following result:

$$a_0 = \frac{1}{2\pi}\int_{-\pi}^{\pi} f(x)\,dx.$$

Thus a_0 is the average value of f on the interval $[-\pi, \pi]$.

To determine the values of any of the other a_k or b_k (for positive k), we use a rather clever method that depends on the following facts. For all integers k and m,

$$\int_{-\pi}^{\pi} \sin(kx)\cos(mx)\,dx = 0,$$

and, provided $k \neq m$,

$$\int_{-\pi}^{\pi} \cos(kx)\cos(mx)\,dx = 0 \quad \text{and} \quad \int_{-\pi}^{\pi} \sin(kx)\sin(mx)\,dx = 0.$$

(See Problems 26–30 on page 558.) In addition, provided $m \neq 0$, we have

$$\int_{-\pi}^{\pi} \cos^2(mx)\,dx = \pi \quad \text{and} \quad \int_{-\pi}^{\pi} \sin^2(mx)\,dx = \pi.$$

To determine a_k, we multiply the Fourier series by $\cos(mx)$, where m is any positive integer:

$$f(x)\cos(mx) = a_0\cos(mx) + \sum_{k=1}^{\infty} a_k\cos(kx)\cos(mx) + \sum_{k=1}^{\infty} b_k\sin(kx)\cos(mx).$$

We integrate this between $-\pi$ and π, term by term:

$$\begin{aligned}\int_{-\pi}^{\pi} f(x)\cos(mx)\,dx &= \int_{-\pi}^{\pi}\left(a_0\cos(mx) + \sum_{k=1}^{\infty} a_k\cos(kx)\cos(mx) + \sum_{k=1}^{\infty} b_k\sin(kx)\cos(mx)\right)dx \\ &= a_0\int_{-\pi}^{\pi}\cos(mx)\,dx + \sum_{k=1}^{\infty}\left(a_k\int_{-\pi}^{\pi}\cos(kx)\cos(mx)\,dx\right) \\ &\quad + \sum_{k=1}^{\infty}\left(b_k\int_{-\pi}^{\pi}\sin(kx)\cos(mx)\,dx\right).\end{aligned}$$

Provided $m \neq 0$, we have $\int_{-\pi}^{\pi}\cos(mx)\,dx = 0$. Since the integral $\int_{-\pi}^{\pi}\sin(kx)\cos(mx)\,dx = 0$, all the terms in the second sum are zero. Since $\int_{-\pi}^{\pi}\cos(kx)\cos(mx)\,dx = 0$ provided $k \neq m$, all the terms in the first sum are zero except where $k = m$. Thus the right-hand side reduces to one term:

$$\int_{-\pi}^{\pi} f(x)\cos(mx)\,dx = a_m\int_{-\pi}^{\pi}\cos(mx)\cos(mx)\,dx = \pi a_m.$$

This leads, for each value of $m = 1, 2, 3 \ldots$, to the following formula:

$$a_m = \frac{1}{\pi}\int_{-\pi}^{\pi} f(x)\cos(mx)\,dx.$$

To determine b_k, we multiply through by $\sin(mx)$ instead of $\cos(mx)$ and eventually obtain, for each value of $m = 1, 2, 3 \ldots$, the following result:

$$b_m = \frac{1}{\pi}\int_{-\pi}^{\pi} f(x)\sin(mx)\,dx.$$

Exercises and Problems for Section 10.5

EXERCISES

Which of the series in Exercises 1–4 are Fourier series?

1. $1+\cos x+\cos^2 x+\cos^3 x+\cos^4 x+\cdots$
2. $\sin x+\sin(x+1)+\sin(x+2)+\cdots$
3. $\dfrac{\cos x}{2}+\sin x-\dfrac{\cos(2x)}{4}-\dfrac{\sin(2x)}{2}+\dfrac{\cos(3x)}{8}+\dfrac{\sin(3x)}{3}-\cdots$
4. $\dfrac{1}{2}-\dfrac{1}{3}\sin x+\dfrac{1}{4}\sin(2x)-\dfrac{1}{5}\sin(3x)+\cdots$
5. Construct the first three Fourier approximations to the square wave function
$$f(x)=\begin{cases}-1 & -\pi\le x<0\\ 1 & 0\le x<\pi.\end{cases}$$
Use a calculator or computer to draw the graph of each approximation.
6. Repeat Exercise 5 with the function
$$f(x)=\begin{cases}-x & -\pi\le x<0\\ x & 0\le x<\pi.\end{cases}$$
7. What fraction of the energy of the function in Problem 6 is contained in the constant term and first three harmonics of its Fourier series?

For Exercises 8–10, find the nth Fourier polynomial for the given functions, assuming them to be periodic with period 2π. Graph the first three approximations with the original function.

8. $f(x)=x^2,\quad -\pi<x\le\pi.$
9. $h(x)=\begin{cases}0 & -\pi<x\le 0\\ x & 0<x\le\pi.\end{cases}$
10. $g(x)=x,\quad -\pi<x\le\pi.$

PROBLEMS

11. Find the constant term of the Fourier series of the triangular wave function defined by $f(x)=|x|$ for $-1\le x\le 1$ and $f(x+2)=f(x)$ for all x.
12. Using your result from Exercise 10, write the Fourier series of $g(x)=x$. Assume that your series converges to $g(x)$ for $-\pi<x<\pi$. Substituting an appropriate value of x into the series, show that
$$\sum_{k=1}^{\infty}(-1)^{k+1}\frac{1}{2k-1}=\frac{\pi}{4}.$$
13. **(a)** For $-2\pi\le x\le 2\pi$, use a calculator to sketch:
i) $y=\sin x+\frac{1}{3}\sin 3x$
ii) $y=\sin x+\frac{1}{3}\sin 3x+\frac{1}{5}\sin 5x$
(b) Each of the functions in part (a) is a Fourier approximation to a function whose graph is a square wave. What term would you add to the right-hand side of the second function in part (a) to get a better approximation to the square wave?
(c) What is the equation of the square wave function? Is this function continuous?
14. **(a)** Find and graph the third Fourier approximation of the square wave $g(x)$ of period 2π:
$$g(x)=\begin{cases}0 & -\pi\le x<-\pi/2\\ 1 & -\pi/2\le x<\pi/2\\ 0 & \pi/2\le x<\pi.\end{cases}$$
(b) How does the result of part (a) differ from that of the square wave in Example 1?
15. Suppose we have a periodic function f with period 1 defined by $f(x)=x$ for $0\le x<1$. Find the fourth-degree Fourier polynomial for f and graph it on the interval $0\le x<1$. [Hint: Remember that since the period is not 2π, you will have to start by doing a substitution. Notice that the terms in the sum are not $\sin(nx)$ and $\cos(nx)$, but instead turn out to be $\sin(2\pi nx)$ and $\cos(2\pi nx)$.]
16. Suppose f has period 2 and $f(x)=x$ for $0\le x<2$. Find the fourth-degree Fourier polynomial and graph it on $0\le x<2$. [Hint: See Problem 15.]
17. Suppose that a spacecraft near Neptune has measured a quantity A and sent it to earth in the form of a periodic signal $A\cos t$ of amplitude A. On its way to earth, the signal picks up periodic noise, containing only second and higher harmonics. Suppose that the signal $h(t)$ actually received on earth is graphed in Figure 10.34. Determine the signal that the spacecraft originally sent and hence the value A of the measurement.

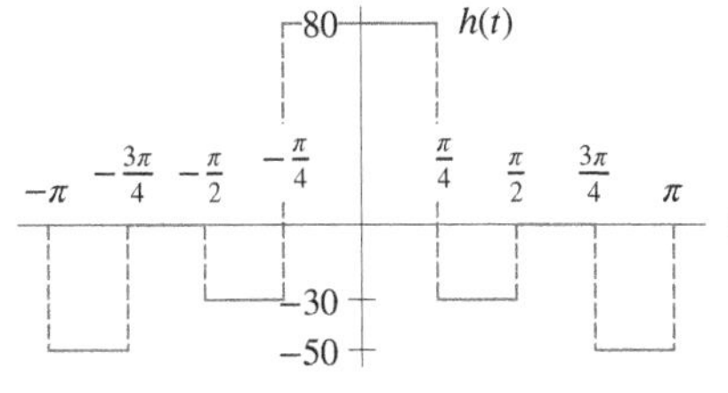

Figure 10.34

18. Figures 10.35 and 10.36 show the waveforms and energy spectra for notes produced by flute and bassoon.[12] Describe the principal differences between the two spectra.

[12] Adapted from C.A. Culver, *Musical Acoustics* (New York: McGraw-Hill, 1956), pp. 200, 213.

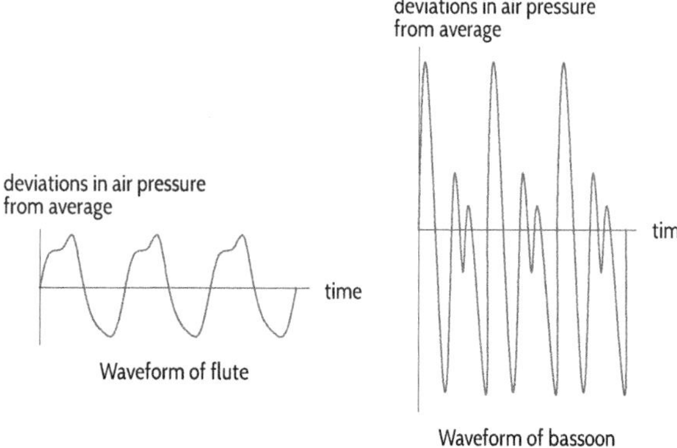

Figure 10.35

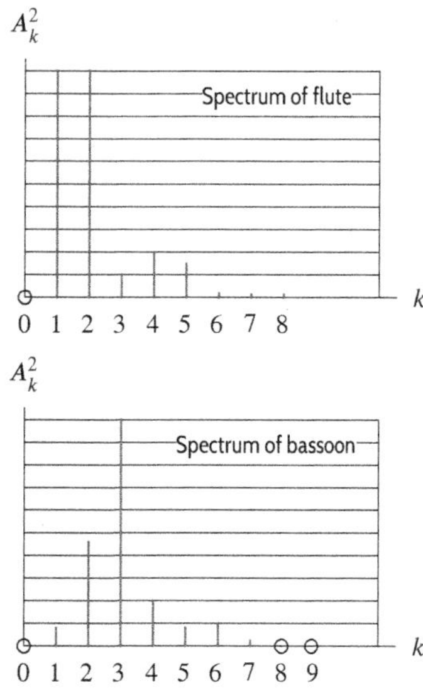

Figure 10.36

19. Show that for positive integers k, the periodic function $f(x) = a_k \cos kx + b_k \sin kx$ of period 2π has energy $a_k^2 + b_k^2$.

20. Given the graph of f in Figure 10.37, find the first two Fourier approximations numerically.

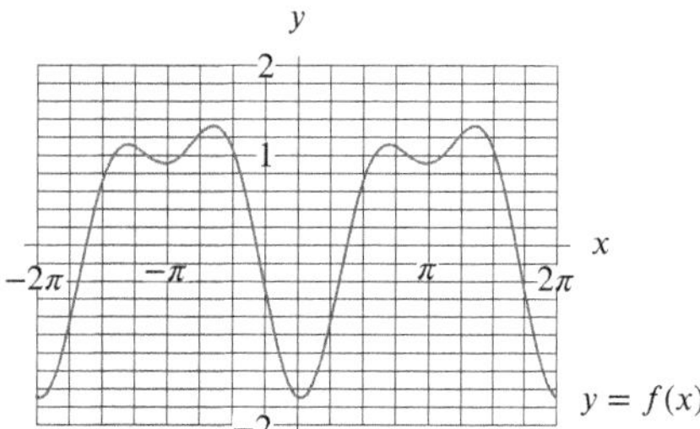

Figure 10.37

21. Justify the formula $b_k = \frac{1}{\pi}\int_{-\pi}^{\pi} f(x)\sin(kx)\,dx$ for the Fourier coefficients, b_k, of a periodic function of period 2π. The argument is similar to that in the text for a_k.

■ In Problems 22–25, the pulse train of width c is the periodic function f of period 2π given by

$$f(x) = \begin{cases} 0 & -\pi \le x < -c/2 \\ 1 & -c/2 \le x < c/2 \\ 0 & c/2 \le x < \pi. \end{cases}$$

22. Suppose that f is the pulse train of width 1.

(a) What fraction of the energy of f is contained in the constant term of its Fourier series? In the constant term and the first harmonic together?
(b) Find a formula for the energy of the kth harmonic of f. Use it to sketch the energy spectrum of f.
(c) How many terms of the Fourier series of f are needed to capture 90% of the energy of f?
(d) Graph f and its fifth Fourier approximation on the interval $[-3\pi, 3\pi]$.

23. Suppose that f is the pulse train of width 0.4.

(a) What fraction of the energy of f is contained in the constant term of its Fourier series? In the constant term and the first harmonic together?
(b) Find a formula for the energy of the kth harmonic of f. Use it to sketch the energy spectrum of f.
(c) What fraction of the energy of f is contained in the constant term and the first five harmonics of f? (The constant term and the first thirteen harmonics are needed to capture 90% of the energy of f.)
(d) Graph f and its fifth Fourier approximation on the interval $[-3\pi, 3\pi]$.

24. Suppose that f is the pulse train of width 2.

(a) What fraction of the energy of f is contained in the constant term of its Fourier series? In the constant term and the first harmonic together?
(b) How many terms of the Fourier series of f are needed to capture 90% of the energy of f?
(c) Graph f and its third Fourier approximation on the interval $[-3\pi, 3\pi]$.

25. After working Problems 22–24, write a paragraph about the approximation of pulse trains by Fourier polynomials. Explain how the energy spectrum of a pulse train of width c changes as c gets closer and closer to 0 and how this affects the number of terms required for an accurate approximation.

■ For Problems 26–30, use the table of integrals inside the back cover to show that the following statements are true for positive integers k and m.

26. $\displaystyle\int_{-\pi}^{\pi} \cos(kx)\cos(mx)\,dx = 0, \quad \text{if } k \ne m.$

27. $\displaystyle\int_{-\pi}^{\pi} \cos^2(mx)\,dx = \pi.$

28. $\int_{-\pi}^{\pi} \sin^2(mx)\,dx = \pi.$

29. $\int_{-\pi}^{\pi} \sin(kx)\cos(mx)\,dx = 0.$

30. $\int_{-\pi}^{\pi} \sin(kx)\sin(mx)\,dx = 0, \quad \text{if } k \neq m.$

31. Suppose that $f(x)$ is a periodic function with period b. Show that

(a) $g(t) = f(bt/2\pi)$ is periodic with period 2π and $f(x) = g(2\pi x/b)$.

(b) The Fourier series for g is given by

$$g(t) = a_0 + \sum_{k=1}^{\infty} \left(a_k \cos(kt) + b_k \sin(kt) \right)$$

where the coefficients a_0, a_k, b_k are given in the box on page 553.

(c) The Fourier series for f is given by

$$f(x) = a_0 + \sum_{k=1}^{\infty} \left(a_k \cos\left(\frac{2\pi kx}{b}\right) + b_k \sin\left(\frac{2\pi kx}{b}\right) \right)$$

where the coefficients are the same as in part (b).

Strengthen Your Understanding

In Problems 32–33, explain what is wrong with the statement.

32. $\int_{-\pi}^{\pi} \sin(kx)\cos(mx)\,dx = \pi$, where k, m are both positive integers.

33. In the Fourier series for $f(x)$ given by $a_0 + \sum_{k=1}^{\infty} a_k \cos(kx) + \sum_{k=1}^{\infty} b_k \sin(kx)$, we have $a_0 = f(0)$.

In Problems 34–35, give an example of:

34. A function, $f(x)$, with period 2π whose Fourier series has no sine terms.

35. A function, $f(x)$, with period 2π whose Fourier series has no cosine terms.

36. True or false? If f is an even function, then the Fourier series for f on $[-\pi, \pi]$ has only cosines. Explain your answer.

37. The graph in Figure 10.38 is the graph of the first three terms of the Fourier series of which of the following functions?

(a) $f(x) = 3(x/\pi)^3$ on $-\pi < x < \pi$ and $f(x + 2\pi) = f(x)$

(b) $f(t) = |x|$ on $-\pi < x < \pi$ and $f(x + 2\pi) = f(x)$

(c) $f(x) = \begin{cases} -3, & -\pi < x < 0 \\ 3, & 0 < x < \pi \end{cases}$ and $f(x + 2\pi) = f(x)$

(d) $f(x) = \begin{cases} \pi + x, & -\pi < x < 0 \\ \pi - x, & 0 < x < \pi \end{cases}$ and $f(x + 2\pi) = f(x)$

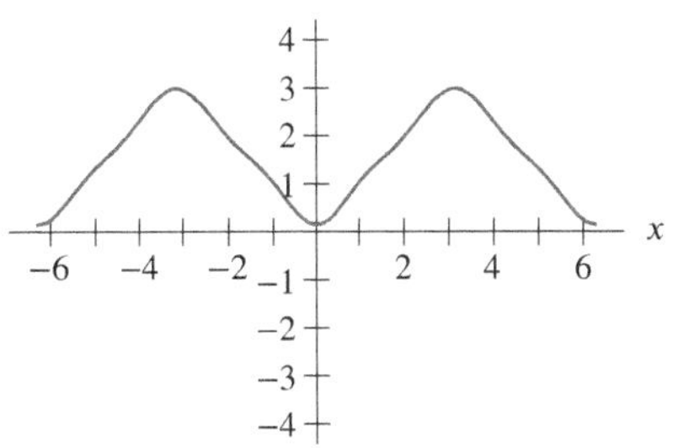

Figure 10.38

Online Resource: Review problems and Projects

Chapter Eleven

DIFFERENTIAL EQUATIONS

Contents

11.1 WHAT IS A DIFFERENTIAL EQUATION?

How Fast Does a Person Learn?

Suppose we are interested in how fast an employee learns a new task. One theory claims that the more the employee already knows of the task, the slower he or she learns. In other words, if $y\%$ is the percentage of the task that has already been mastered and dy/dt the rate at which the employee learns, then dy/dt decreases as y increases.

What can we say about y as a function of time, t? Figure 11.1 shows three graphs whose slope, dy/dt, decreases as y increases. Figure 11.1(a) represents an employee who starts learning at $t = 0$ and who eventually masters 100% of the task. Figure 11.1(b) represents an employee who starts later but eventually masters 100% of the task. Figure 11.1(c) represents an employee who starts learning at $t = 0$ but who does not master the whole task (since y levels off below 100%).

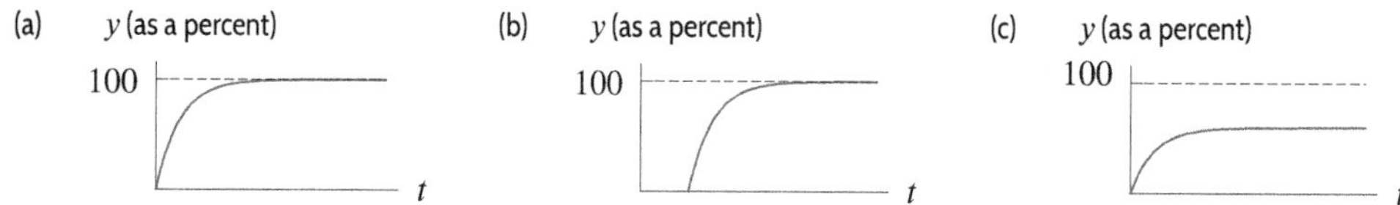

Figure 11.1: Possible graphs showing percentage of task learned, y, as a function of time, t

Setting up a Differential Equation to Model How a Person Learns

To describe more precisely how a person learns, we need more exact information about how dy/dt depends on y. Suppose, if time is measured in weeks, that

$$\text{Rate a person learns} = \text{Percentage of task not yet learned.}$$

Since y is the percentage learned by time t (in weeks), the percentage not yet learned by that time is $100 - y$. So we have

$$\frac{dy}{dt} = 100 - y.$$

Such an equation, which gives information about the rate of change of an unknown function, is called a *differential equation.*

Solving the Differential Equation Numerically

Suppose that the person starts learning at time zero, so $y = 0$ when $t = 0$. Then initially the person is learning at a rate

$$\frac{dy}{dt} = 100 - 0 = 100\% \text{ per week.}$$

In other words, if the person were to continue learning at this rate, the task would be mastered in a week. In fact, however, the rate at which the person learns decreases, so it takes more than a week to get close to mastering the task. Let's assume a five-day work week and that the 100% per week learning rate holds for the whole first day. (It does not, but we assume this for now.) One day is 1/5 of a week, so during the first day the person learns $100(1/5) = 20\%$ of the task. By the end of the first day, the rate at which the person learns has therefore been reduced to

$$\frac{dy}{dt} = 100 - 20 = 80\% \text{ per week.}$$

Thus, during the second day the person learns $80(1/5) = 16\%$, so by the end of the second day, the person knows $20 + 16 = 36\%$ of the task. Continuing in this fashion, we compute the approximate y-values[1] in Table 11.1.

Table 11.1 *Approximate percentage of task learned as a function of time*

Time (working days)	0	1	2	3	4	5	10	20
Percentage learned	0	20	36	48.8	59.0	67.2	89.3	98.8

[1]The values of y after 6, 7, 8, 9, …, 19 days were computed by the same method, but omitted from the table.

A Formula for the Solution to the Differential Equation

A function $y = f(t)$ that satisfies the differential equation is called a *solution*. Figure 11.1(a) shows a possible solution, and Table 11.1 shows approximate numerical values of a solution to the equation

$$\frac{dy}{dt} = 100 - y.$$

Later in this chapter, we see how to obtain a formula for the solution:

$$y = 100 + Ce^{-t},$$

where C is a constant. To check that this formula is correct, we substitute into the differential equation, giving:

$$\text{Left side} = \frac{dy}{dt} = \frac{d}{dt}(100 + Ce^{-t}) = -Ce^{-t}$$
$$\text{Right side} = 100 - y = 100 - (100 + Ce^{-t}) = -Ce^{-t}.$$

Since we get the same result on both sides, $y = 100 + Ce^{-t}$ is a solution of this differential equation.

Finding the Arbitrary Constant: Initial Conditions

To find a value for the arbitrary constant C, we need an additional piece of information—usually the initial value of y. If, for example, we are told that $y = 0$ when $t = 0$, then substituting into

$$y = 100 + Ce^{-t}$$

shows us that

$$0 = 100 + Ce^{0}, \quad \text{so} \quad C = -100.$$

So the function $y = 100 - 100e^{-t}$ satisfies the differential equation *and* the condition that $y = 0$ when $t = 0$.

The Family of Solutions

Any solution to this differential equation is of the form $y = 100 + Ce^{-t}$ for some constant C. Like a family of antiderivatives, this family contains an arbitrary constant, C. We say that the *general solution* to the differential equation $dy/dt = 100 - y$ is the family of functions $y = 100 + Ce^{-t}$. The solution $y = 100 - 100e^{-t}$ that satisfies the differential equation together with the *initial condition* that $y = 0$ when $t = 0$ is called a *particular solution*. The differential equation and the initial condition together are called an *initial value problem*. Several members of the family of solutions are graphed in Figure 11.2. The horizontal solution curve when $C = 0$ is called an *equilibrium solution*.

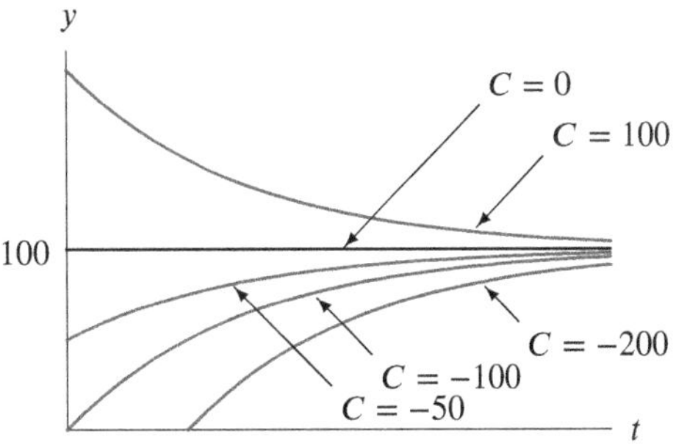

Figure 11.2: Solution curves for $dy/dt = 100 - y$: Members of the family $y = 100 + Ce^{-t}$

First- and Second-Order Differential Equations

First, some more definitions. We often write y' to represent the derivative of y. The differential equation

$$y' = 100 - y$$

is called *first-order* because it involves the first derivative, but no higher derivatives. By contrast, if s is the height (in meters) of a body moving under the force of gravity and t is time (in seconds), then

$$\frac{d^2s}{dt^2} = -9.8.$$

This is a *second-order* differential equation because it involves the second derivative of the unknown function, $s = f(t)$, but no higher derivatives.

Example 1 Show that $y = e^{2t}$ is not a solution to the second-order differential equation

$$\frac{d^2y}{dt^2} + 4y = 0.$$

Solution To decide whether the function $y = e^{2t}$ is a solution, substitute it into the differential equation:

$$\frac{d^2y}{dt^2} + 4y = 2(2e^{2t}) + 4e^{2t} = 8e^{2t}.$$

Since $8e^{2t}$ is not identically zero, $y = e^{2t}$ is not a solution.

How Many Arbitrary Constants Should We Expect in the Family of Solutions?

Since a differential equation involves the derivative of an unknown function, solving it usually involves antidifferentiation, which introduces arbitrary constants. The solution to a first-order differential equation usually involves one antidifferentiation and one arbitrary constant (for example, the C in $y = 100 + Ce^{-t}$). Picking out one particular solution involves knowing one additional piece of information, such as an initial condition. Solving a second-order differential equation generally involves two antidifferentiations and so two arbitrary constants. Consequently, finding a particular solution usually involves two initial conditions.

For example, if s is the height (in meters) of a body above the surface of the earth at time t (in seconds), then

$$\frac{d^2s}{dt^2} = -9.8.$$

Integrating gives

$$\frac{ds}{dt} = -9.8t + C_1,$$

and integrating again gives

$$s = -4.9t^2 + C_1t + C_2.$$

Thus the general solution for s involves the two arbitrary constants C_1 and C_2. We can find C_1 and C_2 if we are told, for example, that the initial velocity at $t = 0$ is 30 meters per second upward and that the initial position is 5 meters above the ground. In this case, $C_1 = 30$ and $C_2 = 5$.

Exercises and Problems for Section 11.1

EXERCISES

1. Is $y = x^3$ a solution to the differential equation

$$xy' - 3y = 0?$$

In Exercises 2–7, which differential equation, (I)–(VI), has the function as a solution?

I. $y' = -2xy$ II. $y' = -xy$ III. $y' = xy$
IV. $y' = -x^2y$ V. $y' = x^{-3}y$ VI. $y' = 2xy$

2. $y = e^{-x^2}$
3. $y = e^{-0.5x^2}$
4. $y = 0.5e^{-x^2}$
5. $y = 0.5e^{x^2}$
6. $y = -e^{-x^3/3}$
7. $y = e^{-0.5x^{-2}}$

8. Determine whether each function is a solution to the differential equation and justify your answer:

$$x\frac{dy}{dx} = 4y.$$

(a) $y = x^4$ (b) $y = x^4 + 3$
(c) $y = x^3$ (d) $y = 7x^4$

9. Determine whether each function is a solution to the differential equation and justify your answer:

$$y\frac{dy}{dx} = 6x^2.$$

(a) $y = 4x^3$ (b) $y = 2x^{3/2}$ (c) $y = 6x^{3/2}$

10. Pick out which functions are solutions to which differential equations. (Note: Functions may be solutions to more than one equation or to none; an equation may have more than one solution.)

(a) $\dfrac{dy}{dx} = -2y$ (I) $y = 2\sin x$

(b) $\dfrac{dy}{dx} = 2y$ (II) $y = \sin 2x$

(c) $\dfrac{d^2y}{dx^2} = 4y$ (III) $y = e^{2x}$

(d) $\dfrac{d^2y}{dx^2} = -4y$ (IV) $y = e^{-2x}$

11. Match solutions and differential equations. (Note: Each equation may have more than one solution.)

(a) $y'' - y = 0$ (I) $y = e^x$

(b) $x^2y'' + 2xy' - 2y = 0$ (II) $y = x^3$

(c) $x^2y'' - 6y = 0$ (III) $y = e^{-x}$

(IV) $y = x^{-2}$

12. Show that, for any constant P_0, the function $P = P_0e^t$ satisfies the equation

$$\frac{dP}{dt} = P.$$

13. Show that $y(x) = Ae^{\lambda x}$ is a solution to the equation $y' = \lambda y$ for any value of A and constant λ.

14. Show that $y = \sin 2t$ satisfies

$$\frac{d^2y}{dt^2} + 4y = 0.$$

15. Is $y(x) = e^{3x}$ the general solution of $y' = 3y$?

16. Use implicit differentiation to show that $x^2 + y^2 = r^2$ is a solution to the differential equation $dy/dx = -x/y$, for any constant r.

17. A quantity Q satisfies the differential equation

$$\frac{dQ}{dt} = \frac{t}{Q} - 0.5.$$

(a) If $Q = 8$ when $t = 2$, use dQ/dt to determine whether Q is increasing or decreasing at $t = 2$.

(b) Use your work in part (a) to estimate the value of Q when $t = 3$. Assume the rate of change stays approximately constant over the interval from $t = 2$ to $t = 3$.

18. Fill in the missing values in the table given that $dy/dt = 0.5y$. Assume the rate of growth given by dy/dt is approximately constant over each unit time interval and that the initial value of y is 8.

t	0	1	2	3	4
y	8				

19. Use the method that generated the data in Table 11.1 on page 562 to fill in the missing y-values for $t = 6, 7, \ldots, 19$ days.

In Exercises 20–23, find the particular solution to a differential equation whose general solution and initial condition are given. (C is the constant of integration.)

20. $x(t) = Ce^{3t}$; $x(0) = 5$

21. $P = C/t$; $P = 5$ when $t = 3$

22. $y = \sqrt{2t + C}$; the solution curve passes through $(1, 3)$

23. $Q = 1/(Ct + C)$; $Q = 4$ when $t = 2$

PROBLEMS

24. Show that $y = A + Ce^{kt}$ is a solution to the equation

$$\frac{dy}{dt} = k(y - A).$$

25. Find the value(s) of ω for which $y = \cos \omega t$ satisfies

$$\frac{d^2y}{dt^2} + 9y = 0.$$

26. Show that any function of the form

$$x = C_1 \cosh \omega t + C_2 \sinh \omega t$$

satisfies the differential equation

$$x'' - \omega^2 x = 0.$$

27. Suppose $Q = Ce^{kt}$ satisfies the differential equation

$$\frac{dQ}{dt} = -0.03Q.$$

What (if anything) does this tell you about the values of C and k?

28. Find the values of k for which $y = x^2 + k$ is a solution to the differential equation

$$2y - xy' = 10.$$

29. For what values of k (if any) does $y = 5 + 3e^{kx}$ satisfy the differential equation

$$\frac{dy}{dx} = 10 - 2y?$$

30. (a) For what values of C and n (if any) is $y = Cx^n$ a solution to the differential equation

$$x\frac{dy}{dx} - 3y = 0?$$

(b) If the solution satisfies $y = 40$ when $x = 2$, what more (if anything) can you say about C and n?

31. (a) Find the value of A so that the equation $y' - xy - x = 0$ has a solution of the form $y(x) = A + Be^{x^2/2}$ for any constant B.
(b) If $y(0) = 1$, find B.

32. In Figure 11.3, the height, y, of the hanging cable above the horizontal line satisfies

$$\frac{d^2y}{dx^2} = k\sqrt{1 + \left(\frac{dy}{dx}\right)^2}.$$

(a) Show that $y = \dfrac{e^x + e^{-x}}{2}$ satisfies this differential equation if $k = 1$.
(b) For general k, one solution to this differential equation is of the form

$$y = \frac{e^{Ax} + e^{-Ax}}{2A}.$$

Substitute this expression for y into the differential equation to find A in terms of k.

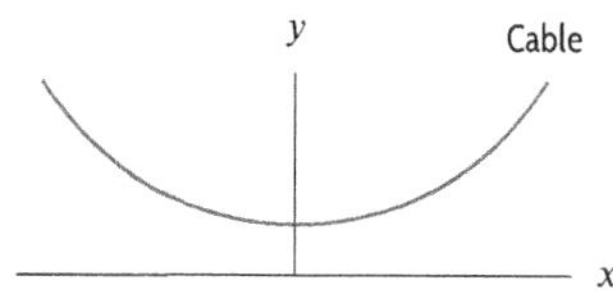

Figure 11.3

33. Families of curves often arise as solutions of differential equations. Match the families of curves with the differential equations of which they are solutions.

(a) $\dfrac{dy}{dx} = \dfrac{y}{x}$ (I) $y = xe^{kx}$

(b) $\dfrac{dy}{dx} = ky$ (II) $y = x^k$

(c) $\dfrac{dy}{dx} = ky + \dfrac{y}{x}$ (III) $y = e^{kx}$

(d) $\dfrac{dy}{dx} = \dfrac{ky}{x}$ (IV) $y = kx$

34. (a) Let $y = A + Be^{-2t}$. For what values of A and B, if any, is y a solution to the differential equation

$$\frac{dy}{dt} = 100 - 2y?$$

Give the general solution to the differential equation.
(b) If the solution satisfies $y = 85$ when $t = 0$, what more (if anything) can you say about A and B? Give the particular solution to the differential equation with this initial condition.

For Problems 35–36, let $y = f(x)$ be a solution to the differential equation. Determine whether $y = 2f(x)$ is a solution.

35. $dy/dx = y$

36. $dy/dx = x$

For Problems 37–38, let $y = f(x)$ and $y = g(x)$ be different solutions to the differential equation. Determine whether $y = f(x) - g(x)$ is a solution to $dy/dx = 0$.

37. $dy/dx = x$

38. $dy/dx = y$

Strengthen Your Understanding

In Problems 39–40, explain what is wrong with the statement.

39. $Q = 6e^{4t}$ is the general solution to the differential equation $dQ/dt = 4Q$.

40. If $dx/dt = 1/x$ and $x = 3$ when $t = 0$, then x is a decreasing function of t.

In Problems 41–47, give an example of:

41. A differential equation with an initial condition.

42. A second-order differential equation.

43. A differential equation and two different solutions to the differential equation.

44. A differential equation that has a trigonometric function as a solution.

45. A differential equation that has a logarithmic function as a solution.

46. A differential equation all of whose solutions are increasing and concave up.

47. A differential equation all of whose solutions have their critical points on the parabola $y = x^2$.

Are the statements in Problems 48–49 true or false? Give an explanation for your answer.

48. If $y = f(t)$ is a particular solution to a first-order differential equation, then the general solution is $y = f(t) + C$, where C is an arbitrary constant.

49. Polynomials are never solutions to differential equations.

■ In Problems 50–57, is the statement true or false? Assume that $y = f(x)$ is a solution to the equation $dy/dx = g(x)$. If the statement is true, explain how you know. If the statement is false, give a counterexample.

50. If $g(x)$ is increasing for all x, then the graph of f is concave up for all x .

51. If $g(x)$ is increasing for $x > 0$, then so is $f(x)$.

52. If $g(0) = 1$ and $g(x)$ is increasing for $x \geq 0$, then $f(x)$ is also increasing for $x \geq 0$.

53. If $g(x)$ is periodic, then $f(x)$ is also periodic.

54. If $\lim_{x\to\infty} g(x) = 0$, then $\lim_{x\to\infty} f(x) = 0$.

55. If $\lim_{x\to\infty} g(x) = \infty$, then $\lim_{x\to\infty} f(x) = \infty$.

56. If $g(x)$ is even, then so is $f(x)$.

57. If $g(x)$ is even, then $f(x)$ is odd.

11.2 SLOPE FIELDS

In this section, we see how to visualize a first-order differential equation. We start with the equation

$$\frac{dy}{dx} = y.$$

Any solution to this differential equation has the property that the slope at any point is equal to the y-coordinate at that point. (That's what the equation $dy/dx = y$ is telling us!) If the solution goes through the point $(0, 0.5)$, its slope there is 0.5; if it goes through a point with $y = 1.5$, its slope there is 1.5. See Figure 11.4.

In Figure 11.4 a small line segment is drawn at each of the marked points showing the slope of the curve there. Imagine drawing many of these line segments, but leaving out the curves; this gives the *slope field* for the equation $dy/dx = y$ in Figure 11.5. From this picture, we can see that above the x-axis, the slopes are all positive (because y is positive there), and they increase as we move upward (as y increases). Below the x-axis, the slopes are all negative and get more so as we move downward. On any horizontal line (where y is constant) the slopes are the same.

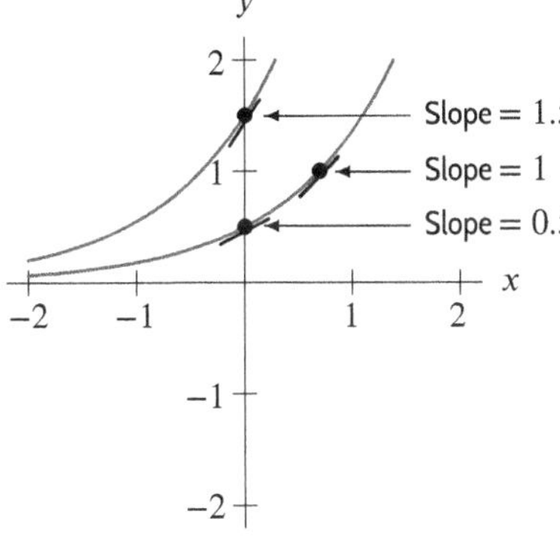

Figure 11.4: Solutions to $dy/dx = y$

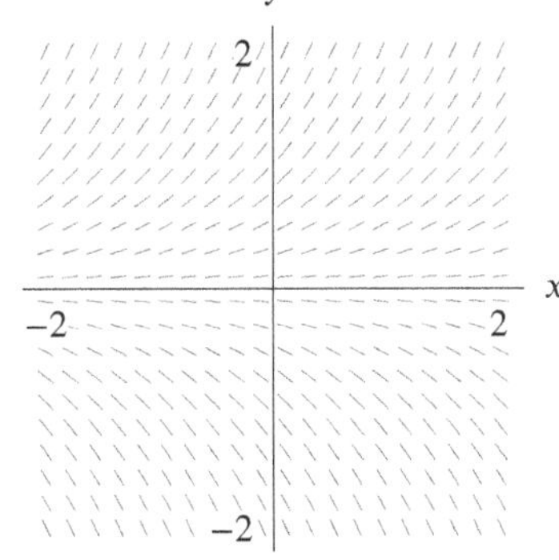

Figure 11.5: Slope field for $dy/dx = y$

In the slope field we can see the ghost of the solution curve lurking. Start anywhere on the plane and move so that the slope lines are tangent to our path; we trace out one of the solution curves. We think of the slope field as a set of signposts pointing in the direction we should go at each point. In this case, the slope field should trace out exponential curves of the form $y = Ce^x$, the solutions to the differential equation $dy/dx = y$.

Example 1 Figure 11.6 shows the slope field of the differential equation $dy/dx = 2x$.

(a) How does the slope field vary as we move around the xy-plane?

(b) Compare the solution curves sketched on the slope field with the formula for the solutions.

Solution

(a) In Figure 11.6 we notice that on a vertical line (where x is constant) the slopes are the same. This is because in this differential equation dy/dx depends on x only. (In the previous example, $dy/dx = y$, the slopes depended on y only.)

(b) The solution curves in Figure 11.7 look like parabolas. By antidifferentiation, we see that the solution to the differential equation $dy/dx = 2x$ is

$$y = \int 2x\,dx = x^2 + C,$$

so the solution curves really are parabolas.

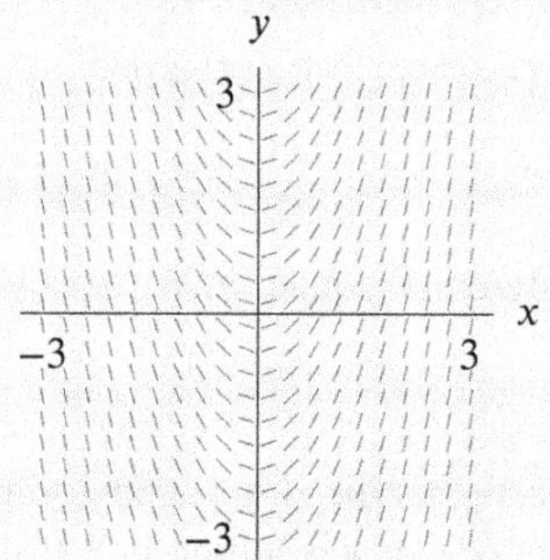

Figure 11.6: Slope field for $dy/dx = 2x$

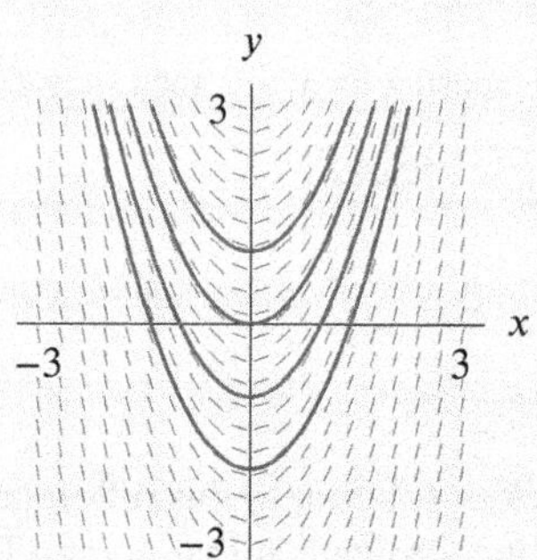

Figure 11.7: Some solutions to $dy/dx = 2x$

Example 2 Using the slope field, guess the form of the solution curves of the differential equation

$$\frac{dy}{dx} = -\frac{x}{y}.$$

Solution The slope field is shown in Figure 11.8. On the y-axis, where x is 0, the slope is 0. On the x-axis, where y is 0, the line segments are vertical and the slope is infinite. At the origin the slope is undefined, and there is no line segment.

The slope field suggests that the solution curves are circles centered at the origin. Later we see how to obtain the solution analytically, but even without this, we can check that the circle is a solution. We take the circle of radius r,

$$x^2 + y^2 = r^2,$$

and differentiate implicitly, thinking of y as a function of x. Using the chain rule, we get

$$2x + 2y \cdot \frac{dy}{dx} = 0.$$

Solving for dy/dx gives our differential equation,

$$\frac{dy}{dx} = -\frac{x}{y}.$$

This tells us that $x^2 + y^2 = r^2$ is a solution to the differential equation.

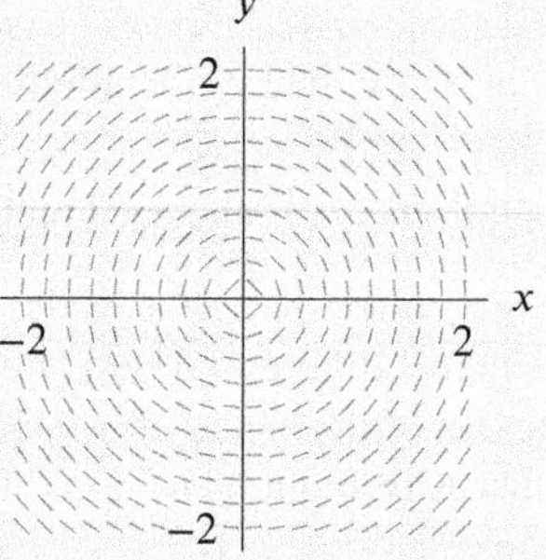

Figure 11.8: Slope field for $dy/dx = -x/y$

The previous example shows that the solution to a differential equation may be an implicit function.

Example 3 The slope fields of $dy/dt = 2 - y$ and $dy/dt = t/y$ are in Figures 11.9 and 11.10.

(a) On each slope field, sketch solution curves with initial conditions
(i) $y = 1$ when $t = 0$ (ii) $y = 0$ when $t = 1$ (iii) $y = 3$ when $t = 0$

(b) For each solution curve, what can you say about the long-run behavior of y? For example, does $\lim_{t\to\infty} y$ exist? If so, what is its value?

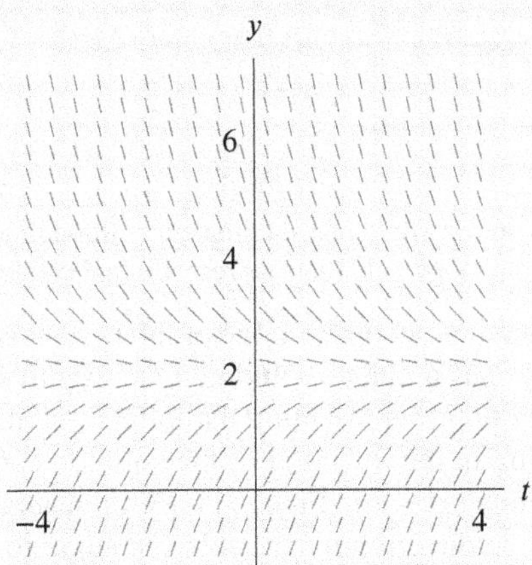

Figure 11.9: Slope field for $dy/dt = 2 - y$

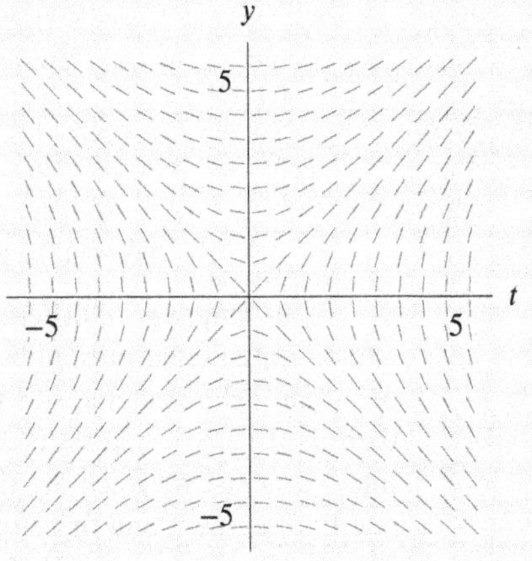

Figure 11.10: Slope field for $dy/dt = t/y$

Solution (a) See Figures 11.11 and 11.12.

(b) For $dy/dt = 2 - y$, all solution curves have $y = 2$ as a horizontal asymptote, so $\lim_{t\to\infty} y = 2$. For $dy/dt = t/y$, as $t \to \infty$, it appears that either $y \to t$ or $y \to -t$.

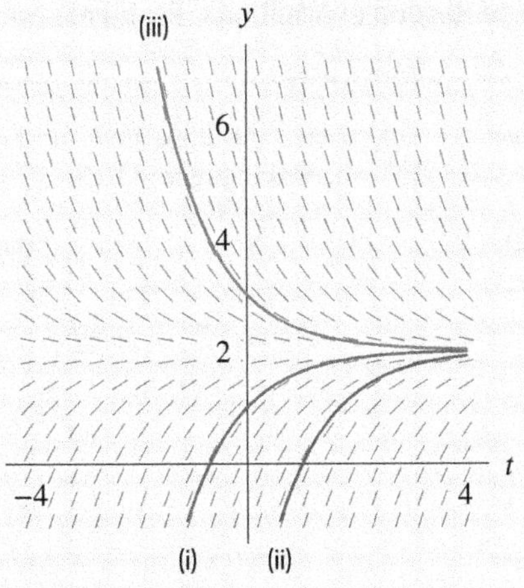

Figure 11.11: Solution curves for $dy/dt = 2 - y$

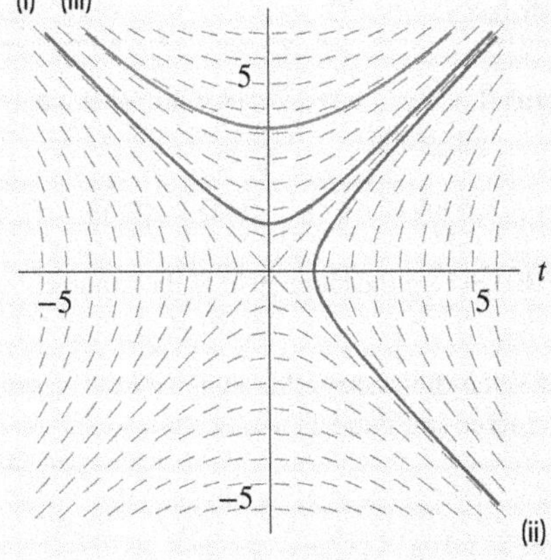

Figure 11.12: Solution curves for $dy/dt = t/y$

Existence and Uniqueness of Solutions

Since differential equations are used to model many real situations, the question of whether a solution is unique can have great practical importance. If we know how the velocity of a satellite is changing, can we know its velocity at any future time? If we know the initial population of a city and we know how the population is changing, can we predict the population in the future? Common sense says yes: if we know the initial value of some quantity and we know exactly how it is changing, we should be able to figure out its future value.

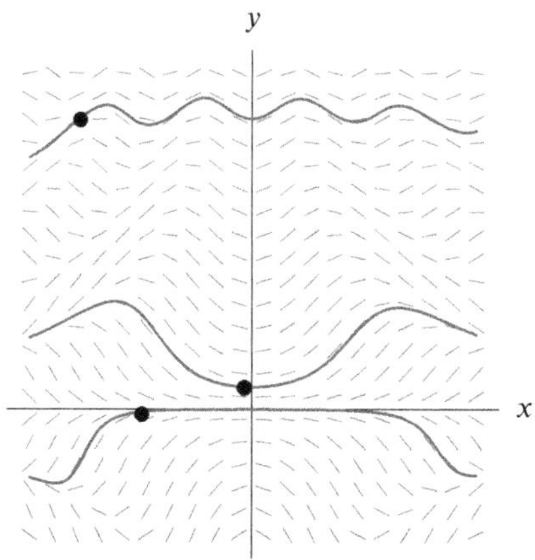

Figure 11.13: There's one and only one solution curve through each point in the plane for this slope field (dots represent initial conditions).

In the language of differential equations, an initial value problem (that is, a differential equation and an initial condition) almost always has a unique solution. One way to see this is by looking at the slope field. Imagine starting at the point representing the initial condition. Through that point there is usually a line segment pointing in the direction of the solution curve. By following these line segments, we trace out the solution curve. See Figure 11.13. In general, at each point there is one line segment and therefore only one direction for the solution curve to go. The solution curve *exists* and is *unique* provided we are given an initial point. Notice that even though we can draw the solution curves, we may have no simple formula for them.

It can be shown that if the slope field is continuous as we move from point to point in the plane, we can be sure that a solution curve exists everywhere. Ensuring that each point has only one solution curve through it requires a slightly stronger condition.

Exercises and Problems for Section 11.2 Online Resource: Additional Problems for Section 11.2

EXERCISES

1. **(a)** For $dy/dx = x^2 - y^2$, find the slope at the following points:

$$(1,0), \quad (0,1), \quad (1,1), \quad (2,1), \quad (1,2), \quad (2,2)$$

(b) Sketch the slope field at these points.

2. Sketch the slope field for $dy/dx = x/y$ at the points marked in Figure 11.14.

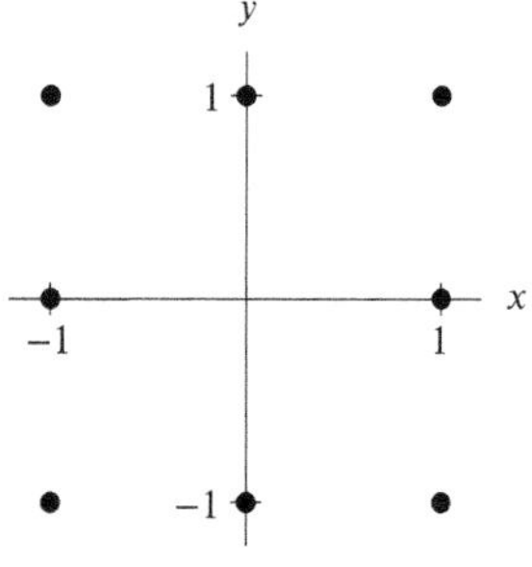

Figure 11.14

3. Sketch the slope field for $dy/dx = y^2$ at the points marked in Figure 11.15.

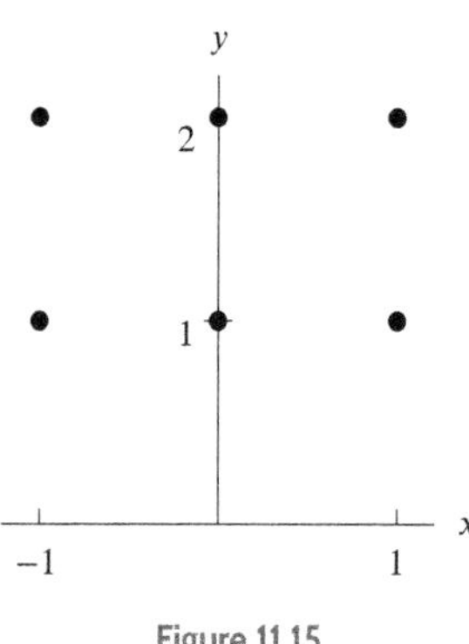

Figure 11.15

4. Match each of the slope field segments in (I)–(VI) with one or more of the differential equations in (a)–(f).

(a) $y' = e^{-x^2}$ **(b)** $y' = \cos y$
(c) $y' = \cos(4 - y)$ **(d)** $y' = y(4 - y)$
(e) $y' = y(3 - y)$ **(f)** $y' = x(3 - x)$

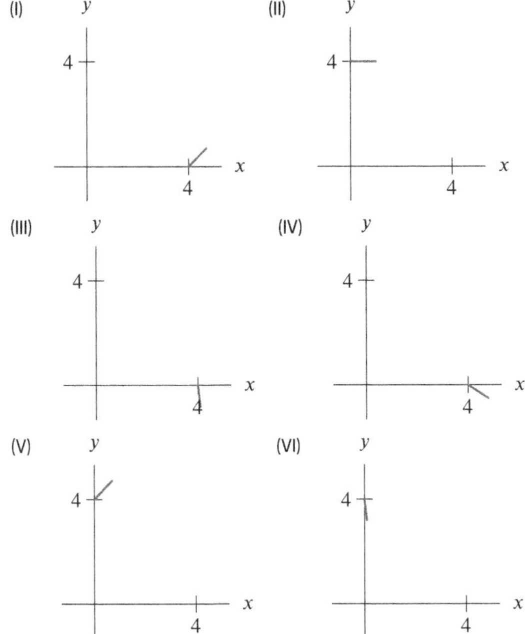

5. Sketch three solution curves for each of the slope fields in Figures 11.16 and 11.17.

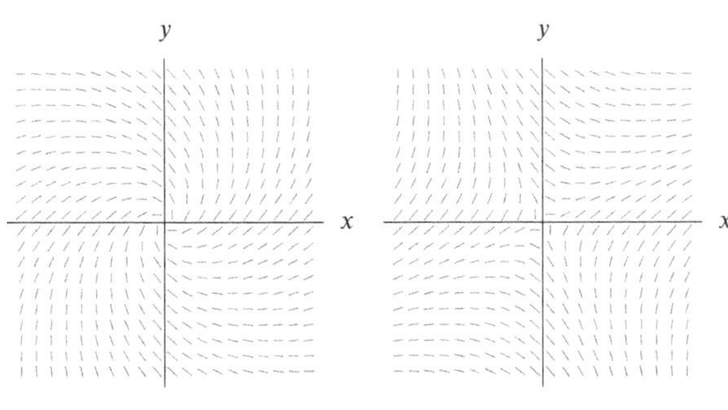

Figure 11.16

Figure 11.17

6. Sketch three solution curves for each of the slope fields in Figure 11.18.

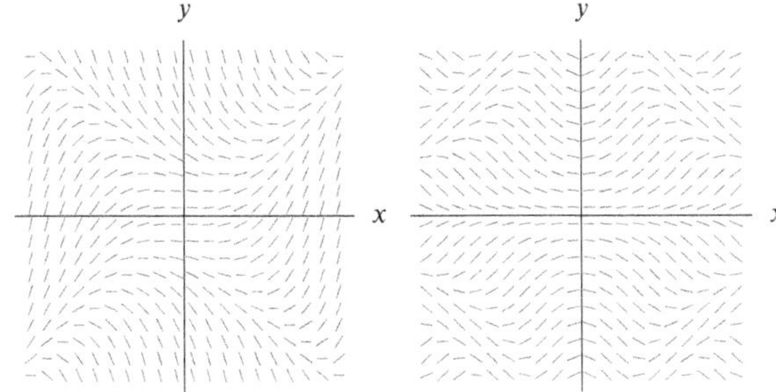

Figure 11.18

■ For Exercises 7–9, at the given point, is the solution curve to $y' = 2y - 3x - 4$ increasing or decreasing?

7. $(1, 4)$ 8. $(2, 4)$ 9. $(0, 3)$

10. One of the slope fields in Figure 11.18 is the slope field for $y' = x^2 - y^2$. Which one? On this field, where is the point $(0, 1)$? The point $(1, 0)$? (Assume that the x and y scales are the same.) Sketch the line $x = 1$ and the solution curve passing through $(0, 1)$ until it crosses $x = 1$.

11. The slope field for the equation $y' = x(y - 1)$ is shown in Figure 11.19.

(a) Sketch the solutions that pass through the points

(i) $(0, 1)$ (ii) $(0, -1)$ (iii) $(0, 0)$

(b) From your sketch, write the equation of the solution with $y(0) = 1$.

(c) Check your solution to part (b) by substituting it into the differential equation.

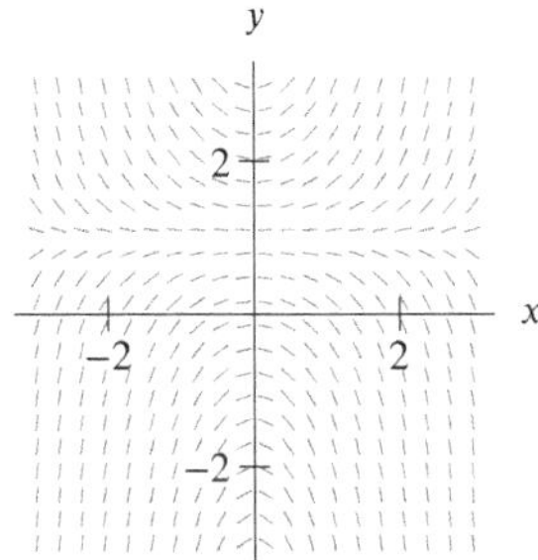

Figure 11.19

12. The slope field for the equation $y' = x + y$ is shown in Figure 11.20.

(a) Sketch the solutions that pass through the points

(i) $(0, 0)$ (ii) $(-3, 1)$ (iii) $(-1, 0)$

(b) From your sketch, write the equation of the solution passing through $(-1, 0)$.

(c) Check your solution to part (b) by substituting it into the differential equation.

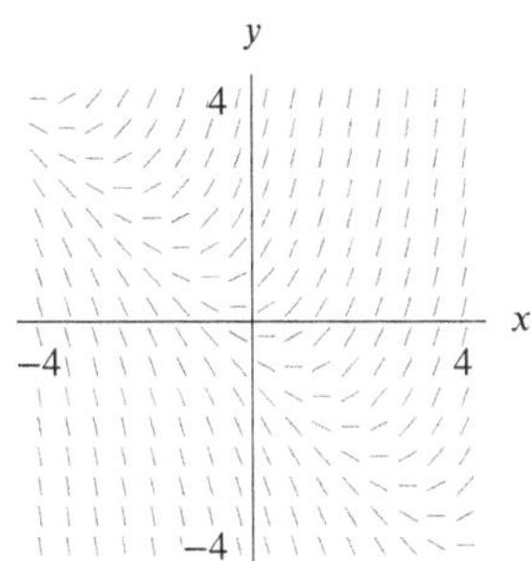

Figure 11.20: Slope field for $y' = x + y$

PROBLEMS

13. **(a)** Match the slope fields (A) and (B) with the following differential equations:

(i) $y' = 0.3y$ (ii) $y' = 0.3t$

(b) For the solutions to each differential equation in part (a), what is $\lim_{t\to\infty} y(t)$ if

(i) $y(0) = 1$ (ii) $y(0) = 0$

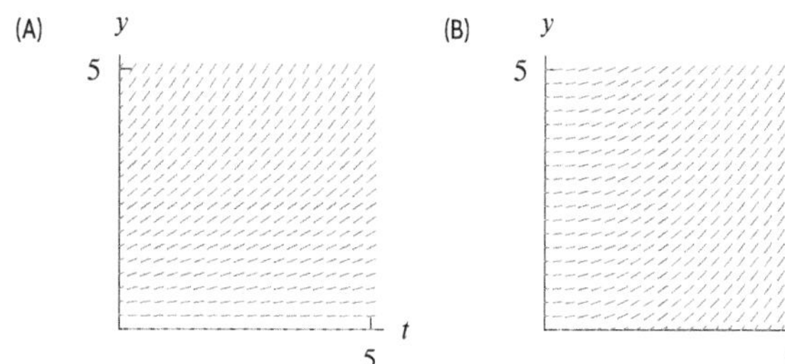

14. A slope field for a differential equation has slopes greater than 0 at every point of the xy-plane.

(a) Explain why any solution curve to this differential equation is increasing everywhere.

(b) Can you conclude that a solution curve to this differential equation is concave up everywhere?

15. Sketch a slope field with the following properties. (Draw at least ten line segments, including some with $x < 0$, with $x > 0$, and with $x = 0$.)

$$\frac{dy}{dx} > 0 \text{ for } x < 0,$$
$$\frac{dy}{dx} < 0 \text{ for } x > 0,$$
$$\frac{dy}{dx} = 0 \text{ for } x = 0.$$

16. Sketch a slope field with the following properties. (Draw at least ten line segments, including some with $P < 2$, with $2 < P < 5$, and with $P > 5$.)

$$\frac{dP}{dt} > 0 \text{ for } 2 < P < 5,$$
$$\frac{dP}{dt} < 0 \text{ for } P < 2 \text{ or } P > 5,$$
$$\frac{dP}{dt} = 0 \text{ for } P = 2 \text{ and } P = 5.$$

17. Is the solution curve to $y' = 2x - 3y - 1$ concave up or concave down at $(3, 2)$?

18. **(a)** Sketch the slope field for the equation $y' = x - y$ in Figure 11.21 at the points indicated.

(b) Find the equation for the solution that passes through the point $(1, 0)$.

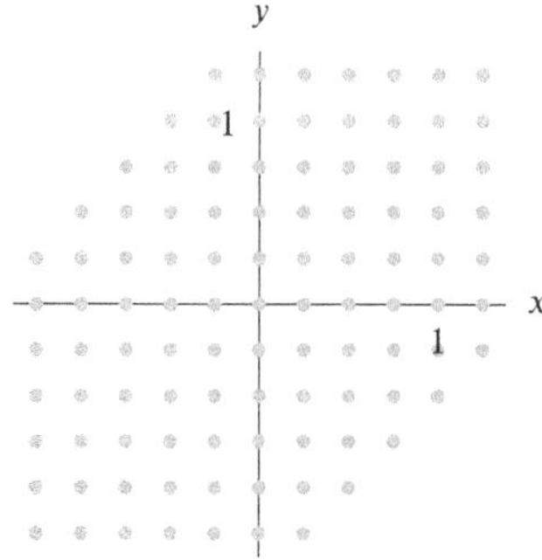

Figure 11.21

19. The slope field for the equation $dP/dt = 0.1P(10-P)$, for $P \geq 0$, is in Figure 11.22.

(a) Sketch the solutions that pass through the points

(i) $(0, 0)$ (ii) $(1, 4)$ (iii) $(4, 1)$
(iv) $(-5, 1)$ (v) $(-2, 12)$ (vi) $(-2, 10)$

(b) For which positive values of P are the solutions increasing? Decreasing? If $P(0) = 5$, what is the limiting value of P as t gets large?

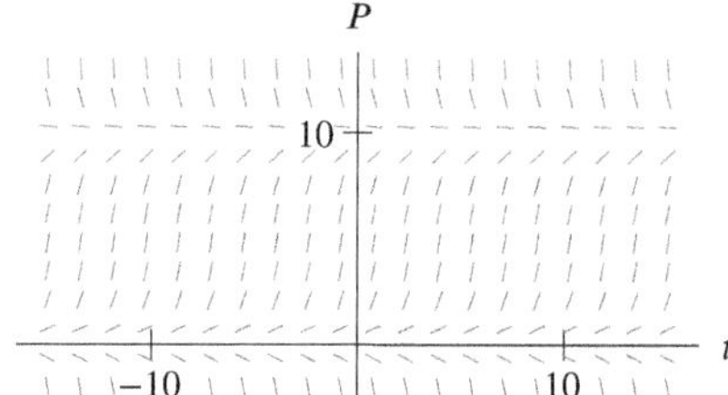

Figure 11.22

20. The slope field for $y' = 0.5(1 + y)(2 - y)$ is shown in Figure 11.23.

(a) Plot the following points on the slope field:

(i) the origin (ii) $(0, 1)$ (iii) $(1, 0)$
(iv) $(0, -1)$ (v) $(0, -5/2)$ (vi) $(0, 5/2)$

(b) Plot solution curves through the points in part (a).

(c) For which regions are all solution curves increasing? For which regions are all solution curves decreasing? When can the solution curves have horizontal tangents? Explain why, using both the slope field and the differential equation.

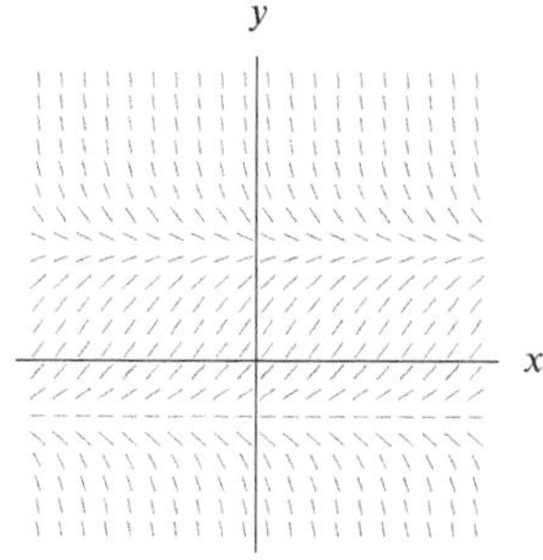

Figure 11.23: Note: x and y scales are equal

21. One of the slope fields in Figure 11.24 is the slope field for $y' = (x + y)/(x - y)$. Which one?

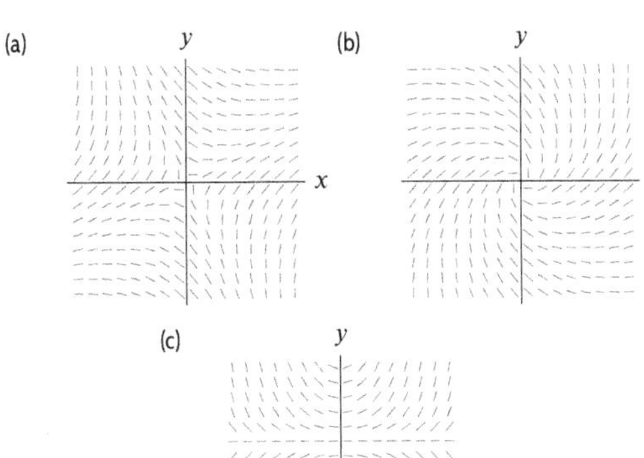

Figure 11.24

22. The slope field for $y' = (\sin x)(\sin y)$ is in Figure 11.25.

(a) Sketch the solutions that pass through the points

(i) $(0, -2)$ (ii) $(0, \pi)$

(b) What is the equation of the solution that passes through $(0, n\pi)$, where n is any integer?

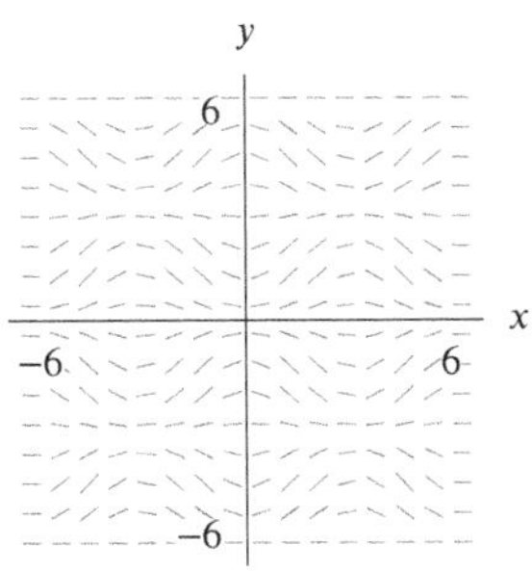

Figure 11.25

23. Match the slope fields in Figure 11.26 with their differential equations. Explain your reasoning.

(a) $y' = -y$ **(b)** $y' = y$ **(c)** $y' = x$

(d) $y' = 1/y$ **(e)** $y' = y^2$

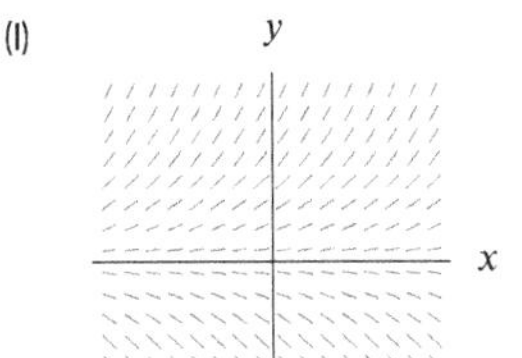

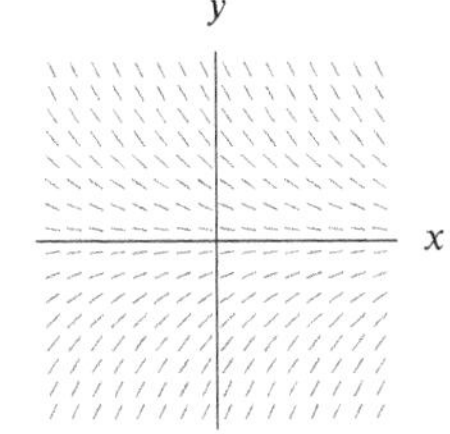

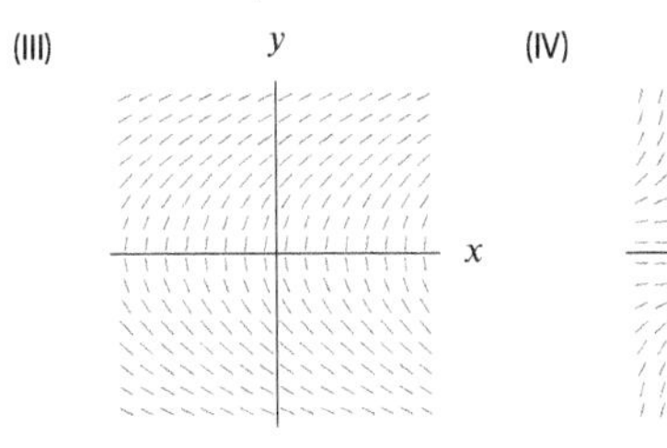

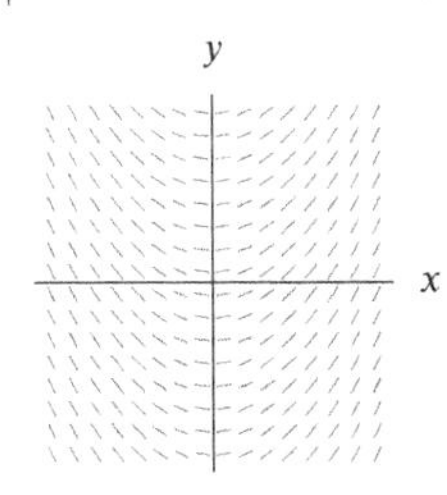

Figure 11.26: Each slope field is graphed for $-5 \le x \le 5, -5 \le y \le 5$

24. Match the slope fields in Figure 11.27 to the corresponding differential equations:

(a) $y' = xe^{-x}$ **(b)** $y' = \sin x$ **(c)** $y' = \cos x$
(d) $y' = x^2e^{-x}$ **(e)** $y' = e^{-x^2}$ **(f)** $y' = e^{-x}$

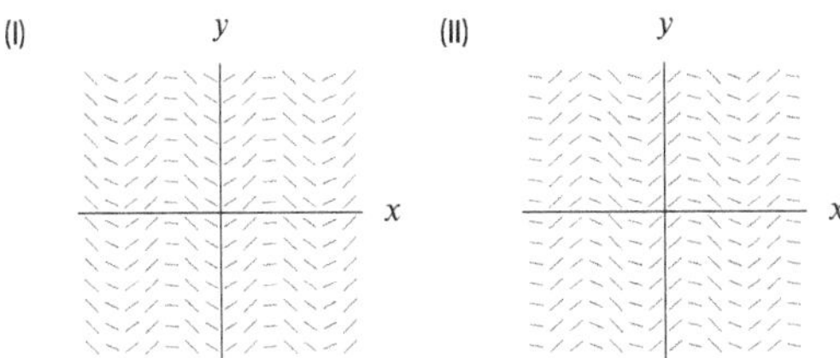

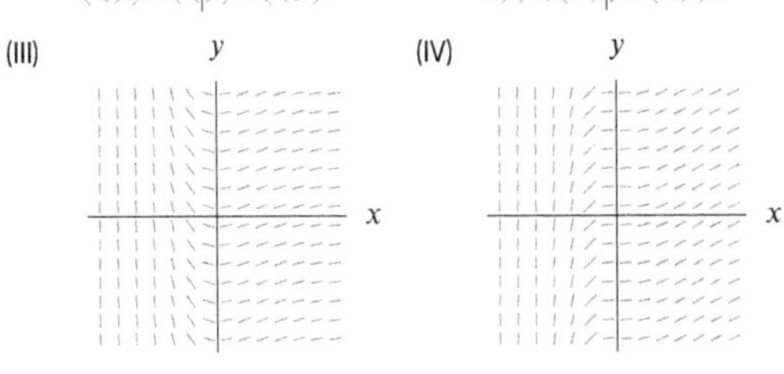

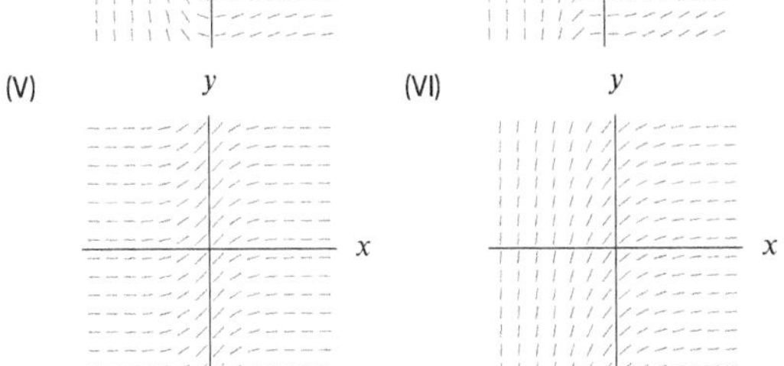

Figure 11.27

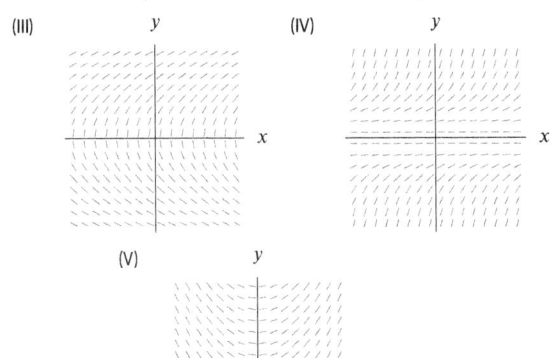

25. Match the following differential equations with the slope fields shown in Figure 11.28.

(a) $\frac{dy}{dx} = e^{x^2}$ (b) $\frac{dy}{dx} = e^{-2x^2}$

(c) $\frac{dy}{dx} = e^{-x^2/2}$ (d) $\frac{dy}{dx} = e^{-0.5x}\cos x$

(e) $\frac{dy}{dx} = \frac{1}{(1+0.5\cos x)^2}$

(f) $\frac{dy}{dx} = -e^{-x^2}$

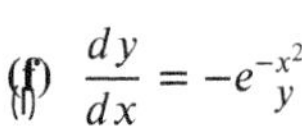

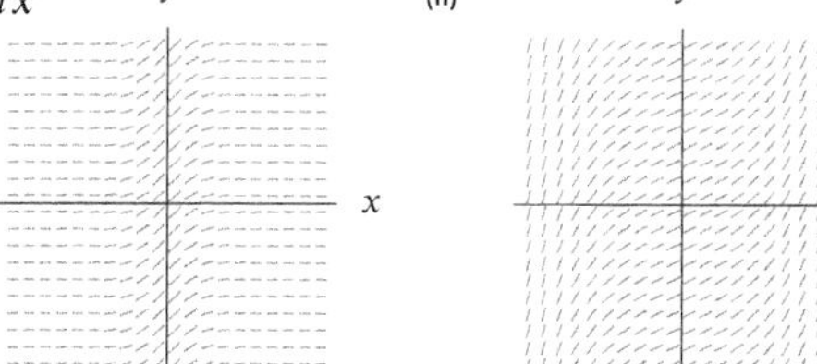

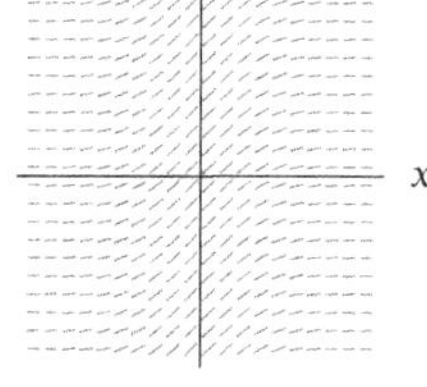

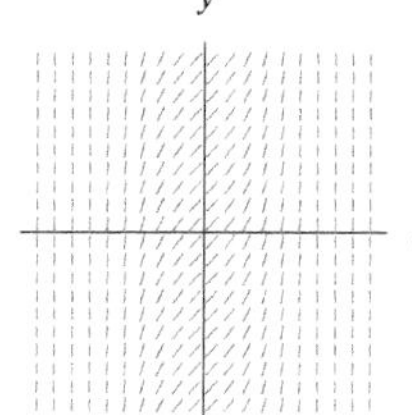

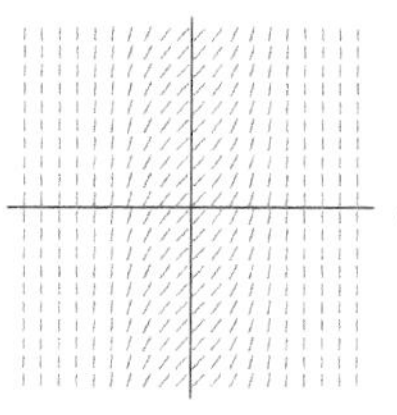

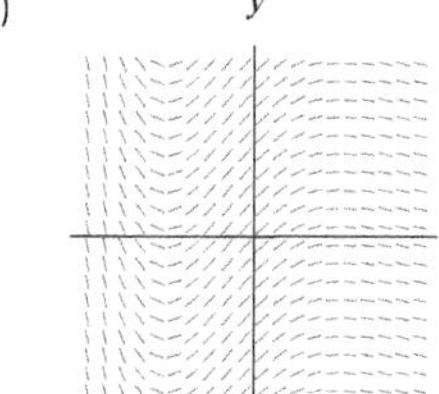

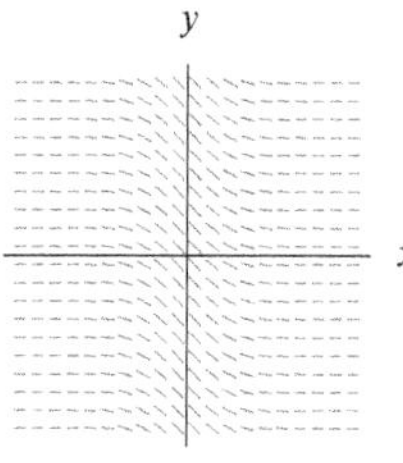

Figure 11.28: Each slope field is graphed for $-3 \le x \le 3$, $-3 \le y \le 3$

In Problems 26–29, match an equation with the slope field.

(a) $y' = 0.05y(10 - y)$ (b) $y' = 0.05x(10 - x)$
(c) $y' = 0.05y(5 - y)$ (d) $y' = 0.05x(5 - x)$
(e) $y' = 0.05y(y - 10)$ (f) $y' = 0.05x(x - 10)$
(g) $y' = 0.05y(y - 5)$ (h) $y' = 0.05x(x - 5)$
(i) $y' = 0.05x(y - 5)$

26.

27.

28.

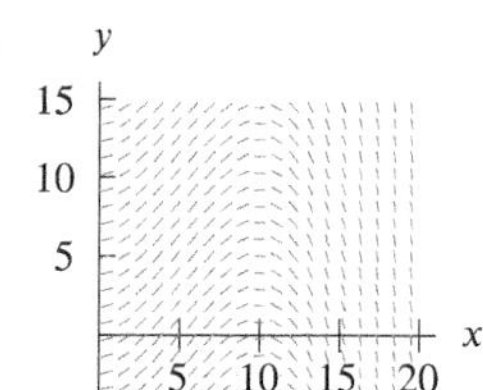

29.

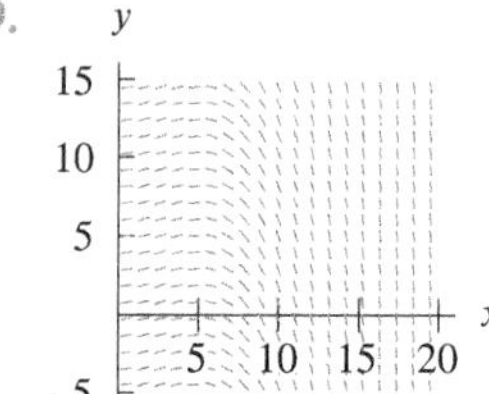

30. Match the slope fields in Figure 11.29 to the differential equations; find a and b assuming $a \neq b$.

(a) $\frac{dy}{dx} = (x-a)(y-b)$ (b) $\frac{dy}{dx} = \frac{1}{(x-a)(y-b)}$

(c) $\frac{dy}{dx} = \frac{x-a}{y-b}$ (d) $\frac{dy}{dx} = \frac{y-b}{x-a}$

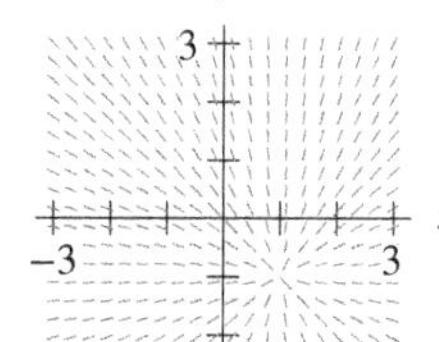

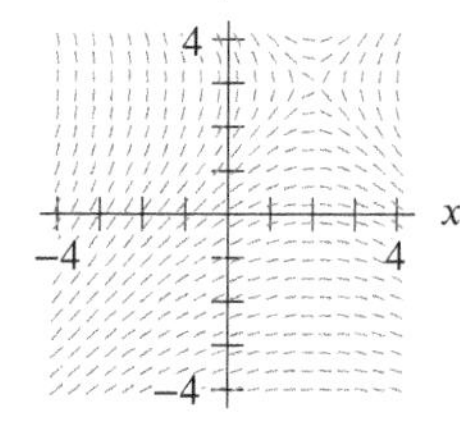

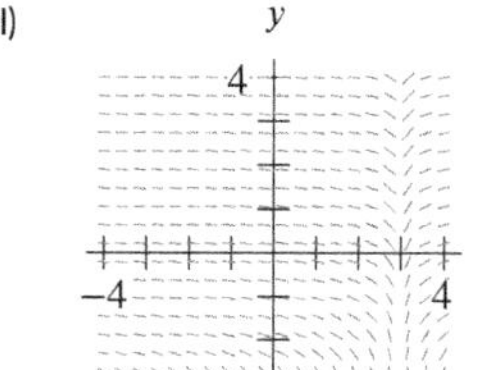

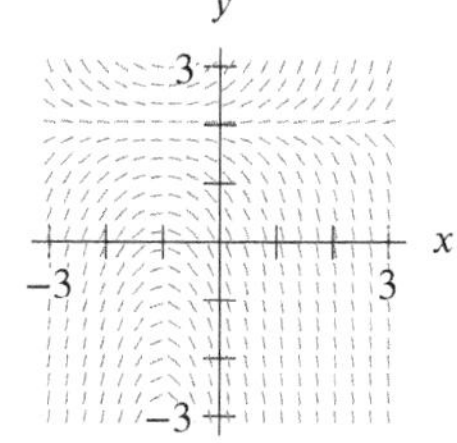

Figure 11.29

Strengthen Your Understanding

In Problems 31–33, explain what is wrong with the statement.

31. There is a differential equation that has $y = x$ as one of its solutions and a slope field with a slope of 0 at the point $(1, 1)$.

32. The differential equation $dy/dx = 0$ has only the solution $y = 0$.

33. Figure 11.30 shows the slope field of $y' = y$.

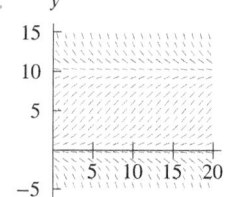

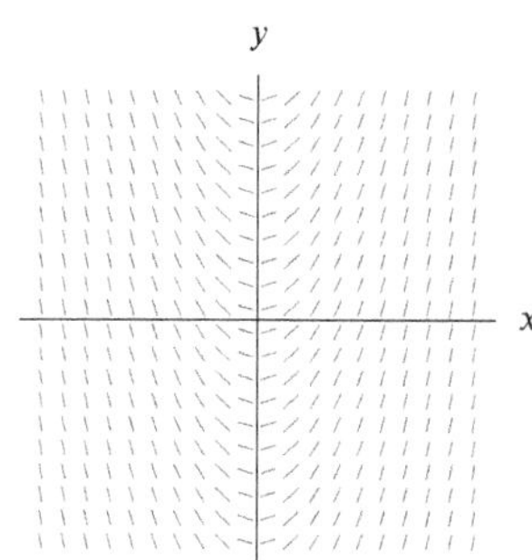

Figure 11.30

■ In Problems 34–37, give an example of:

34. A differential equation whose slope field has all the slopes positive.

35. A differential equation that has a slope field with all the slopes above the x-axis positive and all the slopes below the x-axis negative.

36. A slope field for a differential equation where the formula for dy/dx depends on x but not y.

37. A slope field for a differential equation where the formula for dy/dx depends on y but not x.

■ Are the statements in Problems 38–45 true or false? Give an explanation for your answer.

38. If the slope at the point $(0, 0)$ in a slope field for a differential equation is 0, then the solution of the differential equation passing through that point is the constant $y = 0$.

39. The differential equation $dy/dx = 2$ has straight line solutions with slope 2.

40. The differential equation $dy/dx = 2x$ has only the solution $y = x^2$.

41. All solutions to the differential equation whose slope field is in Figure 11.31 have $\lim_{x\to\infty} y = \infty$.

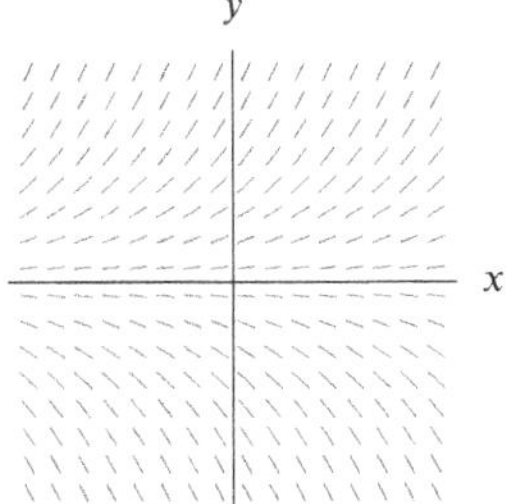

Figure 11.31

42. All solutions to the differential equation whose slope field is in Figure 11.32 have $\lim_{x\to\infty} y = 0$.

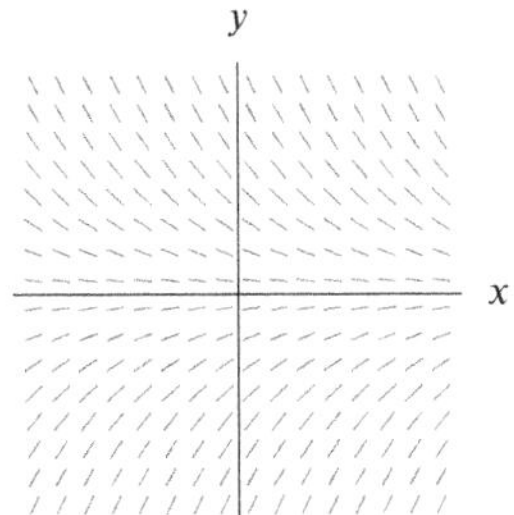

Figure 11.32

43. All solutions to the differential equation whose slope field is in Figure 11.33 have the same limiting value as $x \to \infty$.

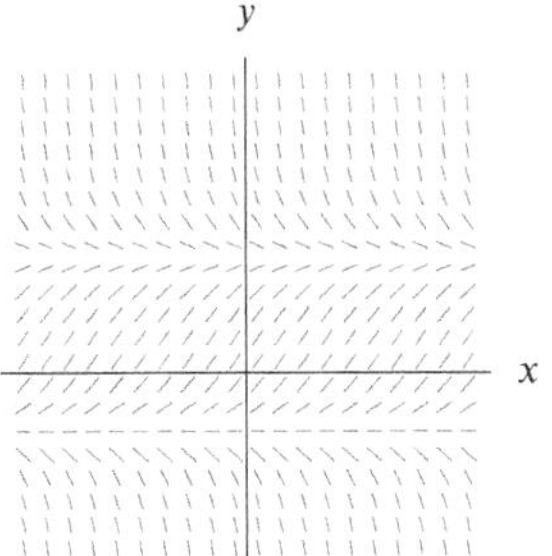

Figure 11.33

44. The solutions of the differential equation $dy/dx = x^2 + y^2 + 1$ are increasing at every point.

45. The solutions of the differential equation $dy/dx = x^2 + y^2 + 1$ are concave up at every point.

■ In Problems 46–54, decide whether the statement is true or false. Assume that $y = f(x)$ is a solution to the equation $dy/dx = 2x - y$. Justify your answer.

46. If $f(a) = b$, the slope of the graph of f at (a, b) is $2a - b$.

47. $f'(x) = 2x - f(x)$.

48. There could be more than one value of x such that $f'(x) = 1$ and $f(x) = 5$.

49. If $y = f(x)$, then $d^2y/dx^2 = 2 - (2x - y)$.

50. If $f(1) = 5$, then $(1, 5)$ could be a critical point of f.

51. The graph of f is decreasing whenever it lies above the line $y = 2x$ and is increasing whenever it lies below the line $y = 2x$.

52. All the inflection points of f lie on the line $y = 2x - 2$.

53. If $g(x)$ is another solution to the differential equation $dy/dx = 2x - y$, then $g(x) = f(x) + C$.

54. If $g(x)$ is a different solution to the differential equation $dy/dx = 2x - y$, then $\lim_{x\to\infty}(g(x) - f(x)) = 0$. [Hint: Show that $w = g(x) - f(x)$ satisfies the differential equation $dw/dx = -w$.]

11.3 EULER'S METHOD

In the preceding section we saw how to sketch a solution curve to a differential equation using its slope field. In this section we compute points on a solution curve numerically using *Euler's method*. (Leonhard Euler was an eighteenth-century Swiss mathematician.) In Section 11.4 we find formulas for some solution curves.

Here's the concept behind Euler's method. Think of the slope field as a set of signposts directing you across the plane. Pick a starting point (corresponding to the initial value), and calculate the slope at that point using the differential equation. This slope is a signpost telling you the direction to take. Go a small distance in that direction. Stop and look at the new signpost. Recalculate the slope from the differential equation, using the coordinates of the new point. Change direction to correspond to the new slope, and move another small distance, and so on.

Example 1 Use Euler's method for $dy/dx = y$. Start at the point $P_0 = (0, 1)$ and take $\Delta x = 0.1$.

Solution The slope at the point $P_0 = (0, 1)$ is $dy/dx = 1$. (See Figure 11.34.) As we move from P_0 to P_1, y increases by Δy, where

$$\Delta y = (\text{slope at } P_0)\Delta x = 1(0.1) = 0.1.$$

So we have

$$y\text{-value at } P_1 = (y \text{ value at } P_0) + \Delta y = 1 + 0.1 = 1.1.$$

Table 11.2 *Euler's method for $dy/dx = y$, starting at (0, 1)*

	x	y	$\Delta y = (\text{Slope})\Delta x$
P_0	0	1	0.1 = (1)(0.1)
P_1	0.1	1.1	0.11 = (1.1)(0.1)
P_2	0.2	1.21	0.121 = (1.21)(0.1)
P_3	0.3	1.331	0.1331 = (1.331)(0.1)
P_4	0.4	1.4641	0.14641 = (1.4641)(0.1)
P_5	0.5	1.61051	0.161051 = (1.61051)(0.1)

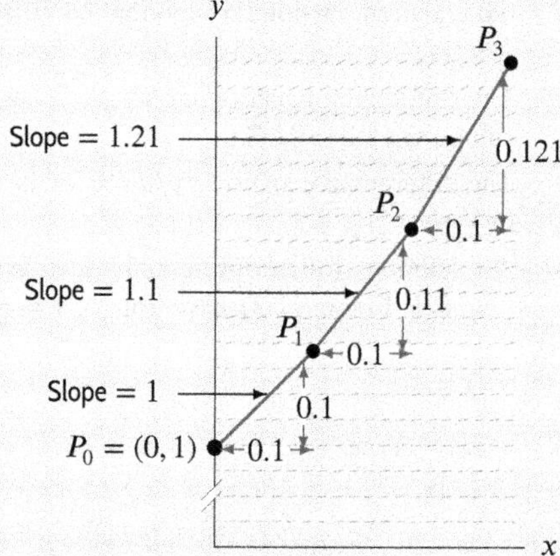

Figure 11.34: Euler's approximate solution to $dy/dx = y$

Thus the point P_1 is (0.1, 1.1). Now, using the differential equation again, we see that

$$\text{slope at } P_1 = 1.1,$$

so if we move to P_2, then y changes by

$$\Delta y = (\text{slope at } P_1)\Delta x = (1.1)(0.1) = 0.11.$$

This means

$$y\text{-value at } P_2 = (y \text{ value at } P_1) + \Delta y = 1.1 + 0.11 = 1.21.$$

Thus P_2 is (0.2, 1.21). Continuing gives the results in Table 11.2.

Since the solution curves of $dy/dx = y$ are exponentials, they are concave up and bend upward away from the line segments of the slope field. Therefore, in this case, Euler's method produces y-values which are too small.

Notice that Euler's method calculates approximate y-values for points on a solution curve; it does not give a formula for y in terms of x.

Example 2 Show that Euler's method for $dy/dx = y$ starting at (0, 1) and using two steps with $\Delta x = 0.05$ gives $y \approx 1.1025$ when $x = 0.1$.

Solution At (0, 1), the slope is 1 and $\Delta y = (1)(0.05) = 0.05$, so new $y = 1 + 0.05 = 1.05$. At (0.05, 1.05), the slope is 1.05 and $\Delta y = (1.05)(0.05) = 0.0525$, so new $y = 1.05 + 0.0525 = 1.1025$ at $x = 0.1$.

In general, dy/dx may be a function of both x and y. Euler's method still works, as the next example shows.

Example 3 Approximate four points on the solution curve to $dy/dx = -x/y$ starting at $(0, 1)$; use $\Delta x = 0.1$. Are the approximate values overestimates or underestimates?

Solution The results from Euler's method are in Table 11.3, along with the y-values (to two decimals) calculated from the equation of the circle $x^2 + y^2 = 1$, which is the solution curve through $(0, 1)$. Since the curve is concave down, the approximate y-values are above the exact ones. (See Figure 11.35.)

Table 11.3 *Euler's method for $dy/dx = -x/y$, starting at* $(0, 1)$

x	Approx. y-value	$\Delta y = (\text{Slope})\Delta x$	True y-value
0	1	$0 = (0)(0.1)$	1
0.1	1	$-0.01 = (-0.1/1)(0.1)$	0.99
0.2	0.99	$-0.02 = (-0.2/0.99)(0.1)$	0.98
0.3	0.97		0.95

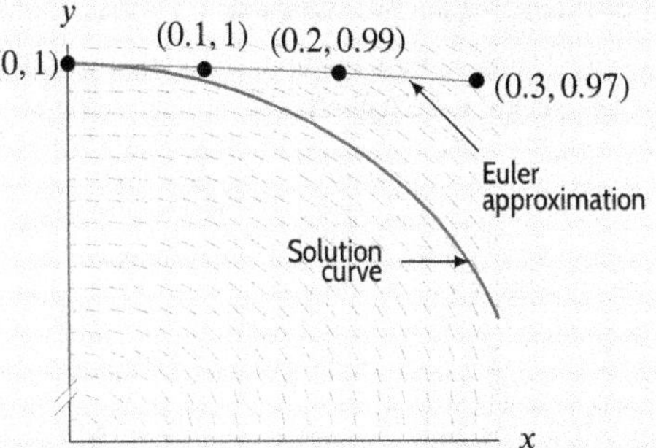

Figure 11.35: Euler's approximate solution to $dy/dx = -x/y$

The Accuracy of Euler's Method

To improve the accuracy of Euler's method, we choose a smaller step size, Δx. Let's go back to the differential equation $dy/dx = y$ and compare the exact and approximate values for different Δx's. The exact solution going through the point $(0, 1)$ is $y = e^x$, so the exact values are calculated using this function. (See Figure 11.36.) Where $x = 0.1$,

$$\text{Exact } y\text{-value } = e^{0.1} \approx 1.1051709.$$

In Example 1 we had $\Delta x = 0.1$, and where $x = 0.1$,

$$\text{Approximate } y\text{-value} = 1.1, \quad \text{so the error} \approx 0.005.$$

In Example 2 we decreased Δx to 0.05. After two steps, $x = 0.1$, and we had

$$\text{Approximate } y\text{-value} = 1.1025, \quad \text{so error} \approx 0.00267.$$

Thus, it appears that halving the step size has approximately halved the error.

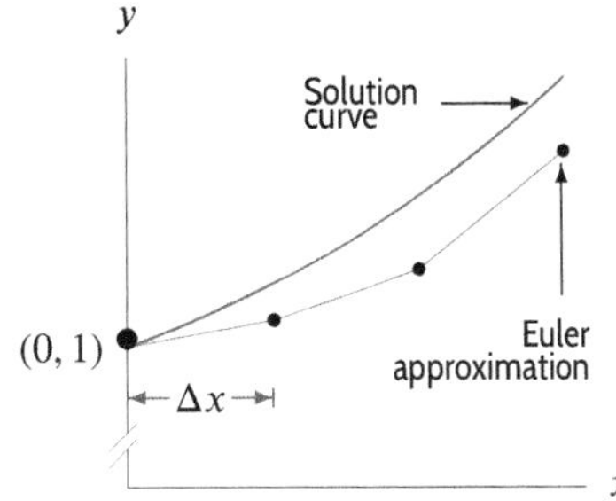

Figure 11.36: Euler's approximate solution to $dy/dx = y$

> The *error* in using Euler's method over a fixed interval is Exact value − Approximate value. If the number of steps used is n, the error is approximately proportional to $1/n$.

Just as there are more accurate numerical integration methods than left-hand and right-hand Riemann sums, there are more accurate methods than Euler's for approximating solution curves. However, Euler's method is all we need in this text.

Exercises and Problems for Section 11.3

EXERCISES

1. Using Euler's method, complete the following table for $y' = (x-2)(y-3)$.

x	y	y'
0.0	4.0	
0.1		
0.2		

2. Using Euler's method, complete the following table for $y' = 4xy$.

x	y	y'
1.00	−3.0	
1.01		
1.02		

3. A population, P, in millions, is 1500 at time $t = 0$ and its growth is governed by
$$\frac{dP}{dt} = 0.00008P(1900 - P).$$
Use Euler's method with $\Delta t = 1$ to estimate P at time $t = 1, 2, 3$.

4. **(a)** Use Euler's method to approximate the value of y at $x = 1$ on the solution curve to the differential equation $dy/dx = 3$ that passes through $(0, 2)$. Use $\Delta x = 0.2$.
(b) What is the solution to the differential equation $dy/dx = 3$ with initial condition $y = 2$ when $x = 0$?
(c) What is the error for the Euler's method approximation at $x = 1$?
(d) Explain why Euler's method is exact in this case.

5. For $y' = 2y - 3x - 4$, use Euler's method with $\Delta x = 0.1$ to estimate y when $x = 0.2$ for the solution curve passing through $(0, 2)$.

6. **(a)** Use five steps in Euler's method to determine an approximate solution for the differential equation $dy/dx = y - x$ with initial condition $y(0) = 10$, using step size $\Delta x = 0.2$. What is the estimated value of y at $x = 1$?
(b) Does the solution to the differential equation appear to be concave up or concave down?
(c) Are the approximate values overestimates or underestimates?

7. **(a)** Use ten steps in Euler's method to determine an approximate solution for the differential equation $y' = x^3$, $y(0) = 0$, using a step size $\Delta x = 0.1$.
(b) What is the exact solution? Compare it to the computed approximation.
(c) Use a sketch of the slope field for this equation to explain the results of part (b).

8. Consider the differential equation $y' = x + y$ whose slope field is in Figure 11.20 on page 571. Use Euler's method with $\Delta x = 0.1$ to estimate y when $x = 0.4$ for the solution curves satisfying

(a) $y(0) = 1$ **(b)** $y(-1) = 0$.

9. **(a)** Using Figure 11.37, sketch the solution curve that passes through $(0, 0)$ for the differential equation
$$\frac{dy}{dx} = x^3 - y^3.$$
(b) Compute the points on the solution curve generated by Euler's method with 5 steps of $\Delta x = 0.2$.
(c) Is your answer to part (b) an overestimate or an underestimate?

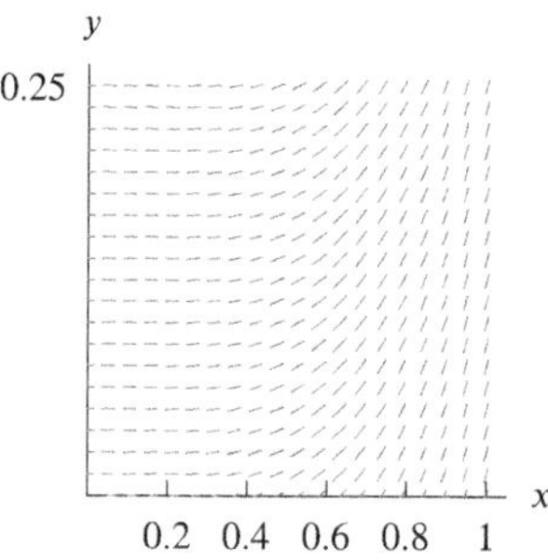

Figure 11.37: Slope field for $dy/dx = x^3 - y^3$

PROBLEMS

10. Consider the differential equation $y' = (\sin x)(\sin y)$.
    (a) Calculate approximate y-values using Euler's method with three steps and $\Delta x = 0.1$, starting at each of the following points:
    (i) $(0, 2)$ (ii) $(0, \pi)$
    (b) Use the slope field in Figure 11.25 on page 573 to explain your solution to part (a)(ii).

11. (a) Use Euler's method with five subintervals to approximate the solution curve to the differential equation $dy/dx = x^2 - y^2$ passing through the point $(0, 1)$ and ending at $x = 1$. (Keep the approximate function values to three decimal places.)
    (b) Repeat this computation using ten subintervals, again ending at $x = 1$.

12. Why are the approximate results you obtained in Problem 11 smaller than the true values? (Note: The slope field for this differential equation is one of those in Figure 11.18 on page 571.)

13. How do the errors of the five-step calculation and the ten-step calculation in Problem 11 compare? Estimate the true value of y on the solution through the point $(0, 1)$ when $x = 1$.

14. (a) Use ten steps of Euler's method to approximate y-values for $dy/dt = 1/t$, starting at $(1, 0)$ and using $\Delta t = 0.1$.
    (b) Using integration, solve the differential equation to find the exact value of y at the end of these ten steps.
    (c) Is your approximate value of y at the end of ten steps bigger or smaller than the exact value? Use a slope field to explain your answer.

15. Consider the differential equation

$$\frac{dy}{dx} = 2x, \quad \text{with initial condition } y(0) = 1.$$

    (a) Use Euler's method with two steps to estimate y when $x = 1$. Now use four steps.
    (b) What is the formula for the exact value of y?
    (c) Does the error in Euler's approximation behave as predicted in the box on page 577?

16. Consider the differential equation

$$\frac{dy}{dx} = \sin(xy), \quad \text{with initial condition } y(1) = 1.$$

Estimate $y(2)$, using Euler's method with step sizes $\Delta x = 0.2, 0.1, 0.05$. Plot the computed approximations for $y(2)$ against Δx. What do you conclude? Use your observations to estimate the exact value of $y(2)$.

17. (a) Use Euler's method to estimate $B(1)$, given that

$$\frac{dB}{dt} = 0.05B$$

    and $B = 1000$ when $t = 0$. Take:
    (i) $\Delta t = 1$ and 1 step (ii) $\Delta t = 0.5$ and 2 steps
    (iii) $\Delta t = 0.25$ and 4 steps
    (b) Suppose B is the balance in a bank account earning interest. Explain why the result of your calculation in part (i) is equivalent to compounding the interest once a year instead of continuously.
    (c) Interpret the result of your calculations in parts (ii) and (iii) in terms of compound interest.

18. With t in years since the start of 2014, copper has been mined worldwide at a rate of $17.9e^{0.025t}$ million tons per year.[2]
    (a) Use Euler's method with $\Delta t = 1$ to estimate values of C, the total quantity of copper extracted worldwide since the start of 2014 until the start of 2018.
    (b) What is the relationship between the value of C from part (a) and $\int_0^4 17.9e^{0.025t}\,dt$?

19. Consider the differential equation $dy/dx = f(x)$ with initial value $y(0) = 0$. Explain why using Euler's method to approximate the solution curve gives the same results as using left-hand Riemann sums to approximate $\int_0^x f(t)\,dt$.

Strengthen Your Understanding

In Problems 20–21, explain what is wrong with the statement.

20. Euler's method never produces an exact solution to a differential equation at a point. There is always some error.

21. If we use Euler's method on the interval $[0, 1]$ to estimate the value of $x(1)$ where $dx/dt = x$, then we get an underestimate.

In Problems 22–23, give an example of:

22. A differential equation for which the approximate values found using Euler's method lie on a straight line.

23. A differential equation and initial condition such that for any step size, the approximate y-value found after one step of Euler's method is an underestimate of the solution value.

Are the statements in Problems 24–29 true or false? Give an

[2]Data from http://minerals.usgs.gov/minerals/pubs/commodity/ Accessed February 8, 2015.

explanation for your answer.

24. Euler's method gives the arc length of a solution curve.

25. Using Euler's method with five steps and $\Delta x = 0.2$ to approximate $y(1)$ when $dy/dx = f(x)$ and $y(0) = 0$ gives the same answer as the left Riemann sum approximation to $\int_0^1 f(x)\,dx$.

26. If $n = 1000$ steps are used in Euler's method, then the error is approximately 0.001.

27. If we increase the steps in Euler's method for a given differential equation from $n = 1000$ to $n = 2000$, then the error decreases by a factor of approximately 0.001.

28. If we use Euler's method to approximate the solution to the differential equation $dy/dx = f(x)$ where $f(x) > 0$ and is increasing, then Euler's method gives an overestimate of the exact solution.

29. If we use Euler's method to approximate the solution to the differential equation $dy/dx = f(x)$ on the interval $[-10, 20]$ with $\Delta x = 0.1$ then we use $n = 30$ steps.

11.4 SEPARATION OF VARIABLES

We have seen how to sketch solution curves of a differential equation using a slope field and how to calculate approximate numerical solutions. Now we see how to solve certain differential equations analytically, finding an equation for the solution curve.

First, we look at a familiar example, the differential equation

$$\frac{dy}{dx} = -\frac{x}{y},$$

whose solution curves are the circles

$$x^2 + y^2 = C.$$

We can check that these circles are solutions by differentiation; the question now is how they were obtained. The method of *separation of variables* works by putting all the x-values on one side of the equation and all the y-values on the other, giving

$$y\,dy = -x\,dx.$$

We then integrate each side separately:

$$\int y\,dy = -\int x\,dx,$$

$$\frac{y^2}{2} = -\frac{x^2}{2} + k.$$

This gives the circles we were expecting:

$$x^2 + y^2 = C \qquad \text{where } C = 2k.$$

You might worry about whether it is legitimate to separate the dx and the dy. The reason it can be done is explained at the end of this section.

Example 1 Using separation of variables, solve the differential equation:

$$\frac{dy}{dx} = ky.$$

Solution Separating variables,

$$\frac{1}{y}dy = k\,dx,$$

and integrating,

$$\int \frac{1}{y}\,dy = \int k\,dx,$$

gives

$$\ln|y| = kx + C \quad \text{for some constant } C.$$

Solving for $|y|$ leads to

$$|y| = e^{kx+C} = e^{kx}e^{C} = Ae^{kx}$$

where $A = e^C$, so A is positive. Thus

$$y = (\pm A)e^{kx} = Be^{kx}$$

where $B = \pm A$, so B is any nonzero constant. Even though there's no C leading to $B = 0$, we can have $B = 0$ because $y = 0$ is a solution to the differential equation. We lost this solution when we divided through by y at the first step. Thus, the general solution is $y = Be^{kx}$ for any B.

The differential equation $dy/dx = ky$ always leads to exponential growth (if $k > 0$) or exponential decay (if $k < 0$). Graphs of solution curves for some fixed $k > 0$ are in Figure 11.38. For $k < 0$, the graphs are reflected across the y-axis.

Example 2 For $k > 0$, find and graph solutions of

$$\frac{dH}{dt} = -k(H - 20).$$

Solution The slope field in Figure 11.39 shows the qualitative behavior of the solutions. To find the equation of the solution curves, we separate variables and integrate:

$$\int \frac{1}{H-20}\, dH = -\int k\, dt.$$

This gives

$$\ln|H - 20| = -kt + C.$$

Solving for H leads to:

$$|H - 20| = e^{-kt+C} = e^{-kt}e^{C} = Ae^{-kt}$$

or

$$H - 20 = (\pm A)e^{-kt} = Be^{-kt}$$

$$H = 20 + Be^{-kt}.$$

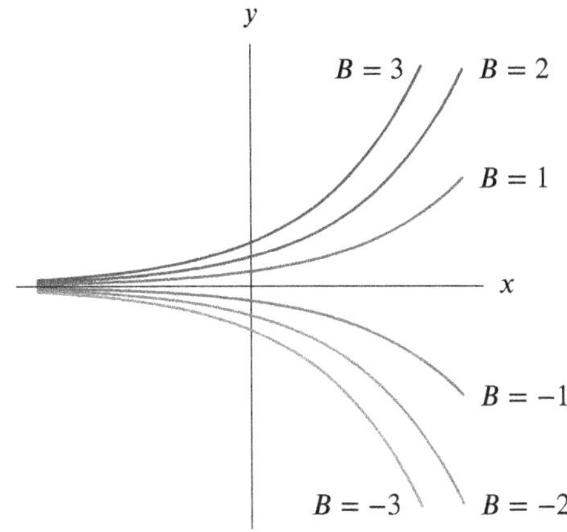

Figure 11.38: Graphs of $y = Be^{kx}$, which are solutions to $dy/dx = ky$, for some fixed $k > 0$

Again, $B = 0$ also gives a solution. Graphs for $k = 1$ and $B = -10, 0, 10$, with $t \geq 0$, are in Figure 11.39.

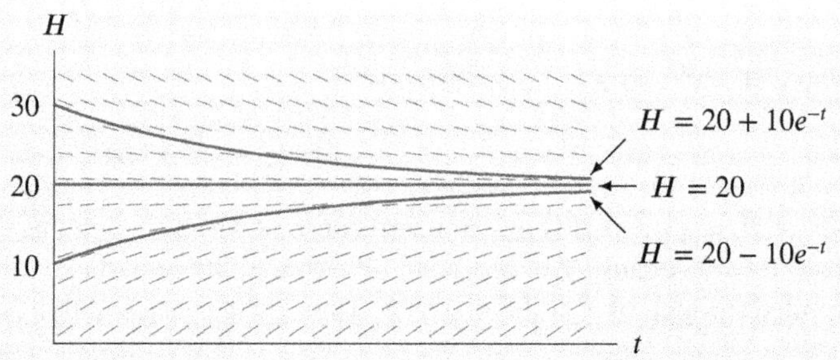

Figure 11.39: Slope field and some solution curves for $dH/dt = -k(H - 20)$, with $k = 1$

This differential equation can be used to represent the temperature, $H(t)$, in °C at time t of a cup of water standing in a room at 20°C. As Figure 11.39 shows, if the initial temperature is 10°C, the water warms up; if the initial temperature is 30°C, the water cools down. If the initial temperature is 20°C, the water remains at 20°C.

Example 3 Find and sketch the solution to

$$\frac{dP}{dt} = 2P - 2Pt \qquad \text{satisfying } P = 5 \text{ when } t = 0.$$

Solution Factoring the right-hand side gives

$$\frac{dP}{dt} = P(2 - 2t).$$

Separating variables, we get

$$\int \frac{dP}{P} = \int (2 - 2t)\, dt,$$

so

$$\ln|P| = 2t - t^2 + C.$$

Solving for P leads to

$$|P| = e^{2t-t^2+C} = e^C e^{2t-t^2} = Ae^{2t-t^2}$$

with $A = e^C$, so $A > 0$. In addition, $A = 0$ gives a solution. Thus the general solution to the differential equation is

$$P = Be^{2t-t^2} \quad \text{for any } B.$$

To find the value of B, substitute $P = 5$ and $t = 0$ into the general solution, giving

$$5 = Be^{2\cdot 0 - 0^2} = B,$$

so

$$P = 5e^{2t-t^2}.$$

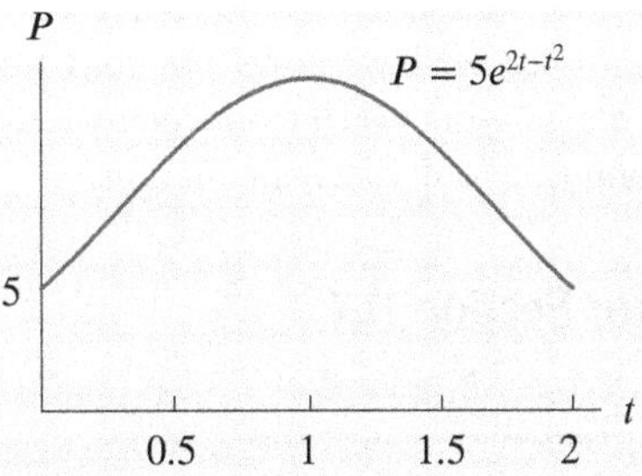

Figure 11.40: Bell-shaped solution curve

The graph of this function is in Figure 11.40. Since the solution can be rewritten as

$$P = 5e^{1-1+2t-t^2} = 5e^1 e^{-1+2t-t^2} = (5e)e^{-(t-1)^2},$$

the graph has the same shape as the graph of $y = e^{-t^2}$, the bell-shaped curve of statistics. Here the maximum, normally at $t = 0$, is shifted one unit to the right to $t = 1$.

Justification for Separation of Variables

A differential equation is called *separable* if it can be written in the form

$$\frac{dy}{dx} = g(x)f(y).$$

Provided $f(y) \neq 0$, we write $f(y) = 1/h(y)$, so the right-hand side can be thought of as a fraction,

$$\frac{dy}{dx} = \frac{g(x)}{h(y)}.$$

If we multiply through by $h(y)$, we get

$$h(y)\frac{dy}{dx} = g(x).$$

Thinking of y as a function of x, so $y = y(x)$, and $dy/dx = y'(x)$, we can rewrite the equation as

$$h(y(x)) \cdot y'(x) = g(x).$$

Now integrate both sides with respect to x:

$$\int h(y(x)) \cdot y'(x)\, dx = \int g(x)\, dx.$$

The form of the integral on the left suggests that we use the substitution $y = y(x)$. Since $dy = y'(x)\, dx$, we get

$$\int h(y)\, dy = \int g(x)\, dx.$$

If we can find antiderivatives of h and g, then this gives the equation of the solution curve.

Note that transforming the original differential equation,

$$\frac{dy}{dx} = \frac{g(x)}{h(y)},$$

into

$$\int h(y)\, dy = \int g(x)\, dx$$

looks as though we have treated dy/dx as a fraction, cross-multiplied, and then integrated. Although that's not exactly what we have done, you may find this a helpful way of remembering the method. In fact, the dy/dx notation was introduced by Leibniz to allow shortcuts like this (more specifically, to make the chain rule look like cancellation).

Exercises and Problems for Section 11.4

EXERCISES

1. Determine which of the following differential equations are separable. Do not solve the equations.

(a) $y' = y$	**(b)** $y' = x + y$
(c) $y' = xy$	**(d)** $y' = \sin(x + y)$
(e) $y' - xy = 0$	**(f)** $y' = y/x$
(g) $y' = \ln(xy)$	**(h)** $y' = (\sin x)(\cos y)$
(i) $y' = (\sin x)(\cos xy)$	**(j)** $y' = x/y$
(k) $y' = 2x$	**(l)** $y' = (x + y)/(x + 2y)$

For Exercises 2–5, determine if the differential equation is separable, and if so, write it in the form $h(y)\,dy = g(x)\,dx$.

2. $y' = xe^y$
3. $y' = xe^y - x$
4. $y' = xe^y - 1$
5. $y' - xy' = y$

In Exercises 6–32, use separation of variables to find the solution to the differential equation subject to the initial condition.

6. $\dfrac{dP}{dt} = -2P, \quad P(0) = 1$
7. $\dfrac{dP}{dt} = 0.02P, \quad P(0) = 20$
8. $\dfrac{dL}{dp} = \dfrac{L}{2}, \quad L(0) = 100$
9. $\dfrac{dQ}{dt} = \dfrac{Q}{5}, \quad Q = 50$ when $t = 0$
10. $P\dfrac{dP}{dt} = 1, \quad P(0) = 1$
11. $\dfrac{dm}{dt} = 3m, \quad m = 5$ when $t = 1$
12. $\dfrac{dI}{dx} = 0.2I, \quad I = 6$ where $x = -1$
13. $\dfrac{1}{z}\dfrac{dz}{dt} = 5, \quad z(1) = 5$
14. $\dfrac{dm}{ds} = m, \quad m(1) = 2$
15. $2\dfrac{du}{dt} = u^2, \quad u(0) = 1$
16. $\dfrac{dz}{dy} = zy, \quad z = 1$ when $y = 0$
17. $\dfrac{dy}{dx} + \dfrac{y}{3} = 0, \quad y(0) = 10$
18. $\dfrac{dy}{dt} = 0.5(y - 200), \quad y = 50$ when $t = 0$
19. $\dfrac{dP}{dt} = P + 4, \quad P = 100$ when $t = 0$
20. $\dfrac{dy}{dx} = 2y - 4$, through $(2, 5)$
21. $\dfrac{dQ}{dt} = 0.3Q - 120, \quad Q = 50$ when $t = 0$
22. $\dfrac{dm}{dt} = 0.1m + 200, \quad m(0) = 1000$
23. $\dfrac{dR}{dy} + R = 1, \quad R(1) = 0.1$
24. $\dfrac{dB}{dt} + 2B = 50, \quad B(1) = 100$
25. $\dfrac{dy}{dt} = \dfrac{y}{3 + t}, \quad y(0) = 1$
26. $\dfrac{dz}{dt} = te^z$, through the origin
27. $\dfrac{dy}{dx} = \dfrac{5y}{x}, \quad y = 3$ where $x = 1$
28. $\dfrac{dy}{dt} = y^2(1 + t), \quad y = 2$ when $t = 1$
29. $\dfrac{dz}{dt} = z + zt^2, \quad z = 5$ when $t = 0$
30. $\dfrac{dw}{d\theta} = \theta w^2 \sin\theta^2, \quad w(0) = 1$
31. $\dfrac{dw}{d\psi} = -w^2 \tan\psi, \quad w(0) = 2$
32. $x(x + 1)\dfrac{du}{dx} = u^2, \quad u(1) = 1$

PROBLEMS

33. **(a)** Solve the differential equation
$$\frac{dy}{dx} = \frac{4x}{y^2}.$$
Write the solution y as an explicit function of x.
(b) Find the particular solution for each initial condition below and graph the three solutions on the same coordinate plane.
$$y(0) = 1, \qquad y(0) = 2, \qquad y(0) = 3.$$

34. **(a)** Solve the differential equation

$$\frac{dP}{dt} = 0.2P - 10.$$

Write the solution P as an explicit function of t.

(b) Find the particular solution for each initial condition below and graph the three solutions on the same coordinate plane.

$$P(0) = 40, \qquad P(0) = 50, \qquad P(0) = 60.$$

35. **(a)** Find the general solution to the differential equation modeling how a person learns:

$$\frac{dy}{dt} = 100 - y.$$

(b) Plot the slope field of this differential equation and sketch solutions with $y(0) = 25$ and $y(0) = 110$.

(c) For each of the initial conditions in part (b), find the particular solution and add it to your sketch.

(d) Which of these two particular solutions could represent how a person learns?

36. A circular oil spill grows at a rate given by the differential equation $dr/dt = k/r$, where r represents the radius of the spill in feet, and time is measured in hours. If the radius of the spill is 400 feet 16 hours after the spill begins, what is the value of k? Include units in your answer.

37. Figure 11.41 shows the slope field for $dy/dx = y^2$.

(a) Sketch the solutions that pass through the points

(i) $(0, 1)$ (ii) $(0, -1)$ (iii) $(0, 0)$

(b) In words, describe the end behavior of the solution curves in part (a).

(c) Find a formula for the general solution.

(d) Show that all solution curves except for $y = 0$ have both a horizontal and a vertical asymptote.

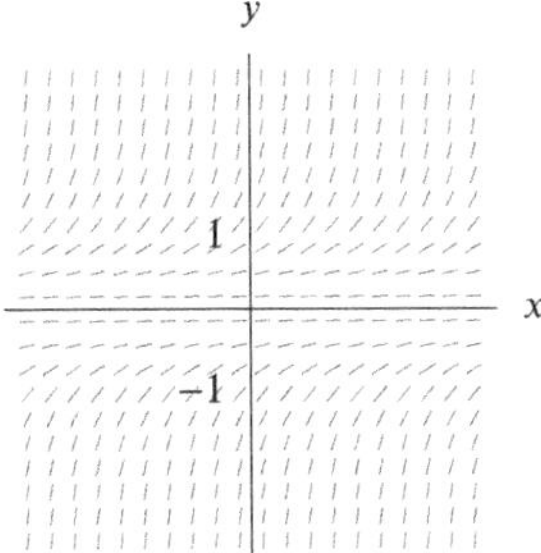

Figure 11.41

■ In Problems 38–47, solve the differential equation. Assume a, b, and k are nonzero constants.

38. $\dfrac{dR}{dt} = kR$

39. $\dfrac{dQ}{dt} - \dfrac{Q}{k} = 0$

40. $\dfrac{dP}{dt} = P - a$

41. $\dfrac{dQ}{dt} = b - Q$

42. $\dfrac{dP}{dt} = k(P - a)$

43. $\dfrac{dR}{dt} = aR + b$

44. $\dfrac{dP}{dt} - aP = b$

45. $\dfrac{dy}{dt} = ky^2(1 + t^2)$

46. $\dfrac{dR}{dx} = a(R^2 + 1)$

47. $\dfrac{dL}{dx} = k(x + a)(L - b)$

■ In Problems 48–51, solve the differential equation. Assume $x, y, t > 0$.

48. $\dfrac{dy}{dt} = y(2 - y), \quad y(0) = 1$

49. $\dfrac{dx}{dt} = \dfrac{x \ln x}{t}$

50. $t\dfrac{dx}{dt} = (1 + 2\ln t)\tan x$

51. $\dfrac{dy}{dt} = -y \ln\left(\dfrac{y}{2}\right), \quad y(0) = 1$

52. Figure 11.42 shows the slope field for the equation

$$\frac{dy}{dx} = \begin{cases} y^2 & \text{if } |y| \ge 1 \\ 1 & \text{if } -1 \le y \le 1. \end{cases}$$

(a) Sketch the solutions that pass through $(0, 0)$.

(b) What can you say about the end behavior of the solution curve in part (a)?

(c) For each of the following regions, find a formula for the general solution

(i) $-1 \le y \le 1$ (ii) $y \le -1$

(iii) $y \ge 1$

(d) Show that each solution curve has two vertical asymptotes.

(e) How far apart are the two asymptotes of a solution curve?

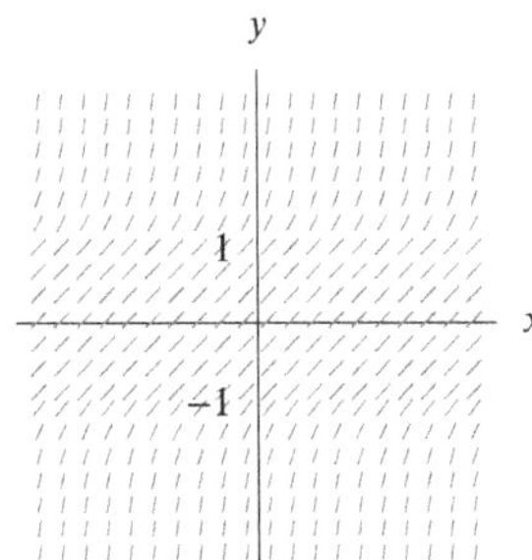

Figure 11.42

53. **(a)** Sketch the slope field for $y' = x/y$.

(b) Sketch several solution curves.

(c) Solve the differential equation analytically.

54. **(a)** Sketch the slope field for $y' = -y/x$.

(b) Sketch several solution curves.

(c) Solve the differential equation analytically.

55. Compare the slope field for $y' = x/y$, Problem 53, with that for $y' = -y/x$, Problem 54. Show that the solution curves of Problem 53 intersect the solution curves of Problem 54 at right angles.

56. Consider the family of differential equations given by $y' = ky^2$ where $k > 0$.

    (a) Explain why the constant function $y = 0$ is a solution.
    (b) Explain why all solutions except $y = 0$ are increasing.
    (c) Use separation of variables to find the general solution of this family, assuming $y \neq 0$. Your answer will have both the parameter k and an arbitrary constant C.
    (d) Does the method of separation of variables give all possible solutions of the family of differential equations? Explain.

Strengthen Your Understanding

In Problems 57–59, explain what is wrong with the statement.

57. Separating variables in $dy/dx = x + y$ gives $-y\,dy = x\,dx$.
58. The solution to $dP/dt = 0.2t$ is $P = Be^{0.2t}$.
59. Separating variables in $dy/dx = e^{x+y}$ gives $-e^y\,dy = e^x\,dx$.

In Problems 60–63, give an example of:

60. A differential equation that is not separable.
61. An expression for $f(x)$ such that the differential equation $dy/dx = f(x) + xy - \cos x$ is separable.
62. A differential equation all of whose solutions form the family of functions $f(x) = x^2 + C$.
63. A differential equation all of whose solutions form the family of hyperbolas $x^2 - y^2 = C$.

Are the statements in Problems 64–69 true or false? Give an explanation for your answer.

64. A differential equation of the form $dy/dx = f(x)$ is separable.
65. A differential equation of the form $dy/dx = 1/g(y)$ is separable.
66. For all constants k, the equation $y' + ky = 0$ has exponential functions as solutions.
67. The differential equation $dy/dx = x + y$ can be solved by separation of variables.
68. The differential equation $dy/dx - xy = x$ can be solved by separation of variables.
69. The only solution to the differential equation $dy/dx = 3y^{2/3}$ passing through the point $(0, 0)$ is $y = x^3$.

11.5 GROWTH AND DECAY

In this section we look at exponential growth and decay equations. Consider the population of a region. If there is no immigration or emigration, the rate at which the population is changing is often proportional to the population. In other words, the larger the population, the faster it is growing because there are more people to have babies. If the population at time t is P and its continuous growth rate is 2% per unit time, then we know

$$\text{Rate of growth of population} = 2\%(\text{Current population}),$$

and we can write this as

$$\frac{dP}{dt} = 0.02P.$$

The 2% growth rate is called the *relative growth rate* to distinguish it from the *absolute growth rate*, dP/dt. Notice they measure different quantities. Since

$$\text{Relative growth rate} = 2\% = \frac{1}{P}\frac{dP}{dt},$$

the relative growth rate is a percent change per unit time, while

$$\text{Absolute growth rate} = \text{Rate of change of population} = \frac{dP}{dt}$$

is a change in population per unit time.

We showed in Section 11.4 that since the equation $dP/dt = 0.02P$ is of the form $dP/dt = kP$ for $k = 0.02$, it has solution

$$P = P_0 e^{0.02t}.$$

Other processes are described by differential equations similar to that for population growth, but with negative values for k. In summary, we have the following result from the preceding section:

> Every solution to the equation
>
> $$\frac{dP}{dt} = kP$$
>
> can be written in the form
>
> $$P = P_0 e^{kt},$$
>
> where P_0 is the value of P at $t = 0$, and $k > 0$ represents growth, whereas $k < 0$ represents decay.

Recall that the *doubling time* of an exponentially growing quantity is the time required for it to double. The *half-life* of an exponentially decaying quantity is the time required for half of it to decay.

Continuously Compounded Interest

At a bank, continuous compounding means that interest is accrued at a rate that is a fixed percentage of the balance at that moment. Thus, the larger the balance, the faster interest is earned and the faster the balance grows.

Example 1 A bank account earns interest continuously at a rate of 5% of the current balance per year. Assume that the initial deposit is \$1000 and that no other deposits or withdrawals are made.

(a) Write the differential equation satisfied by the balance in the account.
(b) Solve the differential equation and graph the solution.

Solution

(a) We are looking for B, the balance in the account in dollars, as a function of t, time in years. Interest is being added to the account continuously at a rate of 5% of the balance at that moment. Since no deposits or withdrawals are made, at any instant,

$$\text{Rate balance increasing} = \text{Rate interest earned} = 5\%(\text{Current balance}),$$

which we write as

$$\frac{dB}{dt} = 0.05B.$$

This is the differential equation that describes the process. It does not involve the initial condition \$1000 because the initial deposit does not affect the process by which interest is earned.

(b) Solving the differential equation by separation of variables gives

$$B = B_0 e^{0.05t},$$

where B_0 is the value of B at $t = 0$, so $B_0 = 1000$. Thus

$$B = 1000e^{0.05t}$$

and this function is graphed in Figure 11.43.

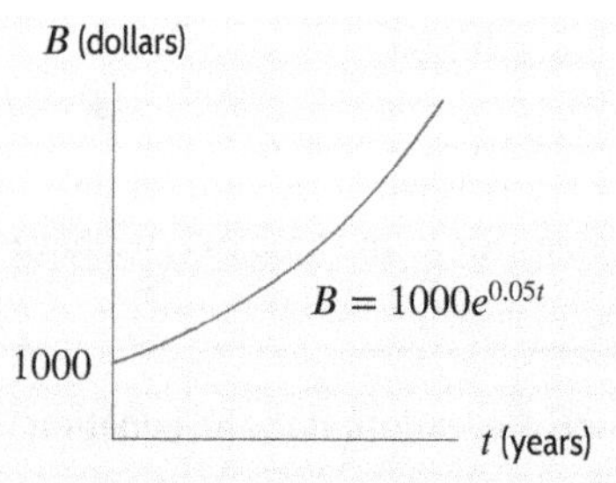

Figure 11.43: Bank balance against time

The Difference Between Continuous and Annual Percentage Growth Rates

If $P = P_0(1+r)^t$ with t in years, we say that r is the *annual* growth rate, while if $P = P_0e^{kt}$, we say that k is the *continuous* growth rate.

The constant k in the differential equation $dP/dt = kP$ is not the annual growth rate, but the continuous growth rate. In Example 1, with a continuous interest rate of 5%, we obtain a balance of $B = B_0e^{0.05t}$, where time, t, is in years. At the end of one year the balance is $B_0e^{0.05}$. In that one year, our balance has changed from B_0 to $B_0e^{0.05}$, that is, by a factor of $e^{0.05} = 1.0513$. Thus the annual growth rate is 5.13%. This is what the bank means when it says "5% compounded continuously for an effective annual yield of 5.13%." Since $P_0e^{0.05t} = P_0(1.0513)^t$, we have two different ways to represent the same function.

Since most growth is measured over discrete time intervals, a continuous growth rate is an idealized concept. A demographer who says a population is growing at the rate of 2% per year usually means that after t years the population is $P = P_0(1.02)^t$. To find the continuous growth rate, k, we express the population as $P = P_0e^{kt}$. At the end of one year $P = P_0e^k$, so $e^k = 1.02$. Thus $k = \ln 1.02 \approx 0.0198$. The continuous growth rate, $k = 1.98\%$, is close to the annual growth rate of 2%, but it is not the same. Again, we have two different representations of the same function since $P_0(1.02)^t = P_0e^{0.0198t}$.

Pollution in the Great Lakes

In the 1960s pollution in the Great Lakes became an issue of public concern. We set up a model for how long it would take for the lakes to flush themselves clean, assuming no further pollutants are being dumped in the lakes.

Suppose Q is the total quantity of pollutant in a lake of volume V at time t. Suppose that clean water is flowing into the lake at a constant rate r and that water flows out at the same rate. Assume that the pollutant is evenly spread throughout the lake and that the clean water coming into the lake immediately mixes with the rest of the water. We investigate how Q varies with time. Since pollutants are being taken out of the lake but not added, Q decreases, and the water leaving the lake becomes less polluted, so the rate at which the pollutants leave decreases. This tells us that Q is decreasing and concave up. In addition, the pollutants are never completely removed from the lake though the quantity remaining becomes arbitrarily small: in other words, Q is asymptotic to the t-axis. (See Figure 11.44.)

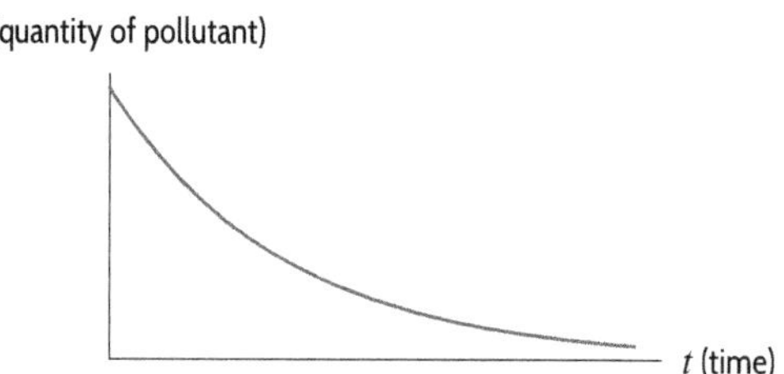

Figure 11.44: Pollutant in lake versus time

Setting Up a Differential Equation for the Pollution

To model exactly how Q changes with time, we write an equation for the rate at which Q changes. We know that

$$\begin{matrix}\text{Rate } Q \\ \text{changes}\end{matrix} = -\begin{pmatrix}\text{Rate pollutants} \\ \text{leave in outflow}\end{pmatrix}$$

where the negative sign represents the fact that Q is decreasing. At time t, the concentration of pollutants is Q/V, and water containing this concentration is leaving at rate r. Thus

$$\begin{matrix}\text{Rate pollutants} \\ \text{leave in outflow}\end{matrix} = \begin{matrix}\text{Rate of} \\ \text{outflow}\end{matrix} \times \text{Concentration} = r \cdot \frac{Q}{V}.$$

So the differential equation is

$$\frac{dQ}{dt} = -\frac{r}{V}Q,$$

and its solution is

$$Q = Q_0 e^{-rt/V}.$$

Table 11.4 contains values of r and V for four of the Great Lakes.[3] We use this data to calculate how long it would take for certain fractions of the pollution to be removed.

Table 11.4 *Volume and outflow in Great Lakes*[4]

	V (thousands of km^3)	r (km^3/year)
Superior	12.2	65.2
Michigan	4.9	158
Erie	0.46	175
Ontario	1.6	209

Example 2 According to this model, how long will it take for 90% of the pollution to be removed from Lake Erie? For 99% to be removed?

Solution Substituting r and V for Lake Erie into the differential equation for Q gives

$$\frac{dQ}{dt} = -\frac{r}{V}Q = \frac{-175}{0.46 \times 10^3}Q = -0.38Q$$

where t is measured in years. Thus Q is given by

$$Q = Q_0 e^{-0.38t}.$$

When 90% of the pollution has been removed, 10% remains, so $Q = 0.1Q_0$. Substituting gives

$$0.1Q_0 = Q_0 e^{-0.38t}.$$

Canceling Q_0 and solving for t, we get

$$t = \frac{-\ln(0.1)}{0.38} \approx 6 \text{ years}.$$

When 99% of the pollution has been removed, $Q = 0.01Q_0$, so t satisfies

$$0.01Q_0 = Q_0 e^{-0.38t}.$$

Solving for t gives

$$t = \frac{-\ln(0.01)}{0.38} \approx 12 \text{ years}.$$

[3] Data from William E. Boyce and Richard C. DiPrima, *Elementary Differential Equations*, 9th Edition (New York: Wiley, 2009), pp. 63–64.

[4] www.epa.gov/greatlakes/

Newton's Law of Heating and Cooling

Newton proposed that the temperature of a hot object decreases at a rate proportional to the difference between its temperature and that of its surroundings. Similarly, a cold object heats up at a rate proportional to the temperature difference between the object and its surroundings.

For example, a hot cup of coffee standing on the kitchen table cools at a rate proportional to the temperature difference between the coffee and the surrounding air. As the coffee cools, the rate at which it cools decreases because the temperature difference between the coffee and the air decreases. In the long run, the rate of cooling tends to zero, and the temperature of the coffee approaches room temperature. See Figure 11.45.

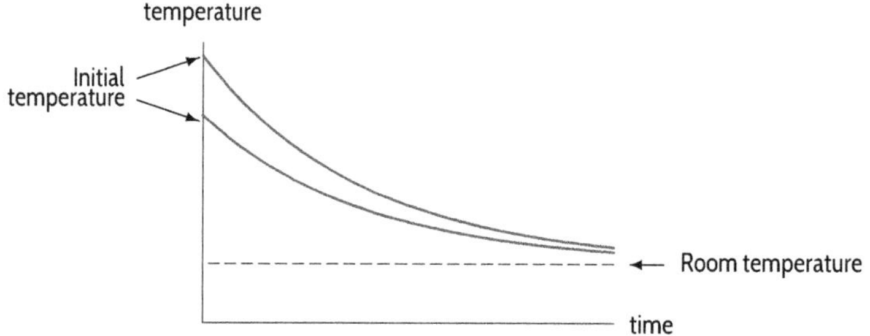

Figure 11.45: Temperature of two cups of coffee with different initial temperatures

Example 3 When a murder is committed, the body, originally at 37°C, cools according to Newton's Law of Cooling. Suppose that after two hours the temperature is 35°C and that the temperature of the surrounding air is a constant 20°C.

(a) Find the temperature, H, of the body as a function of t, the time in hours since the murder was committed.
(b) Sketch a graph of temperature against time.
(c) What happens to the temperature in the long run? Show this on the graph and algebraically.
(d) If the body is found at 4 pm at a temperature of 30°C, when was the murder committed?

Solution (a) We first find a differential equation for the temperature of the body as a function of time. Newton's Law of Cooling says that for some constant α,

$$\text{Rate of change of temperature} = \alpha(\text{Temperature difference}).$$

If H is the temperature of the body, then

$$\text{Temperature difference} = H - 20,$$

so

$$\frac{dH}{dt} = \alpha(H - 20).$$

What about the sign of α? If the temperature difference is positive (that is, $H > 20$), then H is falling, so the rate of change must be negative. Thus α should be negative, so we write:

$$\frac{dH}{dt} = -k(H - 20), \quad \text{for some } k > 0.$$

Separating variables and solving, as in Example 2 on page 581, gives:

$$H - 20 = Be^{-kt}.$$

To find B, substitute the initial condition that $H = 37$ when $t = 0$:

$$37 - 20 = Be^{-k(0)} = B,$$

so $B = 17$. Thus,

$$H - 20 = 17e^{-kt}.$$

To find k, we use the fact that after 2 hours, the temperature is 35°C, so

$$35 - 20 = 17e^{-k(2)}.$$

Dividing by 17 and taking natural logs, we get:

$$\begin{aligned} \ln\left(\frac{15}{17}\right) &= \ln(e^{-2k}) \\ -0.125 &= -2k \\ k &\approx 0.063. \end{aligned}$$

Therefore, the temperature is given by

$$H - 20 = 17e^{-0.063t}$$

or

$$H = 20 + 17e^{-0.063t}.$$

(b) The graph of $H = 20 + 17e^{-0.063t}$ has a vertical intercept of $H = 37$ because the temperature of the body starts at 37°C. The temperature decays exponentially with $H = 20$ as the horizontal asymptote. (See Figure 11.46.)

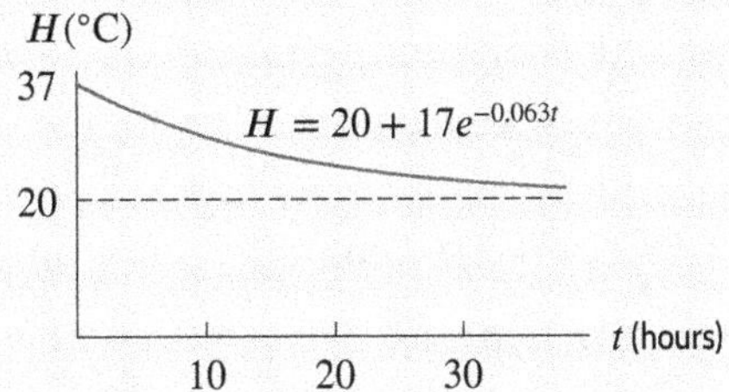

Figure 11.46: Temperature of dead body

(c) "In the long run" means as $t \to \infty$. The graph shows that as $t \to \infty$, $H \to 20$. Algebraically, since $e^{-0.063t} \to 0$ as $t \to \infty$, we have

$$H = 20 + \underbrace{17e^{-0.063t}}_{\text{goes to 0 as } t \to \infty} \longrightarrow 20 \quad \text{as } t \to \infty.$$

(d) We want to know when the temperature reaches 30°C. Substitute $H = 30$ and solve for t:

$$\begin{aligned} 30 &= 20 + 17e^{-0.063t} \\ \frac{10}{17} &= e^{-0.063t}. \end{aligned}$$

Taking natural logs:

$$-0.531 = -0.063t,$$

which gives

$$t \approx 8.4 \text{ hours}.$$

Thus the murder must have been committed about 8.4 hours before 4 pm. Since 8.4 hours = 8 hours 24 minutes, the murder was committed at about 7:30 am.

Equilibrium Solutions

Figure 11.47 shows the temperature of several objects in a 20°C room. One is initially hotter than 20°C and cools down toward 20°C; another is initially cooler and warms up toward 20°C. All these curves are solutions to the differential equation

$$\frac{dH}{dt} = -k(H - 20)$$

for some fixed $k > 0$, and all the solutions have the form

$$H = 20 + Ae^{-kt}$$

for some A. Notice that $H \to 20$ as $t \to \infty$ because $e^{-kt} \to 0$ as $t \to \infty$. In other words, in the long run, the temperature of the object always tends toward 20°C, the temperature of the room. This means that what happens in the long run is independent of the initial condition.

In the special case when $A = 0$, we have the *equilibrium solution*

$$H = 20$$

for all t. This means that if the object starts at 20°C, it remains at 20°C for all time. Notice that such a solution can be found directly from the differential equation by solving $dH/dt = 0$:

$$\frac{dH}{dt} = -k(H - 20) = 0$$

giving

$$H = 20.$$

Regardless of the initial temperature, H always gets closer and closer to 20 as $t \to \infty$. As a result, $H = 20$ is called a *stable* equilibrium[5] for H.

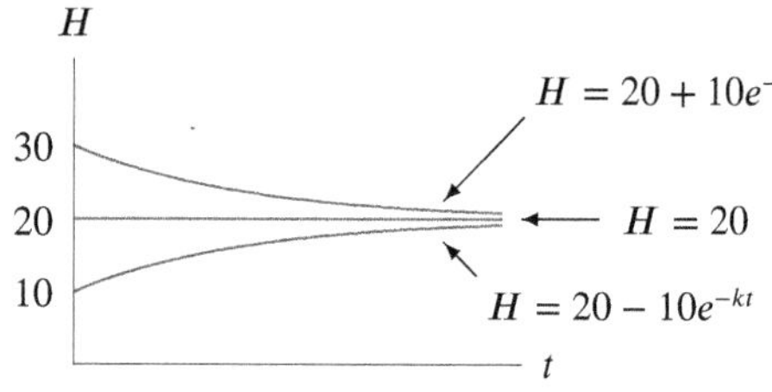

Figure 11.47: $H = 20$ is stable equilibrium ($k > 0$)

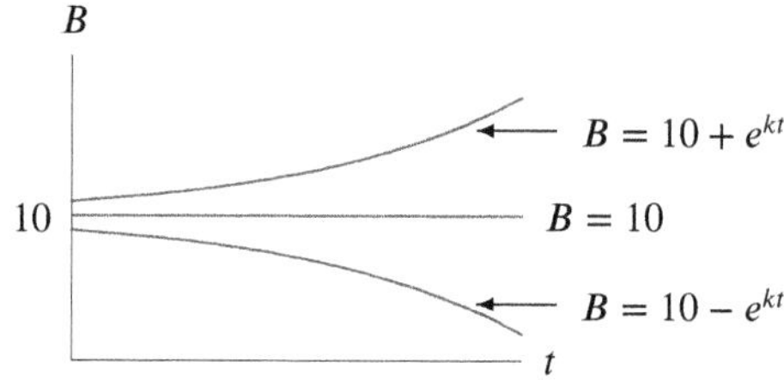

Figure 11.48: $B = 10$ is unstable equilibrium ($k > 0$)

A different situation is displayed in Figure 11.48, which shows solutions to the differential equation

$$\frac{dB}{dt} = k(B - 10)$$

for some fixed $k > 0$. Solving $dB/dt = 0$ gives the equilibrium $B = 10$, which is *unstable* because if B starts near 10, it moves away as $t \to \infty$.

In general, we have the following definitions.

- An **equilibrium solution** is constant for all values of the independent variable. The graph is a horizontal line.
- An equilibrium is **stable** if a small change in the initial conditions gives a solution that tends toward the equilibrium as the independent variable tends to positive infinity.
- An equilibrium is **unstable** if a small change in the initial conditions gives a solution curve that veers away from the equilibrium as the independent variable tends to positive infinity.

[5]In more advanced work, this behavior is described as asymptotic stability.

Solutions that do not veer away from an equilibrium solution are also called stable. If the differential equation is of the form $y' = f(y)$, equilibrium solutions can be found by setting y' to zero.

Example 4 Find the equilibrium solution to the differential equation $dP/dt = -50+4P$ and determine whether the equilibrium is stable or unstable.

Solution In order to find the equilibrium solution, we solve $dP/dt = -50 + 4P = 0$. Thus, we have one equilibrium solution at $P = 50/4 = 12.5$. We can determine whether the equilibrium is stable or unstable by analyzing the behavior of solutions with initial conditions near $P = 12.5$. For example, when $P > 12.5$, the derivative $dP/dt = -50 + 4P > -50 + 4 \cdot 12.5 = 0$, so P is increasing. Thus, solutions above the equilibrium $P = 12.5$ will increase and move away from the equilibrium.

For a solution with initial value $P < 12.5$ we have $dP/dt = -50 + 4P < -50 + 4 \cdot 12.5 = 0$. Thus, solutions that have an initial value less than $P = 12.5$ will decrease and veer away from the equilibrium. Since a small change in the initial condition gives solutions that move away from the equilibrium, $P = 12.5$ is an unstable equilibrium. Notice it is possible to find equilibrium solutions and study their stability without finding a formula for the general solution to the differential equation.

Exercises and Problems for Section 11.5

EXERCISES

1. Match the graphs in Figure 11.49 with the following descriptions.
   - **(a)** The temperature of a glass of ice water left on the kitchen table.
   - **(b)** The amount of money in an interest-bearing bank account into which \$50 is deposited.
   - **(c)** The speed of a constantly decelerating car.
   - **(d)** The temperature of a piece of steel heated in a furnace and left outside to cool.

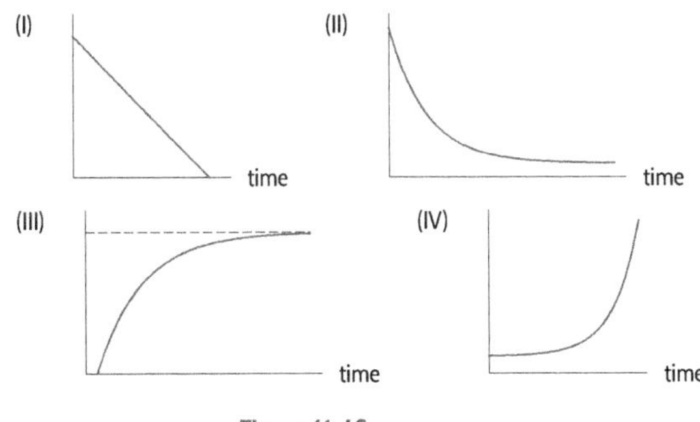

Figure 11.49

2. Each curve in Figure 11.50 represents the balance in a bank account into which a single deposit was made at time zero. Assuming continuously compounded interest, find:
   - **(a)** The curve representing the largest initial deposit.
   - **(b)** The curve representing the largest interest rate.
   - **(c)** Two curves representing the same initial deposit.
   - **(d)** Two curves representing the same interest rate.

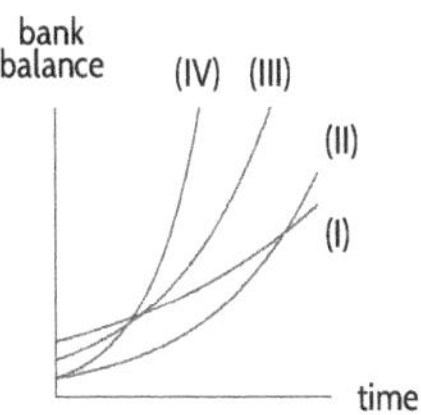

Figure 11.50

3. The slope field for $y' = 0.5(1 + y)(2 - y)$ is given in Figure 11.51.
   - **(a)** List equilibrium solutions and state whether each is stable or unstable.
   - **(b)** Draw solution curves on the slope field through each of the three marked points.

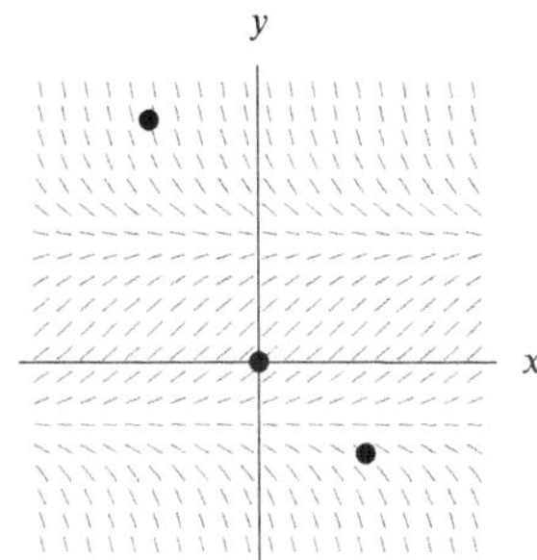

Figure 11.51

4. The slope field for a differential equation is given in Figure 11.52. Estimate all equilibrium solutions for this differential equation, and indicate whether each is stable or unstable.

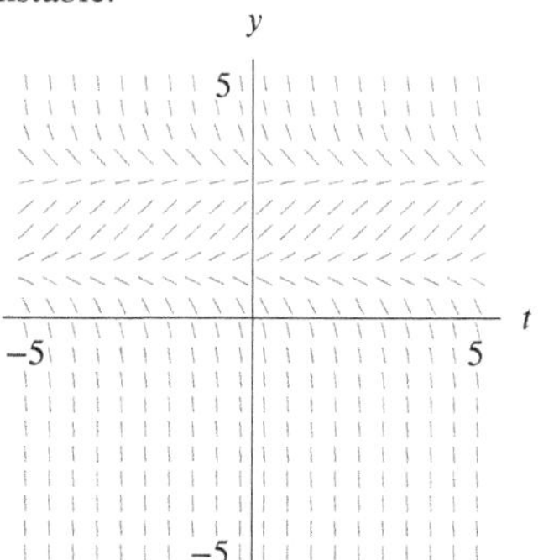

Figure 11.52

For Exercises 5–6, sketch solution curves with a variety of initial values for the differential equations. You do not need to find an equation for the solution.

5. $\dfrac{dy}{dt} = \alpha - y,$
where α is a positive constant.

6. $\dfrac{dw}{dt} = (w-3)(w-7)$

7. A yam is put in a 200°C oven and heats up according to the differential equation

$$\frac{dH}{dt} = -k(H-200), \quad \text{for } k \text{ a positive constant.}$$

(a) If the yam is at 20°C when it is put in the oven, solve the differential equation.
(b) Find k using the fact that after 30 minutes the temperature of the yam is 120°C.

8. (a) Find the equilibrium solution to the differential equation

$$\frac{dy}{dt} = 0.5y - 250.$$

(b) Find the general solution to this differential equation.
(c) Sketch the graphs of several solutions to this differential equation, using different initial values for y.
(d) Is the equilibrium solution stable or unstable?

9. (a) A cup of coffee is made with boiling water and stands in a room where the temperature is 20° C. If $H(t)$ is the temperature of the coffee at time t, in minutes, explain what the differential equation

$$\frac{dH}{dt} = -k(H-20)$$

says in everyday terms. What is the sign of k?
(b) Solve this differential equation. If the coffee cools to 90°C in 2 minutes, how long will it take to cool to 60°C degrees?

In Exercises 10–13, the number of fish, P, in a lake with an initial population of 10,000 satisfies, for constant k and r,

$$\frac{dP}{dt} = kP - r.$$

10. If $k = 0.15$ and $r = 1000$, is the fish population increasing or decreasing at time $t = 0$?

11. If $k = 0.2$ and $r = 3000$, is the fish population increasing or decreasing at time $t = 0$?

12. If $k = 0.05$, what must r be in order for the fish population to remain at a constant level?

13. If $k = 0.10$ and $r = 500$, use a tangent line approximation to estimate P at $t = 0.5$.

For Exercises 14–16, the growth of a tumor with size q is given by

$$q' = kq\ln\left(\frac{L}{q}\right), \quad k, L > 0.$$

14. Will the tumor's size increase or decrease if $q = 30$ and $L = 50$.

15. Will the tumor's size increase or decrease if $q = 25$ and $L = 20$.

16. Let $L = 100$ and $k = 0.1$. If $q = 20$ at time $t = 2$, use the tangent-line approximation to estimate q at $t = 2.1$.

17. In Example 2 on page 589, we saw that it would take about 6 years for 90% of the pollution in Lake Erie to be removed and about 12 years for 99% to be removed. Explain why one time is double the other.

18. Using the model in the text and the data in Table 11.4 on page 589, find how long it would take for 90% of the pollution to be removed from Lake Michigan and from Lake Ontario, assuming no new pollutants are added. Explain how you can tell which lake will take longer to be purified just by looking at the data in the table.

19. Use the model in the text and the data in Table 11.4 on page 589 to determine which of the Great Lakes would require the longest time and which would require the shortest time for 80% of the pollution to be removed, assuming no new pollutants are being added. Find the ratio of these two times.

For Exercises 20–22, is the function a solution to the differential equation $dy/dx = g(y)$ for $g(y)$ in Figure 11.53?

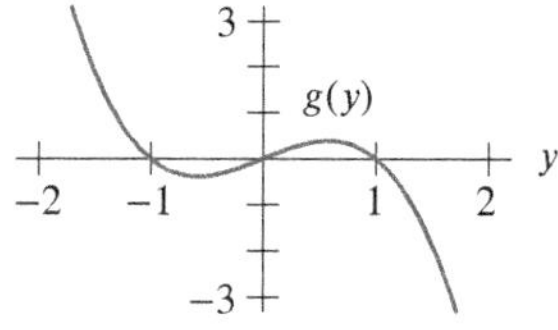

Figure 11.53

20. The constant function $y = 0$.

21. The constant function $y = 1$.

22. The constant function $y = 1/2$.

23. (a) Find all equilibrium solutions for the differential equation
$$\frac{dy}{dx} = 0.5y(y-4)(2+y).$$
(b) Draw a slope field and use it to determine whether each equilibrium solution is stable or unstable.

PROBLEMS

In Problems 24–27,
(a) Define the variables.
(b) Write a differential equation to describe the relationship.
(c) Solve the differential equation.

24. In 2014, the population of India[6] was 1.236 billion people and increasing at a rate proportional to its population. If the population is measured in billions of people and time is measured in years, the constant of proportionality is 0.0125.

25. Nicotine leaves the body at a rate proportional to the amount present, with constant of proportionality 0.347 if the amount of nicotine is in mg and time is in hours. The amount of nicotine in the body immediately after smoking a cigarette is 0.4 mg.

26. By 2014, the cumulative world capacity of solar photovoltaic (PV) reached 178,391 megawatts and was growing exponentially at a continuous rate of 30% per year.[7]

27. In 2007, Grinnell Glacier in Glacier National Park covered 142 acres and was estimated to be shrinking exponentially at a continuous rate of 4.3% per year.[8]

28. Between 2002 and 2013, textbook prices[9] increased at 5.6% per year while inflation was 2.3% per year. Assume both rates are continuous growth rates and let time, t, be in years since the start of 2002.

(a) Write a differential equation satisfied by $B(t)$, the price of a textbook at time t.
(b) Write a differential equation satisfied by $P(t)$, the price at time t of an item growing at the inflation rate.
(c) Solve both differential equations.
(d) What is the doubling time of the price of a textbook?
(e) What is the doubling time of the price of an item growing according to the inflation rate?
(f) How is the ratio of the doubling times related to the ratio of the growth rates? Justify your answer.

29. A bottle of milk is taken out of a 3°C refrigerator and left in a 22°C room.

(a) Write a differential equation whose solution is the temperature of milk as a function of time. (The equation will include a positive constant k.)
(b) Solve this equation.

30. A raw chicken is taken from a 3°C refrigerator and put in a preheated 190°C oven. One hour later, a meat thermometer shows that the internal temperature of the chicken has risen to 40°C. If the safe internal temperature for chicken is 74°C, what is the total cooking time?

31. With time t in years, the population P of fish in a lake (in millions) grows at a continuous rate of r% per year and is subject to continuous harvesting at a constant rate of H million fish per year.

(a) Write a differential equation satisfied by P.
(b) What is the equilibrium level of P if $r = 5$ and $H = 15$?
(c) Without solving the differential equation, explain why the equilibrium you found in part (b) is unstable.
(d) Find a formula for the equilibrium for general positive r and H.
(e) Are there any positive values of r and H that make the equilibrium stable?

32. The population of fish in a lake declines at a continuous rate of 10% per year. To curb this decline, wildlife management services continuously restock the lake at a rate of 5 thousand fish per year. Let P be the fish population (in thousands) as a function of time t in years.

(a) Write a differential equation satisfied by P.
(b) Without solving the differential equation, find the equilibrium fish population.
(c) Is the equilibrium stable or unstable?

33. A radioactive substance decays at a rate proportional to the quantity, Q, present at the time, t. The constant of proportionality is k.

(a) Write a differential equation satisfied by Q.
(b) Find the half-life as a function of k.
(c) Is the half-life an increasing or decreasing function of k?

34. Warfarin is a drug used as an anticoagulant. After administration of the drug is stopped, the quantity remaining in a patient's body decreases at a rate proportional to the quantity remaining. The half-life of warfarin in the body is 37 hours.

(a) Sketch the quantity, Q, of warfarin in a patient's body as a function of the time, t, since stopping administration of the drug. Mark the 37 hours on your graph.

[6] www.indexmundi.com, accessed April 27, 2015.
[7] en.wikipedia.org, accessed 1 April, 2016.
[8] "Warming climate shrinking Glacier Park's glaciers", www.usatoday.com, October 15, 2007.
[9] www.cnbc.com, accessed April 27, 2015.

(b) Write a differential equation satisfied by Q.
(c) How many days does it take for the drug level in the body to be reduced to 25% of the original level?

35. The rate at which a drug leaves the bloodstream and passes into the urine is proportional to the quantity of the drug in the blood at that time. If an initial dose of Q_0 is injected directly into the blood, 20% is left in the blood after 3 hours.

(a) Write and solve a differential equation for the quantity, Q, of the drug in the blood after t hours.
(b) How much of this drug is in a patient's body after 6 hours if the patient is given 100 mg initially?

36. Oil is pumped continuously from a well at a rate proportional to the amount of oil left in the well. Initially there were 1 million barrels of oil in the well; six years later 500,000 barrels remain.

(a) At what rate was the amount of oil in the well decreasing when there were 600,000 barrels remaining?
(b) When will there be 50,000 barrels remaining?

37. The radioactive isotope carbon-14 is present in small quantities in all life forms, and it is constantly replenished until the organism dies, after which it decays to stable carbon-12 at a rate proportional to the amount of carbon-14 present, with a half-life of 5730 years. Let $C(t)$ be the amount of carbon-14 present at time t.

(a) Find the value of the constant k in the differential equation $C' = -kC$.
(b) In 1988 three teams of scientists found that the Shroud of Turin, which was reputed to be the burial cloth of Jesus, contained 91% of the amount of carbon-14 contained in freshly made cloth of the same material.[10] How old was the Shroud of Turin at the time of this data?

38. The amount of radioactive carbon-14 in a sample is measured using a Geiger counter, which records each disintegration of an atom. Living tissue disintegrates at a rate of about 13.5 atoms per minute per gram of carbon. In 1977 a charcoal fragment found at Stonehenge, England, recorded 8.2 disintegrations per minute per gram of carbon. Assuming that the half-life of carbon-14 is 5730 years and that the charcoal was formed during the building of the site, estimate the date when Stonehenge was built.

39. A detective finds a murder victim at 9 am. The temperature of the body is measured at 90.3°F. One hour later, the temperature of the body is 89.0°F. The temperature of the room has been maintained at a constant 68°F.

(a) Assuming the temperature, T, of the body obeys Newton's Law of Cooling, write a differential equation for T.
(b) Solve the differential equation to estimate the time the murder occurred.

[10] *The New York Times*, October 18, 1988.

40. At 1:00 pm one winter afternoon, there is a power failure at your house in Wisconsin, and your heat does not work without electricity. When the power goes out, it is 68°F in your house. At 10:00 pm, it is 57°F in the house, and you notice that it is 10°F outside.

(a) Assuming that the temperature, T, in your home obeys Newton's Law of Cooling, write the differential equation satisfied by T.
(b) Solve the differential equation to estimate the temperature in the house when you get up at 7:00 am the next morning. Should you worry about your water pipes freezing?
(c) What assumption did you make in part (a) about the temperature outside? Given this (probably incorrect) assumption, would you revise your estimate up or down? Why?

41. Before Galileo discovered that the speed of a falling body with no air resistance is proportional to the time since it was dropped, he mistakenly conjectured that the speed was proportional to the distance it had fallen.

(a) Assume the mistaken conjecture to be true and write an equation relating the distance fallen, $D(t)$, at time t, and its derivative.
(b) Using your answer to part (a) and the correct initial conditions, show that D would have to be equal to 0 for all t, and therefore the conjecture must be wrong.

42. (a) An object is placed in a 68°F room. Write a differential equation for H, the temperature of the object at time t.
(b) Find the equilibrium solution to the differential equation. Determine from the differential equation whether the equilibrium is stable or unstable.
(c) Give the general solution for the differential equation.
(d) The temperature of the object is 40°F initially and 48°F one hour later. Find the temperature of the object after 3 hours.

43. Hydrocodone bitartrate is used as a cough suppressant. After the drug is fully absorbed, the quantity of drug in the body decreases at a rate proportional to the amount left in the body. The half-life of hydrocodone bitartrate in the body is 3.8 hours, and the usual oral dose is 10 mg.

(a) Write a differential equation for the quantity, Q, of hydrocodone bitartrate in the body at time t, in hours since the drug was fully absorbed.
(b) Find the equilibrium solution of the differential equation. Based on the context, do you expect the equilibrium to be stable or unstable?
(c) Solve the differential equation given in part (a).
(d) Use the half-life to find the constant of proportionality, k.
(e) How much of the 10 mg dose is still in the body after 12 hours?

44. (a) Let B be the balance at time t of a bank account that earns interest at a rate of $r\%$, compounded continuously. What is the differential equation describing the rate at which the balance changes? What is the constant of proportionality, in terms of r?
(b) Find the equilibrium solution to the differential equation. Determine whether the equilibrium is stable or unstable and explain what this means about the bank account.
(c) What is the solution to this differential equation?
(d) Sketch the graph of B as function of t for an account that starts with \$1000 and earns interest at the following rates:

(i) 4% (ii) 10% (iii) 15%

Strengthen Your Understanding

■ In Problems 45–47, explain what is wrong with the statement.

45. The line $y = 2$ is an equilibrium solution to the differential equation $dy/dx = y^3 - 4xy$.
46. The function $y = x^2$ is an equilibrium solution to the differential equation $dy/dx = y - x^2$.
47. At time $t = 0$, a roast is taken out of a 40°F refrigerator and put in a 350°F oven. If H represents the temperature of the roast at time t minutes after it is put in the oven, we have $dH/dt = k(H - 40)$.

■ In Problems 48–50, give an example of:

48. A differential equation for a quantity that is decaying exponentially at a continuous rate per unit time.
49. A differential equation with an equilibrium solution of $Q = 500$.
50. A graph of three possible solutions, with initial P-values of 20, 25, and 30, respectively, to a differential equation that has an unstable equilibrium solution at $P = 25$.

11.6 APPLICATIONS AND MODELING

Much of this book involves functions that represent real processes, such as how the temperature of a yam or the population of the US is changing with time. You may wonder where such functions come from. In some cases, we fit functions to experimental data by trial and error. In other cases, we take a more theoretical approach, leading to a differential equation whose solution is the function we want. In this section we give examples of the more theoretical approach.

How a Layer of Ice Forms

When ice forms on a lake, the water on the surface freezes first. As heat from the water travels up through the ice and is lost to the air, more ice is formed. The question we will consider is: How thick is the layer of ice as a function of time? Since the thickness of the ice increases with time, the thickness function is increasing. In addition, as the ice gets thicker, it insulates better. Therefore, we expect the layer of ice to form more slowly as time goes on. Hence, the thickness function is increasing at a decreasing rate, so its graph is concave down.

A Differential Equation for the Thickness of the Ice

To get more detailed information about the thickness function, we have to make some assumptions. Suppose y represents the thickness of the ice as a function of time, t. Since the thicker the ice, the longer it takes the heat to get through it, we assume that the rate at which ice is formed is inversely proportional to the thickness. In other words, we assume that for some constant k,

$$\text{Rate thickness is increasing} = \frac{k}{\text{Thickness}},$$

so

$$\frac{dy}{dt} = \frac{k}{y} \quad \text{where} \quad k > 0 \text{ and } y > 0.$$

This differential equation enables us to find a formula for y. Using separation of variables:

$$\int y\,dy = \int k\,dt$$
$$\frac{y^2}{2} = kt + C.$$

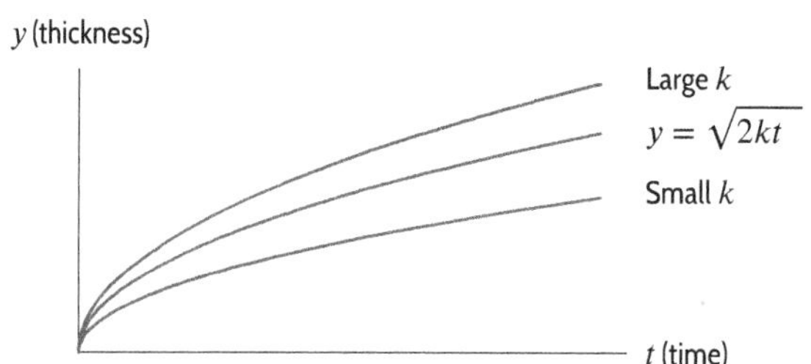

Figure 11.54: Thickness of ice as a function of time

Extending the solution to $y = 0$ and measuring time from then gives $C = 0$. Since y must be non-negative, we have

$$y = \sqrt{2kt}.$$

Graphs of y against t are in Figure 11.54. Notice that the larger y is, the more slowly y increases. In addition, this model suggests that y increases indefinitely as time passes. (Of course, the value of y cannot increase beyond the depth of the lake.)

The Net Worth of a Company

In the preceding section, we saw an example in which money in a bank account was earning interest (Example 1, page 587). Consider a company whose revenues are proportional to its net worth (like interest on a bank account) but that must also make payroll payments. The question is: under what circumstances does the company make money, and under what circumstances does it go bankrupt?

Common sense says that if the payroll exceeds the rate at which revenue is earned, the company will eventually be in trouble, whereas if revenue exceeds payroll, the company should do well. We assume that revenue is earned continuously and that payments are made continuously. (For a large company, this is a good approximation.) We also assume that the only factors affecting net worth are revenue and payroll.

Example 1 A company's revenue is earned at a continuous annual rate of 5% of its net worth. At the same time, the company's payroll obligations amount to \$200 million a year, paid out continuously.

(a) Write a differential equation that governs the net worth of the company, W million dollars.
(b) Solve the differential equation, assuming an initial net worth of W_0 million dollars.
(c) Sketch the solution for $W_0 = 3000$, 4000, and 5000.

Solution First, let's see what we can learn without writing a differential equation. For example, we can ask if there is any initial net worth W_0 that will exactly keep the net worth constant. If there's such an equilibrium, the rate at which revenue is earned must exactly balance the payments made, so

$$\text{Rate revenue is earned} = \text{Rate payments are made.}$$

If net worth is a constant W_0, revenue is earned at a constant rate of $0.05W_0$ per year, so we have

$$0.05W_0 = 200 \quad \text{giving} \quad W_0 = 4000.$$

Therefore, if the net worth starts at \$4000 million, the revenue and payments are equal, and the net worth remains constant. Therefore, \$4000 million is an equilibrium solution.

Suppose, however, the initial net worth is above \$4000 million. Then, the revenue earned is more than the payroll expenses, and the net worth of the company increases, thereby increasing the revenue still further. Thus the net worth increases more and more quickly. On the other hand, if the initial net worth is below \$4000 million, the revenue is not enough to meet the payments, and the net worth of the company declines. This decreases the revenue, making the net worth decrease still more quickly. The net worth will eventually go to zero, and the company goes bankrupt. See Figure 11.55.

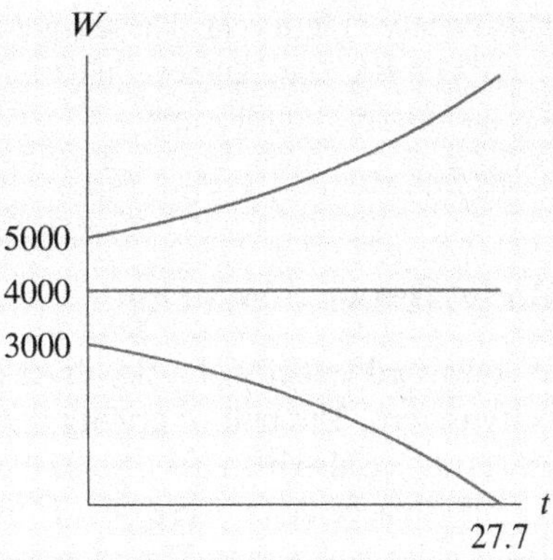

Figure 11.55: Net worth as a function of time: Solutions to $dW/dt = 0.05W - 200$

(a) Now we set up a differential equation for the net worth, using the fact that

$$\begin{array}{c}\text{Rate net worth}\\ \text{is increasing}\end{array} = \begin{array}{c}\text{Rate revenue}\\ \text{is earned}\end{array} - \begin{array}{c}\text{Rate payroll payments}\\ \text{are made}\end{array}.$$

In millions of dollars per year, revenue is earned at a rate of $0.05W$, and payments are made at a rate of 200 per year, so for t in years,

$$\frac{dW}{dt} = 0.05W - 200.$$

The equilibrium solution, $W = 4000$, is obtained by setting $dW/dt = 0$.

(b) We solve this equation by separation of variables. It is helpful to factor out 0.05 before separating, so that the W moves over to the left-hand side without a coefficient:

$$\frac{dW}{dt} = 0.05(W - 4000).$$

Separating and integrating gives

$$\int \frac{dW}{W - 4000} = \int 0.05\, dt,$$

so

$$\ln |W - 4000| = 0.05t + C,$$

or

$$|W - 4000| = e^{0.05t+C} = e^{C}e^{0.05t}.$$

This means

$$W - 4000 = Ae^{0.05t} \qquad \text{where } A = \pm e^{C}.$$

To find A, we use the initial condition that $W = W_0$ when $t = 0$:

$$W_0 - 4000 = Ae^{0} = A.$$

Substituting this value for A into $W = 4000 + Ae^{0.05t}$ gives

$$W = 4000 + (W_0 - 4000)e^{0.05t}.$$

(c) If $W_0 = 4000$, then $W = 4000$, the equilibrium solution.
If $W_0 = 5000$, then $W = 4000 + 1000e^{0.05t}$.
If $W_0 = 3000$, then $W = 4000 - 1000e^{0.05t}$. If $W = 0$, then $t \approx 27.7$, so the company goes bankrupt in its twenty-eighth year. These solutions are shown in Figure 11.55. Notice that if the net worth starts with W_0 near, but not equal to, \$4000 million, then W moves farther away. Thus, $W = 4000$ is an unstable equilibrium.

The Velocity of a Falling Body: Terminal Velocity

Think about the downward velocity of a sky-diver jumping out of a plane. When the sky-diver first jumps, his velocity is zero. The pull of gravity then makes his velocity increase. As the sky-diver speeds up, the air resistance also increases. Since the air resistance partly balances the pull of gravity, the force causing him to accelerate decreases. Thus, the velocity is an increasing function of time, but it is increasing at a decreasing rate. The air resistance increases until it balances gravity, when the sky-diver's velocity levels off. Thus, we expect the the graph of velocity against time to be increasing and concave down with a horizontal asymptote.

A Differential Equation: Air Resistance Proportional to Velocity

In order to compute the velocity function, we need to know exactly how air resistance depends on velocity. To decide whether air resistance is, say, proportional to the velocity, or is some other function of velocity, requires either lab experiments or a theoretical idea of how the air resistance is created. We consider a very small object, such as a dust particle settling on a computer component during manufacturing,[11] and assume that air resistance is proportional to velocity. Thus, the net force on the object is $F = mg - kv$, where mg is the gravitational force, which acts downward, and kv is the air resistance, which acts upward, so $k > 0$. (See Figure 11.56.) Then, by Newton's Second Law of Motion,

$$\text{Force} = \text{Mass} \cdot \text{Acceleration},$$

we have

$$mg - kv = m\frac{dv}{dt}.$$

This differential equation can be solved by separation of variables. It is easier if we factor out $-k/m$ before separating, giving

$$\frac{dv}{dt} = -\frac{k}{m}\left(v - \frac{mg}{k}\right).$$

Separating and integrating gives

$$\int \frac{dv}{v - mg/k} = -\frac{k}{m}\int dt$$

$$\ln\left|v - \frac{mg}{k}\right| = -\frac{k}{m}t + C.$$

Solving for v, we have

$$\left|v - \frac{mg}{k}\right| = e^{-kt/m+C} = e^C e^{-kt/m}$$

$$v - \frac{mg}{k} = Ae^{-kt/m},$$

where A is an arbitrary nonzero constant. We find A from the initial condition that the object starts from rest, so $v = 0$ when $t = 0$. Substituting

$$0 - \frac{mg}{k} = Ae^0$$

[11]Example suggested by Howard Stone.

gives

$$A = -\frac{mg}{k}.$$

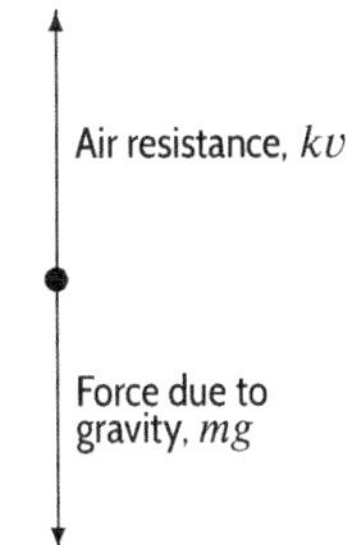

Figure 11.56: Forces acting on a falling object

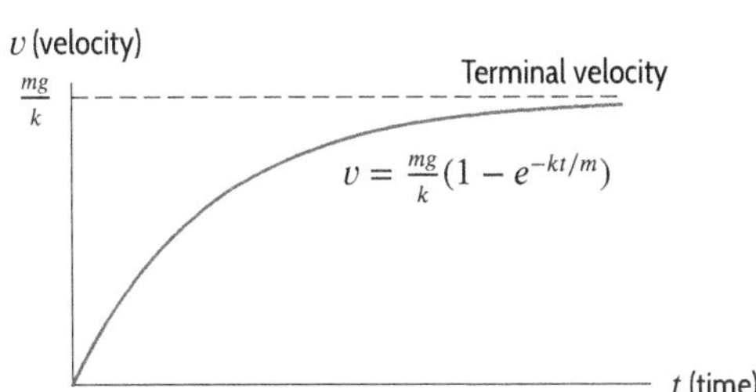

Figure 11.57: Velocity of falling dust particle assuming that air resistance is kv

Thus

$$v = \frac{mg}{k} - \frac{mg}{k}e^{-kt/m} = \frac{mg}{k}(1 - e^{-kt/m}).$$

The graph of this function is in Figure 11.57. The horizontal asymptote represents the *terminal velocity*, mg/k.

Notice that the terminal velocity can also be obtained from the differential equation by setting $dv/dt = 0$ and solving for v:

$$m\frac{dv}{dt} = mg - kv = 0, \quad \text{so} \quad v = \frac{mg}{k}.$$

Compartmental Analysis: A Reservoir

Many processes can be modeled as a container with various solutions flowing in and out—for example, drugs given intravenously or the discharge of pollutants into a lake. We consider a city's water reservoir, fed partly by clean water from a spring and partly by run-off from the surrounding land. In New England and many other areas with much snow in the winter, the run-off contains salt that has been put on the roads to make them safe for driving. We consider the concentration of salt in the reservoir. If there is no salt in the reservoir initially, the concentration builds up until the rate at which the salt is entering into the reservoir balances the rate at which salt flows out. If, on the other hand, the reservoir starts with a great deal of salt in it, then initially, the rate at which the salt is entering is less than the rate at which it is flowing out, and the quantity of salt in the reservoir decreases. In either case, the salt concentration levels off at an equilibrium value.

A Differential Equation for Salt Concentration

A water reservoir holds 100 million gallons of water and supplies a city with 1 million gallons a day. The reservoir is partly refilled by a spring that provides 0.9 million gallons a day, and the rest of the water, 0.1 million gallons a day, comes from run-off from the surrounding land. The spring is clean, but the run-off contains salt with a concentration of 0.0001 pound per gallon. There was no salt in the reservoir initially, and the water is well mixed (that is, the outflow contains the concentration of salt in the tank at that instant). We find the concentration of salt in the reservoir as a function of time.

It is important to distinguish between the total quantity, Q, of salt in pounds and the concentration, C, of salt in pounds/gallon where

$$\text{Concentration} = C = \frac{\text{Quantity of salt}}{\text{Volume of water}} = \frac{Q}{100\text{ million}}\left(\frac{\text{lb}}{\text{gal}}\right).$$

(The volume of the reservoir is 100 million gallons.) We will find Q first, and then C. We know that

$$\begin{array}{c}\text{Rate of change of}\\ \text{quantity of salt}\end{array} = \text{Rate salt entering} - \text{Rate salt leaving}.$$

Salt is entering through the run-off of 0.1 million gallons per day, with each gallon containing 0.0001 pound of salt. Therefore,

$$\begin{aligned}\text{Rate salt entering} &= \text{Concentration} \cdot \text{Volume per day}\\ &= 0.0001\left(\frac{\text{lb}}{\text{gal}}\right)\cdot 0.1\left(\frac{\text{million gal}}{\text{day}}\right)\\ &= 0.00001\left(\frac{\text{million lb}}{\text{day}}\right) = 10\ \text{lb/day}.\end{aligned}$$

Salt is leaving in the million gallons of water used by the city each day . Thus

$$\begin{aligned}\text{Rate salt leaving} &= \text{Concentration} \cdot \text{Volume per day}\\ &= \frac{Q}{100\text{ million}}\left(\frac{\text{lb}}{\text{gal}}\right)\cdot 1\left(\frac{\text{million gal}}{\text{day}}\right) = \frac{Q}{100}\ \text{lb/day}.\end{aligned}$$

Therefore, Q satisfies the differential equation

$$\frac{dQ}{dt} = 10 - \frac{Q}{100}.$$

We factor out $-1/100 = -0.01$ and separate variables, giving

$$\begin{aligned}\frac{dQ}{dt} &= -0.01(Q - 1000)\\ \int \frac{dQ}{Q - 1000} &= -\int 0.01\,dt\\ \ln|Q - 1000| &= -0.01t + k\\ Q - 1000 &= Ae^{-0.01t}.\end{aligned}$$

There is no salt initially, so we substitute $Q = 0$ when $t = 0$:

$$0 - 1000 = Ae^{0} \qquad \text{giving} \qquad A = -1000.$$

Thus

$$Q - 1000 = -1000e^{-0.01t},$$

so

$$Q = 1000(1 - e^{-0.01t}) \quad \text{pounds.}$$

Therefore

$$\text{Concentration} = C = \frac{Q}{100\text{ million}} = \frac{1000}{10^8}(1 - e^{-0.01t}) = 10^{-5}(1 - e^{-0.01t})\ \text{lb/gal}.$$

A sketch of concentration against time is in Figure 11.58.

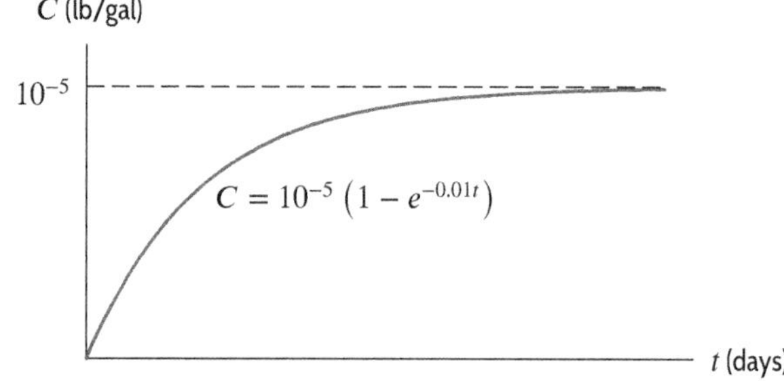

Figure 11.58: Concentration of salt in reservoir

Exercises and Problems for Section 11.6 Online Resource: Additional Problems for Section 11.6

EXERCISES

1. Match the graphs in Figure 11.59 with the following descriptions.
   (a) The population of a new species introduced onto a tropical island.
   (b) The temperature of a metal ingot placed in a furnace and then removed.
   (c) The speed of a car traveling at uniform speed and then braking uniformly.
   (d) The mass of carbon-14 in a historical specimen.
   (e) The concentration of tree pollen in the air over the course of a year.

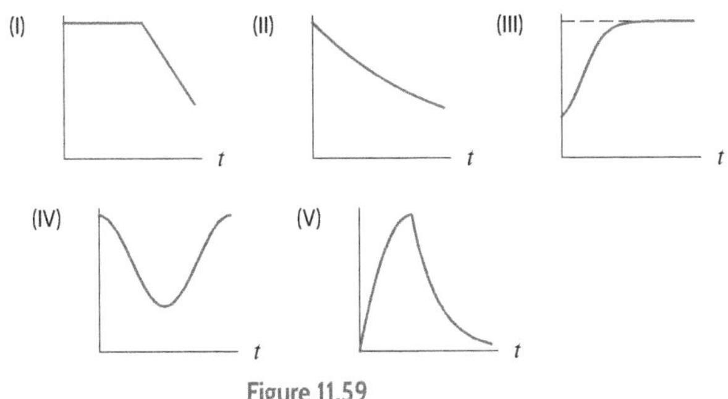

Figure 11.59

In Exercises 2–5, write a differential equation for the balance B in an investment fund with time, t, measured in years.

2. The balance is earning interest at a continuous rate of 5% per year, and payments are being made out of the fund at a continuous rate of \$12,000 per year.

3. The balance is earning interest at a continuous rate of 3.7% per year, and money is being added to the fund at a continuous rate of \$5000 per year.

4. The balance is losing value at a continuous rate of 8% per year, and money is being added to the fund at a continuous rate of \$2000 per year.

5. The balance is losing value at a continuous rate of 6.5% per year, and payments are being made out of the fund at a continuous rate of \$50,000 per year.

6. A bank account that earns 10% interest compounded continuously has an initial balance of zero. Money is deposited into the account at a constant rate of \$1000 per year.
   (a) Write a differential equation that describes the rate of change of the balance $B = f(t)$.
   (b) Solve the differential equation to find the balance as a function of time.

7. At time $t = 0$, a bottle of juice at 90°F is stood in a mountain stream whose temperature is 50°F. After 5 minutes, its temperature is 80°F. Let $H(t)$ denote the temperature of the juice at time t, in minutes.
   (a) Write a differential equation for $H(t)$ using Newton's Law of Cooling.
   (b) Solve the differential equation.
   (c) When will the temperature of the juice have dropped to 60°F?

8. With t in years since the start of 2014, copper has been mined worldwide at a rate of $17.9e^{0.025t}$ million tons per year. At the start of 2014, the world's known copper reserves were 690 million tons.[12]
   (a) Write a differential equation for C, the total quantity of copper mined in t years since the start of 2014.
   (b) Solve the differential equation.
   (c) According to this model, when will the known copper reserves be exhausted?

9. The velocity, v, of a dust particle of mass m and acceleration a satisfies the equation

$$ma = m\frac{dv}{dt} = mg - kv, \quad \text{where } g, k \text{ are constant.}$$

By differentiating this equation, find a differential equation satisfied by a. (Your answer may contain m, g, k, but not v.) Solve for a, given that $a(0) = g$.

PROBLEMS

10. A deposit is made to a bank account paying 2% interest per year compounded continuously. Payments totaling \$2000 per year are made continuously from this account.
    (a) Write a differential equation for the balance, B, in the account after t years.
    (b) Find the equilibrium solution of the differential equation. Is the equilibrium stable or unstable? Explain what happens to an account that begins with slightly more money or slightly less money than the equilibrium value.
    (c) Write the solution to the differential equation.
    (d) How much is in the account after 5 years if the initial deposit is (i) \$80,000? (ii) \$120,000?

11. Dead leaves accumulate on the ground in a forest at a rate of 3 grams per square centimeter per year. At the same time, these leaves decompose at a continuous rate of 75% per year. Write a differential equation for the total quantity of dead leaves (per square centimeter) at time t. Sketch a solution showing that the quantity of

[12]Data from http://minerals.usgs.gov/minerals/pubs/commodity/ Accessed February 8, 2015.

dead leaves tends toward an equilibrium level. What is that equilibrium level?

12. A stream flowing into a lake brings with it a pollutant at a rate of 8 metric tons per year. The river leaving the lake removes the pollutant at a rate proportional to the quantity in the lake, with constant of proportionality -0.16 if time is measured in years.

 (a) Is the quantity of pollutant in the lake increasing or decreasing at a moment at which the quantity is 45 metric tons? At which the quantity is 55 metric tons?
 (b) What is the quantity of pollutant in the lake after a long time?

13. Caffeine is metabolized and excreted at a continuous rate of about 17% per hour. A person with no caffeine in the body starts drinking coffee, containing 130 mg of caffeine per cup, at 7 am. The person drinks coffee continuously all day at the rate of one cup an hour. Write a differential equation for A, the amount of caffeine in the body t hours after 7 am and give the particular solution to this differential equation. How much caffeine is in the person's body at 5 pm?

14. The rate (per foot) at which light is absorbed as it passes through water is proportional to the intensity, or brightness, at that point.

 (a) Find the intensity as a function of the distance the light has traveled through the water.
 (b) If 50% of the light is absorbed in 10 feet, how much is absorbed in 20 feet? 25 feet?

15. In 2012, the world population was 7.052 billion. The birth rate had stabilized to 135 million per year and is projected to remain constant. The death rate is projected to increase from 56 million per year in 2012 to 80 million per year in 2040 and to continue increasing at the same rate.[13]

 (a) Assuming the death rate increases linearly, write a differential equation for $P(t)$, the world population in billions t years from 2012.
 (b) Solve the differential equation.
 (c) Find the population predicted for 2050.

16. A bank account earns 2% annual interest, compounded continuously. Money is deposited in a continuous cash flow at a rate of \$1200 per year into the account.

 (a) Write a differential equation that describes the rate at which the balance $B = f(t)$ is changing.
 (b) Solve the differential equation given an initial balance $B_0 = 0$.
 (c) Find the balance after 5 years.

17. The rate at which barometric pressure decreases with altitude is proportional to the barometric pressure at that altitude. If the barometric pressure is measured in inches of mercury, and the altitude in feet, then the constant of proportionality is $3.7 \cdot 10^{-5}$. The barometric pressure at sea level is 29.92 inches of mercury.

 (a) Calculate the barometric pressure at the top of Mount Whitney, 14,500 feet (the highest mountain in the US outside Alaska), and at the top of Mount Everest, 29,000 feet (the highest mountain in the world).
 (b) People cannot easily survive at a pressure below 15 inches of mercury. What is the highest altitude to which people can safely go?

18. According to a simple physiological model, an athletic adult male needs 20 Calories per day per pound of body weight to maintain his weight. If he consumes more or fewer Calories than those required to maintain his weight, his weight changes at a rate proportional to the difference between the number of Calories consumed and the number needed to maintain his current weight; the constant of proportionality is $1/3500$ pounds per Calorie. Suppose that a particular person has a constant caloric intake of I Calories per day. Let $W(t)$ be the person's weight in pounds at time t (measured in days).

 (a) What differential equation has solution $W(t)$?
 (b) Find the equilibrium solution of the differential equation. Based on the context, do you expect the equilibrium to be stable or unstable?
 (c) Solve this differential equation.
 (d) Graph $W(t)$ if the person starts out weighing 160 pounds and consumes 3000 Calories a day.

19. Morphine is often used as a pain-relieving drug. The half-life of morphine in the body is 2 hours. Suppose morphine is administered to a patient intravenously at a rate of 2.5 mg per hour, and the rate at which the morphine is eliminated is proportional to the amount present.

 (a) Use the half-life to show that, to three decimal places, the constant of proportionality for the rate at which morphine leaves the body (in mg/hour) is $k = -0.347$.
 (b) Write a differential equation for the quantity, Q, of morphine in the blood after t hours.
 (c) Use the differential equation to find the equilibrium solution. (This is the long-term amount of morphine in the body, once the system has stabilized.)

20. Water leaks out of the bottom of a barrel at a rate proportional to the square root of the depth of the water at that time. If the water level starts at 36 inches and drops to 35 inches in 1 hour, how long will it take for all of the water to leak out of the barrel?

21. When a gas expands without gain or loss of heat, the rate of change of pressure with respect to volume is proportional to pressure divided by volume. Find a law connecting pressure and volume in this case.

[13] en.wikipedia.org, accessed 1 April, 2016.

22. A spherical snowball melts at a rate proportional to its surface area.

    (a) Write a differential equation for its volume, V.
    (b) If the initial volume is V_0, solve the differential equation and graph the solution.
    (c) When does the snowball disappear?

23. Water leaks from a vertical cylindrical tank through a small hole in its base at a rate proportional to the square root of the volume of water remaining. If the tank initially contains 200 liters and 20 liters leak out during the first day, when will the tank be half empty? How much water will there be after 4 days?

24. As you know, when a course ends, students start to forget the material they have learned. One model (called the Ebbinghaus model) assumes that the rate at which a student forgets material is proportional to the difference between the material currently remembered and some positive constant, a.

    (a) Let $y = f(t)$ be the fraction of the original material remembered t weeks after the course has ended. Set up a differential equation for y. Your equation will contain two constants; the constant a is less than y for all t.
    (b) Solve the differential equation.
    (c) Describe the practical meaning (in terms of the amount remembered) of the constants in the solution $y = f(t)$.

25. An item is initially sold at a price of $\$p$ per unit. Over time, market forces push the price toward the equilibrium price, $\$p^*$, at which supply balances demand. The Evans Price Adjustment model says that the rate of change in the market price, $\$p$, is proportional to the difference between the market price and the equilibrium price.

    (a) Write a differential equation for p as a function of t.
    (b) Solve for p.
    (c) Sketch solutions for various different initial prices, both above and below the equilibrium price.
    (d) What happens to p as $t \to \infty$?

26. Let L, a constant, be the number of people who would like to see a newly released movie, and let $N(t)$ be the number of people who have seen it during the first t days since its release. The rate that people first go see the movie, dN/dt (in people/day), is proportional to the number of people who would like to see it but haven't yet. Write and solve a differential equation describing dN/dt where t is the number of days since the movie's release. Your solution will involve L and a constant of proportionality, k.

27. A drug is administered intravenously at a constant rate of r mg/hour and is excreted at a rate proportional to the quantity present, with constant of proportionality $\alpha > 0$.

    (a) Solve a differential equation for the quantity, Q, in milligrams, of the drug in the body at time t hours. Assume there is no drug in the body initially. Your answer will contain r and α. Graph Q against t. What is Q_∞, the limiting long-run value of Q?
    (b) What effect does doubling r have on Q_∞? What effect does doubling r have on the time to reach half the limiting value, $\frac{1}{2}Q_\infty$?
    (c) What effect does doubling α have on Q_∞? On the time to reach $\frac{1}{2}Q_\infty$?

28. When people smoke, carbon monoxide is released into the air. In a room of volume 60 m^3, air containing 5% carbon monoxide is introduced at a rate of 0.002 m^3/min. (This means that 5% of the volume of the incoming air is carbon monoxide.) The carbon monoxide mixes immediately with the rest of the air, and the mixture leaves the room at the same rate as it enters.

    (a) Write a differential equation for $c(t)$, the concentration of carbon monoxide at time t, in minutes.
    (b) Solve the differential equation, assuming there is no carbon monoxide in the room initially.
    (c) What happens to the value of $c(t)$ in the long run?

29. (Continuation of Problem 28.) Government agencies warn that exposure to air containing 0.02% carbon monoxide can lead to headaches and dizziness.[14] How long does it take for the concentration of carbon monoxide in the room in Problem 28 to reach this level?

30. An aquarium pool has volume $2 \cdot 10^6$ liters. The pool initially contains pure fresh water. At $t = 0$ minutes, water containing 10 grams/liter of salt is poured into the pool at a rate of 60 liters/minute. The salt water instantly mixes with the fresh water, and the excess mixture is drained out of the pool at the same rate (60 liters/minute).

    (a) Write a differential equation for $S(t)$, the mass of salt in the pool at time t.
    (b) Solve the differential equation to find $S(t)$.
    (c) What happens to $S(t)$ as $t \to \infty$?

31. In 1692, Johann Bernoulli was teaching the Marquis de l'Hopital calculus in Paris. Solve the following problem, which is similar to the one that they did. What is the equation of the curve which has subtangent (distance BC in Figure 11.60) equal to twice its abscissa (distance OC)?

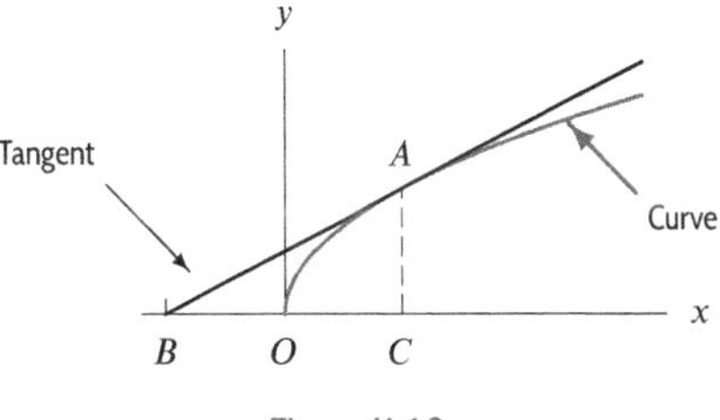

Figure 11.60

[14] www.lni.wa.gov/Safety/Topics/AtoZ/CarbonMonoxide/, accessed on June 14, 2007.

32. An object of mass m is fired vertically upward from the surface of the earth with initial velocity v_0. We will calculate the value of v_0, called the escape velocity, with which the object just escapes the pull of gravity and never returns to earth. Since the object is moving far from the surface of the earth, we must take into account the variation of gravity with altitude. If the acceleration due to gravity at sea level is g, and R is the radius of the earth, the gravitational force, F, on the object of mass m at an altitude h above the surface of the earth is

$$F = \frac{mgR^2}{(R+h)^2}.$$

(a) The velocity of the object (measured upward) is v at time t. Use Newton's Second Law of Motion to show that

$$\frac{dv}{dt} = -\frac{gR^2}{(R+h)^2}.$$

(b) Rewrite this equation with h instead of t as the independent variable using the chain rule $\frac{dv}{dt} = \frac{dv}{dh} \cdot \frac{dh}{dt}$. Hence, show that

$$v\frac{dv}{dh} = -\frac{gR^2}{(R+h)^2}.$$

(c) Solve the differential equation in part (b).

(d) Find the escape velocity, the smallest value of v_0 such that v is never zero.

33. A ball is attached to one end of a rod of length a. A person at the origin holds the other end and moves along the positive x-axis, dragging the ball along a curve, called a *tractrix*, in the plane. See Figure 11.61.

(a) What is the initial position of the ball?

(b) Use the fact that the rod is tangent to the tractrix at every point to find the differential equation that the tractrix satisfies.

(c) Show that the parametric equations $x = a(t - \tanh t)$, $y = a/\cosh t$ satisfy the differential equation and the initial condition.

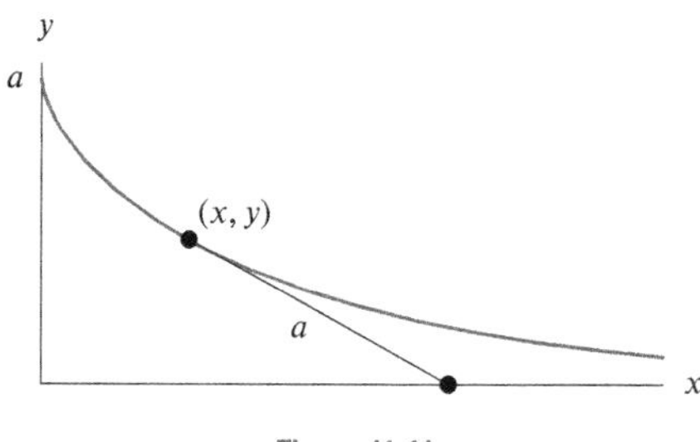

Figure 11.61

Strengthen Your Understanding

In Problems 34–35, explain what is wrong with the statement.

34. At a time when a bank balance $\$B$, which satisfies $dB/dt = 0.08B - 250$, is \$5000, the balance is going down.

35. The differential equation $dQ/dt = -0.15Q + 25$ represents the quantity of a drug in the body if the drug is metabolized at a continuous rate of 15% per day and an IV line is delivering the drug at a constant rate of 25 mg per hour.

In Problems 36–38, give an example of:

36. A differential equation for the quantity of a drug in a patient's body if the patient is receiving the drug at a constant rate through an IV line and is metabolizing the drug at a rate proportional to the quantity present.

37. A differential equation for any quantity which grows in two ways simultaneously: on its own at a rate proportional to the cube root of the amount present and from an external contribution at a constant rate.

38. A differential equation for a quantity that is increasing and grows fastest when the quantity is small and grows more slowly as the quantity gets larger.

11.7 THE LOGISTIC MODEL

Oil prices have a significant impact on the world's economies. In the eighteen months preceding July 2008, the price of oil more than doubled from about \$60 a barrel to about \$140 a barrel.[15] The impact of the increase was significant, from the auto industry, to family budgets, to how people commute. Even the threat of a price hike can send stock markets tumbling.

Many reasons are suggested for the increase, but one fact is inescapable: there is a finite supply of oil in the world. To fuel its expanding economy, the world consumes more oil each succeeding year. This cannot go on indefinitely. Economists and geologists are interested in estimating the remaining oil reserves and the date at which annual oil production is expected to peak (that is, reach a maximum).

[15] http://www.nyse.tv/crude-oil-price-history.htm. Accessed February 2012.

US oil production has already peaked—and the date was predicted in advance. In 1956, geologist M. King Hubbert predicted that annual US oil production would peak some time in the period 1965–1970. Although many did not take his prediction seriously, US oil production did in fact peak in 1970. The economic impact was blunted by the US's increasing reliance on foreign oil.

In this section we introduce the *logistic differential equation* and use it, as Hubbert did, to predict the peak of US oil production.[16] Problems 38–41 investigate the peak of world oil production.

The Logistic Model

The logistic differential equation describes growth subject to a limit. For oil, the limit is the total oil reserves; for a population, the limit is the largest population that the environment can support; for the spread of information or a disease, the limit is the number of people that could be affected. The solution to this differential equation is the family of logistic functions introduced in Section 4.4.

Suppose P is growing logistically toward a limiting value of L, and the relative growth rate, $(1/P)dP/dt$, is k when $P = 0$. In the exponential model, the relative growth rate remains constant at k. But in the logistic model, the relative growth rate decreases linearly to 0 as P approaches L; see Figure 11.62.

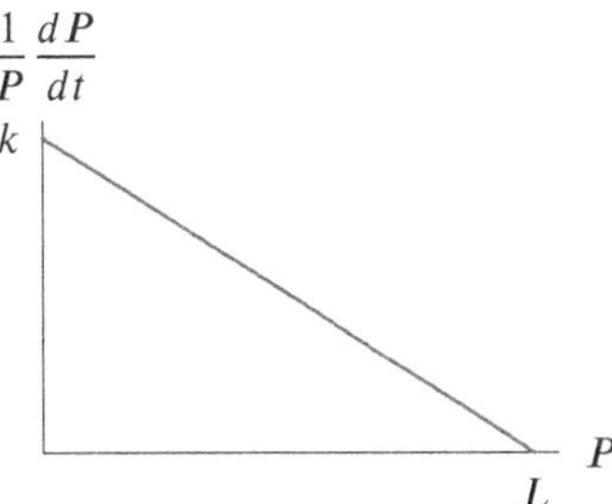

Figure 11.62: Logistic model: Relative growth rate is a linear function of P

So we have

$$\frac{1}{P}\frac{dP}{dt} = k - \frac{k}{L}P = k\left(1 - \frac{P}{L}\right).$$

The logistic differential equation can also be written

$$\frac{dP}{dt} = kP\left(1 - \frac{P}{L}\right).$$

This equation was first proposed as a model for population growth by the Belgian mathematician P. F. Verhulst in the 1830s. In Verhulst's model, L represents the *carrying capacity* of the environment, which is determined by the supply of food and arable land along with the available technology.

Qualitative Solution to the Logistic Equation

Figure 11.63 shows the slope field and characteristic *sigmoid*, or S-shaped, solution curve for the logistic model. Notice that for each fixed value of P, that is, along each horizontal line, the slopes are the same because dP/dt depends only on P and not on t. The slopes are small near $P = 0$ and near $P = L$; they are steepest around $P = L/2$. For $P > L$, the slopes are negative, so if the population is above the carrying capacity, the population decreases.

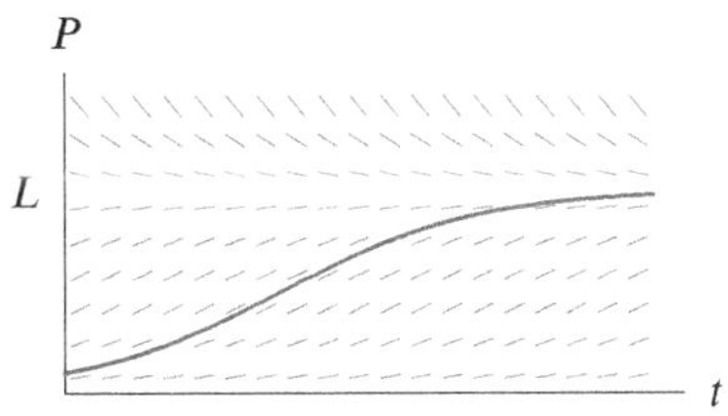

Figure 11.63: Slope field for $dP/dt = kP(1 - P/L)$

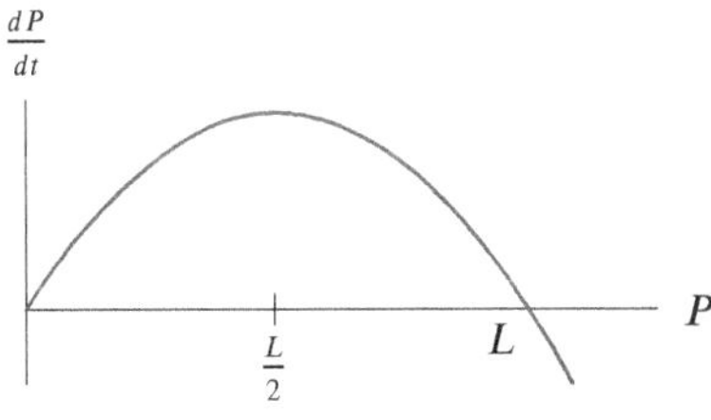

Figure 11.64: $dP/dt = kP(1 - P/L)$

[16]Based on an undergraduate project by Brad Ernst, Colgate University.

We can locate the inflection point where the slope is greatest using Figure 11.64. This graph is a parabola because dP/dt is a quadratic function of P. The horizontal intercepts are at $P = 0$ and $P = L$, so the maximum, where the slope is greatest, is at $P = L/2$. Figure 11.64 also tells us that for $0 < P < L/2$, the slope dP/dt is positive and increasing, so the graph of P against t is concave up. (See Figure 11.65.) For $L/2 < P < L$, the slope dP/dt is positive and decreasing, so the graph of P is concave down. For $P > L$, the slope dP/dt is negative, so P is decreasing.

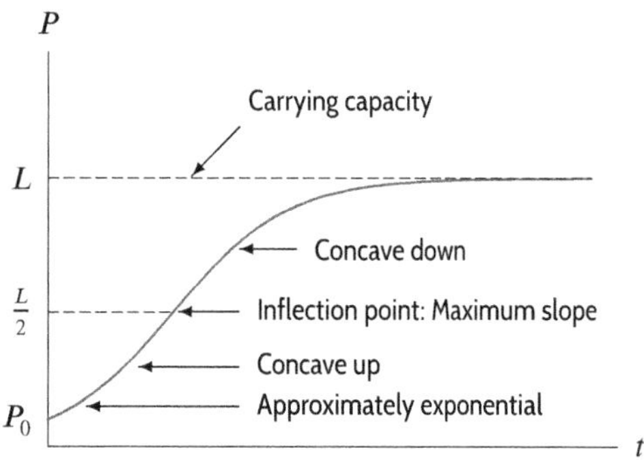

Figure 11.65: Logistic growth with inflection point

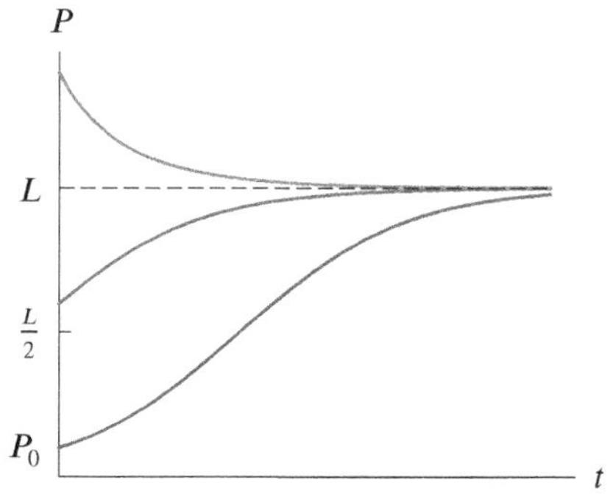

Figure 11.66: Solutions to the logistic equation

If $P = 0$ or $P = L$, there is an equilibrium solution. Figure 11.66 shows that $P = 0$ is an unstable equilibrium because solutions which start near 0 move away from 0. However, $P = L$ is a stable equilibrium.

The Analytic Solution to the Logistic Equation

We have already obtained a lot of information about logistic growth without finding a formula for the solution. However, the equation can be solved analytically by separating variables:

$$\frac{dP}{dt} = kP\left(1 - \frac{P}{L}\right) = kP\left(\frac{L-P}{L}\right)$$

giving

$$\int \frac{dP}{P(L-P)} = \int \frac{k}{L}\,dt.$$

We can integrate the left side using the integral tables (Formula V 26) or by partial fractions:

$$\int \frac{1}{L}\left(\frac{1}{P} + \frac{1}{L-P}\right) dP = \int \frac{k}{L}\,dt.$$

Canceling the constant L, we integrate to get

$$\ln|P| - \ln|L-P| = kt + C.$$

Multiplying through by (-1) and using properties of logarithms, we have

$$\ln\left|\frac{L-P}{P}\right| = -kt - C.$$

Exponentiating both sides gives

$$\left|\frac{L-P}{P}\right| = e^{-kt-C} = e^{-C}e^{-kt}, \qquad \text{so} \qquad \frac{L-P}{P} = \pm e^{-C}e^{-kt}.$$

Then, writing $A = \pm e^{-C}$, we have

$$\frac{L-P}{P} = Ae^{-kt}.$$

We find A by substituting $P = P_0$ when $t = 0$, which, when $P_0 \neq 0$, gives

$$\frac{L - P_0}{P_0} = Ae^0 = A.$$

Since $(L - P)/P = (L/P) - 1$, we have

$$\frac{L}{P} = 1 + Ae^{-kt},$$

which gives the following result:

> The solution to the logistic differential equation:
>
> $$\frac{dP}{dt} = kP\left(1 - \frac{P}{L}\right) \qquad \text{with initial condition } P_0 \text{ when } t = 0$$
>
> is, for $P_0 \neq 0$, the logistic function
>
> $$P = \frac{L}{1 + Ae^{-kt}} \qquad \text{with} \quad A = \frac{L - P_0}{P_0}.$$
>
> The parameter L represents the limiting value. The parameter k represents the relative growth rate when P is small relative to L. The parameter A depends on the initial condition P_0.

Peak Oil: US Production

We apply the logistic model to US oil production as Hubbert did in 1956. To make predictions, we need the values of k and L. We calculate these values from the oil production data Hubbert had available to him in the 1950s; see Table 11.5.

We define P to be the *total* amount of oil, in billions of barrels, produced in the US since 1859, the year the first oil well was built. See Table 11.5. With time, t, in years, dP/dt approximates the *annual* oil production in billions of barrels per year. Peak oil production occurs when dP/dt is a maximum.

Table 11.5 *US oil production[17] for 1931–1950 (billions of barrels)*

Year	dP/dt	P	Year	dP/dt	P	Year	dP/dt	P
1931	0.851	13.8	1938	1.21	21.0	1945	1.71	31.5
1932	0.785	14.6	1939	1.26	22.3	1946	1.73	33.2
1933	0.906	15.5	1940	1.50	23.8	1947	1.86	35.1
1934	0.908	16.4	1941	1.40	25.2	1948	2.02	37.1
1935	0.994	17.4	1942	1.39	26.6	1949	1.84	38.9
1936	1.10	18.5	1943	1.51	28.1	1950	1.97	40.9
1937	1.28	19.8	1944	1.68	29.8			

Figure 11.67 shows a scatterplot of the relative growth rate $(dP/dt)/P$ versus P. If the data follow a logistic differential equation, we see a linear relationship with intercept k and slope $-k/L$:

$$\frac{1}{P}\frac{dP}{dt} = k\left(1 - \frac{P}{L}\right) = k - \frac{k}{L}P.$$

[17]Data from http://www.eia.gov/dnav/pet/hist/LeafHandler.ashx?n=PET&s=MCRFPUS1&f=A. Accessed Feb, 2012.

Figure 11.67 shows a line fitted to the data.[18] The vertical intercept gives the value $k = 0.0649$.

The slope of the line is $-k/L = -0.00036$, so we have $L = 0.0649/0.00036 = 180$ billion barrels of oil. Thus, the model predicts that the total oil reserves in the US (the total amount in the ground before drilling started in 1859) were 180 billion barrels of oil.[19]

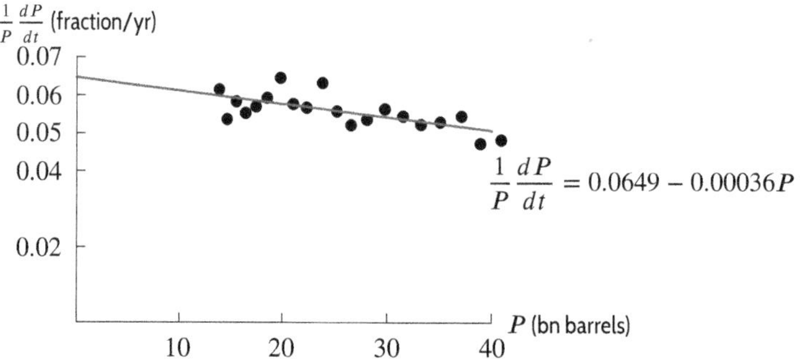

Figure 11.67: US oil production 1931–1950: Scatterplot and line for $1/P(dP/dt)$ versus P

If we let $t = 0$ be 1950, then $P_0 = 40.9$ billion barrels and $A = (180 - 40.9)/40.9 = 3.401$. The logistic function representing US oil production is

$$P = \frac{180}{1 + 3.401e^{-0.0649t}}.$$

Predicting Peak Oil Production

To predict, as Hubbert did, the year when annual US oil production would peak, we use the fact that the maximum value for dP/dt occurs when $P = L/2$. We derive a formula for the peak year (used again in Problems 38–41 to find peak production in world oil). The crucial observation is that, at the peak, the denominator of the expression for P must equal 2. Since

$$P = \frac{L}{2} = \frac{L}{1 + Ae^{-kt}}, \qquad \text{we have} \quad Ae^{-kt} = 1.$$

Using logarithms to solve the equation $Ae^{-kt} = 1$, we get $t = (1/k)\ln A$. Since $A = (L - P_0)/P_0$, we see that the time to peak oil production is an example of the following result:

> For a logistic function, the maximum value of dP/dt occurs when $P = L/2$, and
>
> $$\text{Time to the maximum rate of change} = \frac{1}{k}\ln A = \frac{1}{k}\ln\frac{L - P_0}{P_0}.$$

Thus, for the US,

$$\text{Time to peak oil production} = \frac{1}{0.0649}\ln\frac{180 - 40.9}{40.9} \approx 19 \text{ years};$$

that is, oil production was predicted to peak in the year $1950 + 19 = 1969$. That year, $P = L/2$ and annual production was expected to be

$$\frac{dP}{dt} = kP\left(1 - \frac{P}{L}\right) = 0.0649 \cdot \frac{180}{2}\left(1 - \frac{1}{2}\right) \approx 3 \text{ billion barrels.}$$

[18] The line is a least-squares regression line.

[19] If the same analysis is repeated for other time periods, for example 1900–1950 or 1900–2000, the value for L varies between 120 and 220 billion barrels, while the value of k varies between 0.060 and 0.075.

The actual peak in US oil production was 3.5 billion barrels in 1970. Repeating the analysis using other time periods gives peak oil years in the range 1965-1970, as Hubbert predicted.

Figure 11.68 shows actual annual US production data and the parabola predicting its peak around 1970. Figure 11.69 shows P as a logistic function of t, with the limiting value of $P = 180$ and maximum production at $P = 90$. In fact, the first major oil crisis hit the US in the 1970s, with spiraling gas prices and long lines at service stations. The decline in US oil production since 1970 was partly mitigated by the opening of the Alaskan oil fields, which led to a second but lower peak in 1985. However, the US has increasingly depended on foreign oil.

Although Hubbert's predictions of the peak year proved to be accurate, extrapolation into the future is risky. Figure 11.68 and Figure 11.69 show that since 1970, oil production has slowed, though not as much as predicted.

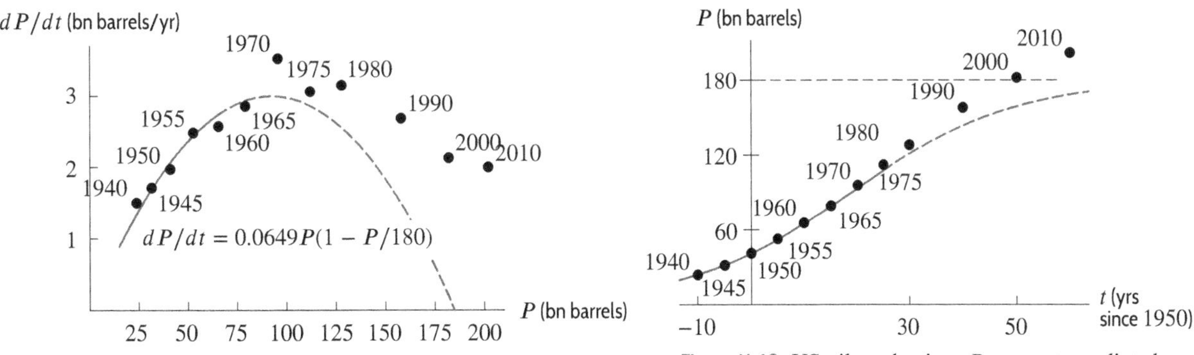

Figure 11.68: US oil production: dP/dt versus P, predicted (parabola) and actual

Figure 11.69: US oil production: P versus t, predicted (logistic) and actual

Interestingly, Hubbert used the logistic model only to estimate k; for L, he relied on geological studies. It is remarkable that using only annual oil production for 1930-1950, we get an estimate for L that is in such close agreement with the geological estimates.

US Population Growth

The logistic equation is often used to model population growth. Table 11.6 gives the annual census figures in millions for the US population from 1790 to 2010.[20] Since we have the population, $P(t)$, of the US at ten-year intervals, we compute the relative growth rate using averages of estimates of the form:

$$\frac{1}{P}\frac{dP}{dt} \text{ in } 1860 = \frac{1}{P(1860)} \cdot \frac{P(1870) - P(1860)}{10}.$$

(See Problem 19 for details.) If we focus on the period 1790–1940, a line fitted to the scatterplot of $(1/P)dP/dt$ versus P has intercept $k = 0.0317$ and slope $-k/L = -0.000165$, so $L \approx 192$. Thus the differential equation modeling the US population during this period is

$$\frac{dP}{dt} = 0.0317P\left(1 - \frac{P}{192}\right).$$

Using 1790 as $t = 0$, we get

$$A = \frac{L - P_0}{P_0} = \frac{192 - 3.9}{3.9} \approx 48.$$

Thus the solution to the logistic differential equation is

$$P = \frac{192}{1 + 48e^{-0.0317t}}.$$

Table 11.6 shows the actual census data for 1790–1940 with projected values from a logistic model; the largest deviations are 4% in 1840 and 1870 (the Civil War accounts for the second one).[21]

[20] www.census.gov.Accessed February 12, 2012.

[21] Calculations were done with more precise values of the population data and the constants k, L, and A than those shown.

Table 11.6 *US population, in millions, for 1790–2010, actual data and logistic predictions*

Year	Actual	Logistic	Year	Actual	Logistic	Year	Actual	Logistic	Year	Actual	Logistic
1790	3.9	3.9	1850	23.2	23.6	1910	92.2	92.9	1970	203.2	165.7
1800	5.3	5.4	1860	31.4	30.9	1920	106.0	108.0	1980	226.5	172.2
1810	7.2	7.3	1870	38.6	40.1	1930	123.2	122.6	1990	248.7	177.2
1820	9.6	9.8	1880	50.2	51.0	1940	132.2	136.0	2000	281.4	181.0
1830	12.9	13.3	1890	63.0	63.7	1950	151.3	147.7	2010	308.7	183.9
1840	17.1	17.8	1900	76.2	77.9	1960	179.3	157.6			

After 1940, the actual figures leave the logistic model in the dust. The model predicts an increase of 9.9 million from 1950 to 1960 versus the actual change of 28 million. By 1970 the actual population of 203.2 million exceeded the predicted limiting population of $L = 192$ million. The unprecedented surge in US population between 1945 and 1965 is referred to as the baby boom.

Exercises and Problems for Section 11.7 Online Resource: Additional Problems for Section 11.7

EXERCISES

1. **(a)** Show that $P = 1/(1+e^{-t})$ satisfies the logistic equation
$$\frac{dP}{dt} = P(1-P).$$
(b) What is the limiting value of P as $t \to \infty$?

2. A quantity P satisfies the differential equation
$$\frac{dP}{dt} = kP\left(1 - \frac{P}{100}\right).$$
Sketch approximate solutions satisfying each of the following initial conditions:
(a) $P_0 = 8$ **(b)** $P_0 = 70$ **(c)** $P_0 = 125$

3. A quantity Q satisfies the differential equation
$$\frac{dQ}{dt} = kQ(1 - 0.0004Q).$$
Sketch approximate solutions satisfying each of the following initial conditions:
(a) $Q_0 = 300$ **(b)** $Q_0 = 1500$ **(c)** $Q_0 = 3500$

4. A quantity P satisfies the differential equation
$$\frac{dP}{dt} = kP\left(1 - \frac{P}{250}\right), \quad \text{with } k > 0.$$
Sketch a graph of dP/dt as a function of P.

5. A quantity A satisfies the differential equation
$$\frac{dA}{dt} = kA(1 - 0.0002A), \quad \text{with } k > 0.$$
Sketch a graph of dA/dt as a function of A.

6. Figure 11.70 shows a graph of dP/dt against P for a logistic differential equation. Sketch several solutions of P against t, using different initial conditions. Include a scale on your vertical axis.

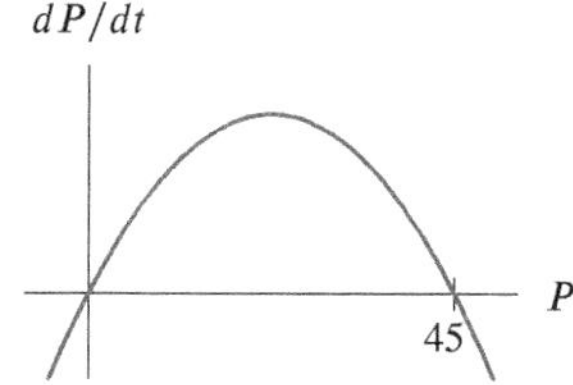

Figure 11.70

7. Figure 11.71 shows a slope field of a differential equation for a quantity Q growing logistically. Sketch a graph of dQ/dt against Q. Include a scale on the horizontal axis.

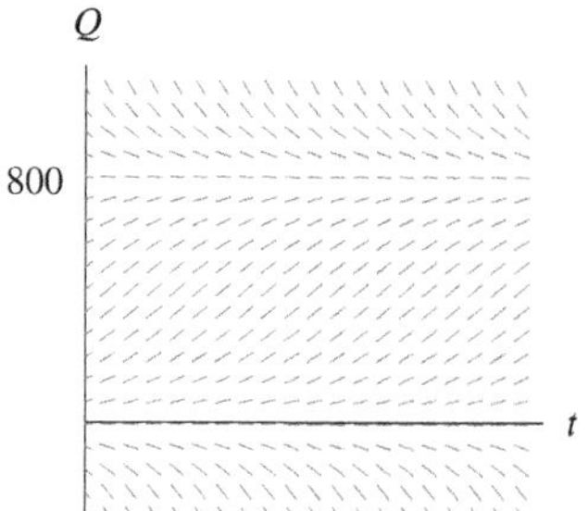

Figure 11.71

8. **(a)** On the slope field for $dP/dt = 3P - 3P^2$ in Figure 11.72, sketch three solution curves showing different types of behavior for the population, P.
(b) Is there a stable value of the population? If so, what is it?
(c) Describe the meaning of the shape of the solution curves for the population: Where is P increasing? Decreasing? What happens in the long run? Are there any inflection points? Where? What do they mean for the population?
(d) Sketch a graph of dP/dt against P. Where is dP/dt positive? Negative? Zero? Maximum? How do your observations about dP/dt explain the shapes of your solution curves?

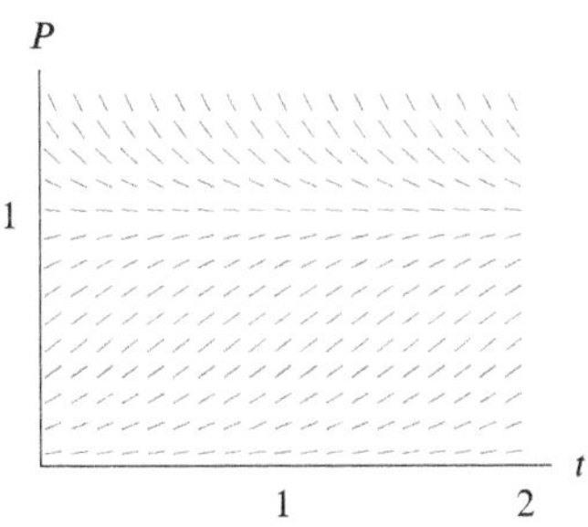

Figure 11.72

■ Exercises 9–10 give a graph of dP/dt against P.

(a) What are the equilibrium values of P?

(b) If $P = 500$, is dP/dt positive or negative? Is P increasing or decreasing?

9. dP/dt 10. dP/dt

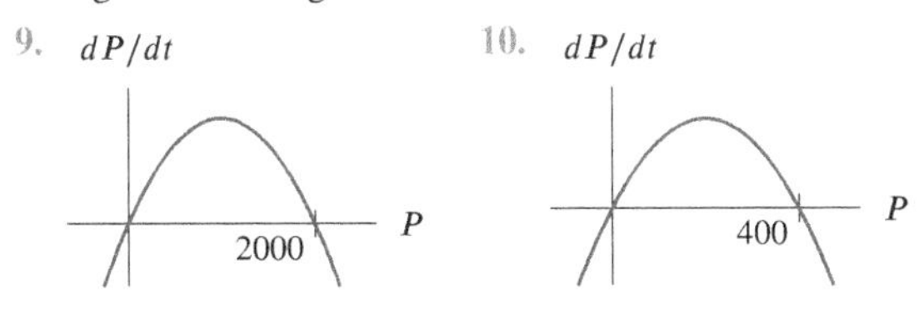

■ For the logistic differential equations in Exercises 11–12,

(a) Give values for k and for L and interpret the meaning of each in terms of the growth of the quantity P.

(b) Give the value of P when the rate of change is at its peak.

11. $\frac{dP}{dt} = 0.035P\left(1 - \frac{P}{6000}\right)$

12. $\frac{dP}{dt} = 0.1P - 0.00008P^2$

■ In Exercises 13–16, give the general solution to the logistic differential equation.

13. $\frac{dP}{dt} = 0.05P\left(1 - \frac{P}{2800}\right)$

14. $\frac{dP}{dt} = 0.012P\left(1 - \frac{P}{5700}\right)$

15. $\frac{dP}{dt} = 0.68P(1 - 0.00025P)$

16. $\frac{dP}{dt} = 0.2P - 0.0008P^2$

■ In Exercises 17–20, give k, L, A, a formula for P as a function of time t, and the time to the peak value of dP/dt.

17. $\frac{dP}{dt} = 10P - 5P^2, \quad P_0 = L/4$

18. $\frac{dP}{dt} = 0.02P - 0.0025P^2, \quad P_0 = 1$

19. $\frac{1}{P}\frac{dP}{dt} = 0.3\left(1 - \frac{P}{100}\right), \quad P_0 = 75$

20. $\frac{1}{10P}\frac{dP}{dt} = 0.012 - 0.002P, \quad P_0 = 2$

■ In Exercises 21–22, give the solution to the logistic differential equation with initial condition.

21. $\frac{dP}{dt} = 0.8P\left(1 - \frac{P}{8500}\right)$ with $P_0 = 500$

22. $\frac{dP}{dt} = 0.04P(1 - 0.0001P)$ with $P_0 = 200$

PROBLEMS

■ Problems 23–27 are about chikungunya, a disease that arrived in the Americas in 2013 and spread rapidly in 2014. While seldom fatal, the disease causes debilitating joint pain and a high fever. In August 2014, a public challenge was issued to predict the number of cases in each of the affected countries. The winners, Joceline Lega and Heidi Brown, used a logistic model to make their predictions.[22] Let N be the total number of cases of chikungunya in a country by week t. From the data given, find approximate values for the country for

(a) The number of cases expected in the long run.

(b) The maximum number of new cases in a week.

23. In Guadeloupe: use the fitted curve in Figure 11.73.

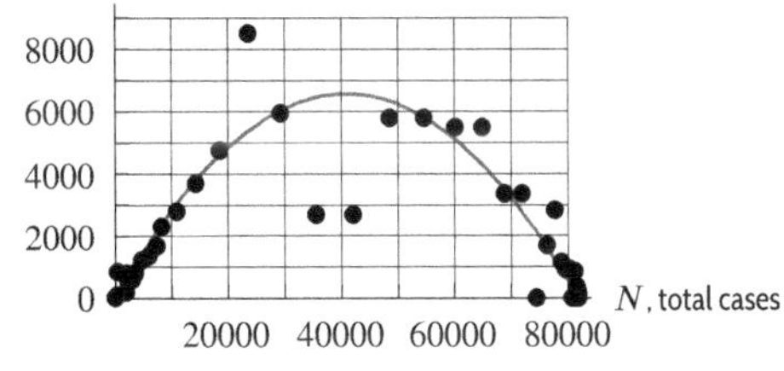

Figure 11.73

24. In Dominica: $\frac{dN}{dt} = 0.31N\left(1 - \frac{N}{3771}\right)$.

25. In the Dominican Republic: $N = \frac{539{,}226}{1 + 177e^{-0.35t}}$.

26. In Haiti: $\frac{1}{N}\frac{dN}{dt} = 0.7 - 0.00001N$.

27. In Ecuador: use the fitted line in Figure 11.74.

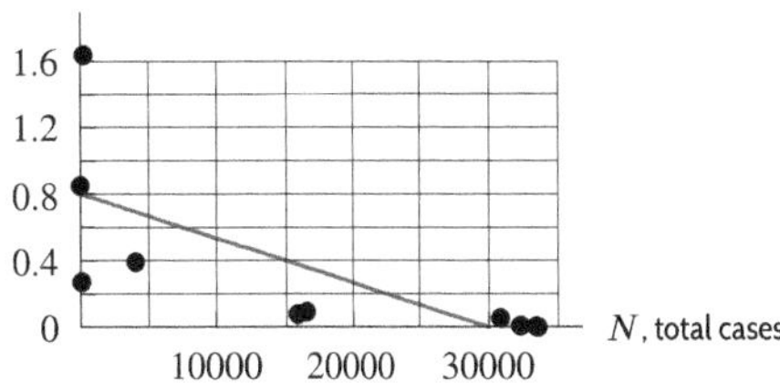

Figure 11.74

■ In Problems 28–29, a population $P(t) = L/(1+10e^{-kt})$ grows logistically, with t in days. Determine how long it takes the population to grow from 10% to 90% of its carrying capacity.

28. $k = 0.1$

29. $k = 0.2$

[22] http://www.darpa.mil/news-events/2015-05-27. Based on preprint "Data Driven Outbreak Forecasting with a Simple Nonlinear Growth Model" J. Lega, H. Brown, April 2016.

30. A rumor spreads among a group of 400 people. The number of people, $N(t)$, who have heard the rumor by time t in hours since the rumor started is approximated by

$$N(t) = \frac{400}{1 + 399e^{-0.4t}}.$$

(a) Find $N(0)$ and interpret it.
(b) How many people will have heard the rumor after 2 hours? After 10 hours?
(c) Graph $N(t)$.
(d) Approximately how long will it take until half the people have heard the rumor? 399 people?
(e) When is the rumor spreading fastest?

31. The total number of people infected with a virus often grows like a logistic curve. Suppose that time, t, is in weeks and that 10 people originally have the virus. In the early stages, the number of people infected is increasing exponentially with $k = 1.78$. In the long run, 5000 people are infected.

(a) Find a logistic function to model the number of people infected.
(b) Sketch a graph of your answer to part (a).
(c) Use your graph to estimate the length of time until the rate at which people are becoming infected starts to decrease. What is the vertical coordinate at this point?

32. The Tojolobal Mayan Indian community in southern Mexico has available a fixed amount of land. The proportion, P, of land in use for farming t years after 1935 is modeled with the logistic function in Figure 11.75:[23]

$$P = \frac{1}{1 + 2.968e^{-0.0275t}}.$$

(a) What proportion of the land was in use for farming in 1935?
(b) What is the long-run prediction of this model?
(c) When was half the land in use for farming?
(d) When is the proportion of land used for farming increasing most rapidly?

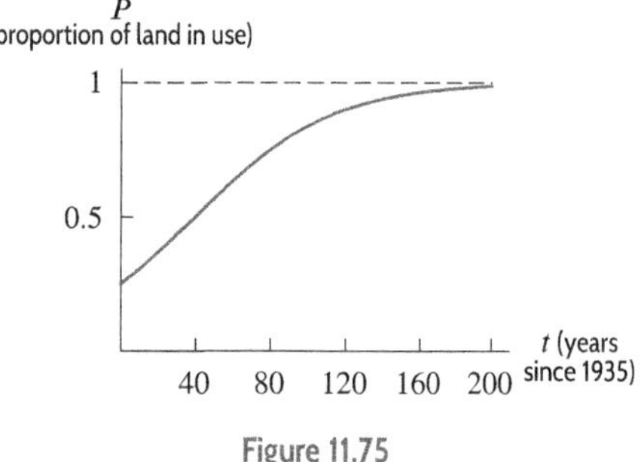

Figure 11.75

33. A model for the population, P, of carp in a landlocked lake at time t is given by the differential equation

$$\frac{dP}{dt} = 0.25P(1 - 0.0004P).$$

(a) What is the long-term equilibrium population of carp in the lake?
(b) A census taken ten years ago found there were 1000 carp in the lake. Estimate the current population.
(c) Under a plan to join the lake to a nearby river, the fish will be able to leave the lake. A net loss of 10% of the carp each year is predicted, but the patterns of birth and death are not expected to change. Revise the differential equation to take this into account. Use the revised differential equation to predict the future development of the carp population.

34. Table 11.7 gives values for a logistic function $P = f(t)$.

(a) Estimate the maximum rate of change of P and estimate the value of t when it occurs.
(b) If P represents the growth of a population, estimate the carrying capacity of the population.

Table 11.7

t	0	10	20	30	40	50	60	70
P	120	125	135	155	195	270	345	385

35. For a population P growing logistically, Table 11.8 gives values of the relative growth rate

$$r = \frac{1}{P}\frac{dP}{dt}.$$

(a) What is the carrying capacity of this population?
(b) What is the maximum value of dP/dt?

Table 11.8

P	1000	2000	3000	4000	5000	6000
r	0.11	0.10	0.09	0.08	0.07	0.06

36. Figure 11.76 shows the spread of the Code-red computer virus during July 2001. Most of the growth took place starting at midnight on July 19; on July 20, the virus attacked the White House, trying (unsuccessfully) to knock its site off-line. The number of computers infected by the virus was a logistic function of time.

(a) Estimate the limiting value of $f(t)$ as t increased. What does this limiting value represent in terms of Code-red?
(b) Estimate the value of t at which $f''(t) = 0$. Estimate the value of n at this time.
(c) What does the answer to part (b) tell us about Code-red?
(d) How are the answers to parts (a) and (b) related?

[23]Adapted from J. S. Thomas and M. C. Robbins, "The Limits to Growth in a Tojolobal Maya Ejido," *Geoscience and Man 26* (Baton Rouge: Geoscience Publications, 1988), pp. 9–16.

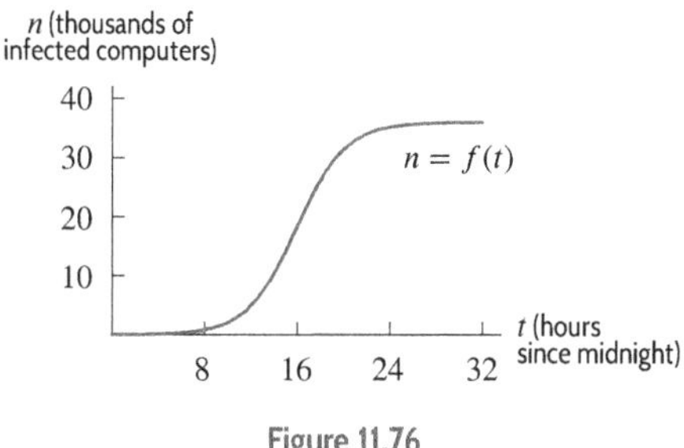

Figure 11.76

37. According to an article in *The New York Times*,[24] pigweed has acquired resistance to the weedkiller Roundup. Let N be the number of acres, in millions, where Roundup-resistant pigweed is found. Suppose the relative growth rate, $(1/N)dN/dt$, was 15% when $N = 5$ and 14.5% when $N = 10$. Assuming the relative growth rate is a linear function of N, write a differential equation to model N as a function of time, and predict how many acres will eventually be afflicted before the spread of Roundup-resistant pigweed halts.

■ In Problems 38–42, we analyze world oil production.[25] When annual world oil production peaks and starts to decline, major economic restructuring will be needed. We investigate when this slowdown is projected to occur.

38. We define P to be the total oil production worldwide since 1859 in billions of barrels. In 1993, annual world oil production was 22.0 billion barrels and the total production was $P = 724$ billion barrels. In 2013, annual production was 27.8 billion barrels and the total production was $P = 1235$ billion barrels. Let t be time in years since 1993.
    - **(a)** Estimate the rate of production, dP/dt, for 1993 and 2013.
    - **(b)** Estimate the relative growth rate, $(1/P)(dP/dt)$, for 1993 and 2013.
    - **(c)** Find an equation for the relative growth rate, $(1/P)(dP/dt)$, as a function of P, assuming that the function is linear.
    - **(d)** Assuming that P increases logistically and that all oil in the ground will ultimately be extracted, estimate the world oil reserves in 1859 to the nearest billion barrels.
    - **(e)** Write and solve the logistic differential equation modeling P.

39. In Problem 38 we used a logistic function to model P, total world oil production since 1859, as a function of time, t, in years since 1993. Use this function to answer the following questions:
    - **(a)** When does peak annual world oil production occur?
    - **(b)** Geologists have estimated world oil reserves to be as high as 3500 billion barrels.[26] When does peak world oil production occur with this assumption? (Assume k and P_0 are unchanged.)

40. As in Problem 38, let P be total world oil production since 1859. In 1998, annual world production was 24.4 billion barrels and total production was $P = 841$ billion barrels. In 2003, annual production was 25.3 billion barrels and total production was $P = 964$ billion barrels. In 2008, annual production was 26.9 billion barrels and the total production was 1100 billion barrels.
    - **(a)** Graph dP/dt versus P from Problem 38 and show the data for 1998, 2003 and 2008. How well does the model fit the data?
    - **(b)** Graph the logistic function modeling worldwide oil production (P versus t) from Problem 38 and show the data for 1998, 2003 and 2008. How well does the model fit the data?

41. Use the logistic function obtained in Problem 38 to model the growth of P, the total oil produced worldwide in billions of barrels since 1859:
    - **(a)** Find the projected value of P for 2014 to the nearest billion barrels.
    - **(b)** Use the derivative to estimate the annual world oil production during 2014.
    - **(c)** How much oil is projected to remain in the ground in 2014?
    - **(d)** Compare the projected production in part (b) with the actual figure of 28.4 billion barrels.

42. With P, the total oil produced worldwide since 1859, in billions of barrels, modeled as a function of time t in years since 1993 as in Problem 38:
    - **(a)** Predict the total quantity of oil produced by 2020.
    - **(b)** In what year does the model predict that only 300 billion barrels remain?

Strengthen Your Understanding

■ In Problems 43–45, explain what is wrong with the statement.

43. The differential equation $dP/dt = 0.08P - 0.0032P^2$ has one equilibrium solution, at $P = 25$.

44. The maximum rate of change occurs at $t = 25$ for a quantity Q growing according to the logistic equation

$$\frac{dQ}{dt} = 0.13Q(1 - 0.02Q).$$

[24] http://www.nytimes.com/2010/05/04/business/energy-environment/04weed.html, accessed May 3, 2010.
[25] cta.ornl.gov, accessed April 29, 2015.
[26] www.hoodriver.k12.or.us, accessed April 29, 2015.

45. Figure 11.77 shows a quantity growing logistically.

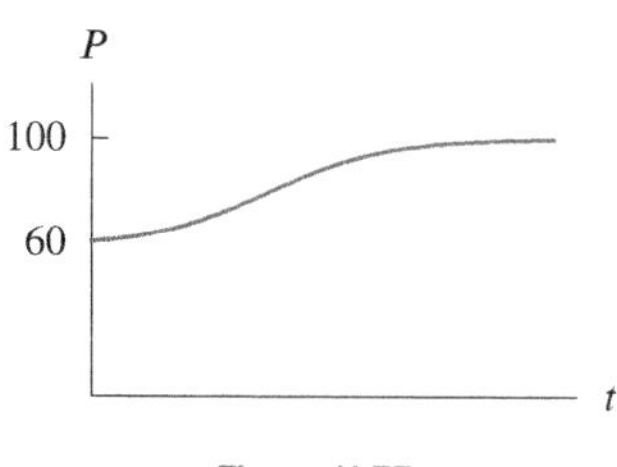

Figure 11.77

In Problems 46–49, give an example of:

46. A quantity that increases logistically.
47. A logistic differential equation for a quantity P such that the maximum rate of change of P occurs when $P = 75$.
48. A graph of dQ/dt against Q if Q is growing logistically and has an equilibrium value at $Q = 500$.
49. A graph of dP/dt against P if P is a logistic function which increases when $0 < P < 20$ and which decreases when $P < 0$ or $P > 20$.

Are the statements in Problems 50–51 true or false? Give an explanation for your answer.

50. There is a solution curve for the logistic differential equation $dP/dt = P(2 - P)$ that goes through the points $(0, 1)$ and $(1, 3)$.
51. For any positive values of the constant k and any positive values of the initial value $P(0)$, the solution to the differential equation $dP/dt = kP(L - P)$ has limiting value L as $t \to \infty$.

11.8 SYSTEMS OF DIFFERENTIAL EQUATIONS

In Section 11.7 we modeled the growth of a single population over time. We now consider the growth of two populations that interact, such as a population of sick people infecting the healthy people around them. This involves not just one differential equation, but a system of two.

Diseases and Epidemics

Differential equations can be used to predict when an outbreak of a disease will become so severe that it is called an *epidemic*[27] and to decide what level of vaccination is necessary to prevent an epidemic. Let's consider a specific example.

Flu in a British Boarding School

In January 1978, 763 students returned to a boys' boarding school after their winter vacation. A week later, one boy developed the flu, followed by two others the next day. By the end of the month, nearly half the boys were sick. Most of the school had been affected by the time the epidemic was over in mid-February.[28]

Being able to predict how many people will get sick, and when, is an important step toward controlling an epidemic. This is one of the responsibilities of Britain's Communicable Disease Surveillance Centre and the US's Center for Disease Control and Prevention.

The S-I-R Model

We apply one of the most commonly used models for an epidemic, called the S-I-R model, to the boarding school flu example. The population of the school is divided into three groups:

S = the number of *susceptibles*, the people who are not yet sick but who could become sick

I = the number of *infecteds*, the people who are currently sick

R = the number of *recovered*, or *removed*, the people who have been sick and can no longer infect others or be reinfected.

The number of susceptibles decreases with time as people become infected. We assume that the rate at which people become infected is proportional to the number of contacts between susceptible

[27] Exactly when a disease should be called an epidemic is not always clear. The medical profession generally classifies a disease an epidemic when the frequency is higher than usually expected—leaving open the question of what is usually expected. See, for example, *Epidemiology in Medicine* by C. H. Hennekens and J. Buring (Boston: Little, Brown, 1987).

[28] Data from the Communicable Disease Surveillance Centre (UK); reported in "Influenza in a Boarding School," *British Medical Journal*, March 4, 1978, and by J. D. Murray in *Mathematical Biology* (New York: Springer Verlag, 1990).

and infected people. We expect the number of contacts between the two groups to be proportional to both S and I. (If S doubles, we expect the number of contacts to double; similarly, if I doubles, we expect the number of contacts to double.) Thus we assume that the number of contacts is proportional to the product, SI. In other words, we assume that for some constant $a > 0$,

$$\frac{dS}{dt} = -\left(\begin{array}{c}\text{Rate susceptibles}\\ \text{get sick}\end{array}\right) = -aSI.$$

(The negative sign is used because S is decreasing.)

The number of infecteds is changing in two ways: newly sick people are added to the infected group, and others are removed. The newly sick people are exactly those people leaving the susceptible group and so accrue at a rate of aSI (with a positive sign this time). People leave the infected group either because they recover (or die), or because they are physically removed from the rest of the group and can no longer infect others. We assume that people are removed at a rate proportional to the number of sick, or bI, where b is a positive constant. Thus,

$$\frac{dI}{dt} = \begin{array}{c}\text{Rate susceptibles}\\ \text{get sick}\end{array} - \begin{array}{c}\text{Rate infecteds}\\ \text{get removed}\end{array} = aSI - bI.$$

Assuming that those who have recovered from the disease are no longer susceptible, the recovered group increases at the rate of bI, so

$$\frac{dR}{dt} = bI.$$

We are assuming that having the flu confers immunity on a person, that is, that the person cannot get the flu again. (This is true for a given strain of flu, at least in the short run.)

In analyzing the flu, we can use the fact that the total population $S + I + R$ is not changing. (The total population, the total number of boys in the school, did not change during the epidemic.) Thus, once we know S and I, we can calculate R. So we restrict our attention to the two equations

$$\frac{dS}{dt} = -aSI$$
$$\frac{dI}{dt} = aSI - bI.$$

The Constants a and b

The constant a measures how infectious the disease is—that is, how quickly it is transmitted from the infecteds to the susceptibles. In the case of the flu, we know from medical accounts that the epidemic started with one sick boy, with two more becoming sick a day later. Thus, when $I = 1$ and $S = 762$, we have $dS/dt \approx -2$, enabling us to roughly[29] approximate a:

$$a = -\frac{dS/dt}{SI} = \frac{2}{(762)(1)} = 0.0026.$$

The constant b represents the rate at which infected people are removed from the infected population. In this case of the flu, boys were generally taken to the infirmary within one or two days of becoming sick. About half the infected population was removed each day, so we take $b \approx 0.5$. Thus, our equations are:

$$\frac{dS}{dt} = -0.0026SI$$
$$\frac{dI}{dt} = 0.0026SI - 0.5I.$$

The Phase Plane

We can get a good idea of the progress of the disease from graphs. You might expect that we would look for graphs of S and I against t, and eventually we will. However, we first look at a graph of I against S. If we plot a point (S, I) representing the number of susceptibles and the number of infecteds at any moment in time, then, as the numbers of

[29]The values of a and b are close to those obtained by J. D. Murray in *Mathematical Biology* (New York: Springer Verlag, 1990).

susceptibles and infecteds change, the point moves. The SI-plane on which the point moves is called the *phase plane*. The path along which the point moves is called the *phase trajectory*, or *orbit*, of the point.

To find the phase trajectory, we need a differential equation relating S and I directly. Thinking of I as a function of S, and S as a function of t, we use the chain rule to get

$$\frac{dI}{dt} = \frac{dI}{dS} \cdot \frac{dS}{dt},$$

giving

$$\frac{dI}{dS} = \frac{dI/dt}{dS/dt}.$$

Substituting for dI/dt and dS/dt, we get

$$\frac{dI}{dS} = \frac{0.0026SI - 0.5I}{-0.0026SI}.$$

Assuming I is not zero, this equation simplifies to approximately

$$\frac{dI}{dS} = -1 + \frac{192}{S}.$$

The slope field of this differential equation is shown in Figure 11.78. The trajectory with initial condition $S_0 = 762$, $I_0 = 1$ is shown in Figure 11.79. Time is represented by the arrow showing the direction that a point moves on the trajectory. The disease starts at the point $S_0 = 762$, $I_0 = 1$. At first, more people become infected and fewer are susceptible. In other words, S decreases and I increases. Later, I decreases as S continues to decrease.

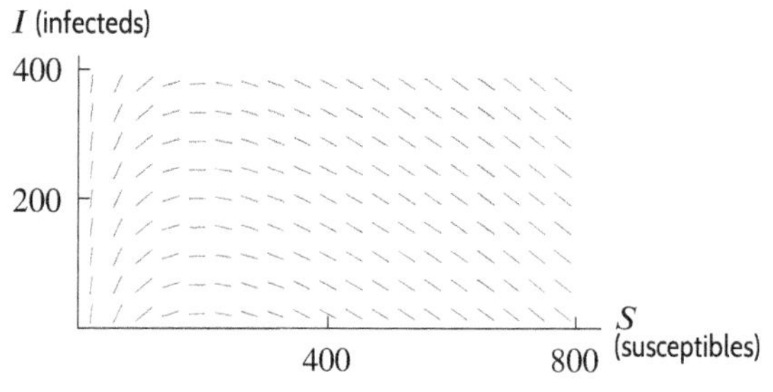

Figure 11.78: Slope field for $dI/dS = -1 + 192/S$

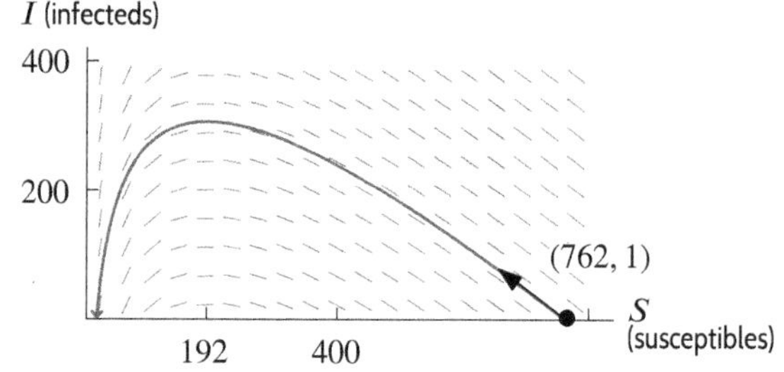

Figure 11.79: Trajectory for $S_0 = 762$, $I_0 = 1$

What Does the SI-Phase Plane Tell Us?

To learn how the disease progresses, look at the shape of the curve in Figure 11.79. The value of I increases to about 300 (the maximum number infected and infectious at any one time); then I decreases to zero. This peak value of I occurs when $S \approx 200$. We can determine exactly when the peak value occurs by solving

$$\frac{dI}{dS} = -1 + \frac{192}{S} = 0,$$

which gives

$$S = 192.$$

Notice that the peak value for I always occurs at the same value of S, namely $S = 192$. The graph shows that if a trajectory starts with $S_0 > 192$, then I first increases and then decreases to zero. On the other hand, if $S_0 < 192$, there is no peak as I decreases right away.

> For this example, the value $S_0 = 192$ is called a *threshold value*. If S_0 is around or below 192, there is no epidemic. If S_0 is significantly greater than 192, an epidemic occurs.[30]

[30]Here we are using J. D. Murray's definition of an epidemic as an outbreak in which the number of infecteds increases from the initial value, I_0. See *Mathematical Biology* (New York: Springer Verlag, 1990).

The phase diagram makes clear that the maximum value of I is about 300. Another question answered by the phase plane diagram is the total number of students who are expected to get sick during the epidemic. (This is not the maximum value reached by I, which gives the maximum number infected at any one time.) The point at which the trajectory crosses the S-axis represents the time when the epidemic has passed (since $I = 0$). The S-intercept shows how many boys never get the flu and thus, how many do get it.

How Many People Should Be Vaccinated?

An epidemic can sometimes be avoided by vaccination. How many boys would have had to be vaccinated to prevent the flu epidemic? To answer this, think of vaccination as removing people from the S category (without increasing I), which amounts to moving the initial point on the trajectory to the left, parallel to the S-axis. To avoid an epidemic, the initial value of S_0 should be at or below the threshold value. Therefore, all but 192 boys would need to be vaccinated.

Graphs of S and I Against t

To find out exactly when I reaches its maximum, we need numerical methods. A modification of Euler's method was used to generate the solution curves of S and I against t in Figure 11.80. Notice that the number of susceptibles drops throughout the disease as healthy people get sick. The number of infecteds peaks after about 6 days and then drops. The epidemic has run its course in 20 days.

Analytical Solution for the SI-Phase Trajectory

The differential equation

$$\frac{dI}{dS} = -1 + \frac{192}{S}$$

can be integrated, giving

$$I = -S + 192 \ln S + C.$$

Using $S_0 = 762$ and $I_0 = 1$ gives $1 = -762 + 192 \ln 762 + C$, so we get $C = 763 - 192 \ln 762$. Substituting this value for C, we get:

$$I = -S + 192 \ln S - 192 \ln 762 + 763$$

$$I = -S + 192 \ln\left(\frac{S}{762}\right) + 763.$$

This is the equation of the solution curve in Figure 11.79.

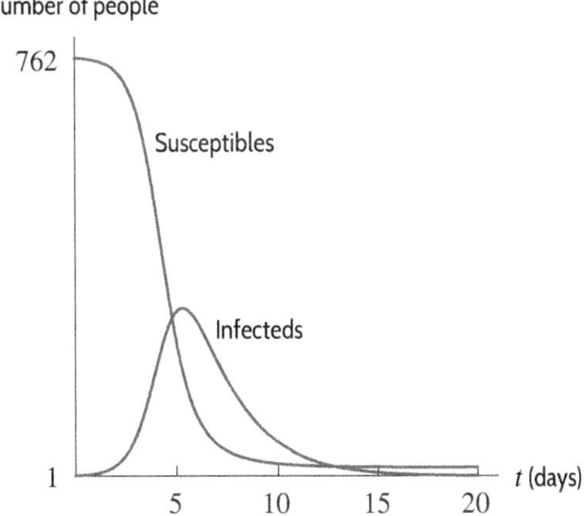

Figure 11.80: Progress of the flu over time

Two Interacting Populations: Predator-Prey

We now consider two populations which interact. They may compete for food, one may prey on the other, or they may enjoy a symbiotic relationship in which each helps the other. We model a predator-prey system using the *Lotka-Volterra equations*.

Robins and Worms

Let's look at an idealized case[31] in which robins are the predators and worms are the prey. There are r thousand robins and w million worms. If there were no robins, the worms would increase exponentially according to the equation

$$\frac{dw}{dt} = aw \quad \text{where } a \text{ is a constant and } a > 0.$$

If there were no worms, the robins would have no food and their population would decrease according to the equation[32]

$$\frac{dr}{dt} = -br \quad \text{where } b \text{ is a constant and } b > 0.$$

Now we account for the effect of the two populations on one another. Clearly, the presence of the robins is bad for the worms, so

$$\frac{dw}{dt} = aw - \text{Effect of robins on worms.}$$

On the other hand, the robins do better with the worms around, so

$$\frac{dr}{dt} = -br + \text{Effect of worms on robins.}$$

How exactly do the two populations interact? Let's assume the effect of one population on the other is proportional to the number of "encounters." (An encounter is when a robin eats a worm.) The number of encounters is likely to be proportional to the product of the populations because the more there are of either population, the more encounters there will be. So we assume

$$\frac{dw}{dt} = aw - cwr \quad \text{and} \quad \frac{dr}{dt} = -br + kwr,$$

where c and k are positive constants.

To analyze this system of equations, let's look at the specific example with $a = b = c = k = 1$:

$$\frac{dw}{dt} = w - wr \quad \text{and} \quad \frac{dr}{dt} = -r + wr.$$

To visualize the solutions to these equations, we look for trajectories in the phase plane. First we use the chain rule,

$$\frac{dr}{dw} = \frac{dr/dt}{dw/dt},$$

to obtain

$$\frac{dr}{dw} = \frac{-r + wr}{w - wr}.$$

The Slope Field and Equilibrium Points

We can get an idea of what solutions of this equation look like from the slope field in Figure 11.81. At the point $(1, 1)$ there is no slope drawn because at this point the rate of change of the worm population with respect to time is zero:

$$\frac{dw}{dt} = 1 - (1)(1) = 0.$$

The rate of change of the robin population with respect to time is also zero:

$$\frac{dr}{dt} = -1 + (1)(1) = 0.$$

[31] Based on ideas from Thomas A. McMahon.

[32] You might criticize this assumption because it predicts that the number of robins will decay exponentially, rather than die out in finite time.

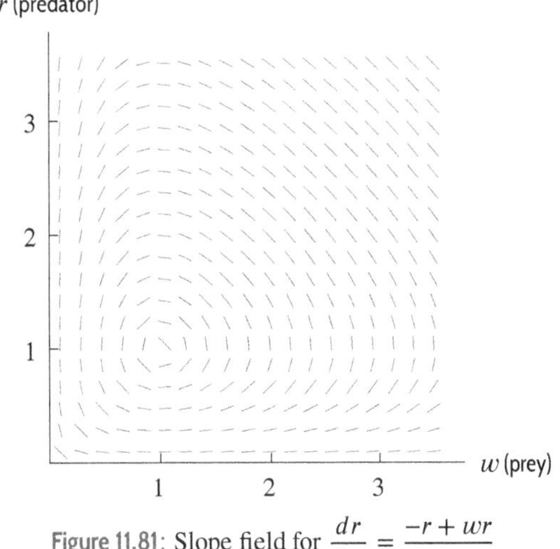

Figure 11.81: Slope field for $\dfrac{dr}{dw} = \dfrac{-r + wr}{w - wr}$

Thus dr/dw is undefined. In terms of worms and robins, this means that if at some moment $w = 1$ and $r = 1$ (that is, there are 1 million worms and 1 thousand robins), then w and r remain constant forever. The point $w = 1, r = 1$ is therefore an equilibrium solution. The slope field suggests that there are no other equilibrium points except the origin.

> At an **equilibrium point**, both w and r are constant, so
> $$\frac{dw}{dt} = 0 \quad \text{and} \quad \frac{dr}{dt} = 0.$$

Therefore, we look for equilibrium points by solving

$$\frac{dw}{dt} = w - wr = 0 \quad \text{and} \quad \frac{dr}{dt} = -r + rw = 0,$$

which has $w = 0, r = 0$ and $w = 1, r = 1$ as the only solutions.

Trajectories in the wr-Phase Plane

Let's look at the trajectories in the phase plane. Remember that a point on a curve represents a pair of populations (w, r) existing at the same time t (though t is not shown on the graph). A short time later, the pair of populations is represented by a nearby point. As time passes, the point traces out a trajectory. The direction is marked on the curve by an arrow. (See Figure 11.82.)

How do we figure out which way to move on the trajectory? Approximating the solution numerically shows that the trajectory is traversed counterclockwise. Alternatively, look at the original pair of differential equations. At the point P_0 in Figure 11.83, where $w > 1$ and $r = 1$,

$$\frac{dr}{dt} = -r + wr = -1 + w > 0.$$

Therefore, r is increasing, so the point is moving counterclockwise around the closed curve.

Now let's think about why the solution curves are closed curves (that is, why they come back and meet themselves). Notice that the slope field is symmetric about the line $w = r$. We can confirm this by observing that interchanging w and r does not alter the differential equation for dr/dw. This means that if we start at point P on the line $w = r$ and travel once around the point $(1, 1)$, we arrive back at the same point P. The reason is that the second half of the path, from Q to P, is the reflection of the first half, from P to Q, about the line $w = r$. (See Figure 11.82.) If we did not end up at P again, the second half of our path would have a different shape from the first half.

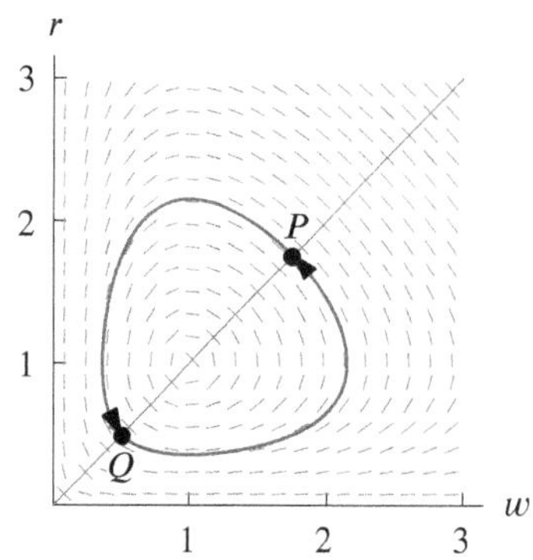

Figure 11.82: Solution curve is closed

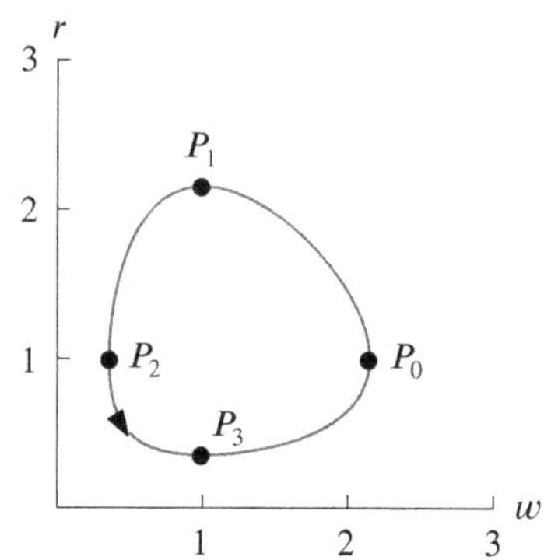

Figure 11.83: Trajectory in the phase plane

The Populations as Functions of Time

The shape of the trajectories tells us how the populations vary with time. We start at $t = 0$ at the point P_0 in Figure 11.83. Then we move to P_1 at time t_1, to P_2 at time t_2, to P_3 at time t_3, and so on. At time t_4 we are back at P_0, and the whole cycle repeats. Since the trajectory is a closed curve, both populations oscillate periodically with the same period. The worms (the prey) are at their maximum a quarter of a cycle before the robins. (See Figure 11.84.)

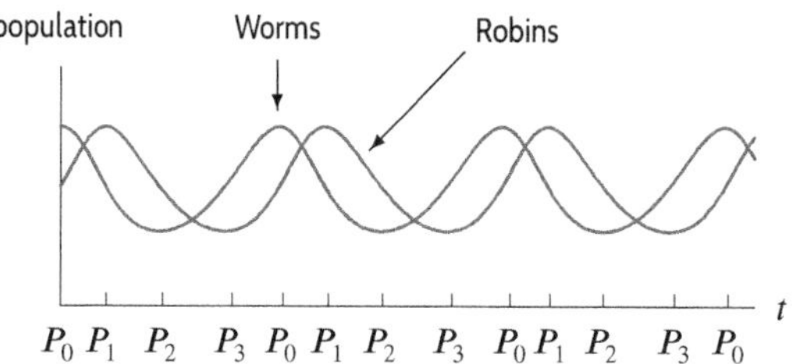

Figure 11.84: Populations of robins (in thousands) and worms (in millions) over time

Exercises and Problems for Section 11.8 Online Resource: Additional Problems for Section 11.8

EXERCISES

For Exercises 1–4, suppose x and y are the populations of two different species. Describe in words how each population changes with time.

1. 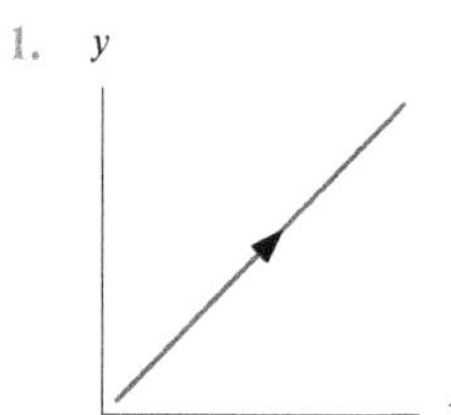

2. y x

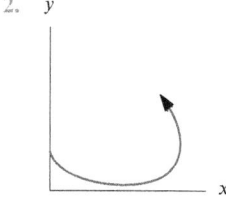

3. 4. 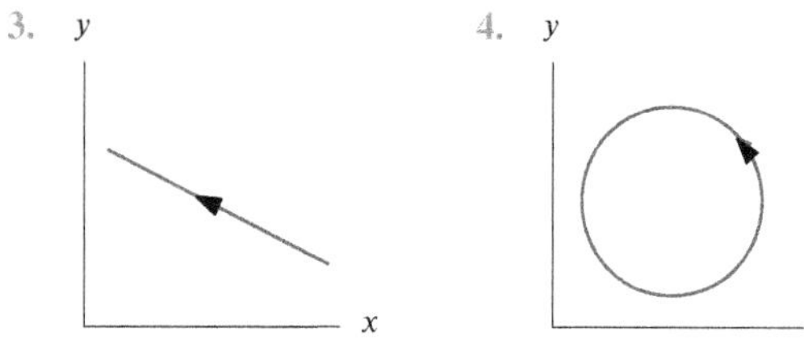

In Exercises 5–8, find all equilibrium points. Give answers as ordered pairs (x, y).

5. $\dfrac{dx}{dt} = -3x + xy$
$\dfrac{dy}{dt} = 5y - xy$

6. $\dfrac{dx}{dt} = -2x + 4xy$
$\dfrac{dy}{dt} = -8y + 2xy$

7. $\dfrac{dx}{dt} = 15x - 5xy$
$\dfrac{dy}{dt} = 10y + 2xy$

8. $\dfrac{dx}{dt} = x^2 - xy$
$\dfrac{dy}{dt} = 15y - 3y^2$

9. Given the system of differential equations

$$\frac{dx}{dt} = 5x - 3xy$$
$$\frac{dy}{dt} = -8y + xy$$

determine whether x and y are increasing or decreasing at the point

(a) $x = 3, y = 2$ **(b)** $x = 5, y = 1$

10. Given the system of differential equations

$$\frac{dP}{dt} = 2P - 10$$
$$\frac{dQ}{dt} = Q - 0.2PQ$$

determine whether P and Q are increasing or decreasing at the point

(a) $P = 2, Q = 3$ **(b)** $P = 6, Q = 5$

For Exercises 11–13, x and y satisfy the system of differential equations

$$\frac{dx}{dt} = x(3 - x + y),$$
$$\frac{dy}{dt} = y(6 - x - y).$$

11. Starting at $P_1 = (3,2)$, as t increases will (x,y) approach $P_2 = (2.90, 1.96)$ or $P_3 = (3.10, 2.03)$?

12. Starting at $Q_1 = (6,2)$, as t increases will (x,y) approach $Q_2 = (5.90, 1.94)$ or $Q_3 = (6.10, 2.07)$?

13. Find all equilibrium points of the system.

For Exercises 14–18, v and w are the number of individuals in two interacting populations with $v, w > 0$ satisfying the system of equations

$$\frac{1}{v}\frac{dv}{dt} = -0.1 + 0.003w$$
$$\frac{1}{w}\frac{dw}{dt} = 0.06 - 0.001v.$$

14. If $v = 20$ and $w = 50$, what is the *relative* growth rate of v?

15. If $v = 20$ and $w = 50$ what is dv/dt?

16. What must v be in order for w to remain constant?

17. Is the presence of the population with w members helpful or harmful to the growth of the population with v members?

18. Is the presence of the population with v members helpful or harmful to the growth of the population with w members?

PROBLEMS

19. Figure 11.85 shows the trajectory through the SI phase plane of a 50-day epidemic.

(a) Make an approximate table of values for the number of susceptibles and infecteds on the days marked on the trajectory.

(b) When is the epidemic at its peak? How many people are infected then?

(c) During the course of the epidemic, how many catch the disease and how many are spared?

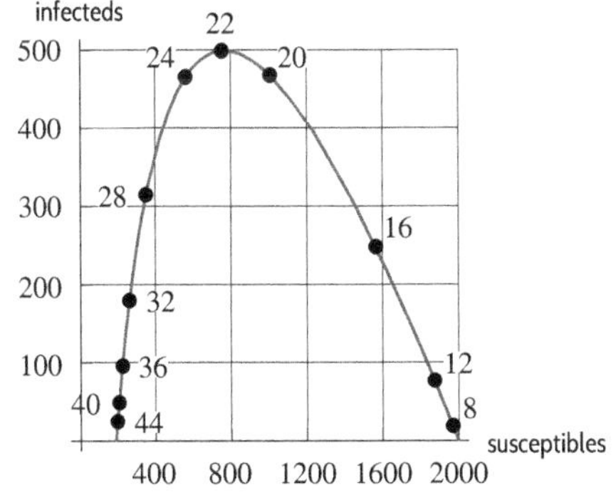

Figure 11.85: Days 8 through 44 of an epidemic

20. Figure 11.86 shows the number of susceptibles and infecteds in a population of 4000 through the course of a 60-day epidemic.

(a) How many are infected on day 20?

(b) How many have had the disease by day 20?

(c) How many have had the disease by the time the epidemic is over?

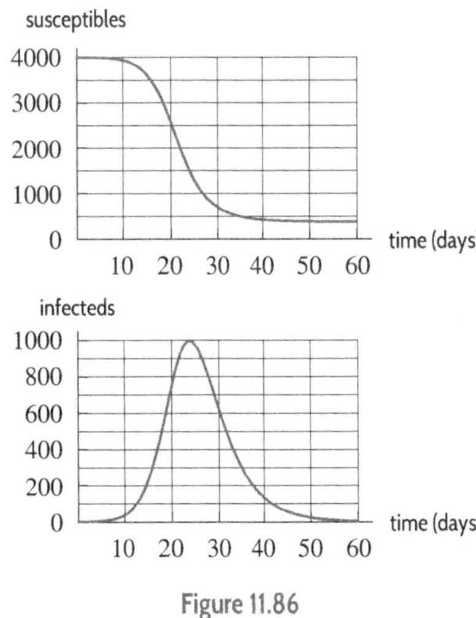

Figure 11.86

21. Humans vs Zombies[33] is a game in which one player starts as a zombie and turns human players into zombies by tagging them. Zombies have to "eat" on a regular basis by tagging human players, or they die of starvation and are out of the game. The game is usually played over a period of about five days. If we let H represent the

[33] http://humansvszombies.org

size of the human population and Z represent the size of the zombie population in the game, then, for constant parameters a, b, and c, we have:

$$\frac{dH}{dt} = aHZ$$
$$\frac{dZ}{dt} = bZ + cHZ$$

(a) Decide whether each of the parameters a, b, c is positive or negative.
(b) What is the relationship, if any, between a and c?

22. Four pairs of species are given, with descriptions of how they interact.

I. Bees/flowers: each needs the other to survive
II. Owls/trees: owls need trees but trees are indifferent
III. Elk/buffalo: in competition and would do fine alone
IV. Fox/hare: fox eats the hare and needs it to survive

Match each system of differential equations with a species pair, and indicate which species is x and which is y.

(a) $\frac{dx}{dt} = -0.2x + 0.03xy$
$\frac{dy}{dt} = 0.4y - 0.08xy$

(b) $\frac{dx}{dt} = 0.18x$
$\frac{dy}{dt} = -0.4y + 0.3xy$

(c) $\frac{dx}{dt} = -0.6x + 0.18xy$
$\frac{dy}{dt} = -0.1y + 0.09xy$

(d) Write a possible system of differential equations for the species pair that does not have a match.

23. Show that if S, I, and R satisfy the differential equations on page 617, the total population, $S + I + R$, is constant.

For Problems 24–32, let w be the number of worms (in millions) and r the number of robins (in thousands) living on an island. Suppose w and r satisfy the following differential equations, which correspond to the slope field in Figure 11.87.

$$\frac{dw}{dt} = w - wr, \qquad \frac{dr}{dt} = -r + wr.$$

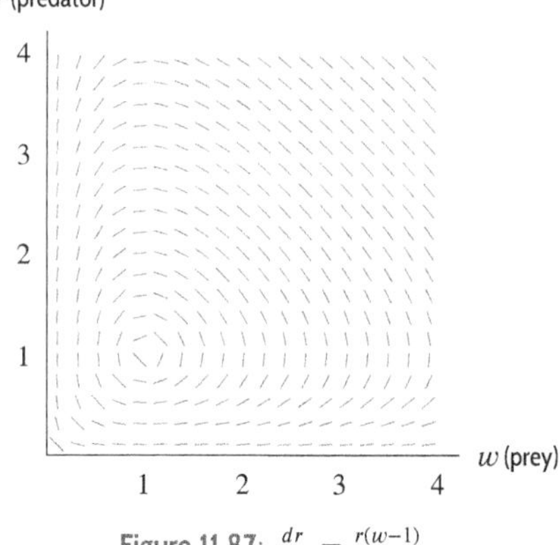

Figure 11.87: $\frac{dr}{dw} = \frac{r(w-1)}{w(1-r)}$

24. Explain why these differential equations are a reasonable model for interaction between the two populations. Why have the signs been chosen this way?

25. Solve these differential equations in the two special cases when there are no robins and when there are no worms living on the island.

26. Describe and explain the symmetry you observe in the slope field. What consequences does this symmetry have for the solution curves?

27. Assume $w = 2$ and $r = 2$ when $t = 0$. Do the numbers of robins and worms increase or decrease at first? What happens in the long run?

28. For the case discussed in Problem 27, estimate the maximum and the minimum values of the robin population. How many worms are there at the time when the robin population reaches its maximum?

29. On the same axes, graph w and r (the worm and the robin populations) against time. Use initial values of 1.5 for w and 1 for r. You may do this without units for t.

30. People on the island like robins so much that they decide to import 200 robins all the way from England, to increase the initial population from $r = 2$ to $r = 2.2$ when $t = 0$. Does this make sense? Why or why not?

31. Assume that $w = 3$ and $r = 1$ when $t = 0$. Do the numbers of robins and worms increase or decrease initially? What happens in the long run?

32. For the case discussed in Problem 31, estimate the maximum and minimum values of the robin population. Estimate the number of worms when the robin population reaches its minimum.

In Problems 33–35, the system of differential equations models the interaction of two populations x and y.
(a) What kinds of interaction (symbiosis,[34] competition, predator-prey) do the equations describe?

[34] Symbiosis takes place when the interaction of two species benefits both. An example is the pollination of plants by insects.

(b) What happens in the long run? Your answer may depend on the initial population. Draw a slope field.

33. $\frac{1}{x}\frac{dx}{dt} = y - 1$
$\frac{1}{y}\frac{dy}{dt} = x - 1$

34. $\frac{1}{x}\frac{dx}{dt} = 1 - \frac{x}{2} - \frac{y}{2}$
$\frac{1}{y}\frac{dy}{dt} = 1 - x - y$

35. $\frac{1}{x}\frac{dx}{dt} = y - 1 - 0.05x$
$\frac{1}{y}\frac{dy}{dt} = 1 - x - 0.05y$

■ For Problems 36–38, p and q are the number of individuals in two interacting populations with $p, q > 0$ satisfying the system of equations

$$\frac{1}{p}\frac{dp}{dt} = 0.01q - 0.3$$
$$\frac{1}{q}\frac{dq}{dt} = 0.02p - 0.2.$$

36. What is the relative rate of change of p if $q = 10$?

37. What populations result in an equilibrium?

38. Give a description of how these populations interact.

■ For Problems 39–42, consider a conflict between two armies of x and y soldiers, respectively. During World War I, F. W. Lanchester assumed that if both armies are fighting a conventional battle within sight of one another, the rate at which soldiers in one army are put out of action (killed or wounded) is proportional to the amount of fire the other army can concentrate on them, which is in turn proportional to the number of soldiers in the opposing army. Thus Lanchester assumed that if there are no reinforcements and t represents time since the start of the battle, then x and y obey the differential equations

$$\frac{dx}{dt} = -ay$$
$$\frac{dy}{dt} = -bx \quad a, b > 0.$$

39. Near the end of World War II a fierce battle took place between US and Japanese troops over the island of Iwo Jima, off the coast of Japan. Applying Lanchester's analysis to this battle, with x representing the number of US troops and y the number of Japanese troops, it has been estimated[35] that $a = 0.05$ and $b = 0.01$.

(a) Using these values for a and b and ignoring reinforcements, write a differential equation involving dy/dx and sketch its slope field.

(b) Assuming that the initial strength of the US forces was 54,000 and that of the Japanese was 21,500, draw the trajectory which describes the battle. What outcome is predicted? (That is, which side do the differential equations predict will win?)

(c) Would knowing that the US in fact had 19,000 reinforcements, while the Japanese had none, alter the outcome predicted?

40. **(a)** For two armies of strengths x and y fighting a conventional battle governed by Lanchester's differential equations, write a differential equation involving dy/dx and the constants of attrition a and b.

(b) Solve the differential equation and hence show that the equation of the phase trajectory is

$$ay^2 - bx^2 = C$$

for some constant C. This equation is called *Lanchester's square law*. The value of C depends on the initial sizes of the two armies.

41. Consider the battle of Iwo Jima, described in Problem 39. Take $a = 0.05$, $b = 0.01$ and assume the initial strength of the US troops to be 54,000 and that of the Japanese troops to be 21,500. (Again, ignore reinforcements.)

(a) Using Lanchester's square law derived in Problem 40, find the equation of the trajectory describing the battle.

(b) Assuming that the Japanese fought without surrendering until they had all been killed, as was the case, how many US troops does this model predict would be left when the battle ended?

42. In this problem we adapt Lanchester's model for a conventional battle to the case in which one or both of the armies is a guerrilla force. We assume that the rate at which a guerrilla force is put out of action is proportional to the product of the strengths of the two armies.

(a) Give a justification for the assumption that the rate at which a guerrilla force is put out of action is proportional to the product of the strengths of the two armies.

(b) Write the differential equations which describe a conflict between a guerrilla army of strength x and a conventional army of strength y, assuming all the constants of proportionality are 1.

(c) Find a differential equation involving dy/dx and solve it to find equations of phase trajectories.

(d) Describe which side wins in terms of the constant of integration. What happens if the constant is zero?

(e) Use your solution to part (d) to divide the phase plane into regions according to which side wins.

[35]See Martin Braun, *Differential Equations and Their Applications*, 2nd ed. (New York: Springer Verlag, 1975).

Strengthen Your Understanding

In Problems 43–44, explain what is wrong with the statement.

43. If $dx/dt = 3x - 0.4xy$ and $dy/dt = 4y - 0.5xy$, then an increase in x corresponds to a decrease in y.

44. For a system of differential equations for x and y, at the point $(2, 3)$, we have $dx/dt < 0$ and $dy/dt > 0$ and $dy/dx > 0$.

In Problems 45–47, give an example of:

45. A system of differential equations for two populations X and Y such that Y needs X to survive and X is indifferent to Y and thrives on its own. Let x represent the size of the X population and y represent the size of the Y population.

46. A system of differential equations for the profits of two companies if each would thrive on its own but the two companies compete for business. Let x and y represent the profits of the two companies.

47. Two diseases D_1 and D_2 such that the parameter a in the S-I-R model on page 616 is larger for disease D_1 than it is for disease D_2. Explain your reasoning.

Are the statements in Problems 48–49 true or false? Give an explanation for your answer.

48. The system of differential equations $dx/dt = -x + xy^2$ and $dy/dt = y - x^2y$ requires initial conditions for both $x(0)$ and $y(0)$ to determine a unique solution.

49. Populations modeled by a system of differential equations never die out.

11.9 ANALYZING THE PHASE PLANE

In the previous section we analyzed a system of differential equations using a slope field. In this section we analyze a system of differential equations using *nullclines*. We consider two species having similar *niches*, or ways of living, and that are in competition for food and space. In such cases, one species often becomes extinct. This phenomenon is called the *Principle of Competitive Exclusion*. We see how differential equations predict this in a particular case.

Competitive Exclusion: Citrus Tree Parasites

The citrus farmers of Southern California are interested in controlling the insects that live on their trees. Some of these insects can be controlled by parasites that live on the trees too. Scientists are, therefore, interested in understanding under what circumstances these parasites flourish or die out. One such parasite was introduced accidentally from the Mediterranean; later, other parasites were introduced from China and India; in each case the previous parasite became extinct over part of its habitat. In 1963 a lab experiment was carried out to determine which one of a pair of species became extinct when they were in competition with each other. The data on one pair of species, called *A. fisheri* and *A. melinus*, with populations P_1 and P_2 respectively, is given in Table 11.9 and shows that *A. melinus* (P_2) became extinct after 8 generations.[36]

Table 11.9 *Population (in thousands) of two species of parasite as a function of time*

Generation number	1	2	3	4	5	6	7	8
Population P_1 (thousands)	0.193	1.093	1.834	5.819	13.705	16.965	18.381	16.234
Population P_2 (thousands)	0.083	0.229	0.282	0.378	0.737	0.507	0.13	0

Data from the same experimenters indicates that, when alone, each population grows logistically. In fact, their data suggests that, when alone, the population of P_1 might grow according to the equation

$$\frac{dP_1}{dt} = 0.05P_1\left(1 - \frac{P_1}{20}\right),$$

and when alone, the population of P_2 might grow according to the equation

$$\frac{dP_2}{dt} = 0.09P_2\left(1 - \frac{P_2}{15}\right).$$

Now suppose both parasites are present. Each tends to reduce the growth rate of the other, so each

[36]Data adapted from Paul DeBach and Ragnhild Sundby, "Competitive Displacement Between Ecological Homologues," *Hilgardia 34*:17 (1963).

differential equation is modified by subtracting a term on the right. The experimental data shows that together P_1 and P_2 can be well described by the equations

$$\frac{dP_1}{dt} = 0.05P_1\left(1 - \frac{P_1}{20}\right) - 0.002P_1P_2$$

$$\frac{dP_2}{dt} = 0.09P_2\left(1 - \frac{P_2}{15}\right) - 0.15P_1P_2.$$

The fact that P_2 dies out with time is reflected in these equations: the coefficient of P_1P_2 is much larger in the equation for P_2 than in the equation for P_1. This indicates that the interaction has a much more devastating effect upon the growth of P_2 than on the growth of P_1.

The Phase Plane and Nullclines

We consider the phase plane with the P_1 axis horizontal and the P_2 axis vertical. To find the trajectories in the P_1P_2 phase plane, we could draw a slope field as in the previous section. Instead, we use a method that gives a good qualitative picture of the behavior of the trajectories even without a calculator or computer. We find the *nullclines* or curves along which $dP_1/dt = 0$ or $dP_2/dt = 0$. At points where $dP_2/dt = 0$, the population P_2 is momentarily constant, so only population P_1 is changing with time. Therefore, at this point the trajectory is horizontal. (See Figure 11.88.) Similarly, at points where $dP_1/dt = 0$, the population P_1 is momentarily constant and population P_2 is the only one changing, so the trajectory is vertical there. A point where both $dP_1/dt = 0$ and $dP_2/dt = 0$ is called an *equilibrium point* because P_1 and P_2 both remain constant if they reach these values.

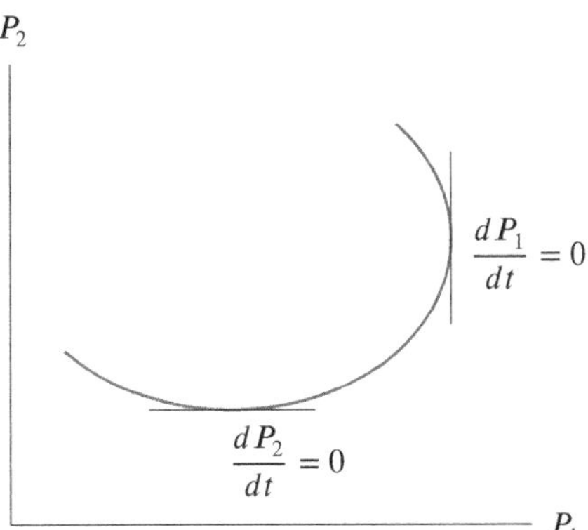

Figure 11.88: Points on a trajectory where $dP_1/dt = 0$ or $dP_2/dt = 0$

> On the P_1P_2 phase plane:
>
> - If $\dfrac{dP_1}{dt} = 0$, the trajectory is vertical.
> - If $\dfrac{dP_2}{dt} = 0$, the trajectory is horizontal.
> - If $\dfrac{dP_1}{dt} = \dfrac{dP_2}{dt} = 0$, there is an equilibrium point.

Using Nullclines to Analyze the Parasite Populations

In order to see where $dP_1/dt = 0$ or $dP_2/dt = 0$, we factor the right side of our differential equations:

$$\frac{dP_1}{dt} = 0.05P_1\left(1 - \frac{P_1}{20}\right) - 0.002P_1P_2 = 0.001P_1(50 - 2.5P_1 - 2P_2)$$

$$\frac{dP_2}{dt} = 0.09P_2\left(1 - \frac{P_2}{15}\right) - 0.15P_1P_2 = 0.001P_2(90 - 150P_1 - 6P_2).$$

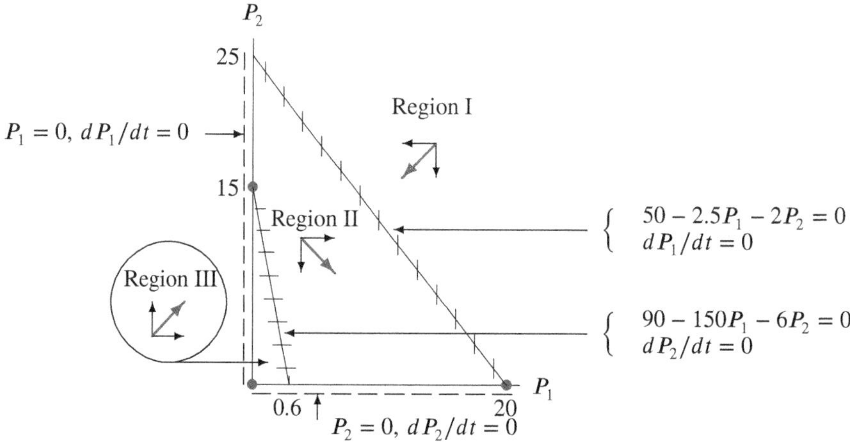

Figure 11.89: Analyzing three regions in the phase plane using nullclines (axes distorted) with equilibrium points represented by dots

Thus $dP_1/dt = 0$ where $P_1 = 0$ or where $50 - 2.5P_1 - 2P_2 = 0$. Graphing these equations in the phase plane gives two lines, which are nullclines. Since the trajectory is vertical where $dP_1/dt = 0$, in Figure 11.89 we draw small vertical line segments on these nullclines to represent the direction of the trajectories as they cross the nullcline. Similarly $dP_2/dt = 0$ where $P_2 = 0$ or where $90 - 150P_1 - 6P_2 = 0$. These equations are graphed in Figure 11.89 with small horizontal line segments on them.

The equilibrium points are where both $dP_1/dt = 0$ and $dP_2/dt = 0$, namely the points $P_1 = 0, P_2 = 0$ (meaning that both species die out); $P_1 = 0, P_2 = 15$ (where P_1 is extinct); and $P_1 = 20, P_2 = 0$ (where P_2 is extinct). Notice that the equilibrium points are located at the intersection of a nullcline with vertical line segments and a nullcline with horizontal line segments.

What Happens in the Regions Between the Nullclines?

Nullclines are useful because they divide the plane into regions in which the signs of dP_1/dt and dP_2/dt are constant. In each region, the direction of every trajectory remains roughly the same.

In Region I, for example, we might try the point $P_1 = 20, P_2 = 25$. Then

$$\frac{dP_1}{dt} = 0.001(20)(50 - 2.5(20) - 2(25)) < 0$$
$$\frac{dP_2}{dt} = 0.001(25)(90 - 150(20) - 6(25)) < 0.$$

Now $dP_1/dt < 0$, so P_1 is decreasing, which can be represented by an arrow in the direction $\leftarrow$. Also, $dP_2/dt < 0$, so P_2 is decreasing, as represented by the arrow $\downarrow$. Combining these directions, we know that the trajectories in this region go approximately in the diagonal direction $\swarrow$ (See Region I in Figure 11.89.)

In Region II, try, for example, $P_1 = 1, P_2 = 1$. Then we have

$$\frac{dP_1}{dt} = 0.001(1)(50 - 2.5 - 2) > 0$$
$$\frac{dP_2}{dt} = 0.001(1)(90 - 150 - 6) < 0.$$

So here, P_1 is increasing while P_2 is decreasing. (See Region II in Figure 11.89.)

In Region III, try $P_1 = 0.1, P_2 = 0.1$:

$$\frac{dP_1}{dt} = 0.001(0.1)(50 - 2.5(0.1) - 2(0.1)) > 0$$
$$\frac{dP_2}{dt} = 0.001(0.1)(90 - 150(0.1) - 6(0.1)) > 0.$$

So here, both P_1 and P_2 are increasing. (See Region III in Figure 11.89.)

Notice that the behavior of the populations in each region makes biological sense. In region I both populations are so large that overpopulation is a problem, so both populations decrease. In Region III both populations are so small that they are effectively not in competition, so both grow. In Region II competition between the species comes into play. The fact that P_1 increases while P_2 decreases in Region II means that P_1 wins.

Solution Trajectories

Suppose the system starts with some of each population. This means that the initial point of the trajectory is not on one of the axes, and so it is in Region I, II, or III. Then the point moves on a trajectory like one of those computed numerically and shown in Figure 11.90. Notice that *all* these trajectories tend toward the point $P_1 = 20, P_2 = 0$, corresponding to a population of 20,000 for P_1 and extinction for P_2. Consequently, this model predicts that no matter what the initial populations are, provided $P_1 \neq 0$, the population of P_2 is excluded by P_1, and P_1 tends to a constant value. This makes biological sense: in the absence of P_2, we would expect P_1 to settle down to the carrying capacity of the niche, which is 20,000.

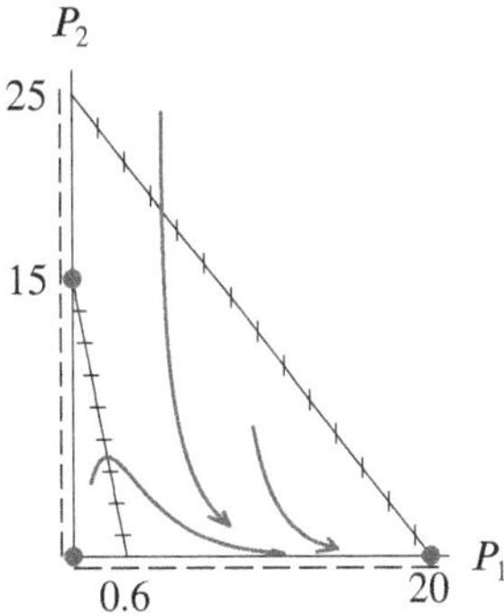

Figure 11.90: Trajectories showing exclusion of population P_2 (not to scale)

Exercises and Problems for Section 11.9

EXERCISES

In Exercises 1–5, use Figure 11.91.

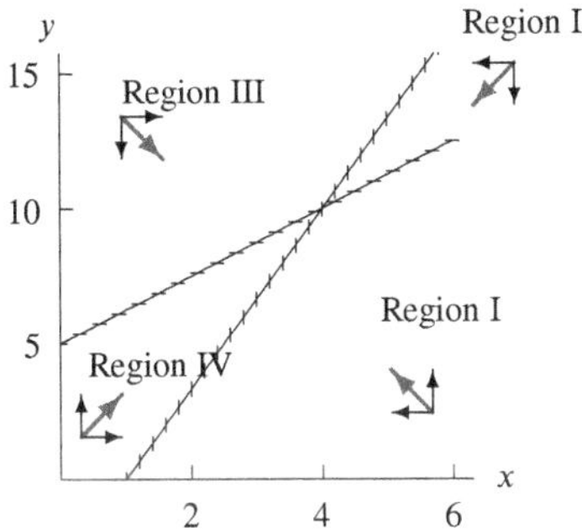

Figure 11.91: Nullclines in the phase plane of a system of differential equations

1. Give the coordinates for each equilibrium point.
2. At each point, give the signs of dx/dt and dy/dt.

   (a) (4, 7) **(b)** (4, 10) **(c)** (6, 15)
3. Draw a possible trajectory that starts at the point (2, 5).
4. Draw a possible trajectory that starts at the point (2, 10).
5. What is the long-run behavior of a trajectory that starts at any point in the first quadrant?

In Exercises 6–10, use Figure 11.92, which shows four nullclines, two of which are on the axes.

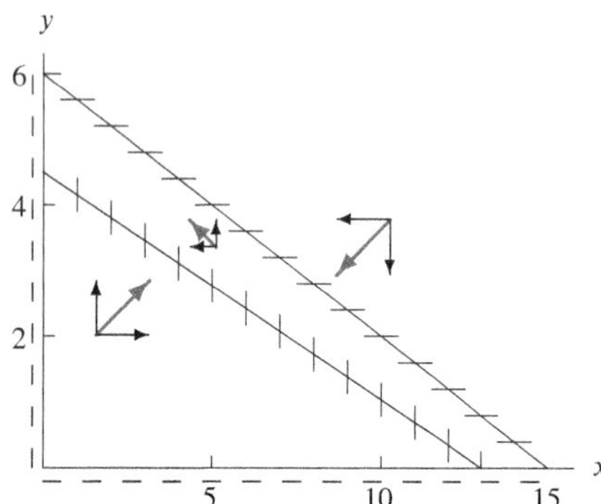

Figure 11.92: Nullclines in the phase plane of a system of differential equations

6. Give the approximate coordinates for each equilibrium point.

7. At each point, give the signs of dx/dt and dy/dt.

   (a) $(5, 2)$ (b) $(10, 2)$ (c) $(10, 1)$

8. Draw a possible trajectory that starts at the point $(2, 5)$.

9. Draw a possible trajectory that starts at the point $(10, 4)$.

10. What is the long-run behavior of a trajectory that starts at any point in the first quadrant?

11. Figure 11.93 shows a phase plane for a system of differential equations. Draw the nullclines.

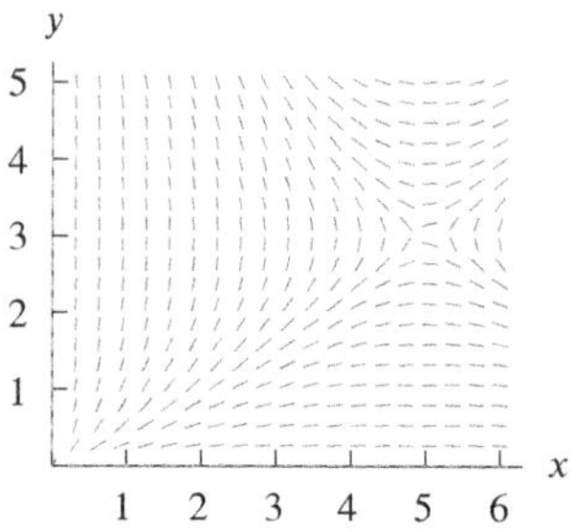

Figure 11.93

12. (a) Find the equilibrium points for the following system of equations

$$\frac{dx}{dt} = 20x - 10xy$$

$$\frac{dy}{dt} = 25y - 5xy.$$

(b) Explain why $x = 2$, $y = 4$ is not an equilibrium point for this system.

PROBLEMS

For Problems 13–18, analyze the phase plane of the differential equations for $x, y \geq 0$. Show the nullclines and equilibrium points, and sketch the direction of the trajectories in each region.

13. $\dfrac{dx}{dt} = x(2 - x - y)$
$\dfrac{dy}{dt} = y(1 - x - y)$

14. $\dfrac{dx}{dt} = x(2 - x - 3y)$
$\dfrac{dy}{dt} = y(1 - 2x)$

15. $\dfrac{dx}{dt} = x(2 - x - 2y)$
$\dfrac{dy}{dt} = y(1 - 2x - y)$

16. $\dfrac{dx}{dt} = x(1 - y - \dfrac{x}{3})$
$\dfrac{dy}{dt} = y(1 - \dfrac{y}{2} - x)$

17. $\dfrac{dx}{dt} = x\left(1 - x - \dfrac{y}{3}\right)$
$\dfrac{dy}{dt} = y\left(1 - y - \dfrac{x}{2}\right)$

18. $\dfrac{dx}{dt} = x\left(1 - \dfrac{x}{2} - y\right)$
$\dfrac{dy}{dt} = y\left(1 - \dfrac{y}{3} - x\right)$

19. The equations describing the flu epidemic in a boarding school are

$$\frac{dS}{dt} = -0.0026SI$$

$$\frac{dI}{dt} = 0.0026SI - 0.5I.$$

(a) Find the nullclines and equilibrium points in the SI phase plane.

(b) Find the direction of the trajectories in each region.

(c) Sketch some typical trajectories and describe their behavior in words.

20. Use the idea of nullclines dividing the plane into sectors to analyze the equations describing the interactions of robins and worms:

$$\frac{dw}{dt} = w - wr$$

$$\frac{dr}{dt} = -r + rw.$$

21. Two companies share the market for a new technology. They have no competition except each other. Let $A(t)$ be the net worth of one company and $B(t)$ the net worth of the other at time t. Suppose that net worth cannot be negative and that A and B satisfy the differential equations

$$A' = 2A - AB$$
$$B' = B - AB.$$

(a) What do these equations predict about the net worth of each company if the other were not present? What effect do the companies have on each other?

(b) Are there any equilibrium points? If so, what are they?

(c) Sketch a slope field for these equations (using a computer or calculator), and hence describe the different possible long-run behaviors.

22. In the 1930s L. F. Richardson proposed that an arms race between two countries could be modeled by a system of differential equations. One arms race that can be reasonably well described by differential equations is the US-Soviet Union arms race between 1945 and 1960. If $\$x$ represents the annual Soviet expenditures on armaments (in billions of dollars) and $\$y$ represents the corresponding US expenditures, it has been suggested[37] that x and y obey the following differential equations:

$$\frac{dx}{dt} = -0.45x + 10.5$$
$$\frac{dy}{dt} = 8.2x - 0.8y - 142.$$

(a) Find the nullclines and equilibrium points for these differential equations. Which direction do the trajectories go in each region?

(b) Sketch some typical trajectories in the phase plane.

(c) What do these differential equations predict will be the long-term outcome of the US-Soviet arms race?

(d) Discuss these predictions in the light of the actual expenditures in Table 11.10.

Table 11.10 *Arms budgets of the United States and the Soviet Union for the years 1945–1960 (billions of dollars)*

	USSR	USA		USSR	USA
1945	14	97	1953	25.7	71.4
1946	14	80	1954	23.9	61.6
1947	15	29	1955	25.5	58.3
1948	20	20	1956	23.2	59.4
1949	20	22	1957	23.0	61.4
1950	21	23	1958	22.3	61.4
1951	22.7	49.6	1959	22.3	61.7
1952	26.0	69.6	1960	22.1	59.6

23. In the 1930s, the Soviet ecologist G. F. Gause performed a series of experiments on competition among two yeasts with populations P_1 and P_2, respectively. By performing population studies at low density in large volumes, he determined what he called the *coefficients of geometric increase* (and we would call continuous exponential growth rates). These coefficients described the growth of each yeast alone:

$$\frac{1}{P_1}\frac{dP_1}{dt} = 0.2$$
$$\frac{1}{P_2}\frac{dP_2}{dt} = 0.06$$

where P_1 and P_2 are measured in units that Gause established.

He also determined that, in his units, the carrying capacity of P_1 was 13 and the carrying capacity of P_2 was 6. He then observed that one P_2 occupies the niche space of 3 P_1 and that one P_1 occupied the niche space of 0.4 P_2. This led him to the following differential equations to describe the interaction of P_1 and P_2:

$$\frac{dP_1}{dt} = 0.2P_1\left(\frac{13-(P_1+3P_2)}{13}\right)$$
$$\frac{dP_2}{dt} = 0.06P_2\left(\frac{6-(P_2+0.4P_1)}{6}\right).$$

When both yeasts were growing together, Gause recorded the data in Table 11.11.

Table 11.11 *Gause's yeast populations*

Time (hours)	6	16	24	29	48	53
P_1	0.375	3.99	4.69	6.15	7.27	8.30
P_2	0.29	0.98	1.47	1.46	1.71	1.84

(a) Carry out a phase plane analysis of Gause's equations.

(b) Mark the data points on the phase plane and describe what would have happened had Gause continued the experiment.

[37]R. Taagepera, G. M. Schiffler, R. T. Perkins and D. L. Wagner, *Soviet-American and Israeli-Arab Arms Races and the Richardson Model* (General Systems, XX, 1975).

Strengthen Your Understanding

In Problems 24–25, explain what is wrong with the statement.

24. A solution trajectory and nullclines for a system of differential equations are shown in Figure 11.94.

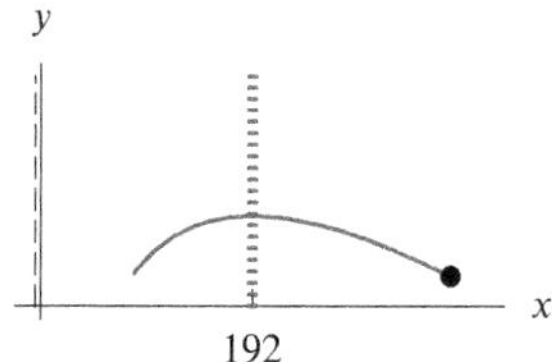

Figure 11.94

25. The nullclines for a system of differential equations are shown in Figure 11.95. The system has an equilibrium at the point $(6, 6)$.

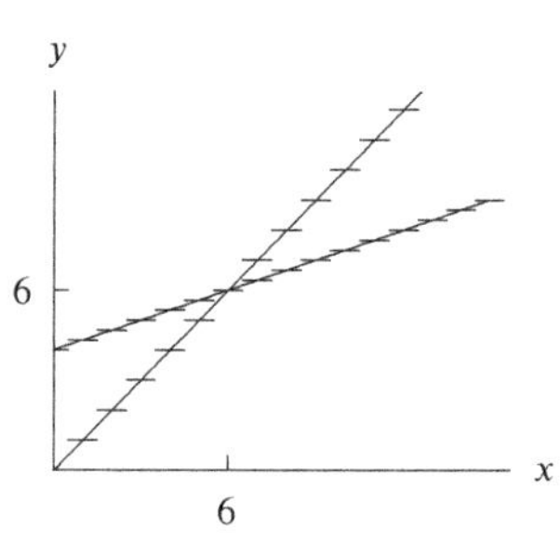

Figure 11.95

In Problems 26–28, give an example of:

26. A graph of the nullclines of a system of differential equations with exactly two equilibrium points in the first quadrant. Label the nullclines to show whether trajectories pass through the nullcline vertically or horizontally.

27. The nullclines of a system of differential equations with the trajectory shown in Figure 11.96.

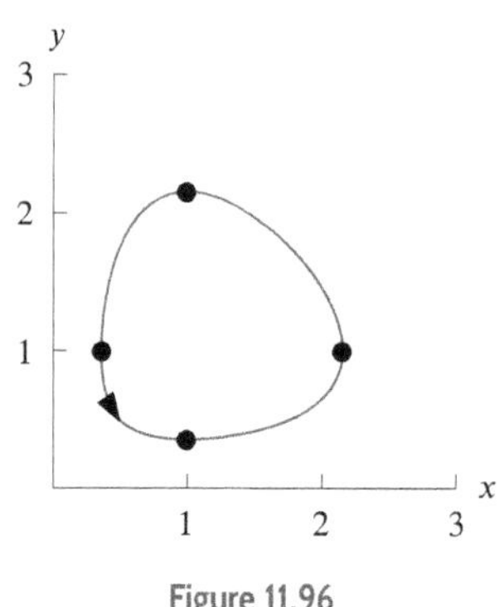

Figure 11.96

28. A trajectory for a system of differential equations with nullcline in Figure 11.97 and initial conditions $x = 1$ and $y = 2$.

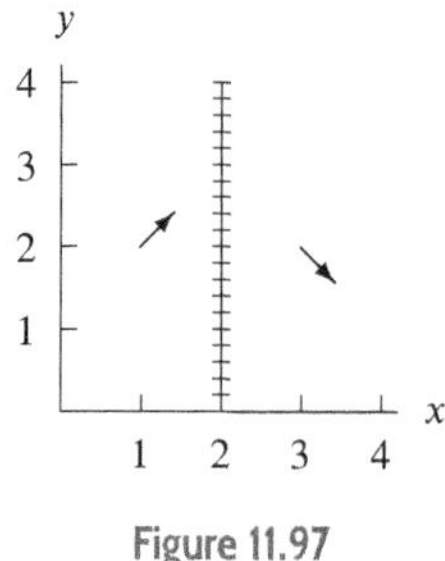

Figure 11.97

Are the statements in Problems 29–32 true or false? Give an explanation for your answer.

29. Nullclines are always straight lines.

30. If two nullclines cross, their intersection point is an equilibrium.

31. A system of two differential equations has at most two nullclines.

32. A system of two differential equations must have at least two nullclines.

11.10 SECOND-ORDER DIFFERENTIAL EQUATIONS: OSCILLATIONS

A Second-Order Differential Equation

When a body moves freely under gravity, we know that

$$\frac{d^2s}{dt^2} = -g,$$

where s is the height of the body above ground at time t and g is the acceleration due to gravity. To solve this equation, we first integrate to get the velocity, $v = ds/dt$:

$$\frac{ds}{dt} = -gt + v_0,$$

where v_0 is the initial velocity. Then we integrate again, giving

$$s = -\frac{1}{2}gt^2 + v_0 t + s_0,$$

where s_0 is the initial height.

The differential equation $d^2s/dt^2 = -g$ is called *second order* because the equation contains a second derivative but no higher derivatives. The general solution to a second-order differential equation is a family of functions with two parameters, here v_0 and s_0. Finding values for the two constants corresponds to picking a particular function out of this family.

A Mass on a Spring

Not every second-order differential equation can be solved simply by integrating twice. Consider a mass m attached to the end of a spring hanging from the ceiling. We assume that the mass of the spring itself is negligible in comparison with the mass m. (See Figure 11.98.)

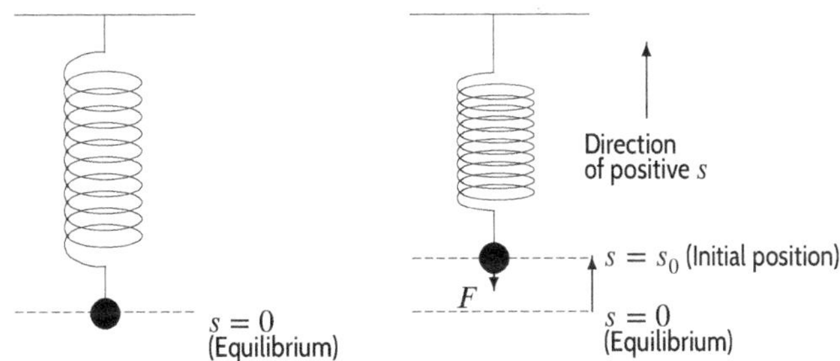

Figure 11.98: Spring and mass in equilibrium position and after upward displacement

When the system is left undisturbed, no net force acts on the mass. The force of gravity is balanced by the force the spring exerts on the mass, and the spring is in the *equilibrium position.* If we pull down on the mass, we feel a force pulling upward. If instead, we push upward on the mass, the opposite happens: a force pushes the mass down.[38]

What happens if we push the mass upward and then release it? We expect the mass to oscillate up and down around the equilibrium position.

Springs and Hooke's Law

In order to figure out how the mass moves, we need to know the exact relationship between its displacement, s, from the equilibrium position and the net force, F, exerted on the mass. (See Figure 11.98.) We would expect that the further the mass is from the equilibrium position, and the more the spring is stretched or compressed, the larger the force. In fact, provided that the displacement is not large enough to deform the spring permanently, experiments show that the net force, F, is approximately proportional to the displacement, s:

$$F = -ks,$$

[38] Pulling down on the mass stretches the spring, increasing the tension, so the combination of gravity and spring force is upward. Pushing up the spring decreases tension in the spring, so the combination of gravity and spring force is downward.

where k is the *spring constant* ($k > 0$) and the negative sign means that the net force is in the opposite direction to the displacement. The value of k depends on the physical properties of the particular spring. This relationship is known as *Hooke's Law.* Suppose we push the mass upward some distance and then release it. After we let go the net force causes the mass to accelerate toward the equilibrium position. By Newton's Second Law of Motion we have

$$\text{Force} = \text{Mass} \times \text{Acceleration}.$$

Since acceleration is d^2s/dt^2 and force is $F = -ks$ by Hooke's law, we have

$$-ks = m\frac{d^2s}{dt^2},$$

which is equivalent to the following differential equation.

Equation for Oscillations of a Mass on a Spring

$$\frac{d^2s}{dt^2} = -\frac{k}{m}s$$

Thus, the motion of the mass is described by a second-order differential equation. Since we expect the mass to oscillate, we guess that the solution to this equation involves trigonometric functions.

Solving the Differential Equation by Guess-and-Check

Guessing is the tried-and-true method for solving differential equations. It may surprise you to learn that there is no systematic method for solving most differential equations analytically, so guesswork is often extremely important.

Example 1 By guessing, find the general solution to the equation

$$\frac{d^2s}{dt^2} = -s.$$

Solution We want to find functions whose second derivative is the negative of the original function. We are already familiar with two functions that have this property: $s(t) = \cos t$ and $s(t) = \sin t$. We check that they are solutions by substituting:

$$\frac{d^2}{dt^2}(\cos t) = \frac{d}{dt}(-\sin t) = -\cos t,$$

and

$$\frac{d^2}{dt^2}(\sin t) = \frac{d}{dt}(\cos t) = -\sin t.$$

The remarkable thing is that starting from these two particular solutions, we can build up *all* the solutions to our equation. Given two constants C_1 and C_2, the function

$$s(t) = C_1\cos t + C_2\sin t$$

satisfies the differential equation, since

$$\frac{d^2}{dt^2}(C_1\cos t + C_2\sin t) = \frac{d}{dt}(-C_1\sin t + C_2\cos t)$$

$$= -C_1 \cos t - C_2 \sin t$$
$$= -(C_1 \cos t + C_2 \sin t).$$

It can be shown (though we will not do so) that $s(t) = C_1 \cos t + C_2 \sin t$ is the most general form of the solution. As expected, it contains two constants, C_1 and C_2.

If the differential equation represents a physical problem, then C_1 and C_2 can often be computed, as shown in the next example.

Example 2 If a mass on a spring is displaced by a distance of s_0 and then released at $t = 0$, find the solution to

$$\frac{d^2s}{dt^2} = -s$$

Solution The position of the mass is given by the equation

$$s(t) = C_1 \cos t + C_2 \sin t.$$

We also know that the initial position is s_0; thus,

$$s(0) = C_1 \cos 0 + C_2 \sin 0 = C_1 \cdot 1 + C_2 \cdot 0 = s_0,$$

so $C_1 = s_0$, the initial displacement. What is C_2? To find it, we use the fact that at $t = 0$, when the mass has just been released, its velocity is 0. Velocity is the derivative of the displacement, so

$$\left.\frac{ds}{dt}\right|_{t=0} = (-C_1 \sin t + C_2 \cos t)\Big|_{t=0} = -C_1 \cdot 0 + C_2 \cdot 1 = 0,$$

so $C_2 = 0$. Therefore the solution is $s = s_0 \cos t$.

Solution to the General Spring Equation

Having found the general solution to the equation $d^2s/dt^2 = -s$, let us return to the equation

$$\frac{d^2s}{dt^2} = -\frac{k}{m}s.$$

We write $\omega = \sqrt{k/m}$ (why will be clear in a moment), so the differential equation becomes

$$\frac{d^2s}{dt^2} = -\omega^2 s.$$

This equation no longer has $\sin t$ and $\cos t$ as solutions, since

$$\frac{d^2}{dt^2}(\sin t) = -\sin t \neq -\omega^2 \sin t.$$

However, since we want a factor of ω^2 after differentiating twice, the chain rule leads us to guess that $\sin \omega t$ may be a solution. Checking this:

$$\frac{d^2}{dt^2}(\sin \omega t) = \frac{d}{dt}(\omega \cos \omega t) = -\omega^2 \sin \omega t,$$

we see that $\sin \omega t$ is a solution, and you can check that $\cos \omega t$ is a solution, too.

The general solution to the equation

$$\frac{d^2s}{dt^2} + \omega^2 s = 0$$

is of the form

$$s(t) = C_1 \cos \omega t + C_2 \sin \omega t,$$

where C_1 and C_2 are arbitrary constants. (We assume $\omega > 0$.) The period of this oscillation is

$$T = \frac{2\pi}{\omega}.$$

Such oscillations are called **simple harmonic motion**.

Thus the solution to our original equation, $\frac{d^2s}{dt^2} + \frac{k}{m}s = 0$, is $s = C_1 \cos\sqrt{\frac{k}{m}}t + C_2 \sin\sqrt{\frac{k}{m}}t$.

Initial Value and Boundary-Value Problems

A problem in which the initial position and the initial velocity are used to determine the particular solution is called an *initial value problem.* (See Example 2.) Alternatively, we may be given the position at two known times. Such a problem is known as a *boundary-value problem.*

Example 3 Find a solution to the differential equation satisfying each set of conditions below

$$\frac{d^2s}{dt^2} + 4s = 0.$$

(a) The boundary conditions $s(0) = 0$, $s(\pi/4) = 20$.
(b) The initial conditions $s(0) = 1$, $s'(0) = -6$.

Solution Since $\omega^2 = 4$, $\omega = 2$, the general solution to the differential equation is

$$s(t) = C_1 \cos 2t + C_2 \sin 2t.$$

(a) Substituting the boundary condition $s(0) = 0$ into the general solution gives

$$s(0) = C_1 \cos(2 \cdot 0) + C_2 \sin(2 \cdot 0) = C_1 \cdot 1 + C_2 \cdot 0 = C_1 = 0.$$

Thus $s(t)$ must have the form $s(t) = C_2 \sin 2t$. The second condition yields the value of C_2:

$$s\left(\frac{\pi}{4}\right) = C_2 \sin\left(2 \cdot \frac{\pi}{4}\right) = C_2 = 20.$$

Therefore, the solution satisfying the boundary conditions is

$$s(t) = 20 \sin 2t.$$

(b) For the initial value problem, we start from the same general solution: $s(t) = C_1 \cos 2t + C_2 \sin 2t$. Substituting 0 for t once again, we find

$$s(0) = C_1 \cos(2 \cdot 0) + C_2 \sin(2 \cdot 0) = C_1 = 1.$$

Differentiating $s(t) = \cos 2t + C_2 \sin 2t$ gives

$$s'(t) = -2 \sin 2t + 2C_2 \cos 2t,$$

and applying the second initial condition gives us C_2:

$$s'(0) = -2\sin(2\cdot 0) + 2C_2\cos(2\cdot 0) = 2C_2 = -6,$$

so $C_2 = -3$ and our solution is

$$s(t) = \cos 2t - 3\sin 2t.$$

What do the Graphs of Our Solutions Look Like?

Since the general solution of the equation $d^2s/dt^2 + \omega^2 s = 0$ is of the form

$$s(t) = C_1\cos\omega t + C_2\sin\omega t$$

it would be useful to know what the graph of such a sum of sines and cosines looks like. We start with the example $s(t) = \cos t + \sin t$, which is graphed in Figure 11.99.

Interestingly, the graph in Figure 11.99 looks like another sine function, and in fact it is one. If we measure the graph carefully, we find that its period is 2π, the same as $\sin t$ and $\cos t$, but the amplitude is approximately 1.414 (in fact, it's $\sqrt{2}$), and the graph is shifted along the t-axis (in fact by $\pi/4$). If we plot $C_1\cos t + C_2\sin t$ for any C_1 and C_2, the resulting graph is always a sine function with period 2π, though the amplitude and shift can vary. For example, the graph of $s = 6\cos t - 8\sin t$ is in Figure 11.100; as expected, it has period 2π.

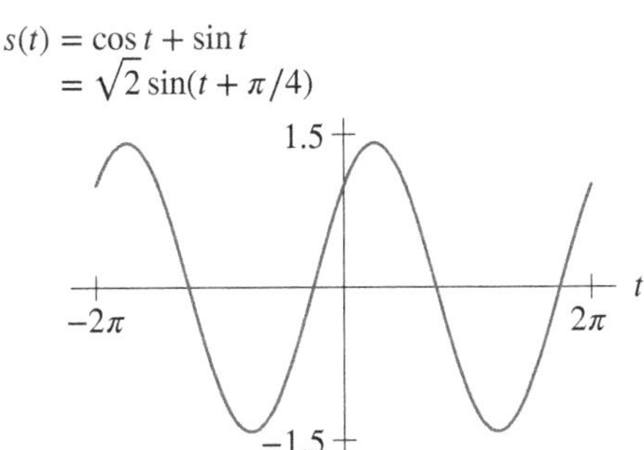

Figure 11.99: Graph of the sum: $s(t) = \cos t + \sin t = \sqrt{2}\sin(t + \pi/4)$

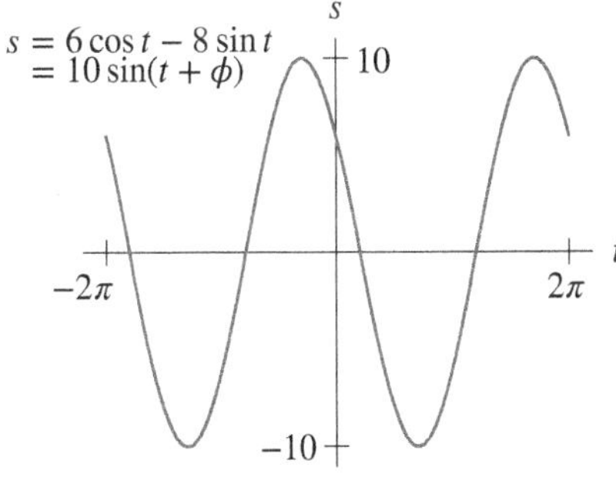

Figure 11.100: Graph of a sum of sine and cosine. Amplitude $A = 10 = \sqrt{6^2 + 8^2}$

These graphs suggest that we can write the sum of a sine and a cosine of the same argument as one single sine function.[39] Problem 27 shows that this can always be done using the following relations.

If $C_1\cos\omega t + C_2\sin\omega t = A\sin(\omega t + \phi)$, the *amplitude*, A, is given by

$$A = \sqrt{C_1^2 + C_2^2}.$$

The angle ϕ is called the *phase shift* and satisfies

$$\tan\phi = \frac{C_1}{C_2}.$$

We choose ϕ in $(-\pi, \pi]$ such that if $C_1 > 0$, ϕ is positive, and if $C_1 < 0$, ϕ is negative.

[39] The sum of a sine and a cosine can also be written as a single cosine function.

The reason that it is often useful to write the solution to the differential equation as a single sine function $A\sin(\omega t+\phi)$, as opposed to the sum of a sine and a cosine, is that the amplitude A and the phase shift ϕ are easier to recognize in this form.

Warning: Phase Shift Is Not the Same as Horizontal Translation

Let's look at $s = \sin(3t + \pi/2)$. If we rewrite this as $s = \sin(3(t + \pi/6))$, we can see from Figure 11.101 that the graph of this function is the graph of $s = \sin 3t$ shifted to the left a distance of $\pi/6$. (Remember that replacing x by $x - 2$ shifts a graph by 2 to the right.) But $\pi/6$ is *not* the phase shift; the phase shift[40] is $\pi/2$. From the point of view of a scientist, the important question is often not the distance the curve has shifted, but the relation between the distance shifted and the period.

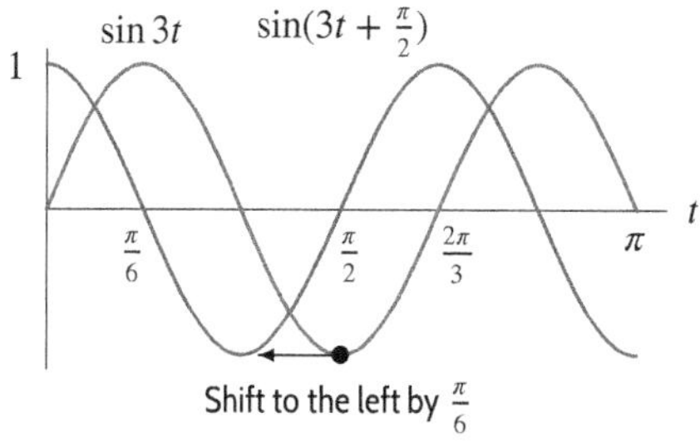

Figure 11.101: Phase shift is $\frac{\pi}{2}$; horizontal translation is $\frac{\pi}{6}$

Exercises and Problems for Section 11.10

EXERCISES

1. Find the amplitude of $y = 3\sin 2t + 4\cos 2t$.
2. Find the amplitude of $3\sin 2t + 7\cos 2t$.
3. Write $5\sin(2t) + 12\cos(2t)$ in the form $A\sin(\omega t + \psi)$.
4. Write $7\sin\omega t + 24\cos\omega t$ in the form $A\sin(\omega t + \phi)$.
5. **(a)** Using a calculator or computer, graph $s = 4\cos t + 3\sin t$ for $-2\pi \le t \le 2\pi$.
   (b) If $4\cos t + 3\sin t = A\sin(t + \phi)$, use your graph to estimate the values of A and ϕ.
   (c) Calculate A and ϕ analytically.
6. Check by differentiation that $y = 2\cos t + 3\sin t$ is a solution to $y'' + y = 0$.
7. Check by differentiation that $y(t) = 3\sin 2t + 2\cos 2t$ is a solution of $y'' + 4y = 0$.
8. Check by differentiation that $y = A\cos t + B\sin t$ is a solution to $y'' + y = 0$ for any constants A and B.
9. Check by differentiation that $y(t) = A\sin 2t + B\cos 2t$ is a solution of $y'' + 4y = 0$ for all values of A and B.
10. Check by differentiation that $y(t) = A\sin\omega t + B\cos\omega t$ is a solution of $y'' + \omega^2 y = 0$ for all values of A and B.
11. What values of α and A make $y = A\cos\alpha t$ a solution to $y'' + 5y = 0$ such that $y'(1) = 3$?

Find a general solution to the differential equations in Exercises 12–13.

12. $\dfrac{d^2z}{dt^2} + \pi^2 z = 0$
13. $9z'' + z = 0$

PROBLEMS

14. A spring has spring constant $k = 0.8$ and mass $m = 10$ grams.
   (a) Write the second-order differential equation for oscillations of the mass on the spring.
   (b) Write the general solution to the differential equation.
15. A spring has spring constant $k = 250$ and mass $m = 100$ grams.
   (a) Write the second-order differential equation for oscillations of the mass on the spring.

[40]This definition of phase shift is the one usually used in the sciences.

(b) Write the solution to the differential equation if the initial position is $s(0) = 5$ and the initial velocity is $s'(0) = -10$.

(c) How far down does the mass go?

The functions in Problems 16–18 describe the motion of a mass on a spring satisfying the differential equation $y'' = -9y$, where y is the displacement of the mass from the equilibrium position at time t, with upward as positive. In each case, describe in words how the motion starts when $t = 0$. For example, is the mass at the highest point, the lowest point, or in the middle? Is it moving up or down or is it at rest?

16. $y = 2\cos 3t$

17. $y = -0.5\sin 3t$

18. $y = -\cos 3t$

19. **(a)** Show that $y = c_1 \sinh wt + c_2 \cosh wt$ is a solution to $y'' - w^2 y = 0$.
   (b) Find a solution of this differential equation such that
   (i) $y(0) = 0,\ y(1) = 6$.
   (ii) $y'(0) = 0,\ y(1) = 6$.

20. **(a)** Find the general solution of the differential equation
$$y'' + 9y = 0.$$
   (b) For each of the following initial conditions, find a particular solution.
   (i) $y(0) = 0,\ y'(0) = 1$
   (ii) $y(0) = 1,\ y'(0) = 0$
   (iii) $y(0) = 1,\ y(1) = 0$
   (iv) $y(0) = 0,\ y(1) = 1$
   (c) Sketch a graph of the solutions found in part (b).

21. Consider the motion described by the differential equations:

(a) $x'' + 16x = 0,\quad x(0) = 5,\quad x'(0) = 0,$
(b) $25x'' + x = 0,\quad x(0) = -1,\quad x'(0) = 2.$

In each case, find a formula for $x(t)$ and calculate the amplitude and period of the motion.

22. Each graph in Figure 11.102 represents a solution to one of the differential equations:
(a) $x'' + x = 0,$ **(b)** $x'' + 4x = 0,$
(c) $x'' + 16x = 0.$

Assuming the t-scales on the four graphs are the same, which graph represents a solution to which equation? Find an equation for each graph.

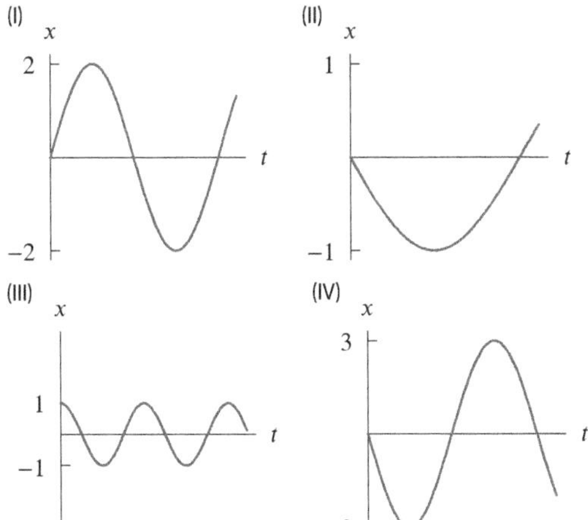

Figure 11.102

23. The following differential equations represent oscillating springs.

(i) $s'' + 4s = 0 \quad s(0) = 5, \quad s'(0) = 0$
(ii) $4s'' + s = 0 \quad s(0) = 10, \quad s'(0) = 0$
(iii) $s'' + 6s = 0 \quad s(0) = 4, \quad s'(0) = 0$
(iv) $6s'' + s = 0 \quad s(0) = 20, \quad s'(0) = 0$

Which differential equation represents:

(a) The spring oscillating most quickly (with the shortest period)?
(b) The spring oscillating with the largest amplitude?
(c) The spring oscillating most slowly (with the longest period)?
(d) The spring with largest maximum velocity?

24. A pendulum of length l makes an angle of x (radians) with the vertical (see Figure 11.103). When x is small, it can be shown that, approximately:
$$\frac{d^2x}{dt^2} = -\frac{g}{l}x,$$
where g is the acceleration due to gravity.

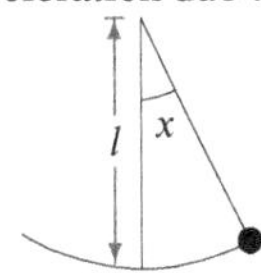

Figure 11.103

(a) Solve this equation assuming that $x(0) = 0$ and $x'(0) = v_0$.
(b) Solve this equation assuming that the pendulum is let go from the position where $x = x_0$. ("Let go" means that the velocity of the pendulum is zero when $x = x_0$. Measure t from the moment when the pendulum is let go.)

25. Look at the pendulum motion in Problem 24. What effect does it have on x as a function of time if:

(a) x_0 is increased? (b) l is increased?

26. A brick of mass 3 kg hangs from the end of a spring. When the brick is at rest, the spring is stretched by 2 cm. The spring is then stretched an additional 5 cm and released. Assume there is no air resistance.

(a) Set up a differential equation with initial conditions describing the motion.
(b) Solve the differential equation.

27. (a) Expand $A\sin(\omega t+\phi)$ using the trigonometric identity $\sin(x+y) = \sin x\cos y + \cos x\sin y$.
(b) Assume $A > 0$. If $A\sin(\omega t+\phi) = C_1\cos\omega t + C_2\sin\omega t$, show that we must have

$$A = \sqrt{C_1^2 + C_2^2} \quad \text{and} \quad \tan\phi = C_1/C_2.$$

■ Problems 28–30 concern the electric circuit in Figure 11.104. A charged capacitor connected to an inductor causes a current to flow through the inductor until the capacitor is fully discharged. The current in the inductor, in turn, charges up the capacitor until the capacitor is fully charged again. If $Q(t)$ is the charge on the capacitor at time t, and I is the current, then

$$I = \frac{dQ}{dt}.$$

If the circuit resistance is zero, then the charge Q and the current I in the circuit satisfy the differential equation

$$L\frac{dI}{dt} + \frac{Q}{C} = 0,$$

where C is the capacitance, and L is the inductance, so

$$L\frac{d^2Q}{dt^2} + \frac{Q}{C} = 0.$$

The unit of charge is the coulomb, the unit of capacitance the farad, the unit of inductance the henry, the unit of current is the ampere, and time is measured in seconds.

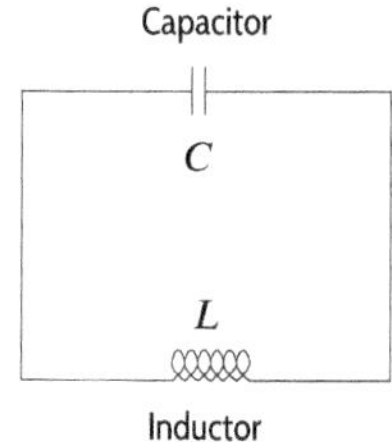

Figure 11.104

28. If $L = 36$ henry and $C = 9$ farad, find a formula for $Q(t)$ if

(a) $Q(0) = 0 \quad I(0) = 2$
(b) $Q(0) = 6 \quad I(0) = 0$

29. Suppose $Q(0) = 0$, $Q'(0) = I(0) = 4$, and the maximum possible charge is $2\sqrt{2}$ coulombs. What is the capacitance if the inductance is 10 henry?

30. What happens to the charge and current as t goes to infinity? What does it mean that the charge and current are sometimes positive and sometimes negative?

Strengthen Your Understanding

■ In Problems 31–32, explain what is wrong with the statement.

31. A solution to the spring equation $d^2s/dt^2 = -s$ with initial displacement 5 is given in Figure 11.105.

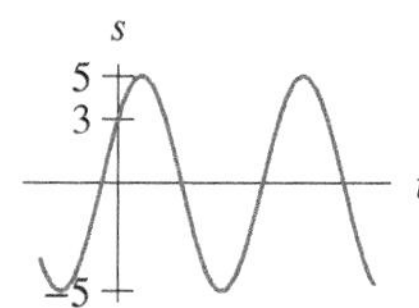

Figure 11.105

32. The general solution to the second-order differential equation $d^2y/dt^2 = k$ with constant k is $y = (k/2)t^2 + t + C$.

■ In Problems 33–34, give an example of:

33. A second-order differential equation which has $y = e^{2x}$ as a solution.

34. Any second-order boundary value problem.

11.11 LINEAR SECOND-ORDER DIFFERENTIAL EQUATIONS

A Spring with Friction: Damped Oscillations

The differential equation $d^2s/dt^2 = -(k/m)s$, which we used to describe the motion of a spring, disregards friction. But there is friction in every real system. For a mass on a spring, the frictional force from air resistance increases with the velocity of the mass. The frictional force is often approximately proportional to velocity, and so we introduce a *damping term* of the form $a(ds/dt)$, where a is a constant called the *damping coefficient* and ds/dt is the velocity of the mass.

Remember that without damping, the differential equation was obtained from

$$\text{Force} = \text{Mass} \cdot \text{Acceleration}.$$

With damping, the spring force $-ks$ is replaced by $-ks - a(ds/dt)$, where a is positive and the $a(ds/dt)$ term is subtracted because the frictional force is in the direction opposite to the motion. The new differential equation is therefore

$$-\underbrace{ks}_{\text{Spring force}} - \underbrace{a\frac{ds}{dt}}_{\text{Frictional force}} = \underbrace{m\frac{d^2s}{dt^2}}_{\text{Mass}\cdot\text{Acceleration}}$$

which is equivalent to the following differential equation:

Equation for Damped Oscillations of a Spring

$$\frac{d^2s}{dt^2} + \frac{a}{m}\frac{ds}{dt} + \frac{k}{m}s = 0$$

We expect the solution to this equation to die away with time, as friction brings the motion to a stop.

The General Solution to a Linear Differential Equation

The equation for damped oscillations is an example of a *linear second-order differential equation with constant coefficients.* This section gives an analytic method of solving the equation,

$$\frac{d^2y}{dt^2} + b\frac{dy}{dt} + cy = 0,$$

for constant b and c. As we saw for the spring equation, if $f_1(t)$ and $f_2(t)$ satisfy the differential equation, then, for any constants C_1 and C_2, another solution is given by

$$y(t) = C_1 f_1(t) + C_2 f_2(t)$$

This is called the *principle of superposition.* It can be shown that the general solution is of this form, provided $f_1(t)$ is not a multiple of $f_2(t)$.

Finding Solutions: The Characteristic Equation

We now use complex numbers to solve the differential equation

$$\frac{d^2y}{dt^2} + b\frac{dy}{dt} + cy = 0.$$

The method is a form of guess-and-check. We ask what kind of function might satisfy a differential equation in which the second derivative d^2y/dt^2 is a sum of multiples of dy/dt and y. One possibility is an exponential function, so we try to find a solution of the form:

$$y = Ce^{rt},$$

where r may be a complex number.[41] To find r, we substitute into the differential equation.

$$\frac{d^2y}{dt^2} + b\frac{dy}{dt} + cy = r^2Ce^{rt} + b \cdot rCe^{rt} + c \cdot Ce^{rt} = Ce^{rt}(r^2 + br + c) = 0.$$

We can divide by $y = Ce^{rt}$ provided $C \neq 0$, because the exponential function is never zero. If $C = 0$, then $y = 0$, which is not a very interesting solution (though it is a solution). So we assume $C \neq 0$. Then $y = Ce^{rt}$ is a solution to the differential equation if

$$r^2 + br + c = 0.$$

This quadratic is called the *characteristic equation* of the differential equation. Its solutions are

$$r = -\frac{1}{2}b \pm \frac{1}{2}\sqrt{b^2 - 4c}.$$

There are three different types of solutions to the differential equation, depending on whether the solutions to the characteristic equation are real and distinct, complex, or repeated. The sign of $b^2 - 4c$ determines the type of solutions.

The Case with $b^2 - 4c > 0$

There are two real solutions r_1 and r_2 to the characteristic equation, and the following two functions satisfy the differential equation:

$$C_1e^{r_1t} \quad \text{and} \quad C_2e^{r_2t}.$$

The sum of these two solutions is the general solution to the differential equation:

If $b^2 - 4c > 0$, the general solution to

$$\frac{d^2y}{dt^2} + b\frac{dy}{dt} + cy = 0$$

is

$$y(t) = C_1e^{r_1t} + C_2e^{r_2t}$$

where r_1 and r_2 are the solutions to the characteristic equation.
If $r_1 < 0$ and $r_2 < 0$, the motion is called **overdamped**.

A physical system satisfying a differential equation of this type is said to be overdamped because it occurs when there is a lot of friction in the system. For example, a spring moving in a thick fluid such as oil or molasses is overdamped: it will not oscillate.

Example 1 A spring is placed in oil, where it satisfies the differential equation

$$\frac{d^2s}{dt^2} + 3\frac{ds}{dt} + 2s = 0.$$

Solve this equation with the initial conditions $s = -0.5$ and $ds/dt = 3$ when $t = 0$.

[41] See Appendix B on complex numbers.

Solution The characteristic equation is

$$r^2 + 3r + 2 = 0,$$

with solutions $r = -1$ and $r = -2$, so the general solution to the differential equation is

$$s(t) = C_1 e^{-t} + C_2 e^{-2t}.$$

We use the initial conditions to find C_1 and C_2. At $t = 0$, we have

$$s = C_1 e^{-0} + C_2 e^{-2(0)} = C_1 + C_2 = -0.5.$$

Furthermore, since $ds/dt = -C_1 e^{-t} - 2C_2 e^{-2t}$, we have

$$\left.\frac{ds}{dt}\right|_{t=0} = -C_1 e^{-0} - 2C_2 e^{-2(0)} = -C_1 - 2C_2 = 3.$$

Solving these equations simultaneously, we find $C_1 = 2$ and $C_2 = -2.5$, so that the solution is

$$s(t) = 2e^{-t} - 2.5e^{-2t}.$$

The graph of this function is in Figure 11.106. The mass is so slowed by the oil that it passes through the equilibrium point only once (when $t \approx 1/4$) and for all practical purposes, it comes to rest after a short time. The motion has been "damped out" by the oil.

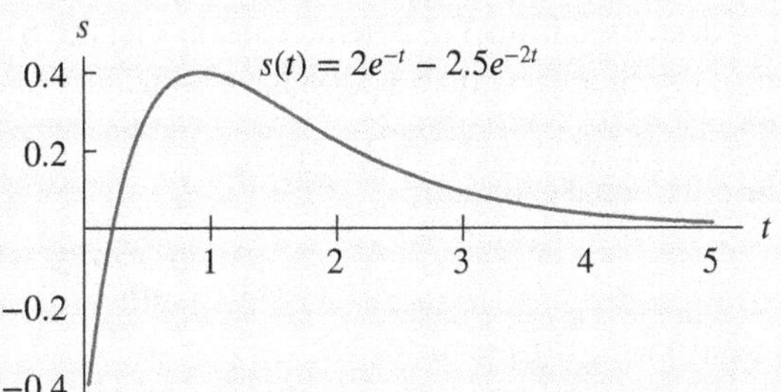

Figure 11.106: Solution to overdamped equation $\frac{d^2s}{dt^2} + 3\frac{ds}{dt} + 2s = 0$

The Case with $b^2 - 4c = 0$

In this case, the characteristic equation has only one solution, $r = -b/2$. By substitution, we can check that both $y = e^{-bt/2}$ and $y = te^{-bt/2}$ are solutions.

> If $b^2 - 4c = 0$,
>
> $$\frac{d^2y}{dt^2} + b\frac{dy}{dt} + cy = 0$$
>
> has general solution
>
> $$y(t) = (C_1 t + C_2)e^{-bt/2}.$$
>
> If $b > 0$, the system is said to be **critically damped**.

The Case with $b^2 - 4c < 0$

In this case, the characteristic equation has complex roots. Using Euler's formula,[42] these complex roots lead to trigonometric functions which represent oscillations.

Example 2 An object of mass $m = 10$ kg is attached to a spring with spring constant $k = 20$ kg/sec^2, and the object experiences a frictional force proportional to the velocity, with constant of proportionality $a = 20$ kg/sec. At time $t = 0$, the object is released from rest 2 meters above the equilibrium position. Write the differential equation that describes the motion.

Solution As we saw on page 641, the differential equation for damped oscillations of a spring is

$$\frac{d^2s}{dt^2} + \frac{a}{m}\frac{ds}{dt} + \frac{k}{m}s = 0.$$

Substituting $m = 10$ kg, $k = 20$ kg/sec^2, and $a = 20$ kg/sec, we obtain the differential equation:

$$\frac{d^2s}{dt^2} + 2\frac{ds}{dt} + 2s = 0.$$

At $t = 0$, the object is at rest 2 meters above equilibrium, so the initial conditions are $s(0) = 2$ and $s'(0) = 0$, where s is in meters and t in seconds.

Notice that this is the same differential equation as in Example 1 except that the coefficient of ds/dt has decreased from 3 to 2, which means that the frictional force has been reduced. This time, the roots of the characteristic equation have imaginary parts which lead to oscillations.

Example 3 Solve the differential equation

$$\frac{d^2s}{dt^2} + 2\frac{ds}{dt} + 2s = 0,$$

subject to $s(0) = 2$, $s'(0) = 0$.

Solution The characteristic equation is

$$r^2 + 2r + 2 = 0 \quad \text{giving} \quad r = -1 \pm i.$$

The solution to the differential equation is

$$s(t) = A_1e^{(-1+i)t} + A_2e^{(-1-i)t},$$

where A_1 and A_2 are arbitrary complex numbers. The initial condition $s(0) = 2$ gives

$$2 = A_1e^{(-1+i)\cdot 0} + A_2e^{(-1-i)\cdot 0} = A_1 + A_2.$$

Also,

$$s'(t) = A_1(-1+i)e^{(-1+i)t} + A_2(-1-i)e^{(-1-i)t},$$

so $s'(0) = 0$ gives

$$0 = A_1(-1+i) + A_2(-1-i).$$

Solving the simultaneous equations for A_1 and A_2 gives (after some algebra)

$$A_1 = 1 - i \quad \text{and} \quad A_2 = 1 + i.$$

[42]See Appendix B on complex numbers.

The solution is therefore

$$s(t) = (1-i)e^{(-1+i)t} + (1+i)e^{(-1-i)t} = (1-i)e^{-t}e^{it} + (1+i)e^{-t}e^{-it}.$$

Using Euler's formula, $e^{it} = \cos t + i\sin t$ and $e^{-it} = \cos t - i\sin t$, we get

$$s(t) = (1-i)e^{-t}(\cos t + i\sin t) + (1+i)e^{-t}(\cos t - i\sin t).$$

Multiplying out and simplifying, all the complex terms drop out, giving

$$\begin{aligned} s(t) &= e^{-t}\cos t + ie^{-t}\sin t - ie^{-t}\cos t + e^{-t}\sin t \\ &\quad + e^{-t}\cos t - ie^{-t}\sin t + ie^{-t}\cos t + e^{-t}\sin t \\ &= 2e^{-t}\cos t + 2e^{-t}\sin t. \end{aligned}$$

The $\cos t$ and $\sin t$ terms tell us that the solution oscillates; the factor of e^{-t} tells us that the oscillations are damped. See Figure 11.107. However, the period of the oscillations does not change as the amplitude decreases. This is why a spring-driven clock can keep accurate time even as it is running down.

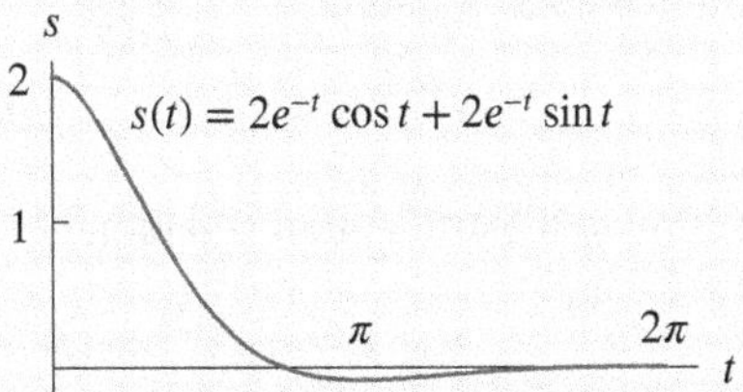

Figure 11.107: Solution to underdamped equation $\frac{d^2s}{dt^2} + 2\frac{ds}{dt} + 2s = 0$

In Example 3, the coefficients A_1 and A_2 are complex, but the solution, $s(t)$, is real. (We expect this, since $s(t)$ represents a real displacement.) In general, provided the coefficients b and c in the original differential equation and the initial values are real, the solution is always real. The coefficients A_1 and A_2 are always complex conjugates (that is, of the form $\alpha \pm i\beta$).

If $b^2 - 4c < 0$, to solve

$$\frac{d^2y}{dt^2} + b\frac{dy}{dt} + cy = 0,$$

- Find the solutions $r = \alpha \pm i\beta$ to the characteristic equation $r^2 + br + c = 0$.
- The general solution to the differential equation is, for some real C_1 and C_2,

$$y = C_1e^{\alpha t}\cos\beta t + C_2e^{\alpha t}\sin\beta t.$$

If $\alpha < 0$, such oscillations are called **underdamped**. If $\alpha = 0$, the oscillations are **undamped**.

Example 4 Find the general solution of the equations (a) $y'' = 9y$ (b) $y'' = -9y$.

Solution (a) The characteristic equation is $r^2 - 9 = 0$, so $r = \pm 3$. Thus the general solution is

$$y = C_1e^{3t} + C_2e^{-3t}.$$

(b) The characteristic equation is $r^2 + 9 = 0$, so $r = 0 \pm 3i$. The general solution is

$$y = C_1e^{0t}\cos 3t + C_2e^{0t}\sin 3t = C_1\cos 3t + C_2\sin 3t.$$

Notice that we have seen the solution to this equation in Section 11.10; it's the equation of undamped simple harmonic motion with $\omega = 3$.

Example 5 Solve the initial value problem

$$y'' + 4y' + 13y = 0, \quad y(0) = 0,\ y'(0) = 30.$$

Solution We solve the characteristic equation

$$r^2 + 4r + 13 = 0, \quad \text{getting} \quad r = -2 \pm 3i.$$

The general solution to the differential equation is

$$y = C_1e^{-2t}\cos 3t + C_2e^{-2t}\sin 3t.$$

Substituting $t = 0$ gives

$$y(0) = C_1 \cdot 1 \cdot 1 + C_2 \cdot 1 \cdot 0 = 0, \quad \text{so} \quad C_1 = 0.$$

Differentiating $y(t) = C_2e^{-2t}\sin 3t$ gives

$$y'(t) = C_2(-2e^{-2t}\sin 3t + 3e^{-2t}\cos 3t).$$

Substituting $t = 0$ gives

$$y'(0) = C_2(-2 \cdot 1 \cdot 0 + 3 \cdot 1 \cdot 1) = 3C_2 = 30 \quad \text{so} \quad C_2 = 10.$$

The solution is therefore

$$y(t) = 10e^{-2t}\sin 3t.$$

Summary of Solutions to $y'' + by' + cy = 0$

If $b^2 - 4c > 0$, then $y = C_1e^{r_1t} + C_2e^{r_2t}$
If $b^2 - 4c = 0$, then $y = (C_1t + C_2)e^{-bt/2}$
If $b^2 - 4c < 0$, then $y = C_1e^{\alpha t}\cos\beta t + C_2e^{\alpha t}\sin\beta t$

Exercises and Problems for Section 11.11

EXERCISES

For Exercises 1–16, find the general solution to the given differential equation.

1. $y'' + 4y' + 3y = 0$
2. $y'' + 4y' + 4y = 0$
3. $y'' + 4y' + 5y = 0$
4. $s'' - 7s = 0$
5. $s'' + 7s = 0$
6. $y'' - 3y' + 2y = 0$
7. $4z'' + 8z' + 3z = 0$
8. $\dfrac{d^2x}{dt^2} + 4\dfrac{dx}{dt} + 8x = 0$
9. $\dfrac{d^2p}{dt^2} + \dfrac{dp}{dt} + p = 0$
10. $z'' + 2z = 0$
11. $z'' + 2z' = 0$
12. $P'' + 2P' + P = 0$
13. $9z'' - z = 0$
14. $y'' + 6y' + 8y = 0$
15. $y'' + 2y' + 3y = 0$
16. $x'' + 2x' + 10x = 0$

For Exercises 17–24, solve the initial value problem.

17. $y'' + 5y' + 6y = 0, \quad y(0) = 1, y'(0) = 0.$
18. $y'' + 5y' + 6y = 0, \quad y(0) = 5, y'(0) = 1.$
19. $y'' - 3y' - 4y = 0, \quad y(0) = 1, y'(0) = 0.$
20. $y'' - 3y' - 4y = 0, \quad y(0) = 0, y'(0) = 0.5.$
21. $y'' + 6y' + 5y = 0, \quad y(0) = 1, \quad y'(0) = 0$
22. $y'' + 6y' + 5y = 0, \quad y(0) = 5, \quad y'(0) = 5$
23. $y'' + 6y' + 10y = 0, \quad y(0) = 0, \quad y'(0) = 2$
24. $y'' + 6y' + 10y = 0, \quad y(0) = 0, \quad y'(0) = 0$

For Exercises 25–28, solve the boundary value problem.

25. $y'' + 5y' + 6y = 0, \quad y(0) = 1, y(1) = 0.$
26. $y'' + 5y' + 6y = 0, \quad y(-2) = 0, y(2) = 3.$
27. $p'' + 2p' + 2p = 0, \quad p(0) = 0, \quad p(\pi/2) = 20$
28. $p'' + 4p' + 5p = 0, \quad p(0) = 1, \quad p(\pi/2) = 5$

PROBLEMS

29. Match the graphs of solutions in Figure 11.108 with the differential equations below.

(a) $x'' + 4x = 0$
(b) $x'' - 4x = 0$
(c) $x'' - 0.2x' + 1.01x = 0$
(d) $x'' + 0.2x' + 1.01x = 0$

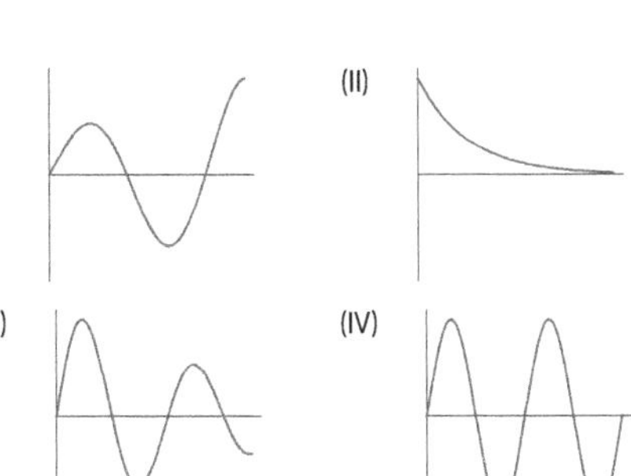

Figure 11.108

30. Match the differential equations to the solution graphs (I)–(IV). Use each graph only once.

(a) $y'' + 5y' + 6y = 0$ **(b)** $y'' + y' - 6y = 0$
(c) $y'' + 4y' + 9y = 0$ **(d)** $y'' = -9y$

Each of the differential equations (i)–(iv) represents the position of a 1 gram mass oscillating on the end of a damped spring. For Problems 31–35, pick the differential equation representing the system which answers the question.

(i) $s'' + s' + 4s = 0$ (ii) $s'' + 2s' + 5s = 0$
(iii) $s'' + 3s' + 3s = 0$ (iv) $s'' + 0.5s' + 2s = 0$

31. Which spring has the largest coefficient of damping?
32. Which spring exerts the smallest restoring force for a given displacement?
33. In which system does the mass experience the frictional force of smallest magnitude for a given velocity?
34. Which oscillation has the longest period?

35. Which spring is the stiffest? [Hint: You need to determine what it means for a spring to be stiff. Think of an industrial strength spring and a slinky.]

■ For each of the differential equations in Problems 36–38, find the values of c that make the general solution:

(a) overdamped, (b) underdamped, (c) critically damped.

36. $s'' + 4s' + cs = 0$

37. $s'' + 2\sqrt{2}s' + cs = 0$

38. $s'' + 6s' + cs = 0$

■ For each of the differential equations in Problems 39–40, find the values of b that make the general solution:

(a) overdamped, (b) underdamped,
(c) critically damped.

39. $s'' + bs' + 5s = 0$

40. $s'' + bs' - 16s = 0$

41. The motion of a mass on the end of a spring satisfies the differential equation

$$\frac{d^2s}{dt^2} + 7\frac{ds}{dt} + 10s = 0.$$

(a) If the mass $m = 10$, what is the spring coefficient k? What is the damping coefficient a?
(b) Solve the differential equation if the initial conditions are $s(0) = -1$ and $s'(0) = -7$.
(c) How low does the mass at the end of the spring go? How high does it go?
(d) How long does it take until the spring stays within 0.1 unit of equilibrium?

42. The motion of a mass on the end of a spring satisfies the differential equation

$$\frac{d^2s}{dt^2} + 2\frac{ds}{dt} + 3s = 0.$$

(a) Give the general solution to the differential equation.
(b) Solve the differential equation if the initial height is +2 and the initial velocity is +5.
(c) How low does the mass at the end of the spring go? How high does it go?
(d) How long does it take until the spring stays within 0.1 unit of equilibrium?

43. If the spring constant $k = 500$ and the mass $m = 100$, what values of the damping coefficient a make the motion

(a) Overdamped?
(b) Critically damped?
(c) Underdamped?

44. If $y = e^{2t}$ is a solution to the differential equation

$$\frac{d^2y}{dt^2} - 5\frac{dy}{dt} + ky = 0,$$

find the value of the constant k and the general solution to this equation.

45. Assuming $b, c > 0$, explain how you know that the solutions of an underdamped differential equation must go to 0 as $t \to \infty$.

46. Could the graph in Figure 11.109 show the position of a mass oscillating at the end of an overdamped spring? Why or why not?

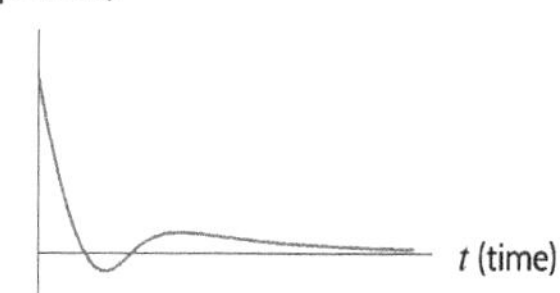

Figure 11.109

47. Find a solution to the following equation which satisfies $z(0) = 3$ and does not tend to infinity as $t \to \infty$:

$$\frac{d^2z}{dt^2} + \frac{dz}{dt} - 2z = 0.$$

48. Consider an overdamped differential equation with $b, c > 0$.

(a) Show that both roots of the characteristic equation are negative.
(b) Show that any solution to the differential equation goes to 0 as $t \to \infty$.

49. Consider the system of differential equations

$$\frac{dx}{dt} = -y \qquad \frac{dy}{dt} = -x.$$

(a) Convert this system to a second order differential equation in y by differentiating the second equation with respect to t and substituting for x from the first equation.
(b) Solve the equation you obtained for y as a function of t; hence find x as a function of t.

50. Juliet is in love with Romeo, who happens (in our version of this story) to be a fickle lover. The more Juliet loves him, the more he begins to dislike her. When she hates him, his feelings for her warm up. On the other hand, her love for him grows when he loves her and withers when he hates her. A model for their ill-fated romance is

$$\frac{dj}{dt} = Ar, \qquad \frac{dr}{dt} = -Bj,$$

where A and B are positive constants, $r(t)$ represents Romeo's love for Juliet at time t, and $j(t)$ represents Juliet's love for Romeo at time t. (Negative love is hate.)

(a) The constant on the right-hand side of Juliet's equation (the one including dj/dt) has a positive sign, whereas the constant in Romeo's equation is negative. Explain why these signs follow from the story.

(b) Derive a second-order differential equation for $r(t)$ and solve it. (Your equation should involve r and its derivatives, but not j and its derivatives.)

(c) Express $r(t)$ and $j(t)$ as functions of t, given $r(0) = 1$ and $j(0) = 0$. Your answer will contain A and B.

(d) As you may have discovered, the outcome of the relationship is a never-ending cycle of love and hate. Find what fraction of the time they both love one another.

51. **(a)** If $r_1 \neq r_2$ and r_1 and r_2 satisfy $r^2+br+c=0$ show that

$$y = \frac{e^{r_1 t} - e^{r_2 t}}{r_1 - r_2}$$

is a solution to $y'' + by' + cy = 0$.

(b) If $r_1 = r_2 + h$, show that the solution in part (a) can be written

$$y = e^{r_2 t}\frac{(e^{ht} - 1)}{h}.$$

(c) Using the Taylor series, show that

$$\frac{(e^{ht} - 1)}{h} = t + \frac{ht^2}{2!} + \frac{h^2t^3}{3!} + \cdots.$$

(d) Use the result of part (c) in the solution from part (b) to show that $\lim_{h\to 0} e^{r_2 t}\frac{(e^{ht} - 1)}{h} = te^{r_2 t}$.

(e) If there is a double root $r_2 = r_1 = -b/2$. By direct substitution, show that $y = te^{-bt/2}$ satisfies $y'' + by' + cy = 0$.

■ Recall the discussion of electric circuits on page 640. Just as a spring can have a damping force which affects its motion, so can a circuit. Problems 52–55 involve a damping force caused by the resistor in Figure 11.110. The charge Q on a capacitor in a circuit with inductance L, capacitance C, and resistance R, in ohms, satisfies the differential equation

$$L\frac{d^2Q}{dt^2} + R\frac{dQ}{dt} + \frac{1}{C}Q = 0.$$

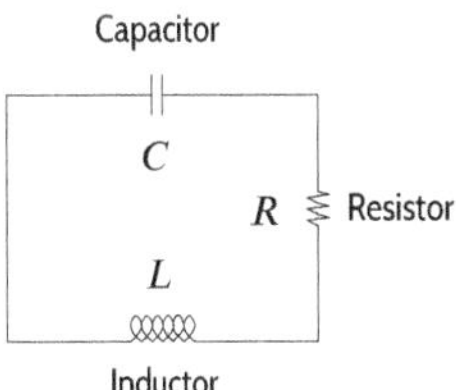

Figure 11.110

52. If $L = 1$ henry, $R = 2$ ohms, and $C = 4$ farads, find a formula for the charge when

(a) $Q(0) = 0, Q'(0) = 2$.
(b) $Q(0) = 2, Q'(0) = 0$.

53. If $L = 1$ henry, $R = 1$ ohm, and $C = 4$ farads, find a formula for the charge when

(a) $Q(0) = 0, Q'(0) = 2$.
(b) $Q(0) = 2, Q'(0) = 0$.
(c) How did reducing the resistance affect the charge? Compare with your solution to Problem 52.

54. If $L = 8$ henry, $R = 2$ ohm, and $C = 4$ farads, find a formula for the charge when

(a) $Q(0) = 0, Q'(0) = 2$.
(b) $Q(0) = 2, Q'(0) = 0$.
(c) How did increasing the inductance affect the charge? Compare with your solution to Problem 52.

55. Given any positive values for R, L and C, what happens to the charge as t goes to infinity?

Strengthen Your Understanding

■ In Problems 56–57, explain what is wrong with the statement.

56. Figure 11.111 represents motion for which the damping coefficient is zero.

Figure 11.111

57. Figure 11.112 represents motion of a mass on a spring with initial conditions $s(0) = 5$ and $s'(0) = 2$.

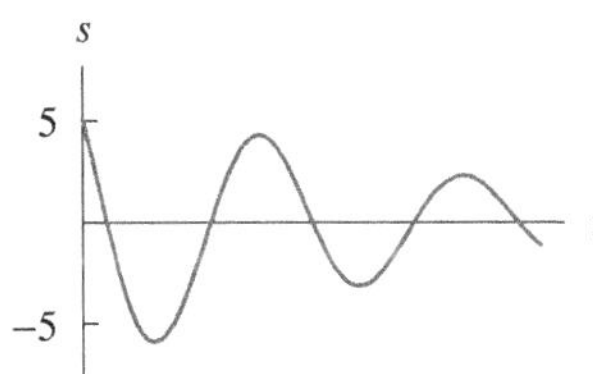

Figure 11.112

In Problems 58–60, give an example of:

58. A linear second-order differential equation representing spring motion that is overdamped.

59. A linear second-order differential equation representing spring motion that is critically damped.

60. Values of the spring constant k, the mass m, and the damping coefficient a so that the motion is underdamped and shows damped oscillations.

Online Resource: Review problems and Projects

Chapter Twelve

FUNCTIONS OF SEVERAL VARIABLES

Contents

12.1 FUNCTIONS OF TWO VARIABLES

Function Notation

Suppose you want to calculate your monthly payment on a five-year car loan; this depends on both the amount of money you borrow and the interest rate. These quantities can vary separately: the loan amount can change while the interest rate remains the same, or the interest rate can change while the loan amount remains the same. To calculate your monthly payment you need to know both. If the monthly payment is $\$m$, the loan amount is $\$L$, and the interest rate is $r\%$, then we express the fact that m is a function of L and r by writing:

$$m = f(L, r).$$

This is just like the function notation of one-variable calculus. The variable m is called the dependent variable, and the variables L and r are called the independent variables. The letter f stands for the *function* or rule that gives the value of m corresponding to given values of L and r.

A function of two variables can be represented graphically, numerically by a table of values, or algebraically by a formula. In this section, we give examples of each.

Graphical Example: A Weather Map

Figure 12.1 shows a weather map from a newspaper. What information does it convey? It displays the predicted high temperature, T, in degrees Fahrenheit (°F), throughout the US on that day. The curves on the map, called *isotherms*, separate the country into zones, according to whether T is in the 60s, 70s, 80s, 90s, or 100s. (*Iso* means same and *therm* means heat.) Notice that the isotherm separating the 80s and 90s zones connects all the points where the temperature is exactly 90°F.

Example 1 Estimate the predicted value of T in Boise, Idaho; Topeka, Kansas; and Buffalo, New York.

Solution Boise and Buffalo are in the 70s region, and Topeka is in the 80s region. Thus, the predicted temperature in Boise and Buffalo is between 70 and 80 while the predicted temperature in Topeka is between 80 and 90. In fact, we can say more. Although both Boise and Buffalo are in the 70s, Boise is quite close to the $T = 70$ isotherm, whereas Buffalo is quite close to the $T = 80$ isotherm. So we estimate the temperature to be in the low 70s in Boise and in the high 70s in Buffalo. Topeka is about halfway between the $T = 80$ isotherm and the $T = 90$ isotherm. Thus, we guess the temperature in Topeka to be in the mid-80s. In fact, the actual high temperatures for that day were 71°F for Boise, 79°F for Buffalo, and 86°F for Topeka.

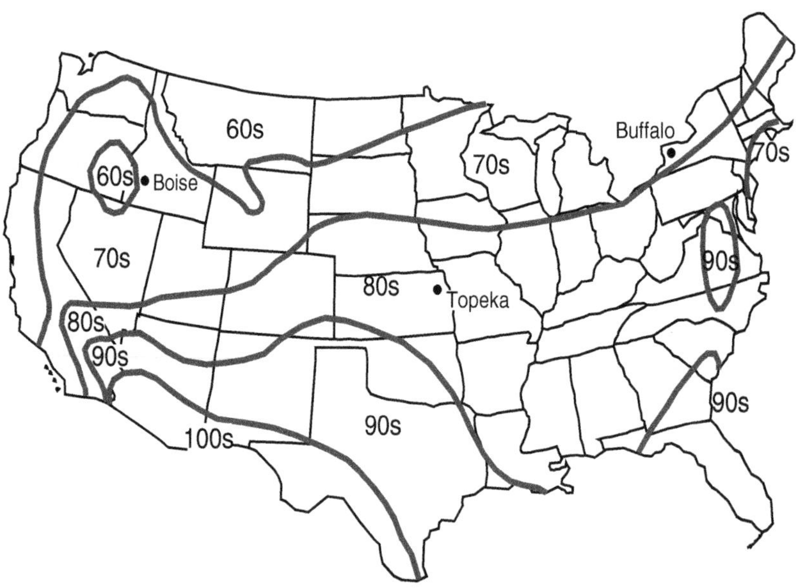

Figure 12.1: Weather map showing predicted high temperatures, T, on a summer day

The predicted high temperature, T, illustrated by the weather map is a function of (that is, depends on) two variables, often longitude and latitude, or miles east-west and miles north-south of a fixed point, say, Topeka. The weather map in Figure 12.1 is called a *contour map* or *contour diagram* of that function. Section 12.2 shows another way of visualizing functions of two variables using surfaces; Section 12.3 looks at contour maps in detail.

Numerical Example: Body Mass Index (BMI)

The body mass index (BMI) is a value that attempts to quantify a person's body fat based on their height h and weight w. In function notation, we write:

$$\text{BMI} = f(h, w).$$

Table 12.1 contains values of this function for h in inches and w in pounds. Values of w are across the top, values of h are down the left side, and corresponding values of $f(h, w)$ are in the table.[1] For example, to find the value of $f(66, 140)$, we look in the row corresponding to $h = 66$ under $w = 140$, where we find the number 22.6. Thus,

$$f(66, 140) = 22.6.$$

This means that if an individual is 66 inches tall and weights 140 lbs, their body mass index is 22.6.

Table 12.1 *Body mass index (BMI)*

Height h (inches) \ Weight w (lbs)	120	140	160	180	200
60	23.4	27.3	31.2	35.2	39.1
63	21.3	24.8	28.3	31.9	35.4
66	19.4	22.6	25.8	29.0	32.3
69	17.7	20.7	23.6	26.6	29.5
72	16.3	19.0	21.7	24.4	27.1
75	15.0	17.5	20.0	22.5	25.0

Notice how this table differs from the table of values of a one-variable function, where one row or one column is enough to list the values of the function. Here many rows and columns are needed because the function has a value for every *pair* of values of the independent variables.

Algebraic Examples: Formulas

In the weather map example there is no formula for the underlying function. That is usually the case for functions representing real-life data. On the other hand, for many models in physics, engineering, and economics, there are exact formulas.

Example 2 Give a formula for the function $M = f(B, t)$ where M is the amount of money in a bank account t years after an initial investment of B dollars, if interest is accrued at a rate of 1.2% per year compounded annually.

Solution Annual compounding means that M increases by a factor of 1.012 every year, so

$$M = f(B, t) = B(1.012)^t.$$

Example 3 A cylinder with closed ends has radius r and height h. If its volume is V and its surface area is A, find formulas for the functions $V = f(r, h)$ and $A = g(r, h)$.

Solution Since the area of the circular base is πr^2, we have

$$V = f(r, h) = \text{Area of base} \cdot \text{Height} = \pi r^2 h.$$

[1] http://www.cdc.gov. Last accessed January 8, 2016.

The surface area of the side is the circumference of the bottom, $2\pi r$, times the height h, giving $2\pi rh$. Thus,

$$A = g(r, h) = 2 \cdot \text{Area of base} + \text{Area of side} = 2\pi r^2 + 2\pi rh.$$

A Tour of 3-Space

In Section 12.2 we see how to visualize a function of two variables as a surface in space. Now we see how to locate points in three-dimensional space (3-space).

Imagine three coordinate axes meeting at the *origin*: a vertical axis, and two horizontal axes at right angles to each other. (See Figure 12.2.) Think of the xy-plane as being horizontal, while the z-axis extends vertically above and below the plane. The labels x, y, and z show which part of each axis is positive; the other side is negative. We generally use *right-handed axes* in which looking down the positive z-axis gives the usual view of the xy-plane. We specify a point in 3-space by giving its coordinates (x, y, z) with respect to these axes. Think of the coordinates as instructions telling you how to get to the point: start at the origin, go x units along the x-axis, then y units in the direction parallel to the y-axis, and finally z units in the direction parallel to the z-axis. The coordinates can be positive, zero or negative; a zero coordinate means "don't move in this direction," and a negative coordinate means "go in the negative direction parallel to this axis." For example, the origin has coordinates $(0, 0, 0)$, since we get there from the origin by doing nothing at all.

Example 4 Describe the position of the points with coordinates $(1, 2, 3)$ and $(0, 0, -1)$.

Solution We get to the point $(1, 2, 3)$ by starting at the origin, going 1 unit along the x-axis, 2 units in the direction parallel to the y-axis, and 3 units up in the direction parallel to the z-axis. (See Figure 12.3.)

To get to $(0, 0, -1)$, we don't move at all in the x- and the y-directions, but move 1 unit in the negative z-direction. So the point is on the negative z-axis. (See Figure 12.4.) You can check that the position of the point is independent of the order of the x, y, and z displacements.

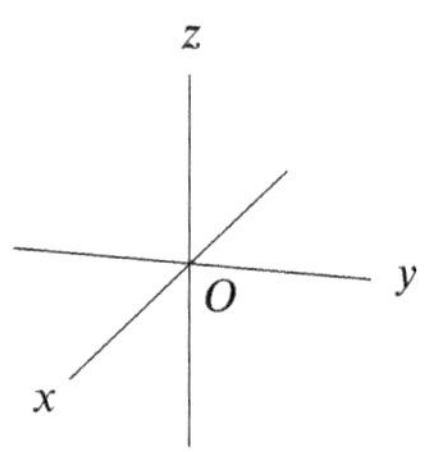

Figure 12.2: Coordinate axes in three-dimensional space

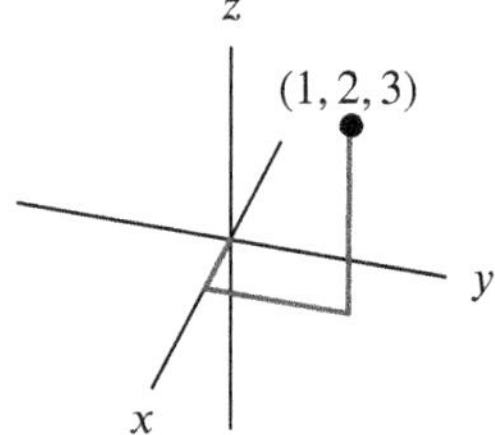

Figure 12.3: The point $(1, 2, 3)$ in 3-space

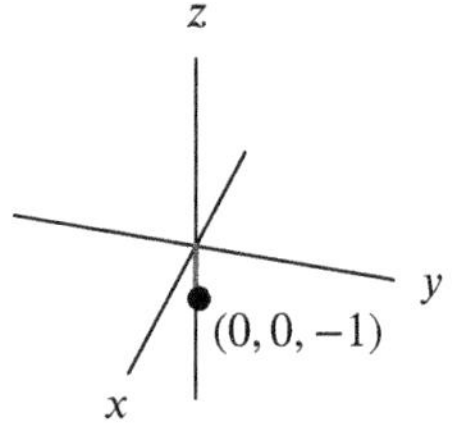

Figure 12.4: The point $(0, 0, -1)$ in 3-space

Example 5 You start at the origin, go along the y-axis a distance of 2 units in the positive direction, and then move vertically upward a distance of 1 unit. What are the coordinates of your final position?

Solution You started at the point $(0, 0, 0)$. When you went along the y-axis, your y-coordinate increased to 2. Moving vertically increased your z-coordinate to 1; your x-coordinate did not change because you did not move in the x-direction. So your final coordinates are $(0, 2, 1)$. (See Figure 12.5.)

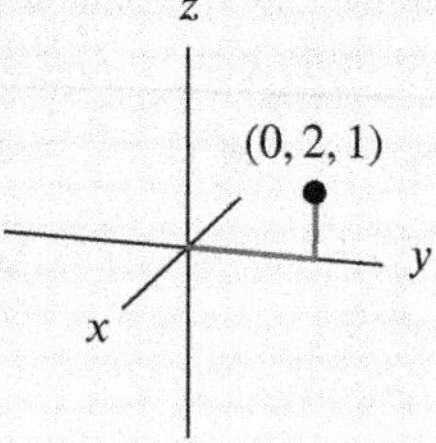

Figure 12.5: The point $(0, 2, 1)$ is reached by moving 2 along the y-axis and 1 upward

It is often helpful to picture a three-dimensional coordinate system in terms of a room. The origin is a corner at floor level where two walls meet the floor. The z-axis is the vertical intersection of the two walls; the x- and the y-axis are the intersections of each wall with the floor. Points with negative coordinates lie behind a wall in the next room or below the floor.

Graphing Equations in 3-Space

We can graph an equation involving the variables x, y, and z in 3-space; such a graph is a picture of all points (x, y, z) that satisfy the equation.

Example 6 What do the graphs of the equations $z = 0$, $z = 3$, and $z = -1$ look like?

Solution To graph $z = 0$, we visualize the set of points whose z-coordinate is zero. If the z-coordinate is 0, then we must be at the same vertical level as the origin; that is, we are in the horizontal plane containing the origin. So the graph of $z = 0$ is the middle plane in Figure 12.6. The graph of $z = 3$ is a plane parallel to the graph of $z = 0$, but three units above it. The graph of $z = -1$ is a plane parallel to the graph of $z = 0$, but one unit below it.

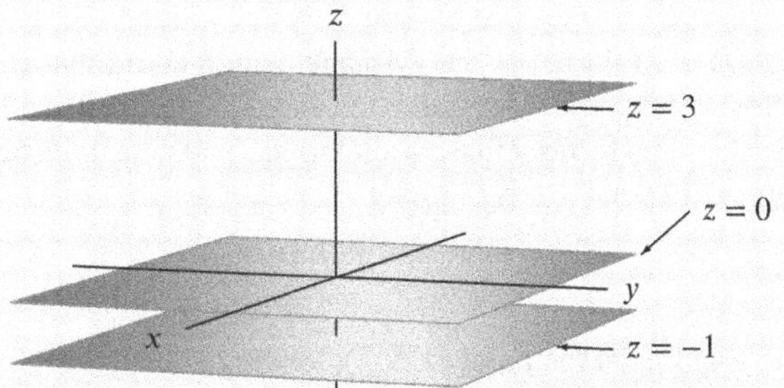

Figure 12.6: The planes $z = -1$, $z = 0$, and $z = 3$

The plane $z = 0$ contains the x- and the y-coordinate axes, and is called the xy-plane. There are two other coordinate planes. The yz-plane contains both the y- and the z-axis, and the xz-plane contains the x- and the z-axis. (See Figure 12.7.)

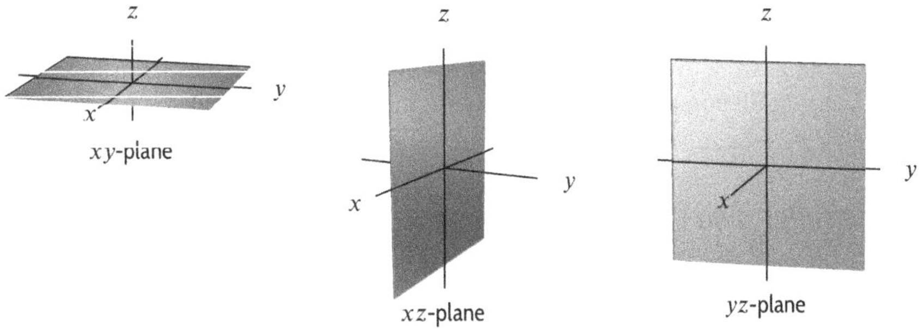

Figure 12.7: The three coordinate planes

Example 7 Which of the points $A = (1, -1, 0)$, $B = (0, 3, 4)$, $C = (2, 2, 1)$, and $D = (0, -4, 0)$ lies closest to the xz-plane? Which point lies on the y-axis?

Solution The magnitude of the y-coordinate gives the distance to the xz-plane. The point A lies closest to that plane, because it has the smallest y-coordinate in magnitude. To get to a point on the y-axis, we move along the y-axis, but we don't move at all in the x- or the z-direction. Thus, a point on the y-axis has both its x- and z-coordinates equal to zero. The only point of the four that satisfies this is D. (See Figure 12.8.)

In general, if a point has one of its coordinates equal to zero, it lies in one of the coordinate planes. If a point has two of its coordinates equal to zero, it lies on one of the coordinate axes.

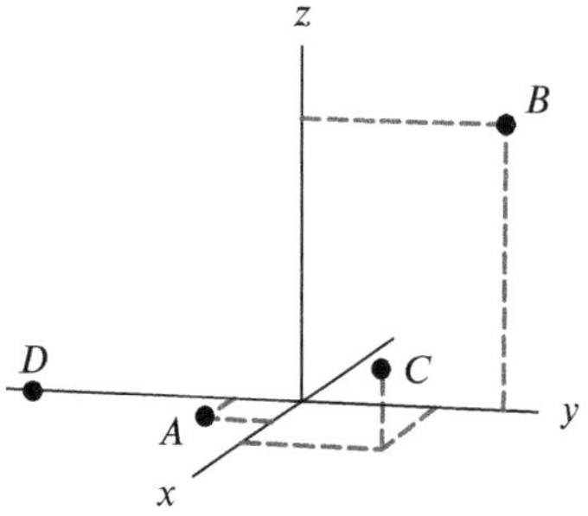

Figure 12.8: Which point lies closest to the xz-plane? Which point lies on the y-axis?

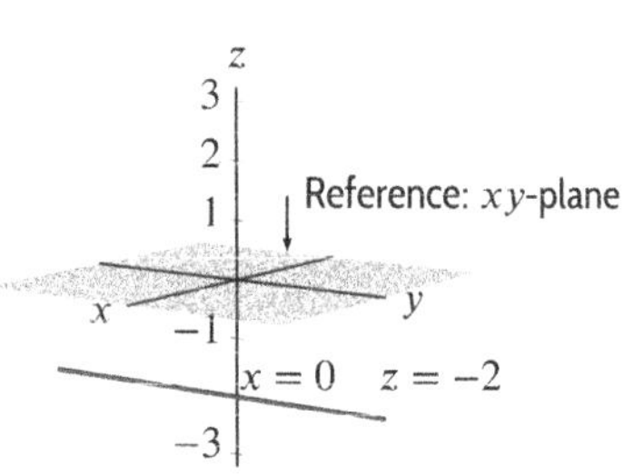

Figure 12.9: The line $x = 0$, $z = -2$

Example 8 You are 2 units below the xy-plane and in the yz-plane. What are your coordinates?

Solution Since you are 2 units below the xy-plane, your z-coordinate is -2. Since you are in the yz-plane, your x-coordinate is 0; your y-coordinate can be anything. Thus, you are at the point $(0, y, -2)$. The set of all such points forms a line parallel to the y-axis, 2 units below the xy-plane, and in the yz-plane. (See Figure 12.9.)

Example 9 You are standing at the point $(4, 5, 2)$, looking at the point $(0.5, 0, 3)$. Are you looking up or down?

Solution The point you are standing at has z-coordinate 2, whereas the point you are looking at has z-coordinate 3; hence you are looking up.

Example 10 Imagine that the yz-plane in Figure 12.7 is a page of this book. Describe the region behind the page algebraically.

Solution The positive part of the x-axis pokes out of the page; moving in the positive x-direction brings you out in front of the page. The region behind the page corresponds to negative values of x, so it is the set of all points in 3-space satisfying the inequality $x < 0$.

Distance Between Two Points

In 2-space, the formula for the distance between two points (x, y) and (a, b) is given by

$$\text{Distance} = \sqrt{(x-a)^2 + (y-b)^2}.$$

The distance between two points (x, y, z) and (a, b, c) in 3-space is represented by PG in Figure 12.10. The side PE is parallel to the x-axis, EF is parallel to the y-axis, and FG is parallel to the z-axis.

Using Pythagoras' theorem twice gives

$$(PG)^2 = (PF)^2 + (FG)^2 = (PE)^2 + (EF)^2 + (FG)^2 = (x-a)^2 + (y-b)^2 + (z-c)^2.$$

Thus, a formula for the distance between the points (x, y, z) and (a, b, c) in 3-space is

$$\boxed{\text{Distance} = \sqrt{(x-a)^2 + (y-b)^2 + (z-c)^2}.}$$

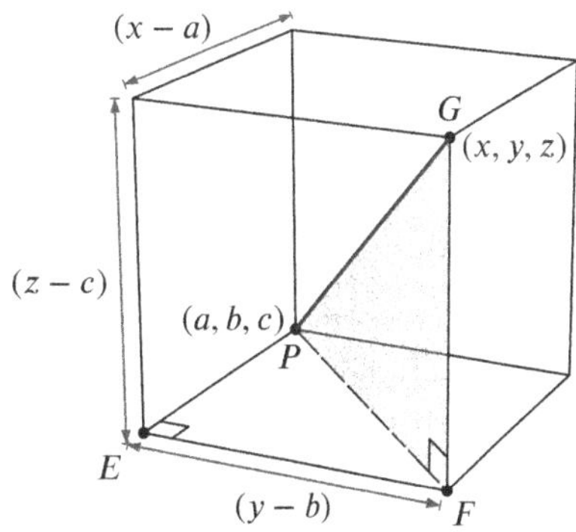

Figure 12.10: The diagonal PG gives the distance between the points (x, y, z) and (a, b, c)

Example 11 Find the distance between $(1, 2, 1)$ and $(-3, 1, 2)$.

Solution $\text{Distance} = \sqrt{(-3-1)^2 + (1-2)^2 + (2-1)^2} = \sqrt{18} = 4.243.$

Example 12 Find an expression for the distance from the origin to the point (x, y, z).

Solution The origin has coordinates $(0, 0, 0)$, so the distance from the origin to (x, y, z) is given by

$$\text{Distance} = \sqrt{(x-0)^2 + (y-0)^2 + (z-0)^2} = \sqrt{x^2 + y^2 + z^2}.$$

Example 13 Find an equation for a sphere of radius 1 with center at the origin.

Solution The sphere consists of all points (x, y, z) whose distance from the origin is 1, that is, which satisfy the equation

$$\sqrt{x^2 + y^2 + z^2} = 1.$$

This is an equation for the sphere. If we square both sides we get the equation in the form

$$x^2 + y^2 + z^2 = 1.$$

Note that this equation represents the *surface* of the sphere. The solid ball enclosed by the sphere is represented by the inequality $x^2 + y^2 + z^2 \le 1$.

Exercises and Problems for Section 12.1 Online Resource: Additional Problems for Section 12.1

EXERCISES

1. Which of the points $P = (1, 2, 1)$ and $Q = (2, 0, 0)$ is closest to the origin?
2. Which two of the three points $P_1 = (1, 2, 3)$, $P_2 = (3, 2, 1)$ and $P_3 = (1, 1, 0)$ are closest to each other?
3. Which of the points $P_1 = (-3, 2, 15)$, $P_2 = (0, -10, 0)$, $P_3 = (-6, 5, 3)$ and $P_4 = (-4, 2, 7)$ is closest to $P = (6, 0, 4)$?
4. Which of the points $A = (1.3, -2.7, 0)$, $B = (0.9, 0, 3.2)$, $C = (2.5, 0.1, -0.3)$ is closest to the yz-plane? Which one lies on the xz-plane? Which one is farthest from the xy-plane?

5. You are at the point $(3, 1, 1)$, standing upright and facing the yz-plane. You walk 2 units forward, turn left, and walk another 2 units. What is your final position? From the point of view of an observer looking at the coordinate system in Figure 12.2 on page 654, are you in front of or behind the yz-plane? To the left or to the right of the xz-plane? Above or below the xy-plane?

6. On a set of x, y and z axes oriented as in Figure 12.5 on page 654, draw a straight line through the origin, lying in the yz-plane and such that if you move along the line with your y-coordinate increasing, your z-coordinate is increasing.

7. What is the midpoint of the line segment joining the points $(-1, 3, 9)$ and $(5, 6, -3)$?

In Exercises 8–11, which of (I)–(IV) lie on the graph of the equation?

I. $(2, 2, 4)$ II. $(-1, 1, 0)$
III. $(-3, -2, -1)$ IV. $(-2, -2, 4)$

8. $z = 4$ **9.** $x + y + z = 0$

10. $x^2 + y^2 + z^2 = 14$ **11.** $x - y = 0$

In Exercises 12–15 sketch graphs of the equations in 3-space.

12. $z = 4$ **13.** $x = -3$

14. $y = 1$ **15.** $z = 2$ and $y = 4$

16. With the z-axis vertical, a sphere has center $(2, 3, 7)$ and lowest point $(2, 3, -1)$. What is the highest point on the sphere?

17. Find an equation of the sphere with radius 5 centered at the origin.

18. Find the equation of the sphere with radius 2 and centered at $(1, 0, 0)$.

19. Find the equation of the vertical plane perpendicular to the y-axis and through the point $(2, 3, 4)$.

Exercises 20–22 refer to the map in Figure 12.1 on page 652.

20. Give the range of daily high temperatures for:

(a) Pennsylvania **(b)** North Dakota
(c) California

21. Sketch a possible graph of the predicted high temperature T on a line north-south through Topeka.

22. Sketch possible graphs of the predicted high temperature on a north-south line and an east-west line through Boise.

For Exercises 23–25, refer to Table 12.1 on page 653 where w is a person's weight (in lbs), and h their height (in inches).

23. Compute a table of values of BMI, with h fixed at 60 inches and w between 120 and 200 lbs at intervals of 20.

24. Medical evidence suggests that BMI values between 18.5 and 24.9 are healthy values.[2] Estimate the range of weights that are considered healthy for a woman who is 6 feet tall.

25. Estimate the BMI of a man who weighs 90 kilograms and is 1.9 meters tall.

PROBLEMS

26. The temperature adjusted for wind chill is a temperature which tells you how cold it feels, as a result of the combination of wind and temperature.[3] See Table 12.2.

Table 12.2 *Temperature adjusted for wind chill (°F) as a function of wind speed and temperature*

		Temperature (°F)							
		35	30	25	20	15	10	5	0
Wind Speed (mph)	5	31	25	19	13	7	1	−5	−11
	10	27	21	15	9	3	−4	−10	−16
	15	25	19	13	6	0	−7	−13	−19
	20	24	17	11	4	−2	−9	−15	−22
	25	23	16	9	3	−4	−11	−17	−24

(a) If the temperature is 0°F and the wind speed is 15 mph, how cold does it feel?

(b) If the temperature is 35°F, what wind speed makes it feel like 24°F?

(c) If the temperature is 25°F, what wind speed makes it feel like 12°F?

(d) If the wind is blowing at 20 mph, what temperature feels like 0°F?

In Problems 27–28, use Table 12.2 to make tables with the given properties.

27. The temperature adjusted for wind chill as a function of wind speed for temperatures of 20°F and 0°F.

28. The temperature adjusted for wind chill as a function of temperature for wind speeds of 5 mph and 20 mph.

For Problems 29–31, refer to Table 12.3 which contains values of beef consumption C (in pounds per week per household) as a function of household income, I (in thousands

[2] http://www.cdc.gov. Accessed January 10, 2016.
[3] www.nws.noaa.gov. Accessed January 10, 2016.

of dollars per year), and the price of beef, p (in dollars per pound). Values of p are shown across the top, values of I are down the left side, and corresponding values of beef consumption $C = f(I, p)$ are given in the table.[4]

Table 12.3 *Quantity of beef bought (pounds/household/week)*

		Price of beef ($/lb)			
		3.00	3.50	4.00	4.50
Household income per year, I ($1000)	20	2.65	2.59	2.51	2.43
	40	4.14	4.05	3.94	3.88
	60	5.11	5.00	4.97	4.84
	80	5.35	5.29	5.19	5.07
	100	5.79	5.77	5.60	5.53

29. Give tables for beef consumption as a function of p, with I fixed at $I = 20$ and $I = 100$. Give tables for beef consumption as a function of I, with p fixed at $p = 3.00$ and $p = 4.00$. Comment on what you see in the tables.
30. Make a table of the proportion, P, of household income spent on beef per week as a function of price and income. (Note that P is the fraction of income spent on beef.)
31. How does beef consumption vary as a function of household income if the price of beef is held constant?

For Problems 32–35, a person's body mass index (BMI) is a function of their weight W (in kg) and height H (in m) given by $B(W, H) = W/H^2$.

32. What is the BMI of a 1.72 m tall man weighing 72 kg?
33. A 1.58 m tall woman has a BMI of 23.2. What is her weight?
34. With a BMI less than 18.5, a person is considered underweight. What is the possible range of weights for an underweight person 1.58 m tall?
35. For weight w in lbs and height h in inches, a persons BMI is approximated using the formula $f(w, h) = 703w/h^2$. Check this approximation by converting the formula $B(W, H)$.
36. A car rental company charges $40 a day and 15 cents a mile for its cars.
    - **(a)** Write a formula for the cost, C, of renting a car as a function, f, of the number of days, d, and the number of miles driven, m.
    - **(b)** If $C = f(d, m)$, find $f(5, 300)$ and interpret it.
37. A cable company charges $100 for a monthly subscription to its services and $5 for each special feature movie that a subscriber chooses to watch.
    - **(a)** Write a formula for the monthly revenue, R in dollars, earned by the cable company as a function of s, the number of monthly subscribers it serves, and m, the total number of special feature movies that its subscribers view.
    - **(b)** If $R = f(s, m)$, find $f(1000, 5000)$ and interpret it in terms of revenue.
38. The gravitational force, F newtons, exerted on an object by the earth depends on the object's mass, m kilograms, and its distance, r meters, from the center of the earth, so $F = f(m, r)$. Interpret the following statement in terms of gravitation: $f(100, 7000000) \approx 820$.
39. A heating element is attached to the center point of a metal rod at time $t = 0$. Let $H = f(d, t)$ represent the temperature in °C of a point d cm from the center after t minutes.
    - **(a)** Interpret the statement $f(2, 5) = 24$ in terms of temperature.
    - **(b)** If d is held constant, is H an increasing or a decreasing function of t? Why?
    - **(c)** If t is held constant, is H an increasing or a decreasing function of d? Why?
40. The pressure, P atmospheres, of 10 moles of nitrogen gas in a steel cylinder depends on the temperature of the gas, T Kelvin, and the volume of the cylinder, V liters, so $P = f(T, V)$. Interpret the following statement in terms of pressure: $f(300, 5) = 49.2$.
41. The monthly payment, m dollars, for a 30-year fixed rate mortgage is a function of the total amount borrowed, P dollars, and the annual interest rate, r%. In other words, $m = f(P, r)$.
    - **(a)** Interpret the following statement in the context of monthly payment: $f(300{,}000, 5) = 1610.46$.
    - **(b)** If P is held constant, is m an increasing or a decreasing function of r? Why?
    - **(c)** If r is held constant, is m an increasing or a decreasing function of P? Why?
42. Consider the acceleration due to gravity, g, at a distance h from the center of a planet of mass m.
    - **(a)** If m is held constant, is g an increasing or decreasing function of h? Why?
    - **(b)** If h is held constant, is g an increasing or decreasing function of m? Why?
43. A cube is located such that its top four corners have the coordinates $(-1, -2, 2)$, $(-1, 3, 2)$, $(4, -2, 2)$ and $(4, 3, 2)$. Give the coordinates of the center of the cube.
44. Describe the set of points whose distance from the x-axis is 2.
45. Describe the set of points whose distance from the x-axis equals the distance from the yz-plane.

[4] Adapted from Richard G. Lipsey, *An Introduction to Positive Economics*, 3rd ed. (London: Weidenfeld and Nicolson, 1971).

46. Find the point on the x-axis closest to the point $(3, 2, 1)$.
47. Does the line parallel to the y-axis through the point $(2, 1, 4)$ intersect the plane $y = 5$? If so, where?
48. Find a formula for the shortest distance between a point (a, b, c) and the y-axis.
49. Find the equations of planes that just touch the sphere $(x-2)^2 + (y-3)^2 + (z-3)^2 = 16$ and are parallel to
    (a) The xy-plane (b) The yz-plane
    (c) The xz-plane
50. Find an equation of the largest sphere contained in the cube determined by the planes $x = 2, x = 6$; $y = 5, y = 9$; and $z = -1, z = 3$.
51. A cube has edges parallel to the axes. One corner is at $A = (5, 1, 2)$ and the corner at the other end of the longest diagonal through A is $B = (12, 7, 4)$.
    (a) What are the coordinates of the other three vertices on the bottom face?
    (b) What are the coordinates of the other three vertices on the top face?
52. An equilateral triangle is standing vertically with a vertex above the xy-plane and its two other vertices at $(7, 0, 0)$ and $(9, 0, 0)$. What is its highest point?
53. (a) Find the midpoint of the line segment joining $A = (1, 5, 7)$ to $B = (5, 13, 19)$.
    (b) Find the point one quarter of the way along the line segment from A to B.
    (c) Find the point one quarter of the way along the line segment from B to A.

Strengthen Your Understanding

In Problems 54–56, explain what is wrong with the statement.

54. In 3-space, $y = 1$ is a line parallel to the x-axis.
55. The xy-plane has equation $xy = 0$.
56. The distance from $(2, 3, 4)$ to the x-axis is 2.

In Problems 57–58, give an example of:

57. A formula for a function $f(x, y)$ that is increasing in x and decreasing in y.
58. A point in 3-space with all its coordinates negative and farther from the xz-plane than from the plane $z = -5$.

Are the statements in Problems 59–72 true or false? Give reasons for your answer.

59. If $f(x, y)$ is a function of two variables defined for all x and y, then $f(10, y)$ is a function of one variable.
60. The volume V of a box of height h and square base of side length s is a function of h and s.
61. If $H = f(t, d)$ is the function giving the water temperature H°C of a lake at time t hours after midnight and depth d meters, then t is a function of d and H.
62. A table for a function $f(x, y)$ cannot have any values of f appearing twice.
63. If $f(x)$ and $g(y)$ are both functions of a single variable, then the product $f(x) \cdot g(y)$ is a function of two variables.
64. The point $(1, 2, 3)$ lies above the plane $z = 2$.
65. The graph of the equation $z = 2$ is a plane parallel to the xz-plane.
66. The points $(1, 0, 1)$ and $(0, -1, 1)$ are the same distance from the origin.
67. The point $(2, -1, 3)$ lies on the graph of the sphere $(x-2)^2 + (y+1)^2 + (z-3)^2 = 25$.
68. There is only one point in the yz-plane that is a distance 3 from the point $(3, 0, 0)$.
69. There is only one point in the yz-plane that is a distance 5 from the point $(3, 0, 0)$.
70. If the point $(0, b, 0)$ has distance 4 from the plane $y = 0$, then b must be 4.
71. A line parallel to the z-axis can intersect the graph of $f(x, y)$ at most once.
72. A line parallel to the y-axis can intersect the graph of $f(x, y)$ at most once.

12.2 GRAPHS AND SURFACES

The weather map on page 652 is one way of visualizing a function of two variables. In this section we see how to visualize a function of two variables in another way, using a surface in 3-space.

Visualizing a Function of Two Variables Using a Graph

For a function of one variable, $y = f(x)$, the graph of f is the set of all points (x, y) in 2-space such that $y = f(x)$. In general, these points lie on a curve in the plane. When a computer or calculator graphs f, it approximates by plotting points in the xy-plane and joining consecutive points by line segments. The more points, the better the approximation.

Now consider a function of two variables.

The **graph** of a function of two variables, f, is the set of all points (x, y, z) such that $z = f(x, y)$. In general, the graph of a function of two variables is a surface in 3-space.

Plotting the Graph of the Function $f(x, y) = x^2 + y^2$

To sketch the graph of f we connect points as for a function of one variable. We first make a table of values of f, such as in Table 12.4.

Table 12.4 *Table of values of* $f(x, y) = x^2 + y^2$

x \ y	−3	−2	−1	0	1	2	3
−3	18	13	10	9	10	13	18
−2	13	8	5	4	5	8	13
−1	10	5	2	1	2	5	10
0	9	4	1	0	1	4	9
1	10	5	2	1	2	5	10
2	13	8	5	4	5	8	13
3	18	13	10	9	10	13	18

Now we plot points. For example, we plot $(1, 2, 5)$ because $f(1, 2) = 5$ and we plot $(0, 2, 4)$ because $f(0, 2) = 4$. Then, we connect the points corresponding to the rows and columns in the table. The result is called a *wire-frame* picture of the graph. Filling in between the wires gives a surface. That is the way a computer drew the graphs in Figures 12.11 and 12.12. As more points are plotted, we get the surface in Figure 12.13, called a *paraboloid.*

You should check to see if the sketches make sense. Notice that the graph goes through the origin since $(x, y, z) = (0, 0, 0)$ satisfies $z = x^2 + y^2$. Observe that if x is held fixed and y is allowed to vary, the graph dips down and then goes back up, just like the entries in the rows of Table 12.4. Similarly, if y is held fixed and x is allowed to vary, the graph dips down and then goes back up, just like the columns of Table 12.4.

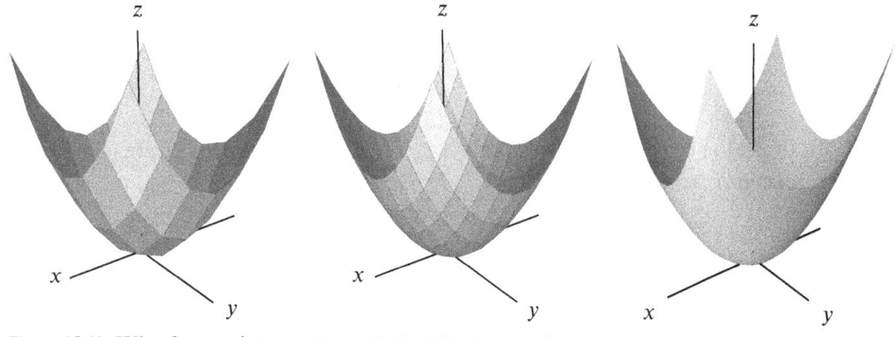

Figure 12.11: Wire frame picture of $f(x, y) = x^2 + y^2$ for $-3 \le x \le 3, -3 \le y \le 3$

Figure 12.12: Wire frame picture of $f(x, y) = x^2 + y^2$ with more points plotted

Figure 12.13: Graph of $f(x, y) = x^2 + y^2$ for $-3 \le x \le 3, -3 \le y \le 3$

New Graphs from Old

We can use the graph of a function to visualize the graphs of related functions.

Example 1 Let $f(x, y) = x^2 + y^2$. Describe in words the graphs of the following functions:
(a) $g(x, y) = x^2 + y^2 + 3$, (b) $h(x, y) = 5 - x^2 - y^2$, (c) $k(x, y) = x^2 + (y - 1)^2$.

Solution We know from Figure 12.13 that the graph of f is a paraboloid, or a bowl, with its vertex at the origin. From this we can work out what the graphs of g, h, and k will look like.

(a) The function $g(x, y) = x^2 + y^2 + 3 = f(x, y) + 3$, so the graph of g is the graph of f, but raised by 3 units. See Figure 12.14.

(b) Since $-x^2 - y^2$ is the negative of $x^2 + y^2$, the graph of $-x^2 - y^2$ is a paraboloid opening downward. Thus, the graph of $h(x, y) = 5 - x^2 - y^2 = 5 - f(x, y)$ looks like a downward-opening paraboloid with vertex at $(0, 0, 5)$, as in Figure 12.15.

(c) The graph of $k(x, y) = x^2 + (y - 1)^2 = f(x, y - 1)$ is a paraboloid with vertex at $x = 0$, $y = 1$, since that is where $k(x, y) = 0$, as in Figure 12.16.

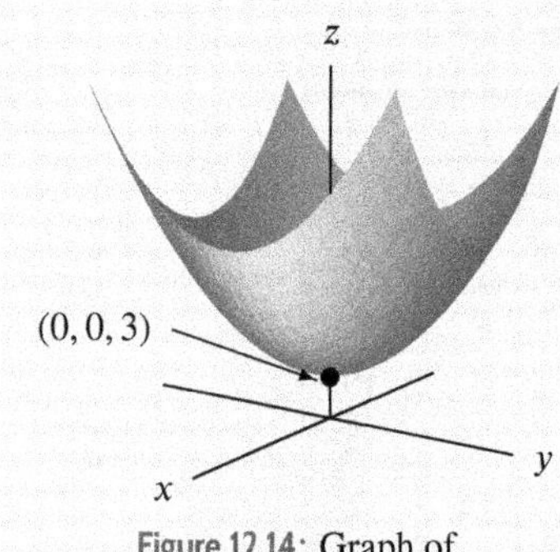

Figure 12.14: Graph of $g(x, y) = x^2 + y^2 + 3$

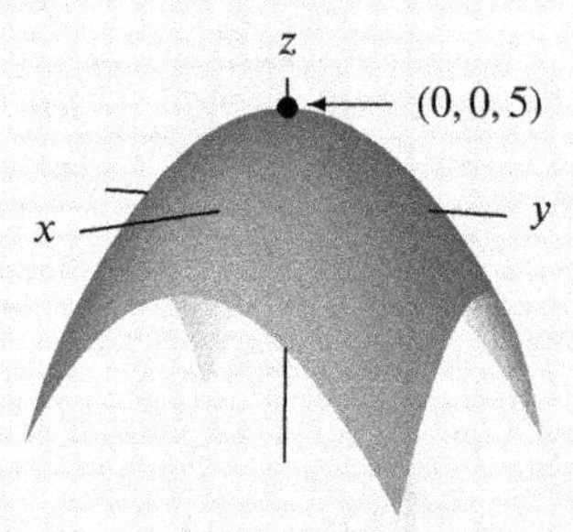

Figure 12.15: Graph of $h(x, y) = 5 - x^2 - y^2$

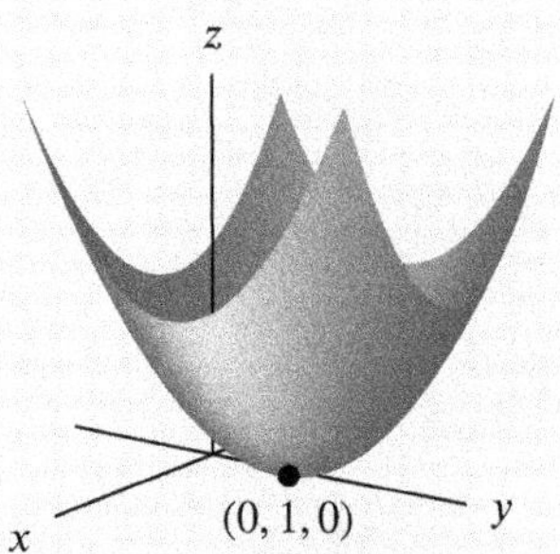

Figure 12.16: Graph of $k(x, y) = x^2 + (y - 1)^2$

Example 2 Describe the graph of $G(x, y) = e^{-(x^2+y^2)}$. What symmetry does it have?

Solution Since the exponential function is always positive, the graph lies entirely above the xy-plane. From the graph of $x^2 + y^2$ we see that $x^2 + y^2$ is zero at the origin and gets larger as we move farther from the origin in any direction. Thus, $e^{-(x^2+y^2)}$ is 1 at the origin, and gets smaller as we move away from the origin in any direction. It can't go below the xy-plane; instead it flattens out, getting closer and closer to the plane. We say the surface is *asymptotic* to the xy-plane. (See Figure 12.17.) Now consider a point (x, y) on the circle $x^2 + y^2 = r^2$. Since

$$G(x, y) = e^{-(x^2+y^2)} = e^{-r^2},$$

the value of the function G is the same at all points on this circle. Thus, we say the graph of G has *circular symmetry*.

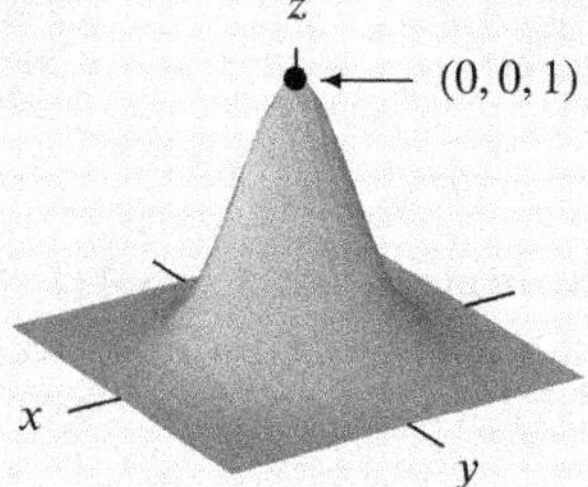

Figure 12.17: Graph of $G(x, y) = e^{-(x^2+y^2)}$

Cross-Sections and the Graph of a Function

We have seen that a good way to analyze a function of two variables is to let one variable vary while the other is kept fixed.

For a function $f(x, y)$, the function we get by holding x fixed and letting y vary is called a **cross-section** of f with x fixed. The graph of the cross-section of $f(x, y)$ with $x = c$ is the curve, or cross-section, we get by intersecting the graph of f with the plane $x = c$. We define a cross-section of f with y fixed similarly.

For example, the cross-section of $f(x, y) = x^2 + y^2$ with $x = 2$ is $f(2, y) = 4 + y^2$. The graph of this cross-section is the curve we get by intersecting the graph of f with the plane perpendicular to the x-axis at $x = 2$. (See Figure 12.18.)

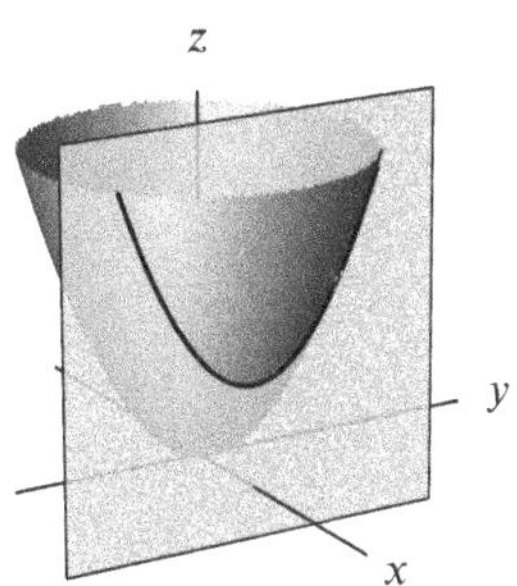

Figure 12.18: Cross-section of the surface $z = f(x, y)$ by the plane $x = 2$

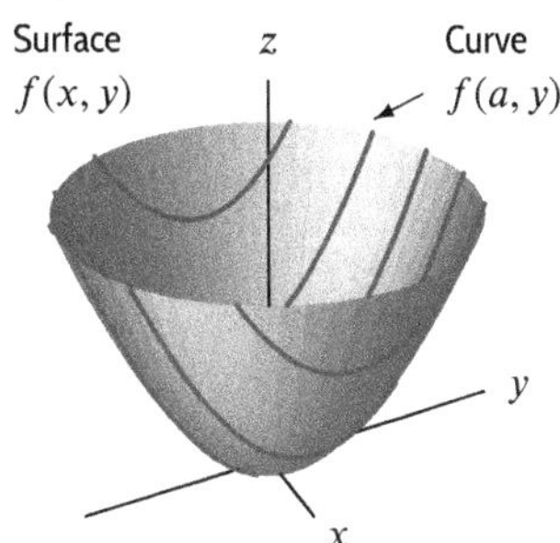

Figure 12.19: The curves $z = f(a, y)$ with a constant: cross-sections with x fixed

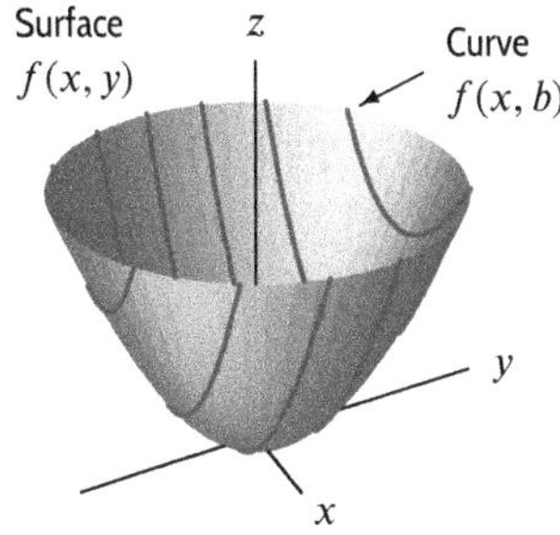

Figure 12.20: The curves $z = f(x, b)$ with b constant: cross-sections with y fixed

Figure 12.19 shows graphs of other cross-sections of f with x fixed; Figure 12.20 shows graphs of cross-sections with y fixed.

Example 3 Describe the cross-sections of the function $g(x, y) = x^2 - y^2$ with y fixed and then with x fixed. Use these cross-sections to describe the shape of the graph of g.

Solution The cross-sections with y fixed at $y = b$ are given by

$$z = g(x, b) = x^2 - b^2.$$

Thus, each cross-section with y fixed gives a parabola opening upward, with minimum $z = -b^2$. The cross-sections with x fixed are of the form

$$z = g(a, y) = a^2 - y^2,$$

which are parabolas opening downward with a maximum of $z = a^2$. (See Figures 12.21 and 12.22.) The graph of g is shown in Figure 12.23. Notice the upward-opening parabolas in the x-direction and the downward-opening parabolas in the y-direction. We say that the surface is *saddle-shaped.*

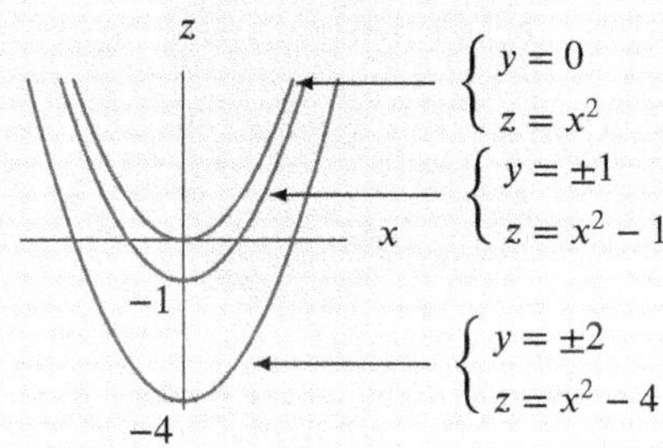

Figure 12.21: Cross-sections of $g(x, y) = x^2 - y^2$ with y fixed

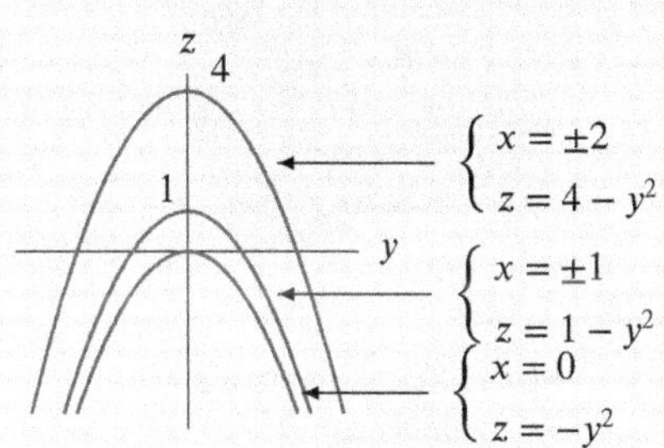

Figure 12.22: Cross-sections of $g(x, y) = x^2 - y^2$ with x fixed

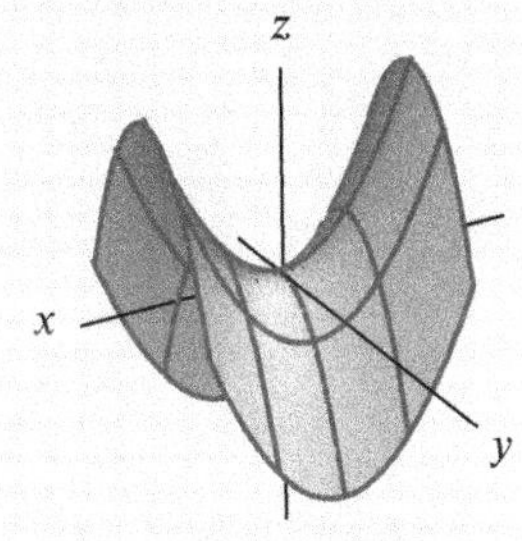

Figure 12.23: Graph of $g(x, y) = x^2 - y^2$ showing cross-sections

Linear Functions

Linear functions are central to single-variable calculus; they are equally important in multivariable calculus. You may be able to guess the shape of the graph of a linear function of two variables. (It's a plane.) Let's look at an example.

Example 4 Describe the graph of $f(x, y) = 1 + x - y$.

Solution The plane $x = a$ is vertical and parallel to the yz-plane. Thus, the cross-section with $x = a$ is the line $z = 1 + a - y$ which slopes downward in the y-direction. Similarly, the plane $y = b$ is parallel to the xz-plane. Thus, the cross-section with $y = b$ is the line $z = 1 + x - b$ which slopes upward in the x-direction. Since all the cross-sections are lines, you might expect the graph to be a flat plane, sloping down in the y-direction and up in the x-direction. This is indeed the case. (See Figure 12.24.)

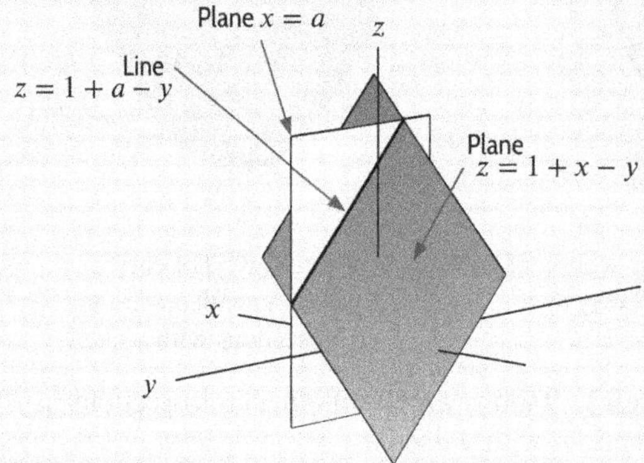

Figure 12.24: Graph of the plane $z = 1 + x - y$ showing cross-section with $x = a$

When One Variable Is Missing: Cylinders

Suppose we graph an equation like $z = x^2$ which has one variable missing. What does the surface look like? Since y is missing from the equation, the cross-sections with y fixed are all the same parabola, $z = x^2$. Letting y vary up and down the y-axis, this parabola sweeps out the trough-shaped surface shown in Figure 12.25. The cross-sections with x fixed are horizontal lines obtained by cutting the surface by a plane perpendicular to the x-axis. This surface is called a *parabolic cylinder*, because it is formed from a parabola in the same way that an ordinary cylinder is formed from a circle; it has a parabolic cross-section instead of a circular one.

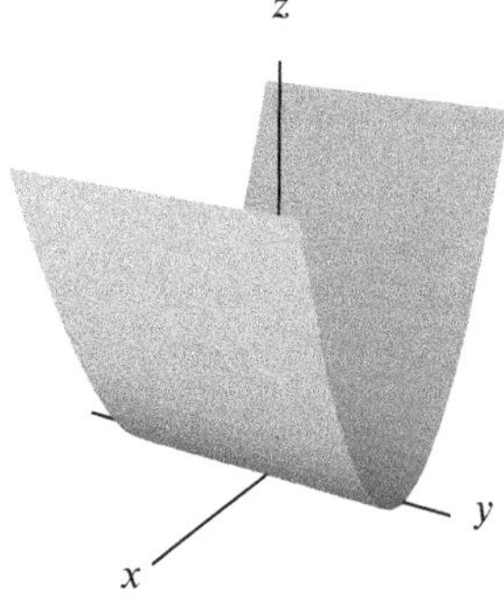

Figure 12.25: A parabolic cylinder $z = x^2$

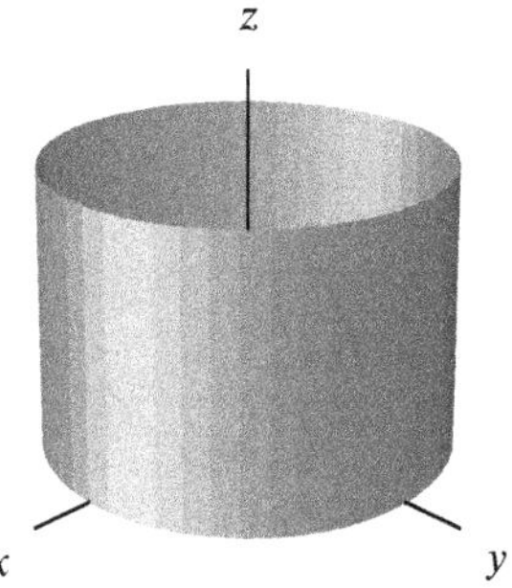

Figure 12.26: Circular cylinder $x^2 + y^2 = 1$

Example 5 Graph the equation $x^2 + y^2 = 1$ in 3-space.

Solution Although the equation $x^2 + y^2 = 1$ does not represent a function, the surface representing it can be graphed by the method used for $z = x^2$. The graph of $x^2 + y^2 = 1$ in the xy-plane is a circle. Since z does not appear in the equation, the intersection of the surface with any horizontal plane will be the same circle $x^2 + y^2 = 1$. Thus, the surface is the cylinder shown in Figure 12.26.

Exercises and Problems for Section 12.2 Online Resource: Additional Problems for Section 12.2

EXERCISES

In Exercises 1–4, which of (I)–(IV) lie on the graph of the function $z = f(x, y)$?

I. $(1, 0, 1)$ II. $(\sqrt{8}, 1, 3)$
III. $(-3, 7, -3)$ IV. $(1, 1, 1/2)$

1. $f(x, y) = -3$
2. $f(x, y) = \sqrt{x^2 + y^2}$
3. $f(x, y) = 1/(x^2 + y^2)$
4. $f(x, y) = 4 - y$

5. Figure 12.27 shows the graph of $z = f(x, y)$.
   (a) Suppose y is fixed and positive. Does z increase or decrease as x increases? Graph z against x.
   (b) Suppose x is fixed and positive. Does z increase or decrease as y increases? Graph z against y.

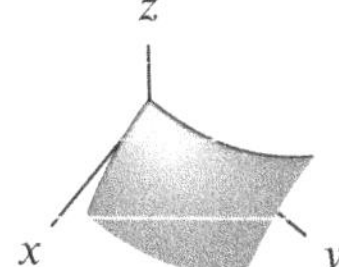

Figure 12.27

6. Without a calculator or computer, match the functions with their graphs in Figure 12.28.
   (a) $z = 2 + x^2 + y^2$ **(b)** $z = 2 - x^2 - y^2$
   (c) $z = 2(x^2 + y^2)$ **(d)** $z = 2 + 2x - y$
   (e) $z = 2$

(I)

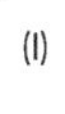

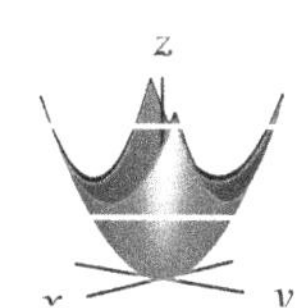

(II)

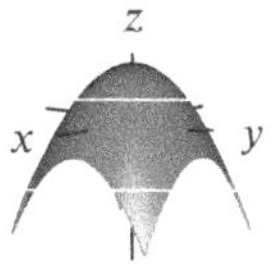

(III)

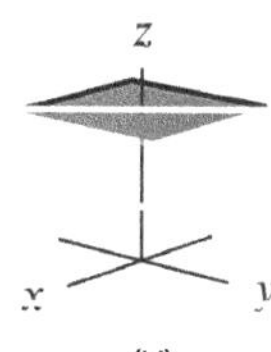

(IV)

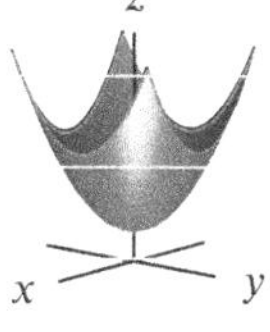

(V)

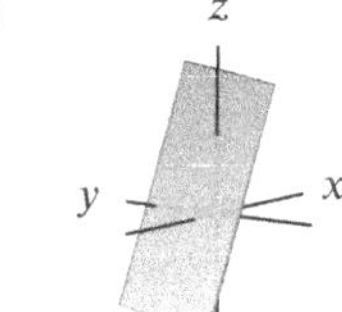

Figure 12.28

7. Without a calculator or computer, match the functions with their graphs in Figure 12.29.
   (a) $z = \dfrac{1}{x^2 + y^2}$ **(b)** $z = -e^{-x^2-y^2}$
   (c) $z = x + 2y + 3$ **(d)** $z = -y^2$
   (e) $z = x^3 - \sin y$.

(I) (II)

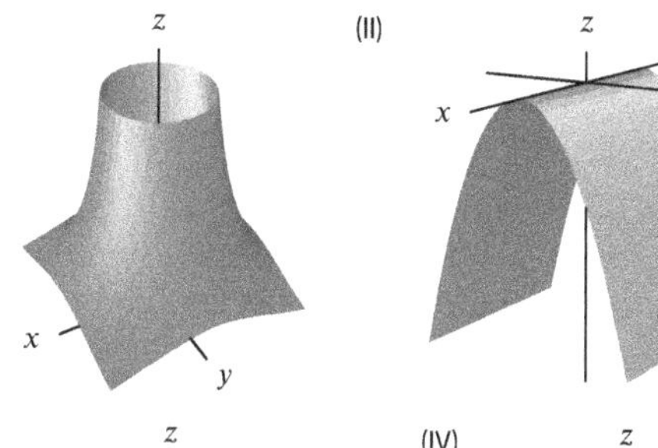

(III)

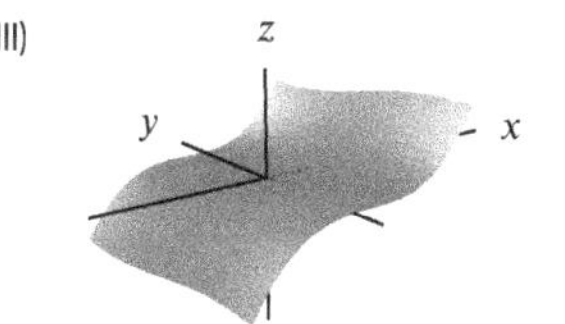

(IV)

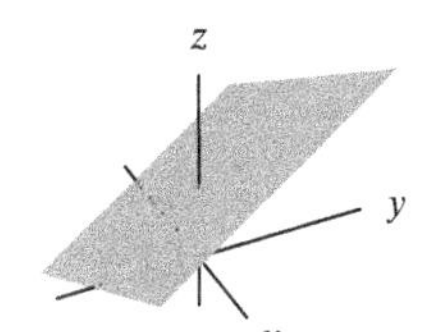

(V)

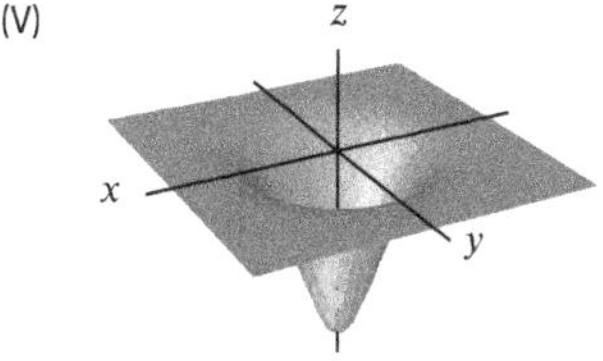

Figure 12.29

In Exercises 8–15, sketch a graph of the surface and briefly describe it in words.

8. $z = 3$
9. $x^2 + y^2 + z^2 = 9$
10. $z = x^2 + y^2 + 4$
11. $z = 5 - x^2 - y^2$
12. $z = y^2$
13. $2x + 4y + 3z = 12$
14. $x^2 + y^2 = 4$
15. $x^2 + z^2 = 4$

In Exercises 16–18, find the equation of the surface.

16. A cylinder of radius $\sqrt{7}$ with its axis along the y-axis.
17. A sphere of radius 3 centered at $\left(0, \sqrt{7}, 0\right)$.
18. The paraboloid obtained by moving the surface $z = x^2 + y^2$ so that its vertex is at $(1, 3, 5)$, its axis is parallel to the x-axis, and the surface opens towards negative x values.

PROBLEMS

19. Consider the function f given by $f(x, y) = y^3 + xy$. Draw graphs of cross-sections with:

    (a) x fixed at $x = -1$, $x = 0$, and $x = 1$.
    (b) y fixed at $y = -1$, $y = 0$, and $y = 1$.

Problems 20–22 concern the concentration, C, in mg per liter, of a drug in the blood as a function of x, the amount, in mg, of the drug given and t, the time in hours since the injection. For $0 \leq x \leq 4$ and $t \geq 0$, we have $C = f(x, t) = te^{-t(5-x)}$.

20. Find $f(3, 2)$. Give units and interpret in terms of drug concentration.

21. Graph the following single-variable functions and explain their significance in terms of drug concentration.

    (a) $f(4, t)$ (b) $f(x, 1)$

22. Graph $f(a, t)$ for $a = 1, 2, 3, 4$ on the same axes. Describe how the graph changes as a increases and explain what this means in terms of drug concentration.

Problems 23–24 concern the kinetic energy, $E = f(m, v) = \frac{1}{2}mv^2$, in joules, of a moving object as a function of its mass $m \geq 0$, in kg, and its speed $v \geq 0$, in m/sec.

23. Find $f(2, 10)$. Give units and interpret this quantity in the context of kinetic energy.

24. Graph the following single-variable functions and explain their significance in terms of kinetic energy.

    (a) $f(6, v)$ (b) $f(m, 20)$

In Problems 25–26, the atmospheric pressure, $P = f(y, t) = (950 + 2t)e^{-y/7}$, in millibars, on a weather balloon, is a function of its height $y \geq 0$, in km above sea level after t hours with $0 \leq t \leq 48$.

25. Find $f(2, 12)$. Give units and interpret this quantity in the context of atmospheric pressure.

26. Graph the following single-variable functions and explain the significance of the shape of the graph in terms of atmospheric pressure.

    (a) $f(3, t)$ (b) $f(y, 24)$

27. Without a computer or calculator, match the equations (a)–(i) with the graphs (I)–(IX).

    (a) $z = xye^{-(x^2+y^2)}$ (b) $z = \cos\left(\sqrt{x^2 + y^2}\right)$
    (c) $z = \sin y$ (d) $z = -\dfrac{1}{x^2 + y^2}$
    (e) $z = \cos^2 x \cos^2 y$ (f) $z = \dfrac{\sin(x^2 + y^2)}{x^2 + y^2}$
    (g) $z = \cos(xy)$ (h) $z = |x||y|$
    (i) $z = (2x^2 + y^2)e^{1-x^2-y^2}$

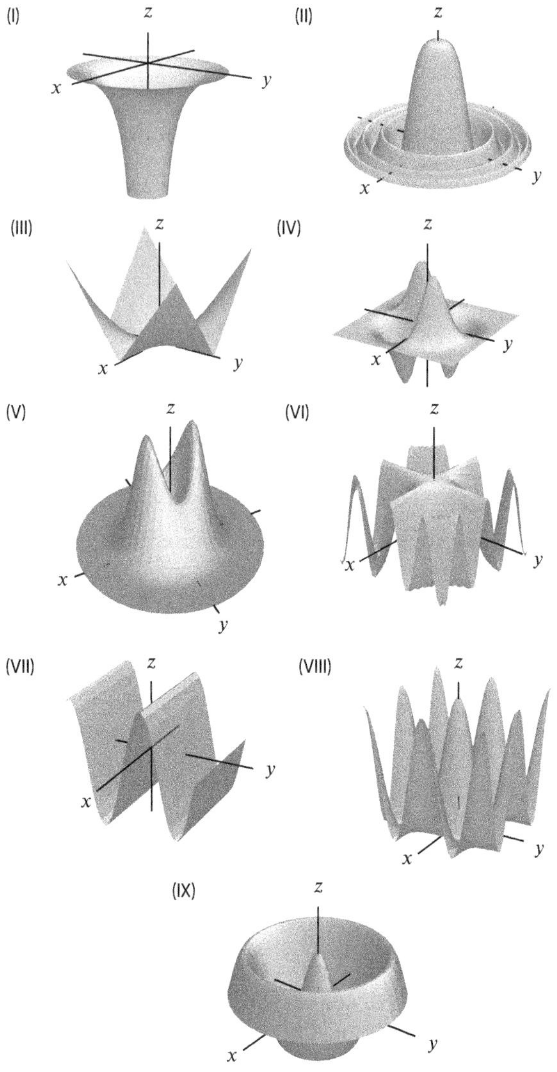

(I) (II) (III) (IV) (V) (VI) (VII) (VIII) (IX)

28. Decide whether the graph of each of the following equations is the shape of a bowl, a plate, or neither. Consider a plate to be any flat surface and a bowl to be anything that could hold water, assuming the positive z-axis is up.

    (a) $z = x^2 + y^2$ (b) $z = 1 - x^2 - y^2$
    (c) $x + y + z = 1$ (d) $z = -\sqrt{5 - x^2 - y^2}$
    (e) $z = 3$

29. Sketch cross-sections for each function in Problem 28.

30. Without a calculator or computer, match the functions with their cross-sections with x fixed in Figure 12.30.

    (a) $z = 1/(1 + x^2 + y^2)$ (b) $z = 1 + x + y$
    (c) $z = e^{-x+y}$ (d) $z = e^{x-y}$
    (e) $z = \sin(xy)$ (f) $z = x^2$.

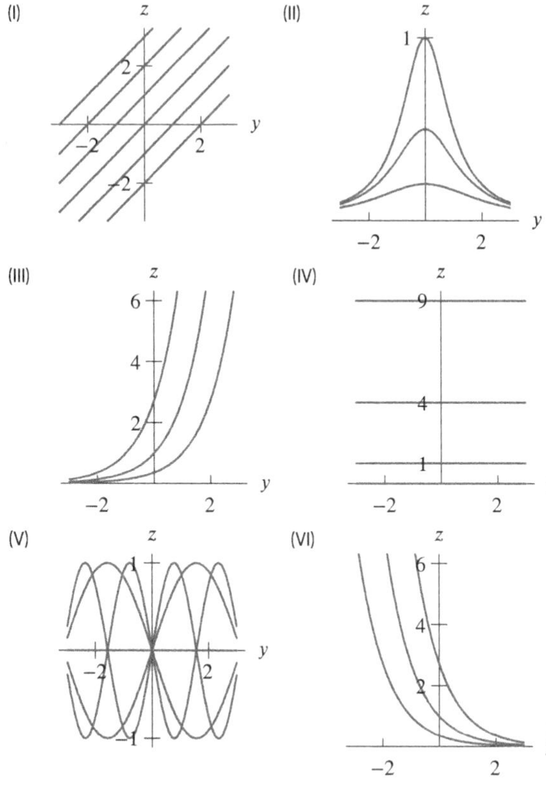

Figure 12.30

31. Without a calculator or computer, for $z = x^2 + 2xy^2$, determine which of (I)–(II) in Figure 12.31 are cross-sections with x fixed and which are cross-sections with y fixed.

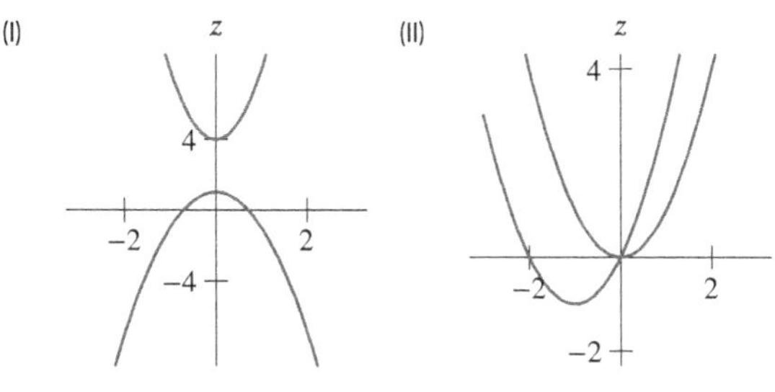

Figure 12.31

32. You like pizza and you like cola. Which of the graphs in Figure 12.32 represents your happiness as a function of how many pizzas and how much cola you have if

(a) There is no such thing as too many pizzas and too much cola?
(b) There is such a thing as too many pizzas or too much cola?
(c) There is such a thing as too much cola but no such thing as too many pizzas?

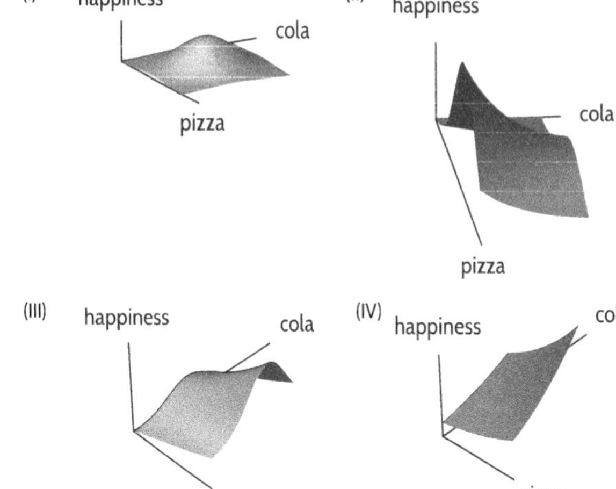

Figure 12.32

33. For each of the graphs I–IV in Problem 32, draw:

(a) Two cross-sections with pizza fixed
(b) Two cross-sections with cola fixed.

For Problems 34–37, give a formula for a function whose graph is described. Sketch it using a computer or calculator.

34. A bowl which opens upward and has its vertex at 5 on the z-axis.

35. A plane which has its x-, y-, and z-intercepts all positive.

36. A parabolic cylinder opening upward from along the line $y = x$ in the xy-plane.

37. A cone of circular cross-section opening downward and with its vertex at the origin.

38. Sketch cross-sections of $f(r, h) = \pi r^2 h$, first keeping h fixed, then keeping r fixed.

39. By setting one variable constant, find a plane that intersects the graph of $z = 4x^2 - y^2 + 1$ in a:

(a) Parabola opening upward
(b) Parabola opening downward
(c) Pair of intersecting straight lines.

40. Sketch cross-sections of the equation $z = y - x^2$ with x fixed and with y fixed and use them to sketch a graph of $z = y - x^2$.

41. A wave travels along a canal. Let x be the distance along the canal, t be the time, and z be the height of the water above the equilibrium level. The graph of z as a function of x and t is in Figure 12.33.

(a) Draw the profile of the wave for $t = -1, 0, 1, 2$. (Put the x-axis to the right and the z-axis vertical.)

(b) Is the wave traveling in the direction of increasing or decreasing x?

(c) Sketch a surface representing a wave traveling in the opposite direction.

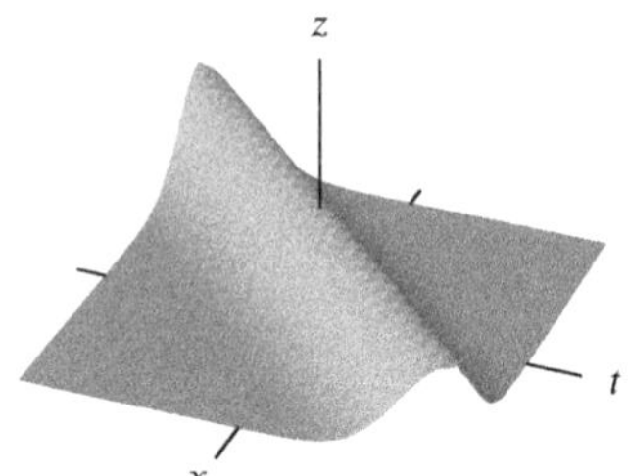

Figure 12.33

42. The pressure of a fixed amount of compressed nitrogen gas in a cylinder is given, in atmospheres, by

$$P = f(T, V) = \frac{10T}{V},$$

where T is the temperature of the gas, in Kelvin, and V is the volume of the cylinder, in liters. Figures 12.34 and 12.35 give cross-sections of the function f.

(a) Which figure shows cross-sections of f with T fixed? What does the shape of the cross-sections tell you about the pressure?

(b) Which figure shows cross-sections of f with V fixed? What does the shape of the cross-sections tell you about the pressure?

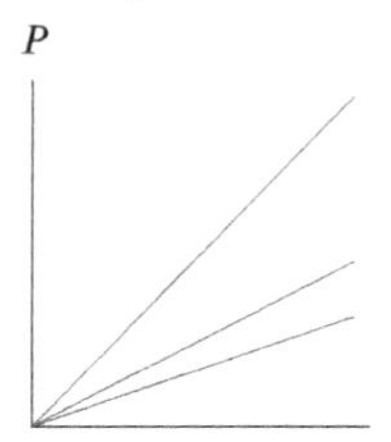

Figure 12.34

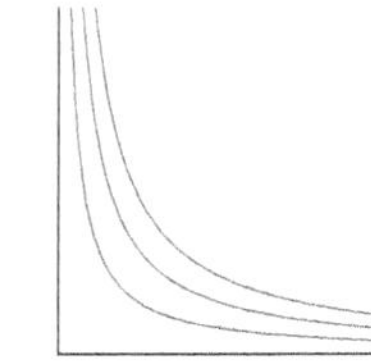

Figure 12.35

Strengthen Your Understanding

In Problems 43–44, explain what is wrong with the statement.

43. The graph of the function $f(x, y) = x^2 + y^2$ is a circle.

44. Cross-sections of the function $f(x, y) = x^2$ with x fixed are parabolas.

In Problems 45–47, give an example of:

45. A function whose graph lies above the xy-plane and intersects the plane $z = 2$ in a single point.

46. A function which intersects the xz-plane in a parabola and the yz-plane in a line.

47. A function which intersects the xy-plane in a circle.

Are the statements in Problems 48–61 true or false? Give reasons for your answer.

48. The function given by the formula $f(v, w) = e^v/w$ is an increasing function of v when w is a nonzero constant.

49. A function $f(x, y)$ can be an increasing function of x with y held fixed, and be a decreasing function of y with x held fixed.

50. A function $f(x, y)$ can have the property that $g(x) = f(x, 5)$ is increasing, whereas $h(x) = f(x, 10)$ is decreasing.

51. The plane $x + 2y - 3z = 1$ passes through the origin.

52. The plane $x + y + z = 3$ intersects the x-axis when $x = 3$.

53. The sphere $x^2+y^2+z^2 = 10$ intersects the plane $x = 10$.

54. The cross-section of the function $f(x, y) = x + y^2$ with $y = 1$ is a line.

55. The function $g(x, y) = 1 - y^2$ has identical parabolas for all cross-sections with x constant.

56. The function $g(x, y) = 1 - y^2$ has lines for all cross-sections with y constant.

57. The graphs of $f(x, y) = \sin(xy)$ and $g(x, y) = \sin(xy) + 2$ never intersect.

58. The graphs of $f(x, y) = x^2+y^2$ and $g(x, y) = 1-x^2-y^2$ intersect in a circle.

59. If all the cross-sections of the graph of $f(x, y)$ with x constant are lines, then the graph of f is a plane.

60. The only point of intersection of the graphs of $f(x, y)$ and $-f(x, y)$ is the origin.

61. The point $(0, 0, 10)$ is the highest point on the graph of the function $f(x, y) = 10 - x^2 - y^2$.

62. The object in 3-space described by $x = 2$ is

(a) A point **(b)** A line
(c) A plane **(d)** Undefined.

12.3 CONTOUR DIAGRAMS

The surface which represents a function of two variables often gives a good idea of the function's general behavior—for example, whether it is increasing or decreasing as one of the variables increases. However, it is difficult to read numerical values off a surface and it can be hard to see all of the function's behavior from a surface. Thus, functions of two variables are often represented by contour diagrams like the weather map on page 652. Contour diagrams have the additional advantage that they can be extended to functions of three variables.

Topographical Maps

One of the most common examples of a contour diagram is a topographical map like that shown in Figure 12.36. It gives the elevation in the region and is a good way of getting an overall picture of the terrain: where the mountains are, where the flat areas are. Such topographical maps are frequently colored green at the lower elevations and brown, red, or white at the higher elevations.

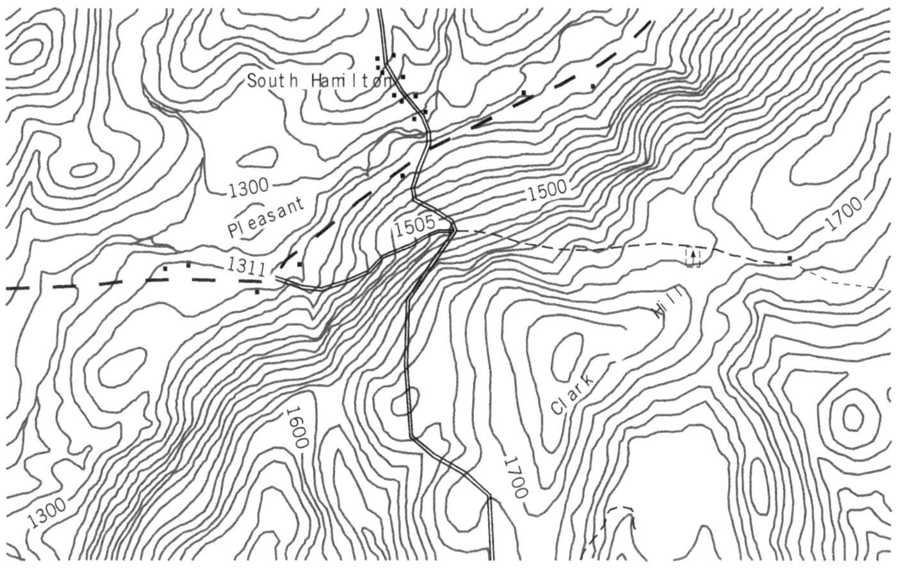

Figure 12.36: A topographical map showing the region around South Hamilton, NY

The curves on a topographical map that separate lower elevations from higher elevations are called *contour lines* because they outline the contour or shape of the land.[5] Because every point along the same contour has the same elevation, contour lines are also called *level curves* or *level sets*. The more closely spaced the contours, the steeper the terrain; the more widely spaced the contours, the flatter the terrain (provided, of course, that the elevation between contours varies by a constant amount). Certain features have distinctive characteristics. A mountain peak is typically surrounded by contour lines like those in Figure 12.37. A pass in a range of mountains may have contours that look like Figure 12.38. A long valley has parallel contour lines indicating the rising elevations on both sides of the valley (see Figure 12.39); a long ridge of mountains has the same type of contour lines, only the elevations decrease on both sides of the ridge. Notice that the elevation numbers on the contour lines are as important as the curves themselves. We usually draw contours for equally spaced values of z.

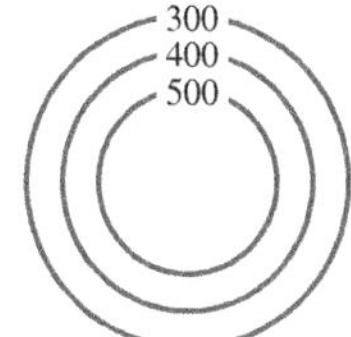

Figure 12.37: Mountain peak

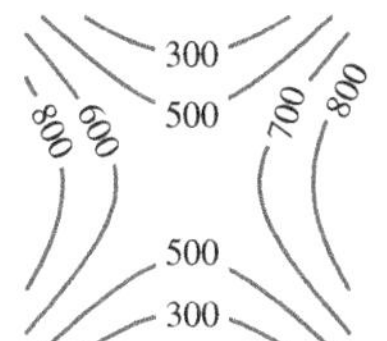

Figure 12.38: Pass between two mountains

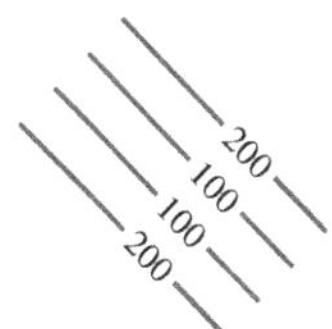

Figure 12.39: Long valley

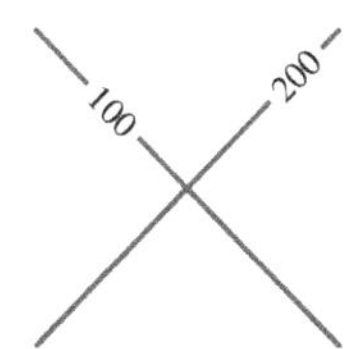

Figure 12.40: Impossible contour lines

Notice that two contours corresponding to different elevations cannot cross each other as shown in Figure 12.40. If they did, the point of intersection of the two curves would have two different elevations, which is impossible (assuming the terrain has no overhangs).

[5]In fact they are usually not straight lines, but curves. They may also be in disconnected pieces.

Corn Production

Contour maps can display information about a function of two variables without reference to a surface. Consider the effect of weather conditions on US corn production. Figure 12.41 gives corn production $C = f(R, T)$ as a function of the total rainfall, R, in inches, and average temperature, T, in degrees Fahrenheit, during the growing season.[6] At the present time, $R = 15$ inches and $T = 76$°F. Production is measured as a percentage of the present production; thus, the contour through $R = 15, T = 76$, has value 100, that is, $C = f(15, 76) = 100$.

Example 1 Use Figure 12.41 to estimate $f(18, 78)$ and $f(12, 76)$ and interpret in terms of corn production.

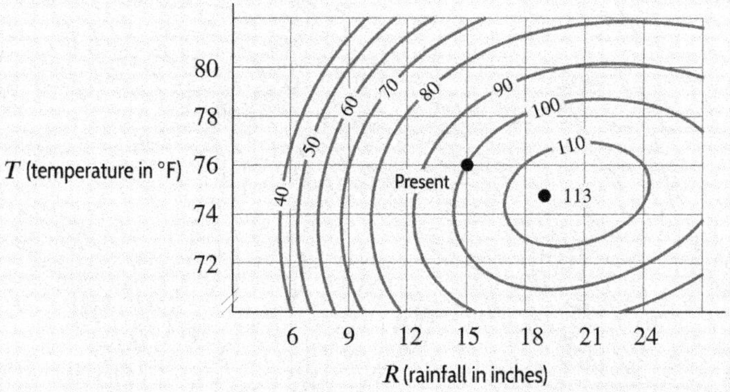

Figure 12.41: Corn production, C, as a function of rainfall and temperature

Solution The point with R-coordinate 18 and T-coordinate 78 is on the contour $C = 100$, so $f(18, 78) = 100$. This means that if the annual rainfall were 18 inches and the temperature were 78°F, the country would produce about the same amount of corn as at present, although it would be wetter and warmer than it is now.

The point with R-coordinate 12 and T-coordinate 76 is about halfway between the $C = 80$ and the $C = 90$ contours, so $f(12, 76) \approx 85$. This means that if the rainfall fell to 12 inches and the temperature stayed at 76°, then corn production would drop to about 85% of what it is now.

Example 2 Use Figure 12.41 to describe in words the cross-sections with T and R constant through the point representing present conditions. Give a common-sense explanation of your answer.

Solution To see what happens to corn production if the temperature stays fixed at 76°F but the rainfall changes, look along the horizontal line $T = 76$. Starting from the present and moving left along the line $T = 76$, the values on the contours decrease. In other words, if there is a drought, corn production decreases. Conversely, as rainfall increases, that is, as we move from the present to the right along the line $T = 76$, corn production increases, reaching a maximum of more than 110% when $R = 21$, and then decreases (too much rainfall floods the fields).

If, instead, rainfall remains at the present value and temperature increases, we move up the vertical line $R = 15$. Under these circumstances corn production decreases; a 2°F increase causes a 10% drop in production. This makes sense since hotter temperatures lead to greater evaporation and hence drier conditions, even with rainfall constant at 15 inches. Similarly, a decrease in temperature leads to a very slight increase in production, reaching a maximum of around 102% when $T = 74$, followed by a decrease (the corn won't grow if it is too cold).

[6]Adapted from S. Beaty and R. Healy, "The Future of American Agriculture," *Scientific American* 248, No. 2, February 1983.

Contour Diagrams and Graphs

Contour diagrams and graphs are two different ways of representing a function of two variables. How do we go from one to the other? In the case of the topographical map, the contour diagram was created by joining all the points at the same height on the surface and dropping the curve into the xy-plane.

How do we go the other way? Suppose we wanted to plot the surface representing the corn production function $C = f(R, T)$ given by the contour diagram in Figure 12.41. Along each contour the function has a constant value; if we take each contour and lift it above the plane to a height equal to this value, we get the surface in Figure 12.42.

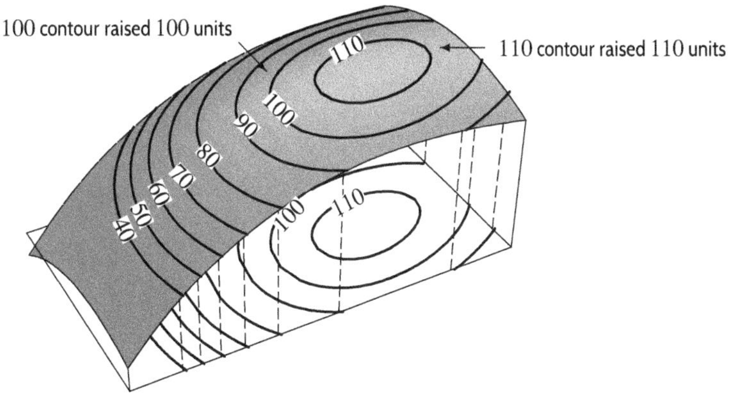

Figure 12.42: Getting the graph of the corn yield function from the contour diagram

Notice that the raised contours are the curves we get by slicing the surface horizontally. In general, we have the following result:

> Contour lines, or level curves, are obtained from a surface by slicing it with horizontal planes. A contour diagram is a collection of level curves labeled with function values.

Finding Contours Algebraically

Algebraic equations for the contours of a function f are easy to find if we have a formula for $f(x, y)$. Suppose the surface has equation

$$z = f(x, y).$$

A contour is obtained by slicing the surface with a horizontal plane with equation $z = c$. Thus, the equation for the contour at height c is given by:

$$f(x, y) = c.$$

Example 3 Find equations for the contours of $f(x, y) = x^2 + y^2$ and draw a contour diagram for f. Relate the contour diagram to the graph of f.

Solution The contour at height c is given by

$$f(x, y) = x^2 + y^2 = c.$$

This is a contour only for $c \geq 0$, For $c > 0$ it is a circle of radius $\sqrt{c}$. For $c = 0$, it is a single point (the origin). Thus, the contours at an elevation of $c = 1, 2, 3, 4, \ldots$ are all circles centered at the origin of radius $1, \sqrt{2}, \sqrt{3}, 2, \ldots$. The contour diagram is shown in Figure 12.43. The bowl–shaped graph of f is shown in Figure 12.44. Notice that the graph of f gets steeper as we move further away from the origin. This is reflected in the fact that the contours become more closely packed as we move further from the origin; for example, the contours for $c = 6$ and $c = 8$ are closer together than the contours for $c = 2$ and $c = 4$.

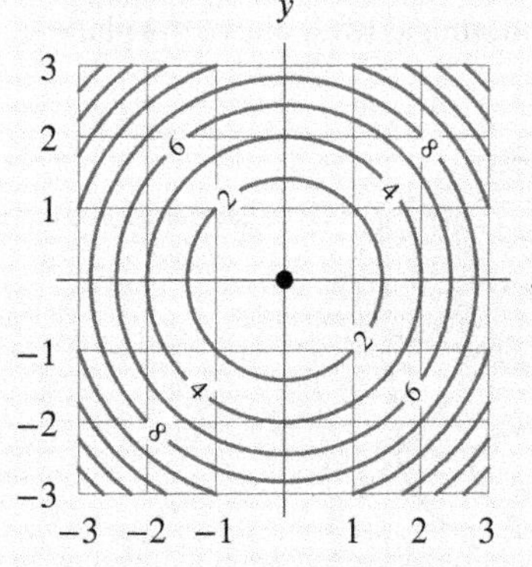

Figure 12.43: Contour diagram for $f(x, y) = x^2 + y^2$ (even values of c only)

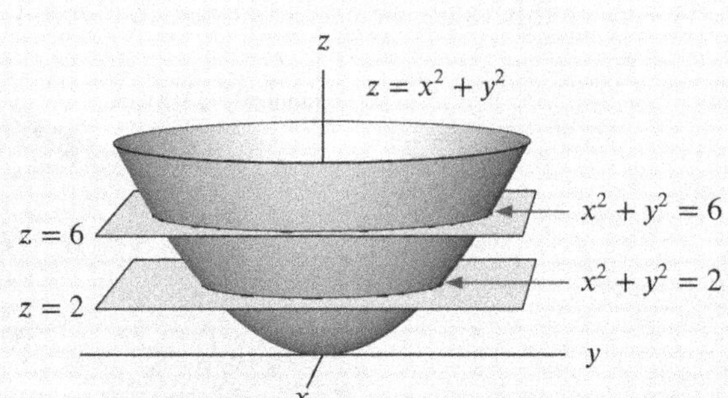

Figure 12.44: The graph of $f(x, y) = x^2 + y^2$

Example 4 Draw a contour diagram for $f(x, y) = \sqrt{x^2 + y^2}$ and relate it to the graph of f.

Solution The contour at level c is given by

$$f(x, y) = \sqrt{x^2 + y^2} = c.$$

For $c > 0$ this is a circle, just as in the previous example, but here the radius is c instead of $\sqrt{c}$. For $c = 0$, it is the origin. Thus, if the level c increases by 1, the radius of the contour increases by 1. This means the contours are equally spaced concentric circles (see Figure 12.45) which do not become more closely packed further from the origin. Thus, the graph of f has the same constant slope as we move away from the origin (see Figure 12.46), making it a cone rather than a bowl.

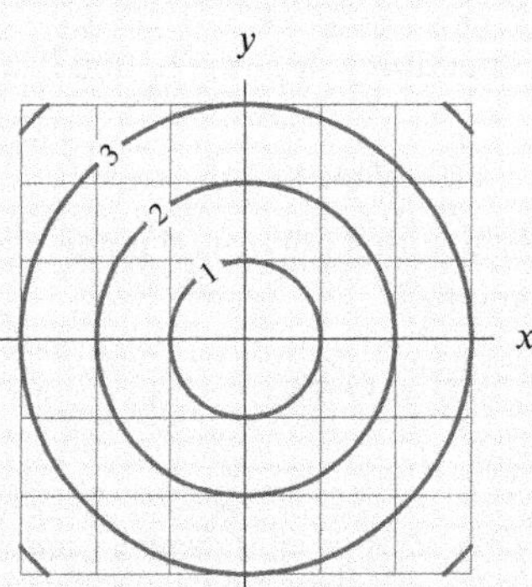

Figure 12.45: A contour diagram for $f(x, y) = \sqrt{x^2 + y^2}$

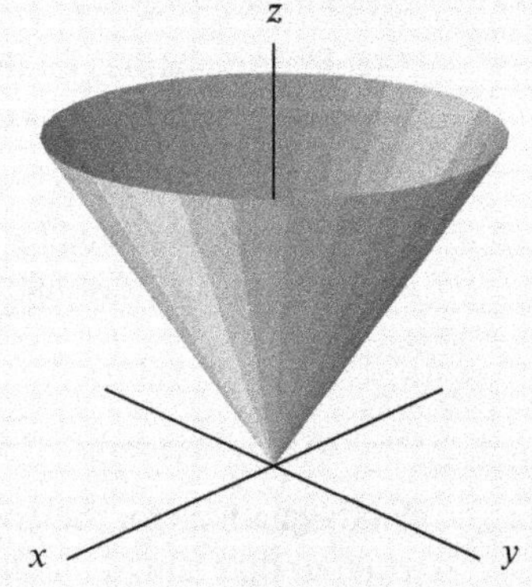

Figure 12.46: The graph of $f(x, y) = \sqrt{x^2 + y^2}$

In both of the previous examples the level curves are concentric circles because the surfaces have circular symmetry. Any function of two variables which depends only on the quantity $(x^2 + y^2)$ has such symmetry: for example, $G(x, y) = e^{-(x^2+y^2)}$ or $H(x, y) = \sin\left(\sqrt{x^2 + y^2}\right)$.

Example 5 Draw a contour diagram for $f(x, y) = 2x + 3y + 1$.

Solution The contour at level c has equation $2x + 3y + 1 = c$. Rewriting this as $y = -(2/3)x + (c - 1)/3$, we see that the contours are parallel lines with slope $-2/3$. The y-intercept for the contour at level c is $(c - 1)/3$; each time c increases by 3, the y-intercept moves up by 1. The contour diagram is shown in Figure 12.47.

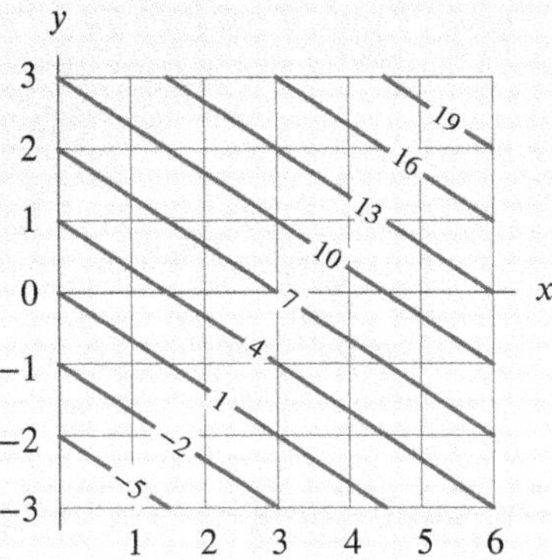

Figure 12.47: A contour diagram for $f(x, y) = 2x + 3y + 1$

Contour Diagrams and Tables

Sometimes we can get an idea of what the contour diagram of a function looks like from its table.

Example 6 Relate the values of $f(x, y) = x^2 - y^2$ in Table 12.5 to its contour diagram in Figure 12.48.

Table 12.5 *Table of values of $f(x, y) = x^2 - y^2$*

y							
3	0	−5	−8	−9	−8	−5	0
2	5	0	−3	−4	−3	0	5
1	8	3	0	−1	0	3	8
0	9	4	1	0	1	4	9
−1	8	3	0	−1	0	3	8
−2	5	0	−3	−4	−3	0	5
−3	0	−5	−8	−9	−8	−5	0
	−3	−2	−1	0	1	2	3
				x			

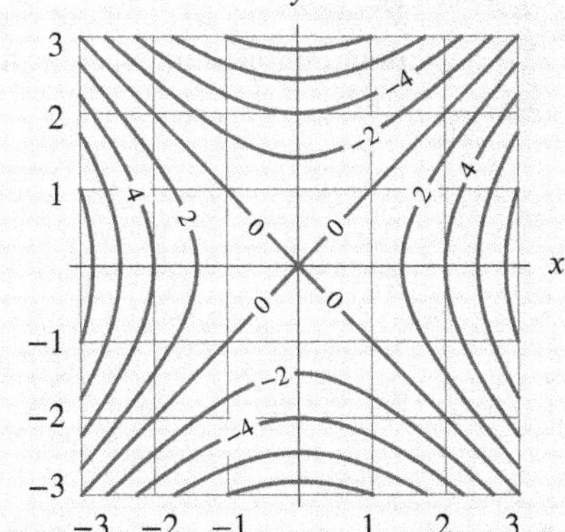

Figure 12.48: Contour map of $f(x, y) = x^2 - y^2$

Solution One striking feature of the values in Table 12.5 is the zeros along the diagonals. This occurs because $x^2 - y^2 = 0$ along the lines $y = x$ and $y = -x$. So the $z = 0$ contour consists of these two lines. In the triangular region of the table that lies to the right of both diagonals, the entries are positive. To the left of both diagonals, the entries are also positive. Thus, in the contour diagram, the positive contours lie in the triangular regions to the right and left of the lines $y = x$ and $y = -x$. Further, the table shows that the numbers on the left are the same as the numbers on the right; thus, each contour has two pieces, one on the left and one on the right. See Figure 12.48. As we move away from the origin along the x-axis, we cross contours corresponding to successively larger values. On the saddle-shaped graph of $f(x, y) = x^2 - y^2$ shown in Figure 12.49, this corresponds to climbing out of the saddle along one of the ridges. Similarly, the negative contours occur in pairs in the top and bottom triangular regions; the values get more and more negative as we go out along the y-axis. This corresponds to descending from the saddle along the valleys that are submerged below the xy-

plane in Figure 12.49. Notice that we could also get the contour diagram by graphing the family of hyperbolas $x^2 - y^2 = 0, \pm 2, \pm 4, \ldots$.

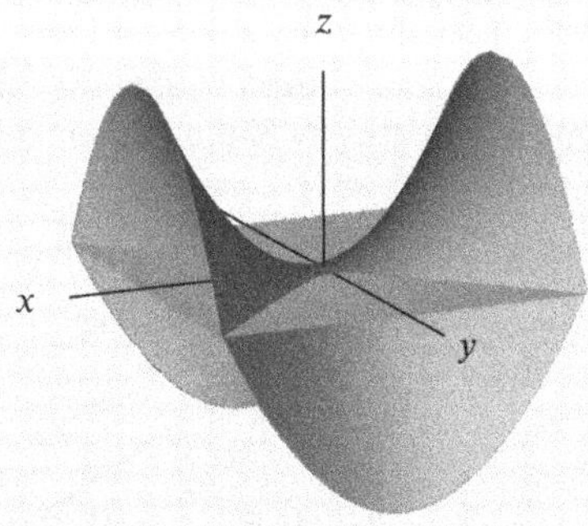

Figure 12.49: Graph of $f(x, y) = x^2 - y^2$ showing plane $z = 0$

Using Contour Diagrams: The Cobb-Douglas Production Function

Suppose you decide to expand your small printing business. Should you start a night shift and hire more workers? Should you buy more expensive but faster computers which will enable the current staff to keep up with the work? Or should you do some combination of the two?

Obviously, the way such a decision is made in practice involves many other considerations—such as whether you could get a suitably trained night shift, or whether there are any faster computers available. Nevertheless, you might model the quantity, P, of work produced by your business as a function of two variables: your total number, N, of workers, and the total value, V, of your equipment. What might the contour diagram of the production function look like?

Example 7 Explain why the contour diagram in Figure 12.50 does not model the behavior expected of the production function, whereas the contour diagram in Figure 12.51 does.

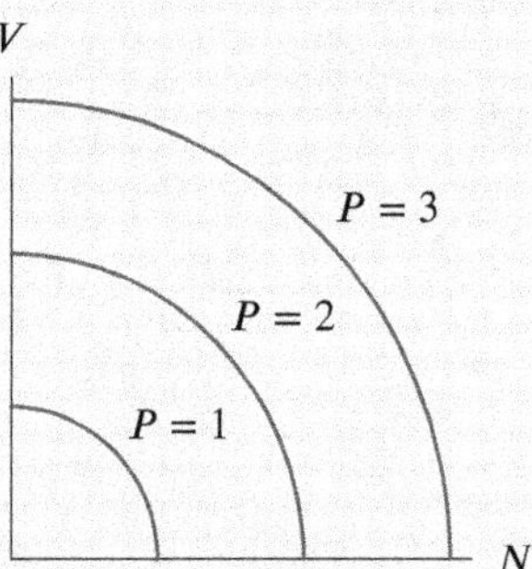

Figure 12.50: Incorrect contours for printing production

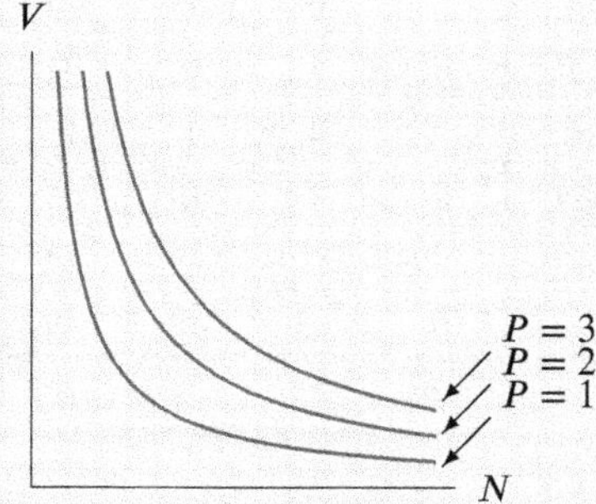

Figure 12.51: Correct contours for printing production

Solution Look at Figure 12.50. Notice that the contour $P = 1$ intersects the N- and the V-axes, suggesting that it is possible to produce work with no workers or with no equipment; this is unreasonable. However, no contours in Figure 12.51 intersect either the N- or the V-axis.

In Figure 12.51, fixing V and letting N increase corresponds to moving to the right, crossing contours less and less frequently. Production increases more and more slowly because hiring additional workers does little to boost production if the machines are already used to capacity.

Similarly, if we fix N and let V increase, Figure 12.51 shows production increasing, but at a decreasing rate. Buying machines without enough people to use them does not increase production much. Thus Figure 12.51 fits the expected behavior of the production function best.

Formula for a Production Function

Production functions are often approximated by formulas of the form

$$P = f(N, V) = cN^{\alpha}V^{\beta}$$

where P is the quantity produced and c, α, and β are positive constants, $0 < \alpha < 1$ and $0 < \beta < 1$.

Example 8 Show that the contours of the function $P = cN^{\alpha}V^{\beta}$ have approximately the shape of the contours in Figure 12.51.

Solution The contours are the curves where P is equal to a constant value, say P_0, that is, where

$$cN^{\alpha}V^{\beta} = P_0.$$

Solving for V, we get

$$V = \left(\frac{P_0}{c}\right)^{1/\beta} N^{-\alpha/\beta}.$$

Thus, V is a power function of N with a negative exponent, so its graph has the general shape shown in Figure 12.51.

The Cobb-Douglas Production Model

In 1928, Cobb and Douglas used a similar function to model the production of the entire US economy in the first quarter of this century. Using government estimates of P, the total yearly production between 1899 and 1922, of K, the total capital investment over the same period, and of L, the total labor force, they found that P was well approximated by the *Cobb-Douglas production function*

$$P = 1.01L^{0.75}K^{0.25}.$$

This function turned out to model the US economy surprisingly well, both for the period on which it was based and for some time afterward.[7]

Exercises and Problems for Section 12.3

EXERCISES

In Exercises 1–4, sketch a possible contour diagram for each surface, marked with reasonable z-values. (Note: There are many possible answers.)

1.

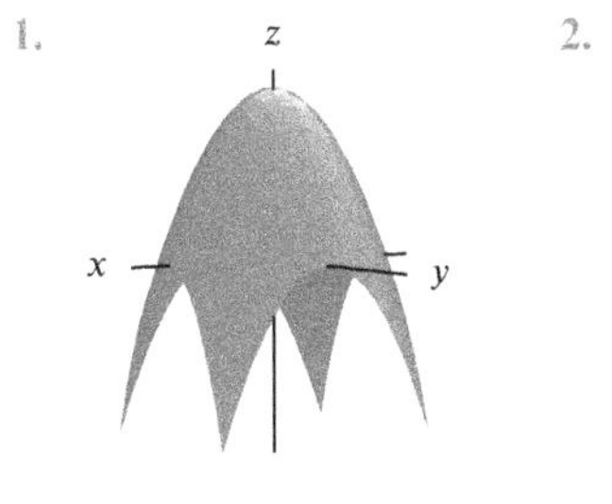

2.

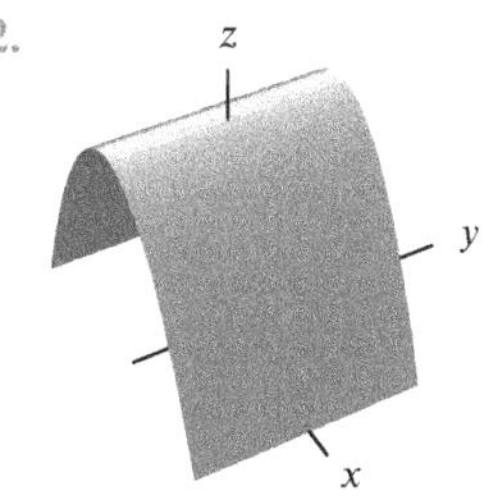

3. 4.

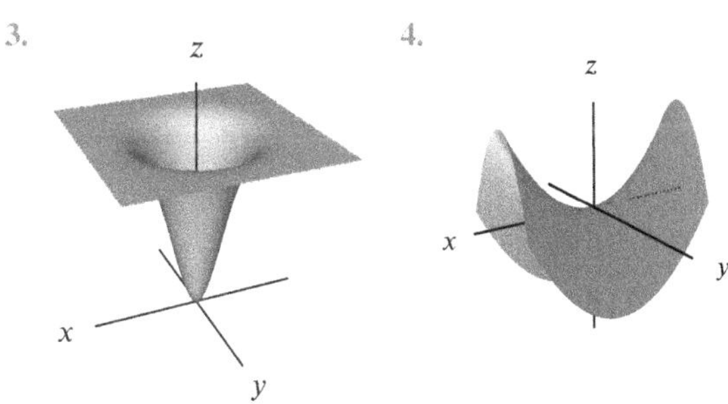

In Exercises 5–13, sketch a contour diagram for the function with at least four labeled contours. Describe in words the contours and how they are spaced.

5. $f(x, y) = x + y$

6. $f(x, y) = 3x + 3y$

7. $f(x, y) = x^2 + y^2$

8. $f(x, y) = -x^2 - y^2 + 1$

9. $f(x, y) = xy$

10. $f(x, y) = y - x^2$

11. $f(x, y) = x^2 + 2y^2$

12. $f(x, y) = \sqrt{x^2 + 2y^2}$

13. $f(x, y) = \cos\sqrt{x^2 + y^2}$

14. Let $f(x, y) = 3x^2y + 7x + 20$. Find an equation for the contour that goes through the point $(5, 10)$.

15. **(a)** For $z = f(x, y) = xy$, sketch and label the level curves $z = \pm 1$, $z = \pm 2$.
(b) Sketch and label cross-sections of f with $x = \pm 1$, $x = \pm 2$.
(c) The surface $z = xy$ is cut by a vertical plane containing the line $y = x$. Sketch the cross-section.

[7] Cobb, C. and Douglas, P., "A Theory of Production", *American Economic Review* 18 (1928: Supplement), pp. 139–165.

16. Match the surfaces (a)–(e) in Figure 12.52 with the contour diagrams (I)–(V) in Figure 12.53.

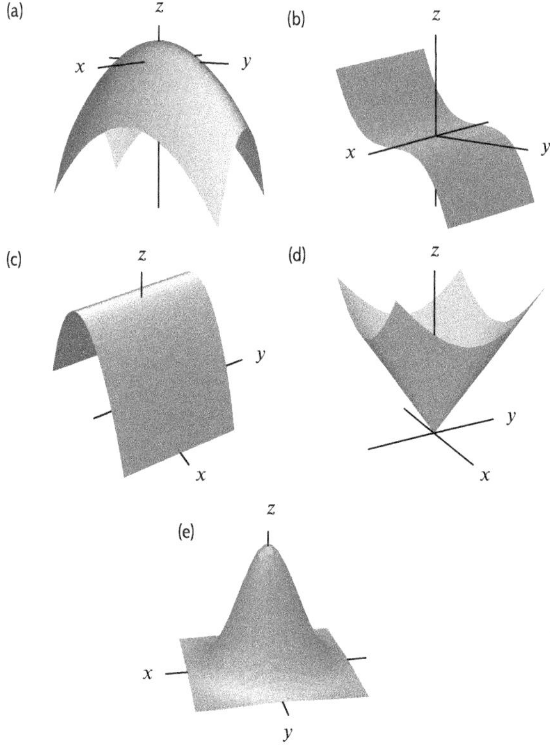

Figure 12.52

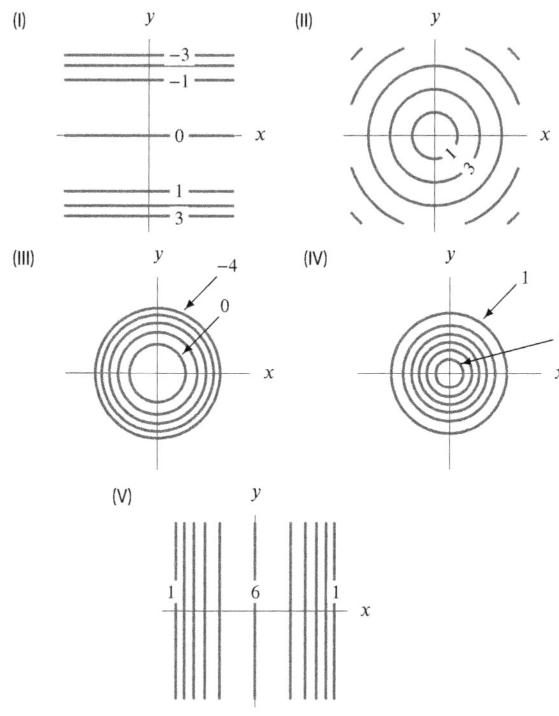

Figure 12.53

17. Figure 12.54 shows the contour diagram of $z = f(x, y)$. Which of the points (I)–(VI) lie on the graph of $z = f(x, y)$?

I. $(1, 0, 2)$ II. $(1, 1, 1)$
III. $(0, -1, -2)$ IV. $(-1, 0, -2)$
V. $(0, 1, 1)$ VI. $(-1, -1, 0)$

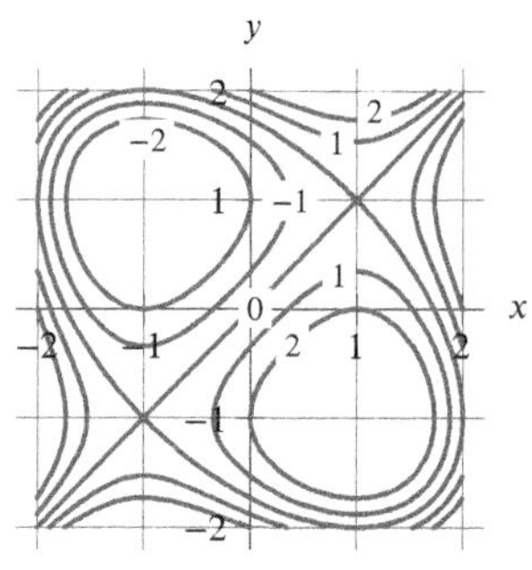

Figure 12.54

18. Match Tables 12.6–12.9 with contour diagrams (I)–(IV) in Figure 12.55.

Table 12.6

$y\backslash x$	-1	0	1
-1	2	1	2
0	1	0	1
1	2	1	2

Table 12.7

$y\backslash x$	-1	0	1
-1	0	1	0
0	1	2	1
1	0	1	0

Table 12.8

$y\backslash x$	-1	0	1
-1	2	0	2
0	2	0	2
1	2	0	2

Table 12.9

$y\backslash x$	-1	0	1
-1	2	2	2
0	0	0	0
1	2	2	2

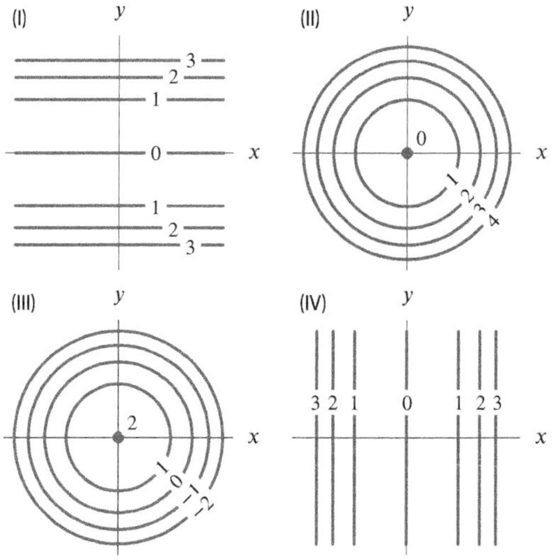

Figure 12.55

PROBLEMS

19. Figure 12.56 shows a graph of $f(x, y) = (\sin x)(\cos y)$ for $-2\pi \le x \le 2\pi$, $-2\pi \le y \le 2\pi$. Use the surface $z = 1/2$ to sketch the contour $f(x, y) = 1/2$.

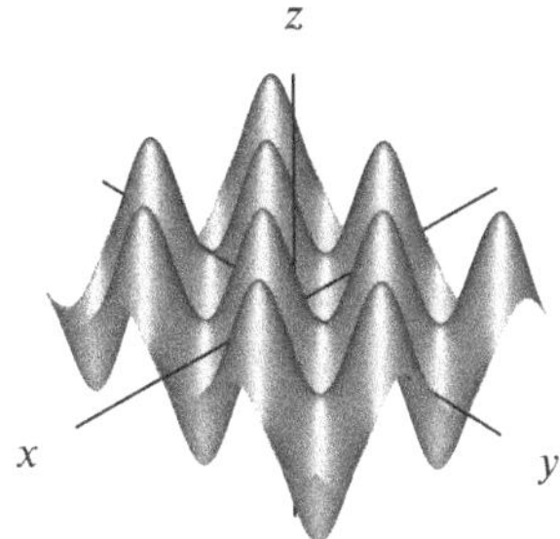

Figure 12.56

20. Total sales, Q, of a product are a function of its price and the amount spent on advertising. Figure 12.57 shows a contour diagram for total sales. Which axis corresponds to the price of the product and which to the amount spent on advertising? Explain.

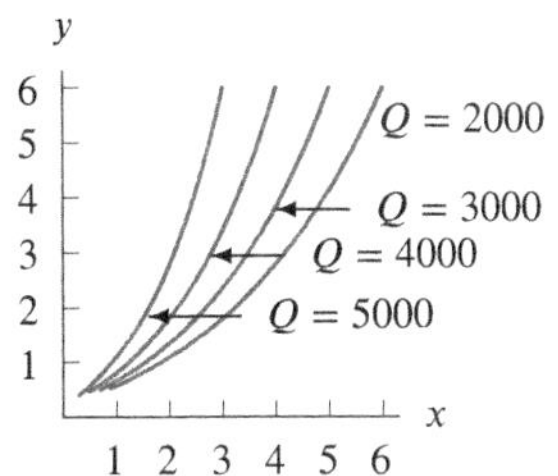

Figure 12.57

21. Each contour diagram (a)–(c) in Figure 12.58 shows satisfaction with quantities of two items X and Y combined. Match (a)–(c) with the items in (I)–(III).

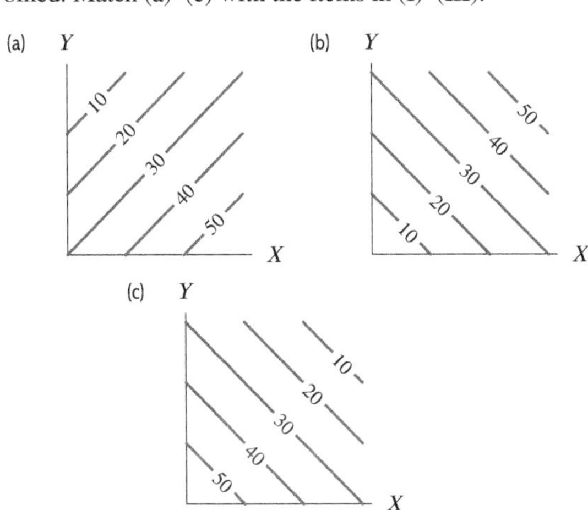

Figure 12.58

(I) X: Income; Y: Leisure time
(II) X: Income; Y: Hours worked
(III) X: Hours worked; Y: Time spent commuting

22. Figure 12.59 shows a contour plot of job satisfaction as a function of the hourly wage and the safety of the workplace (higher values mean safer). Match the jobs at points P, Q, and R with the three descriptions.

(a) The job is so unsafe that higher pay alone would not increase my satisfaction very much.
(b) I could trade a little less safety for a little more pay. It would not matter to me.
(c) The job pays so little that improving safety would not make me happier.

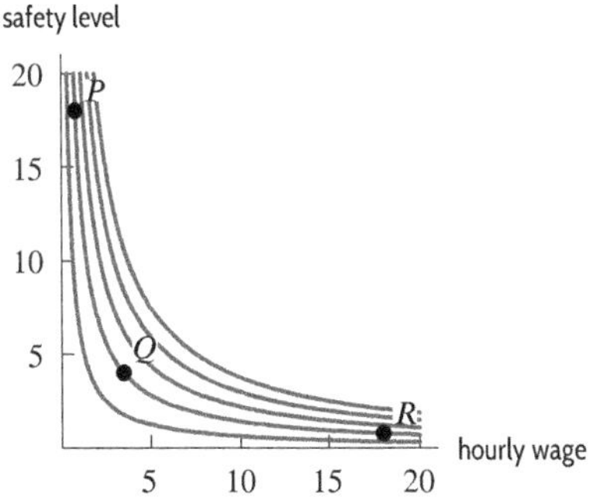

Figure 12.59

23. Figure 12.60 shows a contour diagram of Dan's happiness with snacks of different numbers of cherries and grapes.

(a) What is the slope of the contours?
(b) What does the slope tell you?

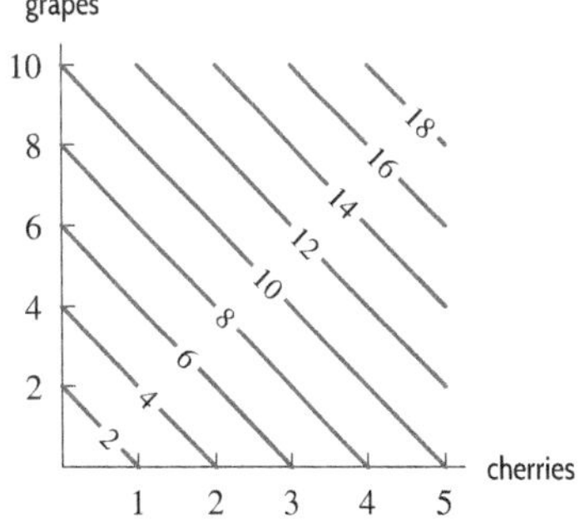

Figure 12.60

24. Figure 12.61 shows contours of $f(x, y) = 100e^x - 50y^2$. Find the values of f on the contours. They are equally spaced multiples of 10.

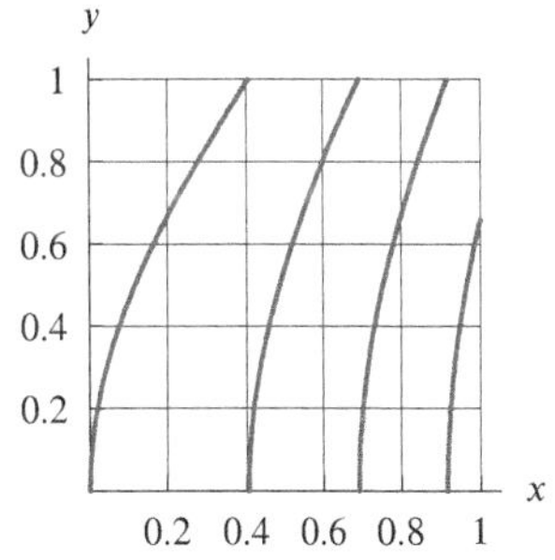

Figure 12.61

25. Figure 12.62 shows contours for a person's body mass index, BMI $= f(w, h) = 703w/h^2$, where w is weight in pounds and h is height in inches. Find the BMI contour values bounding the *underweight* and *normal* regions.

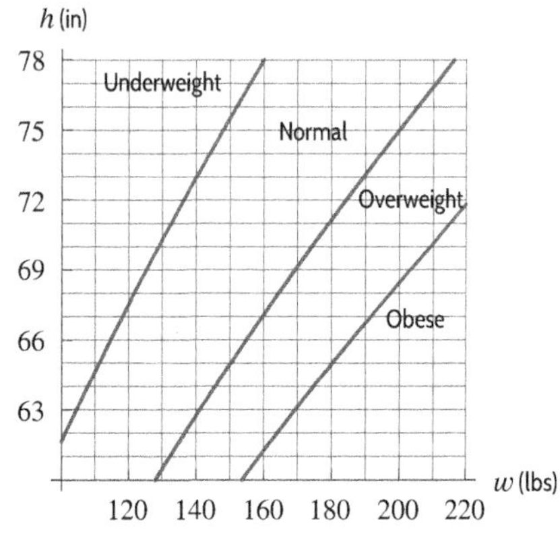

Figure 12.62

26. Match the functions (a)–(f) with the level curves (I)–(VI):

(a) $f(x, y) = x^2 - y^2 - 2x + 4y - 3$
(b) $g(x, y) = x^2 + y^2 - 2x - 4y + 15$
(c) $h(x, y) = -x^2 - y^2 + 2x + 4y - 8$
(d) $j(x, y) = -x^2 + y^2 + 2x - 4y + 3$
(e) $k(x, y) = \sqrt{(x-1)^2 + (y-2)^2}$
(f) $l(x, y) = -\sqrt{(x-1)^2 + (y-2)^2}$

(I)
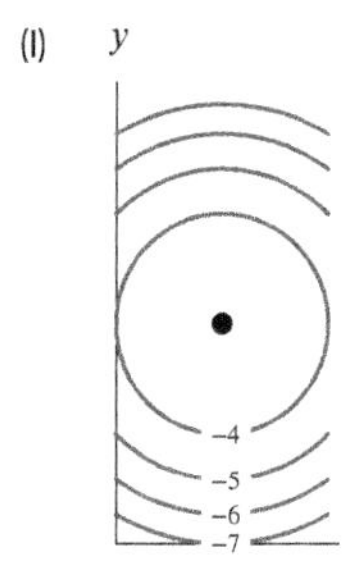

(II)
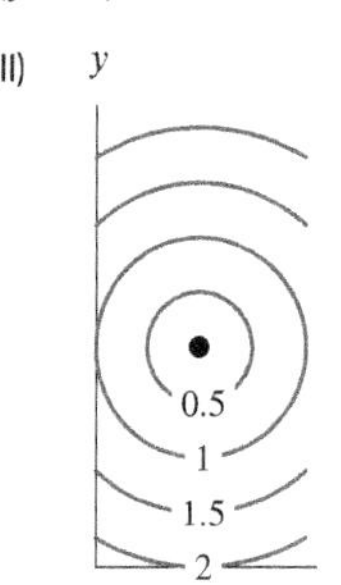

(III)
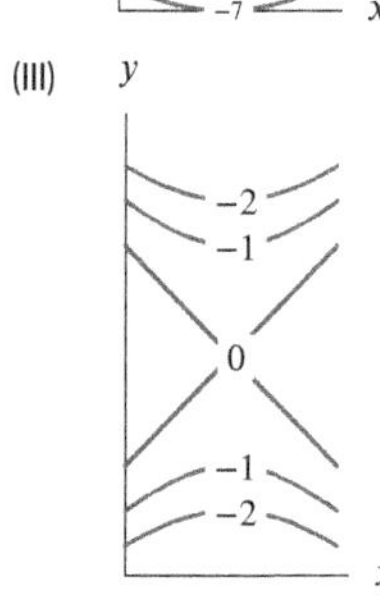

(IV)
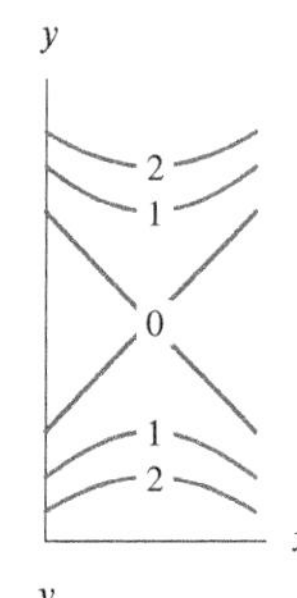

(V)
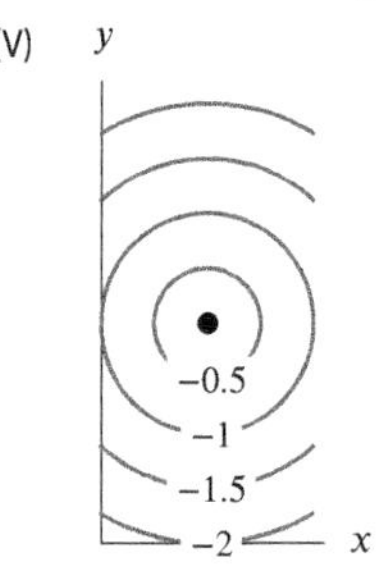

(VI)
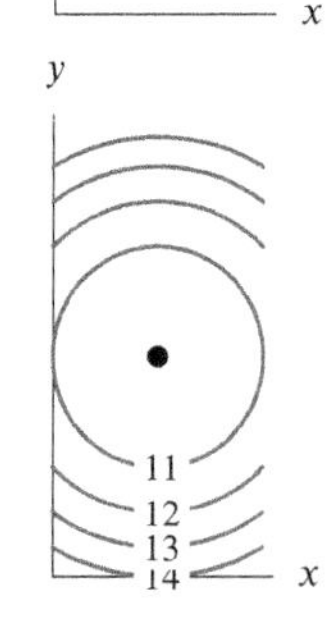

27. The wind chill tells you how cold it feels as a function of the air temperature and wind speed. Figure 12.63 is a contour diagram of wind chill (°F).

(a) If the wind speed is 15 mph, what temperature feels like −20°F?
(b) Estimate the wind chill if the temperature is 0°F and the wind speed is 10 mph.
(c) Humans are at extreme risk when the wind chill is below −50°F. If the temperature is −20°F, estimate the wind speed at which extreme risk begins.
(d) If the wind speed is 15 mph and the temperature drops by 20°F, approximately how much colder do you feel?

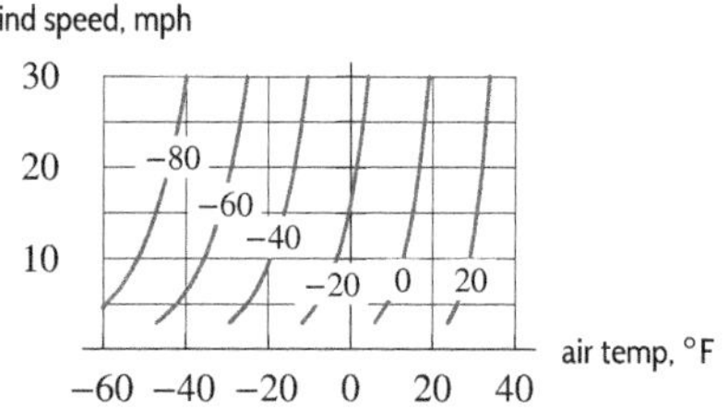

Figure 12.63

28. Figure 12.64 shows contour diagrams of $f(x, y)$ and $g(x, y)$. Sketch the smooth curve with equation $f(x, y) = g(x, y)$.

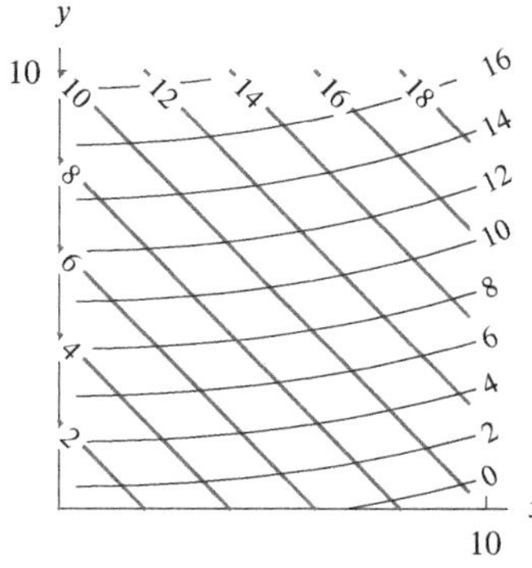

Figure 12.64: Black: $f(x, y)$. Blue: $g(x, y)$

29. Figure 12.65 shows the level curves of the temperature H in a room near a recently opened window. Label the three level curves with reasonable values of H if the house is in the following locations.

(a) Minnesota in winter (where winters are harsh).
(b) San Francisco in winter (where winters are mild).
(c) Houston in summer (where summers are hot).
(d) Oregon in summer (where summers are mild).

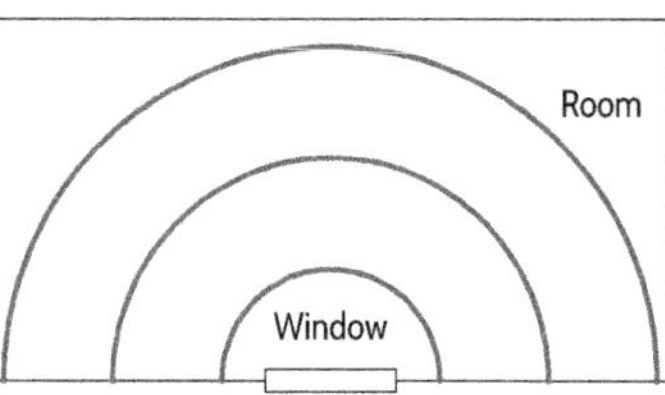

Figure 12.65

30. You are in a room 30 feet long with a heater at one end. In the morning the room is 65°F. You turn on the heater, which quickly warms up to 85°F. Let $H(x, t)$ be the temperature x feet from the heater, t minutes after the heater is turned on. Figure 12.66 shows the contour diagram for H. How warm is it 10 feet from the heater 5 minutes after it was turned on? 10 minutes after it was turned on?

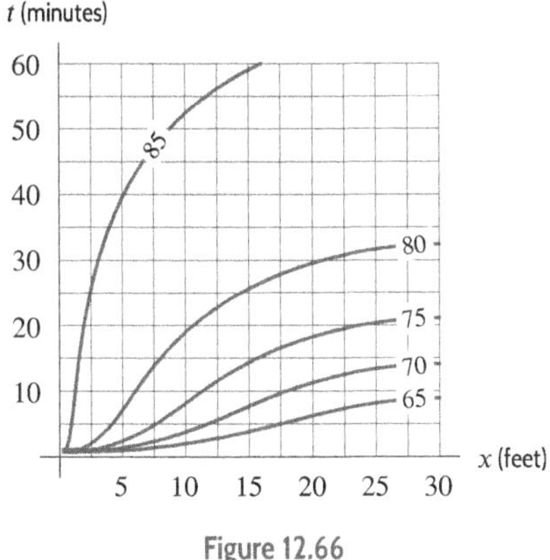

Figure 12.66

31. Using the contour diagram in Figure 12.66, sketch the graphs of the one-variable functions $H(x, 5)$ and $H(x, 20)$. Interpret the two graphs in practical terms, and explain the difference between them.

32. Figure 12.67 shows a contour map of a hill with two paths, A and B.

(a) On which path, A or B, will you have to climb more steeply?
(b) On which path, A or B, will you probably have a better view of the surrounding countryside? (Assume trees do not block your view.)
(c) Alongside which path is there more likely to be a stream?

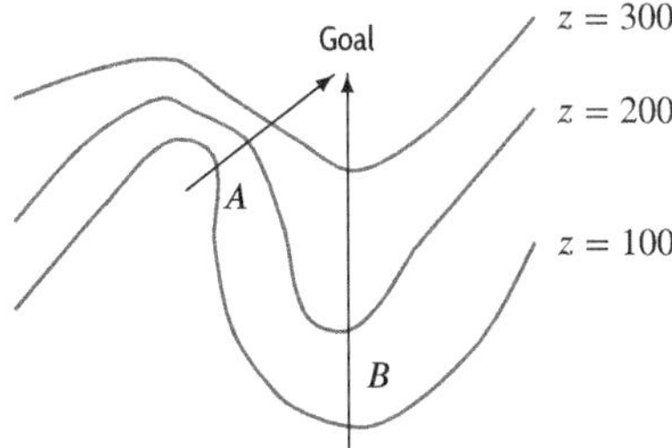

Figure 12.67

In Problems 33–36, for the two given points:
(a) Find the distance h from the first to the second point.
(b) Use Figure 12.68, the contour diagram of $f(x, y)$, to find Δf, the difference between the values of f from the first to the second point.
(c) Find $\Delta f/h$, the average rate of change of $f(x, y)$ from the first to the second point.

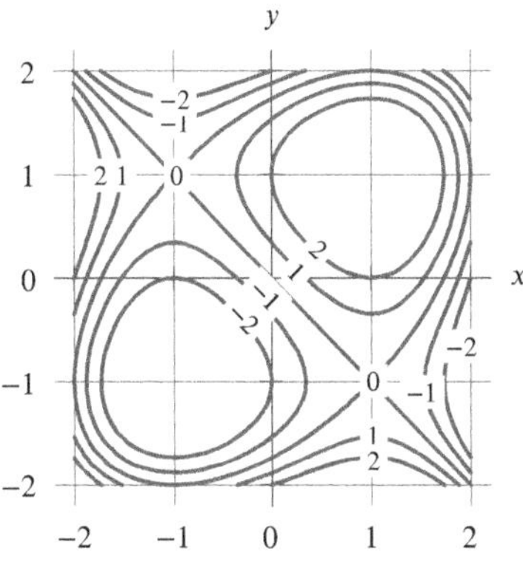

Figure 12.68

33. (0, 0) and (2, 0)

34. (0, −1) and (1, 0)

35. (−1, 1) and (1, 2)

36. (0, −2) and (0, 2)

37. Figure 12.69 is a contour diagram of the monthly payment on a 5-year car loan as a function of the interest rate and the amount you borrow. The interest rate is 13% and you borrow \$6000 for a used car.

(a) What is your monthly payment?
(b) If interest rates drop to 11%, how much more can you borrow without increasing your monthly payment?
(c) Make a table of how much you can borrow without increasing your monthly payment, as a function of the interest rate.

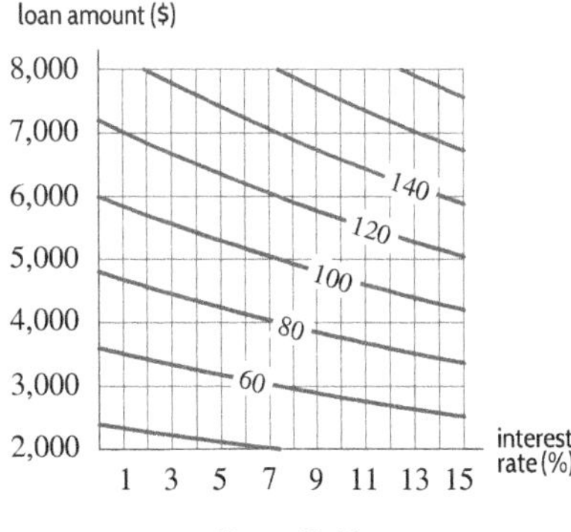

Figure 12.69

38. Hiking on a level trail going due east, you decide to leave the trail and climb toward the mountain on your left. The farther you go along the trail before turning off, the gentler the climb. Sketch a possible topographical map showing the elevation contours.

39. The total productivity $f(n, T)$ of an advertising agency (in ads per day) depends on the number n of workers and the temperature T of the office in degrees Fahrenheit. More workers create more ads, but the farther the temperature from 75°F, the slower they work. Draw a possible contour diagram for the function $f(n, T)$.

40. Match the functions (a)–(d) with the shapes of their level curves (I)–(IV). Sketch each contour diagram.

(a) $f(x, y) = x^2$ **(b)** $f(x, y) = x^2 + 2y^2$
(c) $f(x, y) = y - x^2$ **(d)** $f(x, y) = x^2 - y^2$

I. Lines II. Parabolas
III. Hyperbolas IV. Ellipses

41. Match the functions (a)–(d) with the shapes of their typical level curves (I)–(IV).

(a) $f(x, y) = \dfrac{y}{x^2 + 1}$ **(b)** $f(x, y) = \dfrac{1}{x^2 + 2y^2}$
(c) $f(x, y) = \dfrac{x^2 + 1}{y^2 + 1}$ **(d)** $f(x, y) = \dfrac{x}{x^2 + y^2 + 1}$

I. Circles II. Parabolas
III. Hyperbolas IV. Ellipses

42. Figure 12.70 shows the density of the fox population P (in foxes per square kilometer) for southern England.[8] Draw two different cross-sections along a north-south line and two different cross-sections along an east-west line of the population density P.

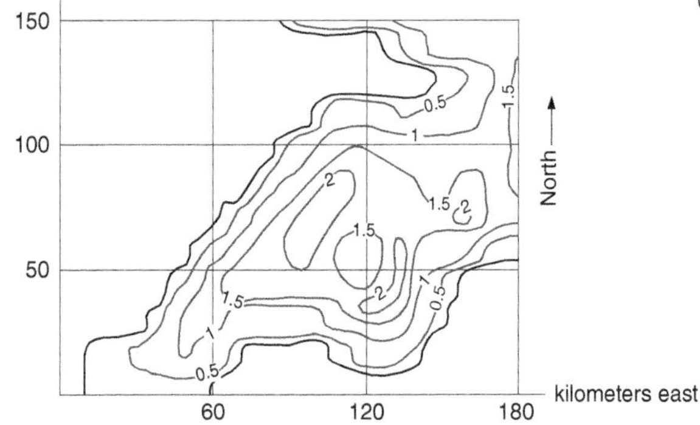

Figure 12.70

43. A manufacturer sells two goods, one at a price of \$3000 a unit and the other at a price of \$12,000 a unit. A quantity q_1 of the first good and q_2 of the second good are sold at a total cost of \$4000 to the manufacturer.

(a) Express the manufacturer's profit, π, as a function of q_1 and q_2.

(b) Sketch curves of constant profit in the q_1q_2-plane for $\pi = 10{,}000$, $\pi = 20{,}000$, and $\pi = 30{,}000$ and the break-even curve $\pi = 0$.

44. A shopper buys x units of item A and y units of item B, obtaining satisfaction $s(x, y)$ from the purchase. (Satisfaction is called *utility* by economists.) The contours $s(x, y) = xy = c$ are called *indifference curves* because they show pairs of purchases that give the shopper the same satisfaction.

(a) A shopper buys 8 units of A and 2 units of B. What is the equation of the indifference curve showing the other purchases that give the shopper the same satisfaction? Sketch this curve.

(b) After buying 4 units of item A, how many units of B must the shopper buy to obtain the same satisfaction as obtained from buying 8 units of A and 2 units of B?

(c) The shopper reduces the purchase of item A by k, a fixed number of units, while increasing the purchase of B to maintain satisfaction. In which of the following cases is the increase in B largest?

- Initial purchase of A is 6 units
- Initial purchase of A is 8 units

45. Match each Cobb-Douglas production function (a)–(c) with a graph in Figure 12.71 and a statement (D)–(G).

(a) $F(L, K) = L^{0.25}K^{0.25}$
(b) $F(L, K) = L^{0.5}K^{0.5}$
(c) $F(L, K) = L^{0.75}K^{0.75}$
(D) Tripling each input triples output.
(E) Quadrupling each input doubles output.
(G) Doubling each input almost triples output.

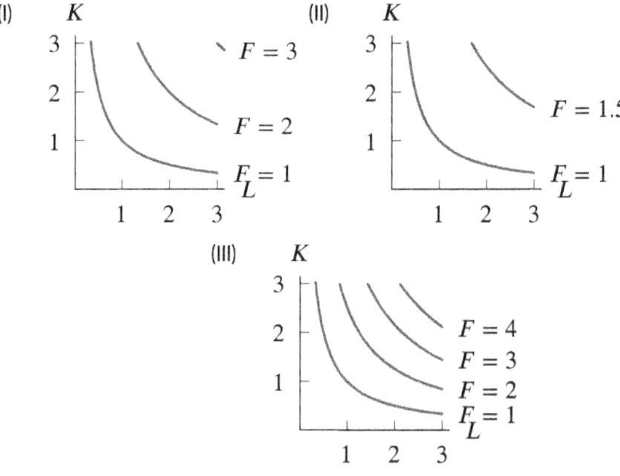

Figure 12.71

46. A Cobb-Douglas production function has the form

$$P = cL^{\alpha}K^{\beta} \quad \text{with } \alpha, \beta > 0.$$

What happens to production if labor and capital are both scaled up? For example, does production double if both labor and capital are doubled? Economists talk about

- *increasing returns to scale* if doubling L and K more than doubles P,
- *constant returns to scale* if doubling L and K exactly doubles P,
- *decreasing returns to scale* if doubling L and K less than doubles P.

What conditions on α and β lead to increasing, constant, or decreasing returns to scale?

[8]From "On the spatial spread of rabies among foxes", Murray, J. D. et al, *Proc. R. Soc. Lond. B*, 229: 111–150, 1986.

47. **(a)** Match $f(x, y) = x^{0.2}y^{0.8}$ and $g(x, y) = x^{0.8}y^{0.2}$ with the level curves in Figures (I) and (II). All scales on the axes are the same.

(b) Figure (III) shows the level curves of $h(x, y) = x^{\alpha}y^{1-\alpha}$ for $0 < \alpha < 1$. Find the range of possible values for α. Again, the scales are the same on both axes.

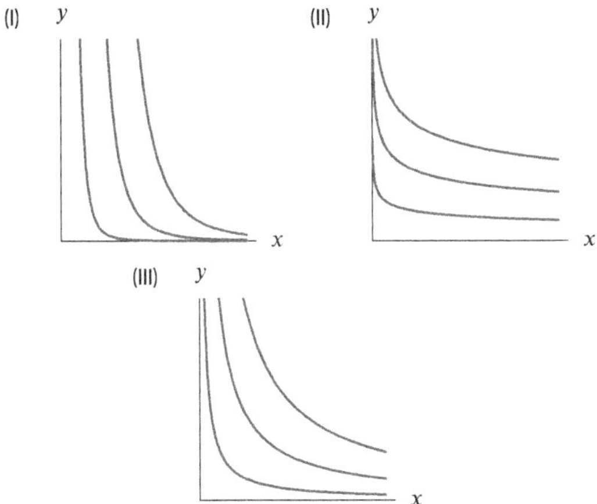

48. Match the functions (a)–(d) with the contour diagrams in Figures I–IV.

(a) $f(x, y) = 0.7\ln x + 0.3\ln y$
(b) $g(x, y) = 0.3\ln x + 0.7\ln y$
(c) $h(x, y) = 0.3x^2 + 0.7y^2$
(d) $j(x, y) = 0.7x^2 + 0.3y^2$

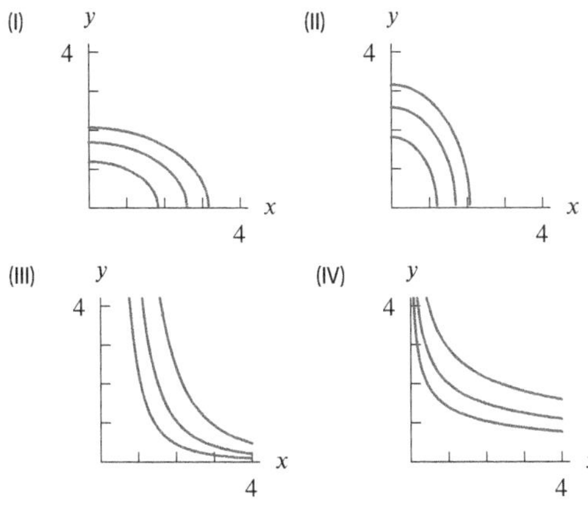

49. Figure 12.72 is the contour diagram of $f(x, y)$. Sketch the contour diagram of each of the following functions.

(a) $3f(x, y)$ **(b)** $f(x, y) - 10$
(c) $f(x - 2, y - 2)$ **(d)** $f(-x, y)$

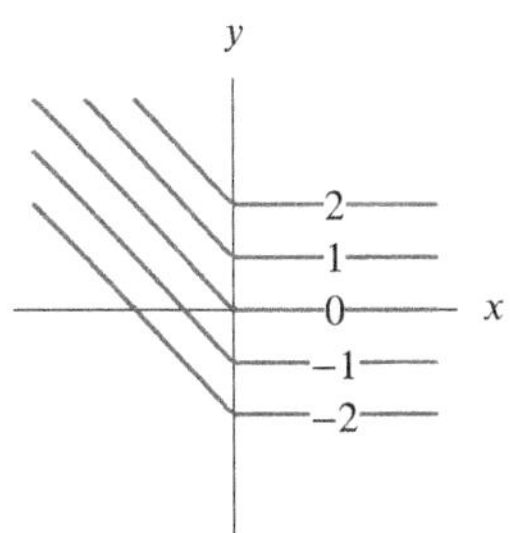

Figure 12.72

50. Figure 12.73 shows part of the contour diagram of $f(x, y)$. Complete the diagram for $x < 0$ if

(a) $f(-x, y) = f(x, y)$ **(b)** $f(-x, y) = -f(x, y)$

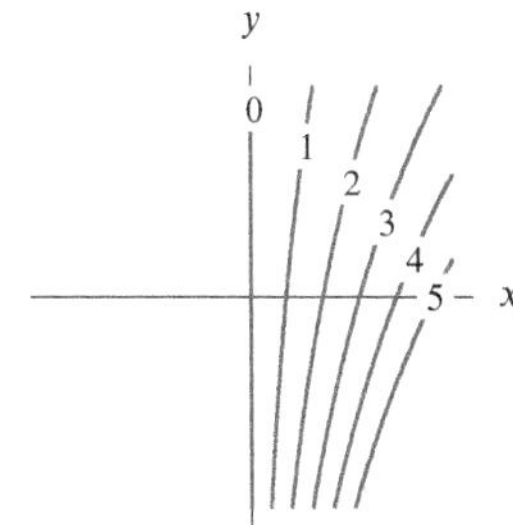

Figure 12.73

51. The contour at level 0 of $f(x, y) = (x+2y)^2 - (3x-4y)^2$ consists of two intersecting lines in the xy-plane. Find equations for the lines.

52. Let $z = f(x, y) = x^2/(x^2 + y^2)$.

(a) Why are there no contours for $z < 0$?
(b) Why are there no contours for $z > 1$?
(c) Sketch a contour diagram for $f(x, y)$ with at least four labeled contours.

53. Let $f(x, y) = x^2 - y^2 = (x - y)(x + y)$. Use the factored form to sketch the contour $f(x, y) = 0$ and to find the regions in the xy-plane where $f(x, y) > 0$ and the regions where $f(x, y) < 0$. Explain how this sketch shows that the graph of $f(x, y)$ is saddle-shaped at the origin.

54. Use Problem 53 to find a formula for a "monkey saddle" surface $z = g(x, y)$ which has three regions with $g(x, y) > 0$ and three with $g(x, y) < 0$.

55. The power P produced by a windmill is proportional to the square of the diameter d of the windmill and to the cube of the speed v of the wind.[9]

(a) Write a formula for P as a function of d and v.
(b) A windmill generates 100 kW of power at a certain wind speed. If a second windmill is built having twice the diameter of the original, what fraction of the original wind speed is needed by the second windmill to produce 100 kW?
(c) Sketch a contour diagram for P.

[9]From www.ecolo.org/documents/documents_in_english/WindmillFormula.htm, accessed October 9, 2011.

Strengthen Your Understanding

In Problems 56–57, explain what is wrong with the statement.

56. A contour diagram for $z = f(x, y)$ is a surface in xyz-space.
57. The functions $f(x, y) = \sqrt{x^2 + y^2}$ and $g(x, y) = x^2+y^2$ have the same contour diagram.

In Problems 58–59, give an example of:

58. A function $f(x, y)$ whose $z = 10$ contour consists of two or more parallel lines.
59. A function whose contours are all parabolas.

Decide if the statements in Problems 60–64 must be true, might be true, or could not be true. The function $z = f(x, y)$ is defined everywhere.

60. The level curves corresponding to $z = 1$ and $z = -1$ cross at the origin.
61. The level curve $z = 1$ consists of the circle $x^2 + y^2 = 2$ and the circle $x^2 + y^2 = 3$, but no other points.
62. The level curve $z = 1$ consists of two lines which intersect at the origin.
63. If $z = e^{-(x^2+y^2)}$, there is a level curve for every value of z.
64. If $z = e^{-(x^2+y^2)}$, there is a level curve through every point (x, y).

Are the statements in Problems 65–72 true or false? Give reasons for your answer.

65. Two isotherms representing distinct temperatures on a weather map cannot intersect.
66. A weather map can have two isotherms representing the same temperature that do not intersect.
67. The contours of the function $f(x, y) = y^2 + (x-2)^2$ are either circles or a single point.
68. If the contours of $g(x, y)$ are concentric circles, then the graph of g is a cone.
69. If the contours for $f(x, y)$ get closer together in a certain direction, then f is increasing in that direction.
70. If all of the contours of $f(x, y)$ are parallel lines, then the graph of f is a plane.
71. If the $f = 10$ contour of the function $f(x, y)$ is identical to the $g = 10$ contour of the function $g(x, y)$, then $f(x, y) = g(x, y)$ for all (x, y).
72. The $f = 5$ contour of the function $f(x, y)$ is identical to the $g = 0$ contour of the function $g(x, y) = f(x, y) - 5$.

12.4 LINEAR FUNCTIONS

What Is a Linear Function of Two Variables?

Linear functions played a central role in one-variable calculus because many one-variable functions have graphs that look like a line when we zoom in. In two-variable calculus, a *linear function* is one whose graph is a plane. In Chapter 14, we see that many two-variable functions have graphs which look like planes when we zoom in.

What Makes a Plane Flat?

What makes the graph of the function $z = f(x, y)$ a plane? Linear functions of *one* variable have straight line graphs because they have constant slope. On a plane, the situation is a bit more complicated. If we walk around on a tilted plane, the slope is not always the same: it depends on the direction in which we walk. However, at every point on the plane, the slope is the same as long as we choose the same direction. If we walk parallel to the x-axis, we always find ourselves walking up or down with the same slope;[10] the same is true if we walk parallel to the y-axis. In other words, the slope ratios $\Delta z/\Delta x$ (with y fixed) and $\Delta z/\Delta y$ (with x fixed) are each constant.

Example 1 A plane cuts the z-axis at $z = 5$ and has slope 2 in the x-direction and slope -1 in the y-direction. What is the equation of the plane?

Solution Finding the equation of the plane means constructing a formula for the z-coordinate of the point on the plane directly above the point (x, y) in the xy-plane. To get to that point start from the point above the origin, where $z = 5$. Then walk x units in the x-direction. Since the slope in the x-direction is 2, the height increases by $2x$. Then walk y units in the y-direction; since the slope in the y-direction is -1, the height decreases by y units. Since the height has changed by $2x - y$ units, the z-coordinate

[10] To be precise, walking in a vertical plane parallel to the x-axis while rising or falling with the plane you are on.

is $5 + 2x - y$. Thus, the equation for the plane is

$$z = 5 + 2x - y.$$

For any linear function, if we know its value at a point (x_0, y_0), its slope in the x-direction, and its slope in the y-direction, then we can write the equation of the function. This is just like the equation of a line in the one-variable case, except that there are two slopes instead of one.

If a **plane** has slope m in the x-direction, has slope n in the y-direction, and passes through the point (x_0, y_0, z_0), then its equation is

$$z = z_0 + m(x - x_0) + n(y - y_0).$$

This plane is the graph of the **linear function**

$$f(x, y) = z_0 + m(x - x_0) + n(y - y_0).$$

If we write $c = z_0 - mx_0 - ny_0$, then we can write $f(x, y)$ in the equivalent form

$$f(x, y) = c + mx + ny.$$

Just as in 2-space a line is determined by two points, so in 3-space a plane is determined by three points, provided they do not lie on a line.

Example 2 Find the equation of the plane passing through the points $(1, 0, 1)$, $(1, -1, 3)$, and $(3, 0, -1)$.

Solution The first two points have the same x-coordinate, so we use them to find the slope of the plane in the y-direction. As the y-coordinate changes from 0 to -1, the z-coordinate changes from 1 to 3, so the slope in the y-direction is $n = \Delta z/\Delta y = (3 - 1)/(-1 - 0) = -2$. The first and third points have the same y-coordinate, so we use them to find the slope in the x-direction; it is $m = \Delta z/\Delta x = (-1 - 1)/(3 - 1) = -1$. Because the plane passes through $(1, 0, 1)$, its equation is

$$z = 1 - (x - 1) - 2(y - 0) \quad \text{or} \quad z = 2 - x - 2y.$$

You should check that this equation is also satisfied by the points $(1, -1, 3)$ and $(3, 0, -1)$.

Example 2 was made easier by the fact that two of the points had the same x-coordinate and two had the same y-coordinate. An alternative method, which works for any three points, is to substitute the x, y, and z-values of each of the three points into the equation $z = c + mx + ny$. The resulting three equations in c, m, n are then solved simultaneously.

Linear Functions from a Numerical Point of View

To avoid flying planes with empty seats, airlines sell some tickets at full price and some at a discount. Table 12.10 shows an airline's revenue in dollars from tickets sold on a particular route, as a function of the number of full-price tickets sold, f, and the number of discount tickets sold, d.

In every column, the revenue jumps by \$40,000 for each extra 200 discount tickets. Thus, each column is a linear function of the number of discount tickets sold. In addition, every column has the same slope, $40{,}000/200 = 200$ dollars/ticket. This is the price of a discount ticket. Similarly, each row is a linear function and all the rows have the same slope, 450, which is the price in dollars of a

Table 12.10 *Revenue from ticket sales (dollars)*

		Full-price tickets (f)			
		100	200	300	400
Discount tickets (d)	200	85,000	130,000	175,000	220,000
	400	125,000	170,000	215,000	260,000
	600	165,000	210,000	255,000	300,000
	800	205,000	250,000	295,000	340,000
	1000	245,000	290,000	335,000	380,000

full-fare ticket. Thus, R is a linear function of f and d, given by:

$$R = 450f + 200d.$$

We have the following general result:

> A **linear function** can be recognized from its table by the following features:
> - Each row and each column is linear.
> - All the rows have the same slope.
> - All the columns have the same slope (although the slope of the rows and the slope of the columns are generally different).

Example 3 The table contains values of a linear function. Fill in the blank and give a formula for the function.

$x\backslash y$	1.5	2.0
2	0.5	1.5
3	−0.5	?

Solution In the first column the function decreases by 1 (from 0.5 to −0.5) as x goes from 2 to 3. Since the function is linear, it must decrease by the same amount in the second column. So the missing entry must be $1.5 - 1 = 0.5$. The slope of the function in the x-direction is -1. The slope in the y-direction is 2, since in each row the function increases by 1 when y increases by 0.5. From the table we get $f(2, 1.5) = 0.5$. Therefore, the formula is

$$f(x, y) = 0.5 - (x - 2) + 2(y - 1.5) = -0.5 - x + 2y.$$

What Does the Contour Diagram of a Linear Function Look Like?

The formula for the airline revenue function in Table 12.10 is $R = 450f + 200d$, where f is the number of full fares and d is the number of discount fares sold.

Notice that the contours of this function in Figure 12.74 are parallel straight lines. What is the practical significance of the slope of these contour lines? Consider the contour $R = 100{,}000$; that means we are looking at combinations of ticket sales that yield \$100,000 in revenue. If we move down and to the right on the contour, the f-coordinate increases and the d-coordinate decreases, so we sell more full fares and fewer discount fares. This is because to receive a fixed revenue of \$100,000, we must sell more full fares if we sell fewer discount fares. The exact trade-off depends on the slope of the contour; the diagram shows that each contour has a slope of about -2. This means that for a fixed revenue, we must sell two discount fares to replace one full fare. This can also be seen by comparing prices. Each full fare brings in \$450; to earn the same amount in discount fares we need to sell $450/200 = 2.25 \approx 2$ fares. Since the price ratio is independent of how many of each type of fare we sell, this slope remains constant over the whole contour map; thus, the contours are all parallel straight lines.

Notice also that the contours are evenly spaced. Thus, no matter which contour we are on, a fixed increase in one of the variables causes the same increase in the value of the function. In terms of revenue, no matter how many fares we have sold, an extra fare, whether full or discount, brings the same revenue as before.

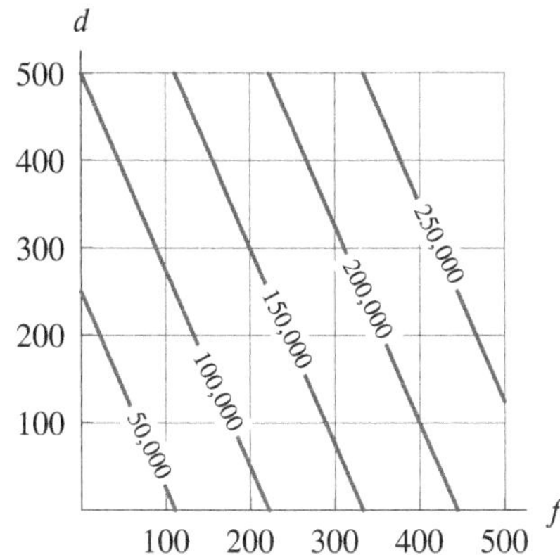

Figure 12.74: Revenue as a function of full and discount fares, $R = 450f + 200d$

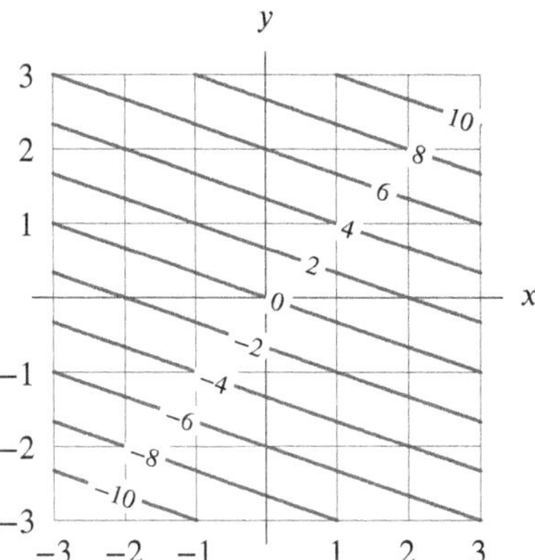

Figure 12.75: Contour map of linear function $f(x, y)$

Example 4 Find the equation of the linear function whose contour diagram is in Figure 12.75.

Solution Suppose we start at the origin on the $z = 0$ contour. Moving 2 units in the y-direction takes us to the $z = 6$ contour, so the slope in the y-direction is $\Delta z/\Delta y = 6/2 = 3$. Similarly, a move of 2 units in the x-direction from the origin takes us to the $z = 2$ contour, so the slope in the x-direction is $\Delta z/\Delta x = 2/2 = 1$. Since $f(0, 0) = 0$, we have $f(x, y) = x + 3y$.

Exercises and Problems for Section 12.4 Online Resource: Additional Problems for Section 12.4

EXERCISES

Exercises 1–2 each contain a partial table of values for a linear function. Fill in the blanks.

1.

$x\backslash y$	0.0	1.0
0.0		1.0
2.0	3.0	5.0

2.

$x\backslash y$	−1.0	0.0	1.0
2.0	4.0		
3.0		3.0	5.0

In Exercises 3–6, could the tables of values represent a linear function?

3.

$x \backslash y$	0	1	2
0	0	1	4
1	1	0	1
2	4	1	0

4.

$x \backslash y$	0	1	2
0	10	13	16
1	6	9	12
2	2	5	8

5.

$x \backslash y$	0	1	2
0	0	5	10
1	2	7	12
2	4	9	14

6.

$x \backslash y$	0	1	2
0	5	7	9
1	6	9	12
2	7	11	15

7. Find the equation of the linear function $z = c+mx+ny$ whose graph contains the points $(0, 0, 0)$, $(0, 2, -1)$, and $(-3, 0, -4)$.

8. Find the linear function whose graph is the plane through the points $(4, 0, 0)$, $(0, 3, 0)$ and $(0, 0, 2)$.

9. Find an equation for the plane containing the line in the xy-plane where $y = 1$, and the line in the xz-plane where $z = 2$.

10. Find the equation of the linear function $z = c+mx+ny$ whose graph intersects the xz-plane in the line $z = 3x+4$ and intersects the yz-plane in the line $z = y+4$.

11. Suppose that z is a linear function of x and y with slope 2 in the x-direction and slope 3 in the y-direction.
    (a) A change of 0.5 in x and -0.2 in y produces what change in z?
    (b) If $z = 2$ when $x = 5$ and $y = 7$, what is the value of z when $x = 4.9$ and $y = 7.2$?

12. **(a)** Find a formula for the linear function whose graph is a plane passing through point $(4, 3, -2)$ with slope 5 in the x-direction and slope -3 in the y-direction.
    (b) Sketch the contour diagram for this function.

In Exercises 13–14, could the contour diagram represent a linear function?

13.

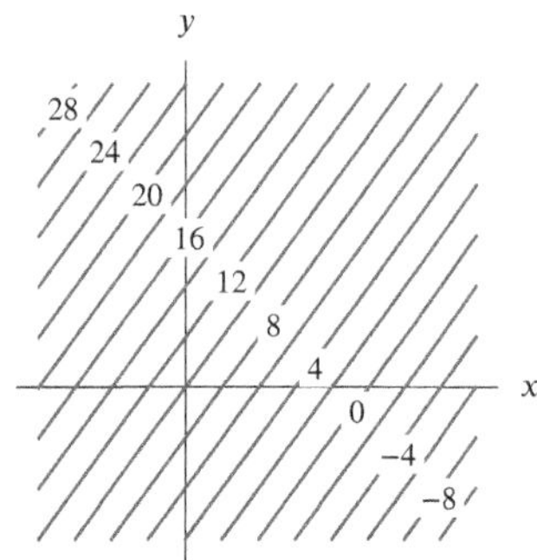

14.

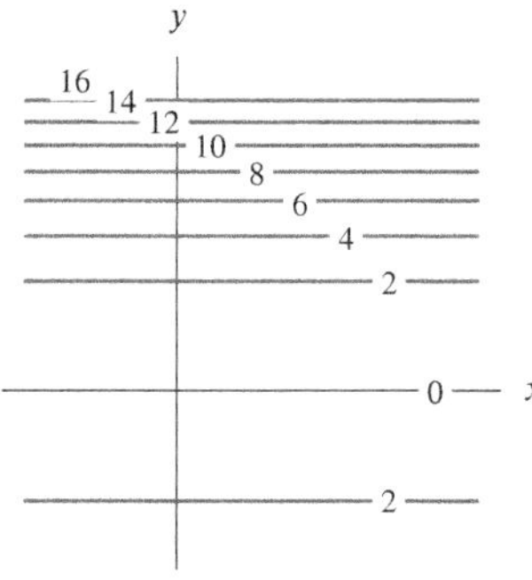

PROBLEMS

15. An internet video streaming company offers a basic and premium monthly streaming subscription package. Figure 12.76 shows the revenue (in dollars per month) of the company as a function of the number, c, of basic subscribers and the number, d, of premium subscribers it has. What is the price of a basic subscription? What is the price of a premium subscription?

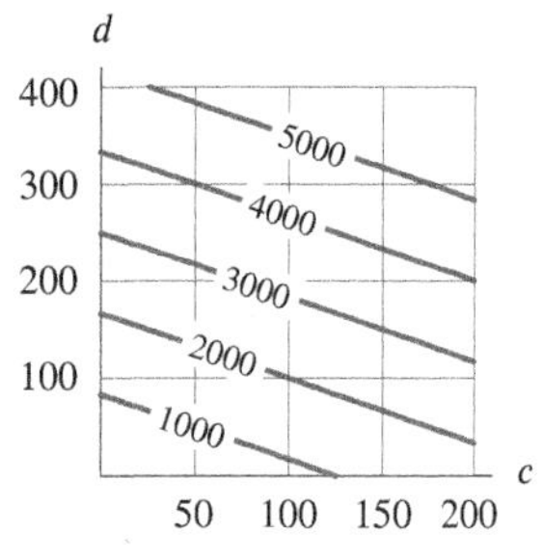

Figure 12.76

16. The charge, C, in dollars, for access to a company's 4G LTE network is a function of m, the number of months of use, and t, the total number of gigabytes used:

$$C = f(m, t) = 99 + 30m + 10t.$$

(a) Is f a linear function?
(b) Give units for the coefficients of m and t, and interpret them as charges.
(c) Interpret the intercept 99 as a charge.
(d) Find $f(3, 8)$ and interpret your answer.

17. A manufacturer makes two products out of two raw materials. Let q_1, q_2 be the quantities sold of the two products, p_1, p_2 their prices, and m_1, m_2 the quantities purchased of the two raw materials. Which of the following functions do you expect to be linear, and why? In each case, assume that all variables except the ones mentioned are held fixed.

(a) Expenditure on raw materials as a function of m_1 and m_2.
(b) Revenue as a function of q_1 and q_2.
(c) Revenue as a function of p_1 and q_1.

Problems 18–20 concern Table 12.11, which gives the number of calories burned per minute for someone roller-blading, as a function of the person's weight and speed.[11]

Table 12.11

	Calories burned per minute			
Weight	8 mph	9 mph	10 mph	11 mph
120 lbs	4.2	5.8	7.4	8.9
140 lbs	5.1	6.7	8.3	9.9
160 lbs	6.1	7.7	9.2	10.8
180 lbs	7.0	8.6	10.2	11.7
200 lbs	7.9	9.5	11.1	12.6

18. Does the data in Table 12.11 look approximately linear? Give a formula for B, the number of calories burned per minute in terms of the weight, w, and the speed, s. Does the formula make sense for all weights or speeds?

19. Who burns more total calories to go 10 miles: A 120-lb person going 10 mph or a 180-lb person going 8 mph? Which of these two people burns more calories per pound for the 10-mile trip?

20. Use Problem 18 to give a formula for P, the number of calories burned per pound, in terms of w and s, for a person weighing w lbs roller-blading 10 miles at s mph.

[11] From the August 28, 1994, issue of *Parade Magazine*.

For Problems 21–22, find a possible equation for a linear function with the given contour diagram.

21.

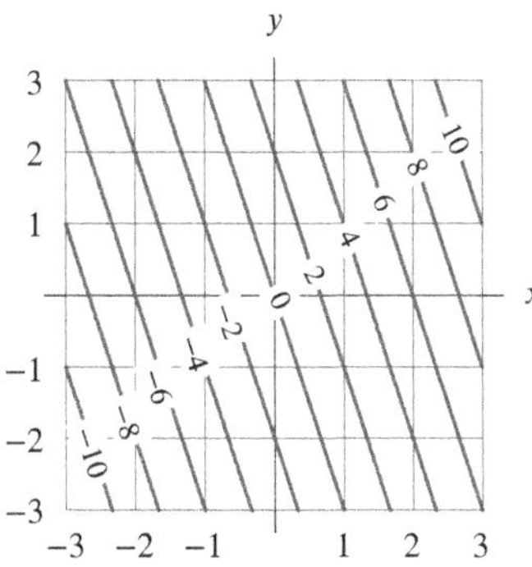

22.

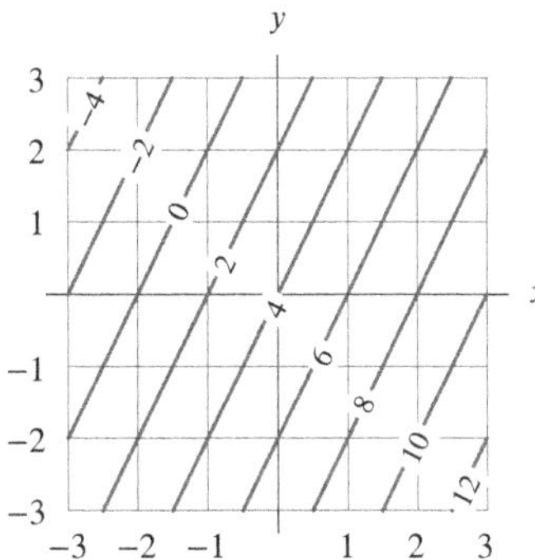

In Problems 23–24, could the contour diagram represent a linear function? If so, find an equation for that function.

23.

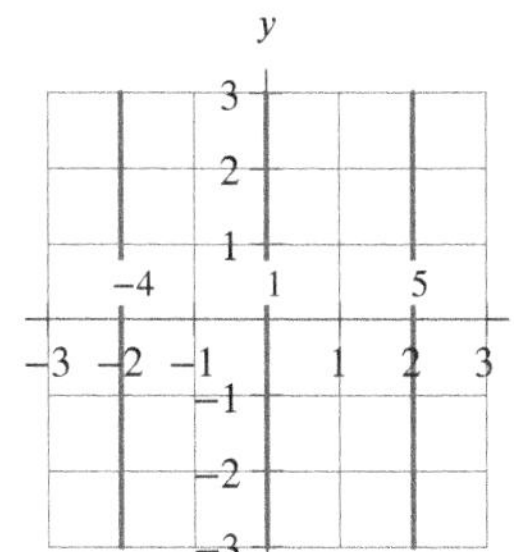

24.

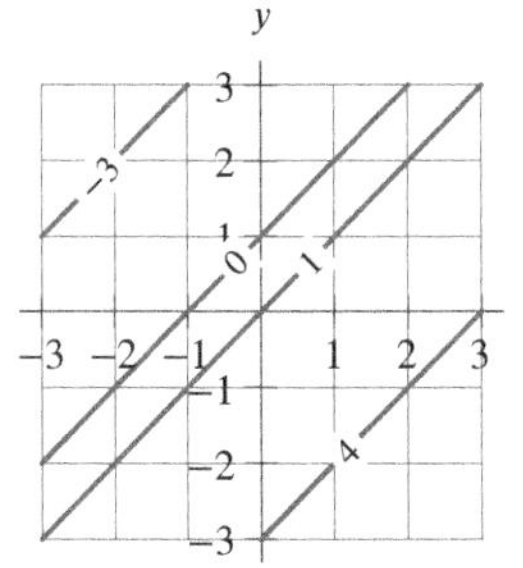

For Problems 25–26, find an equation for the linear function with the given values.

25.

x\y	−1	0	1	2
0	1.5	1	0.5	0
1	3.5	3	2.5	2
2	5.5	5	4.5	4
3	7.5	7	6.5	6

26.

x\y	10	20	30	40
100	3	6	9	12
200	2	5	8	11
300	1	4	7	10
400	0	3	6	9

In Problems 27–34, could the table of values represent a linear function? If so, find a possible formula for the function. If not, give a reason why not.

27.

x\y	1	2	3
1	1	5	9
2	2	6	10
3	3	7	11

28.

x\y	0	1	2
0	1	2	1
1	2	3	2
2	3	4	3

29.

x\y	-2	0	2
-2	2	2	2
0	5	5	5
2	8	8	8

30.

x\y	2	4	6
1	0	3	6
3	1	4	7
5	4	7	10

31.

x\y	0	1	2
0	-5	-7	-9
2	-2	-4	-6
4	1	-1	-3

32.

x\y	1	2	3
1	1	2	3
2	4	5	6
4	7	8	9

33.

x\y	0	2	5
0	3	5	8
1	5	7	10
3	9	11	14

34.

x\y	0	2	5
0	0	4	10
1	1	5	11
2	4	6	14

In Problems 35–38, use the contours of the linear function $z = f(x, y)$ in Figure 12.77 to create possible contour labels for the linear function $z = g(x, y)$ satisfying the given condition.

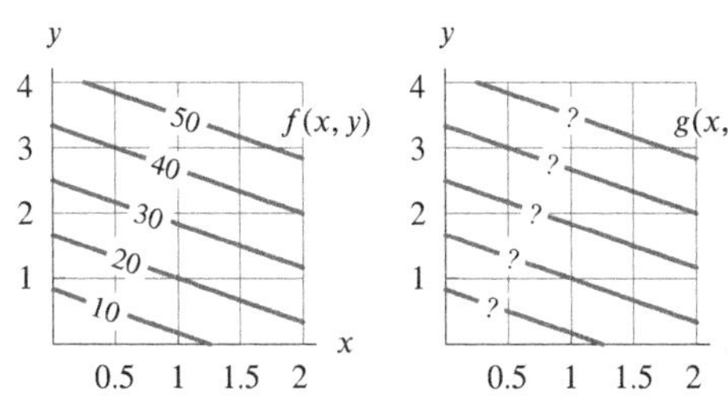

Figure 12.77

35. The graph of g is parallel but different from the graph of f.

36. The graph of g is parallel to the graph of f and passes through the point $(2, 2, 0)$.

37. The graph of g has the same contour as f for the value $z = 30$ but is different from the graph of f.

38. The graph of g has the same contour as f for the value $z = 40$, and a negative slope in the x-direction.

In Problems 39–42, graph the linear function by plotting the x, y, and z-intercepts and joining them by a triangle as in Figure 12.78. This shows the part of the plane in the octant where $x \geq 0$, $y \geq 0$, $z \geq 0$. If the intercepts are not all positive, the same method works if the x, y, and z-axes are drawn from a different perspective.

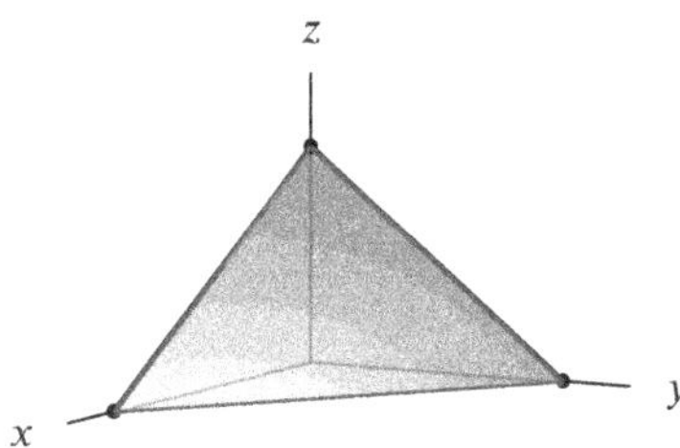

Figure 12.78

39. $z = 2 - 2x + y$

40. $z = 2 - x - 2y$

41. $z = 4 + x - 2y$

42. $z = 6 - 2x - 3y$

43. Figure 12.79 is the contour diagram of a linear function $f(x, y) = mx + 4y + c$. What is the value of m?

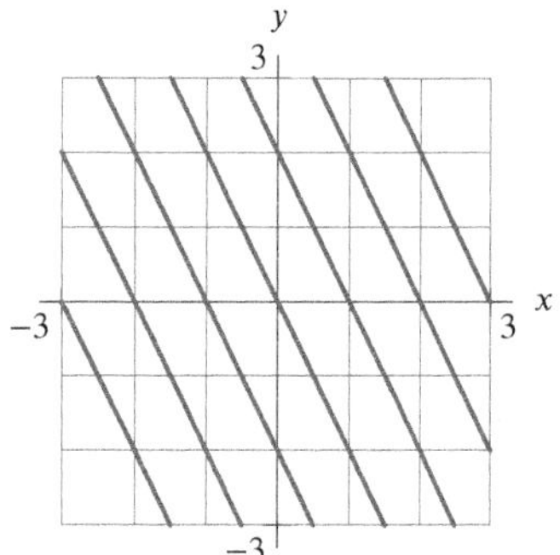

Figure 12.79

44. For the contour diagrams (I)–(IV) on $-2 \leq x \leq 2$, $-2 \leq y \leq 2$, pick the corresponding function.

$f(x, y) = 2x + 3y + 10$ $\quad$ $k(x, y) = -2x + 3y + 12$

$g(x, y) = 2x + 3y + 60$ $\quad$ $m(x, y) = -2x + 3y + 60$

$h(x, y) = 2x - 3y + 12$ $\quad$ $n(x, y) = -2x - 3y + 14$

$j(x, y) = 2x - 3y + 60$ $\quad$ $p(x, y) = -2x - 3y + 60$

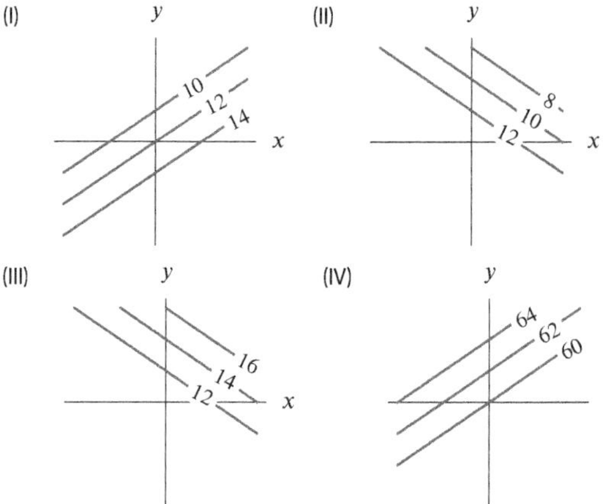

45. A linear function has the formula $f(x, y) = a + 10x - 5y$, but you don't know the value of a. Give a numerical value for the following, if possible.

(a) $f(50, 62)$
(b) $f(51, 60)$
(c) $f(51, 60) - f(50, 62)$

Strengthen Your Understanding

In Problems 46–47, explain what is wrong with the statement.

46. If the contours of f are all parallel lines, then f is linear.

47. A function $f(x, y)$ with linear cross-sections for x fixed and linear cross-sections for y fixed is a linear function.

In Problems 48–49, give an example of:

48. A table of values, with three rows and three columns, for a nonlinear function that is linear in each row and in each column.

49. A linear function whose contours are lines with slope 2.

Are the statements in Problems 50–62 true or false? Give reasons for your answer.

50. The planes $z = 3 + 2x + 4y$ and $z = 5 + 2x + 4y$ intersect.

51. The function represented in Table 12.12 is linear.

Table 12.12

$u \backslash v$	1.1	1.2	1.3	1.4
3.2	11.06	12.06	13.06	14.06
3.4	11.75	12.82	13.89	14.96
3.6	12.44	13.58	14.72	15.86
3.8	13.13	14.34	15.55	16.76
4.0	13.82	15.10	16.38	17.66

52. Contours of $f(x, y) = 3x + 2y$ are lines with slope 3.

53. If f is a non-constant linear function, then the contours of f are parallel lines.

54. If $f(0,0) = 1, f(0,1) = 4, f(0,3) = 5$, then f cannot be linear.

55. The graph of a linear function is always a plane.

56. The cross-section $x = c$ of a linear function $f(x, y)$ is always a line.

57. There is no linear function $f(x, y)$ with a graph parallel to the xy-plane.

58. There is no linear function $f(x, y)$ with a graph parallel to the xz-plane.

59. A linear function $f(x, y) = 2x+3y-5$, has exactly one point (a, b) satisfying $f(a, b) = 0$.

60. In a table of values of a linear function, the columns have the same slope as the rows.

61. There is exactly one linear function $f(x, y)$ whose $f = 0$ contour is $y = 2x + 1$.

62. If the contours of $f(x, y) = c + mx + ny$ are vertical lines, then $n = 0$.

12.5 FUNCTIONS OF THREE VARIABLES

In applications of calculus, functions of any number of variables can arise. The density of matter in the universe is a function of three variables, since it takes three numbers to specify a point in space. Models of the US economy often use functions of ten or more variables. We need to be able to apply calculus to functions of arbitrarily many variables.

One difficulty with functions of more than two variables is that it is hard to visualize them. The graph of a function of one variable is a curve in 2-space, the graph of a function of two variables is a surface in 3-space, so the graph of a function of three variables would be a solid in 4-space. Since we can't easily visualize 4-space, we won't use the graphs of functions of three variables. On the other hand, it is possible to draw contour diagrams for functions of three variables, only now the contours are surfaces in 3-space.

Representing a Function of Three Variables Using a Family of Level Surfaces

A function of two variables, $f(x, y)$, can be represented by a family of level curves of the form $f(x, y) = c$ for various values of the constant, c.

> A **level surface**, or **level set** of a function of three variables, $f(x, y, z)$, is a surface of the form $f(x, y, z) = c$, where c is a constant. The function f can be represented by the family of level surfaces obtained by allowing c to vary.

The value of the function, f, is constant on each level surface.

Example 1 The temperature, in °C, at a point (x, y, z) is given by $T = f(x, y, z) = x^2 + y^2 + z^2$. What do the level surfaces of the function f look like and what do they mean in terms of temperature?

Solution The level surface corresponding to $T = 100$ is the set of all points where the temperature is 100°C. That is, where $f(x, y, z) = 100$, so

$$x^2 + y^2 + z^2 = 100.$$

This is the equation of a sphere of radius 10, with center at the origin. Similarly, the level surface corresponding to $T = 200$ is the sphere with radius $\sqrt{200}$. The other level surfaces are concentric spheres. The temperature is constant on each sphere. We may view the temperature distribution as a set of nested spheres, like concentric layers of an onion, each one labeled with a different temperature, starting from low temperatures in the middle and getting hotter as we go out from the center. (See Figure 12.80.) The level surfaces become more closely spaced as we move farther from the origin because the temperature increases more rapidly the farther we get from the origin.

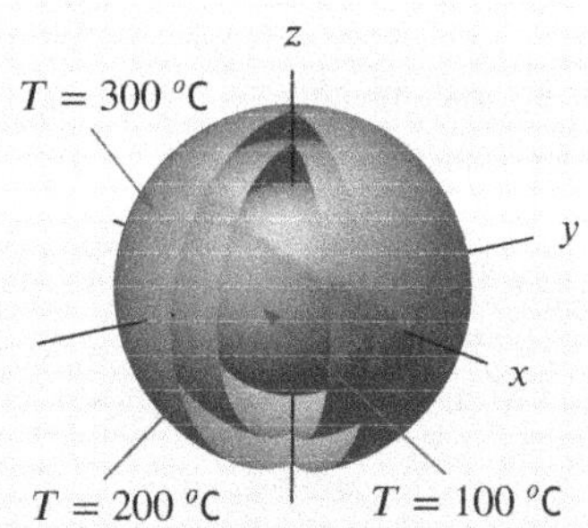

Figure 12.80: Level surfaces of $T = f(x, y, z) = x^2 + y^2 + z^2$, each one having a constant temperature

Example 2 What do the level surfaces of $f(x, y, z) = x^2 + y^2$ and $g(x, y, z) = z - y$ look like?

Solution The level surface of f corresponding to the constant c is the surface consisting of all points satisfying the equation

$$x^2 + y^2 = c.$$

Since there is no z-coordinate in the equation, z can take any value. For $c > 0$, this is a circular cylinder of radius $\sqrt{c}$ around the z-axis. The level surfaces are concentric cylinders; on the narrow ones near the z-axis, f has small values; on the wider ones, f has larger values. See Figure 12.81.

The level surface of g corresponding to the constant c is the plane

$$z - y = c.$$

Since there is no x variable in the equation, these planes are parallel to the x-axis and cut the yz-plane in the line $z - y = c$. See Figure 12.82.

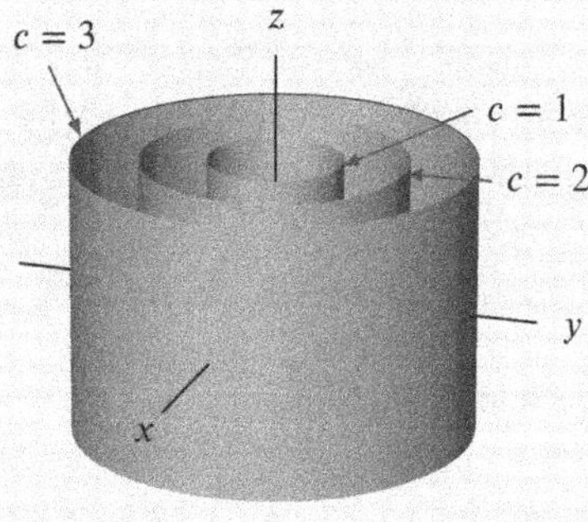

Figure 12.81: Level surfaces of $f(x, y, z) = x^2 + y^2$

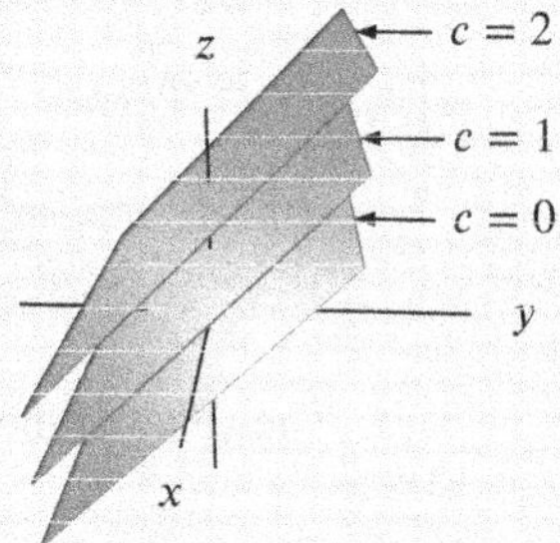

Figure 12.82: Level surfaces of $g(x, y, z) = z - y$

We say $g(x, y, z) = z - y$ in Example 2 is a *linear function* of the three variables x, y, z, whereas $f(x, y) = x^2 + y^2$ and $f(x, y, z) = x^2 + y^2 + z^2$ are *quadratic functions* of three variables.

Example 3 What do the level surfaces of $f(x, y, z) = x^2 + y^2 - z^2$ look like?

Solution In Section 12.3, we saw that the two-variable quadratic function $g(x, y) = x^2 - y^2$ has a saddle-shaped graph and three types of contours. The contour equation $x^2 - y^2 = c$ gives a hyperbola opening right-left when $c > 0$, a hyperbola opening up-down when $c < 0$, and a pair of intersecting lines when $c = 0$. Similarly, the three-variable quadratic function $f(x, y, z) = x^2 + y^2 - z^2$ has three types of level surfaces depending on the value of c in the equation $x^2 + y^2 - z^2 = c$.

Suppose that $c > 0$, say $c = 1$. Rewrite the equation as $x^2 + y^2 = z^2 + 1$ and think of what happens as we cut the surface perpendicular to the z-axis by holding z fixed. The result is a circle, $x^2 + y^2 =$ constant, of radius at least 1 (since the constant $z^2 + 1 \geq 1$). The circles get larger as z gets larger. If we take the $x = 0$ cross-section instead, we get the hyperbola $y^2 - z^2 = 1$. The result is shown in Figure 12.86, with $a = b = c = 1$.

Suppose instead that $c < 0$, say $c = -1$. Then the horizontal cross-sections of $x^2 + y^2 = z^2 - 1$ are again circles except that the radii shrink to 0 at $z = \pm 1$ and between $z = -1$ and $z = 1$ there are no cross-sections at all. The result is shown in Figure 12.87 with $a = b = c = 1$.

When $c = 0$, we get the equation $x^2 + y^2 = z^2$. Again the horizontal cross-sections are circles, this time with the radius shrinking down to exactly 0 when $z = 0$. The resulting surface, shown in Figure 12.88 with $a = b = c = 1$, is the cone $z = \sqrt{x^2 + y^2}$ studied in Section 12.3, together with the lower cone $z = -\sqrt{x^2 + y^2}$.

A Catalog of Surfaces

For later reference, here is a small catalog of the surfaces we have encountered.

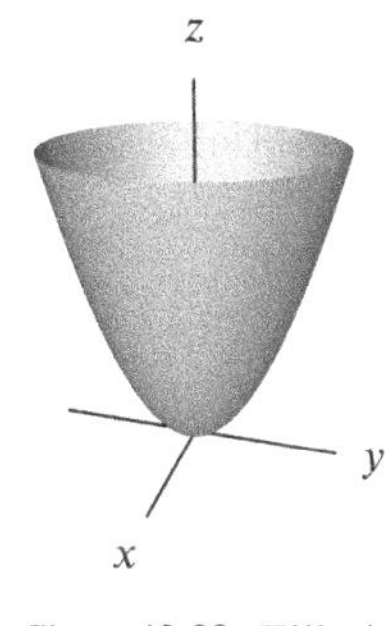

Figure 12.83: Elliptical paraboloid $z = \frac{x^2}{a^2} + \frac{y^2}{b^2}$

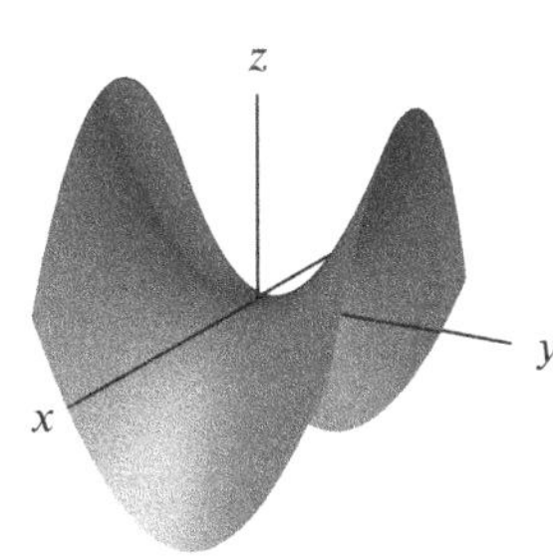

Figure 12.84: Hyperbolic paraboloid $z = -\frac{x^2}{a^2} + \frac{y^2}{b^2}$

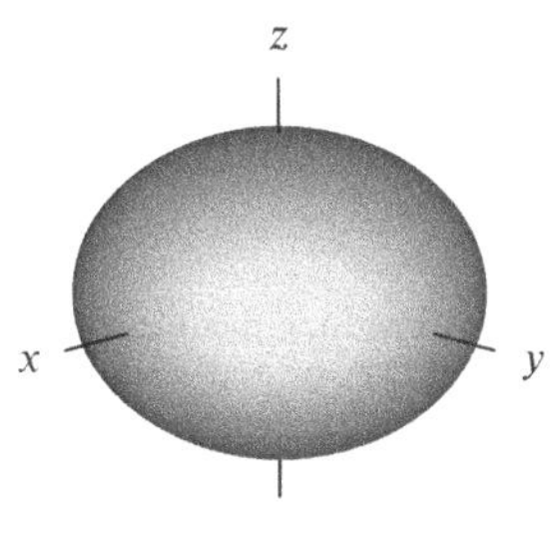

Figure 12.85: Ellipsoid $\frac{x^2}{a^2} + \frac{y^2}{b^2} + \frac{z^2}{c^2} = 1$

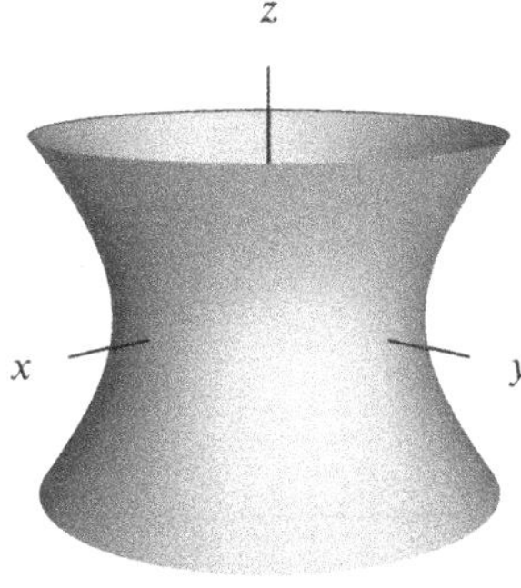

Figure 12.86: Hyperboloid of one sheet $\frac{x^2}{a^2} + \frac{y^2}{b^2} - \frac{z^2}{c^2} = 1$

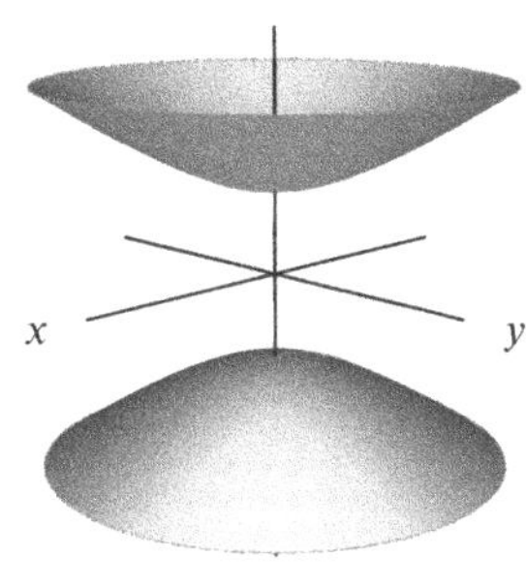

Figure 12.87: Hyperboloid of two sheets $\frac{x^2}{a^2} + \frac{y^2}{b^2} - \frac{z^2}{c^2} = -1$

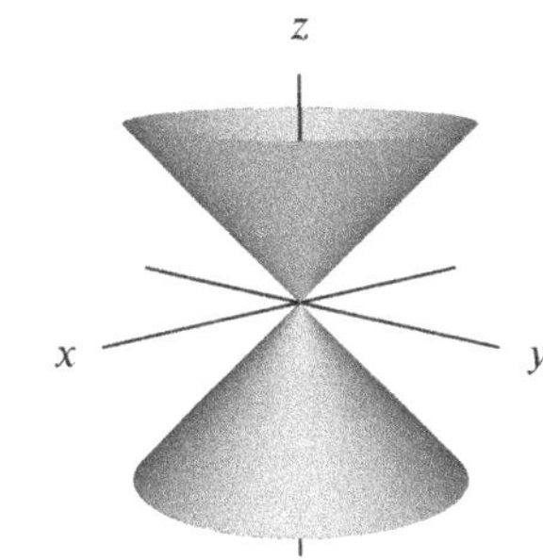

Figure 12.88: Cone $\frac{x^2}{a^2} + \frac{y^2}{b^2} - \frac{z^2}{c^2} = 0$

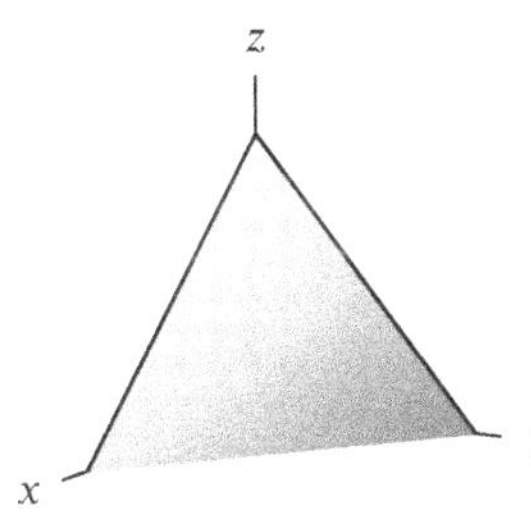

Figure 12.89: Plane $ax + by + cz = d$

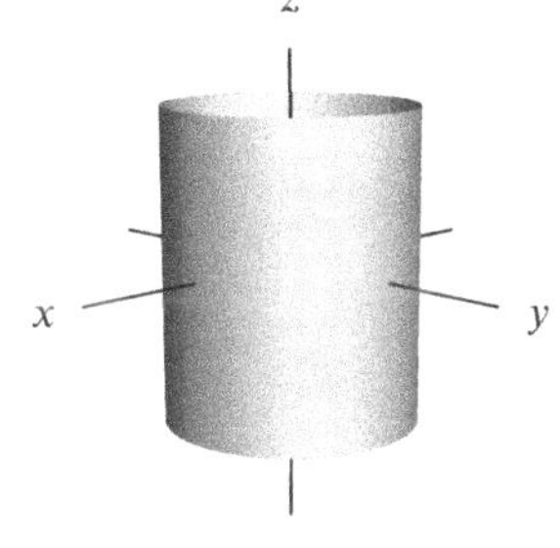

Figure 12.90: Cylindrical surface $x^2 + y^2 = a^2$

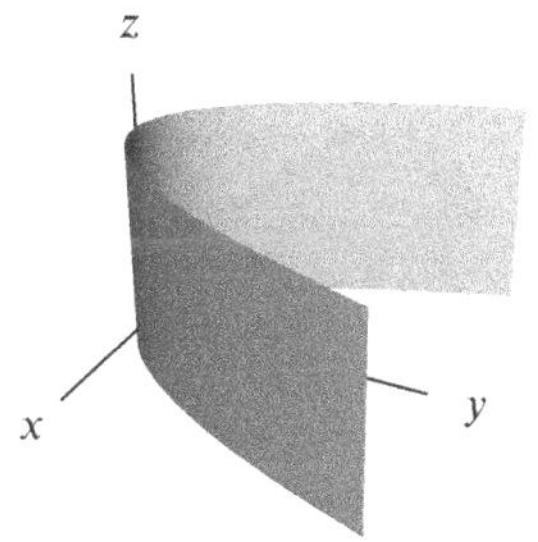

Figure 12.91: Parabolic cylinder $y = ax^2$

(These are viewed as equations in three variables x, y, and z)

How Surfaces Can Represent Functions of Two Variables and Functions of Three Variables

You may have noticed that we have used surfaces to represent functions in two different ways. First, we used a *single* surface to represent a two-variable function $f(x, y)$. Second, we used a *family* of level surfaces to represent a three-variable function $g(x, y, z)$. These level surfaces have equation $g(x, y, z) = c$.

What is the relation between these two uses of surfaces? For example, consider the function

$$f(x, y) = x^2 + y^2 + 3.$$

Define

$$g(x, y, z) = x^2 + y^2 + 3 - z$$

The points on the graph of f satisfy $z = x^2 + y^2 + 3$, so they also satisfy $x^2 + y^2 + 3 - z = 0$. Thus the graph of f is the same as the level surface

$$g(x, y, z) = x^2 + y^2 + 3 - z = 0.$$

In general, we have the following result:

> A single surface that is the graph of a two-variable function $f(x, y)$ can be thought of as one member of the family of level surfaces representing the three-variable function
>
> $$g(x, y, z) = f(x, y) - z.$$
>
> The graph of f is the level surface $g = 0$.

Conversely, a single level surface $g(x, y, z) = c$ can be regarded as the graph of a function $f(x, y)$ if it is possible to solve for z. Sometimes the level surface is pieced together from the graphs of two or more two-variable functions. For example, if $g(x, y, z) = x^2 + y^2 + z^2$, then one member of the family of level surfaces is the sphere

$$x^2 + y^2 + z^2 = 1.$$

This equation defines z implicitly as a function of x and y. Solving it gives two functions

$$z = \sqrt{1 - x^2 - y^2} \quad \text{and} \quad z = -\sqrt{1 - x^2 - y^2}.$$

The graph of the first function is the top half of the sphere and the graph of the second function is the bottom half.

Exercises and Problems for Section 12.5

EXERCISES

1. Match the following functions with the level surfaces in Figure 12.92.

   (a) $f(x, y, z) = y^2 + z^2$ **(b)** $h(x, y, z) = x^2 + z^2$.

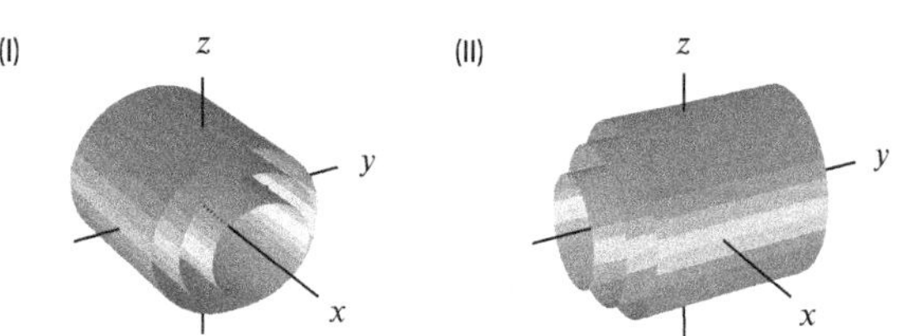

Figure 12.92

2. Match the functions with the level surfaces in Figure 12.93.

   (a) $f(x, y, z) = x^2 + y^2 + z^2$
   (b) $g(x, y, z) = x^2 + z^2$.

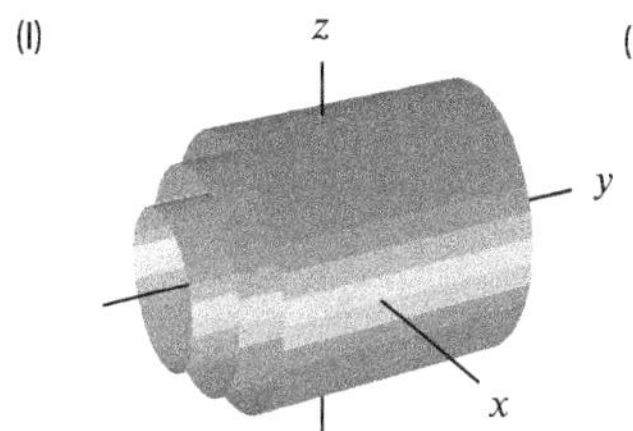

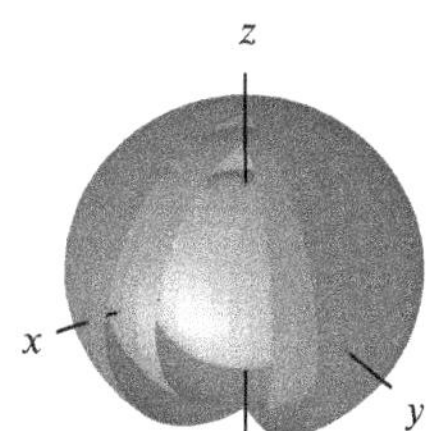

Figure 12.93

3. Write the level surface $x + 2y + 3z = 5$ as the graph of a function $f(x, y)$.
4. Find a formula for a function $f(x, y, z)$ whose level surface $f = 4$ is a sphere of radius 2, centered at the origin.
5. Write the level surface $x^2 + y + \sqrt{z} = 1$ as the graph of a function $f(x, y)$.
6. Find a formula for a function $f(x, y, z)$ whose level surfaces are spheres centered at the point (a, b, c).
7. Which of the graphs in the catalog of surfaces on page 691 is the graph of a function of x and y?

In Exercises 8–11, use the catalog on page 691 to identify the surface.

8. $x^2 + y^2 - z = 0$
9. $-x^2 - y^2 + z^2 = 1$
10. $x + y = 1$
11. $x^2 + y^2/4 + z^2 = 1$

In Exercises 12–15, decide if the given level surface can be expressed as the graph of a function, $f(x, y)$.

12. $z - x^2 - 3y^2 = 0$
13. $2x + 3y - 5z - 10 = 0$
14. $x^2 + y^2 + z^2 - 1 = 0$
15. $z^2 = x^2 + 3y^2$

16. Match the functions (a)–(d) with the descriptions of their level surfaces in I–IV.
   (a) $f(x, y, z) = \sqrt{9 - x^2 - y^2}$
   (b) $f(x, y, z) = \sqrt{x^2 + y^2 + z^2}$
   (c) $f(x, y, z) = \dfrac{1}{x^2 + y^2 + z^2}$
   (d) $f(x, y, z) = 5 + y^2 + z^2$

   I. Cylinders that get larger as the function value increases
   II. Cylinders that get smaller as the function value increases
   III. Spheres that get larger as the function value increases
   IV. Spheres that get smaller as the function value increases

PROBLEMS

In Problems 17–19, represent the surface whose equation is given as the graph of a two-variable function, $f(x, y)$, and as the level surface of a three-variable function, $g(x, y, z) = c$. There are many possible answers.

17. The plane $4x - y - 2z = 6$
18. The top half of the sphere $x^2 + y^2 + z^2 - 10 = 0$
19. The bottom half of the ellipsoid $x^2 + y^2 + z^2/2 = 1$
20. The balance, B, in dollars, in a bank account depends on the amount deposited, A dollars, the annual interest rate, r%, and the time, t, in months since the deposit, so $B = f(A, r, t)$.
   (a) Is f an increasing or decreasing function of A? Of r? Of t?
   (b) Interpret the statement $f(1250, 1, 25) \approx 1276$. Give units.
21. A person's basal metabolic rate (BMR) is the minimal number of daily calories needed to keep their body functioning at rest. The BMR (in kcal/day) of a man of mass m (in kg), height h (in cm) and age a (in years) can be approximated by[12]

$$P = f(m, h, a) = 14m + 5h - 7a + 66$$

and for women by

$$P = g(m, h, a) = 10m + 2h - 5a + 655.$$

   (a) What is the BMR of a 28-year-old man 180 cm tall weighing 59 kg?
   (b) What is the BMR of a 43-year-old woman 162 cm tall weighing 52 kg?
   (c) Describe the level surface $P = 2000$ for a woman and explain what the points on this level surface represent.
   (d) If a 40-year-old man 175 cm tall weighing 77 kg restricts himself to a diet with a daily caloric intake of 1600 kcal, should he expect to lose weight?
22. The monthly payments, P dollars, on a mortgage in which A dollars were borrowed at an annual interest rate of r% for t years is given by $P = f(A, r, t)$. Is f an increasing or decreasing function of A? Of r? Of t?
23. The balance in a bank account, B dollars, is given by $B = f(P, r, t) = P(1 + 0.01r)^t$, where P dollars is the principal amount invested, r% is the annual interest rate, and t years is the time since the investment was made.
   (a) Find a formula for the level surface of f containing the point $(P, r, t) = (1000, 5, 20)$, and explain the significance of this surface in terms of balance.
   (b) Find another point on the level surface in part (a), and explain the significance of this point in terms of balance.
24. The pressure of gas in a storage container, in atmospheres, is given by

$$P = f(n, T, V) = \frac{82nT}{V},$$

where n is the amount of gas, in kilomoles, T is the temperature of the gas, in Kelvin, and V is the volume of the storage container, in liters.
   (a) Find a formula for the level surface of f containing the point $(n, T, V) = (1, 270, 20)$, and explain the significance of this surface in terms of pressure.

[12] www.wikipedia.org, accessed May 11, 2016.

(b) Find another point on the level surface in part (a), and explain the significance of this point in terms of pressure.

25. The mass, in grams, of a rod in the shape of a right circular cylinder, is given by $m = f(r, h, \rho) = \pi r^2 h\rho$, where the rod has a radius of r cm, a height of h cm, and a uniform density of ρ gm/cm^3.

(a) Find a formula for the level surface of f containing the point $(r, h, \rho) = (2, 10, 3)$, and explain the significance of this surface in terms of mass.
(b) Find another point on the level surface in part (a), and explain the significance of this point in terms of mass.

26. Find a function $f(x, y, z)$ whose level surface $f = 1$ is the graph of the function $g(x, y) = x + 2y$.

27. Find two functions $f(x, y)$ and $g(x, y)$ so that the graphs of both together form the ellipsoid $x^2+y^2/4+z^2/9 = 1$.

28. Find a formula for a function $g(x, y, z)$ whose level surfaces are planes parallel to the plane $z = 2x + 3y - 5$.

29. Which of the following functions have planes as level surfaces?

$$f(x, y, z) = e^{x+z} \qquad r(x, y, z) = x^3$$
$$g(x, y, z) = e^x + z \qquad m(x, y, z) = \ln(x + z)$$

30. The surface S is the graph of $f(x, y) = \sqrt{1 - x^2 - y^2}$.

(a) Explain why S is the upper hemisphere of radius 1, with equator in the xy-plane, centered at the origin.
(b) Find a level surface $g(x, y, z) = c$ representing S.

31. The surface S is the graph of $f(x, y) = \sqrt{1 - y^2}$.

(a) Explain why S is the upper half of a circular cylinder of radius 1, centered along the x-axis.
(b) Find a level surface $g(x, y, z) = c$ representing S.

32. A cone C, with height 1 and radius 1, has its base in the xz-plane and its vertex on the positive y-axis. Find a function $g(x, y, z)$ such that C is part of the level surface $g(x, y, z) = 0$. [Hint: The graph of $f(x, y) = \sqrt{x^2 + y^2}$ is a cone which opens up and has vertex at the origin.]

33. Describe the level surface $f(x, y, z) = x^2/4 + z^2 = 1$ in words.

34. Describe the level surface $g(x, y, z) = x^2+y^2/4+z^2 = 1$ in words. [Hint: Look at cross-sections with constant x, y, and z values.]

35. Describe in words the level surfaces of the function $g(x, y, z) = x + y + z$.

36. Describe in words the level surfaces of $f(x, y, z) = \sin(x + y + z)$.

37. Describe the surface $x^2 + y^2 = (2 + \sin z)^2$. In general, if $f(z) \geq 0$ for all z, describe the surface $x^2 + y^2 = (f(z))^2$.

38. What do the level surfaces of $f(x, y, z) = x^2 - y^2 + z^2$ look like? [Hint: Use cross-sections with y constant instead of cross-sections with z constant.]

39. Describe in words the level surfaces of $g(x, y, z) = e^{-(x^2+y^2+z^2)}$.

40. Describe in words the level surfaces of $f(x, y, z) = z/x$.

41. Show that the level surfaces of $g(x, y, z) = ax+by+cz$ where $c \neq 0$ are parallel planes.

42. Sketch and label level surfaces of $h(x, y, z) = e^{z-y}$ for $h = 1, e, e^2$.

43. Sketch and label level surfaces of $f(x, y, z) = 4 - x^2 - y^2 - z^2$ for $f = 0, 1, 2$.

44. Sketch and label level surfaces of $g(x, y, z) = 1-x^2-y^2$ for $g = 0, -1, -2$.

45. What is the relationship between the level surfaces of $g(x, y, z) = f(x, y) - z$ and the graph of $z = f(x, y)$?

46. Describe the level surfaces of $g(x, y, z) = y - f(x)$.

Strengthen Your Understanding

In Problems 47–49, explain what is wrong with the statement.

47. The graph of a function $f(x, y, z)$ is a surface in 3-space.

48. The level surfaces of $f(x, y, z) = x^2 - y^2$ are all saddle-shaped.

49. The level surfaces of $f(x, y, z) = x^2 + y^2$ are paraboloids.

In Problems 50–53, give an example of:

50. A function $f(x, y, z)$ whose level surfaces are equally spaced planes perpendicular to the yz-plane.

51. A function $f(x, y, z)$ whose level sets are concentric cylinders centered on the y-axis.

52. A nonlinear function $f(x, y, z)$ whose level sets are parallel planes.

53. A function $f(x, y, z)$ whose level sets are paraboloids.

Are the statements in Problems 54–64 true or false? Give reasons for your answer.

54. The graph of the function $f(x, y) = x^2 + y^2$ is the same as the level surface $g(x, y, z) = x^2 + y^2 - z = 0$.

55. The graph of $f(x, y) = \sqrt{1 - x^2 - y^2}$ is the same as the level surface $g(x, y, z) = x^2 + y^2 + z^2 = 1$.

56. Any surface which is the graph of a two-variable function $f(x, y)$ can also be represented as the level surface of a three-variable function $g(x, y, z)$.

57. Any surface which is the level surface of a three-variable function $g(x, y, z)$ can also be represented as the graph of a two-variable function $f(x, y)$.

58. The level surfaces of the function $g(x, y, z) = x+2y+z$ are parallel planes.

59. The level surfaces of $g(x, y, z) = x^2 + y + z^2$ are cylinders with axis along the y-axis.

60. A level surface of a function $g(x, y, z)$ cannot be a single point.

61. If $g(x, y, z) = ax + by + cz + d$, where a, b, c, d are nonzero constants, then the level surfaces of g are planes.

62. If the level surfaces of g are planes, then $g(x, y, z) = ax + by + cz + d$, where a, b, c, d are constants.

63. If the level surfaces $g(x, y, z) = k_1$ and $g(x, y, z) = k_2$ are the same surface, then $k_1 = k_2$.

64. If $x^2 + y^2 + z^2 = 1$ is the level surface $g(x, y, z) = 1$, then $x^2 + y^2 + z^2 = 4$ is the level surface $g(x, y, z) = 4$.

12.6 LIMITS AND CONTINUITY

The sheer face of Half Dome, in Yosemite National Park in California, was caused by glacial activity during the Ice Age. (See Figure 12.94.) As we scale the rock from the west, the height of the terrain rises abruptly by nearly 5000 feet from the valley floor, 2000 feet of it vertical.

If we consider the function h giving the height of the terrain above sea level in terms of longitude and latitude, then h has a *discontinuity* along the path at the base of the cliff of Half Dome. Looking at the contour map of the region in Figure 12.95, we see that in most places a small change in position results in a small change in height, except near the cliff. There, no matter how small a step we take, we get a large change in height. (You can see how crowded the contours get near the cliff; some end abruptly along the discontinuity.)

This geological feature illustrates the ideas of continuity and discontinuity. Roughly speaking, a function is said to be *continuous* at a point if its values at places near the point are close to the value at the point. If this is not the case, the function is said to be *discontinuous*.

The property of continuity is one that, practically speaking, we usually assume of the functions we are studying. Informally, we expect (except under special circumstances) that values of a function do not change drastically when making small changes to the input variables. Whenever we model a one-variable function by an unbroken curve, we are making this assumption. Even when functions come to us as tables of data, we usually make the assumption that the missing function values between data points are close to the measured ones.

In this section we study limits and continuity a bit more formally in the context of functions of several variables. For simplicity we study these concepts for functions of two variables, but our discussion can be adapted to functions of three or more variables.

One can show that sums, products, and compositions of continuous functions are continuous,

Figure 12.94: Half Dome in Yosemite National Park

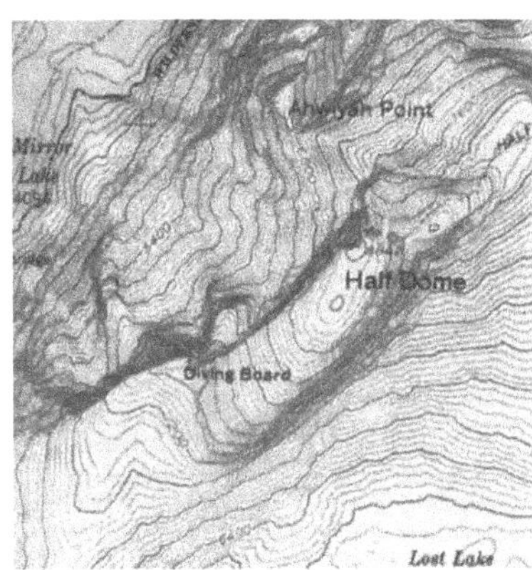

Figure 12.95: A contour map of Half Dome

while the quotient of two continuous functions is continuous everywhere the denominator function is nonzero. Thus, each of the functions

$$\cos(x^2y), \qquad \ln(x^2+y^2), \qquad \frac{e^{x+y}}{x+y}, \qquad \ln(\sin(x^2+y^2))$$

is continuous at all points (x, y) where it is defined. As for functions of one variable, the graph of a continuous function over an unbroken domain is unbroken—that is, the surface has no holes or rips in it.

Example 1 From Figures 12.96–12.99, which of the following functions appear to be continuous at $(0, 0)$?

(a) $f(x,y) = \begin{cases} \dfrac{x^2y}{x^2+y^2}, & (x,y) \neq (0,0), \\ 0, & (x,y) = (0,0). \end{cases}$ (b) $g(x,y) = \begin{cases} \dfrac{x^2}{x^2+y^2}, & (x,y) \neq (0,0), \\ 0, & (x,y) = (0,0). \end{cases}$

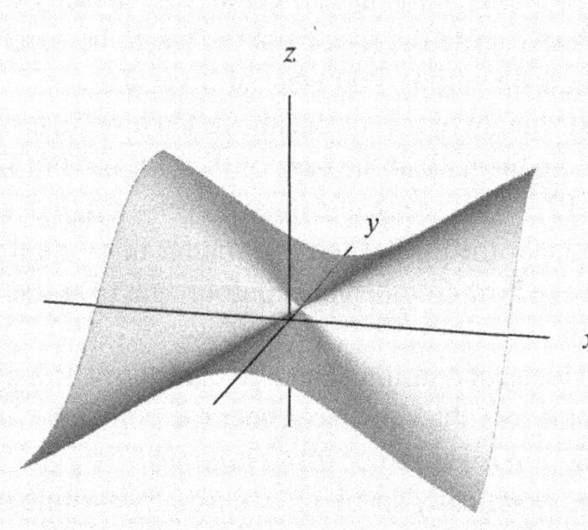

Figure 12.96: Graph of $z = x^2y/(x^2+y^2)$

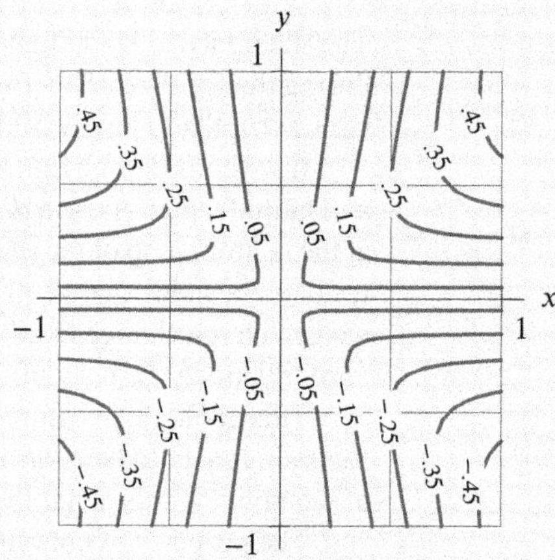

Figure 12.97: Contour diagram of $z = x^2y/(x^2+y^2)$

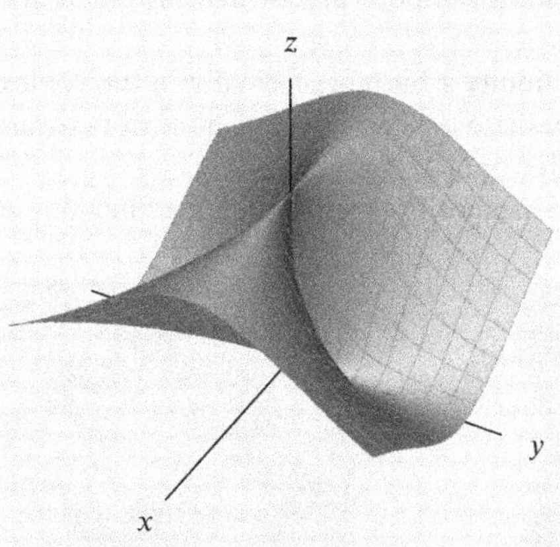

Figure 12.98: Graph of $z = x^2/(x^2+y^2)$

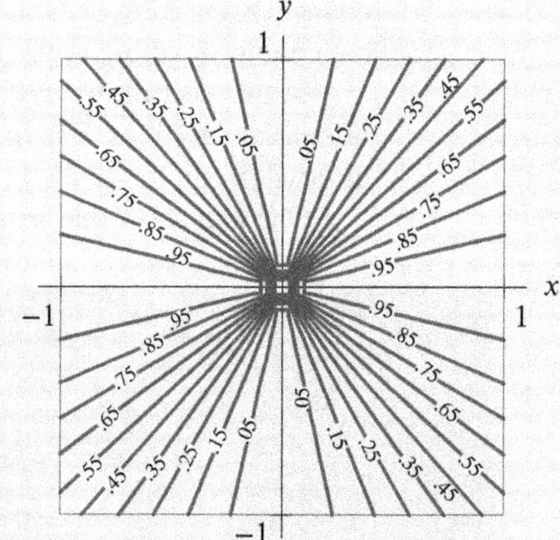

Figure 12.99: Contour diagram of $z = x^2/(x^2+y^2)$

Solution (a) The graph and contour diagram of f in Figures 12.96 and 12.97 suggest that f is close to 0 when (x, y) is close to $(0, 0)$. That is, the figures suggest that f is continuous at the point $(0, 0)$; the graph appears to have no rips or holes there.

However, the figures cannot tell us for sure whether f is continuous. To be certain we must investigate the limit analytically, as is done in Example 2(a) on page 697.

(b) The graph of g and its contours near $(0, 0)$ in Figure 12.98 and 12.99 suggest that g behaves differently from f: The contours of g seem to "crash" at the origin and the graph rises rapidly from 0 to 1 near $(0, 0)$. Small changes in (x, y) near $(0, 0)$ can yield large changes in g, so we expect that g is not continuous at the point $(0, 0)$. Again, a more precise analysis is given in Example 2(b).

The previous example suggests that continuity *at* a point depends on a function's behavior *near* the point. To study behavior near a point more carefully we need the idea of a limit of a function of two variables. Suppose that $f(x, y)$ is a function defined on a set in 2-space, not necessarily containing the point (a, b), but containing points (x, y) arbitrarily close to (a, b); suppose that L is a number.

> The function f has a **limit** L at the point (a, b), written
>
> $$\lim_{(x,y)\to(a,b)} f(x, y) = L,$$
>
> if $f(x, y)$ is as close to L as we please whenever the distance from the point (x, y) to the point (a, b) is sufficiently small, but not zero.

We define continuity for functions of two variables in the same way as for functions of one variable:

> A function f is **continuous at the point** (a, b) if
>
> $$\lim_{(x,y)\to(a,b)} f(x, y) = f(a, b).$$
>
> A function is **continuous on a region** R in the xy-plane if it is continuous at each point in R.

Thus, if f is continuous at the point (a, b), then f must be defined at (a, b) and the limit, $\lim_{(x,y)\to(a,b)} f(x, y)$, must exist and be equal to the value $f(a, b)$. If a function is defined at a point (a, b) but is not continuous there, then we say that f is *discontinuous* at (a, b).

We now apply the definition of continuity to the functions in Example 1, showing that f is continuous at $(0, 0)$ and that g is discontinuous at $(0, 0)$.

Example 2 Let f and g be the functions in Example 1. Use the definition of the limit to show that:
(a) $\lim_{(x,y)\to(0,0)} f(x, y) = 0$ (b) $\lim_{(x,y)\to(0,0)} g(x, y)$ does not exist.

Solution To investigate these limits of f and g, we consider values of these functions near, but not at, the origin, where they are given by the formulas

$$f(x, y) = \frac{x^2y}{x^2 + y^2} \qquad g(x, y) = \frac{x^2}{x^2 + y^2}.$$

(a) The graph and contour diagram of f both suggest that $\lim_{(x,y)\to(0,0)} f(x, y) = 0$. To use the definition of the limit, we estimate $|f(x, y) - L|$ with $L = 0$:

$$|f(x, y) - L| = \left|\frac{x^2y}{x^2 + y^2} - 0\right| = \left|\frac{x^2}{x^2 + y^2}\right| |y| \le |y| \le \sqrt{x^2 + y^2}.$$

Now $\sqrt{x^2+y^2}$ is the distance from (x, y) to $(0, 0)$. Thus, to make $|f(x, y) - 0| < 0.001$, for example, we need only require that (x, y) be within 0.001 of $(0, 0)$. More generally, for any positive number u, no matter how small, we are sure that $|f(x, y) - 0| < u$ whenever (x, y) is no farther than u from $(0, 0)$. This is what we mean by saying that the difference $|f(x, y) - 0|$ can be made as small as we wish by choosing the distance to be sufficiently small. Thus, we conclude that

$$\lim_{(x,y)\to(0,0)} f(x, y) = \lim_{(x,y)\to(0,0)} \frac{x^2y}{x^2+y^2} = 0.$$

Notice that since this limit equals $f(0, 0)$, the function f is continuous at $(0, 0)$.

(b) Although the formula defining the function g looks similar to that of f, we saw in Example 1 that g's behavior near the origin is quite different. If we consider points $(x, 0)$ lying along the x-axis near $(0, 0)$, then the values $g(x, 0)$ are equal to 1, while if we consider points $(0, y)$ lying along the y-axis near $(0, 0)$, then the values $g(0, y)$ are equal to 0. Thus, within any distance (no matter how small) from the origin, there are points where $g = 0$ and points where $g = 1$. Therefore the limit $\lim_{(x,y)\to(0,0)} g(x, y)$ does not exist, and thus g is not continuous at $(0, 0)$.

While the notions of limit and continuity look formally the same for one- and two-variable functions, they are somewhat more subtle in the multivariable case. The reason for this is that on the line (1-space), we can approach a point from just two directions (left or right) but in 2-space there are an infinite number of ways to approach a given point.

Exercises and Problems for Section 12.6

EXERCISES

In Exercises 1–6, is the function continuous at all points in the given region?

1. $\dfrac{1}{x^2+y^2}$ on the square $-1 \le x \le 1, -1 \le y \le 1$

2. $\dfrac{1}{x^2+y^2}$ on the square $1 \le x \le 2, 1 \le y \le 2$

3. $\dfrac{y}{x^2+2}$ on the disk $x^2+y^2 \le 1$

4. $\dfrac{e^{\sin x}}{\cos y}$ on the rectangle $-\frac{\pi}{2} \le x \le \frac{\pi}{2}, 0 \le y \le \frac{\pi}{4}$

5. $\tan(xy)$ on the square $-2 \le x \le 2, -2 \le y \le 2$

6. $\sqrt{2x-y}$ on the disk $x^2+y^2 \le 4$

In Exercises 7–11, find the limit as $(x, y) \to (0, 0)$ of $f(x, y)$. Assume that polynomials, exponentials, logarithmic, and trigonometric functions are continuous.

7. $f(x, y) = e^{-x-y}$

8. $f(x, y) = x^2 + y^2$

9. $f(x, y) = \dfrac{x}{x^2+1}$

10. $f(x, y) = \dfrac{x+y}{(\sin y)+2}$

11. $f(x, y) = \dfrac{\sin(x^2+y^2)}{x^2+y^2}$ [Hint: $\lim_{t\to 0} \dfrac{\sin t}{t} = 1$.]

In Exercises 12–15, use the contour diagram for $f(x, y)$ in Figure 12.100 to suggest an estimate for the limit, or explain why it may not exist.

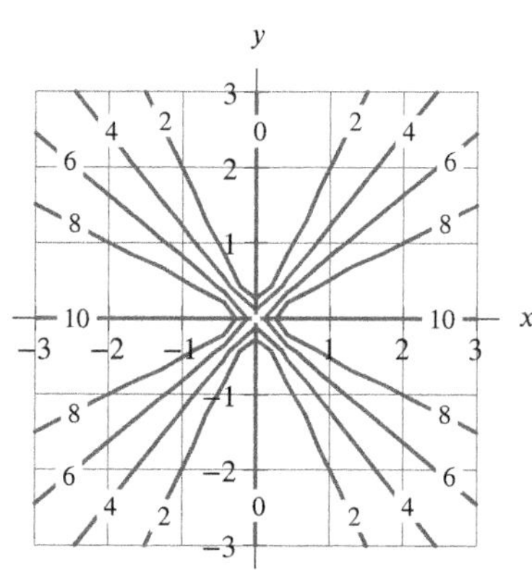

Figure 12.100

12. $\lim_{(x,y)\to(2,1)} f(x, y)$

13. $\lim_{(x,y)\to(-1,2)} f(x, y)$

14. $\lim_{(x,y)\to(-2,0)} f(x, y)$

15. $\lim_{(x,y)\to(0,0)} f(x, y)$

PROBLEMS

In Problems 16–17, show that the function $f(x, y)$ does not have a limit as $(x, y) \to (0, 0)$. [Hint: Use the line $y = mx$.]

16. $f(x, y) = \dfrac{x + y}{x - y}, \quad x \neq y$

17. $f(x, y) = \dfrac{x^2 - y^2}{x^2 + y^2}$

18. By approaching the origin along the positive x-axis and the positive y-axis, show that the following limit does not exist:
$$\lim_{(x,y)\to(0,0)} \frac{2x - y^2}{2x + y^2}.$$

19. Show that $f(x, y)$ has no limit as $(x, y) \to (0, 0)$ if
$$f(x, y) = \frac{xy}{|xy|}, \qquad x \neq 0 \text{ and } y \neq 0.$$

20. Show that the function f does not have a limit at $(0, 0)$ by examining the limits of f as $(x, y) \to (0, 0)$ along the curve $y = kx^2$ for different values of k:
$$f(x, y) = \frac{x^2}{x^2 + y}, \qquad x^2 + y \neq 0.$$

21. Let $f(x, y) = \begin{cases} \dfrac{|x|}{x} y & \text{for } x \neq 0 \\ 0 & \text{for } x = 0. \end{cases}$

Is $f(x, y)$ continuous

(a) On the x-axis? (b) On the y-axis?
(c) At $(0, 0)$?

In Problems 22–23, determine whether there is a value for the constant c making the function continuous everywhere. If so, find it. If not, explain why not.

22. $f(x, y) = \begin{cases} c + y, & x \le 3, \\ 5 - x, & x > 3. \end{cases}$

23. $f(x, y) = \begin{cases} c + y, & x \le 3, \\ 5 - y, & x > 3. \end{cases}$

24. Is the following function continuous at $(0, 0)$?
$$f(x, y) = \begin{cases} x^2 + y^2 & \text{if } (x, y) \neq (0, 0) \\ 2 & \text{if } (x, y) = (0, 0) \end{cases}$$

25. What value of c makes the following function continuous at $(0, 0)$?
$$f(x, y) = \begin{cases} x^2 + y^2 + 1 & \text{if } (x, y) \neq (0, 0) \\ c & \text{if } (x, y) = (0, 0) \end{cases}$$

26. (a) Use a computer to draw the graph and the contour diagram of the following function:
$$f(x, y) = \begin{cases} \dfrac{xy(x^2 - y^2)}{x^2 + y^2}, & (x, y) \neq (0, 0), \\ 0, & (x, y) = (0, 0). \end{cases}$$
(b) Do your answers to part (a) suggest that f is continuous at $(0, 0)$? Explain your answer.

27. The function f, whose graph and contour diagram are in Figures 12.101 and 12.102, is given by
$$f(x, y) = \begin{cases} \dfrac{xy}{x^2 + y^2}, & (x, y) \neq (0, 0), \\ 0, & (x, y) = (0, 0). \end{cases}$$
(a) Show that $f(0, y)$ and $f(x, 0)$ are each continuous functions of one variable.
(b) Show that rays emanating from the origin are contained in contours of f.
(c) Is f continuous at $(0, 0)$?

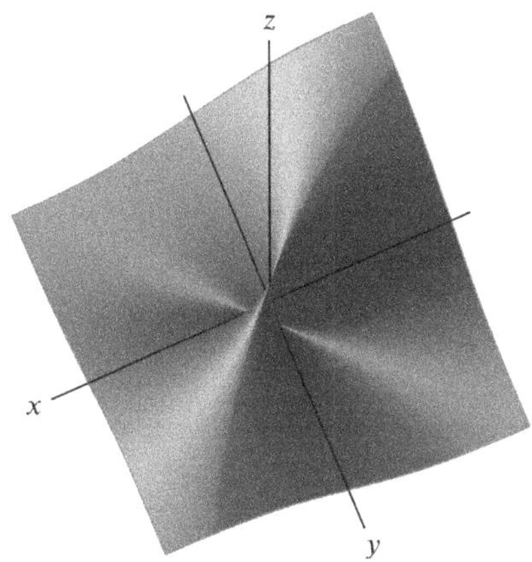

Figure 12.101: Graph of $z = xy/(x^2 + y^2)$

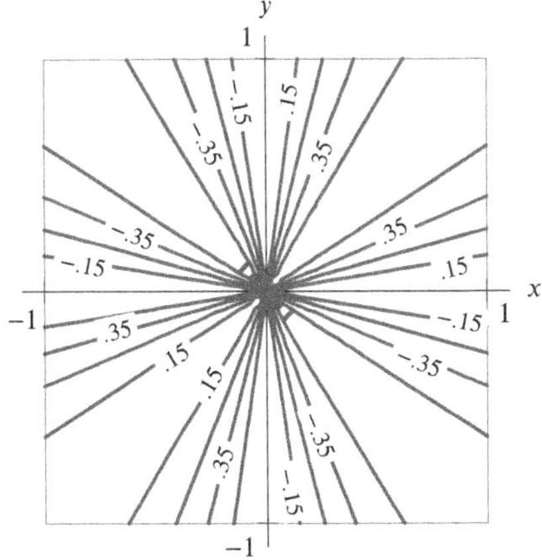

Figure 12.102: Contour diagram of $z = xy/(x^2 + y^2)$

Strengthen Your Understanding

In Problems 28–29, explain what is wrong with the statement.

28. If a function $f(x, y)$ has a limit as (x, y) approaches (a, b), then it is continuous at (a, b).

29. If both f and g are continuous at (a, b), then so are $f + g, fg$ and f/g.

In Problems 30–31, give an example of:

30. A function $f(x, y)$ which is continuous everywhere except at $(0, 0)$ and $(1, 2)$.

31. A function $f(x, y)$ that approaches 1 as (x, y) approaches $(0, 0)$ along the x-axis and approaches 2 as (x, y) approaches $(0, 0)$ along the y-axis.

In Problems 32–34, construct a function $f(x, y)$ with the given property.

32. Not continuous along the line $x = 2$; continuous everywhere else.

33. Not continuous at the point $(2, 0)$; continuous everywhere else.

34. Not continuous along the curve $x^2 + y^2 = 1$; continuous everywhere else.

Are the statements in Problems 35–40 true or false? Give reasons for your answer.

35. If the limit of $f(x, y)$ is 1 as (x, y) approaches $(0, 0)$ along the x-axis, and the limit of $f(x, y)$ is 1 as (x, y) approaches $(0, 0)$ along the y-axis, then

$$\lim_{(x,y)\to(0,0)} f(x, y) \text{ exists.}$$

36. If $f(1, 0) = 2$, then $\lim_{(x,y)\to(1,0)} f(x, y) = 2$.

37. If $f(x, y)$ is continuous and $f(1, 0) = 2$, then

$$\lim_{(x,y)\to(1,0)} f(x, y) = 2.$$

38. If $\lim_{(x,y)\to(0,0)} f(x, y) = 3$, then the limit of $f(x, y)$ is 3 as (x, y) approaches $(0, 0)$ along the x-axis.

39. If $f(x, y)$ is continuous at (a, b), then its limit exists at (a, b).

40. If $\lim_{(x,y)\to(a,b)} f(x, y)$ exists then $f(x, y)$ is continuous at (a, b).

Online Resource: Review problems and Projects

Chapter Thirteen

A FUNDAMENTAL TOOL: VECTORS

Contents

13.1 DISPLACEMENT VECTORS

Suppose you are a pilot planning a flight from Dallas to Pittsburgh. There are two things you must know: the distance to be traveled (so you have enough fuel to make it) and in what direction to go (so you don't miss Pittsburgh). Both these quantities together specify the displacement or *displacement vector* between the two cities.

> The **displacement vector** from one point to another is an arrow with its tail at the first point and its tip at the second. The **magnitude** (or length) of the displacement vector is the distance between the points and is represented by the length of the arrow. The **direction** of the displacement vector is the direction of the arrow.

Figure 13.1 shows a map with the displacement vectors from Dallas to Pittsburgh, from Albuquerque to Oshkosh, and from Los Angeles to Buffalo, SD. These displacement vectors have the same length and the same direction. We say that the displacement vectors between the corresponding cities are the same, even though they do not coincide. In other words,

> Displacement vectors which point in the same direction and have the same magnitude are considered to be the same, even if they do not coincide.

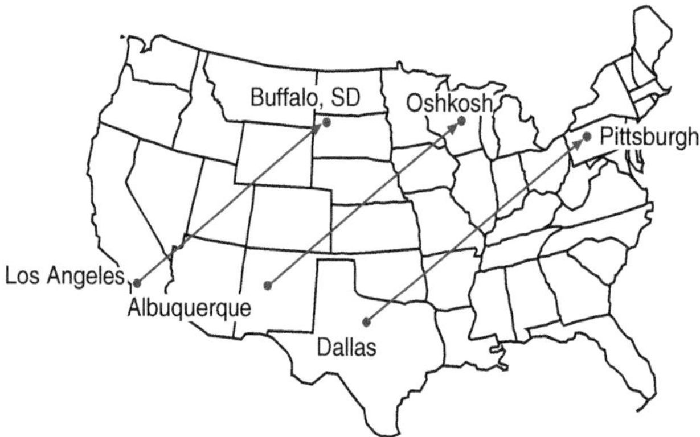

Figure 13.1: Displacement vectors between cities

Notation and Terminology

The displacement vector is our first example of a vector. Vectors have both magnitude and direction; in comparison, a quantity specified only by a number, but no direction, is called a *scalar*.[1] For instance, the time taken by the flight from Dallas to Pittsburgh is a scalar quantity. Displacement is a vector since it requires both distance and direction to specify it.

In this book, vectors are written with an arrow over them, $\vec{v}$, to distinguish them from scalars. Other books use a bold **v** to denote a vector. We use the notation $\overrightarrow{PQ}$ to denote the displacement vector from a point P to a point Q. The magnitude, or length, of a vector $\vec{v}$ is written $\|\vec{v}\|$.

Addition and Subtraction of Displacement Vectors

Suppose NASA commands a robot on Mars to move 75 meters in one direction and then 50 meters in another direction. (See Figure 13.2.) Where does the robot end up? Suppose the displacements are represented by the vectors $\vec{v}$ and $\vec{w}$, respectively. Then the sum $\vec{v} + \vec{w}$ gives the final position.

[1] So named by W. R. Hamilton because they are merely numbers on the *scale* from $-\infty$ to ∞.

The **sum**, $\vec{v} + \vec{w}$, of two vectors $\vec{v}$ and $\vec{w}$ is the combined displacement resulting from first applying $\vec{v}$ and then $\vec{w}$. (See Figure 13.3.) The sum $\vec{w} + \vec{v}$ gives the same displacement.

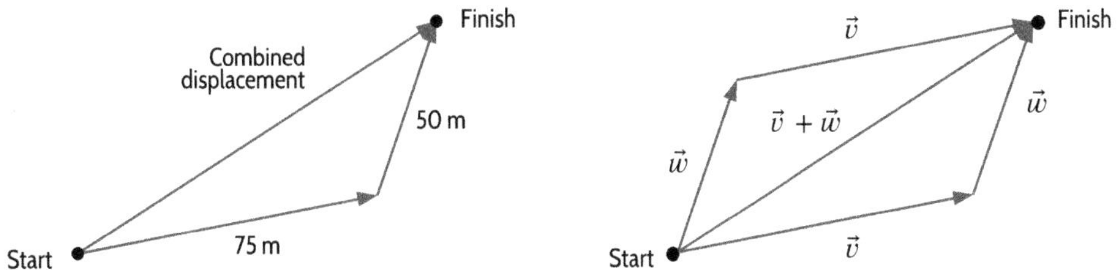

Figure 13.2: Sum of displacements of robots on Mars

Figure 13.3: The sum $\vec{v} + \vec{w} = \vec{w} + \vec{v}$

Suppose two different robots start from the same location. One moves along a displacement vector $\vec{v}$ and the second along a displacement vector $\vec{w}$. What is the displacement vector, $\vec{x}$, from the first robot to the second? (See Figure 13.4.) Since $\vec{v} + \vec{x} = \vec{w}$, we define $\vec{x}$ to be the difference $\vec{x} = \vec{w} - \vec{v}$. In other words, $\vec{w} - \vec{v}$ gets you from the first robot to the second.

The **difference**, $\vec{w} - \vec{v}$, is the displacement vector that, when added to $\vec{v}$, gives $\vec{w}$. That is, $\vec{w} = \vec{v} + (\vec{w} - \vec{v})$. (See Figure 13.4.)

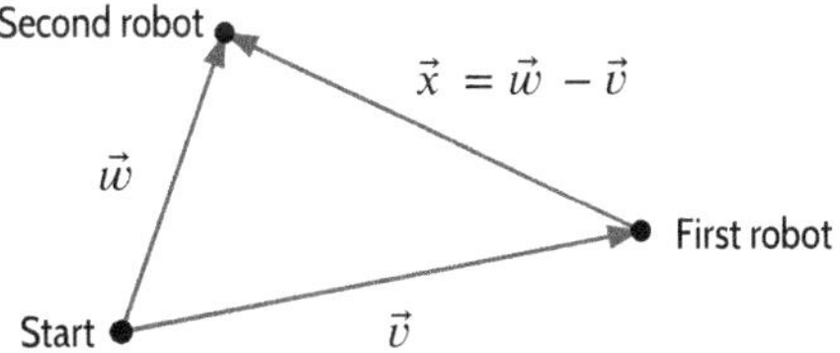

Figure 13.4: The difference $\vec{w} - \vec{v}$

If the robot ends up where it started, then its total displacement vector is the *zero vector*, $\vec{0}$. The zero vector has no direction.

The **zero vector**, $\vec{0}$, is a displacement vector with zero length.

Scalar Multiplication of Displacement Vectors

If $\vec{v}$ represents a displacement vector, the vector $2\vec{v}$ represents a displacement of twice the magnitude in the same direction as $\vec{v}$. Similarly, $-2\vec{v}$ represents a displacement of twice the magnitude in the opposite direction. (See Figure 13.5.)

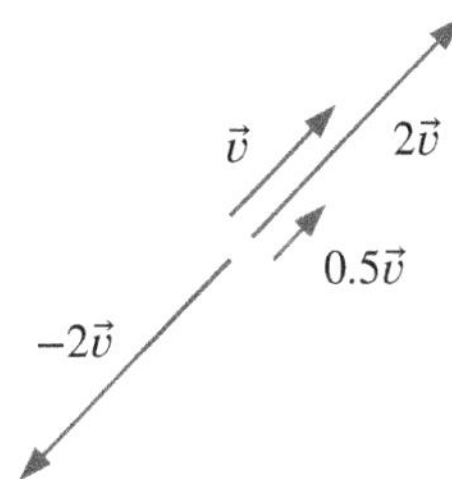

Figure 13.5: Scalar multiples of the vector $\vec{v}$

If λ is a scalar and $\vec{v}$ is a displacement vector, the **scalar multiple of $\vec{v}$ by λ**, written $\lambda\vec{v}$, is the displacement vector with the following properties:

- The displacement vector $\lambda\vec{v}$ is parallel to $\vec{v}$, pointing in the same direction if $\lambda > 0$ and in the opposite direction if $\lambda < 0$.
- The magnitude of $\lambda\vec{v}$ is $|\lambda|$ times the magnitude of $\vec{v}$, that is, $\|\lambda\vec{v}\| = |\lambda|\,\|\vec{v}\|$.

Note that $|\lambda|$ represents the absolute value of the scalar λ while $\|\lambda\vec{v}\|$ represents the magnitude of the vector $\lambda\vec{v}$.

Example 1 Explain why $\vec{w} - \vec{v} = \vec{w} + (-1)\vec{v}$.

Solution The vector $(-1)\vec{v}$ has the same magnitude as $\vec{v}$, but points in the opposite direction. Figure 13.6 shows that the combined displacement $\vec{w} + (-1)\vec{v}$ is the same as the displacement $\vec{w} - \vec{v}$.

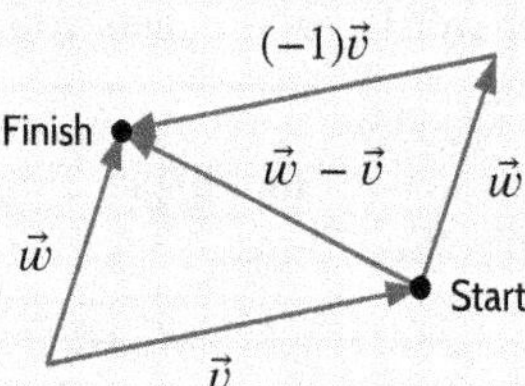

Figure 13.6: Explanation for why $\vec{w} - \vec{v} = \vec{w} + (-1)\vec{v}$

Parallel Vectors

Two vectors $\vec{v}$ and $\vec{w}$ are *parallel* if one is a scalar multiple of the other, that is, if $\vec{w} = \lambda\vec{v}$, for some scalar λ.

Components of Displacement Vectors: The Vectors $\vec{i}$, $\vec{j}$, and $\vec{k}$

Suppose that you live in a city with equally spaced streets running east-west and north-south and that you want to tell someone how to get from one place to another. You'd be likely to tell them how many blocks east-west and how many blocks north-south to go. For example, to get from P to Q in Figure 13.7, we go 4 blocks east and 1 block south. If $\vec{i}$ and $\vec{j}$ are as shown in Figure 13.7, then the displacement vector from P to Q is $4\vec{i} - \vec{j}$.

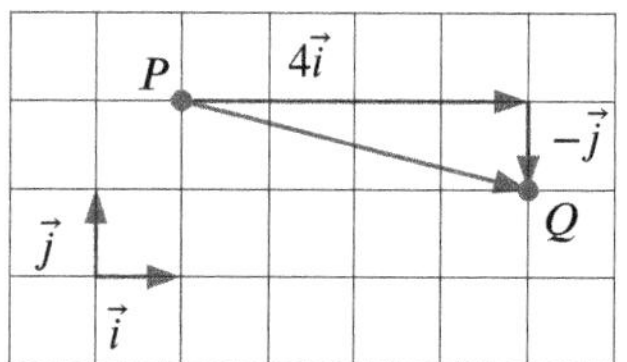

Figure 13.7: The displacement vector from P to Q is $4\vec{i} - \vec{j}$

We extend the same idea to 3 dimensions. First we choose a Cartesian system of coordinate axes. The three vectors of length 1 shown in Figure 13.8 are the vector $\vec{i}$, which points along the positive x-axis, the vector $\vec{j}$, along the positive y-axis, and the vector $\vec{k}$, along the positive z-axis.

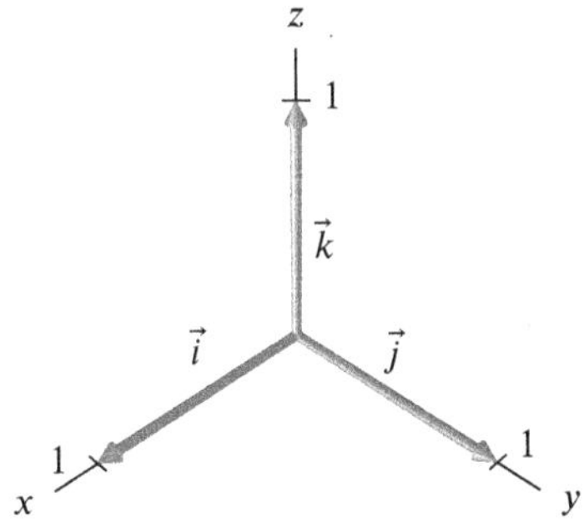

Figure 13.8: The vectors $\vec{i}$, $\vec{j}$ and $\vec{k}$ in 3-space

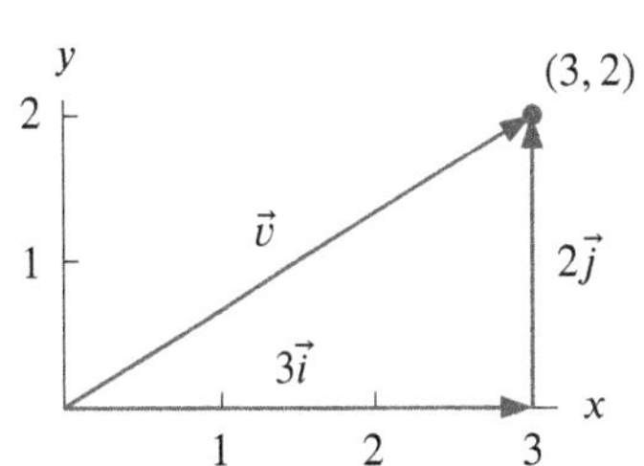

Figure 13.9: We resolve $\vec{v}$ into components by writing $\vec{v} = 3\vec{i} + 2\vec{j}$

Writing Displacement Vectors Using $\vec{i}, \vec{j}, \vec{k}$

Any displacement in 3-space or the plane can be expressed as a combination of displacements in the coordinate directions. For example, Figure 13.9 shows that the displacement vector $\vec{v}$ from the origin to the point $(3, 2)$ can be written as a sum of displacement vectors along the x- and y-axes:

$$\vec{v} = 3\vec{i} + 2\vec{j}.$$

This is called *resolving $\vec{v}$ into components*. In general:

> We **resolve** $\vec{v}$ into components by writing $\vec{v}$ in the form
>
> $$\vec{v} = v_1\vec{i} + v_2\vec{j} + v_3\vec{k},$$
>
> where v_1, v_2, v_3 are scalars. We call $v_1\vec{i}$, $v_2\vec{j}$, and $v_3\vec{k}$ the **components** of $\vec{v}$.

An Alternative Notation for Vectors

Many people write a vector in three dimensions as a string of three numbers, that is, as

$$\vec{v} = (v_1, v_2, v_3) \quad \text{instead of} \quad \vec{v} = v_1\vec{i} + v_2\vec{j} + v_3\vec{k}.$$

Since the first notation can be confused with a point and the second cannot, we usually use the second form.

Example 2 Resolve the displacement vector, $\vec{v}$, from the point $P_1 = (2, 4, 10)$ to the point $P_2 = (3, 7, 6)$ into components.

Solution To get from P_1 to P_2, we move 1 unit in the positive x-direction, 3 units in the positive y-direction, and 4 units in the negative z-direction. Hence $\vec{v} = \vec{i} + 3\vec{j} - 4\vec{k}$.

Example 3 Decide whether the vector $\vec{v} = 2\vec{i} + 3\vec{j} + 5\vec{k}$ is parallel to each of the following vectors:

$$\vec{w} = 4\vec{i} + 6\vec{j} + 10\vec{k}, \quad \vec{a} = -\vec{i} - 1.5\vec{j} - 2.5\vec{k}, \quad \vec{b} = 4\vec{i} + 6\vec{j} + 9\vec{k}.$$

Solution Since $\vec{w} = 2\vec{v}$ and $\vec{a} = -0.5\vec{v}$, the vectors $\vec{v}$, $\vec{w}$, and $\vec{a}$ are parallel. However, $\vec{b}$ is not a multiple of $\vec{v}$ (since, for example, $4/2 \neq 9/5$), so $\vec{v}$ and $\vec{b}$ are not parallel.

In general, Figure 13.10 shows us how to express the displacement vector between two points in components:

Components of Displacement Vectors

The displacement vector from the point $P_1 = (x_1, y_1, z_1)$ to the point $P_2 = (x_2, y_2, z_2)$ is given in components by

$$\overrightarrow{P_1P_2} = (x_2 - x_1)\vec{i} + (y_2 - y_1)\vec{j} + (z_2 - z_1)\vec{k}.$$

Position Vectors: Displacement of a Point from the Origin

A displacement vector whose tail is at the origin is called a *position vector*. Thus, any point (x_0, y_0, z_0) in space has associated with it the position vector $\vec{r}_0 = x_0\vec{i} + y_0\vec{j} + z_0\vec{k}$. (See Figure 13.11.) In general, a position vector gives the displacement of a point from the origin.

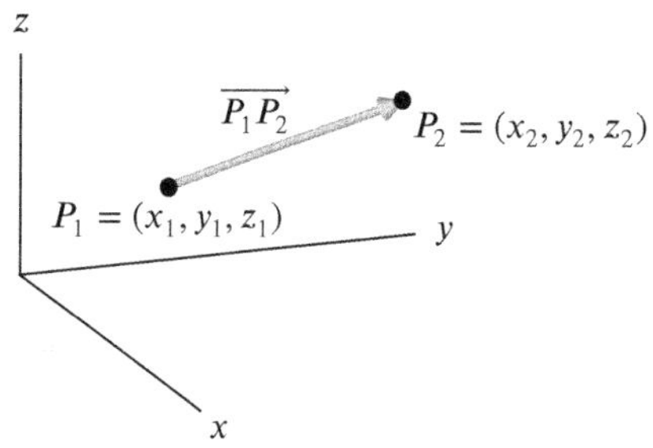

Figure 13.10: The displacement vector $\overrightarrow{P_1P_2} = (x_2 - x_1)\vec{i} + (y_2 - y_1)\vec{j} + (z_2 - z_1)\vec{k}$

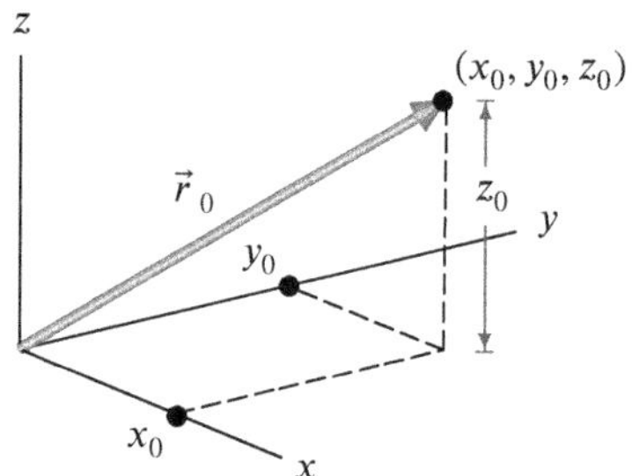

Figure 13.11: The position vector $\vec{r}_0 = x_0\vec{i} + y_0\vec{j} + z_0\vec{k}$

The Components of the Zero Vector

The zero displacement vector has magnitude equal to zero and is written $\vec{0}$. So $\vec{0} = 0\vec{i} + 0\vec{j} + 0\vec{k}$.

The Magnitude of a Vector in Components

For a vector, $\vec{v} = v_1\vec{i} + v_2\vec{j}$, the Pythagorean theorem is used to find its magnitude, $\|\vec{v}\|$. (See Figure 13.12.) The angle θ gives the direction of $\vec{v}$.

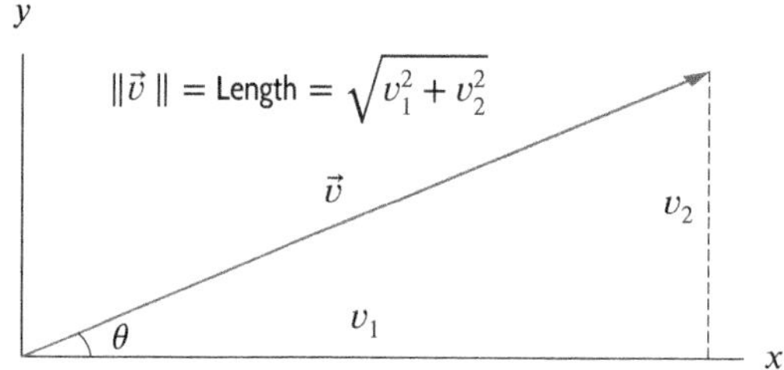

Figure 13.12: Magnitude, $\|\vec{v}\|$, of a 2-dimensional vector, $\vec{v}$

In three dimensions, for a vector $\vec{v} = v_1\vec{i} + v_2\vec{j} + v_3\vec{k}$, we have

$$\text{Magnitude of } \vec{v} = \|\vec{v}\| = \text{Length of the arrow} = \sqrt{v_1^2 + v_2^2 + v_3^2}.$$

For instance, if $\vec{v} = 3\vec{i} - 4\vec{j} + 5\vec{k}$, then $\|\vec{v}\| = \sqrt{3^2 + (-4)^2 + 5^2} = \sqrt{50}$.

Addition and Scalar Multiplication of Vectors in Components

Suppose the vectors $\vec{v}$ and $\vec{w}$ are given in components:

$$\vec{v} = v_1\vec{i} + v_2\vec{j} + v_3\vec{k} \quad \text{and} \quad \vec{w} = w_1\vec{i} + w_2\vec{j} + w_3\vec{k}.$$

Then

$$\boxed{\vec{v} + \vec{w} = (v_1 + w_1)\vec{i} + (v_2 + w_2)\vec{j} + (v_3 + w_3)\vec{k},}$$

and

$$\boxed{\lambda\vec{v} = \lambda v_1\vec{i} + \lambda v_2\vec{j} + \lambda v_3\vec{k}.}$$

Figures 13.13 and 13.14 illustrate these properties in two dimensions. Finally, $\vec{v} - \vec{w} = \vec{v} + (-1)\vec{w}$, so we can write $\vec{v} - \vec{w} = (v_1 - w_1)\vec{i} + (v_2 - w_2)\vec{j} + (v_3 - w_3)\vec{k}$.

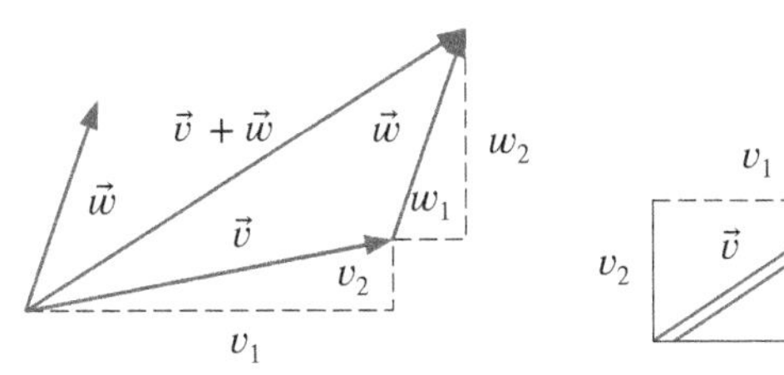

Figure 13.13: Sum $\vec{v} + \vec{w}$ in components

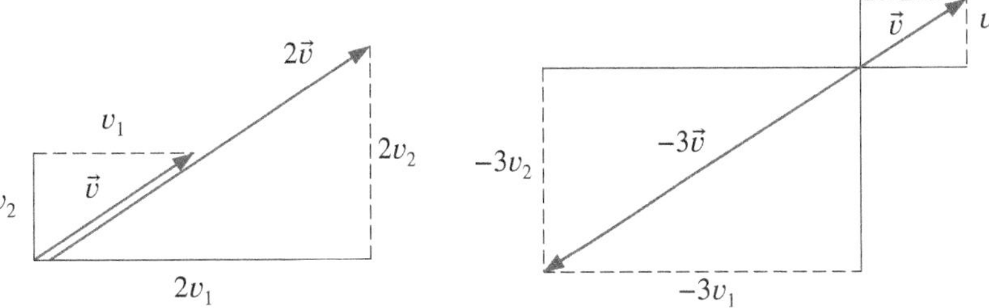

Figure 13.14: Scalar multiples of vectors showing $\vec{v}$, $2\vec{v}$, and $-3\vec{v}$

How to Resolve a Vector into Components

You may wonder how we find the components of a 2-dimensional vector, given its length and direction. Suppose the vector $\vec{v}$ has length v and makes an angle of θ with the x-axis, measured counterclockwise, as in Figure 13.15. If $\vec{v} = v_1\vec{i} + v_2\vec{j}$, Figure 13.15 shows that

$$v_1 = v\cos\theta \quad \text{and} \quad v_2 = v\sin\theta.$$

Thus, we resolve $\vec{v}$ into components by writing

$$\vec{v} = (v\cos\theta)\vec{i} + (v\sin\theta)\vec{j}.$$

Vectors in 3-space are resolved using direction cosines; see Problem 66 (available online).

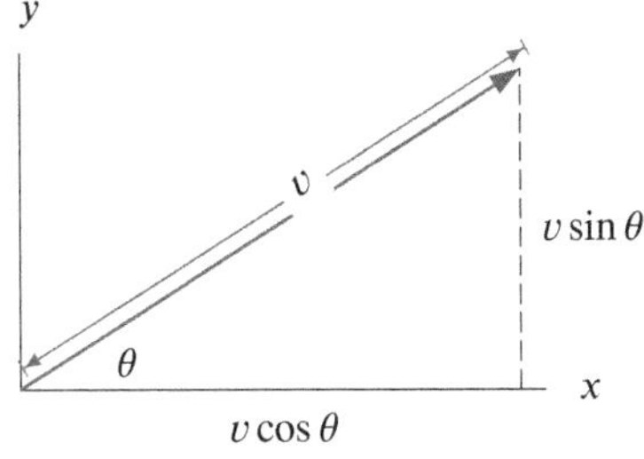

Figure 13.15: Resolving a vector: $\vec{v} = (v\cos\theta)\vec{i} + (v\sin\theta)\vec{j}$

Example 4 Resolve $\vec{v}$ into components if $\|\vec{v}\| = 2$ and $\theta = \pi/6$.

Solution We have $\vec{v} = 2\cos(\pi/6)\vec{i} + 2\sin(\pi/6)\vec{j} = 2\left(\sqrt{3}/2\right)\vec{i} + 2(1/2)\vec{j} = \sqrt{3}\vec{i} + \vec{j}$.

Unit Vectors

A *unit vector* is a vector whose magnitude is 1. The vectors $\vec{i}$, $\vec{j}$, and $\vec{k}$ are unit vectors in the directions of the coordinate axes. It is often helpful to find a unit vector in the same direction as a given vector $\vec{v}$. Suppose that $\|\vec{v}\| = 10$; a unit vector in the same direction as $\vec{v}$ is $\vec{v}/10$. In general, a unit vector in the direction of any nonzero vector $\vec{v}$ is

$$\vec{u} = \frac{\vec{v}}{\|\vec{v}\|}.$$

Example 5 Find a unit vector, $\vec{u}$, in the direction of the vector $\vec{v} = \vec{i} + 3\vec{j}$.

Solution If $\vec{v} = \vec{i} + 3\vec{j}$, then $\|\vec{v}\| = \sqrt{1^2 + 3^2} = \sqrt{10}$. Thus, a unit vector in the same direction is given by

$$\vec{u} = \frac{\vec{v}}{\sqrt{10}} = \frac{1}{\sqrt{10}}(\vec{i} + 3\vec{j}) = \frac{1}{\sqrt{10}}\vec{i} + \frac{3}{\sqrt{10}}\vec{j} \approx 0.32\vec{i} + 0.95\vec{j}.$$

Example 6 Find a unit vector at the point (x, y, z) that points directly outward away from the origin.

Solution The vector from the origin to (x, y, z) is the position vector

$$\vec{r} = x\vec{i} + y\vec{j} + z\vec{k}.$$

Thus, if we put its tail at (x, y, z) it will point away from the origin. Its magnitude is

$$\|\vec{r}\| = \sqrt{x^2 + y^2 + z^2},$$

so a unit vector pointing in the same direction is

$$\frac{\vec{r}}{\|\vec{r}\|} = \frac{x\vec{i} + y\vec{j} + z\vec{k}}{\sqrt{x^2 + y^2 + z^2}} = \frac{x}{\sqrt{x^2 + y^2 + z^2}}\vec{i} + \frac{y}{\sqrt{x^2 + y^2 + z^2}}\vec{j} + \frac{z}{\sqrt{x^2 + y^2 + z^2}}\vec{k}.$$

Exercises and Problems for Section 13.1

Online Resource: Additional Problems for Section 13.1

EXERCISES

In Exercises 1–6, resolve the vectors into components.

1.

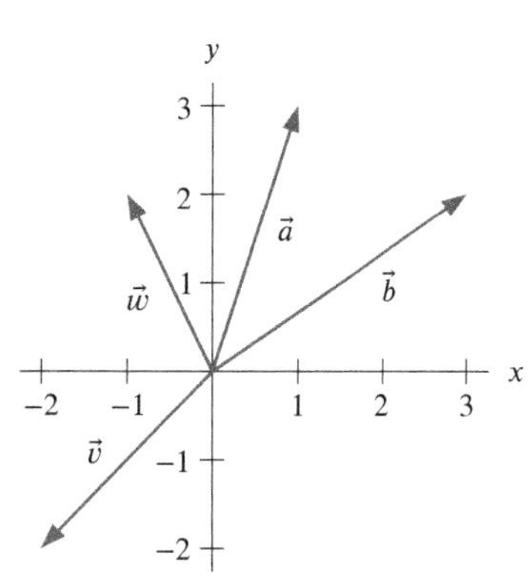

2.

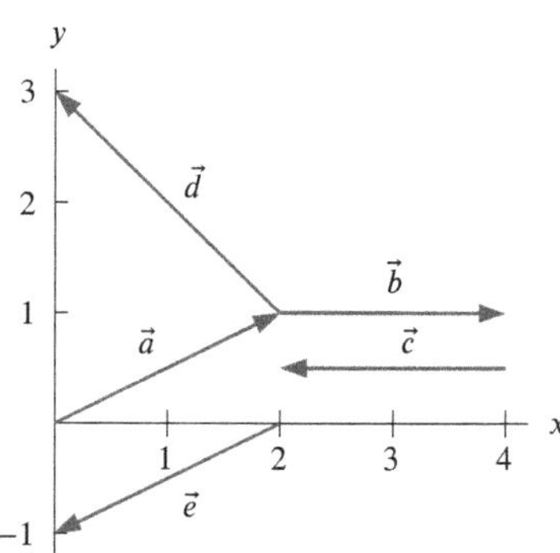

3. A vector starting at the point $Q = (4, 6)$ and ending at the point $P = (1, 2)$.

4. A vector starting at the point $P = (1, 2)$ and ending at the point $Q = (4, 6)$.

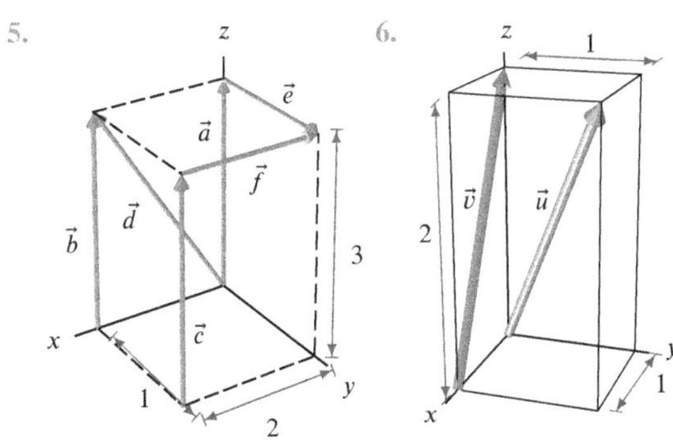

For Exercises 7–14, perform the indicated computation.

7. $(4\vec{i} + 2\vec{j}) - (3\vec{i} - \vec{j})$
8. $(\vec{i} + 2\vec{j}) + (-3)(2\vec{i} + \vec{j})$
9. $-4(\vec{i} - 2\vec{j}) - 0.5(\vec{i} - \vec{k})$
10. $2(0.45\vec{i} - 0.9\vec{j} - 0.01\vec{k}) - 0.5(1.2\vec{i} - 0.1\vec{k})$
11. $(3\vec{i} - 4\vec{j} + 2\vec{k}) - (6\vec{i} + 8\vec{j} - \vec{k})$
12. $(4\vec{i} - 3\vec{j} + 7\vec{k}) - 2(5\vec{i} + \vec{j} - 2\vec{k})$
13. $(0.6\vec{i} + 0.2\vec{j} - \vec{k}) + (0.3\vec{i} + 0.3\vec{k})$
14. $\frac{1}{2}(2\vec{i} - \vec{j} + 3\vec{k}) + 3(\vec{i} - \frac{1}{6}\vec{j} + \frac{1}{2}\vec{k})$

In Exercises 15–19, find the length of the vectors.

15. $\vec{v} = \vec{i} - \vec{j} + 2\vec{k}$
16. $\vec{z} = \vec{i} - 3\vec{j} - \vec{k}$
17. $\vec{v} = \vec{i} - \vec{j} + 3\vec{k}$
18. $\vec{v} = 7.2\vec{i} - 1.5\vec{j} + 2.1\vec{k}$
19. $\vec{v} = 1.2\vec{i} - 3.6\vec{j} + 4.1\vec{k}$

For Exercises 20–25, perform the indicated operations on the following vectors:

$$\vec{a} = 2\vec{j} + \vec{k}, \quad \vec{b} = -3\vec{i} + 5\vec{j} + 4\vec{k}, \quad \vec{c} = \vec{i} + 6\vec{j},$$

$$\vec{x} = -2\vec{i} + 9\vec{j}, \quad \vec{y} = 4\vec{i} - 7\vec{j}, \quad \vec{z} = \vec{i} - 3\vec{j} - \vec{k}.$$

20. $4\vec{z}$
21. $5\vec{a} + 2\vec{b}$
22. $\vec{a} + \vec{z}$
23. $2\vec{c} + \vec{x}$
24. $2\vec{a} + 7\vec{b} - 5\vec{z}$
25. $\|\vec{y} - \vec{x}\|$

26. **(a)** Draw the position vector for $\vec{v} = 5\vec{i} - 7\vec{j}$.
    (b) What is $\|\vec{v}\|$?
    (c) Find the angle between $\vec{v}$ and the positive x-axis.
27. Find the unit vector in the direction of $0.06\vec{i} - 0.08\vec{k}$.
28. Find the unit vector in the opposite direction to $\vec{i} - \vec{j} + \vec{k}$.
29. Find a unit vector in the opposite direction to $2\vec{i} - \vec{j} - \sqrt{11}\vec{k}$.
30. Find a vector with length 2 that points in the same direction as $\vec{i} - \vec{j} + 2\vec{k}$.

PROBLEMS

31. Find the value(s) of a making $\vec{v} = 5a\vec{i} - 3\vec{j}$ parallel to $\vec{w} = a^2\vec{i} + 6\vec{j}$.
32. **(a)** For $a = 1, 2$, and 3, draw position vectors for
    (i) $\vec{v} = a^2\vec{i} + 6\vec{j}$ (ii) $\vec{w} = 5\vec{i} - a^2\vec{j}$
    (b) Explain why there is no value of a that makes $\vec{v}$ and $\vec{w}$ parallel.
33. **(a)** Find a unit vector from the point $P = (1, 2)$ and toward the point $Q = (4, 6)$.
    (b) Find a vector of length 10 pointing in the same direction.
34. If north is the direction of the positive y-axis and east is the direction of the positive x-axis, give the unit vector pointing northwest.
35. Resolve the following vectors into components:
    (a) The vector in 2-space of length 2 pointing up and to the right at an angle of $\pi/4$ with the x-axis.
    (b) The vector in 3-space of length 1 lying in the xz-plane pointing upward at an angle of $\pi/6$ with the positive x-axis.
36. **(a)** From Figure 13.16, read off the coordinates of the five points, A, B, C, D, E, and thus resolve into components the following two vectors: $\vec{u} = (2.5)\overrightarrow{AB} + (-0.8)\overrightarrow{CD}$, $\vec{v} = (2.5)\overrightarrow{BA} - (-0.8)\overrightarrow{CD}$
    (b) What is the relation between $\vec{u}$ and $\vec{v}$? Why was this to be expected?

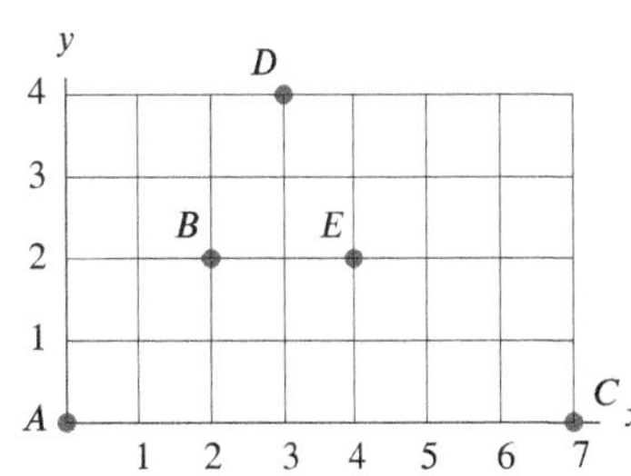

Figure 13.16

37. Find the components of a vector $\vec{p}$ that has the same direction as $\overrightarrow{EA}$ in Figure 13.16 and whose length equals two units.
38. For each of the four statements below, answer the following questions: Does the statement make sense? If yes, is it true for all possible choices of $\vec{a}$ and $\vec{b}$? If no, why not?
    (a) $\vec{a} + \vec{b} = \vec{b} + \vec{a}$
    (b) $\vec{a} + \|\vec{b}\| = \|\vec{a} + \vec{b}\|$
    (c) $\|\vec{b} + \vec{a}\| = \|\vec{a} + \vec{b}\|$
    (d) $\|\vec{a} + \vec{b}\| = \|\vec{a}\| + \|\vec{b}\|$.

39. For each condition, find unit vectors $\vec{a}$ and $\vec{b}$ or explain why no such vectors exist.

(a) $\|\vec{a} + \vec{b}\| = 0$ **(b)** $\|\vec{a} + \vec{b}\| = 1$
(c) $\|\vec{a} + \vec{b}\| = 2$ **(d)** $\|\vec{a} + \vec{b}\| = 3$

40. Two adjacent sides of a regular hexagon are given as the vectors $\vec{u}$ and $\vec{v}$ in Figure 13.17. Label the remaining sides in terms of $\vec{u}$ and $\vec{v}$.

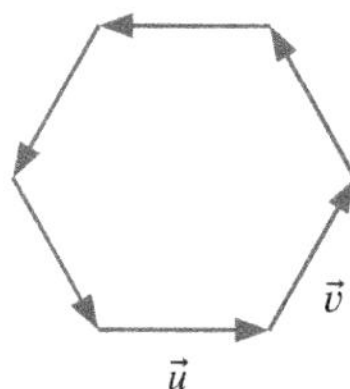

Figure 13.17

41. For what values of t are the following pairs of vectors parallel?

(a) $2\vec{i} + (t^2 + \frac{2}{3}t + 1)\vec{j} + t\vec{k}$, $6\vec{i} + 8\vec{j} + 3\vec{k}$
(b) $t\vec{i} + \vec{j} + (t-1)\vec{k}$, $2\vec{i} - 4\vec{j} + \vec{k}$
(c) $2t\vec{i} + t\vec{j} + t\vec{k}$, $6\vec{i} + 3\vec{j} + 3\vec{k}$.

42. Show that the unit vector $\vec{v} = x\vec{i} + y\vec{j}$ is not parallel to $\vec{w} = y\vec{i} - x\vec{j}$ for any choice of x and y.

43. Find all unit vectors $\vec{v} = x\vec{i} + y\vec{j}$ parallel to $\vec{w} = y\vec{i} + x\vec{j}$.

44. Find all vectors $\vec{v}$ in 2 dimensions having $\|\vec{v}\| = 5$ such that the $\vec{i}$-component of $\vec{v}$ is $3\vec{i}$.

45. **(a)** Find the point on the x-axis closest to the point (a, b, c).
(b) Find a unit vector that points from the point you found in part (a) toward (a, b, c).

46. Figure 13.18 shows a molecule with four atoms at O, A, B and C. Check that every atom in the molecule is 2 units away from every other atom.

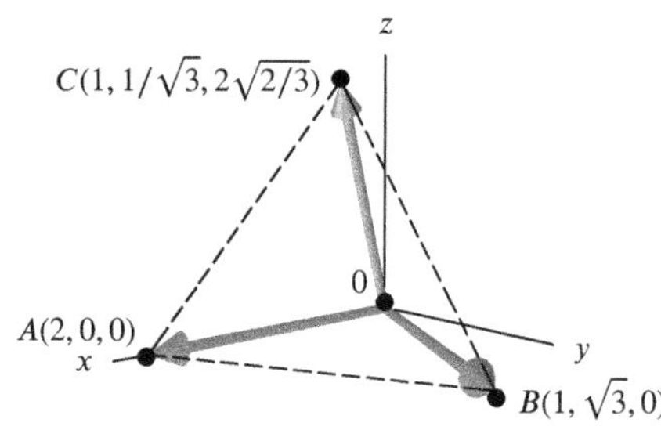

Figure 13.18

Strengthen Your Understanding

In Problems 47–50, explain what is wrong with the statement.

47. If $\|\vec{u}\| = 1$ and $\|\vec{v}\| > 0$, then $\|\vec{u} + \vec{v}\| \geq 1$.

48. The vector $c\vec{u}$ has the same direction as $\vec{u}$.

49. $\|\vec{v} - \vec{u}\|$ is the length of the shorter of the two diagonals of the parallelogram determined by $\vec{u}$ and $\vec{v}$.

50. Given three vectors $\vec{u}$, $\vec{v}$, and $\vec{w}$, if $\vec{u} + \vec{w} = \vec{u}$ then it is possible for $\vec{v} + \vec{w} \neq \vec{v}$.

In Problems 51–53, give an example of:

51. A vector $\vec{v}$ of length 2 with a positive $\vec{k}$-component and lying on a plane parallel to the yz-plane.

52. Two unit vectors $\vec{u}$ and $\vec{v}$ for which $\vec{v} - \vec{u}$ is also a unit vector.

53. Two vectors $\vec{u}$ and $\vec{v}$ that have difference vector $\vec{w} = 2\vec{i} + 3\vec{j}$.

Are the statements in Problems 54–63 true or false? Give reasons for your answer.

54. There is exactly one unit vector parallel to a given nonzero vector $\vec{v}$.

55. The vector $\frac{1}{\sqrt{3}}\vec{i} + \frac{-1}{\sqrt{3}}\vec{j} + \frac{2}{\sqrt{3}}\vec{k}$ is a unit vector.

56. The length of the vector $2\vec{v}$ is twice the length of the vector $\vec{v}$.

57. If $\vec{v}$ and $\vec{w}$ are any two vectors, then $\|\vec{v} + \vec{w}\| = \|\vec{v}\| + \|\vec{w}\|$.

58. If $\vec{v}$ and $\vec{w}$ are any two vectors, then $\|\vec{v} - \vec{w}\| = \|\vec{v}\| - \|\vec{w}\|$.

59. The vectors $2\vec{i} - \vec{j} + \vec{k}$ and $\vec{i} - 2\vec{j} + \vec{k}$ are parallel.

60. The vector $\vec{u} + \vec{v}$ is always larger in magnitude than both $\vec{u}$ and $\vec{v}$.

61. For any scalar c and vector $\vec{v}$ we have $\|c\vec{v}\| = c\|\vec{v}\|$.

62. The displacement vector from $(1, 1, 1)$ to $(1, 2, 3)$ is $-\vec{j} - 2\vec{k}$.

63. The displacement vector from (a, b) to (c, d) is the same as the displacement vector from (c, d) to (a, b).

13.2 VECTORS IN GENERAL

Besides displacement, there are many quantities that have both magnitude and direction and are added and multiplied by scalars in the same way as displacements. Any such quantity is called a *vector* and is represented by an arrow in the same manner we represent displacements. The length of the arrow is the *magnitude* of the vector, and the direction of the arrow is the direction of the vector.

Velocity Versus Speed

The speed of a moving body tells us how fast it is moving, say 80 km/hr. The speed is just a number; it is therefore a scalar. The velocity, on the other hand, tells us both how fast the body is moving and the direction of motion; it is a vector. For instance, if a car is heading northeast at 80 km/hr, then its velocity is a vector of length 80 pointing northeast.

> The **velocity vector** of a moving object is a vector whose magnitude is the speed of the object and whose direction is the direction of its motion.

The velocity vector is the displacement vector if the object moves at constant velocity for one unit of time.

Example 1 A car is traveling north at a speed of 100 km/hr, while a plane above is flying horizontally southwest at a speed of 500 km/hr. Draw the velocity vectors of the car and the plane.

Solution Figure 13.19 shows the velocity vectors. The plane's velocity vector is five times as long as the car's, because its speed is five times as great.

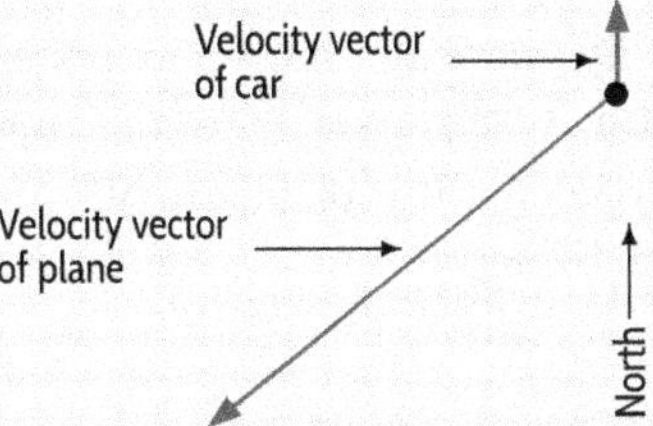

Figure 13.19: Velocity vector of the car is 100 km/hr north and of the plane is 500 km/hr southwest

The next example illustrates that the velocity vectors for two motions add to give the velocity vector for the combined motion, just as displacements do.

Example 2 A riverboat is moving with velocity $\vec{v}$ and a speed of 8 km/hr relative to the water. In addition, the river has a current $\vec{c}$ and a speed of 1 km/hr. (See Figure 13.20.) What is the physical significance of the vector $\vec{v}+\vec{c}$?

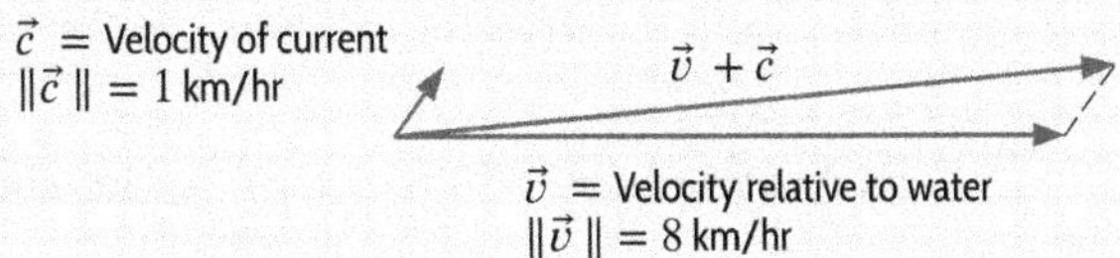

Figure 13.20: Boat's velocity relative to the river bed is the sum $\vec{v}+\vec{c}$

Solution The vector $\vec{v}$ shows how the boat is moving relative to the water, while $\vec{c}$ shows how the water is moving relative to the riverbed. During an hour, imagine that the boat first moves 8 km relative to the water, which remains still; this displacement is represented by $\vec{v}$. Then imagine the water moving 1 km while the boat remains stationary relative to the water; this displacement is represented by $\vec{c}$. The combined displacement is represented by $\vec{v}+\vec{c}$. Thus, the vector $\vec{v}+\vec{c}$ is the velocity of the boat relative to the riverbed.

Note that the effective speed of the boat is not necessarily 9 km/hr unless the boat is moving in the direction of the current. Although we add the velocity vectors, we do not necessarily add their lengths.

Scalar multiplication also makes sense for velocity vectors. For example, if $\vec{v}$ is a velocity vector, then $-2\vec{v}$ represents a velocity of twice the magnitude in the opposite direction.

Example 3 A ball is moving with velocity $\vec{v}$ when it hits a wall at a right angle and bounces straight back, with its speed reduced by 20%. Express its new velocity in terms of the old one.

Solution The new velocity is $-0.8\vec{v}$, where the negative sign expresses the fact that the new velocity is in the direction opposite to the old.

We can represent velocity vectors in components in the same way we did on page 707.

Example 4 Represent the velocity vectors of the car and the plane in Example 1 using components. Take north to be the positive y-axis, east to be the positive x-axis, and upward to be the positive z-axis.

Solution The car is traveling north at 100 km/hr, so the y-component of its velocity is $100\vec{j}$ and the x-component is $0\vec{i}$. Since it is traveling horizontally, the z-component is $0\vec{k}$. So we have

$$\text{Velocity of car} = 0\vec{i} + 100\vec{j} + 0\vec{k} = 100\vec{j}.$$

The plane's velocity vector also has $\vec{k}$ component equal to zero. Since it is traveling southwest, its $\vec{i}$ and $\vec{j}$ components have negative coefficients (north and east are positive). Since the plane is traveling at 500 km/hr, in one hour it is displaced $500/\sqrt{2} \approx 354$ km to the west and 354 km to the south. (See Figure 13.21.) Thus,

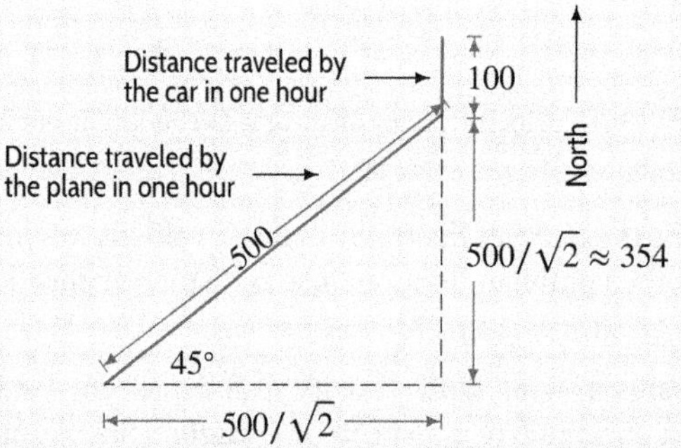

Figure 13.21: Distance traveled by the plane and car in one hour

$$\text{Velocity of plane} = -(500\cos 45°)\vec{i} - (500\sin 45°)\vec{j} \approx -354\vec{i} - 354\vec{j}.$$

Of course, if the car were climbing a hill or if the plane were descending for a landing, then the $\vec{k}$ component would not be zero.

Acceleration

Another example of a vector quantity is acceleration. Acceleration, like velocity, is specified by both a magnitude and a direction — for example, the acceleration due to gravity is 9.81 m/sec^2 vertically downward.

Force

Force is another example of a vector quantity. Suppose you push on an open door. The result depends both on how hard you push and in what direction. Thus, to specify a force we must give its magnitude (or strength) and the direction in which it is acting. For example, the gravitational force exerted on an object by the earth is a vector pointing from the object toward the center of the earth; its magnitude is the strength of the gravitational force.

Example 5 The earth travels around the sun in an ellipse. The gravitational force on the earth and the velocity of the earth are governed by the following laws:

Newton's Law of Gravitation: The gravitational attraction, $\vec{F}$, of a mass m_1 on a mass m_2 at a distance r has magnitude $\|\vec{F}\| = Gm_1m_2/r^2$, where G is a constant, and is directed from m_2 toward m_1.

Kepler's Second Law: The line joining a planet to the sun sweeps out equal areas in equal times.

(a) Sketch vectors representing the gravitational force of the sun on the earth at two different positions in the earth's orbit.

(b) Sketch the velocity vector of the earth at two points in its orbit.

Solution (a) Figure 13.22 shows the earth orbiting the sun. Note that the gravitational force vector always points toward the sun and is larger when the earth is closer to the sun because of the r^2 term in the denominator. (In fact, the real orbit looks much more like a circle than we have shown here.)

(b) The velocity vector points in the direction of motion of the earth. Thus, the velocity vector is tangent to the ellipse. See Figure 13.23. Furthermore, the velocity vector is longer at points of the orbit where the planet is moving quickly, because the magnitude of the velocity vector is the speed. Kepler's Second Law enables us to determine when the earth is moving quickly and when it is moving slowly. Over a fixed period of time, say one month, the line joining the earth to the sun sweeps out a sector having a certain area. Figure 13.23 shows two sectors swept out in two different one-month time-intervals. Kepler's law says that the areas of the two sectors are the same. Thus, the earth must move farther in a month when it is close to the sun than when it is far from the sun. Therefore, the earth moves faster when it is closer to the sun and slower when it is farther away.

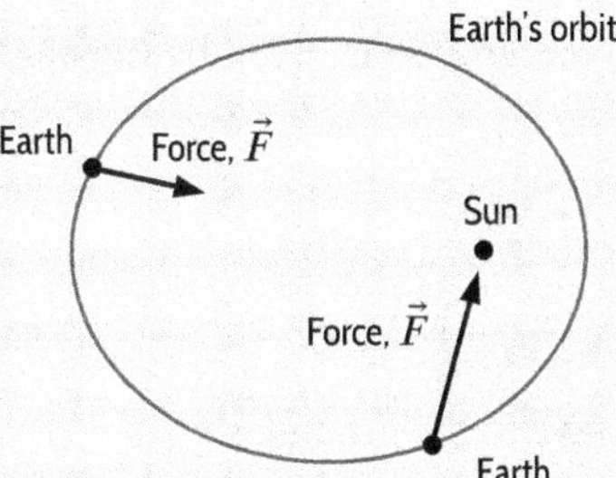

Figure 13.22: Gravitational force, $\vec{F}$, exerted by the sun on the earth: Greater magnitude closer to sun

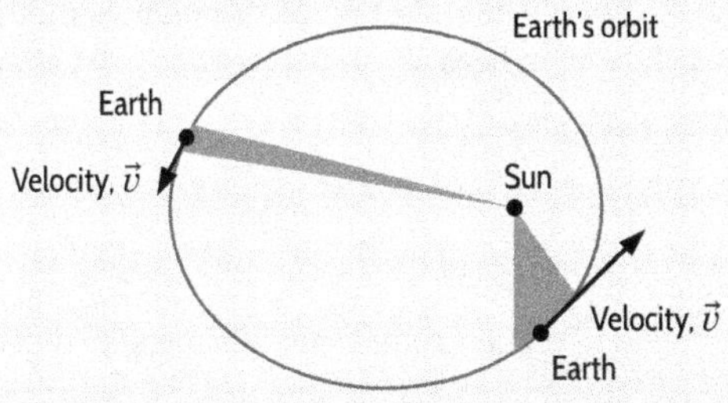

Figure 13.23: The velocity vector, $\vec{v}$, of the earth: Greater magnitude closer to the sun

Properties of Addition and Scalar Multiplication

In general, vectors add, subtract, and are multiplied by scalars in the same way as displacement vectors. Thus, for any vectors $\vec{u}$, $\vec{v}$, and $\vec{w}$ and any scalars α and β, we have the following properties:

Commutativity

1. $\vec{v} + \vec{w} = \vec{w} + \vec{v}$

Associativity

2. $(\vec{u} + \vec{v}) + \vec{w} = \vec{u} + (\vec{v} + \vec{w})$
3. $\alpha(\beta\vec{v}) = (\alpha\beta)\vec{v}$

Distributivity

4. $(\alpha + \beta)\vec{v} = \alpha\vec{v} + \beta\vec{v}$
5. $\alpha(\vec{v} + \vec{w}) = \alpha\vec{v} + \alpha\vec{w}$

Identity

6. $1\vec{v} = \vec{v}$
7. $0\vec{v} = \vec{0}$
8. $\vec{v} + \vec{0} = \vec{v}$
9. $\vec{w} + (-1)\vec{v} = \vec{w} - \vec{v}$

Problems 28–35 at the end of this section ask for a justification of these results in terms of displacement vectors.

Using Components

Example 6 A plane, heading due east at an airspeed of 600 km/hr, experiences a wind of 50 km/hr blowing toward the northeast. Find the plane's direction and ground speed.

Solution We choose a coordinate system with the x-axis pointing east and the y-axis pointing north. See Figure 13.24.

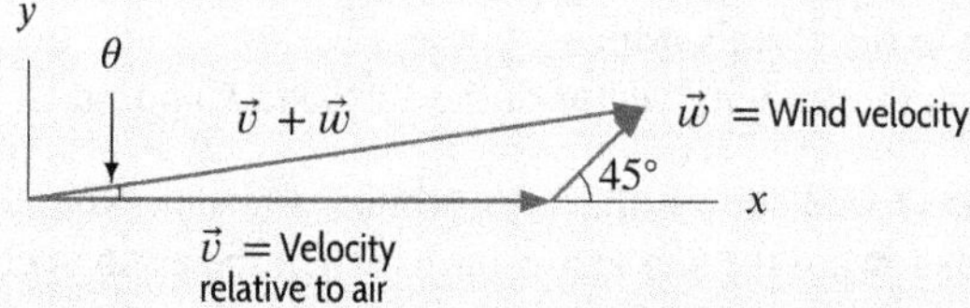

Figure 13.24: Plane's velocity relative to the ground is the sum $\vec{v} + \vec{w}$

The airspeed tells us the speed of the plane relative to still air. Thus, the plane is moving due east with velocity $\vec{v} = 600\vec{i}$ relative to still air. In addition, the air is moving with a velocity $\vec{w}$. Writing $\vec{w}$ in components, we have

$$\vec{w} = (50\cos 45°)\vec{i} + (50\sin 45°)\vec{j} = 35.4\vec{i} + 35.4\vec{j}.$$

The vector $\vec{v} + \vec{w}$ represents the displacement of the plane in one hour relative to the ground. Therefore, $\vec{v} + \vec{w}$ is the velocity of the plane relative to the ground. In components, we have

$$\vec{v} + \vec{w} = 600\vec{i} + \left(35.4\vec{i} + 35.4\vec{j}\right) = 635.4\vec{i} + 35.4\vec{j}.$$

The direction of the plane's motion relative to the ground is given by the angle θ in Figure 13.24, where

$$\tan\theta = \frac{35.4}{635.4}$$

so

$$\theta = \arctan\left(\frac{35.4}{635.4}\right) = 3.2°.$$

The ground speed is the speed of the plane relative to the ground, so

$$\text{Ground speed} = \|\vec{v} + \vec{w}\| = \sqrt{635.4^2 + 35.4^2} = 636.4 \text{ km/hr}.$$

Thus, the speed of the plane relative to the ground has been increased slightly by the wind. (This is as we would expect, as the wind has a positive component in the direction in which the plane is traveling.) The angle θ shows how far the plane is blown off course by the wind.

Vectors in n Dimensions

Using the alternative notation $\vec{v} = (v_1, v_2, v_3)$ for a vector in 3-space, we can define a vector in n dimensions as a string of n numbers. Thus, a vector in n dimensions can be written as

$$\vec{c} = (c_1, c_2, \dots, c_n).$$

Addition and scalar multiplication are defined by the formulas

$$\boxed{\vec{v} + \vec{w} = (v_1, v_2, \dots, v_n) + (w_1, w_2, \dots, w_n) = (v_1 + w_1, v_2 + w_2, \dots, v_n + w_n)}$$

and

$$\boxed{\lambda\vec{v} = \lambda(v_1, v_2, \dots, v_n) = (\lambda v_1, \lambda v_2, \dots, \lambda v_n).}$$

Why Do We Want Vectors in n Dimensions?

Vectors in two and three dimensions can be used to model displacement, velocities, or forces. But what about vectors in n dimensions? There is another interpretation of 3-dimensional vectors (or 3-vectors) that is useful: they can be thought of as listing three different quantities—for example, the displacements parallel to the x-, y-, and z-axes. Similarly, the n-vector

$$\vec{c} = (c_1, c_2, \ldots, c_n)$$

can be thought of as a way of keeping n different quantities organized. For example, a *population* vector $\vec{N}$ shows the number of children and adults in a population:

$$\vec{N} = (\text{Number of children, Number of adults}),$$

or, if we are interested in a more detailed breakdown of ages, we might give the number in each ten-year age bracket in the population (up to age 110) in the form

$$\vec{N} = (N_1, N_2, N_3, N_4, \ldots, N_{10}, N_{11}),$$

where N_1 is the population aged 0–9, and N_2 is the population aged 10–19, and so on.

A *consumption* vector

$$\vec{q} = (q_1, q_2, \ldots, q_n)$$

shows the quantities $q_1, q_2, \ldots, q_n$ consumed of each of n different goods. A *price* vector

$$\vec{p} = (p_1, p_2, \ldots, p_n)$$

contains the prices of n different items.

In 1907, Hermann Minkowski used vectors with four components when he introduced *space-time coordinates*, whereby each event is assigned a vector position $\vec{v}$ with four coordinates, three for its position in space and one for time:

$$\vec{v} = (x, y, z, t).$$

Example 7 Suppose the vector $\vec{I}$ represents the number of copies, in thousands, made by each of four copy centers in the month of December and $\vec{J}$ represents the number of copies made at the same four copy centers during the previous eleven months (the "year-to-date"). If $\vec{I} = (25, 211, 818, 642)$, and $\vec{J} = (331, 3227, 1377, 2570)$, compute $\vec{I} + \vec{J}$. What does this sum represent?

Solution The sum is

$$\vec{I} + \vec{J} = (25 + 331, 211 + 3227, 818 + 1377, 642 + 2570) = (356, 3438, 2195, 3212).$$

Each term in $\vec{I} + \vec{J}$ represents the sum of the number of copies made in December plus those in the previous eleven months, that is, the total number of copies made during the entire year at that particular copy center.

Example 8 The price vector $\vec{p} = (p_1, p_2, p_3)$ represents the prices in dollars of three goods. Write a vector that gives the prices of the same goods in cents.

Solution The prices in cents are $100p_1$, $100p_2$, and $100p_3$ respectively, so the new price vector is

$$(100p_1, 100p_2, 100p_3) = 100\vec{p}.$$

Exercises and Problems for Section 13.2 Online Resource: Additional Problems for Section 13.2

EXERCISES

In Exercises 1–5, say whether the given quantity is a vector or a scalar.

1. The population of the US.
2. The distance from Seattle to St. Louis.
3. The temperature at a point on the earth's surface.
4. The magnetic field at a point on the earth's surface.
5. The populations of each of the 50 states.
6. Give the components of the velocity vector for wind blowing at 10 km/hr toward the southeast. (Assume north is in the positive y-direction.)
7. Give the components of the velocity vector of a boat that is moving at 40 km/hr in a direction 20° south of west. (Assume north is in the positive y-direction.)
8. A car is traveling at a speed of 50 km/hr. The positive y-axis is north and the positive x-axis is east. Resolve the car's velocity vector (in 2-space) into components if the car is traveling in each of the following directions:
   (a) East **(b)** South
   (c) Southeast **(d)** Northwest.
9. Which is traveling faster, a car whose velocity vector is $21\vec{i} + 35\vec{j}$ or a car whose velocity vector is $40\vec{i}$, assuming that the units are the same for both directions?
10. What angle does a force of $\vec{F} = 15\vec{i} + 18\vec{j}$ make with the x-axis?

PROBLEMS

11. The velocity of the current in a river is $\vec{c} = 0.6\vec{i} + 0.8\vec{j}$ km/hr. A boat moves relative to the water with velocity $\vec{v} = 8\vec{i}$ km/hr.
    (a) What is the speed of the boat relative to the riverbed?
    (b) What angle does the velocity of the boat relative to the riverbed make with the vector $\vec{v}$? What does this angle tell us in practical terms?
12. The current in Problem 11 is twice as fast and in the opposite direction. What is the speed of the boat with respect to the riverbed?
13. A boat is heading due east at 25 km/hr (relative to the water). The current is moving toward the southwest at 10 km/hr.
    (a) Give the vector representing the actual movement of the boat.
    (b) How fast is the boat going, relative to the ground?
    (c) By what angle does the current push the boat off of its due east course?
14. A truck is traveling due north at 30 km/hr approaching a crossroad. On a perpendicular road a police car is traveling west toward the intersection at 40 km/hr. Both vehicles will reach the crossroad in exactly one hour. Find the vector currently representing the displacement from the police car to the truck.
15. An airplane heads northeast at an airspeed of 700 km/hr, but there is a wind blowing from the west at 60 km/hr. In what direction does the plane end up flying? What is its speed relative to the ground?
16. Two forces, represented by the vectors $\vec{F}_1 = 8\vec{i} - 6\vec{j}$ and $\vec{F}_2 = 3\vec{i} + 2\vec{j}$, are acting on an object. Give a vector representing the force that must be applied to the object if it is to remain stationary.
17. An airplane is flying at an airspeed of 500 km/hr in a wind blowing at 60 km/hr toward the southeast. In what direction should the plane head to end up going due east? What is the airplane's speed relative to the ground?
18. The current in a river is pushing a boat in direction 25° north of east with a speed of 12 km/hr. The wind is pushing the same boat in a direction 80° south of east with a speed of 7 km/hr. Find the velocity vector of the boat's engine (relative to the water) if the boat actually moves due east at a speed of 40 km/hr relative to the ground.
19. A large ship is being towed by two tugs. The larger tug exerts a force which is 25% greater than the smaller tug and at an angle of 30 degrees north of east. Which direction must the smaller tug pull to ensure that the ship travels due east?
20. An object P is pulled by a force $\vec{F}_1$ of magnitude 15 lb at an angle of 20 degrees north of east. Give the components of a force $\vec{F}_2$ of magnitude 20 lb to ensure that P moves due east.
21. An object is to be moved vertically upward by a crane. As the crane cannot get directly above the object, three ropes are attached to guide the object. One rope is pulled parallel to the ground with a force of 100 newtons in a direction 30° north of east. The second rope is pulled parallel to the ground with a force of 70 newtons in a direction 80° south of east. If the crane is attached to the third rope and can pull with a total force of 3000 newtons, find the force vector for the crane. What is the resulting (total) force on the object? (Assume vector $\vec{i}$ points east, vector $\vec{j}$ points north, and vector $\vec{k}$ points vertically up.)

22. The earth is at the origin, the moon is at the point (384, 0), and a spaceship is at (280, 90), where distance is in thousands of kilometers.

    (a) What is the displacement vector of the moon relative to the earth? Of the spaceship relative to the earth? Of the spaceship relative to the moon?
    (b) How far is the spaceship from the earth? From the moon?
    (c) The gravitational force on the spaceship from the earth is 461 newtons and from the moon is 26 newtons. What is the resulting force?

23. A particle moving with speed v hits a barrier at an angle of 60° and bounces off at an angle of 60° in the opposite direction with speed reduced by 20 percent. See Figure 13.25. Find the velocity vector of the object after impact.

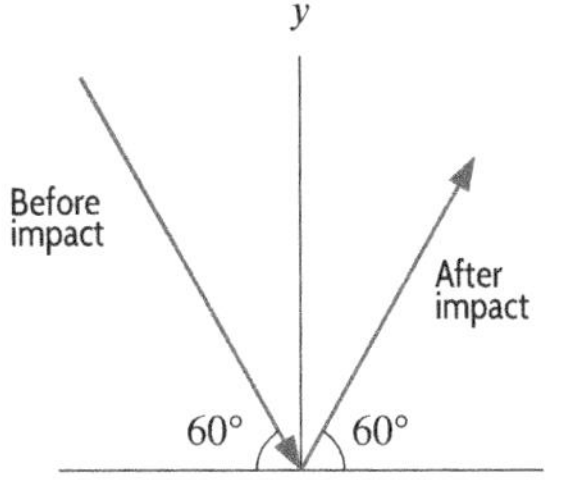

Figure 13.25

24. There are five students in a class. Their scores on the midterm (out of 100) are given by the vector $\vec{v} = (73, 80, 91, 65, 84)$. Their scores on the final (out of 100) are given by $\vec{w} = (82, 79, 88, 70, 92)$. If the final counts twice as much as the midterm, find a vector giving the total scores (as a percentage) of the students.

25. The price vector of beans, rice, and tofu is (1.6, 1.28, 2.60) in dollars per pound. Express it in dollars per ounce.

26. An object is moving counterclockwise at a constant speed around the circle $x^2 + y^2 = 1$, where x and y are measured in meters. It completes one revolution every minute.

    (a) What is its speed?
    (b) What is its velocity vector 30 seconds after it passes the point (1, 0)? Does your answer change if the object is moving clockwise? Explain.

27. An object is attached by a string to a fixed point and rotates 30 times per minute in a horizontal plane. Show that the speed of the object is constant but the velocity is not. What does this imply about the acceleration?

In Problems 28–35, use the geometric definition of addition and scalar multiplication to explain each of the properties.

28. $\vec{w} + \vec{v} = \vec{v} + \vec{w}$

29. $(\alpha + \beta)\vec{v} = \alpha\vec{v} + \beta\vec{v}$

30. $\alpha(\vec{v} + \vec{w}) = \alpha\vec{v} + \alpha\vec{w}$

31. $\alpha(\beta\vec{v}) = (\alpha\beta)\vec{v}$

32. $\vec{v} + \vec{0} = \vec{v}$

33. $1\vec{v} = \vec{v}$

34. $\vec{v} + (-1)\vec{w} = \vec{v} - \vec{w}$

35. $(\vec{u} + \vec{v}) + \vec{w} = \vec{u} + (\vec{v} + \vec{w})$

36. In the game of laser tag, you shoot a harmless laser gun and try to hit a target worn at the waist by other players. Suppose you are standing at the origin of a three-dimensional coordinate system and that the xy-plane is the floor. Suppose that waist-high is 3 feet above floor level and that eye level is 5 feet above the floor. Three of your friends are your opponents. One is standing so that his target is 30 feet along the x-axis, another lying down so that his target is at the point $x = 20$, $y = 15$, and the third lying in ambush so that his target is at a point 8 feet above the point $x = 12$, $y = 30$.

    (a) If you aim with your gun at eye level, find the vector from your gun to each of the three targets.
    (b) If you shoot from waist height, with your gun one foot to the right of the center of your body as you face along the x-axis, find the vector from your gun to each of the three targets.

37. A car drives northeast downhill on a 5° incline at a constant speed of 60 miles per hour. The positive x-axis points east, the y-axis north, and the z-axis up. Resolve the car's velocity into components.

Strengthen Your Understanding

In Problems 38–39, explain what is wrong with the statement.

38. Two vectors in 3-space that have equal $\vec{k}$-components and the same magnitude must be the same vector.

39. A vector $\vec{v}$ in the plane whose $\vec{i}$-component is 0.5 has smaller magnitude than the vector $\vec{w} = 2\vec{i}$.

In Problems 40–41, give an example of:

40. A nonzero vector $\vec{F}$ on the plane that when combined with the force vector $\vec{G} = \vec{i} + \vec{j}$ results in a combined force vector $\vec{R}$ with a positive $\vec{i}$-component and a negative $\vec{j}$-component.

41. Nonzero vectors $\vec{u}$ and $\vec{v}$ such that $\|\vec{u} + \vec{v}\| = \|\vec{u}\| + \|\vec{v}\|$.

In Problems 42–47, is the quantity a vector? Give a reason for your answer.

42. Velocity
43. Speed
44. Force
45. Area
46. Acceleration
47. Volume

13.3 THE DOT PRODUCT

We have seen how to add vectors; can we multiply two vectors together? In the next two sections we will see two different ways of doing so: the *scalar product* (or *dot product*), which produces a scalar, and the *vector product* (or *cross product*), which produces a vector.

Definition of the Dot Product

The dot product links geometry and algebra. We already know how to calculate the length of a vector from its components; the dot product gives us a way of computing the angle between two vectors. For any two vectors $\vec{v} = v_1\vec{i} + v_2\vec{j} + v_3\vec{k}$ and $\vec{w} = w_1\vec{i} + w_2\vec{j} + w_3\vec{k}$, shown in Figure 13.26, we define a scalar as follows:

> The following two definitions of the **dot product**, or **scalar product**, $\vec{v} \cdot \vec{w}$, are equivalent:
> - **Geometric definition**
> $\vec{v} \cdot \vec{w} = \|\vec{v}\|\|\vec{w}\|\cos\theta$ where θ is the angle between $\vec{v}$ and $\vec{w}$ and $0 \le \theta \le \pi$.
> - **Algebraic definition**
> $\vec{v} \cdot \vec{w} = v_1w_1 + v_2w_2 + v_3w_3$.
>
> Notice that the dot product of two vectors is a *number*, not a vector.

Why don't we give just one definition of $\vec{v} \cdot \vec{w}$? The reason is that both definitions are equally important; the geometric definition gives us a picture of what the dot product means and the algebraic definition gives us a way of calculating it.

How do we know the two definitions are equivalent—that is, they really do define the same thing? First, we observe that the two definitions give the same result in a particular example. Then we show why they are equivalent in general.

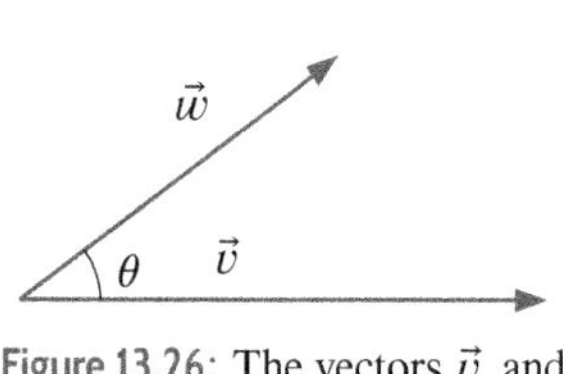

Figure 13.26: The vectors $\vec{v}$ and $\vec{w}$

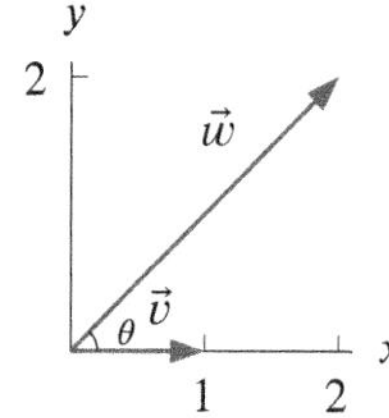

Figure 13.27: Calculating the dot product of the vectors $v = \vec{i}$ and $\vec{w} = 2\vec{i} + 2\vec{j}$ geometrically and algebraically gives the same result

Example 1 Suppose $\vec{v} = \vec{i}$ and $\vec{w} = 2\vec{i} + 2\vec{j}$. Compute $\vec{v} \cdot \vec{w}$ both geometrically and algebraically.

Solution To use the geometric definition, see Figure 13.27. The angle between the vectors is $\pi/4$, or $45°$, and the lengths of the vectors are given by

$$\|\vec{v}\| = 1 \quad\text{and}\quad \|\vec{w}\| = 2\sqrt{2}.$$

Thus,

$$\vec{v} \cdot \vec{w} = \|\vec{v}\|\|\vec{w}\|\cos\theta = 1 \cdot 2\sqrt{2}\cos\left(\frac{\pi}{4}\right) = 2.$$

Using the algebraic definition, we get the same result:

$$\vec{v} \cdot \vec{w} = 1 \cdot 2 + 0 \cdot 2 = 2.$$

Why the Two Definitions of the Dot Product Give the Same Result

In the previous example, the two definitions give the same value for the dot product. To show that the geometric and algebraic definitions of the dot product always give the same result, we must show that, for any vectors $\vec{v} = v_1\vec{i} + v_2\vec{j} + v_3\vec{k}$ and $\vec{w} = w_1\vec{i} + w_2\vec{j} + w_3\vec{k}$ with an angle θ between them:

$$\|\vec{v}\|\|\vec{w}\|\cos\theta = v_1w_1 + v_2w_2 + v_3w_3.$$

One method follows; a method that does not use trigonometry is given in Problem 105 (available online).

Using the Law of Cosines. Suppose that $0 < \theta < \pi$, so that the vectors $\vec{v}$ and $\vec{w}$ form a triangle. (See Figure 13.28.) By the Law of Cosines, we have

$$\|\vec{v} - \vec{w}\|^2 = \|\vec{v}\|^2 + \|\vec{w}\|^2 - 2\|\vec{v}\|\|\vec{w}\|\cos\theta.$$

This result is also true for $\theta = 0$ and $\theta = \pi$. We calculate the lengths using components:

$$\begin{aligned}
\|\vec{v}\|^2 &= v_1^2 + v_2^2 + v_3^2 \\
\|\vec{w}\|^2 &= w_1^2 + w_2^2 + w_3^2 \\
\|\vec{v} - \vec{w}\|^2 &= (v_1 - w_1)^2 + (v_2 - w_2)^2 + (v_3 - w_3)^2 \\
&= v_1^2 - 2v_1w_1 + w_1^2 + v_2^2 - 2v_2w_2 + w_2^2 + v_3^2 - 2v_3w_3 + w_3^2.
\end{aligned}$$

Substituting into the Law of Cosines and canceling, we see that

$$-2v_1w_1 - 2v_2w_2 - 2v_3w_3 = -2\|\vec{v}\|\|\vec{w}\|\cos\theta.$$

Therefore we have the result we wanted, namely that:

$$v_1w_1 + v_2w_2 + v_3w_3 = \|\vec{v}\|\|\vec{w}\|\cos\theta.$$

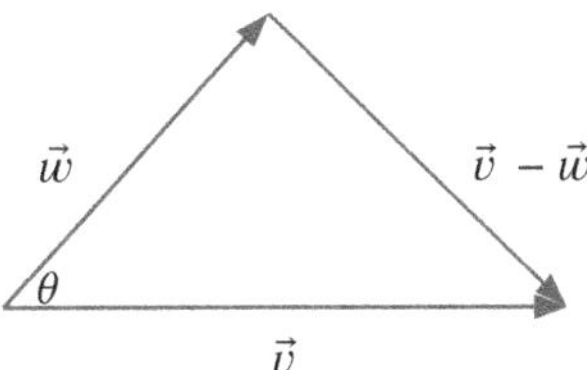

Figure 13.28: Triangle used in the justification of $\|\vec{v}\|\|\vec{w}\|\cos\theta = v_1w_1 + v_2w_2 + v_3w_3$

Properties of the Dot Product

The following properties of the dot product can be justified using the algebraic definition; see Problem 101. For a geometric interpretation of Property 3, see Problem 103 (both available online).

Properties of the Dot Product. For any vectors $\vec{u}$, $\vec{v}$, and $\vec{w}$ and any scalar λ,

1. $\vec{v} \cdot \vec{w} = \vec{w} \cdot \vec{v}$
2. $\vec{v} \cdot (\lambda\vec{w}) = \lambda(\vec{v} \cdot \vec{w}) = (\lambda\vec{v}) \cdot \vec{w}$
3. $(\vec{v} + \vec{w}) \cdot \vec{u} = \vec{v} \cdot \vec{u} + \vec{w} \cdot \vec{u}$

Perpendicularity, Magnitude, and Dot Products

Two vectors are perpendicular if the angle between them is $\pi/2$ or $90°$. Since $\cos(\pi/2) = 0$, if $\vec{v}$ and $\vec{w}$ are perpendicular, then $\vec{v} \cdot \vec{w} = 0$. Conversely, provided that $\vec{v} \cdot \vec{w} = 0$, then $\cos\theta = 0$, so $\theta = \pi/2$ and the vectors are perpendicular. Thus, we have the following result:

> Two nonzero vectors $\vec{v}$ and $\vec{w}$ are **perpendicular**, or **orthogonal**, if and only if
> $$\vec{v} \cdot \vec{w} = 0.$$

For example: $\vec{i} \cdot \vec{j} = 0, \vec{j} \cdot \vec{k} = 0, \vec{i} \cdot \vec{k} = 0$.

If we take the dot product of a vector with itself, then $\theta = 0$ and $\cos\theta = 1$. For any vector $\vec{v}$:

> Magnitude and dot product are related as follows:
> $$\vec{v} \cdot \vec{v} = \|\vec{v}\|^2.$$

For example: $\vec{i} \cdot \vec{i} = 1, \vec{j} \cdot \vec{j} = 1, \vec{k} \cdot \vec{k} = 1$.

Using the Dot Product

Depending on the situation, one definition of the dot product may be more convenient to use than the other. In Example 2, the geometric definition is the only one that can be used because we are not given components. In Example 3, the algebraic definition is used.

Example 2 Suppose the vector $\vec{b}$ is fixed and has length 2; the vector $\vec{a}$ is free to rotate and has length 3. What are the maximum and minimum values of the dot product $\vec{a} \cdot \vec{b}$ as the vector $\vec{a}$ rotates through all possible positions in the plane? What positions of $\vec{a}$ and $\vec{b}$ lead to these values?

Solution The geometric definition gives $\vec{a} \cdot \vec{b} = \|\vec{a}\|\|\vec{b}\|\cos\theta = 3 \cdot 2\cos\theta = 6\cos\theta$. Thus, the maximum value of $\vec{a} \cdot \vec{b}$ is 6, and it occurs when $\cos\theta = 1$ so $\theta = 0$, that is, when $\vec{a}$ and $\vec{b}$ point in the same direction. The minimum value of $\vec{a} \cdot \vec{b}$ is -6, and it occurs when $\cos\theta = -1$ so $\theta = \pi$, that is, when $\vec{a}$ and $\vec{b}$ point in opposite directions. (See Figure 13.29.)

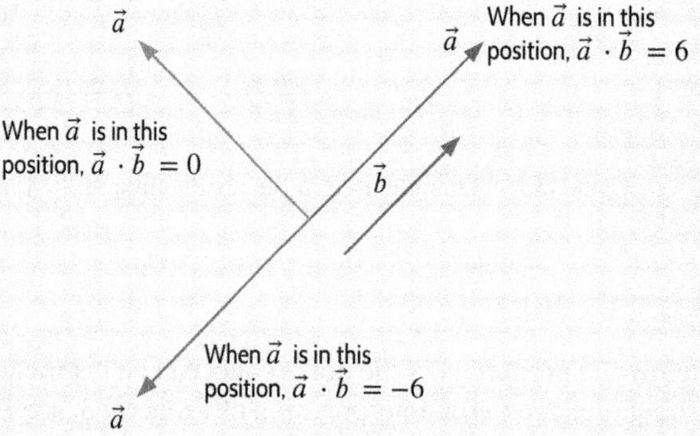

Figure 13.29: Maximum and minimum values of $\vec{a} \cdot \vec{b}$ obtained from a fixed vector $\vec{b}$ of length 2 and rotating vector $\vec{a}$ of length 3

Example 3 Which pairs from the following list of 3-dimensional vectors are perpendicular to one another?

$$\vec{u} = \vec{i} + \sqrt{3}\vec{k}, \quad \vec{v} = \vec{i} + \sqrt{3}\vec{j}, \quad \vec{w} = \sqrt{3}\vec{i} + \vec{j} - \vec{k}.$$

Solution The geometric definition tells us that two vectors are perpendicular if and only if their dot product is zero. Since the vectors are given in components, we calculate dot products using the algebraic definition:

$$\vec{v}\cdot\vec{u} = (\vec{i} + \sqrt{3}\vec{j} + 0\vec{k})\cdot(\vec{i} + 0\vec{j} + \sqrt{3}\vec{k}) = 1\cdot 1 + \sqrt{3}\cdot 0 + 0\cdot\sqrt{3} = 1,$$
$$\vec{v}\cdot\vec{w} = (\vec{i} + \sqrt{3}\vec{j} + 0\vec{k})\cdot(\sqrt{3}\vec{i} + \vec{j} - \vec{k}) = 1\cdot\sqrt{3} + \sqrt{3}\cdot 1 + 0(-1) = 2\sqrt{3},$$
$$\vec{w}\cdot\vec{u} = (\sqrt{3}\vec{i} + \vec{j} - \vec{k})\cdot(\vec{i} + 0\vec{j} + \sqrt{3}\vec{k}) = \sqrt{3}\cdot 1 + 1\cdot 0 + (-1)\cdot\sqrt{3} = 0.$$

So the only two vectors that are perpendicular are $\vec{w}$ and $\vec{u}$.

Example 4 Compute the angle between the vectors $\vec{v}$ and $\vec{w}$ from Example 3.

Solution We know that $\vec{v}\cdot\vec{w} = \|\vec{v}\|\|\vec{w}\|\cos\theta$, so $\cos\theta = \dfrac{\vec{v}\cdot\vec{w}}{\|\vec{v}\|\|\vec{w}\|}$. From Example 3, we know that $\vec{v}\cdot\vec{w} = 2\sqrt{3}$. This gives:

$$\cos\theta = \frac{2\sqrt{3}}{\|\vec{v}\|\|\vec{w}\|} = \frac{2\sqrt{3}}{\sqrt{1^2 + \left(\sqrt{3}\right)^2 + 0^2}\sqrt{\left(\sqrt{3}\right)^2 + 1^2 + (-1)^2}} = \frac{\sqrt{3}}{\sqrt{5}}$$
$$\text{so}\quad \theta = \arccos\left(\frac{\sqrt{3}}{\sqrt{5}}\right) = 39.2315^\circ.$$

Normal Vectors and the Equation of a Plane

In Section 12.4 we wrote the equation of a plane given its x-slope, y-slope and z-intercept. Now we write the equation of a plane using a vector $\vec{n}$ and a point P_0. The key idea is that all the displacement vectors from P_0 that are perpendicular to $\vec{n}$ form a plane. To picture this, imagine a pencil balanced on a table, with other pencils fanned out on the table in different directions. The upright pencil is $\vec{n}$, its base is P_0, the other pencils are perpendicular displacement vectors, and the table is the plane.

More formally, a *normal vector* to a plane is a vector that is perpendicular to the plane, that is, it is perpendicular to every displacement vector between any two points in the plane. Let $\vec{n} = a\vec{i} + b\vec{j} + c\vec{k}$ be a normal vector to the plane, let $P_0 = (x_0, y_0, z_0)$ be a fixed point in the plane, and let $P = (x, y, z)$ be any other point in the plane. Then $\overrightarrow{P_0P} = (x - x_0)\vec{i} + (y - y_0)\vec{j} + (z - z_0)\vec{k}$ is a vector whose head and tail both lie in the plane. (See Figure 13.30.) Thus, the vectors $\vec{n}$ and $\overrightarrow{P_0P}$ are perpendicular, so $\vec{n}\cdot\overrightarrow{P_0P} = 0$. The algebraic definition of the dot product gives $\vec{n}\cdot\overrightarrow{P_0P} = a(x - x_0) + b(y - y_0) + c(z - z_0)$, so we obtain the following result:

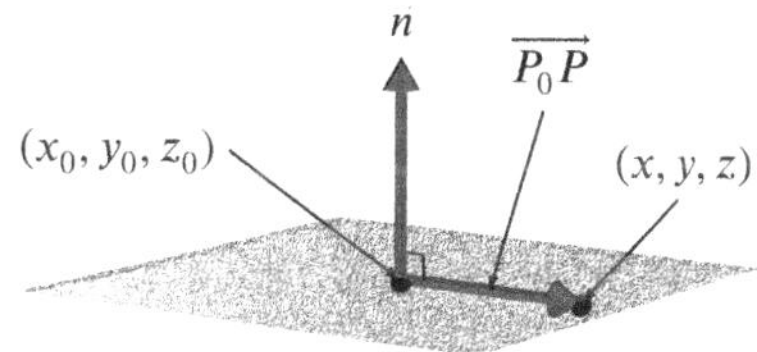

Figure 13.30: Plane with normal $\vec{n}$ and containing a fixed point (x_0, y_0, z_0)

The **equation of the plane** with normal vector $\vec{n} = a\vec{i} + b\vec{j} + c\vec{k}$ and containing the point $P_0 = (x_0, y_0, z_0)$ is

$$a(x - x_0) + b(y - y_0) + c(z - z_0) = 0.$$

Letting $d = ax_0 + by_0 + cz_0$ (a constant), we can write the equation of the plane in the form

$$ax + by + cz = d.$$

Example 5 (a) Find the equation of the plane perpendicular to $\vec{n} = -\vec{i} + 3\vec{j} + 2\vec{k}$ and passing through the point $(1, 0, 4)$.

(b) Find a vector parallel to the plane.

Solution (a) The equation of the plane is

$$-(x - 1) + 3(y - 0) + 2(z - 4) = 0,$$

which can be written as

$$-x + 3y + 2z = 7.$$

(b) Any vector $\vec{v}$ that is perpendicular to n is also parallel to the plane, so we look for any vector satisfying $\vec{v} \cdot \vec{n} = 0$; for example, $\vec{v} = 3\vec{i} + \vec{j}$. There are many other possible vectors.

Example 6 Find a normal vector to the plane with equation (a) $x - y + 2z = 5$ (b) $z = 0.5x + 1.2y$.

Solution (a) Since the coefficients of $\vec{i}$, $\vec{j}$, and $\vec{k}$ in a normal vector are the coefficients of x, y, and z in the equation of the plane, a normal vector is $\vec{n} = \vec{i} - \vec{j} + 2\vec{k}$.

(b) Before we can find a normal vector, we rewrite the equation of the plane in the form

$$0.5x + 1.2y - z = 0.$$

Thus, a normal vector is $\vec{n} = 0.5\vec{i} + 1.2\vec{j} - \vec{k}$.

The Dot Product in n Dimensions

The algebraic definition of the dot product can be extended to vectors in higher dimensions.

If $\vec{u} = (u_1, \ldots, u_n)$ and $\vec{v} = (v_1, \ldots, v_n)$ then the dot product of $\vec{u}$ and $\vec{v}$ is the **scalar**

$$\vec{u} \cdot \vec{v} = u_1v_1 + \cdots + u_nv_n.$$

Example 7 A video store sells videos, tapes, CDs, and computer games. We define the quantity vector $\vec{q} = (q_1, q_2, q_3, q_4)$, where q_1, q_2, q_3, q_4 denote the quantities sold of each of the items, and the price vector $\vec{p} = (p_1, p_2, p_3, p_4)$, where p_1, p_2, p_3, p_4 denote the price per unit of each item. What does the dot product $\vec{p} \cdot \vec{q}$ represent?

Solution The dot product is $\vec{p} \cdot \vec{q} = p_1q_1 + p_2q_2 + p_3q_3 + p_4q_4$. The quantity p_1q_1 represents the revenue received by the store for the videos, p_2q_2 represents the revenue for the tapes, and so on. The dot product represents the total revenue received by the store for the sale of these four items.

Resolving a Vector into Components: Projections

In Section 13.1, we resolved a vector into components parallel to the axes. Now we see how to resolve a vector, $\vec{v}$, into components, called $\vec{v}_{\text{parallel}}$ and $\vec{v}_{\text{perp}}$, which are parallel and perpendicular, respectively, to a given nonzero vector, $\vec{u}$. (See Figure 13.31.)

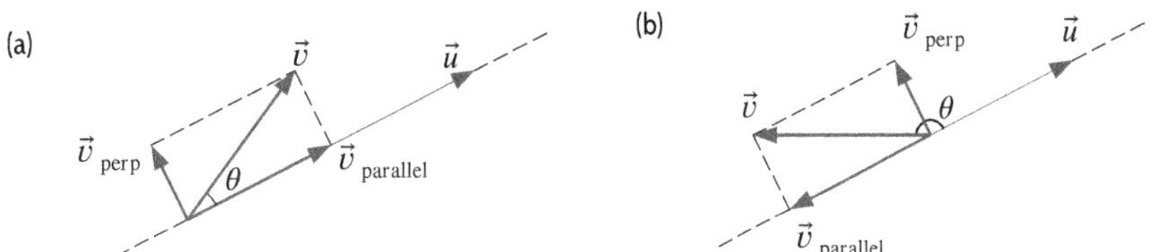

Figure 13.31: Resolving $\vec{v}$ into components parallel and perpendicular to $\vec{u}$
(a) $0 < \theta < \pi/2$ (b) $\pi/2 < \theta < \pi$

The projection of $\vec{v}$ on $\vec{u}$, written $\vec{v}_{\text{parallel}}$, measures (in some sense) how much the vector $\vec{v}$ is aligned with the vector $\vec{u}$. The length of $\vec{v}_{\text{parallel}}$ is the length of the shadow cast by $\vec{v}$ on a line in the direction of $\vec{u}$.

To compute $\vec{v}_{\text{parallel}}$, we assume $\vec{u}$ is a unit vector. (If not, create one by dividing by its length.) Then Figure 13.31(a) shows that, if $0 \le \theta \le \pi/2$:

$$\|\vec{v}_{\text{parallel}}\| = \|\vec{v}\| \cos\theta = \vec{v} \cdot \vec{u} \qquad (\text{since } \|\vec{u}\| = 1).$$

Now $\vec{v}_{\text{parallel}}$ is a scalar multiple of $\vec{u}$, and since $\vec{u}$ is a unit vector,

$$\vec{v}_{\text{parallel}} = (\|\vec{v}\| \cos\theta)\vec{u} = (\vec{v} \cdot \vec{u})\vec{u}.$$

A similar argument shows that if $\pi/2 < \theta \le \pi$, as in Figure 13.31(b), this formula for $\vec{v}_{\text{parallel}}$ still holds. The vector $\vec{v}_{\text{perp}}$ is specified by

$$\vec{v}_{\text{perp}} = \vec{v} - \vec{v}_{\text{parallel}}.$$

Thus, we have the following results:

Projection of $\vec{v}$ on the Line in the Direction of the Unit Vector $\vec{u}$

If $\vec{v}_{\text{parallel}}$ and $\vec{v}_{\text{perp}}$ are components of $\vec{v}$ that are parallel and perpendicular, respectively, to $\vec{u}$, then

$$\text{Projection of } \vec{v} \text{ onto } \vec{u} = \vec{v}_{\text{parallel}} = (\vec{v} \cdot \vec{u})\vec{u} \qquad \text{provided } \|\vec{u}\| = 1$$

$$\text{and} \quad \vec{v} = \vec{v}_{\text{parallel}} + \vec{v}_{\text{perp}} \quad \text{so} \quad \vec{v}_{\text{perp}} = \vec{v} - \vec{v}_{\text{parallel}}.$$

Example 8 Figure 13.32 shows the force the wind exerts on the sail of a sailboat. Find the component of the force in the direction in which the sailboat is traveling.

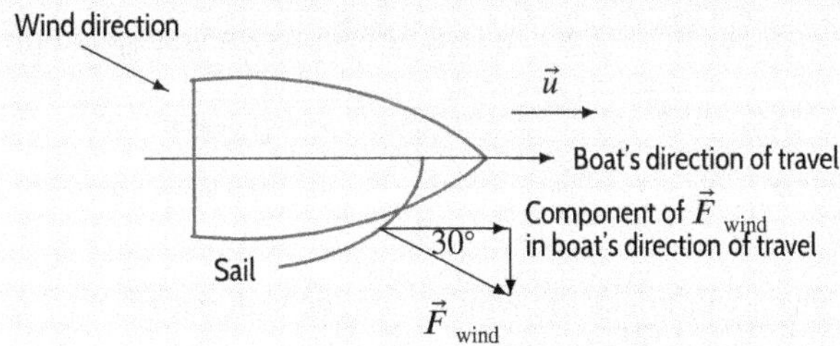

Figure 13.32: Wind moving a sailboat

Solution Let $\vec{u}$ be a unit vector in the direction of travel. The force of the wind on the sail makes an angle of 30° with $\vec{u}$. Thus, the component of this force in the direction of $\vec{u}$ is

$$\vec{F}_{\text{parallel}} = (\vec{F} \cdot \vec{u})\vec{u} = \|\vec{F}\|(\cos 30°)\vec{u} = 0.87\|\vec{F}\|\vec{u}.$$

Thus, the boat is being pushed forward with about 87% of the total force due to the wind. (In fact, the interaction of wind and sail is much more complex than this model suggests.)

A Physical Interpretation of the Dot Product: Work

In physics, the word "work" has a different meaning from its everyday meaning. In physics, when a force of magnitude F acts on an object through a distance d, we say the *work*, W, done by the force is

$$W = Fd,$$

provided the force and the displacement are in the same direction. For example, if a 1 kg body falls 10 meters under the force of gravity, which is 9.8 newtons, then the work done by gravity is

$$W = (9.8 \text{ newtons}) \cdot (10 \text{ meters}) = 98 \text{ joules}.$$

What if the force and the displacement are not in the same direction? Suppose a force $\vec{F}$ acts on an object as it moves along a displacement vector $\vec{d}$. Let θ be the angle between $\vec{F}$ and $\vec{d}$. First, we assume $0 \leq \theta \leq \pi/2$. Figure 13.33 shows how we can resolve $\vec{F}$ into components that are parallel and perpendicular to $\vec{d}$:

$$\vec{F} = \vec{F}_{\text{parallel}} + \vec{F}_{\text{perp}}.$$

Then the work done by $\vec{F}$ is defined to be

$$W = \|\vec{F}_{\text{parallel}}\| \, \|\vec{d}\|.$$

We see from Figure 13.33 that $\vec{F}_{\text{parallel}}$ has magnitude $\|\vec{F}\|\cos\theta$. So the work is given by the dot product:

$$W = (\|\vec{F}\|\cos\theta)\|\vec{d}\| = \|\vec{F}\|\|\vec{d}\|\cos\theta = \vec{F} \cdot \vec{d}.$$

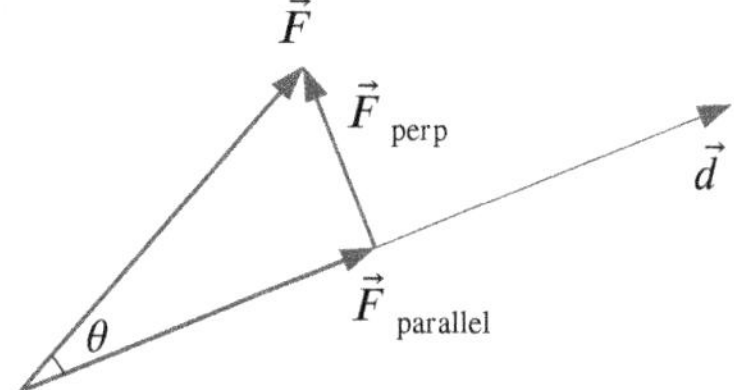

Figure 13.33: Resolving the force $\vec{F}$ into two forces, one parallel to $\vec{d}$, one perpendicular to $\vec{d}$

The formula $W = \vec{F} \cdot \vec{d}$ holds when $\pi/2 < \theta \leq \pi$ also. In that case, the work done by the force is negative and the object is moving against the force. Thus, we have the following definition:

> The **work**, W, done by a force $\vec{F}$ acting on an object through a displacement $\vec{d}$ is given by
>
> $$W = \vec{F} \cdot \vec{d}.$$

Example 9 How much work does the wind do on the sailboat from Example 8 if the boat moves 20 m and the wind's force is 120 newtons?

Solution From Example 8, we know that the force of the wind $\vec{F}$ makes a 30° angle with the boat's displacement $\vec{d}$. Since $\|\vec{F}\| = 120$ and $\|\vec{d}\| = 20$, the work done by the wind on the boat is

$$W = \vec{F} \cdot \vec{d} = \|\vec{F}\|\|\vec{d}\| \cos 30° = 2078.461 \text{ joules}.$$

Notice that if the vectors $\vec{F}$ and $\vec{d}$ are parallel and in the same direction, with magnitudes F and d, then $\cos\theta = \cos 0 = 1$, so $W = \|\vec{F}\|\|\vec{d}\| = Fd$, which is the original definition. When the vectors are perpendicular, $\cos\theta = \cos(\pi/2) = 0$, so $W = 0$ and no work is done in the technical definition of the word. For example, if you carry a heavy box across the room at the same horizontal height, no work is done by gravity because the force of gravity is vertical but the motion is horizontal.

Exercises and Problems for Section 13.3 Online Resource: Additional Problems for Section 13.3

EXERCISES

In Exercises 1–4, evaluate the dot product.

1. $(3\vec{i} + 2\vec{j} - 5\vec{k}) \cdot (\vec{i} - 2\vec{j} - 3\vec{k})$
2. $(\vec{i} + \vec{j} + \vec{k}) \cdot (4\vec{i} + 5\vec{j} + 6\vec{k})$
3. $(3\vec{i} - 2\vec{j} - 4\vec{k}) \cdot (3\vec{i} - 2\vec{j} - 4\vec{k})$
4. $(2i + 5\vec{k}) \cdot 10\vec{j}$

In Exercises 5–6, evaluate $\vec{u} \cdot \vec{w}$.

5. $\|\vec{u}\| = 3$, $\|\vec{w}\| = 5$; the angle between $\vec{u}$ and $\vec{w}$ is 45°.
6. $\|\vec{u}\| = 10$, $\|\vec{w}\| = 20$; the angle between $\vec{u}$ and $\vec{w}$ is 120°.

For Exercises 7–15, perform the following operations on the given 3-dimensional vectors.

$$\vec{a} = 2\vec{j} + \vec{k} \qquad \vec{b} = -3\vec{i} + 5\vec{j} + 4\vec{k} \qquad \vec{c} = \vec{i} + 6\vec{j}$$

$$\vec{y} = 4\vec{i} - 7\vec{j} \qquad \vec{z} = \vec{i} - 3\vec{j} - \vec{k}$$

7. $\vec{a} \cdot \vec{y}$
8. $\vec{c} \cdot \vec{y}$
9. $\vec{a} \cdot \vec{b}$
10. $\vec{a} \cdot \vec{z}$
11. $\vec{c} \cdot \vec{a} + \vec{a} \cdot \vec{y}$
12. $\vec{a} \cdot (\vec{c} + \vec{y})$
13. $(\vec{a} \cdot \vec{b})\vec{a}$
14. $(\vec{a} \cdot \vec{y})(\vec{c} \cdot \vec{z})$
15. $((\vec{c} \cdot \vec{c})\vec{a}) \cdot \vec{a}$

In Exercises 16–20, find a normal vector to the plane.

16. $2x + y - z = 5$
17. $2(x - z) = 3(x + y)$
18. $1.5x + 3.2y + z = 0$
19. $z = 3x + 4y - 7$
20. $\pi(x - 1) = (1 - \pi)(y - z) + \pi$

In Exercises 21–27, find an equation of a plane that satisfies the given conditions.

21. Through $(1, 5, 2)$ perpendicular to $3\vec{i} - \vec{j} + 4\vec{k}$.
22. Through $(2, -1, 3)$ perpendicular to $5\vec{i} + 4\vec{j} - \vec{k}$.
23. Through $(1, 3, 5)$ and normal to $\vec{i} - \vec{j} + \vec{k}$.
24. Perpendicular to $5\vec{i} + \vec{j} - 2\vec{k}$ and passing through $(0, 1, -1)$.
25. Parallel to $2x + 4y - 3z = 1$ and through $(1, 0, -1)$.
26. Through $(-2, 3, 2)$ and parallel to $3x + y + z = 4$.
27. Perpendicular to $\vec{v} = 2\vec{i} - 3\vec{j} + 5\vec{k}$ and through $(4, 5, -2)$.

In Exercises 28–32, compute the angle between the vectors.

28. $\vec{i} + \vec{j} + \vec{k}$ and $\vec{i} - \vec{j} - \vec{k}$.
29. $\vec{i} + \vec{k}$ and $\vec{j} - \vec{k}$.
30. $\vec{i} + \vec{j} - \vec{k}$ and $2\vec{i} + 3\vec{j} + \vec{k}$.
31. $\vec{i} + \vec{j}$ and $\vec{i} + 2\vec{j} - \vec{k}$.
32. $\vec{i}$ and $2\vec{i} + 3\vec{j} - \vec{k}$.

33. Match statements (a)-(c) with diagrams (I)-(III) of vectors $\vec{u}$ and $\vec{w}$ in Figure 13.34.

(a) $\vec{u} \cdot \vec{w} = 0$ **(b)** $\vec{u} \cdot \vec{w} > 0$ **(c)** $\vec{u} \cdot \vec{w} < 0$

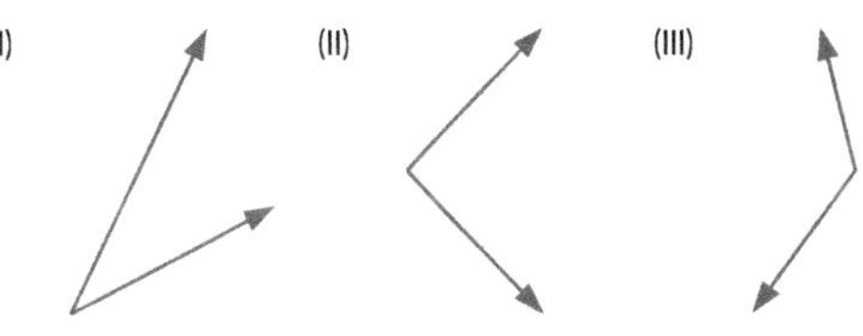

Figure 13.34

PROBLEMS

34. Are the dot products of the two-dimensional vectors in Figure 13.34 positive, negative, or zero?

(a) $\vec{a} \cdot \vec{b}$ **(b)** $\vec{a} \cdot \vec{c}$ **(c)** $\vec{b} \cdot \vec{c}$

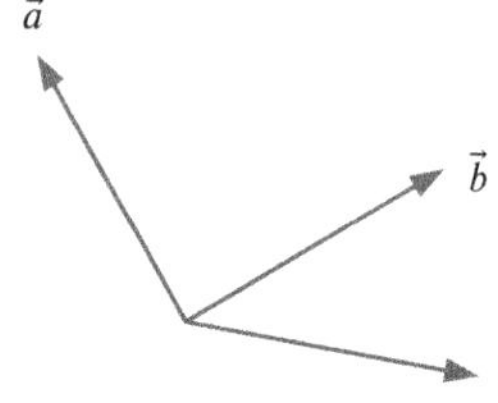

Figure 13.35

35. Give a unit vector

(a) In the same direction as $\vec{v} = 2\vec{i} + 3\vec{j}$.
(b) Perpendicular to $\vec{v}$.

36. A plane has equation $z = 5x - 2y + 7$.

(a) Find a value of λ making the vector $\lambda\vec{i} + \vec{j} + 0.5\vec{k}$ normal to the plane.
(b) Find a value of a so that the point $(a + 1, a, a - 1)$ lies on the plane.

37. Consider the plane $5x - y + 7z = 21$.

(a) Find a point on the x-axis on this plane.
(b) Find two other points on the plane.
(c) Find a vector perpendicular to the plane.
(d) Find a vector parallel to the plane.

38. **(a)** Find a vector perpendicular to the plane $z = 2 + 3x - y$.
(b) Find a vector parallel to the plane.

39. **(a)** Find a vector perpendicular to the plane $z = 2x + 3y$.
(b) Find a vector parallel to the plane.

40. Consider the plane $x + 2y - z = 5$ and the vector $\vec{v} = 2\vec{i} - 5\vec{j} + 3\vec{k}$.

(a) Find a normal vector to the plane.
(b) What is the angle between $\vec{v}$ and the vector you found in part (a)?
(c) What is the angle between $\vec{v}$ and the plane?

41. Match the planes in (a)–(d) with one or more of the descriptions in (I)–(IV). No reasons are needed.

(a) $3x - y + z = 0$ **(b)** $4x + y + 2z - 5 = 0$
(c) $x + y = 5$ **(d)** $x = 5$

I Goes through the origin.
II Has a normal vector parallel to the xy-plane.
III Goes through the point $(0, 5, 0)$.
IV Has a normal vector whose dot products with $\vec{i}$, $\vec{j}$, $\vec{k}$ are all positive.

42. Which pairs (if any) of vectors from the following list

(a) Are perpendicular?
(b) Are parallel?
(c) Have an angle less than $\pi/2$ between them?
(d) Have an angle of more than $\pi/2$ between them?

$$\vec{a} = \vec{i} - 3\vec{j} - \vec{k}, \quad \vec{b} = \vec{i} + \vec{j} + 2\vec{k},$$
$$\vec{c} = -2\vec{i} - \vec{j} + \vec{k}, \quad \vec{d} = -\vec{i} - \vec{j} + \vec{k}.$$

43. List any vectors that are parallel to each other and any vectors that are perpendicular to each other:

$$\vec{v}_1 = \vec{i} - 2\vec{j} \qquad \vec{v}_2 = 2\vec{i} + 4\vec{j}$$
$$\vec{v}_3 = 3\vec{i} + 1.5\vec{j} \qquad \vec{v}_4 = -1.2\vec{i} + 2.4\vec{j}$$
$$\vec{v}_5 = -5\vec{i} - 2.5\vec{j} \qquad \vec{v}_6 = 12\vec{i} - 12\vec{j}$$
$$\vec{v}_7 = 4\vec{i} + 2\vec{j} \qquad \vec{v}_8 = 3\vec{i} - 6\vec{j}$$
$$\vec{v}_9 = 0.70\vec{i} - 0.35\vec{j}$$

44. **(a)** Give a vector that is parallel to, but not equal to, $\vec{v} = 4\vec{i} + 3\vec{j}$.
(b) Give a vector that is perpendicular to $\vec{v}$.

45. For what values of t are $\vec{u} = t\vec{i} - \vec{j} + \vec{k}$ and $\vec{v} = t\vec{i} + t\vec{j} - 2\vec{k}$ perpendicular? Are there values of t for which $\vec{u}$ and $\vec{v}$ are parallel?

46. Let θ be the angle between $\vec{v}$ and $\vec{w}$, with $0 < \theta < \pi/2$. What is the effect on $\vec{v} \cdot \vec{w}$ of increasing each of the following quantities? Does $\vec{v} \cdot \vec{w}$ increase or decrease?

(a) $\|\vec{v}\|$ **(b)** θ

In Problems 47–49, for two-dimensional vectors $\vec{a}$ and $\vec{b}$, if $\|\vec{a}\| = 2$ and $\|\vec{b}\| = 4$, find $\|\vec{a} + \vec{b}\|$ for the given $\vec{a} \cdot \vec{b}$.

47. $\vec{a} \cdot \vec{b} = -8$ **48.** $\vec{a} \cdot \vec{b} = 8$ **49.** $\vec{a} \cdot \vec{b} = 0$

50. For a fixed two-dimensional vector $\vec{a}$ with $\|\vec{a}\| = 2$, determine how many vectors there are with $\|\vec{b}\| = 4$ and $\vec{a} \cdot \vec{b} = 4$.

51. Write $\vec{a} = 3\vec{i} + 2\vec{j} - 6\vec{k}$ as the sum of two vectors, one parallel and one perpendicular to $\vec{d} = 2\vec{i} - 4\vec{j} + \vec{k}$.

52. Find angle BAC if $A = (2, 2, 2)$, $B = (4, 2, 1)$, and $C = (2, 3, 1)$.

53. The points $(5, 0, 0)$, $(0, -3, 0)$, and $(0, 0, 2)$ form a triangle. Find the lengths of the sides of the triangle and each of its angles.

54. Let S be the triangle with vertices $A = (2, 2, 2)$, $B = (4, 2, 1)$, and $C = (2, 3, 1)$.

(a) Find the length of the shortest side of S.
(b) Find the cosine of the angle BAC at vertex A.

In Problems 55–57, find the work done by a force $\vec{F}$ moving an object on the line from point P to point Q. Give answers in joules and foot-pounds, using 1 joule ≈ 0.73756 ft-lb.

55. $\vec{F} = 3\vec{i} + 4\vec{j}$ newtons, $P = (3, 4)$ meters, $Q = (8, 10)$ meters

56. $\vec{F} = 4\vec{i} + 2\vec{j}$ newtons, $P = (10, 9)$ meters, $Q = (12, 2)$ meters

57. $\vec{F} = 20\vec{i} + 30\vec{j}$ pounds, $P = (9, 3)$ feet, $Q = (12, 5)$ feet

In Problems 58–63, given $\vec{v} = 3\vec{i} + 4\vec{j}$ and force vector $\vec{F}$, find:

(a) The component of $\vec{F}$ parallel to $\vec{v}$.

(b) The component of $\vec{F}$ perpendicular to $\vec{v}$.

(c) The work, W, done by force $\vec{F}$ through displacement $\vec{v}$.

58. $\vec{F} = 4\vec{i} + \vec{j}$

59. $\vec{F} = 0.2\vec{i} - 0.5\vec{j}$

60. $\vec{F} = 9\vec{i} + 12\vec{j}$

61. $\vec{F} = -0.4\vec{i} + 0.3\vec{j}$

62. $\vec{F} = -3\vec{i} - 5\vec{j}$

63. $\vec{F} = -6\vec{i} - 8\vec{j}$

In Problems 64–67, the force on an object is $\vec{F} = -20\vec{j}$. For vector $\vec{v}$, find:

(a) The component of $\vec{F}$ parallel to $\vec{v}$.

(b) The component of $\vec{F}$ perpendicular to $\vec{v}$.

(c) The work W done by force $\vec{F}$ through displacement $\vec{v}$.

64. $\vec{v} = 2\vec{i} + 3\vec{j}$

65. $\vec{v} = 5\vec{i} - \vec{j}$

66. $\vec{v} = 3\vec{j}$

67. $\vec{v} = 5\vec{i}$

68. A basketball gymnasium is 25 meters high, 80 meters wide and 200 meters long. For a half-time stunt, the cheerleaders want to run two strings, one from each of the two corners above one basket to the diagonally opposite corners of the gym floor. What is the cosine of the angle made by the strings as they cross?

69. An inner diagonal of a cube runs from one vertex through the center to the opposite vertex. For the cube with vertices $(\pm 1, \pm 1, \pm 1)$, at what acute angle do two distinct inner diagonals intersect?

70. A 100-meter dash is run on a track in the direction of the vector $\vec{v} = 2\vec{i} + 6\vec{j}$. The wind velocity $\vec{w}$ is $5\vec{i} + \vec{j}$ km/hr. The rules say that a legal wind speed measured in the direction of the dash must not exceed 5 km/hr. Will the race results be disqualified due to an illegal wind? Justify your answer.

71. An airplane is flying toward the southeast. Which of the following wind velocity vectors increases the plane's speed the most? Which slows down the plane the most?

$\vec{w}_1 = -4\vec{i} - \vec{j}$ $\quad \vec{w}_2 = \vec{i} - 2\vec{j}$ $\quad \vec{w}_3 = -\vec{i} + 8\vec{j}$

$\vec{w}_4 = 10\vec{i} + 2\vec{j}$ $\quad \vec{w}_5 = 5\vec{i} - 2\vec{j}$

72. A canoe is moving with velocity $\vec{v} = 5\vec{i} + 3\vec{j}$ m/sec relative to the water. The velocity of the current in the water is $\vec{c} = \vec{i} + 2\vec{j}$ m/sec.

(a) What is the speed of the current?

(b) What is the speed of the current in the direction of the canoe's motion?

73. A planet at the point $(30, 60, 90)$ is in a circular orbit about the line through the origin in the direction of the unit vector $\vec{u} = 2/3\vec{i} + 2/3\vec{j} - 1/3\vec{k}$. For the orbit, find the

(a) Center **(b)** Radius

74. Find the shortest distance between the planes $2x - 5y + z = 10$ and $z = 5y - 2x$.

75. A street vendor sells six items, with prices p_1 dollars per unit, p_2 dollars per unit, and so on. The vendor's price vector is $\vec{p} = (p_1, p_2, p_3, p_4, p_5, p_6) = (1.00, 3.50, 4.00, 2.75, 5.00, 3.00)$. The vendor sells q_1 units of the first item, q_2 units of the second item, and so on. The vendor's quantity vector is $\vec{q} = (q_1, q_2, q_3, q_4, q_5, q_6) = (43, 57, 12, 78, 20, 35)$. Find $\vec{p} \cdot \vec{q}$, give its units, and explain its significance to the vendor.

76. A course has four exams, weighted 10%, 15%, 25%, 50%, respectively. The class average on each of these exams is 75%, 91%, 84%, 87%, respectively. What do the vectors $\vec{a} = (0.75, 0.91, 0.84, 0.87)$ and $\vec{w} = (0.1, 0.15, 0.25, 0.5)$ represent, in terms of the course? Calculate the dot product $\vec{w} \cdot \vec{a}$. What does it represent, in terms of the course?

77. A consumption vector of three goods is defined by $\vec{x} = (x_1, x_2, x_3)$, where x_1, x_2 and x_3 are the quantities consumed of the three goods. A budget constraint is represented by the equation $\vec{p} \cdot \vec{x} = k$, where $\vec{p}$ is the price vector of the three goods and k is a constant. Show that the difference between two consumption vectors corresponding to points satisfying the same budget constraint is perpendicular to the price vector $\vec{p}$.

78. What does Property 2 of the dot product in the box on page 719 say geometrically?

79. Show that the vectors $(\vec{b} \cdot \vec{c})\vec{a} - (\vec{a} \cdot \vec{c})\vec{b}$ and $\vec{c}$ are perpendicular.

80. Show that if $\vec{u}$ and $\vec{v}$ are two vectors such that

$$\vec{u} \cdot \vec{w} = \vec{v} \cdot \vec{w}$$

for every vector $\vec{w}$, then

$$\vec{u} = \vec{v}.$$

81. The Law of Cosines for a triangle with side lengths a, b, and c, and with angle C opposite side c, says

$$c^2 = a^2 + b^2 - 2ab\cos C.$$

On page 719, we used the Law of Cosines to show that the two definitions of the dot product are equivalent. In this problem, use the geometric definition of the dot product and its properties in the box on page 719 to prove the Law of Cosines. [Hint: Let $\vec{u}$ and $\vec{v}$ be the displacement vectors from C to the other two vertices, and express c^2 in terms of $\vec{u}$ and $\vec{v}$.]

82. For any vectors $\vec{v}$ and $\vec{w}$, consider the following function of t:

$$q(t) = (\vec{v} + t\vec{w}) \cdot (\vec{v} + t\vec{w}).$$

(a) Explain why $q(t) \geq 0$ for all real t.

(b) Expand $q(t)$ as a quadratic polynomial in t using the properties on page 719.

(c) Using the discriminant of the quadratic, show that

$$|\vec{v} \cdot \vec{w}| \leq \|\vec{v}\|\|\vec{w}\|.$$

Strengthen Your Understanding

In Problems 83–85, explain what is wrong with the statement.

83. For any 3-dimensional vectors $\vec{u}, \vec{v}, \vec{w}$, we have $(\vec{u} \cdot \vec{v}) \cdot \vec{w} = \vec{u} \cdot (\vec{v} \cdot \vec{w})$.

84. If $\vec{u} = \vec{i} + \vec{j}$ and $\vec{v} = 2\vec{i} + \vec{j}$, then the component of $\vec{v}$ parallel to $\vec{u}$ is $\vec{v}_{\text{parallel}} = (\vec{v} \cdot \vec{u})\vec{u} = 3\vec{i} + 3\vec{j}$.

85. A normal vector for the plane $z = 2x + 3y$ is $2\vec{i} + 3\vec{j}$.

In Problems 86–87, give an example of:

86. A point (a, b) such that the displacement vector from $(1, 1)$ to (a, b) is perpendicular to $\vec{i} + 2\vec{j}$.

87. A linear function $f(x, y) = mx + ny + c$ whose graph is perpendicular to $\vec{i} + 2\vec{j} + 3\vec{k}$.

Are the statements in Problems 88–99 true or false? Give reasons for your answer.

88. The quantity $\vec{u} \cdot \vec{v}$ is a vector.

89. The plane $x+2y-3z = 5$ has normal vector $\vec{i} + 2\vec{j} - 3\vec{k}$.

90. If $\vec{u} \cdot \vec{v} < 0$ then the angle between $\vec{u}$ and $\vec{v}$ is greater than $\pi/2$.

91. An equation of the plane with normal vector $\vec{i} + \vec{j} + \vec{k}$ containing the point $(1, 2, 3)$ is $z = x + y$.

92. The triangle in 3-space with vertices $(1, 1, 0)$, $(0, 1, 0)$ and $(0, 1, 1)$ has a right angle.

93. The dot product $\vec{v} \cdot \vec{v}$ is never negative.

94. If $\vec{u} \cdot \vec{v} = 0$ then either $\vec{u} = 0$ or $\vec{v} = 0$.

95. If $\vec{u}, \vec{v}$ and $\vec{w}$ are all nonzero, and $\vec{u} \cdot \vec{v} = \vec{u} \cdot \vec{w}$, then $\vec{v} = \vec{w}$.

96. For any vectors $\vec{u}$ and $\vec{v}$: $(\vec{u} + \vec{v}) \cdot (\vec{u} - \vec{v}) = \|\vec{u}\|^2 - \|\vec{v}\|^2$.

97. If $\|\vec{u}\| = 1$, then the vector $\vec{v} - (\vec{v} \cdot \vec{u})\vec{u}$ is perpendicular to $\vec{u}$.

98. If $\vec{u} \cdot \vec{v} = \|\vec{u}\|\|\vec{v}\|$ then $\|\vec{u} + \vec{v}\| = \|\vec{u}\| + \|\vec{v}\|$.

99. The two nonzero vectors $\vec{v} = x\vec{i} + y\vec{j}$ and $\vec{w} = y\vec{i} - x\vec{j}$ are orthogonal for any choice of x and y.

13.4 THE CROSS PRODUCT

In the previous section we combined two vectors to get a number, the dot product. In this section we see another way of combining two vectors, this time to get a vector, the *cross product*. Any two vectors in 3-space form a parallelogram. We define the cross product using this parallelogram.

The Area of a Parallelogram

Consider the parallelogram formed by the vectors $\vec{v}$ and $\vec{w}$ with an angle of θ between them. Then Figure 13.36 shows

$$\text{Area of parallelogram} = \text{Base} \cdot \text{Height} = \|\vec{v}\|\|\vec{w}\| \sin\theta.$$

How would we compute the area of the parallelogram if we were given $\vec{v}$ and $\vec{w}$ in components, $\vec{v} = v_1\vec{i} + v_2\vec{j} + v_3\vec{k}$ and $\vec{w} = w_1\vec{i} + w_2\vec{j} + w_3\vec{k}$? Project 1 (available online) shows that if $\vec{v}$ and $\vec{w}$ are in the xy-plane so that $v_3 = w_3 = 0$, then

$$\text{Area of parallelogram} = |v_1 w_2 - v_2 w_1|.$$

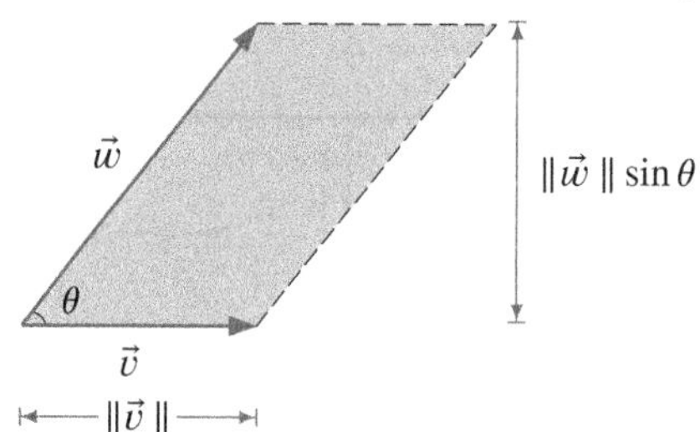

Figure 13.36: Parallelogram formed by $\vec{v}$ and $\vec{w}$ has Area $= \|\vec{v}\|\|\vec{w}\| \sin\theta$

What if $\vec{v}$ and $\vec{w}$ do not lie in the xy-plane? The cross product will enable us to compute the area of the parallelogram formed by any two vectors.

Definition of the Cross Product

We define the cross product of the vectors $\vec{v}$ and $\vec{w}$, written $\vec{v} \times \vec{w}$, to be a vector perpendicular to both $\vec{v}$ and $\vec{w}$. The magnitude of this vector is the area of the parallelogram formed by the two vectors. The direction of $\vec{v} \times \vec{w}$ is given by the normal vector, $\vec{n}$, to the plane defined by $\vec{v}$ and $\vec{w}$. If we require that $\vec{n}$ be a unit vector, there are two choices for $\vec{n}$, pointing out of the plane in opposite directions. We pick one by the following rule (see Figure 13.37):

> **The right-hand rule:** Place $\vec{v}$ and $\vec{w}$ so that their tails coincide and curl the fingers of your right hand through the smaller of the two angles from $\vec{v}$ to $\vec{w}$; your thumb points in the direction of the normal vector, $\vec{n}$.

Like the dot product, there are two equivalent definitions of the cross product:

> The following two definitions of the **cross product** or **vector product** $\vec{v} \times \vec{w}$ are equivalent:
>
> - **Geometric definition**
>   If $\vec{v}$ and $\vec{w}$ are not parallel, then
> $$\vec{v} \times \vec{w} = \begin{pmatrix} \text{Area of parallelogram} \\ \text{with edges } \vec{v} \text{ and } \vec{w} \end{pmatrix} \vec{n} = (\|\vec{v}\|\|\vec{w}\| \sin\theta)\vec{n},$$
>   where $0 \le \theta \le \pi$ is the angle between $\vec{v}$ and $\vec{w}$ and $\vec{n}$ is the unit vector perpendicular to $\vec{v}$ and $\vec{w}$ pointing in the direction given by the right-hand rule. If $\vec{v}$ and $\vec{w}$ are parallel, then $\vec{v} \times \vec{w} = \vec{0}$.
> - **Algebraic definition**
> $$\vec{v} \times \vec{w} = (v_2w_3 - v_3w_2)\vec{i} + (v_3w_1 - v_1w_3)\vec{j} + (v_1w_2 - v_2w_1)\vec{k}$$
>   where $\vec{v} = v_1\vec{i} + v_2\vec{j} + v_3\vec{k}$ and $\vec{w} = w_1\vec{i} + w_2\vec{j} + w_3\vec{k}$.

Problems 79 and 82 (available online) show that the geometric and algebraic definitions of the cross product give the same result.

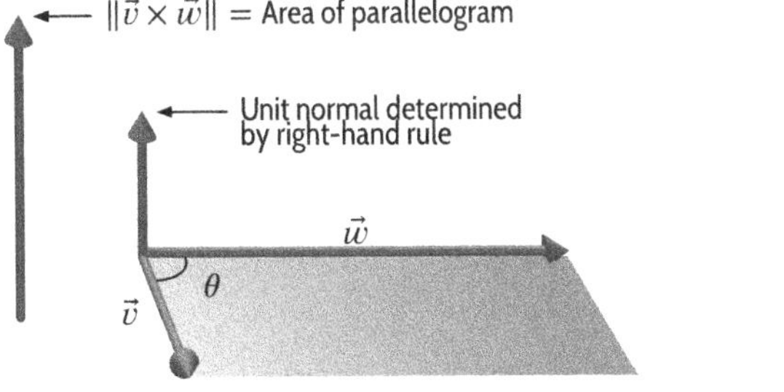

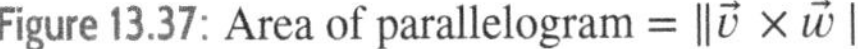

Figure 13.37: Area of parallelogram = $\|\vec{v} \times \vec{w}\|$

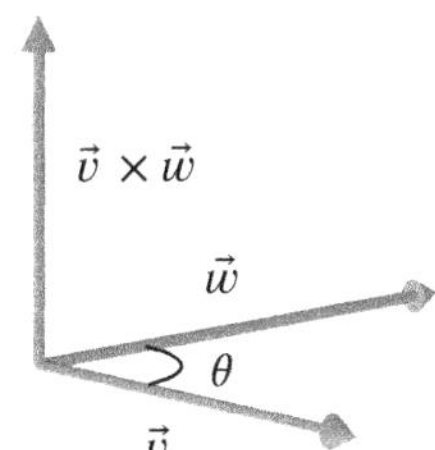

Figure 13.38: The cross product $\vec{v} \times \vec{w}$

The geometric definition shows us that the cross product is *rotation invariant*. Imagine the two vectors $\vec{v}$ and $\vec{w}$ as two metal rods welded together. Attach a third rod whose direction and length correspond to $\vec{v} \times \vec{w}$. (See Figure 13.38.) Then, no matter how we turn this set of rods, the third will still be the cross product of the first two.

The algebraic definition is more easily remembered by writing it as a 3×3 determinant. (See Appendix E.)

$$\vec{v} \times \vec{w} = \begin{vmatrix} \vec{i} & \vec{j} & \vec{k} \\ v_1 & v_2 & v_3 \\ w_1 & w_2 & w_3 \end{vmatrix} = (v_2w_3 - v_3w_2)\vec{i} + (v_3w_1 - v_1w_3)\vec{j} + (v_1w_2 - v_2w_1)\vec{k}.$$

Example 1 Find $\vec{i} \times \vec{j}$ and $\vec{j} \times \vec{i}$.

Solution The vectors $\vec{i}$ and $\vec{j}$ both have magnitude 1 and the angle between them is $\pi/2$. By the right-hand rule, the vector $\vec{i} \times \vec{j}$ is in the direction of $\vec{k}$, so $\vec{n} = \vec{k}$ and we have

$$\vec{i} \times \vec{j} = \left(\|\vec{i}\|\|\vec{j}\| \sin\frac{\pi}{2}\right)\vec{k} = \vec{k}.$$

Similarly, the right-hand rule says that the direction of $\vec{j} \times \vec{i}$ is $-\vec{k}$, so

$$\vec{j} \times \vec{i} = \left(\|\vec{j}\|\|\vec{i}\| \sin\frac{\pi}{2}\right)\left(-\vec{k}\right) = -\vec{k}.$$

Similar calculations show that $\vec{j} \times \vec{k} = \vec{i}$ and $\vec{k} \times \vec{i} = \vec{j}$.

Example 2 For any vector $\vec{v}$, find $\vec{v} \times \vec{v}$.

Solution Since $\vec{v}$ is parallel to itself, $\vec{v} \times \vec{v} = \vec{0}$.

Example 3 Find the cross product of $\vec{v} = 2\vec{i} + \vec{j} - 2\vec{k}$ and $\vec{w} = 3\vec{i} + \vec{k}$ and check that the cross product is perpendicular to both $\vec{v}$ and $\vec{w}$.

Solution Writing $\vec{v} \times \vec{w}$ as a determinant and expanding it into three two-by-two determinants, we have

$$\begin{aligned} \vec{v} \times \vec{w} = \begin{vmatrix} \vec{i} & \vec{j} & \vec{k} \\ 2 & 1 & -2 \\ 3 & 0 & 1 \end{vmatrix} &= \vec{i}\begin{vmatrix} 1 & -2 \\ 0 & 1 \end{vmatrix} - \vec{j}\begin{vmatrix} 2 & -2 \\ 3 & 1 \end{vmatrix} + \vec{k}\begin{vmatrix} 2 & 1 \\ 3 & 0 \end{vmatrix} \\ &= \vec{i}\,(1(1) - 0(-2)) - \vec{j}\,(2(1) - 3(-2)) + \vec{k}\,(2(0) - 3(1)) \\ &= \vec{i} - 8\vec{j} - 3\vec{k}. \end{aligned}$$

To check that $\vec{v} \times \vec{w}$ is perpendicular to $\vec{v}$, we compute the dot product:

$$\vec{v} \cdot (\vec{v} \times \vec{w}) = (2\vec{i} + \vec{j} - 2\vec{k}) \cdot (\vec{i} - 8\vec{j} - 3\vec{k}) = 2 - 8 + 6 = 0.$$

Similarly,

$$\vec{w} \cdot (\vec{v} \times \vec{w}) = (3\vec{i} + 0\vec{j} + \vec{k}) \cdot (\vec{i} - 8\vec{j} - 3\vec{k}) = 3 + 0 - 3 = 0.$$

Thus, $\vec{v} \times \vec{w}$ is perpendicular to both $\vec{v}$ and $\vec{w}$.

Properties of the Cross Product

The right-hand rule tells us that $\vec{v} \times \vec{w}$ and $\vec{w} \times \vec{v}$ point in opposite directions. The magnitudes of $\vec{v} \times \vec{w}$ and $\vec{w} \times \vec{v}$ are the same, so $\vec{w} \times \vec{v} = -(\vec{v} \times \vec{w})$. (See Figure 13.39.)

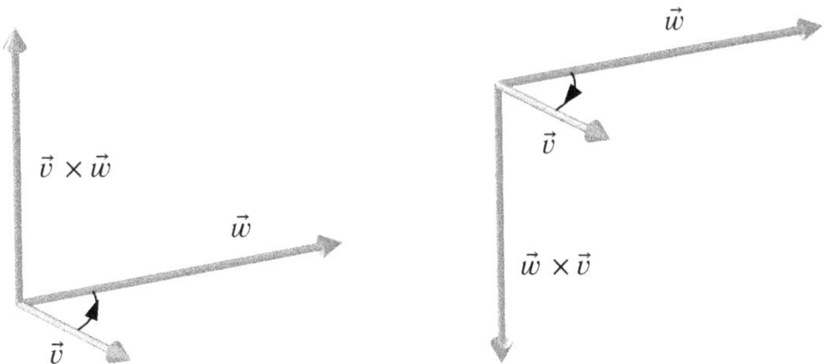

Figure 13.39: Diagram showing $\vec{v} \times \vec{w} = -(\vec{w} \times \vec{v})$

This explains the first of the following properties. The other two are derived in Problems 74, 75, and 82 (available online).

> **Properties of the Cross Product**
>
> For vectors $\vec{u}$, $\vec{v}$, $\vec{w}$ and scalar λ
> 1. $\vec{w} \times \vec{v} = -(\vec{v} \times \vec{w})$
> 2. $(\lambda\vec{v}) \times \vec{w} = \lambda(\vec{v} \times \vec{w}) = \vec{v} \times (\lambda\vec{w})$
> 3. $\vec{u} \times (\vec{v} + \vec{w}) = \vec{u} \times \vec{v} + \vec{u} \times \vec{w}$.

The Equation of a Plane Through Three Points

As we saw on page 721, the equation of a plane is determined by a point $P_0 = (x_0, y_0, z_0)$ on the plane, and a normal vector, $\vec{n} = a\vec{i} + b\vec{j} + c\vec{k}$:

$$a(x - x_0) + b(y - y_0) + c(z - z_0) = 0.$$

However, a plane can also be determined by three points on it (provided they do not lie on the same line). In that case we can find an equation of the plane by first determining two vectors in the plane and then finding a normal vector using the cross product, as in the following example.

Example 4 Find an equation of the plane containing the points $P = (1, 3, 0)$, $Q = (3, 4, -3)$, and $R = (3, 6, 2)$.

Solution Since the points P and Q are in the plane, the displacement vector between them, $\overrightarrow{PQ}$, is in the plane, where

$$\overrightarrow{PQ} = (3-1)\vec{i} + (4-3)\vec{j} + (-3-0)\vec{k} = 2\vec{i} + \vec{j} - 3\vec{k}.$$

The displacement vector $\overrightarrow{PR}$ is also in the plane, where

$$\overrightarrow{PR} = (3-1)\vec{i} + (6-3)\vec{j} + (2-0)\vec{k} = 2\vec{i} + 3\vec{j} + 2\vec{k}.$$

Thus, a normal vector, $\vec{n}$, to the plane is given by

$$\vec{n} = \overrightarrow{PQ} \times \overrightarrow{PR} = \begin{vmatrix} \vec{i} & \vec{j} & \vec{k} \\ 2 & 1 & -3 \\ 2 & 3 & 2 \end{vmatrix} = 11\vec{i} - 10\vec{j} + 4\vec{k}.$$

Since the point $(1, 3, 0)$ is on the plane, the equation of the plane is

$$11(x-1) - 10(y-3) + 4(z-0) = 0,$$

which simplifies to

$$11x - 10y + 4z = -19.$$

You should check that P, Q, and R satisfy this equation, since they lie on the plane.

Areas and Volumes Using the Cross Product and Determinants

We can use the cross product to calculate the area of the parallelogram with sides $\vec{v}$ and $\vec{w}$. We say that $\vec{v} \times \vec{w}$ is the *area vector* of the parallelogram. The geometric definition of the cross product tells us that $\vec{v} \times \vec{w}$ is normal to the parallelogram and gives us the following result:

> **Area of a parallelogram** with edges $\vec{v} = v_1\vec{i} + v_2\vec{j} + v_3\vec{k}$ and $\vec{w} = w_1\vec{i} + w_2\vec{j} + w_3\vec{k}$ is given by
>
> $$\text{Area} = \|\vec{v} \times \vec{w}\|, \qquad \text{where} \quad \vec{v} \times \vec{w} = \begin{vmatrix} \vec{i} & \vec{j} & \vec{k} \\ v_1 & v_2 & v_3 \\ w_1 & w_2 & w_3 \end{vmatrix}.$$

Example 5 Find the area of the parallelogram with edges $\vec{v} = 2\vec{i} + \vec{j} - 3\vec{k}$ and $\vec{w} = \vec{i} + 3\vec{j} + 2\vec{k}$.

Solution We calculate the cross product:

$$\vec{v} \times \vec{w} = \begin{vmatrix} \vec{i} & \vec{j} & \vec{k} \\ 2 & 1 & -3 \\ 1 & 3 & 2 \end{vmatrix} = (2+9)\vec{i} - (4+3)\vec{j} + (6-1)\vec{k} = 11\vec{i} - 7\vec{j} + 5\vec{k}.$$

The area of the parallelogram with edges $\vec{v}$ and $\vec{w}$ is the magnitude of the vector $\vec{v} \times \vec{w}$:

$$\text{Area} = \|\vec{v} \times \vec{w}\| = \sqrt{11^2 + (-7)^2 + 5^2} = \sqrt{195}.$$

Volume of a Parallelepiped

Consider the parallelepiped with sides formed by $\vec{a}$, $\vec{b}$, and $\vec{c}$. (See Figure 13.40.) Since the base is formed by the vectors $\vec{b}$ and $\vec{c}$, we have

$$\text{Area of base of parallelepiped} = \|\vec{b} \times \vec{c}\|.$$

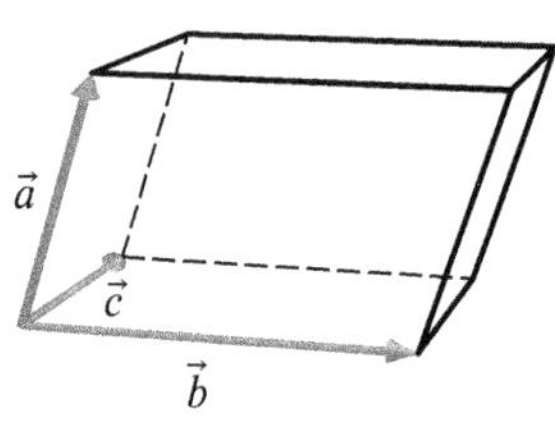

Figure 13.40: Volume of a parallelepiped

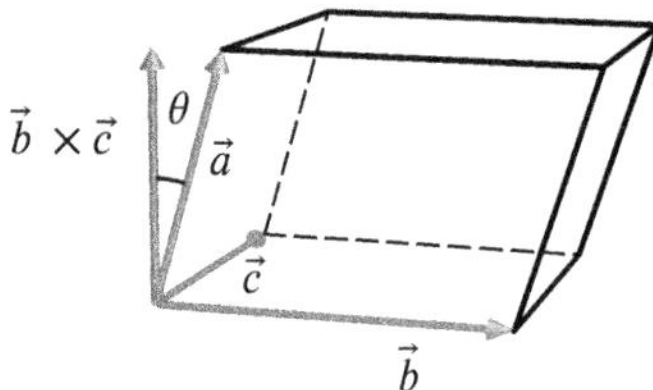

Figure 13.41: The vectors $\vec{a}$, $\vec{b}$, $\vec{c}$ are called a right-handed set

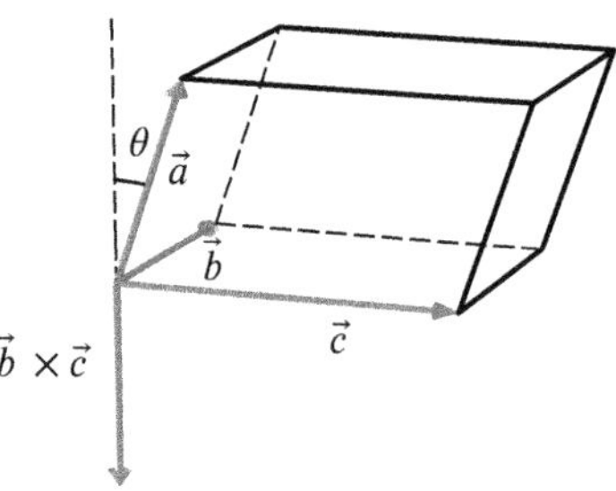

Figure 13.42: The vectors $\vec{a}$, $\vec{b}$, $\vec{c}$ are called a left-handed set

The vectors $\vec{a}$, $\vec{b}$, and $\vec{c}$ can be arranged either as in Figure 13.41 or as in Figure 13.42. In either case,

$$\text{Height of parallelepiped} = \|\vec{a}\| \cos\theta,$$

where θ is the angle shown in the figures. In Figure 13.41 the angle θ is less than $\pi/2$, so the product, $(\vec{b} \times \vec{c}) \cdot \vec{a}$, called the *triple product*, is positive. Thus, in this case

$$\text{Volume of parallelepiped} = \text{Base} \cdot \text{Height} = \|\vec{b} \times \vec{c}\| \cdot \|\vec{a}\| \cos\theta = (\vec{b} \times \vec{c}) \cdot \vec{a}.$$

In Figure 13.42, the angle, $\pi - \theta$, between $\vec{a}$ and $\vec{b} \times \vec{c}$ is more than $\pi/2$, so the product $(\vec{b} \times \vec{c}) \cdot \vec{a}$ is negative. Thus, in this case we have

$$\begin{aligned}\text{Volume} = \text{Base} \cdot \text{Height} &= \|\vec{b} \times \vec{c}\| \cdot \|\vec{a}\| \cos\theta = -\|\vec{b} \times \vec{c}\| \cdot \|\vec{a}\| \cos(\pi - \theta) \\ &= -(\vec{b} \times \vec{c}) \cdot \vec{a} = \left|(\vec{b} \times \vec{c}) \cdot \vec{a}\right|.\end{aligned}$$

Therefore, in both cases the volume is given by $\left|(\vec{b} \times \vec{c}) \cdot \vec{a}\right|$. Using determinants, we can write

Volume of a parallelepiped with edges $\vec{a}$, $\vec{b}$, $\vec{c}$ is given by

$$\text{Volume} = \left|(\vec{b} \times \vec{c}) \cdot \vec{a}\right| = \text{Absolute value of the determinant} \begin{vmatrix} a_1 & a_2 & a_3 \\ b_1 & b_2 & b_3 \\ c_1 & c_2 & c_3 \end{vmatrix}.$$

Angular Velocity

) Angular velocity, which describes rotation about an axis, can be represented by a vector. For example, the angular velocity of the rotating flywheel in Figure 13.43 is represented by the vector $\vec{\omega}$, whose direction is parallel to the axis of rotation in the direction given by the right-hand rule. If the fingers of the right-hand curl around the axis in the direction of the rotation, then the thumb points along the axis in the direction of $\vec{\omega}$. The magnitude $\|\vec{\omega}\|$ is the angular speed of rotation, for example in radians per unit time or revolutions per unit time.

Every point on the flywheel travels a circular orbit around the axis. Since one orbit is 2π radians,

$$\text{Time to complete one orbit} = \frac{\text{Angle traveled}}{\text{Angular speed}} = \frac{2\pi}{\|\vec{\omega}\|}.$$

In Figure 13.43, let $\vec{r}$ be the vector from the center of the orbit to the point P. In one orbit, the point P travels a distance of $2\pi\|\vec{r}\|$ around the circumference of a circle, so

$$\text{Speed} = \frac{\text{Distance}}{\text{Time}} = \frac{2\pi\|\vec{r}\|}{2\pi/\|\vec{\omega}\|} = \|\vec{\omega}\| \cdot \|\vec{r}\|.$$

The velocity vector $\vec{v}$ is tangent to the orbit, so $\vec{v}$ is perpendicular to both the axis and the radius of the orbit. The magnitude of $\vec{v}$ is the speed of P, so $\|\vec{v}\| = \|\vec{\omega}\| \cdot \|\vec{r}\|$. Since the cross product $\vec{\omega} \times \vec{r}$ has the same direction as the velocity (both $\vec{\omega} \times \vec{r}$ and $\vec{v}$ are perpendicular to $\vec{\omega}$ and to $\vec{r}$), and the same magnitude as the velocity (both magnitudes are $\|\vec{\omega}\| \cdot \|\vec{r}\|$), we have

$$\vec{v} = \vec{\omega} \times \vec{r}.$$

The formula $\vec{v} = \vec{\omega} \times \vec{r}$ holds for $\vec{r}$ a vector from any point on the axis of rotation to the point P. This is because $\vec{r}$ can be expressed as the sum of two component vectors, one parallel to the axis and the other a radius vector of the orbit. Only the radial component contributes to the cross product. See Figure 13.44.

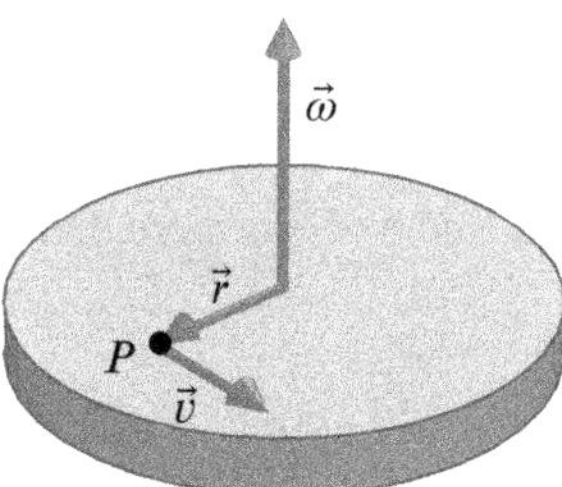

Figure 13.43: Rotating flywheel

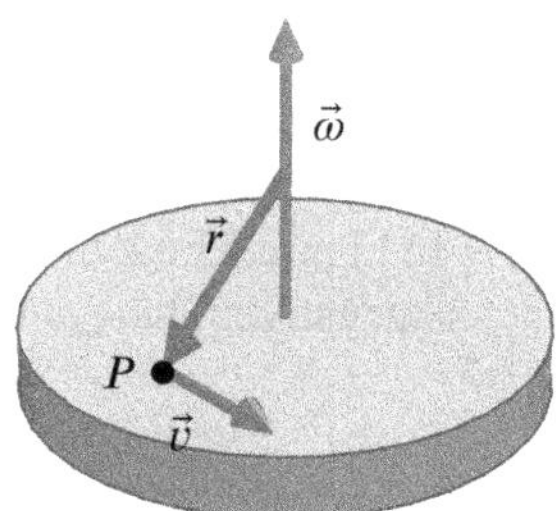

Figure 13.44: Rotating flywheel

Example 6 The world record for the fastest spin by a figure skater, 342 revolutions per minute, is held by Olivia Oliver.[2]

(a) Find Olivia's angular velocity vector, assuming she is vertical and spinning her fastest to her left.
(b) Let her skates touch the ice at the point $(0, 0, 0)$ and her left elbow be at $P = (10, 15, 110)$, where distances are in centimeters. Find the velocity of her elbow.
(c) Find the speed of her elbow in centimeters per minute.

Solution (a) Since she is spinning around a vertical axis, we have $\vec{\omega} = c\vec{k}$ where c, the rate of rotation, is positive because she is spinning to her left. Since 1 revolution corresponds to 2π radians, we have $c = 2\pi \cdot 342 = 2149$. Hence $\vec{\omega} = 2149\vec{k}$ radians per minute.

(b) Her elbow has position vector $\vec{r} = 10\vec{i} + 15\vec{j} + 110\vec{k}$ cm. Her elbow velocity is the vector $\vec{v} = \vec{\omega} \times \vec{r} = -32{,}235\vec{i} + 21{,}490\vec{j}$ cm/min.

(c) Her elbow is moving with speed

$$\|\vec{v}\| = \sqrt{(-32{,}235)^2 + (21{,}490)^2} = 38{,}742 \text{ cm/min.}$$

This is a speed of about 6.5 meters per second.

Exercises and Problems for Section 13.4 Online Resource: Additional Problems for Section 13.4

EXERCISES

In Exercises 1–7, use the algebraic definition to find $\vec{v} \times \vec{w}$.

1. $\vec{v} = \vec{k}, \vec{w} = \vec{j}$
2. $\vec{v} = -\vec{i}, \vec{w} = \vec{j} + \vec{k}$
3. $\vec{v} = \vec{i} + \vec{k}, \vec{w} = \vec{i} + \vec{j}$
4. $\vec{v} = \vec{i} + \vec{j} + \vec{k}, \vec{w} = \vec{i} + \vec{j} + -\vec{k}$
5. $\vec{v} = 2\vec{i} - 3\vec{j} + \vec{k}, \vec{w} = \vec{i} + 2\vec{j} - \vec{k}$
6. $\vec{v} = 2\vec{i} - \vec{j} - \vec{k}, \vec{w} = -6\vec{i} + 3\vec{j} + 3\vec{k}$
7. $\vec{v} = -3\vec{i} + 5\vec{j} + 4\vec{k}, \vec{w} = \vec{i} - 3\vec{j} - \vec{k}$

In Exercises 8–9, use the geometric definition to find:

8. $2\vec{i} \times (\vec{i} + \vec{j})$
9. $(\vec{i} + \vec{j}) \times (\vec{i} - \vec{j})$

In Exercises 10–11, use the properties on page 731 to find:

10. $\left((\vec{i} + \vec{j}) \times \vec{i}\right) \times \vec{j}$
11. $(\vec{i} + \vec{j}) \times (\vec{i} \times \vec{j})$

[2]From www.guinnessworldrecords.com, accessed May 12, 2016.

12. For $\vec{a} = 3\vec{i} + \vec{j} - \vec{k}$ and $\vec{b} = \vec{i} - 4\vec{j} + 2\vec{k}$, find $\vec{a} \times \vec{b}$ and check that it is perpendicular to both $\vec{a}$ and $\vec{b}$.

13. If $\vec{v} = 3\vec{i} - 2\vec{j} + 4\vec{k}$ and $\vec{w} = \vec{i} + 2\vec{j} - \vec{k}$, find $\vec{v} \times \vec{w}$ and $\vec{w} \times \vec{v}$. What is the relation between the two answers?

In Exercises 14–15, find an equation for the plane through the points.

14. $(1,0,0), (0,1,0), (0,0,1)$.

15. $(3,4,2), (-2,1,0), (0,2,1)$.

In Exercises 16–19, find the volume of the parallelepiped with edges $\vec{a}, \vec{b}, \vec{c}$.

16. $\vec{a} = 3\vec{i} + 4\vec{j} + 5\vec{k}, \vec{b} = 5\vec{i} + 4\vec{j} + 3\vec{k}, \vec{c} = \vec{i} + \vec{j} + \vec{k}$.

17. $\vec{a} = -\vec{i} + \vec{j} + \vec{k}, \vec{b} = \vec{i} - \vec{j} + \vec{k}, \vec{c} = \vec{i} + \vec{j} - \vec{k}$.

18. $\vec{a} = -\vec{i} + 8\vec{j} + 7\vec{k}, \vec{b} = 2\vec{j} + 9\vec{k}, \vec{c} = 3\vec{k}$.

19. $\vec{a} = \vec{i} + \vec{j} + 2\vec{k}, \vec{b} = \vec{i} + \vec{k}, \vec{c} = \vec{j} + \vec{k}$.

In Exercises 20–23, the point is rotating around an axis through the origin with angular velocity $\vec{\omega} = 2\vec{i} + \vec{j} - 3\vec{k}$. Find its velocity vector.

20. $(1,2,1)$

21. $(1,0,-1)$

22. $(2,-2,0)$

23. $(4,2,-6)$

PROBLEMS

24. Find a vector parallel to the line of intersection of the planes given by $2y - z = 2$ and $-2x + y = 4$.

25. Find an equation of the plane through the origin that is perpendicular to the line of intersection of the planes in Problem 24.

26. Find an equation of the plane through the point $(4,5,6)$ and perpendicular to the line of intersection of the planes in Problem 24.

27. Find an equation for the plane through the origin containing the points $(1,3,0)$ and $(2,4,1)$.

28. Find a vector parallel to the line of intersection of the two planes $4x - 3y + 2z = 12$ and $x + 5y - z = 25$.

29. Find a vector parallel to the intersection of the planes $2x - 3y + 5z = 2$ and $4x + y - 3z = 7$.

30. Find an equation of the plane through the origin that is perpendicular to the line of intersection of the planes in Problem 29.

31. Find an equation of the plane through the point $(4,5,6)$ that is perpendicular to the line of intersection of the planes in Problem 29.

32. Find the equation of a plane through the origin and perpendicular to $x - y + z = 5$ and $2x + y - 2z = 7$.

33. Given the points $P = (1,2,3)$, $Q = (3,5,7)$, and $R = (2,5,3)$, find:
    - **(a)** A unit vector perpendicular to a plane containing P, Q, R.
    - **(b)** The angle between PQ and PR.
    - **(c)** The area of the triangle PQR.
    - **(d)** The distance from R to the line through P and Q.

34. Let $A = (-1,3,0)$, $B = (3,2,4)$, and $C = (1,-1,5)$.
    - **(a)** Find an equation for the plane that passes through these three points.
    - **(b)** Find the area of the triangle determined by these three points.

35. Consider the plane $z + 2y + x = 4$.
    - **(a)** Find a point on the x-axis on this plane.
    - **(b)** Find a point on the y-axis on this plane.
    - **(c)** Find a point on the z-axis on this plane.
    - **(d)** Find the area of the region of this plane with $x \geq 0$, $y \geq 0$ and $z \geq 0$.

36. If $\vec{v}$ and $\vec{w}$ are both parallel to the xy-plane, what can you conclude about $\vec{v} \times \vec{w}$? Explain.

37. Suppose $\vec{v} \cdot \vec{w} = 5$ and $\|\vec{v} \times \vec{w}\| = 3$, and the angle between $\vec{v}$ and $\vec{w}$ is θ. Find
    - **(a)** $\tan\theta$
    - **(b)** θ.

38. If $\vec{v} \times \vec{w} = 2\vec{i} - 3\vec{j} + 5\vec{k}$, and $\vec{v} \cdot \vec{w} = 3$, find $\tan\theta$ where θ is the angle between $\vec{v}$ and $\vec{w}$.

39. Suppose $\vec{v} \cdot \vec{w} = 8$ and $\vec{v} \times \vec{w} = 12\vec{i} - 3\vec{j} + 4\vec{k}$ and that the angle between $\vec{v}$ and $\vec{w}$ is θ. Find
    - **(a)** $\tan\theta$
    - **(b)** θ

40. If $\vec{v} \cdot (\vec{i} + \vec{j} + \vec{k}) = 6$ and $\vec{v} \times (\vec{i} + \vec{j} + \vec{k}) = \vec{0}$, find $\vec{v}$.

41. Why does a baseball curve? The baseball in Figure 13.45 has velocity $\vec{v}$ meters/sec and is spinning at ω radians per second about an axis in the direction of the unit vector $\vec{n}$. The ball experiences a force, called the Magnus force,[3] $\vec{F}_M$, that is proportional to $\omega\vec{n} \times \vec{v}$.
    - **(a)** What is the effect on $\vec{F}_M$ of increasing ω?
    - **(b)** The ball in Figure 13.45 is moving away from you. What is the direction of the Magnus force?

Figure 13.45: Spinning baseball

[3]Named after German physicist Heinrich Magnus, who first described it in 1853.

42. The London Eye Ferris wheel rotates in a counterclockwise direction when viewed from the east and completes one full rotation in 30 minutes.[4]

(a) Let the x-axis point east, the y-axis north, and the z-axis up. Find the angular velocity, $\vec{\omega}$, of the London Eye.
(b) The passenger capsules of the London Eye are a distance of approximately 200 feet from the center. If the center of the London Eye is at $(0, 0, 0)$, find the velocity vector of a passenger capsule at its highest point.
(c) Find the speed of the capsule.

43. The point P in Figure 13.46 has position vector $\vec{v}$ obtained by rotating the position vector $\vec{r}$ of the point (x, y) by 90° counterclockwise about the origin.

(a) Use the geometric definition of the cross product to explain why $\vec{v} = \vec{k} \times \vec{r}$.
(b) Find the coordinates of P.

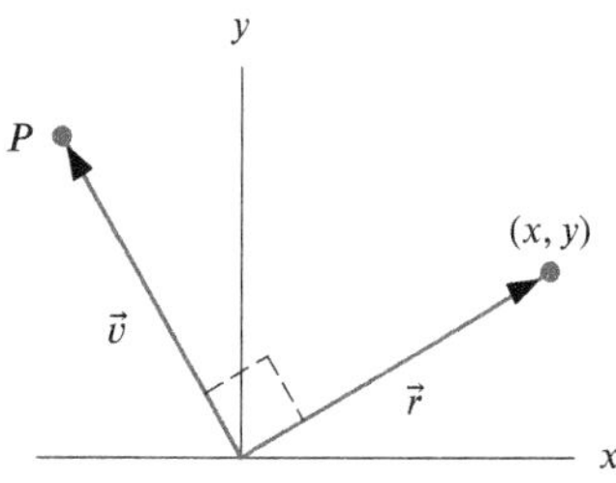

Figure 13.46

44. The points $P_1 = (0, 0, 0)$, $P_2 = (2, 4, 2)$, $P_3 = (3, 0, 0)$, and $P_4 = (5, 4, 2)$ are vertices of a parallelogram.

(a) Find the displacement vectors along each of the four sides. Check that these are equal in pairs.
(b) Find the area of the parallelogram.

In Problems 45–46, find an area vector for the parallelogram with given vertices.

45. $P = (2, 1, 1)$, $Q = (3, 3, 0)$, $R = (4, 0, 2)$, $S = (5, 2, 1)$

46. $P = (-1, -2, 0)$, $Q = (0, -1, 0)$, $R = (-2, -4, 1)$, $S = (-1, -3, 1)$

47. A parallelogram P formed by the vectors $\vec{v} = \vec{i} - 2\vec{j} + \vec{k}$ and $\vec{w} = \vec{i} + 2\vec{j} - 2\vec{k}$ has area vector $\vec{A} = \vec{v} \times \vec{w} = 2\vec{i} + 3\vec{j} + 4\vec{k}$. Find area vectors of each of the following parallelograms and explain how they are related to $\vec{A}$.

(a) The parallelogram obtained by projecting P onto the xy-plane.
(b) The parallelogram obtained by projecting P onto the xz-plane.
(c) The parallelogram obtained by projecting P onto the yz-plane.

[4] http://en.wikipedia.org, accessed May 12, 2016.

48. Using the parallelogram in Problem 44 as a base, create a parallelopiped with side $\overrightarrow{P_1P_5}$ where $P_5 = (1, 0, 4)$. Find the volume of this parallelepiped.

In Problems 49–51, if $0 \le \theta \le \pi$, what are the possible values for the angle, θ, between two nonzero vectors $\vec{v}$ and $\vec{w}$ satisfying the inequality?

49. $|\vec{v} \cdot \vec{w}| = \|\vec{v} \times \vec{w}\|$

50. $|\vec{v} \cdot \vec{w}| < \|\vec{v} \times \vec{w}\|$

51. $|\vec{v} \cdot \vec{w}| > \|\vec{v} \times \vec{w}\|$

52. Use a parallelepiped to show that $\vec{a} \cdot (\vec{b} \times \vec{c}) = (\vec{a} \times \vec{b}) \cdot \vec{c}$ for any vectors $\vec{a}$, $\vec{b}$, and $\vec{c}$.

53. Figure 13.47 shows the tetrahedron determined by three vectors $\vec{a}, \vec{b}, \vec{c}$. The *area vector* of a face is a vector perpendicular to the face, pointing outward, whose magnitude is the area of the face. Show that the sum of the four outward-pointing area vectors of the faces equals the zero vector.

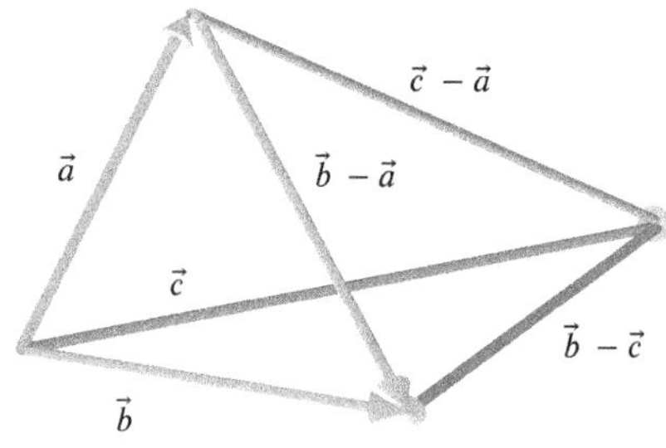

Figure 13.47

In Problems 54–56, find the vector representing the area of a surface. The magnitude of the vector equals the magnitude of the area; the direction is perpendicular to the surface. Since there are two perpendicular directions, we pick one by giving an orientation for the surface.

54. The rectangle with vertices $(0, 0, 0)$, $(0, 1, 0)$, $(2, 1, 0)$, and $(2, 0, 0)$, oriented so that it faces downward.

55. The circle of radius 2 in the yz-plane, facing in the direction of the positive x-axis.

56. The triangle ABC, oriented upward, where $A = (1, 2, 3)$, $B = (3, 1, 2)$, and $C = (2, 1, 3)$.

57. This problem relates the area of a parallelogram S lying in the plane $z = mx + ny + c$ to the area of its projection R in the xy-plane. Let S be determined by the vectors $\vec{u} = u_1\vec{i} + u_2\vec{j} + u_3\vec{k}$ and $\vec{v} = v_1\vec{i} + v_2\vec{j} + v_3\vec{k}$. See Figure 13.48.

(a) Find the area of S.
(b) Find the area of R.
(c) Find m and n in terms of the components of $\vec{u}$ and $\vec{v}$.
(d) Show that

$$\text{Area of } S = \sqrt{1 + m^2 + n^2} \cdot \text{Area of } R.$$

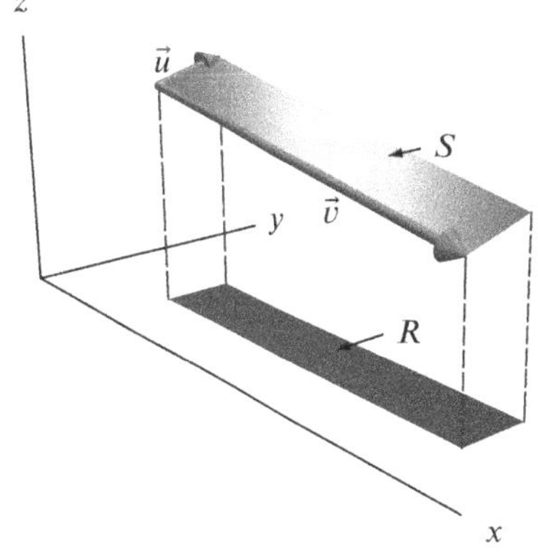

Figure 13.48

Strengthen Your Understanding

In Problems 58–59, explain what is wrong with the statement.

58. There is only one unit vector perpendicular to two non-parallel vectors in 3-space.

59. $\vec{u} \times \vec{v} = \vec{0}$ when $\vec{u}$ and $\vec{v}$ are perpendicular.

In Problems 60–61, give an example of:

60. A vector $\vec{u}$ whose cross product with $\vec{v} = \vec{i} + \vec{j}$ is parallel to $\vec{k}$.

61. A vector $\vec{v}$ such that $\|\vec{u} \times \vec{v}\| = 10$, where $\vec{u} = 3\vec{i} + 4\vec{j}$.

Are the statements in Problems 62–72 true or false? Give reasons for your answer.

62. $\vec{u} \times \vec{v}$ is a vector.

63. $\vec{u} \times \vec{v}$ has direction parallel to both $\vec{u}$ and $\vec{v}$.

64. $\|\vec{u} \times \vec{v}\| = \|\vec{u}\|\|\vec{v}\|$.

65. $(\vec{i} \times \vec{j}) \cdot \vec{k} = \vec{i} \cdot (\vec{j} \times \vec{k})$.

66. If $\vec{v}$ is a nonzero vector and $\vec{v} \times \vec{u} = \vec{v} \times \vec{w}$, then $\vec{u} = \vec{w}$.

67. The value of $\vec{v} \cdot (\vec{v} \times \vec{w})$ is always 0.

68. The value of $\vec{v} \times \vec{w}$ is never the same as $\vec{v} \cdot \vec{w}$.

69. The area of the triangle with two sides given by $\vec{i} + \vec{j}$ and $\vec{j} + 2\vec{k}$ is $3/2$.

70. Given a nonzero vector $\vec{v}$ in 3-space, there is a nonzero vector $\vec{w}$ such that $\vec{v} \times \vec{w} = \vec{0}$.

71. It is never true that $\vec{v} \times \vec{w} = \vec{w} \times \vec{v}$.

72. Two points are circling an axis at a rate of 5 rad/sec. The point closer to the axis has the greater speed.

Online Resource: Review problems and Projects

Chapter Fourteen

DIFFERENTIATING FUNCTIONS OF SEVERAL VARIABLES

Contents

14.1 THE PARTIAL DERIVATIVE

The derivative of a one-variable function measures its rate of change. In this section we see how a two-variable function has two rates of change: one as x changes (with y held constant) and one as y changes (with x held constant).

Rate of Change of Temperature in a Metal Rod: a One-Variable Problem

Imagine an unevenly heated metal rod lying along the x-axis, with its left end at the origin and x measured in meters. (See Figure 14.1.) Let $u(x)$ be the temperature (in °C) of the rod at the point x. Table 14.1 gives values of $u(x)$. We see that the temperature increases as we move along the rod, reaching its maximum at $x = 4$, after which it starts to decrease.

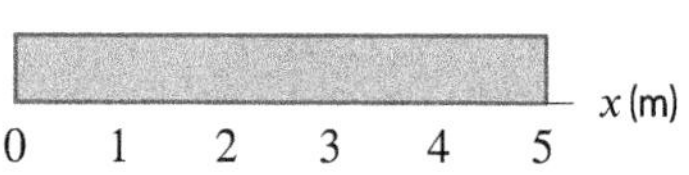

Figure 14.1: Unevenly heated metal rod

Table 14.1 *Temperature $u(x)$ of the rod*

x (m)	0	1	2	3	4	5
$u(x)$ (°C)	125	128	135	160	175	160

Example 1 Estimate the derivative $u'(2)$ using Table 14.1 and explain what the answer means in terms of temperature.

Solution The derivative $u'(2)$ is defined as a limit of difference quotients:

$$u'(2) = \lim_{h\to 0} \frac{u(2+h) - u(2)}{h}.$$

Choosing $h = 1$ so that we can use the data in Table 14.1, we get

$$u'(2) \approx \frac{u(2+1) - u(2)}{1} = \frac{160 - 135}{1} = 25.$$

This means that the temperature increases at a rate of approximately 25°C per meter as we go from left to right, past $x = 2$.

Rate of Change of Temperature in a Metal Plate

Imagine an unevenly heated thin rectangular metal plate lying in the xy-plane with its lower left corner at the origin and x and y measured in meters. The temperature (in °C) at the point (x, y) is $T(x, y)$. See Figure 14.2 and Table 14.2. How does T vary near the point $(2, 1)$? We consider the horizontal line $y = 1$ containing the point $(2, 1)$. The temperature along this line is the cross section, $T(x, 1)$, of the function $T(x, y)$ with $y = 1$. Suppose we write $u(x) = T(x, 1)$.

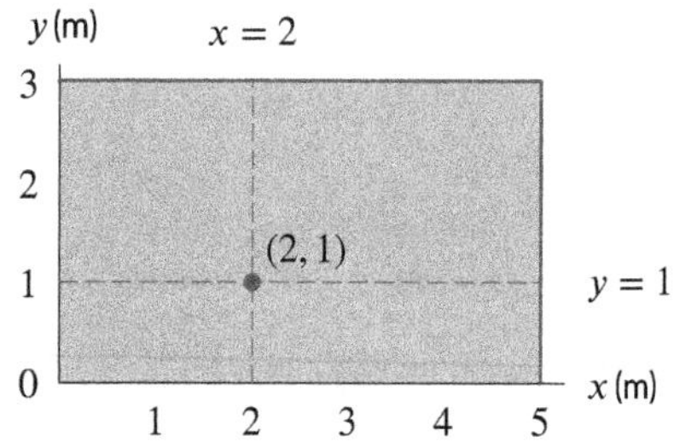

Figure 14.2: Unevenly heated metal plate

Table 14.2 *Temperature (°C) of a metal plate*

y (m) \ x (m)	0	1	2	3	4	5
3	85	90	**110**	135	155	180
2	100	110	**120**	145	190	170
1	**125**	**128**	**135**	**160**	**175**	**160**
0	120	135	**155**	160	160	150

What is the meaning of the derivative $u'(2)$? It is the rate of change of temperature T *in the x-direction* at the point $(2, 1)$, keeping y fixed. Denote this rate of change by $T_x(2, 1)$, so that

$$T_x(2, 1) = u'(2) = \lim_{h\to 0} \frac{u(2+h) - u(2)}{h} = \lim_{h\to 0} \frac{T(2+h, 1) - T(2, 1)}{h}.$$

We call $T_x(2,1)$ the *partial derivative of T with respect to x at the point* $(2, 1)$. Taking $h = 1$, we can read values of T from the row with $y = 1$ in Table 14.2, giving

$$T_x(2,1) \approx \frac{T(3,1) - T(2,1)}{1} = \frac{160 - 135}{1} = 25^\circ\text{C/m}.$$

The fact that $T_x(2,1)$ is positive means that the temperature of the plate is increasing as we move past the point $(2, 1)$ in the direction of increasing x (that is, horizontally from left to right in Figure 14.2).

Example 2 Estimate the rate of change of T in the y-direction at the point $(2, 1)$.

Solution The temperature along the line $x = 2$ is the cross-section of T with $x = 2$, that is, the function $v(y) = T(2, y)$. If we denote the rate of change of T in the y-direction at $(2, 1)$ by $T_y(2,1)$, then

$$T_y(2,1) = v'(1) = \lim_{h\to 0} \frac{v(1+h) - v(1)}{h} = \lim_{h\to 0} \frac{T(2, 1+h) - T(2,1)}{h}.$$

We call $T_y(2,1)$ the *partial derivative of T with respect to y at the point* $(2, 1)$. Taking $h = 1$ so that we can use the column with $x = 2$ in Table 14.2, we get

$$T_y(2,1) \approx \frac{T(2, 1+1) - T(2,1)}{1} = \frac{120 - 135}{1} = -15^\circ\text{C/m}.$$

The fact that $T_y(2,1)$ is negative means that at $(2, 1)$, the temperature decreases as y increases..

Definition of the Partial Derivative

We study the influence of x and y separately on the value of the function $f(x, y)$ by holding one fixed and letting the other vary. This leads to the following definitions.

Partial Derivatives of f with Respect to x and y

For all points at which the limits exist, we define the **partial derivatives at the point (a, b)** by

$$f_x(a,b) = \begin{matrix}\text{Rate of change of } f \text{ with respect to } x \\ \text{at the point } (a,b)\end{matrix} = \lim_{h\to 0} \frac{f(a+h, b) - f(a,b)}{h},$$

$$f_y(a,b) = \begin{matrix}\text{Rate of change of } f \text{ with respect to } y \\ \text{at the point } (a,b)\end{matrix} = \lim_{h\to 0} \frac{f(a, b+h) - f(a,b)}{h}.$$

If we let a and b vary, we have the **partial derivative functions** $f_x(x, y)$ and $f_y(x, y)$.

Just as with ordinary derivatives, there is an alternative notation:

Alternative Notation for Partial Derivatives

If $z = f(x, y)$, we can write

$$f_x(x,y) = \frac{\partial z}{\partial x} \quad \text{and} \quad f_y(x,y) = \frac{\partial z}{\partial y},$$

$$f_x(a,b) = \left.\frac{\partial z}{\partial x}\right|_{(a,b)} \quad \text{and} \quad f_y(a,b) = \left.\frac{\partial z}{\partial y}\right|_{(a,b)}.$$

We use the symbol ∂ to distinguish partial derivatives from ordinary derivatives. In cases where the independent variables have names different from x and y, we adjust the notation accordingly. For example, the partial derivatives of $f(u, v)$ are denoted by f_u and f_v.

Visualizing Partial Derivatives on a Graph

The ordinary derivative of a one-variable function is the slope of its graph. How do we visualize the partial derivative $f_x(a, b)$? The graph of the one-variable function $f(x, b)$ is the curve where the vertical plane $y = b$ cuts the graph of $f(x, y)$. (See Figure 14.3.) Thus, $f_x(a, b)$ is the slope of the tangent line to this curve at $x = a$.

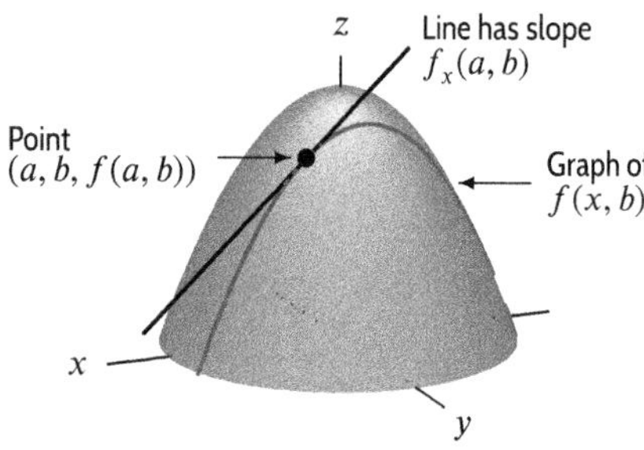

Figure 14.3: The curve $z = f(x, b)$ on the graph of f has slope $f_x(a, b)$ at $x = a$

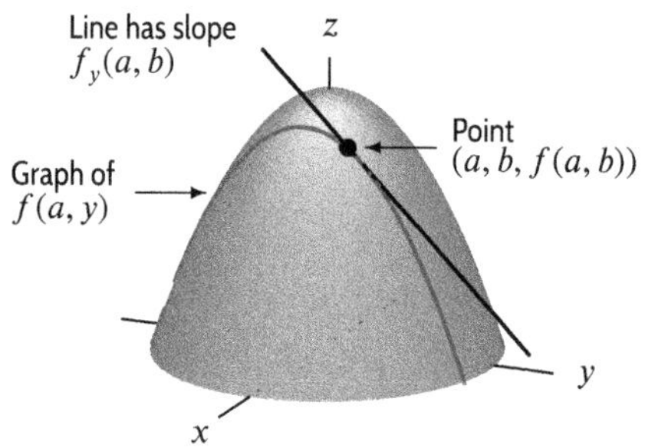

Figure 14.4: The curve $z = f(a, y)$ on the graph of f has slope $f_y(a, b)$ at $y = b$

Similarly, the graph of the function $f(a, y)$ is the curve where the vertical plane $x = a$ cuts the graph of f, and the partial derivative $f_y(a, b)$ is the slope of this curve at $y = b$. (See Figure 14.4.)

Example 3 At each point labeled on the graph of the surface $z = f(x, y)$ in Figure 14.5, say whether each partial derivative is positive or negative.

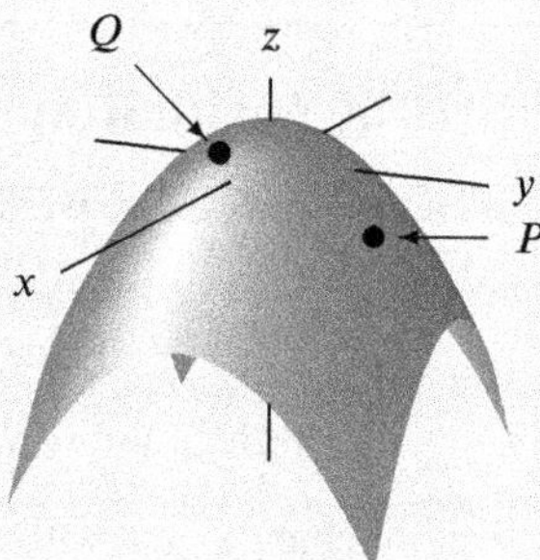

Figure 14.5: Decide the signs of f_x and f_y at P and Q

Solution The positive x-axis points out of the page. Imagine heading off in this direction from the point marked P; we descend steeply. So the partial derivative with respect to x is negative at P, with quite a large absolute value. The same is true for the partial derivative with respect to y at P, since there is also a steep descent in the positive y-direction.

At the point marked Q, heading in the positive x-direction results in a gentle descent, whereas heading in the positive y-direction results in a gentle ascent. Thus, the partial derivative f_x at Q is negative but small (that is, near zero), and the partial derivative f_y is positive but small.

Estimating Partial Derivatives from a Contour Diagram

The graph of a function $f(x, y)$ often makes clear the sign of the partial derivatives. However, numerical estimates of these derivatives are more easily made from a contour diagram than a surface graph. If we move parallel to one of the axes on a contour diagram, the partial derivative is the rate of change of the value of the function on the contours. For example, if the values on the contours are increasing as we move in the positive direction, then the partial derivative must be positive.

Example 4 Figure 14.6 shows the contour diagram for the temperature $H(x, t)$ (in °C) in a room as a function of distance x (in meters) from a heater and time t (in minutes) after the heater has been turned on. What are the signs of $H_x(10, 20)$ and $H_t(10, 20)$? Estimate these partial derivatives and explain the answers in practical terms.

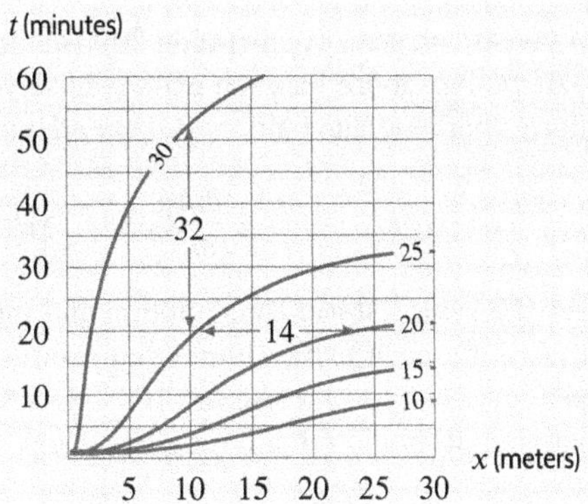

Figure 14.6: Temperature in a heated room: Heater at $x = 0$ is turned on at $t = 0$

Solution The point $(10, 20)$ is nearly on the $H = 25$ contour. As x increases, we move toward the $H = 20$ contour, so H is decreasing and $H_x(10, 20)$ is negative. This makes sense because the $H = 30$ contour is to the left: As we move further from the heater, the temperature drops. On the other hand, as t increases, we move toward the $H = 30$ contour, so H is increasing; as t decreases H decreases. Thus, $H_t(10, 20)$ is positive. This says that as time passes, the room warms up.

To estimate the partial derivatives, use a difference quotient. Looking at the contour diagram, we see there is a point on the $H = 20$ contour about 14 units to the right of the point $(10, 20)$. Hence, H decreases by 5 when x increases by 14, so we find

$$\text{Rate of change of } H \text{ with respect to } x = H_x(10, 20) \approx \frac{-5}{14} \approx -0.36°\text{C/meter}.$$

This means that near the point 10 m from the heater, after 20 minutes the temperature drops about 0.36, or one third, of a degree, for each meter we move away from the heater.

To estimate $H_t(10, 20)$, we notice that the $H = 30$ contour is about 32 units directly above the point $(10, 20)$. So H increases by 5 when t increases by 32. Hence,

$$\text{Rate of change of } H \text{ with respect to } t = H_t(10, 20) \approx \frac{5}{32} = 0.16°\text{C/minute}.$$

This means that after 20 minutes the temperature is going up about 0.16, or 1/6, of a degree each minute at the point 10 m from the heater.

Using Units to Interpret Partial Derivatives

The meaning of a partial derivative can often be explained using units.

Example 5 Suppose that your weight w in pounds is a function $f(c, n)$ of the number c of calories you consume daily and the number n of minutes you exercise daily. Using the units for w, c and n, interpret in everyday terms the statements

$$\frac{\partial w}{\partial c}(2000, 15) = 0.02 \quad \text{and} \quad \frac{\partial w}{\partial n}(2000, 15) = -0.025.$$

Solution The units of $\partial w/\partial c$ are pounds per calorie. The statement

$$\frac{\partial w}{\partial c}(2000, 15) = 0.02$$

means that if you are presently consuming 2000 calories daily and exercising 15 minutes daily, you will weigh 0.02 pounds more for each extra calorie you consume daily, or about 2 pounds for each extra 100 calories per day. The units of $\partial w/\partial n$ are pounds per minute. The statement

$$\frac{\partial w}{\partial n}(2000, 15) = -0.025$$

means that for the same calorie consumption and number of minutes of exercise, you will weigh 0.025 pounds less for each extra minute you exercise daily, or about 1 pound less for each extra 40 minutes per day. So if you eat an extra 100 calories each day and exercise about 80 minutes more each day, your weight should remain roughly steady.

Exercises and Problems for Section 14.1

EXERCISES

1. Given the following table of values for $z = f(x, y)$, estimate $f_x(3, 2)$ and $f_y(3, 2)$, assuming they exist.

$x \backslash y$	0	2	5
1	1	2	4
3	−1	1	2
6	−3	0	0

2. Using difference quotients, estimate $f_x(3, 2)$ and $f_y(3, 2)$ for the function given by

$$f(x, y) = \frac{x^2}{y + 1}.$$

[Recall: A difference quotient is an expression of the form $(f(a + h, b) - f(a, b))/h$.]

3. Use difference quotients with $\Delta x = 0.1$ and $\Delta y = 0.1$ to estimate $f_x(1, 3)$ and $f_y(1, 3)$, where

$$f(x, y) = e^{-x} \sin y.$$

Then give better estimates by using $\Delta x = 0.01$ and $\Delta y = 0.01$.

4. The price P in dollars to purchase a used car is a function of its original cost, C, in dollars, and its age, A, in years.

(a) What are the units of $\partial P/\partial A$?
(b) What is the sign of $\partial P/\partial A$ and why?
(c) What are the units of $\partial P/\partial C$?
(d) What is the sign of $\partial P/\partial C$ and why?

5. Your monthly car payment in dollars is $P = f(P_0, t, r)$, where $\$P_0$ is the amount you borrowed, t is the number of months it takes to pay off the loan, and $r\%$ is the interest rate. What are the units, the financial meaning, and the signs of $\partial P/\partial t$ and $\partial P/\partial r$?

6. A drug is injected into a patient's blood vessel. The function $c = f(x, t)$ represents the concentration of the drug at a distance x mm in the direction of the blood flow measured from the point of injection and at time t seconds since the injection. What are the units of the following partial derivatives? What are their practical interpretations? What do you expect their signs to be?

(a) $\partial c/\partial x$ **(b)** $\partial c/\partial t$

7. You borrow $\$A$ at an interest rate of $r\%$ (per month) and pay it off over t months by making monthly payments of $P = g(A, r, t)$ dollars. In financial terms, what do the following statements tell you?

(a) $g(8000, 1, 24) = 376.59$
(b) $\left.\dfrac{\partial g}{\partial A}\right|_{(8000,1,24)} = 0.047$
(c) $\left.\dfrac{\partial g}{\partial r}\right|_{(8000,1,24)} = 44.83$

8. The sales of a product, $S = f(p, a)$, are a function of the price, p, of the product (in dollars per unit) and the amount, a, spent on advertising (in thousands of dollars).

(a) Do you expect f_p to be positive or negative? Why?
(b) Explain the meaning of the statement $f_a(8, 12) = 150$ in terms of sales.

9. The quantity, Q, of beef purchased at a store, in kilograms per week, is a function of the price of beef, b, and the price of chicken, c, both in dollars per kilogram.

(a) Do you expect $\partial Q/\partial b$ to be positive or negative? Explain.

(b) Do you expect $\partial Q/\partial c$ to be positive or negative? Explain.
(c) Interpret the statement $\partial Q/\partial b = -213$ in terms of quantity of beef purchased.

In Exercises 10–15, a point A is shown on a contour diagram of a function $f(x, y)$.
(a) Evaluate $f(A)$.
(b) Is $f_x(A)$ positive, negative, or zero?
(c) Is $f_y(A)$ positive, negative, or zero?

10.
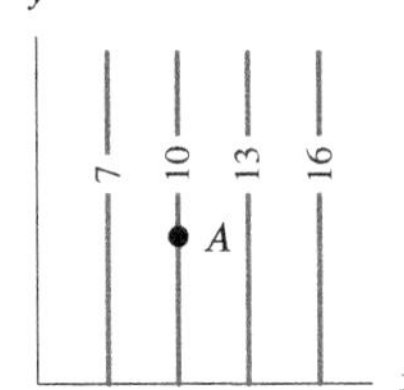

11.
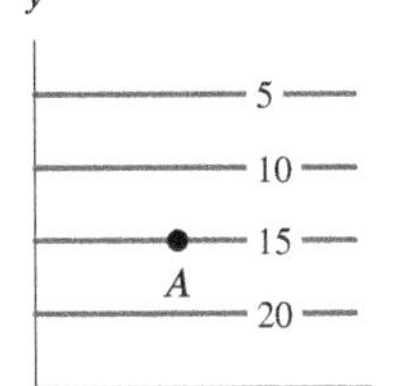

12.
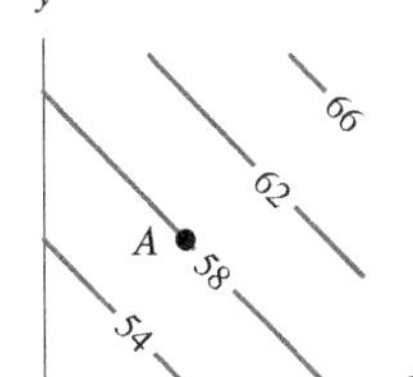

13.
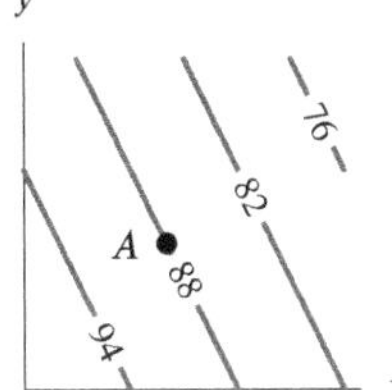

14.
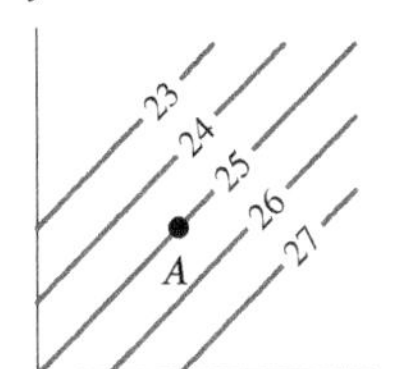

15.
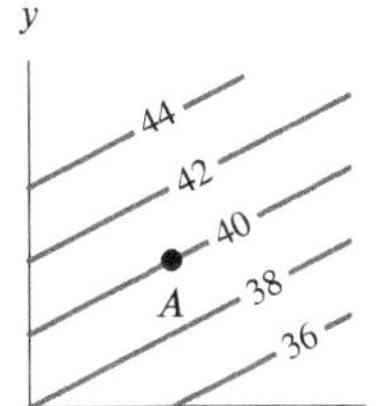

In Exercises 16–19, determine the sign of f_x and f_y at the point using the contour diagram of f in Figure 14.7.

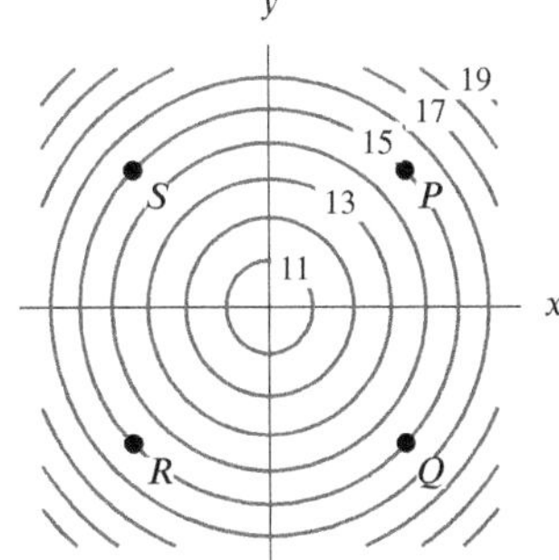

Figure 14.7

16. P
17. Q
18. R
19. S

20. Values of $f(x, y)$ are in Table 14.3. Assuming they exist, decide whether you expect the following partial derivatives to be positive or negative.

(a) $f_x(-2, -1)$
(b) $f_y(2, 1)$
(c) $f_x(2, 1)$
(d) $f_y(0, 3)$

Table 14.3

$x \backslash y$	−1	1	3	5
−2	7	3	2	1
0	8	5	3	2
2	10	7	5	4
4	13	10	8	7

PROBLEMS

21. Figure 14.8 is a contour diagram for $z = f(x, y)$. Is f_x positive or negative? Is f_y positive or negative? Estimate $f(2, 1)$, $f_x(2, 1)$, and $f_y(2, 1)$.

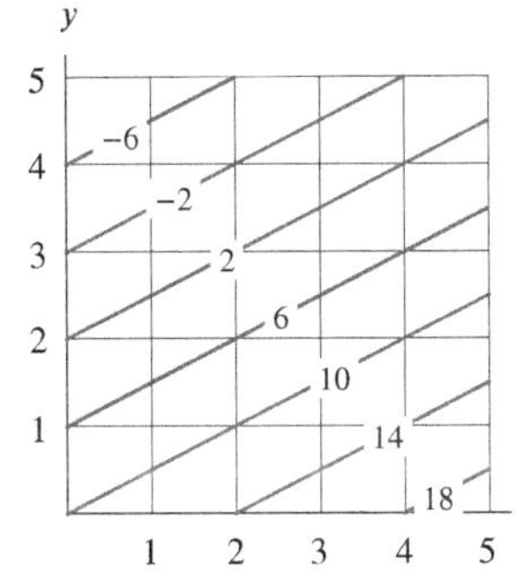

Figure 14.8

22. Approximate $f_x(3, 5)$ using the contour diagram of $f(x, y)$ in Figure 14.9.

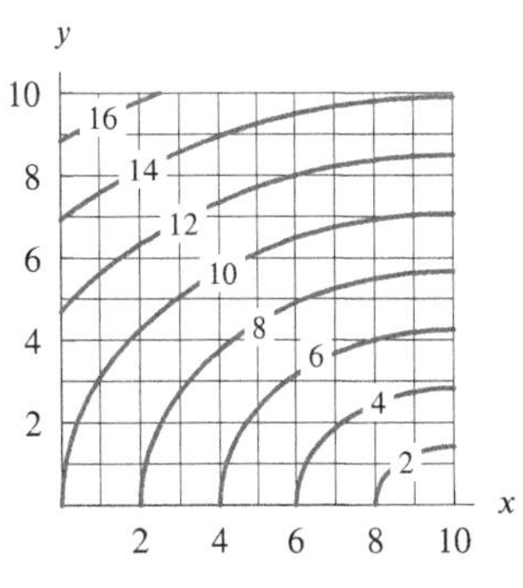

Figure 14.9

23. When riding your bike in winter, the windchill temperature is a measure of how cold you feel as a result of the induced breeze caused by your travel. If W represents windchill temperature (in °F) that you experience, then $W = f(T, v)$, where T is the actual air temperature (in °F) and v is your speed, in meters per second. Match each of the practical interpretations below with a mathematical statement that most accurately describes it be-

low. For the remaining mathematical statement, give a practical interpretation.

(i) "The faster you ride, the colder you'll feel."
(ii) "The warmer the day, the warmer you'll feel."

(a) $f_T(T, v) > 0$ **(b)** $f(0, v) \leq 0$ **(c)** $f_v(T, v) < 0$

24. People commuting to a city can choose to go either by bus or by train. The number of people who choose either method depends in part upon the price of each. Let $f(P_1, P_2)$ be the number of people who take the bus when P_1 is the price of a bus ride and P_2 is the price of a train ride. What can you say about the signs of $\partial f/\partial P_1$ and $\partial f/\partial P_2$? Explain your answers.

25. The average price of large cars getting low gas mileage ("gas guzzlers") is x and the average price of a gallon of gasoline is y. The number, q_1, of gas guzzlers bought in a year, depends on both x and y, so $q_1 = f(x, y)$. Similarly, if q_2 is the number of gallons of gas bought to fill gas guzzlers in a year, then $q_2 = g(x, y)$.

(a) What do you expect the signs of $\partial q_1/\partial x$ and $\partial q_2/\partial y$ to be? Explain.
(b) What do you expect the signs of $\partial q_1/\partial y$ and $\partial q_2/\partial x$ to be? Explain.

For Problems 26–28, refer to Table 12.2 on page 658 giving the temperature adjusted for wind chill, C, in °F, as a function $f(w, T)$ of the wind speed, w, in mph, and the temperature, T, in °F. The temperature adjusted for wind chill tells you how cold it feels, as a result of the combination of wind and temperature.

26. Estimate $f_w(10, 25)$. What does your answer mean in practical terms?

27. Estimate $f_T(5, 20)$. What does your answer mean in practical terms?

28. From Table 12.2 you can see that when the temperature is 20°F, the temperature adjusted for wind-chill drops by an average of about 0.8°F with every 1 mph increase in wind speed from 5 mph to 10 mph. Which partial derivative is this telling you about?

29. An experiment to measure the toxicity of formaldehyde yielded the data in Table 14.4. The values show the percent, $P = f(t, c)$, of rats surviving an exposure to formaldehyde at a concentration of c (in parts per million, ppm) after t months. Estimate $f_t(18, 6)$ and $f_c(18, 6)$. Interpret your answers in terms of formaldehyde toxicity.

Table 14.4

		Time t (months)					
		14	16	18	20	22	24
Conc. c (ppm)	0	100	100	100	99	97	95
	2	100	99	98	97	95	92
	6	96	95	93	90	86	80
	15	96	93	82	70	58	36

30. Figure 14.10 shows contours of $f(x, y)$ with values of f on the contours omitted. If $f_x(P) > 0$, find the sign:

(a) $f_y(P)$ **(b)** $f_y(Q)$ **(c)** $f_x(Q)$

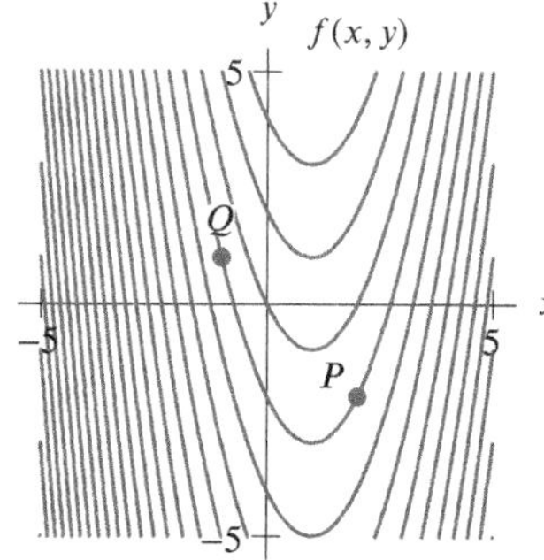

Figure 14.10

31. Figure 14.11 shows the contour diagram of $g(x, y)$. Mark the points on the contours where

(a) $g_x = 0$ **(b)** $g_y = 0$

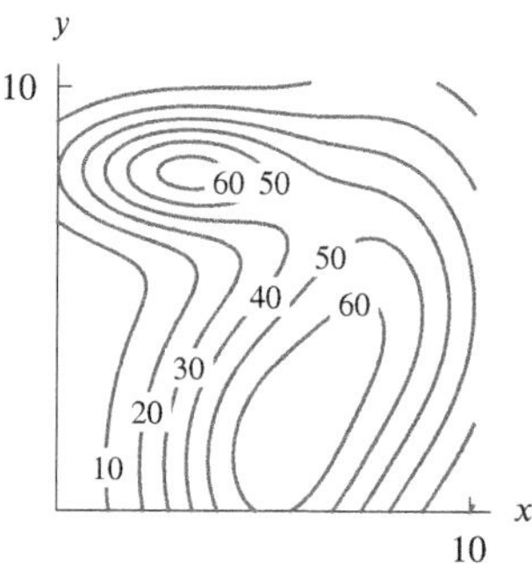

Figure 14.11

32. The surface $z = f(x, y)$ is shown in Figure 14.12. The points A and B are in the xy-plane.

(a) What is the sign of
(i) $f_x(A)$? (ii) $f_y(A)$?
(b) The point P in the xy-plane moves along a straight line from A to B. How does the sign of $f_x(P)$ change? How does the sign of $f_y(P)$ change?

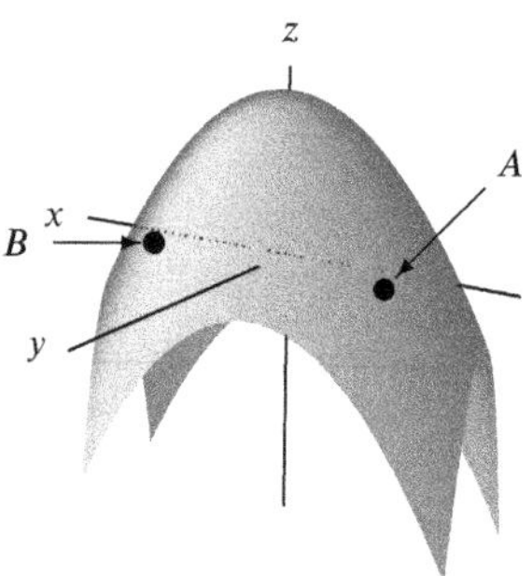

Figure 14.12

33. Figure 14.13 shows the saddle-shaped surface $z = f(x, y)$.

(a) What is the sign of $f_x(0, 5)$?
(b) What is the sign of $f_y(0, 5)$?

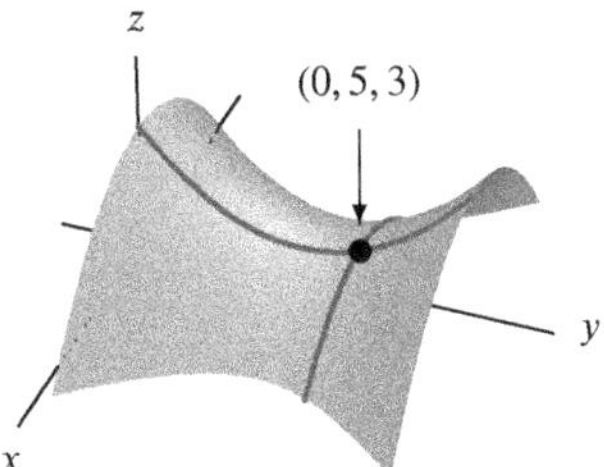

Figure 14.13

34. Figure 14.14 shows the graph of the function $f(x, y)$ on the domain $0 \le x \le 4$ and $0 \le y \le 4$. Use the graph to rank the following quantities in order from smallest to largest: $f_x(3, 2)$, $f_x(1, 2)$, $f_y(3, 2)$, $f_y(1, 2)$, 0.

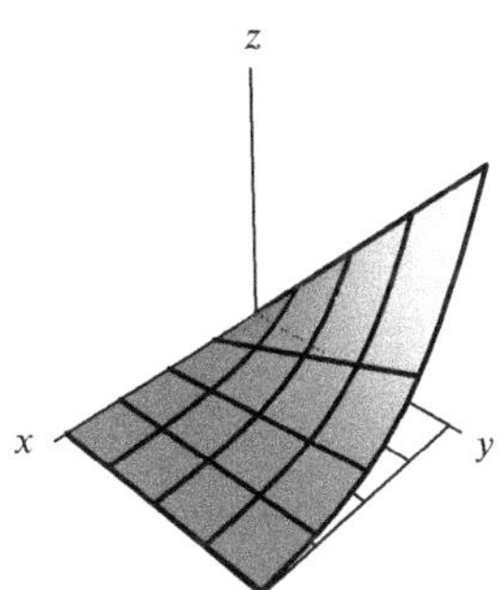

Figure 14.14

35. Figure 14.15 shows a contour diagram for the monthly payment P as a function of the interest rate, $r\%$, and the amount, L, of a 5-year loan. Estimate $\partial P/\partial r$ and $\partial P/\partial L$ at the following points. In each case, give the units and the everyday meaning of your answer.

(a) $r = 8$, $L = 4000$ (b) $r = 8$, $L = 6000$
(c) $r = 13$, $L = 7000$

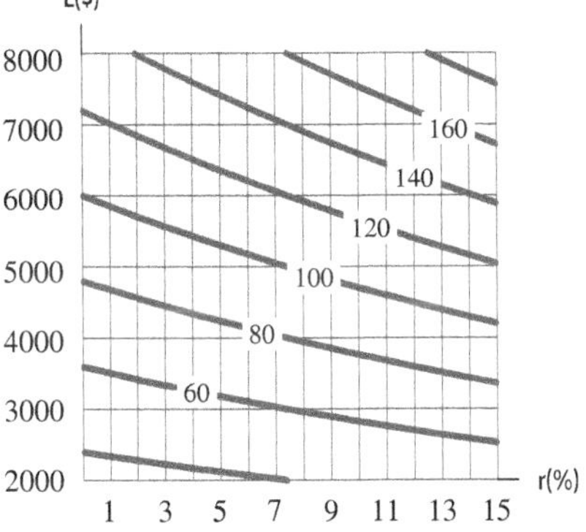

Figure 14.15

36. Figure 14.16 shows a contour diagram for the temperature T (in °C) along a wall in a heated room as a function of distance x along the wall and time t in minutes. Estimate $\partial T/\partial x$ and $\partial T/\partial t$ at the given points. Give units and interpret your answers.

(a) $x = 15, t = 20$ (b) $x = 5, t = 12$

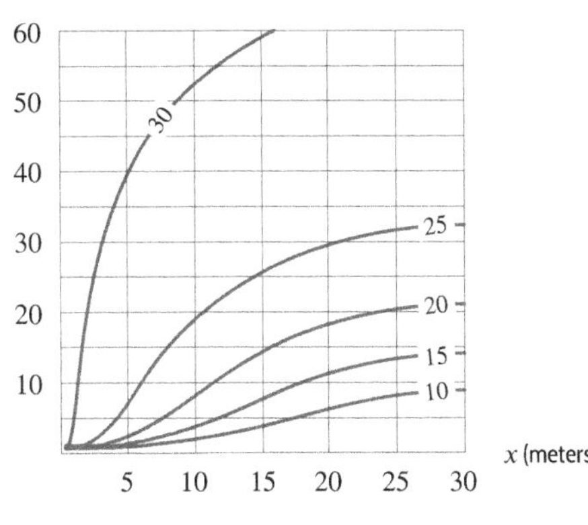

Figure 14.16

In Problems 37–39, we use Figure 14.17 to model the heat required to clear an airport of fog by heating the air. The amount of heat, $H(T, w)$, required (in calories per cubic meter of fog) is a function of the temperature T (in degrees Celsius) and the water content w (in grams per cubic meter of fog). Note that Figure 14.17 is not a contour diagram, but shows cross-sections of H with w fixed at 0.1, 0.2, 0.3, 0.4.

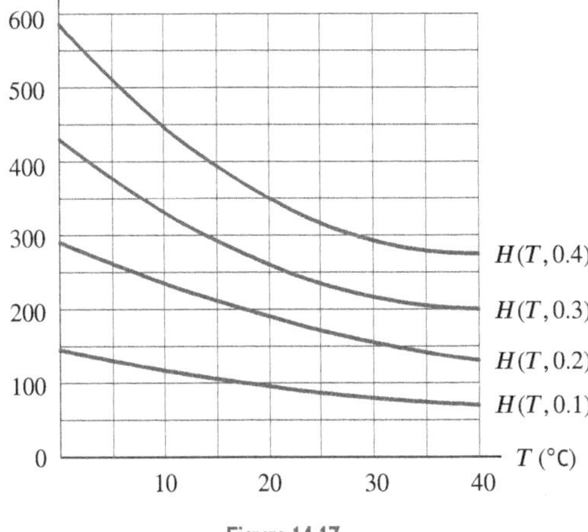

Figure 14.17

37. Use Figure 14.17 to estimate $H_T(10, 0.1)$. Interpret the partial derivative in practical terms.

38. Make a table of values for $H(T, w)$ from Figure 14.17, and use it to estimate $H_T(T, w)$ for $T = 10$, 20, and 30 and $w = 0.1, 0.2$, and 0.3.

39. Repeat Problem 38 for $H_w(T, w)$ at $T = 10$, 20, and 30 and $w = 0.1$, 0.2, and 0.3. What is the practical meaning of these partial derivatives?

40. The cardiac output, represented by c, is the volume of blood flowing through a person's heart per unit time. The systemic vascular resistance (SVR), represented by s, is the resistance to blood flowing through veins and arteries. Let p be a person's blood pressure. Then p is a function of c and s, so $p = f(c, s)$.

    (a) What does $\partial p/\partial c$ represent?

    Suppose now that $p = kcs$, where k is a constant.

    (b) Sketch the level curves of p. What do they represent? Label your axes.
    (c) For a person with a weak heart, it is desirable to have the heart pumping against less resistance, while maintaining the same blood pressure. Such a person may be given the drug nitroglycerine to decrease the SVR and the drug dopamine to increase the cardiac output. Represent this on a graph showing level curves. Put a point A on the graph representing the person's state before drugs are given and a point B for after.
    (d) Right after a heart attack, a patient's cardiac output drops, thereby causing the blood pressure to drop. A common mistake made by medical residents is to get the patient's blood pressure back to normal by using drugs to increase the SVR, rather than by increasing the cardiac output. On a graph of the level curves of p, put a point D representing the patient before the heart attack, a point E representing the patient right after the heart attack, and a third point F representing the patient after the resident has given the drugs to increase the SVR.

41. In each case, give a possible contour diagram for the function $f(x, y)$ if

    (a) $f_x > 0$ and $f_y > 0$ (b) $f_x > 0$ and $f_y < 0$
    (c) $f_x < 0$ and $f_y > 0$ (d) $f_x < 0$ and $f_y < 0$

In Problems 42–45, give a possible contour diagram for the function $f(x, y)$ if

42. $f_x = 0$, $f_y \neq 0$
43. $f_y = 0$, $f_x \neq 0$
44. $f_x = 1$
45. $f_y = -2$

Strengthen Your Understanding

In Problems 46–47, explain what is wrong with the statement.

46. For $f(x, y)$, $\partial f/\partial x$ has the same units as $\partial f/\partial y$.
47. The partial derivative with respect to y is not defined for functions such as $f(x, y) = x^2 + 5$ that have a formula that does not contain y explicitly.

In Problems 48–49, give an example of:

48. A table of values with three rows and three columns of a linear function $f(x, y)$ with $f_x < 0$ and $f_y > 0$.
49. A function $f(x, y)$ with $f_x > 0$ and $f_y < 0$ everywhere.

Are the statements in Problems 50–60 true or false? Give reasons for your answer.

50. If $f(x, y)$ is a function of two variables and $f_x(10, 20)$ is defined, then $f_x(10, 20)$ is a scalar.
51. If $f(x, y) = x^2 + y^2$, then $f_y(1, 1) < 0$.
52. If the graph of $f(x, y)$ is a hemisphere centered at the origin, then $f_x(0, 0) = f_y(0, 0) = 0$.
53. If $P = f(T, V)$ is a function expressing the pressure P (in grams/cm^3) of gas in a piston in terms of the temperature T (in degrees °C) and volume V (in cm^3), then $\partial P/\partial V$ has units of grams.
54. If $f_x(a, b) > 0$, then the values of f decrease as we move in the negative x-direction near (a, b).
55. If $g(r, s) = r^2 + s$, then for fixed s, the partial derivative g_r increases as r increases.
56. Let $P = f(m, d)$ be the purchase price (in dollars) of a used car that has m miles on its engine and originally cost d dollars when new. Then $\partial P/\partial m$ and $\partial P/\partial d$ have the same sign.
57. If $f(x, y)$ is a function with the property that $f_x(x, y)$ and $f_y(x, y)$ are both constant, then f is linear.
58. If $f(x, y)$ has $f_x(a, b) = f_y(a, b) = 0$ at the point (a, b), then f is constant everywhere.
59. If $f_x = 0$ and $f_y \neq 0$, then the contours of $f(x, y)$ are horizontal lines.
60. If the contours of $f(x, y)$ are vertical lines, then $f_y = 0$.

14.2 COMPUTING PARTIAL DERIVATIVES ALGEBRAICALLY

Since the partial derivative $f_x(x, y)$ is the ordinary derivative of the function $f(x, y)$ with y held constant and $f_y(x, y)$ is the ordinary derivative of $f(x, y)$ with x held constant, we can use all the differentiation formulas from one-variable calculus to find partial derivatives.

Example 1 Let $f(x, y) = \dfrac{x^2}{y + 1}$. Find $f_x(3, 2)$ algebraically.

Solution We use the fact that $f_x(3,2)$ equals the derivative of $f(x,2)$ at $x = 3$. Since

$$f(x,2) = \frac{x^2}{2+1} = \frac{x^2}{3},$$

differentiating with respect to x, we have

$$f_x(x,2) = \frac{\partial}{\partial x}\left(\frac{x^2}{3}\right) = \frac{2x}{3}, \qquad \text{and so} \qquad f_x(3,2) = 2.$$

Example 2 Compute the partial derivatives with respect to x and with respect to y for the following functions.
(a) $f(x,y) = y^2e^{3x}$ (b) $z = (3xy+2x)^5$ (c) $g(x,y) = e^{x+3y}\sin(xy)$

Solution (a) This is the product of a function of x (namely e^{3x}) and a function of y (namely y^2). When we differentiate with respect to x, we think of the function of y as a constant, and vice versa. Thus,

$$f_x(x,y) = y^2\frac{\partial}{\partial x}\left(e^{3x}\right) = 3y^2e^{3x},$$
$$f_y(x,y) = e^{3x}\frac{\partial}{\partial y}(y^2) = 2ye^{3x}.$$

(b) Here we use the chain rule:

$$\frac{\partial z}{\partial x} = 5(3xy+2x)^4\frac{\partial}{\partial x}(3xy+2x) = 5(3xy+2x)^4(3y+2),$$
$$\frac{\partial z}{\partial y} = 5(3xy+2x)^4\frac{\partial}{\partial y}(3xy+2x) = 5(3xy+2x)^4 3x = 15x(3xy+2x)^4.$$

(c) Since each function in the product is a function of both x and y, we need to use the product rule for each partial derivative:

$$g_x(x,y) = \left(\frac{\partial}{\partial x}(e^{x+3y})\right)\sin(xy) + e^{x+3y}\frac{\partial}{\partial x}(\sin(xy)) = e^{x+3y}\sin(xy) + e^{x+3y}y\cos(xy),$$
$$g_y(x,y) = \left(\frac{\partial}{\partial y}(e^{x+3y})\right)\sin(xy) + e^{x+3y}\frac{\partial}{\partial y}(\sin(xy)) = 3e^{x+3y}\sin(xy) + e^{x+3y}x\cos(xy).$$

For functions of three or more variables, we find partial derivatives by the same method: Differentiate with respect to one variable, regarding the other variables as constants. For a function $f(x,y,z)$, the partial derivative $f_x(a,b,c)$ gives the rate of change of f with respect to x along the line $y = b, z = c$.

Example 3 Find all the partial derivatives of $f(x,y,z) = \dfrac{x^2y^3}{z}$.

Solution To find $f_x(x,y,z)$, for example, we consider y and z as fixed, giving

$$f_x(x,y,z) = \frac{2xy^3}{z}, \quad \text{and} \quad f_y(x,y,z) = \frac{3x^2y^2}{z}, \quad \text{and} \quad f_z(x,y,z) = -\frac{x^2y^3}{z^2}.$$

Interpretation of Partial Derivatives

Example 4 A vibrating guitar string, originally at rest along the x-axis, is shown in Figure 14.18. Let x be the distance in meters from the left end of the string. At time t seconds the point x has been displaced $y = f(x, t)$ meters vertically from its rest position, where

$$y = f(x,t) = 0.003 \sin(\pi x) \sin(2765t).$$

Evaluate $f_x(0.3, 1)$ and $f_t(0.3, 1)$ and explain what each means in practical terms.

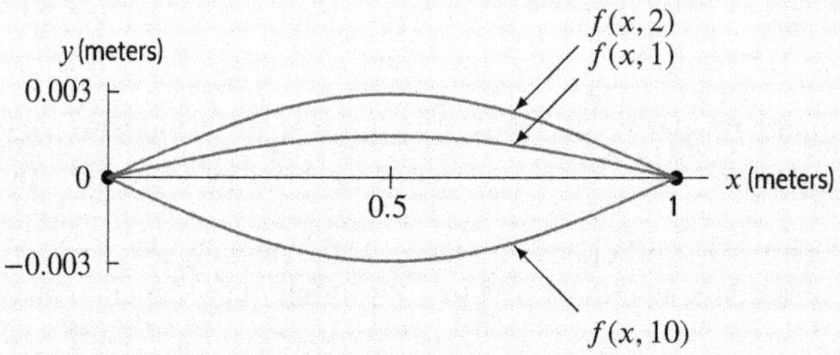

Figure 14.18: The position of a vibrating guitar string at several different times: Graph of $f(x,t)$ for $t = 1, 2, 10$.

Solution Differentiating $f(x,t) = 0.003 \sin(\pi x) \sin(2765t)$ with respect to x, we have

$$f_x(x,t) = 0.003\pi \cos(\pi x) \sin(2765t).$$

In particular, substituting $x = 0.3$ and $t = 1$ gives

$$f_x(0.3, 1) = 0.003\pi \cos(\pi(0.3)) \sin(2765) \approx 0.002.$$

To see what $f_x(0.3, 1)$ means, think about the function $f(x, 1)$. The graph of $f(x, 1)$ in Figure 14.19 is a snapshot of the string at the time $t = 1$. Thus, the derivative $f_x(0.3, 1)$ is the slope of the string at the point $x = 0.3$ at the instant when $t = 1$.

Similarly, taking the derivative of $f(x,t) = 0.003 \sin(\pi x) \sin(2765t)$ with respect to t, we get

$$f_t(x,t) = (0.003)(2765) \sin(\pi x) \cos(2765t) = 8.3 \sin(\pi x) \cos(2765t).$$

Since $f(x,t)$ is in meters and t is in seconds, the derivative $f_t(0.3, 1)$ is in m/sec. Thus, substituting $x = 0.3$ and $t = 1$,

$$f_t(0.3, 1) = 8.3 \sin(\pi(0.3)) \cos(2765(1)) \approx 6 \text{ m/sec}.$$

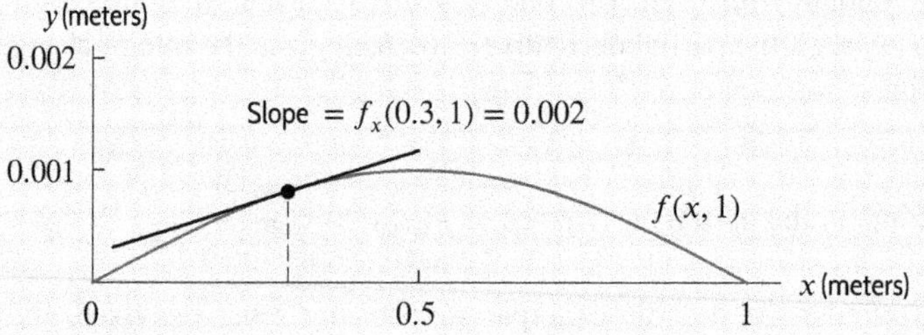

Figure 14.19: Graph of $f(x, 1)$: Snapshot of the shape of the string at $t = 1$ sec

To see what $f_t(0.3, 1)$ means, think about the function $f(0.3, t)$. The graph of $f(0.3, t)$ is a position versus time graph that tracks the up-and-down movement of the point on the string where $x = 0.3$.

(See Figure 14.20.) The derivative $f_t(0.3, 1) = 6$ m/sec is the velocity of that point on the string at time $t = 1$. The fact that $f_t(0.3, 1)$ is positive indicates that the point is moving upward when $t = 1$.

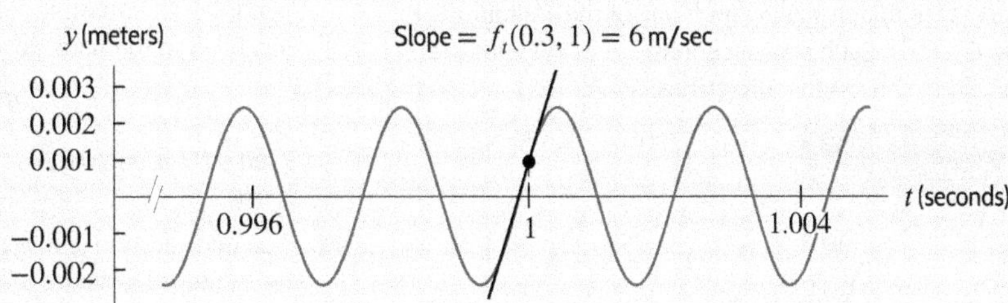

Figure 14.20: Graph of $f(0.3, t)$: Position versus time graph of the point $x = 0.3$ m from the left end of the guitar string

Exercises and Problems for Section 14.2

EXERCISES

1. **(a)** If $f(x, y) = 2x^2 + xy + y^2$, approximate $f_y(3, 2)$ using $\Delta y = 0.01$.
   (b) Find the exact value of $f_y(3, 2)$.

In Exercises 2–40, find the partial derivatives. The variables are restricted to a domain on which the function is defined.

2. f_x and f_y if $f(x, y) = 5x^2y^3 + 8xy^2 - 3x^2$
3. $f_x(1, 2)$ and $f_y(1, 2)$ if $f(x, y) = x^3 + 3x^2y - 2y^2$
4. $\frac{\partial}{\partial y}(3x^5y^7 - 32x^4y^3 + 5xy)$
5. $\frac{\partial z}{\partial x}$ and $\frac{\partial z}{\partial y}$ if $z = (x^2 + x - y)^7$
6. f_x and f_y if $f(x, y) = A^\alpha x^{\alpha+\beta} y^{1-\alpha-\beta}$
7. f_x and f_y if $f(x, y) = \ln(x^{0.6}y^{0.4})$
8. z_x if $z = \frac{1}{2x^2ay} + \frac{3x^5abc}{y}$
9. z_x if $z = x^2y + 2x^5y$
10. $\frac{\partial}{\partial x}(a\sqrt{x})$
11. V_r if $V = \frac{1}{3}\pi r^2 h$
12. $\frac{\partial}{\partial T}\left(\frac{2\pi r}{T}\right)$
13. $\frac{\partial}{\partial x}(xe^{\sqrt{xy}})$
14. $\frac{\partial}{\partial t}e^{\sin(x+ct)}$
15. F_m if $F = mg$
16. a_v if $a = \frac{v^2}{r}$
17. $\frac{\partial A}{\partial h}$ if $A = \frac{1}{2}(a + b)h$
18. $\frac{\partial}{\partial m}\left(\frac{1}{2}mv^2\right)$
19. $\frac{\partial}{\partial B}\left(\frac{1}{u_0}B^2\right)$
20. $\frac{\partial}{\partial r}\left(\frac{2\pi r}{v}\right)$
21. F_v if $F = \frac{mv^2}{r}$
22. $\frac{\partial}{\partial v_0}(v_0 + at)$
23. z_x if $z = \sin(5x^3y - 3xy^2)$
24. $\left.\frac{\partial z}{\partial y}\right|_{(1,0.5)}$ if $z = e^{x+2y}\sin y$
25. g_x if $g(x, y) = \ln(ye^{xy})$
26. $\left.\frac{\partial f}{\partial x}\right|_{(\pi/3,1)}$ if $f(x, y) = x\ln(y\cos x)$
27. z_x and z_y for $z = x^7 + 2^y + x^y$
28. f_x if $f(x, y) = e^{xy}(\ln y)$
29. $\frac{\partial F}{\partial m_2}$ if $F = \frac{Gm_1m_2}{r^2}$
30. $\frac{\partial}{\partial x}\left(\frac{1}{a}e^{-x^2/a^2}\right)$
31. $\frac{\partial}{\partial a}\left(\frac{1}{a}e^{-x^2/a^2}\right)$
32. $\frac{\partial}{\partial t}(v_0t + \frac{1}{2}at^2)$
33. $\frac{\partial}{\partial \theta}(\sin(\pi\theta\phi) + \ln(\theta^2 + \phi))$
34. $\frac{\partial}{\partial M}\left(\frac{2\pi r^{3/2}}{\sqrt{GM}}\right)$
35. f_a if $f(a, b) = e^a\sin(a + b)$
36. F_L if $F(L, K) = 3\sqrt{LK}$
37. $\frac{\partial V}{\partial r}$ and $\frac{\partial V}{\partial h}$ if $V = \frac{4}{3}\pi r^2 h$
38. u_E if $u = \frac{1}{2}\epsilon_0E^2 + \frac{1}{2\mu_0}B^2$
39. $\frac{\partial}{\partial x}\left(\frac{1}{\sqrt{2\pi}\sigma}e^{-(x-\mu)^2/(2\sigma^2)}\right)$
40. $\frac{\partial Q}{\partial K}$ if $Q = c(a_1K^{b_1} + a_2L^{b_2})^\gamma$

PROBLEMS

In Problems 41–43:

(a) Find $f_x(1,1)$ and $f_y(1,1)$.

(b) Use part (a) to match $f(x,y)$ with one of the contour diagrams (I)–(III), each shown centered at $(1,1)$ with the same scale in the x and y directions.

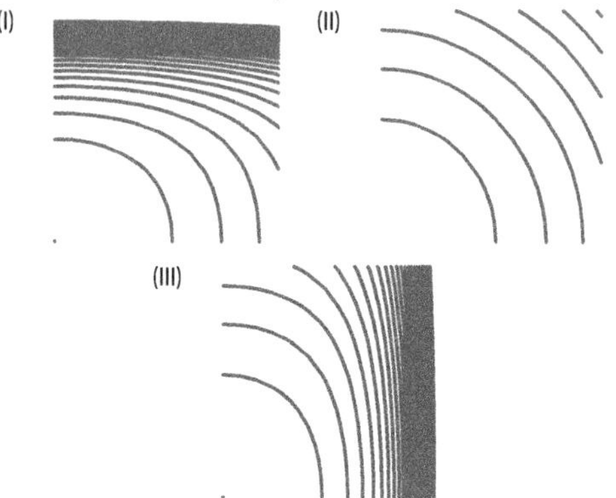

41. $f(x,y) = x^2 + y^2$

42. $f(x,y) = e^{x^2} + y^2$

43. $f(x,y) = x^2 + e^{y^2}$

44. **(a)** Let $f(x,y) = x^2+y^2$. Estimate $f_x(2,1)$ and $f_y(2,1)$ using the contour diagram for f in Figure 14.21.

(b) Estimate $f_x(2,1)$ and $f_y(2,1)$ from a table of values for f with $x = 1.9, 2, 2.1$ and $y = 0.9, 1, 1.1$.

(c) Compare your estimates in parts (a) and (b) with the exact values of $f_x(2,1)$ and $f_y(2,1)$ found algebraically.

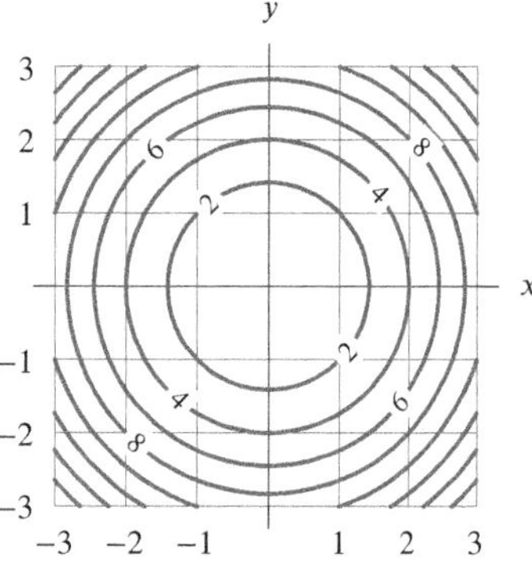

Figure 14.21

45. **(a)** Let $f(w,z) = e^{w \ln z}$. Use difference quotients with $h = 0.01$ to approximate $f_w(2,2)$ and $f_z(2,2)$.

(b) Now evaluate $f_w(2,2)$ and $f_z(2,2)$ exactly.

46. **(a)** The surface S is given, for some constant a, by

$$z = 3x^2 + 4y^2 - axy$$

Find the values of a which ensure that S is sloping upward when we move in the positive x-direction from the point $(1,2)$.

(b) With the values of a from part (a), if you move in the positive y-direction from the point $(1,2)$, does the surface slope up or down? Explain.

47. Money in a bank account earns interest at a continuous rate, r. The amount of money, $\$B$, in the account depends on the amount deposited, $\$P$, and the time, t, it has been in the bank according to the formula

$$B = Pe^{rt}.$$

Find $\partial B/\partial t$ and $\partial B/\partial P$ and interpret each in financial terms.

48. The acceleration g due to gravity, at a distance r from the center of a planet of mass m, is given by

$$g = \frac{Gm}{r^2},$$

where G is the universal gravitational constant.

(a) Find $\partial g/\partial m$ and $\partial g/\partial r$.

(b) Interpret each of the partial derivatives you found in part (a) as the slope of a graph in the plane and sketch the graph.

49. The Dubois formula relates a person's surface area, s, in m^2, to weight, w, in kg, and height, h, in cm, by

$$s = f(w,h) = 0.01w^{0.25}h^{0.75}.$$

Find $f(65,160)$, $f_w(65,160)$, and $f_h(65,160)$. Interpret your answers in terms of surface area, height, and weight.

50. The energy, E, of a body of mass m moving with speed v is given by the formula

$$E = mc^2\left(\frac{1}{\sqrt{1-v^2/c^2}} - 1\right).$$

The speed, v, is nonnegative and less than the speed of light, c, which is a constant.

(a) Find $\partial E/\partial m$. What would you expect the sign of $\partial E/\partial m$ to be? Explain.

(b) Find $\partial E/\partial v$. Explain what you would expect the sign of $\partial E/\partial v$ to be and why.

51. Let $h(x,t) = 5 + \cos(0.5x - t)$ describe a wave. The value of $h(x,t)$ gives the depth of the water in cm at a distance x meters from a fixed point and at time t seconds. Evaluate $h_x(2,5)$ and $h_t(2,5)$ and interpret each in terms of the wave.

52. A one-meter-long bar is heated unevenly, with temperature in °C at a distance x meters from one end at time t given by

$$H(x,t) = 100e^{-0.1t}\sin(\pi x) \qquad 0 \le x \le 1.$$

(a) Sketch a graph of H against x for $t = 0$ and $t = 1$.

(b) Calculate $H_x(0.2,t)$ and $H_x(0.8,t)$. What is the practical interpretation (in terms of temperature) of these two partial derivatives? Explain why each one has the sign it does.

(c) Calculate $H_t(x,t)$. What is its sign? What is its interpretation in terms of temperature?

53. Show that the Cobb-Douglas function

$$Q = bK^{\alpha}L^{1-\alpha} \quad \text{where} \quad 0 < \alpha < 1$$

satisfies the equation

$$K\frac{\partial Q}{\partial K} + L\frac{\partial Q}{\partial L} = Q.$$

In Problems 54–57, find all points where the partial derivatives of $f(x, y)$ are both 0.

54. $f(x, y) = x^2 + y^2$
55. $f(x, y) = xe^y$
56. $f(x, y) = e^{x^2+2x+y^2}$
57. $f(x, y) = x^3 + 3x^2 + y^3 - 3y$
58. Is there a function f which has the following partial derivatives? If so, what is it? Are there any others?

$$f_x(x, y) = 4x^3y^2 - 3y^4,$$
$$f_y(x, y) = 2x^4y - 12xy^3.$$

Strengthen Your Understanding

In Problems 59–60, explain what is wrong with the statement.

59. The partial derivative of $f(x, y) = x^2y^3$ is $2xy^3+3y^2x^2$.
60. For $f(x, y)$, if $\frac{f(0.01, 0)-f(0, 0)}{0.01} > 0$, then $f_x(0, 0) > 0$.

In Problems 61–63, give an example of:

61. A nonlinear function $f(x, y)$ such that $f_x(0, 0) = 2$ and $f_y(0, 0) = 3$.
62. Functions $f(x, y)$ and $g(x, y)$ such that $f_x = g_x$ but $f_y \neq g_y$.
63. A non-constant function $f(x, y)$ such that $f_x = 0$ everywhere.

Are the statements in Problems 64–71 true or false? Give reasons for your answer.

64. There is a function $f(x, y)$ with $f_x(x, y) = y$ and $f_y(x, y) = x$.
65. The function $z(u, v) = u \cos v$ satisfies the equation

$$\cos v\frac{\partial z}{\partial u} - \frac{\sin v}{u}\frac{\partial z}{\partial v} = 1.$$

66. If $f(x, y)$ is a function of two variables and $g(x)$ is a function of a single variable, then

$$\frac{\partial}{\partial y}(g(x)f(x, y)) = g(x)f_y(x, y).$$

67. The function $k(r, s) = rse^s$ is increasing in the s-direction at the point $(r, s) = (-1, 2)$.
68. There is a function $f(x, y)$ with $f_x(x, y) = y^2$ and $f_y(x, y) = x^2$.
69. If $f(x, y)$ has $f_y(x, y) = 0$ then f must be a constant.
70. If $f(x, y) = ye^{g(x)}$ then $f_x = f$.
71. If f is a symmetric two-variable function, that is $f(x, y) = f(y, x)$, then $f_x(x, y) = f_y(x, y)$.
72. Which of the following functions satisfy the following equation (called Euler's Equation):

$$xf_x + yf_y = f?$$

(a) x^2y^3 (b) $x+y+1$ (c) $x^2 + y^2$ (d) $x^{0.4}y^{0.6}$

14.3 LOCAL LINEARITY AND THE DIFFERENTIAL

In Sections 14.1 and 14.2 we studied a function of two variables by allowing one variable at a time to change. We now let both variables change at once to develop a linear approximation for functions of two variables.

Zooming In to See Local Linearity

For a function of one variable, local linearity means that as we zoom in on the graph, it looks like a straight line. As we zoom in on the graph of a two-variable function, the graph usually looks like a plane, which is the graph of a linear function of two variables. (See Figure 14.22.)

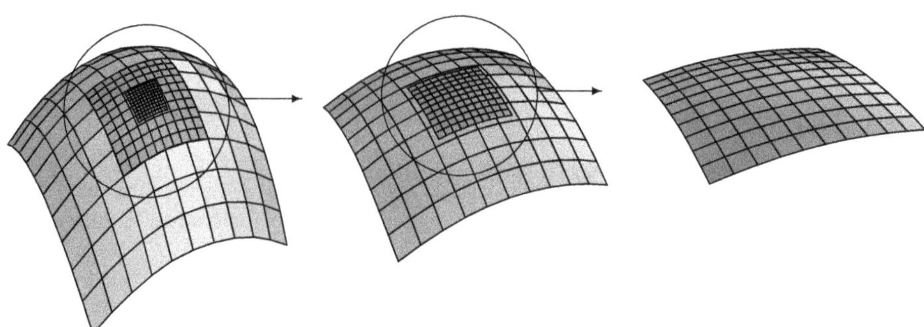

Figure 14.22: Zooming in on the graph of a function of two variables until the graph looks like a plane

Similarly, Figure 14.23 shows three successive views of the contours near a point. As we zoom in, the contours look more like equally spaced parallel lines, which are the contours of a linear function. (As we zoom in, we have to add more contours.)

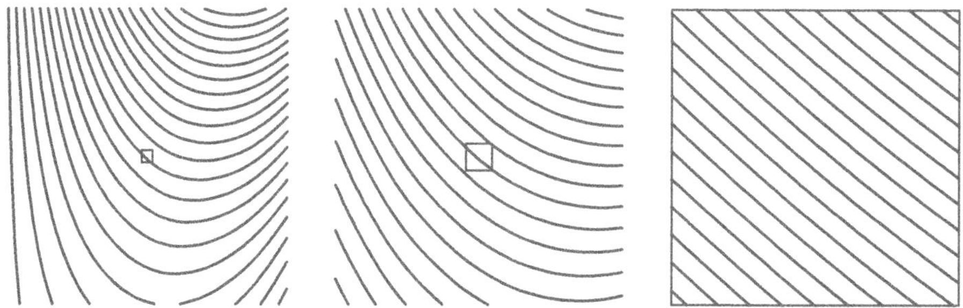

Figure 14.23: Zooming in on a contour diagram until the lines look parallel and equally spaced

This effect can also be seen numerically by zooming in with tables of values. Table 14.5 shows three tables of values for $f(x, y) = x^2 + y^3$ near $x = 2$, $y = 1$, each one a closer view than the previous one. Notice how each table looks more like the table of a linear function.

Table 14.5 *Zooming in on values of $f(x, y) = x^2 + y^3$ near $(2, 1)$ until the table looks linear*

x \ y	0	1	2
1	1	2	9
2	4	5	12
3	9	10	17

x \ y	0.9	1.0	1.1
1.9	4.34	4.61	4.94
2.0	4.73	5.00	5.33
2.1	5.14	5.41	5.74

x \ y	0.99	1.00	1.01
1.99	4.93	4.96	4.99
2.00	4.97	5.00	5.03
2.01	5.01	5.04	5.07

Zooming in Algebraically: Differentiability

Seeing a plane when we zoom in at a point tells us (provided the plane is not vertical) that $f(x, y)$ is closely approximated near that point by a linear function, $L(x, y)$:

$$f(x, y) \approx L(x, y).$$

The Tangent Plane

The graph of the function $z = L(x, y)$ is the tangent plane at that point. See Figure 14.24. Provided the approximation is sufficiently good, we say that $f(x, y)$ is *differentiable* at the point. Section 14.8 on

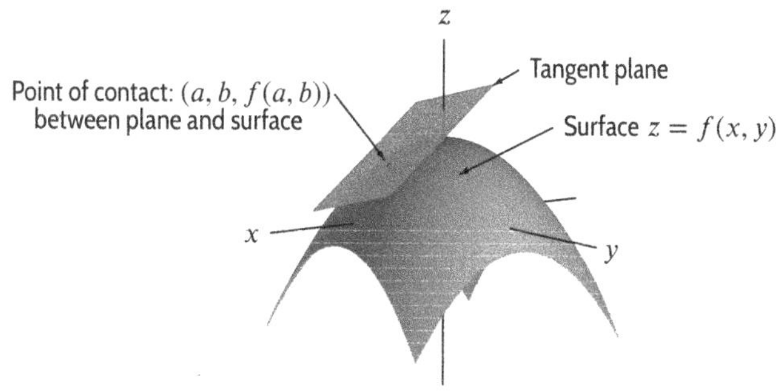

Figure 14.24: The tangent plane to the surface $z = f(x, y)$ at the point (a, b)

page 799 defines precisely what is meant by the approximation being sufficiently good. The functions we encounter are differentiable at most points in their domain.

What is the equation of the tangent plane? At the point (a, b), the x-slope of the graph of f is the partial derivative $f_x(a, b)$ and the y-slope is $f_y(a, b)$. Thus, using the equation for a plane on page 683 of Chapter 12, we have the following result:

Tangent Plane to the Surface $z = f(x, y)$ at the Point (a, b)

Assuming f is differentiable at (a, b), the equation of the tangent plane is

$$z = f(a, b) + f_x(a, b)(x - a) + f_y(a, b)(y - b).$$

Here we are thinking of a and b as fixed, so $f(a, b)$, and $f_x(a, b)$, and $f_y(a, b)$ are constants. Thus, the right side of the equation is a linear function of x and y.

Example 1 Find the equation for the tangent plane to the surface $z = x^2 + y^2$ at the point $(3, 4)$.

Solution We have $f_x(x, y) = 2x$, so $f_x(3, 4) = 6$, and $f_y(x, y) = 2y$, so $f_y(3, 4) = 8$. Also, $f(3, 4) = 3^2 + 4^2 = 25$. Thus, the equation for the tangent plane at $(3, 4)$ is

$$z = 25 + 6(x - 3) + 8(y - 4).$$

Local Linearization

Since the tangent plane lies close to the surface near the point at which they meet, z-values on the tangent plane are close to values of $f(x, y)$ for points near (a, b). Thus, replacing z by $f(x, y)$ in the equation of the tangent plane, we get the following approximation:

Tangent Plane Approximation to $f(x, y)$ for (x, y) Near the Point (a, b)

Provided f is differentiable at (a, b), we can approximate $f(x, y)$:

$$f(x, y) \approx f(a, b) + f_x(a, b)(x - a) + f_y(a, b)(y - b).$$

We are thinking of a and b as fixed, so the expression on the right side is linear in x and y. The right side of this approximation gives the **local linearization** of f near $x = a$, $y = b$.

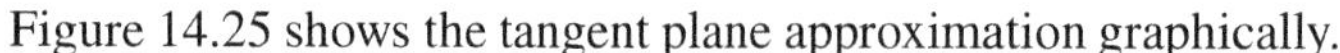

Figure 14.25 shows the tangent plane approximation graphically.

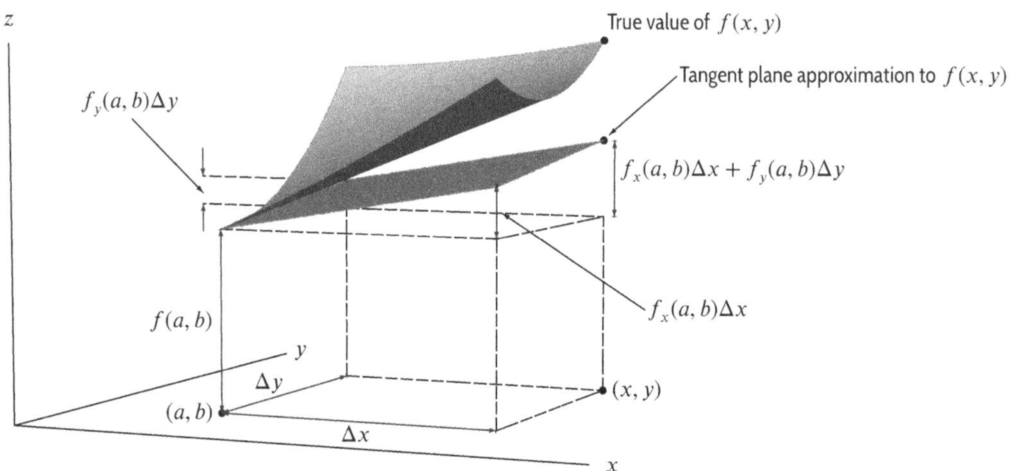

Figure 14.25: Local linearization: Approximating $f(x, y)$ by the z-value from the tangent plane

Example 2 Find the local linearization of $f(x, y) = x^2 + y^2$ at the point $(3, 4)$. Estimate $f(2.9, 4.2)$ and $f(2, 2)$ using the linearization and compare your answers to the true values.

Solution Let $z = f(x, y) = x^2 + y^2$. In Example 1, we found the equation of the tangent plane at $(3, 4)$ to be

$$z = 25 + 6(x - 3) + 8(y - 4).$$

Therefore, for (x, y) near $(3, 4)$, we have the local linearization

$$f(x, y) \approx 25 + 6(x - 3) + 8(y - 4).$$

Substituting $x = 2.9, y = 4.2$ gives

$$f(2.9, 4.2) \approx 25 + 6(-0.1) + 8(0.2) = 26.$$

This compares favorably with the true value $f(2.9, 4.2) = (2.9)^2 + (4.2)^2 = 26.05$.

However, the local linearization does not give a good approximation at points far away from $(3, 4)$. For example, if $x = 2$, $y = 2$, the local linearization gives

$$f(2, 2) \approx 25 + 6(-1) + 8(-2) = 3,$$

whereas the true value of the function is $f(2, 2) = 2^2 + 2^2 = 8$.

Example 3 Designing safe boilers depends on knowing how steam behaves under changes in temperature and pressure. Steam tables, such as Table 14.6, are published giving values of the function $V = f(T, P)$ where V is the volume (in ft^3) of one pound of steam at a temperature T (in °F) and pressure P (in lb/in^2).

(a) Give a linear function approximating $V = f(T, P)$ for T near 500°F and P near 24 lb/in^2.
(b) Estimate the volume of a pound of steam at a temperature of 505°F and a pressure of 24.3 lb/in^2.

Table 14.6 *Volume (in cubic feet) of one pound of steam at various temperatures and pressures*

Temperature T (°F)	Pressure P (lb/in^2) 20	22	24	26
480	27.85	25.31	23.19	21.39
500	28.46	25.86	23.69	21.86
520	29.06	26.41	24.20	22.33
540	29.66	26.95	24.70	22.79

Solution (a) We want the local linearization around the point $T = 500,\ P = 24$, which is

$$f(T, P) \approx f(500, 24) + f_T(500, 24)(T - 500) + f_P(500, 24)(P - 24).$$

We read the value $f(500, 24) = 23.69$ from the table.

Next we approximate $f_T(500, 24)$ by a difference quotient. From the $P = 24$ column, we compute the average rate of change between $T = 500$ and $T = 520$:

$$f_T(500, 24) \approx \frac{f(520, 24) - f(500, 24)}{520 - 500} = \frac{24.20 - 23.69}{20} = 0.0255.$$

Note that $f_T(500, 24)$ is positive, because steam expands when heated.

Next we approximate $f_P(500, 24)$ by looking at the $T = 500$ row and computing the average rate of change between $P = 24$ and $P = 26$:

$$f_P(500, 24) \approx \frac{f(500, 26) - f(500, 24)}{26 - 24} = \frac{21.86 - 23.69}{2} = -0.915.$$

Note that $f_P(500, 24)$ is negative, because increasing the pressure on steam decreases its volume. Using these approximations for the partial derivatives, we obtain the local linearization:

$$V = f(T, P) \approx 23.69 + 0.0255(T - 500) - 0.915(P - 24) \text{ ft}^3 \quad \begin{array}{l}\text{for } T \text{ near } 500\,°\text{F} \\ \text{and } P \text{ near } 24 \text{ lb/in}^2.\end{array}$$

(b) We are interested in the volume at $T = 505°$F and $P = 24.3$ lb/in^2. Since these values are close to $T = 500°$F and $P = 24$ lb/in^2, we use the linear relation obtained in part (a):

$$V \approx 23.69 + 0.0255(505 - 500) - 0.915(24.3 - 24) = 23.54 \text{ ft}^3.$$

Local Linearity with Three or More Variables

Local linear approximations for functions of three or more variables follow the same pattern as for functions of two variables. The local linearization of $f(x, y, z)$ at (a, b, c) is given by

$$f(x, y, z) \approx f(a, b, c) + f_x(a, b, c)(x - a) + f_y(a, b, c)(y - b) + f_z(a, b, c)(z - c).$$

The Differential

We are often interested in the change in the value of the function as we move from the point (a, b) to a nearby point (x, y). We rewrite the tangent plane approximation as

$$\underbrace{f(x, y) - f(a, b)}_{\Delta f} \approx f_x(a, b)\underbrace{(x - a)}_{\Delta x} + f_y(a, b)\underbrace{(y - b)}_{\Delta y},$$

giving us a relationship between Δf, Δx, and Δy:

$$\Delta f \approx f_x(a, b)\Delta x + f_y(a, b)\Delta y.$$

If a and b are fixed, $f_x(a, b)\Delta x + f_y(a, b)\Delta y$ is a linear function of Δx and Δy that can be used to estimate Δf for small Δx and Δy. We introduce new variables dx and dy to represent changes in x and y.

The Differential of a Function $z = f(x, y)$

The **differential**, df (or dz), at a point (a, b) is the linear function of dx and dy given by the formula

$$df = f_x(a, b)\, dx + f_y(a, b)\, dy.$$

The differential at a general point is often written $df = f_x\, dx + f_y\, dy$.

Example 4 Compute the differentials of the following functions.

(a) $f(x, y) = x^2e^{5y}$ (b) $z = x\sin(xy)$ (c) $f(x, y) = x\cos(2x)$

Solution (a) Since $f_x(x, y) = 2xe^{5y}$ and $f_y(x, y) = 5x^2e^{5y}$, we have

$$df = 2xe^{5y}\, dx + 5x^2e^{5y}\, dy.$$

(b) Since $\partial z/\partial x = \sin(xy) + xy\cos(xy)$ and $\partial z/\partial y = x^2\cos(xy)$, we have

$$dz = (\sin(xy) + xy\cos(xy))\, dx + x^2\cos(xy)\, dy.$$

(c) Since $f_x(x, y) = \cos(2x) - 2x\sin(2x)$ and $f_y(x, y) = 0$, we have

$$df = (\cos(2x) - 2x\sin(2x))\, dx + 0\, dy = (\cos(2x) - 2x\sin(2x))\, dx.$$

Example 5 The density ρ (in g/cm^3) of carbon dioxide gas CO_2 depends upon its temperature T (in °C) and pressure P (in atmospheres). The ideal gas model for CO_2 gives what is called the state equation:

$$\rho = \frac{0.5363P}{T + 273.15}.$$

Compute the differential $d\rho$. Explain the signs of the coefficients of dT and dP.

Solution The differential for $\rho = f(T, P)$ is

$$d\rho = f_T(T, P)\, dT + f_P(T, P) dP = \frac{-0.5363P}{(T + 273.15)^2}\, dT + \frac{0.5363}{T + 273.15}\, dP.$$

The coefficient of dT is negative because increasing the temperature expands the gas (if the pressure is kept constant) and therefore decreases its density. The coefficient of dP is positive because increasing the pressure compresses the gas (if the temperature is kept constant) and therefore increases its density.

Where Does the Notation for the Differential Come From?

We write the differential as a linear function of the new variables dx and dy. You may wonder why we chose these names for our variables. The reason is historical: The people who invented calculus thought of dx and dy as "infinitesimal" changes in x and y. The equation

$$df = f_x dx + f_y dy$$

was regarded as an infinitesimal version of the local linear approximation

$$\Delta f \approx f_x \Delta x + f_y \Delta y.$$

In spite of the problems with defining exactly what "infinitesimal" means, some mathematicians, scientists, and engineers think of the differential in terms of infinitesimals.

Figure 14.26 illustrates a way of thinking about differentials that combines the definition with this informal point of view. It shows the graph of f along with a view of the graph around the point $(a, b, f(a, b))$ under a microscope. Since f is locally linear at the point, the magnified view looks like the tangent plane. Under the microscope, we use a magnified coordinate system with its origin at the point $(a, b, f(a, b))$ and with coordinates dx, dy, and dz along the three axes. The graph of the differential df is the tangent plane, which has equation $dz = f_x(a, b)\,dx + f_y(a, b)\,dy$ in the magnified coordinates.

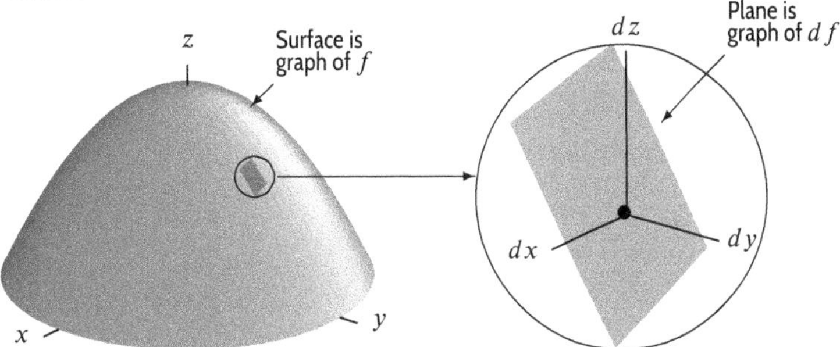

Figure 14.26: The graph of f, with a view through a microscope showing the tangent plane in the magnified coordinate system

Exercises and Problems for Section 14.3

EXERCISES

In Exercises 1–8, find the equation of the tangent plane at the given point.

1. $z = ye^{x/y}$ at the point $(1, 1, e)$
2. $z = \sin(xy)$ at $x = 2$, $y = 3\pi/4$
3. $z = \ln(x^2 + 1) + y^2$ at the point $(0, 3, 9)$
4. $z = e^y + x + x^2 + 6$ at the point $(1, 0, 9)$
5. $z = \frac{1}{2}(x^2 + 4y^2)$ at the point $(2, 1, 4)$
6. $x^2 + y^2 - z = 1$ at the point $(1, 3, 9)$
7. $x^2y^2 + z - 40 = 0$ at $x = 2$, $y = 3$
8. $x^2y + \ln(xy) + z = 6$ at the point $(4, 0.25, 2)$

In Exercises 9–12, find the differential of the function.

9. $f(x, y) = \sin(xy)$
10. $g(u, v) = u^2 + uv$
11. $z = e^{-x}\cos y$
12. $h(x, t) = e^{-3t}\sin(x + 5t)$

In Exercises 13–16, find the differential of the function at the point.

13. $g(x, t) = x^2\sin(2t)$ at $(2, \pi/4)$
14. $f(x, y) = xe^{-y}$ at $(1, 0)$
15. $P(L, K) = 1.01L^{0.25}K^{0.75}$ at $(100, 1)$
16. $F(m, r) = Gm/r^2$ at $(100, 10)$

In Exercises 17–20, assume points P and Q are close. Estimate $\Delta f = f(Q) - f(P)$ using the differential df.

17. $df = 10\,dx - 5\,dy$, $P = (200, 400)$, $Q = (202, 405)$

18. $df = y\,dx + x\,dy$, $P = (10, 5)$, $Q = (9.8, 5.3)$

19. $df = 6\sqrt{1+4x+2y}\,dx + 3\sqrt{1+4x+2y}\,dy$, $P = (1, 2)$, $Q = (1.03, 2.05)$

20. $df = (2x + 2y + 5)\,dx + (2x + 3)\,dy$, $P = (0, 0)$, $Q = (0.1, -0.2)$

In Exercises 21–24, assume points P and Q are close. Estimate $g(Q)$.

21. $P = (60, 80)$, $Q = (60.5, 82)$, $g(P) = 100$, $g_x(P) = 2$, $g_y(P) = -3$.

22. $P = (-150, 200)$, $Q = (-152, 203)$, $g(P) = 2500$, $g_x(P) = 10$, $g_y(P) = 20$.

23. $P = (5, 8)$, $Q = (4.97, 7.99)$, $g(P) = 12$, $g_x(P) = -0.1$, $g_y(P) = -0.2$.

24. $P = (30, 125)$, $Q = (25, 135)$, $g(P) = 840$, $g_x(P) = 4$, $g_y(P) = 1.5$.

PROBLEMS

25. At a distance of x feet from the beach, the price in dollars of a plot of land of area a square feet is $f(a, x)$.
    - **(a)** What are the units of $f_a(a, x)$?
    - **(b)** What does $f_a(1000, 300) = 3$ mean in practical terms?
    - **(c)** What are the units of $f_x(a, x)$?
    - **(d)** What does $f_x(1000, 300) = -2$ mean in practical terms?
    - **(e)** Which is cheaper: 1005 square feet that are 305 feet from the beach or 998 square feet that are 295 feet from the beach? Justify your answer.

26. A student was asked to find the equation of the tangent plane to the surface $z = x^3 - y^2$ at the point $(x, y) = (2, 3)$. The student's answer was

$$z = 3x^2(x - 2) - 2y(y - 3) - 1.$$

    - **(a)** At a glance, how do you know this is wrong?
    - **(b)** What mistake did the student make?
    - **(c)** Answer the question correctly.

27. **(a)** Check the local linearity of $f(x, y) = e^{-x} \sin y$ near $x = 1$, $y = 2$ by making a table of values of f for $x = 0.9, 1.0, 1.1$ and $y = 1.9, 2.0, 2.1$. Express values of f with 4 digits after the decimal point. Then make a table of values for $x = 0.99, 1.00, 1.01$ and $y = 1.99, 2.00. 2.01$, again showing 4 digits after the decimal point. Do both tables look nearly linear? Does the second table look more linear than the first?
    (b) Give the local linearization of $f(x, y) = e^{-x} \sin y$ at $(1, 2)$, first using your tables, and second using the fact that $f_x(x, y) = -e^{-x} \sin y$ and $f_y(x, y) = e^{-x} \cos y$.

28. Find the local linearization of the function $f(x, y) = x^2 y$ at the point $(3, 1)$.

29. The tangent plane to $z = f(x, y)$ at the point $(1, 2)$ is $z = 3x + 2y - 5$.
    - **(a)** Find $f_x(1, 2)$ and $f_y(1, 2)$.
    - **(b)** What is $f(1, 2)$?
    - **(c)** Approximate $f(1.1, 1.9)$.

30. Find an equation for the tangent plane to $z = f(x, y)$ at $(3, -2)$ if the differential at $(3, -2)$ is $df = 5dx + dy$ and $f(3, -2) = 8$.

31. Find df at $(2, -4)$ if the tangent plane to $z = f(x, y)$ at $(2, -4)$ is $z = -3(x - 2) + 2(y + 4) + 3$.

32. Give a linear function approximating $z = f(x, y)$ near $(1, -1)$ using its contour diagram in Figure 14.27.

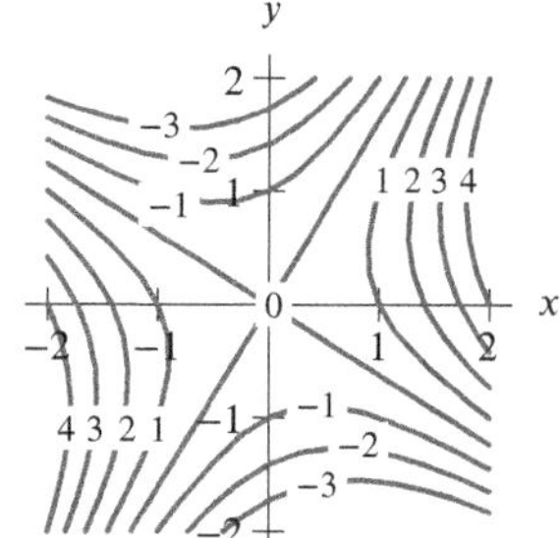

Figure 14.27

33. For the differentiable function $h(x, y)$, we are told that $h(600, 100) = 300$ and $h_x(600, 100) = 12$ and $h_y(600, 100) = -8$. Estimate $h(605, 98)$.

34. **(a)** Find the equation of the plane tangent to the graph of $f(x, y) = x^2 e^{xy}$ at $(1, 0)$.
    (b) Find the linear approximation of $f(x, y)$ for (x, y) near $(1, 0)$.
    (c) Find the differential of f at the point $(1, 0)$.

35. Find the differential of $f(x, y) = \sqrt{x^2 + y^3}$ at the point $(1, 2)$. Use it to estimate $f(1.04, 1.98)$.

36. **(a)** Find the differential of $g(u, v) = u^2 + uv$.
    (b) Use your answer to part (a) to estimate the change in g as you move from $(1, 2)$ to $(1.2, 2.1)$.

37. An unevenly heated plate has temperature $T(x, y)$ in °C at the point (x, y). If $T(2, 1) = 135$, and $T_x(2, 1) = 16$, and $T_y(2, 1) = -15$, estimate the temperature at the point $(2.04, 0.97)$.

38. A right circular cylinder has a radius of 50 cm and a height of 100 cm. Use differentials to estimate the change in volume of the cylinder if its height and radius are both increased by 1 cm.

39. Give the local linearization for the monthly car-loan payment function at each of the points investigated in Problem 35 on page 747.

40. In Example 3 on page 756 we found a linear approximation for $V = f(T, P)$ near $(500, 24)$. Now find a linear approximation near $(480, 20)$.

41. In Example 3 on page 756 we found a linear approximation for $V = f(T, p)$ near $(500, 24)$.

(a) Test the accuracy of this approximation by comparing its predicted value with the four neighboring values in the table. What do you notice? Which predicted values are accurate? Which are not? Explain your answer.
(b) Suggest a linear approximation for $f(T, p)$ near $(500, 24)$ that does not have the property you noticed in part (a). [Hint: Estimate the partial derivatives in a different way.]

42. In a room, the temperature is given by $T = f(x, t)$ degrees Celsius, where x is the distance from a heater (in meters) and t is the elapsed time (in minutes) since the heat has been turned on. A person standing 3 meters from the heater 5 minutes after it has been turned on observes the following: (1) The temperature is increasing by 1.2°C per minute, and (2) As the person walks away from the heater, the temperature decreases by 2°C per meter as time is held constant. Estimate how much cooler or warmer it would be 2.5 meters from the heater after 6 minutes.

43. Van der Waal's equation relates the pressure, P, and the volume, V, of a fixed quantity of a gas at constant temperature T. For a, b, n, R constants, the equation is

$$\left(P + \frac{n^2 a}{V^2}\right)(V - nb) = nRT.$$

(a) Express P as a function of T and V.
(b) Write a linear approximation for the change in pressure, $\Delta P = P - P_0$, resulting from a change in temperature $\Delta T = T - T_0$ and a change in pressure, $\Delta V = V - V_0$.

44. The gas equation for one mole of oxygen relates its pressure, P (in atmospheres), its temperature, T (in K), and its volume, V (in cubic decimeters, dm^3):

$$T = 16.574\frac{1}{V} - 0.52754\frac{1}{V^2} - 0.3879P + 12.187VP.$$

(a) Find the temperature T and differential dT if the volume is 25 dm^3 and the pressure is 1 atmosphere.
(b) Use your answer to part (a) to estimate how much the volume would have to change if the pressure increased by 0.1 atmosphere and the temperature remained constant.

45. The coefficient, β, of thermal expansion of a liquid relates the change in the volume V (in m^3) of a fixed quantity of a liquid to an increase in its temperature T (in °C):

$$dV = \beta V\, dT.$$

(a) Let ρ be the density (in kg/m^3) of water as a function of temperature. (For a mass m of liquid, we have $\rho = m/V$.) Write an expression for $d\rho$ in terms of ρ and dT.
(b) The graph in Figure 14.28 shows density of water as a function of temperature. Use it to estimate β when $T = 20$°C and when $T = 80$°C.

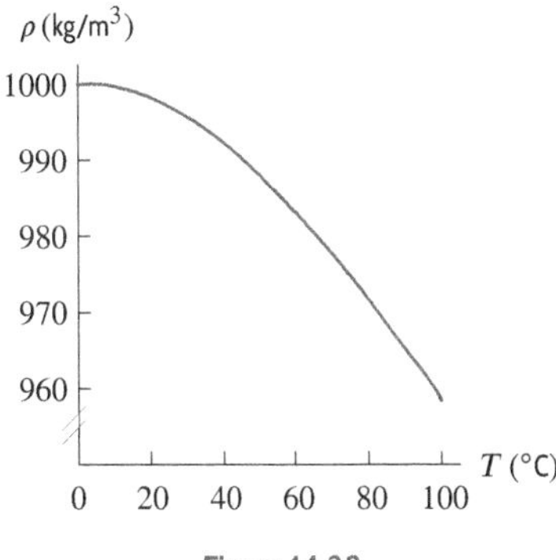

Figure 14.28

46. A fluid moves through a tube of length 1 meter and radius $r = 0.005 \pm 0.00025$ meters under a pressure $p = 10^5 \pm 1000$ pascals, at a rate $v = 0.625 \cdot 10^{-9}$ m^3 per unit time. Use differentials to estimate the maximum error in the viscosity η given by

$$\eta = \frac{\pi}{8}\frac{pr^4}{v}.$$

47. The period, T, of oscillation in seconds of a pendulum clock is given by $T = 2\pi\sqrt{l/g}$, where g is the acceleration due to gravity. The length of the pendulum, l, depends on the temperature, t, according to the formula $l = l_0(1 + \alpha(t - t_0))$ where l_0 is the length of the pendulum at temperature t_0 and α is a constant which characterizes the clock. The clock is set to the correct period at the temperature t_0. How many seconds a day does the clock gain or lose when the temperature is $t_0 + \Delta t$? Show that this gain or loss is independent of l_0.

48. Two functions that have the same local linearization at a point have contours that are tangent at this point.

(a) If $f_x(a, b)$ or $f_y(a, b)$ is nonzero, use the local linearization to show that an equation of the line tangent at (a, b) to the contour of f through (a, b) is $f_x(a, b)(x - a) + f_y(a, b)(y - b) = 0$.
(b) Find the slope of the tangent line if $f_y(a, b) \neq 0$.
(c) Find an equation for the line tangent to the contour of $f(x, y) = x^2 + xy$ at $(3, 4)$.

In Problems 49–52, the point is on the surface in 3-space.
(a) Find the differential of the equation (that is, of each side).

(b) Find dz at the point.
(c) Find an equation of the tangent plane to the surface at the point.

49. $2x^2 + 13 = y^2 + 3z^2$, $(2, 3, 2)$

50. $x^2 + y^2 + z^2 + 1 = xyz + 2x^2 + 3y^2 - 2z^2$, $(1, 1, 1)$

51. $xe^y + z^2 + 1 = \cos(x - 1) + \sqrt{z^2 + 3}$, $(1, 0, 1)$

52. $xz^2 + xy + 5 = x^2 + z^2$, $(2, -1, 1)$

Strengthen Your Understanding

In Problems 53–55, explain what is wrong with the statement.

53. An equation for the tangent plane to the surface $z = f(x, y)$ at the point $(3, 4)$ is

$$z = f(3,4) + f_x(3,4)x + f_y(3,4)y.$$

54. If $f_x(0,0) = g_x(0,0)$ and $f_y(0,0) = g_y(0,0)$, then the surfaces $z = f(x,y)$ and $z = g(x,y)$ have the same tangent planes at the point $(0,0)$.

55. The tangent plane to the surface $z = x^2y$ at the point $(1,2)$ has equation

$$z = 2 + 2xy(x - 1) + x^2(y - 2).$$

In Problems 56–57, give an example of:

56. Two different functions with the same differential.

57. A surface in three space whose tangent plane at $(0, 0, 3)$ is the plane $z = 3$.

Are the statements in Problems 58–65 true or false? Give reasons for your answer.

58. The tangent plane approximation of $f(x, y) = ye^{x^2}$ at the point $(0, 1)$ is $f(x, y) \approx y$.

59. If f is a function with $df = 2y\,dx + \sin(xy)\,dy$, then f changes by about -0.4 between the points $(1, 2)$ and $(0.9, 2.0002)$.

60. The local linearization of $f(x, y) = x^2 + y^2$ at (1,1) gives an overestimate of the value of $f(x, y)$ at the point $(1.04, 0.95)$.

61. If two functions f and g have the same differential at the point $(1, 1)$, then $f = g$.

62. If two functions f and g have the same tangent plane at a point $(1, 1)$, then $f = g$.

63. If $f(x, y)$ is a constant function, then $df = 0$.

64. If $f(x, y)$ is a linear function, then df is a linear function of dx and dy.

65. If you zoom close enough near a point (a, b) on the contour diagram of a differentiable function, the contours are *precisely* parallel and *exactly* equally spaced.

14.4 GRADIENTS AND DIRECTIONAL DERIVATIVES IN THE PLANE

The Rate of Change in an Arbitrary Direction: The Directional Derivative

The partial derivatives of a function f tell us the rate of change of f in the directions parallel to the coordinate axes. In this section we see how to compute the rate of change of f in an arbitrary direction.

Example 1 Figure 14.29 shows the temperature, in °C, at the point (x, y). Estimate the average rate of change of temperature as we walk from point A to point B.

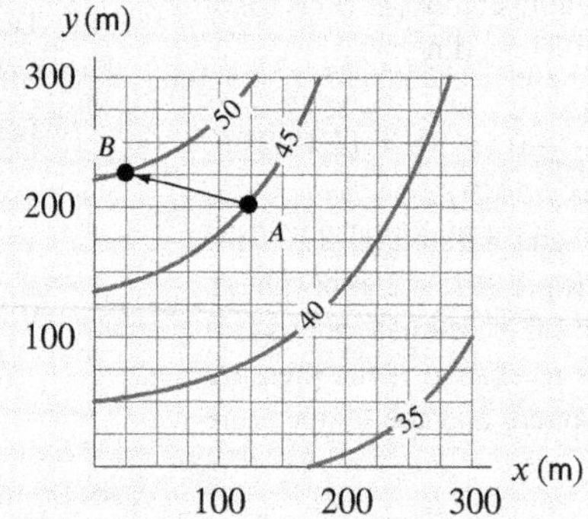

Figure 14.29: Estimating rate of change on a temperature map

Solution At the point A we are on the $H = 45°\text{C}$ contour. At B we are on the $H = 50°\text{C}$ contour. The displacement vector from A to B has x component approximately $-100\vec{i}$ and y component approximately $25\vec{j}$, so its length is $\sqrt{(-100)^2 + 25^2} \approx 103$. Thus, the temperature rises by 5°C as we move 103 meters, so the average rate of change of the temperature in that direction is about $5/103 \approx 0.05°\text{C/m}$.

Suppose we want to compute the rate of change of a function $f(x, y)$ at the point $P = (a, b)$ in the direction of the unit vector $\vec{u} = u_1\vec{i} + u_2\vec{j}$. For $h > 0$, consider the point $Q = (a + hu_1, b + hu_2)$ whose displacement from P is $h\vec{u}$. (See Figure 14.30.) Since $\|\vec{u}\| = 1$, the distance from P to Q is h. Thus,

$$\begin{array}{l}\text{Average rate of change}\\ \text{in } f \text{ from } P \text{ to } Q\end{array} = \frac{\text{Change in } f}{\text{Distance from } P \text{ to } Q} = \frac{f(a + hu_1, b + hu_2) - f(a, b)}{h}.$$

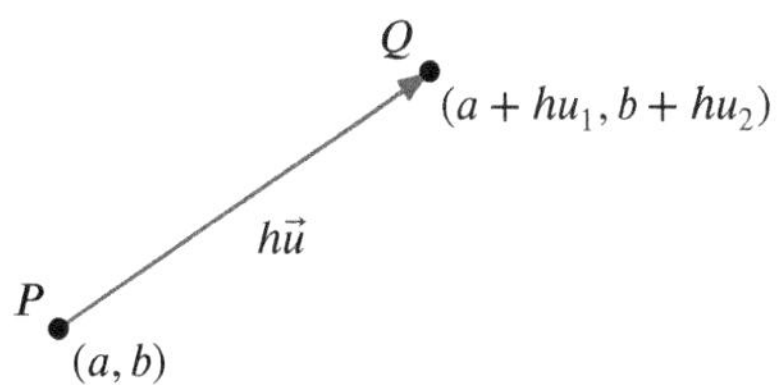

Figure 14.30: Displacement of $h\vec{u}$ from the point (a, b)

Taking the limit as $h \to 0$ gives the instantaneous rate of change and the following definition:

Directional Derivative of f at (a, b) in the Direction of a Unit Vector $\vec{u}$

If $\vec{u} = u_1\vec{i} + u_2\vec{j}$ is a unit vector, we define the directional derivative, $f_{\vec{u}}$, by

$$f_{\vec{u}}(a, b) = \begin{array}{c}\text{Rate of change}\\ \text{of } f \text{ in direction}\\ \text{of } \vec{u} \text{ at } (a, b)\end{array} = \lim_{h \to 0} \frac{f(a + hu_1, b + hu_2) - f(a, b)}{h},$$

provided the limit exists. Note that the directional derivative is a scalar.

Notice that if $\vec{u} = \vec{i}$, so $u_1 = 1, u_2 = 0$, then the directional derivative is f_x, since

$$f_{\vec{i}}(a, b) = \lim_{h \to 0} \frac{f(a + h, b) - f(a, b)}{h} = f_x(a, b).$$

Similarly, if $\vec{u} = \vec{j}$ then the directional derivative $f_{\vec{j}} = f_y$.

What If We Do Not Have a Unit Vector?

We defined $f_{\vec{u}}$ for $\vec{u}$ a unit vector. If $\vec{v}$ is not a unit vector, $\vec{v} \neq \vec{0}$, we construct a unit vector $\vec{u} = \vec{v}/\|\vec{v}\|$ in the same direction as $\vec{v}$ and define the rate of change of f in the direction of $\vec{v}$ as $f_{\vec{u}}$.

Example 2 For each of the functions f, g, and h in Figure 14.31, decide whether the directional derivative at the indicated point is positive, negative, or zero, in the direction of the vector $\vec{v} = \vec{i} + 2\vec{j}$, and in the direction of the vector $\vec{w} = 2\vec{i} + \vec{j}$.

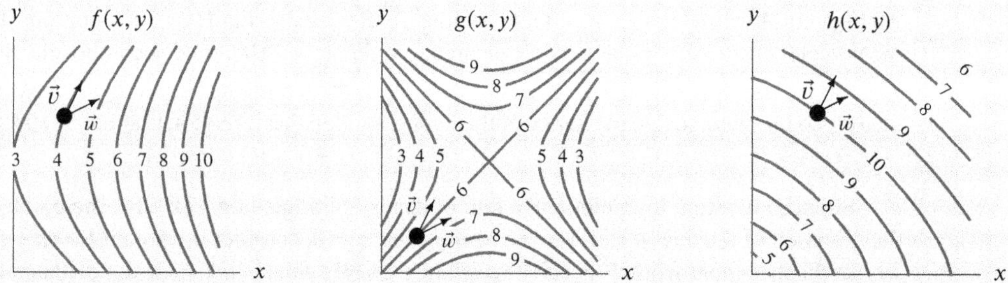

Figure 14.31: Contour diagrams of three functions with direction vectors $\vec{v} = \vec{i} + 2\vec{j}$ and $\vec{w} = 2\vec{i} + \vec{j}$ marked on each

Solution On the contour diagram for f, the vector $\vec{v} = \vec{i} + 2\vec{j}$ appears to be tangent to the contour. Thus, in this direction, the value of the function is not changing, so the directional derivative in the direction of $\vec{v}$ is zero. The vector $\vec{w} = 2\vec{i} + \vec{j}$ points from the contour marked 4 toward the contour marked 5. Thus, the values of the function are increasing and the directional derivative in the direction of $\vec{w}$ is positive.

On the contour diagram for g, the vector $\vec{v} = \vec{i} + 2\vec{j}$ points from the contour marked 6 toward the contour marked 5, so the function is decreasing in that direction. Thus, the rate of change is negative. On the other hand, the vector $\vec{w} = 2\vec{i} + \vec{j}$ points from the contour marked 6 toward the contour marked 7, and hence the directional derivative in the direction of $\vec{w}$ is positive.

Finally, on the contour diagram for h, both vectors point from the $h = 10$ contour to the $h = 9$ contour, so both directional derivatives are negative.

Example 3 Calculate the directional derivative of $f(x, y) = x^2 + y^2$ at $(1, 0)$ in the direction of the vector $\vec{i} + \vec{j}$.

Solution First we have to find the unit vector in the same direction as the vector $\vec{i} + \vec{j}$. Since this vector has magnitude $\sqrt{2}$, the unit vector is

$$\vec{u} = \frac{1}{\sqrt{2}}(\vec{i} + \vec{j}) = \frac{1}{\sqrt{2}}\vec{i} + \frac{1}{\sqrt{2}}\vec{j}.$$

Thus,

$$\begin{aligned} f_{\vec{u}}(1,0) &= \lim_{h\to 0} \frac{f(1 + h/\sqrt{2}, h/\sqrt{2}) - f(1,0)}{h} = \lim_{h\to 0} \frac{(1 + h/\sqrt{2})^2 + (h/\sqrt{2})^2 - 1}{h} \\ &= \lim_{h\to 0} \frac{\sqrt{2}h + h^2}{h} = \lim_{h\to 0}(\sqrt{2} + h) = \sqrt{2}. \end{aligned}$$

Computing Directional Derivatives from Partial Derivatives

If f is differentiable, we will now see how to use local linearity to find a formula for the directional derivative which does not involve a limit. If $\vec{u}$ is a unit vector, the definition of $f_{\vec{u}}$ says

$$f_{\vec{u}}(a, b) = \lim_{h\to 0} \frac{f(a + hu_1, b + hu_2) - f(a, b)}{h} = \lim_{h\to 0} \frac{\Delta f}{h},$$

where $\Delta f = f(a + hu_1, b + hu_2) - f(a, b)$ is the change in f. We write Δx for the change in x, so $\Delta x = (a + hu_1) - a = hu_1$; similarly, $\Delta y = hu_2$. Using local linearity, we have

$$\Delta f \approx f_x(a, b)\Delta x + f_y(a, b)\Delta y = f_x(a, b)hu_1 + f_y(a, b)hu_2.$$

Thus, dividing by h gives

$$\frac{\Delta f}{h} \approx \frac{f_x(a,b)hu_1 + f_y(a,b)hu_2}{h} = f_x(a,b)u_1 + f_y(a,b)u_2.$$

This approximation becomes exact as $h \to 0$, so we have the following formula:

$$f_{\vec{u}}(a,b) = f_x(a,b)u_1 + f_y(a,b)u_2.$$

Example 4 Use the preceding formula to compute the directional derivative in Example 3. Check that we get the same answer as before.

Solution We calculate $f_{\vec{u}}(1,0)$, where $f(x,y) = x^2 + y^2$ and $\vec{u} = \frac{1}{\sqrt{2}}\vec{i} + \frac{1}{\sqrt{2}}\vec{j}$.

The partial derivatives are $f_x(x,y) = 2x$ and $f_y(x,y) = 2y$. So, as before,

$$f_{\vec{u}}(1,0) = f_x(1,0)u_1 + f_y(1,0)u_2 = (2)\left(\frac{1}{\sqrt{2}}\right) + (0)\left(\frac{1}{\sqrt{2}}\right) = \sqrt{2}.$$

The Gradient Vector

Notice that the expression for $f_{\vec{u}}(a,b)$ can be written as a dot product of $\vec{u}$ and a new vector:

$$f_{\vec{u}}(a,b) = f_x(a,b)u_1 + f_y(a,b)u_2 = (f_x(a,b)\vec{i} + f_y(a,b)\vec{j}) \cdot (u_1\vec{i} + u_2\vec{j}).$$

The new vector, $f_x(a,b)\vec{i} + f_y(a,b)\vec{j}$, turns out to be important. Thus, we make the following definition:

The Gradient Vector of a differentiable function f at the point (a,b) is

$$\text{grad}\, f(a,b) = f_x(a,b)\vec{i} + f_y(a,b)\vec{j}$$

The formula for the directional derivative can be written in terms of the gradient as follows:

The Directional Derivative and the Gradient

If f is differentiable at (a,b) and $\vec{u} = u_1\vec{i} + u_2\vec{j}$ is a unit vector, then

$$f_{\vec{u}}(a,b) = f_x(a,b)u_1 + f_y(a,b)u_2 = \text{grad}\, f(a,b) \cdot \vec{u}.$$

The change in f corresponding to a small change $\Delta\vec{r} = \Delta x\vec{i} + \Delta y\vec{j}$ can be estimated using the gradient:

$$\Delta f \approx \text{grad}\, f \cdot \Delta\vec{r}.$$

Example 5 Find the gradient vector of $f(x,y) = x + e^y$ at the point $(1,1)$.

Solution Using the definition, we have

$$\text{grad}\, f = f_x\vec{i} + f_y\vec{j} = \vec{i} + e^y\vec{j},$$

so at the point $(1,1)$

$$\text{grad}\, f(1,1) = \vec{i} + e\vec{j}.$$

Alternative Notation for the Gradient

You can think of $\frac{\partial f}{\partial x}\vec{i} + \frac{\partial f}{\partial y}\vec{j}$ as the result of applying the vector operator (pronounced "del")

$$\nabla = \frac{\partial}{\partial x}\vec{i} + \frac{\partial}{\partial y}\vec{j}$$

to the function f. Thus, we get the alternative notation

$$\operatorname{grad} f = \nabla f.$$

If $z = f(x, y)$, we can write $\operatorname{grad} z$ or ∇z for $\operatorname{grad} f$ or for ∇f.

What Does the Gradient Tell Us?

The fact that $f_{\vec{u}} = \operatorname{grad} f \cdot \vec{u}$ enables us to see what the gradient vector represents. Suppose θ is the angle between the vectors $\operatorname{grad} f$ and $\vec{u}$. At the point (a, b), we have

$$f_{\vec{u}} = \operatorname{grad} f \cdot \vec{u} = \|\operatorname{grad} f\| \underbrace{\|\vec{u}\|}_{1} \cos\theta = \|\operatorname{grad} f\| \cos\theta.$$

Imagine that $\operatorname{grad} f$ is fixed and that $\vec{u}$ can rotate. (See Figure 14.32.) The maximum value of $f_{\vec{u}}$ occurs when $\cos\theta = 1$, so $\theta = 0$ and $\vec{u}$ is pointing in the direction of $\operatorname{grad} f$. Then

$$\text{Maximum } f_{\vec{u}} = \|\operatorname{grad} f\| \cos 0 = \|\operatorname{grad} f\|.$$

The minimum value of $f_{\vec{u}}$ occurs when $\cos\theta = -1$, so $\theta = \pi$ and $\vec{u}$ is pointing in the direction opposite to $\operatorname{grad} f$. Then

$$\text{Minimum } f_{\vec{u}} = \|\operatorname{grad} f\| \cos \pi = -\|\operatorname{grad} f\|.$$

When $\theta = \pi/2$ or $3\pi/2$, so $\cos\theta = 0$, the directional derivative is zero.

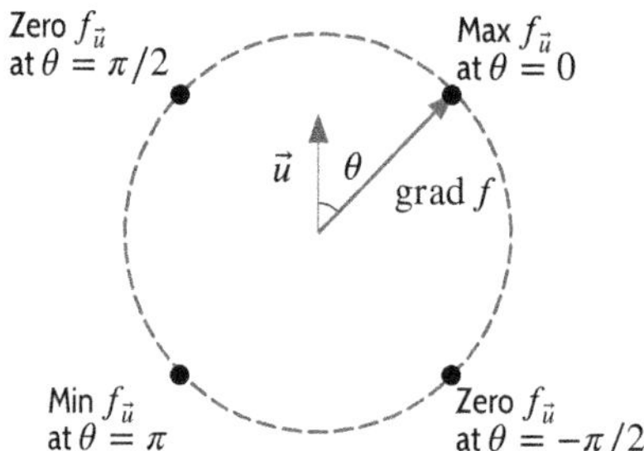

Figure 14.32: Values of the directional derivative at different angles to the gradient

Properties of the Gradient Vector

We have seen that the gradient vector points in the direction of the greatest rate of change at a point and the magnitude of the gradient vector is that rate of change.

Figure 14.33 shows that the gradient vector at a point is perpendicular to the contour through that point. If the contours represent equally spaced f-values and f is differentiable, local linearity tells us that the contours of f around a point appear straight, parallel, and equally spaced. The greatest rate of change is obtained by moving in the direction that takes us to the next contour in the shortest possible distance; that is, perpendicular to the contour. Thus, we have the following:

Geometric Properties of the Gradient Vector in the Plane

If f is a differentiable function at the point (a, b) and grad $f(a, b) \neq \vec{0}$, then:

- The direction of grad $f(a, b)$ is
  - Perpendicular[1] to the contour of f through (a, b);
  - In the direction of the maximum rate of increase of f.
- The magnitude of the gradient vector, $\|$ grad $f \|$, is
  - The maximum rate of change of f at that point;
  - Large when the contours are close together and small when they are far apart.

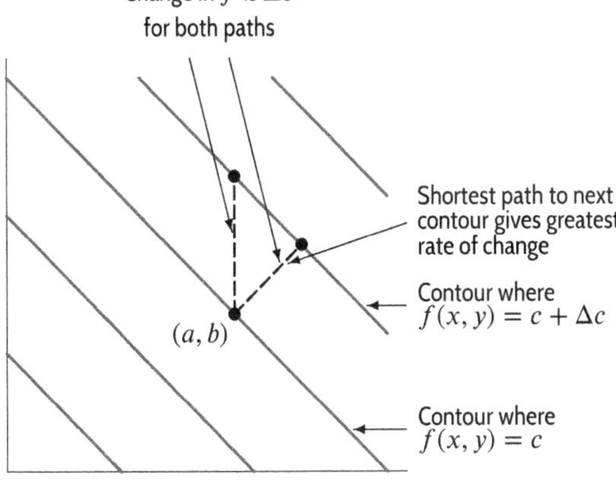

Figure 14.33: Close-up view of the contours around (a, b), showing the gradient is perpendicular to the contours

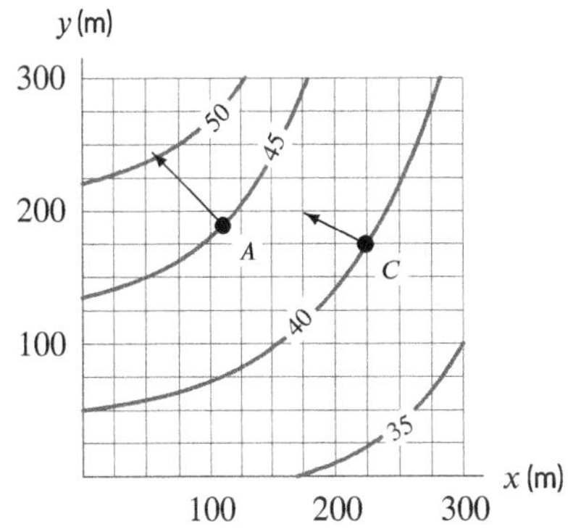

Figure 14.34: A temperature map showing directions and relative magnitudes of two gradient vectors

Examples of Directional Derivatives and Gradient Vectors

Example 6 Explain why the gradient vectors at points A and C in Figure 14.34 have the direction and the relative magnitudes they do.

Solution The gradient vector points in the direction of greatest increase of the function. This means that in Figure 14.34, the gradient points directly toward warmer temperatures. The magnitude of the gradient vector measures the rate of change. The gradient vector at A is longer than the gradient vector at C because the contours are closer together at A, so the rate of change is larger.

Example 2 on page 763 shows how the contour diagram can tell us the sign of the directional derivative. In the next example we compute the directional derivative in three directions, two that are close to that of the gradient vector and one that is not.

Example 7 Use the gradient to find the directional derivative of $f(x, y) = x + e^y$ at the point $(1, 1)$ in the direction of the vectors $\vec{i} - \vec{j}, \vec{i} + 2\vec{j}, \vec{i} + 3\vec{j}$.

Solution In Example 5 we found

$$\text{grad } f(1, 1) = \vec{i} + e\vec{j}.$$

A unit vector in the direction of $\vec{i} - \vec{j}$ is $\vec{s} = (\vec{i} - \vec{j})/\sqrt{2}$, so

$$f_{\vec{s}}(1, 1) = \text{grad } f(1, 1) \cdot \vec{s} = (\vec{i} + e\vec{j}) \cdot \left(\frac{\vec{i} - \vec{j}}{\sqrt{2}}\right) = \frac{1 - e}{\sqrt{2}} \approx -1.215.$$

[1]This assumes that the same scale is used on both axes.

A unit vector in the direction of $\vec{i} + 2\vec{j}$ is $\vec{v} = (\vec{i} + 2\vec{j})/\sqrt{5}$, so

$$f_{\vec{v}}(1,1) = \text{grad}\, f(1,1) \cdot \vec{v} = (\vec{i} + e\vec{j}) \cdot \left(\frac{\vec{i} + 2\vec{j}}{\sqrt{5}}\right) = \frac{1+2e}{\sqrt{5}} \approx 2.879.$$

A unit vector in the direction of $\vec{i} + 3\vec{j}$ is $\vec{w} = (\vec{i} + 3\vec{j})/\sqrt{10}$, so

$$f_{\vec{w}}(1,1) = \text{grad}\, f(1,1) \cdot \vec{w} = (\vec{i} + e\vec{j}) \cdot \left(\frac{\vec{i} + 3\vec{j}}{\sqrt{10}}\right) = \frac{1+3e}{\sqrt{10}} \approx 2.895.$$

Now look back at the answers and compare with the value of $\|\text{grad}\, f\| = \sqrt{1+e^2} \approx 2.896$. One answer is not close to this value; the other two, $f_{\vec{v}} = 2.879$ and $f_{\vec{w}} = 2.895$, are close but slightly smaller than $\|\text{grad}\, f\|$. Since $\|\text{grad}\, f\|$ is the maximum rate of change of f at the point, we have for *any* unit vector $\vec{u}$:

$$f_{\vec{u}}(1,1) \le \|\text{grad}\, f\|.$$

with equality when $\vec{u}$ is in the direction of grad f. Since $e \approx 2.718$, the vectors $\vec{i} + 2\vec{j}$ and $\vec{i} + 3\vec{j}$ both point roughly, but not exactly, in the direction of the gradient vector grad $f(1,1) = \vec{i} + e\vec{j}$. Thus, the values of $f_{\vec{v}}$ and $f_{\vec{w}}$ are both close to the value of $\|\text{grad}\, f\|$. The direction of the vector $\vec{i} - \vec{j}$ is not close to the direction of grad f and the value of $f_{\vec{s}}$ is not close to the value of $\|\text{grad}\, f\|$.

Exercises and Problems for Section 14.4 Online Resource: Additional Problems for Section 14.4

EXERCISES

In Exercises 1–14, find the gradient of the function. Assume the variables are restricted to a domain on which the function is defined.

1. $f(x,y) = \frac{3}{2}x^5 - \frac{4}{7}y^6$
2. $Q = 50K + 100L$
3. $f(m,n) = m^2 + n^2$
4. $z = xe^y$
5. $f(\alpha,\beta) = \sqrt{5\alpha^2 + \beta}$
6. $f(r,h) = \pi r^2 h$
7. $z = (x+y)e^y$
8. $f(K,L) = K^{0.3}L^{0.7}$
9. $f(r,\theta) = r\sin\theta$
10. $f(x,y) = \ln(x^2+y^2)$
11. $z = \sin(x/y)$
12. $z = \tan^{-1}(x/y)$
13. $f(\alpha,\beta) = \dfrac{2\alpha+3\beta}{2\alpha-3\beta}$
14. $z = x\dfrac{e^y}{x+y}$

In Exercises 15–22, find the gradient at the point.

15. $f(x,y) = x^2y + 7xy^3$, at $(1,2)$
16. $f(m,n) = 5m^2 + 3n^4$, at $(5,2)$
17. $f(r,h) = 2\pi rh + \pi r^2$, at $(2,3)$
18. $f(x,y) = e^{\sin y}$, at $(0,\pi)$
19. $f(x,y) = \sin(x^2) + \cos y$, at $(\frac{\sqrt{\pi}}{2}, 0)$
20. $f(x,y) = \ln(x^2+xy)$, at $(4,1)$
21. $f(x,y) = 1/(x^2+y^2)$, at $(-1,3)$
22. $f(x,y) = \sqrt{\tan x + y}$, at $(0,1)$

In Exercises 23–28, which of the following vectors gives the direction of the gradient vector at point A on the contour diagram? The scales on the x- and y-axes are the same.

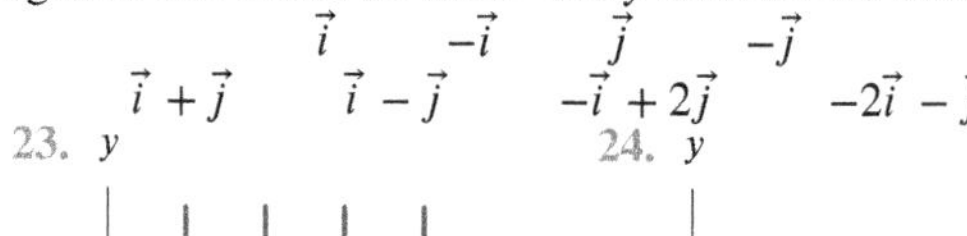

23.
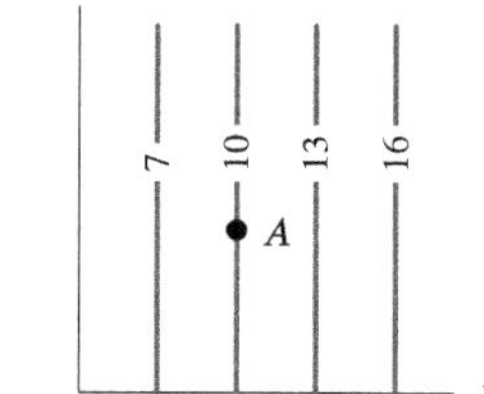

24.
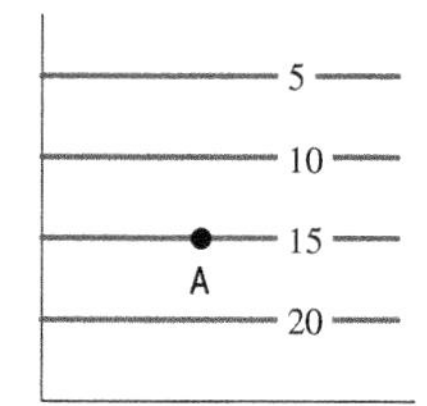

25.

26.
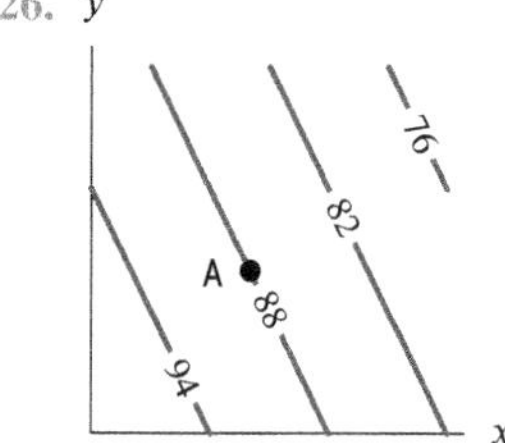

27.
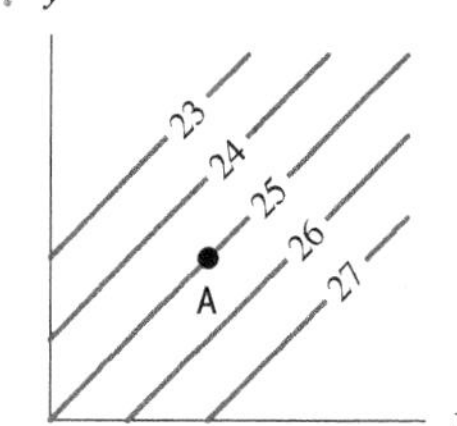

28.
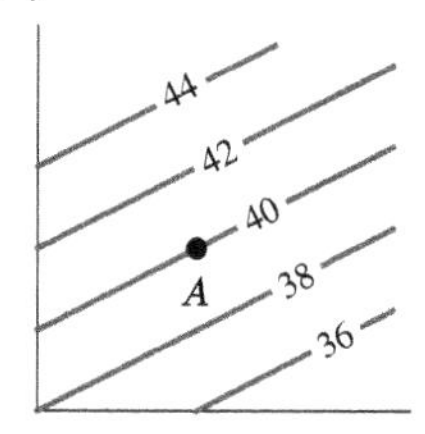

In Exercises 29–34, use the contour diagram of f in Figure 14.35 to decide if the specified directional derivative is positive, negative, or approximately zero.

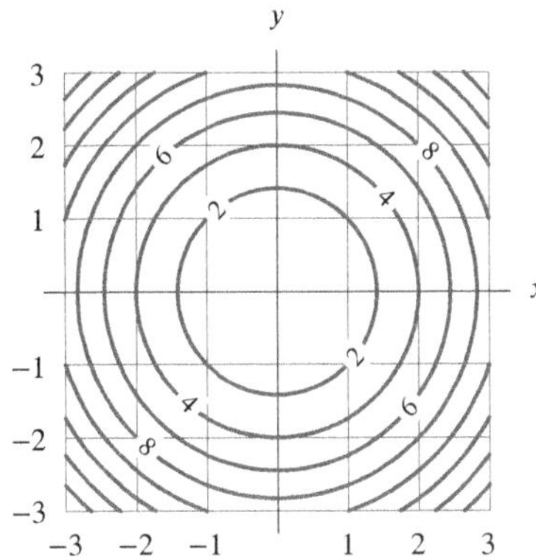

Figure 14.35

29. At point $(-2, 2)$, in direction $\vec{i}$.
30. At point $(0, -2)$, in direction $\vec{j}$.
31. At point $(0, -2)$, in direction $\vec{i} + 2\vec{j}$.
32. At point $(0, -2)$, in direction $\vec{i} - 2\vec{j}$.
33. At point $(-1, 1)$, in direction $\vec{i} + \vec{j}$.
34. At point $(-1, 1)$, in direction $-\vec{i} + \vec{j}$.

In Exercises 35–42, use the contour diagram of f in Figure 14.35 to find the approximate direction of the gradient vector at the given point.

35. $(-2, 0)$
36. $(0, -2)$
37. $(2, 0)$
38. $(0, 2)$
39. $(-2, 2)$
40. $(-2, -2)$
41. $(2, 2)$
42. $(2, -2)$

In Exercises 43–44, approximate the directional derivative of f in the direction from P to Q.

43. $P = (10, 12)$, $Q = (10.3, 12.1)$, $f(P) = 50$, $f(Q) = 52$.
44. $P = (-120, 45)$, $Q = (-122, 47)$, $f(P) = 200$, $f(Q) = 205$.

In Exercises 45–48, find the directional derivative $f_{\vec{u}}(1, 2)$ for the function f with $\vec{u} = (3\vec{i} - 4\vec{j})/5$.

45. $f(x, y) = xy + y^3$
46. $f(x, y) = 3x - 4y$
47. $f(x, y) = x^2 - y^2$
48. $f(x, y) = \sin(2x - y)$

49. If $f(x, y) = x^2 y$ and $\vec{v} = 4\vec{i} - 3\vec{j}$, find the directional derivative at the point $(2, 6)$ in the direction of $\vec{v}$.

In Exercises 50–51, find the differential df from the gradient.

50. $\text{grad } f = y\vec{i} + x\vec{j}$
51. $\text{grad } f = (2x + 3e^y)\vec{i} + 3xe^y\vec{j}$

In Exercises 52–53, find grad f from the differential.

52. $df = 2x\,dx + 10y\,dy$
53. $df = (x + 1)ye^x\,dx + xe^x\,dy$

In Exercises 54–55, assuming P and Q are close, approximate $f(Q)$.

54. $P = (100, 150)$, $Q = (101, 153)$, $f(P) = 2000$, $\text{grad} f(P) = 2\vec{i} - 2\vec{j}$.
55. $P = (10, 10)$, $Q = (10.2, 10.1)$, $f(P) = 50$, $\text{grad} f(P) = 0.5\vec{i} + \vec{j}$.

56. Where is grad$f$ longer: at a point where contour lines of f are far apart or at a point where contour lines of f are close together?

PROBLEMS

57. A student was asked to find the directional derivative of $f(x, y) = x^2 e^y$ at the point $(1, 0)$ in the direction of $\vec{v} = 4\vec{i} + 3\vec{j}$. The student's answer was

$$f_{\vec{u}}(1, 0) = \text{grad } f(1, 0) \cdot \vec{u} = \frac{8}{5}\vec{i} + \frac{3}{5}\vec{j}.$$

(a) At a glance, how do you know this is wrong?
(b) What is the correct answer?

In Problems 58–64, find the quantity. Assume that g is a smooth function and that

$$\nabla g(2, 3) = -2\vec{i} + \vec{j} \quad \text{and} \quad \nabla g(2.4, 3) = 4\vec{i}$$

58. $g_y(2.4, 3)$
59. $g_x(2, 3)$

60. A vector perpendicular to the level curve of g that passes through the point $(2.4, 3)$

61. A vector parallel to the level curve of g that passes through the point $(2, 3)$

62. The slope of the graph of g at the point $(2.4, 3)$ in the direction of the vector $\vec{i} + 3\vec{j}$.

63. The slope of the graph of g at the point $(2, 3)$ in the direction of the vector $\vec{i} + 3\vec{j}$.

64. The greatest slope of the graph of g at the point $(2, 3)$.

65. For $f(x, y) = (x + y)/(1 + x^2)$, find the directional derivative at $(1, -2)$ in the direction of $\vec{v} = 3\vec{i} + 4\vec{j}$.

66. For $g(x, y)$ with $g(5, 10) = 100$ and $g_{\vec{u}}(5, 10) = 0.5$, where $\vec{u}$ is the unit vector in the direction of the vector $\vec{i} + \vec{j}$, estimate $g(5.1, 10.1)$.

67. Let $f(P) = 15$ and $f(Q) = 20$ where $P = (3, 4)$ and $Q = (3.03, 3.96)$. Approximate the directional derivative of f at P in the direction of Q.

68. **(a)** Give Q, the point at a distance of 0.1 from $P = (4, 5)$ in the direction of $\vec{v} = -\vec{i} + 3\vec{j}$. Give five decimal places in your answer.
(b) Use P and Q to approximate the directional derivative of $f(x, y) = \sqrt{x + y}$ in the direction of $\vec{v}$.
(c) Give the exact value for the directional derivative you estimated in part (b).

69. For $f(x, y) = e^x \tan(y) + 2x^2 y$, find the directional derivative at the point $(0, \pi/4)$ in the direction

(a) $\vec{i} - \vec{j}$ **(b)** $\vec{i} + \sqrt{3}\vec{j}$

70. Find the rate of change of $f(x, y) = x^2 + y^2$ at the point $(1, 2)$ in the direction of the vector $\vec{u} = 0.6\vec{i} + 0.8\vec{j}$.

71. **(a)** Let $f(x, y) = (x + y)/(1 + x^2)$. Find the directional derivative of f at $P = (1, -2)$ in the direction of:

(i) $\vec{v} = 3\vec{i} - 2\vec{j}$ (ii) $\vec{v} = -\vec{i} + 4\vec{j}$

(b) What is the direction of greatest increase of f at P?

72. Let $f(5, 10) = 200$ and $f(5.2, 9.9) = 197$.

(a) Approximate the directional derivative at $(5, 10)$ in the direction from $(5, 10)$ toward $(5.2, 9.9)$.
(b) Approximate $f(Q)$ at the point Q that is distance 0.1 from $(5, 10)$ in the direction of $(5.2, 9.9)$.
(c) Give coordinates for the point Q.

73. Let $f(100, 100) = 500$ and $\text{grad} f(100, 100) = 2\vec{i} + 3\vec{j}$.

(a) Find the directional derivative of f at the point $(100, 100)$ in the direction $\vec{i} + \vec{j}$.
(b) Use the directional derivative to approximate $f(102, 102)$.

74. Let $\text{grad} f(50, 60) = 0.3\vec{i} + 0.5\vec{j}$. Approximate the directional derivative of f at the point $(50, 60)$ in the direction of the point $(49.5, 62)$.

75. Let $f(x, y) = x^2 y^3$. At the point $(-1, 2)$, find a vector

(a) In the direction of maximum rate of change.
(b) In the direction of minimum rate of change.
(c) In a direction in which the rate of change is zero.

76. Let $f(x, y) = e^{xy}$. At the point $(1, 1)$, find a unit vector

(a) In the direction of the steepest ascent.
(b) In the direction of the steepest descent.
(c) In a direction in which the rate of change is zero.

For Problems 77–81 use Figure 14.36, showing level curves of $f(x, y)$, to estimate the directional derivatives.

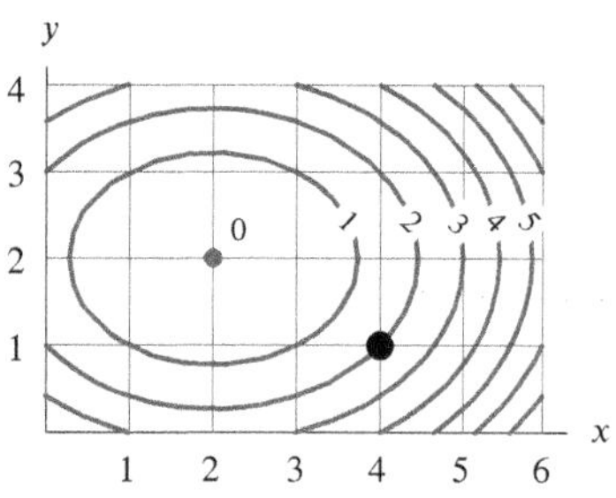

Figure 14.36

77. $f_{\vec{i}}(4, 1)$ **78.** $f_{\vec{j}}(4, 1)$

79. $f_{\vec{u}}(4, 1)$ where $\vec{u} = (\vec{i} - \vec{j})/\sqrt{2}$

80. $f_{\vec{u}}(4, 1)$ where $\vec{u} = (-\vec{i} + \vec{j})/\sqrt{2}$

81. $f_{\vec{u}}(4, 1)$ with $\vec{u} = (-2\vec{i} + \vec{j})/\sqrt{5}$

82. The surface $z = g(x, y)$ is in Figure 14.37. What is the sign of each of the following directional derivatives?

(a) $g_{\vec{u}}(2, 5)$ where $\vec{u} = (\vec{i} - \vec{j})/\sqrt{2}$.
(b) $g_{\vec{u}}(2, 5)$ where $\vec{u} = (\vec{i} + \vec{j})/\sqrt{2}$.

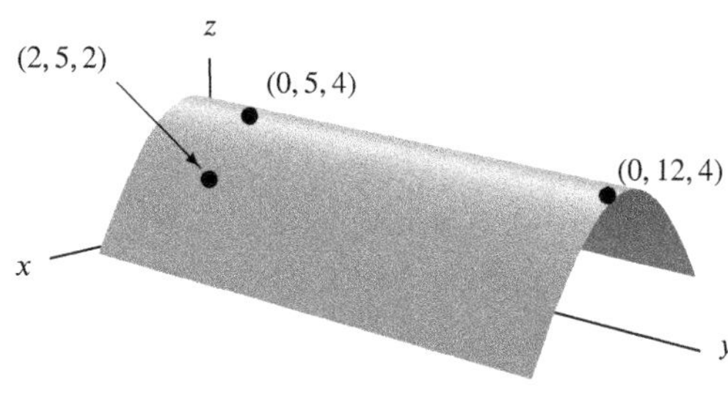

Figure 14.37

83. The table gives values of a differentiable function $f(x, y)$. At the point $(1.2, 0)$, into which quadrant does the gradient vector of f point? Justify your answer.

		y		
		−1	0	1
x	1.0	0.7	0.1	−0.5
	1.2	4.8	4.2	3.6
	1.4	8.9	8.3	7.7

84. The gradient of f at a point P has magnitude 10 and is in the direction of A in Figure 14.38. Find the directional derivatives of f at P in the six directions shown.

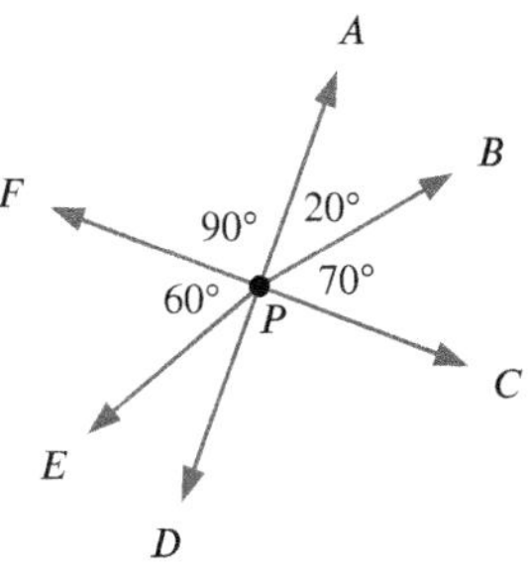

Figure 14.38

85. Figure 14.39 represents the level curves $f(x, y) = c$; the values of f on each curve are marked. In each of the following parts, decide whether the given quantity is positive, negative or zero. Explain your answer.

(a) The value of $\nabla f \cdot \vec{i}$ at P.
(b) The value of $\nabla f \cdot \vec{j}$ at P.
(c) $\partial f/\partial x$ at Q.
(d) $\partial f/\partial y$ at Q.

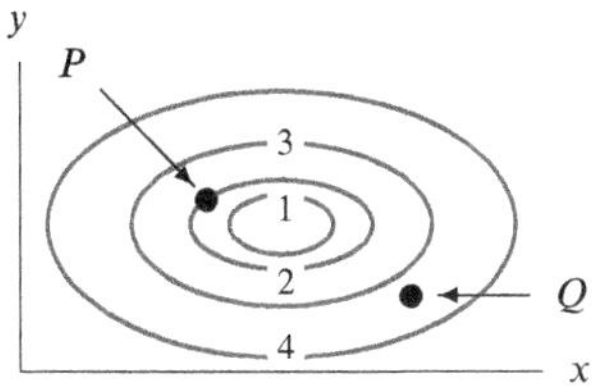

Figure 14.39

86. In Figure 14.39, which is larger: $\|\nabla f\|$ at P or $\|\nabla f\|$ at Q? Explain how you know.

87. Let P, Q, R and S be four distinct points in the plane. Let $\vec{u}$ be the unit vector in the direction from P to Q, $\vec{v}$ the unit vector in the direction from P to R, and $\vec{w}$ the unit vector in the direction from P to S. Let $f(x, y)$ be a linear function with $f(P) = 10$, $f(Q) = 7$, $f(R) = 15$, and $f(S) = 10$. List the directional derivatives $f_{\vec{u}}(P)$, $f_{\vec{v}}(P)$, and $f_{\vec{w}}(P)$ in increasing order.

88. Let $f_x(3, 1) = -5$ and $f_y(3, 1) = 2$. Find a unit vector $\vec{u}$ such that:

(a) $f_{\vec{u}}(3, 1) > 0$ (b) $f_{\vec{u}}(3, 1) < 0$
(c) $f_{\vec{u}}(3, 1) = 0$

89. Let $f(0, 0) = -4$ and $f_{\vec{u}}(0, 0) = 20$ for a unit vector $\vec{u}$. Suppose that points P and Q in Figure 14.40 are close. Find approximate values of $f(P)$ and $f(Q)$.

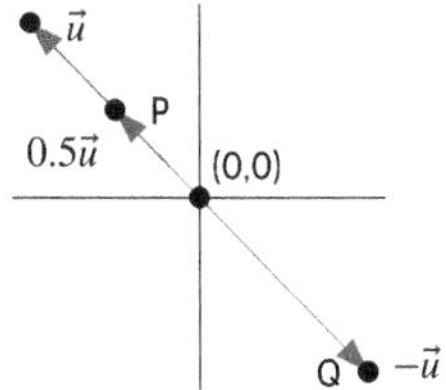

Figure 14.40

In Problems 90–93, check that the point $(2, 3)$ lies on the curve. Then, viewing the curve as a contour of $f(x, y)$, use grad $f(2, 3)$ to find a vector normal to the curve at $(2, 3)$ and an equation for the tangent line to the curve at $(2, 3)$.

90. $x^2 + y^2 = 13$
91. $xy = 6$
92. $y = x^2 - 1$
93. $(y - x)^2 + 2 = xy - 3$

94. The temperature H in °Fahrenheit y miles north of the Canadian border t hours after midnight is given by $H = 30 - 0.05y - 5t$. A moose runs north at a speed of 20 mph. At what rate does the moose perceive the temperature to be changing?

95. At a certain point on a heated plate, the greatest rate of temperature increase, 5° C per meter, is toward the northeast. If an object at this point moves directly north, at what rate is the temperature increasing?

96. An ant is at the point $(1, 1, 3)$ on the surface of a bowl with equation $z = x^2 + 2y^2$, where x and y are in cm. In what two horizontal directions can the ant move away from the point $(1, 1, 3)$ so that its initial rate of ascent is 2 vertical cm for each horizontal cm moved? Give your answers as vectors in the plane.

97. Let $T = f(x, y) = 100e^{-(x^2/2)-y^2}$ represent the temperature, in °C, at the point (x, y) with x and y in meters.

(a) Describe the contours of f, and explain their meaning in the context of this problem.
(b) Find the rate at which the temperature changes as you move away from the point $(1, 1)$ toward the point $(2, 3)$. Give units in your answer.
(c) In what direction would you move away from $(1, 1)$ for the temperature to increase as fast as possible?

98. You are climbing a mountain by the steepest route at a slope of 20° when you come upon a trail branching off at a 30° angle from yours. What is the angle of ascent of the branch trail?

99. You are standing at the point $(1, 1, 3)$ on the hill whose equation is given by $z = 5y - x^2 - y^2$.

(a) If you choose to climb in the direction of steepest ascent, what is your initial rate of ascent relative to the horizontal distance?
(b) If you decide to go straight northwest, will you be ascending or descending? At what rate?
(c) If you decide to maintain your altitude, in what directions can you go?

Strengthen Your Understanding

In Problems 100–102, explain what is wrong with the statement.

100. A function f has a directional derivative given by $f_{\vec{u}}(0, 0) = 3\vec{i} + 4\vec{j}$.

101. A function f has gradient grad $f(0, 0) = 7$.

102. The gradient vector grad $f(x, y)$ is perpendicular to the contours of f, and the closer together the contours for equally spaced values of f, the shorter the gradient vector.

In Problems 103–104, give an example of:

103. A unit vector $\vec{u}$ such that $f_{\vec{u}}(0,0) < 0$, given that $f_x(0,0) = 2$ and $f_y(0,0) = 3$.

104. A contour diagram of a function with two points in the domain where the gradients are parallel but different lengths.

Are the statements in Problems 105–116 true or false? Give reasons for your answer.

105. If the point (a,b) is on the contour $f(x,y) = k$, then the slope of the line tangent to this contour at (a,b) is $f_y(a,b)/f_x(a,b)$.

106. The gradient vector $\operatorname{grad} f(a,b)$ is a vector in 3-space.

107. $\operatorname{grad}(fg) = (\operatorname{grad} f) \cdot (\operatorname{grad} g)$

108. The gradient vector $\operatorname{grad} f(a,b)$ is tangent to the contour of f at (a,b).

109. If you know the gradient vector of f at (a,b) then you can find the directional derivative $f_{\vec{u}}(a,b)$ for any unit vector $\vec{u}$.

110. If you know the directional derivative $f_{\vec{u}}(a,b)$ for all unit vectors $\vec{u}$ then you can find the gradient vector of f at (a,b).

111. The directional derivative $f_{\vec{u}}(a,b)$ is parallel to $\vec{u}$.

112. The gradient $\operatorname{grad} f(3,4)$ is perpendicular to the vector $3\vec{i} + 4\vec{j}$.

113. If $\operatorname{grad} f(1,2) = \vec{i}$, then f decreases in the $-\vec{i}$ direction at $(1,2)$.

114. If $\operatorname{grad} f(1,2) = \vec{i}$, then $f(10,2) > f(1,2)$.

115. At the point $(3,0)$, the function $g(x,y) = x^2 + y^2$ has the same maximal rate of increase as that of the function $h(x,y) = 2xy$.

116. If $f(x,y) = e^{x+y}$, then the directional derivative in any direction $\vec{u}$ (with $\|\vec{u}\| = 1$) at the point $(0,0)$ is always less than or equal to $\sqrt{2}$.

14.5 GRADIENTS AND DIRECTIONAL DERIVATIVES IN SPACE

The Gradient Vector and Directional Derivative of a Function of Three Variables

The gradient of a function of three variables is defined in the same way as for two variables:

> **The gradient vector** of a differentiable function $f(x,y,z)$ is
>
> $$\operatorname{grad} f = f_x\vec{i} + f_y\vec{j} + f_z\vec{k}.$$

As in two dimensions, directional derivatives in space give the rate of change of a function in the direction of a unit vector $\vec{u}$. If a function f of three variables is differentiable at the point (a,b,c) and $\vec{u} = u_1\vec{i} + u_2\vec{j} + u_3\vec{k}$, then the directional derivative $f_{\vec{u}}$ is related to the gradient by

$$f_{\vec{u}}(a,b,c) = f_x(a,b,c)u_1 + f_y(a,b,c)u_2 + f_z(a,b,c)u_3 = \operatorname{grad} f(a,b,c) \cdot \vec{u}.$$

Since $\operatorname{grad} f(a,b,c)\cdot\vec{u} = \|\operatorname{grad} f(a,b,c)\| \cos\theta$, where θ is the angle between $\operatorname{grad} f(a,b,c)$ and $\vec{u}$, the value of $f_{\vec{u}}(a,b,c)$ is largest when $\theta = 0$, that is, when $\vec{u}$ is in the same direction as $\operatorname{grad} f(a,b,c)$. In addition, $f_{\vec{u}}(a,b,c) = 0$ when $\theta = \pi/2$, so $\operatorname{grad} f(a,b,c)$ is perpendicular to the level surface of f. The properties of gradients in space are similar to those in the plane:

> **Properties of the Gradient Vector in Space**
>
> If f is differentiable at (a,b,c) and $\vec{u}$ is a unit vector, then
>
> $$f_{\vec{u}}(a,b,c) = \operatorname{grad} f(a,b,c) \cdot \vec{u}.$$
>
> If, in addition, $\operatorname{grad} f(a,b,c) \neq \vec{0}$, then
> - $\operatorname{grad} f(a,b,c)$ is perpendicular to the level surface of f at (a,b,c)
> - $\operatorname{grad} f(a,b,c)$ is in the direction of the greatest rate of increase of f
> - $\|\operatorname{grad} f(a,b,c)\|$ is the maximum rate of change of f at (a,b,c).

Example 1 Find the directional derivative of $f(x,y,z) = xy + z$ at the point $(-1,0,1)$ in the direction of the vector $\vec{v} = 2\vec{i} + \vec{k}$.

Solution The magnitude of $\vec{v}$ is $\|\vec{v}\| = \sqrt{2^2+1} = \sqrt{5}$, so a unit vector in the same direction as $\vec{v}$ is

$$\vec{u} = \frac{\vec{v}}{\|\vec{v}\|} = \frac{2}{\sqrt{5}}\vec{i} + 0\vec{j} + \frac{1}{\sqrt{5}}\vec{k}.$$

The partial derivatives of f are $f_x(x, y, z) = y$ and $f_y(x, y, z) = x$ and $f_z(x, y, z) = 1$. Thus,

$$\begin{aligned} f_{\vec{u}}(-1, 0, 1) &= f_x(-1, 0, 1)u_1 + f_y(-1, 0, 1)u_2 + f_z(-1, 0, 1)u_3 \\ &= (0)\left(\frac{2}{\sqrt{5}}\right) + (-1)(0) + (1)\left(\frac{1}{\sqrt{5}}\right) = \frac{1}{\sqrt{5}}. \end{aligned}$$

Example 2 Let $f(x, y, z) = x^2 + y^2$ and $g(x, y, z) = -x^2 - y^2 - z^2$. What can we say about the direction of the following vectors?

(a) $\operatorname{grad} f(0, 1, 1)$ (b) $\operatorname{grad} f(1, 0, 1)$ (c) $\operatorname{grad} g(0, 1, 1)$ (d) $\operatorname{grad} g(1, 0, 1)$.

Solution The cylinder $x^2 + y^2 = 1$ in Figure 14.41 is a level surface of f and contains both the points $(0, 1, 1)$ and $(1, 0, 1)$. Since the value of f does not change at all in the z-direction, all the gradient vectors are horizontal. They are perpendicular to the cylinder and point outward because the value of f increases as we move out.

Similarly, the points $(0, 1, 1)$ and $(1, 0, 1)$ also lie on the same level surface of g, namely $g(x, y, z) = -x^2 - y^2 - z^2 = -2$, which is the sphere $x^2 + y^2 + z^2 = 2$. Part of this level surface is shown in Figure 14.42. This time the gradient vectors point inward, since the negative signs mean that the function increases (from large negative values to small negative values) as we move inward.

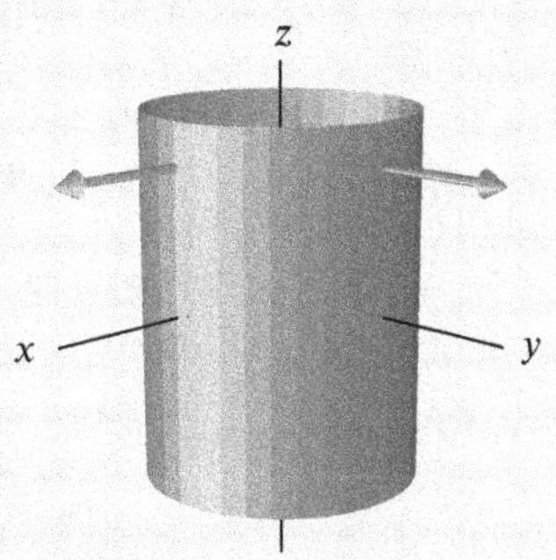

Figure 14.41: The level surface $f(x, y, z) = x^2 + y^2 = 1$ with two gradient vectors

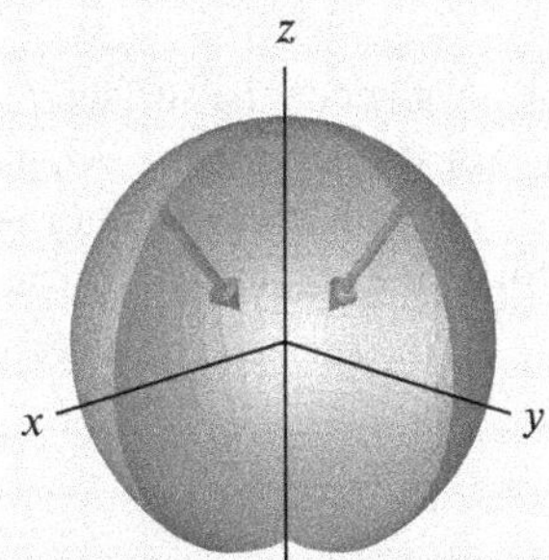

Figure 14.42: The level surface $g(x, y, z) = -x^2 - y^2 - z^2 = -2$ with two gradient vectors

Example 3 Consider the functions $f(x, y) = 4 - x^2 - 2y^2$ and $g(x, y) = 4 - x^2$. Calculate a vector perpendicular to each of the following:

(a) The level curve of f at the point $(1, 1)$ (b) The surface $z = f(x, y)$ at the point $(1, 1, 1)$
(c) The level curve of g at the point $(1, 1)$ (d) The surface $z = g(x, y)$ at the point $(1, 1, 3)$

Solution (a) The vector we want is a 2-vector in the plane. Since $\operatorname{grad} f = -2x\vec{i} - 4y\vec{j}$, we have

$$\operatorname{grad} f(1, 1) = -2\vec{i} - 4\vec{j}.$$

Any nonzero multiple of this vector is perpendicular to the level curve at the point $(1, 1)$.

(b) In this case we want a 3-vector in space. To find it we rewrite $z = 4 - x^2 - 2y^2$ as the level surface of the function F, where

$$F(x, y, z) = 4 - x^2 - 2y^2 - z = 0.$$

Then

$$\operatorname{grad} F = -2x\vec{i} - 4y\vec{j} - \vec{k},$$

so

$$\operatorname{grad} F(1,1,1) = -2\vec{i} - 4\vec{j} - \vec{k},$$

and $\operatorname{grad} F(1,1,1)$ is perpendicular to the surface $z = 4 - x^2 - 2y^2$ at the point $(1,1,1)$. Notice that $-2\vec{i} - 4\vec{j} - \vec{k}$ is not the only possible answer: any multiple of this vector will do.

(c) We are looking for a 2-vector. Since $\operatorname{grad} g = -2x\vec{i} + 0\vec{j}$, we have

$$\operatorname{grad} g(1,1) = -2\vec{i}.$$

Any multiple of this vector is perpendicular to the level curve also.

(d) We are looking for a 3-vector. We rewrite $z = 4 - x^2$ as the level surface of the function G, where

$$G(x, y, z) = 4 - x^2 - z = 0.$$

Then

$$\operatorname{grad} G = -2x\vec{i} - \vec{k}$$

So

$$\operatorname{grad} G(1,1,3) = -2\vec{i} - \vec{k},$$

and any multiple of $\operatorname{grad} G(1,1,3)$ is perpendicular to the surface $z = 4 - x^2$ at this point.

Example 4 (a) A hiker on the surface $f(x, y) = 4 - x^2 - 2y^2$ at the point $(1, -1, 1)$ starts to climb along the path of steepest ascent. What is the relation between the vector $\operatorname{grad} f(1, -1)$ and a vector tangent to the path at the point $(1, -1, 1)$ and pointing uphill?

(b) At the point $(1, -1, 1)$ on the surface $f(x, y) = 4 - x^2 - 2y^2$, calculate a vector, $\vec{n}$, perpendicular to the surface and a vector, $\vec{T}$, tangent to the curve of steepest ascent.

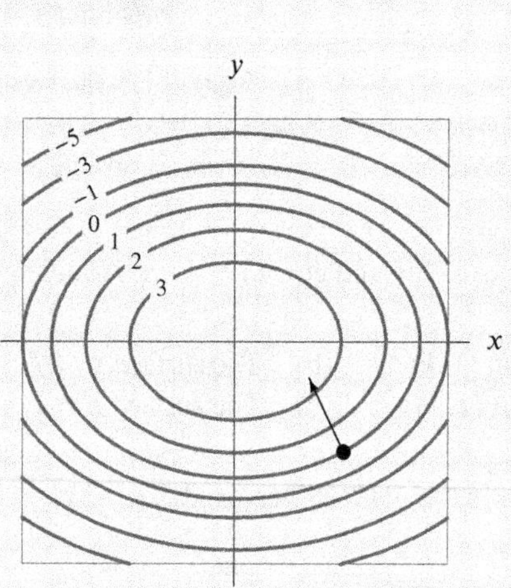

Figure 14.43: Contour diagram for $z = f(x, y) = 4 - x^2 - 2y^2$ showing direction of $\operatorname{grad} f(1, -1)$

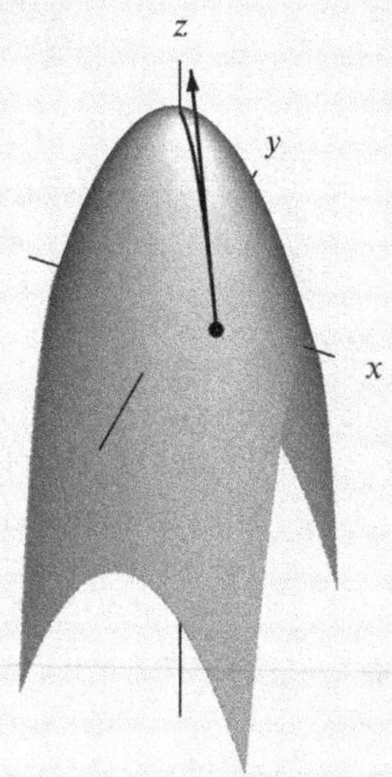

Figure 14.44: Graph of $f(x, y) = 4 - x^2 - 2y^2$ showing path of steepest ascent from the point $(1, -1, 1)$

Solution

(a) The hiker at the point $(1, -1, 1)$ lies directly above the point $(1, -1)$ in the xy-plane. The vector $\operatorname{grad} f(1, -1)$ lies in 2-space, pointing like a compass in the direction in which f increases most rapidly. Therefore, $\operatorname{grad} f(1, -1)$ lies directly under a vector tangent to the hiker's path at $(1, -1, 1)$ and pointing uphill. (See Figures 14.43 and 14.44.)

(b) The surface is represented by $F(x, y, z) = 4 - x^2 - 2y^2 - z = 0$. Since $\operatorname{grad} F = -2x\vec{i} - 4y\vec{j} - \vec{k}$, a normal, $\vec{n}$, to the surface is given by

$$\vec{n} = \operatorname{grad} F(1, -1, 1) = -2(1)\vec{i} - 4(-1)\vec{j} - \vec{k} = -2\vec{i} + 4\vec{j} - \vec{k}.$$

We take the $\vec{i}$ and $\vec{j}$ components of $\vec{T}$ to be the vector $\operatorname{grad} f(1, -1) = -2\vec{i} + 4\vec{j}$. Thus, we have that, for some $a > 0$,

$$\vec{T} = -2\vec{i} + 4\vec{j} + a\vec{k}$$

We want $\vec{n} \cdot \vec{T} = 0$, so

$$\vec{n} \cdot \vec{T} = (-2\vec{i} + 4\vec{j} - \vec{k}) \cdot (-2\vec{i} + 4\vec{j} + a\vec{k}) = 4 + 16 - a = 0$$

So $a = 20$ and hence

$$\vec{T} = -2\vec{i} + 4\vec{j} + 20\vec{k}.$$

Example 5 Find the equation of the tangent plane to the sphere $x^2 + y^2 + z^2 = 14$ at the point $(1, 2, 3)$.

Solution We write the sphere as a level surface as follows:

$$f(x, y, z) = x^2 + y^2 + z^2 = 14.$$

We have

$$\operatorname{grad} f = 2x\vec{i} + 2y\vec{j} + 2z\vec{k},$$

so the vector

$$\operatorname{grad} f(1, 2, 3) = 2\vec{i} + 4\vec{j} + 6\vec{k}$$

is perpendicular to the sphere at the point $(1, 2, 3)$. Since the vector $\operatorname{grad} f(1, 2, 3)$ is normal to the tangent plane, the equation of the plane is

$$2x + 4y + 6z = 2 \cdot 1 + 4 \cdot 2 + 6 \cdot 3 = 28 \quad \text{or} \quad x + 2y + 3z = 14.$$

We could also try to find the tangent plane to the level surface $f(x, y, z) = k$ by solving algebraically for z and using the method of Section 14.3, page 755. (See Problem 47.) Solving for z can be difficult or impossible, however, so the method of Example 5 is preferable.

Tangent Plane to a Level Surface

If $f(x, y, z)$ is differentiable at (a, b, c), then an equation for the tangent plane to the level surface of f at the point (a, b, c) is

$$f_x(a, b, c)(x - a) + f_y(a, b, c)(y - b) + f_z(a, b, c)(z - c) = 0.$$

Caution: Scale on the Axis and the Geometric Interpretation of the Gradient

When we interpreted the gradient of a function geometrically (page 767), we tacitly assumed that the units and scales along the x and y axes were the same. If the scales are not the same, the gradient vector may not look perpendicular to the contours. Consider the function $f(x, y) = x^2 + y$ with gradient vector $\operatorname{grad} f = 2x\vec{i} + \vec{j}$. Figure 14.45 shows the gradient vector at (1, 1) using the same scales in the x and y directions. As expected, the gradient vector is perpendicular to the contour line. Figure 14.46 shows contours of the same function with unequal scales on the two axes. Notice that the gradient vector no longer appears perpendicular to the contour lines. Thus, we see that the geometric interpretation of the gradient vector requires that the same scale be used on both axes.

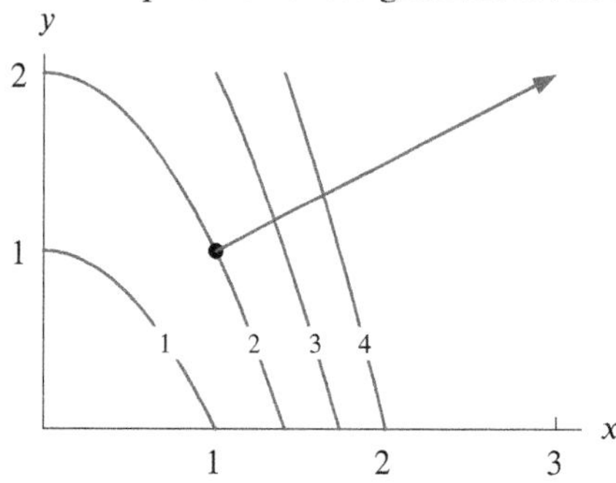

Figure 14.45: The gradient vector with x and y scales equal

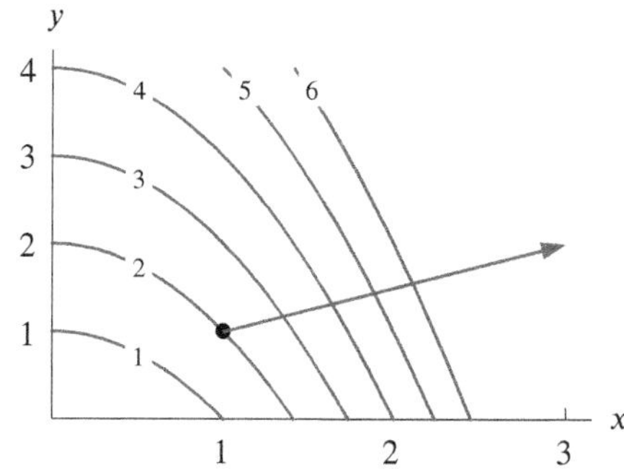

Figure 14.46: The gradient vector with x and y scales unequal

Exercises and Problems for Section 14.5 Online Resource: Additional Problems for Section 14.5

EXERCISES

In Exercises 1–12, find the gradient of the function.

1. $f(x, y, z) = x^2$
2. $f(x, y, z) = x^2 + y^3 - z^4$
3. $f(x, y, z) = e^{x+y+z}$
4. $f(x, y, z) = \cos(x + y) + \sin(y + z)$
5. $f(x, y, z) = yz^2/(1 + x^2)$
6. $f(x, y, z) = 1/(x^2 + y^2 + z^2)$
7. $f(x, y, z) = \sqrt{x^2 + y^2 + z^2}$
8. $f(x, y, z) = xe^y \sin z$
9. $f(x, y, z) = xy + \sin(e^z)$
10. $f(x_1, x_2, x_3) = x_1^2 x_2^3 x_3^4$
11. $f(p, q, r) = e^p + \ln q + e^{r^2}$
12. $f(x, y, z) = e^{z^2} + y\ln(x^2 + 5)$

In Exercises 13–18, find the gradient at the point.

13. $f(x, y, z) = zy^2$, at $(1, 0, 1)$
14. $f(x, y, z) = 2x + 3y + 4z$, at $(1, 1, 1)$
15. $f(x, y, z) = x^2 + y^2 - z^4$, at $(3, 2, 1)$
16. $f(x, y, z) = xyz$, at $(1, 2, 3)$
17. $f(x, y, z) = \sin(xy) + \sin(yz)$, at $(1, \pi, -1)$
18. $f(x, y, z) = x\ln(yz)$, at $(2, 1, e)$

In Exercises 19–24, find the directional derivative using $f(x, y, z) = xy + z^2$.

19. At $(1, 2, 3)$ in the direction of $\vec{i} + \vec{j} + \vec{k}$.
20. At $(1, 1, 1)$ in the direction of $\vec{i} + 2\vec{j} + 3\vec{k}$.
21. As you leave the point $(1, 1, 0)$ heading in the direction of the point $(0, 1, 1)$.
22. As you arrive at $(0, 1, 1)$ from the direction of $(1, 1, 0)$.
23. At the point $(2, 3, 4)$ in the direction of a vector making an angle of $3\pi/4$ with $\operatorname{grad} f(2, 3, 4)$.
24. At the point $(2, 3, 4)$ in the direction of the maximum rate of change of f.

In Exercises 25–30, check that the point $(-1, 1, 2)$ lies on the given surface. Then, viewing the surface as a level surface for a function $f(x, y, z)$, find a vector normal to the surface and an equation for the tangent plane to the surface at $(-1, 1, 2)$.

25. $x^2 - y^2 + z^2 = 4$
26. $z = x^2 + y^2$
27. $y^2 = z^2 - 3$
28. $x^2 - xyz = 3$
29. $\cos(x + y) = e^{xz+2}$
30. $y = 4/(2x + 3z)$

In Exercises 31–32, the gradient of f and a point P on the level surface $f(x, y, z) = 0$ are given. Find an equation for the tangent plane to the surface at the point P.

31. $\operatorname{grad} f = yz\vec{i} + xz\vec{j} + xy\vec{k}$, $P = (1, 2, 3)$
32. $\operatorname{grad} f = 2x\vec{i} + z^2\vec{j} + 2yz\vec{k}$, $P = (10, -10, 30)$

In Exercises 33–37, find an equation of the tangent plane to the surface at the given point.

33. $x^2 + y^2 + z^2 = 17$ at the point $(2, 3, 2)$

34. $x^2 + y^2 = 1$ at the point $(1, 0, 0)$

35. $z = 2x + y + 3$ at the point $(0, 0, 3)$

36. $3x^2 - 4xy + z^2 = 0$ at the point (a, a, a), where $a \neq 0$

37. $z = 9/(x + 4y)$ at the point where $x = 1$ and $y = 2$

38. For $f(x, y, z) = 3x^2y^2+2yz$, find the directional derivative at the point $(-1, 0, 4)$ in the direction of **(a)** $\vec{i} - \vec{k}$ **(b)** $-\vec{i} + 3\vec{j} + 3\vec{k}$

39. If $f(x, y, z) = x^2+3xy+2z$, find the directional derivative at the point $(2, 0, -1)$ in the direction of $2\vec{i} + \vec{j} - 2\vec{k}$.

40. **(a)** Let $f(x, y, z) = x^2 + y^2 - xyz$. Find grad f.
(b) Find the equation for the tangent plane to the surface $f(x, y, z) = 7$ at the point $(2, 3, 1)$.

41. Find the equation of the tangent plane at the point $(3, 2, 2)$ to $z = \sqrt{17 - x^2 - y^2}$.

42. Find the equation of the tangent plane to $z = 8/(xy)$ at the point $(1, 2, 4)$.

43. Find an equation of the tangent plane and of a normal vector to the surface $x = y^3z^7$ at the point $(1, -1, -1)$.

PROBLEMS

44. Let $f(x, y, z)$ represent the temperature in °C at the point (x, y, z) with x, y, z in meters. Let $\vec{v}$ be your velocity in meters per second. Give units and an interpretation of each of the following quantities.

(a) $\|\text{grad } f\|$ **(b)** $\text{grad } f \cdot \vec{v}$ **(c)** $\|\text{grad } f\| \cdot \|\vec{v}\|$

45. Consider the surface $g(x, y) = 4 - x^2$. What is the relation between grad $g(-1, -1)$ and a vector tangent to the path of steepest ascent at $(-1, -1, 3)$? Illustrate your answer with a sketch.

46. Match the functions $f(x, y, z)$ in (a)–(d) with the descriptions of their gradients in (I)–(IV).

(a) $x^2 + y^2 + z^2$ **(b)** $x^2 + y^2$

(c) $\dfrac{1}{x^2 + y^2 + z^2}$ **(d)** $\dfrac{1}{x^2 + y^2}$

I Points radially outward from the z-axis.
II Points radially inward toward the z-axis.
III Points radially outward from the origin.
IV Points radially inward toward the origin.

47. Find the equation of the tangent plane at $(2, 3, 1)$ to the surface $x^2 + y^2 - xyz = 7$. Do this in two ways:

(a) Viewing the surface as the level set of a function of three variables, $F(x, y, z)$.
(b) Viewing the surface as the graph of a function of two variables $z = f(x, y)$.

48. At what point on the surface $z = 1+x^2+y^2$ is its tangent plane parallel to the following planes?

(a) $z = 5$ **(b)** $z = 5 + 6x - 10y$.

49. Let $g_x(2, 1, 7) = 3$, $g_y(2, 1, 7) = 10$, $g_z(2, 1, 7) = -5$. Find the equation of the tangent plane to $g(x, y, z) = 0$ at the point $(2, 1, 7)$.

50. The vector ∇f at point P and four unit vectors $\vec{u}_1, \vec{u}_2, \vec{u}_3, \vec{u}_4$ are shown in Figure 14.47. Arrange the following quantities in ascending order

$$f_{\vec{u}_1}, \quad f_{\vec{u}_2}, \quad f_{\vec{u}_3}, \quad f_{\vec{u}_4}, \quad \text{the number 0.}$$

The directional derivatives are all evaluated at the point P and the function $f(x, y)$ is differentiable at P.

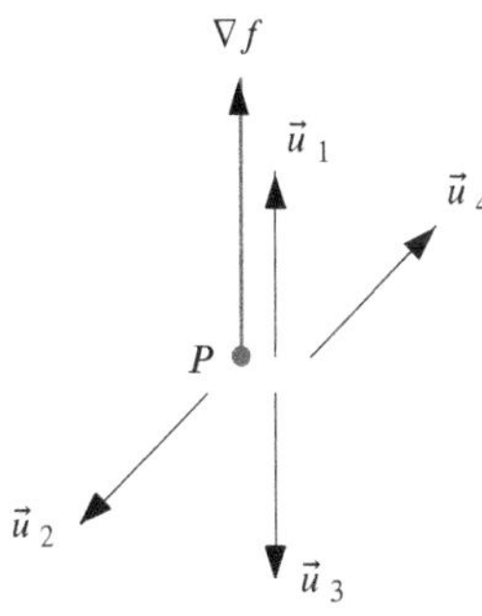

Figure 14.47

51. Let $f(x, y, z) = x^2 + y^2 + z^2$. At the point $(1, 2, 1)$, find the rate of change of f in the direction perpendicular to the plane $x + 2y + 3z = 8$ and moving away from the origin.

52. Let $f(x, y) = \cos x \sin y$ and let S be the surface $z = f(x, y)$.

(a) Find a normal vector to the surface S at the point $(0, \pi/2, 1)$.
(b) What is the equation of the tangent plane to the surface S at the point $(0, \pi/2, 1)$?

53. Let $f(x, y, z) = \sin(x^2 + y^2 + z^2)$.

(a) Describe in words the shape of the level surfaces of f.
(b) Find grad f.
(c) Consider the two vectors $\vec{r} = x\vec{i} + y\vec{j} + z\vec{k}$ and grad f at a point (x, y, z) where $\sin(x^2+y^2+z^2) \neq 0$. What is (are) the possible values(s) of the angle between these vectors?

54. Each diagram (I) – (IV) in Figure 14.48 represents the level curves of a function $f(x, y)$. For each function f, consider the point above P on the surface $z = f(x, y)$ and choose from the lists of vectors and equations that follow:

(a) A vector which could be the normal to the surface at that point;
(b) An equation which could be the equation of the tangent plane to the surface at that point.

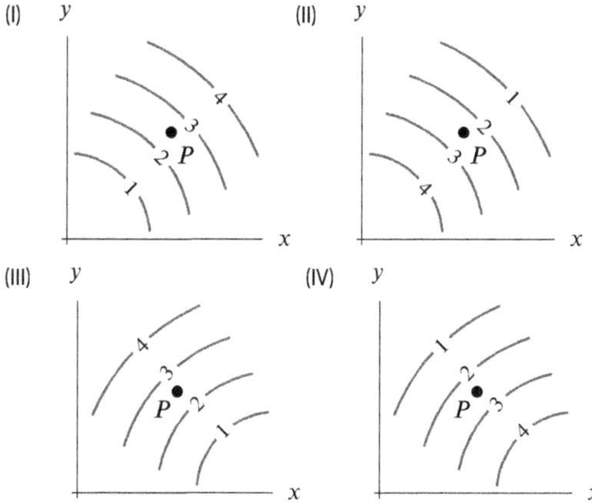

Figure 14.48

Vectors	Equations
(E) $2\vec{i} + 2\vec{j} - 2\vec{k}$	(J) $x + y + z = 4$
(F) $2\vec{i} + 2\vec{j} + 2\vec{k}$	(K) $2x - 2y - 2z = 2$
(G) $2\vec{i} - 2\vec{j} + 2\vec{k}$	(L) $-3x - 3y + 3z = 6$
(H) $-2\vec{i} + 2\vec{j} + 2\vec{k}$	(M) $-\frac{x}{2} + \frac{y}{2} - \frac{z}{2} = -7$

55. (a) What is the shape of the curve in which the following surface cuts the yz-plane:

$$5(x - 1)^2 + 2(y + 1)^2 + 2(z - 3)^2 = 25?$$

(b) Does the curve in part (a) go through the origin?
(c) Find an expression for a vector perpendicular to the surface at the origin.

56. Find the points on the surface $y = 4 + x^2 + z^2$ where the gradient is parallel to $\vec{i} + \vec{j} + \vec{k}$.

57. A particle moves at a speed of 3 units per second perpendicular to the surface $x = 4 + y^2 + z^2$ from the point $(9, 1, 2)$ toward the yz-plane.

(a) What is the particle's velocity vector?
(b) Where is the particle after one second?

58. For the surface $z + 7 = 2x^2 + 3y^2$, where does the tangent plane at the point $(-1, 1, -2)$ meet the three axes?

59. Find a vector perpendicular to the surface $z = 4 - x^2 - y^2$ at the point above the point $(1, 1, 0)$. (The z-axis is vertical.)

60. (a) Where does the surface $x^2 + y^2 - (z - 1)^2 = 0$ cut the xy-plane? What is the shape of the curve?
(b) At the points where the surface cuts the xy-plane, do vectors perpendicular to the surface lie in the xy-plane?

61. A unit vector is perpendicular to the surface $z = x^2 - y^2$. At which point on the surface does this unit vector have the largest dot product with the vector $\vec{i} + 2\vec{j} + 3\vec{k}$?

62. The surface S is represented by the equation $F = 0$ where $F(x, y, z) = x^2 - (y/z^2)$.

(a) Find the unit vectors $\vec{u}_1$ and $\vec{u}_2$ pointing in the direction of maximum increase of F at the points $(0, 0, 1)$ and $(1, 1, 1)$ respectively.
(b) Find the tangent plane to S at the points $(0, 0, 1)$ and $(1, 1, 1)$.
(c) Find all points on S where a normal vector is parallel to the xy-plane.

63. Consider the function $f(x, y) = (e^x - x)\cos y$. Suppose S is the surface $z = f(x, y)$.

(a) Find a vector which is perpendicular to the level curve of f through the point $(2, 3)$ in the direction in which f decreases most rapidly.
(b) Suppose $\vec{v} = 5\vec{i} + 4\vec{j} + a\vec{k}$ is a vector in 3-space which is tangent to the surface S at the point P lying on the surface above $(2, 3)$. What is a?

64. (a) Find the tangent plane to the surface $x^2+y^2+3z^2 = 4$ at the point $(0.6, 0.8, 1)$.
(b) Is there a point on the surface $x^2 + y^2 + 3z^2 = 4$ at which the tangent plane is parallel to the plane $8x + 6y + 30z = 1$? If so, find it. If not, explain why not.

65. Your house lies on the surface $z = f(x, y) = 2x^2 - y^2$ directly above the point $(4, 3)$ in the xy-plane.

(a) How high above the xy-plane do you live?
(b) What is the slope of your lawn as you look from your house directly toward the z-axis (that is, along the vector $-4\vec{i} - 3\vec{j}$)?
(c) When you wash your car in the driveway, on this surface above the point $(4, 3)$, which way does the water run off? (Give your answer as a two-dimensional vector.)
(d) What is the equation of the tangent plane to this surface at your house?

66. (a) Sketch the contours of $z = y - \sin x$ for $z = -1, 0, 1, 2$.
(b) A bug starts on the surface at the point $(\pi/2, 1, 0)$ and walks on the surface $z = y - \sin x$ in the direction parallel to the y-axis, in the direction of increasing y. Is the bug walking in a valley or on top of a ridge? Explain.
(c) On the contour $z = 0$ in your sketch for part (a), draw the gradients of z at $x = 0$, $x = \pi/2$, and $x = \pi$.

67. The function $f(x, y, z) = 2x - 3y + z + 10$ gives the temperature, T, in degrees Celsius, at the point (x, y, z).

(a) In words, describe the isothermal surfaces.
(b) Calculate $f_z(0, 0, 0)$ and interpret in terms of temperature.
(c) If you are standing at the point $(0, 0, 0)$, in what direction should you move to increase your temperature the fastest?
(d) Is $z = -2x + 3y + 17$ an isothermal surface? If so, what is the temperature on this isotherm?

68. The concentration of salt in a fluid at (x, y, z) is given by $F(x, y, z) = x^2 + y^4 + x^2z^2$ mg/cm^3. You are at the point $(-1, 1, 1)$.

(a) In which direction should you move if you want the concentration to increase the fastest?
(b) You start to move in the direction you found in part (a) at a speed of 4 cm/sec. How fast is the concentration changing?

69. The temperature of a gas at the point (x, y, z) is given by $G(x, y, z) = x^2 - 5xy + y^2z$.

(a) What is the rate of change in the temperature at the point $(1, 2, 3)$ in the direction $\vec{v} = 2\vec{i} + \vec{j} - 4\vec{k}$?
(b) What is the direction of maximum rate of change of temperature at the point $(1, 2, 3)$?
(c) What is the maximum rate of change at the point $(1, 2, 3)$?

70. The temperature at the point (x, y, z) in 3-space is given, in degrees Celsius, by $T(x, y, z) = e^{-(x^2+y^2+z^2)}$.

(a) Describe in words the shape of surfaces on which the temperature is constant.
(b) Find grad T.
(c) You travel from the point $(1, 0, 0)$ to the point $(2, 1, 0)$ at a speed of 3 units per second. Find the instantaneous rate of change of the temperature as you leave the point $(1, 0, 0)$. Give units.

71. A spaceship is plunging into the atmosphere of a planet. With coordinates in miles and the origin at the center of the planet, the pressure of the atmosphere at (x, y, z) is

$$P = 5e^{-0.1\sqrt{x^2+y^2+z^2}} \text{ atmospheres.}$$

The velocity, in miles/sec, of the spaceship at $(0, 0, 1)$ is $\vec{v} = \vec{i} - 2.5\vec{k}$. At $(0, 0, 1)$, what is the rate of change with respect to time of the pressure on the spaceship?

72. The earth has mass M and is located at the origin in 3-space, while the moon has mass m. Newton's Law of Gravitation states that if the moon is located at the point (x, y, z) then the attractive force exerted by the earth on the moon is given by the vector

$$\vec{F} = -GMm\frac{\vec{r}}{\|\vec{r}\|^3},$$

where $\vec{r} = x\vec{i} + y\vec{j} + z\vec{k}$. Show that $\vec{F} = \text{grad}\,\varphi$, where φ is the function given by

$$\varphi(x, y, z) = \frac{GMm}{\|\vec{r}\|}.$$

73. Let $\vec{r} = x\vec{i} + y\vec{j} + z\vec{k}$ and $\vec{a}$ be a constant vector. For each of the quantities in (a)–(c), choose the statement in (I)–(V) that describes it. No reasons are needed.

(a) grad$(\vec{r} + \vec{a})$ (b) grad$(\vec{r} \cdot \vec{a})$ (c) grad$(\vec{r} \times \vec{a})$

I Scalar, independent of $\vec{a}$.
II Scalar, depends on $\vec{a}$.
III Vector, independent of $\vec{a}$.
IV Vector, depends on $\vec{a}$.
V Not defined.

Strengthen Your Understanding

In Problems 74–75, explain what is wrong with the statement.

74. The gradient vector grad $f(x, y)$ points in the direction perpendicular to the surface $z = f(x, y)$.

75. The tangent plane at the origin to a surface $f(x, y, z) = 1$ that contains the point $(0, 0, 0)$ has equation

$$f_x(0, 0, 0)x + f_y(0, 0, 0)y + f_z(0, 0, 0)z + 1 = 0.$$

In Problems 76–78, give an example of:

76. A surface $z = f(x, y)$ such that the vector $\vec{i} - 2\vec{j} - \vec{k}$ is normal to the tangent plane at the point where $(x, y) = (0, 0)$.

77. A function $f(x, y, z)$ such that grad $f = 2\vec{i} + 3\vec{j} + 4\vec{k}$.

78. Two nonparallel unit vectors $\vec{u}$ and $\vec{v}$ such that $f_{\vec{u}}(0, 0, 0) = f_{\vec{v}}(0, 0, 0) = 0$, where $f(x, y, z) = 2x - 3y$.

Are the statements in Problems 79–82 true or false? Give reasons for your answer.

79. An equation for the tangent plane to the surface $z = x^2 + y^3$ at $(1, 1)$ is $z = 2 + 2x(x - 1) + 3y^2(y - 1)$.

80. There is a function $f(x, y)$ which has a tangent plane with equation $z = 0$ at a point (a, b).

81. There is a function with $\|\text{grad}\, f\| = 4$ and $f_{\vec{k}} = 5$ at some point.

82. There is a function with $\|\text{grad}\, f\| = 5$ and $f_{\vec{k}} = -3$ at some point.

14.6 THE CHAIN RULE

Composition of Functions of Many Variables and Rates of Change

The chain rule enables us to differentiate *composite functions*. If we have a function of two variables $z = f(x, y)$ and we substitute $x = g(t)$, $y = h(t)$ into $z = f(x, y)$, then we have a composite function in which z is a function of t:

$$z = f(g(t), h(t)).$$

If, on the other hand, we substitute $x = g(u, v)$, $y = h(u, v)$, then we have a different composite function in which z is a function of u and v:

$$z = f(g(u, v), h(u, v)).$$

The next example shows how to calculate the rate of change of a composite function.

Example 1 Corn production, C, depends on annual rainfall, R, and average temperature, T, so $C = f(R,T)$. Global warming predicts that both rainfall and temperature depend on time. Suppose that according to a particular model of global warming, rainfall is decreasing at 0.2 cm per year and temperature is increasing at 0.1°C per year. Use the fact that at current levels of production, $f_R = 3.3$ and $f_T = -5$ to estimate the current rate of change, dC/dt.

Solution By local linearity, we know that changes ΔR and ΔT generate a change, ΔC, in C given approximately by

$$\Delta C \approx f_R \Delta R + f_T \Delta T = 3.3\Delta R - 5\Delta T.$$

We want to know how ΔC depends on the time increment, Δt. A change Δt causes changes ΔR and ΔT, which in turn cause a change ΔC. The model of global warming tells us that

$$\frac{dR}{dt} = -0.2 \quad \text{and} \quad \frac{dT}{dt} = 0.1.$$

Thus, a time increment, Δt, generates changes of ΔR and ΔT given by

$$\Delta R \approx -0.2\Delta t \quad \text{and} \quad \Delta T \approx 0.1\Delta t.$$

Substituting for ΔR and ΔT in the expression for ΔC gives us

$$\Delta C \approx 3.3(-0.2\Delta t) - 5(0.1\Delta t) = -1.16\Delta t.$$

Thus,

$$\frac{\Delta C}{\Delta t} \approx -1.16 \quad \text{and, therefore,} \quad \frac{dC}{dt} \approx -1.16.$$

The relationship between ΔC and Δt, which gives the value of dC/dt, is an example of the *chain rule*. The argument in Example 1 leads to more general versions of the chain rule.

The Chain Rule for $z = f(x, y)$, $x = g(t)$, $y = h(t)$

Since $z = f(g(t), h(t))$ is a function of t, we can consider the derivative dz/dt. The chain rule gives dz/dt in terms of the derivatives of f, g, and h. Since dz/dt represents the rate of change of z with t, we look at the change Δz generated by a small change, Δt.

We substitute the local linearizations

$$\Delta x \approx \frac{dx}{dt}\Delta t \quad \text{and} \quad \Delta y \approx \frac{dy}{dt}\Delta t$$

into the local linearization

$$\Delta z \approx \frac{\partial z}{\partial x}\Delta x + \frac{\partial z}{\partial y}\Delta y,$$

yielding

$$\begin{aligned}\Delta z &\approx \frac{\partial z}{\partial x}\frac{dx}{dt}\Delta t + \frac{\partial z}{\partial y}\frac{dy}{dt}\Delta t \\ &= \left(\frac{\partial z}{\partial x}\frac{dx}{dt} + \frac{\partial z}{\partial y}\frac{dy}{dt}\right)\Delta t.\end{aligned}$$

Thus,

$$\frac{\Delta z}{\Delta t} \approx \frac{\partial z}{\partial x}\frac{dx}{dt} + \frac{\partial z}{\partial y}\frac{dy}{dt}.$$

Taking the limit as $\Delta t \to 0$, we get the following result.

> If f, g, and h are differentiable and if $z = f(x, y)$, and $x = g(t)$, and $y = h(t)$, then
>
> $$\frac{dz}{dt} = \frac{\partial z}{\partial x}\frac{dx}{dt} + \frac{\partial z}{\partial y}\frac{dy}{dt}.$$

Visualizing the Chain Rule with a Diagram

The diagram in Figure 14.49 provides a way of remembering the chain rule. It shows the chain of dependence: z depends on x and y, which in turn depend on t. Each line in the diagram is labeled with a derivative relating the variables at its ends.

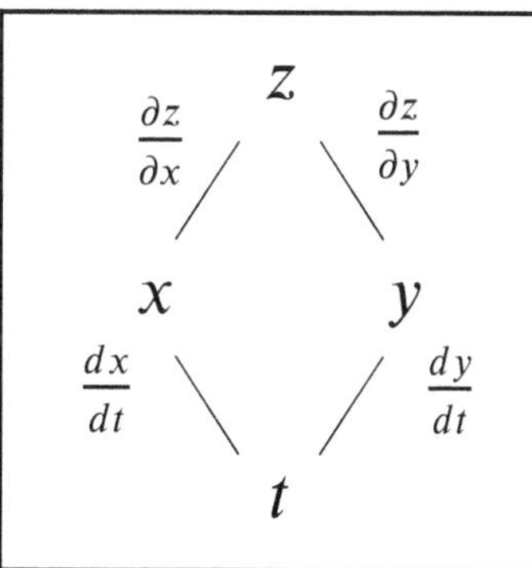

Figure 14.49: Diagram for $z = f(x, y)$, $x = g(t)$, $y = h(t)$. Lines represent dependence of z on x and y, and of x and y on t

The diagram keeps track of how a change in t propagates through the chain of composed functions. There are two paths from t to z, one through x and one through y. For each path, we multiply together the derivatives along the path. Then, to calculate dz/dt, we add the contributions from the two paths.

Example 2 Suppose that $z = f(x, y) = x \sin y$, where $x = t^2$ and $y = 2t + 1$. Let $z = g(t)$. Compute $g'(t)$ directly and using the chain rule.

Solution Since $z = g(t) = f(t^2, 2t+1) = t^2\sin(2t+1)$, it is possible to compute $g'(t)$ directly by one-variable methods:

$$g'(t) = t^2\frac{d}{dt}(\sin(2t+1)) + \left(\frac{d}{dt}(t^2)\right)\sin(2t+1) = 2t^2\cos(2t+1) + 2t\sin(2t+1).$$

The chain rule provides an alternative route to the same answer. We have

$$\frac{dz}{dt} = \frac{\partial z}{\partial x}\frac{dx}{dt} + \frac{\partial z}{\partial y}\frac{dy}{dt} = (\sin y)(2t) + (x\cos y)(2) = 2t\sin(2t+1) + 2t^2\cos(2t+1).$$

Example 3 The capacity, C, of a communication channel, such as a telephone line, to carry information depends on the ratio of the signal strength, S, to the noise, N. For some positive constant k,

$$C = k\ln\left(1 + \frac{S}{N}\right).$$

Suppose that the signal and noise are given as a function of time, t in seconds, by

$$S(t) = 4 + \cos(4\pi t) \qquad N(t) = 2 + \sin(2\pi t).$$

What is dC/dt one second after transmission started? Is the capacity increasing or decreasing at that instant?

Solution By the chain rule

$$\begin{aligned}\frac{dC}{dt} &= \frac{\partial C}{\partial S}\frac{dS}{dt} + \frac{\partial C}{\partial N}\frac{dN}{dt}\\ &= \frac{k}{1+S/N}\cdot\frac{1}{N}(-4\pi\sin 4\pi t) + \frac{k}{1+S/N}\left(-\frac{S}{N^2}\right)(2\pi\cos 2\pi t).\end{aligned}$$

When $t = 1$, the first term is zero, $S(1) = 5$, and $N(1) = 2$, so

$$\frac{dC}{dt} = \frac{k}{1+S(1)/N(1)}\left(-\frac{S(1)}{(N(1))^2}\right)\cdot 2\pi = \frac{k}{1+5/2}\left(-\frac{5}{4}\right)\cdot 2\pi.$$

Since dC/dt is negative, the capacity is decreasing at time $t = 1$ second.

How to Formulate a General Chain Rule

A diagram can be used to write the chain rule for general compositions.

> To find the rate of change of one variable with respect to another in a chain of composed differentiable functions:
>
> - Draw a diagram expressing the relationship between the variables, and label each link in the diagram with the derivative relating the variables at its ends.
> - For each path between the two variables, multiply together the derivatives from each step along the path.
> - Add the contributions from each path.

The diagram keeps track of all the ways in which a change in one variable can cause a change in another; the diagram generates all the terms we would get from the appropriate substitutions into the local linearizations.

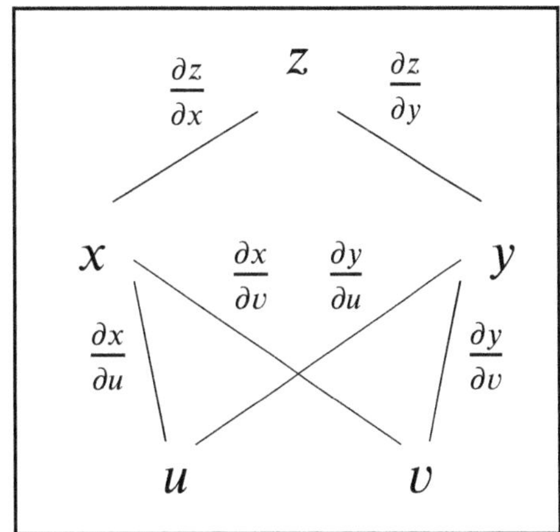

Figure 14.50: Diagram for $z = f(x, y)$, $x = g(u, v)$, $y = h(u, v)$. Lines represent dependence of z on x and y, and of x and y on u and v

For example, we can use Figure 14.50 to find formulas for $\partial z/\partial u$ and $\partial z/\partial v$. Adding the contributions for the two paths from z to u, we get the following results:

If f, g, h are differentiable and if $z = f(x, y)$, with $x = g(u, v)$ and $y = h(u, v)$, then

$$\frac{\partial z}{\partial u} = \frac{\partial z}{\partial x}\frac{\partial x}{\partial u} + \frac{\partial z}{\partial y}\frac{\partial y}{\partial u},$$
$$\frac{\partial z}{\partial v} = \frac{\partial z}{\partial x}\frac{\partial x}{\partial v} + \frac{\partial z}{\partial y}\frac{\partial y}{\partial v}.$$

Example 4 Let $w = x^2e^y$, $x = 4u$, and $y = 3u^2 - 2v$. Compute $\partial w/\partial u$ and $\partial w/\partial v$ using the chain rule.

Solution Using the previous result, we have

$$\frac{\partial w}{\partial u} = \frac{\partial w}{\partial x}\frac{\partial x}{\partial u} + \frac{\partial w}{\partial y}\frac{\partial y}{\partial u} = 2xe^y(4) + x^2e^y(6u) = (8x + 6x^2u)e^y$$
$$= (32u + 96u^3)e^{3u^2-2v}.$$

Similarly,

$$\frac{\partial w}{\partial v} = \frac{\partial w}{\partial x}\frac{\partial x}{\partial v} + \frac{\partial w}{\partial y}\frac{\partial y}{\partial v} = 2xe^y(0) + x^2e^y(-2) = -2x^2e^y$$
$$= -32u^2e^{3u^2-2v}.$$

Example 5 A quantity z can be expressed either as a function of x and y, so that $z = f(x, y)$, or as a function of u and v, so that $z = g(u, v)$. The two coordinate systems are related by

$$x = u + v, \quad y = u - v.$$

(a) Use the chain rule to express $\partial z/\partial u$ and $\partial z/\partial v$ in terms of $\partial z/\partial x$ and $\partial z/\partial y$.
(b) Solve the equations in part (a) for $\partial z/\partial x$ and $\partial z/\partial y$.
(c) Show that the expressions we get in part (b) are the same as we get by expressing u and v in terms of x and y and using the chain rule.

Solution (a) We have $\partial x/\partial u = 1$ and $\partial x/\partial v = 1$, and also $\partial y/\partial u = 1$ and $\partial y/\partial v = -1$. Thus,

$$\frac{\partial z}{\partial u} = \frac{\partial z}{\partial x}(1) + \frac{\partial z}{\partial y}(1) = \frac{\partial z}{\partial x} + \frac{\partial z}{\partial y}$$

and

$$\frac{\partial z}{\partial v} = \frac{\partial z}{\partial x}(1) + \frac{\partial z}{\partial y}(-1) = \frac{\partial z}{\partial x} - \frac{\partial z}{\partial y}.$$

(b) Adding together the equations for $\partial z/\partial u$ and $\partial z/\partial v$, we get

$$\frac{\partial z}{\partial u} + \frac{\partial z}{\partial v} = 2\frac{\partial z}{\partial x}, \quad \text{so} \quad \frac{\partial z}{\partial x} = \frac{1}{2}\frac{\partial z}{\partial u} + \frac{1}{2}\frac{\partial z}{\partial v}.$$

Similarly, subtracting the equations for $\partial z/\partial u$ and $\partial z/\partial v$ yields

$$\frac{\partial z}{\partial y} = \frac{1}{2}\frac{\partial z}{\partial u} - \frac{1}{2}\frac{\partial z}{\partial v}.$$

(c) Alternatively, we can solve the equations

$$x = u + v, \quad y = u - v$$

for u and v, which yields

$$u = \frac{1}{2}x + \frac{1}{2}y, \quad v = \frac{1}{2}x - \frac{1}{2}y.$$

Now we can think of z as a function of u and v, and u and v as functions of x and y, and apply the chain rule again. This gives us

$$\frac{\partial z}{\partial x} = \frac{\partial z}{\partial u}\frac{\partial u}{\partial x} + \frac{\partial z}{\partial v}\frac{\partial v}{\partial x} = \frac{1}{2}\frac{\partial z}{\partial u} + \frac{1}{2}\frac{\partial z}{\partial v}$$

and

$$\frac{\partial z}{\partial y} = \frac{\partial z}{\partial u}\frac{\partial u}{\partial y} + \frac{\partial z}{\partial v}\frac{\partial v}{\partial y} = \frac{1}{2}\frac{\partial z}{\partial u} - \frac{1}{2}\frac{\partial z}{\partial v}.$$

These are the same expressions we got in part (b).

An Application to Physical Chemistry

A chemist investigating the properties of a gas such as carbon dioxide may want to know how the internal energy U of a given quantity of the gas depends on its temperature, T, pressure, P, and volume, V. The three quantities T, P, and V are not independent, however. For instance, according to the ideal gas law, they satisfy the equation

$$PV = kT$$

where k is a constant which depends only upon the quantity of the gas. The internal energy can then be thought of as a function of any two of the three quantities T, P, and V:

$$U = U_1(T, P) = U_2(T, V) = U_3(P, V).$$

The chemist writes, for example, $\left(\frac{\partial U}{\partial T}\right)_P$ to indicate the partial derivative of U with respect to *T holding P constant*, signifying that for this computation U is viewed as a function of T and P. Thus, we interpret $\left(\frac{\partial U}{\partial T}\right)_P$ as

$$\left(\frac{\partial U}{\partial T}\right)_P = \frac{\partial U_1(T, P)}{\partial T}.$$

If U is to be viewed as a function of T and V, the chemist writes $\left(\frac{\partial U}{\partial T}\right)_V$ for the partial derivative of U with respect to T holding V constant: thus, $\left(\frac{\partial U}{\partial T}\right)_V = \frac{\partial U_2(T,V)}{\partial T}$.

Each of the functions U_1, U_2, U_3 gives rise to one of the following formulas for the differential dU:

$$dU = \left(\frac{\partial U}{\partial T}\right)_P dT + \left(\frac{\partial U}{\partial P}\right)_T dP \qquad \text{corresponds to } U_1,$$
$$dU = \left(\frac{\partial U}{\partial T}\right)_V dT + \left(\frac{\partial U}{\partial V}\right)_T dV \qquad \text{corresponds to } U_2,$$
$$dU = \left(\frac{\partial U}{\partial P}\right)_V dP + \left(\frac{\partial U}{\partial V}\right)_P dV \qquad \text{corresponds to } U_3.$$

All the six partial derivatives appearing in formulas for dU have physical meaning, but they are not all equally easy to measure experimentally. A relationship among the partial derivatives, usually derived from the chain rule, may make it possible to evaluate one of the partials in terms of others that are more easily measured.

Example 6 Suppose a gas satisfies the equation $PV = 2T$ and $P = 3$ when $V = 4$. If $\left(\frac{\partial U}{\partial P}\right)_V = 7$ and $\left(\frac{\partial U}{\partial V}\right)_P = 8$, find the values of $\left(\frac{\partial U}{\partial P}\right)_T$ and $\left(\frac{\partial U}{\partial T}\right)_P$.

Solution Since we know the values of $\left(\frac{\partial U}{\partial P}\right)_V$ and $\left(\frac{\partial U}{\partial V}\right)_P$, we think of U as a function of P and V and use the function U_3 to write

$$dU = \left(\frac{\partial U}{\partial P}\right)_V dP + \left(\frac{\partial U}{\partial V}\right)_P dV$$
$$dU = 7dP + 8dV.$$

To calculate $\left(\frac{\partial U}{\partial P}\right)_T$ and $\left(\frac{\partial U}{\partial T}\right)_P$, we think of U as a function of T and P. Thus, we want to substitute for dV in terms of dT and dP. Since $PV = 2T$, we have

$$PdV + VdP = 2dT,$$
$$3dV + 4dP = 2dT.$$

Solving gives $dV = (2dT - 4dP)/3$, so

$$dU = 7dP + 8\left(\frac{2dT - 4dP}{3}\right)$$
$$dU = -\frac{11}{3}dP + \frac{16}{3}dT.$$

Comparing with the formula for dU obtained from U_1,

$$dU = \left(\frac{\partial U}{\partial T}\right)_P dT + \left(\frac{\partial U}{\partial P}\right)_T dP,$$

we have

$$\left(\frac{\partial U}{\partial T}\right)_P = \frac{16}{3} \quad \text{and} \quad \left(\frac{\partial U}{\partial P}\right)_T = -\frac{11}{3}.$$

In Example 6, we could have substituted for dP instead of dV, leading to values of $\left(\frac{\partial U}{\partial T}\right)_V$ and $\left(\frac{\partial U}{\partial V}\right)_T$. See Problem 41.

In general, if for some particular P, V, and T, we can measure two of the six quantities $\left(\frac{\partial U}{\partial P}\right)_V$, $\left(\frac{\partial U}{\partial V}\right)_P$, $\left(\frac{\partial U}{\partial P}\right)_T$, $\left(\frac{\partial U}{\partial T}\right)_P$, $\left(\frac{\partial U}{\partial V}\right)_T$, $\left(\frac{\partial U}{\partial T}\right)_V$, then we can compute the other four using the relationship between dP, dV, and dT given by the gas law. General formulas for each partial derivative in terms of others can be obtained in the same way. See the following example and Problem 41.

Example 7 Express $\left(\frac{\partial U}{\partial T}\right)_P$ in terms of $\left(\frac{\partial U}{\partial T}\right)_V$ and $\left(\frac{\partial U}{\partial V}\right)_T$ and $\left(\frac{\partial V}{\partial T}\right)_P$.

Solution Since we are interested in the derivatives $\left(\frac{\partial U}{\partial T}\right)_V$ and $\left(\frac{\partial U}{\partial V}\right)_T$, we think of U as a function of T and V and use the formula

$$dU = \left(\frac{\partial U}{\partial T}\right)_V dT + \left(\frac{\partial U}{\partial V}\right)_T dV \qquad \text{corresponding to } U_2.$$

We want to find a formula for $\left(\frac{\partial U}{\partial T}\right)_P$, which means thinking of U as a function of T and P. Thus, we want to substitute for dV. Since V is a function of T and P, we have

$$dV = \left(\frac{\partial V}{\partial T}\right)_P dT + \left(\frac{\partial V}{\partial P}\right)_T dP.$$

Substituting for dV into the formula for dU corresponding to U_2 gives

$$dU = \left(\frac{\partial U}{\partial T}\right)_V dT + \left(\frac{\partial U}{\partial V}\right)_T \left(\left(\frac{\partial V}{\partial T}\right)_P dT + \left(\frac{\partial V}{\partial P}\right)_T dP\right).$$

Collecting the terms containing dT and the terms containing dP gives

$$dU = \left(\left(\frac{\partial U}{\partial T}\right)_V + \left(\frac{\partial U}{\partial V}\right)_T \left(\frac{\partial V}{\partial T}\right)_P\right) dT + \left(\frac{\partial U}{\partial V}\right)_T \left(\frac{\partial V}{\partial P}\right)_T dP.$$

But we also have the formula

$$dU = \left(\frac{\partial U}{\partial T}\right)_P dT + \left(\frac{\partial U}{\partial P}\right)_T dP \qquad \text{corresponding to } U_1.$$

We now have two formulas for dU in terms of dT and dP. The coefficients of dT must be identical, so we conclude

$$\left(\frac{\partial U}{\partial T}\right)_P = \left(\frac{\partial U}{\partial T}\right)_V + \left(\frac{\partial U}{\partial V}\right)_T \left(\frac{\partial V}{\partial T}\right)_P.$$

Example 7 expresses $\left(\frac{\partial U}{\partial T}\right)_P$ in terms of three other partial derivatives. Two of them, namely $\left(\frac{\partial U}{\partial T}\right)_V$, the constant-volume heat capacity, and $\frac{1}{V}\left(\frac{\partial V}{\partial T}\right)_P$, the expansion coefficient, can be easily measured experimentally. The third, the internal pressure, $\left(\frac{\partial U}{\partial V}\right)_T$, cannot be measured directly but can be related to $\left(\frac{\partial P}{\partial T}\right)_V$, which is measurable. Thus, $\left(\frac{\partial U}{\partial T}\right)_P$ can be determined indirectly using this identity.

Exercises and Problems for Section 14.6 Online Resource: Additional Problems for Section 14.6

EXERCISES

For Exercises 1–6, find dz/dt using the chain rule. Assume the variables are restricted to domains on which the functions are defined.

1. $z = xy^2,\ x = e^{-t},\ y = \sin t$
2. $z = x\sin y + y\sin x,\ x = t^2,\ y = \ln t$
3. $z = \sin(x/y),\ x = 2t,\ y = 1 - t^2$
4. $z = \ln(x^2 + y^2),\ x = 1/t,\ y = \sqrt{t}$
5. $z = xe^y,\ x = 2t,\ y = 1 - t^2$
6. $z = (x + y)e^y,\ x = 2t,\ y = 1 - t^2$

For Exercises 7–15, find $\partial z/\partial u$ and $\partial z/\partial v$. The variables are restricted to domains on which the functions are defined.

7. $z = \sin(x/y),\ x = \ln u,\ y = v$
8. $z = \ln(xy),\ x = (u^2 + v^2)^2,\ y = (u^3 + v^3)^2$
9. $z = xe^y,\ x = \ln u,\ y = v$
10. $z = (x + y)e^y,\ x = \ln u,\ y = v$
11. $z = xe^y,\ x = u^2 + v^2,\ y = u^2 - v^2$
12. $z = (x + y)e^y,\ x = u^2 + v^2,\ y = u^2 - v^2$
13. $z = xe^{-y} + ye^{-x},\ x = u\sin v,\ y = v\cos u$
14. $z = \cos(x^2 + y^2),\ x = u\cos v,\ y = u\sin v$
15. $z = \tan^{-1}(x/y),\ x = u^2 + v^2,\ y = u^2 - v^2$

PROBLEMS

16. Use the chain rule to find dz/dt, and check the result by expressing z as a function of t and differentiating directly.

$$z = x^3 y^2, \quad x = t^3, \quad y = t^2$$

17. Use the chain rule to find $\partial w/\partial \rho$ and $\partial w/\partial \theta$, given that

$$w = x^2 + y^2 - z^2,$$

and

$$x = \rho \sin\phi \cos\theta, \quad y = \rho \sin\phi \sin\theta, \quad z = \rho \cos\phi.$$

18. Let $z = f(x, y)$ where $x = g(t)$, $y = h(t)$ and f, g, h are all differentiable functions. Given the information in the table, find $\left.\dfrac{\partial z}{\partial t}\right|_{t=1}$.

$f(3,10) = 7$	$f(4,11) = -20$
$f_x(3,10) = 100$	$f_y(3,10) = 0.1$
$f_x(4,11) = 200$	$f_y(4,11) = 0.2$
$f(3,4) = -10$	$f(10,11) = -1$
$g(1) = 3$	$h(1) = 10$
$g'(1) = 4$	$h'(1) = 11$

19. A bison is charging across the plain one morning. His path takes him to location (x, y) at time t where x and y are functions of t and north is in the direction of increasing y. The temperature is always colder farther north. As time passes, the sun rises in the sky, sending out more heat, and a cold front blows in from the east. At time t the air temperature H near the bison is given by $H = f(x, y, t)$. The chain rule expresses the derivative dH/dt as a sum of three terms:

$$\frac{dH}{dt} = \frac{\partial f}{\partial x}\frac{dx}{dt} + \frac{\partial f}{\partial y}\frac{dy}{dt} + \frac{\partial f}{\partial t}.$$

Identify the term that gives the contribution to the change in temperature experienced by the bison that is due to

(a) The rising sun.
(b) The coming cold front.
(c) The bison's change in latitude.

20. The voltage, V (in volts), across a circuit is given by Ohm's law: $V = IR$, where I is the current (in amps) flowing through the circuit and R is the resistance (in ohms). If we place two circuits, with resistance R_1 and R_2, in parallel, then their combined resistance, R, is given by

$$\frac{1}{R} = \frac{1}{R_1} + \frac{1}{R_2}.$$

Suppose the current is 2 amps and increasing at 10^{-2} amp/sec and R_1 is 3 ohms and increasing at 0.5 ohm/sec, while R_2 is 5 ohms and decreasing at 0.1 ohm/sec. Calculate the rate at which the voltage is changing.

21. The air pressure is decreasing at a rate of 2 pascals per kilometer in the eastward direction. In addition, the air pressure is dropping at a constant rate with respect to time everywhere. A ship sailing eastward at 10 km/hour past an island takes barometer readings and records a pressure drop of 50 pascals in 2 hours. Estimate the time rate of change of air pressure on the island. (A pascal is a unit of air pressure.)

22. A steel bar with square cross sections 5 cm by 5 cm and length 3 meters is being heated. For each dimension, the bar expands $13 \cdot 10^{-6}$meters for each 1°C rise in temperature.[2] What is the rate of change in the volume of the steel bar?

23. Corn production, C, is a function of rainfall, R, and temperature, T. (See Example 1 on page 780.) Figures 14.51 and 14.52 show how rainfall and temperature are predicted to vary with time because of global warming. Suppose we know that $\Delta C \approx 3.3\Delta R - 5\Delta T$. Use this to estimate the change in corn production between the year 2020 and the year 2021. Hence, estimate dC/dt when $t = 2020$.

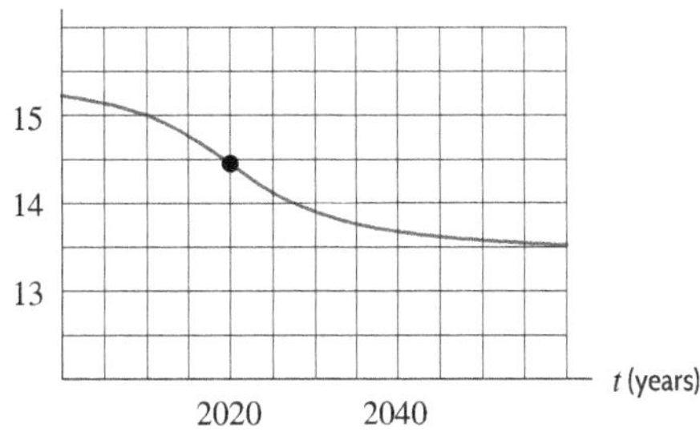

Figure 14.51: Rainfall as a function of time

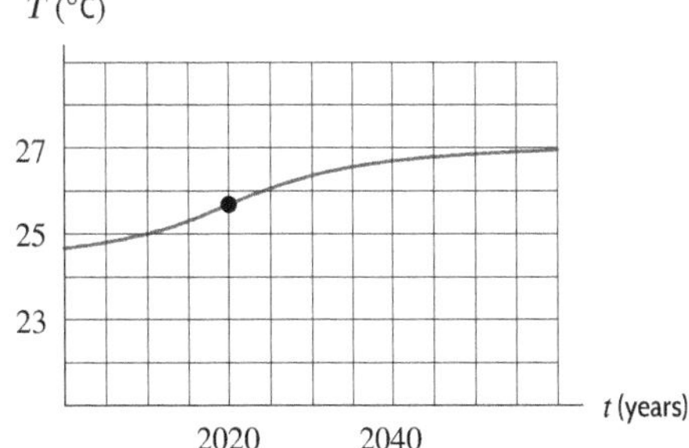

Figure 14.52: Temperature as a function of time

24. At a point x miles east and y miles north of a campground, the height above sea level is $f(x, y)$ feet. Let

$$\nabla f(10, 2) = -2\vec{i} + \vec{j} \text{ and } \vec{v} = \vec{i} + 3\vec{j}.$$

Find the following quantities, including units, as you leave a point 10 miles east and 2 miles north of the campground.

(a) The slope of the land in the direction of $\vec{v}$.

[2] www.engineeringtoolbox.com, accessed January 10th, 2016.

(b) Your vertical speed if you move in the direction of $\vec{v}$ at a speed of 2 miles per hour.
(c) Your vertical speed if you move east at a speed of 2 miles per hour.

25. The function $g(x, y)$ gives the temperature, in degrees Fahrenheit, x miles east and y miles north of a campground. Let $\vec{u}$ be the unit vector in the direction of $\vec{v} = \vec{i} + 3\vec{j}$ and

$$\nabla g(-1, -3) = \vec{i}.$$

A camper located one mile to the west and three miles to the south of the camp starts walking back to camp in the direction $\vec{u}$ at a speed of 2.5 miles/hr. Find the value of the following expressions, and interpret each in everyday terms for the camper.

(a) $g_{\vec{u}}(-1, -3)$ **(b)** $2.5g_{\vec{u}}(-1, -3)$
(c) $2.5g_{\vec{i}}(-1, -3)$

26. Mina's score on her weekly multivariable calculus quiz, S, in points, is a function of the number of hours, H, she spends studying the course materials and the number of problems, P, she solves per week.

- Her score S goes up 2 points for each additional hour spent per week studying the course materials.
- Her score S goes up by 3 points for each additional 5 problems solved during the week.
- The number of weekly hours, H, she spends studying the course materials has been decreasing at a rate of 1.5 hours per week.

Mina's weekly quiz score does not change from week to week.

(a) Find the value of dP/dt, where t is time in weeks, include units.
(b) What can Mina learn from the value of the derivative in part (a)?

27. Let $z = g(u, v, w)$ and $u = u(s, t), v = v(s, t), w = w(s, t)$. How many terms are there in the expression for $\partial z/\partial t$?

28. Suppose $w = f(x, y, z)$ and that x, y, z are functions of u and v. Use a tree diagram to write down the chain rule formula for $\partial w/\partial u$ and $\partial w/\partial v$.

29. Suppose $w = f(x, y, z)$ and that x, y, z are all functions of t. Use a tree diagram to write down the chain rule for dw/dt.

30. Let $z = f(t)g(t)$. Use the chain rule applied to $h(x, y) = f(x)g(y)$ to show that $dz/dt = f'(t)g(t) + f(t)g'(t)$. The one-variable product rule for differentiation is a special case of the two-variable chain rule.

31. Let $F(u, v)$ be a function of two variables. Find $f'(x)$ if

(a) $f(x) = F(x, 3)$ **(b)** $f(x) = F(3, x)$
(c) $f(x) = F(x, x)$ **(d)** $f(x) = F(5x, x^2)$

32. The function $g(\rho)$ is graphed in Figure 14.53. Let $\rho = \sqrt{x^2 + y^2 + z^2}$. Define f, a function of x, y, z by $f(x, y, z) = g\left(\sqrt{x^2 + y^2 + z^2}\right)$. Let $\vec{F} = \text{grad } f$.

(a) Describe precisely in words the level surfaces of f.
(b) Give a unit vector in the direction of $\vec{F}$ at the point $(1, 2, 2)$.
(c) Estimate $\|\vec{F}\|$ at the point $(1, 2, 2)$.
(d) Estimate $\vec{F}$ at the point $(1, 2, 2)$.
(e) The points $(1, 2, 2)$ and $(3, 0, 0)$ are both on the sphere $x^2 + y^2 + z^2 = 9$. Estimate $\vec{F}$ at $(3, 0, 0)$.
(f) If P and Q are any two points on the sphere $x^2 + y^2 + z^2 = k^2$:
(i) Compare the magnitudes of $\vec{F}$ at P and at Q.
(ii) Describe the directions of $\vec{F}$ at P and at Q.

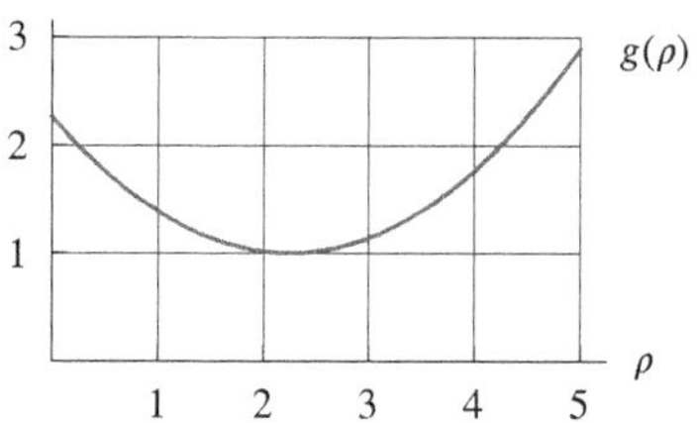

Figure 14.53

In Problems 33–34, let $z = f(x, y)$, $x = x(u, v)$, $y = y(u, v)$ and $x(1, 2) = 5$, $y(1, 2) = 3$, calculate the partial derivative in terms of some of the numbers a, b, c, d, e, k, p, q:

$$\begin{array}{llll} f_x(1,2) = a & f_y(1,2) = c & x_u(1,2) = e & y_u(1,2) = p \\ f_x(5,3) = b & f_y(5,3) = d & x_v(1,2) = k & y_v(1,2) = q \end{array}$$

33. $z_u(1, 2)$ **34.** $z_v(1, 2)$

In Problems 35–36, let $z = f(x, y)$, $x = x(u, v)$, $y = y(u, v)$ and $x(4, 5) = 2$, $y(4, 5) = 3$. Calculate the partial derivative in terms of $a, b, c, d, e, k, p, q, r, s, t, w$:

$$\begin{array}{llll} f_x(4,5) = a & f_y(4,5) = c & x_u(4,5) = e & y_u(4,5) = p \\ f_x(2,3) = b & f_y(2,3) = d & x_v(4,5) = k & y_v(4,5) = q \\ x_u(2,3) = r & y_u(2,3) = s & x_v(2,3) = t & y_v(2,3) = w \end{array}$$

35. $z_u(4, 5)$ **36.** $z_v(4, 5)$

For Problems 37–38, suppose that $x > 0, y > 0$ and that z can be expressed either as a function of Cartesian coordinates (x, y) or as a function of polar coordinates (r, θ), so that $z = f(x, y) = g(r, \theta)$. [Recall that $x = r\cos\theta, y = r\sin\theta, r = \sqrt{x^2 + y^2}$, and, for $x > 0, y > 0, \theta = \arctan(y/x)$.]

37. **(a)** Use the chain rule to find $\partial z/\partial r$ and $\partial z/\partial \theta$ in terms of $\partial z/\partial x$ and $\partial z/\partial y$.
(b) Solve the equations you have just written down for $\partial z/\partial x$ and $\partial z/\partial y$ in terms of $\partial z/\partial r$ and $\partial z/\partial \theta$.
(c) Show that the expressions you get in part (b) are the same as you would get by using the chain rule to find $\partial z/\partial x$ and $\partial z/\partial y$ in terms of $\partial z/\partial r$ and $\partial z/\partial \theta$.

38. Show that

$$\left(\frac{\partial z}{\partial x}\right)^2 + \left(\frac{\partial z}{\partial y}\right)^2 = \left(\frac{\partial z}{\partial r}\right)^2 + \frac{1}{r^2}\left(\frac{\partial z}{\partial \theta}\right)^2.$$

Problems 39–44 are continuations of the physical chemistry example on page 786.

39. Write $\left(\frac{\partial U}{\partial P}\right)_V$ as a partial derivative of one of the functions U_1, U_2, or U_3.

40. Write $\left(\frac{\partial U}{\partial P}\right)_T$ as a partial derivative of one of the functions U_1, U_2, U_3.

41. For the gas in Example 6, find $\left(\frac{\partial U}{\partial T}\right)_V$ and $\left(\frac{\partial U}{\partial V}\right)_T$. [Hint: Use the same method as the example, but substitute for dP instead of dV.]

42. Show that $\left(\frac{\partial T}{\partial V}\right)_P = 1\Big/\left(\frac{\partial V}{\partial T}\right)_P$.

43. Use Example 7 and Problem 42 to show that

$$\left(\frac{\partial U}{\partial V}\right)_P = \left(\frac{\partial U}{\partial V}\right)_T + \frac{\left(\frac{\partial U}{\partial T}\right)_V}{\left(\frac{\partial V}{\partial T}\right)_P}.$$

44. In Example 6, we calculated values of $(\partial U/\partial T)_P$ and $(\partial U/\partial P)_T$ using the relationship $PV = 2T$ for a specific gas. In this problem, you will derive general relationships for these two partial derivatives.

(a) Think of V as a function of P and T and write an expression for dV.

(b) Substitute for dV into the following formula for dU (thinking of U as a function of P and V):

$$dU = \left(\frac{\partial U}{\partial P}\right)_V dP + \left(\frac{\partial U}{\partial V}\right)_P dV.$$

(c) Thinking of U as a function of P and T, write an expression for dU.

(d) By comparing coefficients of dP and dT in your answers to parts (b) and (c), show that

$$\left(\frac{\partial U}{\partial T}\right)_P = \left(\frac{\partial U}{\partial V}\right)_P \cdot \left(\frac{\partial V}{\partial T}\right)_P$$
$$\left(\frac{\partial U}{\partial P}\right)_T = \left(\frac{\partial U}{\partial P}\right)_V + \left(\frac{\partial U}{\partial V}\right)_P \cdot \left(\frac{\partial V}{\partial P}\right)_T.$$

Strengthen Your Understanding

In Problems 45–47, explain what is wrong with the statement.

45. If $z = f(g(t), h(t))$, then $dz/dt = f(g'(t), h(t)) + f(g(t), h'(t))$.

46. If $C = C(R, T), R = R(x, y), T = T(x, y)$ and $R(0, 2) = 5$, $T(0, 2) = 1$, then $C_x(0, 2) = C_R(0, 2)R_x(0, 2) + C_T(0, 2)T_x(0, 2)$.

47. If $z = f(x, y)$ and $x = g(t)$, $y = h(t)$ with $g(0) = 2$ and $h(0) = 3$, then

$$\left.\frac{dz}{dt}\right|_{t=0} = f_x(0, 0)g'(0) + f_y(0, 0)h'(0).$$

In Problems 48–52, give an example of:

48. Functions $x = g(t)$ and $y = h(t)$ such that $(dz/dt)|_{t=0} = 9$, given that $z = x^2y$.

49. A function $z = f(x, y)$ such that $dz/dt|_{t=0} = 10$, given that $x = e^{2t}$ and $y = \sin t$.

50. Functions z, x and y where you need to follow the diagram in order to answer questions about the derivative of z with respect to the other variables.

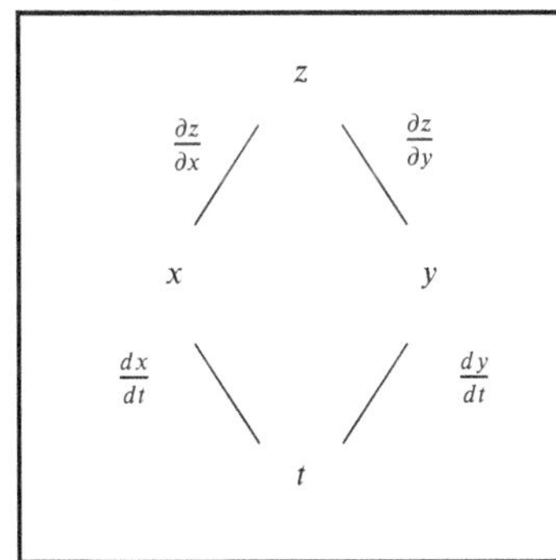

51. Functions w, u and v where you need to follow the diagram in order to answer questions about the derivative of w with respect to the other variables.

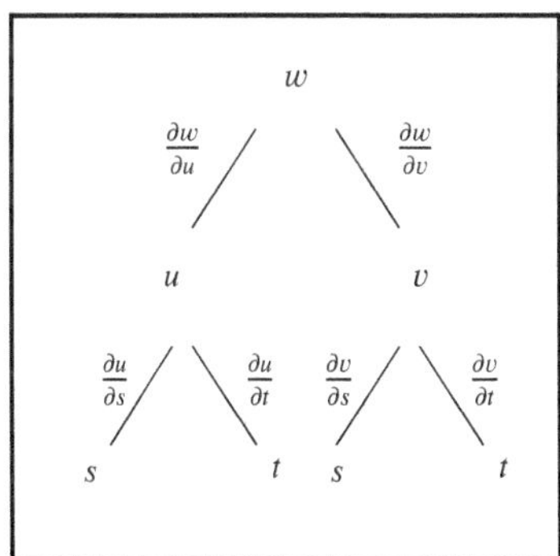

52. Function $z = f(x, y)$ where x and y are functions of one variable, t, for which $\frac{\partial z}{\partial t} = 2$.

53. Let $z = g(u, v)$ and $u = u(x, y, t)$, $v = v(x, y, t)$ and $x = x(t)$, $y = y(t)$. Then the expression for dz/dt has

(a) Three terms (b) Four terms
(c) Six terms (d) Seven terms
(e) Nine terms (f) None of the above

14.7 SECOND-ORDER PARTIAL DERIVATIVES

Since the partial derivatives of a function are themselves functions, we can differentiate them, giving *second-order partial derivatives*. A function $z = f(x, y)$ has two first-order partial derivatives, f_x and f_y, and four second-order partial derivatives.

The Second-Order Partial Derivatives of $z = f(x, y)$

$$\frac{\partial^2 z}{\partial x^2} = f_{xx} = (f_x)_x, \qquad \frac{\partial^2 z}{\partial x \partial y} = f_{yx} = (f_y)_x,$$
$$\frac{\partial^2 z}{\partial y \partial x} = f_{xy} = (f_x)_y, \qquad \frac{\partial^2 z}{\partial y^2} = f_{yy} = (f_y)_y.$$

It is usual to omit the parentheses, writing f_{xy} instead of $(f_x)_y$ and $\dfrac{\partial^2 z}{\partial y\, \partial x}$ instead of $\dfrac{\partial}{\partial y}\left(\dfrac{\partial z}{\partial x}\right)$.

Example 1 Compute the four second-order partial derivatives of $f(x, y) = xy^2 + 3x^2e^y$.

Solution From $f_x(x, y) = y^2 + 6xe^y$ we get

$$f_{xx}(x, y) = \frac{\partial}{\partial x}(y^2 + 6xe^y) = 6e^y \quad \text{and} \quad f_{xy}(x, y) = \frac{\partial}{\partial y}(y^2 + 6xe^y) = 2y + 6xe^y.$$

From $f_y(x, y) = 2xy + 3x^2e^y$ we get

$$f_{yx}(x, y) = \frac{\partial}{\partial x}(2xy + 3x^2e^y) = 2y + 6xe^y \quad \text{and} \quad f_{yy}(x, y) = \frac{\partial}{\partial y}(2xy + 3x^2e^y) = 2x + 3x^2e^y.$$

Observe that $f_{xy} = f_{yx}$ in this example.

Example 2 Use the values of the function $f(x, y)$ in Table 14.7 to estimate $f_{xy}(1, 2)$ and $f_{yx}(1, 2)$.

Table 14.7 *Values of $f(x, y)$*

$y \backslash x$	0.9	1.0	1.1
1.8	4.72	5.83	7.06
2.0	6.48	8.00	9.60
2.2	8.62	10.65	12.88

Solution Since $f_{xy} = (f_x)_y$, we first estimate f_x

$$f_x(1, 2) \approx \frac{f(1.1, 2) - f(1, 2)}{0.1} = \frac{9.60 - 8.00}{0.1} = 16.0,$$
$$f_x(1, 2.2) \approx \frac{f(1.1, 2.2) - f(1, 2.2)}{0.1} = \frac{12.88 - 10.65}{0.1} = 22.3.$$

Thus,

$$f_{xy}(1,2) \approx \frac{f_x(1,2.2) - f_x(1,2)}{0.2} = \frac{22.3 - 16.0}{0.2} = 31.5.$$

Similarly,

$$\begin{aligned} f_{yx}(1,2) \approx \frac{f_y(1.1,2) - f_y(1,2)}{0.1} &\approx \frac{1}{0.1}\left(\frac{f(1.1,2.2) - f(1.1,2)}{0.2} - \frac{f(1,2.2) - f(1,2)}{0.2}\right) \\ &= \frac{1}{0.1}\left(\frac{12.88 - 9.60}{0.2} - \frac{10.65 - 8.00}{0.2}\right) = 31.5. \end{aligned}$$

Observe that in this example also, $f_{xy} = f_{yx}$.

The Mixed Partial Derivatives Are Equal

It is not an accident that the estimates for $f_{xy}(1,2)$ and $f_{yx}(1,2)$ are equal in Example 2, because the same values of the function are used to calculate each one. The fact that $f_{xy} = f_{yx}$ in Examples 1 and 2 corroborates the following general result; Problem 71 (available online) suggests why you might expect it to be true.[3]

Theorem 14.1: Equality of Mixed Partial Derivatives

If f_{xy} and f_{yx} are continuous at (a,b), an interior point of their domain, then

$$f_{xy}(a,b) = f_{yx}(a,b).$$

For most functions f we encounter and most points (a,b) in their domains, not only are f_{xy} and f_{yx} continuous at (a,b), but all their higher-order partial derivatives (such as f_{xxy} or f_{xyyy}) exist and are continuous at (a,b). In that case we say f is *smooth* at (a,b). We say f is smooth on a region R if it is smooth at every point of R.

What Do the Second-Order Partial Derivatives Tell Us?

Example 3 Let us return to the guitar string of Example 4, page 750. The string is 1 meter long and at time t seconds, the point x meters from one end is displaced $f(x,t)$ meters from its rest position, where

$$f(x,t) = 0.003 \sin(\pi x)\sin(2765t).$$

Compute the four second-order partial derivatives of f at the point $(x,t) = (0.3, 1)$ and describe the meaning of their signs in practical terms.

Solution First we compute $f_x(x,t) = 0.003\pi \cos(\pi x)\sin(2765t)$, from which we get

$$f_{xx}(x,t) = \frac{\partial}{\partial x}(f_x(x,t)) = -0.003\pi^2 \sin(\pi x)\sin(2765t), \quad \text{so} \quad f_{xx}(0.3,1) \approx -0.01;$$

and

$$f_{xt}(x,t) = \frac{\partial}{\partial t}(f_x(x,t)) = (0.003)(2765)\pi \cos(\pi x)\cos(2765t), \quad \text{so} \quad f_{xt}(0.3,1) \approx 14.$$

On page 750 we saw that $f_x(x,t)$ gives the slope of the string at any point and time. Therefore,

[3] For a proof, see M. Spivak, *Calculus on Manifolds*, p. 26 (New York: Benjamin, 1965).

$f_{xx}(x,t)$ measures the concavity of the string. The fact that $f_{xx}(0.3, 1) < 0$ means the string is concave down at the point $x = 0.3$ when $t = 1$. (See Figure 14.54.)

On the other hand, $f_{xt}(x,t)$ is the rate of change of the slope of the string with respect to time. Thus, $f_{xt}(0.3, 1) > 0$ means that at time $t = 1$ the slope at the point $x = 0.3$ is increasing. (See Figure 14.55.)

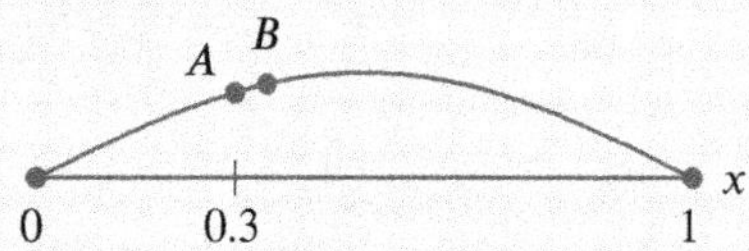

Figure 14.54: Interpretation of $f_{xx}(0.3, 1) < 0$: The concavity of the string at $t = 1$

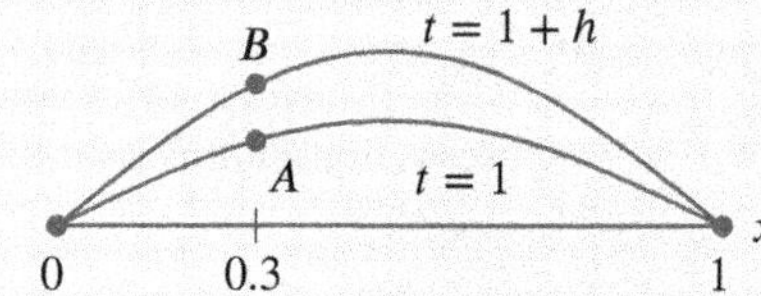

Figure 14.55: Interpretation of $f_{xt}(0.3, 1) > 0$: The slope of one point on the string at two different times

Now we compute $f_t(x,t) = (0.003)(2765)\sin(\pi x)\cos(2765t)$, from which we get

$$f_{tx}(x,t) = \frac{\partial}{\partial x}(f_t(x,t)) = (0.003)(2765)\pi\cos(\pi x)\cos(2765t), \quad \text{so} \quad f_{tx}(0.3, 1) \approx 14$$

and

$$f_{tt}(x,t) = \frac{\partial}{\partial t}(f_t(x,t)) = -(0.003)(2765)^2\sin(\pi x)\sin(2765t), \quad \text{so} \quad f_{tt}(0.3, 1) \approx -7200.$$

On page 750 we saw that $f_t(x,t)$ gives the velocity of the string at any point and time. Therefore, $f_{tx}(x,t)$ and $f_{tt}(x,t)$ will both be rates of change of velocity. That $f_{tx}(0.3, 1) > 0$ means that at time $t = 1$ the velocities of points just to the right of $x = 0.3$ are greater than the velocity at $x = 0.3$. (See Figure 14.56.) That $f_{tt}(0.3, 1) < 0$ means that the velocity of the point $x = 0.3$ is decreasing at time $t = 1$. Thus, $f_{tt}(0.3, 1) = -7200$ m/sec^2 is the acceleration of this point. (See Figure 14.57.)

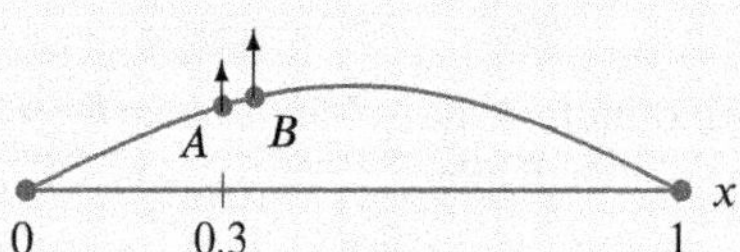

Figure 14.56: Interpretation of $f_{tx}(0.3, 1) > 0$: The velocity of different points on the string at $t = 1$

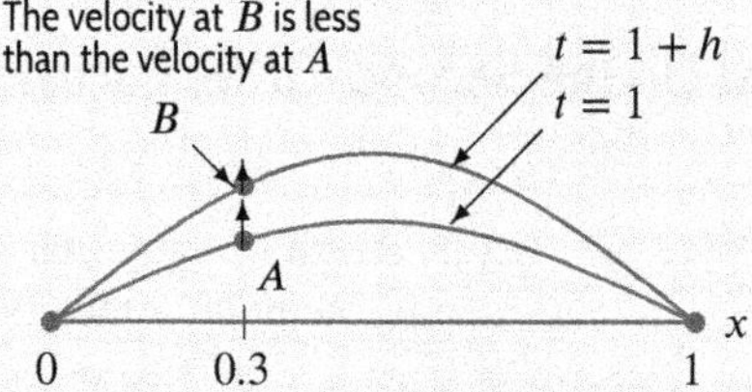

Figure 14.57: Interpretation of $f_{tt}(0.3, 1) < 0$: Negative acceleration. The velocity of one point on the string at two different times

Taylor Approximations

We use second derivatives to construct quadratic Taylor approximations. In Section 14.3, we saw how to approximate $f(x, y)$ by a linear function (its local linearization). We now see how to improve this approximation of $f(x, y)$ using a quadratic function.

Linear and Quadratic Approximations Near (0,0)

For a function of one variable, local linearity tells us that the best *linear* approximation is the degree-1 Taylor polynomial

$$f(x) \approx f(a) + f'(a)(x - a) \quad \text{for } x \text{ near } a.$$

A better approximation to $f(x)$ is given by the degree-2 Taylor polynomial:

$$f(x) \approx f(a) + f'(a)(x-a) + \frac{f''(a)}{2}(x-a)^2 \quad \text{for } x \text{ near } a.$$

For a function of two variables the local linearization for (x, y) near (a, b) is

$$f(x, y) \approx L(x, y) = f(a, b) + f_x(a, b)(x-a) + f_y(a, b)(y-b).$$

In the case $(a, b) = (0, 0)$, we have:

Taylor Polynomial of Degree 1 Approximating $f(x, y)$ for (x, y) near (0,0)
If f has continuous first-order partial derivatives, then

$$f(x, y) \approx L(x, y) = f(0, 0) + f_x(0, 0)x + f_y(0, 0)y.$$

We get a better approximation to f by using a quadratic polynomial. We choose a quadratic polynomial $Q(x, y)$, with the same partial derivatives as the original function f. You can check that the following Taylor polynomial of degree 2 has this property.

Taylor Polynomial of Degree 2 Approximating $f(x, y)$ for (x, y) near (0,0)
If f has continuous second-order partial derivatives, then

$$\begin{aligned} f(x, y) &\approx Q(x, y) \\ &= f(0, 0) + f_x(0, 0)x + f_y(0, 0)y + \frac{f_{xx}(0, 0)}{2}x^2 + f_{xy}(0, 0)xy + \frac{f_{yy}(0, 0)}{2}y^2. \end{aligned}$$

Example 4 Let $f(x, y) = \cos(2x + y) + 3\sin(x + y)$

(a) Compute the linear and quadratic Taylor polynomials, L and Q, approximating f near $(0, 0)$.
(b) Explain why the contour plots of L and Q for $-1 \le x \le 1, -1 \le y \le 1$ look the way they do.

Solution (a) We have $f(0, 0) = 1$. The derivatives we need are as follows:

$$\begin{aligned} f_x(x, y) &= -2\sin(2x + y) + 3\cos(x + y) && \text{so} \quad f_x(0, 0) = 3, \\ f_y(x, y) &= -\sin(2x + y) + 3\cos(x + y) && \text{so} \quad f_y(0, 0) = 3, \\ f_{xx}(x, y) &= -4\cos(2x + y) - 3\sin(x + y) && \text{so} \quad f_{xx}(0, 0) = -4, \\ f_{xy}(x, y) &= -2\cos(2x + y) - 3\sin(x + y) && \text{so} \quad f_{xy}(0, 0) = -2, \\ f_{yy}(x, y) &= -\cos(2x + y) - 3\sin(x + y) && \text{so} \quad f_{yy}(0, 0) = -1. \end{aligned}$$

Thus, the linear approximation, $L(x, y)$, to $f(x, y)$ at $(0, 0)$ is given by

$$f(x, y) \approx L(x, y) = f(0, 0) + f_x(0, 0)x + f_y(0, 0)y = 1 + 3x + 3y.$$

The quadratic approximation, $Q(x, y)$, to $f(x, y)$ near $(0, 0)$ is given by

$$\begin{aligned} f(x, y) &\approx Q(x, y) \\ &= f(0, 0) + f_x(0, 0)x + f_y(0, 0)y + \frac{f_{xx}(0, 0)}{2}x^2 + f_{xy}(0, 0)xy + \frac{f_{yy}(0, 0)}{2}y^2 \\ &= 1 + 3x + 3y - 2x^2 - 2xy - \frac{1}{2}y^2. \end{aligned}$$

Notice that the linear terms in $Q(x, y)$ are the same as the linear terms in $L(x, y)$. The quadratic terms in $Q(x, y)$ can be thought of as "correction terms" to the linear approximation.

(b) The contour plots of $f(x, y)$, $L(x, y)$, and $Q(x, y)$ are in Figures 14.58–14.60.

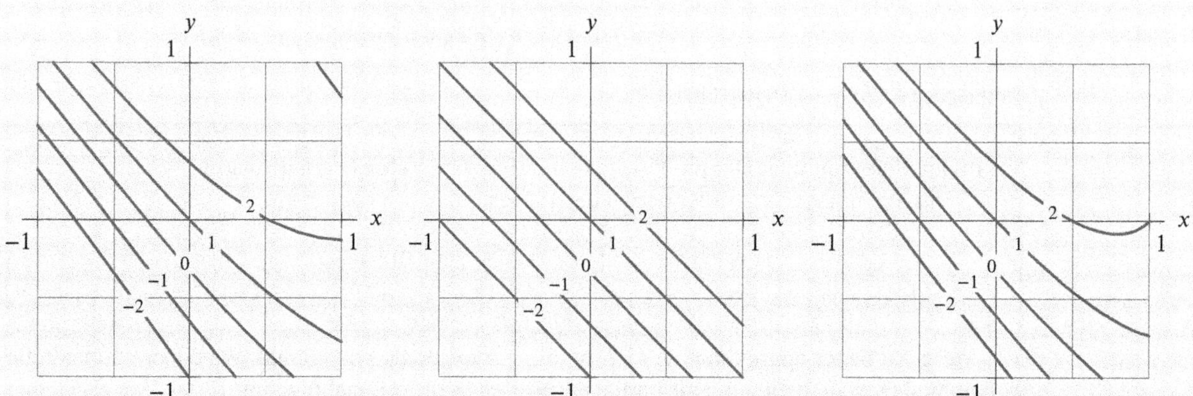

Figure 14.58: Original function, $f(x, y)$

Figure 14.59: Linear approximation, $L(x, y)$

Figure 14.60: Quadratic approximation, $Q(x, y)$

Notice that the contour plot of Q is more similar to the contour plot of f than is the contour plot of L. Since L is linear, the contour plot of L consists of parallel, equally spaced lines.

An alternative, and much quicker, way to find the Taylor polynomial in the previous example is to use the single-variable approximations. For example, since

$$\cos u = 1 - \frac{u^2}{2!} + \frac{u^4}{4!} + \cdots \quad \text{and} \quad \sin v = v - \frac{v^3}{3!} + \cdots,$$

we can substitute $u = 2x + y$ and $v = x + y$ and expand. We discard terms beyond the second (since we want the quadratic polynomial), getting

$$\cos(2x + y) = 1 - \frac{(2x + y)^2}{2!} + \frac{(2x + y)^4}{4!} + \cdots \approx 1 - \frac{1}{2}(4x^2 + 4xy + y^2) = 1 - 2x^2 - 2xy - \frac{1}{2}y^2$$

and

$$\sin(x + y) = (x + y) - \frac{(x + y)^3}{3!} + \cdots \approx x + y.$$

Combining these results, we get

$$\cos(2x + y) + 3\sin(x + y) \approx 1 - 2x^2 - 2xy - \frac{1}{2}y^2 + 3(x + y) = 1 + 3x + 3y - 2x^2 - 2xy - \frac{1}{2}y^2.$$

Linear and Quadratic Approximations Near (a, b)

The local linearization for a function $f(x, y)$ at a point (a, b) is

Taylor Polynomial of Degree 1 Approximating $f(x, y)$ for (x, y) Near (a, b)

If f has continuous first-order partial derivatives, then

$$f(x, y) \approx L(x, y) = f(a, b) + f_x(a, b)(x - a) + f_y(a, b)(y - b).$$

This suggests that a quadratic polynomial approximation $Q(x, y)$ for $f(x, y)$ near a point (a, b) should be written in terms of $(x-a)$ and $(y-b)$ instead of x and y. If we require that $Q(a, b) = f(a, b)$ and that the first- and second-order partial derivatives of Q and f at (a, b) be equal, then we get the following polynomial:

Taylor Polynomial of Degree 2 Approximating $f(x, y)$ for (x, y) Near (a, b)
If f has continuous second-order partial derivatives, then

$$\begin{aligned} f(x, y) &\approx Q(x, y) \\ &= f(a, b) + f_x(a, b)(x-a) + f_y(a, b)(y-b) \\ &\quad + \frac{f_{xx}(a, b)}{2}(x-a)^2 + f_{xy}(a, b)(x-a)(y-b) + \frac{f_{yy}(a, b)}{2}(y-b)^2. \end{aligned}$$

These coefficients are derived in exactly the same way as for $(a, b) = (0, 0)$.

Example 5 Find the Taylor polynomial of degree 2 at the point $(1, 2)$ for the function $f(x, y) = \dfrac{1}{xy}$.

Solution Table 14.8 contains the partial derivatives and their values at the point $(1, 2)$.

Table 14.8 *Partial derivatives of $f(x, y) = 1/(xy)$*

Derivative	Formula	Value at $(1, 2)$	Derivative	Formula	Value at $(1, 2)$
$f(x, y)$	$1/(xy)$	$1/2$	$f_{xx}(x, y)$	$2/(x^3y)$	1
$f_x(x, y)$	$-1/(x^2y)$	$-1/2$	$f_{xy}(x, y)$	$1/(x^2y^2)$	$1/4$
$f_y(x, y)$	$-1/(xy^2)$	$-1/4$	$f_{yy}(x, y)$	$2/(xy^3)$	$1/4$

So, the quadratic Taylor polynomial for f near $(1, 2)$ is

$$\begin{aligned} \frac{1}{xy} &\approx Q(x, y) \\ &= \frac{1}{2} - \frac{1}{2}(x-1) - \frac{1}{4}(y-2) + \frac{1}{2}(1)(x-1)^2 + \frac{1}{4}(x-1)(y-2) + \left(\frac{1}{2}\right)\left(\frac{1}{4}\right)(y-2)^2 \\ &= \frac{1}{2} - \frac{x-1}{2} - \frac{y-2}{4} + \frac{(x-1)^2}{2} + \frac{(x-1)(y-2)}{4} + \frac{(y-2)^2}{8}. \end{aligned}$$

Exercises and Problems for Section 14.7 Online Resource: Additional Problems for Section 14.7

EXERCISES

In Exercises 1–11, calculate all four second-order partial derivatives and check that $f_{xy} = f_{yx}$. Assume the variables are restricted to a domain on which the function is defined.

1. $f(x, y) = (x + y)^2$
2. $f(x, y) = (x + y)^3$
3. $f(x, y) = 3x^2y + 5xy^3$
4. $f(x, y) = e^{2xy}$
5. $f(x, y) = (x + y)e^y$
6. $f(x, y) = xe^y$
7. $f(x, y) = \sin(x/y)$
8. $f(x, y) = \sqrt{x^2 + y^2}$
9. $f(x, y) = 5x^3y^2 - 7xy^3 + 9x^2 + 11$
10. $f(x, y) = \sin(x^2 + y^2)$
11. $f(x, y) = 3\sin 2x \cos 5y$

In Exercises 12–19, find the quadratic Taylor polynomials about $(0,0)$ for the function.

12. $(y-1)(x+1)^2$
13. $(x-y+1)^2$
14. $e^{-2x^2-y^2}$
15. $e^x \cos y$
16. $1/(1+2x-y)$
17. $\cos(x+3y)$
18. $\sin 2x + \cos y$
19. $\ln(1+x^2-y)$

In Exercises 20–21, find the best quadratic approximation for $f(x,y)$ for (x,y) near $(0,0)$.

20. $f(x,y) = \ln(1+x-2y)$
21. $f(x,y) = \sqrt{1+2x-y}$

In Exercises 22–31, use the level curves of the function $z = f(x,y)$ to decide the sign (positive, negative, or zero) of each of the following partial derivatives at the point P. Assume the x- and y-axes are in the usual positions.

(a) $f_x(P)$ **(b)** $f_y(P)$ **(c)** $f_{xx}(P)$
(d) $f_{yy}(P)$ **(e)** $f_{xy}(P)$

22.

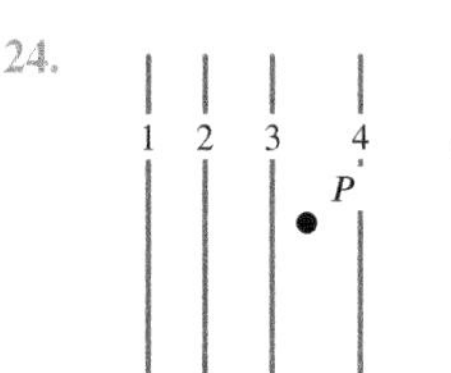

23. 5 4 3 2 1 P

24. 1 2 3 4 5 P

25. 1 2 3 4 5 P

26.

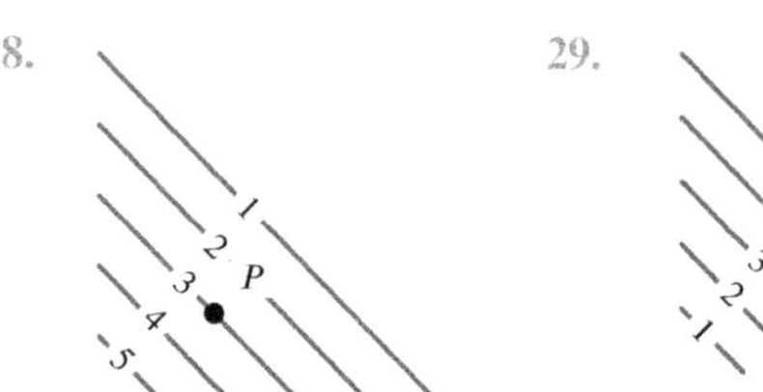

27. 1 2 3 P 4 5

28. 1 2 3 4 5 P

29. 5 4 3 2 1 P

30.

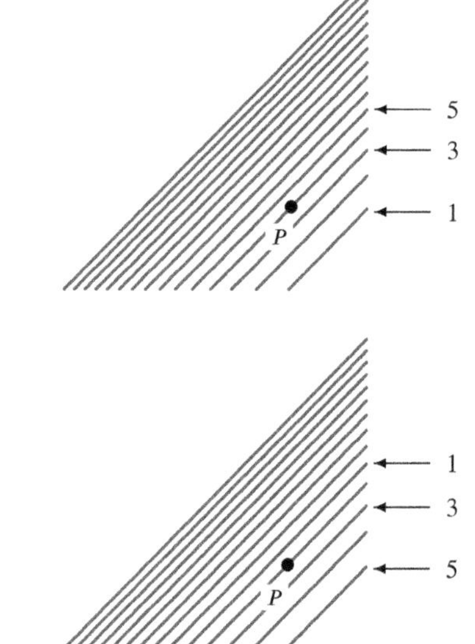

31. 1 3 5 P

PROBLEMS

In Problems 32–35 estimate the quantity, if possible. If it is not possible, explain why. Assume that g is smooth and

$$\nabla g(2,3) = -7\vec{i} + 3\vec{j}$$
$$\nabla g(2.4,3) = -10.2\vec{i} + 4.2\vec{j}$$

32. $g_{yx}(2,3)$ 33. $g_{xx}(2,3)$ 34. $g_{yy}(2,3)$ 35. $g_{xy}(2,3)$

In Problems 36–39 estimate the quantity, if possible. If it is not possible, explain why. Assume that h is smooth and

$$\nabla h(2.4,3) = -20.4\vec{i} + 8.4\vec{j}$$
$$\nabla h(2.4,2.7) = -22.2\vec{i} + 9\vec{j}$$

36. $h_{yy}(2.4,2.7)$
37. $h_{xx}(2.4,3)$
38. $h_{xy}(2.4,2.7)$
39. $h_{yx}(2.4,2.7)$

In Problems 40–44, find the linear, $L(x,y)$, and quadratic, $Q(x,y)$, Taylor polynomials valid near $(1,0)$. Compare the values of the approximations $L(0.9,0.2)$ and $Q(0.9,0.2)$ with the exact value of the function $f(0.9,0.2)$.

40. $f(x,y) = \sqrt{x+2y}$
41. $f(x,y) = x^2 y$
42. $f(x,y) = xe^{-y}$
43. $F(x,y) = e^x \sin y + e^y \sin x$
44. $f(x,y) = \sin(x-1)\cos y$

In Problems 45–46, show that the function satisfies Laplace's equation, $F_{xx} + F_{yy} = 0$.

45. $F(x,y) = e^{-x} \sin y$
46. $F(x,y) = \arctan(y/x)$

47. If $u(x,t) = e^{at}\sin(bx)$ satisfies the heat equation $u_t = u_{xx}$, find the relationship between a and b.

48. **(a)** Check that $u(x,t)$ satisfies the heat equation $u_t = u_{xx}$ for $t > 0$ and all x, where

$$u(x,t) = \frac{1}{2\sqrt{\pi t}}e^{-x^2/(4t)}$$

(b) Graph $u(x,t)$ against x for $t = 0.01, 0.1, 1, 10$. These graphs represent the temperature in an infinitely long insulated rod that at $t = 0$ is 0°C everywhere except at the origin $x = 0$, and that is infinitely hot at $t = 0$ at the origin.

49. Figure 14.61 shows a graph of $z = f(x,y)$. Is $f_{xx}(0,0)$ positive or negative? Is $f_{yy}(0,0)$ positive or negative? Give reasons for your answers.

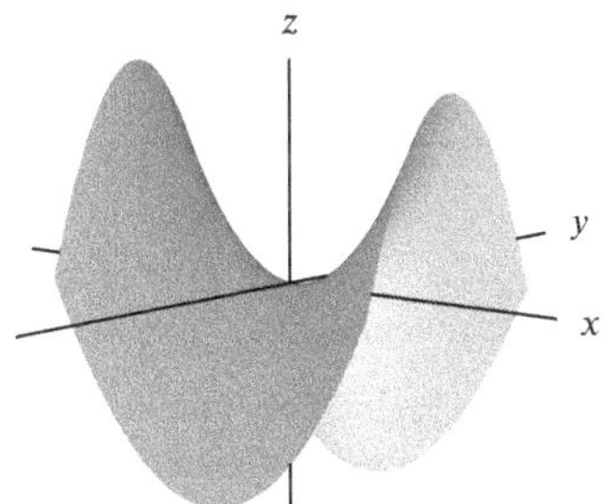

Figure 14.61

50. If $z = f(x) + yg(x)$, what can you say about z_{yy}? Explain your answer.

51. If $z_{xy} = 4y$, what can you say about the value of **(a)** z_{yx}? **(b)** z_{xyx}? **(c)** z_{xyy}?

52. A contour diagram for the smooth function $z = f(x,y)$ is in Figure 14.62.

(a) Is z an increasing or decreasing function of x? Of y?
(b) Is f_x positive or negative? How about f_y?
(c) Is f_{xx} positive or negative? How about f_{yy}?
(d) Sketch the direction of grad f at points P and Q.
(e) Is grad f longer at P or at Q? How do you know?

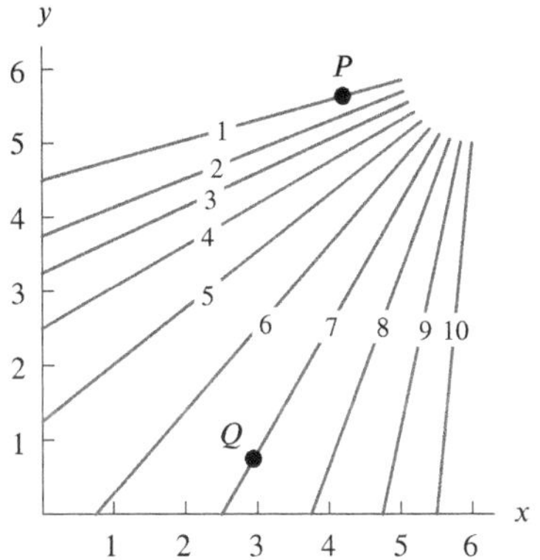

Figure 14.62

Problems 53–56 give tables of values of quadratic polynomials $P(x,y) = a + bx + cy + dx^2 + exy + fy^2$. Determine whether each of the coefficients d, e and f of the quadratic terms is positive, negative, or zero.

53.

		x		
		10	12	14
y	10	35	37	39
	15	45	47	49
	20	55	57	59

54.

		x		
		10	12	14
y	10	26	36	54
	15	31	41	59
	20	36	46	64

55.

		x		
		10	12	14
y	10	90	82	74
	15	75	87	99
	20	10	42	74

56.

		x		
		10	12	14
y	10	13	33	61
	15	28	28	36
	20	93	73	61

57. You are hiking on a level trail going due east and planning to strike off cross country up the mountain to your left. The slope up to the left is too steep now and seems to be gentler the further you go along the trail, so you decide to wait before turning off.

(a) Sketch a topographical contour map that illustrates this story.
(b) What information does the story give about partial derivatives? Define all variables and functions that you use.
(c) What partial derivative influenced your decision to wait before turning?

58. The weekly production, Y, in factories that manufacture a certain item is modeled as a function of the quantity of capital, K, and quantity of labor, L, at the factory. Data shows that hiring a few extra workers increases production. Moreover, for two factories with the same number of workers, hiring a few extra workers increases production more for the factory with more capital. (With more equipment, additional labor can be used more effectively.) What does this tell you about the sign of

(a) $\partial Y/\partial L$?
(b) $\partial^2 Y/(\partial K \partial L)$?

59. Data suggests that human surface area, S, can reasonably be modeled as a function of height, h, and weight, w. In the Dubois model, we have $\partial^2 S/\partial w^2 < 0$ and $\partial^2 S/(\partial h \partial w) > 0$. Two people A and B each gain 1 pound. Which experiences the greater increase in surface area if

(a) They have the same weight but A is taller?
(b) They have the same height, but A is heavier?

60. You plan to buy a used car. You are debating between a 5-year old car and a 10-year old car and thinking about the price. Experts report that the original price matters more when buying a 5-year old car than a 10-year old car. This suggests that we model the average market price, P, in dollars as a function of two variables: the original price, C, in dollars, and the age of the car, A, in years.

(a) Give units for the following partial derivatives and say whether you think they are positive or negative. Explain your reasoning.

(i) $\partial P/\partial A$ (ii) $\partial P/\partial C$

(b) Express the experts' report in terms of partial derivatives.

(c) Using a quadratic polynomial to model P, we have

$$P = a + bC + cA + dC^2 + eCA + fA^2.$$

Which term in this polynomial is most relevant to the experts' report?

61. The tastiness, T, of a soup depends on the volume, V, of the soup in the pot and the quantity, S, of salt in the soup. If you have more soup, you need more salt to make it taste good. Match the three stories (a)–(c) to the three statements (I)–(III) about partial derivatives.

(a) I started adding salt to the soup in the pot. At first the taste improved, but eventually the soup became too salty and continuing to add more salt made it worse.

(b) The soup was too salty, so I started adding unsalted soup. This improved the taste at first, but eventually there was too much soup for the salt, and continuing to add unsalted soup just made it worse.

(c) The soup was too salty, so adding more salt would have made it taste worse. I added a quart of unsalted soup instead. Now it is not salty enough, but I can improve the taste by adding salt.

(I) $\partial^2 T/\partial V^2 < 0$
(II) $\partial^2 T/\partial S^2 < 0$
(III) $\partial^2 T/\partial V \partial S > 0$

62. Figure 14.63 shows the level curves of a function $f(x, y)$ around a maximum or minimum, M. One of the points P and Q has coordinates (x_1, y_1) and the other has coordinates (x_2, y_2). Suppose $b > 0$ and $c > 0$. Consider the two linear approximations to f given by

$$f(x, y) \approx a + b(x - x_1) + c(y - y_1)$$
$$f(x, y) \approx k + m(x - x_2) + n(y - y_2).$$

(a) What is the relationship between the values of a and k?

(b) What are the coordinates of P?

(c) Is M a maximum or a minimum?

(d) What can you say about the sign of the constants m and n?

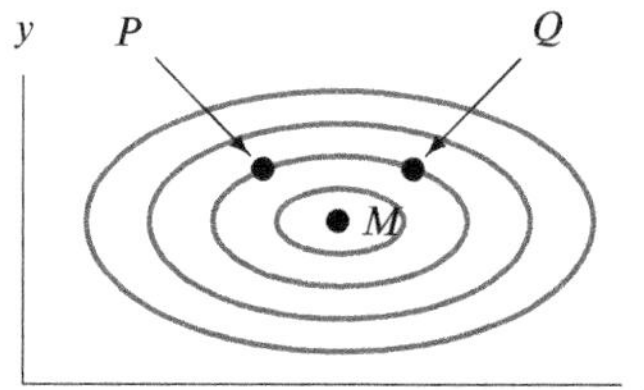

Figure 14.63

63. Consider the function $f(x, y) = (\sin x)(\sin y)$.

(a) Find the Taylor polynomials of degree 2 for f about the points $(0, 0)$ and $(\pi/2, \pi/2)$.

(b) Use the Taylor polynomials to sketch the contours of f close to each of the points $(0, 0)$ and $(\pi/2, \pi/2)$.

64. Let $f(x, y) = \sqrt{x + 2y + 1}$.

(a) Compute the local linearization of f at $(0, 0)$.

(b) Compute the quadratic Taylor polynomial for f at $(0, 0)$.

(c) Compare the values of the linear and quadratic approximations in part (a) and part (b) with the true values for $f(x, y)$ at the points $(0.1, 0.1)$, $(-0.1, 0.1)$, $(0.1, -0.1)$, $(-0.1, -0.1)$. Which approximation gives the closest values?

65. Using a computer and your answer to Problem 64, draw the six contour diagrams of $f(x, y) = \sqrt{x + 2y + 1}$ and its linear and quadratic approximations, $L(x, y)$ and $Q(x, y)$, in the two windows $[-0.6, 0.6]\times[-0.6, 0.6]$ and $[-2, 2]\times[-2, 2]$. Explain the shape of the contours, their spacing, and the relationship between the contours of f, L, and Q.

Strengthen Your Understanding

In Problems 66–67, explain what is wrong with the statement.

66. If $f(x, y) \neq 0$, then the Taylor polynomial of degree 2 approximating $f(x, y)$ near $(0, 0)$ is also nonzero.

67. There is a function $f(x, y)$ with partial derivatives $f_x = xy$ and $f_y = y^2$.

In Problems 68–70, give an example of:

68. A function $f(x, y)$ such that $f_{xx} \neq 0$, $f_{yy} \neq 0$, and $f_{xy} = 0$.

69. Formulas for two different functions $f(x, y)$ and $g(x, y)$ with the same quadratic approximation near $(0, 0)$.

70. Contour diagrams for two different functions $f(x, y)$ and $g(x, y)$ that have the same quadratic approximations near $(0, 0)$.

14.8 DIFFERENTIABILITY

In Section 14.3 we gave an informal introduction to the concept of differentiability. We called a function $f(x, y)$ *differentiable* at a point (a, b) if it is well approximated by a linear function near (a, b). This section focuses on the precise meaning of the phrase "well approximated." By looking at examples, we shall see that local linearity requires the existence of partial derivatives, but they do not tell the whole story. In particular, existence of partial derivatives at a point is not sufficient to guarantee local linearity at that point.

We begin by discussing the relation between continuity and differentiability. As an illustration, take a sheet of paper, crumple it into a ball and smooth it out again. Wherever there is a crease it would be difficult to approximate the surface by a plane—these are points of nondifferentiability of the function giving the height of the paper above the floor. Yet the sheet of paper models a graph which is continuous—there are no breaks. As in the case of one-variable calculus, continuity does not imply differentiability. But differentiability does *require* continuity: there cannot be linear approximations to a surface at points where there are abrupt changes in height.

Differentiability for Functions of Two Variables

For a function of two variables, as for a function of one variable, we define differentiability at a point in terms of the error and the distance from the point. If the point is (a, b) and a nearby point is $(a+h, b+k)$, the distance between them is $\sqrt{h^2+k^2}$. (See Figure 14.64.)

A function $f(x, y)$ is **differentiable at the point** (a, b) if there is a linear function $L(x, y) = f(a, b) + m(x-a) + n(y-b)$ such that if the *error* $E(x, y)$ is defined by

$$f(x, y) = L(x, y) + E(x, y),$$

and if $h = x - a$, $k = y - b$, then the *relative error* $E(a+h, b+k)/\sqrt{h^2+k^2}$ satisfies

$$\lim_{\substack{h\to 0\\ k\to 0}} \frac{E(a+h, b+k)}{\sqrt{h^2+k^2}} = 0.$$

The function f is **differentiable on a region** R if it is differentiable at each point of R. The function $L(x, y)$ is called the *local linearization* of $f(x, y)$ near (a, b).

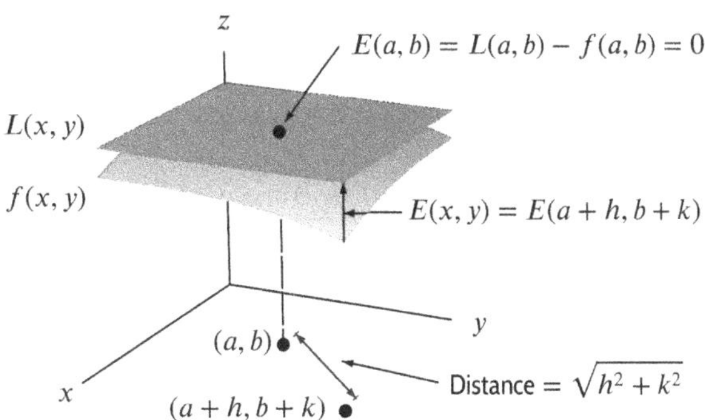

Figure 14.64: Graph of function $z = f(x, y)$ and its local linearization $z = L(x, y)$ near the point (a, b)

Partial Derivatives and Differentiability

In the next example, we show that this definition of differentiability is consistent with our previous notion — that is, that $m = f_x$ and $n = f_y$ and that the graph of $L(x, y)$ is the tangent plane.

Example 1 Show that if f is a differentiable function with local linearization $L(x, y) = f(a, b) + m(x-a) + n(y-b)$, then $m = f_x(a, b)$ and $n = f_y(a, b)$.

Solution Since f is differentiable, we know that the relative error in $L(x, y)$ tends to 0 as we get close to (a, b). Suppose $h > 0$ and $k = 0$. Then we know that

$$\begin{aligned} 0 = \lim_{h\to 0} \frac{E(a+h, b+k)}{\sqrt{h^2+k^2}} = \lim_{h\to 0} \frac{E(a+h, b)}{h} &= \lim_{h\to 0} \frac{f(a+h, b) - L(a+h, b)}{h} \\ &= \lim_{h\to 0} \frac{f(a+h, b) - f(a, b) - mh}{h} \\ &= \lim_{h\to 0} \left(\frac{f(a+h, b) - f(a, b)}{h} \right) - m = f_x(a, b) - m. \end{aligned}$$

A similar result holds if $h < 0$, so we have $m = f_x(a, b)$. The result $n = f_y(a, b)$ is found in a similar manner.

The previous example shows that if a function is differentiable at a point, it has partial derivatives there. Therefore, if any of the partial derivatives fail to exist, then the function cannot be differentiable. This is what happens in the following example of a cone.

Example 2 Consider the function $f(x, y) = \sqrt{x^2 + y^2}$. Is f differentiable at the origin?

Solution If we zoom in on the graph of the function $f(x, y) = \sqrt{x^2 + y^2}$ at the origin, as shown in Figure 14.65, the sharp point remains; the graph never flattens out to look like a plane. Near its vertex, the graph does not look as if is well approximated (in any reasonable sense) by any plane.

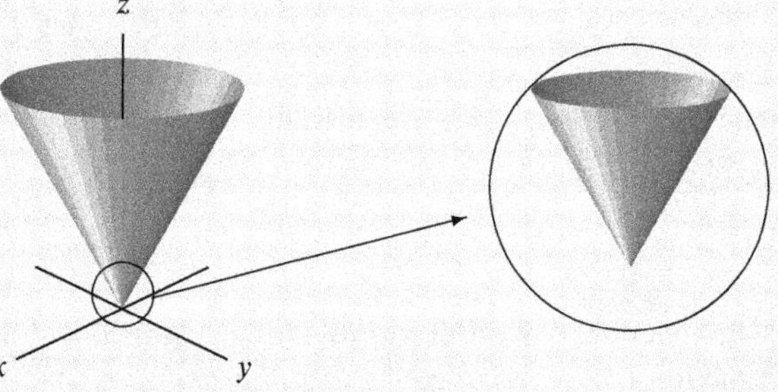

Figure 14.65: The function $f(x, y) = \sqrt{x^2 + y^2}$ is not locally linear at $(0, 0)$: Zooming in around $(0, 0)$ does not make the graph look like a plane

Judging from the graph of f, we would not expect f to be differentiable at $(0, 0)$. Let us check this by trying to compute the partial derivatives of f at $(0, 0)$:

$$f_x(0, 0) = \lim_{h\to 0} \frac{f(h, 0) - f(0, 0)}{h} = \lim_{h\to 0} \frac{\sqrt{h^2 + 0} - 0}{h} = \lim_{h\to 0} \frac{|h|}{h}.$$

Since $|h|/h = \pm 1$, depending on whether h approaches 0 from the left or right, this limit does not exist and so neither does the partial derivative $f_x(0, 0)$. Thus, f cannot be differentiable at the origin. If it were, both of the partial derivatives, $f_x(0, 0)$ and $f_y(0, 0)$, would exist.

Alternatively, we could show directly that there is no linear approximation near $(0, 0)$ that satisfies the small relative error criterion for differentiability. Any plane passing through the point $(0, 0, 0)$ has the form $L(x, y) = mx + ny$ for some constants m and n. If $E(x, y) = f(x, y) - L(x, y)$, then

$$E(x, y) = \sqrt{x^2 + y^2} - mx - ny.$$

Then for f to be differentiable at the origin, we would need to show that

$$\lim_{\substack{h\to 0 \\ k\to 0}} \frac{\sqrt{h^2 + k^2} - mh - nk}{\sqrt{h^2 + k^2}} = 0.$$

Taking $k = 0$ gives

$$\lim_{h\to 0} \frac{|h| - mh}{|h|} = 1 - m \lim_{h\to 0} \frac{h}{|h|}.$$

This limit exists only if $m = 0$ for the same reason as before. But then the value of the limit is 1 and not 0 as required. Thus, we again conclude f is not differentiable.

In Example 2 the partial derivatives f_x and f_y did not exist at the origin and this was sufficient to establish nondifferentiability there. We might expect that if both partial derivatives do exist, then f *is* differentiable. But the next example shows that this not necessarily true: the existence of both partial derivatives at a point is *not* sufficient to guarantee differentiability.

Example 3 Consider the function $f(x, y) = x^{1/3}y^{1/3}$. Show that the partial derivatives $f_x(0, 0)$ and $f_y(0, 0)$ exist, but that f is not differentiable at $(0, 0)$.

Solution See Figure 14.66 for the part of the graph of $z = x^{1/3}y^{1/3}$ when $z \geq 0$. We have $f(0, 0) = 0$ and we compute the partial derivatives using the definition:

$$f_x(0, 0) = \lim_{h\to 0} \frac{f(h, 0) - f(0, 0)}{h} = \lim_{h\to 0} \frac{0 - 0}{h} = 0,$$

and similarly

$$f_y(0, 0) = 0.$$

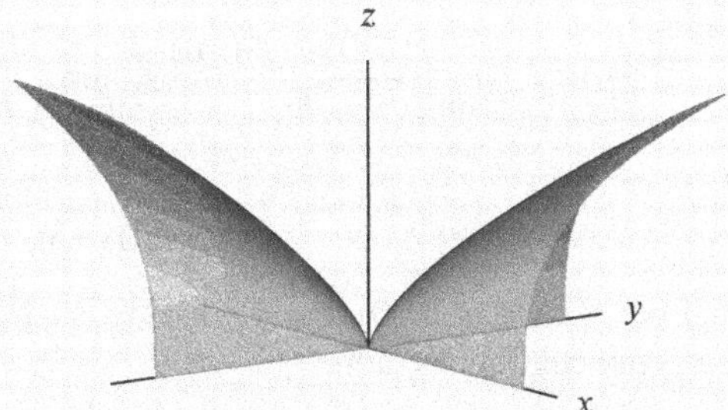

Figure 14.66: Graph of $z = x^{1/3}y^{1/3}$ for $z \geq 0$

So, if there did exist a linear approximation near the origin, it would have to be $L(x, y) = 0$. But we can show that this choice of $L(x, y)$ does not result in the small relative error that is required for differentiability. In fact, since $E(x, y) = f(x, y) - L(x, y) = f(x, y)$, we need to look at the limit

$$\lim_{\substack{h\to 0\\ k\to 0}} \frac{h^{1/3}k^{1/3}}{\sqrt{h^2 + k^2}}.$$

If this limit exists, we get the same value no matter how h and k approach 0. Suppose we take $k = h > 0$. Then the limit becomes

$$\lim_{h\to 0} \frac{h^{1/3}h^{1/3}}{\sqrt{h^2 + h^2}} = \lim_{h\to 0} \frac{h^{2/3}}{h\sqrt{2}} = \lim_{h\to 0} \frac{1}{h^{1/3}\sqrt{2}}.$$

But this limit does not exist, since small values for h will make the fraction arbitrarily large. So the only possible candidate for a linear approximation at the origin does not have a sufficiently small relative error. Thus, this function is *not* differentiable at the origin, even though the partial derivatives $f_x(0, 0)$ and $f_y(0, 0)$ exist. Figure 14.66 confirms that near the origin the graph of $z = f(x, y)$ is not well approximated by any plane.

In summary,

- If a function is differentiable at a point, then both partial derivatives exist there.
- Having both partial derivatives at a point does not guarantee that a function is differentiable there.

Continuity and Differentiability

We know that differentiable functions of one variable are continuous. Similarly, it can be shown that if a function of two variables is differentiable at a point, then the function is continuous there.

In Example 3 the function f was continuous at the point where it was not differentiable. Example 4 shows that even if the partial derivatives of a function exist at a point, the function is not necessarily continuous at that point if it is not differentiable there.

Example 4 Suppose that f is the function of two variables defined by

$$f(x, y) = \begin{cases} \dfrac{xy}{x^2 + y^2}, & (x, y) \neq (0, 0), \\ 0, & (x, y) = (0, 0). \end{cases}$$

Problem 27 on page 699 showed that $f(x, y)$ is not continuous at the origin. Show that the partial derivatives $f_x(0, 0)$ and $f_y(0, 0)$ exist. Could f be differentiable at $(0, 0)$?

Solution From the definition of the partial derivative we see that

$$f_x(0, 0) = \lim_{h\to 0} \frac{f(h, 0) - f(0, 0)}{h} = \lim_{h\to 0} \left(\frac{1}{h} \cdot \frac{0}{h^2 + 0^2} \right) = \lim_{h\to 0} \frac{0}{h} = 0,$$

and similarly

$$f_y(0, 0) = 0.$$

So, the partial derivatives $f_x(0, 0)$ and $f_y(0, 0)$ exist. However, f cannot be differentiable at the origin since it is not continuous there.

In summary,

- If a function is differentiable at a point, then it is continuous there.
- Having both partial derivatives at a point does not guarantee that a function is continuous there.

How Do We Know If a Function Is Differentiable?

Can we use partial derivatives to tell us if a function is differentiable? As we see from Examples 3 and 4, it is not enough that the partial derivatives exist. However, the following theorem gives conditions that *do* guarantee differentiability[4]:

Theorem 14.2: Continuity of Partial Derivatives Implies Differentiability

If the partial derivatives, f_x and f_y, of a function f exist and are continuous on a small disk centered at the point (a, b), then f is differentiable at (a, b).

[4]For a proof, see M. Spivak, *Calculus on Manifolds*, p. 31 (New York: Benjamin, 1965).

We will not prove this theorem, although it provides a criterion for differentiability which is often simpler to use than the definition. It turns out that the requirement of continuous partial derivatives is more stringent than that of differentiability, so there exist differentiable functions which do not have continuous partial derivatives. However, most functions we encounter will have continuous partial derivatives. The class of functions with continuous partial derivatives is given the name C^1.

Example 5 Show that the function $f(x, y) = \ln(x^2 + y^2)$ is differentiable everywhere in its domain.

Solution The domain of f is all of 2-space except for the origin. We shall show that f has continuous partial derivatives everywhere in its domain (that is, the function f is in C^1). The partial derivatives are

$$f_x = \frac{2x}{x^2 + y^2} \quad \text{and} \quad f_y = \frac{2y}{x^2 + y^2}.$$

Since each of f_x and f_y is the quotient of continuous functions, the partial derivatives are continuous everywhere except the origin (where the denominators are zero). Thus, f is differentiable everywhere in its domain.

Most functions built up from elementary functions have continuous partial derivatives, except perhaps at a few obvious points. Thus, in practice, we can often identify functions as being C^1 without explicitly computing the partial derivatives.

Exercises and Problems for Section 14.8

EXERCISES

In Exercises 1–10, list the points in the xy-plane, if any, at which the function $z = f(x, y)$ is not differentiable.

1. $z = -\sqrt{x^2 + y^2}$
2. $z = \sqrt{(x + 1)^2 + y^2}$
3. $z = |x| + |y|$
4. $z = |x + 2| - |y - 3|$
5. $z = e^{-(x^2+y^2)}$
6. $z = x^{1/3} + y^2$
7. $z = |x - 3|^2 + y^3$
8. $z = (\sin x)(\cos |y|)$
9. $z = 4 + \sqrt{(x - 1)^2 + (y - 2)^2}$
10. $z = 1 + \left((x - 1)^2 + (y - 2)^2\right)^2$

PROBLEMS

In Problems 11–14, a function f is given.

(a) Use a computer to draw a contour diagram for f.
(b) Is f differentiable at all points $(x, y) \neq (0, 0)$?
(c) Do the partial derivatives f_x and f_y exist and are they continuous at all points $(x, y) \neq (0, 0)$?
(d) Is f differentiable at $(0, 0)$?
(e) Do the partial derivatives f_x and f_y exist and are they continuous at $(0, 0)$?

11. $f(x, y) = \begin{cases} \dfrac{x}{y} + \dfrac{y}{x}, & x \neq 0 \text{ and } y \neq 0, \\ 0, & x = 0 \text{ or } y = 0. \end{cases}$

12. $f(x, y) = \begin{cases} \dfrac{2xy}{(x^2 + y^2)^2}, & (x, y) \neq (0, 0), \\ 0, & (x, y) = (0, 0). \end{cases}$

13. $f(x, y) = \begin{cases} \dfrac{x^2 y}{x^4 + y^2}, & (x, y) \neq (0, 0), \\ 0, & (x, y) = (0, 0). \end{cases}$

14. $f(x, y) = \begin{cases} \dfrac{xy}{\sqrt{x^2 + y^2}}, & (x, y) \neq (0, 0), \\ 0, & (x, y) = (0, 0). \end{cases}$

15. Consider the function

$$f(x, y) = \begin{cases} \dfrac{xy^2}{x^2 + y^2}, & (x, y) \neq (0, 0), \\ 0, & (x, y) = (0, 0). \end{cases}$$

(a) Use a computer to draw the contour diagram for f.
(b) Is f differentiable for $(x, y) \neq (0, 0)$?
(c) Show that $f_x(0, 0)$ and $f_y(0, 0)$ exist.
(d) Is f differentiable at $(0, 0)$?
(e) Suppose $x(t) = at$ and $y(t) = bt$, where a and b are constants, not both zero. If $g(t) = f(x(t), y(t))$, show that

$$g'(0) = \frac{ab^2}{a^2 + b^2}.$$

(f) Show that

$$f_x(0, 0)x'(0) + f_y(0, 0)y'(0) = 0.$$

Does the chain rule hold for the composite function $g(t)$ at $t = 0$? Explain.

(g) Show that the directional derivative $f_{\vec{u}}(0, 0)$ exists for each unit vector $\vec{u}$. Does this imply that f is differentiable at $(0, 0)$?

16. Consider the function $f(x, y) = \sqrt{|xy|}$.

   (a) Use a computer to draw the contour diagram for f. Does the contour diagram look like that of a plane when we zoom in on the origin?
   (b) Use a computer to draw the graph of f. Does the graph look like a plane when we zoom in on the origin?
   (c) Is f differentiable for $(x, y) \neq (0, 0)$?
   (d) Show that $f_x(0, 0)$ and $f_y(0, 0)$ exist.
   (e) Is f differentiable at $(0, 0)$? [Hint: Consider the directional derivative $f_{\vec{u}}(0, 0)$ for $\vec{u} = (\vec{i} + \vec{j})/\sqrt{2}$.]

17. Consider the function

$$f(x, y) = \begin{cases} \dfrac{xy^2}{x^2 + y^4}, & (x, y) \neq (0, 0), \\ 0, & (x, y) = (0, 0). \end{cases}$$

   (a) Use a computer to draw the contour diagram for f.
   (b) Show that the directional derivative $f_{\vec{u}}(0, 0)$ exists for each unit vector $\vec{u}$.
   (c) Is f continuous at $(0, 0)$? Is f differentiable at $(0, 0)$? Explain.

18. Suppose $f(x, y)$ is a function such that $f_x(0, 0) = 0$ and $f_y(0, 0) = 0$, and $f_{\vec{u}}(0, 0) = 3$ for $\vec{u} = (\vec{i} + \vec{j})/\sqrt{2}$.

   (a) Is f differentiable at $(0, 0)$? Explain.
   (b) Give an example of a function f defined on 2-space which satisfies these conditions. [Hint: The function f does not have to be defined by a single formula valid over all of 2-space.]

19. Consider the following function:

$$f(x, y) = \begin{cases} \dfrac{xy(x^2 - y^2)}{x^2 + y^2}, & (x, y) \neq (0, 0), \\ 0, & (x, y) = (0, 0). \end{cases}$$

   The graph of f is shown in Figure 14.67, and the contour diagram of f is shown in Figure 14.68.

   (a) Find $f_x(x, y)$ and $f_y(x, y)$ for $(x, y) \neq (0, 0)$.
   (b) Show that $f_x(0, 0) = 0$ and $f_y(0, 0) = 0$.
   (c) Are the functions f_x and f_y continuous at $(0, 0)$?
   (d) Is f differentiable at $(0, 0)$?

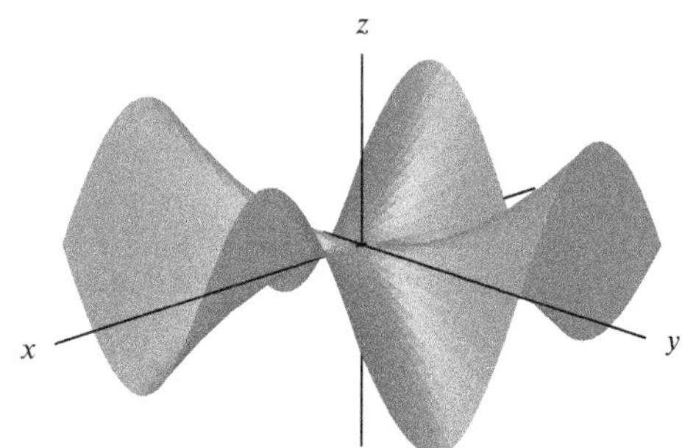

Figure 14.67: Graph of $\dfrac{xy(x^2 - y^2)}{x^2 + y^2}$

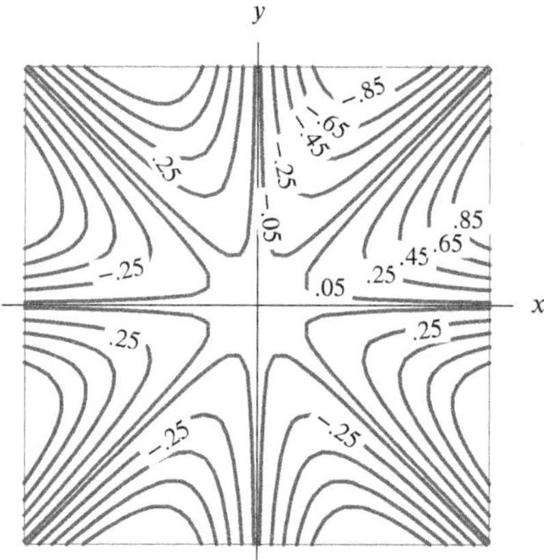

Figure 14.68: Contour diagram of $\dfrac{xy(x^2 - y^2)}{x^2 + y^2}$

20. Suppose a function f is differentiable at the point (a, b). Show that f is continuous at (a, b).

Strengthen Your Understanding

In Problems 21–22, explain what is wrong with the statement.

21. If $f(x, y)$ is continuous at the origin, then it is differentiable at the origin.

22. If the partial derivatives $f_x(0, 0)$ and $f_y(0, 0)$ both exist, then $f(x, y)$ is differentiable at the origin.

In Problems 23–24, give an example of:

23. A continuous function $f(x, y)$ that is not differentiable at the origin.

24. A continuous function $f(x, y)$ that is not differentiable on the line $x = 1$.

25. Which of the following functions $f(x, y)$ is differentiable at the given point?

   (a) $\sqrt{1 - x^2 - y^2}$ at $(0, 0)$ (b) $\sqrt{4 - x^2 - y^2}$ at $(2, 0)$
   (c) $-\sqrt{x^2 + 2y^2}$ at $(0, 0)$ (d) $-\sqrt{x^2 + 2y^2}$ at $(2, 0)$

Online Resource: Review problems and Projects

Chapter Fifteen

OPTIMIZATION: LOCAL AND GLOBAL EXTREMA

Contents

15.1 CRITICAL POINTS: LOCAL EXTREMA AND SADDLE POINTS

Functions of several variables, like functions of one variable, can have *local* and *global* extrema. (That is, local and global maxima and minima.) A function has a local extremum at a point where it takes on the largest or smallest value in a small region around the point. Global extrema are the largest or smallest values anywhere on the domain under consideration. (See Figures 15.1 and 15.2.)

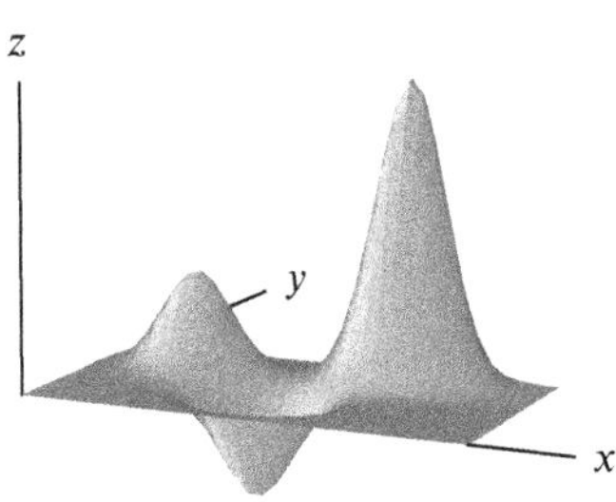

Figure 15.1: Local and global extrema for a function of two variables on $0 \le x \le a$, $0 \le y \le b$

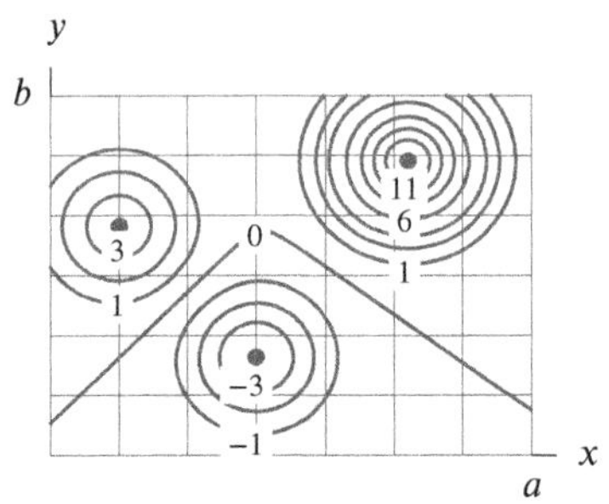

Figure 15.2: Contour map of the function in Figure 15.1

More precisely, considering only points at which f is defined, we say:

- f has a **local maximum** at the point P_0 if $f(P_0) \ge f(P)$ for all points P near P_0.
- f has a **local minimum** at the point P_0 if $f(P_0) \le f(P)$ for all points P near P_0.

For example, the function whose contour map is shown in Figure 15.2 has a local minimum value of -3 and local maximum values of 3 and 11 in the rectangle shown.

How Do We Detect a Local Maximum or Minimum?

Recall that if the gradient vector of a function is defined and nonzero, then it points in a direction in which the function increases. Suppose that a function f has a local maximum at a point P_0 which is not on the boundary of the domain. If the vector grad $f(P_0)$ were defined and nonzero, then we could increase f by moving in the direction of grad $f(P_0)$. Since f has a local maximum at P_0, there is no direction in which f is increasing. Thus, if grad $f(P_0)$ is defined, we must have

$$\text{grad } f(P_0) = \vec{0}.$$

Similarly, suppose f has a local minimum at the point P_0. If grad $f(P_0)$ were defined and nonzero, then we could decrease f by moving in the direction opposite to grad $f(P_0)$, and so we must again have grad $f(P_0) = \vec{0}$. Therefore, we make the following definition:

Points where the gradient is either $\vec{0}$ or undefined are called **critical points** of the function.

If a function has a local maximum or minimum at a point P_0, not on the boundary of its domain, then P_0 is a critical point. For a function of two variables, we can also see that the gradient vector must be zero or undefined at a local maximum by looking at its contour diagram and a plot of its gradient vectors. (See Figures 15.3 and 15.4.) Around the maximum the vectors are all pointing inward, perpendicularly to the contours. At the maximum the gradient vector must be zero or undefined. A similar argument shows that the gradient must be zero or undefined at a local minimum.

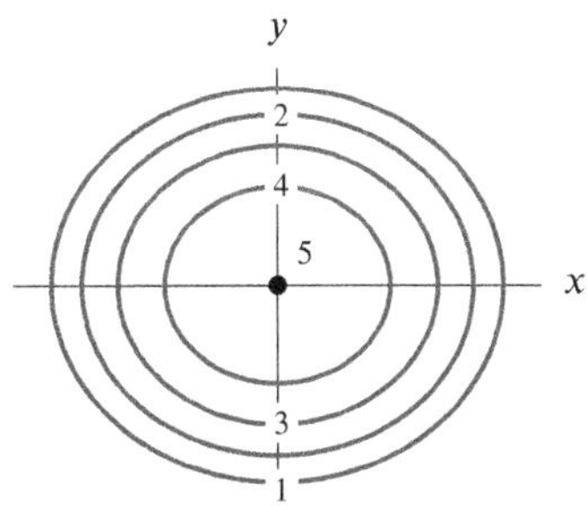

Figure 15.3: Contour diagram around a local maximum of a function

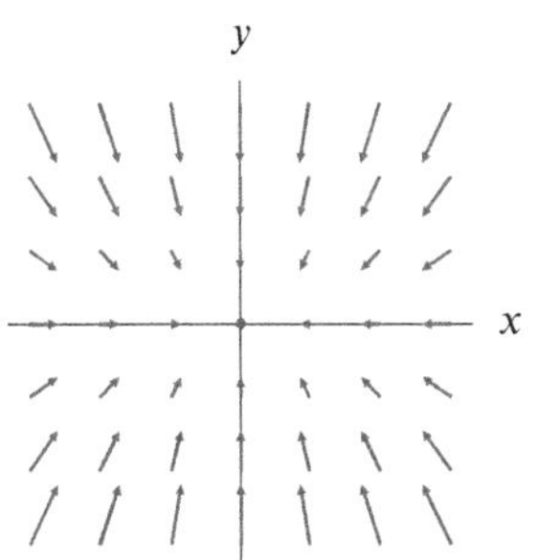

Figure 15.4: Gradients pointing toward the local maximum of the function in Figure 15.3

Finding and Analyzing Critical Points

To find critical points of f we set $\text{grad}\, f = f_x\vec{i} + f_y\vec{j} + f_z\vec{k} = \vec{0}$, which means setting all the partial derivatives of f equal to zero. We must also look for the points where one or more of the partial derivatives is undefined.

Example 1 Find and analyze the critical points of $f(x, y) = x^2 - 2x + y^2 - 4y + 5$.

Solution To find the critical points, we set both partial derivatives equal to zero:

$$f_x(x, y) = 2x - 2 = 0$$
$$f_y(x, y) = 2y - 4 = 0.$$

Solving these equations gives $x = 1$, $y = 2$. Hence, f has only one critical point, namely $(1, 2)$. To see the behavior of f near $(1, 2)$, look at the values of the function in Table 15.1.

Table 15.1 *Values of $f(x, y)$ near the point $(1, 2)$*

		x				
		0.8	0.9	1.0	1.1	1.2
	1.8	0.08	0.05	0.04	0.05	0.08
	1.9	0.05	0.02	0.01	0.02	0.05
y	2.0	0.04	0.01	0.00	0.01	0.04
	2.1	0.05	0.02	0.01	0.02	0.05
	2.2	0.08	0.05	0.04	0.05	0.08

The table suggests that the function has a local minimum value of 0 at $(1, 2)$. We can confirm this by completing the square:

$$f(x, y) = x^2 - 2x + y^2 - 4y + 5 = (x - 1)^2 + (y - 2)^2.$$

Figure 15.5 shows that the graph of f is a paraboloid with vertex at the point $(1, 2, 0)$. It is the same shape as the graph of $z = x^2 + y^2$ (see Figure 12.12 on page 661), except that the vertex has been shifted to $(1, 2)$. So the point $(1, 2)$ is a local minimum of f (as well as a global minimum).

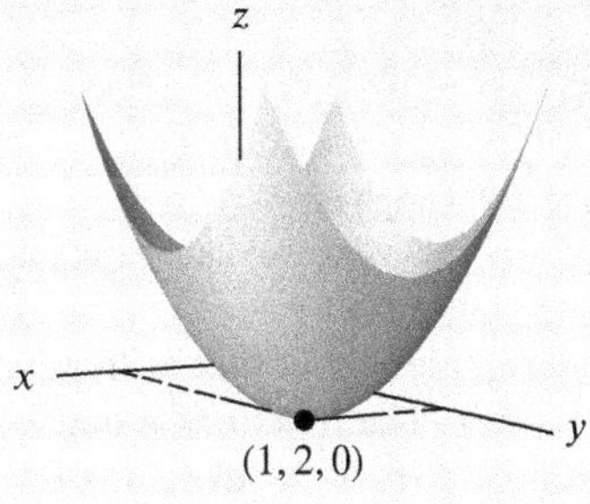

Figure 15.5: The graph of $f(x, y) = x^2 - 2x + y^2 - 4y + 5$ with a local minimum at the point $(1, 2)$

Example 2 Find and analyze any critical points of $f(x, y) = -\sqrt{x^2 + y^2}$.

Solution We look for points where grad $f = \vec{0}$ or is undefined. The partial derivatives are given by

$$f_x(x, y) = -\frac{x}{\sqrt{x^2 + y^2}},$$
$$f_y(x, y) = -\frac{y}{\sqrt{x^2 + y^2}}.$$

These partial derivatives are never simultaneously zero, but they are undefined at $x = 0$, $y = 0$. Thus, $(0, 0)$ is a critical point and a possible extreme point. The graph of f (see Figure 15.6) is a cone, with vertex at $(0, 0)$. So f has a local and global maximum at $(0, 0)$.

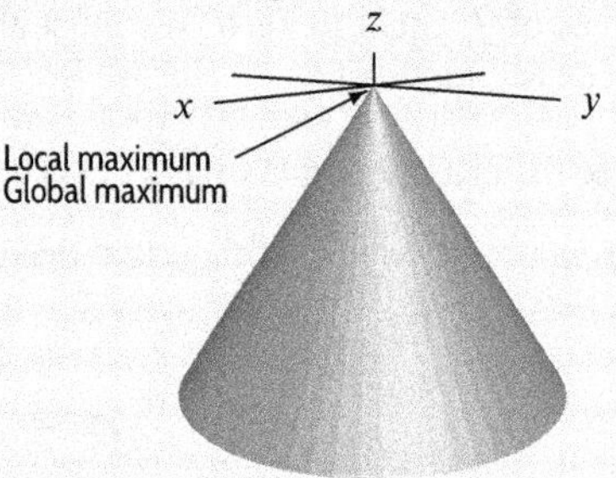

Figure 15.6: Graph of $f(x, y) = -\sqrt{x^2 + y^2}$

Example 3 Find and analyze any critical points of $g(x, y) = x^2 - y^2$.

Solution To find the critical points, we look for points where both partial derivatives are zero:

$$g_x(x, y) = 2x = 0$$
$$g_y(x, y) = -2y = 0.$$

Solving gives $x = 0$, $y = 0$, so the origin is the only critical point.

Figure 15.7 shows that near the origin g takes on both positive and negative values. Since $g(0, 0) = 0$, the origin is a critical point which is neither a local maximum nor a local minimum. The graph of g looks like a saddle.

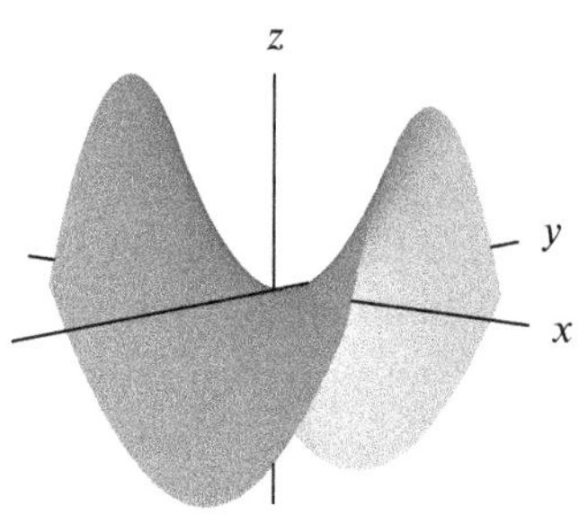

Figure 15.7: Graph of $g(x, y) = x^2 - y^2$, showing saddle shape at the origin

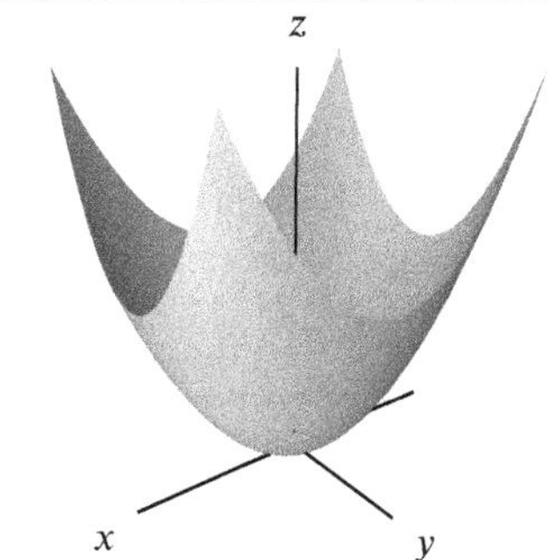

Figure 15.8: Graph of $h(x, y) = x^2 + y^2$, showing minimum at the origin

The previous examples show that critical points can occur at local maxima or minima, or at points which are neither: The functions g and h in Figures 15.7 and 15.8 both have critical points at the origin. Figure 15.9 shows level curves of g. They are hyperbolas showing both positive and negative values of g near $(0, 0)$. Contrast this with the level curves of h near the local minimum in Figure 15.10.

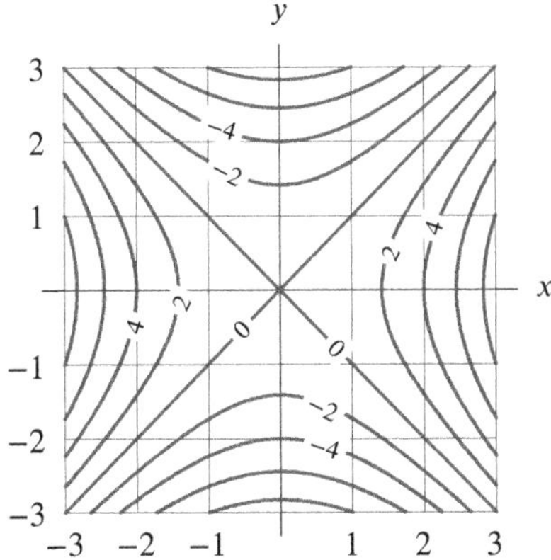

Figure 15.9: Contours of $g(x, y) = x^2 - y^2$, showing a saddle shape at the origin

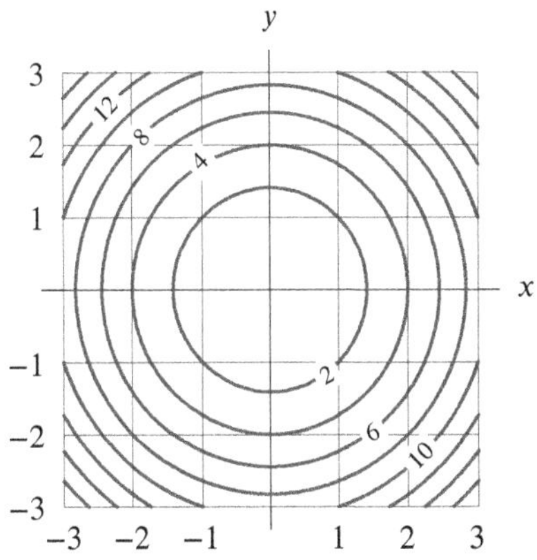

Figure 15.10: Contours of $h(x, y) = x^2 + y^2$, showing a local minimum at the origin

Example 4 Find the local extrema of the function $f(x, y) = 8y^3 + 12x^2 - 24xy$.

Solution We begin by looking for critical points:

$$f_x(x, y) = 24x - 24y,$$
$$f_y(x, y) = 24y^2 - 24x.$$

Setting these expressions equal to zero gives the system of equations

$$x = y, \qquad x = y^2,$$

which has two solutions, $(0, 0)$ and $(1, 1)$. Are these local maxima, local minima or neither? Figure 15.11 shows contours of f near the points. Notice that $f(1, 1) = -4$ and the contours at nearby points have larger function values. This suggests f has a local minimum at $(1, 1)$.

We have $f(0, 0) = 0$ and the contours near $(0, 0)$ show that f takes both positive and negative values nearby. This suggests that $(0, 0)$ is a critical point which is neither a local maximum nor a local minimum.

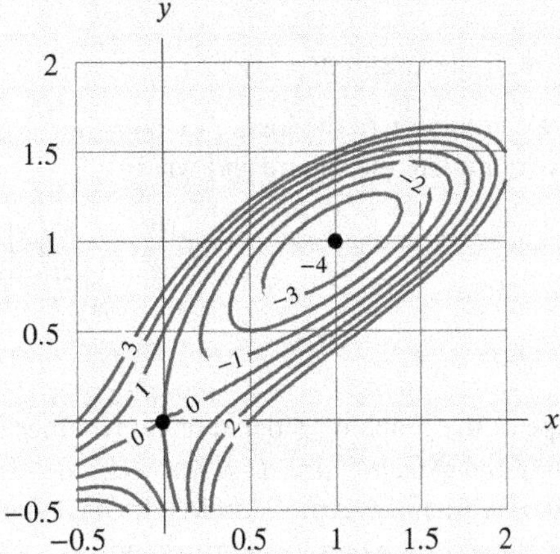

Figure 15.11: Contour diagram of $f(x, y) = 8y^3 + 12x^2 - 24xy$ showing critical points at $(0, 0)$ and $(1, 1)$

Classifying Critical Points

We can see whether a critical point of a function, f, is a local maximum, local minimum, or neither by looking at the contour diagram. There is also an analytic method for making this distinction.

Quadratic Functions of the Form $f(x, y) = ax^2 + bxy + cy^2$

Near most critical points, a function has the same behavior as its quadratic Taylor approximation about that point. Thus, we start by investigating critical points of quadratic functions of the form $f(x, y) = ax^2 + bxy + cy^2$, where a, b and c are constants.

Example 5 Find and analyze the local extrema of the function $f(x, y) = x^2 + xy + y^2$.

Solution To find critical points, we set

$$f_x(x, y) = 2x + y = 0,$$
$$f_y(x, y) = x + 2y = 0.$$

The only critical point is $(0, 0)$, and the value of the function there is $f(0, 0) = 0$. If f is always positive or zero near $(0, 0)$, then $(0, 0)$ is a local minimum; if f is always negative or zero near $(0, 0)$, it is a local maximum; if f takes both positive and negative values, it is neither. The graph in Figure 15.12 suggests that $(0, 0)$ is a local minimum.

How can we be sure that $(0, 0)$ is a local minimum? We complete the square. Writing

$$f(x, y) = x^2 + xy + y^2 = \left(x + \frac{1}{2}y\right)^2 + \frac{3}{4}y^2,$$

shows that $f(x, y)$ is a sum of two nonnegative terms, so it is always greater than or equal to zero. Thus, the critical point is both a local and a global minimum.

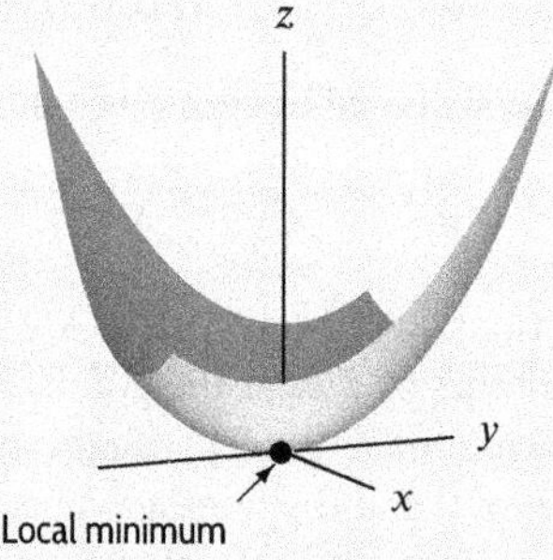

Figure 15.12: Graph of $f(x, y) = x^2 + xy + y^2 = (x + \frac{1}{2}y)^2 + \frac{3}{4}y^2$ showing local minimum at the origin

The Shape of the Graph of $f(x, y) = ax^2 + bxy + cy^2$

In general, a function of the form $f(x, y) = ax^2 + bxy + cy^2$ has one critical point at $(0, 0)$. Assuming $a \neq 0$, we complete the square and write

$$ax^2 + bxy + cy^2 = a\left[x^2 + \frac{b}{a}xy + \frac{c}{a}y^2\right] = a\left[\left(x + \frac{b}{2a}y\right)^2 + \left(\frac{c}{a} - \frac{b^2}{4a^2}\right)y^2\right]$$
$$= a\left[\left(x + \frac{b}{2a}y\right)^2 + \left(\frac{4ac - b^2}{4a^2}\right)y^2\right].$$

The shape of the graph of f depends on whether the coefficient of y^2 is positive, negative, or zero. The sign of the *discriminant*, $D = 4ac - b^2$, determines the sign of the coefficient of y^2.

- If $D > 0$, then the expression inside the square brackets is positive or zero, so the function has a local maximum or a local minimum.
    - If $a > 0$, the function has a local minimum, since the graph is a paraboloid opening upward, like $z = x^2 + y^2$. (See Figure 15.13.)
    - If $a < 0$, the function has a local maximum, since the graph is a paraboloid opening downward, like $z = -x^2 - y^2$. (See Figure 15.14.)
- If $D < 0$, then the function goes up in some directions and goes down in others, like $z = x^2 - y^2$. We say the function has a *saddle point*, that is, a critical point at which the function value increases in some directions but decreases in others. (See Figure 15.15.)
- If $D = 0$, then the quadratic function is $a(x + by/2a)^2$, whose graph is a parabolic cylinder. (See Figure 15.16.)

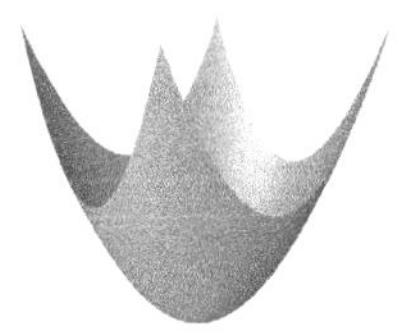
Figure 15.13: Local minimum: $D > 0$ and $a > 0$

Figure 15.14: Local maximum: $D > 0$ and $a < 0$

Figure 15.15: Saddle point: $D < 0$

Figure 15.16: Parabolic cylinder: $D = 0$

More generally, the graph of $g(x, y) = a(x - x_0)^2 + b(x - x_0)(y - y_0) + c(y - y_0)^2$ has the same shape as the graph of $f(x, y) = ax^2 + bxy + cy^2$, except that the critical point is at (x_0, y_0) rather than $(0, 0)$.[1]

Classifying the Critical Points of a Function

Suppose that f is any function with $\text{grad}\, f(0, 0) = \vec{0}$. Its quadratic Taylor polynomial near $(0, 0)$,

$$\begin{aligned} f(x, y) &\approx f(0, 0) + f_x(0, 0)x + f_y(0, 0)y \\ &+ \frac{1}{2} f_{xx}(0, 0)x^2 + f_{xy}(0, 0)xy + \frac{1}{2} f_{yy}(0, 0)y^2, \end{aligned}$$

can be simplified using $f_x(0, 0) = f_y(0, 0) = 0$, which gives

$$f(x, y) - f(0, 0) \approx \frac{1}{2} f_{xx}(0, 0)x^2 + f_{xy}(0, 0)xy + \frac{1}{2} f_{yy}(0, 0)y^2.$$

The discriminant of this quadratic polynomial is

$$D = 4ac - b^2 = 4\left(\frac{1}{2} f_{xx}(0, 0)\right)\left(\frac{1}{2} f_{yy}(0, 0)\right) - \left(f_{xy}(0, 0)\right)^2,$$

which simplifies to

$$D = f_{xx}(0, 0) f_{yy}(0, 0) - (f_{xy}(0, 0))^2.$$

There is a similar formula for D if the critical point is at (x_0, y_0). An analogy with quadratic functions suggests the following test for classifying a critical point of a function of two variables:

Second-Derivative Test for Functions of Two Variables

Suppose (x_0, y_0) is a point where $\text{grad}\, f(x_0, y_0) = \vec{0}$. Let

$$D = f_{xx}(x_0, y_0) f_{yy}(x_0, y_0) - (f_{xy}(x_0, y_0))^2.$$

- If $D > 0$ and $f_{xx}(x_0, y_0) > 0$, then f has a local minimum at (x_0, y_0).
- If $D > 0$ and $f_{xx}(x_0, y_0) < 0$, then f has a local maximum at (x_0, y_0).
- If $D < 0$, then f has a saddle point at (x_0, y_0).
- If $D = 0$, anything can happen: f can have a local maximum, or a local minimum, or a saddle point, or none of these, at (x_0, y_0).

Example 6 Find the local maxima, minima, and saddle points of $f(x, y) = \frac{1}{2}x^2 + 3y^3 + 9y^2 - 3xy + 9y - 9x$.

Solution Setting the partial derivatives of f to zero gives

$$\begin{aligned} f_x(x, y) &= x - 3y - 9 = 0, \\ f_y(x, y) &= 9y^2 + 18y - 3x + 9 = 0. \end{aligned}$$

[1] We assumed that $a \neq 0$. If $a = 0$ and $c \neq 0$, the same argument works. If both $a = 0$ and $c = 0$, then $f(x, y) = bxy$, which has a saddle point.

Eliminating x gives $9y^2 + 9y - 18 = 0$, with solutions $y = -2$ and $y = 1$. The corresponding values of x are $x = 3$ and $x = 12$, so the critical points of f are $(3, -2)$ and $(12, 1)$. The discriminant is

$$D(x, y) = f_{xx}f_{yy} - f_{xy}^2 = (1)(18y + 18) - (-3)^2 = 18y + 9.$$

Since $D(3, -2) = -36 + 9 < 0$, we know that $(3, -2)$ is a saddle point of f. Since $D(12, 1) = 18 + 9 > 0$ and $f_{xx}(12, 1) = 1 > 0$, we know that $(12, 1)$ is a local minimum of f.

The second-derivative test does not give any information if $D = 0$. However, as the following example illustrates, we may still be able to classify the critical points.

Example 7 Classify the critical points of $f(x, y) = x^4 + y^4$, and $g(x, y) = -x^4 - y^4$, and $h(x, y) = x^4 - y^4$.

Solution Each of these functions has a critical point at $(0, 0)$. Since all the second partial derivatives are 0 there, each function has $D = 0$. Near the origin, the graphs of f, g and h look like the surfaces in Figures 15.13–15.15, respectively, so f has a local minimum at $(0, 0)$, and g has a local maximum at $(0, 0)$, and h is saddle-shaped at $(0, 0)$.

We can get the same results algebraically. Since $f(0, 0) = 0$ and $f(x, y) > 0$ elsewhere, f has a local minimum at the origin. Since $g(0, 0) = 0$ and $g(x, y) < 0$ elsewhere, g has a local maximum at the origin. Lastly, h is saddle-shaped at the origin since $h(0, 0) = 0$ and, away from the origin, $h(x, y) > 0$ on the x-axis and $h(x, y) < 0$ on the y-axis.

Exercises and Problems for Section 15.1

EXERCISES

1. Figures (I)–(VI) show level curves of six functions around a critical point P. Does each function have a local maximum, a local minimum, or a saddle point at P?

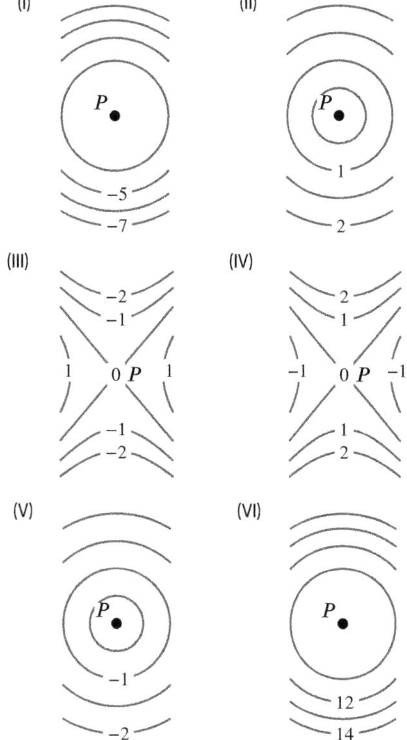

2. Which of the points A, B, C in Figure 15.17 appear to be critical points? Classify those that are critical points.

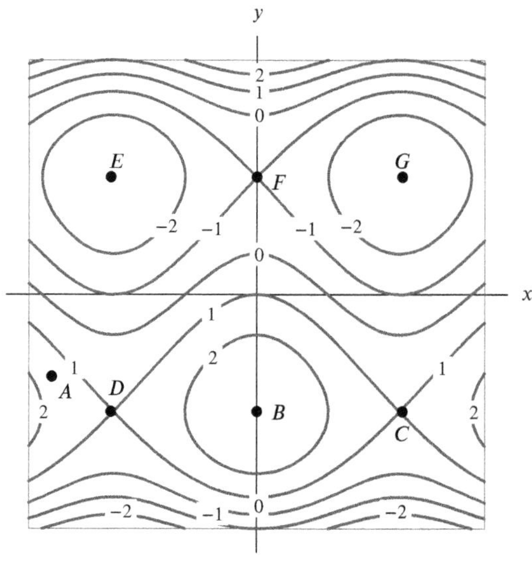

Figure 15.17

3. Which of the points D–G in Figure 15.17 appear to be

(a) Local maxima?
(b) Local minima?
(c) Saddle points?

4. A function $f(x, y)$ has partial derivatives $f_x(1, 2) = 3$, $f_y(1, 2) = 5$. Explain how you know that f does not have a minimum at $(1, 2)$.

5. Assume $f(x, y)$ has a critical point at $(2, 3)$ with $f(2, 3) = 5$. Draw possible cross-sections for $x = 2$ and $y = 3$, and label one value on each axis, if $f(2, 3)$ is:

   (a) A local minimum
   (b) A local maximum
   (c) A saddle point

In Exercises 6–13, the function has a critical point at $(0, 0)$. What sort of critical point is it?

6. $f(x, y) = x^2 - \cos y$
7. $f(x, y) = x \sin y$
8. $g(x, y) = x^4 + y^3$
9. $f(x, y) = x^6 + y^6$
10. $k(x, y) = \sin x \sin y$
11. $h(x, y) = \cos x \cos y$
12. $g(x, y) = (x - e^x)(1 - y^2)$
13. $h(x, y) = x^2 - xy + \sin^2 y$

In Exercises 14–27, find the critical points and classify them as local maxima, local minima, saddle points, or none of these.

14. $f(x, y) = x^2 - 2xy + 3y^2 - 8y$
15. $f(x, y) = 5 + 6x - x^2 + xy - y^2$
16. $f(x, y) = x^2 - y^2 + 4x + 2y$
17. $f(x, y) = 400 - 3x^2 - 4x + 2xy - 5y^2 + 48y$
18. $f(x, y) = 15 - x^2 + 2y^2 + 6x - 8y$
19. $f(x, y) = x^2y + 2y^2 - 2xy + 6$
20. $f(x, y) = 2x^3 - 3x^2y + 6x^2 - 6y^2$
21. $f(x, y) = x^3 - 3x + y^3 - 3y$
22. $f(x, y) = x^3 + y^3 - 3x^2 - 3y + 10$
23. $f(x, y) = x^3 + y^3 - 6y^2 - 3x + 9$
24. $f(x, y) = (x + y)(xy + 1)$
25. $f(x, y) = 8xy - \frac{1}{4}(x + y)^4$
26. $f(x, y) = \sqrt[3]{x^2 + y^2}$
27. $f(x, y) = e^{2x^2+y^2}$

PROBLEMS

28. Let $f(x, y) = 3x^2 + ky^2 + 9xy$. Determine the values of k (if any) for which the critical point at $(0, 0)$ is:

   (a) A saddle point
   (b) A local maximum
   (c) A local minimum

29. Let $f(x, y) = x^3 + ky^2 - 5xy$. Determine the values of k (if any) for which the critical point at $(0, 0)$ is:

   (a) A saddle point
   (b) A local maximum
   (c) A local minimum

30. Find A and B so that $f(x, y) = x^2 + Ax + y^2 + B$ has a local minimum value of 20 at $(1, 0)$.

31. For $f(x, y) = x^2 + xy + y^2 + ax + by + c$, find values of a, b, and c giving a local minimum at $(2, 5)$ and so that $f(2, 5) = 11$.

32. **(a)** Find critical points for $f(x, y) = e^{-(x-a)^2-(y-b)^2}$.
   (b) Find a and b such that the critical point is at $(-1, 5)$.
   (c) For the values of a and b in part (b), is $(-1, 5)$ a local maximum, local minimum, or a saddle point?

33. Let $f(x, y) = kx^2 + y^2 - 4xy$. Determine the values of k (if any) for which the critical point at $(0, 0)$ is:

   (a) A saddle point
   (b) A local maximum
   (c) A local minimum

For Problems 34–36, use the contours of f in Figure 15.18.

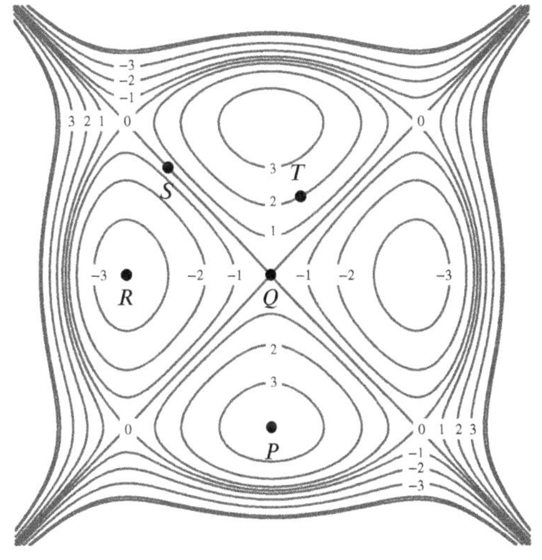

Figure 15.18

34. Decide whether you think each point is a local maximum, local minimum, saddle point, or none of these.

   (a) P **(b)** Q **(c)** R **(d)** S

35. Sketch the direction of ∇f at points surrounding each of P, R, S, and T.

36. At which of P, Q, R, S, or T does $\|\nabla f\|$ seem largest?.

For Problems 37–40, find critical points and classify them as local maxima, local minima, saddle points, or none of these.

37. $f(x, y) = x^3 + e^{-y^2}$

38. $f(x, y) = \sin x \sin y$

39. $f(x, y) = 1 - \cos x + y^2/2$

40. $f(x, y) = e^x(1 - \cos y)$

41. At the point $(1, 3)$, suppose that $f_x = f_y = 0$ and $f_{xx} > 0$, $f_{yy} > 0$, $f_{xy} = 0$.

(a) What can you conclude about the behavior of the function near the point $(1, 3)$?
(b) Sketch a possible contour diagram.

42. At the point (a, b), suppose that $f_x = f_y = 0$, $f_{xx} > 0$, $f_{yy} = 0$, $f_{xy} > 0$.

(a) What can you conclude about the shape of the graph of f near the point (a, b)?
(b) Sketch a possible contour diagram.

43. Let $h(x, y) = f(x)g(y)$ where $f(0) = g(0) = 0$ and $f'(0) \neq 0, g'(0) \neq 0$. Show that $(0, 0)$ is a saddle point of h.

44. Let $h(x, y) = f(x) + g(y)$. Show that h has a critical point at (a, b) if $f'(a) = g'(b) = 0$, and, assuming $f''(a) \neq 0$ and $g''(b) \neq 0$, it is a local maximum or minimum when $f''(a)$ and $g''(b)$ have the same sign and a saddle point when they have opposite signs.

45. Let $h(x, y) = (f(x))^2 + (g(y))^2$. Show that if $f(a) = g(b) = 0$, then (a, b) is a local minimum.

46. Draw a possible contour diagram of f such that $f_x(-1, 0) = 0$, $f_y(-1, 0) < 0$, $f_x(3, 3) > 0$, $f_y(3, 3) > 0$, and f has a local maximum at $(3, -3)$.

47. Draw a possible contour diagram of a function with a saddle point at $(2, 1)$, a local minimum at $(2, 4)$, and no other critical points. Label the contours.

48. For constants a and b with $ab \neq 0$ and $ab \neq 1$, let

$$f(x, y) = ax^2 + by^2 - 2xy - 4x - 6y.$$

(a) Find the x- and y-coordinates of the critical point. Your answer will be in terms of a and b.
(b) If $a = b = 2$, is the critical point a local maximum, a local minimum, or neither? Give a reason for your answer.
(c) Classify the critical point for all values of a and b with $ab \neq 0$ and $ab \neq 1$.

49. (a) Find the critical point of $f(x, y) = (x^2 - y)(x^2 + y)$.
(b) Show that at the critical point, the discriminant $D = 0$, so the second-derivative test gives no information about the nature of the critical point.
(c) Sketch contours near the critical point to determine whether it is a local maximum, a local minimum, a saddle point, or none of these.

50. On a computer, draw contour diagrams for functions

$$f(x, y) = k(x^2 + y^2) - 2xy$$

for $k = -2, -1, 0, 1, 2$. Use these figures to classify the critical point at $(0, 0)$ for each value of k. Explain your observations using the discriminant, D.

51. The behavior of a function can be complicated near a critical point where $D = 0$. Suppose that

$$f(x, y) = x^3 - 3xy^2.$$

Show that there is one critical point at $(0, 0)$ and that $D = 0$ there. Show that the contour for $f(x, y) = 0$ consists of three lines intersecting at the origin and that these lines divide the plane into six regions around the origin where f alternates from positive to negative. Sketch a contour diagram for f near $(0, 0)$. The graph of this function is called a *monkey saddle*.

52. The contour diagrams for four functions $z = f(x, y)$ are in (a)–(d). Each function has a critical point with $z = 0$ at the origin. Graphs (I)–(IV) show the value of z for these four functions on a small circle around the origin, expressed as function of θ, the angle between the positive x-axis and a line through the origin. Match the contour diagrams (a)–(d) with the graphs (I)–(IV). Classify the critical points as local maxima, local minima or saddle points.

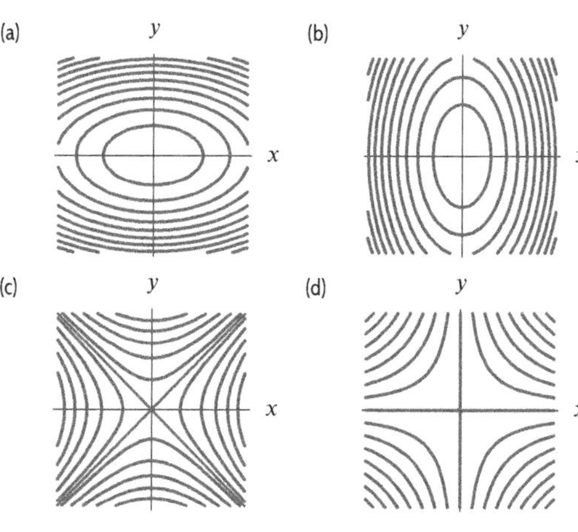

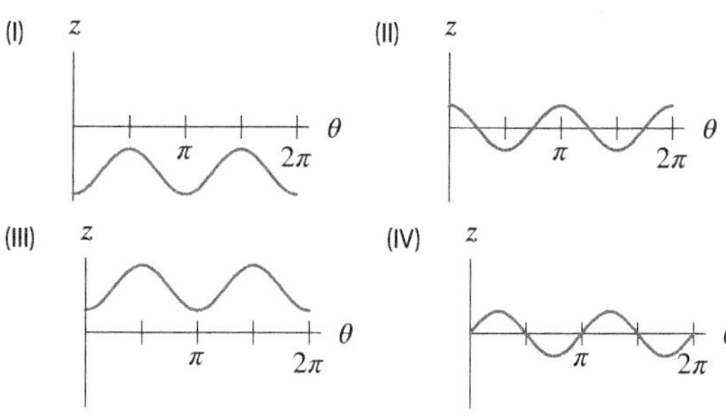

Strengthen Your Understanding

In Problems 53–55, explain what is wrong with the statement.

53. If $f_x = f_y = 0$ at $(1, 3)$, then f has a local maximum or local minimum at $(1, 3)$.

54. For $f(x, y)$, if $D = f_{xx}f_{yy} - (f_{xy})^2 = 0$ at (a, b), then (a, b) is a saddle point.

55. A critical point (a, b) for the function f must be a local minimum if both cross-sections for $x = a$ and $y = b$ are concave up.

In Problems 56–57, give an example of:

56. A nonlinear function having no critical points

57. A function $f(x, y)$ with a local maximum at $(2, -3, 4)$.

Are the statements in Problems 58–69 true or false? Give reasons for your answer.

58. If $f_x(P_0) = f_y(P_0) = 0$, then P_0 is a critical point of f.

59. If $f_x(P_0) = f_y(P_0) = 0$, then P_0 is a local maximum or local minimum of f.

60. If P_0 is a critical point of f, then P_0 is either a local maximum or local minimum of f.

61. If P_0 is a local maximum or local minimum of f, and not on the boundary of the domain of f, then P_0 is a critical point of f.

62. The function whose contour diagram is shown in Figure 15.19 has a saddle point at P.

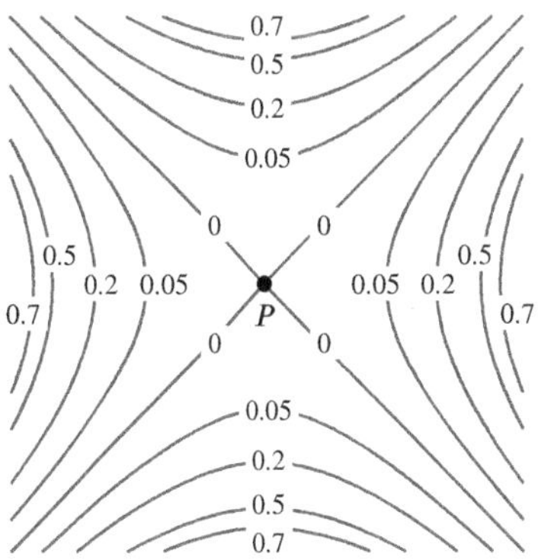

Figure 15.19

63. The function $f(x, y) = \sqrt{x^2 + y^2}$ has a local minimum at the origin.

64. The function $f(x, y) = x^2 - y^2$ has a local minimum at the origin.

65. If f has a local minimum at P_0 then so does the function $g(x, y) = f(x, y) + 5$.

66. If f has a local minimum at P_0 then the function $g(x, y) = -f(x, y)$ has a local maximum at P_0.

67. Every function has at least one local maximum.

68. If P_0 is a local maximum of f, then $f(a, b) \leq f(P_0)$ for all points (a, b) in 2-space.

69. If P_0 is a local maximum of f, then P_0 is also a global maximum of f.

15.2 OPTIMIZATION

Suppose we want to find the highest and the lowest points in Colorado. A contour map is shown in Figure 15.20. The highest point is the top of a mountain peak (point A on the map, Mt. Elbert, 14,440 feet high). What about the lowest point? Colorado does not have large pits without drainage, like Death Valley in California. A drop of rain falling at any point in Colorado will eventually flow out of the state. If there is no local minimum inside the state, where is the lowest point? It must be on the state boundary at a point where a river is flowing out of the state (point B where the Arikaree River leaves the state, 3,315 feet high). The highest point in Colorado is a global maximum for the elevation function in Colorado and the lowest point is the global minimum.

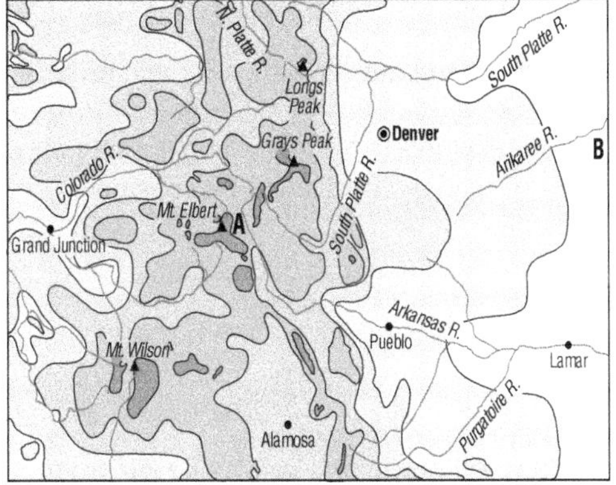

Figure 15.20: The highest and lowest points in the state of Colorado

In general, if we are given a function f defined on a region R, we say:

- f has a **global maximum on** R at the point P_0 if $f(P_0) \geq f(P)$ for all points P in R.
- f has a **global minimum on** R at the point P_0 if $f(P_0) \leq f(P)$ for all points P in R.

The process of finding a global maximum or minimum for a function f on a region R is called *optimization*. If the region R is not stated explicitly, we take it to be the whole xy-plane unless the context of the problem suggests otherwise.

How Do We Find Global Maxima and Minima?

As the Colorado example illustrates, a global extremum can occur either at a critical point inside the region or at a point on the boundary of the region. This is analogous to single-variable calculus, where a function achieves its global extrema on an interval either at a critical point inside the interval or at an endpoint of the interval.

To locate **global maxima and minima** for a function f on a region R:

- Find the critical points of f in the region R.
- Investigate whether the critical points give global maxima or minima.
- If the region R has a boundary, investigate whether f attains a global maximum or minimum on the boundary of R.

Investigating the boundary of a region for possible maxima and minima is the topic of Section 15.1. In this section, we focus on finding global maxima and minima of functions on regions that do not include boundaries.

Not all functions have a global maximum or minimum: it depends on the function and the region. First, we consider applications in which global extrema are expected from practical considerations. At the end of this section, we examine the conditions that lead to global extrema. In general, the fact that a function has a single local maximum or minimum does not guarantee that the point is the global maximum or minimum. (See Problem 38.) An exception is if the function is quadratic, in which case the local maximum or minimum is the global maximum or minimum. (See Example 1 on page 807 and Example 5 on page 810.)

Maximizing Profit and Minimizing Cost

In planning production of an item, a company often chooses the combination of price and quantity that maximizes its profit. We use

$$\text{Profit} = \text{Revenue} - \text{Cost},$$

and, provided the price is constant,

$$\text{Revenue} = \text{Price} \cdot \text{Quantity} = pq.$$

In addition, we need to know how the cost and price depend on quantity.

Example 1 A company manufactures two items which are sold in two separate markets where it has a monopoly. The quantities, q_1 and q_2, demanded by consumers, and the prices, p_1 and p_2 (in dollars), of each item are related by

$$p_1 = 600 - 0.3q_1 \quad \text{and} \quad p_2 = 500 - 0.2q_2.$$

Thus, if the price for either item increases, the demand for it decreases. The company's total production cost is given by

$$C = 16 + 1.2q_1 + 1.5q_2 + 0.2q_1q_2.$$

To maximize its total profit, how much of each product should be produced? What is the maximum profit? [2]

[2] Adapted from M. Rosser and P. Lis, *Basic Mathematics for Economists*, 3rd ed. (New York: Routledge, 2016), p. 351.

Solution The total revenue, R, is the sum of the revenues, p_1q_1 and p_2q_2, from each market. Substituting for p_1 and p_2, we get

$$\begin{aligned} R = p_1q_1 + p_2q_2 &= (600 - 0.3q_1)q_1 + (500 - 0.2q_2)q_2 \\ &= 600q_1 - 0.3q_1^2 + 500q_2 - 0.2q_2^2. \end{aligned}$$

Thus, the total profit P is given by

$$\begin{aligned} P = R - C &= 600q_1 - 0.3q_1^2 + 500q_2 - 0.2q_2^2 - (16 + 1.2q_1 + 1.5q_2 + 0.2q_1q_2) \\ &= -16 + 598.8q_1 - 0.3q_1^2 + 498.5q_2 - 0.2q_2^2 - 0.2q_1q_2. \end{aligned}$$

Since q_1 and q_2 cannot be negative,[3] the region we consider is the first quadrant with boundary $q_1 = 0$ and $q_2 = 0$.

To maximize P, we look for critical points by setting the partial derivatives equal to 0:

$$\frac{\partial P}{\partial q_1} = 598.8 - 0.6q_1 - 0.2q_2 = 0,$$
$$\frac{\partial P}{\partial q_2} = 498.5 - 0.4q_2 - 0.2q_1 = 0.$$

Since grad P is defined everywhere, the only critical points of P are those where grad $P = \vec{0}$. Thus, solving for q_1, and q_2, we find that

$$q_1 = 699.1 \quad \text{and} \quad q_2 = 896.7.$$

The corresponding prices are

$$p_1 = 390.27 \quad \text{and} \quad p_2 = 320.66.$$

To see whether or not we have found a local maximum, we compute second partial derivatives:

$$\frac{\partial^2 P}{\partial q_1^2} = -0.6, \quad \frac{\partial^2 P}{\partial q_2^2} = -0.4, \quad \frac{\partial^2 P}{\partial q_1 \partial q_2} = -0.2,$$

so,

$$D = \frac{\partial^2 P}{\partial q_1^2}\frac{\partial^2 P}{\partial q_2^2} - \left(\frac{\partial^2 P}{\partial q_1 \partial q_2}\right)^2 = (-0.6)(-0.4) - (-0.2)^2 = 0.2.$$

Therefore we have found a local maximum. The graph of P is a paraboloid opening downward, so $(699.1, 896.7)$ is a global maximum. This point is within the region, so points on the boundary give smaller values of P.

The company should produce 699.1 units of the first item priced at \$390.27 per unit, and 896.7 units of the second item priced at \$320.66 per unit. The maximum profit $P(699.1, 896.7) \approx \$433{,}000$.

Example 2 A delivery of 480 cubic meters of gravel is to be made to a landfill. The trucker plans to purchase an open-top box in which to transport the gravel in numerous trips. The total cost to the trucker is the cost of the box plus \$80 per trip. The box must have height 2 meters, but the trucker can choose the length and width. The cost of the box is \$100/m^2 for the ends, \$50/m^2 for the sides and \$200/m^2 for the bottom. Notice the tradeoff: A smaller box is cheaper to buy but requires more trips. What size box should the trucker buy to minimize the total cost?[4]

[3] Restricting prices to be nonnegative further restricts the region but does not alter the solution.
[4] Adapted from Claude McMillan, Jr., *Mathematical Programming*, 2nd ed. (New York: Wiley, 1978), pp. 156–157.

Solution We first get an algebraic expression for the trucker's cost. Let the length of the box be x meters and the width be y meters; the height is 2 meters. (See Figure 15.21.)

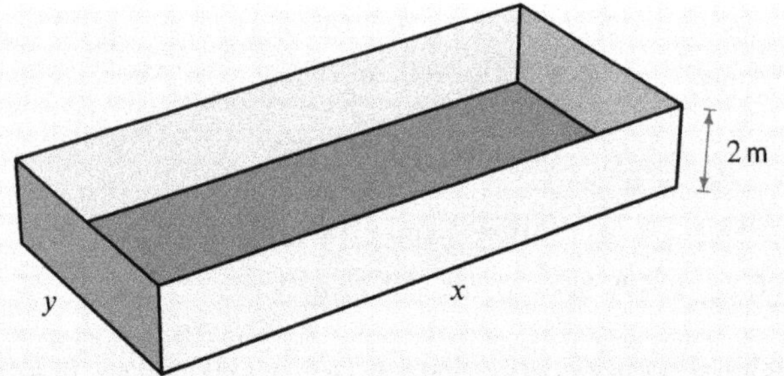

Figure 15.21: The box for transporting gravel

Table 15.2 *Trucker's itemized cost*

Expense	Cost in dollars
Travel: $480/(2xy)$ at \$80/trip	$80 \cdot 480/(2xy)$
Ends: 2 at \$100/m^2 $\cdot\, 2y$ m^2	$400y$
Sides: 2 at \$50/m^2 $\cdot\, 2x$ m^2	$200x$
Bottom: 1 at \$200/m^2 $\cdot\, xy$ m^2	$200xy$

The volume of the box is $2xy$ m^3, so delivery of 480 m^3 of gravel requires $480/(2xy)$ trips. The number of trips is a whole number; however, we treat it as continuous so that we can optimize using derivatives. The trucker's cost is itemized in Table 15.2. The problem is to minimize

$$\text{Total cost} = 80 \cdot \frac{480}{2xy} + 400y + 200x + 200xy = 200\left(\frac{96}{xy} + 2y + x + xy\right).$$

The length and width of the box must be positive. Thus, the region is the first quadrant but it does not contain the boundary, $x = 0$ and $y = 0$.

Our problem is to minimize

$$f(x, y) = \frac{96}{xy} + 2y + x + xy.$$

The critical points of this function occur where

$$f_x(x, y) = -\frac{96}{x^2 y} + 1 + y = 0$$
$$f_y(x, y) = -\frac{96}{xy^2} + 2 + x = 0.$$

We put the $96/(x^2y)$ and $96/(xy^2)$ terms on the other side of the the equation, divide, and simplify:

$$\frac{96/(x^2y)}{96/(xy^2)} = \frac{1+y}{2+x} \quad \text{so} \quad \frac{y}{x} = \frac{1+y}{2+x} \quad \text{giving} \quad 2y = x.$$

Substituting $x = 2y$ in the equation $f_y(x, y) = 0$ gives

$$-\frac{96}{2y \cdot y^2} + 2 + 2y = 0$$
$$y^4 + y^3 - 24 = 0.$$

The only positive solution to this equation is $y = 2$, so the only critical point in the region is $(4, 2)$.

To check that the critical point is a local minimum, we use the second-derivative test. Since

$$D(4, 2) = f_{xx} f_{yy} - (f_{xy})^2 = \frac{192}{4^3 \cdot 2} \cdot \frac{192}{4 \cdot 2^3} - \left(\frac{96}{4^2 \cdot 2^2} + 1\right)^2 = 9 - \frac{25}{4} > 0$$

and $f_{xx}(4, 2) > 0$, the point $(4, 2)$ is a local minimum. Since the value of f increases without bound as x or y increases without bound and as $x \to 0^+$ and $y \to 0^+$, it can be shown that $(4, 2)$ is a global minimum. (See Problem 42.) Thus, the optimal box is 4 meters long and 2 meters wide. With a box of this size, the trucker would need to make 30 trips to haul all of the gravel. This large number lends some credibility to our decision to treat the number of trips as a continuous variable.

Fitting a Line to Data: Least Squares

Suppose we want to fit the "best" line to some data in the plane. We measure the distance from a line to the data points by adding the squares of the vertical distances from each point to the line. The smaller this sum of squares is, the better the line fits the data. The line with the minimum sum of square distances is called the *least squares line*, or the *regression line*. If the data is nearly linear, the least squares line is a good fit; otherwise it may not be. (See Figure 15.22.)

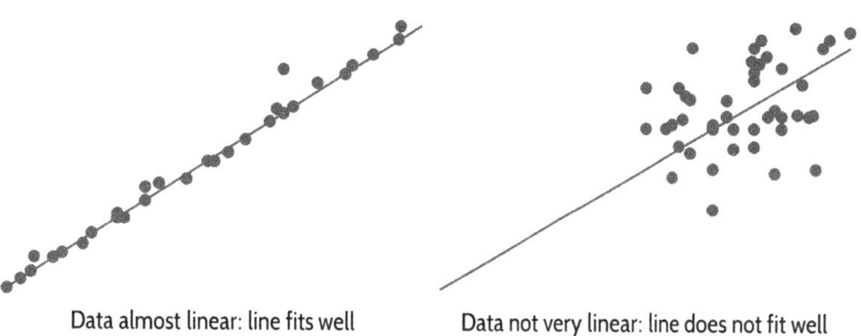

Figure 15.22: Fitting lines to data points

Example 3 Find a least squares line for the following data points: (1, 1), (2, 1), and (3, 3).

Solution Suppose the line has equation $y = b + mx$. If we find b and m then we have found the line. So, for this problem, b and m are the two variables. Any values of m and b are possible, so this is an unconstrained problem. We want to minimize the function $f(b, m)$ that gives the sum of the three squared vertical distances from the points to the line in Figure 15.23.

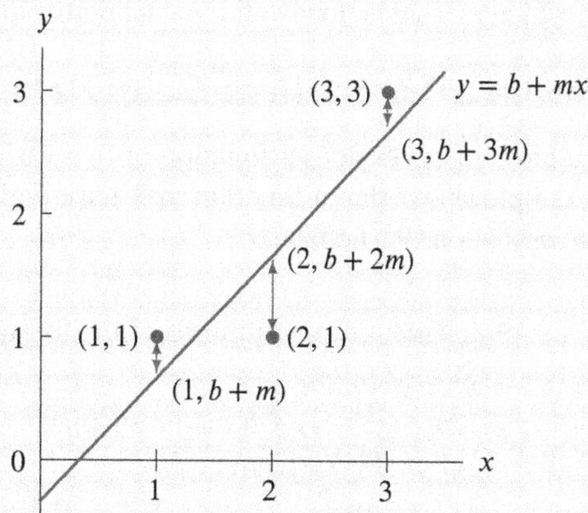

Figure 15.23: The least squares line minimizes the sum of the squares of these vertical distances

The vertical distance from the point (1, 1) to the line is the difference in the y-coordinates $1 - (b + m)$; similarly for the other points. Thus, the sum of squares is

$$f(b, m) = (1 - (b + m))^2 + (1 - (b + 2m))^2 + (3 - (b + 3m))^2.$$

To minimize f we look for critical points. First we differentiate f with respect to b:

$$\begin{aligned}\frac{\partial f}{\partial b} &= -2(1 - (b + m)) - 2(1 - (b + 2m)) - 2(3 - (b + 3m))\\ &= -2 + 2b + 2m - 2 + 2b + 4m - 6 + 2b + 6m\\ &= -10 + 6b + 12m.\end{aligned}$$

Now we differentiate with respect to m:

$$\begin{aligned}\frac{\partial f}{\partial m} &= 2(1-(b+m))(-1)+2(1-(b+2m))(-2)+2(3-(b+3m))(-3)\\ &= -2+2b+2m-4+4b+8m-18+6b+18m\\ &= -24+12b+28m.\end{aligned}$$

The equations $\frac{\partial f}{\partial b}=0$ and $\frac{\partial f}{\partial m}=0$ give a system of two linear equations in two unknowns:

$$\begin{aligned}-10+6b+12m&=0,\\ -24+12b+28m&=0.\end{aligned}$$

The solution to this pair of equations is the critical point $b=-1/3$ and $m=1$. Since

$$D=f_{bb}f_{mm}-(f_{mb})^2=(6)(28)-12^2=24 \quad\text{and}\quad f_{bb}=6>0,$$

we have found a local minimum. The graph of $f(b,m)$ is a parabola opening upward, so the local minimum is the global minimum of f. Thus, the least squares line is

$$y=x-\frac{1}{3}.$$

As a check, notice that the line $y=x$ passes through the points $(1,1)$ and $(3,3)$. It is reasonable that introducing the point $(2,1)$ moves the y-intercept down from 0 to $-1/3$.

The general formulas for the slope and y-intercept of a least squares line are in Project 2 (available online). Many calculators have built-in formulas for b and m, as well as for the *correlation coefficient*, which measures how well the data points fit the least squares line.

How Do We Know Whether a Function Has a Global Maximum or Minimum?

Under what circumstances does a function of two variables have a global maximum or minimum? The next example shows that a function may have both a global maximum and a global minimum on a region, or just one, or neither.

Example 4 Investigate the global maxima and minima of the following functions:

(a) $h(x,y)=1+x^2+y^2$ on the disk $x^2+y^2\le 1$.
(b) $f(x,y)=x^2-2x+y^2-4y+5$ on the xy-plane.
(c) $g(x,y)=x^2-y^2$ on the xy-plane.

Solution

(a) The graph of $h(x,y)=1+x^2+y^2$ is a bowl-shaped paraboloid with a global minimum of 1 at $(0,0)$, and a global maximum of 2 on the edge of the region, $x^2+y^2=1$.
(b) The graph of f in Figure 15.5 on page 807 shows that f has a global minimum at the point $(1,2)$ and no global maximum (because the value of f increases without bound as $x\to\infty$, $y\to\infty$).
(c) The graph of g in Figure 15.7 on page 808 shows that g has no global maximum because $g(x,y)\to\infty$ as $x\to\infty$ if y is constant. Similarly, g has no global minimum because $g(x,y)\to-\infty$ as $y\to\infty$ if x is constant.

Sometimes a function is guaranteed to have a global maximum and minimum. For example, a continuous function, $h(x)$, of one variable has a global maximum and minimum on every closed interval $a\le x\le b$. On a non-closed interval, such as $a\le x<b$ or $a<x<b$, or on an unbounded interval, such as $a<x<\infty$, h may not have a maximum or minimum value.

What is the situation for functions of two variables? As it turns out, a similar result is true for continuous functions defined on regions which are closed and bounded, analogous to the closed and bounded interval $a \leq x \leq b$. In everyday language we say

- A **closed** region is one which contains its boundary;
- A **bounded** region is one which does not stretch to infinity in any direction.

More precise definitions follow. Suppose R is a region in 2-space. A point (x_0, y_0) is a *boundary point* of R if, for every $r > 0$, the circular disk with center (x_0, y_0) and radius r contains both points which are in R and points which are not in R. See Figure 15.24. A point (x_0, y_0) can be a boundary point of the region R without belonging to R. The collection of all the boundary points is the *boundary* of R. The region R is *closed* if it contains its boundary.

A region R in 2-space is *bounded* if the distance between every point (x, y) in R and the origin is less than some constant K. Closed and bounded regions in 3-space are defined in the same way.

Example 5

(a) Consider the square $-1 \leq x \leq 1, -1 \leq y \leq 1$. Every point in this region is within distance $\sqrt{2}$ of the origin, so the region is bounded. The region's boundary consists of four line segments, all of which belong to the region, so the region is closed.

(b) Consider the first quadrant $x \geq 0, y \geq 0$. The boundary of this region consists of the origin, the positive x-axis, and the positive y-axis. All of these belong to the region, so the region is closed. However, the region is not bounded, since there is no upper bound on distances between points in the region and the origin.

(c) The disk $x^2 + y^2 < 1$ is bounded, because each point in the region is within distance 1 of the origin. However, the disk is not closed, because $(1, 0)$ is a boundary point of the region but not included in the region.

(d) The half-plane $y > 0$ is neither closed nor bounded. The origin is a boundary point of this region but is not included in the region.

The reason that closed and bounded regions are useful is the following theorem, which is also true for functions of three or more variables:[5]

Theorem 15.1: Extreme Value Theorem for Multivariable Functions

If f is a continuous function on a closed and bounded region R, then f has a global maximum at some point (x_0, y_0) in R and a global minimum at some point (x_1, y_1) in R.

If f is not continuous or the region R is not closed and bounded, there is no guarantee that f achieves a global maximum or global minimum on R. In Example 4, the function g is continuous but does not achieve a global maximum or minimum in 2-space, a region which is closed but not bounded. Example 6 illustrates what can go wrong when the region is bounded but not closed.

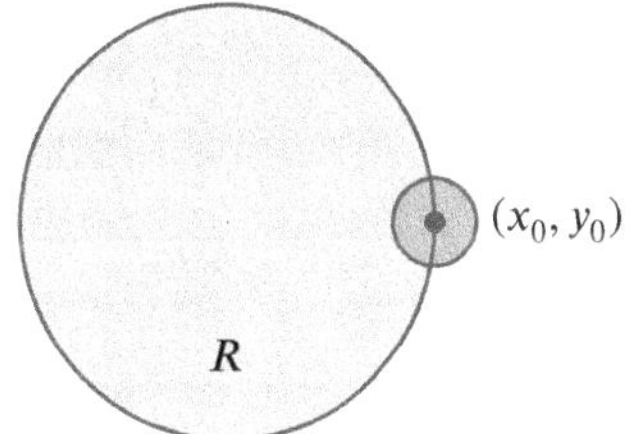

Figure 15.24: Boundary point (x_0, y_0) of R

[5]For a proof, see Walter Rudin, *Principles of Mathematical Analysis*, 3rd ed. (New York: McGraw-Hill, 1976), p. 89.

Example 6 Does the function f have a global maximum or minimum on the region R given by $0 < x^2 + y^2 \le 1$?

$$f(x, y) = \frac{1}{x^2 + y^2}$$

Solution The region R is bounded, but it is not closed since it does not contain the boundary point $(0, 0)$. We see from the graph of $z = f(x, y)$ in Figure 15.25 that f has a global minimum on the circle $x^2 + y^2 = 1$. However, $f(x, y) \to \infty$ as $(x, y) \to (0, 0)$, so f has no global maximum.

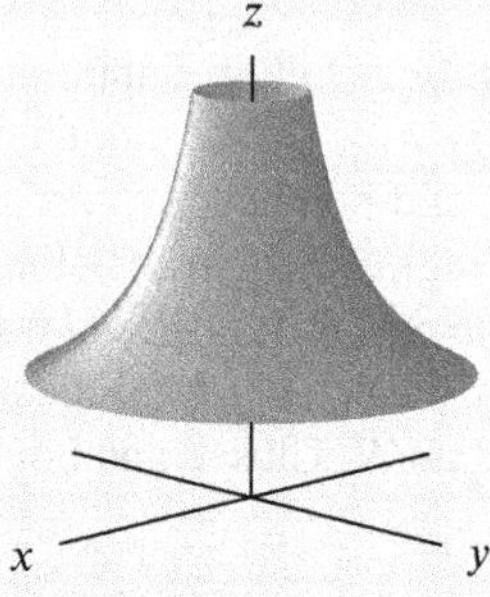

Figure 15.25: Graph showing $f(x, y) = \frac{1}{x^2+y^2}$ has no global maximum on $0 < x^2 + y^2 \le 1$

Exercises and Problems for Section 15.2

EXERCISES

1. By looking at the weather map in Figure 12.1 on page 652, find the maximum and minimum daily high temperatures in the states of Mississippi, Alabama, Pennsylvania, New York, California, Arizona, and Massachusetts.

In Exercises 2–4, estimate the position and approximate value of the global maxima and minima on the closed region shown.

2.

3.

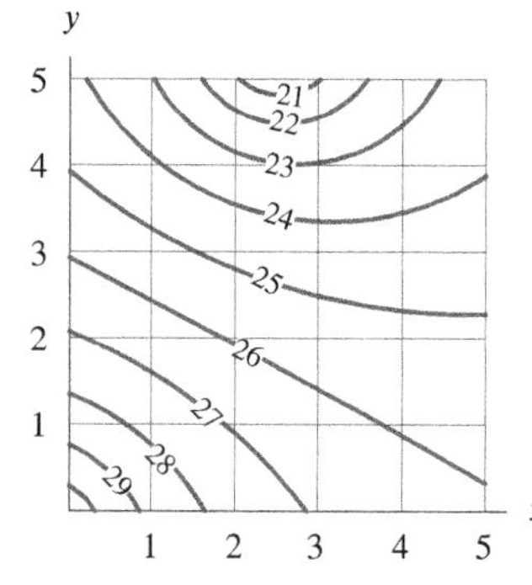

4.

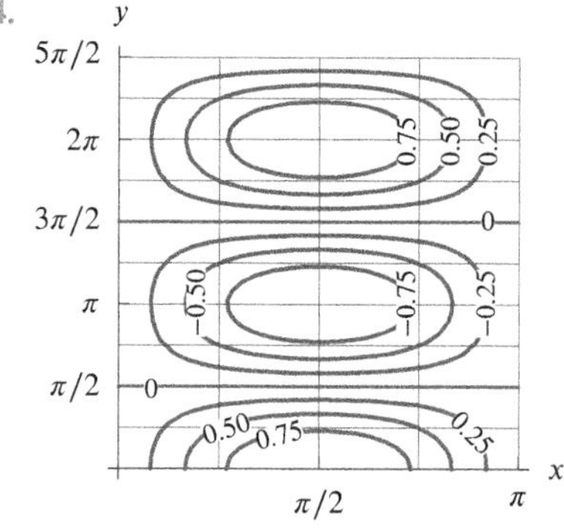

In Problems 5–9, without calculus, find the highest and lowest points (if they exist) on the surface. The z-axis is upward.

5. $x^2 + y^2 + (z - 1)^2 = 49$

6. $(x + 1)^2 + (y - 3)^2 + 2z^2 = 162$

7. $z = (x - 5)^2 + (y - \pi)^2 + 2\pi$

8. $z = 44 - 2x^2 - 2y^2$

9. $x = 4 + y^2 + 2z^2$

10. The surface $z = 27 - x^2 - y^2$ cuts the plane $z = 2$ in a curve. Without calculus, find the point on this curve with the greatest y-coordinate.

In Exercises 11–13, find the global maximum and minimum of the function on $-1 \le x \le 1, -1 \le y \le 1$, and say whether it occurs on the boundary of the square. [Hint: Use graphs.]

11. $z = x^2 + y^2$ 12. $z = -x^2 - y^2$ 13. $z = x^2 - y^2$

In Problems 14–21, does the function have a global maximum? A global minimum?

14. $f(x, y) = x^2 - 2y^2$

15. $g(x, y) = x^2y^2$

16. $h(x, y) = x^3 + y^3$

17. $f(x, y) = -2x^2 - 7y^2$

18. $f(x, y) = e^{x^2+y^2}$

19. $h(x, y) = 1 - y^2e^{xy}$

20. $f(x, y) = x^2/2 + 3y^3 + 9y^2 - 3x$

21. $g(x, y) = x^2 - \cos(x + y)$

PROBLEMS

22. (a) Compute and classify the critical points of $f(x, y) = 2x^2 - 3xy + 8y^2 + x - y$.
    (b) By completing the square, plot the contour diagram of f and show that the local extremum found in part (a) is a global one.

In Problems 23–26, find and classify the critical points of the function. Then decide whether the function has global extrema on the xy-plane, and find them if they exist.

23. $f(x, y) = y^2 + 2xy - y - x^3 - x + 2$
24. $f(x, y) = 2x^2 - 4xy + 5y^2 + 9y + 2$
25. $f(x, y) = -x^4 + 2xy^2 - 2y^3$
26. $f(x, y) = xe^{-x^2-y^2}$
27. A closed rectangular box has volume 32 cm^3. What are the lengths of the edges giving the minimum surface area?
28. A closed rectangular box with faces parallel to the coordinate planes has one bottom corner at the origin and the opposite top corner in the first octant on the plane $3x + 2y + z = 1$. What is the maximum volume of such a box?
29. An international airline has a regulation that each passenger can carry a suitcase having the sum of its width, length and height less than or equal to 135 cm. Find the dimensions of the suitcase of maximum volume that a passenger may carry under this regulation.
30. Design a rectangular milk carton box of width w, length l, and height h which holds 512 cm^3 of milk. The sides of the box cost 1 cent/cm^2 and the top and bottom cost 2 cent/cm^2. Find the dimensions of the box that minimize the total cost of materials used.
31. Find the point on the plane $3x+2y+z = 1$ that is closest to the origin by minimizing the square of the distance.
32. What is the shortest distance from the surface $xy+3x+z^2 = 9$ to the origin?
33. For constants a, b, and c, let $f(x, y) = ax + by + c$ be a linear function, and let R be a region in the xy-plane.
    (a) If R is any disk, show that the maximum and minimum values of f on R occur on the boundary of the disk.
    (b) If R is any rectangle, show that the maximum and minimum values of f on R occur at the corners of the rectangle. They may occur at other points of the rectangle as well.
    (c) Use a graph of the plane $z = f(x, y)$ to explain your answers in parts (a) and (b).
34. Two products are manufactured in quantities q_1 and q_2 and sold at prices of p_1 and p_2, respectively. The cost of producing them is given by
$$C = 2q_1^2 + 2q_2^2 + 10.$$
    (a) Find the maximum profit that can be made, assuming the prices are fixed.
    (b) Find the rate of change of that maximum profit as p_1 increases.
35. A company operates two plants which manufacture the same item and whose total cost functions are
$$C_1 = 8.5 + 0.03q_1^2 \quad \text{and} \quad C_2 = 5.2 + 0.04q_2^2,$$
where q_1 and q_2 are the quantities produced by each plant. The company is a monopoly. The total quantity demanded, $q = q_1 + q_2$, is related to the price, p, by
$$p = 60 - 0.04q.$$
How much should each plant produce in order to maximize the company's profit?[6]
36. The quantity of a product demanded by consumers is a function of its price. The quantity of one product demanded may also depend on the price of other products. For example, if the only chocolate shop in town (a monopoly) sells milk and dark chocolates, the price it sets for each affects the demand of the other. The quantities demanded, q_1 and q_2, of two products depend on their prices, p_1 and p_2, as follows:
$$q_1 = 150 - 2p_1 - p_2$$
$$q_2 = 200 - p_1 - 3p_2.$$
    (a) What does the fact that the coefficients of p_1 and p_2 are negative tell you? Give an example of two products that might be related this way.
    (b) If one manufacturer sells both products, how should the prices be set to generate the maximum possible revenue? What is that maximum possible revenue?
37. A company manufactures a product which requires capital and labor to produce. The quantity, Q, of the product manufactured is given by the Cobb-Douglas function
$$Q = AK^aL^b,$$
where K is the quantity of capital; L is the quantity of labor used; and A, a, and b are positive constants with $0 < a < 1$ and $0 < b < 1$. One unit of capital costs \$$k$ and one unit of labor costs \$$\ell$. The price of the product is fixed at \$$p$ per unit.
    (a) If $a + b < 1$, how much capital and labor should the company use to maximize its profit?
    (b) Is there a maximum profit in the case $a + b = 1$? What about $a + b \geq 1$? Explain.

[6] Adapted from M. Rosser and P. Lis, *Basic Mathematics for Economists*, 3rd ed. (New York: Routledge, 2016), p. 354.

38. Let $f(x, y) = x^2(y + 1)^3 + y^2$. Show that f has only one critical point, namely $(0, 0)$, and that point is a local minimum but not a global minimum. Contrast this with the case of a function with a single local minimum in one-variable calculus.

39. Find the parabola of the form $y = ax^2 + b$ which best fits the points $(1, 0)$, $(2, 2)$, $(3, 4)$ by minimizing the sum of squares, S, given by

$$S = (a + b)^2 + (4a + b - 2)^2 + (9a + b - 4)^2.$$

40. For the data points $(11, 16)$, $(12, 17)$, $(13, 17)$, and $(16, 20)$, find an expression for $f(b, m)$, the sum of squared errors that are minimized on the least squares line $y = b + mx$. (You need not do the minimization.)

41. Find the least squares line for the data points $(0, 4)$, $(1, 3)$, $(2, 1)$.

42. Let $f(x, y) = 80/(xy) + 20y + 10x + 10xy$ in the region R where $x, y > 0$.

(a) Explain why $f(x, y) > f(2, 1)$ at every point in R where

(i) $x > 20$ (ii) $y > 20$

(iii) $x < 0.01$ and $y \leq 20$

(iv) $y < 0.01$ and $x \leq 20$

(b) Explain why f must have a global minimum at a critical point in R.

(c) Explain why f must have a global minimum in R at the point $(2, 1)$.

43. Let $f(x, y) = 2/x + 3/y + 4x + 5y$ in the region R where $x, y > 0$.

(a) Explain why f must have a global minimum at some point in R.

(b) Find the global minimum.

44. **(a)** The energy, E, required to compress a gas from a fixed initial pressure P_0 to a fixed final pressure P_F through an intermediate pressure p is[7]

$$E = \left(\frac{p}{P_0}\right)^2 + \left(\frac{P_F}{p}\right)^2 - 1.$$

How should p be chosen to minimize the energy?

(b) Now suppose the compression takes place in two stages with two intermediate pressures, p_1 and p_2. What choices of p_1 and p_2 minimize the energy if

$$E = \left(\frac{p_1}{P_0}\right)^2 + \left(\frac{p_2}{p_1}\right)^2 + \left(\frac{P_F}{p_2}\right)^2 - 2?$$

45. The Dorfman-Steiner rule shows how a company which has a monopoly should set the price, p, of its product and how much advertising, a, it should buy. The price of advertising is p_a per unit. The quantity, q, of the product sold is given by $q = Kp^{-E}a^{\theta}$, where $K > 0$, $E > 1$, and $0 < \theta < 1$ are constants. The cost to the company to make each item is c.

(a) How does the quantity sold, q, change if the price, p, increases? If the quantity of advertising, a, increases?

(b) Show that the partial derivatives can be written in the form $\partial q/\partial p = -Eq/p$ and $\partial q/\partial a = \theta q/a$.

(c) Explain why profit, π, is given by $\pi = pq - cq - p_a a$.

(d) If the company wants to maximize profit, what must be true of the partial derivatives, $\partial \pi/\partial p$ and $\partial \pi/\partial a$?

(e) Find $\partial \pi/\partial p$ and $\partial \pi/\partial a$.

(f) Use your answers to parts (d) and (e) to show that at maximum profit,

$$\frac{p - c}{p} = \frac{1}{E} \quad \text{and} \quad \frac{p - c}{p_a} = \frac{a}{\theta q}.$$

(g) By dividing your answers in part (f), show that at maximum profit,

$$\frac{p_a a}{pq} = \frac{\theta}{E}.$$

This is the Dorfman-Steiner rule, that the ratio of the advertising budget to revenue does not depend on the price of advertising.

Strengthen Your Understanding

In Problems 46–48, explain what is wrong with the statement.

46. A function having no critical points in a region R cannot have a global maximum in the region.

47. No continuous function has a global minimum on an unbounded region R.

48. If $f(x, y)$ has a local maximum value of 1 at the origin, then the global maximum is 1.

In Problems 49–50, give an example of:

49. A continuous function $f(x, y)$ that has no global maximum and no global minimum on the xy-plane.

50. A function $f(x, y)$ and a region R such that the maximum value of f on R is on the boundary of R.

Are the statements in Problems 51–59 true or false? Give reasons for your answer.

51. If P_0 is a global maximum of f, where f is defined on all of 2-space, then P_0 is also a local maximum of f.

52. Every function has a global maximum.

[7]Adapted from Aris Rutherford, *Discrete Dynamic Programming*, p. 35 (New York: Blaisdell, 1964).

53. The region consisting of all points (x, y) satisfying $x^2 + y^2 < 1$ is bounded.
54. The region consisting of all points (x, y) satisfying $x^2 + y^2 < 1$ is closed.
55. The function $f(x, y) = x^2 + y^2$ has a global minimum on the region $x^2 + y^2 < 1$.
56. The function $f(x, y) = x^2 + y^2$ has a global maximum on the region $x^2 + y^2 < 1$.
57. If P and Q are two distinct points in 2-space, and f has a global maximum at P, then f cannot have a global maximum at Q.
58. The function $f(x, y) = \sin(1 + e^{xy})$ must have a global minimum in the square region $0 \le x \le 1, 0 \le y \le 1$.
59. If P_0 is a global minimum of f on a closed and bounded region, then P_0 need not be a critical point of f.

15.3 CONSTRAINED OPTIMIZATION: LAGRANGE MULTIPLIERS

Many, perhaps most, real optimization problems are constrained by external circumstances. For example, a city wanting to build a public transportation system that will serve the greatest possible number of people has only a limited number of tax dollars it can spend on the project. In this section, we see how to find an optimum value under such constraints.

In Section 15.2, we saw how to optimize a function $f(x, y)$ on a region R. If the region R is the entire xy-plane, we have *unconstrained optimization*; if the region R is not the entire xy-plane, that is, if x or y is restricted in some way, then we have *constrained optimization*.

Graphical Approach: Maximizing Production Subject to a Budget Constraint

Suppose we want to maximize production under a budget constraint. Suppose production, f, is a function of two variables, x and y, which are quantities of two raw materials, and that

$$f(x, y) = x^{2/3}y^{1/3}.$$

If x and y are purchased at prices of p_1 and p_2 thousands of dollars per unit, what is the maximum production f that can be obtained with a budget of c thousand dollars?

To maximize f without regard to the budget, we simply increase x and y. However, the budget constraint prevents us from increasing x and y beyond a certain point. Exactly how does the budget constrain us? With prices of p_1 and p_2, the amount spent on x is p_1x and the amount spent on y is p_2y, so we must have

$$g(x, y) = p_1x + p_2y \le c,$$

where $g(x, y)$ is the total cost of the raw materials and c is the budget in thousands of dollars.

Let's look at the case when $p_1 = p_2 = 1$ and $c = 3.78$. Then

$$x + y \le 3.78.$$

Figure 15.26 shows some contours of f and the budget constraint represented by the line $x + y = 3.78$. Any point on or below the line represents a pair of values of x and y that we can afford. A point on the line completely exhausts the budget, while a point below the line represents values of x and y which can be bought without using up the budget. Any point above the line represents a pair of values that we cannot afford.

To maximize f, we find the point which lies on the level curve with the largest possible value of f *and* which lies within the budget. The point must lie on the budget constraint because production is maximized when we spend all the available money. Unless we are at a point where the budget constraint is tangent to a contour of f, we can increase f by moving in some direction along the line representing the budget constraint in Figure 15.26. For example, if we are on the line to the left of the point of tangency, moving right on the constraint will increase f; if we are on the line to the right of the point of tangency, moving left will increase f. Thus, the maximum value of f on the budget constraint occurs at the point where the budget constraint is tangent to the contour $f = 2$.

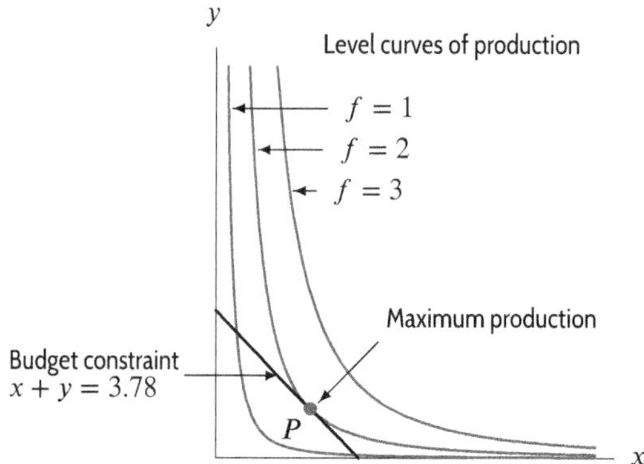

Figure 15.26: Optimal point, P, where budget constraint is tangent to a level of production function

Analytical Solution: Lagrange Multipliers

Figure 15.26 suggests that maximum production is achieved at the point where the budget constraint is tangent to a level curve of the production function. The method of Lagrange multipliers uses this fact in algebraic form. Figure 15.27 shows that at the optimum point, P, the gradient of f and the normal to the budget line $g(x, y) = x + y = 3.78$ are parallel. Thus, at P, grad f and grad g are parallel, so for some scalar λ, called the *Lagrange multiplier,*

$$\operatorname{grad} f = \lambda \operatorname{grad} g.$$

Calculating the gradients, we find that

$$\left(\frac{2}{3}x^{-1/3}y^{1/3}\right)\vec{i} + \left(\frac{1}{3}x^{2/3}y^{-2/3}\right)\vec{j} = \lambda\left(\vec{i} + \vec{j}\right).$$

Equating components gives

$$\frac{2}{3}x^{-1/3}y^{1/3} = \lambda \quad \text{and} \quad \frac{1}{3}x^{2/3}y^{-2/3} = \lambda.$$

Eliminating λ gives

$$\frac{2}{3}x^{-1/3}y^{1/3} = \frac{1}{3}x^{2/3}y^{-2/3}, \quad \text{which leads to} \quad 2y = x.$$

Since the constraint $x + y = 3.78$ must be satisfied, we have $x = 2.52$ and $y = 1.26$. Then

$$f(2.52, 1.26) = (2.52)^{2/3}(1.26)^{1/3} \approx 2.$$

As before, we see that the maximum value of f is approximately 2. Thus, to maximize production on a budget of \$3780, we should use 2.52 units of one raw material and 1.26 units of the other.

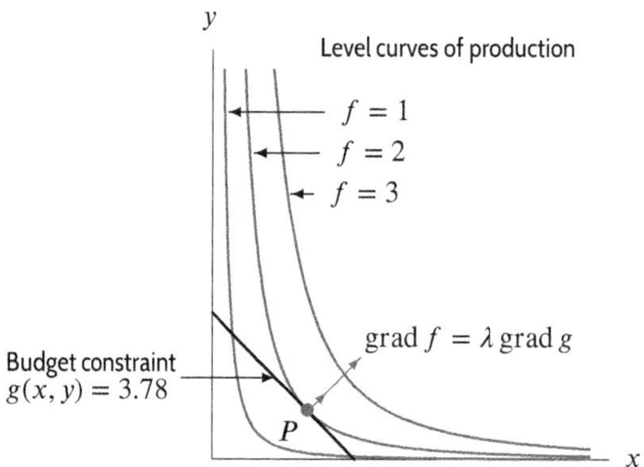

Figure 15.27: At the point, P, of maximum production, the vectors grad f and grad g are parallel

Lagrange Multipliers in General

Suppose we want to optimize an *objective function* $f(x, y)$ subject to a *constraint* $g(x, y) = c$. We look for extrema among the points which satisfy the constraint. We make the following definition.

> Suppose P_0 is a point satisfying the constraint $g(x, y) = c$.
> - f has a **local maximum** at P_0 **subject to the constraint** if $f(P_0) \geq f(P)$ for all points P near P_0 satisfying the constraint.
> - f has a **global maximum** at P_0 **subject to the constraint** if $f(P_0) \geq f(P)$ for all points P satisfying the constraint.
>
> Local and global minima are defined similarly.

As we saw in the production example, constrained extrema occur at points of tangency of contours of f and g; they can also occur at endpoints of constraints. At a point of tangency, grad f is perpendicular to the constraint and so parallel to grad g. At interior points on the constraint where grad f is not perpendicular to the constraint, the value of f can be increased or decreased by moving along the constraint. Therefore constrained extrema occur only at points where grad f and grad g are parallel or at endpoints of the constraint. (See Figure 15.28.) At points where the gradients are parallel, provided grad $g \neq \vec{0}$, there is a constant λ such that grad $f = \lambda$ grad g.

> **Optimizing f Subject to the Constraint $g = c$:**
> If a smooth function, f, has a maximum or minimum subject to a smooth constraint $g = c$ at a point P_0, then either P_0 satisfies the equations
>
> $$\text{grad}\, f = \lambda\, \text{grad}\, g \quad \text{and} \quad g = c,$$
>
> or P_0 is an endpoint of the constraint, or grad $g(P_0) = \vec{0}$. To investigate whether P_0 is a global maximum or minimum, compare values of f at the points satisfying these three conditions. The number λ is called the **Lagrange multiplier**.

If the set of points satisfying the constraint is closed and bounded, such as a circle or line segment, then there must be a global maximum and minimum of f subject to the constraint. If the constraint is not closed and bounded, such as a line or hyperbola, then there may or may not be a global maximum and minimum.

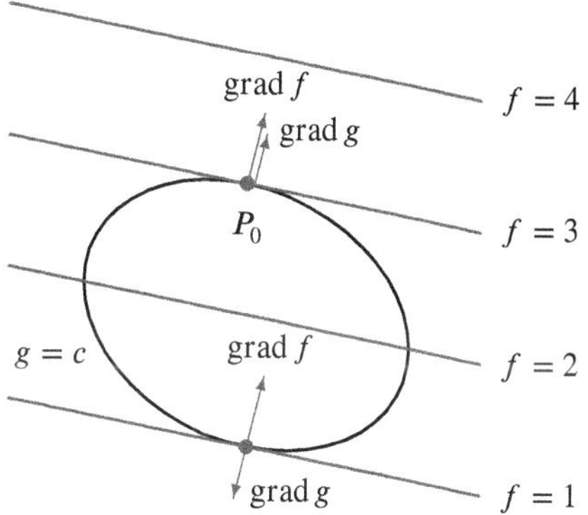

Figure 15.28: Maximum and minimum values of $f(x, y)$ on $g(x, y) = c$ are at points where grad f is parallel to grad g

Example 1 Find the maximum and minimum values of $x + y$ on the circle $x^2 + y^2 = 4$.

Solution The objective function is

$$f(x, y) = x + y,$$

and the constraint is

$$g(x, y) = x^2 + y^2 = 4.$$

Since $\operatorname{grad} f = f_x\vec{i} + f_y\vec{j} = \vec{i} + \vec{j}$ and $\operatorname{grad} g = g_x\vec{i} + g_y\vec{j} = 2x\vec{i} + 2y\vec{j}$, the condition $\operatorname{grad} f = \lambda \operatorname{grad} g$ gives

$$1 = 2\lambda x \quad \text{and} \quad 1 = 2\lambda y,$$

so

$$x = y.$$

We also know that

$$x^2 + y^2 = 4,$$

giving $x = y = \sqrt{2}$ or $x = y = -\sqrt{2}$. The constraint has no endpoints (it's a circle) and $\operatorname{grad} g \neq \vec{0}$ on the circle, so we compare values of f at $(\sqrt{2}, \sqrt{2})$ and $(-\sqrt{2}, -\sqrt{2})$. Since $f(x, y) = x + y$, the maximum value of f is $f(\sqrt{2}, \sqrt{2}) = 2\sqrt{2}$; the minimum value is $f(-\sqrt{2}, -\sqrt{2}) = -2\sqrt{2}$. (See Figure 15.29.)

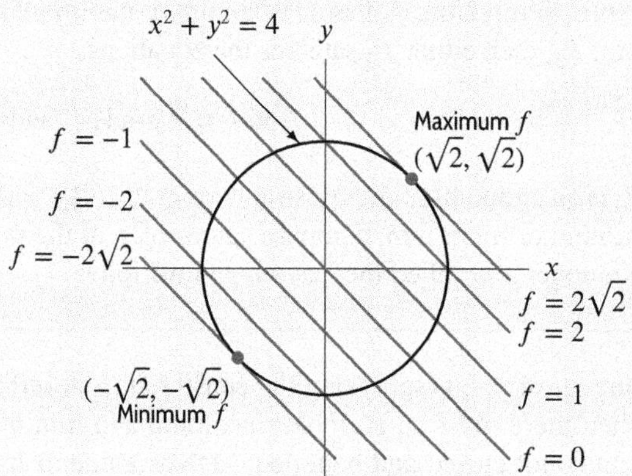

Figure 15.29: Maximum and minimum values of $f(x, y) = x + y$ on the circle $x^2 + y^2 = 4$ are at points where contours of f are tangent to the circle

How to Distinguish Maxima from Minima

There is a second-derivative test[8] for classifying the critical points of constrained optimization problems, but it is more complicated than the test in Section 15.1. However, a graph of the constraint and some contours usually shows which points are maxima, which points are minima, and which are neither.

Optimization with Inequality Constraints

The production problem that we looked at first was to maximize production $f(x, y)$ subject to a budget constraint

$$g(x, y) = p_1x + p_2y \leq c.$$

[8]See J. E. Marsden and A. J. Tromba, *Vector Calculus*, 6th ed. (New York: W.H. Freeman, 2011), p. 220..

Since the inputs are nonnegative, $x \geq 0$ and $y \geq 0$, we have three inequality constraints, which restrict (x, y) to a region of the plane rather than to a curve in the plane. In principle, we should first check to see whether or not $f(x, y)$ has any critical points in the interior:

$$p_1x + p_2y < c, \qquad x > 0 \quad y > 0.$$

However, in the case of a budget constraint, we can see that the maximum of f must occur when the budget is exhausted, so we look for the maximum value of f on the boundary line:

$$p_1x + p_2y = c, \qquad x \geq 0 \quad y \geq 0.$$

Strategy for Optimizing $f(x, y)$ Subject to the Constraint $g(x, y) \leq c$

- Find all points in the region $g(x, y) < c$ where grad f is zero or undefined.
- Use Lagrange multipliers to find the local extrema of f on the boundary $g(x, y) = c$.
- Evaluate f at the points found in the previous two steps and compare the values.

From Section 15.2 we know that if f is continuous on a closed and bounded region, R, then f is guaranteed to attain its global maximum and minimum values on R.

Example 2 Find the maximum and minimum values of $f(x, y) = (x - 1)^2 + (y - 2)^2$ subject to the constraint $x^2 + y^2 \leq 45$.

Solution First, we look for all critical points of f in the interior of the region. Setting

$$f_x(x, y) = 2(x - 1) = 0$$
$$f_y(x, y) = 2(y - 2) = 0,$$

we find f has exactly one critical point at $x = 1$, $y = 2$. Since $1^2 + 2^2 < 45$, that critical point is in the interior of the region.

Next, we find the local extrema of f on the boundary curve $x^2 + y^2 = 45$. To do this, we use Lagrange multipliers with constraint $g(x, y) = x^2 + y^2 = 45$. Setting grad $f = \lambda$ grad g, we get

$$2(x - 1) = \lambda \cdot 2x,$$
$$2(y - 2) = \lambda \cdot 2y.$$

We can't have $x = 0$ since the first equation would become $-2 = 0$. Similarly, $y \neq 0$. So we can solve each equation for λ by dividing by x and y. Setting the expressions for λ equal gives

$$\frac{x - 1}{x} = \frac{y - 2}{y},$$

so

$$y = 2x.$$

Combining this with the constraint $x^2 + y^2 = 45$, we get

$$5x^2 = 45,$$

so

$$x = \pm 3.$$

Since $y = 2x$, we have possible local extrema at $x = 3$, $y = 6$ and $x = -3$, $y = -6$.

We conclude that the only candidates for the maximum and minimum values of f in the region occur at $(1, 2)$, $(3, 6)$, and $(-3, -6)$. Evaluating f at these three points, we find

$$f(1,2) = 0, \qquad f(3,6) = 20, \qquad f(-3,-6) = 80.$$

Therefore, the minimum value of f is 0 at $(1, 2)$ and the maximum value is 80 at $(-3, -6)$.

The Meaning of λ

In the uses of Lagrange multipliers so far, we never found (or needed) the value of λ. However, λ does have a practical interpretation. In the production example, we wanted to maximize

$$f(x, y) = x^{2/3}y^{1/3}$$

subject to the constraint

$$g(x, y) = x + y = 3.78.$$

We solved the equations

$$\frac{2}{3}x^{-1/3}y^{1/3} = \lambda,$$
$$\frac{1}{3}x^{2/3}y^{-2/3} = \lambda,$$
$$x + y = 3.78,$$

to get $x = 2.52$, $y = 1.26$ and $f(2.52, 1.26) \approx 2$. Continuing to find λ gives us

$$\lambda \approx 0.53.$$

Now we do another, apparently unrelated, calculation. Suppose our budget is increased by one, from 3.78 to 4.78, giving a new budget constraint of $x + y = 4.78$. Then the corresponding solution is at $x = 3.19$ and $y = 1.59$ and the new maximum value (instead of $f = 2$) is

$$f = (3.19)^{2/3}(1.59)^{1/3} \approx 2.53.$$

Notice that the amount by which f has increased is 0.53, the value of λ. Thus, in this example, the value of λ represents the extra production achieved by increasing the budget by one—in other words, the extra "bang" you get for an extra "buck" of budget. In fact, this is true in general:

- The value of λ is approximately the increase in the optimum value of f when the budget is increased by 1 unit.

More precisely:

- The value of λ represents the rate of change of the optimum value of f as the budget increases.

An Expression for λ

To interpret λ, we look at how the optimum value of the objective function f changes as the value c of the constraint function g is varied. In general, the optimum point (x_0, y_0) depends on the constraint value c. So, provided x_0 and y_0 are differentiable functions of c, we can use the chain rule to differentiate the optimum value $f(x_0(c), y_0(c))$ with respect to c:

$$\frac{df}{dc} = \frac{\partial f}{\partial x}\frac{dx_0}{dc} + \frac{\partial f}{\partial y}\frac{dy_0}{dc}.$$

At the optimum point (x_0, y_0), we have $f_x = \lambda g_x$ and $f_y = \lambda g_y$, and therefore

$$\frac{df}{dc} = \lambda\left(\frac{\partial g}{\partial x}\frac{dx_0}{dc} + \frac{\partial g}{\partial y}\frac{dy_0}{dc}\right) = \lambda\frac{dg}{dc}.$$

But, as $g(x_0(c), y_0(c)) = c$, we see that $dg/dc = 1$, so $df/dc = \lambda$. Thus, we have the following interpretation of the Lagrange multiplier λ:

The value of λ is the rate of change of the optimum value of f as c increases (where $g(x, y) = c$). If the optimum value of f is written as $f(x_0(c), y_0(c))$, then

$$\frac{d}{dc} f(x_0(c), y_0(c)) = \lambda.$$

Example 3 The quantity of goods produced according to the function $f(x, y) = x^{2/3}y^{1/3}$ is maximized subject to the budget constraint $x + y \leq 3.78$. The budget is increased to allow for a small increase in production. What is the price of the product if the sale of the additional goods covers the budget increase?

Solution We know that $\lambda = 0.53$, which tells us that $df/dc = 0.53$. The constraint corresponds to a budget of \$3.78 thousand. Therefore increasing the budget by \$1000 increases production by about 0.53 units. In order to make the increase in budget profitable, the extra goods produced must sell for more than \$1000. Thus, if p is the price of each unit of the good, then $0.53p$ is the revenue from the extra 0.53 units sold. Thus, we need $0.53p \geq 1000$ so $p \geq 1000/0.53 = \$1890$.

The Lagrangian Function

Constrained optimization problems are frequently solved using a *Lagrangian function*, $\mathcal{L}$. For example, to optimize $f(x, y)$ subject to the constraint $g(x, y) = c$, we use the Lagrangian function

$$\mathcal{L}(x, y, \lambda) = f(x, y) - \lambda(g(x, y) - c).$$

To see how the function $\mathcal{L}$ is used, compute the partial derivatives of $\mathcal{L}$:

$$\frac{\partial \mathcal{L}}{\partial x} = \frac{\partial f}{\partial x} - \lambda \frac{\partial g}{\partial x},$$
$$\frac{\partial \mathcal{L}}{\partial y} = \frac{\partial f}{\partial y} - \lambda \frac{\partial g}{\partial y},$$
$$\frac{\partial \mathcal{L}}{\partial \lambda} = -(g(x, y) - c).$$

Notice that if (x_0, y_0) is an extreme point of $f(x, y)$ subject to the constraint $g(x, y) = c$ and λ_0 is the corresponding Lagrange multiplier, then at the point (x_0, y_0, λ_0) we have

$$\frac{\partial \mathcal{L}}{\partial x} = 0 \quad \text{and} \quad \frac{\partial \mathcal{L}}{\partial y} = 0 \quad \text{and} \quad \frac{\partial \mathcal{L}}{\partial \lambda} = 0.$$

In other words, (x_0, y_0, λ_0) is a critical point for the unconstrained Lagrangian function, $\mathcal{L}(x, y, \lambda)$. Thus, the Lagrangian converts a constrained optimization problem to an unconstrained problem.

Example 4 A company has a production function with three inputs x, y, and z given by

$$f(x, y, z) = 50x^{2/5}y^{1/5}z^{1/5}.$$

The total budget is \$24,000 and the company can buy x, y, and z at \$80, \$12, and \$10 per unit, respectively. What combination of inputs will maximize production?[9]

Solution We need to maximize the objective function

$$f(x, y, z) = 50x^{2/5}y^{1/5}z^{1/5},$$

subject to the constraint

$$g(x, y, z) = 80x + 12y + 10z = 24{,}000.$$

[9] Adapted from M. Rosser and P. Lis, *Basic Mathematics for Economists*, 3rd ed. (New York: Routledge, 2016), p. 360.

The method for functions of two variables works for functions of three variables, so we construct the Lagrangian function

$$\mathcal{L}(x, y, z, \lambda) = 50x^{2/5}y^{1/5}z^{1/5} - \lambda(80x + 12y + 10z - 24{,}000),$$

and solve the system of equations we get from grad $\mathcal{L} = \vec{0}$:

$$\begin{aligned}
\frac{\partial \mathcal{L}}{\partial x} &= 20x^{-3/5}y^{1/5}z^{1/5} - 80\lambda = 0,\\
\frac{\partial \mathcal{L}}{\partial y} &= 10x^{2/5}y^{-4/5}z^{1/5} - 12\lambda = 0,\\
\frac{\partial \mathcal{L}}{\partial z} &= 10x^{2/5}y^{1/5}z^{-4/5} - 10\lambda = 0,\\
\frac{\partial \mathcal{L}}{\partial \lambda} &= -(80x + 12y + 10z - 24{,}000) = 0.
\end{aligned}$$

Simplifying this system gives

$$\begin{aligned}
\lambda &= \frac{1}{4}x^{-3/5}y^{1/5}z^{1/5},\\
\lambda &= \frac{5}{6}x^{2/5}y^{-4/5}z^{1/5},\\
\lambda &= x^{2/5}y^{1/5}z^{-4/5},\\
80x + 12y + 10z &= 24{,}000.
\end{aligned}$$

Eliminating z from the first two equations gives $x = 0.3y$. Eliminating x from the second and third equations gives $z = 1.2y$. Substituting for x and z into $80x + 12y + 10z = 24{,}000$ gives

$$80(0.3y) + 12y + 10(1.2y) = 24{,}000,$$

so $y = 500$. Then $x = 150$ and $z = 600$, and $f(150, 500, 600) = 4{,}622$ units.

The graph of the constraint, $80x + 12y + 10z = 24{,}000$, is a plane. Since the inputs x, y, z must be nonnegative, the graph is a triangle in the first octant, with edges on the coordinate planes. On the boundary of the triangle, one (or more) of the variables x, y, z is zero, so the function f is zero. Thus production is maximized within the budget using $x = 150$, $y = 500$, and $z = 600$.

Exercises and Problems for Section 15.3 Online Resource: Additional Problems for Section 15.3

EXERCISES

In Exercises 1–18, use Lagrange multipliers to find the maximum and minimum values of f subject to the given constraint, if such values exist.

1. $f(x, y) = x + y, \quad x^2 + y^2 = 1$
2. $f(x, y) = x + 3y + 2, \quad x^2 + y^2 = 10$
3. $f(x, y) = (x - 1)^2 + (y + 2)^2, \quad x^2 + y^2 = 5$
4. $f(x, y) = x^3 + y, \quad 3x^2 + y^2 = 4$
5. $f(x, y) = 3x - 2y, \quad x^2 + 2y^2 = 44$
6. $f(x, y) = xy, \quad 4x^2 + y^2 = 8$
7. $f(x, y) = 2xy, \quad 5x + 4y = 100$
8. $f(x_1, x_2) = {x_1}^2 + {x_2}^2, \quad x_1 + x_2 = 1$
9. $f(x, y) = x^2 + y, \quad x^2 - y^2 = 1$
10. $f(x, y, z) = x + 3y + 5z, \quad x^2 + y^2 + z^2 = 1$
11. $f(x, y, z) = x^2 - y^2 - 2z, \quad x^2 + y^2 = z$
12. $f(x, y, z) = xyz, \quad x^2 + y^2 + 4z^2 = 12$
13. $f(x, y) = x^2 + 2y^2, \quad x^2 + y^2 \le 4$
14. $f(x, y) = x + 3y, \quad x^2 + y^2 \le 2$
15. $f(x, y) = xy, \quad x^2 + 2y^2 \le 1$
16. $f(x, y) = x^3 + y, \quad x + y \ge 1$
17. $f(x, y) = (x + 3)^2 + (y - 3)^2, \quad x^2 + y^2 \le 2$
18. $f(x, y) = x^2y + 3y^2 - y, \quad x^2 + y^2 \le 10$

19. For each point marked in Figure 15.30, decide whether:
    (a) The point is a local minimum, maximum, or neither for the function f constrained by the loop.
    (b) The point is a global minimum, maximum, or neither subject to the constraint.

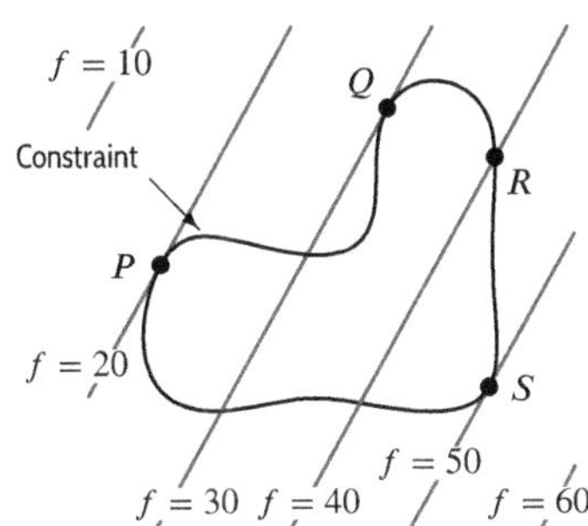

Figure 15.30

In Exercises 20–23, a Cobb-Douglas production function $P(K, L)$ and budget $B(K, L)$ are given, where K represents capital and L represents labor. Use Lagrange multipliers to find the values of K and L that maximize production given a budget constraint or minimize budget given a production constraint. Then give the value for λ and its meaning.

20. Maximize production: $P = K^{1/4}L^{3/4}$
    Budget constraint: $B = 2K + L = 40$

21. Maximize production: $P = K^{2/3}L^{1/3}$
    Budget constraint: $B = 10K + 4L = 60$

22. Maximize production: $P = K^{2/5}L^{3/5}$
    Budget constraint: $B = 4K + 5L = 100$

23. Minimize budget: $B = 4K + L$
    Production constraint: $P = K^{1/2}L^{1/2} = 200$

PROBLEMS

24. Find the maximum value of $f(x, y) = x + y - (x - y)^2$ on the triangular region $x \geq 0$, $y \geq 0$, $x + y \leq 1$.

25. For $f(x, y) = x^2 + 6xy$, find the global maximum and minimum on the closed region in the first quadrant bounded by the line $x + y = 4$ and the curve $xy = 3$.

26. (a) Draw contours of $f(x, y) = 2x + y$ for $z = -7, -5, -3, -1, 1, 3, 5, 7$.
    (b) On the same axes, graph the constraint $x^2+y^2 = 5$.
    (c) Use the graph to approximate the points at which f has a maximum or a minimum value subject to the constraint $x^2 + y^2 = 5$.
    (d) Use Lagrange multipliers to find the maximum and minimum values of $f(x, y) = 2x + y$ subject to $x^2 + y^2 = 5$.

27. Let $f(x, y) = x^{\alpha}y^{1-\alpha}$ for $0 < \alpha < 1$. Find the value of α such that the maximum value of f on the line $2x + 3y = 6$ occurs at $(1.5, 1)$.

28. Figure 15.31 shows contours of f. Does f have a maximum value subject to the constraint $g(x, y) = c$ for $x \geq 0$, $y \geq 0$? If so, approximately where is it and what is its value? Does f have a minimum value subject to the constraint? If so, approximately where and what?

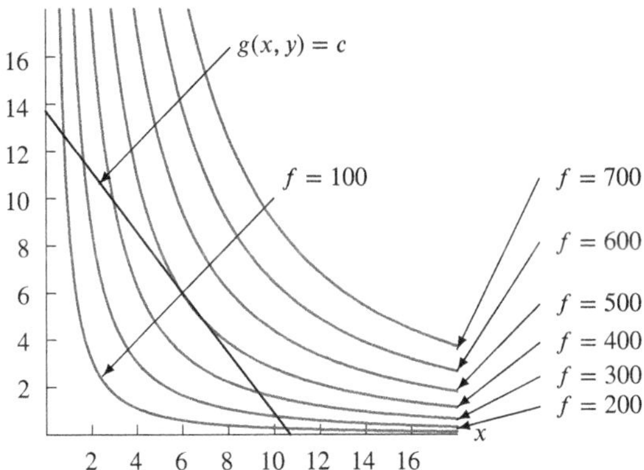

Figure 15.31

29. Each person tries to balance his or her time between leisure and work. The tradeoff is that as you work less your income falls. Therefore each person has *indifference curves* which connect the number of hours of leisure, l, and income, s. If, for example, you are indifferent between 0 hours of leisure and an income of \$1125 a week on the one hand, and 10 hours of leisure and an income of \$750 a week on the other hand, then the points $l = 0$, $s = 1125$, and $l = 10$, $s = 750$ both lie on the same indifference curve. Table 15.3 gives information on three indifference curves, I, II, and III.

Table 15.3

Weekly income			Weekly leisure hours		
I	II	III	I	II	III
1125	1250	1375	0	20	40
750	875	1000	10	30	50
500	625	750	20	40	60
375	500	625	30	50	70
250	375	500	50	70	90

(a) Graph the three indifference curves.
(b) You have 100 hours a week available for work and leisure combined, and you earn \$10/hour. Write an equation in terms of l and s which represents this constraint.
(c) On the same axes, graph this constraint.
(d) Estimate from the graph what combination of leisure hours and income you would choose under these circumstances. Give the corresponding number of hours per week you would work.

30. Figure 15.32 shows ∇f for a function $f(x, y)$ and two curves $g(x, y) = 1$ and $g(x, y) = 2$. Mark the following:

(a) The point(s) A where f has a local maximum.
(b) The point(s) B where f has a saddle point.
(c) The point C where f has a maximum on $g = 1$.
(d) The point D where f has a minimum on $g = 1$.
(e) If you used Lagrange multipliers to find C, what would the sign of λ be? Why?

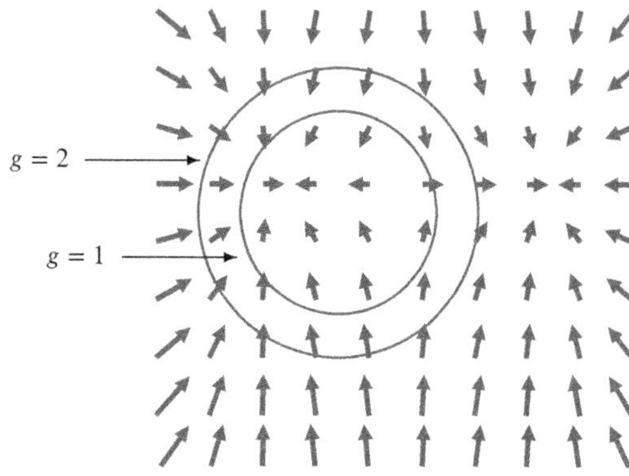

Figure 15.32

31. The point P is a maximum or minimum of the function f subject to the constraint $g(x, y) = x + y = c$, with $x, y \geq 0$. For the graphs (a) and (b), does P give a maximum or a minimum of f? What is the sign of λ? If P gives a maximum, where does the minimum of f occur? If P gives a minimum, where does the maximum of f occur?

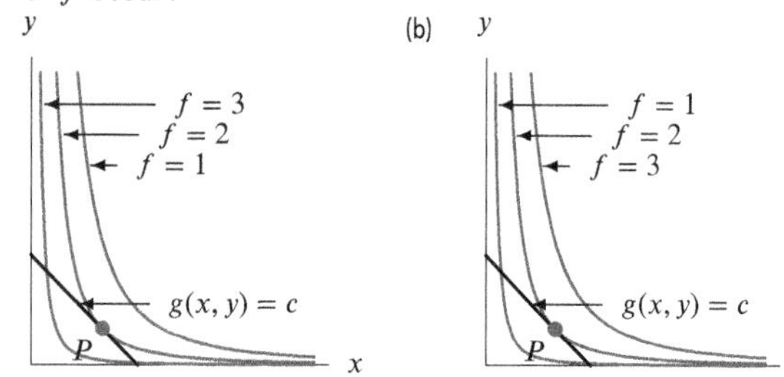

32. Figure 15.33 shows the optimal point (marked with a dot) in three optimization problems with the same constraint. Arrange the corresponding values of λ in increasing order. (Assume λ is positive.)

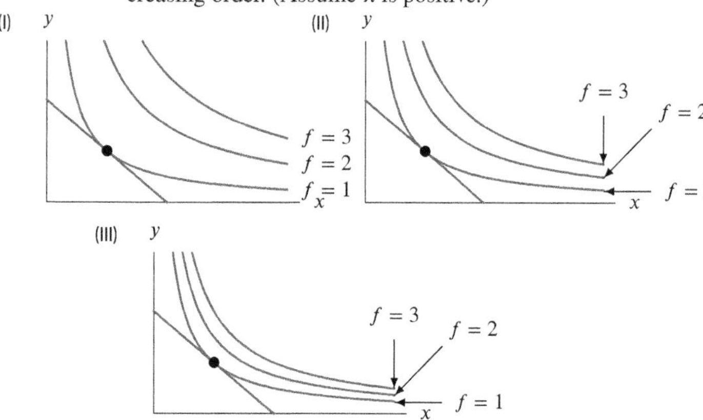

Figure 15.33

33. If the right side of the constraint in Exercise 5 is changed by the small amount Δc, by approximately how much do the maximum and minimum values change?

34. If the right side of the constraint in Exercise 6 is changed by the small amount Δc, by approximately how much do the maximum and minimum values change?

35. The function $P(x, y)$ gives the number of units produced and $C(x, y)$ gives the cost of production.

(a) A company wishes to maximize production at a fixed cost of \$50,000. What is the objective function f? What is the constraint equation? What is the meaning of λ in this situation?
(b) A company wishes to minimize costs at a fixed production level of 2000 units. What is the objective function f? What is the constraint equation? What is the meaning of λ in this situation?

36. Design a closed cylindrical container which holds 100 cm^3 and has the minimal possible surface area. What should its dimensions be?

37. A company manufactures x units of one item and y units of another. The total cost in dollars, C, of producing these two items is approximated by the function

$$C = 5x^2 + 2xy + 3y^2 + 800.$$

(a) If the production quota for the total number of items (both types combined) is 39, find the minimum production cost.
(b) Estimate the additional production cost or savings if the production quota is raised to 40 or lowered to 38.

38. An international organization must decide how to spend the \$2,000,000 they have been allotted for famine relief in a remote area. They expect to divide the money between buying rice at \$38.5/sack and beans at \$35/sack. The number, P, of people who would be fed if they buy x sacks of rice and y sacks of beans is given by

$$P = 1.1x + y - \frac{xy}{10^8}.$$

What is the maximum number of people that can be fed, and how should the organization allocate its money?

39. The quantity, q, of a product manufactured depends on the number of workers, W, and the amount of capital invested, K, and is given by

$$q = 6W^{3/4}K^{1/4}.$$

Labor costs are \$10 per worker and capital costs are \$20 per unit, and the budget is \$3000.

(a) What are the optimum number of workers and the optimum number of units of capital?
(b) Show that at the optimum values of W and K, the ratio of the marginal productivity of labor ($\partial q/\partial W$) to the marginal productivity of capital

$(\partial q/\partial K)$ is the same as the ratio of the cost of a unit of labor to the cost of a unit of capital.

(c) Recompute the optimum values of W and K when the budget is increased by one dollar. Check that increasing the budget by $1 allows the production of λ extra units of the good, where λ is the Lagrange multiplier.

40. A neighborhood health clinic has a budget of $600,000 per quarter. The director of the clinic wants to allocate the budget to maximize the number of patient visits, V, which is given as a function of the number of doctors, D, and the number of nurses, N, by

$$V = 1000D^{0.6}N^{0.3}.$$

A doctor gets $40,000 per quarter; nurses get $10,000 per quarter.

(a) Set up the director's constrained optimization problem.

(b) Describe, in words, the conditions which must be satisfied by $\partial V/\partial D$ and $\partial V/\partial N$ for V to have an optimum value.

(c) Solve the problem formulated in part (a).

(d) Find the value of the Lagrange multiplier and interpret its meaning in this problem.

(e) At the optimum point, what is the marginal cost of a patient visit (that is, the cost of an additional visit)? Will that marginal cost rise or fall with the number of visits? Why?

41. **(a)** In Problem 39, does the value of λ change if the budget changes from $3000 to $4000?

(b) In Problem 40, does the value of λ change if the budget changes from $600,000 to $700,000?

(c) What condition must a Cobb-Douglas production function, $Q = cK^aL^b$, satisfy to ensure that the marginal increase of production (that is, the rate of increase of production with budget) is not affected by the size of the budget?

42. The production function $P(K, L)$ gives the number of pairs of skis produced per week at a factory operating with K units of capital and L units of labor. The contour diagram for P is in Figure 15.34; the parallel lines are budget constraints for budgets, B, in dollars.

(a) On each budget constraint, mark the point that gives the maximum production.

(b) Complete the table, where the budget, B, is in dollars and the maximum production is the number of pairs of skis to be produced each week.

B	2000	4000	6000	8000	10000
M					

(c) Estimate the Lagrange multiplier $\lambda = dM/dB$ at a budget of $6000. Give units for the multiplier.

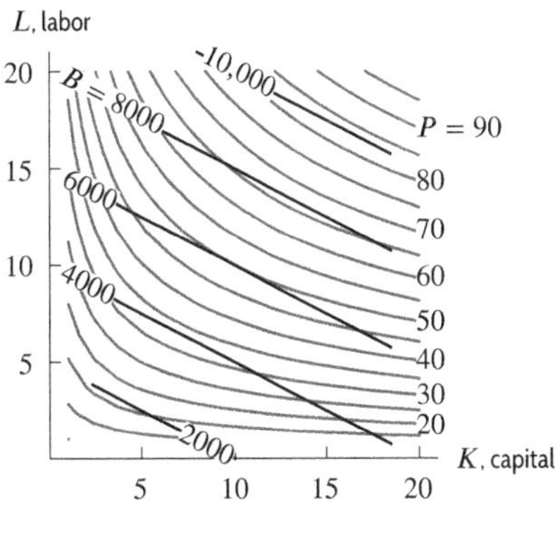

Figure 15.34

43. A doctor wants to schedule visits for two patients who have been operated on for tumors so as to minimize the expected delay in detecting a new tumor. Visits for patients 1 and 2 are scheduled at intervals of x_1 and x_2 weeks, respectively. A total of m visits per week is available for both patients combined.

The recurrence rates for tumors for patients 1 and 2 are judged to be v_1 and v_2 tumors per week, respectively. Thus, $v_1/(v_1+v_2)$ and $v_2/(v_1+v_2)$ are the probabilities that patient 1 and patient 2, respectively, will have the next tumor. It is known that the expected delay in detecting a tumor for a patient checked every x weeks is $x/2$. Hence, the expected detection delay for both patients combined is given by[10]

$$f(x_1, x_2) = \frac{v_1}{v_1+v_2}\cdot\frac{x_1}{2} + \frac{v_2}{v_1+v_2}\cdot\frac{x_2}{2}.$$

Find the values of x_1 and x_2 in terms of v_1 and v_2 that minimize $f(x_1, x_2)$ subject to the fact that m, the number of visits per week, is fixed.

44. What is the value of the Lagrange multiplier in Problem 43? What are the units of λ? What is its practical significance to the doctor?

45. Figure 15.35 shows two weightless springs with spring constants k_1 and k_2 attached between a ceiling and floor without tension or compression. A mass m is placed between the springs which settle into equilibrium as in Figure 15.36. The magnitudes f_1 and f_2 of the forces of the springs on the mass minimize the complementary energy

$$\frac{f_1^2}{2k_1} + \frac{f_2^2}{2k_2}$$

subject to the force balance constraint $f_1 + f_2 = mg$.

(a) Determine f_1 and f_2 by the method of Lagrange multipliers.

[10] Adapted from Daniel Kent, Ross Shachter, *et al.*, "Efficient Scheduling of Cystoscopies in Monitoring for Recurrent Bladder Cancer," *Medical Decision Making* (Philadelphia: Hanley and Belfus, 1989).

(b) If you are familiar with Hooke's law, find the meaning of λ.

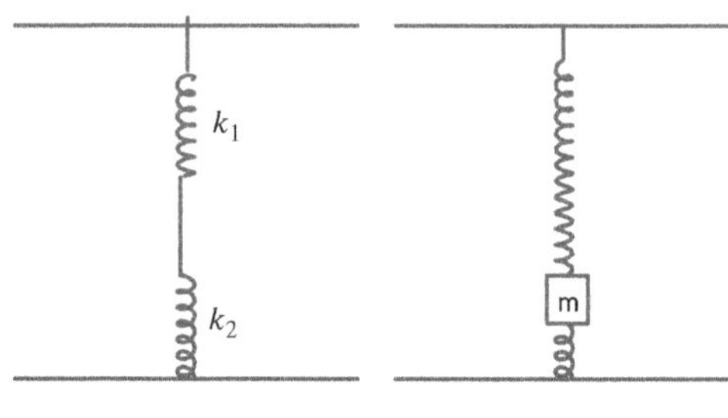

Figure 15.35 Figure 15.36

46. (a) If $\sum_{i=1}^{3} x_i = 1$, find the values of x_1, x_2, x_3 making $\sum_{i=1}^{3} x_i^2$ minimum.
(b) Generalize the result of part (a) to find the minimum value of $\sum_{i=1}^{n} x_i^2$ subject to $\sum_{i=1}^{n} x_i = 1$.

47. Let $f(x, y) = ax^2+bxy+cy^2$. Show that the maximum value of $f(x, y)$ subject to the constraint $x^2 + y^2 = 1$ is equal to λ, the Lagrange multiplier.

48. Find the minimum distance from the point $(1, 2, 10)$ to the paraboloid given by the equation $z = x^2 + y^2$. Give a geometric justification for your answer.

49. A company produces one product from two inputs (for example, capital and labor). Its production function $g(x, y)$ gives the quantity of the product that can be produced with x units of the first input and y units of the second. The *cost function* (or *expenditure function*) is the three-variable function $C(p, q, u)$ where p and q are the unit prices of the two inputs. For fixed p, q, and u, the value $C(p, q, u)$ is the minimum of $f(x, y) = px+qy$ subject to the constraint $g(x, y) = u$.

(a) What is the practical meaning of $C(p, q, u)$?
(b) Find a formula for $C(p, q, u)$ if $g(x, y) = xy$.

50. A *utility function* $U(x, y)$ for two items gives the utility (benefit) to a consumer of x units of item 1 and y units of item 2. The *indirect utility function* is the three-variable function $V(p, q, I)$ where p and q are the unit prices of the two items. For fixed p, q, and I, the value $V(p, q, I)$ is the maximum of $U(x, y)$ subject to the constraint $px + qy = I$.

(a) What is the practical meaning of $V(p, q, I)$?
(b) The Lagrange multiplier λ that arises in the maximization defining V is called the *marginal utility of money*. What is its practical meaning?
(c) Find formulas for $V(p, q, I)$ and λ if $U(x, y) = xy$.

51. The function $h(x, y) = x^2 + y^2 - \lambda(2x + 4y - 15)$ has a minimum value $m(\lambda)$ for each value of λ.

(a) Find $m(\lambda)$.
(b) For which value of λ is $m(\lambda)$ the largest and what is that maximum value?
(c) Find the minimum value of $f(x, y) = x^2 + y^2$ subject to the constraint $2x+4y = 15$ using the method of Lagrange multipliers and evaluate λ.
(d) Compare your answers to parts (b) and (c).

52. Let f be differentiable and grad $f(2, 1) = -3\vec{i} + 4\vec{j}$. You want to see if $(2, 1)$ is a candidate for the maximum and minimum values of f subject to a constraint satisfied by the point $(2, 1)$.

(a) Show $(2, 1)$ is not a candidate if the constraint is $x^2 + y^2 = 5$.
(b) Show $(2, 1)$ is a candidate if the constraint is $(x - 5)^2 + (y + 3)^2 = 25$. From a sketch of the contours for f near $(2, 1)$ and the constraint, decide whether $(2, 1)$ is a candidate for a maximum or minimum.
(c) Do the same as part (b), but using the constraint $(x + 1)^2 + (y - 5)^2 = 25$.

53. A person's satisfaction from consuming a quantity x_1 of one item and a quantity x_2 of another item is given by

$$S = u(x_1, x_2) = a \ln x_1 + (1 - a) \ln x_2,$$

where a is a constant, $0 < a < 1$. The prices of the two items are p_1 and p_2 respectively, and the budget is b.

(a) Express the maximum satisfaction that can be achieved as a function of p_1, p_2, and b.
(b) Find the amount of money that must be spent to achieve a particular level of satisfaction, c, as a function of p_1, p_2, and c.

Strengthen Your Understanding

In Problems 54–55, explain what is wrong with the statement.

54. The function $f(x, y) = xy$ has a maximum of 2 on the constraint $x + y = 2$.

55. If the level curves of $f(x, y)$ and the level curves of $g(x, y)$ are not tangent at any point on the constraint $g(x, y) = c$, $x \geq 0$, $y \geq 0$, then f has no maximum on the constraint.

In Problems 56–60, give an example of:

56. A function $f(x, y)$ whose maximum subject to the constraint $x^2 + y^2 = 5$ is at $(3, 4)$.

57. A function $f(x, y)$ to be optimized with constraint $x^2 + 2y^2 \leq 1$ such that the minimum value does not change when the constraint is changed to $x^2 + y^2 \leq 1 + c$ for $c > 0$.

58. A function $f(x, y)$ with a minimum at $(1, 1)$ on the constraint $x + y = 2$.

59. A function $f(x, y)$ that has a maximum but no minimum on the constraint $x + y = 4$.

60. A contour diagram of a function f whose maximum value on the constraint $x + 2y = 6$, $x \geq 0$, $y \geq 0$ occurs at one of the endpoints.

For Problems 61–62, use Figure 15.37. The grid lines are one unit apart.

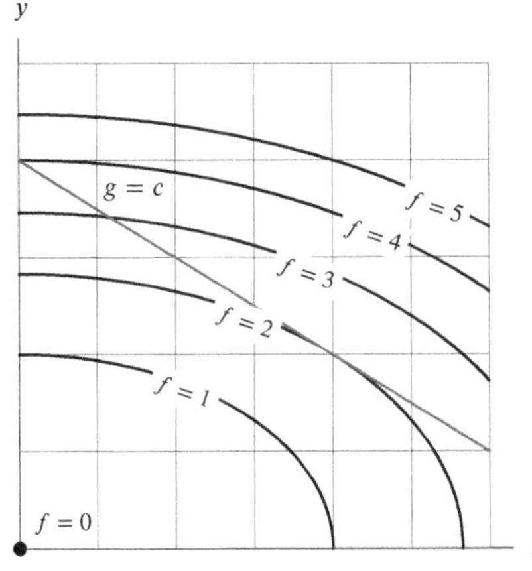

Figure 15.37

61. Find the maximum and minimum values of f on $g = c$. At which points do they occur?

62. Find the maximum and minimum values of f on the triangular region below $g = c$ in the first quadrant.

Are the statements in Problems 63–67 true or false? Give reasons for your answer.

63. If $f(x, y)$ has a local maximum at (a, b) subject to the constraint $g(x, y) = c$, then $g(a, b) = c$.

64. If $f(x, y)$ has a local maximum at (a, b) subject to the constraint $g(x, y) = c$, then $\text{grad} f(a, b) = \vec{0}$.

65. The function $f(x, y) = x + y$ has no global maximum subject to the constraint $x - y = 0$.

66. The point $(2, -1)$ is a local minimum of $f(x, y) = x^2 + y^2$ subject to the constraint $x + 2y = 0$.

67. If grad $f(a, b)$ and grad $g(a, b)$ point in opposite directions, then (a, b) is a local minimum of $f(x, y)$ constrained by $g(x, y) = c$.

In Problems 68–75, suppose that M and m are the maximum and minimum values of $f(x, y)$ subject to the constraint $g(x, y) = c$ and that (a, b) satisfies $g(a, b) = c$. Decide whether the statements are true or false. Give an explanation for your answer.

68. If $f(a, b) = M$, then $f_x(a, b) = f_y(a, b) = 0$.

69. If $f(a, b) = M$, then $f(a, b) = \lambda g(a, b)$ for some value of λ.

70. If grad $f(a, b) = \lambda$ grad $g(a, b)$, then $f(a, b) = M$ or $f(a, b) = m$.

71. If $f(a, b) = M$ and $f_x(a, b)/f_y(a, b) = 5$, then $g_x(a, b)/g_y(a, b) = 5$.

72. If $f(a, b) = m$ and $g_x(a, b) = 0$, then $f_x(a, b) = 0$.

73. Increasing the value of c increases the value of M.

74. Suppose that $f(a, b) = M$ and that grad $f(a, b) = 3$ grad $g(a, b)$. Then increasing the value of c by 0.02 increases the value of M by about 0.06.

75. Suppose that $f(a, b) = m$ and that grad $f(a, b) = 3$ grad $g(a, b)$. Then increasing the value of c by 0.02 decreases the value of m by about 0.06.

Online Resource: Review problems and Projects

Chapter Sixteen

INTEGRATING FUNCTIONS OF SEVERAL VARIABLES

Contents

16.1 THE DEFINITE INTEGRAL OF A FUNCTION OF TWO VARIABLES

The definite integral of a continuous one-variable function, f, is a limit of Riemann sums:

$$\int_a^b f(x)\,dx = \lim_{\Delta x \to 0} \sum_i f(x_i)\,\Delta x,$$

where x_i is a point in the ith subdivision of the interval $[a, b]$. In this section we extend this definition to functions of two variables. We start by considering how to estimate total population from a two-variable population density.

Population Density of Foxes in England

The fox population in parts of England can be important to public health officials because animals can spread diseases, such as rabies. Biologists use a contour diagram to display the fox population density, D; see Figure 16.1, where D is in foxes per square kilometer.[1] The bold contour is the coastline, which may be thought of as the $D = 0$ contour; clearly the density is zero outside it. We can think of D as a function of position, $D = f(x, y)$ where x and y are in kilometers from the southwest corner of the map.

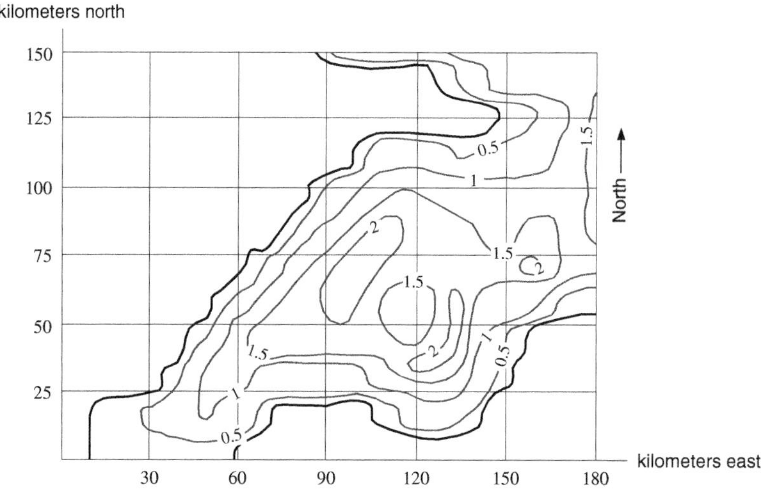

Figure 16.1: Population density of foxes in southwestern England

Example 1 Estimate the total fox population in the region represented by the map in Figure 16.1.

Solution We subdivide the map into the rectangles shown in Figure 16.1 and estimate the population in each rectangle. For simplicity, we use the population density at the northeast corner of each rectangle. For example, in the bottom left rectangle, the density is 0 at the northeast corner; in the next rectangle to the east (right), the density in the northeast corner is 1. Continuing in this way, we get the estimates in Table 16.1. To estimate the population in a rectangle, we multiply the density by the area of the rectangle, $30 \cdot 25 = 750$ km^2. Adding the results, we obtain

$$\begin{aligned}\text{Estimate of population} &= (0.2 + 0.7 + 1.2 + 1.2 + 0.1 + 1.6 + 0.5 + 1.4\\ &\quad + 1.1 + 1.6 + 1.5 + 1.8 + 1.5 + 1.3 + 1.1 + 2.0\\ &\quad + 1.4 + 1.0 + 1.0 + 0.6 + 1.2)750 = 18{,}000 \text{ foxes.}\end{aligned}$$

[1] From "On the spatial spread of rabies among foxes", Murray, J. D. et al, *Proc. R. Soc. Lond. B*, 229: 111–150, 1986.

Taking the upper and lower bounds for the population density on each rectangle enables us to find upper and lower estimates for the population. Using the same rectangles, the upper estimate is approximately 35,000 and the lower estimate is 4,000. There is a wide discrepancy between the upper and lower estimates; we could make them closer by taking finer subdivisions.

Table 16.1 *Estimates of population density (northeast corner)*

0.0	0.0	0.2	0.7	1.2	1.2
0.0	0.0	0.0	0.0	0.1	1.6
0.0	0.0	0.5	1.4	1.1	1.6
0.0	0.0	1.5	1.8	1.5	1.3
0.0	1.1	2.0	1.4	1.0	0.0
0.0	1.0	0.6	1.2	0.0	0.0

Definition of the Definite Integral

The sums used to approximate the fox population are Riemann sums. We now define the definite integral for a function f of two variables on a rectangular region. Given a continuous function $f(x, y)$ defined on a region $a \le x \le b$ and $c \le y \le d$, we subdivide each of the intervals $a \le x \le b$ and $c \le y \le d$ into n and m equal subintervals respectively, giving nm subrectangles. (See Figure 16.2.)

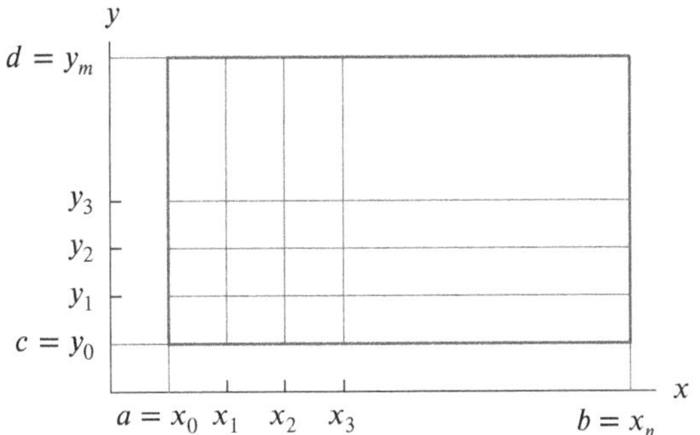

Figure 16.2: Subdivision of a rectangle into nm subrectangles

The area of each subrectangle is $\Delta A = \Delta x\, \Delta y$, where $\Delta x = (b - a)/n$ is the width of each subdivision on the x-axis, and $\Delta y = (d - c)/m$ is the width of each subdivision on the y-axis. To compute the Riemann sum, we multiply the area of each subrectangle by the value of the function at a point in the rectangle and add the resulting numbers. Choosing the maximum value, M_{ij}, of the function on each rectangle and adding for all i, j gives the *upper sum*, $\sum_{i,j} M_{ij}\Delta x\Delta y$.

The *lower sum*, $\sum_{i,j} L_{ij}\Delta x\Delta y$, is obtained by taking the minimum value on each rectangle. If (u_{ij}, v_{ij}) is any point in the ij-th subrectangle, any other Riemann sum satisfies

$$\sum_{i,j} L_{ij}\Delta x\Delta y \le \sum_{i,j} f(u_{ij}, v_{ij})\,\Delta x\,\Delta y \le \sum_{i,j} M_{ij}\Delta x\Delta y.$$

We define the definite integral by taking the limit as the numbers of subdivisions, n and m, tend to infinity. By comparing upper and lower sums, as we did for the fox population, it can be shown that the limit exists when the function, f, is continuous. We get the same limit by letting Δx and Δy tend to 0. Thus, we have the following definition:

Suppose the function f is continuous on R, the rectangle $a \leq x \leq b,\ c \leq y \leq d$. If (u_{ij}, v_{ij}) is any point in the ij-th subrectangle, we define the **definite integral** of f over R

$$\int_R f\,dA = \lim_{\Delta x, \Delta y \to 0} \sum_{i,j} f(u_{ij}, v_{ij}) \Delta x \Delta y.$$

Such an integral is called a **double integral**.

The case when R is not rectangular is considered on page 844. Sometimes we think of dA as being the area of an infinitesimal rectangle of length dx and height dy, so that $dA = dx\,dy$. Then we use the notation[2]

$$\int_R f\,dA = \int_R f(x, y)\,dx\,dy.$$

For this definition, we used a particular type of Riemann sum with equal-sized rectangular subdivisions. In a general Riemann sum, the subdivisions do not all have to be the same size.

Interpretation of the Double Integral as Volume

Just as the definite integral of a positive one-variable function can be interpreted as an area, so the double integral of a positive two-variable function can be interpreted as a volume. In the one-variable case we visualize the Riemann sums as the total area of rectangles above the subdivisions. In the two-variable case we get solid bars instead of rectangles. As the number of subdivisions grows, the tops of the bars approximate the surface better, and the volume of the bars gets closer to the volume under the graph of the function. (See Figure 16.3.)

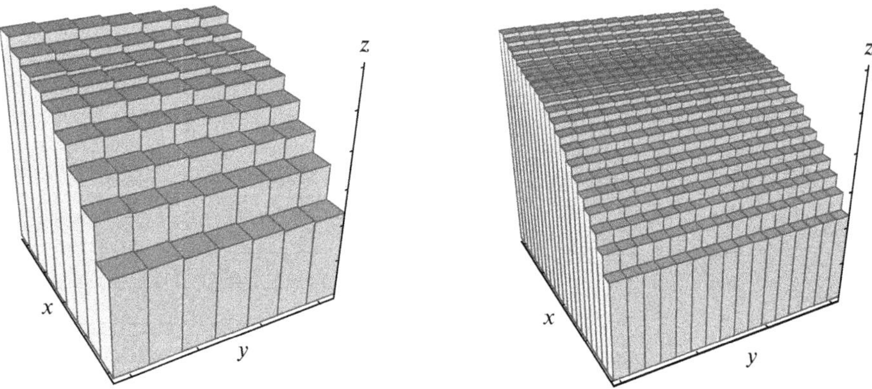

Figure 16.3: Approximating volume under a graph with finer and finer Riemann sums

Thus, we have the following result:

If x, y, z represent length and f is positive, then

$$\text{Volume under graph of } f \text{ above region } R = \int_R f\,dA.$$

Example 2 Let R be the rectangle $0 \leq x \leq 1$ and $0 \leq y \leq 1$. Use Riemann sums to make upper and lower estimates of the volume of the region above R and under the graph of $z = e^{-(x^2+y^2)}$.

[2]Another common notation for the double integral is $\int \int_R f\,dA$.

Solution If R is the rectangle $0 \le x \le 1, 0 \le y \le 1$, the volume we want is given by

$$\text{Volume} = \int_R e^{-(x^2+y^2)}\, dA.$$

We divide R into 16 subrectangles by dividing each edge into four parts. Figure 16.4 shows that $f(x, y) = e^{-(x^2+y^2)}$ decreases as we move away from the origin. Thus, to get an upper sum we evaluate f on each subrectangle at the corner nearest the origin. For example, in the rectangle $0 \le x \le 0.25, 0 \le y \le 0.25$, we evaluate f at $(0, 0)$. Using Table 16.2, we find that

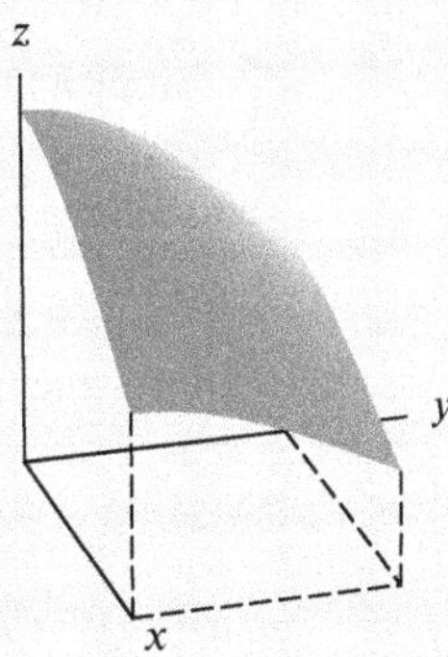

Figure 16.4: Graph of $e^{-(x^2+y^2)}$ above the rectangle R

$$\begin{aligned}\text{Upper sum} = (1 &+ 0.9394 + 0.7788 + 0.5698\\ &+ 0.9394 + 0.8825 + 0.7316 + 0.5353\\ &+ 0.7788 + 0.7316 + 0.6065 + 0.4437\\ &+ 0.5698 + 0.5353 + 0.4437 + 0.3247)(0.0625) = 0.68.\end{aligned}$$

To get a lower sum, we evaluate f at the opposite corner of each rectangle because the surface slopes down in both the x and y directions. This yields a lower sum of 0.44. Thus,

$$0.44 \le \int_R e^{-(x^2+y^2)}\, dA \le 0.68.$$

To get a better approximation of the volume under the graph, we use more subdivisions. See Table 16.3.

Table 16.2 *Values of $f(x, y) = e^{-(x^2+y^2)}$ on the rectangle R*

$x \backslash y$	0.0	0.25	0.50	0.75	1.00
0.0	1	0.9394	0.7788	0.5698	0.3679
0.25	0.9394	0.8825	0.7316	0.5353	0.3456
0.50	0.7788	0.7316	0.6065	0.4437	0.2865
0.75	0.5698	0.5353	0.4437	0.3247	0.2096
1.00	0.3679	0.3456	0.2865	0.2096	0.1353

Table 16.3 *Riemann sum approximations to $\int_R e^{-(x^2+y^2)}\, dA$*

	Number of subdivisions in x and y directions			
	8	16	32	64
Upper	0.6168	0.5873	0.5725	0.5651
Lower	0.4989	0.5283	0.5430	0.5504

The exact value of the double integral, 0.5577 ..., is trapped between the lower and upper sums. Notice that the lower sum increases and the upper sum decreases as the number of subdivisions increases. However, even with 64 subdivisions, the lower and upper sums agree with the exact value of the integral only in the first decimal place.

Interpretation of the Double Integral as Area

In the special case that $f(x, y) = 1$ for all points (x, y) in the region R, each term in the Riemann sum is of the form $1 \cdot \Delta A = \Delta A$ and the double integral gives the area of the region R:

$$\text{Area}(R) = \int_R 1\, dA = \int_R dA$$

Interpretation of the Double Integral as Average Value

As in the one-variable case, the definite integral can be used to define the average value of a function:

$$\begin{array}{c}\text{Average value of } f \\ \text{on the region } R\end{array} = \frac{1}{\text{Area of } R} \int_R f\, dA$$

We can rewrite this as

$$\text{Average value} \times \text{Area of } R = \int_R f\, dA.$$

If we interpret the integral as the volume under the graph of f, then we can think of the average value of f as the height of the box with the same volume that is on the same base. (See Figure 16.5.) Imagine that the volume under the graph is made out of wax. If the wax melted within the perimeter of R, then it would end up box-shaped with height equal to the average value of f.

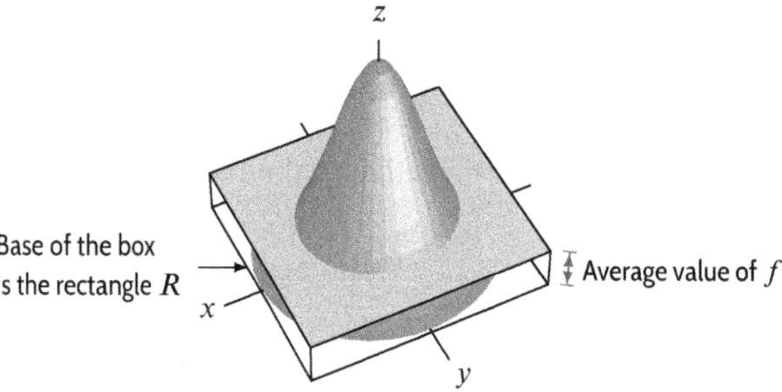

Figure 16.5: Volume and average value

Integral over Regions that Are Not Rectangles

We defined the definite integral $\int_R f(x, y)\, dA$, for a rectangular region R. Now we extend the definition to regions of other shapes, including triangles, circles, and regions bounded by the graphs of piecewise continuous functions.

To approximate the definite integral over a region, R, which is not rectangular, we use a grid of rectangles approximating the region. We obtain this grid by enclosing R in a large rectangle and subdividing that rectangle; we consider just the subrectangles which are inside R.

As before, we pick a point (u_{ij}, v_{ij}) in each subrectangle and form a Riemann sum

$$\sum_{i,j} f(u_{ij}, v_{ij}) \Delta x \Delta y.$$

This time, however, the sum is over only those subrectangles within R. For example, in the case of the fox population we can use the rectangles which are entirely on land. As the subdivisions become finer, the grid approximates the region R more closely. For a function, f, which is continuous on R, we define the definite integral as follows:

$$\int_R f\,dA = \lim_{\Delta x, \Delta y \to 0} \sum_{i,j} f(u_{ij}, v_{ij})\Delta x \Delta y$$

where the Riemann sum is taken over the subrectangles inside R.

You may wonder why we can leave out the rectangles which cover the edge of R—if we included them, might we get a different value for the integral? The answer is that for any region that we are likely to meet, the area of the subrectangles covering the edge tends to 0 as the grid becomes finer. Therefore, omitting these rectangles does not affect the limit.

Convergence of Upper and Lower Sums to Same Limit

We have said that if f is continuous on the rectangle R, then the difference between upper and lower sums for f converges to 0 as Δx and Δy approach 0. In the following example, we show this in a particular case. The ideas in this example can be used in a general proof.

Example 3 Let $f(x, y) = x^2 y$ and let R be the rectangle $0 \leq x \leq 1, 0 \leq y \leq 1$. Show that the difference between upper and lower Riemann sums for f on R converges to 0, as Δx and Δy approach 0.

Solution The difference between the sums is

$$\sum M_{ij} \Delta x \Delta y - \sum L_{ij}\, \Delta x \Delta y = \sum (M_{ij} - L_{ij})\, \Delta x \Delta y,$$

where M_{ij} and L_{ij} are the maximum and minimum of f on the ij-th subrectangle. Since f increases in both the x and y directions, M_{ij} occurs at the corner of the subrectangle farthest from the origin and L_{ij} at the closest. Moreover, since the slopes in the x and y directions don't decrease as x and y increase, the difference $M_{ij} - L_{ij}$ is largest in the subrectangle R_{nm} which is farthest from the origin. Thus,

$$\sum (M_{ij} - L_{ij})\, \Delta x \Delta y \leq (M_{nm} - L_{nm}) \sum \Delta x \Delta y = (M_{nm} - L_{nm})\text{Area}(R).$$

Thus, the difference converges to 0 as long as $(M_{nm} - L_{nm})$ does. The maximum M_{nm} of f on the nm-th subrectangle occurs at $(1, 1)$, the subrectangle's top right corner, and the minimum L_{nm} occurs at the opposite corner, $(1 - 1/n, 1 - 1/m)$. Substituting into $f(x, y) = x^2 y$ gives

$$M_{nm} - L_{nm} = (1)^2(1) - \left(1 - \frac{1}{n}\right)^2 \left(1 - \frac{1}{m}\right) = \frac{2}{n} - \frac{1}{n^2} + \frac{1}{m} - \frac{2}{nm} + \frac{1}{n^2 m}.$$

The right-hand side converges to 0 as $n, m \to \infty$, that is, as $\Delta x, \Delta y \to 0$.

Exercises and Problems for Section 16.1 Online Resource: Additional Problems for Section 16.1

EXERCISES

1. Table 16.4 gives values of the function $f(x, y)$, which is increasing in x and decreasing in y on the region $R : 0 \leq x \leq 6, 0 \leq y \leq 1$. Make the best possible upper and lower estimates of $\int_R f(x, y)\, dA$.

Table 16.4

y \ x	0	3	6
0	5	7	10
0.5	4	5	7
1	3	4	6

2. Values of $f(x, y)$ are in Table 16.5. Let R be the rectangle $1 \leq x \leq 1.2$, $2 \leq y \leq 2.4$. Find Riemann sums which are reasonable over and underestimates for $\int_R f(x, y)\, dA$ with $\Delta x = 0.1$ and $\Delta y = 0.2$.

Table 16.5

y \ x	1.0	1.1	1.2
2.0	5	7	10
2.2	4	6	8
2.4	3	5	4

3. Figure 16.6 shows contours of $g(x, y)$ on the region R, with $5 \le x \le 11$ and $4 \le y \le 10$. Using $\Delta x = \Delta y = 2$, find an overestimate and an underestimate for $\int_R g(x, y)\, dA$.

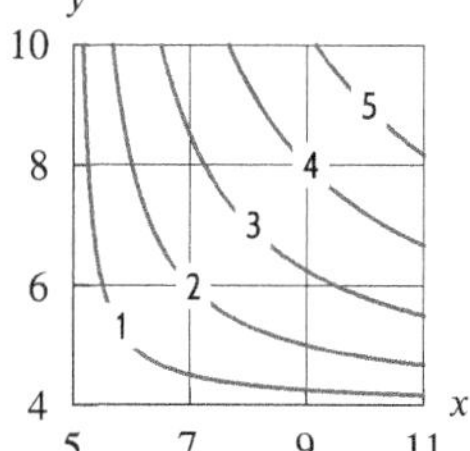

Figure 16.6

4. Figure 16.7 shows contours of $f(x, y)$ on the rectangle R with $0 \le x \le 30$ and $0 \le y \le 15$. Using $\Delta x = 10$ and $\Delta y = 5$, find an overestimate and an underestimate for $\int_R f(x, y)\, dA$.

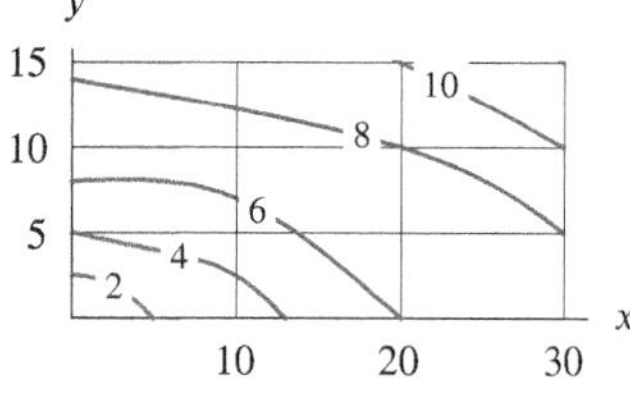

Figure 16.7

5. Figure 16.8 shows a contour plot of population density, people per square kilometer, in a rectangle of land 3 km by 2 km. Estimate the population in the region represented by Figure 16.8.

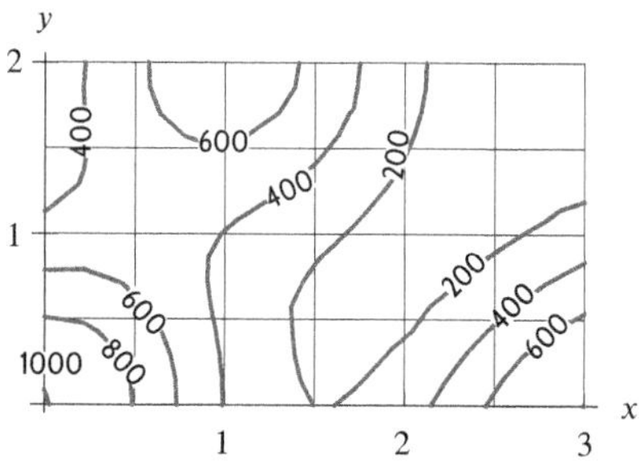

Figure 16.8

In Exercises 6–7, for x and y in meters and R a region on the xy-plane, what does the integral represent? Give units.

6. $\int_R \sigma(x, y)\, dA$, where $\sigma(x, y)$ is bacteria population, in thousands per m^2.

7. $\frac{1}{\text{Area of } R} \int_R h(x, y)\, dA$, where $h(x, y)$ is the height of a tent, in meters.

PROBLEMS

In Problems 8–14, decide (without calculation) whether the integrals are positive, negative, or zero. Let D be the region inside the unit circle centered at the origin, let R be the right half of D, and let B be the bottom half of D.

8. $\int_D 1\, dA$

9. $\int_R 5x\, dA$

10. $\int_B 5x\, dA$

11. $\int_D (y^3 + y^5)\, dA$

12. $\int_B (y^3 + y^5)\, dA$

13. $\int_D (y - y^3)\, dA$

14. $\int_B (y - y^3)\, dA$

15. Figure 16.9 shows contours of $f(x, y)$. Let R be the square $-0.5 \le x \le 1$, $-0.5 \le y \le 1$. Is the integral $\int_R f\, dA$ positive or negative? Explain your reasoning.

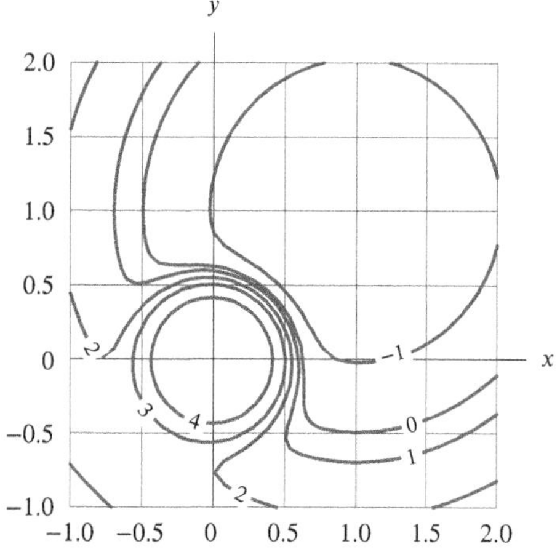

Figure 16.9

16. Table 16.6 gives values of $f(x, y)$, the number of milligrams of mosquito larvae per square meter in a swamp. If x and y are in meters and R is the rectangle $0 \le x \le 8$, $0 \le y \le 6$, estimate $\int_R f(x, y)\, dA$. Give units and interpret your answer.

Table 16.6

y \ x	0	4	8
0	1	3	6
3	2	5	9
6	4	9	15

17. Table 16.7 gives values of $f(x, y)$, the depth of volcanic ash, in meters, after an eruption. If x and y are in kilometers and R is the rectangle $0 \le x \le 100$, $0 \le y \le 100$, estimate the volume of volcanic ash in R in km^3.

Table 16.7

y \ x	0	50	100
0	0.82	0.56	0.43
50	0.63	0.45	0.3
100	0.55	0.44	0.26

18. Table 16.8 gives the density of cacti, $f(x, y)$, in a desert region, in thousands of cacti per km^2. If x and y are in kilometers and R is the square $0 \le x \le 30$, $0 \le y \le 30$, estimate the number of cacti in the region R.

Table 16.8

y \ x	0	10	20	30
0	8.5	8.2	7.9	8.1
10	9.5	10.6	10.5	10.1
20	9.3	10.5	10.4	9.5
30	8.3	8.6	9.3	9.1

19. Use four subrectangles to approximate the volume of the object whose base is the region $0 \le x \le 4$ and $0 \le y \le 6$, and whose height is given by $f(x, y) = x+y$. Find an overestimate and an underestimate and average the two.

20. Figure 16.10 shows the rainfall, in inches, in Tennessee on May 1–2, 2010.[3] Using three contours (red, yellow, and green), make a rough estimate of how many cubic miles of rain fell on the state during this time.

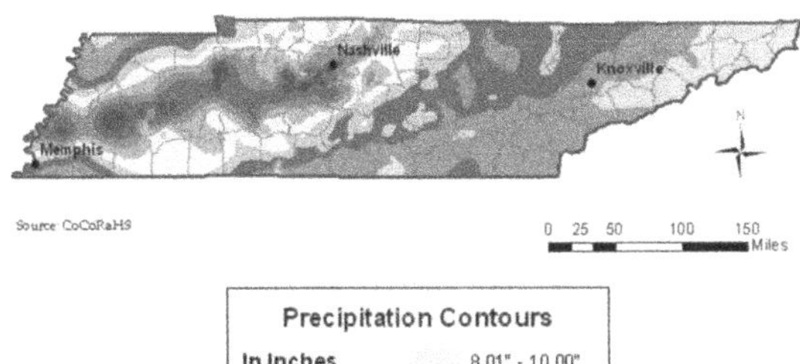

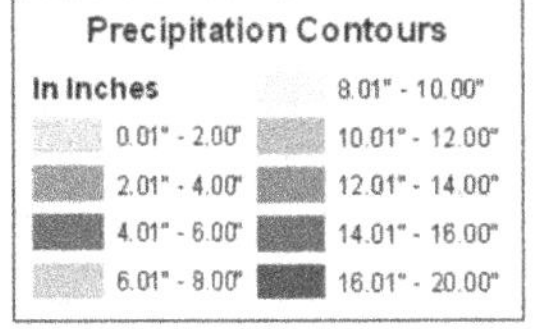

Figure 16.10

Strengthen Your Understanding

In Problems 21–22, explain what is wrong with the statement.

21. For all f, the integral $\int_R f(x, y)\,dA$ gives the volume of the solid under the graph of f over the region R.

22. If R is a region in the third quadrant where $x < 0, y < 0$, then $\int_R f(x, y)\,dA$ is negative.

In Problems 23–24, give an example of:

23. A function $f(x, y)$ and rectangle R such that the Riemann sums obtained using the lower left-hand corner of each subrectangle are an overestimate.

24. A function $f(x, y)$ whose average value over the square $0 \le x \le 1, 0 \le y \le 1$ is negative.

Are the statements in Problems 25–34 true or false? Give reasons for your answer.

25. The double integral $\int_R f\,dA$ is always positive.

26. If $f(x, y) = k$ for all points (x, y) in a region R then $\int_R f\,dA = k \cdot \text{Area}(R)$.

27. If R is the rectangle $0 \le x \le 1, 0 \le y \le 1$ then $\int_R e^{xy}\,dA > 3$.

28. If R is the rectangle $0 \le x \le 2, 0 \le y \le 3$ and S is the rectangle $-2 \le x \le 0, -3 \le y \le 0$, then $\int_R f\,dA = -\int_S f\,dA$.

29. Let $\rho(x, y)$ be the population density of a city, in people per km^2. If R is a region in the city, then $\int_R \rho\,dA$ gives the total number of people in the region R.

30. If $\int_R f\,dA = 0$, then $f(x, y) = 0$ at all points of R.

31. If $g(x, y) = kf(x, y)$, where k is constant, then $\int_R g\,dA = k\int_R f\,dA$.

32. If f and g are two functions continuous on a region R, then $\int_R f \cdot g\,dA = \int_R f\,dA \cdot \int_R g\,dA$.

33. If R is the rectangle $0 \le x \le 1, 0 \le y \le 2$ and S is the square $0 \le x \le 1, 0 \le y \le 1$, then $\int_R f\,dA = 2\int_S f\,dA$.

34. If R is the rectangle $2 \le x \le 4, 5 \le y \le 9$, $f(x, y) = 2x$ and $g(x, y) = x + y$, then the average value of f on R is less than the average value of g on R.

16.2 ITERATED INTEGRALS

In Section 16.1 we approximated double integrals using Riemann sums. In this section we see how to compute double integrals exactly using one-variable integrals.

The Fox Population Again: Expressing a Double Integral as an Iterated Integral

To estimate the fox population, we computed a sum of the form

$$\text{Total population} \approx \sum_{i,j} f(u_{ij}, v_{ij})\Delta x\,\Delta y,$$

where $1 \le i \le n$ and $1 \le j \le m$ and the values $f(u_{ij}, v_{ij})$ can be arranged as in Table 16.9.

[3] www.srh.noaa.gov/images/ohx/rainfall/TN_May2010_rainfall_map.png, accessed June 13, 2016.

Table 16.9 *Estimates for fox population densities for $n = m = 6$*

0.0	0.0	0.2	0.7	1.2	1.2
0.0	0.0	0.0	0.0	0.1	1.6
0.0	0.0	0.5	1.4	1.1	1.6
0.0	0.0	1.5	1.8	1.5	1.3
0.0	1.1	2.0	1.4	1.0	0.0
0.0	1.0	0.6	1.2	0.0	0.0

For any values of n and m, we can either add across the rows first or add down the columns first. If we add rows first, we can write the sum in the form

$$\text{Total population} \approx \sum_{j=1}^{m}\left(\sum_{i=1}^{n} f(u_{ij}, v_{ij})\Delta x\right)\Delta y.$$

The inner sum, $\sum_{i=1}^{n} f(u_{ij}, v_{ij})\,\Delta x$, approximates the integral $\int_0^{180} f(x, v_{ij})\,dx$. Thus, we have

$$\text{Total population} \approx \sum_{j=1}^{m}\left(\int_0^{180} f(x, v_{ij})\,dx\right)\Delta y.$$

The outer Riemann sum approximates another integral, this time with integrand $\int_0^{180} f(x, y)\,dx$, which is a function of y. Thus, we can write the total population in terms of nested, or *iterated*, one-variable integrals:

$$\text{Total population} = \int_0^{150}\left(\int_0^{180} f(x, y)\,dx\right)dy.$$

Since the total population is represented by $\int_R f\,dA$, this suggests the method of computing double integrals in the following theorem:[4]

Theorem 16.1: Writing a Double Integral as an Iterated Integral

If R is the rectangle $a \le x \le b$, $c \le y \le d$ and f is a continuous function on R, then the integral of f over R exists and is equal to the **iterated integral**

$$\int_R f\,dA = \int_{y=c}^{y=d}\left(\int_{x=a}^{x=b} f(x, y)\,dx\right)dy.$$

The expression $\int_{y=c}^{y=d}\left(\int_{x=a}^{x=b} f(x, y)\,dx\right)dy$ can be written $\int_c^d \int_a^b f(x, y)\,dx\,dy$.

To evaluate the iterated integral, first perform the inside integral with respect to x, holding y constant; then integrate the result with respect to y.

Example 1 A building is 8 meters wide and 16 meters long. It has a flat roof that is 12 meters high at one corner and 10 meters high at each of the adjacent corners. What is the volume of the building?

Solution If we put the high corner on the z-axis, the long side along the y-axis, and the short side along the x-axis, as in Figure 16.11, then the roof is a plane with z-intercept 12, and x slope $(-2)/8 = -1/4$, and y slope $(-2)/16 = -1/8$. Hence, the equation of the roof is

$$z = 12 - \tfrac{1}{4}x - \tfrac{1}{8}y.$$

[4]For a proof, see M. Spivak, *Calculus on Manifolds*, pp. 53 and 58 (New York: Benjamin, 1965).

The volume is given by the double integral

$$\text{Volume} = \int_R (12 - \tfrac{1}{4}x - \tfrac{1}{8}y)\,dA,$$

where R is the rectangle $0 \le x \le 8,\ 0 \le y \le 16$. Setting up an iterated integral, we get

$$\text{Volume} = \int_0^{16} \int_0^{8} (12 - \tfrac{1}{4}x - \tfrac{1}{8}y)\,dx\,dy.$$

The inside integral is

$$\int_0^8 (12 - \tfrac{1}{4}x - \tfrac{1}{8}y)\,dx = \left(12x - \tfrac{1}{8}x^2 - \tfrac{1}{8}xy\right)\Big|_{x=0}^{x=8} = 88 - y.$$

Then the outside integral gives

$$\text{Volume} = \int_0^{16} (88 - y)\,dy = (88y - \tfrac{1}{2}y^2)\Big|_0^{16} = 1280.$$

The volume of the building is 1280 cubic meters.

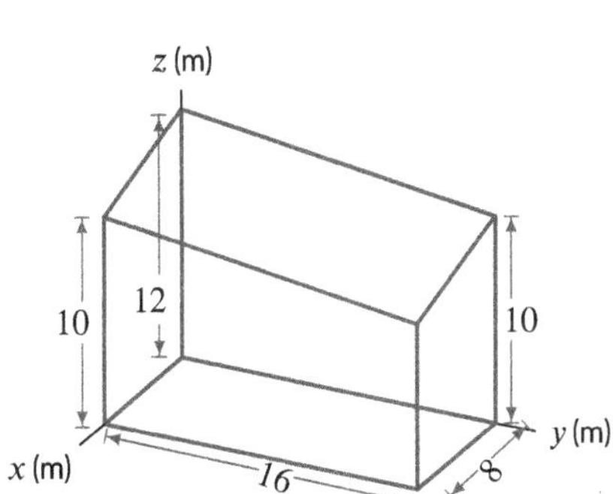

Figure 16.11: A slant-roofed building

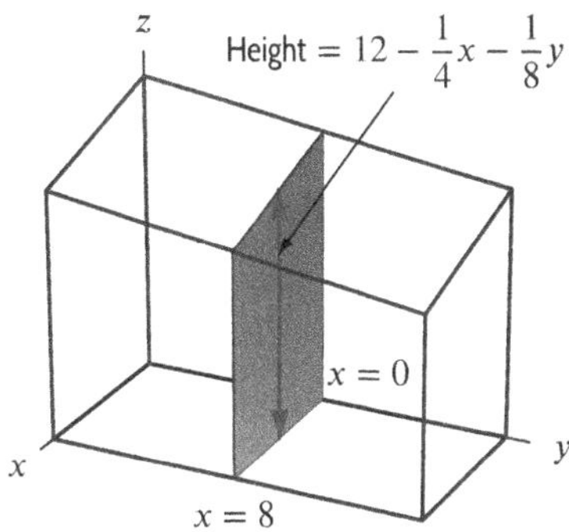

Figure 16.12: Cross-section of a building

Notice that the inner integral $\int_0^8 (12 - \frac{1}{4}x - \frac{1}{8}y)\,dx$ in Example 1 gives the area of the cross section of the building perpendicular to the y-axis in Figure 16.12.

The iterated integral $\int_0^{16} \int_0^8 (12 - \frac{1}{4}x - \frac{1}{8}y)\,dxdy$ thus calculates the volume by adding the volumes of thin cross-sectional slabs.

The Order of Integration

In computing the fox population, we could have chosen to add columns (fixed x) first, instead of the rows. This leads to an iterated integral where x is constant in the inner integral instead of y. Thus,

$$\int_R f(x, y)\,dA = \int_a^b \left(\int_c^d f(x, y)\,dy \right) dx$$

where R is the rectangle $a \le x \le b$ and $c \le y \le d$.

For any function we are likely to meet, it does not matter in which order we integrate over a rectangular region R; we get the same value for the double integral either way.

$$\int_R f\,dA = \int_c^d \left(\int_a^b f(x, y)\,dx \right) dy = \int_a^b \left(\int_c^d f(x, y)\,dy \right) dx$$

Example 2 Compute the volume of Example 1 as an iterated integral by integrating with respect to y first.

Solution Rewriting the integral, we have

$$\text{Volume} = \int_0^8 \left(\int_0^{16} (12 - \tfrac{1}{4}x - \tfrac{1}{8}y)\, dy \right) dx = \int_0^8 \left((12y - \tfrac{1}{4}xy - \tfrac{1}{16}y^2)\Big|_{y=0}^{y=16} \right) dx$$

$$= \int_0^8 (176 - 4x)\, dx = (176x - 2x^2)\Big|_0^8 = 1280 \text{ meter}^3.$$

Iterated Integrals Over Non-Rectangular Regions

Example 3 The density at the point (x, y) of a triangular metal plate, as shown in Figure 16.13, is $\delta(x, y)$. Express its mass as an iterated integral.

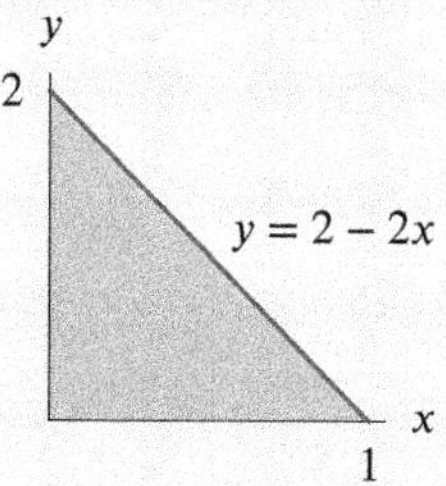

Figure 16.13: A triangular metal plate with density $\delta(x, y)$ at the point (x, y)

Solution Approximate the triangular region using a grid of small rectangles of sides Δx and Δy. The mass of one rectangle is given by

$$\text{Mass of rectangle} \approx \text{Density} \cdot \text{Area} \approx \delta(x, y)\Delta x \Delta y.$$

Summing over all rectangles gives a Riemann sum which approximates the double integral:

$$\text{Mass} = \int_R \delta(x, y)\, dA,$$

where R is the triangle. We want to compute this integral using an iterated integral.

Think about how the iterated integral over the rectangle $a \le x \le b$, $c \le y \le d$ works:

$$\int_a^b \int_c^d f(x, y)\, dy\, dx.$$

The inside integral with respect to y is along vertical strips which begin at the horizontal line $y = c$ and end at the line $y = d$. There is one such strip for each x between $x = a$ and $x = b$. (See Figure 16.14.)

For the triangular region in Figure 16.13, the idea is the same. The only difference is that the individual vertical strips no longer all go from $y = c$ to $y = d$. The vertical strip that starts at the point $(x, 0)$ ends at the point $(x, 2 - 2x)$, because the top edge of the triangle is the line $y = 2 - 2x$. See Figure 16.15. On this vertical strip, y goes from 0 to $2 - 2x$. Hence, the inside integral is

$$\int_0^{2-2x} \delta(x, y)\, dy.$$

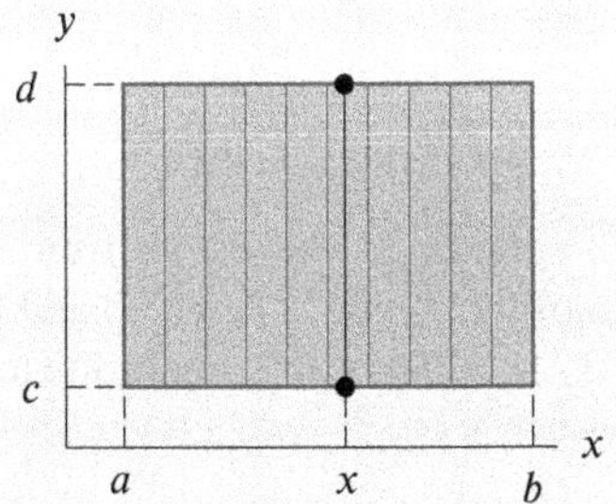

Figure 16.14: Integrating over a rectangle using vertical strips

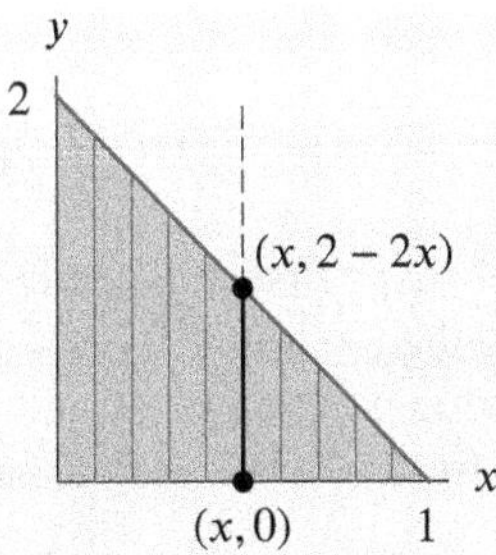

Figure 16.15: Integrating over a triangle using vertical strips

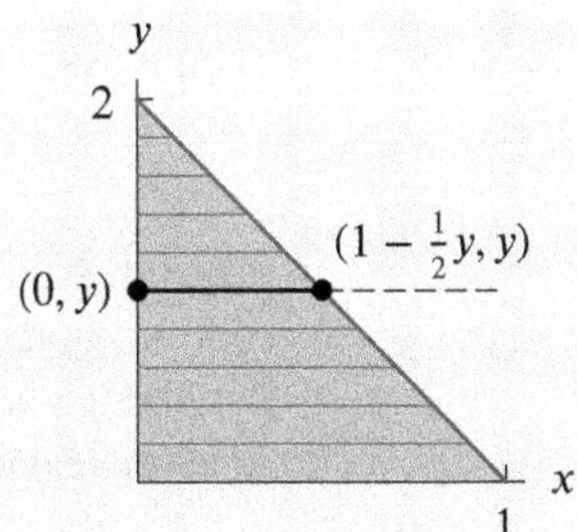

Figure 16.16: Integrating over a triangle using horizontal strips

Finally, since there is a vertical strip for each x between 0 and 1, the outside integral goes from $x = 0$ to $x = 1$. Thus, the iterated integral we want is

$$\text{Mass} = \int_0^1 \int_0^{2-2x} \delta(x, y)\, dy\, dx.$$

We could have chosen to integrate in the opposite order, keeping y fixed in the inner integral instead of x. The limits are formed by looking at horizontal strips instead of vertical ones, and expressing the x-values at the end points in terms of y. See Figure 16.16. To find the right endpoint of the strip, we use the equation of the top edge of the triangle in the form $x = 1 - \frac{1}{2}y$. Thus, a horizontal strip goes from $x = 0$ to $x = 1 - \frac{1}{2}y$. Since there is a strip for every y from 0 to 2, the iterated integral is

$$\text{Mass} = \int_0^2 \int_0^{1-\frac{1}{2}y} \delta(x, y)\, dx\, dy.$$

Limits on Iterated Integrals

- The limits on the outer integral must be constants.
- The limits on the inner integral can involve only the variable in the outer integral. For example, if the inner integral is with respect to x, its limits can be functions of y.

Example 4 Find the mass M of a metal plate R bounded by $y = x$ and $y = x^2$, with density given by $\delta(x, y) = 1 + xy$ kg/meter^2. (See Figure 16.17.)

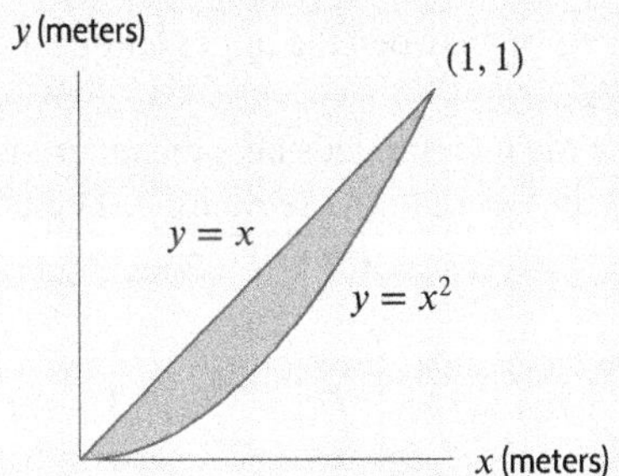

Figure 16.17: A metal plate with density $\delta(x, y)$

Solution The mass is given by

$$M = \int_R \delta(x, y)\, dA.$$

We integrate along vertical strips first; this means we do the y integral first, which goes from the bottom boundary $y = x^2$ to the top boundary $y = x$. The left edge of the region is at $x = 0$ and the right edge is at the intersection point of $y = x$ and $y = x^2$, which is $(1, 1)$. Thus, the x-coordinate of the vertical strips can vary from $x = 0$ to $x = 1$, and so the mass is given by

$$M = \int_0^1 \int_{x^2}^{x} \delta(x, y)\, dy\, dx = \int_0^1 \int_{x^2}^{x} (1 + xy)\, dy\, dx.$$

Calculating the inner integral first gives

$$\begin{aligned} M &= \int_0^1 \int_{x^2}^{x} (1 + xy)\, dy\, dx = \int_0^1 \left(y + x\frac{y^2}{2} \right)\Bigg|_{y=x^2}^{y=x} dx \\ &= \int_0^1 \left(x - x^2 + \frac{x^3}{2} - \frac{x^5}{2} \right) dx = \left(\frac{x^2}{2} - \frac{x^3}{3} + \frac{x^4}{8} - \frac{x^6}{12} \right)\Bigg|_0^1 = \frac{5}{24} = 0.208 \text{ kg}. \end{aligned}$$

Example 5 A semicircular city of radius 3 km borders the ocean on the straight side. Find the average distance from points in the city to the ocean.

Solution Think of the ocean as everything below the x-axis in the xy-plane and think of the city as the upper half of the circular disk of radius 3 bounded by $x^2 + y^2 = 9$. (See Figure 16.18.)

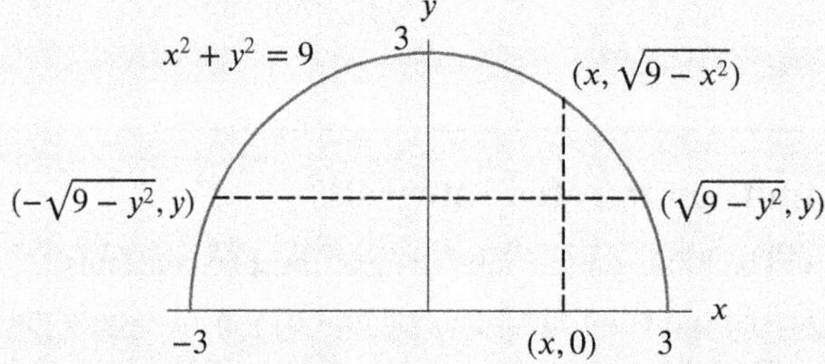

Figure 16.18: The city by the ocean showing a typical vertical strip and a typical horizontal strip

The distance from any point (x, y) in the city to the ocean is the vertical distance to the x-axis, namely y. Thus, we want to compute

$$\text{Average distance} = \frac{1}{\text{Area}(R)} \int_R y\, dA,$$

where R is the region between the upper half of the circle $x^2 + y^2 = 9$ and the x-axis. The area of R is $\pi 3^2/2 = 9\pi/2$.

To compute the integral, let's take the inner integral with respect to y. A vertical strip goes from the x-axis, namely $y = 0$, to the semicircle. The upper limit must be expressed in terms of x, so we solve $x^2 + y^2 = 9$ to get $y = \sqrt{9 - x^2}$. Since there is a strip for every x from -3 to 3, the integral is:

$$\begin{aligned} \int_R y\, dA &= \int_{-3}^{3} \left(\int_0^{\sqrt{9-x^2}} y\, dy \right) dx = \int_{-3}^{3} \left(\frac{y^2}{2}\Bigg|_{y=0}^{y=\sqrt{9-x^2}} \right) dx \\ &= \int_{-3}^{3} \frac{1}{2}(9 - x^2)\, dx = \frac{1}{2}\left(9x - \frac{x^3}{3} \right)\Bigg|_{-3}^{3} = \frac{1}{2}(18 - (-18)) = 18. \end{aligned}$$

Therefore, the average distance is $18/(9\pi/2) = 4/\pi = 1.273$ km.

What if we choose the inner integral with respect to x? Then we get the limits by looking at horizontal strips, not vertical, and we solve $x^2 + y^2 = 9$ for x in terms of y. We get $x = -\sqrt{9-y^2}$ at the left end of the strip and $x = \sqrt{9-y^2}$ at the right. There is a strip for every y from 0 to 3, so

$$\int_R y\,dA = \int_0^3 \left(\int_{-\sqrt{9-y^2}}^{\sqrt{9-y^2}} y\,dx \right) dy = \int_0^3 \left(yx \Big|_{x=-\sqrt{9-y^2}}^{x=\sqrt{9-y^2}} \right) dy = \int_0^3 2y\sqrt{9-y^2}\,dy$$

$$= -\frac{2}{3}(9-y^2)^{3/2}\Big|_0^3 = -\frac{2}{3}(0-27) = 18.$$

We get the same result as before. The average distance to the ocean is $(2/(9\pi))18 = 4/\pi = 1.273$ km.

In the examples so far, a region was given and the problem was to determine the limits for an iterated integral. Sometimes the limits are known and we want to determine the region.

Example 6 Sketch the region of integration for the iterated integral $\displaystyle\int_0^6 \int_{x/3}^2 x\sqrt{y^3+1}\,dy\,dx$.

Solution The inner integral is with respect to y, so we imagine the region built of vertical strips. The bottom of each strip is on the line $y = x/3$, and the top is on the horizontal line $y = 2$. Since the limits of the outer integral are 0 and 6, the whole region is contained between the vertical lines $x = 0$ and $x = 6$. Notice that the lines $y = 2$ and $y = x/3$ meet where $x = 6$. See Figure 16.19.

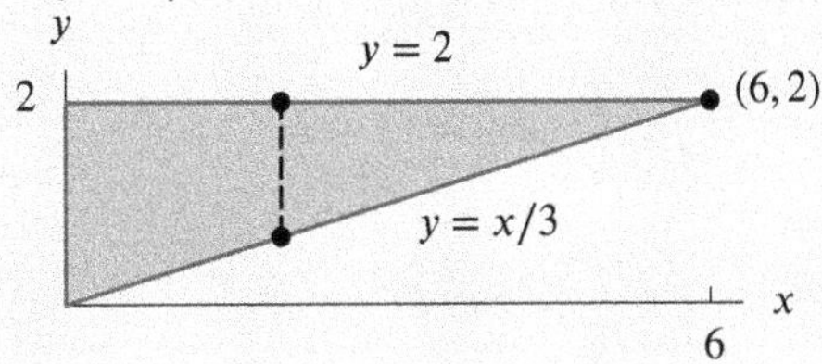

Figure 16.19: The region of integration for Example 6, showing the vertical strip

Reversing the Order of Integration

It is sometimes helpful to reverse the order of integration in an iterated integral. An integral which is difficult or impossible with the integration in one order can be quite straightforward in the other. The next example is such a case.

Example 7 Evaluate $\displaystyle\int_0^6 \int_{x/3}^2 x\sqrt{y^3+1}\,dy\,dx$ using the region sketched in Figure 16.19.

Solution Since $\sqrt{y^3+1}$ has no elementary antiderivative, we cannot calculate the inner integral symbolically. We try reversing the order of integration. From Figure 16.19, we see that horizontal strips go from $x = 0$ to $x = 3y$ and that there is a strip for every y from 0 to 2. Thus, when we change the order of integration we get

$$\int_0^6 \int_{x/3}^2 x\sqrt{y^3+1}\,dy\,dx = \int_0^2 \int_0^{3y} x\sqrt{y^3+1}\,dx\,dy.$$

Now we can at least do the inner integral because we know the antiderivative of x. What about the outer integral?

$$\int_0^2 \int_0^{3y} x\sqrt{y^3+1}\,dx\,dy = \int_0^2 \left(\frac{x^2}{2}\sqrt{y^3+1}\right)\Bigg|_{x=0}^{x=3y} dy = \int_0^2 \frac{9y^2}{2}(y^3+1)^{1/2}\,dy$$
$$= (y^3+1)^{3/2}\Big|_0^2 = 27 - 1 = 26.$$

Thus, reversing the order of integration made the integral in the previous problem much easier. Notice that to reverse the order it is essential first to sketch the region over which the integration is being performed.

Exercises and Problems for Section 16.2 Online Resource: Additional Problems for Section 16.2

EXERCISES

In Exercises 1–4, sketch the region of integration.

1. $\int_0^{\pi} \int_0^{x} y \sin x \, dy\, dx$
2. $\int_0^1 \int_{y^2}^{y} xy \, dx\, dy$
3. $\int_0^2 \int_0^{y^2} y^2 x \, dx\, dy$
4. $\int_0^1 \int_{x-2}^{\cos \pi x} y \, dy\, dx$

For Exercises 5–12, evaluate the integral.

5. $\int_0^3 \int_0^4 (4x+3y)\, dx\, dy$
6. $\int_0^2 \int_0^3 (x^2+y^2)\, dy\, dx$
7. $\int_0^3 \int_0^2 6xy \, dy\, dx$
8. $\int_0^1 \int_0^2 x^2 y \, dy\, dx$
9. $\int_0^1 \int_0^1 ye^{xy} \, dx\, dy$
10. $\int_0^2 \int_0^y y \, dx\, dy$
11. $\int_0^3 \int_0^y \sin x \, dx\, dy$
12. $\int_0^{\pi/2} \int_0^{\sin x} x \, dy\, dx$

For Exercises 13–20, sketch the region of integration and evaluate the integral.

13. $\int_1^3 \int_0^4 e^{x+y} \, dy\, dx$
14. $\int_0^2 \int_0^x e^{x^2} \, dy\, dx$
15. $\int_1^5 \int_x^{2x} \sin x \, dy\, dx$
16. $\int_1^4 \int_{\sqrt{y}}^{y} x^2 y^3 \, dx\, dy$
17. $\int_1^2 \int_y^{3y} xy \, dx\, dy$
18. $\int_0^1 \int_x^{\sqrt{x}} 30x \, dy\, dx$
19. $\int_0^2 \int_0^{2x} xe^{x^3} \, dy\, dx$
20. $\int_0^1 \int_1^{1+x^2} \frac{x}{\sqrt{y}} \, dy\, dx$

In Exercises 21–26, write $\int_R f\,dA$ as an iterated integral for the shaded region R.

21.
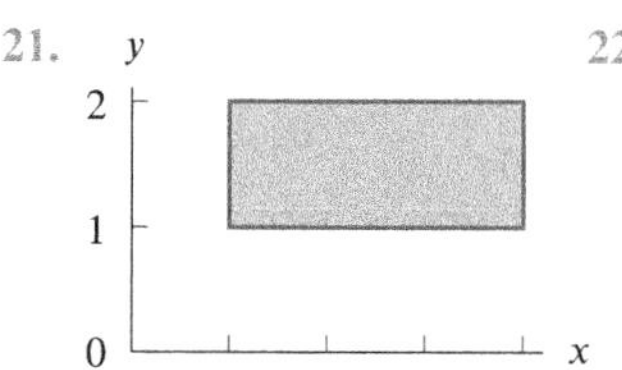

22.
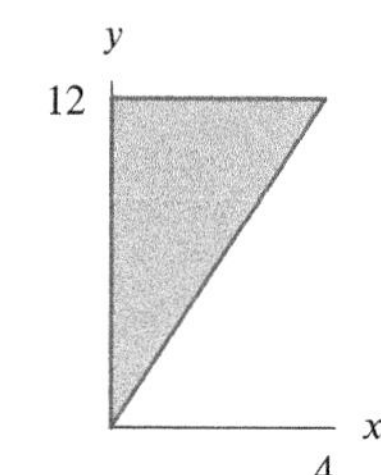

23.
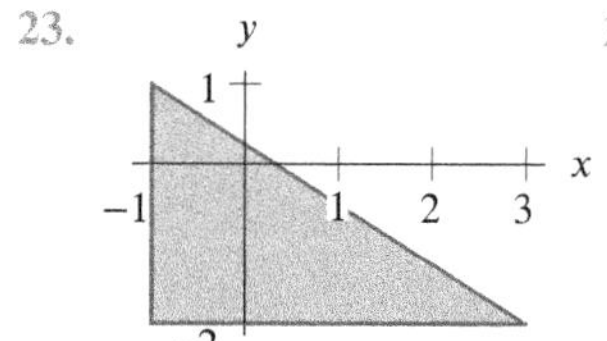

24.
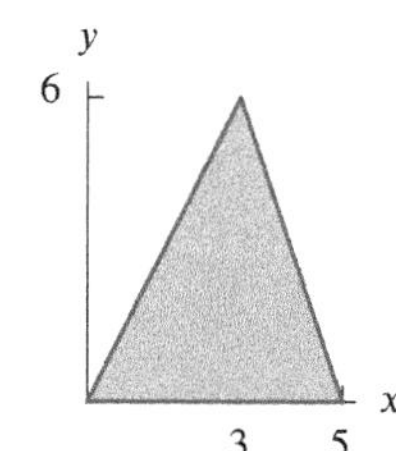

25.
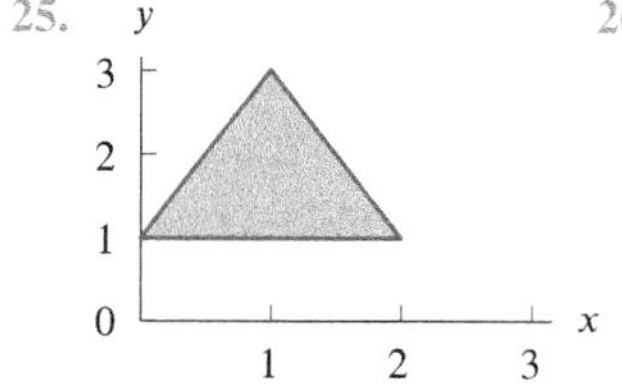

26.
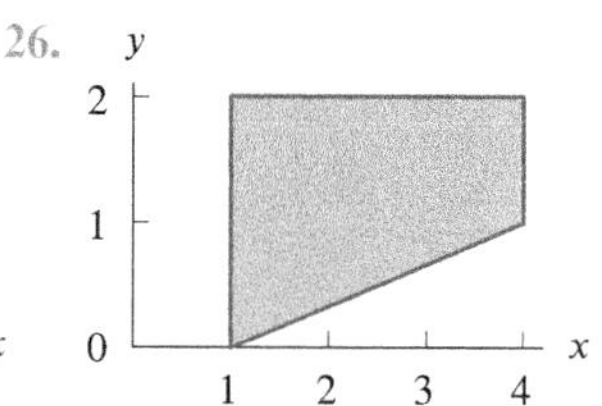

For Exercises 27–28, write $\int_R f\,dA$ as an iterated integral in two different ways for the shaded region R.

27. 28.
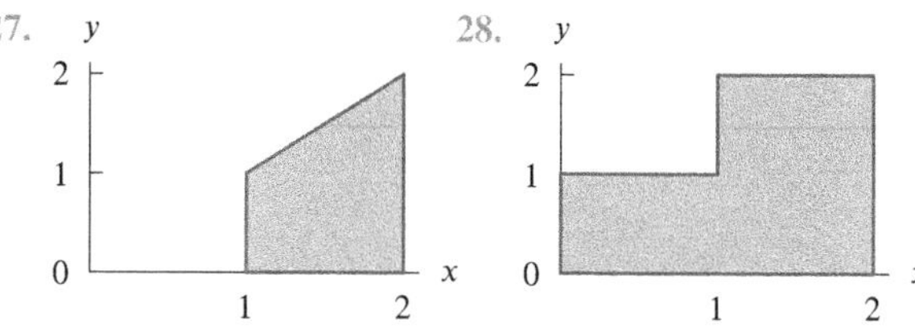

For Exercises 29–33, evaluate the integral.

29. $\int_R \sqrt{x+y}\, dA$, where R is the rectangle $0 \le x \le 1$, $0 \le y \le 2$.

30. The integral in Exercise 29 using the other order of integration.

31. $\int_R (5x^2+1)\sin 3y\, dA$, where R is the rectangle $-1 \le x \le 1, 0 \le y \le \pi/3$.

32. $\int_R xy\, dA$, where R is the triangle $x+y \le 1,\ x \ge 0,\ y \ge 0$.

33. $\int_R (2x+3y)^2\, dA$, where R is the triangle with vertices at $(-1,0)$, $(0,1)$, and $(1,0)$.

PROBLEMS

In Problems 34–37, integrate $f(x,y) = xy$ over the region R.

34.

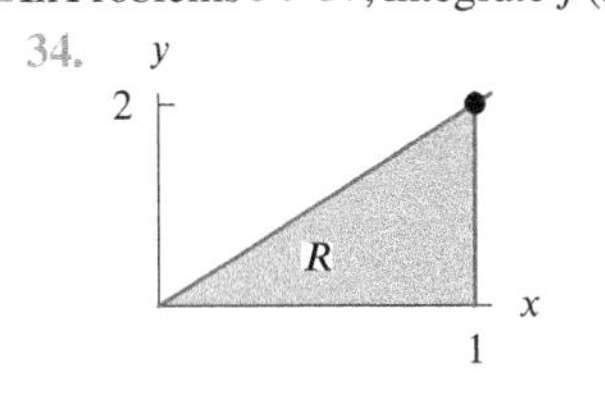

35.

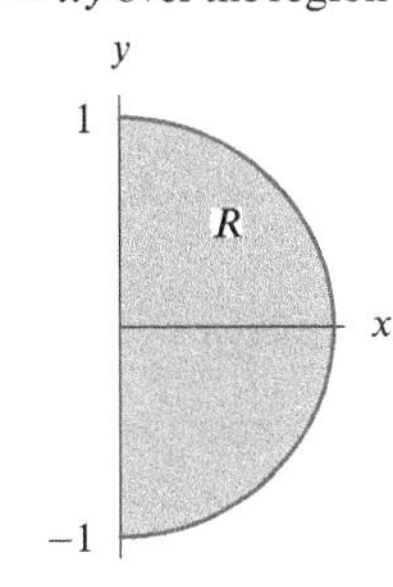

36.

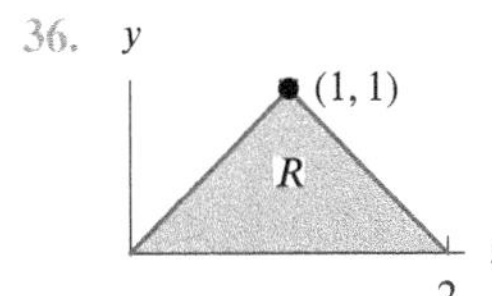

37.

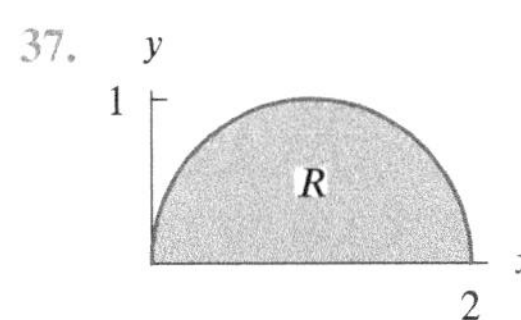

38. **(a)** Use four subrectangles to approximate the volume of the object whose base is the region $0 \le x \le 4$ and $0 \le y \le 6$, and whose height is given by $f(x,y) = xy$. Find an overestimate and an underestimate and average the two.
(b) Integrate to find the exact volume of the three-dimensional object described in part (a).

For Problems 39–42, sketch the region of integration then rewrite the integral with the order of integration reversed.

39. $\displaystyle\int_0^3 \int_{2y}^6 f(x,y)\, dx\, dy$

40. $\displaystyle\int_0^2 \int_0^{\sqrt{4-x^2}} f(x,y)\, dy\, dx$

41. $\displaystyle\int_{-3}^3 \int_0^{9-x^2} f(x,y)\, dy\, dx$

42. $\displaystyle\int_0^2 \int_{y-2}^{2-y} f(x,y)\, dx\, dy$

In Problems 43–50, evaluate the integral by reversing the order of integration.

43. $\displaystyle\int_0^1 \int_y^1 e^{x^2}\, dx\, dy$

44. $\displaystyle\int_0^1 \int_y^1 \sin(x^2)\ dx\, dy$

45. $\displaystyle\int_0^1 \int_{\sqrt{y}}^1 \sqrt{2+x^3}\, dx\, dy$

46. $\displaystyle\int_0^3 \int_{y^2}^9 y\sin(x^2)\, dx\, dy$

47. $\displaystyle\int_0^1 \int_{e^y}^e \frac{x}{\ln x}\, dx\, dy$

48. $\displaystyle\int_0^1 \int_x^1 \cos(y^2)\, dy\, dx$

49. $\displaystyle\int_0^8 \int_{\sqrt[3]{y}}^2 \frac{1}{1+x^4}\, dx\, dy$

50. $\displaystyle\int_0^1 \int_0^x e^{2y-y^2}\, dy\, dx$

51. Each of the integrals (I)–(VI) takes one of two distinct values. Without evaluating, group them by value.

I. $\displaystyle\int_0^5 \int_0^{10} xy^2\, dx\, dy$ II. $\displaystyle\int_0^5 \int_0^{10} xy^2\, dy\, dx$

III. $\displaystyle\int_0^{10} \int_0^5 xy^2\, dx\, dy$ IV. $\displaystyle\int_0^{10} \int_0^5 xy^2\, dy\, dx$

V. $\displaystyle\int_0^5 \int_0^{10} uv^2\, du\, dv$ VI. $\displaystyle\int_0^5 \int_0^{10} uv^2\, dv\, du$

52. Find the volume under the graph of the function $f(x,y) = 6x^2y$ over the region shown in Figure 16.20.

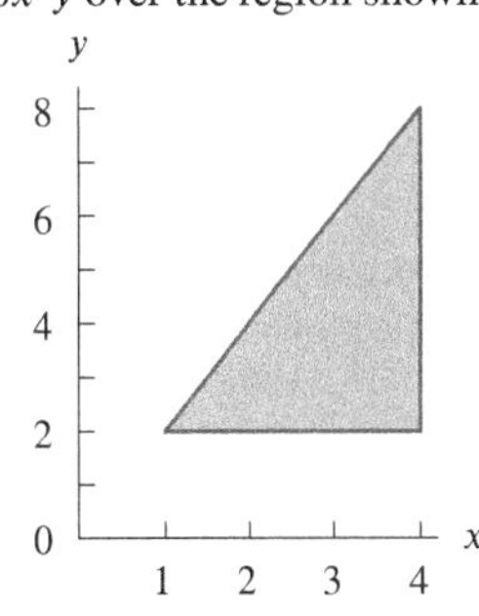

Figure 16.20

53. **(a)** Find the volume below the surface $z = x^2 + y^2$ and above the xy-plane for $-1 \le x \le 1, -1 \le y \le 1$.
(b) Find the volume above the surface $z = x^2 + y^2$ and below the plane $z = 2$ for $-1 \le x \le 1$, $-1 \le y \le 1$.

54. Compute the integral

$$\int\int_R (2x^2 + y)\, dA,$$

where R is the triangular region with vertices at $(0,1)$, $(-2,3)$ and $(2,3)$.

55. **(a)** Sketch the region in the xy-plane bounded by the x-axis, $y = x$, and $x + y = 1$.
(b) Express the integral of $f(x,y)$ over this region in terms of iterated integrals in two ways. (In one, use $dx\,dy$; in the other, use $dy\,dx$.)
(c) Using one of your answers to part (b), evaluate the integral exactly with $f(x,y) = x$.

56. Let $f(x, y) = x^2 e^{x^2}$ and let R be the triangle bounded by the lines $x = 3$, $x = y/2$, and $y = x$ in the xy-plane.

(a) Express $\int_R f\, dA$ as a double integral in two different ways.
(b) Evaluate one of them.

57. Find the average value of $f(x, y) = x^2 + 4y$ on the rectangle $0 \leq x \leq 3$ and $0 \leq y \leq 6$.

58. Find the average value of $f(x, y) = xy^2$ on the rectangle $0 \leq x \leq 4, 0 \leq y \leq 3$.

59. Figure 16.21 shows two metal plates carrying electrical charges. The charge density (in coulombs per square meter) of each at the point (x, y) is $\sigma(x, y) = 6x + 6$ for x, y in meters.

(a) Without calculation, decide which plate carries a greater total charge, and explain your reasoning.
(b) Find the total charge on both plates, and compare to your answer from part (a).

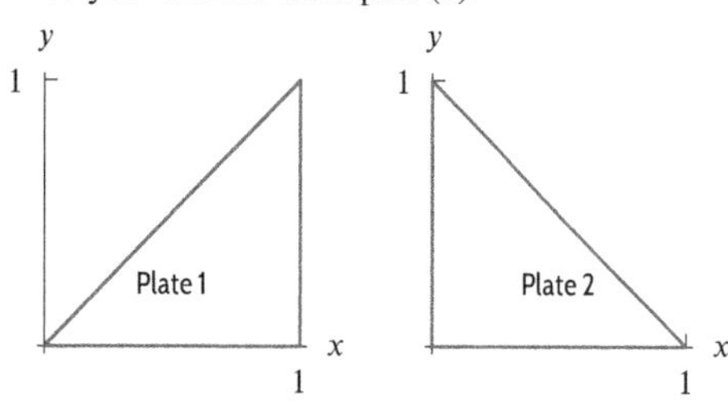

Figure 16.21

60. The population density in people per km^2 for the trapezoid-shaped town in Figure 16.22 for x, y in kilometers is $\delta(x, y) = 100x + 200y$. Find the town's population.

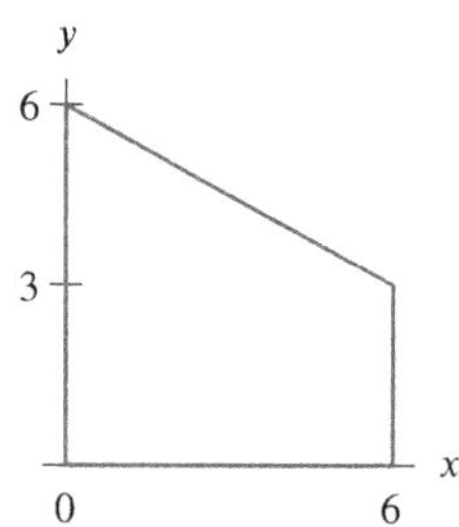

Figure 16.22

61. The quarter-disk-shaped metal plate in Figure 16.23 has radius 3 and density $\sigma(x, y) = 2y$ gm/cm^2, with x, y in cm. Find the mass of the plate.

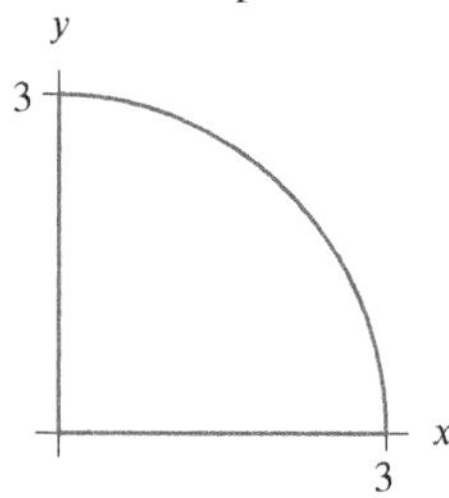

Figure 16.23

In Problems 62–63 set up, but do not evaluate, an iterated integral for the volume of the solid.

62. Under the graph of $f(x, y) = 25 - x^2 - y^2$ and above the xy-plane.

63. Below the graph of $f(x, y) = 25 - x^2 - y^2$ and above the plane $z = 16$.

64. A solid with flat base in the xy-plane is bounded by the vertical planes $y = 0$ and $y - x = 4$, and the slanted plane $2x + y + z = 4$.

(a) Draw the base of the solid.
(b) Set up, but do not evaluate, an iterated integral for the volume of the solid.

In Problems 65–69, find the volume of the solid region.

65. Under the graph of $f(x, y) = xy$ and above the square $0 \leq x \leq 2, 0 \leq y \leq 2$ in the xy-plane.

66. Under the graph of $f(x, y) = x^2 + y^2$ and above the triangle $0 \leq y \leq x, 0 \leq x \leq 1$.

67. Under the graph of $f(x, y) = x + y$ and above the region $y^2 \leq x, 0 \leq x \leq 9, y \geq 0$.

68. Under the graph of $2x + y + z = 4$ in the first octant.

69. The solid region R bounded by the coordinate planes and the graph of $ax + by + cz = 1$. Assume a, b, and $c > 0$.

70. If R is the region $x + y \geq a, x^2 + y^2 \leq a^2$, with $a > 0$, evaluate the integral

$$\int_R xy\, dA.$$

71. The region W lies below the surface $f(x, y) = 2e^{-(x-1)^2 - y^2}$ and above the disk $x^2 + y^2 \leq 4$ in the xy-plane.

(a) Describe in words the contours of f, using $f(x, y) = 1$ as an example.
(b) Write an integral giving the area of the cross-section of W in the plane $x = 1$.
(c) Write an iterated double integral giving the volume of W.

72. Find the average distance to the x-axis for points in the region in the first quadrant bounded by the x-axis and the graph of $y = x - x^2$.

73. Give the contour diagram of a function f whose average value on the square $0 \leq x \leq 1, 0 \leq y \leq 1$ is

(a) Greater than the average of the values of f at the four corners of the square.
(b) Less than the average of the values of f at the four corners of the square.

74. The function $f(x, y) = ax + by$ has an average value of 20 on the rectangle $0 \leq x \leq 2, 0 \leq y \leq 3$.

(a) What can you say about the constants a and b?

(b) Find two different choices for f that have average value 20 on the rectangle, and give their contour diagrams on the rectangle.

75. The function $f(x,y) = ax^2 + bxy + cy^2$ has an average value of 20 on the square $0 \le x \le 2, 0 \le y \le 2$.

(a) What can you say about the constants a, b, and c?

(b) Find two different choices for f that have average value 20 on the square, and give their contour diagrams on the square.

Strengthen Your Understanding

In Problems 76–77, explain what is wrong with the statement.

76. $\int_0^1 \int_0^x f(x,y)\,dy\,dx = \int_0^1 \int_0^y f(x,y)\,dx\,dy$

77. $\int_0^1 \int_0^y xy\,dx\,dy = \int_0^y \int_0^1 xy\,dy\,dx$

In Problems 78–80, give an example of:

78. An iterated double integral, with limits of integration, giving the volume of a cylinder standing vertically with a circular base in the xy-plane.

79. A nonconstant function, f, whose integral is 4 over the triangular region with vertices $(0,0)$, $(1,0)$, $(1,1)$.

80. A double integral representing the volume of a triangular prism of base area 6.

Are the statements in Problems 81–88 true or false? Give reasons for your answer.

81. The iterated integral $\int_0^1 \int_5^{12} f\,dx\,dy$ is computed over the rectangle $0 \le x \le 1, 5 \le y \le 12$.

82. If R is the region inside the triangle with vertices $(0,0)$, $(1,1)$ and $(0,2)$, then the double integral $\int_R f\,dA$ can be evaluated by an iterated integral of the form $\int_0^2 \int_0^1 f\,dx\,dy$.

83. The region of integration of the iterated integral $\int_1^2 \int_{x^2}^{x^3} f\,dy\,dx$ lies completely in the first quadrant (that is, $x \ge 0, y \ge 0$).

84. If the limits a, b, c and d in the iterated integral $\int_a^b \int_c^d f\,dy\,dx$ are all positive, then the value of $\int_a^b \int_c^d f\,dy\,dx$ is also positive.

85. If $f(x,y)$ is a function of y only, then $\int_a^b \int_0^1 f\,dx\,dy = \int_a^b f\,dy$.

86. If R is the region inside a circle of radius a, centered at the origin, then $\int_R f\,dA = \int_{-a}^{a} \int_0^{\sqrt{a^2-x^2}} f\,dy\,dx$.

87. If $f(x,y) = g(x) \cdot h(y)$, where g and h are single-variable functions, then

$$\int_a^b \int_c^d f\,dy\,dx = \left(\int_a^b g(x)\,dx\right) \cdot \left(\int_c^d h(y)\,dy\right).$$

88. If $f(x,y) = g(x) + h(y)$, where g and h are single-variable functions, then

$$\int_a^b \int_c^d f\,dx\,dy = \left(\int_a^b g(x)\,dx\right) + \left(\int_c^d h(y)\,dy\right).$$

16.3 TRIPLE INTEGRALS

A continuous function of three variables can be integrated over a solid region W in 3-space in the same way as a function of two variables is integrated over a flat region in 2-space. Again, we start with a Riemann sum. First we subdivide W into smaller regions, then we multiply the volume of each region by a value of the function in that region, and then we add the results. For example, if W is the box $a \le x \le b$, $c \le y \le d$, $p \le z \le q$, then we subdivide each side into n, m, and l pieces, thereby chopping W into nml smaller boxes, as shown in Figure 16.24.

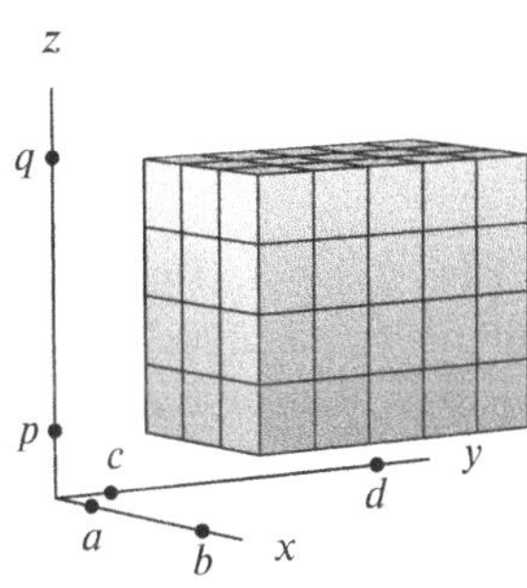

Figure 16.24: Subdividing a three-dimensional box

The volume of each smaller box is

$$\Delta V = \Delta x \Delta y \Delta z,$$

where $\Delta x = (b-a)/n$, and $\Delta y = (d-c)/m$, and $\Delta z = (q-p)/l$. Using this subdivision, we pick a point $(u_{ijk}, v_{ijk}, w_{ijk})$ in the ijk-th small box and construct a Riemann sum

$$\sum_{i,j,k} f(u_{ijk}, v_{ijk}, w_{ijk})\,\Delta V.$$

If f is continuous, as Δx, Δy, and Δz approach 0, this Riemann sum approaches the definite integral, $\int_W f\,dV$, called a *triple integral*, which is defined as

$$\int_W f\,dV = \lim_{\Delta x, \Delta y, \Delta z \to 0} \sum_{i,j,k} f(u_{ijk}, v_{ijk}, w_{ijk})\,\Delta x\,\Delta y\,\Delta z.$$

As in the case of a double integral, we can evaluate this integral as an iterated integral:

Triple integral as an iterated integral

$$\int_W f\,dV = \int_p^q \left(\int_c^d \left(\int_a^b f(x,y,z)\,dx \right) dy \right) dz,$$

where y and z are treated as constants in the innermost (dx) integral, and z is treated as a constant in the middle (dy) integral. Other orders of integration are possible.

Example 1 A cube C has sides of length 4 cm and is made of a material of variable density. If one corner is at the origin and the adjacent corners are on the positive x, y, and z axes, then the density at the point (x,y,z) is $\delta(x,y,z) = 1 + xyz$ gm/cm^3. Find the mass of the cube.

Solution Consider a small piece ΔV of the cube, small enough so that the density remains close to constant over the piece. Then

$$\text{Mass of small piece} = \text{Density} \cdot \text{Volume} \approx \delta(x,y,z)\,\Delta V.$$

To get the total mass, we add the masses of the small pieces and take the limit as $\Delta V \to 0$. Thus, the mass is the triple integral

$$\begin{aligned} M &= \int_C \delta\,dV = \int_0^4 \int_0^4 \int_0^4 (1 + xyz)\,dx\,dy\,dz = \int_0^4 \int_0^4 \left(x + \frac{1}{2}x^2 yz \right)\Bigg|_{x=0}^{x=4} dy\,dz \\ &= \int_0^4 \int_0^4 (4 + 8yz)\,dy\,dz = \int_0^4 \left(4y + 4y^2 z\right)\Bigg|_{y=0}^{y=4} dz = \int_0^4 (16 + 64z)\,dz = 576\,\text{gm}. \end{aligned}$$

Example 2 Express the volume of the building described in Example 1 on page 848 as a triple integral.

Solution The building is given by $0 \le x \le 8$, $0 \le y \le 16$, and $0 \le z \le 12 - x/4 - y/8$. (See Figure 16.25.) To find its volume, divide it into small cubes of volume $\Delta V = \Delta x\,\Delta y\,\Delta z$ and add. First, make a vertical stack of cubes above the point $(x, y, 0)$. This stack goes from $z = 0$ to $z = 12 - x/4 - y/8$, so

$$\text{Volume of vertical stack} \approx \sum_z \Delta V = \sum_z \Delta x\,\Delta y\,\Delta z = \left(\sum_z \Delta z \right) \Delta x\,\Delta y.$$

Next, line up these stacks parallel to the y-axis to form a slice from $y = 0$ to $y = 16$. So

$$\text{Volume of slice} \approx \left(\sum_y \sum_z \Delta z \, \Delta y \right) \Delta x.$$

Finally, line up the slices along the x-axis from $x = 0$ to $x = 8$ and add up their volumes, to get

$$\text{Volume of building} \approx \sum_x \sum_y \sum_z \Delta z \, \Delta y \, \Delta x.$$

Thus, in the limit,

$$\text{Volume of building} = \int_0^8 \int_0^{16} \int_0^{12-x/4-y/8} 1 \, dz \, dy \, dx.$$

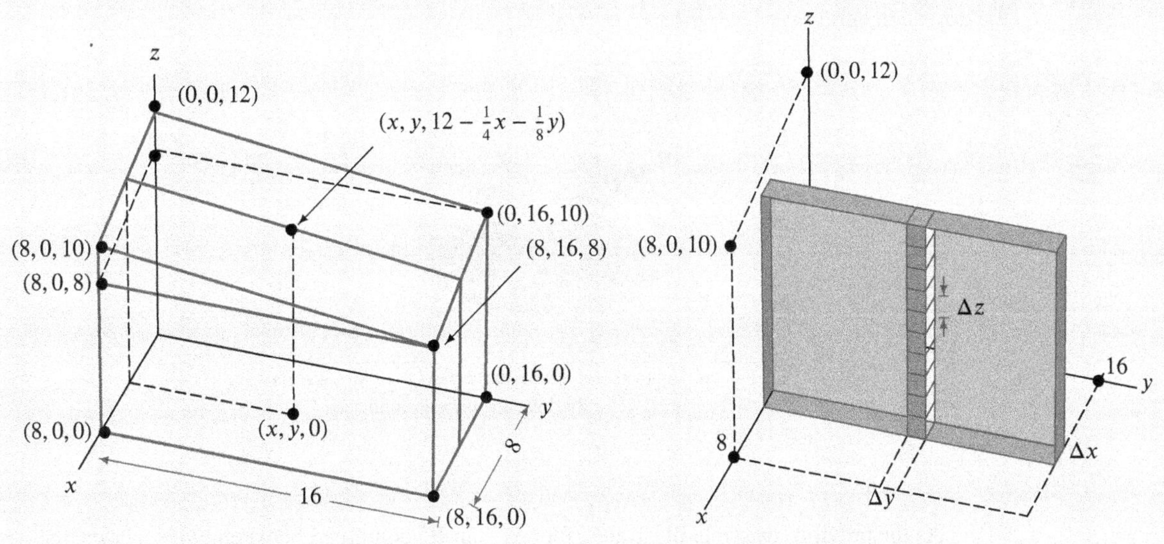

Figure 16.25: Volume of building (shown to left) divided into blocks and slabs for a triple integral

The preceding examples show that the triple integral has interpretations similar to the double integral:

- If $\rho(x, y, z)$ is density, then $\int_W \rho \, dV$ is the total quantity in the solid region W.
- $\int_W 1 \, dV$ is the volume of the solid region W.

Example 3 Set up an iterated integral to compute the mass of the solid cone bounded by $z = \sqrt{x^2 + y^2}$ and $z = 3$, if the density is given by $\delta(x, y, z) = z$.

Solution We break the cone in Figure 16.26 into small cubes of volume $\Delta V = \Delta x \, \Delta y \, \Delta z$, on which the density is approximately constant, and approximate the mass of each cube by $\delta(x, y, z) \, \Delta x \, \Delta y \, \Delta z$. Stacking the cubes vertically above the point $(x, y, 0)$, starting on the cone at height $z = \sqrt{x^2 + y^2}$ and going up to $z = 3$, tells us that the inner integral is

$$\int_{\sqrt{x^2+y^2}}^3 \delta(x, y, z) \, dz = \int_{\sqrt{x^2+y^2}}^3 z \, dz.$$

There is a stack for every point in the xy-plane in the shadow of the cone. The cone $z = \sqrt{x^2+y^2}$ intersects the horizontal plane $z = 3$ in the circle $x^2+y^2 = 9$, so there is a stack for all (x, y) in the region $x^2+y^2 \le 9$. Lining up the stacks parallel to the y-axis gives a slice from $y = -\sqrt{9-x^2}$ to $y = \sqrt{9-x^2}$, for each fixed value of x. Thus, the limits on the middle integral are

$$\int_{-\sqrt{9-x^2}}^{\sqrt{9-x^2}} \int_{\sqrt{x^2+y^2}}^{3} z\,dz\,dy.$$

Finally, there is a slice for each x between -3 and 3, so the integral we want is

$$\text{Mass} = \int_{-3}^{3} \int_{-\sqrt{9-x^2}}^{\sqrt{9-x^2}} \int_{\sqrt{x^2+y^2}}^{3} z\,dz\,dy\,dx.$$

Notice that setting up the limits on the two outer integrals was just like setting up the limits for a double integral over the region $x^2+y^2 \le 9$.

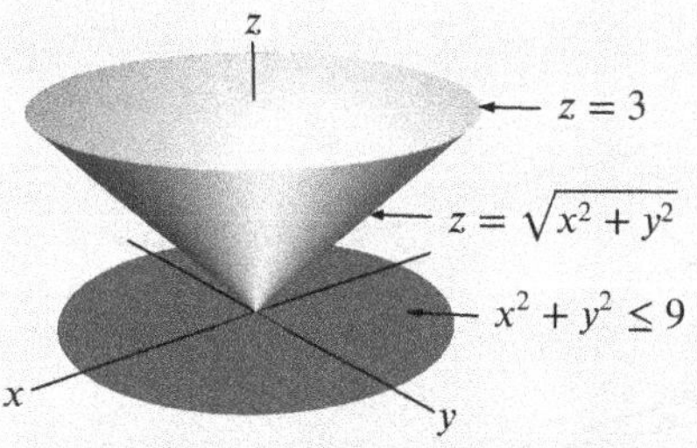

Figure 16.26: The cone $z = \sqrt{x^2+y^2}$ with its shadow on the xy-plane

As the previous example illustrates, for a region W contained between two surfaces, the innermost limits correspond to these surfaces. The middle and outer limits ensure that we integrate over the "shadow" of W in the xy-plane.

Limits on Triple Integrals

- The limits for the outer integral are constants.
- The limits for the middle integral can involve only one variable (that in the outer integral).
- The limits for the inner integral can involve two variables (those on the two outer integrals).

Exercises and Problems for Section 16.3 Online Resource: Additional Problems for Section 16.3

EXERCISES

In Exercises 1–4, find the triple integrals of the function over the region W.

1. $f(x, y, z) = x^2 + 5y^2 - z$, W is the rectangular box $0 \le x \le 2$, $-1 \le y \le 1$, $2 \le z \le 3$.

2. $h(x, y, z) = ax + by + cz$, W is the rectangular box $0 \le x \le 1$, $0 \le y \le 1$, $0 \le z \le 2$.

3. $f(x, y, z) = \sin x \cos(y + z)$, W is the cube $0 \le x \le \pi$, $0 \le y \le \pi$, $0 \le z \le \pi$.

4. $f(x, y, z) = e^{-x-y-z}$, W is the rectangular box with corners at $(0, 0, 0)$, $(a, 0, 0)$, $(0, b, 0)$, and $(0, 0, c)$.

Sketch the region of integration in Exercises 5–13.

5. $\int_0^1\int_{-1}^1\int_0^{\sqrt{1-x^2}} f(x,y,z)\,dz\,dx\,dy$

6. $\int_0^1\int_{-1}^1\int_0^{\sqrt{1-z^2}} f(x,y,z)\,dy\,dz\,dx$

7. $\int_0^1\int_{-1}^1\int_{-\sqrt{1-x^2}}^{\sqrt{1-x^2}} f(x,y,z)\,dz\,dx\,dy$

8. $\int_{-1}^1\int_0^1\int_{-\sqrt{1-z^2}}^{\sqrt{1-z^2}} f(x,y,z)\,dy\,dz\,dx$

9. $\int_{-1}^1\int_{-\sqrt{1-x^2}}^{\sqrt{1-x^2}}\int_0^{\sqrt{1-x^2-z^2}} f(x,y,z)\,dy\,dz\,dx$

10. $\int_0^1\int_{-\sqrt{1-z^2}}^{\sqrt{1-z^2}}\int_0^{\sqrt{1-x^2-z^2}} f(x,y,z)\,dy\,dx\,dz$

11. $\int_0^1\int_0^{\sqrt{1-y^2}}\int_{-\sqrt{1-x^2-y^2}}^{\sqrt{1-x^2-y^2}} f(x,y,z)\,dz\,dx\,dy$

12. $\int_0^1\int_{-\sqrt{1-z^2}}^{\sqrt{1-z^2}}\int_{-\sqrt{1-y^2-z^2}}^{\sqrt{1-y^2-z^2}} f(x,y,z)\,dx\,dy\,dz$

13. $\int_0^1\int_0^{\sqrt{1-z^2}}\int_{-\sqrt{1-x^2-z^2}}^{\sqrt{1-x^2-z^2}} f(x,y,z)\,dy\,dx\,dz$

In Exercises 14–15, for x, y and z in meters, what does the integral over the solid region E represent? Give units.

14. $\int_E 1\,dV$

15. $\int_E \delta(x,y,z)\,dV$, where $\delta(x,y,z)$ is density, in kg/m^3.

PROBLEMS

In Problems 16–20, decide whether the integrals are positive, negative, or zero. Let S be the solid sphere $x^2+y^2+z^2 \le 1$, and T be the top half of this sphere (with $z \ge 0$), and B be the bottom half (with $z \le 0$), and R be the right half of the sphere (with $x \ge 0$), and L be the left half (with $x \le 0$).

16. $\int_T e^z\,dV$

17. $\int_B e^z\,dV$

18. $\int_S \sin z\,dV$

19. $\int_T \sin z\,dV$

20. $\int_R \sin z\,dV$

Let W be the solid cone bounded by $z = \sqrt{x^2+y^2}$ and $z = 2$. For Problems 21–29, decide (without calculating its value) whether the integral is positive, negative, or zero.

21. $\int_W y\,dV$

22. $\int_W x\,dV$

23. $\int_W z\,dV$

24. $\int_W xy\,dV$

25. $\int_W xyz\,dV$

26. $\int_W (z-2)\,dV$

27. $\int_W \sqrt{x^2+y^2}\,dV$

28. $\int_W e^{-xyz}\,dV$

29. $\int_W (z-\sqrt{x^2+y^2})\,dV$

In Problems 30–34, let W be the solid cylinder bounded by $x^2+y^2=1$, $z=0$, and $z=2$. Decide (without calculating its value) whether the integral is positive, negative, or zero.

30. $\int_W x\,dV$

31. $\int_W z\,dV$

32. $\int_W (x^2+y^2-2)\,dV$

33. $\int_W (z-1)\,dV$

34. $\int_W e^{-y}\,dV$

35. Find the volume of the region bounded by the planes $z=3y$, $z=y$, $y=1$, $x=1$, and $x=2$.

36. Find the volume of the region bounded by $z = x^2$, $0 \le x \le 5$, and the planes $y=0$, $y=3$, and $z=0$.

37. Find the volume of the region in the first octant bounded by the coordinate planes and the surface $x+y+z=2$.

38. A trough with triangular cross-section lies along the x-axis for $0 \le x \le 10$. The slanted sides are given by $z=y$ and $z=-y$ for $0 \le z \le 1$ and the ends by $x=0$ and $x=10$, where x, y, z are in meters. The trough contains a sludge whose density at the point (x,y,z) is $\delta = e^{-3x}$ kg per m^3.

   (a) Express the total mass of sludge in the trough in terms of triple integrals.
   (b) Find the mass.

39. Find the volume of the region bounded by $z=x+y$, $z=10$, and the planes $x=0$, $y=0$.

In Problems 40–45, write a triple integral, including limits of integration, that gives the specified volume.

40. Between $z = x+y$ and $z = 1+2x+2y$ and above $0 \le x \le 1, 0 \le y \le 2$.

41. Between the paraboloid $z = x^2+y^2$ and the sphere $x^2+y^2+z^2=4$ and above the disk $x^2+y^2 \le 1$.

42. Between $2x+2y+z=6$ and $3x+4y+z=6$ and above $x+y \le 1, x \ge 0, y \ge 0$.

43. Under the sphere $x^2+y^2+z^2=9$ and above the region between $y=x$ and $y=2x-2$ in the xy-plane in the first quadrant.

44. Between the top portion of the sphere $x^2+y^2+z^2=9$ and the plane $z=2$.

45. Under the sphere $x^2+y^2+z^2=4$ and above the region $x^2+y^2 \le 4, 0 \le x \le 1, 0 \le y \le 2$ in the xy-plane.

In Problems 46–49, write limits of integration for the integral $\int_W f(x, y, z)\,dV$ where W is the quarter or half sphere or cylinder shown.

46.

47.

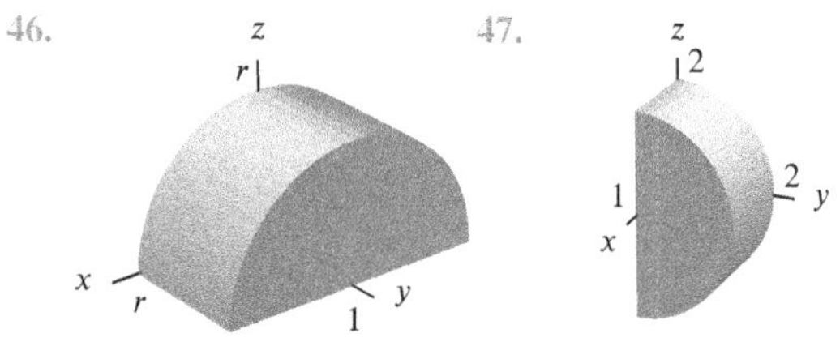

48.

49.

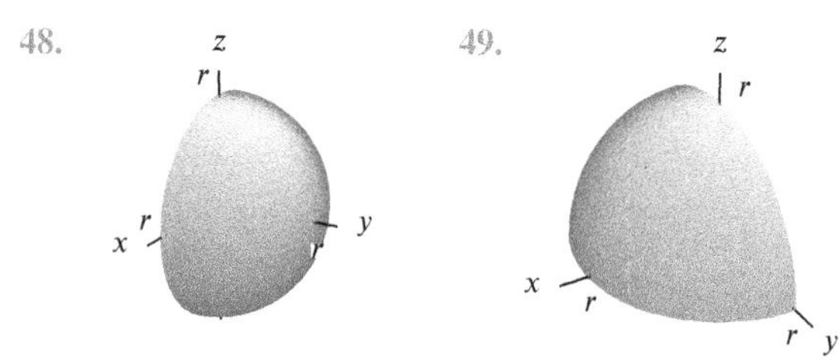

50. Find the volume of the region between the plane $z = x$ and the surface $z = x^2$, and the planes $y = 0$, and $y = 3$.

51. Find the volume of the region bounded by $z = x + y$, $0 \le x \le 5$, $0 \le y \le 5$, and the planes $x = 0$, $y = 0$, and $z = 0$.

52. Find the volume of the pyramid with base in the plane $z = -6$ and sides formed by the three planes $y = 0$ and $y - x = 4$ and $2x + y + z = 4$.

53. Find the volume between the planes $z = 1 + x + y$ and $x + y + z = 1$ and above the triangle $x + y \le 1, x \ge 0$, $y \ge 0$ in the xy-plane.

54. Find the mass of a triangular-shaped solid bounded by the planes $z = 1 + x$, $z = 1 - x$, $z = 0$, and with $0 \le y \le 3$. The density is $\delta = 10 - z$ gm/cm^3, and x, y, z are in cm.

55. Find the mass of the solid bounded by the xy-plane, yz-plane, xz-plane, and the plane $(x/3)+(y/2)+(z/6) = 1$, if the density of the solid is given by $\delta(x, y, z) = x + y$.

56. Find the mass of the pyramid with base in the plane $z = -6$ and sides formed by the three planes $y = 0$ and $y - x = 4$ and $2x + y + z = 4$, if the density of the solid is given by $\delta(x, y, z) = y$.

57. Let E be the solid pyramid bounded by the planes $x + z = 6$, $x - z = 0$, $y + z = 6$, $y - z = 0$, and above the plane $z = 0$ (see Figure 16.27). The density at any point in the pyramid is given by $\delta(x, y, z) = z$ grams per cm^3, where x, y, and z are measured in cm.

(a) Explain in practical terms what the triple integral $\int_E z\,dV$ represents.

(b) In evaluating the integral from part (a), how many separate triple integrals would be required if we chose to integrate in the z-direction first?

(c) Evaluate the triple integral from part (a) by integrating in a well-chosen order.

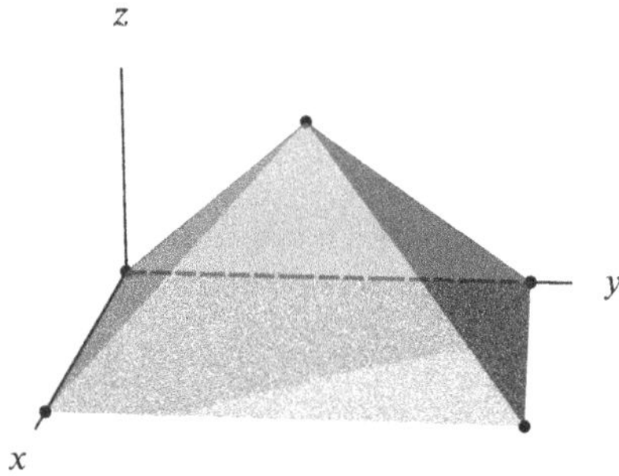

Figure 16.27

58. **(a)** What is the equation of the plane passing through the points $(1, 0, 0)$, $(0, 1, 0)$, and $(0, 0, 1)$?

(b) Find the volume of the region bounded by this plane and the planes $x = 0$, $y = 0$, and $z = 0$.

Problems 59–61 refer to Figure 16.28, which shows triangular portions of the planes $2x+4y+z = 4$, $3x-2y = 0$, $z = 2$, and the three coordinate planes $x = 0$, $y = 0$, and $z = 0$. For each solid region E, write down an iterated integral for the triple integral $\int_E f(x, y, z)\,dV$.

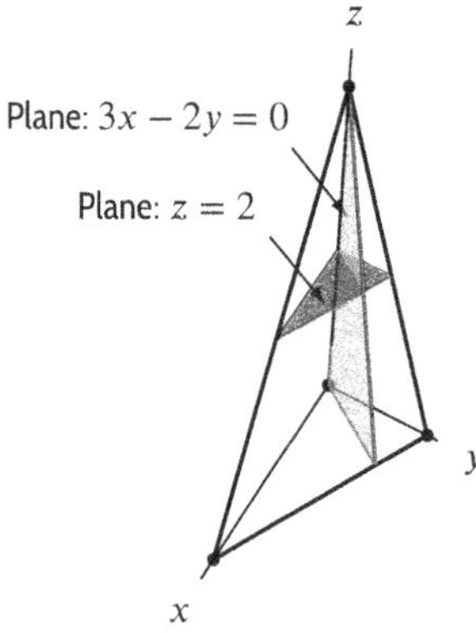

Figure 16.28

59. E is the region bounded by $y = 0$, $z = 0$, $3x - 2y = 0$, and $2x + 4y + z = 4$.

60. E is the region bounded by $x = 0$, $y = 0$, $z = 0$, $z = 2$, and $2x + 4y + z = 4$.

61. E is the region bounded by $x = 0$, $z = 0$, $3x - 2y = 0$, and $2x + 4y + z = 4$.

62. Figure 16.29 shows part of a spherical ball of radius 5 cm. Write an iterated triple integral which represents the volume of this region.

Figure 16.29

63. A solid region D is a half cylinder of radius 1 lying horizontally with its rectangular base in the xy-plane and its axis along the y-axis from $y = 0$ to $y = 10$. (The region is above the xy-plane.)

(a) What is the equation of the curved surface of this half cylinder?

(b) Write the limits of integration of the integral $\int_D f(x, y, z)\, dV$ in Cartesian coordinates.

64. Set up, but do not evaluate, an iterated integral for the volume of the solid formed by the intersections of the cylinders $x^2 + z^2 = 1$ and $y^2 + z^2 = 1$.

Problems 65–67 refer to Figure 16.30, which shows E, the region in the first octant bounded by the parabolic cylinder $z = 6y^2$ and the elliptical cylinder $x^2 + 3y^2 = 12$. For the given order of integration, write an iterated integral equivalent to the triple integral $\int_E f(x, y, z)\, dV$.

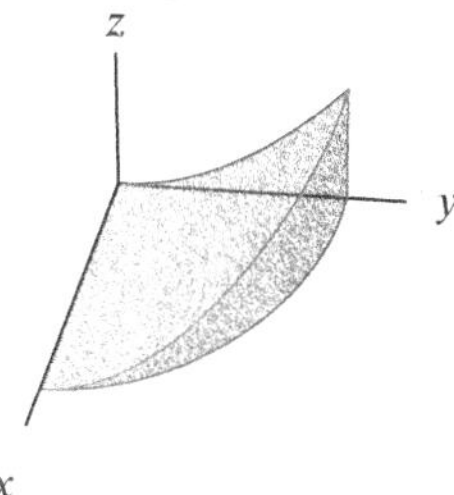

Figure 16.30

65. $dz\,dx\,dy$ 66. $dx\,dz\,dy$ 67. $dy\,dz\,dx$

Problems 68–71 refer to Figure 16.31, which shows E, the region in the first octant bounded by the planes $z = 5$ and $5x + 3z = 15$ and the elliptical cylinder $4x^2 + 9y^2 = 36$. For the given order of integration, write an iterated integral equivalent to the triple integral $\int_E f(x, y, z)\, dV$.

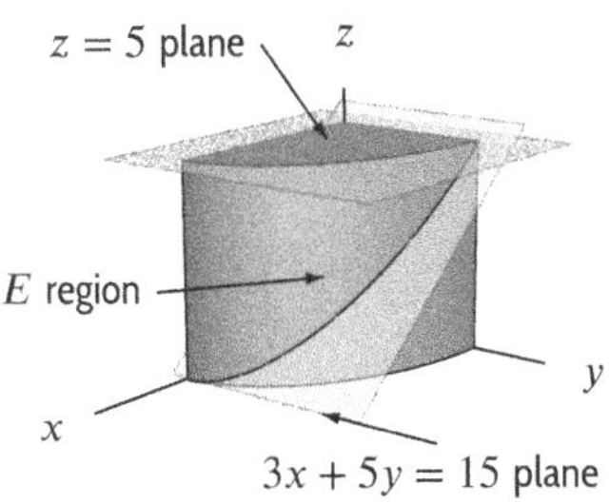

Figure 16.31

68. $dz\,dy\,dx$ 69. $dz\,dx\,dy$

70. $dy\,dz\,dx$ 71. $dy\,dx\,dz$

Problems 72–74 refer to Figure 16.32, which shows E, the region in the first octant bounded by the plane $x + y = 2$ and the parabolic cylinder $z = 4 - x^2$. For the given order of integration, write an iterated integral, or sum of integrals, equivalent to the triple integral $\int_E f(x, y, z)\, dV$.

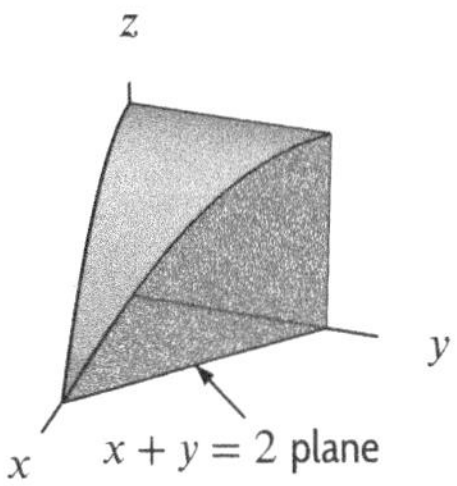

Figure 16.32

72. $dz\,dy\,dx$ 73. $dy\,dz\,dx$ 74. $dy\,dx\,dz$

Problems 75–76 concern the *center of mass*, the point at which the mass of a solid body in motion can be considered to be concentrated. If the object has density $\rho(x, y, z)$ at the point (x, y, z) and occupies a region W, then the coordinates $(\bar{x}, \bar{y}, \bar{z})$ of the center of mass are given by

$$\bar{x} = \frac{1}{m}\int_W x\rho\, dV \quad \bar{y} = \frac{1}{m}\int_W y\rho\, dV \quad \bar{z} = \frac{1}{m}\int_W z\rho\, dV$$

where $m = \int_W \rho\, dV$ is the total mass of the body.

75. A solid is bounded below by the square $z = 0, 0 \le x \le 1, 0 \le y \le 1$ and above by the surface $z = x + y + 1$. Find the total mass and the coordinates of the center of mass if the density is 1 gm/cm^3 and x, y, z are measured in centimeters.

76. Find the center of mass of the tetrahedron that is bounded by the xy, yz, xz planes and the plane $x + 2y + 3z = 1$. Assume the density is 1 gm/cm^3 and x, y, z are in centimeters.

Strengthen Your Understanding

In Problems 77–78, explain what is wrong with the statement.

77. Let S be the solid sphere $x^2 + y^2 + z^2 \le 1$ and let U be the upper half of S where $z \ge 0$. Then $\int_S f(x, y, z)\, dV = 2\int_U f(x, y, z)\, dV$.

78. $\int_0^1 \int_0^x \int_0^y f(x, y, z)\, dz\, dy\, dx = \int_0^1 \int_y^1 \int_0^x f(x, y, z)\, dz\, dx\, dy$

In Problems 79–80, give an example of:

79. A function f such that $\int_R f\, dV = 7$, where R is the cylinder $x^2 + y^2 \le 4, 0 \le z \le 3$.

80. A nonconstant function $f(x, y, z)$ such that if B is the region enclosed by the sphere of radius 1 centered at the origin, the integral $\int_B f(x, y, z)\, dx\, dy\, dz$ is zero.

Are the statements in Problems 81–90 true or false? Give reasons for your answer.

81. If $\rho(x, y, z)$ is mass density of a material in 3-space, then $\int_W \rho(x, y, z)\,dV$ gives the volume of the solid region W.

82. The region of integration of the triple iterated integral $\int_0^1 \int_0^1 \int_0^x f\,dz\,dy\,dx$ lies above a square in the xy-plane and below a plane.

83. If W is the unit ball $x^2 + y^2 + z^2 \le 1$ then an iterated integral over W is $\int_0^1 \int_0^{\sqrt{1-x^2}} \int_0^{\sqrt{1-x^2-y^2}} f\,dz\,dy\,dx$.

84. The iterated integrals $\int_0^1 \int_0^{1-x} \int_0^{1-x-y} f\,dz\,dy\,dx$ and $\int_0^1 \int_0^{1-z} \int_0^{1-y-z} f\,dx\,dy\,dz$ are equal.

85. The iterated integrals $\int_{-1}^1 \int_0^1 \int_0^{1-x^2} f\,dz\,dy\,dx$ and $\int_0^1 \int_0^1 \int_{-\sqrt{1-z}}^{\sqrt{1-z}} f\,dx\,dy\,dz$ are equal.

86. If W is a rectangular solid in 3-space, then $\int_W f\,dV = \int_a^b \int_c^d \int_e^k f\,dz\,dy\,dx$, where a, b, c, d, e, and k are constants.

87. If W is the unit cube $0 \le x \le 1, 0 \le y \le 1, 0 \le z \le 1$ and $\int_W f\,dV = 0$, then $f = 0$ everywhere in the unit cube.

88. If $f > g$ at all points in the solid region W, then $\int_W f\,dV > \int_W g\,dV$.

89. If W_1 and W_2 are solid regions with volume$(W_1) >$ volume(W_2) then $\int_{W_1} f\,dV > \int_{W_2} f\,dV$.

90. Both double and triple integrals can be used to compute volume.

16.4 DOUBLE INTEGRALS IN POLAR COORDINATES

Integration in Polar Coordinates

We started this chapter by putting a rectangular grid on the fox population density map, to estimate the total population using a Riemann sum. However, sometimes a polar grid is more appropriate.

Example 1 A biologist studying insect populations around a circular lake divides the area into the polar sectors of radii 2, 3, and 4 km in Figure 16.33. The approximate population density in each sector is shown in millions per square km. Estimate the total insect population around the lake.

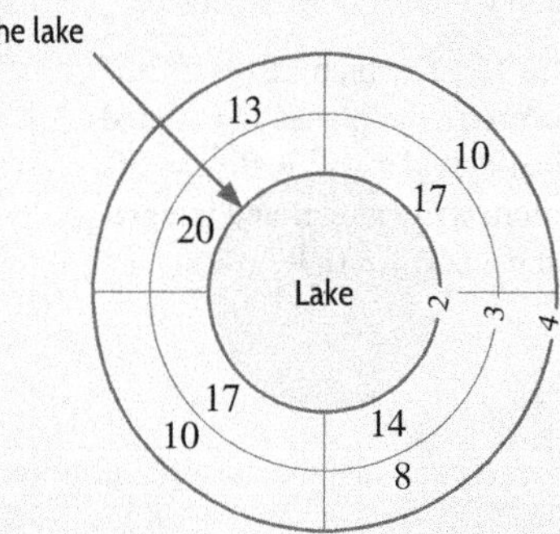

Figure 16.33: An insect-infested lake showing the insect population density by sector

Solution To get the estimate, we multiply the population density in each sector by the area of that sector. Unlike the rectangles in a rectangular grid, the sectors in this grid do not all have the same area. The inner sectors have area

$$\frac{1}{4}(\pi 3^2 - \pi 2^2) = \frac{5\pi}{4} \approx 3.93 \text{ km}^2,$$

and the outer sectors have area

$$\frac{1}{4}(\pi 4^2 - \pi 3^2) = \frac{7\pi}{4} \approx 5.50 \text{ km}^2,$$

so we estimate

$$\begin{aligned} \text{Population} &\approx (20)(3.93) + (17)(3.93) + (14)(3.93) + (17)(3.93) \\ &\quad + (13)(5.50) + (10)(5.50) + (8)(5.50) + (10)(5.50) \\ &= 492.74 \text{ million insects.} \end{aligned}$$

What Is dA in Polar Coordinates?

The previous example used a polar grid rather than a rectangular grid. A rectangular grid is constructed from vertical and horizontal lines of the form $x = k$ (a constant) and $y = l$ (another constant). In polar coordinates, $r = k$ gives a circle of radius k centered at the origin and $\theta = l$ gives a ray emanating from the origin (at angle l with the x-axis). A polar grid is built out of these circles and rays. Suppose we want to integrate $f(r, \theta)$ over the region R in Figure 16.34.

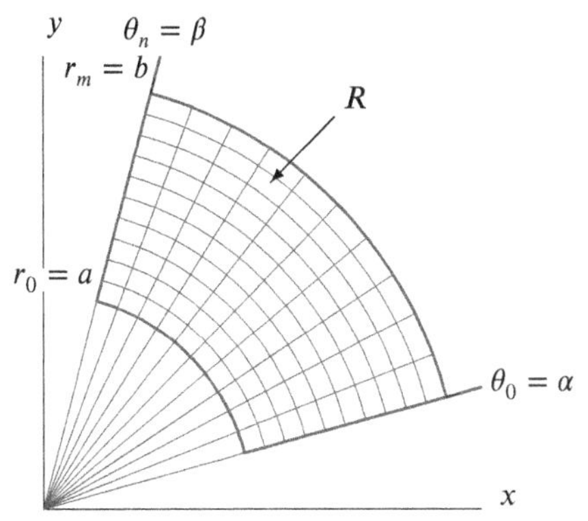

Figure 16.34: Dividing up a region using a polar grid

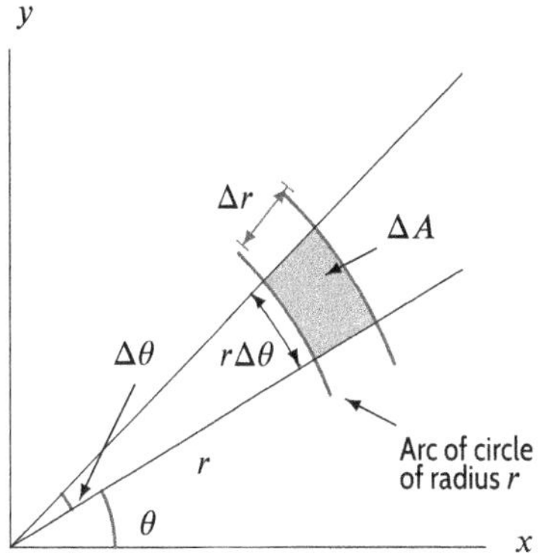

Figure 16.35: Calculating area ΔA in polar coordinates

Choosing (r_{ij}, θ_{ij}) in the ij-th bent rectangle in Figure 16.34 gives a Riemann sum:

$$\sum_{i,j} f(r_{ij}, \theta_{ij})\,\Delta A.$$

To calculate the area ΔA, look at Figure 16.35. If Δr and $\Delta\theta$ are small, the shaded region is approximately a rectangle with sides $r\,\Delta\theta$ and Δr, so

$$\Delta A \approx r\Delta\theta\Delta r.$$

Thus, the Riemann sum is approximately

$$\sum_{i,j} f(r_{ij}, \theta_{ij})\, r_{ij}\,\Delta\theta\,\Delta r.$$

If we take the limit as Δr and $\Delta\theta$ approach 0, we obtain

$$\int_R f\,dA = \int_\alpha^\beta \int_a^b f(r, \theta)\, r\,dr\,d\theta.$$

> When computing integrals in polar coordinates, use $x = r\cos\theta$, $y = r\sin\theta$, $x^2 + y^2 = r^2$. Put $dA = r\,dr\,d\theta$ or $dA = r\,d\theta\,dr$.

Example 2 Compute the integral of $f(x, y) = 1/(x^2 + y^2)^{3/2}$ over the region R shown in Figure 16.36.

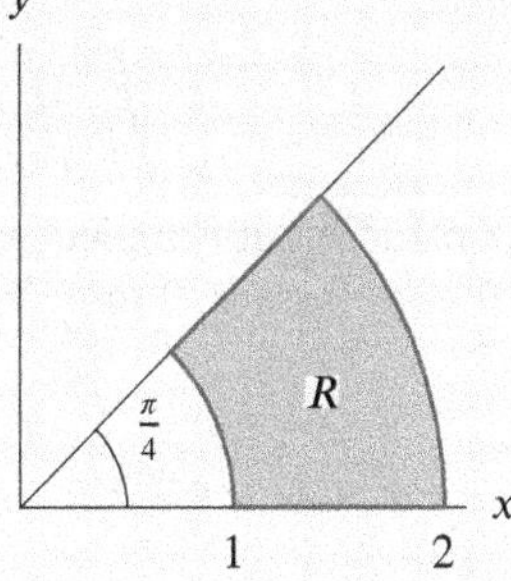

Figure 16.36: Integrate f over the polar region

Solution The region R is described by the inequalities $1 \le r \le 2$, $0 \le \theta \le \pi/4$. In polar coordinates, $r = \sqrt{x^2 + y^2}$, so we can write f as

$$f(x, y) = \frac{1}{(x^2 + y^2)^{3/2}} = \frac{1}{(r^2)^{3/2}} = \frac{1}{r^3}.$$

Then

$$\int_R f \, dA = \int_0^{\pi/4} \int_1^2 \frac{1}{r^3} r \, dr \, d\theta = \int_0^{\pi/4} \left(\int_1^2 r^{-2} \, dr \right) d\theta$$

$$= \int_0^{\pi/4} -\frac{1}{r}\bigg|_{r=1}^{r=2} d\theta = \int_0^{\pi/4} \frac{1}{2} \, d\theta = \frac{\pi}{8}.$$

Example 3 For each region in Figure 16.37, decide whether to integrate using polar or Cartesian coordinates. On the basis of its shape, write an iterated integral of an arbitrary function $f(x, y)$ over the region.

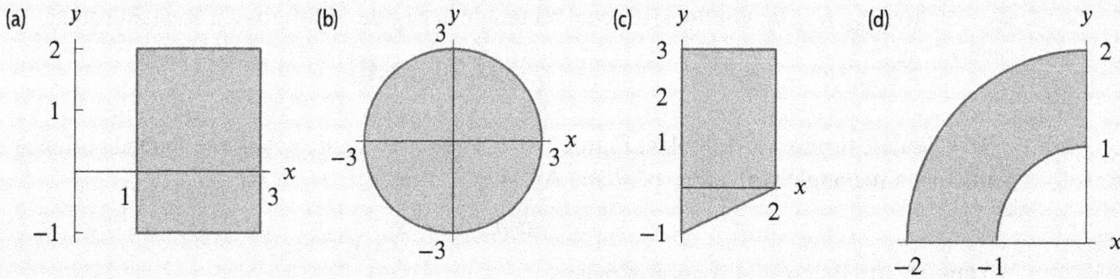

Figure 16.37

Solution (a) Since this is a rectangular region, Cartesian coordinates are likely to be a better choice. The rectangle is described by the inequalities $1 \le x \le 3$ and $-1 \le y \le 2$, so the integral is

$$\int_{-1}^{2} \int_1^3 f(x, y) \, dx \, dy.$$

(b) A circle is best described in polar coordinates. The radius is 3, so r goes from 0 to 3, and to describe the whole circle, θ goes from 0 to 2π. The integral is

$$\int_0^{2\pi} \int_0^3 f(r\cos\theta, r\sin\theta) \, r \, dr \, d\theta.$$

(c) The bottom boundary of this trapezoid is the line $y = (x/2) - 1$ and the top is the line $y = 3$, so we use Cartesian coordinates. If we integrate with respect to y first, the lower limit of the integral is $(x/2) - 1$ and the upper limit is 3. The x limits are $x = 0$ to $x = 2$. So the integral is

$$\int_0^2 \int_{(x/2)-1}^3 f(x, y) \, dy \, dx.$$

(d) This is another polar region: it is a piece of a ring in which r goes from 1 to 2. Since it is in the second quadrant, θ goes from $\pi/2$ to π. The integral is

$$\int_{\pi/2}^{\pi} \int_1^2 f(r\cos\theta, r\sin\theta) \, r \, dr \, d\theta.$$

Exercises and Problems for Section 16.4 Online Resource: Additional Problems for Section 16.4

EXERCISES

For the regions R in Exercises 1–4, write $\int_R f\,dA$ as an iterated integral in polar coordinates.

1.

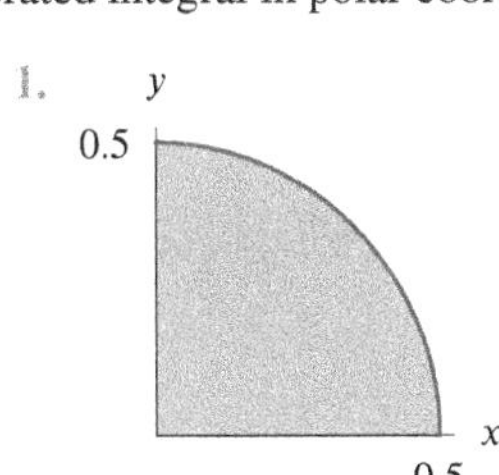

2.

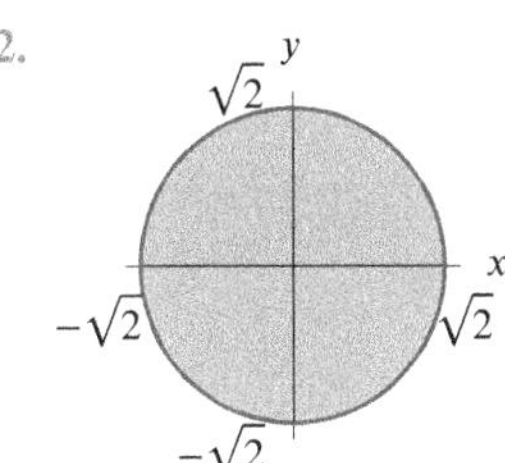

3.

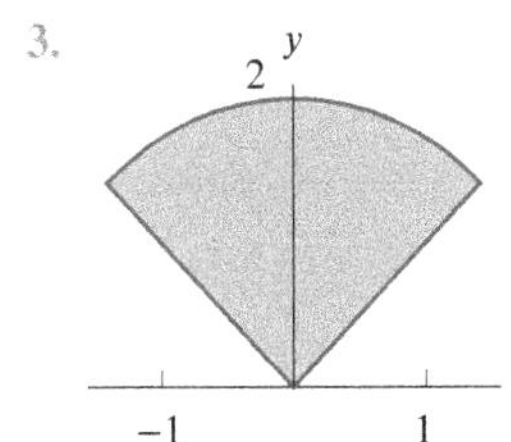

4.

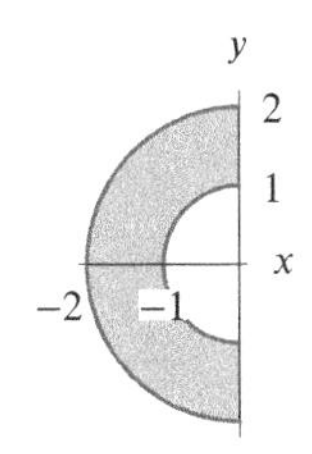

In Exercises 5–8, choose rectangular or polar coordinates to set up an iterated integral of an arbitrary function $f(x, y)$ over the region.

5.

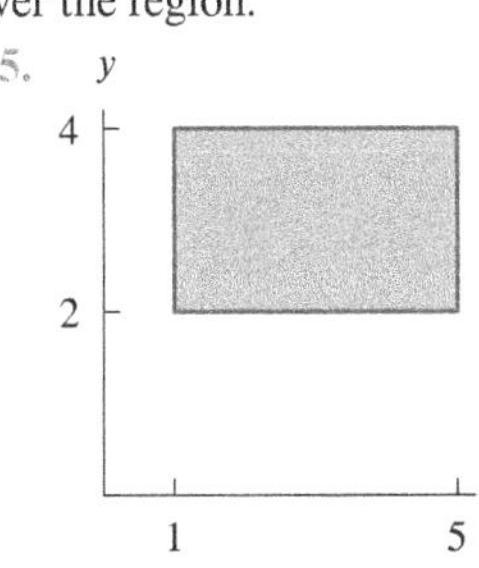

6.

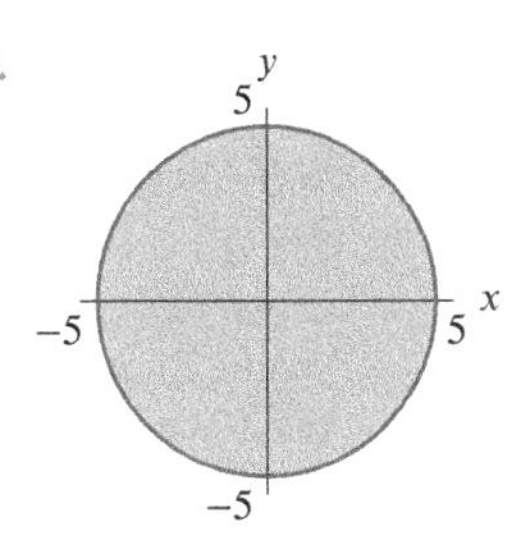

7. 8.

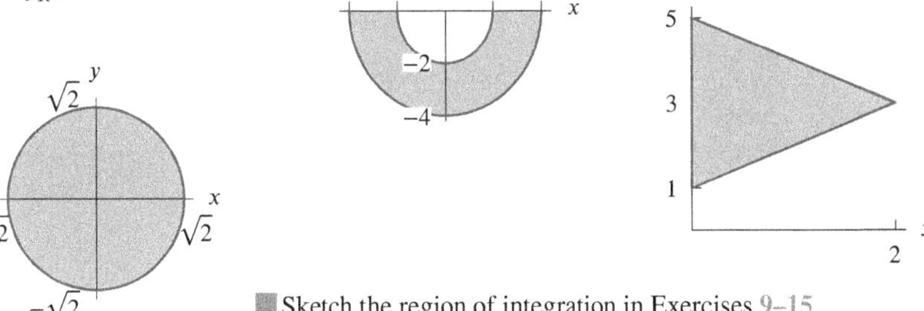

Sketch the region of integration in Exercises 9–15.

9. $\displaystyle\int_0^4 \int_{-\pi/2}^{\pi/2} f(r, \theta)\, r\, d\theta\, dr$

10. $\displaystyle\int_{\pi/2}^{\pi} \int_0^1 f(r, \theta)\, r\, dr\, d\theta$

11. $\displaystyle\int_0^{2\pi} \int_1^2 f(r, \theta)\, r\, dr\, d\theta$

12. $\displaystyle\int_{\pi/6}^{\pi/3} \int_0^1 f(r, \theta)\, r\, dr\, d\theta$

13. $\displaystyle\int_0^{\pi/4} \int_0^{1/\cos\theta} f(r, \theta)\, r\, dr\, d\theta$

14. $\displaystyle\int_3^4 \int_{3\pi/4}^{3\pi/2} f(r, \theta)\, r\, d\theta\, dr$

15. $\displaystyle\int_{\pi/4}^{\pi/2} \int_0^{2/\sin\theta} f(r, \theta)\, r\, dr\, d\theta$

PROBLEMS

In Problems 16–18, evaluate the integral.

16. $\int_R \sqrt{x^2 + y^2}\, dx dy$ where R is $4 \le x^2 + y^2 \le 9$.

17. $\int_R \sin(x^2 + y^2)\, dA$, where R is the disk of radius 2 centered at the origin.

18. $\int_R (x^2 - y^2)\, dA$, where R is the first quadrant region between the circles of radius 1 and radius 2.

Convert the integrals in Problems 19–21 to polar coordinates and evaluate.

19. $\displaystyle\int_{-1}^{0} \int_{-\sqrt{1-x^2}}^{\sqrt{1-x^2}} x\, dy\, dx$

20. $\displaystyle\int_0^{\sqrt{6}} \int_{-x}^{x} dy\, dx$

21. $\displaystyle\int_0^{\sqrt{2}} \int_y^{\sqrt{4-y^2}} xy\, dx\, dy$

Problems 22–26 concern Figure 16.38, which shows regions R_1, R_2, and R_3 contained in the semicircle $x^2 + y^2 = 4$ with $y \ge 0$.

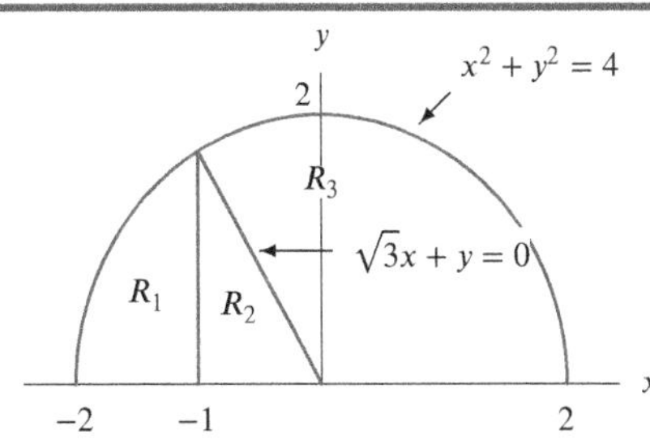

Figure 16.38

22. In Cartesian coordinates, write $\int_{R_1} 2y\, dA$ as an iterated integral in two different ways and then evaluate it.

23. In Cartesian coordinates, write $\int_{R_2} 2y\, dA$ as an iterated integral in two different ways.

24. Evaluate $\int_{R_3} (x^2 + y^2)\, dA$.

25. Evaluate $\int_R 12y\, dA$, where R is the region formed by combining the regions R_1 and R_2.

26. Evaluate $\int_S x\,dA$, where S is the region formed by combining the regions R_2 and R_3.

27. Consider the integral $\int_0^3 \int_{x/3}^1 f(x, y)\,dy\,dx$.
    (a) Sketch the region R over which the integration is being performed.
    (b) Rewrite the integral with the order of integration reversed.
    (c) Rewrite the integral in polar coordinates.

28. Describe the region of integration for $\int_{\pi/4}^{\pi/2} \int_{1/\sin\theta}^{4/\sin\theta} f(r, \theta) r\,dr\,d\theta$.

29. Evaluate the integral by converting it into Cartesian coordinates:
$$\int_0^{\pi/6} \int_0^{2/\cos\theta} r\,dr\,d\theta.$$

30. **(a)** Sketch the region of integration of
$$\int_0^1 \int_{\sqrt{1-x^2}}^{\sqrt{4-x^2}} x\,dy\,dx + \int_1^2 \int_0^{\sqrt{4-x^2}} x\,dy\,dx$$
    (b) Evaluate the quantity in part (a).

31. Find the volume of the region between the graph of $f(x, y) = 25 - x^2 - y^2$ and the xy plane.

32. Find the volume of an ice cream cone bounded by the hemisphere $z = \sqrt{8 - x^2 - y^2}$ and the cone $z = \sqrt{x^2 + y^2}$.

33. **(a)** For $a > 0$, find the volume under the graph of $z = e^{-(x^2+y^2)}$ above the disk $x^2 + y^2 \leq a^2$.
    (b) What happens to the volume as $a \to \infty$?

34. A circular metal disk of radius 3 lies in the xy-plane with its center at the origin. At a distance r from the origin, the density of the metal per unit area is $\delta = \dfrac{1}{r^2 + 1}$.
    (a) Write a double integral giving the total mass of the disk. Include limits of integration.
    (b) Evaluate the integral.

35. A city surrounds a bay as shown in Figure 16.39. The population density of the city (in thousands of people per square km) is $\delta(r, \theta)$, where r and θ are polar coordinates and distances are in km.
    (a) Set up an iterated integral in polar coordinates giving the total population of the city.
    (b) The population density decreases the farther you live from the shoreline of the bay; it also decreases the farther you live from the ocean. Which of the following functions best describes this situation?
    (i) $\delta(r, \theta) = (4 - r)(2 + \cos\theta)$
    (ii) $\delta(r, \theta) = (4 - r)(2 + \sin\theta)$
    (iii) $\delta(r, \theta) = (r + 4)(2 + \cos\theta)$
    (c) Estimate the population using your answers to parts (a) and (b).

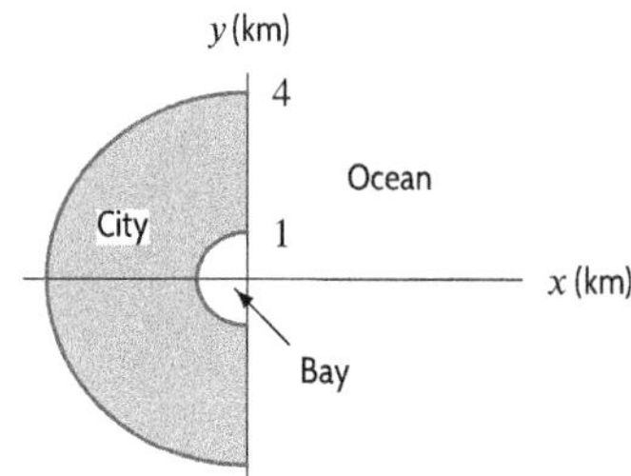

Figure 16.39

36. A disk of radius 5 cm has density 10 gm/cm^2 at its center and density 0 at its edge, and its density is a linear function of the distance from the center. Find the mass of the disk.

37. Electric charge is distributed over the xy-plane, with density inversely proportional to the distance from the origin. Show that the total charge inside a circle of radius R centered at the origin is proportional to R. What is the constant of proportionality?

38. **(a)** Graph $r = 1/(2\cos\theta)$ for $-\pi/2 \leq \theta \leq \pi/2$ and $r = 1$.
    (b) Write an iterated integral representing the area inside the curve $r = 1$ and to the right of $r = 1/(2\cos\theta)$. Evaluate the integral.

39. **(a)** Sketch the circles $r = 2\cos\theta$ for $-\pi/2 \leq \theta \leq \pi/2$ and $r = 1$.
    (b) Write an iterated integral representing the area inside the circle $r = 2\cos\theta$ and outside the circle $r = 1$. Evaluate the integral.

Strengthen Your Understanding

In Problems 40–44, explain what is wrong with the statement.

40. If R is the region bounded by $x = 1, y = 0, y = x$, then in polar coordinates $\int_R x\,dA = \int_0^{\pi/4} \int_0^1 r^2 \cos\theta\,dr\,d\theta$.

41. If R is the region $x^2 + y^2 \leq 4$, then $\int_R (x^2 + y^2)\,dA = \int_0^{2\pi} \int_0^2 r^2\,dr\,d\theta$.

42. $\int_0^1 \int_0^1 \sqrt{x^2 + y^2}\,dy\,dx = \int_0^{\pi/2} \int_0^1 r^2\,dr\,d\theta$

43. $\int_1^2 \int_0^{\sqrt{4-x^2}} 1\,dy\,dx = \int_0^{\pi/2} \int_1^2 r\,dr\,d\theta$

44. $\int_0^1 \int_0^{\pi} r\,dr\,d\theta = \int_0^{\pi} \int_0^1 r\,dr\,d\theta$

In Problems 45–48, give an example of:

45. A region R of integration in the first quadrant which suggests the use of polar coordinates.

46. An integrand $f(x, y)$ that suggests the use of polar coordinates.

47. A function $f(x, y)$ such that $\int_R f(x, y)\, dy\, dx$ in polar coordinates has an integrand without a factor of r.

48. A region R such that $\int_R f(x, y)\, dA$ must be broken into two integrals in Cartesian coordinates, but only needs one integral in polar coordinates.

49. Which of the following integrals give the area of the unit circle?

(a) $\int_{-1}^{1} \int_{-\sqrt{1-x^2}}^{\sqrt{1-x^2}} dy\, dx$ (b) $\int_{-1}^{1} \int_{-\sqrt{1-x^2}}^{\sqrt{1-x^2}} x\, dy\, dx$

(c) $\int_{0}^{2\pi} \int_{0}^{1} r\, dr\, d\theta$ (d) $\int_{0}^{2\pi} \int_{0}^{1} dr\, d\theta$

(e) $\int_{0}^{1} \int_{0}^{2\pi} r\, d\theta\, dr$ (f) $\int_{0}^{1} \int_{0}^{2\pi} d\theta\, dr$

Are the statements in Problems 50–55 true or false? Give reasons for your answer.

50. The integral $\int_0^{2\pi} \int_0^1 dr\, d\theta$ gives the area of the unit circle.

51. The quantity $8 \int_5^7 \int_0^{\pi/4} r\, d\theta\, dr$ gives the area of a ring with radius between 5 and 7.

52. Let R be the region inside the semicircle $x^2 + y^2 = 9$ with $y \geq 0$. Then $\int_R (x + y)\, dA = \int_0^{\pi} \int_0^3 r\, dr\, d\theta$

53. The integrals $\int_0^{\pi} \int_0^1 r^2 \cos\theta\, dr\, d\theta$ and $2 \int_0^{\pi/2} \int_0^1 r^2 \cos\theta\, dr\, d\theta$ are equal.

54. The integral $\int_0^{\pi/4} \int_0^{1/\cos\theta} r\, dr\, d\theta$ gives the area of the region $0 \leq x \leq 1, 0 \leq y \leq x$.

55. The integral $\int_0^{2\pi} \int_0^1 r^3\, dr\, d\theta$ gives the area of the unit circle.

16.5 INTEGRALS IN CYLINDRICAL AND SPHERICAL COORDINATES

Some double integrals are easier to evaluate in polar, rather than Cartesian, coordinates. Similarly, some triple integrals are easier in non-Cartesian coordinates.

Cylindrical Coordinates

The cylindrical coordinates of a point (x, y, z) in 3-space are obtained by representing the x and y coordinates in polar coordinates and letting the z-coordinate be the z-coordinate of the Cartesian coordinate system. (See Figure 16.40.)

Relation Between Cartesian and Cylindrical Coordinates

Each point in 3-space is represented using $0 \leq r < \infty, 0 \leq \theta \leq 2\pi, -\infty < z < \infty$.

$$x = r\cos\theta,$$
$$y = r\sin\theta,$$
$$z = z.$$

As with polar coordinates in the plane, note that $x^2 + y^2 = r^2$.

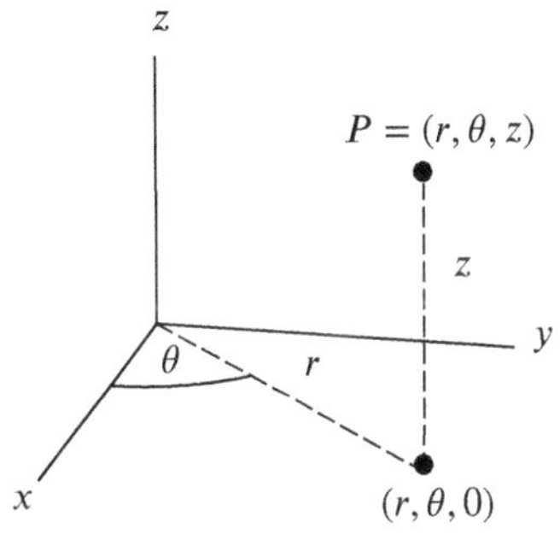

Figure 16.40: Cylindrical coordinates: (r, θ, z)

A useful way to visualize cylindrical coordinates is to sketch the surfaces obtained by setting one of the coordinates equal to a constant. See Figures 16.41–16.43.

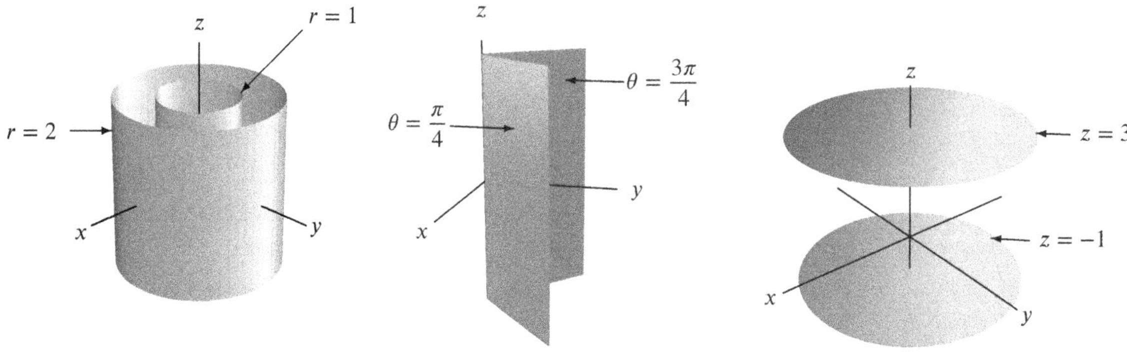

Figure 16.41: The surfaces $r = 1$ and $r = 2$

Figure 16.42: The surfaces $\theta = \pi/4$ and $\theta = 3\pi/4$

Figure 16.43: The surfaces $z = -1$ and $z = 3$

Setting $r = c$ (where c is constant) gives a cylinder around the z-axis whose radius is c. Setting $\theta = c$ gives a half-plane perpendicular to the xy plane, with one edge along the z-axis, making an angle c with the x-axis. Setting $z = c$ gives a horizontal plane $|c|$ units from the xy-plane. We call these *fundamental surfaces*.

The regions that can most easily be described in cylindrical coordinates are those regions whose boundaries are such fundamental surfaces. (For example, vertical cylinders, or wedge-shaped parts of vertical cylinders.)

Example 1 Describe in cylindrical coordinates a wedge of cheese cut from a cylinder 4 cm high and 6 cm in radius; this wedge subtends an angle of $\pi/6$ at the center. (See Figure 16.44.)

Solution The wedge is described by the inequalities $0 \le r \le 6$, and $0 \le z \le 4$, and $0 \le \theta \le \pi/6$.

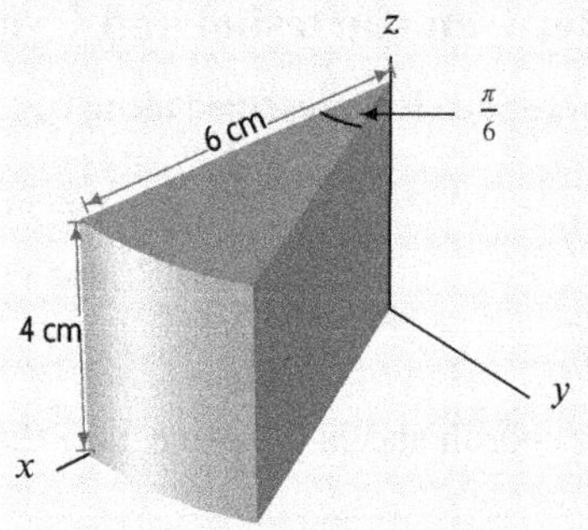

Figure 16.44: A wedge of cheese

Integration in Cylindrical Coordinates

To integrate a double integral $\int_R f\,dA$ in polar coordinates, we had to express the area element dA in terms of polar coordinates: $dA = r\,dr\,d\theta$. To evaluate a triple integral $\int_W f\,dV$ in cylindrical coordinates, we need to express the volume element dV in cylindrical coordinates.

In Figure 16.45, consider the volume element ΔV bounded by fundamental surfaces. The area of the base is $\Delta A \approx r\,\Delta r\,\Delta\theta$. Since the height is Δz, the volume element is given approximately by $\Delta V \approx r\,\Delta r\,\Delta\theta\,\Delta z$.

When computing integrals in cylindrical coordinates, put $dV = r\,dr\,d\theta\,dz$. Other orders of integration are also possible.

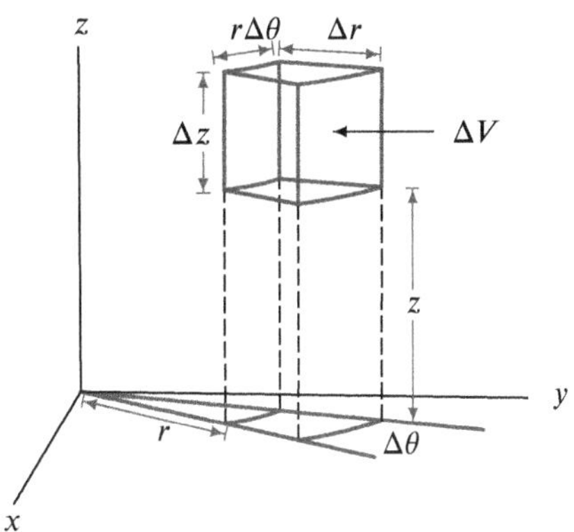

Figure 16.45: Volume element in cylindrical coordinates

Example 2 Find the mass of the wedge of cheese in Example 1, if its density is 1.2 grams/cm^3.

Solution If the wedge is W, its mass is

$$\int_W 1.2\,dV.$$

In cylindrical coordinates this integral is

$$\int_0^4 \int_0^{\pi/6} \int_0^6 1.2\,r\,dr\,d\theta\,dz = \int_0^4 \int_0^{\pi/6} 0.6r^2 \bigg|_0^6 d\theta\,dz = 21.6 \int_0^4 \int_0^{\pi/6} d\theta\,dz$$
$$= 21.6\left(\frac{\pi}{6}\right)4 = 45.239 \text{ grams.}$$

Example 3 A water tank in the shape of a hemisphere has radius a; its base is its plane face. Find the volume, V, of water in the tank as a function of h, the depth of the water.

Solution In Cartesian coordinates, a sphere of radius a has the equation $x^2 + y^2 + z^2 = a^2$. (See Figure 16.46.) In cylindrical coordinates, $r^2 = x^2 + y^2$, so this becomes

$$r^2 + z^2 = a^2.$$

Thus, if we want to describe the amount of water in the tank in cylindrical coordinates, we let r go from 0 to $\sqrt{a^2 - z^2}$, we let θ go from 0 to 2π, and we let z go from 0 to h, giving

$$\text{Volume of water} = \int_W 1\,dV = \int_0^{2\pi} \int_0^h \int_0^{\sqrt{a^2-z^2}} r\,dr\,dz\,d\theta = \int_0^{2\pi} \int_0^h \frac{r^2}{2}\bigg|_{r=0}^{r=\sqrt{a^2-z^2}} dz\,d\theta$$
$$= \int_0^{2\pi} \int_0^h \frac{1}{2}(a^2 - z^2)\,dz\,d\theta = \int_0^{2\pi} \frac{1}{2}\left(a^2 z - \frac{z^3}{3}\right)\bigg|_{z=0}^{z=h} d\theta$$
$$= \int_0^{2\pi} \frac{1}{2}\left(a^2 h - \frac{h^3}{3}\right) d\theta = \pi\left(a^2 h - \frac{h^3}{3}\right).$$

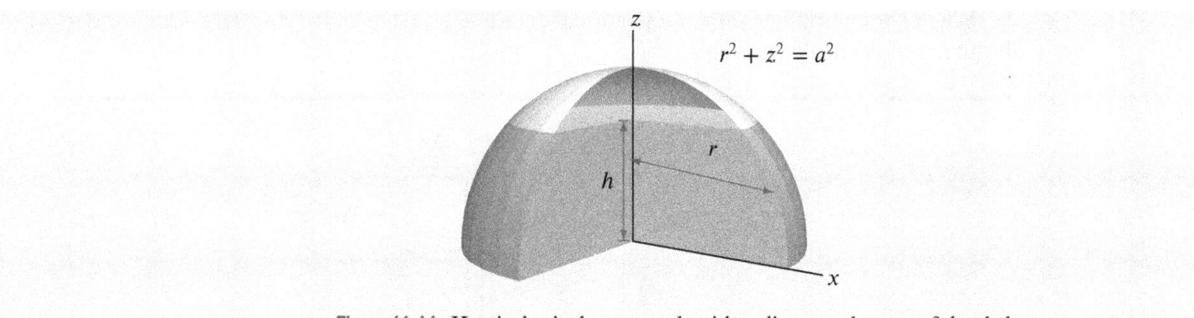

Figure 16.46: Hemispherical water tank with radius a and water of depth h

Spherical Coordinates

In Figure 16.47, the point P has coordinates (x, y, z) in the Cartesian coordinate system. We define spherical coordinates ρ, ϕ, and θ for P as follows: $\rho = \sqrt{x^2 + y^2 + z^2}$ is the distance of P from the origin; ϕ is the angle between the positive z-axis and the line through the origin and the point P; and θ is the same as in cylindrical coordinates.

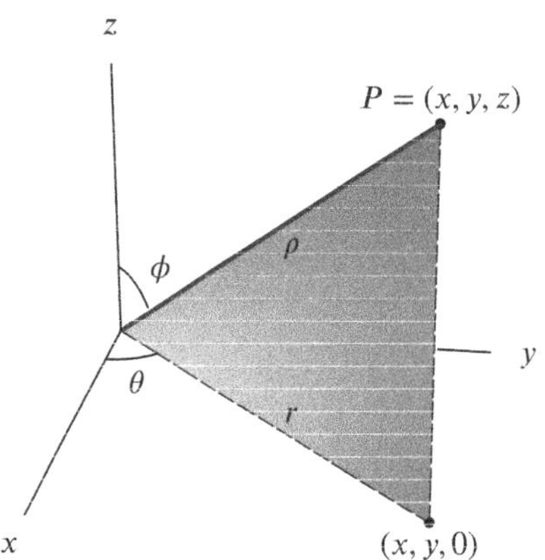

Figure 16.47: Spherical coordinates: (ρ, ϕ, θ)

In cylindrical coordinates,

$$x = r\cos\theta, \quad \text{and} \quad y = r\sin\theta, \quad \text{and} \quad z = z.$$

From Figure 16.47 we have $z = \rho\cos\phi$ and $r = \rho\sin\phi$, giving the following relationship:

Relation Between Cartesian and Spherical Coordinates

Each point in 3-space is represented using $0 \le \rho < \infty, 0 \le \phi \le \pi$, and $0 \le \theta \le 2\pi$.

$$x = \rho\sin\phi\cos\theta$$
$$y = \rho\sin\phi\sin\theta$$
$$z = \rho\cos\phi.$$

Also, $\rho^2 = x^2 + y^2 + z^2$.

This system of coordinates is useful when there is spherical symmetry with respect to the origin, either in the region of integration or in the integrand. The fundamental surfaces in spherical coordinates are $\rho = k$ (a constant), which is a sphere of radius k centered at the origin, $\theta = k$ (a constant), which is the half-plane with its edge along the z-axis, and $\phi = k$ (a constant), which is a cone if $k \ne \pi/2$ and the xy-plane if $k = \pi/2$. (See Figures 16.48–16.50.)

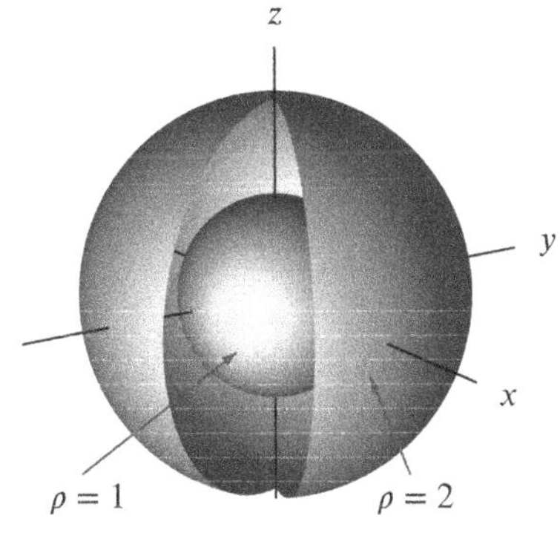

Figure 16.48: The surfaces $\rho = 1$ and $\rho = 2$

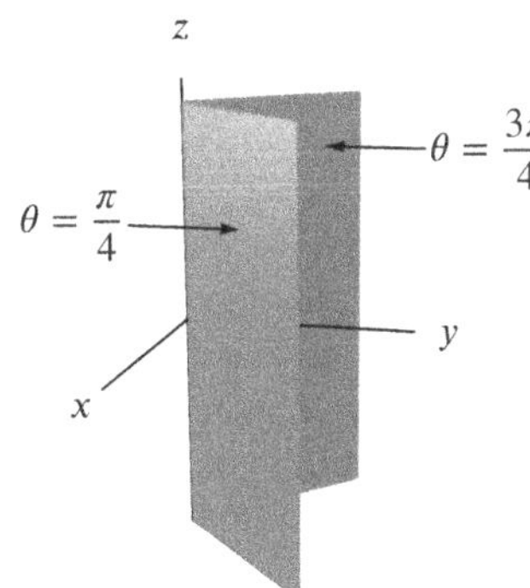

Figure 16.49: The surfaces $\theta = \pi/4$ and $\theta = 3\pi/4$

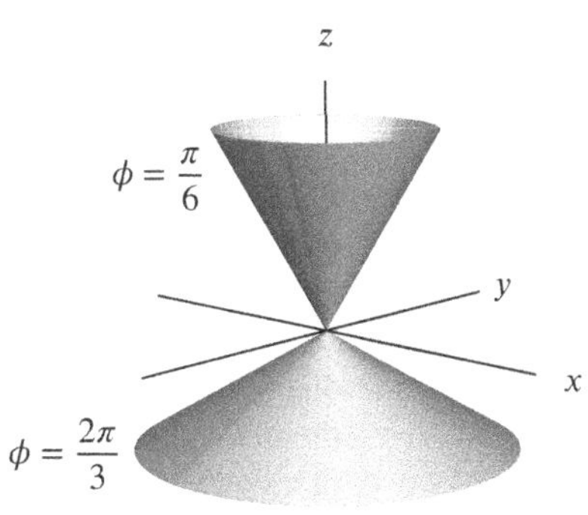

Figure 16.50: The surfaces $\phi = \pi/6$ and $\phi = 2\pi/3$

Integration in Spherical Coordinates

To use spherical coordinates in triple integrals we need to express the volume element, dV, in spherical coordinates. From Figure 16.51, we see that the volume element can be approximated by a box with curved edges. One edge has length $\Delta\rho$. The edge parallel to the xy-plane is an arc of a circle made from rotating the cylindrical radius r ($= \rho\sin\phi$) through an angle $\Delta\theta$, and so has length $\rho\sin\phi\,\Delta\theta$. The remaining edge comes from rotating the radius ρ through an angle $\Delta\phi$, and so has length $\rho\,\Delta\phi$. Therefore, $\Delta V \approx \Delta\rho(\rho\,\Delta\phi)(\rho\sin\phi\,\Delta\theta) = \rho^2\sin\phi\,\Delta\rho\,\Delta\phi\,\Delta\theta$.

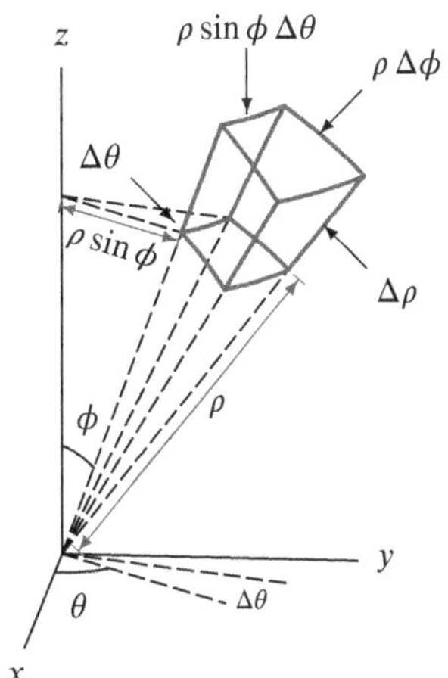

Figure 16.51: Volume element in spherical coordinates

Thus:

> When computing integrals in spherical coordinates, put $dV = \rho^2\sin\phi\,d\rho\,d\phi\,d\theta$. Other orders of integration are also possible.

Example 4 Use spherical coordinates to derive the formula for the volume of a ball of radius a.

Solution In spherical coordinates, a ball of radius a is described by the inequalities $0 \le \rho \le a$, $0 \le \theta \le 2\pi$, and $0 \le \phi \le \pi$. Note that θ goes from 0 to 2π, whereas ϕ goes from 0 to π. We find the volume by integrating the constant density function 1 over the ball:

$$\begin{aligned}\text{Volume} = \int_R 1\,dV &= \int_0^{2\pi}\int_0^{\pi}\int_0^{a}\rho^2\sin\phi\,d\rho\,d\phi\,d\theta = \int_0^{2\pi}\int_0^{\pi}\frac{1}{3}a^3\sin\phi\,d\phi\,d\theta \\ &= \frac{1}{3}a^3\int_0^{2\pi} -\cos\phi\Big|_0^{\pi}\,d\theta = \frac{2}{3}a^3\int_0^{2\pi}d\theta = \frac{4\pi a^3}{3}.\end{aligned}$$

Example 5 Find the magnitude of the gravitational force exerted by a solid hemisphere of radius a and constant density δ on a unit mass located at the center of the base of the hemisphere.

Solution Assume the base of the hemisphere rests on the xy-plane with center at the origin. (See Figure 16.52.) Newton's law of gravitation says that the force between two masses m_1 and m_2 at a distance r apart is $F = Gm_1m_2/r^2$, where G is the gravitational constant.

In this example, symmetry shows that the net component of the force on the particle at the origin due to the hemisphere is in the z direction only. Any force in the x or y direction from some part of the hemisphere is canceled by the force from another part of the hemisphere directly opposite the first.

To compute the net z-component of the gravitational force, we imagine a small piece of the hemisphere with volume ΔV, located at spherical coordinates (ρ, θ, ϕ). This piece has mass $\delta\Delta V$ and exerts a force of magnitude F on the unit mass at the origin. The z-component of this force is given by its projection onto the z-axis, which can be seen from the figure to be $F\cos\phi$. The distance from the mass $\delta\Delta V$ to the unit mass at the origin is the spherical coordinate ρ. Therefore, the z-component of the force due to the small piece ΔV is

$$\begin{matrix}z\text{-component}\\ \text{of force}\end{matrix} = \frac{G(\delta\Delta V)(1)}{\rho^2}\cos\phi.$$

Adding the contributions of the small pieces, we get a vertical force with magnitude

$$\begin{aligned}F &= \int_0^{2\pi}\int_0^{\pi/2}\int_0^a \left(\frac{G\delta}{\rho^2}\right)(\cos\phi)\rho^2\sin\phi\, d\rho\, d\phi\, d\theta = \int_0^{2\pi}\int_0^{\pi/2} G\delta(\cos\phi\sin\phi)\rho\Big|_{\rho=0}^{\rho=a} d\phi d\theta\\ &= \int_0^{2\pi}\int_0^{\pi/2} G\delta a\cos\phi\sin\phi\, d\phi\, d\theta = \int_0^{2\pi} G\delta a\left(-\frac{(\cos\phi)^2}{2}\right)\Bigg|_{\phi=0}^{\phi=\pi/2} d\theta\\ &= \int_0^{2\pi} G\delta a\left(\frac{1}{2}\right) d\theta = G\delta a\pi.\end{aligned}$$

The integral in this example is improper because the region of integration contains the origin, where the force is undefined. However, it can be shown that the result is nevertheless correct.

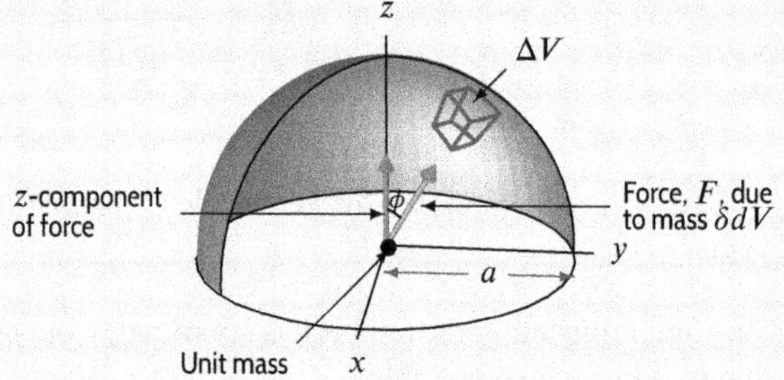

Figure 16.52: Gravitational force of hemisphere on mass at origin

Exercises and Problems for Section 16.5 Online Resource: Additional Problems for Section 16.5

EXERCISES

1. Match the equations in (a)–(f) with one of the surfaces in (I)–(VII).

(a) $x = 5$ (b) $x^2 + z^2 = 7$ (c) $\rho = 5$
(d) $z = 1$ (e) $r = 3$ (f) $\theta = 2\pi$

(I) Cylinder, centered on x-axis.
(II) Cylinder, centered on y-axis.
(III) Cylinder, centered on z-axis.
(IV) Plane, perpendicular to the x-axis.
(V) Plane, perpendicular to the y-axis.
(VI) Plane, perpendicular to the z-axis.
(VII) Sphere.

In Exercises 2–7, find an equation for the surface.

2. The vertical plane $y = x$ in cylindrical coordinates.
3. The top half of the sphere $x^2+y^2+z^2 = 1$ in cylindrical coordinates.
4. The cone $z = \sqrt{x^2 + y^2}$ in cylindrical coordinates.
5. The cone $z = \sqrt{x^2 + y^2}$ in spherical coordinates.
6. The plane $z = 10$ in spherical coordinates.
7. The plane $z = 4$ in spherical coordinates.

In Exercises 8–9, evaluate the triple integrals in cylindrical coordinates over the region W.

8. $f(x, y, z) = \sin(x^2 + y^2)$, W is the solid cylinder with height 4 and with base of radius 1 centered on the z axis at $z = -1$.
9. $f(x, y, z) = x^2 + y^2 + z^2$, W is the region $0 \leq r \leq 4$, $\pi/4 \leq \theta \leq 3\pi/4$, $-1 \leq z \leq 1$.

In Exercises 10–11, evaluate the triple integrals in spherical coordinates.

10. $f(\rho, \theta, \phi) = \sin\phi$, over the region $0 \leq \theta \leq 2\pi$, $0 \leq \phi \leq \pi/4$, $1 \leq \rho \leq 2$.
11. $f(x, y, z) = 1/(x^2 + y^2 + z^2)^{1/2}$ over the bottom half of the sphere of radius 5 centered at the origin.

For Exercises 12–18, choose coordinates and set up a triple integral, including limits of integration, for a density function f over the region.

12.

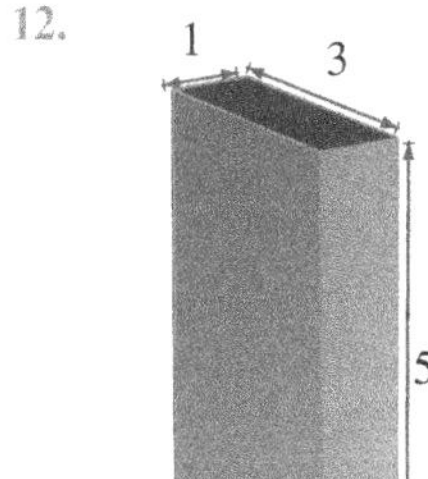

13.

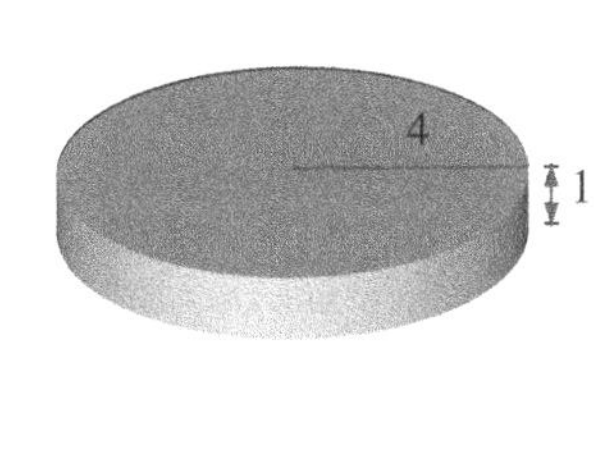

14. 15.

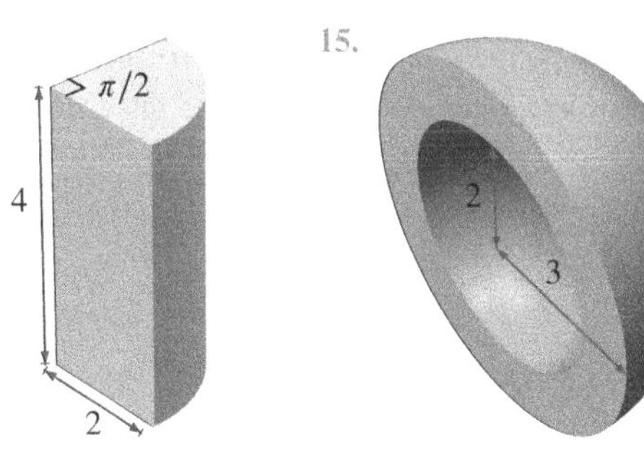

16. A piece of a sphere; angle at the center is $\pi/3$.

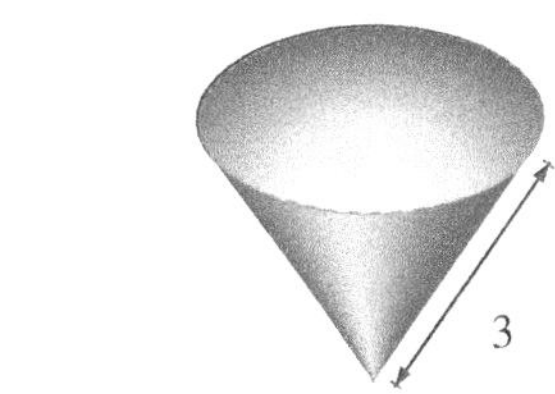

17.

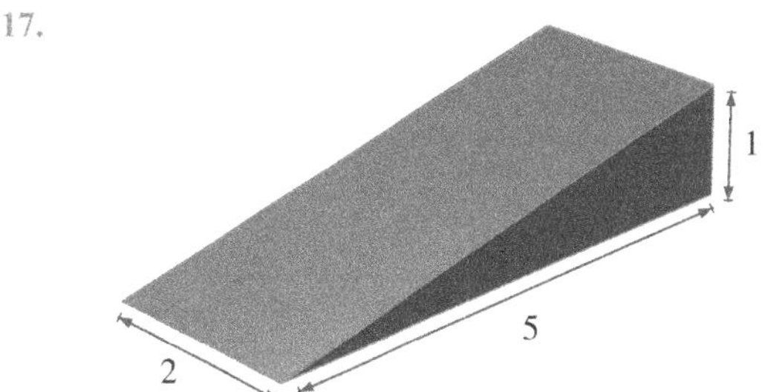

18.

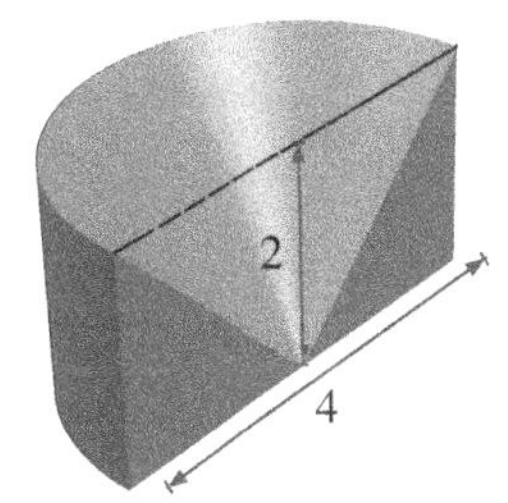

PROBLEMS

In Problems 19–21, if W is the region in Figure 16.53, what are the limits of integration?

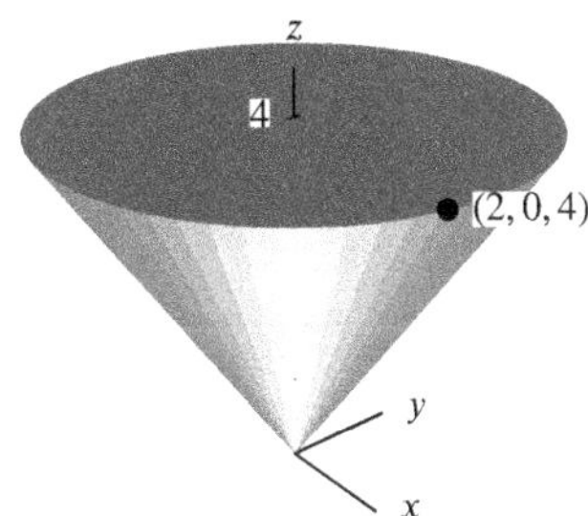

Figure 16.53: Cone with flat top, symmetric about z-axis

19. $\int_?^? \int_?^? \int_?^? f(r, \theta, z) r\, dz\, dr\, d\theta$

20. $\int_?^? \int_?^? \int_?^? g(\rho, \phi, \theta)\rho^2 \sin\phi\, d\rho\, d\phi\, d\theta$

21. $\int_?^? \int_?^? \int_?^? h(x, y, z)\, dz\, dy\, dx$

22. Write a triple integral in cylindrical coordinates giving the volume of a sphere of radius K centered at the origin. Use the order $dz\, dr\, d\theta$.

23. Write a triple integral in spherical coordinates giving the volume of a sphere of radius K centered at the origin. Use the order $d\theta\, d\rho\, d\phi$.

In Problems 24–26, for the regions W shown, write the limits of integration for $\int_W dV$ in the following coordinates:

(a) Cartesian **(b)** Cylindrical **(c)** Spherical

24.

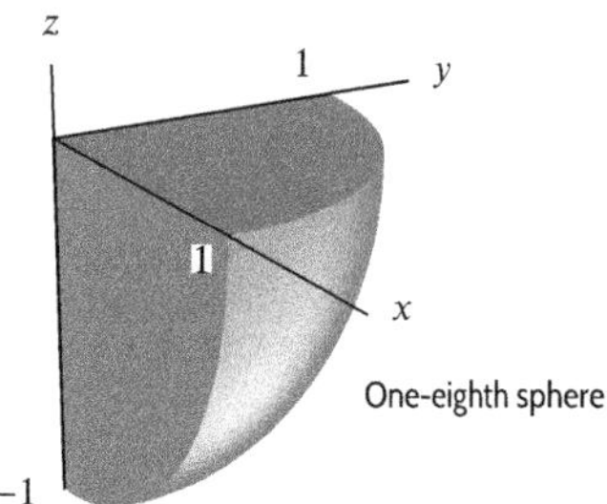

25.

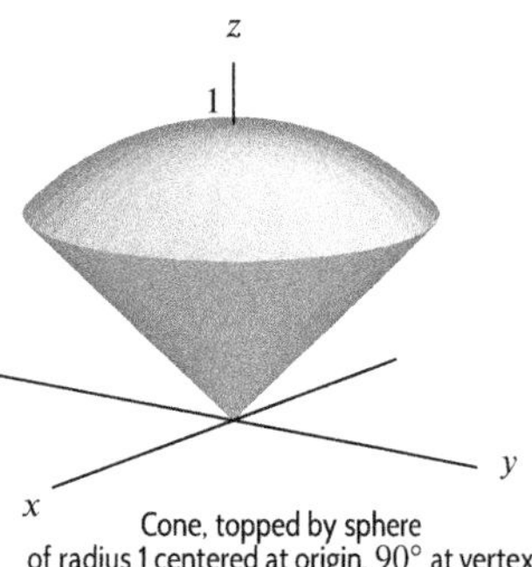

26.

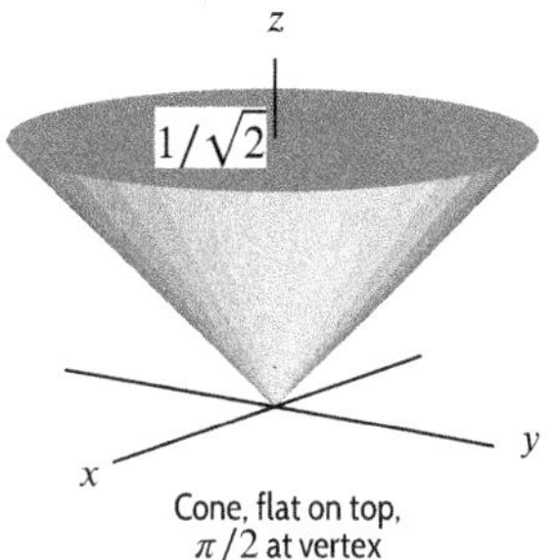

27. Write a triple integral representing the volume above the cone $z = \sqrt{x^2 + y^2}$ and below the sphere of radius 2 centered at the origin. Include limits of integration but do not evaluate. Use:

(a) Cylindrical coordinates
(b) Spherical coordinates

28. Write a triple integral representing the volume of the region between spheres of radius 1 and 2, both centered at the origin. Include limits of integration but do not evaluate. Use:

(a) Spherical coordinates.
(b) Cylindrical coordinates. Write your answer as the difference of two integrals.

In Problems 29–34, write a triple integral including limits of integration that gives the specified volume.

29. Under $\rho = 3$ and above $\phi = \pi/3$.

30. Under $\rho = 3$ and above $z = r$.

31. The region between $z = 5$ and $z = 10$, with $2 \le x^2 + y^2 \le 3$ and $0 \le \theta \le \pi$.

32. Between the cone $z = \sqrt{x^2 + y^2}$ and the first quadrant of the xy-plane, with $x^2 + y^2 \le 7$.

33. The cap of the solid sphere $x^2 + y^2 + z^2 \le 10$ cut off by the plane $z = 1$.

34. Below the cone $z = r$, above the xy-plane, and inside the sphere $x^2 + y^2 + z^2 = 8$.

35. **(a)** Write an integral (including limits of integration) representing the volume of the region inside the cone $z = \sqrt{3(x^2 + y^2)}$ and below the plane $z = 1$.
(b) Evaluate the integral.

36. Find the volume between the cone $z = \sqrt{x^2 + y^2}$ and the plane $z = 10 + x$ above the disk $x^2 + y^2 \le 1$.

37. Find the volume between the cone $x = \sqrt{y^2 + z^2}$ and the sphere $x^2 + y^2 + z^2 = 4$.

38. The sphere of radius 2 centered at the origin is sliced horizontally at $z = 1$. What is the volume of the cap above the plane $z = 1$?

39. Suppose W is the region outside the cylinder $x^2 + y^2 = 1$ and inside the sphere $x^2 + y^2 + z^2 = 2$. Calculate

$$\int_W (x^2 + y^2)\, dV.$$

40. Write and evaluate a triple integral representing the volume of a slice of the cylindrical cake of height 2 and radius 5 between the planes $\theta = \pi/6$ and $\theta = \pi/3$.

41. Write a triple integral representing the volume of the cone in Figure 16.54 and evaluate it.

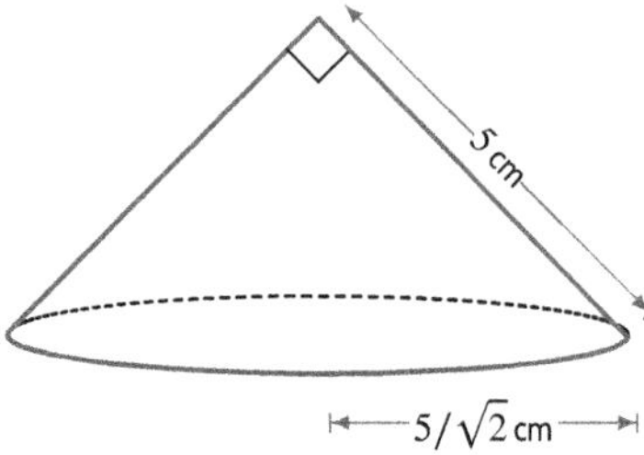

Figure 16.54

42. Find the average distance from the origin of

(a) The points in the interval $|x| \le 12$.
(b) The points in the plane in the disc $r \le 12$.
(c) The points in space in the ball $\rho \le 12$.

In Problems 43–44, without performing the integration, decide whether the integral is positive, negative, or zero.

43. W_1 is the unit ball, $x^2 + y^2 + z^2 \leq 1$.

(a) $\int_{W_1} \sin\phi \, dV$ (b) $\int_{W_1} \cos\phi \, dV$

44. W_2 is $0 \leq z \leq \sqrt{1 - x^2 - y^2}$, the top half of the unit ball.

(a) $\int_{W_2} (z^2 - z) \, dV$ (b) $\int_{W_2} (-xz) \, dV$

45. The insulation surrounding a pipe of length l is the region between two cylinders with the same axis. The inner cylinder has radius a, the outer radius of the pipe, and the insulation has thickness h. Write a triple integral, including limits of integration, giving the volume of the insulation. Evaluate the integral.

46. Assume p, q, r are positive constants. Find the volume contained between the coordinate planes and the plane

$$\frac{x}{p} + \frac{y}{q} + \frac{z}{r} = 1.$$

47. A cone stands with its flat base on a table. The cone's circular base has radius a; the vertex (tip) is at a height of h above the center of the base. Write a triple integral, including limits of integration, representing the volume of the cone. Evaluate the integral.

48. A half-melon is approximated by the region between two concentric spheres, one of radius a and the other of radius b, with $0 < a < b$. Write a triple integral, including limits of integration, giving the volume of the half-melon. Evaluate the integral.

49. A bead is made by drilling a cylindrical hole of radius 1 mm through a sphere of radius 5 mm. See Figure 16.55.

(a) Set up a triple integral in cylindrical coordinates representing the volume of the bead.
(b) Evaluate the integral.

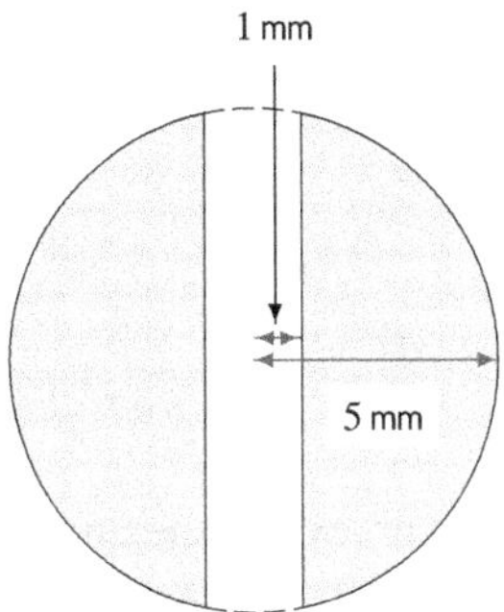

Figure 16.55

50. A pile of hay is in the region $0 \leq z \leq 2 - x^2 - y^2$, where x, y, z are in meters. At height z, the density of the hay is $\delta = (2 - z)$ kg/m^3.

(a) Write an integral representing the mass of hay in the pile.
(b) Evaluate the integral.

51. Find the mass M of the solid region W given in spherical coordinates by $0 \leq \rho \leq 3, 0 \leq \theta < 2\pi, 0 \leq \phi \leq \pi/4$. The density, $\delta(P)$, at any point P is given by the distance of P from the origin.

52. Write an integral representing the mass of a sphere of radius 3 if the density of the sphere at any point is twice the distance of that point from the center of the sphere.

53. A sphere has density at each point proportional to the square of the distance of the point from the z-axis. The density is 2 gm/cm^3 at a distance of 2 cm from the axis. What is the mass of the sphere if it is centered at the origin and has radius 3 cm?

54. The density of a solid sphere at any point is proportional to the square of the distance of the point to the center of the sphere. What is the ratio of the mass of a sphere of radius 1 to a sphere of radius 2?

55. A spherical shell centered at the origin has an inner radius of 6 cm and an outer radius of 7 cm. The density, δ, of the material increases linearly with the distance from the center. At the inner surface, $\delta = 9$ gm/cm^3; at the outer surface, $\delta = 11$ gm/cm^3.

(a) Using spherical coordinates, write the density, δ, as a function of radius, ρ.
(b) Write an integral giving the mass of the shell.
(c) Find the mass of the shell.

56. (a) Write an iterated integral which represents the mass of a solid ball of radius a. The density at each point in the ball is k times the distance from that point to a fixed plane passing through the center of the ball.
(b) Evaluate the integral.

57. In the region under $z = 4 - x^2 - y^2$ and above the xy-plane the density of a gas is $\delta = e^{-x-y}$gm/cm^3, where x, y, z are in cm. Write an integral, with limits of integration, representing the mass of the gas.

58. The density, δ, of the cylinder $x^2 + y^2 \leq 4, 0 \leq z \leq 3$ varies with the distance, r, from the z-axis:

$$\delta = 1 + r \text{ gm/cm}^3.$$

Find the mass of the cylinder if x, y, z are in cm.

59. The density of material at a point in a solid cylinder is proportional to the distance of the point from the z-axis. What is the ratio of the mass of the cylinder $x^2 + y^2 \leq 1$, $0 \leq z \leq 2$ to the mass of the cylinder $x^2 + y^2 \leq 9$, $0 \leq z \leq 2$?

60. Electric charge is distributed throughout 3-space, with density proportional to the distance from the xy-plane. Show that the total charge inside a cylinder of radius R and height h, sitting on the xy-plane and centered along the z-axis, is proportional to R^2h^2.

61. Electric charge is distributed throughout 3-space with density inversely proportional to the distance from the origin. Show that the total charge inside a sphere of radius R is proportional to R^2.

For Problems 62–65, use the definition of center of mass given on page 863. Assume x, y, z are in cm.

62. Let C be a solid cone with both height and radius 1 and contained between the surfaces $z = \sqrt{x^2+y^2}$ and $z = 1$. If C has constant mass density of 1 gm/cm^3, find the z-coordinate of C's center of mass.

63. The density of the cone C in Problem 62 is given by $\delta(z) = z^2$ gm/cm^3. Find

(a) The mass of C.
(b) The z-coordinate of C's center of mass.

64. For $a > 0$, consider the family of solids bounded below by the paraboloid $z = a(x^2+y^2)$ and above by the plane $z = 1$. If the solids all have constant mass density 1 gm/cm^3, show that the z-coordinate of the center of mass is $2/3$ and so independent of the parameter a.

65. Find the location of the center of mass of a hemisphere of radius a and density b gm/cm^3.

Strengthen Your Understanding

In Problems 66–68, explain what is wrong with the statement.

66. The integral $\int_0^{2\pi}\int_0^{\pi}\int_0^1 1\,d\rho\,d\phi\,d\theta$ gives the volume inside the sphere of radius 1.

67. Changing the order of integration gives

$$\int_0^{2\pi}\int_0^{\pi/4}\int_0^{2/\cos\phi} \rho^2\sin\phi\,d\rho\,d\phi\,d\theta = \int_0^{2/\cos\phi}\int_0^{\pi/4}\int_0^{2\pi} \rho^2\sin\phi\,d\theta\,d\phi\,d\rho.$$

68. The volume of a cylinder of height and radius 1 is $\int_0^{2\pi}\int_0^{1}\int_0^1 1\,dz\,dr\,d\theta$.

In Problems 69–70, give an example of:

69. An integral in spherical coordinates that gives the volume of a hemisphere.

70. An integral for which it is more convenient to use spherical coordinates than to use Cartesian coordinates.

71. Which of the following integrals give the volume of the unit sphere?

(a) $\int_0^{2\pi}\int_0^{2\pi}\int_0^1 1\,d\rho\,d\theta\,d\phi$

(b) $\int_0^{\pi}\int_0^{2\pi}\int_0^1 1\,d\rho\,d\theta\,d\phi$

(c) $\int_0^{\pi}\int_0^{2\pi}\int_0^1 \rho^2\sin\phi\,d\rho\,d\theta\,d\phi$

(d) $\int_0^{\pi}\int_0^{2\pi}\int_0^1 \rho^2\sin\phi\,d\rho\,d\phi\,d\theta$

(e) $\int_0^{\pi}\int_0^{2\pi}\int_0^1 \rho\,d\rho\,d\phi\,d\theta$

16.6 APPLICATIONS OF INTEGRATION TO PROBABILITY

To represent how a quantity such as height or weight is distributed throughout a population, we use a density function. To study two or more quantities at the same time and see how they are related, we use a multivariable density function.

Density Functions

Distribution of Weight and Height in Expectant Mothers

Table 16.10 shows the distribution of weight and height in a survey of expectant mothers. The histogram in Figure 16.56 is constructed so that the volume of each bar represents the percentage in the corresponding weight and height range. For example, the bar representing the mothers who weighed 60–70 kg and were 160–165 cm tall has base of area 10 kg · 5 cm = 50 kg cm. The volume of this bar is 12%, so its height is 12%/50 kg cm = 0.24%/ kg cm. Notice that the units on the vertical axis are % per kg cm, so the volume of a bar is a %. The total volume is 100% = 1.

Table 16.10 *Distribution of weight and height in a survey of expectant mothers, in %*

	45-50 kg	50-60 kg	60-70 kg	70-80 kg	80-105 kg	Totals by height
150-155 cm	2	4	4	2	1	13
155-160 cm	0	12	8	2	1	23
160-165 cm	1	7	12	4	3	27
165-170 cm	0	8	12	6	2	28
170-180 cm	0	1	3	4	1	9
Totals by weight	3	32	39	18	8	100

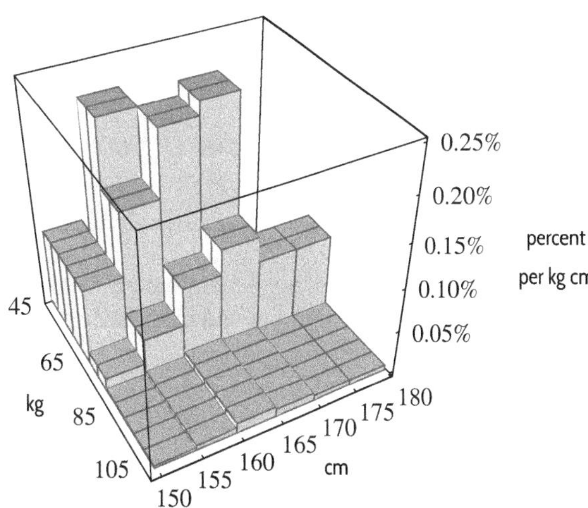

Figure 16.56: Histogram representing the data in Table 16.10

Example 1 Find the percentage of mothers in the survey with height between 170 and 180 cm.

Solution We add the percentages across the row corresponding to the 170–180 cm height range; this is equivalent to adding the volumes of the corresponding rectangular solids in the histogram.

$$\text{Percentage of mothers} = 0 + 1 + 3 + 4 + 1 = 9\%.$$

Smoothing the Histogram

If we group the data using narrower weight and height groups (and a larger sample), we can draw a smoother histogram and get finer estimates. In the limit, we replace the histogram with a smooth surface, in such a way that the volume under the surface above a rectangle is the percentage of mothers in that rectangle. We define a *density function*, $p(w, h)$, to be the function whose graph is the smooth surface. It has the property that

$$\begin{array}{c}\text{Fraction of sample with}\\ \text{weight between } a \text{ and } b \text{ and}\\ \text{height between } c \text{ and } d\end{array} = \begin{array}{c}\text{Volume under graph of } p\\ \text{over the rectangle}\\ a \le w \le b,\ c \le h \le d\end{array} = \int_a^b \int_c^d p(w, h)\, dh\, dw.$$

This density also gives the probability that a mother is in these height and weight groups.

Joint Probability Density Functions

We generalize this idea to represent any two characteristics, x and y, distributed throughout a population.

> A function $p(x, y)$ is called a **joint probability density function**, or **pdf**, for x and y if
>
> $$\begin{array}{c}\text{Probability that member of}\\ \text{population has } x \text{ between } a \text{ and } b\\ \text{and } y \text{ between } c \text{ and } d\end{array} = \begin{array}{c}\text{Volume under graph of } p\\ \text{above the rectangle}\\ a \le x \le b,\ c \le y \le d\end{array} = \int_a^b \int_c^d p(x, y)\,dy\,dx,$$
>
> where
>
> $$\int_{-\infty}^{\infty} \int_{-\infty}^{\infty} p(x, y)\,dy\,dx = 1 \quad \text{and} \quad p(x, y) \ge 0 \text{ for all } x \text{ and } y.$$
>
> The probability that x falls in an interval of width Δx around x_0 and y falls in an interval of width Δy around y_0 is approximately $p(x_0, y_0)\Delta x\Delta y$.

A joint density function need not be continuous, as in Example 2. In addition, as in Example 4, the integrals involved may be improper and must be computed by methods similar to those used for improper one-variable integrals.

Example 2 Let $p(x, y)$ be defined on the square $0 \le x \le 1, 0 \le y \le 1$ by $p(x, y) = x + y$; let $p(x, y) = 0$ if (x, y) is outside this square. Check that p is a joint density function. In terms of the distribution of x and y in the population, what does it mean that $p(x, y) = 0$ outside the square?

Solution First, we have $p(x, y) \ge 0$ for all x and y. To check that p is a joint density function, we show that the total volume under the graph is 1:

$$\int_{-\infty}^{\infty} \int_{-\infty}^{\infty} p(x, y)\,dy\,dx = \int_0^1 \int_0^1 (x + y)\,dy\,dx$$

$$= \int_0^1 \left(xy + \frac{y^2}{2}\right)\bigg|_0^1 dx = \int_0^1 \left(x + \frac{1}{2}\right) dx = \left(\frac{x^2}{2} + \frac{x}{2}\right)\bigg|_0^1 = 1.$$

The fact that $p(x, y) = 0$ outside the square means that the variables x and y never take values outside the interval $[0, 1]$; that is, the value of x and y for any individual in the population is always between 0 and 1.

Example 3 Two variables x and y are distributed in a population according to the density function of Example 2. Find the fraction of the population with $x \le 1/2$, the fraction with $y \le 1/2$, and the fraction with both $x \le 1/2$ and $y \le 1/2$.

Solution The fraction with $x \le 1/2$ is the volume under the graph to the left of the line $x = 1/2$:

$$\int_0^{1/2} \int_0^1 (x + y)\,dy\,dx = \int_0^{1/2} \left(xy + \frac{y^2}{2}\right)\bigg|_0^1 dx = \int_0^{1/2} \left(x + \frac{1}{2}\right) dx$$

$$= \left(\frac{x^2}{2} + \frac{x}{2}\right)\bigg|_0^{1/2} = \frac{1}{8} + \frac{1}{4} = \frac{3}{8}.$$

Since the function and the regions of integration are symmetric in x and y, the fraction with $y \le 1/2$

is also $3/8$. Finally, the fraction with both $x \le 1/2$ and $y \le 1/2$ is

$$\int_0^{1/2}\int_0^{1/2}(x+y)\,dy\,dx = \int_0^{1/2}\left(xy+\frac{y^2}{2}\right)\Big|_0^{1/2}dx = \int_0^{1/2}\left(\frac{1}{2}x+\frac{1}{8}\right)dx$$
$$= \left(\frac{1}{4}x^2+\frac{1}{8}x\right)\Big|_0^{1/2} = \frac{1}{16}+\frac{1}{16} = \frac{1}{8}.$$

Recall that a one-variable density function $p(x)$ is a function such that $p(x) \ge 0$ for all x, and $\int_{-\infty}^{\infty} p(x)\,dx = 1$.

Example 4 Let p_1 and p_2 be one-variable density functions for x and y, respectively. Check that $p(x, y) = p_1(x)p_2(y)$ is a joint density function.

Solution Since both p_1 and p_2 are density functions, they are nonnegative everywhere. Thus, their product $p_1(x)p_2(x) = p(x, y)$ is nonnegative everywhere. Now we must check that the volume under the graph of p is 1. Since $\int_{-\infty}^{\infty} p_2(y)\,dy = 1$ and $\int_{-\infty}^{\infty} p_1(x)\,dx = 1$, we have

$$\int_{-\infty}^{\infty}\int_{-\infty}^{\infty} p(x,y)\,dy\,dx = \int_{-\infty}^{\infty}\int_{-\infty}^{\infty} p_1(x)p_2(y)\,dy\,dx = \int_{-\infty}^{\infty} p_1(x)\left(\int_{-\infty}^{\infty} p_2(y)\,dy\right)dx$$
$$= \int_{-\infty}^{\infty} p_1(x)(1)\,dx = \int_{-\infty}^{\infty} p_1(x)\,dx = 1.$$

Example 5 A machine in a factory is set to produce components 10 cm long and 5 cm in diameter. In fact, there is a slight variation from one component to the next. A component is usable if its length and diameter deviate from the correct values by less than 0.1 cm. With the length, x, in cm and the diameter, y, in cm, the probability density function is

$$p(x, y) = \frac{50\sqrt{2}}{\pi}e^{-100(x-10)^2}e^{-50(y-5)^2}.$$

What is the probability that a component is usable? (See Figure 16.57.)

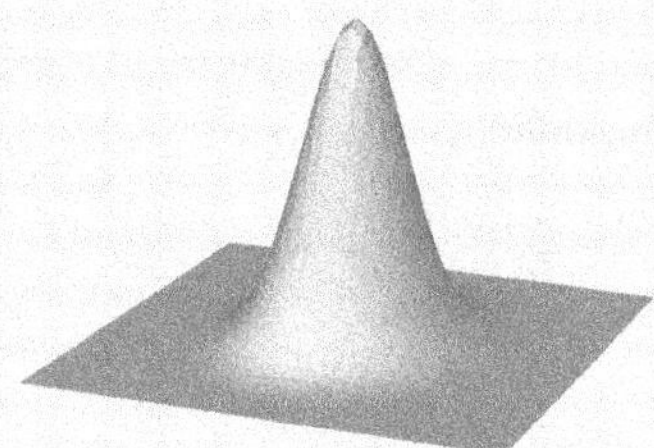

Figure 16.57: The density function $p(x, y) = \frac{50\sqrt{2}}{\pi}e^{-100(x-10)^2}e^{-50(y-5)^2}$

Solution We know that

$$\begin{array}{l}\text{Probability that } x \text{ and } y \text{ satisfy}\\ x_0-\Delta x \le x \le x_0+\Delta x\\ y_0-\Delta y \le y \le y_0+\Delta y\end{array} = \frac{50\sqrt{2}}{\pi}\int_{y_0-\Delta y}^{y_0+\Delta y}\int_{x_0-\Delta x}^{x_0+\Delta x} e^{-100(x-10)^2}e^{-50(y-5)^2}\,dx\,dy.$$

Thus,

$$\begin{array}{c}\text{Probability that}\\ \text{component is usable}\end{array} = \frac{50\sqrt{2}}{\pi}\int_{4.9}^{5.1}\int_{9.9}^{10.1} e^{-100(x-10)^2} e^{-50(y-5)^2}\,dx\,dy.$$

The double integral must be evaluated numerically. This yields

$$\begin{array}{c}\text{Probability that}\\ \text{component is usable}\end{array} = \frac{50\sqrt{2}}{\pi}(0.02556) = 0.57530.$$

Thus, there is a 57.530% chance that the component is usable.

Exercises and Problems for Section 16.6

EXERCISES

In Exercises 1–6, check whether p is a joint density function. Assume $p(x, y) = 0$ outside the region R.

1. $p(x, y) = 1/2$, where R is $4 \le x \le 5, -2 \le y \le 0$
2. $p(x, y) = 1$, where R is $0 \le x \le 1, 0 \le y \le 2$
3. $p(x, y) = x + y$, where R is $-1 \le x \le 1, 0 \le y \le 1$
4. $p(x, y) = 6(y - x)$, where R is $0 \le x \le y \le 2$
5. $p(x, y) = (2/\pi)(1 - x^2 - y^2)$, where R is $x^2 + y^2 \le 1$
6. $p(x, y) = xye^{-x-y}$, where R is $x \ge 0, y \ge 0$

In Exercises 7–10, a joint probability density function is given by $p(x, y) = xy/4$ in R, the rectangle $0 \le x \le 2$, $0 \le y \le 2$, and $p(x, y) = 0$ else. Find the probability that a point (x, y) satisfies the given conditions.

7. $x \le 1$ and $y \le 1$
8. $x \ge 1$ and $y \ge 1$
9. $x \ge 1$ and $y \le 1$
10. $1/3 \le x \le 1$

In Exercises 11–14, a joint probability density function is given by $p(x, y) = 0.005x + 0.025y$ in R, the rectangle $0 \le x \le 10$, $0 \le y \le 2$, and $p(x, y) = 0$ else. Find the probability that a point (x, y) satisfies the given conditions.

11. $x \le 4$
12. $y \ge 1$
13. $x \le 4$ and $y \ge 1$
14. $x \ge 5$ and $y \ge 1$

In Exercises 15–22, let p be the joint density function such that $p(x, y) = xy$ in R, the rectangle $0 \le x \le 2, 0 \le y \le 1$, and $p(x, y) = 0$ outside R. Find the fraction of the population satisfying the given constraints.

15. $x \ge 3$
16. $x = 1$
17. $x + y \le 3$
18. $-1 \le x \le 1$
19. $x \ge y$
20. $x + y \le 1$
21. $0 \le x \le 1, 0 \le y \le 1/2$
22. Within a distance 1 from the origin

PROBLEMS

23. Let x and y have joint density function

$$p(x, y) = \begin{cases} \frac{2}{3}(x + 2y) & \text{for } 0 \le x \le 1, 0 \le y \le 1,\\ 0 & \text{otherwise.}\end{cases}$$

Find the probability that

(a) $x > 1/3$. **(b)** $x < (1/3) + y$.

24. The joint density function for x, y is given by

$$f(x, y) = \begin{cases} kxy & \text{for } 0 \le x \le y \le 1,\\ 0 & \text{otherwise.}\end{cases}$$

(a) Determine the value of k.
(b) Find the probability that (x, y) lies in the shaded region in Figure 16.58.

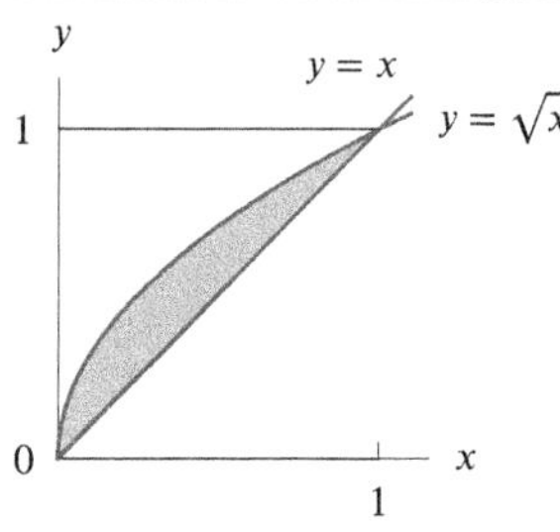

Figure 16.58

25. A joint density function is given by

$$f(x, y) = \begin{cases} kx^2 & \text{for } 0 \le x \le 2 \text{ and } 0 \le y \le 1, \\ 0 & \text{otherwise.} \end{cases}$$

(a) Find the value of the constant k.
(b) Find the probability that (x, y) satisfies $x + y \le 2$.
(c) Find the probability that (x, y) satisfies $x \le 1$ and $y \le 1/2$.

26. A point is chosen at random from the region S in the xy-plane containing all points (x, y) such that $-1 \le x \le 1, -2 \le y \le 2$ and $x - y \ge 0$ ("at random" means that the density function is constant on S).

(a) Determine the joint density function for x and y.
(b) If T is a subset of S with area α, then find the probability that a point (x, y) is in T.

27. A probability density function on a square has constant values in different triangular regions as shown in Figure 16.59. Find the probability that

(a) $x \ge 2$
(b) $y \ge x$
(c) $y \ge x$ and $x \ge 2$

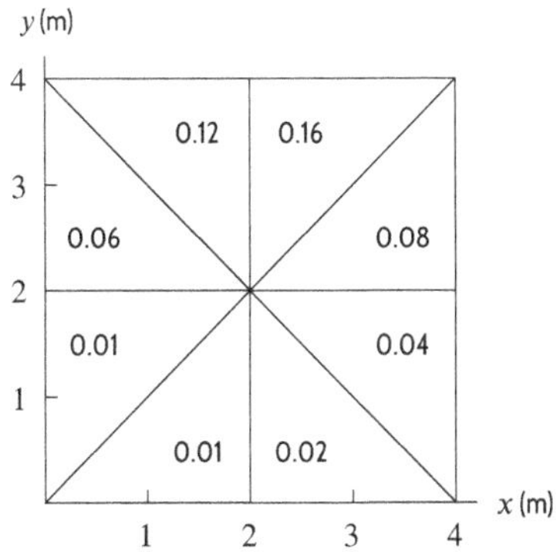

Figure 16.59: Probability density on a square (per m^2)

28. A health insurance company wants to know what proportion of its policies are going to cost the company a lot of money because the insured people are over 65 and sick. In order to compute this proportion, the company defines a *disability index*, x, with $0 \le x \le 1$, where $x = 0$ represents perfect health and $x = 1$ represents total disability. In addition, the company uses a density function, $f(x, y)$, defined in such a way that the quantity

$$f(x, y)\,\Delta x\,\Delta y$$

approximates the fraction of the population with disability index between x and $x+\Delta x$, and aged between y and $y+\Delta y$. The company knows from experience that a policy no longer covers its costs if the insured person is over 65 and has a disability index exceeding 0.8. Write an expression for the fraction of the company's policies held by people meeting these criteria.

29. The probability that a radioactive substance will decay at time t is modeled by the density function

$$p(t) = \lambda e^{-\lambda t}$$

for $t \ge 0$, and $p(t) = 0$ for $t < 0$. The positive constant λ depends on the material, and is called the decay rate.

(a) Check that p is a density function.
(b) Two materials with decay rates λ and μ decay independently of each other; their joint density function is the product of the individual density functions. Write the joint density function for the probability that the first material decays at time t and the second at time s.
(c) Find the probability that the first substance decays before the second.

30. Figure 16.60 represents a baseball field, with the bases at $(1, 0), (1, 1), (0, 1)$, and home plate at $(0, 0)$. The outer bound of the outfield is a piece of a circle about the origin with radius 4. When a ball is hit by a batter we record the spot on the field where the ball is caught. Let $p(r, \theta)$ be a function in the plane that gives the density of the distribution of such spots. Write an expression that represents the probability that a hit is caught in

(a) The right field (region R).
(b) The center field (region C).

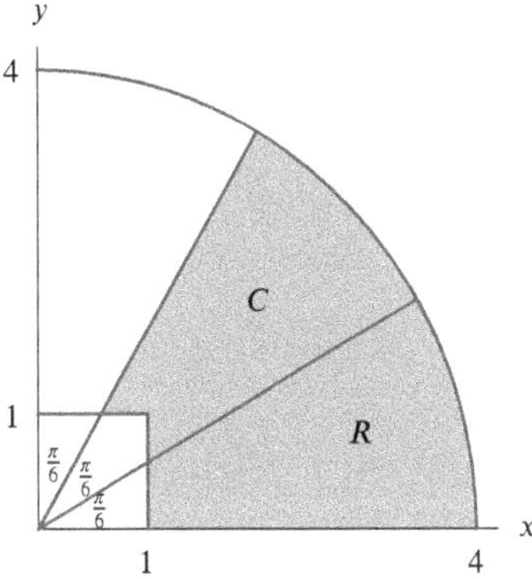

Figure 16.60

31. Two independent random numbers x and y between 0 and 1 have joint density function

$$p(x, y) = \begin{cases} 1 & \text{if } 0 \le x, y \le 1 \\ 0 & \text{otherwise.} \end{cases}$$

This problem concerns the average $z = (x+y)/2$, which has a one-variable probability density function of its own.

(a) Find $F(t)$, the probability that $z \le t$. Treat separately the cases $t \le 0$, $0 < t \le 1/2$, $1/2 < t \le 1$, $1 < t$. Note that $F(t)$ is the cumulative distribution function of z.
(b) Find and graph the probability density function of z.
(c) Are x and y more likely to be near 0, 1/2, or 1? What about z?

Strengthen Your Understanding

In Problems 32–33, explain what is wrong with the statement.

32. If $p_1(x, y)$ and $p_2(x, y)$ are joint density functions, then $p_1(x, y) + p_2(x, y)$ is a joint density function.

33. If $p(w, h)$ is the probability density function of the weight and height of mothers discussed in Section 16.6, then the probability that a mother weighs 60 kg and has a height of 170 cm is $p(60, 170)$.

In Problems 34–35, give an example of:

34. Values for a, b, c and d such that f is a joint density function:

$$f(x, y) = \begin{cases} 1 & \text{for } a \le x \le b \text{ and } c \le y \le d, \\ 0 & \text{otherwise} \end{cases}$$

35. A one-variable function $g(y)$ such that f is a joint density function:

$$f(x, y) = \begin{cases} g(y) & \text{for } 0 \le x \le 2 \text{ and } 0 \le y \le 1, \\ 0 & \text{otherwise} \end{cases}$$

For Problems 36–39, let $p(x, y)$ be a joint density function for x and y. Are the following statements true or false?

36. $\int_a^b \int_{-\infty}^{\infty} p(x, y)\, dy\, dx$ is the probability that $a \le x \le b$.

37. $0 \le p(x, y) \le 1$ for all x.

38. $\int_a^b p(x, y)\, dx$ is the probability that $a \le x \le b$.

39. $\int_{-\infty}^{\infty} \int_{-\infty}^{\infty} p(x, y)\, dy\, dx = 1$.

Online Resource: Review problems and Projects

Chapter Seventeen

PARAMETERIZATION AND VECTOR FIELDS

Contents

17.1 PARAMETERIZED CURVES

A curve in the plane may be parameterized by a pair of equations of the form $x = f(t)$, $y = g(t)$. As the parameter t changes, the point (x, y) traces out the curve. In this section we find parametric equations for curves in three dimensions, and we see how to write parametric equations using position vectors.

Parametric Equations in Three Dimensions

We describe motion in the plane by giving parametric equations for x and y in terms of t. To describe a motion in 3-space parametrically, we need a third equation giving z in terms of t.

Example 1 Find parametric equations for the curve $y = x^2$ in the xy-plane.

Solution A possible parameterization in two dimensions is $x = t$, $y = t^2$. Since the curve is in the xy-plane, the z-coordinate is zero, so a parameterization in three dimensions is

$$x = t, \quad y = t^2, \quad z = 0.$$

Example 2 Find parametric equations for a particle that starts at $(0, 3, 0)$ and moves around a circle as shown in Figure 17.1.

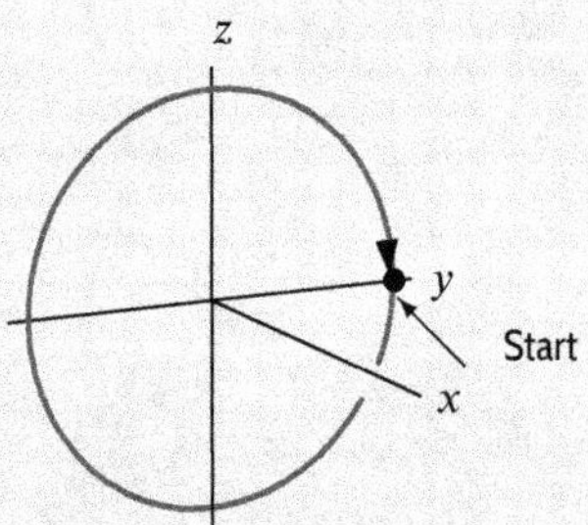

Figure 17.1: Circle of radius 3 in the yz-plane, centered at origin

Solution Since the motion is in the yz-plane, we have $x = 0$ at all times t. Looking at the yz-plane from the positive x-direction we see motion around a circle of radius 3 in the clockwise direction. Thus,

$$x = 0, \quad y = 3\cos t, \quad z = -3\sin t.$$

Example 3 Describe in words the motion given parametrically by

$$x = \cos t, \quad y = \sin t, \quad z = t.$$

Solution The particle's x- and y-coordinates give circular motion in the xy-plane, while the z-coordinate increases steadily. Thus, the particle traces out a rising spiral, like a coiled spring. (See Figure 17.2.) This curve is called a *helix*.

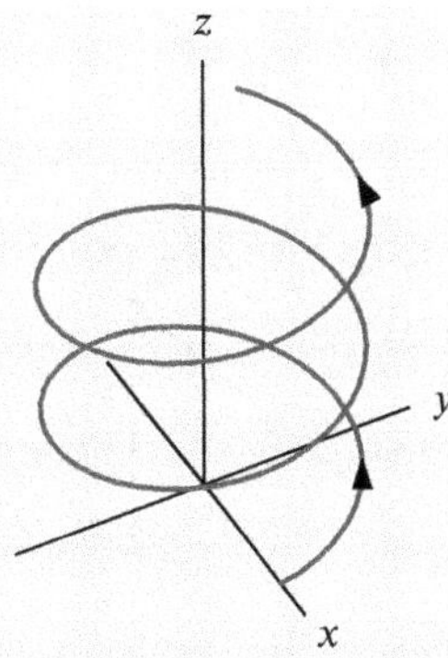

Figure 17.2: The helix $x = \cos t, y = \sin t, z = t$

Example 4 Find parametric equations for the line parallel to the vector $2\vec{i} + 3\vec{j} + 4\vec{k}$ and through the point $(1, 5, 7)$.

Solution Let's imagine a particle at the point $(1, 5, 7)$ at time $t = 0$ and moving through a displacement of $2\vec{i} + 3\vec{j} + 4\vec{k}$ for each unit of time, t. When $t = 0$, $x = 1$ and x increases by 2 units for every unit of time. Thus, at time t, the x-coordinate of the particle is given by

$$x = 1 + 2t.$$

Similarly, the y-coordinate starts at $y = 5$ and increases at a rate of 3 units for every unit of time. The z-coordinate starts at $y = 7$ and increases by 4 units for every unit of time. Thus, the parametric equations of the line are

$$x = 1 + 2t, \quad y = 5 + 3t, \quad z = 7 + 4t.$$

We can generalize the previous example as follows:

> **Parametric Equations of a Line** through the point (x_0, y_0, z_0) and parallel to the vector $a\vec{i} + b\vec{j} + c\vec{k}$ are
>
> $$x = x_0 + at, \quad y = y_0 + bt, \quad z = z_0 + ct.$$

Notice that the coordinates x, y, and z are linear functions of the parameter t.

Example 5
(a) Describe in words the curve given by the parametric equations $x = 3 + t$, $y = 2t$, $z = 1 - t$.
(b) Find parametric equations for the line through the points $(1, 2, -1)$ and $(3, 3, 4)$.

Solution
(a) The curve is a line through the point $(3, 0, 1)$ and parallel to the vector $\vec{i} + 2\vec{j} - \vec{k}$.
(b) The line is parallel to the vector between the points $P = (1, 2, -1)$ and $Q = (3, 3, 4)$.

$$\overrightarrow{PQ} = (3 - 1)\vec{i} + (3 - 2)\vec{j} + (4 - (-1))\vec{k} = 2\vec{i} + \vec{j} + 5\vec{k}.$$

Thus, using the point P, the parametric equations are

$$x = 1 + 2t, \quad y = 2 + t, \quad z = -1 + 5t.$$

Using the point Q gives the equations $x = 3 + 2t$, $y = 3 + t$, $z = 4 + 5t$, which represent the same line. The point where $t = 0$ in the second equations is given by $t = 1$ in the first equations.

Using Position Vectors to Write Parameterized Curves as Vector-Valued Functions

A point in the plane with coordinates (x, y) can be represented by the position vector $\vec{r} = x\vec{i} + y\vec{j}$ in Figure 17.3. Similarly, in 3-space we write $\vec{r} = x\vec{i} + y\vec{j} + z\vec{k}$. (See Figure 17.4.)

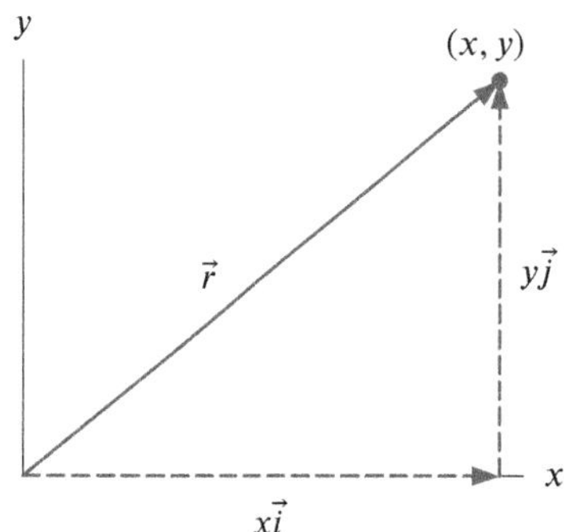

Figure 17.3: Position vector $\vec{r}$ for the point (x, y)

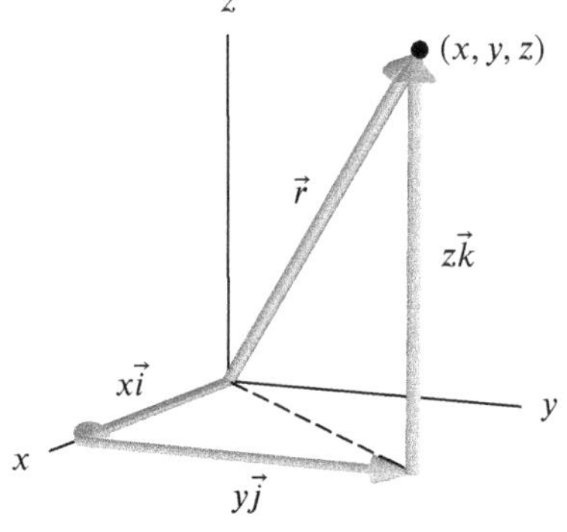

Figure 17.4: Position vector $\vec{r}$ for the point (x, y, z)

We can write the parametric equations $x = f(t)$, $y = g(t)$, $z = h(t)$ as a single vector equation

$$\vec{r}(t) = f(t)\vec{i} + g(t)\vec{j} + h(t)\vec{k}$$

called a *parameterization*. As the parameter t varies, the point with position vector $\vec{r}(t)$ traces out a curve in 3-space. For example, the circular motion in the plane

$$x = \cos t, y = \sin t \quad \text{can be written as} \quad \vec{r} = (\cos t)\vec{i} + (\sin t)\vec{j}$$

and the helix in 3-space

$$x = \cos t, y = \sin t, z = t \quad \text{can be written as} \quad \vec{r} = (\cos t)\vec{i} + (\sin t)\vec{j} + t\vec{k}.$$

See Figure 17.5.

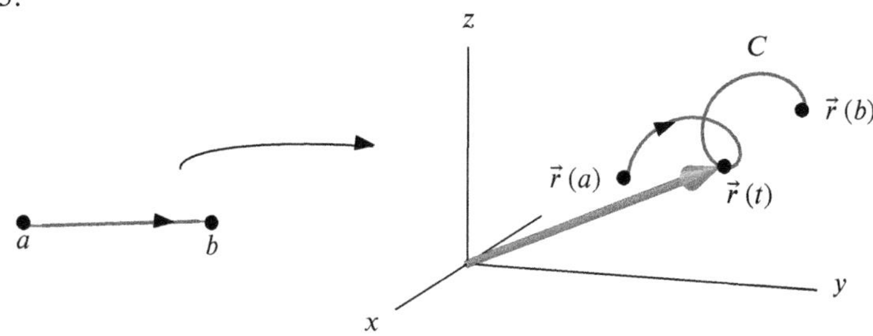

Figure 17.5: The parameterization sends the interval, $a \le t \le b$, to the curve, C, in 3-space

Example 6 Use vectors to give a parameterization for the circle of radius $\frac{1}{2}$ centered at the point $(-1, 2)$.

Solution The circle of radius 1 centered at the origin is parameterized by the vector-valued function

$$\vec{r}_1(t) = \cos t\vec{i} + \sin t\vec{j}, \quad 0 \le t \le 2\pi.$$

The point $(-1, 2)$ has position vector $\vec{r}_0 = -\vec{i} + 2\vec{j}$. The position vector, $\vec{r}(t)$, of a point on the circle of radius $\frac{1}{2}$ centered at $(-1, 2)$ is found by adding $\frac{1}{2}\vec{r}_1$ to $\vec{r}_0$. (See Figures 17.6 and 17.7.) Thus,

$$\vec{r}(t) = \vec{r}_0 + \tfrac{1}{2}\vec{r}_1(t) = -\vec{i} + 2\vec{j} + \tfrac{1}{2}(\cos t\vec{i} + \sin t\vec{j}) = (-1 + \tfrac{1}{2}\cos t)\vec{i} + (2 + \tfrac{1}{2}\sin t)\vec{j},$$

or, equivalently,

$$x = -1 + \tfrac{1}{2}\cos t, \quad y = 2 + \tfrac{1}{2}\sin t, \quad 0 \le t \le 2\pi.$$

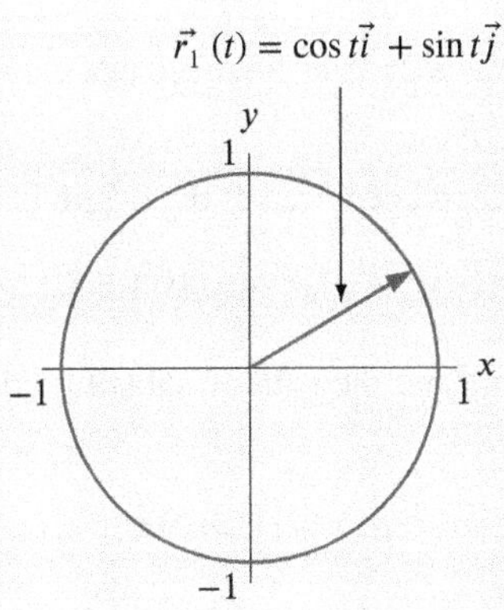

Figure 17.6: The circle $x^2 + y^2 = 1$ parameterized by $\vec{r}_1(t) = \cos t\vec{i} + \sin t\vec{j}$

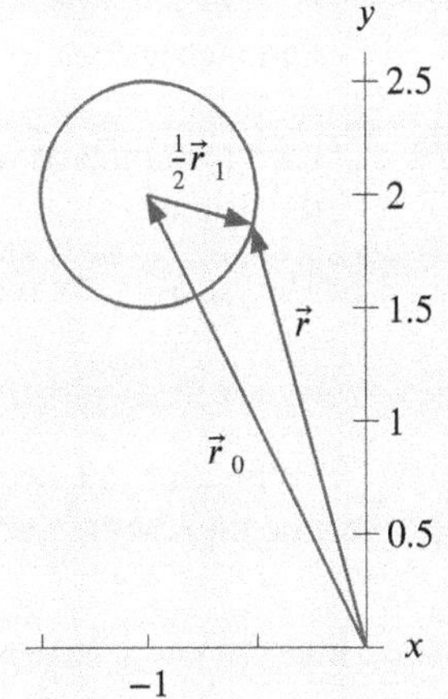

Figure 17.7: The circle of radius $\frac{1}{2}$ and center $(-1, 2)$ parameterized by $\vec{r}(t) = \vec{r}_0 + \frac{1}{2}\vec{r}_1(t)$

Parametric Equation of a Line

Consider a straight line in the direction of a vector $\vec{v}$ passing through the point (x_0, y_0, z_0) with position vector $\vec{r}_0$. We start at $\vec{r}_0$ and move up and down the line, adding different multiples of $\vec{v}$ to $\vec{r}_0$. (See Figure 17.8.)

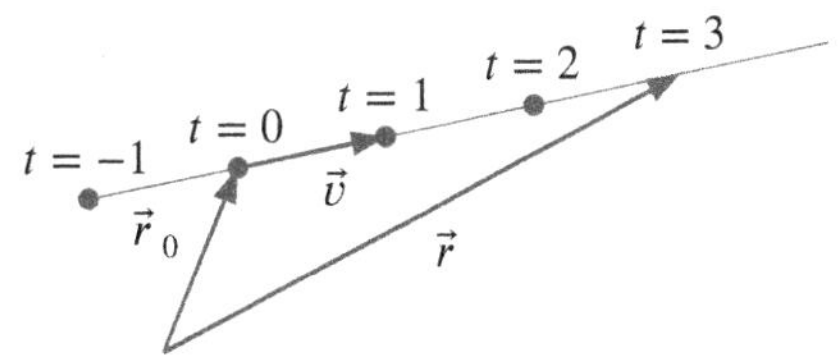

Figure 17.8: The line $\vec{r}(t) = \vec{r}_0 + t\vec{v}$

In this way, every point on the line can be written as $\vec{r}_0 + t\vec{v}$, which yields the following:

Parametric Equation of a Line in Vector Form

The line through the point with position vector $\vec{r}_0 = x_0\vec{i} + y_0\vec{j} + z_0\vec{k}$ in the direction of the vector $\vec{v} = a\vec{i} + b\vec{j} + c\vec{k}$ has parametric equation

$$\vec{r}(t) = \vec{r}_0 + t\vec{v}.$$

Example 7

(a) Find parametric equations for the line passing through the points $(2, -1, 3)$ and $(-1, 5, 4)$.
(b) Represent the line segment from $(2, -1, 3)$ to $(-1, 5, 4)$ parametrically.

Solution

(a) The line passes through $(2, -1, 3)$ and is parallel to the displacement vector $\vec{v} = -3\vec{i} + 6\vec{j} + \vec{k}$ from $(2, -1, 3)$ to $(-1, 5, 4)$. Thus, the parametric equation is

$$\vec{r}(t) = 2\vec{i} - \vec{j} + 3\vec{k} + t(-3\vec{i} + 6\vec{j} + \vec{k}).$$

(b) In the parameterization in part (a), $t = 0$ corresponds to the point $(2, -1, 3)$ and $t = 1$ corresponds to the point $(-1, 5, 4)$. So the parameterization of the segment is

$$\vec{r}(t) = 2\vec{i} - \vec{j} + 3\vec{k} + t(-3\vec{i} + 6\vec{j} + \vec{k}), \qquad 0 \le t \le 1.$$

Intersection of Curves and Surfaces

Parametric equations for a curve enable us to find where a curve intersects a surface.

Example 8 Find the points at which the line $x = t$, $y = 2t$, $z = 1 + t$ pierces the sphere of radius 10 centered at the origin.

Solution The equation for the sphere of radius 10 and centered at the origin is

$$x^2 + y^2 + z^2 = 100.$$

To find the intersection points of the line and the sphere, substitute the parametric equations of the line into the equation of the sphere, giving

$$t^2 + 4t^2 + (1 + t)^2 = 100,$$

so

$$6t^2 + 2t - 99 = 0,$$

which has the two solutions at approximately $t = -4.23$ and $t = 3.90$. Using the parametric equation for the line, $(x, y, z) = (t, 2t, 1 + t)$, we see that the line cuts the sphere at the two points

$$(x, y, z) = (-4.23, 2(-4.23), 1 + (-4.23)) = (-4.23, -8.46, -3.23),$$

and

$$(x, y, z) = (3.90, 2(3.90), 1 + 3.90) = (3.90, 7.80, 4.90).$$

We can also use parametric equations to find the intersection of two curves.

Example 9 Two particles move through space, with equations $\vec{r}_1(t) = t\vec{i} + (1 + 2t)\vec{j} + (3 - 2t)\vec{k}$ and $\vec{r}_2(t) = (-2 - 2t)\vec{i} + (1 - 2t)\vec{j} + (1 + t)\vec{k}$. Do the particles ever collide? Do their paths cross?

Solution To see if the particles collide, we must find out if they pass through the same point at the same time t. So we must find a solution to the vector equation $\vec{r}_1(t) = \vec{r}_2(t)$, which is the same as finding a common solution to the three scalar equations

$$t = -2 - 2t, \qquad 1 + 2t = 1 - 2t, \qquad 3 - 2t = 1 + t.$$

Separately, the solutions are $t = -2/3$, $t = 0$, and $t = 2/3$, so there is no common solution, and the particles don't collide. To see if their paths cross, we find out if they pass through the same point at two possibly different times, t_1 and t_2. So we solve the equations

$$t_1 = -2 - 2t_2, \qquad 1 + 2t_1 = 1 - 2t_2, \qquad 3 - 2t_1 = 1 + t_2.$$

We solve the first two equations simultaneously and get $t_1 = 2$, $t_2 = -2$. Since these values also satisfy the third equation, the paths cross. The position of the first particle at time $t = 2$ is the same as the position of the second particle at time $t = -2$, namely the point $(2, 5, -1)$.

Example 10 Are the lines $x = -1 + t$, $y = 1 + 2t$, $z = 5 - t$ and $x = 2 + 2t$, $y = 4 + t$, $z = 3 + t$ parallel? Do they intersect?

Solution In vector form the lines are parameterized by

$$\vec{r} = -\vec{i} + \vec{j} + 5\vec{k} + t(\vec{i} + 2\vec{j} - \vec{k})$$
$$\vec{r} = 2\vec{i} + 4\vec{j} + 3\vec{k} + t(2\vec{i} + \vec{j} + \vec{k})$$

Their direction vectors $\vec{i} + 2\vec{j} - \vec{k}$ and $2\vec{i} + \vec{j} + \vec{k}$ are not multiples of each other, so the lines are not parallel. To find out if they intersect, we see if they pass through the same point at two possibly different times, t_1 and t_2:

$$-1 + t_1 = 2 + 2t_2, \qquad 1 + 2t_1 = 4 + t_2, \qquad 5 - t_1 = 3 + t_2.$$

The first two equations give $t_1 = 1, t_2 = -1$. Since these values do not satisfy the third equation, the paths do not cross, and so the lines do not intersect.

The next example shows how to tell if two different parameterizations give the same line.

Example 11 Show that the following two lines are the same:

$$\vec{r} = -\vec{i} - \vec{j} + \vec{k} + t(3\vec{i} + 6\vec{j} - 3\vec{k})$$
$$\vec{r} = \vec{i} + 3\vec{j} - \vec{k} + t(-\vec{i} - 2\vec{j} + \vec{k})$$

Solution The direction vectors of the two lines, $3\vec{i} + 6\vec{j} - 3\vec{k}$ and $-\vec{i} - 2\vec{j} + \vec{k}$, are multiples of each other, so the lines are parallel. To see if they are the same, we pick a point on the first line and see if it is on the second line. For example, the point on the first line with $t = 0$ has position vector $-\vec{i} - \vec{j} + \vec{k}$. Solving

$$\vec{i} + 3\vec{j} - \vec{k} + t(-\vec{i} - 2\vec{j} + \vec{k}) = -\vec{i} - \vec{j} + \vec{k},$$

we get $t = 2$, so the two lines have a point in common. Thus, they are the same line, parameterized in two different ways.

Exercises and Problems for Section 17.1

Online Resource: Additional Problems for Section 17.1

EXERCISES

In Exercises 1–6, find a parameterization for the curve.

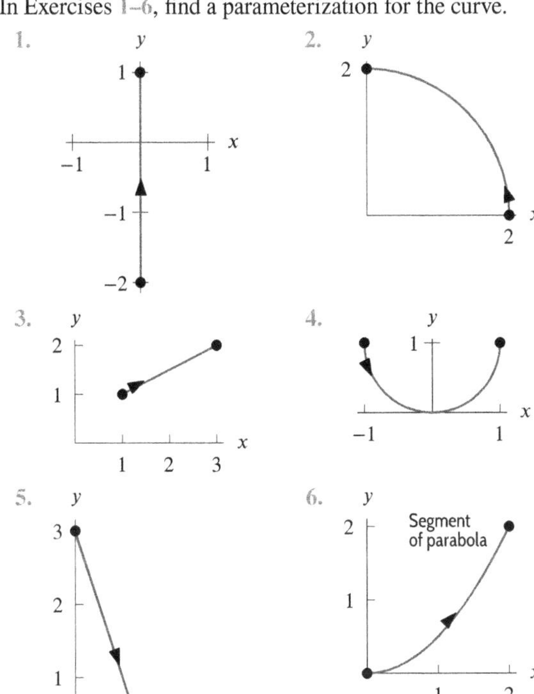

In Exercises 7–17, find parametric equations for the line.

7. The line in the direction of the vector $\vec{i} - \vec{k}$ and through the point $(0, 1, 0)$.

8. The line in the direction of the vector $\vec{i} + 2\vec{j} - \vec{k}$ and through the point $(3, 0, -4)$.

9. The line parallel to the z-axis passing through the point $(1, 0, 0)$.

10. The line in the direction of the vector $5\vec{j} + 2\vec{k}$ and through the point $(5, -1, 1)$.

11. The line in the direction of the vector $3\vec{i} - 3\vec{j} + \vec{k}$ and through the point $(1, 2, 3)$.

12. The line in the direction of the vector $2\vec{i} + 2\vec{j} - 3\vec{k}$ and through the point $(-3, 4, -2)$.

13. The line through $(-3, -2, 1)$ and $(-1, -3, -1)$.

14. The line through the points $(1, 5, 2)$ and $(5, 0, -1)$.

15. The line through the points $(2, 3, -1)$ and $(5, 2, 0)$.

16. The line through $(3, -2, 2)$ and intersecting the y-axis at $y = 2$.

17. The line intersecting the x-axis at $x = 3$ and the z-axis at $z = -5$.

In Exercises 18–34, find a parameterization for the curve.

18. A line segment between $(2, 1, 3)$ and $(4, 3, 2)$.

19. A circle of radius 3 centered on the z-axis and lying in the plane $z = 5$.

20. A line perpendicular to the plane $z = 2x - 3y + 7$ and through the point $(1, 1, 6)$.

21. The circle of radius 2 in the xy-plane, centered at the origin, clockwise.

22. The circle of radius 2 parallel to the xy-plane, centered at the point $(0, 0, 1)$, and traversed counterclockwise when viewed from below.

23. The circle of radius 2 in the xz-plane, centered at the origin.

24. The circle of radius 3 parallel to the xy-plane, centered at the point $(0, 0, 2)$.

25. The circle of radius 3 in the yz-plane, centered at the point $(0, 0, 2)$.

26. The circle of radius 5 parallel to the yz-plane, centered at the point $(-1, 0, -2)$.

27. The curve $x = y^2$ in the xy-plane.

28. The curve $y = x^3$ in the xy-plane.

29. The curve $x = -3z^2$ in the xz-plane.

30. The curve in which the plane $z = 2$ cuts the surface $z = \sqrt{x^2 + y^2}$.

31. The curve $y = 4 - 5x^4$ through the point $(0, 4, 4)$, parallel to the xy-plane.

32. The ellipse of major diameter 5 parallel to the y-axis and minor diameter 2 parallel to the z-axis, centered at $(0, 1, -2)$.

33. The ellipse of major diameter 6 along the x-axis and minor diameter 4 along the y-axis, centered at the origin.

34. The ellipse of major diameter 3 parallel to the x-axis and minor diameter 2 parallel to the z-axis, centered at $(0, 1, -2)$.

In Exercises 35–42, find a parametric equation for the curve segment.

35. Line from $(-1, 2, -3)$ to $(2, 2, 2)$.

36. Line from $P_0 = (-1, -3)$ to $P_1 = (5, 2)$.

37. Line from $P_0 = (1, -3, 2)$ to $P_1 = (4, 1, -3)$.

38. Semicircle from $(0, 0, 5)$ to $(0, 0, -5)$ in the yz-plane with $y \geq 0$.

39. Semicircle from $(1, 0, 0)$ to $(-1, 0, 0)$ in the xy-plane with $y \geq 0$.

40. Graph of $y = \sqrt{x}$ from $(1, 1)$ to $(16, 4)$.

41. Arc of a circle of radius 5 from $P = (0, 0)$ to $Q = (10, 0)$.

42. Quarter-ellipse from $(4, 0, 3)$ to $(0, -3, 3)$ in the plane $z = 3$.

In Exercises 43–46, find parametric equations for a helix satisfying the given conditions.

43. Centered on the z-axis, with radius 10.

44. Centered on the x-axis, with radius 5.

45. Centered on the y-axis, with radius 2.

46. Centered on the vertical line passing through $(3, 5, 0)$, with radius 1.

PROBLEMS

In Problems 47–51, parameterize the line through $P = (2, 5)$ and $Q = (12, 9)$ so that the points P and Q correspond to the given parameter values.

47. $t = 0$ and 1

48. $t = 0$ and 5

49. $t = 20$ and 30

50. $t = 10$ and 11

51. $t = 0$ and -1

52. At the point where $t = -1$, find an equation for the plane perpendicular to the line

$$x = 5 - 3t, \quad y = 5t - 7, \quad \frac{z}{t} = 6.$$

53. Determine whether the following line is parallel to the plane $2x - 3y + 5z = 5$:

$$x = 5 + 7t, \quad y = 4 + 3t, \quad z = -3 - 2t.$$

54. Show that the equations $x = 3 + t$, $y = 2t$, $z = 1 - t$ satisfy the equations $x + y + 3z = 6$ and $x - y - z = 2$. What does this tell you about the curve parameterized by these equations?

55. **(a)** Explain why the line of intersection of two planes must be parallel to the cross product of a normal vector to the first plane and a normal vector to the second.

(b) Find a vector parallel to the line of intersection of the two planes $x + 2y - 3z = 7$ and $3x - y + z = 0$.

(c) Find parametric equations for the line in part (b).

56. Find an equation for the plane containing the point $(2, 3, 4)$ and the line $x = 1 + 2t$, $y = 3 - t$, $z = 4 + t$.

57. **(a)** Find an equation for the line perpendicular to the plane $2x - 3y = z$ and through the point $(1, 3, 7)$.

(b) Where does the line cut the plane?

(c) What is the distance between the point $(1, 3, 7)$ and the plane?

58. Consider two points P_0 and P_1 in 3-space.

(a) Show that the line segment from P_0 to P_1 can be parameterized by

$$\vec{r}(t) = (1-t)\overrightarrow{OP_0} + t\overrightarrow{OP_1}, \quad 0 \le t \le 1.$$

(b) What is represented by the parametric equation

$$\vec{r}(t) = t\overrightarrow{OP_0} + (1-t)\overrightarrow{OP_1}, \quad 0 \le t \le 1?$$

59. **(a)** Find a vector parallel to the line of intersection of the planes $2x - y - 3z = 0$ and $x + y + z = 1$.
(b) Show that the point $(1, -1, 1)$ lies on both planes.
(c) Find parametric equations for the line of intersection.

60. Find the intersection of the line $x = 5 + 7t$, $y = 4 + 3t$, $z = -3 - 2t$ and the plane $2x - 3y + 5z = -7$.

In Problems 61–64, are the lines L_1 and L_2 the same line?

61. L_1: $x = 5 + t$, $y = 3 - 2t$, $z = 5t$
L_2: $x = 5 + 2t$, $y = 3 - 4t$, $z = 10t$

62. L_1: $x = 2 + 3t$, $y = 1 + 4t$, $z = 6 - t$
L_2: $x = 2 + 6t$, $y = 4 + 3t$, $z = 3 - 2t$

63. L_1: $x = 2 + 3t$, $y = 1 + 4t$, $z = 6 - t$
L_2: $x = 5 + 6t$, $y = 5 + 8t$, $z = 5 - 2t$

64. L_1: $x = 1 + 2t$, $y = 1 - 3t$, $z = 1 + t$
L_2: $x = 1 - 4t$, $y = 6t$, $z = 4 - 2t$

In Problems 65–67 two parameterized lines are given. Are they the same line?

65. $\vec{r}_1(t) = (5 - 3t)\vec{i} + 2t\vec{j} + (7 + t)\vec{k}$
$\vec{r}_2(t) = (5 - 6t)\vec{i} + 4t\vec{j} + (7 + 3t)\vec{k}$

66. $\vec{r}_1(t) = (5 - 3t)\vec{i} + (1 + t)\vec{j} + 2t\vec{k}$
$\vec{r}_2(t) = (2 + 6t)\vec{i} + (2 - 2t)\vec{j} + (2 - 4t)\vec{k}$

67. $\vec{r}_1(t) = (5 - 3t)\vec{i} + (1 + t)\vec{j} + 2t\vec{k}$
$\vec{r}_2(t) = (2 + 6t)\vec{i} + (2 - 2t)\vec{j} + (3 - 4t)\vec{k}$

68. If it exists, find the value of c for which the lines $l(t) = (c + t, 1 + t, 5 + t)$ and $m(s) = (s, 1 - s, 3 + s)$ intersect.

69. **(a)** Where does the line $\vec{r} = 2\vec{i} + 5\vec{j} + t(3\vec{i} + \vec{j} + 2\vec{k})$ cut the plane $x + y + z = 1$?
(b) Find a vector perpendicular to the line and lying in the plane.
(c) Find an equation for the line that passes through the point of intersection of the line and plane, is perpendicular to the line, and lies in the plane.

In Problems 70–73, find parametric equations for the line.

70. The line of intersection of the planes $x - y + z = 3$ and $2x + y - z = 5$.

71. The line of intersection of the planes $x + y + z = 3$ and $x - y + 2z = 2$.

72. The line perpendicular to the surface $z = x^2 + y^2$ at the point $(1, 2, 5)$.

73. The line through the point $(-4, 2, 3)$ and parallel to a line in the yz-plane which makes a 45° angle with the positive y-axis and the positive z-axis.

74. Is the point $(-3, -4, 2)$ visible from the point $(4, 5, 0)$ if there is an opaque ball of radius 1 centered at the origin?

75. Two particles are traveling through space. At time t the first particle is at the point $(-1 + t, 4 - t, -1 + 2t)$ and the second particle is at $(-7 + 2t, -6 + 2t, -1 + t)$.

(a) Describe the two paths in words.
(b) Do the two particles collide? If so, when and where?
(c) Do the paths of the two particles cross? If so, where?

76. Match the parameterizations with their graphs in Figure 17.9.

(a) $x = 2\cos 4\pi t$, $y = 2\sin 4\pi t$, $z = t$
(b) $x = 2\cos 4\pi t$, $y = \sin 4\pi t$, $z = t$
(c) $x = 0.5t\cos 4\pi t$, $y = 0.5t\sin 4\pi t$, $z = t$
(d) $x = 2\cos 4\pi t$, $y = 2\sin 4\pi t$, $z = 0.5t^3$

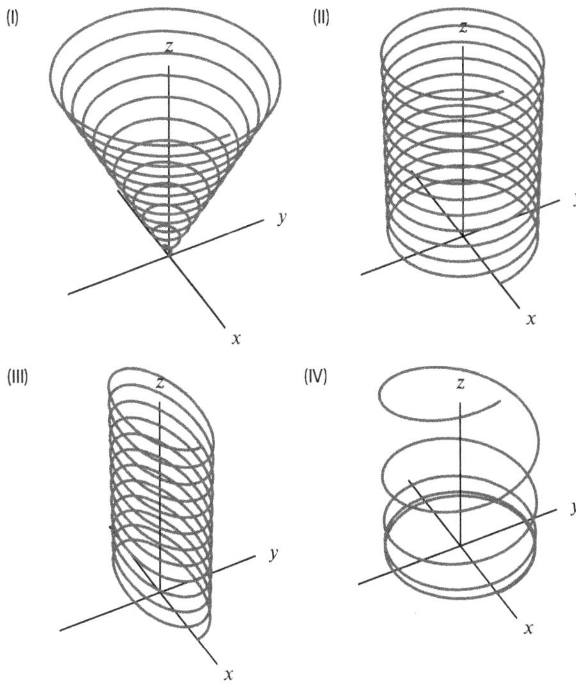

Figure 17.9

In Problems 77–80 find c so that one revolution about the z-axis of the helix gives an increase of Δz in the z-coordinate.

77. $x = 2\cos t$, $y = 2\sin t$, $z = ct$, $\Delta z = 15$

78. $x = 2\cos t,\ y = 2\sin t,\ z = ct,\ \Delta z = 50$

79. $x = 2\cos 3t,\ y = 2\sin 3t,\ z = ct,\ \Delta z = 10$

80. $x = 2\cos \pi t,\ y = 2\sin \pi t,\ z = ct,\ \Delta z = 20$

81. For $t > 0$, a particle moves along the curve $x = a + b\sin kt$, $y = a + b\cos kt$, where a, b, k are positive constants.

(a) Describe the motion in words.
(b) What is the effect on the curve of the following changes?
  (i) Increasing b
  (ii) Increasing a
  (iii) Increasing k
  (iv) Setting a and b equal

82. In the Atlantic Ocean off the coast of Newfoundland, Canada, the temperature and salinity (saltiness) vary throughout the year. Figure 17.10 shows a parametric curve giving the average temperature, T (in °C) and salinity (in grams of salt per kg of water) for t in months, with $t = 1$ corresponding to mid-January.[1]

(a) Why does the parameterized curve form a loop?
(b) When is the water temperature highest?
(c) When is the water saltiest?
(d) Estimate dT/dt at $t = 6$, and give the units. What is the meaning of your answer for seawater?

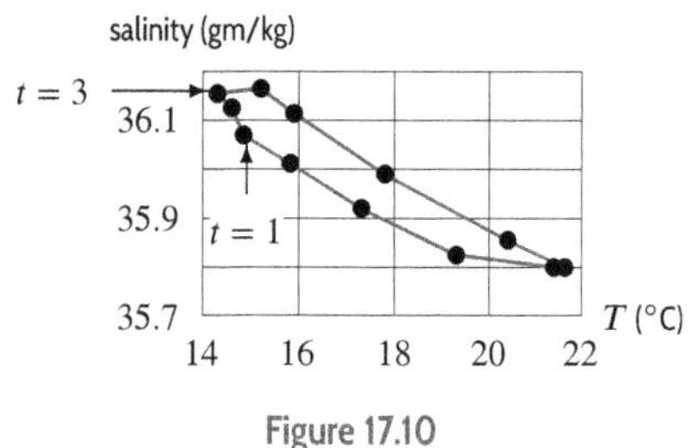

Figure 17.10

83. A light shines on the helix of Example 3 on page 886 from far down each axis. Sketch the shadow the helix casts on each of the coordinate planes: xy, xz, and yz.

84. The paraboloid $z = x^2+y^2$ and the plane $z = 2x+4y+4$ intersect in a curve in 3-space.

(a) Show that the shadow of the intersection in the xy-plane is a circle and find its center and radius.
(b) Parameterize the circle in the xy-plane.
(c) Parameterize the intersection of the paraboloid and the plane in 3-space.

85. For a positive constant a and $t \geq 0$, the parametric equations I-V represent the curves described in (a)-(e). Match each description (a)-(e) with its parametric equations and write an equation involving only x and y for the curve.

(a) Line through the origin.
(b) Line not through the origin.
(c) Hyperbola opening along x-axis.
(d) Circle traversed clockwise.
(e) Circle traversed counterclockwise.

I. $x = a\sin t,\ y = a\cos t$ II. $x = a\sin t,\ y = a\sin t$
III. $x = a\cos t,\ y = a\sin t$ IV. $x = a\cos^2 t,\ y = a\sin^2 t$
V. $x = a/\cos t,\ y = a\tan t$

86. **(a)** Find a parametric equation for the line through the point $(2, 1, 3)$ and in the direction of $a\vec{i} + b\vec{j} + c\vec{k}$.
(b) Find conditions on a, b, c so that the line you found in part (a) goes through the origin. Give a reason for your answer.

87. Consider the line $x = 5 - 2t$, $y = 3 + 7t$, $z = 4t$ and the plane $ax + by + cz = d$. All the following questions have many possible answers. Find values of a, b, c, d such that:

(a) The plane is perpendicular to the line.
(b) The plane is perpendicular to the line and through the point $(5, 3, 0)$.
(c) The line lies in the plane.

88. Explain the significance of the constants $\alpha > 0$ and $\beta > 0$ in the family of helices given by $\vec{r} = \alpha\cos t\vec{i} + \alpha\sin t\vec{j} + \beta t\vec{k}$.

89. Find parametric equations of the line passing through the points $(1, 2, 3)$, $(3, 5, 7)$ and calculate the shortest distance from the line to the origin.

90. Show that for a fixed value of θ, the line parameterized by $x = \cos\theta + t\sin\theta$, $y = \sin\theta - t\cos\theta$ and $z = t$ lies on the graph of the hyperboloid $x^2 + y^2 = z^2 + 1$.

91. A line has equation $\vec{r} = \vec{a} + t\vec{b}$ where $\vec{r} = x\vec{i} + y\vec{j} + z\vec{k}$ and $\vec{a}$ and $\vec{b}$ are constant vectors such that $\vec{a} \neq \vec{0}$, $\vec{b} \neq \vec{0}$, $\vec{b}$ not parallel or perpendicular to $\vec{a}$. For each of the planes (a)–(c), pick the equation (i)–(ix) which represents it. Explain your choice.

(a) A plane perpendicular to the line and through the origin.
(b) A plane perpendicular to the line and not through the origin.
(c) A plane containing the line.

(i) $\vec{a} \cdot \vec{r} = \|\vec{b}\|$ (ii) $\vec{b} \cdot \vec{r} = \|\vec{a}\|$
(iii) $\vec{a} \cdot \vec{r} = \vec{b} \cdot \vec{r}$ (iv) $(\vec{a} \times \vec{b}) \cdot (\vec{r} - \vec{a}) = 0$
(v) $\vec{r} - \vec{a} = \vec{b}$ (vi) $\vec{a} \cdot \vec{r} = 0$
(vii) $\vec{b} \cdot \vec{r} = 0$ (viii) $\vec{a} + \vec{r} = \vec{b}$
(ix) $(\vec{a} \times \vec{b}) \cdot (\vec{r} - \vec{b}) = \|\vec{a}\|$

92. **(a)** Find a parametric equation for the line through the point $(1, 5, 2)$ and in the direction of the vector $2\vec{i} + 3\vec{j} - \vec{k}$.
(b) By minimizing the square of the distance from a point on the line to the origin, find the exact point on the line which is closest to the origin.

[1]Based on http://www.vub.ac.be. Accessed November, 2011.

93. A plane from Denver, Colorado, (altitude 1650 meters) flies to Bismark, North Dakota (altitude 550 meters). It travels at 650 km/hour at a constant height of 8000 meters above the line joining Denver and Bismark. Bismark is about 850 km in the direction 60° north of east from Denver. Find parametric equations describing the plane's motion. Assume the origin is at sea level beneath Denver, that the x-axis points east and the y-axis points north, and that the earth is flat. Measure distances in kilometers and time in hours.

94. The vector $\vec{n}$ is perpendicular to the plane P_1. The vector $\vec{v}$ is parallel to the line L.

(a) If $\vec{n} \cdot \vec{v} = 0$, what does this tell you about the directions of P_1 and L? (Are they parallel? Perpendicular? Or is it impossible to tell?)

(b) Suppose $\vec{n} \times \vec{v} \neq \vec{0}$. The plane P_2 has normal $\vec{n} \times \vec{v}$. What can you say about the directions of

(i) P_1 and P_2? (ii) L and P_2?

95. Figure 17.11 shows the parametric curve $x = x(t), y = y(t)$ for $a \leq t \leq b$.

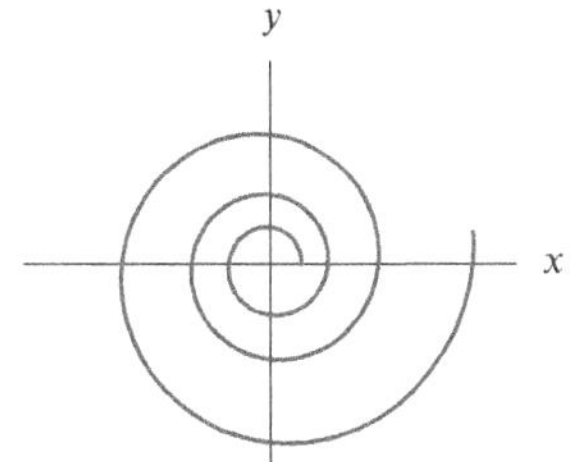

Figure 17.11

(a) Match a graph to each of the parametric curves given, for the same t values, by

(i) $(-x(t), -y(t))$ (ii) $(-x(t), y(t))$

(iii) $(x(t) + 1, y(t))$ (iv) $(x(t) + 1, y(t) + 1)$

(A)

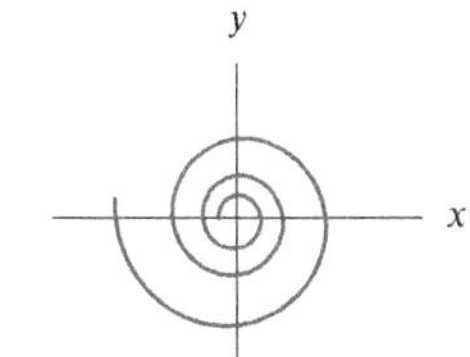

(B)

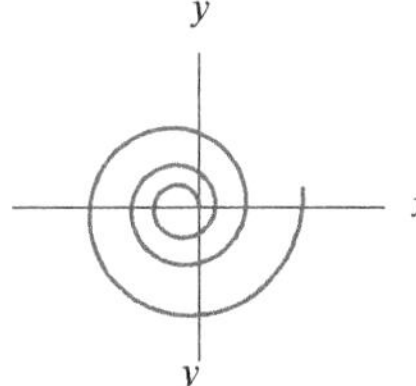

(C)

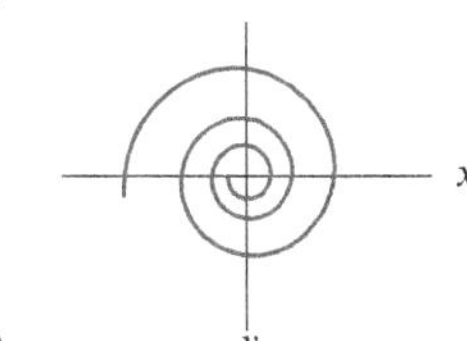

(D)

(E)

(F)

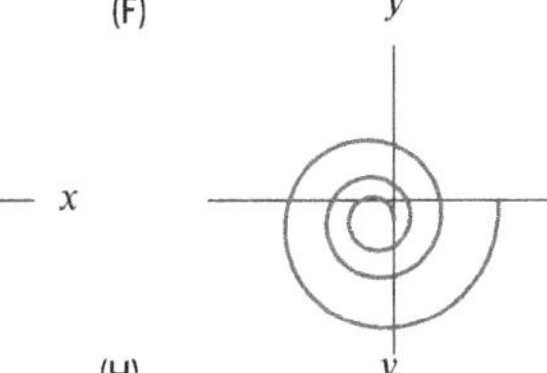

(G)

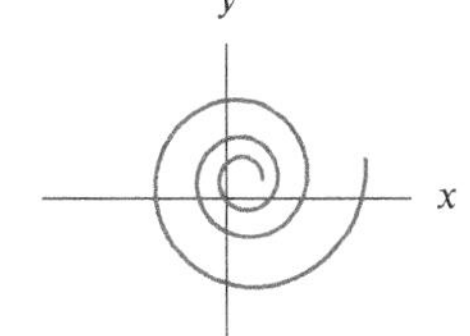

(H)

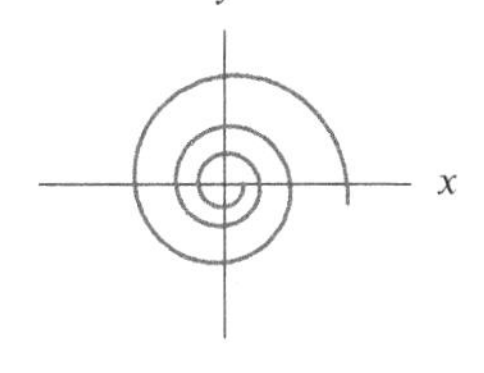

(b) Which of the following could be the formulas for the functions $x(t), y(t)$?

(i) $x = 10\cos t \quad y = 10\sin t$

(ii) $x = (10 + 8t)\cos t \quad y = (10 + 8t)\sin t$

(iii) $x = e^{t^2/200}\cos t \quad y = e^{t^2/200}\sin t$

(iv) $x = (10 - 8t)\cos t \quad y = (10 - 8t)\sin t$

(v) $x = 10\cos(t^2 + t) \quad y = 10\sin(t^2 + t)$

Strengthen Your Understanding

In Problems 96–97, explain what is wrong with the statement.

96. The curve parameterized by $\vec{r}_1(t) = \vec{r}(t - 2)$, defined for all t, is a shift in the $\vec{i}$-direction of the curve parameterized by $\vec{r}(t)$.

97. All points of the curve $\vec{r}(t) = R\cos t\vec{i} + R\sin t\vec{j} + t\vec{k}$ are the same distance, R, from the origin.

In Problems 98–100, give an example of:

98. Parameterizations of two different circles that have the same center and equal radii.

99. Parameterizations of two different lines that intersect at the point $(1, 2, 3)$.

100. A parameterization of the line $x = t, y = 2t, z = 3 + 4t$ that is not given by linear functions.

Are the statements in Problems 101–112 true or false? Give reasons for your answer.

101. The parametric curve $x = 3t + 2, y = -2t$ for $0 \leq t \leq 5$ passes through the origin.

102. The parametric curve $x = t^2, y = t^4$ for $0 \leq t \leq 1$ is a parabola.

103. A parametric curve $x = g(t), y = h(t)$ for $a \leq t \leq b$ is always the graph of a function $y = f(x)$.

104. The parametric curve $x = (3t + 2)^2, y = (3t + 2)^2 - 1$ for $0 \leq t \leq 3$ is a line.

105. The parametric curve $x = -\sin t, y = -\cos t$ for $0 \leq t \leq 2\pi$ traces out a unit circle counterclockwise as t increases.

106. A parameterization of the graph of $y = \ln x$ for $x > 0$ is given by $x = e^t, y = t$ for $-\infty < t < \infty$.

107. Both $x = -t + 1, y = 2t$ and $x = 2s, y = -4s + 2$ describe the same line.

108. The line of intersection of the two planes $z = x + y$ and $z = 1 - x - y$ can be parameterized by $x = t, y = \frac{1}{2} - t, z = \frac{1}{2}$.

109. The two lines given by $x = t, y = 2 + t, z = 3 + t$ and $x = 2s, y = 1 - s, z = s$ do not intersect.

110. The line parameterized by $x = 1, y = 2t, z = 3 + t$ is parallel to the x-axis.

111. The equation $\vec{r}(t) = 3t\vec{i} + (6t + 1)\vec{j}$ parameterizes a line.

112. The lines parameterized by $\vec{r}_1(t) = t\vec{i} + (-2t + 1)\vec{j}$ and $\vec{r}_2(t) = (2t + 5)\vec{i} + (-t)\vec{j}$ are parallel.

17.2 MOTION, VELOCITY, AND ACCELERATION

In this section we see how to find the vector quantities of velocity and acceleration from a parametric equation for the motion of an object.

The Velocity Vector

The velocity of a moving particle can be represented by a vector with the following properties:

The **velocity vector** of a moving object is a vector $\vec{v}$ such that:

- The magnitude of $\vec{v}$ is the speed of the object.
- The direction of $\vec{v}$ is the direction of motion.

Thus, the speed of the object is $\|\vec{v}\|$ and the velocity vector is tangent to the object's path.

Example 1 A child is sitting on a Ferris wheel of diameter 10 meters, making one revolution every 2 minutes. Find the speed of the child and draw velocity vectors at two different times.

Solution The child moves at a constant speed around a circle of radius 5 meters, completing one revolution every 2 minutes. One revolution around a circle of radius 5 is a distance of 10π, so the child's speed is $10\pi/2 = 5\pi \approx 15.7$ m/min. Hence, the magnitude of the velocity vector is 15.7 m/min. The direction of motion is tangent to the circle, and hence perpendicular to the radius at that point. Figure 17.12 shows the direction of the vector at two different times.

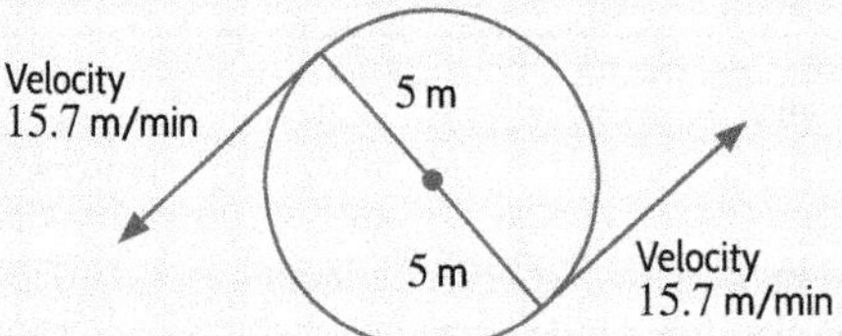

Figure 17.12: Velocity vectors of a child on a Ferris wheel (note that vectors would be in opposite direction if viewed from the other side)

Computing the Velocity

We find the velocity, as in one-variable calculus, by taking a limit. If the position vector of the particle is $\vec{r}(t)$ at time t, then the displacement vector between its positions at times t and $t + \Delta t$ is $\Delta\vec{r} = \vec{r}(t + \Delta t) - \vec{r}(t)$. (See Figure 17.13.) Over this interval,

$$\text{Average velocity} = \frac{\Delta\vec{r}}{\Delta t}.$$

In the limit as Δt goes to zero we have the instantaneous velocity at time t:

The **velocity vector**, $\vec{v}(t)$, of a moving object with position vector $\vec{r}(t)$ at time t is

$$\vec{v}(t) = \lim_{\Delta t \to 0} \frac{\Delta \vec{r}}{\Delta t} = \lim_{\Delta t \to 0} \frac{\vec{r}(t + \Delta t) - \vec{r}(t)}{\Delta t},$$

whenever the limit exists. We use the notation $\vec{v} = \dfrac{d\vec{r}}{dt} = \vec{r}\,'(t)$.

Notice that the direction of the velocity vector $\vec{r}\,'(t)$ in Figure 17.13 is approximated by the direction of the vector $\Delta\vec{r}$ and that the approximation gets better as $\Delta t \to 0$.

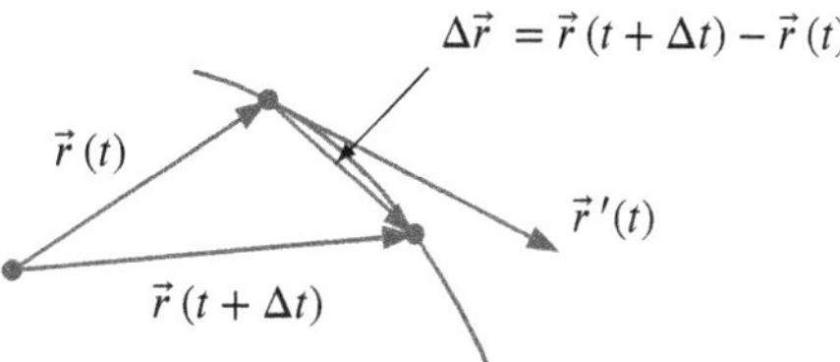

Figure 17.13: The change, $\Delta\vec{r}$, in the position vector for a particle moving on a curve and the velocity vector $\vec{v} = \vec{r}\,'(t)$

The Components of the Velocity Vector

If we represent a curve parametrically by $x = f(t)$, $y = g(t)$, $z = h(t)$, then we can write its position vector as: $\vec{r}(t) = f(t)\vec{i} + g(t)\vec{j} + h(t)\vec{k}$. Now we can compute the velocity vector:

$$\begin{aligned}
\vec{v}(t) &= \lim_{\Delta t \to 0} \frac{\vec{r}(t + \Delta t) - \vec{r}(t)}{\Delta t} \\
&= \lim_{\Delta t \to 0} \frac{(f(t+\Delta t)\vec{i} + g(t+\Delta t)\vec{j} + h(t+\Delta t)\vec{k}) - (f(t)\vec{i} + g(t)\vec{j} + h(t)\vec{k})}{\Delta t} \\
&= \lim_{\Delta t \to 0} \left(\frac{f(t+\Delta t) - f(t)}{\Delta t}\vec{i} + \frac{g(t+\Delta t) - g(t)}{\Delta t}\vec{j} + \frac{h(t+\Delta t) - h(t)}{\Delta t}\vec{k} \right) \\
&= f'(t)\vec{i} + g'(t)\vec{j} + h'(t)\vec{k} \\
&= \frac{dx}{dt}\vec{i} + \frac{dy}{dt}\vec{j} + \frac{dz}{dt}\vec{k}.
\end{aligned}$$

Thus, we have the following result:

The **components of the velocity vector** of a particle moving in space with position vector $\vec{r}(t) = f(t)\vec{i} + g(t)\vec{j} + h(t)\vec{k}$ at time t are given by

$$\vec{v}(t) = f'(t)\vec{i} + g'(t)\vec{j} + h'(t)\vec{k} = \frac{dx}{dt}\vec{i} + \frac{dy}{dt}\vec{j} + \frac{dz}{dt}\vec{k}.$$

Example 2 Find the components of the velocity vector for the child on the Ferris wheel in Example 1 using a coordinate system which has its origin at the center of the Ferris wheel and which makes the rotation counterclockwise.

Solution The Ferris wheel has radius 5 meters and completes 1 revolution counterclockwise every 2 minutes. The motion is parameterized by an equation of the form

$$\vec{r}(t) = 5\cos(\omega t)\vec{i} + 5\sin(\omega t)\vec{j},$$

where ω is chosen to make the period 2 minutes. Since the period of $\cos(\omega t)$ and $\sin(\omega t)$ is $2\pi/\omega$, we must have

$$\frac{2\pi}{\omega} = 2, \quad \text{so} \quad \omega = \pi.$$

Thus, the motion is described by the equation

$$\vec{r}(t) = 5\cos(\pi t)\vec{i} + 5\sin(\pi t)\vec{j},$$

where t is in minutes. The velocity is given by

$$\vec{v} = \frac{dx}{dt}\vec{i} + \frac{dy}{dt}\vec{j} = -5\pi\sin(\pi t)\vec{i} + 5\pi\cos(\pi t)\vec{j}.$$

To check, we calculate the magnitude of $\vec{v}$,

$$\|\vec{v}\| = \sqrt{(-5\pi)^2\sin^2(\pi t) + (5\pi)^2\cos^2(\pi t)} = 5\pi\sqrt{\sin^2(\pi t) + \cos^2(\pi t)} = 5\pi \approx 15.7,$$

which agrees with the speed we calculated in Example 1. To see that the direction is correct, we must show that the vector $\vec{v}$ at any time t is perpendicular to the position vector of the child at time t. To do this, we compute the dot product of $\vec{v}$ and $\vec{r}$:

$$\begin{aligned}\vec{v}\cdot\vec{r} &= (-5\pi\sin(\pi t)\vec{i} + 5\pi\cos(\pi t)\vec{j}) \cdot (5\cos(\pi t)\vec{i} + 5\sin(\pi t)\vec{j}) \\ &= -25\pi\sin(\pi t)\cos(\pi t) + 25\pi\cos(\pi t)\sin(\pi t) = 0.\end{aligned}$$

So the velocity vector, $\vec{v}$, is perpendicular to $\vec{r}$ and hence tangent to the circle. The direction is counterclockwise, since in the first quadrant, x is decreasing while y is increasing. (See Figure 17.14.)

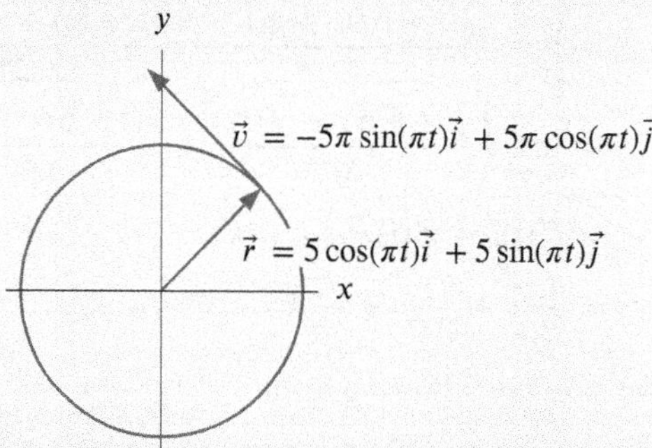

Figure 17.14: Velocity and radius vector of motion around a circle

Velocity Vectors and Tangent Lines

Since the velocity vector is tangent to the path of motion, it can be used to find parametric equations for the tangent line, if there is one.

Example 3 Find the tangent line at the point $(1, 1, 2)$ to the curve defined by the parametric equation

$$\vec{r}(t) = t^2\vec{i} + t^3\vec{j} + 2t\vec{k}.$$

Solution At time $t = 1$ the particle is at the point $(1, 1, 2)$ with position vector $\vec{r}_0 = \vec{i} + \vec{j} + 2\vec{k}$. The velocity vector at time t is $\vec{r}\,'(t) = 2t\vec{i} + 3t^2\vec{j} + 2\vec{k}$, so at time $t = 1$ the velocity is $\vec{v} = \vec{r}\,'(1) = 2\vec{i} + 3\vec{j} + 2\vec{k}$. The tangent line passes through $(1, 1, 2)$ in the direction of $\vec{v}$, so it has the parametric equation

$$\vec{r}(t) = \vec{r}_0 + t\vec{v} = (\vec{i} + \vec{j} + 2\vec{k}) + t(2\vec{i} + 3\vec{j} + 2\vec{k}).$$

The Acceleration Vector

Just as the velocity of a particle moving in 2-space or 3-space is a vector quantity, so is the rate of change of the velocity of the particle, namely its acceleration. Figure 17.15 shows a particle at time t with velocity vector $\vec{v}(t)$ and then a little later at time $t + \Delta t$. The vector $\Delta\vec{v} = \vec{v}(t + \Delta t) - \vec{v}(t)$ is the change in velocity and points approximately in the direction of the acceleration. So,

$$\text{Average acceleration} = \frac{\Delta\vec{v}}{\Delta t}.$$

In the limit as $\Delta t \to 0$, we have the instantaneous acceleration at time t:

> The **acceleration vector** of an object moving with velocity $\vec{v}(t)$ at time t is
>
> $$\vec{a}(t) = \lim_{\Delta t \to 0} \frac{\Delta\vec{v}}{\Delta t} = \lim_{\Delta t \to 0} \frac{\vec{v}(t + \Delta t) - \vec{v}(t)}{\Delta t},$$
>
> if the limit exists. We use the notation $\vec{a} = \dfrac{d\vec{v}}{dt} = \dfrac{d^2\vec{r}}{dt^2} = \vec{r}\,''(t)$.

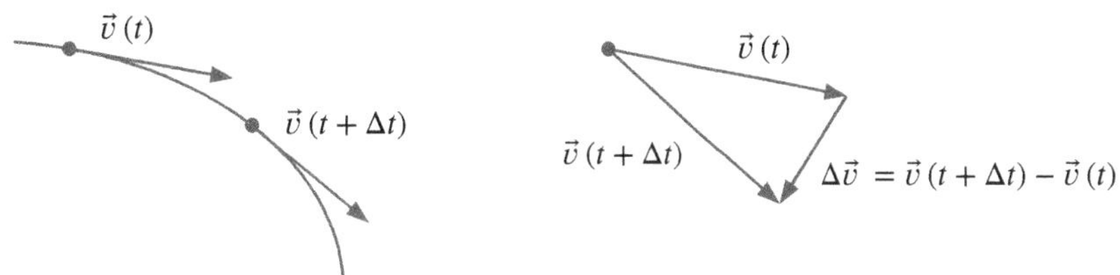

Figure 17.15: Computing the difference between two velocity vectors

Components of the Acceleration Vector

If we represent a curve in space parametrically by $x = f(t)$, $y = g(t)$, $z = h(t)$, we can express the acceleration in components. The velocity vector $\vec{v}(t)$ is given by

$$\vec{v}(t) = f'(t)\vec{i} + g'(t)\vec{j} + h'(t)\vec{k}.$$

From the definition of the acceleration vector, we have

$$\vec{a}(t) = \lim_{\Delta t \to 0} \frac{\vec{v}(t + \Delta t) - \vec{v}(t)}{\Delta t} = \frac{d\vec{v}}{dt}.$$

Using the same method to compute $d\vec{v}/dt$ as we used to compute $d\vec{r}/dt$ on page 897, we obtain

> The **components of the acceleration vector**, $\vec{a}(t)$, at time t of a particle moving in space with position vector $\vec{r}(t) = f(t)\vec{i} + g(t)\vec{j} + h(t)\vec{k}$ at time t are given by
>
> $$\vec{a}(t) = f''(t)\vec{i} + g''(t)\vec{j} + h''(t)\vec{k} = \frac{d^2x}{dt^2}\vec{i} + \frac{d^2y}{dt^2}\vec{j} + \frac{d^2z}{dt^2}\vec{k}.$$

Motion in a Circle and Along a Line

We now consider the velocity and acceleration vectors for two basic motions: uniform motion around a circle, and motion along a straight line.

Example 4 Find the acceleration vector for the child on the Ferris wheel in Examples 1 and 2.

Solution The child's position vector is given by $\vec{r}(t) = 5\cos(\pi t)\vec{i} + 5\sin(\pi t)\vec{j}$. In Example 2 we saw that the velocity vector is

$$\vec{v}(t) = \frac{dx}{dt}\vec{i} + \frac{dy}{dt}\vec{j} = -5\pi\sin(\pi t)\vec{i} + 5\pi\cos(\pi t)\vec{j}.$$

Thus, the acceleration vector is

$$\begin{aligned}\vec{a}(t) = \frac{d^2x}{dt^2}\vec{i} + \frac{d^2y}{dt^2}\vec{j} &= -(5\pi)\cdot\pi\cos(\pi t)\vec{i} - (5\pi)\cdot\pi\sin(\pi t)\vec{j} \\ &= -5\pi^2\cos(\pi t)\vec{i} - 5\pi^2\sin(\pi t)\vec{j}.\end{aligned}$$

Notice that $\vec{a}(t) = -\pi^2\vec{r}(t)$. Thus, the acceleration vector is a multiple of $\vec{r}(t)$ and points toward the origin.

The motion of the child on the Ferris wheel is an example of uniform circular motion, whose properties follow. (See Problem 45.)

Uniform Circular Motion: For a particle whose motion is described by

$$\vec{r}(t) = R\cos(\omega t)\vec{i} + R\sin(\omega t)\vec{j}$$

- Motion is in a circle of radius R with period $2\pi/|\omega|$.
- Velocity, $\vec{v}$, is tangent to the circle and speed is constant $\|\vec{v}\| = |\omega|R$.
- Acceleration, $\vec{a}$, points toward the center of the circle with $\|\vec{a}\| = \|\vec{v}\|^2/R$.

In uniform circular motion, the acceleration vector is perpendicular to the velocity vector, $\vec{v}$, because $\vec{v}$ does not change in magnitude, only in direction. There is no acceleration in the direction of $\vec{v}$.

We now look at straight-line motion in which the velocity vector always has the same direction but its magnitude changes. In straight-line motion, the acceleration vector points in the same direction as the velocity vector if the speed is increasing and in the opposite direction to the velocity vector if the speed is decreasing.

Example 5 Consider the motion given by the vector equation

$$\vec{r}(t) = 2\vec{i} + 6\vec{j} + (t^3 + t)(4\vec{i} + 3\vec{j} + \vec{k}).$$

Show that this is straight-line motion in the direction of the vector $4\vec{i} + 3\vec{j} + \vec{k}$ and relate the acceleration vector to the velocity vector.

Solution The velocity vector is

$$\vec{v} = (3t^2 + 1)(4\vec{i} + 3\vec{j} + \vec{k}).$$

Since $(3t^2 + 1)$ is a positive scalar, the velocity vector $\vec{v}$ always points in the direction of the vector $4\vec{i} + 3\vec{j} + \vec{k}$. In addition,

$$\text{Speed} = \|\vec{v}\| = (3t^2 + 1)\sqrt{4^2 + 3^2 + 1^2} = \sqrt{26}(3t^2 + 1).$$

Notice that the speed is decreasing until $t = 0$, then starts increasing. The acceleration vector is

$$\vec{a} = 6t(4\vec{i} + 3\vec{j} + \vec{k}).$$

For $t > 0$, the acceleration vector points in the same direction as $4\vec{i} + 3\vec{j} + \vec{k}$, which is the same direction as $\vec{v}$. This makes sense because the object is speeding up. For $t < 0$, the acceleration vector $6t(4\vec{i} + 3\vec{j} + \vec{k})$ points in the opposite direction to $\vec{v}$ because the object is slowing down.

Motion in a Straight Line: For a particle whose motion is described by

$$\vec{r}(t) = \vec{r}_0 + f(t)\vec{v}$$

- Motion is along a straight line through the point with position vector $\vec{r}_0$ parallel to $\vec{v}$.
- Velocity, $\vec{v}$, and acceleration, $\vec{a}$, are parallel to the line.

If $f(t) = t$, then we have $\vec{r}(t) = \vec{r}_0 + t\vec{v}$, the equation of a line obtained on page 889.

The Length of a Curve

The speed of a particle is the magnitude of its velocity vector:

$$\text{Speed} = \|\vec{v}\| = \sqrt{\left(\frac{dx}{dt}\right)^2 + \left(\frac{dy}{dt}\right)^2 + \left(\frac{dz}{dt}\right)^2}.$$

As in one dimension, we can find the distance traveled by a particle along a curve by integrating its speed. Thus,

$$\text{Distance traveled} = \int_a^b \|\vec{v}(t)\|\,dt.$$

If the particle never stops or reverses its direction as it moves along the curve, the distance it travels will be the same as the length of the curve. This suggests the following formula, which is justified in Problem 71 (available online):

If the curve C is given parametrically for $a \le t \le b$ by smooth functions and if the velocity vector $\vec{v}$ is not $\vec{0}$ for $a < t < b$, then

$$\text{Length of } C = \int_a^b \|\vec{v}\|\,dt.$$

Example 6 Find the circumference of the ellipse given by the parametric equations

$$x = 2\cos t, \quad y = \sin t, \quad 0 \le t \le 2\pi.$$

Solution The circumference of this curve is given by an integral which must be calculated numerically:

$$\begin{aligned}\text{Circumference} &= \int_0^{2\pi} \sqrt{\left(\frac{dx}{dt}\right)^2 + \left(\frac{dy}{dt}\right)^2}\,dt = \int_0^{2\pi} \sqrt{(-2\sin t)^2 + (\cos t)^2}\,dt \\ &= \int_0^{2\pi} \sqrt{4\sin^2 t + \cos^2 t}\,dt = 9.69.\end{aligned}$$

Since the ellipse is inscribed in a circle of radius 2 and circumscribes a circle of radius 1, we would expect the length of the ellipse to be between $2\pi(2) \approx 12.57$ and $2\pi(1) \approx 6.28$, so the value of 9.69 is reasonable.

Exercises and Problems for Section 17.2 Online Resource: Additional Problems for Section 17.2

EXERCISES

In Exercises 1–6, find the velocity and acceleration vectors.

1. $x = 2 + 3t$, $y = 4 + t$, $z = 1 - t$
2. $x = 2 + 3t^2$, $y = 4 + t^2$, $z = 1 - t^2$
3. $x = t$, $y = t^2$, $z = t^3$
4. $x = t$, $y = t^3 - t$
5. $x = 3\cos t$, $y = 4\sin t$
6. $x = 3\cos(t^2)$, $y = 3\sin(t^2)$, $z = t^2$

In Exercises 7–12, find the velocity $\vec{v}(t)$ and speed $\|\vec{v}(t)\|$. Find any times at which the particle stops.

7. $x = t$, $y = t^2$, $z = t^3$
8. $x = \cos 3t$, $y = \sin 5t$
9. $x = 3t^2$, $y = t^3 + 1$
10. $x = (t-1)^2$, $y = 2$, $z = 2t^3 - 3t^2$
11. $x = 3\sin(t^2) - 1$, $y = 3\cos(t^2)$
12. $x = 3\sin^2 t$, $y = \cos t - 1$, $z = t^2$

In Exercises 13–16, find the length of the curve.

13. $x = 3 + 5t$, $y = 1 + 4t$, $z = 3 - t$ for $1 \le t \le 2$. Check by calculating the length by another method.
14. $x = \cos 3t$, $y = \sin 5t$ for $0 \le t \le 2\pi$.
15. $x = \cos(e^t)$, $y = \sin(e^t)$ for $0 \le t \le 1$. Check by calculating the length by another method.
16. $\vec{r}(t) = 2t\vec{i} + \ln t\vec{j} + t^2\vec{k}$ for $1 \le t \le 2$.

In Exercises 17–18, find the velocity and acceleration vectors of the uniform circular motion and check that they are perpendicular. Check that the speed and magnitude of the acceleration are constant.

17. $x = 3\cos(2\pi t)$, $y = 3\sin(2\pi t)$, $z = 0$
18. $x = 2\pi$, $y = 2\sin(3t)$, $z = 2\cos(3t)$

In Exercises 19–20, find the velocity and acceleration vectors of the straight-line motion. Check that the acceleration vector points in the same direction as the velocity vector if the speed is increasing and in the opposite direction if the speed is decreasing.

19. $x = 2 + t^2$, $y = 3 - 2t^2$, $z = 5 - t^2$
20. $x = -2t^3 - 3t + 1$, $y = 4t^3 + 6t - 5$, $z = 6t^3 + 9t - 2$
21. Find parametric equations for the tangent line at $t = 2$ for Exercise 10.

PROBLEMS

22. A particle passes through the point $P = (5, 4, -2)$ at time $t = 4$, moving with constant velocity $\vec{v} = 2\vec{i} - 3\vec{j} + \vec{k}$. Find a parametric equation for its motion.

In Problems 23–24, find all values of t for which the particle is moving parallel to the x-axis and to the y-axis. Determine the end behavior and graph the particle's path.

23. $x = t^2 - 6t$, $y = t^3 - 3t$
24. $x = t^3 - 12t$, $y = t^2 + 10t$
25. The table gives x and y coordinates of a particle in the plane at time t. Assuming that the particle moves smoothly and that the points given show all the major features of the motion, estimate the following quantities:

 (a) The velocity vector and speed at time $t = 2$.
 (b) Any times when the particle is moving parallel to the y-axis.
 (c) Any times when the particle has come to a stop.

t	0	0.5	1.0	1.5	2.0	2.5	3.0	3.5	4.0
x	1	4	6	7	6	3	2	3	5
y	3	2	3	5	8	10	11	10	9

26. A particle starts at the point $P = (3, 2, -5)$ and moves along a straight line toward $Q = (5, 7, -2)$ at a speed of 5 cm/sec. Let x, y, z be measured in centimeters.

 (a) Find the particle's velocity vector.
 (b) Find parametric equations for the particle's motion.

27. A particle moves at a constant speed along a line from the point $P = (2, -1, 5)$ at time $t = 0$ to the point $Q = (5, 3, -1)$. Find parametric equations for the particle's motion if:

 (a) The particle takes 5 seconds to move from P to Q.
 (b) The speed of the particle is 5 units per second.

28. A particle travels along the line $x = 1 + t$, $y = 5 + 2t$, $z = -7 + t$, where t is in seconds and x, y, z are in meters.

 (a) When and where does the particle hit the plane $x + y + z = 1$?
 (b) How fast is the particle going when it hits the plane? Give units.

29. A stone is thrown from a rooftop at time $t = 0$ seconds. Its position at time t is given by

$$\vec{r}(t) = 10t\vec{i} - 5t\vec{j} + (6.4 - 4.9t^2)\vec{k}.$$

The origin is at the base of the building, which is standing on flat ground. Distance is measured in meters. The vector $\vec{i}$ points east, $\vec{j}$ points north, and $\vec{k}$ points up.

(a) How high is the rooftop above the ground?
(b) At what time does the stone hit the ground?
(c) How fast is the stone moving when it hits the ground?
(d) Where does the stone hit the ground?
(e) What is the stone's acceleration when it hits the ground?

30. A child wanders slowly down a circular staircase from the top of a tower. With x, y, z in feet and the origin at the base of the tower, her position t minutes from the start is given by

$$x = 10\cos t, \quad y = 10\sin t, \quad z = 90 - 5t.$$

(a) How tall is the tower?
(b) When does the child reach the bottom?
(c) What is her speed at time t?
(d) What is her acceleration at time t?

31. The origin is on flat ground and the z-axis points upward. For time $0 \le t \le 10$ in seconds and distance in centimeters, a particle moves along a path given by

$$\vec{r} = 2t\vec{i} + 3t\vec{j} + (100 - (t-5)^2)\vec{k}.$$

(a) When is the particle at the highest point? What is that point?
(b) When in the interval $0 \le t \le 10$ is the particle moving fastest? What is its speed at that moment?
(c) When in the interval $0 \le t \le 10$ is the particle moving slowest? What is its speed at that moment?

32. The function $w = f(x, y, z)$ has grad $f(7, 2, 5) = 4\vec{i} - 3\vec{j} + \vec{k}$. A particle moves along the curve $\vec{r}(t)$, arriving at the point $(7, 2, 5)$ with velocity $2\vec{i} + 3\vec{j} + 6\vec{k}$ when $t = 0$. Find the rate of change of w with respect to time at $t = 0$.

33. Suppose x measures horizontal distance in meters, and y measures distance above the ground in meters. At time $t = 0$ in seconds, a projectile starts from a point h meters above the origin with speed v meters/sec at an angle θ to the horizontal. Its path is given by

$$x = (v\cos\theta)t, \quad y = h + (v\sin\theta)t - \frac{1}{2}gt^2.$$

Using this information about a general projectile, analyze the motion of a ball which travels along the path

$$x = 20t, \quad y = 2 + 25t - 4.9t^2.$$

(a) When does the ball hit the ground?
(b) Where does the ball hit the ground?
(c) At what height above the ground does the ball start?
(d) What is the value of g, the acceleration due to gravity?
(e) What are the values of v and θ?

34. A particle is moving on a path in the xz-plane given by $x = 20t$, $z = 5t - 0.5t^2$, where z is the height of the particle above the ground in meters, x is the horizontal distance in meters, and t is time in seconds.

(a) What is the equation of the path in terms of x and z only?
(b) When is the particle at ground level?
(c) What is the velocity of the particle at time t?
(d) What is the speed of the particle at time t?
(e) Is the speed ever 0?
(f) When is the particle at the highest point?

35. The base of a 20-meter tower is at the origin; the base of a 20-meter tree is at $(0, 20, 0)$. The ground is flat and the z-axis points upward. The following parametric equations describe the motion of six projectiles each launched at time $t = 0$ in seconds.

(I) $\vec{r}(t) = (20 + t^2)\vec{k}$
(II) $\vec{r}(t) = 2t^2\vec{j} + 2t^2\vec{k}$
(III) $\vec{r}(t) = 20\vec{i} + 20\vec{j} + (20 - t^2)\vec{k}$
(IV) $\vec{r}(t) = 2t\vec{j} + (20 - t^2)\vec{k}$
(V) $\vec{r}(t) = (20 - 2t)\vec{i} + 2t\vec{j} + (20 - t)\vec{k}$
(VI) $\vec{r}(t) = t\vec{i} + t\vec{j} + t\vec{k}$

(a) Which projectile is launched from the top of the tower and goes downward? When and where does it hit the ground?
(b) Which projectile hits the top of the tree? When? From where is it launched?
(c) Which projectile is not launched from somewhere on the tower and hits the tree? Where and when does it hit the tree?

36. A particle moves on a circle of radius 5 cm, centered at the origin, in the xy-plane (x and y measured in centimeters). It starts at the point $(0, 5)$ and moves counterclockwise, going once around the circle in 8 seconds.

(a) Write a parameterization for the particle's motion.
(b) What is the particle's speed? Give units.

37. A particle moves along a curve with velocity vector $\vec{v}(t) = -\sin t\vec{i} + \cos t\vec{j}$. At time $t = 0$ the particle is at $(2, 3)$.

(a) Find the displacement vector for the particle from time $t = 0$ to $t = \pi$.
(b) Find the position of the particle at time $t = \pi$.
(c) Find the distance traveled by the particle from time $t = 0$ to time $t = \pi$.

38. Determine the position vector $\vec{r}(t)$ for a rocket which is launched from the origin at time $t = 0$ seconds, reaches its highest point of $(x, y, z) = (1000, 3000, 10{,}000)$, where x, y, z are in meters, and after the launch is subject only to the acceleration due to gravity, 9.8 m/sec^2.

39. Emily is standing on the outer edge of a merry-go-round, 10 meters from the center. The merry-go-round completes one full revolution every 20 seconds. As Emily passes over a point P on the ground, she drops a ball from 3 meters above the ground.

(a) How fast is Emily going?
(b) How far from P does the ball hit the ground? (The acceleration due to gravity is 9.8 m/sec^2.)
(c) How far from Emily does the ball hit the ground?

40. A point P moves in a circle of radius a. Show that $\vec{r}(t)$, the position vector of P, and its velocity vector $\vec{r}\,'(t)$ are perpendicular.

41. A wheel of radius 1 meter rests on the x-axis with its center on the y-axis. There is a spot on the rim at the point $(1, 1)$. See Figure 17.16. At time $t = 0$ the wheel starts rolling on the x-axis in the direction shown at a rate of 1 radian per second.

(a) Find parametric equations describing the motion of the center of the wheel.
(b) Find parametric equations describing the motion of the spot on the rim. Plot its path.

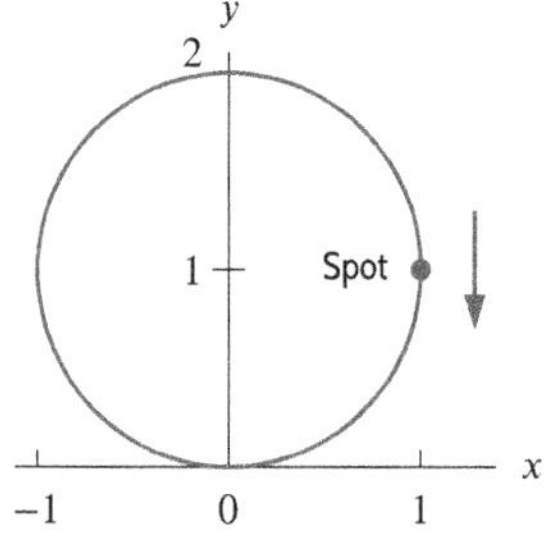

Figure 17.16

42. Suppose $\vec{r}(t) = \cos t\,\vec{i} + \sin t\,\vec{j} + 2t\,\vec{k}$ represents the position of a particle on a helix, where z is the height of the particle above the ground.

(a) Is the particle ever moving downward? When?
(b) When does the particle reach a point 10 units above the ground?
(c) What is the velocity of the particle when it is 10 units above the ground?
(d) When it is 10 units above the ground, the particle leaves the helix and moves along the tangent. Find parametric equations for this tangent line.

43. Show that the helix $\vec{r} = \alpha\cos t\vec{i} + \alpha\sin t\vec{j} + \beta t\vec{k}$ is parameterized with constant speed.

44. An ant crawls along the radius from the center to the edge of a circular disk of radius 1 meter, moving at a constant rate of 1 cm/sec. Meanwhile, the disk is turning counterclockwise about its center at 1 revolution per second.

(a) Parameterize the motion of the ant.
(b) Find the velocity and speed of the ant.
(c) Determine the acceleration and magnitude of the acceleration of the ant.

45. The motion of a particle is given by $\vec{r}(t) = R\cos(\omega t)\vec{i} + R\sin(\omega t)\vec{j}$, with $R > 0$, $\omega > 0$.

(a) Show that the particle moves on a circle and find the radius, direction, and period.
(b) Determine the velocity vector of the particle and its direction and speed.
(c) What are the direction and magnitude of the acceleration vector of the particle?

46. You bicycle along a straight flat road with a safety light attached to one foot. Your bike moves at a speed of 25 km/hr and your foot moves in a circle of radius 20 cm centered 30 cm above the ground, making one revolution per second.

(a) Find parametric equations for x and y which describe the path traced out by the light, where y is distance (in cm) above the ground.
(b) Sketch the light's path.
(c) How fast (in revolutions/sec) would your foot have to be rotating if an observer standing at the side of the road sees the light moving backward?

47. How do the motions of objects A and B differ, if A has position vector $\vec{r}_A(t)$ and B has position vector $\vec{r}_B(t) = \vec{r}_A(2t)$ for $t \geq 0$. Illustrate your answer with $\vec{r}_A(t) = t\vec{i} + t^2\vec{j}$.

48. At time $t = 0$ an object is moving with velocity vector $\vec{v} = 2\vec{i} + \vec{j}$ and acceleration vector $\vec{a} = \vec{i} + \vec{j}$. Can it be in uniform circular motion about some point in the plane?

49. Figure 17.17 shows the velocity and acceleration vectors of an object in uniform circular motion about a point in the plane at a particular moment. Is it moving round the circle in the clockwise or counterclockwise direction?

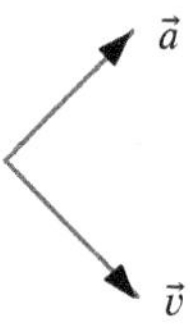

Figure 17.17

50. Let $\vec{v}(t)$ be the velocity of a particle moving in the plane. Let $s(t)$ be the magnitude of $\vec{v}$ and let $\theta(t)$ be the angle of $\vec{v}(t)$ with the positive x-axis at time t, so that $\vec{v} = s\cos\theta\vec{i} + s\sin\theta\vec{j}$.

Let $\vec{T}$ be the unit vector in the direction of $\vec{v}$, and let $\vec{N}$ be the unit vector in the direction of $\vec{k} \times \vec{v}$, perpendicular to $\vec{v}$. Show that the acceleration $\vec{a}(t)$ is given by

$$\vec{a} = \frac{ds}{dt}\vec{T} + s\frac{d\theta}{dt}\vec{N}.$$

This shows how to separate the acceleration into the sum of one component, $\frac{ds}{dt}\vec{T}$, due to changing speed and a perpendicular component, $s\frac{d\theta}{dt}\vec{N}$, due to changing direction of the motion.

Strengthen Your Understanding

In Problems 51–53, explain what is wrong with the statement.

51. When a particle moves around a circle its velocity and acceleration are always orthogonal.

52. A particle with position $\vec{r}(t)$ at time t has acceleration equal to 3 m/sec^2 at time $t = 0$.

53. A parameterized curve $\vec{r}(t)$, $A \le t \le B$, has length $B - A$.

In Problems 54–55, give an example of:

54. A function $\vec{r}(t)$ such that the particle with position $\vec{r}(t)$ at time t has velocity $\vec{v} = \vec{i} + 2\vec{j}$ and acceleration $\vec{a} = 4\vec{i} + 6\vec{k}$ at $t = 0$.

55. An interval $a \le t \le b$ corresponding to a piece of the helix $\vec{r}(t) = \cos t\vec{i} + \sin t\vec{j} + t\vec{k}$ of length 10.

Are the statements in Problems 56–63 true or false? Give reasons for your answer.

56. A particle whose motion in the plane is given by $\vec{r}(t) = t^2\vec{i} + (1-t)\vec{j}$ has the same velocity at $t = 1$ and $t = -1$.

57. A particle whose motion in the plane is given by $\vec{r}(t) = t^2\vec{i} + (1-t)\vec{j}$ has the same speed at $t = 1$ and $t = -1$.

58. If a particle is moving along a parameterized curve $\vec{r}(t)$ then the acceleration vector at any point is always perpendicular to the velocity vector at that point.

59. If a particle is moving along a parameterized curve $\vec{r}(t)$ then the acceleration vector at a point cannot be parallel to the velocity vector at that point.

60. If $\vec{r}(t)$ for $a \le t \le b$ is a parameterized curve, then $\vec{r}(-t)$ for $a \le t \le b$ is the same curve traced backward.

61. If $\vec{r}(t)$ for $a \le t \le b$ is a parameterized curve C and the speed $\|\vec{v}(t)\| = 1$, then the length of C is $b - a$.

62. If a particle moves with motion $\vec{r}(t) = 3t\vec{i} + 2t\vec{j} + t\vec{k}$, then the particle stops at the origin.

63. If a particle moves with constant speed, the path of the particle must be a line.

For Problems 64–67, decide if the statement is true or false for all smooth parameterized curves $\vec{r}(t)$ and all values of t for which $\vec{r}\,'(t) \neq \vec{0}$.

64. The vector $\vec{r}\,'(t)$ is tangent to the curve at the point with position vector $\vec{r}(t)$.

65. $\vec{r}\,'(t) \times \vec{r}(t) = \vec{0}$

66. $\vec{r}\,'(t) \cdot \vec{r}(t) = 0$

67. $\vec{r}\,''(t) = -\omega^2\vec{r}(t)$

17.3 VECTOR FIELDS

Introduction to Vector Fields

A *vector field* is a function that assigns a vector to each point in the plane or in 3-space. One example of a vector field is the gradient of a function $f(x, y)$; at each point (x, y) the vector $\operatorname{grad} f(x, y)$ points in the direction of maximum rate of increase of f. In this section we look at other vector fields representing velocities and forces.

Velocity Vector Fields

Figure 17.18 shows the flow of a part of the Gulf Stream, a current in the Atlantic Ocean.[2] It is an example of a *velocity vector field*: each vector shows the velocity of the current at that point. The current is fastest where the velocity vectors are longest in the middle of the stream. Beside the stream are eddies where the water flows round and round in circles.

[2]Based on data supplied by Avijit Gangopadhyay of the Jet Propulsion Laboratory.

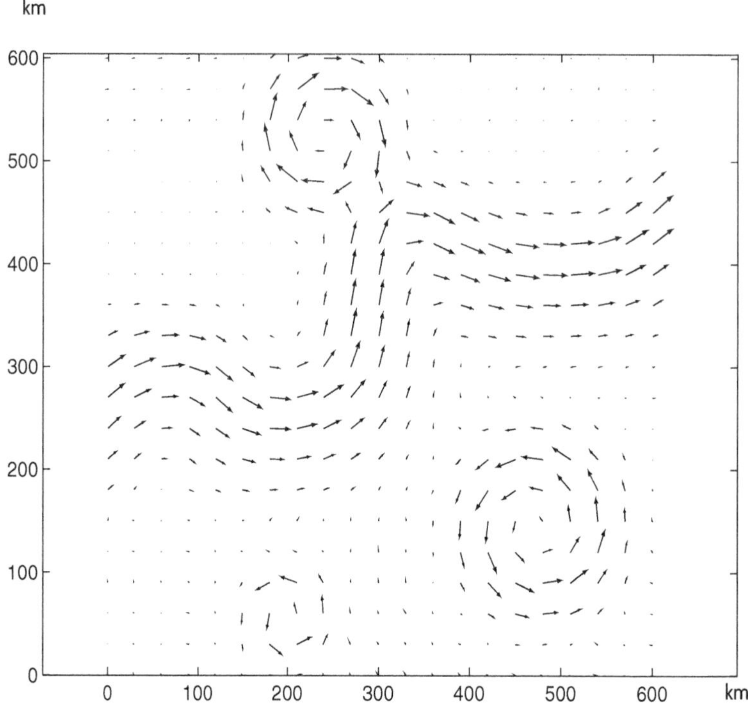

Figure 17.18: The velocity vector field of the Gulf Stream

Force Fields

Another physical quantity represented by a vector is force. When we experience a force, sometimes it results from direct contact with the object that supplies the force (for example, a push). Many forces, however, can be felt at all points in space. For example, the earth exerts a gravitational pull on all other masses. Such forces can be represented by vector fields.

Figure 17.19 shows the gravitational force exerted by the earth on a mass of one kilogram at different points in space. This is a sketch of the vector field in 3-space. You can see that the vectors all point toward the earth (which is not shown in the diagram) and that the vectors farther from the earth are smaller in magnitude.

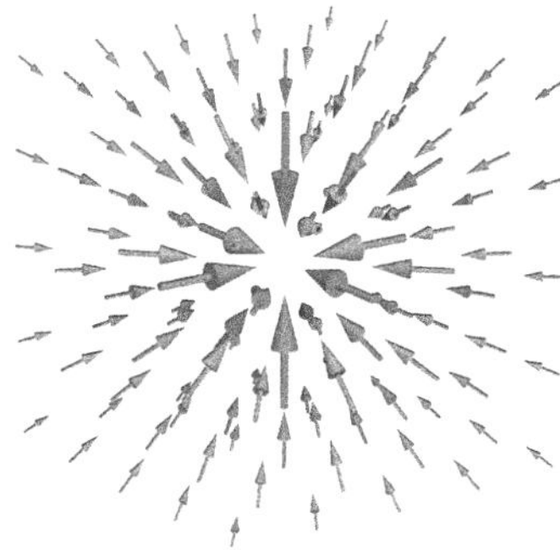

Figure 17.19: The gravitational field of the earth

Definition of a Vector Field

Now that you have seen some examples of vector fields, we give a more formal definition.

> A **vector field** in 2-space is a function $\vec{F}(x, y)$ whose value at a point (x, y) is a 2-dimensional vector. Similarly, a vector field in 3-space is a function $\vec{F}(x, y, z)$ whose values are 3-dimensional vectors.

Notice the arrow over the function, $\vec{F}$, indicating that its value is a vector, not a scalar. We often represent the point (x, y) or (x, y, z) by its position vector $\vec{r}$ and write the vector field as $\vec{F}(\vec{r})$.

Visualizing a Vector Field Given by a Formula

Since a vector field is a function that assigns a vector to each point, a vector field can often be given by a formula.

Example 1 Sketch the vector field in 2-space given by $\vec{F}(x, y) = -y\vec{i} + x\vec{j}$.

Solution Table 17.1 shows the value of the vector field at a few points. Notice that each value is a vector. To plot the vector field, we plot $\vec{F}(x, y)$ with its tail at (x, y). (See Figure 17.20.)

Table 17.1 *Values of $\vec{F}(x, y) = -y\vec{i} + x\vec{j}$*

		y: -1	0	1
x	-1	$\vec{i} - \vec{j}$	$-\vec{j}$	$-\vec{i} - \vec{j}$
	0	$\vec{i}$	$\vec{0}$	$-\vec{i}$
	1	$\vec{i} + \vec{j}$	$\vec{j}$	$-\vec{i} + \vec{j}$

Now we look at the formula. The magnitude of the vector at (x, y) is the distance from (x, y) to the origin since

$$\|\vec{F}(x, y)\| = \| -y\vec{i} + x\vec{j}\,\| = \sqrt{x^2 + y^2}.$$

Therefore, all the vectors at a fixed distance from the origin (that is, on a circle centered at the origin) have the same magnitude. The magnitude gets larger as we move farther from the origin.

What about the direction? Figure 17.20 suggests that at each point (x, y) the vector $\vec{F}(x, y)$ is perpendicular to the position vector $\vec{r} = x\vec{i} + y\vec{j}$. We confirm this using the dot product:

$$\vec{r} \cdot \vec{F}(x, y) = (x\vec{i} + y\vec{j}) \cdot (-y\vec{i} + x\vec{j}) = 0.$$

Thus, the vectors are tangent to circles centered at the origin and get longer as we go out. In Figure 17.21, the vectors have been scaled so that they do not obscure each other.

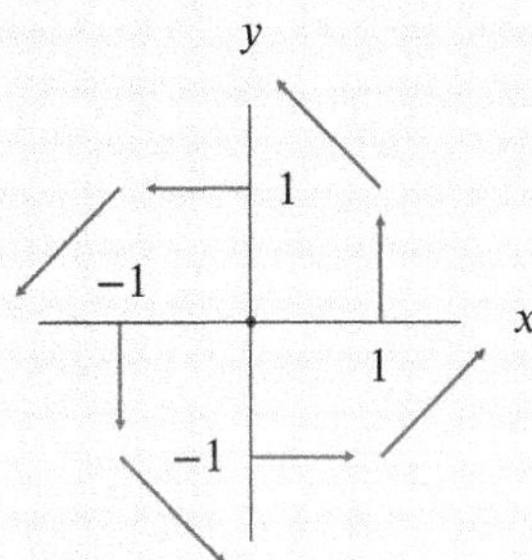

Figure 17.20: The value $\vec{F}(x, y)$ is placed at the point (x, y)

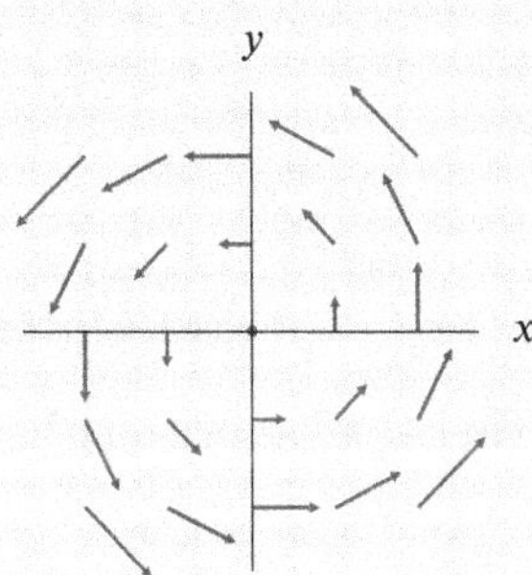

Figure 17.21: The vector field $\vec{F}(x, y) = -y\vec{i} + x\vec{j}$, vectors scaled smaller to fit in diagram

Example 2 Sketch the vector fields in 2-space given by (a) $\vec{F}(x, y) = x\vec{j}$ (b) $\vec{G}(x, y) = x\vec{i}$.

Solution (a) The vector $x\vec{j}$ is parallel to the y-direction, pointing up when x is positive and down when x is negative. Also, the larger $|x|$ is, the longer the vector. The vectors in the field are constant along vertical lines since the vector field does not depend on y. (See Figure 17.22.)

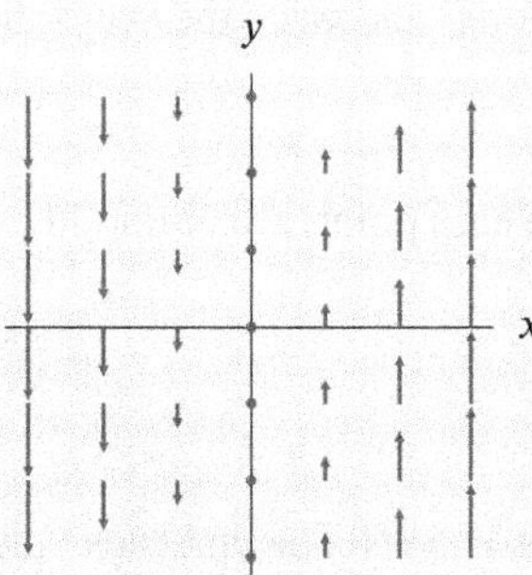

Figure 17.22: The vector field $\vec{F}(x, y) = x\vec{j}$

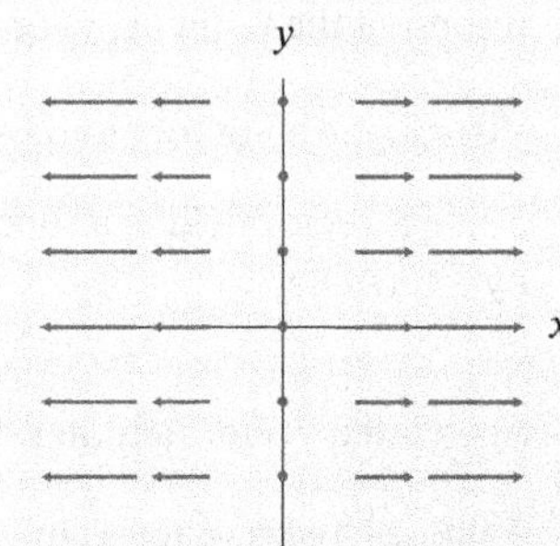

Figure 17.23: The vector field $\vec{F}(x, y) = x\vec{i}$

(b) This is similar to the previous example, except that the vector $x\vec{i}$ is parallel to the x-direction, pointing to the right when x is positive and to the left when x is negative. Again, the larger $|x|$ is the longer the vector, and the vectors are constant along vertical lines, since the vector field does not depend on y. (See Figure 17.23.)

Example 3 Describe the vector field in 3-space given by $\vec{F}(\vec{r}) = \vec{r}$, where $\vec{r} = x\vec{i} + y\vec{j} + z\vec{k}$.

Solution The notation $\vec{F}(\vec{r}) = \vec{r}$ means that the value of $\vec{F}$ at the point (x, y, z) with position vector $\vec{r}$ is the vector $\vec{r}$ with its tail at (x, y, z). Thus, the vector field points outward. See Figure 17.24. Note that the lengths of the vectors have been scaled down so as to fit into the diagram.

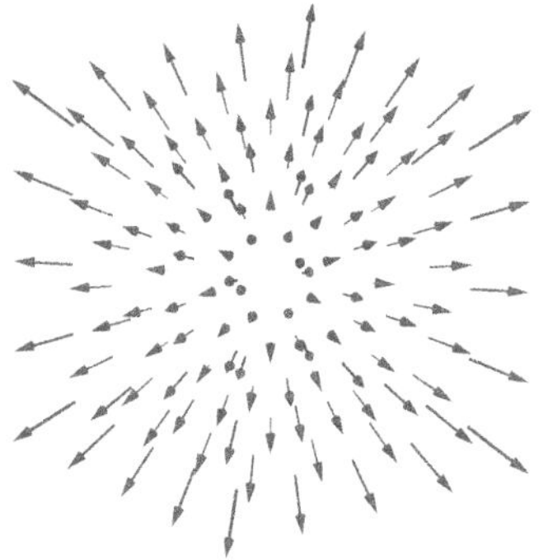

Figure 17.24: The vector field $\vec{F}(\vec{r}) = \vec{r}$

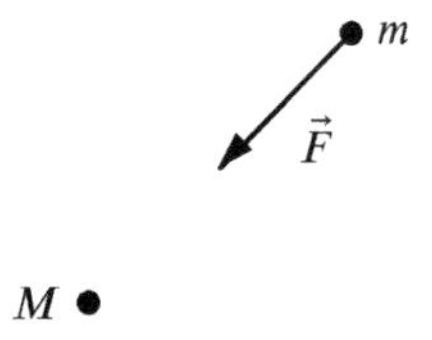

Figure 17.25: Force exerted on mass m by mass M

Finding a Formula for a Vector Field

Example 4 Newton's Law of Gravitation states that the magnitude of the gravitational force exerted by an object of mass M on an object of mass m is proportional to M and m and inversely proportional to the square of the distance between them. The direction of the force is from m to M along the line connecting them. (See Figure 17.25.) Find a formula for the vector field $\vec{F}(\vec{r})$ that represents the gravitational force, assuming M is located at the origin and m is located at the point with position vector $\vec{r}$.

Solution Since the mass m is located at $\vec{r}$, Newton's law says that the magnitude of the force is given by

$$\|\vec{F}(\vec{r})\| = \frac{GMm}{\|\vec{r}\|^2},$$

where G is the universal gravitational constant. A unit vector in the direction of the force is $-\vec{r}/\|\vec{r}\|$, where the negative sign indicates that the direction of force is toward the origin (gravity is attractive). By taking the product of the magnitude of the force and a unit vector in the direction of the force, we obtain an expression for the force vector field:

$$\vec{F}(\vec{r}) = \frac{GMm}{\|\vec{r}\|^2}\left(-\frac{\vec{r}}{\|\vec{r}\|}\right) = \frac{-GMm\vec{r}}{\|\vec{r}\|^3}.$$

We have already seen a picture of this vector field in Figure 17.19.

Gradient Vector Fields

The gradient of a scalar function f is a function that assigns a vector to each point, and is therefore a vector field. It is called the *gradient field* of f. Many vector fields in physics are gradient fields.

Example 5 Sketch the gradient field of the functions in Figures 17.26–17.28.

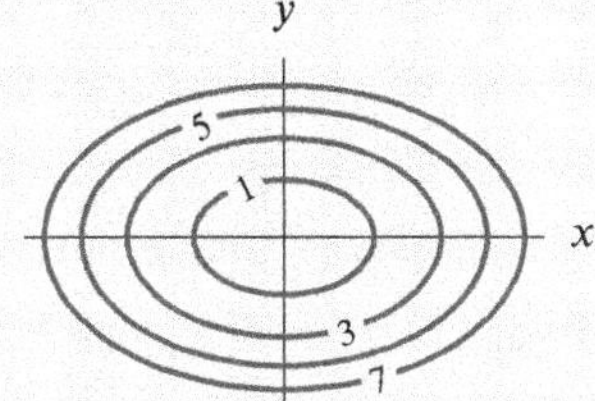

Figure 17.26: The contour map of $f(x, y) = x^2 + 2y^2$

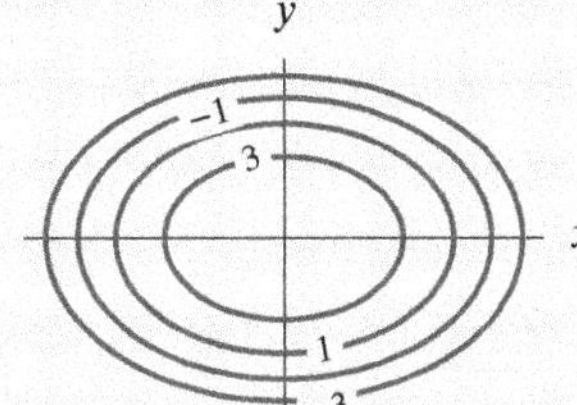

Figure 17.27: The contour map of $g(x, y) = 5 - x^2 - 2y^2$

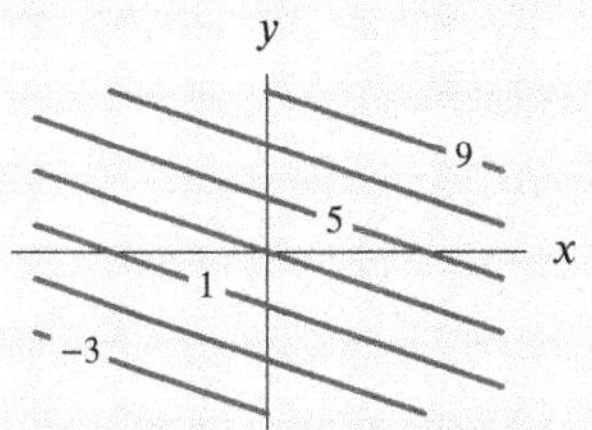

Figure 17.28: The contour map of $h(x, y) = x + 2y + 3$

Solution See Figures 17.29–17.31. For a function $f(x, y)$, the gradient vector of f at a point is perpendicular to the contours in the direction of increasing f and its magnitude is the rate of change in that direction. The rate of change is large when the contours are close together and small when they are far apart. Notice that in Figure 17.29 the vectors all point outward, away from the local minimum of f, and in Figure 17.30 the vectors of grad g all point inward, toward the local maximum of g. Since h is a linear function, its gradient is constant, so grad h in Figure 17.31 is a constant vector field.

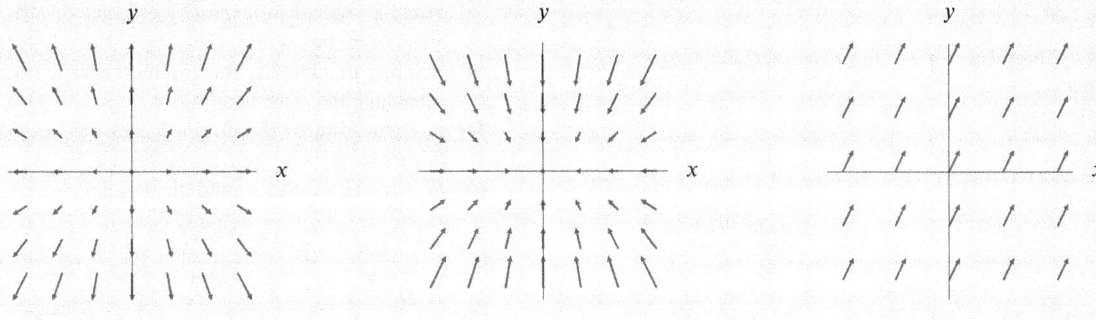

Figure 17.29: grad f **Figure 17.30**: grad g **Figure 17.31**: grad h

Exercises and Problems for Section 17.3

EXERCISES

For Exercises 1–6, find formulas for the vector fields. (There are many possible answers.)

1. 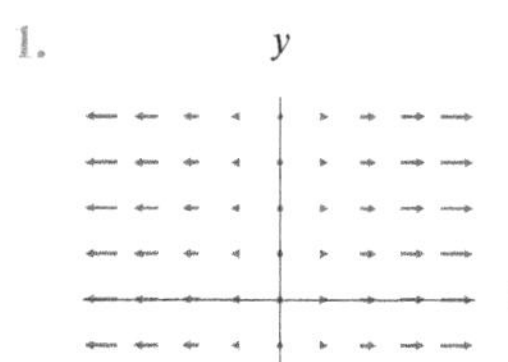

2. 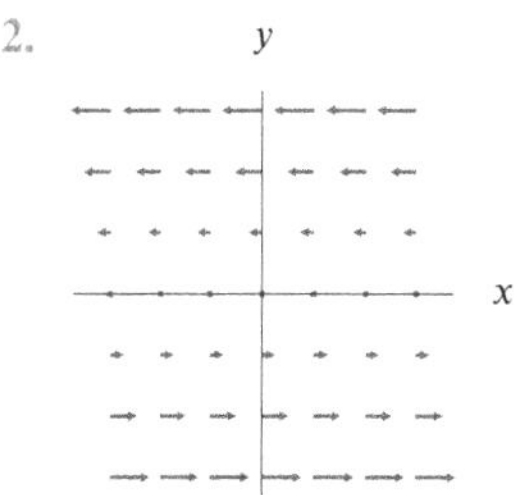

3. 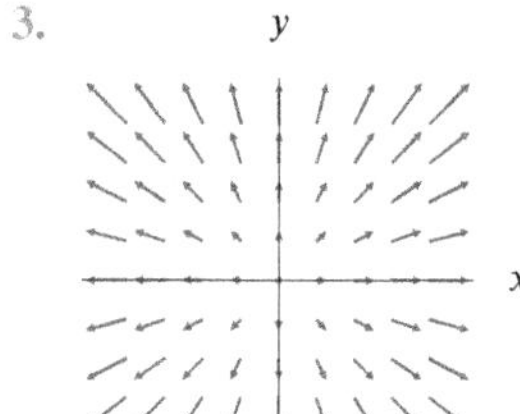

4. 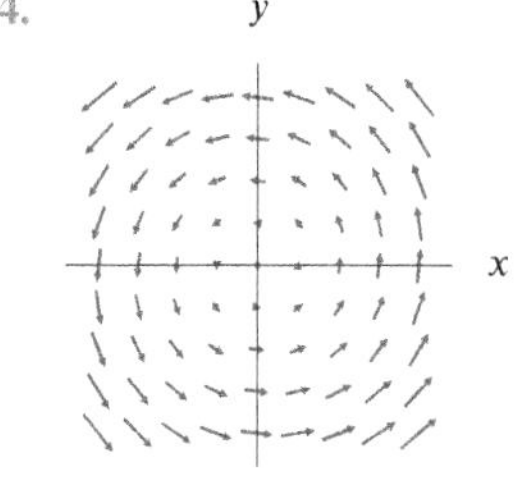

5. 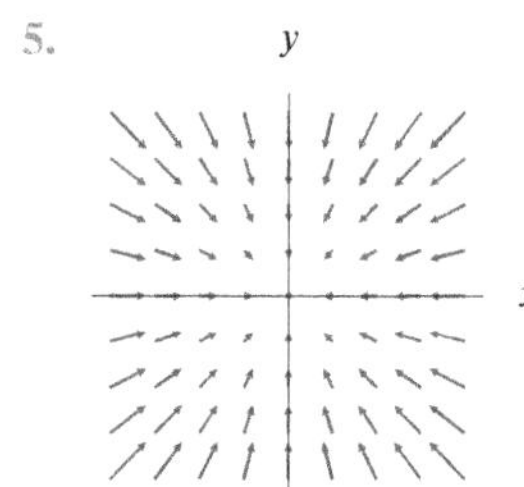

6. 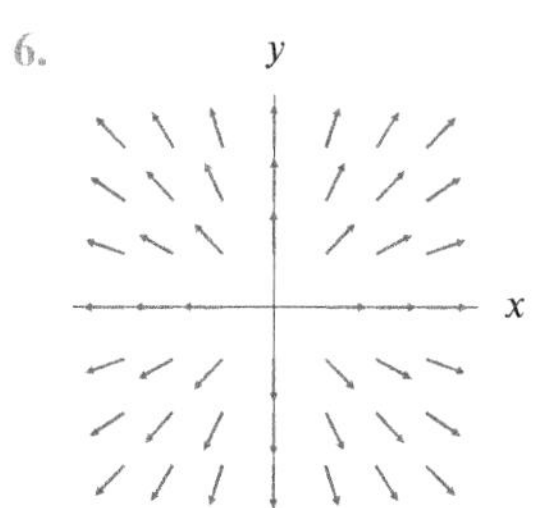

In Exercises 7–10, assume $x, y > 0$ and decide if

(a) The vector field is parallel to the x-axis, parallel to the y-axis, or neither.

(b) As x increases, the length increases, decreases, or neither.

(c) As y increases, the length increases, decreases, or neither.

Assume $x, y > 0$.

7. $\vec{F} = x\vec{j}$

8. $\vec{F} = y\vec{i} + \vec{j}$

9. $\vec{F} = (x + e^{1-y})\vec{i}$

10. $\text{grad}(x^4 + e^{3y})$

Sketch the vector fields in Exercises 11–19 in the xy-plane.

11. $\vec{F}(x, y) = 2\vec{i} + 3\vec{j}$

12. $\vec{F}(x, y) = y\vec{i}$

13. $\vec{F}(x, y) = -y\vec{j}$

14. $\vec{F}(\vec{r}) = 2\vec{r}$

15. $\vec{F}(\vec{r}) = \vec{r}/\|\vec{r}\|$

16. $\vec{F}(\vec{r}) = -\vec{r}/\|\vec{r}\|^3$

17. $\vec{F} = y\vec{i} - x\vec{j}$

18. $\vec{F}(x, y) = 2x\vec{i} + x\vec{j}$

19. $\vec{F}(x, y) = (x + y)\vec{i} + (x - y)\vec{j}$

20. Match vector fields $\vec{A}$–$\vec{D}$ in the tables with vector fields (I)–(IV) in Figure 17.32.

Vector field $\vec{A}$

$y\backslash x$	-1	1
-1	$\vec{i} + \vec{j}$	$\vec{i} + \vec{j}$
1	$\vec{i} + \vec{j}$	$\vec{i} + \vec{j}$

Vector field $\vec{B}$

$y\backslash x$	-1	1
-1	$-\vec{i} - \vec{j}$	$-\vec{i} - \vec{j}$
1	$\vec{i} + \vec{j}$	$\vec{i} + \vec{j}$

Vector field $\vec{C}$

$y\backslash x$	-1	1
-1	$-2\vec{i} + \vec{j}$	$2\vec{i} + \vec{j}$
1	$-2\vec{i} + \vec{j}$	$2\vec{i} + \vec{j}$

Vector field $\vec{D}$

$y\backslash x$	-1	1
-1	$\vec{i} + \vec{j}$	$-\vec{i} - \vec{j}$
1	$-\vec{i} + \vec{j}$	$\vec{i} - \vec{j}$

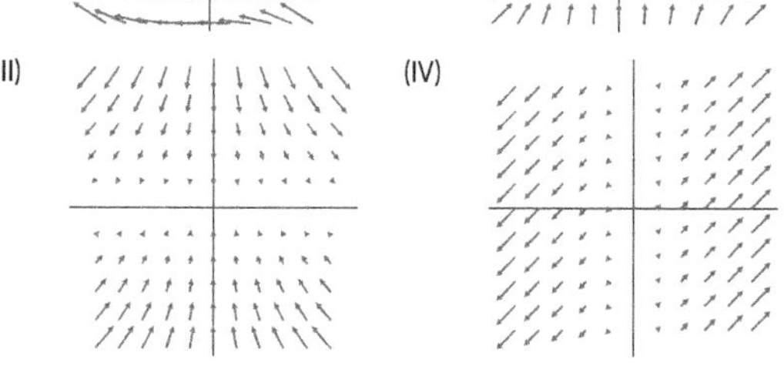

Figure 17.32

21. For each description of a vector field in (a)-(d), choose one or more of the vector fields I-IX.

(a) Pointing radially outward, increasing in length away from the origin.

(b) Pointing in a circular direction around the origin, remaining the same length.

(c) Pointing towards the origin, increasing in length farther from the origin.

(d) Pointing clockwise around the origin.

I. $\dfrac{x\vec{i} + y\vec{j}}{\sqrt{x^2 + y^2}}$ II. $\dfrac{-y\vec{i} + x\vec{j}}{\sqrt{x^2 + y^2}}$ III. $\vec{r}$

IV. $-\vec{r}$ V. $-y\vec{i} + x\vec{j}$ VI. $y\vec{i} - x\vec{j}$

VII. $y\vec{i} + x\vec{j}$ VIII. $\dfrac{\vec{r}}{\|\vec{r}\|^3}$ IX. $-\dfrac{\vec{r}}{\|\vec{r}\|^3}$

22. Each vector field in Figures (I)–(IV) represents the force on a particle at different points in space as a result of another particle at the origin. Match up the vector fields with the descriptions below.

(a) A repulsive force whose magnitude decreases as distance increases, such as between electric charges of the same sign.
(b) A repulsive force whose magnitude increases as distance increases.
(c) An attractive force whose magnitude decreases as distance increases, such as gravity.
(d) An attractive force whose magnitude increases as distance increases.

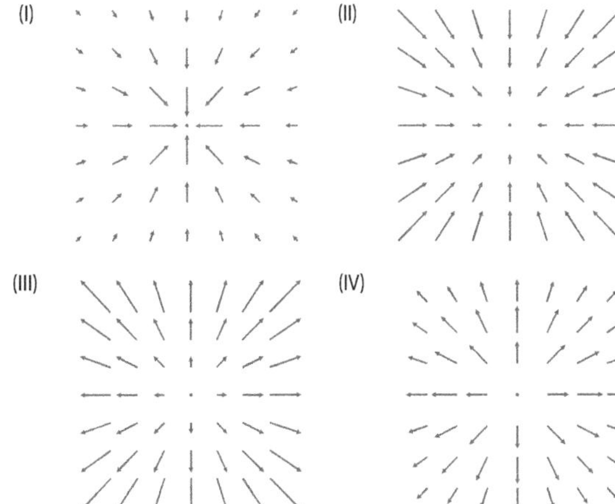

PROBLEMS

In Problems 23–27, give an example of a vector field $\vec{F}(x, y)$ in 2-space with the stated properties.

23. $\vec{F}$ is constant

24. $\vec{F}$ has a constant direction but $\|\vec{F}\|$ is not constant

25. $\|\vec{F}\|$ is constant but $\vec{F}$ is not constant

26. Neither $\|\vec{F}\|$ nor the direction of $\vec{F}$ is constant

27. $\vec{F}$ is perpendicular to $\vec{G} = (x+y)\vec{i} + (1+y^2)\vec{j}$ at every point

28. Match the level curves in (I)–(IV) with the gradient fields in (A)–(D). All figures use the same square window.

(I)

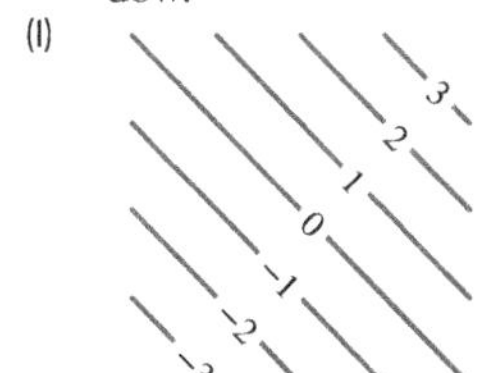

(II)

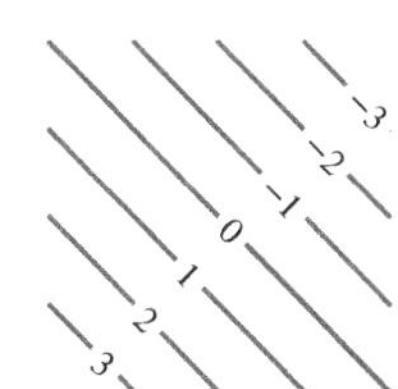

(III)

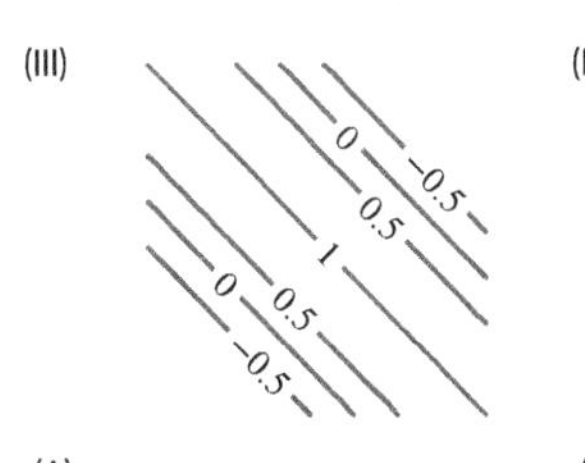

(IV)

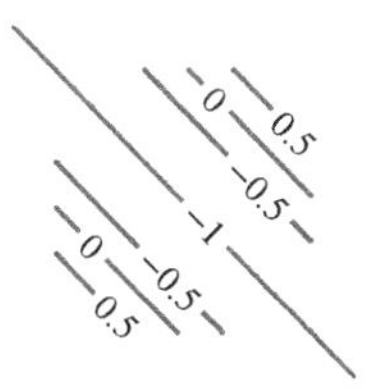

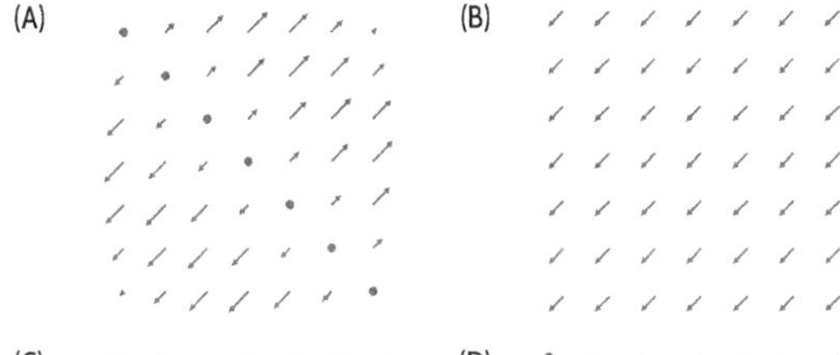

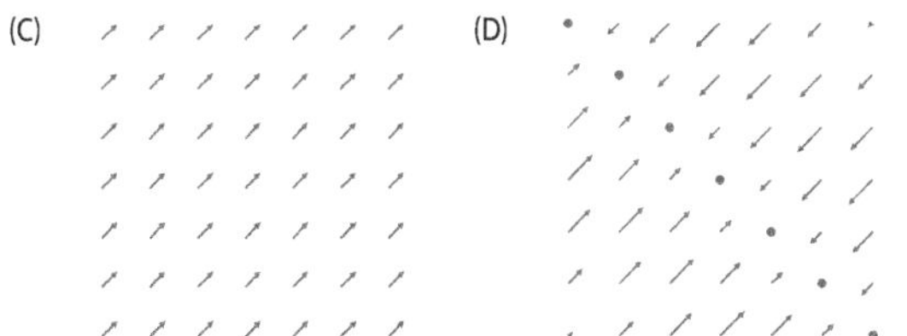

Problems 29–30 concern the vector fields $\vec{F} = x\vec{i} + y\vec{j}$, $\vec{G} = -y\vec{i} + x\vec{j}$, and $\vec{H} = x\vec{i} - y\vec{j}$.

29. Match $\vec{F}$, $\vec{G}$, $\vec{H}$ with their sketches in (I)–(III).

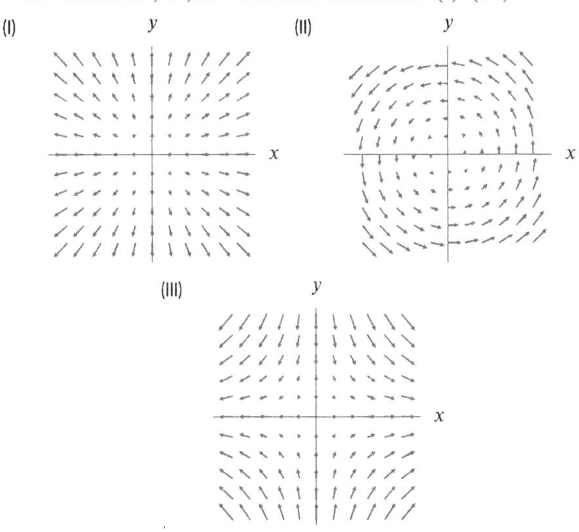

30. Match the vector fields with their sketches, (I)–(IV).

(a) $\vec{F} + \vec{G}$ **(b)** $\vec{F} + \vec{H}$ **(c)** $\vec{G} + \vec{H}$ **(d)** $-\vec{F} + \vec{G}$

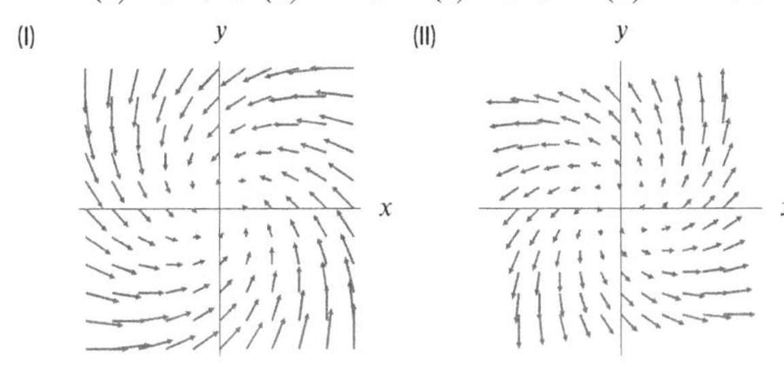

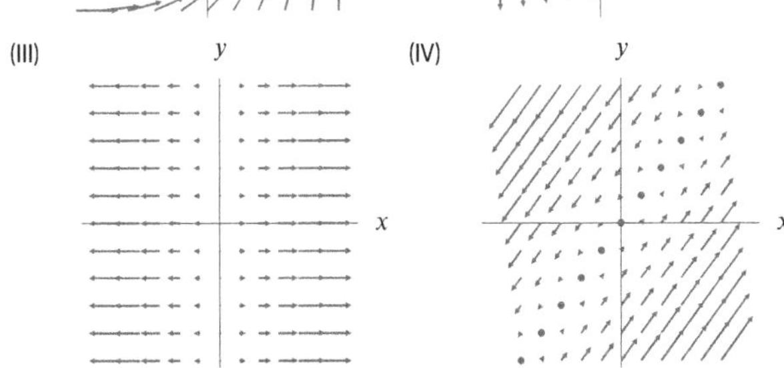

31. Match vector fields (a)–(f) with their graphs (I)–(VI).

(**a**) $-y\vec{i} + x\vec{j}$ (**b**) $x\vec{i}$
(**c**) $y\vec{j}$ (**d**) $z\vec{k}$
(**e**) $2\vec{k}$ (**f**) $\vec{r}$

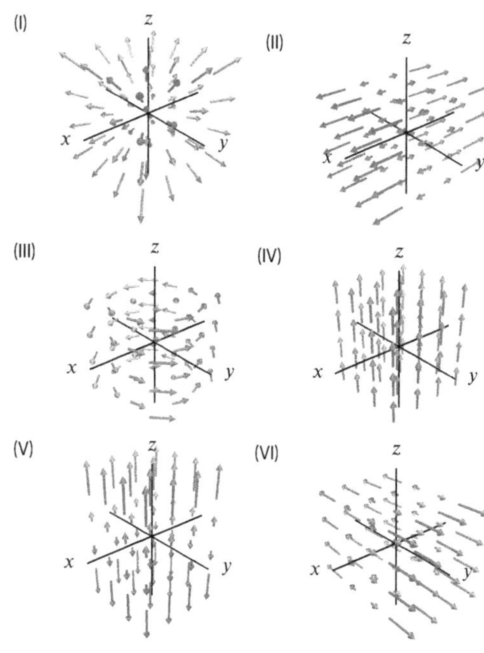

In Problems 32–34, write formulas for vector fields with the given properties.

32. All vectors are parallel to the x-axis; all vectors on a vertical line have the same magnitude.

33. All vectors point toward the origin and have constant length.

34. All vectors are of unit length and perpendicular to the position vector at that point.

35. (**a**) Let $\vec{F} = x\vec{i} + (x+y)\vec{j} + (x-y+z)\vec{k}$. Find a point at which $\vec{F}$ is parallel to l, the line $x = 5+t$, $y = 6-2t$, $z = 7-3t$.
(**b**) Find a point at which $\vec{F}$ and l are perpendicular.
(**c**) Give an equation for and describe in words the set of all points at which $\vec{F}$ and l are perpendicular.

In Problems 36–37, let $\vec{F} = x\vec{i} + y\vec{j}$ and $\vec{G} = -y\vec{i} + x\vec{j}$.

36. Sketch the vector field $\vec{L} = a\vec{F} + \vec{G}$ if:

(**a**) $a = 0$ (**b**) $a > 0$ (**c**) $a < 0$

37. Sketch the vector field $\vec{L} = \vec{F} + b\vec{G}$ if:

(**a**) $b = 0$ (**b**) $b > 0$ (**c**) $b < 0$

38. In the middle of a wide, steadily flowing river there is a fountain that spouts water horizontally in all directions. The river flows in the $\vec{i}$-direction in the xy-plane and the fountain is at the origin.

(**a**) If $A > 0$, $K > 0$, explain why the following expression could represent the velocity field for the combined flow of the river and the fountain:

$$\vec{v} = A\vec{i} + K(x^2+y^2)^{-1}(x\vec{i} + y\vec{j}).$$

(**b**) What is the significance of the constants A and K?
(**c**) Using a computer, sketch the vector field $\vec{v}$ for $K = 1$ and $A = 1$ and $A = 2$, and for $A = 0.2$, $K = 2$.

39. Figures 17.33 and 17.34 show the gradient of the functions $z = f(x, y)$ and $z = g(x, y)$.

(**a**) For each function, draw a rough sketch of the level curves, showing possible z-values.
(**b**) The xz-plane cuts each of the surfaces $z = f(x, y)$ and $z = g(x, y)$ in a curve. Sketch each of these curves, making clear how they are similar and how they are different from one another.

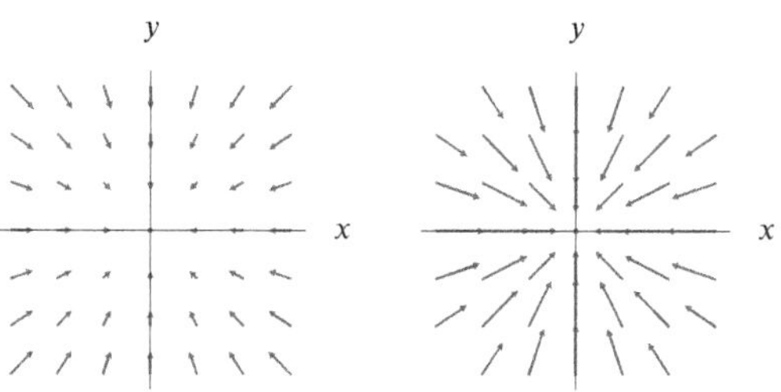

Figure 17.33: Gradient of $z = f(x, y)$

Figure 17.34: Gradient of $z = g(x, y)$

40. Let $\vec{F} = u\vec{i} + v\vec{j}$ be a vector field in 2-space with magnitude $F = \|\vec{F}\|$.

(**a**) Let $\vec{T} = (1/F)\vec{F}$. Show that $\vec{T}$ is the unit vector in the direction of $\vec{F}$. See Figure 17.35.
(**b**) Let $\vec{N} = (1/F)(\vec{k} \times \vec{F}) = (1/F)(-v\vec{i} + u\vec{j})$. Show that $\vec{N}$ is the unit vector pointing to the left of and at right angles to $\vec{F}$. See Figure 17.35.

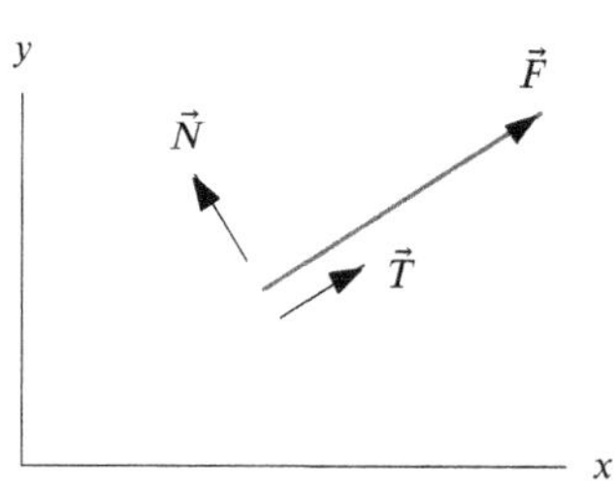

Figure 17.35

Strengthen Your Understanding

In Problems 41–42, explain what is wrong with the statement.

41. A plot of the vector field $\vec{G}(x, y, z) = \vec{F}(2x, 2y, 2z)$ can be obtained from a plot of the vector field $\vec{F}(x, y, z)$ by doubling the lengths of all the arrows.

42. A vector field $\vec{F}$ is defined by the formula $\vec{F}(x, y, z) = x^2 - yz$.

In Problems 43–44, give an example of:

43. A nonconstant vector field that is parallel to $\vec{i} + \vec{j} + \vec{k}$ at every point.

44. A nonconstant vector field with magnitude 1 at every point.

17.4 THE FLOW OF A VECTOR FIELD

When an iceberg is spotted in the North Atlantic, it is important to be able to predict where the iceberg is likely to be a day or a week later. To do this, one needs to know the velocity vector field of the ocean currents, that is, how fast and in what direction the water is moving at each point.

In this section we use differential equations to find the path of an object in a fluid flow. This path is called a flow line. Figure 17.36 shows several flow lines for the Gulf Stream velocity vector field in Figure 17.18 on page 906. The arrows on each flow line indicate the direction of flow.

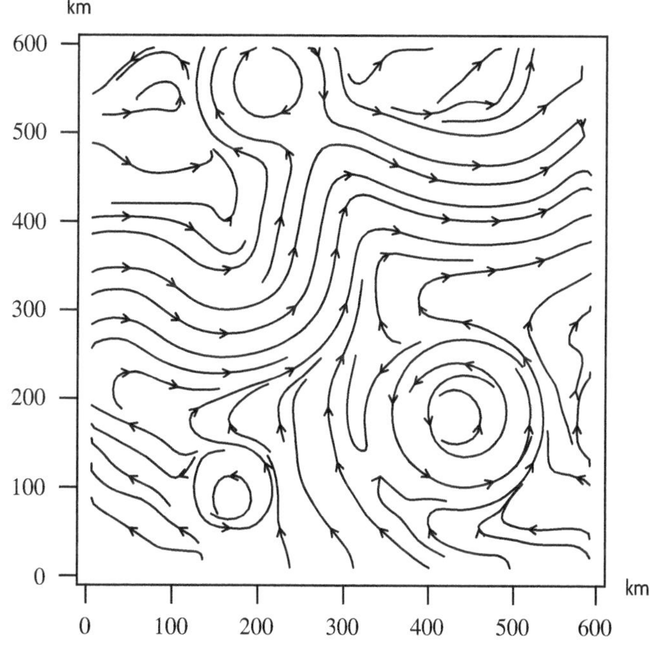

Figure 17.36: Flow lines for objects in the Gulf Stream with different starting points

How Do We Find a Flow Line?

Suppose that $\vec{F}$ is the velocity vector field of water on the surface of a creek and imagine a seed being carried along by the current. We want to know the position vector $\vec{r}(t)$ of the seed at time t. We know

$$\text{Velocity of seed at time } t = \text{Velocity of current at seed's position at time } t;$$

that is,

$$\vec{r}\,'(t) = \vec{F}(\vec{r}(t)).$$

We make the following definition:

> A **flow line** of a vector field $\vec{v} = \vec{F}(\vec{r})$ is a path $\vec{r}(t)$ whose velocity vector equals $\vec{v}$. Thus,
>
> $$\vec{r}\,'(t) = \vec{v} = \vec{F}(\vec{r}(t)).$$
>
> The **flow** of a vector field is the family of all of its flow lines.

A flow line is also called an *integral curve* or a *streamline*. We define flow lines for any vector field, as it turns out to be useful to study the flow of fields (for example, electric and magnetic) that are not velocity fields.

After resolving $\vec{F}$ and $\vec{r}$ into components, $\vec{F} = F_1\vec{i} + F_2\vec{j}$ and $\vec{r}(t) = x(t)\vec{i} + y(t)\vec{j}$, the definition of a flow line tells us that $x(t)$ and $y(t)$ satisfy the system of differential equations

$$x'(t) = F_1(x(t), y(t)) \quad \text{and} \quad y'(t) = F_2(x(t), y(t)).$$

Solving these differential equations gives a parameterization of the flow line.

Example 1 Find the flow line of the constant velocity field $\vec{v} = 3\vec{i} + 4\vec{j}$ cm/sec that passes through the point $(1, 2)$ at time $t = 0$.

Solution Let $\vec{r}(t) = x(t)\vec{i} + y(t)\vec{j}$ be the position in cm of a particle at time t, where t is in seconds. We have

$$x'(t) = 3 \quad \text{and} \quad y'(t) = 4.$$

Thus,

$$x(t) = 3t + x_0 \quad \text{and} \quad y(t) = 4t + y_0.$$

Since the path passes the point $(1, 2)$ at $t = 0$, we have $x_0 = 1$ and $y_0 = 2$ and so

$$x(t) = 3t + 1 \qquad \text{and} \qquad y(t) = 4t + 2.$$

Thus, the path is the line given parametrically by

$$\vec{r}(t) = (3t + 1)\vec{i} + (4t + 2)\vec{j}.$$

(See Figure 17.37.) To find an explicit equation for the path, eliminate t between these expressions to get

$$\frac{x-1}{3} = \frac{y-2}{4} \qquad \text{or} \qquad y = \frac{4}{3}x + \frac{2}{3}.$$

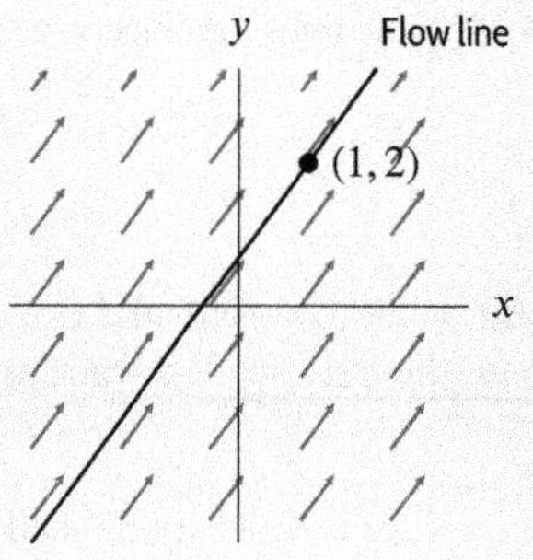

Figure 17.37: Vector field $\vec{F} = 3\vec{i} + 4\vec{j}$ with the flow line through $(1, 2)$

Example 2 The velocity of a flow at the point (x, y) is $\vec{F}(x, y) = \vec{i} + x\vec{j}$. Find the path of motion of an object in the flow that is at the point $(-2, 2)$ at time $t = 0$.

Solution Figure 17.38 shows this field. Since $\vec{r}\,'(t) = \vec{F}(\vec{r}(t))$, we are looking for the flow line that satisfies the system of differential equations

$$x'(t) = 1, \quad y'(t) = x(t) \qquad \text{satisfying } x(0) = -2 \text{ and } y(0) = 2.$$

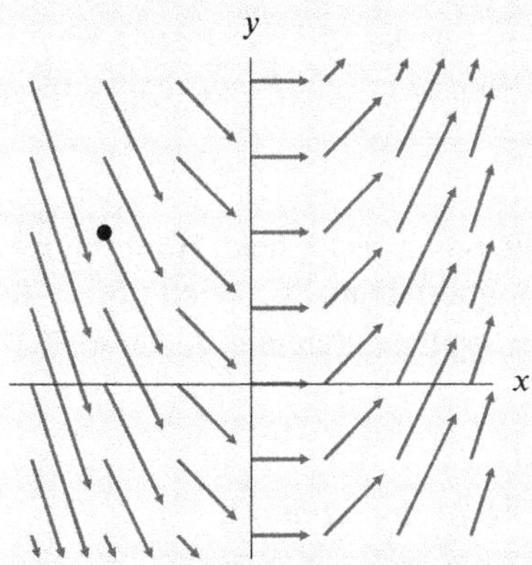

Figure 17.38: The velocity field $\vec{v} = \vec{i} + x\vec{j}$

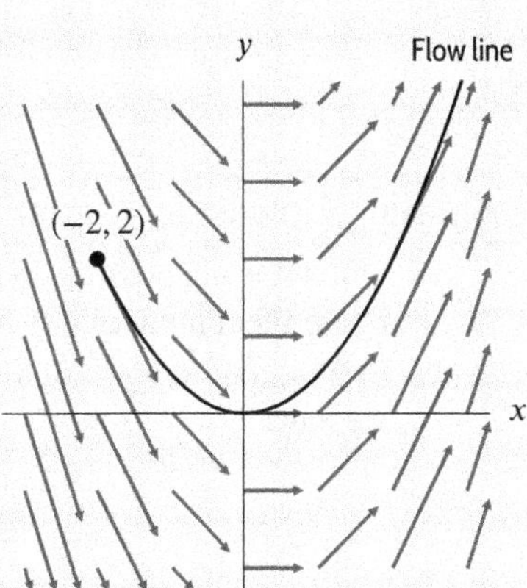

Figure 17.39: A flow line of the velocity field $\vec{v} = \vec{i} + x\vec{j}$

Solving for $x(t)$ first, we get $x(t) = t + x_0$, where x_0 is a constant of integration. Thus, $y'(t) = t + x_0$, so $y(t) = \frac{1}{2}t^2 + x_0 t + y_0$, where y_0 is also a constant of integration. Since $x(0) = x_0 = -2$ and $y(0) = y_0 = 2$, the path of motion is given by

$$x(t) = t - 2, \quad y(t) = \frac{1}{2}t^2 - 2t + 2,$$

or, equivalently,

$$\vec{r}(t) = (t-2)\vec{i} + (\tfrac{1}{2}t^2 - 2t + 2)\vec{j}.$$

The graph of this flow line in Figure 17.39 looks like a parabola. We check this by seeing that an explicit equation for the path is $y = \frac{1}{2}x^2$.

Example 3 Determine the flow of the vector field $\vec{v} = -y\vec{i} + x\vec{j}$.

Solution Figure 17.40 suggests that the flow consists of concentric counterclockwise circles, centered at the origin. The system of differential equations for the flow is

$$x'(t) = -y(t) \qquad y'(t) = x(t).$$

The equations $(x(t), y(t)) = (a\cos t,\ a\sin t)$ parameterize a family of counterclockwise circles of radius a, centered at the origin. We check that this family satisfies the system of differential equations:

$$x'(t) = -a\sin t = -y(t) \quad \text{and} \quad y'(t) = a\cos t = x(t).$$

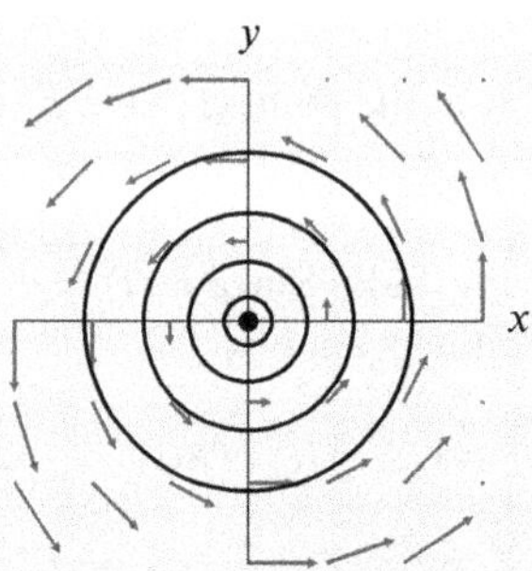

Figure 17.40: The flow of the vector field $\vec{v} = -y\vec{i} + x\vec{j}$

Approximating Flow Lines Numerically

Often it is not possible to find formulas for the flow lines of a vector field. However, we can approximate them numerically by Euler's method for solving differential equations. Since the flow lines $\vec{r}(t) = x(t)\vec{i} + y(t)\vec{j}$ of a vector field $\vec{v} = \vec{F}(x, y)$ satisfy the differential equation $\vec{r}\,'(t) = \vec{F}(\vec{r}(t))$, we have

$$\begin{aligned}\vec{r}(t + \Delta t) &\approx \vec{r}(t) + (\Delta t)\vec{r}\,'(t) \\ &= \vec{r}(t) + (\Delta t)\vec{F}(\vec{r}(t)) \quad \text{for } \Delta t \text{ near } 0.\end{aligned}$$

To approximate a flow line, we start at a point $\vec{r}_0 = \vec{r}(0)$ and estimate the position $\vec{r}_1$ of a particle at time Δt later:

$$\begin{aligned}\vec{r}_1 = \vec{r}(\Delta t) &\approx \vec{r}(0) + (\Delta t)\vec{F}(\vec{r}(0)) \\ &= \vec{r}_0 + (\Delta t)\vec{F}(\vec{r}_0).\end{aligned}$$

We then repeat the same procedure starting at $\vec{r}_1$, and so on. The general formula for getting from one point to the next is

$$\vec{r}_{n+1} = \vec{r}_n + (\Delta t)\vec{F}(\vec{r}_n).$$

The points with position vectors $\vec{r}_0, \vec{r}_1, \ldots$ trace out the path, as shown in the next example.

Example 4 Use Euler's method to approximate the flow line through $(1, 2)$ for the vector field $\vec{v} = y^2\vec{i} + 2x^2\vec{j}$.

Solution The flow is determined by the differential equations $\vec{r}\,'(t) = \vec{v}$, or equivalently

$$x'(t) = y^2, \qquad y'(t) = 2x^2.$$

We use Euler's method with $\Delta t = 0.02$, giving

$$\begin{aligned}\vec{r}_{n+1} &= \vec{r}_n + 0.02\,\vec{v}(x_n, y_n) \\ &= x_n\vec{i} + y_n\vec{j} + 0.02(y_n^2\vec{i} + 2x_n^2\vec{j}),\end{aligned}$$

or equivalently

$$x_{n+1} = x_n + 0.02{y_n}^2, \qquad y_{n+1} = y_n + 0.02 \cdot 2{x_n}^2.$$

When $t = 0$, we have $(x_0, y_0) = (1, 2)$. Then

$$\begin{aligned}x_1 &= x_0 + 0.02 \cdot {y_0}^2 = 1 + 0.02 \cdot 2^2 = 1.08, \\ y_1 &= y_0 + 0.02 \cdot 2x_0^2 = 2 + 0.02 \cdot 2 \cdot 1^2 = 2.04.\end{aligned}$$

So after one step $x(0.02) \approx 1.08$ and $y(0.02) \approx 2.04$. Similarly, $x(0.04) = x(2\Delta t) \approx 1.16$, $y(0.04) = y(2\Delta t) \approx 2.08$ and so on. Farther values along the flow line are given in Table 17.2 and plotted in Figure 17.41.

Table 17.2 *Approximated flow line starting at* $(1, 2)$ *for the vector field* $\vec{v} = y^2\vec{i} + 2x^2\vec{j}$

t	0	0.02	0.04	0.06	0.08	0.1	0.12	0.14	0.16	0.18
x	1	1.08	1.16	1.25	1.34	1.44	1.54	1.65	1.77	1.90
y	2	2.04	2.08	2.14	2.20	2.28	2.36	2.45	2.56	2.69

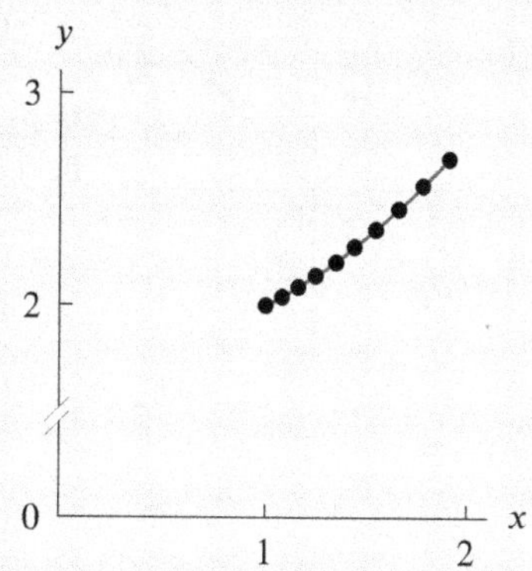

Figure 17.41: Euler's method solution to $x' = y^2$, $y' = 2x^2$

Exercises and Problems for Section 17.4

EXERCISES

In Exercises 1–3, sketch the vector field and its flow.

1. $\vec{v} = 2\vec{j}$
2. $\vec{v} = 3\vec{i}$
3. $\vec{v} = 3\vec{i} - 2\vec{j}$

In Exercises 4–9, sketch the vector field and the flow. Then find the system of differential equations associated with the vector field and check that the flow satisfies the system.

4. $\vec{v} = x\vec{i}$; $x(t) = ae^t$, $y(t) = b$
5. $\vec{v} = x\vec{j}$; $x(t) = a$, $y(t) = at + b$
6. $\vec{v} = x\vec{i} + y\vec{j}$; $x(t) = ae^t$, $y(t) = be^t$
7. $\vec{v} = x\vec{i} - y\vec{j}$; $x(t) = ae^t$, $y(t) = be^{-t}$
8. $\vec{v} = y\vec{i} - x\vec{j}$; $x(t) = a\sin t$, $y(t) = a\cos t$
9. $\vec{v} = y\vec{i} + x\vec{j}$; $x(t) = a(e^t + e^{-t})$, $y(t) = a(e^t - e^{-t})$
10. Use a computer or calculator with Euler's method to approximate the flow line through $(1, 2)$ for the vector field $\vec{v} = y^2\vec{i} + 2x^2\vec{j}$ using 5 steps with $\Delta t = 0.1$.

PROBLEMS

For Problems 11–14, find the region of the Gulf Stream velocity field in Figure 17.18 on page 906 represented by the given table of velocity vectors (in cm/sec).

11.

$35\vec{i} + 131\vec{j}$	$48\vec{i} + 92\vec{j}$	$47\vec{i} + \vec{j}$
$-32\vec{i} + 132\vec{j}$	$-44\vec{i} + 92\vec{j}$	$-42\vec{i} + \vec{j}$
$-51\vec{i} + 73\vec{j}$	$-119\vec{i} + 84\vec{j}$	$-128\vec{i} + 6\vec{j}$

12.

$10\vec{i} - 3\vec{j}$	$11\vec{i} + 16\vec{j}$	$20\vec{i} + 75\vec{j}$
$53\vec{i} - 7\vec{j}$	$58\vec{i} + 23\vec{j}$	$64\vec{i} + 80\vec{j}$
$119\vec{i} - 8\vec{j}$	$121\vec{i} + 31\vec{j}$	$114\vec{i} + 66\vec{j}$

13.

$97\vec{i} - 41\vec{j}$	$72\vec{i} - 24\vec{j}$	$54\vec{i} - 10\vec{j}$
$134\vec{i} - 49\vec{j}$	$131\vec{i} - 44\vec{j}$	$129\vec{i} - 18\vec{j}$
$103\vec{i} - 36\vec{j}$	$122\vec{i} - 30\vec{j}$	$131\vec{i} - 17\vec{j}$

14.

$-95\vec{i} - 60\vec{j}$	$18\vec{i} - 48\vec{j}$	$82\vec{i} - 22\vec{j}$
$-29\vec{i} + 48\vec{j}$	$76\vec{i} + 63\vec{j}$	$128\vec{i} - 16\vec{j}$
$26\vec{i} + 105\vec{j}$	$49\vec{i} + 119\vec{j}$	$88\vec{i} + 13\vec{j}$

15. $\vec{F}(x, y)$ and $\vec{G}(x, y) = 2\vec{F}(x, y)$ are two vector fields. Illustrating your answer with $\vec{F}(x, y) = -y\vec{i} + x\vec{j}$, describe the graphical difference between:

(a) The vector fields **(b)** Their flows

16. Match the vector fields (a)–(f) with their flow lines (I)–(VI). Put arrows on the flow lines indicating the direction of flow.

(a) $y\vec{i} + x\vec{j}$
(b) $-y\vec{i} + x\vec{j}$
(c) $x\vec{i} + y\vec{j}$
(d) $-y\vec{i} + (x + y/10)\vec{j}$
(e) $-y\vec{i} + (x - y/10)\vec{j}$
(f) $(x - y)\vec{i} + (x - y)\vec{j}$

(I)
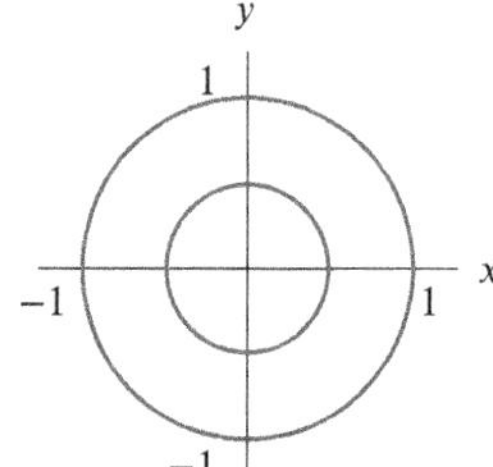

(II)
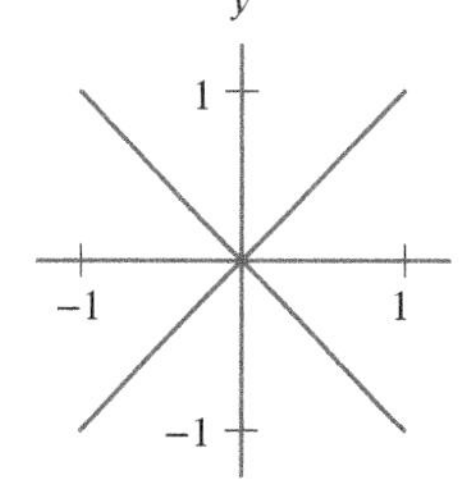

(III)
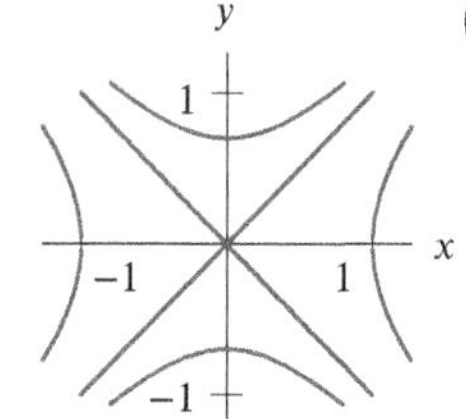

(IV)
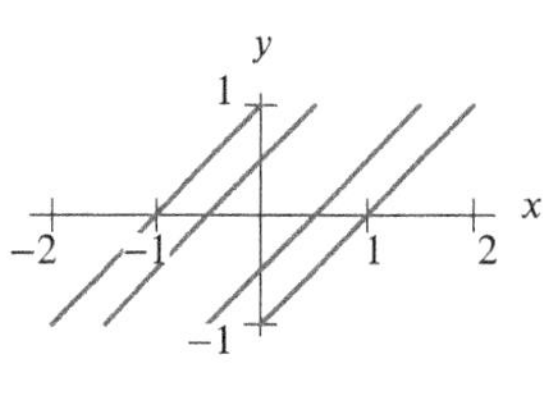

(V)
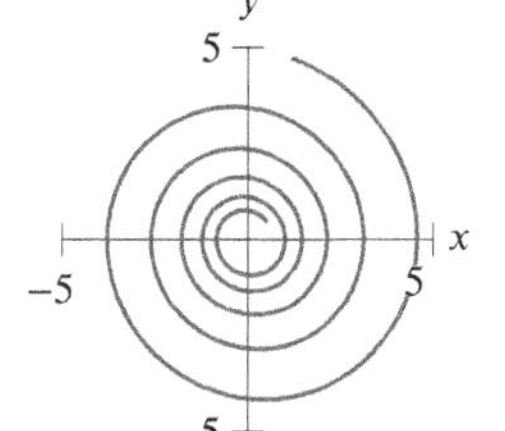

(VI)
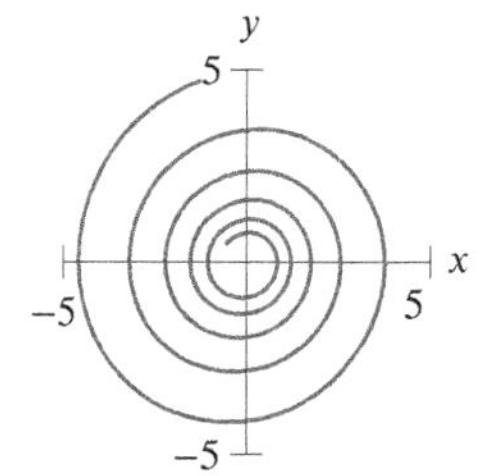

17. Show that the acceleration $\vec{a}$ of an object flowing in a velocity field $\vec{F}(x, y) = u(x, y)\vec{i} + v(x, y)\vec{j}$ is given by $\vec{a} = (u_x u + u_y v)\vec{i} + (v_x u + v_y v)\vec{j}$.

18. A velocity vector field $\vec{v} = -H_y\vec{i} + H_x\vec{j}$ is based on the partial derivatives of a smooth function $H(x, y)$. Explain why

(a) $\vec{v}$ is perpendicular to grad H.
(b) the flow lines of $\vec{v}$ are along the level curves of H.

In Problems 19–21, show that every flow line of the vector field $\vec{v}$ lies on a level curve for the function $f(x, y)$.

19. $\vec{v} = x\vec{i} - y\vec{j}$, $f(x, y) = xy$

20. $\vec{v} = y\vec{i} + x\vec{j}$, $f(x, y) = x^2 - y^2$

21. $\vec{v} = ay\vec{i} + bx\vec{j}$, $f(x, y) = bx^2 - ay^2$

22. A velocity vector field, $\vec{F}(x, y) = (x + 2y)\vec{i} + xy\vec{j}$, in meters per sec, has x and y in meters. For an object starting at (2, 1), use Euler's method with $\Delta t = 0.01$ sec to approximate its position 0.01 sec later.

23. A solid metal ball has its center at the origin of a fixed set of axes. The ball rotates once every 24 hours around the z-axis. The direction of rotation is counterclockwise when viewed from above. Let $\vec{v}(x, y, z)$ be the velocity vector of the particle of metal at the point (x, y, z) inside the ball. Time is in hours and x, y, z are in meters.

(a) Find a formula for the vector field $\vec{v}$. Give units for your answer.
(b) Describe in words the flow lines of $\vec{v}$.

24. (a) Show that $h(t) = e^{-2at}(x^2 + y^2)$ is constant along any flow line of $\vec{v} = (ax - y)\vec{i} + (x + ay)\vec{j}$.
(b) Show that points moving with the flow that are on the unit circle centered at the origin at time 0 are on the circle of radius e^{at} centered at the origin at time t.

Strengthen Your Understanding

In Problems 25–26, explain what is wrong with the statement.

25. The flow lines of a vector field whose components are linear functions are all straight lines.

26. If the flow lines of a vector field are all straight lines with the same slope pointing in the same direction, then the vector field is constant.

In Problems 27–28, give an example of:

27. A vector field $\vec{F}(x, y, z)$ such that the path $\vec{r}(t) = t\vec{i} + t^2\vec{j} + t^3\vec{k}$ is a flow line.

28. A vector field whose flow lines are rays from the origin.

Are the statements in Problems 29–38 true or false? Give reasons for your answer.

29. The flow lines for $\vec{F}(x, y) = x\vec{j}$ are parallel to the y-axis.

30. The flow lines of $\vec{F}(x, y) = y\vec{i} - x\vec{j}$ are hyperbolas.

31. The flow lines of $\vec{F}(x, y) = x\vec{i}$ are parabolas.

32. The vector field in Figure 17.42 has a flow line which lies in the first and third quadrants.

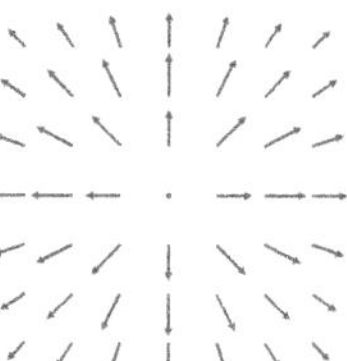

Figure 17.42

33. The vector field in Figure 17.42 has a flow line on which both x and y tend to infinity.

34. If $\vec{F}$ is a gradient vector field, $\vec{F}(x, y) = \nabla f(x, y)$, then the flow lines for $\vec{F}$ are the contours for f.

35. If the flow lines for the vector field $\vec{F}(\vec{r})$ are all concentric circles centered at the origin, then $\vec{F}(\vec{r}) \cdot \vec{r} = 0$ for all $\vec{r}$.

36. If the flow lines for the vector field $\vec{F}(x, y)$ are all straight lines parallel to the constant vector $\vec{v} = 3\vec{i} + 5\vec{j}$, then $\vec{F}(x, y) = \vec{v}$.

37. No flow line for the vector field $\vec{F}(x, y) = x\vec{i} + 2\vec{j}$ has a point where the y-coordinate reaches a relative maximum.

38. The vector field $\vec{F}(x, y) = e^x\vec{i} + y\vec{j}$ has a flow line that crosses the x-axis.

Online Resource: Review problems and Projects

Chapter Eighteen

LINE INTEGRALS

Contents

18.1 THE IDEA OF A LINE INTEGRAL

Imagine that you are rowing on a river with a noticeable current. At times you may be working against the current and at other times you may be moving with it. At the end you have a sense of whether, overall, you were helped or hindered by the current. The line integral, defined in this section, measures the extent to which a curve in a vector field is, overall, going with the vector field or against it.

Orientation of a Curve

A curve can be traced out in two directions, as shown in Figure 18.1. We need to choose one direction before we can define a line integral.

> A curve is said to be **oriented** if we have chosen a direction of travel on it.

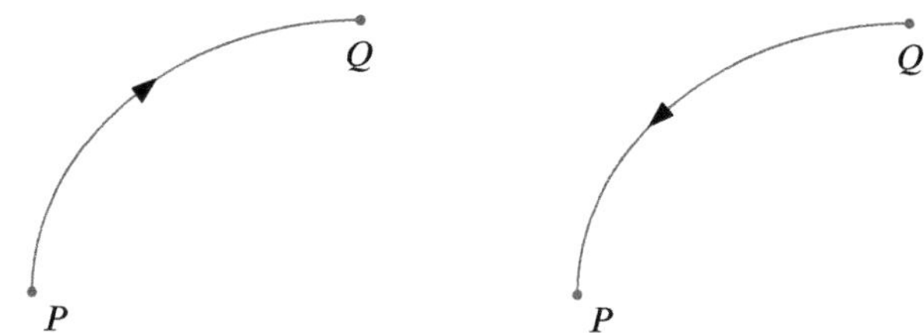

Figure 18.1: A curve with two different orientations represented by arrowheads

Definition of the Line Integral

Consider a vector field $\vec{F}$ and an oriented curve C. We begin by dividing C into n small, almost straight pieces along which $\vec{F}$ is approximately constant. Each piece can be represented by a displacement vector $\Delta\vec{r}_i = \vec{r}_{i+1} - \vec{r}_i$ and the value of $\vec{F}$ at each point of this small piece of C is approximately $\vec{F}(\vec{r}_i)$. See Figures 18.2 and 18.3.

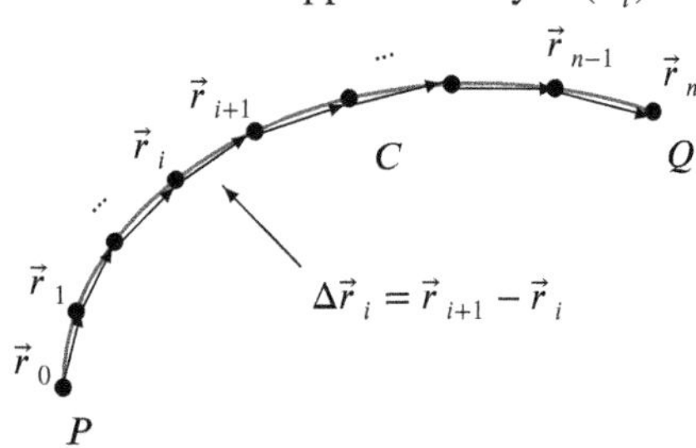

Figure 18.2: The curve C, oriented from P to Q, approximated by straight line segments represented by displacement vectors $\Delta\vec{r}_i = \vec{r}_{i+1} - \vec{r}_i$

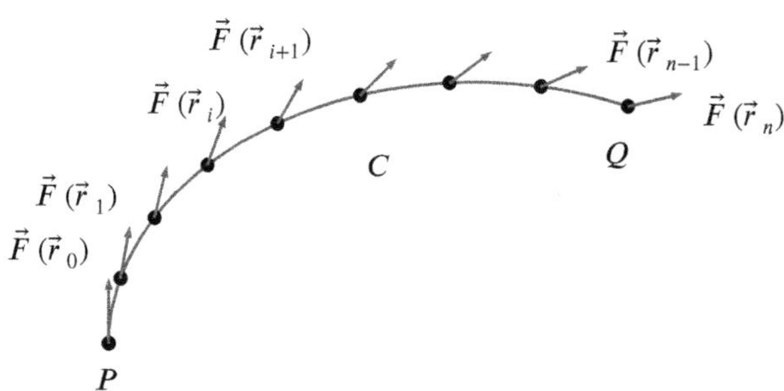

Figure 18.3: The vector field $\vec{F}$ evaluated at the points with position vector $\vec{r}_i$ on the curve C oriented from P to Q

Returning to our initial example, the vector field $\vec{F}$ represents the current and the oriented curve C is the path of the person rowing the boat. We wish to determine to what extent the vector field $\vec{F}$ helps or hinders motion along C. Since the dot product can be used to measure to what extent two vectors point in the same or opposing directions, we form the dot product $\vec{F}(\vec{r}_i) \cdot \Delta\vec{r}_i$ for each point with position vector $\vec{r}_i$ on C. Summing over all such pieces, we get a Riemann sum:

$$\sum_{i=0}^{n-1} \vec{F}(\vec{r}_i) \cdot \Delta\vec{r}_i.$$

We define the line integral, written $\int_C \vec{F} \cdot d\vec{r}$, by taking the limit as $\|\Delta\vec{r}_i\| \to 0$. Provided the limit exists, we make the following definition:

The **line integral** of a vector field $\vec{F}$ along an oriented curve C is

$$\int_C \vec{F} \cdot d\vec{r} = \lim_{\|\Delta \vec{r}_i\| \to 0} \sum_{i=0}^{n-1} \vec{F}(\vec{r}_i) \cdot \Delta \vec{r}_i.$$

How Does the Limit Defining a Line Integral Work?

The limit in the definition of a line integral exists if $\vec{F}$ is continuous on the curve C and if C is made by joining end to end a finite number of smooth curves. (A vector field $\vec{F} = F_1\vec{i} + F_2\vec{j} + F_3\vec{k}$ is *continuous* if F_1, F_2, and F_3 are continuous, and a *smooth curve* is one that can be parameterized by smooth functions.) We subdivide the curve using a parameterization that goes from one end of the curve to the other, in the forward direction, without retracing any portion of the curve. A subdivision of the parameter interval gives a subdivision of the curve. All the curves we consider in this book are *piecewise smooth* in this sense. Section 18.2 shows how to use a parameterization to compute a line integral.

Example 1 Find the line integral of the constant vector field $\vec{F} = \vec{i} + 2\vec{j}$ along the path from (1, 1) to (10, 10) shown in Figure 18.4.

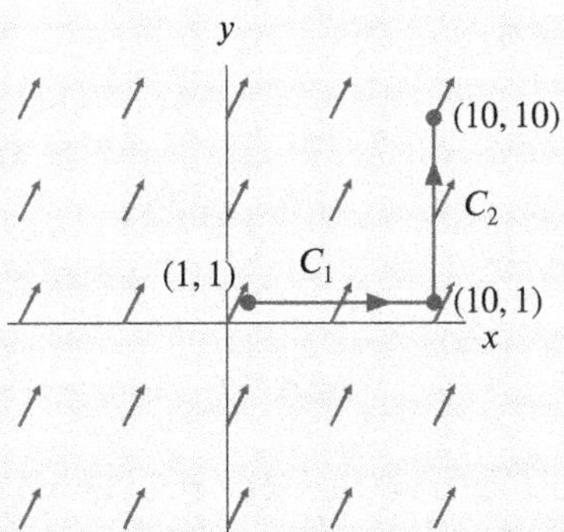

Figure 18.4: The constant vector field $\vec{F} = \vec{i} + 2\vec{j}$ and the path from (1, 1) to (10, 10)

Solution Let C_1 be the horizontal segment of the path going from (1, 1) to (10, 1). When we break this path into pieces, each piece $\Delta\vec{r}$ is horizontal, so $\Delta\vec{r} = \Delta x\vec{i}$ and $\vec{F} \cdot \Delta\vec{r} = (\vec{i} + 2\vec{j}) \cdot \Delta x\vec{i} = \Delta x$. Hence,

$$\int_{C_1} \vec{F} \cdot d\vec{r} = \int_{x=1}^{x=10} dx = 9.$$

Similarly, along the vertical segment C_2, we have $\Delta\vec{r} = \Delta y\vec{j}$ and $\vec{F} \cdot \Delta\vec{r} = (\vec{i} + 2\vec{j}) \cdot \Delta y\vec{j} = 2\Delta y$, so

$$\int_{C_2} \vec{F} \cdot d\vec{r} = \int_{y=1}^{y=10} 2\,dy = 18.$$

Thus,

$$\int_C \vec{F} \cdot d\vec{r} = \int_{C_1} \vec{F} \cdot d\vec{r} + \int_{C_2} \vec{F} \cdot d\vec{r} = 9 + 18 = 27.$$

What Does the Line Integral Tell Us?

Remember that for any two vectors $\vec{u}$ and $\vec{v}$, the dot product $\vec{u} \cdot \vec{v}$ is positive if $\vec{u}$ and $\vec{v}$ point roughly in the same direction (that is, if the angle between them is less than $\pi/2$). The dot product is zero if $\vec{u}$ is perpendicular to $\vec{v}$ and is negative if they point roughly in opposite directions (that is, if the angle between them is greater than $\pi/2$).

The line integral of $\vec{F}$ adds the dot products of $\vec{F}$ and $\Delta\vec{r}$ along the path. If $||\vec{F}||$ is constant, the line integral gives a positive number if $\vec{F}$ is mostly pointing in the same direction as C, and a negative number if $\vec{F}$ is mostly pointing in the opposite direction to C. The line integral is zero if $\vec{F}$ is perpendicular to the path at all points or if the positive and negative contributions cancel out. In general, the line integral of a vector field $\vec{F}$ along a curve C measures the extent to which C is going with $\vec{F}$ or against it.

Example 2 The vector field $\vec{F}$ and the oriented curves C_1, C_2, C_3, C_4 are shown in Figure 18.5. The curves C_1 and C_3 are the same length. Which of the line integrals $\int_{C_i} \vec{F} \cdot d\vec{r}$, for $i = 1, 2, 3, 4$, are positive? Which are negative? Arrange these line integrals in ascending order.

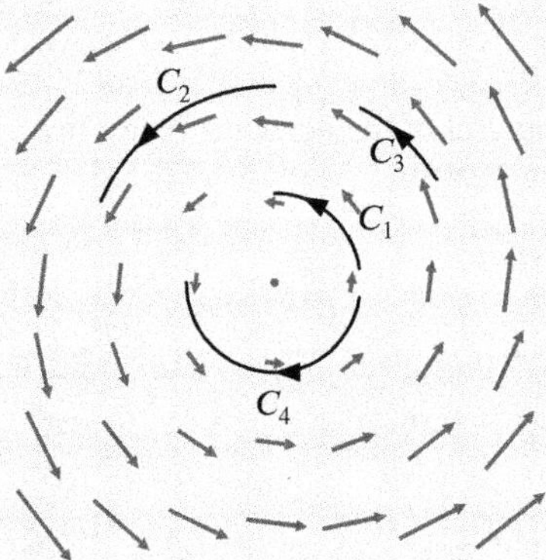

Figure 18.5: Vector field and paths C_1, C_2, C_3, C_4

Solution The vector field $\vec{F}$ and the line segments $\Delta\vec{r}$ are approximately parallel and in the same direction for the curves C_1, C_2, and C_3. So the contributions of each term $\vec{F} \cdot \Delta\vec{r}$ are positive for these curves. Thus, $\int_{C_1} \vec{F} \cdot d\vec{r}$, $\int_{C_2} \vec{F} \cdot d\vec{r}$, and $\int_{C_3} \vec{F} \cdot d\vec{r}$ are each positive. For the curve C_4, the vector field and the line segments are in opposite directions, so each term $\vec{F} \cdot \Delta\vec{r}$ is negative, and therefore the integral $\int_{C_4} \vec{F} \cdot d\vec{r}$ is negative.

Since the magnitude of the vector field is smaller along C_1 than along C_3, and these two curves are the same length, we have

$$\int_{C_1} \vec{F} \cdot d\vec{r} < \int_{C_3} \vec{F} \cdot d\vec{r}.$$

In addition, the magnitude of the vector field is the same along C_2 and C_3, but the curve C_2 is longer than the curve C_3. Thus,

$$\int_{C_3} \vec{F} \cdot d\vec{r} < \int_{C_2} \vec{F} \cdot d\vec{r}.$$

Putting these results together with the fact that $\int_{C_4} \vec{F} \cdot d\vec{r}$ is negative, we have

$$\int_{C_4} \vec{F} \cdot d\vec{r} < \int_{C_1} \vec{F} \cdot d\vec{r} < \int_{C_3} \vec{F} \cdot d\vec{r} < \int_{C_2} \vec{F} \cdot d\vec{r}.$$

Interpretations of the Line Integral

Work

Recall from Section 13.3 that if a constant force $\vec{F}$ acts on an object while it moves along a straight line through a displacement $\vec{d}$, the work done by the force on the object is

$$\text{Work done} = \vec{F} \cdot \vec{d}.$$

Now suppose we want to find the work done by gravity on an object moving far above the surface of the earth. Since the force of gravity varies with distance from the earth and the path may not be straight, we can't use the formula $\vec{F} \cdot \vec{d}$. We approximate the path by line segments which are small enough that the force is approximately constant on each one. Suppose the force at a point with position vector $\vec{r}$ is $\vec{F}(\vec{r})$, as in Figures 18.2 and 18.3. Then

$$\begin{array}{c}\text{Work done by force } \vec{F}(\vec{r}_i) \\ \text{over small displacement } \Delta\vec{r}_i\end{array} \approx \vec{F}(\vec{r}_i) \cdot \Delta\vec{r}_i,$$

and so,

$$\begin{array}{c}\text{Total work done by force} \\ \text{along oriented curve } C\end{array} \approx \sum_i \vec{F}(\vec{r}_i) \cdot \Delta\vec{r}_i.$$

Taking the limit as $\|\Delta\vec{r}_i\| \to 0$, we get

$$\begin{array}{c}\text{Work done by force } \vec{F}(\vec{r}) \\ \text{along curve } C\end{array} = \lim_{\|\Delta\vec{r}_i\|\to 0} \sum_i \vec{F}(\vec{r}_i) \cdot \Delta\vec{r}_i = \int_C \vec{F} \cdot d\vec{r}.$$

Example 3 A mass lying on a flat table is attached to a spring whose other end is fastened to the wall. (See Figure 18.6.) The spring is extended 20 cm beyond its rest position and released. If the axes are as shown in Figure 18.6, when the spring is extended by a distance of x, by Hooke's Law, the force exerted by the spring on the mass is given by

$$\vec{F}(x) = -kx\vec{i},$$

where k is a positive constant that depends on the properties of the particular spring.

Suppose the mass moves back to the rest position. How much work is done by the force exerted by the spring?

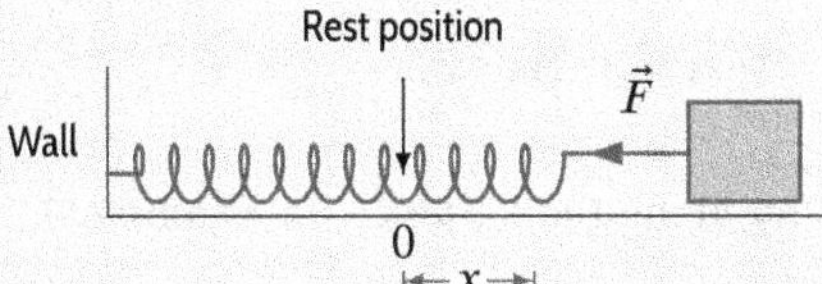

Figure 18.6: Force on mass due to an extended spring

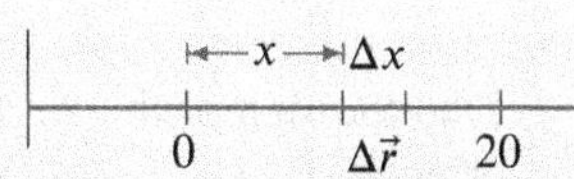

Figure 18.7: Dividing up the interval $0 \le x \le 20$ in order to calculate the work done

Solution The path from $x = 20$ to $x = 0$ is divided as shown in Figure 18.7, with a typical segment represented by

$$\Delta\vec{r} = \Delta x\vec{i}.$$

Since we are moving from $x = 20$ to $x = 0$, the quantity Δx will be negative. The work done by the force as the mass moves through this segment is approximated by

$$\text{Work done} \approx \vec{F} \cdot \Delta\vec{r} = (-kx\vec{i}\,) \cdot (\Delta x\vec{i}\,) = -kx\,\Delta x.$$

Thus, we have

$$\text{Total work done} \approx \sum -kx\,\Delta x.$$

In the limit, as $\|\Delta x\| \to 0$, this sum becomes an ordinary definite integral. Since the path starts at $x = 20$, this is the lower limit of integration; $x = 0$ is the upper limit. Thus, we get

$$\text{Total work done} = \int_{x=20}^{x=0} -kx\,dx = -\frac{kx^2}{2}\bigg|_{20}^{0} = \frac{k(20)^2}{2} = 200k.$$

Note that the work done is positive, since the force acts in the direction of motion.

Example 3 shows how a line integral over a path parallel to the x-axis reduces to a one-variable integral. Section 18.2 shows how to convert *any* line integral into a one-variable integral.

Example 4 A particle with position vector $\vec{r}$ is subject to a force, $\vec{F}$, due to gravity. What is the *sign* of the work done by $\vec{F}$ as the particle moves along the path C_1, a radial line through the center of the earth, starting 8000 km from the center and ending 10,000 km from the center? (See Figure 18.8.)

Solution We divide the path into small radial segments, $\Delta\vec{r}$, pointing away from the center of the earth and parallel to the gravitational force. The vectors $\vec{F}$ and $\Delta\vec{r}$ point in opposite directions, so each term $\vec{F} \cdot \Delta\vec{r}$ is negative. Adding all these negative quantities and taking the limit results in a negative value for the total work. Thus, the work done by gravity is negative. The negative sign indicates that we would have to do work *against* gravity to move the particle along the path C_1.

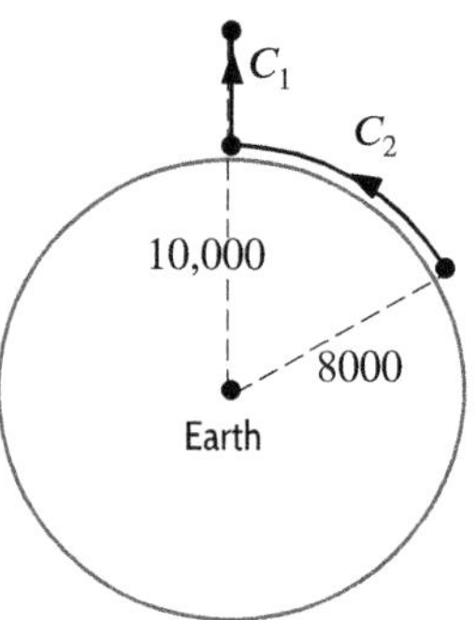

Figure 18.8: The earth

Example 5 Find the sign of the work done by gravity along the curve C_1 in Example 4, but with the opposite orientation.

Solution Tracing a curve in the opposite direction changes the sign of the line integral because all the segments $\Delta\vec{r}$ change direction, and so every term $\vec{F} \cdot \Delta\vec{r}$ changes sign. Thus, the result will be the negative of the answer found in Example 4. Therefore, the work done by gravity as a particle moves along C_1 toward the center of the earth is positive.

Example 6 Find the work done by gravity as a particle moves along C_2, an arc of a circle 8000 km long at a distance of 8000 km from the center of the earth. (See Figure 18.8.)

Solution Since C_2 is everywhere perpendicular to the gravitational force, $\vec{F}\cdot\Delta\vec{r}=0$ for all $\Delta\vec{r}$ along C_2. Thus,

$$\text{Work done} = \int_{C_2} \vec{F}\cdot d\vec{r} = 0,$$

so the work done is zero. This is why satellites can remain in orbit without expending any fuel, once they have attained the correct altitude and velocity.

Circulation

The velocity vector field for the Gulf Stream on page 906 shows distinct eddies or regions where the water circulates. We can measure this circulation using a *closed curve*, that is, one that starts and ends at the same point.

> If C is an oriented closed curve, the line integral of a vector field $\vec{F}$ around C is called the **circulation** of $\vec{F}$ around C.

Circulation is a measure of the net tendency of the vector field to point around the curve C. To emphasize that C is closed, the circulation is sometimes denoted $\oint_C \vec{F}\cdot d\vec{r}$, with a small circle on the integral sign.

Example 7 Describe the rotation of the vector fields in Figures 18.9 and 18.10. Find the sign of the circulation of the vector fields around the indicated paths.

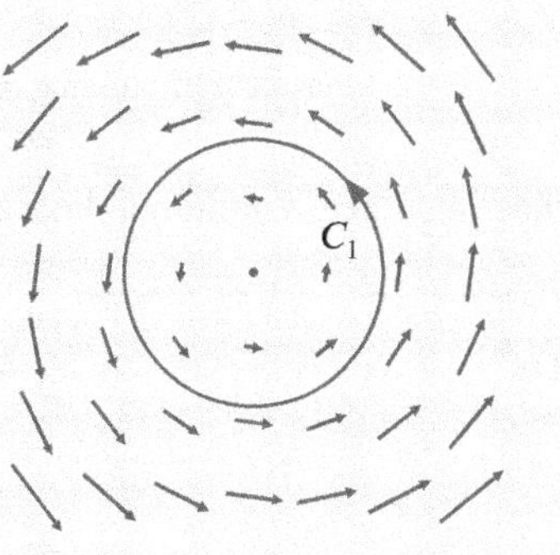

Figure 18.9: A circulating flow

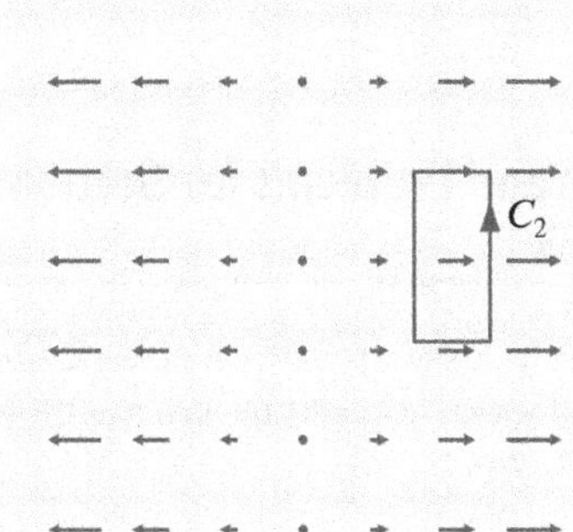

Figure 18.10: A flow with zero circulation

Solution Consider the vector field in Figure 18.9. If you think of this as representing the velocity of water flowing in a pond, you see that the water is circulating. The line integral around C_1, measuring the circulation around C_1, is positive, because the vectors of the field are all pointing in the direction of the path. By way of contrast, look at the vector field in Figure 18.10. Here the line integral around C_2 is zero because the vertical portions of the path are perpendicular to the field and the contributions from the two horizontal portions cancel out. This means that there is no net tendency for the water to circulate around C_2.

It turns out that the vector field in Figure 18.10 has the property that its circulation around *any* closed path is zero. Water moving according to this vector field has no tendency to circulate around any point, and a leaf dropped into the water will not spin. We'll look at such special fields again later when we introduce the notion of the *curl* of a vector field.

Properties of Line Integrals

Line integrals share some basic properties with ordinary one-variable integrals:

> For a scalar constant λ, vector fields $\vec{F}$ and $\vec{G}$, and oriented curves C, C_1, and C_2,
>
> 1. $\displaystyle\int_C \lambda\vec{F}\cdot d\vec{r} = \lambda\int_C \vec{F}\cdot d\vec{r}.$
> 2. $\displaystyle\int_C (\vec{F}+\vec{G})\cdot d\vec{r} = \int_C \vec{F}\cdot d\vec{r} + \int_C \vec{G}\cdot d\vec{r}.$
> 3. $\displaystyle\int_{-C} \vec{F}\cdot d\vec{r} = -\int_C \vec{F}\cdot d\vec{r}.$
> 4. $\displaystyle\int_{C_1+C_2} \vec{F}\cdot d\vec{r} = \int_{C_1} \vec{F}\cdot d\vec{r} + \int_{C_2} \vec{F}\cdot d\vec{r}.$

Properties 3 and 4 are concerned with the curve C over which the line integral is taken. If C is an oriented curve, then $-C$ is the same curve traversed in the opposite direction, that is, with the opposite orientation. (See Figure 18.11.) Property 3 holds because if we integrate along $-C$, the vectors $\Delta\vec{r}$ point in the opposite direction and the dot products $\vec{F}\cdot\Delta\vec{r}$ are the negatives of what they were along C.

If C_1 and C_2 are oriented curves with C_1 ending where C_2 begins, we construct a new oriented curve, called $C_1 + C_2$, by joining them together. (See Figure 18.12.) Property 4 is the analogue for line integrals of the property for definite integrals which says that

$$\int_a^b f(x)\,dx = \int_a^c f(x)\,dx + \int_c^b f(x)\,dx.$$

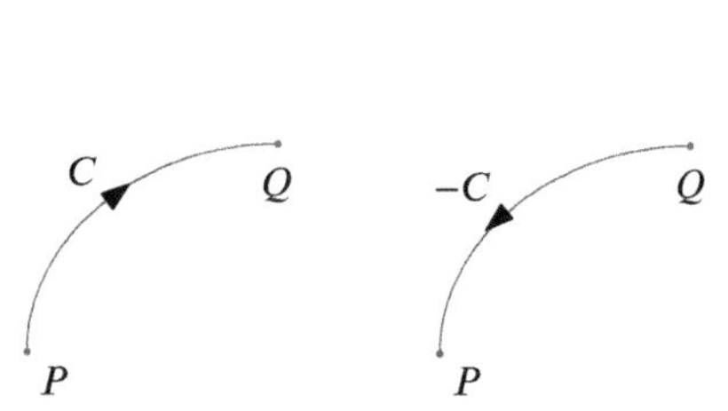

Figure 18.11: A curve, C, and its opposite, $-C$

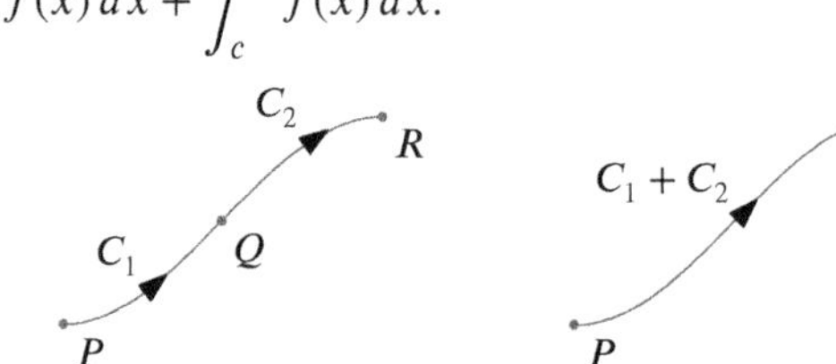

Figure 18.12: Joining two curves, C_1, and C_2, to make a new one, $C_1 + C_2$

Exercises and Problems for Section 18.1 Online Resource: Additional Problems for Section 18.1

EXERCISES

In Exercises 1–6, say whether you expect the line integral of the pictured vector field over the given curve to be positive, negative, or zero.

1.

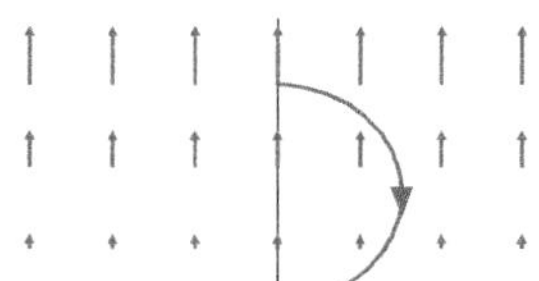

2.

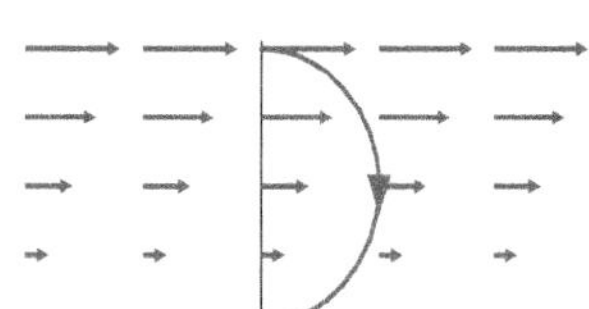

3.

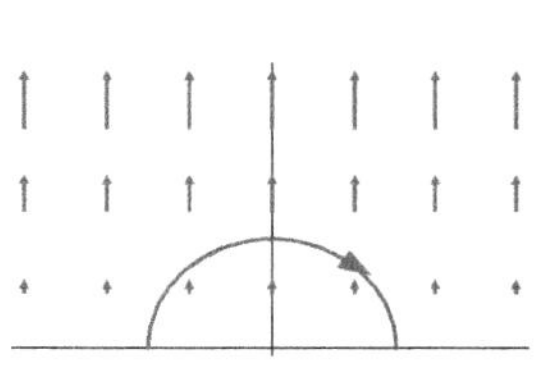

4.

5.

6.

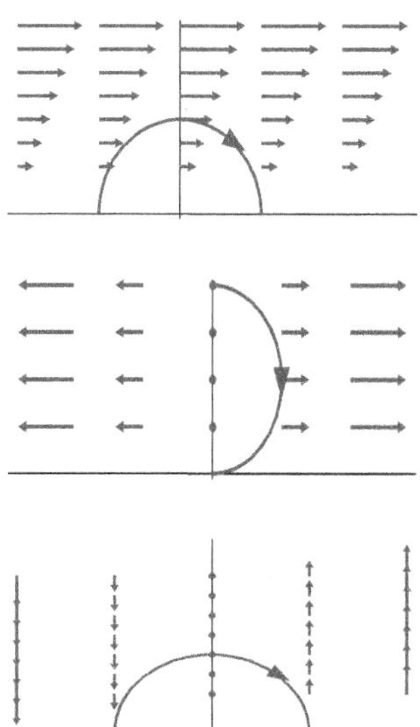

In Exercises 7–15, calculate the line integral of the vector field along the line between the given points.

7. $\vec{F} = x\vec{j}$, from (1, 0) to (3, 0)
8. $\vec{F} = x\vec{j}$, from (2, 0) to (2, 5)
9. $\vec{F} = 6\vec{i} - 7\vec{j}$, from (0, 0) to (7, 6)
10. $\vec{F} = 6\vec{i} + y^2\vec{j}$, from (3, 0) to (7, 0)
11. $\vec{F} = 3\vec{i} + 4\vec{j}$, from (0, 6) to (0, 13)
12. $\vec{F} = x\vec{i}$, from (2, 0) to (6, 0)
13. $\vec{F} = x\vec{i} + y\vec{j}$, from (2, 0) to (6, 0)
14. $\vec{F} = \vec{r} = x\vec{i} + y\vec{j}$, from (2, 2) to (6, 6)
15. $\vec{F} = x\vec{i} + 6\vec{j} - \vec{k}$, from (0, −2, 0) to (0, −10, 0)

In Exercises 16–18, find $\int_C \vec{F} \cdot d\vec{r}$ for the given $\vec{F}$ and C.

16. $\vec{F} = 5\vec{i} + 7\vec{j}$, and C is the x-axis from (−1, 0) to (−9, 0).
17. $\vec{F} = x^2\vec{i} + y^2\vec{j}$, and C is the x-axis from (2, 0) to (3, 0).
18. $\vec{F} = 6x\vec{i} + (x + y^2)\vec{j}$; C is the y-axis from (0, 3) to (0, 5).

In Exercises 19–23, calculate the line integral.

19. $\int_C (2\vec{j} + 3\vec{k}) \cdot d\vec{r}$ where C is the y-axis from the origin to the point (0, 10, 0).
20. $\int_C (2x\vec{i} + 3y\vec{j}) \cdot d\vec{r}$, where C is the line from (1, 0, 0) to (1, 0, 5).
21. $\int_C ((2y + 7)\vec{i} + 3x\vec{j}) \cdot d\vec{r}$, where C is the line from (1, 0, 0) to (5, 0, 0).
22. $\int_C (x\vec{i} + y\vec{j} + z\vec{k}) \cdot d\vec{r}$ where C is the unit circle in the xy-plane, oriented counterclockwise.
23. $\int_C (3z\vec{i} + 4x^2\vec{j} - xy\vec{k}) \cdot d\vec{r}$, where C is the line from (2, 1, 3) to (2, 1, 8).

In Exercises 24–27, find the work done by the force $\vec{F}$ along the curve C.

24. $\vec{F} = 3\vec{i} - x\vec{j}$, C is the line from (2, 6) to (9, 6).
25. $\vec{F} = y^3\vec{i} + 2xy\vec{j}$, C is the line from (−1, 0) to (−1, 3).
26. $\vec{F} = 7\vec{i} - 5\vec{j}$, C is the line from (2, −2) to (1, 6).
27. $\vec{F} = -x\vec{i} - y\vec{j}$, C is the upper half of the unit circle from (1, 0) to (−1, 0).

PROBLEMS

In Problems 28–31, let C_1 be the line from (0, 0) to (0, 1); let C_2 be the line from (1, 0) to (0, 1); let C_3 be the semicircle in the upper half plane from (−1, 0) to (1, 0). Do the line integrals of the vector field along each of the paths C_1, C_2, and C_3 appear to be positive, negative, or zero?

28.

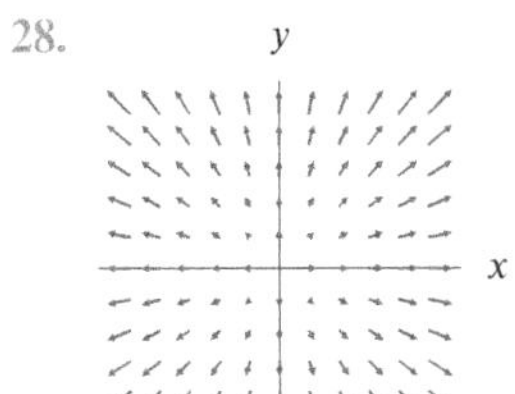

29.

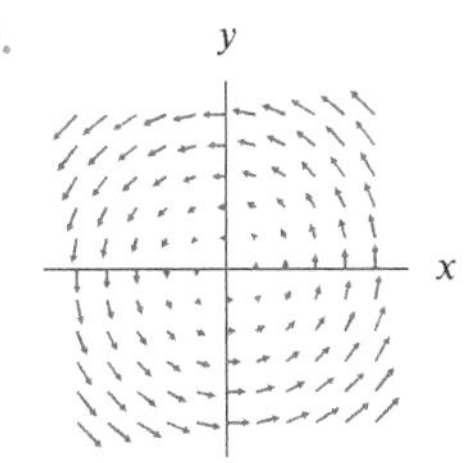

30.

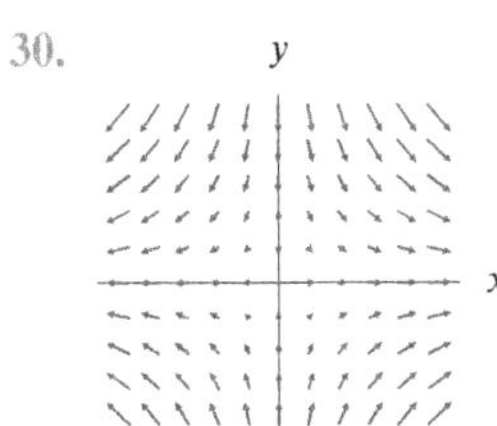

31.

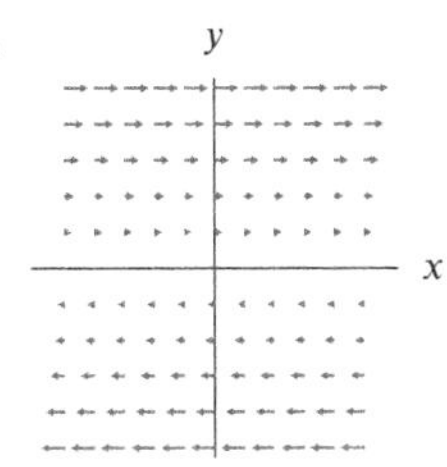

32. Consider the vector field $\vec{F}$ shown in Figure 18.13, together with the paths C_1, C_2, and C_3. Arrange the line integrals $\int_{C_1} \vec{F} \cdot d\vec{r}$, $\int_{C_2} \vec{F} \cdot d\vec{r}$ and $\int_{C_3} \vec{F} \cdot d\vec{r}$ in ascending order.

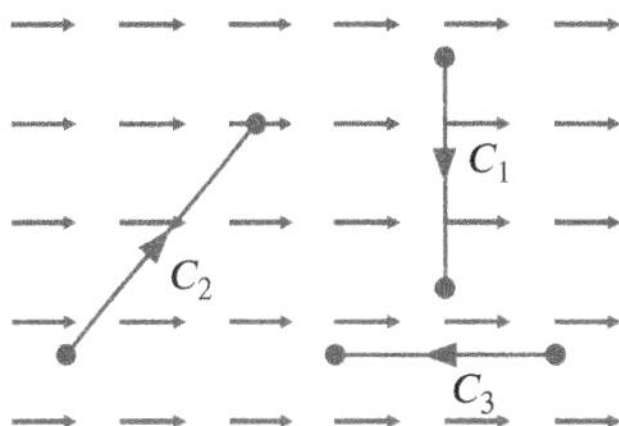

Figure 18.13

33. Compute $\int_C \vec{F} \cdot d\vec{r}$, where C is the oriented curve in Figure 18.14 and $\vec{F}$ is a vector field constant on each of the three straight segments of C:

$$\vec{F} = \begin{cases} \vec{i} & \text{on } PQ \\ 2\vec{i} - \vec{j} & \text{on } QR \\ 3\vec{i} + \vec{j} & \text{on } RS. \end{cases}$$

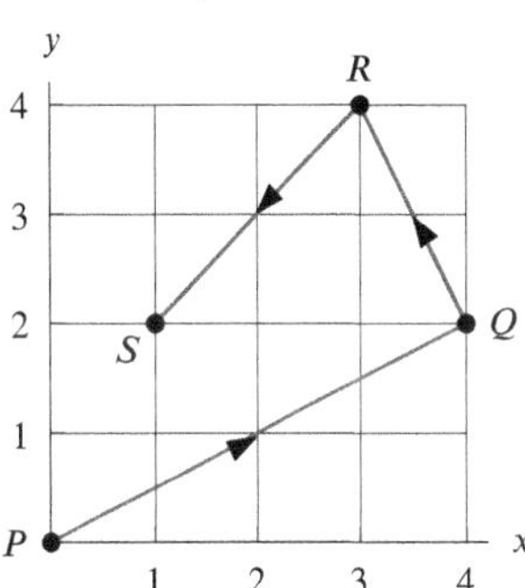

Figure 18.14

34. An object moves along the curve C in Figure 18.15 while being acted on by the force field $\vec{F}(x, y) = y\vec{i} + x^2\vec{j}$.

(a) Evaluate $\vec{F}$ at the points $(0, -1)$, $(1, -1)$, $(2, -1)$, $(3, -1)$, $(4, -1)$, $(4, 0)$, $(4, 1)$, $(4, 2)$, $(4, 3)$.
(b) Make a sketch showing the force field along C.
(c) Find the work done by $\vec{F}$ on the object.

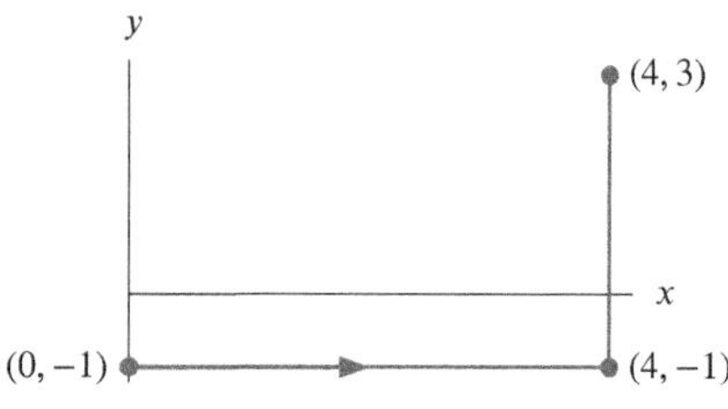

Figure 18.15

35. Let $\vec{F}$ be the constant force field $\vec{j}$ in Figure 18.16. On which of the paths C_1, C_2, C_3 is zero work done by $\vec{F}$? Explain.

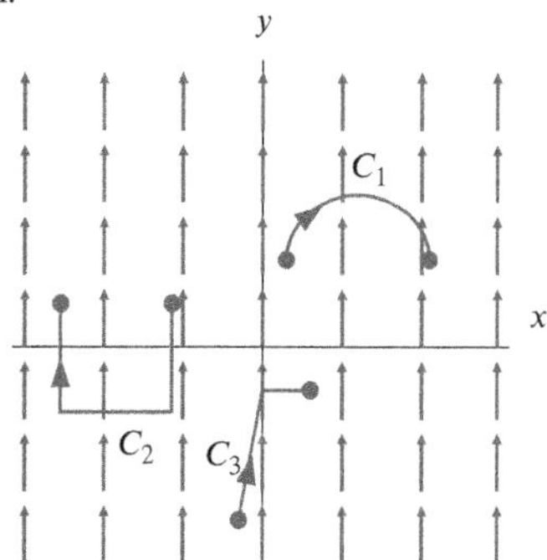

Figure 18.16

In Problems 36–40, give conditions on one or more of the constants a, b, c to ensure that the line integral $\int_C \vec{F} \cdot d\vec{r}$ has the given sign.

36. Positive for $\vec{F} = a\vec{i} + b\vec{j} + c\vec{k}$ and C is the line from the origin to $(10, 0, 0)$.

37. Positive for $\vec{F} = ay\vec{i} + c\vec{k}$ and C is the unit circle in the xy-plane, centered at the origin and oriented counterclockwise when viewed from above.

38. Negative for $\vec{F} = b\vec{j} + c\vec{k}$ and C is the parabola $y = x^2$ in the xy-plane from the origin to $(3, 9, 0)$.

39. Positive for $\vec{F} = ay\vec{i} - ax\vec{j} + (c - 1)\vec{k}$ and C is the line segment from the origin to $(1, 1, 1)$.

40. Negative for $\vec{F} = a\vec{i} + b\vec{j} - \vec{k}$ and C is the line segment from $(1, 2, 3)$ to $(1, 2, c)$.

41. (a) For each of the vector fields, $\vec{F}$, shown in Figure 18.17, sketch a curve for which the integral $\int_C \vec{F} \cdot d\vec{r}$ is positive.
(b) For which of the vector fields is it possible to make your answer to part (a) a closed curve?

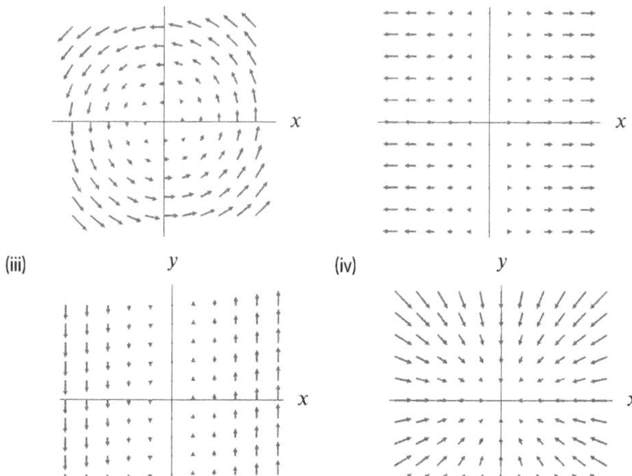

Figure 18.17

For Problems 42–46, say whether you expect the given vector field to have positive, negative, or zero circulation around the closed curve $C = C_1 + C_2 + C_3 + C_4$ in Figure 18.18. The segments C_1 and C_3 are circular arcs centered at the origin; C_2 and C_4 are radial line segments. You may find it helpful to sketch the vector field.

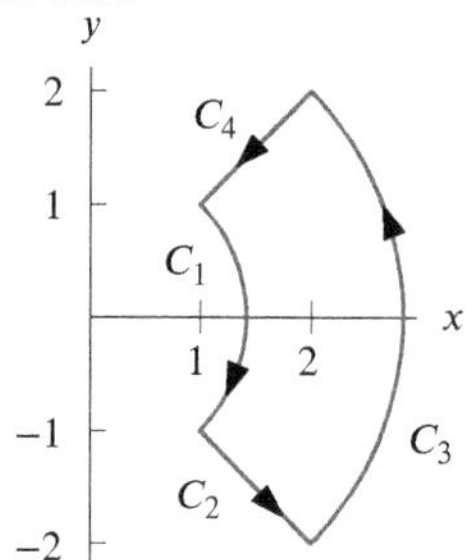

Figure 18.18

42. $\vec{F}(x, y) = x\vec{i} + y\vec{j}$

43. $\vec{F}(x, y) = -y\vec{i} + x\vec{j}$

44. $\vec{F}(x, y) = y\vec{i} - x\vec{j}$

45. $\vec{F}(x, y) = x^2\vec{i}$

46. $\vec{F}(x, y) = -\dfrac{y}{x^2 + y^2}\vec{i} + \dfrac{x}{x^2 + y^2}\vec{j}$

In Problems 47–50, C_1 and C_2 are oriented curves, and C_1 ends where C_2 begins. Find the integral given that $\int_{C_1} \vec{F} \cdot d\vec{r} = 8$, $\int_{C_1} \vec{G} \cdot d\vec{r} = 3$, $\int_{C_2} \vec{F} \cdot d\vec{r} = -5$, and $\int_{C_2} \vec{G} \cdot d\vec{r} = 15$.

47. $\displaystyle\int_{C_1} \left(\vec{F} + \vec{G}\right) \cdot d\vec{r}$

48. $\displaystyle\int_{C_2} 3\vec{G} \cdot d\vec{r}$

49. $\displaystyle\int_{C_1 + C_2} 2\vec{F} \cdot d\vec{r}$

50. $\displaystyle\int_{C_1 + C_2} \left(\vec{G} - \vec{F}\right) \cdot d\vec{r}$

51. A force $\vec{F}$ moves an object in a line from $(1, 1)$ to $(2, 4)$ with force $\vec{F} = 2\vec{i} + 3\vec{j}$, and then along a line from $(2, 4)$ to $(3, 3)$ with force $\vec{F} = \vec{i} - \vec{j}$. How much work does the force do on the object in total?

52. Find the work done by a constant force $\vec{F}$ moving an object through straight line displacement $\vec{r}$ if

(a) $\vec{F}$ is in the same direction as $\vec{r}$, $\|\vec{F}\| = 5$ newtons and $\|\vec{r}\| = 3$ meters.

(b) $\vec{F}$ and $\vec{r}$ are perpendicular.

(c) $\vec{F} = 4\vec{i} + \vec{j} + 4\vec{k}$ pounds and $\vec{r} = \vec{i} + \vec{j} + \vec{k}$ foot.

53. A horizontal square has sides of 1000 km running north-south and east-west. A wind blows from the east and decreases in magnitude toward the north at a rate of 6 meter/sec for every 500 km. Compute the circulation of the wind counterclockwise around the square.

54. Let $\vec{F} = x\vec{i} + y\vec{j}$ and let C_1 be the line joining $(1,0)$ to $(0,2)$ and let C_2 be the line joining $(0,2)$ to $(-1,0)$. Is $\int_{C_1} \vec{F} \cdot d\vec{r} = -\int_{C_2} \vec{F} \cdot d\vec{r}$? Explain.

55. The vector field $\vec{F}$ has $\|\vec{F}\| \leq 7$ everywhere and C is the circle of radius 1 centered at the origin. What is the largest possible value of $\int_C \vec{F} \cdot d\vec{r}$? The smallest possible value? What conditions lead to these values?

56. Along a curve C, a vector field $\vec{F}$ is everywhere tangent to C in the direction of orientation and has constant magnitude $\|\vec{F}\| = m$. Use the definition of the line integral to explain why

$$\int_C \vec{F} \cdot d\vec{r} = m \cdot \text{Length of } C.$$

57. Explain why the following statement is true: Whenever the line integral of a vector field around every closed curve is zero, the line integral along a curve with fixed endpoints has a constant value independent of the path taken between the endpoints.

58. Explain why the converse to the statement in Problem 57 is also true: Whenever the line integral of a vector field depends only on endpoints and not on paths, the circulation around every closed curve is zero.

In Problems 59–60, use the fact that the force of gravity on a particle of mass m at the point with position vector $\vec{r}$ is

$$\vec{F} = -\frac{GMm\vec{r}}{\|\vec{r}\|^3}$$

where G is a constant and M is the mass of the earth.

59. Calculate the work done by the force of gravity on a particle of mass m as it moves radially from 8000 km to 10,000 km from the center of the earth.

60. Calculate the work done by the force of gravity on a particle of mass m as it moves radially from 8000 km from the center of the earth to infinitely far away.

Strengthen Your Understanding

In Problems 61–62, explain what is wrong with the statement.

61. If $\vec{F}$ is a vector field and C is an oriented curve, then $\int_{-C} \vec{F} \cdot d\vec{r}$ must be less than zero.

62. It is possible that for a certain vector field $\vec{F}$ and oriented path C, we have $\int_C \vec{F} \cdot d\vec{r} = 2\vec{i} - 3\vec{j}$.

In Problems 63–64, give an example of:

63. A nonzero vector field $\vec{F}$ such that $\int_C \vec{F} \cdot d\vec{r} = 0$, where C is the straight line curve from $(0,0)$ to $(1,1)$.

64. Two oriented curves C_1 and C_2 in the plane such that, for $\vec{F}(x,y) = x\vec{j}$, we have $\int_{C_1} \vec{F} \cdot d\vec{r} > 0$ and $\int_{C_2} \vec{F} \cdot d\vec{r} < 0$.

Are the statements in Problems 65–67 true or false? Explain why or give a counterexample.

65. $\int_C \vec{F} \cdot d\vec{r}$ is a vector.

66. Suppose C_1 is the unit square joining the points $(0,0)$, $(1,0)$, $(1,1)$, $(0,1)$ oriented clockwise and C_2 is the same square but traversed twice in the opposite direction. If $\int_{C_1} \vec{F} \cdot d\vec{r} = 3$, then $\int_{C_2} \vec{F} \cdot d\vec{r} = -6$.

67. The line integral of $\vec{F} = x\vec{i} + y\vec{j} = \vec{r}$ along the semicircle $x^2 + y^2 = 1$, $y \geq 0$, oriented counterclockwise, is zero.

Are the statements in Problems 68–74 true or false? Give reasons for your answer.

68. The line integral $\int_C \vec{F} \cdot d\vec{r}$ is a scalar.

69. If C_1 and C_2 are oriented curves and C_1 is longer than C_2, then $\int_{C_1} \vec{F} \cdot d\vec{r} > \int_{C_2} \vec{F} \cdot d\vec{r}$.

70. If C is an oriented curve and $\int_C \vec{F} \cdot d\vec{r} = 0$, then $\vec{F} = \vec{0}$.

71. If $\vec{F} = \vec{i}$ is a vector field in 2-space, then $\int_C \vec{F} \cdot d\vec{r} > 0$, where C is the oriented line from $(0,0)$ to $(1,0)$.

72. If $\vec{F} = \vec{i}$ is a vector field in 2-space, then $\int_C \vec{F} \cdot d\vec{r} > 0$, where C is the oriented line from $(0,0)$ to $(0,1)$.

73. If C_1 is the upper semicircle $x^2 + y^2 = 1$, $y \geq 0$ and C_2 is the lower semicircle $x^2+y^2 = 1$, $y \leq 0$, both oriented counterclockwise, then for any vector field $\vec{F}$, we have $\int_{C_1} \vec{F} \cdot d\vec{r} = -\int_{C_2} \vec{F} \cdot d\vec{r}$.

74. The work done by the force $\vec{F} = -y\vec{i} + x\vec{j}$ on a particle moving clockwise around the boundary of the square $-1 \leq x \leq 1, -1 \leq y \leq 1$ is positive.

18.2 COMPUTING LINE INTEGRALS OVER PARAMETERIZED CURVES

The goal of this section is to show how to use a parameterization of a curve to convert a line integral into an ordinary one-variable integral.

Using a Parameterization to Evaluate a Line Integral

Recall the definition of the line integral,

$$\int_C \vec{F} \cdot d\vec{r} = \lim_{\|\Delta \vec{r}_i\| \to 0} \sum \vec{F}(\vec{r}_i) \cdot \Delta \vec{r}_i,$$

where the $\vec{r}_i$ are the position vectors of points subdividing the curve into short pieces. Now suppose we have a smooth parameterization, $\vec{r}(t)$, of C for $a \le t \le b$, so that $\vec{r}(a)$ is the position vector of the starting point of the curve and $\vec{r}(b)$ is the position vector of the end. Then we can divide C into n pieces by dividing the interval $a \le t \le b$ into n pieces, each of size $\Delta t = (b-a)/n$. See Figures 18.19 and 18.20.

At each point $\vec{r}_i = \vec{r}(t_i)$ we want to compute

$$\vec{F}(\vec{r}_i) \cdot \Delta \vec{r}_i.$$

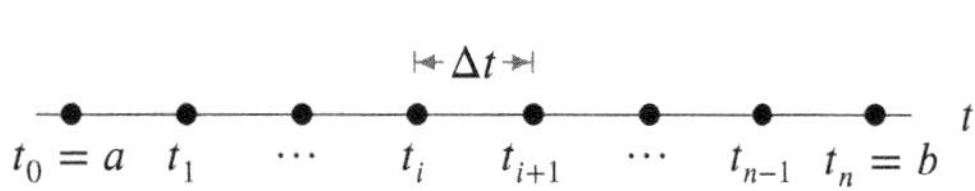

Figure 18.19: Subdivision of the interval $a \le t \le b$

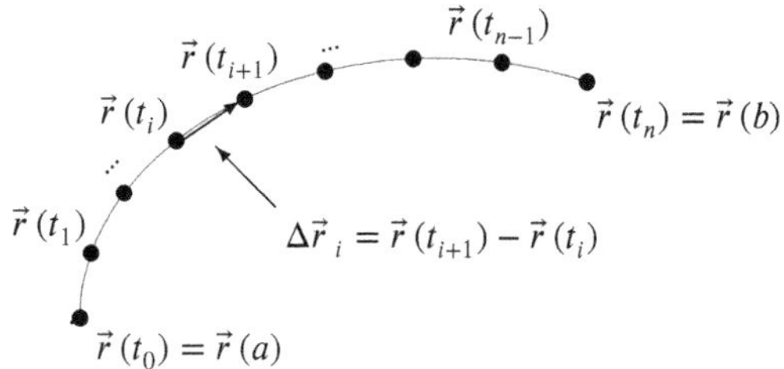

Figure 18.20: Corresponding subdivision of the parameterized path C

Since $t_{i+1} = t_i + \Delta t$, the displacement vectors $\Delta \vec{r}_i$ are given by

$$\begin{aligned}
\Delta \vec{r}_i &= \vec{r}(t_{i+1}) - \vec{r}(t_i) \\
&= \vec{r}(t_i + \Delta t) - \vec{r}(t_i) \\
&= \frac{\vec{r}(t_i + \Delta t) - \vec{r}(t_i)}{\Delta t} \cdot \Delta t \\
&\approx \vec{r}\,'(t_i)\Delta t,
\end{aligned}$$

where we use the facts that Δt is small and that $\vec{r}(t)$ is differentiable to obtain the last approximation.

Therefore,

$$\int_C \vec{F} \cdot d\vec{r} \approx \sum \vec{F}(\vec{r}_i) \cdot \Delta \vec{r}_i \approx \sum \vec{F}(\vec{r}(t_i)) \cdot \vec{r}\,'(t_i)\,\Delta t.$$

Notice that $\vec{F}(\vec{r}(t_i)) \cdot \vec{r}\,'(t_i)$ is the value at t_i of a one-variable function of t, so this last sum is really a one-variable Riemann sum. In the limit as $\Delta t \to 0$, we get a definite integral:

$$\lim_{\Delta t \to 0} \sum \vec{F}(\vec{r}(t_i)) \cdot \vec{r}\,'(t_i)\,\Delta t = \int_a^b \vec{F}(\vec{r}(t)) \cdot \vec{r}\,'(t)\,dt.$$

Thus, we have the following result:

> If $\vec{r}(t)$, for $a \le t \le b$, is a smooth parameterization of an oriented curve C and $\vec{F}$ is a vector field which is continuous on C, then
>
> $$\int_C \vec{F} \cdot d\vec{r} = \int_a^b \vec{F}(\vec{r}(t)) \cdot \vec{r}\,'(t)\,dt.$$
>
> In words: To compute the line integral of $\vec{F}$ over C, take the dot product of $\vec{F}$ evaluated on C with the velocity vector, $\vec{r}\,'(t)$, of the parameterization of C, then integrate along the curve.

Even though we assumed that C is smooth, we can use the same formula to compute line integrals over curves which are *piecewise smooth*, such as the boundary of a rectangle. If C is piecewise smooth, we apply the formula to each one of the smooth pieces and add the results by applying property 4 on page 928.

Example 1 Compute $\int_C \vec{F} \cdot d\vec{r}$ where $\vec{F} = (x+y)\vec{i} + y\vec{j}$ and C is the quarter unit circle, oriented counterclockwise as shown in Figure 18.21.

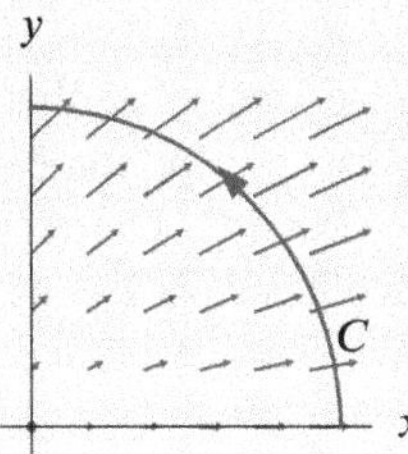

Figure 18.21: The vector field $\vec{F} = (x+y)\vec{i} + y\vec{j}$ and the quarter circle C

Solution Since most of the vectors in $\vec{F}$ along C point generally in a direction opposite to the orientation of C, we expect our answer to be negative. The first step is to parameterize C by

$$\vec{r}(t) = x(t)\vec{i} + y(t)\vec{j} = \cos t\vec{i} + \sin t\vec{j}, \quad 0 \le t \le \frac{\pi}{2}.$$

Substituting the parameterization into $\vec{F}$, we get $\vec{F}(x(t), y(t)) = (\cos t + \sin t)\vec{i} + \sin t\vec{j}$. The vector $\vec{r}\,'(t) = x'(t)\vec{i} + y'(t)\vec{j} = -\sin t\vec{i} + \cos t\vec{j}$. Then

$$\begin{aligned}\int_C \vec{F} \cdot d\vec{r} &= \int_0^{\pi/2} ((\cos t + \sin t)\vec{i} + \sin t\vec{j}) \cdot (-\sin t\vec{i} + \cos t\vec{j})dt \\ &= \int_0^{\pi/2} (-\cos t \sin t - \sin^2 t + \sin t \cos t)dt \\ &= \int_0^{\pi/2} -\sin^2 t\, dt = -\frac{\pi}{4} \approx -0.7854.\end{aligned}$$

So the answer is negative, as expected.

Example 2 Consider the vector field $\vec{F} = x\vec{i} + y\vec{j}$.

(a) Suppose C_1 is the line segment joining $(1,0)$ to $(0,2)$ and C_2 is a part of a parabola with its vertex at $(0,2)$, joining the same points in the same order. (See Figure 18.22.) Verify that

$$\int_{C_1} \vec{F} \cdot d\vec{r} = \int_{C_2} \vec{F} \cdot d\vec{r}.$$

(b) If C is the triangle shown in Figure 18.23, show that $\int_C \vec{F} \cdot d\vec{r} = 0$.

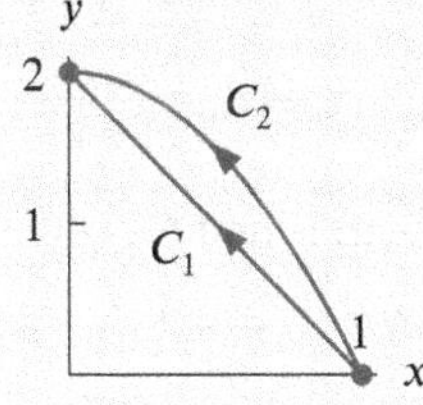

Figure 18.22

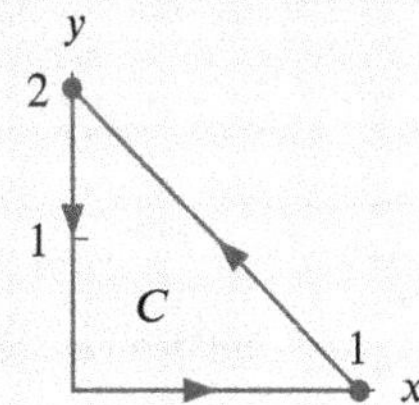

Figure 18.23

Solution (a) We parameterize C_1 by $\vec{r}(t) = (1-t)\vec{i} + 2t\vec{j}$ with $0 \le t \le 1$. Then $\vec{r}\,'(t) = -\vec{i} + 2\vec{j}$, so

$$\int_{C_1} \vec{F} \cdot d\vec{r} = \int_0^1 \vec{F}(1-t, 2t) \cdot (-\vec{i} + 2\vec{j})\, dt = \int_0^1 ((1-t)\vec{i} + 2t\vec{j}) \cdot (-\vec{i} + 2\vec{j})\, dt$$
$$= \int_0^1 (5t-1)\, dt = \frac{3}{2}.$$

To parameterize C_2, we use the fact that it is part of a parabola with vertex at $(0,2)$, so its equation is of the form $y = -kx^2 + 2$ for some k. Since the parabola crosses the x-axis at $(1,0)$, we find that $k = 2$ and $y = -2x^2 + 2$. Therefore, we use the parameterization $\vec{r}(t) = t\vec{i} + (-2t^2 + 2)\vec{j}$ with $0 \le t \le 1$, which has $\vec{r}\,' = \vec{i} - 4t\vec{j}$. This traces out C_2 in reverse, since $t = 0$ gives $(0, 2)$, and $t = 1$ gives $(1, 0)$. Thus, we make $t = 0$ the upper limit of integration and $t = 1$ the lower limit:

$$\int_{C_2} \vec{F} \cdot d\vec{r} = \int_1^0 \vec{F}(t, -2t^2+2) \cdot (\vec{i} - 4t\vec{j})\, dt = -\int_0^1 (t\vec{i} + (-2t^2+2)\vec{j}) \cdot (\vec{i} - 4t\vec{j})\, dt$$
$$= -\int_0^1 (8t^3 - 7t)\, dt = \frac{3}{2}.$$

So the line integrals of $\vec{F}$ along C_1 and C_2 have the same value.

(b) We break $\int_C \vec{F} \cdot d\vec{r}$ into three pieces, one of which we have already computed (namely, the piece connecting $(1, 0)$ to $(0, 2)$, where the line integral has value $3/2$). The piece running from $(0, 2)$ to $(0, 0)$ can be parameterized by $\vec{r}(t) = (2-t)\vec{j}$ with $0 \le t \le 2$. The piece running from $(0, 0)$ to $(1, 0)$ can be parameterized by $\vec{r}(t) = t\vec{i}$ with $0 \le t \le 1$. Then

$$\int_C \vec{F} \cdot d\vec{r} = \frac{3}{2} + \int_0^2 \vec{F}(0, 2-t) \cdot (-\vec{j})\, dt + \int_0^1 \vec{F}(t, 0) \cdot \vec{i}\, dt$$
$$= \frac{3}{2} + \int_0^2 (2-t)\vec{j} \cdot (-\vec{j})\, dt + \int_0^1 t\vec{i} \cdot \vec{i}\, dt$$
$$= \frac{3}{2} + \int_0^2 (t-2)\, dt + \int_0^1 t\, dt = \frac{3}{2} + (-2) + \frac{1}{2} = 0.$$

Example 3 Let C be the closed curve consisting of the upper half-circle of radius 1 and the line forming its diameter along the x-axis, oriented counterclockwise. (See Figure 18.24.) Find $\int_C \vec{F} \cdot d\vec{r}$ where $\vec{F}(x, y) = -y\vec{i} + x\vec{j}$.

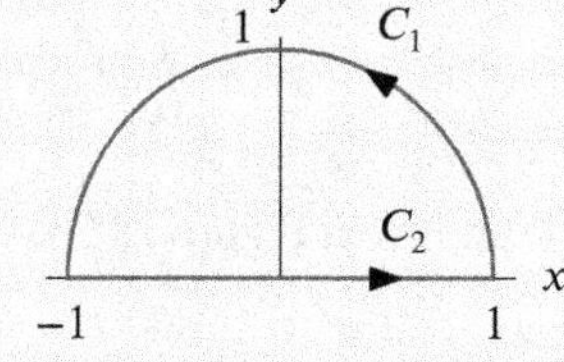

Figure 18.24: The curve $C = C_1 + C_2$ for Example 3

Solution We write $C = C_1 + C_2$ where C_1 is the half-circle and C_2 is the line, and compute $\int_{C_1} \vec{F} \cdot d\vec{r}$ and $\int_{C_2} \vec{F} \cdot d\vec{r}$ separately. We parameterize C_1 by $\vec{r}(t) = \cos t\vec{i} + \sin t\vec{j}$, with $0 \le t \le \pi$. Then

$$\int_{C_1} \vec{F} \cdot d\vec{r} = \int_0^\pi (-\sin t\vec{i} + \cos t\vec{j}) \cdot (-\sin t\vec{i} + \cos t\vec{j})\, dt$$

$$= \int_0^{\pi} (\sin^2 t + \cos^2 t)\, dt = \int_0^{\pi} 1\, dt = \pi.$$

For C_2, we have $\int_{C_2} \vec{F} \cdot d\vec{r} = 0$, since the vector field $\vec{F}$ has no $\vec{i}$ component along the x-axis (where $y = 0$) and is therefore perpendicular to C_2 at all points.

Finally, we can write

$$\int_C \vec{F} \cdot d\vec{r} = \int_{C_1} \vec{F} \cdot d\vec{r} + \int_{C_2} \vec{F} \cdot d\vec{r} = \pi + 0 = \pi.$$

It is no accident that the result for $\int_{C_1} \vec{F} \cdot d\vec{r}$ is the same as the length of the curve C_1. See Problem 56 on page 931 and Problem 41 on page 938.

The next example illustrates the computation of a line integral over a path in 3-space.

Example 4 A particle travels along the helix C given by $\vec{r}(t) = \cos t\vec{i} + \sin t\vec{j} + 2t\vec{k}$ and is subject to a force $\vec{F} = x\vec{i} + z\vec{j} - xy\vec{k}$. Find the total work done on the particle by the force for $0 \le t \le 3\pi$.

Solution The work done is given by a line integral, which we evaluate using the given parameterization:

$$\begin{aligned}
\text{Work done} = \int_C \vec{F} \cdot d\vec{r} &= \int_0^{3\pi} \vec{F}(\vec{r}(t)) \cdot \vec{r}\,'(t)\, dt \\
&= \int_0^{3\pi} (\cos t\vec{i} + 2t\vec{j} - \cos t \sin t\vec{k}) \cdot (-\sin t\vec{i} + \cos t\vec{j} + 2\vec{k})\, dt \\
&= \int_0^{3\pi} (-\cos t \sin t + 2t \cos t - 2\cos t \sin t)\, dt \\
&= \int_0^{3\pi} (-3\cos t \sin t + 2t \cos t)\, dt = -4.
\end{aligned}$$

The Differential Notation $\int_C P\,dx + Q\,dy + R dz$

There is an alternative notation for line integrals that is often useful. For the vector field $\vec{F} = P(x, y, z)\vec{i} + Q(x, y, z)\vec{j} + R(x, y, z)\vec{k}$ and an oriented curve C, if we write $d\vec{r} = dx\vec{i} + dy\vec{j} + dz\vec{k}$ we have

$$\int_C \vec{F} \cdot d\vec{r} = \int_C P(x, y, z)dx + Q(x, y, z)dy + R(x, y, z)dz.$$

Example 5 Evaluate $\int_C xy\,dx - y^2\,dy$ where C is the line segment from $(0, 0)$ to $(2, 6)$.

Solution We parameterize C by $x = t, y = 3t,\ 0 \le t \le 2$. Thus, $dx = dt, dy = 3dt$, so

$$\int_C xy\,dx - y^2\,dy = \int_0^2 t(3t)dt - (3t)^2(3dt) = \int_0^2 (-24t^2)\, dt = -64.$$

Line integrals can be expressed either using vectors or using differentials. If the independent variables are distances, then visualizing a line integral in terms of dot products can be useful. However, if the independent variables are, for example, temperature and volume, then the dot product does not have physical meaning, so differentials are more natural.

Independence of Parameterization

Since there are many different ways of parameterizing a given oriented curve, you may be wondering what happens to the value of a given line integral if you choose another parameterization. The answer is that the choice of parameterization makes no difference. Since we initially defined the line integral without reference to any particular parameterization, this is exactly as we would expect.

Example 6 Consider the oriented path which is a straight-line segment L running from $(0,0)$ to $(1,1)$. Calculate the line integral of the vector field $\vec{F} = (3x-y)\vec{i} + x\vec{j}$ along L using each of the parameterizations

(a) $A(t) = (t,t), \quad 0 \le t \le 1,$ (b) $D(t) = (e^t-1, e^t-1), \quad 0 \le t \le \ln 2.$

Solution The line L has equation $y = x$. Both $A(t)$ and $D(t)$ give a parameterization of L: each has both coordinates equal and each begins at (0,0) and ends at (1,1). Now let's calculate the line integral of the vector field $\vec{F} = (3x-y)\vec{i} + x\vec{j}$ using each parameterization.

(a) Using $A(t)$, we get

$$\int_L \vec{F} \cdot d\vec{r} = \int_0^1 ((3t-t)\vec{i} + t\vec{j}) \cdot (\vec{i} + \vec{j})\, dt = \int_0^1 3t\, dt = \frac{3t^2}{2}\bigg|_0^1 = \frac{3}{2}.$$

(b) Using $D(t)$, we get

$$\begin{aligned}\int_L \vec{F} \cdot d\vec{r} &= \int_0^{\ln 2} \left((3(e^t-1)-(e^t-1))\vec{i} + (e^t-1)\vec{j}\right) \cdot (e^t\vec{i} + e^t\vec{j})\, dt \\ &= \int_0^{\ln 2} 3(e^{2t}-e^t)\, dt = 3\left(\frac{e^{2t}}{2} - e^t\right)\bigg|_0^{\ln 2} = \frac{3}{2}.\end{aligned}$$

The fact that both answers are the same illustrates that the value of a line integral is independent of the parameterization of the path. Problems 59–61 (available online) give another way of seeing this.

Exercises and Problems for Section 18.2 Online Resource: Additional Problems for Section 18.2

EXERCISES

In Exercises 1–3, write $\int_C \vec{F} \cdot d\vec{r}$ in the form $\int_a^b g(t)\, dt$. (Give a formula for g and numbers for a and b. You do not need to evaluate the integral.)

1. $\vec{F} = y\vec{i} + x\vec{j}$ and C is the semicircle from $(0,1)$ to $(0,-1)$ with $x > 0$.
2. $\vec{F} = x\vec{i} + z^2\vec{k}$ and C is the line from $(0,1,0)$ to $(2,3,2)$.
3. $\vec{F} = (\cos x)\vec{i} + (\cos y)\vec{j} + (\cos z)\vec{k}$ and C is the unit circle in the plane $z = 10$, centered on the z-axis and oriented counterclockwise when viewed from above.

In Exercises 4–8, find the line integral.

4. $\int_C (3\vec{i} + (y+5)\vec{j}) \cdot d\vec{r}$ where C is the line from $(0,0)$ to $(0,3)$.
5. $\int_C (2x\vec{i} + 3y\vec{j}) \cdot d\vec{r}$ where C is the line from $(1,0,0)$ to $(5,0,0)$.
6. $\int_C (2y^2\vec{i} + x\vec{j}) \cdot d\vec{r}$ where C is the line segment from $(3,1)$ to $(0,0)$.
7. $\int_C (x\vec{i} + y\vec{j}) \cdot d\vec{r}$ where C is the semicircle with center at $(2,0)$ and going from $(3,0)$ to $(1,0)$ in the region $y > 0$.
8. Find $\int_C ((x^2+y)\vec{i} + y^3\vec{j}) \cdot d\vec{r}$ where C consists of the three line segments from $(4,0,0)$ to $(4,3,0)$ to $(0,3,0)$ to $(0,3,5)$.

In Exercises 9–23, find $\int_C \vec{F} \cdot d\vec{r}$ for the given $\vec{F}$ and C.

9. $\vec{F} = 2\vec{i} + \vec{j}$; C is the x-axis from $x = 10$ to $x = 7$.
10. $\vec{F} = 3\vec{j} - \vec{i}$; C is the line $y = x$ from $(1,1)$ to $(5,5)$.
11. $\vec{F} = x\vec{i} + y\vec{j}$ and C is the line from $(0,0)$ to $(3,3)$.
12. $\vec{F} = y\vec{i} - x\vec{j}$ and C is the right-hand side of the unit circle, starting at $(0,1)$.
13. $\vec{F} = x^2\vec{i} + y^2\vec{j}$ and C is the line from the point $(1,2)$ to the point $(3,4)$.
14. $\vec{F} = -y\sin x\vec{i} + \cos x\vec{j}$ and C is the parabola $y = x^2$ between $(0,0)$ and $(2,4)$.
15. $\vec{F} = y^3\vec{i} + x^2\vec{j}$ and C is the line from $(0,0)$ to $(3,2)$.

16. $\vec{F} = 2y\vec{i} - (\sin y)\vec{j}$ counterclockwise around the unit circle C starting at the point $(1, 0)$.

17. $\vec{F} = \ln y\vec{i} + \ln x\vec{j}$ and C is the curve given parametrically by $(2t, t^3)$, for $2 \le t \le 4$.

18. $\vec{F} = x\vec{i} + 6\vec{j} - \vec{k}$, and C is the line $x = y = z$ from $(0, 0, 0)$ to $(2, 2, 2)$.

19. $\vec{F} = (2x-y+4)\vec{i} + (5y+3x-6)\vec{j}$ and C is the triangle with vertices $(0, 0), (3, 0), (3, 2)$ traversed counterclockwise.

20. $\vec{F} = x\vec{i} + 2zy\vec{j} + x\vec{k}$ and C is $\vec{r} = t\vec{i} + t^2\vec{j} + t^3\vec{k}$ for $1 \le t \le 2$.

21. $\vec{F} = x^3\vec{i} + y^2\vec{j} + z\vec{k}$ and C is the line from the origin to the point $(2, 3, 4)$.

22. $\vec{F} = -y\vec{i} + x\vec{j} + 5\vec{k}$ and C is the helix $x = \cos t, y = \sin t, z = t$, for $0 \le t \le 4\pi$.

23. $\vec{F} = e^y\vec{i} + \ln(x^2+1)\vec{j} + \vec{k}$ and C is the circle of radius 2 centered at the origin in the yz-plane in Figure 18.25.

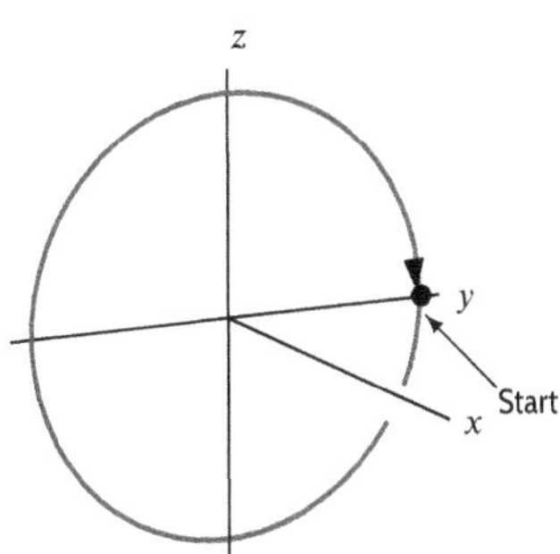

Figure 18.25

24. Every line integral can be written in both vector notation and differential notation. For example,

$$\int_C (2x\vec{i} + (x+y)\vec{j}) \cdot d\vec{r} = \int_C 2x\,dx + (x+y)\,dy.$$

(a) Express $\int_C 3\,dx + xy\,dy$ in vector notation.
(b) Express $\int_C (100\cos x\vec{i} + e^y \sin x\vec{j}) \cdot d\vec{r}$ in differential notation.

In Exercises 25–26, express the line integral $\int_C \vec{F} \cdot d\vec{r}$ in differential notation.

25. $\vec{F} = 3x\vec{i} - y\sin x\vec{j}$

26. $\vec{F} = y^2\vec{i} + z^2\vec{j} + (x^2-5)\vec{k}$

In Exercises 27–28, find $\vec{F}$ so that the line integral equals $\int_C \vec{F} \cdot d\vec{r}$.

27. $\int_C (x+2y)dx + x^2y\,dy$

28. $\int_C e^{-3y}\,dx - yz(\sin x)\,dy + (y+z)\,dz$

Evaluate the line integrals in Exercises 29–34.

29. $\int_C y\,dx + x\,dy$ where C is the parameterized path $x = t^2, y = t^3, 1 \le t \le 5$.

30. $\int_C dx + y\,dy + z\,dz$ where C is one turn of the helix $x = \cos t,\ y = \sin t,\ z = 3t,\ 0 \le t \le 2\pi$.

31. $\int_C 3y\,dx + 4x\,dy$ where C is the straight-line path from $(1, 3)$ to $(5, 9)$.

32. $\int_C x\,dx + z\,dy - y\,dz$ where C is the circle of radius 3 in the yz-plane centered at the origin, oriented counterclockwise when viewed from the positive y-axis.

33. $\int_C (x+y)\,dx + x^2\,dy$ where C is the path $x = t^2, y = t, 0 \le t \le 10$.

34. $\int_C x\,dy$ where C is the quarter circle centered at the origin going counterclockwise from $(2, 0)$ to $(0, 2)$.

PROBLEMS

35. Evaluate the line integral of $\vec{F} = (3x-y)\vec{i} + x\vec{j}$ over two different paths from $(0, 0)$ to $(1, 1)$.

(a) The path (t, t^2), with $0 \le t \le 1$
(b) The path (t^2, t), with $0 \le t \le 1$

36. Curves C_1 and C_2 are parameterized as follows:

$$C_1 \text{ is } (x(t), y(t)) = (0, t) \quad \text{for } -1 \le t \le 1$$
$$C_2 \text{ is } (x(t), y(t)) = (\cos t, \sin t) \quad \text{for } \frac{\pi}{2} \le t \le \frac{3\pi}{2}.$$

(a) Sketch C_1 and C_2 with arrows showing their orientation.
(b) Suppose $\vec{F} = (x+3y)\vec{i} + y\vec{j}$. Calculate $\int_C \vec{F} \cdot d\vec{r}$, where C is the curve given by $C = C_1 + C_2$.

37. Calculate the line integral of $\vec{F} = (3x-y)\vec{i} + x\vec{j}$ along the line segment L from $(0, 0)$ to $(1, 1)$ using each of the parameterizations

(a) $B(t) = (2t, 2t), \quad 0 \le t \le 1/2$
(b) $C(t) = \left(\frac{t^2-1}{3}, \frac{t^2-1}{3}\right), \quad 1 \le t \le 2$

In Problems 38–39 evaluate the line integral using a shortcut available in two special situations.

- If $\vec{F}$ is perpendicular to C at every point of C, then

$$\int_C \vec{F} \cdot d\vec{r} = 0.$$

- If $\vec{F}$ is tangent to C at every point of C and has constant magnitude on C, then

$$\int_C \vec{F} \cdot d\vec{r} = \pm\|\vec{F}\| \times \text{Length of } C.$$

Choose the + sign if $\vec{F}$ points in the direction of integration and choose the − sign if $\vec{F}$ points in the direction opposite to the direction of integration.

38. $\int_C (x\vec{i} + y\vec{j}) \cdot d\vec{r}$, where C is the circle of radius 10 centered at the origin, oriented counterclockwise.

39. $\int_C (-y\vec{i} + x\vec{j}) \cdot d\vec{r}$, where C is the circle of radius 10 centered at the origin, oriented counterclockwise.

40. Justify the following without parameterizing the paths.

 (a) $\int_C (x^2 + \cos y)\, dy = 0$ where C is the straight line path from $(10, 5)$ to $(20, 5)$.
 (b) $\int_C (x^2 + \cos y)\, dx = 0$ where C is the straight line path from $(10, 5)$ to $(10, 50)$.

41. Let $\vec{F} = -y\vec{i} + x\vec{j}$ and let C be the unit circle oriented counterclockwise.

 (a) Show that $\vec{F}$ has a constant magnitude of 1 on C.
 (b) Show that $\vec{F}$ is always tangent to the circle C.
 (c) Show that $\int_C \vec{F} \cdot d\vec{r}$ = Length of C.

42. A spiral staircase in a building is in the shape of a helix of radius 5 meters. Between two floors of the building, the stairs make one full revolution and climb by 4 meters. A person carries a bag of groceries up two floors. The combined mass of the person and the groceries is $m = 70$ kg and the gravitational force is mg downward, where $g = 9.8$ m/sec^2 is the acceleration due to gravity. Calculate the work done by the person against gravity.

43. If C is $\vec{r} = (t+1)\vec{i} + 2t\vec{j} + 3t\vec{k}$ for $0 \le t \le 1$, we know $\int_C \vec{F}(\vec{r}) \cdot d\vec{r} = 5$. Find the value of the integrals:

 (a) $\int_1^0 \vec{F}((t+1)\vec{i} + 2t\vec{j} + 3t\vec{k}) \cdot (\vec{i} + 2\vec{j} + 3\vec{k})\, dt$
 (b) $\int_0^1 \vec{F}((t^2+1)\vec{i} + 2t^2\vec{j} + 3t^2\vec{k}) \cdot (2t\vec{i} + 4t\vec{j} + 6t\vec{k})\, dt$
 (c) $\int_{-1}^1 \vec{F}((t^2+1)\vec{i} + 2t^2\vec{j} + 3t^2\vec{k}) \cdot (2t\vec{i} + 4t\vec{j} + 6t\vec{k})\, dt$

Strengthen Your Understanding

In Problems 44–45, explain what is wrong with the statement.

44. For the vector field $\vec{F} = x\vec{i} - y\vec{j}$ and oriented path C parameterized by $x = \cos t$, $y = \sin t$, $0 \le t \le \pi/2$, we have

$$\int_C \vec{F} \cdot d\vec{r} = \int_0^{\pi/2} (\cos t\vec{i} - \sin t\vec{j}) \cdot (\cos t\vec{i} + \sin t\vec{j})\, dt.$$

45. If $\int_C \vec{F} \cdot d\vec{r} = 0$, then $\vec{F}$ is perpendicular to C at every point on C.

In Problems 46–47, give an example of:

46. A vector field $\vec{F}$ such that, for the parameterized path $\vec{r}(t) = 3\cos t\vec{i} + 3\sin t\vec{j}$, $-\pi/2 \le t \le \pi/2$, the integral $\int_C \vec{F} \cdot d\vec{r}$ can be computed geometrically, without using the parameterization.

47. A parameterized path C such that, for the vector field $\vec{F}(x, y) = \sin y\vec{i}$, the integral $\int_C \vec{F} \cdot d\vec{r}$ is nonzero and can be computed geometrically, without using the parameterization.

Are the statements in Problems 48–56 true or false? Give reasons for your answer.

48. If C_1 and C_2 are oriented curves with C_2 beginning where C_1 ends, then $\int_{C_1+C_2} \vec{F} \cdot d\vec{r} > \int_{C_1} \vec{F} \cdot d\vec{r}$.

49. The line integral $\int_C 4\vec{i} \cdot d\vec{r}$ over the curve C parameterized by $\vec{r}(t) = t\vec{i} + t^2\vec{j}$, for $0 \le t \le 2$, is positive.

50. If C_1 is the curve parameterized by $\vec{r}_1(t) = \cos t\vec{i} + \sin t\vec{j}$, with $0 \le t \le \pi$, and C_2 is the curve parameterized by $\vec{r}_2(t) = \cos t\vec{i} - \sin t\vec{j}$, $0 \le t \le \pi$, then for any vector field $\vec{F}$ we have $\int_{C_1} \vec{F} \cdot d\vec{r} = \int_{C_2} \vec{F} \cdot d\vec{r}$.

51. If C_1 is the curve parameterized by $\vec{r}_1(t) = \cos t\vec{i} + \sin t\vec{j}$, with $0 \le t \le \pi$, and C_2 is the curve parameterized by $\vec{r}_2(t) = \cos(2t)\vec{i} + \sin(2t)\vec{j}$, $0 \le t \le \frac{\pi}{2}$, then for any vector field $\vec{F}$ we have $\int_{C_1} \vec{F} \cdot d\vec{r} = \int_{C_2} \vec{F} \cdot d\vec{r}$.

52. If C is the curve parameterized by $\vec{r}(t)$, for $a \le t \le b$ with $\vec{r}(a) = \vec{r}(b)$, then $\int_C \vec{F} \cdot d\vec{r} = 0$ for any vector field $\vec{F}$. (Note that C starts and ends at the same place.)

53. If C_1 is the line segment from $(0, 0)$ to $(1, 0)$ and C_2 is the line segment from $(0, 0)$ to $(2, 0)$, then for any vector field $\vec{F}$, we have $\int_{C_2} \vec{F} \cdot d\vec{r} = 2\int_{C_1} \vec{F} \cdot d\vec{r}$.

54. If C is a circle of radius a, centered at the origin and oriented counterclockwise, then $\int_C (2x\vec{i} + y\vec{j}) \cdot d\vec{r} = 0$.

55. If C is a circle of radius a, centered at the origin and oriented counterclockwise, then $\int_C (2y\vec{i} + x\vec{j}) \cdot d\vec{r} = 0$.

56. If C_1 is the curve parameterized by $\vec{r}_1(t) = t\vec{i} + t^2\vec{j}$, with $0 \le t \le 2$, and C_2 is the curve parameterized by $\vec{r}_2(t) = (2-t)\vec{i} + (2-t)^2\vec{j}$, $0 \le t \le 2$, then for any vector field $\vec{F}$ we have $\int_{C_1} \vec{F} \cdot d\vec{r} = -\int_{C_2} \vec{F} \cdot d\vec{r}$.

57. If C_1 is the path parameterized by $\vec{r}_1(t) = (t, t)$, $0 \le t \le 1$, and if C_2 is the path parameterized by $\vec{r}_2(t) = (1-t, 1-t)$, $0 \le t \le 1$, and if $\vec{F} = x\vec{i} + y\vec{j}$, which of the following is true?

 (a) $\int_{C_1} \vec{F} \cdot d\vec{r} > \int_{C_2} \vec{F} \cdot d\vec{r}$
 (b) $\int_{C_1} \vec{F} \cdot d\vec{r} < \int_{C_2} \vec{F} \cdot d\vec{r}$
 (c) $\int_{C_1} \vec{F} \cdot d\vec{r} = \int_{C_2} \vec{F} \cdot d\vec{r}$

58. If C_1 is the path parameterized by $\vec{r}_1(t) = (t, t)$, for $0 \le t \le 1$, and if C_2 is the path parameterized by $\vec{r}_2(t) = (\sin t, \sin t)$, for $0 \le t \le 1$, and if $\vec{F} = x\vec{i} + y\vec{j}$, which of the following is true?

 (a) $\int_{C_1} \vec{F} \cdot d\vec{r} > \int_{C_2} \vec{F} \cdot d\vec{r}$
 (b) $\int_{C_1} \vec{F} \cdot d\vec{r} < \int_{C_2} \vec{F} \cdot d\vec{r}$
 (c) $\int_{C_1} \vec{F} \cdot d\vec{r} = \int_{C_2} \vec{F} \cdot d\vec{r}$

18.3 GRADIENT FIELDS AND PATH-INDEPENDENT FIELDS

For a function, f, of one variable, the Fundamental Theorem of Calculus tells us that the definite integral of a rate of change, f', gives the total change in f:

$$\int_a^b f'(t)\,dt = f(b) - f(a).$$

What about functions of two or more variables? The quantity that describes the rate of change is the gradient vector field. If we know the gradient of a function f, can we compute the total change in f between two points? The answer is yes, using a line integral.

Finding the Total Change in f from grad f: The Fundamental Theorem

To find the change in f between two points P and Q, we choose a smooth path C from P to Q, then divide the path into many small pieces. See Figure 18.26.

First we estimate the change in f as we move through a displacement $\Delta\vec{r}_i$ from $\vec{r}_i$ to $\vec{r}_{i+1}$. Suppose $\vec{u}$ is a unit vector in the direction of $\Delta\vec{r}_i$. Then the change in f is given by

$$\begin{aligned} f(\vec{r}_{i+1}) - f(\vec{r}_i) &\approx \text{Rate of change of } f \times \text{Distance moved in direction of } \vec{u} \\ &= f_{\vec{u}}(\vec{r}_i)\|\Delta\vec{r}_i\| \\ &= \operatorname{grad} f \cdot \vec{u}\,\|\Delta\vec{r}_i\| \\ &= \operatorname{grad} f \cdot \Delta\vec{r}_i. \qquad \text{since } \Delta\vec{r}_i = \|\Delta\vec{r}_i\|\vec{u} \end{aligned}$$

Therefore, summing over all pieces of the path, the total change in f is given by

$$\text{Total change} = f(Q) - f(P) \approx \sum_{i=0}^{n-1} \operatorname{grad} f(\vec{r}_i)\cdot\Delta\vec{r}_i.$$

In the limit as $\|\Delta\vec{r}_i\|$ approaches zero, this suggests the following result:

Theorem 18.1: The Fundamental Theorem of Calculus for Line Integrals

Suppose C is a piecewise smooth oriented path with starting point P and ending point Q. If f is a function whose gradient is continuous on the path C, then

$$\int_C \operatorname{grad} f \cdot d\vec{r} = f(Q) - f(P).$$

Notice that there are many different paths from P to Q. (See Figure 18.27.) However, the value of the line integral $\int_C \operatorname{grad} f \cdot d\vec{r}$ depends only on the endpoints of C; it does not depend on where C goes in between. Problem 88 (available online) shows how the Fundamental Theorem for Line Integrals can be derived from the one-variable Fundamental Theorem of Calculus.

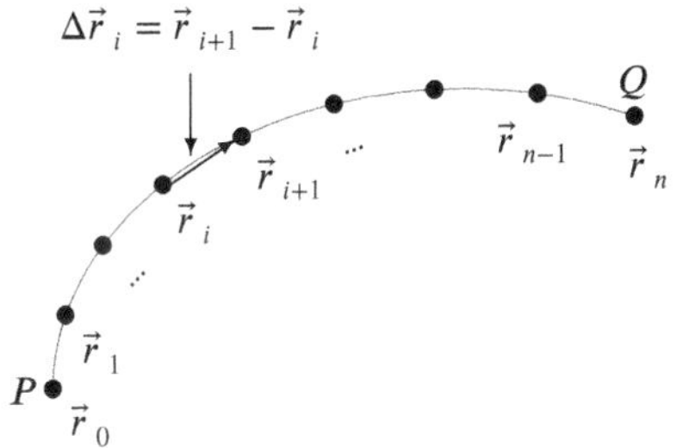

Figure 18.26: Subdivision of the path from P to Q. We estimate the change in f along $\Delta\vec{r}_i$

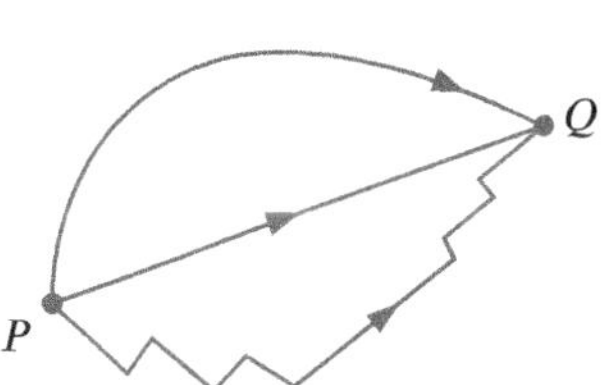

Figure 18.27: There are many different paths from P to Q: all give the same value of $\int_C \operatorname{grad} f \cdot d\vec{r}$

Example 1 Suppose that grad f is everywhere perpendicular to the curve joining P and Q shown in Figure 18.28.

(a) Explain why you expect the path joining P and Q to be a contour.

(b) Using a line integral, show that $f(P) = f(Q)$.

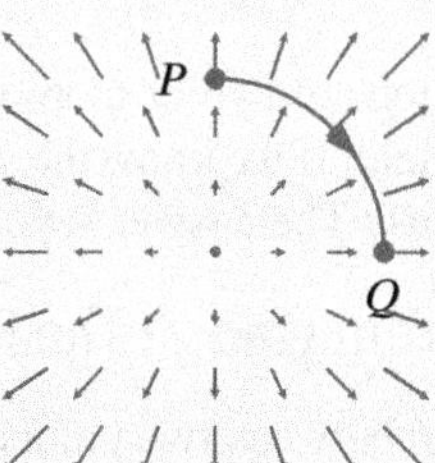

Figure 18.28: The gradient vector field of the function f

Solution (a) The gradient of f is everywhere perpendicular to the path from P to Q, as you expect along a contour.

(b) Consider the path from P to Q shown in Figure 18.28 and evaluate the line integral

$$\int_C \text{grad}\, f \cdot d\vec{r} = f(Q) - f(P).$$

Since grad f is everywhere perpendicular to the path, the line integral is 0. Thus, $f(Q) = f(P)$.

Example 2 Consider the vector field $\vec{F} = x\vec{i} + y\vec{j}$. In Example 2 on page 933 we calculated $\int_{C_1} \vec{F} \cdot d\vec{r}$ and $\int_{C_2} \vec{F} \cdot d\vec{r}$ over the oriented curves shown in Figure 18.29 and found they were the same. Find a scalar function f with grad $f = \vec{F}$. Hence, find an easy way to calculate the line integrals, and explain why we could have expected them to be the same.

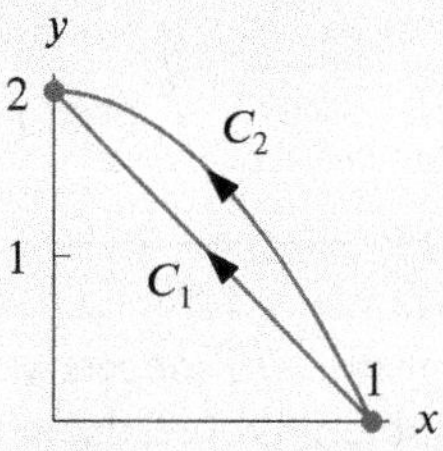

Figure 18.29: Find the line integral of $\vec{F} = x\vec{i} + y\vec{j}$ over the curves C_1 and C_2

Solution We want to find a function $f(x, y)$ for which $f_x = x$ and $f_y = y$. One possibility for f is

$$f(x, y) = \frac{x^2}{2} + \frac{y^2}{2}.$$

You can check that grad $f = x\vec{i} + y\vec{j}$. Now we can use the Fundamental Theorem to compute the line integral. Since $\vec{F} = \text{grad}\, f$ we have

$$\int_{C_1} \vec{F} \cdot d\vec{r} = \int_{C_1} \text{grad}\, f \cdot d\vec{r} = f(0, 2) - f(1, 0) = \frac{3}{2}.$$

Notice that the calculation looks exactly the same for C_2. Since the value of the integral depends only on the values of f at the endpoints, it is the same no matter what path we choose.

Path-Independent, or Conservative, Vector Fields

In the previous example, the line integral was independent of the path taken between the two (fixed) endpoints. We give vector fields whose line integrals have this property a special name.

> A vector field $\vec{F}$ is said to be **path-independent**, or **conservative**, if for any two points P and Q, the line integral $\int_C \vec{F} \cdot d\vec{r}$ has the same value along any piecewise smooth path C from P to Q lying in the domain of $\vec{F}$.

If, on the other hand, the line integral $\int_C \vec{F} \cdot d\vec{r}$ does depend on the path C joining P to Q, then $\vec{F}$ is said to be a *path-dependent* vector field.

Now suppose that $\vec{F}$ is any continuous gradient field, so $\vec{F} = \operatorname{grad} f$. If C is a path from P to Q, the Fundamental Theorem for Line Integrals tells us that

$$\int_C \vec{F} \cdot d\vec{r} = f(Q) - f(P).$$

Since the right-hand side of this equation does not depend on the path, but only on the endpoints of the path, the vector field $\vec{F}$ is path-independent. Thus, we have the following important result:

> If $\vec{F}$ is a continuous gradient vector field, then $\vec{F}$ is path-independent.

Why Do We Care About Path-Independent, or Conservative, Vector Fields?

Many of the fundamental vector fields of nature are path-independent—for example, the gravitational field and the electric field of particles at rest. The fact that the gravitational field is path-independent means that the work done by gravity when an object moves depends only on the starting and ending points and not on the path taken. For example, the work done by gravity (computed by the line integral) on a bicycle being carried to a sixth floor apartment is the same whether it is carried up the stairs in a zig-zag path or taken straight up in an elevator.

When a vector field is path-independent, we can define the *potential energy* of a body. When the body moves to another position, the potential energy changes by an amount equal to the work done by the vector field, which depends only on the starting and ending positions. If the work done had not been path-independent, the potential energy would depend both on the body's current position *and* on how it got there, making it impossible to define a useful potential energy.

Project 1 (available online) explains why path-independent force vector fields are also called *conservative* vector fields: When a particle moves under the influence of a conservative vector field, the *total energy* of the particle is *conserved*. It turns out that the force field is obtained from the gradient of the potential energy function.

Path-Independent Fields and Gradient Fields

We have seen that every gradient field is path-independent. What about the converse? That is, given a path-independent vector field $\vec{F}$, can we find a function f such that $\vec{F} = \operatorname{grad} f$? The answer is yes, provided that $\vec{F}$ is continuous.

How to Construct f from $\vec{F}$

First, notice that there are many different choices for f, since we can add a constant to f without changing $\operatorname{grad} f$. If we pick a fixed starting point P, then by adding or subtracting a constant to f, we can ensure that $f(P) = 0$. For any other point Q, we define $f(Q)$ by the formula

$$f(Q) = \int_C \vec{F} \cdot d\vec{r}, \quad \text{where } C \text{ is any path from } P \text{ to } Q.$$

Since $\vec{F}$ is path-independent, it does not matter which path we choose from P to Q. On the other hand, if $\vec{F}$ is not path-independent, then different choices might give different values for $f(Q)$, so f would not be a function (a function has to have a single value at each point).

We still have to show that the gradient of the function f really is $\vec{F}$; we do this on page 943. However, by constructing a function f in this manner, we have the following result:

Theorem 18.2: Path-independent Fields Are Gradient Fields

If $\vec{F}$ is a continuous path-independent vector field on an open region R, then $\vec{F} = \text{grad} f$ for some f defined on R.

Combining Theorems 18.1 and 18.2, we have

A continuous vector field $\vec{F}$ defined on an open region is path-independent if and only if $\vec{F}$ is a gradient vector field.

The function f is sufficiently important that it is given a special name:

If a vector field $\vec{F}$ is of the form $\vec{F} = \text{grad} f$ for some scalar function f, then f is called a **potential function** for the vector field $\vec{F}$.

Warning

Physicists use the convention that a function ϕ is a potential function for a vector field $\vec{F}$ if $\vec{F} = -\text{grad}\,\phi$. See Problem 89 (available online).

Example 3 Show that the vector field $\vec{F}(x, y) = y\cos x\vec{i} + (\sin x + y)\vec{j}$ is path-independent.

Solution Let's suppose $\vec{F}$ does have a potential function f, so that $\vec{F} = \text{grad} f$. This means

$$\frac{\partial f}{\partial x} = y\cos x \quad \text{and} \quad \frac{\partial f}{\partial y} = \sin x + y.$$

Integrating the expression for $\partial f/\partial x$ with respect to x shows that

$$f(x, y) = y\sin x + C(y) \qquad \text{where } C(y) \text{ is a function of } y \text{ only.}$$

The constant of integration here is an arbitrary function $C(y)$ of y, since $\partial(C(y))/\partial x = 0$. Differentiating this expression for $f(x, y)$ with respect to y and using $\partial f/\partial y = \sin x + y$ gives

$$\frac{\partial f}{\partial y} = \sin x + C'(y) = \sin x + y.$$

Thus, we must have $C'(y) = y$, so $g(y) = y^2/2 + A$, where A is some constant. Thus,

$$f(x, y) = y\sin x + \frac{y^2}{2} + A$$

is a potential function for $\vec{F}$. Therefore, $\vec{F}$ is path-independent.

Example 4 The gravitational field, $\vec{F}$, of an object of mass M is given by

$$\vec{F} = -\frac{GM}{r^3}\vec{r}.$$

Show that $\vec{F}$ is a gradient field by finding f, a potential function for $\vec{F}$.

Solution The vector $\vec{F}$ points directly in toward the origin. If $\vec{F} = \operatorname{grad} f$, then $\vec{F}$ must be perpendicular to the level surfaces of f, so the level surfaces of f must be spheres. Also, if $\operatorname{grad} f = \vec{F}$, then $\|\operatorname{grad} f\| = \|\vec{F}\| = GM/r^2$ is the rate of change of f in the direction toward the origin. Now, differentiating with respect to r gives the rate of change in a radially outward direction. Thus, if we write $w = f(x, y, z)$, we have

$$\frac{dw}{dr} = -\frac{GM}{r^2} = GM\left(-\frac{1}{r^2}\right) = GM\frac{d}{dr}\left(\frac{1}{r}\right).$$

So for the potential function, let's try

$$w = \frac{GM}{r} \quad \text{or} \quad f(x, y, z) = \frac{GM}{\sqrt{x^2 + y^2 + z^2}}.$$

We check that f is the potential function by calculating

$$f_x = \frac{\partial}{\partial x}\frac{GM}{\sqrt{x^2 + y^2 + z^2}} = \frac{-GMx}{(x^2 + y^2 + z^2)^{3/2}},$$

$$f_y = \frac{\partial}{\partial y}\frac{GM}{\sqrt{x^2 + y^2 + z^2}} = \frac{-GMy}{(x^2 + y^2 + z^2)^{3/2}},$$

$$f_z = \frac{\partial}{\partial z}\frac{GM}{\sqrt{x^2 + y^2 + z^2}} = \frac{-GMz}{(x^2 + y^2 + z^2)^{3/2}}.$$

So

$$\operatorname{grad} f = f_x\vec{i} + f_y\vec{j} + f_z\vec{k} = \frac{-GM}{(x^2 + y^2 + z^2)^{3/2}}(x\vec{i} + y\vec{j} + z\vec{k}) = \frac{-GM}{r^3}\vec{r} = \vec{F}.$$

Our computations show that $\vec{F}$ is a gradient field and that $f = GM/r$ is a potential function for $\vec{F}$.

Path-independent vector fields are rare, but often important. Section 18.4 gives a method for determining whether a vector field has the property.

Why Path-Independent Vector Fields Are Gradient Fields: Showing grad $f = \vec{F}$

Suppose $\vec{F}$ is a path-independent vector field. On page 941 we defined the function f, which we hope will satisfy $\operatorname{grad} f = \vec{F}$, as follows:

$$f(x_0, y_0) = \int_C \vec{F} \cdot d\vec{r},$$

where C is a path from a fixed starting point P to a point $Q = (x_0, y_0)$. This integral has the same value for any path from P to Q because $\vec{F}$ is path-independent. Now we show why $\operatorname{grad} f = \vec{F}$. We consider vector fields in 2-space; the argument in 3-space is essentially the same.

First, we write the line integral in terms of the components $\vec{F}(x, y) = F_1(x, y)\vec{i} + F_2(x, y)\vec{j}$ and the components $d\vec{r} = dx\vec{i} + dy\vec{j}$:

$$f(x_0, y_0) = \int_C F_1(x, y)\,dx + F_2(x, y)\,dy.$$

We want to compute the partial derivatives of f, that is, the rate of change of f at (x_0, y_0) parallel to the axes. To do this easily, we choose a path which reaches the point (x_0, y_0) on a horizontal or vertical line segment. Let C' be a path from P which stops short of Q at a fixed point (a, b) and let

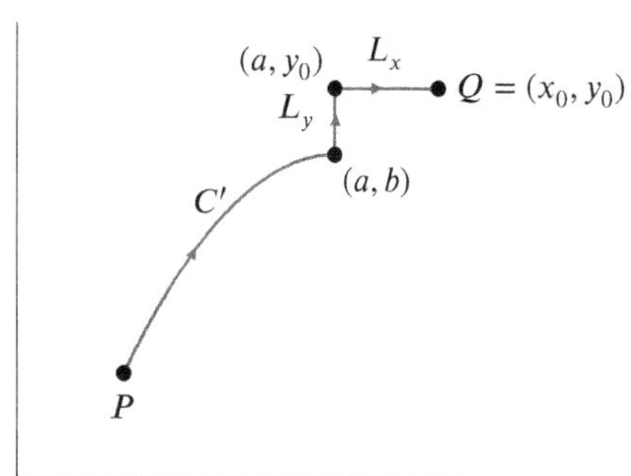

Figure 18.30: The path $C' + L_y + L_x$ is used to show $f_x = F_1$

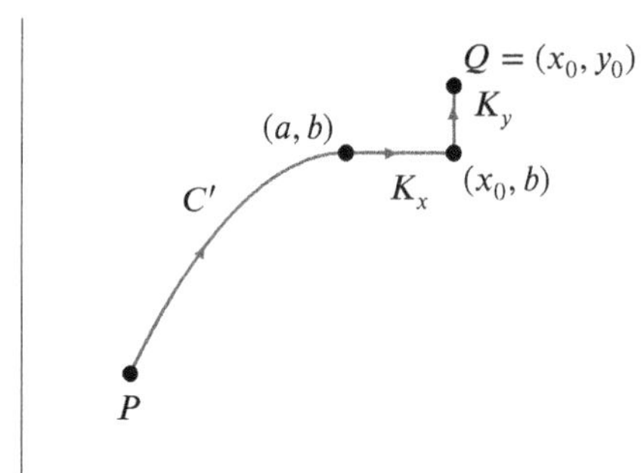

Figure 18.31: The path $C' + K_x + K_y$ is used to show $f_y = F_2$

L_x and L_y be the paths shown in Figure 18.30. Then we can split the line integral into three pieces. Since $d\vec{r} = \vec{j}\,dy$ on L_y and $d\vec{r} = \vec{i}\,dx$ on L_x, we have:

$$f(x_0, y_0) = \int_{C'} \vec{F}\cdot d\vec{r} + \int_{L_y} \vec{F}\cdot d\vec{r} + \int_{L_x} \vec{F}\cdot d\vec{r} = \int_{C'} \vec{F}\cdot d\vec{r} + \int_b^{y_0} F_2(a, y)\,dy + \int_a^{x_0} F_1(x, y_0)\,dx.$$

The first two integrals do not involve x_0. Thinking of x_0 as a variable and differentiating with respect to it gives

$$\begin{aligned} f_{x_0}(x_0, y_0) &= \frac{\partial}{\partial x_0}\int_{C'} \vec{F}\cdot d\vec{r} + \frac{\partial}{\partial x_0}\int_b^{y_0} F_2(a, y)dy + \frac{\partial}{\partial x_0}\int_a^{x_0} F_1(x, y_0)dx \\ &= 0 + 0 + F_1(x_0, y_0) = F_1(x_0, y_0), \end{aligned}$$

and thus

$$f_x(x, y) = F_1(x, y).$$

A similar calculation for y using the path from P to Q shown in Figure 18.31 gives

$$f_{y_0}(x_0, y_0) = F_2(x_0, y_0).$$

Therefore, as we claimed,

$$\text{grad}\, f = f_x\vec{i} + f_y\vec{j} = F_1\vec{i} + F_2\vec{j} = \vec{F}.$$

Exercises and Problems for Section 18.3 Online Resource: Additional Problems for Section 18.3

EXERCISES

1. If $\vec{F} = \text{grad}(x^2 + y^4)$, find $\int_C \vec{F}\cdot d\vec{r}$ where C is the quarter of the circle $x^2 + y^2 = 4$ in the first quadrant, oriented counterclockwise.

2. If $\vec{F} = \text{grad}(\sin(xy) + e^z)$, find $\int_C \vec{F}\cdot d\vec{r}$ where C consists of a line from $(0, 0, 0)$ to $(0, 0, 1)$ followed by a line to $(0, \sqrt{2}, 3)$, followed by a line to $(\sqrt{2}, \sqrt{5}, 2)$.

In Exercises 3–6, let C be the curve consisting of a square of side 2, centered at the origin with sides on the lines $x = \pm 1$, $y = \pm 1$ and traversed counterclockwise. What is the sign of the line integrals of the vector fields around the curve C? Indicate whether each vector field is path-independent.

3.

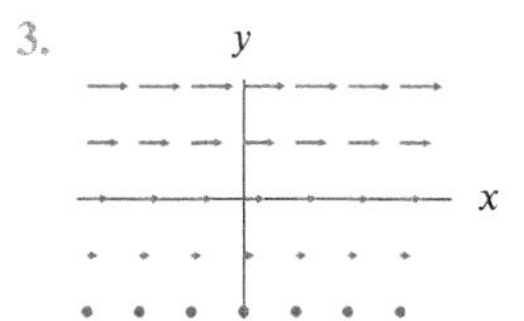

4.

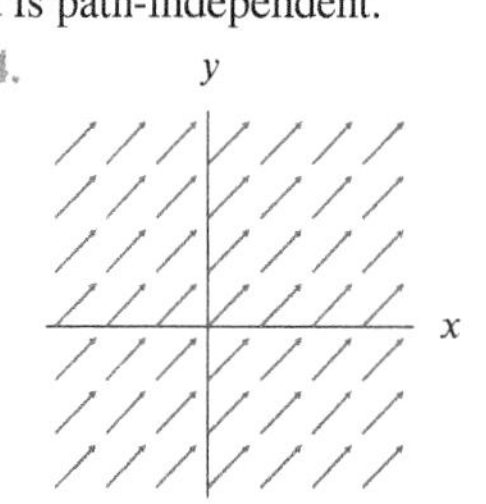

5.

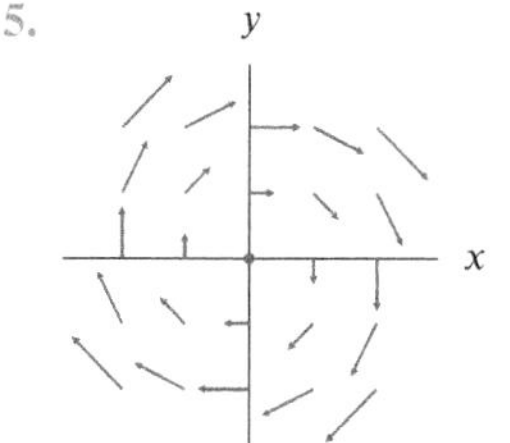

6.

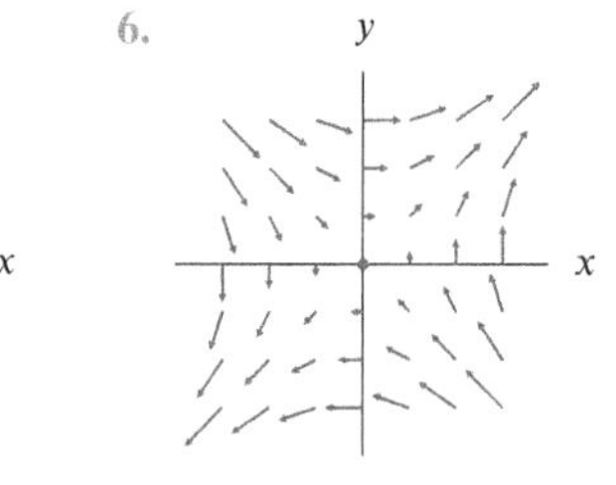

In Exercises 7–12, does the vector field appear to be path-independent (conservative)?

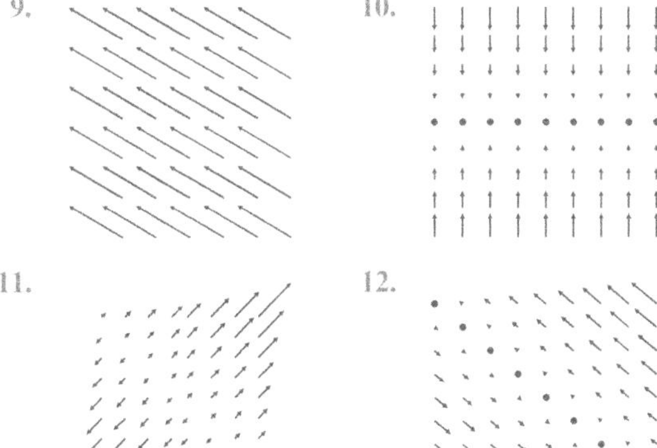

13. Find f if grad $f = 2xy\vec{i} + x^2\vec{j}$.

14. Find f if grad $f = 2xy\vec{i} + (x^2 + 8y^3)\vec{j}$.

15. Find f if grad $f = (yze^{xyz} + z^2\cos(xz^2))\vec{i} + xze^{xyz}\vec{j} + (xye^{xyz} + 2xz\cos(xz^2))\vec{k}$.

16. Let $f(x, y, z) = x^2 + 2y^3 + 3z^4$ and $\vec{F} = \text{grad } f$. Find $\int_C \vec{F} \cdot d\vec{r}$ where C consists of four line segments from $(4, 0, 0)$ to $(4, 3, 0)$ to $(0, 3, 0)$ to $(0, 3, 5)$ to $(0, 0, 5)$.

In Exercises 17–25, use the Fundamental Theorem of Line Integrals to calculate $\int_C \vec{F} \cdot d\vec{r}$ exactly.

17. $\vec{F} = 3x^2\vec{i} + 4y^3\vec{j}$ around the top of the unit circle from $(1, 0)$ to $(-1, 0)$.

18. $\vec{F} = (x + 2)\vec{i} + (2y + 3)\vec{j}$ and C is the line from $(1, 0)$ to $(3, 1)$.

19. $\vec{F} = 2\sin(2x + y)\vec{i} + \sin(2x + y)\vec{j}$ along the path consisting of a line from $(\pi, 0)$ to $(2, 5)$ followed by a line to $(5\pi, 0)$ followed by a quarter circle to $(0, 5\pi)$.

20. $\vec{F} = 2x\vec{i} - 4y\vec{j} + (2z - 3)\vec{k}$ and C is the line from $(1, 1, 1)$ to $(2, 3, -1)$.

21. $\vec{F} = x^{2/3}\vec{i} + e^{7y}\vec{j}$ and C is the unit circle oriented clockwise.

22. $\vec{F} = x^{2/3}\vec{i} + e^{7y}\vec{j}$ and C is the quarter of the unit circle in the first quadrant, traced counterclockwise from $(1, 0)$ to $(0, 1)$.

23. $\vec{F} = ye^{xy}\vec{i} + xe^{xy}\vec{j} + (\cos z)\vec{k}$ along the curve consisting of a line from $(0, 0, \pi)$ to $(1, 1, \pi)$ followed by the parabola $z = \pi x^2$ in the plane $y = 1$ to the point $(3, 1, 9\pi)$.

24. $\vec{F} = y\sin(xy)\vec{i} + x\sin(xy)\vec{j}$ and C is the parabola $y = 2x^2$ from $(1, 2)$ to $(3, 18)$.

25. $\vec{F} = 2xy^2ze^{x^2y^2z}\vec{i} + 2x^2yze^{x^2y^2z}\vec{j} + x^2y^2e^{x^2y^2z}\vec{k}$ and C is the circle of radius 1 in the plane $z = 1$, centered on the z-axis, starting at $(1, 0, 1)$ and oriented counterclockwise viewed from above.

PROBLEMS

26. Let $\vec{v} = \text{grad}(x^2 + y^2)$. Consider the path C which is a line between any two of the following points: $(0, 0)$; $(5, 0)$; $(-5, 0)$; $(0, 6)$; $(0, -6)$; $(5, 4)$; $(-3, -5)$. Suppose you want to choose the path C in order to maximize $\int_C \vec{v} \cdot d\vec{r}$. What point should be the start of C? What point should be the end of C? Explain your answer.

27. Let $\vec{F} = \text{grad}(2x^2 + 3y^2)$. Which one of the three paths PQ, QR, and RS in Figure 18.32 should you choose as C in order to maximize $\int_C \vec{F} \cdot d\vec{r}$?

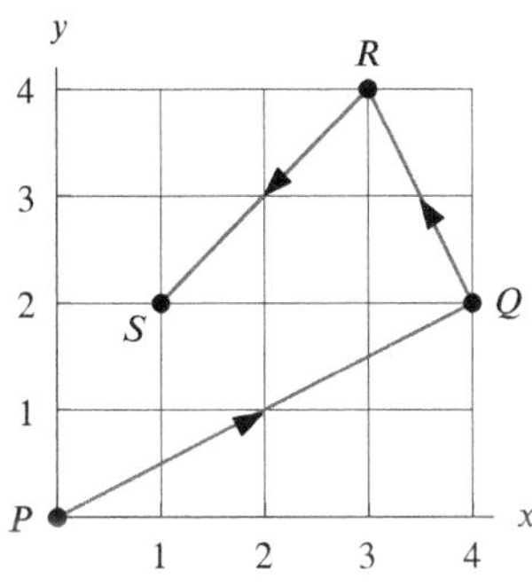

Figure 18.32

28. Compute $\displaystyle\int_C \left(\cos(xy)e^{\sin(xy)}(y\vec{i} + x\vec{j}) + \vec{k}\right) \cdot d\vec{r}$ where C is the line from $(\pi, 2, 5)$ to $(0.5, \pi, 7)$.

29. The vector field $\vec{F}(x, y) = x\vec{i} + y\vec{j}$ is path-independent. Compute geometrically the line integrals over the three paths A, B, and C shown in Figure 18.33 from $(1, 0)$ to $(0, 1)$ and check that they are equal. Here A is a portion of a circle, B is a line, and C consists of two line segments meeting at a right angle.

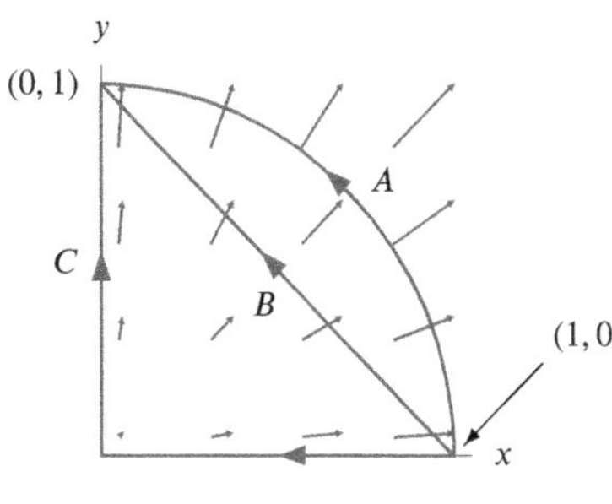

Figure 18.33

30. The vector field $\vec{F}(x, y) = x\vec{i} + y\vec{j}$ is path-independent. Compute algebraically the line integrals over the three paths A, B, and C shown in Figure 18.34 from $(0, 0)$ to $(1, 1)$ and check that they are equal. Here A is a line segment, B is part of the graph of $f(x) = x^2$, and C consists of two line segments meeting at a right angle.

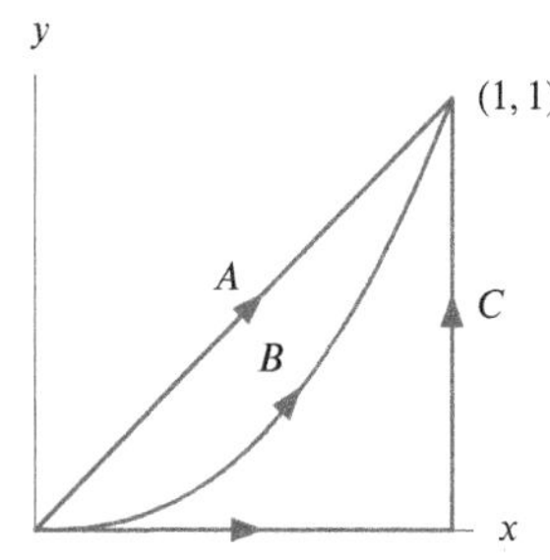

Figure 18.34

In Problems 31–34, decide whether the vector field could be a gradient vector field. Justify your answer.

31. $\vec{F}(x, y) = x\vec{i}$

32. $\vec{G}(x, y) = (x^2 - y^2)\vec{i} - 2xy\vec{j}$

33. $\vec{F}(\vec{r}) = \vec{r}/||\vec{r}||^3$, where $\vec{r} = x\vec{i} + y\vec{j} + z\vec{k}$

34. $\vec{F}(x, y, z) = \dfrac{-z}{\sqrt{x^2+z^2}}\vec{i} + \dfrac{y}{\sqrt{x^2+z^2}}\vec{j} + \dfrac{x}{\sqrt{x^2+z^2}}\vec{k}$

35. Find a potential function for $\vec{F} = 5y\vec{i} + (5x + y)\vec{j}$. Use it to evaluate the integral $\int_C \vec{F} \cdot d\vec{r}$ on the path C if

(a) C runs from $(10, 0)$ to $(0, -10)$ along the circle of radius 10 centered at the origin.
(b) C runs from $(10, 0)$ to $(0, -10)$ along a straight line.

36. Suppose C is a path that begins and ends at the same point $P = (15, 20)$. What, if anything, can you say about $\int_C (p(x, y)\vec{i} + q(x, y)\vec{j}) \cdot d\vec{r}$?

(a) With no assumptions about p and q.
(b) If $p(x, y)\vec{i} + q(x, y)\vec{j}$ has a potential function.

37. Let $\vec{F} = -y\vec{i} + x\vec{j}$ and let C be the circle of radius 5 centered at the origin, oriented counterclockwise.

(a) Evaluate $\int_C \vec{F} \cdot d\vec{r}$.
(b) Give a potential function for $\vec{F}$ or explain why there are none.

38. Let $\vec{F} = y\vec{i}$.

(a) Evaluate $\int_{C_1} \vec{F} \cdot d\vec{r}$ if C_1 is the straight line path from $(0, 0)$ to $(1, 0)$.
(b) Evaluate $\int_{C_2} \vec{F} \cdot d\vec{r}$ if C_2 is the path along three edges of a square, from $(0, 0)$ to $(0, 1)$ to $(1, 1)$ to $(1, 0)$.
(c) Does $\vec{F}$ have a potential function? Either give one or explain why there are none.

39. If $df = p\,dx + q\,dy$ for smooth f, explain why

$$\frac{\partial p}{\partial y} = \frac{\partial q}{\partial x}.$$

40. If $\vec{F}(x, y, z) = 2xe^{x^2+yz}\vec{i} + ze^{x^2+yz}\vec{j} + ye^{x^2+yz}\vec{k}$, find exactly the line integral of $\vec{F}$ along the curve consisting of the two half circles in the plane $z = 0$ in Figure 18.35.

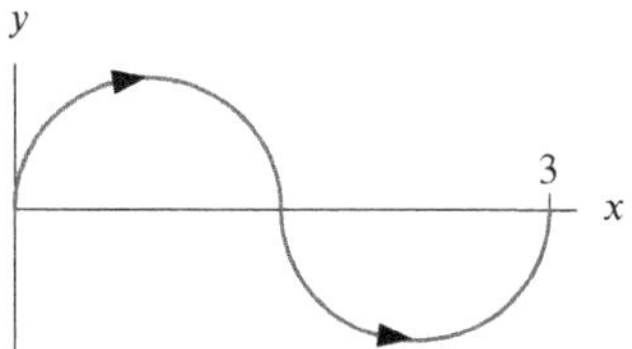

Figure 18.35

41. Let grad $f = 2xe^{x^2} \sin y\vec{i} + e^{x^2} \cos y\vec{j}$. Find the change in f between $(0, 0)$ and $(1, \pi/2)$:

(a) By computing a line integral.
(b) By computing f.

42. Let C be the quarter of the unit circle centered at the origin, traversed counterclockwise starting on the negative x-axis. Find the exact values of

(a) $\displaystyle\int_C (2\pi x\vec{i} + y^2\vec{j}) \cdot d\vec{r}$ (b) $\displaystyle\int_C (-2y\vec{i} + x\vec{j}) \cdot d\vec{r}$

For the vector fields in Problems 43–46, find the line integral along the curve C from the origin along the x-axis to the point $(3, 0)$ and then counterclockwise around the circumference of the circle $x^2 + y^2 = 9$ to the point $(3/\sqrt{2}, 3/\sqrt{2})$.

43. $\vec{F} = x\vec{i} + y\vec{j}$

44. $\vec{H} = -y\vec{i} + x\vec{j}$

45. $\vec{F} = y(x+1)^{-1}\vec{i} + \ln(x+1)\vec{j}$

46. $\vec{G} = (ye^{xy} + \cos(x+y))\vec{i} + (xe^{xy} + \cos(x+y))\vec{j}$

47. Let C be the helix $x = \cos t$, $y = \sin t$, $z = t$ for $0 \le t \le 1.25\pi$. Find $\int_C \vec{F} \cdot d\vec{r}$ exactly for

$$\vec{F} = yz^2e^{xyz^2}\vec{i} + xz^2e^{xyz^2}\vec{j} + 2xyze^{xyz^2}\vec{k}.$$

48. Let $\vec{F} = 2x\vec{i} + 2y\vec{j} + 2z\vec{k}$ and $\vec{G} = (2x + y)\vec{i} + 2y\vec{j} + 2z\vec{k}$. Let C be the line from the origin to the point $(1, 5, 9)$. Find $\int_C \vec{F} \cdot d\vec{r}$ and use the result to find $\int_C \vec{G} \cdot d\vec{r}$.

49. (a) If $\vec{F} = ye^x\vec{i} + e^x\vec{j}$, explain how the Fundamental Theorem of Calculus for Line Integrals enables you to calculate $\int_C \vec{F} \cdot d\vec{r}$ where C is any curve going from the point $(1, 2)$ to the point $(3, 7)$. Explain why it does not matter how the curve goes.
(b) If C is the line from the point $(1, 2)$ to the point $(3, 7)$, calculate the line integral in part (a) without using the Fundamental Theorem.

50. Calculate the line integral $\int_C \vec{F} \cdot d\vec{r}$ exactly, where C is the curve from P to Q in Figure 18.36 and

$$\vec{F} = \sin\left(\frac{x}{2}\right)\sin\left(\frac{y}{2}\right)\vec{i} - \cos\left(\frac{x}{2}\right)\cos\left(\frac{y}{2}\right)\vec{j}.$$

The curves PR, RS and SQ are trigonometric functions of period 2π and amplitude 1.

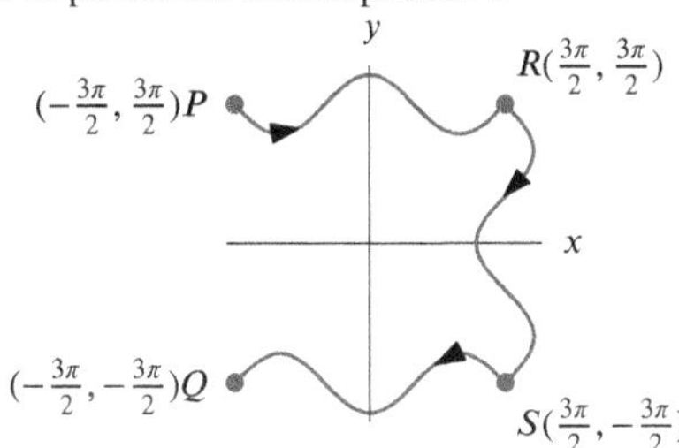

Figure 18.36

51. The domain of $f(x, y)$ is the xy-plane; values of f are in Table 18.1. Find $\int_C \text{grad } f \cdot d\vec{r}$, where C is

(a) A line from (0, 2) to (3, 4).
(b) A circle of radius 1 centered at (1, 2) traversed counterclockwise.

Table 18.1

$y \backslash x$	0	1	2	3	4
0	53	57	59	58	56
1	56	58	59	59	57
2	57	58	59	60	59
3	59	60	61	62	61
4	62	63	65	66	69

52. Figure 18.37 shows the vector field $\vec{F}(x, y) = x\vec{j}$.

(a) Find paths C_1, C_2, and C_3 from P to Q such that

$$\int_{C_1} \vec{F} \cdot d\vec{r} = 0, \quad \int_{C_2} \vec{F} \cdot d\vec{r} > 0, \quad \int_{C_3} \vec{F} \cdot d\vec{r} < 0.$$

(b) Is $\vec{F}$ a gradient field? Explain.

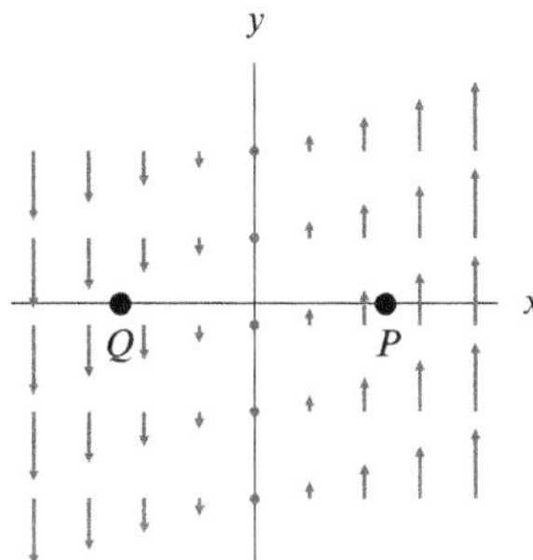

Figure 18.37

53. **(a)** Figure 18.38 shows level curves of $f(x, y)$. Sketch a vector at P in the direction of grad f.
(b) Is the length of grad f at P longer, shorter, or the same length as the length of grad f at Q?
(c) If C is a curve going from P to Q, find $\int_C \text{grad } f \cdot d\vec{r}$.

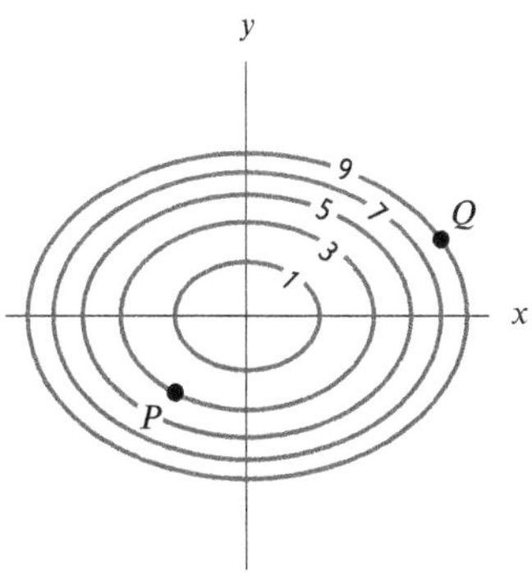

Figure 18.38

54. Consider the line integrals, $\int_{C_i} \vec{F} \cdot d\vec{r}$, for $i = 1, 2, 3, 4$, where C_i is the path from P_i to Q_i shown in Figure 18.39 and $\vec{F} = \text{grad } f$. Level curves of f are also shown in Figure 18.39.

(a) Which of the line integral(s) is (are) zero?
(b) Arrange the four line integrals in ascending order (from least to greatest).
(c) Two of the nonzero line integrals have equal and opposite values. Which are they? Which is negative and which is positive?

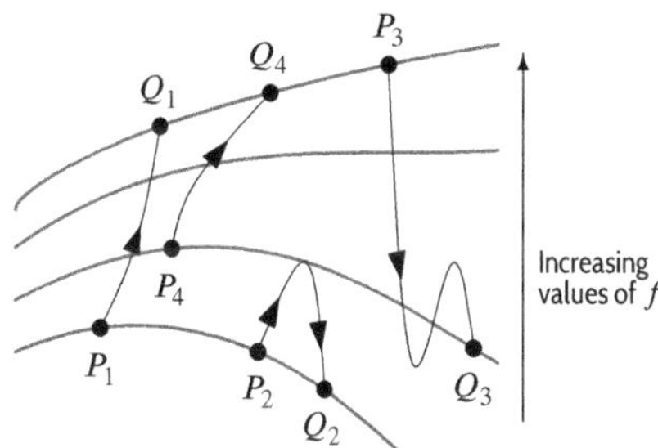

Figure 18.39

55. Consider the vector field $\vec{F}$ shown in Figure 18.40.

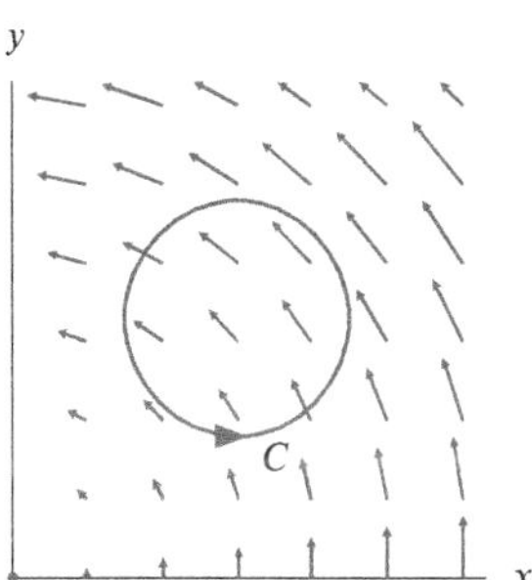

Figure 18.40

(a) Is $\int_C \vec{F} \cdot d\vec{r}$ positive, negative, or zero?

(b) From your answer to part (a), can you determine whether or not $\vec{F} = \operatorname{grad} f$ for some function f?

(c) Which of the following formulas best fits $\vec{F}$?

$$\vec{F}_1 = \frac{x}{x^2+y^2}\vec{i} + \frac{y}{x^2+y^2}\vec{j},$$
$$\vec{F}_2 = -y\vec{i} + x\vec{j},$$
$$\vec{F}_3 = \frac{-y}{(x^2+y^2)^2}\vec{i} + \frac{x}{(x^2+y^2)^2}\vec{j}.$$

56. If $\vec{F}$ is a path-independent vector field, with $\int_{(0,0)}^{(1,0)} \vec{F} \cdot d\vec{r} = 5.1$ and $\int_{(1,0)}^{(1,1)} \vec{F} \cdot d\vec{r} = 3.2$ and $\int_{(0,1)}^{(1,1)} \vec{F} \cdot d\vec{r} = 4.7$, find

$$\int_{(0,1)}^{(0,0)} \vec{F} \cdot d\vec{r}.$$

57. The path C is a line segment of length 10 in the plane starting at $(2, 1)$. For $f(x, y) = 3x + 4y$, consider

$$\int_C \operatorname{grad} f \cdot d\vec{r}.$$

(a) Where should the other end of the line segment C be placed to maximize the value of the integral?

(b) What is the maximum value of the integral?

58. Let $\vec{r} = x\vec{i} + y\vec{j} + z\vec{k}$ and $\vec{a} = a_1\vec{i} + a_2\vec{j} + a_3\vec{k}$, a constant vector.

(a) Find $\operatorname{grad}(\vec{r} \cdot \vec{a})$.

(b) Let C be a path from the origin to the point with position vector $\vec{r}_0$. Find $\int_C \operatorname{grad}(\vec{r} \cdot \vec{a}) \cdot d\vec{r}$.

(c) If $||\vec{r}_0|| = 10$, what is the maximum possible value of $\int_C \operatorname{grad}(\vec{r} \cdot \vec{a}) \cdot d\vec{r}$? Explain.

59. The force exerted by gravity on a refrigerator of mass m is $\vec{F} = -mg\vec{k}$.

(a) Find the work done against this force in moving from the point $(1, 0, 0)$ to the point $(1, 0, 2\pi)$ along the curve $x = \cos t, y = \sin t, z = t$ by calculating a line integral.

(b) Is $\vec{F}$ conservative (that is, path independent)? Give a reason for your answer.

60. A particle subject to a force $\vec{F}(x, y) = y\vec{i} - x\vec{j}$ moves clockwise along the arc of the unit circle, centered at the origin, that begins at $(-1, 0)$ and ends at $(0, 1)$.

(a) Find the work done by $\vec{F}$. Explain the sign of your answer.

(b) Is $\vec{F}$ path-independent? Explain.

Strengthen Your Understanding

In Problems 61–63, explain what is wrong with the statement.

61. If $\vec{F}$ is a gradient field and C is an oriented path from point P to point Q, then $\int_C \vec{F} \cdot d\vec{r} = \vec{F}(Q) - \vec{F}(P)$.

62. Given any vector field $\vec{F}$ and a point P, the function $f(Q) = \int_C \vec{F} \cdot d\vec{r}$, where C is a path from P to Q, is a potential function for $\vec{F}$.

63. If a vector field $\vec{F}$ is not a gradient vector field, then $\int_C \vec{F} \cdot d\vec{r}$ can't be evaluated.

In Problems 64–65, give an example of:

64. A vector field $\vec{F}$ such that $\int_C \vec{F} \cdot d\vec{r} = 100$, for every oriented path C from $(0, 0)$ to $(1, 2)$.

65. A path-independent vector field.

In Problems 66–69, each of the statements is *false*. Explain why or give a counterexample.

66. If $\int_C \vec{F} \cdot d\vec{r} = 0$ for one particular closed path C, then $\vec{F}$ is path-independent.

67. $\int_C \vec{F} \cdot d\vec{r}$ is the total change in $\vec{F}$ along C.

68. If the vector fields $\vec{F}$ and $\vec{G}$ have $\int_C \vec{F} \cdot d\vec{r} = \int_C \vec{G} \cdot d\vec{r}$ for a particular path C, then $\vec{F} = \vec{G}$.

69. If the total change of a function f along a curve C is zero, then C must be a contour of f.

Are the statements in Problems 70–80 true or false? Give reasons for your answer.

70. The fact that the line integral of a vector field $\vec{F}$ is zero around the unit circle $x^2 + y^2 = 1$ means that $\vec{F}$ must be a gradient vector field.

71. If C is the line segment that starts at $(0, 0)$ and ends at (a, b) then $\int_C (x\vec{i} + y\vec{j}) \cdot d\vec{r} = \frac{1}{2}(a^2 + b^2)$.

72. The circulation of any vector field $\vec{F}$ around any closed curve C is zero.

73. If $\vec{F} = \operatorname{grad} f$, then $\vec{F}$ is path-independent.

74. If $\vec{F}$ is path-independent, then $\int_{C_1} \vec{F} \cdot d\vec{r} = \int_{C_2} \vec{F} \cdot d\vec{r}$, where C_1 and C_2 are any paths.

75. The line integral $\int_C \vec{F} \cdot d\vec{r}$ is the total change of $\vec{F}$ along C.

76. If $\vec{F}$ is path-independent, then there is a potential function for $\vec{F}$.

77. If $f(x, y) = e^{\cos(xy)}$, and C_1 is the upper semicircle $x^2+y^2 = 1$ from $(-1, 0)$ to $(1, 0)$, and C_2 is the line from $(-1, 0)$ to $(1, 0)$, then $\int_{C_1} \operatorname{grad} f \cdot d\vec{r} = \int_{C_2} \operatorname{grad} f \cdot d\vec{r}$.

78. If $\vec{F}$ is path-independent, and C is any closed curve, then $\int_C \vec{F} \cdot d\vec{r} = 0$.

79. The vector field $\vec{F}(x, y) = y^2\vec{i} + k\vec{j}$, where k is constant, is a gradient field.

80. If $\int_C \vec{F} \cdot d\vec{r} = 0$, where C is any circle of the form $x^2 + y^2 = a^2$, then $\vec{F}$ is path-independent.

18.4 PATH-DEPENDENT VECTOR FIELDS AND GREEN'S THEOREM

Suppose we are given a vector field but are not told whether it is path-independent. How can we tell if it has a potential function, that is, if it is a gradient field?

How to Tell If a Vector Field Is Path-Dependent Using Line Integrals

One way to decide if a vector field is path-dependent is to find two paths with the same endpoints such that the line integrals of the vector field along the two paths have different values.

Example 1 Is the vector field $\vec{G}$ shown in Figure 18.41 path-independent? At any point $\vec{G}$ has magnitude equal to the distance from the origin and direction perpendicular to the line joining the point to the origin.

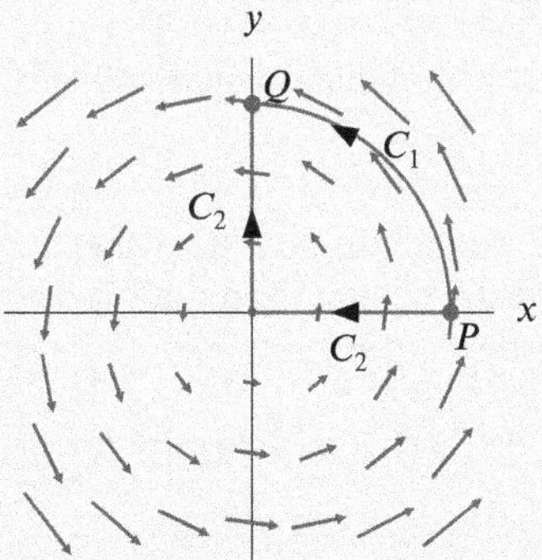

Figure 18.41: Is this vector field path-independent?

Solution We choose $P = (1, 0)$ and $Q = (0, 1)$ and two paths between them: C_1, a quarter circle of radius 1, and C_2, formed by parts of the x- and y-axes. (See Figure 18.41.)

Along C_1, the line integral $\int_{C_1} \vec{G} \cdot d\vec{r} > 0$, since $\vec{G}$ points in the direction of the curve.

Along C_2, however, we have $\int_{C_2} \vec{G} \cdot d\vec{r} = 0$, since $\vec{G}$ is perpendicular to C_2 everywhere.

Thus, $\vec{G}$ is not path-independent.

Path-Dependent Fields and Circulation

Notice that the vector field in the previous example has nonzero circulation around the origin. What can we say about the circulation of a general path-independent vector field $\vec{F}$ around a closed curve, C? Suppose C is a *simple* closed curve, that is, a closed curve that does not cross itself. If P and Q are any two points on the path, then we can think of C (oriented as shown in Figure 18.42) as made up of the path C_1 followed by $-C_2$. Since $\vec{F}$ is path-independent, we know that

$$\int_{C_1} \vec{F} \cdot d\vec{r} = \int_{C_2} \vec{F} \cdot d\vec{r}.$$

Thus, we see that the circulation around C is zero:

$$\int_{C} \vec{F} \cdot d\vec{r} = \int_{C_1} \vec{F} \cdot d\vec{r} + \int_{-C_2} \vec{F} \cdot d\vec{r} = \int_{C_1} \vec{F} \cdot d\vec{r} - \int_{C_2} \vec{F} \cdot d\vec{r} = 0.$$

If the closed curve C does cross itself, we break it into simple closed curves as shown in Figure 18.43 and apply the same argument to each one.

Now suppose we know that the line integral around any closed curve is zero. For any two points, P and Q, with two paths, C_1 and C_2, between them, create a closed curve, C, as in Figure 18.42. Since the circulation around this closed curve, C, is zero, the line integrals along the two paths, C_1 and C_2, are equal.[1] Thus, $\vec{F}$ is path-independent.

[1]A similar argument is used in Problems 57 and 58 on page 931.

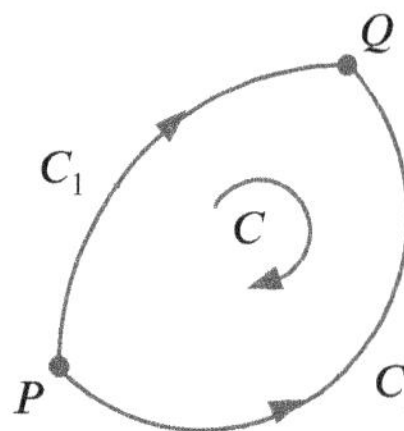

Figure 18.42: A simple closed curve C broken into two pieces, C_1 and C_2

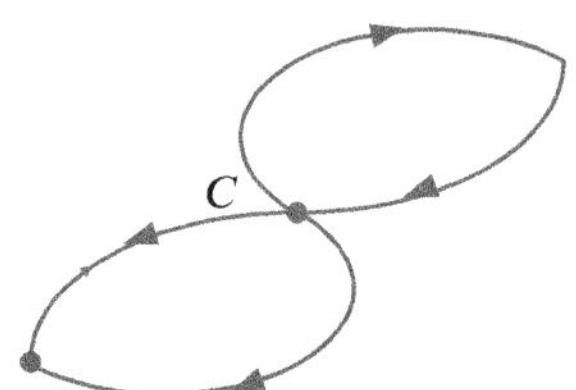

Figure 18.43: A curve C which crosses itself can be broken into simple closed curves

Thus, we have the following result:

> A vector field is path-independent if and only if $\int_C \vec{F} \cdot d\vec{r} = 0$ for every closed curve C.

Hence, to see if a field is *path-dependent*, we look for a closed path with nonzero circulation. For instance, the vector field in Example 1 has nonzero circulation around a circle around the origin, showing it is path-dependent.

How to Tell If a Vector Field Is Path-Dependent Algebraically: The Curl

Example 2 Does the vector field $\vec{F} = 2xy\vec{i} + xy\vec{j}$ have a potential function? If so, find it.

Solution Let's suppose $\vec{F}$ does have a potential function, f, so $\vec{F} = \operatorname{grad} f$. This means that

$$\frac{\partial f}{\partial x} = 2xy \quad \text{and} \quad \frac{\partial f}{\partial y} = xy.$$

Integrating the expression for $\partial f/\partial x$ shows that we must have

$$f(x, y) = x^2y + C(y) \qquad \text{where } C(y) \text{ is a function of } y.$$

Differentiating this expression for $f(x, y)$ with respect to y and using the fact that $\partial f/\partial y = xy$, we get

$$\frac{\partial f}{\partial y} = x^2 + C'(y) = xy.$$

Thus, we must have

$$C'(y) = xy - x^2.$$

But this expression for $C'(y)$ is impossible because $C'(y)$ is a function of y alone. This argument shows that there is no potential function for the vector field $\vec{F}$.

Is there an easier way to see that a vector field has no potential function, other than by trying to find the potential function and failing? The answer is yes. First we look at a 2-dimensional vector field $\vec{F} = F_1\vec{i} + F_2\vec{j}$. If $\vec{F}$ is a gradient field, then there is a potential function f such that

$$\vec{F} = F_1\vec{i} + F_2\vec{j} = \frac{\partial f}{\partial x}\vec{i} + \frac{\partial f}{\partial y}\vec{j}.$$

Thus,

$$F_1 = \frac{\partial f}{\partial x} \quad \text{and} \quad F_2 = \frac{\partial f}{\partial y}.$$

Let us assume that f has continuous second partial derivatives. Then, by the equality of mixed partial derivatives,

$$\frac{\partial F_1}{\partial y} = \frac{\partial^2 f}{\partial y \partial x} = \frac{\partial^2 f}{\partial x \partial y} = \frac{\partial F_2}{\partial x}.$$

Thus, we have the following result:

> If $\vec{F}(x, y) = F_1\vec{i} + F_2\vec{j}$ is a gradient vector field with continuous partial derivatives, then
>
> $$\frac{\partial F_2}{\partial x} - \frac{\partial F_1}{\partial y} = 0.$$
>
> If $\vec{F}(x, y) = F_1\vec{i} + F_2\vec{j}$ is an arbitrary vector field, then we define the 2-dimensional or scalar **curl** of the vector field $\vec{F}$ to be
>
> $$\frac{\partial F_2}{\partial x} - \frac{\partial F_1}{\partial y}.$$

Notice that we now know that if $\vec{F}$ is a gradient field, then its curl is 0. We do not (yet) know whether the converse is true. (That is: If the curl is 0, does $\vec{F}$ have to be a gradient field?) However, the curl already enables us to show that a vector field is *not* a gradient field.

Example 3 Show that $\vec{F} = 2xy\vec{i} + xy\vec{j}$ cannot be a gradient vector field.

Solution We have $F_1 = 2xy$ and $F_2 = xy$. Since $\partial F_1/\partial y = 2x$ and $\partial F_2/\partial x = y$, in this case

$$\partial F_2/\partial x - \partial F_1/\partial y \neq 0$$

so $\vec{F}$ cannot be a gradient field.

Green's Theorem

We now have two ways of seeing that a vector field $\vec{F}$ in the plane is path-dependent. We can evaluate $\int_C \vec{F} \cdot d\vec{r}$ for some closed curve and find it is not zero, or we can show that $\partial F_2/\partial x - \partial F_1/\partial y \neq 0$. It's natural to think that

$$\int_C \vec{F} \cdot d\vec{r} \quad \text{and} \quad \frac{\partial F_2}{\partial x} - \frac{\partial F_1}{\partial y}$$

might be related. The relation is called Green's Theorem.

> ### Theorem 18.3: Green's Theorem
>
> Suppose C is a piecewise smooth simple closed curve that is the boundary of a region R in the plane and oriented so that the region is on the left as we move around the curve. See Figure 18.44. Suppose $\vec{F} = F_1\vec{i} + F_2\vec{j}$ is a smooth vector field on a region[2] containing R and C. Then
>
> $$\int_C \vec{F} \cdot d\vec{r} = \int_R \left(\frac{\partial F_2}{\partial x} - \frac{\partial F_1}{\partial y} \right) dx\, dy.$$

The online supplement at www.wiley.com/college/hughes-hallett contains a proof of Green's Theorem with different, but equivalent, conditions on the region R.

[2]The region is an open region containing R and C.

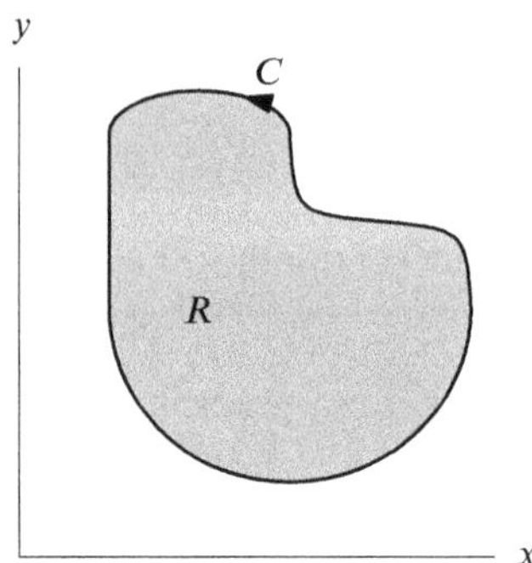

Figure 18.44: Boundary C oriented with R on the left

We first prove Green's Theorem in the case where the region R is the rectangle $a \leq x \leq b, c \leq y \leq d$. Figure 18.45 shows the boundary of R divided into four curves.

On C_1, where $y = c$ and $dy = 0$, we have $d\vec{r} = dx\vec{i}$ and thus

$$\int_{C_1} \vec{F} \cdot d\vec{r} = \int_a^b F_1(x, c)\, dx.$$

Similarly, on C_3 where $y = d$ we have

$$\int_{C_3} \vec{F} \cdot d\vec{r} = \int_b^a F_1(x, d)\, dx = -\int_a^b F_1(x, d)\, dx.$$

Hence

$$\int_{C_1+C_3} \vec{F} \cdot d\vec{r} = \int_a^b F_1(x, c)\, dx - \int_a^b F_1(x, d)\, dx = -\int_a^b (F_1(x, d) - F_1(x, c))\, dx.$$

By the Fundamental Theorem of Calculus,

$$F_1(x, d) - F_1(x, c) = \int_c^d \frac{\partial F_1}{\partial y}\, dy$$

and therefore

$$\int_{C_1+C_3} \vec{F} \cdot d\vec{r} = -\int_a^b \int_c^d \frac{\partial F_1}{\partial y}\, dy\, dx = -\int_c^d \int_a^b \frac{\partial F_1}{\partial y}\, dx\, dy.$$

Along the curve C_2, where $x = b$, and the curve C_4, where $x = a$, we get, by a similar argument,

$$\int_{C_2+C_4} \vec{F} \cdot d\vec{r} = \int_c^d (F_2(b, y) - F_2(a, y))\, dy = \int_c^d \int_a^b \frac{\partial F_2}{\partial x}\, dx\, dy.$$

Adding the line integrals over $C_1 + C_3$ and $C_2 + C_4$, we get

$$\int_C \vec{F} \cdot d\vec{r} = \int_R \left(\frac{\partial F_2}{\partial x} - \frac{\partial F_1}{\partial y} \right) dx\, dy.$$

If R is not a rectangle, we subdivide it into small rectangular pieces as shown in Figure 18.46. The contribution to the integral of the non-rectangular pieces can be made as small as we like by making the subdivision fine enough. The double integrals over each piece add up to the double integral over the whole region R. Figure 18.47 shows how the circulations around adjacent pieces cancel along the common edge, so the circulations around all the pieces add up to the circulation around the boundary C. Since Green's Theorem holds for the rectangular pieces, it holds for the whole region R.

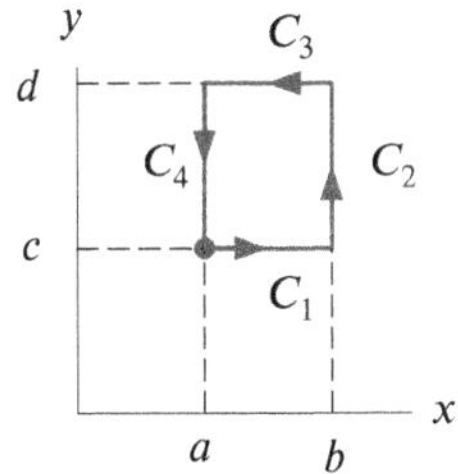

Figure 18.45: The boundary of a rectangle broken into C_1, C_2, C_3, C_4

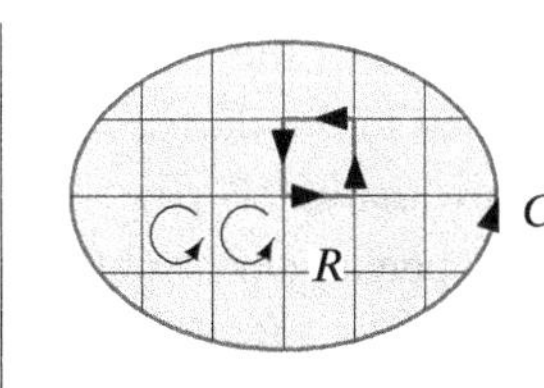

Figure 18.46: Region R bounded by a closed curve C and split into many small regions, ΔR

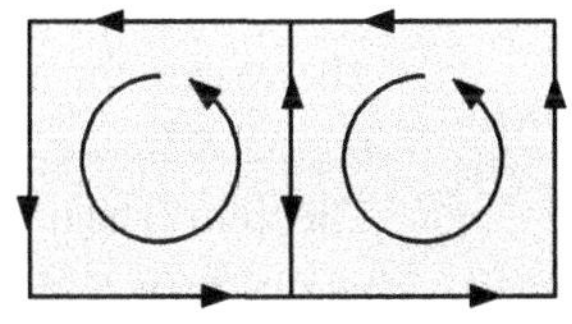

Figure 18.47: Two adjacent small closed curves

Example 4 Use Green's Theorem to evaluate $\int_C \left(y^2\vec{i} + x\vec{j}\right)\cdot d\vec{r}$ where C is the counterclockwise path around the perimeter of the rectangle $0 \leq x \leq 2, 0 \leq y \leq 3$.

Solution We have $F_1 = y^2$ and $F_2 = x$. By Green's Theorem,

$$\int_C \left(y^2\vec{i} + x\vec{j}\right)\cdot d\vec{r} = \int_R \left(\frac{\partial F_2}{\partial x} - \frac{\partial F_1}{\partial y}\right) dx\,dy = \int_0^3 \int_0^2 (1-2y)\,dx\,dy = -12.$$

The Curl Test for Vector Fields in the Plane

We already know that if $\vec{F} = F_1\vec{i} + F_2\vec{j}$ is a gradient field with continuous partial derivatives, then

$$\frac{\partial F_2}{\partial x} - \frac{\partial F_1}{\partial y} = 0.$$

Now we show that the converse is true if the domain of $\vec{F}$ has no holes in it. This means that we assume that

$$\frac{\partial F_2}{\partial x} - \frac{\partial F_1}{\partial y} = 0$$

and show that $\vec{F}$ is path-independent. If C is any oriented simple closed curve in the domain of $\vec{F}$ and R is the region inside C, then

$$\int_R \left(\frac{\partial F_2}{\partial x} - \frac{\partial F_1}{\partial y}\right) dx\,dy = 0$$

since the integrand is identically 0. Therefore, by Green's Theorem,

$$\int_C \vec{F}\cdot d\vec{r} = \int_R \left(\frac{\partial F_2}{\partial x} - \frac{\partial F_1}{\partial y}\right) dxdy = 0.$$

Thus, $\vec{F}$ is path-independent and therefore a gradient field. This argument is valid for every closed curve, C, provided the region R is entirely in the domain of $\vec{F}$. Thus, we have the following result:

The Curl Test for Vector Fields in 2-Space

Suppose $\vec{F} = F_1\vec{i} + F_2\vec{j}$ is a vector field with continuous partial derivatives such that

- The domain of $\vec{F}$ has the property that every closed curve in it encircles a region that lies entirely within the domain. In particular, the domain of $\vec{F}$ has no holes.
- $\dfrac{\partial F_2}{\partial x} - \dfrac{\partial F_1}{\partial y} = 0.$

Then $\vec{F}$ is path-independent, so $\vec{F}$ is a gradient field and has a potential function.

Why Are Holes in the Domain of the Vector Field Important?

The reason for assuming that the domain of the vector field $\vec{F}$ has no holes is to ensure that the region R inside C is actually contained in the domain of $\vec{F}$. Otherwise, we cannot apply Green's Theorem. The next two examples show that if $\partial F_2/\partial x - \partial F_1/\partial y = 0$ but the domain of $\vec{F}$ contains a hole, then $\vec{F}$ can either be path-independent or be path-dependent.

Example 5 Let $\vec{F}$ be the vector field given by $\vec{F}(x, y) = \dfrac{-y\vec{i} + x\vec{j}}{x^2 + y^2}$.

(a) Calculate $\dfrac{\partial F_2}{\partial x} - \dfrac{\partial F_1}{\partial y}$. Does the curl test imply that $\vec{F}$ is path-independent?

(b) Calculate $\int_C \vec{F} \cdot d\vec{r}$, where C is the unit circle centered at the origin and oriented counterclockwise. Is $\vec{F}$ a path-independent vector field?

(c) Explain why the answers to parts (a) and (b) do not contradict Green's Theorem.

Solution (a) Taking partial derivatives, we have

$$\frac{\partial F_2}{\partial x} = \frac{\partial}{\partial x}\left(\frac{x}{x^2+y^2}\right) = \frac{1}{x^2+y^2} - \frac{x \cdot 2x}{(x^2+y^2)^2} = \frac{y^2-x^2}{(x^2+y^2)^2}.$$

Similarly,

$$\frac{\partial F_1}{\partial y} = \frac{\partial}{\partial y}\left(\frac{-y}{x^2+y^2}\right) = \frac{-1}{x^2+y^2} + \frac{y \cdot 2y}{(x^2+y^2)^2} = \frac{y^2-x^2}{(x^2+y^2)^2}.$$

Thus,

$$\frac{\partial F_2}{\partial x} - \frac{\partial F_1}{\partial y} = 0.$$

Since $\vec{F}$ is undefined at the origin, the domain of $\vec{F}$ contains a hole. Therefore, the curl test does not apply.

(b) See Figure 18.49. On the unit circle, $\vec{F}$ is tangent to the circle and $\|\vec{F}\| = 1$. Thus,[3]

$$\int_C \vec{F} \cdot d\vec{r} = \|\vec{F}\| \cdot \text{Length of curve} = 1 \cdot 2\pi = 2\pi.$$

Since the line integral around the closed curve C is nonzero, $\vec{F}$ is not path-independent. We observe that $\vec{F} = \text{grad}(\arctan(y/x))$ and $\arctan(y/x)$ is θ from polar coordinates, for $-\pi/2 < \theta < \pi/2$. The fact that θ increases by 2π each time we wind once around the origin counterclockwise explains why $\vec{F}$ is not path-independent.

(c) The domain of $\vec{F}$ is the "punctured plane," as shown in Figure 18.48. Since $\vec{F}$ is not defined at the origin, which is inside C, Green's Theorem does not apply. In this case

$$2\pi = \int_C \vec{F} \cdot d\vec{r} \neq \int_R \left(\frac{\partial F_2}{\partial x} - \frac{\partial F_1}{\partial y}\right) dx\,dy = 0.$$

[3] See Problem 56 on page 931.

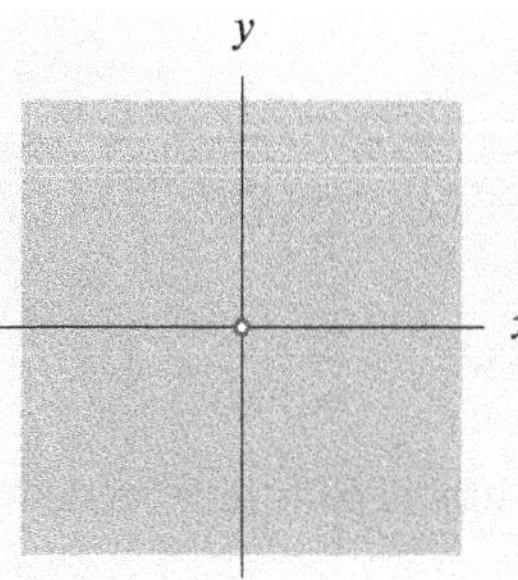

Figure 18.48: The domain of $\vec{F}(x, y) = \frac{-y\vec{i} + x\vec{j}}{x^2+y^2}$ is the plane minus the origin

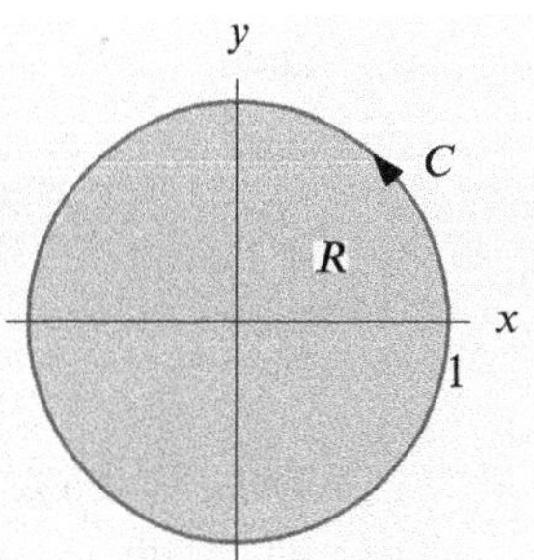

Figure 18.49: The region R is *not* contained in the domain of $\vec{F}(x, y) = \frac{-y\vec{i} + x\vec{j}}{x^2+y^2}$

Although the vector field $\vec{F}$ in the last example was not defined at the origin, this by itself does not prevent the vector field from being path-independent, as we see in the following example.

Example 6 Consider the vector field $\vec{F}$ given by $\vec{F}(x, y) = \dfrac{x\vec{i} + y\vec{j}}{x^2 + y^2}$.

(a) Calculate $\dfrac{\partial F_2}{\partial x} - \dfrac{\partial F_1}{\partial y}$. Does the curl test imply that $\vec{F}$ is path-independent?

(b) Explain how we know that $\int_C \vec{F} \cdot d\vec{r} = 0$, where C is the unit circle centered at the origin and oriented counterclockwise. Does this imply that $\vec{F}$ is path-independent?

(c) Check that $f(x, y) = \frac{1}{2}\ln(x^2 + y^2)$ is a potential function for $\vec{F}$. Does this imply that $\vec{F}$ is path-independent?

Solution (a) Taking partial derivatives, we have

$$\frac{\partial F_2}{\partial x} = \frac{\partial}{\partial x}\left(\frac{y}{x^2 + y^2}\right) = \frac{-2xy}{(x^2 + y^2)^2}, \quad \text{and} \quad \frac{\partial F_1}{\partial y} = \frac{\partial}{\partial y}\left(\frac{x}{x^2 + y^2}\right) = \frac{-2xy}{(x^2 + y^2)^2}.$$

Therefore,

$$\frac{\partial F_2}{\partial x} - \frac{\partial F_1}{\partial y} = 0.$$

This does *not* imply that $\vec{F}$ is path-independent: The domain of $\vec{F}$ contains a hole since $\vec{F}$ is undefined at the origin. Thus, the curl test does not apply.

(b) Since $\vec{F}(x, y) = x\vec{i} + y\vec{j} = \vec{r}$ on the unit circle C, the field $\vec{F}$ is everywhere perpendicular to C. Thus

$$\int_C \vec{F} \cdot d\vec{r} = 0.$$

The fact that $\int_C \vec{F} \cdot d\vec{r} = 0$ when C is the unit circle does *not* imply that $\vec{F}$ is path-independent. To be sure that $\vec{F}$ is path-independent, we would have to show that $\int_C \vec{F} \cdot d\vec{r} = 0$ for *every* closed curve C in the domain of $\vec{F}$, not just the unit circle.

(c) To check that grad $f = \vec{F}$, we differentiate f:

$$f_x = \frac{1}{2}\frac{\partial}{\partial x}\ln(x^2 + y^2) = \frac{1}{2}\frac{2x}{x^2 + y^2} = \frac{x}{x^2 + y^2},$$

and

$$f_y = \frac{1}{2}\frac{\partial}{\partial y}\ln(x^2+y^2) = \frac{1}{2}\frac{2y}{x^2+y^2} = \frac{y}{x^2+y^2},$$

so that

$$\text{grad } f = \frac{x\vec{i} + y\vec{j}}{x^2+y^2} = \vec{F}.$$

Thus, $\vec{F}$ is a gradient field and therefore is path-independent—even though $\vec{F}$ is undefined at the origin.

The Curl Test for Vector Fields in 3-Space

The curl test is a convenient way of deciding whether a 2-dimensional vector field is path-independent. Fortunately, there is an analogous test for 3-dimensional vector fields, although we cannot justify it until Chapter 20.

If $\vec{F}(x, y, z) = F_1\vec{i} + F_2\vec{j} + F_3\vec{k}$ is a vector field on 3-space we define a new vector field, curl $\vec{F}$, on 3-space by

$$\text{curl } \vec{F} = \left(\frac{\partial F_3}{\partial y} - \frac{\partial F_2}{\partial z}\right)\vec{i} + \left(\frac{\partial F_1}{\partial z} - \frac{\partial F_3}{\partial x}\right)\vec{j} + \left(\frac{\partial F_2}{\partial x} - \frac{\partial F_1}{\partial y}\right)\vec{k}.$$

The vector field curl $\vec{F}$ can be used to determine whether the vector field $\vec{F}$ is path-independent.

The Curl Test for Vector Fields in 3-Space

Suppose $\vec{F}$ is a vector field on 3-space with continuous partial derivatives such that

- The domain of $\vec{F}$ has the property that every closed curve in it can be contracted to a point in a smooth way, staying at all times within the domain.
- curl $\vec{F} = \vec{0}$.

Then $\vec{F}$ is path-independent, so $\vec{F}$ is a gradient field and has a potential function.

For the 2-dimensional curl test, the domain of $\vec{F}$ must have no holes. This meant that if $\vec{F}$ was defined on a simple closed curve C, then it was also defined at all points inside C. One way to test for holes is to try to "lasso" them with a closed curve. If every closed curve in the domain can be pulled to a point without hitting a hole, that is, without straying outside the domain, then the domain has no holes. In 3-space, we need the same condition to be satisfied: we must be able to pull every closed curve to a point, like a lasso, without straying outside the domain.

Example 7 Decide if the following vector fields are path-independent and whether or not the curl test applies.

(a) $\vec{F} = \dfrac{x\vec{i} + y\vec{j} + z\vec{k}}{(x^2+y^2+z^2)^{3/2}}$ (b) $\vec{G} = \dfrac{-y\vec{i} + x\vec{j}}{x^2+y^2} + z^2\vec{k}$

Solution

(a) Suppose $f = -(x^2+y^2+z^2)^{-1/2}$. Then $f_x = x(x^2+y^2+z^2)^{-3/2}$ and f_y and f_z are similar, so grad $f = \vec{F}$. Thus, $\vec{F}$ is a gradient field and therefore path-independent. Calculations show curl $\vec{F} = \vec{0}$. The domain of $\vec{F}$ is all of 3-space minus the origin, and any closed curve in the domain can be pulled to a point without leaving the domain. Thus, the curl test applies.

(b) Let C be the circle $x^2+y^2 = 1$, $z = 0$ traversed counterclockwise when viewed from the positive

z-axis. Since $z = 0$ on the curve C, the vector field $\vec{G}$ reduces to the vector field in Example 5 and is everywhere tangent to C and of magnitude 1, so

$$\int_C \vec{G} \cdot d\vec{r} = \|\vec{G}\| \cdot \text{Length of curve} = 1 \cdot 2\pi = 2\pi.$$

Since the line integral around this closed curve is nonzero, $\vec{G}$ is path-dependent. Computations show $\text{curl}\,\vec{G} = \vec{0}$. However, the domain of $\vec{G}$ is all of 3-space minus the z-axis, and it does not satisfy the curl test domain criterion. For example, the circle, C, is lassoed around the z-axis, and cannot be pulled to a point without hitting the z-axis. Thus, the curl test does not apply.

Exercises and Problems for Section 18.4

EXERCISES

In Exercises 1–10, decide if the given vector field is the gradient of a function f. If so, find f. If not, explain why not.

1. $y\vec{i} - x\vec{j}$
2. $2xy\vec{i} + x^2\vec{j}$
3. $y\vec{i} + y\vec{j}$
4. $2xy\vec{i} + 2xy\vec{j}$
5. $(x^2 + y^2)\vec{i} + 2xy\vec{j}$
6. $(2xy^3 + y)\vec{i} + (3x^2y^2 + x)\vec{j}$
7. $\dfrac{\vec{i}}{x} + \dfrac{\vec{j}}{y} + \dfrac{\vec{k}}{z}$
8. $\dfrac{\vec{i}}{x} + \dfrac{\vec{j}}{y} + \dfrac{\vec{k}}{xy}$
9. $2x\cos(x^2 + z^2)\vec{i} + \sin(x^2 + z^2)\vec{j} + 2z\cos(x^2 + z^2)\vec{k}$
10. $\dfrac{y}{x^2 + y^2}\vec{i} - \dfrac{x}{x^2 + y^2}\vec{j}$

In Exercises 11–14, use Green's Theorem to calculate the circulation of $\vec{F}$ around the curve, oriented counterclockwise.

11. $\vec{F} = y\vec{i} - x\vec{j}$ around the unit circle.
12. $\vec{F} = xy\vec{j}$ around the square $0 \le x \le 1, 0 \le y \le 1$.
13. $\vec{F} = (2x^2+3y)\vec{i} + (2x+3y^2)\vec{j}$ around the triangle with vertices $(2, 0)$, $(0, 3)$, $(-2, 0)$.
14. $\vec{F} = 3y\vec{i} + xy\vec{j}$ around the unit circle.
15. Use Green's Theorem to evaluate $\int_C \left(y^2\vec{i} + x\vec{j}\right) \cdot d\vec{r}$ where C is the counterclockwise path around the perimeter of the rectangle $0 \le x \le 2, 0 \le y \le 3$.
16. If C goes counterclockwise around the perimeter of the rectangle R with vertices $(10, 10)$, $(30, 10)$, $(30, 20)$, and $(10, 20)$, use Green's theorem to evaluate
$$\int_C -y\,dx + x\,dy.$$
17. Calculate $\int_C ((3x + 5y)\vec{i} + (2x + 7y)\vec{j}) \cdot d\vec{r}$ where C is the circular path with center (a, b) and radius m, oriented counterclockwise. Use Green's Theorem.

PROBLEMS

18. Find the line integral of $\vec{F} = e^{x^2}\vec{i} + (7x + 1)\vec{j}$ around the closed curve C consisting of the two line segments and the circular arc in Figure 18.50.

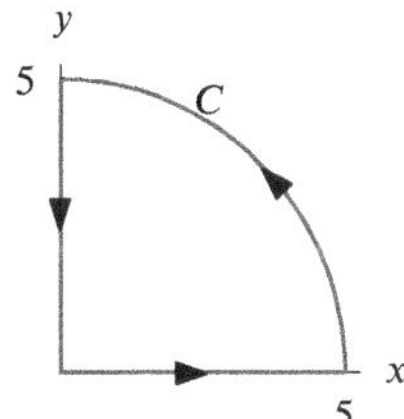

Figure 18.50

19. **(a)** Sketch $\vec{F} = y\vec{i}$ and determine the sign of the circulation of $\vec{F}$ around the unit circle centered at the origin and oriented counterclockwise.
    (b) Use Green's Theorem to compute the circulation in part (a) exactly.
20. Let $\vec{F} = (\sin x)\vec{i} + (x + y)\vec{j}$. Find the line integral of $\vec{F}$ around the perimeter of the rectangle with corners $(3, 0)$, $(3, 5)$, $(-1, 5)$, $(-1, 0)$, traversed in that order.
21. Find $\displaystyle\int_C ((\sin(x^2)\cos y)\vec{i} + (\sin(y^2) + e^x)\vec{j}) \cdot d\vec{r}$ where C is the square of side 1 in the first quadrant of the xy-plane, with one vertex at the origin and sides along the axes, and oriented counterclockwise when viewed from above.

In Problems 22–23 find the line integral of $\vec{F}$ around the closed curve in Figure 18.51. The arc is part of a circle.

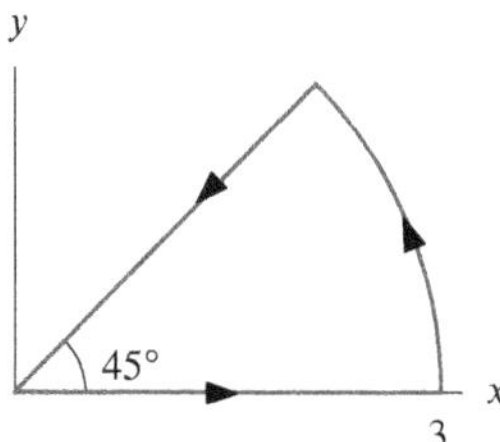

Figure 18.51

22. $\vec{F} = (x - y)\vec{i} + x\vec{j}$
23. $\vec{F} = (x + y)\vec{i} + \sin y\vec{j}$
24. Let $\vec{F} = 2xe^y\vec{i} + x^2e^y\vec{j}$ and $\vec{G} = (x - y)\vec{i} + (x + y)\vec{j}$. Let C be the line from $(0, 0)$ to $(2, 4)$. Find exactly:
    (a) $\int_C \vec{F} \cdot d\vec{r}$ (b) $\int_C \vec{G} \cdot d\vec{r}$
25. Let $\vec{F} = y\vec{i} + x\vec{j}$ and $\vec{G} = 3y\vec{i} - 3x\vec{j}$. In Figure 18.52, the curve C_2 is the semicircle centered at the origin from $(-1, 1)$ to $(1, -1)$ and C_1 is the line segment from $(-1, 1)$ to $(1, -1)$, and $C = C_2 - C_1$. Find the following line integrals:
    (a) $\int_{C_1} \vec{F} \cdot d\vec{r}$ (b) $\int_C \vec{F} \cdot d\vec{r}$
    (c) $\int_{C_2} \vec{F} \cdot d\vec{r}$ (d) $\int_{C_2} \vec{G} \cdot d\vec{r}$
    (e) $\int_C \vec{G} \cdot d\vec{r}$ (f) $\int_{C_1} \vec{G} \cdot d\vec{r}$
    (g) $\int_C (\vec{F} + \vec{G}) \cdot d\vec{r}$

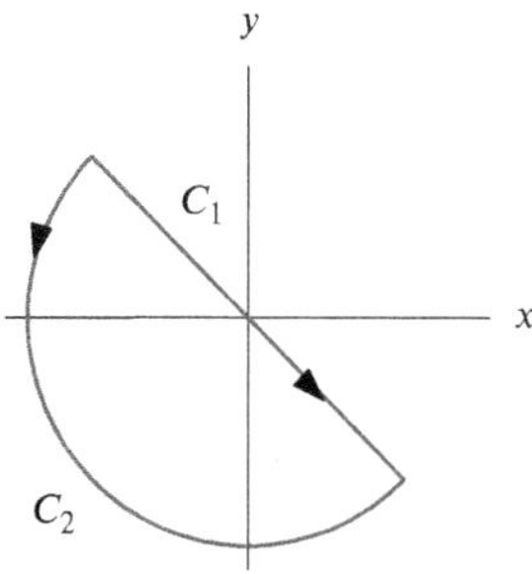

Figure 18.52

26. Calculate $\int_C \left((x^2 - y)\vec{i} + (y^2 + x)\vec{j}\right) \cdot d\vec{r}$ if:
    (a) C is the circle $(x - 5)^2 + (y - 4)^2 = 9$ oriented counterclockwise.
    (b) C is the circle $(x - a)^2 + (y - b)^2 = R^2$ in the xy-plane oriented counterclockwise.
27. Let C_1 be the curve consisting of the circle of radius 2, centered at the origin and oriented counterclockwise, and C_2 be the curve consisting of the line segment from $(0, 0)$ to $(1, 1)$ followed by the line segment from $(1, 1)$ to $(3, 1)$. Let $\vec{F} = 2xy^2\vec{i} + (2yx^2 + 2y)\vec{j}$ and let $\vec{G} = (y + x)\vec{i} + (y - x)\vec{j}$. Compute the following line integrals.
    (a) $\int_{C_1} \vec{F} \cdot d\vec{r}$ (b) $\int_{C_2} \vec{F} \cdot d\vec{r}$
    (c) $\int_{C_1} \vec{G} \cdot d\vec{r}$ (d) $\int_{C_2} \vec{G} \cdot d\vec{r}$
28. Prove that Green's theorem is true when the integrand of the line integral has a potential function.
29. Consider the following parametric equations:
$$C_1 : \vec{r}(t) = t\cos(2\pi t)\vec{i} + t\sin(2\pi t)\vec{k}, 0 \le t \le 2$$
$$C_2 : \vec{r}(t) = t\cos(2\pi t)\vec{i} + t\vec{j} + t\sin(2\pi t)\vec{k}, 0 \le t \le 2$$
    (a) Describe, in words, the motion of a particle moving along each of the paths.
    (b) Evaluate $\int_{C_2} \vec{F} \cdot d\vec{r}$, for the vector field $\vec{F} = yz\vec{i} + z(x + 1)\vec{j} + (xy + y + 1)\vec{k}$.
    (c) Find a nonzero vector field $\vec{G}$ such that:
$$\int_{C_1} \vec{G} \cdot d\vec{r} = \int_{C_2} \vec{G} \cdot d\vec{r}.$$
    Explain how you reasoned to find $\vec{G}$.
    (d) Find two different, nonzero vector fields $\vec{H}_1$, $\vec{H}_2$ such that:
$$\int_{C_1} \vec{H}_1 \cdot d\vec{r} = \int_{C_1} \vec{H}_2 \cdot d\vec{r}.$$
    Explain how you reasoned to find the two fields.
30. Show that the line integral of $\vec{F} = x\vec{j}$ around a closed curve in the xy-plane, oriented as in Green's Theorem, measures the area of the region enclosed by the curve.

In Problems 31–33, use the result of Problem 30 to calculate the area of the region within the parameterized curves. In each case, sketch the curve.

31. The ellipse $x^2/a^2 + y^2/b^2 = 1$ parameterized by $x = a\cos t$, $y = b\sin t$, for $0 \le t \le 2\pi$.
32. The hypocycloid $x^{2/3} + y^{2/3} = a^{2/3}$ parameterized by $x = a\cos^3 t$, $y = a\sin^3 t$, $0 \le t \le 2\pi$.
33. The folium of Descartes, $x^3 + y^3 = 3xy$, parameterized by $x = \dfrac{3t}{1 + t^3}$, $y = \dfrac{3t^2}{1 + t^3}$, for $0 \le t < \infty$.
34. The vector field $\vec{F}$ is defined on the disk D of radius 5 centered at the origin in the plane:
$$\vec{F} = (-y^3 + y\sin(xy))\vec{i} + (4x(1 - y^2) + x\sin(xy))\vec{j}.$$
    Consider the line integral $\int_C \vec{F} \cdot d\vec{r}$, where C is some closed curve contained in D. For which C is the value of this integral the largest? [Hint: Assume C is a closed curve, made up of smooth pieces and never crossing itself, and oriented counterclockwise.]

35. Example 1 on page 949 showed that the vector field in Figure 18.53 could not be a gradient field by showing that it is not path-independent. Here is another way to see the same thing. Suppose that the vector field were the gradient of a function f. Draw and label a diagram showing what the contours of f would have to look like, and explain why it would not be possible for f to have a single value at any given point.

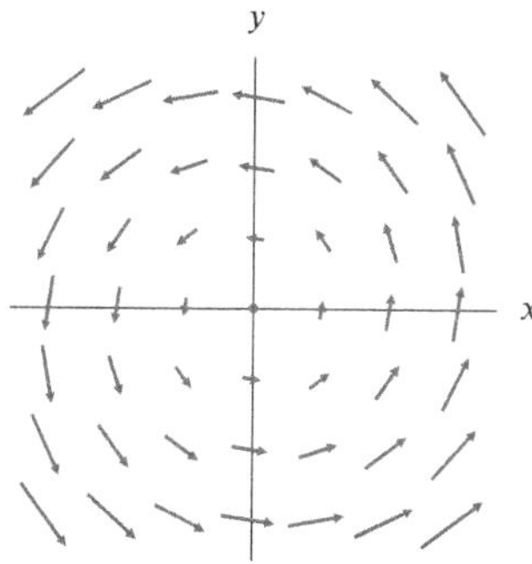

Figure 18.53

36. Repeat Problem 35 for the vector field in Problem 52 on page 947.

37. **(a)** By finding potential functions, show that each of the vector fields $\vec{F}, \vec{G}, \vec{H}$ is a gradient field on some domain (not necessarily the whole plane).

(b) Find the line integrals of $\vec{F}, \vec{G}, \vec{H}$ around the unit circle in the xy-plane, centered at the origin, and traversed counterclockwise.

(c) For which of the three vector fields can Green's Theorem be used to calculate the line integral in part (b)? Why or why not?

$$\vec{F} = y\vec{i} + x\vec{j}, \ \vec{G} = \frac{y\vec{i} - x\vec{j}}{x^2 + y^2}, \ \vec{H} = \frac{x\vec{i} + y\vec{j}}{(x^2 + y^2)^{1/2}}$$

38. **(a)** For which of the following can you use Green's Theorem to evaluate the integral? Explain.

I $\int_C (x^2 + y^2)\,dx + (x^2 + y^2)\,dy$ where C is the curve defined by $y = x$, $y = x^2, 0 \le x \le 1$ with counterclockwise orientation.

II $\int_C \frac{1}{\sqrt{x^2 + y^2}}\,dx - \frac{1}{\sqrt{x^2 + y^2}}\,dy$ where C is the unit circle centered at the origin, oriented counterclockwise.

III $\int_C \vec{F} \cdot d\vec{r}$ where $\vec{F} = x\vec{i} + y\vec{j}$ where C is the line segment from the origin to $(1, 1)$.

(b) Use Green's Theorem to evaluate the integrals in part (a) that can be done that way.

39. Arrange the line integrals L_1, L_2, L_3 in ascending order, where

$$L_i = \int_{C_i} (-x^2 y\vec{i} + (xy^2 - x)\vec{j}) \cdot d\vec{r}.$$

The points A, B, D lie on the unit circle and C_i is one of the curves shown in Figure 18.54.

C_1: Line segment A to B

C_2: Line segment A to D followed by line segment D to B

C_3: Semicircle ADB

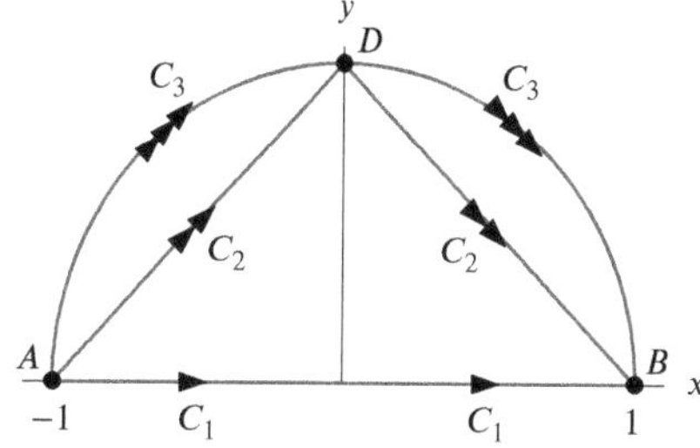

Figure 18.54

40. For all x, y, let $\vec{F} = F_1(x, y)\vec{i} + F_2(x, y)\vec{j}$ satisfy

$$\frac{\partial F_2}{\partial x} - \frac{\partial F_1}{\partial y} = 3.$$

(a) Calculate $\int_{C_1} \vec{F} \cdot d\vec{r}$ where C_1 is the unit circle in the xy-plane centered at the origin, oriented counterclockwise.

(b) Calculate $\int_{C_2} \vec{F} \cdot d\vec{r}$ where C_2 is the boundary of the rectangle of $4 \le x \le 7, 5 \le y \le 7$, oriented counterclockwise.

(c) Let C_3 be the circle of radius 7 centered at the point $(10, 2)$; let C_4 be the circle of radius 8 centered at the origin; let C_5 be the square of side 14 centered at $(7, 7)$ with sides parallel to the axes; C_3, C_4, C_5 are all oriented counterclockwise. Arrange the integrals $\int_{C_3} \vec{F} \cdot d\vec{r}$, $\int_{C_4} \vec{F} \cdot d\vec{r}$, $\int_{C_5} \vec{F} \cdot d\vec{r}$ in increasing order.

41. Let $\vec{F} = (3x^2y + y^3 + e^x)\vec{i} + (e^{y^2} + 12x)\vec{j}$. Consider the line integral of $\vec{F}$ around the circle of radius a, centered at the origin and traversed counterclockwise.

(a) Find the line integral for $a = 1$.

(b) For which value of a is the line integral a maximum? Explain.

42. Let

$$\vec{F}(x, y) = \frac{-y\vec{i} + x\vec{j}}{x^2 + y^2}$$

and let oriented curves C_1 and C_2 be as in Figure 18.55. The curve C_2 is an arc of the unit circle centered at the origin. Show that

(a) The curl of $\vec{F}$ is zero.

(b) $\int_{C_1} \vec{F} \cdot d\vec{r} = \int_{C_2} \vec{F} \cdot d\vec{r}$.

(c) $\int_{C_1} \vec{F} \cdot d\vec{r} = \theta$, the angle at the origin subtended by the oriented curve C_1.

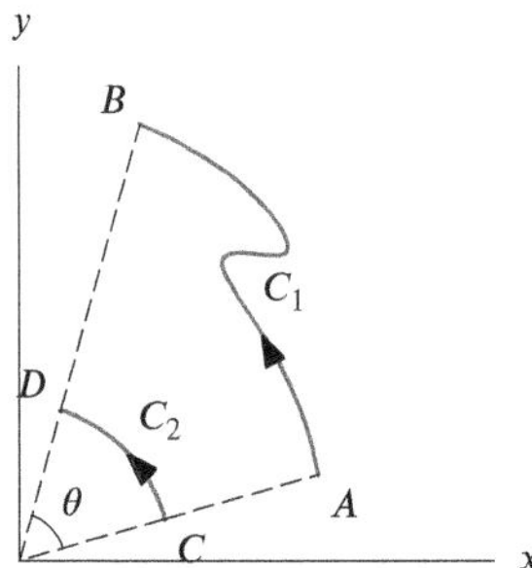

Figure 18.55

43. The electric field $\vec{E}$, at the point with position vector $\vec{r}$ in 3-space, due to a charge q at the origin is given by

$$\vec{E}(\vec{r}) = q\frac{\vec{r}}{\|\vec{r}\|^3}.$$

(a) Compute curl $\vec{E}$. Is $\vec{E}$ a path-independent vector field? Explain.

(b) Find a potential function φ for $\vec{E}$, if possible.

Strengthen Your Understanding

In Problems 44–45, explain what is wrong with the statement.

44. If $\int_C \vec{F} \cdot d\vec{r} = 0$ for a specific closed path C, then $\vec{F}$ must be path-independent.

45. Let $\vec{F} = F_1(x,y)\vec{i} + F_2(x,y)\vec{j}$ with

$$\frac{\partial F_2}{\partial x} - \frac{\partial F_1}{\partial y} = 3$$

and let C be the path consisting of line segments from $(0,0)$ to $(1,1)$ to $(2,0)$. Then

$$\int_C \vec{F} \cdot d\vec{r} = 3.$$

In Problems 46–48, give an example of:

46. A function $Q(x,y)$ such that $\vec{F} = xy\vec{i} + Q(x,y)\vec{j}$ is a gradient field.

47. Two oriented curves, C_1 and C_2, from $(1,0)$ to $(0,1)$ such that if

$$\vec{F}(x,y) = \frac{-y\vec{i} + x\vec{j}}{x^2+y^2},$$

then

$$\int_{C_1} \vec{F} \cdot d\vec{r} \neq \int_{C_2} \vec{F} \cdot d\vec{r}.$$

[Note that the scalar curl of $\vec{F}$ is 0 where $\vec{F}$ is defined.]

48. A vector field that is not a gradient field.

Are the statements in Problems 49–56 true or false? Give reasons for your answer.

49. If $f(x)$ and $g(y)$ are continuous one-variable functions, then the vector field $\vec{F} = f(x)\vec{i} + g(y)\vec{j}$ is path-independent.

50. If $\vec{F} = \operatorname{grad} f$, and C is the perimeter of a square of side length a oriented counterclockwise and surrounding the region R, then

$$\int_C \vec{F} \cdot d\vec{r} = \int_R f\,dA.$$

51. If $\vec{F}$ and $\vec{G}$ are both path-independent vector fields, then $\vec{F} + \vec{G}$ is path-independent.

52. If $\vec{F}$ and $\vec{G}$ are both path-dependent vector fields, then $\vec{F} + \vec{G}$ is path-dependent.

53. The vector field $\vec{F}(\vec{r}) = \vec{r}$ in 3-space is path-independent.

54. A constant vector field $\vec{F} = a\vec{i} + b\vec{j}$ is path-independent.

55. If $\vec{F}$ is path-independent and k is a constant, then the vector field $k\vec{F}$ is path-independent.

56. If $\vec{F}$ is path-independent and $h(x,y)$ is a scalar function, then the vector field $h(x,y)\vec{F}$ is path-independent.

Online Resource: Review problems and Projects

Chapter Nineteen

FLUX INTEGRALS AND DIVERGENCE

Contents

19.1 THE IDEA OF A FLUX INTEGRAL

Flow Through a Surface

Imagine water flowing through a fishing net stretched across a stream. Suppose we want to measure the flow rate of water through the net, that is, the volume of fluid that passes through the surface per unit time.

Example 1 A flat square surface of area A, in m^2, is immersed in a fluid. The fluid flows with constant velocity $\vec{v}$, in m/sec, perpendicular to the square. Write an expression for the rate of flow in m^3/sec.

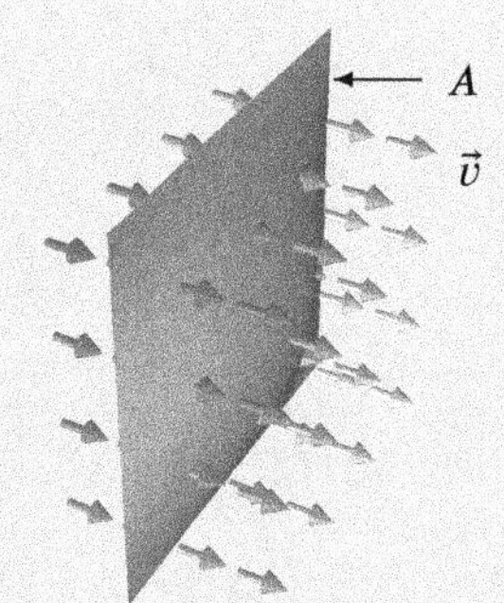

Figure 19.1: Fluid flowing perpendicular to a surface

Solution In one second a given particle of water moves a distance of $\|\vec{v}\,\|$ in the direction perpendicular to the square. Thus, the entire body of water moving through the square in one second is a box of length $\|\vec{v}\,\|$ and cross-sectional area A. So the box has volume $\|\vec{v}\,\|A$ m^3, and

$$\text{Flow rate} = \|\vec{v}\,\|A \text{ m}^3/\text{sec}.$$

This flow rate is called the *flux* of the fluid through the surface. We can also compute the flux of vector fields, such as electric and magnetic fields, where no flow is actually taking place. If the vector field is constant and perpendicular to the surface, and if the surface is flat, as in Example 1, the flux is obtained by multiplying the speed by the area.

Next we find the flux of a constant vector field through a flat surface that is not perpendicular to the vector field, using a dot product. In general, we break a surface into small pieces which are approximately flat and where the vector field is approximately constant, leading to a flux integral.

Orientation of a Surface

Before computing the flux of a vector field through a surface, we need to decide which direction of flow through the surface is the positive direction; this is described as choosing an orientation.[1]

> At each point on a smooth surface there are two unit normals, one in each direction. **Choosing an orientation** means picking one of these normals at every point of the surface in a continuous way. The unit normal vector in the direction of the orientation is denoted by $\vec{n}$. For a closed surface (that is, the boundary of a solid region), we choose the **outward orientation** unless otherwise specified.

We say the flux through a piece of surface is positive if the flow is in the direction of the orientation and negative if it is in the opposite direction. (See Figure 19.2.)

[1] Although we will not study them, there are a few surfaces for which this cannot be done. See page 968.

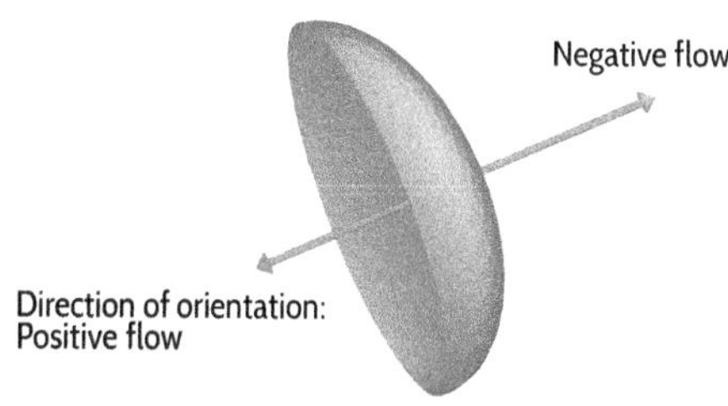

Figure 19.2: An oriented surface showing directions of positive and negative flow

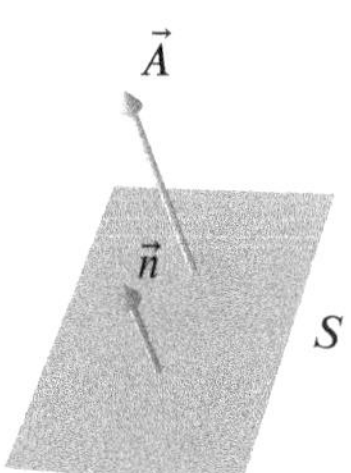

Figure 19.3: Area vector $\vec{A} = \vec{n} A$ of flat surface with area A and orientation $\vec{n}$

The Area Vector

The flux through a flat surface depends both on the area of the surface and its orientation. Thus, it is useful to represent its area by a vector as shown in Figure 19.3.

> The **area vector** of a flat, oriented surface is a vector $\vec{A}$ such that
> - The magnitude of $\vec{A}$ is the area of the surface.
> - The direction of $\vec{A}$ is the direction of the orientation vector $\vec{n}$.

The Flux of a Constant Vector Field Through a Flat Surface

Suppose the velocity vector field, $\vec{v}$, of a fluid is constant and $\vec{A}$ is the area vector of a flat surface. The flux through this surface is the volume of fluid that flows through in one unit of time. The skewed box in Figure 19.4 has cross-sectional area $\|\vec{A}\|$ and height $\|\vec{v}\|\cos\theta$, so its volume is $(\|\vec{v}\|\cos\theta)\|\vec{A}\| = \vec{v}\cdot\vec{A}$. Thus, we have the following result:

> If $\vec{v}$ is constant and $\vec{A}$ is the area vector of a flat surface, then
>
> $$\text{Flux through surface} = \vec{v}\cdot\vec{A}.$$

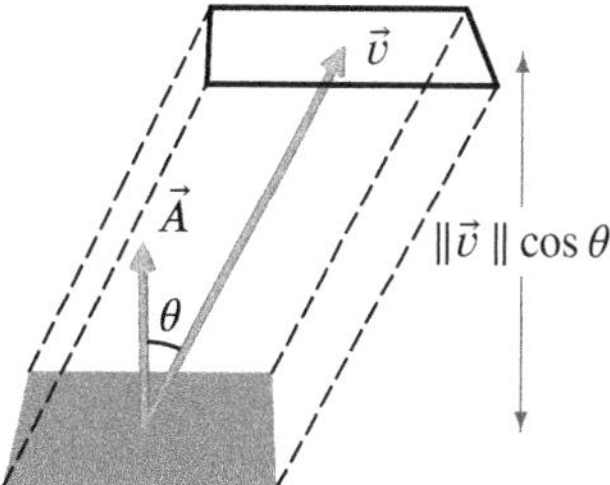

Figure 19.4: Flux of $\vec{v}$ through a surface with area vector $\vec{A}$ is the volume of this skewed box

Example 2 Water is flowing down a cylindrical pipe 2 cm in radius with a velocity of 3 cm/sec. Find the flux of the velocity vector field through the ellipse-shaped region shown in Figure 19.5. The normal to the ellipse makes an angle of θ with the direction of flow and the area of the ellipse is $4\pi/(\cos\theta)$ cm^2.

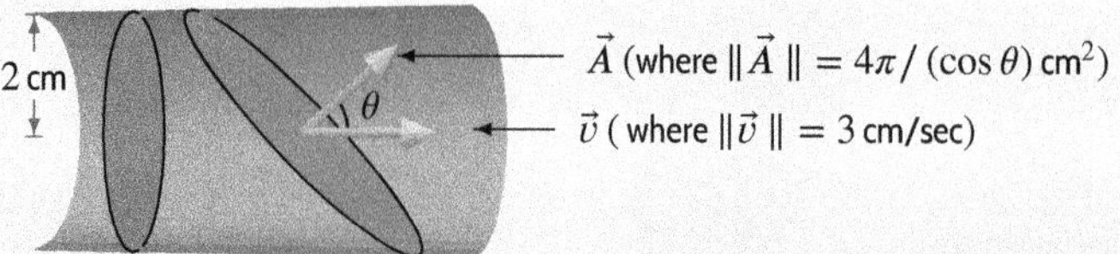

Figure 19.5: Flux through ellipse-shaped region across a cylindrical pipe

Solution There are two ways to approach this problem. One is to use the formula we just derived, which gives

$$\begin{aligned}\text{Flux through ellipse} &= \vec{v}\cdot\vec{A} = \|\vec{v}\|\|\vec{A}\|\cos\theta = 3(\text{Area of ellipse})\cos\theta\\ &= 3\left(\frac{4\pi}{\cos\theta}\right)\cos\theta = 12\pi\ \text{cm}^3/\text{sec}.\end{aligned}$$

The second way is to notice that the flux through the ellipse is equal to the flux through the circle perpendicular to the pipe in Figure 19.5. Since the flux is the rate at which water is flowing down the pipe, we have

$$\text{Flux through circle} = \begin{matrix}\text{Velocity}\\ \text{of water}\end{matrix} \times \begin{matrix}\text{Area of}\\ \text{circle}\end{matrix} = \left(3\,\frac{\text{cm}}{\text{sec}}\right)(\pi 2^2\ \text{cm}^2) = 12\pi\ \text{cm}^3/\text{sec}.$$

The Flux Integral

If the vector field, $\vec{F}$, is not constant or the surface, S, is not flat, we divide the surface into a patchwork of small, almost flat pieces. (See Figure 19.6.) For a particular patch with area ΔA, we pick a unit orientation vector $\vec{n}$ at a point on the patch and define the area vector of the patch, $\Delta\vec{A}$, as

$$\Delta\vec{A} = \vec{n}\,\Delta A.$$

(See Figure 19.7.) If the patches are small enough, we can assume that $\vec{F}$ is approximately constant on each piece. Then we know that

$$\text{Flux through patch} \approx \vec{F}\cdot\Delta\vec{A},$$

so, adding the fluxes through all the small pieces, we have

$$\text{Flux through whole surface} \approx \sum \vec{F}\cdot\Delta\vec{A},$$

As each patch becomes smaller and $\|\Delta\vec{A}\| \to 0$, the approximation gets better and we get

$$\text{Flux through } S = \lim_{\|\Delta\vec{A}\|\to 0} \sum \vec{F}\cdot\Delta\vec{A}.$$

Thus, provided the limit exists, we make the following definition:

> The **flux integral** of the vector field $\vec{F}$ through the oriented surface S is
>
> $$\int_S \vec{F}\cdot d\vec{A} = \lim_{\|\Delta\vec{A}\|\to 0} \sum \vec{F}\cdot\Delta\vec{A}.$$
>
> If S is a closed surface oriented outward, we describe the flux through S as the flux out of S.

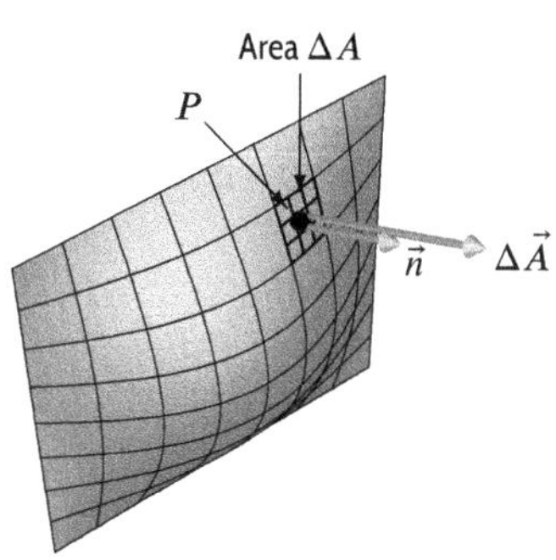

Figure 19.6: Surface S divided into small, almost flat pieces, showing a typical orientation vector $\vec{n}$ and area vector $\Delta \vec{A}$

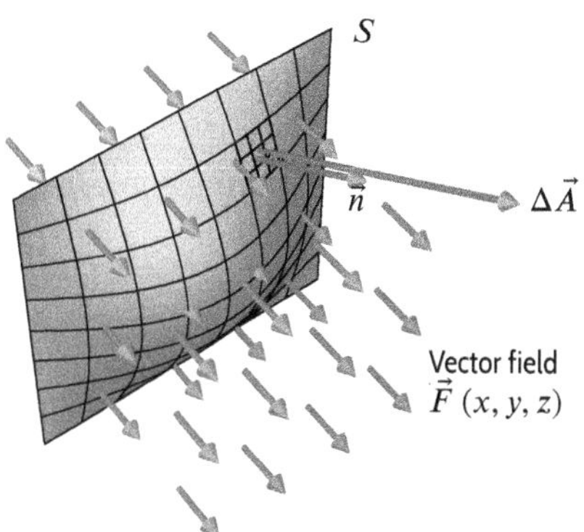

Figure 19.7: Flux of a vector field through a curved surface S

In computing a flux integral, we have to divide the surface up in a reasonable way, or the limit might not exist. In practice this problem seldom arises; however, one way to avoid it is to define flux integrals by the method used to compute them shown in Section 21.3.

Flux and Fluid Flow

If $\vec{v}$ is the velocity vector field of a fluid, we have

$$\text{Rate fluid flows through surface } S = \text{Flux of } \vec{v} \text{ through } S = \int_S \vec{v} \cdot d\vec{A}$$

The rate of fluid flow is measured in units of volume per unit time.

Example 3 Find the flux of the vector field $\vec{B}(x, y, z)$ shown in Figure 19.8 through the square S of side 2 shown in Figure 19.9, oriented in the $\vec{j}$ direction, where

$$\vec{B}(x, y, z) = \frac{-y\vec{i} + x\vec{j}}{x^2 + y^2}.$$

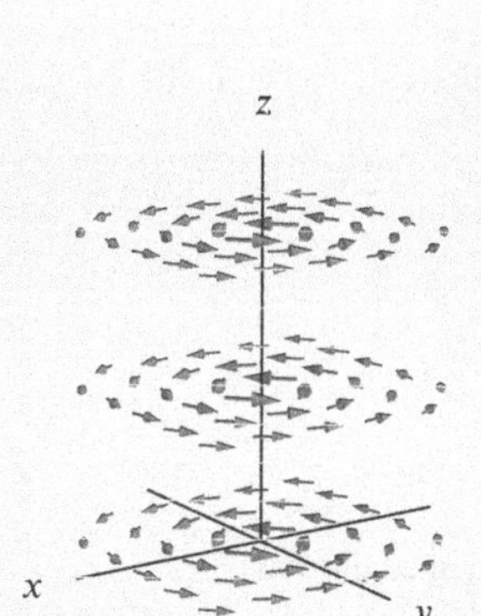

Figure 19.8: The vector field $\vec{B}$ in planes $z = 0$, $z = 1$, $z = 2$, where $\vec{B}(x, y, z) = \dfrac{-y\vec{i} + x\vec{j}}{x^2 + y^2}$

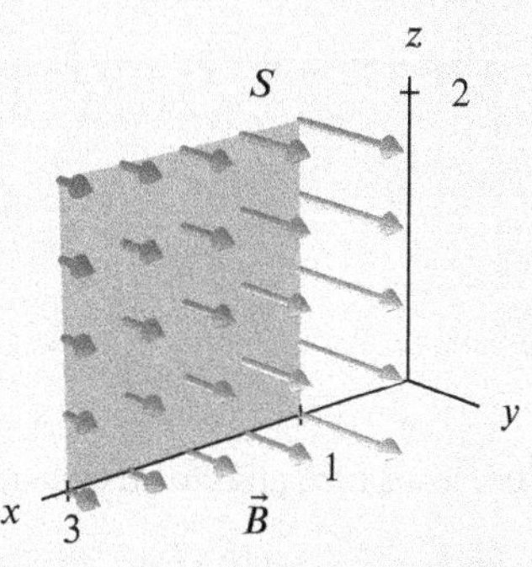

Figure 19.9: Flux of $\vec{B}$ through the square S of side 2 in xy-plane and oriented in $\vec{j}$ direction

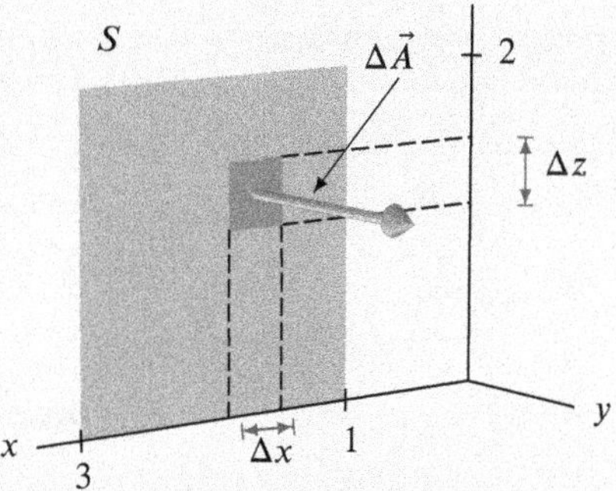

Figure 19.10: A small patch of surface with area $\|\Delta \vec{A}\| = \Delta x \Delta z$

Solution Consider a small rectangular patch with area vector $\Delta \vec{A}$ in S, with sides Δx and Δz so that $\|\Delta \vec{A}\| = \Delta x\, \Delta z$. Since $\Delta \vec{A}$ points in the $\vec{j}$ direction, we have $\Delta \vec{A} = \vec{j}\, \Delta x \Delta z$. (See Figure 19.10.)

At the point $(x, 0, z)$ in S, substituting $y = 0$ into $\vec{B}$ gives $\vec{B}(x, 0, z) = (1/x)\vec{j}$. Thus, we have

$$\text{Flux through small patch} \approx \vec{B} \cdot \Delta \vec{A} = \left(\frac{1}{x}\vec{j}\right) \cdot (\vec{j}\, \Delta x \Delta z) = \frac{1}{x}\, \Delta x\, \Delta z.$$

Therefore,

$$\text{Flux through surface} = \int_S \vec{B} \cdot d\vec{A} = \lim_{\|\Delta \vec{A}\| \to 0} \sum \vec{B} \cdot \Delta \vec{A} = \lim_{\substack{\Delta x \to 0 \\ \Delta z \to 0}} \sum \frac{1}{x} \Delta x \, \Delta z.$$

This last expression is a Riemann sum for the double integral $\int_R \frac{1}{x} \, dA$, where R is the square $1 \leq x \leq 3,\ 0 \leq z \leq 2$. Thus,

$$\text{Flux through surface} = \int_S \vec{B} \cdot d\vec{A} = \int_R \frac{1}{x} \, dA = \int_0^2 \int_1^3 \frac{1}{x} \, dx \, dz = 2 \ln 3.$$

The result is positive since the vector field is passing through the surface in the positive direction.

Example 4 Each of the vector fields in Figure 19.11 consists entirely of vectors parallel to the xy-plane, and is constant in the z direction (that is, the vector field looks the same in any plane parallel to the xy-plane). For each one, say whether you expect the flux through a closed surface surrounding the origin to be positive, negative, or zero. In part (a) the surface is a closed cube with faces perpendicular to the axes; in parts (b) and (c) the surface is a closed cylinder. In each case we choose the outward orientation. (See Figure 19.12.)

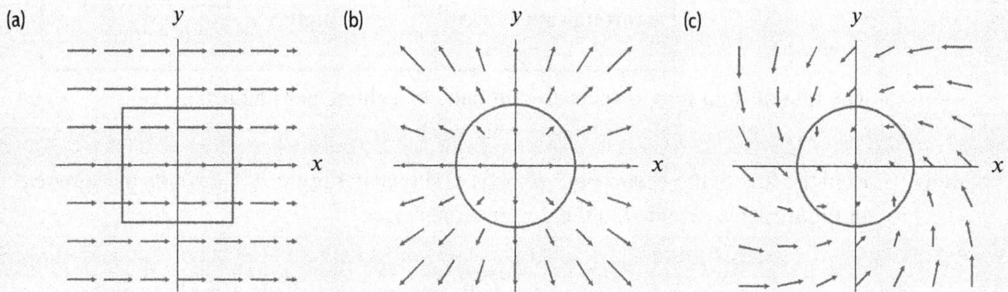

Figure 19.11: Flux of a vector field through the closed surfaces whose cross-sections are shown in the xy-plane

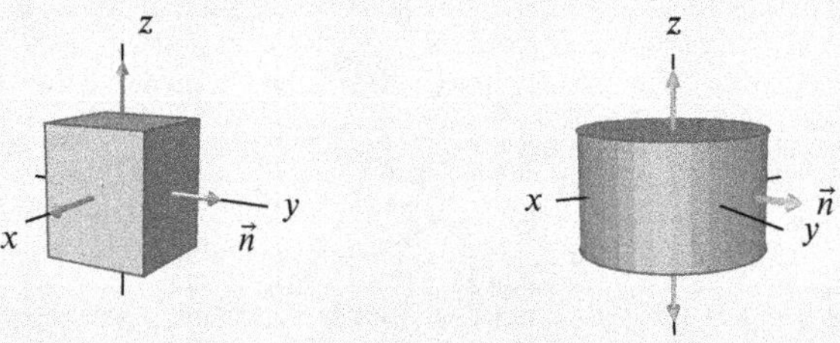

Figure 19.12: The closed cube and closed cylinder, both oriented outward

Solution

(a) Since the vector field appears to be parallel to the faces of the cube which are perpendicular to the y- and z-axes, we expect the flux through these faces to be zero. The fluxes through the two faces perpendicular to the x-axis appear to be equal in magnitude and opposite in sign, so we expect the net flux to be zero.

(b) Since the top and bottom of the cylinder are parallel to the flow, the flux through them is zero. On the curved surface of the cylinder, $\vec{v}$ and $\Delta \vec{A}$ appear to be everywhere parallel and in the same direction, so we expect each term $\vec{v} \cdot \Delta \vec{A}$ to be positive, and therefore the flux integral $\int_S \vec{v} \cdot d\vec{A}$ to be positive.

(c) As in part (b), the flux through the top and bottom of the cylinder is zero. In this case $\vec{v}$ and $\Delta \vec{A}$ are not parallel on the round surface of the cylinder, but since the fluid appears to be flowing inward as well as swirling, we expect each term $\vec{v} \cdot \Delta \vec{A}$ to be negative, and therefore the flux integral to be negative.

Calculating Flux Integrals Using $d\vec{A} = \vec{n}\, dA$

For a small patch of surface ΔS with unit normal $\vec{n}$ and area ΔA, the area vector is $\Delta\vec{A} = \vec{n}\,\Delta A$. The next example shows how we can use this relationship to compute a flux integral.

Example 5 An electric charge q is placed at the origin in 3-space. The resulting electric field $\vec{E}(\vec{r})$ at the point with position vector $\vec{r}$ is given by

$$\vec{E}(\vec{r}) = q\frac{\vec{r}}{\|\vec{r}\|^3}, \qquad \vec{r} \neq \vec{0}.$$

Find the flux of $\vec{E}$ out of the sphere of radius R centered at the origin. (See Figure 19.13.)

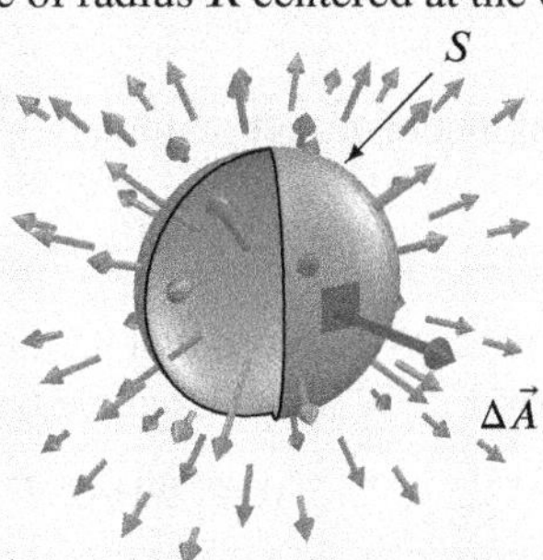

Figure 19.13: Flux of $\vec{E} = q\vec{r}/\|\vec{r}\|^3$ through the surface of a sphere of radius R centered at the origin

Solution This vector field points radially outward from the origin in the same direction as $\vec{n}$. Thus, since $\vec{n}$ is a unit vector,

$$\vec{E}\cdot\Delta\vec{A} = \vec{E}\cdot\vec{n}\,\Delta A = \|\vec{E}\|\,\Delta A.$$

On the sphere, $\|\vec{E}\| = q/R^2$, so

$$\int_S \vec{E}\cdot d\vec{A} = \lim_{\|\Delta\vec{A}\|\to 0}\sum \vec{E}\cdot\Delta\vec{A} = \lim_{\Delta A\to 0}\sum \frac{q}{R^2}\Delta A = \frac{q}{R^2}\lim_{\Delta A\to 0}\sum \Delta A.$$

The last sum approximates the surface area of the sphere. In the limit as the subdivisions get finer we have

$$\lim_{\Delta A\to 0}\sum \Delta A = \text{Surface area of sphere}.$$

Thus, the flux is given by

$$\int_S \vec{E}\cdot d\vec{A} = \frac{q}{R^2}\lim_{\Delta A\to 0}\sum \Delta A = \frac{q}{R^2}\cdot(\text{Surface area of sphere}) = \frac{q}{R^2}(4\pi R^2) = 4\pi q.$$

This result is known as Gauss's law.

To compute a flux with an integral instead of Riemann sums, we often write $d\vec{A} = \vec{n}\, dA$, as in the next example.

Example 6 Let S be the surface of the cube bounded by the six planes $x = \pm 1$, $y = \pm 1$, and $z = \pm 1$. Compute the flux of the electric field $\vec{E}$ of the previous example outward through S.

Solution It is enough to compute the flux of $\vec{E}$ through a single face, say the top face S_1 defined by $z = 1$, where $-1 \le x \le 1$ and $-1 \le y \le 1$. By symmetry, the flux of $\vec{E}$ through the other five faces of S must be the same.

On the top face, S_1, we have $d\vec{A} = \vec{n}\, dA = \vec{k}\, dx\, dy$ and

$$\vec{E}(x, y, 1) = q\frac{x\vec{i} + y\vec{j} + \vec{k}}{(x^2 + y^2 + 1)^{3/2}}.$$

The corresponding flux integral is given by

$$\int_{S_1} \vec{E} \cdot d\vec{A} = q\int_{-1}^{1}\int_{-1}^{1} \frac{x\vec{i} + y\vec{j} + \vec{k}}{(x^2+y^2+1)^{3/2}} \cdot \vec{k}\, dx\, dy = q\int_{-1}^{1}\int_{-1}^{1} \frac{1}{(x^2+y^2+1)^{3/2}}\, dx\, dy.$$

Computing this integral numerically shows that

$$\text{Flux through top face} = \int_{S_1} \vec{E} \cdot d\vec{A} \approx 2.0944q.$$

Thus,

$$\text{Total flux of } \vec{E} \text{ out of cube} = \int_{S} \vec{E} \cdot d\vec{A} \approx 6(2.0944q) = 12.5664q.$$

Example 5 on page 967 showed that the flux of $\vec{E}$ through a sphere of radius R centered at the origin is $4\pi q$. Since $4\pi \approx 12.5664$, Example 6 suggests that

$$\text{Total flux of } \vec{E} \text{ out of cube} = 4\pi q.$$

By computing the flux integral in Example 6 exactly, it is possible to verify that the flux of $\vec{E}$ through the cube and the sphere are exactly equal. When we encounter the Divergence Theorem in Chapter 20 we will see why this is so.

Notes on Orientation

Two difficulties can occur in choosing an orientation. The first is that if the surface is not smooth, it may not have a normal vector at every point. For example, a cube does not have a normal vector along its edges. When we have a surface, such as a cube, which is made of a finite number of smooth pieces, we choose an orientation for each piece separately. The best way to do this is usually clear. For example, on the cube we choose the outward orientation on each face. (See Figure 19.14.)

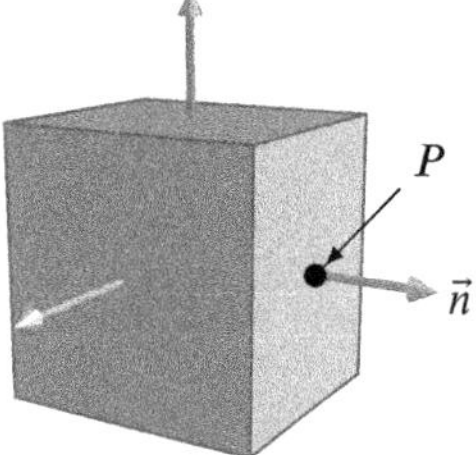

Figure 19.14: The orientation vector field $\vec{n}$ on the cube surface determined by the choice of unit normal vector at the point P

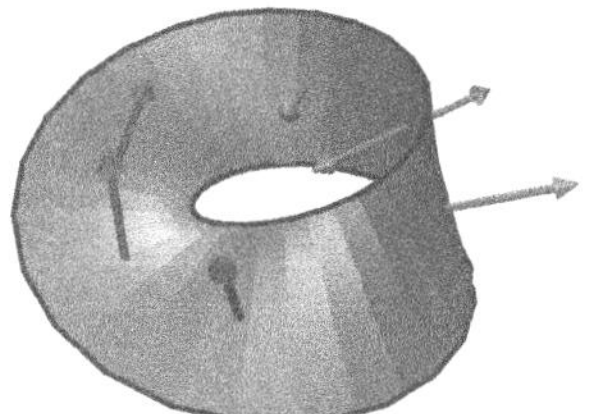

Figure 19.15: The Möbius strip is an example of a non-orientable surface

The second difficulty is that there are some surfaces which cannot be oriented at all, such as the *Möbius strip* in Figure 19.15.

Exercises and Problems for Section 19.1 Online Resource: Additional Problems for Section 19.1

EXERCISES

In Exercises 1–4, find the area vector of the oriented flat surface.

1. The triangle with vertices $(0,0,0)$, $(0,2,0)$, $(0,0,3)$ oriented in the negative x direction.

2. The circular disc of radius 5 in the xy-plane, oriented upward.

3. $y = 10$, $0 \le x \le 5$, $0 \le z \le 3$, oriented away from the xz-plane.

4. $y = -10$, $0 \le x \le 5$, $0 \le z \le 3$, oriented away from the xz-plane.

5. Find an oriented flat surface with area vector $150\vec{j}$.

6. Let S be the disk of radius 3 perpendicular the the y-axis, centered at $(0, 6, 0)$ and oriented away from the origin. Is $\int_S (x\vec{i} + y\vec{j}) \cdot d\vec{A}$ a vector or a scalar?

7. Compute $\int_S (4\vec{i} + 5\vec{k}) \cdot d\vec{A}$, where S is the square of side length 3 perpendicular to the z-axis, centered at $(0, 0, -2)$ and oriented

   (a) Toward the origin. **(b)** Away from the origin.

8. Compute $\int_S (2\vec{i} + 3\vec{k}) \cdot d\vec{A}$, where S is the disk of radius 4 perpendicular to the x-axis, centered at $(5, 0, 0)$ and oriented

   (a) Toward the origin. **(b)** Away from the origin.

In Exercises 9–12, for each of the surfaces in (a)–(e), say whether the flux of $\vec{F}$ through the surface is positive, negative, or zero. The normal vector shows the orientation.

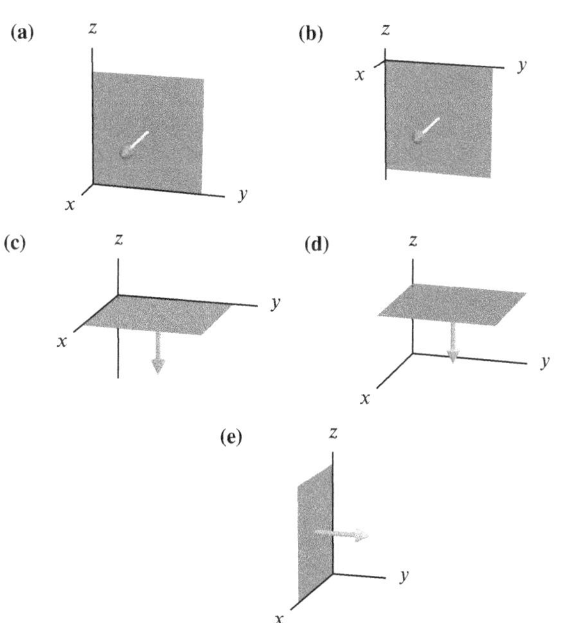

9. $\vec{F}(x, y, z) = \vec{i} + 2\vec{j} + \vec{k}$.

10. $\vec{F}(x, y, z) = z\vec{i}$.

11. $\vec{F}(x, y, z) = -z\vec{i} + x\vec{k}$.

12. $\vec{F}(\vec{r}) = \vec{r}$.

In Exercises 13–16, compute the flux of $\vec{v} = \vec{i} + 2\vec{j} - 3\vec{k}$ through the rectangular region with the orientation shown.

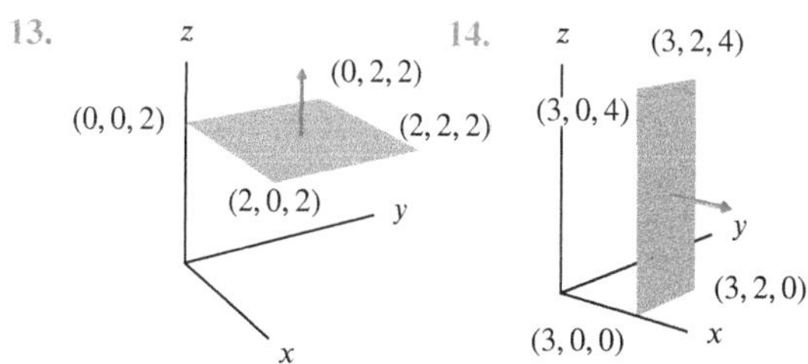

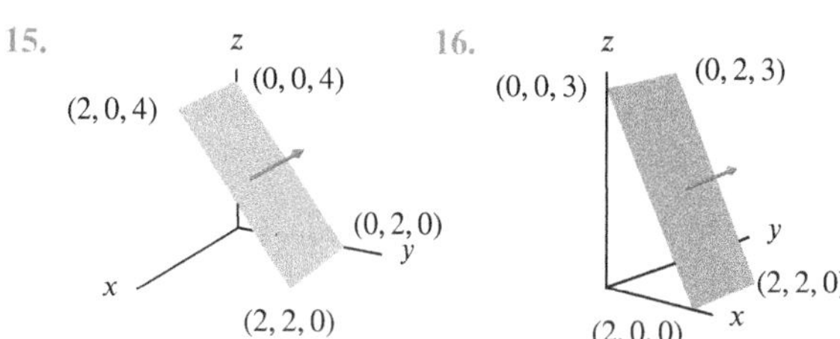

For Exercises 17–20 find the flux of the constant vector field $\vec{v} = \vec{i} - \vec{j} + 3\vec{k}$ through the given surface.

17. A disk of radius 2 in the xy-plane oriented upward.

18. A triangular plate of area 4 in the yz-plane oriented in the positive x-direction.

19. A square plate of area 4 in the yz-plane oriented in the positive x-direction.

20. The triangular plate with vertices $(1, 0, 0)$, $(0, 1, 0)$, $(0, 0, 1)$, oriented away from the origin.

In Exercises 21–23, find the flux of $\vec{H} = 2\vec{i} + 3\vec{j} + 5\vec{k}$ through the surface S.

21. S is the cylinder $x^2 + y^2 = 1$, closed at the ends by the planes $z = 0$ and $z = 1$ and oriented outward.

22. S is the disk of radius 1 in the plane $x = 2$ oriented in the positive x-direction.

23. S is the disk of radius 1 in the plane $x + y + z = 1$ oriented in upward.

Find the flux of the vector fields in Exercises 24–26 out of the closed box $0 \leq x \leq 1, 0 \leq y \leq 2, 0 \leq z \leq 3$.

24. $\vec{F} = 3\vec{i} + 2\vec{j} + \vec{k}$

25. $\vec{G} = x\vec{i}$

26. $\vec{H} = zx\vec{k}$

In Exercises 27–30, calculate the flux integral.

27. $\int_S (x\vec{i} + 4\vec{j}) \cdot d\vec{A}$ where S is the disk of radius 5 perpendicular to the x-axis, centered at $(3, 0, 0)$ and oriented toward the origin.

28. $\int_S \vec{r} \cdot d\vec{A}$ where S is the sphere of radius 3 centered at the origin.

29. $\int_S (\sin x\vec{i} + (y^2 + z^2)\vec{j} + y^2\vec{k}) \cdot d\vec{A}$ where S is a disk of radius π in the plane $x = 3\pi/2$, oriented in the positive x-direction.

30. $\int_S (5\vec{i} + 5\vec{j} + 5\vec{k}) \cdot d\vec{A}$ where S is a disk of radius 3 in the plane $x + y + z = 1$, oriented upward.

In Exercises 31–34, calculate the flux integral using a shortcut arising from two special cases:

- If $\vec{F}$ is tangent at every point of S, then $\int_S \vec{F} \cdot d\vec{A} = 0$.
- If $\vec{F}$ is perpendicular at every point of S and has constant magnitude on S, then

$$\int_S \vec{F} \cdot d\vec{A} = \pm\|\vec{F}\| \cdot \text{Area of } S.$$

Choose the positive sign if $\vec{F}$ points in the same direction as the orientation of S; choose the negative sign if $\vec{F}$ points in the direction opposite the orientation of S.

31. $\int_S (x\vec{i} + y\vec{j}) \cdot d\vec{A}$, where S is the cylinder of radius 10, centered on the z-axis between $z = 0$ and $z = 10$ and oriented away from the z-axis.

32. $\int_S (-y\vec{i} + x\vec{j}) \cdot d\vec{A}$, where S is the cylinder of radius 10, centered on the z-axis between $z = 0$ and $z = 10$ and oriented away from the z-axis.

33. $\int_S (x\vec{i} + y\vec{j} + z\vec{k}) \cdot d\vec{A}$, where S is the sphere of radius 20, centered at the origin and oriented outward.

34. $\int_S (-y\vec{i} + x\vec{j}) \cdot d\vec{A}$, where S is the sphere of radius 20, centered at the origin and oriented outward.

In Exercises 35–57, calculate the flux of the vector field through the surface.

35. $\vec{F} = 2\vec{i} + 3\vec{j}$ through the square of side π in the xy-plane, oriented upward.

36. $\vec{F} = 2\vec{i} + 3\vec{j}$ through the unit disk in the yz-plane, centered at the origin and oriented in the positive x-direction.

37. $\vec{F} = x\vec{i} + y\vec{j} + z\vec{k}$ through the square of side 1.6 centered at $(2, 5, 8)$, parallel to the xz-plane and oriented away from the origin.

38. $\vec{F} = z\vec{k}$ through a square of side $\sqrt{14}$ in a horizontal plane 2 units below the xy-plane and oriented downward.

39. $\vec{F} = -y\vec{i} + x\vec{j}$ and S is the square plate in the yz-plane with corners at $(0, 1, 1)$, $(0, -1, 1)$, $(0, 1, -1)$, and $(0, -1, -1)$, oriented in the positive x-direction.

40. $\vec{F} = 7\vec{i} + 6\vec{j} + 5\vec{k}$ and S is a disk of radius 2 in the yz-plane, centered at the origin and oriented in the positive x-direction.

41. $\vec{F} = x\vec{i} + 2y\vec{j} + 3z\vec{k}$ and S is a square of side 2 in the plane $y = 3$, oriented in the positive y-direction.

42. $\vec{F} = 7\vec{i} + 6\vec{j} + 5\vec{k}$ and S is a sphere of radius π centered at the origin.

43. $\vec{F} = -5\vec{r}$ through the sphere of radius 2 centered at the origin.

44. $\vec{F} = x\vec{i} + y\vec{j} + (z^2 + 3)\vec{k}$ and S is the rectangle $z = 4$, $0 \le x \le 2$, $0 \le y \le 3$, oriented in the positive z-direction.

45. $\vec{F} = 6\vec{i} + 7\vec{j}$ through a triangle of area 10 in the plane $x + y = 5$, oriented in the positive x-direction.

46. $\vec{F} = 6\vec{i} + x^2\vec{j} - \vec{k}$, through the square of side 4 in the plane $y = 3$, centered on the y-axis, with sides parallel to the x and z axes, and oriented in the positive y-direction.

47. $\vec{F} = (x+3)\vec{i} + (y+5)\vec{j} + (z+7)\vec{k}$ through the rectangle $x = 4$, $0 \le y \le 2$, $0 \le z \le 3$, oriented in the positive x-direction.

48. $\vec{F} = 7\vec{r}$ through the sphere of radius 3 centered at the origin.

49. $\vec{F} = -3\vec{r}$ through the sphere of radius 2 centered at the origin.

50. $\vec{F} = 2z\vec{i} + x\vec{j} + x\vec{k}$ through the rectangle $x = 4$, $0 \le y \le 2$, $0 \le z \le 3$, oriented in the positive x-direction.

51. $\vec{F} = \vec{i} + 2\vec{j}$ through a square of side 2 lying in the plane $x + y + z = 1$, oriented away from the origin.

52. $\vec{F} = (x^2 + y^2)\vec{k}$ through the disk of radius 3 in the xy-plane, centered at the origin and oriented upward.

53. $\vec{F} = \cos(x^2 + y^2)\vec{k}$ through the disk $x^2 + y^2 \le 9$ oriented upward in the plane $z = 1$.

54. $\vec{F} = e^{y^2+z^2}\vec{i}$ through the disk of radius 2 in the yz-plane, centered at the origin and oriented in the positive x-direction.

55. $\vec{F} = -y\vec{i} + x\vec{j}$ through the disk in the xy-plane with radius 2, oriented upward and centered at the origin.

56. $\vec{F} = \vec{r}$ through the disk of radius 2 parallel to the xy-plane oriented upward and centered at $(0, 0, 2)$.

57. $\vec{F} = (2 - x)\vec{i}$ through the cube whose vertices include the points $(0, 0, 0)$, $(3, 0, 0)$, $(0, 3, 0)$, $(0, 0, 3)$, and oriented outward.

PROBLEMS

58. Let B be the surface of a box centered at the origin, with edges parallel to the axes and in the planes $x = \pm 1$, $y = \pm 1$, $z = \pm 1$, and let S be the sphere of radius 1 centered at origin.

(a) Indicate whether the following flux integrals are positive, negative, or zero. No reasons needed.

(a) $\int_B x\vec{i} \cdot d\vec{A}$ **(b)** $\int_B y\vec{i} \cdot d\vec{A}$

(c) $\int_S |x|\vec{i} \cdot d\vec{A}$ **(d)** $\int_S (y - x)\vec{i} \cdot d\vec{A}$

(b) Explain with reasons how you know which flux integral is greater:

$$\int_S x\vec{i} \cdot d\vec{A} \quad \text{or} \quad \int_B x\vec{i} \cdot d\vec{A}\,?$$

59. Suppose that $\vec{E}$ is a uniform electric field on 3-space, so $\vec{E}(x,y,z) = a\vec{i} + b\vec{j} + c\vec{k}$, for all points (x,y,z), where a,b,c are constants. Show, with the aid of symmetry, that the flux of $\vec{E}$ through each of the following closed surfaces S is zero:

(a) S is the cube bounded by the planes $x = \pm 1$, $y = \pm 1$, and $z = \pm 1$.
(b) S is the sphere $x^2 + y^2 + z^2 = 1$.
(c) S is the cylinder bounded by $x^2 + y^2 = 1$, $z = 0$, and $z = 2$.

60. Water is flowing down a cylindrical pipe of radius 2 cm; its speed is $(3 - (3/4)r^2)$ cm/sec at a distance r cm from the center of the pipe. Find the flux through the circular cross section of the pipe, oriented so that the flux is positive.

61. **(a)** What do you think will be the electric flux through the cylindrical surface that is placed as shown in the constant electric field in Figure 19.16? Why?
(b) What if the cylinder is placed upright, as shown in Figure 19.17? Explain.

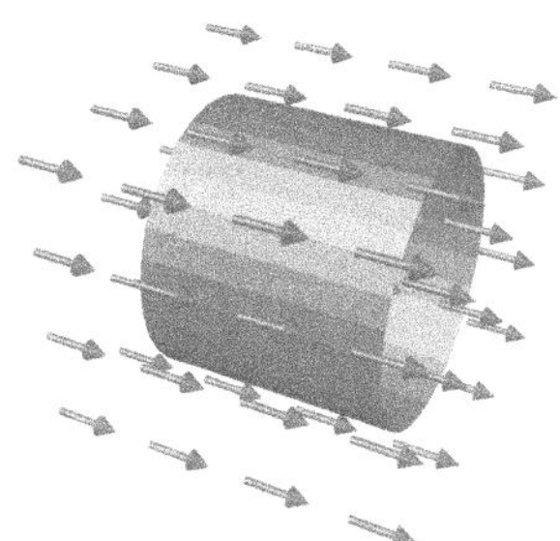

Figure 19.16

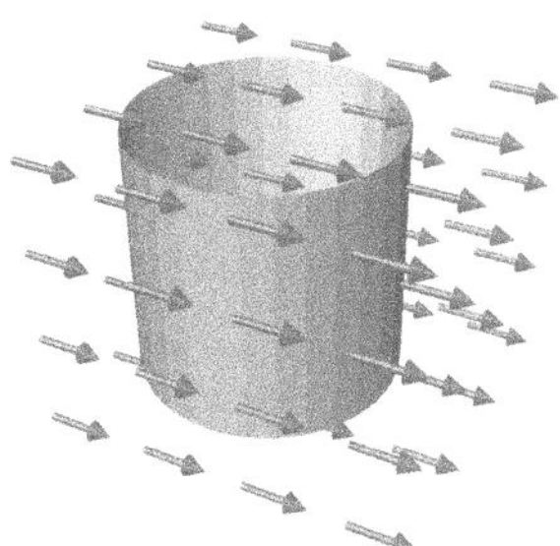

Figure 19.17

62. Let S be part of a cylinder centered on the y-axis. Explain why the three vectors fields $\vec{F}$, $\vec{G}$, and $\vec{H}$ have the same flux through S. Do not compute the flux.

$$\vec{F} = x\vec{i} + 2yz\vec{k}$$
$$\vec{G} = x\vec{i} + y\sin x\vec{j} + 2yz\vec{k}$$
$$\vec{H} = x\vec{i} + \cos(x^2+z)\vec{j} + 2yz\vec{k}$$

63. Find the flux of $\vec{F} = \vec{r}/\|\vec{r}\|^3$ out of the sphere of radius R centered at the origin.

64. Find the flux of $\vec{F} = \vec{r}/||r||^2$ out of the sphere of radius R centered at the origin.

65. Consider the flux of the vector field $\vec{F} = \vec{r}/||\vec{r}||^p$ for $p \geq 0$ out of the sphere of radius 2 centered at the origin.

(a) For what value of p is the flux a maximum?
(b) What is that maximum value?

66. Let S be the cube with side length 2, faces parallel to the coordinate planes, and centered at the origin.

(a) Calculate the total flux of the constant vector field $\vec{v} = -\vec{i} + 2\vec{j} + \vec{k}$ out of S by computing the flux through each face separately.
(b) Calculate the flux out of S for any constant vector field $\vec{v} = a\vec{i} + b\vec{j} + c\vec{k}$.
(c) Explain why the answers to parts (a) and (b) make sense.

67. Let S be the tetrahedron with vertices at the origin and at $(1,0,0)$, $(0,1,0)$ and $(0,0,1)$.

(a) Calculate the total flux of the constant vector field $\vec{v} = -\vec{i} + 2\vec{j} + \vec{k}$ out of S by computing the flux through each face separately.
(b) Calculate the flux out of S in part (a) for any constant vector field $\vec{v}$.
(c) Explain why the answers to parts (a) and (b) make sense.

68. Let $P(x,y,z)$ be the pressure at the point (x,y,z) in a fluid. Let $\vec{F}(x,y,z) = P(x,y,z)\vec{k}$. Let S be the surface of a body submerged in the fluid. If S is oriented inward, show that $\int_S \vec{F} \cdot d\vec{A}$ is the buoyant force on the body, that is, the force upward on the body due to the pressure of the fluid surrounding it. [Hint: $\vec{F} \cdot d\vec{A} = P(x,y,z)\vec{k} \cdot d\vec{A} = (P(x,y,z)\,d\vec{A}) \cdot \vec{k}$.]

69. A region of 3-space has a temperature which varies from point to point. Let $T(x,y,z)$ be the temperature at a point (x,y,z). Newton's law of cooling says that grad T is proportional to the heat flow vector field, $\vec{F}$, where $\vec{F}$ points in the direction in which heat is flowing and has magnitude equal to the rate of flow of heat.

(a) Suppose $\vec{F} = k \operatorname{grad} T$ for some constant k. What is the sign of k?
(b) Explain why this form of Newton's law of cooling makes sense.
(c) Let W be a region of space bounded by the surface S. Explain why

$$\text{Rate of heat loss from } W = k\int_S (\operatorname{grad} T) \cdot d\vec{A}.$$

70. The z-axis carries a constant electric charge density of λ units of charge per unit length, with $\lambda > 0$. The resulting electric field is $\vec{E}$.

(a) Sketch the electric field, $\vec{E}$, in the xy-plane, given

$$\vec{E}(x,y,z) = 2\lambda\frac{x\vec{i} + y\vec{j}}{x^2+y^2}.$$

(b) Compute the flux of $\vec{E}$ outward through the cylinder $x^2 + y^2 = R^2$, for $0 \le z \le h$.

71. An infinitely long straight wire lying along the z-axis carries an electric current I flowing in the $\vec{k}$ direction. Ampère's Law in magnetostatics says that the current gives rise to a magnetic field $\vec{B}$ given by

$$\vec{B}(x, y, z) = \frac{I}{2\pi} \frac{-y\vec{i} + x\vec{j}}{x^2 + y^2}.$$

(a) Sketch the field $\vec{B}$ in the xy-plane.
(b) Let S_1 be the disk with center at $(0, 0, h)$, radius a, and parallel to the xy-plane, oriented in the $\vec{k}$ direction. What is the flux of $\vec{B}$ through S_1? Does your answer seem reasonable?
(c) Let S_2 be the rectangle given by $x = 0$, $a \le y \le b$, $0 \le z \le h$, and oriented in the $-\vec{i}$ direction. What is the flux of $\vec{B}$ through S_2? Does your answer seem reasonable?

Strengthen Your Understanding

In Problems 72–73, explain what is wrong with the statement.

72. For a certain vector field $\vec{F}$ and oriented surface S, we have $\int_S \vec{F} \cdot d\vec{A} = 2\vec{i} - 3\vec{j} + \vec{k}$.

73. If S is a region in the xy-plane oriented upwards then $\int_S \vec{F} \cdot d\vec{A} > 0$.

In Problems 74–75, give an example of:

74. A nonzero vector field $\vec{F}$ such that $\int_S \vec{F} \cdot d\vec{A} = 0$, where S is the triangular surface with corners $(1, 0, 0)$, $(0, 1, 0)$, $(0, 0, 1)$, oriented away from the origin.

75. A nonconstant vector field $\vec{F}(x, y, z)$ and an oriented surface S such that $\int_S \vec{F} \cdot d\vec{A} = 1$.

Are the statements in Problems 76–85 true or false? Give reasons for your answer.

76. The value of a flux integral is a scalar.

77. The area vector $\vec{A}$ of a flat, oriented surface is parallel to the surface.

78. If S is the unit sphere centered at the origin, oriented outward and the flux integral $\int_S \vec{F} \cdot d\vec{A}$ is zero, then $\vec{F} = \vec{0}$.

79. The flux of the vector field $\vec{F} = \vec{i}$ through the plane $x = 0$, with $0 \le y \le 1, 0 \le z \le 1$, oriented in the $\vec{i}$ direction is positive.

80. If S is the unit sphere centered at the origin, oriented outward and $\vec{F} = x\vec{i} + y\vec{j} + z\vec{k} = \vec{r}$, then the flux integral $\int_S \vec{F} \cdot d\vec{A}$ is positive.

81. If S is the cube bounded by the six planes $x = \pm 1, y = \pm 1, z = \pm 1$, oriented outward, and $\vec{F} = \vec{k}$, then $\int_S \vec{F} \cdot d\vec{A} = 0$.

82. If S is an oriented surface in 3-space, and $-S$ is the same surface, but with the opposite orientation, then $\int_S \vec{F} \cdot d\vec{A} = -\int_{-S} \vec{F} \cdot d\vec{A}$.

83. If S_1 is a rectangle with area 1 and S_2 is a rectangle with area 2, then $2\int_{S_1} \vec{F} \cdot d\vec{A} = \int_{S_2} \vec{F} \cdot d\vec{A}$.

84. If $\vec{F} = 2\vec{G}$, then $\int_S \vec{F} \cdot d\vec{A} = 2\int_S \vec{G} \cdot d\vec{A}$.

85. If $\int_S \vec{F} \cdot d\vec{A} > \int_S \vec{G} \cdot d\vec{A}$ then $\|\vec{F}\| > \|\vec{G}\|$ at all points on the surface S.

86. For each of the surfaces in (a)–(e), pick the vector field $\vec{F}_1, \vec{F}_2, \vec{F}_3, \vec{F}_4, \vec{F}_5$, with the largest flux through the surface. The surfaces are all squares of the same size. Note that the orientation is shown.

$$\begin{aligned}
\vec{F}_1 &= 2\vec{i} - 3\vec{j} - 4\vec{k} \\
\vec{F}_2 &= \vec{i} - 2\vec{j} + 7\vec{k} \\
\vec{F}_3 &= -7\vec{i} + 5\vec{j} + 6\vec{k} \\
\vec{F}_4 &= -11\vec{i} + 4\vec{j} - 5\vec{k} \\
\vec{F}_5 &= -5\vec{i} + 3\vec{j} + 5\vec{k}
\end{aligned}$$

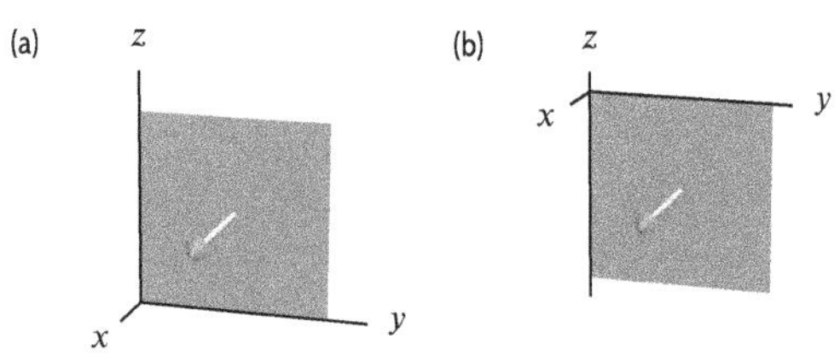

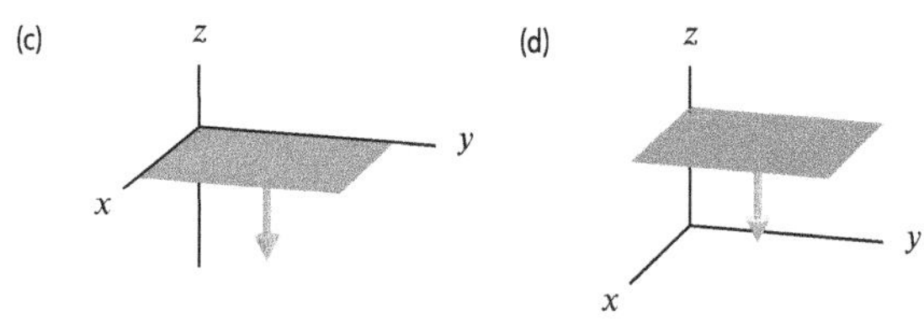

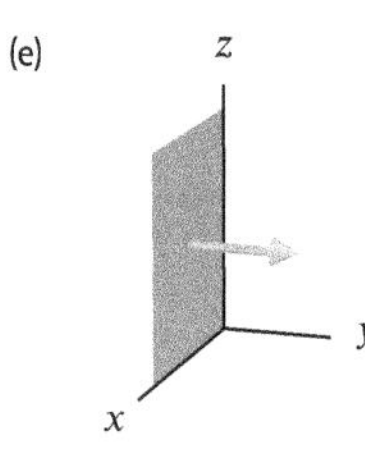

19.2 FLUX INTEGRALS FOR GRAPHS, CYLINDERS, AND SPHERES

In Section 19.1 we computed flux integrals in certain simple cases. In this section we see how to compute flux through surfaces that are graphs of functions, through cylinders, and through spheres.

Flux of a Vector Field Through the Graph of $z = f(x, y)$

Suppose S is the graph of the differentiable function $z = f(x, y)$, oriented upward, and that $\vec{F}$ is a smooth vector field. In Section 19.1 we subdivided the surface into small pieces with area vector $\Delta\vec{A}$ and defined the flux of $\vec{F}$ through S as follows:

$$\int_S \vec{F} \cdot d\vec{A} = \lim_{\|\Delta\vec{A}\|\to 0} \sum \vec{F} \cdot \Delta\vec{A}.$$

How do we divide S into small pieces? One way is to use the cross sections of f with x or y constant and take the patches in a wire frame representation of the surface. So we must calculate the area vector of one of these patches, which is approximately a parallelogram.

The Area Vector of a Coordinate Patch

According to the geometric definition of the cross product on page 729, the vector $\vec{v} \times \vec{w}$ has magnitude equal to the area of the parallelogram formed by $\vec{v}$ and $\vec{w}$ and direction perpendicular to this parallelogram and determined by the right-hand rule. Thus, we have

$$\text{Area vector of parallelogram} = \vec{A} = \vec{v} \times \vec{w}.$$

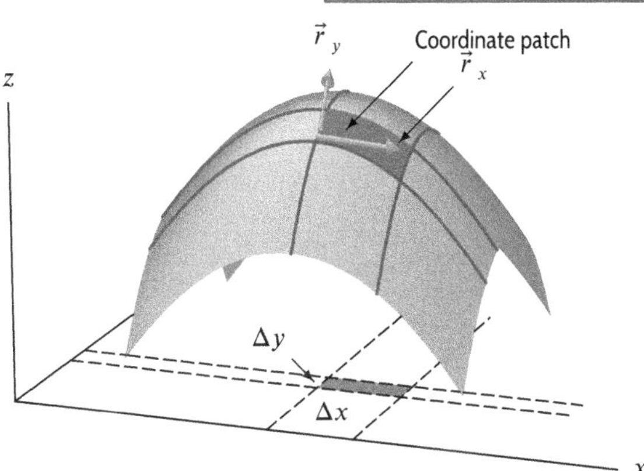

Figure 19.18: Surface showing coordinate patch and tangent vectors $\vec{r}_x$ and $\vec{r}_y$

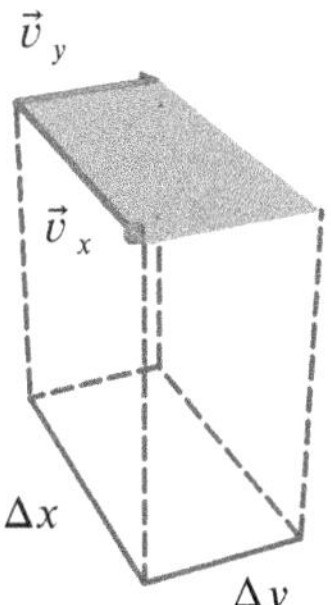

Figure 19.19: Parallelogram-shaped patch in the tangent plane to the surface

Consider the patch of surface above the rectangular region with sides Δx and Δy in the xy-plane shown in Figure 19.18. We approximate the area vector, $\Delta\vec{A}$, of this patch by the area vector of the corresponding patch on the tangent plane to the surface. See Figure 19.19. This patch is the parallelogram determined by the vectors $\vec{v}_x$ and $\vec{v}_y$, so its area vector is given by

$$\Delta\vec{A} \approx \vec{v}_x \times \vec{v}_y.$$

To find $\vec{v}_x$ and $\vec{v}_y$, notice that a point on the surface has position vector $\vec{r} = x\vec{i} + y\vec{j} + f(x, y)\vec{k}$. Thus, a cross section of S with y constant has tangent vector

$$\vec{r}_x = \frac{\partial\vec{r}}{\partial x} = \vec{i} + f_x\vec{k},$$

and a cross section with x constant has tangent vector

$$\vec{r}_y = \frac{\partial\vec{r}}{\partial y} = \vec{j} + f_y\vec{k}.$$

The vectors $\vec{r}_x$ and $\vec{v}_x$ are parallel because they are both tangent to the surface and parallel to the xz-plane. Since the x-component of $\vec{r}_x$ is $\vec{i}$ and the x-component of $\vec{v}_x$ is $(\Delta x)\vec{i}$, we have $\vec{v}_x = (\Delta x)\vec{r}_x$. Similarly, we have $\vec{v}_y = (\Delta y)\vec{r}_y$. So the upward-pointing area vector of the parallelogram is

$$\Delta\vec{A} \approx \vec{v}_x \times \vec{v}_y = \left(\vec{r}_x \times \vec{r}_y\right)\Delta x\,\Delta y = \left(-f_x\vec{i} - f_y\vec{j} + \vec{k}\right)\Delta x\,\Delta y.$$

This is our approximation for the area vector $\Delta\vec{A}$ on the surface. Replacing $\Delta\vec{A}$, Δx, and Δy by $d\vec{A}$, dx and dy, we write

$$d\vec{A} = \left(-f_x\vec{i} - f_y\vec{j} + \vec{k}\right)dx\,dy.$$

The Flux of $\vec{F}$ Through a Surface Given by a Graph of $z = f(x, y)$

Suppose the surface S is the part of the graph of $z = f(x, y)$ above[2] a region R in the xy-plane, and suppose S is oriented upward. The flux of $\vec{F}$ through S is

$$\int_S \vec{F}\cdot d\vec{A} = \int_R \vec{F}(x, y, f(x, y))\cdot\left(-f_x\vec{i} - f_y\vec{j} + \vec{k}\right)dx\,dy.$$

Example 1 Compute $\int_S \vec{F}\cdot d\vec{A}$ where $\vec{F}(x, y, z) = z\vec{k}$ and S is the rectangular plate with corners $(0,0,0)$, $(1,0,0)$, $(0,1,3)$, $(1,1,3)$, oriented upward. See Figure 19.20.

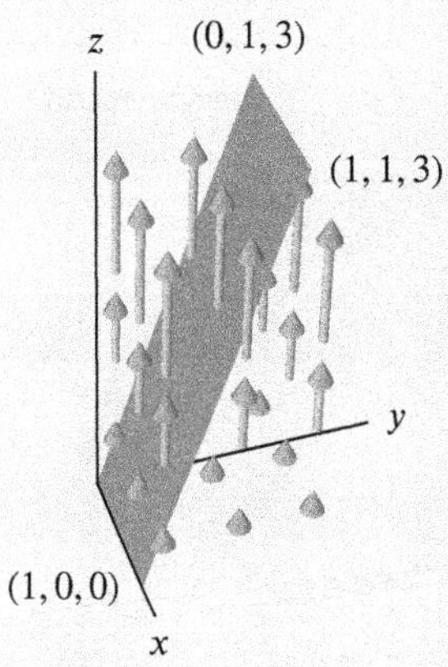

Figure 19.20: The vector field $\vec{F} = z\vec{k}$ on the rectangular surface S

Solution We find the equation for the plane S in the form $z = f(x, y)$. Since f is linear, with x-slope equal to 0 and y-slope equal to 3, and $f(0, 0) = 0$, we have

$$z = f(x, y) = 0 + 0x + 3y = 3y.$$

Thus, we have

$$d\vec{A} = (-f_x\vec{i} - f_y\vec{j} + \vec{k})\,dx\,dy = (0\vec{i} - 3\vec{j} + \vec{k})\,dx\,dy = (-3\vec{j} + \vec{k})\,dx\,dy.$$

The flux integral is therefore

$$\int_S \vec{F}\cdot d\vec{A} = \int_0^1\int_0^1 3y\vec{k}\cdot(-3\vec{j} + \vec{k})\,dx\,dy = \int_0^1\int_0^1 3y\,dx\,dy = 1.5.$$

[2]The formula is also correct when the graph is below the region R.

Surface Area of a Graph

Since the magnitude of an area vector is area, we can find area of a surface by integrating the magnitude $\|d\vec{A}\|$. If a surface is the graph of a function $z = f(x, y)$, we have

$$\|d\vec{A}\| = \| - f_x\vec{i} - f_y\vec{j} + \vec{k}\| \, dx\, dy = \sqrt{(f_x)^2 + (f_y)^2 + 1}\, dx\, dy.$$

Thus we have the following result:

> Suppose a surface S is the part of the graph $z = f(x, y)$ where (x, y) is in a region R in the xy-plane. Then
>
> $$\text{Area of } S = \int_R \sqrt{(f_x)^2 + (f_y)^2 + 1}\, dx\, dy.$$

Example 2 Find the area of the surface $z = f(x, y)$ where $0 \le x \le 4$, $0 \le y \le 5$, when:

(a) $f(x, y) = 2x + 3y + 4$ (b) $f(x, y) = x^2 + y^2$

Solution (a) Since $f_x = 2$ and $f_y = 3$, we have

$$\text{Area} = \int_0^5 \int_0^4 \sqrt{2^2 + 3^2 + 1}\, dx\, dy = 20\sqrt{14}.$$

(b) Since $f_x = 2x$ and $f_y = 2y$. we have

$$\text{Area} = \int_0^5 \int_0^4 \sqrt{4x^2 + 4y^2 + 1}\, dx\, dy = 140.089.$$

Surface area integrals can often only be evaluated numerically.

Flux of a Vector Field Through a Cylindrical Surface

Consider the cylinder of radius R centered on the z-axis illustrated in Figure 19.21 and oriented away from the z-axis. The coordinate patch in Figure 19.22 has surface area given by

$$\Delta A \approx R\,\Delta\theta\,\Delta z.$$

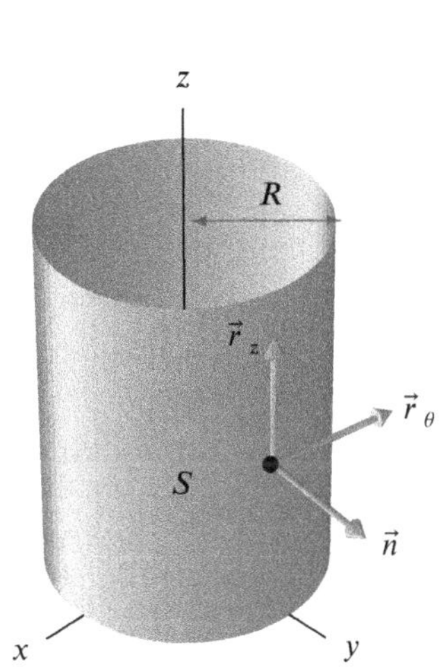

Figure 19.21: Outward-oriented cylinder

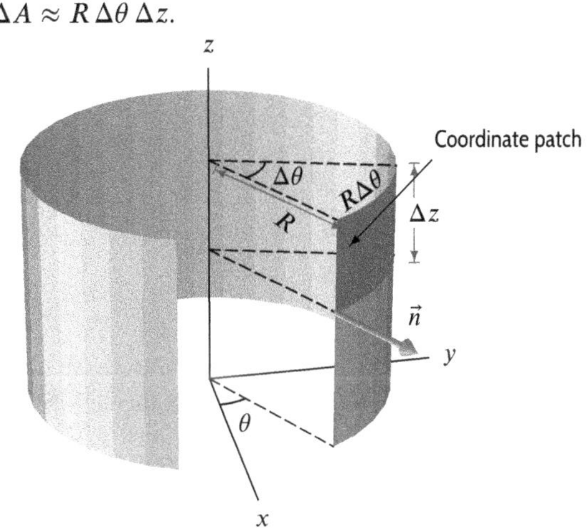

Figure 19.22: Coordinate patch with area $\Delta\vec{A}$ on surface of a cylinder

The outward unit normal $\vec{n}$ points in the direction of $x\vec{i} + y\vec{j}$, so

$$\vec{n} = \frac{x\vec{i} + y\vec{j}}{\|x\vec{i} + y\vec{j}\|} = \frac{R\cos\theta\vec{i} + R\sin\theta\vec{j}}{R} = \cos\theta\vec{i} + \sin\theta\vec{j}.$$

Therefore, the area vector of the coordinate patch is approximated by

$$\Delta\vec{A} = \vec{n}\,\Delta A \approx \left(\cos\theta\vec{i} + \sin\theta\vec{j}\right) R\,\Delta z\,\Delta\theta.$$

Replacing $\Delta\vec{A}$, Δz, and $\Delta\theta$ by $d\vec{A}$, dz, and $d\theta$, we write

$$d\vec{A} = \left(\cos\theta\vec{i} + \sin\theta\vec{j}\right) R\,dz\,d\theta.$$

This gives the following result:

The Flux of a Vector Field Through a Cylinder

The flux of $\vec{F}$ through the cylindrical surface S, of radius R and oriented away from the z-axis, is given by

$$\int_S \vec{F} \cdot d\vec{A} = \int_T \vec{F}(R,\theta,z) \cdot \left(\cos\theta\vec{i} + \sin\theta\vec{j}\right) R\,dz\,d\theta,$$

where T is the θz-region corresponding to S.

Example 3 Compute $\int_S \vec{F} \cdot d\vec{A}$ where $\vec{F}(x,y,z) = y\vec{j}$ and S is the part of the cylinder of radius 2 centered on the z-axis with $x \geq 0$, $y \geq 0$, and $0 \leq z \leq 3$. The surface is oriented toward the z-axis.

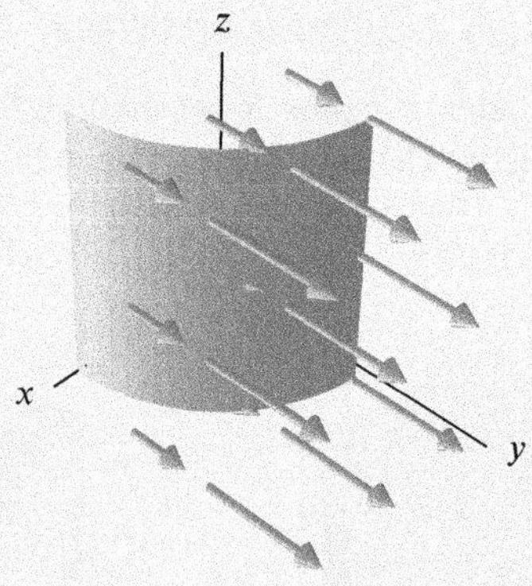

Figure 19.23: The vector field $\vec{F} = y\vec{j}$ on the surface S

Solution In cylindrical coordinates, we have $R = 2$ and $\vec{F} = y\vec{j} = 2\sin\theta\vec{j}$. Since the orientation of S is toward the z-axis, the flux through S is given by

$$\int_S \vec{F} \cdot d\vec{A} = -\int_T 2\sin\theta\vec{j} \cdot (\cos\theta\vec{i} + \sin\theta\vec{j})2\,dz\,d\theta = -4\int_0^{\pi/2}\int_0^3 \sin^2\theta\,dz\,d\theta = -3\pi.$$

Flux of a Vector Field Through a Spherical Surface

Consider the piece of the sphere of radius R centered at the origin, oriented outward, as illustrated in Figure 19.24. The coordinate patch in Figure 19.24 has surface area given by

$$\Delta A \approx R^2 \sin\phi \, \Delta\phi \, \Delta\theta.$$

The outward unit normal $\vec{n}$ points in the direction of $\vec{r} = x\vec{i} + y\vec{j} + z\vec{k}$, so

$$\vec{n} = \frac{\vec{r}}{\|\vec{r}\|} = \sin\phi\cos\theta\vec{i} + \sin\phi\sin\theta\vec{j} + \cos\phi\vec{k}.$$

Therefore, the area vector of the coordinate patch is approximated by

$$\Delta\vec{A} \approx \vec{n}\,\Delta A = \frac{\vec{r}}{\|\vec{r}\|}\Delta A = \left(\sin\phi\cos\theta\vec{i} + \sin\phi\sin\theta\vec{j} + \cos\phi\vec{k}\right) R^2 \sin\phi\,\Delta\phi\,\Delta\theta.$$

Replacing $\Delta\vec{A}$, $\Delta\phi$, and $\Delta\theta$ by $d\vec{A}$, $d\phi$, and $d\theta$, we write

$$d\vec{A} = \frac{\vec{r}}{\|\vec{r}\|} dA = \left(\sin\phi\cos\theta\vec{i} + \sin\phi\sin\theta\vec{j} + \cos\phi\vec{k}\right) R^2 \sin\phi\, d\phi\, d\theta.$$

Thus, we obtain the following result:

The Flux of a Vector Field Through a Sphere

The flux of $\vec{F}$ through the spherical surface S, with radius R and oriented away from the origin, is given by

$$\int_S \vec{F}\cdot d\vec{A} = \int_S \vec{F}\cdot\frac{\vec{r}}{\|\vec{r}\|} dA$$

$$= \int_T \vec{F}(R,\theta,\phi)\cdot\left(\sin\phi\cos\theta\vec{i} + \sin\phi\sin\theta\vec{j} + \cos\phi\vec{k}\right) R^2 \sin\phi\, d\phi\, d\theta,$$

where T is the $\theta\phi$-region corresponding to S.

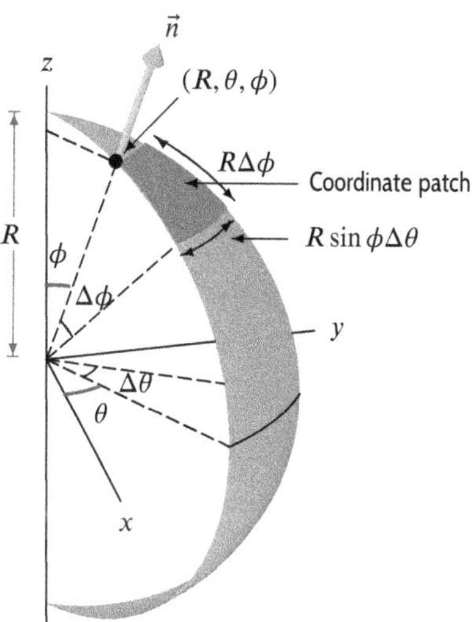

Figure 19.24: Coordinate patch with area $\Delta\vec{A}$ on surface of a sphere

Example 4 Find the flux of $\vec{F} = z\vec{k}$ through S, the upper hemisphere of radius 2 centered at the origin, oriented outward.

Solution The hemisphere S is parameterized by spherical coordinates θ and ϕ, with $0 \le \theta \le 2\pi$ and $0 \le \phi \le \pi/2$. Since $R = 2$ and $\vec{F} = z\vec{k} = 2\cos\phi\vec{k}$, the flux is

$$\int_S \vec{F} \cdot d\vec{A} = \int_S 2\cos\phi\vec{k} \cdot (\sin\phi\cos\theta\vec{i} + \sin\phi\sin\theta\vec{j} + \cos\phi\vec{k})4\sin\phi\, d\phi\, d\theta$$

$$= \int_0^{2\pi}\int_0^{\pi/2} 8\sin\phi\cos^2\phi\, d\phi\, d\theta = 2\pi\left(8\left(\frac{-\cos^3\phi}{3}\right)\Big|_{\phi=0}^{\pi/2}\right) = \frac{16\pi}{3}.$$

Example 5 The magnetic field $\vec{B}$ due to an *ideal magnetic dipole*, $\vec{\mu}$, located at the origin is a multiple of

$$\vec{B}(\vec{r}) = -\frac{\vec{\mu}}{\|\vec{r}\|^3} + \frac{3(\vec{\mu}\cdot\vec{r})\vec{r}}{\|\vec{r}\|^5}.$$

Figure 19.25 shows a sketch of $\vec{B}$ in the plane $z = 0$ for the dipole $\vec{\mu} = \vec{i}$. Notice that $\vec{B}$ is similar to the magnetic field of a bar magnet with its north pole at the tip of the vector $\vec{i}$ and its south pole at the tail of the vector $\vec{i}$.

Compute the flux of $\vec{B}$ outward through the sphere S with center at the origin and radius R.

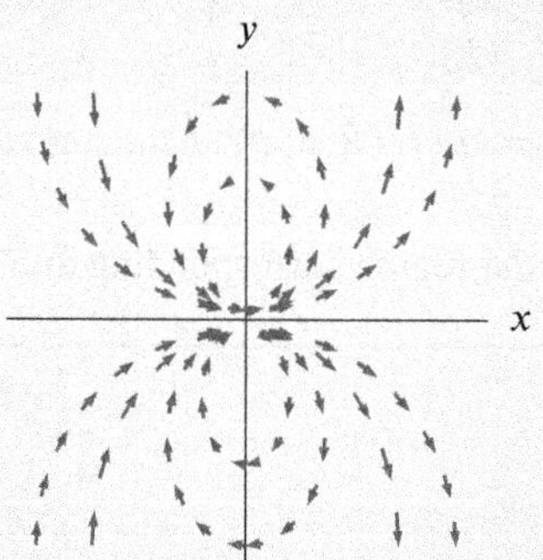

Figure 19.25: The magnetic field of a dipole, $\vec{i}$, at the origin: $\vec{B} = -\frac{\vec{i}}{\|\vec{r}\|^3} + \frac{3(\vec{i}\cdot\vec{r})\vec{r}}{\|\vec{r}\|^5}$

Solution Since $\vec{i}\cdot\vec{r} = x$ and $\|\vec{r}\| = R$ on the sphere of radius R, we have

$$\int_S \vec{B}\cdot d\vec{A} = \int_S\left(-\frac{\vec{i}}{\|\vec{r}\|^3} + \frac{3(\vec{i}\cdot\vec{r})\vec{r}}{\|\vec{r}\|^5}\right)\cdot\frac{\vec{r}}{\|\vec{r}\|}\,dA = \int_S\left(-\frac{\vec{i}\cdot\vec{r}}{\|\vec{r}\|^4} + \frac{3(\vec{i}\cdot\vec{r})\|\vec{r}\|^2}{\|\vec{r}\|^6}\right)dA$$

$$= \int_S \frac{2\vec{i}\cdot\vec{r}}{\|\vec{r}\|^4}\,dA = \int_S \frac{2x}{\|\vec{r}\|^4}\,dA = \frac{2}{R^4}\int_S x\,dA.$$

But the sphere S is centered at the origin. Thus, the contribution to the integral from each positive x-value is canceled by the contribution from the corresponding negative x-value; so $\int_S x\,dA = 0$. Therefore,

$$\int_S \vec{B}\cdot d\vec{A} = \frac{2}{R^4}\int_S x\,dA = 0.$$

Exercises and Problems for Section 19.2

EXERCISES

In Exercises 1–4, find the area vector $d\vec{A}$ for the surface $z = f(x, y)$, oriented upward.

1. $f(x, y) = 3x - 5y$
2. $f(x, y) = 8x + 7y$
3. $f(x, y) = 2x^2 - 3y^2$
4. $f(x, y) = xy + y^2$

In Exercises 5–8, write an iterated integral for the flux of $\vec{F}$ through the surface S, which is the part of the graph of $z = f(x, y)$ corresponding to the region R, oriented upward. Do not evaluate the integral.

5. $\vec{F}(x, y, z) = 10\vec{i} + 20\vec{j} + 30\vec{k}$
   $f(x, y) = 2x - 3y$
   R: $-2 \le x \le 3, 0 \le y \le 5$
6. $\vec{F}(x, y, z) = z\vec{i} + x\vec{j} + y\vec{k}$
   $f(x, y) = 50 - 4x + 10y$
   R: $0 \le x \le 4, 0 \le y \le 8$
7. $\vec{F}(x, y, z) = yz\vec{i} + xy\vec{j} + xy\vec{k}$
   $f(x, y) = \cos x + \sin 2y$
   R: Triangle with vertices $(0, 0), (0, 5), (5, 0)$
8. $\vec{F}(x, y, z) = \cos(x + 2y)\vec{j}$
   $f(x, y) = xe^{3y}$
   R: Quarter disk of radius 5 centered at the origin, in quadrant I

In Exercises 9–12, compute the flux of $\vec{F}$ through the surface S, which is the part of the graph of $z = f(x, y)$ corresponding to region R, oriented upward.

9. $\vec{F}(x, y, z) = 3\vec{i} - 2\vec{j} + 6\vec{k}$
   $f(x, y) = 4x - 2y$
   R: $0 \le x \le 5, 0 \le y \le 10$
10. $\vec{F}(x, y, z) = \vec{i} - 2\vec{j} + z\vec{k}$
    $f(x, y) = xy$
    R: $0 \le x \le 10, 0 \le y \le 10$
11. $\vec{F}(x, y, z) = \cos y\vec{i} + z\vec{j} + \vec{k}$
    $f(x, y) = x^2 + 2y$
    R: $0 \le x \le 1, 0 \le y \le 1$
12. $\vec{F}(x, y, z) = x\vec{i} + z\vec{k}$
    $f(x, y) = x + y + 2$
    R: Triangle with vertices $(-1, 0), (1, 0), (0, 1)$

In Exercises 13–16, write an iterated integral for the flux of $\vec{F}$ through the cylindrical surface S centered on the z-axis, oriented away from the z-axis. Do not evaluate the integral.

13. $\vec{F}(x, y, z) = \vec{i} + 2\vec{j} + 3\vec{k}$
    S: radius 10, $x \ge 0, y \ge 0, 0 \le z \le 5$
14. $\vec{F}(x, y, z) = x\vec{i} + 2y\vec{j} + 3z\vec{k}$
    S: radius 10, $0 \le z \le 5$
15. $\vec{F}(x, y, z) = z^2\vec{i} + e^x\vec{j} + \vec{k}$
    S: radius 6, inside sphere of radius 10
16. $\vec{F}(x, y, z) = x^2yz\vec{j} + z^3\vec{k}$
    S: radius 2, between the xy-plane and the paraboloid $z = x^2 + y^2$

In Exercises 17–20, compute the flux of $\vec{F}$ through the cylindrical surface S centered on the z-axis, oriented away from the z-axis.

17. $\vec{F}(x, y, z) = z\vec{j} + 6x\vec{k}$
    S: radius 5, $y \ge 0, 0 \le z \le 20$
18. $\vec{F}(x, y, z) = y\vec{i} + xz\vec{k}$
    S: radius 10, $x \ge 0, y \ge 0, 0 \le z \le 3$
19. $\vec{F}(x, y, z) = xyz\vec{j} + xe^z\vec{k}$
    S: radius 2, $0 \le y \le x, 0 \le z \le 10$
20. $\vec{F}(x, y, z) = xy\vec{i} + 2z\vec{j}$
    S: radius 1, $x \ge 0, 0 \le y \le 1/2, 0 \le z \le 2$

In Exercises 21–24, write an iterated integral for the flux of $\vec{F}$ through the spherical surface S centered at the origin, oriented away from the origin. Do not evaluate the integral.

21. $\vec{F}(x, y, z) = \vec{i} + 2\vec{j} + 3\vec{k}$
    S: radius 10, $z \ge 0$
22. $\vec{F}(x, y, z) = x\vec{i} + 2y\vec{j} + 3z\vec{k}$
    S: radius 5, entire sphere
23. $\vec{F}(x, y, z) = z^2\vec{i}$
    S: radius 2, $x \ge 0$
24. $\vec{F}(x, y, z) = e^x\vec{k}$
    S: radius 3, $y \ge 0, z \le 0$

In Exercises 25–27, compute the flux of $\vec{F}$ through the spherical surface S centered at the origin, oriented away from the origin.

25. $\vec{F}(x, y, z) = z\vec{i}$
    S: radius 20, $x \ge 0, y \ge 0, z \ge 0$
26. $\vec{F}(x, y, z) = y\vec{i} - x\vec{j} + z\vec{k}$
    S: radius 4, entire sphere

27. $\vec{F}(x,y,z) = x\vec{i} + y\vec{j}$
S: radius 1, above the cone $\phi = \pi/4$.

In Exercises 28–29, compute the flux of $\vec{v} = z\vec{k}$ through the rectangular region with the orientation shown.

28. 29.

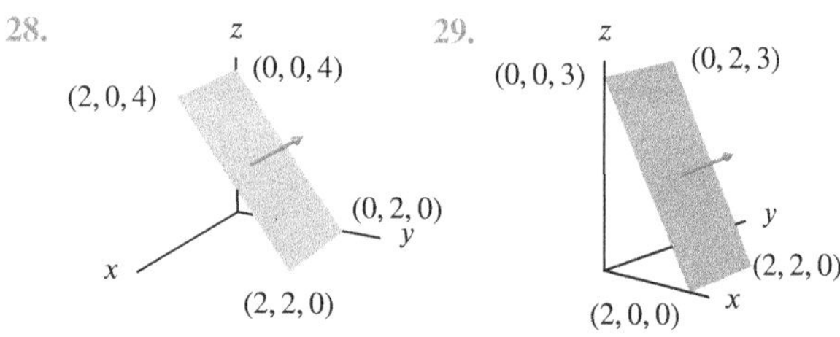

PROBLEMS

In Problems 30–46 compute the flux of the vector field $\vec{F}$ through the surface S.

30. $\vec{F} = z\vec{k}$ and S is the portion of the plane $x+y+z=1$ that lies in the first octant, oriented upward.

31. $\vec{F} = (x-y)\vec{i} + z\vec{j} + 3x\vec{k}$ and S is the part of the plane $z = x + y$ above the rectangle $0 \le x \le 2$, $0 \le y \le 3$, oriented upward.

32. $\vec{F} = 2x\vec{j} + y\vec{k}$ and S is the part of the surface $z = -y + 1$ above the square $0 \le x \le 1$, $0 \le y \le 1$, oriented upward.

33. $\vec{F} = -y\vec{j} + z\vec{k}$ and S is the part of the surface $z = y^2 + 5$ over the rectangle $-2 \le x \le 1$, $0 \le y \le 1$, oriented upward.

34. $\vec{F} = \ln(x^2)\vec{i} + e^x\vec{j} + \cos(1-z)\vec{k}$ and S is the part of the surface $z = -y + 1$ above the square $0 \le x \le 1$, $0 \le y \le 1$, oriented upward.

35. $\vec{F} = 5\vec{i} + 7\vec{j} + z\vec{k}$ and S is a closed cylinder of radius 3 centered on the z-axis, with $-2 \le z \le 2$, and oriented outward.

36. $\vec{F} = x\vec{i} + y\vec{j} + z\vec{k}$ and S is a closed cylinder of radius 2 centered on the y-axis, with $-3 \le y \le 3$, and oriented outward.

37. $\vec{F} = 3x\vec{i} + y\vec{j} + z\vec{k}$ and S is the part of the surface $z = -2x - 4y + 1$, oriented upward, with (x, y) in the triangle R with vertices $(0,0)$, $(0,2)$, $(1,0)$.

38. $\vec{F} = x\vec{i} + y\vec{j}$ and S is the part of the surface $z = 25 - (x^2 + y^2)$ above the disk of radius 5 centered at the origin, oriented upward.

39. $\vec{F} = \cos(x^2 + y^2)\vec{k}$ and S is as in Exercise 38.

40. $\vec{F} = \vec{r}$ and S is the part of the plane $x + y + z = 1$ above the rectangle $0 \le x \le 2$, $0 \le y \le 3$, oriented downward.

41. $\vec{F} = \vec{r}$ and S is the part of the surface $z = x^2 + y^2$ above the disk $x^2 + y^2 \le 1$, oriented downward.

42. $\vec{F} = xz\vec{i} + y\vec{k}$ and S is the hemisphere $x^2 + y^2 + z^2 = 9$, $z \ge 0$, oriented upward.

43. $\vec{F} = -xz\vec{i} - yz\vec{j} + z^2\vec{k}$ and S is the cone $z = \sqrt{x^2 + y^2}$ for $0 \le z \le 6$, oriented upward.

44. $\vec{F} = yz^4\vec{i} - xz^4\vec{j} + e^{z^2}\vec{k}$ and S is the cone $z = \sqrt{x^2 + y^2}$ for $1 \le z \le 2$, oriented upward.

45. $\vec{F} = y\vec{i} + \vec{j} - xz\vec{k}$ and S is the surface $y = x^2 + z^2$, with $x^2 + z^2 \le 1$, oriented in the positive y-direction.

46. $\vec{F} = x^2\vec{i} + y^2\vec{j} + z^2\vec{k}$ and S is the oriented triangular surface shown in Figure 19.26.

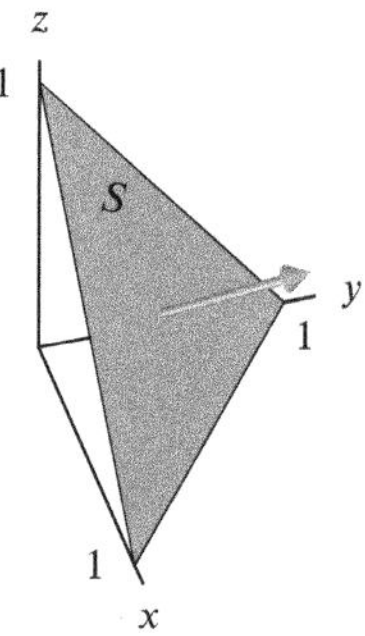

Figure 19.26

In Problems 47–50 find the area of the surface $z = f(x, y)$ over the region R in the xy-plane.

47. $f(x,y) = 50 + 5x - y$, R: $-5 \le x \le 5$, $0 \le y \le 10$

48. $f(x,y) = 50 + 5x - y$, R: circle of radius 3 centered at the origin

49. $f(x,y) = xe^y$, R: $0 \le x \le 1$, $0 \le y \le 1$

50. $f(x,y) = (\sin x)(\sin y)$, R: $0 \le x \le \pi/2$, $0 \le y \le \pi/2$

51. Let S be the hemisphere $x^2 + y^2 + z^2 = a^2$ of radius a, where $z \ge 0$.

    (a) Express the surface area of S as an integral in Cartesian coordinates.

(b) Change variables to express the area integral in polar coordinates.
(c) Find the area of S.

In Problems 52–53, compute the flux of $\vec{F}$ through the cylindrical surface in Figure 19.27, oriented away from the z-axis.

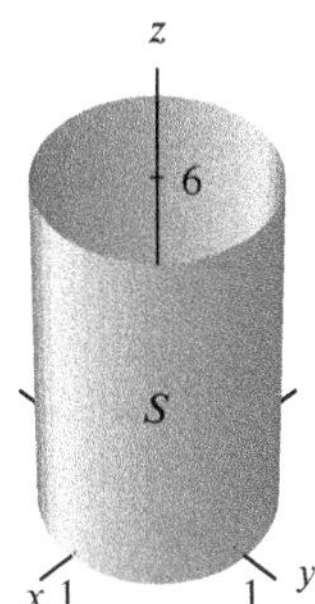

Figure 19.27

52. $\vec{F} = x\vec{i} + y\vec{j}$
53. $\vec{F} = xz\vec{i} + yz\vec{j} + z^3\vec{k}$

In Problems 54–57, compute the flux of $\vec{F}$ through the spherical surface, S.

54. $\vec{F} = z\vec{k}$ and S is the upper hemisphere of radius 2 centered at the origin, oriented outward.
55. $\vec{F} = y\vec{i} - x\vec{j} + z\vec{k}$ and S is the spherical cap given by $x^2 + y^2 + z^2 = 1$, $z \geq 0$, oriented upward.
56. $\vec{F} = z^2\vec{k}$ and S is the upper hemisphere of the sphere $x^2 + y^2 + z^2 = 25$, oriented away from the origin.
57. $\vec{F} = x\vec{i} + y\vec{j} + z\vec{k}$ and S is the surface of the sphere $x^2 + y^2 + z^2 = a^2$, oriented outward.
58. Compute the flux of $\vec{F} = x\vec{i} + y\vec{j} + z\vec{k}$ over the quarter cylinder S given by $x^2 + y^2 = 1, 0 \leq x \leq 1, 0 \leq y \leq 1$, $0 \leq z \leq 1$, oriented outward.
59. Compute the flux of $\vec{F} = x\vec{i} + \vec{j} + \vec{k}$ through the surface S given by $x = \sin y \sin z$, with $0 \leq y \leq \pi/2$, $0 \leq z \leq \pi/2$, oriented in the direction of increasing x.
60. Compute the flux of $\vec{F} = (x + z)\vec{i} + \vec{j} + z\vec{k}$ through the surface S given by $y = x^2 + z^2$, with $0 \leq y \leq 1$, $x \geq 0$, $z \geq 0$, oriented toward the xz-plane.
61. Let $\vec{F} = (xze^{yz})\vec{i} + xz\vec{j} + (5+x^2+y^2)\vec{k}$. Calculate the flux of $\vec{F}$ through the disk $x^2 + y^2 \leq 1$ in the xy-plane, oriented upward.
62. Let $\vec{H} = (e^{xy}+3z+5)\vec{i} + (e^{xy}+5z+3)\vec{j} + (3z+e^{xy})\vec{k}$. Calculate the flux of $\vec{H}$ through the square of side 2 with one vertex at the origin, one edge along the positive y-axis, one edge in the xz-plane with $x > 0$, $z > 0$, and the normal $\vec{n} = \vec{i} - \vec{k}$.
63. The vector field, $\vec{F}$, in Figure 19.28 depends only on z; that is, it is of the form $g(z)\vec{k}$, where g is an increasing function. The integral $\int_S \vec{F} \cdot d\vec{A}$ represents the flux of $\vec{F}$ through this rectangle, S, oriented upward. In each of the following cases, how does the flux change?
(a) The rectangle is twice as wide in the x-direction, with new corners at the origin, $(2, 0, 0)$, $(2, 1, 3)$, $(0, 1, 3)$.
(b) The rectangle is moved so that its corners are at $(1, 0, 0)$, $(2, 0, 0)$, $(2, 1, 3)$, $(1, 1, 3)$.
(c) The orientation is changed to downward.
(d) The rectangle is tripled in size, so that its new corners are at the origin, $(3, 0, 0)$, $(3, 3, 9)$, $(0, 3, 9)$.

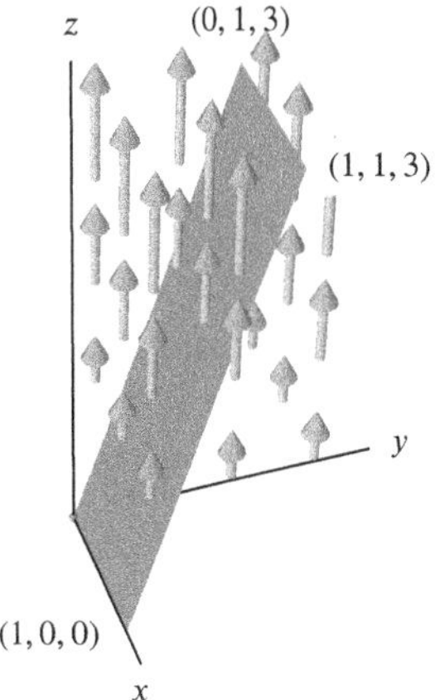

Figure 19.28

64. Electric charge is distributed in space with density (in coulomb/m^3) given in spherical coordinates by
$$\delta(\rho, \phi, \theta) = \begin{cases} \delta_0 \text{ (a constant)} & \rho \leq a \\ 0 & \rho > a. \end{cases}$$
(a) Describe the charge distribution in words.
(b) Find the electric field $\vec{E}$ due to δ. Assume that $\vec{E}$ can be written in spherical coordinates as $\vec{E} = E(\rho)\vec{e}_\rho$, where $\vec{e}_\rho$ is the unit outward normal to the sphere of radius ρ. In addition, $\vec{E}$ satisfies Gauss's Law for any simple closed surface S enclosing a volume W:
$$\int_S \vec{E} \cdot d\vec{A} = k \int_W \delta \, dV, \quad k \text{ a constant.}$$
65. Electric charge is distributed in space with density (in coulomb/m^3) given in cylindrical coordinates by
$$\delta(r, \theta, z) = \begin{cases} \delta_0 \text{ (a constant)} & \text{if } r \leq a \\ 0 & \text{if } r > a \end{cases}$$
(a) Describe the charge distribution in words.
(b) Find the electric field $\vec{E}$ due to δ. Assume that $\vec{E}$ can be written in cylindrical coordinates as $\vec{E} = E(r)\vec{e}_r$, where $\vec{e}_r$ is the unit outward vector to the cylinder of radius r, and that $\vec{E}$ satisfies Gauss's Law (see Problem 64).

Strengthen Your Understanding

In Problems 66–67, explain what is wrong with the statement.

66. Flux outward through the cone, given in cylindrical coordinates by $z = r$, can be computed using the formula $d\vec{A} = \left(\cos\theta\vec{i} + \sin\theta\vec{j}\right) R\,dz\,d\theta$.

67. For the surface $z = f(x, y)$ oriented upward, the formula

$$d\vec{A} = \vec{n}\,dA = \left(-f_x\vec{i} - f_y\vec{j} + \vec{k}\right) dx\,dy$$

gives $\vec{n} = -f_x\vec{i} - f_y\vec{j} + \vec{k}$ and $dA = dx\,dy$.

In Problems 68–69, give an example of:

68. A function $f(x, y)$ such that, for the surface $z = f(x, y)$ oriented upwards, we have $d\vec{A} = (\vec{i} + \vec{j} + \vec{k})\,dx\,dy$.

69. An oriented surface S on the cylinder of radius 10 centered on the z-axis such that $\int_S \vec{F} \cdot d\vec{A} = 600$, where $\vec{F} = x\vec{i} + y\vec{j}$.

Are the statements in Problems 70–72 true or false? Give reasons for your answer.

70. If S is the part of the graph of $z = f(x, y)$ above $a \le x \le b$, $c \le y \le d$, then S has surface area $\int_a^b \int_c^d \sqrt{(f_x)^2 + (f_y)^2 + 1}\,dx\,dy$.

71. If $\vec{A}(x, y)$ is the area vector for $z = f(x, y)$ oriented upward and $\vec{B}(x, y)$ is the area vector for $z = -f(x, y)$ oriented upward, then $\vec{A}(x, y) = -\vec{B}(x, y)$.

72. If S is the sphere $x^2 + y^2 + z^2 = 1$ oriented outward and $\int_S \vec{F} \cdot d\vec{A} = 0$, then $\vec{F}(x, y, z)$ is perpendicular to $x\vec{i} + y\vec{j} + z\vec{k}$ at every point of S.

19.3 THE DIVERGENCE OF A VECTOR FIELD

Imagine that the vector fields in Figures 19.29 and 19.30 are velocity vector fields describing the flow of a fluid.[3] Figure 19.29 suggests outflow from the origin; for example, it could represent the expanding cloud of matter in the big-bang theory of the origin of the universe. We say that the origin is a *source*. Figure 19.30 suggests flow into the origin; in this case we say that the origin is a *sink*.

In this section we use the flux out of a closed surface surrounding a point to measure the outflow per unit volume there, also called the *divergence*, or *flux density*.

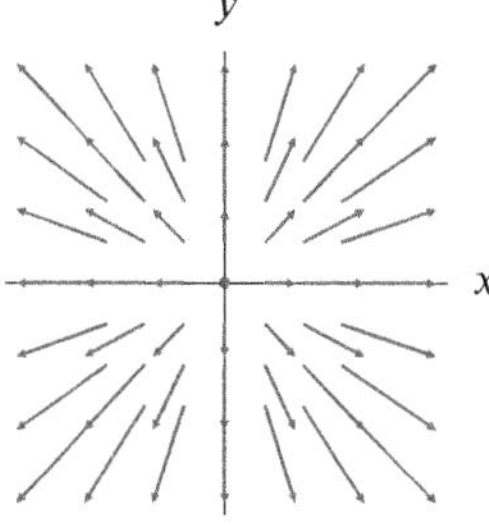

Figure 19.29: Vector field showing a source

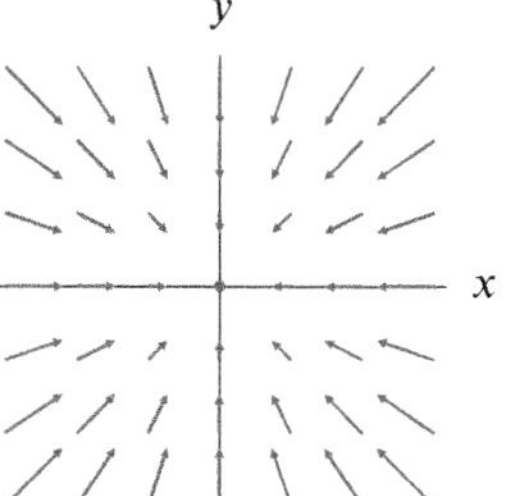

Figure 19.30: Vector field showing a sink

Definition of Divergence

To measure the outflow per unit volume of a vector field at a point, we calculate the flux out of a small sphere centered at the point, divide by the volume enclosed by the sphere, then take the limit of this flux-to-volume ratio as the sphere contracts around the point.

Geometric Definition of Divergence

The **divergence**, or **flux density**, of a smooth vector field $\vec{F}$, written $\mathbf{div}\vec{F}$, is a scalar-valued function defined by

$$\operatorname{div}\vec{F}(x, y, z) = \lim_{\text{Volume}\to 0} \frac{\int_S \vec{F} \cdot d\vec{A}}{\text{Volume of } S}.$$

Here S is a sphere centered at (x, y, z), oriented outward, that contracts down to (x, y, z) in the limit. The limit can be computed using other shapes as well, such as the cubes in Example 2.

[3]Although not all vector fields represent physically realistic fluid flows, it is useful to think of them in this way.

In Cartesian coordinates, the divergence can also be calculated using the following formula. We show these definitions are equivalent later in the section.

Cartesian Coordinate Definition of Divergence

If $\vec{F} = F_1\vec{i} + F_2\vec{j} + F_3\vec{k}$, then

$$\operatorname{div} \vec{F} = \frac{\partial F_1}{\partial x} + \frac{\partial F_2}{\partial y} + \frac{\partial F_3}{\partial z}.$$

The dot product formula gives an easy way to remember the Cartesian coordinate definition, and suggests another common notation for div $\vec{F}$, namely $\nabla \cdot \vec{F}$. Using $\nabla = \frac{\partial}{\partial x}\vec{i} + \frac{\partial}{\partial y}\vec{j} + \frac{\partial}{\partial z}\vec{k}$, we can write

$$\operatorname{div} \vec{F} = \nabla \cdot \vec{F} = \left(\frac{\partial}{\partial x}\vec{i} + \frac{\partial}{\partial y}\vec{j} + \frac{\partial}{\partial z}\vec{k}\right) \cdot (F_1\vec{i} + F_2\vec{j} + F_3\vec{k}) = \frac{\partial F_1}{\partial x} + \frac{\partial F_2}{\partial y} + \frac{\partial F_3}{\partial z}.$$

Example 1 Calculate the divergence of $\vec{F}(\vec{r}) = \vec{r}$ at the origin

(a) Using the geometric definition.
(b) Using the Cartesian coordinate definition.

Solution (a) Using the method of Example 5 on page 967, we can calculate the flux of $\vec{F}$ out of the sphere of radius a, centered at the origin; it is $4\pi a^3$. So we have

$$\operatorname{div} \vec{F}(0,0,0) = \lim_{a\to 0} \frac{\text{Flux}}{\text{Volume}} = \lim_{a\to 0} \frac{4\pi a^3}{\frac{4}{3}\pi a^3} = \lim_{a\to 0} 3 = 3.$$

(b) In Cartesian coordinates, $\vec{F}(x, y, z) = x\vec{i} + y\vec{j} + z\vec{k}$, so

$$\operatorname{div} \vec{F} = \frac{\partial}{\partial x}(x) + \frac{\partial}{\partial y}(y) + \frac{\partial}{\partial z}(z) = 1 + 1 + 1 = 3.$$

The next example shows that the divergence can be negative if there is net inflow to a point.

Example 2 (a) Using the geometric definition, find the divergence of $\vec{v} = -x\vec{i}$ at: (i) $(0,0,0)$ (ii) $(2,2,0)$.
(b) Confirm that the coordinate definition gives the same results.

Solution (a) (i) The vector field $\vec{v} = -x\vec{i}$ is parallel to the x-axis and is shown in the xy-plane in Figure 19.31. To compute the flux density at $(0,0,0)$, we use a cube S_1, centered at the origin with edges parallel to the axes, of length $2c$. Then the flux through the faces perpendicular to the y- and z-axes is zero (because the vector field is parallel to these faces). On the faces perpendicular to the x-axis, the vector field and the outward normal are parallel but point in opposite directions. On the face at $x = c$, where $\vec{v} = -c\vec{i}$ and $\Delta\vec{A} = \|\vec{A}\|\vec{i}$, we have

$$\vec{v} \cdot \Delta\vec{A} = -c\,\|\Delta\vec{A}\|.$$

On the face at $x = -c$, where $\vec{v} = c\vec{i}$ and $\Delta\vec{A} = -\|\vec{A}\|\vec{i}$, the dot product is still negative:

$$\vec{v} \cdot \Delta\vec{A} = -c\,\|\Delta\vec{A}\|.$$

Therefore, the flux through the cube is given by

$$\int_{S_1} \vec{v} \cdot d\vec{A} = \int_{\text{Face } x=-c} \vec{v} \cdot d\vec{A} + \int_{\text{Face } x=c} \vec{v} \cdot d\vec{A}$$

$$= -c \cdot \text{Area of one face} + (-c) \cdot \text{Area of other face} = -2c(2c)^2 = -8c^3.$$

Thus,

$$\operatorname{div} \vec{v}\,(0,0,0) = \lim_{\text{Volume}\to 0} \frac{\int_S \vec{v} \cdot d\vec{A}}{\text{Volume of cube}} = \lim_{c\to 0}\left(\frac{-8c^3}{(2c)^3}\right) = -1.$$

Since the vector field points inward toward the yz-plane, it makes sense that the divergence is negative at the origin.

(ii) Take S_2 to be a cube as before, but centered this time at the point $(2,2,0)$. See Figure 19.31. As before, the flux through the faces perpendicular to the y- and z-axes is zero. On the face at $x = 2 + c$,

$$\vec{v} \cdot \Delta\vec{A} = -(2+c)\,\|\Delta\vec{A}\,\|.$$

On the face at $x = 2 - c$ with outward normal, the dot product is positive, and

$$\vec{v} \cdot \Delta\vec{A} = (2-c)\,\|\Delta\vec{A}\,\|.$$

Therefore, the flux through the cube is given by

$$\int_{S_2} \vec{v} \cdot d\vec{A} = \int_{\text{Face } x=2-c} \vec{v} \cdot d\vec{A} + \int_{\text{Face } x=2+c} \vec{v} \cdot d\vec{A}$$

$$= (2-c) \cdot \text{Area of one face} - (2+c) \cdot \text{Area of other face} = -2c(2c)^2 = -8c^3.$$

Then, as before,

$$\operatorname{div} \vec{v}\,(2,2,0) = \lim_{\text{Volume}\to 0} \frac{\int_S \vec{v} \cdot d\vec{A}}{\text{Volume of cube}} = \lim_{c\to 0}\left(\frac{-8c^3}{(2c)^3}\right) = -1.$$

Although the vector field is flowing away from the point $(2,2,0)$ on the left, this outflow is smaller in magnitude than the inflow on the right, so the net outflow is negative.

(b) Since $\vec{v} = -x\vec{i} + 0\vec{j} + 0\vec{k}$, the formula gives

$$\operatorname{div} \vec{v} = \frac{\partial}{\partial x}(-x) + \frac{\partial}{\partial y}(0) + \frac{\partial}{\partial z}(0) = -1 + 0 + 0 = -1.$$

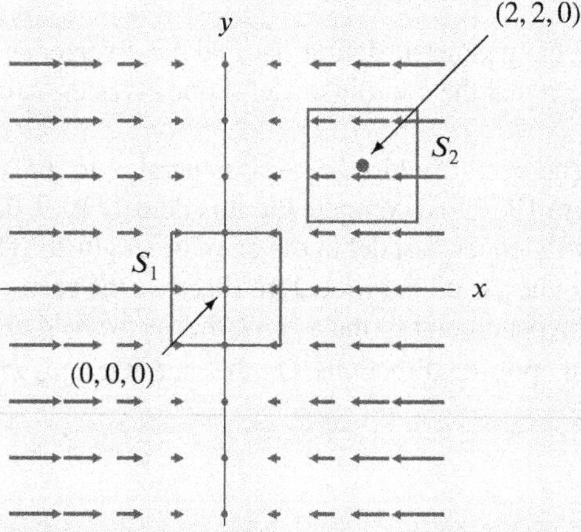

Figure 19.31: Vector field $\vec{v} = -x\vec{i}$ in the xy-plane

Why Do the Two Definitions of Divergence Give the Same Result?

The geometric definition defines div $\vec{F}$ as the flux density of $\vec{F}$. To see why the coordinate definition is also the flux density, imagine computing the flux out of a small box-shaped surface S at (x_0, y_0, z_0), with sides of length Δx, Δy, and Δz parallel to the axes. On S_1 (the back face of the box shown in Figure 19.32, where $x = x_0$), the outward normal is in the negative x-direction, so $d\vec{A} = -dy\,dz\,\vec{i}$. Assuming $\vec{F}$ is approximately constant on S_1, we have

$$\begin{aligned}\int_{S_1} \vec{F} \cdot d\vec{A} &= \int_{S_1} \vec{F} \cdot (-\vec{i})\,dy\,dz \approx -F_1(x_0, y_0, z_0) \int_{S_1} dy\,dz \\ &= -F_1(x_0, y_0, z_0) \cdot \text{Area of } S_1 = -F_1(x_0, y_0, z_0)\,\Delta y\,\Delta z.\end{aligned}$$

On S_2, the face where $x = x_0 + \Delta x$, the outward normal points in the positive x-direction, so $d\vec{A} = dy\,dz\,\vec{i}$. Therefore,

$$\begin{aligned}\int_{S_2} \vec{F} \cdot d\vec{A} &= \int_{S_2} \vec{F} \cdot \vec{i}\,dy\,dz \approx F_1(x_0 + \Delta x, y_0, z_0) \int_{S_2} dy\,dz \\ &= F_1(x_0 + \Delta x, y_0, z_0) \cdot \text{Area of } S_2 = F_1(x_0 + \Delta x, y_0, z_0)\,\Delta y\,\Delta z.\end{aligned}$$

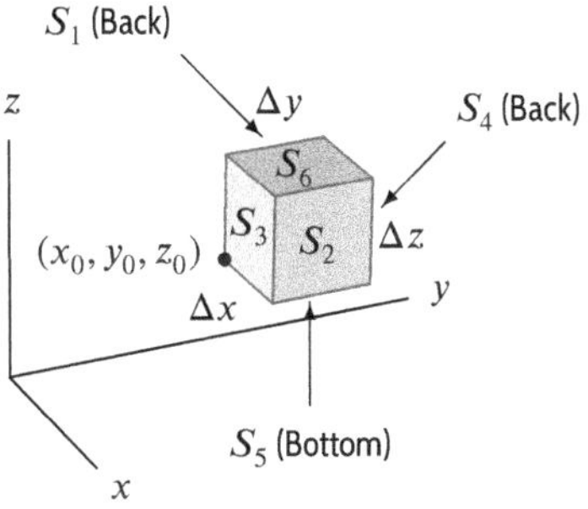

Figure 19.32: Box used to find div $\vec{F}$ at (x_0, y_0, z_0)

Thus,

$$\begin{aligned}\int_{S_1} \vec{F} \cdot d\vec{A} + \int_{S_2} \vec{F} \cdot d\vec{A} &\approx F_1(x_0 + \Delta x, y_0, z_0)\Delta y \Delta z - F_1(x_0, y_0, z_0)\Delta y \Delta z \\ &= \frac{F_1(x_0 + \Delta x, y_0, z_0) - F_1(x_0, y_0, z_0)}{\Delta x}\Delta x \Delta y \Delta z \\ &\approx \frac{\partial F_1}{\partial x}\Delta x \Delta y \Delta z.\end{aligned}$$

By an analogous argument, the contribution to the flux from S_3 and S_4 (the surfaces perpendicular to the y-axis) is approximately

$$\frac{\partial F_2}{\partial y}\,\Delta x\,\Delta y\,\Delta z,$$

and the contribution to the flux from S_5 and S_6 is approximately

$$\frac{\partial F_3}{\partial z}\,\Delta x\,\Delta y\,\Delta z.$$

Thus, adding these contributions, we have

$$\text{Total flux through } S \approx \frac{\partial F_1}{\partial x}\,\Delta x\,\Delta y\,\Delta z + \frac{\partial F_2}{\partial y}\,\Delta x\,\Delta y\,\Delta z + \frac{\partial F_3}{\partial z}\,\Delta x\,\Delta y\,\Delta z.$$

Since the volume of the box is $\Delta x\,\Delta y\,\Delta z$, the flux density is

$$\frac{\text{Total flux through } S}{\text{Volume of box}} \approx \frac{\frac{\partial F_1}{\partial x}\Delta x\Delta y\Delta z + \frac{\partial F_2}{\partial y}\Delta x\Delta y\Delta z + \frac{\partial F_3}{\partial z}\Delta x\Delta y\Delta z}{\Delta x\Delta y\Delta z} = \frac{\partial F_1}{\partial x} + \frac{\partial F_2}{\partial y} + \frac{\partial F_3}{\partial z}.$$

Divergence-Free Vector Fields

A vector field $\vec{F}$ is said to be *divergence free* or *solenoidal* if $\operatorname{div}\vec{F} = 0$ everywhere that $\vec{F}$ is defined.

Example 3 Figure 19.33 shows, for three values of the constant p, the vector field

$$\vec{E} = \frac{\vec{r}}{\|\vec{r}\|^p} \qquad \vec{r} = x\vec{i} + y\vec{j} + z\vec{k},\ \vec{r} \neq \vec{0}.$$

(a) Find a formula for $\operatorname{div}\vec{E}$.
(b) Is there a value of p for which $\vec{E}$ is divergence-free? If so, find it.

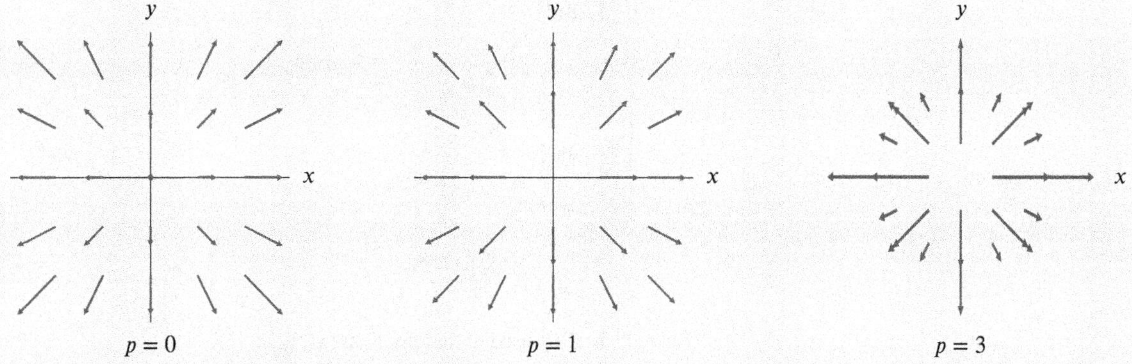

Figure 19.33: The vector field $\vec{E}(\vec{r}) = \vec{r}/\|\vec{r}\|^p$ for $p = 0$, 1, and 3 in the xy-plane

Solution (a) The components of $\vec{E}$ are

$$\vec{E} = \frac{x}{(x^2+y^2+z^2)^{p/2}}\vec{i} + \frac{y}{(x^2+y^2+z^2)^{p/2}}\vec{j} + \frac{z}{(x^2+y^2+z^2)^{p/2}}\vec{k}.$$

We compute the partial derivatives

$$\frac{\partial}{\partial x}\left(\frac{x}{(x^2+y^2+z^2)^{p/2}}\right) = \frac{1}{(x^2+y^2+z^2)^{p/2}} - \frac{px^2}{(x^2+y^2+z^2)^{(p/2)+1}}$$

$$\frac{\partial}{\partial y}\left(\frac{y}{(x^2+y^2+z^2)^{p/2}}\right) = \frac{1}{(x^2+y^2+z^2)^{p/2}} - \frac{py^2}{(x^2+y^2+z^2)^{(p/2)+1}}$$

$$\frac{\partial}{\partial z}\left(\frac{z}{(x^2+y^2+z^2)^{p/2}}\right) = \frac{1}{(x^2+y^2+z^2)^{p/2}} - \frac{pz^2}{(x^2+y^2+z^2)^{(p/2)+1}}.$$

So

$$\operatorname{div}\vec{E} = \frac{3}{(x^2+y^2+z^2)^{p/2}} - \frac{p(x^2+y^2+z^2)}{(x^2+y^2+z^2)^{(p/2)+1}} = \frac{3-p}{(x^2+y^2+z^2)^{p/2}} = \frac{3-p}{\|\vec{r}\|^p}.$$

(b) The divergence is zero when $p = 3$, so $\vec{F}(\vec{r}) = \vec{r}/\|\vec{r}\|^3$ is a divergence-free vector field. Notice that the divergence is zero even though the vectors point outward from the origin.

Magnetic Fields

An important class of divergence-free vector fields is the magnetic fields. One of Maxwell's Laws of Electromagnetism is that the magnetic field $\vec{B}$ satisfies

$$\operatorname{div} \vec{B} = 0.$$

Example 4 An infinitesimal current loop, similar to that shown in Figure 19.34, is called a *magnetic dipole*. Its magnitude is described by a constant vector $\vec{\mu}$, called the dipole moment. The magnetic field due to a magnetic dipole with moment $\vec{\mu}$ is a multiple of

$$\vec{B} = -\frac{\vec{\mu}}{\|\vec{r}\|^3} + \frac{3(\vec{\mu} \cdot \vec{r})\vec{r}}{\|\vec{r}\|^5}, \qquad \vec{r} \neq \vec{0}.$$

Show that $\operatorname{div} \vec{B} = 0$.

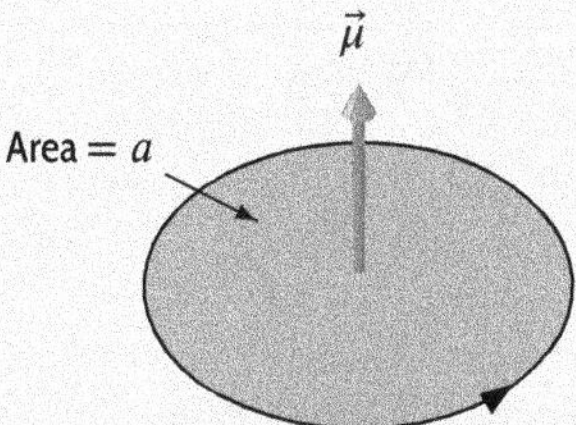

Figure 19.34: A current loop

Solution To show that $\operatorname{div} \vec{B} = 0$ we can use the following version of the product rule for the divergence: if g is a scalar function and $\vec{F}$ is a vector field, then

$$\operatorname{div}(g\vec{F}) = (\operatorname{grad} g) \cdot \vec{F} + g \operatorname{div} \vec{F}.$$

(See Problem 37 on page 990.) Thus, since $\vec{\mu}$ is constant and $\operatorname{div} \vec{\mu} = 0$, we have

$$\operatorname{div}\left(\frac{\vec{\mu}}{\|\vec{r}\|^3}\right) = \operatorname{div}\left(\frac{1}{\|\vec{r}\|^3}\vec{\mu}\right) = \operatorname{grad}\left(\frac{1}{\|\vec{r}\|^3}\right) \cdot \vec{\mu} + \left(\frac{1}{\|\vec{r}\|^3}\right)0$$

and

$$\operatorname{div}\left(\frac{(\vec{\mu} \cdot \vec{r})\vec{r}}{\|\vec{r}\|^5}\right) = \operatorname{div}\left(\vec{\mu} \cdot \vec{r}\ \frac{\vec{r}}{\|\vec{r}\|^5}\right) = \operatorname{grad}(\vec{\mu} \cdot \vec{r}) \cdot \frac{\vec{r}}{\|\vec{r}\|^5} + (\vec{\mu} \cdot \vec{r}) \operatorname{div}\left(\frac{\vec{r}}{\|\vec{r}\|^5}\right).$$

From Problems 83 and 84 of Section 14.5 (available online) and Example 3 on page 986, we have

$$\operatorname{grad}\left(\frac{1}{\|\vec{r}\|^3}\right) = \frac{-3\vec{r}}{\|\vec{r}\|^5}, \qquad \operatorname{grad}(\vec{\mu} \cdot \vec{r}) = \vec{\mu}, \qquad \operatorname{div}\left(\frac{\vec{r}}{\|\vec{r}\|^5}\right) = \frac{-2}{\|\vec{r}\|^5}.$$

Putting these results together gives

$$\begin{aligned}
\operatorname{div} \vec{B} &= -\operatorname{grad}\left(\frac{1}{\|\vec{r}\|^3}\right) \cdot \vec{\mu} + 3 \operatorname{grad}(\vec{\mu} \cdot \vec{r}) \cdot \frac{\vec{r}}{\|\vec{r}\|^5} + 3(\vec{\mu} \cdot \vec{r}) \operatorname{div}\left(\frac{\vec{r}}{\|\vec{r}\|^5}\right) \\
&= \frac{3\vec{r} \cdot \vec{\mu}}{\|\vec{r}\|^5} + \frac{3\vec{\mu} \cdot \vec{r}}{\|\vec{r}\|^5} - \frac{6\vec{\mu} \cdot \vec{r}}{\|\vec{r}\|^5} \\
&= 0.
\end{aligned}$$

Exercises and Problems for Section 19.3 Online Resource: Additional Problems for Section 19.3

EXERCISES

Are the quantities in Exercises 1–2 vectors or scalars? Calculate them.

1. $\operatorname{div}\left((x^2+y)\vec{i}+(xye^z)\vec{j}-\ln(x^2+y^2)\vec{k}\right)$

2. $\operatorname{div}\left((2\sin(xy)+\tan z)\vec{i}+(\tan y)\vec{j}+(e^{x^2+y^2})\vec{k}\right)$

3. Which of the following two vector fields, sketched in the xy-plane, appears to have the greater divergence at the origin? The scales are the same on each.

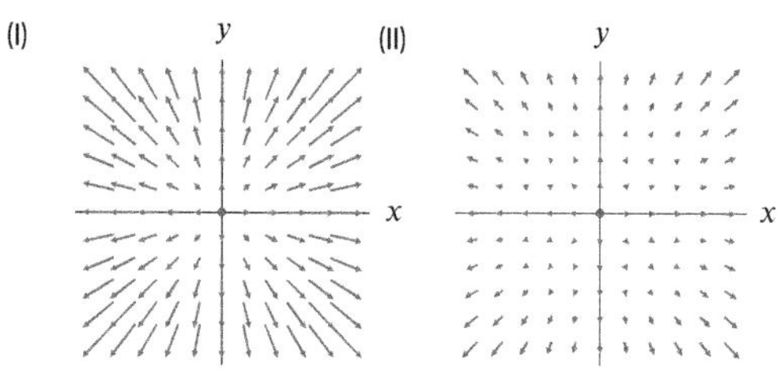

In Exercises 4–12, find the divergence of the vector field. (Note: $\vec{r}=x\vec{i}+y\vec{j}+z\vec{k}$.)

4. $\vec{F}(x,y)=-y\vec{i}+x\vec{j}$

5. $\vec{F}(x,y)=-x\vec{i}+y\vec{j}$

6. $\vec{F}(x,y,z)=(-x+y)\vec{i}+(y+z)\vec{j}+(-z+x)\vec{k}$

7. $\vec{F}(x,y)=(x^2-y^2)\vec{i}+2xy\vec{j}$

8. $\vec{F}(x,y,z)=3x^2\vec{i}-\sin(xz)(\vec{i}+\vec{k})$

9. $\vec{F}=\left(\ln(x^2+1)\vec{i}+(\cos y)\vec{j}+(xye^z)\vec{k}\right)$

10. $\vec{F}(\vec{r})=\vec{a}\times\vec{r}$

11. $\vec{F}(x,y)=\dfrac{-y\vec{i}+x\vec{j}}{x^2+y^2}$

12. $\vec{F}(\vec{r})=\dfrac{\vec{r}-\vec{r}_0}{\|\vec{r}-\vec{r}_0\|},\quad \vec{r}\neq\vec{r}_0$

13. For each of the following vector fields, sketched in the xy-plane, decide if the divergence is positive, zero, or negative at the indicated point.

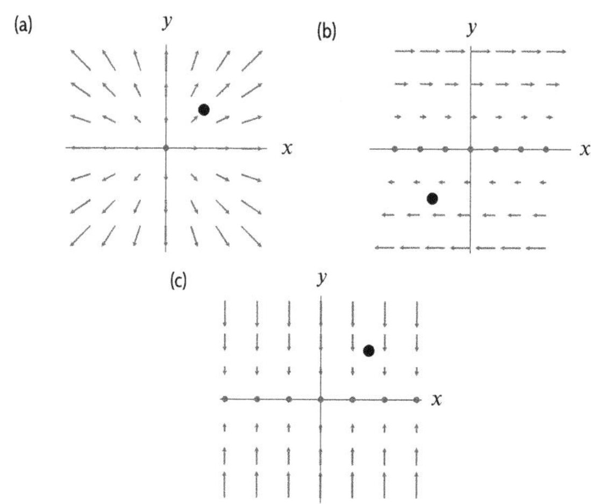

PROBLEMS

14. Draw two vector fields that have positive divergence everywhere.

15. Draw two vector fields that have negative divergence everywhere.

16. Draw two vector fields that have zero divergence everywhere.

17. A small sphere of radius 0.1 surrounds the point $(2,3,-1)$. The flux of a vector field $\vec{G}$ into this sphere is 0.00004π. Estimate $\operatorname{div}\vec{G}$ at the point $(2,3,-1)$.

18. A smooth vector field $\vec{F}$ has $\operatorname{div}\vec{F}(1,2,3)=5$. Estimate the flux of $\vec{F}$ out of a small sphere of radius 0.01 centered at the point $(1,2,3)$.

19. Let $\vec{F}$ be a vector field with $\operatorname{div}\vec{F}=x^2+y^2-z$.
   (a) Estimate $\int_S \vec{F}\cdot d\vec{A}$ where S is
      (i) A sphere of radius 0.1 centered at $(2,0,0)$.
      (ii) A box of side 0.2 with edges parallel to the axes and centered at $(0,0,10)$.
   (b) The point $(2,0,0)$ is called a *source* for the vector field $\vec{F}$; the point $(0,0,10)$ is called a *sink*. Explain the reason for these names using your answer to part (a).

20. The flux of $\vec{F}$ out of a small sphere of radius 0.1 centered at $(4,5,2)$ is 0.0125. Estimate:
   (a) $\operatorname{div}\vec{F}$ at $(4,5,2)$
   (b) The flux of $\vec{F}$ out of a sphere of radius 0.2 centered at $(4,5,2)$.

21. (a) Find the flux of $\vec{F}=2x\vec{i}-3y\vec{j}+5z\vec{k}$ through a cube with four of its corners at the points (a,b,c), $(a+w,b,c)$, $(a,b+w,c)$, $(a,b,c+w)$ and edge length w. See Figure 19.35.
   (b) Use the geometric definition and part (a) to find $\operatorname{div}\vec{F}$ at the point (a,b,c).
   (c) Find $\operatorname{div}\vec{F}$ using partial derivatives.

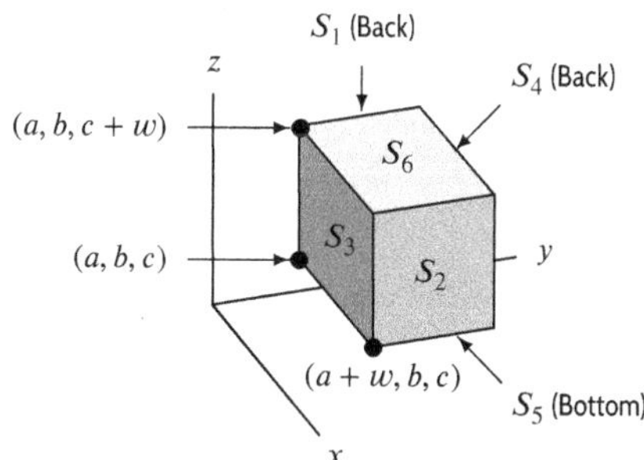

Figure 19.35

22. Suppose $\vec{F} = (3x + 2)\vec{i} + 4x\vec{j} + (5x + 1)\vec{k}$. Use the method of Exercise 21 to find div $\vec{F}$ at the point (a, b, c) by two different methods.

23. Use the geometric definition of divergence to find div $\vec{v}$ at the origin, where $\vec{v} = -2\vec{r}$. Check that you get the same result using the definition in Cartesian coordinates.

24. **(a)** Let $f(x, y) = axy + ax^2y + y^3$. Find div grad f.
    (b) Choose a so that div grad $f = 0$ for all x, y.

25. Let $\vec{F} = (9a^2x + 10ay^2)\vec{i} + (10z^3 - 6ay)\vec{j} - (3z + 10x^2 + 10y^2)\vec{k}$. Find the value(s) of a making div $\vec{F}$
    (a) 0 **(b)** A minimum

26. Let $\vec{F}(\vec{r}) = \vec{r}/\|\vec{r}\|^3$ (in 3-space), $\vec{r} \neq \vec{0}$.
    (a) Calculate div $\vec{F}$.
    (b) Sketch $\vec{F}$. Does $\vec{F}$ appear to have nonzero divergence? Does this agree with your answer to part (a)?

27. The vector field $\vec{F}(\vec{r}) = \vec{r}/\|\vec{r}\|^3$ is not defined at the origin. Nevertheless, we can attempt to use the flux definition to compute div $\vec{F}$ at the origin. What is the result?

28. Let $\vec{F}(x, y, z) = z\vec{k}$.
    (a) Calculate div $\vec{F}$.
    (b) Sketch $\vec{F}$. Does $\vec{F}$ appear to have nonzero divergence? Does this agree with your answer to part (a)?

29. The divergence of a magnetic vector field $\vec{B}$ must be zero everywhere. Which of the following vector fields cannot be a magnetic vector field?
    (a) $\vec{B}(x, y, z) = -y\vec{i} + x\vec{j} + (x + y)\vec{k}$
    (b) $\vec{B}(x, y, z) = -z\vec{i} + y\vec{j} + x\vec{k}$
    (c) $\vec{B}(x, y, z) = (x^2 - y^2 - x)\vec{i} + (y - 2xy)\vec{j}$

Problems 30–31 involve electric fields. Electric charge produces a vector field $\vec{E}$, called the electric field, which represents the force on a unit positive charge placed at the point. Two positive or two negative charges repel one another, whereas two charges of opposite sign attract one another. The divergence of $\vec{E}$ is proportional to the density of the electric charge (that is, the charge per unit volume), with a positive constant of proportionality.

30. A certain distribution of electric charge produces the electric field shown in Figure 19.36. Where are the charges that produced this electric field concentrated? Which concentrations are positive and which are negative?

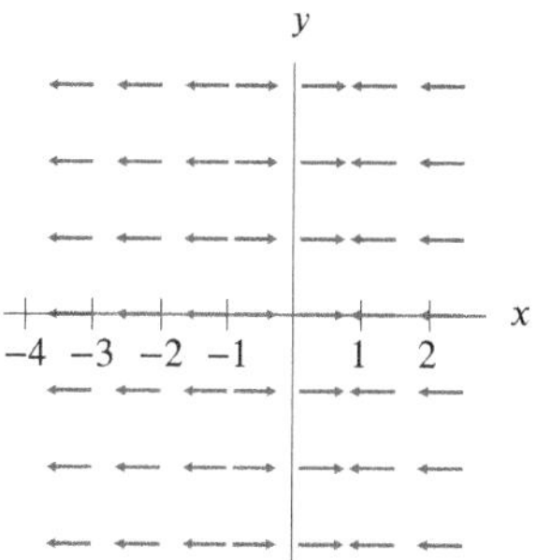

Figure 19.36

31. The electric field at the point $\vec{r}$ as a result of a point charge at the origin is $\vec{E}(\vec{r}) = k\vec{r}/\|\vec{r}\|^3$.
    (a) Calculate div $\vec{E}$ for $\vec{r} \neq \vec{0}$.
    (b) Calculate the limit suggested by the geometric definition of div $\vec{E}$ at the point $(0, 0, 0)$.
    (c) Explain what your answers mean in terms of charge density.

32. Due to roadwork ahead, the traffic on a highway slows linearly from 55 miles/hour to 15 miles/hour over a 2000-foot stretch of road, then crawls along at 15 miles/hour for 5000 feet, then speeds back up linearly to 55 miles/hour in the next 1000 feet, after which it moves steadily at 55 miles/hour.
    (a) Sketch a velocity vector field for the traffic flow.
    (b) Write a formula for the velocity vector field $\vec{v}$ (miles/hour) as a function of the distance x feet from the initial point of slowdown. (Take the direction of motion to be $\vec{i}$ and consider the various sections of the road separately.)
    (c) Compute div $\vec{v}$ at $x = 1000, 5000, 7500, 10{,}000$. Be sure to include the proper units.

33. The velocity field $\vec{v}$ in Problem 32 does not give a complete description of the traffic flow, for it takes no account of the spacing between vehicles. Let ρ be the density (cars/mile) of highway, where we assume that ρ depends only on x.
    (a) Using your highway experience, arrange in ascending order: $\rho(0), \rho(1000), \rho(5000)$.
    (b) What are the units and interpretation of the vector field $\rho\vec{v}$?
    (c) Would you expect $\rho\vec{v}$ to be constant? Why? What does this mean for div$(\rho\vec{v})$?
    (d) Determine $\rho(x)$ if $\rho(0) = 75$ cars/mile and $\rho\vec{v}$ is constant.
    (e) If the highway has two lanes, find the approximate number of feet between cars at $x = 0$, 1000, and 5000.

34. For $\vec{r} = x\vec{i} + y\vec{j} + z\vec{k}$, an arbitrary function $f(x, y, z)$, and an arbitrary vector field $\vec{F}(x, y, z)$, which of the following is a vector field and which is a constant vector field?

(a) grad f **(b)** $(\text{div}\,\vec{F})\vec{i}$ **(c)** $(\text{div}\,\vec{r})\vec{i}$
(d) $(\text{div}\,\vec{i})\vec{F}$ **(e)** grad(div $\vec{F}$)

35. Let $\vec{r} = x\vec{i} + y\vec{j} + z\vec{k}$ and $\vec{c} = c_1\vec{i} + c_2\vec{j} + c_3\vec{k}$, a constant vector; let S be a sphere of radius R centered at the origin. Find

(a) $\text{div}(\vec{r} \times \vec{c})$ **(b)** $\int_S (\vec{r} \times \vec{c}) \cdot d\vec{A}$

36. Show that if $\vec{a}$ is a constant vector and $f(x, y, z)$ is a function, then $\text{div}(f\vec{a}) = (\text{grad}\, f) \cdot \vec{a}$.

37. Show that if $g(x, y, z)$ is a scalar-valued function and $\vec{F}(x, y, z)$ is a vector field, then

$$\text{div}(g\vec{F}) = (\text{grad}\, g) \cdot \vec{F} + g\,\text{div}\,\vec{F}.$$

38. If $f(x, y, z)$ and $g(x, y, z)$ are functions with continuous second partial derivatives, show that

$$\text{div}(\text{grad}\, f \times \text{grad}\, g) = 0.$$

In Problems 39–41, use Problems 37 and 38 to find the divergence of the vector field. The vectors $\vec{a}$ and $\vec{b}$ are constant.

39. $\vec{F} = \dfrac{1}{\|\vec{r}\|^p}\vec{a} \times \vec{r}$ 40. $\vec{B} = \dfrac{1}{x^a}\vec{r}$

41. $\vec{G} = (\vec{b} \cdot \vec{r})\vec{a} \times \vec{r}$

42. A vector field, $\vec{v}$, in the plane is a *point source* at the origin if its direction is away from the origin at every point, its magnitude depends only on the distance from the origin, and its divergence is zero away from the origin.

(a) Explain why a point source at the origin must be of the form $\vec{v} = \left(f(x^2 + y^2)\right)(x\vec{i} + y\vec{j})$ for some positive function f.
(b) Show that $\vec{v} = K(x^2 + y^2)^{-1}(x\vec{i} + y\vec{j})$ is a point source at the origin if $K > 0$.
(c) What is the magnitude $\|\vec{v}\|$ of the source in part (b) as a function of the distance from its center?
(d) Sketch the vector field $\vec{v} = (x^2 + y^2)^{-1}(x\vec{i} + y\vec{j})$.
(e) Show that $\phi = \frac{K}{2}\log(x^2 + y^2)$ is a potential function for the source in part (b).

43. A vector field, $\vec{v}$, in the plane is a *point sink* at the origin if its direction is toward the origin at every point, its magnitude depends only on the distance from the origin, and its divergence is zero away from the origin.

(a) Explain why a point sink at the origin must be of the form $\vec{v} = \left(f(x^2 + y^2)\right)(x\vec{i} + y\vec{j})$ for some negative function f.
(b) Show that $\vec{v} = K(x^2 + y^2)^{-1}(x\vec{i} + y\vec{j})$ is a point sink at the origin if $K < 0$.
(c) Determine the magnitude $\|\vec{v}\|$ of the sink in part (b) as a function of the distance from its center.
(d) Sketch $\vec{v} = -(x^2 + y^2)^{-1}(x\vec{i} + y\vec{j})$.
(e) Show that $\phi = \frac{K}{2}\log(x^2 + y^2)$ is a potential function for the sink in part (b).

Strengthen Your Understanding

In Problems 44–46, explain what is wrong with the statement.

44. $\text{div}(2x\vec{i}) = 2\vec{i}$.
45. For $\vec{F}(x, y, z) = (x^2 + y)\vec{i} + (2y + z)\vec{j} - z^2\vec{k}$ we have $\text{div}\,\vec{F} = 2x\vec{i} + 2\vec{j} - 2z\vec{k}$.
46. The divergence of $f(x, y, z) = x^2 + yz$ is given by $\text{div}\, f(x, y, z) = 2x + z + y$.

In Problems 47–49, give an example of:

47. A vector field $\vec{F}(x, y, z)$ whose divergence is a nonzero constant.
48. A nonzero vector field $\vec{F}(x, y, z)$ whose divergence is zero.
49. A vector field that is not divergence free.

Are the statements in Problems 50–62 true or false? Give reasons for your answer.

50. $\text{div}(\vec{F} + \vec{G}) = \text{div}\,\vec{F} + \text{div}\,\vec{G}$
51. $\text{grad}(\vec{F} \cdot \vec{G}) = \vec{F}(\text{div}\,\vec{G}) + (\text{div}\,\vec{F})\vec{G}$
52. div $\vec{F}$ is a scalar whose value can vary from point to point.
53. If $\vec{F}$ is a vector field in 3-space, then div$\vec{F}$ is also a vector field.
54. A constant vector field $\vec{F} = a\vec{i} + b\vec{j} + c\vec{k}$ has zero divergence.
55. If a vector field $\vec{F}$ in 3-space has zero divergence then $\vec{F} = a\vec{i} + b\vec{j} + c\vec{k}$ where a, b and c are constants.
56. If $\vec{F}$ is a vector field in 3-space, and f is a scalar function, then $\text{div}(f\vec{F}) = f\text{div}\vec{F}$.
57. If $\vec{F}$ is a vector field in 3-space, and $\vec{F} = \text{grad}\, f$, then $\text{div}\,\vec{F} = 0$.
58. If $\vec{F}$ is a vector field in 3-space, then $\text{grad}(\text{div}\,\vec{F}) = \vec{0}$.
59. The field $\vec{F}(\vec{r}) = \vec{r}$ is divergence free.
60. If $f(x, y, z)$ is any given continuous scalar function, then there is at least one vector field $\vec{F}$ such that $\text{div}\vec{F} = f$.
61. If $\vec{F}$ and $\vec{G}$ are vector fields satisfying $\text{div}\vec{F} = \text{div}\vec{G}$ then $\vec{F} = \vec{G}$.
62. There exist a scalar function f and a vector field $\vec{F}$ satisfying $\text{div}(\text{grad}\, f) = \text{grad}(\text{div}\,\vec{F})$.

19.4 THE DIVERGENCE THEOREM

The Divergence Theorem is a multivariable analogue of the Fundamental Theorem of Calculus; it says that the integral of the flux density over a solid region equals the flux integral through the boundary of the region.

The Boundary of a Solid Region

The boundary, S, of a solid region, W, may be thought of as the skin between the interior of the region and the space around it. For example, the boundary of a solid ball is a spherical surface, the boundary of a solid cube is its six faces, and the boundary of a solid cylinder is a tube sealed at both ends by disks. (See Figure 19.37). A surface which is the boundary of a solid region is called a *closed surface*.

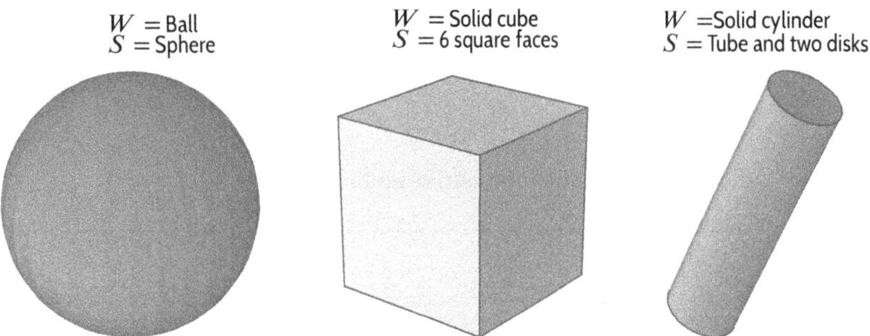

Figure 19.37: Several solid regions and their boundaries

Calculating the Flux from the Flux Density

Consider a solid region W in 3-space whose boundary is the closed surface S. There are two ways to find the total flux of a vector field $\vec{F}$ out of W. One is to calculate the flux of $\vec{F}$ through S:

$$\text{Flux out of } W = \int_S \vec{F} \cdot d\vec{A}.$$

Another way is to use div $\vec{F}$, which gives the flux density at any point in W. We subdivide W into small boxes, as shown in Figure 19.38. Then, for a small box of volume ΔV,

$$\text{Flux out of box} \approx \text{Flux density} \cdot \text{Volume} = \operatorname{div} \vec{F}\ \Delta V.$$

What happens when we add the fluxes out of all the boxes? Consider two adjacent boxes, as shown in Figure 19.39. The flux through the shared wall is counted twice, once out of the box on each side. When we add the fluxes, these two contributions cancel, so we get the flux out of the solid region formed by joining the two boxes. Continuing in this way, we find that

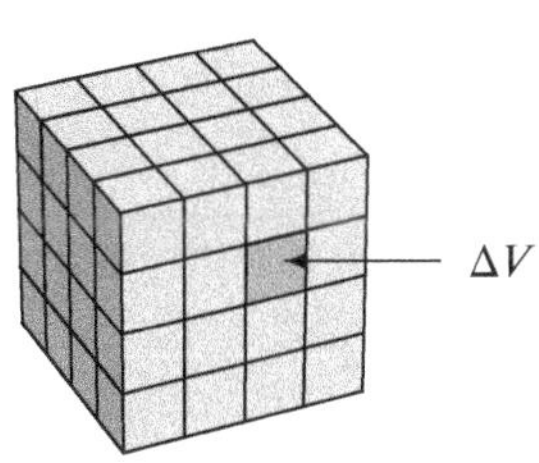

Figure 19.38: Subdivision of region into small boxes

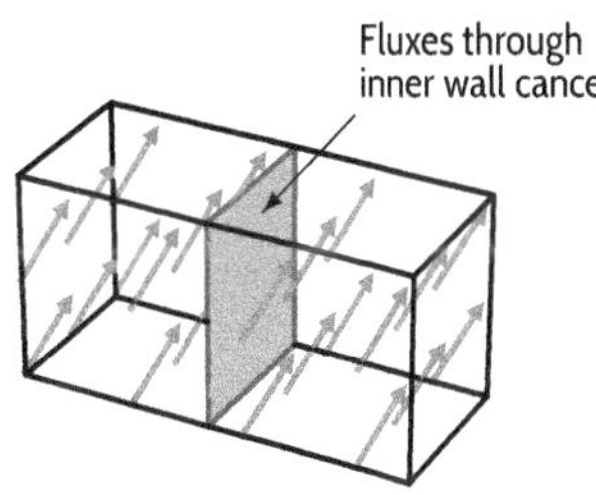

Figure 19.39: Adding the flux out of adjacent boxes

$$\text{Flux out of } W = \sum \text{Flux out of small boxes} \approx \sum \operatorname{div} \vec{F}\ \Delta V.$$

We have approximated the flux by a Riemann sum. As the subdivision gets finer, the sum approaches an integral, so

$$\text{Flux out of } W = \int_W \operatorname{div} \vec{F}\ dV.$$

We have calculated the flux in two ways, as a flux integral and as a volume integral. Therefore, these two integrals must be equal. This result holds even if W is not a rectangular solid. Thus, we have the following result.[4]

Theorem 19.1: The Divergence Theorem

If W is a solid region whose boundary S is a piecewise smooth surface, and if $\vec{F}$ is a smooth vector field on a solid region[5] containing W and S, then

$$\int_S \vec{F} \cdot d\vec{A} = \int_W \operatorname{div} \vec{F}\ dV,$$

where S is given the outward orientation.

Example 1 Use the Divergence Theorem to calculate the flux of the vector field $\vec{F}(\vec{r}) = \vec{r}$ through the sphere of radius a centered at the origin.

Solution In Example 5 on page 967 we computed the flux using the definition of a flux integral, giving

$$\int_S \vec{r} \cdot d\vec{A} = 4\pi a^3.$$

Now we use $\operatorname{div} \vec{F} = \operatorname{div}(x\vec{i} + y\vec{j} + z\vec{k}) = 3$ and the Divergence Theorem:

$$\int_S \vec{r} \cdot d\vec{A} = \int_W \operatorname{div} \vec{F}\ dV = \int_W 3\, dV = 3 \cdot \frac{4}{3}\pi a^3 = 4\pi a^3.$$

Example 2 Use the Divergence Theorem to calculate the flux of the vector field

$$\vec{F}(x, y, z) = (x^2 + y^2)\vec{i} + (y^2 + z^2)\vec{j} + (x^2 + z^2)\vec{k}$$

through the cube in Figure 19.40.

[4] A proof of the Divergence Theorem using the coordinate definition of the divergence can be found in the online supplement at www.wiley.com/college/hughes-hallett.

[5] The region containing W and S is open.

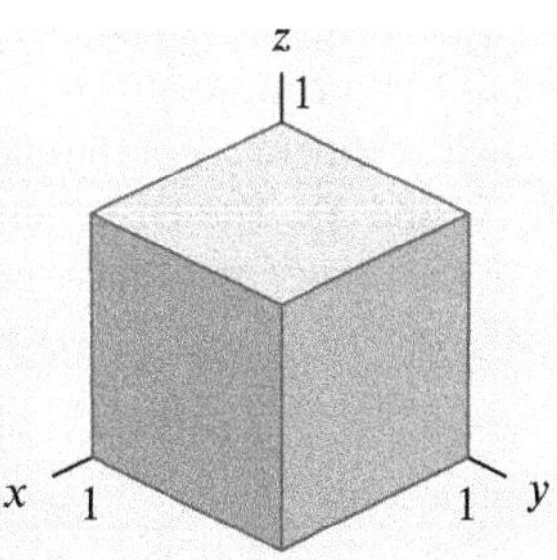

Figure 19.40

Solution The divergence of $\vec{F}$ is $\operatorname{div}\vec{F} = 2x + 2y + 2z$. Since $\operatorname{div}\vec{F}$ is positive everywhere in the first quadrant, the flux through S is positive. By the Divergence Theorem,

$$\begin{aligned}\int_S \vec{F}\cdot d\vec{A} &= \int_0^1\int_0^1\int_0^1 2(x+y+z)\,dx\,dy\,dz = \int_0^1\int_0^1 (x^2+2x(y+z))\Big|_0^1\,dy\,dz\\ &= \int_0^1\int_0^1 1+2(y+z)\,dy\,dz = \int_0^1 (y+y^2+2yz)\Big|_0^1\,dz\\ &= \int_0^1 (2+2z)\,dz = (2z+z^2)\Big|_0^1 = 3.\end{aligned}$$

The Divergence Theorem and Divergence-Free Vector Fields

An important application of the Divergence Theorem is the study of divergence-free vector fields.

Example 3 In Example 3 on page 986 we saw that the following vector field is divergence free:

$$\vec{F}(\vec{r}) = \frac{\vec{r}}{\|\vec{r}\|^3}, \qquad \vec{r} \neq \vec{0}.$$

Calculate $\int_S \vec{F}\cdot d\vec{A}$, using the Divergence Theorem if possible, for the following surfaces:

(a) S_1 is the sphere of radius a centered at the origin.

(b) S_2 is the sphere of radius a centered at the point $(2a, 0, 0)$.

Solution (a) We cannot use the Divergence Theorem directly because $\vec{F}$ is not defined everywhere inside the sphere (it is not defined at the origin). Since $\vec{F}$ points outward everywhere on S_1, the flux out of S_1 is positive. On S_1,

$$\vec{F}\cdot d\vec{A} = \|\vec{F}\|dA = \frac{a}{a^3}dA,$$

so

$$\int_{S_1}\vec{F}\cdot d\vec{A} = \frac{1}{a^2}\int_{S_1} dA = \frac{1}{a^2}(\text{Area of } S_1) = \frac{1}{a^2}4\pi a^2 = 4\pi.$$

Notice that the flux is not zero, although $\operatorname{div}\vec{F}$ is zero everywhere it is defined.

(b) Let W be the solid region enclosed by S_2. Since $\operatorname{div}\vec{F} = 0$ everywhere in W, we can use the Divergence Theorem in this case, giving

$$\int_{S_2}\vec{F}\cdot d\vec{A} = \int_W \operatorname{div}\vec{F}\,dV = \int_W 0\,dV = 0.$$

The Divergence Theorem applies to any solid region W and its boundary S, even in cases where the boundary consists of two or more surfaces. For example, if W is the solid region between the sphere S_1 of radius 1 and the sphere S_2 of radius 2, both centered at the same point, then the boundary of W consists of both S_1 and S_2. The Divergence Theorem requires the outward orientation, which on S_2 points away from the center and on S_1 points toward the center. (See Figure 19.41.)

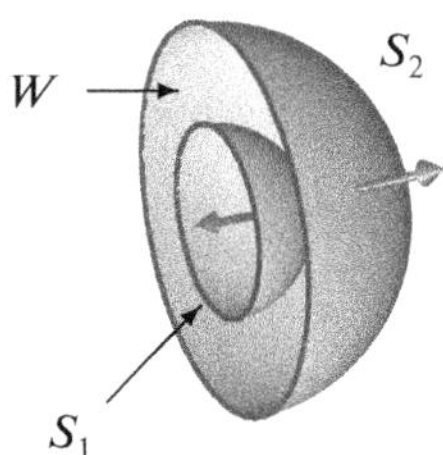

Figure 19.41: Cutaway view of the region W between two spheres, showing orientation vectors

Example 4 Let S_1 be the sphere of radius 1 centered at the origin and let S_2 be the ellipsoid $x^2+y^2+4z^2=16$, both oriented outward. For

$$\vec{F}(\vec{r})=\frac{\vec{r}}{\|\vec{r}\|^3},\qquad \vec{r}\neq\vec{0},$$

show that

$$\int_{S_1}\vec{F}\cdot d\vec{A}=\int_{S_2}\vec{F}\cdot d\vec{A}.$$

Solution The ellipsoid contains the sphere; let W be the solid region between them. Since W does not contain the origin, $\operatorname{div}\vec{F}$ is defined and equal to zero everywhere in W. Thus, if S is the boundary of W, then

$$\int_S\vec{F}\cdot d\vec{A}=\int_W\operatorname{div}\vec{F}\,dV=0.$$

But S consists of S_2 oriented outward and S_1 oriented inward, so

$$0=\int_S\vec{F}\cdot d\vec{A}=\int_{S_2}\vec{F}\cdot d\vec{A}-\int_{S_1}\vec{F}\cdot d\vec{A},$$

and thus

$$\int_{S_2}\vec{F}\cdot d\vec{A}=\int_{S_1}\vec{F}\cdot d\vec{A}.$$

In Example 3 we showed that $\int_{S_1}\vec{F}\cdot d\vec{A}=4\pi$, so $\int_{S_2}\vec{F}\cdot d\vec{A}=4\pi$ also. Note that it would have been more difficult to compute the integral over the ellipsoid directly.

Electric Fields

The electric field produced by a positive point charge q placed at the origin is

$$\vec{E}=q\frac{\vec{r}}{\|\vec{r}\|^3}.$$

Using Example 3, we see that the flux of the electric field through any sphere centered at the origin is $4\pi q$. In fact, using the idea of Example 4, we can show that the flux of $\vec{E}$ through any closed surface containing the origin is $4\pi q$. See Problems 37 and 38 on page 997. This is a special case of Gauss's Law, which states that the flux of an electric field through any closed surface is proportional to the total charge enclosed by the surface. Carl Friedrich Gauss (1777–1855) also discovered the Divergence Theorem, which is sometimes called Gauss's Theorem.

Exercises and Problems for Section 19.4

EXERCISES

For Exercises 1–5, compute the flux integral $\int_S \vec{F} \cdot d\vec{A}$ in two ways, if possible, directly and using the Divergence Theorem. In each case, S is closed and oriented outward.

1. $\vec{F}(\vec{r}) = \vec{r}$ and S is the cube enclosing the volume $0 \le x \le 2, 0 \le y \le 2$, and $0 \le z \le 2$.
2. $\vec{F} = y\vec{j}$ and S is a closed vertical cylinder of height 2, with its base a circle of radius 1 on the xy-plane, centered at the origin.
3. $\vec{F} = x^2\vec{i} + 2y^2\vec{j} + 3z^2\vec{k}$ and S is the surface of the box with faces $x = 1, x = 2, y = 0, y = 1, z = 0, z = 1$.
4. $\vec{F} = (z^2 + x)\vec{i} + (x^2 + y)\vec{j} + (y^2 + z)\vec{k}$ and S is the closed cylinder $x^2 + z^2 = 1$, with $0 \le y \le 1$, oriented outward.
5. $\vec{F} = -z\vec{i} + x\vec{k}$ and S is a square pyramid with height 3 and base on the xy-plane of side length 1.

In Exercises 6–12, find the flux of the vector field out of the closed box $0 \le x \le 2, 0 \le y \le 3, 0 \le z \le 4$.

6. $\vec{F} = 4\vec{i} + 7\vec{j} - \vec{k}$
7. $\vec{G} = y\vec{i} + z\vec{k}$
8. $\vec{H} = xy\vec{i} + z\vec{j} + y\vec{k}$
9. $\vec{J} = xy^2\vec{j} + x\vec{k}$
10. $\vec{N} = e^z\vec{i} + \sin(xy)\vec{k}$
11. $\vec{M} = (3x + 4y)\vec{i} + (4y + 5z)\vec{j} + (5z + 3x)\vec{k}$
12. $\vec{M}$ where $\operatorname{div}\vec{M} = xy + 5$

PROBLEMS

13. Find the flux of $\vec{F} = z\vec{i} + y\vec{j} + x\vec{k}$ out of a sphere of radius 3 centered at the origin.
14. Find the flux of $\vec{F} = xy\vec{i} + yz\vec{j} + zx\vec{k}$ out of a sphere of radius 1 centered at the origin.
15. Find the flux of $\vec{F} = x^3\vec{i} + y^3\vec{j} + z^3\vec{k}$ through the closed surface bounding the solid region $x^2 + y^2 \le 4$, $0 \le z \le 5$, oriented outward.
16. The region W lies between the spheres $x^2 + y^2 + z^2 = 4$ and $x^2 + y^2 + z^2 = 9$ and within the cone $z = \sqrt{x^2 + y^2}$ with $z \ge 0$; its boundary is the closed surface, S, oriented outward. Find the flux of $\vec{F} = x^3\vec{i} + y^3\vec{j} + z^3\vec{k}$ out of S.
17. For $\vec{F} = (2x + \sin z)\vec{i} + (xz - y)\vec{j} + (e^x + 2z)\vec{k}$, find the flux of $\vec{F}$ out of the closed silo-shaped region within the cylinder $x^2 + y^2 = 1$, below the hemisphere $z = 1 + \sqrt{1 - x^2 - y^2}$, and above the xy-plane.
18. Find the flux of $\vec{F}$ through the closed cylinder of radius 2, centered on the z-axis, with $3 \le z \le 7$, if $\vec{F} = (x + 3e^{yz})\vec{i} + (\ln(x^2z^2 + 1) + y)\vec{j} + z\vec{k}$.
19. Find the flux of $\vec{F} = e^{y^2z^2}\vec{i} + (\tan(0.001x^2z^2) + y^2)\vec{j} + (\ln(1 + x^2y^2) + z^2)\vec{k}$ out of the closed box $0 \le x \le 5$, $0 \le y \le 4, 0 \le z \le 3$.
20. Find the flux of $\vec{F} = x^2\vec{i} + z\vec{j} + y\vec{k}$ out of the closed cone $x = \sqrt{y^2 + z^2}$, with $0 \le x \le 1$.
21. Suppose $\vec{F}$ is a vector field with $\operatorname{div}\vec{F} = 10$. Find the flux of $\vec{F}$ out of a cylinder of height a and radius a, centered on the z-axis and with base in the xy-plane.
22. Let $\vec{F} = (5x+7y)\vec{i} + (7y+9z)\vec{j} + (9z+11x)\vec{k}$, and let Q_i be the flux of $\vec{F}$ through the surfaces S_i for i =1–4. Arrange Q_i in ascending order, where
    - **(a)** S_1 is the sphere of radius 2 centered at the origin
    - **(b)** S_2 is the cube of side 2 centered at the origin and with sides parallel to the axes
    - **(c)** S_3 is the sphere of radius 1 centered at the origin
    - **(d)** S_4 is a pyramid with all four corners lying on S_3
23. A cone has its tip at the point $(0, 0, 5)$, and its base is the disk $D, x^2 + y^2 \le 1$, in the xy-plane. The surface of the cone is the curved and slanted face, S, oriented upward, and the flat base, D, oriented downward. The flux of the constant vector field $\vec{F} = a\vec{i} + b\vec{j} + c\vec{k}$ through S is given by
$$\int_S \vec{F} \cdot d\vec{A} = 3.22.$$
Is it possible to calculate $\int_D \vec{F} \cdot d\vec{A}$? If so, give the answer. If not, explain what additional information you would need to be able to make this calculation.
24. If V is a volume surrounded by a closed surface S, show that $\frac{1}{3}\int_S \vec{r} \cdot d\vec{A} = V$.
25. A vector field $\vec{F}$ satisfies $\operatorname{div}\vec{F} = 0$ everywhere. Show that $\int_S \vec{F} \cdot d\vec{A} = 0$ for every closed surface S.

26. Let S be the cube in the first quadrant with side 2, one corner at the origin and edges parallel to the axes. Let

$$\vec{F}_1 = (xy^2 + 3xz^2)\vec{i} + (3x^2y + 2yz^2)\vec{j} + 3zy^2\vec{k}$$
$$\vec{F}_2 = (xy^2 + 5e^{yz})\vec{i} + (yz^2 + 7\sin(xz))\vec{j} + (x^2z + \cos(xy))\vec{k}$$
$$\vec{F}_3 = \left(xz^2 + \frac{x^3}{3}\right)\vec{i} + \left(yz^2 + \frac{y^3}{3}\right)\vec{j} + \left(zy^2 + \frac{z^3}{3}\right)\vec{k}.$$

Arrange the flux integrals of $\vec{F}_1$, $\vec{F}_2$, $\vec{F}_3$ out of S in increasing order.

27. Let $\operatorname{div}\vec{F} = 2(6 - x)$ and $0 \le a, b, c \le 10$.
   (a) Find the flux of $\vec{F}$ out of the rectangular box given by $0 \le x \le a, 0 \le y \le b, 0 \le z \le c$.
   (b) For what values of a, b, c is the flux largest? What is that largest flux?

28. (a) Find $\operatorname{div}(\vec{r}/||\vec{r}||^2)$ where $\vec{r} = x\vec{i} + y\vec{j}$ for $\vec{r} \ne \vec{0}$.
   (b) Can you use the Divergence Theorem to compute the flux of $\vec{r}/||\vec{r}||^2$ out of a closed cylinder of radius 1, length 2, centered at the origin, and with its axis along the z-axis?
   (c) Compute the flux of $\vec{r}/||\vec{r}||^2$ out of the cylinder in part (b).
   (d) Find the flux of $\vec{r}/||\vec{r}||^2$ out of a closed cylinder of radius 2, length 2, centered at the origin, and with its axis along the z-axis.

29. Let $\vec{r} = x\vec{i} + y\vec{j} + z\vec{k}$ and let $\vec{F}$ be the vector field given by

$$\vec{F} = \frac{\vec{r}}{||\vec{r}||^3}.$$

   (a) Find the flux of $\vec{F}$ out of the sphere $x^2+y^2+z^2 = 1$ oriented outward.
   (b) Calculate $\operatorname{div}\vec{F}$. Show your work and simplify your answer completely.
   (c) Use your answers to parts (a) and (b) to calculate the flux out of a box of side 10 centered at the origin and with sides parallel to the coordinate planes. (The box is also oriented outward.)

In Problems 30–31, find the flux of $\vec{F} = \vec{r}/||\vec{r}||^3$ through the surface. [Hint: Use the method of Problem 29.]

30. S is the ellipsoid $x^2 + 2y^2 + 3z^2 = 6$.

31. S is the closed cylinder $y^2 + z^2 = 4, -2 \le x \le 2$.

32. (a) Let $\operatorname{div}\vec{F} = x^2+y^2+z^2+3$. Calculate $\int_{S_1}\vec{F}\cdot d\vec{A}$ where S_1 is the sphere of radius 1 centered at the origin.
   (b) Let S_2 be the sphere of radius 2 centered at the origin; let S_3 be the sphere of radius 3 centered at the origin; let S_4 be the box of side 6 centered at the origin with edges parallel to the axes. Without calculating them, arrange the following integrals in increasing order:

$$\int_{S_2}\vec{F}\cdot d\vec{A}, \quad \int_{S_3}\vec{F}\cdot d\vec{A}, \quad \int_{S_4}\vec{F}\cdot d\vec{A}.$$

33. Suppose $\operatorname{div}\vec{F} = xyz^2$.
   (a) Find $\operatorname{div}\vec{F}$ at the point $(1, 2, 1)$. [Note: You are given $\operatorname{div}\vec{F}$, not $\vec{F}$.]
   (b) Using your answer to part (a), but no other information about the vector field $\vec{F}$, estimate the flux out of a small box of side 0.2 centered at the point $(1, 2, 1)$ and with edges parallel to the axes.
   (c) Without computing the vector field $\vec{F}$, calculate the exact flux out of the box.

34. Suppose $\operatorname{div}\vec{F} = x^2+y^2+3$. Find a surface S such that $\int_S \vec{F}\cdot d\vec{A}$ is negative, or explain why no such surface exists.

35. As a result of radioactive decay, heat is generated uniformly throughout the interior of the earth at a rate of 30 watts per cubic kilometer. (A watt is a rate of heat production.) The heat then flows to the earth's surface where it is lost to space. Let $\vec{F}(x, y, z)$ denote the rate of flow of heat measured in watts per square kilometer. By definition, the flux of $\vec{F}$ across a surface is the quantity of heat flowing through the surface per unit of time.
   (a) What is the value of $\operatorname{div}\vec{F}$? Include units.
   (b) Assume the heat flows outward symmetrically. Verify that $\vec{F} = \alpha\vec{r}$, where $\vec{r} = x\vec{i} + y\vec{j} + z\vec{k}$ and α is a suitable constant, satisfies the given conditions. Find α.
   (c) Let $T(x, y, z)$ denote the temperature inside the earth. Heat flows according to the equation $\vec{F} = -k \operatorname{grad} T$, where k is a constant. Explain why this makes sense physically.
   (d) If T is in °C, then $k = 30{,}000$ watts/km°C. Assuming the earth is a sphere with radius 6400 km and surface temperature 20°C, what is the temperature at the center?

36. If a surface S is submerged in an incompressible fluid, a force $\vec{F}$ is exerted on one side of the surface by the pressure in the fluid. If the z-axis is vertical, with the positive direction upward and the fluid level at $z = 0$, then the component of force in the direction of a unit vector $\vec{u}$ is given by the following:

$$\vec{F}\cdot\vec{u} = -\int_S z\delta g\vec{u}\cdot d\vec{A},$$

where δ is the density of the fluid (mass/volume), g is the acceleration due to gravity, and the surface is oriented away from the side on which the force is exerted. In this problem we consider a totally submerged closed surface enclosing a volume V. We are interested in the force of the liquid on the external surface, so S is oriented inward. Use the Divergence Theorem to show that:
   (a) The force in the $\vec{i}$ and $\vec{j}$ directions is zero.
   (b) The force in the $\vec{k}$ direction is δgV, the weight of the volume of fluid with the same volume as V. This is *Archimedes' Principle.*

37. According to Coulomb's Law, the electrostatic field $\vec{E}$ at the point $\vec{r}$ due to a charge q at the origin is given by

$$\vec{E}(\vec{r}) = q\frac{\vec{r}}{\|\vec{r}\|^3}.$$

(a) Compute $\operatorname{div} \vec{E}$.
(b) Let S_a be the sphere of radius a centered at the origin and oriented outward. Show that the flux of $\vec{E}$ through S_a is $4\pi q$.
(c) Could you have used the Divergence Theorem in part (b)? Explain why or why not.
(d) Let S be an arbitrary, closed, outward-oriented surface surrounding the origin. Show that the flux of $\vec{E}$ through S is again $4\pi q$. [Hint: Apply the Divergence Theorem to the solid region lying between a small sphere S_a and the surface S.]

38. According to Coulomb's Law, the electric field $\vec{E}$ at the point $\vec{r}$ due to a charge q at the point $\vec{r}_0$ is given by

$$\vec{E}(\vec{r}) = q\frac{(\vec{r}-\vec{r}_0)}{\|\vec{r}-\vec{r}_0\|^3}.$$

Suppose S is a closed, outward-oriented surface and that $\vec{r}_0$ does not lie on S. Use Problem 37 to show that

$$\int_S \vec{E}\cdot d\vec{A} = \begin{cases} 4\pi q & \text{if } q \text{ lies inside } S, \\ 0 & \text{if } q \text{ lies outside } S. \end{cases}$$

Strengthen Your Understanding

In Problems 39–40, explain what is wrong with the statement.

39. The flux integral $\int_S \vec{F}\cdot d\vec{A}$ can be evaluated using the Divergence Theorem, where $\vec{F} = 2x\vec{i} - 3\vec{j}$ and S is the triangular surface with corners $(1,0,0)$, $(0,1,0)$, $(0,0,1)$ oriented away from the origin.

40. If S is the boundary of a solid region W, where S is oriented outward, and $\vec{F}$ is a vector field, then

$$\int_S \operatorname{div} \vec{F}\, d\vec{A} = \int_W \vec{F}\, dV.$$

In Problems 41–42, give an example of:

41. A surface S that is the boundary of a solid region such that $\int_S \vec{F}\cdot d\vec{A} = 0$ if $\vec{F}(x,y,z) = y\vec{i} + xz\vec{j} + y^2\vec{k}$.

42. A vector field $\vec{F}$ such that the flux of $\vec{F}$ out of a sphere of radius 1 centered at the origin is 3.

Are the statements in Problems 43–47 true or false? The smooth vector field $\vec{F}$ is defined everywhere in 3-space and has constant divergence equal to 4.

43. The field $\vec{F}$ has a net inflow per unit volume at the point $(-3,4,0)$.

44. The vector field $\vec{F}$ could be $\vec{F} = x\vec{i} + (3y)\vec{j} + (y-5x)\vec{k}$.

45. The vector field $\vec{F}$ could be a constant field.

46. The flux of $\vec{F}$ through a circle of radius 5 lying anywhere on the xy-plane and oriented upward is $4(\pi 5^2)$.

47. The flux of $\vec{F}$ through a closed cylinder of radius 1 centered along the y-axis, $0 \le y \le 3$ and oriented outward is $4(3\pi)$.

Are the statements in Problems 48–55 true or false? Give reasons for your answer.

48. $\int_S \vec{F}\cdot d\vec{A} = \operatorname{div} \vec{F}$.

49. If $\vec{F}$ is a divergence-free vector field in 3-space and S is a closed surface oriented inward, then $\int_S \vec{F}\cdot d\vec{A} = 0$.

50. If $\vec{F}$ is a vector field in 3-space satisfying $\operatorname{div} \vec{F} = 1$, and S is a closed surface oriented outward, then $\int_S \vec{F}\cdot d\vec{A}$ is equal to the volume enclosed by S.

51. Let W be the solid region between the sphere S_1 of radius 1 and S_2 of radius 2, both centered at the origin and oriented outward. If $\vec{F}$ is a vector field in 3-space, then $\int_W \operatorname{div} \vec{F}\, dV = \int_{S_2} \vec{F}\cdot d\vec{A} - \int_{S_1} \vec{F}\cdot d\vec{A}$.

52. Let S_1 be the square $0 \le x \le 1, 0 \le y \le 1, z = 0$ oriented downward and let S_2 be the square $0 \le x \le 1, 0 \le y \le 1, z = 1$ oriented upward. If $\vec{F}$ is a vector field, then $\int_W \operatorname{div} \vec{F}\, dV = \int_{S_2} \vec{F}\cdot d\vec{A} + \int_{S_1} \vec{F}\cdot d\vec{A}$, where W is the solid cube $0 \le x \le 1, 0 \le y \le 1, 0 \le z \le 1$.

53. Let S_1 be the square $0 \le x \le 1, 0 \le y \le 1, z = 0$ oriented downward and let S_2 be the square $0 \le x \le 1, 0 \le y \le 1, z = 1$ oriented upward. If $\vec{F} = \cos(xyz)\vec{k}$, then $\int_W \operatorname{div} \vec{F}\, dV = \int_{S_2} \vec{F}\cdot d\vec{A} + \int_{S_1} \vec{F}\cdot d\vec{A}$, where W is the solid cube $0 \le x \le 1, 0 \le y \le 1, 0 \le z \le 1$.

54. If S is a sphere of radius 1, centered at the origin, oriented outward, and $\vec{F}$ is a vector field satisfying $\int_S \vec{F}\cdot d\vec{A} = 0$, then $\operatorname{div} \vec{F} = 0$ at all points inside S.

55. Let S_h be the surface consisting of a cylinder of height h, closed at the top. The curved sides are $x^2 + y^2 = 1$, for $0 \le z \le h$, and the top $x^2 + y^2 \le 1$, for $z = h$, oriented outward. If $\vec{F}$ is divergence free, then $\int_{S_h} \vec{F}\cdot d\vec{A}$ is independent of the height h.

Online Resource: Review problems and Projects

Chapter Twenty

THE CURL AND STOKES' THEOREM

Contents

20.1 THE CURL OF A VECTOR FIELD

The divergence is a scalar derivative which measures the outflow of a vector field per unit volume. Now we introduce a vector derivative, the curl, which measures the circulation of a vector field.

Imagine holding the paddle-wheel in Figure 20.1 in the flow shown by Figure 20.2. The speed at which the paddle-wheel spins measures the strength of circulation. Notice that the angular velocity depends on the direction in which the stick is pointing. If the stick is pointing horizontally the paddle-wheel does not spin; if the stick is vertical, the paddle wheel spins.

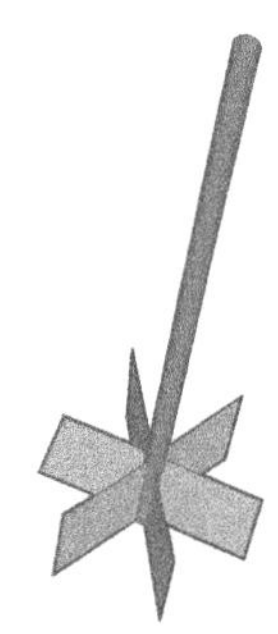

Figure 20.1: A device for measuring circulation

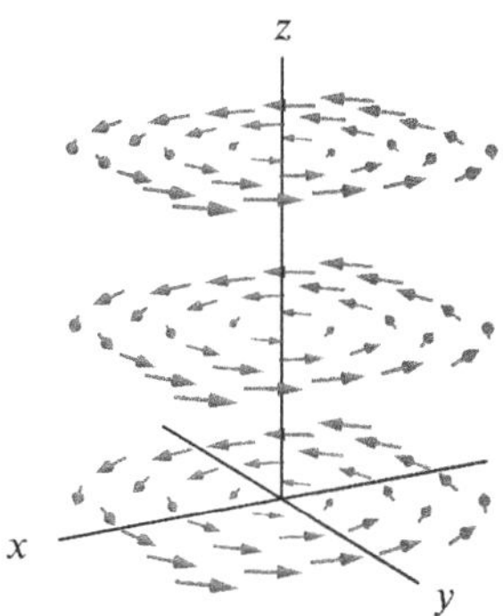

Figure 20.2: A vector field (in the planes $z = 1$, $z = 2$, $z = 3$) with circulation about the z-axis

Circulation Density

We measure the strength of the circulation using a closed curve. Suppose C is a circle with center $P = (x, y, z)$ in the plane perpendicular to $\vec{n}$, traversed in the direction determined from $\vec{n}$ by the right-hand rule. (See Figures 20.3 and 20.4.)

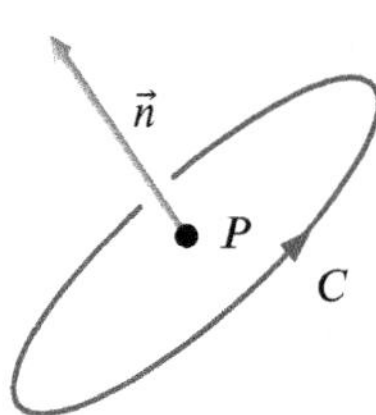

Figure 20.3: Direction of C relates to direction of $\vec{n}$ by the right-hand rule

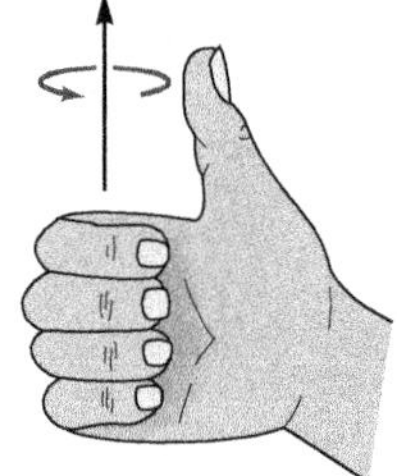

Figure 20.4: When the thumb points in the direction of $\vec{n}$, the fingers curl in the forward direction around C

We make the following definition:

The **circulation density** of a smooth vector field $\vec{F}$ at (x, y, z) around the direction of the unit vector $\vec{n}$ is defined, provided the limit exists, to be

$$\text{circ}_{\vec{n}}\, \vec{F}(x, y, z) = \lim_{\text{Area}\to 0} \frac{\text{Circulation around } C}{\text{Area inside } C} = \lim_{\text{Area}\to 0} \frac{\int_C \vec{F} \cdot d\vec{r}}{\text{Area inside } C},$$

The circle C is in the plane perpendicular to $\vec{n}$ and oriented by the right-hand rule.

We can use other closed curves for C, such as rectangles, that lie in a plane perpendicular to $\vec{n}$ and include the point (x, y, z).

The circulation density determines the angular velocity of the paddle-wheel in Figure 20.1 provided you could make one sufficiently small and light and insert it without disturbing the flow.

Example 1 Consider the vector field $\vec{F}$ in Figure 20.2. Suppose that $\vec{F}$ is parallel to the xy-plane and that at a distance r from the z-axis it has magnitude $2r$. Calculate $\text{circ}_{\vec{n}}\, \vec{F}$ at the origin for

(a) $\vec{n} = \vec{k}$ (b) $\vec{n} = -\vec{k}$ (c) $\vec{n} = \vec{i}$.

Solution (a) Take a circle C of radius a in the xy-plane, centered at the origin, traversed in a direction determined from $\vec{k}$ by the right-hand rule. Then, since $\vec{F}$ is tangent to C everywhere and points in the forward direction around C, we have

$$\text{Circulation around } C = \int_C \vec{F} \cdot d\vec{r} = \|\vec{F}\| \cdot \text{Circumference of } C = 2a(2\pi a) = 4\pi a^2.$$

Thus, the circulation density is

$$\text{circ}_{\vec{k}}\, \vec{F} = \lim_{a\to 0} \frac{\text{Circulation around } C}{\text{Area inside } C} = \lim_{a\to 0} \frac{4\pi a^2}{\pi a^2} = 4.$$

(b) If $\vec{n} = -\vec{k}$ the circle is traversed in the opposite direction, so the line integral changes sign. Thus,

$$\text{circ}_{-\vec{k}}\, \vec{F} = -4.$$

(c) The circulation around $\vec{i}$ is calculated using circles in the yz-plane. Since $\vec{F}$ is everywhere perpendicular to such a circle C,

$$\int_C \vec{F} \cdot d\vec{r} = 0.$$

Thus, we have

$$\text{circ}_{\vec{i}}\, \vec{F} = \lim_{a\to 0} \frac{\int_C \vec{F} \cdot d\vec{r}}{\pi a^2} = \lim_{a\to 0} \frac{0}{\pi a^2} = 0.$$

Definition of the Curl

Example 1 shows that the circulation density of a vector field can be positive, negative, or zero, depending on the direction. We assume that there is one direction in which the circulation density is greatest and define a single vector quantity that incorporates all these different circulation densities. We give two definitions, one geometric and one algebraic, which turn out to lead to the same result.

Geometric Definition of Curl

The curl of a smooth vector field $\vec{F}$, written $\text{curl}\, \vec{F}$, is the vector field with the following properties:

- The direction of $\text{curl}\, \vec{F}(x, y, z)$ is the direction $\vec{n}$ for which $\text{circ}_{\vec{n}}\, \vec{F}(x, y, z)$ is the greatest.
- The magnitude of $\text{curl}\, \vec{F}(x, y, z)$ is the circulation density of $\vec{F}$ around that direction.

If the circulation density is zero around every direction, then we define the curl to be $\vec{0}$.

Cartesian Coordinate Definition of Curl

If $\vec{F} = F_1\vec{i} + F_2\vec{j} + F_3\vec{k}$, then

$$\operatorname{curl}\vec{F} = \left(\frac{\partial F_3}{\partial y} - \frac{\partial F_2}{\partial z}\right)\vec{i} + \left(\frac{\partial F_1}{\partial z} - \frac{\partial F_3}{\partial x}\right)\vec{j} + \left(\frac{\partial F_2}{\partial x} - \frac{\partial F_1}{\partial y}\right)\vec{k}.$$

The cross-product formula gives an easy way to remember the Cartesian coordinate definition and suggests another common notation for curl $\vec{F}$, namely $\nabla \times \vec{F}$. Using $\nabla = \frac{\partial}{\partial x}\vec{i} + \frac{\partial}{\partial y}\vec{j} + \frac{\partial}{\partial z}\vec{k}$, we can write

$$\operatorname{curl}\vec{F} = \nabla \times \vec{F} = \begin{vmatrix} \vec{i} & \vec{j} & \vec{k} \\ \frac{\partial}{\partial x} & \frac{\partial}{\partial y} & \frac{\partial}{\partial z} \\ F_1 & F_2 & F_3 \end{vmatrix}.$$

Example 2 For each field in Figure 20.5, use the sketch and the geometric definition to decide whether the curl at the origin appears to point up or down, or to be the zero vector. Then check your answer using the coordinate definition of curl and the formulas in the caption of Figure 20.5. Note that the vector fields have no $\vec{k}$-components and are independent of z.

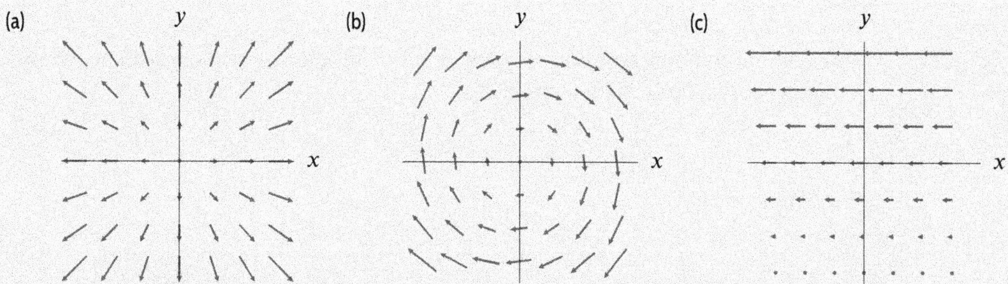

Figure 20.5: Sketches in the xy-plane of (a) $\vec{F} = x\vec{i} + y\vec{j}$ (b) $\vec{F} = y\vec{i} - x\vec{j}$ (c) $\vec{F} = -(y+1)\vec{i}$

Solution (a) This vector field shows no rotation, and the circulation around any circle in the xy-plane centered at the origin appears to be zero, so we suspect that the circulation density around $\vec{k}$ is zero. The coordinate definition of curl gives

$$\operatorname{curl}\vec{F} = \left(\frac{\partial(0)}{\partial y} - \frac{\partial y}{\partial z}\right)\vec{i} + \left(\frac{\partial x}{\partial z} - \frac{\partial(0)}{\partial x}\right)\vec{j} + \left(\frac{\partial y}{\partial x} - \frac{\partial x}{\partial y}\right)\vec{k} = \vec{0}.$$

(b) This vector field appears to be rotating around the z-axis. By the right-hand rule, the circulation density around $\vec{k}$ is negative, so we expect the z-component of the curl to point down. The coordinate definition gives

$$\operatorname{curl}\vec{F} = \left(\frac{\partial(0)}{\partial y} - \frac{\partial(-x)}{\partial z}\right)\vec{i} + \left(\frac{\partial y}{\partial z} - \frac{\partial(0)}{\partial x}\right)\vec{j} + \left(\frac{\partial(-x)}{\partial x} - \frac{\partial y}{\partial y}\right)\vec{k} = -2\vec{k}.$$

(c) At first glance, you might expect this vector field to have zero curl, as all the vectors are parallel to the x-axis. However, if you find the circulation around the curve C in Figure 20.6, the sides contribute nothing (they are perpendicular to the vector field), the bottom contributes a negative quantity (the curve is in the opposite direction to the vector field), and the top contributes a larger positive quantity (the curve is in the same direction as the vector field and the magnitude of the vector field is larger at the top than at the bottom). Thus, the circulation around C is positive and hence we expect the curl to be nonzero and point up. The coordinate definition gives

$$\operatorname{curl}\vec{F} = \left(\frac{\partial(0)}{\partial y} - \frac{\partial(0)}{\partial z}\right)\vec{i} + \left(\frac{\partial(-(y+1))}{\partial z} - \frac{\partial(0)}{\partial x}\right)\vec{j} + \left(\frac{\partial(0)}{\partial x} - \frac{\partial(-(y+1))}{\partial y}\right)\vec{k} = \vec{k}.$$

Another way to see that the curl is nonzero in this case is to imagine the vector field representing the velocity of moving water. A boat sitting in the water tends to rotate, as the water moves faster on one side than the other.

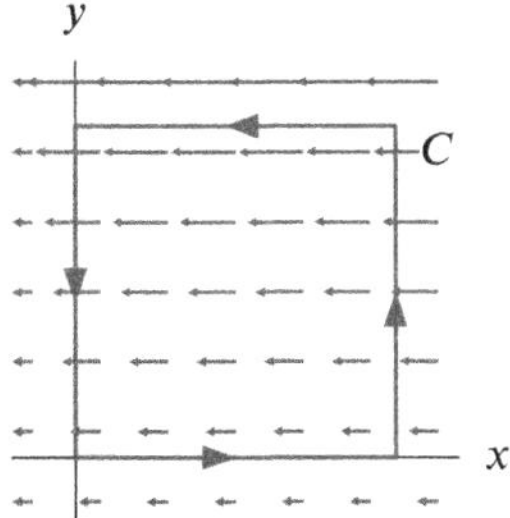

Figure 20.6: Rectangular curve in xy-plane

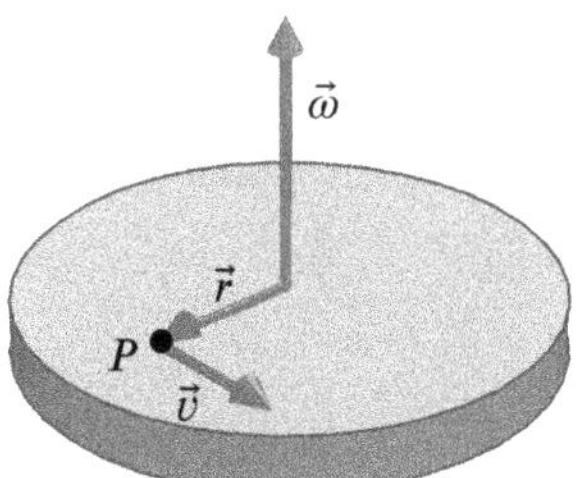

Figure 20.7: Rotating flywheel

Example 3 A flywheel is rotating with angular velocity $\vec{\omega}$ and the velocity of a point P with position vector $\vec{r}$ is given by $\vec{v} = \vec{\omega} \times \vec{r}$. (See Figure 20.7.) Calculate $\operatorname{curl}\vec{v}$.

Solution If $\vec{\omega} = \omega_1\vec{i} + \omega_2\vec{j} + \omega_3\vec{k}$, using the determinant notation introduced in Section 13.4, we have

$$\vec{v} = \vec{\omega} \times \vec{r} = \begin{vmatrix} \vec{i} & \vec{j} & \vec{k} \\ \omega_1 & \omega_2 & \omega_3 \\ x & y & z \end{vmatrix} = (\omega_2 z - \omega_3 y)\vec{i} + (\omega_3 x - \omega_1 z)\vec{j} + (\omega_1 y - \omega_2 x)\vec{k}.$$

The curl formula can also be written using a determinant:

$$\begin{aligned}
\operatorname{curl}\vec{v} &= \begin{vmatrix} \vec{i} & \vec{j} & \vec{k} \\ \frac{\partial}{\partial x} & \frac{\partial}{\partial y} & \frac{\partial}{\partial z} \\ \omega_2 z - \omega_3 y & \omega_3 x - \omega_1 z & \omega_1 y - \omega_2 x \end{vmatrix} \\
&= \left(\frac{\partial}{\partial y}(\omega_1 y - \omega_2 x) - \frac{\partial}{\partial z}(\omega_3 x - \omega_1 z)\right)\vec{i} + \left(\frac{\partial}{\partial z}(\omega_2 z - \omega_3 y) - \frac{\partial}{\partial x}(\omega_1 y - \omega_2 x)\right)\vec{j} \\
&\quad + \left(\frac{\partial}{\partial x}(\omega_3 x - \omega_1 z) - \frac{\partial}{\partial y}(\omega_2 z - \omega_3 y)\right)\vec{k} \\
&= 2\omega_1\vec{i} + 2\omega_2\vec{j} + 2\omega_3\vec{k} = 2\vec{\omega}.
\end{aligned}$$

Thus, as we would expect, $\operatorname{curl}\vec{v}$ is parallel to the axis of rotation of the flywheel (namely, the direction of $\vec{\omega}$) and the magnitude of $\operatorname{curl}\vec{v}$ is larger the faster the flywheel is rotating (that is, the larger the magnitude of $\vec{\omega}$).

Why Do the Two Definitions of Curl Give the Same Result?

Using Green's Theorem in Cartesian coordinates, we can show that for curl $\vec{F}$ defined in Cartesian coordinates

$$\boxed{\operatorname{curl}\vec{F}\cdot\vec{n} = \operatorname{circ}_{\vec{n}}\vec{F}.}$$

This shows that curl $\vec{F}$ defined in Cartesian coordinates satisfies the geometric definition, since the left-hand side takes its maximum value when $\vec{n}$ points in the same direction as curl $\vec{F}$, and in that case its value is $\|\operatorname{curl}\vec{F}\|$.

The following example justifies this formula in a specific case.

Example 4 Use the definition of curl in Cartesian coordinates and Green's Theorem to show that

$$\left(\operatorname{curl}\vec{F}\right)\cdot\vec{k} = \operatorname{circ}_{\vec{k}}\vec{F}.$$

Solution Using the definition of curl in Cartesian coordinates, the left-hand side of the formula is

$$\left(\operatorname{curl}\vec{F}\right)\cdot\vec{k} = \frac{\partial F_2}{\partial x} - \frac{\partial F_1}{\partial y}.$$

Now let's look at the right-hand side. The circulation density around $\vec{k}$ is calculated using circles perpendicular to $\vec{k}$; hence, the $\vec{k}$-component of $\vec{F}$ does not contribute to it; that is, the circulation density of $\vec{F}$ around $\vec{k}$ is the same as the circulation density of $F_1\vec{i} + F_2\vec{j}$ around $\vec{k}$. But in any plane perpendicular to $\vec{k}$, z is constant, so in that plane F_1 and F_2 are functions of x and y alone. Thus, $F_1\vec{i} + F_2\vec{j}$ can be thought of as a two-dimensional vector field on the horizontal plane through the point (x, y, z) where the circulation density is being calculated. Let C be a circle in this plane, with radius a and centered at (x, y, z), and let R be the region enclosed by C. Green's Theorem says that

$$\int_C (F_1\vec{i} + F_2\vec{j})\cdot d\vec{r} = \int_R \left(\frac{\partial F_2}{\partial x} - \frac{\partial F_1}{\partial y}\right) dA.$$

When the circle is small, $\partial F_2/\partial x - \partial F_1/\partial y$ is approximately constant on R, so

$$\int_R \left(\frac{\partial F_2}{\partial x} - \frac{\partial F_1}{\partial y}\right) dA \approx \left(\frac{\partial F_2}{\partial x} - \frac{\partial F_1}{\partial y}\right)\cdot \text{Area of } R = \left(\frac{\partial F_2}{\partial x} - \frac{\partial F_1}{\partial y}\right)\pi a^2.$$

Thus, taking a limit as the radius of the circle goes to zero, we have

$$\operatorname{circ}_{\vec{k}}\vec{F}(x, y, z) = \lim_{a\to 0}\frac{\int_C (F_1\vec{i} + F_2\vec{j})\cdot d\vec{r}}{\pi a^2} = \lim_{a\to 0}\frac{\int_R \left(\frac{\partial F_2}{\partial x} - \frac{\partial F_1}{\partial y}\right) dA}{\pi a^2} = \frac{\partial F_2}{\partial x} - \frac{\partial F_1}{\partial y}.$$

Curl-Free Vector Fields

A vector field is said to be *curl free* or *irrotational* if curl $\vec{F} = \vec{0}$ everywhere that $\vec{F}$ is defined.

Example 5 Figure 20.8 shows the vector field $\vec{B}$ for three values of the constant p, where $\vec{B}$ is defined on 3-space by

$$\vec{B} = \frac{-y\vec{i} + x\vec{j}}{(x^2 + y^2)^{p/2}}.$$

(a) Find a formula for curl $\vec{B}$.
(b) Is there a value of p for which $\vec{B}$ is curl free? If so, find it.

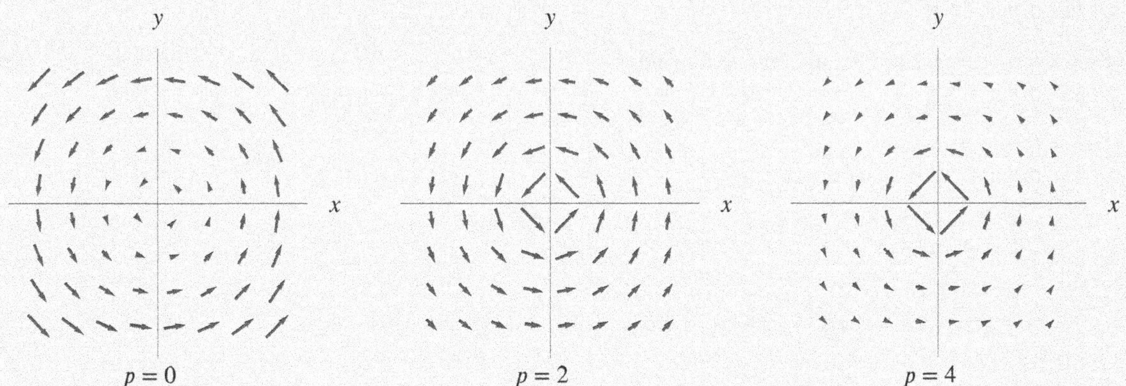

Figure 20.8: The vector field $\vec{B}(\vec{r}) = (-y\vec{i} + x\vec{j})/(x^2 + y^2)^{p/2}$ for $p = 0, 2$, and 4

Solution (a) We can use the following version of the product rule for curl. If ϕ is a scalar function and $\vec{F}$ is a vector field, then

$$\text{curl}(\phi\vec{F}) = \phi \,\text{curl}\,\vec{F} + (\text{grad}\,\phi) \times \vec{F}.$$

(See Problem 32 on page 1007.) We write $\vec{B} = \phi\vec{F} = \frac{1}{(x^2 + y^2)^{p/2}}(-y\vec{i} + x\vec{j})$. Then

$$\text{curl}\,\vec{F} = \text{curl}(-y\vec{i} + x\vec{j}) = 2\vec{k}$$
$$\text{grad}\,\phi = \text{grad}\left(\frac{1}{(x^2 + y^2)^{p/2}}\right) = \frac{-p}{(x^2 + y^2)^{(p/2)+1}}(x\vec{i} + y\vec{j}).$$

Thus, we have

$$\begin{aligned}
\text{curl}\,\vec{B} &= \frac{1}{(x^2 + y^2)^{p/2}}\,\text{curl}(-y\vec{i} + x\vec{j}) + \text{grad}\left(\frac{1}{(x^2 + y^2)^{p/2}}\right) \times (-y\vec{i} + x\vec{j}) \\
&= \frac{1}{(x^2 + y^2)^{p/2}}2\vec{k} + \frac{-p}{(x^2 + y^2)^{(p/2)+1}}(x\vec{i} + y\vec{j}) \times (-y\vec{i} + x\vec{j}) \\
&= \frac{1}{(x^2 + y^2)^{p/2}}2\vec{k} + \frac{-p}{(x^2 + y^2)^{(p/2)+1}}(x^2 + y^2)\vec{k} \\
&= \frac{2 - p}{(x^2 + y^2)^{p/2}}\vec{k}.
\end{aligned}$$

(b) The curl is zero when $p = 2$. Thus, when $p = 2$ the vector field is curl free:

$$\vec{B} = \frac{-y\vec{i} + x\vec{j}}{x^2 + y^2}.$$

Exercises and Problems for Section 20.1 Online Resource: Additional Problems for Section 20.1

EXERCISES

In Exercises 1–5, is the quantity a vector or a scalar? Calculate it.

1. $\operatorname{curl}(z\vec{i} - x\vec{j} + y\vec{k})$
2. $\operatorname{circ}_{\vec{i}}(z\vec{i} - x\vec{j} + y\vec{k})$
3. $\operatorname{curl}(-2z\vec{i} - z\vec{j} + xy\vec{k})$
4. $\operatorname{circ}_{\vec{k}}(-2z\vec{i} - z\vec{j} + xy\vec{k})$
5. $\operatorname{curl}(x\vec{i} + y\vec{j} + z\vec{k})$

In Exercises 6–13, compute the curl of the vector field.

6. $\vec{F} = 3x\vec{i} - 5z\vec{j} + y\vec{k}$
7. $\vec{F} = (x^2 - y^2)\vec{i} + 2xy\vec{j}$
8. $\vec{F} = (-x + y)\vec{i} + (y + z)\vec{j} + (-z + x)\vec{k}$
9. $\vec{F} = 2yz\vec{i} + 3xz\vec{j} + 7xy\vec{k}$
10. $\vec{F} = x^2\vec{i} + y^3\vec{j} + z^4\vec{k}$
11. $\vec{F} = e^x\vec{i} + \cos y\vec{j} + e^{z^2}\vec{k}$
12. $\vec{F} = (x + yz)\vec{i} + (y^2 + xzy)\vec{j} + (zx^3y^2 + x^7y^6)\vec{k}$
13. $\vec{F}(\vec{r}) = \vec{r}/\|\vec{r}\|$

In Exercises 14–17, does the vector field appear to have nonzero curl at the origin? The vector field is shown in the xy-plane; it has no z-component and is independent of z.

14.
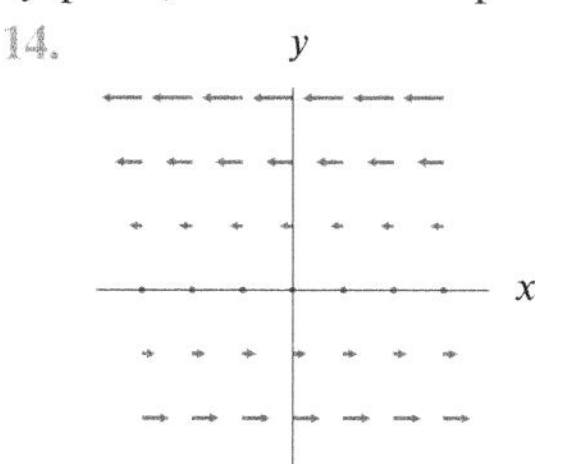

15.
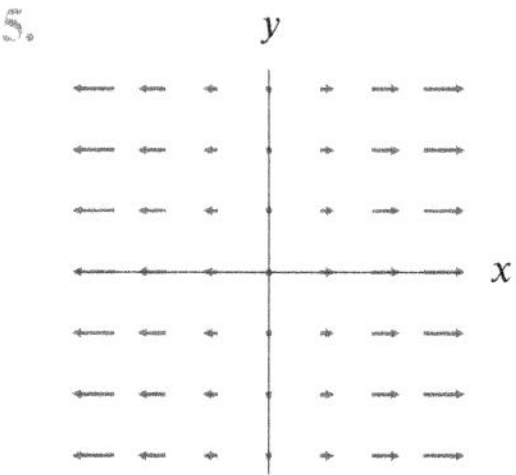

16.
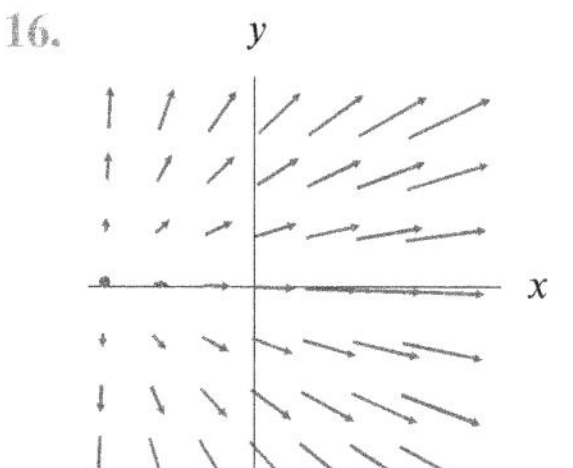

17.
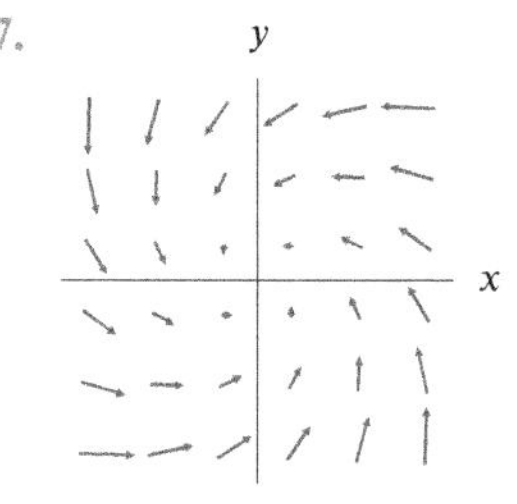

PROBLEMS

18. Let $\vec{F}$ be the vector field in Figure 20.2 on page 1000. It is rotating counterclockwise around the z-axis when viewed from above. At a distance r from the z-axis, $\vec{F}$ has magnitude $2r$.
    (a) Find a formula for $\vec{F}$.
    (b) Find $\operatorname{curl}\vec{F}$ using the coordinate definition and relate your answer to circulation density.

19. Use the geometric definition to find the curl of the vector field $\vec{F}(\vec{r}) = \vec{r}$. Check your answer using the coordinate definition.

20. A smooth vector field $\vec{G}$ has $\operatorname{curl}\vec{G}(0,0,0) = 2\vec{i} - 3\vec{j} + 5\vec{k}$. Estimate the circulation around a circle of radius 0.01 centered at the origin in each of the following planes:
    (a) xy-plane, oriented counterclockwise when viewed from the positive z-axis.
    (b) yz-plane, oriented counterclockwise when viewed from the positive x-axis.
    (c) xz-plane, oriented counterclockwise when viewed from the positive y-axis.

21. Three small circles, C_1, C_2, and C_3, each with radius 0.1 and centered at the origin, are in the xy-, yz-, and xz-planes, respectively. The circles are oriented counterclockwise when viewed from the positive z-, x-, and y-axes, respectively. A vector field, $\vec{F}$, has circulation around C_1 of 0.02π, around C_2 of 0.5π, and around C_3 of 3π. Estimate $\operatorname{curl}\vec{F}$ at the origin.

22. Using your answers to Exercises 10–11, make a conjecture about a particular form of the vector field $\vec{F} \neq \vec{0}$ that has $\operatorname{curl}\vec{F} = \vec{0}$. What form? Show why your conjecture is true.

23. (a) Find $\operatorname{curl}\vec{G}$ if $\vec{G} = (ay^3 + be^z)\vec{i} + (cz + dx^2)\vec{j} + (e\sin x + fy)\vec{k}$ and a, b, c, d, e, f are constants.
    (b) If $\operatorname{curl}\vec{G}$ is everywhere parallel to the yz-plane, what can you say about the constants a–f?
    (c) If $\operatorname{curl}\vec{G}$ is everywhere parallel to the z-axis, what can you say about the constants a–f?

24. Figure 20.9 gives a sketch of the velocity vector field $\vec{F} = y\vec{i} + x\vec{j}$ in the xy-plane.
    (a) What is the direction of rotation of a thin twig placed at the origin along the x-axis?
    (b) What is the direction of rotation of a thin twig placed at the origin along the y-axis?
    (c) Compute $\operatorname{curl}\vec{F}$.

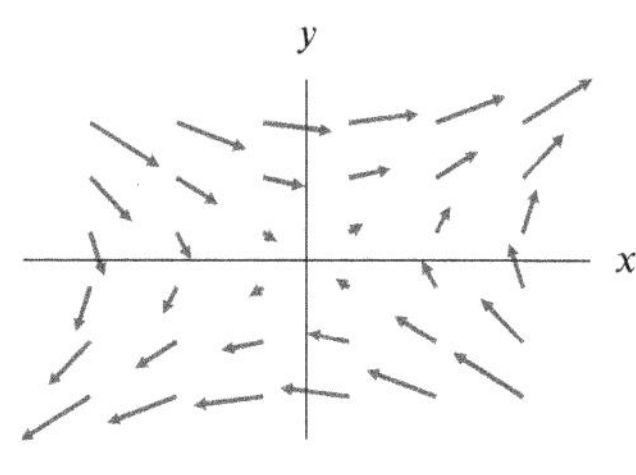

Figure 20.9

25. A tornado is formed when a tube of air circling a horizontal axis is tilted up vertically by the updraft from a thunderstorm. If t is time, this process can be modeled by the wind velocity field

$$\vec{F}(t,x,y,z) = (\cos t\vec{j} + \sin t\vec{k}) \times \vec{r} \quad \text{and} \quad 0 \le t \le \frac{\pi}{2}.$$

Determine the direction of curl $\vec{F}$:

(a) At $t = 0$ **(b)** At $t = \pi/2$
(c) For $0 < t < \pi/2$

26. A large fire becomes a fire-storm when the nearby air acquires a circulatory motion. The associated updraft has the effect of bringing more air to the fire, causing it to burn faster. Records show that a fire-storm developed during the Chicago Fire of 1871 and during the Second World War bombing of Hamburg, Germany, but there was no fire-storm during the Great Fire of London in 1666. Explain how a fire-storm could be identified using the curl of a vector field.

27. A vortex that rotates at constant angular velocity ω about the z-axis has velocity vector field $\vec{v} = \omega(-y\vec{i} + x\vec{j})$.

(a) Sketch the vector field with $\omega = 1$ and the vector field with $\omega = -1$.
(b) Determine the speed $\|\vec{v}\|$ of the vortex as a function of the distance from its center.
(c) Compute div $\vec{v}$ and curl $\vec{v}$.
(d) Compute the circulation of $\vec{v}$ counterclockwise about the circle of radius R in the xy-plane, centered at the origin.

28. A central vector field is one of the form $\vec{F} = f(r)\vec{r}$ where f is any function of $r = \|\vec{r}\|$. Show that any central vector field is irrotational.

29. Show that curl $(\vec{F} + \vec{C}) =$ curl $\vec{F}$ for a constant vector field $\vec{C}$.

30. If $\vec{F}$ is any vector field whose components have continuous second partial derivatives, show div curl $\vec{F} = 0$.

31. We have seen that the Fundamental Theorem of Calculus for Line Integrals implies $\int_C \text{grad } f \cdot d\vec{r} = 0$ for any smooth closed path C and any smooth function f.

(a) Use the geometric definition of curl to deduce that curl grad $f = \vec{0}$.
(b) Show that curl grad $f = \vec{0}$ using the coordinate definition.

32. Show that curl $(\phi\vec{F}) = \phi$ curl $\vec{F} + (\text{grad } \phi) \times \vec{F}$ for a scalar function ϕ and a vector field $\vec{F}$.

33. Show that if $\vec{F} = f$ grad g for some scalar functions f and g, then curl $\vec{F}$ is everywhere perpendicular to $\vec{F}$.

Strengthen Your Understanding

In Problems 34–35, explain what is wrong with the statement.

34. A vector field $\vec{F}$ has curl given by curl $\vec{F} = 2x - 3y$.
35. If all the vectors of a vector field $\vec{F}$ are parallel, then curl $\vec{F} = \vec{0}$.

In Problems 36–37, give an example of:

36. A vector field $\vec{F}(x,y,z)$ such that curl $\vec{F} = \vec{0}$.
37. A vector field $\vec{F}(x,y,z)$ such that curl $\vec{F} = \vec{j}$.

In Problems 38–46, is the statement true or false? Assume $\vec{F}$ and $\vec{G}$ are smooth vector fields and f is a smooth function on 3-space. Explain.

38. The circulation density, $\text{circ}_{\vec{n}}\vec{F}(x,y,z)$, is a scalar.
39. curl grad $f = 0$.
40. If $\vec{F}$ is a vector field with div$\vec{F} = 0$ and curl$\vec{F} = \vec{0}$, then $\vec{F} = \vec{0}$.
41. If $\vec{F}$ and $\vec{G}$ are vector fields, then curl$(\vec{F} + \vec{G}) =$ curl$\vec{F}$ + curl$\vec{G}$.
42. If $\vec{F}$ and $\vec{G}$ are vector fields, then curl$(\vec{F} \cdot \vec{G}) =$ curl$\vec{F} \cdot$ curl$\vec{G}$.
43. If $\vec{F}$ and $\vec{G}$ are vector fields, then curl$(\vec{F} \times \vec{G}) =$ (curl$\vec{F}$) $\times$ (curl$\vec{G}$).
44. curl$(f\vec{G}) = (\text{grad } f) \times \vec{G} + f(\text{curl } \vec{G})$.
45. For any vector field $\vec{F}$, the curl of $\vec{F}$ is perpendicular at every point to $\vec{F}$.
46. If $\vec{F}$ is as shown in Figure 20.10, then curl $\vec{F} \cdot \vec{j} > 0$.

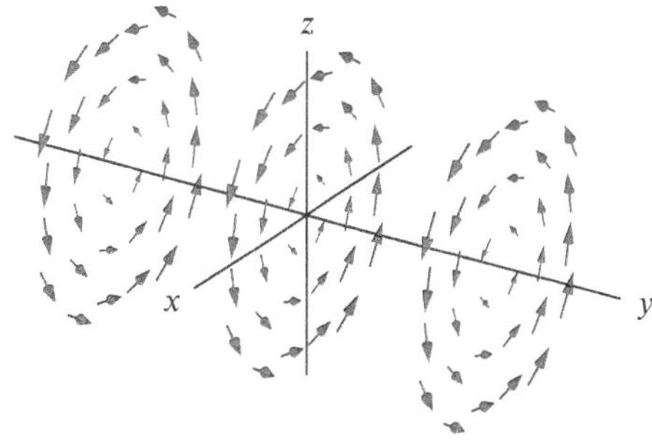

Figure 20.10

47. Of the following vector fields, which ones have a curl which is parallel to one of the axes? Which axis?

(a) $y\vec{i} - x\vec{j} + z\vec{k}$ **(b)** $y\vec{i} + z\vec{j} + x\vec{k}$ **(c)** $-z\vec{i} + y\vec{j} + x\vec{k}$
(d) $x\vec{i} + z\vec{j} - y\vec{k}$ **(e)** $z\vec{i} + x\vec{j} + y\vec{k}$

20.2 STOKES' THEOREM

The Divergence Theorem says that the integral of the flux density over a solid region is equal to the flux through the surface bounding the region. Similarly, Stokes' Theorem says that the integral of the circulation density over a surface is equal to the circulation around the boundary of the surface.

The Boundary of a Surface

The *boundary* of a surface S is the curve or curves running around the edge of S (like the hem around the edge of a piece of cloth). An orientation of S determines an orientation for its boundary, C, as follows. Pick a positive normal vector $\vec{n}$ on S, near C, and use the right-hand rule to determine a direction of travel around $\vec{n}$. This in turn determines a direction of travel around the boundary C. See Figure 20.11. Another way of describing the orientation on C is that someone walking along C in the forward direction, body upright in the direction of the positive normal on S, would have the surface on their left. Notice that the boundary can consist of two or more curves, as the surface on the right in Figure 20.11 shows.

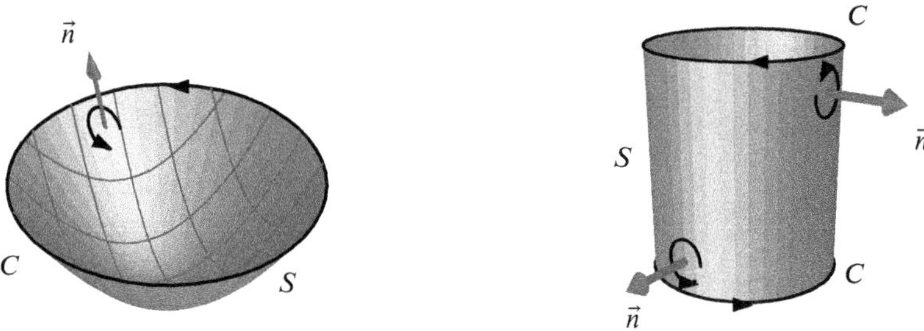

Figure 20.11: Two oriented surfaces and their boundaries

Calculating the Circulation from the Circulation Density

Consider a closed, oriented curve C in 3-space. We can find the circulation of a vector field $\vec{F}$ around C by calculating the line integral:

$$\begin{array}{c}\text{Circulation}\\ \text{around } C\end{array} = \int_C \vec{F} \cdot d\vec{r}.$$

If C is the boundary of an oriented surface S, there is another way to calculate the circulation using $\operatorname{curl} \vec{F}$. We subdivide S into pieces as shown on the surface on the left in Figure 20.11. If $\vec{n}$ is a positive unit normal vector to a piece of surface with area ΔA, then $\Delta \vec{A} = \vec{n}\,\Delta A$. In addition, $\operatorname{circ}_{\vec{n}} \vec{F}$ is the circulation density of $\vec{F}$ around $\vec{n}$, so

$$\begin{array}{c}\text{Circulation of } \vec{F} \text{ around}\\ \text{boundary of the piece}\end{array} \approx \left(\operatorname{circ}_{\vec{n}} \vec{F}\right) \Delta A = ((\operatorname{curl} \vec{F}) \cdot \vec{n})\Delta A = (\operatorname{curl}\vec{F}) \cdot \Delta\vec{A}.$$

Next we add up the circulations around all the small pieces. The line integral along the common edge of a pair of adjacent pieces appears with opposite sign in each piece, so it cancels out. (See Figure 20.12.) When we add up all the pieces the internal edges cancel and we are left with the circulation around C, the boundary of the entire surface. Thus,

$$\begin{array}{c}\text{Circulation}\\ \text{around } C\end{array} = \sum \begin{array}{c}\text{Circulation around}\\ \text{boundary of pieces}\end{array} \approx \sum \operatorname{curl} \vec{F} \cdot \Delta\vec{A}.$$

Taking the limit as $\Delta A \to 0$, we get

$$\begin{array}{c}\text{Circulation}\\ \text{around } C\end{array} = \int_S \operatorname{curl} \vec{F} \cdot d\vec{A}.$$

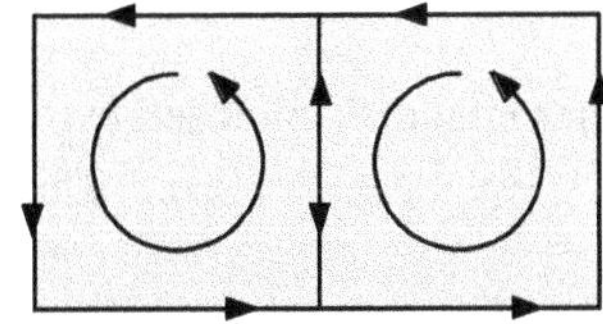

Figure 20.12: Two adjacent pieces of the surface

We have expressed the circulation as a line integral around C and as a flux integral over S; thus, the two integrals must be equal. Hence we have[1]

Theorem 20.1: Stokes' Theorem

If S is a smooth oriented surface with piecewise smooth, oriented boundary C, and if $\vec{F}$ is a smooth vector field on a solid region[2] containing S and C, then

$$\int_C \vec{F} \cdot d\vec{r} = \int_S \operatorname{curl} \vec{F} \cdot d\vec{A}.$$

The orientation of C is determined from the orientation of S according to the right-hand rule.

Example 1 Let $\vec{F}(x, y, z) = -2y\vec{i} + 2x\vec{j}$. Use Stokes' Theorem to find $\int_C \vec{F} \cdot d\vec{r}$, where C is a circle.

(a) Parallel to the yz-plane, of radius a, centered at a point on the x-axis, with either orientation.

(b) Parallel to the xy-plane, of radius a, centered at a point on the z-axis, oriented counterclockwise as viewed from a point on the z-axis above the circle.

Solution We have $\operatorname{curl} \vec{F} = 4\vec{k}$. Figure 20.13 shows sketches of $\vec{F}$ and $\operatorname{curl} \vec{F}$.

(a) Let S be the disk enclosed by C. Since S lies in a vertical plane and $\operatorname{curl} \vec{F}$ points vertically everywhere, the flux of $\operatorname{curl} \vec{F}$ through S is zero. Hence, by Stokes' Theorem,

$$\int_C \vec{F} \cdot d\vec{r} = \int_S \operatorname{curl} \vec{F} \cdot d\vec{A} = 0.$$

It makes sense that the line integral is zero. If C is parallel to the yz-plane (even if it is not lying in the plane), the symmetry of the vector field means that the line integral of $\vec{F}$ over the top half of the circle cancels the line integral over the bottom half.

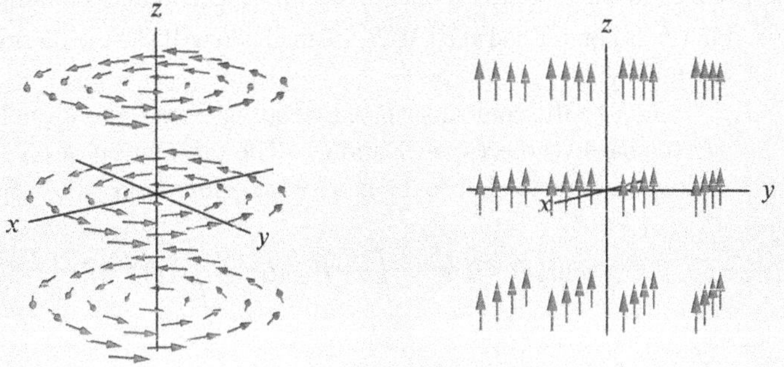

Figure 20.13: The vector fields $\vec{F}$ and $\operatorname{curl} \vec{F}$ (in the planes $z = -1$, $z = 0$, $z = 1$)

[1]A proof of Stokes' Theorem using the coordinate definition of curl can be found in the online supplement at www.wiley.com/college/hughes-hallett.

[2]The region containing S and C is open.

(b) Let S be the horizontal disk enclosed by C. Since curl $\vec{F}$ is a constant vector field pointing in the direction of $\vec{k}$, we have, by Stokes' Theorem,

$$\int_C \vec{F} \cdot d\vec{r} = \int_S \text{curl}\,\vec{F} \cdot d\vec{A} = \|\text{curl}\,\vec{F}\| \cdot \text{Area of } S = 4\pi a^2.$$

Since $\vec{F}$ is circling around the z-axis in the same direction as C, we expect the line integral to be positive. In fact, in Example 1 on page 1001, we computed this line integral directly.

Curl-Free Vector Fields

Stokes' Theorem applies to any oriented surface S and its boundary C, even in cases where the boundary consists of two or more curves. This is useful in studying curl-free vector fields.

Example 2 A current I flows along the z-axis in the $\vec{k}$ direction. The induced magnetic field $\vec{B}(x, y, z)$ is

$$\vec{B}(x, y, z) = \frac{2I}{c}\left(\frac{-y\vec{i} + x\vec{j}}{x^2 + y^2}\right),$$

where c is the speed of light. Example 5 on page 1005 shows that curl $\vec{B} = \vec{0}$.

(a) Compute the circulation of $\vec{B}$ around the circle C_1 in the xy-plane of radius a, centered at the origin, and oriented counterclockwise when viewed from above.

(b) Use part (a) and Stokes' Theorem to compute $\int_{C_2} \vec{B} \cdot d\vec{r}$, where C_2 is the ellipse $x^2 + 9y^2 = 9$ in the plane $z = 2$, oriented counterclockwise when viewed from above.

Solution (a) On the circle C_1, we have $\|\vec{B}\| = 2I/(ca)$. Since $\vec{B}$ is tangent to C_1 everywhere and points in the forward direction around C_1,

$$\int_{C_1} \vec{B} \cdot d\vec{r} = \|\vec{B}\| \cdot \text{Length of } C_1 = \frac{2I}{ca} \cdot 2\pi a = \frac{4\pi I}{c}.$$

(b) We cannot use Stokes' Theorem on the elliptical disk bounded by C_2 in the plane $z = 2$ because curl $\vec{B}$ is not defined at $(0, 0, 2)$. Instead, we will use the theorem on a conical surface connecting C_1 and C_2.

Let S be the conical surface extending from C_1 to C_2 in Figure 20.14. The boundary of this surface has two pieces, $-C_2$ and C_1. The orientation of C_1 leads to the outward normal on S, which forces us to choose the clockwise orientation on C_2. By Stokes' Theorem,

$$\int_S \text{curl}\,\vec{B} \cdot d\vec{A} = \int_{-C_2} \vec{B} \cdot d\vec{r} + \int_{C_1} \vec{B} \cdot d\vec{r} = -\int_{C_2} \vec{B} \cdot d\vec{r} + \int_{C_1} \vec{B} \cdot d\vec{r}.$$

Since curl $\vec{B} = \vec{0}$, we have $\int_S \text{curl}\,\vec{B} \cdot d\vec{A} = 0$, so the two line integrals must be equal:

$$\int_{C_2} \vec{B} \cdot d\vec{r} = \int_{C_1} \vec{B} \cdot d\vec{r} = \frac{4\pi I}{c}.$$

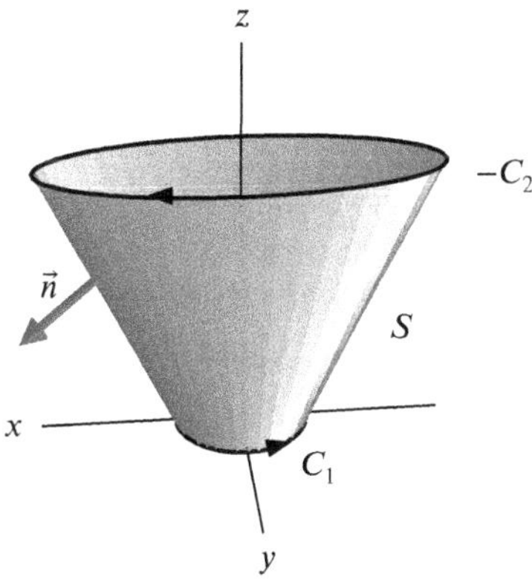

Figure 20.14: Surface joining C_1 to C_2, oriented to satisfy the conditions of Stokes' Theorem

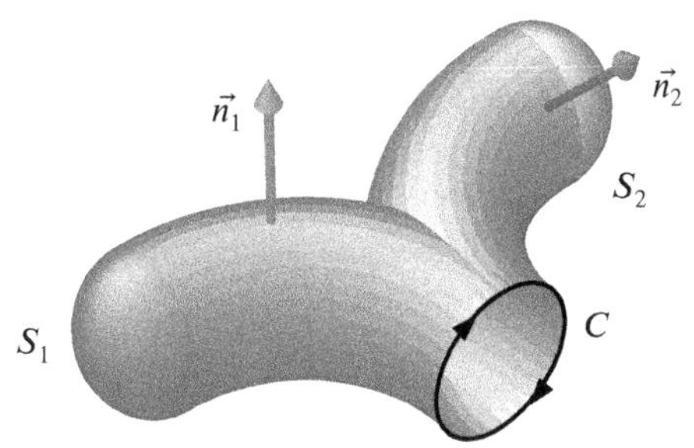

Figure 20.15: The flux of a curl is the same through the two surfaces S_1 and S_2 if they determine the same orientation on the boundary, C

Curl Fields

A vector field $\vec{F}$ is called a *curl field* if $\vec{F} = \operatorname{curl}\vec{G}$ for some vector field $\vec{G}$. Recall that if $\vec{F} = \operatorname{grad} f$, then f is called a potential function. By analogy, if a vector field $\vec{F} = \operatorname{curl}\vec{G}$, then $\vec{G}$ is called a *vector potential* for $\vec{F}$. The following example shows that the flux of a curl field through a surface depends only on the boundary of the surface. This is analogous to the fact that the line integral of a gradient field depends only on the endpoints of the path.

Example 3 Suppose $\vec{F} = \operatorname{curl}\vec{G}$, and that S_1 and S_2 are two oriented surfaces with the same boundary C. Show that, if S_1 and S_2 determine the same orientation on C (as in Figure 20.15), then

$$\int_{S_1} \vec{F} \cdot d\vec{A} = \int_{S_2} \vec{F} \cdot d\vec{A}.$$

If S_1 and S_2 determine opposite orientations on C, then

$$\int_{S_1} \vec{F} \cdot d\vec{A} = -\int_{S_2} \vec{F} \cdot d\vec{A}.$$

Solution If S_1 and S_2 determine the same orientation on C, then since $\vec{F} = \operatorname{curl}\vec{G}$, by Stokes' Theorem we have

$$\int_{S_1} \vec{F} \cdot d\vec{A} = \int_{S_1} \operatorname{curl}\vec{G} \cdot d\vec{A} = \int_C \vec{G} \cdot d\vec{r}$$

and

$$\int_{S_2} \vec{F} \cdot d\vec{A} = \int_{S_2} \operatorname{curl}\vec{G} \cdot d\vec{A} = \int_C \vec{G} \cdot d\vec{r}.$$

In each case the line integral on the right must be computed using the orientation determined by the surface. Thus, the two flux integrals of $\vec{F}$ are the same if the orientations are the same and they are opposite if the orientations are opposite.

Exercises and Problems for Section 20.2

EXERCISES

1. If $\operatorname{curl}\vec{F} = \vec{k}$, find the circulation of $\vec{F}$ around C, a circle of radius 1, centered at the origin, with
   (a) C in the xy-plane, oriented counterclockwise when viewed from above.
   (b) C in the yz-plane, oriented counterclockwise when viewed from the positive x-axis.

In Exercises 2–3, for the circle C and curl $\vec{F}$ as described, is the circulation of $\vec{F}$ around C positive, negative, or zero?

2. C is in the xy-plane oriented counterclockwise when viewed from above, and curl $\vec{F}$ points parallel to $\vec{j}$ everywhere.
3. C is in the yz-plane oriented clockwise when viewed from the positive x-axis, and curl $\vec{F}$ points parallel to and in the direction of $-\vec{i}$.

In Exercises 4–8, calculate the circulation, $\int_C \vec{F} \cdot d\vec{r}$, in two ways, directly and using Stokes' Theorem.

4. $\vec{F} = (x+z)\vec{i} + x\vec{j} + y\vec{k}$ and C is the upper half of the circle $x^2 + z^2 = 9$ in the plane $y = 0$, together with the x-axis from $(3, 0, 0)$ to $(-3, 0, 0)$, traversed counterclockwise when viewed from the positive y-axis.
5. $\vec{F} = y\vec{i} - x\vec{j}$ and C is the boundary of S, the part of the surface $z = 4 - x^2 - y^2$ above the xy-plane, oriented upward.
6. $\vec{F} = (x-y+z)(\vec{i} + \vec{j})$ and C is the triangle with vertices $(0, 0, 0)$, $(5, 0, 0)$, $(5, 5, 0)$, traversed in that order.
7. $\vec{F} = xy\vec{i} + yz\vec{j} + xz\vec{k}$ and C is the boundary of S, the surface $z = 1 - x^2$ for $0 \le x \le 1$ and $-2 \le y \le 2$, oriented upward. Sketch S and C.
8. $\vec{F} = y\vec{i} + z\vec{j} + x\vec{k}$ and C is the boundary of S, the paraboloid $z = 1 - (x^2 + y^2)$, $z \ge 0$ oriented upward. [Hint: Use polar coordinates.]

In Exercises 9–12, use Stokes' Theorem to calculate the integral.

9. $\int_C \vec{F} \cdot d\vec{r}$ where $\vec{F} = x^2\vec{i} + y^2\vec{j} + z^2\vec{k}$ and C is the unit circle in the xz-plane, oriented counterclockwise when viewed from the positive y-axis.
10. $\int_C \vec{F} \cdot d\vec{r}$ where $\vec{F} = (y-x)\vec{i} + (z-y)\vec{j} + (x-z)\vec{k}$ and C is the circle $x^2 + y^2 = 5$ in the xy-plane, oriented counterclockwise when viewed from above.
11. $\int_S \operatorname{curl} \vec{F} \cdot d\vec{A}$ where $\vec{F} = -y\vec{i} + x\vec{j} + (xy + \cos z)\vec{k}$ and S is the disk $x^2 + y^2 \le 9$, oriented upward in the xy-plane.
12. $\int_S \operatorname{curl} \vec{F} \cdot d\vec{A}$ where $\vec{F} = (x+7)\vec{j} + e^{x+y+z}\vec{k}$ and S is the rectangle $0 \le x \le 3$, $0 \le y \le 2$, $z = 0$, oriented upward.

13. Let $\vec{F} = y\vec{i} - x\vec{j}$ and let C be the unit circle in the xy-plane centered at the origin and oriented counterclockwise when viewed from above.
    - **(a)** Calculate $\int_C \vec{F} \cdot d\vec{r}$ by parameterizing the circle.
    - **(b)** Calculate curl $\vec{F}$.
    - **(c)** Calculate $\int_C \vec{F} \cdot d\vec{r}$ using your result from part (b).
    - **(d)** What theorem did you use in part (c)?
14. **(a)** If $\vec{F} = (\cos x)\vec{i} + e^y\vec{j} + (x - y - z)\vec{k}$, find curl $\vec{F}$.
    (b) Find $\int_C \vec{F} \cdot d\vec{r}$ where C is the circle of radius 3 in the plane $x+y+z = 1$, centered at $(1, 0, 0)$ oriented counterclockwise when viewed from above.
15. Can you use Stokes' Theorem to compute the line integral $\int_C (2x\vec{i} + 2y\vec{j} + 2z\vec{k}) \cdot d\vec{r}$ where C is the straight line from the point $(1, 2, 3)$ to the point $(4, 5, 6)$? Why or why not?

PROBLEMS

16. At all points in 3-space curl $\vec{F}$ points in the direction of $\vec{i} - \vec{j} - \vec{k}$. Let C be a circle in the yz-plane, oriented clockwise when viewed from the positive x-axis. Is the circulation of $\vec{F}$ around C positive, zero, or negative?
17. If curl $\vec{F} = (x^2 + z^2)\vec{j} + 5\vec{k}$, find $\int_C \vec{F} \cdot d\vec{r}$, where C is a circle of radius 3, centered at the origin, with
    - **(a)** C in the xy-plane, oriented counterclockwise when viewed from above.
    - **(b)** C in the xz-plane, oriented counterclockwise when viewed from the positive y-axis.
18. **(a)** Find $\operatorname{curl}(y\vec{i} + z\vec{j} + x\vec{k})$.
    (b) Find $\int_C (y\vec{i} + z\vec{j} + x\vec{k}) \cdot d\vec{r}$ where C is the boundary of the triangle with vertices $(2, 0, 0)$, $(0, 3, 0)$, $(-2, 0, 0)$, traversed in that order.
19. **(a)** Let $\vec{F} = y\vec{i} + z\vec{j} + x\vec{k}$. Find curl $\vec{F}$.
    (b) Calculate $\int_C \vec{F} \cdot d\vec{r}$ where C is
    - (i) A circle of radius 2 centered at $(1, 1, 3)$ in the plane $z = 3$, oriented counterclockwise when viewed from above.
    - (ii) The triangle obtained by tracing out the path $(2, 0, 0)$ to $(2, 0, 5)$ to $(2, 3, 5)$ to $(2, 0, 0)$.
20. **(a)** Find $\operatorname{curl}(z\vec{i} + x\vec{j} + y\vec{k})$.
    (b) Find $\int_C (z\vec{i} + x\vec{j} + y\vec{k}) \cdot d\vec{r}$ where C is a square of side 2 lying in the plane $x + y + z = 5$, oriented counterclockwise when viewed from the origin.

In Problems 21–26, find $\int_C \vec{F} \cdot d\vec{r}$ where C is a circle of radius 2 in the plane $x + y + z = 3$, centered at $(1, 1, 1)$ and oriented clockwise when viewed from the origin.

21. $\vec{F} = \vec{i} + \vec{j} + 3\vec{k}$
22. $\vec{F} = -y\vec{i} + x\vec{j} + z\vec{k}$
23. $\vec{F} = y\vec{i} - x\vec{j} + (y - x)\vec{k}$
24. $\vec{F} = (2y + e^x)\vec{i} + ((\sin y) - x)\vec{j} + (2y - x + \cos z^2)\vec{k}$
25. $\vec{F} = -z\vec{j} + y\vec{k}$
26. $\vec{F} = (z - y)\vec{i} + (x - z)\vec{j} + (y - x)\vec{k}$
27. For positive constants a, b, and c, let

$$f(x, y, z) = \ln(1 + ax^2 + by^2 + cz^2).$$

    (a) What is the domain of f?

(b) Find grad f.
(c) Find curl(grad f).
(d) Find $\int_C \vec{F} \cdot d\vec{r}$ where C is the helix $x = \cos t$, $y = \sin t$, $z = t$ for $0 \le t \le 13\pi/2$ and

$$\vec{F} = \frac{2x\vec{i} + 4y\vec{j} + 6z\vec{k}}{1 + x^2 + 2y^2 + 3z^2}.$$

28. Figure 20.16 shows an open cylindrical can, S, standing on the xy-plane. (S has a bottom and sides, but no top.)

(a) Give equation(s) for the rim, C.
(b) If S is oriented outward and downward, find $\int_S \operatorname{curl}(-y\vec{i} + x\vec{j} + z\vec{k}) \cdot d\vec{A}$.

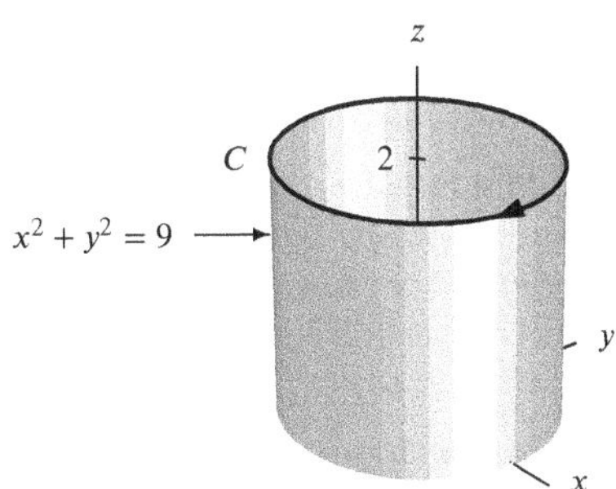

Figure 20.16

29. Evaluate $\int_C (-z\vec{i} + y\vec{j} + x\vec{k}) \cdot d\vec{r}$, where C is a circle of radius 2, parallel to the xz-plane and around the y-axis with the orientation shown in Figure 20.17.

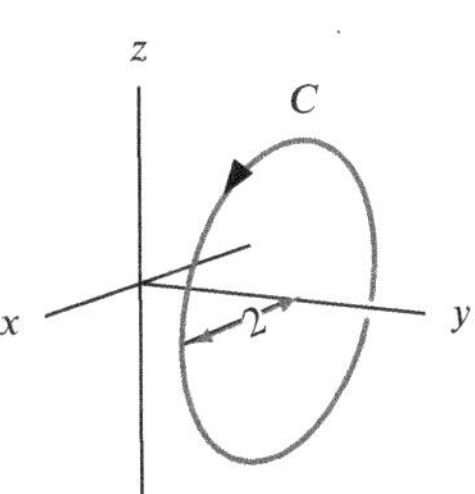

Figure 20.17

30. Evaluate the circulation of $\vec{G} = xy\vec{i} + z\vec{j} + 3y\vec{k}$ around a square of side 6, centered at the origin, lying in the yz-plane, and oriented counterclockwise viewed from the positive x-axis.

31. Find the flux of $\vec{F} = \operatorname{curl}((x^3 + \cos(z^2))\vec{i} + (x + \sin(y^2))\vec{j} + (y^2 \sin(x^2))\vec{k})$ through the upper half of the sphere of radius 2, with center at the origin and oriented upward.

32. Suppose that C is a closed curve in the xy-plane, oriented counterclockwise when viewed from above. Show that $\frac{1}{2}\int_C (-y\vec{i} + x\vec{j}) \cdot d\vec{r}$ equals the area of the region R in the xy-plane enclosed by C.

33. In the region between the circles $C_1 : x^2 + y^2 = 4$ and $C_2 : x^2 + y^2 = 25$ in the xy-plane, the vector field $\vec{F}$ has curl $\vec{F} = 3\vec{k}$. If C_1 and C_2 are both oriented counterclockwise when viewed from above, find the value of

$$\int_{C_2} \vec{F} \cdot d\vec{r} - \int_{C_1} \vec{F} \cdot d\vec{r}.$$

34. Let curl $\vec{F} = 3x\vec{i} + 3y\vec{j} - 6z\vec{k}$ and let C_1 and C_2 be the closed curves in the planes $z = 0$ and $z = 5$ in Figure 20.18. Find

$$\int_{C_1} \vec{F} \cdot d\vec{r} + \int_{C_2} \vec{F} \cdot d\vec{r}.$$

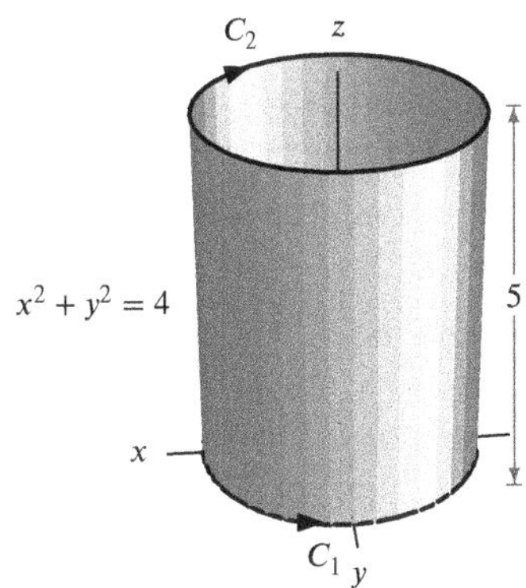

Figure 20.18

35. **(a)** Find $\operatorname{curl}(x^3\vec{i} + \sin(y^3)\vec{j} + e^{z^3}\vec{k})$.
(b) What does your answer to part (a) tell you about $\int_C (x^3\vec{i} + \sin(y^3)\vec{j} + e^{z^3}\vec{k}) \cdot d\vec{r}$ where C is the circle $(x-10)^2 + (y-20)^2 = 1$ in the xy-plane, oriented clockwise?
(c) If C is any closed curve, what can you say about $\int_C (x^3\vec{i} + \sin(y^3)\vec{j} + e^{z^3}\vec{k}) \cdot d\vec{r}$?

36. For C, the intersection of the cylinder $x^2 + y^2 = 9$ and the plane $z = -2 - x + 2y$ oriented counterclockwise when viewed from above, use Stokes' Theorem to find

$$\int_C \left((x^2 - 3y^2)\vec{i} + (\frac{z^2}{2} + y)\vec{j} + (x + 2z^2)\vec{k} \right) \cdot d\vec{r}.$$

37. Let $\vec{F}(x, y, z) = F_1(x, y)\vec{i} + F_2(x, y)\vec{j}$, where F_1 and F_2 are continuously differentiable for all x, y.

(a) Describe in words how $\vec{F}$ varies through space.
(b) Find an expression for curl $\vec{F}$ in terms of F_1 and F_2.
(c) Let C be a closed curve in the xy-plane, oriented counterclockwise when viewed from above, and let S be the region enclosed by C. Use your answer to part (b) to simplify the statement of Stokes' Theorem for this $\vec{F}$ and C.
(d) The result in part (c) is usually known by another name. What is it?

38. Water in a bathtub has velocity vector field near the drain given, for x, y, z in cm, by

$$\vec F = -\frac{y+xz}{(z^2+1)^2}\vec i - \frac{yz-x}{(z^2+1)^2}\vec j - \frac{1}{z^2+1}\vec k \text{ cm/sec.}$$

(a) Rewriting $\vec F$ as follows, describe in words how the water is moving:

$$\vec F = \frac{-y\vec i + x\vec j}{(z^2+1)^2} + \frac{-z(x\vec i + y\vec j)}{(z^2+1)^2} - \frac{\vec k}{z^2+1}.$$

(b) The drain in the bathtub is a disk in the xy-plane with center at the origin and radius 1 cm. Find the rate at which the water is leaving the bathtub. (That is, find the rate at which water is flowing through the disk.) Give units for your answer.

(c) Find the divergence of $\vec F$.

(d) Find the flux of the water through the hemisphere of radius 1, centered at the origin, lying below the xy-plane and oriented downward.

(e) Find $\int_C \vec G \cdot d\vec r$ where C is the edge of the drain, oriented clockwise when viewed from above, and where

$$\vec G = \frac{1}{2}\left(\frac{y}{z^2+1}\vec i - \frac{x}{z^2+1}\vec j - \frac{x^2+y^2}{(z^2+1)^2}\vec k\right).$$

(f) Calculate curl $\vec G$.

(g) Explain why your answers to parts (d) and (e) are equal.

Strengthen Your Understanding

In Problems 39–40, explain what is wrong with the statement.

39. The line integral $\int_C \vec F \cdot d\vec r$ can be evaluated using Stokes' Theorem, where $\vec F = 2x\vec i - 3\vec j + \vec k$ and C is an oriented curve from $(0,0,0)$ to $(3,4,5)$.

40. If S is the unit circular disc $x^2+y^2 \le 1$, $z=0$, in the xy-plane, oriented downward, C is the unit circle in the xy-plane oriented counterclockwise, and $\vec F$ is a vector field, then

$$\int_C \vec F \cdot d\vec r = \int_S \text{curl}\,\vec F \cdot d\vec A.$$

In Problems 41–42, give an example of:

41. An oriented closed curve C such that $\int_C \vec F \cdot d\vec r = 0$, where $\vec F(x,y,z) = x\vec i + y^2\vec j + z^3\vec k$.

42. A surface S, oriented appropriately to use Stokes' Theorem, which has as its boundary the circle C of radius 1 centered at the origin, lying in the xy-plane, and oriented counterclockwise when viewed from above.

In Problems 43–51, is the statement true or false? Give a reason for your answer.

43. If curl $\vec F$ is everywhere perpendicular to the z-axis, and C is a circle in the xy-plane, then the circulation of $\vec F$ around C is zero.

44. If S is the upper unit hemisphere $x^2+y^2+z^2=1$, $z \ge 0$, oriented upward, then the boundary of S used in Stokes' Theorem is the circle $x^2+y^2=1$, $z=0$, with orientation counterclockwise when viewed from the positive z-axis.

45. Let S be the cylinder $x^2+z^2=1, 0 \le y \le 2$, oriented with inward-pointing normal. Then the boundary of S consists of two circles C_1 $(x^2+z^2=1, y=0)$ and C_2 $(x^2+z^2=1, y=2)$, both oriented clockwise when viewed from the positive y-axis.

46. If C is the boundary of an oriented surface S, oriented by the right-hand rule, then $\int_C \text{curl}\,\vec F \cdot d\vec r = \int_S \vec F \cdot d\vec A$.

47. Let S_1 be the disk $x^2+y^2 \le 1, z=0$ and let S_2 be the upper unit hemisphere $x^2+y^2+z^2=1, z \ge 0$, both oriented upward. If $\vec F$ is a vector field then $\int_{S_1} \text{curl}\vec F \cdot d\vec A = \int_{S_2} \text{curl}\vec F \cdot d\vec A$.

48. Let S be the closed unit sphere $x^2+y^2+z^2=1$, oriented outward. If $\vec F$ is a vector field, then $\int_S \text{curl}\vec F \cdot d\vec A = 0$.

49. If $\vec F$ and $\vec G$ are vector fields satisfying curl$\vec F$ = curl$\vec G$, then $\int_C \vec F \cdot d\vec r = \int_C \vec G \cdot d\vec r$, where C is any oriented circle in 3-space.

50. If $\vec F$ is a vector field satisfying curl$\vec F = \vec 0$, then $\int_C \vec F \cdot d\vec r = 0$, where C is any oriented path around a rectangle in 3-space.

51. Let S be an oriented surface, with oriented boundary C, and suppose that $\vec F$ is a vector field such that $\int_C \vec F \cdot d\vec r = 0$. Then curl $\vec F = \vec 0$ everywhere on S.

52. The circle C has radius 3 and lies in a plane through the origin. Let $\vec F = (2z+3y)\vec i + (x-z)\vec j + (6y-7x)\vec k$. What is the equation of the plane and what is the orientation of the circle that make the circulation, $\int_C \vec F \cdot d\vec r$, a maximum? [Note: You should specify the orientation of the circle by saying that it is clockwise or counterclockwise when viewed from the positive or negative x- or y- or z-axis.]

20.3 THE THREE FUNDAMENTAL THEOREMS

We have now seen three multivariable versions of the Fundamental Theorem of Calculus. In this section we will examine some consequences of these theorems.

Fundamental Theorem of Calculus for Line Integrals

$$\int_C \operatorname{grad} f \cdot d\vec{r} = f(Q) - f(P).$$

Stokes' Theorem

$$\int_S \operatorname{curl} \vec{F} \cdot d\vec{A} = \int_C \vec{F} \cdot d\vec{r}.$$

Divergence Theorem

$$\int_W \operatorname{div} \vec{F}\, dV = \int_S \vec{F} \cdot d\vec{A}.$$

Notice that, in each case, the region of integration on the right is the boundary of the region on the left (except that for the first theorem we simply evaluate f at the boundary points); the integrand on the left is a sort of derivative of the integrand on the right; see Figure 20.19.

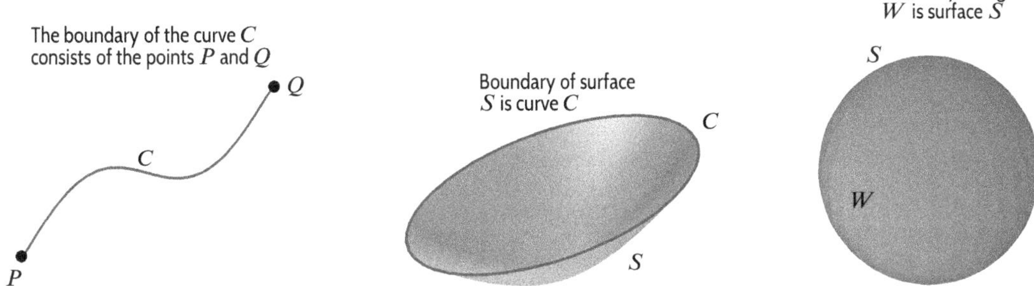

Figure 20.19: Regions and their boundaries for the three fundamental theorems

The Gradient and the Curl

Suppose that $\vec{F}$ is a smooth gradient field, so $\vec{F} = \operatorname{grad} f$ for some function f. Using the Fundamental Theorem for Line Integrals, we saw in Chapter 18 that

$$\int_C \vec{F} \cdot d\vec{r} = 0$$

for any closed curve C. Thus, for any unit vector $\vec{n}$

$$\operatorname{circ}_{\vec{n}} \vec{F} = \lim_{\text{Area}\to 0} \frac{\int_C \vec{F} \cdot d\vec{r}}{\text{Area of } C} = \lim_{\text{Area}\to 0} \frac{0}{\text{Area}} = 0,$$

where the limit is taken over circles C in a plane perpendicular to $\vec{n}$, and oriented by the right-hand rule. Thus, the circulation density of $\vec{F}$ is zero in every direction, so $\operatorname{curl} \vec{F} = \vec{0}$, that is,

$$\boxed{\operatorname{curl} \operatorname{grad} f = \vec{0}.}$$

(This formula can also be verified using the coordinate definition of curl. See Problem 31 on page 1007.)

Is the converse true? Is any vector field whose curl is zero a gradient field? Suppose that curl $\vec{F} = \vec{0}$ and let us consider the line integral $\int_C \vec{F} \cdot d\vec{r}$ for a closed curve C contained in the domain of $\vec{F}$. If C is the boundary curve of an oriented surface S that lies wholly in the domain of curl $\vec{F}$, then Stokes' Theorem asserts that

$$\int_C \vec{F} \cdot d\vec{r} = \int_S \operatorname{curl} \vec{F} \cdot d\vec{A} = \int_S \vec{0} \cdot d\vec{A} = 0.$$

If we knew that $\int_C \vec{F} \cdot d\vec{r} = 0$ for every closed curve C, then $\vec{F}$ would be path-independent, and hence a gradient field. Thus, we need to know whether every closed curve in the domain of $\vec{F}$ is the boundary of an oriented surface contained in the domain. It can be quite difficult to determine if a given curve is the boundary of a surface (suppose, for example, that the curve is knotted in a complicated way). However, if the curve can be contracted smoothly to a point, remaining all the time in the domain of $\vec{F}$, then it is the boundary of a surface, namely, the surface it sweeps through as it contracts. Thus, we have proved the test for a gradient field that we stated in Chapter 18.

The Curl Test for Vector Fields in 3-Space

Suppose $\vec{F}$ is a smooth vector field on 3-space such that

- The domain of $\vec{F}$ has the property that every closed curve in it can be contracted to a point in a smooth way, staying at all times within the domain.
- curl $\vec{F} = \vec{0}$.

Then $\vec{F}$ is path-independent, and thus is a gradient field.

Example 7 on page 956 shows how the curl test is applied.

The Curl and the Divergence

In this section we will use the second two fundamental theorems to get a test for a vector field to be a curl field, that is, a field of the form $\vec{F} = \operatorname{curl} \vec{G}$ for some $\vec{G}$.

Example 1 Suppose that $\vec{F}$ is a smooth curl field. Use Stokes' Theorem to show that for any closed surface, S, contained in the domain of $\vec{F}$,

$$\int_S \vec{F} \cdot d\vec{A} = 0.$$

Solution Suppose $\vec{F} = \operatorname{curl} \vec{G}$. Draw a closed curve C on the surface S, thus dividing S into two surfaces S_1 and S_2 as shown in Figure 20.20. Pick the orientation for C corresponding to S_1; then the orientation of C corresponding to S_2 is the opposite. Thus, using Stokes' Theorem,

$$\int_{S_1} \vec{F} \cdot d\vec{A} = \int_{S_1} \operatorname{curl} \vec{G} \cdot d\vec{A} = \int_C \vec{G} \cdot d\vec{r} = -\int_{S_2} \operatorname{curl} \vec{G} \cdot d\vec{A} = -\int_{S_2} \vec{F} \cdot d\vec{A}.$$

Thus, for any closed surface S, we have

$$\int_S \vec{F} \cdot d\vec{A} = \int_{S_1} \vec{F} \cdot d\vec{A} + \int_{S_2} \vec{F} \cdot d\vec{A} = 0.$$

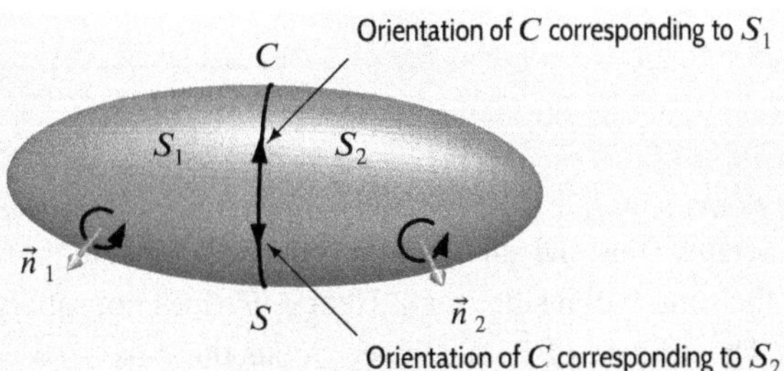

Figure 20.20: The closed surface S divided into two surfaces S_1 and S_2

Thus, if $\vec{F} = \operatorname{curl}\vec{G}$, we use the result of Example 1 to see that

$$\operatorname{div}\vec{F} = \lim_{\text{Volume}\to 0} \frac{\int_S \vec{F}\cdot d\vec{A}}{\text{Volume enclosed by } S} = \lim_{\text{Volume}\to 0} \frac{0}{\text{Volume}} = 0,$$

where the limit is taken over spheres S contracting down to a point. So we conclude that:

$$\boxed{\operatorname{div}\operatorname{curl}\vec{G} = 0.}$$

(This formula can also be verified using coordinates. See Problem 30 on page 1007.)

Is every vector field whose divergence is zero a curl field? It turns out that we have the following analogue of the curl test, though we will not prove it.

The Divergence Test for Vector Fields in 3-Space

Suppose $\vec{F}$ is a smooth vector field on 3-space such that

- The domain of $\vec{F}$ has the property that every closed surface in it is the boundary of a solid region completely contained in the domain.
- $\operatorname{div}\vec{F} = 0$.

Then $\vec{F}$ is a curl field.

Example 2 Consider the vector fields $\vec{E} = q\dfrac{\vec{r}}{\|\vec{r}\|^3}$ and $\vec{B} = \dfrac{2I}{c}\left(\dfrac{-y\vec{i} + x\vec{j}}{x^2+y^2}\right)$.

(a) Calculate $\operatorname{div}\vec{E}$ and $\operatorname{div}\vec{B}$.
(b) Do $\vec{E}$ and $\vec{B}$ satisfy the divergence test?
(c) Is either $\vec{E}$ or $\vec{B}$ a curl field?

Solution (a) Example 3 on page 986 shows that $\operatorname{div}\vec{E} = 0$. The following calculation shows $\operatorname{div}\vec{B} = 0$ also:

$$\operatorname{div}\vec{B} = \frac{2I}{c}\left(\frac{\partial}{\partial x}\left(\frac{-y}{x^2+y^2}\right) + \frac{\partial}{\partial y}\left(\frac{x}{x^2+y^2}\right) + \frac{\partial}{\partial z}(0)\right)$$

$$= \frac{2I}{c}\left(\frac{2xy}{(x^2+y^2)^2} + \frac{-2yx}{(x^2+y^2)^2}\right) = 0.$$

(b) The domain of $\vec{E}$ is 3-space minus the origin, so a region is contained in the domain if it misses the origin. Thus, the surface of a sphere centered at the origin is contained in the domain of E, but the solid ball inside is not. Hence, $\vec{E}$ does not satisfy the divergence test.

The domain of $\vec{B}$ is 3-space minus the z-axis, so a region is contained in the domain if it avoids the z-axis. If S is a surface bounding a solid region W, then the z-axis cannot pierce W without piercing S as well. Hence, if S avoids the z-axis, so does W. Thus, $\vec{B}$ satisfies the divergence test.

(c) In Example 3 on page 993 we computed the flux of $\vec{r}/\|\vec{r}\|^3$ through a sphere centered at the origin, and found it was 4π, so the flux of $\vec{E}$ through this sphere is $4\pi q$. Thus, $\vec{E}$ cannot be a curl field, because by Example 1, the flux of a curl field through a closed surface is zero.

On the other hand, $\vec{B}$ satisfies the divergence test, so it must be a curl field. In fact, Problem 26 shows that

$$\vec{B} = \operatorname{curl}\left(\frac{-I}{c}\ln(x^2+y^2)\vec{k}\right).$$

Exercises and Problems for Section 20.3 Online Resource: Additional Problems for Section 20.3

EXERCISES

In Exercises 1–6, is the vector field a gradient field?

1. $\vec{F} = 2x\vec{i} + z\vec{j} + y\vec{k}$
2. $\vec{F} = y\vec{i} + z\vec{j} + x\vec{k}$
3. $\vec{F} = (y+2z)\vec{i} + (x+z)\vec{j} + (2x+y)\vec{k}$
4. $\vec{F} = (y-2z)\vec{i} + (x-z)\vec{j} + (2x-y)\vec{k}$
5. $\vec{G} = -y\vec{i} + x\vec{j}$
6. $\vec{F} = yz\vec{i} + (xz+z^2)\vec{j} + (xy+2yz)\vec{k}$

In Exercises 7–12, is the vector field a curl field?

7. $\vec{F} = z\vec{i} + x\vec{j} + y\vec{k}$
8. $\vec{F} = z\vec{i} + y\vec{j} + x\vec{k}$
9. $\vec{F} = 2x\vec{i} - y\vec{j} - z\vec{k}$
10. $\vec{F} = (x+y)\vec{i} + (y+z)\vec{j} + (x+z)\vec{k}$
11. $\vec{F} = (-xy)\vec{i} + (2yz)\vec{j} + (yz-z^2))\vec{k}$
12. $\vec{F} = (xy)\vec{i} + (xy)\vec{j} + (xy)\vec{k}$

13. Let $\vec{F}$ be a vector field defined everywhere except the z-axis and with $\operatorname{curl}\vec{F} = 0$ at all points of its domain. Determine whether Stokes' Theorem implies that $\int_C \vec{F}\cdot d\vec{r} = 0$ for a circle C of radius 1, where
(a) C is parallel to the xy-plane with center $(0,0,1)$.
(b) C is parallel to the yz-plane with center $(1,0,0)$.

14. Let $\vec{F}$ be a vector field defined everywhere except the origin and with $\operatorname{div}\vec{F} = 0$ at all points of its domain. Determine whether the Divergence Theorem implies that $\int_S \vec{F}\cdot d\vec{A} = 0$ for a sphere S of radius 1, where
(a) S is centered at $(0,1,1)$.
(b) S is centered at $(0.5,0,0)$.

In Exercises 15–18, can the curl test and the divergence test be applied to a vector field whose domain is the given region?

15. All points (x,y,z) such that $z > 0$.
16. All points (x,y,z) not on the y-axis.
17. All points (x,y,z) not on the positive z-axis.
18. All points (x,y,z) except the x-axis with $0 \le x \le 1$.

PROBLEMS

19. Let $\vec{B} = b\vec{k}$, for some constant b. Show that the following are all possible vector potentials for $\vec{B}$:
(a) $\vec{A} = -by\vec{i}$ **(b)** $\vec{A} = bx\vec{j}$
(c) $\vec{A} = \frac{1}{2}\vec{B}\times\vec{r}$.

20. Find a vector field $\vec{F}$ such that $\operatorname{curl}\vec{F} = 2\vec{i} - 3\vec{j} + 4\vec{k}$. [Hint: Try $\vec{F} = \vec{v}\times\vec{r}$ for some vector $\vec{v}$.]

21. Find a vector potential for the constant vector field $\vec{B}$ whose value at every point is $\vec{b}$.

22. Express $(3x+2y)\vec{i}+(4x+9y)\vec{j}$ as the sum of a curl-free vector field and a divergence-free vector field.

In Problems 23–24, does a vector potential exist for the vector field given? If so, find one.

23. $\vec{G} = x^2\vec{i} + y^2\vec{j} + z^2\vec{k}$

24. $\vec{F} = 2x\vec{i} + (3y - z^2)\vec{j} + (x - 5z)\vec{k}$

25. An electric charge q at the origin produces an electric field $\vec{E} = q\vec{r}/\|\vec{r}\|^3$.
   (a) Does $\operatorname{curl}\vec{E} = \vec{0}$?
   (b) Does $\vec{E}$ satisfy the curl test?
   (c) Is $\vec{E}$ a gradient field?

26. Show that $\vec{A} = \dfrac{-I}{c}\ln(x^2 + y^2)\vec{k}$ is a vector potential for
$$\vec{B} = \frac{2I}{c}\left(\frac{-y\vec{i} + x\vec{j}}{x^2 + y^2}\right).$$

27. Suppose c is the speed of light. A thin wire along the z-axis carrying a current I produces a magnetic field
$$\vec{B} = \frac{2I}{c}\left(\frac{-y\vec{i} + x\vec{j}}{x^2 + y^2}\right).$$
   (a) Does $\operatorname{curl}\vec{B} = \vec{0}$?
   (b) Does $\vec{B}$ satisfy the curl test?
   (c) Is $\vec{B}$ a gradient field?

28. Use Stokes' Theorem to show that if $u(x, y)$ and $v(x, y)$ are two functions of x and y and C is a closed curve in the xy-plane oriented counterclockwise, then
$$\int_C (u\vec{i} + v\vec{j}) \cdot d\vec{r} = \int_R \left(\frac{\partial v}{\partial x} - \frac{\partial u}{\partial y}\right) dx\,dy$$
where R is the region in the xy-plane enclosed by C. This is Green's Theorem.

29. Suppose that $\vec{A}$ is a vector potential for $\vec{B}$.
   (a) Show that $\vec{A} + \operatorname{grad}\psi$ is also a vector potential for $\vec{B}$, for any function ψ with continuous second-order partial derivatives. (The vector potentials $\vec{A}$ and $\vec{A} + \operatorname{grad}\psi$ are called *gauge equivalent* and the transformation, for any ψ, from $\vec{A}$ to $\vec{A} + \operatorname{grad}\psi$ is called a *gauge transformation*.)
   (b) What is the divergence of $\vec{A} + \operatorname{grad}\psi$? How should ψ be chosen such that $\vec{A} + \operatorname{grad}\psi$ has zero divergence? (If $\operatorname{div}\vec{A} = 0$, the magnetic vector potential $\vec{A}$ is said to be in *Coulomb gauge*.)

Strengthen Your Understanding

In Problems 30–31, explain what is wrong with the statement.

30. The curl of a vector field $\vec{F}$ is given by $\operatorname{curl}\vec{F} = x\vec{i}$.

31. For a certain vector field $\vec{F}$, we have $\operatorname{curl}\operatorname{div}\vec{F} = y\vec{i}$.

In Problems 32–33, give an example of:

32. A vector field $\vec{F}$ that is not the curl of another vector field.

33. A function f such that $\operatorname{div}\operatorname{grad} f \neq 0$.

In Problems 34–37, is the statement true or false? Give a reason for your answer.

34. There exists a vector field $\vec{F}$ with $\operatorname{curl}\vec{F} = \vec{i}$.

35. There exists a vector field $\vec{F}$ (whose components have continuous second partial derivatives) satisfying $\operatorname{curl}\vec{F} = x\vec{i}$.

36. Let S be an oriented surface, with oriented boundary C, and suppose that $\vec{F}$ is a vector field such that $\int_S \operatorname{curl}\vec{F} \cdot d\vec{A} = 0$. Then $\vec{F}$ is a gradient field.

37. If $\vec{F}$ is a gradient field, then $\int_S \operatorname{curl}\vec{F} \cdot d\vec{A} = 0$, for any smooth oriented surface, S, in 3-space.

38. Let $f(x, y, z)$ be a scalar function with continuous second partial derivatives. Let $\vec{F}(x, y, z)$ be a vector field with continuous second partial derivatives. Which of the following quantities are identically zero?
   (a) $\operatorname{curl}\operatorname{grad} f$
   (b) $\vec{F} \times \operatorname{curl}\vec{F}$
   (c) $\operatorname{grad}\operatorname{div}\vec{F}$
   (d) $\operatorname{div}\operatorname{curl}\vec{F}$
   (e) $\operatorname{div}\operatorname{grad} f$

Online Resource: Review problems and Projects

Chapter Twentyone

PARAMETERS, COORDINATES, AND INTEGRALS

Contents

21.1 COORDINATES AND PARAMETERIZED SURFACES

In Chapter 17 we parameterized curves in 2- and 3-space, and in Chapter 16 we used polar, cylindrical, and spherical coordinates to simplify iterated integrals. We now take a second look at parameterizations and coordinate systems, and see that they are the same thing in different disguises: functions from one space to another.

We have already seen this with parameterized curves, which we view as a function from an interval $a \leq t \leq b$ to a curve in xyz-space. See Figure 21.1.

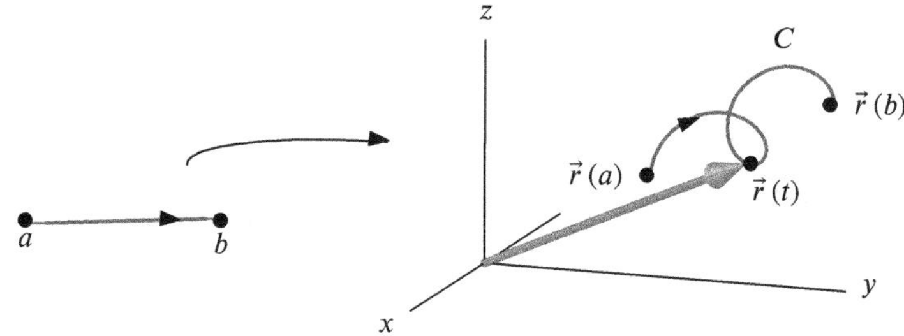

Figure 21.1: The parameterization is a function from the interval, $a \leq t \leq b$, to 3-space, whose image is the curve, C

Polar, Cylindrical, and Spherical Coordinates Revisited

The equations for polar coordinates,

$$x = r\cos\theta$$
$$y = r\sin\theta,$$

can also be viewed as defining a function from the $r\theta$-plane into the xy-plane. This function transforms the rectangle on the left of Figure 21.2 into the quarter disk on the right. We need two parameters to describe this disk because it is a two-dimensional object.

Polar Coordinates as Families of Parameterized Curves

Polar coordinates give two families of parameterized curves, which form the polar coordinate grid. The lines $r =$ Constant in the $r\theta$-plane correspond to circles in the xy-plane, each circle parameterized by θ; the lines $\theta =$ Constant correspond to rays in the xy-plane, each ray parameterized by r.

Cylindrical and Spherical Coordinates

Similarly, cylindrical and spherical coordinates may be viewed as functions from 3-space to 3-space. Cylindrical coordinates take rectangular boxes in $r\theta z$-space and map them to cylindrical regions in xyz-space; spherical coordinates take rectangular boxes in $\rho\phi\theta$-space and map them to spherical regions in xyz-space.

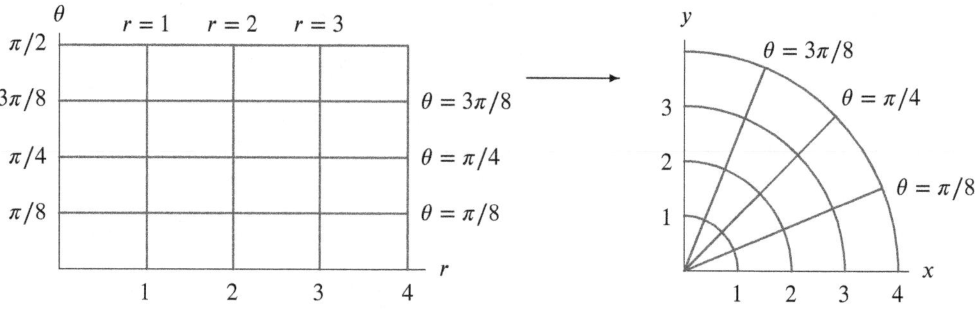

Figure 21.2: A grid in the $r\theta$-plane and the corresponding curved grid in the xy-plane

General Parameterizations

In general, a parameterization or coordinate system provides a way of representing a curved object by means of a simple region in the *parameter space* (an interval, rectangle, or rectangular box), along with a function mapping that region into the curved object. In the next section, we use this idea to parameterize curved surfaces in 3-space.

How Do We Parameterize a Surface?

In Section 17.1 we parameterized a circle in 2-space using the equations

$$x = \cos t, \quad y = \sin t.$$

In 3-space, the same circle in the xy-plane has parametric equations

$$x = \cos t, \quad y = \sin t, \quad z = 0.$$

We add the equation $z = 0$ to specify that the circle is in the xy-plane. If we wanted a circle in the plane $z = 3$, we would use the equations

$$x = \cos t, \quad y = \sin t, \quad z = 3.$$

Suppose now we let z vary freely, as well as t. We get circles in every horizontal plane, forming a cylinder as in the left of Figure 21.3. Thus, we need two parameters, t and z, to parameterize the cylinder.

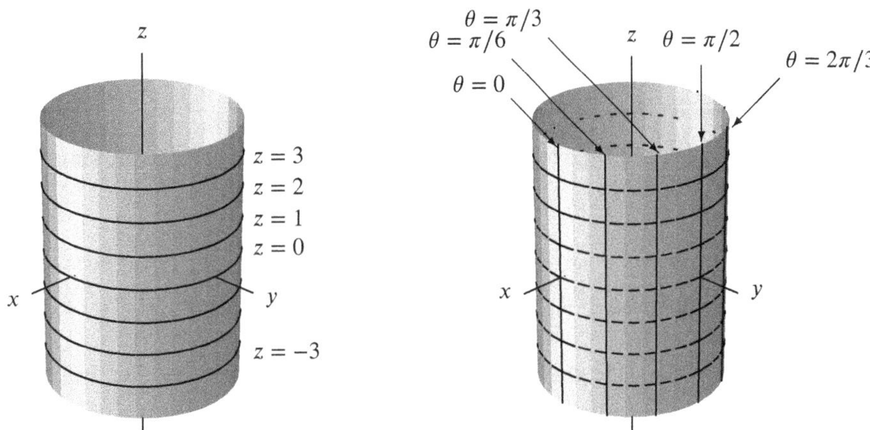

Figure 21.3: The cylinder $x = \cos t,\ y = \sin t,\ z = z$

We can contrast curves and surfaces. A curve, though it may live in two or three dimensions, is itself one-dimensional; if we move along it we can only move backward and forward in one direction. Thus, it only requires one parameter to trace out a curve.

A surface is 2-dimensional; at any given point there are two independent directions we can move. For example, on the cylinder we can move vertically, or we can circle around the z-axis horizontally. So we need *two* parameters to describe it. We can think of the parameters as map coordinates, like longitude and latitude on the surface of the earth. Just as polar coordinates give a polar grid on a circular region, so the parameters for a surface give a grid on the surface. See Figure 21.3 on the right.

In the case of the cylinder our parameters are t and z, so

$$x = \cos t, \quad y = \sin t, \quad z = z, \qquad 0 \le t < 2\pi, \quad -\infty < z < \infty.$$

The last equation, $z = z$, looks strange, but it reminds us that we are in three dimensions, not two, and that the z-coordinate on our surface is allowed to vary freely.

In general, we express the coordinates, (x, y, z), of a point on a surface S in terms of two parameters, s and t:

$$x = f_1(s,t), \quad y = f_2(s,t), \quad z = f_3(s,t).$$

As the values of s and t vary, the corresponding point (x, y, z) sweeps out the surface, S. (See Figure 21.4.) The function which sends the point (s, t) to the point (x, y, z) is called the *parameterization of the surface*.

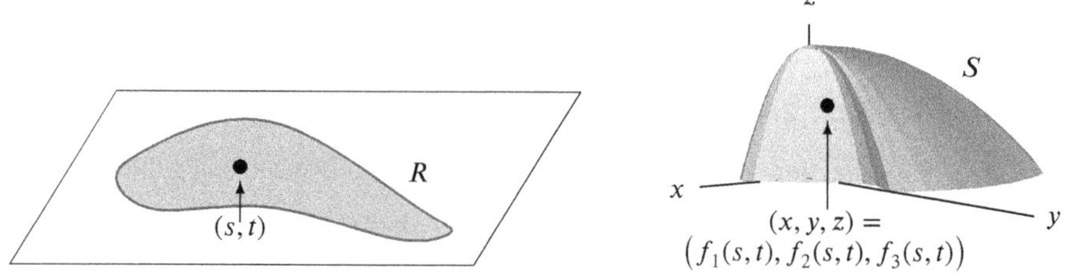

Figure 21.4: The parameterization sends each point (s, t) in the parameter region, R, to a point $(x, y, z) = (f_1(s, t), f_2(s, t), f_3(s, t))$ in the surface, S

Using Position Vectors

We can use the position vector $\vec{r} = x\vec{i} + y\vec{j} + z\vec{k}$ to combine the three parametric equations for a surface into a single vector equation. For example, the parameterization of the cylinder $x = \cos t$, $y = \sin t$, $z = z$ can be written as

$$\vec{r}(t, z) = \cos t\vec{i} + \sin t\vec{j} + z\vec{k} \qquad 0 \le t < 2\pi, \quad -\infty < z < \infty.$$

For a general parameterized surface S, we write

$$\vec{r}(s, t) = f_1(s, t)\vec{i} + f_2(s, t)\vec{j} + f_3(s, t)\vec{k}.$$

Parameterizing a Surface of the Form $z = f(x, y)$

The graph of a function $z = f(x, y)$ can be given parametrically simply by letting the parameters s and t be x and y:

$$x = s, \quad y = t, \quad z = f(s, t).$$

Example 1 Give a parametric description of the lower hemisphere of the sphere $x^2 + y^2 + z^2 = 1$.

Solution The surface is the graph of the function $z = -\sqrt{1 - x^2 - y^2}$ over the region $x^2 + y^2 \le 1$ in the plane. Then parametric equations are $x = s$, $y = t$, $z = -\sqrt{1 - s^2 - t^2}$, where the parameters s and t vary inside and on the unit circle.

In practice we often think of x and y as parameters rather than introduce new parameters s and t. Thus, we may write $x = x$, $y = y$, $z = f(x, y)$.

Parameterizing Planes

Consider a plane containing two nonparallel vectors $\vec{v}_1$ and $\vec{v}_2$ and a point P_0 with position vector $\vec{r}_0$. We can get to any point on the plane by starting at P_0 and moving parallel to $\vec{v}_1$ or $\vec{v}_2$, adding multiples of them to $\vec{r}_0$. (See Figure 21.5.)

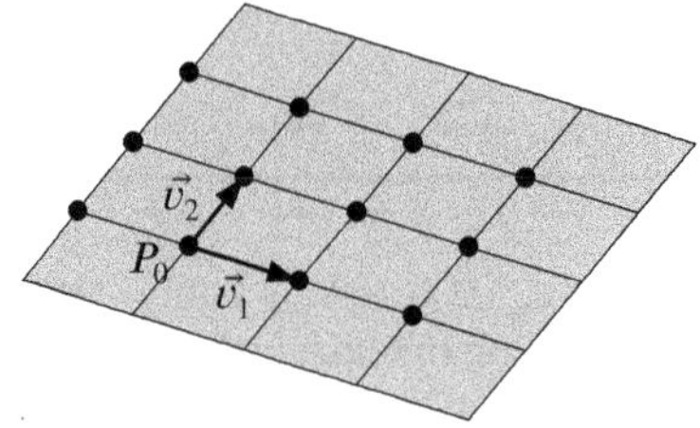

Figure 21.5: The plane $\vec{r}(s, t) = \vec{r}_0 + s\vec{v}_1 + t\vec{v}_2$ and some points corresponding to various choices of s and t

Since $s\vec{v}_1$ is parallel to $\vec{v}_1$ and $t\vec{v}_2$ is parallel to $\vec{v}_2$, we have the following result:

Parameterizing a Plane

The plane through the point with position vector $\vec{r}_0$ and containing the two nonparallel vectors $\vec{v}_1$ and $\vec{v}_2$ has parameterization

$$\vec{r}(s,t) = \vec{r}_0 + s\vec{v}_1 + t\vec{v}_2.$$

If $\vec{r}_0 = x_0\vec{i} + y_0\vec{j} + z_0\vec{k}$, and $\vec{v}_1 = a_1\vec{i} + a_2\vec{j} + a_3\vec{k}$, and $\vec{v}_2 = b_1\vec{i} + b_2\vec{j} + b_3\vec{k}$, then the parameterization of the plane can be expressed with the parametric equations

$$x = x_0 + sa_1 + tb_1, \quad y = y_0 + sa_2 + tb_2, \quad z = z_0 + sa_3 + tb_3.$$

Notice that the parameterization of the plane expresses the coordinates x, y, and z as linear functions of the parameters s and t.

Example 2 Write a parameterization for the plane through the point $(2,-1,3)$ and containing the vectors $\vec{v}_1 = 2\vec{i} + 3\vec{j} - \vec{k}$ and $\vec{v}_2 = \vec{i} - 4\vec{j} + 5\vec{k}$.

Solution A possible parameterization is

$$\begin{aligned}\vec{r}(s,t) = \vec{r}_0 + s\vec{v}_1 + t\vec{v}_2 &= 2\vec{i} - \vec{j} + 3\vec{k} + s(2\vec{i} + 3\vec{j} - \vec{k}) + t(\vec{i} - 4\vec{j} + 5\vec{k}) \\ &= (2 + 2s + t)\vec{i} + (-1 + 3s - 4t)\vec{j} + (3 - s + 5t)\vec{k},\end{aligned}$$

or equivalently,

$$x = 2 + 2s + t, \quad y = -1 + 3s - 4t, \quad z = 3 - s + 5t.$$

Parameterizations Using Spherical Coordinates

Recall the spherical coordinates ρ, ϕ, and θ introduced on page 872 of Chapter 16. On a sphere of radius $\rho = a$ we can use ϕ and θ as coordinates, similar to latitude and longitude on the surface of the earth. (See Figure 21.6.) The latitude, however, is measured from the equator, whereas ϕ is measured from the north pole. If the positive x-axis passes through the Greenwich meridian, the longitude and θ are equal for $0 \leq \theta \leq \pi$.

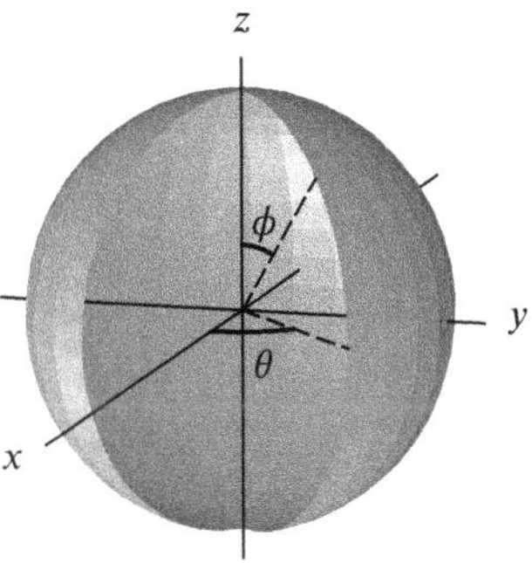

Figure 21.6: Parameterizing the sphere by ϕ and θ

Example 3 You are at a point on a sphere with $\phi = 3\pi/4$. Are you in the northern or southern hemisphere? If ϕ decreases, do you move closer to or farther from the equator?

Solution The equator has $\phi = \pi/2$. Since $3\pi/4 > \pi/2$, you are in the southern hemisphere. If ϕ decreases, you move closer to the equator.

Example 4 On a sphere, you are standing at a point with coordinates θ_0 and ϕ_0. Your *antipodal* point is the point on the other side of the sphere on a line through you and the center. What are the θ, ϕ coordinates of your antipodal point?

Solution Figure 21.7 shows that the coordinates are $\theta = \theta_0 + \pi$ if $\theta_0 < \pi$ or $\theta = \theta_0 - \pi$ if $\pi \le \theta_0 \le 2\pi$, and $\phi = \pi - \phi_0$. Notice that if you are on the equator, then so is your antipodal point.

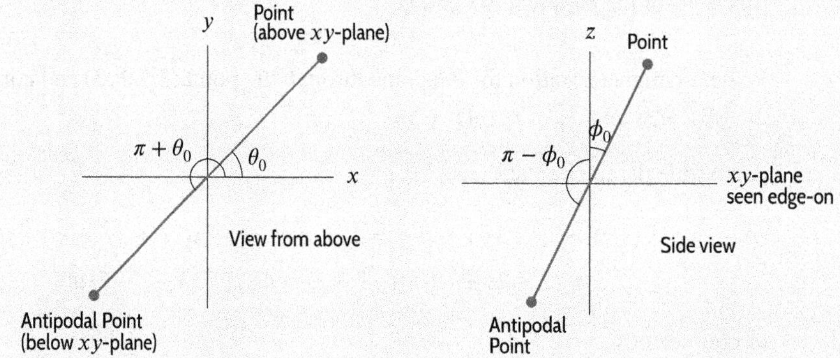

Figure 21.7: Two views of the xyz-coordinate system showing coordinates of antipodal points

Parameterizing a Sphere Using Spherical Coordinates

The sphere with radius 1 centered at the origin is parameterized by

$$x = \sin\phi\cos\theta, \qquad y = \sin\phi\sin\theta, \qquad z = \cos\phi,$$

where $0 \le \theta \le 2\pi$ and $0 \le \phi \le \pi$. (See Figure 21.8.)

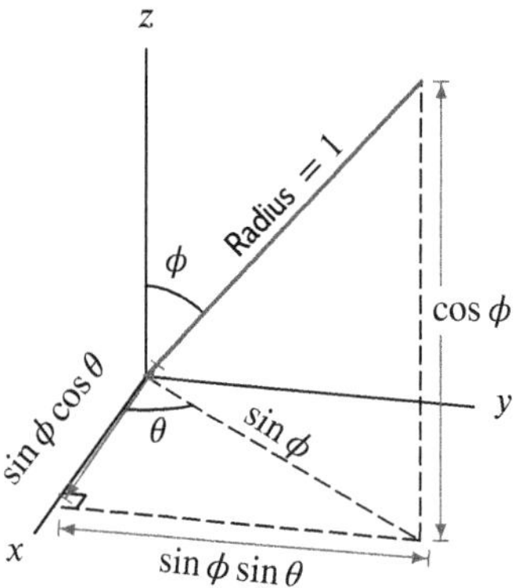

Figure 21.8: The relationship between x, y, z and ϕ, θ on a sphere of radius 1

We can also write these equations in vector form:

$$\vec{r}(\theta, \phi) = \sin\phi\cos\theta\,\vec{i} + \sin\phi\sin\theta\,\vec{j} + \cos\phi\,\vec{k}.$$

Since $x^2 + y^2 + z^2 = \sin^2\phi(\cos^2\theta + \sin^2\theta) + \cos^2\phi = \sin^2\phi + \cos^2\phi = 1$, this verifies that the point with position vector $\vec{r}(\theta, \phi)$ does lie on the sphere of radius 1. Notice that the z-coordinate depends only on the parameter ϕ. Geometrically, this means that all points on the same latitude have the same z-coordinate.

Example 5 Find parametric equations for the following spheres:

(a) Center at the origin and radius 2.
(b) Center at the point with Cartesian coordinates $(2, -1, 3)$ and radius 2.

Solution (a) We must scale the distance from the origin by 2. Thus, we have

$$x = 2\sin\phi\cos\theta, \qquad y = 2\sin\phi\sin\theta, \qquad z = 2\cos\phi,$$

where $0 \leq \theta \leq 2\pi$ and $0 \leq \phi \leq \pi$. In vector form, this is written

$$\vec{r}(\theta, \phi) = 2\sin\phi\cos\theta\vec{i} + 2\sin\phi\sin\theta\vec{j} + 2\cos\phi\vec{k}.$$

(b) To shift the center of the sphere from the origin to the point $(2, -1, 3)$, we add the vector parameterization we found in part (a) to the position vector of $(2, -1, 3)$. (See Figure 21.9.) This gives

$$\begin{aligned}\vec{r}(\theta, \phi) &= 2\vec{i} - \vec{j} + 3\vec{k} + (2\sin\phi\cos\theta\vec{i} + 2\sin\phi\sin\theta\vec{j} + 2\cos\phi\vec{k}) \\ &= (2 + 2\sin\phi\cos\theta)\vec{i} + (-1 + 2\sin\phi\sin\theta)\vec{j} + (3 + 2\cos\phi)\vec{k},\end{aligned}$$

where $0 \leq \theta \leq 2\pi$ and $0 \leq \phi \leq \pi$. Alternatively,

$$x = 2 + 2\sin\phi\cos\theta, \qquad y = -1 + 2\sin\phi\sin\theta, \qquad z = 3 + 2\cos\phi.$$

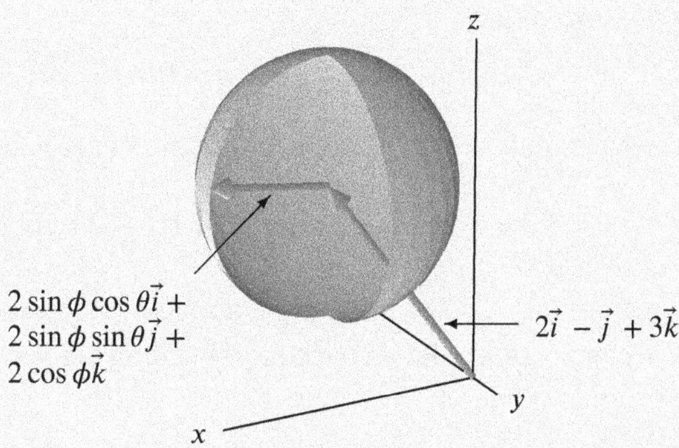

Figure 21.9: Sphere with center at the point $(2, -1, 3)$ and radius 2

Note that the same point can have more than one value for θ or ϕ. For example, points with $\theta = 0$ also have $\theta = 2\pi$, unless we restrict θ to the range $0 \leq \theta < 2\pi$. Also, the north pole, at $\phi = 0$, and the south pole, at $\phi = \pi$, can have any value of θ.

Parameterizing Surfaces of Revolution

Many surfaces have an axis of rotational symmetry and circular cross sections perpendicular to that axis. These surfaces are referred to as *surfaces of revolution.*

Example 6 Find a parameterization of the cone whose base is the circle $x^2 + y^2 = a^2$ in the xy-plane and whose vertex is at height h above the xy-plane. (See Figure 21.10.)

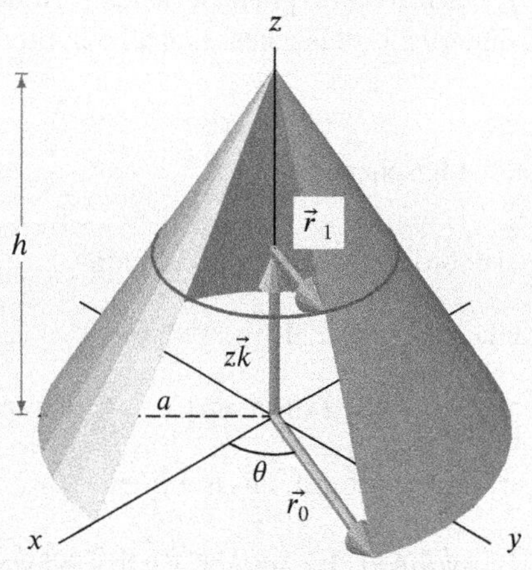

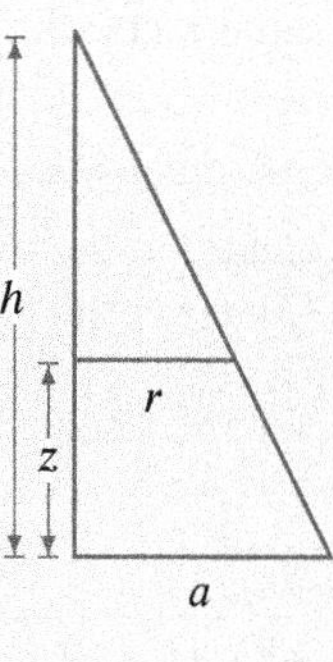

Figure 21.10: The cone whose base is the circle $x^2 + y^2 = a^2$ in the xy-plane and whose vertex is at the point $(0, 0, h)$ and the vertical cross section through the cone

Solution We use cylindrical coordinates, r, θ, z. (See Figure 21.10.) In the xy-plane, the radius vector, $\vec{r}_0$, from the z-axis to a point on the cone in the xy-plane is

$$\vec{r}_0 = a\cos\theta\vec{i} + a\sin\theta\vec{j}.$$

Above the xy-plane, the radius of the circular cross section, r, decreases linearly from $r = a$ when $z = 0$ to $r = 0$ when $z = h$. From the similar triangles in Figure 21.10,

$$\frac{a}{h} = \frac{r}{h - z}.$$

Solving for r, we have

$$r = \left(1 - \frac{z}{h}\right)a.$$

The horizontal radius vector, $\vec{r}_1$, at height z has components similar to $\vec{r}_0$, but with a replaced by r:

$$\vec{r}_1 = r\cos\theta\vec{i} + r\sin\theta\vec{j} = \left(1 - \frac{z}{h}\right)a\cos\theta\vec{i} + \left(1 - \frac{z}{h}\right)a\sin\theta\vec{j}.$$

As θ goes from 0 to 2π, the vector $\vec{r}_1$ traces out the horizontal circle in Figure 21.10. We get the position vector, $\vec{r}$, of a point on the cone by adding the vector $z\vec{k}$, so

$$\vec{r} = \vec{r}_1 + z\vec{k} = a\left(1 - \frac{z}{h}\right)\cos\theta\vec{i} + a\left(1 - \frac{z}{h}\right)\sin\theta\vec{j} + z\vec{k}, \quad \text{for } 0 \le z \le h \text{ and } 0 \le \theta \le 2\pi.$$

These equations can be written as

$$x = \left(1 - \frac{z}{h}\right)a\cos\theta, \quad y = \left(1 - \frac{z}{h}\right)a\sin\theta, \quad z = z.$$

The parameters are θ and z.

Example 7 Consider the bell of a trumpet. A model for the radius $z = f(x)$ of the horn (in cm) at a distance x cm from the large open end is given by the function

$$f(x) = \frac{6}{(x+1)^{0.7}}.$$

The bell is obtained by rotating the graph of f about the x-axis. Find a parameterization for the first 24 cm of the bell. (See Figure 21.11.)

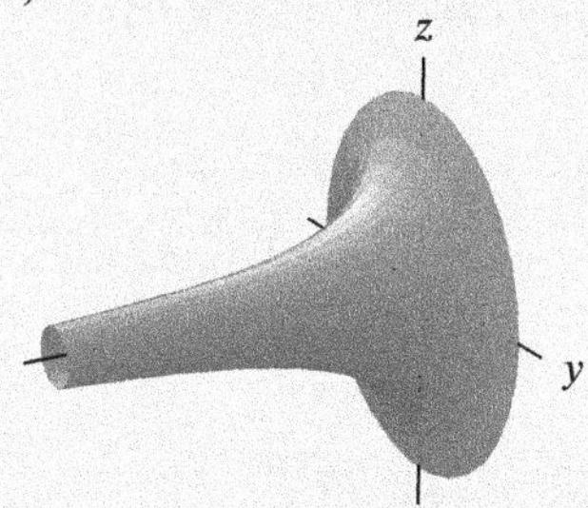

Figure 21.11: The bell of a trumpet obtained by rotating the graph of $z = f(x)$ about the x-axis

Solution At distance x from the large open end of the horn, the cross section parallel to the yz-plane is a circle of radius $f(x)$, with center on the x-axis. Such a circle can be parameterized by $y = f(x)\cos\theta$, $z = f(x)\sin\theta$. Thus, we have the parameterization

$$x = x, \quad y = \left(\frac{6}{(x+1)^{0.7}}\right)\cos\theta, \quad z = \left(\frac{6}{(x+1)^{0.7}}\right)\sin\theta, \quad 0 \le x \le 24, \quad 0 \le \theta \le 2\pi.$$

The parameters are x and θ.

Parameter Curves

On a parameterized surface, the curve obtained by setting one of the parameters equal to a constant and letting the other vary is called a *parameter curve*. If the surface is parameterized by

$$\vec{r}(s,t) = f_1(s,t)\vec{i} + f_2(s,t)\vec{j} + f_3(s,t)\vec{k},$$

there are two families of parameter curves on the surface, one family with t constant and the other with s constant.

Example 8 Consider the vertical cylinder

$$x = \cos t, \quad y = \sin t, \quad z = z.$$

(a) Describe the two parameter curves through the point $(0, 1, 1)$.
(b) Describe the family of parameter curves with t constant and the family with z constant.

Solution (a) Since the point $(0, 1, 1)$ corresponds to the parameter values $t = \pi/2$ and $z = 1$, there are two parameter curves, one with $t = \pi/2$ and the other with $z = 1$. The parameter curve with $t = \pi/2$ has the parametric equations

$$x = \cos\left(\frac{\pi}{2}\right) = 0, \quad y = \sin\left(\frac{\pi}{2}\right) = 1, \quad z = z,$$

with parameter z. This is a line through the point $(0, 1, 1)$ parallel to the z-axis.

The parameter curve with $z = 1$ has the parametric equations

$$x = \cos t, \quad y = \sin t, \quad z = 1,$$

with parameter t. This is a unit circle parallel to and one unit above the xy-plane centered on the z-axis.

(b) First, fix $t = t_0$ for t and let z vary. The curves parameterized by z have equations

$$x = \cos t_0, \quad y = \sin t_0, \quad z = z.$$

These are vertical lines on the cylinder parallel to the z-axis. (See Figure 21.12.)

The other family is obtained by fixing $z = z_0$ and varying t. Curves in this family are parameterized by t and have equations

$$x = \cos t, \quad y = \sin t, \quad z = z_0.$$

They are circles of radius 1 parallel to the xy-plane centered on the z-axis. (See Figure 21.13.)

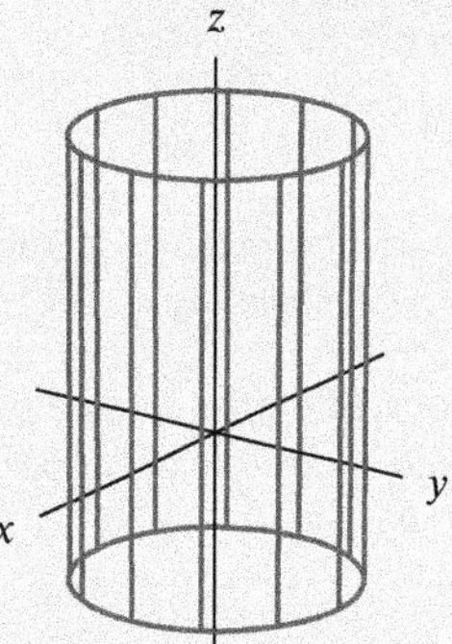

Figure 21.12: The family of parameter curves with $t = t_0$ for the cylinder $x = \cos t, y = \sin t, z = z$

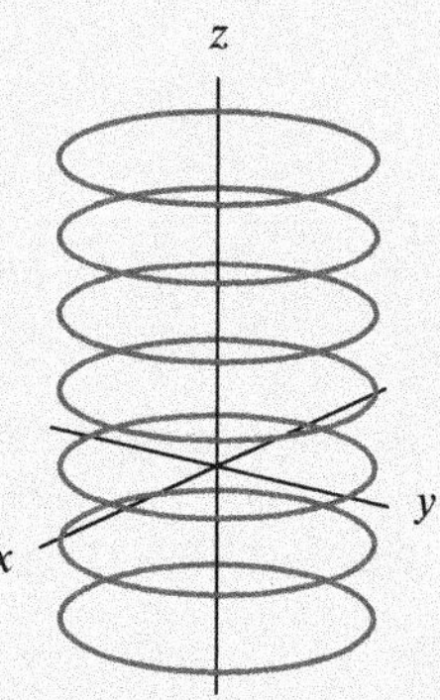

Figure 21.13: The family of parameter curves with $z = z_0$ for the cylinder $x = \cos t, y = \sin t, z = z$

Example 9 Describe the families of parameter curves with $\theta = \theta_0$ and $\phi = \phi_0$ for the sphere

$$x = \sin\phi\cos\theta, \qquad y = \sin\phi\sin\theta, \qquad z = \cos\phi,$$

where $0 \le \theta \le 2\pi, 0 \le \phi \le \pi$.

Solution Since ϕ measures latitude, the family with ϕ constant consists of the circles of constant latitude. (See Figure 21.14.) Similarly, the family with θ constant consists of the meridians (semicircles) running between the north and south poles. (See Figure 21.15.)

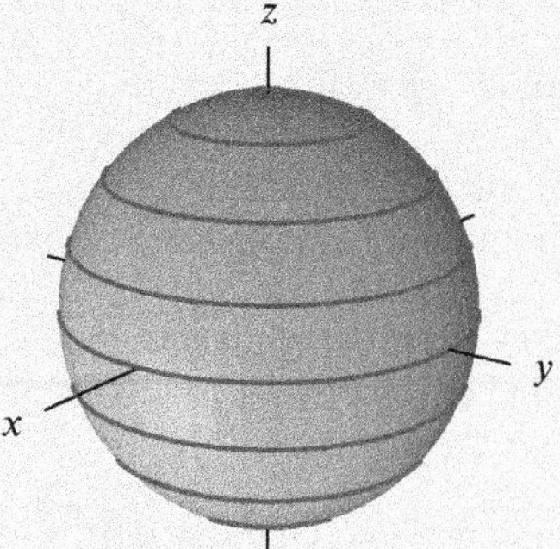

Figure 21.14: The family of parameter curves with $\phi = \phi_0$ for the sphere parameterized by (θ, ϕ)

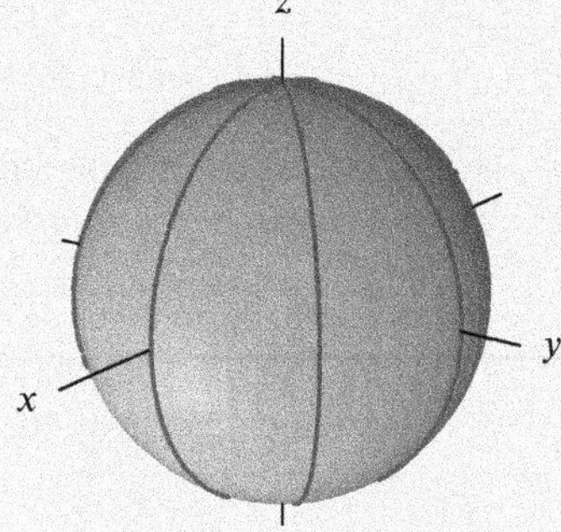

Figure 21.15: The family of parameter curves with $\theta = \theta_0$ for the sphere parameterized by (θ, ϕ)

We have seen parameter curves before on pages 661-663 of Section 12.2: The cross sections with $x = a$ or $y = b$ on a surface $z = f(x, y)$ are examples of parameter curves. So are the grid lines on a computer sketch of a surface. The small regions shaped like parallelograms surrounded by nearby pairs of parameter curves are called *parameter rectangles*. See Figure 21.16.

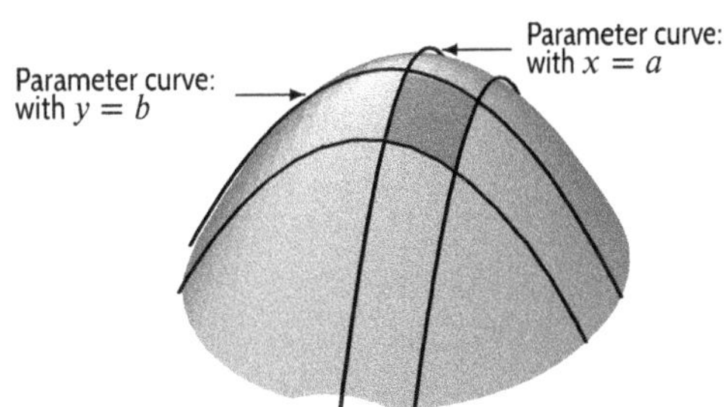

Figure 21.16: Parameter curves $x = a$ or $y = b$ on a surface $z = f(x, y)$; the darker region is a parameter rectangle

Exercises and Problems for Section 21.1

EXERCISES

In Exercises 1–4 decide if the parameterization describes a curve or a surface.

1. $\vec{r}(s) = s\vec{i} + (3 - s)\vec{j} + s^2\vec{k}$
2. $\vec{r}(s,t) = (s + t)\vec{i} + (3 - s)\vec{j}$
3. $\vec{r}(s,t) = \cos s\,\vec{i} + \sin s\,\vec{j} + t^2\vec{k}$
4. $\vec{r}(s) = \cos s\,\vec{i} + \sin s\,\vec{j} + s^2\vec{k}$

Describe in words the objects parameterized by the equations in Exercises 5–8. (Note: r and θ are cylindrical coordinates.)

5. $x = r\cos\theta \quad y = r\sin\theta \quad z = 7$
   $0 \le r \le 5 \quad 0 \le \theta \le 2\pi$
6. $x = 5\cos\theta \quad y = 5\sin\theta \quad z = z$
   $0 \le \theta \le 2\pi \quad 0 \le z \le 7$
7. $x = 5\cos\theta \quad y = 5\sin\theta \quad z = 5\theta$
   $0 \le \theta \le 2\pi$
8. $x = r\cos\theta \quad y = r\sin\theta \quad z = r$
   $0 \le r \le 5 \quad 0 \le \theta \le 2\pi$

In Exercises 9–12, for a sphere parameterized using the spherical coordinates θ and ϕ, describe in words the part of the sphere given by the restrictions.

9. $0 \le \theta < 2\pi, \quad 0 \le \phi \le \pi/2$
10. $\pi \le \theta < 2\pi, \quad 0 \le \phi \le \pi$
11. $\pi/4 \le \theta < \pi/3, \quad 0 \le \phi \le \pi$
12. $0 \le \theta \le \pi, \quad \pi/4 \le \phi < \pi/3$

PROBLEMS

In Problems 13–16, give parametric equations for the plane through the point with position vector $\vec{r}_0$ and containing the vectors $\vec{v}_1$ and $\vec{v}_2$.

13. $\vec{r}_0 = \vec{i}, \vec{v}_1 = \vec{j}, \vec{v}_2 = \vec{k}$
14. $\vec{r}_0 = \vec{j}, \vec{v}_1 = \vec{k}, \vec{v}_2 = \vec{i}$
15. $\vec{r}_0 = \vec{i} + \vec{j}, \vec{v}_1 = \vec{j} + \vec{k}, \vec{v}_2 = \vec{i} + \vec{k}$
16. $\vec{r}_0 = \vec{i} + \vec{j} + \vec{k}, \vec{v}_1 = \vec{i} - \vec{k}, \vec{v}_2 = -\vec{j} + \vec{k}$

In Problems 17–18, parameterize the plane that contains the three points.

17. $(0, 0, 0), (1, 2, 3), (2, 1, 0)$
18. $(1, 2, 3), (2, 5, 8), (5, 2, 0)$

In Problems 19–20, give two nonparallel vectors and the coordinates of a point in the plane with given parametric equations

19. $x = 2s + 3t, \quad y = s - 5t, \quad z = -s + 2t$
20. $x = 2 + s + t, \quad y = s - t, \quad z = -1 + s + t$

In Problems 21–22, parameterize the plane through the point with the given normal vector.

21. $(3, 5, 7), \vec{i} + \vec{j} + \vec{k}$
22. $(5, 1, 4), \vec{i} + 2\vec{j} + 3\vec{k}$

23. Does the plane $\vec{r}(s,t) = (2 + s)\vec{i} + (3 + s + t)\vec{j} + 4t\vec{k}$ contain the following points?
    (a) $(4, 8, 12)$ **(b)** $(1, 2, 3)$

24. Are the following two planes parallel?

$$x = 2 + s + t, \quad y = 4 + s - t, \quad z = 1 + 2s, \quad \text{and}$$
$$x = 2 + s + 2t, \quad y = t, \quad z = s - t.$$

In Problems 25–28, describe the families of parameter curves with $s = s_0$ and $t = t_0$ for the parameterized surface.

25. $x = s, \quad y = t, \quad z = 1$ for $-\infty < s < \infty, -\infty < t < \infty$

26. $x = s, \quad y = \cos t, \quad z = \sin t$ for $-\infty < s < \infty$, $0 \le t \le 2\pi$

27. $x = s \quad y = t, \quad z = s^2 + t^2$ for $-\infty < s < \infty$, $-\infty < t < \infty$

28. $x = \cos s \sin t, \quad y = \sin s \sin t, \quad z = \cos t$ for $0 \le s \le 2\pi, 0 \le t \le \pi$

29. A city is described parametrically by the equation

$$\vec{r} = (x_0\vec{i} + y_0\vec{j} + z_0\vec{k}) + s\vec{v}_1 + t\vec{v}_2$$

where $\vec{v}_1 = 2\vec{i} - 3\vec{j} + 2\vec{k}$ and $\vec{v}_2 = \vec{i} + 4\vec{j} + 5\vec{k}$. A city block is a rectangle determined by $\vec{v}_1$ and $\vec{v}_2$. East is in the direction of $\vec{v}_1$ and north is in the direction of $\vec{v}_2$. Starting at the point (x_0, y_0, z_0), you walk 5 blocks east, 4 blocks north, 1 block west and 2 blocks south. What are the parameters of the point where you end up? What are your x, y and z coordinates at that point?

30. You are at a point on the earth with longitude 80° West of Greenwich, England, and latitude 40° North of the equator.

(a) If your latitude decreases, have you moved nearer to or farther from the equator?
(b) If your latitude decreases, have you moved nearer to or farther from the north pole?
(c) If your longitude increases (say, to 90° West), have you moved nearer to or farther from Greenwich?

31. Describe in words the curve $\phi = \pi/4$ on the surface of the globe.

32. Describe in words the curve $\theta = \pi/4$ on the surface of the globe.

33. A decorative oak post is 48″ long and is turned on a lathe so that its profile is sinusoidal, as shown in Figure 21.17.

(a) Describe the surface of the post parametrically using cylindrical coordinates.
(b) Find the volume of the post.

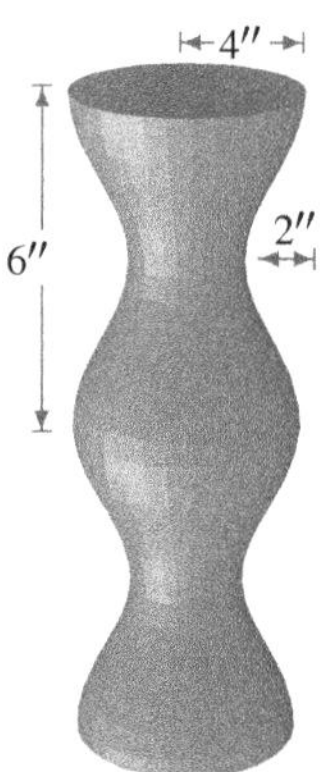

Figure 21.17

34. Find parametric equations for the sphere $(x - a)^2 + (y - b)^2 + (z - c)^2 = d^2$.

35. You are standing at a point on the equator of a sphere parameterized by spherical coordinates θ_0 and ϕ_0. If you go halfway around the equator and halfway up toward the north pole along a longitude, what are your new θ and ϕ coordinates?

36. Find parametric equations for the cone $x^2 + y^2 = z^2$.

37. Parameterize the cone in Example 6 on page 1028 in terms of r and θ.

38. Give a parameterization of the circle of radius a centered at the point (x_0, y_0, z_0) and in the plane parallel to two given unit vectors $\vec{u}$ and $\vec{v}$ such that $\vec{u} \cdot \vec{v} = 0$.

For Problems 39–41,
(a) Write an equation in x, y, z and identify the parametric surface.
(b) Draw a picture of the surface.

39. $x = 2s \quad y = s + t \quad z = 1 + s - t$
$0 \le s \le 1 \quad 0 \le t \le 1$

40. $x = s \quad y = t \quad z = \sqrt{1 - s^2 - t^2}$
$s^2 + t^2 \le 1 \quad s, t \ge 0$

41. $x = s + t \quad y = s - t \quad z = s^2 + t^2$
$0 \le s \le 1 \quad 0 \le t \le 1$

Strengthen Your Understanding

In Problems 42–43, explain what is wrong with the statement.

42. The parameter curves of a parameterized surface intersect at right angles.

43. The parameter curves for constant ϕ on the sphere $\vec{r}(\theta, \phi) = R\sin\phi\cos\theta\vec{i} + R\sin\phi\sin\theta\vec{j} + R\cos\phi\vec{k}$ are circles of radius R.

In Problems 44–46, give an example of:

44. A parameterization $\vec{r}(s, t)$ of the plane tangent to the unit sphere at the point where $\theta = \pi/4$ and $\phi = \pi/4$.

45. An equation of the form $f(x, y, z) = 0$ for the plane

$$\vec{r}(s,t) = (s+1)\vec{i} + (t+2)\vec{j} + (s+t)\vec{k}.$$

46. A parameterized curve on the sphere $\vec{r}(\theta, \phi) = \sin\phi\cos\theta\vec{i} + \sin\phi\sin\theta\vec{j} + \cos\phi\vec{k}$ that is not a parameter curve.

Are the statements in Problems 47–53 true or false? Give reasons for your answer.

47. The equations $x = s+1$, $y = t-2$, $z = 3$ parameterize a plane.
48. The equations $x = 2s-1$, $y = -s+3$, $z = 4+s$ parameterize a plane.
49. If $\vec{r} = \vec{r}(s,t)$ parameterizes the upper hemisphere $x^2+y^2+z^2 = 1$, $z \geq 0$, then $\vec{r} = -\vec{r}(s,t)$ parameterizes the lower hemisphere $x^2+y^2+z^2 = 1$, $z \leq 0$.
50. If $\vec{r} = \vec{r}(s,t)$ parameterizes the upper hemisphere $x^2+y^2+z^2 = 1$, $z \geq 0$, then $\vec{r} = \vec{r}(-s,-t)$ parameterizes the lower hemisphere $x^2+y^2+z^2 = 1$, $z \leq 0$.
51. If $\vec{r}_1(s,t)$ parameterizes a plane then $\vec{r}_2(s,t) = \vec{r}_1(s,t) + 2\vec{i} - 3\vec{j} + \vec{k}$ parameterizes a parallel plane.
52. Every point on a parameterized surface has a parameter curve passing through it.
53. If $s_0 \neq s_1$, then the parameter curves $\vec{r}(s_0,t)$ and $\vec{r}(s_1,t)$ do not intersect.

54. Match the parameterizations (I)–(IV) with the surfaces (a)–(d). In all cases $0 \leq s \leq \pi/2$, $0 \leq t \leq \pi/2$. Note that only part of the surface may be described by the given parameterization.

(a) Cylinder
(b) Plane
(c) Sphere
(d) Cone

I. $x = \cos s$, $y = \sin t$, $z = \cos s + \sin t$
II. $x = \cos s$, $y = \sin s$, $z = \cos t$
III. $x = \sin s\cos t$, $y = \sin s\sin t$, $z = \cos s$
IV. $x = \cos s$, $y = \sin t$, $z = \sqrt{\cos^2 s + \sin^2 t}$

21.2 CHANGE OF COORDINATES IN A MULTIPLE INTEGRAL

In Chapter 16 we used polar, cylindrical, and spherical coordinates to simplify iterated integrals. In this section, we discuss more general changes of coordinate. In the process, we see where the factors r and $\rho^2\sin\phi$ come from when we convert to polar, cylindrical, or spherical coordinates (see pages 865, 871, and 873).

Polar Change of Coordinates Revisited

Consider the integral $\int_R (x+y)\,dA$ where R is the region in the first quadrant bounded by the circle $x^2+y^2 = 16$ and the x and y-axes. Writing the integral in Cartesian and polar coordinates, we have

$$\int_R (x+y)\,dA = \int_0^4 \int_0^{\sqrt{16-x^2}} (x+y)\,dy\,dx = \int_0^{\pi/2} \int_0^4 (r\cos\theta + r\sin\theta) r\,dr\,d\theta.$$

The integral on the right is over the rectangle in the $r\theta$-plane given by $0 \leq r \leq 4$, $0 \leq \theta \leq \pi/2$. The conversion from polar to Cartesian coordinates changes this rectangle into a quarter-disk. Figure 21.18 shows how a typical rectangle (shaded) in the $r\theta$-plane with sides of length Δr and $\Delta\theta$ corresponds to a curved rectangle in the xy-plane with sides of length Δr and $r\Delta\theta$. The extra r is needed because the correspondence between r, θ and x, y not only curves the lines $r = 1, 2, 3\ldots$ into circles, it also stretches those lines around larger and larger circles.

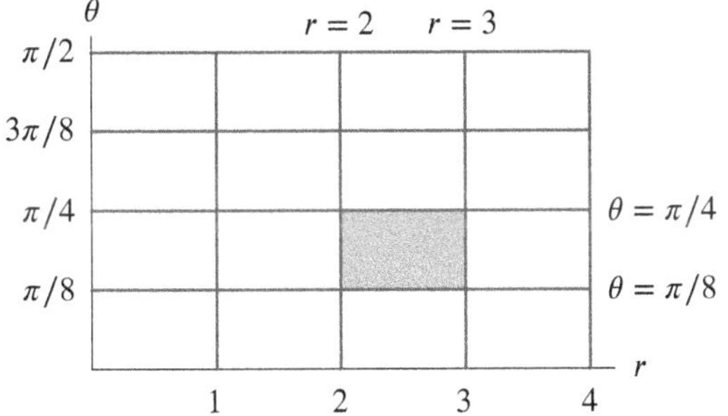

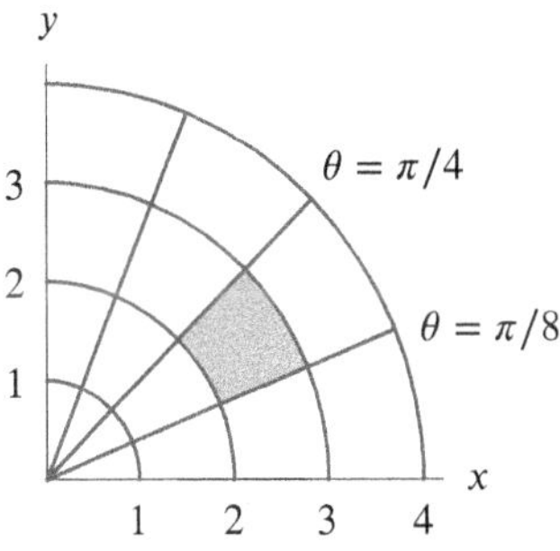

Figure 21.18: A grid in the $r\theta$-plane and the corresponding curved grid in the xy-plane

General Change of Coordinates

We now consider a general change of coordinates, where x, y coordinates are related to s, t coordinates by the differentiable functions

$$x = x(s,t) \quad y = y(s,t).$$

Just as a rectangular region in the $r\theta$-plane corresponds to a region in the xy-plane, a rectangular region, T, in the st-plane corresponds to a region, R, in the xy-plane. We assume that the change of coordinates is one-to-one, that is, that each point in R corresponds to only one point in T.

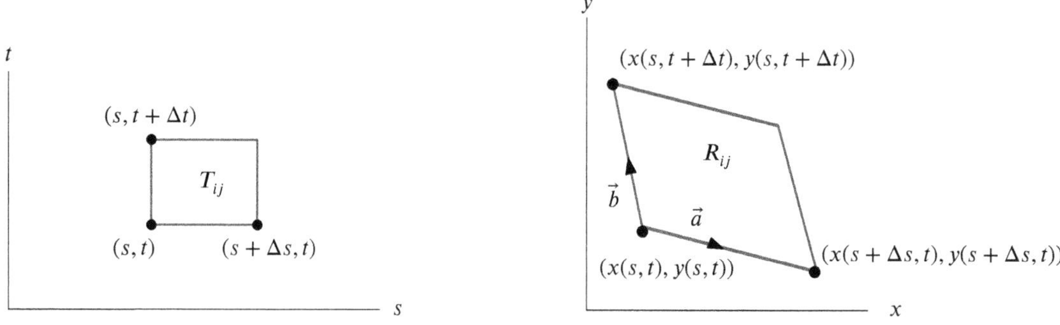

Figure 21.19: A small rectangle T_{ij} in the st-plane and the corresponding region R_{ij} of the xy-plane

We divide T into small rectangles T_{ij} with sides of length Δs and Δt. (See Figure 21.19.) The corresponding piece R_{ij} of the xy-plane is a quadrilateral with curved sides. If we choose Δs and Δt small, then by local linearity of $x(s,t)$ and $y(s,t)$, we know R_{ij} is approximately a parallelogram.

Recall from Chapter 13 that the area of the parallelogram with sides $\vec{a}$ and $\vec{b}$ is $\|\vec{a} \times \vec{b}\|$. Thus, we need to find the sides of R_{ij} as vectors. The side of R_{ij} corresponding to the bottom side of T_{ij} has endpoints $(x(s,t), y(s,t))$ and $(x(s+\Delta s, t), y(s+\Delta s, t))$, so in vector form that side is

$$\vec{a} = (x(s+\Delta s,t) - x(s,t))\vec{i} + (y(s+\Delta s,t) - y(s,t))\vec{j} \approx \left(\frac{\partial x}{\partial s}\Delta s\right)\vec{i} + \left(\frac{\partial y}{\partial s}\Delta s\right)\vec{j}.$$

Similarly, the side of R_{ij} corresponding to the left edge of T_{ij} is given by

$$\vec{b} \approx \left(\frac{\partial x}{\partial t}\Delta t\right)\vec{i} + \left(\frac{\partial y}{\partial t}\Delta t\right)\vec{j}.$$

Computing the cross product, we get

$$\begin{aligned}\text{Area } R_{ij} \approx \|\vec{a} \times \vec{b}\| &\approx \left|\left(\frac{\partial x}{\partial s}\Delta s\right)\left(\frac{\partial y}{\partial t}\Delta t\right) - \left(\frac{\partial x}{\partial t}\Delta t\right)\left(\frac{\partial y}{\partial s}\Delta s\right)\right| \\ &= \left|\frac{\partial x}{\partial s}\cdot\frac{\partial y}{\partial t} - \frac{\partial x}{\partial t}\cdot\frac{\partial y}{\partial s}\right|\Delta s\Delta t.\end{aligned}$$

Using determinant notation,[1] we define the *Jacobian*, $\dfrac{\partial(x,y)}{\partial(s,t)}$, as follows:

$$\frac{\partial(x,y)}{\partial(s,t)} = \frac{\partial x}{\partial s}\cdot\frac{\partial y}{\partial t} - \frac{\partial x}{\partial t}\cdot\frac{\partial y}{\partial s} = \begin{vmatrix} \frac{\partial x}{\partial s} & \frac{\partial x}{\partial t} \\ \frac{\partial y}{\partial s} & \frac{\partial y}{\partial t} \end{vmatrix}.$$

[1]See Appendix E.

Thus, we can write

$$\text{Area } R_{ij} \approx \left|\frac{\partial(x,y)}{\partial(s,t)}\right| \Delta s\, \Delta t.$$

To compute $\int_R f(x,y)\,dA$, where f is a continuous function, we look at the Riemann sum obtained by dividing the region R into the small curved regions R_{ij}, giving

$$\int_R f(x,y)\,dA \approx \sum_{i,j} f(u_{ij}, v_{ij}) \cdot \text{Area of } R_{ij} \approx \sum_{i,j} f(u_{ij}, v_{ij}) \left|\frac{\partial(x,y)}{\partial(s,t)}\right| \Delta s\, \Delta t.$$

Each point (u_{ij}, v_{ij}) in R_{ij} corresponds to a point (s_{ij}, t_{ij}) in T_{ij}, so the sum can be written in terms of s and t:

$$\sum_{i,j} f(x(s_{ij}, t_{ij}), y(s_{ij}, t_{ij})) \left|\frac{\partial(x,y)}{\partial(s,t)}\right| \Delta s\, \Delta t.$$

This is a Riemann sum in terms of s and t, so as Δs and Δt approach 0, we get

$$\int_R f(x,y)\,dA = \int_T f(x(s,t), y(s,t)) \left|\frac{\partial(x,y)}{\partial(s,t)}\right| ds\, dt.$$

> To convert an integral from x, y to s, t coordinates we make three changes:
>
> 1. Substitute for x and y in the integrand in terms of s and t.
> 2. Change the xy region R into an st region T.
> 3. Use the absolute value of the Jacobian to change the area element by making the substitution $dx\,dy = \left|\frac{\partial(x,y)}{\partial(s,t)}\right| ds\, dt$.

Example 1 Check that the Jacobian $\dfrac{\partial(x,y)}{\partial(r,\theta)} = r$ for polar coordinates $x = r\cos\theta,\ y = r\sin\theta$.

Solution We have $$\frac{\partial(x,y)}{\partial(r,\theta)} = \begin{vmatrix} \frac{\partial x}{\partial r} & \frac{\partial x}{\partial \theta} \\ \frac{\partial y}{\partial r} & \frac{\partial y}{\partial \theta} \end{vmatrix} = \begin{vmatrix} \cos\theta & -r\sin\theta \\ \sin\theta & r\cos\theta \end{vmatrix} = r\cos^2\theta + r\sin^2\theta = r.$$

Example 2 Find the area of the ellipse $\dfrac{x^2}{a^2} + \dfrac{y^2}{b^2} = 1$.

Solution Let $x = as,\ y = bt$. Then the ellipse $x^2/a^2 + y^2/b^2 = 1$ in the xy-plane corresponds to the circle $s^2 + t^2 = 1$ in the st-plane. The Jacobian is $\begin{vmatrix} a & 0 \\ 0 & b \end{vmatrix} = ab$. Thus, if R is the ellipse in the xy-plane and T is the unit circle in the st-plane, we get

$$\text{Area of } xy\text{-ellipse} = \int_R 1\,dA = \int_T 1\,ab\,ds\,dt = ab\int_T ds\,dt = ab \cdot \text{ Area of } st\text{-circle} = \pi ab.$$

Change of Coordinates in Triple Integrals

For triple integrals, there is a similar formula. Suppose the differentiable functions

$$x = x(s,t,u), \quad y = y(s,t,u), \quad z = z(s,t,u)$$

define a one-to-one change of coordinates from a region S in stu-space to a region W in xyz-space. Then, the Jacobian of this change of coordinates is given by the determinant[2]

$$\frac{\partial(x,y,z)}{\partial(s,t,u)} = \begin{vmatrix} \frac{\partial x}{\partial s} & \frac{\partial x}{\partial t} & \frac{\partial x}{\partial u} \\ \frac{\partial y}{\partial s} & \frac{\partial y}{\partial t} & \frac{\partial y}{\partial u} \\ \frac{\partial z}{\partial s} & \frac{\partial z}{\partial t} & \frac{\partial z}{\partial u} \end{vmatrix}.$$

Just as the Jacobian in two dimensions gives us the change in the area element, the Jacobian in three dimensions represents the change in the volume element. Thus, we have

$$\int_W f(x,y,z)\,dx\,dy\,dz = \int_S f(x(s,t,u), y(s,t,u), z(s,t,u)) \left|\frac{\partial(x,y,z)}{\partial(s,t,u)}\right| ds\,dt\,du.$$

Problem 11 at the end of this section asks you to check that the Jacobian for the change of coordinates to spherical coordinates is $\rho^2 \sin\phi$. The next example generalizes Example 2 to ellipsoids.

Example 3 Find the volume of the ellipsoid $\frac{x^2}{a^2} + \frac{y^2}{b^2} + \frac{z^2}{c^2} = 1$.

Solution Let $x = as,\ y = bt,\ z = cu$. The Jacobian is computed to be abc. The xyz-ellipsoid corresponds to the stu-sphere $s^2 + t^2 + u^2 = 1$. Thus, as in Example 2,

$$\text{Volume of } xyz\text{-ellipsoid} = abc \cdot \text{Volume of } stu\text{-sphere} = abc\frac{4}{3}\pi = \frac{4}{3}\pi abc.$$

Exercises and Problems for Section 21.2

EXERCISES

In Exercises 1–4, find the absolute value of the Jacobian, $\left|\frac{\partial(x,y)}{\partial(s,t)}\right|$, for the given change of coordinates.

1. $x = 5s + 2t,\ y = 3s + t$
2. $x = s^2 - t^2,\ y = 2st$
3. $x = e^s \cos t,\ y = e^s \sin t$
4. $x = s^3 - 3st^2,\ y = 3s^2t - t^3$

In Exercises 5–7, find positive numbers a and b so that the change of coordinates $s = ax, t = by$ transforms the integral $\int\int_R dx\,dy$ into

$$\int\int_T \left|\frac{\partial(x,y)}{\partial(s,t)}\right| ds\,dt$$

for the given regions R and T.

5. R is the rectangle $0 \le x \le 10$, $0 \le y \le 1$ and T is the square $0 \le s,t \le 1$.
6. R is the rectangle $0 \le x \le 1$, $0 \le y \le 1/4$ and T is the square $0 \le s,t \le 1$.
7. R is the rectangle $0 \le x \le 50$, $0 \le y \le 10$ and T is the square $0 \le s,t \le 1$.

In Exercises 8–9, find a number a so that the change of coordinates $s = x + ay, t = y$ transforms the integral $\int\int_R dx\,dy$ over the parallelogram R in the xy-plane into an integral

$$\int\int_T \left|\frac{\partial(x,y)}{\partial(s,t)}\right| ds\,dt$$

over a rectangle T in the st-plane.

8. R has vertices $(0,0), (10,0), (12,3), (22,3)$
9. R has vertices $(0,0), (10,0), (-15,5), (-5,5)$

[2] See Appendix E.

PROBLEMS

10. Find the region R in the xy-plane corresponding to the region T consisting of points (s,t) with $0 \le s \le 3$, $0 \le t \le 2$ for the change of coordinates $x = 2s-3t$, $y = s-2t$. Check that

$$\int_R dx\,dy = \int_T \left|\frac{\partial(x,y)}{\partial(s,t)}\right| ds\,dt.$$

11. Compute the Jacobian for the change of coordinates into spherical coordinates:

$$x = \rho\sin\phi\cos\theta, \quad y = \rho\sin\phi\sin\theta, \quad z = \rho\cos\phi.$$

12. For the change of coordinates $x = 3s-4t$, $y = 5s+2t$, show that

$$\frac{\partial(x,y)}{\partial(s,t)}\cdot\frac{\partial(s,t)}{\partial(x,y)} = 1$$

13. Use the change of coordinates $x = 2s+t$, $y = s-t$ to compute the integral $\int_R (x+y)\,dA$, where R is the parallelogram formed by $(0,0)$, $(3,-3)$, $(5,-2)$, and $(2,1)$.

14. Use the change of coordinates $s = x+y$, $t = y$ to find the area of the ellipse $x^2+2xy+2y^2 = 1$.

15. Use the change of coordinates $s = y$, $t = y-x^2$ to evaluate $\int\int_R x\,dx\,dy$ over the region R in the first quadrant bounded by $y = 0$, $y = 16$, $y = x^2$, and $y = x^2-9$.

16. If R is the triangle bounded by $x+y = 1$, $x = 0$, and $y = 0$, evaluate the integral $\int_R \cos\left(\frac{x-y}{x+y}\right) dx\,dy$.

17. Two independent random numbers x and y from a normal distribution with mean 0 and standard deviation σ have joint density function $p(x,y) = (1/(2\pi\sigma^2)e^{-(x^2+y^2)/(2\sigma^2)}$. The average $z = (x+y)/2$ has a one-variable probability density function of its own.

(a) Give a double integral expression for $F(t)$, the probability that $z \le t$.

(b) Give a single integral expression for $F(t)$. To do this, make the change of coordinates: $u = (x+y)/2$, $v = (x-y)/2$ and then do the integral on dv. Use the fact that $\int_{-\infty}^{\infty} e^{-x^2/a^2}\,dx = a\sqrt{\pi}$.

(c) Find the probability density function $F'(t)$ of z.

(d) What is the name of the distribution of z?

18. A river follows the path $y = f(x)$ where x, y are in kilometers. Near the sea, it widens into a lagoon, then narrows again at its mouth. See Figure 21.20. At the point (x,y), the depth, $d(x,y)$, of the lagoon is given by

$$d(x,y) = 40 - 160(y-f(x))^2 - 40x^2 \text{ meters.}$$

The lagoon itself is described by $d(x,y) \ge 0$. What is the volume of the lagoon in cubic meters? [Hint: Use new coordinates $u = x/2$, $v = y-f(x)$ and Jacobians.]

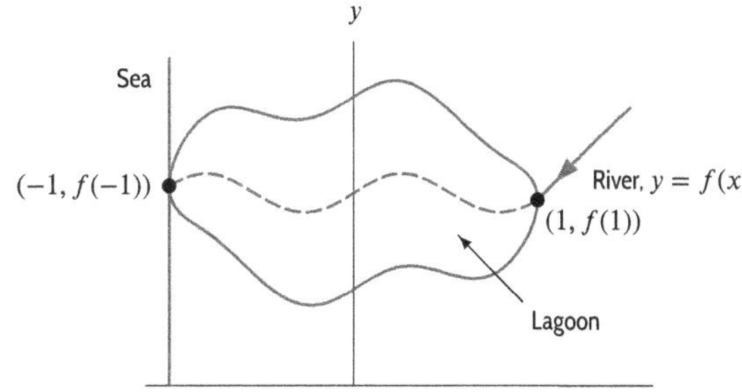

Figure 21.20

Strengthen Your Understanding

In Problems 19–20, explain what is wrong with the statement.

19. If R is the region $0 \le x \le 1$, $0 \le y \le 4$ and T is the region $0 \le s \le 1$, $-2 \le t \le 2$, using the formulas $x = s$, $y = t^2$, we have

$$\int_R f(x,y)\,dx\,dy = \int_T f(s,t^2)\left|\frac{\partial(x,y)}{\partial(s,t)}\right| ds\,dt.$$

20. If R and T are corresponding regions of the xy- and st-planes, the change of coordinates $x = t^3$, $y = s$ leads to the formula

$$\int_R (x+2y)\,dx\,dy = \int_T \left(t^3+2s\right)\left(-3t^2\right) ds\,dt.$$

In Problems 21–22, give an example of:

21. A change of coordinates $x = x(s,t)$, $y = y(s,t)$ where the rectangle $0 \le s \le 1$, $0 \le t \le 1$ in the st-plane corresponds to a different rectangle in the xy-plane.

22. A change of coordinates $x = x(s,t)$, $y = y(s,t)$ where every region in the st-plane corresponds to a region in the xy-plane with twice the area.

In Problems 23–24, consider a change of variable in the integral $\int_R f(x,y)\,dA$ from x, y to s, t. Are the following statements true or false?

23. If the Jacobian $\left|\frac{\partial(x,y)}{\partial(s,t)}\right| > 1$, the value of the s,t-integral is greater than the original x,y-integral.

24. The Jacobian cannot be negative.

21.3 FLUX INTEGRALS OVER PARAMETERIZED SURFACES

Most of the flux integrals we are likely to encounter can be computed using the methods of Sections 19.1 and 19.2. In this section, we briefly consider the general case: how to compute the flux of a smooth vector field $\vec{F}$ through a smooth oriented surface, S, parameterized by

$$\vec{r} = \vec{r}(s, t),$$

for (s, t) in some region R of the parameter space. The method is similar to the one used for graphs in Section 19.2. We consider a parameter rectangle on the surface S corresponding to a rectangular region with sides Δs and Δt in the parameter space. (See Figure 21.21.)

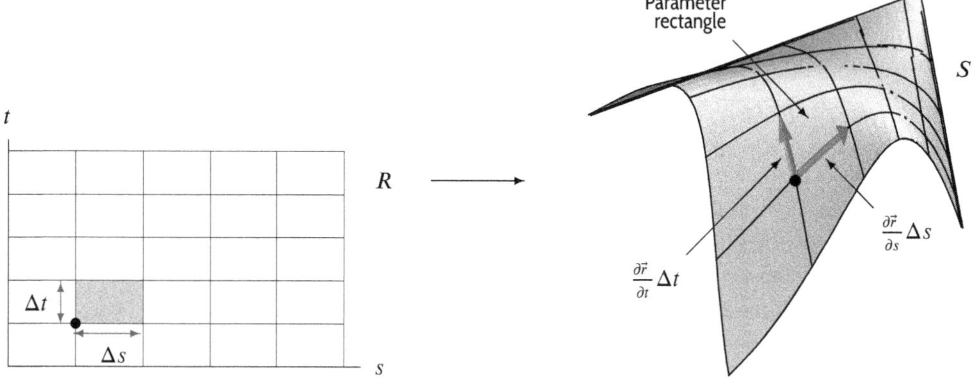

Figure 21.21: Parameter rectangle on the surface S corresponding to a small rectangular region in the parameter space, R

If Δs and Δt are small, the area vector, $\Delta \vec{A}$, of the patch is approximately the area vector of the parallelogram defined by the vectors

$$\vec{r}(s + \Delta s, t) - \vec{r}(s, t) \approx \frac{\partial \vec{r}}{\partial s}\Delta s, \qquad \text{and} \qquad \vec{r}(s, t + \Delta t) - \vec{r}(s, t) \approx \frac{\partial \vec{r}}{\partial t}\Delta t.$$

Thus,

$$\Delta \vec{A} \approx \frac{\partial \vec{r}}{\partial s} \times \frac{\partial \vec{r}}{\partial t}\Delta s\,\Delta t.$$

We assume that the vector $\partial \vec{r}/\partial s \times \partial \vec{r}/\partial t$ is never zero and points in the direction of the unit normal orientation vector $\vec{n}$. If the vector $\partial \vec{r}/\partial s \times \partial \vec{r}/\partial t$ points in the opposite direction to $\vec{n}$, we reverse the order of the cross product. Replacing $\Delta \vec{A}$, Δs, and Δt by $d\vec{A}$, ds, and dt, we write

$$d\vec{A} = \left(\frac{\partial \vec{r}}{\partial s} \times \frac{\partial \vec{r}}{\partial t}\right) ds\,dt.$$

The Flux of a Vector Field Through a Parameterized Surface

The flux of a smooth vector field $\vec{F}$ through a smooth oriented surface S parameterized by $\vec{r} = \vec{r}(s, t)$, where (s, t) varies in a parameter region R, is given by

$$\int_S \vec{F} \cdot d\vec{A} = \int_R \vec{F}(\vec{r}(s, t)) \cdot \left(\frac{\partial \vec{r}}{\partial s} \times \frac{\partial \vec{r}}{\partial t}\right) ds\,dt.$$

We choose the parameterization so that $\partial \vec{r}/\partial s \times \partial \vec{r}/\partial t$ is never zero and points in the direction of $\vec{n}$ everywhere.

Example 1 Find the flux of the vector field $\vec{F} = x\vec{i} + y\vec{j}$ through the surface S, oriented downward and given by

$$x = 2s, \quad y = s + t, \quad z = 1 + s - t, \qquad \text{where } 0 \le s \le 1, \quad 0 \le t \le 1.$$

Solution Since S is parameterized by

$$\vec{r}(s,t) = 2s\vec{i} + (s+t)\vec{j} + (1+s-t)\vec{k},$$

we have

$$\frac{\partial \vec{r}}{\partial s} = 2\vec{i} + \vec{j} + \vec{k} \quad \text{and} \quad \frac{\partial r}{\partial t} = \vec{j} - \vec{k},$$

so

$$\frac{\partial \vec{r}}{\partial s} \times \frac{\partial \vec{r}}{\partial t} = \begin{vmatrix} \vec{i} & \vec{j} & \vec{k} \\ 2 & 1 & 1 \\ 0 & 1 & -1 \end{vmatrix} = -2\vec{i} + 2\vec{j} + 2\vec{k}.$$

Since the vector $-2\vec{i} + 2\vec{j} + 2\vec{k}$ points upward, we use $2\vec{i} - 2\vec{j} - 2\vec{k}$ for downward orientation. Thus, the flux integral is given by

$$\begin{aligned}\int_S \vec{F} \cdot d\vec{A} &= \int_0^1 \int_0^1 (2s\vec{i} + (s+t)\vec{j}) \cdot (2\vec{i} - 2\vec{j} - 2\vec{k})\, ds\, dt \\ &= \int_0^1 \int_0^1 (4s - 2s - 2t)\, ds dt = \int_0^1 \int_0^1 (2s - 2t)\, ds\, dt \\ &= \int_0^1 \left(s^2 - 2st \Big|_{s=0}^{s=1} \right) dt = \int_0^1 (1 - 2t)\, dt = t - t^2 \Big|_0^1 = 0.\end{aligned}$$

Area of a Parameterized Surface

The area ΔA of a small parameter rectangle is the magnitude of its area vector $\Delta \vec{A}$. Therefore,

$$\text{Area of } S = \sum \Delta A = \sum \|\Delta \vec{A}\| \approx \sum \left\| \frac{\partial \vec{r}}{\partial s} \times \frac{\partial \vec{r}}{\partial t} \right\| \Delta s\, \Delta t.$$

Taking the limit as the area of the parameter rectangles tends to zero, we are led to the following expression for the area of S.

The Area of a Parameterized Surface

The area of a surface S which is parameterized by $\vec{r} = \vec{r}(s,t)$, where (s,t) varies in a parameter region R, is given by

$$\int_S dA = \int_R \left\| \frac{\partial \vec{r}}{\partial s} \times \frac{\partial \vec{r}}{\partial t} \right\| ds\, dt.$$

Example 2 Compute the surface area of a sphere of radius a.

Solution We take the sphere S of radius a centered at the origin and parameterize it with the spherical coordinates ϕ and θ. The parameterization is

$$x = a\sin\phi\cos\theta, \quad y = a\sin\phi\sin\theta, \quad z = a\cos\phi, \qquad \text{for} \quad 0 \le \theta \le 2\pi, \quad 0 \le \phi \le \pi.$$

We compute

$$\begin{aligned}\frac{\partial\vec{r}}{\partial\phi}\times\frac{\partial\vec{r}}{\partial\theta} &= (a\cos\phi\cos\theta\vec{i} + a\cos\phi\sin\theta\vec{j} - a\sin\phi\vec{k})\times(-a\sin\phi\sin\theta\vec{i} + a\sin\phi\cos\theta\vec{j})\\ &= a^2(\sin^2\phi\cos\theta\vec{i} + \sin^2\phi\sin\theta\vec{j} + \sin\phi\cos\phi\vec{k})\end{aligned}$$

and so

$$\left\|\frac{\partial\vec{r}}{\partial\phi}\times\frac{\partial\vec{r}}{\partial\theta}\right\| = a^2\sin\phi.$$

Thus, we see that the surface area of the sphere S is given by

$$\text{Surface area} = \int_S dA = \int_R \left\|\frac{\partial\vec{r}}{\partial\phi}\times\frac{\partial\vec{r}}{\partial\theta}\right\| d\phi d\theta = \int_{\phi=0}^{\pi}\int_{\theta=0}^{2\pi} a^2\sin\phi\, d\theta\, d\phi = 4\pi a^2.$$

Exercises and Problems for Section 21.3

EXERCISES

In Exercises 1–4 compute $d\vec{A}$ for the given parameterization for one of the two orientations.

1. $x = s+t, \quad y = s-t, \quad z = st$
2. $x = \sin t, \quad y = \cos t, \quad z = s+t$
3. $x = e^s, \quad y = \cos t, \quad z = \sin t$
4. $x = 0, \quad y = u+v, \quad z = u-v$

In Exercises 5–9 compute the flux of the vector field $\vec{F}$ through the parameterized surface S.

5. $\vec{F} = z\vec{k}$ and S is oriented upward and given, for $0 \le s \le 1, \ 0 \le t \le 1$, by
$$x = s+t, \quad y = s-t, \quad z = s^2+t^2.$$
6. $\vec{F} = x\vec{i} + y\vec{j}$ and S is oriented downward and given, for $0 \le s \le 1, 0 \le t \le 1$, by
$$x = 2s, \quad y = s+t, \quad z = 1+s-t.$$
7. $\vec{F} = x\vec{i}$ through the surface S oriented downward and parameterized for $0 \le s \le 4, 0 \le t \le \pi/6$ by
$$x = e^s, \quad y = \cos(3t), \quad z = 6s.$$
8. $\vec{F} = y\vec{i} + x\vec{j}$ and S is oriented away from the z-axis and given, for $0 \le s \le \pi, \ 0 \le t \le 1$, by
$$x = 3\sin s, \quad y = 3\cos s, \quad z = t+1.$$
9. $\vec{F} = x^2y^2z\vec{k}$ and S is the cone $\sqrt{x^2+y^2} = z$, with $0 \le z \le R$, oriented downward. Parameterize the cone using cylindrical coordinates. (See Figure 21.22.)

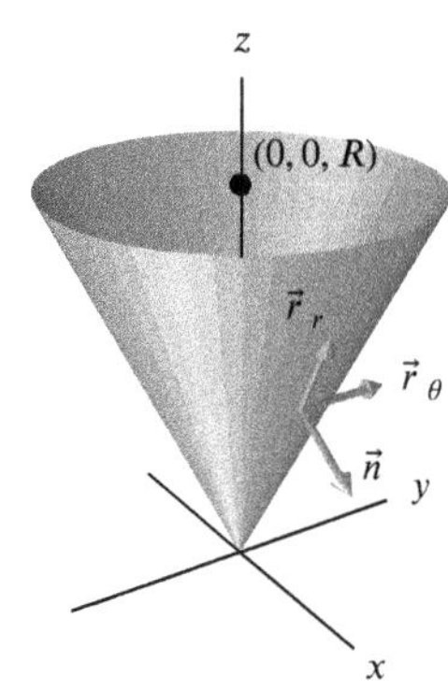

Figure 21.22

In Exercises 10–11, find the surface area.

10. A cylinder of radius a and length L.
11. The region S in the plane $z = 3x+2y$ such that $0 \le x \le 10$ and $0 \le y \le 20$.

PROBLEMS

12. Compute the flux of the vector field $\vec{F} = (x+z)\vec{i} + \vec{j} + z\vec{k}$ through the surface S given by $y = x^2 + z^2$, $1/4 \le x^2 + z^2 \le 1$ oriented away from the y-axis.

13. Find the area of the ellipse S on the plane $2x+y+z = 2$ cut out by the circular cylinder $x^2 + y^2 = 2x$. (See Figure 21.23.)

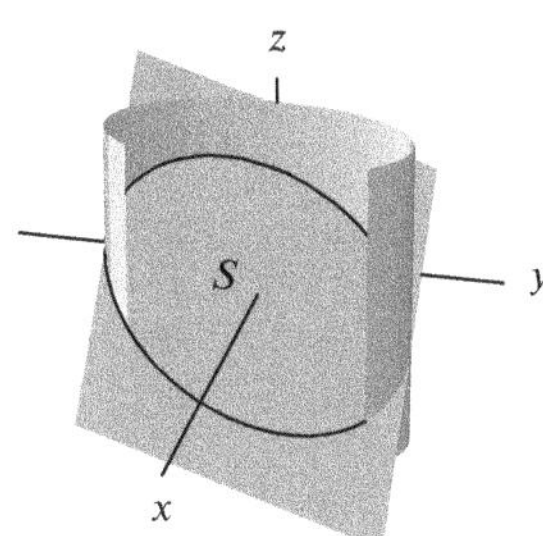

Figure 21.23

14. Consider the surface S formed by rotating the graph of $y = f(x)$ around the x-axis between $x = a$ and $x = b$. Assume that $f(x) \ge 0$ for $a \le x \le b$. Show that the surface area of S is $2\pi \int_a^b f(x)\sqrt{1 + f'(x)^2}\, dx$.

15. A rectangular channel of width w and depth h meters lies in the $\vec{j}$ direction. At a point d_1 meters from one side and d_2 meters from the other side, the velocity vector of fluid in the channel is $\vec{v} = kd_1d_2\vec{j}$ meters/sec. Find the flux through a rectangle stretching the full width and depth of the channel, and perpendicular to the flow.

16. The base of a cone is the unit circle centered at the origin in the xy-plane and vertex $P = (a,b,c)$, where $c > 0$.
    (a) Parameterize the cone.
    (b) Express the surface area of the cone as an integral.
    (c) Use a numerical method to find the surface area of the cone with vertex $P = (2,0,1)$.

As we remarked in Section 19.1, the limit defining a flux integral might not exist if we subdivide the surface in the wrong way. One way to get around this is to take the formula for a flux integral over a parameterized surface that we have developed in this section and to use it as the *definition* of the flux integral. In Problems 17–20 we explore how this works.

17. Use a parameterization to verify the formula for a flux integral over a surface graph on page 974.

18. Use a parameterization to verify the formula for a flux integral over a cylindrical surface on page 976.

19. Use a parameterization to verify the formula for a flux integral over a spherical surface on page 977.

20. One problem with defining the flux integral using a parameterization is that the integral appears to depend on the choice of parameterization. However, the flux through a surface ought not to depend on how the surface is parameterized. Suppose that the surface S has two parameterizations, $\vec{r} = \vec{r}(s,t)$ for (s,t) in the region R of st-space, and also $\vec{r} = r(u,v)$ for (u,v) in the region T in uv-space, and suppose that the two parameterizations are related by the change of coordinates
$$u = u(s,t) \quad v = v(s,t).$$
Suppose that the Jacobian determinant $\partial(u,v)/\partial(s,t)$ is positive at every point (s,t) in R. Use the change of coordinates formula for double integrals on page 1035 to show that computing the flux integral using either parameterization gives the same result.

Strengthen Your Understanding

In Problems 21–22, explain what is wrong with the statement.

21. The area of the surface parameterized by $x = s, y = t, z = f(s,t)$ above the square $0 \le x \le 1, 0 \le y \le 1$ is given by the integral
$$\text{Area} = \int_0^1 \int_0^1 f(s,t)\, ds\, dt.$$

22. The surface S parameterized by $x = f(s,t)$, $y = g(s,t)$, $z = h(s,t)$, where $0 \le s \le 2, 0 \le t \le 3$, has area 6.

In Problems 23–24, give an example of:

23. A parameterization $\vec{r} = \vec{r}(s,t)$ of the xy-plane such that $dA = 2\, ds\, dt$.

24. A vector field $\vec{F}$ such that $\int_S \vec{F} \cdot d\vec{A} > 0$, where S is the surface $\vec{r} = (s-t)\vec{i} + t^2\vec{j} + (s+t)\vec{k}$, $0 \le s \le 1$, $0 \le t \le 1$, oriented in the direction of $\dfrac{\partial \vec{r}}{\partial s} \times \dfrac{\partial \vec{r}}{\partial t}$.

Are the statements in Problems 25–27 true or false? Give reasons for your answer.

25. If $\vec{r}(s,t), 0 \le s \le 1, 0 \le t \le 1$ is an oriented parameterized surface S, and $\vec{F}$ is a vector field that is everywhere tangent to S, then the flux of $\vec{F}$ through S is zero.

26. For any parameterization of the surface $x^2 - y^2 + z^2 = 6$, $d\vec{A}$ at $(1,2,3)$ is a multiple of $(2\vec{i} - 4\vec{j} + 6\vec{k})dx\,dy$.

27. If you parameterize the plane $3x + 4y + 5z = 7$, then there is a constant c such that, at any point (x,y,z), $d\vec{A} = c(3\vec{i} + 4\vec{j} + 5\vec{k})dx\,dy$.

28. Let S be the hemisphere $x^2 + y^2 + z^2 = 1$ with $x \leq 0$, oriented away from the origin. Which of the following integrals represents the flux of $\vec{F}(x, y, z)$ through S?

(a) $\int_R \vec{F}(x, y, z(x, y)) \cdot \frac{\partial \vec{r}}{\partial x} \times \frac{\partial \vec{r}}{\partial y} \, dx \, dy$

(b) $\int_R \vec{F}(x, y, z(x, y)) \cdot \frac{\partial \vec{r}}{\partial y} \times \frac{\partial \vec{r}}{\partial x} \, dy \, dx$

(c) $\int_R \vec{F}(x, y(x, z), z) \cdot \frac{\partial \vec{r}}{\partial x} \times \frac{\partial \vec{r}}{\partial z} \, dx \, dz$

(d) $\int_R \vec{F}(x, y(x, z), z) \cdot \frac{\partial \vec{r}}{\partial z} \times \frac{\partial \vec{r}}{\partial x} \, dz \, dx$

(e) $\int_R \vec{F}(x(y, z), y, z) \cdot \frac{\partial \vec{r}}{\partial y} \times \frac{\partial \vec{r}}{\partial z} \, dy \, dz$

(f) $\int_R \vec{F}(x(y, z), y, z) \cdot \frac{\partial \vec{r}}{\partial z} \times \frac{\partial \vec{r}}{\partial y} \, dz \, dy$

Online Resource: Review problems and Projects

Ready Reference

READY REFERENCE

A Library of Functions

Linear functions (p. 4) have the form $y = f(x) = b + mx$, where m is the **slope**, or rate of change of y with respect to x (p. 5) and b is the **vertical intercept**, or value of y when x is zero (p. 5). The slope is

$$m = \frac{\text{Rise}}{\text{Run}} = \frac{\Delta y}{\Delta x} = \frac{f(x_2) - f(x_1)}{x_2 - x_1} \quad \text{(p. 5).}$$

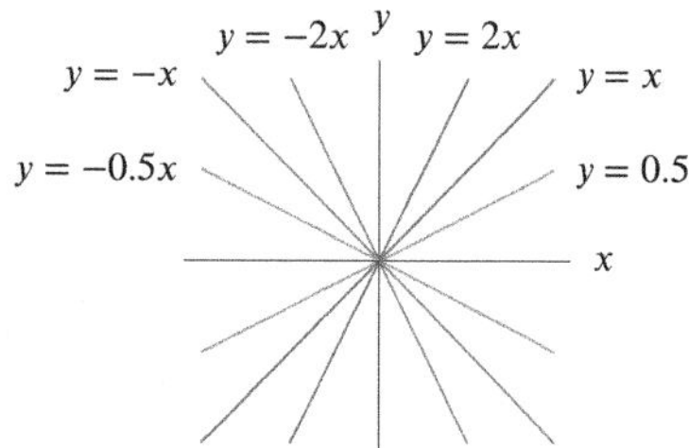

Figure R.1: The family $y = mx$ (with $b = 0$) (p. 6)

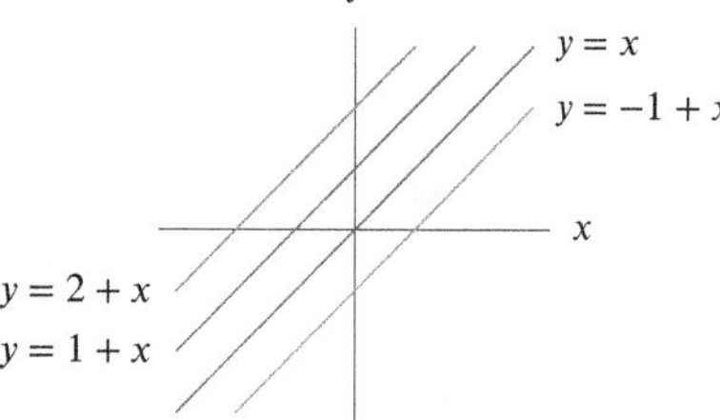

Figure R.2: The family $y = b + x$ (with $m = 1$) (p. 6)

Exponential functions have the form $P = P_0 a^t$ (p. 15) or $P = P_0 e^{kt}$ (p. 17), where P_0 is the initial quantity (p. 15), a is the growth (decay) factor per unit time (p. 15), $|k|$ is the continuous growth (decay) rate (pp. 17, 587), and $r = |a-1|$ is the growth (decay) rate per unit time (p. 15).

Suppose $P_0 > 0$. If $a > 1$ or $k > 0$, we have **exponential growth**; if $0 < a < 1$ or $k < 0$, we have **exponential decay** (p. 16). The **doubling time** (for growth) is the time required for P to double (p. 16). The **half-life** (for decay) is the time required for P to be reduced by a factor of one half (p. 16). The continuous growth rate $k = \ln(1 + r)$ is slightly less than, but very close to, r, provided r is small (p. 588).

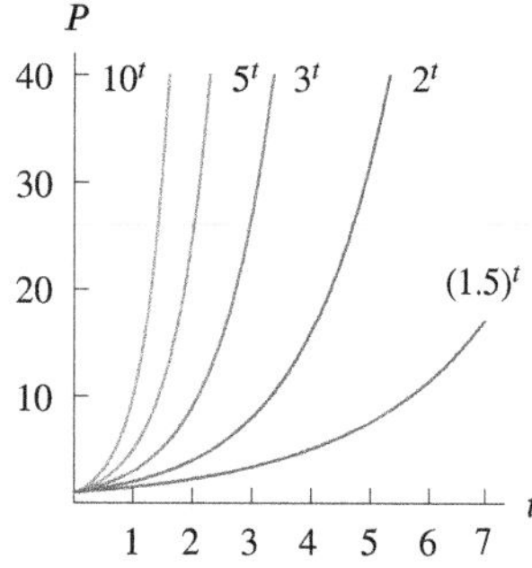

Figure R.3: Exponential growth: $P = a^t$, for $a > 1$ (p. 17)

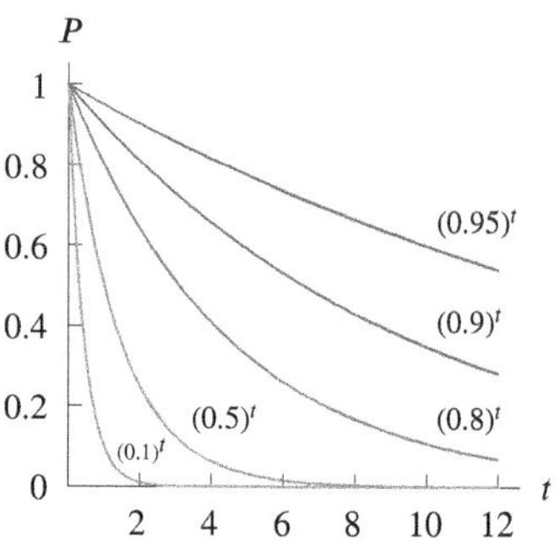

Figure R.4: Exponential decay: $P = a^t$, for $0 < a < 1$ (p. 17)

Common Logarithm and Natural Logarithm

$\log_{10} x = \log x =$ power of 10 that gives x (p. 32)
$\log_{10} x = c$ means $10^c = x$ (p. 32)
$\ln x =$ power of e that gives x (p. 32)
$\ln x = c$ means $e^c = x$ (p. 32)
$\log x$ and $\ln x$ are not defined if x is negative or 0 (p. 32).

Properties of Logarithms (p. 33)

1. $\log(AB) = \log A + \log B$
2. $\log\left(\frac{A}{B}\right) = \log A - \log B$
3. $\log(A^p) = p \log A$
4. $\log(10^x) = x$
5. $10^{\log x} = x$
6. $\log 1 = 0$

The natural logarithm satisfies properties 1, 2, and 3, and $\ln e^x = x$, $e^{\ln x} = x$, $\ln 1 = 0$ (p. 33).

Trigonometric Functions The sine and cosine are defined in Figure R.5 (see also p. 40). The tangent is $\tan t = \frac{\sin t}{\cos t} =$ slope of the line through the origin (0, 0) and P if $\cos t \neq 0$ (p. 43). The period of sin and cos is 2π (p. 40), the period of tan is π (p. 44).

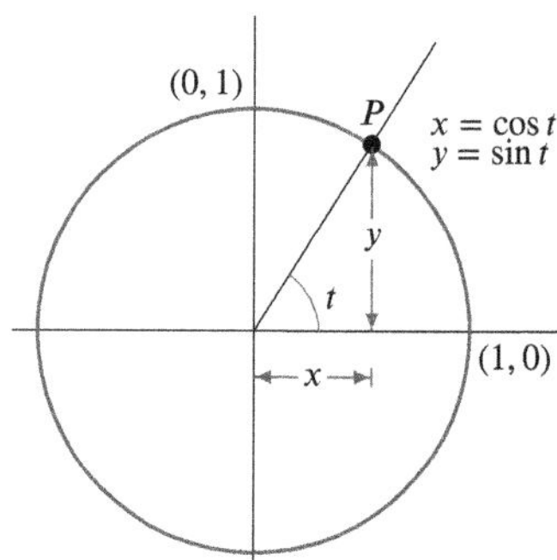

Figure R.5: The definitions of $\sin t$ and $\cos t$ (p. 40)

A **sinusoidal function** (p. 41) has the form
$y = C + A\sin(B(t + h))$ or $y = C + A\cos(B(t + h))$.

The **amplitude** is $|A|$, (p. 41), and the **period** is $2\pi/|B|$ (p. 41).

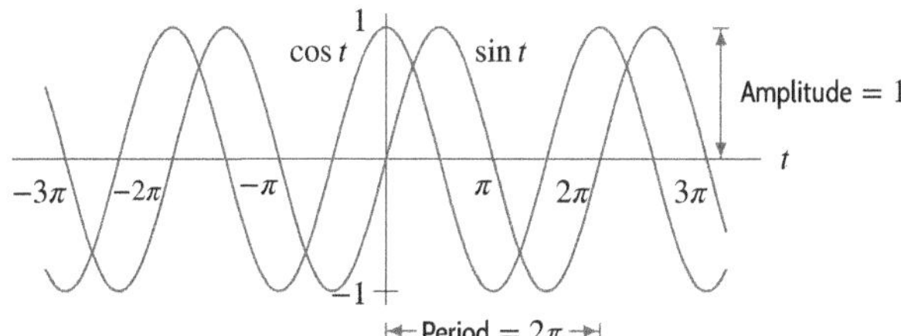

Figure R.6: Graphs of $\cos t$ and $\sin t$ (p. 41)

Trigonometric Identities

$$\sin^2 x + \cos^2 x = 1$$
$$\sin(2x) = 2\sin x \cos x$$
$$\cos(2x) = \cos^2 x - \sin^2 x = 2\cos^2 x - 1 = 1 - 2\sin^2 x$$
$$\cos(a+b) = \cos a \cos b - \sin a \sin b$$
$$\sin(a+b) = \sin a \cos b + \cos a \sin b$$

Inverse Trigonometric Functions: $\arcsin y = x$ means $\sin x = y$ with $-(\pi/2) \leq x \leq (\pi/2)$ (p. 44), $\arccos y = x$ means $\cos x = y$ with $0 \leq x \leq \pi$ (Problem 74, p. 49), $\arctan y = x$ means $\tan x = y$ with $-(\pi/2) < x < (\pi/2)$ (p. 45). The domain of arcsin and arccos is $[-1, 1]$ (p. 44), the domain of arctan is all numbers (p. 45). **Power Functions** have the form $f(x) = kx^p$ (p. 49). Graphs for positive powers:

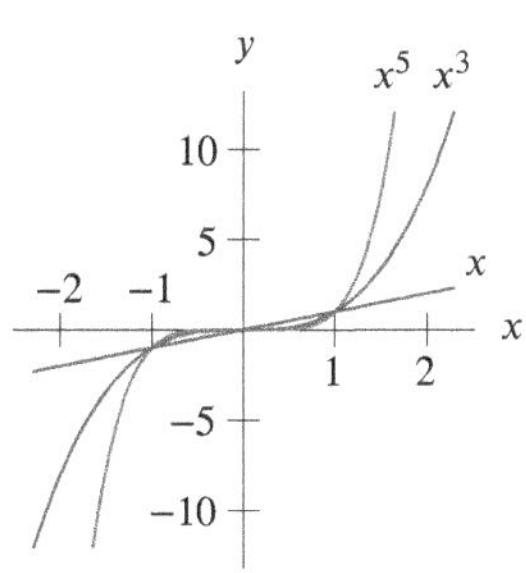

Figure R.7: Odd integer powers of x: "Seat" shaped for $k > 1$ (p. 50)

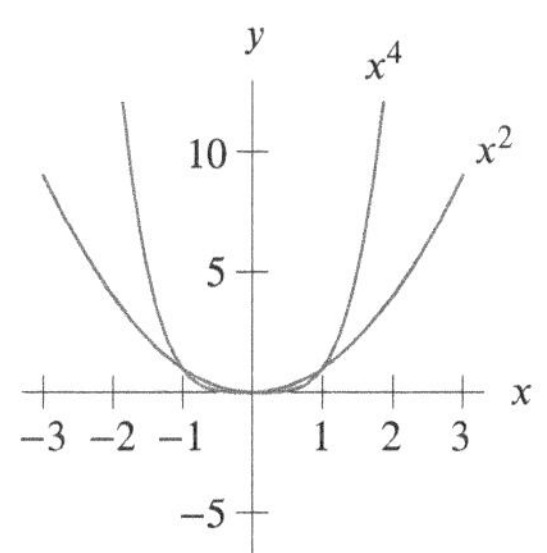

Figure R.8: Even integer powers of x: ⋃-shaped (p. 50)

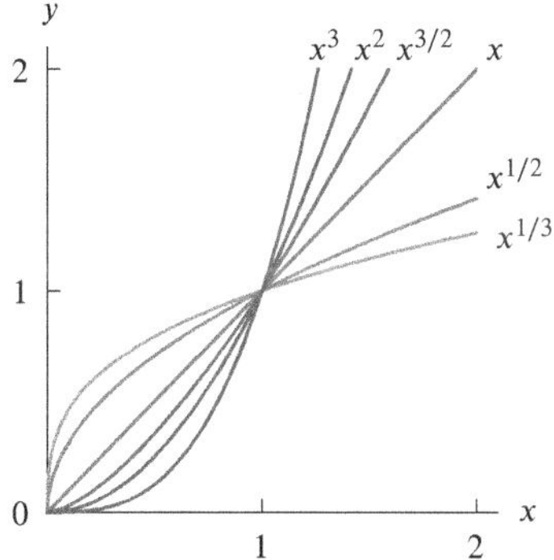

Figure R.9: Comparison of some fractional powers of x

Graphs for zero and negative powers:

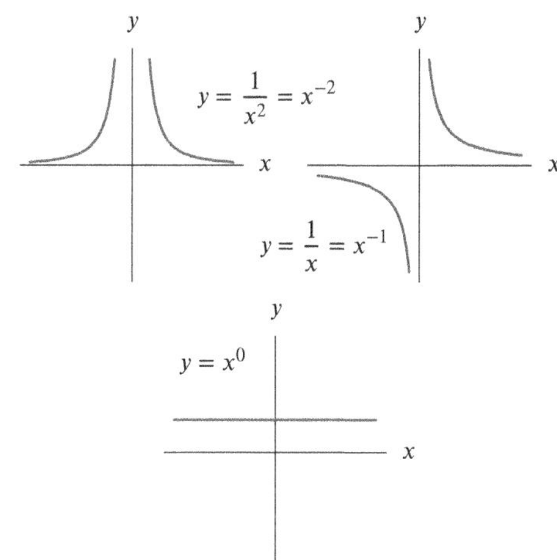

Figure R.10: Comparison of zero and negative powers of x

Polynomials have the form

$$f(x) = a_n x^n + a_{n-1}x^{n-1} + \cdots + a_1 x + a_0, \ a_n \neq 0 \ \text{(p. 51)}.$$

The **degree** is n (p. 51) and the **leading coefficient** is a_n (p. 51).

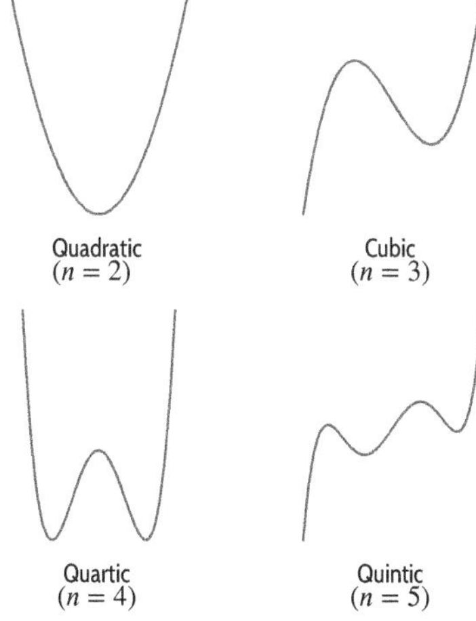

Figure R.11: Graphs of typical polynomials of degree n (p. 51)

Rational Functions have the form $f(x) = \dfrac{p(x)}{q(x)}$, where p and q are polynomials (p. 53). There is usually a vertical asymptote at $x = a$ if $q(a) = 0$ and a horizontal asymptote at $y = L$ if $\lim_{x\to\infty} f(x) = L$ or $\lim_{x\to-\infty} f(x) = L$ (p. 53).

Hyperbolic Functions:

$$\cosh x = \frac{e^x + e^{-x}}{2} \qquad \sinh x = \frac{e^x - e^{-x}}{2} \qquad \text{(p. 174)}.$$

Relative Growth Rates of Functions

Power Functions As $x \to \infty$, higher powers of x dominate, as $x \to 0$, smaller powers dominate.

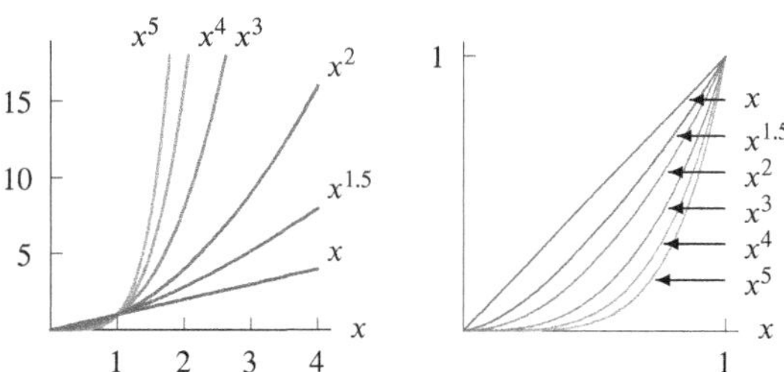

Figure R.12: For large x: Large powers of x dominate

Figure R.13: For $0 \le x \le 1$: Small powers of x dominate

Power Functions Versus Exponential Functions Every exponential growth function eventually dominates every power function (p. 50).

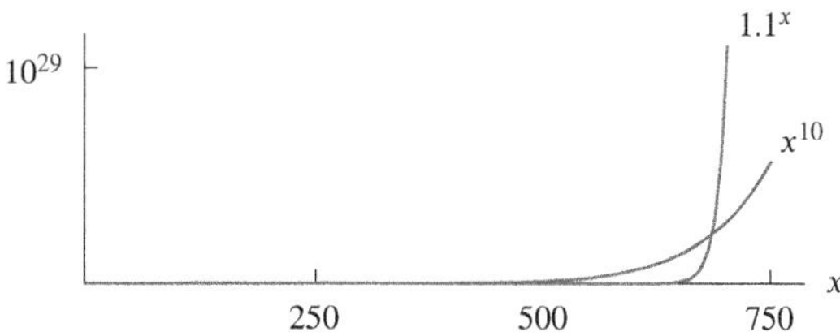

Figure R.14: Exponential function eventually dominates power function

Power Functions Versus Logarithm Functions The power function x^p dominates $A \log x$ for large x for all values of $p > 0$ and $A > 0$.

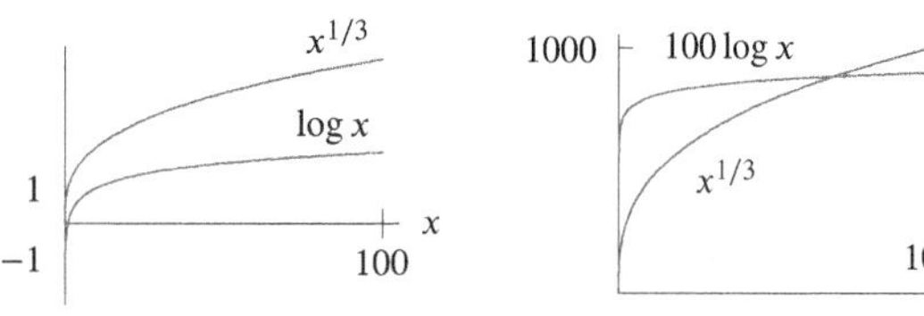

Figure R.15: Comparison of $x^{1/3}$ and $\log x$

Figure R.16: Comparison of $x^{1/3}$ and $100 \log x$

Numerical comparisons of growth rates:

Table R.1 *Comparison of $x^{0.001}$ and $1000 \log x$*

x	$x^{0.001}$	$1000 \log x$
10^{5000}	10^5	$5 \cdot 10^6$
10^{6000}	10^6	$6 \cdot 10^6$
10^{7000}	10^7	$7 \cdot 10^6$

Table R.2 *Comparison of x^{100} and 1.01^x*

x	x^{100}	1.01^x
10^4	10^{400}	$1.6 \cdot 10^{43}$
10^5	10^{500}	$1.4 \cdot 10^{432}$
10^6	10^{600}	$2.4 \cdot 10^{4321}$

Operations on Functions

Shifts, Stretches, and Composition Multiplying by a constant, c, stretches (if $c > 1$) or shrinks (if $0 < c < 1$) the graph vertically. A negative sign (if $c < 0$) reflects the graph about the x-axis, in addition to shrinking or stretching (p. 24). Replacing y by $(y - k)$ moves a graph up by k (down if k is negative) (p. 24). Replacing x by $(x - h)$ moves a graph to the right by h (to the left if h is negative) (p. 24). The composite of f and g is the function $f(g(x))$; f is the outside function, g the inside function (p. 24).

Symmetry We say f is an **even** function if $f(-x) = f(x)$ (p. 25) and f is an **odd** function if $f(-x) = -f(x)$ (p. 25).

Inverse Functions A function f has an inverse if (and only if) its graph intersects any horizontal line at most once (p. 27). If f has an inverse, it is written f^{-1}, and $f^{-1}(x) = y$ means $f(y) = x$ (p. 27). Provided the x and y scales are equal, the graph of f^{-1} is the reflection of the graph of f about the line $y = x$ (p. 28).

Proportionality We say y is proportional to x if $y = kx$ for k a nonzero constant. We say y is inversely proportional to x if $y = k(1/x)$ (p. 6).

Limits and Continuity

Idea of Limit (p. 60) If there is a number L such that $f(x)$ is as close to L as we please whenever x is sufficiently close to c (but $x \neq c$), then $\lim_{x \to c} f(x) = L$.

Definition of Limit()Section 1.10 online) If there is a number L such that for any $\epsilon > 0$, there exists a $\delta > 0$ such that if $|x - c| < \delta$ and $x \neq c$, then $|f(x) - L| < \epsilon$, then $\lim_{x \to c} f(x) = L$.

One-sided Limits (p. 67) If $f(x)$ approaches L as x approaches c through values greater than c, then $\lim_{x \to c^+} f(x) = L$. If $f(x)$ approaches L as x approaches c through values less than c, then $\lim_{x \to c^-} f(x) = L$.

Limits at Infinity (p. 68) If $f(x)$ gets as close to L as we please when x gets sufficiently large, then $\lim_{x \to \infty} f(x) = L$. Similarly, if $f(x)$ approaches L as x gets more and more negative, then $\lim_{x \to -\infty} f(x) = L$.

Limits of Continuous Functions (p. 62) If a function $f(x)$ is continuous at $x = c$, the limit is the value of $f(x)$ there: $\lim_{x \to c} f(x) = f(c)$. Thus, to find limits for a continuous function: Substitute c.

Theorem: Properties of Limits (p. 71) Assuming all the limits on the right-hand side exist:

1. If b is a constant, then $\lim_{x \to c} (bf(x)) = b \left(\lim_{x \to c} f(x) \right)$.
2. $\lim_{x \to c} (f(x) + g(x)) = \lim_{x \to c} f(x) + \lim_{x \to c} g(x)$.
3. $\lim_{x \to c} (f(x)g(x)) = \left(\lim_{x \to c} f(x) \right) \left(\lim_{x \to c} g(x) \right)$.
4. $\lim_{x \to c} \dfrac{f(x)}{g(x)} = \dfrac{\lim_{x \to c} f(x)}{\lim_{x \to c} g(x)}$, provided $\lim_{x \to c} g(x) \neq 0$.
5. For any constant k, $\lim_{x \to c} k = k$.
6. $\lim_{x \to c} x = c$.

Idea of Continuity (p. 58) A function is **continuous on an interval** if its graph has no breaks, jumps, or holes in that interval.

Definition of Continuity (p. 60) The function f is **continuous** at $x = c$ if f is defined at $x = c$ and

$$\lim_{x \to c} f(x) = f(c).$$

The function is **continuous on an interval** if it is continuous at every point in the interval.

Theorem: Continuity of Sums, Products, Quotients (p. 71) Suppose that f and g are continuous on an interval and that b is a constant. Then, on that same interval, the following functions are also continuous: $bf(x)$, $f(x) + g(x)$, $f(x)g(x)$. Further, $f(x)/g(x)$ is continuous provided $g(x) \neq 0$ on the interval.

Theorem: Continuity of Composite Functions (p. 71) Suppose f and g are continuous and $f(g(x))$ is defined on an interval. Then on that interval $f(g(x))$ is continuous.

Intermediate Value Theorem (p. 61) Suppose f is continuous on a closed interval $[a, b]$. If k is any number between $f(a)$ and $f(b)$, then there is at least one number c in $[a, b]$ such that $f(c) = k$.

The Extreme Value Theorem (p. 203) If f is continuous on the interval $[a, b]$, then f has a global maximum and a global minimum on that interval.

The Squeeze Theorem (p. 79) If $b(x) \leq f(x) \leq a(x)$ for all x close to $x = c$ except possibly at $x = c$, and $\lim_{x \to c} b(x) = L = \lim_{x \to c} a(x)$, then $\lim_{x \to c} f(x) = L$.

The Derivative

The slope of the secant line of $f(x)$ over an interval $[a, b]$ gives:

$$\text{Average rate of change of } f \text{ over } [a,b] = \frac{f(b) - f(a)}{b - a} \quad \text{(p. 91)}.$$

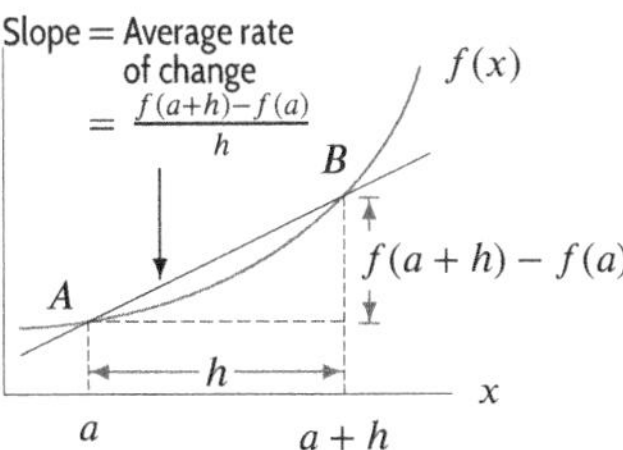

Figure R.17: Visualizing the average rate of change of f (p. 93)

The **derivative** of f at a is the slope of the line tangent to the graph of f at the point $(a, f(a))$:

$$f'(a) = \lim_{h \to 0} \frac{f(a+h) - f(a)}{h},$$

and gives the **instantaneous rate of change** of f at a (p. 92). The function f is **differentiable** at a if this limit exists (p. 92). The **second derivative** of f, denoted f'', is the derivative of f' (p. 115).

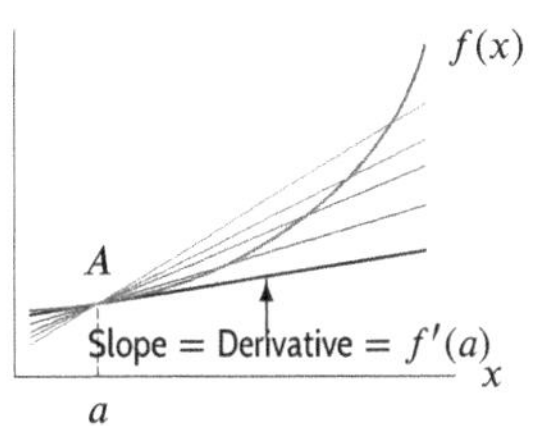

Figure R.18: Visualizing the instantaneous rate of change of f (p. 93)

The **units** of $f'(x)$ are: $\dfrac{\text{Units of } f(x)}{\text{Units of } x}$ (p. 109). If $f' > 0$ on an interval, then f is **increasing** over that interval (p. 101). If $f' < 0$ on an interval, then f is **decreasing** over that interval (p. 101). If $f'' > 0$ on an interval, then f is **concave up** over that interval (p. 115). If $f'' < 0$ on an interval, then f is **concave down** over that interval (p. 115).

The **tangent line** at $(a, f(a))$ is the graph of $y = f(a) + f'(a)(x-a)$ (p. 178). The **tangent line approximation** says that for values of x near a, $f(x) \approx f(a) + f'(a)(x - a)$. The expression $f(a) + f'(a)(x - a)$ is called the **local linearization** of f near $x = a$ (p. 178).

Derivatives of elementary functions

$$\frac{d}{dx}(x^n) = nx^{n-1} \text{ (p. 132)} \qquad \frac{d}{dx}(e^x) = e^x \text{ (p. 141)}$$

$$\frac{d}{dx}(a^x) = (\ln a)a^x \text{ (p. 142)} \qquad \frac{d}{dx}(\ln x) = \frac{1}{x} \text{ (p. 165)}$$

$$\frac{d}{dx}(\sin x) = \cos x \qquad \frac{d}{dx}(\cos x) = -\sin x \text{ (p. 160)}$$

$$\frac{d}{dx}(\arctan x) = \frac{1}{1+x^2} \text{ (p. 166)}$$

$$\frac{d}{dx}(\arcsin x) = \frac{1}{\sqrt{1-x^2}} \text{ (p. 166)}$$

Derivatives of sums, differences, and constant multiples

$$\frac{d}{dx}[f(x) \pm g(x)] = f'(x) \pm g'(x) \text{ (p. 131)}$$

$$\frac{d}{dx}[cf(x)] = cf'(x) \text{ (p. 130)}$$

Product and quotient rules

$$(fg)' = f'g + fg' \text{ (p. 146)}$$

$$\left(\frac{f}{g}\right)' = \frac{f'g - fg'}{g^2} \text{ (p. 147)}$$

Chain rule

$$\frac{d}{dx} f(g(x)) = f'(g(x)) \cdot g'(x) \text{ (p. 152)}$$

Derivative of an inverse function (p. 166). If f has a differentiable inverse, f^{-1}, then

$$\frac{d}{dx}(f^{-1}(x)) = \frac{1}{f'(f^{-1}(x))}.$$

Implicit differentiation (p. 171) If y is implicitly defined as a function of x by an equation, then, to find dy/dx, differentiate the equation (remembering to apply the chain rule).

Applications of the Derivative

A function f has a **local maximum** at p if $f(p)$ is greater than or equal to the values of f at points near p, and a **local minimum** at p if $f(p)$ is less than or equal to the values of f at points near p (p. 193). It has a **global maximum** at p if $f(p)$ is greater than or equal to the value of f at any point in the interval, and a **global minimum** at p if $f(p)$ is less than or equal to the value of f at any point in the interval (p. 203).

A **critical point** of a function $f(x)$ is a point p in the domain of f where $f'(p) = 0$ or $f'(p)$ is undefined (p. 193).

Theorem: Local maxima and minima which do not occur at endpoints of the domain occur at critical points (pp. 194, 198).

The First-Derivative Test for Local Maxima and Minima (p. 194):

- If f' changes from negative to positive at p, then f has a local minimum at p.
- If f' changes from positive to negative at p, then f has a local maximum at p.

The Second-Derivative Test for Local Maxima and Minima (p. 196):

- If $f'(p) = 0$ and $f''(p) > 0$ then f has a local minimum at p.
- If $f'(p) = 0$ and $f''(p) < 0$ then f has a local maximum at p.
- If $f'(p) = 0$ and $f''(p) = 0$ then the test tells us nothing.

To find the **global maximum and minimum** of a function on an interval we compare values of f at all critical points in the interval and at the endpoints of the interval (or $\lim_{x\to\pm\infty} f(x)$ if the interval is unbounded) (p. 204).

An **inflection point** of f is a point at which the graph of f changes concavity (p. 196); f'' is zero or undefined at an inflection point (p. 196).

L'Hopital's rule (p. 252) If f and g are continuous, $f(a) = g(a) = 0$, and $g'(a) \neq 0$, then

$$\lim_{x\to a}\frac{f(x)}{g(x)} = \frac{f'(a)}{g'(a)}.$$

Parametric equations (p. 259) If a curve is given by the parametric equations $x = f(t)$, $y = g(t)$, the slope of the curve as a function of t is $dy/dx = (dy/dt)/(dx/dt)$.

Theorems About Derivatives

Theorem: Local Extrema and Critical Points (pp. 194, 198) Suppose f is defined on an interval and has a local maximum or minimum at the point $x = a$, which is not an endpoint of the interval. If f is differentiable at $x = a$, then $f'(a) = 0$.

The Mean Value Theorem (p. 186) If f is continuous on $[a, b]$ and differentiable on (a, b), then there exists a number c, with $a < c < b$, such that

$$f'(c) = \frac{f(b) - f(a)}{b - a}.$$

The Increasing Function Theorem (p. 187) Suppose that f is continuous on $[a, b]$ and differentiable on (a, b).

- If $f'(x) > 0$ on (a, b), then f is increasing on $[a, b]$.
- If $f'(x) \geq 0$ on (a, b), then f is nondecreasing on $[a, b]$.

The Constant Function Theorem (p. 187) Suppose that f is continuous on $[a, b]$ and differentiable on (a, b). If $f'(x) = 0$ on (a, b), then f is constant on $[a, b]$.

The Racetrack Principle (p. 188) Suppose that g and h are continuous on $[a, b]$ and differentiable on (a, b), and that $g'(x) \leq h'(x)$ for $a < x < b$.

- If $g(a) = h(a)$, then $g(x) \leq h(x)$ for $a \leq x \leq b$.
- If $g(b) = h(b)$, then $g(x) \geq h(x)$ for $a \leq x \leq b$.

Theorem: Differentiability and Local Linearity (p. 179) Suppose f is differentiable at $x = a$ and $E(x)$ is the error in the tangent line approximation, that is: $E(x) = f(x) - f(a) - f'(a)(x - a)$. Then $\lim_{x\to a}\frac{E(x)}{x - a} = 0$.

Theorem: A Differentiable Function Is Continuous (p. 125) If $f(x)$ is differentiable at a point $x = a$, then $f(x)$ is continuous at $x = a$.

The Definite Integral

The **definite integral of f from a to b** (p. 284), denoted $\int_a^b f(x)\,dx$, is the limit of the left and the right sums as the width of the rectangles is shrunk to 0, where

$$\textbf{Left-hand sum} = \sum_{i=0}^{n-1} f(x_i)\Delta x \quad \text{(p. 277)}$$
$$= f(x_0)\Delta x + f(x_1)\Delta x + \cdots + f(x_{n-1})\Delta x$$

$$\textbf{Right-hand sum} = \sum_{i=1}^{n} f(x_i)\Delta x \quad \text{(p. 283)}$$
$$= f(x_1)\Delta x + f(x_2)\Delta x + \cdots + f(x_n)\Delta x$$

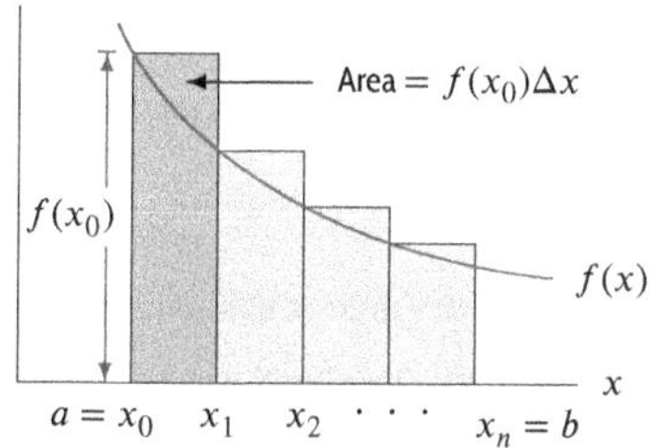

Figure R.19: Left-hand sum (p. 277)

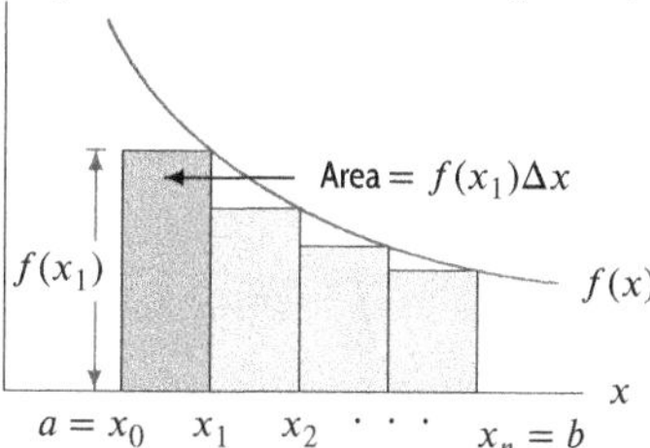

Figure R.20: Right-hand sum (p. 277)

If f is nonnegative, $\int_a^b f(x)\,dx$ represents the area under the curve between $x = a$ and $x = b$ (p. 285). If f has any sign, $\int_a^b f(x)\,dx$ is the sum of the areas above the x-axis, counted positively, and the areas below the x-axis, counted negatively (p. 286). If $F'(t)$ is the rate of change of some quantity $F(t)$, then $\int_a^b F'(t)\,dt$ is the **total change** in $F(t)$ between $t = a$ and $t = b$ (p. 294). The **average value** of f on the interval $[a, b]$ is given by $\frac{1}{b-a}\int_a^b f(x)\,dx$ (p. 308).

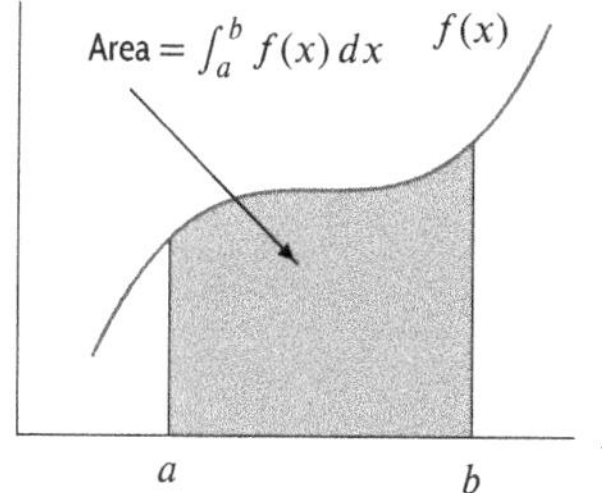

Figure R.21: The definite integral $\int_a^b f(x)\,dx$ (p. 285)

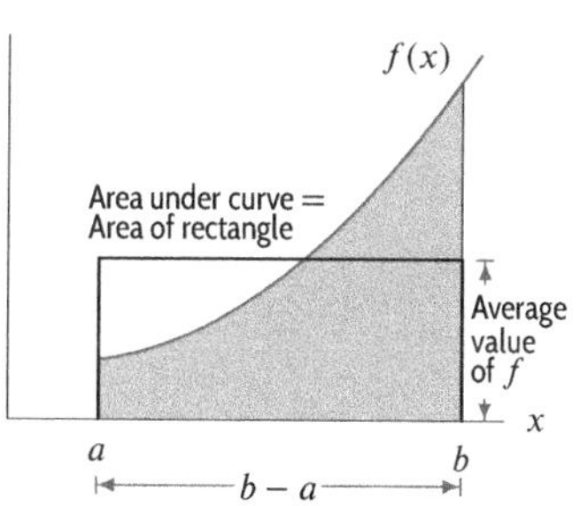

Figure R.22: Area and average value (p. 309)

The **units** of $\int_a^b f(x)\,dx$ are Units of $f(x) \times$ Units of x (p. 292).

The Fundamental Theorem of Calculus If f is continuous on $[a, b]$ and $f(x) = F'(x)$, then

$$\int_a^b f(x)\,dx = F(b) - F(a) \quad \text{(p. 293).}$$

Properties of Definite Integrals (pp. 302, 304)
If a, b, and c are any numbers and f, g are continuous functions, then

$$\int_b^a f(x)\,dx = -\int_a^b f(x)\,dx$$

$$\int_a^b (f(x) \pm g(x))\,dx = \int_a^b f(x)\,dx \pm \int_a^b g(x)\,dx$$

$$\int_a^c f(x)\,dx + \int_c^b f(x)\,dx = \int_a^b f(x)\,dx$$

$$\int_a^b cf(x)\,dx = c\int_a^b f(x)\,dx$$

Antiderivatives

An **antiderivative** of a function $f(x)$ is a function $F(x)$ such that $F'(x) = f(x)$ (p. 316). There are infinitely many antiderivatives of f since $F(x) + C$ is an antiderivative of f for any constant C, provided $F'(x) = f(x)$ (p. 316). The **indefinite integral** of f is the family of antiderivatives $\int f(x)\,dx = F(x) + C$ (p. 322).

Construction Theorem (Second Fundamental Theorem of Calculus) If f is a continuous function on an interval and a is any number in that interval, then $F(x) = \int_a^x f(t)\,dt$ is an antiderivative of f (p. 336).

Properties of Indefinite Integrals (p. 325)

$$\int (f(x) \pm g(x))\,dx = \int f(x)\,dx \pm \int g(x)\,dx$$

$$\int cf(x)\,dx = c\int f(x)\,dx$$

Some antiderivatives:

$$\int k\,dx = kx + C \quad \text{(p. 323)}$$

$$\int x^n\,dx = \frac{x^{n+1}}{n+1} + C, \quad n \neq -1 \quad \text{(p. 323)}$$

$$\int \frac{1}{x}\,dx = \ln|x| + C \quad \text{(p. 324)}$$

$$\int e^x\,dx = e^x + C \quad \text{(p. 324)}$$

$$\int \cos x\,dx = \sin x + C \quad \text{(p. 324)}$$

$$\int \sin x\,dx = -\cos x + C \quad \text{(p. 324)}$$

$$\int \frac{dx}{1+x^2} = \arctan x + C \quad \text{(p. 372)}$$

$$\int \frac{dx}{\sqrt{1-x^2}} = \arcsin x + C \quad \text{(p. 369)}$$

Substitution (p. 342) For integrals of the form $\int f(g(x))g'(x)\,dx$, let $w = g(x)$. Choose w, find dw/dx and substitute for x and dx. Convert limits of integration for definite integrals.

Integration by Parts (p. 353) Used mainly for products; also for integrating $\ln x$, $\arctan x$, $\arcsin x$.

$$\int uv'\,dx = uv - \int u'v\,dx$$

Partial Fractions (p. 366) To integrate a rational function, $P(x)/Q(x)$, express as a sum of a polynomial and terms of the form $A/(x-c)^n$ and $(Ax+B)/q(x)$, where $q(x)$ is an unfactorable quadratic.

Trigonometric Substitutions (p. 369) To simplify $\sqrt{x^2 - a^2}$, try $x = a\sin\theta$ (p. 370). To simplify $a^2 + x^2$ or $\sqrt{a^2 + x^2}$, try $x = a\tan\theta$ (p. 372).

Numerical Approximations for Definite Integrals (p. 376)
Riemann sums (left, right, midpoint) (p. 377), trapezoid rule (p. 378), Simpson's rule (p. 381)

Approximation Errors (p. 379)
f concave up: midpoint underestimates, trapezoid overestimates (p. 379)
f concave down: trapezoid underestimates, midpoint overestimates (p. 379)

Evaluating Improper Integrals (p. 385)

- Infinite limit of integration: $\int_a^\infty f(x)\,dx = \lim_{b\to\infty}\int_a^b f(x)\,dx$ (p. 385)
- For $a < b$, if integrand is unbounded at $x = b$, then: $\int_a^b f(x)\,dx = \lim_{c\to b^-}\int_a^c f(x)\,dx$ (p. 389)

Testing Improper Integrals for Convergence by Comparison (p. 395):

- If $0 \le f(x) \le g(x)$ and $\int_a^\infty g(x)\,dx$ converges, then $\int_a^\infty f(x)\,dx$ converges
- If $0 \le g(x) \le f(x)$ and $\int_a^\infty g(x)\,dx$ diverges, then $\int_a^\infty f(x)\,dx$ diverges

Applications of Integration

Total quantities can be approximated by slicing them into small pieces and summing the pieces. The limit of this sum is a definite integral which gives the exact total quantity.

Applications to Geometry
To calculate the volume of a solid, slice the volume into pieces whose volumes you can estimate (pp. 402, 410). Use this method to calculate volumes of revolution (p. 411) and volumes of solids with known cross sectional area (p. 412). Curve $f(x)$ from $x = a$ to $x = b$ has

$$\text{Arc length} = \int_a^b \sqrt{1 + (f'(x))^2}\,dx \quad \text{(p. 414)}.$$

Mass and Center of Mass from Density, δ

$$\text{Total mass} = \int_a^b \delta(x)\,dx \text{ (p. 430)}$$

$$\text{Center of mass} = \frac{\int_a^b x\delta(x)\,dx}{\int_a^b \delta(x)\,dx} \text{ (p. 434)}$$

To find the center of mass of two- and three-dimensional objects, use the formula separately on each coordinate (p. 435).

Applications to Physics

$$\text{Work done} = \text{Force} \times \text{Distance} \text{ (p. 440)}$$
$$\text{Pressure} = \text{Density} \times g \times \text{Depth} \text{ (p. 444)}$$
$$\text{Force} = \text{Pressure} \times \text{Area} \text{ (p. 444)}$$

Applications to Economics Present and future value of income stream, $P(t)$ (p. 450); consumer and producer surplus (p. 453).

$$\text{Present value} = \int_0^M P(t)e^{-rt}\,dt \text{ (p. 450)}$$
$$\text{Future value} = \int_0^M P(t)e^{r(M-t)}\,dt \text{ (p. 450)}$$

Applications to Probability Given a density function $p(x)$, the fraction of the population for which x is between a and b is the area under the graph of p between a and b (p. 459). The cumulative distribution function $P(t)$ is the fraction having values of x below t (p. 460). The median is the value T such that half the population has values of x less than or equal to T (p. 466). The mean (p. 468) is defined by

$$\text{Mean value} = \int_{-\infty}^{\infty} xp(x)dx.$$

Polar Coordinates

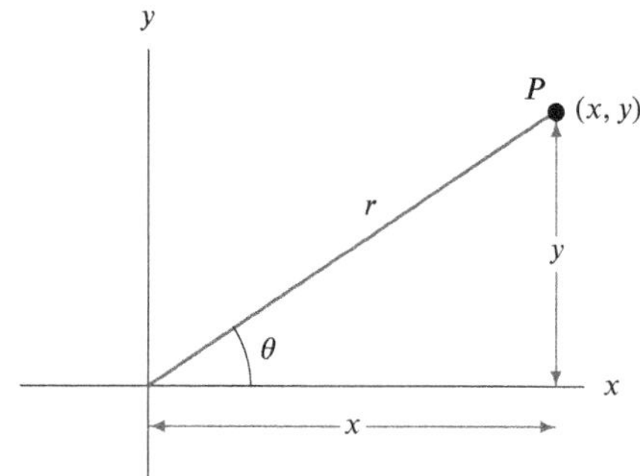

Figure R.23: Cartesian and polar coordinates for the point P

The **polar coordinates** (p. 421) of a point are related to its Cartesian coordinates by

- $x = r\cos\theta$ and $y = r\sin\theta$
- $r = \sqrt{x^2 + y^2}$ and $\tan\theta = \dfrac{y}{x}$, $x \ne 0$

For a constant a, the equation $r = a$ gives a circle of radius a, and the equation $\theta = a$ gives a ray from the origin making an angle of θ with the positive x-axis. The equation $r = \theta$ gives an **Archimedean spiral** (p. 423).

Area in Polar Coordinates (p. 426)
For a curve $r = f(\theta)$, with $\alpha \le \theta \le \beta$, which does not cross itself,

$$\text{Area of region enclosed } = \frac{1}{2}\int_{\alpha}^{\beta} f(\theta)^2\, d\theta.$$

Slope and Arclength in Polar Coordinates (p. 427)
For a curve $r = f(\theta)$, we can express x and y in terms of θ as a parameter, giving

$$x = r\cos\theta = f(\theta)\cos\theta \quad \text{and} \quad y = r\sin\theta = f(\theta)\sin\theta.$$

Then

$$\text{Slope} = \frac{dy}{dx} = \frac{dy/d\theta}{dx/d\theta}$$

and

$$\text{Arc length } = \int_{\alpha}^{\beta}\sqrt{\left(\frac{dx}{d\theta}\right)^2 + \left(\frac{dy}{d\theta}\right)^2}\, d\theta.$$

Alternatively (p. 429),

$$\text{Arc length } = \int_{\alpha}^{\beta}\sqrt{(f'(\theta))^2 + (f(\theta))^2}\, d\theta.$$

Sequences and Series

A **sequence** $s_1, s_2, s_3, \ldots, s_n, \ldots$ has a **limit** L, written $\lim_{n\to\infty} s_n = L$, if we can make s_n as close to L as we please by choosing a sufficiently large n. The sequence **converges** if a limit exists, **diverges** if no limit exists (see p. 476). Limits of sequences satisfy the same properties as limits of functions stated in Theorem 1.2 (p. 71) and

$$\lim_{n\to\infty} x^n = 0 \quad \text{if } |x| < 1 \qquad \lim_{n\to\infty} 1/n = 0 \quad \text{(p. 476)}$$

A sequence s_n is **bounded** if there are constants K and M such that $K \le s_n \le M$ for all n (p. 477). A convergent sequence is bounded. A sequence is monotone if it is either increasing, that is $s_n < s_{n+1}$ for all n, or decreasing, that is $s_n > s_{n+1}$ for all n (p. 477).

Theorem: Convergence of a Monotone, Bounded Sequence (p. 477): If a sequence s_n is bounded and monotone, it converges.

A **series** is an infinite sum $\sum a_n = a_1 + a_2 + \cdots$. The nth **partial sum** is $S_n = a_1 + a_2 + \cdots + a_n$ (p. 488). If $S = \lim_{n\to\infty} S_n$ exists, then the series $\sum a_n$ **converges**, and its sum is S (p. 488). If a series does not converge, we say that it **diverges** (p. 488). The sum of a **finite geometric series** is (p. 482):

$$a + ax + ax^2 + \cdots + ax^{n-1} = \frac{a(1-x^n)}{1-x}, \qquad x \ne 1.$$

The sum of an **infinite geometric series** is (p. 483):

$$a + ax + ax^2 + \cdots + ax^n + \cdots = \frac{a}{1-x}, \qquad |x| < 1.$$

The p-**series** $\sum 1/n^p$ converges if $p > 1$ and diverges if $p \le 1$ (p. 493). The **harmonic series** $\sum 1/n$ diverges (p. 491), the **alternating harmonic series** $\sum(-1)^{n-1}(1/n)$ converges (p. 500). An **alternating series** can be absolutely or conditionally convergent (p. 500).

Convergence Tests

Theorem: Convergence Properties of Series (p. 490)

1. If $\sum a_n$ and $\sum b_n$ converge and if k is a constant, then
   - $\sum(a_n + b_n)$ converges to $\sum a_n + \sum b_n$.
   - $\sum ka_n$ converges to $k\sum a_n$.
2. Changing a finite number of terms in a series does not change whether or not it converges, although it may change the value of its sum if it does converge.
3. If $\lim_{n\to\infty} a_n \ne 0$ or $\lim_{n\to\infty} a_n$ does not exist, then $\sum a_n$ diverges.
4. If $\sum a_n$ diverges, then $\sum ka_n$ diverges if $k \ne 0$.

Theorem: Integral Test (p. 492) Suppose $c \ge 0$ and $f(x)$ is a decreasing positive function, defined for all $x \ge c$, with $a_n = f(n)$ for all n.

- If $\int_c^{\infty} f(x)\,dx$ converges, then $\sum a_n$ converges.
- If $\int_c^{\infty} f(x)\,dx$ diverges, then $\sum a_n$ diverges.

Theorem: Comparison Test (p. 494) Suppose $0 \le a_n \le b_n$ for all n.

- If $\sum b_n$ converges, then $\sum a_n$ converges.
- If $\sum a_n$ diverges, then $\sum b_n$ diverges.

Theorem: Limit Comparison Test (p. 496) Suppose $a_n > 0$ and $b_n > 0$ for all n. If

$$\lim_{n\to\infty} \frac{a_n}{b_n} = c \qquad \text{where } c > 0,$$

then the two series $\sum a_n$ and $\sum b_n$ either both converge or both diverge.

Theorem: Convergence of Absolute Values Implies Convergence (p. 497): If $\sum |a_n|$ converges, then so does $\sum a_n$.
We say $\sum a_n$ is **absolutely convergent** if $\sum a_n$ and $\sum |a_n|$ both converge and **conditionally convergent** if $\sum a_n$ converges but $\sum |a_n|$ diverges (p. 500).

Theorem: The Ratio Test (p. 498) For a series $\sum a_n$, suppose the sequence of ratios $|a_{n+1}|/|a_n|$ has a limit:

$$\lim_{n\to\infty} \frac{|a_{n+1}|}{|a_n|} = L.$$

- If $L < 1$, then $\sum a_n$ converges.
- If $L > 1$, or if L is infinite, then $\sum a_n$ diverges.
- If $L = 1$, the test does not tell us anything about the convergence of $\sum a_n$.

Theorem: Alternating Series Test (p. 499) A series of the form

$$\sum_{n=1}^{\infty}(-1)^{n-1}a_n = a_1 - a_2 + a_3 - a_4 + \cdots + (-1)^{n-1}a_n + \cdots$$

converges if $0 < a_{n+1} < a_n$ for all n and $\lim_{n\to\infty} a_n = 0$.

Theorem: Error Bounds for Alternating Series (p. 500) Let $S_n = \sum_{i=1}^{n}(-1)^{i-1}a_i$ be the nth partial sum of an alternating series and let $S = \lim_{n\to\infty} S_n$. Suppose that $0 < a_{n+1} < a_n$ for all n and $\lim_{n\to\infty} a_n = 0$. Then $|S - S_n| < a_{n+1}$.

Absolutely and conditionally convergent series behave differently when their **terms are rearranged** (p. 501)

- **Absolutely** convergent series: Rearranging terms does not change the sum.
- **Conditionally** convergent series: Rearranging terms can change the sum to any number.

Power Series

$$\begin{aligned}P(x) &= C_0 + C_1(x-a) + C_2(x-a)^2 + \cdots \\ &\quad + C_n(x-a)^n + \cdots \\ &= \sum_{n=0}^{\infty} C_n(x-a)^n \quad \text{(p. 504).}\end{aligned}$$

The **radius of convergence** is 0 if the series converges only for $x = a$, ∞ if it converges for all x, and the positive number R if it converges for $|x-a| < R$ and diverges for $|x-a| > R$ (p. 506). The **interval of convergence** is the interval between $a - R$ and $a + R$, including any endpoint where the series converges (p. 506).

Theorem: Method for Computing Radius of Convergence (p. 507) To calculate the radius of convergence, R, for the power series $\sum_{n=0}^{\infty} C_n(x-a)^n$, use the ratio test with $a_n = C_n(x-a)^n$.

- If $\lim_{n\to\infty} |a_{n+1}|/|a_n|$ is infinite, then $R = 0$.
- If $\lim_{n\to\infty} |a_{n+1}|/|a_n| = 0$, then $R = \infty$.
- If $\lim_{n\to\infty} |a_{n+1}|/|a_n| = K|x-a|$, where K is finite and nonzero, then $R = 1/K$.

Theorem: A Power Series is its Own Taylor Series (p. 527) If a power series about $x = a$ converges to $f(x)$ for $|x-a| < R$, then the power series is the Taylor series for $f(x)$ about $x = a$.

Approximations

The **Taylor polynomial of degree** n approximating $f(x)$ for x near a is:

$$\begin{aligned}P_n(x) &= f(a) + f'(a)(x-a) + \frac{f''(a)}{2!}(x-a)^2 + \cdots\cdots \\ &\quad + \frac{f^{(n)}(a)}{n!}(x-a)^n \quad \text{(p. 517)}\end{aligned}$$

The **Maclaurin polynomial of degree** n is the Taylor polynomial of degree n near $x = 0$ (p. 517). The **Taylor series** approximating $f(x)$ for x near a is:

$$\begin{aligned}f(x) &= f(a) + f'(a)(x-a) + \frac{f''(a)}{2!}(x-a)^2 + \cdots \\ &\quad + \frac{f^{(n)}(a)}{n!}(x-a)^n + \cdots \quad \text{(p. 525)}\end{aligned}$$

The **Maclaurin series** is the Taylor series approximating $f(x)$ near $x = 0$ (p. 524).

Theorem: The Lagrange Error Bound for $P_n(x)$ (p. 540) Suppose f and all its derivatives are continuous. If $P_n(x)$ is the nth Taylor polynomial for to $f(x)$ about a, then

$$|E_n(x)| = |f(x) - P_n(x)| \le \frac{M}{(n+1)!}|x-a|^{n+1},$$

where $\max \left|f^{(n+1)}\right| \le M$ on the interval between a and x.

Taylor Series for sin x, cos x, e^x (p. 524):

$$\begin{aligned}\sin x &= x - \frac{x^3}{3!} + \frac{x^5}{5!} - \frac{x^7}{7!} + \frac{x^9}{9!} - \cdots \\ \cos x &= 1 - \frac{x^2}{2!} + \frac{x^4}{4!} - \frac{x^6}{6!} + \frac{x^8}{8!} - \cdots \\ e^x &= 1 + x + \frac{x^2}{2!} + \frac{x^3}{3!} + \frac{x^4}{4!} + \cdots\end{aligned}$$

Taylor Series for ln x about $x = 1$ converges for $0 < x \le 2$ (p. 526):

$$(x-1) - \frac{(x-1)^2}{2} + \frac{(x-1)^3}{3} - \frac{(x-1)^4}{4} + \cdots$$

The Binomial Series for $(1+x)^p$ converges for $-1 < x < 1$ (p. 525):

$$1 + px + \frac{p(p-1)}{2!}x^2 + \frac{p(p-1)(p-2)}{3!}x^3 + \cdots$$

The **Fourier Series** of $f(x)$ is (p. 550)

$$\begin{aligned}f(x) &= a_0 + a_1\cos x + a_2\cos 2x + a_3\cos 3x + \cdots \\ &\quad + b_1\sin x + b_2\sin 2x + b_3\sin 3x + \cdots,\end{aligned}$$

where

$$\begin{aligned}a_0 &= \frac{1}{2\pi}\int_{-\pi}^{\pi} f(x)\,dx, \\ a_k &= \frac{1}{\pi}\int_{-\pi}^{\pi} f(x)\cos(kx)\,dx \quad \text{for } k \text{ a positive integer,} \\ b_k &= \frac{1}{\pi}\int_{-\pi}^{\pi} f(x)\sin(kx)\,dx \quad \text{for } k \text{ a positive integer.}\end{aligned}$$

Differential Equations

A **differential equation** for the function $y(x)$ is an equation involving x, y and the derivatives of y (p. 562). The **order** of a differential equation is the order of the highest-order derivative appearing in the equation (p. 563). A **solution** to a differential equation is any function y that satisfies the equation (p. 563). The **general solution** to a differential equation is the **family of functions** that satisfies the equation (p. 563). An **initial value problem** is a differential equation together with an **initial condition**; a solution to an initial value problem is called a **particular solution** (p. 563). An **equilibrium solution** to a differential equation is a particular solution where y is constant and $dy/dx = 0$ (p. 592).

First-order equations: methods of solution. A **slope field** corresponding to a differential equation is a plot in the xy-plane of small line segments with slope given by the differential equation (p. 567). **Euler's method** approximates the solution of an initial value problem with a string of small line segments (p. 575). Differential equations of the form $dy/dx = g(x)f(y)$ can be solved analytically by **separation of variables** (p. 580).

First-order equations: applications. The differential equation for **exponential growth and decay** is of the form

$$\frac{dP}{dt} = kP.$$

The **solution** is of the form $P = P_0 e^{kt}$, where P_0 is the initial value of P, and positive k represents growth while negative k represents decay (p. 587). Applications of growth and decay include **continuously compounded interest (p. 587), population growth (p. 606),** and **Newton's law of heating and cooling (p. 590)**. The **logistic equation** for population growth is of the form

$$\frac{dP}{dt} = kP\left(1 - \frac{P}{L}\right),$$

where L is the **carrying capacity** of the population (p. 607). The **solution to the logistic equation** is of the form $P = \dfrac{L}{1 + Ae^{-kt}}$, where $A = \dfrac{L - P_0}{P_0}$ (p. 608).

Systems of differential equations. Two interacting populations, w and r, can be modeled by two equations

$$\frac{dw}{dt} = aw - cwr \quad \text{and} \quad \frac{dr}{dt} = -br + kwr \quad \text{(p. 620).}$$

Solutions can be visualized as **trajectories** in the wr-**phase plane** (p. 620). An **equilibrium point** is one at which $dw/dt = 0$ and $dr/dt = 0$ (p. 621). A **nullcline** is a curve along which $dw/dt = 0$ or $dr/dt = 0$ (p. 627).

Multivariable Functions

Points in 3-space are represented by a system of **Cartesian coordinates** (p. 654). The **distance** between (x, y, z) and (a, b, c) is $\sqrt{(x-a)^2 + (y-b)^2 + (z-c)^2}$ (p. 656).

Functions of two variables can be represented by **graphs** (p. 661), **contour diagrams** (p. 668), **cross-sections** (p. 663), and **tables** (p. 653).

Functions of three variables can be represented by the family of **level surfaces** $f(x, y, z) = c$ for various values of the constant c (p. 689).

A **linear function** $f(x, y)$ has equation

$$\begin{aligned} f(x, y) &= z_0 + m(x - x_0) + n(y - y_0) \text{ (p. 683)} \\ &= c + mx + ny, \text{ where } c = z_0 - mx_0 - ny_0. \end{aligned}$$

Its **graph** is a plane with slope m in the x-direction, slope n in the y-direction, through (x_0, y_0, z_0) (p. 683). Its **table of values** has linear rows (of same slope) and linear columns (of same slope) (p. 684). Its **contour diagram** is equally spaced parallel straight lines (p. 684).

The **limit** of f at the point (a, b), written $\lim_{(x,y)\to(a,b)} f(x, y)$, is the number L, if one exists, such that $f(x, y)$ is as close to L as we please whenever the distance from the point (x, y) to the point (a, b) is sufficiently small, but not zero. (p. 697).

A function f is **continuous at the point** (a, b) if $\lim_{(x,y)\to(a,b)} f(x, y) = f(a, b)$. A function is **continuous on a region** R if it is continuous at each point of R (p. 697).

Vectors

A **vector** $\vec{v}$ has **magnitude** (denoted $\|\vec{v}\|$) and **direction**. Examples are **displacement vectors** (p. 702), **velocity and acceleration vectors** (pp. 711, 712). and **force** (p. 712). We can add vectors, and multiply a vector by a scalar (p. 703). Two non-zero vectors, $\vec{v}$ and $\vec{w}$, are **parallel** if one is a scalar multiple of the other (p. 704).

A **unit vector** has magnitude 1. The vectors $\vec{i}$, $\vec{j}$, and $\vec{k}$ are unit vectors in the directions of the coordinate axes. A unit vector in the direction of any nonzero vector $\vec{v}$ is $\vec{u} = \vec{v}/\|\vec{v}\|$ (p. 708). We **resolve** $\vec{v}$ **into components** by writing $\vec{v} = v_1\vec{i} + v_2\vec{j} + v_3\vec{k}$ (p. 705).

If $\vec{v} = v_1\vec{i} + v_2\vec{j} + v_3\vec{k}$ and $\vec{w} = w_1\vec{i} + w_2\vec{j} + w_3\vec{k}$ then

$$\|\vec{v}\| = \sqrt{v_1^2 + v_2^2 + v_3^2} \text{ (p. 706)}$$

$$\vec{v} + \vec{w} = (v_1 + w_1)\vec{i} + (v_2 + w_2)\vec{j} + (v_3 + w_3)\vec{k} \text{ (p. 707),}$$

$$\lambda\vec{v} = \lambda v_1\vec{i} + \lambda v_2\vec{j} + \lambda v_3\vec{k} \text{ (p. 707).}$$

The **displacement vector** from $P_1 = (x_1, y_1, z_1)$ to $P_2 = (x_2, y_2, z_2)$ is

$$\overrightarrow{P_1P_2} = (x_2 - x_1)\vec{i} + (y_2 - y_1)\vec{j} + (z_2 - z_1)\vec{k} \quad \text{(p. 706).}$$

The **position vector** of $P = (x, y, z)$ is $\overrightarrow{OP}$ (p. 706). A **vector in n dimensions** is a string of numbers $\vec{v} = (v_1, v_2, \ldots, v_n)$ (p. 714).

Dot Product (Scalar Product) (p. 718).
Geometric definition: $\vec{v} \cdot \vec{w} = \|\vec{v}\|\|\vec{w}\| \cos\theta$ where θ is the angle between $\vec{v}$ and $\vec{w}$ and $0 \le \theta \le \pi$.
Algebraic definition: $\vec{v} \cdot \vec{w} = v_1w_1 + v_2w_2 + v_3w_3$.

Two nonzero vectors $\vec{v}$ and $\vec{w}$ are **perpendicular** if and only if $\vec{v} \cdot \vec{w} = 0$ (p. 720). Magnitude and dot product are related by $\vec{v} \cdot \vec{v} = \|\vec{v}\|^2$ (p. 720). If $\vec{u} = (u_1, \dots, u_n)$ and $\vec{v} = (v_1, \dots, v_n)$ then the dot product of $\vec{u}$ and $\vec{v}$ is $\vec{u} \cdot \vec{v} = u_1 v_1 + \dots + u_n v_n$ (p. 722).

The **equation of the plane** with **normal vector** $\vec{n} = a\vec{i} + b\vec{j} + c\vec{k}$ and containing the point $P_0 = (x_0, y_0, z_0)$ is $\vec{n} \cdot (\vec{r} - \vec{r}_0) = a(x - x_0) + b(y - y_0) + c(z - z_0) = 0$ or $ax + by + cz = d$, where $d = ax_0 + by_0 + cz_0$ (p. 721).

If $\vec{v}_{\text{parallel}}$ and $\vec{v}_{\text{perp}}$ are components of $\vec{v}$ which are parallel and perpendicular, respectively, to a unit vector $\vec{u}$, then $\vec{v}_{\text{parallel}} = (\vec{v} \cdot \vec{u})\vec{u}$ and $\vec{v}_{\text{perp}} = \vec{v} - \vec{v}_{\text{parallel}}$ (p. 723).

The **work**, W, done by a force $\vec{F}$ acting on an object through a displacement $\vec{d}$ is $W = \vec{F} \cdot \vec{d}$ (p. 724).

Cross Product (Vector Product) (p. 729, 729)
Geometric definition

$$\vec{v} \times \vec{w} = \begin{pmatrix}\text{Area of parallelogram} \\ \text{with edges } \vec{v} \text{ and } \vec{w}\end{pmatrix} \vec{n}$$
$$= (\|\vec{v}\|\|\vec{w}\| \sin\theta)\vec{n},$$

where $0 \le \theta \le \pi$ is the angle between $\vec{v}$ and $\vec{w}$ and $\vec{n}$ is the unit vector perpendicular to $\vec{v}$ and $\vec{w}$ pointing in the direction given by the **right-hand rule.**
Algebraic definition

$$\vec{v} \times \vec{w} = (v_2 w_3 - v_3 w_2)\vec{i} + (v_3 w_1 - v_1 w_3)\vec{j} + (v_1 w_2 - v_2 w_1)\vec{k}$$

$$\vec{v} = v_1\vec{i} + v_2\vec{j} + v_3\vec{k},\ \vec{w} = w_1\vec{i} + w_2\vec{j} + w_3\vec{k}.$$

To find the **equation of a plane through three points** that do not lie on a line, determine two vectors in the plane and then find a normal vector using the cross product (p. 731). The **area of a parallelogram** with edges $\vec{v}$ and $\vec{w}$ is $\|\vec{v} \times \vec{w}\|$. The **volume of a parallelepiped** with edges $\vec{a}, \vec{b}, \vec{c}$ is $|(\vec{b} \times \vec{c}) \cdot \vec{a}|$ (p. 733).

The **angular velocity** (p. 733) of a flywheel can be represented by a vector $\vec{\omega}$ whose direction is parallel to the axis of rotation and magnitude is the angular speed of rotation. The velocity vector $\vec{v}$ of a point P on the flywheel is $\vec{v} = \vec{\omega} \times \vec{r}$ where $\vec{r}$ is a vector from the axis of rotation to P.

Differentiation of Multivariable Functions

Partial derivatives of f (p. 741).

$$f_x(a,b) = \text{Rate of change of } f \text{ with respect to } x \text{ at the point } (a,b)$$
$$= \lim_{h\to 0} \frac{f(a+h,b) - f(a,b)}{h},$$
$$f_y(a,b) = \text{Rate of change of } f \text{ with respect to } y \text{ at the point } (a,b)$$
$$= \lim_{h\to 0} \frac{f(a,b+h) - f(a,b)}{h}.$$

On the graph of f, the partial derivatives $f_x(a,b)$ and $f_y(a,b)$ give the slope in the x and y directions, respectively (p. 742). The **tangent plane** to $z = f(x,y)$ at (a,b) is

$$z = f(a,b) + f_x(a,b)(x-a) + f_y(a,b)(y-b) \quad \text{(p. 755).}$$

Partial derivatives can be estimated from a contour diagram or table of values using difference quotients (p. 742), and can be computed algebraically using the same rules of differentiation as for one-variable calculus (p. 748). Partial derivatives for functions of three or more variables are defined and computed in the same way (p 749).

The gradient vector $\operatorname{grad} f$ of f is $\operatorname{grad} f(a,b) = f_x(a,b)\vec{i} + f_y(a,b)\vec{j}$ (2 variables) (p. 765) or $\operatorname{grad} f(a,b,c) = f_x(a,b,c)\vec{i} + f_y(a,b,c)\vec{j} + f_z(a,b,c)\vec{k}$ (3 variables) (p. 772). The gradient vector at P: Points in the direction of increasing f; is perpendicular to the level curve or level surface of f through P; and has magnitude $\|\operatorname{grad} f\|$ equal to the maximum rate of change of f at P (pp. 767, 772). The magnitude is large when the level curves or surfaces are close together and small when they are far apart.

The **directional derivative** of f at P in the direction of a unit vector $\vec{u}$ is (pp. 763, 765)

$$f_{\vec{u}}(P) = \begin{matrix}\text{Rate of change} \\ \text{of } f \text{ in direction} \\ \text{of } \vec{u} \text{ at } P\end{matrix} = \operatorname{grad} f(P) \cdot \vec{u}$$

The **tangent plane approximation** to $f(x,y)$ for (x,y) near the point (a,b) is

$$f(x,y) \approx f(a,b) + f_x(a,b)(x-a) + f_y(a,b)(y-b).$$

The right-hand side is the **local linearization** (p. 755). The **differential of** $z = f(x,y)$ at (a,b) is the linear function of dx and dy

$$df = f_x(a,b)\,dx + f_y(a,b)\,dy \quad \text{(p. 758).}$$

Local linearity with three or more variables follows the same pattern as for functions of two variables (p. 757).

The **tangent plane to a level surface** of a function of three-variables f at (a,b,c) is (p. 775)

$$f_x(a,b,c)(x-a) + f_y(a,b,c)(y-b) + f_z(a,b,c)(z-c) = 0.$$

The Chain Rule for the partial derivative of one variable with respect to another in a chain of composed functions (p. 782):

- Draw a diagram expressing the relationship between the variables, and label each link in the diagram with the derivative relating the variables at its ends.
- For each path between the two variables, multiply together the derivatives from each step along the path.

- Add the contributions from each path.

If $z = f(x, y)$, and $x = g(t)$, and $y = h(t)$, then

$$\frac{dz}{dt} = \frac{\partial z}{\partial x}\frac{dx}{dt} + \frac{\partial z}{\partial y}\frac{dy}{dt} \quad \text{(p. 781).}$$

If $z = f(x, y)$, with $x = g(u, v)$ and $y = h(u, v)$, then

$$\frac{\partial z}{\partial u} = \frac{\partial z}{\partial x}\frac{\partial x}{\partial u} + \frac{\partial z}{\partial y}\frac{\partial y}{\partial u},$$
$$\frac{\partial z}{\partial v} = \frac{\partial z}{\partial x}\frac{\partial x}{\partial v} + \frac{\partial z}{\partial y}\frac{\partial y}{\partial v} \quad \text{(p. 783).}$$

Second-order partial derivatives (p. 790)

$$\frac{\partial^2 z}{\partial x^2} = f_{xx} = (f_x)_x, \qquad \frac{\partial^2 z}{\partial x \partial y} = f_{yx} = (f_y)_x,$$
$$\frac{\partial^2 z}{\partial y \partial x} = f_{xy} = (f_x)_y, \qquad \frac{\partial^2 z}{\partial y^2} = f_{yy} = (f_y)_y.$$

Theorem: Equality of Mixed Partial Derivatives. If f_{xy} and f_{yx} are continuous at (a, b), an interior point of their domain, then $f_{xy}(a, b) = f_{yx}(a, b)$ (p. 791).

Taylor Polynomial of Degree 1 Approximating $f(x, y)$ for (x, y) near (a, b) (p. 794)

$$f(x, y) \approx L(x, y) = f(a, b) + f_x(a, b)(x - a) + f_y(a, b)(y - b).$$

Taylor Polynomial of Degree 2 (p. 795)

$$\begin{aligned} f(x, y) &\approx Q(x, y) \\ &= f(a, b) + f_x(a, b)(x - a) + f_y(a, b)(y - b) \\ &\quad + \frac{f_{xx}(a, b)}{2}(x - a)^2 + f_{xy}(a, b)(x - a)(y - b) \\ &\quad + \frac{f_{yy}(a, b)}{2}(y - b)^2. \end{aligned}$$

Definition of Differentiability (p. 799). A function $f(x, y)$ is **differentiable at the point** (a, b) if there is a linear function $L(x, y) = f(a, b) + m(x - a) + n(y - b)$ such that if the **error** $E(x, y)$ is defined by

$$f(x, y) = L(x, y) + E(x, y),$$

and if $h = x - a, k = y - b$, then the **relative error** $E(a + h, b + k)/\sqrt{h^2 + k^2}$ satisfies

$$\lim_{\substack{h \to 0 \\ k \to 0}} \frac{E(a + h, b + k)}{\sqrt{h^2 + k^2}} = 0.$$

Theorem: Continuity of Partial Derivatives Implies Differentiability (p. 802). If the partial derivatives, f_x and f_y, of a function f exist and are continuous on a small disk centered at the point (a, b), then f is differentiable at (a, b).

Optimization

A function f has a **local maximum** at the point P_0 if $f(P_0) \geq f(P)$ for all points P near P_0, and a **local minimum** at the point P_0 if $f(P_0) \leq f(P)$ for all points P near P_0 (p. 806). A **critical point** of a function f is a point where grad f is either $\vec{0}$ or undefined. If f has a local maximum or minimum at a point P_0, not on the boundary of its domain, then P_0 is a critical point (p. 806). A **quadratic function** $f(x, y) = ax^2 + bxy + cz^2$ generally has one critical point, which can be a local maximum, a local minimum, or a **saddle point** (p. 809).

Second derivative test for functions of two variables (p. 811). Suppose grad $f(x_0, y_0) = \vec{0}$. Let $D = f_{xx}(x_0, y_0)f_{yy}(x_0, y_0) - (f_{xy}(x_0, y_0))^2$.

- If $D > 0$ and $f_{xx}(x_0, y_0) > 0$, then f has a local minimum at (x_0, y_0).
- If $D > 0$ and $f_{xx}(x_0, y_0) < 0$, then f has a local maximum at (x_0, y_0).
- If $D < 0$, then f has a saddle point at (x_0, y_0).
- If $D = 0$, anything can happen.

Unconstrained optimization
A function f defined on a region R has a **global maximum on** R at the point P_0 if $f(P_0) \geq f(P)$ for all points P in R, and a **global minimum on** R at the point P_0 if $f(P_0) \leq f(P)$ for all points P in R (p. 816). For an **unconstrained optimization problem**, find the critical points and investigate whether the critical points give global maxima or minima (p. 816).

A **closed** region is one which contains its boundary; a **bounded** region is one which does not stretch to infinity in any direction (p. 821).

Extreme Value Theorem for Multivariable Functions. If f is a continuous function on a closed and bounded region R, then f has a global maximum at some point (x_0, y_0) in R and a global minimum at some point (x_1, y_1) in R (p. 821).

Constrained optimization
Suppose P_0 is a point satisfying the constraint $g(x, y) = c$. A function f has a **local maximum** at P_0 **subject to the constraint** if $f(P_0) \geq f(P)$ for all points P near P_0 satisfying the constraint (p. 827). It has a **global maximum** at P_0 **subject to the constraint** if $f(P_0) \geq f(P)$ for all points P satisfying the constraint (p. 827). Local and global minima are defined similarly (p. 827). A local maximum or minimum of $f(x, y)$ subject to a constraint $g(x, y) = c$ occurs at a point where the constraint is tangent to a level curve of f, and thus where grad g is parallel to grad f (p. 827).

To optimize f subject to the constraint $g = c$ (p. 827), find the points satisfying the equations

$$\text{grad}\, f = \lambda\, \text{grad}\, g \quad \text{and} \quad g = c.$$

Then compare values of f at these points, at points on the constraint where grad $g = \vec{0}$, and at the endpoints of the constraint. The number λ is called the **Lagrange multiplier**.

To optimize f subject to the constraint $g \le c$ (p. 829), find all points in the interior $g(x, y) < c$ where grad f is zero or undefined; then use Lagrange multipliers to find the local extrema of f on the boundary $g(x, y) = c$. Evaluate f at the points found and compare the values.

The value of λ is the rate of change of the optimum value of f as c increases (where $g(x, y) = c$) (p. 831). The **Lagrangian function** $\mathcal{L}(x, y, \lambda) = f(x, y) - \lambda(g(x, y) - c)$ can be used to convert a constrained optimization problem for f subject the constraint $g = c$ into an unconstrained problem for $\mathcal{L}$ (p. 831).

Multivariable Integration

The **definite integral** of f, a continuous function of two variables, over R, the rectangle $a \le x \le b, c \le y \le d$, is called a **double integral**, and is a limit of **Riemann sums**

$$\int_R f\,dA = \lim_{\Delta x, \Delta y \to 0} \sum_{i,j} f(u_{ij}, v_{ij})\Delta x \Delta y \quad \text{(p. 842).}$$

The Riemann sum is constructed by subdividing R into subrectangles of width Δx and height Δy, and choosing a point (u_{ij}, v_{ij}) in the ij-th rectangle.

A **triple integral** of f, a continuous function of three variables, over W, the box $a \le x \le b, c \le y \le d, p \le z \le q$ in 3-space, is defined in a similar way using three-variable Riemann sums (p. 858).

Interpretations

If $f(x, y)$ is positive, $\int_R f\,dA$ is the **volume** under graph of f above the region R (p. 842). If $f(x, y) = 1$ for all x and y, then the area of R is $\int_R 1\,dA = \int_R dA$ (p. 844). If $f(x, y)$ is a **density**, then $\int_R f\,dA$ is the **total quantity** in the region R (p. 840). The **average value** of $f(x, y)$ on the region R is $\frac{1}{\text{Area of } R}\int_R f\,dA$ (p. 844). In probability, if $p(x, y)$ is a **joint density function** then $\int_a^b \int_c^d p(x, y)\,dy\,dx$ is the fraction of population with $a \le x \le b$ and $c \le y \le d$ (p. 880).

Iterated integrals

Double and triple integrals can be written as **iterated integrals**

$$\int_R f\,dA = \int_c^d \int_a^b f(x, y)\,dx\,dy \text{ (p. 848)}$$

$$\int_W f\,dV = \int_p^q \int_c^d \int_a^b f(x, y, z)\,dx\,dy\,dz \text{ (p. 858)}$$

Other orders of integration are possible. For iterated integrals over **non-rectangular regions** (p. 850), limits on outer integral are constants and limits on inner integrals involve only the variables in the integrals further out (pp. 851, 860).

Integrals in other coordinate systems

When computing double integrals in polar coordinates, put $dA = r\,dr\,d\theta$ or $dA = r\,d\theta\,dr$ (p. 864). Cylindrical coordinates are given by $x = r\cos\theta$, $y = r\sin\theta$, $z = z$, for $0 \le r < \infty, 0 \le \theta \le 2\pi, -\infty < z < \infty$ (p. 869). Spherical coordinates are given by $x = \rho\sin\phi\cos\theta$, $y = \rho\sin\phi\sin\theta$, $z = \rho\cos\phi$, for $0 \le \rho < \infty, 0 \le \phi \le \pi, 0 \le \theta \le 2\pi$ (p. 872). When computing triple integrals in cylindrical or spherical coordinates, put $dV = r\,dr\,d\theta\,dz$ for cylindrical coordinates (p. 871), $dV = \rho^2\sin\phi\,d\rho\,d\phi\,d\theta$ for spherical coordinates (p. 873). Other orders of integration are also possible.

For a **change of variables** $x = x(s, t)$, $y = y(s, t)$, the **Jacobian** is

$$\frac{\partial(x, y)}{\partial(s, t)} = \frac{\partial x}{\partial s}\cdot\frac{\partial y}{\partial t} - \frac{\partial x}{\partial t}\cdot\frac{\partial y}{\partial s} = \begin{vmatrix} \frac{\partial x}{\partial s} & \frac{\partial x}{\partial t} \\ \frac{\partial y}{\partial s} & \frac{\partial y}{\partial t} \end{vmatrix} \quad \text{(p. 1034).}$$

To convert an integral from x, y to s, t coordinates (p. 1035): Substitute for x and y in terms of s and t, change the xy region R into an st region T, and change the area element by making the substitution $dxdy = \left|\frac{\partial(x,y)}{\partial(s,t)}\right| dsdt$. For triple integrals, there is a similar formula (p. 1036).

Parameterizations and Vector Fields

Parameterized curves

The motion of a particle is described by **parametric equations** $x = f(t), y = g(t)$ (2-space) or $x = f(t), y = g(t), z = h(t)$ (3-space). The path of the particle is a **parameterized curve** (p. 886). Parameterizations are also written in **vector form** $\vec{r}(t) = f(t)\vec{i} + g(t)\vec{j} + h(t)\vec{k}$ (p. 888). For a **curve segment** we restrict the parameter to to a closed interval $a \le t \le b$ (p. 889). **Parametric equations for the graph** of $y = f(x)$ are $x = t, y = f(t)$.

Parametric equations for a line through (x_0, y_0) in the direction of $\vec{v} = a\vec{i} + b\vec{j}$ are $x = x_0 + at$, $y = y_0 + bt$. In 3-space, the line through (x_0, y_0, z_0) in the direction of $\vec{v} = a\vec{i} + b\vec{j} + c\vec{k}$ is $x = x_0 + at$, $y = y_0 + bt$, $z = z_0 + ct$ (p. 887). In vector form, the equation for a line is $\vec{r}(t) = \vec{r}_0 + t\vec{v}$, where $\vec{r}_0 = x_0\vec{i} + y_0\vec{j} + z_0\vec{k}$ (p. 889).

Parametric equations for a circle of radius R in the plane, centered at the origin are $x = R\cos t$, $y = R\sin t$ (counterclockwise), $x = R\cos t$, $y = -R\sin t$ (clockwise).

To find the **intersection points** of a curve $\vec{r}(t) = f(t)\vec{i} + g(t)\vec{j} + h(t)\vec{k}$ with a surface $F(x, y, z) = c$, solve $F(f(t), g(t), h(t)) = c$ for t (p. 890). To find the intersection points of two curves $\vec{r}_1(t)$ and $\vec{r}_2(t)$, solve $\vec{r}_1(t_1) = \vec{r}_2(t_2)$ for t_1 and t_2 (p. 890).

The **length of a curve segment** C given parametrically for $a \le t \le b$ with velocity vector $\vec{v}$ is $\int_a^b \|\vec{v}\|dt$ if $\vec{v} \ne \vec{0}$ for $a < t < b$ (p. 901).

The **velocity** and **acceleration** of a moving object with position vector $\vec{r}(t)$ at time t are

$$\vec{v}(t) = \lim_{\Delta t \to 0}\frac{\Delta\vec{r}}{\Delta t} \quad \text{(p. 897)}$$

$$\vec{a}(t) = \lim_{\Delta t \to 0}\frac{\Delta\vec{v}}{\Delta t} \quad \text{(p. 899)}$$

We write $\vec{v} = \dfrac{d\vec{r}}{dt} = \vec{r}\,'(t)$ and $\vec{a} = \dfrac{d\vec{v}}{dt} = \dfrac{d^2\vec{r}}{dt^2} = \vec{r}\,''(t)$.

The **components of the velocity and acceleration vectors** are

$$\vec{v}(t) = \frac{dx}{dt}\vec{i} + \frac{dy}{dt}\vec{j} + \frac{dz}{dt}\vec{k} \quad \text{(p. 897)}$$

$$\vec{a}(t) = \frac{d^2x}{dt^2}\vec{i} + \frac{d^2y}{dt^2}\vec{j} + \frac{d^2z}{dt^2}\vec{k} \quad \text{(p. 899)}$$

The **speed** is $\|\vec{v}\| = \sqrt{(dx/dt)^2 + (dy/dt)^2 + (dz/dt)^2}$ (p. 901). Analogous formulas for velocity, speed, and acceleration hold in 2-space.

Uniform Circular Motion (p. 900) For a particle $\vec{r}(t) = R\cos(\omega t)\vec{i} + R\sin(\omega t)\vec{j}$: motion is in a circle of radius R with period $2\pi/\omega$; velocity, $\vec{v}$, is tangent to the circle and speed is constant $\|\vec{v}\| = \omega R$; acceleration, $\vec{a}$, points toward the center of the circle with $\|\vec{a}\| = \|\vec{v}\|^2/R$.

Motion in a Straight Line (p. 901) For a particle $\vec{r}(t) = \vec{r}_0 + f(t)\vec{v}_0$: Motion is along a straight line through the point with position vector $\vec{r}_0$ parallel to $\vec{v}_0$; velocity, $\vec{v}$, and acceleration, $\vec{a}$, are parallel to the line.

Vector fields

A **vector field** in 2-space is a function $\vec{F}(x, y)$ whose value at a point (x, y) is a 2-dimensional vector (p. 905). Similarly, a vector field in 3-space is a function $\vec{F}(x, y, z)$ whose values are 3-dimensional vectors (p. 905). Examples are the **gradient** of a differentiable function f, the **velocity field** of a fluid flow, and **force fields** (p. 905). A **flow line** of a vector field $\vec{v} = \vec{F}(\vec{r})$ is a path $\vec{r}(t)$ whose velocity vector equals $\vec{v}$, thus $\vec{r}\,'(t) = \vec{v} = \vec{F}(\vec{r}(t))$ (p. 914). The **flow** of a vector field is the family of all of its flow line (p. 914). Flow lines can be approximated numerically using Euler's method (p. 916).

Parameterized surfaces

We **parameterize a surface** with two parameters, $x = f_1(s,t)$, $y = f_2(s,t)$, $z = f_3(s,t)$ (p. 1024). We also use the vector form $\vec{r}(s,t) = f_1(s,t)\vec{i} + f_2(s,t)\vec{j} + f_3(s,t)\vec{k}$ (p. 1024). **Parametric equations for the graph** of $z = f(x, y)$ are $x = s$, $y = t$, and $z = f(s,t)$ (p. 1024). **Parametric equation for a plane** through the point with position vector $\vec{r}_0$ and containing the two nonparallel vectors $\vec{v}_1$ and $\vec{v}_2$ is $\vec{r}(s,t) = \vec{r}_0 + s\vec{v}_1 + t\vec{v}_2$ (p. 1025). **Parametric equation for a sphere** of radius R centered at the origin is $\vec{r}(\theta, \phi) = R\sin\phi\cos\theta\,\vec{i} + R\sin\phi\sin\theta\,\vec{j} + \cos\phi\,\vec{k}$, $0 \le \theta \le 2\pi, 0 \le \phi \le \pi$ (p. 1025). **Parametric equation for a cylinder** of radius R along the z-axis is $\vec{r}(\theta, z) = R\cos\theta\vec{i} + R\sin\theta\vec{j} + z\vec{k}$, $0 \le \theta \le 2\pi, -\infty < z < \infty$ (p. 1023). A **parameter curve** is the curve obtained by holding one of the parameters constant and letting the other vary (p. 1029).

Line Integrals

The **line integral** of a vector field $\vec{F}$ along an **oriented curve** C (p. 922) is

$$\int_C \vec{F} \cdot d\vec{r} = \lim_{\|\Delta\vec{r}_i\| \to 0} \sum_{i=0}^{n-1} \vec{F}(\vec{r}_i) \cdot \Delta\vec{r}_i,$$

where the direction of $\Delta\vec{r}_i$ is the direction of the orientation (p. 923).

The line integral measures the extent to which C is going with $\vec{F}$ or against it (p. 924). For oriented curves C, C_1, and C_2, $\int_{-C}\vec{F} \cdot d\vec{r} = -\int_C \vec{F} \cdot d\vec{r}$, where $-C$ is the curve C parameterized in the opposite direction, and $\int_{C_1+C_2}\vec{F} \cdot d\vec{r} = \int_{C_1}\vec{F} \cdot d\vec{r} + \int_{C_2}\vec{F} \cdot d\vec{r}$, where C_1+C_2 is the curve obtained by joining the endpoint of C_1 to the starting point of C_2 (p. 928).

The **work done by a force** $\vec{F}$ along a curve C is $\int_C \vec{F} \cdot d\vec{r}$ (p. 925). The **circulation** of $\vec{F}$ around an oriented closed curve is $\int_C \vec{F} \cdot d\vec{r}$ (p. 927).

Given a parameterization of C, $\vec{r}(t)$, for $a \le t \le b$, the line integral can be calculated as

$$\int_C \vec{F} \cdot d\vec{r} = \int_a^b \vec{F}(\vec{r}(t)) \cdot \vec{r}\,'(t)\,dt \quad \text{(p. 932).}$$

Fundamental Theorem for Line Integrals (p. 939):
Suppose C is a piecewise smooth oriented path with starting point P and endpoint Q. If f is a function whose gradient is continuous on the path C, then

$$\int_C \operatorname{grad} f \cdot d\vec{r} = f(Q) - f(P).$$

Path-independent fields and gradient fields

A vector field $\vec{F}$ is said to be **path-independent**, or **conservative**, if for any two points P and Q, the line integral $\int_C \vec{F} \cdot d\vec{r}$ has the same value along any piecewise smooth path C from P to Q lying in the domain of $\vec{F}$ (p. 941). A **gradient field** is a vector field of the form $\vec{F} = \operatorname{grad} f$ for some scalar function f, and f is called a **potential function** for the vector field $\vec{F}$ (p. 942). A vector field $\vec{F}$ is path-independent if and only if $\vec{F}$ is a gradient vector field (p. 942). A vector field $\vec{F}$ is path-independent if and only if $\int_C \vec{F} \cdot d\vec{r} = 0$ for every closed curve C (p. 950). If $\vec{F}$ is a gradient field, then $\frac{\partial F_2}{\partial x} - \frac{\partial F_1}{\partial y} = 0$ (p. 951). The quantity $\frac{\partial F_2}{\partial x} - \frac{\partial F_1}{\partial y}$ is called the 2-dimensional or scalar **curl** of $\vec{F}$.

Green's Theorem (p. 951):
Suppose C is a piecewise smooth simple closed curve that is the boundary of an open region R in the plane and oriented so that the region is on the left as we move around the curve. Suppose $\vec{F} = F_1\vec{i} + F_2\vec{j}$ is a smooth vector field defined at every point of the region R and boundary C. Then

$$\int_C \vec{F} \cdot d\vec{r} = \int_R \left(\frac{\partial F_2}{\partial x} - \frac{\partial F_1}{\partial y}\right) dx\,dy.$$

Curl test for vector fields in 2-space: If $\frac{\partial F_2}{\partial x} - \frac{\partial F_1}{\partial y} = 0$ and the domain of $\vec{F}$ has no holes, then $\vec{F}$ is path-independent, and hence a gradient field (p. 953). The condition that the domain have no holes is important. It is not always true that if the scalar curl of $\vec{F}$ is zero then $\vec{F}$ is a gradient field (p. 954).

Surface Integrals

A surface is **oriented** if a unit normal vector $\vec{n}$ has been chosen at every point on it in a continuous way (p. 962). For a closed surface, we usually choose the outward orientation (p. 962). The **area vector** of a flat, oriented surface is a vector $\vec{A}$ whose magnitude is the area of the surface, and whose direction is the direction of the orientation vector $\vec{n}$ (p. 963). If $\vec{v}$ is the velocity vector of a constant fluid flow and $\vec{A}$ is the area vector of a flat surface, then the total flow through the surface in units of volume per unit time is called the **flux** of $\vec{v}$ through the surface and is given by $\vec{v} \cdot \vec{A}$ (p. 963).

The **surface integral** or **flux integral** of the vector field $\vec{F}$ through the oriented surface S is

$$\int_S \vec{F} \cdot d\vec{A} = \lim_{\|\Delta \vec{A}\| \to 0} \sum \vec{F} \cdot \Delta \vec{A},$$

where the direction of $\Delta \vec{A}$ is the direction of the orientation (p. 964). If $\vec{v}$ is a variable vector field and then $\int_S \vec{v} \cdot d\vec{A}$ is the flux through the surface S (p. 965).

Simple flux integrals can be calculated by putting $d\vec{A} = \vec{n}\, dA$ and using geometry or converting to a double integral (p. 967).

The **flux through a graph** of $z = f(x, y)$ above a region R in the xy-plane, oriented upward, is

$$\int_R \vec{F}(x, y, f(x, y)) \cdot \left(-f_x \vec{i} - f_y \vec{j} + \vec{k}\right) dx\, dy \quad \text{(p. 974)}.$$

The **area of the part of the graph** of $z = f(x, y)$ above a region R in the xy-plane is

$$\text{Area of } S = \int_R \sqrt{(f_x)^2 + (f_y)^2 + 1}\, dx\, dy \quad \text{(p. 975)}.$$

The **flux through a cylindrical surface** S of radius R and oriented away from the z-axis is

$$\int_T \vec{F}(R, \theta, z) \cdot \left(\cos\theta \vec{i} + \sin\theta \vec{j}\right) R\, dz\, d\theta \quad \text{(p. 976)},$$

where T is the θz-region corresponding to S.

The **flux through a spherical surface** S of radius R and oriented away from the origin is

$$\int_T \vec{F}(R, \theta, \phi) \cdot \left(\sin\phi\cos\theta \vec{i} + \sin\phi\sin\theta \vec{j} + \cos\phi \vec{k}\right) R^2 \sin\phi\, d\phi\, d\theta, \quad \text{(p. 977)}$$

where T is the $\theta\phi$-region corresponding to S.

The **flux through a parameterized surface** S, parameterized by $\vec{r} = \vec{r}(s, t)$, where (s, t) varies in a parameter region R, is

$$\int_R \vec{F}(\vec{r}(s, t)) \cdot \left(\frac{\partial \vec{r}}{\partial s} \times \frac{\partial \vec{r}}{\partial t}\right) ds\, dt \quad \text{(p. 1038)}.$$

We choose the parameterization so that $\partial\vec{r}/\partial s \times \partial\vec{r}/\partial t$ is never zero and points in the direction of $\vec{n}$ everywhere.

The **area of a parameterized surface** S, parameterized by $\vec{r} = \vec{r}(s, t)$, where (s, t) varies in a parameter region R, is

$$\int_S dA = \int_R \left\| \frac{\partial \vec{r}}{\partial s} \times \frac{\partial \vec{r}}{\partial t} \right\| ds\, dt \quad \text{(p. 1039)}.$$

Divergence and Curl

Divergence

Definition of Divergence (p. 982).
Geometric definition: The **divergence** of $\vec{F}$ is

$$\operatorname{div} \vec{F}(x, y, z) = \lim_{\text{Volume} \to 0} \frac{\int_S \vec{F} \cdot d\vec{A}}{\text{Volume of } S}.$$

Here S is a sphere centered at (x, y, z), oriented outwards, that contracts down to (x, y, z) in the limit.

Cartesian coordinate definition: If $\vec{F} = F_1\vec{i} + F_2\vec{j} + F_3\vec{k}$, then

$$\operatorname{div} \vec{F} = \frac{\partial F_1}{\partial x} + \frac{\partial F_2}{\partial y} + \frac{\partial F_3}{\partial z}.$$

The divergence can be thought of as the outflow per unit volume of the vector field. A vector field $\vec{F}$ is said to be **divergence free** or **solenoidal** if $\operatorname{div}\vec{F} = 0$ everywhere that $\vec{F}$ is defined. Magnetic fields are divergence free (p. 986).

The Divergence Theorem (p. 992). If W is a solid region whose boundary S is a piecewise smooth surface, and if $\vec{F}$ is a smooth vector field which is defined everywhere in W and on S, then

$$\int_S \vec{F} \cdot d\vec{A} = \int_W \operatorname{div} \vec{F}\, dV,$$

where S is given the outward orientation. In words, the Divergence Theorem says that the total flux out of a closed surface is the integral of the flux density over the volume it encloses.

Curl

The **circulation density** of a smooth vector field $\vec{F}$ at (x, y, z) around the direction of the unit vector $\vec{n}$ is defined to be

$$\operatorname{circ}_{\vec{n}} \vec{F}(x, y, z) = \lim_{\text{Area} \to 0} \frac{\text{Circulation around } C}{\text{Area inside } C} = \lim_{\text{Area} \to 0} \frac{\int_C \vec{F} \cdot d\vec{r}}{\text{Area inside } C} \quad \text{(p. 1000)}.$$

Circulation density is calculated using the **right-hand rule** (p. 1000).

Definition of curl (p. 1001).
Geometric definition The curl of $\vec{F}$, written curl $\vec{F}$, is the vector field with the following properties

- The direction of curl $\vec{F}(x, y, z)$ is the direction $\vec{n}$ for which $\text{circ}_{\vec{n}}(x, y, z)$ is greatest.
- The magnitude of curl $\vec{F}(x, y, z)$ is the circulation density of $\vec{F}$ around that direction.

Cartesian coordinate definition If $\vec{F} = F_1\vec{i} + F_2\vec{j} + F_3\vec{k}$, then

$$\text{curl}\,\vec{F} = \left(\frac{\partial F_3}{\partial y} - \frac{\partial F_2}{\partial z}\right)\vec{i} + \left(\frac{\partial F_1}{\partial z} - \frac{\partial F_3}{\partial x}\right)\vec{j} + \left(\frac{\partial F_2}{\partial x} - \frac{\partial F_1}{\partial y}\right)\vec{k}.$$

Curl and circulation density are related by $\text{circ}_{\vec{n}}\,\vec{F} = \text{curl}\,\vec{F} \cdot \vec{n}$ (p. 1004). A vector field is said to be **curl free** or **irrotational** if curl $\vec{F} = \vec{0}$ everywhere that $\vec{F}$ is defined (p. 1004).

Given an oriented surface S with a boundary curve C we use the right-hand rule to determine the orientation of C (p. 1008).

Stokes' Theorem (p. 1009). If S is a smooth oriented surface with piecewise smooth, oriented boundary C, and if $\vec{F}$ is a smooth vector field which is defined on S and C, then

$$\int_C \vec{F} \cdot d\vec{r} = \int_S \text{curl}\,\vec{F} \cdot d\vec{A}.$$

Stokes' Theorem says that the total circulation around C is the integral over S of the circulation density. A **curl field** is a vector field $\vec{F}$ that can be written as $\vec{F} = \text{curl}\,\vec{G}$ for some vector field $\vec{G}$, called a **vector potential** for $\vec{F}$ (p. 1011).

Relation between divergence, gradient, and curl
The curl and gradient are related by curl grad $f = 0$ (p. 1015). Divergence and curl are related by div curl $\vec{F} = 0$ (p. 1017).

The curl test for vector fields in 3-space (p. 1016) Suppose that curl $\vec{F} = \vec{0}$, and that the domain of $\vec{F}$ has the property that every closed curve in it can be contracted to a point in a smooth way, staying at all times within the domain. Then $\vec{F}$ is path-independent, so $\vec{F}$ is a gradient field and has a potential function.

The divergence test for vector fields in 3-space (p. 1017) Suppose that div $\vec{F} = 0$, and that the domain of $\vec{F}$ has the property that every closed surface in it is the boundary of a solid region completely contained in the domain. Then $\vec{F}$ is a curl field.

ANSWERS TO ODD NUMBERED PROBLEMS

Section 1.1

1 Pop 7 million in 2015

5 $y = (1/2)x + 2$

7 $y = 2x + 2$

9 Slope: $-12/7$
Vertical intercept: $2/7$

11 Slope: 2
Vertical intercept: $-2/3$

13 (a) (V)
(b) (VI)
(c) (I)
(d) (IV)
(e) (III)
(f) (II)

15 We use the point-slope form, getting $y - c = m(x - a)$. Alternatively, we can substitute the point (a, c) into the equation $y = b + mx$ and solve for b:

$$c = ma + b$$
$$b = c - ma.$$

Substituting our value of b into $y = b + mx$ gives:

$$y = c - ma + mx = c + m(x - a).$$

17 $y = -\frac{1}{5}x + \frac{7}{5}$

19 Parallel: $y = m(x - a) + b$
Perpendicular:
$y = (-1/m)(x - a) + b$

21 Domain: $1 \le x \le 5$
Range: $1 \le y \le 6$

23 Domain: $0 \le x \le 5$
Range: $0 \le y \le 4$

25 Domain: all x
Range: $0 < y \le 1/2$

27 $V = kr^3$

29 $S = kh^2$

31 $N = k/l^2$

33 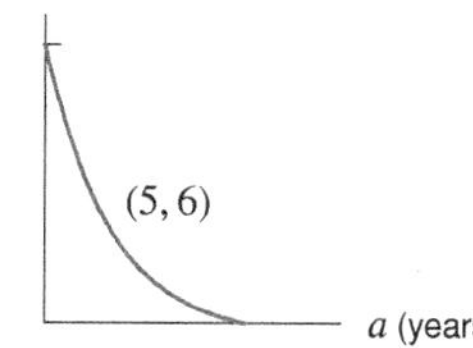

35 $f(12) = 7.049$

37 $f(63) = f(12)$

39 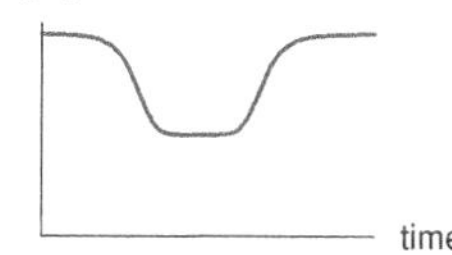

41 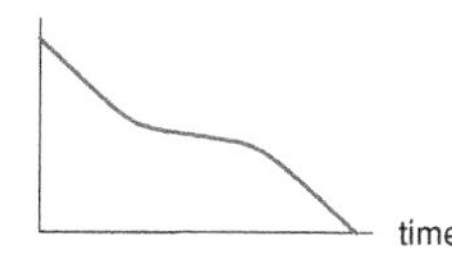

45 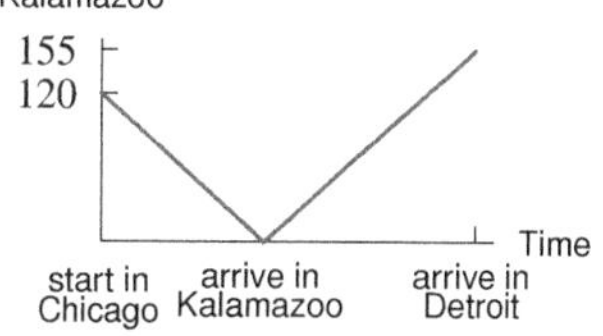

47 (a) About 10.4 kg/hectare per mm
(b) Additional 1 mm annual rainfall corresponds to additional 10.4 kg grass per hectare
(c) $Q = -440 + 10.4r$

49 1939: more addl grass/addl rainfall

51 (a) About 10 days
(b) About 0.775 days per day
(c) Flower later when snow cover lasts longer
(d) $y = 41.25 + 0.775x$. Other answers are possible

53 (a) $C = 4.16 + 0.12w$
(b) 0.12 \$/gal
(c) \$4.16

55 (a) \$0.025/cubic foot
(b) $c = 15 + 0.025w$
c = cost of water
w = cubic feet of water
(c) 3400 cubic feet

57 (a) (i) $f(2001) = 272$
(ii) $f(2014) = 525$
(b) $(f(2014) - f(2001))/(2014 - 2001) =$ 19.46 billionaires/yr
(c) $f(t) = 19.46t - 38{,}667.5$

59 $Q(m) = T + L + Pm$
T = fuel for take-off
L = fuel for landing
P = fuel per mile in the air
m = length of the trip (miles)

61 (a) -0.01
(b) -0.2

63 (a) 2007–2009
(b) 2012–2014

65 (a) 7.094 meters
(b) 1958, 1883

67 (a) $C_1 = 40 + 0.15m$
$C_2 = 50 + 0.10m$
(b) 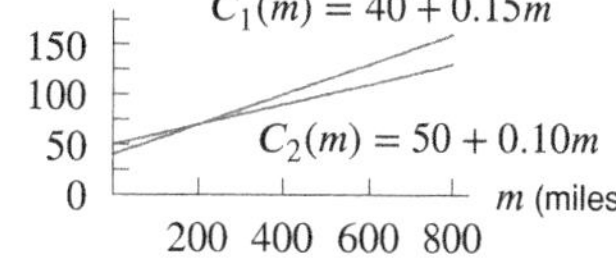

(c) For distances less than 200 miles, C_1 is cheaper.
For distances more than 200 miles, C_2 is cheaper.

69 (a) Higher price: customers want less but owner wants to sell more
(b) (60, 18): No
(120, 12): Yes
(c) Shaded region is all possible quantities of cakes owner is willing to sell and customers willing to buy

p
25
20
15
10
5
$d(q) = 20 - q/20$
(120, 14)
$s(q) = 11 + q/40$
q
40 80 120 160 200

(d) (120, 14): equilibrium price.

73 $x = 3$ not a function of x

75 $y = 0.5 - 3x$ is decreasing

77 $y = 2x + 3$

79 False

81 True

83 False; $y = x + 1$ at points (1, 2) and (2, 3)

85 (b), (c)

Section 1.2

1 Concave up

3 Neither

5 5; 7%

7 3.2; 3% (continuous)

9 $P = 15(1.2840)^t$; growth

11 $P = P_0(1.2214)^t$; growth

13 (a) 1.5
(b) 50%

15 (a) $P = 1000 + 50t$
(b) $P = 1000(1.05)^t$

17 (a) D to E, H to I
(b) A to B, E to F
(c) C to D, G to H
(d) B to C, F to G

19 (a) $h(x) = 31 - 3x$
(b) $g(x) = 36(1.5)^x$

21 Table D

23 $f(s) = 2(1.1)^s$
$g(s) = 3(1.05)^s$
$h(s) = (1.03)^s$

25 (a) $g(x)$
(b) $h(x)$
(c) $f(x)$

27 $y = 3(2^x)$

29 $y = 2(3^x)$

31 $y = 4\left(\frac{1}{2}\right)^x$

33 (a) 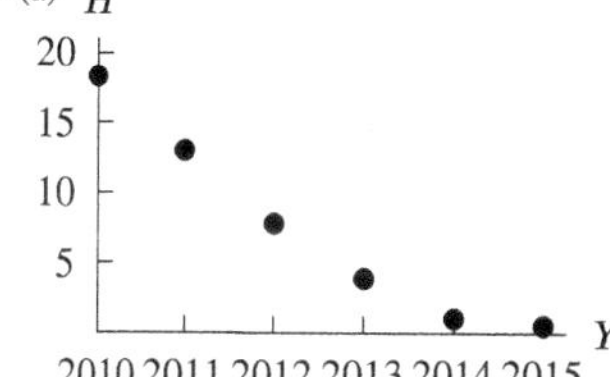

(b) No
(c) No

35 (a)

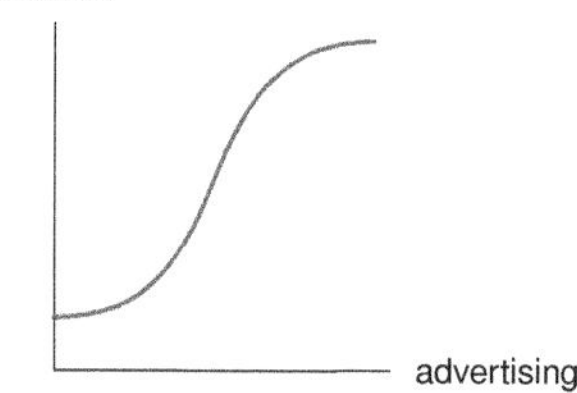

(b)

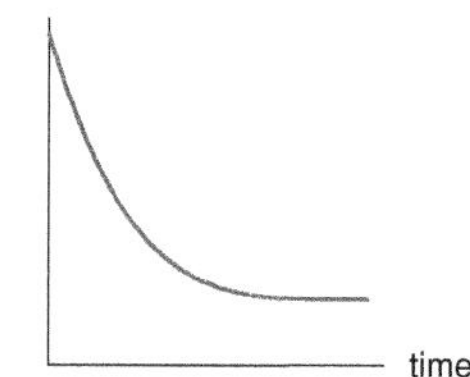

37 17.882 mn barrels/day

39 $d = 670(1.096)^{h/1000} = 670(1.00009124)^h$

41 (a) II
(b) 18.9% growth; 10.9% loss; 20.2% loss

43 (a) $S(t) = 219(1.05946)^t$
(b) 5.946%

45 (a) $2P_0, 4P_0, 8P_0$
(b) $t/50$; $P = P_0 2^{t/50}$

47 (b) 80.731%

49 (a) Users in 2006, thousands
(b) 81% per year
(c) 124.8% per year
(d) Less

51 (a) False
(b) 2011

53 (a) 923 trillion BTUs, 1163 trillion BTUs
(b)

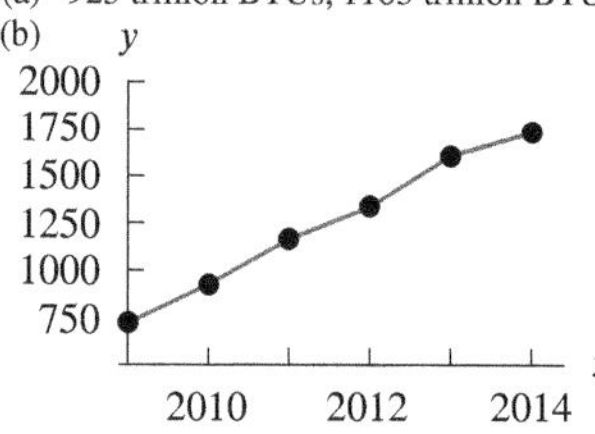

(c) 2013, 268 trillion BTUs

55 (a) a, c, p
(b) a, d, q
(c) $c = p$
(d) a and b are reciprocals
p and q are reciprocals

57 $y = e^{-0.25x}$ concave up

59 $(2, 2e)$, $(3, 3e)$ not on the graph

61 $q = 2.2(0.97)^t$

63 $y = e^{-x} - 5$

65 True

67 True

69 True

71 True

Section 1.3

1 (a)

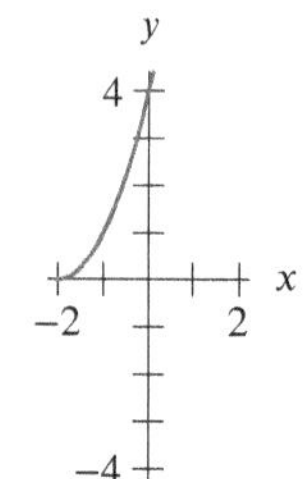

(b)

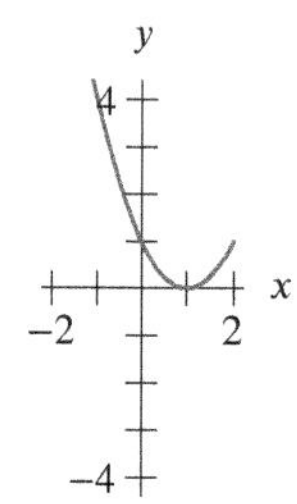

(c)

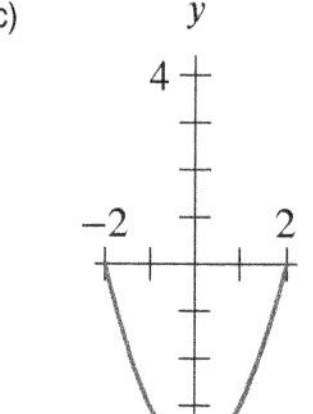

(d)

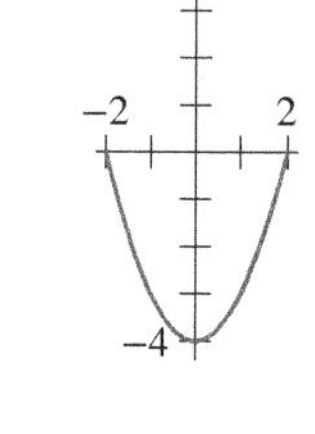

(e)

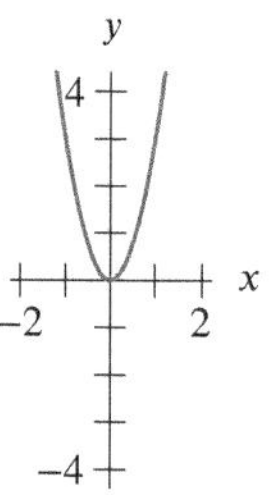

(f)

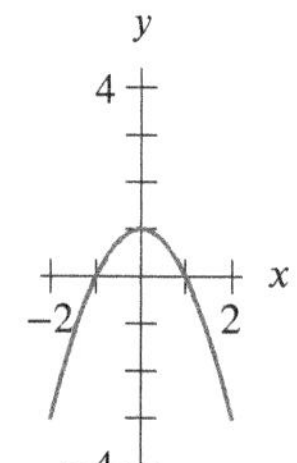

3 (a)

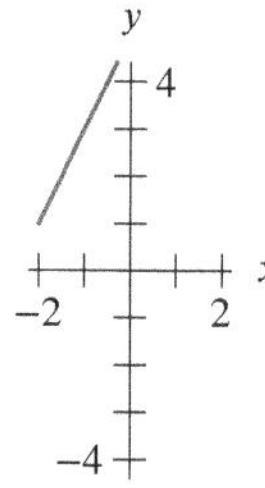

(b)

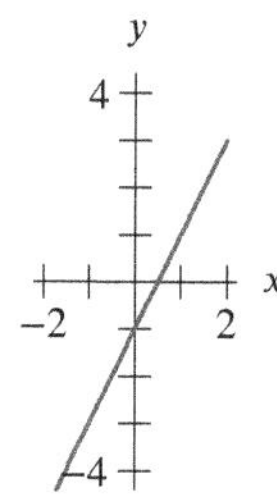

(c)

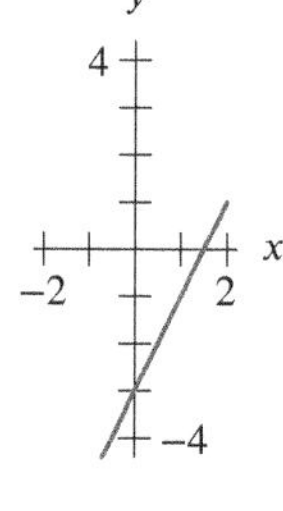

(d)

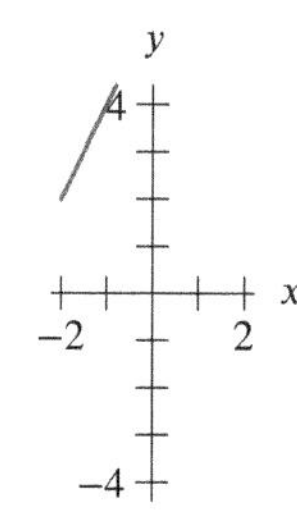

(e)

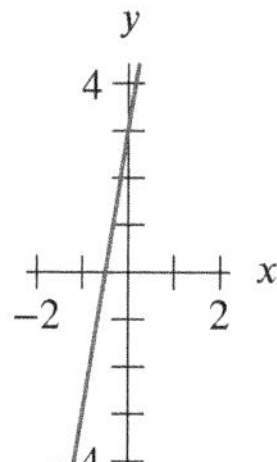

(f)

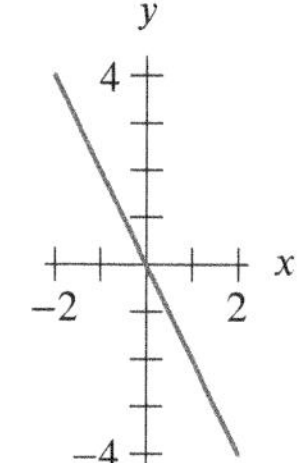

5

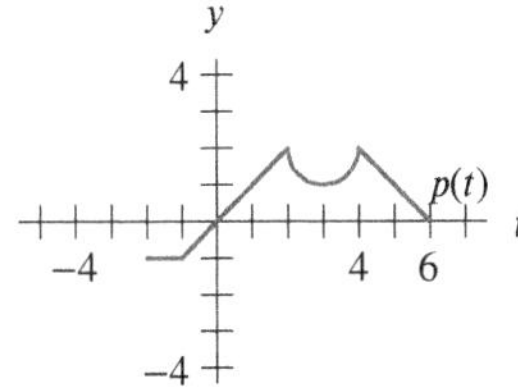

7

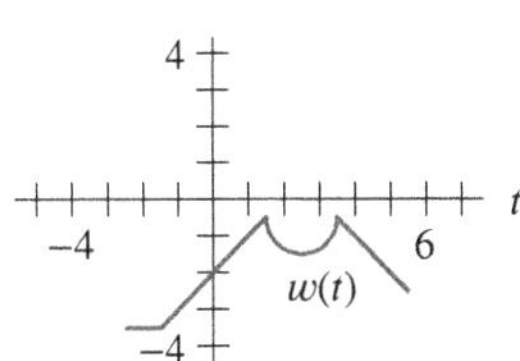

9 (a) 4
(b) 2
(c) $(x+1)^2$
(d) x^2+1
(e) $t^2(t+1)$

11 (a) e
(b) e^2
(c) e^{x^2}
(d) e^{2x}
(e) $e^t t^2$

13 (a) t^2+2t+2
(b) t^4+2t^2+2
(c) 5
(d) $2t^2+2$
(e) t^4+2t^2+2

15 $2z+1$

17 $2zh-h^2$

19 Neither

21 Even

23 odd

25 Odd

27 Not invertible

29 not invertible

31 Invertible

35 (a) $y=2x^2+1$

(b) $y=2(x^2+1)$
(c) No

37 $y=-(x+1)^2+3$

39 Can't be done

41 About 60

43 0.4

45 −0.9

47

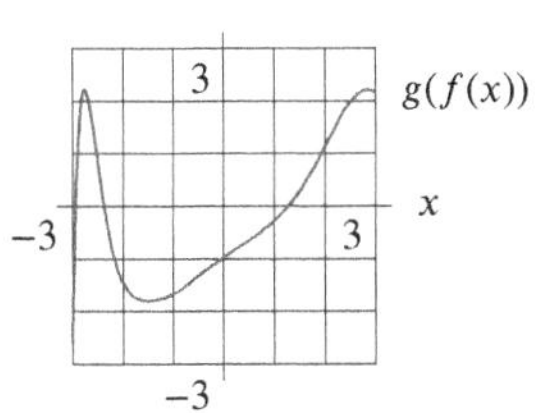

49 $f(x)=x^3$
$g(x)=x+1$

51 $f(x)=\sqrt{x}$
$g(x)=x^2+4$

53 $2(y-1)^3-(y-1)^2$

57 (a) −1
(b)

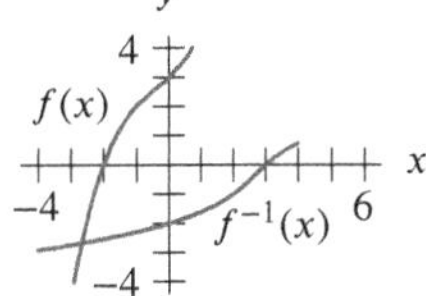

59 Not invertible

61 Not invertible

67 $g(f(t))$ ft^3

69 $f^{-1}(30)$ min

71 (a) $q=\frac{C-100}{2}=f^{-1}(C)$
(b) Number of articles produced at given cost

73 (a) $f(15)\approx 48$
(b) Yes
(c) $f^{-1}(120)\approx 35$
Rock is 35 millions yrs old at depth of 120 meters

75 Shift left

77 $f(g(x))\neq x$

79 $f^{-1}(x)=x$

81 $f(x)=x^2+2$

83 $f(x)=1.5x$, $g(x)=1.5x+3$

85 True

87 True

89 False

91 False

93 True

95 Impossible

97 Impossible

Section 1.4

1 1/2

3 $5A^2$

5 $-1+\ln A+\ln B$

7 $(\log 11)/(\log 3)=2.2$

9 $(\log(2/5))/(\log 1.04)=-23.4$

11 1.68

13 6.212

15 0.26

17 1

19 $(\log a)/(\log b)$

21 $(\log Q-\log Q_0)/(n\log a)$

23 $\ln(a/b)$

25 $P=15e^{0.4055t}$

27 $P=174e^{-0.1054t}$

29 $p^{-1}(t)\approx 58.708\log t$

31 $f^{-1}(t)=e^{t-1}$

33 0.0693

35 13,500 bacteria

37 (a) 2024
(b) 336.49 million people

39 19.807 mn barrels per day

41 (a) $Q_0(1.0033)^x$
(b) 210.391 microgm/cubic m

43 (a) $A=100e^{-0.17t}$
(b) $t\approx 4$ hours

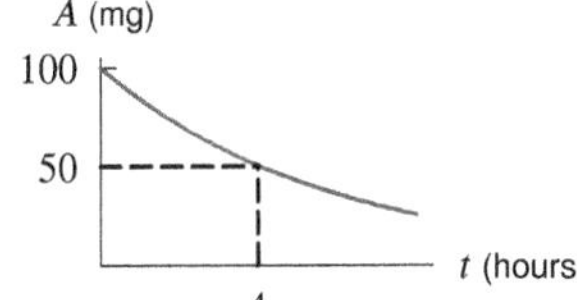

(c) $t=4.077$ hours

45 (a) 10 mg
(b) 18%
(c) 3.04 mg
(d) 11.60 hours

47 7.925 hours

49 (a) $B(t)=B_0e^{0.067t}$
(b) $P(t)=P_0e^{0.033t}$
(c) $t=20.387$; in 2000

51 2023

53 (a) 0.00664
(b) $t=22.277$; April 11, 2032

55 2054

57 It is a fake

59 (a) 63.096 million
(b) 3.162

61 A: e^x
B: x^2
C: $x^{1/2}$
D: $\ln x$

63 (a) Increases
(b) Moves to the left

65 (a) No effect
(b) Moves toward the origin

67 Moves toward the origin

69 Function even

71 Only for $x>1$

73 Correct property is $\ln(AB)=\ln A+\ln B$

75 $y=0.7\log x$

77 $\ln(x-3)$

79 False

81 False

Section 1.5

1 Negative
0
Undefined

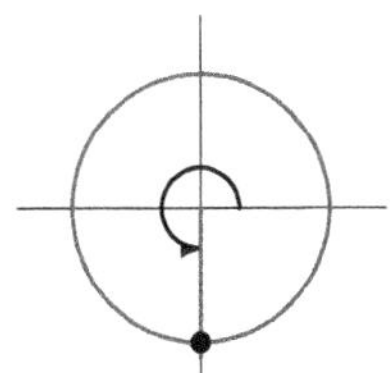

3 Positive
Positive
Positive

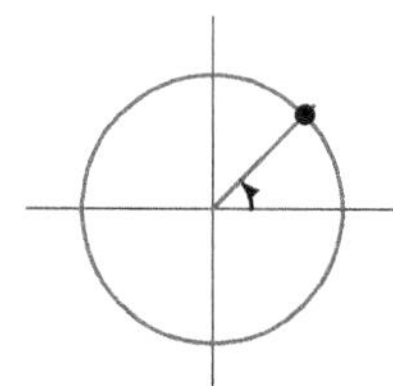

5 Positive
Positive
Positive

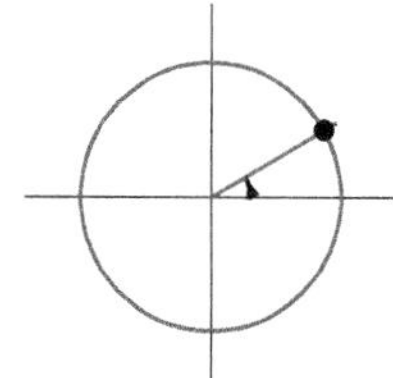

7 Positive
Negative
Negative

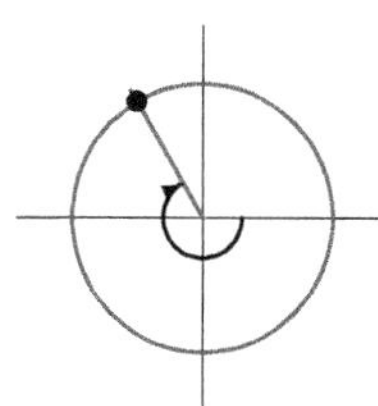

9 Negative
Positive
Negative

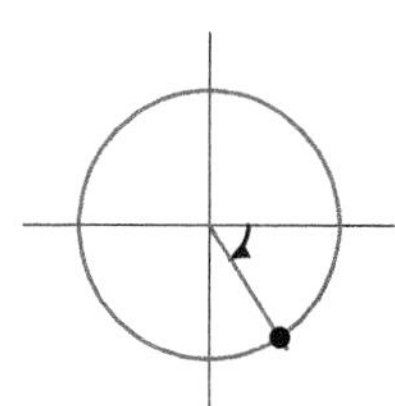

11 $8\pi; 3$

13 $2; 0.1$

15 $f(x) = 5\cos(x/3)$

17 $f(x) = -8\cos(x/10)$

19 $f(x) = 2\cos(5x)$

21 $f(x) = 3\sin(\pi x/9)$

23 $f(x) = 3 + 3\sin((\pi/4)x)$

25 0.588

27 −0.259

29 0.727

31 $(\cos^{-1}(0.5) - 1)/2 \approx 0.0236$

33 $(\arctan(0.5) - 1)/2 = -0.268$

35 One year

37 0.3 seconds

39 (a) 4.76°
(b) 7.13°
(c) 2.86°

41 (a) $h(t)$
(b) $f(t)$
(c) $g(t)$

43

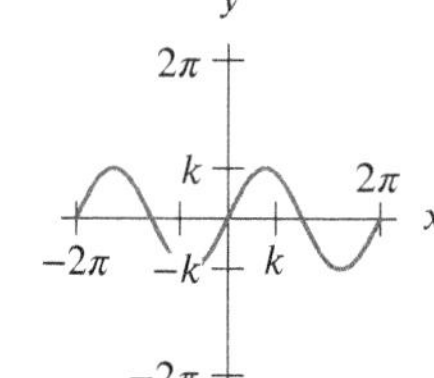

45

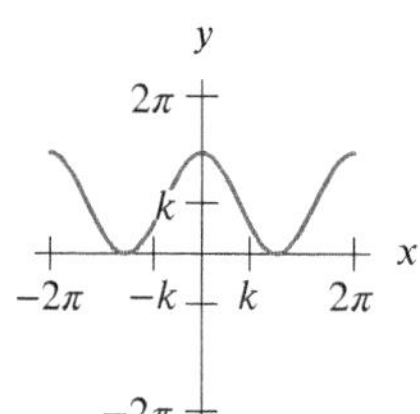

47

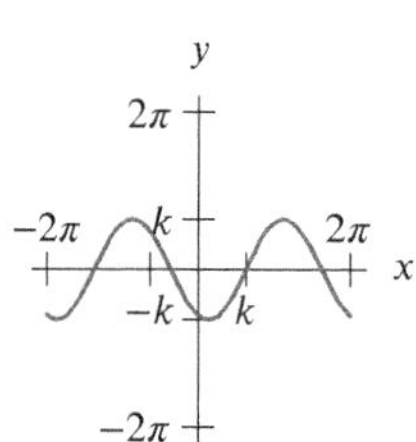

49

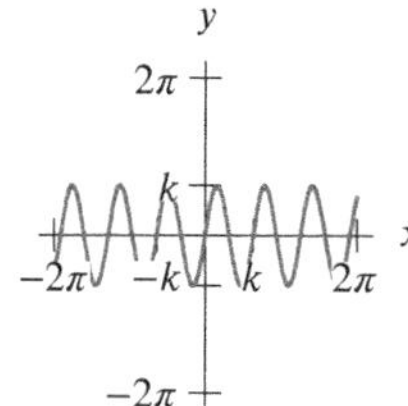

51 12°C

53 7.5 months

55 275 thousand ft^3/sec

57 1 month

59 If $f(x) = \sin x$ and $g(x) = x^2$ then
$\sin x^2 = f(g(x))$
$\sin^2 x = g(f(x))$
$\sin(\sin x) = f(f(x))$

61 (a)

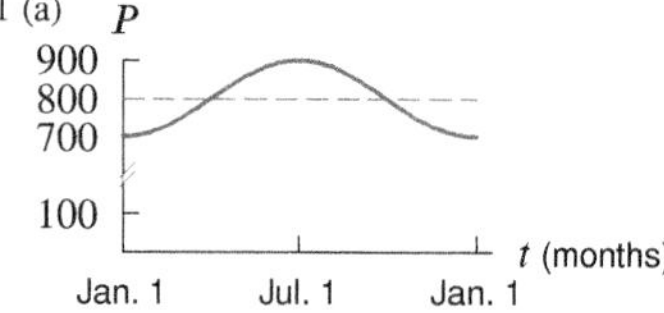

(b) $P = 800 - 100\cos(\pi t/6)$

63 Depth $= 7 + 1.5\sin(\pi t/3)$

65 (a) $f(t) = 10\sin(\pi t/14)$
(b)

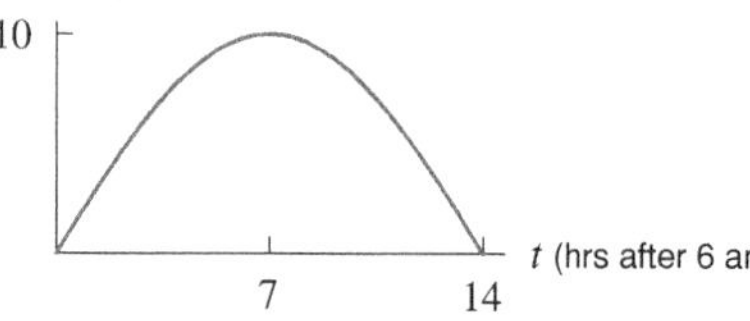

(c) 1 pm; 10 watts
(d) $g(t) = 10\sin(\pi t/9)$

67 (a) $f(t) = -0.5 + \sin t$
$g(t) = 1.5 + \sin t$
$h(t) = -1.5 + \sin t$
$k(t) = 0.5 + \sin t$
(b) $g(t) = 1 + k(t)$

69 (a) Average depth of water
(b) $A = 7.5$
(c) $B = 0.507$
(d) The time of a high tide

71 (a) About 10 ppm
(b) Estimating coefficients $y = (1/6) + 381$
(c) Period: about 12 months; amplitude: about 3.5 ppm
$y = 3.5\sin(\pi t/6)$
(d) $h(t) = 3.5\sin(\pi t/6) + (t/6) + 381$

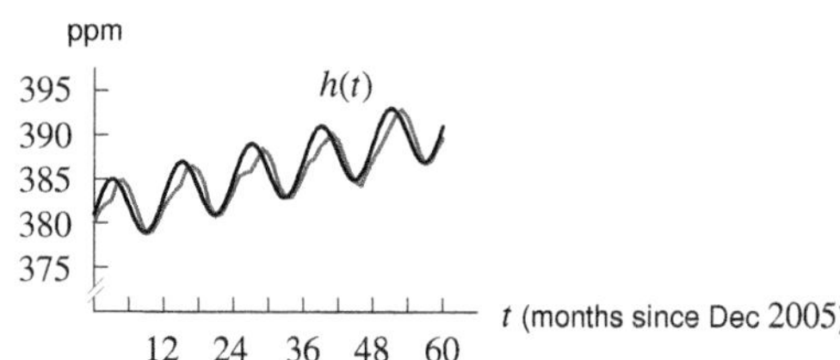

73 (a) 0.4 and 2.7
(b) $\arcsin(0.4) \approx 0.4$
$\pi - \arcsin(0.4) \approx 2.7$
(c) −0.4 and −2.7.
(d) $-0.4 \approx -\arcsin(0.4)$
$-2.7 \approx \arcsin(0.4) - \pi$

75 f has period 2π, but g has different period

77 Period is less than 2π

79 $\sin(2\pi x/23)$

81 True

83 False

85 True

87 False

89 False

91 False

93 False

95 True

97 True

Section 1.6

1 As $x \to \infty$, $y \to \infty$
As $x \to -\infty$, $y \to -\infty$

3 $f(x) \to -\infty$ as $x \to +\infty$
$f(x) \to -\infty$ as $x \to -\infty$

5 $f(x) \to +\infty$ as $x \to +\infty$
$f(x) \to +\infty$ as $x \to -\infty$

7 $f(x) \to 3$ as $x \to +\infty$
$f(x) \to 3$ as $x \to -\infty$

9 $f(x) \to 0$ as $x \to +\infty$
$f(x) \to 0$ as $x \to -\infty$

11 $0.2x^5$

13 1.05^x

15 $25 - 40x^2 + x^3 + 3x^5$

17 (I) **(a)** 3 **(b)** Negative
(II) **(a)** 4 **(b)** Positive
(III) **(a)** 4 **(b)** Negative
(IV) **(a)** 5 **(b)** Negative
(V) **(a)** 5 **(b)** Positive

19 $y = -\frac{1}{2}(x+2)^2(x-2)$

21 $f(x) = kx(x+3)(x-4)$
$(k < 0)$

23 $f(x) =$
$k(x+2)(x-2)^2(x-5)$
$(k < 0)$

25 r

27 (III), (IV)

29 (II), (III), (IV)

31 (II), (IV)

33 IV

35 I, II, V

37 V

39 II

41 Logarithmic

43 Rational

45 1, 2, 3, 4, or 5 roots

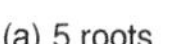

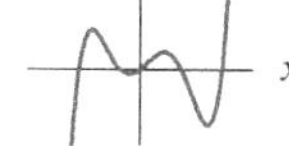

(b) 4 roots

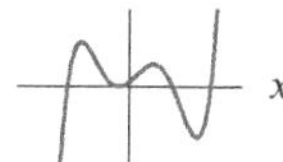

(c) 3 roots

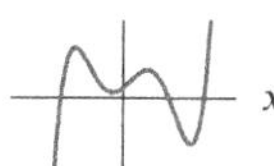

(d) 2 roots

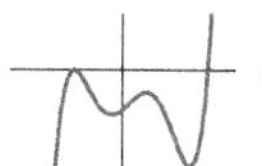

(e) 1 root

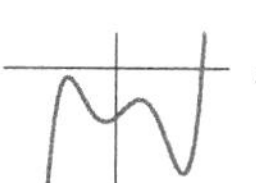

47 3 zeros:
$x \approx -1, x \approx 3,$
$x > 10$

49 $h = V/x^2$

h

$h = V/x^2$

x

51 (a) 1.3 m^2
(b) 86.8 kg
(c) $h = 112.6s^{4/3}$

53 (a) $R = kr^4$
(k is a constant)
(b) $R = 4.938r^4$
(c) $3086.42 \text{ cm}^3/\text{sec}$

55 (a) 0
(b) $t = 2v_0/g$
(c) $t = v_0/g$
(d) $(v_0)^2/(2g)$

57 Yes

59 No

61 No

63 Horizontal: $y = 1$;
Vertical: $x = -2, x = 2$

65 (a) $-\infty, -\infty$
(b) $3/2, 3/2$
(c) $0, +\infty$

67 $(3/x) + 6/(x-2)$
Horizontal asymptote: x-axis
Vertical asymptote: $x = 0$ and $x = 2$

69 $h(t) = ab^t$
$g(t) = kt^3$
$f(t) = ct^2$

71 $y = x$

73 $y = 100e^{-0.2z}$.

75 May tend to negative infinity

77 Only for $x > 1$

79 $f(x) = x/(x^2+1)$ crosses its horizontal asymptote $y = 0$

81 $f(x) = 3x/(x-10)$

83 $f(x) = 1/(x+7\pi)$

85 $f(x) = (x-1)/(x-2)$

87 True

89 True.

91 (a) (I)
(b) (III)
(c) (IV)
(d) (II)

Section 1.7

1 (a) $x = -1, 1$
(b) $-3 < x < -1, -1 < x < 1, 1 < x < 3$

3 (a) 1
(b) 2
(c) Does not exist
(d) 1
(e) 0
(f) 2

5 (a) $x = -2, 3$
(b) $\lim_{x\to-2} f(x) = -3$; $\lim_{x\to3} f(x)$ does not exist

7 1

9 (b) 3

11 Yes

13 Yes

15 No

17 (a) Continuous
(b) Not continuous

23 4

25 6

27 5/9

29 $k = 2$

31 $k = 5/3$

33 $k = 4$

35 (a) Continuous
(b) Not continuous
(c) Not continuous

37 (a) No
(b) Yes

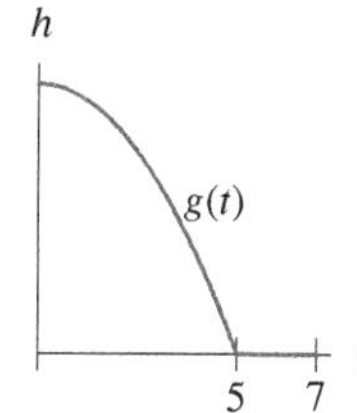

39 Not continuous

41 Does not exist

43 2

45 1

47 0.693

49 $k = \pi - 12$

51 No value

53 $k = \pi/6$

55 (a) $x = 1, 3, 5$
(b) $x = 3$

57 $Q = \begin{cases} 1.2t & 0 \le t \le 0.5 \\ 0.6e^{0.001}e^{-.002t} & 0.5 < t \end{cases}$

59 $x = 5$; $g(5) = 6$

61 $t = \pm 3$; $q(3) = -3, q(-3) = 3$

63 No. Limit does not exist at 0

65 No. $f(0) \neq$ limit at 0

67 Three zeros: one between 5 and 10, one between 10 and 12, the third either less than 5 or greater than 12

69 (a) No
(b) $k = 3$; other answers possible

71 e

73 (a) $x = 1/(n\pi)$,
$n = 1, 2, 3, \ldots$
(b) $x = 2/(n\pi)$,
$n = 1, 5, 9, \ldots$
(c) $x = 2/(n\pi)$,
$n = 3, 7, 11, \ldots$

75 $f(x) = 5$ for some x, not necessarily for $x = 1$

77 $f(x) = \begin{cases} 1 & x \ge 15 \\ -1 & x < 15 \end{cases}$

79 $f(x) = \begin{cases} 1 & x \le 2 \\ x & x > 2 \end{cases}$

81 False, $f(x) = \begin{cases} 1 & x \le 3 \\ 2 & x > 3 \end{cases}$

83 False

Section 1.8

1 (a) 1
(b) 0
(c) 0
(d) 1
(e) Does not exist
(f) 1

3 (a) 4
(b) 2
(c) −4
(d) 2
(e) 4
(f) 0
(g) Does not exist
(h) 2

5 (a) 3
(b) Does not exist or $-\infty$

7 16

9 (a) 8
(b) 6
(c) 15
(d) 4

11 Other answers are possible

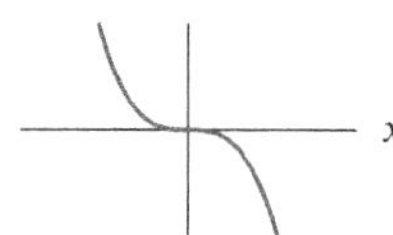

13 Other answers are possible

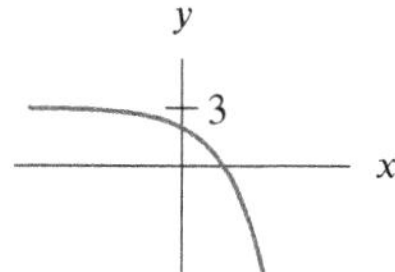

15 Other answers are possible

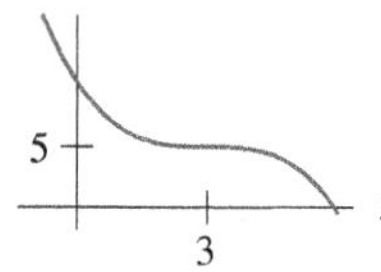

17 ∞

19 ∞

21 0

23 ∞

25 0

27 0

29 $\lim_{x\to-\infty} f(x) = -\infty$;
$\lim_{x\to\infty} f(x) = -\infty$

31 $\lim_{x\to-\infty} f(x) = -\infty$;
$\lim_{x\to\infty} f(x) = \infty$

33 $\lim_{x\to-\infty} f(x) = 0$;
$\lim_{x\to+\infty} f(x) = 0$

35 Yes; not on any interval containing 0

37 $\lim_{x\to2^+} f(x) = \lim_{x\to2^-} f(x) = \lim_{x\to2} f(x) = 0$

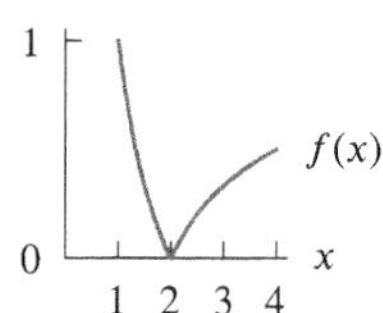

39 e

41 (a)

(b) Yes

43 −1

45 0

47 ∞

49 0

51 5/7

53 (a)

(b) No

55 Other answers are possible

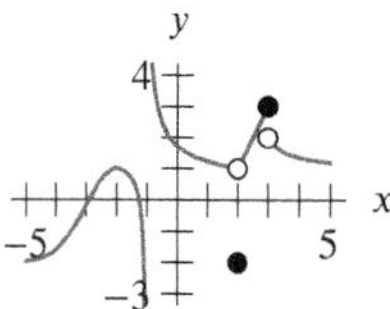

57 (a) (i) 50 mg
(ii) 150 mg
(b) $t = 1, 2, 3, 4$ sec

59 2

61 (a) $\dfrac{x-5}{(x+5)(x-5)}$
(b) 1/10
(c) Neither $\lim_{x\to5} \dfrac{1}{x-5}$ nor $-\frac{10}{x^2-25}$ exist.

65 $P(x)/Q(x)$ is not necessarily defined at all x

67 $f(x) = (x+3)(x-1)/(x-1)$

69 True

71 True

73 True

75 True

77 False

79 True

81 False

83 1

85 3

87 (a) Follows
(b) Does not follow (although true)
(c) Follows
(d) Does not follow

Section 1.9

1 3

3 Does not exist

5 −1/9

7 −1/11

9 −4

11 6/7

13 6

15 5/6

17 −1/9

19 −1/16

21 1/6

23 4

25 (a) Yes
(b) 1

27 0

29 4/5

31 ∞

33 28

35 10

37 4

39 5

41 6

43 7 or −7

45 Any k

47 $k \leq 5$

49 $k \geq 0$

51 5

53 1/3

55 2/5

59 0

61 0

63 There are many possibilities, one being $g(x) = (x^3 - 3x^2 + x - 3)/(x-3)$; (3, 10)

65 There are many possibilities, one being $g(x) = (x \ln x - \ln x)/(x-1)$; (1, 0)

67 0

69 1

71 Does not exist

73 $c = \pm\infty, L = 0$

75 The limit of a function does not depend on the value of a function at a point

77 True

79 False

81 False

83 False

Section 1.10

1 (a) No
(b) No
(c) Yes
(d) Yes
(e) Yes

3 0.1, 0.05, 0.01, 0.005, 0.001, 0.0005

5 0.025

7 0.02

11 0.45, 0.0447, 0.00447

13 $\delta = 1 - e^{-0.1}$

15 $\delta = 1/3$

17 (b) 0

(c) 0.01, −0.01, −0.015, 0.015

(d) $-0.015 < x < 0.015$,
$-0.01 < y < 0.01$

35 False

37 False

39 False

Chapter 1 Review

1 $y = 2x - 10$

3 $y = -3x^2 + 3$

5 $y = 2 + \sqrt{9 - (x+1)^2}$

7 $y = -5x/(x-2)$

9

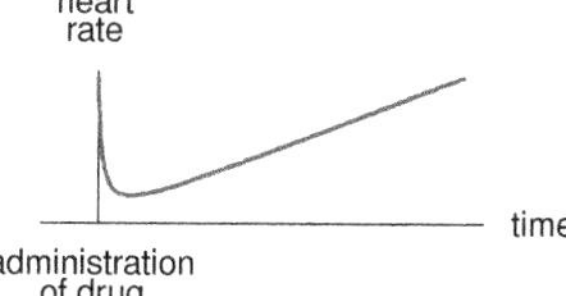

11 (a) $[0, 7]$
(b) $[-2, 5]$
(c) 5
(d) $(1, 7)$
(e) Concave up
(f) 1
(g) No

13 (a) Min gross income for \$100,000 loan
(b) Size of loan for income of \$75,000

15 $t \approx 35.003$

17 $t = \log(12.01/5.02)/\log(1.04/1.03) = 90.283$

19 $P = 5.23e^{-1.6904t}$

21 $f(x) = x^3$
$g(x) = \ln x$

23 Amplitude: 2
Period: $2\pi/5$

25 (a) $f(x) \to \infty$ as $x \to \infty$; $f(x) \to -\infty$ as as $x \to -\infty$
(b) $f(x) \to -\infty$ as as $x \to \pm\infty$
(c) $f(x) \to 0$ as $x \to \pm\infty$
(d) $f(x) \to 6$ as $x \to \pm\infty$

27 $0.25\sqrt{x}$

29 $y = e^{0.4621x}$ or $y = (1.5874)^x$

31 $y = -k(x^2 + 5x)$
$(k > 0)$

33 $z = 1 - \cos\theta$

35 $x = k(y^2 - 4y)$
$(k > 0)$

37 $y = -(x+5)(x+1)(x-3)^2$

39 Simplest is $y = 1 - e^{-x}$

41 $f(x) = \sin(2(\pi/5)x)$

43 Not continuous

45 $\lim_{x\to 3^+} f(x) = 54$, $\lim_{x\to 3^-} f(x) = -54$, $\lim_{x\to 3} f(x)$ does not exist

100, 50, 0, −50, −100; $f(x)$; 1, 2, 3, 4, x

47 (a) −5
(b) −1
(c) Does not exist
(d) Does not exist

49 Yes

51 No

53 0

55 1.098

57 (b) 1

59 0

61 $-5/3$

65 (a) 1
(b) −1
(c) Does not exist
(d) −5

67 10

69 −1

71 9

73 (b) 200 bushels
(c) 80 lbs
(d) About $0 \le Y \le 550$
(e) Decreasing
(f) Concave down

75 (a) (i) $q = 320 - (2/5)p$
(ii) $p = 800 - (5/2)q$
(b)

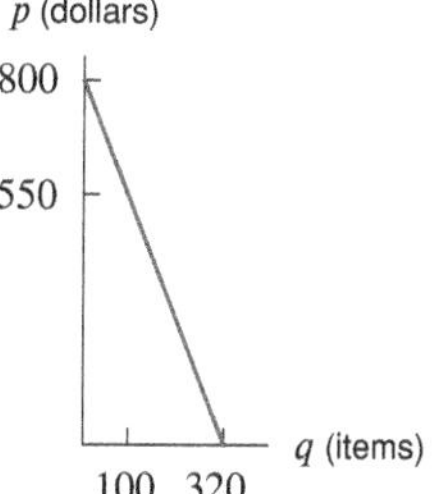

77 (a) (i) Attractive force, pulling atoms together
(ii) Repulsive force, pushing atoms apart
(b) Yes

79 2024

81 (a) Increased: 2011; decreased: 2012
(b) False
(c) True

83 Parabola opening downward

85 10 hours

87 About 14.21 years

89 (a) 81%
(b) 32.9 hours
(c)

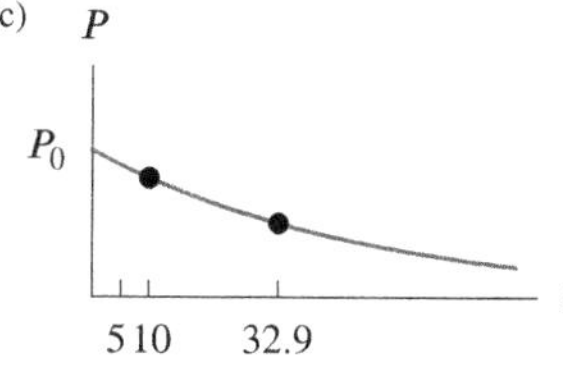

91 One hour

93 US: 156 volts max, 60 cycles/sec
Eur: 339 volts max, 50 cycles/sec

95 (a) (i) $V = 3\pi r^2$
(ii) $V = \pi r^2 h$
(b) (i)

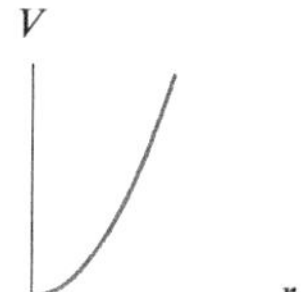

(ii) V, k

97 (a)

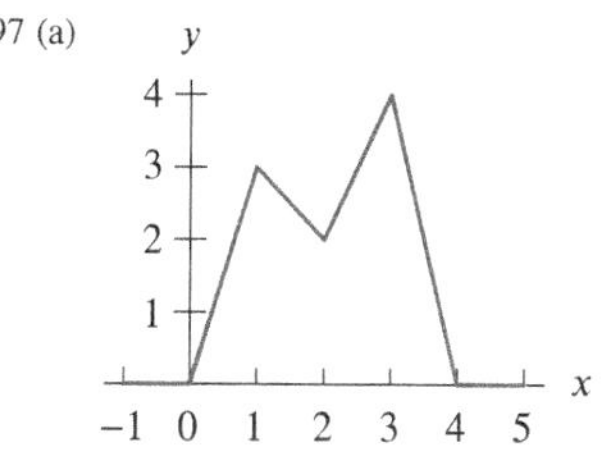

99 (a) III
(b) IV
(c) I
(d) II

101 20

103 Velocity: Not continuous
Distance: Continuous

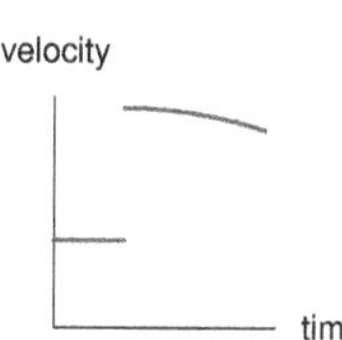

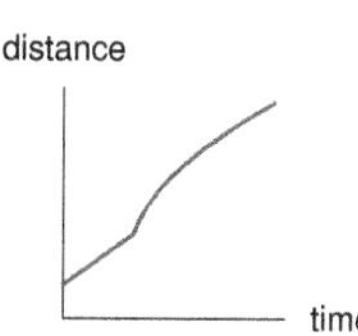

105 (a) 5
(b) 5
(c) 5

107 (a)

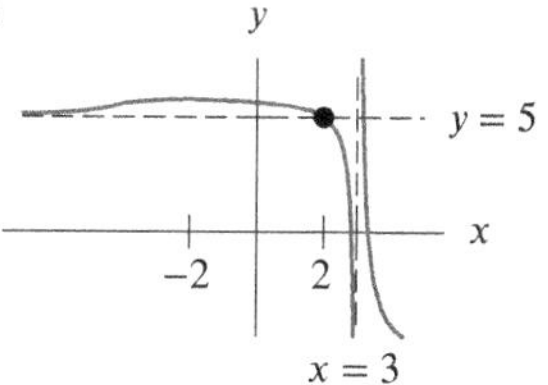

(b) No

109 False

111 Cannot be determined

113 (a) 3.5
(b) Yes

115 $1/(2\sqrt{x})$

117 48/7

121 11

123 25

125 The limit appears to be 1

127 (a) 1
(b) $\lim_{x\to 0} \frac{1}{x}$ does not exist

129 (b) -1
(c)

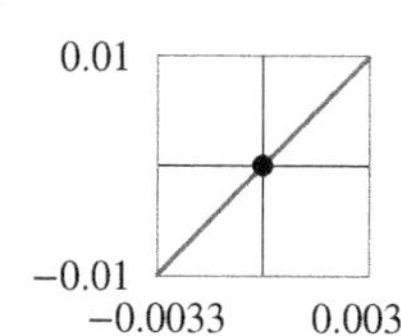

(d) $-0.099 < x < 0.099$, $-1.01 < y < -0.99$

131 (b) 0
(c)

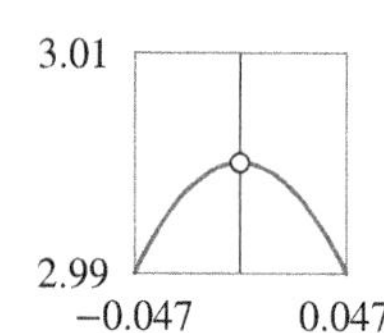

(d) $-0.0033 < x < 0.0033$, $-0.01 < y < 0.01$

133 (b) 3
(c)

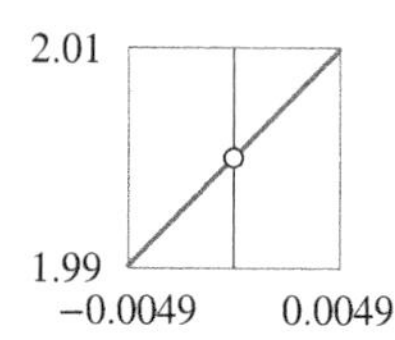

(d) $-0.047 < x < 0.047$, $2.99 < y < 3.01$

135 (b) 2
(c)

2.01

1.99

−0.0049 0.0049

(d) $-0.0049 < x < 0.0049$, $1.99 < y < 2.01$

137 (a) $f(x) = -(x-1)^2(x-3)^3$
(b)

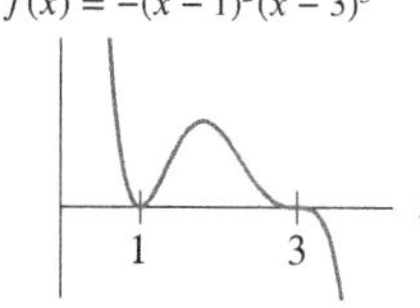

139 (a) $x - 2x^3 + x^4$
$-x+x^2+2x^3-5x^4+2x^5+4x^6-4x^7+x^8$

(b) 64

141 $5\sin x - 20\sin^3 x + 16\sin^5 x$

Section 2.1

1 $265/3$ km/hr

3 -3 angstroms/sec

5 2 meters/sec

7 23.605 μm/sec

9 (a) (i) 6.3 m/sec
(ii) 6.03 m/sec
(iii) 6.003 m/sec
(b) 6 m/sec

11 (a) (i) -1.00801 m/sec
(ii) -0.8504 m/sec
(iii) -0.834 m/sec
(b) -0.83 m/sec

13

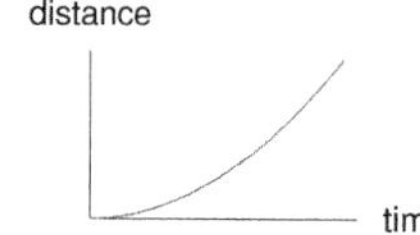

15 27

17 1.9

19 1

21

Slope	−3	−1	0	1/2	1	2
Point	F	C	E	A	B	D

23 From smallest to largest:
0, slope at C, slope at B
slope of AB, 1, slope at A

25 $v_{avg} = 2.5$ ft/sec
$v(0.2) = 4.5$ ft/sec

27

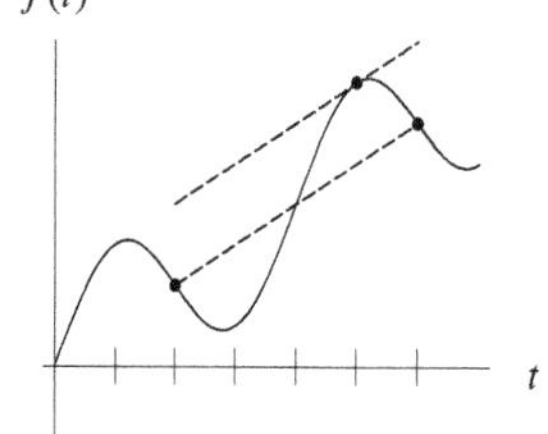

29 (a) 36 feet
(b) 34 ft/sec
(c) 18 ft/sec
(d) 75 ft, 0 ft/sec
(e)

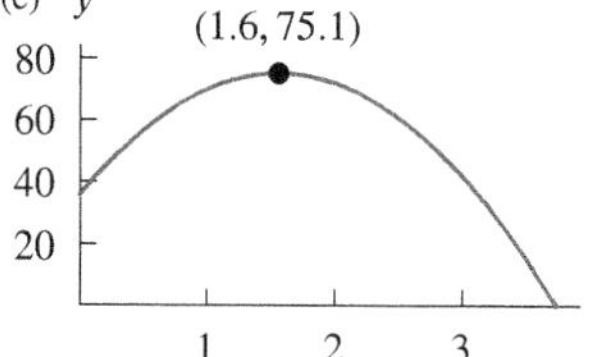

(f) $t = 1.6$ sec

31 3

33 6

35 Expand and simplify first

37 $f(t) = t^2$

39 False

41 True

43 True

Section 2.2

1 12

3 (a) 70 \$/kg; 50 \$/kg
(b) About 60 \$/kg

5 (b) 0.24
(c) 0.22

7 negative

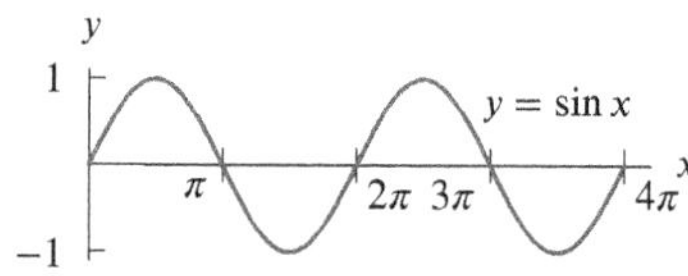

9 $f'(2) \approx 9.89$

11 (a) $x = 2$ and $x = 4$
(b) $g(4)$
(c) $g'(4)$

13

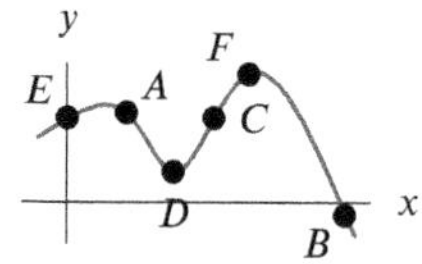

15 16 million km^2 covered by ice in Feb 1983

17 14 million km^2 covered by ice on Feb 1, 2015

19 About 41

21 (a) $f(4)$
(b) $f(2) - f(1)$
(c) $(f(2) - f(1))/(2-1)$
(d) $f'(1)$

23 (4, 25); (4.2, 25.3); (3.9, 24.85)

25

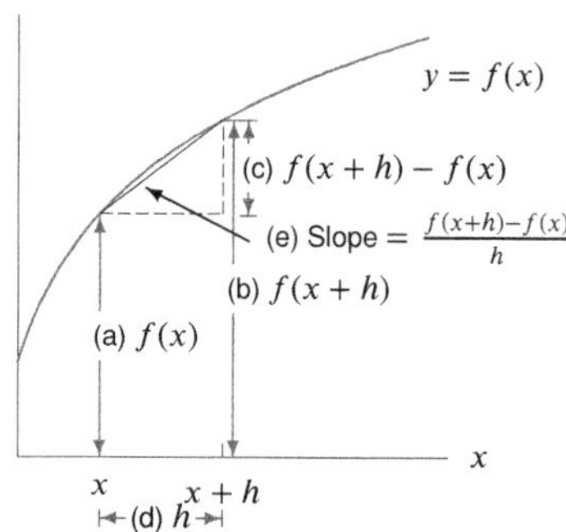

27 (a) $(f(b) - f(a))/(b-a)$
(b) Slopes same
(c) Yes

29 $g'(-4) = 5$

31 $y = 7x - 9$

33 $f'(2) \approx 6.77$

35 $f'(1) \approx -1.558$
$f'(\frac{\pi}{4}) \approx -1$

37 300 million people;
2.867 million people/yr

39 6

41 -4

45 -6

47 -1
49 $1/4$
51 3
53 $1/2$
55 1
57 12
59 3
61 $-1/4$
63 $y = 12x + 16$
65 $y = -2x + 3$
67 Derivative at a point is slope of tangent line
69 $f(x) = 2x + 1$
71 True
73 (a)

Section 2.3

1 (a) About 3
(b) Positive: $0 < x < 4$
Negative: $4 < x < 12$

3 Between -9 and -6
Between 6 and 9

5
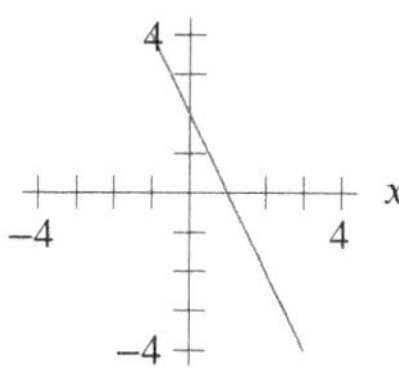

7
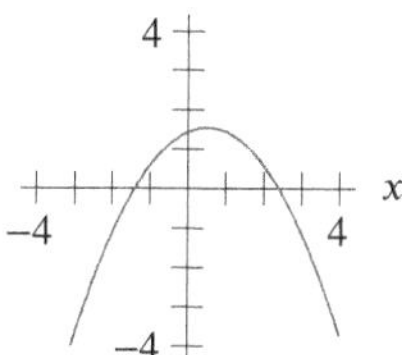

9
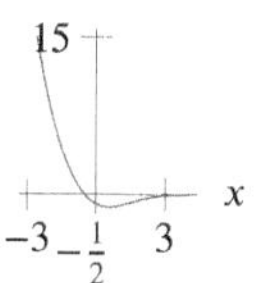

11
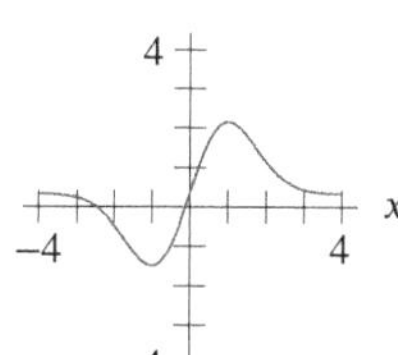

13
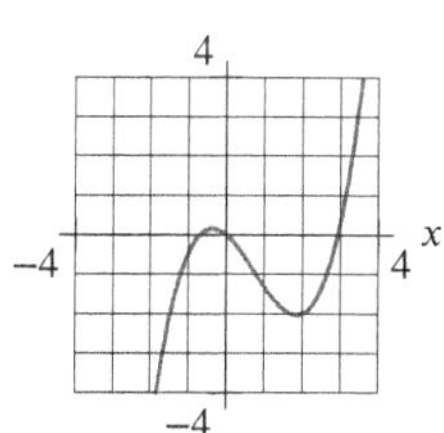

15
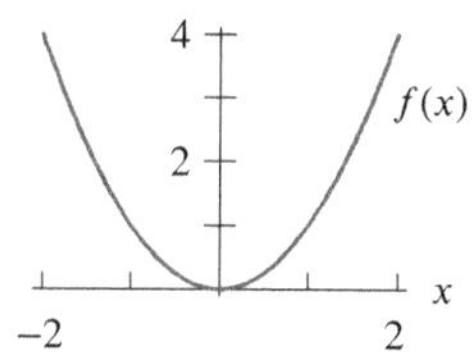

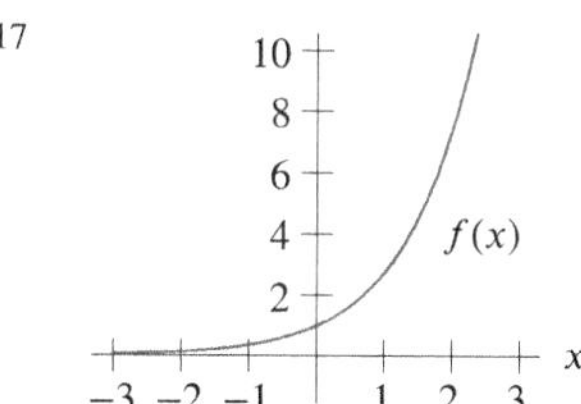

17

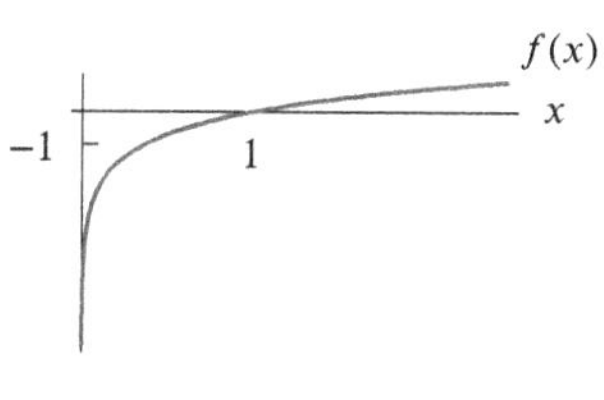

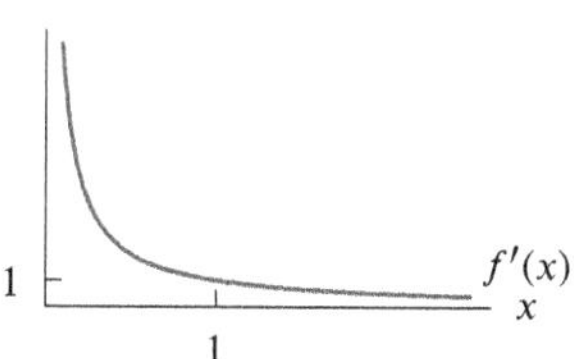

19
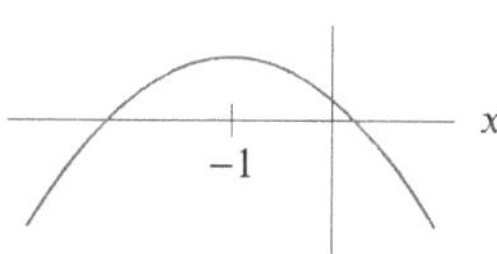

21 $-2/x^3$
23 $-1/(x+1)^2$
25 $2x + 2$
27 $-1/(2x^{3/2})$
29

Other answers possible

31 $f'(x)$ positive: $4 \le x \le 8$
$f'(x)$ negative: $0 \le x \le 3$
$f'(x)$ greatest: at $x \approx 8$

33 (a) x_3
(b) x_4
(c) x_5
(d) x_3

35
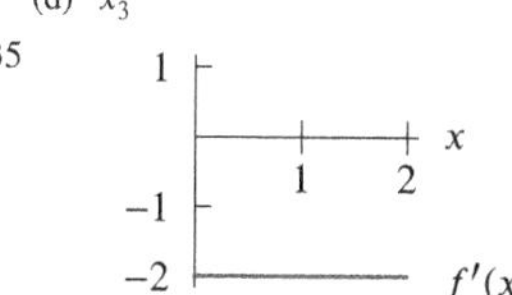

37
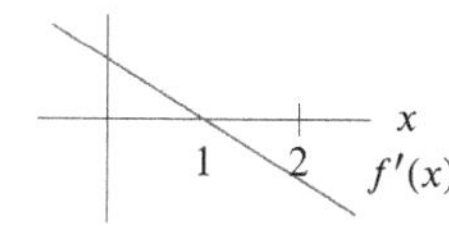

39
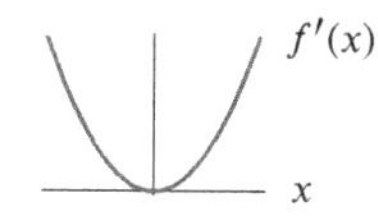

41
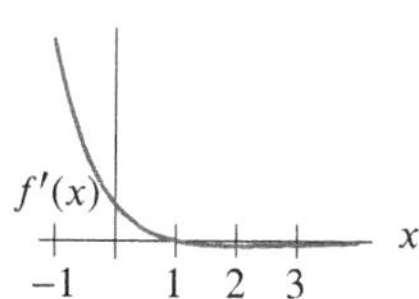

43
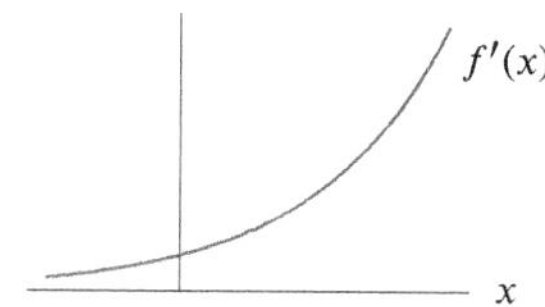

45 (V)
47 (II)
49 (a) Graph II
(b) Graph I
(c) Graph III

51 current

57
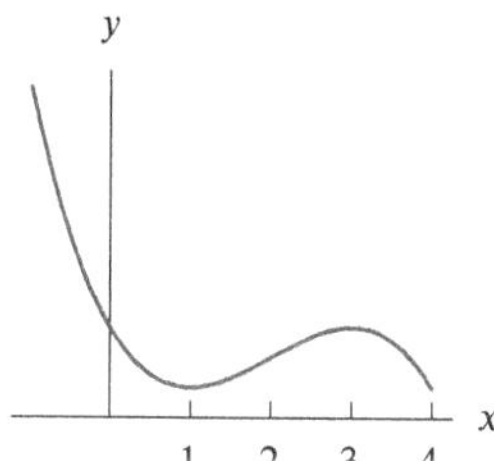

Other answers possible

63 counterexample: $f(x) = 1, g(x) = 2$
65 $f(x) = b + 2x$
67 True
69 False

Section 2.4

1 (a) Costs \$1300 for 200 gallons
(b) Costs about \$6 for 201st gallon

3 (a) Positive
(b) °F/min

5 (a) Quarts; dollars.
(b) Quarts; dollars/quart

7 (a) acre-feet per week
(b) water storage increases; recent rainfall increases inflow

9 Dollars/percent; positive

11 Ft/sec; negative

13 Dollars per percent; positive

15 Joule/cm; positive

17 Dollars/year

19 (a) $°C/\text{km}$
(b) Air temp decreases about 6.5° for a 1 km increase in altitude

21 (a) $\Delta R \approx 3\Delta S$
(b) 0.6
(c) 13.6

23 About 1.445 billion people in 2020; growing at about 8.64 million people per year

25 (a) f
(b) g and h

27 (b) Pounds/(Calories/day)

29 $f(\text{points}) =$ Revenue
$f'(4.3) \approx 55$ million dollars/point

31 (a) In 2008 pop Mexico increase 1.35 m people/yr
(b) Pop 113.2 m in 2009
(c) 113.2 m: about 0.74 yrs for pop increase of 1 m

33 (a) Depth 3 ft at $t = 5$ hrs
(b) Depth increases 0.7 ft/hr
(c) Time 7 hrs when depth 5 ft
(d) Depth at 5 ft increases 1 ft in 1.2 hrs

35 (a) The weight of the object, in Newtons, at a distance of 80 kilometers above the surface of the earth; positive
(b) For an object at a distance of 80 km from the surface of the earth, $f'(80)$ represents the approximate number of Newtons by which the weight of an object will change for each additional kilometer that the object moves away from the earth's surface; negative
(c) The distance of the object from the surface of the earth, in kilometers, that would be necessary for the object to weigh 200 Newtons; positive
(d) For an object that weighs 200 Newtons, $(f^{-1})'(200)$ represents the approximate amount by which the object's distance from the earth's surface must change in order for its weight to increase by 1 Newton; negative

39 mpg/mph

41 Barrels/year; negative

43 (a) Gal/minute
(b) (i) 0
(ii) Negative
(iii) 0

45 (a) Mn km^2/day
Growing 73,000 km^2/day
(b) 0.365 mn km^2
Grew by 0.365 mn km^2

47 (a) mm (of rain)
(b) mm (of leaf width) per mm (of rain)
(c) 4.36 mm

49 (a) Cell divisions per hour per 10^{-4} M of concentration
(b) 10^{-4} M
(c) 0.02 cell divisions per hour

51 (a) 0.0729 people/sec
(b) 14 seconds

53 Number of people 65.5–66.5 inches
Units: People per inch
$P'(66)$ between 17 and 34 million people/in
$P'(x)$ is never negative

55 No units are given

57 Minutes/page

61 True

63 False

65 (b), (d)

Section 2.5

1 (a) Increasing, concave up
(b) Decreasing, concave down

3 B

5 (a)

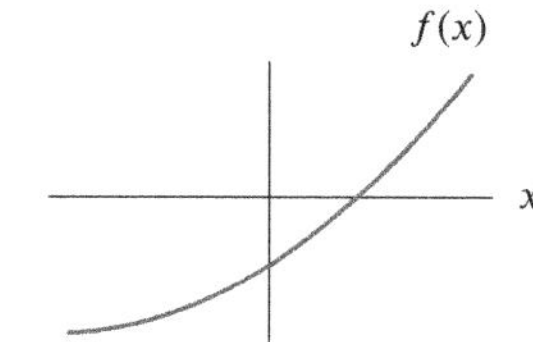

(b)

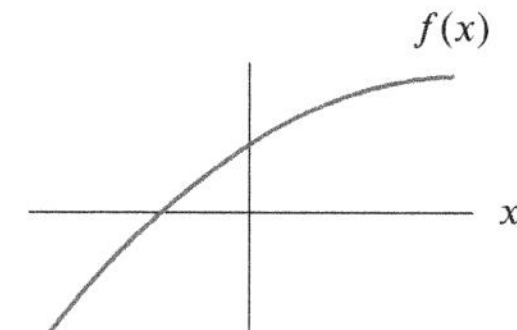

(c)

$f(x)$

(d)

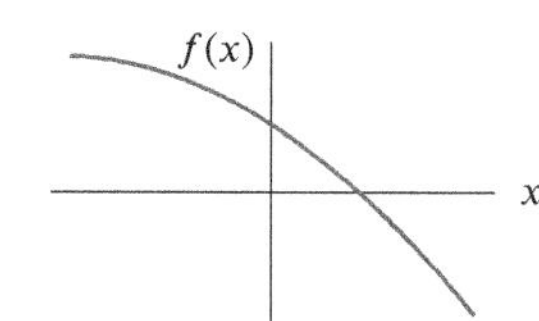

7

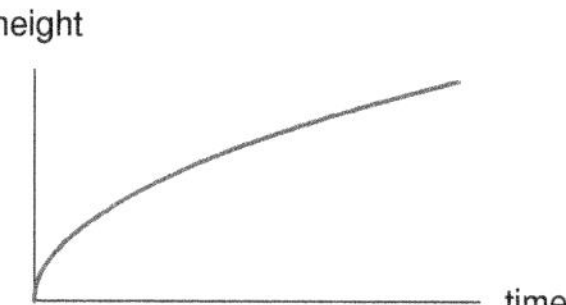

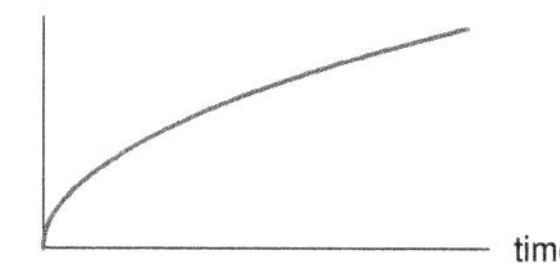

9 $f'(x) < 0$
$f''(x) = 0$

11 $f'(x) < 0$
$f''(x) > 0$

13 $f'(x) < 0$
$f''(x) < 0$

15 (a) Positive; negative
(b) Neither; positive
(c) Number of cars increasing at about 600,000 million cars per year in 2005

17 (a)

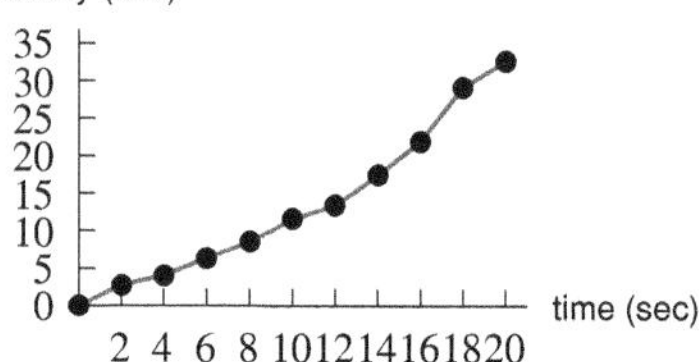

(b) 3.6

19

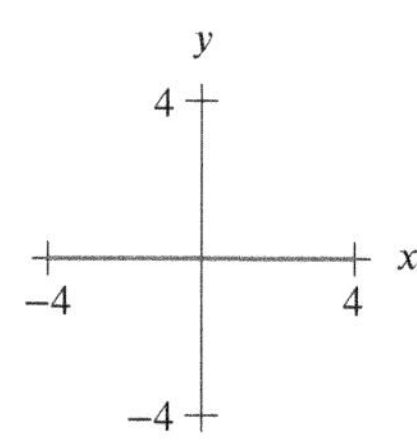

21

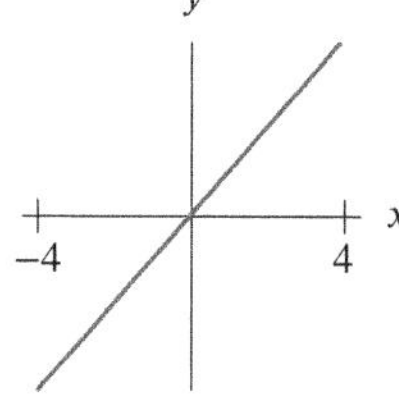

23

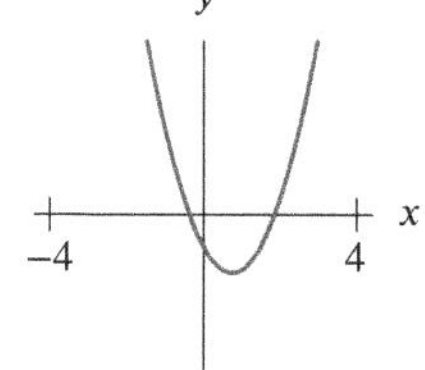

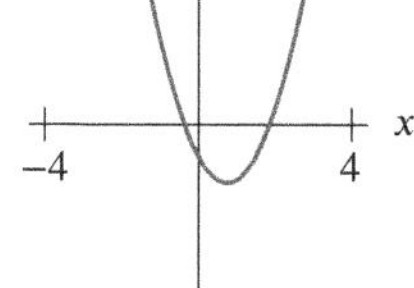

25 (a) $dP/dt > 0, d^2P/dt^2 > 0$
(b) $dP/dt < 0, d^2P/dt^2 > 0$ (but dP/dt is close to zero)

27 (a)

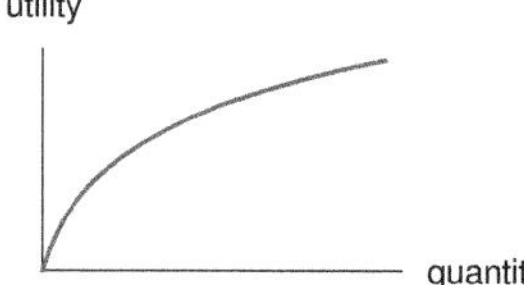

(b) Derivative of utility is positive
2nd derivative of utility is negative

31 No

33 No

35 Yes

37 (a) t_3, t_4, t_5
(b) t_2, t_3
(c) t_1, t_2, t_5
(d) t_1, t_4, t_5
(e) t_3, t_4

39

41 22 only possible value

43 (a) All positive x
(b) All positive x
(c) All positive x

47 $f(x) = x^2$

49 True

51 True

Section 2.6

1 (a) $x = 1$
(b) $x = 1, 2, 3$

3 No

5 Yes

7 Yes

9 No

11 (a)

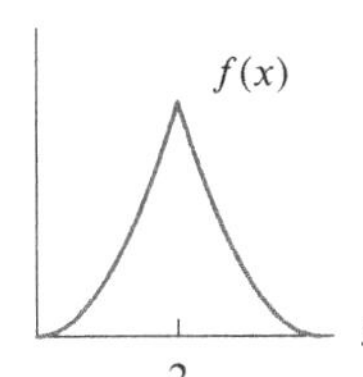

(b)

$f(x)$

15 (a) $B = B_0$ at $r = r_0$
B_0 is max

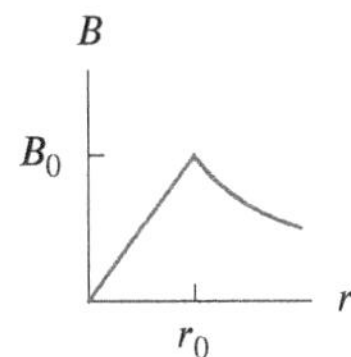

(b) Yes
(c) No

17 (a) Yes
(b) Yes

19 (a) No; yes.
(b) $f'(x)$ not differentiable at $x = 0$; $f''(x)$ neither differentiable nor continuous at $x = 0$

21 Other cases are possible

23 $f(x) = |x - 2|$

25 $f(x) = (x^2 - 1)/(x^2 - 4)$

27 True; $f(x) = x^2$

29 True; $f(x) = \begin{cases} 1 & x \geq 0 \\ -1 & x < 0 \end{cases}$

31 (a) Not a counterexample
(b) Counterexample
(c) Not a counterexample
(d) Not a counterexample

Chapter 2 Review

1 $72/7 = 10.286$ cm/sec

3 $(\ln 3)/2 = 0.549$ mm/sec

5 0 mm/sec

7 0 mm/sec

9

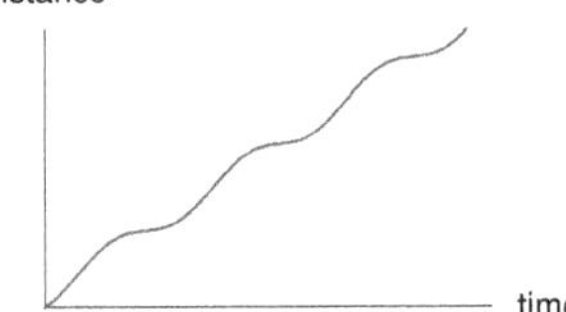

11 (a)

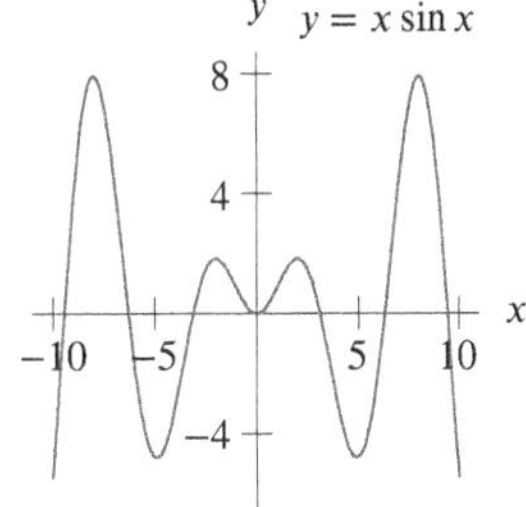

(b) Seven
(c) Increasing at $x = 1$
Decreasing at $x = 4$
(d) $6 \leq x \leq 8$
(e) $x = -9$

13

15

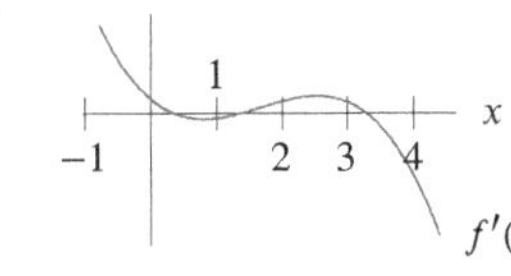

17

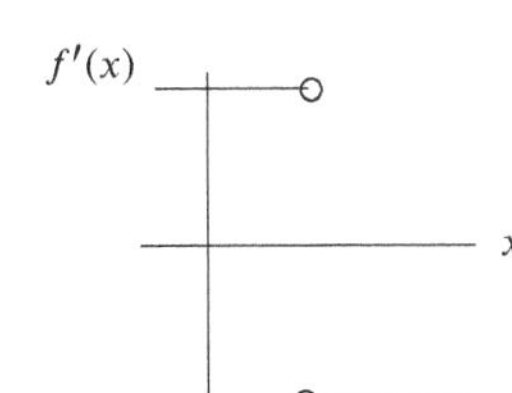

19

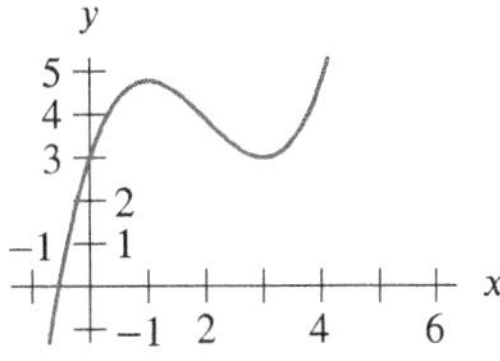

21

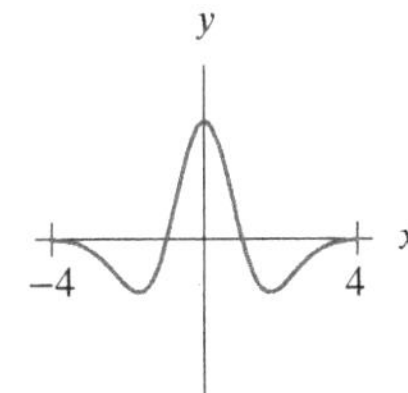

23 $-1/x^2$

25 (a) (i) C and D
(ii) B and C
(b) A and B, C and D

27 $10x$

29 $6x$

31 $2a$

33 $-2/a^3$

35 $-1/(2(\sqrt{a})^3)$

37 From smallest to largest: 0, Vel at C, Vel at B, Av vel between A and B, 1, Vel at A

39 (a) A, C, F, and H.
(b) Acceleration 0

41 (a) $f'(t) > 0$: depth increasing
$f'(t) < 0$: depth decreasing
(b) Depth increasing 20 cm/min
(c) 12 meters/hr

43 (a) Thousand \$ per \$ per gallon
Rate change revenue with price/gal
(b) \$ per gal/thousand \$
Rate change price/gal with revenue

47 Concave up

49 (a)

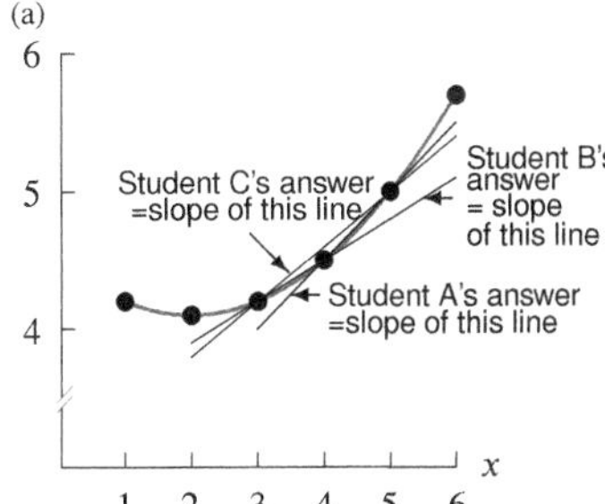

(b) Student C's
(c) $f'(x) = \dfrac{f(x+h) - f(x-h)}{2h}$

51 $x_1 = 0.9, x_2 = 1, x_3 = 1.1$
$y_1 = 2.8, y_2 = 3, y_3 = 3.2$

53 (a) Negative
(b) $dw/dt = 0$
(c) $|dw/dt|$ increases; dw/dt decreases

55 (a) Dose for 140 lbs is 120 mg
Dose increases by 3mg/lb
(b) About 135 mg

57 $P'(t) = 0.008P(t)$

59 (a) $f'(0.6) \approx 0.5$
$f'(0.5) \approx 2$
(b) $f''(0.6) \approx -15$
(c) Maximum: near $x = 0.8$
minimum: near $x = 0.3$

61 (a) At $(0, \sqrt{19})$: slope = 0
At $(\sqrt{19}, 0)$: slope is undefined
(b) slope $\approx 1/2$
(c) At $(-2, \sqrt{15})$: slope $\approx 1/2$
At $(-2, -\sqrt{15})$: slope $\approx -1/2$
At $(2, \sqrt{15})$: slope $\approx -1/2$

63 (a) Period 12 months

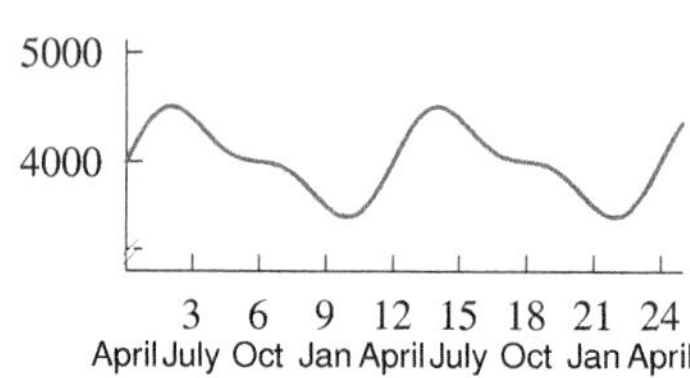

(b) Max of 4500 on June 1st
(c) Min of 3500 on Feb 1st
(d) Growing fastest:
April 1st
Decreasing fastest:
July 15 and Dec 15
(e) About 400 deer/month

65 (a) Concave down

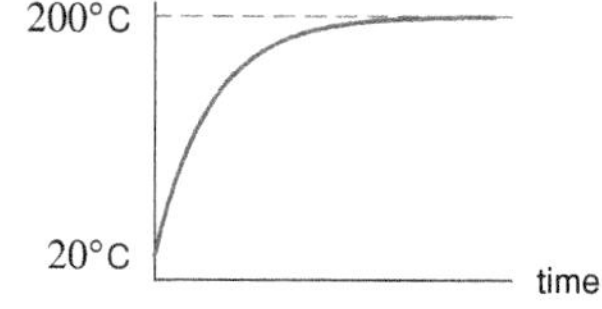

(b) $120° < T < 140°$
(c) $135° < T < 140°$
(d) $45 < t < 50$

67 (a) $f'(0) = 1.00000$
$f'(0.3) = 1.04534$
$f'(0.7) = 1.25521$
$f'(1) = 1.54314$
(b) They are about the same

69 0, because $f(x)$ constant

71 (a) $-2a/e^{ax^2} + 4a^2x^2/e^{ax^2}$
(b)

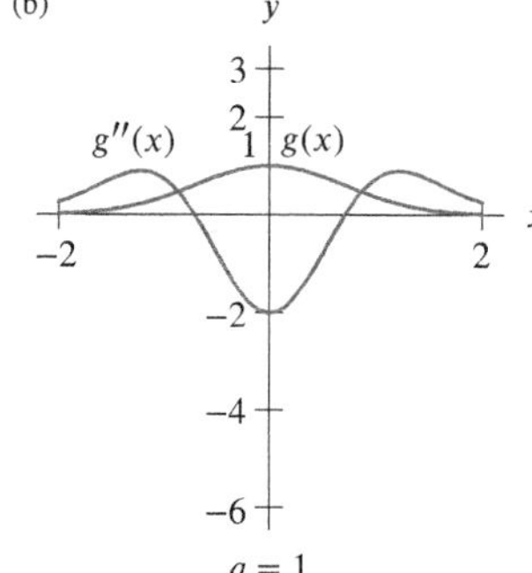

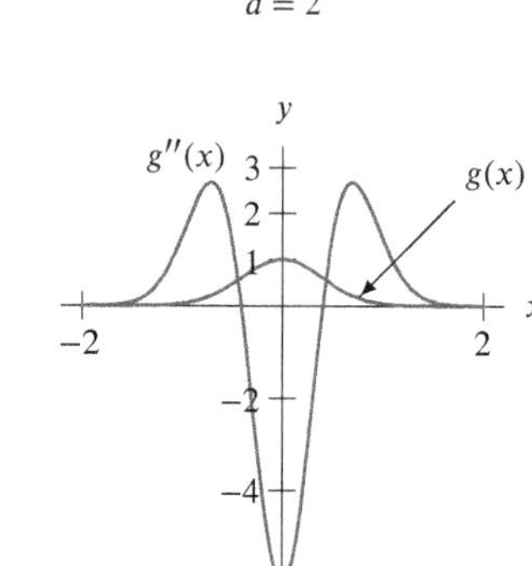

(graph for $a = 3$)

73 (a) $4x(x^2+1)$, $6x(x^2+1)^2$, $8x(x^2+1)^3$
(b) $2nx(x^2+1)^{n-1}$

Section 3.1

3 Do not apply
5 $y' = \pi x^{\pi-1}$ (power rule)
7 $11x^{10}$
9 $-12x^{-13}$
11 $-3x^{-7/4}/4$
13 $3x^{-1/4}/4$
15 $3t^2 - 6t + 8$
17 $-5t^{-6}$
19 $-(7/2)r^{-9/2}$
21 $x^{-3/4}/4$
23 $-(3/2)x^{-5/2}$
25 $6x^{1/2} - \frac{5}{2}x^{-1/2}$
27 $17 + 12x^{-1/2}$
29 $20x^3 - 2/x^3$
31 $-12x^3 - 12x^2 - 6$
33 $6t - 6/t^{3/2} + 2/t^3$
35 $3t^{1/2} + 2t$
37 $(1/2)\theta^{-1/2} + \theta^{-2}$
39 $(z^2 - 1)/3z^2$
41 $1/(2\sqrt{\theta}) + 1/(2\theta^{3/2})$
43 $3x^2/a + 2ax/b - c$
45 a/c
47 $3ab^2$
49 $b/(2\sqrt{t})$
51 4.44
53 −3.85
55 −3.98
57 −4.15

59 (a) $f'(x) = 7x^6$; $f''(x) = 42x^5$; $f'''(x) = 210x^4$
(b) $n = 8$.

61 $d^2w/dx^2 = (-1/4)x^{-3/2} + (3/4)x^{-5/2}$; $d^3w/dx^3 = (3/8)x^{-5/2} - (15/8)x^{-7/2}$

63 Rules of this section do not apply

65 $6x$
(power rule)

67 $-2/3z^3$ (power rule)

69 $y = 2x - 1$

71 $y = -9.333 + 6.333x$

73 (a) $f'(1) < f'(0) < f'(-1) < f'(4)$
(b) $f'(1) = -1$, $f'(0) = 2$, $f'(-1) = 11$, $f'(4) = 26$

75 For $x < 0$ or $2 < x < 3$

77 (a) 0
(b) 5040

79 (a) 0
(b) −6
(c) (IV)

81 (a) 0
(b) 6
(c) (I)

83 (a) 12.5 liters
(b) −6.25 liters per atmosphere
(c) 1 atmosphere

85 (a) 15.2 m/sec
(b) 5.4 m/sec
(c) -9.8 m/sec^2
(d) 34.9 m
(e) 5.2 sec

87 (a) $5159v^{-0.33}$
(b) 11030 m^3/sec per km^3

89 Approximately 10%

91 Approximately 20%

93 (a) $0.0467x^{-1/3}$
(b) 0.015 mm per meter
(c) 0.09 mm

95 (a) $dT/dl = \pi/\sqrt{gl}$
(b) Positive, so period increases as length increases

99 $n = 4$, $a = 3/32$

101 (a) $f'(x) = g'(x) = 3x^2 + 6x - 2$
(b) One is a vertical shift of the other, so they have the same slopes everywhere
(c) Any vertical shift has the same derivative

103 $x^2 + a$

105 $f(x) = x^2$, $g(x) = 3x$

107 $f(x) = 3x^2$

109 False

111 True

113 $y = 2x$ and $y = -6x$

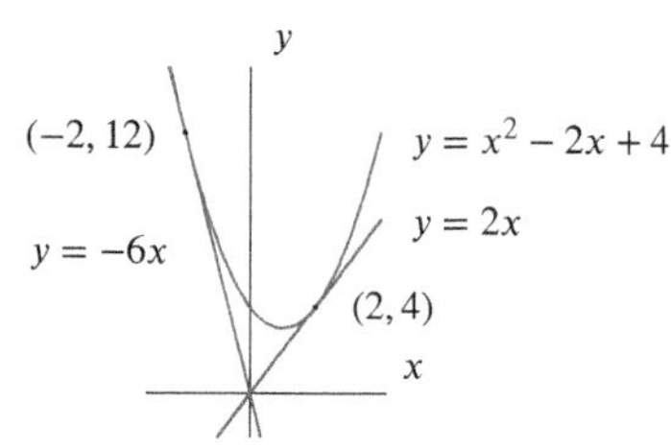

115 (a) $x < 1/2$
(b) $x > 1/4$

(c) $x < 0$ or $x > 0$

117 $n = 1/13$

119 $a = 2, b = 2$

121 Infinitely many, each of the form $f(x) = C$

Section 3.2

1 $2e^x + 2x$

3 $5\ln(a)a^{5x}$

5 $10x + (\ln 2)2^x$

7 $4(\ln 10)10^x - 3x^2$

9 $((\ln 3)3^x)/3 - (33x^{-3/2})/2$

11 $(\ln 4)^2 4^x$

13 $5 \cdot 5^t \ln 5 + 6 \cdot 6^t \ln 6$

15 ex^{e-1}

17 $(\ln \pi)\pi^x$

19 $(\ln k)k^x$

21 e^{t+2}

23 $a^x \ln a + ax^{a-1}$

25 $2 + 1/(3x^{4/3}) + 3^x \ln 3$

27 $f''(t) = (\ln 5)^2 5^{t+1}$; $f'''(t) = (\ln 5)^3 5^{t+1}$

29 $2x + (\ln 2)2^x$

31 Rules do not apply

33 e^{x+5}

35 Rules do not apply

37 Rules do not apply

39 Rules do not apply

41 16.021; pop incr $\approx$ 16 animals/yr when $t = 5$

43 1013.454 mb; increasing at 2.261 mb/hr

45 (a) $17.9 \cdot \ln 1.025\left(1.025^t\right)$ m. tonnes/yr
(b) 0.580 m. tonnes/year
(c) 2.9 m. tonnes
(d) Smaller

47 (a) 1.247 million people per year
(b) 2.813 million people per year
US, since $dU/dt > dM/dt$ at $t = 0$

49 $g(x) = x^2/2 + x + 1$

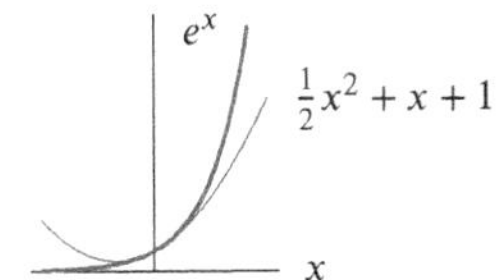

51 $f'(x) = 0$

53 $f(x) = e^x$

55 False

59 (a) $f'(0) = -1$
(b) $y = -x$
(c) $y = x$

61 e

63 $0 < a < 1$ and $C < 0$

65 $a > 1$ and $C < 0$

Section 3.3

1 $5x^4 + 10x$

3 $e^x(x + 1)$

5 $2^x/(2\sqrt{x}) + \sqrt{x}(\ln 2)2^x$

7 $3^x[(\ln 3)(x^2 - x^{\frac{1}{2}}) + (2x - 1/(2\sqrt{x}))]$

9 $(1 - x)/e^x$

11 $(1 - (t + 1)\ln 2)/2^t$

13 $6/(5r + 2)^2$

15 $1/(5t + 2)^2$

17 $2e^t + 2te^t + 1/(2t^{3/2})$

19 $2y - 6,\ y \neq 0$

21 $\sqrt{z}(3 - z^{-2})/2$

23 $2r(r + 1)/(2r + 1)^2$

25 $17e^x(1 - \ln 2)/2^x$

27 $1, x \neq -1$

29 $(-4 - 6x)(6x^e - 3\pi) + (2 - 4x - 3x^2)(6ex^{e-1})$

31 (a) 4
(b) Does not exist
(c) -4

33 (a) -2
(b) Does not exist
(c) 0

35 Approx 0.4

37 Approx 0.7

39 Approx -21.2

41 $f'(x) = 2e^{2x}$

43 $x > -2$

45 $y = 5x$

47 $y = 7x - 5$

49 $4^x(\ln 4 \cdot f(x) + \ln 4 \cdot g(x) + f'(x) + g'(x))$

51 $(f'(x)g(x)h(x) + f(x)g'(x)h(x) - f(x)g(x)h'(x))/(h(x))^2$

53 (a) 19
(b) -11

55 $f(x) = x^{10}e^x$

57 (a) $f(140) = 15{,}000$:
If the cost \$140 per board then 15,000 skateboards are sold
$f'(140) = -100$:
Every dollar of increase from \$140 will decrease the total sales by about 100 boards
(b) $dR/dp|_{p=140} = 1000$
(c) Positive
Increase by about \$1000

59 (a) $g(v) = 1/f(v)$
$g(80) = 20$ km/liter
$g'(80) = -(1/5)$ km/liter for each 1 km/hr increase in speed
(b) $h(v) = v \cdot f(v)$
$h(80) = 4$ liters/hr
$h'(80) = 0.09$ liters/hr for each 1 km/hr inc. in speed

61 (a) $P(t) = A(t)/N(t)$
(b) Purchase more shares to offset loss
(c) Purchase 28,750 additional shares

63 0

65 0

67 Product rule gives two terms

69 Write $f(x) = e^x(x + 1)$ and apply the product rule

71 $f(x) = x^2 = x \cdot x$

73 True

75 (a) Not a counterexample
(b) Not a counterexample
(c) Not a counterexample
(d) Counterexample

77 False, choose $g(2) = -1/0.7$ and $h(2) = 2$

79 True

81 (a) $\frac{d}{dx}\left(\frac{e^x}{x}\right) = \frac{e^x}{x} - \frac{e^x}{x^2}$
$\frac{d}{dx}\left(\frac{e^x}{x^2}\right) = \frac{e^x}{x^2} - \frac{2e^x}{x^3}$
$\frac{d}{dx}\left(\frac{e^x}{x^3}\right) = \frac{e^x}{x^3} - \frac{3e^x}{x^4}$
(b) $\frac{d}{dx}\left(\frac{e^x}{x^n}\right) = \frac{e^x}{x^n} - \frac{ne^x}{x^{n+1}}$

83 (a) $g'(a) = -f'(a)/(f(a))^2 < 0$ since $f'(a) > 0$
(b) If $g'(a) = -f'(a)/(f(a))^2 = 0$, so must $f'(a)$
(c) Not if $f(a) = 0$, too

87 $f'(x) = f(x)\left(\frac{1}{x - r_1} + \frac{1}{x - r_2} + \cdots + \frac{1}{x - r_n}\right)$

Section 3.4

1 $99(x + 1)^{98}$

3 $56x(4x^2 + 1)^6$

5 $e^x/(2\sqrt{e^x + 1})$

7 $5(w^4 - 2w)^4(4w^3 - 2)$

9 $2r^3/\sqrt{r^4 + 1}$

11 $e^{2x}\left[2x^2 + 2x + (\ln 5 + 2)5^x\right]$

13 $\pi e^{\pi x}$

15 $-200xe^{-x^2}$

17 $(\ln \pi)\pi^{(x+2)}$

19 $e^{5-2t}(1 - 2t)$

21 $(2t - ct^2)e^{-ct}$

23 $(e^{\sqrt{s}})/(2\sqrt{s})$

25 $3s^2/(2\sqrt{s^3 + 1})$

27 $(e^{-z})/(2\sqrt{z}) - \sqrt{z}e^{-z}$

29 $5 \cdot \ln 2 \cdot 2^{5t-3}$

31 $-(\ln 10)(10^{\frac{5}{2} - \frac{y}{2}})/2$

33 $(1 - 2z\ln 2)/(2^{z+1}\sqrt{z})$

35 $\dfrac{\sqrt{x + 3}(x^2 + 6x - 9)}{2\sqrt{x^2 + 9}(x + 3)^2}$

37 $-(3e^{3x} + 2x)/(e^{3x} + x^2)^2$

39 $-1.5x^2(x^3 + 1)^{-1.5}$

41 $(2t + 3)(1 - e^{-2t}) + (t^2 + 3t)(2e^{-2t})$

43 $30e^{5x} - 2xe^{-x^2}$

45 $2we^{w^2}(5w^2 + 8)$

47 $-3te^{-3t^2}/\sqrt{e^{-3t^2} + 5}$

49 $2ye^{[e^{(y^2)} + y^2]}$

51 $6ax(ax^2 + b)^2$

53 $ae^{-bx} - abxe^{-bx}$

55 $abce^{-cx}e^{-be^{-cx}}$

57 $6x\left(x^2 + 5\right)^2\left(3x^3 - 2\right)\left(6x^3 + 15x - 2\right)$

59 (a) -2
(b) Chain rule does not apply
(c) 2

61 (a) 1
(b) 1
(c) 1

63 0

65 1/2

67 $y = 4451.66x - 3560.81$

69 $-1/\sqrt{2} < x < 1/\sqrt{2}$

71 $f'(x) =$
$[(2x+1)^9(3x-1)^6] \times (102x+1)$
$f''(x) =$
$[9(2x+1)^8(2)(3x-1)^6 + (2x+1)^9(6)(3x-1)^5(3)] \times (102x+1) + (2x+1)^9(3x-1)^6(102)$

73 (a) $H(4) = 1$
(b) $H'(4) = 30$
(c) $H(4) = 4$
(d) $H'(4) = 56$
(e) $H'(4) = -1$

75 (a) $g'(2) = 42$
(b) $h'(2) = -8$

77 (a) 13,394 fish
(b) 8037 fish/month

79 $f(3) = 2.51$ billion dollars;
$f'(3) = 0.301$ billion dollars of 4th quarter sales per year

81 (a) \$5000; 2% compounded continuously
(b) $f(10) = 6107.01$ dollars;
$f'(10) = 122.14$ dollars per year

83 $b = 1/40$ and $a = 169.36$

85 (a) $P(1 + r/100)^t \ln(1 + r/100)$
(b) $Pt(1 + r/100)^{t-1}/100$

87 (a) $f(t)$: linear
$g(t)$: exponential
$h(t)$: quadratic polynomial
(b) 1.3 ppm/yr; 1.451 ppm/yr; 2.133 ppm/yr
(c) Linear < Exp < Quad
(d) No; Exponential eventually largest

89 Decreasing

91 Increasing

93 $g'(x) = 5e^x(e^x + 2)^4$

95 Not necessarily equal to 0

97 $f(x) = (x^2 + 1)^2$

99 False; $f(x) = 6, g(x) = 10$

101 False; $f(x) = e^{-x}, g(x) = x^2$

103 Zero; decreasing

105 Positive; positive

109 (a) 4
(b) 2
(c) $1/2$

111 (a) $g'(1) = 3/4$
(b) $h'(1) = 3/2$

115 $f''(x)(g(x))^{-1} - 2f'(x)(g(x))^{-2}g'(x) + 2f(x)(g(x))^{-3}(g'(x))^2 - f(x)(g(x))^{-2}g''(x)$

Section 3.5

3 $\cos^2\theta - \sin^2\theta = \cos 2\theta$

5 $3\cos(3x)$

7 $-8\sin(2t)$

9 $3\pi\sin(\pi x)$

11 $3\pi\cos(\pi t)(2 + \sin(\pi t))^2$

13 $e^t\cos(e^t)$

15 $(\cos y)e^{\sin y}$

17 $3\cos(3\theta)e^{\sin(3\theta)}$

19 $2x/\cos^2(x^2)$

21 $4\cos(8x)(3 + \sin(8x))^{-0.5}$

23 $\cos x/\cos^2(\sin x)$

25 $2\sin(3x) + 6x\cos(3x)$

27 $e^{-2x}[\cos x - 2\sin x]$

29 $5\sin^4\theta\cos\theta$

31 $-3e^{-3\theta}/\cos^2(e^{-3\theta})$

33 $-2e^{2x}\sin(e^{2x})$

35 $-\sin\alpha + 3\cos\alpha$

37 $3\theta^2\cos\theta - \theta^3\sin\theta$

39 $\cos(\cos x + \sin x) \cdot (\cos x - \sin x)$

41 $(-t\sin t - 3\cos t)/t^4$

43 $\dfrac{\sqrt{1-\cos x}(1 - \cos x - \sin x)}{2\sqrt{1-\sin x}(1-\cos x)^2}$

45 $(6\sin x\cos x)/(\cos^2 x + 1)^2$

47 $-ab\sin(bt + c)$

49 $2x^3e^{5x}\cos(2x) + 5x^3e^{5x}\sin(2x) + 3x^2e^{5x}\sin(2x)$

51 $f''(\theta) = -(\theta\cos\theta + 2\sin\theta)$; $f'''(\theta) = \theta\sin\theta - 3\cos\theta$

53 Decreasing, concave up

55 $y = 3x + 1$

57 $d^{50}y/dx^{50} = -\cos x$

59 (a) 0
(b) $\sin^2 x + \cos^2 x = 1$, so $f'(x) = 0$

61 (a) $v(t) = 2\pi\cos(2\pi t)$
(b)

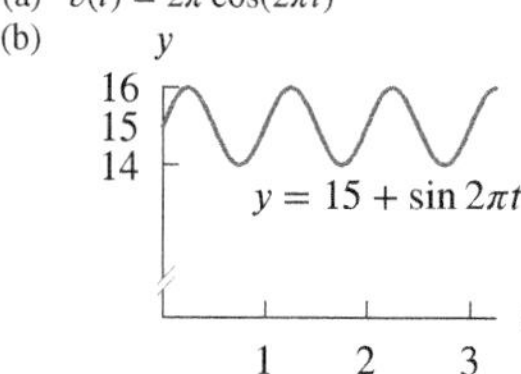

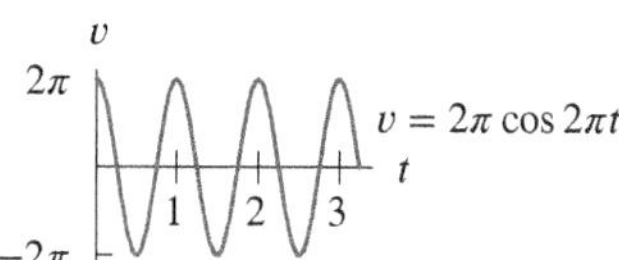

63 (a) $t = (\pi/2)(m/k)^{1/2}$;
$t = 0$;
$t = (3\pi/2)(m/k)^{1/2}$
(b) $T = 2\pi(m/k)^{1/2}$
(c) $dT/dm = \pi/\sqrt{km}$;
Positive sign means an increase in mass causes the period to increase

65 (a) $0 \le t \le 2$
(b) No, not at $t = 2$

67 CO_2 increasing at 1.999 ppm/month on Dec 1, 2008

69 CO_2 decreasing at 1.666 ppm/month on June 1, 2008

71 (a) $d'(t) = -2e^{-t}\sin t$
(b) $t = 0, \pi, 2\pi, 3\pi, \ldots$
(c) 0; the wave stops

75 Cannot use product rule

77 $\sin x$

79 True

81 $(\sin x + x\cos x)x\sin x$

83 $2\sin(x^4) + 8x^4\cos(x^4)$

85 $k = 7.46$, $(3\pi/4, 1/\sqrt{2})$

Section 3.6

1 $2t/(t^2 + 1)$

3 $10x/(5x^2 + 3)$

5 $1/\sqrt{1 - (x+1)^2}$

7 $(6x + 15)/(x^2 + 5x + 3)$

9 2

11 $e^{-x}/(1 - e^{-x})$

13 $e^x/(e^x + 1)$

15 $ae^{ax}/(e^{ax} + b)$

17 $3w^2\ln(10w) + w^2$

19 e

21 $1/t$

23 $-1/(1 + (2-x)^2)$

25 $e^{\arctan(3t^2)}(6t)/(1 + 9t^4)$

27 $(\ln 2)z^{(\ln 2 - 1)}$

29 k

31 $-x/\sqrt{1 - x^2}$

33 $-1/z(\ln z)^2$

35 $(\cos x - \sin x)/(\sin x + \cos x)$

37 $3w^{-1/2} - 2w^{-3} + 5/w$

39 $-(x+1)/(\sqrt{1 - (x+1)^2})$

41 $1/(1 + 2u + 2u^2)$

43 $-1 < x < 1$

45 $\frac{d}{dx}(\log x) = \frac{1}{(\ln 10)x}$

47 $g(5000) = 32.189$ years
$g'(5000) = 0.004$ years per dollar

49 (a) $f'(x) = 32.7/x$
(b) 0.01635 mm of leaf width per mm of rain
(c) 2.4525 mm

51 (a) $f'(x) = 0$
(b) f is a constant function

53 (a) $10x^4 + 9x^2 + 1$
(b) $f' > 0$ so f increasing everywhere
(c) 6
(d) 20
(e) $1/20$

55 Any x with $25 < x < 50$

57 Any x with $75 < x < 100$

59 2.8

61 1.4

63 −0.12

65 $1/5$

67 (a) Pop is 319 m in 2014
(b) 2014
(c) Pop incr by 2.44 m/yr
(d) 0.41 yr/million

69 $1/3$

71 (a) 1
(b) 1
(c) 1

73 $(f^{-1})'$ does not change sign

75 $w'(x) = 4x^3/(1 + x^4)$

77 Formula for $(f^{-1})'$ incorrect

79 $y = c\ln x$ for constant c.

81 $f(x) = x$

83 False; $f(x) = x^3$

85 (b)

Section 3.7

1 $dy/dx = -x/y$

3 $dy/dx = (y^2 - y - 2x)/(x - 3y^2 - 2xy)$

5 $-(1 + y)/(1 + x)$

7 $dy/dx = (y - 2xy^3)/(3x^2y^2 - x)$

9 $dy/dx = -\sqrt{y/x}$

11 $-3x/2y$

13 $-y/(2x)$

15 $dy/dx = (2 - y\cos(xy))/(x\cos(xy))$

17 $(y^2 + x^4y^4 - 2xy)/(x^2 - 2xy - 2x^5y^3)$

19 $(a - x)/y$

21 $(y + b\sin(bx))/$
$(a\cos(ay) - x)$

23 Slope is infinite

25 $-23/9$

27 $y = e^2x$

29 $y = x/a$

31 (a) $(4 - 2x)/(2y + 7)$
(b) Horizontal if $x = 2$,
Vertical if $y = -7/2$

33 (a) $(1, -1)$ and $(1, 0)$.
(b) $dy/dx = -(y + 2x)/(x + 2y)$
(c) 1 at $(1, -1)$ and -2 at $(1, 0)$

35 (a) $y - 3 = -4(x - 4)/3$
and
$y + 3 = 4(x - 4)/3$
(b) $y = 3x/4$
and
$y = -3x/4$
(c) $(0, 0)$

39 (a) $-(1/2P)f(1 - f^2)$

41 Need to use implicit differentiation

43 $dy/dx = (x^2 - 4)/(y - 2)$

45 True

47 $(-1/3, 2\sqrt{2}/3)$; $(7/3, 4\sqrt{2}/3)$

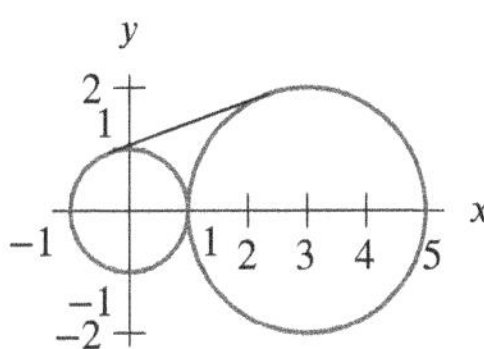

Section 3.8

1 $3\cosh(3z + 5)$

3 $2\cosh t \cdot \sinh t$

5 $3t^2 \sinh t + t^3 \cosh t$

7 $18/\cosh^2(12 + 18x)$

9 $\tanh(1 + \theta)$

11 0

15 $(t^2 + 1)/2t$

19 $\sinh(2x) = 2\sinh x \cosh x$

23 0

25 1

27 $|k| \le 3$

29 (a) $0.54\,T/w$

31 (a)

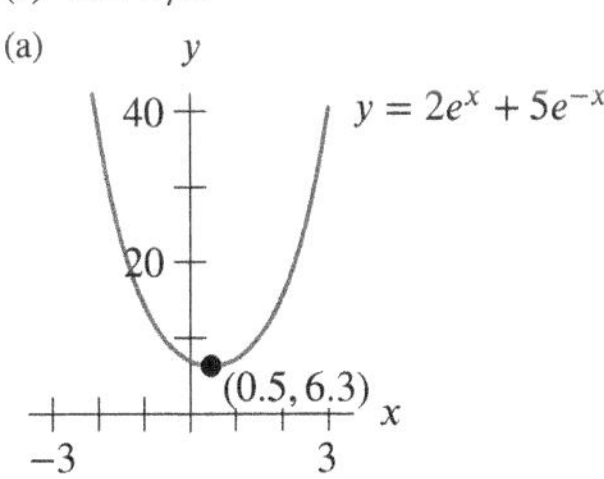

(b) $A = 6.325$
(stretch factor)
$c = 0.458$
(horizontal shift)

33 (a) 0
(b) Positive for $x > 0$
Negative for $x < 0$
Zero for $x = 0$
(c) Increasing everywhere
(d) $1, -1$

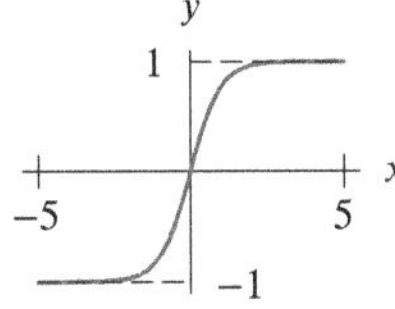

(e) Yes; derivative positive everywhere

35 $f'(x) = \sinh x$

37 $\tanh x \to 1$ as $x \to \infty$

39 $k = 0$

41 True

43 True

45 False

Section 3.9

1 $\sqrt{1 + x} \approx 1 + x/2$

3 $1/x \approx 2 - x$

5 $e^{x^2} \approx 2ex - e$

9 $|\text{Error}| < 0.2$
Overestimate, $x > 0$
Underestimate, $x < 0$

11 (a) $L(x) = 1 + x$
(b) Positive for $x \neq 0$
(c) 0.718

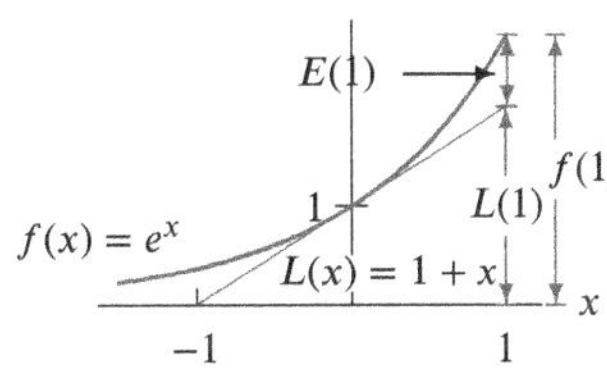

(d) $E(1)$ larger
(e) 0.005

13 (a) $99.5 + 0.2(x - 50)$
(b) 52.5
(c) Approx straight near $x = 50$

15 (a) and (c)

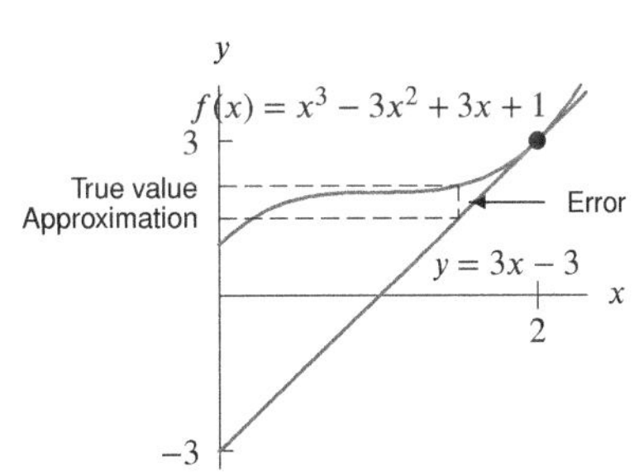

(b) $y = 3x - 3$

17 $a = 1$; $f(a) = 1$
Underestimate
$f(1.2) \approx 1.4$

19 0.1

21 (a) 2.5
(b) 7.25
(c) Underestimate

23 $331.3 + 0.606T$ m/sec

25 (a) $23.681L^{-0.421}$
(b) 23.681 days per cm
(c) 4.736 days
(d) $40.9 + 23.681(L - 1)$ days

27 (a) 10.034 mn users/yr
(b) 5.7 mn users/yr
(c) $t = 1.6$; mid 2011
(d) Rates of change constant

29 (a) 9.14, 4.21, 1.51
(b) $-134.16s^{-2.118}$
(c) -0.0338 attacks per bug per hour per number of bugs
(d) $a \approx 1.51 - 0.0338(s - 50)$ attacks per bug per hour
(e) Linear: -0.169, Power function: -0.166

31 (b) 1% increase

33 $f(1 + \Delta x) \ge f(1) + f'(1)\Delta x$

35 (a) 16,398 m
(b) $16{,}398 + 682(\theta - 20)$ m
(c) True: 17,070 m
Approx: 17,080 m

37 (a) 1492 m
(b) $1492 + 143(\theta - 20)$ m
(c) True: 1638 m
Approx: 1635 m

39 $f(x) \approx 1 + kx$
$e^{0.3} \approx 1.3$

41 $E(x) = x^4 - (1 + 4(x - 1))$
$k = 6$; $f''(1) = 12$
$E(x) \approx 6(x - 1)^2$

43 $E(x) = e^x - (1 + x)$
$k = 1/2$; $f''(0) = 1$
$E(x) \approx (1/2)x^2$

45 $E(x) = \ln x - (x - 1)$
$k = -1/2$; $f''(1) = -1$
$E(x) \approx -(1/2)(x - 1)^2$

47 (c) 0

53 Only near $x = 0$

55 $f(x) = x^3 + 1$, $g(x) = x^4 + 1$

57 $f(x) = |x + 1|$

Section 3.10

1 False

3 False

5 True

7 No; no

9 No; no

11 $f'(c) = -0.5$, $f'(x_1) > -0.5$,
$f'(x_2) < -0.5$

13 6 distinct zeros

19 Racetrack

21 Constant Function

23 $21 \le f(2) \le 25$

31 Not continuous

33 $1 \le x \le 2$

35 Possible answer: $f(x) = |x|$

37 Possible answer

$$f(x) = \begin{cases} x^2 & \text{if } 0 \le x < 1 \\ 1/2 & \text{if } x = 1 \end{cases}$$

39 False

41 False

Chapter 3 Review

1 $200t(t^2+1)^{99}$

3 $(t^2+2t+2)/(t+1)^2$

5 $-8/(4+t)^2$

7 $x^2 \ln x$

9 $(\cos\theta)e^{\sin\theta}$

11 $-\tan(w-1)$

13 $kx^{k-1} + k^x \ln k$

15 $3\sin^2\theta\cos\theta$

17 $6\tan(2+3\alpha)\cos^{-2}(2+3\alpha)$

19 $(-e^{-t}-1)/(e^{-t}-t)$

21 $1/\sin^2\theta - 2\theta\cos\theta/\sin^3\theta$

23 $-(2^w \ln 2 + e^w)/(2^w+e^w)^2$

25 $1/(\sqrt{\sin(2z)}\sqrt{\cos^3(2z)})$

27 $2^{-4z}[-4\ln(2)\sin(\pi z) + \pi\cos(\pi z)]$

29 $e^{(e^\theta+e^{-\theta})}(e^\theta - e^{-\theta})$

31 $e^{\tan(\sin\alpha)}\cos\alpha/\cos^2(\sin\alpha)$

33 $e(\tan 2 + \tan r)^{e-1}/\cos^2 r$

35 $2e^{2x}\sin(3x)$
$(\sin(3x)+3\cos(3x))$

37 $2^{\sin x}\left((\ln 2)\cos^2 x - \sin x\right)$

39 $e^{\theta-1}$

41 $(-cat^2+2at-bc)e^{-ct}$

43 $(\ln 5)5^x$

45 $(2abr - ar^4)/(b+r^3)^2$

47 $-2/(x^2+4)$

49 $20w/(a^2-w^2)^3$

51 $ae^{au}/(a^2+b^2)$

53 $\ln x/(1+\ln x)^2$

55 $e^t\cos\sqrt{e^t+1}/(2\sqrt{e^t+1})$

57 $18x^2+8x-2$

59 $(2\ln 3)z + (\ln 4)e^z$

61 $3x^2 + 3^x \ln 3$

63 $-\frac{5}{2}\frac{\sin(5\theta)}{\sqrt{\cos(5\theta)}} + 12\sin(6\theta)\cos(6\theta)$

65 $4s^3-1$

67 $ke^{kt}(\sin at + \cos bt) +$
$e^{kt}(a\cos at - b\sin bt)$

69 $f'(t) = 4(\sin(2t)-\cos(3t))^3$
$(2\cos(2t)+3\sin(3t))$

71 $-16-12x+48x^2-32x^3+$
$28x^6-9x^8+20x^9$

73 $f'(z) = \frac{5}{2}(5z)^{-1/2} + \frac{5}{2}z^{-1/2} - \frac{5}{2}z^{-3/2} + \frac{\sqrt{5}}{2}z^{-3/2}$

75 $dy/dx = 0$

77 $dy/dx = -3$

79 $f'(t) = 6t^2-8t+3$
$f''(t) = 12t-8$

81 $x=-1$ and $x=5$

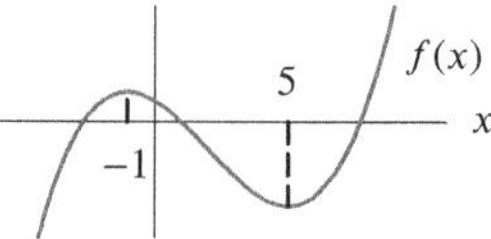

83 4

85 $x \approx 1, -0.4$, or 2.4

87 $y = -4x+6$

89 Approx 1

91 Approx 1.9

93 (a) $H'(2) = 11$
(b) $H'(2) = -1/4$
(c) $H'(2) = r'(1)\cdot 3$
(we don't know $r'(1)$)
(d) $H'(2) = -3$

95 (a) $y = 20x-48$
(b) $y = 11x/9 - 16/9$
(c) $y = -4x+20$
(d) $y = -24x+57$
(e) $y = 8.06x - 15.84$
(f) $y = -0.94x + 6.27$

97 1.909 radians (109.4°) or 1.231 radians (70.5°)

99 Not perpendicular; $x \approx 1.3$

101 0

103 1

105 (a) $2, 1, -3$
(b) $f(2.1) \approx 0.7$, overestimate
$f(1.98) \approx 1.06$, underestimate
Second better

107 (a) 9.801 million
(b) 0.0196 million/year

109 (a) $v(t) = 10e^{t/2}$
(b) $v(t) = s(t)/2$

111 (a) $v = -2\pi\omega y_0\sin(2\pi\omega t)$
$a = -4\pi^2\omega^2 y_0\cos(2\pi\omega t)$
(b) Amplitudes: different
$(y_0,\ 2\pi\omega y_0,\ 4\pi^2\omega^2 y_0)$
Periods $= 1/\omega$

113 (a) $0.33268/k$ years
(b) 30.24 years
(c) $30.24 - 2749.42(k-0.011)$ years
(d) Exponential: 33.3 years
Approx 32.99 years

115 1/6

119 (a) $y^{(n)} =$
$(-1)^{n+1}(n-1)!x^{-n}$
(b) $y^{(n)} = xe^x + ne^x$

121 $(f/g)'/(f/g) = (f'/f) - (g'/g)$

123 (a) 1
(b) 1
(c) $\sin(\arcsin x) = x$

125 (a) 0
(b) 0
(c) $\ln(1-1/t) + \ln(t/(t-1)) = \ln 1$

Section 4.1

1

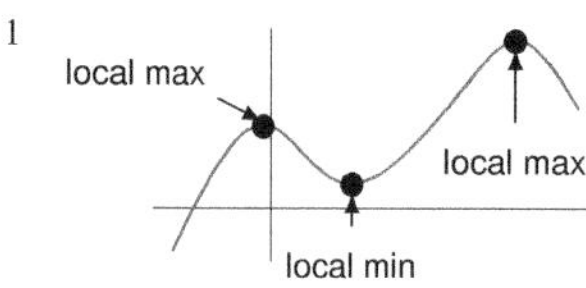

3

Critical point
Not local max or min
Local min
x

5 Critical points: $x=0, x=-\sqrt{6}, x=\sqrt{6}$
Inflection points: $x=0, x=-\sqrt{3}, x=\sqrt{3}$

7 Critical point: $x = 3/5$
No inflection points

9 Critical points:
$x=0$ and $x=1$
Extrema:
$f(1)$ local minimum
$f(0)$ not a local extremum

11 Critical points:
$x=0$ and $x=2$
Extrema:
$f(2)$ local minimum
$f(0)$ not a local extremum.

13 Critical points: $x=0, x=4$
$x=0$: local minimum
$x=4$: local maximum

15 Critical point: $x=0$
$x=0$: local maximum

17 Critical point: $x=1/3$, local maximum

19 Crit pts: max: $x = \frac{\pi}{3} + 2n\pi$,
min: $x = \frac{5\pi}{3} + 2n\pi$

21 (a) Critical point $x \approx 0$;
Inflection points between -1 and 0 and between 0 and 1
(b) Critical point at $x=0$,
Inflection points at $x = \pm 1/\sqrt{2}$

23 $f'(2) < f'(6)$

25 (a) Increasing for $x>0$
Decreasing for $x<0$
(b) Local and global min: $f(0)$

27 (a) Increasing for $0<x<4$
Decreasing for $x<0$ and $x>4$
(b) Local max: $f(4)$
Local min: $f(0)$

29 (a) $\pm\sqrt{a/3}$
(b) $a=12$

31 (a) $x=b$
(b) Local minimum

33 (a) $f'(x)=0$ has no solutions
(b) $f'(x)>0, f''(x)<0$

35

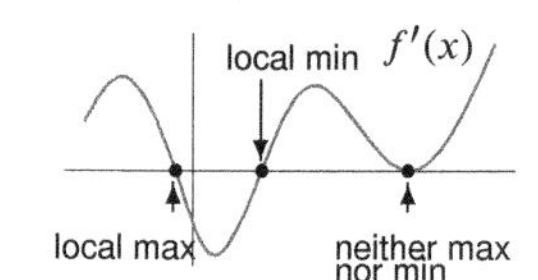

37

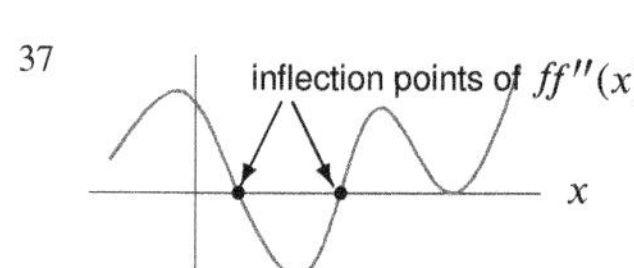

39 (a) D, G
(b) D, F
(c) One; Zero

41

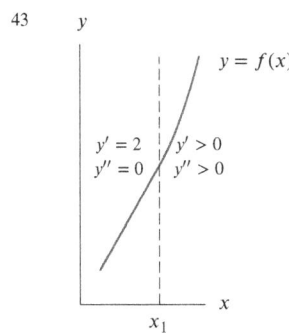

$y' < 0$ everywhere
$y'' = 0$ $y'' = 0$ $y'' = 0$
$y'' > 0$ $y'' < 0$ $y'' > 0$ $y'' < 0$
$y = f(x)$
x_1 x_2 x_3

43

$y = f(x)$
$y' = 2$ $y'' = 0$
$y' > 0$ $y'' > 0$
x_1

45 (a) Local maxima at $x \approx 1$, $x \approx 8$,
Local minimum at $x \approx 4$
(b) Local maxima at $x \approx 2.5$, $x \approx 9.5$
Local minimum at $x \approx 6.5$,

47

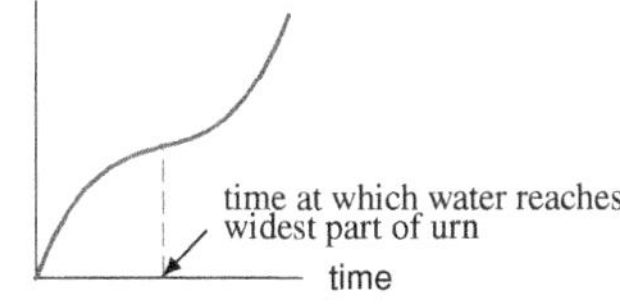

49 $x = 0$, local min if m even, inflect if m odd;
$x = m/(m+n)$, local max;
$x = 1$, local min if n even, inflect if n odd

51 (a) Positive

53 $a = -1/3$

55 (a) 0
(b) Two
(c) Local maximum

59 $B = f, A = f', C = f''$

61 III even; I, II odd
I is f'', II is f, III is f'

63 Consider $f(x) = x^3$

65 Consider $f(x) = x^4$

67 $f(x) = x$

69 $f(x) = \cos x$ or $f(x) = \sin x$

71 True

73 False

75 False

77 True

79 True

81 Impossible

83 Impossible

85 True

87 False

89 Yes

91 Not enough information.

93 (a) $x = \pm 2$
(b) $x = 0, \pm\sqrt{5}$

95 (a) No critical points
(b) Decreasing everywhere

97 Impossible

Section 4.2

1

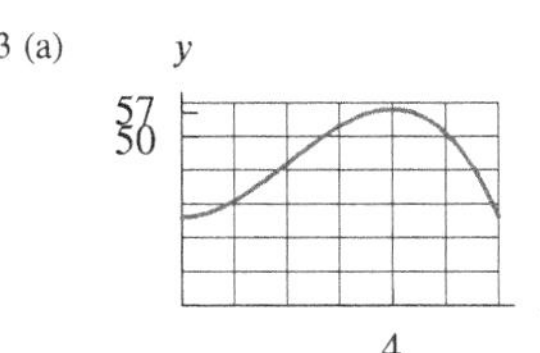

3 (a)

y
57
50
4
x

(b) $x = 4$, $y = 57$

5 Max: 9 at $x = -3$;
Min: -16 at $x = -2$

7 Max: 2 at $x = 1$;
Min: -2 at $x = -1, 8$

9 Max: 8 at $x = 4$;
Min: -1 at $x = -1, 1$

11 (a) $f(1)$ local minimum;
$f(0)$, $f(2)$ local maxima
(b) $f(1)$ global minimum
$f(2)$ global maximum

13 (a) $f(2\pi/3)$ local maximum
$f(0)$ and $f(\pi)$ local minima
(b) $f(2\pi/3)$ global maximum
$f(0)$ global minimum

15 Global min $= 2$ at $x = 1$
No global max

17 Global min $= 1$ at $x = 1$
No global max

19 Global max $= 2$ at $t = 0, \pm 2\pi, \ldots$
Global min $= -2$ at $t = \pm\pi, \pm 3\pi, \ldots$

21 Global min $-4/27$ at $x = \ln(2/3)$

23 $0.91 < y \le 1.00$

25 $0 \le y \le 2\pi$

27 $0 \le y < 1.61$

29 Yes, at G; No

31 $x = -b/2a$,
Max if $a < 0$, min if $a > 0$

33 $T = S/2$

35 $t = 2/b$

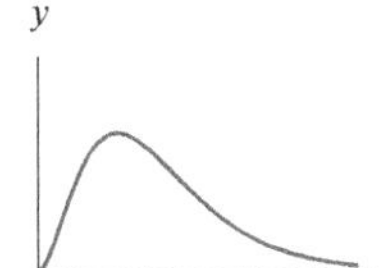

37 (a) Ordering: a/q
Storage: bq
(b) $\sqrt{a/b}$

39 (a) $0 \le y \le a$

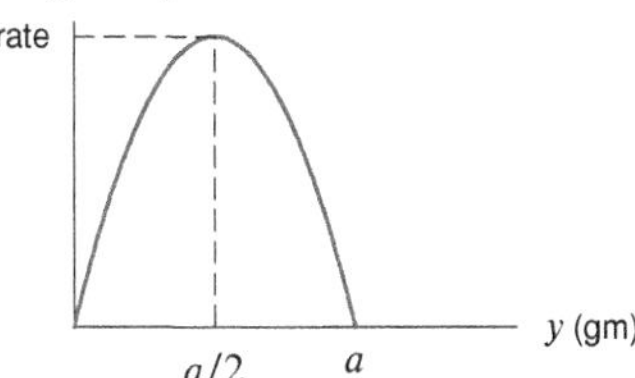

(b) $y = a/2$

41 $r = 3B/(2A)$

43 $x = L/2$

45 $(-1, -1/2)$; $(1, 1/2)$

47 (b) Yes, at $x = 0$
(c) Max: $x = -2$, Min: $x = 2$
(d) $5 > g(0) > g(2)$

49 (a) f is always increasing; g is always decreasing
(b) 5

51 $x = \left(\sum_{i=1}^{n} a_i\right)/n$

53 On $1 \le x \le 2$, global minimum at $x = 1$

55 $f(x) = 1 - x$

57 $\sqrt{2} \le x \le \sqrt{5}$

59 True

61 True

63 True

65 False

67 True

69 True

71 (a) $x = a/2$
(b) $x = a/2$

Section 4.3

1 2500

3 9000

5 Square of side 16 cm

7 $x = h = 8^{1/3}$ cm

9 $r = (4/\pi)^{1/3}$ cm, $h = 2(4/\pi)^{1/3}$ cm

11 11.270

13 $(\pm 1, 0)$, $(\pm 1, 1/2)$

15 (a) $1/(2e)$
(b) $(\ln 2) + 1$

17 When the rectangle is a square

19 1/2

21 (a) $xy + \pi y^2/8$
(b) $2x + y + \pi y/2$
(c) $x = 100/(4+\pi)$, $y = 200/(4+\pi)$

23 (a) $xy + \pi y^2/4 + \pi x^2/4$
(b) $\pi x + \pi y$
(c) $x = 0, y = 100/\pi$; $x = 100/\pi, y = 0$

25 0, 10

27 $10 + 5\sqrt{3}$, $10\sqrt{5}$

29 $(1/2, 1/\sqrt{2})$

31 $(-2.5, \sqrt{2.5})$;
Minimum distance is 2.958

33 Radius 3.854 cm
Height 7.708 cm
Volume 359.721 cm^3

35 13.13 mi from first smokestack

37 Min $v = \sqrt{2k}$; no max

39 Minimum: $0.148mg$ newtons
Maximum: $1.0mg$ newtons

41 Maximum revenue = \$27,225
Minimum = \$0

43 (a) q/r months
(b) $(ra/q) + rb$ dollars
(c) $C = (ra/q) + rb + kq/2$ dollars
(d) $q = \sqrt{2ra/k}$

45 0.8 miles from Town 1

47 65.1 meters

49 (a) $x = e$
(b) $n = 3$
(c) $3^{1/3} > \pi^{1/\pi}$

51 Max slope $= 1/(3e)$ at $x = 1/3$

53 (a) 10
(b) 9

55 (a)

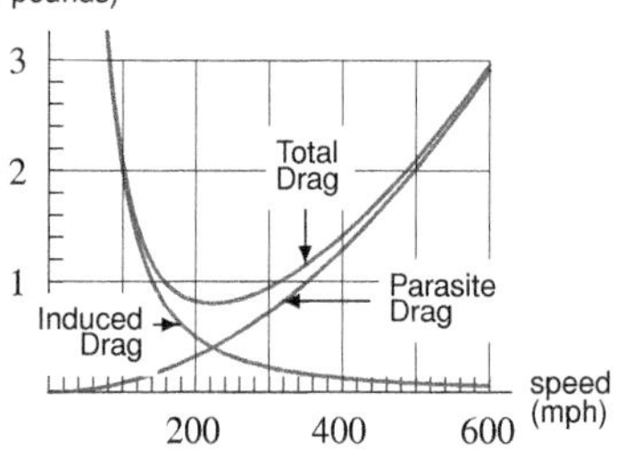

(b) 160 mph or 320 mph; no; yes
(c) 220 mph

57 $y = mg/k$

59 Max at $x = 10$

61 Optimum may occur at an endpoint

Section 4.4

1 (a)

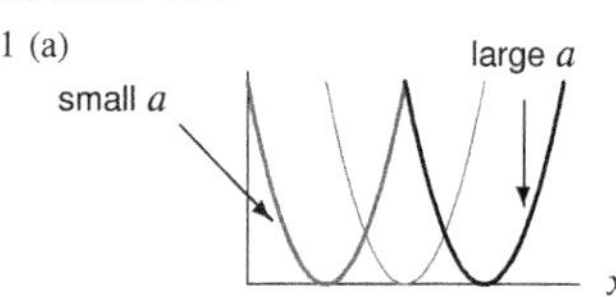

(b) Critical point moves right
(c) $x = a$

3 (a)

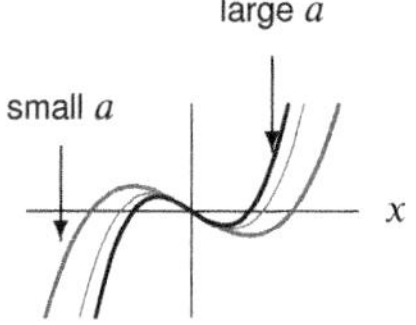

(b) 2 critical points move closer to origin
(c) $x = \pm\sqrt{1/(3a)}$

5 (a)

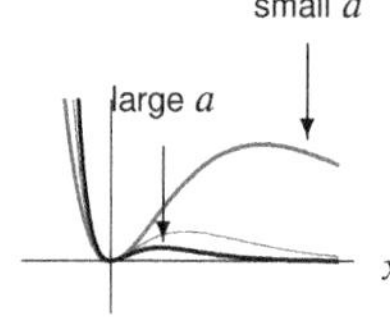

(b) Nonzero critical point moves down to the left
(c) $x = 0, 2/a$

7 A has $a = 1$, B has $a = 2$, C has $a = 5$

9 (a) Larger $|A|$, steeper
(b) Shifted horizontally by B
Left for $B > 0$; right for $B < 0$
Vertical asymptote $x = -B$
(c)

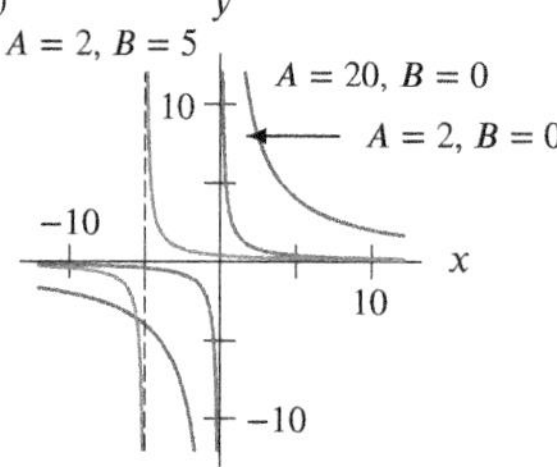

11 (a)

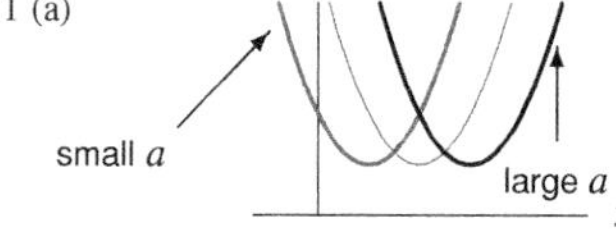

(b)

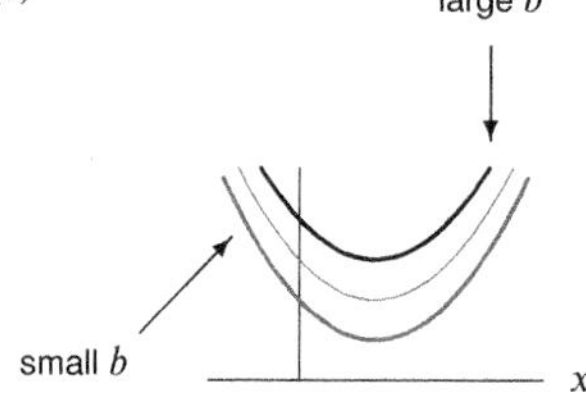

(c) a moves critical point right;
b moves critical point up
(d) $x = a$

13 (a)

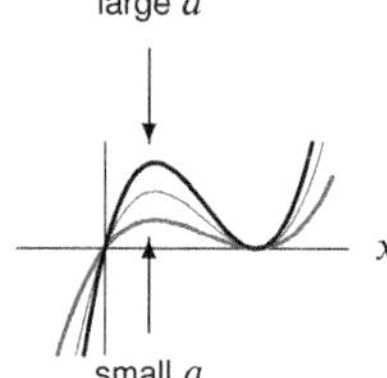

(b)

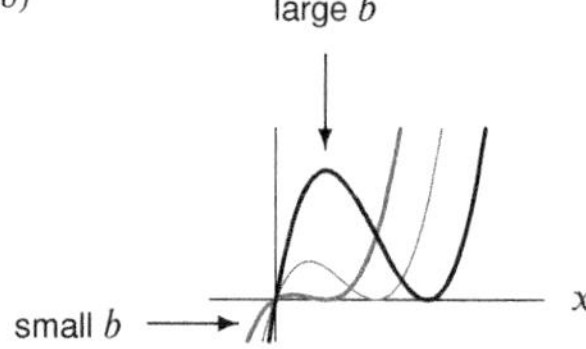

(c) a moves one critical point up, does not move the other;
b moves one critical point up to right, moves the other right

(d) $x = b/3, b$

15 (a)

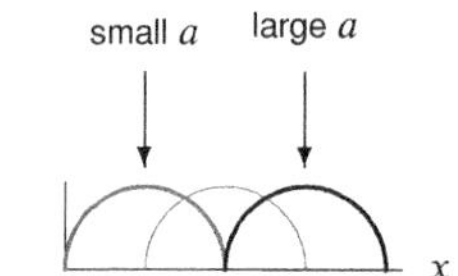

(b)

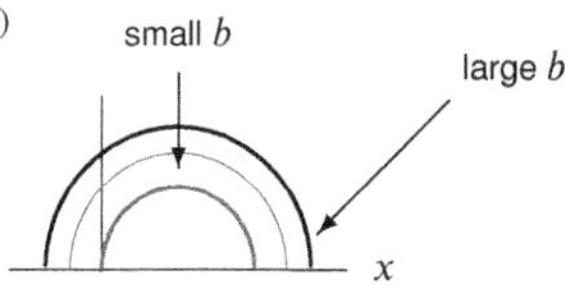

(c) a moves critical points to the right;
b moves one critical point left, one up, one right

(d) $x = a, a \pm \sqrt{b}$

17 C has $a = 1$, B has $a = 2$, A has $a = 3$

19 (a)

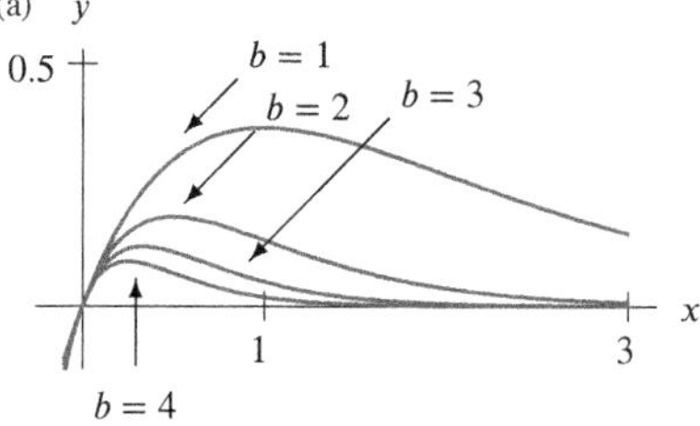

(b) $(1/b, 1/be)$

21 Two critical points: $x = b$, $x = b/(n + 1)$

23 (a)

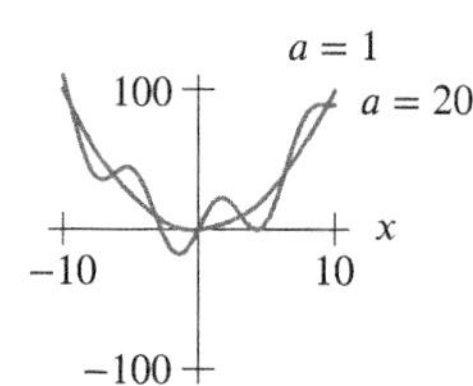

(b) $-2 < a < 2$

25 (a) $x = -1/b$
(b) Local minimum

27 $k > 0$

29 (a) $x = e^a$
(b) $a = -1$:

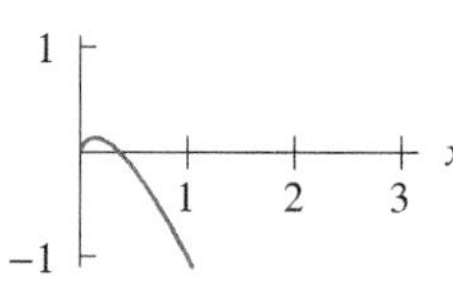

$a = 1$:

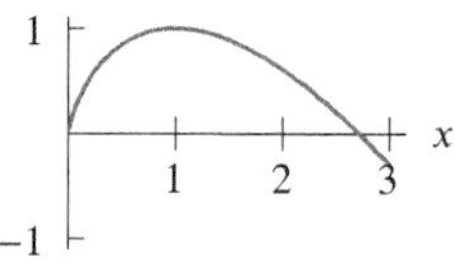

(c) Max at (e^{a-1}, e^{a-1}) for any a

33 $f''(x) = a^2e^{-ax} + b^2e^{bx}$; always positive

35 $f'(x) = a/x - b$; positive for $x < a/b$, negative for $x > a/b$

37 $f''(x) = 12ax^2 - 2b$,
graph f'' is upward parabola with neg vert intercept

39 (a) A: value for large negative x
B: value for large positive x

(c) (I) $(8, -4)$

(II) $(-2, 5)$

(III) $(7, 0)$

(IV) $(2, 6)$.

41 (a) Local max: $x = 1/b$
No local minima
Inflection point: $x = 2/b$

(b) Varying a stretches or flattens the graph vertically.
Incr b shifts critical, inflection points to left; lowers max

(c)

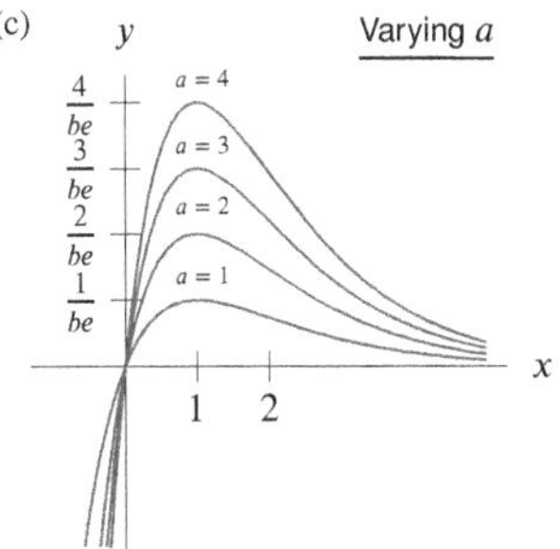

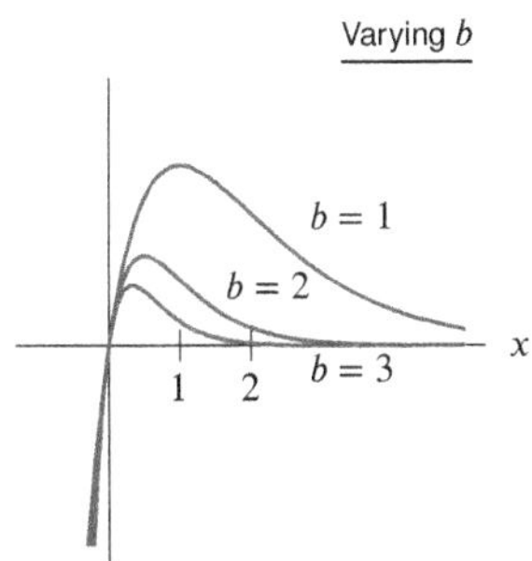

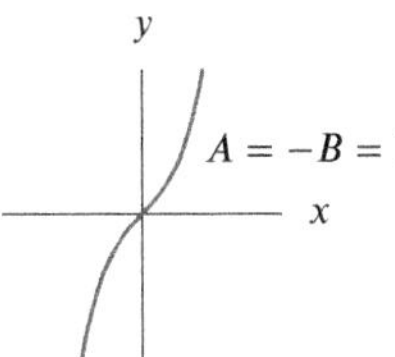

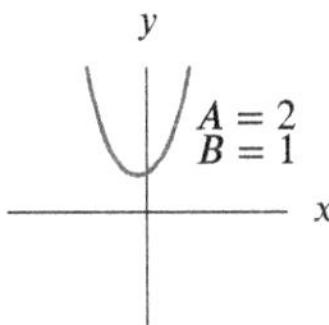

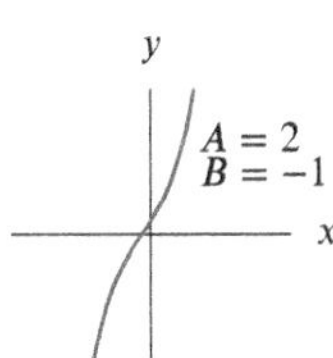

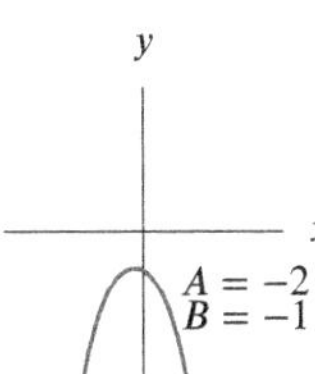

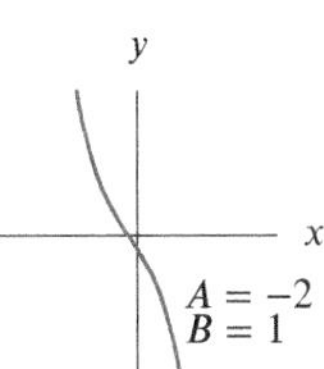

45 $Ax^3 + Cx$, Two

47 $y = 5(1 - e^{-bx})$

49 $y = e^{-(x-2)^2/2}$

51 $y = (2 + 2e)/(1 + e^{1-t})$

53 $y = -2x^4 + 4x^2$

55 $y = (\sqrt{2}/\pi)\cos(\pi t^2)$

57 $y = 2xe^{1-x/3}$

59 (a) For $a \leq 0$, all b; $x = b \pm \sqrt{-a}$

(b) $a = 1/5$, $b = 3$

61 (a)

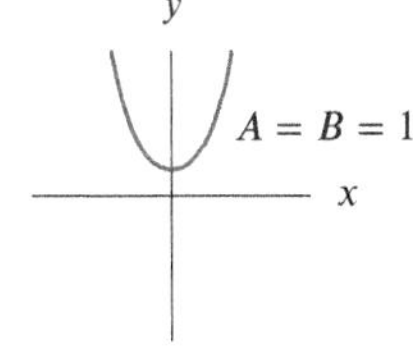

(b) U-shaped

(c) Incr ($A > 0$)
or Decr ($A < 0$)

(d) Max: $A < 0, B < 0$
Min: $A > 0, B > 0$

63 (a) Intercept: $x = a$
Asymptotes: $x = 0$,
$U = 0$

(b) Local min: $(2a, -b/4)$
Local max: none

(c)

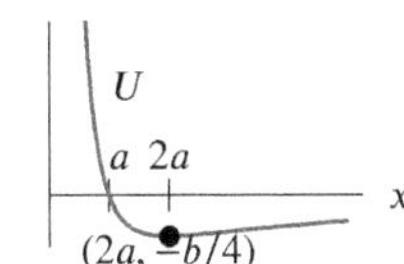

65 (a) $+\infty$

(b) $r = (2A/B)^{1/6}$; local minimum

(c) $r = (2A/B)^{1/6}$

67 (a) Vertical intercept: $W = Ae^{-e^b}$, Horizontal asymptote $W = A$

(b) No critical points, inflection point at $t = b/c$, $W = Ae^{-1}$

(c)

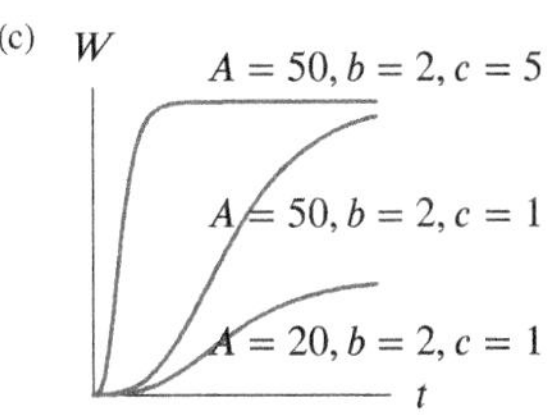

(d) Yes

69 Two critical points only if a and b have the same sign.

71 $f(x) = x^3 + x$

73 One possibility:
$g(x) = ax^3 + bx^2$, $a, b \neq 0$

75 (b), (c)

Section 4.5

1 \$5000, \$2.40, \$4

3 Below 4500

5 About \$1.1 m, 70, \$1.2 m

7 $C(q) = 500+6q$, $R(q) = 12q$, $\pi(q) = 6q-500$

9 $C(q) = 5000 + 15q$, $R(q) = 60q$,
$\pi(q) = 45q - 5000$

11 $\pi(q) = 490q - q^2 - 150$
Max at $q = 245$

13 \$0.20/item

15 (a) Increase production

(b) $q = 8000$

17 Above 2000

19 Greater than 100

21 (a) \$0

(b) \$96.56

(c) Raise the price by \$5

23 (a)

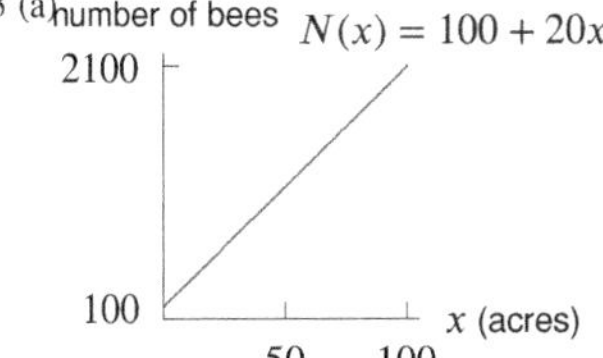

(b) (i) $N'(x) = 20$

(ii) $N(x)/x = (100/x) + 20$

25 (a)

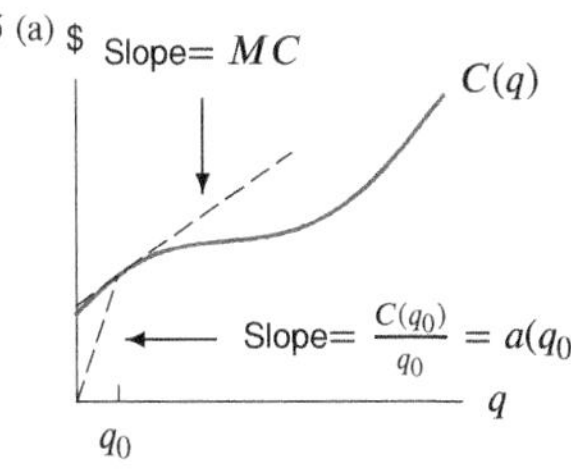

27 (a)

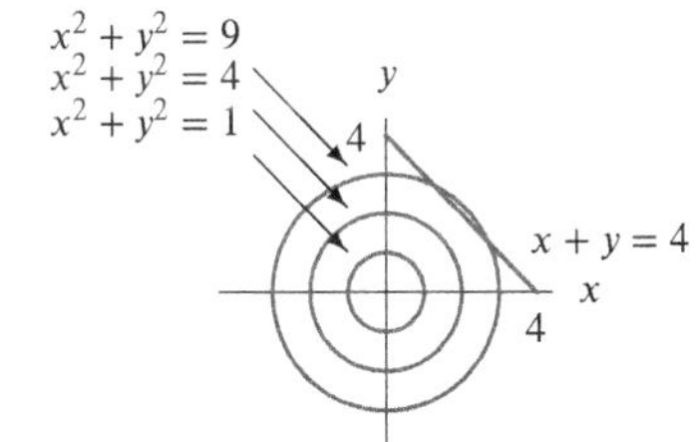

(c) 8

29 (a)

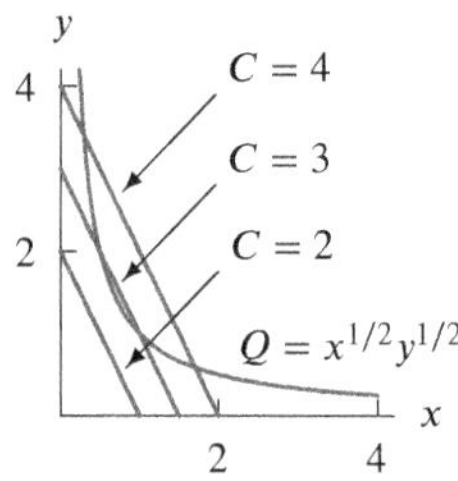

(c) $2\sqrt{2}$

31 Cannot determine whether local or global, max or min

33 2 or 15

35 (e)

37 (a) Positive
(b) Cost = wx, Revenue = $pf(x)$
(c) w/p
(d) Negative
(e) $1/(pf''(x^*))$, negative
(f) Less

Section 4.6

1 $-0.32°$C/min; $-0.262°$C/min

3 (a) 0
(b) -4.5
(c) -8

5 (a) $\pi/\sqrt{9.8l}$
(b) Decr

7 $-1°C$ minute

9 (a) 5644 ft
(b) -124 ft/min

11 $-81/R^2$

13 0.211 units/sec

15 24 meters^3/yr

17 0 atmospheres/hour

19 -1.440 atmospheres/hour

21 30 cm^2/min

23 -7 cm^2/sec

25 Moving closer at 4 mph

27 (a) $CD - D^2$
(b) $D < C$

29 (a) (i) $5r$ \$/yr
(ii) $5re^{0.02r}$ \$/yr
(b) 24.916 \$/yr

31 5.655 ft^3/sec

33 $180\pi = 565.487$ cm^3/hr

35 2/9 ohms/min

37 g; $ge^{-k/m}$; acceleration decreases from g at $t = 0$

39 (a) $-92.8V^{-2.4}$ atm/cm^3
(b) Decreasing at 0.0529 atm/min

41 (a) 94.248 m^2/min
(b) 0.0000267 m/min

43 (a) $80\pi = 251.327$ sec
(b) $V = 3\pi h^3/25$ cm^3
(c) 0.0207 cm/sec

45 $8/\pi$ meters/min

47 0.253 meters/second

49 Between 12°F/min and 20°F/min

51 (a) $z = \sqrt{0.25 + x^2}$
(b) 0.693 km/min

(c) 0.4 radians/min

53 $1/3$

55 (a) $h(t) = 300 - 30t$
$0 \le t \le 10$
(b) $\theta = \arctan((200 - 30t)/150)$
$d\theta/dt =$
$-\left(\frac{1}{5}\right)\left(\frac{150^2}{150^2+(200-30t)^2}\right)$
(c) When elevator is at level of observer

57 $dD/dt = 2 \cdot dR/dt$

59 $y = f(x) = 2x + 1$ and $x = g(t) = 5t$

61 True

63 (c)

65 $(1/V)dV/dt$

67 (a) 900
(b) 34 ships
(c) End of battle
(d) 30 ships
(e) $y'(t) = kx$; negative
(g) $k = 2$, 68 ships/hour

Section 4.7

1 Yes

3 Yes

5 No

7 No

9 Yes

11 No

13 $1/4$

15 1.5

17 0

19 0

21 $(1/3)a^{-2/3}$

23 -0.7

25 $10/9$

27 0.75

29 Does not exist

31 Does not exist

33 $7/9$

35 0

37 0

39 $0.1x^7$

41 $x^{0.2}$

43 -2

45 Positive

47 Negative

49 0

51 0

53 none, no

55 $\infty - \infty$, yes

57 ∞^0, yes

59 0

61 0

63 0

65 0

67 $1/2$

69 -2

71 0

73 Does not exist

75 e

77 e^{kt}

79 $e^{-\lambda}$

81 5

83 2

85 $-25/2$

87 1

89 -2

91 Limit of derivative does not exist

93 $f(x) = x$

95 (b)

Section 4.8

1 The particle moves on straight lines from $(0, 1)$ to $(1, 0)$ to $(0, -1)$ to $(-1, 0)$ and back to $(0, 1)$

3 Two diamonds meeting at $(1, 0)$

5 $x = 3\cos t$, $y = -3\sin t$, $0 \le t \le 2\pi$

7 $x = 2 + 5\cos t$, $y = 1 + 5\sin t$, $0 \le t \le 2\pi$

9 $x = t$, $y = -4t + 7$

11 $x = -3\cos t$, $y = -7\sin t$, $0 \le t \le 2\pi$

13 Clockwise for all t

15 Clockwise: $-\sqrt{1/3} < t < \sqrt{1/3}$
Counterclockwise: $t < -\sqrt{1/3}$ or $t > \sqrt{1/3}$

17 Clockwise: $2k\pi < t < (2k + 1)\pi$
Counterclockwise: $(2k - 1)\pi < t < 2k\pi$

19 Line segment:
$y + x = 4$, $1 \le x \le 3$

21 Line $y = (x - 13)/3$, left to right

23 Parabola $y = x^2 - 8x + 13$, left to right

25 Circle $x^2 + y^2 = 9$, counterclockwise

27 $(y - 4) = (4/11)(x - 6)$

29 $y = -(4/3)x$

31 Speed $= 2|t|$, stops at $t = 0$

33 Speed $= ((2t - 4)^2 + (3t^2 - 12)^2)^{1/2}$, Stops at $t = 2$

35

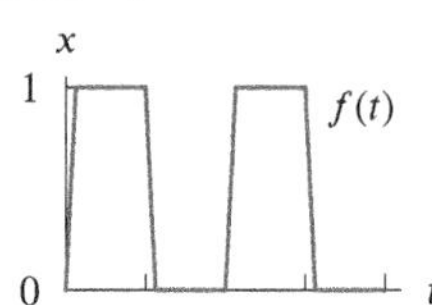

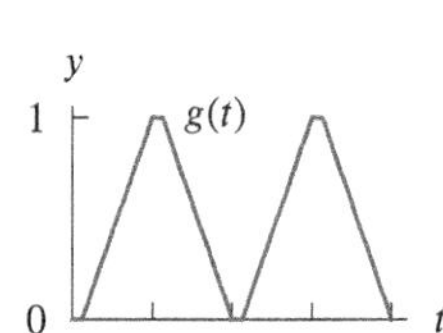

37 (a) The part of the line with $x < 10$ and $y < 0$
(b) The line segment between $(10, 0)$ and $(11, 2)$

39 (a) Both parameterize line
$y = 3x - 2$
(b) Slope = 3
y-intercept = -2

41 (a) $a = b = 0, k = 5$ or -5
(b) $a = 0, b = 5, k = 5$ or -5
(c) $a = 10, b = -10, k = \sqrt{200}$ or $-\sqrt{200}$

43 A straight line through the point $(3, 5)$

45 (a) $dy/dx = 4e^t$
(b) $y = 2x^2$
(c) $dy/dx = 4x$

47 (a) $(11, -2)$
(b) $1/12$
(c) $\sqrt{145}$

49 (a) Yes, $t = 1$, $(x, y) = (-2, -1)$
(b) Yes, $t = -1$, $(x, y) = (2, 3)$
(c) Never

51 (a) Increasing
(b) P to Q
(c) Concave down

53 (a) No
(b) $k = 1$
(c) Particle B

55 (a) $y + \frac{1}{2} = -\frac{\sqrt{3}}{3}(x - \pi)$
(b) $t = \pi$
(c) 0.291, concave up

57 (a) $t = \pi/4$; at that time, speed $= \sqrt{9/2}$
(b) Yes, when $t = \pi/2$ or $t = 3\pi/2$
(c) Concave down everywhere

59 Line traversed in the wrong direction

61 $x = 2\cos t$, $y = 2\sin t$, $0 \le t \le \frac{\pi}{2}$

63 False

67

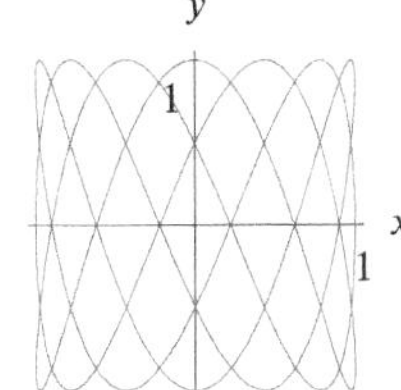

69

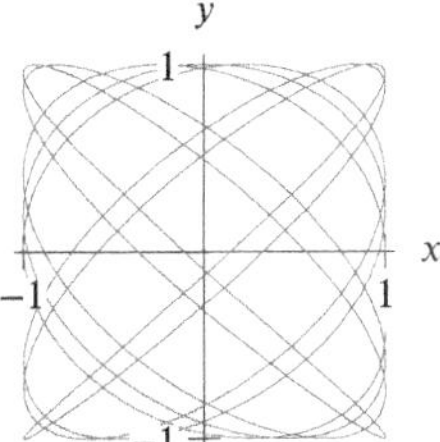

Chapter 4 Review

1

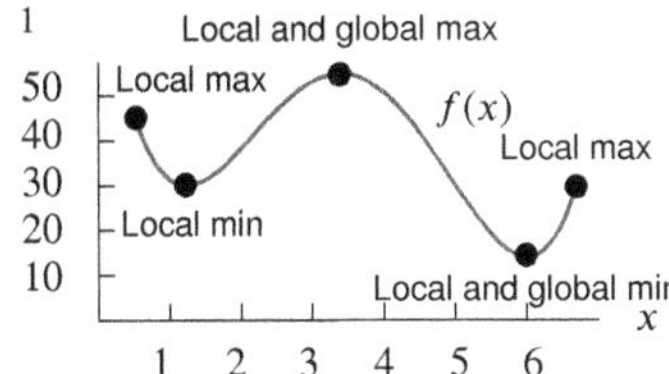

3 (a) $f'(x) = 3x(x - 2)$
$f''(x) = 6(x - 1)$
(b) $x = 0$
$x = 2$
(c) Inflection point: $x = 1$
(d) Endpoints:
$f(-1) = -4$ and $f(3) = 0$
Critical points:
$f(0) = 0$ and $f(2) = -4$
Global max:
$f(0) = 0$ and $f(3) = 0$
Global min:
$f(-1) = -4$ and $f(2) = -4$
(e) f increasing:
for $x < 0$ and $x > 2$
f decreasing:
for $0 < x < 2$
f concave up:
for $x > 1$
f concave down:
for $x < 1$

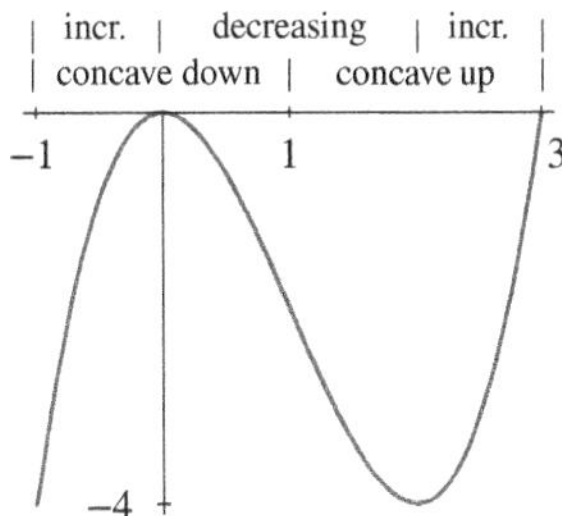

5 (a) $f'(x) =$
$-e^{-x}\sin x + e^{-x}\cos x$
$f''(x) = -2e^{-x}\cos x$
(b) Critical points:
$x = \pi/4$ and $5\pi/4$
(c) Inflection points:
$x = \pi/2$ and $3\pi/2$
(d) Endpoints:
$f(0) = 0$ and $f(2\pi) = 0$
Global max:
$f(\pi/4) = e^{-\pi/4}(\sqrt{2}/2)$
Global min:
$f(5\pi/4) = -e^{-5\pi/4}(\sqrt{2}/2)$
(e) f increasing:
$0 < x < \pi/4$ and
$5\pi/4 < x < 2\pi$
f decreasing:
$\pi/4 < x < 5\pi/4$
f concave down:
for $0 \le x < \pi/2$
and $3\pi/2 < x \le 2\pi$
f concave up:
for $\pi/2 < x < 3\pi/2$

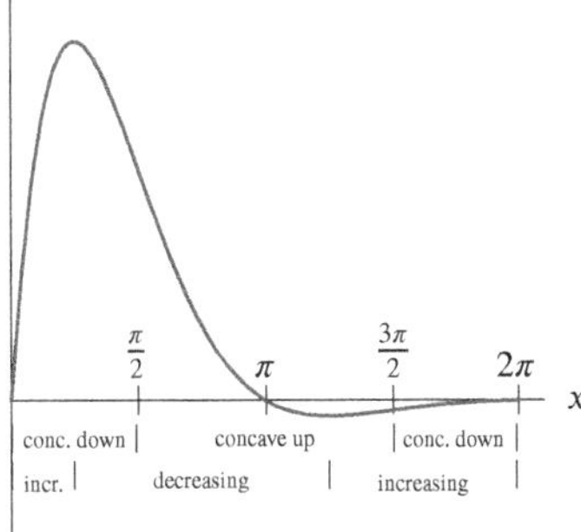

7 $\lim_{x\to\infty} f(x) = \infty$
$\lim_{x\to-\infty} f(x) = -\infty$
(a) $f'(x) = 6(x - 2)(x - 1)$
$f''(x) = 6(2x - 3)$
(b) $x = 1$ and $x = 2$
(c) $x = 3/2$
(d) Critical points:
$f(1) = 6$, $f(2) = 5$
Local max: $f(1) = 6$
Local min: $f(2) = 5$
Global max and min: none
(e) f increasing: $x < 1$ and $x > 2$
Decreasing: $1 < x < 2$
f concave up: $x > 3/2$
f concave down: $x < 3/2$

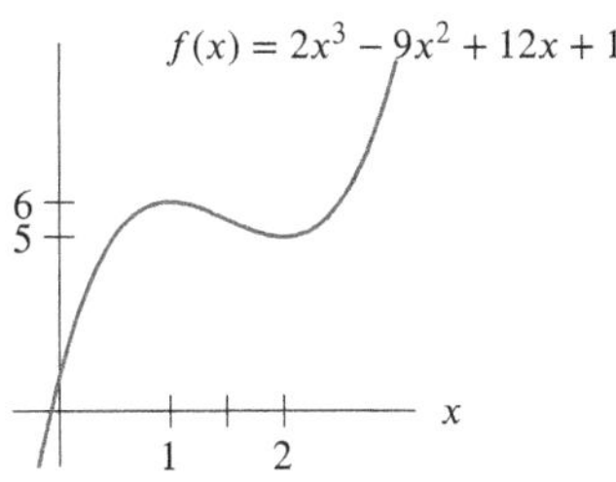

9 $\lim_{x\to-\infty} f(x) = -\infty$
$\lim_{x\to\infty} f(x) = 0$
(a) $f'(x) = (1 - x)e^{-x}$
$f''(x) = (x - 2)e^{-x}$
(b) Only critical point is at $x = 1$
(c) Inflection point: $f(2) = 2/e^2$
(d) Global max: $f(1) = 1/e$
Local and global min: none
(e) f increasing: $x < 1$
f decreasing: $x > 1$
f concave up: $x > 2$
f concave down: $x < 2$

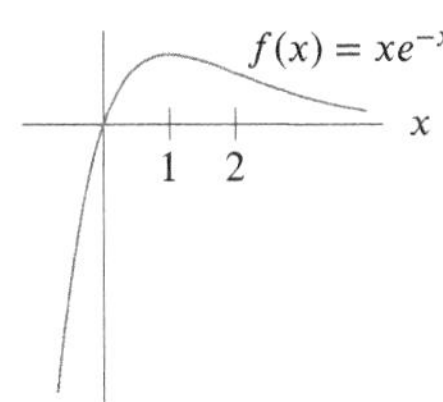

11 D

13 Max: 22.141 at $x = \pi$;
Min: 2 at $x = 0$

15 Max: 1 at $x = 0$;
Min: Approx 0 at $x = 10$

17 Global max $= 1$ at $t = 0$
No global min

19 Local max: $f(-3) = 12$
Local min: $f(1) = -20$
Inflection pt: $x = -1$
Global max and min: none

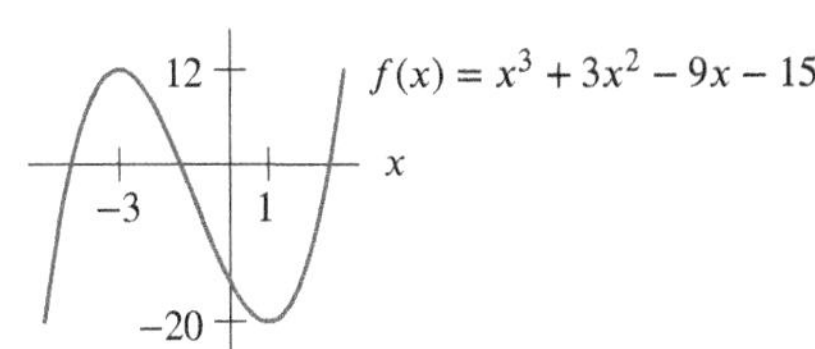

21 Global and local min: $x = 2$
Global and local max: none
Inflection pts: none

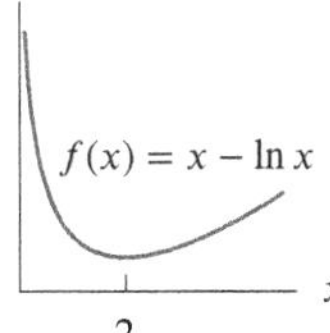

23 Local max: $f(-2/5)$
Global and local min: $f(0)$
Inflection pts: $x = (-2 \pm \sqrt{2})/5$
Global max: none

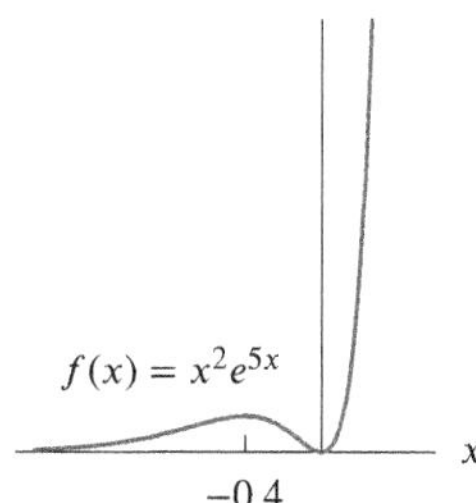

25 Max: $r = \frac{2}{3}a$; Min: $r = 0, a$

27 A has $a = 1$, B has $a = 2$, C has $a = 3$

29 (a) Critical points: $x = 0$, $x = \sqrt{a}$, $x = -\sqrt{a}$
Inflection points: $x = \sqrt{a/3}$, $x = -\sqrt{a/3}$
(b) $a = 4, b = 21$
(c) $x = \sqrt{4/3}, x = -\sqrt{4/3}$

31 Domain: All real numbers except $x = b$;
Critical points: $x = 0$, $x = 2b$

33 18

35 (a) $C(q) = 3 + 0.4q$ m
(b) $R(q) = 0.5q$ m
(c) $\pi(q) = 0.1q - 3$ m

37 2.5 gm/hr

39 Line $y = -2x + 9$, right to left

41 (a) x_3
(b) x_1, x_5
(c) x_2
(d) 0

43

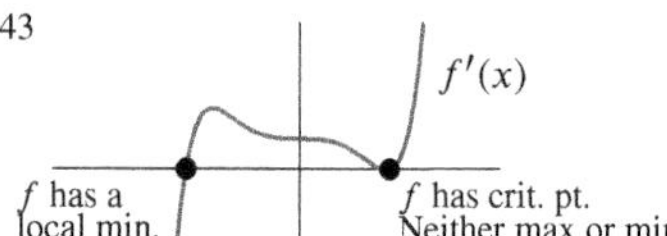

45 $y = x^3 - 6x^2 + 9x + 5$

47 $y = 3x^{-1/2} \ln x$

49 $y = 2xe^{(1-x^2)/2}$

51 $x = \sqrt{8/3}$ cm, $h = \sqrt{2/3}$ cm

53 $r = h = \sqrt{8/3\pi}$ cm

55 $-4.81 \le f(x) \le 1.82$

57 Max: $m = -5k/(12j)$

59 (a) $t = 0$ and $t = 2/b$
(b) $b = 0.4$; $a = 3.547$
(c) Local min: $t = 0$; Local max: $t = 5$

61

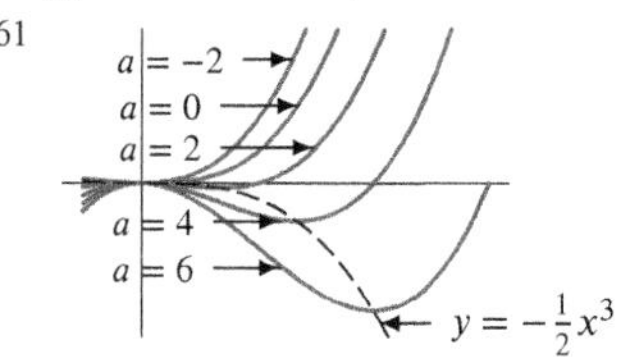

63 1.688 mm/sec away from lens

65 Minimum: −2 amps
Maximum: 2 amps

67 Max area = 1
$\left(\pm\frac{1}{\sqrt{2}}, 0\right)$; $\left(\pm\frac{1}{\sqrt{2}}, \frac{1}{\sqrt{2}}\right)$

69 (0.59, 0.35)

71 $r = \sqrt{2A/(4+\pi)}$;
$h = \frac{A}{2} \cdot \frac{\sqrt{4+\pi}}{\sqrt{2A}} - \frac{\pi}{4} \cdot \frac{\sqrt{2A}}{\sqrt{4+\pi}}$

73 40 feet by 80 feet

75 (a) $\pi(q)$ max when $R(q) > C(q)$ and R and Q are farthest apart

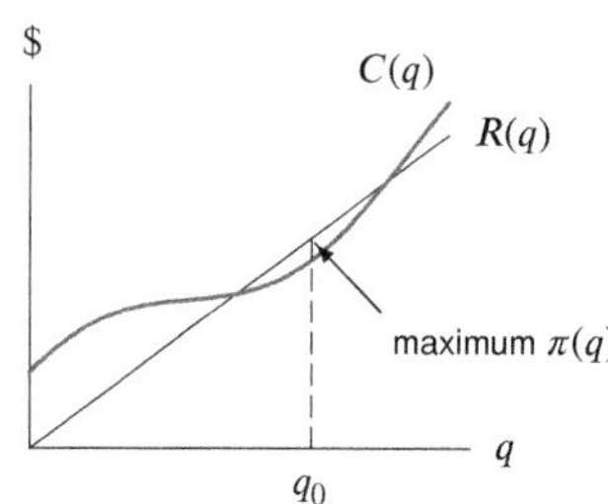

(b) $C'(q_0) = R'(q_0) = p$
(c)

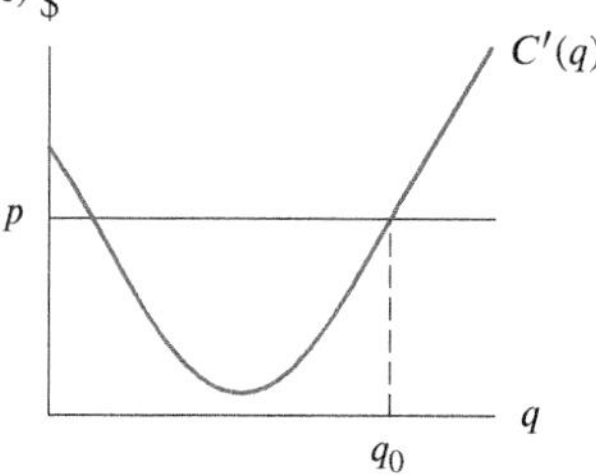

77 About 331 meters, Maintain course

79 (a) $g(e)$ is a global maximum; there is no minimum
(b) There are exactly two solutions
(c) $x = 5$ and $x \approx 1.75$

81 (a) depth of water

(b) depth of water

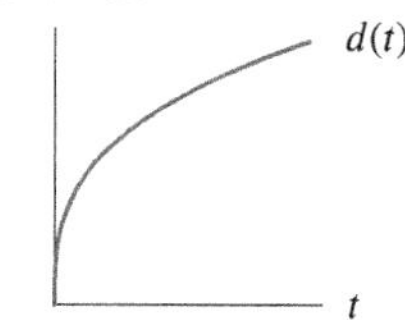

83 -0.0545 m^3/min

85 0/0, yes

87 $-1/2$

89 $-1/6$

91 $1/\pi \approx 0.32\mu$m/day

93 (a) Angular velocity
(b) $v = a(d\theta/dt)$

95 $dM/dt = K(1 - 1/(1+r))dr/dt$

97 Height: $0.3/(\pi h^2)$ meters/hour
Radius: $(0.3/\sqrt{3})/(\pi h^2)$ meters/hour

99 (a) Increasing
(b) Not changing
(c) Decreasing

101 398.103 mph

103 (a) Decreases
(b) -0.25 cm^3/min

105 (a)

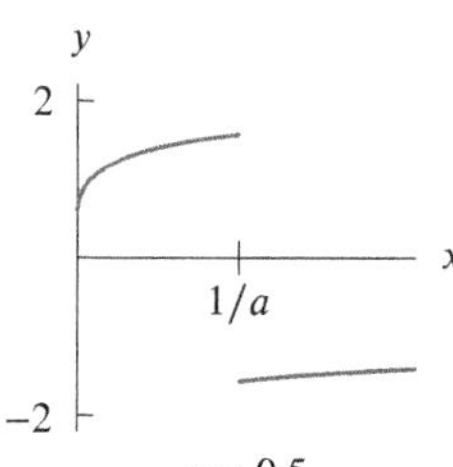

$a = 0.5$

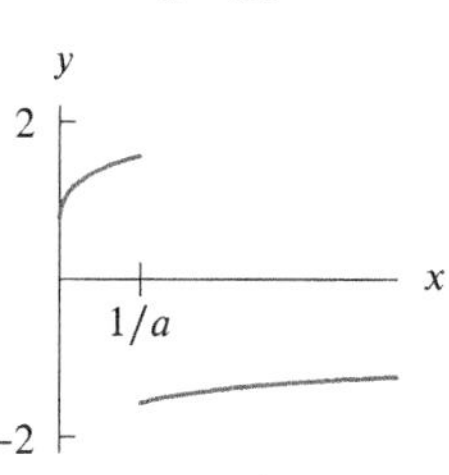

$a = 1$

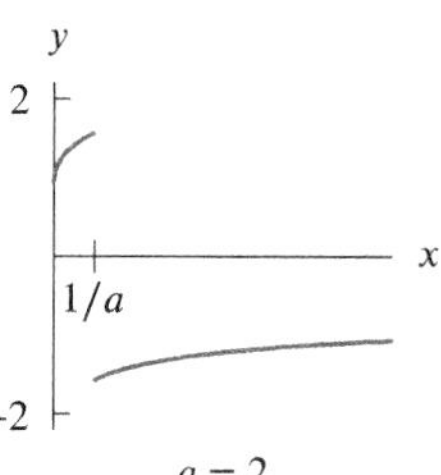

$a = 2$

(b) Same graph all a

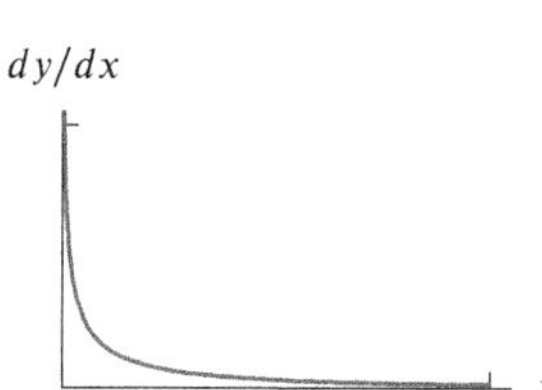

(c) dy/dx simplifies to $\sqrt{x}/(2x(1+x))$

107 (a) $1/\sqrt{x^2-1}$

109 (a) $dy/dx = \frac{\tan(x/2)}{2\sqrt{(1-\cos x)/(1+\cos x)}}$
(b) $dy/dx = 1/2$ on $0 < x < \pi$
$dy/dx = -1/2$ on $\pi < x < 2\pi$

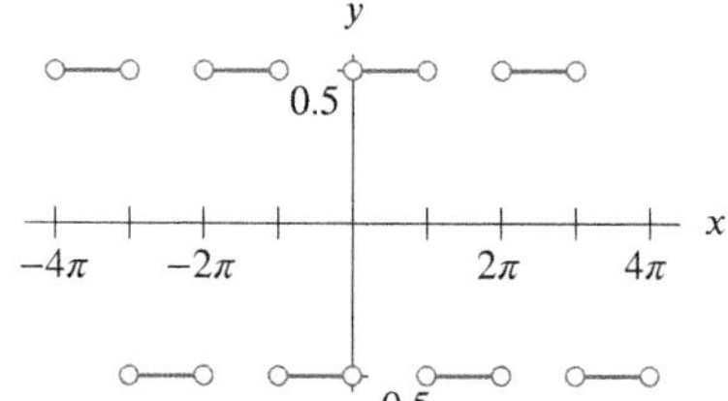

Section 5.1

1 (a) Left sum
(b) Upper estimate
(c) 6
(d) $\Delta t = 2$
(e) Upper estimate ≈ 24.4

3 (a) 54 meters; upper
(b) 30 meters; lower

5 Between 140 and 150 meters

7 (a) $b = 15$
(b) 75 feet

9 (a) $a = 20,\ b = 90,\ c = 12$
(b) 660 feet

11 (a) 408
(b) 390

13 lower: 235.2 meters; upper: 295.0 meters

15 Lower est = 46 m
Upper est = 118 m
Average = 82 m

17 Change: 0 cm, no change in position
Total distance: 12 cm

19 Change: 15 cm to the left
Total distance: 15 cm

21 (a) 120, 240, 180 ft
(b) 180 ft

23 (a) 2; 15, 17, 19, 21, 23; 10, 13, 18, 20, 30
(b) 122; 162
(c) 4; 15, 19, 23; 10, 18, 30
(d) 112; 192

25 (a) Lower estimate = 5.25 mi
Upper estimate = 5.75 mi
(b) Lower estimate = 11.5 mi
Upper estimate = 14.5 mi
(c) Every 30 seconds

27 1/5

29 0.0157

31 (a)

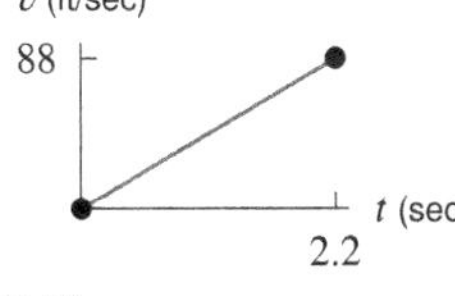

(b) 96.8 ft

33 About 65 km from home
3 hours
About 90 km

35 (a) Up from 0 to 90 seconds; down from 100 to 190 seconds
(b) 870 ft

37 (a) A: 8 hrs
B: 4 hrs
(b) 100 km/hr
(c) A: 400 km
B: 100 km

39 (a)

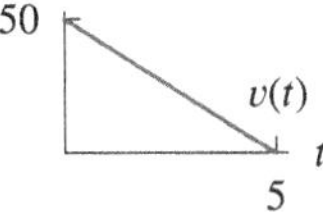

(b) 125 feet
(c) 4 times as far

41 20 minutes

43 Only true if car accelerates at a constant rate

45 $f(x) = x^2$, $[a, b] = [-2, -1]$

47 True

49 False

51 About 0.0635 miles or 335 feet

Section 5.2

1 (a) Right
(b) Upper
(c) 3
(d) 2

3 (a) Left; smaller
(b) 0, 2, 6, 1/3

5 (a) 224

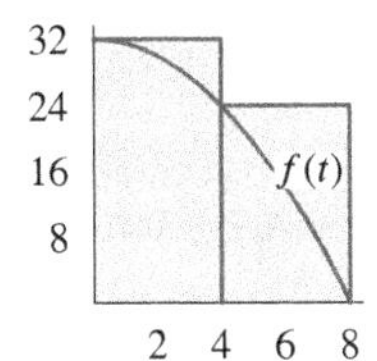

(b) 96

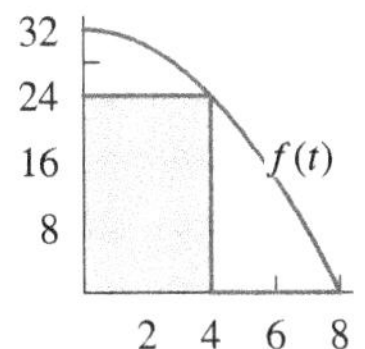

(c) About 200

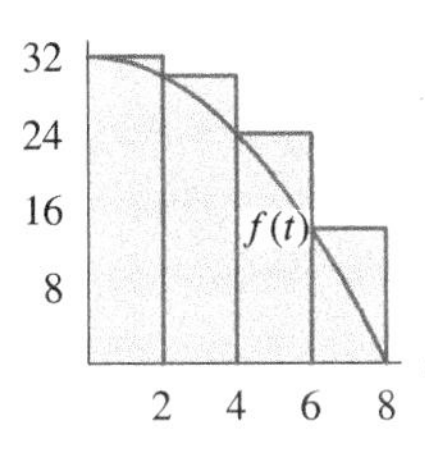

(d) About 136

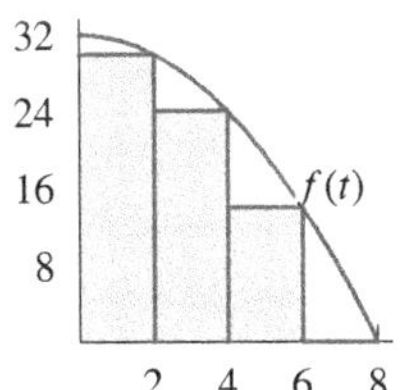

7 205.5

11 About 350

13 28.5

15 1.4936

17 0.3103

19 (a) 160
(b) 219
(c) −18.5

21 (a) 80
(b) 123
(c) 88

23 IV: 93.47

25 7.667

27 14,052.5

29 2.545

31 1.106

33 21.527

35 (a) −4
(b) 0
(c) 8

37 (a) 1
(b) 2π
(c) $2\pi - 1/2$
(d) $\pi - 3/2$

39 −16

41 22/3

43 0.80

45 (a) 0.808
(b) 0.681

47 Left-hand sum = 792; Right-hand sum = 2832.

49 Left-hand sum = 50.5; Right-hand sum = 18.625.

51 Left-hand sum = $(\pi/4)(\sqrt{2} + 1)$; Right-hand sum = $(\pi/4)(\sqrt{2} + 1)$.

53 Left-hand sum = $1 + \sqrt{2} + \sqrt{3}$; Right-hand sum = $\sqrt{2} + \sqrt{3} + 2$.

55 (a)

sin x

x

$-\pi$ π

(b)

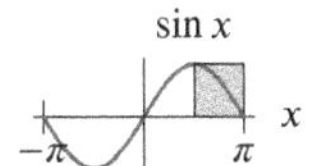

(c) 0

57 (a)

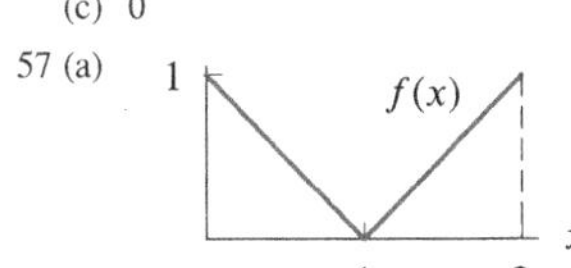

(b) 1
(c) 1; same value

59 2/125

61 $\int_0^3 2t\,dt$

63 $\int_0^{1/2} (7t^2 + 3)\,dt$

65 $\int_1^4 \sqrt{t^2 + t}\,dt$

67 $\int_0^4 \sqrt{t}\,dt = 16/3$

69 $\int_1^2 (8t - 8)\,dt = 4$

71 $a = 1$

73 $a = 1, b = 5$

75 Too many terms in sum

77 $f(x) = -1$, $[a, b] = [0, 1]$

79 False

81 False

83 True

85

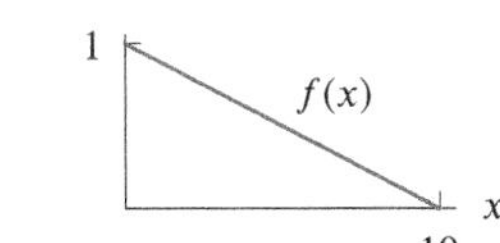

87

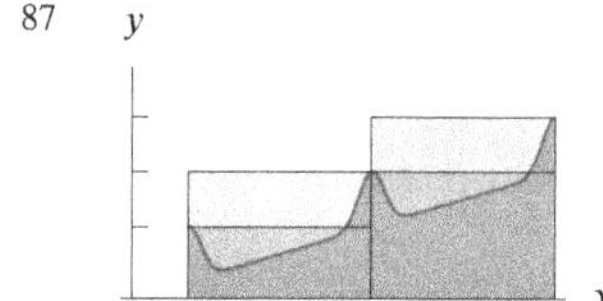

89 (c) $n \geq 10^5$

Section 5.3

1 Dollars

3 Foot-pounds

5 Change in velocity; km/hr

7 Change in salinity; gm/liter

9 $\int_1^3 2t\,dt = 8$

11 $\int_1^5 1/t\,dt = \ln 5$

13 $\int_2^3 7\ln(4)\cdot 4^t\,dt = 336$

15 (a) $3x^2 + 1$
(b) 10

17 (a) $\sin t \cos t$
(b) (i) ≈ 0.056
(ii) $\frac{1}{2}(\sin^2(0.4) - \sin^2(0.2))$

19 75

21 (a) Removal rate 500 kg/day on day 12
(b) Days, days, kilograms
(c) 4000 kg removed between day 5 and day 15

23 (a) $\int_0^2 R(t)\,dt$
(b)

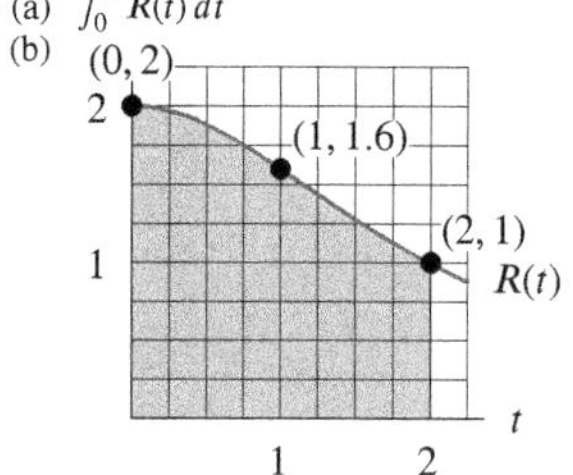

(c) lower estimate: 2.81
upper estimate: 3.38

25 (a) $\int_0^5 f(t)\,dt$
(b) 177.270 billion barrels
(c) Lower estimate of year's oil consumption

27 (a) Emissions 2000–2012, m. tons CO_2 equiv
(b) 1912.7 m. tons CO_2 equiv
(Possible answers from 1898 to 1928)

29 13.295 billion tons
$\int_0^{12} f(t)\,dt$

31 About $11,600

33 8.9064 kwh

35 45.8°C.

37 (a) Rate is 50 ng/ml per hour after 1 hour
(b) Concentration is 730 mg/ml after 3 hours

39 $f(3) - f(2)$,
$(f(4) - f(2))/2$,
$f(4) - f(3)$

41 $7.1024 < f(2) < 7.1724$

43 (a) V
(b) IV
(c) III
(d) II
(e) III
(f) I

45 Amount oil pumped from well from 0 to t_0

47 (a) Boys: black curve; girls: colored curve
(b) About 43 cm
(c) Boys: about 23 cm; girls: about 18 cm
(d) About 13 cm taller

49 (a) $S(0) = 182{,}566$ acre-feet; $S(3) = 171{,}770$ acre-feet
(b) Max in April; min in November
(c) Between June and July

51 −4.3

53 5; costs $5 mil more to plow 24 in than 15 in

55 0.4; cost increases by 0.4 million/inch after 24 in fallen

57 $\int_0^4 r(t)\,dt < 4r(4)$

59 (II), (III), (I)

61 (III), (II), (I)

63 $\frac{d}{dx}(\sqrt{x}) \neq \sqrt{x}$

65

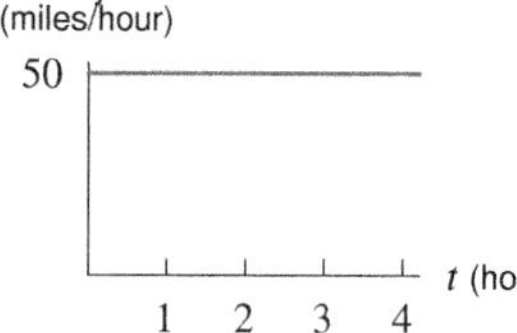

67 0.732 gals, 1.032 gals;
Better estimates possible

Section 5.4

1 10

3 9

5 $2c_1 + 12c_2^2$

7 2

9 2

11 (a) 8.5
(b) 1.7

13 4.389

15 0.083

17 7.799

19 0.172

21 $\int_0^1 x^2\,dx$

23 $\int_2^3 \cos(x)\,dx$

25 (a) 4
(b) 16
(c) 12

27 $\int_0^3 f(x)\,dx =$
$\frac{1}{2}\int_{-1}^1 f(x)\,dx + \int_1^3 f(x)\,dx$

29 (a) −2
(b) 6
(c) 1

31 (a) (i) 14
(ii) −7
(b) $a = 9, b = 11$

33 (a) $\int_a^b x\,dx = (b^2 - a^2)/2$
(b) (i) $\int_2^5 x\,dx = 21/2$
(ii) $\int_{-3}^8 x\,dx = 55/2$
(iii) $\int_1^3 5x\,dx = 20$

35 10

37 42

39 (a) 0.375 thousand/hour
(b) 1.75 thousands

41 (a) 22°C
(b) 183°C
(c) Smaller

43 (a) Less
(b) 0.856

45 Lower 12, upper 15

47 Lower 12, upper 20

49

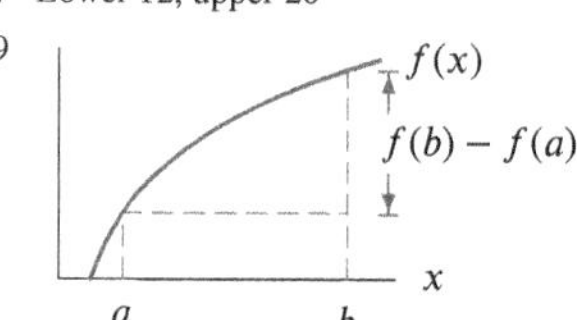

51

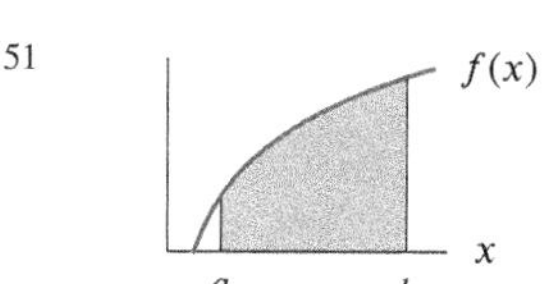

53 (a) Negative
(b) Negative
(c) Positive
(d) Negative

55 (a) (i) 1/4
(ii) 1/4
(iii) 0
(b) Not true

57 (a) 0.1574
(b) 0.9759

59 12

61 Not enough information

63 39

65

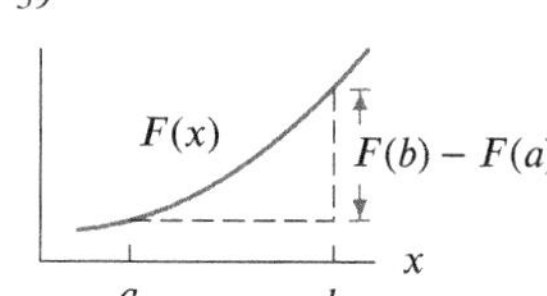

67 $f(x)$ could have both positive and negative values

69 Time should be in days

71 $f(x) = 2 - x$

73 True

75 False

77 False

79 False

81 False

83 True

85 True

87 False

89 (a) Does not follow
(b) Follows
(c) Follows

Chapter 5 Review

1 (a) Lower estimate = 122 ft
Upper estimate = 298 ft

(b)

3 20

5 396

7 (a) Upper estimate = 34.16 m/sec
Lower estimate = 27.96 m/sec
(b) 31.06 m/sec;
It is too high

9 $\int_{-2}^{1}(12t^3 - 15t^2 + 5)\,dt = -75$

11 36

13 1.142

15 13.457

17 $\int_{-1}^{1} |x|\,dx = 1$

19 (a) 260 ft
(b) Every 0.5 sec

21 0.399 miles

23 (a) Emissions 1970–2000, m. metric tons
(b) 772.8 million metric tons

25 (a) Overestimate = 7 tons
Underestimate = 5 tons
(b) Overestimate = 74 tons
Underestimate = 59 tons
(c) Every 2 days

27 (a) $(1/5)\int_0^5 f(x)\,dx$
(b) $(1/5)(\int_0^2 f(x)\,dx - \int_2^5 f(x)\,dx)$

29 (a) $f(1)$, $f(2)$
(b) 2, 2.31, 2.80, 2.77

31 (a) $F(0) = 0$
(b) F increases
(c) $F(1) \approx 0.7468$
$F(2) \approx 0.8821$
$F(3) \approx 0.8862$

33 4

35 3

37 3

39 (a) Odd integrand; areas cancel
(b) 0.4045
(c) −0.4049
(d) No. Different sized rectangles

41 (a) 318,000 megawatts; 711,556.784 megawatts
(b) 11.5%
(c) 488,730 megawatts

43 $\int_0^{0.5T} r(t)\,dt > \int_{0.5T}^{T} r(t)\,dt$

45 $\int_0^{T_h} r(t)\,dt < \int_0^{0.5T} r(t)\,dt$

47 $30/7$

49 V < IV < II < III < I
I, II, III positive
IV, V negative

51 2.5; ice 2.5 in thick at 4 am

53 (a) $2\int_0^2 f(x)\,dx$
(b) $\int_0^5 f(x)\,dx - \int_2^5 f(x)\,dx$
(c) $\int_{-2}^{5} f(x)\,dx - \frac{1}{2}\int_{-2}^{2} f(x)\,dx$

55 About 554 feet

57 (a) Positive: $0 \le t < 40$
and just before $t = 60$;
negative: from $t = 40$
to just before $t = 60$
(b) About 500 feet
at $t = 42$ seconds
(c) Just before $t = 60$
(d) Just after $t = 40$
(e) A catastrophe
(f) Total area under curve is positive
About 280 feet

59 $(a) < (c) < (b) < (d)$

61 Exact

63 Underestimate

67 (a) $\sum_{i=1}^{n} i^5/n^6$
(b) $(2n^4 + 6n^3 + 5n^2 - 1)/12n^4$
(c) $1/6$

69 (a) $\sum_{i=0}^{n-1}(n+i)^2/n^3$
(b) $7/3 + 1/(6n^2) - 3/(2n)$
(c) 7/3
(d) 7/3

71 (a) $\cos(ac)/c - \cos(bc)/c$
(b) $-\cos(cx)/c$

Section 6.1

1 (a) Increasing
(b) Concave up

3 $1, 0, -1/2, 0, 1$

5

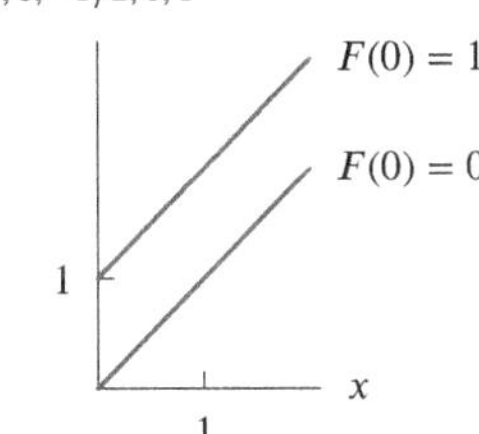

7

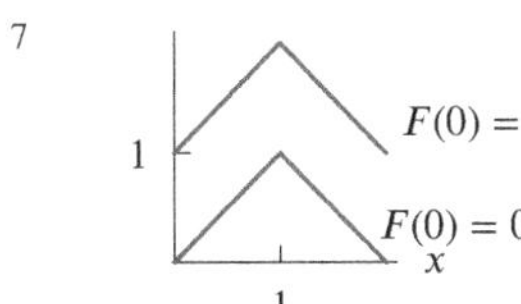

9

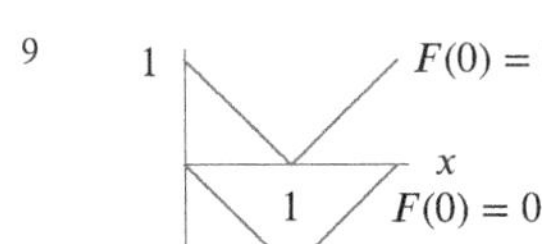

11

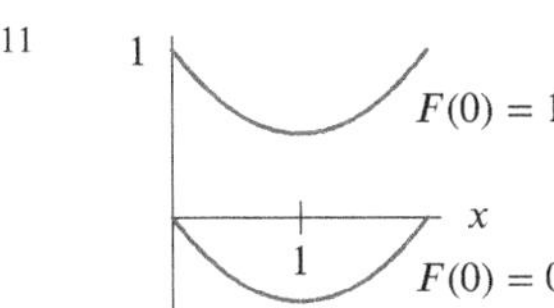

13 62

15 (a) $F(2) = 6$
(b) $F(100) = F(0)$

17 (a) −16
(b) 84

19 52.545

21 82, 107, 119

23 (0, 1); (2, 3); (6, −4); (8, 0)

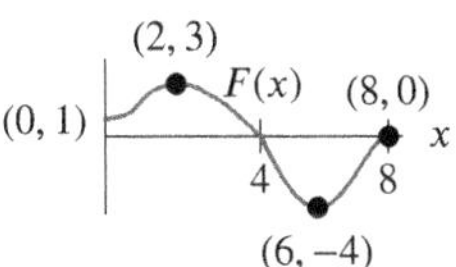

25 (a) 0; 3000; 12,000; 21,000; 27,000; 30,000
(b)

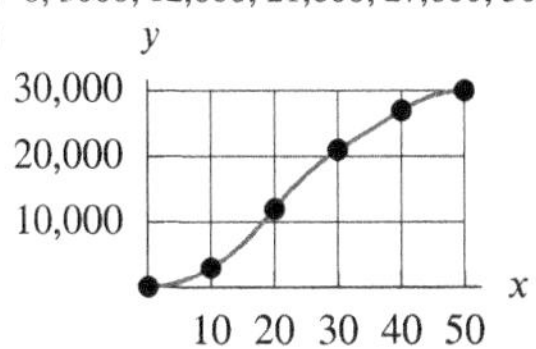

27 (a) $-1 < x < 1$
(b) $x < 0$

29 x_1 local min;
x_2 inflection pt;
x_3 local max

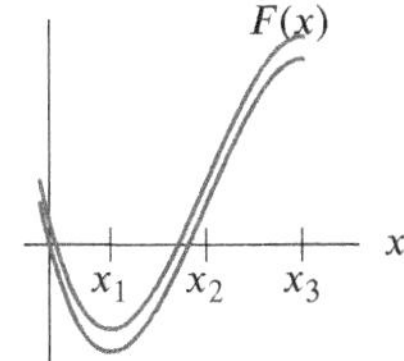

31 x_1 local max;
x_2 inflection pt;
x_3 local min

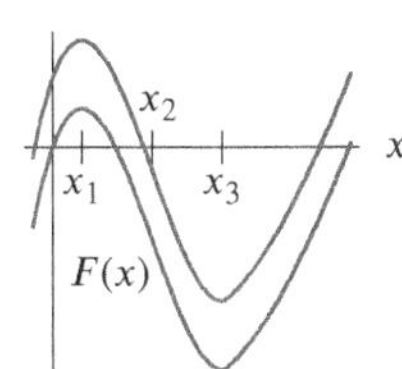

33 (a) $f(3) = 1$; $f(7) = 0$
(b) $x = 0, 5.5, 7$
(c)

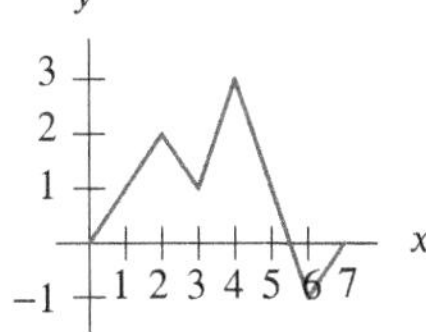

35 (b) Maximum in July 2016
Minimum in Jan 2017
(c) Increasing fastest in May 2016
Decreasing fastest in Oct 2016

37 Statement has $f(x)$ and $F(x)$ reversed

39 $f(x) = 1 - x$

41 True

43 Maximum = 6.17 at $x = 1.77$

45

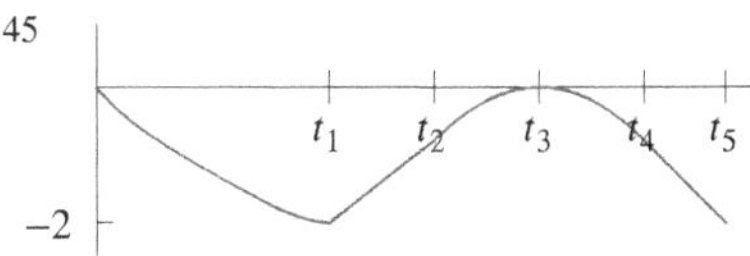

Section 6.2

1 $\int q(x)\,dx = p(x) + C$

3 (II) and (IV) are antiderivatives

5 (III), (IV) and (V) are antiderivatives

7 $5t^2/2$

9 $t^3/3 + t^2/2$

11 $\ln|z|$

13 $\sin t$

15 $y^5/5 + \ln|y|$

17 $-\cos t$

19 $5x^2/2 - 2x^{3/2}/3$

21 $t^4/4 - t^3/6 - t^2/2$

23 $x^4/4 - x^2/2 + C$

25 $2t^{3/2}/3 + C$

27 $z^2/2 + e^z + C$

29 $x^4 - 7x + C$

31 $2t - \cos t + C$

33 $7\tan t + C$

35 $F(x) = x^2$
(only possibility)

37 $F(x) = 2x + 2x^2 + (5/3)x^3$
(only possibility)

39 $F(x) = x^3/3$
(only possibility)

41 $F(x) = -\cos x + 1$
(only possibility)

43 $2t^2 + \ln|t| + C$

45 $7e^x + C$

47 $4e^x + 3\cos x + C$

49 $x^3/3 + 3x^2 + 9x + C$

51 $3\ln|t| + \dfrac{2}{t} + C$

53 $t^6/6 + t^4/4$

55 $x + \ln|x| + C$

57 $\ln 3 \approx 1.0986$

59 $2e - 2 \approx 3.437$

61 $609/4 - 39\pi \approx 29.728$

63 $\ln 2 + 3/2 \approx 2.193$

65 1

67 False

69 True

71 False

73 True

75 False

77 $1/2$

79 22

81 $e^2 + 4\ln 2 - 7$

83 $1/2$

85 $b = 7$

87 $2 + \pi^3/6$

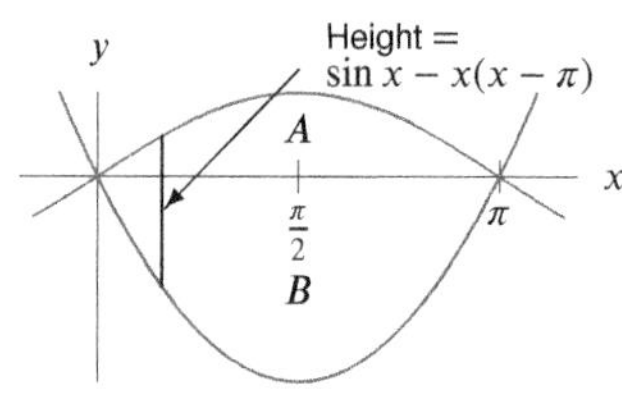

89 (a) 0

(b) $2/\pi$

91 5.500 ft

93 $C(x) =$
$5x^2 + 4000x + 1{,}000{,}000$ riyals

95 26

97 (II)

99 None

101 All $n \neq -1$

103 $f(x) = -1$

105 True

107 False

109 True

111 False

113 80

117 5 to 1

Section 6.3

3 $y = x^2 + C$

5 $y = x^4/4 + x^5 + C$

7 $y = \sin x + C$

9 $y = x^3 + 5$

11 $y = e^x + 6$

13 80.624 ft/sec downward

15 10 sec

17 (a) $R(p) = 25p - p^2$
(b) Increasing for $p < 12.5$
Decreasing for $p > 12.5$

19 (a) $a(t) = -9.8$ m/sec^2
$v(t) = -9.8t + 40$ m/sec
$h(t) = -4.9t^2 + 40t + 25$ m
(b) 106.633 m; 4.082 sec
(c) 8.747 sec

21 (a) 32 ft/sec^2
(b) Constant rate of change
(c) 5 sec
(d) 10 sec
(e)

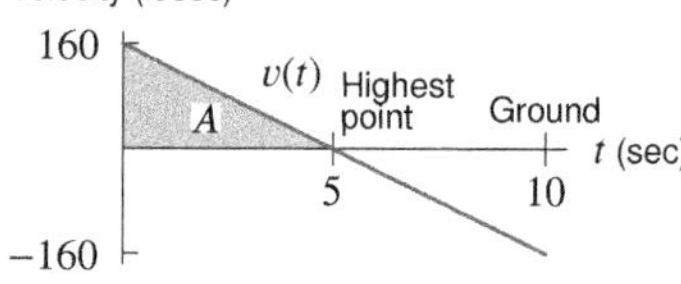

(f) Height = 400 feet
(g) $v(t) = -32t + 160$
Height = 400 feet

23 5/6 miles

25 −33.56 ft/sec^2

27 (a) $y = -\cos x + 2x + C$

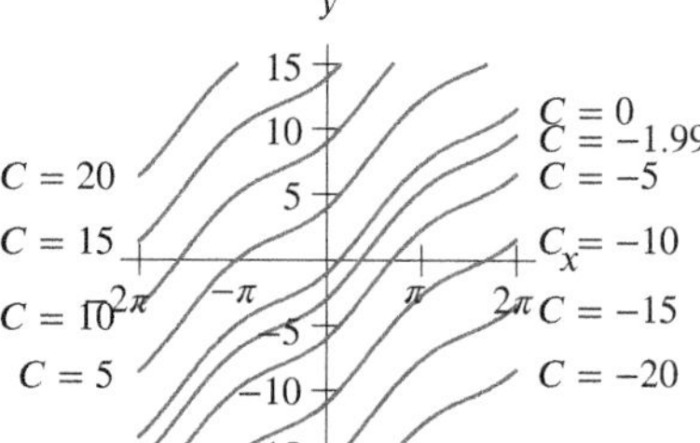

(b) $y = -\cos x + 2x - 1.99$

29 (a) 80 ft/sec
(b) 640 ft

31 128 ft/sec^2

33 10 ft; 4 sec

35 Positive, zero, negative,
positive, zero

37 If $y = \cos(t^2)$, $dy/dt \neq -\sin(t^2)$

39 $dy/dx = 0$

41 $dy/dx = -5\sin(5x)$

43 True

45 True

47 False

49 True

51 True

53 (a) $t = s/(\frac{1}{2}v_{\max})$

55 (a) First second: $-g/2$
Second: $-3g/2$
Third: $-5g/2$
Fourth: $-7g/2$
(b) Galileo seems to have been correct

Section 6.4

3 (a) $\mathrm{Si}(4) \approx 1.76$
$\mathrm{Si}(5) \approx 1.55$
(b) $(\sin x)/x$ is negative on that interval

5 $f(x) = 5 + \int_1^x (\sin t)/t\,dt$

7

$F(x)$

x

9

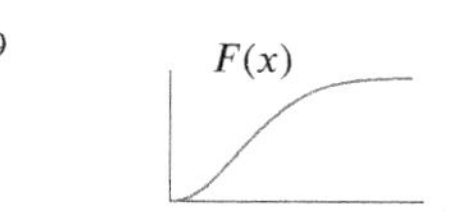

11 $\cos(x^2)$

13 $(1 + x)^{200}$

15 $\arctan(x^2)$

17 Concave up $x < 0$, concave down $x > 0$

19 $F(0) = 0$
$F(0.5) = 0.041$
$F(1) = 0.310$
$F(1.5) = 0.778$
$F(2) = 0.805$
$F(2.5) = 0.431$

21 Max at $x = \sqrt{\pi}$;
$F(\sqrt{\pi}) = 0.895$

23 500

25 (a) 0
(b) −2
(c) $1 \leq x \leq 6$
(d) 8

27 (a) $F'(x) = 1/(\ln x)$
(b) Increasing, concave down

29 −3.905

31 −1.4; exact

33 −8, an overestimate; or −6, an underestimate

35 $\cos(\sin^2 t)(\cos t)$

37 $4xe^{x^4}$

39 $e^{-x}/\sqrt{\pi x}$

41 $3x^2e^{-x^6} - e^{-x^2}$

43 (a) -5
(b) 0
(c) 12
(d) -5

45 $F(x)$ is non-decreasing; min at $x = -2$

47 $F(x) = \int_0^x t^2\,dt$

49 True

51 True

53 True

55 Increasing; concave up

57 (a) $R(0) = 0$, R is an odd function.
(b) Increasing everywhere
(c) Concave up for $x > 0$, concave down for $x < 0$.
(d)

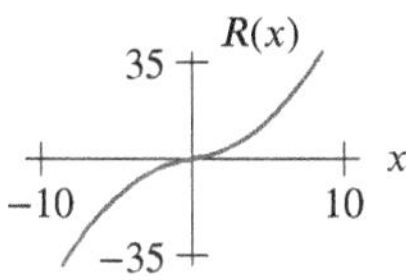

(e) $1/2$

Chapter 6 Review

3

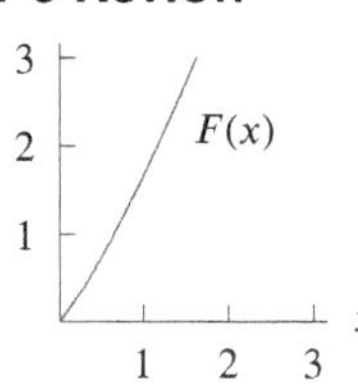

5 (a) About -19
(b) About 6

7 $x^4/4 + C$

9 $x^4/4 - 2x + C$

11 $-4/t + C$

13 $8w^{3/2}/3 + C$

15 $\sin\theta + C$

17 $x^2/2 + 2x^{1/2} + C$

19 $3\sin t + 2t^{3/2} + C$

21 $\tan x + C$

23 $(1/2)x^2 + x + \ln|x| + C$

25 $2^x/\ln 2 + C$

27 $2e^x - 8\sin x + C$

29 4

31 $x^4/4 + +2x^3 - 4x + 4.$

33 $e^x + 3$

35 $\sin x + 4$

39 $4x^2 + \ln|x| + C$

41 $-3\cos p + C$

43 $10e^t + 15$

45 $2z - \cos z + 6$

47 $-\ln x$

49 Acceleration is zero at points A and C

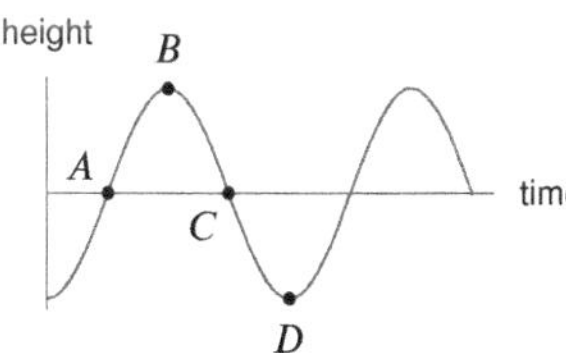

51 (a) $x = -1, x = 1, x = 3$
(b) Local min at $x = -1, x = 3$; local max at $x = 1$
(c)

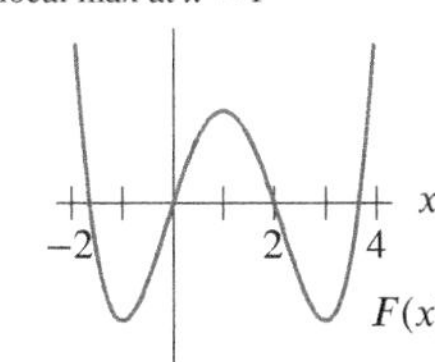

53 (a)

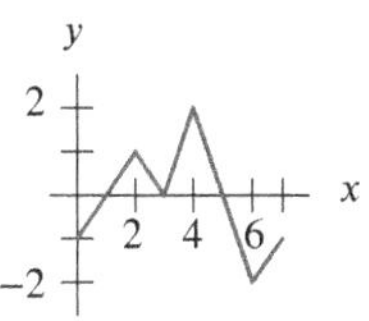

(b) $f(0) = f(7) = -1$
(c) 0

55 36

57 $1/30$

59 $\sqrt{2} - 1$

61 $e - 1 - \sin 1$

63 $c = 3$

65 $c = 6$

67 0.4

69 $16/3$

71 $3x^2\sin(x^6)$

73 $2e^{-x^4}$

75 (a) $f(t) = Q - \frac{Q}{A}t$
(b) $Q/2$

77 (a) $t = 2, t = 5$
(b) $f(2) \approx 55$, $f(5) \approx 40$
(c) -10

79 (a) 14,000 revs/min^2
(b) 180 revolutions

81 19.55 ft/sec^2

83 (b) Highest pt: $t = 2.5$ sec
Hits ground: $t = 5$ sec
(c) Left sum:
136 ft (an overest.)
Right sum:
56 ft (an underest.)
(d) 100 ft

85 $V(r) = (4/3)\pi r^3$

87

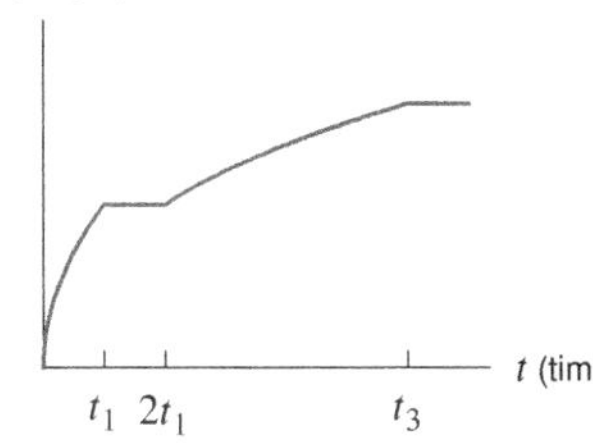

89 Increasing; concave down

91 (a) Zeros: $x = 0, 5$; Critical points: $x = 3$
(b) Zeros: $x = 1$; Critical points: $x = 0, 5$

93 (a) $\Delta x = (b - a)/n$
$x_i = a + i(b - a)/n$
(b) $(b^4 - a^4)/4$

95 (a) $-(1/3)\cos(3x)$, $-(1/4)\cos(4x)$, $-(1/3)\cos(3x + 5)$
(b) $-(1/a)\cos(ax + b)$

97 (a) $(1/2)(\ln|x - 3| - \ln|x - 1|)$, $(1/3)(\ln|x - 4| - \ln|x - 1|)$, $(1/4)(\ln|x + 3| - \ln|x - 1|)$
(b) $(1/(b - a))(\ln|x - b| - \ln|x - a|)$

Section 7.1

1 (a) $(\ln 2)/2$
(b) $(\ln 2)/2$

3 $(1/3)e^{3x} + C$

5 $-e^{-x} + C$

7 $-0.5\cos(2x) + C$

9 $\cos(3 - t) + C$

11 $(r + 1)^4/4 + C$

13 $(1/18)(1 + 2x^3)^3 + C$

15 $(1/6)(x^2 + 3)^3 + C$

17 $(1/5)y^5 + (1/2)y^4 + (1/3)y^3 + C$

19 $(1/3)e^{x^3+1} + C$

21 $-2\sqrt{4 - x} + C$

23 $-10e^{-0.1t+4} + C$

25 $-(1/8)(\cos\theta + 5)^8 + C$

27 $(1/7)\sin^7\theta + C$

29 $(1/35)\sin^7 5\theta + C$

31 $(1/3)(\ln z)^3 + C$

33 $t + 2\ln|t| - 1/t + C$

35 $(1/\sqrt{2})\arctan(\sqrt{2}x) + C$

37 $2\sin\sqrt{x} + C$

39 $2\sqrt{x + e^x} + C$

41 $(1/2)\ln(x^2 + 2x + 19) + C$

43 $\ln(e^x + e^{-x}) + C$

45 $(1/3)\cosh 3t + C$

47 $(1/2)\sinh(2w + 1) + C$

49 $\frac{1}{3}\cosh^3 x + C$

51 $(\pi/4)t^4 + 2t^2 + C$

53 $\sin x^2 + C$

55 $(1/5)\cos(2 - 5x) + C$

57 $(1/2)\ln(x^2 + 1) + C$

59 0

61 $1 - (1/e)$

63 $2e(e - 1)$

65 $2(\sin 2 - \sin 1)$

67 40

69 $\ln 3$

71 $14/3$

73 $(2/5)(y + 1)^{5/2} - (2/3)(y + 1)^{3/2} + C$

75 $(2/5)(t + 1)^{5/2} - (2/3)(t + 1)^{3/2} + C$

77 $(2/7)(x - 2)^{7/2} + (8/5)(x - 2)^{5/2} + (8/3)(x - 2)^{3/2} + C$

79 $(2/3)(t + 1)^{3/2} - 2(t + 1)^{1/2} + C$

81 $\ln(1 + e^t) + C$

83 $t = s^2$

85 Let $t = \pi - x$

87 $1/w^2$

89 $e^{11} - e^7$

91 3

93 218/3

95 Substitute $w = \sin x$

97 Substitute $w = \sin x$, $w = \arcsin x$

99 Substitute $w = x + 1$, $w = 1 + \sqrt{x}$

101 $e^{g(x)} + C$

103 $2(1 + g(x))^{3/2}/3 + C$

105 $w = \sin t, k = 1, n = -1$

107 $k = 0.3, n = -1/2, w_0 = 12, w_1 = 132$

109 $k = 0.5, n = 7, w_0 = 0.5, w_1 = 1$

111 $w = \sin\phi, k = 1$

113 $w = z^3, k = 1/3$

115 $a = 3, b = 11, A = 0.5, w = 2t - 3$

117 (a) $2\sqrt{x} + C$
(b) $2\sqrt{x+1} + C$
(c) $2\sqrt{x} - 2\ln(\sqrt{x} + 1) + C$

119 $(1/2)(e^4 - 1)$

121 $2\cosh 1 - 2$

123 $e^3 - e^2 - e + 1$

125 $2V_0/\omega$

127 $(1/2)\ln 3$

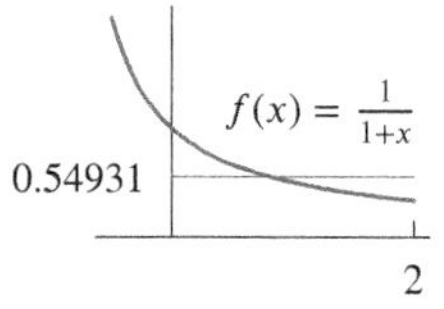

129 (a) 10
(b) 5

131 (a) 0
(b) 2/3

133 (a) $(\sin^2\theta)/2 + C$
(b) $-(\cos^2\theta)/2 + C$
(c) $-(\cos 2\theta)/4 + C$
(d) Functions differ by a constant

135 $w = \cos x$
$k = -1, a = 1, b = -1$ or
$k = 1, a = -1, b = 1$

137 $w = \ln(x^2 + 1), k = 1/2, a = \ln 5, b = \ln 26$

139 13

141 0.661

143 One possible answer
$g(x) = e^{\sqrt{x}} f(x) = 0.5\cos x,$

147 114.414 million

149 (a) $\int_0^{72} f(t)\,dt$
(b) $\int_0^3 f(24T)\,24\,dT$

151 (a) 50 thousand liters/minute
15.06 thousand liters/minute
(b) 1747 thousand liters

153 (a) $E(t) = 1.4e^{0.07t}$
(b) $0.2(e^7 - 1) \approx 219$ million megawatt-hours
(c) 1972
(d) Graph $E(t)$ and estimate t such that $E(t) = 219$

155 (a) First case: 19,923
Second case: 1.99 billion
(b) In both cases, 6.47 yrs
(c) 3.5 yrs

157 Integrand needs extra factor of $f'(x)$

159 Change limits of integration for substitution in definite integral

161 $\int \sin(x^3 - 3x)(x^2 - 1)\,dx$

163 False

165 $(mg/k)t - (m^2g/k^2)(1 - e^{-kt/m}) + h_0$

Section 7.2

1 (a) $x^3e^x/3 - (1/3)\int x^3e^x\,dx$
(b) $x^2e^x - 2\int xe^x\,dx$

3 $-t\cos t + \sin t + C$

5 $\frac{1}{5}te^{5t} - \frac{1}{25}e^{5t} + C$

7 $-10pe^{(-0.1)p} - 100e^{(-0.1)p} + C$

9 $(1/2)x^2\ln x - (1/4)x^2 + C$

11 $(1/6)q^6\ln 5q - (1/36)q^6 + C$

13 $-(1/2)\sin\theta\cos\theta + \theta/2 + C$

15 $t(\ln t)^2 - 2t\ln t + 2t + C$

17 $(2/3)y(y+3)^{3/2} - (4/15)(y+3)^{5/2} + C$

19 $-(\theta+1)\cos(\theta+1) + \sin(\theta+1) + C$

21 $-x^{-1}\ln x - x^{-1} + C$

23 $-2t(5-t)^{1/2} - (4/3)(5-t)^{3/2} - 14(5-t)^{1/2} + C$

25 $(2/3)x^{3/2}\ln x - (4/9)x^{3/2} + C$

27 $r^2[(\ln r)^2 - \ln r + (1/2)]/2 + C$

29 $z\arctan 7z - \frac{1}{14}\ln(1 + 49z^2) + C$

31 $(1/2)x^2e^{x^2} - (1/2)e^{x^2} + C$

33 $x\cosh x - \sinh x + C$

35 $2\sqrt{x}e^{\sqrt{x}} - 2e^{\sqrt{x}} + C$

37 $(1/18)(2x+1)^3(3\ln(2x+1) - 1) + C$

39 $5\ln 5 - 4 \approx 4.047$

41 $-11e^{-10} + 1 \approx 0.9995$

43 $\pi/4 - (1/2)\ln 2 \approx 0.439$

45 $\pi/2 - 1 \approx 0.571$

47 (a) Parts
(b) Substitution
(c) Substitution
(d) Substitution
(e) Substitution
(f) Parts
(g) Parts

49 $w = 5 - 3x, k = -2/3$

51 $w = \ln x, k = 3$

53 $2\arctan 2 - (\ln 5)/2$

55 $2\ln 2 - 1$

57 π

59 Integration by parts gives:
$(1/2)\sin\theta\cos\theta + (1/2)\theta + C$
The identity for $\cos^2\theta$ gives:
$(1/2)\theta + (1/4)\sin 2\theta + C$

61 $(1/2)e^\theta(\sin\theta + \cos\theta) + C$

63 $(1/2)\theta e^\theta(\sin\theta + \cos\theta) - (1/2)e^\theta\sin\theta + C$

65 $xf'(x) - f(x) + C$

67 Integrate by parts choosing $u = x^n$, $v' = \cos ax$

69 Integrate by parts choosing $u = \cos^{n-1}x$, $v' = \cos x$

71 125

73 $f(x) = x^3, g(x) = (1/3)\cos x$

75 $h(x) - 0.5g(x) + C$

77 (a) $-a^2e^{-a} - 2ae^{-a} - 2e^{-a} + 2$
(b) Increasing
(c) Concave up

79 $h(x) - g(x) + C$

81 $\mathrm{Li}(x)\ln x - x + C$

83 Write $\arctan x = (1)(\arctan x)$

85 $\int \theta^2\sin\theta\,d\theta$

87 $\int e^x\sin x\,dx$

89 True

91 Approximately 77

93 (a) $-(1/a)Te^{-aT} + (1/a^2)(1 - e^{-aT})$
(b) $\lim_{T\to\infty} E = 1/a^2$

Section 7.3

1 No formula

3 IV-17; $n = 4$

5 No formula

7 V-26; $a = 4, b = -1$ or $a = -1, b = 4$

9 VI-31, then VI-29; $a = 3$

11 No formula

13 V-25; $a = 3, b = 4, c = -2$

15 $(1/6)x^6\ln x - (1/36)x^6 + C$

17 $-(1/5)x^3\cos 5x + (3/25)x^2\sin 5x + (6/125)x\cos 5x - (6/625)\sin 5x + C$

19 $(1/7)x^7 + (5/2)x^4 + 25x + C$

21 $-(1/4)\sin^3 x\cos x - (3/8)\sin x\cos x + (3/8)x + C$

23 $((1/3)x^2 - (2/9)x + 2/27)e^{3x} + C$

25 $((1/3)x^4 - (4/9)x^3 + (4/9)x^2 - (8/27)x + (8/81))e^{3x} + C$

27 $(1/\sqrt{3})\arctan(y/\sqrt{3}) + C$

29 $(1/4)\arcsin(4x/5) + C$

31 $(5/16)\sin 3\theta\sin 5\theta + (3/16)\cos 3\theta\cos 5\theta + C$

33 $\frac{1}{2}\frac{\sin x}{\cos^2 x} + \frac{1}{4}\ln\left|\frac{\sin x+1}{\sin x-1}\right| + C$

35 $(1/34)e^{5x}(5\sin 3x - 3\cos 3x) + C$

37 $-(1/2)y^2\cos 2y + (1/2)y\sin 2y + (1/4)\cos 2y + C$

39 $\frac{1}{21}(\tan 7x/\cos^2 7x) + \frac{2}{21}\tan 7x + C$

41 $-\frac{1}{2\tan 2\theta} + C$

43 $\frac{1}{2}(\ln|x+1| - \ln|x+3|) + C$

45 $-(1/3)(\ln|z| - \ln|z-3|) + C$

47 $\arctan(z+2) + C$

49 $-(1/3)\sin^2 x\cos x - (2/3)\cos x + C$

51 $\frac{1}{5}\cosh^5 x - \frac{1}{3}\cosh^3 x + C$

53 $-(1/9)(\cos^3 a\theta) + (1/15)(\cos^5 a\theta) + C$

55 $\frac{1}{3}(1 - \frac{\sqrt{2}}{2})$

57 $9/8 - 6\ln 2 = -3.034$

59 1/2

61 $\pi/12$

63 0.5398

65 $k = 0.5, w = 2x + 1, n = 3$

67 Form (i) with $a = -1/6, b = -1/4, c = 5$

69 Form (iii) with $a = 1, b = -5, c = 6, n = 7$

71 $a = 5, b = 4, \lambda = 2$

75 (a) $r'(t) = t/(t^2+1)^{3/2} > 0$; $r(t)$ increases from 0 to 1
(b) Increasing, concave up
(c) $v(t) = t - \ln|t + \sqrt{t^2+1}|$

77 If $a = 3$, denominator factors; answer involves ln, not arctan

79 $\int \sin x \cos x\, dx = (\sin^2 x)/2 + C$

81 $\int 1/\sqrt{2x - x^2}\, dx$

83 True

85 False

87 (a) 0
(b) $V_0/\sqrt{2}$
(c) 156 volts

Section 7.4

1 $(1/6)/(x) + (5/6)/(6 + x)$

3 $1/(w - 1) - 1/w - 1/w^2 - 1/w^3$

5 $-2/y + 1/(y - 2) + 1/(y + 2)$

7 $1/(2(s - 1)) - 1/(2(s + 1)) - 1/(s^2 + 1)$

9 $-2 \ln|5 - x| + 2 \ln|5 + x| + C$

11 $\ln|y - 1| + \arctan y - \frac{1}{2} \ln|y^2 + 1| + C$

13 $\ln|s| - \ln|s + 2| + C$

15 $\ln|x - 2| + \ln|x + 1| + \ln|x - 3| + K$

17 $2 \ln|x - 5| - \ln|x^2 + 1| + K$

19 $x^3/3 + \ln|x + 1| - \ln|x + 2| + C$

21 $\arcsin(x - 2) + C$

23 (a) Yes; $x = 3 \sin\theta$
(b) No

25 $w = 3x^2 - x - 2, k = 2$

27 $a = -5/3, b = 4/5, c = -2/15, d = -3/5$

29 (a) $2 \ln|x| + \ln|x + 3| + C$

31 $x = (4 \tan\theta) - 3$

33 $w = (x + 1)^2$

35 $w = 1 - (z - 1)^2$

37 $w = (\theta - 2)^2 - 4$

39 $\ln|x + 2| - \ln|x + 3| + C$

41 $-\ln|x - 1| + 2 \ln|x - 2| + C$

43 $(\ln|x + 1| - \ln|x + 4|)/3 + C$

45 $-4 \ln|x - 1| + 7 \ln|x - 2| + C$

47 $\ln|x| - (1/2) \ln|x^2 + 1| + \arctan x + K$

49 $y - 5 \arctan(y/5) + C$

51 $2 \ln|s + 2| - \ln|s^2 + 1| + 4 \arctan s + K$

53 $(1/3)(\ln|e^x - 1| - \ln|e^x + 2|) + C$

55 $-x\sqrt{9 - x^2}/2 + (9/2) \arcsin(x/3) + C$

57 $(5/2) \ln|(5 - \sqrt{25 - 9x^2})/(5 + \sqrt{25 - 9x^2})| + \sqrt{25 - 9x^2} + C$

59 $(1/2) \ln|(1 - \sqrt{1 + 16x^2})/(1 + \sqrt{1 + 16x^2})| + C$

61 $(1/25)(x/\sqrt{25 + 4x^2}) + C$

63 $(1/27)(\frac{1}{2} \ln|(3x + \sqrt{1 + 9x^2})(3x - \sqrt{1 + 9x^2})| + 3x/\sqrt{1 + 9x^2}) + C$

65 $-4\sqrt{4 - x^2} + \frac{1}{3}(4 - x^2)^{3/2} + C$

67 $2 \ln 2 - \pi/4$

69 $8(2/3 - 5/(6\sqrt{2}))$

71 $(1/6) \ln((3 - 2\sqrt{2})(7 + 4\sqrt{3}))$

73 $\ln|x + 1| + \ln|x - 1| + C$

75 (b) $-\sqrt{5 - y^2}/(5y) + C$

77 (a) $(a \ln|x - a| - b \ln|x - b|)/(a - b) + C$
(b) $\ln|x - a| - a/(x - a) + C$

79 (a) $t(p) = b \ln(p(1 - a))/(a(1 - p))$
(b) $a = 0.01$
(c) $b = 0.218$
(d) 1.478

81 Use $x = 2 \tan\theta$

83 $\int 1/(x^3 - x)\, dx$

85 $\int (1/\sqrt{9 - 4x^2})\, dx$

87 True

89 (c), (b)

91 (a) $(k/(b - a)) \ln|(2b - a)/b|$
(b) $T \to \infty$

Section 7.5

1 (a) Underestimate

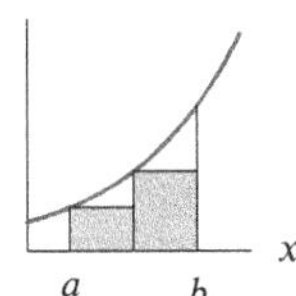

(b) Overestimate

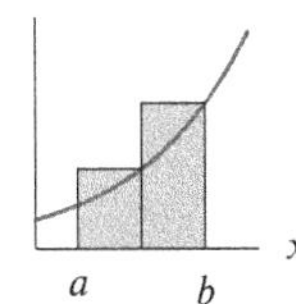

(c) Overestimate

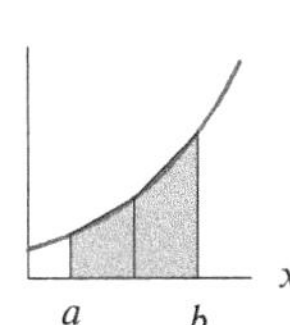

(d) Underestimate

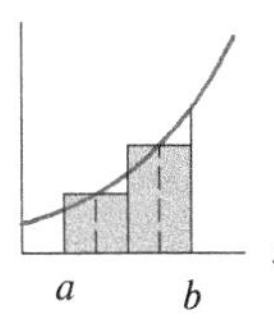

3 (a) Underestimate

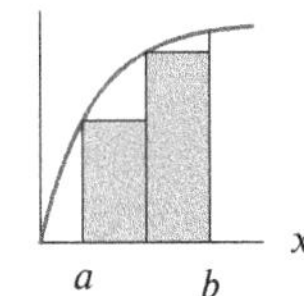

(b) Overestimate

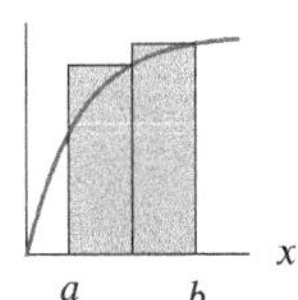

(c) Underestimate

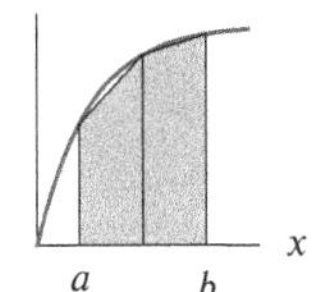

(d) Overestimate

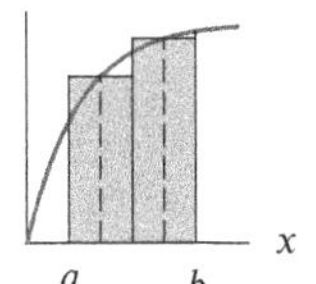

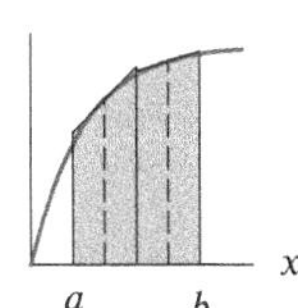

5 (a) Underestimate

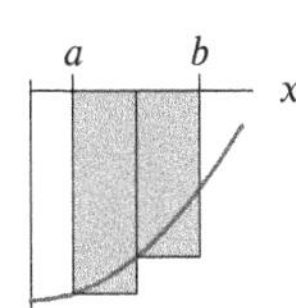

(b) Overestimate

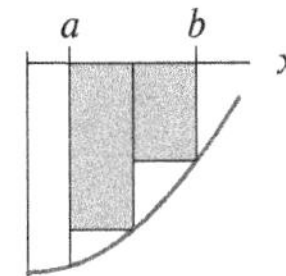

(c) Overestimate

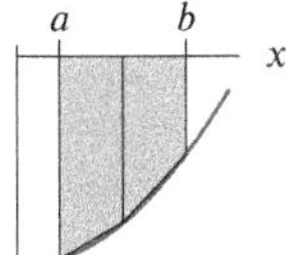

(d) Underestimate

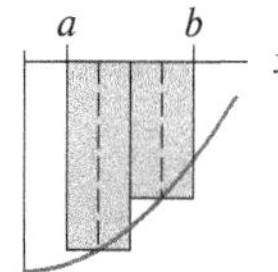

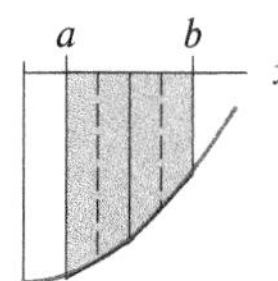

7 (a) 27
(b) 135
(c) 81
(d) 67.5

9 (a) LEFT(2)= 12;
RIGHT(2)= 44
(b) LEFT(2) underestimate;
RIGHT(2) overestimate

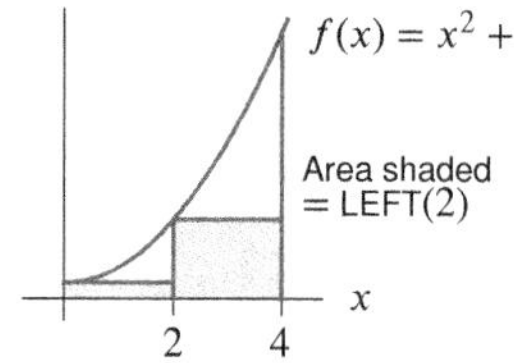

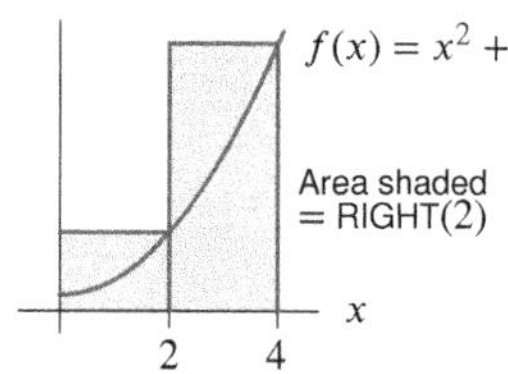

11 (a) $\pi/2$
(b) $\pi/2$
(c) $\pi/2$
(d) $\sqrt{2}\pi/2$

13 18.1

15 4.863

17 I, IV

19 0.1649

21 LEFT(6) = 31
RIGHT(6) = 39
TRAP(6) = 35

23 (a) 12
(b) SIMP(2) = 12
SIMP(4) = 12
SIMP(100) = 12

25 (a) RIGHT = 0.368, LEFT = 1, TRAP = 0.684, MID =0.882, $\int_0^1 e^{-x^2/2}dx = 0.856$
(b) LEFT = 0.368, RIGHT = 1, TRAP = 0.684, MID =0.882, $\int_{-1}^0 e^{-x^2/2}dx = 0.856$

27 (a) TRAP(5) = 0.3846
(b) Concave down
(c) 0.3863

29 MID: over; TRAP: under

31 TRAP: over; MID: under

33 (a) $f(x)$
(b) $f(x)$

35 (a) Increasing; concave down

37 (a) 1
(d) Ratio of errors:
LEFT = 1.78,
RIGHT = 2.18,
TRAP = 3.96,
MID = 3.93,
SIMP = 15.91

39 (a) 0
(c) MID(n) = 0 for all n

41 Large $|f''|$ gives large error

43 445 lbs. of fertilizer

49 Midpoint rule exact for linear functions

51 If f concave down, TRAP(n) $\leq$ MID(n)

53 $f(x) = 1 - x^2$

55 True

57 False

59 False

61 True

63 False

65 False

67 True

69 (a) II
(b) III
(c) V
(d) V
(e) III

Section 7.6

1 (a)

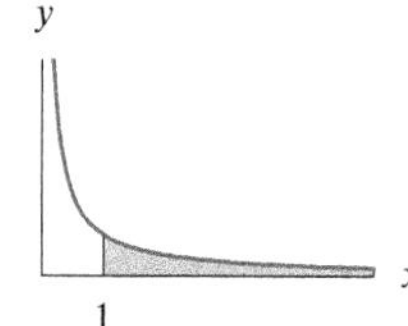

(b)

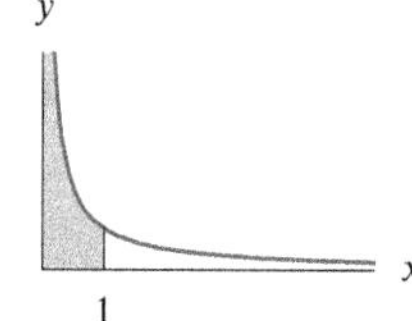

3 (a) 0.9596, 0.9995, 0.99999996
(b) 1.0

5 Diverges

7 −1

9 Converges to 1/2

11 1

13 ln 2

15 Converges to 1/16

17 4

19 Does not converge

21 $\pi/4$

23 Diverges

25 Does not converge

27 Does not converge

29 0.01317

31 $1/\ln 3$

33 1/3

35 Does not converge

37 Converges to 1/4

39 Does not converge

41 (a) 0.421
(b) 0.500

43 The area is infinite

45 1/2

47 2/27

49 9×10^9 joules

51 $\sqrt{b\pi}$

53 $\pi^4/120$

55 1/4

57 $f(x) = 1/x$

59 $f(x) = 1/(x-3)^2$

61 True

63 False

65 False

67 True

69 False.

71 (a) $\Gamma(1) = 1$
$\Gamma(2) = 1$
(c) $\Gamma(n) = (n-1)!$

Section 7.7

1 Converges; behaves like $1/x^2$

3 Diverges; behaves like $1/x$

5 Diverges; behaves like $1/x$

7 Converges; behaves like $5/x^3$

9 Converges; behaves like $1/x^2$

11 Inconclusive

13 Converges

15 Does not converge

17 Converges

19 Does not converge

21 Converges

23 Converges

25 Converges

27 Converges

29 Does not converge

31 (a) Converges
(b) Either
(c) Converges
(d) Either
(e) Converges
(f) Diverges
(g) Converges

33 Converges for $p > 1$
Diverges for $p \leq 1$

35 (a) $\int_3^\infty e^{-x^2}\,dx \leq e^{-9}/3$
(b) $\int_n^\infty e^{-x^2}\,dx \leq (1/n)e^{-n^2}$

37 Cannot compare, first integrand sometimes larger than second

39 Comparison test not applicable

41 $f(x) = 3/(2x^2 + 1)$

43 True

45 True

47 False

Chapter 7 Review

1 $(t+1)^3/3$

3 $5^x/\ln 5$

5 $3w^2/2+7w+C$

7 $-\cos t+C$

9 $(1/5)e^{5z}+C$

11 $(-1/2)\cos 2\theta+C$

13 $2x^{5/2}/5+3x^{5/3}/5+C$

15 $-e^{-z}+C$

17 $x^2/2+\ln|x|-x^{-1}+C$

19 $(1/2)e^{t^2}+C$

21 $((1/2)x^2-(1/2)x+1/4)e^{2x}+C$

23 $(1/2)y^2\ln y-(1/4)y^2+C$

25 $x(\ln x)^2-2x\ln x$
$+2x+C$

27 $(1/3)\sin^3\theta+C$

29 $(1/2)u^2+3u+3\ln|u|-1/u+C$

31 $\tan z+C$

33 $(1/12)t^{12}-(10/11)t^{11}+C$

35 $(1/3)(\ln x)^3+C$

37 $(1/3)x^3+x^2+\ln|x|+C$

39 $(1/2)e^{t^2+1}+C$

41 $(1/10)\sin^2(5\theta)+C$ (other forms of answer are possible)

43 $\arctan z+C$

45 $-(1/8)\cos^4 2\theta+C$

47 $(-1/4)\cos^4 z+$
$(1/6)\cos^6 z+C$

49 $(2/3)(1+\sin\theta)^{3/2}+C$

51 $t^3e^t-3t^2e^t+6te^t-6e^t+C$

53 $(3z+5)^4/12+C$

55 $\arctan(\sin w)+C$

57 $-\cos(\ln x)+C$

59 $y-(1/2)e^{-2y}+C$

61 $\ln|\ln x|+C$

63 $\sin\sqrt{x^2+1}+C$

65 $ue^{ku}/k-e^{ku}/k^2+C$

67 $(1/\sqrt{2})e^{\sqrt{2}+3}+C$

69 $(u^3\ln u)/3-u^3/9+C$

71 $-\cos(2x)/(4\sin^2(2x))$
$+\frac{1}{8}\ln\left|\frac{(\cos(2x)-1)}{(\cos(2x)+1)}\right|+C$

73 $-y^2\cos(cy)/c+2y\sin(cy)/c^2$
$+2\cos(cy)/c^3+C$

75 $\frac{1}{34}e^{5x}(5\cos(3x)+3\sin(3x))+C$

77 $(\sqrt{3}/4)(2x\sqrt{1+4x^2}$
$+\ln|2x+\sqrt{1+4x^2}|)+C$

79 $x^2/2-3x-\ln|x+1|$
$+8\ln|x+2|+C$

81 $(1/b)(\ln|x|-\ln|x+b/a|)+C$

83 $(x^3/27)+2x-(9/x)+C$

85 $-(1/\ln 10)10^{1-x}+C$

87 $((1/2)v^2-1/4)\arcsin v+$
$(1/4)v\sqrt{1-v^2}+K$

89 $z^3/3+5z^2/2+25z$
$+125\ln|z-5|+C$

91 $(1/3)\ln|\sin(3\theta)|+C$

93 $(2/3)x^{3/2}+2\sqrt{x}+C$

95 $(4/3)(\sqrt{x}+1)^{3/2}+C$

97 $-1/(4(z^2-5)^2)+C$

99 $(1+\tan x)^4/4+C$

101 $e^{x^2+x}+C$

103 $-(2/9)(2+3\cos x)^{3/2}+C$

105 $\sin^3(2\theta)/6-\sin^5(2\theta)/10+C$

107 $(x+\sin x)^4/4+C$

109 $\frac{1}{3}\sinh^3 x+C$

111 $49932\frac{1}{6}$

113 201,760

115 $3/2+\ln 2$

117 $e-2\approx 0.71828$

119 $-11e^{-10}+1$

121 $3(e^2-e)$

123 0

125 $\frac{1}{2}\ln|x-1|-\frac{1}{2}\ln|x+1|+C$

127 $(\ln|x|-\ln|L-x|)/L+C$

129 $\arcsin(x/5)+C$

131 $(1/3)\arcsin(3x)+C$

133 $2\ln|x-1|-\ln|x+1|-\ln|x|+K$

135 $\arctan(x+1)+C$

137 $(1/b)\arcsin(bx/a)+C$

139 $(1/2)(\ln|e^x-1|-\ln|e^x+1|)+C$

141 Does not converge

143 6

145 Does not converge

147 Does not converge

149 Converges to $\pi/8$

151 Converges to $2\sqrt{15}$

153 Does not converge

155 Converges to value between 0 and $\pi/2$

157 Does not converge

159 $e^9/3-e^6/2+1/6$

161 Area $=2\sqrt{2}$

163 $w=3x^5+2, p=1/2, k=1/5$

165 $A=2/5, B=-3/5$

167 $\int 0.5ue^u\,du$;
$k=0.5, u=-x^2$

169 Substitute $w=x^2$ into second integral

171 Substitute $w=1-x^2$, $w=\ln x$

173 $w=2x$

175 5/6

177 (a) 3.5
(b) 35

179 11/9

181 Wrong; improper integral treated as proper integral, integral diverges

183 RT(5) < RT(10) < TR(10) < Exact < MID(10) < LF(10) < LF(5)

185 (a) 4 places: 2 seconds
8 places: ≈ 6 hours
12 places: ≈ 6 years
20 places: ≈ 600 million years
(b) 4 places: 2 seconds
8 places: ≈ 3 minutes
12 places: ≈ 6 hours
20 places: ≈ 6 years

187 (a) $(k/L)\ln 3$
(b) $T\to\infty$

189 (a) 0.5 ml
(b) 99.95%

191 (a) 360 degree-days
(b)

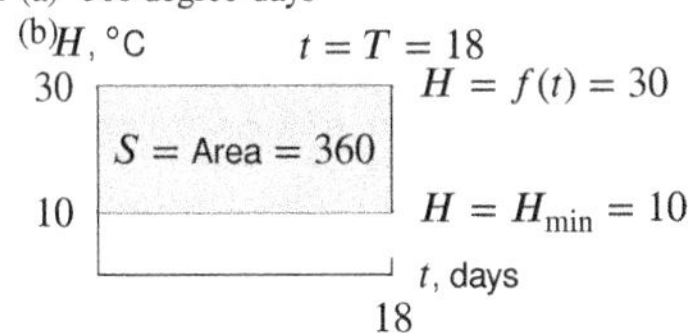

(c) $T_2=36$

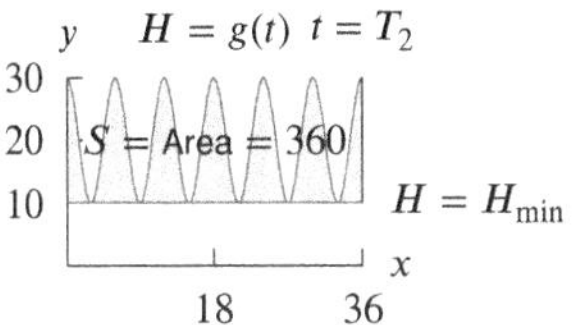

193 (a) $(\ln x)^2/2$, $(\ln x)^3/3$, $(\ln x)^4/4$
(b) $(\ln x)^{n+1}/(n+1)$

195 (a) $(-9\cos x+\cos(3x))/12$
(b) $(3\sin x-\sin(3x))/4$

197 (a) $x+x/(2(1+x^2))-(3/2)\arctan x$
(b) $1-(x^2/(1+x^2)^2)-1/(1+x^2)$

Section 8.1

1 (a) $\sum 2x\,\Delta x$
(b) 9

3 (a) $\sum\sqrt{y}\,\Delta y$
(b) 18

5 15

7 15/2

9 $(5/2)\pi$

11 1/6

13 (a) I:C; II:B; III:A; IV:D
(b) II:$\int_0^1 x\,dx+\int_1^2(2-x)\,dx$
III:$\int_1^2(2x-2)\,dx$

15 $\int_0^6\pi(x^2/9)\,dx=8\pi$ cm^3

17 $\int_0^7 20\sqrt{7^2-y^2}\,dy=245\pi$ m^3

19 $\int_0^2(2-y)^2dy=8/3$ m^3

21 $\int_0^{15}6\,dx=90$ cm^3

23 Semicircle $r=9$

25 Triangle; $b, h=5, 7$

27 (a) $\int_0^3(3x-x^2)\,dx=4.5$
(b) $\int_0^9(\sqrt{y}-(y/3))\,dy=4.5$

29 (a) $\int_0^2 x^2\,dx+\int_2^6(6-x)\,dx=2.667+8=10.667$
(b) $\int_0^4((6-y)-\sqrt{y})\,dy=10.667$

31 Hemisphere, $r=12$

33 Sphere, $r=8$

35 Hemisphere $r=2$

37 (a) 0.2.
(b) $0.2\pi\left(16-h^2\right)$

39 $36\pi=113.097$ m^3

41 (a) $3\Delta x$;
$\int_0^4 3\,dx=12$ cm^3

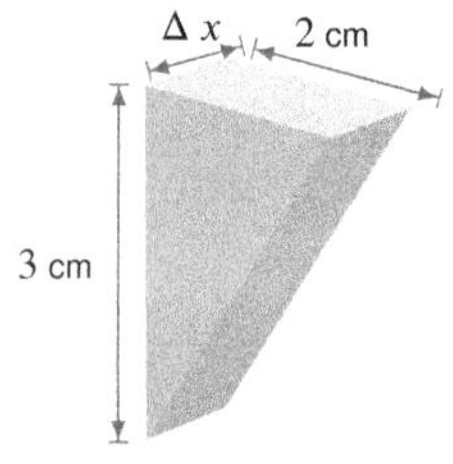

(b) $8(1 - h/3)\Delta h$;
$\int_0^3 8(1 - h/3)\,dh = 12\text{ cm}^3$

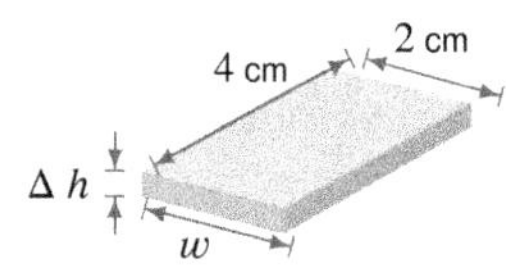

43 $\int_0^{150} 1400(160 - h)\,dh = 1.785 \cdot 10^7\text{ m}^3$

45 $36\pi = 113.097\text{ m}^3$

47 III

49 IV

51 Change to $\int_{-10}^{10} \pi\left(10^2 - x^2\right)\,dx$

53 Region between positive x-axis, positive y-axis, and $y = 1 - x$

55 False

57 True

59 True

Section 8.2

1 (a) $4\pi\int_0^3 x^2\,dx$
(b) 36π

3 (a) $\int_0^9 \pi y\,dy$
(b) 40.5π

5 $\pi/5$

7 $256\pi/15$

9 $\pi(e^2 - e^{-2})/2$

11 $\pi/2$

13 $2\pi/15$

15 $V = \int_0^6 \pi y^2/9\,dy = 8\pi$

17 $V = \int_0^{\ln 2} \pi\left(4 - e^{2y}\right)dy = 4\pi\ln 2 - 3\pi/2$

19 2.958

21 2.302

23 π

25 $\sqrt{41}$

27 ≈ 24.6

29 $V = \int_0^5 \pi((5x)^2 - (x^2)^2)\,dx$

31 $V = \int_0^5 \pi((4 + 5x)^2 - (4 + x^2)^2)\,dx$

33 $V = \int_0^2 [\pi(9 - y^3)^2 - \pi(9 - 4y)^2]\,dy$

35 $V = \int_0^9 [\pi(2 + \frac{1}{3}x)^2 - \pi 2^2]\,dx$

37 3.820

39 (a) $8a/3$
(b) $32a^2\pi/5$

41 $V = (16/7)\pi \approx 7.18$

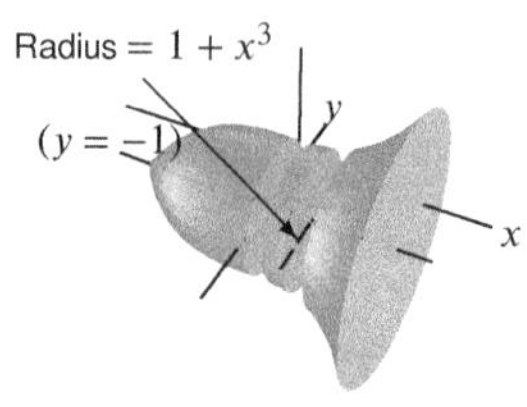

43 $V = (\pi^2/2) \approx 4.935$

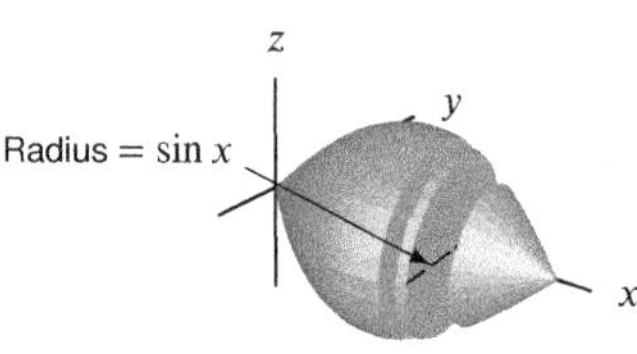

45 $4\pi/5$

47 $8/15$

49 $\sqrt{3}/8$

51 $V \approx 42.42$

53 $V = (e^2 - 1)/2$

55 $V = \int_0^4 \pi\left(3y^2/4 - 12y + 36\right)dy = 64\pi$

57 $V = \int_0^1 \pi\left((2 - y)^4 - y^4\right)\,dy = 6\pi$

59 (a) $16\pi/3$
(b) 1.48

61 $V = 32\pi a^3/105$

63 $40{,}000LH^{3/2}/(3\sqrt{a})$

65 3509 cubic inches

67 $\int_0^4 \sqrt{1 + (4 - 2x)^2}\,dx$

69 (a) $(\pi/2)(a^2 - x^2)\,\Delta x$
(b) $(\pi/2)\int_{-a}^{a}(a^2 - x^2)\,dx$
(c) $2\pi a^3/3$

71 $64\sqrt{3}/15$

73 $k = 6$

75 122.782 m

77 80

79 (a) 3π
(c) 2

81 (a) Change in x is 27
Change in y is 54
(b) $(20, 65)$
(c) 61

83 $V = 2\pi$

85 2595 cubic feet

87 $y = 0.8x^{5/4}$

89 $\int_1^3 \sqrt{1 + \left(4x^3 - 24x^2 + 36x + 3\right)^2}\,dx$

91 $\int_{-5}^5 \sqrt{1 + \left((e^{x/a} - e^{-x/a})/2\right)^2}\,dx$

93 Change to $\int_0^{\pi/4} \sqrt{1 + \cos^2 x}\,dx$

95 Change to $\int_0^1 \sqrt{(-2\pi\sin(2\pi t))^2 + (2\pi\cos(2\pi t))^2}\,dt$

97 Region bounded by $y = 2x$, x-axis, $0 \le x \le 1$

99 $f(x) = x^2$

101 False

103 False

105 (c) $f(x) = \sqrt{3}x$

Section 8.3

1 $(-1/2, \sqrt{3}/2)$

3 $(3, -\sqrt{3})$

5 $(\sqrt{2}, \pi/4)$

7 $(2\sqrt{2}, -\pi/6)$

9 (b)

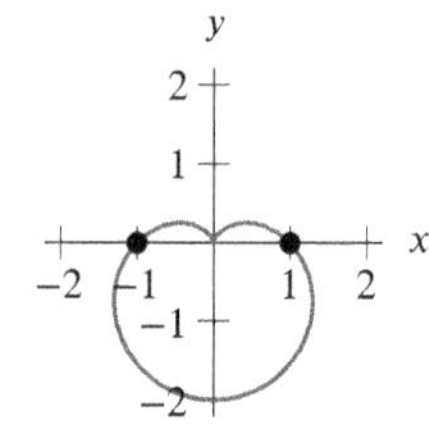

(c) Cartesian:
$(\sqrt{3}/4, 1/4)$;
$(-\sqrt{3}/4, 1/4)$ or polar:
$r = 1/2$, $\theta = \pi/6$ or $5\pi/6$

(d)

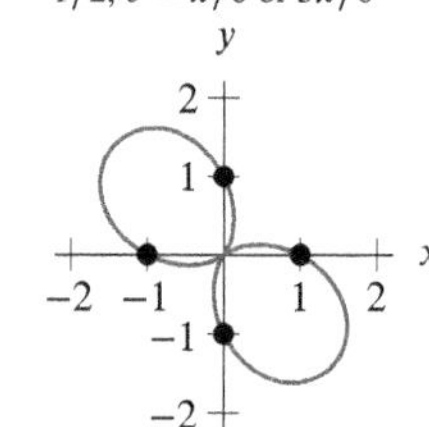

11 Looks the same

13 Rotated by 90° clockwise

15 $\pi/4 \le \theta \le 5\pi/4$;
$0 \le \theta \le \pi/4$ and $5\pi/4 \le \theta \le 2\pi$

17 $\sqrt{8} \le r \le \sqrt{18}$ and $\pi/4 \le \theta \le \pi/2$

19 $0 \le \theta \le \pi/2$ and $1 \le r \le 2/\cos\theta$

21 -1

23 $\sqrt{2}(e^{\pi} - e^{\pi/2})$

25

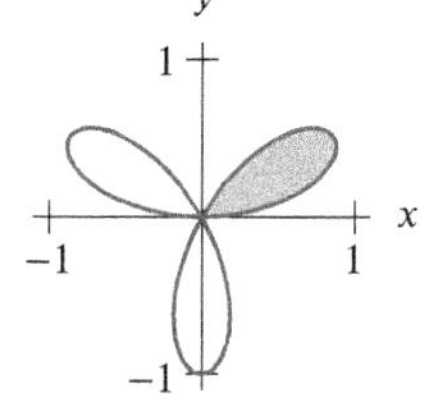

27 $4\pi^3$

29 (a) $r = 1/\cos\theta$; $r = 2$
(b) $\frac{1}{2}\int_{-\pi/3}^{\pi/3}(2^2 - (1/\cos\theta)^2)\,d\theta$
(c) $(4\pi/3) - \sqrt{3}$

31 $2\pi a$

33 $(5\pi/6) + 7\sqrt{3}/8$

35 (a)

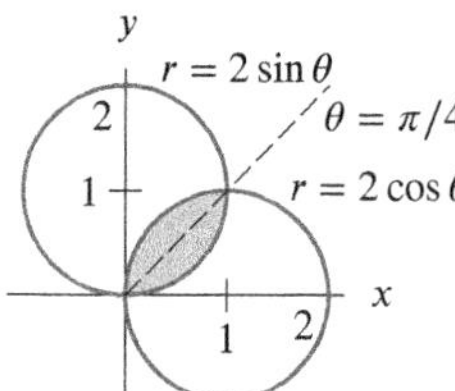

(b) $(\pi/2) - 1$

37 (a)

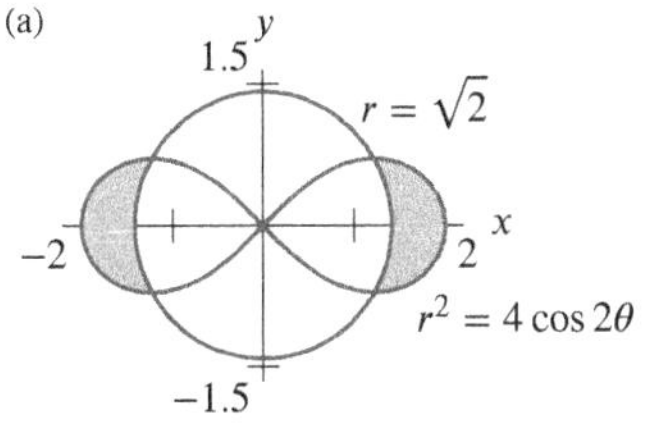

(b) $2\sqrt{3} - 2\pi/3$

39 Horiz: $(\pm 1.633, \pm 2.309)$; $(0, 0)$
Vert: $(\pm 2.309, \pm 1.633)$; $(0, 0)$

41 (a) $y = (2/\pi)x + (2/\pi)$
(b) $y = 1$

43 21.256

47 2.828

49 Points are in quadrant IV

51

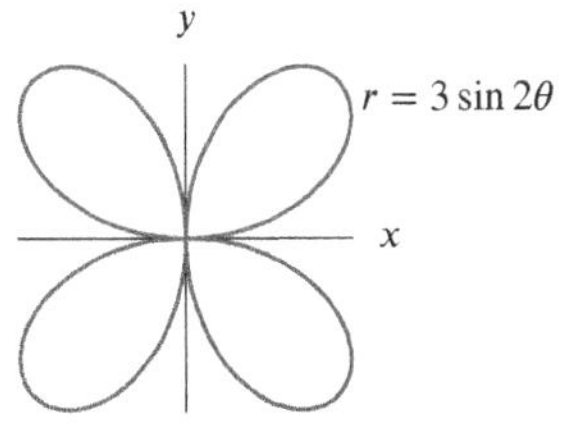

53 $r = 100$

55 $r = 1 + \cos\theta$

Section 8.4

1 $1 - e^{-10}$ gm

3 (a) $\sum_{i=1}^{N}(2 + 6x_i)\Delta x$
(b) 16 grams

5 (b) $\sum_{i=1}^{N}[600 + 300\sin(4\sqrt{x_i + 0.15})](20/N)$
(c) ≈ 11513

7 2 cm to right of origin

9 45

11 186,925

13 1 gm

15 $M = \int_a^b \delta(x)[f(x) - g(x)]dx$

17 (a) $\sum_{i=1}^{N} 2\pi r_i\,(50/(1 + r_i))\,\Delta r$
(b) $3.14 \cdot 10^6$ kg
(c) 5009 meters

19 (a) $C(s) = 10 + 2s$; 30 g/m^3
(b) 157.080 g; overestimate
(c) $(10 + 2s)\pi\Delta s$
(d) 1,884.956 g

21 Total mass = 12; $\bar{x} = 2.06$

23 (a) $\bar{x} = (3/4)((2 + k)/(3 + k))$

25 $\bar{x} = 4/(3\pi)$, $\bar{y} = 0$

27 (a) 10/3 gm
(b) $\bar{x} = 3/5$ cm; $\bar{y} = 3/8$ cm

29 1.25 cm from center of base

31 (a) $16000\delta/3$ gm
(b) 2.5 cm above center of base

33 $\int_0^{60}(1/144)g(t)\,dt$ ft^3

35 $M_U \approx 6.50 \times 10^{27}$ g
$M_L \approx 5.46 \times 10^{27}$ g

37 $\left(\int_0^{10} x^3\,dx\right) / \left(\int_0^{10} x^2\,dx\right)$

39 Mass < 27π gm

41 $\delta(x) = x$

43 False

45 False

47 False

49 False

Section 8.5

1 30 ft-lb

3 1.333 ft-lb

5 27/2 joules

7 $1.176 \cdot 10^7$ lb

9 20 ft-lb

11 $1.489 \cdot 10^{10}$ joules

13 3437.5 ft-lbs

15 180,000 ft-lb

17 3,088,800 ft-lbs

19 2,822,909.50 ft-lbs

21 661,619.41 ft-lb

23 (a) $62.4\pi(8y - 6y^2 + y^3)\Delta y$
(b) $\int_0^1 62.4\pi(8y - 6y^2 + y^3)\,dy = 140.4\pi$

25 354,673 ft-lbs

27 1,170,000 lbs

29 4,992,000 lbs

31 (a) $1.76 \cdot 10^6$ nt/m^2
(b) $1.96 \cdot 10^7 \int_0^{180} h\,dh = 3.2 \cdot 10^{11}$ nt

33 (a) 780,000 lb/ft^2
About 5400 lb/in^2
(b) $124.8\int_{-3}^{3}(12{,}500 - h)\sqrt{9 - h^2}\,dh = 2.2 \cdot 10^7$ lb

35 $\int_0^{7000} dh/v(h)$ seconds

37 (a) $14 + 0.2h$ kg
(b) $(137.2 + 1.96h)\,\Delta h$ joules
(c) 6 m
(d) 858.48 joules
(e) 1387.68 joules

39 Potential $= 2\pi\sigma(\sqrt{R^2 + a^2} - R)$

41 Not all of rope raised 20 m

43 196 joules

45 *A* tall, thin; *B* short, fat

47 True

49 True

51 True

53 $\dfrac{GM_1M_2}{l_1l_2}\ln\left(\dfrac{(a + l_1)(a + l_2)}{a(a + l_1 + l_2)}\right)$

55 60 joules

Section 8.6

1 $C(1 + 0.02)^{20}$ dollars

3 $C/e^{0.02(10)}$ dollars

5 $\int_0^{15} C\,e^{0.02(15-t)}\,dt$ dollars

7 $\ln(25{,}000/C)/30$ per year

9 \$22,658.65, \$27,675.34

11 (a) Future value = \$6389.06
Present value = \$864.66
(b) 17.92 years

13 3.466%/ yr

15 \$4000/ yr; \$21,034.18

17 \$1000/ yr; \$24,591.23

19 (a) \$300 million
(b) 15 years

21 (a) Lump sum better at 6%
Continuous payments better at 3%
(b) Interest rates remain high, above 5.3% per year

23 (a) \$5716.59 per year
(b) \$74,081.82

25 Accept installments

27 3.641%/ yr

29 9.519%/ yr

31 (a) Option 1
(b) Option 1: \$10.929 million;
Option 2: \$10.530 million

33 In 10 years

35 (a) $\sum_{i=0}^{n-1}(2000 - 100t_i)e^{-0.03t_i}\,\Delta t$
(b) $\int_0^M e^{-0.03t}(2000 - 100t)\,dt$
(c) After 20 years
\$16,534.63

39 (a)price

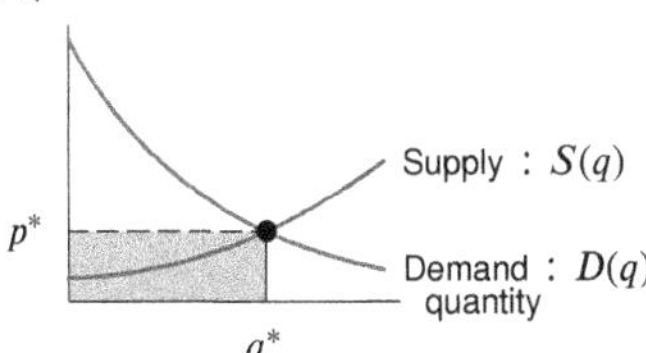

(b)price

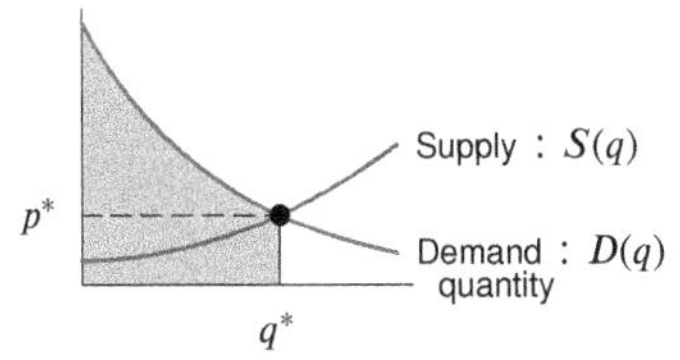

(c)price

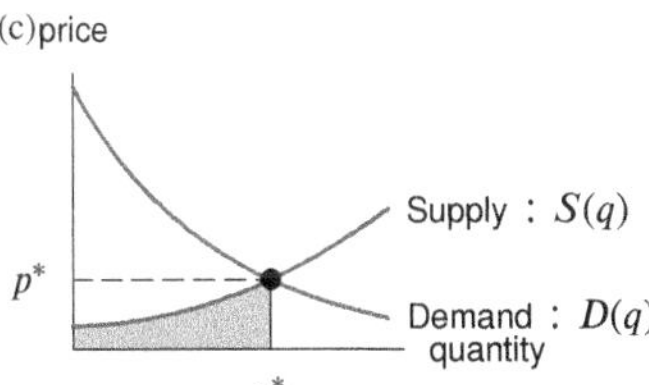

(d)price

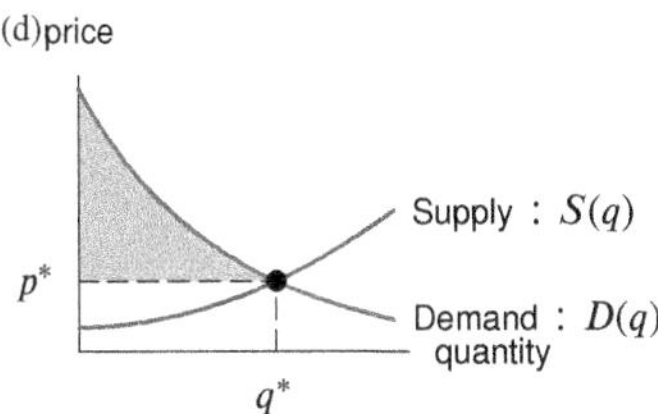

(e)price

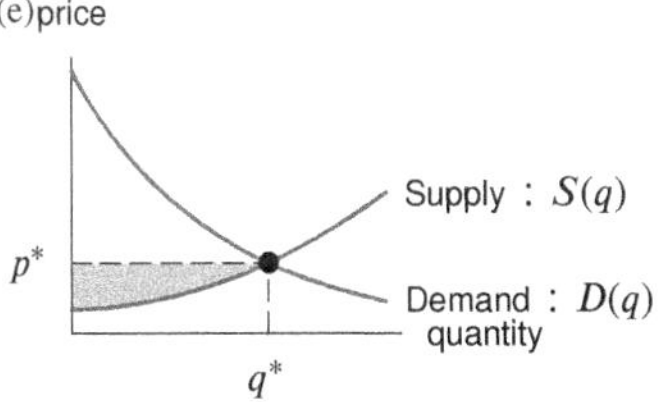

(f)

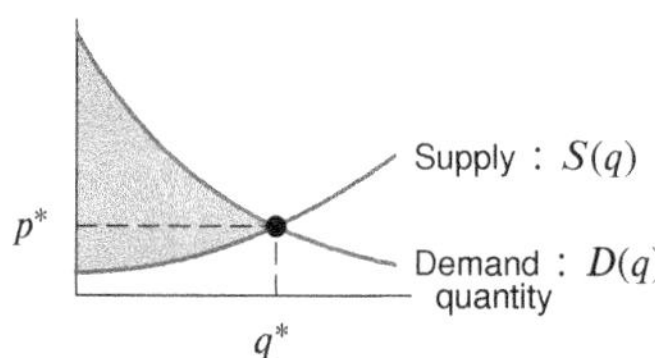

41 (a) Less
(b) Cannot tell
(c) Less

43 Smaller with 4%

45 Dollars

49 Assuming 10%/yr and t in yrs from now:

t	0	1	2	3	4
\$	3855	4241	4665	5131	5644

Section 8.7

1 (a)-(II), (b)-(I), (c)-(III)

3

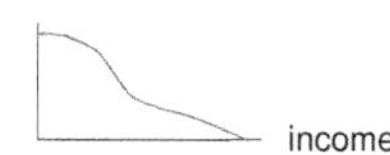

5 pdf; 1/4

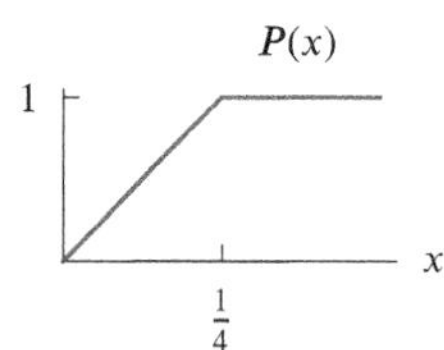

7 cdf; 1

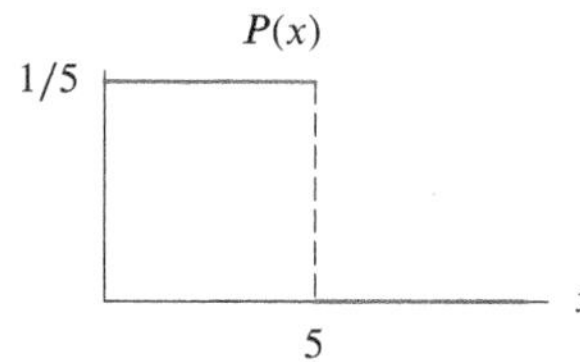

9 cdf; 1/3

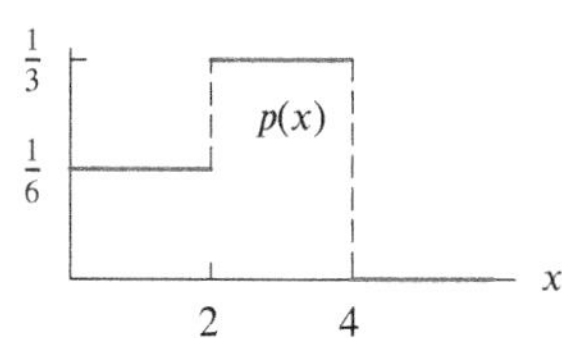

11 For small Δx around 70, fraction of families with incomes in that interval about $0.05\Delta x$

13 A is 300 kelvins;
B is 500 kelvins

15 (a) About 0.9 m–1.1 m

19 (a) Cumulative distribution increasing
(b) Vertical 0.2, horizontal 2

21 (a) 22.1%
(b) 33.0%
(c) 30.1%
(d) $C(h) = 1 - e^{-0.4h}$

23 (b) About 3/4

25 Prob $0.02\Delta x$ in interval around 1

27 $\int_{-\infty}^{\infty} p(t)\,dt \neq 1$

29 $P(x)$ grows without bound as $x \to \infty$, instead of approaching 1

31 Cumulative distribution increasing

33 $P(t) = \begin{cases} 0, & t < 0 \\ t, & 0 \le t \le 1 \\ 1, & t > 1 \end{cases}$

35 $P(x) = (x-3)/4,\ 3 \le x \le 7$

37 False

Section 8.8

5 2.65 m

7 (a) $A = 0.015$, $B = 0.005$
(b) 33.33
(c) 37.5
(d)

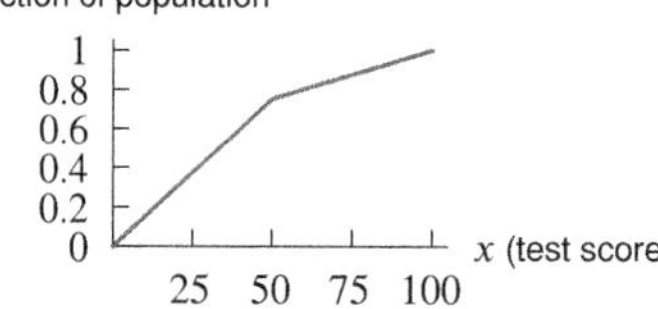

9 20%

11 22%

13 Mean 2/3; Median $2 - \sqrt{2} = 0.586$

15 (a) 0.684 : 1
(b) 1.6 hours
(c) 1.682 hours

17 (a) $P(t)$ = Fraction of population who survive up to t years after treatment
(b) $S(t) = e^{-Ct}$
(c) 0.178

19 (a) $p(x) = \frac{1}{15\sqrt{2\pi}} e^{-\frac{1}{2}\left(\frac{x-100}{15}\right)^2}$
(b) 6.7% of the population

21 (c) μ represents the mean of the distribution, while σ is the standard deviation

23 (b)

25 (a) $p(r) = 4r^2 e^{-2r}$

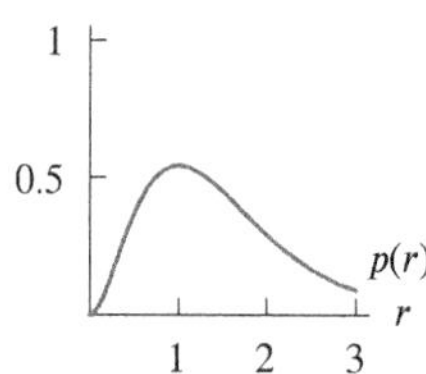

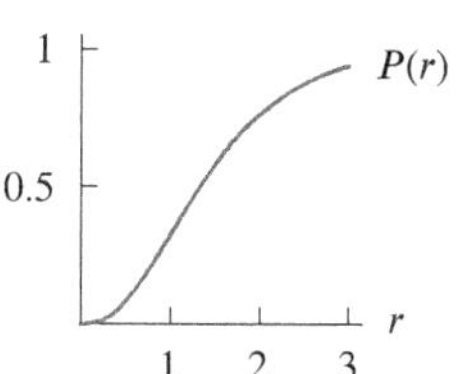

(b) Mean: 1.5 Bohr radii
Median: 1.33 Bohr radii
Most likely:
1 Bohr radius

27 Median cannot be 1 since all of the area is to the left of $x = 1$

29 $p(x) = \begin{cases} 0 & \text{for } x < 0 \\ 1 & \text{for } 0 \le x \le 1 \\ 0 & \text{for } x > 1 \end{cases}$

31 True

33 False

Chapter 8 Review

1

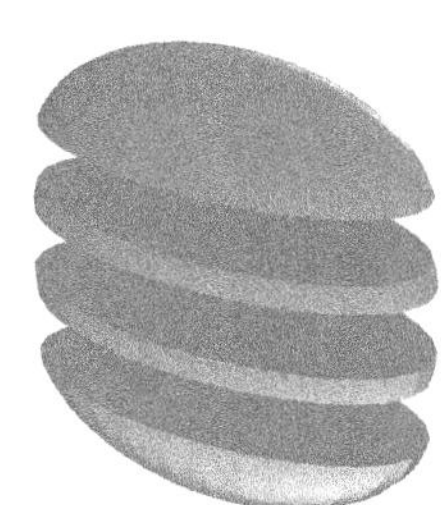

3 $2\int_{-r}^{r} \sqrt{r^2 - x^2}\,dx = \pi r^2$

5 (a) $\sum 2\sqrt{9-y}\,\Delta y$
(b) 36

7 $3\pi/2 = 4.712$

9 $512\pi/15 = 107.233$

11 (a) $12\pi \int_0^9 \sqrt{9-y}\,dy$
(b) 216π

13 $V = \int_0^2 \pi (y^3)^2\,dy$

15 $V = \int_0^2 \left[\pi(10-4y)^2 - \pi(2)^2\right] dy$

17 $V = \int_0^2 \left[\pi(4y+3)^2 - \pi(y^3+3)^2\right] dx$

19 $V = \pi$

21 $2\int_{-a}^{a} \sqrt{1 + b^2x^2/(a^2(a^2-x^2))}\,dx$

23 45.230

25 4.785

27 15.865

29 $\int_0^1 f(x)\,dx < 1/2$

31 $\int_0^1 f^{-1}(x)\,dx > 1/2$

33 $\int_0^1 \sqrt{1 + (f'(x))^2}\,dx > \sqrt{2}$

35

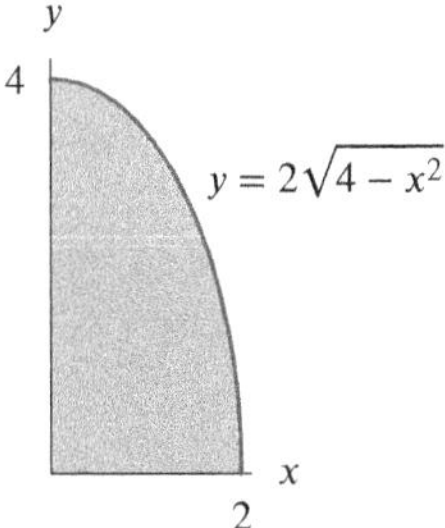

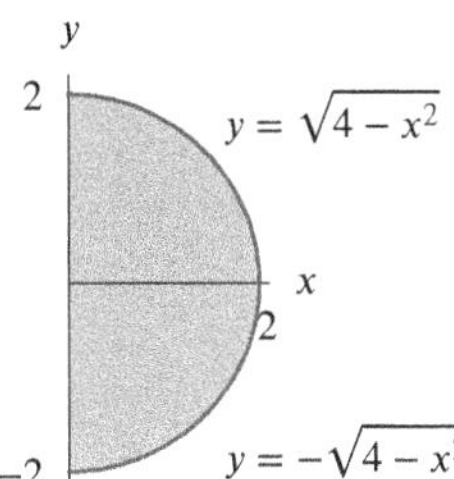

37 $V = \int_0^{25} \pi((8 - y/5)^2 - (8 - \sqrt{y})^2)\, dy$

39 (a) $V = 5\pi/6$
(b) $V = 4\pi/5$

41 $736\pi/3 = 770.737$

43 $2\pi/3$

45 $\pi^2/2 - 2\pi/3 = 2.840$

47 $\pi/24$

49 Area under graph of f from $x = 0$ to $x = 3$

51 $e - e^{-1}$

53 (a) $a = b/l$
(b) $(1/3)\pi b^2 l$

55 Volume $= 6\pi^2$

57 $y = (1/3)e^{3t}$ from $t = 3$ to $t = 8$

59 $(x - a)^2 + y^2 = a^2$

61 $(\pi/3 + \sqrt{3}/2)a^2$; 61%

63 $2\pi a$

65 $1/4$

67 4400 ft-lb

69 1000 ft-lb

71 518,363 ft-lbs

73 (a) Force on dam $\approx \sum_{i=0}^{N-1} 1000(62.4h_i)\Delta h$
(b) $\int_0^{50} = 1000(62.4h)\, dh =$ 78,000,000 pounds

75 $0.366(k + 1.077)g\pi$ joules

77 33 billion m^3/year

79 (a) $\sum_{i=1}^{N} \pi \left((3.5 \cdot 10^5)/\sqrt{h + 600}\right)^2 \Delta h$
(b) $1.042 \cdot 10^{12}$ cubic feet

81 92% of pop younger than 70

83 $\int_0^R 2\pi N r S(r) dr$

85 (a) $\pi P R^4/(8\eta l)$

87 (a) $\pi h^2/(2a)$
(b) $\pi h/a$
(c) $dh/dt = -k$
(d) h_0/k

89 The thin spherical shell

91 (c) $\pi^2 a^3/8$

93 (a) $\frac{1}{2}\sqrt{t}\sqrt{1 + 4t} + \frac{1}{4}\text{arcsinh}(2\sqrt{t})$
(b) t

Section 9.1

1 3, 5, 9, 17, 33

3 2/3, 4/5, 6/7, 8/9, 10/11

5 1, −1/2, 1/4, −1/8, 1/16

7 $s_n = 2^{n+1}$

9 $s_n = n^2 + 1$

11 $s_n = n/(2n + 1)$

13 (a) (IV)
(b) (III)
(c) (II)
(d) (I)

15 (a) (II)
(b) (III)
(c) (I)
(d) (IV)
(e) (V)

17 Converges to 0

19 Converges to 0

21 Converges to 0

23 Converges to 0

25 Converges to 1/2

27 Diverges

29 1, 3, 6, 10, 15, 21

31 1, 5, 7, 17, 31, 65

33 13, 27, 50, 85

35 $s_n = s_{n-1} + 2$, $s_1 = 1$

37 $s_n = 2s_{n-1} - 1$, $s_1 = 3$

39 $s_n = s_{n-1} + n$, $s_1 = 1$

45 45

47 2

49 0.878, 0.540, 0.071, −0.416, −0.801, −0.990

51 0, 6, −6, 6, −6, 6, … and 3, 0, 2, −2, 2, …

53 1.5, 2, 3, 4, 5, 6, 7 … and 1.75, 2.17, 3, 4, 5, 6, …

55 Converges to 0.5671

57 Converges to 1

59 (a) 2, 4, 2^n
(b) 33 generations; overlap

61 (a) k rows; $a_n = n$
(b) $T_n = T_{n-1} + n, n > 1; T_1 = 1$

65 Converges to 3/7

67 $s_n = 2 - 1/n$

69 $s_n = n$

71 False

73 True

75 False

77 False

79 (b)

Section 9.2

1 Series

3 Sequence

5 Series

7 Series

9 Yes, $a = 2$, ratio $= 1/2$

11 No

13 No

15 Yes, $a = 1$, ratio $= -y^2$

17 No

19 Geometric; diverges

21 Not geometric

23 Not geometric

25 10 terms; 0.222

27 14 terms; 15.999

29 −486

31 $1/(1 - 5x)$, $-1/5 < x < 1/5$

33 $1/x$, $0 < x < 4$

35 $1/(1 - z/2)$, $-2 < z < 2$

37 $y/(1 + y)$, $-1 < y < 1$

39 $3 + x/(1 - x)$, $-1 < x < 1$

41 $8/(1 - (x^2 - 5))$
$-\sqrt{6} < x < -2$ and $2 < x < \sqrt{6}$

43 Does not converge

45 Does not converge

47 32 tons

49 (a) $0.232323\ldots = 0.23 + 0.23(0.01) + 0.23(0.01)^2 + \cdots$
(b) $0.23/(1 - 0.01) = (23)/(99)$

51 (a) (i) \$16.43 million
(ii) \$24.01 million
(b) \$16.87 million

53 (a) $17.9(1.025) + 17.9(1.025)^2 + \ldots + 17.9(1.025)^n$;
$733.9(1.025^n - 1)$
(b) 2040
(c) Right Riemann sum with n subdivisions

55 (a) $P_n = 250(0.04) + 250(0.04)^2 + 250(0.04)^3 + \cdots + 250(0.04)^{n-1}$
(b) $P_n = 250 \cdot 0.04(1 - (0.04)^{n-1})/(1 - 0.04)$
(c) $\lim_{n\to\infty} P_n \approx 10.42$
Difference between them is 250 mg

57

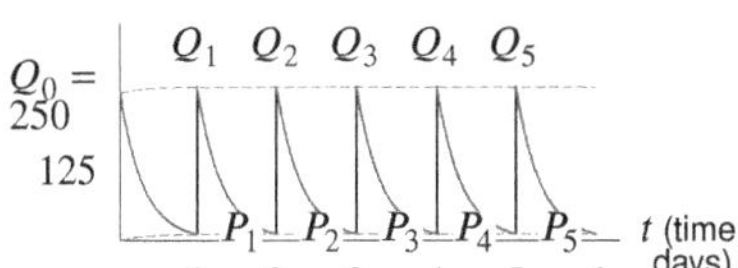

59 \$400 million

61 (a) \$1000
(b) When the interest rate is 5%, the present value equals the principal
(c) The principal
(d) Because the present value is more than the principal

65 $|x| \geq 1$

67 $3 + 6 + 12 + 24 + \cdots$

69 $16/3 + 8/3 + 4/3 + 2/3$

71 $a_n = 2r^n$, $b_n = 3r^n$, $|r| < 1$

73 (c)

Section 9.3

1 1, 3, 6, 10, 15

3 1/2, 2/3, 3/4, 4/5, 5/6

5 Diverges

7 Converges

11 $f(x) = (-1)^x/x$ undefined
13 Diverges
15 Diverges
17 Converges
19 Converges
21 Converges
23 Diverges
25 Converges
27 Diverges
29 Diverges
31 Converges
35 (a) $\ln(n+1)$
(b) Diverges
37 (b) $S_3 = 1 - 1/4; S_{10} = 1 - 1/11; S_n = 1 - 1/(n+1)$
45 Terms approaching zero does not guarantee convergence
47 Converges for $k < -1$ and diverges for $k \geq -1$
49 $\sum_{n=1}^{\infty} 1/n$
51 True
53 False
55 True
57 False
63 (a) 1.596
(b) 3.09
(c) 1.635; 3.13
(d) 0.05; 0.01
67 (d)

Section 9.4

5 Behaves like $\sum_{n=1}^{\infty} 1/n$; Diverges
7 Behaves like $\sum_{n=1}^{\infty} 1/n^4$; Converges
9 Converges
11 Converges
13 Converges
15 Converges
17 Converges
19 Converges
21 Diverges
23 Converges
25 Converges
27 Diverges
29 Diverges
31 Diverges
33 Alternating
35 Not alternating
37 Converges
39 Converges
41 Converges
43 (a) No
(b) Converges
(c) Converges
45 Conditionally convergent
47 Divergent
49 Conditionally convergent
51 Conditionally convergent
53 Terms not positive
55 Limit of ratios is 1
57 $a_{n+1} > a_n$ or $\lim_{n\to\infty} a_n \neq 0$
59 $\lim_{n\to\infty} a_n \neq 0$
61 Does not converge
63 Converges to approximately 0.3679
65 Diverges
67 Converges
69 Diverges
71 Converges
73 Converges
75 Diverges
77 Converges
79 Diverges
81 Diverges
83 Diverges
85 Diverges
87 Diverges
89 Converges
91 Converges
93 (a) Converges
(b) Converges
95 (a) Converges
(b) Diverges
97 (a) Diverges
(b) Diverges
99 Unaffected
101 Affected
103 Converges for $a > 2$ and diverges for $a \leq 2$
105 Converges for $a > 1$ and diverges for $a \leq 1$
107 99 or more terms
109 2 or more terms
113 $(-1)^{2n} = 1$ so series is not alternating
115 Need $1/n^{3/2} \leq 1/n^2$ for all n to make comparison
117 $\sum(-1)^n n$
119 False
121 True
123 False
125 True
127 False
129 False
131 True
133 False
135 (b)
139 Converges

Section 9.5

1 Yes
3 No
5 $1 \cdot 3 \cdot 5 \cdots (2n-1)x^n/(2^n \cdot n!)$; $n \geq 1$
7 $(-1)^k(x-1)^{2k}/(2k)!$; $k \geq 0$
9 $(x-a)^n/(2^{n-1} \cdot n!)$; $n \geq 1$
11 1
13 1
15 2
17 2
19 1
21 1/4
23 1
25 (a) $R = 1$
(b) $-1 < x \leq 1$
27 $-3 < x < 3$
29 $-2 < x < 2$
31 $-\infty < x < \infty$
33 $-1/5 \leq x < 1/5$
35 $\sum_{n=0}^{\infty}(-2z)^n, -1/2 < z < 1/2$
37 $\sum_{n=0}^{\infty} 3(z/2)^n, -2 < z < 2$
39 (a) 5.833
(b) $-4 < z < 10$
41 (a) False
(b) True
(c) False
(d) Cannot determine
43 $5 \leq R \leq 7$
45 $k = 0.1, C_0 = 3, C_1 = 1, C_2 = 4, C_3 = 1$
49 (a) Odd; $g(0) = 0$
(c) $\sin x$
51 Limit may be > 2
53 $\sum_{n=0}^{\infty}(x-2)^n/n$
55 $\sum_{n=1}^{\infty}(1/n)(x-10)^n$
57 False
59 False
61 True
63 False
65 True
67 True
69 True
71 (a) All real numbers
(b) $J(0) = 1$
(c) $S_4(x) = 1 - \frac{x^2}{4} + \frac{x^4}{64} - \frac{x^6}{2304} + \frac{x^8}{147456}$
(d) 0.765
(e) 0.765

Chapter 9 Review

1 $3(2^{11} - 1)/2^{10}$
3 619.235
5 $b^5(1 - b^6)/(1 - b)$ when $b \neq 1$; 6 when $b = 1$
7 $(3^{17} - 1)/(2 \cdot 3^{20})$
9 36, 48, 52, 53.333; $S_n = 36(1 - (1/3)^n)/(1 - 1/3)$; $S = 54$
11 $-810, -270, -630, -390$; $S_n = -810(1 - (-2/3)^n)/(1 + 2/3)$; $S = -486$
13 Converges 4/7
15 Diverges
17 Converges
19 Converges
21 Converges
23 Diverges
25 Converges
27 Divergent
29 Divergent
33 Converges
35 Converges
37 Converges
39 Converges
41 Diverges
43 Diverges
45 Convergent
47 Converges
49 Converges

51 Diverges

53 Converges

55 Diverges

57 Diverges

59 ∞

61 1

63 $-3 < x < 7$

65 $-\infty < x < \infty$

67 $s_n = 2n + 3$

69 9, 15, 23

71 Converges if $r > 1$, diverges if $0 < r \leq 1$

73 (a) All real numbers
(b) Even

75 240 million tons

77 \$926.40

79 (a) \$124.50
(b) \$125

81 £250

83 (a) (i) $B_0 4^n$
(ii) $B_0 2^{n/10}$
(iii) $(2^{1.9})^n$
(b) 10.490 hours

85 5/6

87 $\sum(3/2)^n$, $\sum(1/2)^n$, other answers possible

89 Converges

91 Not enough information

93 Converges

95 (a) C_0 ————————
C_1 ——— ———
C_2 — — — —
C_3 -- -- -- --

(b) $1/3 + 2/9 + 4/27 + \cdots + (1/3)(2/3)^{n-1}$
(c) 1

97 12

Section 10.1

1 $P_3(x) = 1 + x + x^2 + x^3$
$P_5(x) = 1 + x + x^2 + x^3 + x^4 + x^5$
$P_7(x) = 1 + x + x^2 + x^3 + x^4 + x^5 + x^6 + x^7$

3 $P_2(x) = 1 + (1/2)x - (1/8)x^2$
$P_3(x) = 1 + (1/2)x - (1/8)x^2 + (1/16)x^3$
$P_4(x) = 1 + (1/2)x - (1/8)x^2 + (1/16)x^3 - (5/128)x^4$

5 $P_2(x) = 1 - x^2/2!$
$P_4(x) = 1 - x^2/2! + x^4/4!$
$P_6(x) = 1 - x^2/2! + x^4/4! - x^6/6!$

7 $P_3(x) = P_4(x) = x - (1/3)x^3$

9 $P_2(x) = 1 - (1/2)x + (3/8)x^2$
$P_3(x) = 1 - (1/2)x + (3/8)x^2 - (5/16)x^3$
$P_4(x) = 1 - (1/2)x + (3/8)x^2 - (5/16)x^3 + (35/128)x^4$

11 $P_3(x) = 1 - \frac{x}{2} - \frac{x^2}{8} - \frac{x^3}{16}$

13 $P_4(x) = \frac{1}{3}[1 - \frac{x-2}{3} + \frac{(x-2)^2}{3^2} - \frac{(x-2)^3}{3^3} + \frac{(x-2)^4}{3^4}]$

15 $P_3(x) = \frac{\sqrt{2}}{2}[-1 + \left(x + \frac{\pi}{4}\right) + \frac{1}{2}\left(x + \frac{\pi}{4}\right)^2 - \frac{1}{6}\left(x + \frac{\pi}{4}\right)^3]$

17 (a) 2
(b) −1
(c) −2/3
(d) 12

19 $P_3(x) = 1 - 5x + 7x^2 + x^3$
$f(x) = P_3(x)$

21 $1 - x/3 + 5x^2/7 + 8x^3$

23 $-3 + 5x - x^2 - x^4/24 + x^5/30$

25 (a) Positive
(b) $P_1(x) = \ln 4 + (1/2)x$
(c) $P_2(x) = \ln 4 + (1/2)x - (1/8)x^2$
$P_3(x) = \ln 4 + (1/2)x - (1/8)x^2 + (1/24)x^3$

27 $a > 0, b < 0, c < 0$

29 $a < 0, b < 0, c > 0$

31 $f'''(1) = 3$

35 −2.616

39 (a) 1/2
(b) 1/6

41 (a) fs are Figure 10.9
gs are Figure 10.8
(b) $A = (0, 1)$
$B = (0, 2)$
(c) $f_1 = \text{III}, f_2 = \text{I}, f_3 = \text{II}$
$g_1 = \text{III}, g_2 = \text{II}, g_3 = \text{I}$

43 (a) $1 + 3x + 2x^2$
(b) $1 + 3x + 7x^2$
(c) No

45 (a) $P_4(x) = 1 + x^2 + (1/2)x^4$
(b) If we substitute x^2 for x in the Taylor polynomial for e^x of degree 2, we will get $P_4(x)$, the Taylor polynomial for e^{x^2} of degree 4
(c) $P_{20}(x) = 1 + x^2/1! + x^4/2! + \cdots + x^{20}/10!$
(d) $e^{-2x} \approx 1 - 2x + 2x^2 - (4/3)x^3 + (2/3)x^4 - (4/15)x^5$

47 (a) Infinitely many; 3
(b) That near $x = 0$
Taylor poly only accurate near 0

49 (b) 0, 0.2

53 $f'(0) = 1$

55 $f(x) = \sin x$

57 $f(x) = 1 + x$; $g(x) = 1 + x + x^2$

59 False

61 False

63 False

65 False

Section 10.2

1 $f(x) = 1 + 3x/2 + 3x^2/8 - x^3/16 + \cdots$

3 $f(x) = -x + x^3/3! - x^5/5! + x^7/7! + \cdots$

5 $f(x) = 1 + x + x^2 + x^3 + \cdots$

7 $f(y) = 1 - y/3 - y^2/9 - 5y^3/81 - \cdots$

9 $\ln 5 + (2/5)x - (2/25)x^2 + (8/475)x^3$

11 $\cos\theta = \sqrt{2}/2 - (\sqrt{2}/2)(\theta - \pi/4) - (\sqrt{2}/4)(\theta - \pi/4)^2 + (\sqrt{2}/12)(\theta - \pi/4)^3 - \cdots$

13 $\sin\theta = -\sqrt{2}/2 + (\sqrt{2}/2)(\theta + \pi/4) + (\sqrt{2}/4)(\theta + \pi/4)^2 - (\sqrt{2}/12)(\theta + \pi/4)^3 + \cdots$

15 $1/x = 1 - (x - 1) + (x - 1)^2 - (x - 1)^3 + \cdots$

17 $\frac{1}{x} = -1 - (x + 1) - (x + 1)^2 - (x + 1)^3 - \cdots$

19 $(-1)^n x^n$; $n \geq 0$

21 $(-1)^{n-1} x^n / n$; $n \geq 1$

23 $(-1)^k x^{2k+1} / (2k + 1)$; $k \geq 0$

25 $(-1)^k x^{4k+2} / (2k)!$; $k \geq 0$

27 (a) 0
(b) Increasing
(c) Concave down

29 1

31 $\frac{d}{dx}(x^2 e^{x^2})|_{x=0} = 0$
$\frac{d^6}{dx^6}(x^2 e^{x^2})|_{x=0} = \frac{6!}{2} = 360$

33 $f^{(n)}(0) > 0$ for all n

35 $f^{(n)}(0) = 0$ for even n

37 $-1 < x < 1$

39 (a) $-1 < x < 1$
(b) 1

41 1

43 e^2

45 4/3

47 $\ln(3/2)$

49 e^3

51 $e^{-0.1}$

53 4/5

55 216

57 1

61 Only converges for $-1 < x < 1$

63 $f(x) = \cos x$

65 False

67 True

69 True

71 True

Section 10.3

1 $e^{-x} = 1 - x + x^2/2! - x^3/3! + \cdots$

3 $\cos(\theta^2) = 1 - \theta^4/2! + \theta^8/4! - \theta^{12}/6! + \cdots$

5 $\arcsin x = x + (1/6)x^3 + (3/40)x^5 + (5/112)x^7 + \cdots$

7 $1/\sqrt{1 - z^2} = 1 + (1/2)z^2 + (3/8)z^4 + (5/16)z^6 + \cdots$

9 $\phi^3 \cos(\phi^2) = \phi^3 - \phi^7/2! + \phi^{11}/4! - \phi^{15}/6! + \cdots$

11 $1 + t^2/2! + t^4/4! + t^6/6! + \cdots$

13 $1 + 3x + 3x^2 + x^3$
$0 \cdot x^n$ for $n \geq 4$

15 $1 + (1/2)y^2 + (3/8)y^4 + \cdots + ((1/2)(3/2)\cdots(1/2 + n - 1)y^{2n})/n! + \cdots$; $n \geq 1$

17 $\ln(3/2)$

19 $\sqrt{T + h} = \sqrt{T}(1 + (1/2)(h/T) - (1/8)(h/T)^2 + (1/16)(h/T)^3 + \cdots)$

21 $1/(a + r)^2 = (1/a^2)(1 - 2(r/a) + 3(r/a)^2 - 4(r/a)^3 + \cdots)$

23 $a/\sqrt{a^2 + x^2} = 1 - (1/2)(x/a)^2 + (3/8)(x/a)^4 - (5/16)(x/a)^6 + \cdots$

25 $e^t \cos t = 1 + t - t^3/3 - t^4/6 + \cdots$

27 $1 + t + t^2/2 + t^3/3$

29 (a) $1/2 - (x-1)/4 + (x-1)^2/8 - (x-1)^3/16 + \cdots + (-1)^n (x - 1)^n / 2^{n+1} + \cdots$

(b) $(x-1)^2/2-(x-1)^3/4+(x-1)^4/8-(x-1)^5/16+\cdots+(-1)^n(x-1)^{n+2}/2^{n+1}+\cdots$

31 $\sinh 2x = \sum_{m=0}^{\infty}(2x)^{2m+1}/(2m+1)!$
$\cosh 2x = \sum_{m=0}^{\infty}(2x)^{2m}/(2m)!$

33 (a) $2x + x^3/3 + x^5/60 + \cdots$
(b) $P_3(x) = 2x + x^3/3$

35 $1/(1-x)^2$

37 $1 + \sin\theta \le e^\theta \le 1/\sqrt{1-2\theta}$

39 (a) I
(b) IV
(c) III
(d) II

41 $f(x) = (1/3)x^3 - (1/42)x^7 + (1/1320)x^{11} - (1/75{,}600)x^{15} + \cdots$

43 1

45 (a) $f(x) = 1 + (a-b)x + (b^2 - ab)x^2 + \cdots$
(b) $a = 1/2, b = -1/2$

47 (a) If $M >> m$, then $\mu \approx mM/M = m$
(b) $\mu = m[1 - m/M + (m/M)^2 - (m/M)^3 + \cdots]$
(c) -0.0545%

49 (a) $\theta'' + (g/l)\theta = 0$
(b) 0.76%

51 (a) $((l_1 + l_2)/c)\cdot(v^2/c^2 + (5/4)v^4/c^4)$
(b) $v^2, (l_1 + l_2)/c^3$

53 (a) Set $dV/dr = 0$, solve for r. Check for max or min.
(b) $V(r) = -V_0 + 72V_0r_0^{-2}\cdot(r - r_0)^2(1/2) + \cdots$
(d) $F = 0$ when $r - r_0$

55 (a) $1 + x^2/2! + x^4/4! + x^6/6! + x^8/8!$
(b) 1.54308
(c) $x + x^3/3! + x^5/5! + x^7/7!$

57 -720

61 $1 - x + \frac{x^2}{2!} - \frac{x^3}{3!} + \cdots$

63 $|x|$

65 True

67 True

69 (c)

Section 10.4

1 $|E_3| \le 0.00000460, E_3 = 0.00000425$

3 $|E_3| \le 0.000338, E_3 = 0.000336$

5 $|E_4| \le 0.0156, E_3 = -0.0112$

7 $|E_3| \le 16.5, E_3 = 0.224$

9 $1/5!$

11 $1/5!$

13 $1/50$

15 (c)

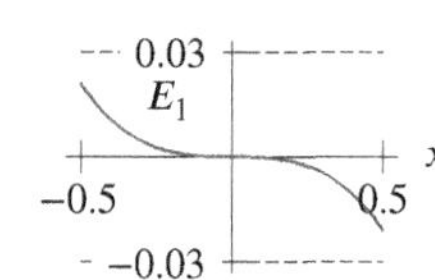

17 (a) 5
(b) 1.06

19 (a) Overestimate: $0 < \theta \le 1$
Underestimate: $-1 \le \theta < 0$
(b) $|E_2| \le 0.17$

21 (a) Underestimate
(b) 1

23 (a) (i) 4, 0.2
(ii) 1
(b) $-4 \le x \le 4$

25 Four decimal places: $n = 7$
Six decimal places: $n = 9$

29 (a) $\pi \approx 2.67$
(b) $\pi \approx 2.33$
(c) $|E_n| \le 0.78$
(d) Derivatives unbounded near $x = 1$

31 (a) $(1/2)^{n+1}/(n+1)!$
(b) $(1/2)^{n+1}/(n+1)!$
(c) Equal

33 $|f(a) - P_n(a)| = 0$

35 $f(x) = \sin x$

37 e^x on $[-1, 1]$

41 True

Section 10.5

1 Not a Fourier series

3 Fourier series

5 $F_1(x) = F_2(x) = (4/\pi)\sin x$
$F_3(x) = (4/\pi)\sin x + (4/3\pi)\sin 3x$

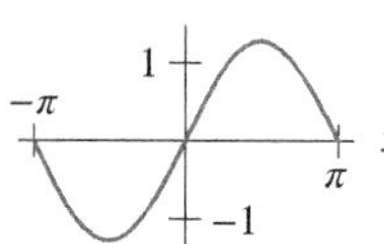

$F_1(x) = F_2(x) = \frac{4}{\pi}\sin x$

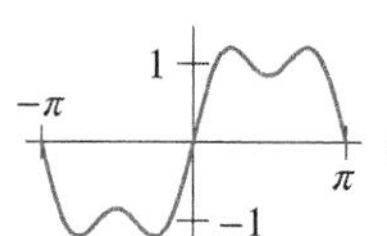

$F_3(x) = \frac{4}{\pi}\sin x + \frac{4}{3\pi}\sin 3x$

7 99.942% of the total energy

9 $H_n(x) = \pi/4 + \sum_{i=1}^{n}((-1)^{i+1}\sin(ix))/i + \sum_{i=1}^{[n/2]}(-2/((2i-1)^2\pi))\cos((2i-1)x)$, where $[n/2]$ denotes the biggest integer smaller than or equal to $n/2$

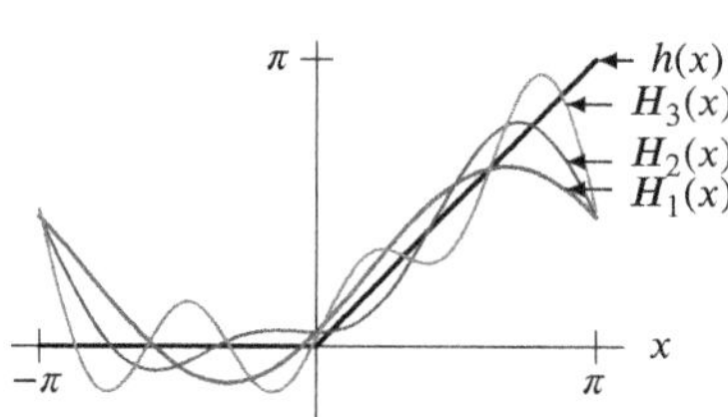

11 $a_0 = 1/2$

13 (b) $(1/7)\sin 7x$
(c) $f(x) = 1$ for $2\pi \le x < -\pi$
$f(x) = -1$ for $-\pi \le x < 0$
$f(x) = 1$ for $0 \le x < \pi$
$f(x) = -1$ for $\pi \le x < 2\pi\ldots$
Not continuous

15 $F_4(x) = 1/2 - (2/\pi)\sin(2\pi x) - (1/\pi)\sin(4\pi x) - (2/3\pi)\sin(6\pi x) - (1/2\pi)\sin(8\pi x)$

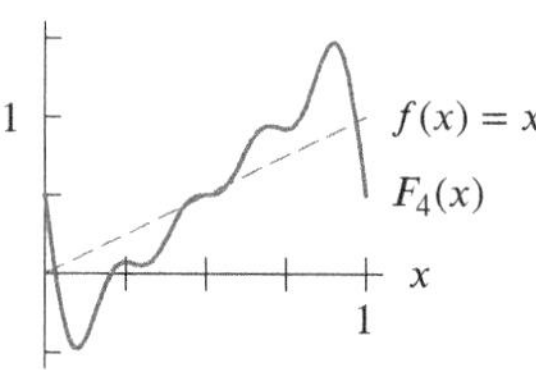

17 52.93

23 (a) 6.3662%, 18.929%
(b) $(4\sin^2(k/5))/(k^2\pi^2)$
(c) 61.5255%
(d) $F_5(x) = 1/(5\pi) + (2\sin(1/5)/\pi)\cos x + (\sin(2/5)/\pi)\cos 2x + (2\sin(3/5)/(3\pi))\cos 3x + (\sin(4/5)/(2\pi))\cos 4x + (2\sin 1/(5\pi))\cos 5x$

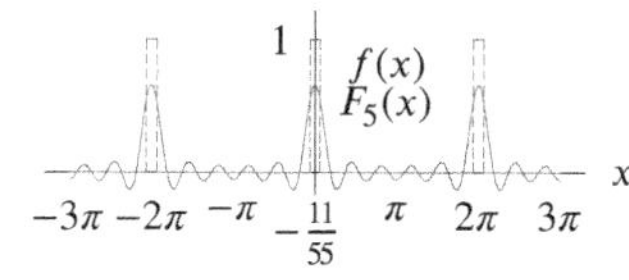

25 The energy of the pulse train is spread out over more harmonics as c gets closer to 0

33 a_0 is average of $f(x)$ on interval of approximation

35 Any odd function with period 2π

37 (b)

Chapter 10 Review

1 $e^x \approx 1 + e(x-1) + (e/2)(x-1)^2$

3 $\sin x \approx -1/\sqrt{2} + (1/\sqrt{2})(x + \pi/4) + (1/2\sqrt{2})(x + \pi/4)^2$

5 $P_3(x) = 4 + 12(x-1) + 10(x-1)^2 + (x-1)^3$

7 $P_3(x) = \sqrt{2}[1 + (x-1)/4 - (x-1)^2/32 + (x-1)^3/128]$

9 $P_7 = 3x - 9x^3 + (27/2)x^5 - (27/2)x^7$

11 $t^2 + t^3 + (1/2)t^4 + (1/6)t^5 + \cdots$

13 $\theta^2\cos\theta^2 = \theta^2 - \theta^6/2! + \theta^{10}/4! - \theta^{14}/6! + \cdots$

15 $t/(1+t) = t - t^2 + t^3 - t^4 + \cdots$

17 $1/\sqrt{4-x} = 1/2 + (1/8)x + (3/64)x^2 + (5/256)x^3 + \cdots$

19 $a/(a+b) = 1 - b/a + (b/a)^2 - (b/a)^3 \cdots$

21 $(B^2 + y^2)^{3/2} = B^3(1 + (3/2)(y/B)^2 + (3/8)(y/B)^4 - (1/16)(y/B)^6 + \cdots)$

23 1.45

25 3/4

27 $\sin 2$

29 (a) $7(1.02^{104} - 1)/(0.02(1.02)^{100})$
(b) $7e^{0.01}$

31 Smallest to largest:
$1 - \cos x,\ x\sqrt{1-x},$
$\ln(1+x), \arctan x,\ \sin x,$
$x,\ e^x - 1$

33 $1/2$

35 0.10008

37 0.9046

39 (a) $\arcsin x = x + (1/6)x^3 + (3/40)x^5 + (5/112)x^7 + (35/1152)x^9 + \cdots$
(b) 1

41 (a)

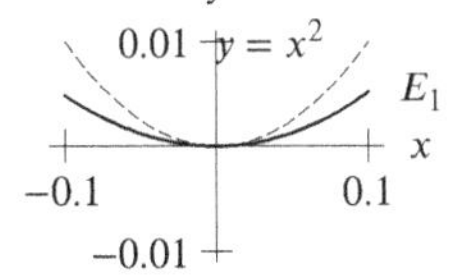

(b)
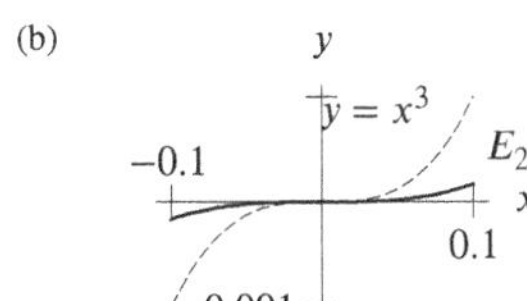

43 (a) $V(x) \approx V(0) + V''(0)x^2/2$
$V''(0) > 0$
(b) Force $\approx -V''(0)x$
$V''(0) > 0$
Toward origin

45 $x - x^2 + (3/2)x^3 - (8/3)x^4$

47 $P_3(t) = t + t^2/4 + (1/18)t^3$

49 (a) $F = GM/R^2 + Gm/(R+r)^2$
(b) $F = GM/R^2 + (Gm/R^2)(1 - 2(r/R) + 3(r/R)^2 - \cdots)$

51 (b) $F = mg(1 - 2h/R + 3h^2/R^2 - 4h^3/R^3 + \cdots)$
(c) 300 km

57 (b) If the amplitude of the kth harmonic of f is A_k, then the amplitude of the kth harmonic of f' is kA_k
(c) The energy of the kth harmonic of f' is k^2 times the energy of the kth harmonic of f

61 (a) $P_7(x) = x - x^3/6 + x^5/120 - x^7/5040$
$Q_7(x) = x - 2x^3/3 + 2x^5/15 - 4x^7/315$
(b) For n odd, ratio of coefficients of x^n is 2^{n-1}

63 (a) $P_{10}(x) = 1 + x^2/12 - x^4/720 + x^6/30240 - x^8/1209600 + x^{10}/47900160$
(b) All even powers
(c) f even

Section 11.1

1 Yes

3 (II)

5 (VI)

7 (V)

9 (a) Not a solution
(b) Solution
(c) Not a solution

11 (a) (I), (III)
(b) (IV)
(c) (II), (IV)

15 No

17 (a) Decreasing
(b) 7.75

21 $P = 15/t$

23 $Q = 12/(t+1)$

25 $\omega = \pm 3$

27 $k = -0.03$ and C is any number, or $C = 0$ and k is any number

29 -2

31 (a) $A = -1$
(b) $B = 2$

33 (a) (IV)
(b) (III)
(c) (I)
(d) (II)

35 Is a solution

37 Is a solution

39 $Q = 6e^{4t}$ is particular solution, not general

41 $dy/dx = x/y$ and $y = 100$ when $x = 0$

43 $dy/dx = 2x$ with solutions $y = x^2$ and $y = x^2 + 5$

45 $dy/dx = 1/x$

47 $dy/dx = y - x^2$

49 False

51 False

53 False

55 True

57 False

Section 11.2

1 (a) 1; −1; 0; 3; −3; 0
(b)
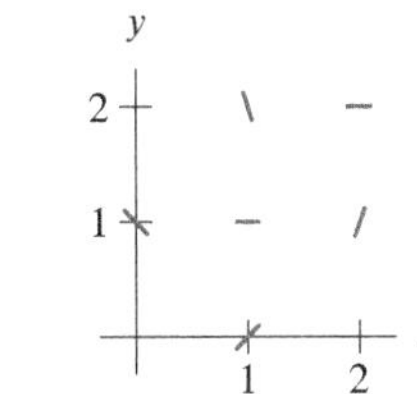

3
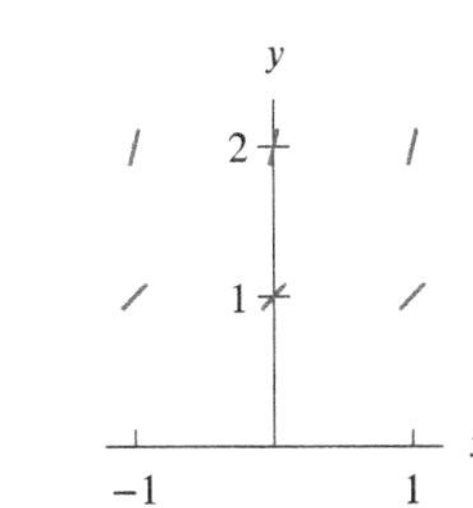

5 Possible curves:
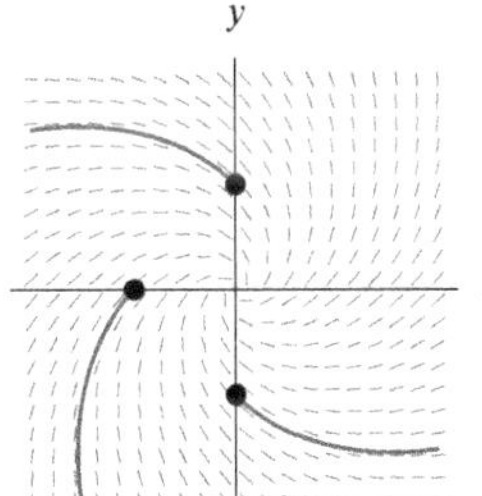

7 Increasing

9 Increasing

11 (a)
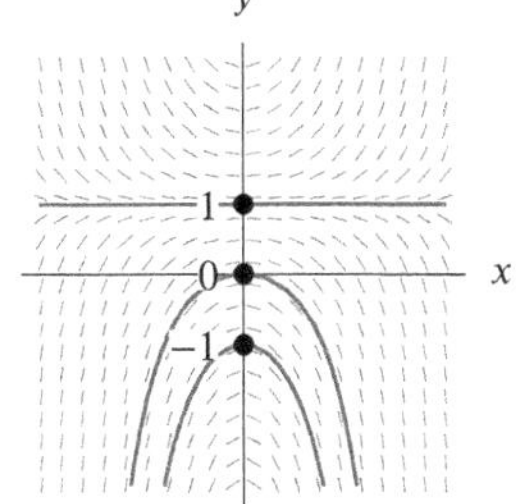

(b) $y(x) = 1$

13 (a) (A) is $y' = 0.3y$
(B) is $y' = 0.3t$
(b) For $y' = 0.3y$:
(i) ∞
(ii) 0
For $y' = 0.3t$:
(i) ∞
(ii) ∞

15
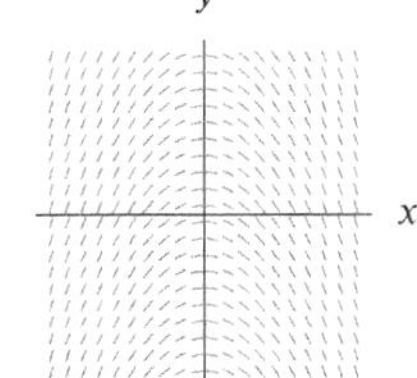

17 Concave up

19 (a)
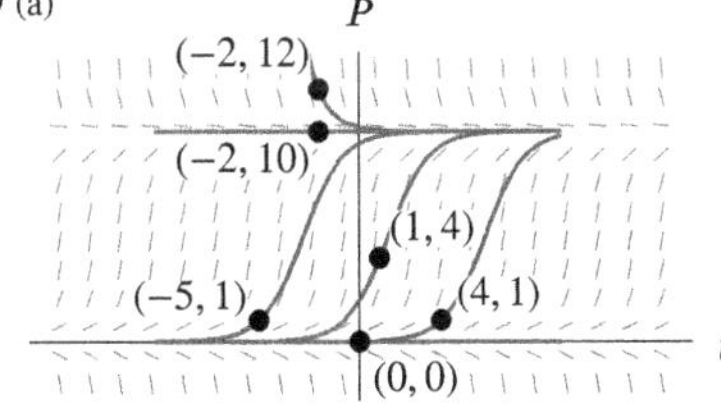

(b) Incr: $0 < P < 10$; decr: $P > 10$; P tends to 10

21 Slope field (b)

23 (a) (II)
(b) (I)
(c) (V)
(d) (III)
(e) (IV)

25 (a) IV
(b) I
(c) III
(d) V
(e) II
(f) VI

27 (a) $y' = 0.05y(10 - y)$.

29 (d) $y' = 0.05x(5 - x)$.

31 Graph of $y = x$ not tangent to slope field at (1, 1)

33 Slopes of $y' = y$ in quadrant II are positive

35 $dy/dx = y$

37

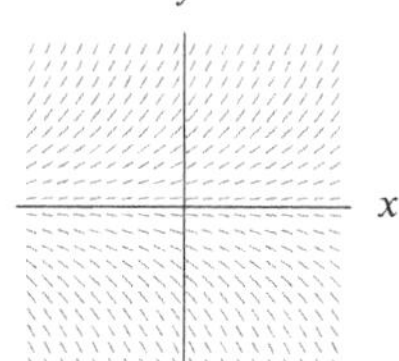

39 True

41 False

43 False

45 False

47 True

49 True

51 True

53 False

Section 11.3

3 1548, 1591.5917, 1630.860

5 1.97

7 (b) $y(x) = x^4/4$

9 (a)

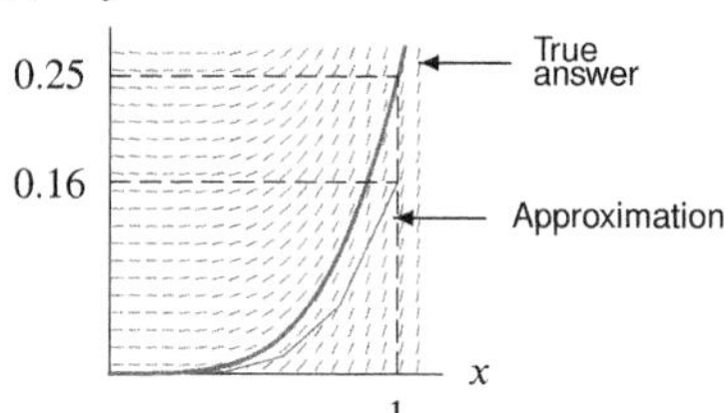

(b) At $x = 1$, $y \approx 0.16$
(c) Underestimate

11 (a) $y = 0.667$
(b) $y = 0.710$

13 Error in ten-step is half error ins five-step; $y = 0.753$

15 (a) $\Delta x = 0.5$, $y(1) \approx 1.5$
$\Delta x = 0.25$, $y \approx 1.75$
(b) $y = x^2 + 1$, so $y(1) = 2$
(c) Yes

17 (a) (i) $B \approx 1050$
(ii) $B \approx 1050.63$
(iii) $B \approx 1050.94$

21 Overestimate if $x(0) < 0$

23 $dy/dx = y$, $y(0) = 1$

25 True

27 Error decreases by factor of approx 1/2

29 We would need $n = 300$

Section 11.4

1 (a) Yes	(b) No	(c) Yes
(d) No	(e) Yes	(f) Yes
(g) No	(h) Yes	(i) No
(j) Yes	(k) Yes	(l) No

3 Separable; $1/(e^y - 1)dy = x\,dx$

5 Separable; $(1/y)dy = 1/(1 - x)dx$

7 $P = 20e^{0.02t}$

9 $Q = 50e^{(1/5)t}$

11 $m = 5e^{3t-3}$

13 $z = 5e^{5t-5}$

15 $u = 1/(1 - (1/2)t)$

17 $y = 10e^{-x/3}$

19 $P = 104e^t - 4$

21 $Q = 400 - 350e^{0.3t}$

23 $R = 1 - 0.9e^{1-y}$

25 $y = (1/3)(3 + t)$

27 $y = 3x^5$

29 $z = 5e^{t+t^3/3}$

31 $w = -1/(\ln|\cos\psi| - 1/2)$

33 (a) $y = \sqrt[3]{6x^2 + B}$
(b) $y = \sqrt[3]{6x^2 + 1}$; $y = \sqrt[3]{6x^2 + 8}$; $y = \sqrt[3]{6x^2 + 27}$

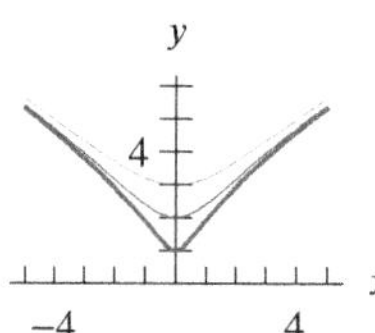

35 (a) $y(t) = 100 - Ae^{-t}$
(b)

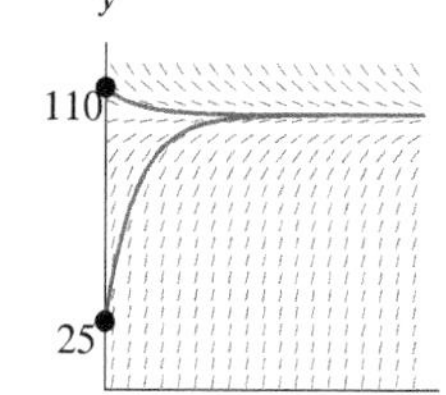

(c) $y(t) = 100 - 75e^{-t}$
$y(t) = 100 + 10e^{-t}$
(d) $y(t) = 100 - 75e^{-t}$

37 (a)

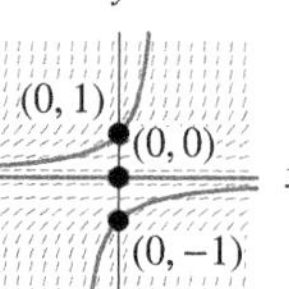

(b) One end asymptotic to $y = 0$, other end unbounded
(c) $y = -1/(x + C)$, $x \neq -C$

39 $Q = Ae^{t/k}$

41 $Q = b - Ae^{-t}$

43 $R = -(b/a) + Ae^{at}$

45 $y = -1/\left(k(t + t^3/3) + C\right)$

47 $L = b + Ae^{k((1/2)x^2 + ax)}$

49 $x = e^{At}$

51 $y = 2(2^{-e^{-t}})$

53 (a) and (b)

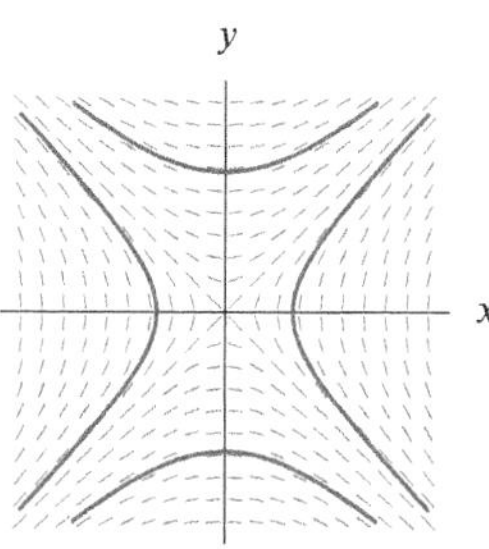

(c) $y^2 - x^2 = 2C$

57 Impossible to separate the variables in $dy/dx = x + y$

59 $e^{-y}\,dy = e^x\,dx$

61 $f(x) = \cos x$

63 $dy/dx = x/y$

65 True

67 False

69 False

Section 11.5

1 (a) (III)
(b) (IV)
(c) (I)
(d) (II)

3 (a) $y = 2$: stable; $y = -1$: unstable
(b)

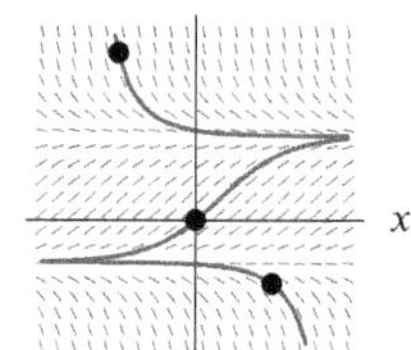

5 y α t

7 (a) $H = 200 - 180e^{-kt}$
(b) $k \approx 0.027$ (if t is in minutes)

9 (a) Positive
(b) Approximately 10 minutes

11 Decreasing

13 10,250

15 Decrease

19 Longest: Lake Superior
Shortest: Lake Erie
The ratio is about 75

21 Is a solution

23 (a) $y = 0$, $y = 4$, $y = -2$
(b) $y = 0$ stable, $y = 4$ and $y = -2$ unstable

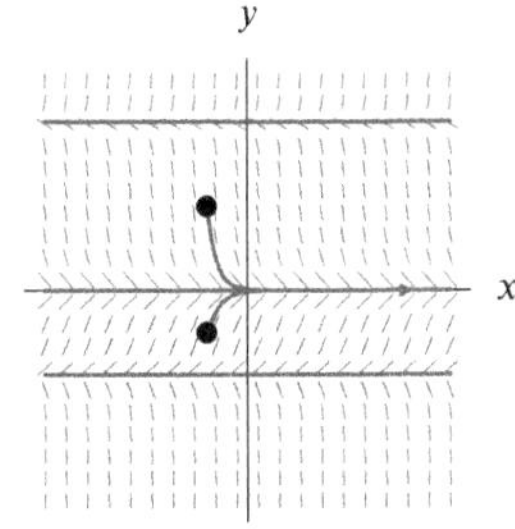

25 (a) N = amount of nicotine in mg at time t; t = number of hours since smoking a cigarette
(b) $dN/dt = -0.347N$
(c) $N = 0.4e^{-0.347t}$

27 (a) G = size in acres of Grinnell Glacier in year t; t = number of years since 2007
(b) $dG/dt = -0.043G$
(c) $G = 142e^{-0.043t}$

29 (a) $dH/dt = -k(H - 22)$
(b) $H = 22 - 19e^{-kt}$

31 (a) $dP/dt = 0.01rP - H$
(b) $P = 300$ million fish
(d) $P = 100H/r$
(e) No

33 (a) $dQ/dt = -kQ$
(b) Half-life is $(\ln 2)/k$
(c) Decreasing

35 (a) $dQ/dt = -0.5365Q$
$Q = Q_0e^{-0.5365t}$
(b) 4 mg

37 (a) $k \approx 0.000121$
(b) 779.4 years

39 (a) $dT/dt = -k(T - 68)$
(b) $T = 68 + 22.3e^{-0.06t}$; 3:45 am.

41 (a) $dD/dt = kD$

43 (a) $dQ/dt = kQ$
(b) $Q = 0$, stable
(c) $Q = Ce^{kt}$
(d) $k \approx -0.182$
(e) $Q(12) \approx 1.126$ mg

45 $dy/dx \neq 0$ when $y = 2$

47 Surrounding air is 350°F not 40°F

49 $dQ/dt = Q - 500$

Section 11.6

1 (a) (III)
(b) (V)
(c) (I)
(d) (II)
(e) (IV)

3 $dB/dt = 0.037B + 5000$

5 $dB/dt = -0.065B - 50{,}000$

7 (a) $H' = k(H - 50)$; $H(0) = 90$
(b) $H(t) = 50 + 40e^{-0.05754t}$
(c) 24 minutes

9 $da/dt = -ka/m$
$a = ge^{-kt/m}$

11 $dD/dt = -0.75(D - 4)$
Equilibrium = 4 g/cm^2

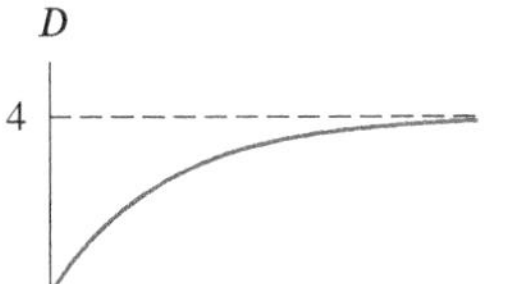

13 $dA/dt = -0.17A + 130$; $A = 764.7 - 764.7e^{-0.17t}$; 625 mg

15 (a) $dP/dt = 0.079 - 0.000857t$
(b) $P = 0.079t - 0.000857t^2/2 + 7.052$
(c) 9.435 bn

17 (a) Mt. Whitney: 17.50 inches
Mt. Everest: 10.23 inches
(b) 18,661.5 feet

19 (b) $dQ/dt = -0.347Q + 2.5$
(c) $Q = 7.2$ mg

21 $P = AV^k$

23 5.7 days, 126.32 liters

25 (a) $dp/dt = -k(p - p^*)$
(b) $p = p^* + (p_0 - p^*)e^{-kt}$
(c)

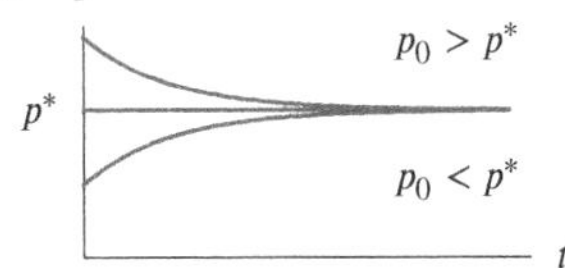

(d) As $t \to \infty$, $p \to p^*$

27 (a) $Q = (r/\alpha)(1 - e^{-\alpha t})$
$Q_\infty = r/\alpha$

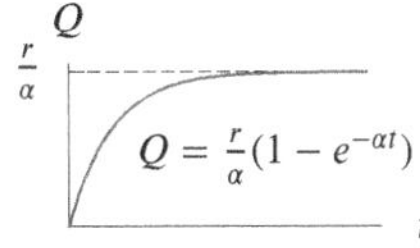

(b) Doubling r doubles Q_∞;
Altering r does not alter the time it takes to reach $(1/2)Q_\infty$
(c) Both Q_∞ and the time to reach $(1/2)Q_0$ are halved by doubling α

29 $t = 133.601$ min

31 $y = A\sqrt{x}$

33 (a) $(0, a)$
(b) $dy/dx = -y/\sqrt{a^2 - y^2}$

35 Units of time are not consistent

37 $dQ/dt = 0.2\sqrt[3]{Q} + 100$

39 (a) $B_{60} = \sum_{k=0}^{59} 100e^{(0.02)k/12}$
(b) $B_{60} = 6304.998$
(c) Smaller because deposits are made later

Section 11.7

1 (b) 1

3

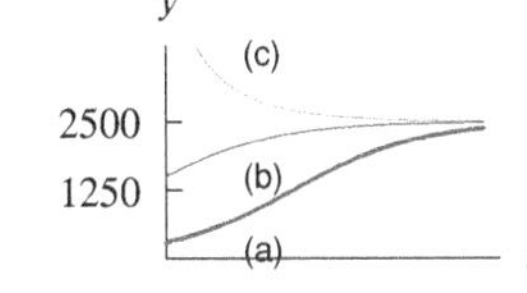

5

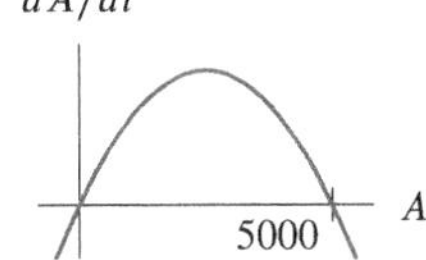

7

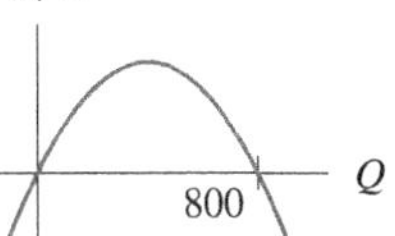

9 (a) $P = 0$, $P = 2000$
(b) dP/dt positive; P increasing

11 (a) $k = 0.035$; gives relative growth rate when P is small
$L = 6000$; gives limiting value on the size of P
(b) $P = 3000$

13 $P = 2800/(1 + Ae^{-0.05t})$

15 $P = 4000/(1 + Ae^{-0.68t})$

17 $k = 10, L = 2, A = 3, P = 2/(1+3e^{-10t}), t = \ln(3)/10$

19 $k = 0.3, L = 100, A = 1/3, P = 100/(1 + e^{-0.3t}/3), t = -\ln(3)/0.3$

21 $P = 8500/(1 + 16e^{-0.8t})$

23 (a) Between 80,000 and 85,000 cases
(b) About 6500 cases/week

25 (a) 539,226 cases
(b) 47,182 cases/week

27 (a) 30,000 cases
(b) 6000 cases/week

29 22 days

31 (a) $P = 5000/(1 + 499e^{-1.78t})$
(b)

y
5000
2500
3.5
t

(c) $t \approx 3.5$; $P \approx 2500$

33 (a) 2500 fish
(b) 2230 fish
(c) 1500 fish

35 (a) $L = 12{,}000$
(b) 360

37 $dN/dt = N(0.155 - 0.001N)$; 155 million

39 (a) 2017
(b) 2025

41 (a) 1261 bn barrels
(b) About 28 bn barrels
(c) 1434 bn barrels
(d) Very close

43 $P = 0$ also equilibrium

45 Inflection point should be at $P = 50$

47 $dP/dt = 0.2P(1 - P/150)$

49

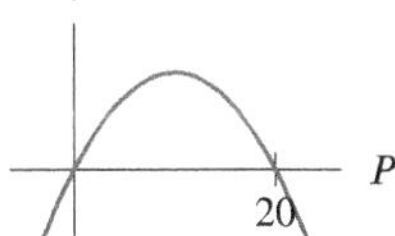

51 True

53 (a) $L \approx 13$
(b) $k \approx 0.30$

55 $dI/dt = 0.001I(1 - 10(I/M))$; $M/10$

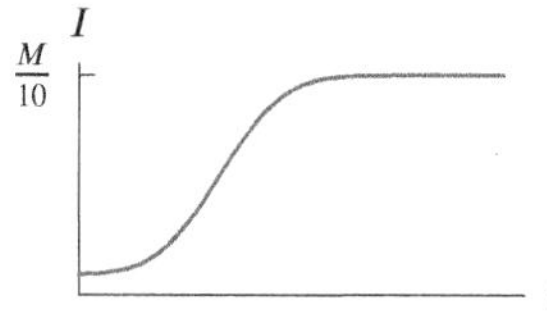

Section 11.8

1 x and y increase, about same rate

3 x decreases quickly while y increases more slowly

5 $(0, 0)$ and $(5, 3)$

7 $(0, 0)$ and $(-5, 3)$

9 (a) Both x and y are decreasing
(b) x is increasing; y is decreasing

11 P_3

13 $(0, 0), (3, 0), (0, 6), (4.5, 1.5)$

15 1

17 Helpful

19 (a) Susceptibles: 1950, 1850, 1550, 1000, 750, 550, 350, 250, 200, 200, 200
Infecteds: 20, 80, 240, 460, 500, 460, 320, 180, 100, 40, 20
(b) Day 22, 500
(c) 1800, 200

21 (a) a is negative; b is negative; c is positive
(b) $a = -c$

25 $r = r_0e^{-t}$, $w = w_0e^t$

27 Worms decrease, robins increase. Long run: populations oscillate

29

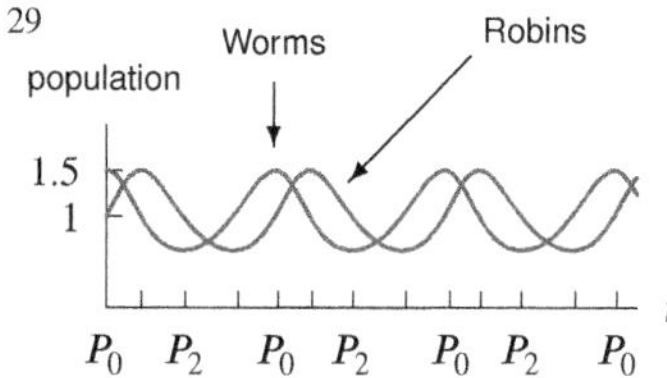

31 Robins increase;
Worms constant, then decrease;
Both oscillate in long run

33 (a) Symbiosis
(b) Both $\to \infty$ or both $\to 0$

35 (a) Predator-prey
(b) x, y tend to ≈ 1

37 $(10, 30)$

39 (a) $dy/dx = x/5y$
(b) US victory
(c) No

41 (a) $x^2 - 5y^2 = 604.75$
(b) About 25,000 troops

43 dx/dt, dy/dt can both be positive

45 $dx/dt = 0.5x$; $dy/dt = -0.1y + 0.3xy$

47 D_1 is flu and D_2 is HIV

49 False

51 $a = 0.00001, k = 0.06$

53 (a) $w = 0$, $r = 0$ and $w = b/k$, $r = a/c$
(b) Worm equilibrium unchanged, robin equilibrium reduced

Section 11.9

1 $(4, 10)$

3

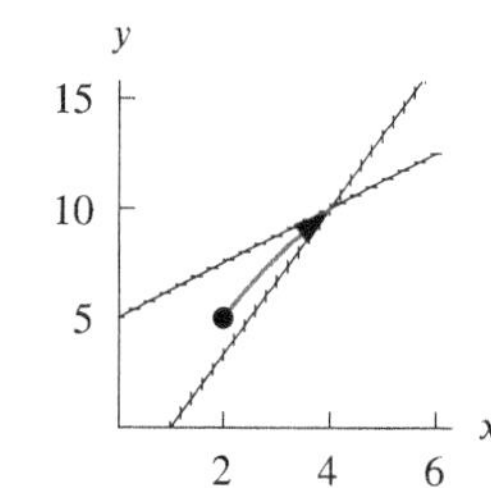

5 Tends towards point $(4, 10)$

7 (a) $dx/dt > 0$ and $dy/dt > 0$
(b) $dx/dt < 0$ and $dy/dt = 0$
(c) $dx/dt = 0$ and $dy/dt > 0$

9

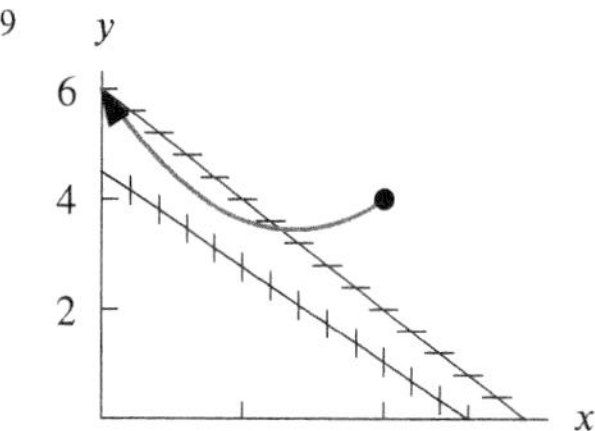

11

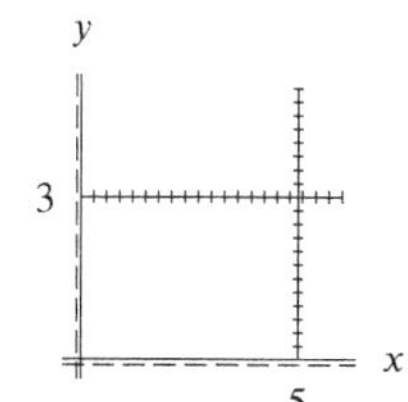

13 Vertical nullclines:
$x = 0, x + y = 2$
Horizontal nullclines:
$y = 0, x + y = 1$
Equilibrium points:
$(0, 0), (0, 1), (2, 0)$

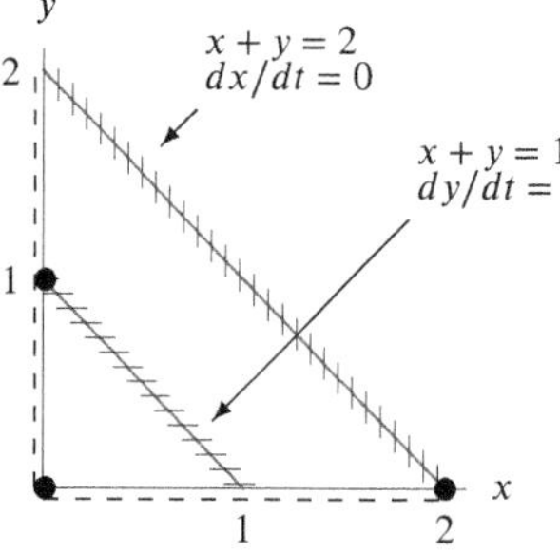

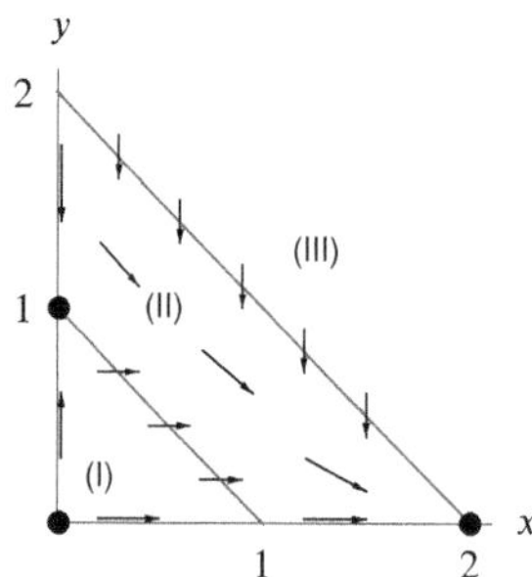

15 Horizontal nullclines;
$y = 0, y = 1 - 2x$
Vertical nullclines;
$x = 0, y = (1/2)(2 - x)$
Equilibrium points;
$(0, 0), (0, 1), (2, 0)$

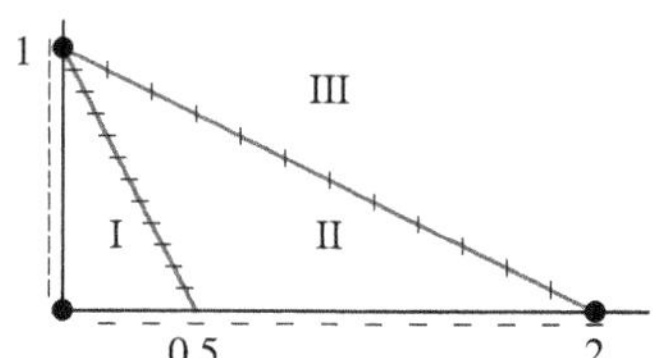

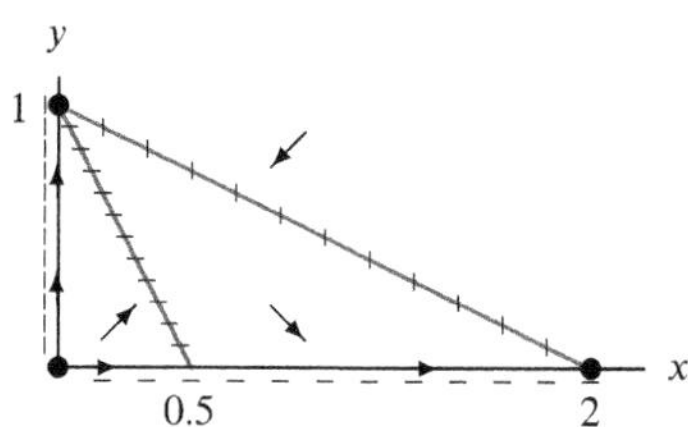

17 $dx/dt = 0$ when
$x = 0$ or $x + y/3 = 1$
$dy/dt = 0$ when
$y = 0$ or $y + x/2 = 1$
Equilibrium points:
$(0, 0), (0, 1), (1, 0), (4/5, 3/5)$

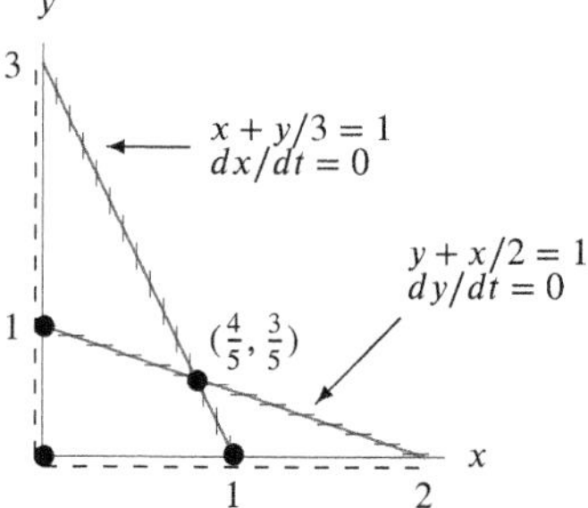

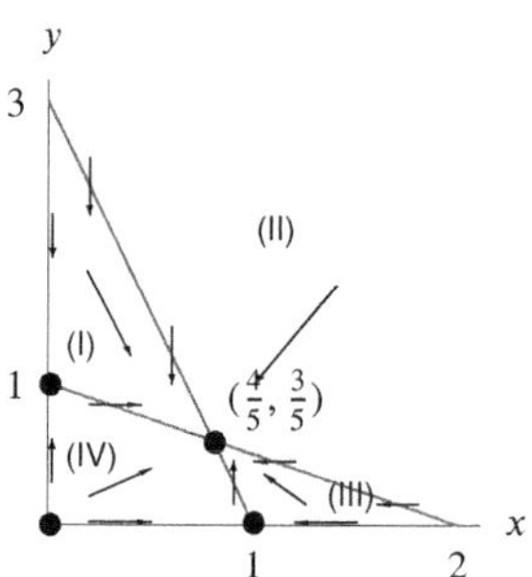

19 (a) $dS/dt = 0$ where $S = 0$ or $I = 0$
$dI/dt = 0$ where $I = 0$ or $S = 192$

(b) Where $S > 192$,
$dS/dt < 0$ and $dI/dt > 0$
Where $S < 192$,
$dS/dt < 0$ and $dI/dt < 0$

(c)

21 (a) In the absence of the other, each company grows exponentially
The two companies restrain each other's growth if they are both present
(b) $(0, 0)$ and $(1, 2)$
(c) In the long run, one of the companies will go out of business

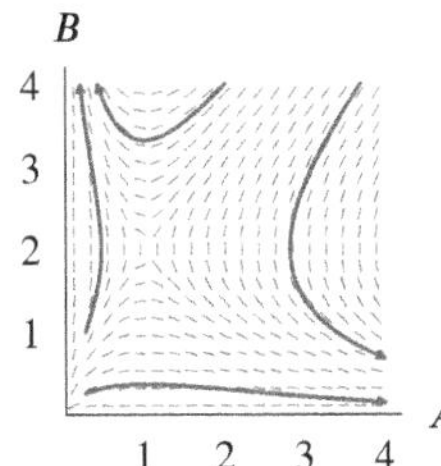

23 (a) $dP_1/dt = 0$ where $P = 0$ or $P_1 + 3P_2 = 13$
$dP_2/dt = 0$ where $P = 0$ or $P_2 + 0.4P_1 = 6$

25 $dx/dt \neq 0$

27

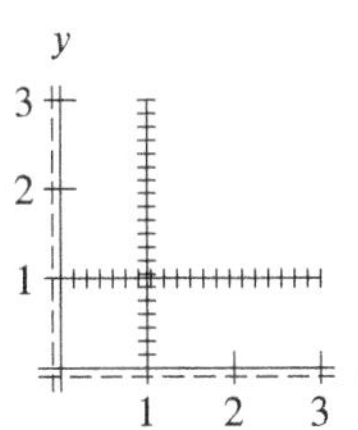

29 False

31 False

Section 11.10

1 $A = \sqrt{3^2 + 4^2} = 5$

3 $\omega = 2$, $A = 13$, $\psi = tan^{-1} 12/5$

5 (a)

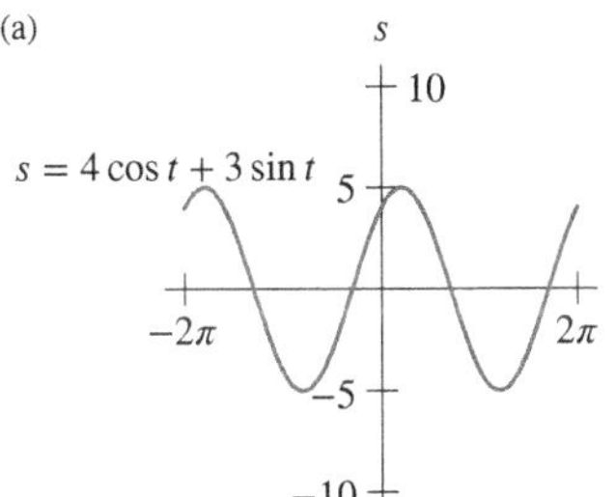

(c) $A = 5, \phi = 0.93$

11 $A \approx -1.705$
$\alpha = \pm\sqrt{5}$

13 $y(t) =$
$C_1 \cos(t/3) + C_2 \sin(t/3)$

15 (a) $d^2s/dt^2 = -2.5s$
(b) $s(t) = 5\cos(1.581t) - 6.325\cos(1.581t)$
(c) -8.06

17 Middle point, moving down

19 (b) (i) $y = 6 \sinh wt/ \sinh w$
(ii) $y = 6 \cosh wt/ \cosh w$

21 (a) $x(t) = 5\cos 4t$
$A = 5$, period $= \pi/2$
(b) $x(t) = -\cos(t/5) + 10\sin(t/5)$
$A = \sqrt{101}$, period $= 10\pi$

23 (a) Spring (iii)
(b) Spring (iv)
(c) Spring (iv)
(d) Spring (i)

25 (a) Starts higher; period same
(b) Period increases

27 (a) $A\sin(\omega t + \phi) =$
$(A \sin\phi)\cos\omega t +$
$(A\cos\phi)\sin\omega t$

29 $C = 1/20$ farads

31 Initial displacement is 3

33 $y'' = 4y$

Section 11.11

1 $y(t) = C_1e^{-t} + C_2e^{-3t}$

3 $y(t) =$
$C_1e^{-2t}\cos t + C_2e^{-2t}\sin t$

5 $s(t) =$
$C_1 \cos\sqrt{7}t + C_2 \sin\sqrt{7}t$

7 $z(t) = C_1e^{-t/2} + C_2e^{-3t/2}$

9 $p(t) = C_1e^{-t/2}\cos\frac{\sqrt{3}}{2}t+$
$C_2e^{-t/2}\sin\frac{\sqrt{3}}{2}t$

11 $y(t) = A + Be^{-2t}$

13 $y(t) = C_1e^{t/3} + C_2e^{-t/3}$

15 $y(t) = e^{-t}(A\sin\sqrt{2}t + B\cos\sqrt{2}t)$

17 $y(t) = -2e^{-3t} + 3e^{-2t}$

19 $y(t) = \frac{1}{5}e^{4t} + \frac{4}{5}e^{-t}$

21 $y(t) = \frac{5}{4}e^{-t} - \frac{1}{4}e^{-5t}$

23 $y(t) = 2e^{-3t}\sin t$

25 $y(t) = \frac{1}{1-e}e^{-2t} + \frac{-e}{1-e}e^{-3t}$

27 $p(t) = 20e^{(\pi/2)-t}\sin t$

29 (a) (IV)
(b) (II)
(c) (I)
(d) (III)

31 (iii)

33 (iv)

35 (ii)

37 Overdamped: $c < 2$
Critically damped: $c = 2$
Underdamped: $c > 2$

39 Overdamped if:
$b > 2\sqrt{5}$
Critically damped if:
$b = 2\sqrt{5}$
Underdamped if:
$0 < b < 2\sqrt{5}$

41 (a) $k = 100$; $a = 70$

(b) $s(t) = -4e^{-2t} + 3e^{-5t}$
(c) $s = -1.58$; $s = 0$
(d) $t \geq 1.843$

43 (a) $a > 447.216$
(b) $a = 447.216$
(c) $0 < a < 447.216$

47 $z(t) = 3e^{-2t}$

49 (a) $d^2y/dt^2 - y = 0$
(b) $y = C_1e^t + C_2e^{-t}$,
so $x = C_2e^{-t} - C_1e^t$

53 (a) $Q(t) = 2te^{-t/2}$
(b) $Q(t) = (2 + t)e^{-t/2}$

55 As $t \to \infty$, $Q(t) \to 0$

57 Curve shows $s'(0) < 0$

59 $d^2s/dt^2 + 2ds/dt + s = 0$

Chapter 11 Review

1 (I) satisfies equation (d)
(II) satisfies equation (c)

3 (a) (I)
(b) (IV)
(c) (III)

5 $y(x) = 40 + Ae^{0.2x}$

7 $H = Ae^{0.5t} - 20$

9 $P = 100Ae^{0.4t}/(1 + Ae^{0.4t})$

11 $P = (40000/3)(e^{0.03t} - 1)$

13 $y = \sqrt[3]{33 - 6\cos x}$

15 $\frac{1}{2}y - 4\ln|y| = 3\ln|x| - x$
$+ \frac{7}{2} - 4\ln 5$

17 $z(t) = 1/(1 - 0.9e^t)$

19 $20\ln|y| - y =$
$100\ln|x| - x + 20\ln 20 - 19$

21 $y = \ln(e^x + e - 1)$

23 $z = \sin(-e^{\cos\theta} + \pi/6 + e)$

25 $y = -\ln(\frac{\ln 2}{3}\sin^2 t\cos t$
$+ \frac{2\ln 2}{3}\cos t - \frac{2\ln 2}{3} + 1)/\ln 2$

27 (a) $y = 3$ and $y = -2$
(b) $y = 3$ unstable;
$y = -2$ stable

29 No

31 (a) $y(1) \approx 3.689$
(b) Overestimate

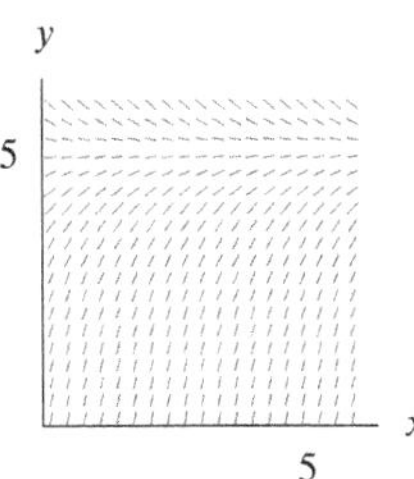

(c) $y = 5 - 4e^{-x}$
$y(1) = 5 - 4e^{-1} \approx$
3.528
(d) ≈ 3.61

33 (b) 2070

35 (a) $dM/dt = rM$
(b) $M = 2000e^{rt}$

(c)

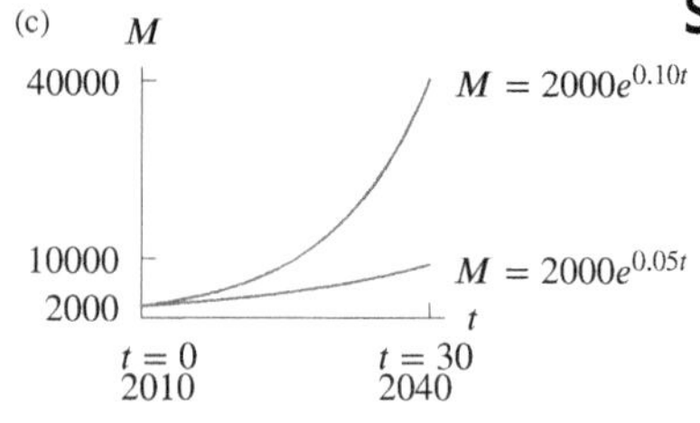

37

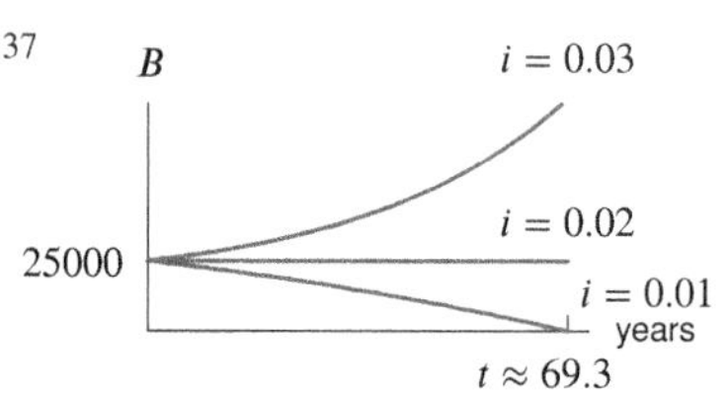

39 (a) $dB/dt = 0.05B - 12{,}000$
(b) $B = (B_0 - 240{,}000)e^{0.05t}$ $+ 240{,}000$
(c) \$151,708.93

41 $C(t) = 30e^{-100.33t}$

43 No, no

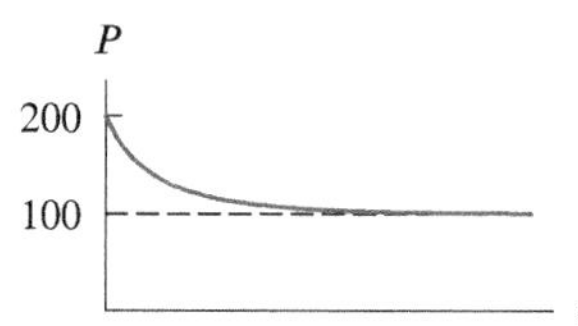

45 (a) $dx/dt = k(a-x)(b-x)$
(b) $x = ab(e^{bkt} - e^{akt})/(be^{bkt} - ae^{akt})$

47 (a) $x \to \infty$ exponentially
$y \to 0$ exponentially
(b) Predator-prey

49 (a) $x \to \infty$ exponentially
$y \to 0$ exponentially

(b) y is helped by the presence of x

51 (a) $dy/dx = y(x+3)/(x(y+2))$
$y^2e^y = Ax^3e^x$
(d) $y^2e^y = x^3e^{x+4}$
(e) $y = x = e^{-4} \approx 0.0183$
(f) $y \approx e^{-13}$

53 (a) $x \to \infty$ if $y = 0$
$y \to 0$ if $x = 0$
(b) (0, 0) and (10, 000, 150)
(c) $dy/dx = y(-10+0.001x)/(x(3-0.02y))$
$3\ln y - 0.02y =$
$-10\ln x + 0.001x + 94.1$
(f) $A \to B \to C \to D$
(h) At points A and C:
$dy/dx = 0$
At points B and D:
$dx/dy = 0$

55 P decreases to 0

57 P decreases towards $2L$

59 (c)

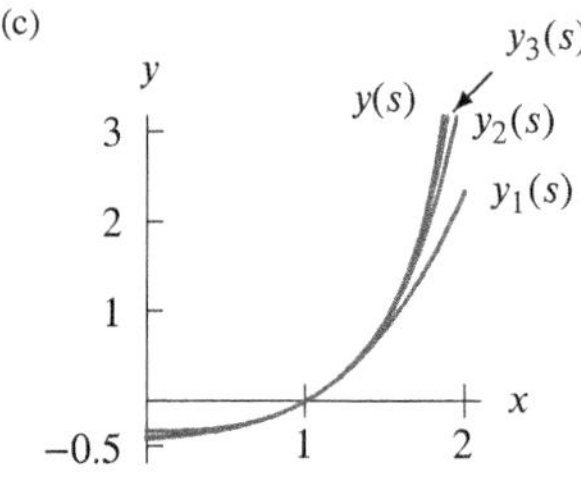

Section 12.1

1 Q

3 $(-4, 2, 7)$

5 $(1, -1, 1)$; Front, left, above

7 $(2, 4.5, 3)$

9 $(-1, 1, 0)$, $(-2, -2, 4)$

11 $(2, 2, 4)$, $(-2, -2, 4)$

13

31

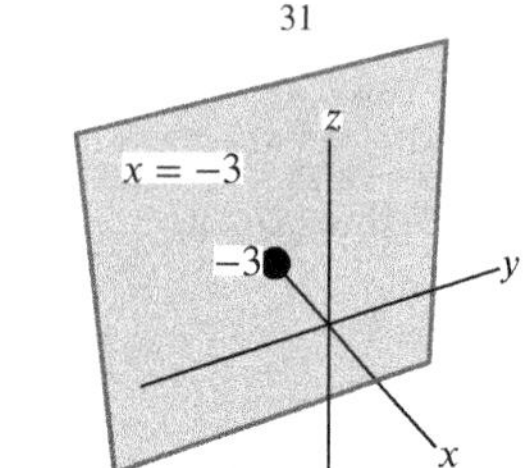

15

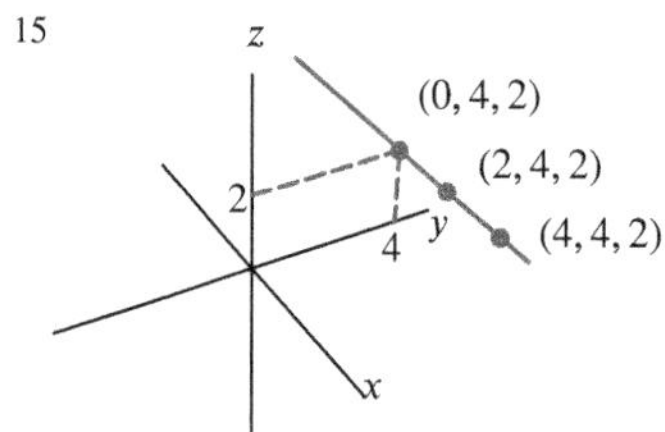

17 $x^2 + y^2 + z^2 = 25$

19 $y = 3$

21

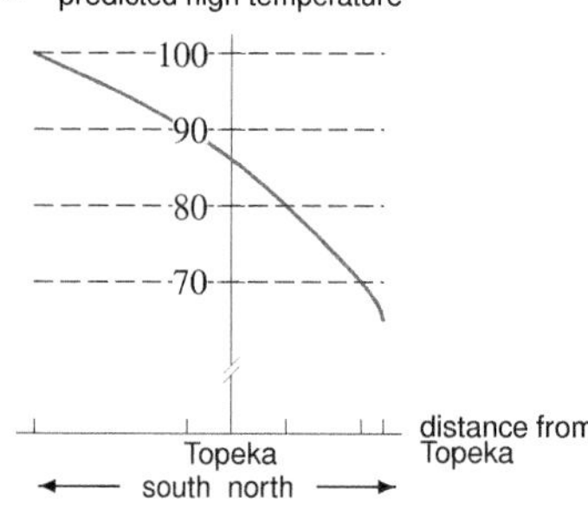

23 $f(w, 60)$: 23.4, 27.3, 31.2, 35.2, 39.1

25 25

29 $f(20, p)$: 2.65, 2.59, 2.51, 2.43
$f(100, p)$: 5.79, 5.77, 5.60, 5.53
$f(I, 3.00)$: 2.65, 4.14, 5.11, 5.35, 5.79
$f(I, 4.00)$: 2.51, 3.94, 4.97, 5.19, 5.60

31 Increasing function

33 57.9 kg

37 (a) $R = 100s + 5m$
(b) 125,000 dollars

39 (b) Increasing
(c) Decreasing

41 (b) Increasing
(c) Increasing

43 $(1.5, 0.5, -0.5)$

45 Cone, tip at origin, along x-axis with slope of 1

47 Yes; (2, 5, 4)

49 (a) $z = 7$, $z = -1$
(b) $x = 6$, $x = -2$
(c) $y = 7$, $y = -1$

51 (a) (12, 7, 2); (5, 7, 2); (12, 1, 2)
(b) (5, 1, 4); (5, 7, 4); (12, 1, 4)

53 (a) (3, 9, 13)
(b) (2, 7, 10)
(c) (4, 11, 16)

55 xy-plane is $z = 0$
$xy = 0$ is yz-plane and xz-plane

57 $f(x, y) = x - y$

59 True

61 False

63 True

65 False

67 False

69 False

71 True

73 (a) yz-plane: circle $(y+3)^2 + (z-2)^2 = 3$
xz-plane: none
xy-plane: point $(1, -3, 0)$

(b) Does not intersect

Section 12.2

1 (III)

3 (I), (IV)

5 (a) Decreases

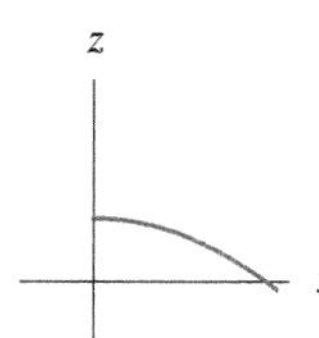

(b) Increases

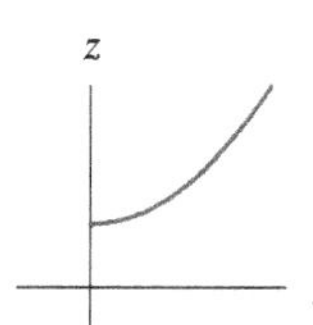

7 (a) I
(b) V
(c) IV
(d) II
(e) III

9 Sphere, radius 3

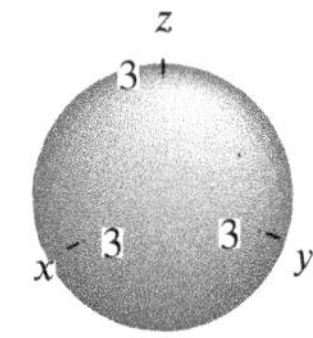

11 Upside-down bowl, vertex (0, 0, 5)

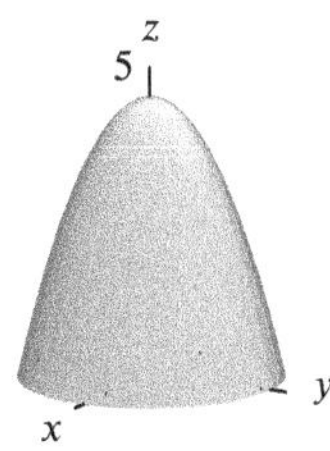

13 Plane, x-intercept 6, y-intercept 3, z-intercept 4

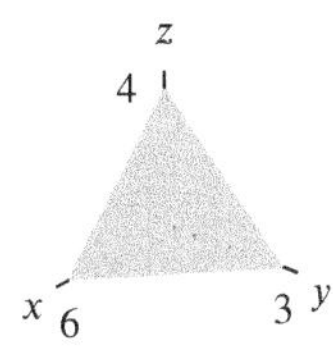

15 Circular cylinder extended in the y-direction

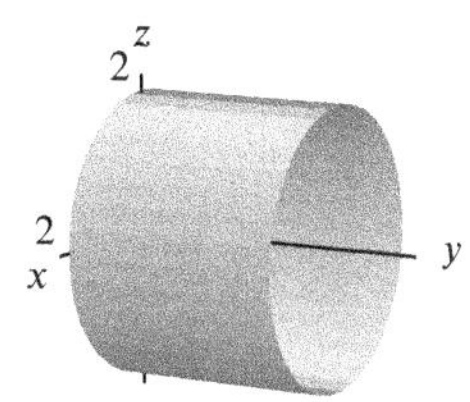

17 $x^2 + (y - \sqrt{7})^2 + z^2 = 9$

19 (a)

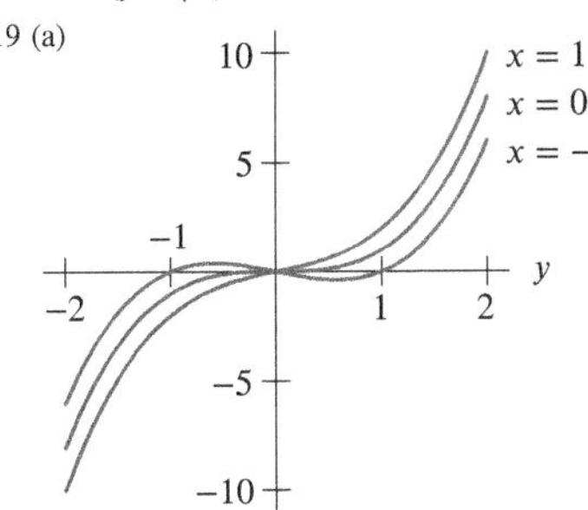

(b)

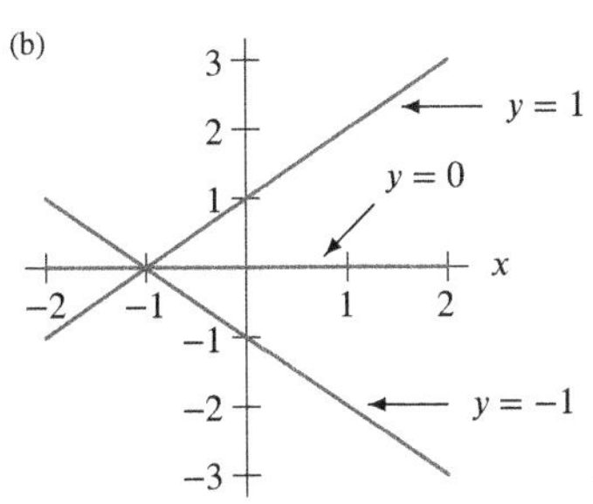

21 (a)

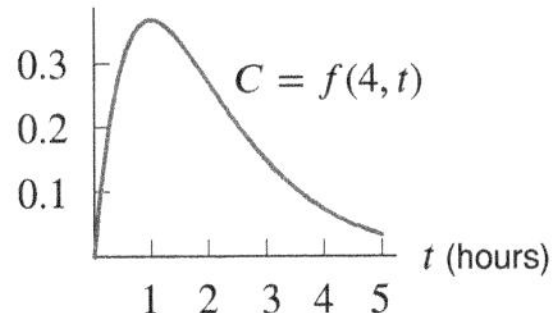

(b)

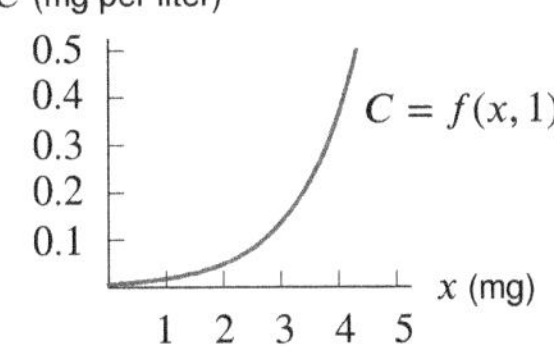

23 $f(2, 10) = 100$ joules

25 $f(2, 12) = 731.9$ millibars.

27 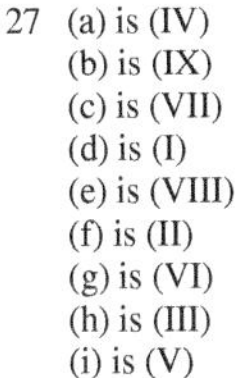
(a) is (IV)
(b) is (IX)
(c) is (VII)
(d) is (I)
(e) is (VIII)
(f) is (II)
(g) is (VI)
(h) is (III)
(i) is (V)

29 (a) (i)

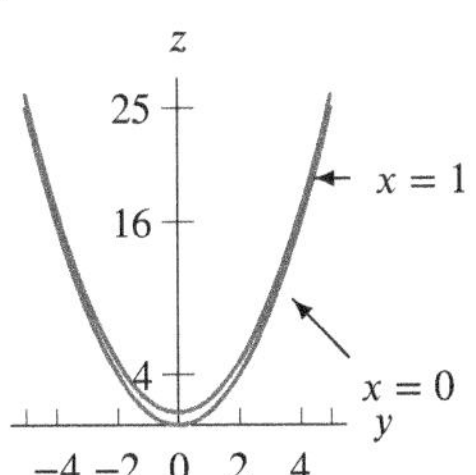

(ii)

(b) (i)

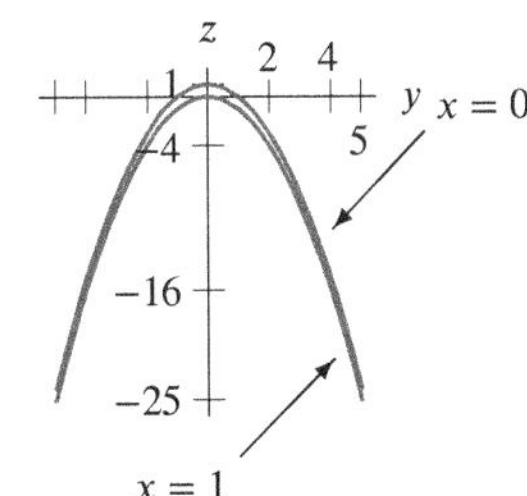

(ii)

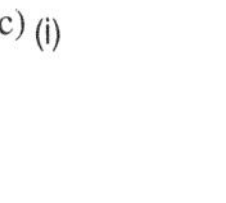
(c) (i)

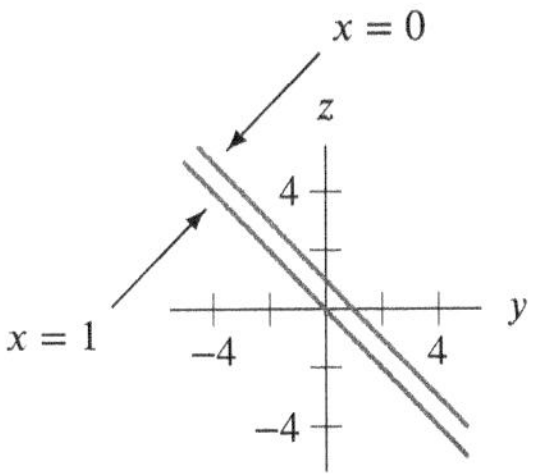

(ii)

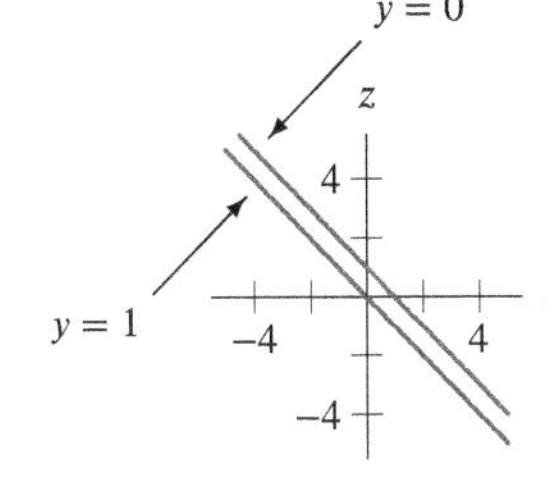

(d) (i)

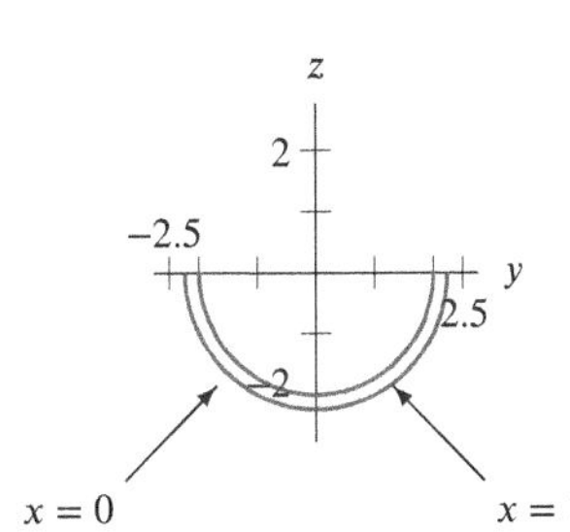

(ii)

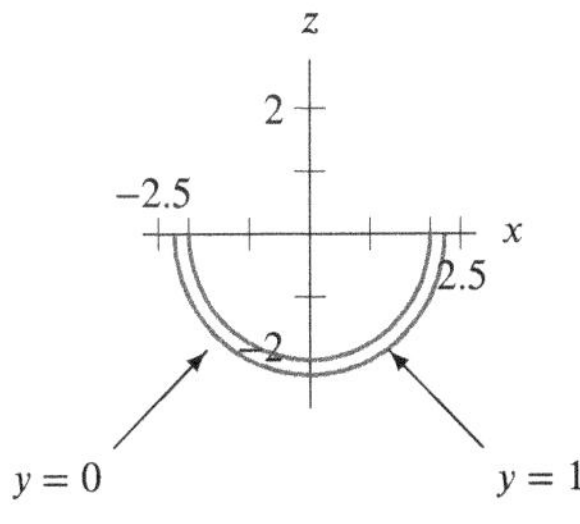

(e) (i)

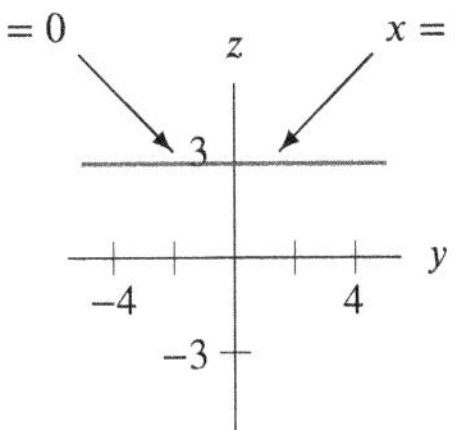

(ii)

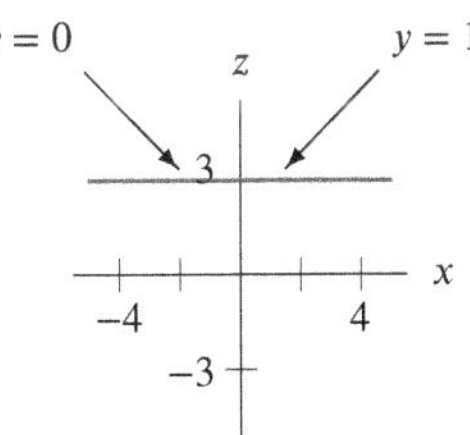

31 (I) cross-sections with x fixed, (II) cross-sections with y fixed

33 Cross-sections graph I:

(a)

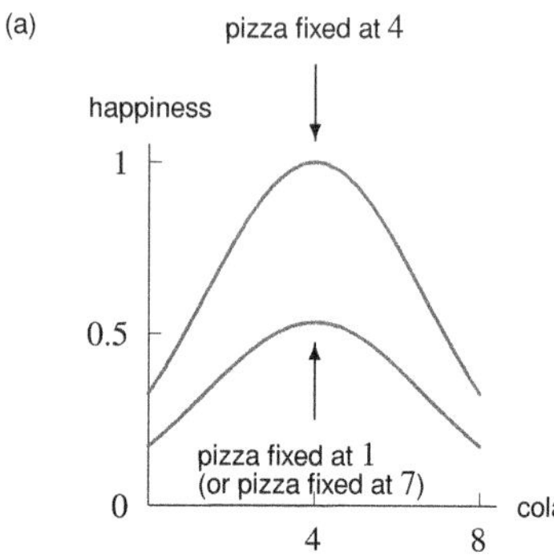

(b)

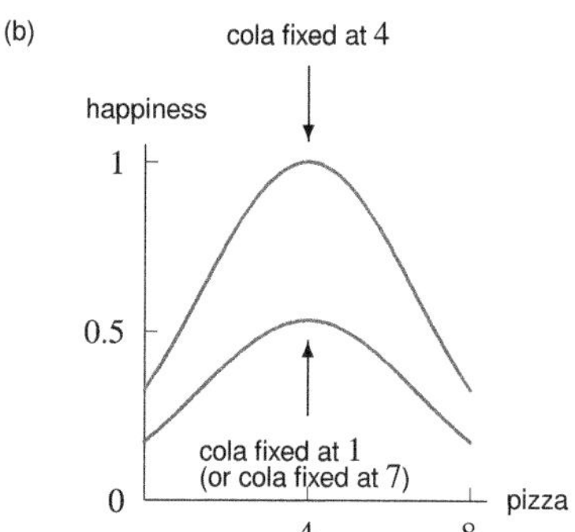

Cross-sections graph II:

(a)

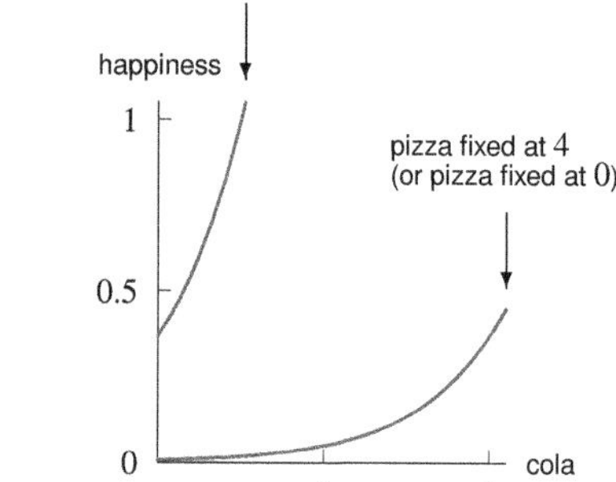

(b)

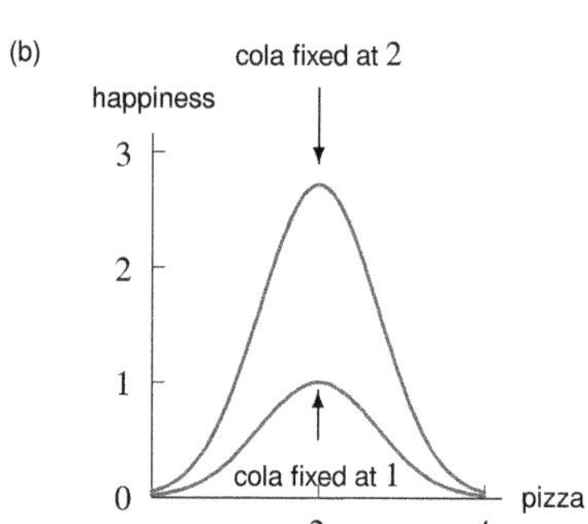

Cross-sections graph III:

(a)

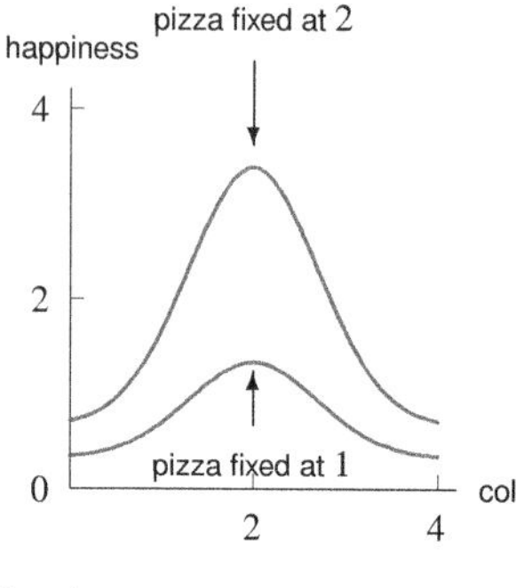

(b)

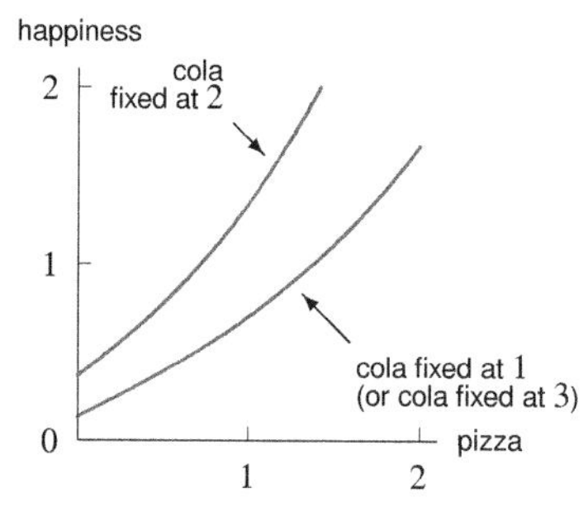

Cross-sections graph IV:

(a)

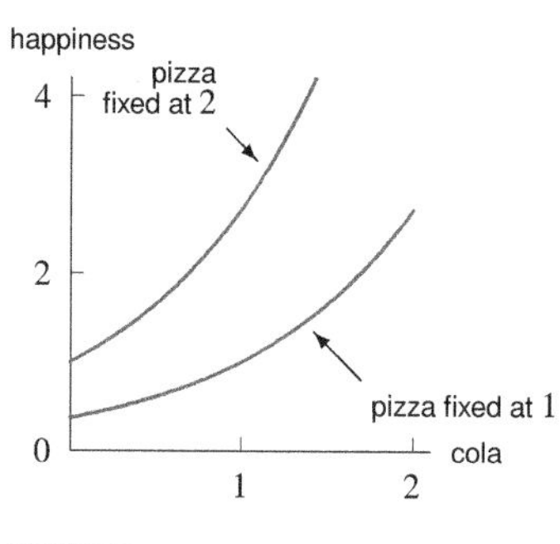

(b)

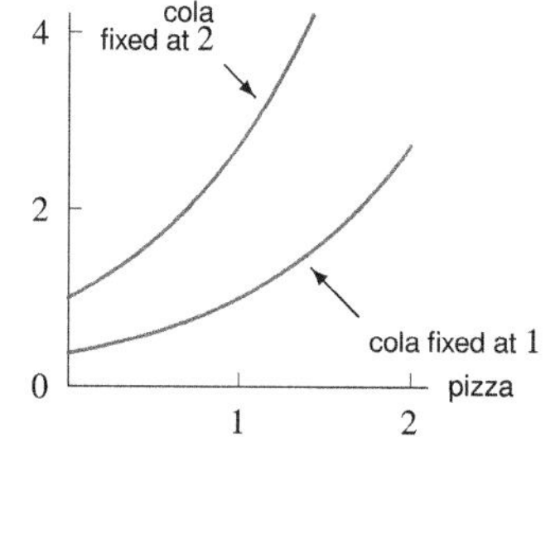

35

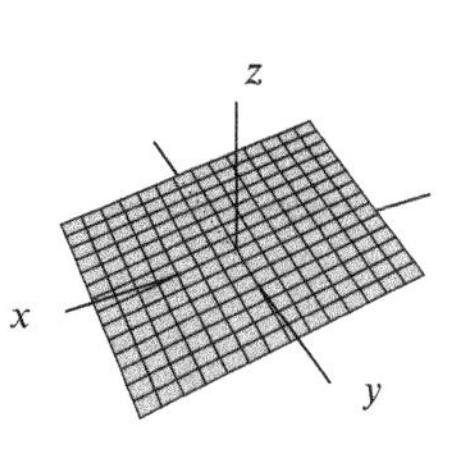

37

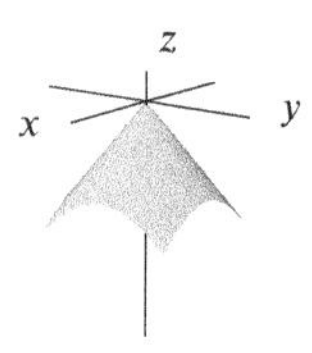

39 (a) $y = 0$
(b) $x = 0$
(c) $z = 1$

41 (a)

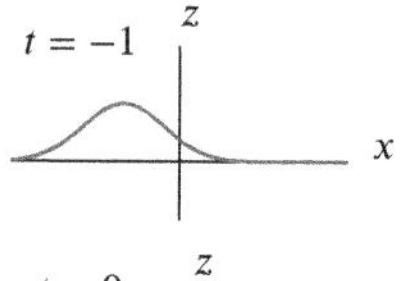

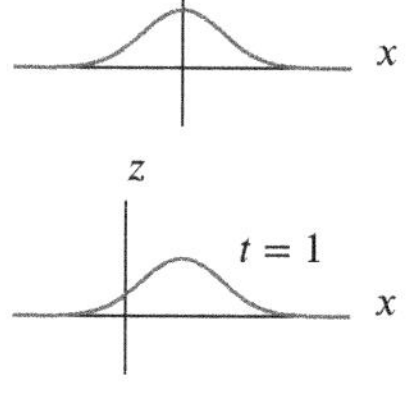

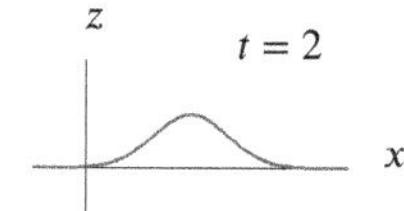

(b) Increasing x
(c)

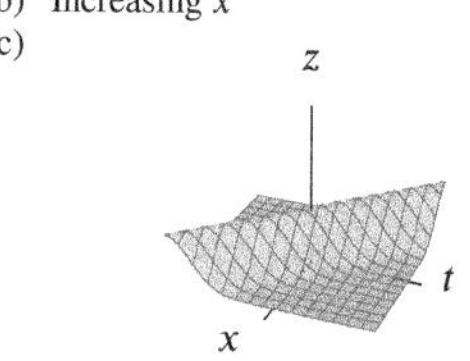

43 Graph is surface in 3-space

45 $f(x, y) = x^2 + y^2 + 2$

47 $f(x, y) = 1 - x^2 - y^2$

49 True

51 False

53 False

55 True

57 True

59 False

61 True

63 (a)

$t = 0$ $t = \pi/4$
1
−1
x
π

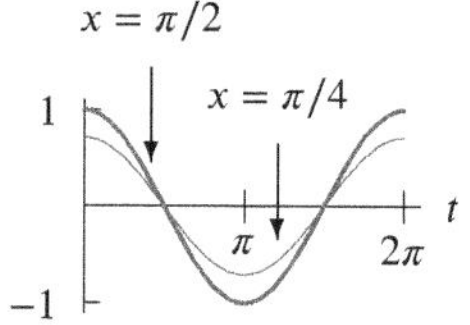

(b) $f = 0$; ends of string don't move

Section 12.3

1

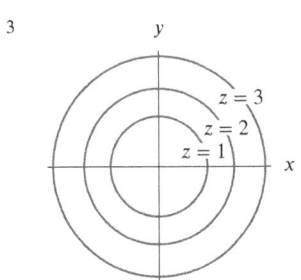

3

5 Contours evenly spaced parallel lines

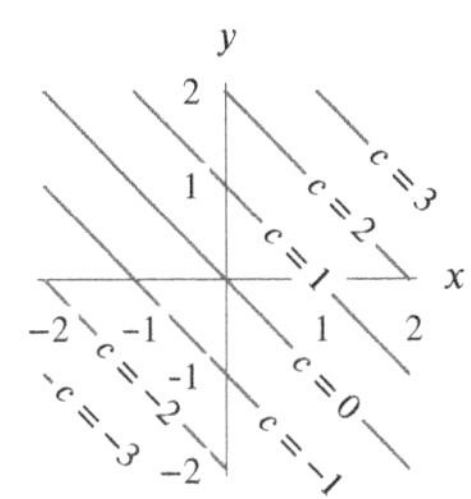

7

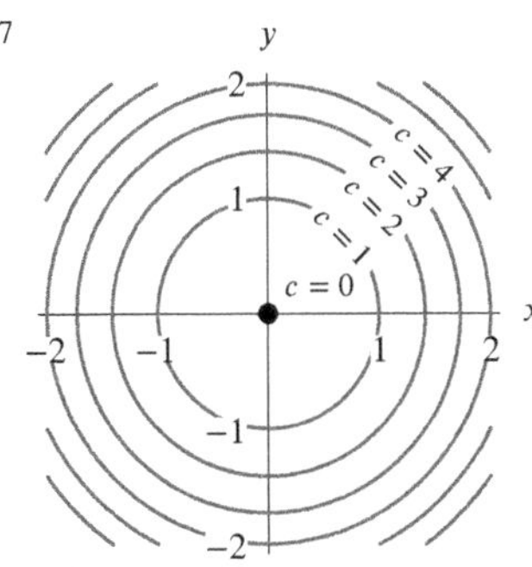

9

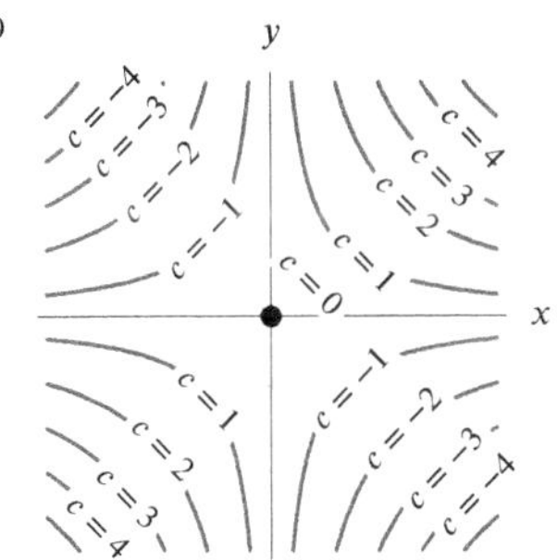

11

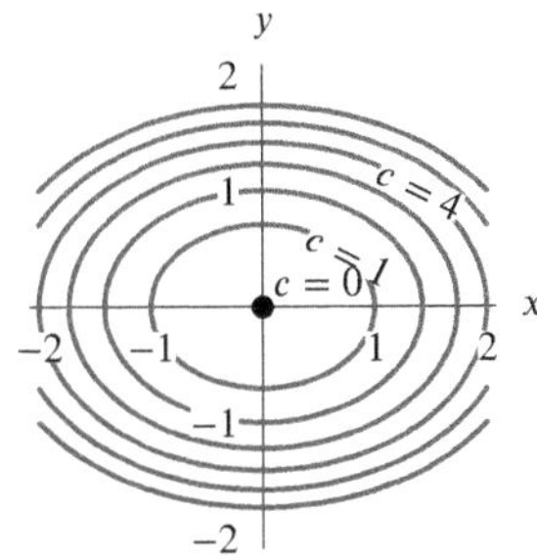

15 (a)

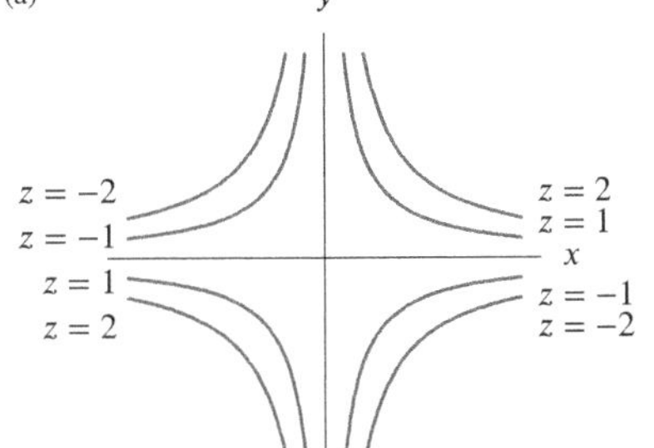

(b)

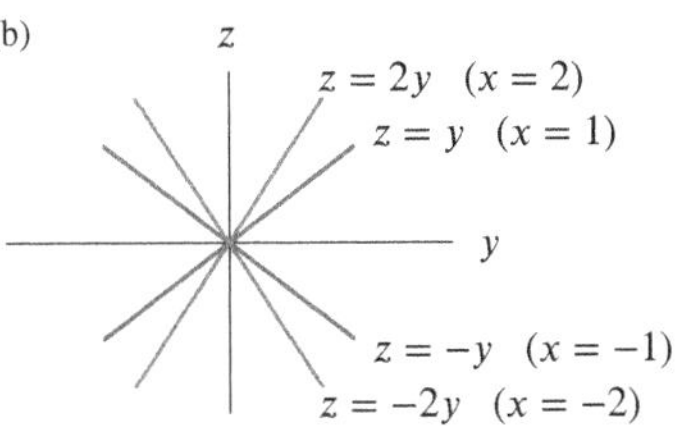

(c)

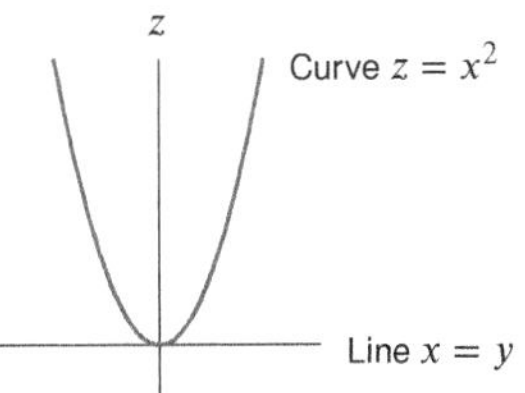

17 (I), (IV), (VI)

19

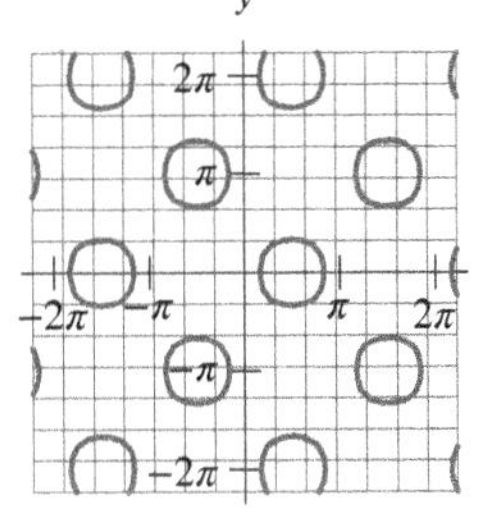

21 (a) is (II)
(b) is (I)
(c) is (III)

23 (a) −2 Grapes/Cherry
(b) No change in happiness when replacing 2 grapes with one cherry

25 Underweight: below 18.5, Normal: 18.5-25

27 (a) About 0°F
(b) About −16°F
(c) About 23 mph
(d) About 25°F

29 Answers in °C:

(a)

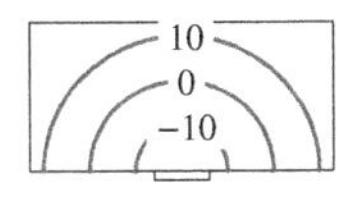

(b)

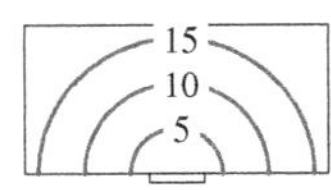

(c)

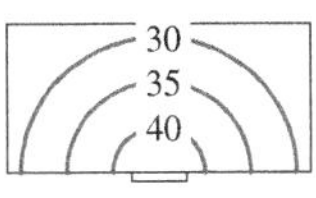

(d)

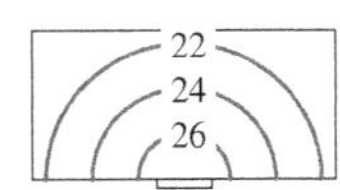

31

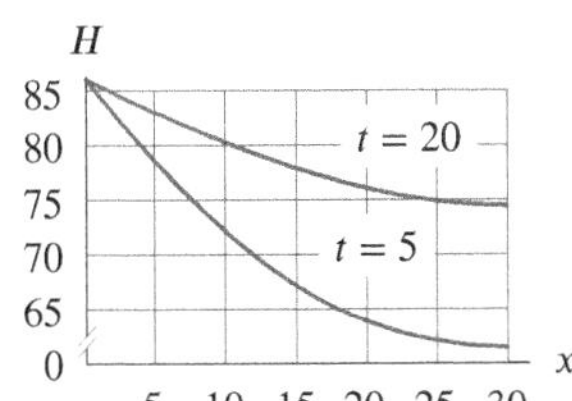

33 (a) 2
(b) −2
(c) −1

35 (a) $\sqrt{5}$
(b) 0
(c) 0

37 (a) About \$137
(b) About \$250

(a) About \$122
(b) About \$350

39 Other answers possible

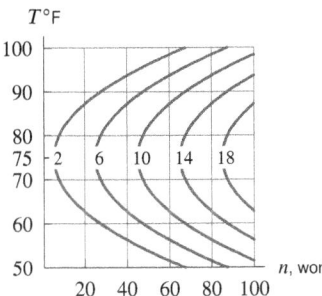

41 (a) II
(b) IV

(c) III
(d) I

43 (a) $\pi = 3q_1 + 12q_2 - 4$ (thousands)
(b)

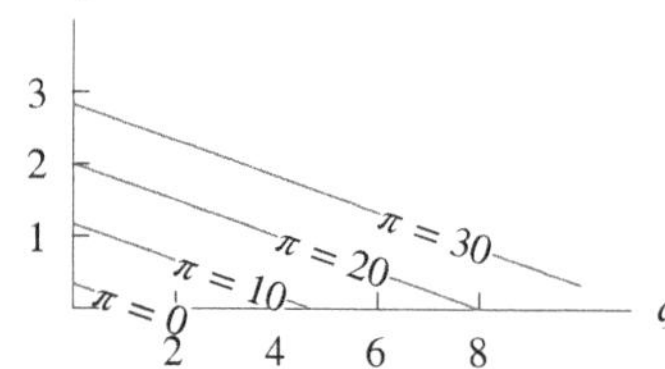

45 (a) (II) (E)
(b) (I) (D)
(c) (III) (G)

47 (a) (I) g
(II) f
(b) $0.2 < \alpha < 0.8$

49 (a)

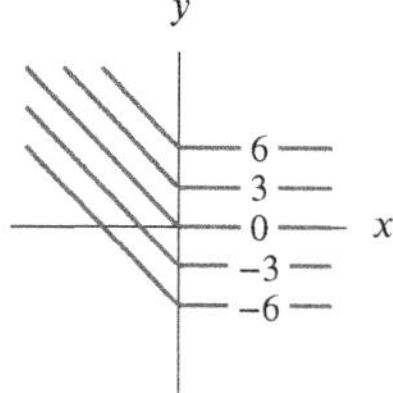

(b)

(c)

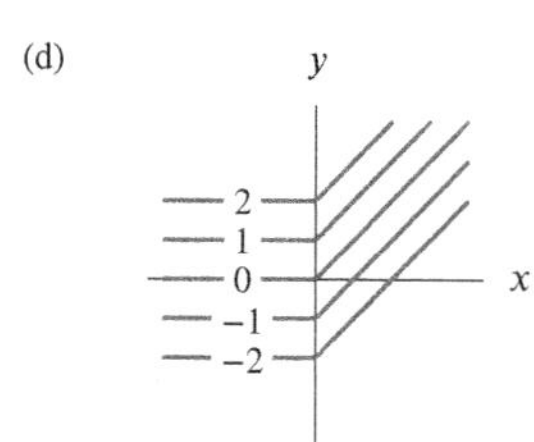

(d)

51 $y = 2x$, $y = (1/3)x$

55 (a) $P(d, v) = kd^2v^3$
(b) $1/\sqrt[3]{4}$
(c)

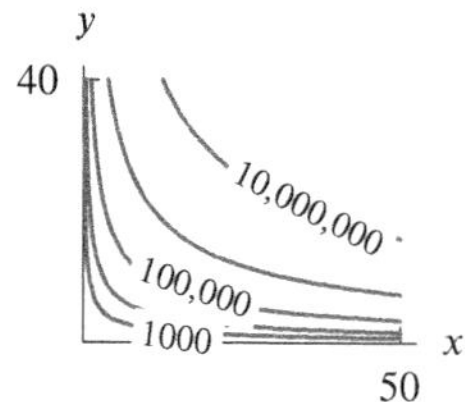

57 Spacing of the contours of f and g are different

59 $f(x, y) = y - x^2$

61 Might be true

63 Not true

65 True

67 True

69 False

71 False

Section 12.4

1 −1.0

3 Not a linear function

5 Linear function

7 $z = \frac{4}{3}x - \frac{1}{2}y$

9 $z = -2y + 2$

11 $\Delta z = 0.4$; $z = 2.4$

13 Linear

15 Basic subscription cost \$8
Premium subscription cost \$12

17 (a) Linear
(b) Linear
(c) Not linear

19 180 lb person at 8 mph
120 lb person at 10 mph

21 $g(x, y) = 3x + y$

23 Not linear

25 $f(x, y) = 2x - 0.5y + 1$

27 Could be linear; $z = -4 + x + 4y$

29 Could be linear; $z = 5 + (3/2)x$

31 Could be linear; $z = -5 + (3/2)x - 2y$

33 Could be linear; $z = 2x + y + 3$

35

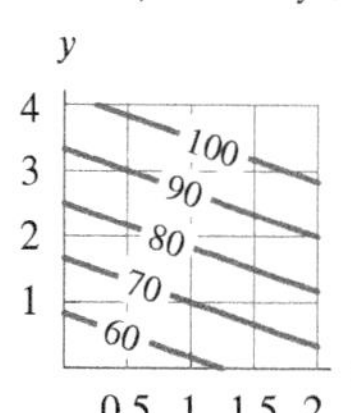

Other answers possible

37

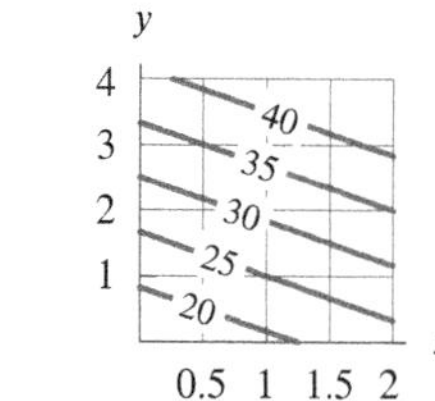

Other answers are possible

39

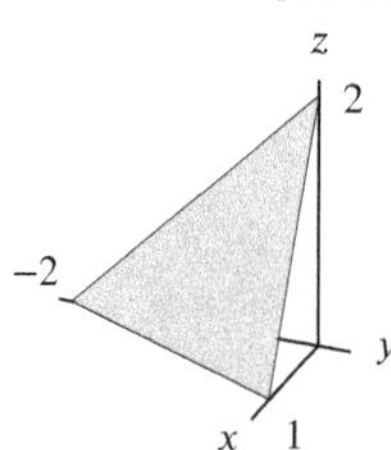

41

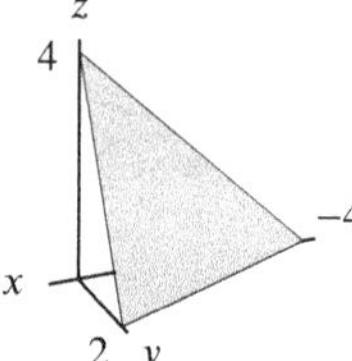

43 8

45 (a) Impossible
(b) Impossible
(c) 20

47 $f(x, y) = xy$ has linear cross-sections

49 $z = -2x + y$

51 False

53 True

55 True

57 False

59 False

61 False

65 (a) $7/\sqrt{29}$
(b) $-5/\sqrt{104}$

Section 12.5

1 (a) I
(b) II

3 $f(x, y) = \frac{1}{3}(5 - x - 2y)$

5 $f(x, y) = (1 - x^2 - y)^2$

7 Elliptical and hyperbolic paraboloid, plane

9 Hyperboloid of two sheets

11 Ellipsoid

13 Yes, $f(x, y) = (2x + 3y - 10)/5$

15 No

17 $f(x, y) = 2x - (y/2) - 3$
$g(x, y, z) = 4x - y - 2z = 6$

19 $f(x, y) = -\sqrt{2(1 - x^2 - y^2)}$
$g(x, y, z) = x^2 + y^2 + z^2/2 = 1$

21 (a) 1596 kcal/day
(b) 1284 kcal/day
(c) Plane; weight, height, age combinations of woman whose BMR is 2000 kcal/day
(d) Lose weight

23 (a) $P(1 + 0.01r)^t = 2653.3$
(b) $(P, r, t) = (1628.9, 5, 10)$; other answers possible

25 (a) $r^2h\rho = 120$
(b) $(r, h, \rho) = (2, 5, 6)$; other answers possible

27 $f(x, y) = 3\sqrt{1 - x^2 - y^2/4}$;
$g(x, y) = -3\sqrt{1 - x^2 - y^2/4}$

29 $f(x, y, z)$, $r(x, y, z)$, $m(x, y, z)$

31 (a) Graph of f is the graph of $y^2 + z^2 = 1, z \geq 0$
(b) $\sqrt{1 - y^2} - z = 0$

33 Elliptical cylinder along y-axis

35 Parallel planes

37 Surface of rotation

39 Spheres

43

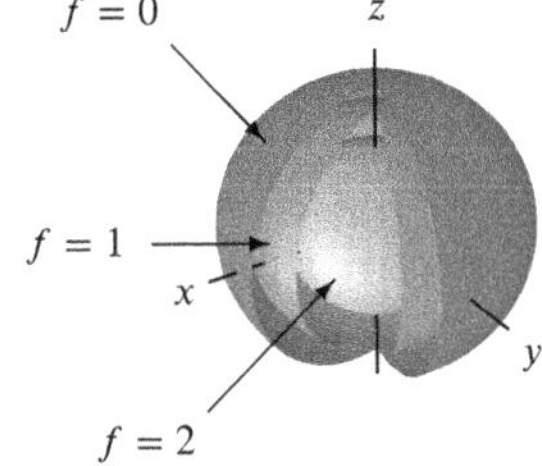

45 Vertical shifts

47 Graph of $f(x, y, z)$ needs 4 dimensions

49 Level surfaces cylinders

51 $f(x, y, z) = x^2 + z^2$

53 $f(x, y, z) = x^2 + y^2 - z$

55 False

57 False

59 False

61 True

63 True

Section 12.6

1 Not continuous

3 Continuous

5 Not continuous

7 1

9 0

11 1

13 2

15 Does not exist

23 No

25 $c = 1$

27 (c) No

29 For quotient, need $g(a, b) \neq 0$

31 $f(x, y) = (x^2 + 2y^2)/(x^2 + y^2)$

33 $f(x, y) = 1/((x - 2)^2 + y^2)$

35 False

37 True

39 True

Chapter 12 Review

1 A, B, C

3

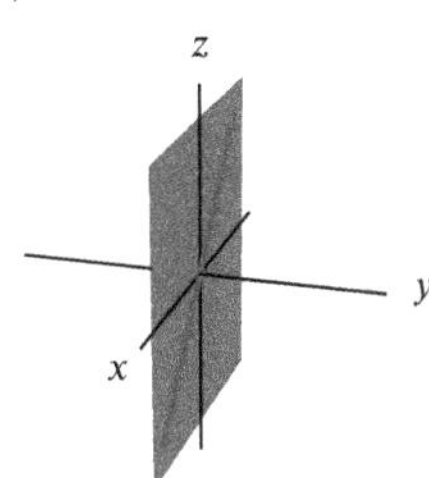

5 Not a function

7 IV

9 (a) is (I)
(b) is (IV)
(c) is (II)
(d) is (III)

11 $y = k$

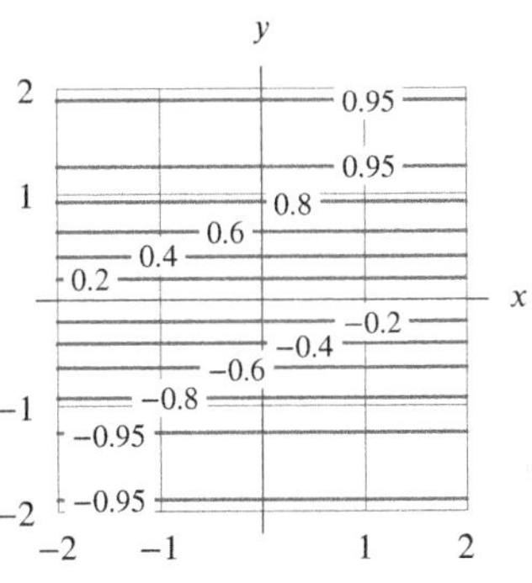

13 $2x^2 + y^2 = k$

15 $(x - 1)^2 + (y - 2)^2 + (z - 3)^2 = 25$

17 $(-2, 3, -6)$; 7

19 Linear

21 $f(x, y, z) = x^2 + y^2$

23 $z = f(x, y) = 4 - 2x - 4y/3$
$g(x, y, z) = (x/2) + (y/3) + (z/4) = 1$

25 $z = f(x, y) = -\sqrt{4 - (x - 3)^2 - y^2}$
$g(x, y, z) = (x - 3)^2 + y^2 + z^2 = 4$

27 Cylinder

29 $g(x, y) = 200x + 100y$

31 Doubles production

37 (a) No
(b) No
(c) No
(e) 0.5 of GPA

39 (a)

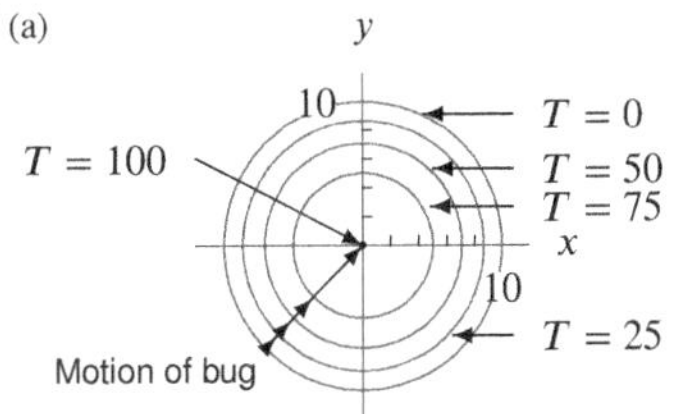

(b) Toward origin

41 (a)

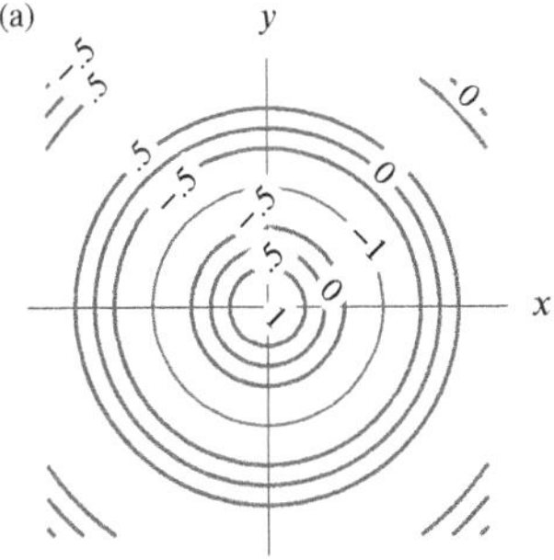

(b)

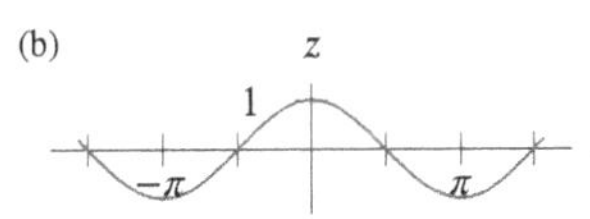

(c)

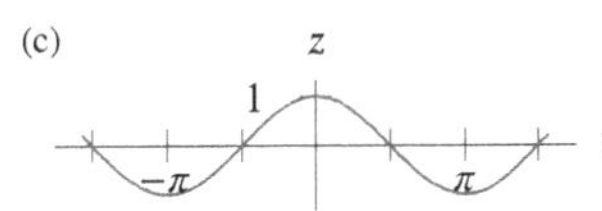

45 (a) For $t = 0$:

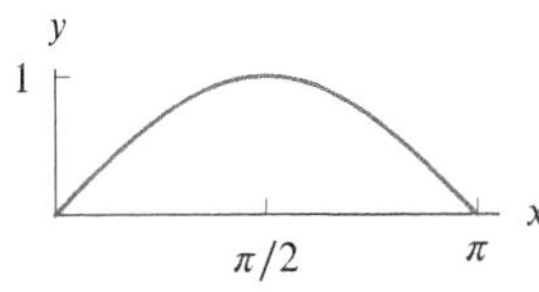

For $t = \pi/4$:

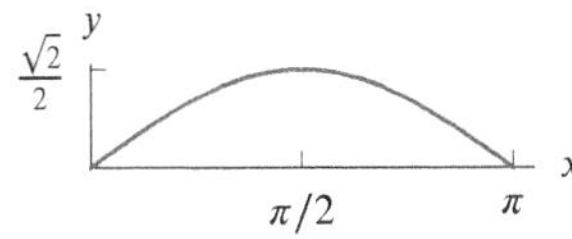

For $t = \pi/2$:

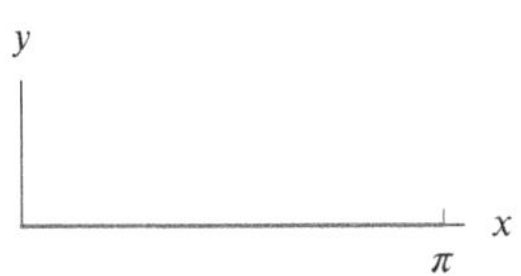

For $t = 3\pi/4$:

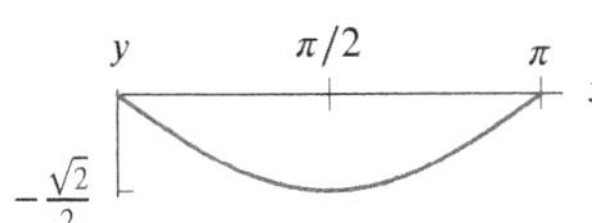

For $t = \pi$:

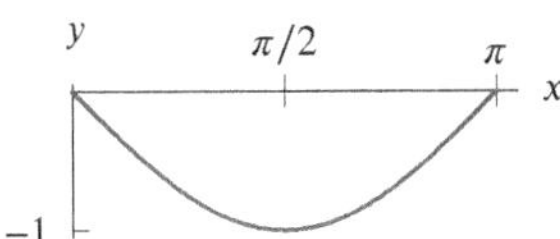

47 (a) $(1, \sqrt{3}, 0)$
(b) $(1, 1/\sqrt{3}, 2\sqrt{2}/\sqrt{3})$
(c) Tetrahedron

49 (a) $f(1, 1, 3) \approx 26$, $f(1, 2, 1) \approx 12$
(b) $f(1, 1, 3)$ exact, $f(1, 2, 1)$ not exact
(c) $5x^2 + 2yz + 3zx^3 + 6 \cdot 2^{x-y}$
(d) $f(1, 1, 3) = 26$, $f(1, 2, 1) = 15$

Section 13.1

1 $\vec{a} = \vec{i} + 3\vec{j}$
$\vec{b} = 3\vec{i} + 2\vec{j}$
$\vec{v} = -2\vec{i} - 2\vec{j}$
$\vec{w} = -\vec{i} + 2\vec{j}$

3 $-3\vec{i} - 4\vec{j}$

5 $\vec{a} = \vec{b} = \vec{c} = 3\vec{k}$
$\vec{d} = 2\vec{i} + 3\vec{k}$
$\vec{e} = \vec{j}$
$\vec{f} = -2\vec{i}$

7 $\vec{i} + 3\vec{j}$

9 $-4.5\vec{i} + 8\vec{j} + 0.5\vec{k}$

11 $-3\vec{i} - 12\vec{j} + 3\vec{k}$

13 $0.9\vec{i} + 0.2\vec{j} - 0.7\vec{k}$

15 $\sqrt{6}$

17 $\sqrt{11}$

19 5.6

21 $-6\vec{i} + 20\vec{j} + 13\vec{k}$

23 $21\vec{j}$

25 $2\sqrt{73}$

27 $0.6\vec{i} - 0.8\vec{k}$

29 $-\vec{i}/2 + \vec{j}/4 + \sqrt{11}\vec{k}/4$

31 $0, -10$

33 (a) $(3/5)\vec{i} + (4/5)\vec{j}$
(b) $6\vec{i} + 8\vec{j}$

35 (a) $\sqrt{2}\vec{i} + \sqrt{2}\vec{j}$
(b) $(\sqrt{3}/2)\vec{i} + \vec{k}/2$

37 $\vec{p} = -\frac{4\sqrt{5}}{5}\vec{i} - \frac{2\sqrt{5}}{5}\vec{j}$

39 (a) $\vec{a} = \vec{i}$ and $\vec{b} = -\vec{i}$; other answers possible
(b) $\vec{a} = (1/\sqrt{2})\vec{i} - (1/\sqrt{2})\vec{j}$ and $\vec{b} = (1/\sqrt{2})\vec{j} + (1/\sqrt{2})\vec{k}$; other answers possible
(c) $\vec{a} = \vec{i}$ and $\vec{b} = \vec{i}$; other answers possible
(d) Not possible

41 (a) $t = 1$
(b) No t values
(c) Any t values

43 $(\vec{i} + \vec{j})/\sqrt{2}$, $(\vec{i} - \vec{j})/\sqrt{2}$, $(-\vec{i} + \vec{j})/\sqrt{2}$, $(-\vec{i} - \vec{j})/\sqrt{2}$

45 (a) $(a, 0, 0)$
(b) $\left(b/\sqrt{b^2+c^2}\right)\vec{j} + \left(c/\sqrt{b^2+c^2}\right)\vec{k}$

47 $\|\vec{u} + \vec{v}\|$ could be less than 1

49 Longer diagonal if angle between $\vec{u}$ and $\vec{v}$ more than 90°

51 $\vec{v} = \vec{j} + \sqrt{3}\vec{k}$

53 $\vec{u} = \vec{i}$, $\vec{v} = 3\vec{i} + 3\vec{j}$

55 False

57 False

59 False

61 False

63 False

Section 13.2

1 Scalar

3 Scalar

5 Vector

7 $-37.59\vec{i}, -13.68\vec{j}$

9 $21\vec{i} + 35\vec{j}$

11 (a) 8.64 km/hr
(b) 0.093 radian or about 5° off course

13 (a) $17.93\vec{i} - 7.07\vec{j}$
(b) 19.27 km/hr
(c) 21.52° south of east

15 48.3° east of north
744 km/hr

17 4.87° north of east
540.63 km/hr

19 38.7° south of east

21 $-98.76\vec{i} + 18.94\vec{j} + 2998.31\vec{k}$
2998.31 newtons directly up

23 $0.4v\vec{i} + 0.7v\vec{j}$

25 $0.1\vec{i} + 0.08\vec{j} + 0.1625\vec{k}$ or $(0.1, 0.08, 0.1625)$

37 $42.265\vec{i} + 42.265\vec{j} - 5.229\vec{k}$ mph

39 Not if $|\vec{j}$-component$| \geq 2$

41 Let $\vec{v} = c\vec{v}$ for any $c > 0$

43 No

45 No

47 No

49 3.4° north of east

Section 13.3

1 14

3 29

5 $7.5\sqrt{2}$

7 -14

9 14

11 -2

13 $28\vec{j} + 14\vec{k}$

15 185

17 $\vec{i} + 3\vec{j} + 2\vec{k}$ (multiples of)

19 $3\vec{i} + 4\vec{j} - \vec{k}$ (multiples of)

21 $3x - y + 4z = 6$

23 $x - y + z = 3$

25 $2x + 4y - 3z = 5$

27 $2x - 3y + 5z = -17$

29 $2\pi/3$ radians (120°)

31 $\pi/6$ radians (30°)

33 (a) - (II)
(b) - (I)
(c) - (III)

35 (a) $(2/\sqrt{13})\vec{i} + (3/\sqrt{13})\vec{j}$
(b) Multiples of $3\vec{i} - 2\vec{j}$

37 (a) $(21/5, 0, 0)$
(b) $(0, -21, 0)$; $(0, 0, 3)$ (for example)
(c) $\vec{n} = 5\vec{i} - \vec{j} + 7\vec{k}$ (for example)
(d) $21\vec{j} + 3\vec{k}$ (for example)

39 Possible answers:
(a) $2\vec{i} + 3\vec{j} - \vec{k}$
(b) $3\vec{i} - 2\vec{j}$

41 (a) is (I); (b) is (III), (IV); (c) is (II), (III); (d) is (II)

43 $\vec{v}_1, \vec{v}_4, \vec{v}_8$ all parallel
$\vec{v}_3, \vec{v}_5, \vec{v}_7$ all parallel
$\vec{v}_1, \vec{v}_4, \vec{v}_8$ perpendicular to $\vec{v}_3, \vec{v}_5, \vec{v}_7$
$\vec{v}_2$ and $\vec{v}_9$ perpendicular

45 $\vec{u} \perp \vec{v}$ for $t = 2$ or -1.
No values of t make $\vec{u}$ parallel to $\vec{v}$

47 2

49 $\sqrt{20}$

51 $\vec{a} = -\frac{8}{21}\vec{d} + (\frac{79}{21}\vec{i} + \frac{10}{21}\vec{j} - \frac{118}{21}\vec{k})$

53 Lengths: $\sqrt{34}$, $\sqrt{29}$, $\sqrt{13}$
Angles: 37.235°, 64.654°, 78.111°

55 39 joules; 28.765 foot-pounds

57 120 foot-pounds; 162.698 joules

59 (a) $\vec{F}_{\text{parallel}} = -0.168\vec{i} - 0.224\vec{j}$
(b) $\vec{F}_{\text{perp}} = 0.368\vec{i} - 0.276\vec{j}$
(c) $W = -1.4$

61 (a) $\vec{F}_{\text{parallel}} = \vec{0}$
(b) $\vec{F}_{\text{perp}} = \vec{F}$
(c) $W = 0$

63 (a) $\vec{F}_{\text{parallel}} = \vec{F}$
(b) $\vec{F}_{\text{perp}} = \vec{0}$
(c) $W = -50$

65 (a) $\vec{F}_{\text{parallel}} = 3.846\vec{i} - 0.769\vec{j}$
(b) $\vec{F}_{\text{perp}} = -3.846\vec{i} - 19.231\vec{j}$
(c) $W = 20$

67 (a) $\vec{F}_{\text{parallel}} = \vec{0}$
(b) $\vec{F}_{\text{perp}} = \vec{F}$
(c) $W = 0$

69 70.529°

71 $\vec{w}_4$ increases most
$\vec{w}_3$ decreases most

73 (a) $(20, 20, -10)$
(b) 108.167

75 \$710 revenue

83 Can't take dot product of a scalar and a vector

85 Normal vector is $2\vec{i} + 3\vec{j} - \vec{k}$

87 $f(x, y) = (-1/3)x + (-2/3)y$

89 True

91 False

93 True

95 False

97 True

99 True

Section 13.4

1 $-\vec{i}$

3 $-\vec{i} + \vec{j} + \vec{k}$

5 $\vec{i} + 3\vec{j} + 7\vec{k}$

7 $7\vec{i} + \vec{j} + 4\vec{k}$

9 $-2\vec{k}$

11 $\vec{i} - \vec{j}$

13 $\vec{v} \times \vec{w} = -6\vec{i} + 7\vec{j} + 8\vec{k}$
$\vec{w} \times \vec{v} = 6\vec{i} - 7\vec{j} - 8\vec{k}$
$\vec{v} \times \vec{w} = -(\vec{w} \times \vec{v})$

15 $x - y - z = -3$

17 4

19 0

21 $-\vec{i} - \vec{j} - \vec{k}$

23 $\vec{0}$

25 $x + 2y + 2z = 0$

27 $3x - y - 2z = 0$

29 $4\vec{i} + 26\vec{j} + 14\vec{k}$

31 $4(x - 4) + 26(y - 5) + 14(z - 6) = 0$

33 (a) $\vec{u}$ and $-\vec{u}$ where
$\vec{u} = \frac{12}{13}\vec{i} - \frac{4}{13}\vec{j} - \frac{3}{13}\vec{k}$
(b) $\theta \approx 49.76°$
(c) $13/2$
(d) $13/\sqrt{29}$

35 (a) $(4, 0, 0)$
(b) $(0, 2, 0)$
(c) $(0, 0, 4)$
(d) 9.798

37 (a) 0.6
(b) 0.540

39 (a) 1.625
(b) 1.019

41 (a) Increases force
(b) Ball moves down and to the left

43 (b) $(-y, x)$

45 $\vec{i} - 3\vec{j} - 5\vec{k}$

47 (a) $4\vec{k}$
(b) $3\vec{j}$
(c) $2\vec{i}$

49 $\theta = \pi/4$ or $3\pi/4$

51 $0 \le \theta < \pi/4$ or $3\pi/4 < \theta \le \pi$

55 $4\pi\vec{i}$

57 (a) $((u_2v_3 - u_3v_2)^2 + (u_3v_1 - u_1v_3)^2 + (u_1v_2 - u_2v_1)^2)^{1/2}$
(b) $|u_1v_2 - u_2v_1|$
(c) $m = (u_2v_3 - u_3v_2)/(u_2v_1 - u_1v_2)$, $n = (u_3v_1 - u_1v_3)/(u_2v_1 - u_1v_2)$

59 Parallel, not perpendicular

61 $\vec{v} = (8\vec{i} - 6\vec{j})/5$

63 False

65 True

67 True

69 True

71 False

Chapter 13 Review

1 Scalar; −1

3 −1

5 $\vec{a} = -2\vec{j}$, $\vec{b} = 3\vec{i}$, $\vec{c} = \vec{i} + \vec{j}$, $\vec{d} = 2\vec{j}$, $\vec{e} = \vec{i} - 2\vec{j}$, $\vec{f} = -3\vec{i} - \vec{j}$

7 $5\vec{i} + 30\vec{j}$

9 $3\sqrt{2}$

11 $3\vec{i} + 7\vec{j} - 4\vec{k}$

13 −3

15 $\vec{0}$

17 0

19 $\vec{0}$

21 $-5\vec{i} + 3\vec{j} + \vec{k}$ (multiples of)

23 (a) 4
(b) $-4\vec{i} - 11\vec{j} - 17\vec{k}$
(c) $3.64\vec{i} + 2.43\vec{j} - 2.43\vec{k}$
(d) 79.0°
(e) 0.784.
(f) $2\vec{i} - 2\vec{j} + \vec{k}$ (many answers possible)
(g) $-4\vec{i} - 11\vec{j} - 17\vec{k}$.

25 $\pm(-\vec{i} + \vec{j} - 2\vec{k})/\sqrt{6}$

27 $\vec{n} = 4\vec{i} + 6\vec{k}$

29 $-3\vec{i} + 4\vec{j}$

31 $\vec{u}$ and $\vec{w}$; $\vec{v}$ and $\vec{q}$.

33 $\vec{F}_{\text{parallel}} = \vec{F}$
$\vec{F}_{\text{perp}} = \vec{0}$
$W = -10$

35 $\vec{F}_{\text{parallel}} = -(6/5)\vec{i} + (8/5)\vec{j}$
$\vec{F}_{\text{perp}} = (16/5)\vec{i} + (12/5)\vec{j}$
$W = -10$

37 $\vec{F}_{\text{parallel}} = 2\vec{j}$
$\vec{F}_{\text{perp}} = 5\vec{i}$
$W = 6$

39 (a) True
(b) False
(c) False
(d) True
(e) True
(f) False

41 548.6 km/hr

43

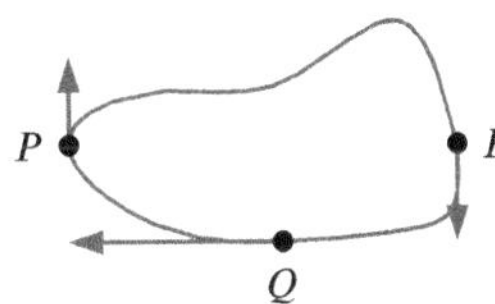

45 Parallel:
$3\vec{i} + \sqrt{3}\vec{j}$ and $\sqrt{3}\vec{i} + \vec{j}$
Perpendicular:
$\sqrt{3}\vec{i} + \vec{j}$ and $\vec{i} - \sqrt{3}\vec{j}$
$3\vec{i} + \sqrt{3}\vec{j}$ and $\vec{i} - \sqrt{3}\vec{j}$

47 2, 8

49 $2x - 3y + 7z = 19$

51 $\sqrt{6}/2$

53 (a) 1.5
(b) $y = 1$

55 $9x - 16y + 12z = 5$
0.23

57 $7.0710\vec{i} + 2.5882\vec{j} + 6.580\vec{k}$

59 $0, \vec{0}$

61 (a) $\vec{c} = -\vec{i} - 4\vec{j} + 3\vec{k}$ or $\vec{c} = \vec{i} + 4\vec{j} - 3\vec{k}$
(b) $\vec{i} + 4\vec{j} - 3\vec{k}$

63 (a) $(tv - sw - ty + wy + sz - vz)\vec{i} + (-tu + rw + tx - wx - rz + uz)\vec{j} + (su - rv - sx + vx + ry - uy)\vec{k}$
(b) $((s(u - x) + vx - uy + r(-v + y))/c)(a\vec{i} + b\vec{j} + c\vec{k})$

Section 14.1

1 $f_x(3, 2) \approx -2/5$; $f_y(3, 2) \approx 3/5$

3 −0.0493, −0.3660
−0.0501, −0.3629

5 $\partial P/\partial t$:
dollars/month
Rate of change in payments with time
negative
$\partial P/\partial r$:
dollars/percentage point
Rate of change in payments with interest rate
positive

7 (a) Payment \$376.59/mo at 1% for 24 mos
(b) 4.7¢ extra/mo for \$1 increase
(c) Approx \$44.83 increase for 1% interest increase

9 (a) Negative
(b) Positive

11 (a) $f(A) = 15$
(b) Zero
(c) Negative

13 (a) $f(A) = 88$
(b) Negative
(c) Negative

15 (a) $f(A) = 40$
(b) Negative
(c) Positive

17 $f_x > 0$, $f_y < 0$

19 $f_x < 0$, $f_y > 0$

21 Positive, Negative, 10, 2, −4

23 (i)(c); (ii)(a)

25 (a) Both negative
(b) Both negative

27 $f_T(5, 20) \approx 1.2$°F/°F

29 −1.5 and −1.22

31 (a)

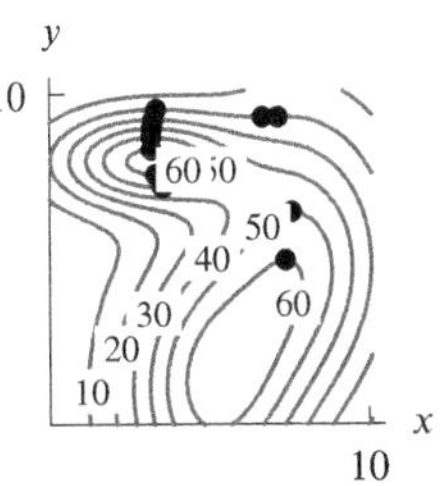

(b)

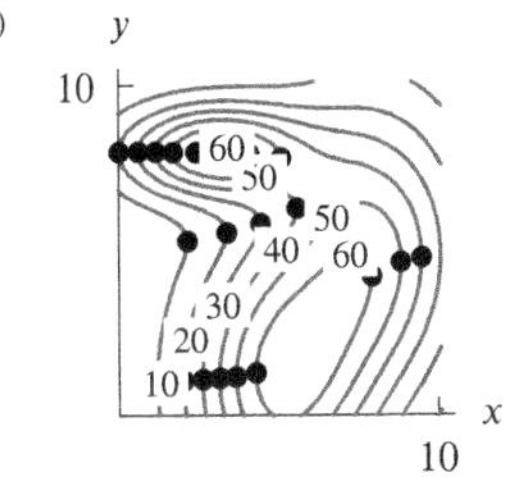

33 (a) Negative
(b) Positive

35 (a) 2.5, 0.02
(b) 3.33, 0.02
(c) 3.33, 0.02

37 −2.5

39

		w (gm/m^3)		
		0.1	0.2	0.3
T (°C)	10	1300	900	1200
	20	800	800	900
	30	800	700	800

41 (a)

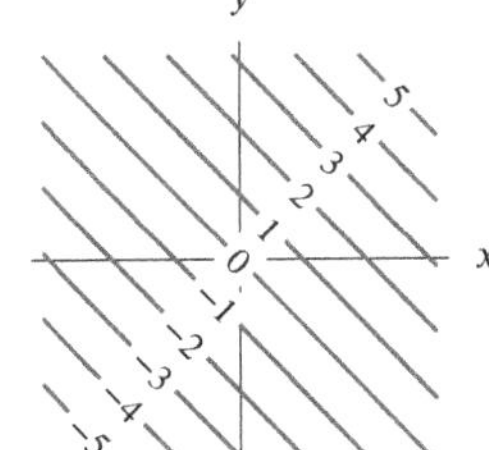

(b)

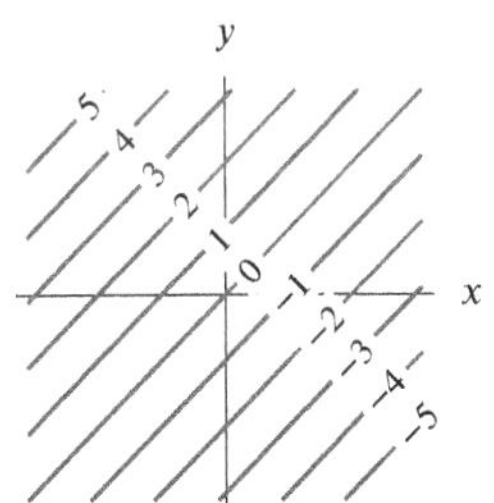

(c)

(d)

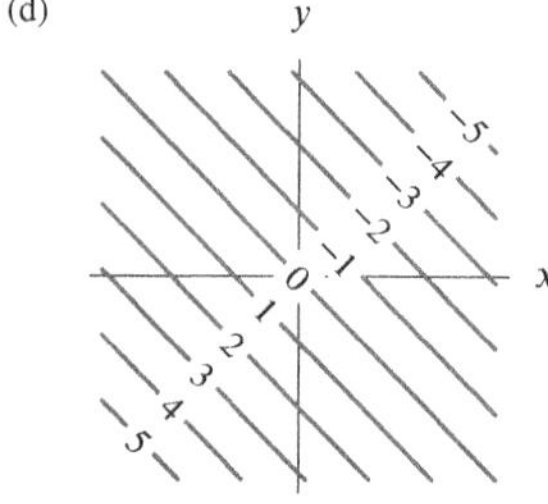

43 There are many possibilities.

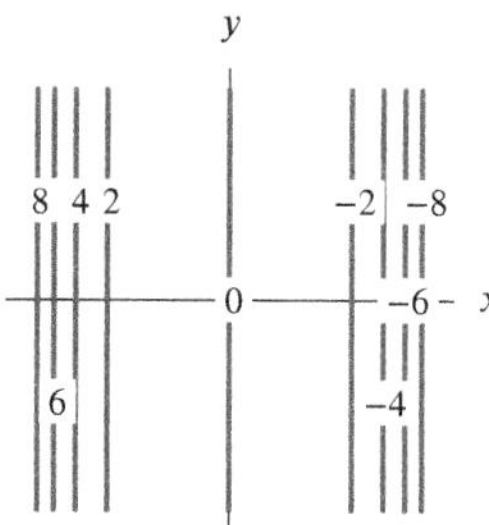

45 There are many possibilities.

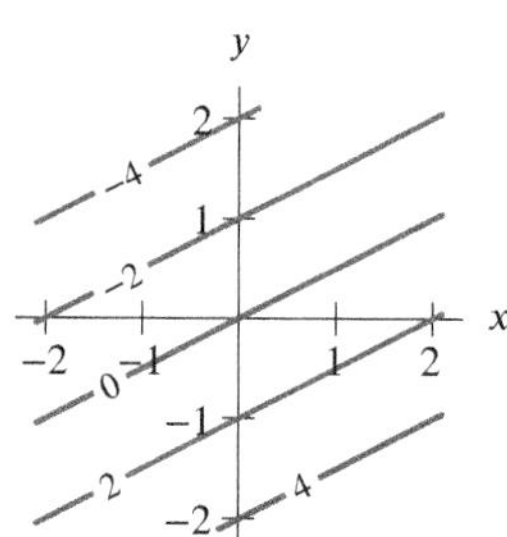

47 $f_y = 0$

49 $f(x, y) = 4x - y$

51 False

53 False

55 True

57 True

59 True

Section 14.2

1 (a) 7.01
(b) 7

3 $f_x(1, 2) = 15$,
$f_y(1, 2) = -5$

5 $\partial z/\partial x = \frac{14x+7}{(x^2+x-y)^{-6}}$
$\partial z/\partial y = -7(x^2 + x - y)^6$

7 $f_x = 0.6/x$
$f_y = 0.4/y$

9 $2xy + 10x^4 y$

11 $V_r = \frac{2}{3}\pi rh$

13 $e^{\sqrt{xy}}(1 + \sqrt{xy}/2)$

15 g

17 $(a + b)/2$

19 $2B/u_0$

21 $2mv/r$

23 $(15x^2y - 3y^2)\cos(5x^3y - 3xy^2)$

25 y

27 $z_x = 7x^6 + yx^{y-1}$
$z_y = 2^y \ln 2 + x^y \ln x$

29 Gm_1/r^2

31 $-e^{-x^2/a^2}(a^2 - 2x^2)/a^4$

33 $\cos(\pi\theta\phi) \cdot \pi\phi + \dfrac{1}{\theta^2 + \phi} \cdot 2\theta$

35 $f_a = e^a \sin(a + b) + e^a \cos(a + b)$

37 $\partial V/\partial r = \frac{8}{3}\pi rh$,
$\partial V/\partial h = \frac{4}{3}\pi r^2$

39 $-(x - \mu)e^{-(x-\mu)^2/(2\sigma^2)}/(\sqrt{2\pi}\sigma^3)$

41 (a) $f_x(1, 1) = 2$; $f_y(1, 1) = 2$
(b) (II)

43 (a) $f_x(1, 1) = 2$; $f_y(1, 1) = 2e$
(b) (I)

45 (a) $f_w(2, 2) \approx 2.78$
$f_z(2, 2) \approx 4.01$
(b) $f_w(2, 2) \approx 2.773$
$f_z(2, 2) = 4$

47 (a) Pre^{rt}
(b) e^{rt}

49 1.277 m^2, 0.005 m^2/kg, 0.006 m^2/cm

51 $h_x(2, 5) \approx -0.38$ cm/meter
$h_t(2, 5) \approx 0.76$ cm/second

55 No such points exist

57 $(0, 1)$, $(0, -1)$, $(-2, 1)$ and $(-2, -1)$

59 $\partial f/\partial x$ or $\partial f/\partial y$?

61 $f(x, y) = 2x + 3y + x^2$

63 $f(x, y) = y^2 + 1$

65 True

67 False

69 False

71 False

Section 14.3

1 $z = ex$

3 $z = 6y - 9$

5 $z = -4 + 2x + 4y$

7 $z = -36x - 24y + 148$

9 $df = y\cos(xy)\,dx + x\cos(xy)\,dy$

11 $dz = -e^{-x}\cos(y)dx - e^{-x}\sin(y)dy$

13 $dg = 4\,dx$

15 $dP \approx 2.395\,dK + 0.008\,dL$

17 -5

19 0.99

21 95

23 12.005

25 (a) Dollars/Square foot
(b) Larger plots at same distance \$3/ft^2 more
(c) Dollars/Foot
(d) Farther from beach but same area \$2/ft less
(e) 998 ft^2

27 (b) $f(x, y) \approx$
$0.3345 - 0.33(x - 1) - 0.15(y - 2)$
(c) $f(x, y) \approx 0.3345 -$
$0.3345(x - 1) - 0.1531(y - 2)$

29 (a) $f_x(1, 2) = 3$; $f_y(1, 2) = 2$
(b) 2
(c) 2.1

31 $df = -3dx + 2dy$ at $(2, -4)$

33 376

35 $df = \frac{1}{3}dx + 2dy$
$f(1.04, 1.98) \approx 2.973$

37 136.09°C

39 $P(r, L) \approx$
$80 + 2.5(r - 8) + 0.02(L - 4000)$
$P(r, L) \approx$
$120 + 3.33(r - 8) + 0.02(L - 6000)$
$P(r, L) \approx$
$160 + 3.33(r - 13) + 0.02(L - 7000)$

43 (a) $nRT/(V - nb) - n^2a/V^2$
(b) $\Delta P \approx (nR/(V_0 - nb))\Delta T + (2n^2a/V_0^3 - nRT_0/((V_0 - nb)^2))\Delta V$

45 (a) $d\rho = -\beta\rho\,dT$
(b) 0.00015, $\beta \approx 0.0005$

47 $-43200\Delta t$
Slow if $\Delta t > 0$; fast if $\Delta t < 0$

49 (a) $4x\,dx = 2y\,dy + 6z\,dz$
(b) $dz = \frac{2}{3}dx - \frac{1}{2}dy$
(c) $z = 2 + \frac{2}{3}(x - 2) - \frac{1}{2}(y - 3)$

51 (a) $e^y\,dx + xe^y\,dy + 2z\,dz = -\sin(x-1)\,dx + \frac{z}{\sqrt{z^2+3}}\,dz$
(b) $dz = -\frac{2}{3}dx - \frac{2}{3}dy$
(c) $z = 1 - \frac{2}{3}(x - 1) - \frac{2}{3}(y - 0)$

53 $z = f(3, 4) + f_x(3, 4)(x - 3) + f_y(3, 4)(y - 4)$

55 Equation not linear

57 sphere of radius 3 centered at the origin

59 True

61 False

63 True

65 False

Section 14.4

1 $(\frac{15}{2}x^4)\vec{i} - (\frac{24}{7}y^5)\vec{j}$

3 $2m\vec{i} + 2n\vec{j}$

5 $\left(\frac{5\alpha}{\sqrt{5\alpha^2+\beta}}\right)\vec{i} + \left(\frac{1}{2\sqrt{5\alpha^2+\beta}}\right)\vec{j}$

7 $\nabla z = e^y\vec{i} + e^y(1 + x + y)\vec{j}$

9 $\sin\theta\vec{i} + r\cos\theta\vec{j}$

11 $\nabla z = \dfrac{1}{y}\cos\left(\dfrac{x}{y}\right)\vec{i} - \dfrac{x}{y^2}\cos\left(\dfrac{x}{y}\right)\vec{j}$

13 $\left(\dfrac{-12\beta}{(2\alpha - 3\beta)^2}\right)\vec{i} + \left(\dfrac{12\alpha}{(2\alpha - 3\beta)^2}\right)\vec{j}$

15 $60\vec{i} + 85\vec{j}$

17 $10\pi\vec{i} + 4\pi\vec{j}$

19 $(\pi/2)^{1/2}\,\vec{i}$

21 $\frac{1}{100}(2\vec{i} - 6\vec{j})$

23 $\vec{i}$

25 $\vec{i} + \vec{j}$

27 $\vec{i} - \vec{j}$

29 Negative

31 Negative

33 Approximately zero

35 $-\vec{i}$

37 $\vec{i}$

39 $-\vec{i} + \vec{j}$

41 $\vec{i} + \vec{j}$

43 6.325

45 $-46/5$

47 $22/5$

49 $84/5$

51 $(2x + 3e^y)dx + 3xe^y dy$

53 $(x + 1)ye^x\vec{i} + xe^x\vec{j}$

55 50.2

57 (a) Should be number
(b) $11/5$

59 -2

61 $\vec{i} + 2\vec{j}$ or any multiple

63 0.316

65 1

67 100

69 (a) $-\sqrt{2}/2$
(b) $\sqrt{3} + 1/2$

71 (a) $2/\sqrt{13}$
(b) $1/\sqrt{17}$
(c) $\vec{i} + \frac{1}{2}\vec{j}$

73 (a) $5/\sqrt{2}$
(b) 510

75 (a) $-16\vec{i} + 12\vec{j}$
(b) $16\vec{i} - 12\vec{j}$
(c) $12\vec{i} + 16\vec{j}$; answers may vary

77 1.7; closer estimate is 1.35

79 2.5; better estimate is 1.8

81 -0.9; better estimate is -1.8

83 Fourth quadrant

85 (a) Negative
(b) Positive
(c) Positive
(d) Negative

87 $f_{\vec{u}}(P) < f_{\vec{w}}(P) < f_{\vec{v}}(P)$

89 $f(P) \approx 6$, $f(Q) \approx -24$

91 $3\vec{i} + 2\vec{j}$; $3(x - 2) + 2(y - 3) = 0$

93 $-5\vec{i}$; $x = 2$

95 $5/\sqrt{2}$

97 (a) ellipses centered at $(0, 0)$
(b) decreasing at 49.9°C per meter
(c) $-\vec{i} - 2\vec{j}$. Other answers possible

99 (a) $\sqrt{13}$ meters ascended/horizontal meter
(b) 3.54 meters ascended/horizontal meter
(c) $\vec{u} = 3\vec{i} + 2\vec{j}$; $\vec{u} = -3\vec{i} - 2\vec{j}$

101 grad $f(0, 0)$ is vector, not scalar

103 $-\vec{i}$

105 False

107 False

109 True

111 False

113 True

115 True

117 (a) Perpendicular to contour of f at P
(b) Maximum directional derivative of f at P
(c) Directional derivative $f_{\vec{u}}(P)$

119 19.612

121 356.5

123 (a) -3.268
(b) -4.919

125 Yes

127 Yes

129 (a)

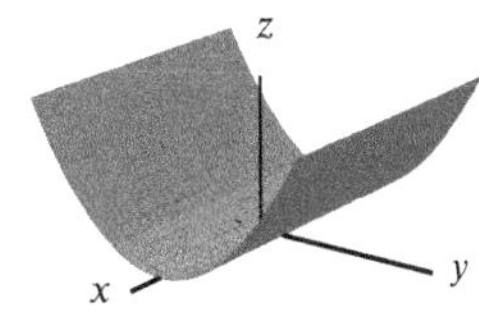

(b)

y
4
$z = 16$
$z = 9$
$z = 4$
$z = 1$
$z = 0$
x
$z = 1$
$z = 4$
-2

(c) $\vec{j}$

131 (a) Circles centered at P
(b) away from P
(c) 1

133 $(3\sqrt{5} - 2\sqrt{2})\vec{i} + (4\sqrt{2} - 3\sqrt{5})\vec{j}$

135 $4\sqrt{2}$,
$6\vec{i} + 2\vec{j}$

139 (a) $\sqrt{m^2 + n^2}$
(b) $(C_2 - C_1)/\sqrt{m^2 + n^2}$

141 (a) $a\vec{i} + 2b\vec{j}$
(b) $\sqrt{(2 + a)/(2 - a)}$
(c) $\sqrt{(2 - a)/(2 + a)}$

143 True

145 True

147 False

Section 14.5

1 $2x\vec{i}$

3 $e^x e^y e^z(\vec{i} + \vec{j} + \vec{k})$

5 $\dfrac{-2xyz^2}{(1 + x^2)^2}\vec{i} + \dfrac{z^2}{1 + x^2}\vec{j} + \dfrac{2yz}{1 + x^2}\vec{k}$

7 $(x\vec{i} + y\vec{j} + z\vec{k})/\sqrt{x^2 + y^2 + z^2}$

9 $y\vec{i} + x\vec{j} + e^z \cos(e^z)\vec{k}$

11 $e^p\vec{i} + (1/q)\vec{j} + 2re^{r^2}\vec{k}$

13 $\vec{0}$

15 $6\vec{i} + 4\vec{j} - 4\vec{k}$

17 $-\pi\vec{i} - \pi\vec{k}$

19 $9/\sqrt{3}$

21 $-1/\sqrt{2}$

23 $-\sqrt{77/2}$

25 $-2\vec{i} - 2\vec{j} + 4\vec{k}$;
$-2(x + 1) - 2(y - 1) + 4(z - 2) = 0$

27 $2\vec{j} - 4\vec{k}$; $2(y - 1) - 4(z - 2) = 0$

29 $-2\vec{i} + \vec{k}$; $-2(x + 1) + (z - 2) = 0$

31 $6(x - 1) + 3(y - 2) + 2(z - 1) = 0$

33 $2x + 3y + 2z = 17$

35 $z = 2x + y + 3$

37 $x + 4y + 10z = 18$

39 $10/3$

41 $2z + 3x + 2y = 17$

43 $x + 3y + 7z = -9$
$\vec{i} + 3\vec{j} + 7\vec{k}$

45 grad $g(-1, -1)$ lies directly under path of steepest descent

47 (a) $(x - 2) + 4(y - 3) - 6(z - 1) = 0$
(b) $z = 1 + (1/6)(x - 2) + (2/3)(y - 3)$

49 $3x + 10y - 5z + 19 = 0$

51 $16/\sqrt{14}$

53 (a) Spheres centered at the origin
(b) $2x\sin(x^2 + y^2 + z^2)\vec{i} + 2y\sin(x^2 + y^2 + z^2)\vec{j} + 2z\sin(x^2 + y^2 + z^2)\vec{k}$
(c) 0, 180°

55 (a) Circle: $(y + 1)^2 + (z - 3)^2 = 10$
(b) Yes
(c) Multiples of $-10\vec{i} + 4\vec{j} - 12\vec{k}$

57 (a) $(-3\vec{i} + 6\vec{j} + 12\vec{k})/\sqrt{21}$
(b) $(8.345, 2.309, 4.619)$

59 Any multiple of $2\vec{i} + 2\vec{j} + \vec{k}$

61 $(-1/6, 1/3, -1/12)$

63 (a) $6.33\vec{i} + 0.76\vec{j}$
(b) -34.69

65 (a) 23
(b) -9.2
(c) $-16\vec{i} + 6\vec{j}$
(d) $16x - 6y - z = 23$

67 (a) Parallel planes: $2x - 3y + z = T - 10$
(b) $f_z(0, 0, 0) = 1$, temp increases 1°C per unit in z-direction
(c) $2\vec{i} - 3\vec{j} + \vec{k}$
(d) Yes; 27°C

69 (a) $-25/\sqrt{21}$
(b) $-8\vec{i} + 7\vec{j} + 4\vec{k}$
(c) $\sqrt{129}$

71 1.131 atm/sec

73 (a) is (V); (b) is (IV); (c) is (V)

75 $f_x(0, 0, 0)x + f_y(0, 0, 0)y + f_z(0, 0, 0)z = 0$

77 $f(x, y, z) = 2x + 3y + 4z + 100$

79 False

81 False

Section 14.6

1 $\dfrac{dz}{dt} = e^{-t}\sin(t)(2\cos t - \sin t)$

3 $2\cos\left(\frac{2t}{1-t^2}\right)\frac{1+t^2}{(1-t^2)^2}$

5 $2e^{1-t^2}(1 - 2t^2)$

7 $\dfrac{\partial z}{\partial u} = \dfrac{1}{vu}\cos\left(\dfrac{\ln u}{v}\right)$
$\dfrac{\partial z}{\partial v} = -\dfrac{\ln u}{v^2}\cos\left(\dfrac{\ln u}{v}\right)$

9 $\dfrac{\partial z}{\partial u} = \dfrac{e^v}{u}$
$\dfrac{\partial z}{\partial v} = e^v \ln u$

11 $\dfrac{\partial z}{\partial u} = 2ue^{(u^2-v^2)}(1 + u^2 + v^2)$
$\dfrac{\partial z}{\partial v} = 2ve^{(u^2-v^2)}(1 - u^2 - v^2)$

13 $\dfrac{\partial z}{\partial u} =$
$(e^{-v\cos u} - v(\cos u)e^{-u\sin v})\sin v$
$-(-u(\sin v)e^{-v\cos u} + e^{-u\sin v})v\sin u$

$\frac{\partial z}{\partial v} =$
$(e^{-v\cos u} - v(\cos u)e^{-u\sin v})u\cos v$
$+ (-u(\sin v)e^{-v\cos u} + e^{-u\sin v})\cos u$

15 $\frac{\partial z}{\partial u} = \frac{-2uv^2}{u^4+v^4}$
$\frac{\partial z}{\partial v} = \frac{2vu^2}{u^4+v^4}$

17 $-2\rho\cos 2\phi, 0$

19 (a) $\partial f/\partial t$
(b) $(\partial f/\partial x)(dx/dt)$
(c) $(\partial f/\partial y)(dy/dt)$

21 -5 pascal/hour

23 -0.6

25 (a) $1/\sqrt{10} = 0.316$ °F/mile
(b) $2.5/\sqrt{10} = 0.791$°F/hr
(c) 2.5 °F/hr

27 Three

29 $\frac{dw}{dt} = \frac{\partial w}{\partial x}\frac{dx}{dt} + \frac{\partial w}{\partial y}\frac{dy}{dt} + \frac{\partial w}{\partial z}\frac{dz}{dt}$

31 (a) $F_u(x,3)$
(b) $F_v(3,x)$
(c) $F_u(x,x) + F_v(x,x)$
(d) $F_u(5x,x^2)(5) + F_v(5x,x^2)(2x)$

33 $b\cdot e + d\cdot p$

35 $b\cdot e + d\cdot p$

37 (a) $\frac{\partial z}{\partial r} = \cos\theta\frac{\partial z}{\partial x} + \sin\theta\frac{\partial z}{\partial y}$
$\frac{\partial z}{\partial \theta} = r(\cos\theta\frac{\partial z}{\partial y} - \sin\theta\frac{\partial z}{\partial x})$
(b) $\frac{\partial z}{\partial y} = \sin\theta\frac{\partial z}{\partial r} + \frac{\cos\theta}{r}\frac{\partial z}{\partial\theta}$
$\frac{\partial z}{\partial x} = \cos\theta\frac{\partial z}{\partial r} - \frac{\sin\theta}{r}\frac{\partial z}{\partial\theta}$

39 $(\frac{\partial U_3}{\partial P})_V$

41 $(\frac{\partial U}{\partial T})_V = 7/2$
$(\frac{\partial U}{\partial V})_T = 11/4$

45 $dz/dt = f_x(g(t),h(t))g'(t) + f_y(g(t),h(t))h'(t)$

47 $dz/dt|_{t=0} = f_x(2,3)g'(0) + f_y(2,3)h'(0)$

49 $f(x,y) = 4x + 2y$

51 $w = uv$, $u = 2s^2 + t$ and $v = e^{st}$, many other answers are possible

53 (c)

57 $\int_0^b F_u(x,y)\,dy$

Section 14.7

1 $f_{xx} = 2$
$f_{yy} = 2$
$f_{yx} = 2$
$f_{xy} = 2$

3 $f_{xx} = 6y$
$f_{xy} = 6x + 15y^2$
$f_{yx} = 6x + 15y^2$
$f_{yy} = 30xy$

5 $f_{xx} = 0$
$f_{yx} = e^y = f_{xy}$
$f_{yy} = e^y(x+2+y)$

7 $f_{xx} = -\left(\sin\left(\frac{x}{y}\right)\right)\left(\frac{1}{y^2}\right)$
$f_{xy} = -\left(\sin\left(\frac{x}{y}\right)\right)\left(\frac{-x}{y^2}\right)\left(\frac{1}{y}\right)$
$+ \left(\cos\left(\frac{x}{y}\right)\right)\left(\frac{-1}{y^2}\right) = f_{yx}$
$f_{yy} = -\left(\sin\left(\frac{x}{y}\right)\right)\left(\frac{-x}{y^2}\right)^2$
$+ \left(\cos\left(\frac{x}{y}\right)\right)\left(\frac{2x}{y^3}\right)$

9 $f_{xx} = 30xy^2 + 18$
$f_{xy} = 30x^2y - 21y^2$
$f_{yx} = 30x^2y - 21y^2$
$f_{yy} = 10x^3 - 42xy$

11 $f_{xx} = -12\sin 2x\cos 5y$
$f_{xy} = -30\cos 2x\sin 5y$
$f_{yx} = -30\cos 2x\sin 5y$
$f_{yy} = -75\sin 2x\cos 5y$

13 $Q(x,y) = 1 + 2x - 2y + x^2 - 2xy + y^2$

15 $Q(x,y) = 1 + x + x^2/2 - y^2/2$

17 $Q(x,y) = 1 - x^2/2 - 3xy - (9/2)y^2$

19 $Q(x,y) = -y + x^2 - y^2/2$

21 $1 + x - y/2 - x^2/2 + xy/2 - y^2/8$

23 (a) Negative
(b) Zero
(c) Negative
(d) Zero
(e) Zero

25 (a) Positive
(b) Zero
(c) Positive
(d) Zero
(e) Zero

27 (a) Zero
(b) Negative
(c) Zero
(d) Negative
(e) Zero

29 (a) Positive
(b) Positive
(c) Zero
(d) Zero
(e) Zero

31 (a) Positive
(b) Negative
(c) Negative
(d) Negative
(e) Positive

33 -8

35 3

37 Not possible

39 6

41 $L(x,y) = y$
$Q(x,y) = y + 2(x-1)y$
$L(0.9, 0.2) = 0.2$
$Q(0.9, 0.2) = 0.16$
$f(0.9, 0.2) = 0.162$

47 $a = -b^2$

49 Positive, negative

51 (a) $z_{yx} = 4y$
(b) $z_{xyx} = 0$
(c) $z_{xyy} = 4$

53 $d = e = f = 0$

55 $d = 0, e > 0, f < 0$

57 (a)

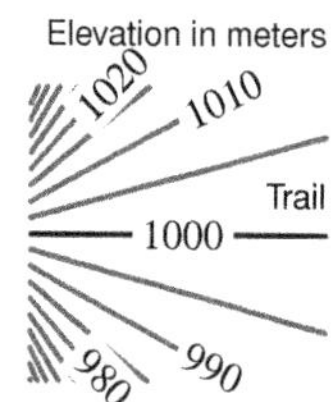

(b) $\partial h/\partial x = 0$, $\partial h/\partial y > 0$, $(\partial^2 h)/(\partial x\partial y) < 0$
(c) $(\partial^2 h)/(\partial x\partial y)$

59 (a) A
(b) B

61 (a) (II)
(b) (I)
(c) (III)

63 (a) xy
$1 - \frac{1}{2}(x - \frac{\pi}{2})^2 - \frac{1}{2}(y - \frac{\pi}{2})^2$
(b)

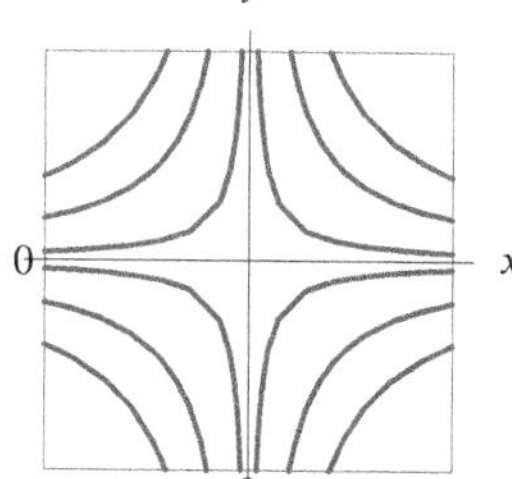

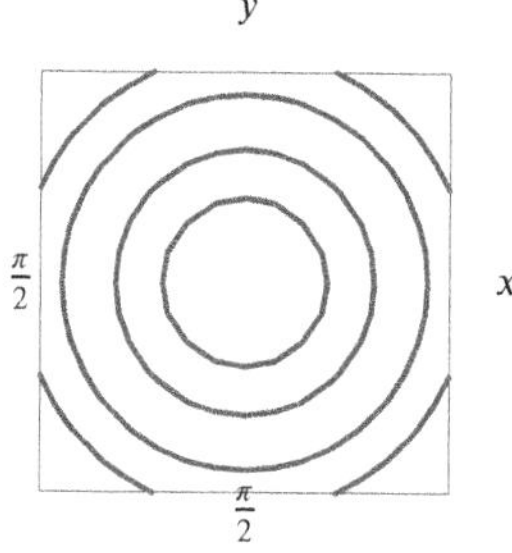

65 $f(x,y)$:

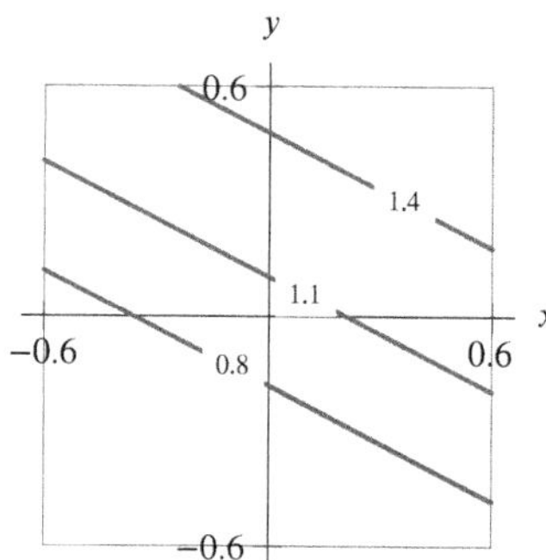

$L(x,y)$:

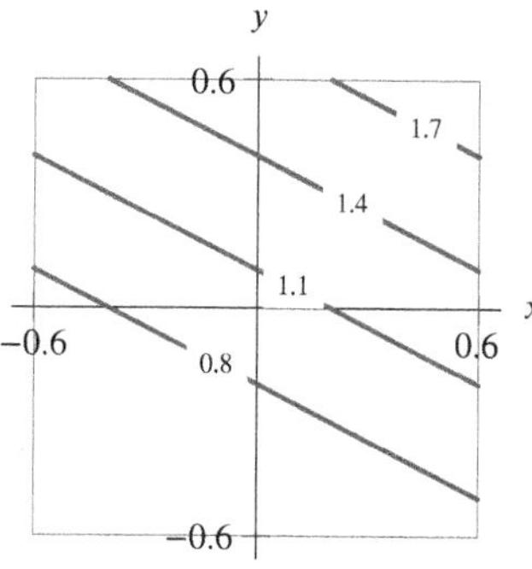

$Q(x,y)$:

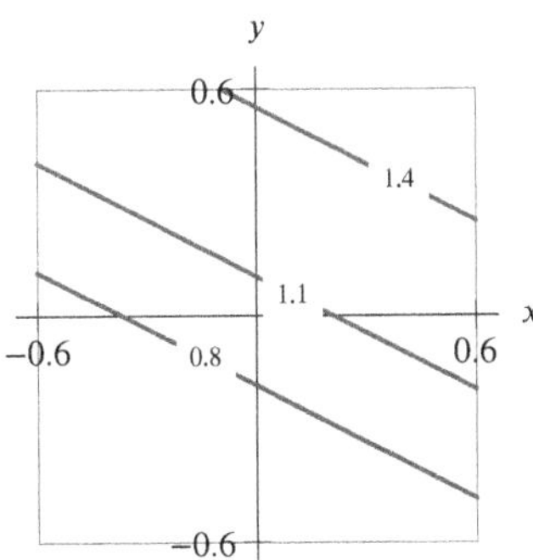

$f(x, y)$:

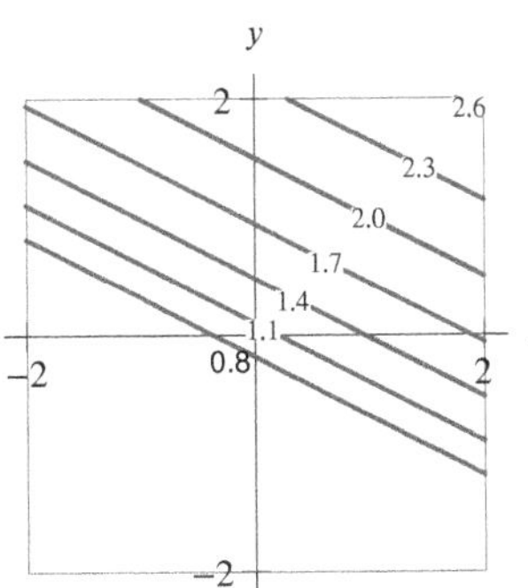

$L(x, y)$:

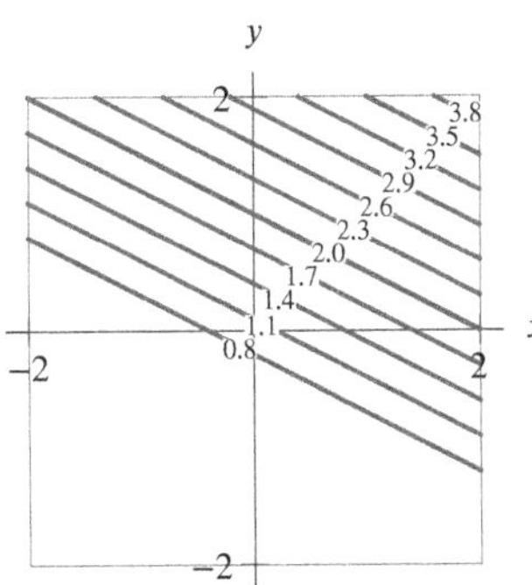

$Q(x, y)$:

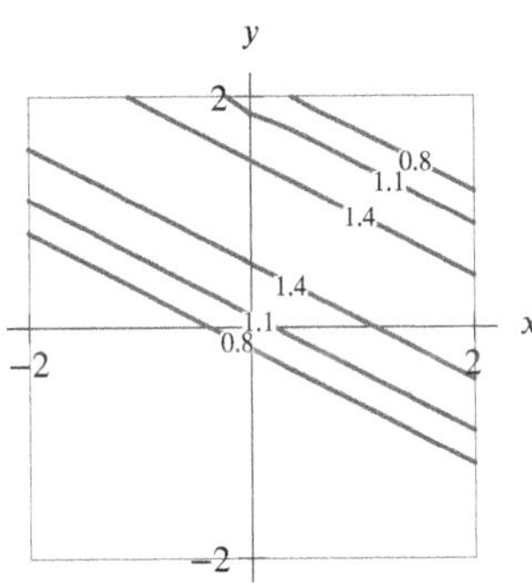

67 None since $f_{xy} \neq f_{yx}$

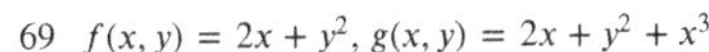
69 $f(x, y) = 2x + y^2$, $g(x, y) = 2x + y^2 + x^3$

Section 14.8

1 $(0, 0)$

3 x-axis and y-axis

5 None

7 None

9 $(1, 2)$

11 (a)

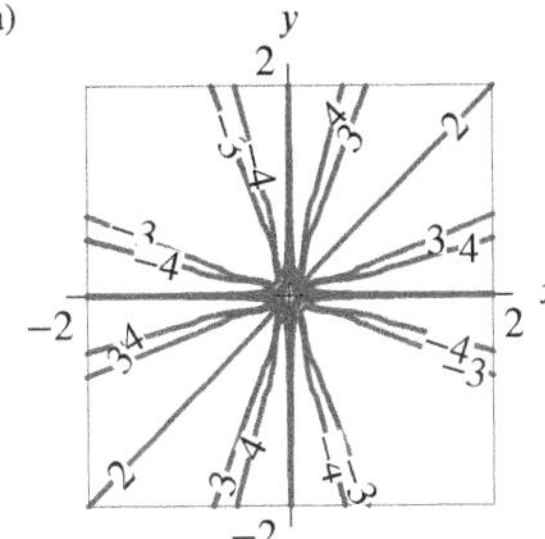

(b) No
(c) No
(d) No
(e) Exist, not continuous

13 (a)

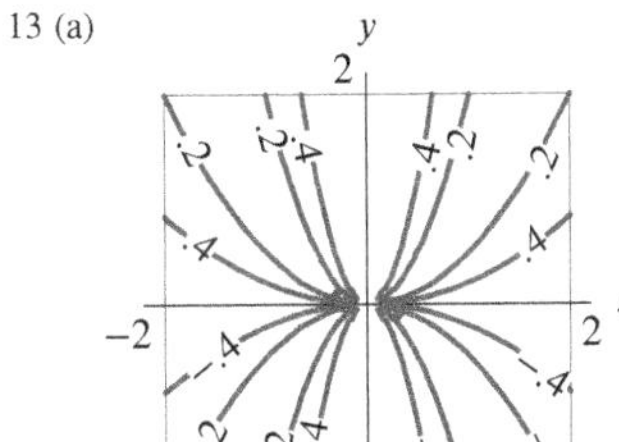

(b) Yes
(c) Yes
(d) No
(e) Exist, not continuous

15 (a)

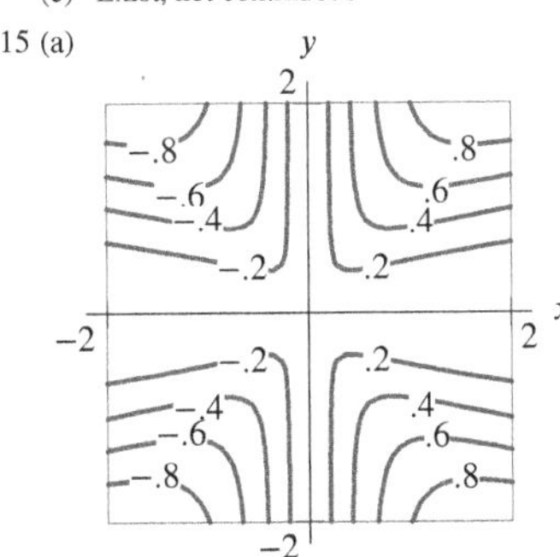

(b) Yes
(d) No
(f) No

17 (a)

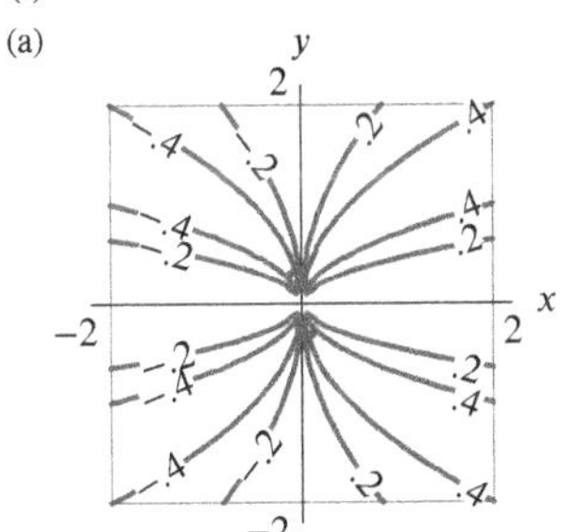

(c) No, no

19 (a) $f_x(x, y) = (x^4 y + 4x^2 y^3 - y^5)/(x^2 + y^2)^2$
$f_y(x, y) = (x^5 - 4x^3 y^2 - xy^4)/(x^2 + y^2)^2$
(c) Yes
(d) Yes

21 Counterexample: $\sqrt{x^2 + y^2}$

23 $f(x, y) = \sqrt{x^2 + y^2}$

25 (a) Differentiable
(b) Not differentiable
(c) Not differentiable
(d) Differentiable

Chapter 14 Review

1 Vector; $3e^{-1}\vec{i} - \frac{1}{2}e^{-1}\vec{j}$

3 Vector; $-(\sin x)e^y\vec{i} + (\cos x)e^y\vec{j} + \vec{k}$

5 $f_x = 2xy + 3x^2 - 7y^6$
$f_y = x^2 - 42xy^5$

7 $\pi/\sqrt{lg}$

9 $f_x = \frac{2xy^3}{(x^2+y^2)^2}$, $f_y = \frac{x^4 - x^2y^2}{(x^2+y^2)^2}$

11 $\partial f/\partial p = (1/q)e^{p/q}$
$\partial f/\partial q = -(p/q^2)e^{p/q}$

13 $f_N = c\alpha N^{\alpha-1}V^\beta$

15 $x/\left(2\sqrt{\omega x}\cos^2\left(\sqrt{\omega x}\right)\right)$

17 $\dfrac{270x^3y^7 - 168x^2y^6 - 15xy^2 + 16y}{(15xy - 8)^2}$

19 $\pi xy/\sqrt{2\pi xyw - 13x^7y^3v}$

21 $\frac{7}{2}\left(\frac{w-1}{x^2yw - xy^3w^7}\right)^{-9/2} \left(\frac{x^2y + 6xy^3w^7 - 7xy^3w^6}{(w-1)^2}\right)$

23 $-1/(4\pi L\sqrt{LC})$

25 $u_{xx} = e^x \sin y$, $u_{yy} = -e^x \sin y$

27 $f_{xxy} = f_{yxx} = 2\cos(x - 2y)$

29 $2x\vec{i} + (2y + 3y^2)\vec{j}$

31 $-((1/x)\vec{i} + (1/y)\vec{j} + (1/z)\vec{k})/(xyz)$

33 $\nabla z = 2x\cos(x^2 + y^2)\vec{i} + 2y\cos(x^2 + y^2)\vec{j}$

35 $\cos(x^2 + y^2 + z^2)\left(2x\vec{i} + 2y\vec{j} + 2z\vec{k}\right)$

37 $-\frac{(t^2-2t+4)}{(2s\sqrt{s})}\vec{i} + \frac{(2t-2)}{\sqrt{s}}\vec{j}$

39 $y[\cos(xy) - \sin(xy)]\vec{i} + x[\cos(xy) - \sin(xy)]\vec{j}$

41 $2\vec{i} + \vec{k}$

43 -1

45 0

47 $2/\sqrt{3}$

49 $5\vec{i} + 4\vec{j} + 3\vec{k}$

51 $-4x - 3y + 4z = 9$

53 $x + y + z = 3$

55 $\cos t \sin(\cos t) - \sin^2 t \cos(\cos t)$

57 $100t^3$

59 $3/t + 2t/(t^2 + 1)$

61 $Q(x, y) = 2 + 6x + y + 6x^2 + 3xy$

63 $Q(x, y) = 1 + (x-3) - \frac{1}{2}(y-5)$
$-\frac{1}{2}(x-3)^2 + \frac{1}{2}(x-3)(y-5) - \frac{1}{8}(y-5)^2$

65 (a) $2x - 4y + az = a - 2$
(b) $a = 2$

67 (a) Q, R
(b) Q, P
(c) P, Q, R, S
(d) None

71 $F = 684$ newtons,
$\partial F/\partial m = 9.77$ newtons/kg,
$\partial F/\partial r = -0.000214$ newtons/meter

73 (a) 20 hours per day
(b) 18.615 hours per day

75 (a) P, S
(b) R, S
(c) P, Q, R, S
(d) None

77 0.3

79 0.8

81 0

83 (a) $-5\sqrt{2}/2$
(b) $4\vec{i} + \vec{j}$

85 (a) 98.387 ft/mile
(b) 295.161 ft/hour

87 (a) $-4e^{-81}$ °C/meter
(b) $-40e^{-81}$ °C/sec
(c) $\sqrt{932}e^{-81}$ °C/meter

89 $-2x\vec{i} - 2y\vec{j}$

91 Yes

95 (a) $F_u(x, y, 3)$
(b) $F_w(3, y, x)$
(c) $F_u(x, y, x) + F_w(x, y, x)$
(d) $F_u(x, y, xy) + yF_w(x, y, xy)$

97 $dP \approx 47.6\,dL + 17.8\,dK$

101 (a) $-3\vec{i} + 4\vec{j} - \vec{k}$
(b) $-3\vec{i} + 4\vec{j}$

105 15°C/minute

107 Approx 7.5 at (1.94, 1.08)

109 $x - y$

111 (a) Negative, positive,
Up if positive, down if negative
(b) $\pi < t < 2\pi$
(c) $0 < x < 3\pi/2$ and
$0 < t < \pi/2$ or $3\pi/2 < t < 5\pi/2$

113 (a) $A_0 + A_1 + 2A_2 + A_3 + 2A_4 + 4A_5 + (A_1 + 2A_3 + 2A_4)(x-1) + (A_2 + A_4 + 4A_5)(y-2)$,
$1 + B_1 t,\ 2 + C_1 t$
(b) $A_1B_1 + 2A_3B_1 + 2A_4B_1 + A_2C_1 + A_4C_1 + 4A_5C_1$

115 $\frac{\partial w}{\partial x}\frac{\partial x}{\partial u}\frac{du}{dt} + \frac{\partial w}{\partial y}\frac{\partial y}{\partial u}\frac{du}{dt} + \frac{\partial w}{\partial x}\frac{\partial x}{\partial v}\frac{dv}{dt} + \frac{\partial w}{\partial y}\frac{\partial y}{\partial v}\frac{dv}{dt} + \frac{\partial w}{\partial z}\frac{dz}{dt}$

Section 15.1

1 (I) and (V) Local maximum, (II) and (VI) Local minimum, (III) and (IV) Saddle point

3 (a) None
(b) E, G
(c) D, F

5 (a)

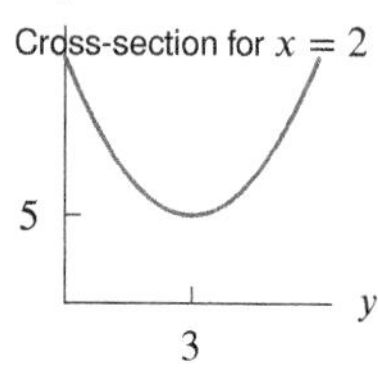

z
Cross-section for $y = 3$
5
x
2

(b)

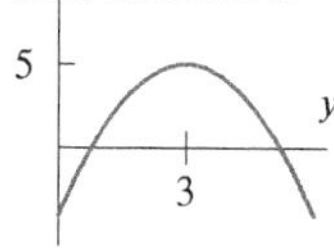

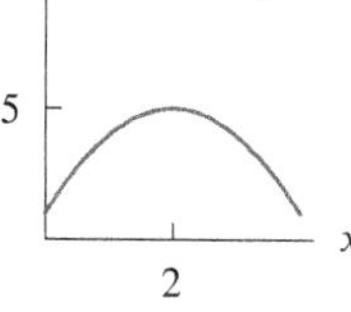

(c)

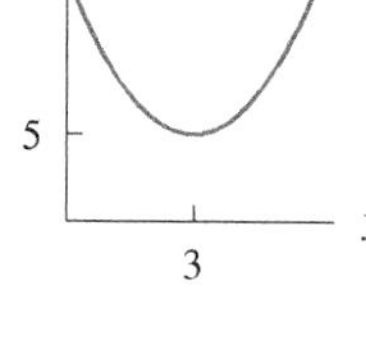

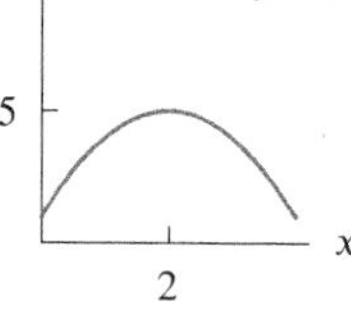

7 Saddle point

9 Local minimum

11 Local maximum

13 Local minimum

15 Local max: (4, 2)

17 Local max: (1, 5)

19 Saddle point: (0, 0)
Saddle point: (2, 0)
Local min: (1, 0.25)

21 Saddle pts: (1, −1), (−1, 1)
Local max: (−1, −1)
Local min: (1, 1)

23 Local max: (−1, 0)
Saddle pts: (1, 0), (−1, 4)
Local min: (1, 4)

25 Saddle point: (0, 0)
Local max: (1, 1), (−1, −1)

27 Local min: (0, 0)

29 (a) All values of k
(b) None
(c) None

31 $a = -9, b = -12, c = 50$

33 (a) $k < 4$
(b) None
(c) $k \geq 4$

35

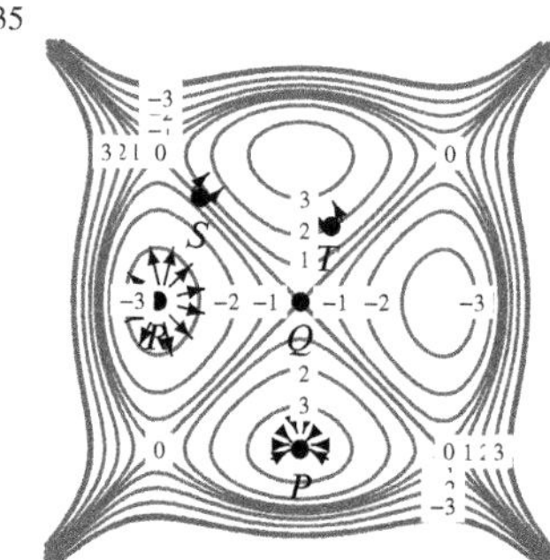

37 Saddle point: (0, 0).

39 Critical points: $(0, 0), (\pm\pi, 0)$, $(\pm 2\pi, 0), (\pm 3\pi, 0), \cdots$
Local minima: $(0, 0)$, $(\pm 2\pi, 0), \pm 4\pi, 0), \cdots$
Saddle points: $(\pm\pi, 0)$, $(\pm 3\pi, 0), (\pm 5\pi, 0), \cdots$

41 (a) (1, 3) is a local minimum
(b)

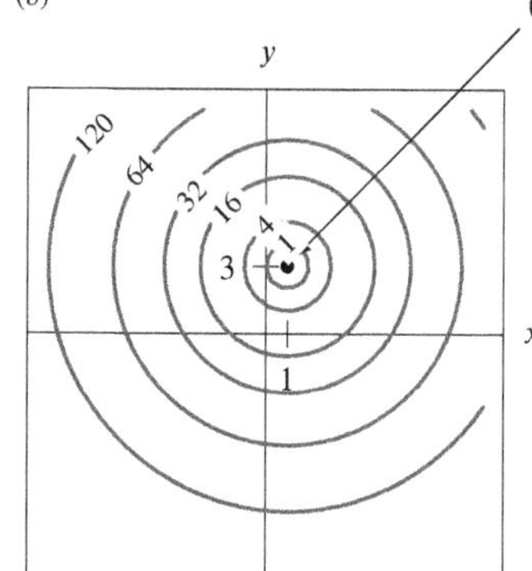

47

y
6
5
4
3
2
1
−4
−3
−2
−1
0 1 2 3 4
−1
x
−1 1 2 3 4 5 6

49 (a) (0, 0)
(b) $D = -24x^2$
(c) Saddle point

51

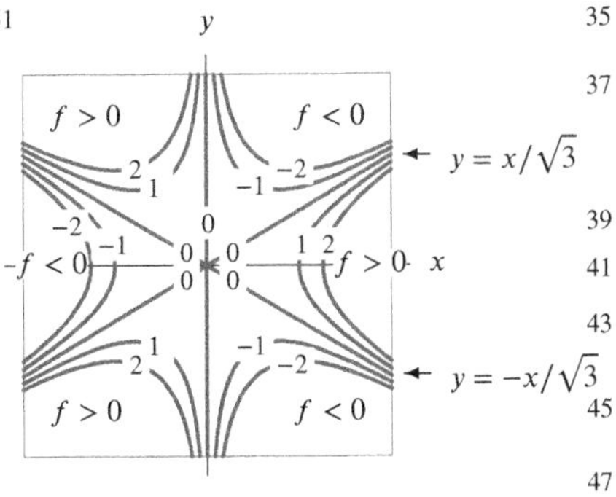

53 (1, 3) could be saddle point

55 Can be saddle if f_{xy} large

57 $f(x, y) = 4 - (x - 2)^2 - (y + 3)^2$

59 False

61 True

63 True

65 True

67 False

69 False

Section 15.2

1 Mississippi:
 87 − 88 (max), 83 − 87 (min)
 Alabama:
 88 − 89 (max), 83 − 87 (min)
 Pennsylvania:
 89 − 90 (max), 80 (min)
 New York:
 81 − 84 (max), 74 − 76 (min)
 California:
 100 − 101 (max), 65 − 68 (min)
 Arizona:
 102 − 107 (max), 85 − 87 (min)
 Massachusetts:
 81 − 84 (max), 70 (min)

3 Max: 30.5 at (0, 0)
 Min: 20.5 at (2.5, 5)

5 High: (0, 0, 8)
 Low: (0, 0, −6)

7 High: None
 Low: $(5, \pi, 2\pi)$

9 None

11 Min = 0 at (0, 0)
 (not on boundary)
 Max = 2 at (1, 1), (1, −1),
 (−1, −1) and (−1, 1)
 (on boundary)

13 max= 1 at (1, 0) and (−1, 0)
 (on boundary)
 min= −1 at (0, 1), (0, −1)
 (on boundary)

15 Global min; no global max

17 Global max; no global min

19 Global max; no global min

21 Global min at $(0, 2\pi n)$, all n
 No global max

23 Saddle at (0, 1/2); local min at (−2/3, 7/6); no global max or min

25 Saddle at (0, 0); local max at (2/9, 4/27); no global max or min

27 All edges $(32)^{1/3}$ cm

29 $l = w = h = 45$ cm

31 (3/14, 1/7, 1/14)

35 $q_1 = 300, q_2 = 225$.

37 (a) $L = \left[pA\left(\frac{a}{k}\right)^a\left(\frac{l}{b}\right)^{(a-1)}\right]^{1/(1-a-b)}$
 $K = \frac{la}{kb}L$
 (b) No

39 $y = 24x^2/49 - 2/7$

41 $y = \frac{25}{6} - \frac{3}{2}x$

43 (b) $f(\sqrt{1/2}, \sqrt{3/5}) = 4\sqrt{2} + 2\sqrt{15} \approx 13.403$

45 (a) Decrease; increase
 (d) Both zero

47 Some do, like $f(x, y) = x^2 + y^2$; some don't

49 $f(x, y) = x + y$

51 True

53 True

55 True

57 False

59 True

Section 15.3

1 Min $= -\sqrt{2}$, max $= \sqrt{2}$

3 Max: 20 at (−1, 2);
 Min: 0 at (1, −2)

5 Min = −22, max = 22

7 Maximum $f(10, 12.5) = 250$;
 No minimum

9 Min $= \frac{3}{4}$, no max

11 Max = 0, no min

13 Max: $f(0, 2) = f(0, -2) = 8$
 Min: $f(0, 0) = 0$

15 Max $= \frac{\sqrt{2}}{4}$, min $= -\frac{\sqrt{2}}{4}$

17 Max: 32 at (1, −1);
 Min: 8 at (−1, 1)

19 (a) P minimum, Q minimum, R neither, S maximum
 (b) P minimum, Q neither, R neither, S maximum

21 $K = 4, L = 5, \lambda = 0.072$

23 $K = 100, L = 400, \lambda = 4$

25 Global max (12/5, 8/5); global min (1, 3)

27 0.5

29 (a)

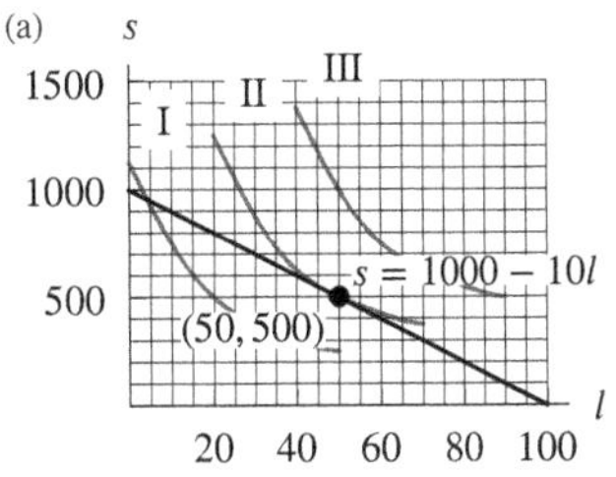

(b) $s = 1000 - 10l$

31 (a) Min; max at endpt of constraint
 λ neg
 (b) Max; min at endpt of constraint
 λ pos

33 $\Delta c/4$; $-\Delta c/4$

37 (a) $C = \$4349$
 (b) $182

39 (a) $W = 225$
 $K = 37.5$
 (c) $W = 225$
 $K = 37.5$
 $\lambda = 0.29$

41 (a) No
 (b) Yes
 (c) $a + b = 1$

43 $x_1 = ((v_1)^{1/2} + (v_2)^{1/2})/(m(v_1)^{1/2})$
 $x_2 = ((v_1)^{1/2} + (v_2)^{1/2})/(m(v_2)^{1/2})$

45 (a) $f_1 = \frac{k_1}{k_1+k_2}mg$, $f_2 = \frac{k_2}{k_1+k_2}mg$
 (b) Distance the mass stretches the top spring and compresses the lower spring

49 (a) Cost of producing quantity u when prices are p, q
 (b) $2\sqrt{pqu}$

51 (a) $-5\lambda^2 + 15\lambda$
 (b) 1.5, 11.25
 (c) 11.25, 1.5
 (d) same

53 (a) $S = \ln(a^a(1 - a)^{(1-a)}) + \ln b - a\ln p_1 - (1 - a)\ln p_2$
 (b) $b = e^c p_1^a p_2^{(1-a)}/(a^a(1 - a)^{(1-a)})$

55

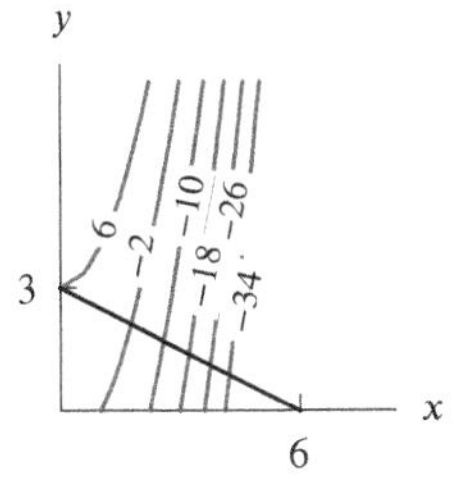

57 $f(x, y) = x^2 + y^2$

59 $f(x, y) = 10 - x^2 - y^2$

63 True

65 True

67 False

69 False

71 True

73 False

75 False

77 (a)

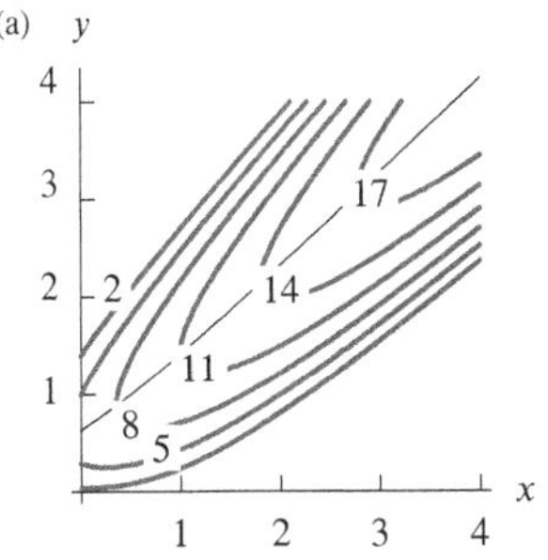

(b)

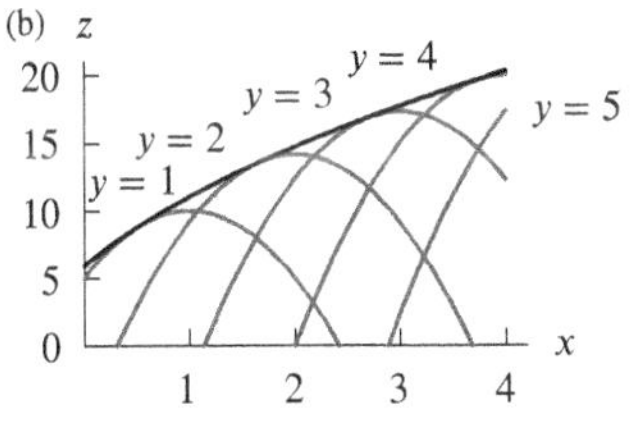

Chapter 15 Review

1 $(3, -1)$, Saddle point

3 Saddle pt: $(0, -5)$
Local min: $(2, -5)$

5 $(2, 1)$; local min

7 $(2, 3)$: Local and global max

9 Local minimum: $(1, 2)$
Local maximum: $(-1, -2)$
Saddle points: $(1, -2)$ and $(-1, 2)$
No global maximum or minimum

11 Minimum $f(-3/\sqrt{5}, 4/\sqrt{5}) = -5\sqrt{5}$,
Maximum $f(3/\sqrt{5}, -4/\sqrt{5}) = 5\sqrt{5}$

13 Min $= 11.25$; no max

15 Minimum $f(27.907, 23.256) = 1860.484$;
No maximum

17 Max $= 5/2$, min $= -2$

19 Maxima: $(-1, 1)$ and $(1, -1)$
Minimum: $(0, 0)$

21 Max $= 1$
Min $= -1$

23 Minimum

25 Neither

27 (a) $\sqrt{(x-3)^2 + (y-4)^2}$
(b) 4 at $(0.6, 0.8)$
(c) 6 at $(-0.6, -0.8)$

29 $y = 2/3 - x/2$

31 $A = 10, B = 4, C = -2$

33 $q_1 = 50$ units
$q_2 = 150$ units

35 6340

37 (a) Reduce K by 1/2 unit, increase L by 1 unit.

39 Along line $x = 2y$

41 $p_1 = 110, p_2 = 115$.

43 $x \approx 23.47$, $y \approx 23.47$, $z \approx 75.1$

45 (a) $i_1 = R_2 I/(R_1 + R_2)$,
$i_2 = R_1 I/(R_1 + R_2)$
(b) $\lambda = 2 \cdot$ Voltage

49 $d \approx 5.37$ m, $w \approx 6.21$ m,
$\theta = \pi/3$ radians

51 Student B correct. Local max, not global

Section 16.1

1 24; 43.5

3 Over: Approx 137
Under: Approx 60

5 about 2300

7 Average height of a tent in meters

9 Positive

11 Zero

13 Zero

15 Positive

17 About 4.888 km^3

19 210

21 Need f nonnegative everywhere

23 $f(x, y) = 5 - x - y$; R is square with vertices $(\pm 1, \pm 1)$

25 False

27 False

29 True

31 True

33 False

35 25.2°C

Section 16.2

1

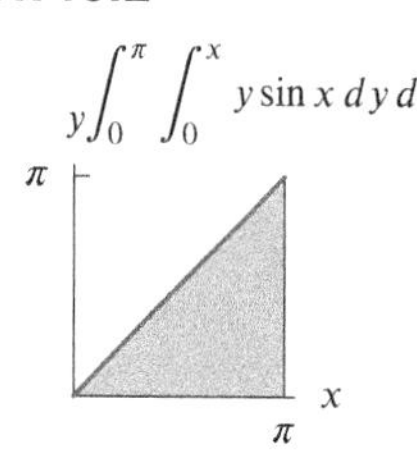

3

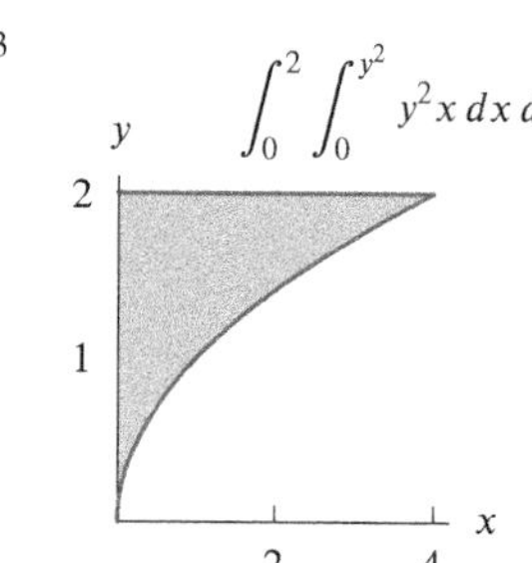

5 150

7 54

9 $e - 2$

11 $3 - \sin 3$

13 $(e^4 - 1)(e^2 - 1)e$

15 -2.678

17 15

19 $\frac{2}{3}(e^8 - 1)$

21 $\int_1^4 \int_1^2 f \, dy \, dx$ or $\int_1^2 \int_1^4 f \, dx \, dy$

23 $\int_{-1}^3 \int_{-2}^{(1-3x)/4} f \, dy \, dx$
or $\int_{-2}^1 \int_{-1}^{(1-4y)/3} f \, dx \, dy$

25 $\int_1^3 \int_{\frac{1}{2}(y-1)}^{-\frac{1}{2}(y-5)} f \, dx \, dy$ or
$\int_0^1 \int_1^{2x+1} f \, dy dx +$
$\int_1^2 \int_1^{-2x+5} f \, dy dx$

27 $\int_1^2 \int_0^x f \, dy \, dx$ or
$\int_0^1 \int_1^2 f \, dx dy +$
$\int_1^2 \int_y^2 f \, dx dy$

29 $\frac{4}{15}(9\sqrt{3} - 4\sqrt{2} - 1) = 2.38176$

31 $32/9$

33 $13/6$

35 0

37 2/3

39 $\int_0^6 \int_0^{x/2} f(x, y) \, dy \, dx$

41 $\int_0^9 \int_{-\sqrt{9-y}}^{\sqrt{9-y}} f(x, y) \, dx \, dy$

43 $(e - 1)/2$

45 $\frac{2}{9}(3\sqrt{3} - 2\sqrt{2})$

47 $\frac{1}{2}(e^2 - 1)$

49 $\ln(17)/4$

51 $\{(I),(IV),(V)\}$, $\{(II),(III),(VI)\}$

53 (a) $8/3$
(b) $16/3$

55 (a) $\int_0^{1/2} \int_y^{1-y} f(x, y) \, dx \, dy$
$\int_0^{1/2} \int_0^x f(x, y) \, dy \, dx$ +
$\int_{1/2}^1 \int_0^{1-x} f(x, y) \, dy \, dx$
(b) 1/8

57 15

59 (a) Plate 1
(b) Plate 1: 5 coulombs; Plate 2: 4 coulombs

61 18 gm

63 $\int_{-3}^3 \int_{-\sqrt{9-y^2}}^{\sqrt{9-y^2}} (9 - x^2 - y^2) \, dx \, dy$

65 4

67 117.45

69 Volume $= 1/(6abc)$

71 (a) Circles centered at $(1, 0)$
(b) $\int_{-\sqrt{3}}^{\sqrt{3}} e^{-y^2} \, dy$
(c) $\int_{-2}^2 \int_{-\sqrt{4-x^2}}^{\sqrt{4-x^2}} e^{-(x-1)^2 - y^2} \, dy \, dx$

73 (a)

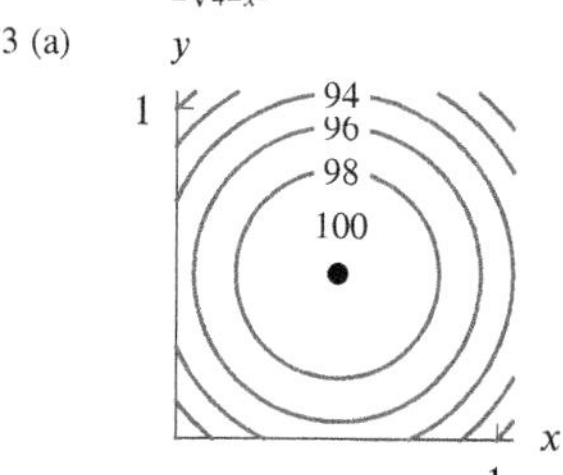

(b)

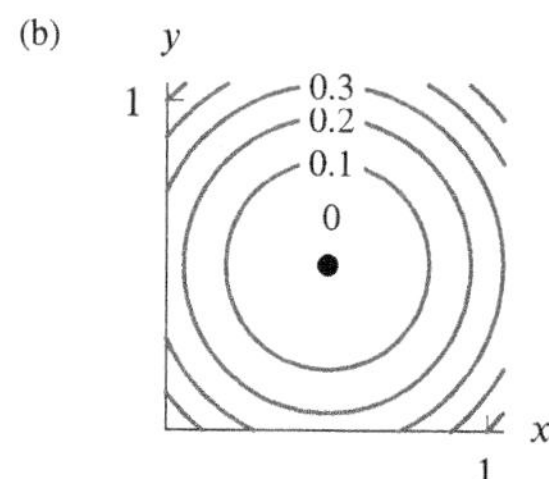

75 (a) $(4/3)a + b + (4/3)c = 20$
(b) $f(x, y) = x^2 + \frac{44}{3}xy + 3y^2$:

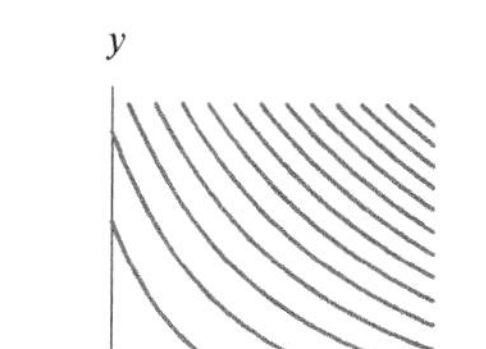

$f(x, y) = -3x^2 + 24xy$:

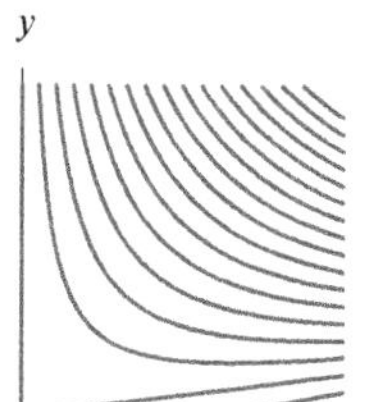

77 Outside limits on right should be constants

79 $f(x, y) = 12x$

81 False

83 True

85 True

87 True

89 Volume = 6

91 $k(a^3b + ab^3)/3$

Section 16.3

1 2

3 −8

5

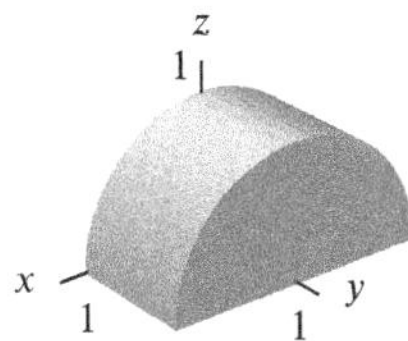

7

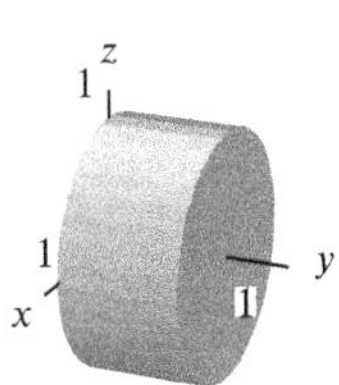

9

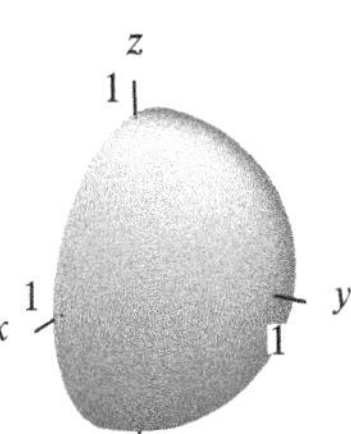

11

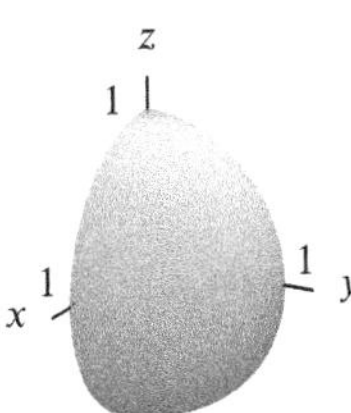

13

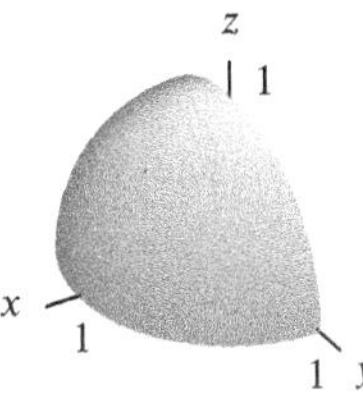

15 Mass of E in kg

17 Positive

19 Positive

21 Zero

23 Positive

25 Zero

27 Positive

29 Positive

31 Positive

33 Zero

35 1

37 4/3

39 500/3

41 $V = \int_{-1}^{1} \int_{-\sqrt{1-x^2}}^{\sqrt{1-x^2}} \int_{x^2+y^2}^{\sqrt{4-x^2-y^2}} 1\,dz\,dy\,dx$
Can reverse order x, y

43 $V = \int_0^2 \int_y^{(y+2)/2} \int_0^{\sqrt{9-x^2-y^2}} 1\,dz\,dx\,dy$

45 $V = \int_0^1 \int_0^{\sqrt{4-x^2}} \int_0^{\sqrt{4-x^2-y^2}} 1\,dz\,dy\,dx$

47 $\int_0^1 \int_{-2}^2 \int_0^{\sqrt{4-z^2}} f(x, y, z)\,dy\,dz\,dx$

49 $\int_0^r \int_{-\sqrt{r^2-x^2}}^{\sqrt{r^2-x^2}} \int_0^{\sqrt{r^2-x^2-y^2}} f(x, y, z)\,dz\,dy\,dx$

51 125/3

53 2/3

55 15/2

57 (a) Mass of pyramid in grams
(b) Four
(c) 27 grams

59 $\int_0^{3/4} \int_{\frac{2y}{3}}^{2-2y} \int_0^{4-2x-4y} f(x, y, z)\,dz\,dx\,dy$

61 $\int_0^{\frac{1}{2}} \int_0^{4-8x} \int_{\frac{3x}{2}}^{1-\frac{x}{2}-\frac{z}{4}} f(x, y, z)\,dy\,dz\,dx$

63 (a) $z = \sqrt{1-x^2}, 0 \le y \le 10$
(b) $\int_0^{10} \int_{-1}^1 \int_0^{\sqrt{1-x^2}} f(x, y, z)\,dz\,dx\,dy$

65 $\int_0^2 \int_0^{\sqrt{12-3y^2}} \int_0^{6y^2} f(x, y, z)\,dz\,dx\,dy$

67 $\int_0^{\sqrt{12}} \int_0^{24-2x^2} \int_{\sqrt{\frac{z}{6}}}^{\sqrt{\frac{12-x^2}{3}}} f(x, y, z)\,dy\,dz\,dx$

69 $\int_0^2 \int_0^{(3/2)\sqrt{4-y^2}} \int_{(15-5x)/3}^5 f(x, y, z)\,dz\,dx\,dy$

71 $\int_0^5 \int_0^{(15-3z)/5} \int_0^{(2/3)\sqrt{9-x^2}} f(x, y, z)\,dy\,dx\,dz$

73 $\int_0^2 \int_0^{4-x^2} \int_0^{2-x} f(x, y, z)\,dy\,dz\,dx$

75 $m = 2$;
$(\bar{x}, \bar{y}, \bar{z}) = (13/24, 13/24, 25/24)$

77 Not true for $f(x, y, z) = z$

79 $f(x, y, z) = 7/(12\pi)$

81 False

83 False

85 True

87 False

89 False

91 4

93 1

95 (a) $\int_0^2 \int_{\sqrt{\frac{x}{2}}}^{\frac{4-x}{2}} \int_0^{4-x-2y} f(x, y, z)\,dz\,dy\,dx$
(b) $\int_0^2 \int_0^{4-x-\sqrt{2x}} \int_{\sqrt{\frac{x}{2}}}^{\frac{4-x-z}{2}} f(x, y, z)\,dy\,dz\,dx$

97 $m(b^2 + c^2)/3$

Section 16.4

1 $\int_0^{\pi/2} \int_0^{1/2} f\,r\,dr\,d\theta$

3 $\int_{\pi/4}^{3\pi/4} \int_0^2 f\,r\,dr\,d\theta$

5 $\int_1^5 \int_2^4 f(x, y)\,dy\,dx$

7 $\int_\pi^{2\pi} \int_2^4 f(r\cos\theta, r\sin\theta)\,r\,dr\,d\theta$

9

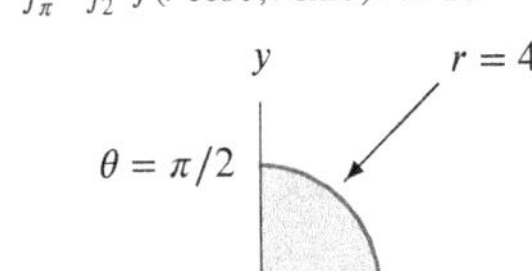

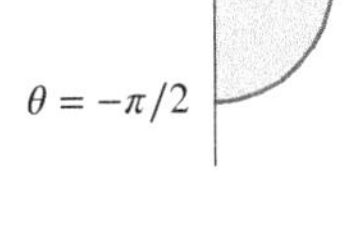

11

y
r = 2
r = 1
x
1
2

13

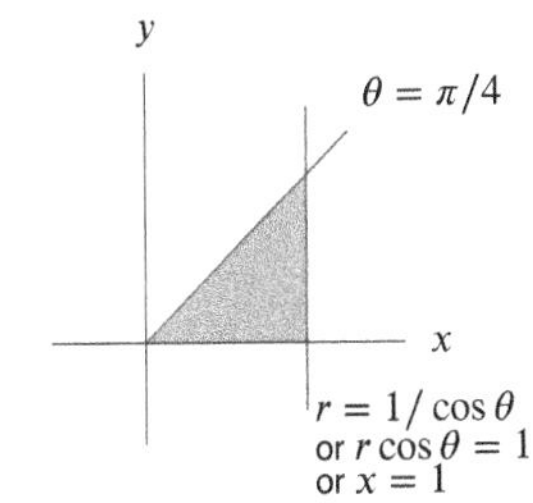

15

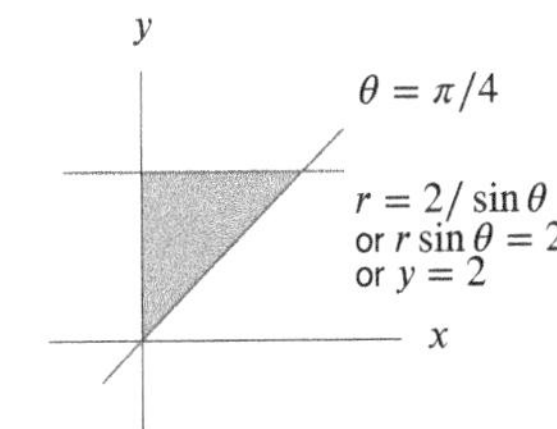

17 $\pi(1 - \cos 4)$

19 $-2/3$

21 1.

23 $\int_{-1}^0 \int_0^{-\sqrt{3}x} 2y\,dy\,dx$; $\int_0^{\sqrt{3}} \int_{-1}^{-y/\sqrt{3}} 2y\,dx\,dy$

25 16

27 (a)

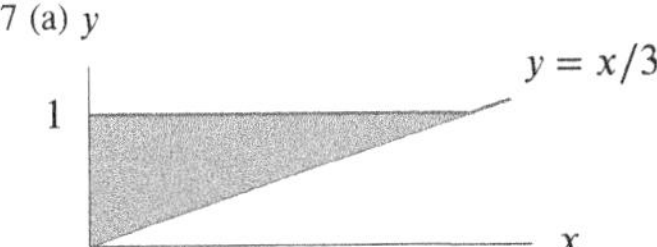

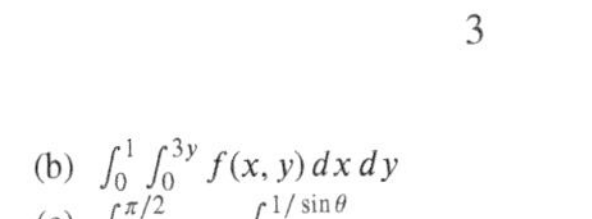

(b) $\int_0^1 \int_0^{3y} f(x, y)\,dx\,dy$
(c) $\int_{\tan^{-1}(1/3)}^{\pi/2} \int_0^{1/\sin\theta}$
$f(r\cos\theta, r\sin\theta)r\,dr\,d\theta$

29 $2/\sqrt{3}$

31 $625\pi/2$

33 (a) $\pi(1 - e^{-a^2})$
(b) Volume tends to π

35 (a) $\int_{\pi/2}^{3\pi/2} \int_1^4 \delta(r, \theta)\, r\, dr\, d\theta$
(b) (i)
(c) About 39,000

37 Total charge $= 2k\pi R$

39 (a)

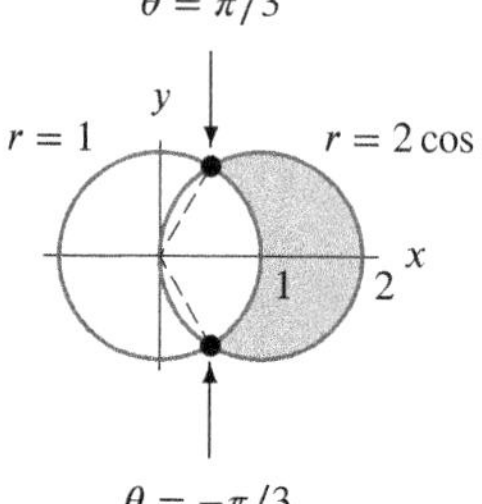

(b) $\sqrt{3}/2 + \pi/3$

41 Integrand r^3 instead of r^2

43 Regions of integration are not the same

45 Quarter disk $0 \le x \le 1, 0 \le y \le \sqrt{1 - x^2}$

47 $f(x, y) = 1/\sqrt{x^2 + y^2}$

49 (a), (c), (e)

51 True

53 False

55 False

57 (a) $\int_{-\sqrt{3}/2}^{\sqrt{3}/2} \int_{1-\sqrt{1-y^2}}^{\sqrt{1-y^2}} dx\, dy$
(b) $\int_0^1 \int_{-\arccos(r/2)}^{\arccos(r/2)} r\, d\theta\, dr$

Section 16.5

1 (a) is (IV); (b) is (II); (c) is (VII); (d) is (VI); (e) is (III); (f) is (V)

3 $z = \sqrt{1 - r^2}$

5 $\phi = \pi/4$

7 $\rho = 4/\cos\phi$

9 $200\pi/3$

11 25π

13 $\int_0^1 \int_0^{2\pi} \int_0^4 f \cdot r\, dr\, d\theta\, dz$

15 $\int_0^{\pi} \int_0^{\pi} \int_2^3 f \cdot \rho^2 \sin\phi\, d\rho\, d\phi\, d\theta$

17 $\int_0^5 \int_0^2 \int_0^{x/5} f\, dz\, dy\, dx$

19 $\int_0^{2\pi} \int_0^2 \int_{2r}^4 f(r, \theta, z) r\, dz\, dr\, d\theta$

21 $\int_{-2}^{2} \int_{-\sqrt{4-x^2}}^{\sqrt{4-x^2}} \int_{2\sqrt{x^2+y^2}}^{4} h(x, y, z)\, dz\, dy\, dx$

23 $\int_0^{\pi} \int_0^{K} \int_0^{2\pi} \rho^2 \sin\phi\, d\theta\, d\rho\, d\phi$

25 (a) $\int_{-1/\sqrt{2}}^{1/\sqrt{2}} \int_{-\sqrt{(1/2)-x^2}}^{\sqrt{(1/2)-x^2}} \int_{\sqrt{x^2+y^2}}^{\sqrt{1-x^2-y^2}} dz\, dy\, dx$
(b) $\int_0^{2\pi} \int_0^{1/\sqrt{2}} \int_r^{\sqrt{1-r^2}} r\, dz\, dr\, d\theta$
(c) $\int_0^{2\pi} \int_0^{\pi/4} \int_0^1 \rho^2 \sin\phi\, d\rho\, d\phi\, d\theta$

27 (a) $\int_0^{2\pi} \int_0^{\sqrt{2}} \int_r^{\sqrt{4-r^2}} r\, dz\, dr\, d\theta$
(b) $\int_0^{2\pi} \int_0^{\pi/4} \int_0^2 \rho^2 \sin\phi\, d\rho\, d\phi\, d\theta$

29 $V = \int_0^{2\pi} \int_0^{\pi/3} \int_0^3 \rho^2 \sin\phi\, d\rho\, d\phi\, d\theta$
Order of integration can be altered; other coordinates can be used

31 $V = \int_0^{\pi} \int_{\sqrt{2}}^{\sqrt{3}} \int_5^{10} r\, dz\, dr\, d\theta$;
Order of integration can be altered; other coordinates can be used

33 $V = \int_0^{2\pi} \int_0^3 \int_1^{\sqrt{10-r^2}} r\, dz\, dr\, d\theta$
or
$V = \int_0^{2\pi} \int_1^{\sqrt{10}} \int_0^{\sqrt{10-z^2}} r\, dr\, dz\, d\theta$ Order of integration can be altered; other coordinates can be used

35 (a) $\int_0^{2\pi} \int_0^{1/\sqrt{3}} \int_{\sqrt{3}r}^{1} r\, dz\, dr\, d\theta$
(b) $\pi/9$

37 $16\pi(\sqrt{2} - 1)/(3\sqrt{2})$

39 $28\pi/15$

41 $\int_0^{2\pi} \int_0^{5/\sqrt{2}} \int_r^{5/\sqrt{2}} r\, dz\, dr\, d\theta = 125\pi/(6\sqrt{2}) = 46.28\ \text{cm}^3$

43 (a) Positive
(b) Zero

45 $\int_0^{2\pi} \int_0^{l} \int_a^{a+h} r\, dr dz d\theta = \pi l((a + h)^2 - a^2)$

47 $\int_0^{2\pi} \int_0^{a} \int_{hr/a}^{h} r\, dz dr d\theta = \pi h a^2/3$

49 (a) $\int_0^{2\pi} \int_1^5 \int_{-\sqrt{25-r^2}}^{\sqrt{25-r^2}} r\, dz\, dr\, d\theta$
(b) $64\sqrt{6}\pi = 492.5\ \text{mm}^3$

51 $81\pi(-\sqrt{2} + 2)/4$

53 $324\pi/5$ gm

55 1702π gm

57 Mass $= \int_{-2}^{2} \int_{-\sqrt{4-x^2}}^{\sqrt{4-x^2}} \int_0^{4-x^2-y^2} e^{-x-y}\, dz\, dy\, dx$ gm

59 1/27

61 Total charge $= 2\pi k R^2$

63 (a) $\pi/5$
(b) $5/6$

65 $3a/8b$ above center of base

67 Limits of outer integral not constant

69 $\int_0^{2\pi} \int_0^{\pi/2} \int_0^5 \rho^2 \sin\phi\, d\rho\, d\phi\, d\theta$

71 (c)

73 $W = \int_0^1 \int_0^{2\pi} \int_{\sqrt{1-r^2}}^{(\sqrt{9-r^2})-1} r\, dz\, d\theta dr + \int_1^{2\sqrt{2}} \int_0^{2\pi} \int_0^{(\sqrt{9-r^2})-1} r\, dz\, d\theta\, dr$

75 $3I = \frac{6}{5}a^2$; $I = \frac{2}{5}a^2$

77 $(q^2/8\pi\epsilon)((1/a) - (1/b))$

79 $r^2 \sin\theta\, dr\, d\theta\, d\phi$

Section 16.6

1 Is a joint density function

3 Not a joint density function

5 Is joint density function

7 1/16

9 3/16

11 0.28

13 0.19

15 0

17 1

19 7/8

21 1/16

23 (a) 20/27
(b) 199/243

25 (a) $k = 3/8$
(b) 15/32
(c) 1/16

27 (a) 0.60
(b) 0.70
(c) 0.32

29 (a) $\lambda/(\lambda + \mu)$

31 (a) 0 if $t \le 0$, $2t^2$ if $0 < t \le 1/2$, $1 - 2(1 - t)^2$ if $1/2 < t \le 1$, 1 if $1 < t$
(b) 0 if $t \le 0$, $4t$ if $0 < t \le 1/2$, $4 - 4t$ if $1/2 < t \le 1$, 0 if $1 < t$

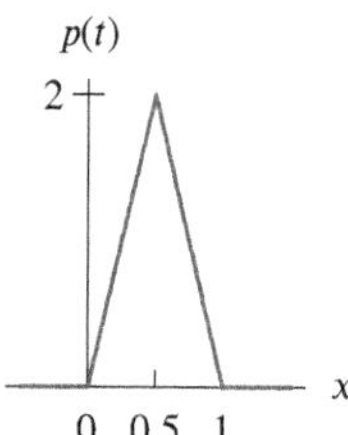

(c) x, y: All equally likely
z: Near 1/2

33 $p(60, 170)$ not a probability

35 $g(y) = y$

37 False

39 True

Chapter 16 Review

1 94.5

3 14

5

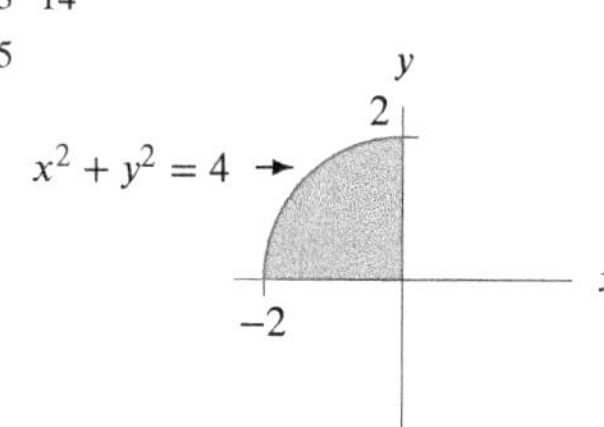

7

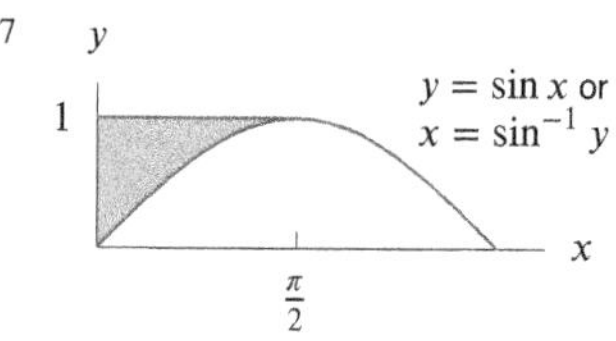

11 $\int_0^{2\pi} \int_0^2 \int_0^3 f\, r\, dr dz d\theta$

13 $\int_0^{\pi/2} \int_0^{\pi} \int_0^5 f \rho^2 \sin\phi\, d\rho d\phi d\theta$

15 (a)

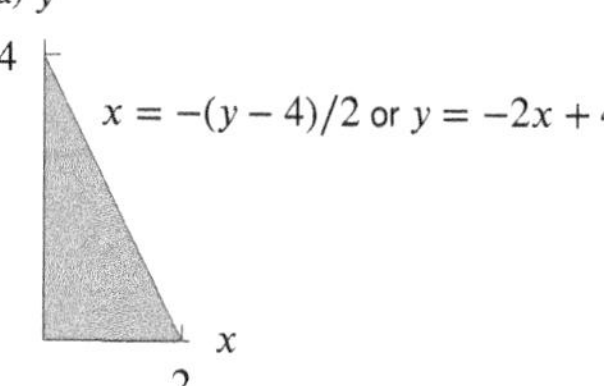

(b) $\int_0^2 \int_0^{-2x+4} g(x, y)\, dy\, dx$

17 $10(e-2)$

19 $(1/20)\sin^5 1$

21 $(\pi/2)(1-e^{-1})$

23 1/48

25 $\int_0^{2\pi}\int_{3\pi/4}^{\pi}\int_0^{\sqrt{2}} f(\rho,\phi,\theta)\rho^2\sin\phi\,d\rho\,d\phi\,d\theta$

27 $\int_{-1}^{1}\int_{-\sqrt{1-x^2}}^{\sqrt{1-x^2}}\int_{-\sqrt{2-x^2-y^2}}^{-\sqrt{x^2+y^2}} h(x,y,z)\,dz\,dy\,dx$

29 Positive

31 Positive

33 Negative

35 Zero

37 Positive

39 1.571

41 (a) $\int_0^{2\pi}\int_{\pi/4}^{\pi/2}\int_0^{3/\sin\theta}\rho^2\sin\phi\,d\rho\,d\phi\,d\theta$
(b) 18π

43

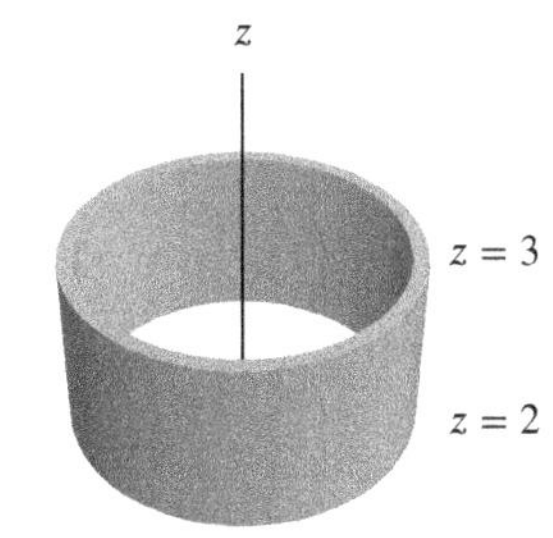

$\int_0^{2\pi}\int_2^3\int_{\sqrt{5}}^{\sqrt{6}} r\,dr\,dz\,d\theta$
Order of integration can be changed

45

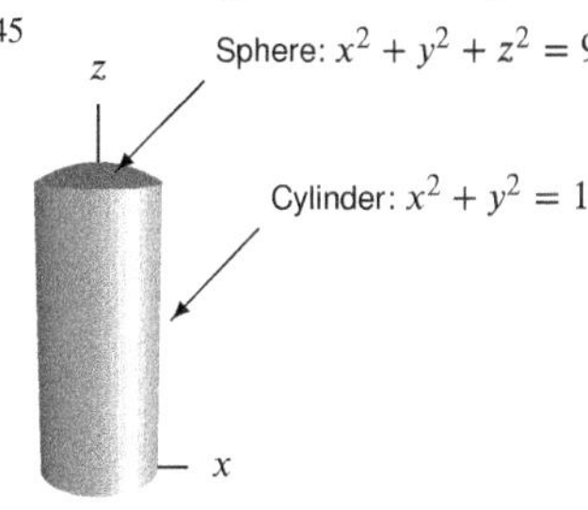

$\int_0^{2\pi}\int_0^1\int_0^{\sqrt{9-r^2}} r\,dz\,dr\,d\theta$

47 Negative

49 Can't tell

51 Zero

53 Zero

55 Zero

57 Negative

59 $\pi(3-2\ln 2)$

61 $162\pi/5$

63 (a)

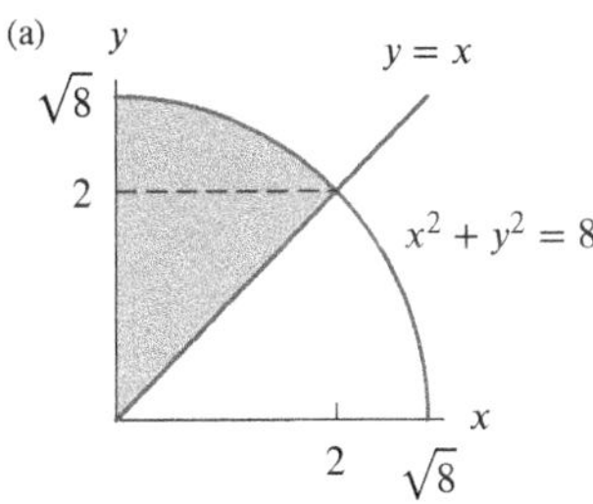

(b) $\pi(1-e^{-8})/8$

65 (a) $z=\sqrt{1-(y-1)^2}\quad 0\le y\le 2\quad 0\le x\le 10$
(b) $\int_0^{10}\int_0^2\int_0^{\sqrt{1-(y-1)^2}} f(x,y,z)\,dz\,dy\,dx$

67 (a) Half cylinder, radius 1, along x-axis, $y\ge 0$, $-1\le x\le 1$
(b) $2\pi/5$

69 $\int_0^{2\pi}\int_0^4\int_3^{\sqrt{25-r^2}} r\,dz\,dr\,d\theta$
$=52\pi/3=54.45\text{ cm}^3$

71 $\int_0^{2\pi}\int_0^{\arccos(3/5)}\int_{3/\cos\phi}^{5}\rho^2\sin\phi\,d\rho\,d\phi\,d\theta$
$=52\pi/3=54.45\text{ cm}^3$

73 8π gm

75 $\int_{-1}^{1}\int_{-\sqrt{1-x^2}}^{\sqrt{1-x^2}}\int_{-1+\sqrt{2-x^2-y^2}}^{\sqrt{1-x^2-y^2}} dz\,dy\,dx$
$=2\pi((7/6)-2\sqrt{2/3})$

77 $8\pi R^5/15$

79 (a) 3/2000
(b) 343/1000

81 $6\pi/7$

83 $a, b+c, 0$

Section 17.1

1 $x=0, y=t, -2\le t\le 1$

3 $x=1+2t,\quad y=1+t,\quad 0\le t\le 1$

5 $x=t,\quad y=3-3t,\quad 0\le t\le 1$

7 $x=t, y=1, z=-t$

9 $x=1,\quad y=0,\quad z=t$

11 $x=1+3t,\quad y=2-3t,\quad z=3+t$

13 $x=-3+2t, y=-2-t, z=1-2t$

15 $x=2+3t,\quad y=3-t,\quad z=-1+t$

17 $x=3-3t, y=0, z=-5t$

19 $x=3\cos t, y=3\sin t, z=5, 0\le t<2\pi$

21 $x=2\cos t, y=-2\sin t, z=0$

23 $x=2\cos t, y=0, z=2\sin t$

25 $x=0, y=3\cos t, z=2+3\sin t$

27 $x=t^2, y=t, z=0$

29 $x=-3t^2, y=0, z=t$

31 $x=t, y=4-5t^4, z=4$

33 $x=3\cos t, y=2\sin t, z=0$

35 $x=-1+3t, y=2, z=-3+5t$

37 $\vec r(t)=\vec i-3\vec j+2\vec k+t(3\vec i+4\vec j-5\vec k)$, $0\le t\le 1$, $x=1+3t$, $y=-3+4t$, $z=2-5t$, $0\le t\le 1$

39 $x=\cos t, y=\sin t, z=0, 0\le t\le\pi$

41 Two arcs:
$\vec r(t)=5\vec i+5(-\cos t\vec i+\sin t\vec j)$, $0\le t\le\pi$ or
$\vec r(t)=5\vec i+5(\cos t\vec i+\sin t\vec j)$, $\pi\le t\le 2\pi$

43 $x=10\cos t, y=10\sin t, z=t$

45 $x=2\cos t, y=t, z=2\sin t$

47 $\vec r(t)=(2+10t)\vec i+(5+4t)\vec j$

49 $\vec r(t)=(2+((t-20)/10)10)\vec i$
$+(5+((t-20)/10)4)\vec j$

51 $\vec r(t)=(2-10t)\vec i+(5-4t)\vec j$

53 No

55 (b) $-\vec i-10\vec j-7\vec k$
(c) $\vec r=(1-t)\vec i+(3-10t)\vec j-7t\vec k$

57 (a) $\vec r=(\vec i+3\vec j+7\vec k)+t(2\vec i-3\vec j-\vec k)$
(b) $(3,0,6)$
(c) $\sqrt{14}$

59 (a) $2\vec i-5\vec j+3\vec k$
(c) $x=1+2t, y=-1-5t, z=1+3t$

61 Same

63 Same

65 Different lines

67 Different lines

69 (a) $(-1,4,-2)$
(b) $-\vec i-\vec j+2\vec k$; other answers possible
(c) $\vec r=-\vec i+4\vec j-2\vec k+t(-\vec i-\vec j+2\vec k)$

71 $x=-\frac{3}{2}t+\frac{5}{2}, y=\frac{1}{2}t+\frac{1}{2}, z=t.$

73 $x=-4, y=2+t, z=3+t$

75 (a) Straight lines
(b) No
(c) $(1,2,3)$

77 $15/(2\pi)$

79 $15/\pi$

81 (a) Circle, center (a,a), rad b, per $2\pi/k$
(b) (i) Increases radius
(ii) Center moves outward on $y=x$
(iii) Speeds up
(iv) Touches axes

83 Circle, cosine, sine

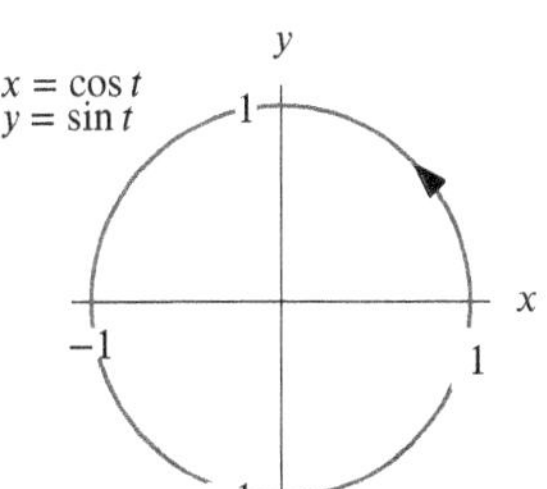

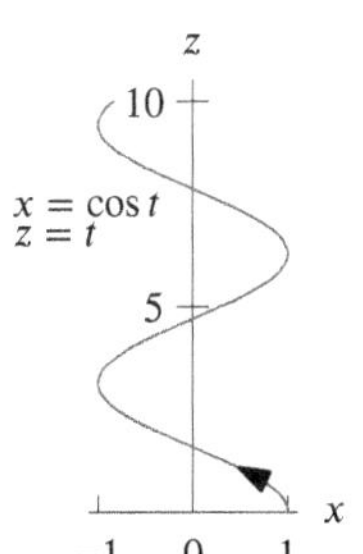

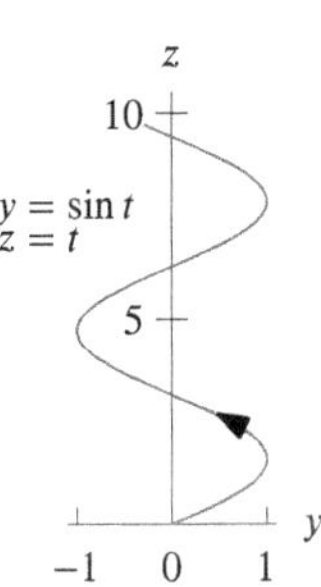

85 (a) II, $y=x$
(b) IV, $x+y=a$
(c) V, $x^2-y^2=a^2$
(d) I, $x^2+y^2=a^2$

(e) III, $x^2 + y^2 = a^2$

87 Many possible answers

(a) $a = -2, b = 7, c = 4, d = 0$
(b) $a = -2, b = 7, c = 4, d = 11$
(c) $a = 7, b = 2, c = 0, d = 41$

89 Line Equation:
$x = 1 + 2t$
$y = 2 + 3t$
$z = 3 + 4t$
Shortest distance: $\sqrt{6/29}$

91 (a) (vii)
(b) (ii)
(c) (iv)

93 $\vec{r} = 9.65\vec{k} + t(325\vec{i} + 563\vec{j} - 0.84\vec{k})$

95 (a) (i) is (C); (ii) is (A); (iii) is (D); (iv) is (G)
(b) (iii)

97 Distance $|R|$ from z-axis
Distance $\sqrt{R^2 + t^2}$ from origin

99 $\vec{i} + 2\vec{j} + 3\vec{k} + t\left(\vec{i} + 2\vec{j}\right)$
$\vec{i} + 2\vec{j} + 3\vec{k} + t\left(\vec{i} - \vec{k}\right)$

101 False

103 False

105 False

107 True

109 True

111 True

113 (a) Center: $(a/2, b/2)$, Radius: $\sqrt{c + (a^2 + b^2)/4}$.
(b) $x = a/2 + \sqrt{c + (a^2 + b^2)/4}\cos t$
$y = b/2 + \sqrt{c + (a^2 + b^2)/4}\sin t$
$0 \le t \le 2\pi$
(c) $x = a/2 + \sqrt{c + (a^2 + b^2)/4}\cos t$
$y = b/2 + \sqrt{c + (a^2 + b^2)/4}\sin t$
$z = (a^2 + b^2)/2 + c + a\sqrt{c + (a^2 + b^2)/4}\cos t + b\sqrt{c + (a^2 + b^2)/4}\sin t$
$0 \le t \le 2\pi$

115 (a) $-2e^{-1}/3$ μg/m^3/m
(b) $t = \pm\sqrt{3}/2$ sec

Section 17.2

1 $\vec{v} = 3\vec{i} + \vec{j} - \vec{k}$, $\vec{a} = \vec{0}$

3 $\vec{v} = \vec{i} + 2t\vec{j} + 3t^2\vec{k}$, $\vec{a} = 2\vec{j} + 6t\vec{k}$

5 $\vec{v} = -3\sin t\vec{i} + 4\cos t\vec{j}$,
$\vec{a} = -3\cos t\vec{i} - 4\sin t\vec{j}$

7 $\vec{v} = \vec{i} + 2t\vec{j} + 3t^2\vec{k}$,
Speed $= \sqrt{1 + 4t^2 + 9t^4}$,
Particle never stops

9 $\vec{v} = 6t\vec{i} + 3t^2\vec{j}$,
$\|\vec{v}\| = 3|t| \cdot \sqrt{4 + t^2}$,
Stops when $t = 0$

11 $\vec{v} = 6t\cos(t^2)\vec{i} - 6t\sin(t^2)\vec{j}$,
$\|\vec{v}\| = 6|t|$,
Stops when $t = 0$

13 Length $= \sqrt{42}$

15 Length $= e - 1$

17 $\vec{v} = -6\pi\sin(2\pi t)\vec{i} + 6\pi\cos(2\pi t)\vec{j}$,
$\vec{a} = -12\pi^2\cos(2\pi t)\vec{i} - 12\pi^2\sin(2\pi t)\vec{j}$,
$\vec{v} \cdot \vec{a} = 0$, $\|\vec{v}\| = 6\pi$, $\|\vec{a}\| = 12\pi^2$

19 Line through $(2, 3, 5)$ in direction of $\vec{i} - 2\vec{j} - \vec{k}$,
$\vec{v} = 2t(\vec{i} - 2\vec{j} - \vec{k})$, $\vec{a} = 2(\vec{i} - 2\vec{j} - \vec{k})$

21 $x = 1 + 2(t - 2)$, $y = 2$, $z = 4 + 12(t - 2)$

23 Vertical: $t = 3$
Horizontal: $t = \pm 1$
As $t \to \infty$, $x \to \infty$, $y \to \infty$
As $t \to -\infty$, $x \to \infty$, $y \to -\infty$

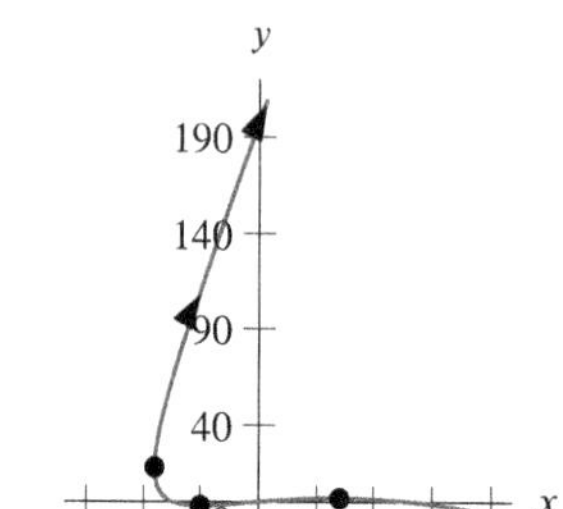

25 (a) $\vec{v}(2) \approx -4\vec{i} + 5\vec{j}$,
Speed $\approx \sqrt{41}$
(b) About $t = 1.5$
(c) About $t = 3$

27 (a) $x = 2 + 0.6t$, $y = -1 + 0.8t$, $z = 5 - 1.2t$, $0 \le t \le 5$
(b) $x = 2 + 1.92t$, $y = -1 + 2.56t$, $z = 5 - 3.84t$, $0 \le t \le 1.56$

29 (a) 6.4 meters
(b) 1.14 sec
(c) 15.81 m/sec
(d) $(11.4, -5.7, 0)$
(e) -9.8 m/sec^2

31 (a) 5 secs; $(10, 15, 100)$
(b) $t = 0, 10$ secs, $\sqrt{113}$ cm/sec
(c) 5 secs, $\sqrt{13}$ cm/sec

33 (a) $t = 5.181$ sec
(b) $x = 103.616$ meters
(c) 2 meters
(d) 9.8 meters/sec^2
(e) $\theta = 0.896$; $v = 32.016$ meters/sec

35 (a) (IV); 4.5 sec; (0, 8.9 m, 0)
(b) (II); 3.2 sec; base of tower
(c) (V); 10 sec; halfway up

37 (a) $-2\vec{i}$
(b) $(0, 3)$
(c) π

39 (a) π m/sec
(b) 2.45 m
(c) 3.01 m

41 (a) $(x, y) = (t, 1)$
(b) $(x, y) = (t + \cos t, 1 - \sin t)$

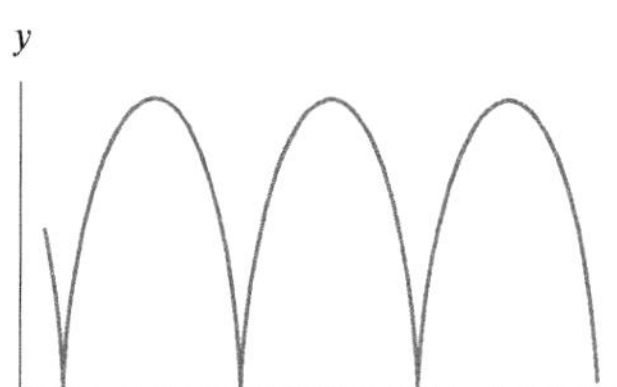

43 SHORT ANSWER NOT WRITTEN

45 (a) R, counterclockwise, $2\pi/\omega$
(b) $\vec{v} = -\omega R\sin(\omega t)\vec{i} + \omega R\cos(\omega t)\vec{j}$
(c) $\vec{a} = -\omega^2\vec{r}$

47 Same path, B moves twice as fast

49 Counterclockwise

51 Orthogonal only if speed is constant

53 Length $= \int_A^B \|\vec{v}(t)\|\, dt$

55 $0 \le t \le 10/\sqrt{2}$

57 True

59 False

61 True

63 False

65 False

67 False

69 (a) $x - \sqrt{6}y + z = 3 - 7\sqrt{6}i$
(b) $\pi/3$
(c) 4 ppm/sec

Section 17.3

1 $\vec{V} = x\vec{i}$

3 $\vec{V} = x\vec{i} + y\vec{j} = \vec{r}$

5 $\vec{V} = -x\vec{i} - y\vec{j} = -\vec{r}$

7 (a) y-axis
(b) Increasing
(c) Neither

9 (a) x-axis
(b) Increases
(c) Decreases

11

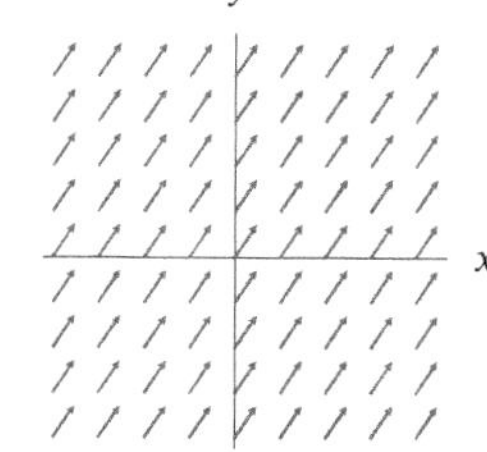

13

y

x

15

y

x

17

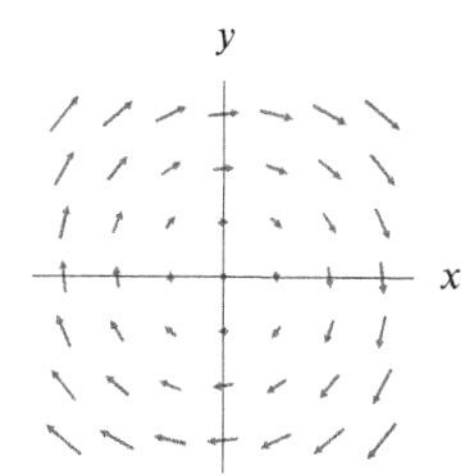

19

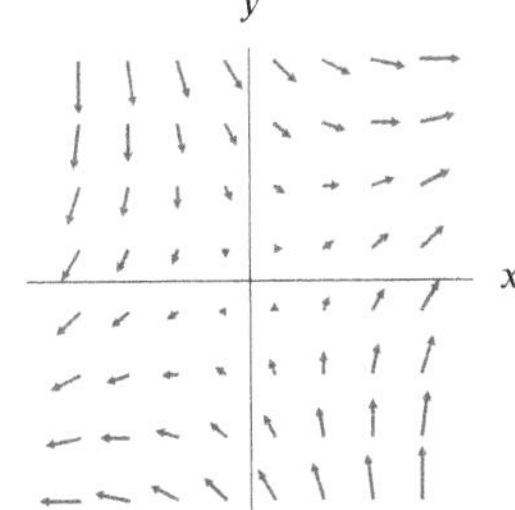

21 (a) III
(b) II
(c) IV
(d) VI

23 $3\vec{i} - 4\vec{j}$, other answers possible

25 $(1/\sqrt{1+x^2})(\vec{i} - x\vec{j})$, other answers possible

27 $\vec{F}(x, y) = (y + \cos x)((1 + y^2)\vec{i} - (x + y)\vec{j})$, other answers possible

29 I, II, III

31 (a) (III)
(b) (II)
(c) (VI)
(d) (V)
(e) (IV)
(f) (I)

33 $\vec{F}(x, y) = \frac{-x\vec{i} - y\vec{j}}{\sqrt{x^2+y^2}}$ (for example)

35 (a) $(1, -3, -7)$; other answers possible
(b) $(0, 0, 0)$; other answers possible
(c) $-4x + y - 3z = 0$; plane through origin

37 (a) Radiates out from origin

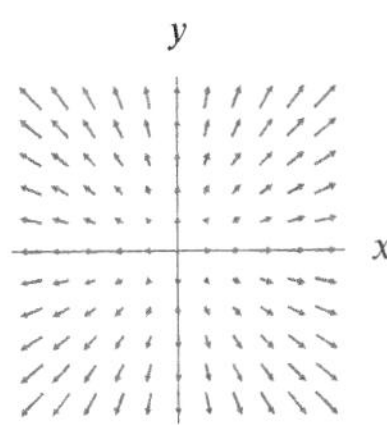

(b) Spirals outward counterclockwise around origin

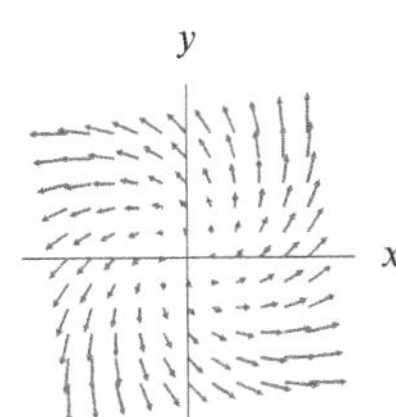

(c) Spirals outward clockwise around origin

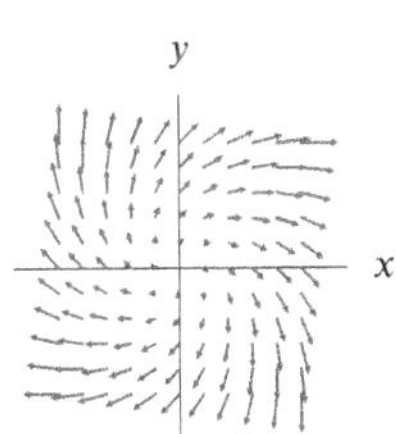

39 (a) $z = f(x, y)$:

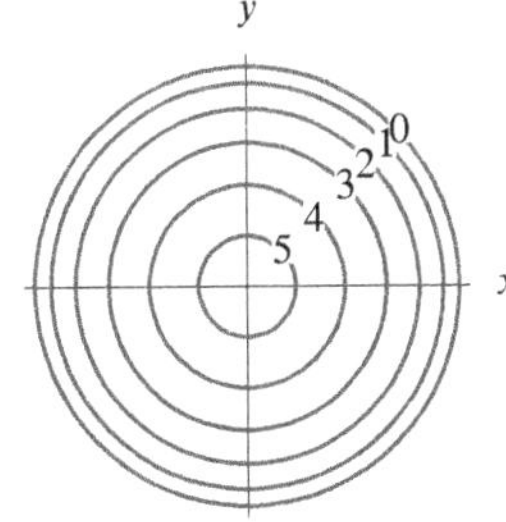

$z = g(x, y)$:

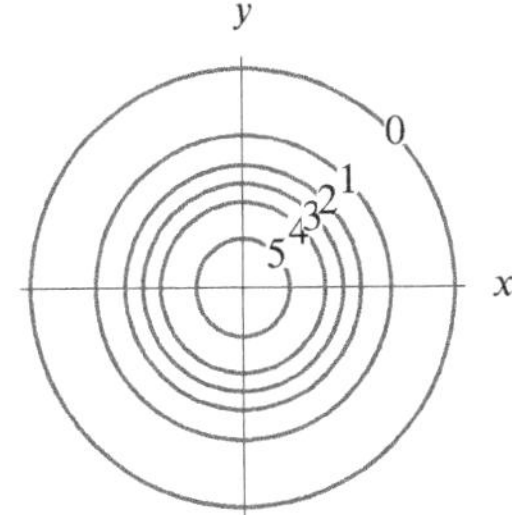

(b)

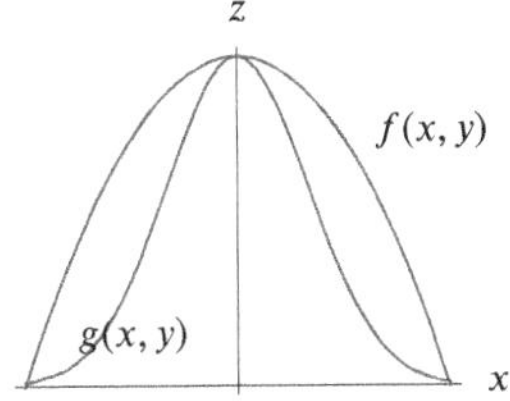

41 To plot $\vec{G}(x, y, z)$ move arrows of $\vec{F}(2x, 2y, 2z)$ halfway to origin

43 $(x^2 + 1)\left(\vec{i} + \vec{j} + \vec{k}\right)$

Section 17.4

1 Field:

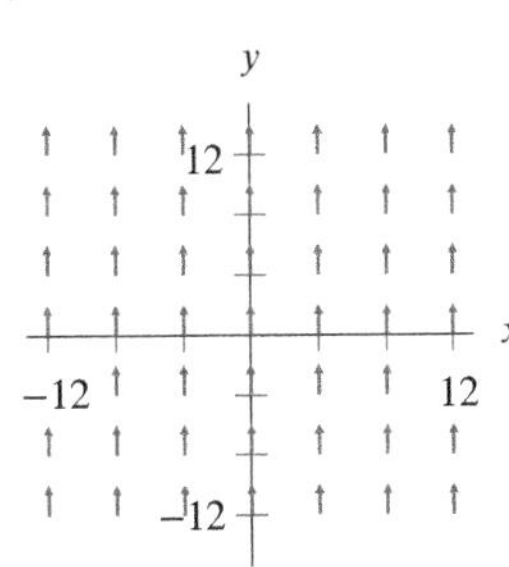

Flow, x = constant:

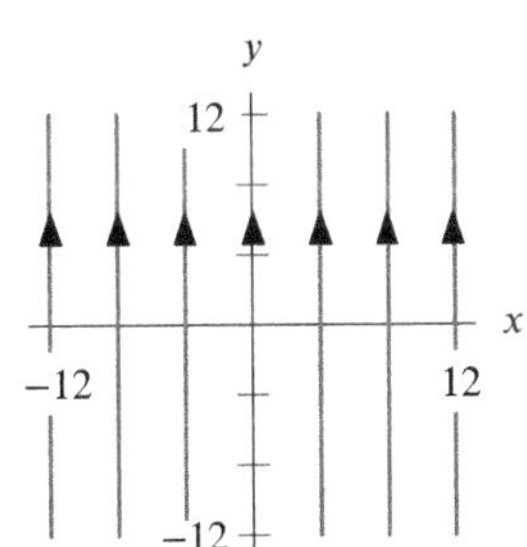

3 Field:

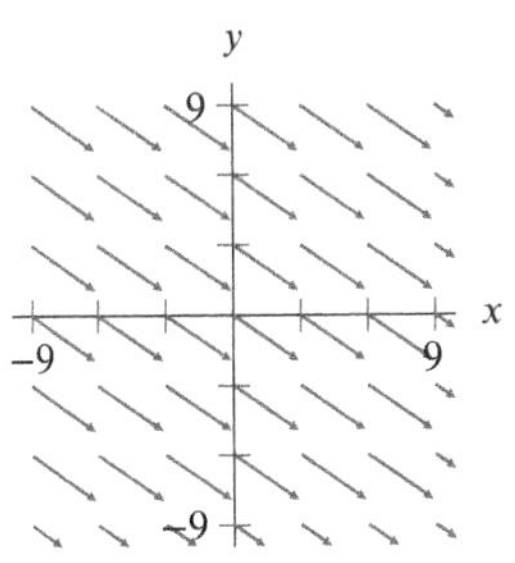

Flow, $y = -(2/3)x + c$:

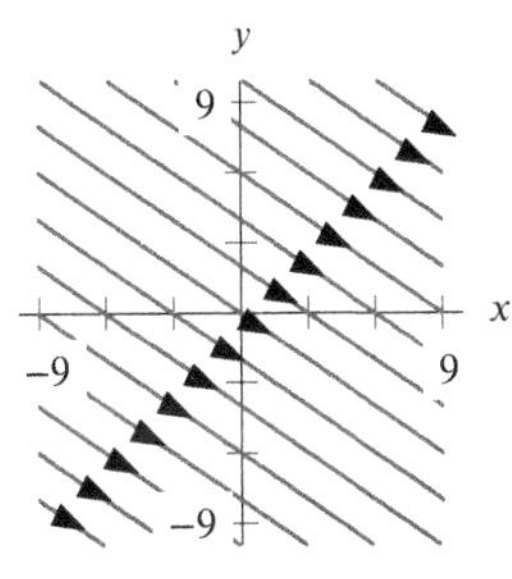

5 Field:

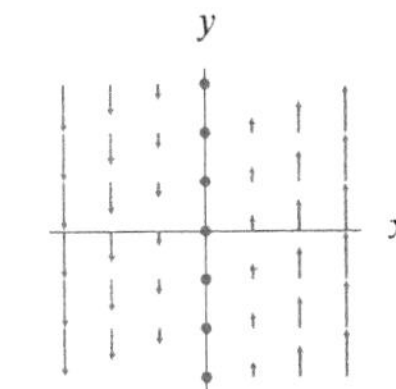

Flow:

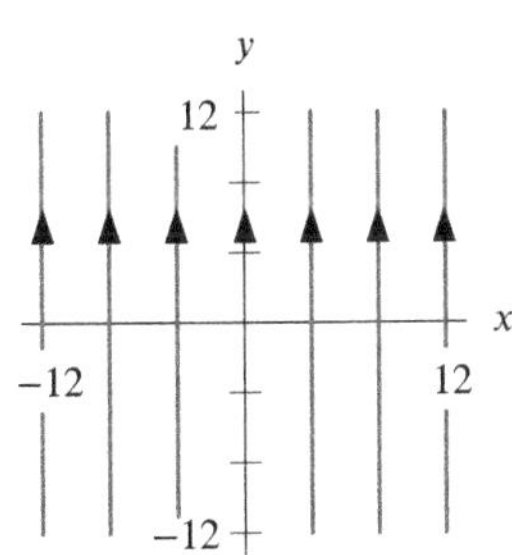

7 Field:

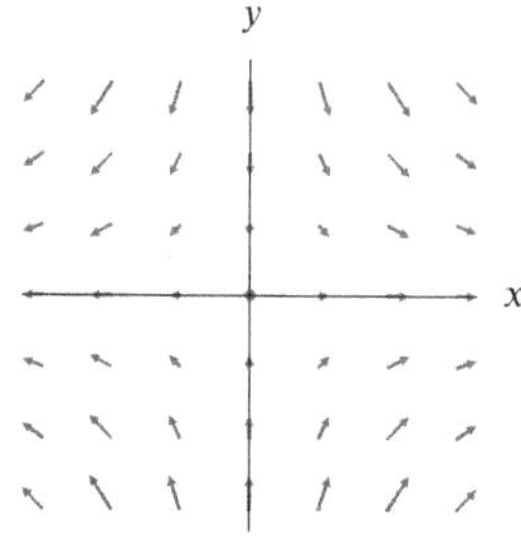

Flow:

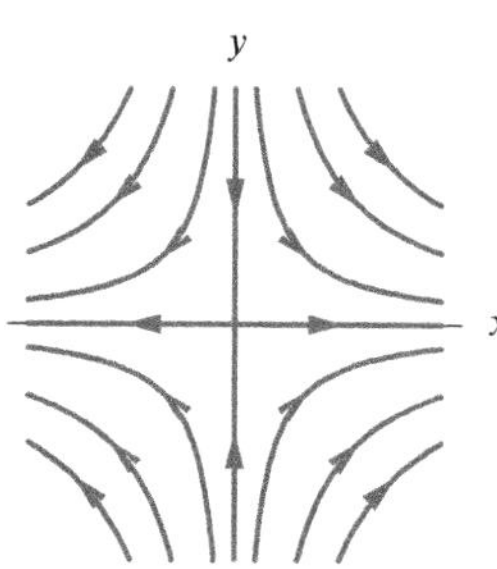

9 Field:

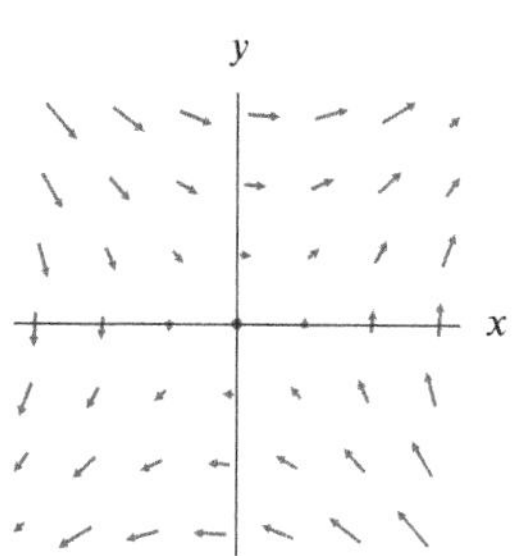

Flow:

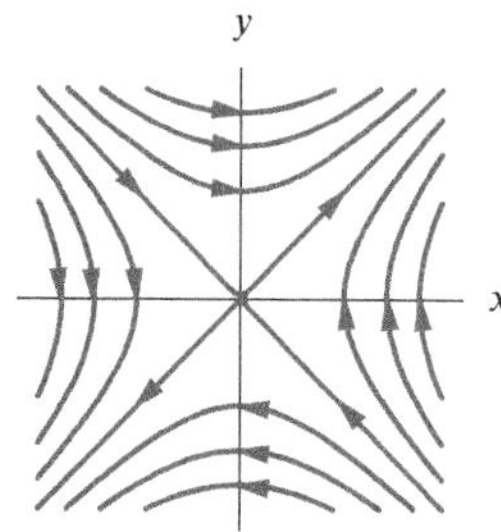

15 (a) Same directions, different magnitudes
(b) Same curves, different parameterizations

23 (a) $\vec{v} = \pi(-y\vec{i} + x\vec{j})/12$
(b) Horizontal circles

25 Counterexample: $\vec{F} = -y\vec{i} + x\vec{j}$

27 $\vec{F}(x, y, z) = \vec{i} + 2x\vec{j} + 3y\vec{k}$

29 True

31 False

33 True

35 True

37 True

Chapter 17 Review

1 $\vec{r} = 2\vec{i} - \vec{j} + 3\vec{k} + t(5\vec{i} + 4\vec{j} - \vec{k})$

3 $x = t,\ y = 5$

5 $x = 4 + 4\sin t,\ y = 4 - 4\cos t$

7 $x = 2 - t, y = -1 + 3t, z = 4 + t.$

9 $x = 1 + 2t, y = 1 - 3t, z = 1 + 5t.$

11 $x = 3\cos t$
$y = 5$
$z = -3\sin t$

13 $\vec{r} = 10\cos(2\pi t/30)\vec{i} - 10\sin(2\pi t/30)\vec{j} + 7\vec{k}$

15 $\vec{v} = \vec{i} + (3t^2 - 1)\vec{j}$

17 $\vec{v} = 6t\vec{i} + 2t\vec{j} - 2t\vec{k}$

19 Vector; $\left((3\cos\sqrt{2t+1})\vec{i} - (3\sin\sqrt{2t+1})\vec{j} + \vec{k}\right)/\sqrt{2t+1}$

21 Vector; $-(\cos t/(2\sqrt{3 + \sin t}))\vec{i} - (\sin t/(2\sqrt{3 + \cos t}))\vec{j}$

23 No

25 Same direction $-\vec{i} + 4\vec{j} - 2\vec{k}$, point $(3, 3, -1)$ in common

27

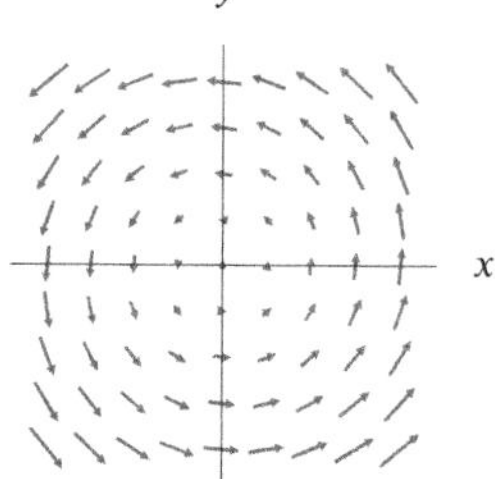

29

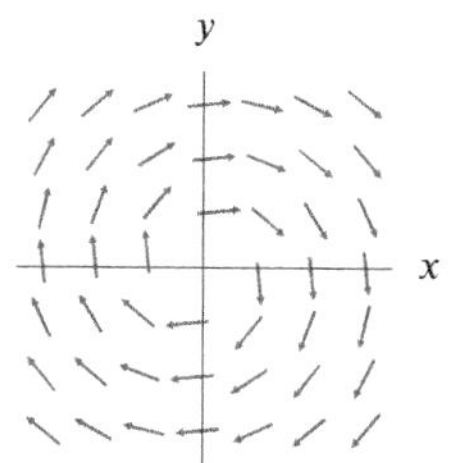

31 (a) $(2, 3, 0)$
(b) 2
(c) No; not on line

33 (b) (i) Yes
(ii) No
(iii) Yes
(iv) No

35 $\vec{E}$: (IV); $\vec{F}$: (I);
$\vec{G}$: (II); $\vec{H}$: (III)

37 $\vec{v} = 0.2\vec{i} - 0.4\vec{j} + 0.4\vec{k}$ m/sec
$\vec{a} = \vec{0}$

39 (a) Max rate change temp with distance; °C/cm
(b) $\sqrt{(g'(t))^2 + (k'(t))^2}$; cm/sec
(c) $f_x \cdot g'(t) + f_y \cdot k'(t)$; °C/sec

41 (a) Yes, $t = 1$, $(x, y) = (-2, -1)$
(b) Yes, $t = -1$, $(x, y) = (2, 3)$
(c) No

43 (a) $x(t) = 5\sin t,\ y(t) = 5\cos t,\ z(t) = 8$
(b) $\vec{v} = -5\vec{j},\ \vec{a} = -5\vec{i}$
(c) $x_{tt}(t) = y_{tt}(t) = 0,$
$z_{tt}(t) = -g,$
$x_t(0) = z_t(0) = 0,$
$y_t(0) = -5, x_t(0) = 5, y_t(0) = 0, z_t(0) = 8$

45 (a) In the direction given by the vector: $\vec{i} - \vec{j}$
(b) Directions given by unit vectors:
$\frac{1}{\sqrt{2}}\vec{i} + \frac{1}{\sqrt{2}}\vec{j}$
$-\frac{1}{\sqrt{2}}\vec{i} - \frac{1}{\sqrt{2}}\vec{j}$
(c) -4

47 $-(2 + 4t^2)/(t^3(1 + t^2)^2)$

49 (a) $52/\sqrt{13}, (8, 12)$
(b) $52/\sqrt{13}, (18, 62)$

51 No, since the point $(0, 1)$ is not on the curve

53 ω: Rate of change of polar angle θ of particle, a: Rate of change of particle's distance from origin

55 (a) $\frac{1}{x^2+y^2+z^2}$
(b) $\frac{1}{\sqrt{x^2+y^2+z^2}}$
(c) $\frac{x}{\sqrt{x^2+y^2+z^2}}\vec{i} + \frac{y}{\sqrt{x^2+y^2+z^2}}\vec{j} + \frac{z}{\sqrt{x^2+y^y+z^2}}\vec{k}$
(d) $\frac{-x}{\sqrt{x+y^2+z^2}}\vec{i} + \frac{-y}{\sqrt{x^2+y^2+z^2}}\vec{j} + \frac{-z}{\sqrt{x^2+y^y+z^2}}\vec{k}$
(e) $\frac{\cos t}{2\sqrt{2}}\vec{i} + \frac{\sin t}{2\sqrt{2}}\vec{j} + \frac{1}{2\sqrt{2}}\vec{k}$
(f) $\frac{1}{\sqrt{2}}$

57 No

59 (b) $\vec{i} + 2\vec{j} + 3\vec{k} + 5\cos t((1/\sqrt{2})\vec{i} - (1/\sqrt{2})\vec{j}) + 5\sin t((1/\sqrt{6})\vec{i} + (1/\sqrt{6})\vec{j} - (\sqrt{2}/\sqrt{3})\vec{k})$

61 (b) $e^{-t}\vec{i} - 2e^{-t}\vec{j}$,
$(0.0025e^{3t} + 0.9975e^{-t})\vec{i} + (0.005e^{3t} - 1.995e^{-t})\vec{j}$,
$(-0.0025e^{3t} + 1.0025e^{-t})\vec{i} + (-0.005e^{3t} + 2.005e^{-t})\vec{j}$

Section 18.1

1 Negative

3 Zero

5 Zero

7 0

9 0

11 28

13 16

15 −48

17 19/3

19 20

21 28

23 −10

25 −9

27 0

29 C_1 is zero; C_2 is pos; C_3 is neg

31 C_1 is 0; C_2 is neg; C_3 is pos

33 −8

35 C_1, C_2

37 $a < 0$

39 $c > 1$

41 (a) (i)

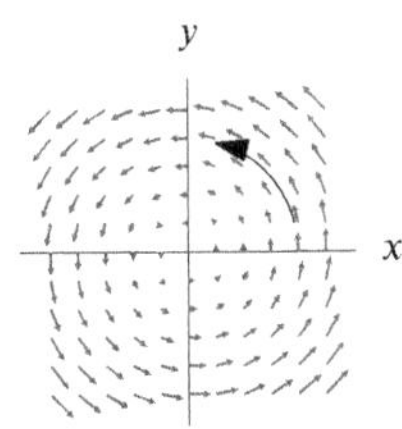

(ii)

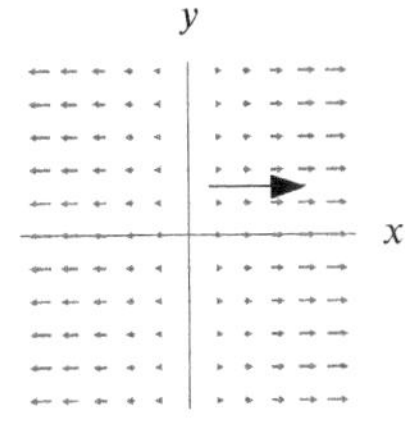

(iii)

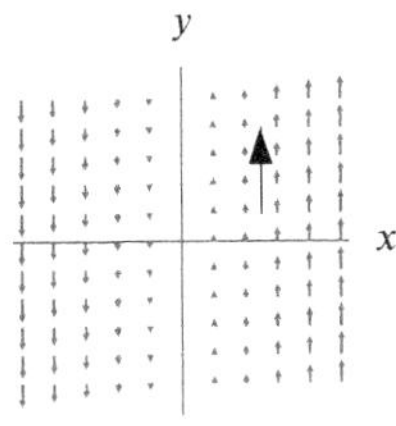

(iv)

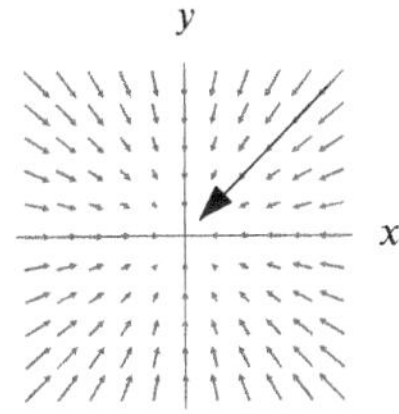

(b) (i), (iii)

43 Positive

45 0

47 11

49 6

51 13

53 $-1.2 \cdot 10^7$ meter^2/sec

55 14π; tangent to C
Same direction, $||\vec{F}|| = 7$
-14π; tangent to C
Opposite direction, $||\vec{F}|| = 7$

59 $-2.5 \cdot 10^{-5} GMm$

61 If $\int_C \vec{F} \cdot d\vec{r} < 0$, then $\int_{-C} \vec{F} \cdot d\vec{r} > 0$

63 $\vec{F} = \vec{i} - \vec{j}$

65 False

67 True

69 False

71 True

73 False

75 $-\int_C \vec{E} \cdot d\vec{r}$

77 Spheres centered at origin

Section 18.2

1 $\int_0^{\pi} (\cos^2 t - \sin^2 t)\, dt$
Other answers are possible

3 $\int_0^{2\pi} (-\sin t \cos(\cos t) + \cos t \cos(\sin t)) dt$

5 24

7 -4

9 -6

11 9

13 82/3

15 12

17 116.28

19 12

21 21

23 0

25 $\int_C 3x dx - y \sin x dy$

27 $(x + 2y)\vec{i} + x^2 y\vec{j}$

29 3124

31 144

33 77,000/3

35 (a) $11/6$
(b) $7/6$

37 (a) $3/2$
(b) $3/2$

39 200π

43 (a) -5
(b) 5
(c) 0

45 F could point with C at some points and against C at others

47 $y = \pi/2$, $x = t$, $0 \le t \le 3$, $\int_C \vec{F} \cdot d\vec{r} = 3$

49 True

51 True

53 False

55 False

57 (a)

Section 18.3

1 12

3 Negative, not path-independent

5 Negative, not path-independent

7 Path-independent

9 Path-independent

11 Path-independent

13 $f(x, y) = x^2 y + K$

15 $f(x, y, z) = e^{xyz} + \sin(xz^2) + C$
C = constant

17 -2

19 2

21 0

23 $e^3 - 1$

25 0

27 PQ

31 Yes

33 Yes.

35 $5xy + y^2/2$
(a) 50
(b) 50

37 (a) 50π
(b) No, integral over closed path not zero

39 Use $f_{xy} = f_{yx}$

41 (a) e
(b) e

43 9/2

45 $\frac{3}{\sqrt{2}} \ln(\frac{3}{\sqrt{2}} + 1)$

47 $e^{(1.25\pi)^2/2} - 1$

49 (a) $7e^3 - 2e$
(b) $7e^3 - 2e$

51 (a) 9
(b) 0

53 (a)

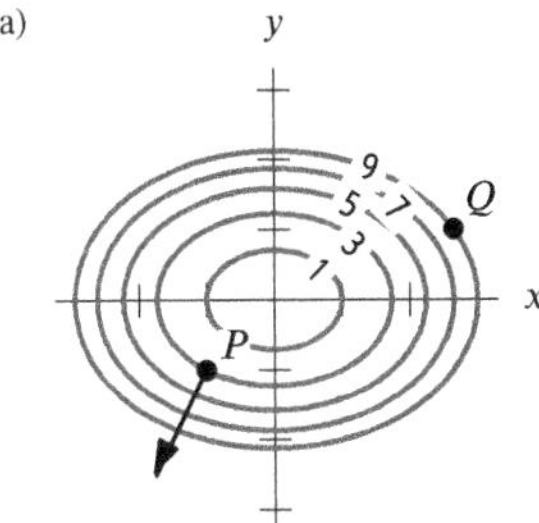

(b) Shorter
(c) 6

55 (a) Positive
(b) Not gradient
(c) $\vec{F_2}$

57 (a) (8, 9)
(b) 50

59 (a) $2\pi mg$
(b) Yes

61 $f(Q) - f(P)$ where $\vec{F} = \text{grad} f$

63 Methods other than Theorem 18.1 can be used

65 Gradient of any function

71 True

73 True

75 False

77 True

79 False

83 If $A'(x) = a(x)$, then $f(x, y) = A(x)$ is potential function
$x + x^2/2 + x^3/3 + C$, any C

85 (a) $\vec{F} - \text{grad}\,\phi = -y\,\text{grad}\,h$
(b) 30

87 (a) $\vec{F} - \text{grad}\,\phi = -(x + 2y)\,\text{grad}\,h$
(b) -50

89 (a) Increases

Section 18.4

1 No

3 No

5 $f(x, y) = x^3/3 + xy^2 + C$

7 Yes, $f = \ln A|xyz|$ where $A > 0$

9 No

11 -2π

13 -6

15 -12

17 $-3\pi m^2$

19 (a)

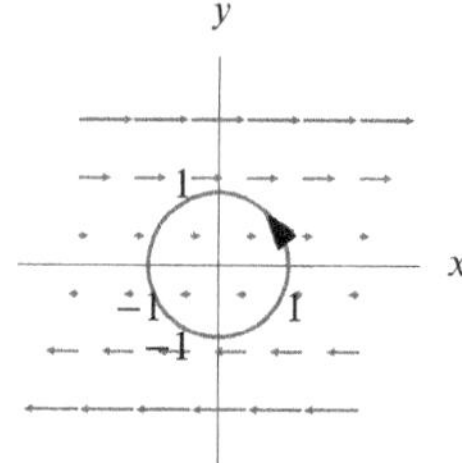

(b) $-\pi$

21 $e - \cos 1$

23 $-9\pi/8$

25 (a) 0
(b) 0
(c) 0
(d) -6π
(e) -6π
(f) 0
(g) -6π

27 (a) 0
(b) 10
(c) -8π
(d) 7

29 (b) 0
(c) $\vec{G} = \nabla(xyz + zy + z)$
(d) $\vec{H}_1 = \nabla(yx^2), \vec{H}_2 = \nabla(y(x+z))$

31 πab

33 3/2

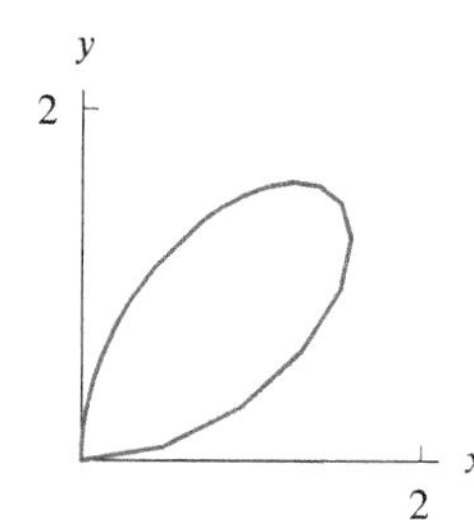

35

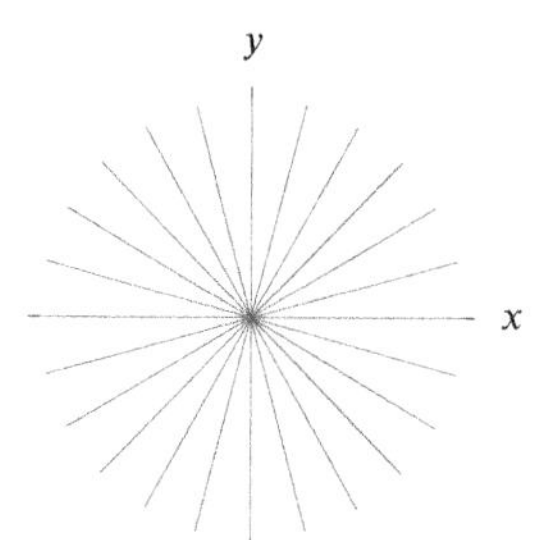

37 (a) Possible answers are:
$\vec{F} = \text{grad}(xy)$
$\vec{G} = \text{grad}(\arctan(x/y)),\ y \neq 0$
$\vec{H} = \text{grad}\left((x^2 + y^2)^{1/2}\right),\ (x, y) \neq (0, 0)$
(b) $0, -2\pi, 0$
(c) Does not apply to $\vec{G}$, $\vec{H}$; holes in domain

39 $L_1 < L_2 < L_3$

41 (a) $21\pi/2$
(b) 2

43 (a) $\vec{0}$
(b) $q/\|\vec{r}\|$

45 Green's Theorem does not apply;
Line integral depends on $\vec{F}$

47

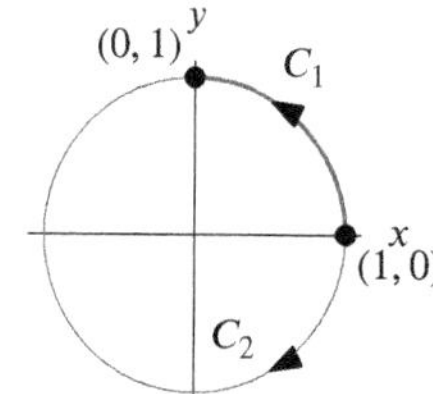

49 True

51 True

53 True

55 True

Chapter 18 Review

1 Positive

3 (a) Zero
(b) C_1, C_3: Zero
C_2: Negative
C_4: Positive
(c) Zero

5 Scalar; 12

7 −58

9 50

11 18

13 1372

15 Not path-independent

17 Path-independent

19 Path-independent

21 Path-independent

23 350

25 $27\pi/2$

27 45

29 0, 100

31 (ii), (iv)

33 (a) 24
(b) 12
(c) −12

37 18

39 −36

41 (a) 0
(b) 24

43 (a) 0
(b) 6
(c) $75\pi/2$
(d) 14

45 (a) 9/2
(b) −9/2

47

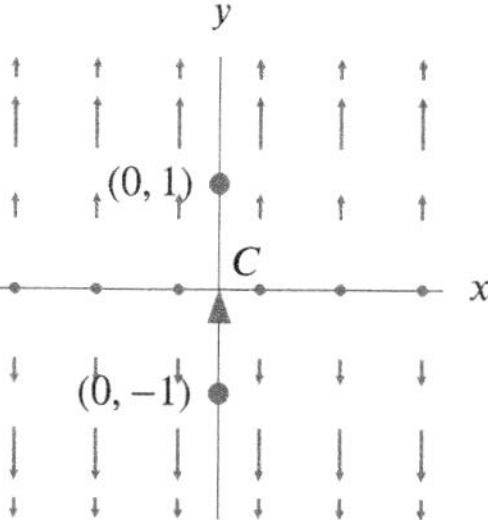

49 (a) Closed curve oriented counterclockwise
(b) Closed curve oriented clockwise with $y > 0$ or
Closed curve oriented counterclockwise if $y < 0$
(Other answers are possible)

51 (a)

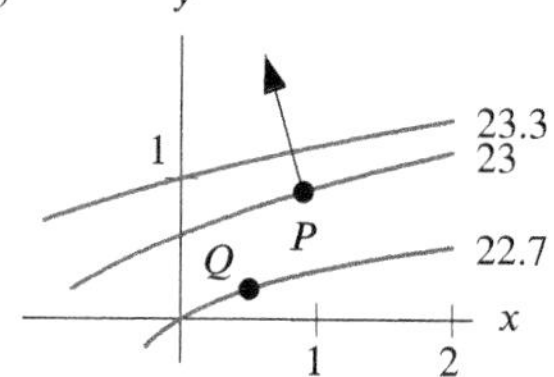

(b) Longer
(c) −0.3

53 (a) $\pi/2$
(b) 0

57 (a) $\omega = 3000$ rad/hr
$K = 3 \cdot 10^7$ m^2·rad/hr
(b) Inside tornado:

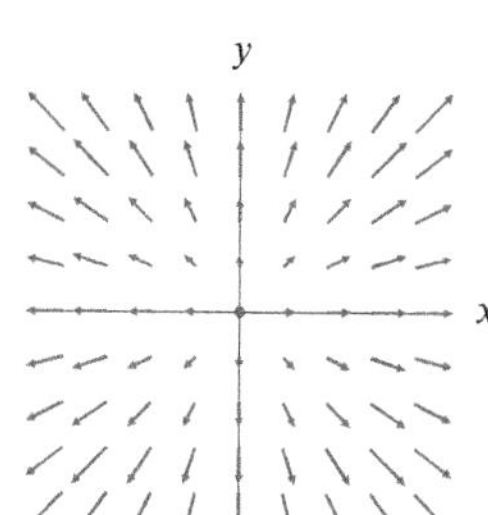

View from great distance:

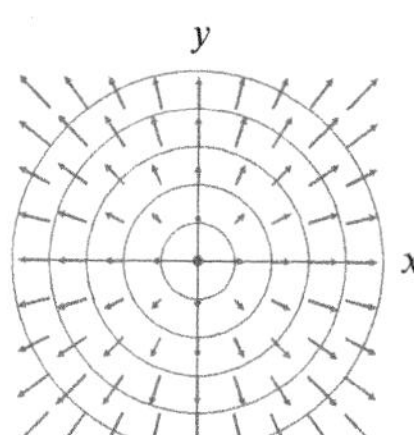

(c) $r < 100$ m, circulation is $2\omega\pi r^2$
$r \geq 100$ m, circulation is $2K\pi$

59 (a) $-(\pi/2)(-6a^2 + a^4)$, $a = \sqrt{3}$
(b) Integrand $3 - x^2 - y^2$ positive inside disk of radius $\sqrt{3}$, negative outside

61 $18a + 18b + 36c + (81d/2)$,
$-18a - 18b - 36c - (81d/2)$,
curves go in opposite directions

Section 19.1

1 $-3\vec{i}$

3 $15\vec{j}$

5 Rectangle in xz plane with area 150, oriented pos y direction

7 (a) 45
(b) -45

9 (a) Positive
(b) Positive
(c) Negative
(d) Negative
(e) Positive

11 (a) Negative
(b) Positive
(c) Negative
(d) Negative
(e) Zero

13 -12

15 4

17 12π

19 4

21 0

23 $10\pi/\sqrt{3}$

25 6

27 -75π

29 $-\pi^3$

31 2000π

33 32000π

35 0

37 12.8

39 Zero

41 24

43 -160π

45 $130/\sqrt{2}$

47 42

49 -96π

51 $4\sqrt{3}$

53 $\pi \sin 9$

55 0

57 -27

61 (a) Zero
(b) Zero

63 4π

65 (a) 0
(b) 32π

67 (a) Zero
(b) Zero

71 (a)

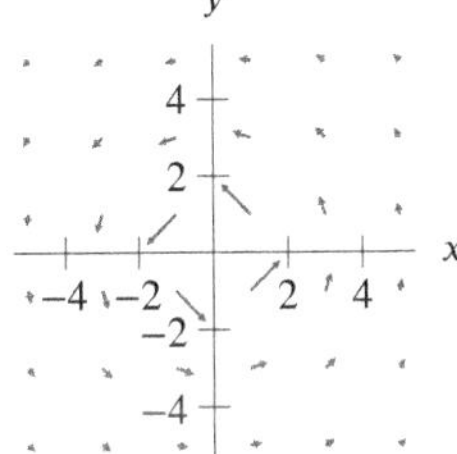

(b) 0
(c) $Ih \ln |b/a|/2\pi$

73 Sign of $\int_S \vec{F} \cdot d\vec{A}$ depends on both $\vec{F}$ and S

75 $\vec{F} = z\vec{k}$
S: $0 \le x \le 1$, $0 \le y \le 1$, $z = 1$, oriented upwards

77 False

79 True

81 True

83 False

85 False

87 (a) 0
(b) 0

Section 19.2

1 $\left(-3\vec{i} + 5\vec{j} + \vec{k}\right) dx\,dy$

3 $\left(-4x\vec{i} + 6y\vec{j} + \vec{k}\right) dx\,dy$

5 $\int_{-2}^{3} \int_0^5 70\,dy\,dx$

7 $\int_0^5 \int_0^{5-x} (yz \sin x - 2xy \cos 2y + xy)\,dy\,dx$

9 -500

11 $-5/3 - \sin 1 = -2.508$

13 $\int_0^{\pi/2} \int_0^5 10(\cos\theta + 2\sin\theta)\,dz\,d\theta$

15 $\int_0^{2\pi} \int_{-8}^{8} \left(6z^2 \cos\theta + 6 \sin\theta e^{6\cos\theta}\right) dz\,d\theta$

17 2000

19 $100\sqrt{2}/3$

21 $\int_0^{2\pi} \int_0^{\pi/2} 100(\sin\phi\cos\theta + 2\sin\phi\sin\theta + 3\cos\phi)\sin\phi\,d\phi\,d\theta$

23 $\int_{-\pi/2}^{\pi/2} \int_0^{\pi} 16\cos^2\phi \sin^2\phi \cos\theta\,d\phi\,d\theta$

25 $8000/3$

27 $(8 - 5\sqrt{2})\pi/6 = 0.486$

29 6

31 6

33 18

35 36π

37 $7/3$

39 $\pi \sin 25$

41 $\pi/2$

43 1296π

45 π

47 $100\sqrt{27}$

49 2.228

51 (a) $\int_R a/\sqrt{a^2 - x^2 - y^2}\,dx\,dy$
(b) $\int_0^{2\pi} \int_0^a ar/\sqrt{a^2 - r^2}\,dr\,d\theta$.
(c) $2\pi a^2$

53 36π

55 $2\pi/3$

57 $4\pi a^3$

59 -1

61 $11\pi/2$

65 (a) Constant inside cylinder radius a
(b) $\vec{E} = \begin{cases} \frac{1}{2}k\delta_0 r\vec{e}_r & \text{if } r \le a \\ \frac{1}{2}k\delta_0 \frac{a^2}{r}\vec{e}_r & \text{if } r > a \end{cases}$

67 $\vec{n} = \left(-f_x\vec{i} - f_y\vec{j} + \vec{k}\right)/\sqrt{f_x^2 + f_y^2 + 1}$
$dA = \sqrt{f_x^2 + f_y^2 + 1}\,dx\,dy$

69 $r = 10$, $0 \le \theta \le 2$, $0 \le z \le 3$, oriented outwards

71 False

Section 19.3

1 $2x + xe^z$

3 (I)

5 0

7 $4x$

9 $2x/(x^2+1) - \sin y + xye^z$

11 0

13 (a) Positive
(b) Zero
(c) Negative

15

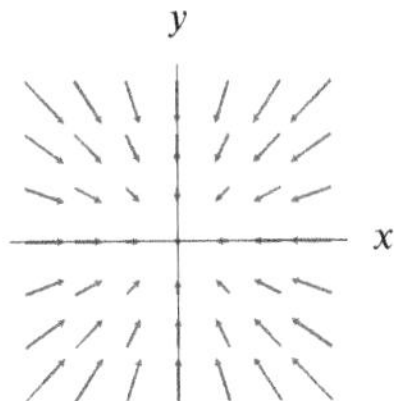

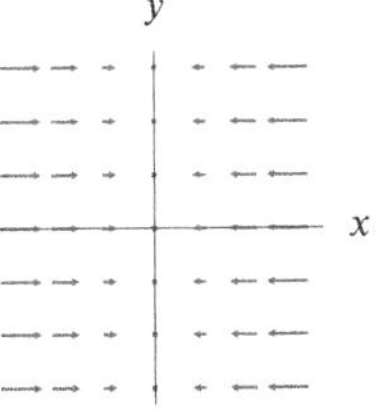

17 -0.030

19 (a) (i) $0.016\pi/3$
(ii) -0.08
(b) Flux positive at $(2, 0, 0)$ and negative at $(0, 0, 10)$

21 (a) $4w^3$
(b) 4
(c) 4

23 $\operatorname{div}\vec{v} = -6$

25 (a) $-1/3, 1$
(b) $1/3$

27 Undefined

29 (b)

31 (a) 0
(b) Undefined

33 (a) $\rho(0) < \rho(1000) < \rho(5000)$
(b) cars/hour
(d) $\rho(x) = 4125/(55 - x/50)$
if $0 \le x < 2000$
$\rho(x) = 4125/15 = 275$
if $2000 \le x < 7000$
$\rho(x) = \dfrac{4125}{(15 + (x - 7000)/25)}$
if $7000 \le x < 8000$
$\rho(x) = 4125/55 = 75$
if $x \ge 8000$
(e) 139 ft. at $x = 0$
89 ft. at $x = 1000$
38 ft. at $x = 5000$

35 (a) 0
(b) 0

39 0

41 $\vec{b} \cdot (\vec{a} \times \vec{r})$

43 (d)

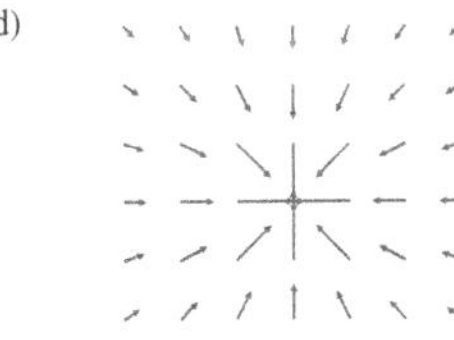

45 div $\vec{F} = 2x + 2 - 2z$

47 $\vec{F}(x, y, z) = 2x\vec{i} + 3y\vec{j} + 4z\vec{k}$

49 $\vec{F}(x, y) = 2x\vec{i}$

51 False

53 False

55 False

57 False

59 False

61 False

Section 19.4

1 24

3 8

5 Zero

7 24

9 72

11 288

13 36π

15 620π

17 5π

19 420

21 $10\pi a^3$

23 Yes; −3.22

25 $\int_S \vec{F} \cdot d\vec{A} = \int_W \operatorname{div} \vec{F}\, dV = 0$

27 (a) $cb(12a - a^2)$
(b) 6, 10, 10; 3600

29 (a) 4π
(b) 0
(c) 4π

31 4π

33 (a) 2
(b) 0.016
(c) 0.016053 ⋯

35 (a) 30 watts/km^3
(b) $\alpha = 10$ watts/km^3
(d) 6847°C

37 (a) 0
(c) No

39 S not the boundary of a solid region

41 Any sphere

43 False

45 False.

47 True.

49 True

51 True

53 True

55 True

Chapter 19 Review

1 Scalar; -100π

3 0

5 0

7 75π

9 80

11 −12

13 0

15 Zero

17 $3(8 + \sin 4)$

19 20

21 2π

23 $b > 0$

25 $a > 0$

27 (a) 3/2
(b) 3/2

29 32π

31 162

33 120π

35 Flux through S_1 = Flux through S_2 < Flux through S_3 < Flux through S_4

37 $a = 6$
Cannot say anything about b and c

39 (a) Negative
(b) Positive
(c) Zero

41 (a) Flux = c^3
(b) div $\vec{F} = 1$
(c) div $\vec{F} = 1$

43 (a) $2c^3$
(b) 2
(c) 2

45 -2π

47 42π

49 $-7(\vec{i} + \vec{k})/(16\pi)$

51 (a) div $\vec{F} = 0$
(b) 12

53 $(20{,}000\pi/3) - 128$

55 (a) 35π
(b) 105π
(c) 70π

57 (a) 0, except at origin
(b) 4π
(c) 0
(d) 4π
(e) 4π if origin inside
0 if origin outside

61 (a) (i) Total charge inside W
(ii) Total current out of S

63 (b) $\|\vec{v}\| = K/r^2$
(c) Flux = $4\pi K/3$
(d) Zero

65 (a)

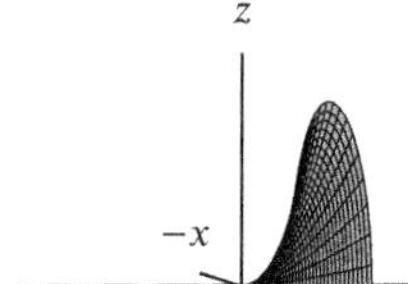

(b) $32\pi/5$

Section 20.1

1 Vector; $\vec{i} + \vec{j} - \vec{k}$

3 Vector; $(x + 1)\vec{i} - (y + 2)\vec{j}$

5 Vector; $\vec{0}$

7 $4y\vec{k}$

9 $4x\vec{i} - 5y\vec{j} + z\vec{k}$

11 $\vec{0}$

13 $\vec{0}$

15 Zero curl

17 Nonzero curl

19 0

21 $50\vec{i} + 300\vec{j} + 2\vec{k}$

23 (a) $(f-c)\vec{i} + (be^z - e\cos x)\vec{j} + (2dx - 3ay^2)\vec{k}$
(b) $f = c$
(c) $f = c$, $b = e = 0$

25 (a) Horizontal
(b) Vertical
(c) Parallel to the yz-plane, making angle t with horizontal

27 (a) $w = 1$

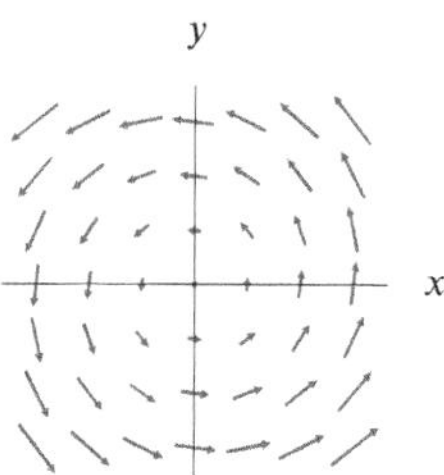

$w = -1$

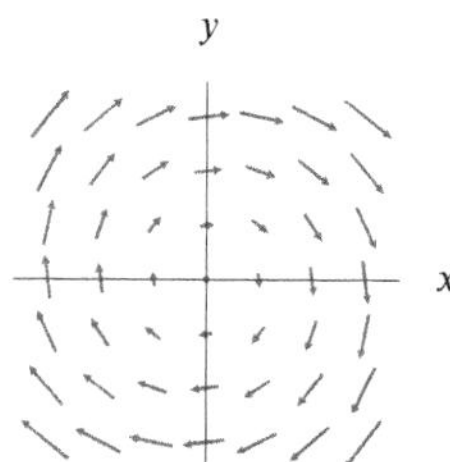

(b) $|\omega| \cdot \sqrt{x^2 + y^2}$
(c) div $\vec{v} = 0$
curl $\vec{v} = 2\omega\vec{k}$
(d) $2\pi\omega R^2$

35 Counterexample: $\vec{F} = y\vec{i}$

37 $\vec{F} = z\vec{i}$

39 True.

41 True

43 False

45 False

47 (a)-z, (c)-y, (d)-x

Section 20.2

1 (a) π
(b) 0

3 Positive

5 -8π

7 −2

9 0

11 18π

13 (a) -2π
(b) $-2\vec{k}$
(c) -2π
(d) Stokes' Theorem

15 No

17 (a) 45π
(b) $81\pi/2$

19 (a) $-\vec{i} - \vec{j} - \vec{k}$
(b) (i) -4π
(ii) $15/2$

21 0

23 0

25 $8\pi/\sqrt{3}$

27 (a) All 3-space
(b) $\frac{2ax\vec{i} + 2by\vec{j} + 2cz\vec{k}}{1+ax^2+by^2+cz^2}$
(c) 0
(d) $\ln(3 + 507\pi^2/4) - \ln(2)$

29 -8π

31 4π

33 63π

35 (a) $\vec{0}$
(b) 0
(c) 0

37 (a) Parallel to xy-plane; same in all horizontal planes
(b) $(\partial F_2/\partial x - \partial F_1/\partial y)\vec{k}$
(d) Green's Theorem

39 C not the boundary of a surface

41 Any oriented circle

43 True

45 False

47 True

49 True

51 False

Section 20.3

1 Yes

3 Yes

5 No

7 Yes

9 Yes

11 Yes

13 (a) No
(b) Yes

15 Curl yes; Divergence yes

17 Curl yes; Divergence yes

21 $(1/2)\vec{b} \times \vec{r}$

23 No

25 (a) Yes
(b) Yes
(c) Yes

27 (a) Yes
(b) No
(c) No

29 (b) $\nabla^2\psi = -\operatorname{div}\vec{A}$

31 Curl of scalar function not defined

33 $f(x, y, z) = x^2$

35 False

37 True

39 (a) $\operatorname{curl}\vec{E} = \vec{0}$
(b) 3-space minus a point if $p > 0$
3-space if $p \le 0$.
(c) Satisfies test for all p.
$\phi(r) = r^{2-p}$ if $p \neq 2$.
$\phi(r) = \ln r$ if $p = 2$.

Chapter 20 Review

1 $\vec{i} - \vec{j} - \vec{k}$

3 $(c), (d), (f)$

5 C_2, C_3, C_4, C_6

7 Defined; scalar

9 Defined; vector

11 Nonzero curl

13 $\operatorname{div}\vec{F} = y + z + x$
$\operatorname{curl}\vec{F} = -y\vec{i} - z\vec{j} - x\vec{k}$
$\vec{F}$ not solenoidal; not irrotational

15 $\operatorname{div}\vec{F} = 0$
$\operatorname{curl}\vec{F} = (2y - \cos(x+z))\vec{i}$
$- \left(2x - e^{y+z}\right)\vec{j} + \left(\cos(x+z) - e^{y+z}\right)\vec{k}$
$\vec{F}$ is solenoidal; not irrotational

17 (a) 18π
(b) 18π

19 20π

21 0

23 (a) is (I); (b) is (I); (c) is (V); (d) is (III); (e) is (IV); (f) is (V)

25 (a) -1125π
(b) Not defined
(c) 0
(d) Not defined
(e) $384\pi/5$
(f) $353/4$
(g) Not defined
(h) 54

27 (a) 23
(b) Approx $0.0003\pi/\sqrt{3}$

29 (a) $\operatorname{div}\vec{F} < 0$, $\operatorname{div}\vec{G} < 0$
(b) $\operatorname{curl}\vec{F} = 0$, $\operatorname{curl}\vec{G} = 0$
(c) Yes
(d) Yes
(e) No
(f) No

31 0

33 4π

35 210

37 $12\pi/5$

39 150π

41 $e - 1$

43 (a) 18
(b) $81/2$

45 (a) No
(b) No

47 (a) 2π
(b) $\vec{0}$ except on z-axis
(c) No
(d) Yes; 0
(e) No

49 (a) 50
(b) 30π
(c) 8
(d) -18π

51 (a) $(4(p+q)\pi R^3/3)$
(b) $(4(p+q)\pi R^3/3)$

Section 21.1

1 Curve

3 Surface

5 Horizontal disk of radius 5 in plane $z = 7$

7 Helix radius 5 about z-axis

9 Top hemisphere

11 Vertical segment

13 $x = 1$
$y = s$
$z = t$

15 $x = 1 + t$
$y = 1 + s$
$z = s + t$

17 $\vec{r}(s,t) = (s + 2t)\vec{i} + (2s + t)\vec{j} + 3s\vec{k}$, other answers possible

19 $(0,0,0)$, $2\vec{i} + \vec{j} - \vec{k}$, $3\vec{i} - 5\vec{j} + 2\vec{k}$

21 $\vec{r}(s,t) = (3 + s + t)\vec{i} + (5 - s)\vec{j} + (7 - t)\vec{k}$, other answers possible

23 (a) Yes
(b) No

25 $s = s_0$: lines parallel to y-axis with $z = 1$
$t = t_0$: lines parallel to x-axis with $z = 1$

27 $s = s_0$: parabolas in planes parallel to yz-plane
$t = t_0$: parabolas in planes parallel to xz-plane

29 $s = 4, t = 2$
$(x, y, z) = (x_0 + 10, y_0 - 4, z_0 + 18)$

31 Horizontal circle

33 (a) $x = \left(\cos\left(\frac{\pi}{3}t\right) + 3\right)\cos\theta$
$y = \left(\cos\left(\frac{\pi}{3}t\right) + 3\right)\sin\theta$
$z = t \quad 0 \le \theta \le 2\pi,\ 0 \le t \le 48$
(b) 456π in^3

35 If $\theta < \pi$, then $(\theta + \pi, \pi/4)$
If $\theta \ge \pi$, then $(\theta - \pi, \pi/4)$

37 $x = r\cos\theta, \quad 0 \le r \le a$
$y = r\sin\theta, \quad 0 \le \theta \le 2\pi$
$z = (1 - r/a)h$

39 (a) $-x + y + z = 1$,
$0 \le x \le 2$,
$-1 \le y - z \le 1$
(b)

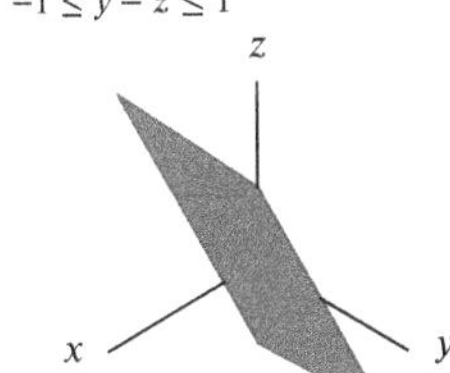

41 (a) $z = (x^2/2) + (y^2/2)$
$0 \le x + y \le 2$
$0 \le x - y \le 2$
(b)

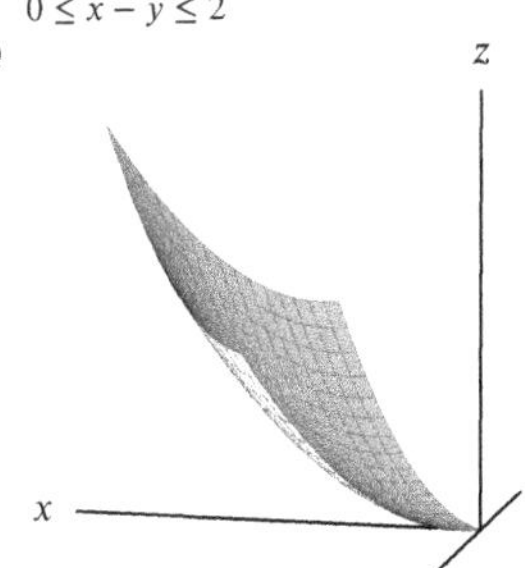

43 Radius: $R\sin\phi$

45 $x + y - z - 3 = 0$

47 True

49 True

51 True

53 False

Section 21.2

1 1

3 e^{2s}

5 $a = 1/10, b = 1$

7 $a = 1/50, b = 1/10$

9 3

11 $\rho^2 \sin\phi$

13 13.5

15 72

17 (a) $(1/(2\pi\sigma^2)) \int_{-\infty}^{\infty} \int_{-\infty}^{2t-x} e^{-(x^2+y^2)/(2\sigma^2)} dy\, dx$
(b) $(1/(\sqrt{\pi}\sigma)) \int_{-\infty}^{t} e^{-u^2/\sigma^2} du$
(c) $(1/(\sqrt{\pi}\sigma)) e^{-t^2/\sigma^2}$
(d) Normal, mean 0, standard deviation $\sigma/\sqrt{2}$

19 R does not correspond to T

21 $x = 2s$, $y = 3t$

23 False

Section 21.3

1 $((s+t)\vec{i} - (s-t)\vec{j} - 2\vec{k})\, ds\, dt$

3 $-e^s(\cos t\,\vec{j} + \sin t\,\vec{k})\, ds\, dt$

5 4/3

7 $6(e^4 - 1)$

9 $-\pi R^7/28$

11 $200\sqrt{14}$

13 $\sqrt{6}\pi$

15 $khw^3/6$ meter^3/sec.

21 Integral gives volume

23 $\vec{r}(s,t) = 2s\vec{i} + t\vec{j}$

25 True

27 False

Chapter 21 Review

1 Cone, height 7 and radius 14

3 $a = 1/15, b = 1/15$

5 $-1/2$

7 0

9 $x = 2 + 5\sin\phi\cos\theta$
$y = -1 + 5\sin\phi\sin\theta$
$z = 3 + 5\cos\phi$

11 (a) Cylinder
(b) Helices

13 $x = a\sin\phi\cos\theta \quad 0 \le \phi \le \pi$
$y = b\sin\phi\sin\theta \quad 0 \le \theta \le 2\pi$
$z = c\cos\phi$

15 (a) $x^2 + y^2 = 9, x \ge 0, 1 \le z \le 2.$
(b)

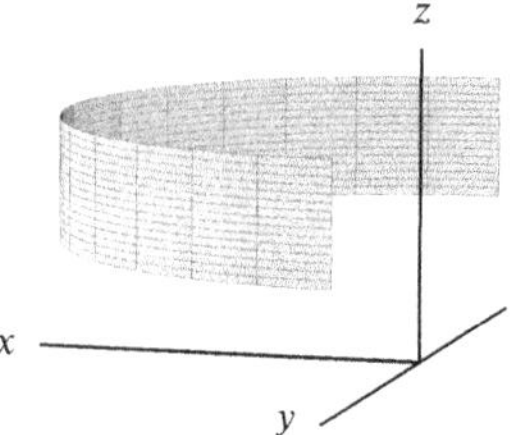

17 0

19 $2\pi c(a^2 + b^2)$

Appendix A

1 (a) $y \le 30$
(b) two zeros

3 -1.05

5 2.5

7 $x = -1.1$

9 0.45

11 1.3

13 (a) $x = -1.15$
(b) $x = 1, x = 1.41$, and $x = -1.41$

15 (a) $x \approx 0.7$
(b) $x \approx 0.4$

17 (a) 4 zeros
(b) [0.65, 0.66], [0.72, 0.73], [1.43, 1.44], [1.7, 1.71]

19 (b) $x \approx 5.573$

21 Bounded $-5 \le f(x) \le 4$

23 Not bounded

Appendix B

1 $2e^{i\pi/2}$

3 $\sqrt{2}e^{i\pi/4}$

5 $0e^{i\theta}$, for any θ.

7 $\sqrt{10}e^{i(\arctan(-3)+\pi)}$

9 $-3 - 4i$

11 $-5 + 12i$

13 $1/4 - 9i/8$

15 $-1/2 + i\sqrt{3}/2$

17 $-125i$

19 $\sqrt{2}/2 + i\sqrt{2}/2$

21 $\sqrt{3}/2 + i/2$

23 -2^{50}

25 $2i\sqrt[3]{4}$

27 $(1/\sqrt{2})\cos(-\pi/12) + (i/\sqrt{2})\sin(-\pi/12)$

29 $-i, -1, i, 1$
$i^{-36} = 1, i^{-41} = -i$

31 $A_1 = 1 + i$
$A_2 = 1 - i$

37 True

39 False

41 True

Appendix C

1 (a) $f'(x) = 3x^2 + 6x + 3$
(b) At most one
(c) [0, 1]
(d) $x \approx 0.913$

3 $\sqrt[4]{100} \approx 3.162$

5 $x \approx 0.511$

7 $x \approx 1.310$

9 $x \approx 1.763$

11 $x \approx 0.682328$

Appendix D

1 3, 0 radians

3 2, $3\pi/4$ radians

5 $7\vec{j}$

7 $\|3\vec{i} + 4\vec{j}\| = \|-5\vec{i}\| = \|5\vec{j}\|$, $\|\vec{i} + \vec{j}\| = \|\sqrt{2}\vec{j}\|$

9 $5\vec{j}$ and $-6\vec{j}$; $\sqrt{2}\vec{j}$ and $-6\vec{j}$

11 (a) $(-3/5)\vec{i} + (4/5)\vec{j}$
(b) $(3/5)\vec{i} + (-4/5)\vec{j}$

13 $8\vec{i} - 6\vec{j}$

15 $\vec{i} + 2\vec{j}$

17 Equal

19 Equal

21 $\vec{i} + \vec{j}$, $\sqrt{2}$, $\vec{i} - \vec{j}$

23 Pos: $(1/\sqrt{2})\vec{i} + (1/\sqrt{2})\vec{j}$
Vel: $(-1/\sqrt{2})\vec{i} + (1/\sqrt{2})\vec{j}$
Speed: 1

INDEX

For "online" material, please see www.wiley.com/college/hughes-hallett.

Differentiation Formulas

1. $(f(x) \pm g(x))' = f'(x) \pm g'(x)$
2. $(kf(x))' = kf'(x)$
3. $(f(x)g(x))' = f'(x)g(x) + f(x)g'(x)$
4. $\left(\dfrac{f(x)}{g(x)}\right)' = \dfrac{f'(x)g(x) - f(x)g'(x)}{(g(x))^2}$
5. $(f(g(x)))' = f'(g(x)) \cdot g'(x)$
6. $\dfrac{d}{dx}(x^n) = nx^{n-1}$
7. $\dfrac{d}{dx}(e^x) = e^x$
8. $\dfrac{d}{dx}(a^x) = a^x \ln a \quad (a > 0)$
9. $\dfrac{d}{dx}(\ln x) = \dfrac{1}{x}$
10. $\dfrac{d}{dx}(\sin x) = \cos x$
11. $\dfrac{d}{dx}(\cos x) = -\sin x$
12. $\dfrac{d}{dx}(\tan x) = \dfrac{1}{\cos^2 x}$
13. $\dfrac{d}{dx}(\arcsin x) = \dfrac{1}{\sqrt{1-x^2}}$
14. $\dfrac{d}{dx}(\arctan x) = \dfrac{1}{1+x^2}$

A Short Table of Indefinite Integrals

I. Basic Functions

1. $\displaystyle\int x^n\,dx = \frac{1}{n+1}x^{n+1} + C, \quad n \neq -1$
2. $\displaystyle\int \frac{1}{x}\,dx = \ln|x| + C$
3. $\displaystyle\int a^x\,dx = \frac{1}{\ln a}a^x + C, \quad a > 0$
4. $\displaystyle\int \ln x\,dx = x\ln x - x + C$
5. $\displaystyle\int \sin x\,dx = -\cos x + C$
6. $\displaystyle\int \cos x\,dx = \sin x + C$
7. $\displaystyle\int \tan x\,dx = -\ln|\cos x| + C$

II. Products of e^x, $\cos x$, and $\sin x$

8. $\displaystyle\int e^{ax}\sin(bx)\,dx = \frac{1}{a^2+b^2}e^{ax}[a\sin(bx) - b\cos(bx)] + C$
9. $\displaystyle\int e^{ax}\cos(bx)\,dx = \frac{1}{a^2+b^2}e^{ax}[a\cos(bx) + b\sin(bx)] + C$
10. $\displaystyle\int \sin(ax)\sin(bx)\,dx = \frac{1}{b^2-a^2}[a\cos(ax)\sin(bx) - b\sin(ax)\cos(bx)] + C, \quad a \neq b$
11. $\displaystyle\int \cos(ax)\cos(bx)\,dx = \frac{1}{b^2-a^2}[b\cos(ax)\sin(bx) - a\sin(ax)\cos(bx)] + C, \quad a \neq b$
12. $\displaystyle\int \sin(ax)\cos(bx)\,dx = \frac{1}{b^2-a^2}[b\sin(ax)\sin(bx) + a\cos(ax)\cos(bx)] + C, \quad a \neq b$

III. Product of Polynomial $p(x)$ with $\ln x$, e^x, $\cos x$, $\sin x$

13. $\displaystyle\int x^n\ln x\,dx = \frac{1}{n+1}x^{n+1}\ln x - \frac{1}{(n+1)^2}x^{n+1} + C, \quad n \neq -1$
14. $\displaystyle\int p(x)e^{ax}\,dx = \frac{1}{a}p(x)e^{ax} - \frac{1}{a}\int p'(x)e^{ax}\,dx$

$$= \frac{1}{a}p(x)e^{ax} - \frac{1}{a^2}p'(x)e^{ax} + \frac{1}{a^3}p''(x)e^{ax} - \cdots$$

$(+ - + - \ldots)$

(signs alternate)

15. $\displaystyle\int p(x)\sin ax\,dx = -\frac{1}{a}p(x)\cos ax + \frac{1}{a}\int p'(x)\cos ax\,dx$

$\displaystyle\quad = -\frac{1}{a}p(x)\cos ax + \frac{1}{a^2}p'(x)\sin ax + \frac{1}{a^3}p''(x)\cos ax - \cdots$

$(-++--++\ldots)$

(signs alternate in pairs after first term)

16. $\displaystyle\int p(x)\cos ax\,dx = \frac{1}{a}p(x)\sin ax - \frac{1}{a}\int p'(x)\sin ax\,dx$

$\displaystyle\quad = \frac{1}{a}p(x)\sin ax + \frac{1}{a^2}p'(x)\cos ax - \frac{1}{a^3}p''(x)\sin ax - \cdots$

$(++--++--\ldots)$ (signs alternate in pairs)

IV. Integer Powers of $\sin x$ and $\cos x$

17. $\displaystyle\int \sin^n x\,dx = -\frac{1}{n}\sin^{n-1} x\cos x + \frac{n-1}{n}\int \sin^{n-2} x\,dx,$ n positive

18. $\displaystyle\int \cos^n x\,dx = \frac{1}{n}\cos^{n-1} x\sin x + \frac{n-1}{n}\int \cos^{n-2} x\,dx,$ n positive

19. $\displaystyle\int \frac{1}{\sin^m x}\,dx = \frac{-1}{m-1}\frac{\cos x}{\sin^{m-1} x} + \frac{m-2}{m-1}\int \frac{1}{\sin^{m-2} x}\,dx,$ $m \neq 1$, m positive

20. $\displaystyle\int \frac{1}{\sin x}\,dx = \frac{1}{2}\ln\left|\frac{(\cos x)-1}{(\cos x)+1}\right| + C$

21. $\displaystyle\int \frac{1}{\cos^m x}\,dx = \frac{1}{m-1}\frac{\sin x}{\cos^{m-1} x} + \frac{m-2}{m-1}\int \frac{1}{\cos^{m-2} x}\,dx,$ $m \neq 1$, m positive

22. $\displaystyle\int \frac{1}{\cos x}\,dx = \frac{1}{2}\ln\left|\frac{(\sin x)+1}{(\sin x)-1}\right| + C$

23. $\displaystyle\int \sin^m x\cos^n x\,dx$: If m is odd, let $w = \cos x$. If n is odd, let $w = \sin x$. If both m and n are even and positive, convert all to $\sin x$ or all to $\cos x$ (using $\sin^2 x + \cos^2 x = 1$), and use IV-17 or IV-18. If m and n are even and one of them is negative, convert to whichever function is in the denominator and use IV-19 or IV-21. If both m and n are even and negative, substitute $w = \tan x$, which converts the integrand to a rational function that can be integrated by the method of partial fractions.

V. Quadratic in the Denominator

24. $\displaystyle\int \frac{1}{x^2+a^2}\,dx = \frac{1}{a}\arctan\frac{x}{a} + C,\quad a \neq 0$

25. $\displaystyle\int \frac{bx+c}{x^2+a^2}\,dx = \frac{b}{2}\ln|x^2+a^2| + \frac{c}{a}\arctan\frac{x}{a} + C,\quad a \neq 0$

26. $\displaystyle\int \frac{1}{(x-a)(x-b)}\,dx = \frac{1}{a-b}(\ln|x-a| - \ln|x-b|) + C,\quad a \neq b$

27. $\displaystyle\int \frac{cx+d}{(x-a)(x-b)}\,dx = \frac{1}{a-b}[(ac+d)\ln|x-a| - (bc+d)\ln|x-b|] + C,\quad a \neq b$

VI. Integrands Involving $\sqrt{a^2+x^2}$, $\sqrt{a^2-x^2}$, $\sqrt{x^2-a^2}$, $a > 0$

28. $\displaystyle\int \frac{1}{\sqrt{a^2-x^2}}\,dx = \arcsin\frac{x}{a} + C$

29. $\displaystyle\int \frac{1}{\sqrt{x^2 \pm a^2}}\,dx = \ln\left|x + \sqrt{x^2 \pm a^2}\right| + C$

30. $\displaystyle\int \sqrt{a^2 \pm x^2}\,dx = \frac{1}{2}\left(x\sqrt{a^2 \pm x^2} + a^2\int \frac{1}{\sqrt{a^2 \pm x^2}}\,dx\right) + C$

31. $\displaystyle\int \sqrt{x^2 - a^2}\,dx = \frac{1}{2}\left(x\sqrt{x^2 - a^2} - a^2\int \frac{1}{\sqrt{x^2 - a^2}}\,dx\right) + C$